Proceedings of the 20th International Symposium on

SHOCK WAVES

Proceedings of the 20th International Symposium on

SHOCK WAVES

Volume I

Pasadena, California, USA July 1995

Editors

B Sturtevant

J E Shepherd

H G Hornung

Graduate Aeronautical Laboratories
California Institute of Technology

World Scientific
Singapore • New Jersey • London • Hong Kong

Published by

World Scientific Publishing Co. Pte. Ltd.

P O Box 128, Farrer Road, Singapore 912805

USA office: Suite 1B, 1060 Main Street, River Edge, NJ 07661

UK office: 57 Shelton Street, Covent Garden, London WC2H 9HE

This work relates to Department of Navy Grant N00014-95-1-0301 issued by the Office of Naval Research. The U.S. Government has a royalty-free license throughout the world in all copyrightable material contained herein.

This work is sponsored in part by the Defense Nuclear Agency under Grant DNA001-95-1-0007. This manuscript is submitted for publication with the understanding that the U.S. Government is authorized to reproduce and distribute reprints for Government purposes notwithstanding any copyright notions thereon.

British Library Cataloguing-in-Publication Data
A catalogue record for this book is available from the British Library.

PROCEEDINGS OF THE 20TH INTERNATIONAL SYMPOSIUM ON SHOCK WAVES (Vol. I)

ISBN 981-02-2593-8 (Set)
ISBN 981-02-2957- 7 (Vol. I)
ISBN 981-02-2958-5 (Vol. II)

Printed in Singapore.

Preface

The International Symposia on Shock Waves are the definitive meetings of the scientific community devoted to the study and use of shock-wave phenomena of all kinds. They take place every two years, usually at places close to centers of activity in the field. The format of the meetings has grown over the first thirty years of their existence, but the number of presentations has deliberately been held approximately constant since then. Because of steadily increasing activity in the field, this requires fairly stringent selection of the material to be presented, and consequently also leads to the high standard and good reputation that the proceedings enjoy. Although much of the work published there has also reached the archival literature in more detailed form, these volumes are a valuable resource because they bring the shock-wave information together in one place.

The broad range of topics covered by the presentations may be divided into applications such as combustion, high-speed flight, detonations, physical chemistry, medicine, and geology; the use and development of devices that produce shock waves, such as shock tubes and shock tunnels, with the associated technology advances including diagnostic techniques; and fundamental studies such as the refraction and reflection of shocks, their interaction with vortices, and chemical kinetics.

Four hundred fifty abstracts were submitted to the 20th Symposium. Each abstract was judged by three members of the Program Committee; 320 abstracts were selected for presentation, 165 as oral and 155 as poster presentations. The presentations included 12 memorable invited lectures. Two of these, the Paul Vieille and Otto Laporte lectures, were delivered by Paul Roth of Duisburg and Heinz Reichenbach of the Ernst Mach Institut in Freiburg, respectively. Both were high-quality and rich presentations that will be remembered as highlights of the Symposium.

Three hundred thirty four participants and 30 companions came to Pasadena from 18 countries: USA, 110; Japan, 57; Germany, 36; France, 32; Australia, 18; Russia, 17; UK, 16; Israel, 13; Canada, 11; China, 9; South Africa and Netherlands, 4 each; Korea, 2; and Belarus, Brazil, Czech Republic, India, and Poland, 1 each.

Two hundred fifty eight papers are published in the Proceedings. This number is smaller than the number of presentations at the Symposium for a number of reasons. For example, some authors had already published similar material elsewhere, as in the case of one of the invited talks (Kanamori), and some wished to publish in other places, where the usual limitation of the ISSW Proceedings to six pages did not constrain them.

The Proceedings is organized such that papers on related topics are grouped together, without making distinctions between poster and oral presentations. It was necessary to break the material into two volumes, which should, however, be seen as parts of a set. Accordingly, the page numbers do not start again in Volume II, but continue from the end of Volume I. For convenience, the Table of Contents in each volume includes the entries of both volumes. Similarly, complete author and subject indexes may be found at the end of each volume.

The papers here were selected from contributions by the Scientific Committee. We thank these individuals for their service to the shock-wave community:

T. Ahrens	R. Akbar	D. Baganoff	W. Behrens
J. Bélanger	R.F. Benjamin	D. Bershader	R. Bonazza
C. Brennen	M. Brouillette	A. Bruckner	J. Budzinski
G. Candler	P. Cassady	J. Cates	B.M. Cetegen
W. Christiansen	P. Colella	M.S. Cramer	E. Cummings
J. Davis	J.M. Dewey	J.L. Ellzey	R.J. Emrich
D. Frost	H. Glaz	J.J. Gottlieb	R.K. Hanson
B. Hartenbaum	H. Hornung	J.W. Jacobs	J.H. Kiefer
D. Knight	J.K. Knowles	J.H.S. Lee	P.V. Marrone
J.R. Maus	S. McCahan	J.C. McDaniel	R. Menikoff
R.B. Miles	P. Miller	S. Molder	F.K. Owen
E.G. Puckett	D. Pullin	G. Ravichandran	V. Rupert
A. Roshko	D.A. Russell	R. Samtaney	S.R. Sanderson
D.W. Schwendeman	R.E. Setchell	S. Sharma	J.E. Shepherd
M. Sichel	B. Sturtevant	A. Teodorczyk	P. Thibault
W.P. Thompson	W. Tsang	T. Vreeland	G. Wilson
N.J. Zabusky			

The International Advisory Committee of the Symposium contributed many suggestions. We recognize these individuals for their assistance:

T.V. Bazhenova	G. Ben-Dor	D. Bershader	M. Brouillette
R. Brun	F. Demmig	J.M. Dewey	L. Dumitrescu
R. East	J.L. Ellzey	N.A. Fomin	B. Forestier
B. Gelfand	J.J. Gottlieb	H. Grönig	Z. Han
R.K. Hanson	L.F. Henderson	A. Hertzberg	F. Higashino
R. Hillier	H. Honma	H. Hornung	L. Houas
F. Houwing	O. Igra	V.K. Kedrinskii	Y.W. Kim
A. Lifshitz	H. Matsui	B.E. Milton	H. Miyajima
F. Obermeier	N.M. Reddy	H.W. Reichenbach	P. Roth
D.A. Russell	J.E. Shepherd	B. Skews	R.J. Stalker
J.L. Stollery	B. Sturtevant	K. Takayama	M.E.H. Van Dongen
Z.A. Walenta	H.-Y. Yu		

The following organizations contributed financial support essential to the operation of the Symposium:

Air Force Office of Scientific Research under Grant F49610-92-J-0110
Defense Nuclear Agency under Grant DNA001-95-1-0007
National Science Foundation under Grant CTS-9509360
Office of Naval Research under Grant N00014-95-1-0301
Division of Engineering and Applied Science,
California Institute of Technology
Seismological Laboratory at the California Institute of Technology
The Estate of Ralph D. Baker, M.S., Ph.D., Caltech, 1938

We gratefully acknowledge the efforts of Jamie Guthrie, Donna Shepherd, Joe Cates, James Quirk, the GALCIT, Mechanical Engineering and Geophysics graduate students, and Caltech staff in organizing the Symposium. We also wish to thank Marionne Kirk for performing the arduous task of copy editing the Proceedings.

The Executive Committee

H. G. HORNUNG	Chairman
T. J. AHRENS	
G. RAVICHANDRAN	
A. ROSHKO	
J. E. SHEPHERD	
B. STURTEVANT	Editor
K. CHEETHAM	Secretary

Contents

Volume I

Preface v

Paul Vielle Lecture

Shock-tube kinetics in homogeneous and heterogeneous reaction systems
Roth, P. 1

Otto Laporte Lecture

120 years of shock-wave visualization — a fascinating activity
Reichenbach, H. 17

Plenary Lectures

Transport and relaxation behind shock waves: old and new problems
Brun, R. 33

Tuned operation of a free-piston shock tunnel
Itoh, K. 43

High-resolution imaging of hypervelocity metal jets using advanced high-speed photographic techniques
Shaw, L.L., Muelder, S.A. 53

Numerical simulation of high-enthalpy impulse facilities
Wilson, G.J. 67

Blast effects of confined and unconfined explosions
Dorofeev, S.B. 77

Anomalous physical effects from artificial numerical length scales
Menikoff, R., Lackner, K.S. 87

Curved detonation fronts in solid explosives: collisions and boundary interactions
Bdzil, J.B., Aslam, T.D., Stewart, D.S. 97

Bubbly cavitation in intense rarefaction waves and its effects
Kedrinskii, V.K. 107

Wave propagation in vapor explosions
Frost, D.L. 121

Irving Glass Memorial Session

Wave machinery for chemical processing and high-efficiency heat engines
Christiansen, W.H., Hertzberg, A. 135

Holographic interferometric animation of shock diffraction and reflection using a highly repeatable shock tube
Yang, J.M., Babinsky, H., Takayama, K., Saito, T., Sislian, J.P. 141

Universal attenuation law of planar shock waves propagating in dust-gas suspensions
Aizik, F., Ben-Dor, G., Elperin, T., Igra, O., Mond, M., Grönig, H. 147

Development of a detonation driver for a shock tunnel
Lenartz, M., Wang, B., Grönig, H. 153

Behavior of laser-induced bubbles and pressure waves in liquid nitrogen
Sato, H., Odagawa, M., Sun, X.W., Maeno, K., Honma, H. 159

The interaction of a two-dimensional strong compressible vortex with a shock wave
Kao, C.T, von Ellenrieder, K., Bershader, D. 165

Hypersonic Flow

Interference heating effects caused by a 3D transverse jet in hypersonic flow
Mudford, N.R., Roberts, G.T., Schuricht, P.H., Ball, G.J., East, R.A. 173

Comparison between shock-tunnel measurements on a planetary-probe configuration and DSMC predictions
Holden, M.S., Chadwick, K.M., Gallis, M.A., Harvey, J.K. 179

Hyperboloid flare experiments in the HEG
Krek, R.M., Eitelberg, G., Beck, W.H. 185

Experiments on the Halis axisymmetric configuration in the T5 shock tunnel
Davis, J., Sturtevant, B., Muylaert, J. 191

Shock-tube measurements for aero-thermodynamic gas-wall interaction
Maul, J. 197

Computational analysis of experiments in T5 and HHK for the study of flap-induced separation on an axisymmetric configuration
Muylaert, J., Walpot, L., Hironde, N., Tumino, G., Simeonides, G., Koppenwallner, G., Davis, J. 203

Shock-interference heating in hypervelocity flow
Sanderson, S.R., Sturtevant, B. 209

Investigation of shock-shock interactions in hypersonic reentry flows
Brück, S. 215

Recompression of nonequilibrium flow in the wake of a blunted cone
Kastell, D., Hannemann, K., Eitelberg, G., Horvath, T.J. 221

Flow in the TCM2-Marseille hypersonic facility: some new experimental results
Brun, R., Burtschell, Y., Chaix, A., Druguet, M.-C., Dumitrescu, L.Z., Dumitrescu, M.-P., Martin, R., Zeitoun, D. 227

Leading edge bluntness effects in hypervelocity compression corner flow
Mallinson, S.G., Gai, S.L., Mudford, N.R. 233

Hypersonic high-enthalpy flow over a blunt-stepped cone
Gai, S.L., Milthorpe, J.F. 239

Surface heat flux to the Viking aeroshell in hypersonic reacting flow
Mudford, N.R., Gai, S.L. 245

Heat transfer in a superorbital chemically frozen boundary layer
Neely, A., Morgan, R. 251

Standard model experiments in high-enthalpy facilities
Ueda, S., Hannemann, K., Eitelberg, G. 257

Radiation around a hypersonic ablative flight model
Komurasaki, K., Yano, S., Fujiwara, T. 263

Laser-induced fluorescence measurements in high-enthalpy facilities: application to the study of air plasma/surface interaction
Robin, L., Honore, D., Vervisch, P. 269

Use of an expansion tube for imaging studies of hypersonic reactive flows around blunt bodies
Kamel, M.R., Morris, C.I., Thurber, M.C., Wehe, S.D., Hanson, R.K. 275

Spatially and spectrally resolved laser-induced fluorescence measurements in the HEG
Wollenhaupt, M., Rosenhauer, M., Müller, T., Beck, W.H., Scheer, M. 281

Bow-shock/disturbance interaction for hypersonic flow over a cylinder
Zhong, X., Joubert, X. 287

Monte Carlo study of threshold line models for dissociation in hypersonic shock waves
Boyd, I.D. 293

Analytical hypersonic waverider configurations
Merlen, A., Fabre, E., Cieslak, S. 299

A more precise method to determine stagnation-point heat-transfer rates
Olivier, H., Grönig, H. 305

Millisecond aerodynamic force measurement using a free-flying side-jet model with built-in gas reservoir
Naumann, K.W., Ende, H., Mathieu, G., George, A. 311

Effects of catalycity and boundary layer reactions on surface heat flux in hypersonic high enthalpy flows
Gai, S.L., Mudford, N.R., Roberts, G.T., Mallinson, S.G. 317

Study of experiments sensitive to vibration-dissociation coupling models
Olejniczak, J., Candler, G.V. 323

Supersonic Flow

Disturbances on the shock front ahead of a concave body due to resonant excitement of vortices
Serov, A.O., Shtemenko, L.S., Shugaev, F.V. 331

An inverse method for calculating nonequilibrium flow behind curved shock waves
Mölder, S. 337

Shock-cell structure in a jet at the critical pressure ratio
Makhsud, A., Brocher, E., Chambarel, A., Cordonnier, P. 343

Mach disc and shock structure in coflowing dual axisymmetric jets
Kurian, J., Ramana Rao, T.V., Ramesh Kumar, R. 349

A study of three-dimensional supersonic mixing flow phenomena induced by ramp injectors
Umii, K., Aso, S., Tannou, M., Yamane, Y., Ando, Y., Fukuda, M. 355

Effects of cavity configurations on supersonic internal flow
Watanabe, Y., Kudo, S., Higashino, F. 361

Wave Reflection and Diffraction

Experimental reconsideration of the state of the art of oblique shock-wave reflections in steady flows
Chpoun, A., Passerel, D., Lengrand, J.C., Li, H., Ben-Dor, G. 369

Experimental and theoretical study of stationary reflection of shock waves
Gvoždeva, L.G., Sherbak, N.B., Chpoun, A., Passerel, D. 375

Mach-stem growth in von Neumann reflection
Virgona, R.J., Higashino, F. 381

Mach-reflection flow at hypersonic speeds
Le, J., Ni, H., Hu, G. 387

Wave configuration of Mach reflection in steady flows: analytical solution and dependence on downstream influences
Li, H., Schotz, M., Ben-Dor, G. 393

Non-self-similarity of the initial stages of Mach reflection
Dewey, J.M., van Netten A.A. 399

Numerical confirmation of the hysteresis phenomenon in the regular to Mach reflection transition in steady flows
Chpoun, A., Ben-Dor, G. 405

The existence of reflected finite-compression waves in weak pseudo-steady Mach reflections over not-"small" wedge angles
Liu, J.J. 411

Shock-wave reflections from a cone with attack angles in a symmetric plane
Suzuki, T., Sakamura, Y., Yin, X.-.Z 417

Calculation of focusing of weak shocks using geometrical shock dynamics
Cates, J., Sturtevant, B. 423

The shock reflection process over a non-straight surface
Kobayashi, S., Adachi, T., Suzuki, T. 429

Oblique shock interactions and the von Neumann paradox
Čanić, S., Keyfitz, B.L. 435

Shock-wave interactions at impingement of impulsive supersonic jet upon an obstacle
Bazhenova, T.V., Golub, V.V., Shulmeister, A.M., Bormotova, T.A., Bazarov, S.B. 441

Application of the principle of minimum entropy production to the regular-to-Mach-reflection transition in steady flows
Li, H., Ben-Dor, G. 447

Evolution of Courant-Friedrichs structure: self-collimating outflows behind reflected shocks
Chernin, A., Voinovich, P. 453

Numerical investigations of the prediction of the Mach-stem height in steady flows
Vuillon, J., Zeitoun, D., Ben-Dor, G. 459

Interaction of shock waves emitted from a shock tube with a supersonic projectile
Jiang, Z., Takayama, K. 465

Numerical investigation of shock-wave reflection problems in steady flows
Ivanov, M.S., Gimelshein, S.F., Markelov, G.N., Beylich, A.E. 471

The interaction of a shock wave discharged from the open end of a shock tube with a flat plate
Bazhenova, T.V., Golub, V.V., Shulmeister, A.M., Bormotova, T.A., Bazarov, S.B., Pichugin, S.V. 477

Analyses of reflection and diffraction of weak shock waves
Takano, Y., Iwamoto, N., Goto, T. 483

Computation and validation of shock-wave interaction with fluid inhomogeneities
Falcovitz, J., Yamada, K., Nagoya, H., Takayama, K. 489

The influence of geometric variations on shock focusing in cylindrical cavities
Babinsky, H., Onodera, O., Takayama, K., Timofeev, E., Voinovich, P. 495

An experimental/numerical approach to shock-wave diffraction
Kleine, H., Yu, Q., Ritzerfeld, E., Grönig, H. 501

Compressibility effects on the propagation and interaction of shock-generated vortices
Hébert, C., Brouillette, M. 507

Shock-wave diffraction by an elastic sphere in water
Merlen, A., Pernod, P., Ahyi, A., Kemmou, A. 513

Physics of a shock-generated freely moving compressible vortex
von Ellenrieder, K., Kao, C.T., Bershader, D. 519

Study of the propagation of shock waves over bodies by holographic interferometry
Le, J., Ye, X., Wu, X., Yang, H. 525

Accuracy of shock-wave propagation using geometrical shock dynamics
Schwendeman, D.W. 531

A locally adaptive structured/unstructured Navier-Stokes solver for essentially unsteady shock-obstacle interactions at high Reynolds numbers
Ofengeim, D., Timofeev, E., Voinovich, P. 537

Wave Refraction

Vortex-ring impingement on a wall with shock formation
Minota, T., Nishida, M., Lee, M.G. 545

Oblique reflection of shock waves in a slow sound-speed fluorocarbon liquid
Mortimer, B.J.P., Skews, B.W., Felthun, L.T. 551

Refraction of a blast wave propagating parallel to a contact discontinuity: a Free-Lagrange numerical simulation
Ball, G.J., East, R.A. 557

Shock waves generated by the interaction of two vortex rings
Minota, T., Nishida, M., Lee, M.G. 563

Mechanisms of shock-wave generation resulting from shock-wave/bubble interaction in liquids
Yamada, K., Kodama, T., Takayama, K. 569

Passage of two shocks through an interface
Zaytsev, S.G., Chebotareva, E.I., Aleshin, A.N., Lazareva, E.V., Titov, S.N. 575

Shock-hydrodynamics experiments on the Nova laser
Miller, P.L., Peyser, T.A., Stry, P.E., Budil, K.S. 581

Shock-accelerated heavy-gas layers
Budzinski, J.M., Benjamin, R.F., Jacobs, J.W. 587

Experiments on the shock-excited interfacial instability at an air/SF_6 interface
Vetter, M., Sturtevant, B. 593

Mathematical modeling of gravitational turbulent mixing in non-self-similar problems
Neuvazhayev, V.Y. 599

Numerical simulation of Richtmyer-Meshkov instability
Lebo, I.G., Rozanov, V.B., Tishkin, V.F., Nikishin, V.V., Popov, I.V., Favorsky, A.P. 605

The effect of refraction regime on the evolution of perturbation for Richtmyer-Meshkov instability
Zaytsev, S.G., Aleshin, A.N., Lazareva, E.V. 611

A numerical study of shock acceleration of a diffuse helium cylinder
Greenough, J.A., Bell, J.B., Colella, P., Puckett, E.G., Jacobs, J.W. 617

Hypervelocity Richtmyer-Meshkov instability
Samtaney, R., Meiron, D.I., Pullin, D.I. 623

Numerical investigation of Richtmyer-Meshkov instability using front tracking
Grove, J.W., Holmes, R.L., Sharp, D.H. 629

Richtmyer-Meshkov mixing-zone deformation induced by shock-tube wall boundary layers
Jourdan, G., Houas, L. 635

Vortex methods for early and intermediate time Richtmyer-Meshkov environments
Ray, J., Zabusky, N.J., Samtaney, R. 641

Visualization of shocked mixing zones using differential interferometry and X-rays
Rodriguez, G., Galametz, I., Thorembey, M.-H., Rayer, C., Haas, J.-F. 647

Interaction of a turbulent shear flow with a Mach 1.5 normal shock wave
Mahesh, K., Lele, S.K., Moin, P. 653

Waves, General

Converging shock waves: modeling and analysis of structures
Wallus, H., Demmig, F. 661

Propagation and stability of strong shocks in gases
Mond, M., Rutkevich, I., Toffin, E. 667

Blast parameters from unconfined gaseous detonations
Dorofeev, S.B., Sidorov, V.P., Dvoinishnikov, A.E. 673

Numerical simulation of the blast in the World Trade Center
Baum, J.D., Luo, H., Löhner, R. 679

Blast-wave interaction with multiple structures
Absil, L.H.J., van den Berg, A.C., Weerheijm, J. 685

Efficient 2D and 3D locally adaptive unstructured Euler solvers for unsteady shocked gas flows
Galyukov, A., Fursenko, A., Voinovich, P., Saito, T., Timofeev, E. 691

Viscous Effects

Boundary-layer transition on slender blunt cones at hypersonic speeds
Zanchetta, M.A., Hillier, R. 699

Reflected-shock/side-boundary-layer interaction in a shock tube
Nishida, M., Lee, M.G. 705

Effects of shear-layer transition and leading-edge bluntness on control-flap effectiveness
Kumar, D., Stollery, J.L. 711

Statistical simulation of nonequilibrium hypersonic flows over concave bodies
Ivanov, M.S., Gimelshein, S.F., Markelov, G.N., Gorbachev, Y.E. 717

Transition of unsteady boundary layer induced by propagating compression wave
Aoki, T., Kondoh, N., Matsuo, K., Mashimo, S. 723

Structure of multiple shock-wave/turbulent boundary-layer interactions in a rectangular duct
Sugiyama, H., Arai, T., Hongo, S., Kawase, T., Uno, N. 729

Asymptotic defect boundary layer applied to nonequilibrium laminar hypersonic air flows
Séror, S., Schall, E., Zeitoun, D., Brazier, J.P. 735

Analysis of nonequilibrium, hypersonic blunt-body flow with the local-similarity method
Klomfass, A., Müller, S., Ballmann, J. 741

Formation of standing shock waves due to evaporation from a spherical condensed phase
Onishi, Y., Fuchs, J.W. 747

Knudsen-number effects in high-speed microchannel flows
Nguyen, T.X., Oh, C.K., Saint-Martin-Tillet, X.N., Piana, J.V., Oran, E.S. 753

Studies of separated flow structure and oscillating separation shock waves in shock-wave/turbulent boundary-layer interaction regions induced by a blunt protuberance
Maekawa, S., Aso, S., Karashima, K., Sato, K. 759

A numerical study of shock-wave reflections in nonequilibrium chemically reacting flows
Aso, S., Mitomo, T., Fujiwara, T., Hayashi, M. 765

Turbulent spots in boundary layers in a free-piston shock-tunnel flow
Mee, D.J., Goyne, C.P. 771

Influence of reflected-shock/boundary-layer interaction on driver-gas contamination in high-enthalpy shock tunnels
Chue, R.S.M., Itoh, K. 777

Suppression of shock-induced separation in dense gases
Cramer, M.S., Park, S., Watson, L.T. 783

Shock waves diffracting over rigid ramps in a viscous, heat-conducting gas (argon)
Virgona, R.J., Crutchfield, W.Y., Henderson, L.F. 789

Comparison of continuum and molecular approaches for the flow around a perpendicular disk
Elizarova, T.G., Graur, I.A., Chpoun, A., Lengrand, J.C. 795

The numerical simulation of unsteady shock-wave/boundary-layer interactions in reacting gas mixtures
Weber, Y.S., Oran, E.S., Boris, J.P., Anderson, J.D. 801

Author Index 807

Keyword Index 815

Volume II

Chemical Physics

UV- and IR-absorption measurements in a shock tube using multiple reflections
Thoma, M.L., Hindelang, F.J. 827

High-temperature thermolysis of ethylene-d_4
Kern, R.D., Chen, H., Yao, J. 833

Single-pulse shock-tube study of tetraethoxysilane decomposition
Herzler, J., Manion, J.A., Tsang, W. 839

Bimolecular S-atom reactions studied in pyrolysis and photolysis shock-wave experiments
Woiki, D., Roth, P. 845

Laser-flash-photolysis study of the decomposition of vibrationally excited CO_2
Eremin, A.V., Shumova, V.V., Ziborov, V.S., Mick, H.-J., Roth, P. 881

Formation of C_2 during high-temperature pyrolysis of fullerene C_{60} in shock waves
Sommer, T., Kruse, T., Roth, P. 887

Thermal decomposition of gaseous hydrazine behind a reflected shock wave
Halat-Augier, C., Dupré, G., Paillard, C.E. 893

Shock-tube study of acetone pyrolysis
Hidaka, Y., Sato, K., Iwamoto, R., Koike, T. 899

Measurements of rotational-vibrational excitation behind incident shock waves by tunable diode laser absorption spectrometry
Klenk, W., Widdecke, N., Frohn, A. 905

Absorption lineshape measurement of atomic oxygen at 130 nm
Meyer, S., Bershader, D., Gilmore, J., Sharma, S., Exberger, R., Whiting, E. 911

Spectroscopic detection of highly excited molecular-state transformations behind shock waves and master-equation implementation
Zaslonko, I.S., Eremin, A.V., Shumova, V.V. 917

Measurement of the absorption spectrum of the OH A-X (0,0) 306-nm band-head at high pressures
Davidson, D.F., Röhrig, M., Petersen, E.L., Di Rosa, M.D., Hanson, R.K. 923

Self-ignition of n-heptane/iso-octane mixtures and n-decane with air under engine conditions
Fieweger, K., Pfahl, U., Adomeit, G. 929

Self-ignition of H_2-air mixtures at high pressure and low temperature
Blumenthal, R., Fieweger, K., Komp, K.H., Adomeit, G., Gelfand, B.E. 935

High-pressure shock-tube measurements of ignition times in stoichiometric $H_2/O_2/Ar$ mixtures
Petersen, E.L., Davidson, D.F., Röhrig, M., Hanson, R.K. 941

Soot formation from rich hydrocarbon oxidation under elevated pressure conditions
Kellerer, H., Bauer, H.-J., Wittig, S. 947

Single-pulse shock-tube study on the decomposition of C_3F_6
He, Y. 953

Direct measurements of rate coefficients for thermal decomposition of methyl halides using shock-tube ARAS technique
Takahashi, K., Inoue, A., Inomata, T. 959

Examination of the oxidation mechanism of H_2S
Tsuchiya, K., Oya, M., Matsui, H. 965

Thermal decomposition of furonitrile: studies with a single-pulse shock tube
Laskin, A., Lifshitz, A. 971

Comparison of relaxation models with shock-tube data
Stuhler, H., Frohn, A. 977

Various models of coupling of vibrational-dissociational kinetics behind shock waves
Nagnibeda, E.A., Méolans, J.G., Mouti, M., Chauvin, A.H. 983

Combustion

Optical diagnostic measurements in a supersonic-combustion ramjet
McIntyre, T.J., Houwing, A.F.P., Rabbath, P.A.B., Palma, P.C. 991

Numerical investigation of induction-zone development in hydrogen-oxygen mixtures
Figura, E.S., Cullis, I.G. 997

Shock-tube investigations of H_2 combustion in a high-temperature supersonic air flow
Smeets, G., Quenett, C. 1003

Numerical simulation of reactive shock diffraction around an object containing a cavity
Jiang, B., Ingram, D.M., Causon, D.M., Saunders, R. 1009

Modeling of explosion in explosion behind a reactive shock wave by the joint velocity-composition PDF method
Frolov, S.M., Noskov, M.A., Tatschl, R., Wolanski, P. 1015

Spontaneous ignition of a reactive gas mixture by shock-vortex interaction
Takayama, F., Kambe, T., Taki, T., Sakurai, A. 1021

Shock-tube investigations of atomization, evaporation, and ignition of n-decane and α-methylnaphthalene droplets
Pfahl, U., Fieweger, K., Adomeit, G., Gelfand, B.E. 1027

Numerical modeling of chemistry and gas dynamics during shock-induced ethylene combustion
Clifford, L., Milne, A., Boulton, D. 1033

Shock-tube combustion of liquid sprays of ethanol: effect of additives on ignition delays and combustion characteristics
Cadman, P. 1039

The effect of halocarbons on methane ignition
Burcat, A., Dvinyaninov, M., Lifshitz, A. 1045

LIPF measurements and modeling of jet ignition through an orifice between two separated rooms
Mori, S., Yoshida, T., Hayashi, A.K., Fujiwara, T. 1051

Gasdynamic limits of ignition and combustion of a gas mixture in ISL's RAMAC 30 scram accelerator
Seiler, F., Patz, G., Smeets, G., Srulijes, J. 1057

Ram-accelerator operation with simplified geometries
Sasoh, A., Knowlen, C., Higgins, A.J., Bruckner, A.P. 1063

Detonation

Triple-shock structure in dusty gas detonations
Hayashi, A.K., Fuyuto, T., Fujiwara, T. 1071

Detailed modeling of detonation in heterogeneous flows
Uphoff, U., Hänel, D., Roth, P. 1077

Detonation-tube studies on aluminum dust-air mixtures under weak ignition conditions
Pu, Y.K., Li, K.Y. 1083

Detonation propagation in a one-dimensional heterogeneous condensed-phase explosive
Lee, J.J., Brouillette, M. 1089

Transmission of a detonation across an air gap
Teodorczyk, A., Thomas, G.O., Ward, S.M. 1095

Large-scale confined hydrogen-air detonation experiments and their numerical simulation
Breitung, W., Dorofeev, S.B., Efimenko, A.A., Kochurko, A.S., Redlinger, R., Sidorov, V.P. 1101

Transition from deflagration to detonation with a dense aluminum-dust slug in a combustible gas
Zhang, F., Thibault, P. A. 1107

Effect of confinement on the propagation of a gaseous detonation in a porous medium
Makris, A., Frost, D.L., Abdul-Baki, I., Yoshinaka, A., Lee, J.H.S. 1113

Detonation initiation by hypervelocity projectiles
Bélanger, J., Kaneshige, M., Shepherd, J.E. 1119

Numerical investigation of pulsating detonations
Quirk, J.J., Short, M. 1125

Initiation of detonation by gasdynamic methods in a half-limited space
Achasov, O.V., Labuda, S.A., Penyazkov, O.G., Ragozin, D.S. 1131

Characterization of heterogeneous detonations of powders dispersed in nitrogen
Tulis, A.J., Sumida, W.K., Norikan, K., Dihu, R., Patel, D.L., Heberlein, D.C. 1137

Numerical simulation of the reflection of detonation waves
Yu, Q., Grönig, H. 1143

Analysis of detonation on a hypervelocity projectile in the in-tube chemical accelerator
Rom, J. 1149

Plasma and Radiation

Regular-shock-reflection phenomena in magnetohydrodynamics
Ogawa, H., Fujiwara, T., Hayashi, A.K. 1157

Nonlinear magnetohydrodynamic waves
Tarkenton, G.M., Cramer, M.S. 1163

Mechanisms of the formation of highly excited ArI and ArII level populations in shock-heated jets
Bazhenova, T.V., Eremin, A.V., Emelyanov, A.V., Shumova, V.V. 1169

Interaction between shock-generated plasma and electromagnetic field in a reentry vehicle
Leblanc, J.E., Fujiwara, T. 1175

Time-resolved imaging spectroscopy of nonequilibrium shock-front radiation in air
Koreeda, J., Ohama, Y., Maeno, K., Honma, H. 1181

Multi-phase Flows

Measurements of void fraction in a vapor explosion by a light-extinction technique
El-Beshbeeshy, M., Bonazza, R., Shamoun, B. 1189

A hybrid front-tracking scheme for shock wave propagation in gas/water media
Causon, D.M. 1195

Shock waves in dilute/non-dilute bubbly flows
Kameda, M., Shimaura, N., Higashino, F., Matsumoto, Y. 1201

Hydraulic shock loading due to supercavitating projectiles
Davies, G.A.O, Flanagan, O., Hillier, R., Hitchings, D., Lord, S.J. 1207

Shock waves and noise in the collapse of a cloud of cavitation bubbles
Wang, Y.-C., Brennen, C.E. 1213

Shock-tube-deduced condensation coefficient
Fujikawa, S., Kotani, M., Sato, H., Takasugi, N., Matsumoto, M. 1219

Unsteady shock formation in transonic flow with nonequilibrium phase transition
Schnerr, G.H., Adam, S. 1225

Condensation phenomena behind incident and reflected shock waves in heavy vapors at low Mach numbers
Garen, W., Becker, M., Paál, G. 1231

On the formation of embedded shock waves in rarefaction waves by homogeneous condensation
Delale, C.F., Maršík, F., Schnerr, G.H. 1237

Condensation induced by rarefaction waves and reflected rarefaction waves
Fu, Y., Han, Z., Gong, M. 1243

Condensation shock wave as the consequence of the principle of minimum entropy production
Maršík, F., Sopuch, P., Blaha, J. 1249

Shock-wave reflection around a two-dimensional wedge in low-temperature gases with phase change
Maeno, K., Nozawa, H., Nishimori, Y., Honma, H. 1255

Complex droplet deformation and breakup modes at intermediate Weber numbers from laser holography
Zerf, G., Hornung, K. 1261

Entrainment of liquid droplets into shock-induced airstreams
Archer, R.D., Milton, B.E., Behnia, M., Casey, R.T. 1267

Combined sublimation and thermal decomposition of NH_4Cl/Ar aerosols in shock waves
Schading, G.N., Roth, P. 1273

Effective sphere drag coefficient for unsteady gas-particle flows
Kurian, J., Das, H.K. 1279

Acceleration of a sphere behind planar shock waves
Britan, A., Elperin, T., Igra, O., Jiang, J.P. 1285

Dusty shock reflection and diffraction in aligned baffle systems
Wang, B.-Y., Xu, Y.-H., Xu, Y.-Q., Wu, Q.-S. 1291

Numerical analysis of wave patterns of shock propagation in a dusty gas along convergent-divergent nozzles
Wu, Q.-S., Wang, D.-Z., Xu, Y.-H., Wang, B.-Y. 1297

Dusty gas flow in a nozzle
Elperin, I., Igra, O., Ben-Dor, G., Vasiliev, E. 1303

Head-on collision of a planar shock wave with a dusty gas layer
Jiang, J.P., Elperin, T., Igra, O., Falcovitz, J. 1309

An experimental and theoretical study of shock-wave induced particle coagulation
Strecker, J.J.F., Roth, P. 1315

Experimental verification of the concept of amplification of shock loading by means of compressible covering as applied to the propagation of blast waves over a dust deposit
Medvedev, S.P., Polenov, A.N., Khomik, S.V., Gelfand, B.E. 1321

Effect of thermal expansion on the shock speeds of first, second, and fourth sound in He II
Braun, S., Kluwick, A., Cramer, M.S. 1327

Analytical prediction of regular reflection over porous surfaces
Levy, A., Li, H., Ben-Dor, G. 1333

Experimental observation of shock-wave evolution in an elastomeric foam
Zaretsky, E., Igra, O. 1339

Experimental investigation of interaction of shock wave with closed cellular compressible foam
Lagutov, Y.P., Sharov, Y.L., Gvozdeva, L.G. 1345

Blast-wave propagation in foam
Kleine, H., Diaconescu, G., Lee, J.H.S. 1351

Experimental and numerical investigation of the head-on collision of planar shock waves with rigid porous materials
Levy, A., Sorek, S., Ben-Dor, G., Skews, B.W. 1357

Shock-induced wave phenomena in open-cell permeable and flexible foams
van Dongen, M.E.H., Smeets, G.V.R., Hoeijmakers, H.W.M., Smeulders, D.M.J. 1363

Equilibrium gas dynamics and shock relations for flexible foams
van Dongen, M.E.H., van Gool, C.J., Pruiksma, J.P., Smeulders, D.M.J. 1369

Axisymmetric tests of shock-wave impact on PUR foam
Skews, B.W., Sellschop, R.M., Legoete, T. 1375

Shock impact on porous plugs with a fixed gap between the plug and a wall
Seitz, M.W., Skews, B.W. 1381

Attenuation of a blast wave with a cellular material
Makris, A., Frost, D.L., Nerenberg, J., Lee, J. H.S. 1387

Structure of deformation waves in porous compressible materials — influence of percolation
Gvozdeva, L.G., Lagutov, Y.P., Sharov, Y.L., Sherbak, N.B. 1393

Condensed Matter

Ultra-high-pressure thermal-conductivity measurements of griceite and corundum
Gallagher, K.G., Ahrens, T.J. 1401

Combustion synthesis of advanced ceramics assisted by explosive compaction
Timnat, Y.M., Rosenband, V. 1407

Dynamic compaction of granular materials in a tube with wall friction, applied to deflagration-to-detonation transition
Hill, L.G., Kapila, A.K. 1413

Investigation of HE-driven cylindrical liner implosions
Tan, T.-H., Marsh, S.P. 1419

Modeling shock-wave propagation in obliquely loaded unidirectional composites
Resnyansky, A.D., Romensky, E.I., Belov, V.Y., Williams, G.H. 1425

Shock consolidation of powders by utilizing a converged underwater shock-wave assembly
Kubota, S., Nagano, S., Fujita, M., Itoh, S. 1431

Thermodynamic relations, Hugoniot adiabatics, and stability of strong shocks in solids
Rutkevich, I., Zaretsky, E., Mond, M. 1437

Numerical simulation of shock-wave generation with KALIF
Vorobiev, O.Y., Fortov, V.E., Goel, B. 1443

Effects of solid-solid phase changes on impact mechanics
Knowles, J.K., Winfree, N.A., Ahrens, T.J. 1449

Comet Shoemaker-Levy-9 impact on Jupiter
Ahrens, T.J., Takata, T., O'Keefe, J.D. 1455

Impact jetting of geological materials
Yang, W., Ahrens, T.J. 1461

Numerical computation of jetting impacts
Puckett, E.G., Miller, G.H. 1467

Compositional changes of the Moon and Earth by impacts
Miura, Y., Okamoto, M. 1473

Technology and Diagnostics

Experimental investigation of tunnel sonic-boom suppression
Sasoh, A., Onodera, O., Takayama, K., Kaneko, R., Matsui, Y. 1481

Discarding the tuned-piston conditions for Stalker tubes: possible gains and drawbacks
Dumitrescu, M.-P. 1487

Compact entropy-raising driver for impulse facilities
Kendall, M.A., Morgan, R.G., Jacobs, P.A. 1491

Temperature distribution in free-piston compressors
Griffin, D.K, Stalker, R.J., Page, N.W. 1497

Soft-landing operation for free-piston driver by adjusting piston-stopper height
Tanno, H., Itoh, K., Komuro, T., Sato, K. 1503

Preliminary experimental results of the detonation-driven shock tube
Yu, H.-R., Zhao, W., Li, Z.-F., Gu, J.-H., Zhang, X.-Y. 1509

Operational characteristics of a free-piston shock tunnel with a light-gas gun as the shock tube
Gou, G.X., Gao, D.Z. 1515

Characteristics of shock waves in exhaust systems and exhaust gas flows of automobile engines
Sekine, N., Onodera, O., Takayama, K. 1521

Influence of the curvature of the diaphragm on flow quality in shock tubes
Daru, V., Damion, J.P. 1527

Modifications to the AEDC shock-tunnel code and comparisons with data
Maus, J.R., McKenzie, N.R. 1533

Numerical simulation of diaphragm rupture
Petrie, P.J., Jacobs, P.A. 1539

Starting-phase numerical simulation of the TCM2 wind-tunnel nozzle
Cardoso, M., Burtschell, Y., Druguet, M.C., Zeitoun, D., Abgrall, R. 1545

Combustion driver with interface compression and computer code for hypersonic shock tunnel to produce stagnation temperatures to 20,000 K and Mach numbers of 35
Nagamatsu, H.T., Messitt, D.G., Myrabo, L.N., Minucci, M.A.S. 1551

A simple shock-tunnel driver-gas detector
Paull, A. 1557

Boundary-layer blockage in expansion-tube nozzles
Sudnitsin, O., Morgan, R.G. 1563

Throat design for hypersonic nozzles
Dumitrescu, L.Z., Burtschell, Y., Zeitoun, D. 1569

Latest calibration results of the HEG contoured nozzle
Hannemann, K., Krek, R., Eitelberg, G. 1575

Trapping the boundary layer: a method to diminish flow contamination in shock tunnels
Dumitrescu, M.-P. 1581

High-enthalpy nozzle flow sensitivity study and effects on heat transfer
Walpot, L.M.G., Muylaert, J., Bakker, P.G., Simeonides, G. 1587

Wave processes in a pressure-wave refrigerator
Galyukov, A., Voinovich, P., Timofeev, E. 1593

Quantitative holographic interferometry of shock-wave flows using Fourier transform fringe analysis
Babinsky, H., Takayama, K. 1599

Single-pulse spontaneous Raman-scattering measurements in an expanding nitrogen/oxygen admixture
Gilmore, J.O., Sharma, S., Fletcher, D., Bershader, D. 1605

Laser-induced thermal acoustics (LITA) measurement of gas properties
Cummings, E.B., Hornung, H.G., Brown, M.S. 1611

PLIF imaging of nitric oxide in high enthalpy flows
Palma, P.C., McIntyre, T.J., Houwing, A.F.P. 1617

A skin-friction gauge for impulsive facilities
Goyne, C.P., Paull, A., Stalker, R.J. 1623

Imaging and quantitative measurement of transverse injection in supersonic flow using LIF
Shida, H., Nagayama, T., Deguchi, Y. 1629

On the role of laser absorption in planar laser-induced fluorescence imaging of high-enthalpy flows
Ruyten, W.M., Smith, M.S., Williams, W.D. 1635

High-speed image processing of interferograms in experiments on weak shock waves
Kurokui, Y., Aoyama, T., Watanabe, N., Maeno, K., Honma, H. 1641

Medical Applications

Experimental and mathematical model of living-tissue cell damage induced by plane shock waves
Tamagawa, M., Akamatsu, T., Yoshida, K. 1649

Application of extracorporeal shock wave on bone
Ikeda, K., Nakatani, A., Tomita, K., Takayama, K. 1655

Author Index 1661

Keyword Index 1669

Volume I

Shock-tube kinetics in homogeneous and heterogeneous reaction systems

P. Roth
Institut für Verbrennung und Gasdynamik
Gerhard-Mercator-Universität Duisburg
D-47048 Duisburg, Germany

Abstract: The shock tube used as a high-temperature wave reactor has dominated high-temperature kinetics for more than 45 years. The nearly instantaneous heating to high temperatures, the accessible wide temperature and pressure ranges, and the diffusion-free reaction conditions are the main advantages of this technique for measuring rate coefficients at high temperatures. In this paper some applications of the shock-tube technique for kinetic studies in homogeneous and heterogeneous reaction systems will be discussed. The examples to be presented were obtained in the author's laboratory. They include thermally and photolytically induced chemical reactions, which were studied by applying different optical absorption techniques.

Key words: Shock-tube kinetics

1. Introduction

The modern applications of shock-tube technology for kinetic studies are characterized by two interrelated facts: very high dilution of the reactants in an inert bath gas (mostly argon) and high sensitivity of the optical diagnostics used. The big advantage of highly diluted reactants is that the endothermicity or exothermicity of the chemical reactions has no effect on the structure and stability of the wave. Furthermore, the number of elementary reactions, which determine progress in the shock-induced reaction system, is limited, thus facilitating the kinetic interpretation of measured properties. On the other hand, access to intermediate chemical species by the use of modern optical diagnostics (see Hanson 1995) provides an insight into details of chemical reactions which are highly influenced by radical species.

Compared to research on homogeneous chemical reactions in shock waves, which was strongly stimulated by the Göttingen researchers (see Wagner 1971), research on heterogeneous kinetics in shock waves is at an early stage. The reasons are that both the degree of reaction complexity and the difficulties in the diagnostics are significantly higher. A precondition for this type of experiment is that the solid phase must be homogeneously distributed in the gas phase. The kinetic problems interfere to a high degree with the two-phase flow difficulties. Detailed access to the adsorption-desorption surface reactions is not possible in shock-tube experiments so that rate parameters obtained are limited to global expressions. On the diagnostics side, sensitive optical methods for gas-phase properties like laser schlieren or laser absorption are significantly disturbed by the dispersed particles.

In the present paper, some examples of different types of shock-wave-induced chemical reactions all obtained in the author's laboratory will be illustrated: simple thermal decomposition reactions, thermally or photolytically initiated radical reactions, and reactions in dispersed systems (aerosols). In all cases, optical absorption ranging from vacuum ultra-violet (VUV) to the infrared region was used, together with the Lambert-Beer Law, to measure time-dependent concentrations of reactive species.

2. Thermal decomposition reactions

The classic application of shock-tube technology, which reached a certain state of completeness many years ago, is to study the above type of reaction by observing the relaxation towards chemical equilibrium after shock heating of the test gas. The decomposition of molecules A can in those cases be described by the Lindemann mechanism, which considers a collisional activation or deactivation followed by the dissociation:

$$\mathrm{A} + \mathrm{M} \underset{2}{\overset{1}{\rightleftharpoons}} \mathrm{A}^* + \mathrm{M}$$
$$\mathrm{A}^* \overset{3}{\rightarrow} \text{products}$$

M is the collision partner. Besides temperature, the progress in the disappearance of A depends on pressure. Two limiting cases are possible: at low pressure the collisional deactivation of A^* is much rarer than its chemical reaction and the whole reaction obeys a second-order rate law (low-pressure limit). At high pressure, the rate of disappearance of A becomes independent of the concentration of the collision partner M and obeys a first-order rate law (high-pressure limit). The reality of the above type of reaction is normally much more complicated. The energy required for dissociation is transferred to the molecule by a large number of collisions which can have different efficiencies. Also the internal dynamics of the energy transfer is of importance, i.e., the coupling between vibrational relaxation and dissociation can play a significant role resulting in an apparent delayed dissociation.

The thermal decomposition of diatomic molecules proceeds for most applications under conditions close to the low-pressure limit. A good example is the dissociation of N_2. Because of its importance in hypersonic flows, its rate coefficient was determined in numerous studies, all following the disappearance rate of N_2. For this and other types of decomposition reactions it seems useful to measure the rate of appearance of the atomic dissociation product and to determine in this way the dissociation rate coefficient of the mother molecule. In the case of N_2, the low-pressure decomposition is simply:

$$N_2 + M \xrightarrow{4} N + N + M$$

It is not an easy task to follow N-atom formation under shock-tube conditions. Since the middle of the 1970s atomic resonance absorption spectroscopy (ARAS) for measuring H- and O-atom concentrations has been made available for shock-tube research (see Just 1981), which was later extended to other species like N, C, Si, and S atoms. The diagnostic method is basically a spectroscopic line-emission line-absorption technique. The respective atomic species have strong resonance absorption lines in the vacuum UV, which makes it possible to measure their ground-state concentration with very high sensitivity.

A schematic diagram of the experimental arrangement employed for ARAS is shown in Fig. 1 together with other devices, which will be of interest later on. Microwave power is coupled into a low-pressure flow of a He/N_2 mixture. The

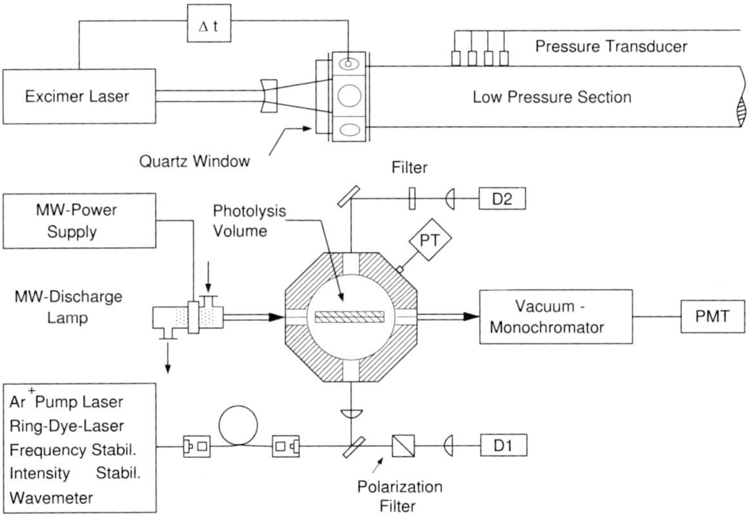

Figure 1. Experimental arrangement for measurements in a shock tube: atomic resonance absorption spectroscopy (ARAS), ring-dye-laser spectroscopy (RDLS), and laser photolysis (LP).

high-temperature plasma formed contains the NI triplet radiation at 119.9 nm, which is single-passed through the shock tube about 10 mm from the end wall.

A one-meter VUV monochromator separates the strongest of the N-atom lines from the radiation of the plasma lamp. N atoms formed in the shock tube during high-temperature thermal decomposition absorb the resonance radiation very effectively. The absorption measured by a solar-blind photomultiplier is related to the N-atom concentration. For a detailed evaluation of the signal calibration is needed, see Thielen and Roth (1986).

An example of a data trace for N absorption/concentration in a reflected shock experiment in a mixture of 200 ppm N_2/Ar at 5500 K and 1.43 bar is shown in Fig. 2 (left). The arrival time of the reflected shock is marked. After

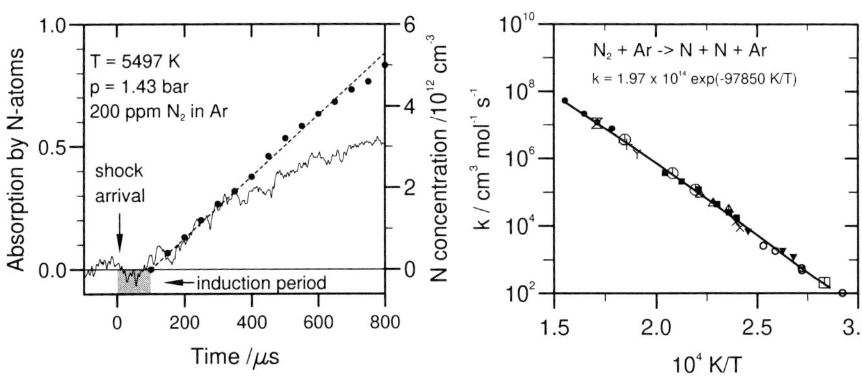

Figure 2. N-atom ARAS signal and corresponding concentration during thermal decomposition of N_2 (left). Rate coefficient of N_2 dissociation (right).

a certain delay the N-atom concentration increases linearly with time. From the N-atom slope the rate coefficient for the N_2 dissociation can be determined. Results of a series of shock-wave experiments in N_2/Ar mixtures are summarized in Arrhenius form in Fig. 2 (right). Because of the high sensitivity of the optical diagnostics, the low temperature end is very much extended compared to other experimental results, the data points are spread over several orders of magnitude, and the rms deviation is quite small.

The thermal decomposition of triatomic molecules like H_2O, COS, or N_2O is more complicated. They can dissociate immediately out of the ground state or can dissociate in a spin-forbidden way from an electronically excited state, which requires a singlet-triplet transition. Decomposition reactions of molecules with four or more atoms are more and more influenced by secondary and consecutive reactions, which must — even under high dilution conditions — be described with kinetic mechanisms.

3. Precursor-initiated radical reactions

The second type of elementary reaction is motivated by the strong interest in obtaining rate-coefficient data of radical–molecule or radical–radical reactions. A chemical precursor undergoing a very fast "priming" reaction after shock heating seeds the test gas with reactive intermediate species at known concentrations. The reaction of interest begins to proceed at a measurable rate only after the radical formation process is nearly finished. This idea of using the shock tube as a chemical reactor goes back to Gardiner et al. (1971). It is clear that the radical formation reaction must be fast compared to the reaction of interest. Through the years, many precursors were tested in our and in other laboratories. The application and successful use of the precursors depend on the detailed reaction conditions.

A good example for a precursor-initiated atom reaction, which also illustrates again the high sensitivity of ARAS, is the reaction of Si atoms with CO.

$$\mathrm{Si} + \mathrm{CO} \xrightarrow{7} \mathrm{SiO} + \mathrm{C} \qquad \Delta^{\mathrm{R}}\mathrm{h} = 276.1\ \mathrm{kJ/mol}$$

This reaction was studied in mixtures containing 0.75 to 20 ppm SiH_4 and 0.2 to 5% CO diluted in Ar at temperatures 2720 K$\leq T \leq$ 5190 K using the Si-atom ARAS as given in Fig. 1. An individual example is represented in Fig. 3 (left)

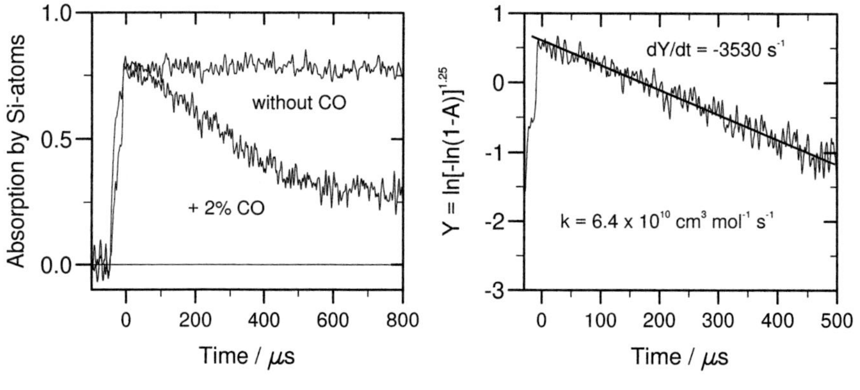

Figure 3. Si-atom absorption measured in mixtures of Ar and 1 ppm SiH_4 without and with CO addition, T = 3660 K, p = 0.8 bar (left). Evaluation of the absorption signal by the first-order method (right).

showing Si-atom absorption measurements in mixtures without and with CO. Immediately behind the reflected shock wave the precursor SiH_4 decomposes nearly instantaneously to form Si atoms (and H_2) at a constant level. In the mixture with CO, the silicon atoms in turn react as given in the above equation, resulting in a decrease in the Si absorption.

The kinetic evaluation of the above type of absorption signal is facilitated by the fact that the disappearance of Si proceeds under first-order conditions, because the CO concentration is practically constant.

$$d\ln[\mathrm{Si}]/dt = -\mathrm{k}_7[\mathrm{CO}]_0$$

The silicon atom concentration can, via the modified Lambert-Beer Law with a concentration exponent of n = 0.8, be directly related to the measured absorption. The above equation rewritten yields:

$$\frac{d\ln\{-\ln(1-\mathrm{A})^{1/n}\}}{dt} \equiv \frac{d\mathrm{Y}}{dt} = -\mathrm{k}_7[\mathrm{CO}]_0$$

with A being the measured fractional Si absorption. The diagram on the right-hand side of Fig. 3 shows the property Y as a function of the reaction time. The linear behavior of Y confirms the first-order assumption of reaction 7 and allows determination of the rate coefficient from the slope. A summary of all rate coefficients obtained is shown in the Arrhenius diagram of Fig. 4. It also

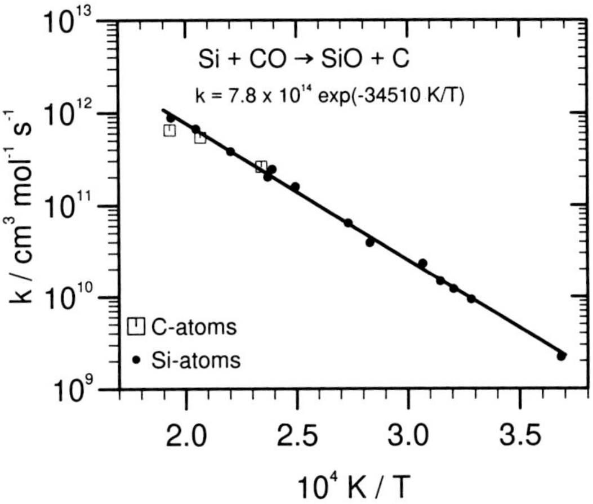

Figure 4. Rate coefficient of the reaction Si + CO ⟶ SiO + C.

includes results of C-atom measurements. The standard deviation of the data due to experimental scatter is, at ±10%, very low. Rate coefficients of several silicon reactions obtained in the author's laboratory are summarized in Table 1.

Table 1. Rate coefficients of various Si-containing reactions obtained by ARAS; $k_i = A_i \exp(-C_i/T)$ cm^3 mol^{-1}s^{-1}.

REACTION			A_i	C_i
Si + CO	$\longrightarrow$	SiO + C	7.8×10^{14}	34510
Si + CO_2	$\longrightarrow$	SiO + CO	6.0×10^{14}	9420
Si + NO	$\longrightarrow$	SiO + N	3.2×10^{13}	1775
Si + O_2	$\longrightarrow$	products	2.7×10^{14}	1765
Si + N_2O	$\longrightarrow$	SiN + NO	5.0×10^{14}	8100
	$\longrightarrow$	SiO + N_2	8.0×10^{13}	—

4. Photolytically initiated radical reactions

The method of generating radical species by photolysis and following their further reaction is an old and well-established technique in chemical kinetics, mostly used in room-temperature static cells. In the shock-wave community, flash lamps were first used to photolyze molecules in shock-heated gases. With the availability of high-intensity pulsed-UV lasers, which have a higher spectral intensity and a greater monochromaticity compared to flashlamps, these light sources were also used in shock-tube experiments, see e.g. Hanson (1995), Yoshimura et al. (1992). Compared to the chemically formed radicals, described in the previous section, pulsed-laser photolysis is an effective and instantaneous source allowing direct preparation of radical species widely independent of the properties of the shocked gases. An extension to low temperatures is therefore unproblematic. A disadvantage is that the energy transfer or the absorption coefficient during photolysis is not known in detail, rendering prediction of the identity and the energy states of the photofragments produced more difficult.

To illustrate a photolytically initiated radical reaction behind the reflected shock wave, I have chosen the reaction:

$$\mathrm{S} + \mathrm{H_2} \xrightarrow{9} \mathrm{HS} + \mathrm{H} \qquad \Delta^R h = 80.3 \text{ kJ/mol}$$

for which no experimental high-temperature data are available. Sulfur atoms were photolytically produced by an excimer-laser pulse at 193 nm in mixtures of 25 or 30 ppm CS_2 and 1000 to 4000 ppm H_2 diluted in Ar. The laser was coupled into the measurement plane of the shock tube through an end plate made of quartz glass, see Fig. 1. The originally rectangular beam was expanded by a cylindrical lens allowing the illumination of the whole ARAS diagnostic path; for more details see Woiki and Roth (1993). The temperature range of the present experiments was 1260 to 1840 K. The progress in the chemical reaction studied was followed by applying S(^{3}P) ARAS at 147.4 nm.

The absorption profile of a typical experiment is shown on the left-hand side of Fig. 5. The arrival of the reflected shock, which is associated with a

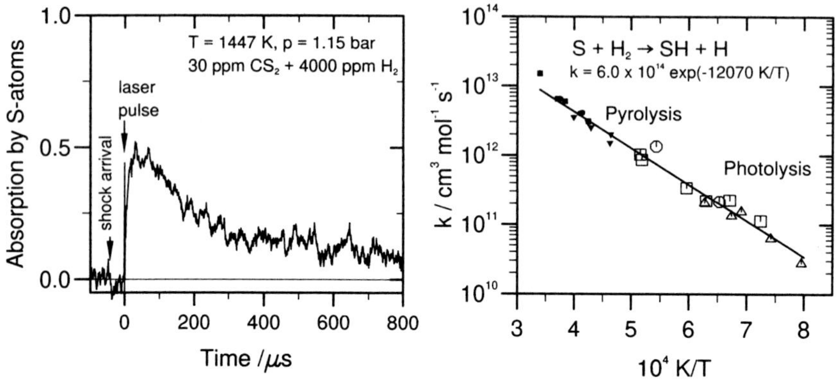

Figure 5. S-atom ARAS signal after photolysis of a shock-heated $CS_2/H_2/Ar$ mixture (left). Rate coefficient of the reaction $S+H_2 \longrightarrow HS + H$ (right).

temperature discontinuity does not initiate any measurable S-atom formation in the mixture. Only the laser photolysis pulse of about 13-ns duration, which was delayed with respect to the reflected wave by about 50 μs, causes a quasi-step-like peak in S-atom absorption. Subsequently the signal decreases due to S consumption by the above reaction with H_2. Data reduction could again be achieved by assuming first-order decay of S, as illustrated in the last section. Results of all the experiments are summarized in the Arrhenius diagram of Fig. 5, right-hand side. The data points can be approximated by the Arrhenius expression

$$k_9 = 6.0 \times 10^{14} \exp(-12070 \text{ K}/T) \text{ cm}^3\text{mol}^{-1}\text{s}^{-1}$$

In this diagram also the results of another series of experiments are included, where the S atoms were chemically formed by pyrolysis of COS. Both data sets agree quite nicely and illustrate the high potential of photolysis experiments at low temperatures, and the good applicability of pyrolysis-generated reactive species at high-temperature reaction conditions.

5. Thermally initiated reactions in dispersed systems

The reactions of solid particles homogeneously dispersed in a gas mixture at high temperature is of interest in many technical applications, for example, soot oxidation, pulverized fuel combustion, or SO_2 reduction by limestone particles. The degree of complexity of such heterogeneous reactions is much higher than in the case of gas-phase reactions, and the availability of kinetic data, even for global processes, is poor. The idea of applying the shock-tube technique to reactive aerosol systems seemed to suggest itself, because a dilute suspension of particles behaves essentially as a gas. The initially unreacted mixture can be heated by a shock wave thus initiating the post-shock chemical reactions, which must be recorded by optical methods. But there are some difficulties compared to experiments with gas mixtures:

- the preparation of the gas/particle mixture (aerosol) and its homogeneous distribution along the low-pressure part of the shock tube,
- the difficulties to be expected during shock-wave heating of the two-phase system,
- the necessity to measure the reactive surface of the aerosol system,
- the application of an optical diagnostic for gas-phase species in a particle-loaded system.

The first problem was solved by developing an expansion-wave-driven aerosol generator allowing dispersion of small amounts of powders, see Roth and von Gersum (1993). The second difficulty can be overcome by choosing sub-μm particles having low overall particle load. In this case the effect of particles on the fluid-flow properties is very weak, i.e. gas temperature and pressure increase quasi-instantaneously behind a shock wave to nearly constant levels. Because

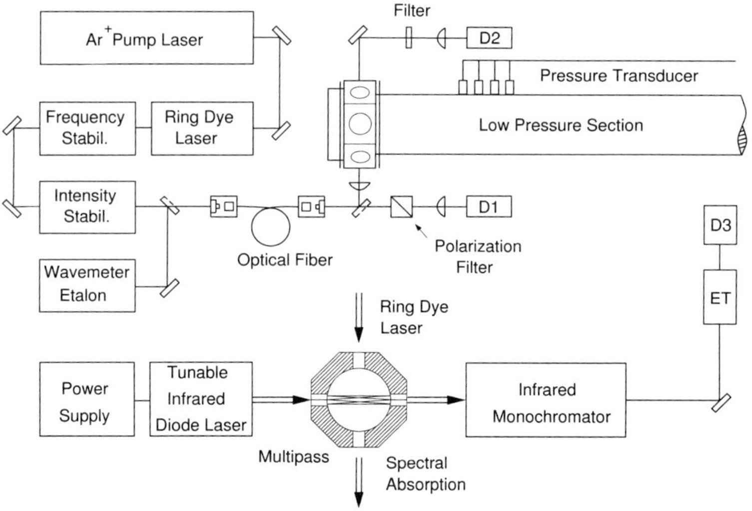

Figure 6. Experimental arrangement for measurement in a shock tube: ring-dye-laser spectroscopy (RDLS) and infrared-diode-laser spectroscopy (IRDS).

of the small size of the particles, the slip-flow region behind the shock wave, in which the heat and momentum transfer from the gas to the particle phase occurs, is very small compared to the extension of the heterogeneous reaction zone, i.e. the suspended particles are also nearly instantaneously heated up behind the shock wave to a constant level. If gas-phase reactants are either initially available in the mixture or chemically formed during a fast homogeneous rate process, their reactions with the hot solid particle surface can proceed. The rate of a heterogeneous reaction depends on, besides other surface properties, the size or the reacting area, i.e. the particles must be characterized with respect to these properties. This can be done by applying laser-light scattering or laser-light extinction at different wavelengths, which will not be discussed here; for details see Brandt and Roth (1989).

The remaining problem of the above list is to determine the progress of the heterogeneous reaction by measuring the formation of gas-phase products or the reduction of gas-phase educts. A well-suited technique is tunable IR-diode laser absorption spectroscopy, because interference by the particles is small, as light extinction by small particles is weak in the infrared region. The optical arrangement used is shown in Fig. 6. It consists of a pulsed, tunable IR-diode laser, the shock tube arranged as a multipass optical cell, a 0.5-m infrared monochromator for mode filtering, and an InSb detection system. The laser is pulsed with a frequency of 25 kHz and can additionally be tuned in each pulse over one or two lines thus producing a series of highly resolved absorption lines, from which time-dependent species concentration can be determined, see Brandt and Roth (1989).

An example of a thermally initiated reaction in a dispersed system is the oxidation of soot particles by O atoms. The expected oxidation products are CO and CO_2, but the only clearly detectable product in the time scale of the present experiments was CO, i.e. the oxidation proceeds in the present case mainly via the global heterogeneous reaction:

$$\mathrm{C_s} + \mathrm{O} \longrightarrow \mathrm{CO}$$

C_s represents the solid carbon in the soot particles. The O atoms were generated by the fast thermal decomposition of N_2O that was present in the initial gas/particle mixture. According to the above reaction, the rate of formation of CO is:

$$\frac{d[\mathrm{CO}]}{dt} = \alpha_0 Z_0 a_p \qquad Z_0 = \frac{1}{4}\,\overline{c_0}\,[\mathrm{O}]$$

In this equation α_0, Z_0, and a_p are the reaction probability of O atoms with the soot surface, the collision number of O atoms per time and unit area, and the reacting particle surface per unit volume of suspension, respectively. The collision number Z_0 can be calculated from the Hertz-Knudsen equation with $\overline{c_0}$ being the mean thermal velocity of O atoms. The desirable kinetic property is the reaction probability α_0, which can for known [O] be determined from measured properties [CO] and a_p.

A typical example of an individual shock-tube experiment in soot/N_2O/Ar mixture at 2490 K is shown in Fig. 7. The laser was tuned in a wavelength range close to the (0,1) P8 line of CO, the pulse frequency was about 25 kHz. The arrival of the shock front is marked. In the lower left part of Fig. 7 selected scans of the same experiment are shown in more detail with an enlarged time scale. The strong absorption lines in scan A were caused by spectral absorption of N_2O under pre-shock conditions. Beginning with the first post-shock scans, sharp spectral lines of CO appear showing increasing absorption. These illustrate the formation of an increasing amount of CO during the heterogeneous soot oxidation.

The reduction of the spectroscopic data in terms of CO concentration can be performed based on the Lambert-Beer Law. The necessary spectral-absorption coefficient depends on the line strength and the line-shape factor. The latter was determined by fitting a Voigt profile to the measured absorption line. Taking this into account, the sequence of absorption obtained can be transformed into a time-dependent CO concentration, as illustrated on the right-hand side of Fig. 7. From the initial slope together with the above equation or from computer simulation of this and other oxidation experiments, reaction probabilities α_0 were determined. A summary is given in the Arrhenius diagram of Fig. 8,

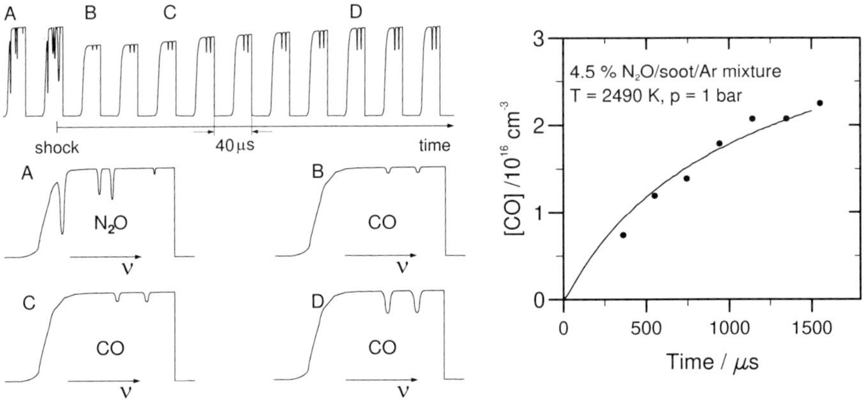

Figure 7. Example of pulsed-IR-diode-laser scans in a shock heated soot/N_2O/Ar mixture showing increasing CO spectral absorptions (left); corresponding CO concentration (right).

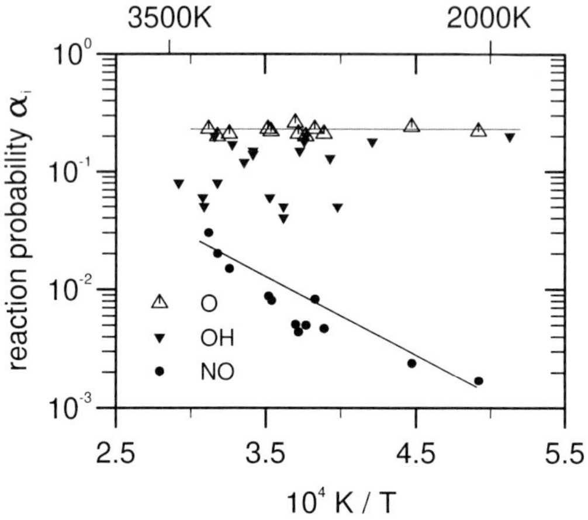

Figure 8. Reaction probabilities of soot oxidation by O, OH, and NO obtained from shock-tube experiments.

which also contains results of soot oxidation by OH and NO. The α_0 values obtained are nearly independent of temperature and can be approximated by a mean value of $\alpha_0 = 0.23$. For more details see Roth and von Gersum (1993).

6. Conclusion

Modern applications of shock-tube technology for kinetic studies in both homogeneous and heterogeneous reaction systems were illustrated in several examples. It could be shown that the trends are in the direction of studying the rate processes in highly diluted systems, partly as low as 1 ppm of reactant. The availability of highly sensitive optical, mostly laser-based diagnostics make such experiments possible. Care is required concerning the influence of impurities in the test gases as well as in the shock tubes, which are in our case constructed and used as ultra-high vacuum apparatuses. A problem can also be wall adsorption of strongly polar reactants in the highly diluted systems. Nevertheless the high-temperature wave reactor will also in the future be of great importance in high-temperature kinetics.

Acknowledgement. The financial support of the German Science Foundation is gratefully acknowledged.

References

Brandt O, Roth P (1989) A tunable IR-diode laser technique for measuring reaction rates of high-temperature aerosols. Comb and Flame 77:69.

Gardiner WC, McFarland M, Morinaga K, Takeyama T, Walkes BF (1971) The shock tube as a chemical reactor. In: Proceedings of the 8th Int. Shock Tube Symposium, Chapman and Hall, London, p. 23/1.

Hanson RK (1995) The rate of lasers in shock tube studies of chemical kinetics. In: Proceedings of the 19th Int. Shock Wave Symposium, R Brun and LZ Dumitrescu (eds), Springer-Verlag.

Just T (1981) Atomic resonance absorption spectrometry in shock waves. In: Shock Waves in Chemistry, A Lifshitz (ed), Marcel Dekker Inc., New York, p. 279.

Roth P, von Gersum S (1993) High temperature oxidation of soot particles by O, OH, and NO. In: Turbulence and Molecular Processes in Combustion, T Takeno (ed), Elsevier, Amsterdam, p. 149

Thielen K, Roth P (1986) N atom measurements in high-temperature N_2 dissociation kinetics. AIAA Journal 24:1102–1105.

Wagner HG (1971) Chemical reaction of shock waves. In: Proceedings of the 8th Int. Shock Tube Symp., Chapman and Hall, London, p. 4/1.

Woiki D, Roth P (1993) ARAS measurments on $S(^1D)$ formation in shock heated COS/Ar mixtures. Ber Bunsenges Phys Chem 97:996.

Yoshimura M, Koshi M, Matsui H, Kamia K, Hideaki U (1992) Non-arrhenius temperature dependance of the rate constant for the reaction $H+H_2S$. Chem Phys Lett 189:195–204.

Otto Laporte Lecture

120 years of shock-wave visualization — a fascinating activity

H. Reichenbach
Fraunhofer-Institut für Kurzzeitdynamik
Ernst-Mach Instach Institut
Freiburg, Germany

Abstract: After a brief chronology of Otto Laporte's life and a few personal remembrances, contributions are described of some pioneers in shock-wave visualization. August Toepler (1836–1912) observed shock waves created by electric sparks only with his eye. Ernst Mach (1838–1916), philosopher and originator of instationary gasdynamics and shock-wave research, found the irregular shock-wave reflection and invented various high-speed diagnostic methods. Cranz (1858–1946) and Schardin (1902–1965) contributed significantly to the development of shock visualization. From countless applications only a few are picked out to demonstrate the fascinating development to the state of the art nowadays.

Key words: Shock waves, Visualization, Historical report

1. Introduction

Assuming that not all participants know exactly who Otto Laporte was, in Table 1 a brief chronology is presented, quoted from a bibliographical memoir (Crane and Dennison 1979), where also many details of his life[1] are given. Otto Laporte is well known as an excellent spectroscopist. From his youth he was interested in this subject, starting when his father gave him a prism of an old periscope. Otto Laporte spoke German and English fluently, and he was also perfect in French and Japanese. Now and then he liked to apply these languages in the same sentence. He managed perfectly to mix in such a manner that nobody had any difficulty in understanding him.

With such a talent for languages, it is not surprising that Otto Laporte was given the task of interviewing German scientists concerning their research work during the Second World War. In this way he met Hubert Schardin who was Director of the Ballistics Institute in Berlin until 1945.

During the interview with Schardin, Otto Laporte did not act like an inquisitor — he was too much a scientist for that. On the contrary, he spoke passionately about shock waves and the manifold applications of shock tubes.

[1]I gratefully acknowledge the help of Prof. Y. W. Kim, Lehigh University, Bethlehem PA, who made the Otto Laporte bibliographical memoir available to me.

Table 1. Chronology: Otto Laporte.

1902	Born July 23, Mainz, Germany. Parents: Wilhelm Laporte, heavy artillery officer in the Imperial German Army; Anna Geyl
1923	Analyzed iron and vanadium spectra. Enunciated Laporte rule
1924	Ph.D. University of Munich, under Arnold Sommerfeld. Thesis: "Exact Treatment of Scattering of EM Waves by a Sphere"
1924-26	International Education Fellow; studied in Europe, Japan, and finally in the USA in the spectroscopy section of the National Bureau of Standards
1927	Assistant Professor
1935	Became naturalized US citizen
1944	Changed field of activity from atomic spectroscopy to fluid mechanics
1945	Promoted to full professor
1946	Assumed charge of shock-tube laboratory at University of Michigan
1949-50	Scientific Intelligence Analyst, US Army of Occupation, Heidelberg, Germany
1951	Initiated research on reflected shock waves
1954-55	Scientific attaché at US Embassy in Tokyo
1956	Cited by US State Department as instrumental in securing atomic-energy agreement between USA and Japan
1965	Chairman, Division of Fluid Dynamics, American Physical Society
1971	Died March 28, Ann Arbor, Michigan
1971	Elected, posthumously, to the National Academy of Sciences

As ballistician, Schardin was of course very well acquainted with shock waves and high-speed measurement techniques. Remember, together with his academic teacher and mentor, he developed the famous Cranz-Schardin Camera in 1928. Of course, Schardin immediately realized the advantage of combining a shock tube with a Cranz-Schardin Camera for shock wave research. Even today, nearly 50 years later, this combination is still an important tool in this field.

As shock-wave research flourished in Germany, France, and other European countries after the Second World War, Otto Laporte was one of those who gave a significant impetus to the development of shock-wave visualization.

If I dare now to reflect on the development of the last 120 or more years, I must restrict myself to a small, modest, and subjective selection. I apologize ahead of time if I cannot mention all important contributions in this field.

2. First visualizations of shock waves

Although Ernst Mach is now recognized as the pioneer and originator of instationary gasdynamics and shock-wave research, there was another scientist before him who carefully investigated shock-wave phenomena from spark discharges. Since no sensitive photographic plates were available at that time, he observed the phenomena in a *subjective* manner, namely, with his eye by means of a telescope. The name of this great scientist was August Toepler[2].

[2]P. Krehl from Ernst-Mach Institut, Freiburg, has investigated intensively the literature on Toepler's publications (see Krehl and Engemann 1995). I gratefully acknowledge Krehl's

2.1. August Toepler (1836–1912)

Toepler is best known as the originator of the schlieren method. But his contributions to shock-wave phenomena and other fields of physics and chemistry have fallen into oblivion. His pioneering work was at first not noticed by experts. He was forced to print it at his own expense in 1864. Besides the spherical pressure waves that he first called shock waves (Stoßwellen), he also investigated the reflection of shock waves.

Toepler was not only an excellent experimentalist but also a talented artist. The shock fronts he saw were afterwards drawn by hand using pen and ink. Figure 1 is a reproduction of a woodcut of such a drawing, published 1864. He found that the wave fronts reflected in such a way that it appeared as if a second source (c') was present and symmetrically orientated to the reflecting surface. He noticed that the reflected wave front met the incident front at the surface, but he also mentioned that in some tests the reflected wave seemed to *precede* the incident wave near the reflection point. What he saw was probably the pattern of a Mach reflection! Such a reflection was unimaginable for him, so he assumed that it must be an optical illusion. Toepler also studied shock refraction phenomena, and investigated gas combinations with sound velocities larger and smaller than air.

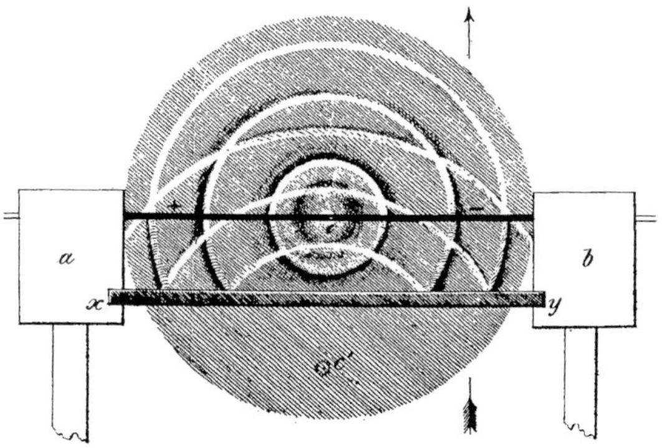

Figure 1. Reflection of spark-generated shock waves (woodcut, 1864).

Nowadays it sounds unbelievable that Toepler observed the instationary gasdynamic effects only with his eye. While he had unfortunately no appropriate method at his disposal for an *objective* registration, another scientist was much more successful in this respect.

help and information he gave me concerning Toepler's scientific contributions.

2.2. Ernst Mach (1838–1916)

Ernst Mach was an inventive scientist, not only skilled in experimental techniques, but also in theoretical interpretation of experimental observations. He invented various high-speed diagnostic methods. His life and his contributions to philosophy and physics are discussed in numerous publications (e.g., Merzkirch 1970; Blackmore 1972; Reichenbach 1983; Wolters 1987; Prosser and Folta 1988).

The discovery of an irregular reflection pattern at shock waves — today called Mach reflection — is certainly Mach's most important and spectacular contribution to fluid dynamics. Even today this phenomenon is not completely understood. The era of objective visualization started when E. Mach and J. Wosyka published in June 1875 (exactly 120 years ago) the result of their ingenious experiments. A spark discharge along a line, drawn with a conducting paint on a glass plate, was used as the source of a strong acoustic wave. The line

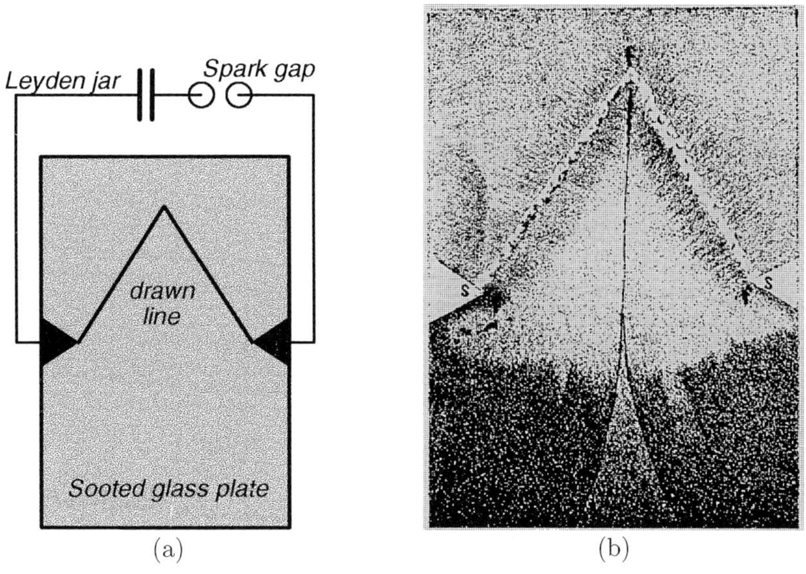

Figure 2. Mach's study of the V-phenomenon (1875): (a) test arrangement; (b) visualization of Mach-V by means of a sooted plate.

formed an angle as illustrated in Fig. 2a. A Leyden jar capacitor was charged by an hand-operated electrostatic machine until the spark gap fired, thus closing the circuit. At discharge, two cylindrical shocks start out from the angular line. To visualize the intersection Mach applied the soot-plate technique developed by Antolik (1875). The glass plate with the drawn circuit was sooted. The soot was partly blown away by the shock waves and revealed their intersection, with the so-called Mach V into which the line of intersection bifurcated. This can be seen in Fig. 2b which is a reproduction of an original woodcut of an exposed soot plate.

Mach assumed immediately that a new wave must be the reason for the bifurcation. To visualize this wave, he used in another ingenious experiment two line systems in opposite positions. If this new wave exists, the soot must remain at rest at the intersection line inside the Mach Vs. That is exactly what the experiment showed. This wave is called today Mach wave or Mach stem. Only a few years later in 1878, Mach formulated a criterion for Mach reflection (Mach 1878). This formula corresponds to what we call today: the "Mechanical Equilibrium Condition." It was Ernst Mach who clearly emphasized that the experiment must be the basis and the touchstone of theory in natural science.

3. Progress in experimental techniques

Serious shock-wave research at the end of the last century was just barely possible, because the required experimental techniques were not available. New precise tools were necessary if Mach's idea of the control or validation mechanisms of experiments should be translated into action.

3.1. Carl Cranz (1858–1946)

Carl Cranz was one of those who fully accepted Mach's concept. In 1903 he took the Chair of Ballistics of the Military-Technical Academy in Berlin; he received enough funding to make his Institute a premier institution for more than three decades.

Figure 3, taken before 1914 (Cranz et al. 1914), gives proof of the excellent quality of the high-speed pictures. It shows a projectile propagating inside a tube with slots. The bow wave can escape through the openings. It was shown that the bow wave is the envelope of Huygens wavelets.

But the most famous technical development of Cranz and his student Hubert Schardin was the invention of a multiple-spark camera in 1928, today known as Cranz Schardin Camera. In the classical paper "Cinematography on Fixed Film with Extremely High Picture Frequency" numerous applications were treated; picture frequencies as high as 3 million frames per second were achieved (Cranz and Schardin 1929).

3.2. Hubert Schardin (1902–1965)

The work of Cranz was continued by his successor and former student Hubert Schardin. Like Cranz, he was an excellent scientist, with an ability to grasp the principle underlying complicated events and develop a simple approach for their study. Schardin studied for instance the question whether rarefaction shocks exist. When an evacuated container implodes, a bang is heard. Is this caused by the expansion wave or by the pressure wave propagating after the implosion? Schardin's set-up was indeed a shock tube where the outer atmosphere represented the pressure chamber. The shock-wave studies, undertaken so hopefully at first, were not followed up. They did not find a receptive audience at the time, since people had not yet realized the numerous applications of a shock tube. One must remember that in those days ballistic problems seemed more

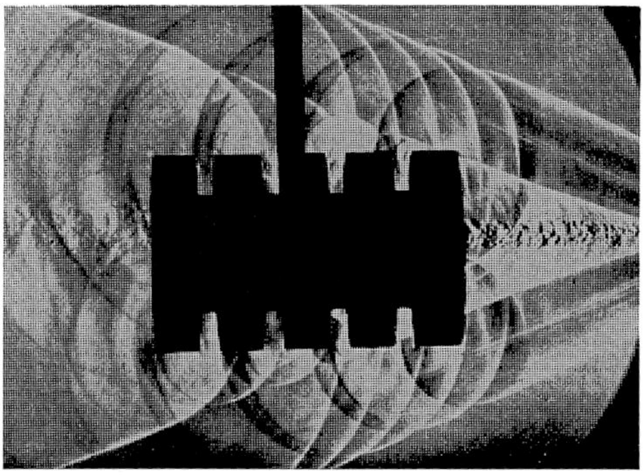

Figure 3. Projectile transversing a tube with slots (Cranz, before 1914).

important. In retrospect one could say that all the knowledge of visualization techniques was developed in Cranz's institute and available to Schardin in the thirties; only the idea of carrying out shock-wave experiments in a shock tube was lacking. As often happens in science, all information and methods are available, but a new impetus is necessary to create a new era.

4. Applications

Two military problems spawned rapidly increasing activities in shock-wave research: nuclear explosions and protection against their effects, and the forces and heat transfer acting on intercontinental missiles at reentry into the atmosphere. During and after the Second World War, experiments were performed to solve these problems and to understand the physics of the numerous new phenomena. As examples, I would like to mention Bleakney's group in Princeton and Glass's group in Toronto, and also Schardin in St. Louis (and later on in Freiburg). New visualization principles were developed, such as phase-contrast systems, differential interferometer, color-schlieren systems, and color interferometry (e.g. Holder and North 1952; Wolters 1956).

4.1. Reflection and diffraction

Let me present a few examples of reflection and diffraction. The wave pattern around a triangular prism is shown in Fig. 4. Most remarkable are the spiral vortices at the corners, Mach effects at the crossing diffracted shock fronts, rarefaction waves centered at the corners, and slip lines starting at the triple points.

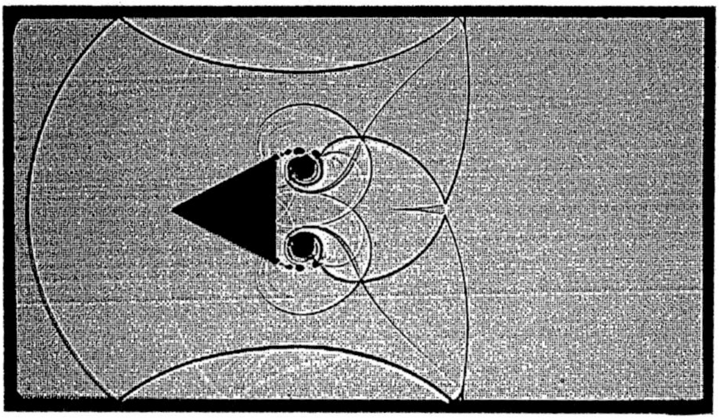

Figure 4. Diffraction around a triangular prism.

To understand the complicated wave interactions, a systematic study of simple cases was necessary, such as reflection of a plane shock front at oblique solid surfaces. Regular reflection and Mach reflection were expected. But at higher shock Mach numbers a new type of reflection appeared; the double Mach reflection, with two triple points and two slip lines. In the meantime various kinds of shock-wave reflection were isolated and transition criteria were formulated. Ben-Dor recently published a book dealing only with shock-wave reflection phenomena (Ben-Dor 1992). Is it not fascinating to find that there are so many solutions to such a simple problem as the reflection of a plane shock from a flat rigid surface? Even today it is not clear whether all possible solutions have been discovered (Ben-Dor and Takayama 1992).

Very early in the fifties, investigations were started on the diffraction at corners and edges. Around that time, color films became available and color-schlieren systems were developed. (Holder and North 1952; North 1954). Schardin noticed that color-schlieren photography is an ideal method for visualizing diffraction and vortex formation at corners. One of his first attempts is shown in Fig. 5. It is a good example to demonstrate the advantages of color photography. Color interferometry, applied by us since the sixties, is also useful for visualizing diffraction phenomena.

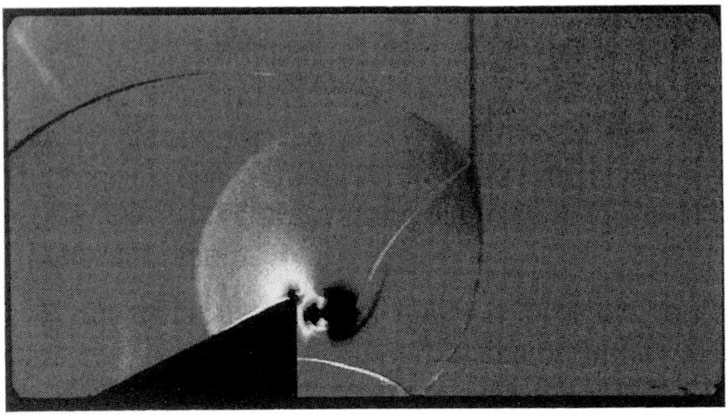

Figure 5. Color-schlieren photograph: diffraction at a wedge (Schardin, about 1953).

4.2. Influence of surface properties

Surface properties (e.g., roughness, porosity, and elasticity) can influence the propagation and reflection of shock waves. Extensive tests were performed to investigate the physical mechanisms. If a shock wave propagates along a rough surface, it is reflected and diffracted at each roughness center. To visualize this effect, a surface was chosen with periodic grooves perpendicular to the propagation direction of the shock wave. The result is given in Fig. 6 for a groove depth of 2.5 mm. The elementary waves emanating from each step (representing two-dimensional roughness centers) overlap each other, causing distinct bending of the original vertical front, starting where the first elementary wave hits the incident shock front.

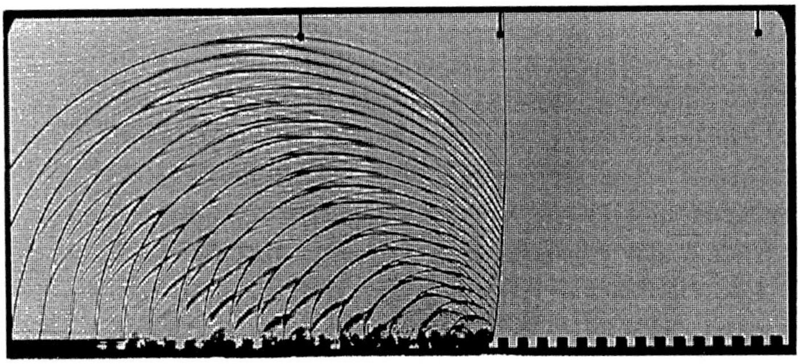

Figure 6. Shock propagation along a surface with periodic grooves; $M_S = 1.38$.

The same effect must also result from oblique reflections at a rough surface. This implies that, at the reflection point, the angle between shock front and the wedge surface does not exactly correspond to the wedge angle. The transition from regular to Mach reflection on rough surfaces must occur at smaller wedge angles than on a smooth surface. This is indeed observed. But no comprehensive theory has yet been found.

4.3. Precursors and decursors

Another field where further effects are involved is the shock-wave interaction with thermal layers, or layers with gases of different sound speeds. We have developed several methods to produce such layers in a shock tube: the most important ones are: (a) electric heating of the surface, (b) diffusion of a gas through a porous ceramic flat plate (Reichenbach and Kuhl 1987). The sound speed of helium for instance is much higher than that of air.

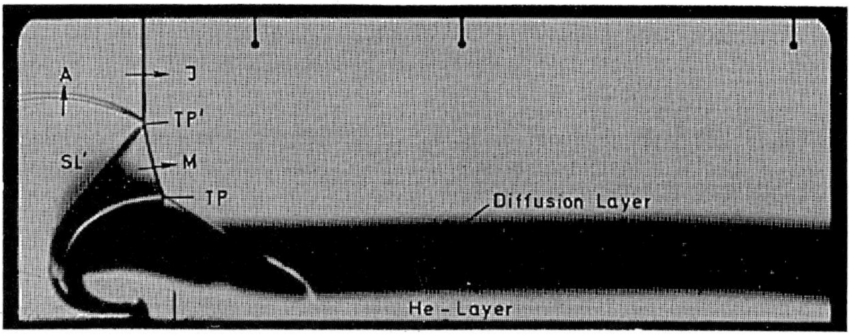

Figure 7. Precursor start-up period, $M_S = 1.77$.

If a shock wave interacts with such a layer a precursor develops. Figure 7 shows the start-up period of the interaction with a helium layer. Between the helium layer (white zone at the bottom) and the pure air in the upper part of the schlieren picture, a mixing layer is visible as a black zone. The start-up of the precursor proceeds through a transition phase to the final phase.

A completely different behavior occurs with a freon layer; a decursor develops. The sound speed of freon R12 is 150 m/s (at $T = 293$ K). Therefore a shock wave propagates more slowly in the Freon layer than in air, Fig. 8. This causes the shock front to be curved backward near the wall. The oblique shock compresses the layer and reflects from the floor as a regular reflection. The reflected shock reflects off the layer interface as a rarefaction wave which stops the compression of the layer. The rarefaction wave reflects back and forth within the layer, causing a series of expansion and compression waves that radiate from the layer. These waves strongly perturb the interface, which

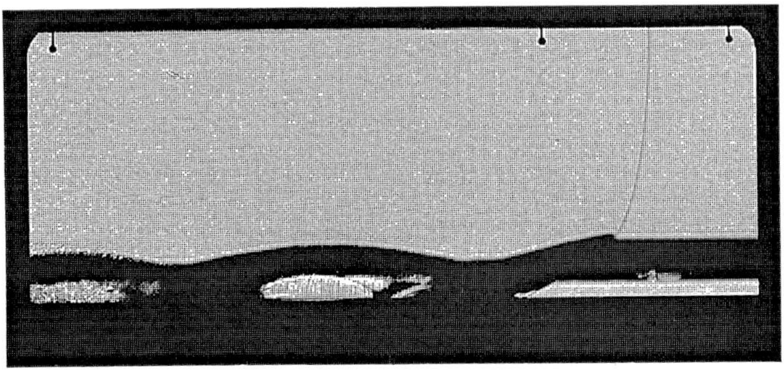

Figure 8. Decursor formation, $M_S = 1.29$.

acquires an undulated shape and transitions rapidly to a turbulent boundary-layer flow. Wave-dynamics effects damp out and mixing processes begin to dominate. Coherent structures and striations appear (Kuhl et al. 1992). We call this a baroclinic boundary layer.

4.4. Shock-induced turbulent flow

When visualization techniques developed into a practical tool, investigations of shock-induced flow and interactions with small-scale structures increased rapidly. As only one example I choose the propagation of a planar shock through an aligned baffle system. Figure 9a shows a single-frame color interferogram in the infinite-fringe mode of the same shock-induced flow. The undisturbed region of this aligned 4-baffle system appears in a uniform color. In this case, the fringes are lines of constant density (isopycnics).

(a) (b)

Figure 9. Shock wave passing a baffle system: (a) experimental color interferogram; (b) numerical interferogram.

The last figure demonstrates a classical visualization method. Numerous improvements have been made in the last 50 years, and new principles were developed like phase-contrast and differential-interferometry methods. The advent of the laser in 1960 provided a revolution in optics, and opened up applications such as holographic visualization techniques. There is not enough time to describe these methods. Professor Takayama from Sendai for instance and many others have presented excellent results again and again. I would now like to pass onto another type of visualization that is applied more and more frequently nowadays.

5. Numerical simulation

As a consequence of the Nuclear Atmospheric Test Ban Treaty of 1962, multidimensional hydrocodes have been developed for shock-wave research, thus allowing visualization by numerical simulation. With sophisticated dataprocessing programs, it is possible to present the results of numerically simulated shock-wave phenomena in form of shadow-, schlieren- or interferograms etc. A new era of visualization possibilities has started.

Figure 9b demonstrates an example. The shock-induced flow in the baffle system under the same conditions as for the previously presented experiments was calculated with the Adaptive Mesh Refinement code AMR (Bell et al. 1994; Berger and Colella 1989). At first glance, it is not easy to distinguish between the experimental and the numerical interferogram. However, in small details differences can be detected, especially in the vortex region where some instabilities seem to be more apparent in the calculated interferogram. Altogether, the agreement between experiment and simulation is excellent.

Numerical methods also provide information on quantities which cannot be measured or visualized directly, like entropy, internal energy, vorticity etc. In order to avoid misinterpretations, it becomes important to validate computer codes, under the assumption that a good agreement between calculated and measured quantities also implies that non-directly measurable data are reproduced in a trustworthy manner. Experiments will therefore never become unnecessary, since an experiment is essentially nothing but an analog computer.

6. Final remarks

It was the aim of my presentation to trace in rough outline the historic development of shock-wave visualization. It is fascinating to see what has resulted from the scientific work of a few pioneers of our research discipline. Shock-wave research leads often to surprising results, and visualized effects do not only help create a better physical understanding, but sometimes also fulfill artistic and aesthetic ambitions. Also, for the future we can certainly assume further significant progress in shock-wave research by sophisticated visualization methods.

References

Antolik K (1875) Das Gleiten elektrischer Funken. Pogg Ann Phys Chem 1:14.

Bell JB, Berger M, Saltzman J, Welcome M (1994) Three-dimensional adaptive mesh refinement for hyperbolic conservation laws. SIAM J Stat Comp 15(1):127.

Ben-Dor G (1992) Shock Wave Reflection Phenomena, Springer, Berlin, Heidelberg, New York.

Ben-Dor G, Takayama K (1992) The phenomena of shock wave reflection — a review of unsolved problems and future research needs. Shock Waves 2:211.

Berger MJ, Colella P (1989) Local adaptive refinement for shock hydrodynamics. J Comp Phys 82:64.

Blackmore JT (1972) Ernst Mach — His Work, Life and Influence, Univ Calif Press, Berkley, Los Angeles, London.

Crane HR, Dennison DM (1979) Otto Laporte 1902–1971; A Bibliographical Memoir, Bibliogr Memoirs, Vol 50, Nat Acad of Sciences, USA.

Cranz C, Günther PA, Külp F (1914) Photographische Aufnahmen von sehr rasch verlaufenden Vorgängen, insbesondere von Schußvorgängen mittels Vorderbelichtung durch das Licht elektrischer Funken. Z.f.d. Gesamte Schieß-und Sprengstoffwesen 4:42–59.

Cranz C, Schardin H (1929) Kinematographie auf ruhendem Film und mit extrem hoher Bildfrequenz. Z f Physik 56:147.

Holder DW, North RJ (1952) A schlieren apparatus giving an image in colour. Nature 15:466.

Krehl P, Engemann S (1995) August Toepler — the first who visualized shock waves. Shock Waves 5:1.

Kuhl AL, Reichenbach H, Ferguson RE (1992) Shock interaction with dense-gas wall layers. In: Proc 18th Int Symp on Shock Waves, pp 159-166.

Mach E, Wosyka J (1875) Über einige mechanischen Wirkungen des elektrischen Funkens. Sitzungsber Akad Wiss Wien 72:44.

Mach E (1878) Über den Verlauf der Funkenwellen in der Ebene und im Raum. Sitzungsber Akad Wiss Wien 77:819.

Merzkirch W (1970) Mach's contribution to the development of gasdynamics. Boston studies in philosophy and science 6:42.

North RJ (1954) A colour schlieren system using multi-colour filters of simple construction. NPL/Aero 266.

Prosser V, Folta J (1988) Ernst Mach and the development of physics. In: Int Conf Prague, Karolinum (ed), (conference paper), Prague 1991.

Reichenbach H (1969) Contribution of H. Schardin to the theory and applications of the shock tube. Phys Fluids, Shock Tube Symp 12(5):I3–I8.

Reichenbach H (1983) Contribution of Ernst Mach to fluid mechanics. Ann Rev Fluid Mech 1–28.

Reichenbach H, Kuhl AL (1987) Techniques for creating precursors in shock tubes. In: Proc 16th Int Symp on Shock Waves, pp 847–853.

Reichenbach H (1992a) In the footsteps of Ernst Mach — a historical review of shock wave research at the Ernst Mach Institut. Shock Waves 2:65.

Reichenbach H (1992b) Hubert Schardin (1902–1965) — his life and his work. Proc 20th Int Congr on High Speed Photogr and Photonics.

Schardin H (1932) Bemerkungen zum Druckausgleichsvorgang in einer Rohrleitung. Phys Z 2:60.

Schardin H (1959) Carl Cranz als Mitbegründer der Kurzzeitphotographie. Proc 4th Int Congr on High Speed Photogr and Cinematogr.

Toepler A (1864) Beobachtungen nach einer neuen Methode — Ein Beitrag zur Experimentalphysik, Max Cohen & Sohn, Bonn.

Wolters G (1987) Mach I, Mach II, Einstein und die Relativitätstheorie - Eine Fälschung und ihre Folgen, Walter de Gruyter, Berlin, New York.

Wolters H (1956) Schlieren-, Phasenkontrast- und Lichtschnittverfahren. Handb Phys XXIV, Grundlagen der Optik, Springer, Berlin, Göttingen, Heidelberg.

Plenary Lectures

Transport and relaxation behind shock waves: old and new problems

R. Brun
Laboratoire IUSTI, MHEQ Université de Provence
Centre Saint Jérôme
13397 Marseille, Cedex 20 France

Abstract: An analysis of possible ways of modeling nonequilibrium flow behind shock waves is presented: the corresponding Navier-Stokes systems differ by their transport expressions and relaxation equations. Then, a general model is exposed which avoids misunderstandings and covers all nonequilibrium situations: particular emphasis is laid on vibrational nonequilibrium and examples of computation of transport terms are given. Furthermore, vibrational relaxation models and corresponding experimental results are examined, particularly the evolution of individual vibrational populations behind shocks with and without dissociation; models for the vibration-dissociation coupling are also presented. Finally, emission measurements of CN behind strong shock waves moving in the reconstituted Titan atmosphere of a free-piston shock tube are discussed.

Key words: Nonequilibrium flow models, Transport terms, Vibrational populations, Dissociation

1. Introduction

Transport and relaxation are two aspects of the same phenomenon, characterizing the nonequilibrium of parameters representing physico-chemical processes having a time scale larger than those of perturbations in the considered medium. The case of a shock wave, representing an intense, brief, and well-defined perturbation, allows one to analyze simply the evolution of these processes behind the shock, where the flow time and process time have the same order of magnitude. From a theoretical point of view, the modeling of these phenomena may give rise to different formulations and it is necessary to know the precise starting assumptions and definitions, particularly about the time scales, known or assumed, of the different processes. In the same way, the interpretation of experiments in nonequilibrium regions necessitates the use of more or less sophisticated models with results to be carefully examined.

At the macroscopic level in the continuum regime, the description of the flow behind shock waves is essentially based on the Euler or Navier-Stokes equations more or less simplified: Viscous Shock Layer (VSL), Parabolized Navier-Stokes (PNS), Boundary Layer (BL). These equations result from successive approximations of Boltzmann equations specific to different species and (or) quantum levels. If we except the N-moment methods and Gross-Jackson procedures, which are valid outside the collisional regime but difficult to apply, the most popular method of solving the Boltzmann equations remains the Chapman-Enskog method (CE) first established for pure structureless gases, then extended to mixtures and to gases with internal modes in nonequilibrium situations or eventually reactive (Brun et al. 1984). Based on an expansion of the distribution function in a series of a "small" parameter chosen in most but not all cases as the ratio of the relaxation translation-rotation time to a reference flow time, the method requires one to define the time scale of the different processes as compared to this reference time: depending on the considered cases, frozen, equilibrium, and nonequilibrium flows may be defined. The Chapman-Enskog expansion differs for each case and it is the same for the corresponding Navier-Stokes systems. There is also no matching between the solutions which therefore cannot describe a complete flow passing for example from a frozen to an equilibrium situation (shock) or the inverse (expansion). A general matching method or generalized Chapman-Enskog method (GCE) is proposed hereafter which can cover all situations, whatever the degree of nonequilibrium.

Applications to vibrational nonequilibrium are treated for pure gases (Brun 1988) and then for mixtures like N_2/O_2 (Pascal & Brun 1993). Furthermore, the collision integrals which unavoidably appear in the formal solutions may be expressed as functions of transport terms insensitive to the internal content of molecules (shear viscosity, mass diffusion) or of global phenomenological relaxation times (rotation, vibration) (Mason & Monchick 1962), so that explicit solutions are presented. Connected to the CE method, the problem of vibration-dissociation coupling is globally treated and allows one to explicitly compute the influence of the vibrational nonequilibrium on the dissociation rate constants.

As for the relaxation itself, the availability of CW lasers allows one to carry out absorption measurements of characteristic rotational-vibrational lines and to follow the evolution of vibrational populations (NO and CO) (Hanson et al. 1976). Thus, the non-Boltzmann character of these populations is evidenced and incubation times for high levels may be measured. A comparison of different models including TV and VV transitions is also presented without and with dissociation and popular models of vibration-dissociation coupling are discussed.

Finally, regarding chemical kinetics, some recent results obtained in the free-piston shock tube of Marseille are analyzed. They consist of emission measurements of CN lines behind an intense shock propagating in a $N_2/CH_4/Ar$ mixture (Titan atmosphere).

2. Theoretical models for transport

2.1. Boltzmann equation and Chapman-Enskog method

In a reactive flow, the Boltzmann equation for a component p and/or an internal level i, with i = ir, iv, where ir and iv are respectively rotational and vibrational levels, is symbolically writen:

$$\frac{d\,f_{\rm ip}}{d\,t} = J \tag{1}$$

where J is the collisional balance of the corresponding particles and f is the corresponding distribution function. The CE method consists in expanding f in a series of a small parameter ε and here, in stopping the expansion at the first order, that is $f_{\rm ip} = f^0_{\rm ip}\,(1+\phi_{\rm ip})$ where $f^0_{\rm ip}$ is the zeroth-order term and $\phi_{\rm ip}$ the perturbation brought at the first order. Then the collisions are classified in two types depending on their probability or macroscopically on the characteristic relaxation times, that is $\tau_{\rm I}$ and $\tau_{\rm II}$. Noting $\varepsilon_{\rm I} = \tau_{\rm I}/\vartheta$ and $\varepsilon_{\rm II} = \tau_{\rm II}/\vartheta$, where ϑ represents a characteristic flow time, (1) may be written under the nondimensional form:

$$\frac{d\,f^*_{\rm ip}}{d\,t^*} = \frac{1}{\varepsilon_{\rm I}} J^*_{\rm I} + \frac{1}{\varepsilon_{\rm II}} J^*_{\rm II} \tag{2}$$

In the continuum regime $\varepsilon_{\rm I} = \varepsilon \ll 1$ and represents the "small parameter." As for $\varepsilon_{\rm II}$, three cases are possible, depending on its order of magnitude. The first case, for which $\varepsilon_{\rm I} = 1/\varepsilon \ll 1$, may be left out, since it represents the frozen case, for which the processes of type II have no influence on the first two orders of f and therefore on the description of the flow. The two other cases: $\varepsilon_{\rm II} \ll 1$ and $\varepsilon_{\rm II} \approx 1$, correspond respectively to the so-called Weak Non-Equilibrium (WNE) and Strong Non-Equilibrium (SNE) cases, examined below.

2.2. WNE and SNE cases

In the first case ($\varepsilon_{\rm II} \ll 1$), the expansion allows one to obtain the following system:

$$J^0_{\rm I} + J^0_{\rm II} = 0 \qquad\qquad \frac{d\,f^0_{\rm ip}}{d\,t} = J^1_{\rm I} + J^1_{\rm II} \tag{3}$$

The corresponding zeroth-order solution is therefore an equilibrium solution (Euler level) and the first order one brings a weak nonequilibrium solution (Navier-Stokes level) giving transport terms in which collisions of type I and II are included. Thus, this is called the Weak Non-Equilibrium (WNE) case.

In the second case ($\varepsilon_{\rm II} \approx 1$), the system is written:

$$J^0_{\rm I} = 0 \qquad\qquad \frac{d\,f^0_{\rm ip}}{d\,t} = J^1_{\rm I} + J^0_{\rm II} \tag{4}$$

The zeroth-order solution is therefore a nonequilibrium solution and at the corresponding Euler level, relaxation (or kinetic) equations are present, as well as at the first-order Navier-Stokes level in which the transport terms include only collisions of type I. This is called the Strong Non-Equilibrium case (SNE).

The application of both methods has been made for polyatomic gases or mixtures possessing vibrational modes eventually in nonequilibrium. The extension to the case of chemical reactions is in principle similar and a particular example proposed hereafter allows us to treat the problem of vibration-dissociation interaction.

WNE and SNE cases have been compared in the case of a steady nozzle expansion (Fig. 1). They give different results for the freezing level of the vibrational temperature. This is not surprising because in the SNE case the vibrational nonequilibrium is calculated through relaxation equations and in the WNE case the vibrational energy (or temperature) is calculated like a transport term. Of course, in concrete cases, the flow may be treatable successively by both methods and it is therefore necessary to use a universal method (presented below).

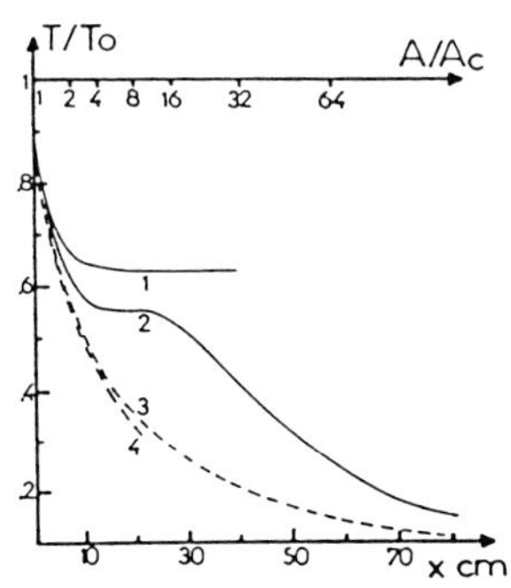

Figure 1. Temperature evolution along a nozzle, N_2, 2500 K; 10^7 Pa; 1: T_V/T_0 (SNE), 2: T_V/T_0 (WNE) 3: T/T_0 (WNE), 4: T/T_0 (SNE). Divergent angle: 4°.

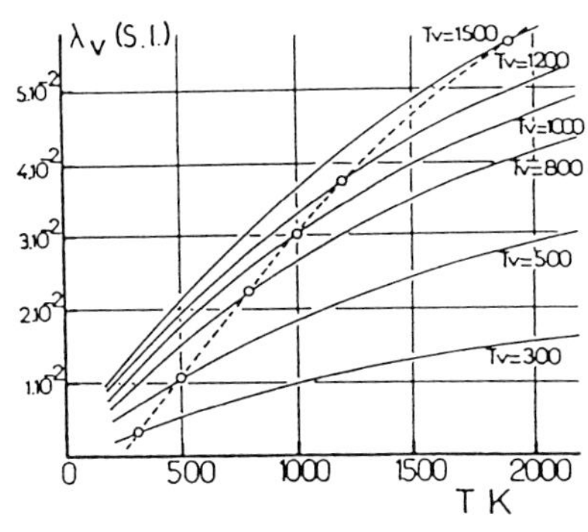

Figure 2. Vibrational conductivity for carbon dioxide; —o— is equilibrium curve.

2.3. Generalized CE method

The matching of WNE and SNE cases is realized by simply adding the term J^1_{II} in the linearized equation (4), so that the system becomes (Kogan et al. 1979,

Table 1. Comparison of WNE and SNE methods for gases in vibrational nonequilibrium.

Method	WNE	SNE
Zeroth order	• Rot. and vib. modes in equilibrium • No relaxation equations	• Rot. mode in equilibrium vib. mode in nonequilibrium • Vib. relaxation equations
First order	• Navier-Stokes with one temperature T • Bulk viscosity due to rot. and vib. nonequilibrium • One global conductivity coefficient • Transport terms functions of collision integrals I and II • T_{trans}, T_{rot}, T_{vib} computed as transport terms	• Navier-Stokes + vib. relaxation equations • Bulk viscosity due to rot. nonequilibrium • Relaxation pressure • Cross terms for each conductivity coefficient • Transport terms functions of collision integrals I • T_{trans}, T_{rot}, computed as transport terms T_{vib} with relaxation eq.

Brun et al. 1984):

$$J^1_{\text{II}} = 0 \qquad \frac{d\,f^0_{\text{ip}}}{d\,t} = J^1_{\text{I}} + J^0_{\text{II}} + J^1_{\text{II}} \tag{5}$$

Of course, solving (5) becomes more complex because of the structure of the operator J^1_{II}. In the same way, approximations of the collision integrals are often necessary. The results concerning the transport terms are formally identical to the SNE case but collisions I and II are present, which allows the matching with the WNE case. The nonequilibrium explicitly appears in these terms but they are not all affected in the same way, the most sensitive being the vibrational conductivity (Fig. 2). Another problem lies in the calculation of inelastic collision integrals; this represents a formidable task but an alternative solution consists in expressing the most affected terms as functions of those which are not too much affected (viscosity, diffusion, etc.) and of global phenomenological relaxation times τ_{R}, τ_{V}, τ_{VV}, τ_{TV}, which are, in principle, easy to determine.

2.4. Vibration-dissociation coupling

The CE method may also be successfully used for the description of nonequilibrium flows behind shocks involving vibrational relaxation and dissociation kinetics simultaneously.

In a wide range of temperatures, the WNE method may be applied because of the different orders of magnitude of the vibration and dissociation times. Thus, at the zeroth order, the dissociative molecules remain in an equilibrium distribution, but the computation of the first-order distribution function leads to an explicit expression of the dissociation rate constant K_f depending on both temperature translation-rotation T and vibration T_v (Méolans et al. 1995). In the same way, the vibrational nonequilibrium depends on K_f. An example of the computation of K_f is presented in Fig. 3 for nitrogen and compared with well-known correlations of experimental results (Park 1993).

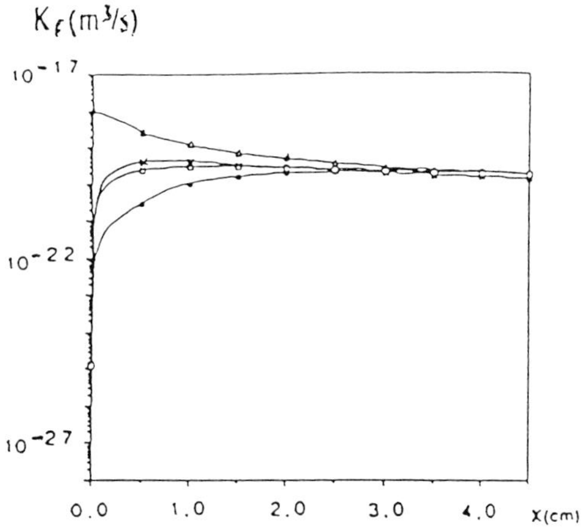

Figure 3. Nitrogen dissociation rate constant evolution behind a strong shock wave: $M = 19$, 235 K, 10 Pa. $\triangle$: K_f (zeroth order); $\times$: K_f (first order); •: Park (first correlation); □: Park (second correlation).

3. Experimental data on relaxation and kinetics

3.1. Vibrational populations

The development of laser diagnostic techniques, particularly in the IR domain, allows one to carry out absorption measurements of individual RV lines and therefore to determine the evolution of individual vibrational populations for polar gases (NO,CO...), (Hanson et al. 1976). In spite of difficulties in the interpretation of results, the example shown in Fig. 4 gives a proof of the non-

Boltzmann behavior of these populations in the relaxation zone (Méolans & Brun 1986): the importance of the VV transitions reducing the TV transfers and the existence of an incubation time specific to high vibrational levels are also evidenced. These results contribute to questioning the validity of LT-type equations, as was already the case for the nozzle expansions and the low-temperature regimes. The extension to higher temperatures for which multiple quantum transitions are possible contributes also to the elaboration of new models (Adamovich et al. 1995).

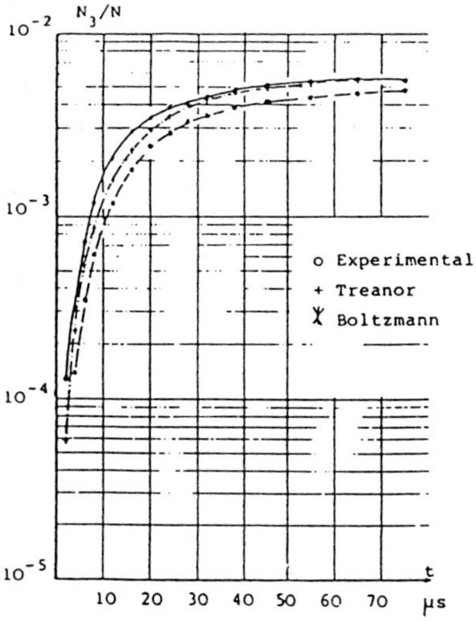

Figure 4. Relative population of the 3rd vibrational level: evolution behind a shock in CO. $M_S = 5.6$. Initial pressure 2.0×10^3 Pa. Initial temperature 291 K.

In the same way, the influence of the dissociation being important at high temperatures, the vibration-dissociation coupling influences the vibrational populations themselves. Thus, a complete analysis of nonequilibrium phenomena is presented with O_2 as a test gas. First, the relaxation is described by the master equations of each level coupled with conservation equations. The transition probabilities appearing in the relaxation equations are computed taking into account the anharmonicity and the potential angular dependence (Méolans et al. 1995). The bound-free transition probabilities of Marrone & Treanor (1963) (MT model) are used in the master equations. The population evolution is not monotonic and a time lag appears, as shown experimentally, and for high levels, the influence of the anharmonicity and VV processes is significant and strongly related (Fig. 5); these VV processes reduce the effects of the TV transfer. Finally, it is shown that the value of the "characteristic probability" U of the MT model has no great importance on the evolution of the level populations.

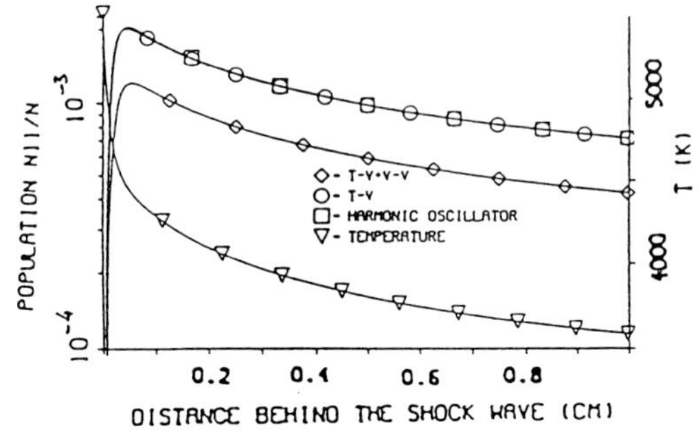

Figure 5. Relative population of the 11th level as a function of the distance from a shock in O_2; 4.5×10^4 Pa, 5525 K, 550 m/s.

However it is important to know that a "universal" curve $U(T)$ may be obtained from several runs when excluding absolutely the strong nonequilibrium part of each run and keeping only the common part of each curve. In Fig. 6, one sees that U may considerably vary around the value T_{diss} for the considered temperature range. Experiments with other gases than O_2 are in progress.

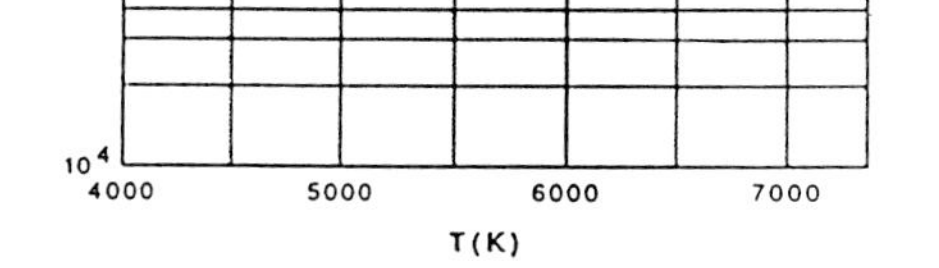

Figure 6. Intrinsic curve $U_{O_2 \to O_2}$ versus the temperature T.

3.2. Chemical kinetics: an example

Large varieties of physico-chemical processes behind shock waves have been experimentally examined, but just one single example will be given here. It concerns one of the last processes studied in the Marseille TCM2 shock tube at large Mach numbers (Dumitrescu 1995).

CN spontaneous emission is recorded behind a shock wave propagating along the shock tube with a velocity of about 6 km/s, in a reconstituted Titan atmosphere $N_2/CH_4/Ar$. At each run, time-resolved emission intensity measurements of three lines belonging to the (0,0) and (1,1) band system of CN violet spectrum (electronic transition) are carried out with three photomultiplier tubes. The results are compared to theoretical calculations and to previous measurements (Park & Bershader 1992), including a systematic analysis of the influence of argon concentration. From emission recordings, the evolution of rotational and vibrational temperatures may also be obtained (Fig. 7). Presently, these measurements are extended to CO_2/N_2 mixtures by recording CN and C_2 emission lines.

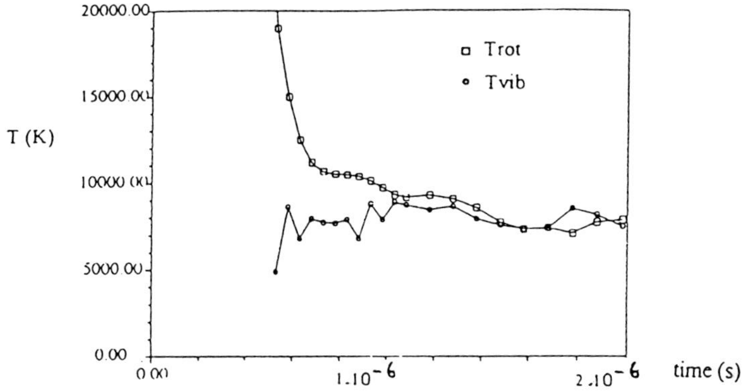

Figure 7. Rotational and vibrational temperatures determined from emission intensities behind a shock in the TCM2 shock tube, $N_2/CH_4/Ar$, 2×10^{-3} bar, $M_S = 18$.

4. Summary and conclusions

An analysis of different possible models of nonequilibrium flows behind shock waves is presented which leads to Navier-Stokes systems differing by their transport and relaxation terms or equations. However, a general model is proposed which avoids misunderstandings and covers all nonequilibrium situations. In the same way, vibrational relaxation models and corresponding experimental results are examined concerning particularly the evolution of individual vibrational populations: the influence of different types of transitions is analyzed and new interpretations and models for the vibration-dissociation coupling are also presented.

For the future, the development of computational techniques will certainly provide quantitative data on the inelastic collision frequencies necessary for precise computation of transport coefficients. The planned use of sophisticated experimental techniques — CCD cameras are a good example — will also provide considerable extension and convenience for recording time-dependent spectra.

Acknowledgement. The author is greatly indebted to his coworkers, particularly to Dr. J. G. Méolans.

References

Adamovich IV, Macheret SO, Rich JW, Treanor CE (1995) Vibrational relaxation and dissociation behind shock waves. AIAA J 33(6):1064.

Brun R (1988) Transport Properties in Reactive Gas Flows, AIAA Paper 88-2655.

Brun R, Villa MP, Méolans SG (1984) Generalized transport terms in vibrationally relaxing flows. In: 14th RGD Symp, H Oguchi (ed), Tokyo.

Dumitrescu MP (1995) Operation and development of the TCM2 free piston hypersonic facility in Marseille. In: 2nd Europ Symp Aerothermod, ESTEC, Noordwijk ESA S P 367:341.

Hanson RK, Monat JP, Kruger CH (1976) Absorption of CO laser radiation by NO. Jour Quant Spectros Rad Transf 16:705.

Kogan MN, Galkin VS, Makashev NK (1979) Generalized Chapman-Enskog method: derivation of the nonequilibrium gasdynamic equations. In: 11th RGD Symp, R Campargue (ed), Paris.

Marrone PV, Treanor CE (1963) Chemical relaxation with preferential dissociation from excited vibrational levels. Phys Fluids 6:1215.

Mason EA, Monchick L (1962) Heat conductivity of polyatomic and polar gases. J Chem Phys 36(6):1622.

Méolans JG, Brun R (1986) Nonequilibrium vibrational population behind shock waves: comparison theory-experiments. In: 15th RGD Symp, V Boffi, C Cercignani (eds), Teubner, Stuttgart.

Méolans JG, Brun R, Mouti M, Llorca M, Chauvin A (1995) Vibration-dissociation coupling in high temperature nonequilibrium flows In: 2nd Europ Symp Aerothermod, ESTEC, Noordwijk ESA SP 367:293.

Park C (1993) Review of chemical-kinetic problems of future NASA mission I: Earth entries. J Therm and Heat Transf 7(3):385.

Park CS, Bershader D (1992) Studies of radiative emission from the simulated shock layer of the Huygens probe. In: Shock Waves, K Takayama (ed), Sendai, p 671.

Pascal S, Brun R (1993) Transport properties of nonequilibrium gas mixtures. Phys Review 47(5):3251.

Tuned operation of a free-piston shock tunnel

K. Itoh
National Aerospace Laboratory
Kakuda Research Center
1 Koganesawa, Kimigaya, Kakuda, Miyagi 981-15 Japan

Abstract: The National Aerospace Laboratory (NAL) has been planning to build a large free-piston shock tunnel, named HIEST, for the reentry aerothermodynamic tests of HOPE. Recently, the HIEST project proceeded to the construction phase, which will be completed in 1997. To support this project NAL has done various experimental and theoretical studies and applied the results to the basic design. In this paper, an experimental study with the medium-sized free-piston shock tunnel HEK in terms of tuned operation is presented. The detailed piston-landing motion was measured with a simple method and it was verified that tuned operation was successfully achieved. The effects of high-efficiency driving on facility performance were also studied. The results suggest that the use of a light piston is expected to be effective to achieve tuned operation when aiming for a more powerful shock-tube driver.

Key words: Free-piston shock tunnel, Tuned operation, Free-piston driver

1. Introduction

A free-piston shock tunnel has great capability to produce high-enthalpy flow, by applying adiabatic compression with the free piston to increase driver-gas temperature and pressure. However, since the driver gas is highly compressed to a very small volume, it is difficult to maintain a useful duration time for a test.

To overcome this essential problem, the concept of high-speed piston operation was given by Stalker (1967) and Hornung & Belanger (1990). According to the concept, a piston is operated to travel at sufficiently high speed even at diaphragm rupture so that the driver-gas pressure can be held nearly constant for a certain duration. Thus, the size of the facility can be drastically reduced. Here, the driver based on this concept is referred to as a high-efficiency driver.

However, because the required piston speed is very high, it is essential to achieve the soft landing operation of the driver, or tuned operation, to avoid damaging the facility. In order to find an effective way to achieve tuned operation, a tuning principle was formulated by Itoh (1994) based on the eigen-value theory. In the present study, an attempt was made at achieving tuned operation with the medium-sized free-piston shock tunnel HEK by following this tuned-operation formula. The detailed piston landing motion was measured with a simple method and it was verified that tuned operation was successfully achieved. The effect of high-efficiency driving on facility performance was also studied. The results suggest that the use of a light piston is expected to be effective to achieve tuned operation when aiming for a more powerful shock-tube driver.

2. Tuned operation of free-piston shock tunnel

During a piston trajectory after diaphragm rupture, there exists a point where the pressures ahead of and behind the piston are balanced and thus minimum piston velocity occurs. This interesting piston motion can be characterized by $\omega \equiv \sqrt{c(\gamma)P_0V_0/W_p(A^*/A)^2a_0^2}$, which acts as a deceleration factor. Here, W_p is piston mass, P_0, V_0, a_0 are initial pressure, volume, and sound speed of the driver gas at piston launch. The quantity A^*/A is the area ratio of the exit to the main part of the compression tube.

When ω is too large, the piston is too much decelerated and rebounds before impact. In contrast, when ω is too small, the piston can not be decelerated enough and impacts directly. Between them, there is good value for which the piston decelerates and comes to rest at the point where the pressures ahead of and behind the piston are just balanced. Then, a soft landing is possible by adjusting the piston stopper height to this point. Therefore, the tuned-operation problem can be recognized as the eigen-value problem to find ω which satisfies $u = 0$ at $du/dt = 0$. In the previous study (Itoh 1994), the formula which represents the relation of ω and β satisfying this condition was derived; β is defined as the ratio of piston speed at diaphragm rupture to reference speed needed to hold the driver-gas pressure constant, thus the high-efficiency driver is characterized by $\beta > 1$.

The relation of β and ω representing tuned operation is shown in Fig. 1. It should be remarked that a sufficiently large ω is needed to achieve tuned operation, and β should be greater than unity.

For given facility and major test conditions, there are various possibilities to achieve a sufficiently large ω, i.e., use of a lower sound-speed driver gas or a smaller area ratio. However, both of these methods cause a reduction in performance. Therefore, we proposed use of a light piston to achieve a more powerful shock-tube driver; and this concept was incorporated into the basic design of HIEST. However, a light piston has difficulty maintaining enough strength against high pressures. Thus, the mechanical design of the piston becomes critical.

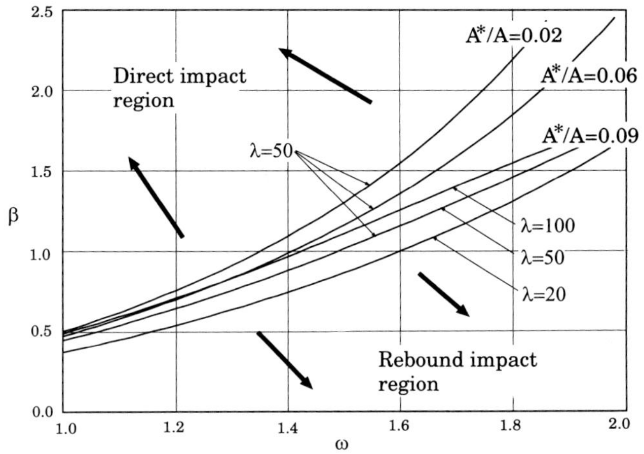

Figure 1. Eigen values for tuned operation.

3. HEK experiments

The above-mentioned theoretical prediction for achieving tuned operation was examined by experiments with the medium-sized free-piston shock tunnel HEK, of which the compression tube is 16 m long and 210 mm in diameter; the shock tube is 6.5 m long and 72 mm in diameter; the orifice between the compression tube and the shock tube is 50 mm to 64 mm in diameter; and the piston mass is 30 kg. The purposes of the HEK experiments were to attempt to carry out tuned operation and to investigate the effect of tuned operation on facility performance.

3.1. Piston-motion measurement system

The unique feature of the HEK is the simple measurement system for detailed piston motion. On the piston surface, black stripes are made with fine pitch using an anode oxidation method. When black stripe comes in contact with a laser beam, the reflection signal of the laser beam disappears. By using this method, detailed piston motion can be measured with a few piston sensors placed in the high-pressure segment of the compression tube. The piston is made of Duralmin and consists of an 240-mm long main body and an 180-mm long skirt which acts as the valve for reservoir gas. The width of the stripes and intervals are 6 mm on the main body and 10 mm on the skirt. The o-rings on the skirt were also detected by the photo sensors.

3.2. Operating conditions

The operating conditions in shots 10–20 (excluding shot 14 in which the data sampling failed) are summarized in Table 1. In all shots, the diaphragm-rupture pressures were uniformly 52 MPa, and the same piston was used. Therefore, ω and the compression ratio are no longer independent and commonly changed by changing the initial pressure of the driver gas. According to the tuned-operation formula, the β values were decided with respect to ω. Then, the reservoir pressures were adjusted to obtain the intended β values.

Table 1. Operating conditions.

Shot #	Orifice	P_0 (kPa)	P_{R_0} (MPa)
10	ϕ50	108.0	5.216
11	ϕ50	108.9	5.100
12	ϕ50	87.8	4.262
13	ϕ50	98.5	4.706
15	ϕ54	132.5	6.240
16	ϕ52	113.0	5.460
17	ϕ52	94.5	4.420
18,19	ϕ52	113.0	4.888
20	ϕ52	94.5	3.954

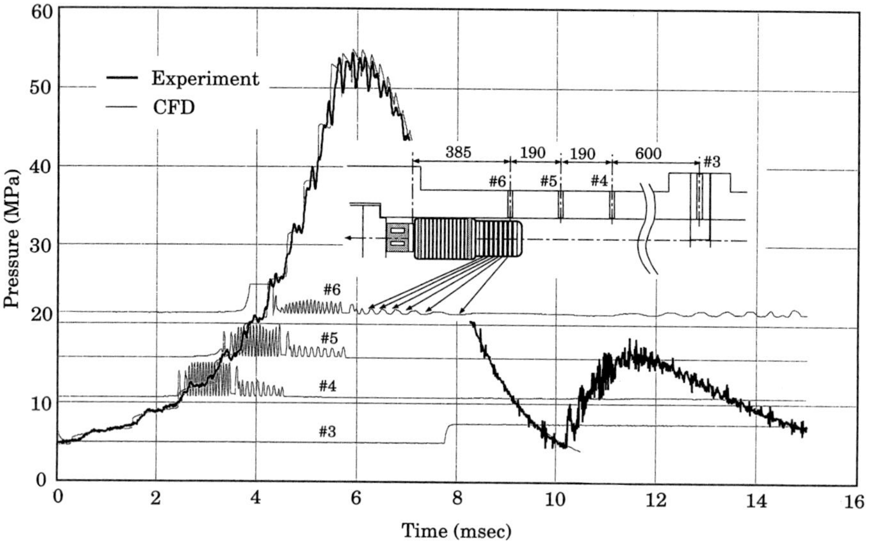

Figure 2. Compression tube pressure and piston motion in shot 16.

3.3. Characteristics of compression-tube pressure

Figure 2 shows an example of the compression-tube pressure as predicted by CFD and as signals detected by the piston sensors obtained in shot 16. As seen in the figure, the detailed piston motion could be successfully detected. The signal from sensor No. 3 was used as a starting point in this measurement. The signal from No. 6 showed that the piston was rapidly decelerated just before landing and almost came to rest at landing. When the reflected shock arrived, the piston was pushed back. In this case, the measured β value was 1.23 so that the driver gas pressure was held nearly constant for about 0.8 msec. Figure 3

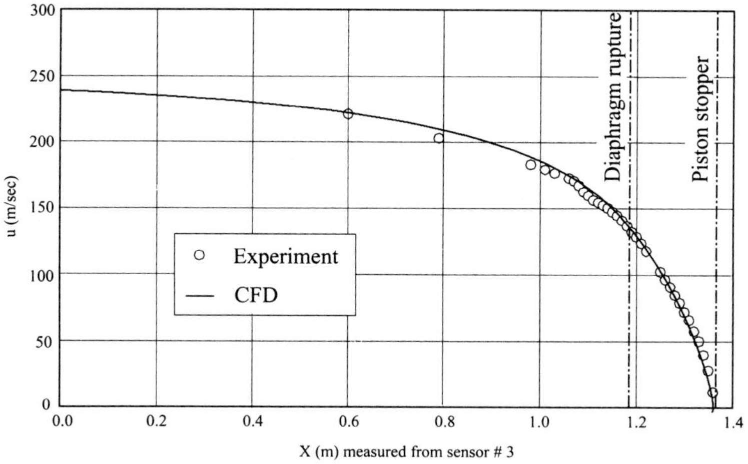

Figure 3. Piston velocity in shot 16.

shows the piston velocity calculated from piston trajectory versus the distance from sensor No. 3. It is seen that the operation was very well tuned so that piston came to rest just at landing.

The characteristics of the compression tube were estimated by the piston-motion measurements and compression-tube pressures and they are summarized in Table 2 and also represented in the $\omega - \beta$ operating map in Fig. 4.

According to the tuned-operation formula, the goal was to carry out tuned operation in all shots except 13, 18, and 20. In shots 13, 18, and 20, the intention was to try to slightly rebound the piston. The results are quite consistent with the theoretical prediction, and it was verified that tuned operation could be successfuly achieved by using the tuned-operation formula.

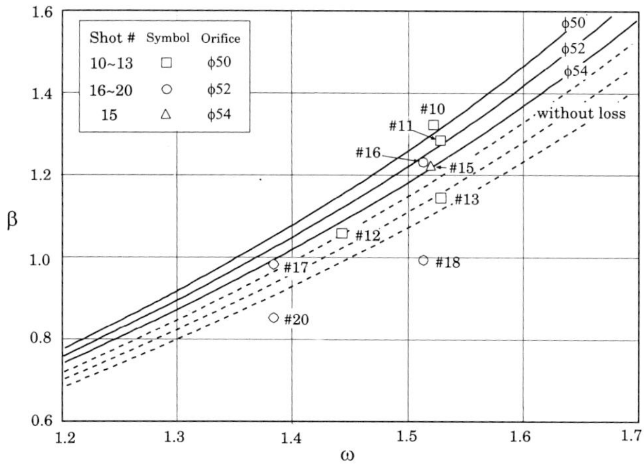

Figure 4. The $\omega - \beta$ operating map.

Table 2. Characteristics of compression tube.

Shot #	Orifice	λ	u_r (m/sec)	ω	β	τ (msec)
10	$\phi 50$	46.2	135.6	1.522	1.324	1.30
11	$\phi 50$	46.0	131.4	1.528	1.285	1.20
12	$\phi 50$	52.2	113.1	1.443	1.058	0.27
13	$\phi 50$	50.2	117.9	1.528	1.146	0.54
15	$\phi 54$	40.8	140.6	1.520	1.225	0.84
16	$\phi 52$	45.6	134.3	1.513	1.232	0.76
17	$\phi 52$	51.1	110.6	1.384	0.982	–
18	$\phi 52$	45.8	108.0	1.513	0.993	–
20	$\phi 52$	54.4	93.0	1.384	0.853	–

3.4. Effect of high-efficiency driving on facility performance

Next, the effect of high-efficiency driving on facility performance was studied. As indicated in Table 2, the useful duration in which the pressure is held nearly constant can be obtained when β is greater than unity; and, when β is less than unity there is no useful duration. This difference in driver-gas condition clearly appeared in shock-tube performance. As shown in Fig. 5, decreasing β led to the pressure level going down; moreover, the pressure profile deviated from the plateau one. The effect of changing orifice size was also investigated with the aim of obtaining a more powerful shock-tube driver. As shown in Fig. 6, the pressure level is clearly increased when orifice size is increased. Both β and orifice-size effects are summarized in terms of the pressure recovery factor in Fig. 7. In order to increase β and orifice size, ω had to be increased by decreasing

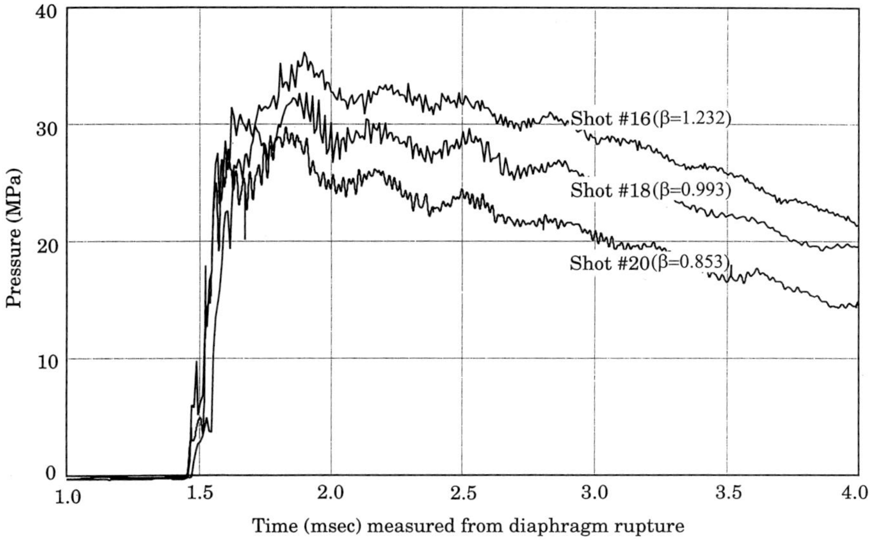

Figure 5. Effect of β on shock-tube pressure (end wall).

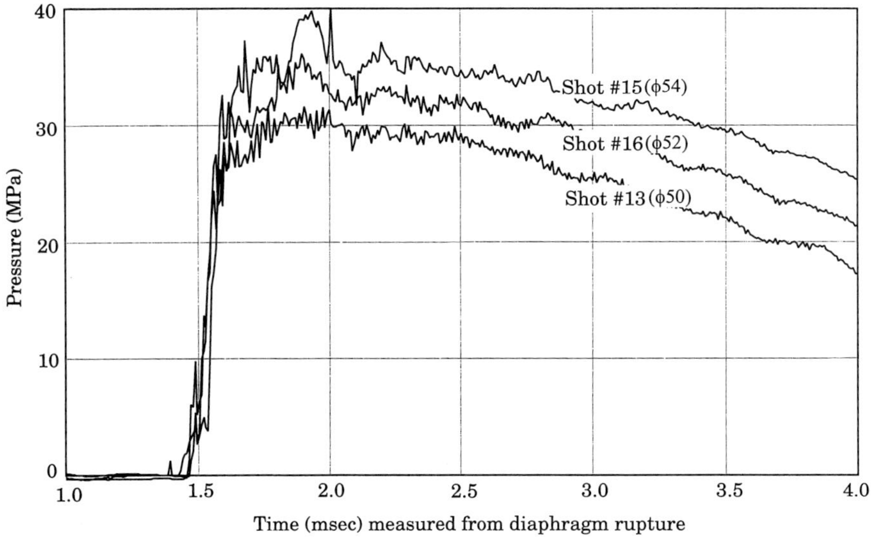

Figure 6. Effect of orifice size on shock-tube pressure (end wall).

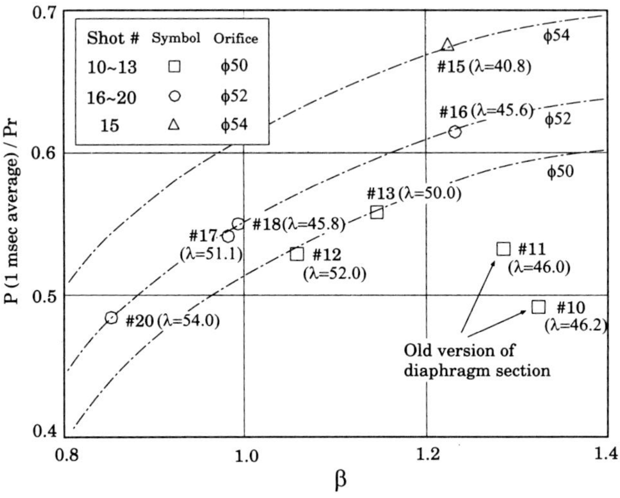

Figure 7. Effect of β and orifice size on pressure recovery factor.

the compression ratio, as indicated by the values in Fig. 7, because the piston mass and diaphragm-rupture pressure were fixed in this series. Therefore, if in practice the piston mass is reduced, even greater effect on ω is expected to be obtained.

4. Conclusions

In the present study, an attempt was made at achieving tuned operation with the medium-sized free-piston shock tunnel HEK by following the tuned-operation formula. Conclusions are as follows:

- Tuned operation was successfully achieved by following the tuned operation formula.
- Effects of high-efficiency driving on facility performance were verified.
- Use of a light piston is expected to be a more effective way to get a more powerful shock-tube driver.

References

Eitelberg G (1994) First results of calibration and use of the HEG, AIAA Paper 94-2525.

Hornung HG (1988) The piston motion in a free-piston driver for shock tubes and tunnels, GALCIT FM88-1, California Institute of Technology.
Hornung HG, Belanger J (1990) Role and techniques of ground testing for simulation of flows up to orbital speed, AIAA Paper 90-1377.
Itoh K (1994) Tuned operation theory for free-piston driver of high-enthalpy shock tunnel. In: Symposium on Shock Waves Japan '94, pp 29–32.
Stalker RJ (1967) A study of the free-piston shock tunnel. AIAA J 5:2160–2165.

High-resolution imaging of hypervelocity metal jets using advanced high-speed photographic techniques

L. L. Shaw, S. A. Muelder
Lawrence Livermore National Laboratory
Box 808, L-281, Livermore, California 94550 USA

Abstract: It is now possible to obtain high-resolution sequential photographs of the initial formation and evolution of hypervelocity metal jets formed by shaped-charge devices fired in air. Researchers have been frustrated by the high velocity of the jet material and the luminous sheath of hot gases cloaking the jet that made detailed observation of the jet body extremely difficult. The camera system that provides the photographs is a large-format multi-frame electro-optic camera, referred to as an IC camera (IC stands for image converter), that utilizes electro-optic shuttering, monochromatic pulsed-laser illumination, and band-pass filtering to provide sequential pictures (in 3D if desired) with minimal degradation due to luminous air shocks or motion blur. The large format (75-mm image plane), short exposure (15 ns minimum), ruby-laser illumination, and band-pass filtering (monochromatic illumination while excluding extraneous light) produces clear, sharp, images of the detailed surface structure of a metal shaped-charge jet during early jet formation, elongation of the jet body, jet-tip evolution, and subsequent particulation (breakup) of the jet body. By utilizing the new camera system in conjunction with the more traditional rotating-mirror high-speed cameras, pulsed radiography, and electrical sensors, a maximum amount of often unique data can be extracted from a single experiment. This paper was intended primarily as an oral presentation. For purposes of continuity and simplicity in these proceedings, we have chosen to concentrate on the development of the IC-camera system and its impact on the photography of high-speed shaped-charge jets.

Key words: Image-converter camera, Laser-illuminated photography, Electro-optic shutter, Multi-pulsed ruby laser, High-speed photography, Shaped-charge jet

1. Introduction

High-speed photography is a major diagnostic tool in the study of fast phenomena such as the acceleration and deformation of high-explosive driven metal surfaces. However, multiple-frame high-speed photography has been of limited benefit in the study of the jets formed by shaped-charge devices fired in air. One problem is poor spatial resolution due to the motion blur caused by the movement of the jet material during the exposure time of the camera frame. A second problem is the shock-induced luminosity cloaking the tip and body of the jet that blocks direct observation of the jet surface. In order to minimize the effect of object motion on picture resolution, exposure times as short as 10^{-8} sec must be possible. To photograph the surface of the jet, the camera must "see through" the shock-induced luminosity and exclude it from the camera while at the same time photographing an adequately illuminated jet surface. The use of the recently developed multi-frame electro-optic shuttered image-converter camera (IC camera) with pulsed-ruby laser illumination meets the performance parameters just stated and provides the means to obtain high-resolution sequential photographs of these jets. A nominal 20-nanosecond exposure time greatly reduces the effect of material motion, even for velocities in the range of 10 kilometers per second. The monochromatic 694-nanometer pulsed-ruby laser illumination with 1.5-nanometer band-pass filtering allows the self-luminosity and other extraneous light to be excluded from the photographs. With this system, sequential pictures of the surface of a metal shaped charge have been obtained that show the detailed metallurgy as the jet proceeds from initial formation through particulation.

2. IC-camera system

The IC-camera system described in this paper is a direct outgrowth of the development work in proximity focused image-converter tubes and in pulsed solid-state lasers that has occurred over the past 40 years. Experimenters and diagnosticians realized that the ultra-short exposure time possible using the image-converter tube as a shutter permitted the photographing of fast phenomena minimizing motion blur, and a number of framing cameras utilizing these tubes were built (Chancel and Pelte 1972, Owren 1959). During the same time period, experimenters realized that the exposure time of mechanical cameras could be dramatically reduced by using pulses from Q-switched or cavity-dumped lasers as light sources. There were also two other desirable attributes of using lasers: extraneous light could be excluded from the camera by the use of band-pass filtering that transmits only the wavelength of the laser light, and the technique proved very adaptable for stereo photography (Carpluk 1970, Paisley 1976). Over the past fifteen years, we have combined the best attributes of the electro-optic framing camera and pulsed-laser illumination into a high-speed framing-photography system (Shaw et al. 1984).

2.1. IC camera

The IC camera has a modular design as shown in Fig. 1. Each module is a complete two-frame camera with its own objective optics and mechanical shutter. An internal 50% mirror creates two image planes from a single entrance optic. Individually triggered 75-mm diameter proximity focused image-converter tubes are positioned at each image plane; these have adjustable exposure times with a minimum of 12 to 15 ns. A detailed description of the camera timing

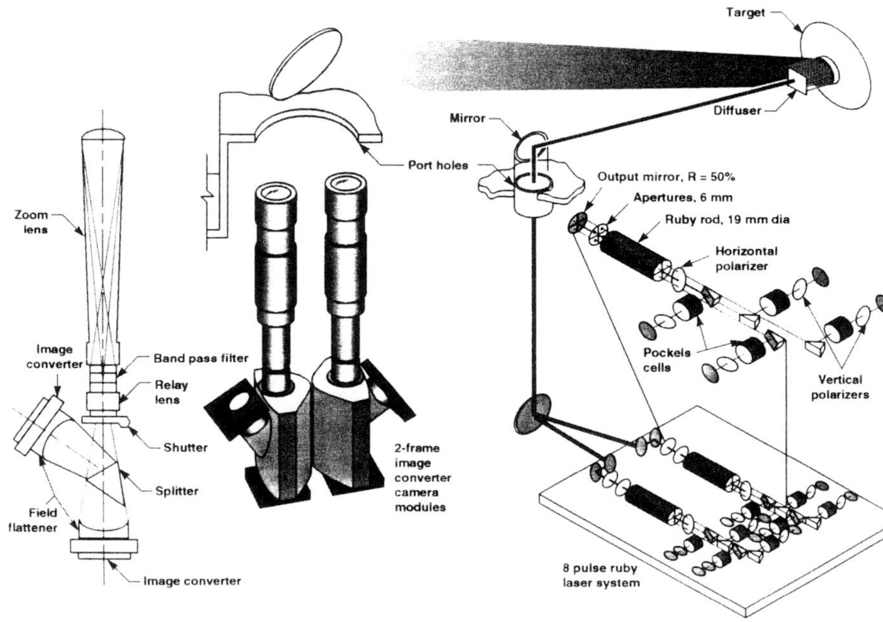

Figure 1. Diagram of the eight-frame IC (image-converter) camera and the eight-pulse ruby-laser illuminator. Note the modular camera construction with two frames per objective optic. The laser is configured to provide four independent pulses from a single rod. Both the camera and the laser are located in a concrete reinforced underground bunker below and to the side of the experimental firing area.

and control system is provided in Shaw et al. (1984, 1993). The optical system for each module consists of a 200-mm to 1140-mm zoom objective lens with a 35-mm image format. A five-to-one magnifying relay lens transmits and enlarges a portion of the image onto the two 75-mm image-converter tubes. The original configuration of eight frames is made up of four modules designed to be clustered as a unit on a single hydraulic stand, but it is often more convenient to use just two modules per stand as shown in Fig. 2. The modular construction and this type of mounting introduce between one and two degrees of parallax between modules. This is acceptable for monocular photography, and proves very useful and quite adequate for stereo photography.

The film used in this camera is Kodak T-MAX 3200 4×5-inch sheet film. The speed, resolution, and low fog level of this film have been major factors in the high quality of the photographs taken with this system. The film is routinely developed at ISO 6400 and quite often pushed to ISO 12800. This allows significant leeway in the choice of illumination light levels and in the adjustment of the aperture to manipulate film exposure and the camera depth of field. Extraneous light from air shocks, explosive products, and other sources is prevented from entering the camera by a 1.5-nm band-pass filter centered at 694 nm (ruby-laser wavelength) and positioned at the output of the objective optics.

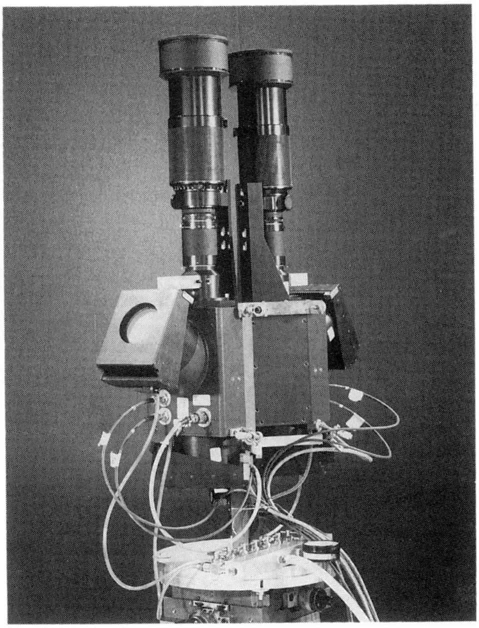

Figure 2. A pair of IC-camera modules mounted on a vertical hoist in the firing bunker camera room; the modules view the experimental firing table through a 30-cm glass porthole in the camera-room ceiling. The 200–1140-mm focal length zoom lens on the camera and the magnification of five by the internal relay lens yield an image on film up to nearly full size at an object distance of 10 meters. Object details as small as 100 microns can be resolved in the resultant photographs.

2.2. Eight-pulse ruby-laser illuminator

The eight-pulse Q-switched ruby laser has a novel design that permits the extraction of four separate pulses from a single ruby rod during a single pumping of the rod. This is illustrated in cartoon fashion in Fig. 1; and the equipment is shown in Fig. 3. A mask with four 6-mm diameter holes allows each quadrant

Figure 3. Shown is the eight-pulse ruby-laser system using two ruby rods located in the flashlamp heads seen at the right center of the photograph. Also clearly visible at the lower left are the four white Pockels cells used to Q-switch the four cavities of the nearer ruby rod. The beams from the two ruby rods are brought together and guided to the firing table through the vertical tube just visible at the top of the photograph.

of a 19-mm diameter rod to be used in a separate laser cavity. The nominal energy of each pulse is 0.5 joules and is approximately Gaussian with a full width at half maximum of 50 ns. Although this design was developed independently, we quickly discovered the approach had previously been developed by others (Jones 1974). Figure 4 is a graphic example of the IC camera's ability to produce detailed photographs in the presence of intense ambient light by using monochromatic laser illumination coupled with a band-pass filter.

The timing and control system, not addressed in this paper, consists of programmed delays and triggers that initiate the individual image-converter tube drivers and the laser Pockels-cell drivers. The jitter of this system is less than 5 ns allowing the 15–20-ns exposure time of each frame to be well centered in the 50-ns laser-illumination pulse reflected from the experiment. Transient digitizers measure laser-to-camera simultaneity and reference timing to the main experimental timing.

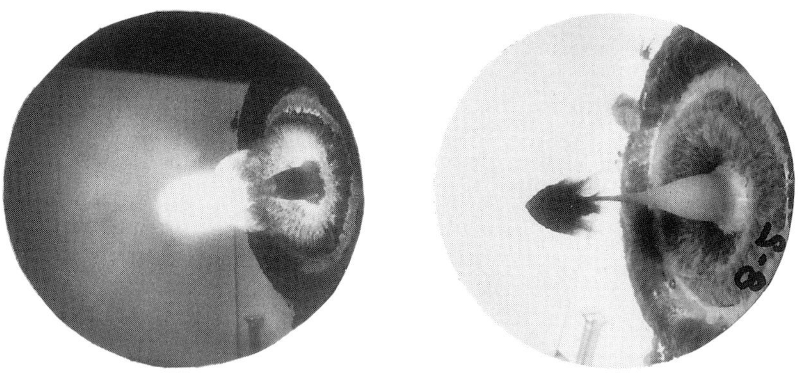

(a) (b)

Figure 4. IC-camera photographs of a jet from a shaped charge with a hemispherical liner. The two pictures were taken 200 ns apart on the same experiment using two different camera modules — one module without the 1.5-nm band-pass filter, and one with it. The exposure time for both photographs was 15 ns. In (a), the only illumination is from the luminous air shock surrounding the jet. In (b), the illumination is from the pulsed ruby laser where the band-pass filter excludes the luminous air shock allowing observation of the underlying metal structure.

3. Experiments

The illustrations used in this paper are of experiments evaluating the performance of explosive shaped charges. Figure 5a is a cartoon of a representative experimental layout. Figure 5b is a cross-sectional view of a typical shaped charge. In diagnosing experiments, we place many diverse diagnostics on the same experiment. These include low-energy pulsed radiography, electrical signal pins, and rotating-mirror framing and streak cameras. These additional diagnostics have been excluded from this paper but are discussed in some detail in Shaw et al. (1979, 1993). The IC camera and laser are located in an underground bunker adjacent to and below the level of the gravel test area known as the firing table. Front surface mirrors direct the incident laser beam to the experiment and the reflected laser light to the camera. The laser pulses strike a ground-glass diffuser placed close to the experiment that scatters the laser light over the experiment. Matte white poster board placed around the experiment provides a light background and helps to "splash" the laser light onto the target.

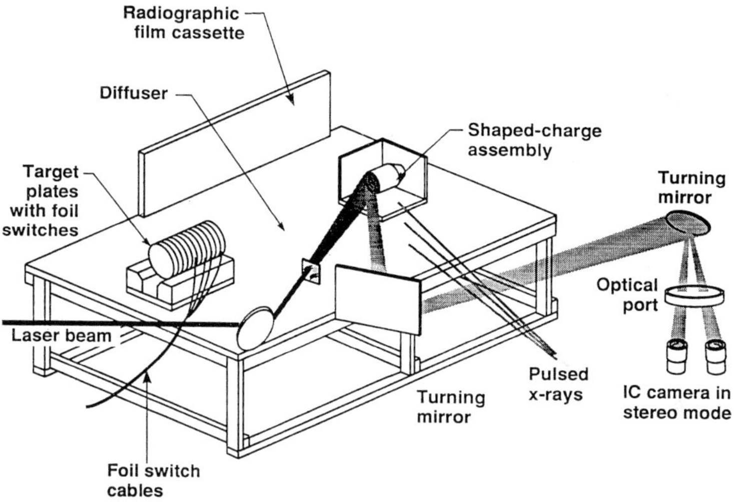

(a)

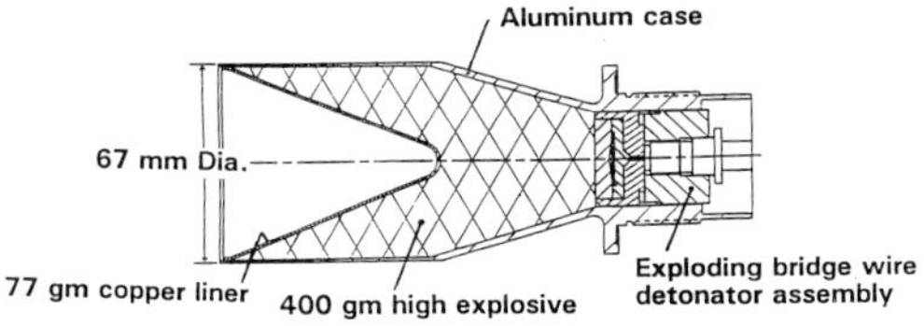

(b)

Figure 5. (a) Diagram of a typical test setup used for shaped-charge experiments. For simplicity, only one IC-camera station is shown. Not shown are some of the other diagnostics often fielded in these experiments such as rotating-mirror framing cameras. (b) Section view of the "Viper" shaped charge used in some of these experiments. The electrically initiated exploding bridge wire on the right end launches a detonation wave through the high explosive. The detonation causes the copper liner to collapse, turn inside out, and "squirt" to the left at velocities approaching 10 km/sec.

4. Results

In the oral presentation, we discussed photographing the entire evolution of the shaped-charge hypervelocity jet. In this paper, we have concentrated on the

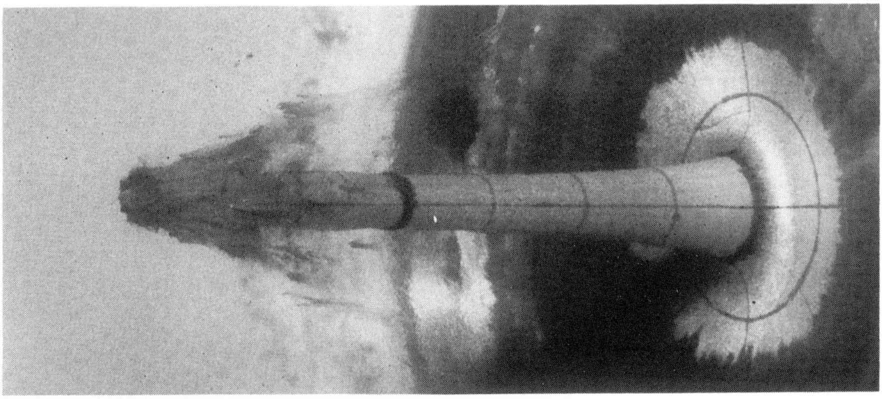

(a)

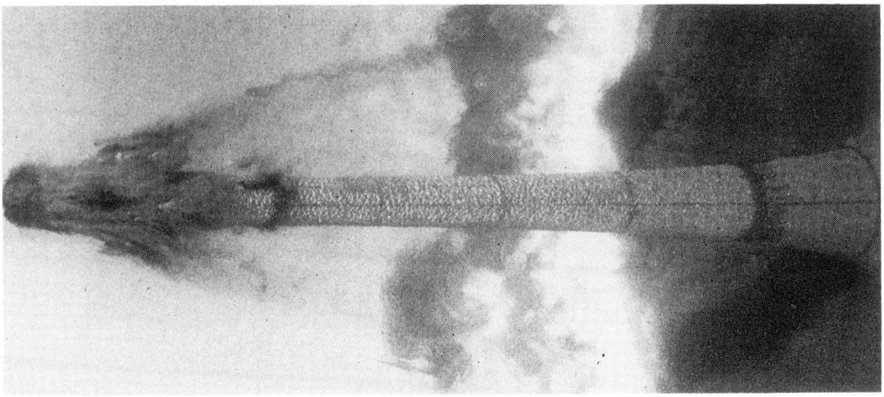

(b)

Figure 6. Laser-illuminated photographs of a shaped charge with a conical liner. Grid lines are made with a fine felt-tipped permanent marker on bare metal. The lines thicken and stretch with the liner deformation providing valuable quantitative strain and timing data. (a) 15-ns exposure taken 40.8 μs after shaped-charge detonator initiation. The jet diameter behind the tip is approximately 10 mm. (b) Same exposure taken 6.7 μs later. Note the ablated copper and the turbulent shock waves in the air behind the tip, and the texture change of the jet surface as it begins to stretch. The jet diameter is now about 6 mm.

high-resolution detail the IC camera is able to record as the metal surface of the jet stretches and cracks during initial elongation and then pulls apart into what are eventually discrete pieces as the jet particulates. In Figs. 6, 7, 10, and

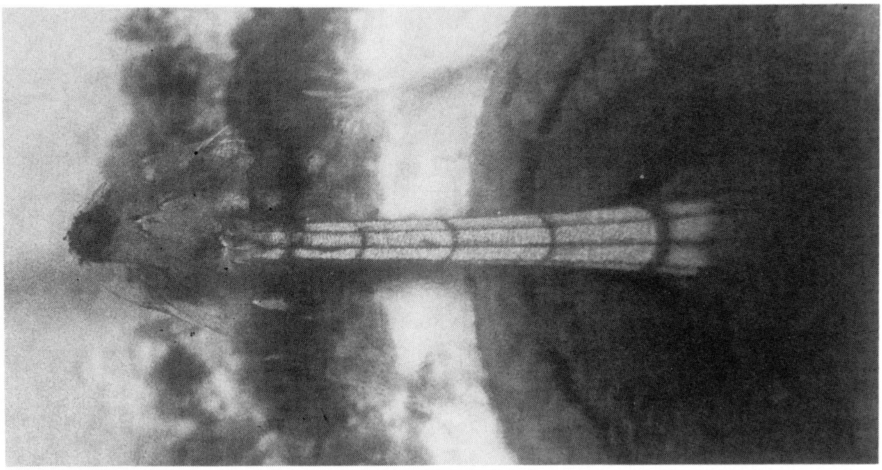

(a)

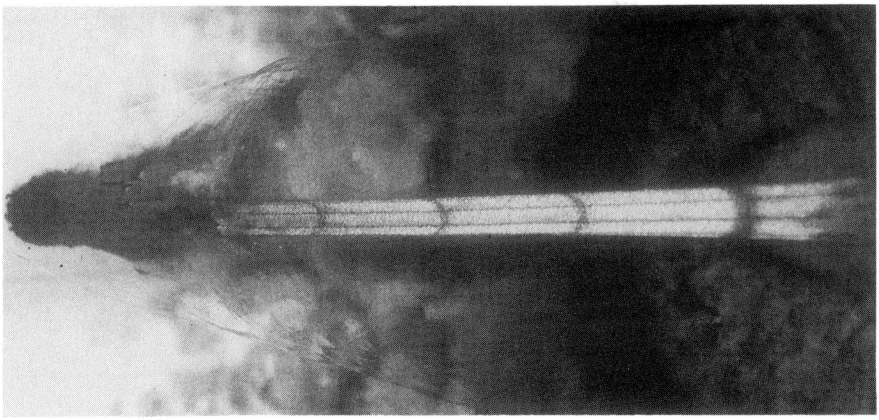

(b)

Figure 7. These are IC photographs of the same type of shaped charge as shown in Fig. 6. The two photographs were taken 10.0 μs apart. The jets tips extend out from the base of the conical liner by approximately 150 mm (top) and 235 mm (bottom). A total of eight photographs were taken of this experiment, and were used to verify that the fabrication technique used for this liner imparts a spin to the resultant jet. An examination of the ink lines along the jet axis show the rotation as the liner collapses and the jet forms.

11 we have chosen to show IC-camera pictures of jets which, while illustrative of the ability to record overall jet characteristics with unique tip structure and air-shock formations, were chosen primarily to illustrate the surface characteristics of the elongating jet.

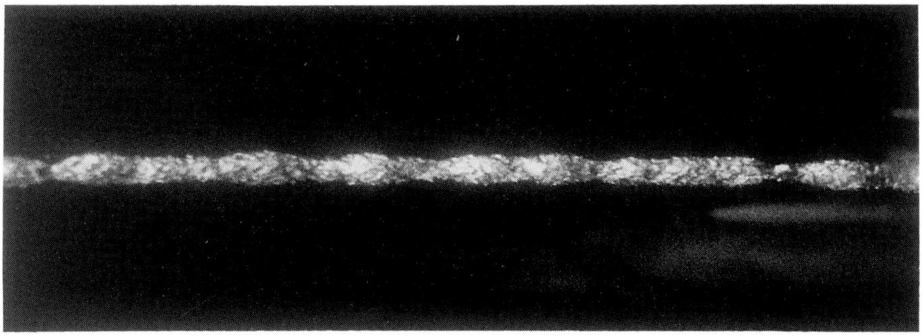

Figure 8. Main body of a jet 320 mm behind the tip. The jet has traveled 870 mm from the exit orifice of the shaped charge. The jet tip has a higher velocity than the tail causing the jet to stretch and thin during flight. This jet had a nominal diameter of 1.4 mm at the time of this photograph, and is beginning to show signs of breaking into separate particles. Note the fine structure on the jet surface including sections that appear cross-hatched.

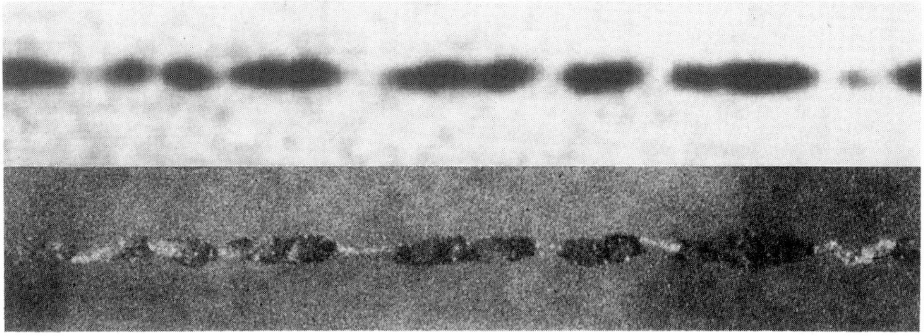

Figure 9. Body of shaped-charge jet 200 mm behind jet tip and approximately 1100 mm from the exit orifice of the shaped charge. The jet has apparently broken into 2-mm diameter particles. Above is a section of a 450-KeV X-ray taken of this jet at the same time as the IC photograph below. Note that the IC photograph clearly shows reflective material between the larger dark particles whereas the radiograph shows little or none. One explanation currently being explored is that there is a thread of liquid copper in the center of the jet that becomes exposed when the jet breaks apart, and that the thread is reflective to the laser illuminator, but too transparent to show in the radiograph.

The particulating jets shown in Figs. 8 and 9 are of unique and recent interest. Particulated jet fragments had previously been observed only in silhouette with low-energy pulsed radiography and occasional rotating-mirror framing photography. The IC-camera photographs show surface crazing and stretching not previously observable. Noted, and not unexpected, are slip dislocation lines running at 45 degrees with respect to the jet axis (Fig. 8).

The similarity to solid metal rods undergoing static tension leads to the conclusion that the Viper jet surface is solid rather than liquid. Totally surprising, however, is the fact that the IC-camera pictures show the particulated segments to be connected by highly reflective threads where the radiographs show clear gaps (Fig. 9). The threads seem to be liquid copper that are radiographically transparent. Since the physical state of the metal is critical to the understanding and computer modeling of these jets, current efforts are being directed towards dynamic measurements of the temperature profiles of the jets in flight.

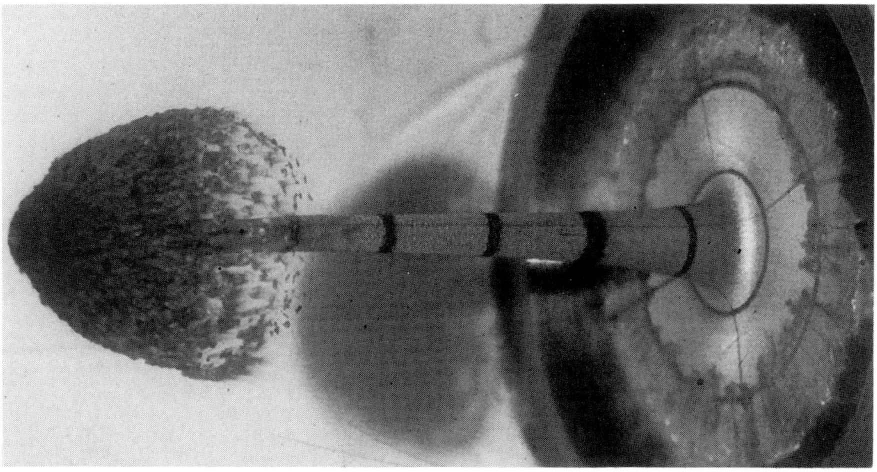

Figure 10. Example of a Viper shaped charge (see Fig. 4) taken when the tip has traveled 80 mm from the exit orifice (31.5 μs after detonator initiation). Tip velocity is 9.2 km/s creating a clearly visible shock structure in the surrounding air and a shadowgraph on the background surface. The "umbrella"-shaped cloud of debris surrounding the tip is a result of ablation and of the rounded apex of the liner cone. The jets shown in Figs. 6 and 7 were from larger shaped charges with conical liners having sharp apexes which produce different debris clouds.

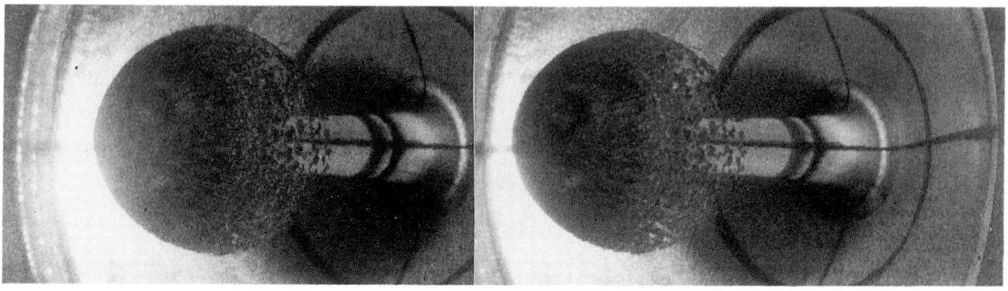

Figure 11. Representative stereo pair of pictures of a Viper taken with the IC camera. To view these images in three dimensions, place the photographs at normal reading distance. Relax your eyes so that you see double images and allow the center two (of the four) to overlap. If you experience difficulty in this free-focus technique, try placing a card between the two images from the page towards your nose so that each eye can see only one image. Relax your eyes until the images overlap to see the three-dimensional effect.

5. Conclusions

The IC-camera system provides a unique capability to photograph very fast phenomena such as the metal jets from shaped-charge devices. This system provides the resolution to observe the minute surface and metallurgical details of the liner deformation and subsequent jet formation and evolution. It has proven invaluable in providing a detailed picture of the breakup process and surface characteristics during the process of jet particulation. Data from the IC-camera pictures have proven indispensable for accurate computer simulations. As was shown in the oral presentation, such data have also has enhanced the quality of the information obtained from more conventional rotating-mirror high-speed cameras viewing the same event, by providing a highly detailed reference when evaluating the data contained in the color frames from these cameras. We continue to improve this system and are currently building the third iteration of the camera that will incorporate a new higher-resolution image-converter tube, greatly improved electronics for pulsing the new tubes, and improved objective optics. Because of our desire to emphasize the IC-camera system's pictures, we have kept the text to a minimum in this paper. We encourage the reader to review the list of references or contact the authors for more detailed descriptions of the camera system and its applications.

Acknowledgement. The authors acknowledge the programmatic support of LLNL's B-Division, the technical expertise on the experiments of R. Don Breithaupt, the leadership in shaped-charge technology of Dennis W. Baum, Leonard C. Haselman Jr., Christian Simonson III, and Kris Winer of the Advanced Conventional Weapons Program in B-Division, and the technical staff of bunkers 851 and 801 at LLNL Site 300. The authors extend a special acknowledgment to Anthony T. Rivera, James T. Wade, and James L. Dunmire for their outstanding contributions to the IC-camera-system fabrication, maintenance, and use.

This work was performed under the auspices of the U.S. Department of Energy by the University of California, Lawrence Livermore National Laboratory under contract W-7405-Eng-48; and was supported in part by the Office of Munitions, U.S. Department of Defense, Washington, D.C.

References

Baum DW, Shaw LL, Honodel CA (1992) Application of IC camera with laser illumination to shaped charge metal jet formation. In: Proc. 13th Int. Symposium on Ballistics, Vol. 2, ADPA, Arlington, VA, USA, pp 575–583.

Baum DW, Shaw LL, Simonson SC, Winer KA (1993) Shaped charge liner collapse and early jet formation. In: Proc. 14th Int. Symposium on Ballistics, Vol. 2, ADPA, Arlington, VA, USA, pp 13–22.

Carpluk GT (1970) Laser stereo photography. In: Proc. 9th Int. Congress on High Speed Photography, SPIE, Bellingham, WA, USA, p 64.

Chancel C, Pelte C (1972) Multi-frame image tube camera. In: Proc. 10th Int. Congress on High Speed Photography, SPIE, Bellingham, WA, USA.

Chavailler, C (1990) High speed photographic systems. In: Proc. 19th Int. Congress on High Speed Photography, Videography, and Photonics, SPIE 1358, Bellingham, WA, USA, pp 210–226.

Jones PA (1974) Four pulse Q-switched ruby laser. J Phys E 7:704.

Muelder SA, Huen T, Firpo CS, Hein DF, Steinmetz LL, Shaw LL (1987) Microchannel plate intensifier version of the IC camera. In: High Speed Photography, Videography, and Photonics V, SPIE 832, Bellingham, WA, USA, pp 57–62.

Owren HM (1959) Product bulletin for single frame image converter camera. Abtronics Inc., Livermore, CA, USA.

Paisley DL (1976) Laser stereo photography. In: Proc. 12th Int. Congress on High Speed Photography, SPIE 97, Bellingham, WA, USA, pp 184–188.

Parker VC, Crews JL (1987) Laser shadowgraphs with a rotating mirror camera. In: High Speed Photography, Videography, and Photonics V, SPIE 832, Bellingham, WA, USA, pp 88–95.

Shaw LL, Sonderman JB, Seppala LG (1979) 70 mm format rotating mirror framing camera. In: Proc. 13th Int. Congress on High Speed Photography, Videography and Photonics, SPIE 189, Bellingham, WA, USA, pp 218–221.

Shaw LL, Steinmetz LL, Behrendt WC, Sonderman JB, Mikkelson K, Beer GK, Seppala LG, Romero E (1984) Eight frame electro-optic camera with laser illumination. In: Proc. 16th Int. Congress on High Speed Photography, SPIE

491, Bellingham, WA, USA, pp 267–273.

Shaw LL, Muelder SA, Rivera AT, Dunmire JL, Breithaupt R (1993) Continuing evaluation of shaped charge metal jet characteristics using multi-frame electro-optic camera with laser illumination. In: Proc. 20th Int. Congress on High Speed Photography, SPIE 1801, Bellingham, WA, USA, pp 92–105.

Winer KA, Shaw L, Muelder S, Breithaupt D, Baum D (1993) Comparison of dynamic behavior of shear-formed shaped charge liners to computer simulations. In: Propellants, Explosives, Pyrotechnics, 18, Weinheim, Germany, pp 345–351.

Winer KA, Breithaupt D, Shaw L, Muelder S, Baum D (1995) High speed, high resolution observations of shaped charge metal jets undergoing particulation. In: Proc. 15th Int. Symposium on Ballistics, ADPA, Arlington, VA, USA, pp 89–96.

Numerical simulation of high-enthalpy impulse facilities

G. J. Wilson*
Thermosciences Institute
Moffett Field, CA 94035 USA

Abstract: Axisymmetric flows inside shock tubes and expansion tubes including the effects of finite-rate chemistry and both laminar and turbulent boundary layers are computed. A simulation of a low-pressure shock tube filled with nitrogen is carried out until the limiting flow regime is reached. Grid-refinement studies are made and the results are compared to Mirels' correlations. Simulations in the acceleration-tube section of the HYPULSE expansion tube are presented for two operating conditions, and the computations are compared with experimental data. One of these operating conditions is characterized by a laminar boundary layer and the other by a turbulent boundary layer. The simulations compare well with experiment and reveal that the nonuniformity in pressure observed during the test time is related to variations in the boundary-layer displacement thickness.

Key words: Shock tube, Expansion tube, Unsteady flow, Boundary layer

1. Introduction

One-dimensional analysis is a valuable tool for understanding and predicting the flow within shock tubes, shock tunnels, and expansion tubes. However, there are many non-ideal multi-dimensional phenomena which occur within these facilities which cause performance to deviate from one-dimensional theory. The result of these non-ideal processes is often temporally and spatially nonuniform test conditions and significantly reduced test times. Examples of these processes include the time-dependent growth of the boundary layer behind the incident shock, diaphragm-rupture mechanics, gas-interface instabilities, the propagation of transverse waves created by an area change or the diaphragm rupture, and the interaction of the reflected shock with the boundary layer. Understanding and quantifying these phenomena are important for interpreting experimental data and form a prerequisite for improving the design and operation of impulse facilities. Improvements in numerical methods and computer power provide opportunities for computational fluid dynamics (CFD) to add to the understanding of impulse-facility flows.

---

*Mailing address: MS 230-2, NASA Ames Research Center, Moffett Field, CA 94035.

Unsteady simulations of the viscous flow in shock tubes and expansion tubes are computationally expensive because they must include a large physical domain (i.e., the facility length) and simultaneously resolve small length scales associated with features such as the boundary layer or a gas interface. The modeling of a real gas adds to the computational expense by requiring multiple species and energy equations, and chemical kinetics. Recently, researchers have begun to explore the modeling requirements for detailed multi-dimensional simulations of shock tubes and expansion tubes. A few examples of these works include Weber et al. (1994), Daru and Damion (1993), Wilson et al. (1993), and Jacobs (1991). The literature suggests the research community is primarily investigating separate multi-dimensional phenomena in isolation in order to reduce the problem complexity. This approach will probably continue until more experience in simulating these flows is gained and computers become faster.

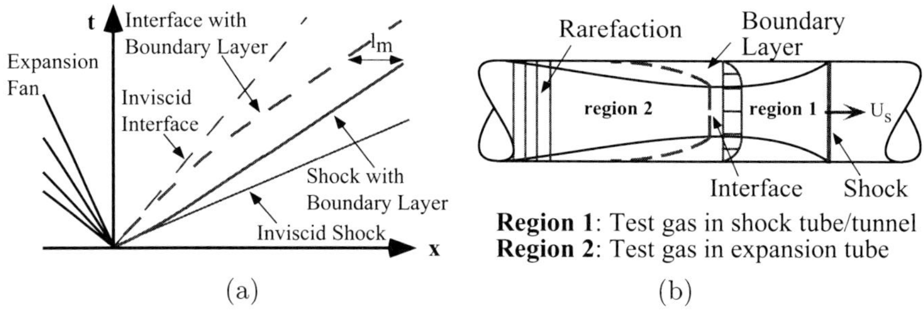

Figure 1. Shock-tube flow including the boundary layer (a) x-t diagram, (b) schematic diagram in laboratory coordinates.

The simulations presented herein focus on the influence of the boundary layer on shock-tube and expansion-tube flow. The x-t diagram of shock-tube flow in Fig. 1a illustrates the effect of the boundary layer on the shock and interface speeds. Shock speed decreases (attenuates) compared to the inviscid shock speed and the interface speed increases. After traveling a sufficient distance, the shock and interface reach the same speed and the separation distance remains constant at a limiting value, l_m. Figure 1b contains a schematic diagram of the major features of shock-tube flow with a boundary layer in laboratory coordinates. There is a large body of analysis, mostly due to Mirels (1963, 1966, 1971), which quantifies the nonuniformity of the gas between the shock and the interface (region 1) due to the boundary layer and which also quantifies the effects of the shock attenuation. However, this theory offers little detailed information about the effect of the boundary-layer growth on the expanded driver gas (region 2). In the acceleration section of an expansion tube, region 2 (expanded intermediate tube gas) is the test gas and the primary region of interest. Since CFD can provide detailed information about both regions, it can be an important tool for studying expansion-tube flow. In addition, CFD simulations can incorporate the effects of thermochemical nonequilibrium and turbulent transition which are not included in Mirels' analysis.

2. Numerical method

The axisymmetric flow through a shock tube or expansion tube is modeled using the thin-layer Navier-Stokes equations for a chemically reacting ideal-gas mixture. The present gas model includes three gas species (He, N_2, and N). A vibrational energy equation is included so that vibrational nonequilibrium can be modeled. Species diffusion is not included. The gas-dynamic equations are solved using a semi-implicit finite-volume form of the Harten-Yee upwind TVD scheme (Yee 1989). Grid points are clustered at the shock and interface and translated with these features to minimize numerical errors. The viscous terms are treated implicitly so that the allowable time-step size is dictated by the explicit stability criteria of the inviscid terms. The source terms representing the finite-rate chemical kinetics and vibrational relaxation are also implicit. The moving grid and single-step implicit algorithm reduce the formal time accuracy to first order. A single-equation algebraic model proposed by Cebeci and Smith (1974) is employed for turbulent-flow cases.

3. Shock-tube flow

Flow conditions within a shock tube are chosen so that the assumptions of Mirels' laminar analysis are valid (Sharma and Wilson 1995). The shock-tube diameter is chosen to be 0.05 m with a driver length of 0.76 m and a driven length of 12.0 m. A helium driver is used and the driven tube is filled with nitrogen at 70 Pa. The incident shock speed is approximately 2.9 km/sec so that there is little dissociation. The simulations are started by instantly removing a planar diaphragm and are run until the distance between the shock and interface has become nearly constant. The axial direction is divided into 700 grid points and the radius into 63 grid points. To assess the accuracy of the simulations, a grid-refinement study is performed with the number of grid points kept constant and minimum spacing at the wall reduced by half. Figure 2 is a plot of the computed distance between the shock and the interface versus time on four grids. The predicted separation distance at the end of the driven tube decreases with each reduction in wall-grid spacing and changes by less than 3% percent on the two finest grids. A computation with 126 points normal to the wall shows little difference compared to a solution with the same wall spacing and 63 points. For these conditions, both Mirels' (1963) analysis and the finest grid solution closely agree and predict a separation distance of 0.151 m.

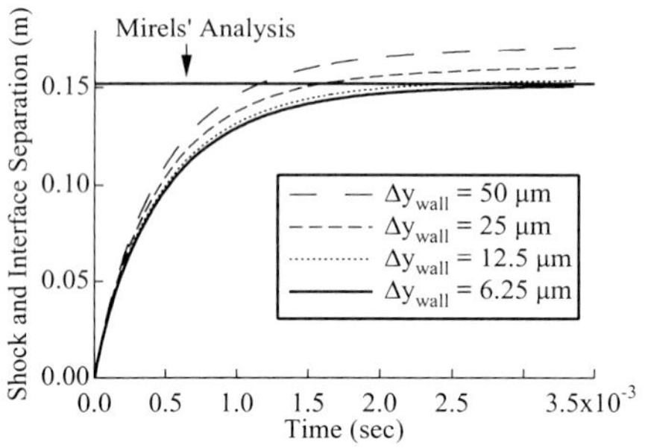

Figure 2. Time history of the separation distance between shock wave and interface for grid-refinement study.

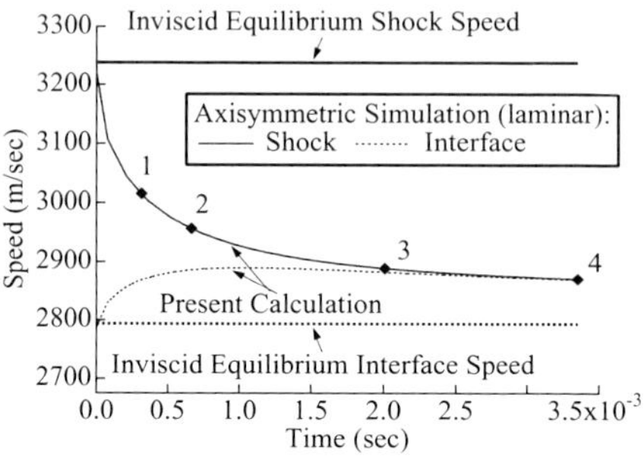

Figure 3. Time history of shock-wave and interface speed starting at diaphragm rupture.

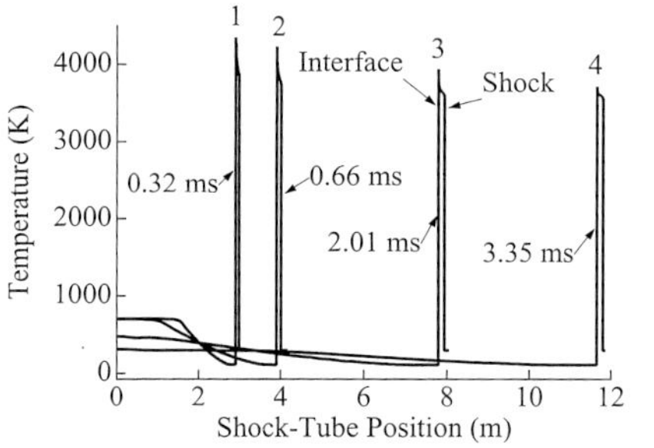

Figure 4. Shock-tube temperature profiles at four times corresponding to the markers on Fig. 3.

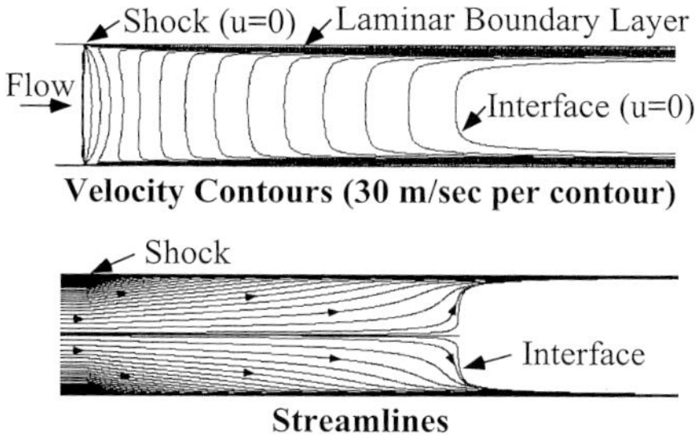

Figure 5. Limiting flow velocity contours and streamlines in shock-fixed coordinates.

Figure 3 shows the variation of the computed shock and interface speed in the driven tube compared with the speeds predicted from inviscid theory. The nearly identical shock and interface speed at the end of the shock tube shows that the limiting flow regime is reached. Figure 4 contains temperature profiles in the shock tube at the four times corresponding to the markers in Fig. 3. The temperature profiles show significant nonuniformity between the shock and interface due to shock attenuation, especially at early times when the rate of change of the shock speed is highest. The highest temperature gas occurs near the interface because this gas is processed by the highest speed shock. The temperature profile of the gas near the end of the tube is the most uniform because the shock speed is nearly uniform and the gas initially processed by the shock has been entrained into the boundary layer. At this point the variations are primarily due to the quasi-steady shape of the boundary layer.

Figure 5 shows computed velocity contours and streamlines in shock-fixed coordinates at the limiting flow regime. The velocity contours show the size and shape of the boundary layer and illustrate the nonuniformity in velocity due to the effective area change created by the boundary layer. The velocity decelerates from the post-shock velocity to zero velocity at the interface so that the shock and interface have the same speed. The streamlines show the gas crossing the shock entering the boundary layer before the interface. At the limiting flow regime, the mass flow across the shock is equal to amount of gas being entrained by the boundary layer. Figure 6 shows a comparison of

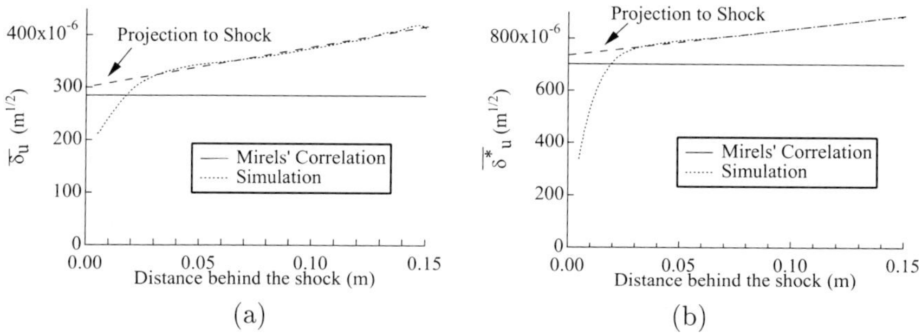

Figure 6. Comparison between Mirels' correlations and simulation (a) boundary-layer-thickness parameter, (b) boundary-layer-displacement-thickness parameter.

the computed boundary-layer-thickness and displacement-thickness parameters with values predicted with Mirels' (1966) correlations. Mirels' values for these quantities are constant because constant post-shock conditions are assumed. Agreement between the correlations and the simulations is expected just behind the shock. One reason for the discrepancy at this location is the curved incident shock captured by the simulations (see Fig. 5). Another possible reason is the lack of grid resolution in the very thin boundary layer immediately behind the shock. If values for the parameters at the shock front are predicted based on the trends observed farther behind the shock, as illustrated in Fig. 6, then the computed values are found to be in good agreement with Mirels' analysis. Differences between the two methods farther behind the shock presumably give an indication of the error introduced by the assumption of constant post-shock conditions made in Mirels' analysis.

4. Acceleration-tube flow

Simulations of the HYPULSE expansion tube are presented for two operating conditions and the computations are compared with experimental data (Wilson et al. 1995). The operating gas for both cases is nitrogen. The first test condition is at a total enthalpy of 15.2 MJ/kg and a pressure of 2 kPa and is referred to as the Mach 17 condition. This case is characterized by a laminar boundary layer and significant chemical nonequilibrium in the acceleration gas. The second test condition is at a total enthalpy of 10.2 MJ/kg and a pressure of 38 kPa and is referred to as the Mach 14 HP (high pressure) condition. This case is characterized by a turbulent boundary layer. These simulations begin with a shock wave in the intermediate tube which travels to the location of the secondary diaphragm. The secondary diaphragm is planar and is broken instantly.

4.1. Laminar boundary layer

A 14.62-m acceleration tube is used for the Mach 17 test condition giving a length-to-diameter (L/D) ratio of approximately 96. A 400×63 grid (63 points normal to the wall) is used and the grid spacing at the wall is varied to assess the required grid resolution in the boundary layer. For this case, the high post-shock temperature and the cold wall create an extremely small thermal boundary layer which is difficult to resolve. The finest grid-wall spacing attempted was 12.5 μm, which gave a value of l_m 5% smaller than the value obtained with a grid-wall spacing of 25 μm. A solution of the flow through the facility on the finest grid requires approximately 700,000 iterations and 70 hours of single-processor CPU time on a Cray C90. The value of l_m predicted by Mirels' (1963) analysis is calculated to be approximately 0.098 m, which compares to the value of 0.103 m from the simulation. Close agreement between the computation and analysis is not necessarily expected, as it was in the shock-tube case, because Mirels' assumption of chemical equilibrium behind the shock is not satisfied for this case.

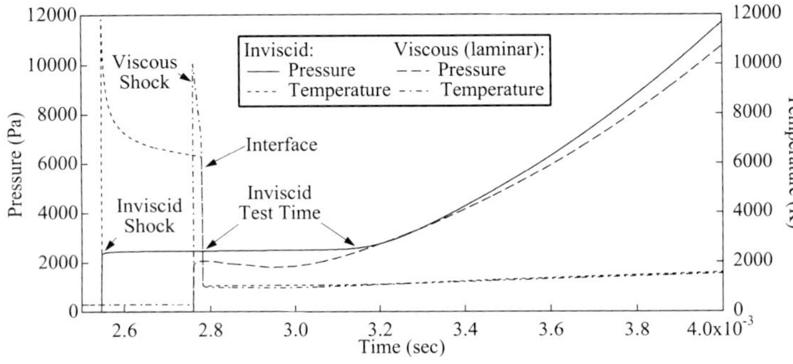

Figure 7. Computed pressure and temperature traces at the end of the acceleration for the Mach 17 condition (inviscid and laminar).

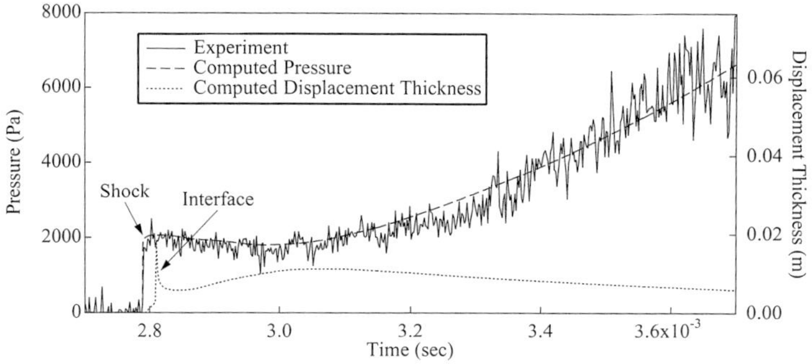

Figure 8. Computed and experimental pressure traces and computed displacement thickness for the Mach 17 condition.

Figure 7 presents pressure and temperature traces at the end of the acceleration tube for the inviscid and viscous solutions. The time of arrival of the interface for the two solutions has been aligned. The large temperature variation behind the acceleration-tube shock is primarily due to chemical nonequilibrium. The pressures and temperatures behind the shock for the viscous solution are lower than the values predicted by the inviscid solution because of shock attenuation. The laminar solution has a much smaller separation distance between the shock and the interface.

A comparison of experimental and computed pressure traces at the end of the acceleration tube is presented in Fig. 8, which shows good agreement for the post-shock pressure level and the variation of pressure during the test time. The computed displacement thickness at the exit of the tube is also presented. The height of the right axis is set equal to the radius of the acceleration tube to allow an easy comparison of the displacement thickness to the radius of the tube. The sharp peak in the displacement thickness at the interface results from the integration path crossing the curved interface. The minimum pressure during the test time is associated with a maximum in the displacement thickness.

For this condition, the agreement between experiment and computation for the temperature and pitot pressure is not as good as it is for the static pressure. Pitot pressure is overpredicted by 10–20% and temperature is similarly underpredicted. Simulations with varied initial conditions and including vibrational nonequilibrium do not account for the observed differences. A possible cause of the discrepancy may be the partial reflection of the intermediate-tube shock off of the secondary diaphragm which is observed in the experimental data. Simulations which delay (10 μsec) the opening of the secondary diaphragm support this conjecture. The processing of the test gas upstream of the secondary diaphragm by the reflected shock decreases the pitot pressure during the test time and increases the static temperature bringing the simulations in better agreement with the data. The static pressure, which is in good agreement with experiment, is found to be relatively insensitive to small delays in the opening of the diaphragm.

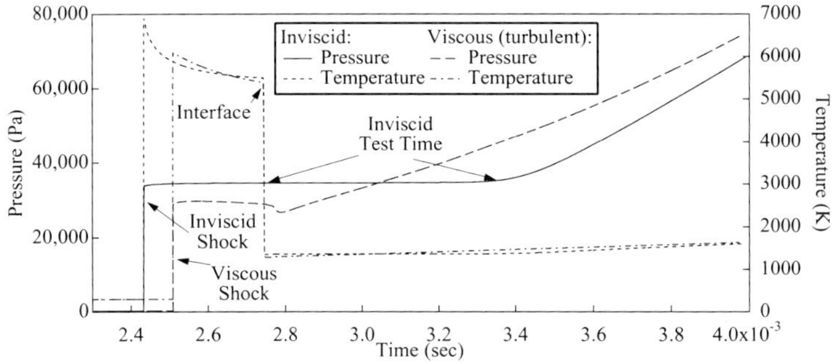

Figure 9. Computed pressure and temperature traces at the end of the acceleration for the Mach 14 HP condition (inviscid and turbulent).

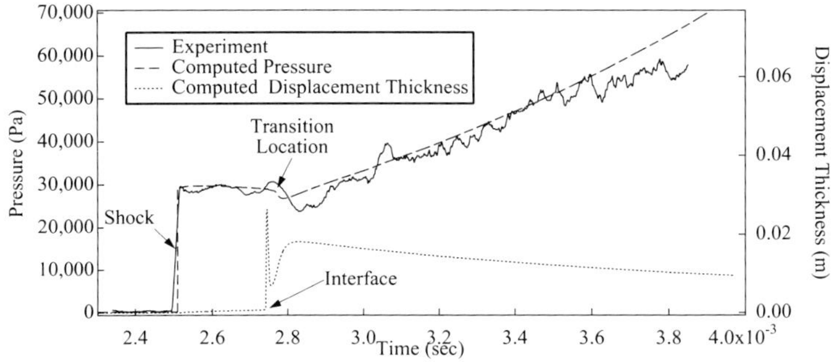

Figure 10. Computed and experimental pressure traces and computed displacement thickness for the Mach 14 HP condition.

4.2. Turbulent boundary layer

The Mach 14 HP condition uses a 10-m acceleration tube resulting in an acceleration-tube L/D ratio of approximately 66. The acceleration-tube fill pressure for the Mach 14 HP case is much higher than it is for the Mach 17 condition (193 Pa versus 7.2 Pa) and the acceleration-tube shock speed is lower (approximately 4 km/sec versus 5 km/sec). As a result, this case has a significantly higher post-shock pressure and a lower post-shock temperature with less dissociation, and the gas is much closer to chemical equilibrium. The flow does not reach the limiting flow regime by the end of the acceleration tube. Experimental heat-transfer measurements (Erdos et al. 1994) suggest that the boundary layer in the region between the shock and interface is laminar, and the transition to turbulence occurs at the start of the test time just behind the interface. For this reason, the simulations specify a turbulent boundary layer starting 2.54 cm behind the interface. A 450×53 grid is used with the wall grid spacing on the finest grid of 25 μm. It is known from the grid-resolution study that this grid accurately captures the time evolution of the shock and interface but slightly underpredicts the boundary-layer thickness in the test gas.

Figure 9 shows the pressure and temperature traces from inviscid and viscous solutions. A comparison of the experimental and computed pressure traces at the end of the acceleration tube is presented in Fig. 10. Both the experiment and simulation show similar characteristics and are in good agreement. There is a drop in pressure near the transition point and a steady rise in pressure during the test time. Figure 10 also shows the computed value of the displacement thickness at the exit of the tube with the plot axis scaled to the tube diameter. As with the Mach 17 simulation, the minimum pressure over the test time is associated with a maximum in the displacement thickness. The decrease in the displacement thickness during the test time seems to be the cause of the rise in pressure during the test time; however, it is difficult to distinguish between this effect and the arrival of the rarefaction. Additional studies of the effect of Reynolds number and acceleration-tube length on the turbulent boundary layer for these conditions are found in Wilson et al. (1995).

5. Conclusions and future work

The present work makes a detailed comparison between axisymmetric numerical simulations and Mirels' correlations for shock-tube flow with a boundary layer. It also represents an initial attempt to characterize laminar and turbulent flow in an expansion tube. These simulations are shown to provide valuable physical understanding about impulse-facility flow. The present numerical approach is computationally expensive and is impractical at higher Reynolds numbers where finer grid spacing at the walls is required to resolve the boundary layer. One goal of future numerical work should be to apply more advanced numerical methods such as higher-order schemes and fully-implicit time-accurate algorithms. There also is a need to identify or develop appropriate turbulence models for these hypersonic internal flows.

In addition to investigations of the boundary layer, numerical studies of the diaphragm-opening processes and pressure disturbances (noise) within shock tubes would be valuable. These topics are related because the diaphragm-rupture process is the likely cause of much of the noise observed in these facilities. Daru and Damion (1993) investigated flow perturbations in a shock tube due to diaphragm curvature. These computations also seem to show a mechanism of driver-gas contamination caused by the diaphragm curvature which has also been observed at higher pressure ratios by the author. Further investigations of these topics may lead to a better understanding of test-gas contamination, as well as facility noise.

Acknowledgement. Support to the author was provided by NASA contract NAS2-14031 to Eloret. The author thanks Dr. M. A. Sussman, Dr. S. P. Sharma, and Dr. R. J. Bakos for their significant contributions. Computer time was provided by the National Aerodynamic Simulation facility at NASA Ames Research Center.

References

Cebeci T, Smith AMO (1974) Analysis of Turbulent Boundary Layers, Academic Press, New York.

Daru V, Damion JP (1993) Analysis of the flow perturbations in a shock tube due to the curvature of the diaphragm. In: Shock Waves @ Marseille, Brun R and Dumitrescu LZ (eds), Springer, Vol 1, pp 451–456.

Erdos J, Calleja J, Tamagno J (1994) Increase in the Hypervelocity Test Envelope of the HYPULSE Shock-Expansion Tube, AIAA Paper 94-2524.

Jacobs PA (1991) Numerical Simulation of Transient Hypervelocity Flow in an Expansion Tube, NASA CR 189615.

Mirels H (1963) Test time in low pressure shock tubes. Phys Fluids 6:1201.

Mirels H (1966) Correlation formulas for laminar shock tube boundary layer. Physics of Fluids 9:1265–1272.

Mirels H (1971) Boundary growth effects in shock tubes. In: 8th Symp (Intl) on Shock Waves, London, pp 6/1–6/30.

Sharma SP, Wilson GJ (1995) Test Times in Hypersonic Shock Tubes, AIAA Paper 95-0713.

Weber YS, Anderson JD, Oran ES, Boris JP (1994) The Numerical Simulation of the Shock Bifurcation Near the End Wall in a Shock Tube, AIAA Paper 94-2307.

Wilson GJ, Sharma SP, Gillespie WD (1993) Time-dependent simulation of the reflected-shock/boundary layer interaction in a shock tube. In: Shock Waves @ Marseille, Brun R and Dumitrescu LZ (eds), Springer, Vol 1, pp 439–444.

Wilson GJ, Sussman MA, Bakos RJ (1995) Numerical Simulations of the Flow in the HYPULSE Expansion Tube, NASA TM 110357.

Yee HC (1989) A Class of High-Resolution Explicit and Implicit Shock-Capturing Methods, NASA TM 101088.

Blast effects of confined and unconfined explosions

S. B. Dorofeev
Russian Research Centre 'Kurchatov Institute'
Moscow, 123182 Russia

Abstract: A brief review of current and past results on the blast effect of confined and unconfined explosions is presented. The processes considered include unconfined gaseous and heterogeneous detonations, deflagrations, and confined explosions. Scaling of unconfined deflagration, detonation, and numerical simulations of confined explosions are discussed.

Key words: Blast waves, Detonation, Deflagration, Explosions, Scaling

1. Introduction

Confined and unconfined fuel-air explosions represent one of the most hazardous events in industrial accidents. A characteristic feature of these events is a distinguishable blast effect. At the same time, these explosions can vary widely in the physical processes involved and in initial and boundary conditions. The physical processes involved include a variety of explosion modes from rather slow deflagrations to detonations. Gaseous and heterogeneous mixtures can form an explosive. The mixtures can differ in stoichiometry, specific energy content, and chemical reaction rate. Shapes of explosive clouds can be different also. For confined conditions, these clouds can occupy different parts of an enclosure. These types of accidental explosions are the subject of much continuing research. The aim of this presentation is a brief review of results and current activity in this field at Russian Research Centre 'Kurchatov Institute.'

2. Unconfined fuel-air explosions

2.1. Gaseous detonations

The effects of unconfined gaseous detonations have been studied in many experiments and calculations (Kogarko et al. 1965; Fishburn 1976; Baker et al. 1983; Zhdan 1983; Brossard et al. 1984; Lannoy 1984; Borisov et al. 1985; Gelfand et al. 1985; Dorofeev et al. 1991, Dorofeev et al. 1993). The main objective of these studies was to obtain relationships for description of air blast, similar to those for high explosives (HE). The parameters of interest in these studies are maximum overpressure P^+, impulse I^+ of the positive phase of the wave, corresponding parameters of the negative phase P^-, I^-, and also duration of these phases. Sachs dimensionless variables are usually used for scaling, similar to those for HE explosions. Corresponding functions for overpressure and impulse of the positive phase are presented in Fig. 1.

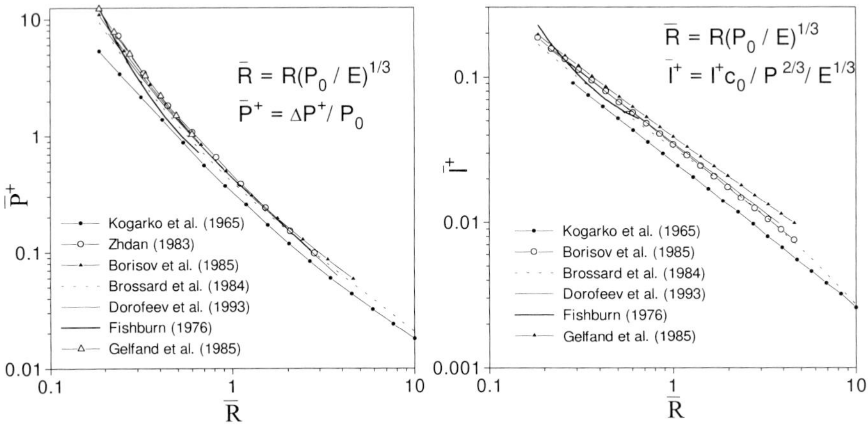

Figure 1. Positive overpressure and impulse in air blast wave versus distance (Sachs' variables) for gaseous detonation.

After about 30 years of experimentation and analysis, we have come to the conclusion that the description of the air blast from gaseous detonation is at the same level as that for HE explosions. The description of the negative phase, however, needs to be improved. In this situation, there is no reason to use TNT equivalents estimating the blast effect from gaseous detonation (Dorofeev et al. 1996a). The dimensionless overpressure and impulse for the positive phase can be calculated from the following expressions (Dorofeev et al. 1993):

$$\bar{P}^+ = 0.34/\bar{R}^{4/3} + 0.062/\bar{R}^2 + 0.033/\bar{R}^3, \tag{1}$$

$$\bar{I}^+ = 0.0353/\bar{R}^{0.968}, \qquad 0.21 < \bar{R} < 3.77. \tag{2}$$

Here and in the following discussion all parameters correspond to spherical explosion in air.

2.2. Heterogeneous detonations

The description of the air blast from detonation of heterogeneous clouds can be found in a similar form as that for gaseous explosives. Generally, the efficiency of these explosions is not equal to gaseous detonation because of incomplete combustion of fuel droplets in the detonation wave. The parameters of the blast wave depend on the fuel type, droplet size, vapor fraction, etc. Nevertheless, in a variety of applications, there is a need to have some approximate description. Such a description has been suggested (Dorofeev et al. 1991) on the basis of numerous experiments with air sprays of motor fuel. Corresponding approximations for near-stoichiometric mixtures are given by:

$$\bar{P}^{+} = 0.125/\bar{R}^{1} + 0.137/\bar{R}^{2} + 0.023/\bar{R}^{3}, \tag{3}$$

$$\bar{I}^{+} = 0.022/\bar{R}, \qquad 0.3 < \bar{R} < 2.0. \tag{4}$$

Some of the experimental results are presented in Fig. 2. The data presented show that the impulse depends on the vapor fraction, which is very different in gasoline and kerosene sprays.

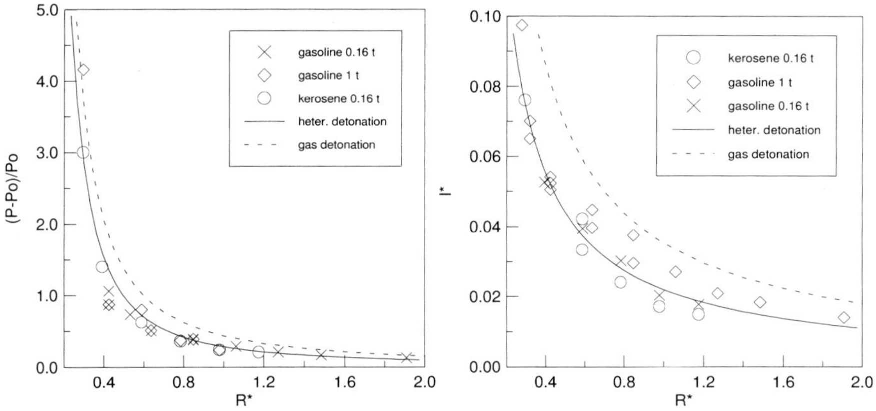

Figure 2. Positive overpressure and impulse in air blast wave versus distance (Sachs' variables) for detonation of heterogeneous clouds.

2.3. Fuel-rich clouds

The previous discussion addressed near-stoichiometric mixtures, while fuel-rich clouds are very typical in industrial accidents. In the case of a small additional amount of fuel (compared with the stoichiometric concentration), it is possible to find the blast parameters using the description for stoichiometric mixtures. It is necessary to calculate only the amount of energy release according to oxygen content or cloud volume. Systematic experiments on the effect of fuel concentration (Dorofeev et al. 1993) have shown that this is the case for all detonable concentrations. Corresponding experimental data for gaseous and heterogeneous mixtures are presented in Figs. 3 and 4. Variables scaled according to the mixture volume V are used in Fig. 4. This type of scaling is used because the clouds were not hemispherical in these experiments. Some data spread for the heterogeneous mixture is defined by difference in cloud shapes and by different vapor fractions.

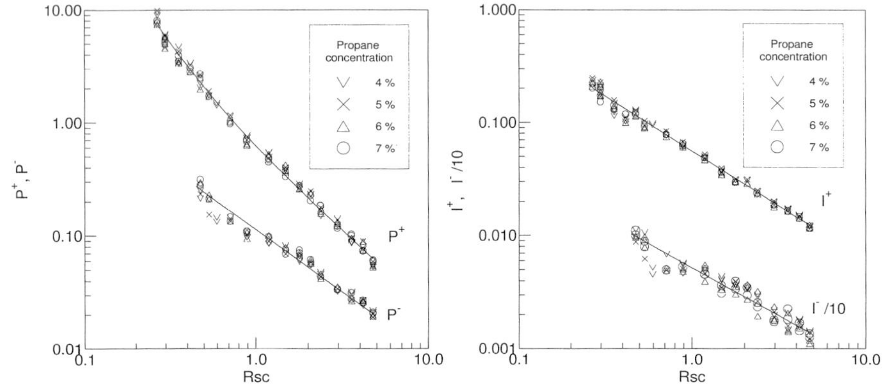

Figure 3. Positive and negative overpressures and impulses in air blast wave versus distance (Sachs' variables) for stoichiometric and fuel-rich gaseous mixtures.

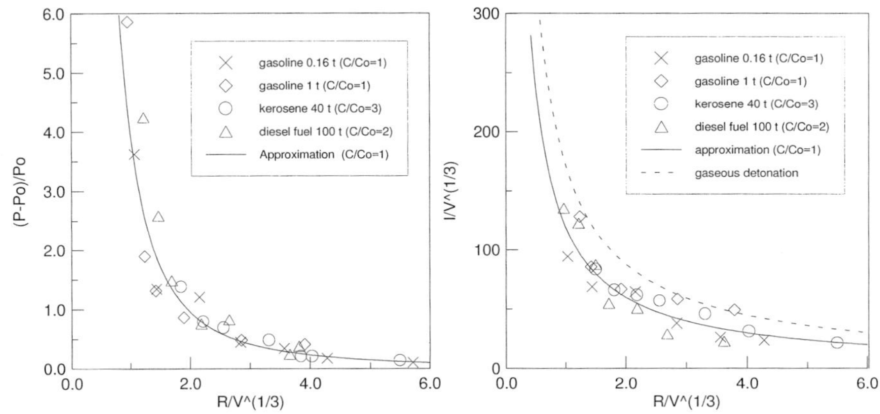

Figure 4. Positive overpressure and impulse in air blast wave versus distance for detonation of stoichiometric (C_o) and fuel-rich ($C/C_0 > 1$) heterogeneous clouds.

2.4. Unconfined deflagrations

While an unconfined detonation defines an upper bound of the explosion effect, this is not a very probable event in industrial accidents. An appropriate description for deflagration should necessarily include the flame speed. In this case, energetic Sachs' scaling can not be used without some modifications. Nevertheless, an estimate for the maximum overpressure and impulse at the cloud boundary as a function of the flame speed can be obtained by quasi-acoustic approximation. Corresponding analysis of the data for different flame speeds available in the literature (Kuhl et al. 1973; Strehlow RA 1975; Desbordes et al. 1978; Geisbrecht et al. 1981; and other studies) has given the following description of the blast parameters (Dorofeev et al. 1991):

$$\bar{P}^{+} = u^2 \frac{\sigma - 1}{\sigma}(0.83/\bar{R} - 0.14/\bar{R}^2), \tag{5}$$

$$\bar{I}^{+} = \frac{u}{c_0}\frac{\sigma - 1}{\sigma}(1 - 0.4\frac{u}{c_0}\frac{\sigma - 1}{\sigma})(0.06/\bar{R} + 0.01/\bar{R}^2 - 0.0025/\bar{R}^3), \tag{6}$$

where u denotes the apparent flame speed, c_0 the sound speed, and σ the expansion ratio. The value of the expansion ratio σ is about 7 for gaseous mixtures and decreases down to 4 for heterogeneous mixtures. The functions (5) and (6) were received for apparent flame speeds $u < 300$ m/s. The experiments conducted with higher flame speeds (Dorofeev et al. 1991) have shown that these functions are applicable also up to about 500 m/s. In this case, the blast parameters are given by:

$$\bar{P}^{+} = \min(\bar{P}_1, \bar{P}_2), \tag{7}$$

$$\bar{I}^{+} = \min(\bar{I}_1, \bar{I}_2), \tag{8}$$

where the values of $\bar{P}_1$, $\bar{P}_2$, $\bar{I}_1$, $\bar{I}_2$ are defined by (1), (5), (2), and (6) for gaseous, and (3), (5), (4), and (6) for heterogeneous mixtures, correspondingly. Figure 5 shows examples of experimental and calculated blast parameters for high-speed deflagrations. Although the prediction of turbulent flame speeds for some particular initial and boundary conditions is a rather complicated problem, an approximate description of the air blast for known flame speed can be suggested.

3. Confined explosions

A description of the blast effect of confined explosions should be different in form from that for unconfined ones. First, a simple engineering approach to find functions of blast parameters on distance is inappropriate, since 3D effects are of prime importance. Second, it becomes more important to predict the explosion mode, because both the effect and the mode appear to be strongly interconnected. Third, details of boundary conditions have a strong influence, and some universal description similar to dimensionless functions for air-blast parameters seems hardly possible. The only useful tool to describe confined conditions is multi-dimensional simulation of explosion processes. A simulation of all possible explosion modes in the frame of one model is not really practical at

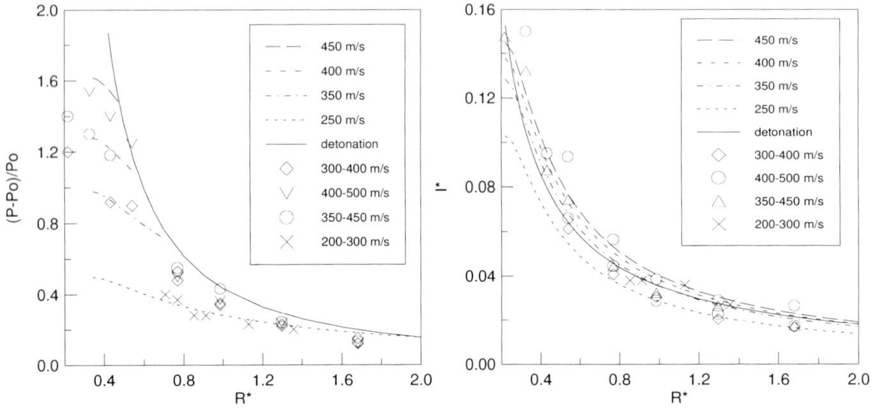

Figure 5. Positive overpressure and impulse in air blast wave versus distance (Sachs' variables) for deflagrations with different flame speeds.

present, since it requires computation of unsteady, turbulent, multi-dimensional compressible reacting flows. Simplified models are necessary, but they should be validated against appropriate experimental data.

3.1. Confined detonations

Confined detonations were studied in a series of large-scale experiments and 3D numerical simulations (Breitung et al. 1994; Breitung et al. 1996). Experiments have been carried out with hydrogen-air mixtures in a large confined volume of 263 m^3 (Fig. 6). The data obtained include all important features of confined 3D detonations, such as curved and oblique fronts, reflections, cellular structure, detonation decay, and reinitiation. Results of 3D simulation show good prediction of detailed load parameters for the case of stable developed detonation waves (Breitung et al. 1996).

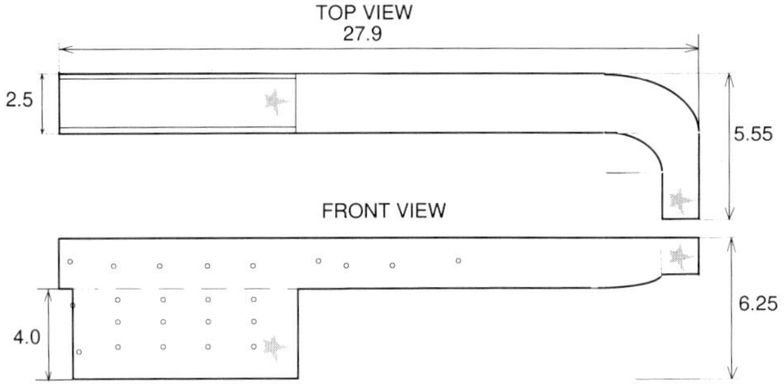

Figure 6. Scheme of the RUT facility. Stars show two initiation positions.

3.2. Transient explosion processes

The case of unstable detonations near the limit of propagation requires much more sophisticated models. First of all, an appropriate description of chemical reaction is necessary. Direct application of detailed kinetic schemes for detonation modeling has many problems (Efimenko et al. 1994). Turbulent effects seem to play a role in the complex 3D structure of detonations. Some simplified models of the rate of energy release for near-limit mixtures show rather reasonable agreement with experimental results (Breitung et al. 1996). An example of such a calculation is shown in Fig. 7. The pressure field for confined turbulent

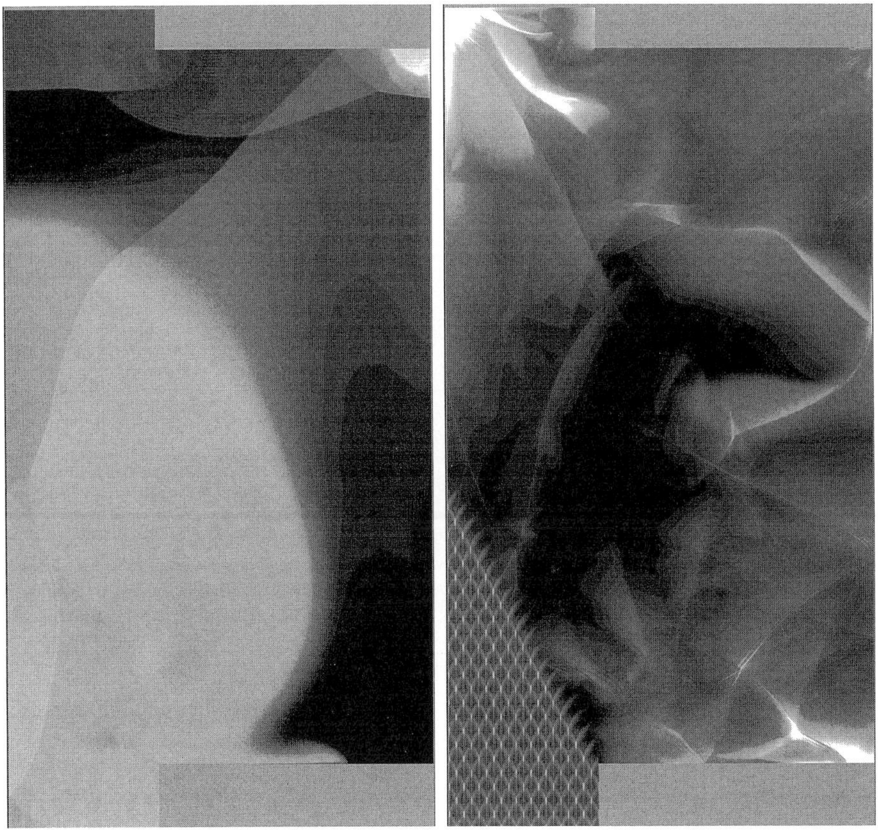

Figure 7. Two-dimensional simulation of marginal detonation in RUT facility. Detonation propagates upward from bottom of the figure. Reinitiation is observed in upper right corner. Temperature field (left) at 18.9 ms ranged from 388 to 2248 K. Maximum pressures (right) ranged from 1.2 to 9 MPa.

deflagration varies greatly in the range of processes from slow deflagration to DDT events (Fig. 8). A predictive modeling of turbulent deflagrations is not possible at present. Nevertheless, corresponding computer models are being developed. Some useful information for the estimation of an explosion effect can be obtained experimentally. An example is that impulse appeared to be almost independent of explosion mode, including fast turbulent deflagrations, detonations, and DDT events (Fig. 9). Another example is an estimation of conditions for spontaneous onset of detonations (Dorofeev et al. 1996b). An approximate scale limitation representing necessary condition for the onset of detonation can be suggested.

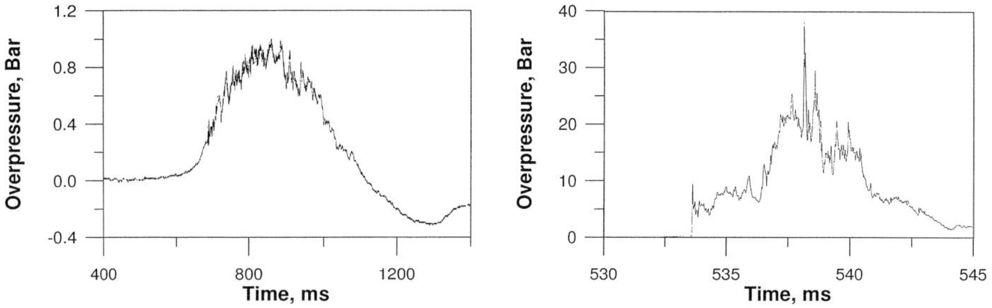

Figure 8. Overpressures versus time for slow deflagration (left) and DDT (right) in large confined RUT volume.

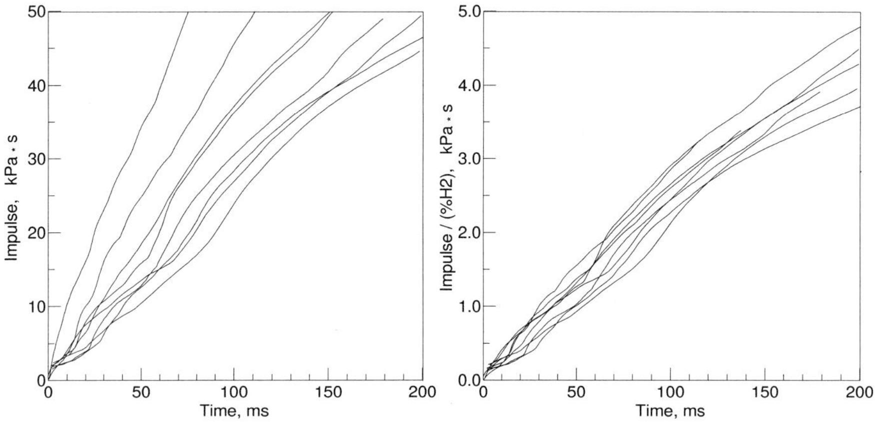

Figure 9. Positive impulse (left) and impulse normalized to hydrogen concentration (right) versus time for different explosion processes in RUT facility. Data range from slow deflagration to fully developed detonations.

4. Concluding remarks

In conclusion, the blast effect of confined and unconfined explosions can be described, with different degrees of confidence, for different types of processes. Confined and unconfined detonations seem to be the most-studied cases. Appropriate numerical tools and simple engineering descriptions of blast parameters are available. More problems are still to be solved for confined and unconfined turbulent deflagrations and for transient explosion processes.

Acknowledgement. The author is thankful to the staff of the Laboratory of Induced Chemical Reactions RRC-KI, and especially to V. Alekseev, A. Dvoinishnikov, A. Efimenko, A. Kochurko, M. Kuznetsov, and V. Sidorov for their technical excellency and dedicated work. This work was sponsored by Russian State Research Program 'Safety,' Project of Nuclear Safety Research (PSF) Forschungszentrum Karlsruhe, Germany, and the U.S. Nuclear Regulatory Commission.

References

Baker WE, Cox PA, Westine PS, Kulesz JJ, Strehlow RA (1983) Explosion Hazards and Evaluation, Elsivier, Amsterdam, Oxford, New York.

Borisov AA, Gelfand BE, Tsyganov SA (1985) On modelling of pressure waves formed by detonation and combustion of gas mixtures. J Phys Combust and Explos (Russian Acad of Sci) 2:90.

Breitung W, Dorofeev SB, Efimenko AA, Kochurko AS, Redlinger R, Sidorov VP (1994) Large scale experiments on hydrogen-air detonation loads and their numerical simulation. In: Proceedings of ARS'94 Int Topical Meeting on Advanced Reactor Safety, Pittsburg, PA, Vol 2, pp 733–745.

Breitung W, Dorofeev SB, Efimenko AA, Kochurko AS, Redlinger R, Sidorov VP (1996) Large-scale hydrogen-air detonation experiments and their numerical simulation. In: Shock Waves, Proc of the 20th ISSW, Sturtevant B, Shepherd JE, Hornung HG (eds), World Scientific, Vol 2, pp 1101–1106.

Brossard J, Leyer JC, Desbordes D, Saint Clouds JP, Hendrickx S, Garnier JL, Lannoy A, Perrot J (1984) Air blast unconfined gaseous detonations. Prog Astron Aeron 94:556.

Desbordes D, Manson N, Brossard J (1978) Explosion dans l'air de charges spheriques non confenees de melanges reactifs gazeux. Acta Astronautica 5:1009.

Dorofeev SB, Sidorov VP, Alekseev VI (1991) Air-Blast Parameters from Detonation of Gaseous, Heterogeneous and Fuel Rich Fuel-Air Explosives, State Research Program 'Safety' Report SRDO IRIS-91/1.6.08, Moscow.

Dorofeev SB, Sidorov VP, Dvoinishnikov AE, Alekseev VI, Kuznetsov MS (1993) Experimental study of air blast parameters from fuel rich mixtures detonation. In: 14th Int Colloquium on Dynamics of Explosions and Reactive Systems (ICDERS), 2:E1.4.1.

Dorofeev SB, Sidorov VP, Dvoinishnikov AE (1996a) Blast parameters from unconfined gaseous detonations. In: Shock Waves, Proc of the 20th ISSW, Sturtevant B, Shepherd JE, Hornung HG (eds), World Scientific, Vol 1, pp 673–678.

Dorofeev SB, Sidorov VP, Dvoinishnikov AE, Breitung W (1996b) Deflagration to detonation transition in large confined volume of hydrogen-air mixture. Comb and Flame (to appear).

Efimenko AA, Dorofeev SB, Shepherd J (1994) Chemical reaction kinetics and detonation modeling. Presented at 25th International Symposium on Combustion, W-I-P Abstracts The University of California, Irvine, p 251.

Fishburn BD (1976) Some aspects of blast from fuel-air explosives. Acta Astronautica. 3:1049.

Gelfand BE, Gubin CA, Mikhalkin VN, Shargatov VA (1985) Computation of shock wave parameters by detonation of combustible gaseous mixtures of a variable composition. J Phys Combust Explos (Russian Acad of Sci) 3:92.

Geisbrecht H, Hemmer G, Hess K, Leuckel W, Stoeckel A (1981) Analysis of explosion hazards on spontaneous release of inflammable gases into the atmosphere. Ger Chem Eng 4:315.

Kogarko SM, Adushkin VV, Liamin AG (1965) Investigation of spherical detonation in gas mixtures. Combust Explos Shock Waves, USSR 1-2:22.

Kuhl AL, Kamel MM, Oppenheim AK (1973) On flame generated self-similar blast waves. In: 14th Symposium International on Combustion, The Combustion Institute, pp 1201–1214.

Lannoy A (1984) Analyse des explosions air-hydrocarbures en milieu libre: Etudes deterministe et probabiliste du scenario d'accident. Prevision des effets de surpression. Bulletin Direct, Etudes et Recherches EDF, A4.

Strehlow RA (1975) Blast waves generated by constant velocity flames. Comb and Flame 24:257.

Zhdan SA (1983) Calculation of gas mixture explosion with regard to shift of chemical equilibrium products. J Phys Combust Explos (Russian Acad of Sci) 1:131.

Anomalous physical effects from artificial numerical length scales

R. Menikoff, K. S. Lackner
Theoretical Division
Los Alamos National Laboratory
Los Alamos, NM 87545 USA

Abstract: Shock-capturing algorithms are widely used for simulations of compressible fluid flow. Though these algorithms resolve a shock wave within a couple of grid points, the artificial length scale from the numerical shock profile can have side effects. The side effects are similar to physical effects that occur when a relaxation process gives rise to fully or partly dispersed shock waves. Two anomalies due to a non-zero shock width are discussed: (i) in one dimension, a non-decaying entropy spike results from a transient when a shock profile is formed or changed; (ii) in higher dimensions, front curvature affects the propagation of a shock wave. We show that both the entropy anomaly and the curvature effect are natural consequences of the conservation laws. The same analysis applies both to the continuum equations and to their finite-difference approximations in conservation form. Consequently, the artificial length scale inherent in a shock-capturing algorithm can mimic real physical effects that are associated with partly dispersed shock waves.

Key words: Shock-capturing, Wave propagation, Shock interaction, Curvature effect, Hugoniot jump conditions

1. Introduction

Fluid flow is described by the Euler equations. When transport effects are neglected, the Euler equations are a system of hyperbolic partial differential equations (PDEs) modeling the conservation of mass, momentum, and energy. An understanding of the wave structure of solutions to the PDEs plays a crucial role in developing numerical algorithms for fluid flow.

In the early days of computer simulations it was found that straightforward finite-difference algorithms of the Euler equations were adequate for smooth solutions but were unstable for shock waves. There are two reasons for the numerical instability: (i) the truncation error associated with the finite-difference discretization is large at a shock discontinuity; (ii) the hyperbolic PDEs have no dissipative terms yet entropy must increase across a physical shock. To cure the numerical difficulty, von Neumann & Richtmyer (1950) introduced the concept of artificial viscosity. The idea is to add dissipation to the equations in such a way that (i) smooth flows are minimally affected, (ii) the state behind a shock is correct, and (iii) the time step for an explicit algorithm is not unduly restricted. This is achieved by smoothing out the discontinuity and giving a profile to a shock wave. The profile is analogous to what occurs for real viscosity. However, the numerical profile must be a few cells wide, and hence is artificially large compared to the real physical shock width.

This numerical procedure works because (i) an artificial profile is a localized error that typically does not affect the global solution, and (ii) the conservation laws give rise to the Hugoniot jump conditions which determine the state behind a shock independent of the form of dissipation generating the entropy. This led to the development of "shock-capturing" algorithms. By modifying the equations such that a shock has a narrow profile, a simple uniform algorithm can be used for the entire flow rather than employing a specialized treatment to detect and track a shock discontinuity. This was especially important in the days when computers were limited in both speed and memory.

With current computers, much more elaborate algorithms and much higher resolution are possible. Shock-capturing algorithms have evolved to use Riemann solvers or approximate Riemann solvers to generate the dissipation necessary for a shock and to account properly for the flow of information along characteristics. They generally give accurate solutions and spread a shock wave over only a couple of grid points. However, calculated solutions with shock-capturing algorithms sometimes display "glitches" which usually manifest themselves as localized errors; see e.g., Quirk (1994). We show that some of these numerical anomalies mimic real physical effects and are a consequence of the artificial shock width.

Two types of physical effects are described. In Section 2, we analyze in one dimension an entropy anomaly that arises from the interaction of fully or partly dispersed shock waves. The entropy anomaly explains a wide class of common numerical errors such as excessive wall heating when a shock reflects from a rigid wall, startup errors when a shock is initialized as a discontinuity, errors from an impedance mismatch when a shock impacts a material interface, and errors from a shock passing through an abrupt change in mesh spacing. In Section 3, we show in higher dimensions that front curvature leads to a modification of the Hugoniot jump conditions. A consequence of the modified jump conditions is the diameter effect in a rate stick of explosive. The curvature effect also explains the artificially large numerical diameter effect that occurs when the reaction zone of a detonation wave is captured and not resolved. Our results are summarized in Section 4.

2. Transient effects in one dimension

The Euler equations for fluid flow in one dimension can be expressed in conservation form as

$$\frac{\partial}{\partial t}\begin{pmatrix}\rho\\ \rho u\\ \rho E\end{pmatrix}+\frac{\partial}{\partial x}\begin{pmatrix}\rho u\\ \rho u^2+P\\ \rho u(E+PV)\end{pmatrix}=\begin{pmatrix}0\\0\\0\end{pmatrix}, \tag{1}$$

where ρ, e, and u are the density, specific internal energy, and particle velocity. $E = \frac{1}{2}u^2 + e$ is the total specific energy, $V = 1/\rho$ is the specific volume, and the pressure $P(V, e)$ characterizes the gas or fluid of interest. For a steady-state shock wave, the conservation form of the equations leads to the Hugoniot jump conditions. The shock speed and the state behind a shock are independent of the dissipative mechanism that generates the necessary shock entropy.

The jump conditions do not apply to transients when a shock profile is formed or changed. We show that the amount of entropy generated in a transient depends on the dissipative mechanism. When heat conduction is neglected, outside a shock profile entropy is convected along particle trajectories. Consequently, the entropy generated in a transient is frozen into the particle trajectories and does not diffuse. When shock waves interact, a localized entropy anomaly originates in the region in which the shock profiles of the incoming waves overlap.

As an example, we consider the simple case of a partly dispersed shock wave reflecting from a rigid wall. Because this example is equivalent to the symmetric collision of two shock waves, it also serves to illustrate the entropy anomaly caused by the interaction of shock waves. Moreover, this example has been studied in shock-tube experiments. The gas and incoming shock strength can be selected such that a relaxation process causes the shock to have a measurable width. Experiments and numerical simulations with relaxation processes included show that the reflection of a partly dispersed shock wave results in an entropy layer at the wall. This is nicely summarized in a review article by Becker (1972), see in particular his Fig. 9.

Sometime later an idealized problem was formulated by Noh (1987) to test the effect of artificial viscosity in calculations using the Euler equations. In the Noh problem, an ideal γ-law gas is initialized to have a uniform state consisting of a cold gas and a velocity directed towards a rigid wall. The problem has a simple analytic solution consisting of a strong shock wave propagating away from the wall. His calculations showed that a numerical error occurred near the wall. The particle velocity and pressure were correct but the specific energy was too large and was compensated for by a low density. Noh called this numerical anomaly "excess wall heating." It is qualitatively similar to the entropy layer at the wall that had previously been measured in experiments with partly dispersed shock waves.

Both the physical effect and the numerical effect can be explained based on the conservation laws. We outline the analysis given by Menikoff (1994). The solution with dispersed shock waves is compared to the corresponding solution of the Euler equations with discontinuous shock waves. We start by defining the front position for a dispersed shock wave. Let the subscripts a and b denote the states on the shock Hugoniot ahead and behind the shock front. The shock position x_s is defined such that the mass of the dispersed wave matches that of the idealized discontinuous shock. For a right-facing wave,

$$\int_{x<x_s} [\rho_b - \rho(x)]dx = \int_{x>x_s} [\rho(x) - \rho_a]dx \ . \tag{2}$$

Matching the momentum would give the same position. But the position based on matching the total energy is frame dependent. However, the relative energy between the dispersed wave and the discontinuous shock

$$\delta\mathcal{E} = \int_{x<x_s} [\rho(x)E(x) - \rho_b E_b]dx + \int_{x>x_s} [\rho(x)E(x) - \rho_a E_a]dx \tag{3}$$

is Galilean invariant. The key observation is that the relative energy depends on shock strength.

The asymptotic outgoing waves from a shock interaction are the solution to the Riemann problem with initial data corresponding to the states behind the incoming waves. Typically, a shock profile forms and equilibrates on a very fast time scale. The asymptotics set in after the outgoing waves are separated by a distance on the order of the sum of the widths of the equilibrium shock profiles. Otherwise, the dissipative mechanism would affect smooth flows as well as generating the entropy needed for a shock wave.

We consider the case of a symmetric shock collision. This is equivalent to the reflection from a rigid wall. Let the position of the wall be $x = 0$ and let a left-facing shock wave impact the wall at $t = 0$. Suppose at time $t > 0$, the outgoing shock wave has reached its asymptotic steady-state profile. The flux of mass and energy into a fixed region that contains the outgoing wave is given by

$$\hat{m} = -\rho_a u_a \qquad \text{and} \qquad \hat{E} = -\rho_a u_a(E_a + P_a V_a) \ .$$

A uniform state behind the outgoing shock would lead to the equations for conservation of mass and total energy

$$\begin{aligned} (\rho_b - \rho_a) \cdot x_s &= \hat{m}_L t \ , \\ \Big[(\rho E)_b - (\rho E)_a\Big] \cdot x_s &= \hat{E}t + \delta E_{\text{in}} - \delta E_{\text{out}} \ . \end{aligned}$$

Since we are considering a symmetric collision, the total momentum is always zero and momentum conservation is identically satisfied.

The conservation laws reduce to two equations for the one unknown shock position. The constraints from the Hugoniot jump condition imply that there is a solution for the hyperbolic problem when $x_s = \sigma t$, where σ is the outgoing shock speed. The difference in the relative energies of the incoming and outgoing shock waves, $\delta E_{\text{in}} - \delta E_{\text{out}}$, is not zero. This upsets the degeneracy of the equations. Therefore, the states behind the outgoing shock wave can not be uniform. Consequently, an anomaly must occur when the incoming and outgoing wave profiles overlap.

Without heat conduction, the entropy anomaly is frozen into the particle trajectories. Because of the wall, the velocity behind the outgoing shock must be zero. Therefore, the entropy anomaly appears as a boundary layer at the wall. For weak waves the entropy anomaly is small because the entropy change across a shock is third order in the shock strength. The anomaly is large for the Noh problem because the outgoing wave is a strong shock. Thus, by isolating and exaggerating a single numerical effect, the Noh problem is a good test problem for shock-capturing algorithms.

The argument based on the conservation laws can be extended to a non-symmetric wave interaction. In general there are three degrees of freedom corresponding to the positions of the three outgoing waves, and three equations for conservation of mass, momentum, and total energy. Typically, the contact is smeared out and there are small offsets or phase shifts in the asymptotic wave positions compared to the hyperbolic problem. Entropy is an important variable for the transient when wave profiles overlap because of its role as an adiabatic invariant.

A pronounced entropy anomaly at the contact occurs in some cases. A shock wave impacting a material interface is an important example. In this case, there is a separate conservation law for the mass of each material. This leads to four equations for the three unknown wave positions. Again a constant state behind the outgoing waves is not possible. The entropy anomaly can be large when the equations of state of the two materials are very different. When the equations of state are similar, for an Eulerian algorithm, mass diffusion from advection through a mixed cell can cause the entropy anomaly at the contact to artificially diffuse away. A pronounced entropy anomaly also occurs when a shock wave impacts an abrupt change in mesh, and when a shock wave is initialized as a discontinuity. In the last case, in addition to an entropy error which convects along the particle trajectories, there can be a weak wave of the opposite family. The initial data are not compatible with an additional shock. Instead there is an N-wave or discrete diffusion wave, see Chern (1991).

A numerical shock width may have additional side effects. The mass, momentum, and energy within the discrete profile oscillate as the shock front propagates through a cell. The oscillation acts as a source of acoustic noise. For upwind schemes the noise radiates behind the shock front. Numerical stability requires sufficient dissipation to damp short acoustic wavelengths which the mesh can not resolve. However, for a nearly stationary shock in an Eulerian mesh, the acoustic noise has a long wavelength. This is indistinguishable from acoustics waves that the grid is supposed to resolve. Consequently, it is difficult to damp the noise without affecting real signals. This example shows how the discretization can break the Galilean invariance of the continuum equations.

3. Wave propagation with curved front

In higher dimensions, a discontinuous shock wave obeys the Hugoniot jump conditions in the direction normal to the front. However, when the wave has a width, there is a correction to the jump conditions from the front curvature. This gives rise to a curvature effect on a propagating partly dispersed shock wave. A good example of the combined effect of a non-zero wave width and front curvature is the diameter effect in an explosive as measured in rate-stick experiments.

A rate stick is a cylinder of explosive. When detonated at one end, the flow rapidly approaches a steady-state detonation wave. The front is curved due to a rarefaction from the side boundary. Moreover, the asymptotic detonation velocity decreases as the diameter of the rate stick decreases. The diameter effect of the rate-stick experiments is a consequence of the curvature effect; see e.g., Menikoff (1989) and references contained therein.

In the model of Zel'dovich, von Neumann, and Doering (ZND), a detonation wave is a shock followed by a thin reaction zone. It is similar to a partly dispersed shock wave. Because of the chemical energy released, there exists a self-sustaining detonation wave; the lead shock heats the explosive and triggers a chemical reaction which releases energy and drives the lead shock. Moreover, an important effect of the energy release on the Hugoniot locus is that the self-sustaining detonation has the minimum possible wave speed. The curvature effect is an inescapable conclusion of the fact that the detonation speed measured in a rate-stick experiment is less than the minimum value for a planar detonation wave obtained from the Hugoniot jump conditions.

Rate-stick experiments can be simulated using the reactive Euler equations. A systematic numerical study by Donguy & Legrand (1981) used a two-dimensional Eulerian code with a uniform grid and varied both the charge diameter and the cell size. With a fine grid, the calculated diameter effect agreed with experiments. However, on a coarse grid, the reaction zone was not resolved and the numerical diameter effect exceeded the physical effect. Furthermore, the numerical effect increased with cell size. On a coarse grid, the numerical reaction dynamics serves as an artificial process to give the detonation wave a profile, and plays an analogous role to artificial viscosity for a shock wave. In effect, there is a numerical curvature effect that increases with the artificial reaction-zone width. The numerical variation in the curvature effect with cell size is qualitatively similar to experimental results on heterogeneous explosives that used the distribution of grain sizes to vary the reaction zone width; see Aleksandrov et al. (1967) and Moulard (1989).

The analysis by Menikoff et al. (1995) explains both the physical and numerical curvature effects in terms of the conservation laws. We assume that the detonation front is smooth and the reaction zone width w is sufficiently small. Then within the reaction zone the derivatives of the flow variables tangent to the front are small compared to those normal to the front, and the transit time for a particle in the reaction zone is small compared to the time scale for variations in the flow outside the reaction zone. These assumptions imply a long-wavelength, low-frequency approximation that describes the average behavior of a detonation wave and also is consistent with the typical mesh resolution used in numerical simulations. Locally the reaction zone can be approximated as one dimensional and quasi-steady. However, for a curved front, the normal directions are not parallel to each other. This effect can not be neglected. We show that the leading-order effect in higher dimensions can be accounted for by a geometric source term.

Consider the local flow tube generated by the streak lines from a small patch on the detonation front. Let A be the cross-sectional area of the flow tube. The reaction zone is modeled by the reactive flow equations in a duct of variable cross-sectional area

$$\frac{\partial}{\partial t}\begin{pmatrix}\rho A\\ \rho A u\\ \rho A E\\ \rho A\lambda\end{pmatrix}+\frac{\partial}{\partial x}\begin{pmatrix}\rho A u\\ \rho A[u^2+PV]\\ \rho A u(E+PV)\\ \rho A u\lambda\end{pmatrix}=\begin{pmatrix}0\\ P\partial_x A\\ -P\partial_t A\\ \rho A\mathcal{R}\end{pmatrix}, \tag{4}$$

where λ is the mass fraction of the reaction products and $\mathcal{R}$ is the specific reaction rate.

The flow tube is not Galilean invariant. There are two special frames in which to leading order $\partial_t A = 0$ and the equations simplify: (i) the shock-attached frame, and (ii) the normal frame. In these frames the only geometric term that enters the duct equations is the local divergence of the streak lines, $\kappa = \lim_{A\to 0}(\partial_x A/A)$. The shock-attached frame is unsuitable because $\kappa(x)$ is not *a priori* known. The normal frame is defined by the condition that the velocity ahead of the front is on average normal and has magnitude given by

$$\kappa_f u_{\text{ahead}} = \nabla_T \cdot \vec{u} \ , \tag{5}$$

where κ_f is the front curvature and $\nabla_T \cdot \vec{u}$ is the velocity divergence in the plane tangent to the front. In two dimensions the normal coordinates correspond to the streak lines. In three dimensions the streak lines are neither normal to the front nor do they lead to a stationary flow tube. Nevertheless, in this distinguished frame $\kappa = \kappa_f$ is constant to $O(\kappa_f w)$ (Menikoff et al. 1995).

To leading order the reaction zone is quasi-steady and adjusts adiabatically to its local front curvature. The PDEs then reduce to ODEs for the reaction-zone profile. The ODEs extend the ZND model of a planar detonation wave to the case of a curved detonation wave. Integrating the ODEs across the reaction zone leads to the modified jump conditions

$$\Delta[\rho(D-u)] = \kappa_f w \, \langle \rho u \rangle \ , \tag{6}$$

$$\Delta\left[(\rho(D-u))^2 V + P\right] = \kappa_f w \, \langle \rho(D-u)u \rangle \ , \tag{7}$$

$$\Delta\left[E + PV + \frac{1}{2}(D-u)^2\right] = 0 \ , \tag{8}$$

where D is the detonation velocity, $\Delta[f] = f(x_a) - f(x_b)$ is the change of variable f across the detonation wave, and

$$\langle f \rangle = \frac{1}{w} \int_{x_b}^{x_a} dx \, f$$

is the average value of f in the reaction zone. The left-hand sides correspond to the standard jump conditions, and the correction terms on the right-hand sides are proportional to the dimensionless product $\kappa_f w$. Because these relations are derived from the quasi-steady approximation, they are only valid in the normal frame and are only accurate to $O\Big((\kappa_f w)^2\Big)$.

In contrast to a planar detonation in which the jump conditions depend solely on the equation of state of the detonation products, for a curved detonation the jump conditions are also affected by the reaction-zone dynamics, in particular by the competition between the source term for the heat release and the geometric source term due to the front curvature. The averages are slowly varying functions. The dominant factor in the correction term is the reaction-zone width. Consequently, a large numerical reaction-zone width leads to an artificially large curvature effect. Similarly, changing the grain size of an explosive varies the reaction-zone width and has a corresponding effect on the physical curvature effect.

The one-dimensional approximation breaks down in regions of high front curvature when the tangential flow gradients are comparable to the normal gradients. This also occurs in the interaction region in which wave profiles overlap. When the interaction of discontinuous waves does not have a unique solution based on a shock polar analysis then the dynamics within the wave profiles would resolve the non-uniqueness. For example, the transition between regular and Mach reflection is a threshold phenomenon and could be affected by the dissipative mechanism that gives shock waves a profile. The experiment by Lambourn & Wright (1965) shows that the von Neumann interaction identified by Colella & Henderson (1990) occurs when Mach reflection of detonation waves is expected.

Since the physical dissipative mechanism is difficult to vary experimentally, numerical calculations in which the wave profile is resolved would be very useful for developing an understanding of how the bifurcation of two-dimensional wave patterns can be affected by dissipative mechanisms that take place on short length scales and that for a shock-capturing algorithm represent sub-grid physics. Transitions can depend on dimensionless ratios of length scales associated with different dissipative mechanisms even in the limit as the shock width vanishes. In assessing the accuracy of numerical solutions, it is important to know when localized errors can trigger phenomena that change the global flow.

4. Summary

The conservation form of the fluid equations leads to strong constraints on the wave structure of fluid flow. In one dimension, when a shock wave has a non-zero width, transients can cause an entropy anomaly. This is because of the mass, momentum, and energy contained within a shock profile that must be accounted for when a profile is formed or changed. An entropy anomaly commonly results from interactions involving shock waves. The effect has been measured in experiments with partly dispersed shock waves. In this case, the shock width is due to a physical relaxation process.

In higher dimensions, the leading-order effect on a shock profile is an extra geometric source term due to front curvature. The competition between source terms results in a modification of the Hugoniot jump conditions proportional to the dimensionless product of the shock width and the front curvature. This leads to a curvature effect on the propagation of a partly dispersed shock wave. The curvature effect is most noticeable for a detonation because of the minimum speed at which a steady-state detonation wave can propagate. The effect has been measured in rate-stick experiments.

In order to simulate correctly the propagation of a steady-state shock wave, finite-difference shock-capturing algorithms are written in conservation form.

Numerical solutions are subject to the same analysis as those for the continuum PDEs. A numerical shock width has qualitatively the same effect as the width of a partly dispersed shock wave from a relaxation phenomenon or a chemical reaction. Hence, an artificial numerical length scale can mimic real physical effects. This analogy provides a unifying explanation for a large class of diverse and seemingly unrelated numerical errors.

Acknowledgement. This research supported by the U.S. Department of Energy, Los Alamos National Laboratory.

References

Aleksandrov EN, Veretennikov VA, Dremin AN, Shvedov KK (1967) The detonation mechanism in porous explosives. Fizika Goreniya i Vzryva 3:471–484. (English translation in Combustion Explosion and Shock Wave 3:285–293).

Becker E (1972) Chemically reacting flow. Ann Rev of Fluid Mech 4:155–194.

Chern I-L (1991) Large-time behavior of solutions of Lax-Friedrichs finite difference equations for hyperbolic systems of conservation laws. Mathematics of Computation 56:107–118.

Colella P, Henderson LF (1990) The von Neumann paradox for the diffraction of weak shock waves. J Fluid Mech 213:71–94.

Donguy P, Legrand N (1981) Numerical simulations of non-ideal detonations of a heterogeneous explosive with the two dimensional Eulerian code C.E.E. In: Seventh Symposium (Int) on Detonation, pp 695–702.

Lambourn BD, Wright PW (1965) Mach interaction of two plane detonation waves. In: Fourth Symposium (Int) on Detonation, pp 142–152.

Menikoff R (1989) Determining curvature effect on detonation velocity from rate stick experiments. Impact of Computing in Science and Engineering 1:168–179.

Menikoff R (1994) Errors when shock waves interact due to numerical shock width. SIAM J Sci Comp 15:1227–1242.

Menikoff R, Lackner KS, Bukiet BG (1995) Modeling flows with curved detonation waves. Combustion and Flame (to appear).

Moulard H (1989) Particular aspect of the explosive particle size effect on shock sensitivity of cast PBX formulations. In: Ninth Symposium (Int) on Detonation, pp 18–24.

Noh WF (1987) Errors for calculations of strong shocks using artificial viscosity and an artificial heat flux. J Comp Phys 72:78–120.

Quirk JJ (1994) A contribution to the great Riemann solver debate. Int J for Numerical Methods in Fluids 18:555–574.

von Neumann J, Richtmyer RD (1950) A method for the numerical calculation of hydrodynamic shocks. J Appl Phys 21:232–237.

Curved detonation fronts in solid explosives: collisions and boundary interactions

J. B. Bdzil
Los Alamos National Laboratory
Los Alamos, NM 87545 USA

T. D. Aslam, D. S. Stewart
TAM Department, University of Illinois
Urbana, IL 61801 USA

Abstract: Detonation Shock Dynamics (DSD) can be used to model the effects that shock curvature κ has on detonation speed $D_n(\kappa)$. At the edges of the explosive, $D_n(\kappa)$ is supplemented with boundary conditions. By direct numerical simulation (DNS), we study how the reaction zone interacts with the edge. DSD theory has been integrated with the level-set method of Osher & Sethian and the Los Alamos DNS code Mesa to create a powerful tool for simulating complex explosive-containing systems.

Key words: Detonation, Curvature effect, Edge interactions, Mach reflection

1. Introduction

To accurately predict the propagation of detonation through an explosive, one needs to model the physics that occurs on the chemical reaction-zone scale η_r. In sharp contrast to the structured shocks observed for gaseous detonation, those for heterogeneous solid explosives are broadly curved on the η_r scale. The speed of the detonation is strongly influenced by the curvature of the shock κ; with reductions of speed of 40% in strongly divergent flows.

Safety concerns have led to the use of explosives that satisfy $\mathcal{L}/\eta_r = O(10^3)$, where $\mathcal{L}$ is a representative dimension of the explosive. Curvature effects have a strong influence on detonation in these systems, and highly resolved multidimensional simulations are "expensive." A body of theory and supporting experiments, called Detonation Shock Dynamics (DSD) (Bdzil & Stewart 1989), has been developed that treats these curvature effects (Aslam et al. 1996). The DSD front theory derives a speed function $D_n(\kappa)$ based on a weakly divergent, quasi-one-dimensional (1D) model of the detonation reaction zone. This function can also be determined directly from experiments. The regions of strongest flow divergence are found near the explosives' boundaries. Boundary conditions (BC) must be supplied in addition to $D_n(\kappa)$ to treat these interactions. The dynamics of broadly curved fronts, described by DSD, interacts with the edge through a narrow boundary layer a few η_r thick, where the flow is both reactive and fully two dimensional (2D).

In Section 2 we give a brief review of DSD. Front theories require a front propagation algorithm. In Section 3 we describe an engineering implementation of DSD that uses a level-set (LS) algorithm (Osher & Sethian 1988) to propagate the DSD front (Aslam et al. 1996). To integrate DSD front theory with hydrodynamic simulations, an accurate method is needed to quickly burn the explosive and capture the detonation state consistent with $D_n(\kappa)$. In Section 4 we describe our new burn model and show a full DSD-based simulation. In Section 5 we discuss results obtained from high-resolution simulations of two edge problems: (1) the sudden loss of confinement, and (2) oblique interaction of detonation with a rigid wall. Owing to reaction-zone effects, we find that detonations exhibit von Neumann-like reflection (Colella & Henderson 1990).

2. DSD theory

DSD is the name given to the body of multidimensional detonation theory and experiments that is used to describe the dynamics of detonation with broadly curved shocks on the reaction-zone scale η_r. The model equations used to describe this limit derive from the 2D, reactive Euler equations transformed to shock-attached, intrinsic coordinates. Shown in Fig. 1 is our coordinate net of straight lines normal and curves locally parallel to the shock, all moving with the shock normal speed D_n.

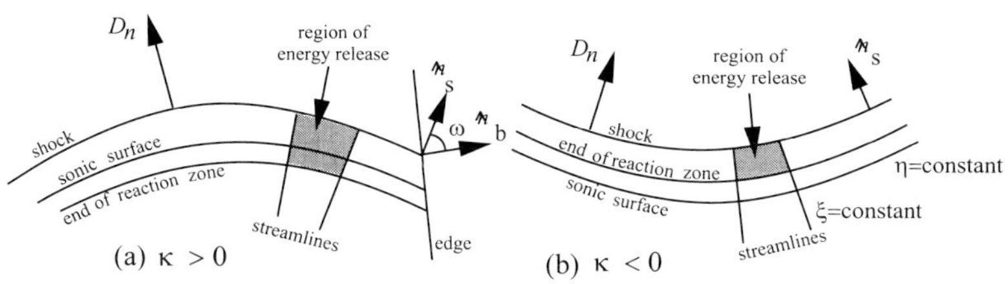

Figure 1. A snapshot showing the intrinsic coordinates and the definition of the boundary-edge angle ω; in (a) for a diverging detonation and in (b) for a converging detonation. The speed of the wave is influenced by two factors, the convergence/divergence and the location of the sonic surface.

2.1. Interior flow

The weak shock-curvature limit defined by $\kappa = O(\epsilon)$, where $\epsilon \equiv (reaction - zone\ scale)/(shock\ radius\ of\ curvature)$, is the basis of most theoretical analysis of 2D detonation. This limit envisions that $O(1)$ changes in the field variables occur over $O(\epsilon^{-1})$ distances in the ξ-direction, and the flow velocity in the ξ-direction is no greater than $O(\epsilon)$. Under these assumptions, the DSD-related time derivatives are $O(\epsilon)$, the flow is principally in the η-direction, and is "nozzle"-like (see Fig. 1), and 2D effects enters the flow only parametrically, via $\kappa(\xi, t)$. This limit allows for time dependence on the $O(1)$ reaction scale, provided no $O(1)$ velocities are generated in the ξ-direction. The various streamtubes then communicate with one another only through the shock-surface compatibility condition (i.e., through the LS equation or whichever propagation method we use). The DSD reaction zone equations are:

$$(\dot{\rho}) + \rho(D_n - U)_\eta = -\rho U \kappa \,, \tag{1}$$

$$\rho(\dot{U}) - P_\eta = 0 \,, \tag{2}$$

$$\rho^2(\dot{e}) - P(\dot{\rho}) = 0 \,, \tag{3}$$

where $(\dot{\ }) \equiv (\partial/\partial t)_{\eta,\xi} + (D_n - U)(\partial/\partial\eta)_{t,\xi}$ and ρ, U, P, $e(P, \rho, \lambda q)$, λ, and q are the density, laboratory particle velocity in the η- direction, pressure, specific internal energy, degree of reaction, and heat of detonation, respectively. Henceforth, when t, η, ξ, etc. are used as subscripts they denote partial derivatives. The *master equation*, derived from Eqs. (1) – (3) helps us analyze the 2D reaction zone

$$P_t + \rho(D_n - U)U_t = -\rho[(D_n - U)^2 - C^2]U_\eta + \sigma \,, \tag{4}$$

where $\sigma \equiv -(e_\lambda/e_P)R - \rho C^2 U\kappa$ and C, $-(e_\lambda/e_P)R \geq 0$, and $(\dot{\lambda}) \equiv R \geq 0$ are the sound speed, heat-release rate, and chemical rate law, respectively.

In its simplest realization, DSD theory assumes that $D_n/D_{CJ} = 1 + O(\epsilon)$, that the departures from the Zeldovich-von Neumann-Doring (ZND) limit are small, and neglects the inertia of the reaction zone (setting t derivatives in Eqs. 1 – 4 to zero). The reaction-zone structure is then obtained by solving the steady form of Eqs. (1) – (3), subject to the shock conditions and generalized Chapman-Jouguet (CJ) condition obtained from Eq. (4) (i.e., $(D_n - U)^2 - C^2 = 0$ when $\sigma = 0$, provided that $\kappa \geq 0$). This defines an eigenvalue problem that obtains $D_n(\kappa)$. Figure 1 shows that two effects contribute to $D_n \leq D_{CJ}$: (1) the divergence and (2) the movement of the sonic surface. For weakly convergent systems, effect (1) dominates and $D_n(\kappa) \geq D_{CJ}$ is derived by requiring $(D_n - U)^2 - C^2 = 0$ at the end of the reaction zone (i.e., $\lambda = 1$). Details can be found in (Bdzil 1981), (Bdzil & Stewart 1986), (Stewart et al. 1995), (Aslam et al. 1996) and references therein. Importantly for $D_n(\kappa)$, the shock evolves by parabolic dynamics and so is smooth. For real heterogeneous solid explosives, R is not well known. Then $D_n(\kappa)$ is determined directly from experiment. The $D_n(\kappa)$ for the explosive PBX 9502 is shown in Fig. 2a. Note the $O(1)$ variations of $(D_{CJ} - D_n)/D_{CJ}$.

2.2. Boundary conditions

The slow variations in the ξ-direction that are a part of DSD can break down at explosive boundaries. There the flow can be fully 2D and time-dependent on the $O(1)$ reaction-zone scale (Bdzil & Stewart 1986). On the slow-time DSD scale, this region can appear to be steady. Then the DSD and boundary regions are coupled as follows: (1) the DSD region drives the boundary with information on the shock slope and D_n, and in turn (2) the boundary region uses this data to return a (possibly) modified shock slope, etc. (Bdzil 1981). Using the boundary angle ω defined in Fig. 1, DSD supplies ω_{in} to the boundary flow, which then returns ω_{out} to the DSD region. The boundary is itself characterized by two angles that depend on the explosive/inert pair being considered: a critical angle ω_s and a confinement angle ω_c.

We've distilled these interactions into the following recipe. If $\omega_{in} < \omega_s$, then $\omega_{out} = \omega_{in}$; otherwise, $\omega_{out} = \omega_c$. The boundary dynamics are considered in more detail later in this paper.

3. Front propagators

Together, the $D_n(\kappa)$ function and the boundary conditions provide a complete dynamical description of the detonation front at the DSD level. A second element is needed to propagate the front: a shock compatibility condition that relates how changes in the speed of the front affect its shape. One of the most widely used forms of this condition is Whitham's ray method (Whitham 1973). Every section of the detonation front advances along a system of rays that resembles the bicharacteristic rays of geometrical optics. Although physically appealing, such methods can be logically complex. The front-attached rays converge and diverge with the front, which can lead to numerical problems. This is one member of the family of marker-particle or Lagrangian methods. Such methods are not well suited for engineering applications where the problem geometries are complex.

3.1. Level-set method

Osher and Sethian (1988) devised a powerful algorithm for propagating fronts with curvature-dependent speed. Their method obviates the need for the complex logic to treat collisions and avoids the problem of marker-particle methods. They consider the shock as a level curve $C(x, y, t) = 0$ embedded in a higher-dimensional LS function $\psi(x, y, t)$ that's defined on an Eulerian grid. The evolution equation for the LS function is derived by using the property $\psi(x, y, t) = \mathit{constant}$ along a level curve to obtain

$$\psi_t + \vec{D_n}(\kappa) \cdot \vec{\nabla}\psi = 0 \ . \tag{5}$$

Equation (5) is the LS-method surface compatibility condition.

The precise form selected for $\psi(x,y,0)$ is unimportant, however, $\psi(x,y,0)$ must be single valued. This insensitivity requires that; (1) D_n depend only on data from a single curve and (2) that the level curves not cross with time. The first property is automatic, since $D_n(\kappa)$. The second property follows from the fact that the distance between level curves, as measured by $d(x,y,t) = |\vec{\nabla}\psi(x,y,0)|/|\vec{\nabla}\psi|$, does not go to zero. By way of example, for $D_n = D_{CJ} - \alpha\kappa$ with $\alpha \geq 0$, the d-equation (which is derived from Eq. 5) is

$$d_t + D_{CJ}\vec{n} \cdot \vec{\nabla}(d) = \alpha\kappa^2 d \ , \tag{6}$$

for a cylindrically symmetric system. Since $\alpha\kappa^2 \geq 0$ and $d(x,y,0) = 1$, d increases with time. The result $d(x,y,t) > 0$ continues to hold for fully 2D systems with strong divergence at the boundaries (Aslam et al. 1996). In practice,

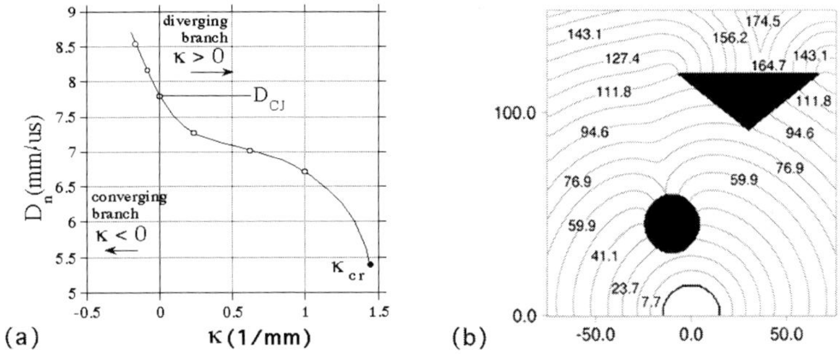

Figure 2. The $D_n(\kappa)$ for a typical condensed phase explosive, PBX 9502 is shown in Fig. 2a. Burn time contours for an engineering-style problem obtained with the LS-method are shown in Fig. 2b. The dark areas are obstacles.

we assign $\psi = 0$ to the shock, with $\psi > 0$ in unburnt material and $\psi < 0$ in burnt material. The burn time $t_b(x,y)$ is taken as the first time that $\psi < 0$ at a point. Figure 2b shows the result of a LS calculation of the burn times for an engineering-style problem. Initially, the detonation is a semi-circle with origin (0,0) and obeys $D_n = 8$ mm/μs $- \kappa \times 66$ mm^2/μs everywhere. Bifurcation, merging, convergence, and divergence of the wave are all well captured. All of these ideas easily carryover to 3D.

4. Hydrodynamic simulations using DSD

The work that has gone into DSD was motivated by the need to capture reaction-zone effects in numerical simulations of engineering systems. Two elements are required to accomplish this goal: (1) accurate front evolution and (2) an ability to deposit the proper detonation state (i.e., $P(D_n)$, $\rho(D_n)$, $U(D_n)$) at the front.

To address element (2), we've developed a numerical reaction-zone model that uses a "pseudo" divergence Ω in place of κ in simulations, so that a detonation state consistent with D_n is captured. To eliminate precursors to the $D_n(\kappa)$ wave, we built a model around the reaction zone equation of state (EOS), $e(P/\lambda, \rho, q)$, which together with the trigger provided by the DSD burn times $t_b(x, y)$, causes the numerical "reaction zone" to be on the weak detonation branch. Starting with $P = 0$ and $\lambda = 0$, at $t_b(x, y)$ the explosive is rapidly burned (i.e., $\lambda \to 1$ in about the time it takes the wave to traverse one computational zone). From Eq. (4) it follows that the ratio κ/R determines D_n. Since this nonphysical numerical reaction rate $\tilde{R}$ is associated with the size of the computational zones, we replace $U\kappa$ in Eqs. (3) – (4), by

$$U\kappa \Rightarrow \Omega\tilde{R}\,, \qquad R \Rightarrow \tilde{R}\,, \tag{7}$$

and then solve for $D_n(\Omega)$. Elimination then yields $\Omega(\kappa)$, the "pseudo" divergence required to get the numerical detonation to be compatible with $D_n(\kappa)$. This 2-part model is implemented in a second-order, 2D, multi-material, Eulerian grid hydrodynamics code at Los Alamos called Mesa (Holian et al. 1989). The results of two calculations of the detonation Cu-cylinder test (one using the standard $D_n = D_{CJ}$ model and the other using DSD) are displayed in Fig. 3. The difference in the timing and wave shapes is striking.

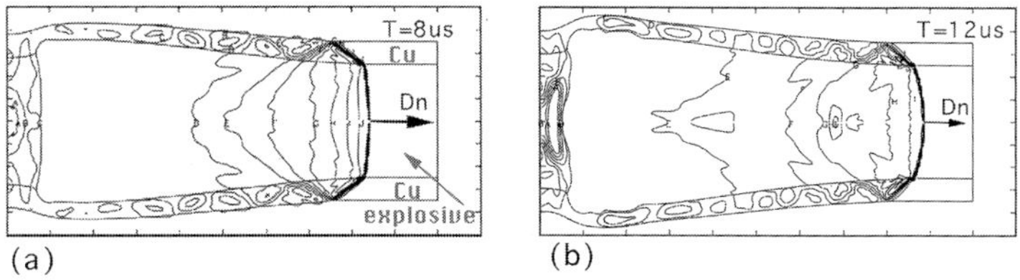

Figure 3. Pressure contours for Mesa calculations of the detonation cylinder test. The detonation wave is moving to the right. Frame (a) shows the results at $t = 8\ \mu$s from the $D_n = D_{CJ}$ model, (b) shows the results for the $D_n(\kappa)$ model at $t = 12\ \mu$s.

5. Boundary-condition study

A complete theoretical analysis of fully 2D, reactive edge flows is out of reach. Here we present some results obtained using high-resolution numerical simulations. A limited amount of supporting theory is offered to help with the interpretation of these results.

We study a simplified Euler fluid model, and use a polytropic EOS with a large γ to mimic condensed phase explosive, where

$$e = \frac{P}{(\gamma - 1)\rho} - q\lambda \; , \tag{8}$$

with $\gamma = 3$, an initial density $\rho_0 = 2000$ kg/m^3, and a heat of detonation $q = 4 \times 10^6$ m^2/s^2, so that $P_{CJ} = 3.2 \times 10^{10}$ N/m^2 is large. A simplified state-independent rate law is used ($k = 2.51 \times 10^6$ s^{-1})

$$(\dot{\lambda}) = R = k\sqrt{1 - \lambda} \; , \tag{9}$$

for which the 1D steady-state reaction-zone length is 4×10^{-3} m and the particle reaction time is 0.8 μs. The numerical simulations were done using a second-order Godunov code called Caveat (Addessio et al. 1990). The grid size used in the calculations was 2×10^{-4} m in the streamwise direction. Two types of boundary interactions are studied here. For both, the detonation is initially a plane ZND wave whose direction of propagation is colinear with a *flat rigid* wall (i.e., $\omega = 90°$). Problem (1) considers the response of the detonation to a sudden loss of confinement, while the interaction of the detonation with a converging, rigid wedge is studied in problem (2).

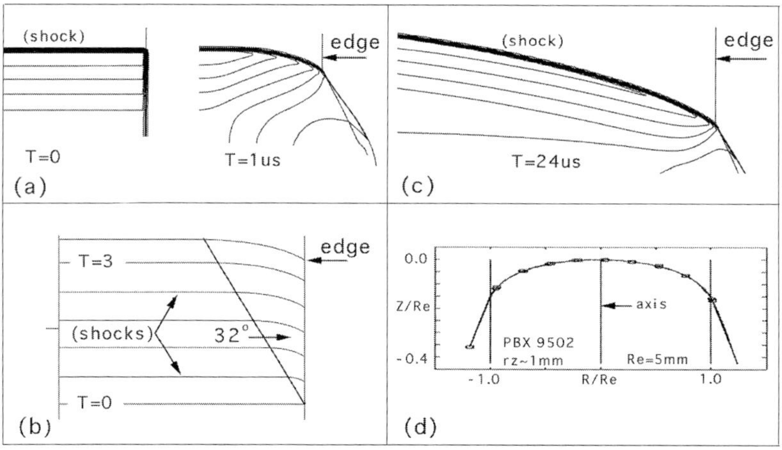

Figure 4. Response of a detonation reaction zone to a sudden loss of confinement. Frames (a), (b), (c), and (d) show the reaction-zone pressure contours, the evolving shock-front shape, late-time pressure contours, and an experimentally measured shock-front shape, respectively.

5.1. Loss of confinement

A collection of the results from problem (1) is shown in Fig. 4. Propagating upwards initially, the detonation loses confinement along its right boundary at $t > 0$. Figure 4a shows strong 2D flow at the edge as evidenced by $P_\xi = O(1)$ and the rapid change in the shock slope (ω changes from 90° to 55°). The Prandtl-Meyer expansion that develops limits the decrease in pressure at the shock which results in a fixed shock-edge angle after a very short transient, as shown in Fig. 4b. The rarefaction moves along the shock at $(\sqrt{C^2 - (D_n - U)^2})_{\eta=0} = 0.7 D_{CJ}$. Figure 4c shows a broadly curved DSD wave at $t = 24$ μs. The value $\omega_c = 55°$, corresponds to a flow that is exactly sonic as measured with respect to the shock-edge intersection point. A discussion of the effect that increasing confinement and changes in ω_{in} have on these results is given elsewhere (Bdzil & Stewart 1986, Aslam et al. 1996). The experimentally measured shock-arrival trace for a $Re = 5$ mm radius detonating cylinder of PBX 9502 is shown in Fig. 4d. The $\omega_c = 45°$, corresponds to the sonic angle for this material. For this explosive, $\eta_r = 1$ mm. Measured on this scale, the shock for this 10-mm diameter explosive is *broadly curved.*

5.2. Converging wedge

A collection of results from problem (2) is shown in Fig. 5. A ZND detonation propagating to the left meets a rigid, converging wedge at $x = 190$ mm ($x = 0$ is at the left). Standard three-shock theory for a CJ detonation predicts Mach reflection for $\omega_{in} \geq 45°$. A simulation of the 50° wedge problem shows regular reflection. The evolving shock and D_n for both a 40° and 20° wedge shown in Fig. 5a display irregular reflection. The pressure contours at $t = 24$ μs shown in Fig. 5b reveal that the irregular reflection grows slowly for the 40° case. Although the structure looks somewhat classical, the reflected wave is clearly. dispersed.

The pressure contours at $t = 24$ μs for the 20° wedge reveal a totally non-classical von Neumann-like reflection (Colella & Henderson 1990). The leading shock is broadly curved with no evidence of a reflected shock. This is a consequence of the diffraction of the reflected wave by the reaction-zone flow gradient. By contrast, the Mach reflection of an inert shock of comparable strength, shown in Fig. 5d, is classical. Figure 5e shows "Mach" reflection during the collision of two PBX 9502 detonations. These measurements were made at Los Alamos by Larry Hull (Hull 1995). With the progress of time (time advances to the left), the wave-interaction region (the left-most part of each trace) becomes wider and more "rounded."

We present a qualitative theoretical argument to help understand these observations. Equation (4) is exact along the shock:

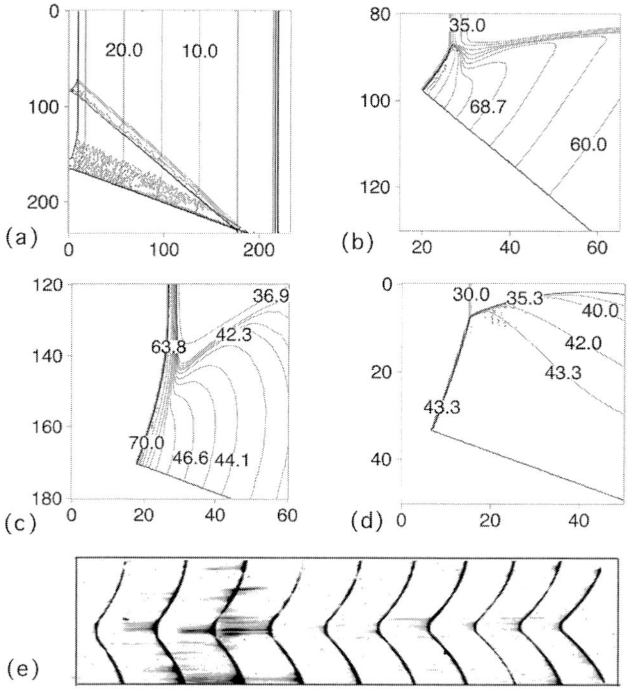

Figure 5. The oblique interaction of a resolved reaction zone detonation with a rigid wedge. Frames (a), (b), (c), (d), and (e) show the evolving lead shocks for a 40° and 20° wedge, pressure profiles at $t = 24$ μs for the 40° and 20° wedge, a strong inert shock over a 20° wedge and the curved "Mach" interaction for PBX 9502, respectively.

$$(D_n^2)_t = \frac{(\gamma+1)^2}{3}\left(\frac{(\gamma-1)}{(\gamma+1)}D_n^2\right)(U_\eta)_{\eta=0} + \frac{(\gamma+1)}{(\gamma-1)}\frac{D_{CJ}^2}{6}(R)_{\eta=0} - \frac{4\gamma}{3(\gamma+1)}D_n^3\kappa\ . \tag{10}$$

Differentiating Eq. (10) with respect to ξ, using the shock-compatibility condition, and then transforming the resulting equation to a reference frame moving at acoustic speed along the shock ($\nu = -(\sqrt{C^2 - (D_n - U)^2})_{\eta=0}$ with $d\omega \equiv d\xi - \nu dt$), allows us to derive an equation for the amplitude of the leading edge of a weak 2D disturbance moving into a 1D ZND detonation

$$4\tilde{K}_\tau = -\frac{2(\gamma+9)}{3(\gamma+1)}\tilde{K}^2 - \frac{(\gamma+1)}{6(\gamma-1)}(1-\sigma)\tilde{K} + \frac{2(\gamma-3)}{3\sqrt{\gamma^2-1}}\tilde{K}_{\tilde{\omega}} + O((U_{\eta\omega})_{\eta=0})\ . \tag{11}$$

The variable $\tilde{K} \equiv (\kappa D_{CJ}/k)_{wavehead}$ is the scaled shock curvature, $\tau = kt$

is the scaled time, and σ measures the state dependence of the rate (e.g., $R = k\sqrt{1-\lambda}\exp(\sigma D_n/D_{CJ})$). The strength of the 2D disturbance is $\tilde{K}$, with $|\tilde{K}| \to \infty$ denoting triple-point formation on the shock. Neglecting the $O((U_{\eta\omega})_{\eta=0})$ term in Eq. (11), setting $\gamma = 3$, and solving, one finds that for $\sigma < 1$, $\tilde{K} \to 0$ when $\tilde{K}_{t=0} > -(1-\sigma)/6$ and $\tilde{K} \to -\infty$ otherwise. Thus, sufficiently weak convergence does not lead to triple points. This is what we see in our simulations. Suppressing the reaction-related terms in Eq. (11), one finds that all levels of convergence lead to triple points for inert flows. The DSD boundary conditions derived from this example are: (1) when $\omega_{in} < \omega_s = 35°$, then $\omega_{out} = \omega_{in}$ and (2) when $\omega_{in} > \omega_s = 35°$, then $\omega_{out} = \omega_c = 90°$.

Acknowledgement. We thank Rudy Henninger for Mesa-DSD. John Bdzil was supported by the U.S. Department of Energy. Tariq Aslam and Scott Stewart were supported by the U.S. Air Force, Wright Laboratory, F08630-92-K0057.

References

Addessio FL, Baumgardner JR, Dukowicz JK, Johnson NL, Kashiwa BA, Rauenzahn RM (1990) Caveat: A Computer Code for Fluid Dynamics with Large Distortions and Internal Slip, Los Alamos Report, LA-10613.

Aslam TD, Bdzil JB, Stewart DS (1996) Level set methods applied to modeling detonation shock dynamics. J Computational Physics (accepted).

Bdzil JB (1981) Steady-state two-dimensional detonation. J Fluid Mech 108:195–226.

Bdzil JB, Stewart DS (1986) Time-dependent two-dimensional detonation: the interaction of edge rarefactions with finite-length reaction zones. J Fluid Mech 171:1–26.

Bdzil JB, Stewart DS (1989) Modeling two-dimensional detonation with detonation shock dynamics. Phys Fluids A 1:1261–1267.

Colella P, Henderson LF (1990) The von Neumann paradox for the diffraction of weak shock waves. J Fluid Mech 213:71–94.

Holian KS, Mandell DA, Adams TF, Addessio FL, Baumgardner JR, Mosso SJ (1989) Mesa: A 3-D computer code for armor/anti-armor applications. In: Supercomputing World Conference, Oct 17–20, 1989, San Francisco, CA.

Hull LM (1995) Los Alamos National Laboratory, private communication.

Osher S, Sethian JA (1988) Fronts propagating with curvature-dependent speed: algorithms based on Hamilton-Jacobi formulations. J Computational Physics 79:12–49.

Stewart DS, Aslam TD, Yao J, Bdzil JB (1995) Level-set techniques applied to unsteady detonation propagation. In: Modeling in Combustion Science, Buckmaster J and Takeno T (eds), Springer-Verlag, Berlin.

Whitham GB (1973) Linear and Nonlinear Waves, Wiley, New York.

Bubbly cavitation in intense rarefaction waves and its effects

V. K. Kedrinskii
Lavrentyev Institute of Hydrodynamics
Novosibirsk 630090 Russia

Abstract: The paper presents a review of experimental results, mathematical models, and numerical investigations of wave processes and the dynamics of liquid media with microinhomogeneities under impulse loading. The presence of microinhomogeneities can substantially change the state of the liquid and the structure of the applied wave field as well as the mechanisms of physical processes that take place in liquid media.

Key words: Microinhomogeneities, Cavitation structure, Rarefaction focusing, ESWL, Shock loading, Detonation initiation mechanism, Vapor explosion

1. Introduction

Microinhomogeneities are microbubbles of free gas, solid microparticles, and their combinations (Besov et al. 1992), which are called the cavitation nuclei.

According to modern data for distilled water, based on the experimental studies of the dynamics of scattering indicatrix of a laser beam on microinhomogeneities and on the analysis of the tracks of diffraction spots of nuclei illuminated by a laser beam, the spectrum of nuclei distribution can be described by a simple dependence (Kedrinskii 1989) where $N_o \simeq 10^5 - 10^6$ cm^{-3} is the total density of nuclei (see Fig. 1). When pulsed tensile stresses appear in the fluid, the nuclei begin to grow, forming a cluster whose form and dynamics are determined by a spectrum of initial nuclei sizes, by the character of the stress applied, and by the boundary conditions.

The above data enable us to make conclusions regarding the two-phase structure of a real liquid and to use a two-phase mathematical model for average characteristics to describe the wave processes in a liquid cavitating behind a rarefaction wave front (Kedrinskii 1975). This model can be supplemented by the conditions of heat and mass exchange (for instance, this can be the energy equation with phase transition, Nakoryakov et al. 1990) and by the equation of chemical kinetics for bimolecular reactions (Todes 1933). When numerically analyzing wave processes in chemically active media one usually considers the

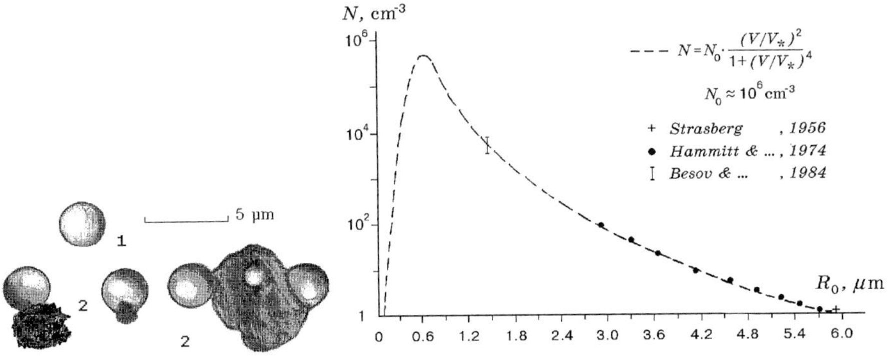

Figure 1. Structure of microinhomogeneities (left): 1 – free gas bubble, 2 – combination structure; spectrum of cavitation nuclei (right).

model of the so-called adiabatic explosion (Kedrinskii and Zamaraev 1989). The study of different stages of a chemical reaction occurring under explosion was carried out by Hasegava and Fujiwara (1982).

As is obvious from practice, the physics of cavitation has direct relation to: the problem of detonation initiation in liquid High Explosive (HE); the problem of the mechanics of fragmentation of kidney stones in lithotriptors; the fundamental problem of liquid strength under explosive loading; as well as to well-known but not completely understood effects of large-scale explosions of fuel containers during transport. Some approaches to the problems mentioned for which cavitation effects can be of principal importance will be considered.

2. Disintegration mechanism of target at shock-wave focusing in ESWL systems

Experimental studies carried out by many researches confirm that the role of the cavitation zone developing around the target when a shock wave focuses on it can appear essential for the process of the target disintegration in lithotriptor systems. However, in laboratory studies of wave focusing in a liquid, as a rule, the rarefaction phase receives little attention, although it naturally arises when using open segments of ellipsoids and spheres as reflectors or sources.

Grönig (1989) in his survey gives a short review of ESWL (External Shock Wave Lithotripsy), distinguishing the works of Jutkin, Goldberg, Forssman, Chaussy et al., Reichenberger and Naser, Ziegler and Wurster, Muller, Delius and Brendel, Takayama et al., and Coleman and Saunders. As a rule, in these papers cavitation effects are simulated primarily as the results of the interaction of shock waves with artificially produced gas bubbles and have no connection to cavitation, as a new medium state, which considerably varies the wave-field parameters (Kedrinskii 1976) and the conditions at which bubble collapse with cumulative jet formation is possible.

Physical aspects of wave focusing in the context of ESWL were analyzed by Sturtevant (1989). He reports that the role of rarefaction waves in the process is to cause the sharply peaked wave form and to limit the amplitude of the shock wave.

Kuwahara (1989) and Kitayama et al. (1987) noted that after the shock wave hits the stone, cavitation phenomenon is seen around the stone; thereafter, the stone deforms and disintegrates. Delius (1990) related the damage effect to the formation of cumulative microjets. Prat (1993) concluded that a shock-induced cavitation has a potential for selective and noninvasive destruction of deep tumors through an extra-corporeal approach. Nagoya et al. (1993) discovered a hysteresis effects in studying human blood. This effect was also described and explained by Besov et al. (1992). Its nature appeared to be determined by the two-phase state of the medium containing microinhomogeneities which played the role of cavitation nuclei and changed their structure under the effect of the wave field.

Sato et al. (1993) suggested the use of a soft-flash radiography system to visualize shock-wave-induced cavitation bubbles in living tissue. Earlier, the technique of pulse X-ray was used for the same purpose in studying distilled water (Bayikov et al. 1984, Berngardt et al. 1984). Church and Crum (1990) suggested the idealized pressure-wave forms produced by commercially available lithotriptors.

The references could be continued, but the explanations of the reasons for fracture would remain the same as mentioned above: cumulative jets arising under collapse of cavitation bubbles and shear stresses inside the target resulting in formation of microcracks. On the basis of the above statements and known experimental facts, we shall try to analyze other possible mechanisms of the evolution of the process of target disintegration.

Experimental simulation. First of all we will note the statements of Delius and Philipp (Philipp 1991). They have carried out experiments on the process of formation of a bubble cluster on the surface of a gall stone as a result of focusing a shock wave generated by Dornier Lithotriptor XLI with amplitude in the front of 65 MPa and a rarefaction phase of maximum amplitude 6 MPa. In these experiments a gall stone was positioned onto foil in the geometrical focus of the Lithotriptor. High-speed photo registration has shown that during the process of cavitation development a gall stone (target) was completely surrounded by a dense bubble cluster which then was transformed probably into a gas-vapor layer on the surface. Analysis and prediction of further behavior of the cluster and the effect of its dynamics on the target are impossible to state directly.

However, principal moments of the process can be simulated by the example of cavitation effects arising at the bottom of a tube filled with liquid, the tube being accelerated vertically downward by the impact (as in the experiments of Hansson et al. 1982). The amplitude of acceleration was about $2 \cdot 10^4$ ms^{-2} with the width of the pulse near the base of the acceleration peak about 300 μs. The high-speed photos show that a dense cavitation zone transformed over 4 ms into a gas-vapor layer.

One can consider that the pressure in this cluster is of the order of the pressure of saturated vapors (for water 2.34 kPa), and the pressure at the outer boundary of the cluster is equal to atmospheric pressure ($p_\infty = 10^2$ kPa). Then under the effect of pressure difference, the layer starts collapsing and over the time 7 – 8 ms disappears. In this time the gauges placed on the tube bottom record a sharp jump in acceleration (Fig. 2, $t = 7.735$ ms) with the amplitude exceeding the initial value and pressure pulse (Fig. 3b); with the amplitude of about 1.5 MPa and duration 1.5 ms (unpublished data received by the author together with I. Hansson and K. Morch). It is clear that the effect was caused by hydraulic impact. Under the effect of the second acceleration pulse a new cavitation cluster develops near the bottom and the process repeats. The oscillograms of the acceleration (Fig. 2) record repeated intense hydraulic impacts ($t = 13.69$ and 18.425 ms) and thus the hydraulic impacts can be considered as the principal mechanism of failure. The pressure oscillograms (Fig. 3c) confirm this conclusion. One should not exclude the possibility of the generation of a series of intermediate weaker hydraulic impacts (Fig. 3b) that are probably connected with the collapse of individual large bubbles or their clusters near the bottom. The digits (Fig. 3) show the parameters of $p(t)$-curve in the location points of the marks.

3. Some problems of pulse fracture of liquid

The process of fracturing a free-boundary liquid under explosive loading is called cavitation fracture, since all features of its development are completely determined by the phenomena of bubble cavitation. This is a complex of phenomena whose sequence can be presented as the stages of bubble cluster formation, its transformation into a foam-like structure, and then into a droplet system. The first two stages were studied in detail using the following methods and measuring techniques (Kedrinskii 1993): (1) two-diaphragm hydrodynamic and electromagnetic shock tubes, (2) flash radiography, (3) scattering indicatrix (statics

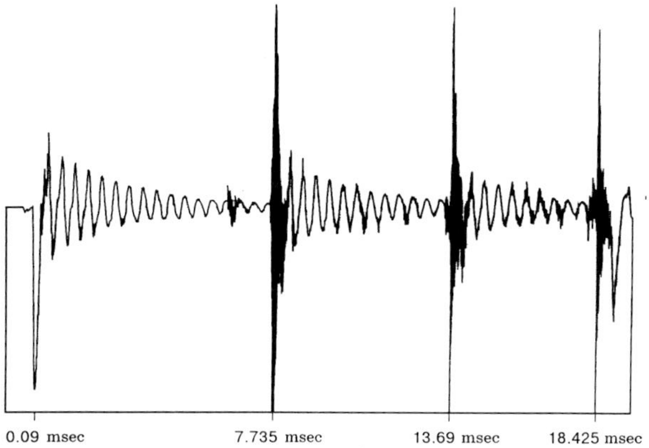

Figure 2. Oscillogram of acceleration: repeated hydraulic impacts.

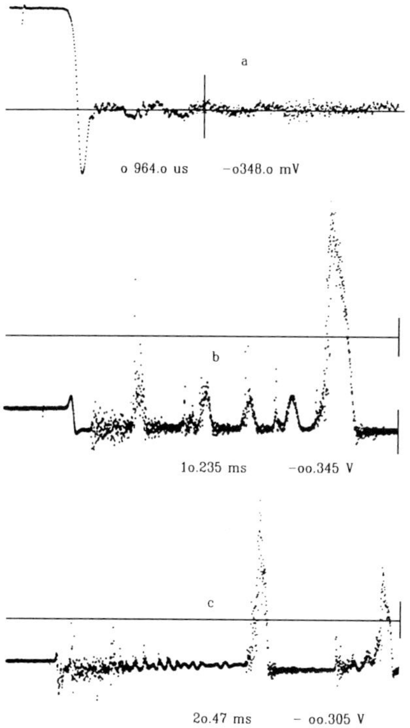

Figure 3. Pressure oscillograms: the tube bottom accelerated by an impact.

and dynamics), (4) electromagnetic capacitance and pressure gauges, the dynamic thrust and electric potential methods, and (5) computer processing.

The mechanism of development of cavitation zones has been formulated (Kedrinskii 1986): in weak rarefaction waves a cluster of visible bubbles is gradually formed depending on the initial position of the nuclei in the spectrum; during the high-intensity rarefaction phase all nuclei attain visible size simultaneously; the liquid becomes a monodisperse medium that is very essential for the mathematical model.

Application of the two-phase model made it possible to calculate the dynamics of a visible cavitation zone in the problems of underwater explosion and rarefaction wave profiles, as well as to show that the steepness of the rarefaction wave front plays a major role in the problem of extreme tensile stresses admitted by a real liquid (Kedrinskii 1976).

A number of criteria of the development process have been obtained (Kedrinskii 1968, 1989; Berngardt et al. 1995): (1) characteristic distance of wave attenuation, (2) relaxation time of tensile stresses and ultimate tensile stresses, and (3) relaxation time of cavitating medium to the state of "bulk density." For example, for $k_o = 10^{-10}$, $R_o = 10^{-4}$ cm, and $p = -30$ MPa in the rarefaction wave, the relaxation time equals $t_{cav} \simeq 8.3 \cdot 10^{-8}$ s, and the liquid withstands only $p_* \simeq -3.8$ MPa at $\tau = 1$ μs.

The third stage poses the principal problems in the investigation of the process of cavitation fracture and consists in revealing a physical transition mechanism of a cavitating liquid state from foam to droplet phase. The problem was solved using the process of the disintegration of a liquid drop as a result of ultra-short impact. A drop of distilled water (a few millimeters in size) was placed on the diaphragm of the electromagnetic shock tube which under the action of a pulse magnetic field impacted on the drop. As a result a shock wave of amplitude 5–6 MPa and a duration of the order of 3–4 μs was formed in the drop.

We could observe a circular cumulative jet as a thin sheet and the formation of a cavitation zone. By 70 μs the whole drop "boils," acquiring a clear cellular structure. The base of this structure consists of large clusters formed by junction of cavitation nuclei growing in the tensile stress field.

Figure 4 shows a frame (left) and detail (right) of a high-speed photo of the last stage of the fracture (original drop diameter was 5 mm) when the characteristic size of the structure reached a few centimeters at $t \simeq 1.5$ ms. The drop transforms into a peculiar foam phase with clear boundaries, as a mesh of liquid bundles in which the main mass of the original drop is likely to be concentrated. The elements of the liquid mesh gradually decompose into

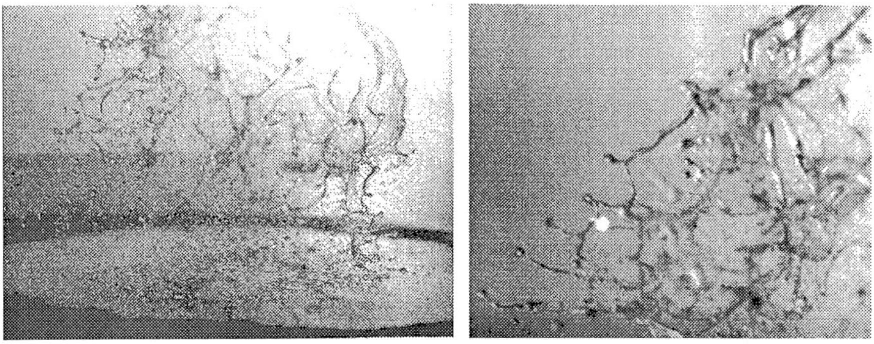

Figure 4. Cellular structure of fracturing drop at pulse impact.

separate fragments under tension, transforming into jets that later disintegrate into separate drops. Notice that even in the drop state the flow conserves the cellular structure pattern for a rather long time.

4. Cavitation effects in chemically active liquid media

In this survey we select from a large number of papers only those dealing with the problem of the effect of microinhomogeneities. This survey does not claim to span the information completely, but it is intended to analyze possible mechanisms of bubbly initiation proposed in the literature.

Bowden and Yoffe (1958) pointed to the thermal nature of explosion when initiating liquid HE by an impact. Small gas bubbles that are usually available in a liquid were recognized as one of the possible seats of heating up. To initiate the explosion the temperature in the seat should be of the order of 800 K with heating duration in the range of 1–100 μs.

Andreev (1957) and Johansson (1957) formulated a model of initiation seat according to which the possible availability of small particles or gaseous phase of HE in bubbles is more important. They concluded that the fast increase of pressure in gaseous combustion products of suspended particles or vapors of HE inside the bubble is responsible for the initiation of detonation.

Campbell et al. (1961) experimentally proved that the mechanism of shock initiation has its origin in the interaction of the wave with microinhomogeneities resulting in the formation of hot spots, and its successive amplification. Dubovik and Bobolev (1978) consider that, in spite of high adiabatic temperatures, a heat exchange in the system "gas bubble-liquid" is unlikely to explain the mechanism of liquid HE initiation. They analyzed possible mechanisms of the occurrence of the smallest droplets of HE in a bubble and estimated their critical value when the droplets could be quickly heated at the time of bubble collapse as well as the value of adiabatic period of induction of mixture explosion inside the bubble.

The experimental facts of increasing the mechanical sensitivity of HE as a result of a successive series of loading pulses (a so-called fractional impact) are known. The first shock wave in a specimen of liquid HE interacts with its free surface and transforms into a rarefaction wave. The subsequent interaction of successive shock waves with cavitation bubbles results in HE initiation (Dubovik and Bobolev 1978).

According to the results of many experiments the shock sensitivity of liquid HE really depends on the quantitative characteristics and the character of inclusions of different nature (Winning 1956, Campbell et al. 1961, Dremin et al. 1970) and the detonation processes in liquid HE develop only on the condition that the liquid becomes nonuniform owing to the above-mentioned loadings (Dubovik et al. 1967, Amster et al. 1967).

Field et al. (1992) and Bourne and Field (1991) considered the effects of stronger shocks of 3 GPa passing over single cavities and cavity arrays, as well as the collapse of cavities in reactive emulsion systems. Roberts and Field (1993) showed that voids of diameter 1.5 mm introduced into sample of octol and cyclotol in the vicinity of the point of projectile impact were the main source of ignition.

This question has not been adequately investigated, in particular, because the numerical models of wave phenomena in cavitating liquid HE have not been seriously studied. Only in recent years the wave processes in a chemically active bubble medium and a model of bubbly detonation have attracted consideration (Hasegawa, Fujiwara, Sychev, Pinaev, Kedrinskii, Mader, Beylich, Gulhan, Lee, Frost, etc.).

From the viewpoint of hot-spot mechanisms, the gas-phase heating realized in a bubbly medium can be a dominant mechanism, according to Field et al. (1992), as large bubbles will be collapsed relatively slowly by an incident shock wave. Therefore a model of an initiation of a cavitating liquid HE by single or fractional impacts is rather close to the approaches developed at the analysis of the wave processes in a liquid medium with the natural inhomogeneities considered above. Unfortunately, there is practically no information on the initial parameters of microinhomogeneities in liquid HE or combustible liquids. But we can assume that the possible range of their volumetric concentrations is from fractions to several percent, and radii are of the order of tens and hundreds of microns.

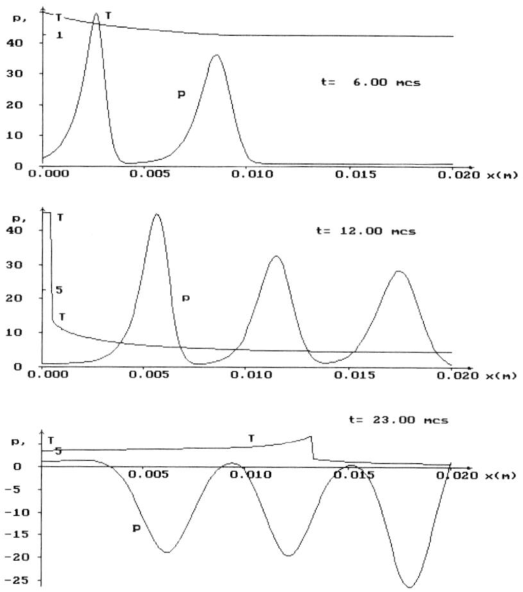

Figure 5. Simulation of a fractional impact. Exponential shock wave sequence propagation, $k_o = 10^{-4}$, $R_o = 0.05$ cm, $l = 2$ cm, $p_{1,2,3} = 300$ bar, $\tau = 0.5$ μs.

Two problems of initiation (a fractional impact and the influence of a rarefaction phase) were considered within the framework of a two-phase model of bubbly detonation (the system of equations is analogous to the system used by Kedrinskii and Zamaraev 1989).

Figure 5 shows the process of bubbly detonation development due to the propagation of a sequence (in 4 μs) of three exponential shock waves with amplitudes 30 MPa and $\tau = 0.5$ μs ($k_o = 10^{-4}$, $R_o = 0.05$ cm) along the sample with a 2-cm length.

The main reason for initiation turned out to be not related to the reflection of the first shock wave. The energy of the incident first and second shock waves was not enough to initiate the process (Fig. 5, $t = 6$ μ s) and the explosion of the gas mixture appeared only after the third wave ($t = 12$ μs, the temperature T of the gas mixture increased tenfold in the vicinity of left boundary of the sample). The influence of the right free boundary of the sample turned out to be contrary: the reflection of the shock-wave sequence interrupted a process of bubbly detonation propagation ($t = 23$ μs) at the point with coordinate $x \simeq 13$ mm. At least, in 10 μs a front of a temperature jump remained at the same place.

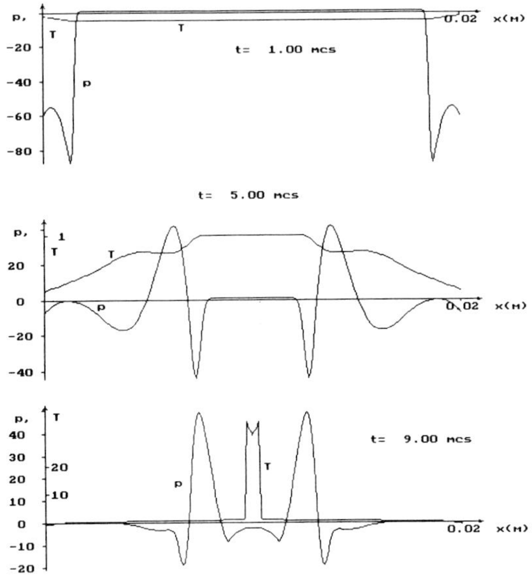

Figure 6. Ignition of a chemical reaction in bubbles at rarefaction wave focusing, $k_o = 10^{-3}$, $R_o = 0.01$ cm, $d = 2$ cm, $p_{l,r} = 100$ bar, $\tau = 3$ μs.

An unexpected effect was discovered upon analysis of the second statement regarding the free-boundary influence. Clearly, as a result of the impact on the plane of a thin sample (one-dimensional case) two exponential rarefaction waves will propagate from its free boundaries to the center (Fig. 6, $t = 1$, 5, and 9 μs). Under the collision of these waves the bubbly detonation process turned out to be excited at the center of the sample and then propagate to the periphery (Fig. 6, $t = 9$ μs, the temperature jump T). The length of the sample is 2 cm, $k_o = 10^{-3}$, $R_o = 0.01$ cm, the amplitudes of rarefaction waves are -10 MPa, and $\tau = 3$ μs. The reason for this effect is the transformation of an exponential profile of a rarefaction wave in the oscillating structure ($t = 5$ μs, Fig. 6) with an intense positive phase behind the front. In both cases the microbubbles in liquid samples contained the acetylene-oxygen mixture. The calculations were carried out using a program developed by Dr. V. A. Vshivkov.

5. Conclusion

The above experimental data, statements, calculation results, and survey have shown that many physical phenomena appearing under impulsive loading of a liquid and the mechanisms of their development can be considered from the viewpoint of wave processes in cavitating media.

Application of the two-phase mathematical model and model experiments have shown that the mechanism of the destruction of kidney stones in lithotriptor systems with high probability is determined by hydraulic impact of the liquid under collapse of dense cavitation clusters. The calculations have shown that the same effect can be achieved when focusing rarefaction waves without initial shock positive phase.

It was found that at the last stage of liquid fracture under pulse loading, the initial liquid volume is transformed into the structure of the liquid mesh type. The decomposition of the mesh cells into separate jets and further into drops forms the basis for the process of inversion of the two-phase state of the liquid.

In the problem of initiating the detonation of liquid HE the mechanisms concerned with the effect of microinhomogeneities of the type of bubbles of inert or chemically active gas play a special role. Application of two-phase models to describe and to estimate the wave-field parameters in such systems makes it possible to extend the spectrum of mechanisms governing the development of the phenomena.

Acknowledgement. This work was supported by the Russian Foundation for Fundamental Studies, Grant 93-01-16383, and by ISF, Grant J7N100.

References

Amster AB, Woolfolk RW (1967) Anomalies in detonation initiation of liquid explosives. Nature 215(5108):1370–1371.

Andreev KK (1958) Some considerations on the mechanism of initiation of detonation in explosives. Proc R Soc London A 246(1245):257–267.

Bayikov I, Berngardt A, Kedrinskii V, Pal'chikov E (1984) Experimental methods of study of cavitation cluster dynamics. J Appl Mech and Tech Phys (translated from Russian by Plenum Publ Corp) 5:30–34.

Berngardt A, Bichenkov E, Kedrinskii V, Pal'chikov E (1984) Optic and x-ray investigation of water fracture in rarefaction wave at later stages. In: Proc of IUTAM Symposium on Optical Methods in the Dynamics of Fluids and Solids, Prague, Czechoslovakia, M Pichal (ed), Springer-Verlag, pp 137–142.

Berngardt AR, Kedrinskii VK, Pal'chikov EI (1995) Evolution of inner structure of a liquid fracture zone under impulsive loading. J Appl Mech and Tech Phys (translated from Russian by Plenum Publ Corp) 36(2):99–105.

Besov A, Kedrinskii VK, Pal'chikov E, Matsumoto Y, Ohashi H (1992) Microinhomogeneity structures and hysteresis effects in cavitating liquid. In: Proc of 14th Int Congress on Acoustics, 3–10 September, Beijing, China.

Bourne NK, Field JE (1991) Bubble collapse and the initiation of explosion. Proc R Soc London A 435:423–435.

Bowden FP, Yoffe AD (1958) Fast Reaction in Solids, Butterworths Scientific Publications, London.

Campbell A, Davis W, Ramsay J, Travis J (1961) Shock initiation of solid explosives. Phys of Fluids 4:511.

Campbell AW, Davis WC, Travis JR (1961) Shock initiation of detonation in liquid explosives. Phys of Fluids 4(4):498.

Church C, Crum L (1990) A theoretical study of cavitation generated by four commercially available ESWL. In: Frontiers of Nonlinear Acoustics, Proc of 12th ISNA, M Hamilton, D Blackstock (eds), ESP Ltd, London, pp 433–438.

Delius M (1990) Effect of lithotripter shock waves on tissues and materials. In: Frontiers of Nonlinear Acoustics, Proc of 12th ISNA, M Hamilton, D Blackstock (eds), ESP Ltd, London, pp 31–46.

Dubovik AV, Bobolev VK (1978) Sensitivity of Liquid Systems to an Impact, Moscow, Nauka Publishers.

Dremin AN, Savrov SD, Trophimov VS, Shvedov KK (1970) Detonation Waves in Condensed Media, Moscow, Nauka Publishers.

Dubovic AV, Bobolev VK (1967) Exciting and propagation of detonation processes at weak initiation of liquid explosives. Fizika Goreniya i Vzryva 3(4):493–501.

Field JE, Bourne NK, Palmer SJ, Walley SM (1992) Hot-spot ignition mechanisms for explosives and propellants. Phil Trans R Soc London 339:269–283.

Grönig H (1989) Past, present and future of shock focusing research. In: Proc of the Int Workshop on Shock Wave Focusing, Inst of High Speed Mechanics, Sendai, Japan, March 22–23, pp 1–38.

Hasegawa T, Fujiwara T (1982) Detonation in oxyhydrogen bubbled liquids. In: Proc of 19th Int Symp on Combustion, Haifa, pp 675–683.

Hansson I, Kedrinskii V, Morch K (1982) On the dynamics of cavity clusters. J Physics D: Appl Phys 15:1725–1734.

Johansson CH (1958) The initiation of liquid explosives by shock and the importance of liquid break-up. Proc R Soc London A 246(1245):160–166.

Kedrinskii VK (1989) On relaxation of tensile stresses in cavitating liquid. In: Proc of 13th Int Congress on Acoustics, Belgrade, Yugoslavia, Vol 1, pp 327–330.

Kedrinskii VK (1968) Wave propagation in a liquid with gas bubbles. J of Appl Mechanics and Technical Physics (in Russian) 4:29–34.

Kedrinskii VK (1975) Dynamics of cavitation zone at underwater explosion near free surface. J of Appl Mechanics and Technical Physics (translated from Russian by Plenum Publ Corp) 5:68–78.

Kedrinskii V (1976) Negative pressure profile in cavitation zone at underwater explosion near free surface. Acta Astronautica, July-August 3(7-8):623–632.

Kedrinskii VK, Zamaraev FN (1989) Wave amplification in chemically active

bubble medium. In: AIP Conf Proc, YW Kim (ed), Current Topics in Shock Waves, 17th ISSWST, Bethlehem, pp 51–62.
Kedrinskii VK (1986) On multiplication mechanism of cavitation nuclei. In: Proc of 12th Int Congress on Acoustics, Toronto, Canada.
Kedrinskii VK (1993) Nonlinear problems of cavitation fracture at explosive loading. J of Appl Mechanics and Technical Physics (translated from Russian by Plenum Publ Corp) 3:74–91.
Kitayama O, Ise H, Sato T, Takayama K (1987) Non-invasive gallstone disintegration by underwater shock focusing. In: Proc of the 16th Int Symposium on Shock Tubes and Waves, H. Grönig (ed), Aachen, pp 897–904.
Kuwahara M (1989) Extracorporeal shock wave lithotripsy. In: Proc of the Int Workshop on Shock Wave Focusing, Inst of High Speed Mechanics, Sendai, Japan, March 22–23, pp 65–89.
Nakoryakov VE, Pokusaev BG, Shreiber IR (1990) Wave Dynamics of Gas- and Vapour-Liquid Media, Moscow, Energoatomizdat.
Nagoya H, Obara T, Takayama K (1993) Underwater shock wave propagation and focusing in inhomogeneous media. In: Shock Waves @ Marseille, Brun and Dumitrescu (eds), Springer, Vol III, pp 439–444.
Philipp A (1991) Private communication.
Prat F, Arefiev A (1993) Cellural and tissular effects of shock waves-induced cavitation: potential application to digestive cancers. In: Shock Waves @ Marseille, Brun and Dumitrescu (eds), Springer, Vol III, pp 21–28.
Roberts P, Field JE (1993) Simulated fragment attack on cased munition. In: Proc of 10th Int Det Symp, Boston, MA, 12–16 July.
Sato E, Takahashi K, Sagae M, Shikoda A, Oizumi T, Ojima H, Takayama K, Sasaki K, Tamakawa Y, Yanagisawa T (1993) Soft flash x-ray system for shock wave research. In: Shock Waves @ Marseille, Brun and Dumitrescu (eds), Springer, Vol III, pp 449–454.
Sturtevant B (1989) The physics of shock focusing in the context of ESWL. In: Proc of the Int Workshop on Shock Wave Focusing, Inst of High Speed Mechanics, Sendai, Japan, March 22–23, pp 39–64.
Todes OM (1933) Adiabatic term explosion. In: J of Phys Chem 4(1):71.
Winning CH (1956) Enhanced initiating effect of multiple shock intersection. Nature 177(4497):33–34.

Wave propagation in vapor explosions

D. L. Frost
McGill University
Mechanical Engineering Department
817 Sherbrooke St. W., Montreal
Quebec, Canada H3A 2K6

Abstract: If a hot liquid such as molten metal contacts a cooler volatile liquid, rapid phase transformation of the volatile liquid may lead to a violent vapor explosion. Propagating interactions in melt-coolant mixtures are typically inefficient in converting the available thermal energy to mechanical energy and highly nonideal in comparison with combustion processes in premixed gases. The propagation of the interaction is closely tied to the rates of melt fragmentation and energy transfer to the coolant. The characteristics of vapor explosion waves that have been observed in melt-water mixtures are discussed with reference to recent experimental results. The relevance of wave solutions calculated using equilibrium thermodynamics to the experimental results is described together with a summary of current outstanding issues that need to be resolved to provide a more realistic description of vapor explosion waves.

Key words: Vapor explosion, Fuel-coolant interaction, Rapid phase transformation

1. Introduction

If a hot liquid contacts a cooler volatile liquid, the rapid phase transition of the volatile liquid can have an explosive character. Such explosions are called vapor explosions (or thermal or physical, in contrast with chemical explosions) or steam explosions if the cooler liquid is water, and are a hazard in a wide variety of technological situations. Explosions involving spills of molten metal into water are a concern in the metallurgical, paper, and nuclear industries and often are referred to as fuel- or melt-coolant interactions in this context. Reviews of the various aspects of vapor explosions can be found in the papers by Corradini et al. (1988) and Fletcher & Anderson (1990) as well as the proceedings of two recent specialist conferences (e.g., Park et al. 1993, Hohmann et al. 1994).

In a vapor explosion, the rapid conversion of thermal to mechanical energy may occur across a propagating wave front. Phase-transition waves have been observed in *homogeneous* liquids that have been superheated by heat transfer from a hot liquid (e.g., Shepherd & Sturtevant 1982) or by a sudden drop in the ambient pressure (e.g., Hill 1990). The focus of the present paper is

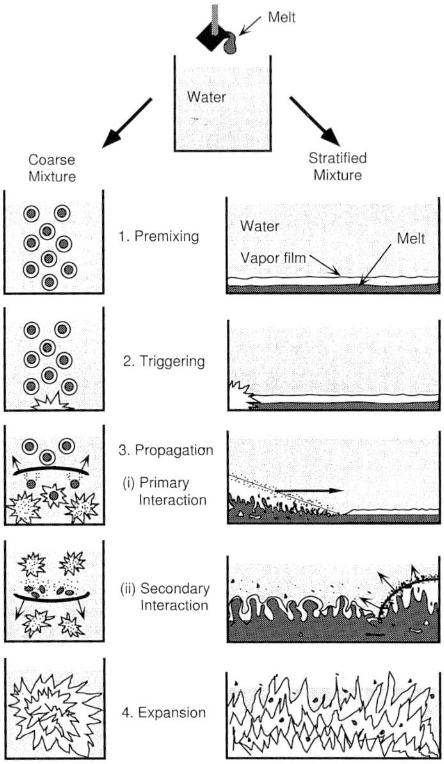

Figure 1. Schematic of the stages of a vapor explosion for two different contact modes following a spill of melt into water.

on the characteristics of propagating interactions in *heterogeneous* melt-coolant mixtures. This is analogous to the combustion-wave propagation in heterogeneous combustible mixtures such as liquid-fuel droplets or a dust suspension in air in contrast to homogeneous premixed gaseous mixtures. Figure 1 shows a schematic of a common accident scenario in which a heterogeneous mixture is formed when a quantity of melt is spilled into water. As the melt jet breaks up, a coarse mixture will be formed consisting of hot-liquid fragments dispersed in the cold liquid. A stratified system will be produced if the melt settles to the bottom of the vessel. Propagating interactions can occur in both of the contact modes shown in Fig. 1. In the initial premixing stage, stable vapor films separate the two liquids physically and thermally. The interaction is triggered by some disturbance which initiates local collapse of the vapor film, resulting in rapid heat transfer and in vaporization of the cold liquid forming a high-pressure vapor bubble. Subsequent expansion of the bubble causes pressurization of the surrounding liquid causing adjacent vapor films to collapse and leading to spatial propagation of the interaction. Multiple interactions may occur if the melt is not completely quenched during the initial interaction.

The dynamics of the propagating interactions shown in Fig. 1 are closely tied to the rate of thermal energy transfer to the coolant which is governed by the rate at which the melt is fragmented and dispersed in the coolant. The fine fragmentation process is the most important process in a melt-coolant interaction and determines if a local interaction will escalate to form a coherent propagating wave which transports momentum and energy in the direction of propagation. For the case in which melt drops are dispersed in a coolant, a global wave phenomenon will only exist if the local explosions of individual drops can couple to form a macroscopic front with associated global pressure and velocity fields. If the drops are too far apart, or too few in number to provide sufficient confinement, sequential explosions of the drops may still occur, but a macroscopic wave will not be formed.

In this paper, the dynamics of propagating vapor explosions in melt-coolant mixtures will be discussed with reference to recent experimental results (using molten tin-water or molten alumina-water mixtures) and compared with ideal wave solutions calculated using equilibrium thermodynamics. To characterize the propagating interactions observed experimentally, it is necessary to examine the characteristic length- and time-scales associated with the local explosion of single drops within the reaction zone and compare them with the appropriate scales associated with the macroscopic mixture.

2. Experimental results

2.1. Explosion of a single molten drop

If a hot melt drop is immersed in water, an interaction may be initiated by disrupting the vapor film around the drop with a pressure or flow disturbance. If the vapor film is collapsed with a weak underwater shock wave, the fragmentation of the drop will be dominated by so-called “thermal-fragmentation” mechanisms in which the energy dissipated in the melt breakup is derived from the internal energy of the melt itself. Conversely, if the drop is subjected to a large relative velocity between the drop and the surrounding coolant, direct “hydrodynamic fragmentation” of the drop will occur due to the acceleration of the drop.

Figure 2, taken from the work of Ciccarelli & Frost (1994), shows the characteristics of the explosion of 0.5-g drops of molten tin at 700°C triggered by a shock wave. The drop breakup is driven by thermal energy transfer to the

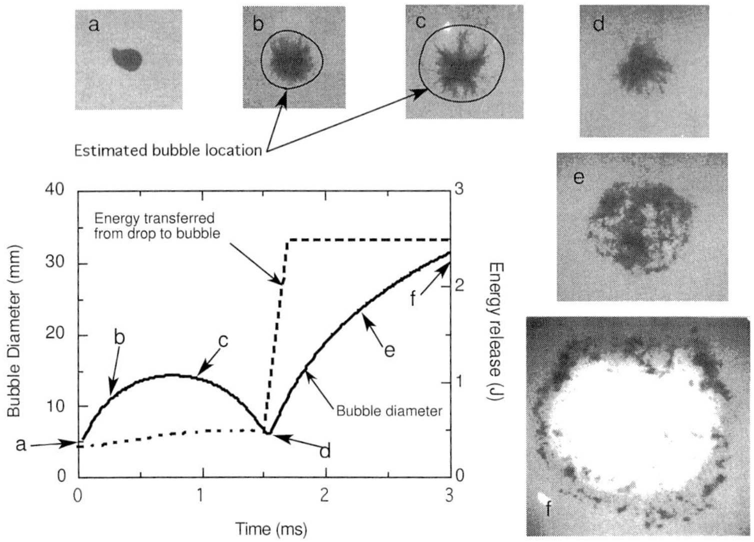

Figure 2. Dynamics of vapor explosion of single 0.5-g drop of tin at 700°C immersed in water at 65°C. X-ray radiographs show fragmentation of the melt drop at various stages of the interaction. Each radiograph corresponds to a different experiment.

water and the most significant feature of the interaction is that the thermal energy is released in several bubble-growth and collapse cycles. Following the initial vapor-film collapse, a steam bubble is formed and the drop surface becomes highly convoluted as shown in the X-ray radiographs. A simple model for the bubble dynamics indicates that about 3% of the thermal energy of the drop is transferred to the thermal energy of the vapor bubble, with the majority transferred at the time of bubble collapse (at $t = 1.5$ ms) when the drop is completely fragmented (Ciccarelli and Frost 1994).

Figure 3 contains X-ray radiographs showing a comparison between the fragmentation of hot and cold drops subjected to an ambient water flow (moving from bottom to top) of 40 m/s. The drops at the top of the figure are cold liquid metal drops (a low melting-point alloy consisting of 45% bismuth, 23% lead, 19% indium, 8% tin and 5% cadmium) and the drop fragmentation is strictly due to the relative flow. For the hot drops shown at the bottom of the figure, a vapor bubble is still generated, but the fragmentation within the bubble is enhanced by the internal turbulent vapor flow. At later times, the bubble is displaced and condensed, again bringing the melt in direct contact with the water. It is conventional to define a nondimensional time $T = t/t^*$,

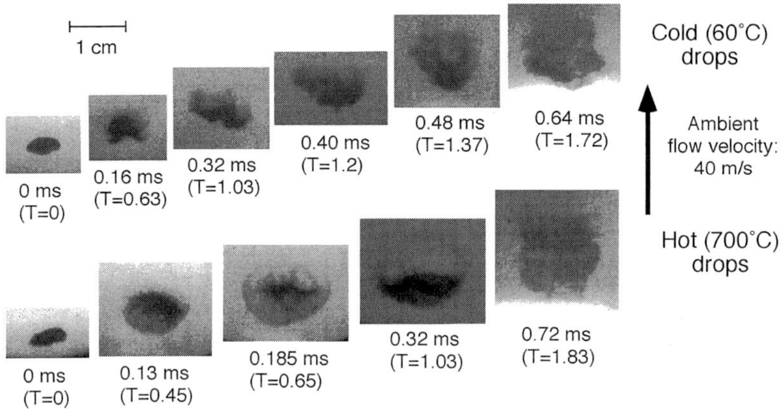

Figure 3. X-ray radiographs showing a comparison of the effect of ambient flow velocity on the fragmentation of *cold* 0.5-g liquid metal drops (low melting-point alloy) at 60°C shown on top and *hot* 0.5-g tin drops at 700°C shown on the bottom. Each picture corresponds to a different experiment. The vertical position of the drops shown is approximate. T corresponds to the nondimensional time as defined in the text.

where the characteristic time $t^* = (d/v_{rel})\sqrt{\rho_d/\rho_c}$, where d is the initial drop diameter, v_{rel} is the initial relative velocity between the drop and coolant, ρ is the density, and the subscripts d and c refer to the drop and coolant, respectively. Physically t^* corresponds to the time for the drop to move a distance approximately equal to its initial diameter due to the acceleration by the flow. Note that for both the hot and cold drops, complete fragmentation of the drop occurs at a time of about $T \approx 2$.

2.2. Propagating interactions in melt-coolant mixtures

The characteristic length scale for the fragmentation of a tin drop in a propagating interaction will depend on the pressure and particle velocity fields within the reaction zone following the leading pressure front. If the flow velocities are small (e.g., < 5 m/s), then the characteristic dimension will correspond to the size of the steam bubble generated, which is of the order of 3–4 cm for the explosion of 0.5-g molten tin drops. If the propagating interaction escalates to generate high particle velocities behind the wavefront, the characteristic reaction-zone thickness $\delta \sim D\tau$, where D is the propagation velocity and τ is the time for complete fragmentation of a drop. If we take a typical propagation velocity of 200 m/s for well-confined tin-water mixtures (e.g., Bürger et al. 1991), a relative flow velocity of 40 m/s and a nondimensional breakup time of 2 based on

Fig. 3, then $\delta \approx 15$ cm. For both of these cases, propagating interactions will be strongly influenced by the boundary conditions unless the scale of the mixture is much greater than the reaction-zone thickness (otherwise expansion waves due to the lateral expansion of the flow will attenuate the pressure within the reaction zone). This implies that mixture scales of greater than 1 m are required for the propagating interaction to be relatively unaffected by the confinement of the mixture.

For an unconfined linear drop array, an incoherent propagating interaction can occur, although little kinetic energy is transferred in the direction of propagation. This is shown by the series of photographs in Fig. 4a which illustrates the propagation of an interaction through an array of 20 5-g drops of tin at 800°C dropped into a water tank. The interaction propagates from one drop to the next due to the blast wave (duration ≈ 10 μs) generated during bubble collapse. The time for the shock to propagate from one drop to another (~ 10 μs) is negligible compared with the characteristic time for bubble growth and collapse, t_{bubble}, so the explosion propagates in a pulsating manner through the drops with an average propagation speed of 8 m/s with the speed proportional to Δ/t_{bubble}, where Δ is the characteristic drop spacing for which the explosion of one drop is sufficient to trigger the explosion of the adjacent drop.

The experiment described above was repeated, with the drops confined laterally by two glass plates forming a channel with a spacing of 1.25 cm, and the results are shown in Fig. 4b. Since the channel width is smaller than the size of the steam bubbles generated, spherical bubble growth is prevented and the flow velocity generated by an expanding bubble is enhanced. This has the effect of reducing the induction time for the explosion of one drop to trigger the next and the interaction sweeps through the drops to the left with a velocity of 50–60 m/s, generating a pressure pulse (with a rise time of ≈ 1 ms and peak pressure of ≈ 0.5 MPa measured at the tank base). Although the expansion is constrained in one dimension, the vapor generated is free to expand vertically which limits the overpressure within the reaction zone. Due to the relatively low pressure amplitudes attained and the high compressibility of the upstream mixture, the velocity of the propagating interaction is of the same magnitude as the homogeneous two-phase sound speed for the upstream mixture (the sound speed for homogeneous water-steam mixtures varies from 39 to 54 m/s for void fractions from 10 to 5%, respectively). A similar propagation velocity is obtained if the molten tin is allowed to fall to the base of the channel, forming a stratified mixture (see Frost et al. 1995). The energy yield (i.e., the conversion of thermal to mechanical energy) per surface area of the tin as well as the overpressures generated are also of the same order for both the dispersed- and stratified-geometry experiments.

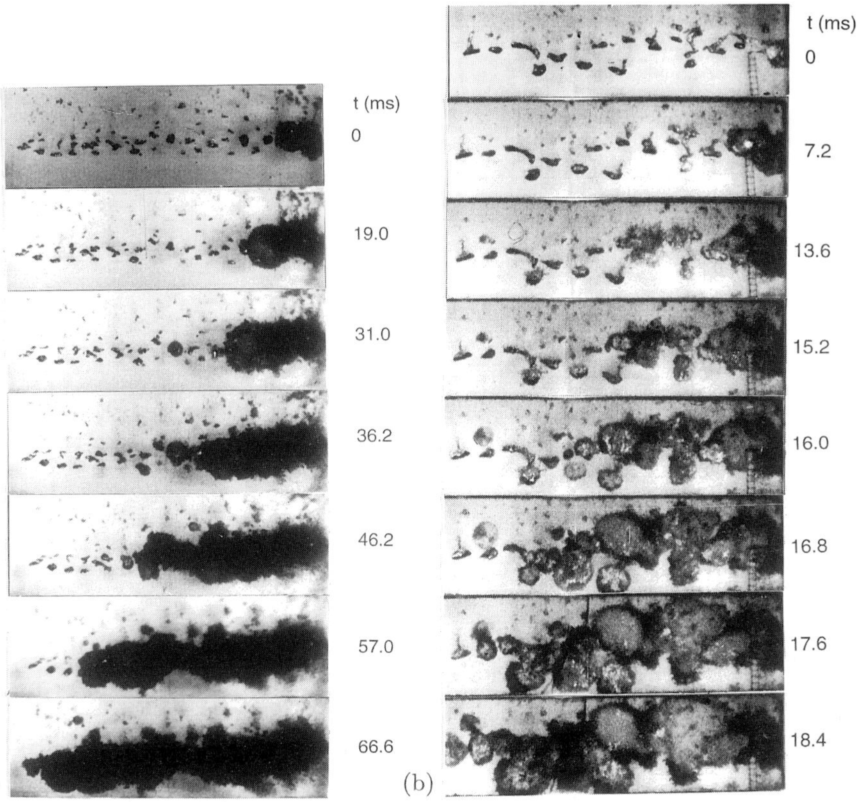

(a)

Figure 4. Effect of confinement on the propagation of an interaction in 20 5-g drops of tin at 800°C in water at 60°C. Drops on the left (a) are within a tank with a width of 20 cm. Drops on the right (b) are confined laterally within glass plates 1.25 cm apart. Explosion is initiated on the right by either the contact with a local region of cold water (unconfined drops) or by the impact with a wire mesh (confined drops).

A molten tin-water mixture may be more fully confined by placing it within a narrow tube. If the interaction is initiated at the closed end, the expansion of the high-pressure vapor in the reaction zone will act like a moving piston, driving a pressure wave to supersonic speeds into the upstream melt-coolant mixture. Experiments with this geometry have been carried out in the KROTOS (Bürger et al. 1991) and WFCI (Park et al. 1993) facilities. In both cases, a quantity of tin (between 5 and 7.5 kg) was heated to between 1000°C and 1100°C and poured into a vertical tube filled with a quantity (between 7.5 and 8.7 liters) of water. Propagation velocities and peak pressures recorded were typically on the order of 200 m/s and 5 MPa, respectively. Since the tube diameter (in both cases about 9 cm) is of the same order of magnitude as the reaction-zone thickness estimated earlier, these propagating interactions are strongly influenced by the boundary conditions (i.e., the confinement provided by the tube walls). The reaction-zone thickness is likely even longer than the earlier estimate since the interference between neighboring drops will reduce the relative flow velocity between the drop and coolant and hence the rate of fragmentation.

The KROTOS facility has also been used to study the propagation of vapor explosions in high-temperature molten alumina-water mixtures that represent a better simulant of the molten fuel produced in a hypothetical loss-of-coolant scenario in a nuclear reactor. For example, in the KROTOS-28 test (Hohmann et al. 1994) 1.45 kg of aluminum oxide melt at 2400°C was released into 7.5 liters of water at 87°C. An interaction propagated in the mixture with a velocity between 650 and 1000 m/s. The supercritical pressures attained were poorly characterized due to saturation of the transducers, but were greater than 50 MPa. Due to the high pressures, the corresponding high density of the fluid in the reaction zone will enhance the hydrodynamic fragmentation rate and rate of energy transfer to the coolant. An order-of-magnitude estimate of the reaction-zone thickness can be made by taking a propagation velocity of 1000 m/s, an average relative velocity between the flow and a 1-cm drop of 100 m/s, and a nondimensional total fragmentation time of 2, giving a value for the reaction-zone length of about 10 cm. Since this value is of the same order as the tube diameter, the dynamic parameters of the propagating interaction will again be strongly dependent on boundary conditions.

3. Wave propagation models

3.1. Steady models

The nature of possible steady wave solutions supported by a melt-coolant mixture can be investigated without regard to the complex dynamical processes involved. The analysis involves solving the one-dimensional conservation laws of mass, momentum, and energy with appropriate equations of state and mixture rules across a steady control volume that contains the wave. The downstream locus-of-equilibrium states calculated in this manner define the Hugoniot curve as in a chemically reacting medium. A unique Chapman-Jouguet (CJ) detonation solution can be determined by using the classical construction of the Rayleigh line tangent to the Hugoniot curve. This procedure was first carried out by Board et al. (1975) for mixtures of molten tin and water. For a low-temperature melt-coolant mixture, the Hugoniot curve has two distinct sections separated by a discontinuity or kink at the saturation boundary. As a result, classical CJ detonation solutions can only be obtained for initial void fractions below a certain value. The analysis was refined by McCahan & Shepherd (1993) by incorporating a compressible equation of state for the tin.

Two assumptions in the ideal Hugoniot analysis are not satisfied, in general, for propagating interactions observed in melt-coolant mixtures. First, although the reaction-zone length does not come into the Hugoniot analysis, since boundary conditions do not influence the solution, the dimensions of the mixture must be much greater than the reaction-zone thickness, a condition clearly not satisfied for the propagation experiments. Secondly, the attainment of thermal equilibrium at the downstream state is hindered by the low thermal conductivity of the coolant and the presence of vapor which isolates the coolant from hot melt fragments.

Despite the limited scale of current propagation experiments, it is of interest to determine the properties of the ideal wave solutions for a high-temperature melt-coolant mixture, assuming that the mixture dimension is large enough to support a thermal detonation wave. If a molten alumina-water mixture is considered, with an initial melt temperature (2400°C) and melt, coolant, and vapor volume fractions (0.08, 0.88, and 0.04, respectively) corresponding to the KROTOS-28 test (Hohmann et al. 1994), and a compressible equation of state for molten alumina is used (following McCahan & Shepherd 1994), the equilibrium Hugoniot may be computed and is shown in Fig. 5a. The CJ detonation pressure and velocity (400 MPa and 2600 m/s, respectively) are much higher than the values observed experimentally. A second ideal wave solution, first proposed by McCahan & Shepherd (1993) for wave propagation in a closed-end tube, may be computed which consists of a leading shock wave that precompresses the mixture followed by a deflagration wave that brings the flow to rest downstream. To obtain a unique solution we can use the condition that the deflagration satisfies the CJ criterion. The results for the same initial conditions as for the detonation solution are shown in Fig. 5b. State 0 is the upstream state, state 1 is downstream of the shock wave, and the final state 2 is downstream of the CJ deflagration wave. The structure of this wave solution differs from that of the detonation solution in that it is explicitly assumed that a shock wave is present. To determine the state downstream of the shock, the shock is assumed to be isentropic with the melt and coolant downstream of the shock in mechanical but not thermal equilibrium. This ad-hoc assumption leads to a deficit in the conservation of energy across the shock by about 17% in this case. In this solution, the shock and deflagration waves are not coupled as in a detonation wave and move at different velocities (1690 m/s and 1208 m/s, respectively). Following a high pressure of 474 MPa across the shock, the CJ deflagration pressure is 77 MPa. Although the choice of the CJ criterion cannot be justified on a fundamental basis, the predicted velocities and pressure at the CJ plane are closer to the experimental values suggesting that the shock/deflagration wave complex may be more representative of the observed propagating interaction than a fully coupled detonation wave. However, in this model, uncertainties remain concerning the validity of the assumptions used in calculating the jump conditions across the shock in the multiphase mixture.

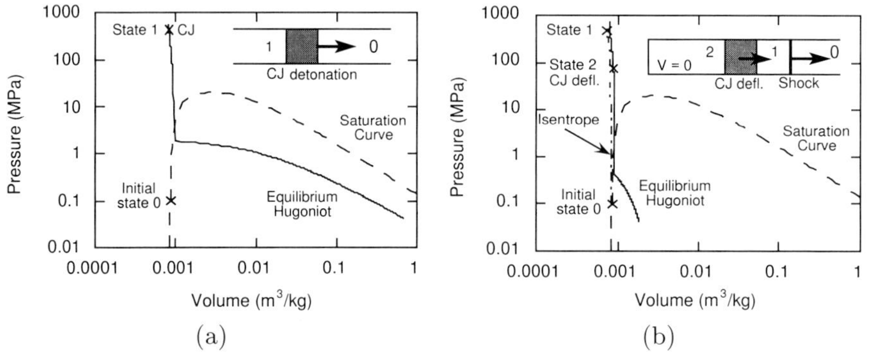

Figure 5. Hugoniot curves showing (a) ideal CJ-detonation and (b) isentropic shock/CJ-deflagration solutions for a mixture of molten alumina at 2400°C and water and steam.

The Hugoniot analysis can be modified to include the effect of momentum losses due to friction at tube walls. For example, Gelfand et al. (1992) included frictional effects by using the traditional method based on the ZND model for detonations in combustible gas mixtures to find the singularity or eigenvalue conditions at the CJ plane that represent valid wave solutions. However, this adds an additional complication because if downstream equilibrium is not assumed, a model for the rate of energy release (involving the rates of fragmentation of the melt and heat transfer to the coolant) must be included in the model. However, even if a realistic fragmentation model is incorporated into the analysis, there is no guarantee that such a wave exists in reality or in fact can be attained following the transient development of the explosion.

3.2. Transient models

A number of multiphase models have been proposed to address the transient development of a propagating interaction by considering the complex processes within the reaction zone. The current generation of models (e.g., CULDESAC, Fletcher & Anderson 1990; ESPROSE, Yuen et al. 1992; IDEMO, Bürger et al. 1991; TEXAS-III, Tang & Corradini 1994) is useful for studying the parametric dependence of various properties on the propagation dynamics. However, uncertainties in the constitutive physical models prevent quantitative predictions from being made with any confidence. The lack of well-characterized experimental data is the limiting factor in the validation of the transient propagation models. This is due largely to two factors: first, it is difficult experimentally, particularly at large scale, to produce a well-defined stable homogeneous mixture of melt and coolant (other than a stratified mixture) and secondly, it is very difficult to obtain time- and space-resolved measurements of temperature, pressure, and the distribution of phases within the reaction zone of a propagating interaction sufficient to resolve the important fine fragmentation process.

4. Outstanding issues

A detailed description of the fine fragmentation process within a propagating melt-coolant interaction remains the most important outstanding factor that must be addressed to improve our ability to estimate the dynamic parameters of vapor explosion waves. In particular, the relative importance of thermal- and hydrodynamic-fragmentation mechanisms and the effect of interference between adjacent fragmenting drops must be determined as a function of the transient pressure and velocity fields within the reaction zone. Realistic models for other processes at the micro-level, such as for the transient vaporization of the nonuniformly heated coolant following melt-coolant contact, are not available. A second major challenge is the theoretical description of how the explosion of discrete melt drops interact and couple to form a coherent macroscopic wave. The critical energy to directly initiate the propagation of a coherent wave and the propagation limits (analagous to the corresponding dynamic parameters for combustible mixtures) are poorly understood for melt-coolant mixtures. The scale of current propagation experiments is typically of the same order as the fragmentation zone length and therefore the interaction is strongly dependent on the boundary conditions, unlike the ideal wave solutions that may be determined using equilibrium thermodynamics. Although progress is being made on the description of the constitutive physical processes within a vapor explosion, complete models for propagating vapor explosion waves with a truly predictive capability will likely remain elusive for some time to come.

Acknowledgement. The author would like to acknowledge the work of his former graduate students Gaby Ciccarelli, Phil Watts, and Barbara Bruckert as well as the assistance of Walter Lei and Patton Chan in the Hugoniot calculations. Valuable comments were also provided by John Lee. This work has been supported by the Natural Sciences and Engineering Research Council of Canada, Atomic Energy Control Board of Canada, and Sandia National Labs.

References

Board SJ, Hall RW, Hall RS (1975) Detonation of fuel coolant explosions. Nature 254:319.

Bürger M, Müller K, Buck M, Cho SH, Schatz A, Schins H, Zeyen R, Hohmann H (1991) Examination of thermal detonation codes and included fragmentation models by means of triggered propagation experiments in a tin/water mixture. Nucl Eng Design 131:61.

Ciccarelli G, Frost DL (1994) Fragmentation mechanisms based on single drop steam explosion experiments using flash X-ray radiography. Nucl Eng Des 146:109.

Corradini ML, Kim BJ, Oh MD (1988) Vapor explosions in light water reactors: a review of theory and modeling. Prog Nucl Energy 22(1):1.

Fletcher DF, Anderson RP (1990) A review of pressure-induced propagation models of the vapour explosion process. Prog in Nucl Energy 23:137.

Frost DL, Bruckert B, Ciccarelli G (1995) Effect of boundary conditions on the propagation of a vapor explosion in stratified molten tin/water systems. Nucl Eng Des 155:311.

Gelfand BE, Bartenev AM, Frolov SM, Tsyganov SA (1992) Thermal detonation in molten Sn-water suspension. Archivum Combustionis 12(1):5.

Hill LG (1990) An Experimental Study of Evaporation Waves in a Superheated Liquid, Ph.D. Thesis, California Institute of Technology.

Hohmann H, Magallon D, Schins H, Yerkes A (1994) FCI experiments in the aluminumoxide/water system. In: Proc CSNI Specialist Meeting on Fuel Coolant Interactions, NUREG/CP-0127, Washington, p. 193.

McCahan S, Shepherd JE (1993) Models of rapid evaporation in nonequilibrium mixtures of tin and water. AIAA Prog in Astr and Aero 154:432.

McCahan S, Shepherd JE (1994) A thermodynamic model for aluminum-water interactions. In: Proc CSNI Specialist Meeting on Fuel Coolant Interactions, NUREG/CP-0127, Washington, p. 134.

Park HS, Yoon C, Corradini ML, Bang KH (1993) Vapor explosion escalation/propagation experiments and possible fragmentation mechanisms. In: Int Sem on the Physics of Vapor Explosions, Tomakomai, Japan.

Shepherd JE, Sturtevant B (1982) Rapid evaporation at the superheat limit. J Fluid Mech 121:379.

Tang J, Corradini ML (1994) Modelling of the complete process of one-dimensional vapour explosions. In: Proc CSNI Specialist Meeting on Fuel Coolant Interactions, NUREG/CP-0127, Washington, p. 204.

Yuen WW, Chen X, Theofanous TG (1992) On the fundamental microinteractions that support the propagation of steam explosions. In: Proc NURETH-5, Vol. 2, p. 627.

Irving Glass Memorial Session

Wave machinery for chemical processing and high-efficiency heat engines

W. H. Christiansen, A. Hertzberg
Department of Aeronautics and Astronautics
P.O. Box 352400
University of Washington
Seattle, WA 98195-2400 USA

Abstract: Wave machines using nonsteady gasdynamic processes, i.e., mechanizations of shock tubes to produce steady gas flows, have been used in energy and chemical processes. The potential of this class of machinery is investigated here using on-rotor combustion as a means of extending the usefulness of this type of machinery. On-rotor combustion adds yet another variable in the complex gasdynamic process on the rotor, but permits the possibility of using the device as a stand-alone heat engine.

Key words: Wave machinery, Heat engine

1. Introduction

Wave machinery involving direct pressure exchange between hot and cold gases has always been an appealing concept, since the average temperature of the pressure-exchange tubes can be maintained at a relatively low level, allowing extremely high temperatures to be explored. However, it was not until after World War II that Brown-Boveri (now Asea Brown-Boveri) under the direction of Claude Seippel (Seippel 1946) built wave machinery in a topping unit using a comprex (trade name, but meaning compression and expansion in the same device, used by Brown-Boveri). Kantrowitz at Cornell University in the late 1940s also studied the comprex concept as a heat engine component (Kantrowitz et al. 1948). In 1950 at Cornell Aero Lab (now Calspan Corporation), A. Hertzberg laid out the design of a device which could, in principle, produce NO efficiently (Glick et al. 1953). Also the design was capable of very efficient pyrolysis and production of acetylene. Al Flax of Cornell Lab adapted these ideas and pushed forward to the construction of a machine capable of heating air to about $3,500^\circ$K for periods of seconds, to be used as a hypersonic nose-cone testing device (Hertzberg et al. 1959). The machine was successful and proved the high temperature capability of wave machinery. In 1960 using the knowledge gained from these efforts in wave machinery, Hertzberg and Weatherston (Hertzberg et al. 1959) developed a concept for a high-temperature gas-turbine device which promised significant increases in efficiency over existing gas turbines of that period. During this period Brown-Boveri solved the mechanical problems of the comprex and developed the concept into an effective supercharger for diesel trucks. Wave machinery is now currently being studied by Asea Brown-Boveri as a method of increasing the efficiency of gas turbines. NASA is supporting efforts at Cornell University under Dr. Ed Resler (Resler et al. 1994) and Dr. Jakob Keller (private communication) at the University of Washington for cycles that would be suitable for aircraft gas turbines. These machines are attractive because they can operate at high temperatures and higher efficiency.

In this paper we would like to discuss a study on alternate cycles which promises relative simplicity of operation, for either high-temperature chemical pyrolysis, or gas-turbine power cycles. A "fly-wheel"-type of driver system was suggested (Hertzberg and Christiansen 1976) in which the compression energy inside the cycle is returned to keep the cycle operating, permitting a stand-alone wave device without auxiliary equipment. Losses in the circulating system were originally anticipated to be made up by an overall pressure drop through the machine with careful adjustment of the wave system, when used as a pyrolysis-inducing device for ethylene, acetylene, or NO production. The term "fly-wheel" is used representing the motion of a massless imaginary piston, whose operating motion can be adapted at will to the cycle. Thus, there is basically one inlet duct (and one exit duct) to let in the fresh unprocessed gas, and to exhaust the spent processed gas from the machine.

2. Details of the fly-wheel driver concept

Figure 1 is the wave diagram (x-t diagram) Hertzberg and Christiansen (1976) used to illustrate the fly-wheel driver concept. In this diagram the reactant and driver gas are idealized, for illustration only, to be perfect air, thus ensuring tailored operation and minimum interface mixing. The temperatures are given in degrees Kelvin, and the pressures in atm — in this case chosen to be in the appropriate range for NO production. At the bottom of the figure high-pressure reactant air is taken into the device from the left at a Mach number of 0.68 and static pressure of 3.48 atm. To start this cycle, the reactant gas is brought to rest by a weak hammer shock. Additional isentropic pressurization is carried out by driver gas coming in from the right through wave action until the reaction conditions are met — in this case 50 atm and 3300°K. The dashed line in this drawing represents the interface between the reactant and driver gas as this process is carried out. After a suitable dwell time, the pressurization process is reversed and the driver gas again leaves to the right, but at pressures slightly higher than those at which it came in. This is carefully carried out until the original driver gas has left the rotor or until the interface reaches the right-hand edge of the rotor. The remaining reactant gas is further removed with the help of new reactant gas from the left, thus arriving at the same gas conditions in order to repeat the cycle.

Further close scrutiny of the total pressures, total enthalpies, and mass flows indicated that it was possible to exhaust the driver gas with excess total pressure so that it could be collected via suitable manifolds and used again without the need of pumps on the driver side of the wave machine. Of course the energy gain in the driver gas was made up by a corresponding energy loss in the reactant gas loop. This illustrates the concept of the "fly-wheel driver," a driver-gas system not requiring any external pumping device. But observation of this drawing shows that the driver over pressures are very small O (3% max) and the excess total enthalpy is also O (1% max). This is a marginal pumping situation, particularly in view of any calculational errors present and external mixing processes, and one which may not be changed easily. Thus, a method to improve on this notion is needed.

3. Modification of concept via on-rotor combustion

This cycle has been modified recently to use the addition of either a layer of combustible gas at the interface, or by modifying the inlet mixture, so that a certain amount of combustion on the rotor in the manner of a reciprocating combustion engine can be employed to provide significant driving energy for the system consistent with the needs of the system. The combustible gas mixture undergoes compression on the rotor. At the appropriate point in the cycle, say at maximum compression, it is ignited. If combustion takes place on the rotor, there are two limiting cases. In one, the combustion time is slow compared to the acoustic transit time, L/a, so that pressure continuity is maintained at the interfaces. While the overall volume, indicated by L, is constant, the individual volumes of the driver gas and the combustible mixture are not constant during combustion. The second possibility is that the combustion time is short compared to L/a, which results in pressure discontinuities at the interfaces because the individual gas volumes remain constant during the process. This situation will produce a shock-wave relaxation process in addition to the wave processes already present in the wave machine. For the rest of the study, it is assumed that only slow combustion takes place.

Following the details in Fig. 1, we imagine that the reactant gas is burnable. At maximum compression, i.e., $p = 50$, there is no flow and the wave system is at rest. One can add significant dwell time, or in this case, combustion time here without affecting the subsequent wave processes necessary to close the cycle. The schematic of the wave diagram under this scenario is shown in Fig. 2, wherein after combustion the interface has moved to the right. Since there is no outflow during this part of the cycle, the pressure of the system

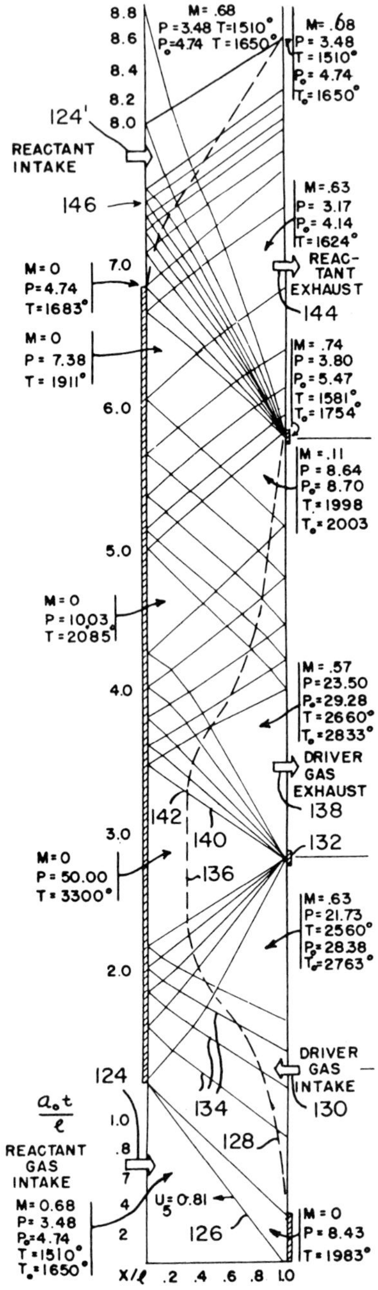

Figure 1. Wave diagram.

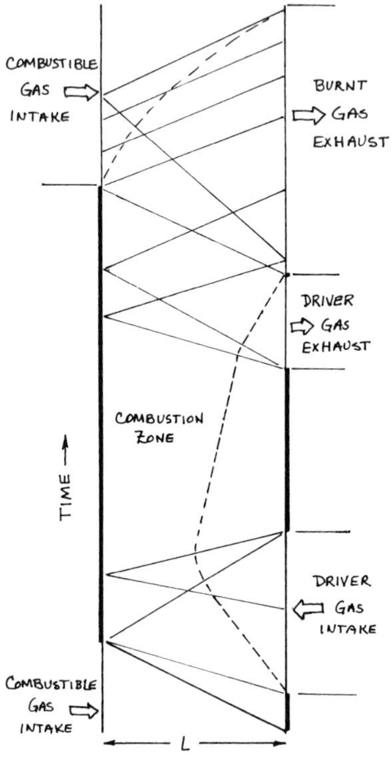

Figure 2. Wave diagram with combustion.

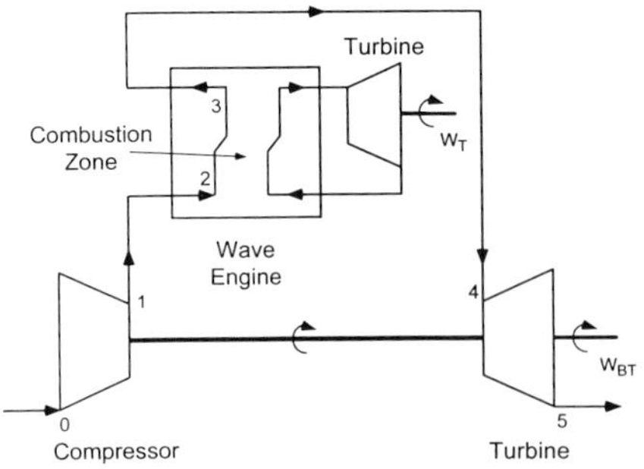

Figure 3. Block diagram of topping unit.

increases, including the driver gas which is being compressed by the movement of the interface. After combustion there is pressure continuity and a temperature discontinuity at the interface, which results in a slight off-tailoring condition. But the point to notice is that significant compression occurs in the inert driver gas, depending on the degree of combustion on-rotor. Examining this cycle more carefully, we can see that by using a combustible mixture as the process gas, the machine becomes in itself a self-driving hot-gas generator with considerable pre-compression. More than enough energy can easily be drained off to continue the fly-wheel action, and in some cases shaft energy could be withdrawn from the driver outflow.

4. Example use with a topping unit

Figure 3 shows a sketch of a wave machine used as a high-temperature combustor in a normal Brayton cycle. In this schematic, the combustor is imagined to run along the lines of the wave diagrams depicted earlier. First, reactant gas is compressed by an external compressor which is then led to the wave engine. It is further compressed in the wave engine and burned there. The driver gas is tailored, i.e., by using a lower molecular-weight gas, to operate at lower driver-cycle temperatures so that current turbine technology can be used. The driver gas is also sufficiently compressed by the combustion process so that additional work can be extracted by a topping turbine. The normal high temperatures involved in combustion are mitigated in the rotor because of the work exchanged in the process. The resulting reactant gas is drawn off the rotor, still at modest temperature and relatively high pressure, to be used in the bottoming turbine. The resulting thermodynamic cycle is more efficient than a stand-alone Brayton cycle as shown in Fig. 3. Of course, other versions including additional external burning off the rotor are also possible.

References

Glick H, Squire W, Hertzberg A (1953) A New Shock Tube Technique for the Study of High-Temperature Gas-Phase Reactions, Project SQUID Report.

Hertzberg A, Glick HS, Squire W, Wetherston RC (1959) Methods and Apparatus for Carrying Out Gas-Phase Reactions Which Require a High Temperature to Promote the Reaction and Rapid Cooling to Preserve the Reaction Product, U.S. Patent Numbers 2,832,665 and 2,832,666 (1958) and 2,902,337.

Hertzberg A, Christiansen W (1976) Method for Continuously Carrying Out a Gas Phase Reaction and Apparatus Therefore, U.S. Patent Number 3,998,711.

Kantrowitz A, Hertzberg A, McDonald E, Resler E (1948) Heat Engines Based on Wave Processes, Cornell University Graduate School of Aeronautical Engineering Report.
Resler EL Jr, Mocsari JC, Nalim MR (1994) Analytic design methods for wave rotor cycles. J Propulsion and Power 10(5): 683–689.
Seippel C (1946) Pressure Exchanger, U.S. Patent Number 2,399,394.

Holographic interferometric animation of shock diffraction and reflection using a highly repeatable shock tube

J. M. Yang*, H. Babinsky†, K. Takayama
Shock Wave Research Center, Institute of Fluid Science, Tohoku University
Katahira 2-1-1, Aoba, Sendai 980, Japan
T. Saito
Cray Research Japan Ltd.
2-5-5 Shinyokohama, Minatokita-ku, Yokohama, Japan
J. P. Sislian
Institute for Aerospace Studies, University of Toronto, Canada

Abstract: An animated holographic observation was made of shock diffraction and reflection in a 60 mm × 150 mm diaphragmless shock tube which has a high degree of repeatability. Two animations were made. One is a shock diffraction over a 90° sharp corner; the other examines the variation of the shock-wave reflection configuration with changing inclination of a wedge. It was found that as the inclination of the wedge increases, the reflection pattern continuously changes from von Neumann reflection through simple Mach reflection to regular reflection. The animation of shock-wave diffraction showed some non-self-similarity near the corner.

Some new results of 3-dimensional shock-wave reflection over an oblique cylinder are also reported. This experiment was conducted in the 100 mm × 180 mm shock tube (Glass 1991) which was recently moved from the Institute for Aerospace Studies, University of Toronto to the Shock Wave Research Center, Institute of Fluid Science, Tohoku University.

Key words: Reflection, Diffraction, Animation, Holographic interferometry, Diaphragmless shock tube

1. Introduction

The shock-wave reflection over a straight wedge or shock-wave diffraction past a 90° corner are two of the benchmark tests of shock dynamics (Skews 1967; Ben-Dor 1992; Han and Yin 1993). Motion of such shock waves are usually vi-

---

*Present address: Department of Modern Mechanics, University of Science and Technology of China, Hefei, Anhui, 230026, China.

†Present address: Department of Mechanical Engineering, University of Cambridge, England.

sualized by means of high-speed cameras. Unfortunately, the spatial resolution of existing high-speed cameras is generally at most 20 lines/mm. So far, very precise sequential observations have been recognized to be impossible. Double-exposure holographic interferometry has a very high spatial resolution of more than 100 lines/mm, however, this method can only supply a single image at each test. By using the combination of holographic interferometry and a shock tube which has very high repeatability, it is possible to record a large number of interferograms under nearly identical initial conditions. Images obtained in such a fashion are equivalent to individual frames in a high-speed movie. Through image processing, these frames can be combined into an animation sequence with little degradation of the spatial resolution (Babinsky et al. 1995). Two configurations have been investigated with this technique. One is shock-wave diffraction over a 90° sharp corner, the other examines shock-reflection patterns on a wedge. The wedge inclination was increased by a small amount between each test, while the shock position was kept constant. The thus obtained animation shows the change of the shock-wave reflection configuration with increasing wedge inclination.

2. Experiment

The experiments were conducted in a diaphragmless shock tube at the Shock Wave Research Center, Tohoku University. The shock tube consisted of a low-pressure channel of 60 mm × 150 mm, and a high-pressure chamber whose diameter is 290 mm. Upon the quick movement of a rubber membrane which seals the high-pressure and low-pressure sections, a shock wave can be generated with high repeatability. The scatter of the generated shock Mach number was found to be within ±0.25 % for more than 500 runs (Yang et al. 1994).

Double-exposure holographic interferometry was used to visualize the flow field. The basic optical arrangement is similar to a conventional shadowgraph system (Takayama 1983). A double-pulse ruby laser (Apollo Laser Inc. 22DH, 25-ns pulse duration and 2 J/pulse) was used as the light source. In the experiments of shock-wave reflection over wedges, a 60-mm wide wedge whose inclination angle can be adjusted continuously was installed in the 60 mm × 150 mm test section, and the inclination angles of the wedge varied from 3° to 55° in steps of about 0.7°. Nitrogen was used as the test gas in the reflection experiments. The shock Mach number was 1.20 and its scatter was found to be less than ±0.15% over a total of 156 runs. For the experiments of shock-wave diffraction over a 90° corner, a sudden backward-facing step from 60 mm × 150 mm to 60 mm × 360 mm was used as the test section. Such a large test section allowed an extension of the observation time of up to 800 μs after the shock wave reached the corner. Two shock Mach numbers, $M_s = 1.20$ and 1.50, were tested for this case and the scatter in both cases was below ±0.2 %.

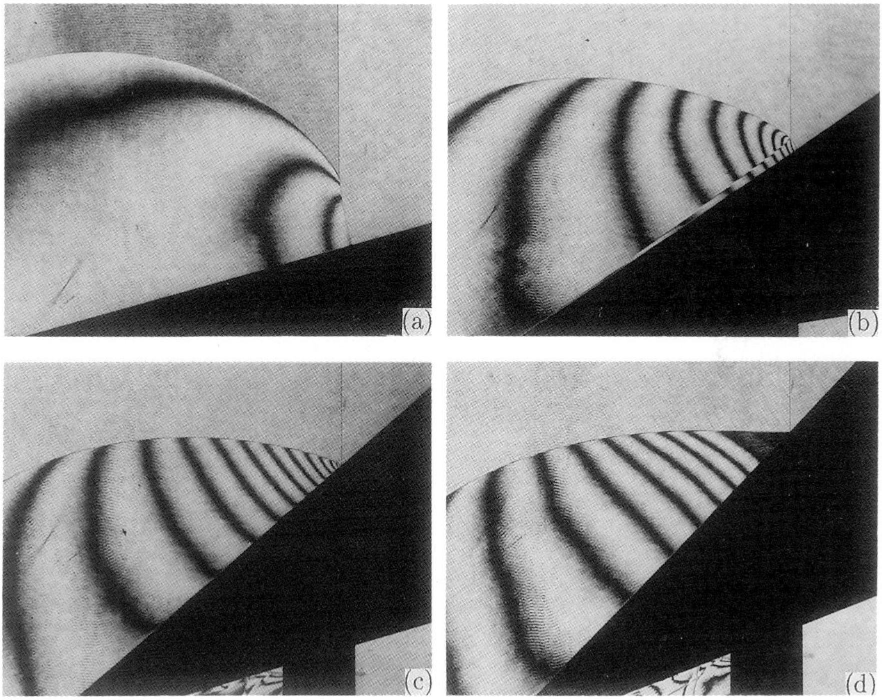

Figure 1. Holographic interferograms of shock wave reflection over wedges of different inclination, $M_s = 1.20$: (a) $\theta = 15°$; (b) $\theta = 34°$; (c) $\theta = 41°$; (d) $\theta = 45°$.

3. Results and discussion

Figure 1 shows some holographic interferograms of the animation result. For the the case of shock-wave reflection over a wedge, as shown in Fig. 1a, when the inclination angle of the wedge is small, the pattern of shock reflection appears to be what is known as von Neumann reflection (Sasoh and Takayama 1994), in which the incident shock wave and Mach stem appear to intersect with zero angle and no slip line can be recognized near the triple point. With increasing inclination angle, a discontinuity at the triple point appears and the slip line becomes clearly observable; this configuration is known as simple Mach reflection and is shown in Fig. 1b. With further increase of the inclination angle, the Mach stem reduces in length and finally the Mach reflection terminates and regular reflection occurs. Since the shock tube is highly repeatable, it is easy to observe the shock-wave reflection pattern near the transition between Mach reflection and regular reflection. Figure 1c shows the configuration near the transition point and the transition angle was found to be about 42° for shock Mach number 1.20. Figure 1d shows regular reflection. The animation sequence clearly shows that as the inclination angle of the wedge increases, the reflection pattern almost continuously changes from von Neumann reflection through simple Mach reflection to regular reflection, with no sudden jump observable. It should be pointed out that this animation cannot be obtained with a high-speed camera because the motion animated here never exists in the real world. It was found to be helpful for the understanding of shock-wave reflection phenomena.

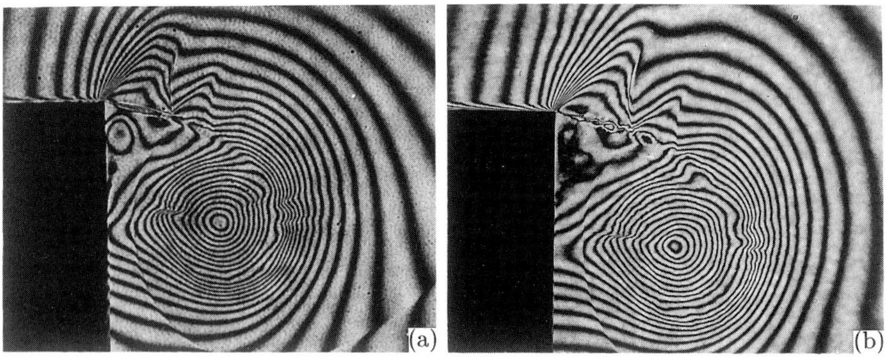

Figure 2. Holographic interferograms of shock wave diffraction over a 90° corner, $M_s = 1.50$: (a) $t = 300\mu s$, (b) $t = 865\mu s$.

The animation of shock diffraction shows the dynamic development of the shock wave diffracting around the corner. Not only the shock front but also the flow field behind the shock wave can be observed. Especially in the case of shock Mach number 1.50, the secondary shock-wave system, the vortex, the slip stream, and the contact surface, nearly self-similarly changing with time, can be clearly seen. With the aid of the high resolution of the hologram, a careful study can be made of the flow field near the corner. It is not difficult to see that there is a secondary vortex near the corner and that this vortex becomes weaker and disappears with time. Figure 2a shows the shock-wave configuration at 300 μs after the shock wave reaches the corner, while Fig. 2b is the visualization result at 865 μs. The secondary vortex can be easily observed in Fig. 2a but is almost invisible in Fig. 2b.

4. Shock reflection over an oblique cylinder

For observation of three-dimensional shock-wave reflection, the situation becomes much more complicated. A shock reflection over an oblique cylinder was visualized by using diffuse holography. This experiment was carried out in a

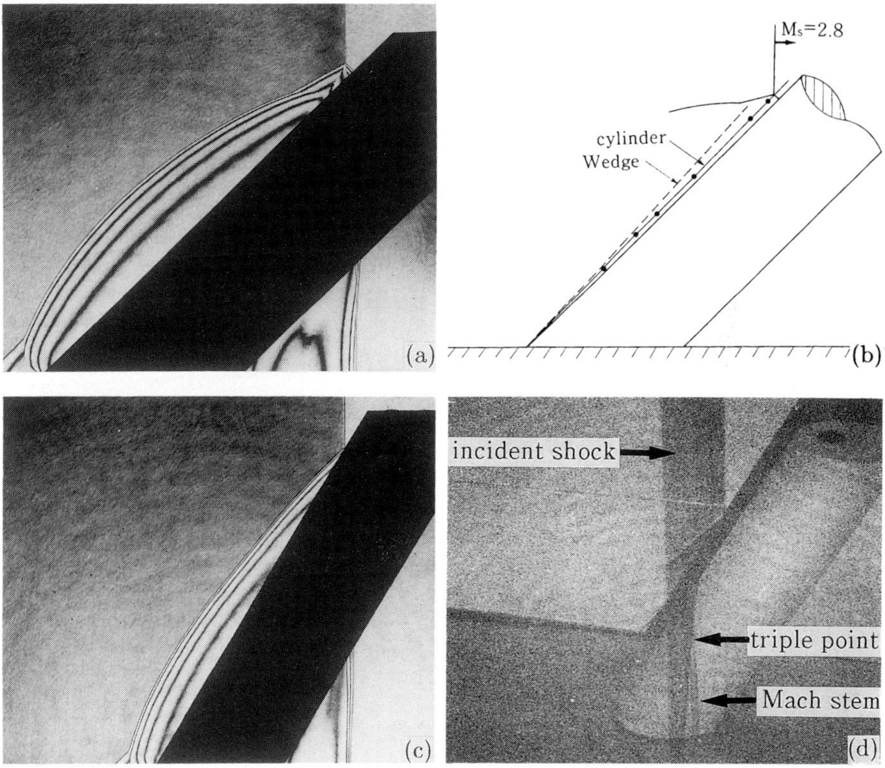

Figure 3. Visualization results of shock-wave reflection over an oblique cylinder, $M_s = 2.8$: (a) $\theta = 45°$, interferogram; (b) $\theta = 45°$, triple-point trajectory; (c) $\theta = 60°$, 2D interferogram; (d) $\theta = 60°$, diffuse (3D) interferogram.

100 mm × 180 mm shock tube which has been used for about three decades at the Institute for Aerospace Studies, University of Toronto, and was recently moved to the Shock Wave Research Center, Tohoku University.

A preliminary result is shown in Fig. 3. The incident-shock Mach number was 2.8 in air. At an inclination angle of the oblique cylinder of 45°, double Mach reflection was observed, as shown in Fig. 3a, and this reflection pattern is different from the two-dimensional one. The triple-point trajectory, unlike the straight line in two-dimensional cases, bends towards the cylinder surface and becomes parallel to the wall (Fig. 3b). When the inclination angle of the cylinder is 60°, regular reflection occurs at the front surface (Fig. 3c), whereas the diffuse interferogram shows that Mach reflection starts on the side wall (Fig. 3d).

Further investigations will be carried out in the near future and experimental results will be used to verify the three-dimensional numerical scheme.

5. Conclusions

By using the combination of holographic interferometry and a shock tube which has very high repeatability, an animated observation was successfully made for two-dimensional shock-wave reflection and diffraction. It was found that for $M_s = 1.20$, the reflection pattern continuously changes from von Neumann reflection through simple Mach reflection to regular reflection. In the case of shock diffraction over a 90° corner, dynamic motion of the complicated flow field behind the diffracted shock is clearly visible and some non-self-similarity was also observed. A three-dimensional shock-wave reflection over an oblique cylinder has been visualized at $M_s = 2.8$.

Acknowledgement. The authors would like to express their thanks to Professor I. I. Glass for his encouragement of shock-wave research at the Institute of Fluid Science, Tohoku University. The authors also express thanks to Messrs. O. Onodera, H. Ojima, and T. Ogawa of the Shock Wave Research Center, Institute of Fluid Science, Tohoku University, for their help in conducting the experiments.

References

Babinsky H, Yang JM, Takayama K (1995) Animated visualization of shock wave flowfields for dynamic comparison between experiment and computational prediction. In: Proc. SPIE 2410:101.

Ben-Dor G (1992) Shock Wave Reflection Phenomena, Springer-Verlag, New York.

Glass II (1991) Over forty years of continuous research at UTIAS on nonstationary flows and shock waves. Shock Waves 1:75.

Han ZY, Yin XZ (1993) Shock Dynamics, Kluwer Acadamic Publishers/Science Press, Dordrecht, The Netherlands.

Sasoh A, Takayama K (1994) Characterization of disturbance propagation in weak shock-wave reflections. J Fluid Mech 277:331.

Skews GW (1967) The perturbed region behind a diffracting shock wave. J Fluid Mech 29:705.

Takayama K (1983) Application of holographic interferometry into shock wave research. In: Proc. SPIE 398:174.

Yang JM, Onodera O, Takayama K (1994) Design and performance of a quick opening shock tube using a rubber membrane for weak shock wave generation. J JSME (in Japanese) 60-570:473.

Universal attenuation law of planar shock waves propagating in dust-gas suspensions

F. Aizik, G. Ben-Dor, T. Elperin, O. Igra, M. Mond
Pearlstone Center for Aeronautical Engineering Studies
Department of Mechanical Engineering
Ben-Gurion University of the Negev, Beer Sheva, Israel
H. Grönig
Stosswellenlabor, RWTH Aachen, Aachen, Germany

Abstract: In a previous study, Olim et al. (1990) proposed an attenuation law for weak to moderate planar shock waves propagating through dust-gas suspensions with relatively high loading ratios of the solid particles. In the present study, a more general attenuation law has been developed. It is valid for shock waves having an initial Mach number in the range $1 < M_0 < 5$ and dust-loading ratios in the range $0.25 < \eta < 10$.

Key words: Attenuation law, Dust-gas suspensions

1. Introduction

A few years ago, Olim et al. (1990) proposed a general law for describing the instantaneous shock-wave Mach number as it attenuates while propagating through dust-gas suspensions. Their proposed attenuation law was

$$M_s = (M_0 - 1)\exp(-x/\chi) + 1 \tag{1}$$

where M_0 is the initial shock-wave Mach number, i.e., the shock-wave Mach number when it first encounters the dust-gas suspension whose edge is at $x = 0$. The attenuation coefficient χ was found to be linearly related to the diameter of the dust particles, D, and inversely to their loading ratio, η, through the expression:

$$\chi = \frac{1}{\eta}\left[\alpha(M_0) + \beta(M_0)\,D\right] \tag{2}$$

In this expression α and β are constants for a given initial shock-wave Mach number. Olim et al. (1990) also calculated the appropriate values of these two constants for $M_0 = 1.5$. The appropriate values as obtained by them were $\alpha = 2.275$ m and $\beta = 3.07e4$.

The correlation proposed by Olim et al. (1990) implies that when $x \to \infty$ the instantaneous shock-wave Mach number degenerates to a sound wave, i.e.,

$M \to 1$. This was found experimentally by Sommerfeld (1984, 1985) to be the case only if the initial shock wave is weak to moderate and the dust-loading ratio is relatively high. If, however, the initial shock wave is strong and the dust-loading ratio is low to moderate, the shock wave does not attenuate to a sound wave, but to a finite-strength shock wave whose Mach number, M_e, is larger than unity.

Owing to this observation, it was suggested by Olim et al. (1990) that Eq. (1) be replaced by the following more general expression

$$M_s = (M_0 - M_e) \exp(-x/\chi) + M_e \tag{3}$$

Finding the functional dependence of both χ and M_e on the initial shock-wave Mach number, M_0, the diameter of the solid particles, D, and the dust-loading ratio, η, was the purpose of the study presented in the following sections.

2. Present study

In order to find the functional dependence of both χ and M_e on M_0, D, and η, the governing equations describing the propagation of a planar shock wave through a dust-gas suspension, which were formulated by Olim et al. (1990), were solved numerically for various combinations of M_0, D, and η.

Unlike Olim et al. (1990), whose numerical simulation was based on the flux-corrected-transport technique (FCT), the numerical simulation in the present study was based on the random-choice method (RCM).

The computer code was validated by comparing its predictions to all the experimental results which were reported by Sommerfeld (1984). Good to very good agreement was evident. Details of the comparison can be found in Aizik et al. (1995), where a detailed derivation of the governing equations and their final form are also given. In the following only the assumptions upon which the governing equations were based are presented.

The assumptions

1. The flow field is one dimensional and unsteady.
2. The gaseous phase behaves as a perfect gas.
3. The solid particles are rigid, spherical, and inert. They are identical in all their physical properties including their diameter.
4. The solid particles are uniformly distributed in the suspension.

5. The number density of the solid particles is high enough so that the solid phase could be considered as a pseudo-gas.
6. The solid particles do not interact with each other. As a result their partial pressure in the suspension is negligibly small.
7. The volume occupied by the solid particles is negligibly small.
8. Beside the momentum and energy exchange between the solid and the gaseous phases, the gaseous phase is assumed to be an ideal fluid, i.e., $\mu_g = 0$ and $k_g = 0$, where μ_g and k_g are the dynamic viscosity and the thermal conductivity, respectively.
9. The weight of the solid particles and the buoyancy force are negligibly small compared to the drag force acting on them.
10. The solid particles are too large to experience Brownian motion.
11. The temperature within the solid particles is uniform.
12. The dynamic viscosity, the thermal conductivity, and the specific-heat capacity at constant pressure of the gaseous phase depend solely on its temperature.
13. The heat capacity of the solid particles is constant.
14. The diameter of the solid particles is much larger than the shock-wave thickness.

The governing equations were solved for the following ranges of parameters: $1 < M_0 < 5$, $25 < D < 500$ μm, and $0.25 < \eta < 10$. The upper limit of the initial shock-wave Mach-number value, M_0, is due to assumption 2, namely, the gas behaves as a perfect gas. The diameter, D, is limited from above by assumptions 6, 7, 9, and 11, while its lower limit is governed by assumptions 10 and 14. The highest loading-ratio value, η, is limited by assumption 7. In order to quantify the latter, the volume of the solid particles was limited to maximum 1% of the suspension volume. All the one hundred ($4 \times 5 \times 5$) possible combinations of the following values of the above-mentioned three parameters were used as initial conditions for the numerical simulation undertaken during the course of this study.

$$D = 25, 50, 100, 500\ \mu\text{m}; \quad \eta = 0.25, 0.75, 2, 6, 10; \quad M_0 = 1.1, 1.5, 2, 4, 5$$

As mentioned earlier, the purpose of the present study was to find the functional dependence of both χ and M_e, which are the free parameters in Eq. (3) on M_0, D, and η. The process of evaluating this functional dependence is given in detail in Aizik et al. (1995). In the following only the final results are provided.

3. The general attenuation law

For the reader's convenience the general attenuation law, which was developed during the course of this study, is presented in the following

$$M_s(x) = (M_0 - M_e)\exp(-x/\chi) + M_e \tag{4}$$

where the attenuation coefficient, χ, as a function of the parameters affecting the shock-wave attenuation, is given by

$$\chi(M_0, D, \eta) = (b_1 D/M_0)\,\eta^{-\left(b_2/M_0^{b_3}+b_4\right)} \tag{5}$$

and the equilibrium shock-wave Mach number, M_e, which is obtained at $x \to \infty$ is expressed by

$$M_e = A(M_0) + [M_0 - A(M_0)]\exp[-B(M_0)\,\eta] \tag{6}$$

where

$$A(M_0) = b_5 + b_6 M_0 + b_7 M_0^2 \tag{7}$$

and

$$B(M_0) = b_8/\ln(b_9 M_0) \tag{8}$$

The values of $b_1, b_2, \ldots, b_9$ as obtained in the course of the present study and which ensure a 99% correlation that the curve fits with the numerical data are:

$b_1 = 0.02728$	$b_2 = 1.19194$	$b_3 = 6.63244$	$b_4 = 0.36737$	$b_5 = 1.15934$
$b_6 = -0.23607$	$b_7 = 0.10080$	$b_8 = 0.52311$	and	$b_9 = 0.92796$

It could be seen that the general attenuation law given by Eq. (4) contains all the major flow parameters and dust physical properties which are known to affect the shock-wave attenuation, namely, the initial shock-wave Mach number, M_0, the dust-loading ratio, η, and its diameter, D.

The general attenuation law enables one to calculate the instantaneous shock-wave Mach number within a 5% accuracy. As a result, the jump conditions across the attenuating shock wave could be readily obtained without the need to numerically solve the entire flow field.

4. Validation of the general attenuation law

The shock-wave attenuation, as predicted by the above-developed general attenuation law, were compared to the experimental results of Sommerfeld (1984, 1985). A typical comparison is shown in Fig. 1 for $M_0 = 1.7$, $D = 27\ \mu$m, and $\eta = 1.4$. As can be seen, the agreement between the attenuation predicted by the general attenuation law (Eq. 4) and the actual attenuation is very good.

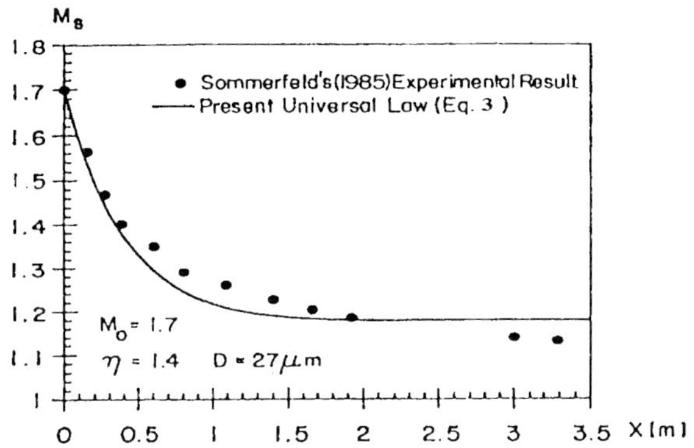

Figure 1. Comparison between the attenuation predicted by the "General Attenuation Law" developed in the course of the present study and the experimental results of Sommerfeld (1984) for $M_0 = 1.7$, $\eta = 1.4$, and $D = 27\mu$m.

A comparison between the general attenuation law (Eq. 4) and the attenuation law developed by Olim et al. (1990) (Eq. 1), which as mentioned earlier is applicable only to weak to moderate planar shock waves and relatively high loading ratios, is shown in Fig. 2 for $M_0 = 1.5$, $D = 200$ mm, and $\eta = 4.0$. It is evident that the general attenuation law predicts an attenuation similar to that predicted by Olim et al.'s (1990) limited law. The difference of about 3 to 4% which is developed at large values of x is due to the fact that while the present general attenuation law dictates a realistic equilibrium shock-wave Mach number, $M_e > 1$, Olim et al.'s (1990) specific attenuation law inherently dictates $M_e = 1$ when $x \to \infty$.

5. Conclusions

The governing equations describing the flow field which develops when a planar shock wave propagates inside a dust-gas suspension were solved numerically using the random-choice method.

The shock-wave attenuation was investigated numerically. Based on the numerical results, a general law for describing the shock-wave attenuation as it propagates inside the dust-gas suspension was developed. It is an improvement of a simplified specific atenuation law which was developed a few years ago by Olim et al. (1990).

The general attenuation law was validated by comparing its predictions to actual experimental results. Very good agreement was evident.

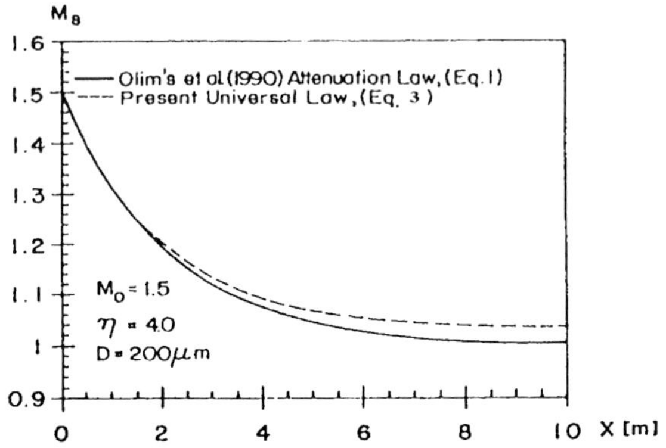

Figure 2. Comparison between the predictions of the general attenuation laws developed during the course of the present study (dashed line) and the limited attenuation law developed by Olim et al. (1990).

The general attenuation law enables one to simply and cheaply calculate the instantaneous shock-wave Mach number, and as a result to immediately obtain the jump conditions across it without the need to conduct a tedious numerical simulation.

Acknowledgement. This research was supported by a grant from the German-Israeli Foundation for Scientific Research and Development. Their support is acknowledged with thanks.

References

Aizik F, Ben-Dor G, Elperin T, Igra O, Mond M, Grönig H (1995) Attenuation law of planar shock waves propagating through dusty-gases suspensions. AIAA J 33(5):953–955.

Olim M, Ben-Dor G, Mond M, Igra O (1990) A general attenuation law of moderate planar shock waves propagating into dusty gases with relatively high loading ratios of solid particles. Fluid Dynamics Research 6:185–200.

Sommerfeld M (1984) Instationare Stosswellenausbreitung in Gas-teilchengemiscen, Ph.D. Thesis, Faculty of Mechanical Engineering, RWTH Aachen, Aachen, Germany.

Sommerfeld M (1985) The unsteadiness of shock waves propagating through gas-particle mixtures. Experiments in Fluids 3:197–206.

Development of a detonation driver for a shock tunnel

M. Lenartz, B. Wang*, H. Grönig
Shock Wave Laboratory
Templergraben 55, 52056 Aachen, Germany

Abstract: The detonation driver represents a new type of high-pressure section for a conventional shock tunnel to increase its performance range. The detonation driver consists of a detonation section which is comparable to the conventional high-pressure section and an additional damping tube. The high driver energy is provided by an oxyhydrogen detonation initiated at the main diaphragm. Simulation calculations show that it is theoretically possible to obtain a high stagnation enthalpy in front of the nozzle and to adjust tailored interface conditions by the proper choice of the pressure ratio between the driver and the driven section and of the composition of the oxyhydrogen. A detonation-driven shock tube was built to investigate the real operation performance (total length about 19 m, inner diameter 140 mm).

Key words: Detonation driver, Shock tunnel, Detonation

1. Introduction

A detonation-driven shock tunnel consists of the "normal" low-pressure side (i.e., driven section, nozzle, test section) of a conventional shock tunnel, a detonation section including an initiation tube at the main diaphragm, and an additional damping section (see Fig. 1). The damping and detonation section,

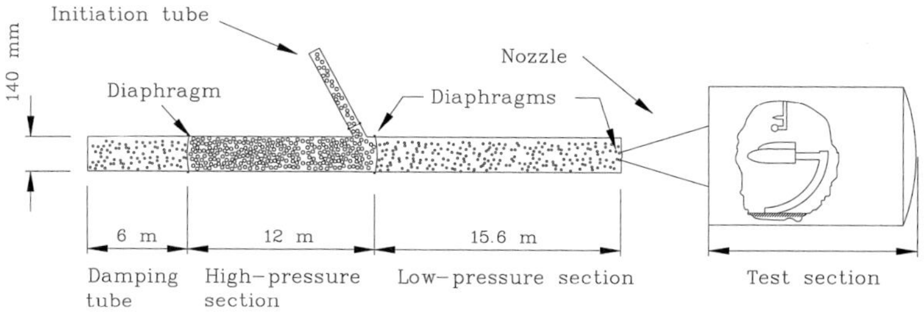

Figure 1. Set-up of a detonation-driven shock tunnel.

*On leave of absence from College of Chemical Engineering, Nanjing University of Science and Technology, Nanjing, China.

and the low-pressure section and the nozzle are separated by diaphragms. The detonation section is filled with oxyhydrogen, the damping section is evacuated, and the medium in the low-pressure section acts as test gas.

The experiment starts with the ignition of the oxyhydrogen at the end of the initiation tube. After a certain induction distance a fully developed detonation wave is established which propagates through the initiation tube into the main tube. There, the detonation front moves towards the damping section, breaks the diaphragm nearly without being reflected, and expands into the damping section which acts to reduce the load on the structure and to increase the duration of the constant stagnation state in front of the nozzle. The velocity of the detonation front is stable and depends mainly on the initial oxyhydrogen composition. The detonation velocity and the state of the burned gas can be predicted by the classical one-dimensional Chapman-Jouguet theory. The planar detonation front is followed by a Taylor expansion fan (Nettleton 1987) which reduces the flow velocity to zero. This state is comparable to the initial state of the high-pressure section of a conventional shock tube. Its pressure level is about 5–8 times higher than the initial pressure and its temperature lies between 3500–4200 K depending on the initial oxyhydrogen composition and its pressure. The wave phenomena inside the low-pressure section are comparable to those of a conventional shock tunnel.

2. Numerical simulation

Numerical simulations of the detonation-driven shock tube have been performed using a program (KASIMIR) which was developed at the Shock Wave Laboratory (Esser 1992, Kanira 1990) and has been extended by a routine to calculate the Chapman-Jouguet detonation for a given oxyhydrogen mixture. The computer code is based on a one-dimensional exact Riemann solver, takes into account high-temperature effects (e.g., dissociation, vibrational excitation), and uses the method of characteristics to calculate the wave diagram of a shock tube with given initial conditions; but it does not consider any losses (e.g., shock attenuation, diaphragm rupture). Figure 2 shows the wave diagram of a detonation-driven shock tube with the same lengths as given in Fig. 1. Two results obtained by the numerical simulation are particularly important (Yu et al. 1992): it is possible to obtain a high stagnation enthalpy in front of the nozzle (see Fig. 3) and to adjust the tailored interface condition for a wide range of pressure ratios p_4/p_1 by matching the oxyhydrogen composition. The dashed line in Fig. 3 determines the initial conditions in the high- and low-pressure sections for a maximum test time under tailored interface conditions. Thus, for a stagnation enthalpy of 10 MJ/kg, a pressure ratio p_4/p_1 of 170 is required with an initial ratio H_2/O_2 of about 1.

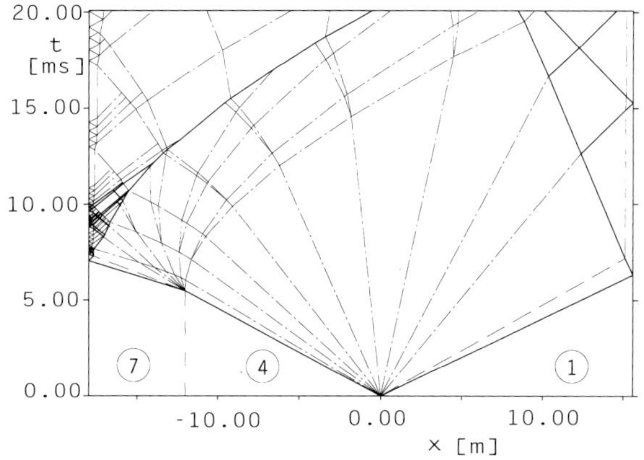

Figure 2. Wave diagram of a detonation-driven shock tube for the following initial conditions: p_4 = 4 MPa, p_1 = 4 kPa, p_7 = 5 kPa; composition of the oxyhydrogen: 60 vol% H_2 and 40 vol% O_2; the temperature of all sections is 293 K; the damping gas is N_2 and the test gas is air.

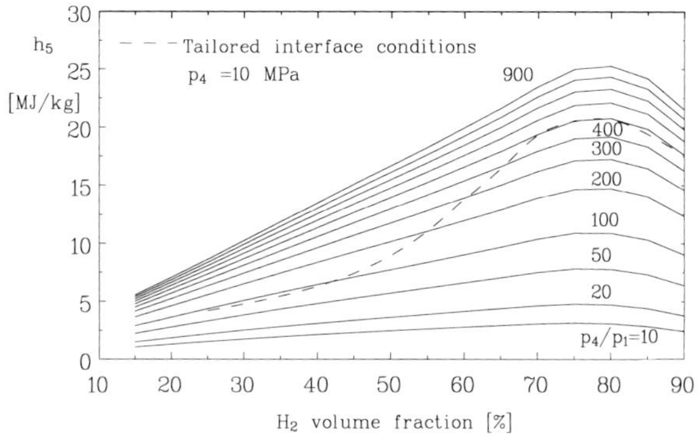

Figure 3. Specific enthalpy h_5 of a detonation-driven shock tube depending on the oxyhydrogen composition and the pressure ratio p_4/p_1; p_4 = 10 MPa.

3. Set-up of a detonation-driven shock tube

The detonation-driven shock tube which was built to check the numerical results and to study its performance is sketched in Fig. 4. Its total length is about 19 m; the lengths of the damping, the detonation, and the driven sections are about 6 m each; the inner diameter of all sections is 140 mm, separated by hydraulically clamped diaphragms. The length of the initiation tube is about 900 mm with an inner diameter of 30 mm. Mass-flow controllers adjust the

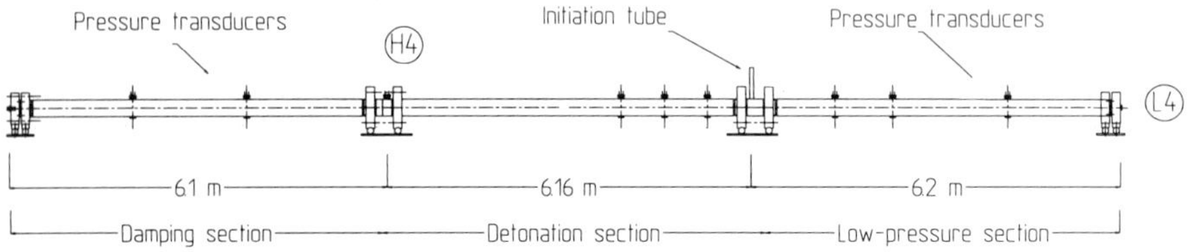

Figure 4. Set-up of the detonation-driven shock tube.

oxyhydrogen composition up to a maximum filling pressure of about 7.5 MPa. There are several locations for piezoelectric pressure transducers in all three sections (detonation section: KISTLER 601H, 6205, 6215; damping and driven section: KISTLER 603B, 601H). The signals of the pressure transducers were recorded by a transient memory system with a maximum sampling rate of 10 MHz.

4. Shock-tube experiments

In order to answer the question if any quenching and reignition processes take place, the first experiments in the shock tube dealt with the transition of the detonation from the initiation tube into the main tube.

These experiments were performed with the detonation and damping sections only. The main diaphragm was replaced by an end plate where a pressure transducer was located in the center. The measurements yielded the direct initiation of an oxyhydrogen detonation in the main tube and proved that the ignition equipment (exploding wire, initiation-tube design) provided a well-working system.

The following experiments concern the complete detonation-driven shock tube to investigate its operational performance. Figure 5 shows the wave diagram based on the following initial conditions providing a nearly tailored interface: $p_4 = 1.5$ MPa, $H_2/O_2 = 1$, $p_1 = 8$ kPa, $p_7 = 5$ kPa; the initial temperature of all sections is 288 K; the damping gas is nitrogen and the driven gas air. Measured and calculated pressure curves at two positions along the tube are presented in Figs. 6 and 7. The agreement between the simulation and the experiment is very good for the detonation itself (see Fig. 6 between 3 and 10 ms) because the state of the burned gas is determined by the CJ condition and is not very sensitive to disturbing influences (Bartlmä 1975). The shift between the measured and simulated pressure curves increases with time because all losses are neglected in the calculation. Figure 7 shows the signal of the pressure transducer mounted on the end plate of the low-pressure section and confirms the tailored operation mode: the pressure level behind the reflected shock has

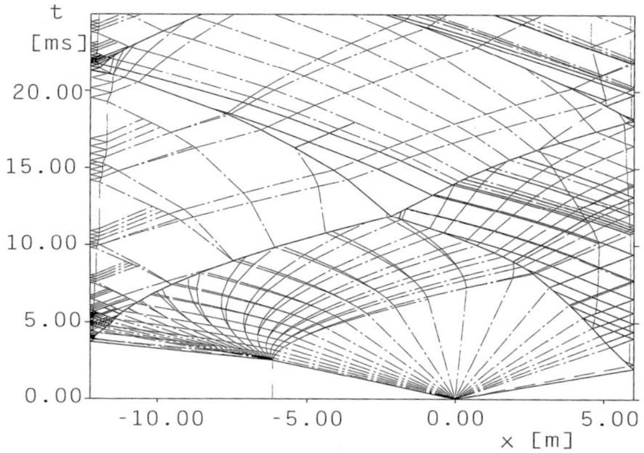

Figure 5. Wave diagram: $p_4 = 1.5$ MPa, $H_2/O_2 = 1$, $p_1 = 8$ kPa, $p_7 = 5$ kPa; the initial temperature of all sections is 288 K; the damping gas is nitrogen and the driven gas air.

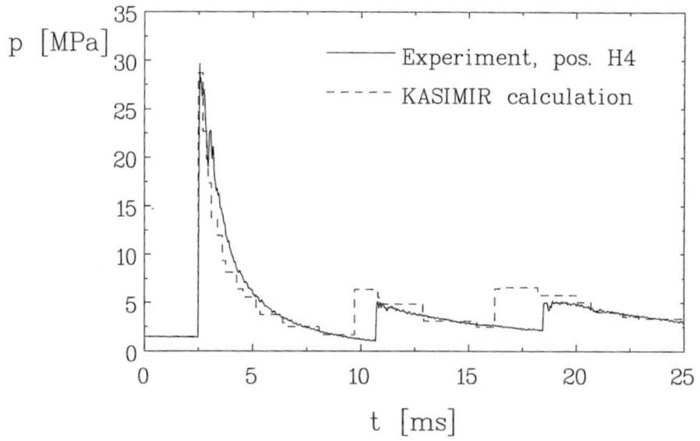

Figure 6. Measured and calculated pressure in the detonation section 6 m from the main diaphragm. Same initial conditions as given in Fig. 5.

a constant value of about 6 MPa during 2 ms; the corresponding stagnation enthalpy is about 8 MJ/kg.

5. Conclusion

The simulation calculations and the experiments performed show that using a detonation driver increases the performance range of a conventional shock tunnel. It is possible to increase its specific enthalpy and to obtain longer test times by operating the tube in the tailored interface mode. The initiation system presented (initiation tube, exploding wire) provides a direct initiation

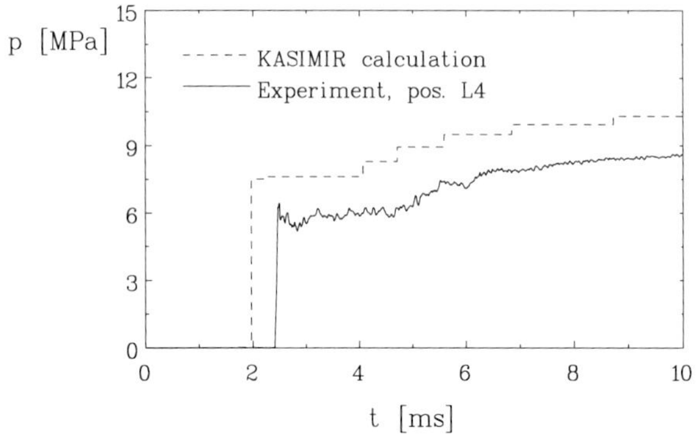

Figure 7. Measured and calculated pressure at the end plate of the low-pressure section. Same initial conditions as given in Fig. 5.

of the oxyhydrogen detonation inside the main tube. The construction of the whole tube is comparably simple, only an additional damping tube has to be added. Compressors to fill the high-pressure section are not needed. But the requirements regarding the safety equipment (e.g., design of the tube structure, leakage detection, gas-mixing system) are much higher. The water generated by the detonation is also a problem (corrosion of the materials and the operation of the facility itself) which has to be treated very carefully.

Acknowledgement. This research project was partially sponsored by the Federal Ministry of Science and Technology, Bonn, by the Ministry of Science of the State of North Rhine-Westfalia, Düsseldorf, and by ESA/ESTEC, Noordwijk.

References

Bartlmä F (1975) Gasdynamik der Verbrennung, Springer, Wien, New York, pp 97–101.

Esser B (1992) Die Zustandsgrößen im Stoßwellenkanal als Ergebnis eines exakten Riemannlösers, Doctoral Thesis, RWTH Aachen, Shock Wave Laboratory.

Kanira I (1990) Simulation der Wellenausbreitung im Stoßwellenrohr, Master Thesis, RWTH Aachen, Shock Wave Laboratory.

Nettleton MA (1987) Gaseous Detonations, Chapman and Hall, London, Chapter 2.

Yu H-R, Esser B, Lenartz M, Grönig H (1992) Gaseous detonation driver for a shock tunnel. Shock Waves 2:245–254.

Behavior of laser-induced bubbles and pressure waves in liquid nitrogen

H. Sato, M. Odagawa, X. W. Sun, K. Maeno, H. Honma
Dept of Mechanical Engineering, Faculty of Engineering, Chiba University
1-33 Yayoi, Inage-ku, Chiba 263 Japan

Abstract: This paper deals with the behavior of laser-induced vapor bubbles in liquid nitrogen. The pulsed irradiation of a ruby laser beam is focused in a special cryostat to generate bubbles. The laser-induced bubbles are visualized by high-speed photography. The pressure signal from bubble motion in liquid nitrogen is measured under near-equilibrium and subcooled conditions. Furthermore, numerical simulation of the dynamics of a single spherical bubble is also performed under the conditions corresponding to our experiments.

Key words: Cryogenic liquid, Cavitation, Bubble dynamics, Pulse laser

1. Introduction

When the pressure in a liquid decreases locally below the saturated value of its temperature due to high flow velocity, bubbles are generated, which expand and collapse, and pressure waves from the bubbles are generated and propagate into the liquid. These phenomena taken together are called cavitation. Cavitation phenomena, mainly in water, have been investigated and widely reported. Recently, in various fields such as aerospace engineering, superconductivity technology, and liquefaction plants for LPG and LNG, cryogenic liquids have been very important objectives of research especially regarding cavitation phenomena in relation to their transportation and preservation. There have been, however, only a few experimental reports which treat cavitation phenomena and two-phase flow in cryogenic liquids, and in particular laser-induced cavitation bubbles (Maeno et al. 1990, Sato et al. 1995).

This paper deals with an investigation of the behavior of laser-induced vapor bubbles in liquid nitrogen. The bubbles produced by pulsed irradiation of a ruby laser beam and pressure waves combined with the bubble behavior are visualized by using both the framing and streak modes of an image-converter camera. The pressure variation is measured by a piezoelectric transducer under equilibrium and pressurized subcool conditions. Furthermore, a numerical simulation of the dynamics of a single spherical bubble including phase-change phenomena at the bubble wall is performed. The results obtained from both experiments and numerical simulations are compared.

2. Experimental setup

In the present experiment, a special cryostat (shown in Fig. 1) has been developed for a cavitation experiment in cryogenic liquids. It is possible to use liquid nitrogen and argon as the test fluid, and by using liquid helium for the coolant, we can also do the experiment with liquid hydrogen. In the present report, liquid nitrogen is employed as the test fluid. The inner vessel to be filled with the test liquid is 150 mm in height and 38 mm in diameter, and two pairs of quartz windows of 20-mm diameter and 8-mm thickness are equipped for visualization and laser-beam introduction. It is also possible to pressurize the test liquid by the gas for an experiment under nonequilibrium conditions. The pressure variation in liquid nitrogen is measured by a transducer (PCB-102A11B) and the temperature in liquid nitrogen is monitored by a silicon diode thermometer (Scientific Instruments, Inc., Si-400).

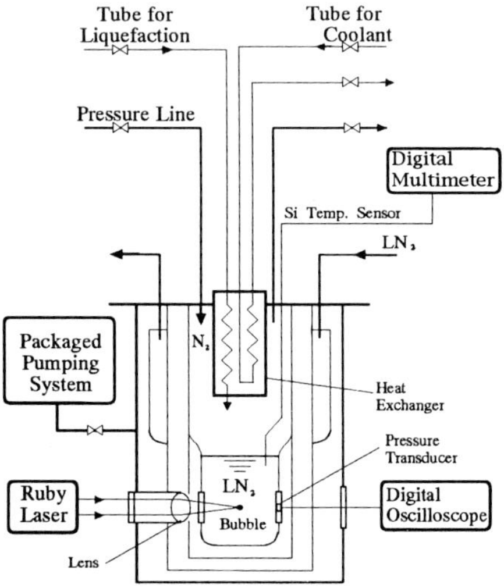

Figure 1. Experimental apparatus (cryostat).

Figure 2 shows a schematic diagram for bubble generation and visualization. A pulsed ruby laser (Nihon Kagaku Engng. Ltd., NAL-707P1) is introduced for bubble formation through the lens system. Iris and stop are used to adjust the light intensity and aberration of the focusing lens. The behavior of the laser-induced bubble and pressure waves are visualized by using the framing or streak mode of an image-converter camera (Hadland Photonics, Imacon-790).

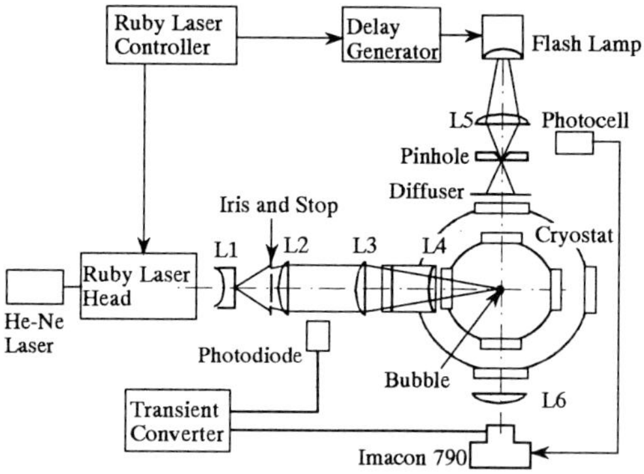

Figure 2. Laser focus and visualization.

3. Numerical analysis

A numerical analysis of a single spherical vapor bubble in liquid nitrogen is carried out to compare with our experimental results. Let us consider a single spherical bubble containing saturated vapor with initial radius R_0 in an infinite volume of liquid nitrogen. The surrounding pressure increases instantaneously to p_∞ at time zero, and the bubble starts to collapse with mass transfer at the interface. The following equations are employed to determine the behavior of the bubble

$$R\ddot{R} + \frac{3}{2}\dot{R}^2 = \frac{1}{\rho_\infty}\left(p_v - p_\infty - \frac{2\sigma}{R} - \frac{4\mu\dot{R}}{R}\right), \tag{1}$$

$$\frac{d}{dt}\left(\frac{4}{3}\pi R^3 \rho_v \hat{U}_v\right) = -p_v \frac{d}{dt}\left(\frac{4}{3}\pi R^3\right) - 4\pi R^2 \Delta U_{v\to L}, \tag{2}$$

$$\frac{d}{dt}\left(\frac{4}{3}\pi R^3 \rho_v\right) = 4\pi R^2 \dot{m}, \tag{3}$$

where $\mu, \sigma, \hat{U}_v$, and $\Delta U_{v\to L}$ denote viscosity, surface tension, specific internal energy, and net rate of the flux of internal energy from the vapor to liquid, respectively. Subscripts L and v denote liquid and vapor, respectively. It is assumed that the vapor within the bubble is described by the equation of state for a perfect gas. In Eq. (3), $\dot{m}$ denotes mass transfer per unit time and area, and is given as follows

$$\dot{m} = \frac{\alpha_M}{\sqrt{2\pi R_v}}\left\{\frac{p_L^*(T_L)}{\sqrt{T_L}} - \frac{p_v}{\sqrt{T_v}}\right\}, \tag{4}$$

where α_M indicates the coefficient of condensation or evaporation. The following energy equation for the liquid

$$\frac{\partial T}{\partial t} + \frac{R^2\dot{R}}{r^2}\frac{\partial T}{\partial r} = D\left(\frac{\partial^2 T}{\partial r^2} + \frac{2}{r}\frac{\partial T}{\partial r}\right), \tag{5}$$

where D denotes thermal diffusivity of the liquid, is solved by means of an integral method (Tomita and Sima 1979) to estimate the liquid temperature at the interface. In this method we assume that the temperature boundary layer in the liquid has a parabolic temperature profile. For easier calculation, a simulation is also performed with the assumption that the liquid temperature at the interface has the same value as at infinity and both results are compared with each other.

4. Results and discussion

4.1. Measurement of the pressure signal and behavior of the bubble

The time history of the pressure variation in liquid nitrogen under atmospheric conditions is shown in Fig. 3. In the figure, compression waves generated by laser irradiation and bubble formation appear on the positive side, and rarefaction waves generated by reflection at the liquid surface appear on the negative side. These pressure waves reflect randomly at the liquid surface and at the wall of the test section. The pressure signal from rebound is not detected clearly, because the distance between the pressure transduser and focal point of the laser is 26 mm in length.

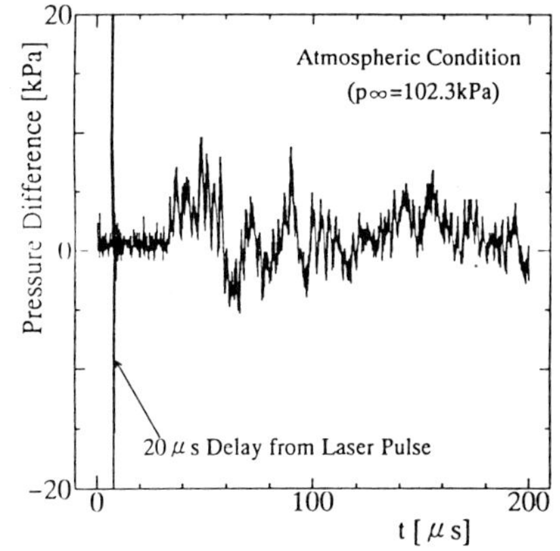

Figure 3. Signal of pressure transducer.

Figure 4 shows a typical series of diffusive shadowgraphs of a laser-induced bubble in liquid nitrogen taken under atmospheric conditions. The frame speed is 5×10^4 frames/s and first frame is taken 5 μs before the ruby-laser pulse. The induced bubble is not spherical but ellipsoidal in its initial stage of bubble expansion because of the small angle for laser focusing. In these shadowgraphs, the bubble expands to maximum radius in the seventh frame. Then it shrinks by the twelveth frame and rebounds. The bubble oscillation period measured from the shadowgraphs is about 200 μs. In the case of pressurized conditions, the bubble radii are smaller and the period of bubble oscillation is shorter than for the case of atmospheric conditions (Sato et al. 1995). The instability of the bubble surface is clearly observed in Fig. 5, this is a characteristic phenomenon of bubble motion in cryogenic liquid. In the present experiment the generated bubble is shaped as an ellipsoidal sphere in the initial stage of bubble motion, as discussed before. So the generated bubble originates with a second-order mode of shape oscillation. But the bubble and its surface in cryogenic liquid are strongly influenced by thermal effects, and higher-order modes of shape oscillation can be excited.

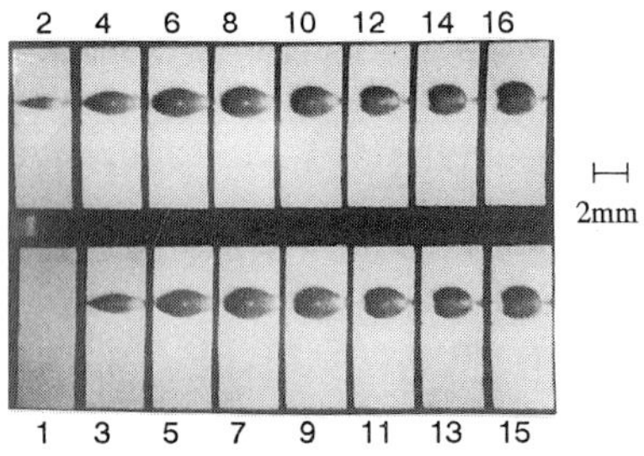

Figure 4. Shadowgraph of induced bubble.

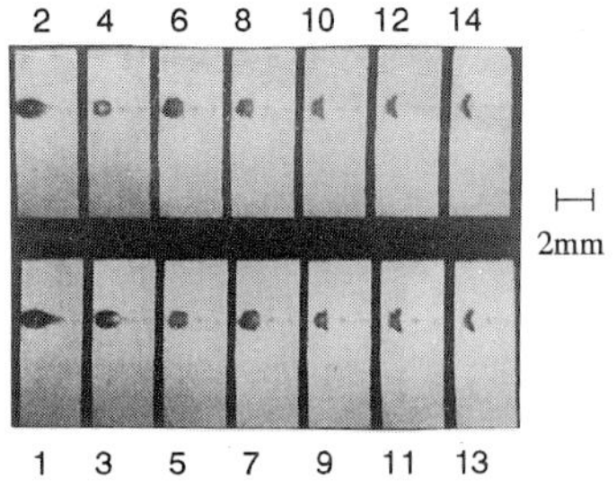

Figure 5. The instability of bubble surface.

Taking notice of the bubble formation, the initial deposit of energy results in the optical breakdown of the cryogenic liquid nitrogen under high-power laser irradiation. This breakdown is characterized by an ionization of the liquid nitrogen by absorption of the incident laser pulse leading to the formation of luminous plasma. Consequently, a shock wave is emitted and a small volume of liquid nitrogen is heated up to the bulk vaporization.

A streak shadowgraph taken 0.39 μs before the laser irradiation under atmospheric conditions is shown in Fig. 6. Parallel light of a Xe flash lamp is employed for visualization of pressure waves. The streak speed is 50 ns/mm. The black line in the middle of the frame indicates the bubble, and a shock wave caused by the laser irradiation is emitted outward from the bubble center. The wave velocity and shock Mach number measured from the shadowgraph are 1187.5 m/s and $M_s = 1.26$, respectively.

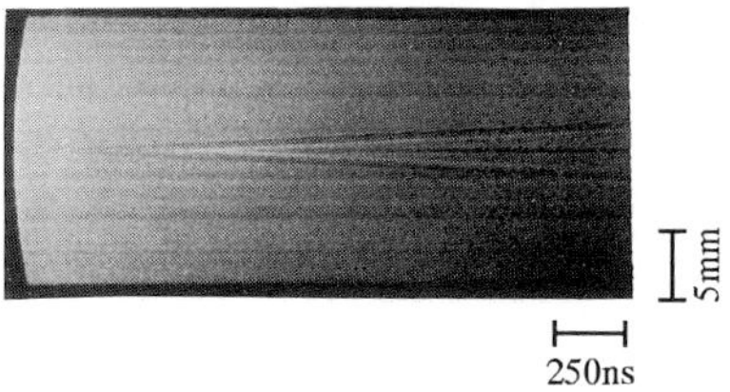

Figure 6. Streak shadowgraph of bubble initiation.

4.2. Results of numerical simulation

Figure 7 shows numerical results with nondimensionalized radius for a single spherical bubble in liquid nitrogen under pressurized conditions with $\alpha_M = 0.01$. The solid line and broken line indicate the result when $T_L = T_\infty$ and the result obtained by means of the integral method, respectively. Nondimensional radii of the bubble obtained by the integral method rebound faster than when $T_L = T_\infty$. It is possible to know the bubble behavior by the integral method up to its rebound point assuming the temperature distribution in the thermal boundary layer is not valid after the rebound point.

In the present experiments we consider the shapes of the laser-induced bubbles, such as the series of shadowgraphs in Fig. 4 shows, to be ellipsoids of revolution, and then we estimate an equivalent radius that the bubble would have if it were a sphere of the same volume. The experimental results for temporal motion of a laser-induced bubble in cryogenic liquid nitrogen are compared with numerical results, as shown in Fig. 8. The Rayleigh solution for an empty cavity is also plotted in Fig. 8, where from $t = 0$ to the maximum radius the Rayleigh curve is folded at the maximum point. This solution is for only the inertia-dominant case in liquid nitrogen. In this figure, note that the dynamics in the growth phase of bubble are in good agreement with the folded Rayleigh solution. In the collapse stage, however, a significant retardation appears in the bubble motion due to thermal effects and phase change.

The initial inflating phase of the bubble is assumed as a modified adiabatic change, i.e., the temperature within bubble is obtained when the bubble radius is 1/3 of the maximum radius from the adiabatic relation, and pressure is adjusted. The numerical results with these assumption are shown in Fig. 8 with initial conditions. Present numerical results are in good agreement with our experimental results from the bubble inflating phase to the first rebound of the bubble. Therefore, the effects of phase change explain well the behavior of a laser-induced bubble in cryogenic liquid nitrogen.

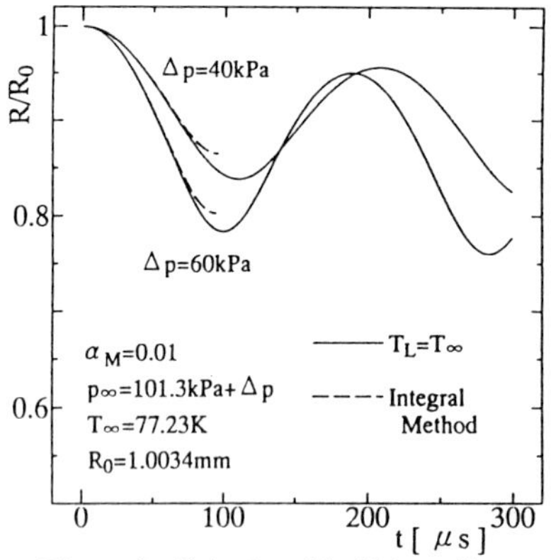

Figure 7. Calculated bubble radius.

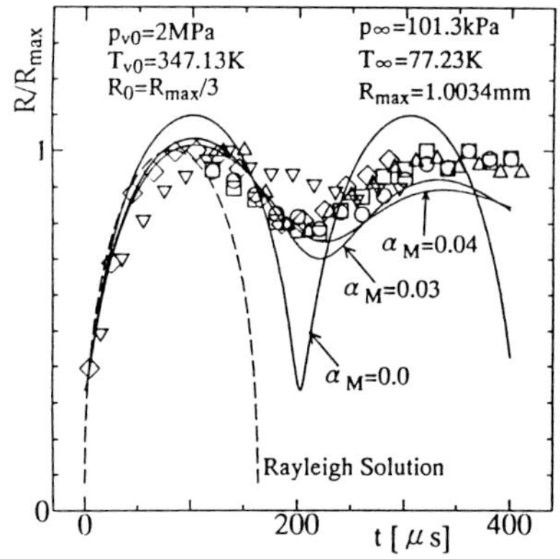

Figure 8. Measured and calculated bubble radius.

5. Conclusion

The behavior of a laser-induced vapor bubble in liquid nitrogen is studied both experimentally and numerically. The induced bubble and initial shock wave combined with pulsed laser irradiation are visualized by high-speed photography. Numerical results are in good agreement with experiments from the bubble inflating phase to the first rebound, and phase-change phenomena influence the bubble motion significantly in liquid nitrogen.

References

Maeno K, Yokoyama S, Hanaoka Y (1990) Study on laser-induced bubble in cryogenic liquids. In: Proc 17th Int Symp on Shock Waves and Shock Tubes, AIP-208:383–388.

Sato H, Sun XW, Odagawa M, Maeno K, Honma H (1995) An investigation on the behavior of laser induced bubble in cryogenic liquid nitrogen. In: Proc Int Symp on Cavitation, CAV'95, Deauville, France, pp 391–397.

Tomita Y, Shima A (1979) The effects of heat transfer on the behavior of a bubble and the impulse pressure in a viscous compressible liquid. ZAMM 59:297–306.

The interaction of a two-dimensional strong compressible vortex with a shock wave

C. T. Kao, K. von Ellenrieder, D. Bershader*
Department of Aeronautics and Astronautics
Stanford University
Stanford, CA 94305 USA

Abstract: The two-dimensional compressible vortex-shock interaction is studied both experimentally and numerically. Shadow and schlieren pictures indicate that the flow field of the strong compressible vortex has a significant effect on the shock configuration in the interaction. A numerical simulation utilizing a first-order shock-capturing scheme shows good qualitative agreement with the present experimental results and previous research. The generation and propagation of diffracted and/or reflected waves due to the interaction are also revealed in the numerical simulation.

Key words: Shock-vortex interaction, Flow visualization, Shock tube

1. Introduction

This investigation is aimed at obtaining a fuller physical insight into the familiar but complex problem of the head-on collision of a shock wave and a two-dimensional strong compressible freely moving vortex of essentially a circular cross section. The obvious questions are: what happens to the shock, what happens to the vortex, and what secondary phenomena are generated, if any? The existence of continuous density, pressure, and velocity variations across the vortex means that the immediate upstream conditions also vary continuously from point to point along the shock, and they vary with time as the inter-penetration proceeds. Also, it is evident that there is a point-wise angular variation of the flow incident on the shock. Clearly, it is expected that an initially plane uniform shock will undergo distortion in both strength and direction. Corresponding variations take place in the downstream flow, including changes in vorticity and entropy.

The density and temperature variations within the vortex imply that the shock is traversing a medium of variable acoustic impedance, a feature which results in the modification of wave properties. The latter has been observed in experimental studies (Dosanjh and Weeks 1965, Naumann and Hermanns 1973)

*In memoriam, deceased on May 30, 1995.

and in numerical simulations (Pao and Salas 1981, Meadows et al. 1991, Ellzey et al. 1995). Also, Ribner (1985) provided a theoretical analysis of the problem. It is evident that the nature of the distorted wave configuration and alteration of the vortex structure will depend on shock Mach number, vortex strength, density level, as well as the structure of the external flow field. However, proper scaling of the overall problem needs further examination. An improved understanding of the shock-vortex interaction is warranted both in connection with the interest in reducing wing-tip noise generation in the transonic flow regime of rotary wing aircraft, and the enhancement of sound radiation from supersonic jets.

2. The compressible freely moving vortex

The method of vortex generation in this study has been described by Bershader (1988), in which the detailed physical configuration of the free vortex prior to shock interaction is also illustrated. Thus, for example, the density as a function of radius and time is expressed by

$$\frac{\rho}{\rho_\infty} = 1 - \frac{\rho^* e^{-\frac{t}{\tau}}}{1 + \frac{r^2}{(r_o + mt)^2}}, \tag{1}$$

where ρ^*, r_o, m, τ are empirical constants equal to 0.898, 0.143 cm, 57.2 cm/s, 1325 μs respectively, while ρ_∞ is the free-stream density. The pressure distribution is measured by a fast-response pressure transducer. The tangential velocity profile, which is especially significant for the shock-vortex interaction, was obtained by insertion of the density and pressure data into the simplified radial momentum equation:

$$\rho \frac{{V_\theta}^2}{r} = \frac{\partial P}{\partial r}. \tag{2}$$

Note that the terms involving radial velocities have been ignored since the radial velocity is at least two orders of magnitude smaller than the tangential velocity. The tangential-velocity profile shown in Fig. 1a is quite similar to that of the classical Rankine vortex, normally linear with radius, r, (solid body rotation) in the core and decaying as $1/r$ outside the core, characteristic of a potential vortex. It is seen that the maximum tangential velocity of the vortex relative to its center is $\sim$ 230 m/s near the edge of the core. Figure 1b gives the corresponding vorticity distribution indicating that the vorticity is concentrated within the vortex core, and the overall circulation is close to 2.5 m^2/s. The vortex is considered strong in the sense that the radial density and pressure distributions exhibit in excess of a 60% drop from free stream to the vortex center.

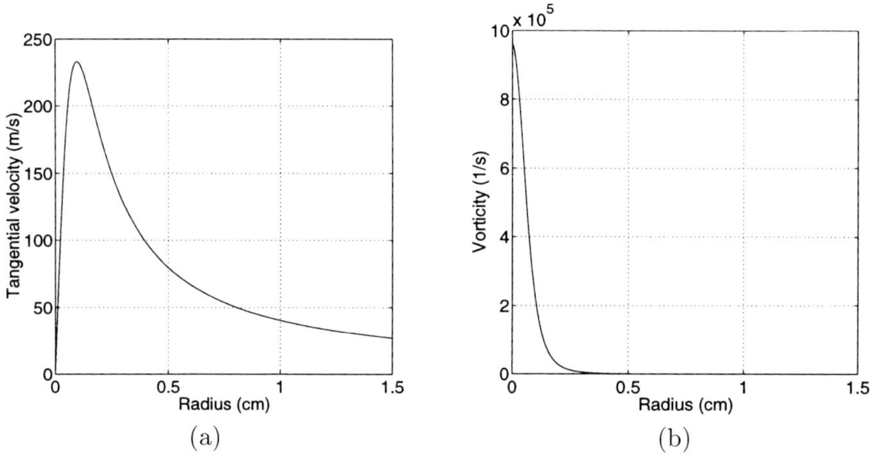

Figure 1. (a) Tangential velocity and (b) vorticity profiles of the free vortex.

3. Experimental observation

The shock-vortex interaction experiments reported here are two dimensional, taking place in a test section bounded by two lucite windows but open to the laboratory atmosphere in the other dimensions. This is in contrast to earlier studies where the interaction occurred in a closed shock tube or channel. The interacting shock originates in the incident shock reflection from a downstream shock-tube model; the portion interacting with the vortex shows only slight curvature. Figure 2 shows both shadow and schlieren pictures of one stage of the interaction. In these photos, the vortex moves to the right approximately with the free-stream velocity of $\sim$ 180 m/s, while the shock moves to the left with a Mach number of $\sim$ 1.2 relative to the incoming free stream. (Other features visible on both photos represent contact surfaces resulting from triple points occurring during earlier shock progress.) The obvious asymmetry of the shock structure occurs because the normal components of the velocity vectors face toward the shock on the lower half of the counter-clockwise rotating vortex, while the opposite is true along the upper half. It is seen that the portion of the shock emerging from the upper half of the vortex is relatively weak; the part diffracting into the lower region of the picture undergoes substantial further curvature and additional weakening. On the other hand, the lower shock portion is stronger, and its progress with respect to the laboratory is diminished due to the opposing velocity of the counter-clockwise vortex fluid rotation. The increased strength corresponds to an increased effective Mach number relative to the opposing rotational vortex velocity. The latter reaches a maximum with respect to the vortex center at the edge of the core, a radius of about 1 mm. It is seen from the shading on both the shadow and schlieren pictures that the density derivatives become especially large within the core, a feature that is also true for the pressure and tangential velocity profiles (see Bershader 1988). Figure 2 also shows an apparent termination of the edge of the lower shock

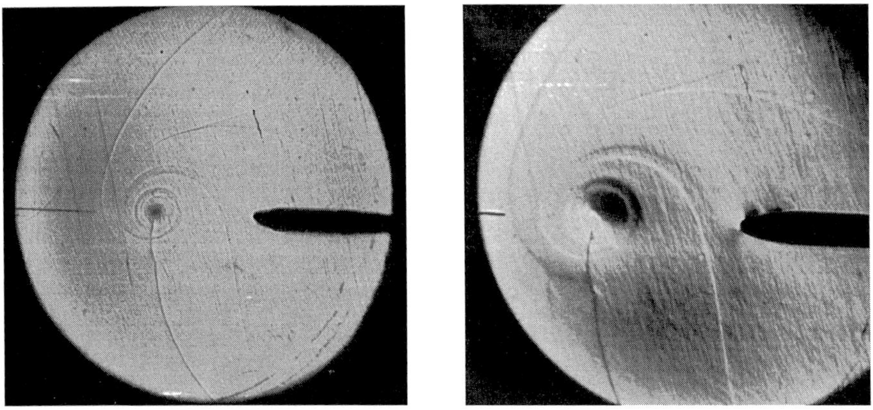

(a) (b)

Figure 2. (a) Shadow and (b) schlieren pictures of the shock-vortex interaction.

portion within the vortex core. That indicates the possible existence of an instability due to the presence of high gradients of physical properties parallel to the shock front. Finally, one may say qualitatively that in the present shock-vortex interaction, the shock "gets the worst of it" in the sense that it is more drastically affected than the vortex. Presumably, the torque generated by the shock during the interaction is not very successful in modifying the angular momentum of the vortex.

4. Numerical simulation

A preliminary numerical study has been performed utilizing the finite-volume formulation to simulate the physical phenomena observed in the experiment. The unsteady two-dimensional Euler equations in conservative form are solved explicitly using the modified Steger-Warming flux-splitting technique. This method is essentially a first-order upwind scheme, in which the fluxes between two adjacent cells are evaluated by averaging the flow properties of the two cells. In the present numerical simulation, the experimentally obtained vortex structure is used as initial conditions, and the shock is assumed initially to be planar. Pressure distributions at two time stages of the interaction are shown in Figs. 3 and 4. The distorted shock structure shown in Fig. 2 is clearly resolved in Fig. 3 at a slightly earlier time stage. Figure 4 illustrates the flow field at a later stage, in which the so-called "acoustic waves" can be identified especially in the 3D plot. The somewhat "thick" shock waves shown in Figs. 3 and 4 are due to the numerical dissipation of the first-order discretization in space. The calculated flow-field features agree with those of the present experiments and previous numerical calculations (for example, Ellzey et al. 1995).

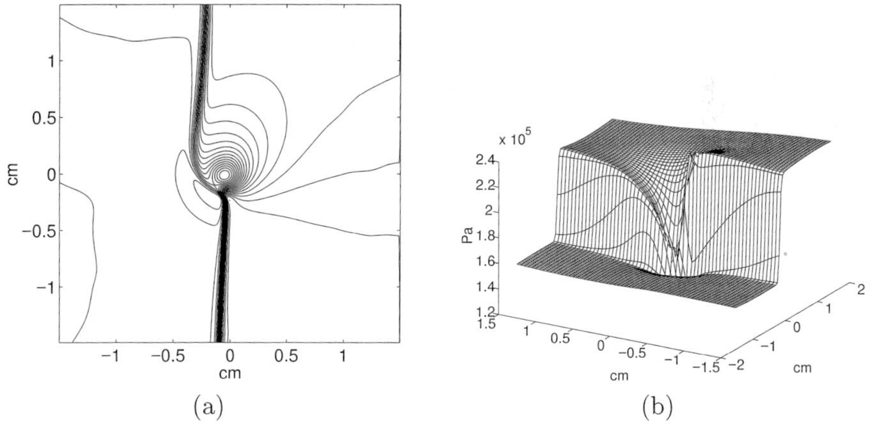

Figure 3. (a) Contour and (b) 3D plot of pressure distributions.

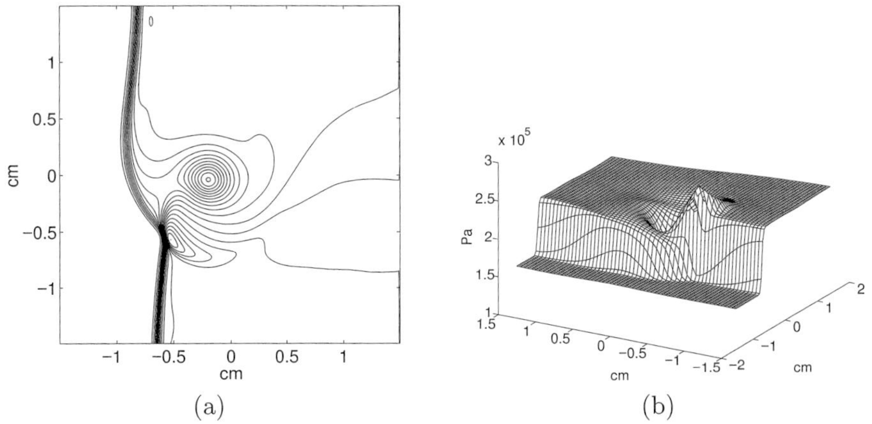

Figure 4. (a) Contour and (b) 3D plot of pressure distributions (approximately 15 μs later than those in Fig. 3).

5. Discussion and conclusions

Diffracted and/or reflected waves due to shock-vortex interaction have been described by some authors, e.g., Dosanjh and Weeks (1965), as *acoustic* waves. Of course, the acoustic wave is a limiting case of a shock of near-zero strength. However, the term "acoustic" can be misleading in that it implies a train of waves of either narrow or broad frequency spread. The strength of the pressure pulse shown in Fig. 4 is approximately one third of the pressure increase across the shock wave, corresponding to a sound pressure level over 180 dB. In other words, the pressure pulse is about one hundred times stronger than the pressure waves one would receive near a jet engine, typically 140 dB.

The present investigation has increased understanding of the physical basis for most of the admittedly complex phenomena associated with the shock-vortex interaction. The vortex flow field has a drastic effect on the configuration of the shock, including point-wise shock-strength variations. Also, the apparent termination of the edge of the lower shock portion in the vortex core raises a basic question of the nature of non-linear wave behavior in the presence of strong physical gradients parallel to the wave surface. A higher order shock-capturing scheme, e.g., a TVD (total variation diminishing) scheme, including viscous effects is a logical extension of the present numerical work to yield quantitative data for further specification of the phenomenon.

Acknowledgement. The authors are grateful for partial support of this work by NASA Consortium agreement number NCC2-5002 and the NASA Graduate Student Researcher's program. Valuable suggestions from Professor Robert W. MacCormack at Stanford University are also highly appreciated.

References

Bershader D (1988) Shock tube studies of vortex structure and behavior. In: Proc 16th Int Symp on Shock Tubes and Waves, Aachen, Germany, pp 5–18.

Dosanjh DS, Weeks TM (1965) Interaction of a starting vortex as well as a vortex street with a traveling shock wave. AIAA J 3:216.

Ellzey JL, Henneke MR, Picone JM, Oran ES (1995) The interaction of a shock with a vortex: shock distortion and the production of acoustic wave. Phys Fluids A 7:172.

Meadows KR, Kumar A, Hussaini MY (1991) Computational study on the interaction between a vortex and a shock wave. AIAA J 29:174.

Naumann A, Hermanns E (1973) On the Interaction Between a Shock Wave and a Vortex Field, AGARD-CP-131.

Pao SP, Salas MD (1981) A Numerical Study of Two-Dimensional Shock Vortex Interaction, AIAA Paper 81-1205.

Ribner HS (1985) Cylindrical sound wave generated by a shock-vortex interaction. AIAA J 23:1708.

Hypersonic Flow

Interference heating effects caused by a 3D transverse jet in hypersonic flow

N. R. Mudford
Department of Aerospace and Mechanical Engineering
Australian Defence Force Academy
Canberra ACT 2604 Australia

G. T. Roberts, P. H. Schuricht, G. J. Ball, R. A. East
Department of Aeronautics and Astronautics
University of Southampton
Hants SO17 1BJ UK

Abstract: Experiments have been performed in which the heating distribution near a single, circular, sonic jet issuing normally from a flat plate into a hypersonic freestream flow has been quantitatively measured using thermochromic liquid crystals. The results indicate that the heating distribution is sensitive to both injectant gas composition and jet stagnation pressure.

Key words: Control jets, Interference heating, Hypersonic flow

1. Introduction

When a jet is directed transversely into a hypersonic flow, the jet interaction with the freestream creates an enhanced surface pressure field near the jet, resulting in a reaction force which may be several times greater than that of the jet itself. Therefore, such jets are particularly effective for controlling vehicle attitude and stability or creating large side forces for rapid maneuvering.

Powrie et al. (1992, 1993) have recently mapped the surface-pressure distributions on a zero-incidence flat plate with 2D and axisymmetric jets issuing normally into a hypersonic (Mach 6.7) freestream. Qualitative surface visualization using thermochromic (i.e. temperature-sensitive) liquid crystals also showed that large variations in local surface heating were associated with the jet interaction. Reported here are the results of further experiments with a single, circular, sonic transverse jet on a flat plate in which the surface-heating distributions have been quantitatively measured using a recent development of the liquid-crystal thermographic technique (Roberts and East 1995). As with the former work, the influence of injectant-gas composition and jet stagnation pressure was explored under conditions where the upstream (undisturbed) boundary-layer flow on the plate was laminar.

2. Experimental details

2.1. Experimental facility and model design

The experiments were conducted in the Southampton free-piston hypersonic wind tunnel. This facility provides a uniform, steady low-enthalpy flow for a period of $\sim$ 0.5 s; its design and operation have been described elsewhere (Qaswari 1977). For the present work the run conditions were: stagnation pressure $\sim$ 13 bar; stagnation temperature $\sim$ 630 K; freestream Mach number $\sim$ 6.7; unit Reynolds number $\sim 4.3 \times 10^6$ m^{-1}. The freestream gas was nitrogen.

The surface of the 200-mm $\times$ 100-mm flat-plate model, onto which the liquid crystals were sprayed, was made from a black plastic (Acetal). The depth of the plastic (3 mm) was chosen so that it remained effectively semi-infinite (as far as heat conduction is concerned) throughout the duration of the run. The initial temperature of the model (measured by platinum-resistance thermometers) was regulated by an integral heating/cooling system; this also allowed in situ calibrations of the liquid crystals to be performed. A single, 3-mm diameter, sonic jet orifice was located $\sim$ 140 mm downstream of the leading edge of the plate at the mid-span. The injectant gases (Ar, He, N_2, CH_4) were fed to the nozzle via a large-volume plenum chamber; the plenum pressures (p_j) used here were approximately 0.6, 1.0, and 1.6 bar (corresponding to jet-to-freestream stagnation pressure ratios of 0.048, 0.083, and 0.127) and the jet orifice remained choked (and under-expanded) under all conditions.

2.2. Heat-flux measurement technique

The liquid-crystal response to the suddenly applied aerodynamic heating was captured by a video recorder operating at 50 frames/second. Each of the 20–25 frames recorded was analyzed digitally (via a PC and frame grabber) to produce a map of the surface-temperature distribution. To do this, the grabbed red, green, and blue (RGB) values were converted to hue, saturation, and intensity. By comparing the measured hue values (which contain all the color information) with those obtained from an in situ static calibration of hue vs. temperature, the surface-temperature distribution on the model in each frame was obtained. From this, the convective heat-transfer coefficient was determined at all locations on the plate by employing the 1D non-steady heat-conduction equation (see Roberts and East 1995 for details). The above procedure was carried out for both jet-on and jet-off (i.e. undisturbed flat plate) flows. For the latter the local heat-transfer coefficients deduced (h_u) were shown to be in good agreement with laminar boundary-layer predictions using the reference enthalpy correlation of Eckert (1955) (Fig. 1). This correlation was then used to normalize the data to obtain the local ratio of jet-on to jet-off heating, h/h_u. The heat fluxes measured were $O(10^4)$ Wm^{-2}.

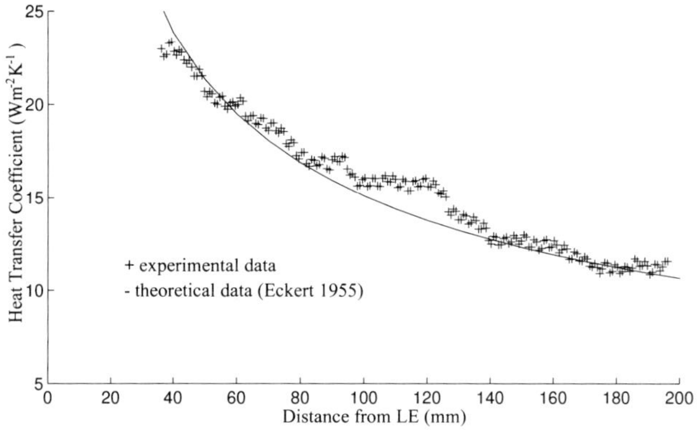

Figure 1. Variation of heat-transfer coefficient h_u with distance along plate.

3. Results and discussion

Figure 2 is an isometric plot of normalized heat-transfer coefficient in the region surrounding an argon jet with $p_j = 1.0$ bar. Data from only one side of the centerline are presented and the region covered extends just upstream of the zone of influence of the jet and to the side and rear of the plate. The spatial resolution of the data (corresponding to the pixel size in the original video image) is approximately 1 mm, although the pixels have been 3×3 neighborhood-averaged to reduce noise. Evidently, the liquid-crystal mapping technique is well able to resolve and measure the large-amplitude features present in the flow field.

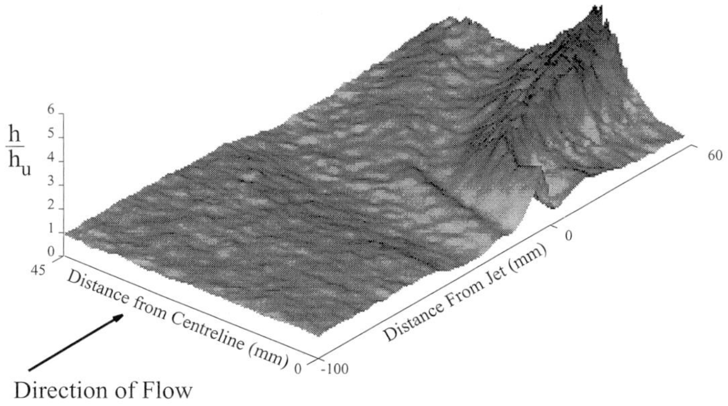

Figure 2. Map of heat-transfer coefficient near jet: argon injectant, $p_j = 1$ bar.

Figure 3 shows the normalized heat-transfer variation along the centerline for the same conditions. From Figs. 2 and 3, and the oil-flow experiments

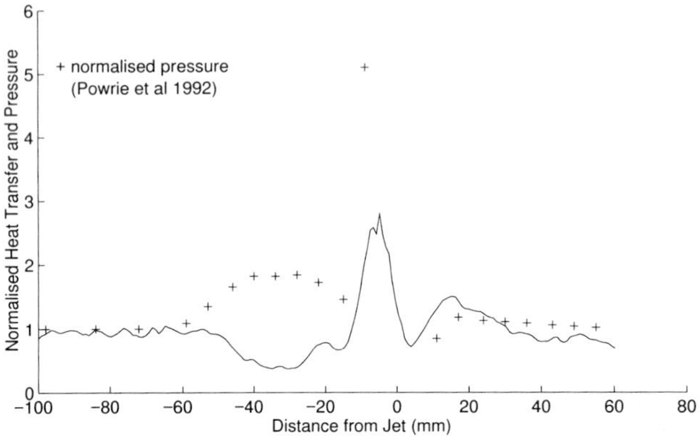

Figure 3. Variation of normalized heat-transfer and pressure along centerline.

reported in Powrie et al. (1993), several features can be identified: the pressure rise associated with the jet interaction (as noted by the centerline pressure measurements, also illustrated in Fig. 3) causes the oncoming boundary layer to separate and roll up into a series of horseshoe vortices. The initial reduction in heat transfer to a minimum value and rise in pressure to a plateau level upstream of the jet are characteristic of a laminar separation (Holden 1978), whereas the peaks in heat transfer correspond to the positions of reattachment of the vortical flows on the plate surface. Evidently, there are at least two of these vortices within the separation zone upstream of the jet and it is probable that the local trough and peak in heating just downstream of the jet are also caused by a relatively small separation zone which is confined close to the jet itself.

As, on the centerline (Fig. 3), the peak heating associated with the reattached flow of the inner vortex approximately coincides with the measured peak pressure, it is tempting to invoke the pressure-interaction theory in which the peak heat transfer is related to the measured peak pressure. Although this theory has been applied reasonably successfully to 2D laminar shock-boundary layer interactions (e.g., Hung 1973), we consider it inappropriate for this type of reattaching flow field as it does not take into account the inherent three dimensionality of the flow nor the differing thermodynamic properties (e.g., stagnation temperatures and heat capacities) of the two gases.

Unlike the blunt-fin problem (Haq et al. 1991), with which the flow is otherwise qualitatively similar, the heat flux along the reattachment line increases as the vortices are swept round the jet; indeed, it is probable that the normalized heat flux would have continued to increase had the plate been longer. Table 1 shows the maximum normalized heating values (measured at the trailing edge of the plate) along the reattachment lines of the outer and inner horseshoe vortices ($(h_p/h_u)_1$ and $(h_p/h_u)_2$, respectively) for argon jets with $p_j = 0.6$, 1.0, and 1.6 bar. These indicate that the amplitude of heating associated with the outer vortex is relatively insensitive to jet pressure whereas that associated with the inner vortex increases with injectant pressure. It has also been observed that the size of the separation zone (evident by the onset of reduction in heat flux from the undisturbed flat-plate value) increases with jet pressure. This agrees with the oil-flow data of Powrie et al. (1993) who showed a strong dependence of the separation length on the jet-to-freestream (stagnation) pressure ratio and a rather weaker dependence on the specific-heat ratio of the jet gas: both influence the effective size of obstruction the jet presents to the oncoming flow.

Table 1. Variation in peak heating (h_p/h_u) with p_j and injectant gas.

Gas	Ar			He	N_2	CH_4
p_j (bar)	0.6	1.0	1.6	1.0	1.0	1.0
$(h_p/h_u)_1$	2.45	2.32	2.53	2.17	1.67	1.37
$(h_p/h_u)_2$	4.33	5.34	7.25	3.22	3.73	2.64

Similar heating features are also observed with different injectant gases. Table 1 also shows the corresponding maximum heating values measured with He, N_2, and CH_4 injectants at $p_j = 1$ bar. Evidently, for a given jet pressure, the heating associated with the reattaching vortical flows is sensitive to the injectant gas composition. Although force data for reaction control jets are often correlated in terms of the jet-to-freestream momentum ratio (e.g., Spaid and Cassel 1973) the data do not incorporate the relevant thermodynamic properties of the injectant gas and cannot, therefore, correlate the present heat-transfer data. Further work is required to identify the appropriate heat-flux correlation parameter.

With helium injectant there are regions where the jet interaction has actually cooled the flat-plate surface. The mechanism for this is as yet unknown, but presumably results from the relatively cool injectant gas (the jet-to-freestream stagnation temperature ratio was ~ 0.48 in these experiments) being brought into contact with the plate surface before it has had an opportunity to mix and exchange energy with the freestream gas. This could be related to the relatively high heat capacity of helium compared with the other injectant gases. Again this warrants further study.

4. Conclusions

The heating distribution on a flat plate near a circular, sonic, transverse jet in a hypersonic freestream flow has been quantitatively measured using thermochromic liquid crystals. The results show that the jet interaction causes the heating to be locally amplified by an amount which is dependent on both jet-injectant pressure and composition. However, further work is necessary in order to identify the relevant correlation parameters for this type of interaction.

Acknowledgement. The work has been carried out with the financial support of the UK Defence Research Agency. P. H. Schuricht is a CASE student supported by the EPSRC and British Aerospace (Dynamics) Ltd. The authors would like to acknowledge the help of Mr. R. P. Clarke in carrying out the experiments and Ms. H. Powrie for her helpful suggestions and advice.

References

Eckert ERG (1955) Engineering relations for skin friction and heat transfer to surfaces in high velocity flows. J Aerospace Sciences 22:585.

Haq ZU, Roberts GT, East RA (1991) Interference heating near fin/body junctions on hypersonic vehicles. In: Proc 1st European Symposium on Aerothermodynamics of Space Vehicles, ESTEC, Netherlands, pp 171–176.

Holden MS (1978) A Study of Flow Separation in Regions of Shock Wave - Boundary Layer Interaction in Hypersonic Flow, AIAA Paper 78-1169.

Hung FT (1973) Interference Heating Due to Shock Wave Impingement on Laminar Boundary Layers, AIAA Paper 73-678.

Powrie HEG, Ball GJ, East RA (1992) Experimental study of the interaction of a two dimensional reaction control jet with a hypersonic flow. In: Proc IUTAM Symposium on Aerothermochemistry of Spacecraft and Associated Hypersonic Flows, Marseille, pp 362–367.

Powrie HEG, Ball GJ, East RA (1993) Comparison of the interaction of two and three dimensional transverse jets with a hypersonic freestream. In: Proc Agard Conf on Computational and Experimental Assessment of Jets in Cross Flow, Winchester, UK, Agard CP534, pp 20.1–20.7.

Qaswari AMS (1977) Measurement of Hypersonic Dynamic Stability of Pitching Blunt Nosed Bodies in a Short Duration Facility, Ph.D. Thesis, University of Southampton.

Roberts GT, East RA (1995) The Use of Thermochromic Liquid Crystals for Heat Transfer Measurement in Short Duration Hypersonic Wind Tunnel Facilities, AIAA Paper 95-6115.

Spaid FW, Cassel LA (1973) Aerodynamic Interference Induced by Reaction Controls, AGARD-AG-173.

Comparison between shock-tunnel measurements on a planetary-probe configuration and DSMC predictions

M. S. Holden, K. M. Chadwick
Calspan-University of Buffalo
New York, New York 14225 USA
M. A. Gallis, J. K. Harvey
Imperial College
London SW7 2BY UK

Abstract: This paper describes an experimental and numerical study of the flow field of a sting-mounted planetary-probe configuration. Two issues are addressed: the flow establishment time of the base flow and the effects of transition to turbulence in the shear layer and reattachment flows. The experimental studies have been conducted at enthalpies up to 10 MJ/kg in the Calspan LENS facility where heat-transfer and pressure measurements have been made for the complete body. Transient Direct Simulation Monte Carlo (DSMC) computations have been made which confirm the wind-tunnel findings.

Key words: 70-deg cone, Hypersonic, Base flow, Transition, High-enthalpy effects, Flow establishment times

1. Introduction

The experimental studies conducted in the LENS facility (Holden 1994) and HEG high-enthalpy shock tunnels with the 70-deg cone planetary-probe configuration coupled with the numerical simulations conducted with the DSMC and Navier-Stokes solvers have resulted in a number of issues that must be resolved to provide an understanding of the key fluid dynamic and chemical modeling of these flows, particularly in the wake. The flow establishment time (FET) for steady attached and separated flows in short-duration facilities is an important factor in the design of an experimental investigation. During the past 30 years, a number of studies have been conducted to determine suitable criteria to judge FET for simple and complex configurations. Typically, from 3 to 4 flow lengths (L_{fl}) based on body diameter are required to establish a steady attached flow, while from 20 to 50 are required for complex shock-induced separated regions and base flows. However, most of these studies were conducted in relatively low-temperature flows where real-gas effects were unimportant. The measurements in the base flow of the planetary-probe configuration in continuum and non-continuum flows for energy levels of 5 MJ/kg and above may be significantly influenced by non-equilibrium flow effects. Thus, the recent high-enthalpy experiments made in the ground tests on the planetary-probe are of great interest.

Another question of central importance, which has arisen during our analysis, is whether transition to turbulent flow occurs in the shear layer or just downstream of the reattachment point in the base-flow region. It is interesting to note in these hypervelocity base flows that there is a relatively narrow laminar-flow regime between the transition associated with non-continuum to continuum flow and the transition to turbulence. It is important that the experimental studies be conducted over a range of Reynolds numbers sufficient to ensure that the transition to turbulence occurs well downstream of the rear stagnation point to provide confidence in the comparisons between experimental and numerical laminar-flow simulations.

2. Time to establish steady flow in the base region

The FET for the base region of blunt bodies has been examined in earlier shock-tunnel studies by Holden (1971) for a sphere supported by fine wires. In these studies, heat-transfer and pressure measurements were made in continuum low-Re flow which in the base flow indicated that this region remained laminar. The pressure measurements showed that the base-flow structure was set-up within $25L_{fl}$ while the heat-transfer measurements suggested that the flow took close to $50L_{fl}$ to become steady.

The determination of FET is complicated by the start time of the tunnel which is controlled principally by establishing the steady flow in the nozzle. In these latter studies, nozzle-flow starting times were less than one ms; nearly an order of magnitude smaller than to establish a steady base flow. In the present studies, covering a wide range of Re and total enthalpies, uniform flow in the test section was established within the first ms of the typically 6-ms flow duration. The steady test duration gave from 15 to 25 m of flow or from 100 to $160L_{fl}$ in which to establish a steady flow over the base region. A typical set of time histories obtained under the 5-MJ/kg conditions was presented in Holden (1994). These measurements suggested that the region of flow close to the shear-layer reattachment point on the sting is the most sensitive indication of FET, and both the heat-transfer and pressure measurements suggest that close to $50L_{fl}$ are required for the position of the shear layer to stabilize in this region. Typically, the recirculation flow in the base region, which is established during tunnel start-up, changes in size causing the pressure and heat transfer at the base of the planetary probe to rise slightly and along the sting to increase or decrease depending on the position in the reattachment region. This adjustment requires about 2 ms to occur. Figure 1 shows the time-history traces for pressure and heat transfer obtained at the 10-MJ/kg condition. We again observe that the tunnel flow is established in one ms, and that while there is relatively minor adjustment to the pressure and heat-transfer levels close to the base of the body, the heat transfer and pressure in the recompression region are sensitive to small movements of the reattaching shear layer as it stabilizes. A steady base flow is reached in 2 ms (or $50L_{fl}$) after the flow stabilizes over the forebody leaving approximately 3 ms over which the flow in the base region is steady.

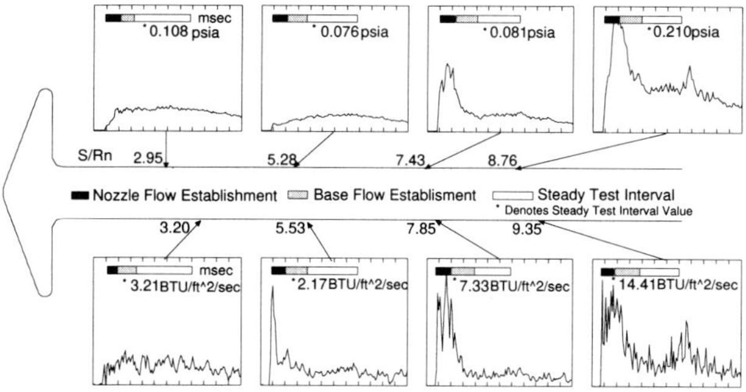

Figure 1. LENS Case E, base flow heat-transfer and pressure time establishment.

3. High-enthalpy non-continuum and continuum flows

Measurements of heat transfer and pressure were made over a range of unit Re spanning the continuum to non-continuum flow regime. Experiments were conducted at stagnation enthalpies of 5 and 10 MJ/kg to investigate the effects of oxygen dissociation. The initial series of studies was designed to provide information for fully laminar base flows over the sting-mounted planetary-probe configuration for direct comparison with solutions from DSMC and Navier-Stokes codes, and was selected so there could be some overlapping in the range of the applicability of these two prediction techniques. Measurements were then made at higher Re to examine when transition to turbulence occurred in the base region. Figure 2 (left) shows the heat-transfer and pressure measurements on the forebody and in the base region compared with DSMC solutions obtained independently by Moss et al. (1995)and Gallis & Harvey (1995) for a rarefied flow condition.

Both theory and experiment indicate that the pressures and heat-transfer rates in the base-flow region are close to 1% of the cone value, rising to 8% at the end of the reattachment compression process. Increasing the Reynolds number to obtain close to continuum conditions in the base-flow region allows us to compare the measurements with solutions by Gallis & Harvey (1995) and Moss et al. (1995) (see Fig. 2, right) employing the DSMC and Navier-Stokes codes, respectively. Again, we see the heat-transfer and pressure levels in the recirculation region at the level of 1% of the cone value, with the heating values downstream of reattachment approaching 10% of the cone value. The DSMC solution is only an instant in the evolution of the flow corresponding to 0.3 ms and shows an overprediction of the heat transfer near the base of the body.

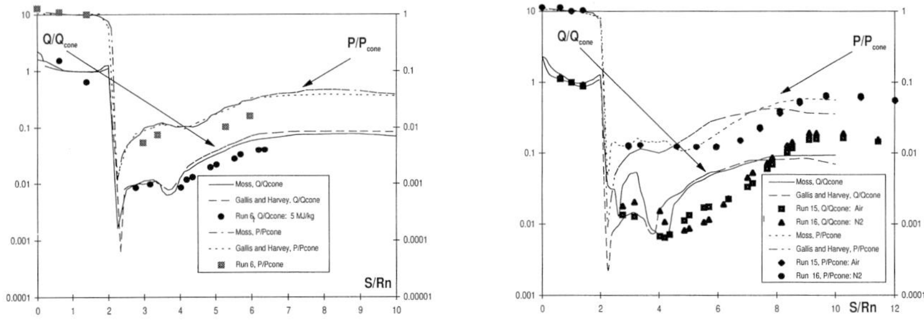

Figure 2. Comparisons between heat-transfer and pressure measurements and computations with DSMC and Navier-Stokes codes.

Transient computations have been performed for the 5-MJ/kg flow to determine the FET using the Imperial College Maximum Entropy DSMC code (Gallis & Harvey 1995). Density contours, non-dimensionalized by the free-stream value, for 0.5, 1.1, 3.0, and 5.0 ms are given in Figs. 3 and 4. From these figures we can identify from the 0.3 contour the changes in the approximate extent of the recirculating region immediately behind the body which has not reached steady state by 3.0 ms. The reattachment shock can be identified by the cluster of 0.4 to 0.6 contours emanating from the sting. It is diffuse because of the low densities in the wake and it plays a dominant role in the formation of the sting boundary and hence on the heat transfer at and downstream of the

Figure 3. Density contours at: 0.5 ms (left) and 1.1 ms (right).

reattachment point. The wake is a low-density region and flow entering it from the forebody is far from equilibrium due to the rapid expansion around the rim. The FET is thus determined principally by the (a) convection of particles as

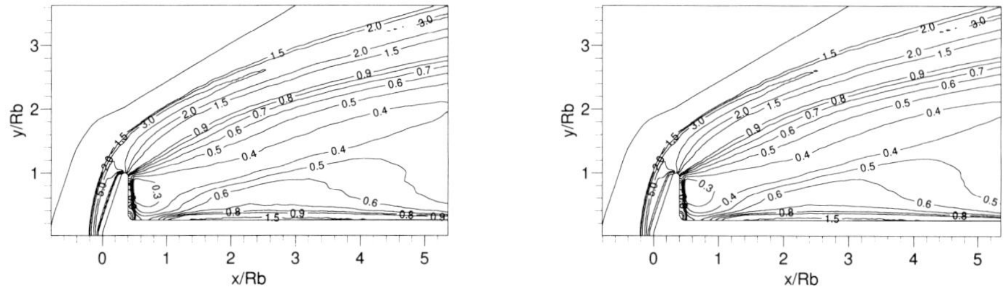

Figure 4. Density contours at: 3.0 ms (left) and (b) 5.0 ms (right).

they recirculate and (b) by the relatively slow real-gas relaxation processes. The evolution of the heat transfer on the sting is presented in Fig. 5 (left). While C_p converges after 1 ms we note that C_h does not reach steady state until 2 ms. This observation is supported by the predicted evolution of the wake.

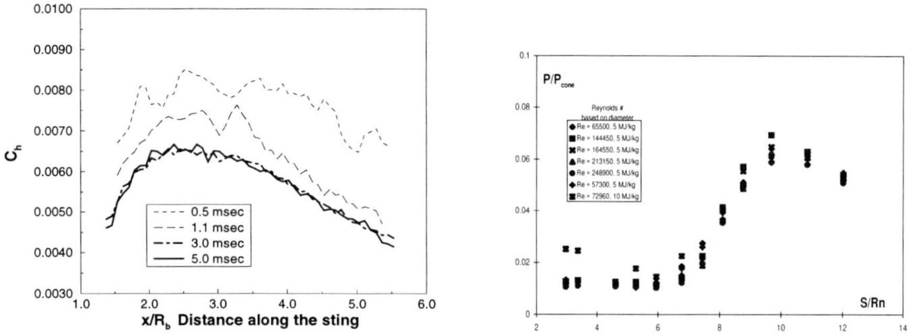

Figure 5. Left: C_h evolution along the sting. Stagnation point at $x/R_b = 0$. Right: LENS planetary-probe data, Reynolds number effect on pressure.

To investigate whether transition to turbulence influenced the pressure and heat transfer in and downstream of the base recirculation region, measurements were conducted at a series of higher Re conditions. These indicated that while the shape and magnitude of the pressure distribution remains roughly independent of Re, above 10×10^4, there is an increase in the maximum heat transfer in the reattachment region accompanied by a forward movement of the peak value (see Figs. 5, right and 6, left). Both these effects suggest that transition has occurred in the recompression region which moves forward with increasing Re, but has yet failed to influence significantly the properties in the recirculation region. For measurements made at similar Re, doubling the total enthalpy of the flow results in a significant decrease in the heating levels in the recompres-

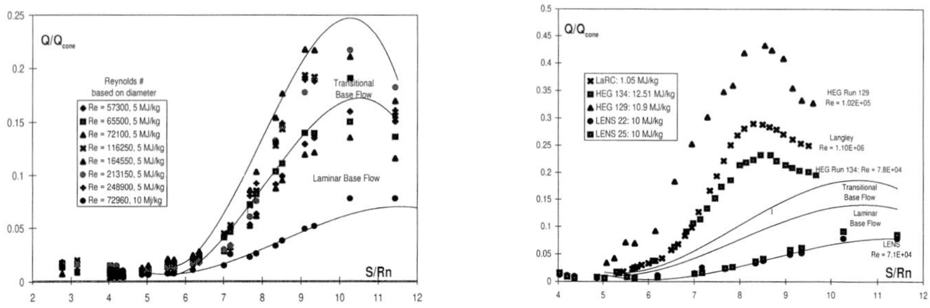

Figure 6. Left: Reynolds number and total enthalpy effects on planetary probe sting pressure and heat-transfer distribution. Right: Comparisons between heat-transfer distribution from LENS, HEG, and NASA LaRC experimental studies.

sion process coupled with an enlarging of the base-flow interaction region. Such a trend was observed in the HEG studies (Kastell et al. 1994), however, in these studies, heating levels in the reattachment region were found to be significantly larger, as shown in Fig. 6, right. Also shown are measurements made in the NASA-Langley Mach 10 tunnel for very large Re conditions ($Re_D = 10^6$) where the base flow is surely turbulent. Plotting our measurements for fully laminar and transition base flows (Fig. 6, right), it could be speculated that the differences between the measurements at the 10-MJ/kg condition could result from differences in the turbulent structure of the shear layer. However, at these high-enthalpy conditions where non-equilibrium flow effects may play an important role in controlling flow-field structure and surface heating, a more detailed examination of these effects and the measurement techniques employed is required to provide a definitive explanation of the differences in heating values observed.

References

Holden MS (1994) Development and Code Evaluation Studies in Hypervelocity Flows in the LENS Facility, AIAA Paper 95–0291.

Holden MS (1971) The establishment time of laminar-separated flows. AIAA J 9(11):2296–2298.

Gallis MA, Harvey JK (1995) Comparison of the Maximum Entropy DSMC Code with Flowfield Measurements, AIAA Paper 95–0413.

Kastell D, Horvath T, Eitelberg G (1994) Nonequilibrium flow expansion experiment around a blunted cone. In: 2nd European Symp on Aerothermodynamics for Space Vehicles, ESTEC, The Netherlands.

Moss JN, Price JM, Dogra VK, Hash DB (1995) Comparison of DSMC and Experimental Results for Hypersonic External Flows, AIAA Paper 95–2028.

Hyperboloid flare experiments in the HEG

R. M. Krek, G. Eitelberg, W. H. Beck
DLR – Institute for Fluid Mechanics, Bunsenstraße 10
37073 Göttingen, Germany

Abstract: Tests on a hyperboloid flare model have been undertaken in the HEG facility. Pressure and heat-flux profiles were measured and the flow field was visualized by a schlieren system. Transitional behavior of the separation shear layer was exhibited by the two high Reynolds number conditions. The experimental heat-flux profile for a low Reynolds number, high total enthalpy condition compares well with Navier-Stokes calculations, assuming a non-catalytic wall. The model forebody pressure and heat-flux profiles for the six conditions tested are self-similar until separation occurs. It is argued, that at constant viscous interaction parameter, which is also at constant relaxation length, the separation bubble size decreases with increasing enthalpy.

Key words: HEG shock tunnel, Hyperboloid flare, Separation, Hypervelocity

1. Introduction

Control surfaces such as body flaps, elevons, and rudders on hypersonic vehicles can produce separation regions that reduce the efficiency of these surfaces. To investigate the flow about deflected surfaces, an axisymmetric body with a flare extension, provided by ESA as a standard model, was utilized. The hyperboloid flare has been tested in many cold wind-tunnel facilities (Krogmann 1994, Koppenewallner et al. 1994, Niezgodka & Henckels 1994, Sagnier 1994a) and one moderate enthalpy facility F4, where the flow is chemically frozen (Sagnier 1994b). The hyperboloid flare model was tested in the High Enthalpy shock tunnel located in Göttingen, HEG, that produces flows with a total enthalpy range of $h_0 \approx 10 - 24$ MJ/kg, with highly nonequilibrium chemistry. Pressure and heat-flux measurements were made as well as flow visualization using a schlieren arrangement.

2. Experimental

The hyperboloid flare is an axisymmetric model whose profile describes the windward centerline of the Hermes re-entry vehicle, at an incidence angle $\alpha = 30°$ and a flap deflection $\delta \approx 20°$. The length of the model is $L_{ref} = 111.4$ mm and the base diameter is $d = 150$ mm. The model was instrumented with 20

Kulite pressure transducers, type XCS-093-50D, and 17 ONERA-made thermocouples. The HEG is a free-piston shock tunnel that can produce flows to simulate the binary scaling parameter of re-entry to Earth's atmosphere (Eitelberg et al. 1992). The model was tested at the HEG operating conditions given in Table 1.

Table 1. Summary of flow properties of the six HEG standard operating conditions.

Condition	p_0 MPa	T_0 K	h_0 MJ/kg	p_∞ kPa	T_∞ K	ρ_∞ g/m^3	u_∞ m/s	M_∞	Re_∞ /m ($\times 10^{-6}$)
I	39	9050	21.1	0.43	790	1.64	5920	9.7	0.27
II	91	9730	22.3	1.21	1040	3.59	6160	9.0	0.52
III	45	7280	13.2	0.47	550	2.83	4810	10.0	0.48
IV	110	8110	14.8	1.32	720	6.15	5150	9.5	0.93
V	51	6340	10.6	0.55	480	3.91	4370	9.8	0.65
VI	94	6870	11.5	1.02	540	6.57	4570	9.8	1.07

Here p is pressure, T is temperature, h is enthalpy, ρ is density, u is velocity, M is Mach number, and Re is Reynolds number.

3. Flow establishment and transition

The flow about the hyperboloid flare is complicated by the shock/shock and shock/boundary layer interactions that occur in the region of the flare, Fig. 1. Flow establishment was monitored by sensors located near the flare hinge line. For all conditions, the model forebody boundary layer is expected to be laminar. For the two higher Reynolds number conditions, the separation bubble displays an oscillatory behavior which was not detected in the other conditions. Figure 2 displays the heat-flux signals taken from sensors located upstream, on the edge of, and inside the separation bubble. The signal produced by the middle sensor displays in an alternating manner the separated and the attached boundary layer heat flux. This is possibly due to shear-layer transition. At the shear layer Reynolds numbers $\approx 1.2 \times 10^5$, determined for the higher pressure conditions, according to the criteria by Holden & Kolly (1995), the shear layer becomes transitional. The transition of the separation shear layer also causes an oscillation in the reattachment point, but to a greater spatial extent than the separation. Figure 4 shows plots of the normalized pressure and heat-flux profiles along the model. The arrows on the symbols indicate the oscillation of the signals during the test time and the symbols are the average of the oscillations. For both the pressure and heat-flux results, the oscillation is many times larger for condition VI than for condition I. The signals on the forebody are steady. It should also be noted that the oscillation amplitudes in the pressure signals are less than those in the heat-flux signals.

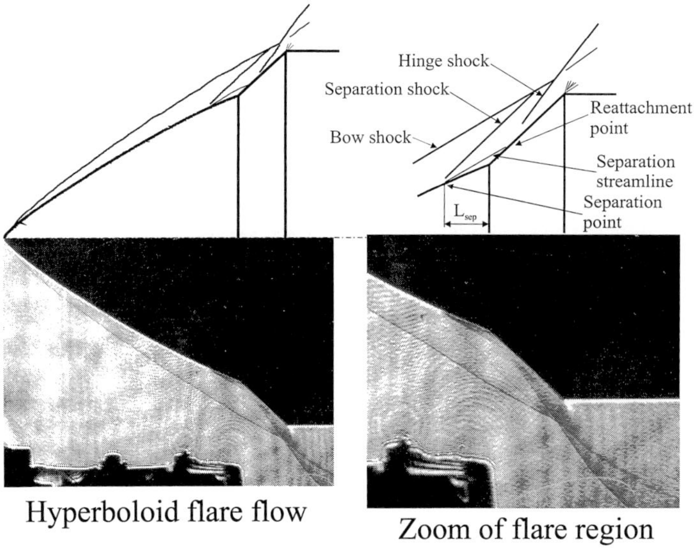

Figure 1. Schlieren photographs and sketch of the flow structure about the hyperboloid flare and close up of the flare region for flow condition III.

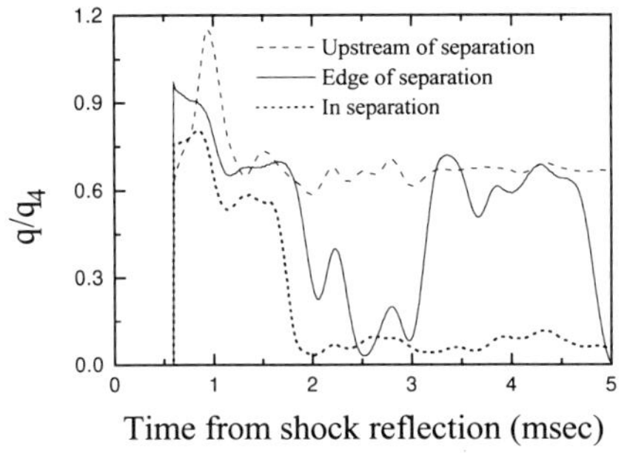

Figure 2. Heat-flux histories for sensors located upstream, on the edge of, and inside the separation bubble for flow condition VI.

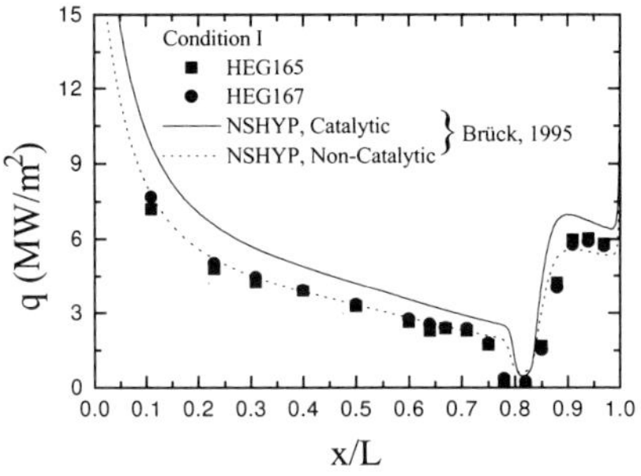

Figure 3. Comparison of the measured and calculated catalytic/non-catalytic heat-flux rates on the hyperboloid flare model for flow condition I.

4. Results

Figure 3 shows the heat-transfer rate distribution measured on the model, compared with Navier-Stokes calculations for catalytic and non-catalytic surfaces. The numerical calculations (Brück 1995) use the NSHYP code (Brenner 1994).

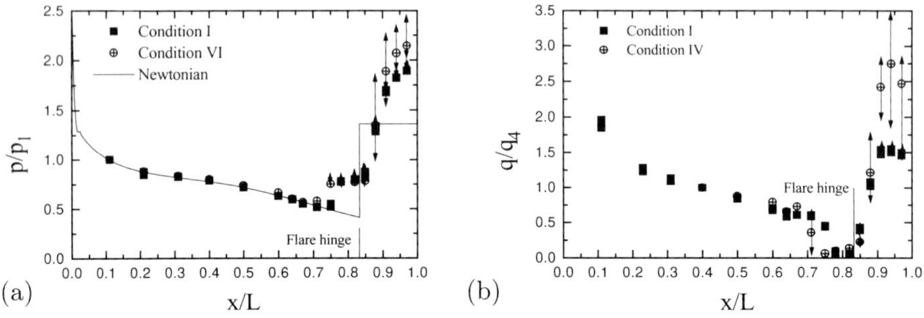

Figure 4. Oscillation of the signals during the test time. Comparison of conditions I and VI: (a) p/p_1 versus x/L_{ref}; (b) q/q_4 versus x/L_{ref}.

From Fig. 3, the measured heat-transfer rates follow the non-catalytic calculations on the forebody. This is in agreement with the prediction by Inger (1994) for large sweepback angles. The experimental results on the flare display slightly higher heat fluxes than the non-catalytic calculations. Due to the complex, unsteady flow in the reattachment region, no conclusions about catalycity effects can be drawn at present. Figure 4 shows examples of pressure and heat-flux distributions for two of the HEG conditions. The results are normalized with respect to measured model quantities to eliminate the influence of non-uniformities in the nozzle flow. The pressure and heat-transfer profiles are self-similar on the forebody of the model until separation of the flow occurs. The Newtonian pressure profile is an excellent approximation for the experimental results.

5. Discussion

Figure 5 shows the separation length versus the viscous interaction parameter for the six HEG conditions and for other hypersonic wind tunnels. The separation length L_{sep} is measured from the separation point, determined from the schlieren images, to the flare hinge line (refer to Fig. 1). The viscous interaction parameter is defined by

$$\bar{V}_\infty = \frac{M_\infty}{\sqrt{Re_\infty}}. \tag{1}$$

All the wind tunnels, except for the HEG and F4, are cold, with the F4 facility being operated from cold conditions to conditions with similar enthalpies as the HEG. Figure 5 shows that all the cold wind-tunnel results, including the cold F4 results, follow the same curve that has a maximum at $\bar{V} \approx 0.015$. The hot F4 and HEG results deviate from this curve, suggesting that the enthalpy is affecting the separation size. The size of the separation bubble is also affected

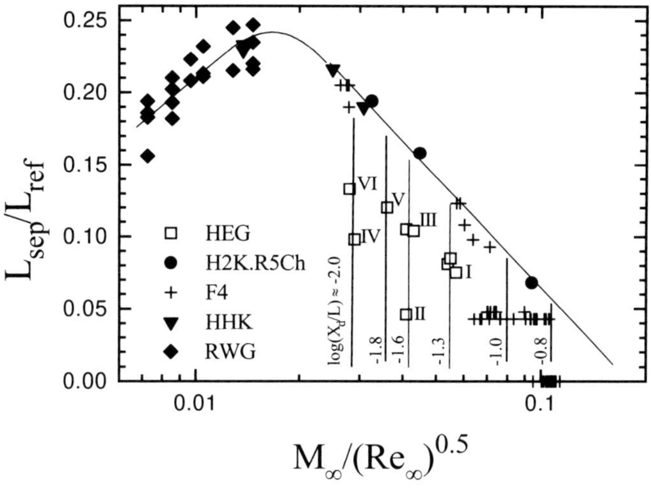

Figure 5. Separation length L_{sep}/L_{ref} versus the viscous interaction parameter $\bar{V}$ for the six HEG conditions and for various other wind tunnels.

by the extent of flow nonequilibrium. The extent of nonequilibrium can be described by the characteristic length for dissociation, calculated assuming a strong shock and ideal dissociating gas, which is given by Eitelberg et al. (1992)

$$X_d = \frac{u_\infty}{C\rho_\infty \Theta_d^\eta}, \tag{2}$$

where Θ_d is the characteristic dissociation temperature, and C and η are constants. The dissociation length X_d is then normalized by a characteristic length scale, in this case, the length of the model L_{ref}. Assuming Sutherland's law for viscosity and frozen freestream, combining Eqs. (1) and (2), the following relationship can be developed:

$$\bar{V}_\infty \propto \sqrt{\frac{X_d}{L_{ref}} \frac{1}{\sqrt[4]{T_\infty}}}. \tag{3}$$

Figure 5 plots lines of constant X_d/L_{ref} calculated for the six HEG conditions and the hot F4 data. The curve representing the cold wind-tunnel data is the frozen limit. For constant $\bar{V}$, and therefore for constant X_d/L_{ref}, as the enthalpy increases, the separation length decreases. Conditions IV and VI are exceptions, but this is possibly due to these conditions being transitional or the uncertainty in determining the freestream conditions. Also for constant $\bar{V}$, e.g., conditions II and III, as the enthalpy increases and the separation point moves downstream, the reattachment point moves upstream.

6. Conclusions

Tests on a hyperboloid flare have been conducted in the HEG at six conditions. The two highest Reynolds number conditions exhibit transitional behavior of the separation shear layer. The experimental results for condition I were compared with numerical calculations and yielded good agreement. All pressure and heat-flux profiles on the model forebody are self-similar for the six HEG conditions. For the HEG range of total enthalpies, it was shown that the viscous interaction parameter is proportional to the square root of the dissociation length scale for the freestream. The separation size was found to be affected by not only the total enthalpy of the flow, but also by the relaxation process.

Acknowledgement. The authors wish to thank ESA for supplying the hyperboloid flare model and for partial funding of the tests. Thanks also go to D. Kastell for the optical visualization, J. Lenz for the HEG operation, and D. Garbe for the installation and interfacing of the model.

References

Brenner G (1994) Numerische Simulation von Wechselwirkungen zwischen Stößen und Grenzschicten in chemisch reagierenden Hyperschallströmungen, DLR FB 94-04.

Brück S (1995) Private communication.

Eitelberg G, McIntyre TJ, Beck WH, Lacey J (1992) The High Enthalpy Shock Tunnel in Göttingen, AIAA Paper 92-3942.

Holden M, Kolly J (1995) Measurements of Heating in Regions of Shock/Shock Interaction in Hypersonic Flow, AIAA Paper 95-0640.

Inger GR (1994) Nonequilibrium-Dissociated Stagnation Boundary Layer on an Arbitrarily Swept Wing, AIAA Paper 94-2072.

Koppenwallner G, Müller-Eigmer R, Johannsmier D, Holze K (1994) Hyperboloid Flare Tests in the HHK, HTG Report November 18 1994.

Krogmann P (1994) Hyperboloid Flare Experiments at Mach 6.8 in RWG, DLR IB 223-94 C 44.

Niezgodka J, Henckels A (1994) Pressure Distribution and Stanton Number Distribution Measurements on a Hyperboloid Flare Model at $M_\infty = 8.7$, DLR IB 39113 94 C 07.

Sagnier P (1994a) Synthesis of Measurement Results Over a Hyperboloid Flare Configuration in ONERA R5Ch Wind Tunnel, ONERA RT No. 64/6121SY.

Sagnier P (1994b) Synthesis of the Hyperboloid Flare Campaign in F4, ONERA RT No. 70/6121SY.

Experiments on the Halis axisymmetric configuration in the T5 shock tunnel

J. Davis, B. Sturtevant
Graduate Aeronautical Laboratories, California Institute of Technology
Pasadena, CA 91125 USA

J. Muylaert
Aerothermodynamics Section YPA, ESA-ESTEC
2200AG, Noordwijk, The Netherlands

Abstract: A model of the Halis Axisymmetric Configuration (HAC) has been built and tested in the T5 hypervelocity shock tunnel, a high-enthalpy impulse facility. Experiments were performed at four test conditions using both air and nitrogen. Static pressures and heat-transfer rates were measured on the model surface. Flow visualization was performed using infinite-fringe differential interferometry. A separated region at the flare root was observed under all test conditions. Pressure and heat-transfer results showed reasonable repeatability, but exhibited unsteadiness in the region downstream of reattachment. Transition to turbulence was found to occur in the separation region. Separation length was seen to increase with viscous interaction parameter $\bar{V}_\infty$, with a possible high-enthalpy effect decreasing the separation length at constant $\bar{V}_\infty$.

Key words: Hypersonic, Hypervelocity, Real gas, Separation, Shock tunnel, Measurements, T5, High enthalpy, Halis, Shock/boundary-layer interaction

1. Introduction

The T5 facility at GALCIT is a free-piston driven shock tunnel capable of reaching very high stagnation temperatures and pressures at relatively high densities, simulating real gas flows around vehicles flying at sub-orbital speeds through the atmosphere. The HAC is an axisymmetric geometry extracted from the windward centerline profile of the STS orbiter. The particular configuration used in the present experiments, shown in Fig. 1a, corresponds to an angle of attack $\alpha = 35°$ and a flap deflection angle $\delta = 20°$. The HAC is a generic axisymmetric form that reproduces flow phenomenon found on 3D flight geometries like the orbiter. While not an exact duplication of the orbiter's geometry (the HAC flare length is much longer than the orbiter flap length), the HAC does incorporate the essential curvature just upstream of the flap, which is expected to decrease the size of the separation region and increase the sensitivity of the separation region to high-enthalpy effects. Separation lengths can be measured from experiments at different enthalpy levels in T5 to look for real-gas effects. A separate paper in this symposium (Muylaert et al. 1996) uses the data obtained from these experiments in conjunction with computational results to study real-gas effects on flap-induced separation.

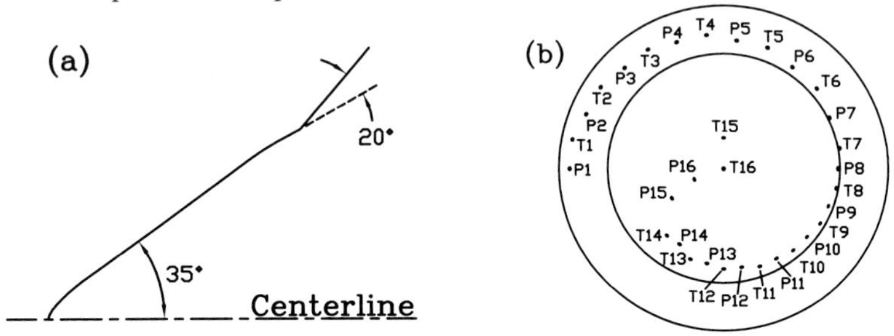

Figure 1. (a) Profile of the HAC geometry; (b) front view of the HAC model.

2. Test model and instrumentation

The stainless steel HAC model of 17-cm base diameter holds 16 pressure sensors and 16 thermocouples, arrayed in a spiral pattern as shown in Fig. 1b. The axial positions of pressure- and temperature-measurement locations are coincident except near the nose, where the pressure sensors and thermocouples are staggered. A thermocouple is used at the stagnation point because the pressure there could be inferred from previous measurements of the free-stream pitot pressure measured for different test conditions.

Surface-temperature measurements were made using type-E coaxial thermocouples with a 1-μs response time. The amplified voltage traces were converted to temperature using a fourth-order polynomial fit for type-E thermocouples. Then time-resolved heat fluxes were calculated using a one-dimensional semi-infinite slab theory for unsteady heat transfer via an indirect method due to Schultz & Jones (1973) and detailed in Germain (1994).

Static-pressure measurements were made using high-frequency, acceleration-compensated, piezoelectric pressure transducers. The recessed mounting of these transducers was designed for fast response time and incorporates several features that protect the sensors from the harsh environment in T5. The possible error in pressure measurement due to the relatively large hole size on the HAC model was estimated to be on the order of 1–2% (Davis 1994); this error was ignored when computing uncertainties for the pressure measurements because it is small and cannot be accurately predicted.

3. Experiments

A detailed description of the T5 facility and its performance envelope can be found in Hornung (1992). The test conditions for the HAC experiments form a matrix of two stagnation pressures and two stagnation enthalpies. Table 1 shows a typical run for each of the four test conditions. Reservoir enthalpy h_0 was computed using ESTC (McIntosh 1970), an equilibrium shock-tube calculation. Free-stream conditions were computed using SURF (Rein 1989), a 2D nonequilibrium, inviscid nozzle-flow code. Re_∞ is based on the model diameter (17 cm) and $\bar{V}_\infty = M_\infty/\sqrt{Re_\infty}$ is the viscous interaction parameter.

Table 1. Typical reservoir and free-stream conditions for HAC tests.

Test Cond.	Run No.	p_0 (MPa)	h_0 (MJ/kg)	ρ_∞ (kg/m^3)	u_∞ (m/s)	T_∞ (K)	M_∞	Re_∞	$\bar{V}_\infty$
1	662	58.5	10.8	0.050	4140	1320	5.7	7.33×10^5	0.0067
2	730	59.3	19.1	0.034	5310	2560	5.0	3.96×10^5	0.0079
3	720	23.5	10.9	0.020	4120	1240	5.8	3.08×10^5	0.0105
4	719	27.2	17.2	0.017	5050	1880	5.4	2.24×10^5	0.0114

Each test condition had four possible types of runs: (1) the original run with test gas of air; (2) a run to show repeatability; (3) a run with the model rotated 180° in roll to show axisymmetric flow uniformity; and (4) a run with N_2 test gas to look for chemical effects. Typical repeatability is shown in Fig. 2 by comparing all four types of runs at test condition 2. Error bars and spline fit are shown only on Run 726. The error bars are based on the variance due to flow unsteadiness during the test time. For pressure measurements they also include uncertainties in sensor calibration and free-stream pitot pressure p_{t2}. Previous measurements of the T5 nozzle flow (Rousset 1995) showed $p_{t2} \propto p_0$ within 5% for air, thus normalization by p_{t2} accounts for variation of p_0 between runs of the same nominal test condition. Figure 2 shows no significant effects due to roll angle or use of N_2 test gas. The unsteadiness and scatter in the data are both larger on the flare than on the forebody.

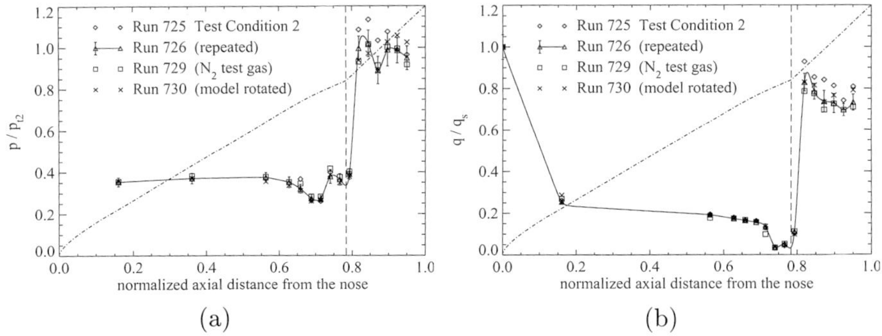

Figure 2. Repeatability of measurements for test condition 2: (a) normalized static pressure, (b) normalized heat flux. Vertical dashed line is hinge-line location, dash-dot line represents model profile.

4. Results

The pressure and heat-flux distributions in Fig. 2 and Fig. 3 have a shape typical of all runs. The pressure is relatively constant over the conical forebody, drops slowly at the curvature upstream of the flare, rises to an intermediate plateau in the separated region, and jumps to stagnation level at reattachment. The drop and subsequent rise in pressure downstream of reattachment appeared in all runs and does not correspond to any flow structure visible in the interferograms; it may be due to the curvature upstream of the flare, which produces a local expansion and causes the reattachment shock to be slightly curved. The heat flux falls to a very low value in the separated region, then overshoots at reattachment and falls off over the flare without a subsequent rise. The experiments were originally intended to encompass fully laminar flow, but preliminary computations (Muylaert et al. 1996) indicate that transition to a turbulent boundary layer at or near reattachment is required to attain the observed heat flux on the flare. This is also consistent with the increased unsteadiness in measurements on the flare.

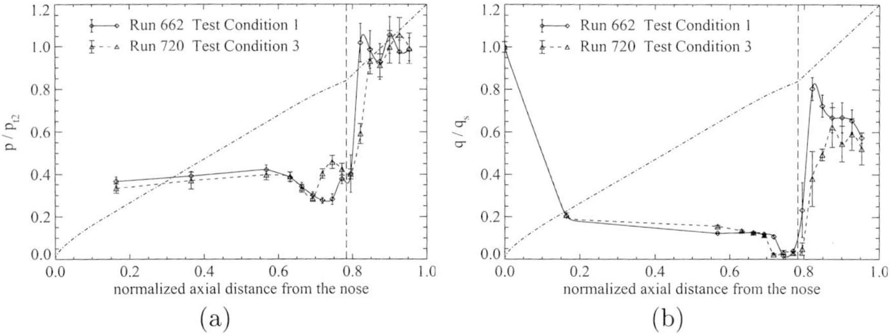

Figure 3. Comparison of data for test conditions 1 and 3: (a) normalized static pressure, (b) normalized heat flux.

The separated region was observed for all runs. Comparison of data between test conditions 1 and 3 in Fig. 3 shows a large difference in separation length due to variation in Re_∞. Figure 4a and Fig. 4b are two differential interferograms, with beam shear aligned at 35° and 135° respectively (measured counterclockwise from the flow direction). The interferograms allow higher resolution measurement of separation length than obtainable with pressure data. The separated shear layer and a corresponding weak compression wave are visible in the interferograms. The separation point is taken where this compression wave intersects the body.

The separation length was measured from interferograms for each of the four runs in Table 1 as the axial distance from the flare root to the separation

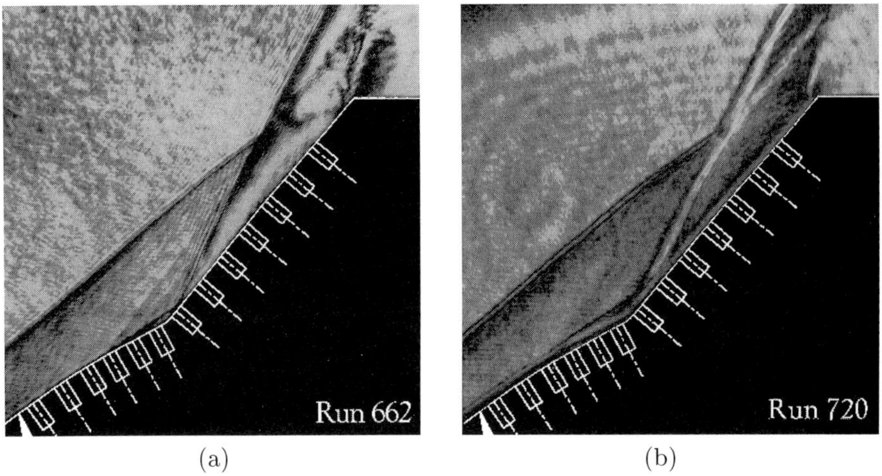

Figure 4. Differential interferograms for: (a) test condition 1, and (b) test condition 3. Flow is from left to right. The overlay shows axial locations for measurements, and the arrow indicates start of curvature downstream of conical forebody.

point. These lengths are normalized by the HAC profile axial length (0.1 m) and plotted in Fig. 5 using two different correlations. Figure 5a is an empirical correlation from experiments on a 2D compression corner in cold hypersonic flow (Needham & Stollery 1966), where p_{flare} and p_{sep} are measured surface pressures averaged over the flare and separation regions respectively. M_0 and Re_{x0} at the boundary-layer edge just upstream of separation have been taken from computational results (Muylaert et al. 1996) for laminar flow on the forebody. The decreasing trend with Re_{x0} is consistent with Needham's results for transitional flows (laminar separation and turbulent reattachment) and is opposite the trend for fully laminar or fully turbulent flows. This correlation does not, however, show any clear enthalpy effect with the present data.

The second correlation, Fig. 5b, has been used recently for the hyperboloid flare, an axisymmetric configuration similar to the HAC (Krek et al. 1996). The upward trend with $\bar{V}_\infty$ is the same found for perfect-gas experiments on the hyperboloid flare with $\bar{V}_\infty < 0.15$ and opposite that found for low-density, high-enthalpy experiments with $\bar{V}_\infty > 0.15$, again indicating transitional flows. The high-enthalpy test conditions 2 and 4 show increased separation length compared to the low-enthalpy conditions 1 and 3. This is possibly an enthalpy effect on the separation length, though the evidence is weak considering there are only four data points with significant error bars. Further experiments are planned for T5 using a high-area-ratio conical nozzle to achieve lower density (and fully laminar) test conditions of reasonable free-stream uniformity and to increase the number of data points for correlating separation length.

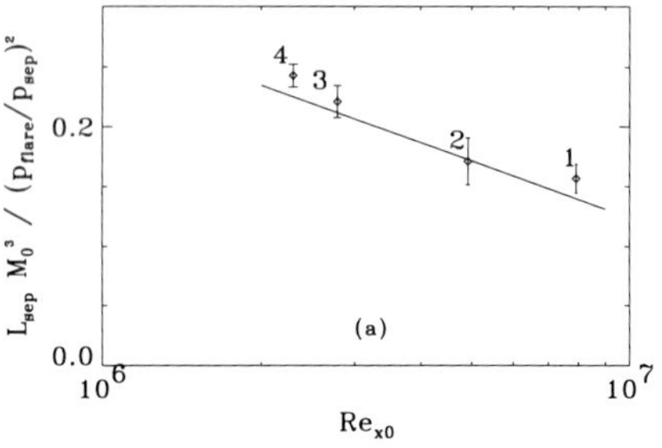

(a)

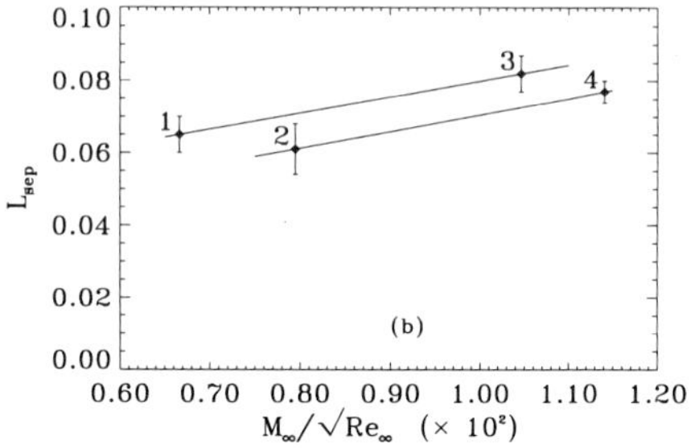

(b)

Figure 5. Two correlations for separation length: (a) empirical correlation due to Needham, and (b) against interaction parameter $\bar{V}_\infty$. Labels indicate test condition.

References

Davis J (1994) Tests on the HALIS Axisymmetric Configuration in the T5 Hypervelocity Shock Tunnel, GALCIT Report FM 94-4, Pasadena, CA.

Germain P (1994) The Boundary Layer on a Sharp Cone in High-Enthalpy Flow, PhD Thesis, California Institute of Technology, Pasadena, CA.

Hornung HG (1992) Performance Data of the New Free-Piston Shock Tunnel at GALCIT, AIAA Paper 92-3943.

Krek RM, Eitelberg G, Beck WH (1996) Hyperboloid flare experiments in the HEG. In: Shock Waves, Proc of the 20th ISSW, Sturtevant B, Shepherd JE, Hornung HG (eds), World Scientific, Vol 1, pp 185–190.

McIntosh MK (1970) A Computer Program for the Numerical Calculation of Equilibrium and Perfect Gas Conditions in Shock Tunnels, Tech Note CPD 169, Australian Defence Scientific Service, Salisbury, South Australia.

Muylaert J, Walpot L, Hironde N, Tumino G, Simeonides G, Koppenwallner G, Davis J (1996) Computational analysis of experiments in T5 and HHK for the study of flap induced separation on an axisymmetric configuration. In: Shock Waves, Proc of the 20th ISSW, Sturtevant B, Shepherd JE, Hornung HG (eds), World Scientific, Vol 1, pp 203–208.

Needham DA, Stollery JL (1966) Boundary Layer Separation in Hypersonic Flow, AIAA Paper 66-455.

Rein M (1989) Surf: A Program for Calculating Inviscid Supersonic Reacting Flows in Nozzles, GALCIT Report FM 89-1.

Rousset B (1995) Calibration and Study of the Contoured Nozzle of the T5 Free-Piston Hypervelocity Shock Tunnel, Engineer Degree Thesis, California Institute of Technology, Pasadena, CA.

Schultz DL, Jones TV (1973) Heat Transfer Measurements in Short Duration Facilities, Technical Report 165, AGARD.

Shock-tube measurements for aero-thermodynamic gas-wall interaction

J. Maul
Institut für Strömungslehre und Strömungsmaschinen
Universität Karlsruhe (TH), Kaiserstrasse 12
D - 76128 Karlsruhe, Germany

Abstract: The motivation for this paper about aerothermodynamic gas-wall interaction comes from the re-entry of a capsule. During the re-entry, the flow is in a thermochemical non-equilibrium state in the higher atmosphere. As a result, the surface catalycity becomes important for the total heat flux into the capsule's heat shield. Due to the surface recombination process an additional heat flux occurs. To determine the influence of the surface catalycity, experiments were performed with oxygen in a shock tube using flow conditions which differed from re-entry conditions. Heat flux and the partial density distribution of atomic and molecular oxygen within the boundary layer of a flat plate coated with aluminium-oxide (Al_2O_3) and silicium-dioxide (SiO_2) were measured. A newly developed two-wavelength laser differential interferometer was applied to measure the partial density distribution. With these investigations, the parameters of a gas-wall interaction model, the atom recombination coefficient γ'_O, the recombination velocity k_{WO}, and the energy accommodation coefficient β_{Oi} were determined.

Key words: Gas-wall interaction, Optical measuring technique, Shock tube, Experiment

1. Introduction

Catalytic recombination is a complex process which includes the absorption of atoms, the chemical reaction at the wall, and the desorption of molecules. Because catalytic recombination cannot be calculated in detail with numerical programs, it is described by a simplified model (Oertel et al. 1994). A common model used in numerical programs is explained by Scott (1990). It consists of three parameters. The first parameter is the atom recombination coefficient γ'_i, which is the ratio of the mass flux of the recombined atoms, $\dot{m}_{i,recomb}$, divided by the mass flux of the atoms hitting the surface, $\dot{m}_i$. The second parameter is the energy accommodation coefficient β_i, which describes the amount of the available energy of formation transformed in an additional heat flux into the structure. The third parameter is the recombination velocity k_{Wi}. The equations for these parameters are:

$$\gamma_i' = \frac{\dot{m}_{recomb}}{\dot{m}_i} \tag{1}$$

$$\beta_i = \frac{\dot{q}_{recomb}}{\dot{m}_{recomb} h_{Di}} \tag{2}$$

$$k_{Wi} = \sqrt{2\pi T_{tr}/k M_i} \tag{3}$$

where $\dot{q}_{recomb}$ is the heat flux due to the catalytic recombination, h_{Di} is the energy of formation, and i specifies species i. T_{tr} is the translational temperature at the wall, k is the Stefan-Boltzmann constant, and M_i is the mass of species i.

To determine these parameters the following terms have to be measured (Scott 1990): the density ρ, the mass fraction c_i, the partial density ρ_i of the species i at the wall, the gradient of the mass fraction of the species i $\partial c_i/\partial z$ at the wall, and the heat flux due to the catalytic recombination $\dot{q}_{recomb}$.

2. Experiments

The experiments were performed in the new shock tube at the Institute for Fluid Mechanics of the Technical University of Braunschweig (Maul and Oertel 1995). This shock tube has a total length of 20 m, an inner diameter of $d_i = 270$ mm, and a test section with a squared cross section of $A = 170 \times 170$ mm^2. In this facility $M_S \approx 11$ can be reached.

A flat-plate boundary layer was investigated. The test gas was oxygen since numerical sensitivity studies have shown that the heat conduction into the structure of a re-entry capsule reacts sensitively to the dissociation of oxygen (Holthoff 1995). Two surface materials were tested, Al_2O_3 and SiO_2. Heat fluxes for different flow conditions from $M_S = 7$ to $M_S = 11$ have been measured. The partial density distribution of atomic and molecular oxygen within the boundary layer of the flat plate has been measured for the following flow conditions: $M_S = 10$, $p_0 = 0.4$ mbar, $M_1 = 3.0$, and $T_1 = 2900$ K (Lieseberg 1994, Maul 1994).

2.1. Two-wavelength laser differential interferometer

To determine the partial density distribution of atomic and molecular oxygen, a two-wavelength laser differential interferometer which works with two different wavelengths simultaneously (Lieseberg 1994) has been developed. The principle is based on laser differential interferometry. It uses the relationship between the density ρ and the refraction index n, the Gladstone-Dale relation, and the assumption that for a gas mixture the Gladstone-Dale constant $G(\lambda)$ consists of the sum of the Gladstone-Dale constant of the single species i weighted with the mass fractions of species i:

$$n - 1 = \sum_i \frac{\rho_i}{\rho} \cdot G_i \cdot \rho \tag{4}$$

Equation 4 can also be regarded as the difference of the refraction index Δn between the light beam 1 and 2 on the basis of a integral density difference $\Delta\rho$:

$$\Delta n_{12} = G(\lambda) \cdot \Delta\rho \tag{5}$$

For two different wavelengths, λ_1 and λ_2, the following systems of equations for the unknown partial densities can be developed:

$$\underbrace{\begin{pmatrix} \Delta n_{12}(\lambda_1) \\ \Delta n_{12}(\lambda_2) \end{pmatrix}}_{\text{measured}} = \underbrace{\begin{pmatrix} G_{O_2}(\lambda_1) & G_O(\lambda_1) \\ G_{O_2}(\lambda_2) & G_O(\lambda_2) \end{pmatrix}}_{\text{given}} \cdot \underbrace{\begin{pmatrix} \Delta\rho_{O_2} \\ \Delta\rho_O \end{pmatrix}}_{\text{unknown}} \tag{6}$$

This system is explicitly solvable when the determinant $||G|| \neq 0$. To guarantee this, one wavelength has to operate in the vicinity of an absorption line of the investigated gas, where $G(\lambda)$ changes drastically. With $\lambda = 760.8$ nm, a specific wavelength for molecular oxygen exists. The choice of the second wavelength is uncritical. For these experiments $\lambda_2 = 632.8$ nm was chosen.

2.2. Thin-film thermometer

Thin-film thermometers (Jessen and Grönig 1992) were used for the determination of the heat flux. They were coated to a thickness of 100 nm with the investigated materials. An estimation shows that the temporal signal is changed only slightly by the coating.

For these flow conditions, the recombination process changes the temperature profile unessentially. Because SiO_2 has non-catalytic behavior, the heat flux due to the catalytic recombination can be obtained. It is the difference between the heat fluxes for the two materials.

3. Experimental results

One necessity for the evaluation of the oxygen partial densities is the knowledge of the Gladstone-Dale constants. Experiments have been performed to determine the Gladstone-Dale constants for the two wavelengths. The results are shown in Fig. 1.

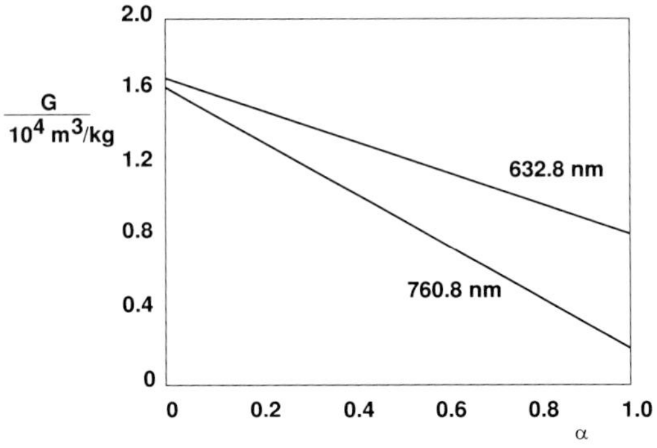

Figure 1. Gladstone-Dale constants for $\lambda_1 = 760.8$ nm and $\lambda_2 = 632.8$ nm.

The diagram shows $G(\lambda)$ plotted against the degree of dissociation α. Obviously the curves have different gradients. Therefore, Eq. (6) has a non-ambigous solution and the partial densities can be determined.

The partial densities for atomic and molecular oxygen have been measured for differences between $z = 0.4$ mm and $z = 2.37$ mm distant from the surface. Figure 2 shows the atomic mass fraction of oxygen c_O for both materials plotted against the dimensionless boundary layer coordinate z/δ.

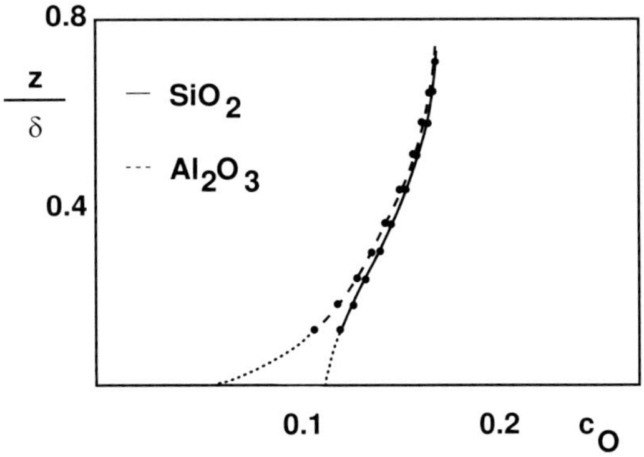

Figure 2. Atomic mass fraction for SiO_2 and Al_2O_3.

For a height above $z/\delta \approx 0.5$ the curves are nearly identical; the surface material has no influence. Below $z/\delta \approx 0.5$, the difference grows. Due to the finite extension of the laser beam, the last measuring point was $z = 0.4$ mm above the surface. To determine c_O and $\partial c_O/\partial z$ at the wall the partial densities have been extrapolated to the wall. The extrapolation is dotted in

Fig. 2. In spite of a detailed inspection, the extrapolation represents the greatest experimental error in the determination of the gas-wall interaction parameters. The error for these investigations is 20%.

Figure 3 shows the heat fluxes for both materials plotted against M_S. For $M_S \leq 7$ the difference between the two curves vanishes because no recombinable oxygen exists in the flow behind the shock. With increasing M_S the difference between the two investigated materials grows due to the increasing degree of dissociation α_O. For $M_S = 10$ the difference is approximately 20%. In Table 1 the experimental results and the evaluated parameters are summarized.

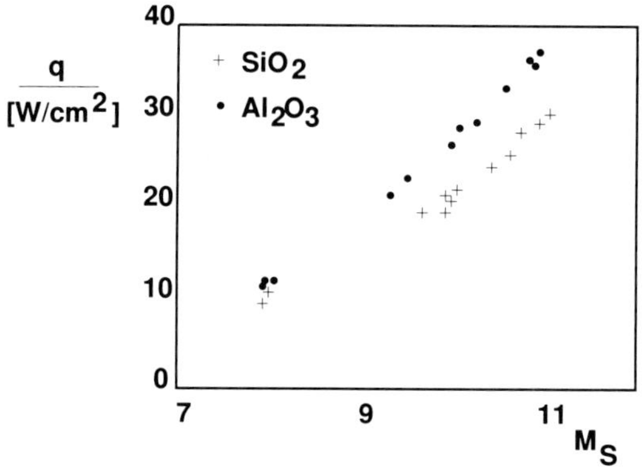

Figure 3. Heat fluxes for SiO_2 and Al_2O_3.

Table 1. Experimental results and evaluated parameters.

	c_O	$\partial c_O/\partial z$ $\left[\frac{1}{mm}\right]$	γ'_O	k_{WO} $\left[\frac{cm}{s}\right]$	β_O
SiO_2	0.117	0.01	0.0003	5.0	—
Al_2O_3	0.067	0.15	0.009	139.0	0.78

The results show that SiO_2 behaves non-catalytically. In comparison with values known from the literature, γ'_O and k_{WO} agree within the measuring error. Al_2O_3 behaves partly catalytically, but γ'_O and k_{WO} obtained by these experiments are greater than literature values. But the advantages of these experiments were the exact knowledge of the flow conditions and that no assumptions for the evaluation of the experimental data needed to be made. Contrary to previous assumptions, the experimental results show that Al_2O_3 does not emit the entire available energy of formation during the recombination. That means the recombined molecules leave the surface either in an excited state or with a higher velocity than that corresponding to the wall temperature. To determine β_O for SiO_2, numerical calculations for the experimental conditions have to be carried out. By a comparison between experiment and numerical results β_O can be obtained.

4. Summary

Experiments have been carried out to determine the parameters of a model describing the interaction of a hot dissociated gas with a catalytic (Al_2O_3) and non-catalytic (SiO_2) surface. A two-wavelength laser differential interferometer has been developed and applied to measure the partial density distribution of atomic and molecular oxygen in a boundary layer. The results show that Al_2O_3 behaves partly catalytically while SiO_2 is non-catalytic. Furthermore, Al_2O_3 does not emit the entire available energy of formation during the recombination. The molecules leave the surface either in an excited state or with a higher velocity than that corresponding to the wall temperature.

Acknowledgement. This research was supported by the Deutsche Forschungsgemeinschaft, Germany. The author wants to thank Dr. A. Lieseberg, who has carried out the optical experiments.

References

Holthoff H (1995) Numerische Simulation des Wärmeübergangs in Hyperschallströmungen, ZLR-Bericht der TU Braunschweig.

Jessen C, Grönig H (1992) Sensoren zur Messung von Wärmestrom und Wandschubspannung innerhalb von Millisekunden, Stosswellenlabor, RWTH Aachen.

Lieseberg A (1994) Interferometrische Messung der aerothermodynamischen Wand-Wechselwirkung an einer längsangeströmten Platte, ZLR-Bericht der TU Braunschweig.

Maul J (1994) Stossrohrmessungen zur aerothermodynamischen Wand-Wechselwirkung, ZLR-Bericht 94-09, TU Braunschweig.

Maul J, Oertel H (1995) Stossrohrmessungen zur aerothermodynamischen Wand-Wechselwirkung (revised manuscript submitted to ZFW).

Oertel H, Böhle M, Delfs J, Hafermann D, Holthoff H (1994) Aerothermodynamik, Springer-Verlag.

Scott CD (1990) Wall Catalytic Recombination and Boundary Conditions in Nonequilibrium Hypersonic Flows — With Application, Third Joint Europe/US Short Course in Hypersonics, RWTH Aachen.

Computational analysis of experiments in T5 and HHK for the study of flap-induced separation on an axisymmetric configuration

J. Muylaert, L. Walpot, N. Hironde, G. Tumino, G. Simeonides
Aerothermodynamic Section YPA, ESA-ESTEC
P.O. Box 299, NL-2200 AG, Noordwijk, The Netherlands
G. Koppenwallner
Hypershall Technology Göttingen
Max Planck Strasse 1, D-37191 Kattlenburg Lindau, Germany
J. Davis
Graduate Aeronautical Laboratories
California Institute of Technology, Pasadena, California, 91125 USA

Abstract: Experiments on an axisymmetric configuration to investigate the influence of real-gas and viscous interaction effects on flap-induced separation and heating were carried out in T5 and the HHK. Nonequilibrium Navier-Stokes computations were performed in conditions including nozzle-flow nonuniformity. For T5, the influence of the pitot-pressure nonuniformity lays within the experimental error. Experiments performed in the perfect-gas Mach 9 facility HHK, with the same viscous interaction parameter as for T5, show a larger separation bubble. For all the T5 experiments, transition to a turbulent boundary layer occurred at the point where the separated shear layer reattaches on the flare. This was confirmed by comparing the experimental heating levels on the flare with Navier-Stokes computations and simpler boundary-layer methods assuming that the flow transitions to turbulence at reattachment.

Key words: Shock tunnel, Shock/boundary-layer interaction, Real-gas effects, Nonequilibrium Navier-Stokes, T5, HHK

1. Introduction

Flap efficiency and associated peak-heating assessment for stability and trimming is a critical design point for space transportation systems. Weilmuenster et al. (1993) showed that the space shuttle STS1 pitching-moment anomaly was due to real-gas effects. However, a detailed systematic analysis of real-gas and viscous interaction effects on flap efficiency and heating has not been done. It is the purpose of the present work to analyze these effects on an "affordable" axisymmetric geometry which incorporates all relevant physical effects associated with the shuttle geometry i.e., the entropy layer from the nose at incidence, the expansion on the windward side upstream of the body-flap hinge, and the compression due to flap deflection. We refer to this configuration as the Halis Axisymmetric Configuration (HAC).

2. Model design and experimental campaigns

The axisymmetric configuration was generated by the windward centerline of the Orbiter at high incidence with body flap deflected. The incidence and flare angles were designed to maximize the laminar separation extent to obtain the largest resolution. Different configurations were tested in the Ludwieg tube HHK in Lindau, Germany. Incidence was varied from 30° to 40° and flap angle from 10° to 20°. The aim was to freeze a configuration which is free from base effects and which produces a steady shock/boundary-layer interaction. The frozen configuration has an equivalent incidence of 35° and a flare angle of 20°. The T5 HAC model was designed and instrumented at Caltech; experiments were performed by Davis et al. (1996). The stainless steel T5 model with a 17-cm base diameter holds 16 pressure sensors and 16 thermocouples. The HHK model contains 20 pressure sensors and is exactly half the size of the T5 model. The T5 and HHK conditions are shown in Table 1 with reference length $L = 0.1$ m for T5 and 0.05 m for HHK, where $\chi_L = M_\infty/\sqrt{Re_L}$.

Table 1. Free-stream conditions.

Test Cond.	Run No.	P_0 (MPa)	H_0 ($\frac{\text{MJ}}{\text{kg}}$)	ρ_∞ ($\frac{\text{kg}}{\text{m}^3}$)	u_∞ ($\frac{\text{m}}{\text{s}}$)	T_∞ (K)	Tv_∞ (K)	M_∞	Re_L ($\times 10^5$)	χ_L
T5 1	662	58.5	10.8	0.049	4216	1185	1769	6.0	4.5	0.009
T5 2	730	59.3	19.1	0.026	5433	2076	2342	5.6	2.2	0.012
T5 3	723	22.7	10.8	0.017	4170	969	2161	6.5	1.8	0.015
T5 4	719	27.2	17.2	0.013	5143	1528	2580	6.1	1.3	0.017
HHK	–	4.26	–*	0.014	1105	37	–	9.06	4.3	0.015

*T_0=645K

3. Computational analysis of experiments

3.1. Codes and free-stream conditions

To have the proper nonequilibrium free-stream data for T5, the LORE 1D code (nonequilibrium multi-temperature, with Edenfield boundary-layer correction) was used (Walpot et al. 1996). The computed free-stream conditions were obtained in such a way that the computed pitot pressures match the measured ones. The matching was done by reducing accordingly the displacement thickness in Edenfield. Table 1 shows the free-stream conditions; they correspond well with those computed by Davis et al. (1996) using the SURF code. The 2D nonequilibrium computations on the HAC geometry were done with the TINA code developed by Netterfield (1992). For the HHK, the perfect-gas code Euranus was used.

3.2. Influence of nozzle nonuniformity

First the influence of the nonuniformity of the pitot-pressure distribution at nozzle exit was assessed. A series of quasi-1D computations were performed along equivalent streamlines, all having the same reservoir pressure and enthalpy; pitot pressures at the exit corresponded to those obtained by measurements at the respective radial positions. These exit conditions were used as 2D inlet boundary conditions for the laminar HAC computations. The influence of the nonuniform pitot was verified for all 4 conditions. Figure 1 shows the extent of pitot-pressure nonuniformity compared to HAC model size. Figure 2 shows the influence of the pitot nonuniformity on pressure distribution. For this configuration the influence of pitot nonuniformity lays within experimental error.

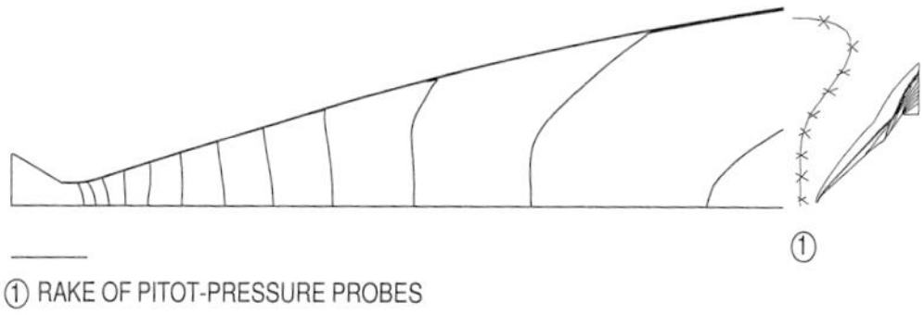

Figure 1. Numerical reconstruction of the HAC model installed in the T5 nozzle.

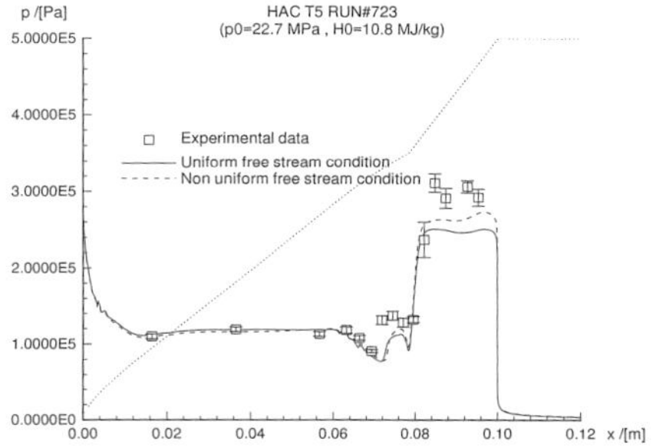

Figure 2. Influence of pitot-pressure nonuniformity.

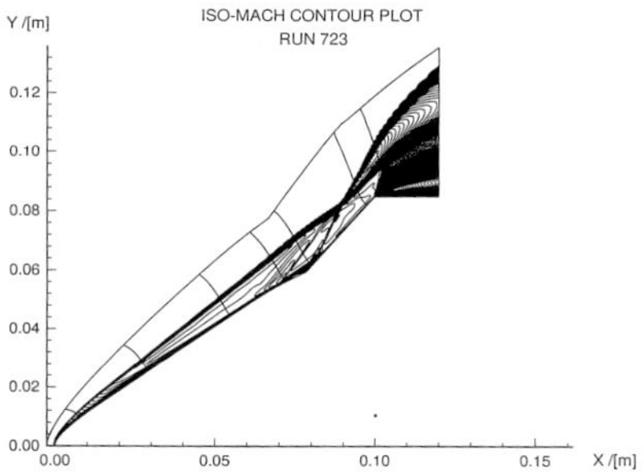

Figure 3. Iso-mach contour for 723.

3.3. Flap-induced separation and reattachment

Figure 3 shows a typical iso-mach plot where all features of the interaction are clearly visible. When comparing the pressure and heat-flux data in Figs. 4 and 5 we notice excellent agreement between experiments and computations on the attached forward part. The computations show only a small influence of catalycity. Large discrepancies occur between experimental and computed heat fluxes on the flare region. It was postulated that the flow at the reattachment point

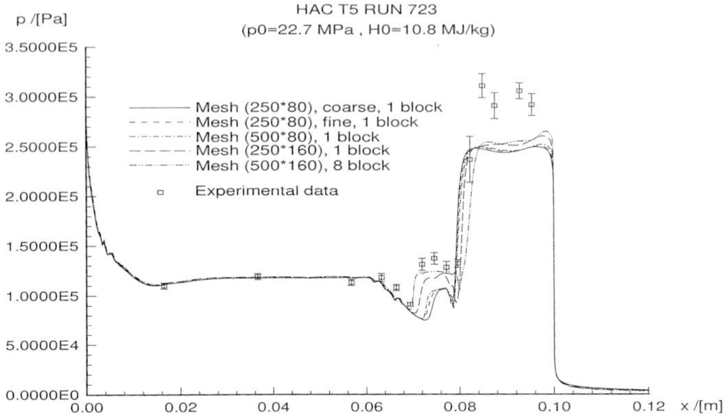

Figure 4. Pressure distribution: TINA results with various computational meshes.

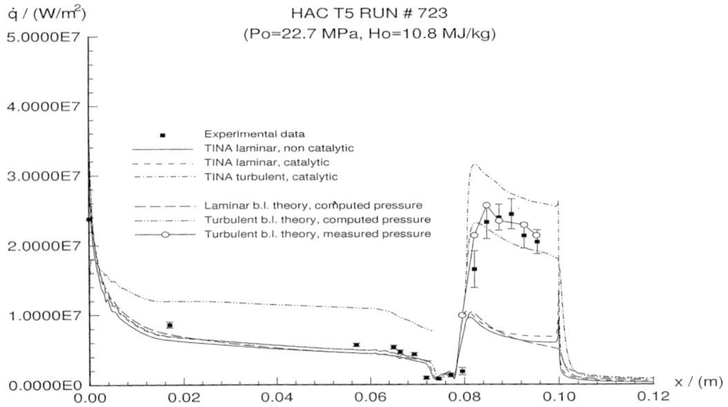

Figure 5. Heat flux distribution: TINA vs. boundary-layer method.

was turbulent; different ways to prove this were considered. First, TINA computations were carried out where turbulence, using the Baldwin-Lomax model, was switched on at reattachment. Second, a boundary-layer method was applied where the computed boundary-layer edge pressure and velocity served as input for the computation of the laminar or turbulent heating. Note that the figure shows an even better match when the measured pressures on the flare are taken instead of the computed ones. And third, the measured and computed peak heating were compared to a correlation derived from earlier work by Simeonides (1995). Figure 6 shows this correlation when accounting for the conicity of the flow on the forebody and on the flare. It shows that for T5, the peak-heating ratio corresponds to a critical Re* in the vicinity of the intersection between laminar and turbulent regimes. The critical Re* is the Re at the location of the peak heating using Eckert's reference enthalpy and free-stream pressure at equilibrium conditions. In Fig. 8 the separation length

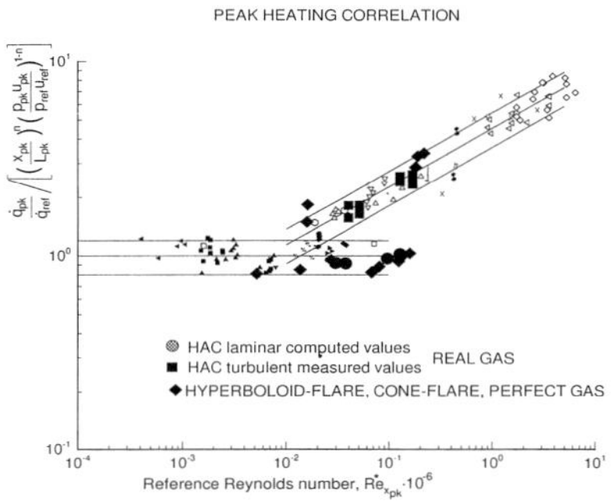

Figure 6. Peak-heating correlation from ramp flows: comparison with transformed HAC data.

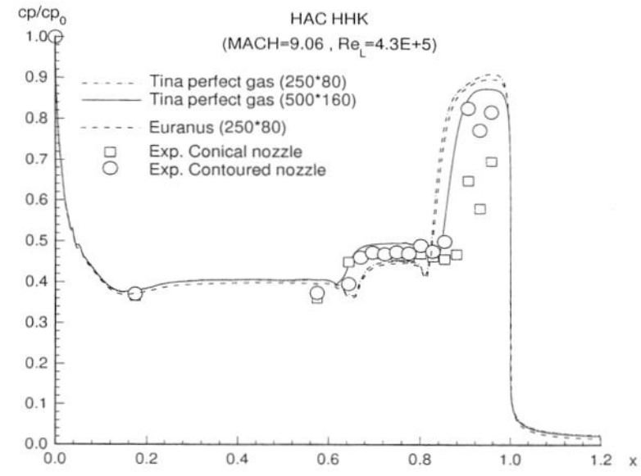

Figure 7. Measured and computed pressure distribution on the HHK HAC model.

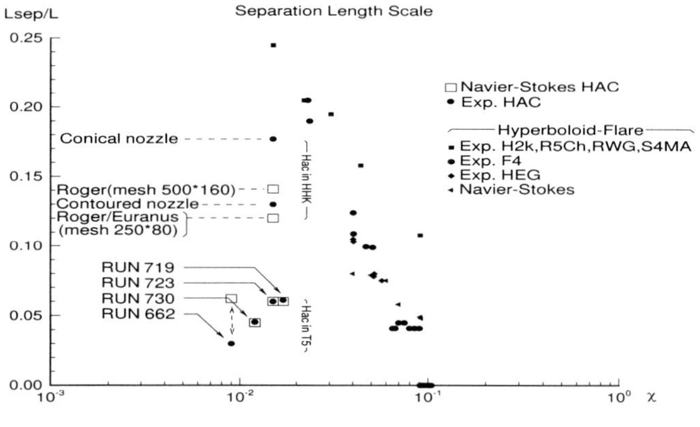

Figure 8. Separation length vs. viscous interaction parameter.

upstream of the hinge-line is plotted as a function of the viscous interaction parameter for the HAC model and compared with hyperboloid-flare data from Sagnier (1994). The trend of the computations with Re follows that of similar hyperboloid-flare laminar data, but the experimental data follow a different trend, confirming again the fact that the flow became turbulent at reattachment. To obtain the influence of real gas at a constant viscous interaction parameter, HHK tests were carried out with reservoir conditions such that for both HHK and T5 the viscous interaction parameter is 0.015. The HAC geometry is much more sensitive to real-gas effects than the hyperboloid flare due to the increased nose bluntness and the expansion upstream of the flare. Figure 7 shows the HHK pressure distribution for conditions as shown in Table 1 for the conical and contoured nozzles and a comparison with the perfect-gas Euranus and TINA results using a 250×80 grid. The perfect-gas computations nearly match the contoured-nozzle data, however, the latter differ considerably from the conical experimental data. T5 HAC experiments show a separation-length reduction with increasing enthalpy. More grid refinement studies are needed for the nonequilibrium laminar computations to detect a similar reduction in separation length.

4. Conclusion and recommendation

An experimental campaign took place in T5 and in the HHK to study the influence of real-gas and viscous interaction effects on flap efficiency and heating. An axisymmetric configuration was designed to enhance the sensitivity to real-gas effects and ease computational analysis. Transition to turbulence occurred at reattachment, making the analysis much more complicated than initially envisioned. It is recommended that the analysis be completed by studying reference laminar interactions at high enthalpy to understand the influence of real-gas effects at constant viscous interaction without the additional complexity of having transition at reattachment.

Acknowledgement. The authors wish to acknowledge H. Wong and M. Netterfield for their support in running TINA and EURANUS.

References

Davis J, Sturtevant B, Muylaert J (1996) Experiments on the Halis axisymmetric configuration in the T5 shock tunnel. In: Shock Waves, Proc of the 20th ISSW, Sturtevant B, Shepherd JE, Hornung HG (eds), World Scientific, Vol 1, pp 191–196.

Netterfield M (1992) Validation of a Navier Stokes code for thermo-chemical non equilibrium flows. In: Proc. 27th Thermo-physics Conf., AIAA 92-2878.

Sagnier P (1994) Synthesis of the Hyperboloid Flare Campaign in F4, Onera RT 70/6121 SY December 1994.

Simeonides G, Haase W (1995) Experimental and computational investigations of hypersonic flow about ramp. J Fluid Mech 283(1):17–47.

Walpot L, Muylaert J, Bakker P, Simeonides G (1996) High-enthalpy nozzle flow sensitivity study and effects on heat transfer. In: Shock Waves, Proc of the 20th ISSW, Sturtevant B, Shepherd JE, Hornung HG (eds), World Scientific, Vol 2, pp 1587–1592.

Weilmuenster KJ, Gnoffo PA, Greene FA (1993) Navier Stokes Simulations of the Shuttle Orbiter Aerodynamic Characteristics with Emphasis on Pitch Trim and Body Flap, AIAA 93-2814.

Shock-interference heating in hypervelocity flow

S. R. Sanderson*, B. Sturtevant
Graduate Aeronautical Laboratories
California Institute of Technology
Pasadena, CA 91125 USA

Abstract: The interaction of a weak oblique shock with the strong bow shock ahead of a blunt body in supersonic flow produces extreme heat-transfer rates and surface pressures. Experiments are conducted in a free-piston shock tunnel to determine the effects of thermochemistry on the problem at high enthalpy. The flow topology is simplified by studying the nominally two-dimensional flow about a cylinder with a coplanar impinging shock wave. Increases in shock-interference heating rates due to real-gas effects are observed to be significantly less than previously expected. A model is discussed that describes the thermochemical processes occurring in the interaction region. The model accurately reproduces the experimental observations.

Key words: Shock interference, Shock impingement, Shock-on-shock

1. Introduction

Although the shock-interference heating problem has been studied extensively in low-enthalpy flows (Edney 1968), the influence of high-enthalpy real-gas effects is poorly understood. Edney (1968) observed and classified six distinct interaction regimes that are known as types I–VI. Severe heating occurs for the type-IV flow where the interaction causes the formation of a supersonic jet that impinges on the surface of the body (Fig. 1). By mapping the shock-wave interactions that occur in the supersonic jet into the pressure-flow deflection angle (p-δ) plane, Edney (1968) studied the variation of the pressure at the jet-impingement point for a perfect gas. It was demonstrated that reductions of the ratio of specific heats, γ, at high enthalpy produce greatly increased surface pressures. The jet-impingement pressure also increases strongly with Mach number. Although the ratio of specific heats is only a crude measure of the influence of equilibrium thermochemistry, the magnitude of the effect predicted by Edney (1968) motivated the current experiments. The objective of the present work was to utilize the capabilities of the GALCIT T5 free-piston

---

*Present address: GE Corporate Research & Development, P.O. Box 8, Schenectady, NY 12301.

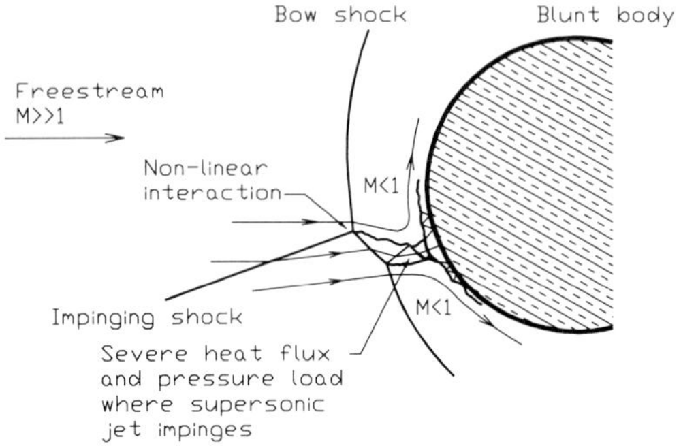

Figure 1. Schematic representation of the type-IV interaction.

shock tunnel to determine the quantitative effects of thermochemistry on the shock-impingement phenomenon.

2. Experimental investigations

The experimental arrangement is shown in Fig. 2. High-temperature and pres-

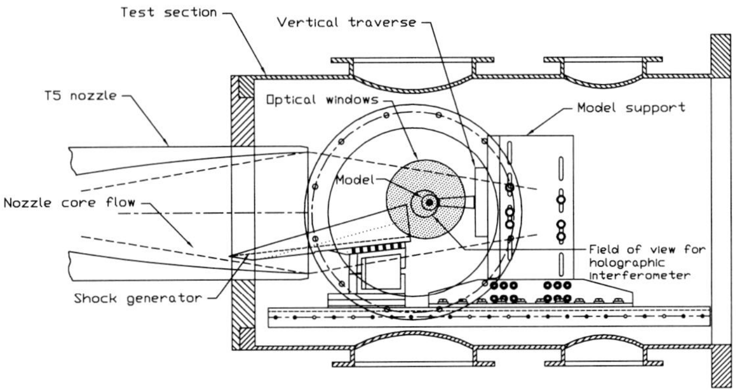

Figure 2. Layout of apparatus in the test section of T5.

sure gas was generated in the nozzle reservoir by the T5 hypervelocity shock tunnel. Nitrogen test gas was used for all experiments (Table 1).

The cylindrical model (ϕ40 mm $\times$ 180 mm) was positioned in the free-jet test section and a large shock generator was partially inserted into the uniform nozzle core flow upstream of the nozzle exit. Holographic interferometry was used to investigate variations in the shock-impingement flow field as the model was vertically translated with respect to the exit of the shock genera-

Table 1. Summary of the free-stream conditions.

Test condition		A	B	C
Test section conditions	Velocity (m/s)	2540	4450	5350
	Density (kg/m^3)	0.0218	0.0155	0.0157
	Pressure (kPa)	1.03	5.48	11.4
	N concentration (mole/kg)	9.9×10^{-5}	6.9×10^{-1}	3.65×10^{0}
	M_∞	9.9	6.3	5.3
	Re per mm	5500	1600	1350

tor. Surface-junction thermocouple gauges provided time-resolved heat-transfer measurements at 24 circumferential locations around the model forebody. Figure 3 shows a holographic interferogram of a type-IV interaction at the high-

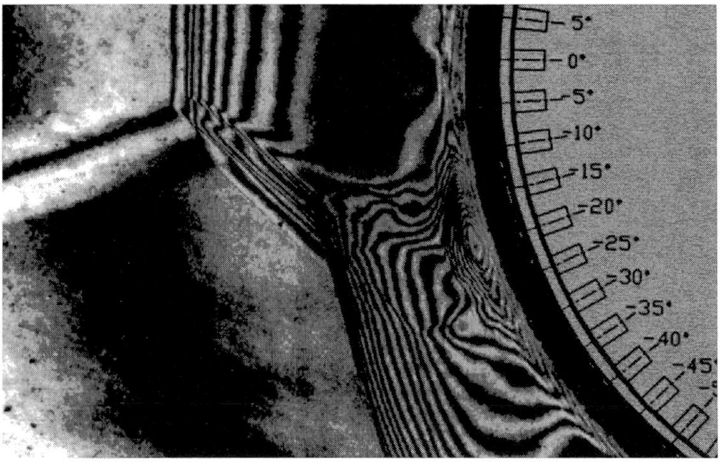

Figure 3. Holographic interferogram of type-IV interaction; condition C.

enthalpy condition, C. Two mutually inverted λ-shocks are observed along with a terminating strong shock and the region of elevated density behind it. The peak heat fluxes recorded during the entire sequence of shots at each test condition are listed in Table 2. The high bandwidth of the heat-flux gauges used for these experiments permitted examination of the temporal characteristics of the heat-transfer signals. Sanderson (1995) discusses observations of coherent unsteady fluctuations of the type-IV jet that are consistent with oscillations observed in recent numerical simulations (Gaitonde 1993, Zhong 1994). Additional data are reported in detail by Sanderson (1995).

Table 2. Measured and predicted peak heat-transfer intensification for type-IV flows.

Test condition		A	B	C
Peak measured $\frac{\dot{q}_{jet}}{\dot{q}_{body}}$	Instantaneous	15.2	10.8	7.9
	100 μs averaged	13.5	7.6	6.0
Predicted $\frac{\dot{q}_{jet}}{\dot{q}_{body}}$	Frozen	14.1	7.2	5.1
	Nonequilibrium	14.1	6.7	4.4
	Equilibrium	14.1	7.4	7.2

The conclusion of the experimental investigation was that the stagnation density and heat-flux intensification decreased at high enthalpy. Care must be exercised in the interpretation of this trend since the Mach number of the flow in T5 decreased as the stagnation enthalpy was increased (Table 1). The net result was that further increases in the large heat fluxes produced by shock impingement, due to real-gas effects, were observed to be much lower than those predicted by existing models (Section 1).

3. Model of the influence of real-gas effects

Modeling of the influence of real-gas effects followed the pattern of the existing perfect-gas model of the type-IV flow (Edney 1968). Normal shock solutions were obtained using the ideal dissociating gas model of Lighthill (1957). The normal shock solutions were extended to oblique waves and p-δ plane methods were used to obtain locally valid solutions for the λ-shock points. These inviscid shock-interaction solutions determine the state of the gas in the type-IV jet. Heating due to impingement of the type-IV jet was modeled using the equilibrium reacting-flow similarity solution of Fay and Riddell (1958). Sanderson (1995) shows that the heat-transfer intensification at the jet-impingement point has the following functional dependence:

$$\frac{\dot{q}_{jet}}{\dot{q}_{body}}\sqrt{\frac{\Delta x}{D}\frac{\zeta^2_{1_{body}}\zeta_{2_{body}}}{\zeta^2_{1_{jet}}\zeta_{2_{jet}}}} = \mathrm{func}(P_\infty, H_{0_\infty}, \hat{\rho}_d, \alpha_\infty, \beta_1). \tag{1}$$

The parameters are $P_\infty = p_\infty/\rho_\infty u^2_\infty \sim 1/M^2_\infty$, H_{0_∞} is the stagnation enthalpy normalized with respect to the dissociation energy of the gas, $\hat{\rho}_d$ is the equilibrium constant, α_∞ is the dissociated mass fraction of the free stream, and β_1 is the angle of the impinging wave. The geometrical terms on the left-hand side specialize the result for a particular configuration. The similarity coefficients, ζ, are determined from the work of Fay and Riddell (1958) and the Newtonian solution for the velocity gradient at the outer edge of the boundary layer. The jet-width-to-body-diameter ratio, $\Delta x/D$, is empirically scaled with the undisturbed shock standoff distance, χ/D, and the divergence angle of the oblique wave and shear layer at the shock-impingement point, δ;

$$\frac{\Delta x}{D} = \frac{\chi}{D} \sin \delta. \tag{2}$$

All of the quantities appearing in Eqs. (1) and (2) may be predicted a priori.

The behavior of the local solution (1) with stagnation enthalpy, H_{0_∞}, is illustrated in Fig. 4 for typical values of the remaining parameters. Three solutions are obtained using different thermodynamic models to demonstrate the influence of nonequilibrium thermochemistry. The dotted lines represent chemically frozen solutions that apply when the reaction rates are slow with respect to the rate of flow over the body. The dash-dot curves are equilibrium solutions that apply when the reaction rates are fast and are obtained using the Lighthill (1957) model for the shock-jump computations. The dashed curves represent an approximate nonequilibrium solution that is motivated by the disparate shock strengths produced at the λ-points. The weak oblique waves in the supersonic jet are assumed to remain frozen whilst the strong bow shock and terminating jet shock relax to equilibrium. The validity of the nonequilibrium approximation is assessed by computing the reaction rates (Freeman 1958) immediately downstream of the translational discontinuities. Figure 4 shows that equilibrium real-gas effects increase the heat-transfer intensification with respect to the frozen solution. This occurs because of the increased density ratio across the weak oblique jet shocks due to dissociation. Dissociation across the jet shock waves is precluded in the nonequilibrium approximation and the heat-transfer intensification decreases slightly with respect to the frozen solution. Within the scope of the approximation, non-monotonic behavior with the reaction rate is indicated. The significance of real-gas effects declines at higher Mach numbers because the oblique jet shocks become weaker relative to the bow shock and terminating jet shock. This causes reduced dissociation levels and reaction rates in the supersonic jet. Heating rates initially increase with impinging shock strength before reaching a maximum for modest-strength waves. Further analysis (Sanderson 1995) demonstrates that the relative influence of real-gas effects is minimized at this maximum heating condition.

Specific predictions of the model for the test conditions listed in Table 1 are compared with the experimental data in Table 2. The extent of chemical reactions at the low-enthalpy condition A was negligible and so only the frozen solution is shown. Reaction rates for condition C were sufficiently high that departures from equilibrium were expected to be slight. The calculated values accurately reproduce the experimental heat-transfer measurements.

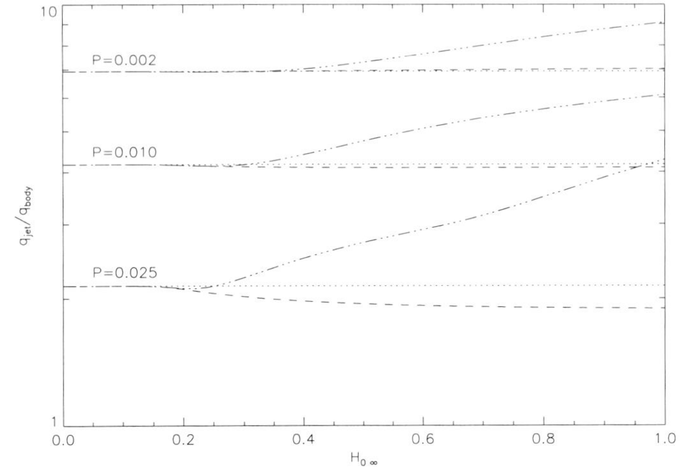

Figure 4. Influence of dimensionless stagnation enthalpy, H_{0_∞}, Mach number, $\sim 1/P_\infty^{0.5}$, and reaction rate on type-IV heat-transfer intensification, $\frac{\dot{q}_{jet}}{\dot{q}_{body}}(\Delta x \zeta_{1_{body}}^2 \zeta_{2_{body}} / D \zeta_{1_{jet}}^2 \zeta_{2_{jet}})^{0.5}$. The parameters are $\hat{\rho}_d = 1 \times 10^7$, $\alpha_\infty = 0$, and $\beta_1 = 16°$.

Acknowledgement. This work was supported by the Air Force Office of Scientific Research under Grant Nos. F 49620-92-J-0110 and F 49620-93-1-0338.

References

Edney BE (1968) Anomalous Heat Transfer and Pressure Distributions on Blunt Bodies at Hypersonic Speeds in the Presence of an Impinging Shock, FFA Report 115.

Fay JA, Riddell FR (1958) Theory of stagnation point heat transfer in dissociated air. Journal of the Aeronautical Sciences 25:73.

Freeman NC (1958) Non-equilibrium flow of an ideal dissociating gas. J Fl Mech 4:407.

Gaitonde D (1993) Calculations on Unsteady Type IV Interaction at Mach 8, Wright Laboratory, WL-TR-93-3002.

Lighthill MJ (1957) Dynamics of a dissociating gas. Part I. Equilibrium flow. J Fl Mech 2:1.

Sanderson SR (1995) Shock Wave Interaction in Hypervelocity Flow, Ph.D. Thesis, California Institute of Technology.

Zhong X (1994) Application of essentially non-oscillatory schemes to unsteady hypersonic shock-shock interference heating problems. AIAA J 32:1606.

Investigation of shock-shock interactions in hypersonic reentry flows

S. Brück
DLR Institute of Fluid Mechanics
Bunsenstraße 10
D-37073 Göttingen, Germany

Abstract: Numerical simulations of shock-shock interactions in front of a circular cylinder have been performed. The numerical results of a validation test case are compared with experimental data. Models of a perfect gas and a mixture of chemically reacting perfect gases have been applied to reentry flow conditions. Differences of the results of the two models are demonstrated in view of the calculated shock configurations and the peak loads on the surface.

Key words: Shock-shock interactions, Viscous nonequilibrium flows, TVD method, Cylinder flow

1. Introduction

The understanding of the physics of complex flow phenomena in hypersonic flow is important, because the phenomena may result in strong local pressure and temperature and/or heat-flux peaks on the considered body surface. One example of these phenomena is the shock-shock interaction. The physics of such flow phenomena was experimentally investigated in a two-dimensional supersonic model flow by Edney (1968). A weak shock generated by means of a wedge interacts with the bow shock in front of a cylinder. Based on these experiments, six different types of shock-shock interactions were classified for cold hypersonic flows, depending on the position of the interaction location and the strength of the impinging shock wave. In chemically reacting flows, the typical shock configuration of the various interaction types may change due to dissociation effects. Therefore, it is necessary to investigate such flows to predict the influence of shock-shock interactions at free-flight conditions.

2. Numerical method

The numerical code NSHYP (Brenner 1994) used for the investigation of the flow phenomena is based on the solution of the time-dependent thin-layer Navier-Stokes equations. The numerical algorithm employs an implicit finite-difference scheme which is solved by using a line Gauß-Seidel relaxation method. The spatial order of the evaluation of the inviscid fluxes is formally second order and based on a TVD upwind discretization and a non-MUSCL approach. For the simulation of flows in chemical nonequilibrium, a mixture of monatomic and diatomic electrically neutral gases is considered. It is assumed that the density of the gas is sufficiently low. Thus, each component can be regarded as a perfect gas. The model under consideration includes the conservation of the partial densities of the chemical components N_2, O_2, NO, N, and O for air, the momentum in the three spatial coordinates, the total energy per volume, and the vibrational energy per volume of the diatomic species NO excluded. The rates of the chemical reactions used here as well as the modeling of the vibrational relaxation can be found in Park (1989). For the nonequilibrium calculations, a fully catalytic wall is assumed.

3. Validation

At first, numerical investigations should be compared with standard wind-tunnel tests before extrapolating to free-flight conditions. A test case of shock-shock interactions is considered in order to validate the numerical code for defined conditions. The two-dimensional flow field in front of a circular cylinder is calculated where an impinging shock is present at the inflow boundary of the computational domain at a defined position. The interaction with the detached bow shock produces a complex flow field which requires a good local resolution. The test case was experimentally investigated by Wieting & Holden (1987) in a cold wind tunnel ($Ma_\infty = 8.03$, $T_\infty = 122$ K) and is widely used for comparative studies in the literature. For the presented test case where only perfect

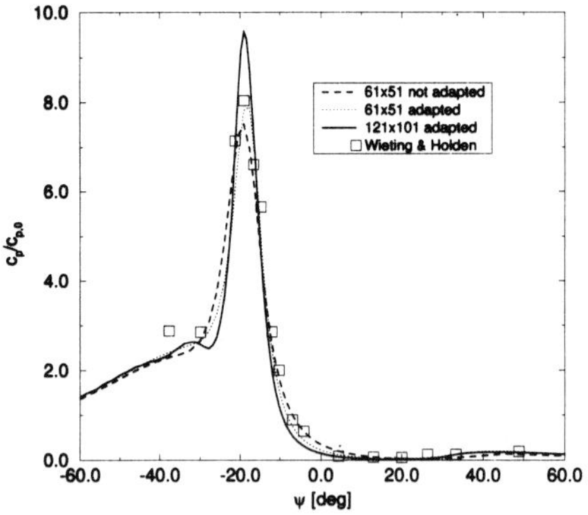

Figure 1. Pressure distribution of several simulations and the experiment, flow conditions of Wieting & Holden (1987).

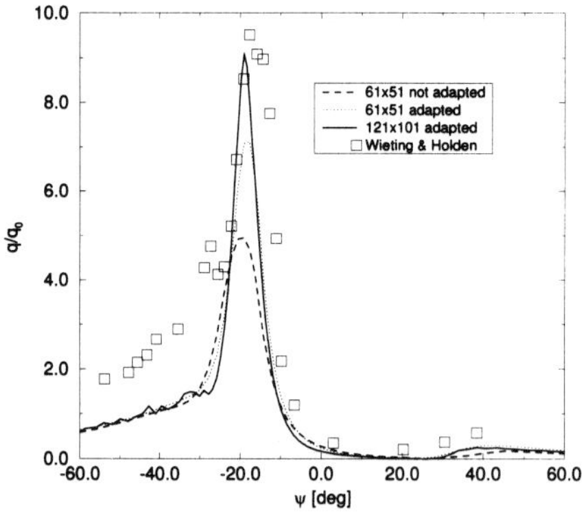

Figure 2. Heat-flux distribution of several simulations and the experiment, flow conditions Wieting & Holden (1987): $q_0 = 681,\ 395\ \text{W/m}^2$.

gas has to be considered, a type-IV interaction is achieved: a supersonic jet impinges on the body surface and produces a high peak in the pressure and heat-flux distribution, as shown in Figs. 1 and 2. The results produced by the finest grid show good agreement with the experimental data. The code has also been applied to similar flow fields including the modeling of chemical and thermal nonequilibrium for pure nitrogen. Results have been presented by Brück (1993) and Hannemann et al. (1993).

4. Reentry-flow simulation

For the flow around a cylinder with the presence of an impinging shock, computations have been performed to investigate the influence of the shock position and of nonequilibrium effects due to dissociation. The free-stream conditions were taken from COESA (1976) and correspond to an altitude of 53.5 km. The free-stream velocity is 4.0 km/s. The incident-shock positions at the inflow boundary of the computational domain are obtained by using the Rankine-Hugoniot relations. For small shock angle, i.e. weak shocks, the composition and the vibrational relaxation can be assumed to be frozen. The shock positions are chosen such that shock configurations of type-III and type-IV interaction occur. Another point of interest is the location of the incident shock for which the jet turns around the body surface, named here type-IVa interaction.

In Fig. 3, Mach-number contours and contours of O_2 concentration (only for chemically reacting flow) are presented for several interaction types and chemistry modeling (perfect gas, thermal and chemical nonequilibrium). For the comparison between perfect gas and a chemically reacting flow field, the vertical position of the incident shock is chosen such that the stagnation point has nearly the same location. Three flow patterns of type-IV interaction are shown to demonstrate the effects of the different flow modeling. The pictures in the first row (Figs. 3a–c) show the flow field of a type-IVa interaction where the supersonic jet turns around the surface and remains supersonic. The dissociation rate within the nonequilibrium flow field indicates that the flow inside the jet is only weakly dissociated, compared with the areas behind the strong bow shock where the oxygen is nearly completely dissociated. This behavior can be observed in all nonequilibrium flow fields. In the next case shown in Figs. 3d–f, a typical type-IV interaction is achieved where the supersonic jets impinge on the body surface. Just in front of the surface, the flow becomes subsonic through a strong normal shock. Downstream of the stagnation point, the flow is accelerated to supersonic. Behind the shock at the end of the jet, the concentration of molecular oxygen decreases due to the higher temperature. Only a small jet is present in the last flow pattern shown in Figs. 3g–i, which can also be classified as a type-IV interaction. In this case the supersonic jet flow also becomes subsonic by a shock just in front of the stagnation point. Downstream

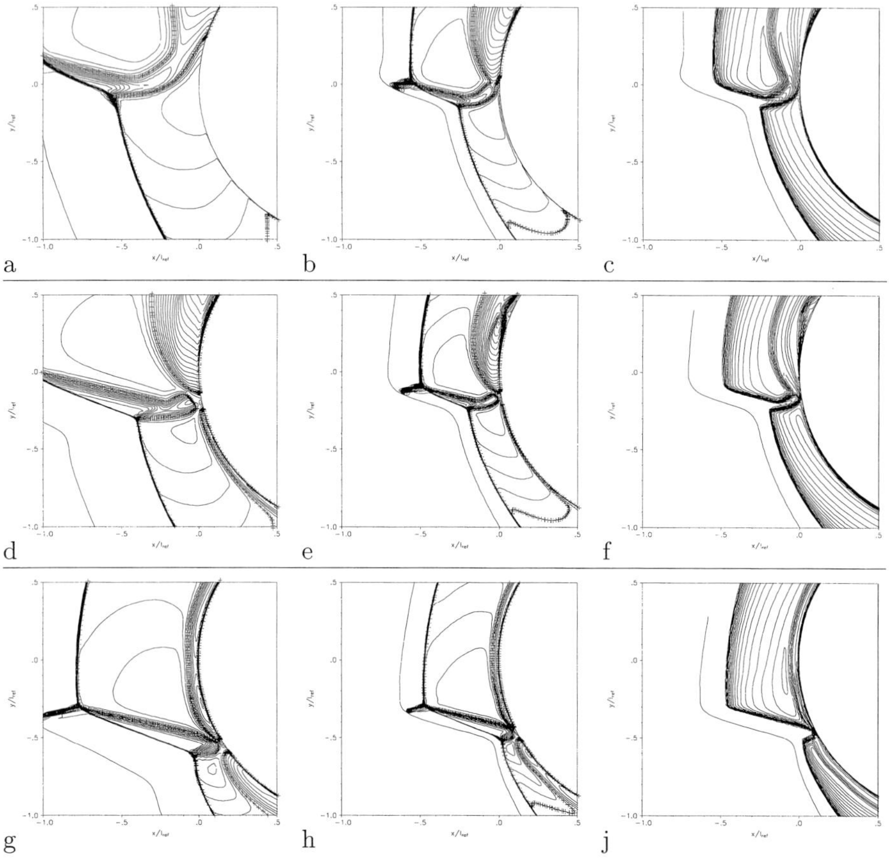

Figure 3. Calculated free-flight flow field for various incident shock positions, $u_\infty = 4$ km/s, H = 53.5 km (left: Mach-number perfect gas; center: Mach-number thermal and chemical nonequilibrium, +++ sonic line; right: concentration of O_2; top: type-IVa interaction; middle and bottom: type-IV interaction).

of the stagnation point, the flow expands and interacts in the upper part with the jet.

Starting from a type-III interaction, the flow field develops towards type IV by shifting up the incident shock, where in the transitional region the highest peak value is observed in the pressure distributions in Fig. 4. The typical peak-pressure levels vanish with the development of a type-IVa interaction where the supersonic jet runs along the surface. The heat-flux distributions in Fig. 6

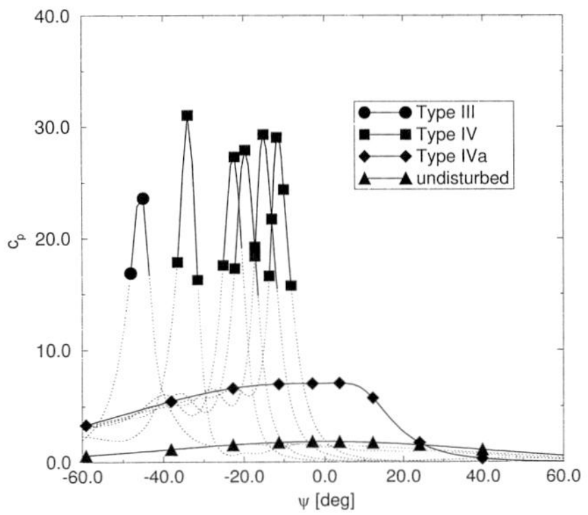

Figure 4. Pressure distributions for different shock positions (perfect gas).

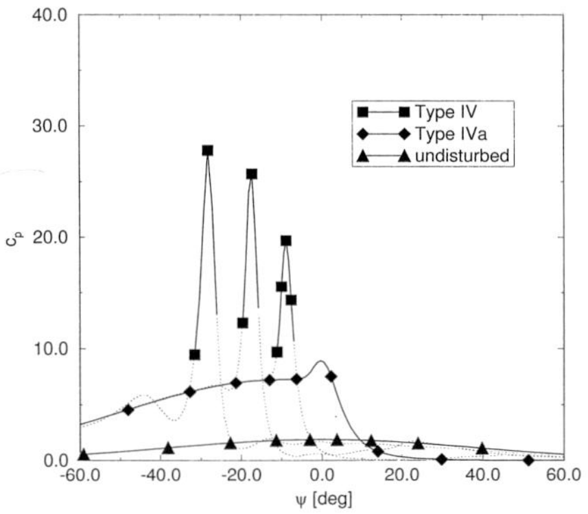

Figure 5. Pressure distributions for different shock positions (nonequilibrium).

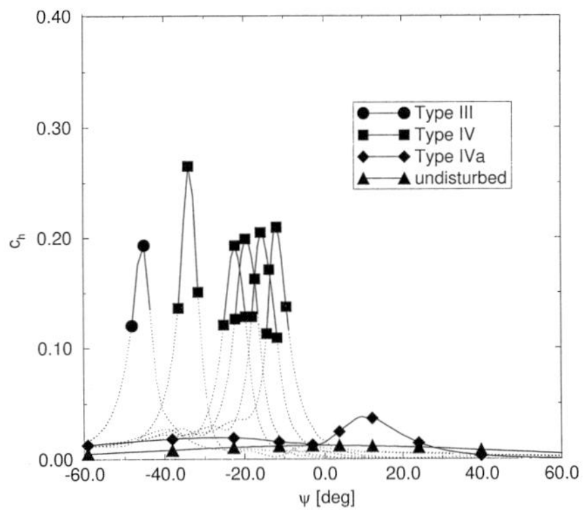

Figure 6. Heat-flux distributions for different shock positions (perfect gas).

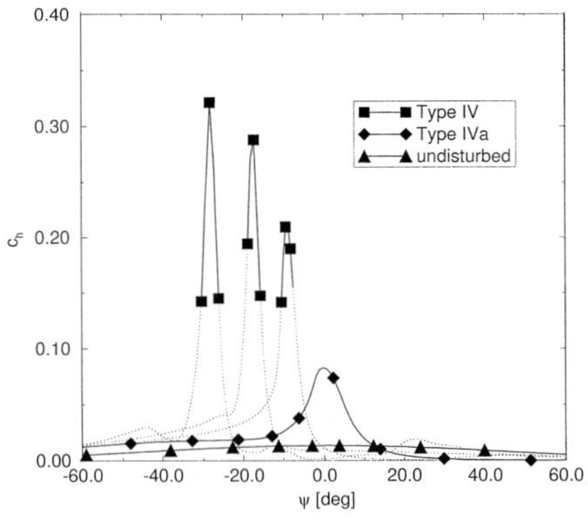

Figure 7. Heat-flux distributions for different shock positions (nonequilibrium).

show the same behavior. Only for type IVa can an increase of the heat-flux be observed where the jet comes close to the wall. The transition between type IV and IVa is sharp for the perfect gas flow, while the nonequilibrium results are showing a continuous transition (Figs. 5 and 7). The maximum loads for the nonequilibrium results are achieved when the impinging shock location for a type-IV interaction is close to the impinging shock location for type III. The peak values in the pressure distribution have nearly the same levels as observed for the perfect-gas solutions, while the heat-flux level is increased due to recombination.

5. Conclusion

The results of the validation test case indicate that the chosen grid resolution is sufficient to capture the most important effects of a typical shock-shock interaction. The predicted surface values agree well with the experimental data. The results of the reentry-flow simulations do not show qualitative changes of the shock configuration due to dissociation, while the shock stand-off distance is decreased. Only small dissociation effects are found in the jet-flow regions. Differences can be seen in the distribution of the surface loads, where the nonequilibrium results show lower peak values in the transitional region from type IV to type IVa. On closer examination we see that for nonequilibrium flows the range of shock positions is decreased when type-IV interactions are observed. The highest peak values are predicted for a jet impingement interacting with an expanding flow.

Acknowledgement. This work is partially funded by the Deutsche Forschungsgemeinschaft and ESA/CNES. Furthermore, I wish to thank K. Hannemann and G. Eitelberg for helpful discussions and their interest.

References

Brenner G (1994) Numerische Simulation von Wechselwirkungen zwischen Stößen und Grenzschichten in chemisch reagierenden Hyperschallströmungen, DLR-FB 94-04, Göttingen.

Brück S (1993) Numerical Simulation of Shock-Shock Interactions in Transitional/Turbulent Flow, DLR IB 221-93 A 23, Göttingen.

Edney B (1968) Anomalous Heat Transfer and Pressure Distributions on Blunt Bodies at Hypersonic Speeds in the Presence of an Impinging Shock, FFA Report 115, Stockholm.

Hannemann K, Brück S, Brenner G (1993) Numerical Simulation of Reacting Flows Related to the HEG. In: Shock Waves @ Marseille, Brun R, Dumitrescu LZ (eds), Springer, pp 251–256.

Park C (1989) Nonequilibrium Hypersonic Aerothermodynamics, John Wiley & Sons, New York.

United States Committee on Extension to the U.S. Standard Atmosphere (COESA) (1976), U.S. Standard Atmosphere, NASA Technical Memorandum X-74335, Washington, D.C.

Wieting AR, Holden MS (1987) Experimental Study of Shock Wave Interference Heating on a Cylindrical Leading Edge at Mach 6 and 8. In: AIAA 22nd Thermophysics Conference, AIAA-87-1511.

Recompression of nonequilibrium flow in the wake of a blunted cone

D. Kastell, K. Hannemann, G. Eitelberg
DLR, Institut für Strömungsmechanik, Bunsenstr. 10, 37073 Göttingen, FRG
T. J. Horvath
NASA Langley RC, GDD AB, Hampton, VA 23681 USA

Abstract: The entry of a planetary probe, which is a 140° total-included-angle blunted cone with a spherical nose, rounded edges, and a cylindrical afterbody is significantly influenced by nonequilibrium flow effects. Therefore its aerodynamic characteristics have been investigated in the continuum flow of the High Enthalpy Shock Tunnel at the DLR in Göttingen (HEG). Calculations with a Thin-Layer Navier-Stokes code which is capable of simulating chemical and thermal nonequilibrium flows have been carried out and compared to measured heat-transfer and pressure distributions. Depending on the operating condition of the tunnel, up to 63% of the stagnation heat-transfer rate was reached on the front-body corner junction and up to 13% was measured on the afterbody sting due to the reattachment of the separated shear layer. The numerical prediction on the forebody compares well with the experimental data, whereas the maximum heat loads on the sting were underestimated by the calculations.

Key words: HEG, Nonequilibrium flow, Hypersonic, Experiment, Numerical simulation, Blunted cone, Wake flow, Shear layer

1. Introduction

Almost any flight mission that involves changes in orbital altitude and/or inclination near an atmosphere-bearing planet is a candidate for the application of aeroassisted transfer vehicles (ASTVs). These vehicles take advantage of the aerodynamic forces generated during an atmospheric pass to change orbit with a much smaller expenditure of energy than is required for an all-propulsive maneuver. Due to the presence of high-temperature effects, which dominate the main flight regime of an ASTV, the influence of chemical and thermal nonequilibrium on the flow field has to be quantified. Experimental data and validated numerical codes are not available for this particular problem at the moment (Deiwert 1993). To overcome this deficiency, experiments have been carried out in the HEG along with numerical simulations with a three-dimensional Thin-Layer Navier-Stokes code including chemical reactions at the DLR in Göttingen, Germany. This investigation was part of the working group 18 activity of the AGARD fluid dynamics panel.

2. Description of the experiments

The HEG is a 60-m total-length free-piston-driven shock tunnel producing a hypervelocity continuum flow with a flow core diameter of 450 mm. At the maximum reservoir enthalpies of 23 MJ/kg it is capable of dissociating oxygen as well as nitrogen behind the bow shock of a model. An extended description of the facility is given by Eitelberg et al. (1992). The model used in the present investigation is a blunted cone with a base radius of $R_B = 76.2$ mm, a nose radius of $R_N = 38.1$ mm, a corner edge radius of $R_C = 3.81$ mm, and a total angle of 140° ($R_N/R_B = 0.5$, $R_c/R_B = 0.05$). The model is supported in the test section with an instrumented cylindrical sting (diameter 38.1 mm) having a length of 457.32 mm. Two models were built, each individually instrumented for three different test campaigns. The first model was instrumented with 15 thermocouples, most of them located on the sting, and two pressure gauges on the forebody. The second model had a forebody of chromel to match the thermal properties of its coaxial sensors and was instrumented with 72 thin-film gauges, located on the sting and the back of the body. For the latest experiments the first model was redesigned with a thin-film sensor insert in the sting. Pressure gauges were added to the sting and to the back of the body. Holographic interferometry and laser schlieren were conducted. A more detailed description of instrumentation, data-reduction procedures, test conditions, and criteria used to determine near-wake flow establishment can be found in Kastell et al. (1994).

3. Numerical analysis

The numerical simulations were performed using the implicit upwind-TVD Euler/Thin-Layer Navier-Stokes code NSHYP (Brenner et al. 1993). The three-dimensional, time-dependent Thin-Layer Navier-Stokes (TLNS) equations in conservation form for a multicomponent mixture of perfect gases are considered. For the simulations discussed in this paper a fully laminar air flow is considered. The isothermal wall ($T_{wall} = 300$ K) of the blunted cone is taken to be fully catalytic. The relevant chemical reactions were taken from Park (1989). The determination of the vibrational energy of the molecular species is based on the assumption of a harmonic oscillator. The transition of energy between the vibrational modes and the translational modes is approximated using the Landau-Teller formulation. The vibrational relaxation times were obtained from correlations of Millikan and White and Park (1989). Table 1 shows the free-stream conditions used for the computations. The data were calculated with NSHYP from nominal reservoir conditions for the axisymmetric contoured nozzle of the HEG (Hannemann et al. 1996). The actual run-to-run conditions vary and have to be calculated individually. Free-stream data presented in Kastell et al. (1994) and Legge (1995) have been calculated with a one-dimensional nozzle code.

Table 1. Nominal flow conditions used for the TLNS calculations.

	u_∞ $[\frac{m}{s}]$	ρ_∞ $[\frac{g}{m^3}]$	Re_∞ $[\frac{1}{m}]$	M_∞	T_∞ [K]	p_o [MPa]	h_o $[\frac{MJ}{kg}]$
Condition I	5700	2.19	$2.7 \cdot 10^4$	9.2	797	38.6	20.35
Condition II	6037	4.45	$4.7 \cdot 10^4$	8.5	1073	90.8	22.27
Condition III	4737	2.99	$3.7 \cdot 10^4$	9.5	574	44.9	13.33

4. Results and comparison

Most experiments and numerical calculations related to ASTVs have only been conducted in continiuum flow at low reservoir enthalpy. They also concentrated only on forebody measurements (see e.g. Stewart & Chen 1994). None of the investigators have focused their attention on the influence of high-temperature effects nor have they considered the flow in the near wake of the aeroshell-shaped vehicle. This region is of interest for the design of aeroassisted flight vehicles because shielding of the payload on its back might have to be considered. The rapid expansion at the shoulder turns the flow towards the centerline of the body where it impinges onto the sting, leading to a local peak in heat transfer and the formation of a recompression shock. The present experiments and numerical investigations were focused on the above-described topics. For three different flow conditions, with varying reservoir enthalpy and free-stream Reynolds number, the heat-flux and pressure distributions on the forebody, back, and sting of

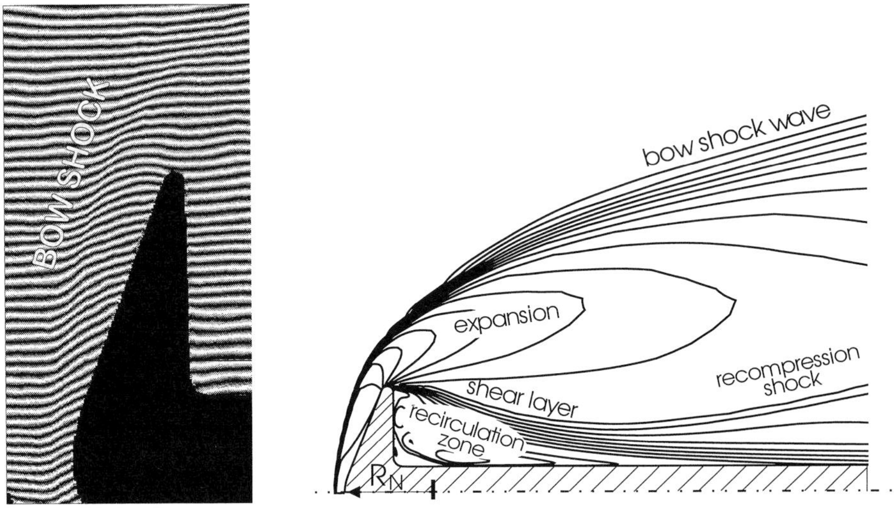

Figure 1. Interferogram (left) and Mach-number distribution for condition II (right).

the model are presented. Figure 1 (right) shows the calculated Mach-number distribution for flow condition II in comparison with an experimental interferogram in Fig. 1 (left). The primary flow features like the bow shock wave, the separated shear layer with the enclosed recirculation zone, and the recompression shock are evident in the calculations. In the interferogram, only the bow shock wave with a stand-off distance of $\Delta/R_N \approx 0.21$ can be identified.

Figure 2 shows the measured pressure on the blunt body together with the numerical calculations as a function of the normalized geometric arc-length S/R_N. To compare the numerical calculations with the experiments, the data were normalized by a value close to the stagnation point ($S/R_N \approx 0.68$), because this point was instrumented with a thermocouple. As can be seen in the plot the calculated pressure distributions on the forebody are equal for all conditions, and only on the sting a small variation is present. These discrepancies are more obvious for the heating distribution shown in Fig. 3. The data were normalized by the values measured at $S/R_N \approx 0.34$. This was done to minimize the influence of stagnation-point heat-transfer measurement and because of possible uncertanties due to grid singularity in the numerical code at this point. On the forebody and on the back of the model surface, the experimental data taken with two different models during three different runs follow the numerical curves well. On the sting, the location of the maximum heat-transfer load is predicted correctly, but the maximum heat-transfer rate of 20% of the reference value is underestimated by the calculations. This discrepancy appears for all conditions, but is more pronounced at lower reservoir enthalpies (cond. III).

Forebody-surface-heating distributions for conditions I, II, and III are shown in Fig. 4. The numerical curves fall on top of each other except at the stagnation point. This indicates that different total enthalpies affect mainly the heat flux on the spherical nose of the model. The experimental data follow the trend of the curves well. The situation on the back and on the sting is different, as shown

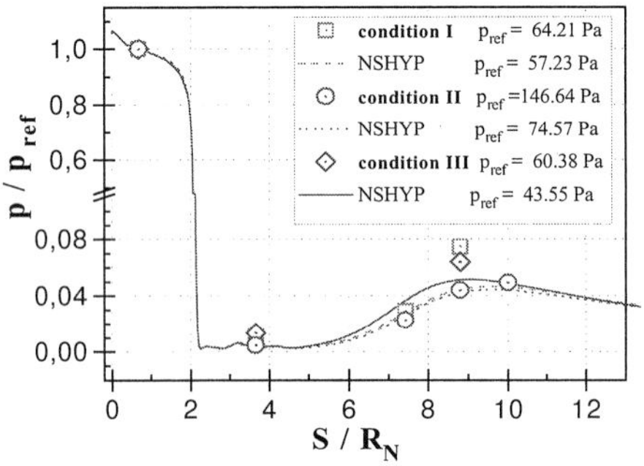

Figure 2. Surface-pressure distribution.

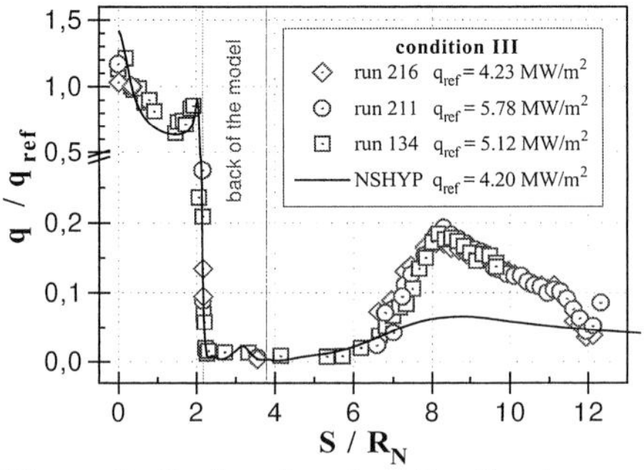

Figure 3. Surface-heating distribution.

in Fig. 5. The experimental data are taken from two different test series. On the back of the model and on the beginning of the sting the experimental data follow the numerical curves. The discrepancy between measured and predicted data becomes evident at the point of maximum heat transfer, where the calculated values are lower than the measured ones. Also, the measured peaks are narrower than the numerical ones. The higher measured heating indicates that shear-layer transition might be a significant feature of the near wake flow field. Tests recently conducted with schlieren visualization and intrusive hot-wire, hot-film, and pitot-probe surveys in a perfect-gas Mach-6 flow at NASA LaRC (with the same model) by Horvath (1995) support this possibility. A similar transition effect has also been noted by Baker & Martin (1966) for a backward-facing step. Significant influence might also cause change of the continuum behavior of the wake flow by the strong shoulder expansion. However the potential effects of high-temperature real-gas behavior on shear-layer transition remain coupled.

Experimentally, the location and magnitude of the peak heating associated with the reattaching shear layer appear to be very sensitive to free-stream conditions, whereas the numerical simulations indicate very little effect. Additional calculations with an adapted grid have shown almost no or little influence. For a noncatalytic-wall-model solution the normalized values are approximately half the magnitude of the presented ones. The calculated and the experimental data show both a shift towards the model base and an increase of the maximum heat-transfer load with increasing Reynolds number and decreasing the total enthalpy respectively (cond. III relative to cond. I). This shift is believed to be a viscous phenomenon whereas the heat-transfer change is influenced by the high temperature. At high-reservoir enthalpy and Reynolds number (cond. II) the high-temperature effects predominate leading to a low maximum heat transfer.

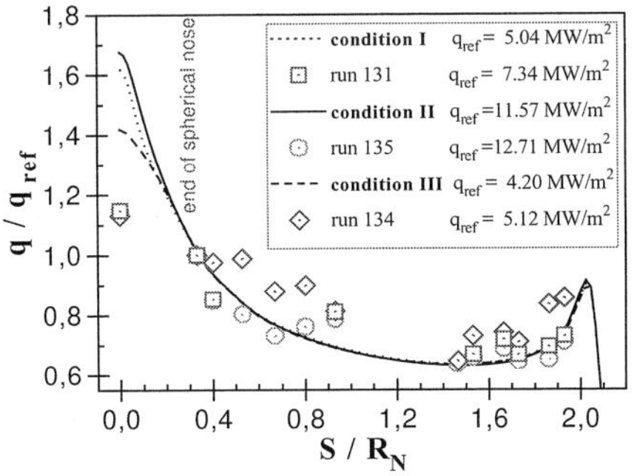

Figure 4. Forebody surface heating.

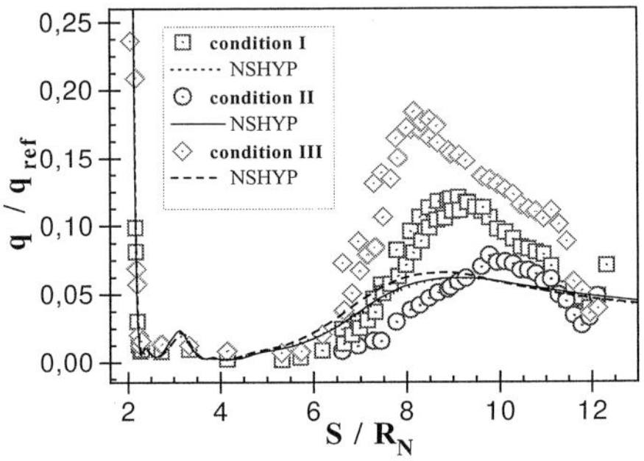

Figure 5. Sting-heating distribution.

5. Summary and conclusions

The heating and pressure distributions on the forebody, back, and afterbody of a 140° total-included-angle blunted cone have been measured in nonequilibrium flow. The reservoir enthalpies were varied between 14 and 23 MJ/kg. The measured data were compared with the results of a Thin-Layer Navier-Stokes code with embedded chemical reactions. If normalized by a value close to the stagnation point, the calculated heat flux on the forebody, except on the spherical nose, is not and on the sting is little affected by altered flow conditions. The same holds for the experimentally measured forebody heating. Nevertheless, the heat-transfer on the sting is strongly influenced by the different free-stream flow conditions. Increasing the Reynolds number moves the maximum heat load on the sting upstream and lowering the total enthalpies shrinks the peak. After all, the maximum heat load on the sting due to the recompression of the flow is underestimated by the numerical code. These discrepancies between the calculated and the experimental data in the wake flow of the blunted cone are not fully understood at this time and are the main issue of continued research.

References

Baker PJ, Martin BW (1966) Heat transfer in supersonic separated flow over a two dimensional backward facing step. JHM Transfer 9:1081–1088.

Brenner G, Gerhold T, Hannemann K, Rues D (1993) Numerical simulation of shock/shock and shock-wave/boundary-layer interactions in hypersonic flows. Computers Fluids. 22(4/5):427–439.

Deiwert GS (1993) Issues and Approach to Develop Validated Analysis Tools for Hypersonic Flows: One Perspective, AGARD CP-514, Ref 24.

Eitelberg G, McIntyre TJ, Beck WH, Lacey J (1992) The High Enthalpy Shock Tunnel in Göttingen, AIAA Paper 92-3942.

Hannemann K, Krek R, Eitelberg G (1995) Latest calibration results of the HEG contoured nozzle. In: Shock Waves, Proc of the 20th ISSW, Sturtevant B, Shepherd JE, Hornung HG (eds), World Scientific, Vol 2, pp 1575–1580.

Horvath TJ (1995) Private communication. NASA Langley RC, VA USA.

Kastell D, Horvath TJ, Eitelberg G (1994) Nonequilibrium flow expansion around a blunted cone. In: 2nd European Symposium on Aerothermodynamic for Space Vehicles, Nordwijk, NL, ESA SP-367 (1995), pp 383–389.

Legge H (1995) Experiments on a 70 Degree Blunted Cone in Rarefied Hypersonic Flow, AIAA Paper 95-2140.

Park C (1989) Nonequilibrium Hypersonic Aerothermodynamics, J Wiley, NY.

Stewart DA, Chen YK (1994) Hypersonic convective heat transfer over 140-deg blunt cones in different gases. J Spacecraft 31(5):735–743.

Flow in the TCM2-Marseille hypersonic facility: some new experimental results

R. Brun, Y. Burtschell, A. Chaix, M.-C. Druguet, L. Z. Dumitrescu,
M.-P. Dumitrescu, R. Martin, D. Zeitoun
Lab. I.U.S.T.I. (U.R.A. CNRS 1168) Dept. MHEQ
Université de Provence, Marseille, France

Abstract: Some questions arising during the development of a new free-piston hypersonic facility (TCM2 Marseille) are discussed. It is shown that, to ensure a disturbance-free flow, in addition to having a tuned piston velocity (Hornung 1988), one must also keep this velocity practically constant for a duration of about 1.6 times the desired test duration. A method to ascertain the achievement of these conditions is presented. Then, the use of argon addition for fine adjustment of interface tailoring is advocated. If the above requirements are fulfilled, it is shown that the actual reservoir pressure, behind the reflected shock, coincides with the computed value (from M_S), and no supplementary loss mechanism has to be invoked. Disturbance-free operation of TCM2 for over one msec is demonstrated. Attempts to determine the reservoir enthalpy from the end-wall heat flux yielded much higher values than expected. No explanation is yet available for these aberrant results.

Key words: Hypersonic facility, Free-piston shock tunnel

1. Introduction

The TCM2 facility (Burtschell et al. 1991) is a medium-size free-piston shock tunnel designed for basic studies in hypersonics, especially on the interconnection between fluid-dynamic and physico-chemical phenomena at high enthalpies. The present paper will report on the work carried out since the start of the project: a number of novel developments will be presented, as well as some unexpected findings, for which an explanation is still uncertain.

2. The facility

A schematic of TCM2 is shown in Fig. 1. No attempt has been made by the design to duplicate free-flight conditions, in order to avoid going to exceedingly high pressures: instead, the emphasis was on producing well-controlled flow conditions, at moderate enthalpies (up to 12 MJ/kg), reasonably free of impurities.

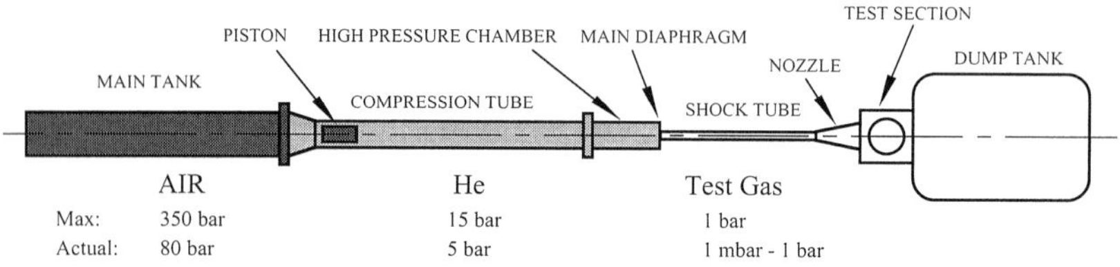

Figure 1. Schematic of the TCM2 free-piston shock tunnel.

Following erection of the facility, a systematic study of the gasdynamic processes taking place in the various compartments has been undertaken, with the aim of a full characterization of the operation of the machine.

3. Gas-dynamic processes in the compression tube and the "tuned-piston" problem

The first problem encountered was the discovery that the pressure build-up in the compression tube does not proceed smoothly, but as a series of successive pressure steps, on which strong oscillations are superimposed (Dumitrescu et al. 1993). Later is was found that in other facilities the pressure displays the same behavior (see, e.g. Fig. 10 in Maus et al. 1993). The cause having been uncovered (acoustic resonance of a cavity near the diaphragm), a remedy has been found: by a special design of the piston (Dumitrescu et al. 1993), the pressure rise in the compression tube of TCM2 is made smooth, while the main diaphragm bursting pressure, p_R, is controlled to within a few percent.

A second problem is that of tuning the piston motion (Hornung 1988). This term designates the setting up of the operating parameters (piston mass, initial pressures, etc.) so that, at diaphragm burst, the piston velocity, V_R, is such as to push the driver gas through the shock-tube entrance with exactly the required mass-flow rate, thus preventing the generation of an expansion wave that would then travel downstream the shock tube and catch up with the primary shock. Upon closer investigation it became clear, however, that the condition on the velocity is not sufficient: it is furthermore required that the piston's inertia be sufficiently great, so that the piston could continue to travel, with almost constant velocity (say, $\pm 10 - 20\%$), during a certain time, t_{**}, necessary to ensure a given test duration, t_* in the facility. To derive a relationship between t_* and t_{**}, one may examine the wave diagram of the shock tube (Fig. 2): it may be seen that, to avoid disturbing the constant regime at the end wall, behind the reflected shock, at least for a duration t_*, the head of the expansion wave coming from the piston should not reach the end wall earlier than t_B (point B). For TCM2, at a nominal M_s of about 10, one gets: $t_{**} \geq 1.6 \times t_*$ (for $t_* \approx 1$ msec). Inspection of Fig. 2 reveals also that the useful test time disappears altogether if

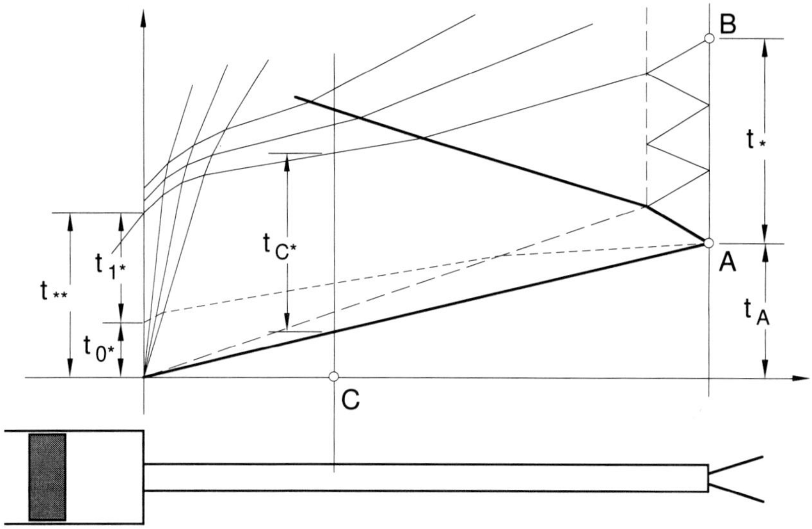

Figure 2. Wave diagram of the shock-tube processes.

the expansion wave generated by the slowing-down motion of the piston arrives at the end wall at the same time as the primary shock (point A); for TCM2, one finds $t_{0*} \approx 0.6$ msec. This means that the relationship between t_{**} and t_* is nonlinear, and that any reduction in the constant-velocity stroke of the piston may cut dramatically the available test duration. The holding time t_{0*} has, in addition, to be increased by about 10 – 20%, and t_{**} has to be supplemented with this quantity, to account for the primary-shock viscous attenuation.

Now, if the required constant-velocity piston stroke is allowed for, the piston will come very close to the end of the compression tube; this raises the question of finding a means to stop the piston without damage over the remaining very short stroke. The problem has been solved at TCM2 by a special piston design (Dumitrescu M-P 1996a).

4. Gas-dynamic processes in the shock tube

While the compression tube acts as a driver for the machine, it is the correct evolution of the gas-dynamic processes in the shock tube that has to be controlled. One important question is then how to verify that the required piston velocity (and its constancy) have been achieved. Recording the pressure in the compression tube provides an indication; but one would like to have a direct check. The evolution of the reflected-shock pressure cannot be used for this purpose, since it is affected by other phenomena (see below). However, an unambiguous way to ascertain the correct piston tuning is simply to inspect the pattern of the pressure trace at a station sufficiently far upstream along the shock tube (point C, Fig. 1).

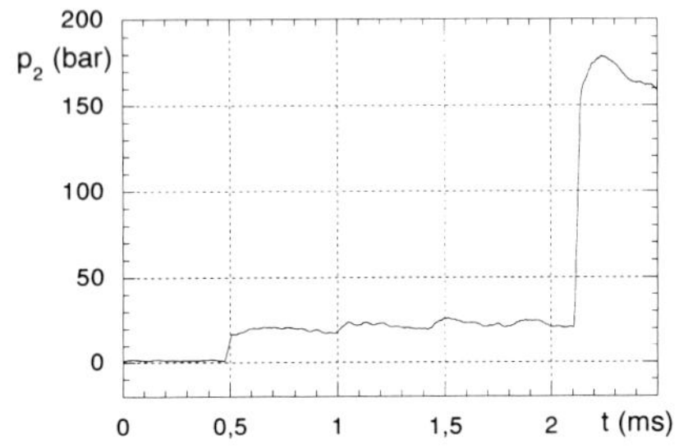

Figure 3. Pressure behind the primary shock at station C.

It is known that, behind a shock wave being attenuated only by viscous effects, the pressure p_2 registers a gradual increase; whereas it would start to fall if an expansion wave came from upstream. Therefore, if the duration of non-decreasing pressure p_2 at station C, t_{C*}, is at least equal to the required t_{**} (for a desired test duration t_*), then the proof is made. Such a recording in TCM2 is presented in Fig. 3, showing at least 1.6 msec of quasi-constant p_2, and thus giving about 1 msec of test time.

Fine tuning the piston motion is nevertheless insufficient to ensure that constant reservoir conditions are obtained: one has also to produce so-called tailored interface operations, that is, to have equal acoustic impedances across the contact surface when it is struck by the reflected shock. Viscous shock attenuation tends to shift conditions towards undertailoring at a given M_S. So, a convenient way to adjust finely the degree of tailoring is to add a small percentage of heavy gas (argon) into the helium driver, in the compression tube, thus increasing its impedance. Tailoring can be verified on the end-wall pressure trace, when a quasi-constant level, lasting the required test duration t_*, is seen. In the undertailored condition, the pressure evolves as a series of plateaus (progressively smeared), whose levels alternate above and below that of the first (just after the arrival of the primary shock) (Fig. 4); whereas overtailoring is signaled by a progression of increasing pressure steps.

Of course, adding argon reduces somewhat the shock Mach number and the enthalpy, but this is the price to pay for having well-defined test conditions. As an example, Fig. 5 shows what can be achieved by careful tuning of the machine, in terms of pressure steadiness.

A final comment is due: if all is set properly, the measured pressure, p_5, should coincide with the value computed (keeping account of real-gas effects) from the *measured* shock Mach number M_S at a station close to the end wall: no supplementary losses need to be invoked to explain any discrepancy. For instance, in the test of Figs. 4 and 5 the actual pressure level is within 3% of the computed value. We maintain, therefore, that, were a significant pressure

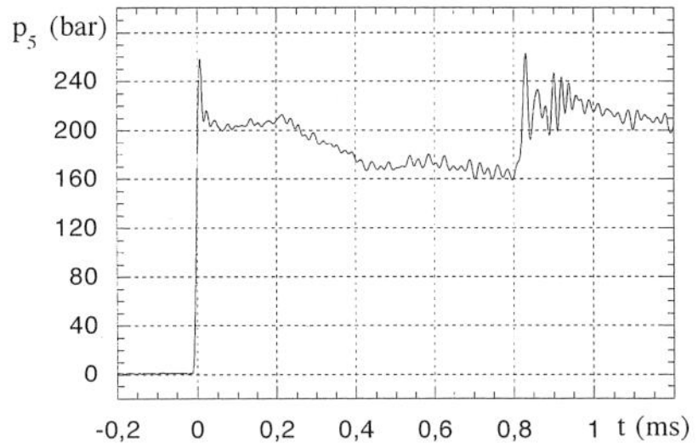

Figure 4. Undertailored end-wall pressure, driver pressure is 170 bar.

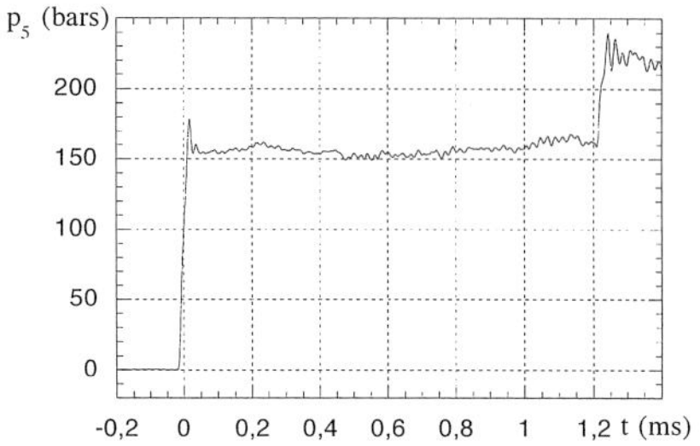

Figure 5. End-wall pressure for tuned and tailored operation, $M_S = 5.0$.

loss to be found, the cause should be pinned down to a poor tuning of the piston velocity and/or decceleration.

5. Determination of the reservoir enthalpy

All data in a hypersonic test are referred to reservoir conditions (p_5, h_5). But, whereas p_5 is readily measured, determination of the reservoir enthalpy remains problematic. Knowing the actual shock Mach number, M_S, real-gas computations give T_5 (and p_5); now, if the measured pressure coincides with the computed value, one can say that the phenomena are really taking place as expected, and infer that the actual temperature should also be equal to that deduced from M_S. Of course, this reasoning fails if significant discrepancies persist between measured and computed p_5; and an independent determination of T_5 (or h_5) would be welcome.

One way to achieve this is by measuring the heat flux from the hot, stagnant gas to the end wall. In the absence of any gas motion, a thermal boundary layer develops near the wall after shock reflection. The relevant equations are readily written and solved, using, as for an ordinary boundary layer, the von Mises transform. Equilibrium real-gas properties are assumed. Theory shows that the wall temperature should display a step-like evolution, consistent with a heat flux varying essentially as the inverse square root of time.

We have placed a number of coaxial thermocouples flush with the end wall of the facility, and recorded the signals: the expected step-like behavior was obtained, but the resulting heat flux (and gas temperature), as computed by the procedure described, was several times higher than expected. In the presence of such aberrant results, all common-sense checks were made, and the tests were repeated, using thin-film platinum gauges and different electronic hardware (but similar results); unfortunately, no definite explanation can yet be advanced, and further tests are now in progress. Anyway, as explained above, provided the machine is properly tuned, the stagnation temperature can confidently be inferred from the shock Mach number, correlated with p_5. The survey of the heat flux at several points across the end-wall surface has produced other surprising results, which are discussed in a companion paper (Dumitrescu M-P 1996b).

6. The nozzle

The TCM2 facility has been equipped with a conical hypersonic nozzle, designed for a free-stream Mach number of about 6. Thorough tests are planned; presently a discussion of the most relevant points concerning the nozzle design is reported (Dumitrescu LZ et al. 1996).

Acknowledgement. This research was carried out within the framework of a project sponsored by the European Space Agency, aimed at a better understanding of the operation and use of ground test facilities for hypersonic flow. The authors wish to acknowledge the continuing support and valuable contributions of G. Durand, J. Muylaert, and D. Vanmol, from ESA.

References

Burtschell Y, Brun R, Zeitoun D (1991) Two-dimensional numerical simulation of the Marseille University free-piston shock tunnel TCM2. In: Shock Waves, Proc 18th Int Symp on Shock Waves, Takayama K (ed), Sendai, Japan, Springer-Verlag, Heidelberg, pp 583–590.

Dumitrescu LZ, Burtschell Y, Zeitoun D (1996) Throat design for hypersonic nozzles. In: Shock Waves, Proc of the 20th ISSW, Sturtevant B, Shepherd JE, Hornung HG (eds), World Scientific, Vol 2, pp 1569–1574.

Dumitrescu M-P, Brun R, Billiotte M, Bertoni JM, Burtschell Y, Canova A, Dumitrescu LZ, Houas L, Labracherie L, Zeitoun D (1993) Dealing with pressure oscillations in Stalker tubes. In: Shock Waves @ Marseille, Proc 19th Int Symp on Shock Waves, Brun R, Dumitrescu LZ (eds), Marseille, France, Springer-Verlag, Heidelberg, pp 247–250.

Dumitrescu M-P (1996a) Discarding the tuned-piston condition for Stalker tubes: possible gains and drawbacks. In: Shock Waves, Proc of the 20th ISSW, Sturtevant B, Shepherd JE, Hornung HG (eds), World Scientific, Vol 2, pp 1487–1490.

Dumitrescu M-P (1996b) Trapping the boundary layer: a method to diminish flow contamination in shock tunnels. In: Shock Waves, Proc of the 20th ISSW, Sturtevant B, Shepherd JE, Hornung HG (eds), World Scientific, Vol 2, pp 1581–1586.

Hornung HG (1988) The Piston Motion in a Free-Piston Driver for Shock Tubes and Tunnels, GALCIT Report FM 88-1.

Maus J, Blanks J, Dewitt J (1993) Calibration and Tests of a New Free-Piston Shock Tunnel, AIAA Paper 93-5003.

Leading edge bluntness effects in hypervelocity compression corner flow

S. G. Mallinson, S. L. Gai, N. R. Mudford
Department of Aerospace and Mechanical Engineering, University College
University of New South Wales, Australian Defence Force Academy
Canberra ACT 2600 Australia

Abstract: Compression corner flow is examined under the combined effects of real gas behavior and leading edge bluntness. Pressure and heat transfer distributions have been obtained using a free-piston shock tunnel at two enthalpies: 2.8 MJ kg^{-1} and 18.0 MJ kg^{-1}. On the flat plate and upstream of the corner, the data for the blunt leading edge configuration were found to lie above the corresponding sharp leading values, whilst on the ramp, there was a considerable reduction in the pressure and heat transfer when the leading edge was blunt. Also, the extent of the interaction was seen to be smaller with the blunt leading edge. These results are similar to those from perfect gas studies. The differences between the sharp and blunt leading edge data seem to be less pronounced at the higher enthalpy and this is thought likely to be due to the reduced shock stand-off as a result of dissociation.

Key words: Compression corner, Hypervelocity, Leading edge bluntness

1. Introduction

The compression corner is a convenient configuration for examining shock wave / boundary layer interactions. It has been widely studied with both sharp and blunt leading edges (see Délery 1989; Kumar & Stollery 1994). The blunt leading edge has been found to increase the pressure and heat transfer upstream of the corner, and to decrease them downstream of the corner, in comparison to the sharp leading edge flows. Also, for small leading edge bluntness, the transition location is moved upstream and separation is delayed, whilst for large leading edge bluntness, the transition location is moved downstream and separation is hastened. Small and large bluntness may be conveniently delineated in terms of β, the combined bluntness-viscous interaction parameter (Holden 1971). The leading edge bluntness is considered small if $0.5 > \beta > 0.1$ and large if $\beta \leq 0.1$. An examination of leading edge bluntness effects using triple-deck theory arrived at similar criteria (Lagrée 1992).

The majority of investigations of the compression corner have considered perfect gas flow with only a few experimental studies (Stalker & Rayner 1985;

Mallinson et al. 1995) and theoretical studies (Ikawa 1979; Grasso & Leone 1992) examining the real gas effects. Certainly, there has been little if any attention paid to the combined effects of leading edge bluntness and real gas behavior on compression corner flow. This paper therefore examines the high enthalpy, hypersonic flow over a compression corner with blunt leading edges. The data are compared with that of a sharp leading edge obtained previously in the same facility (Mallinson et al. 1995).

2. Experimental details

The experiments have been performed using the free-piston shock tunnel T3, located at the Australian National University (Stalker 1972). Table 1 shows typical values of the reservoir and free-stream conditions. The surface pressure and heat transfer were measured using PCB 113M165 pressure transducers and co-axial chromel-alumel (Type K) thermocouples, respectively. The uncertainty in the measurements from these gauges is ±10% and ±15%, respectively (Mallinson et al. 1995; Gai & Joe 1992).

Table 1. Reservoir and free-stream values for conditions B and G; h = enthalpy, p = pressure, T = temperature, ρ = density, U = velocity, M = Mach number, γ_{fr} = frozen ratio of specific heats, subscripts 0 and ∞ refer to reservoir and free-stream conditions, respectively.

	h_0 MJ kg^{-1}	p_0 MPa	T_0 K	p_∞ kPa	T_∞ K	ρ_∞ g m^{-3}	U_∞ m s^{-1}	M_∞	γ_{fr_∞}	β_{cyl}	β_{sqr}
B	18.0	20.0	8100	0.85	1080	2.5	5300	7.5	1.45	0.07	0.05
G	2.78	22.0	2300	0.65	150	16.0	2300	9.1	1.40	0.03	0.02

The model employed in the present experiments is shown in Fig. 1. It consisted of a flat plate with a detachable leading edge and a ramp plate. These plates rested upon gauge housings. The corner angle was adjusted by inserting wedges beneath the housings. In the present paper, we will discuss results for corner angles of 0°, 15°, 18° and 24°.

Two blunt leading edges were employed: hemi-cylindrical and square. These have nose drag coefficients of $C_D = 4/3$ and 2, respectively (Anderson 1989). The sharp leading edge, the results of which are used to compare the present data, is described by Mallinson et al. (1995). The three leading edges are shown in Fig. 1.

3. The flat plate flow

The flat plate pressure and heat transfer distributions for conditions B and G with the hemi-cylindrical and square leading edges are compared in Figs. 2 and 3.

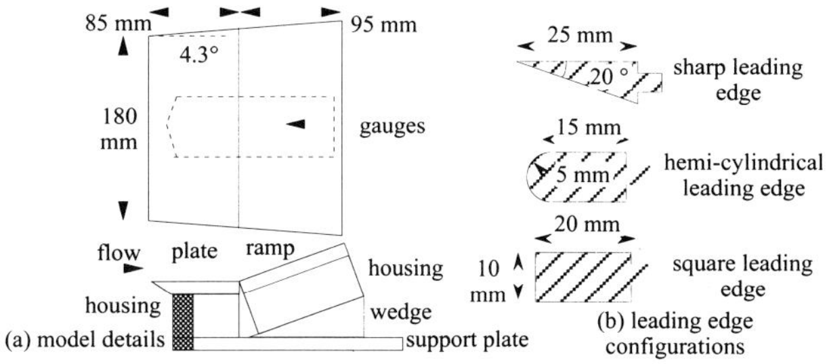

Figure 1. Schematic of the model and leading edges.

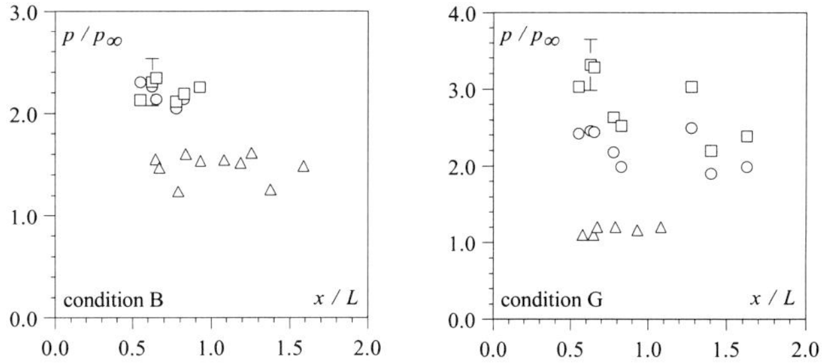

Figure 2. Flat plate pressure distributions. ○ = hemi-cylindrical leading edge; □ = square leading edge; △ = sharp leading edge.

Also shown are the data obtained with a sharp leading edge (Mallinson et al. 1995).

Firstly, consider the pressure distributions. For the high enthalpy condition B, the blunt leading edge data seem to lie closer to the sharp leading edge data than is the case for low enthalpy condition G. Also, the differences between the hemi-cylindrical and square leading edge data are less for condition B than for condition G.

It is seen that the heat transfer for the square leading edge lies above both the hemi-cylindrical and sharp leading edge data. Again, this difference is more pronounced for condition G. In fact, the sharp leading edge data could be said to be in good agreement with the hemi-cylindrical leading edge data for condition B. This may be due to the reduced shock stand-off that is experienced under high enthalpy conditions (Hornung 1972). This would lead to a thinner entropy layer and hence a smaller effect of the blunt leading edge.

4. The compression corner flow

The pressure and heat transfer distributions for conditions B and G with corner angles of 15°, 18° and 24° are shown in Figs. 4 and 5, respectively for both hemi-

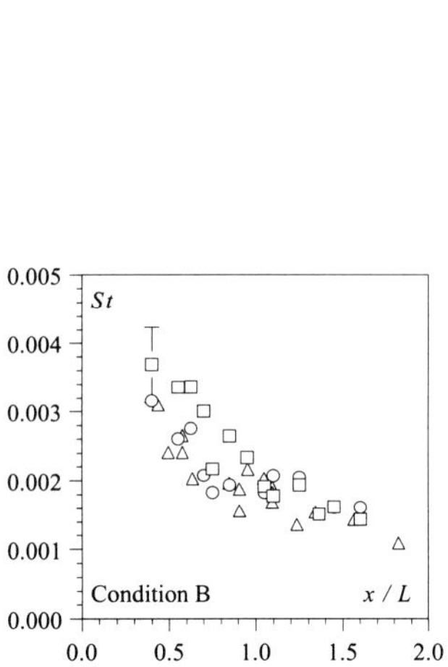

Figure 3. Flat plate heat transfer distributions. Refer to Fig. 2 for explanation of symbols.

cylindrical and square leading edge compression corners. Also shown are the pressure and heat transfer distributions from the sharp leading edge experiments (Mallinson et al. 1995).

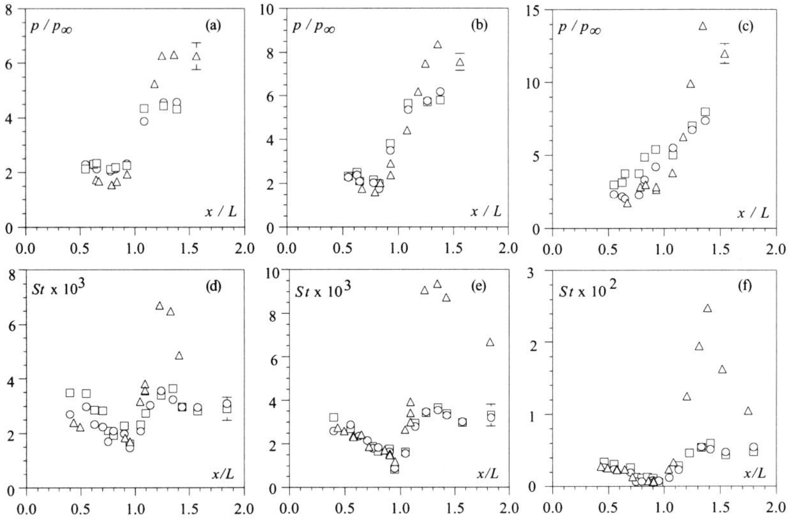

Figure 4. Compression corner pressure (a–c) and heat transfer (d–f) distributions, condition B. (a, d) = 15°, (b, e) = 18°, (c, f) = 24°. Refer to Fig. 2 for explanation of symbols.

We note that on the ramp face, the pressure and heat transfer for the blunt leading edge configuration are lower than that for the sharp leading edge. Overall, there does not seem to be much difference between the hemi-cylindrical and

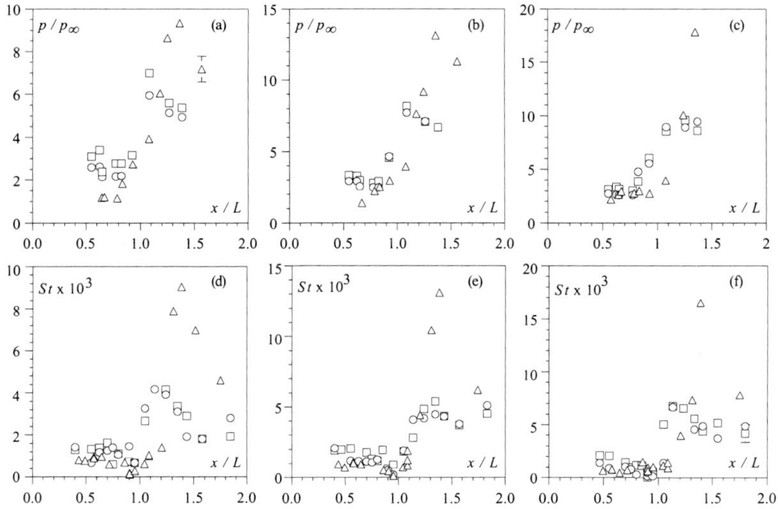

Figure 5. Compression corner pressure (a–c) and heat transfer (d–f) distributions, condition G. (a, d) = 15°, (b, e) = 18°, (c, f) = 24°. Refer to Fig. 2 for explanation of symbols.

square leading edge flows with regards to the levels of pressure and heat transfer on the ramp face.

The values of β for conditions B and G, as given in Table 1, indicate that the leading edges employed here have large bluntness. From the distributions presented, it would seem that the scale of the interaction is reduced for the blunt leading edges which is, therefore, in agreement with previous perfect gas studies (see Kumar & Stollery 1994).

5. Conclusions

The combined effects of leading edge bluntness and real gas behavior on flat plate and compression corner flow have been examined.

On the flat plate, the pressure and heat transfer lie above the sharp leading edge values, which is similar to the previous studies performed under perfect gas flow conditions. The difference between sharp and blunt leading edge flows seems to be more pronounced at low enthalpy. The smaller differences observed at the high enthalpy flow condition are attributed to the smaller bow shock stand-off due to dissociation.

For the compression corner, the pressure and heat transfer on the ramp face and the scale of interaction were reduced, which again is similar to the previous studies performed under perfect gas flow conditions with large leading edge bluntness.

Acknowledgement. The authors would like to acknowledge Mr. P.M. Walsh of the Department of Physics and Theoretical Physics, The Faculties, Australian National University, for assistance with the experiments.

References

Anderson JD Jr (1989) Hypersonic and High Temperature Gas Dynamics, McGraw-Hill.

Délery J (1989) Shock/Shock and Shock-Wave/Boundary-Layer Interactions in Hypersonic Flows, AGARD Rept 761, Part 9.

Gai SL, Joe WS (1992) Laminar heat transfer to blunt cones in high-enthalpy hypervelocity flows. J Thermophysics 6:433.

Grasso F, Leone G (1992) Chemistry effects in shock wave boundary layer interaction problems. In: Aerothermochemistry of Spacecraft and Associated Hypersonic Flows, Brun R, Chikhaoui AA (eds), Proc IUTAM Symp, Marseille, p 220.

Holden MS (1971) Boundary-layer displacement and leading-edge bluntness effects on attached and separated laminar boundary layers in a compression corner. Part II: Experimental study. AIAA J 9:84.

Hornung HG (1972) Non-equilibrium dissociating nitrogen flow over spheres and circular cylinders. J Fluid Mech 53:149.

Ikawa H (1979) Real Gas Laminar Boundary Layer Separation Methodology as Applied to Orbiter Control Surface Effectiveness Prediction, AIAA Paper 79-0212.

Kumar D, Stollery JL (1994) Hypersonic control flap effectiveness. ICAS Paper 94-4.4.3 In: 19th Congress of the International Council of Aeronautical Sciences.

Lagrée PY (1992) Influence of the entropy layer on viscous triple deck hypersonic scales. In: Aerothermochemistry of Spacecraft and Associated Hypersonic Flows, Brun R, Chikhaoui AA (eds), Proc IUTAM Symp, Marseille, France, p 358.

Mallinson SG, Gai SL, Mudford NR (1995) High Enthalpy, Hypersonic Compression Corner Flow, AIAA Paper 95-0155.

Stalker RJ (1972) Development of a hypervelocity wind tunnel. Aeronautical J 76:374.

Stalker RJ, Rayner JP (1985) Shock wave - laminar boundary layer interaction at finite span compression corners. In: 15th Int Symp on Shock Waves and Shock Tubes, Bershader D, Hanson R (eds), Stanford University, p 509.

Hypersonic high-enthalpy flow over a blunt-stepped cone

S. L. Gai, J. F. Milthorpe
Department of Aerospace and Mechanical Engineering, University College
University of New South Wales, Australian Defence Force Academy
Canberra ACT 2600 Australia

Abstract: The paper describes some experimental and computational results of a high-enthalpy flow over a blunt-stepped cone. The enthalpy range covered in the experiments was from 2.6 MJ/kg to 19.8 MJ/kg with equivalent velocities ranging from 2.1 km/s to 5.6 km/s. The results are of particular interest to proposed hypersonic aerospace planes.

Key words: Blunt cone, Backwards step, Hypervelocity

1. Introduction

There has been a resurgence of interest in the study of aero-thermodynamic problems of hypervelocity flight with the proposed aerospace planes and aeroassisted space transfer vehicles (ASTVs). This new generation of spacecraft will be subjected to prolonged radiative and convective heating during the trajectory through low-density upper atmosphere at speeds ranging from 5 km/s to 10 km/s. It is important, therefore, to understand the physical and chemical gas-dynamic processes in the flow over such vehicle configurations.

A simple configuration amenable to both theoretical and experimental studies is a spherically blunted cone. Flow around sphere cones has been studied for a long time (Lees 1956, Griffith and Lewis 1964, Gai and Joe 1992). However, data is sparse regarding configurations which may contain separating and reattaching flows as part of the total flow around the vehicle. These separating and reattaching flow phenomena are important because they may affect the control effectiveness and maneuverability of the vehicle.

In the present investigation the flow over a blunt stepped cone under high-enthalpy hypersonic conditions is studied using a free-piston shock tunnel T3 at the Australian National University in Canberra, Australia (Stalker 1985).

2. Model, facility, and experimental conditions

The configuration studied is shown in Fig. 1. It was a spherically blunted cone of nose radius (RN) 21.7 mm and cone half angle of 12°. The cone had a recessed step at a distance of 101 mm from the nose and was then extended parallel to the fore-cone surface for another 120mm. An axisymmetric rearward-facing step of height either 6 mm (0.30 RN) or 3 mm (0.15 RN) was thus formed where the flow separated at the end of the fore-cone and reattached on the aft-cone.

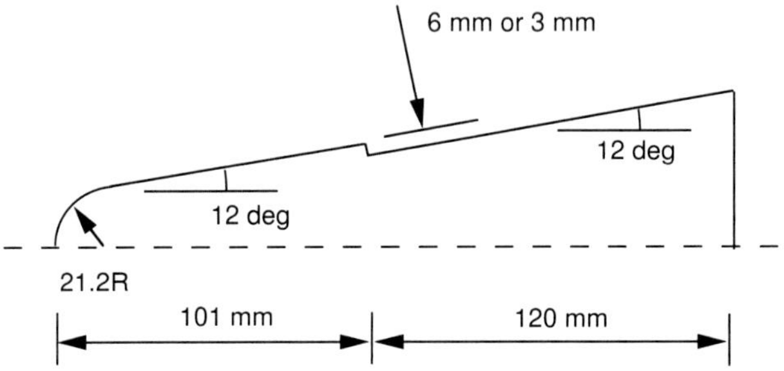

Figure 1. Schematic of the model.

The tests were conducted at the Australian National University free-piston shock tunnel T3 (Stalker 1985) with air as test gas and helium as driver gas. The flow was generated through a conical nozzle of 305 mm exit diameter and a 12.7 mm throat with a 7.5° half angle. This provided a test gas of frozen composition in the Mach number range 7.7 to 10 depending on the stagnation enthalpy and pressure. The tunnel was operated at three different conditions. These conditions, designated as B, D, and G, were typical of high, medium, and low-enthalpy flow, respectively. The flow conditions are summarized in Gai and Joe (1992).

3. Numerical method

The algorithm used to calculate the flow values is based on conservation of mass, momentum, and energy as the simulation proceeds over discrete time-steps. The flow domain is divided into rectangular cells. Each cell contains, at any given time, a certain block of material which possesses a certain momentum and energy. In the course of the next time-step, this block, with its associated mass, momentum, and energy, is convected a small distance. It may also gain or lose mass, momentum or energy by diffusion and the action of pressure gradients. The algorithm evaluates the changes in these quantities for the material block and proportionally redistributes the mass, momentum, and energy to the cells over which the block is located at the end of the time-step.

The mechanism of the algorithm is illustrated in Fig. 2 for convection with positive u and v velocities. The letters A, B, C, D represent the proportion of the material block located in a particular cell at the end of a time-step. These proportions of the mass, the momentum, and the energy in the material

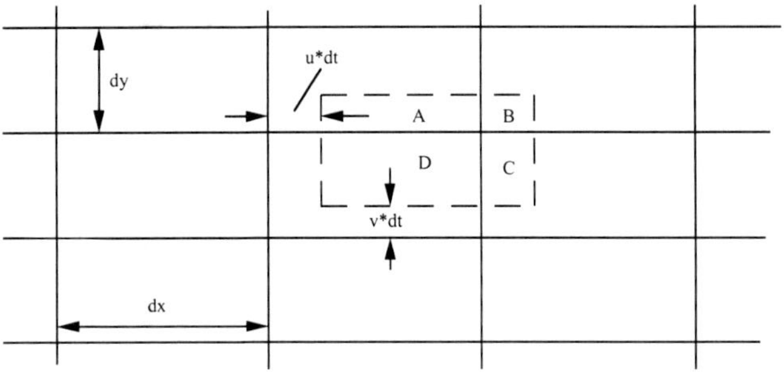

Figure 2. Numerical convection scheme.

block are allocated to their respective cells. A cell is regarded as moving with unchanged dimensions relative to the mesh of cell boundaries, so that it overlaps adjacent cells. The letter A indicates the area of the cell which remains in the original cell at the end of the time-step, while B, C, and D represent the areas of a cell which are located in adjacent cells. The quantities of mass, momentum, and energy in the cell are divided in proportion to the areas A, B, C, and D. Those proportions represented by B, C, and D are subtracted from the quantities in the original cell and added to the appropriate adjacent cells.

When the adjustment of mass, momentum, and energy has been completed for all cells, the flow variables ρ, u, v, p, and t can be calculated for all cells.

4. Stability limits

It is more economical to use the largest possible time-step in the computations, but there are a number of limits imposed by accuracy and stability considerations. The first is that, as convection is only calculated into the adjacent cells, the distance traveled in a time-step must not exceed the cell dimension. The second limit is that due to over-prediction of diffusion, as described by Roache (1976), which restricts the time-step to

$$\Delta t < \frac{\rho \Delta x^2}{2k}, \tag{1}$$

where Δx is the step size, ρ is the density, and k is the diffusion coefficient for any quantity.

5. Results and discussion

Figure 3 shows heat-transfer results for flow condition G normalized with the stagnation-point heat transfer measured on the nose. Also shown on the figure is a numerical result calculated for a similar flow condition. The calculations were made at Reynolds numbers similar to those in the experiments. It is seen that the code is able to predict the fore-cone heat transfer reasonably well.

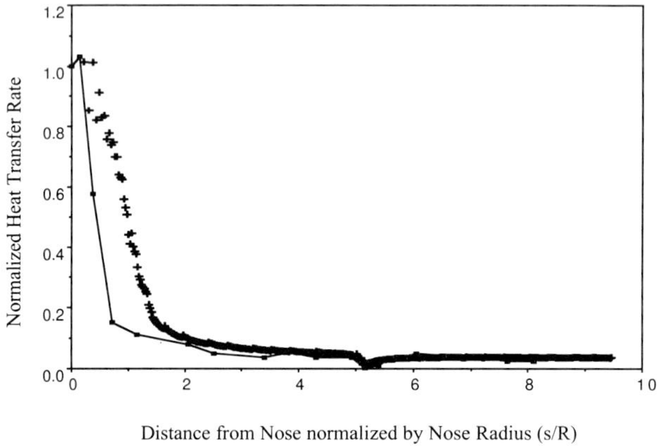

Figure 3. Normalized heat-transfer rate on blunt-stepped cone — measured and computed (condition G; $h = 6$ mm).

Downstream of the step the heat transfer (see Fig. 4) is typical of that in separated flow, viz. a sudden fall in heat transfer very near the step and then a gradual increase. While the experimental data show a decrease in heat transfer after reattachment, the numerical prediction exhibits a plateau for a considerable distance.

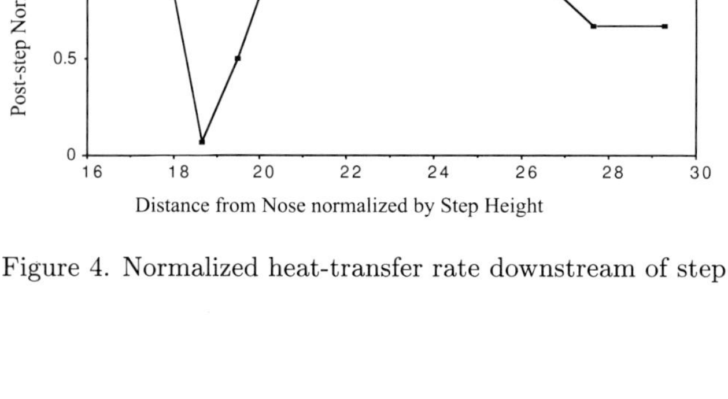

Figure 4. Normalized heat-transfer rate downstream of step.

Figure 5. Schlieren photograph of flow behind the step (condition G; $h = 6$ mm).

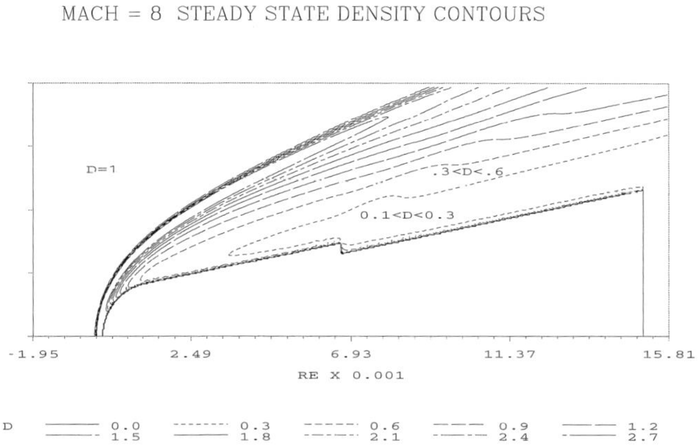

Figure 6. Non-dimensional density — computation.

Figure 5 shows an example of a schlieren photograph. We can clearly see an expansion at the corner, a separated shear layer behind the step, a wake neck, and a clear recompression shock. A notable feature is the strong curvature of the separated shear layer which is in contrast to that observed in a two-dimensional rearward-facing step flow (Gai et al. 1989). This is possibly due to the axisymmetric nature of the flow configuration.

Figure 6 shows the non-dimensional density field which shows a strong expansion, a clearly separated flow behind the step, a wake neck, and a strong recompression shock, which are consistent with the schlieren data.

The results are generally similar to those behind a two-dimensional rearward-facing step (Gai et al. 1989) but the shear layer appears to have a rather strong curvature which is unlike the two-dimensional case. This is attributed to the axisymmetric nature of the flow.

Acknowledgement. The authors would like to acknowledge the Australian Research Council for their generous support. We would also like to thank Dr. S. G. Mallinson for his help in gathering and analyzing the data, and Mr. P. Walsh for his assistance with the experiments.

References

Gai SL, Joe WS (1992) Laminar heat transfer to blunt cones in high enthalpy hypervelocity flows. J Thermo Phys Heat Transfer AIAA 6:433.

Gai SL, Reynolds NT, Ross C, Baird JP (1989) J Fluid Mech 199:541.

Griffith BJ, Lewis CH (1964) AIAA Journal 2:438.

Lees L (1956) Laminar heat transfer over blunt-nosed bodies at hypersonic speeds. Jet Propulsion 26:259–269, 274.

Milthorpe JF (1993) Numerical simulation of two-dimensional compressible flows. Int J Num Methods for Heat & Fluid Flow 3:223.

Roache PJ (1976) Computational Fluid Dynamics, Hermosa Publishers, Albuquerque.

Stalker RJ (1985) Free Piston Shock Tunnel T3 Facility Handbook, University of Queensland, St Lucia.

Surface heat flux to the Viking aeroshell in hypersonic reacting flow

N. R. Mudford, S. L. Gai
Department of Aerospace and Mechanical Engineering, University College
University of New South Wales, Australian Defence Force Academy
Canberra ACT 2600 Australia

Abstract: Measurements are reported of surface heat flux on the windward face of a Viking aeroshell model at zero angle of attack in high-enthalpy, dissociating, hypersonic flow. The reduction of surface heat flux due to dissociation is evaluated and used to infer the extent of recombination in the boundary layer.

Key words: Heat flux, Dissociation, Boundary layer, Hypersonic

1. Introduction

Surface heat flux is an important consideration in the design of atmospheric entry vehicles particularly for Aeroassisted Space Transfer and single stage to orbit vehicles. At Earth orbital speeds and above, molecular dissociation in the body shock layer causes a significant proportion of the total flow enthalpy to be present as chemical enthalpy which may contribute to surface heat flux only if recombination reactions proceed in the gas phase in the boundary layer or via surface catalyzed reaction at the vehicle surface. A thorough understanding of these recombination mechanisms is thus required.

The theory of chemical enthalpy recovery is well developed, e.g. Lees (1956), Fay and Riddell (1958), but experimental tests are rare and have yet to cover the full range of simulated flight conditions. In the present work, surface heat-flux measurements are reported for the windward surface of a Viking aeroshell model at zero angle of attack in hypersonic high-enthalpy, chemically reacting air flows. These are compared with measurements in low-enthalpy flow and with the results of calculations of expected surface heat flux under conditions of equilibrium and frozen boundary-layer chemistry. The effects of molecular dissociation are then assessed.

2. Facility and experimental conditions

The test flows were produced in the Australian National University's (ANU) T3 Free-Piston Shock Tunnel (FPST) (Stalker 1985) operating in the reflected mode with a 12.7-mm diameter throat and a 305-mm diameter exit conical nozzle.

The free-stream flow conditions for the tests are shown in Table 1. They are a subset of the air flows used in the flat plate experiments of East et al. (1980) and Gai et al. (1996). Conditions B and D have high total enthalpies and high levels of dissociation behind the bow shock around the model. Condition G has low total enthalpy and negligible dissociation behind the bow shock.

Table 1. Reservoir and free-stream values for conditions B, D, and G.

	h_0 MJ kg^{-1}	p_0 MPa	T_0 K	T_∞ K	ρ_∞ g m^{-3}	U_∞ m s^{-1}	M_∞	$\alpha_{N\infty}$	$\alpha_{O\infty}$
B	19.0	22.0	8400	1200	2.5	5500	7.5	10^{-7}	0.82
D	13.0	21.0	7100	930	3.5	4600	7.4	10^{-9}	0.43
G	2.78	22.0	2300	150	16.0	2300	9.1	0.00	0.00

Here h = enthalpy, p = pressure, T = temperature, ρ = density, U = velocity, M = Mach number, α_N = nitrogen dissociation fraction, α_O = oxygen dissociation fraction, subcripts 0 and ∞ refer to reservoir and free-stream conditions, respectively.

3. Model and instrumentation

The two mild steel models, A and B, are 70° half-angle sphere cones with base radius r_b = 25.0 mm and nose radius r_n = 12.5 mm. The conical skirt terminates in a sharp corner. In experimental series 1, models A and B were placed in the test section at the same downstream station, with model A on the centerline. Sufficient lateral separation existed to prevent interference. In experimental series 2, only model A was present on the flow centerline. Two shots were fired at each flow condition in each series.

The measurements were made using type-K surface junction coaxial thermocouples manufactured in-house to an ANU design (see Gai and Mudford 1992).

The thermocouple thermal product $(\rho ck)^{1/2}$ was obtained using the 'dipping' technique of Jessen et al. (1993) with distilled water as the fluid. This produced a value of $9500 \pm 500 \mathrm{Ws}^{1/2}\mathrm{m}^{-2}\mathrm{K}^{-1}$ (Mallinson 1994) which compares well with the $9900\mathrm{Ws}^{1/2}\mathrm{m}^{-2}\mathrm{K}^{-1}$ obtained by Lyons and Gai (1988) for type-K thermocouples.

4. Theory

East et al. (1980) investigated the surface heat flux on flat plates at 0° and 12° incidence under the flow conditions of the present study. They argue that the appropriate recovery enthalpy for surface heat flux under conditions of equilibrium chemistry in the boundary layer is given by

$$(h_r)_{eq} = c_{pf}T_e + h_{c,e} + h_{v,e} + \sigma^{1/2}(u_e^2/2) \tag{1}$$

where $h_{c,e}$ is the chemical enthalpy in molecular dissociation and the remaining terms on the RHS are, respectively, the translational and rotational enthalpy, the vibrational enthalpy, and the kinetic energy contributions. The subscript e denotes conditions at the boundary-layer edge. Their corresponding expression for recovery enthalpy for the case of frozen boundary-layer chemistry and a non-catalytic wall is

$$(h_r)_{fr} = (h_r)_{eq} - h_{c,e} \tag{2}$$

Lees (1956) derives an expression (his Eqs. 12 and 12a) for q_E/q_{sE} for a bluff body in hypersonic, high-enthalpy flow. q_E is the surface heat flux with equilibrium boundary-layer chemistry and unity Lewis number, Le, and q_{sE} is the corresponding quantity at the stagnation point of a sphere. The surface heat flux q_{FN} with frozen boundary-layer chemistry and non-catalytic wall, is given by $q_{FN} = q_E(h_r)_{fr}/(h_r)_{eq}$. With $Le = 1$, q_E is the upper limit in a dissociated flow as all of $h_{c,e}$ is recovered and contributes to surface heat flux. Conversely, q_{FN} is the lower limit, as no part of $h_{c,e}$ contributes to surface heat flux.

The boundary-layer edge conditions for the present calculations of q_E/q_{sE} and q_{FN}/q_{sE} were obtained using the BSHOCK program which calculates the full inviscid flow field using the inverse method (Garr and Marrone 1963). It includes nonequilibrium chemistry and equilibrium vibrational excitation using a single temperature model. BSHOCK has successfully predicted the density fields ahead of circular cylinders in reacting flows in T3 (Hornung 1972). The reaction rates employed were those of Park (1990).

5. Results and discussion

The measurements are presented here in the form q/q_{sE} against r/r_b where r is the radius from the model symmetry axis; q_{sE} is for a sphere of radius equal to the model nose radius of curvature r_n, and its value is obtained from the correlation of Fay and Riddell (1958).

The results for condition G are shown in Fig. 1 together with the BSHOCK prediction, shown as a solid line, and the results of Shih and Gay (1985), who measured surface heat flux on a 70° half angle sphere cone in low-enthalpy hypersonic flows. The Shih and Gay data are shown as $*$ on the figures. The r_n/r_b for their model is not known and the trailing edge of the conical skirt is radiused. Also shown are the average values of the present q/q_{sE} at each station, excluding outlying points. The figure shows that the experimental q/q_{sE} is close to 0.8 at the stagnation point. Both the BSHOCK prediction and the Shih and Gay (1985) results are in good agreement with the present results from $r/r_b = 0.2$ to the limit of the instrumented region of the present model. The high q/q_{sE} in the Shih and Gay data for $r/r_b < 0.15$ may be due to $r_n/r_b < 0.5$ for their model. Interestingly, the BSHOCK prediction follows

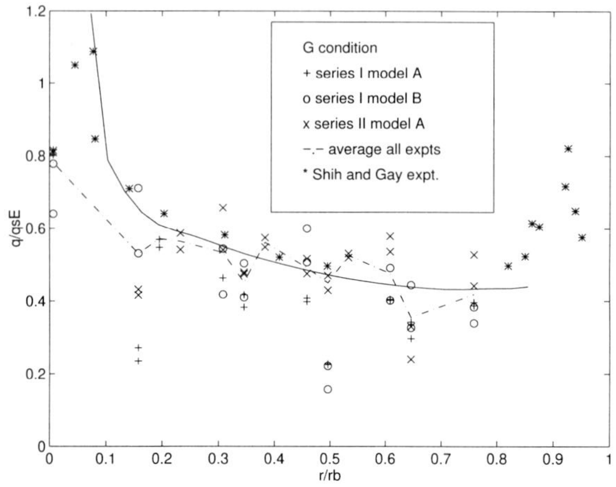

Figure 1. Condition G.

the Shih and Gay data closely in this region. Note that BSHOCK is inaccurate close to the stagnation point.

Figures 2 and 3 show the reacting flow-field results with the BSHOCK predictions for q_E/q_{sE} and q_{FN}/q_{sE} labeled E and FN respectively. Note that for both these conditions $h_{c,e}$, and consequently q_{FN}/q_E, are almost constant throughout the flow field. The negligible difference between the q_E/q_{sE} for the three flow conditions is expected on dimensional grounds (Fay and Riddell 1958).

It can be seen from the figures that q/q_{sE}, at the stagnation point and for $r/r_b > 0.5$, is unaffected by $h_{c,e} \neq 0$ indicating full recovery of $h_{c,e}$ in these regions. On the conical skirt in the region $0.1 < r_n/r_b < 0.5$, the q/q_{sE} for conditions D and B are lower than for condition G and lie between the E and FN limits. Consideration of Figs. 2 and 3 indicates that in this latter region approximately 60% of $h_{c,e}$ is recovered in the D condition flow and 70% in the B condition flow.

6. Conclusions

Measurements in non-reacting flow in the T3 FPST have been shown to agree well with those performed elsewhere. The present theoretical predictions agree well with the non-reacting flow results from all sources. Significant reductions below the value corresponding to full recovery of chemical enthalpy were detected in the inner regions of the conical skirt in the reacting flows. It is not possible from these experiments to determine whether gas-phase or surface-catalyzed reaction is responsible for the observed chemical-enthalpy recovery.

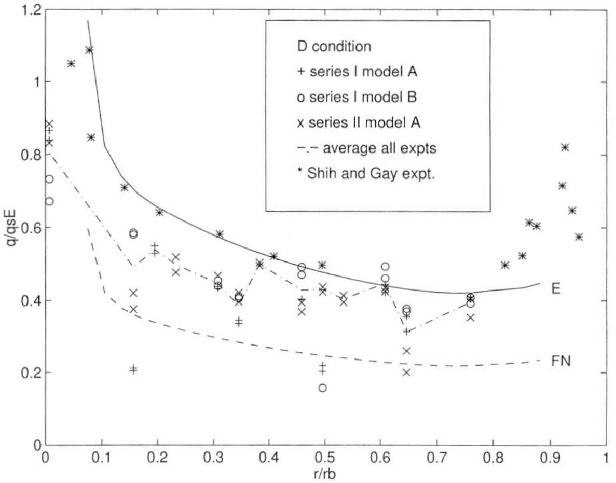

Figure 2. Condition D.

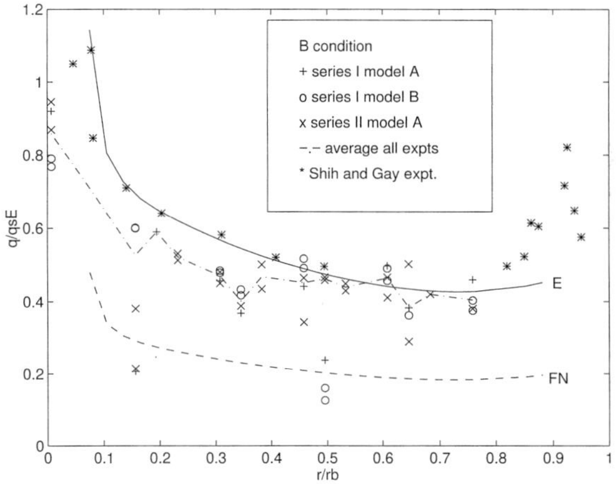

Figure 3. Condition B.

The high levels of chemical-enthalpy recovery observed here contrast with zero recovery found by East et al. (1980) and Gai et al. (1996) on sharp-nosed flat plates at low incidence under the same flow conditions. Clearly, the regime over which the boundary-layer chemistry moves from frozen to equilibrium lies between the two extremes of slender and bluff bodies in the FPST flows. This merits further investigation.

Acknowledgement. The authors would like to thank Mr. G. F. Foppoli of the Department of Aerospace and Mechanical Engineering, ADFA for making the models and the thermocouples to such a high standard. They would also like to thank Mr. P. M. Walsh and Mr. W. S. Joe of the Department of Physics and Theoretical Physics, the Faculties, ANU for their invaluable assistance with the experiments.

References

East RA, Stalker RJ, Baird JP (1980) Measurement of heat transfer to a flat plate in a dissociated high enthalpy laminar air flow. J Fluid Mech 97:673.

Fay JA, Riddell FR (1958) Theory of stagnation point heat transfer in dissociated air. J Aero Sci 25:73–122.

Gai SL, Mudford NR (1992) Stagnation point heat flux in hypersonic high enthalpy flow. Shock Waves 2:43–47.

Gai SL, Mudford NR, Roberts GT, Mallinson SG (1996) Effects of catalycity and boundary layer reactions on surface heat flux in hypersonic high enthalpy flows. In: Shock Waves, Proc of the 20th ISSW, Sturtevant B, Shepherd JE, Hornung HG (eds), World Scientific, Vol 1, pp 317–322.

Garr LJ, Marrone PV (1963) Cornell Aeron Lab Rep QM-1626-A-12(II).

Hornung HG (1972) Nonequilibrium dissociating nitrogen flows over spheres and circular cylinders. J Fluid Mech 53:149–176.

Jessen C, Vetter M, Grönig H (1993) Experimental studies in the Aachen hypersonic shock tunnel. Z Flugwiss Weltraumforsch 17(2):73–81.

Lees L (1956) Laminar heat transfer over blunt nosed bodies at hypersonic flight speeds. Jet Prop 26:259–269, 274.

Lordi JA, Mates RE, Moselle JR (1966) Computer program for the numerical solution of nonequilibrium expansions of reacting gas mixtures. NACA Computer Rept CR-472.

Lyons PRA, Gai SL (1988) A method for the accurate determination of the thermal product $(\rho ck)^{1/2}$ for thin film heat transfer or thermocouple gauges. J Phys E: Sci Instrum 21(5):445–448.

Mallinson SG (1994) Shock wave/boundary layer interaction at a compression corner in hypervelocity flows. PhD Thesis, Mech Eng Dept, University College, UNSW, Australian Defence Force Academy, Canberra, Australia.

Park C (1990) Nonequilibrium Hypersonic Aerothermodynamics, Wiley Interscience, New York.

Shih PK-T, Gay A (1985) Low lift-to-drag aerobrake heat transfer test at Mach 10. In: Thermal Design of Aeroassisted Orbital Transfer Vehicles (Nelson HF, ed), Progress in Astronautics and Aeronautics, AIAA, New York.

Schultz DL, Jones TV (1973) Heat-Transfer Measurements in Short-Duration Hypersonic Facilities, AGARD Rept No 165.

Stalker RJ (1985) Free Piston Shock Tunnel T3, A Facility Handbook, Univ of Qld, Australia.

Heat transfer in a superorbital chemically frozen boundary layer

A. Neely*, R. Morgan
Department of Mechanical Engineering
The University of Queensland
Brisbane, Queensland 4072 Australia

Abstract: The chemical behavior of flat-plate boundary layers and the associated heat-transfer rates have been investigated using 104-MJkg^{-1} and 108-MJkg^{-1} flows produced in the X1 superorbital expansion tube. The Damkohler numbers for these flows indicate that the chemical composition of the boundary layer will remain frozen along a plate at zero incidence, in agreement with measurements. Inclination of the surface to the flow raises the density in the boundary layer and increases the likelihood of the recombination reactions proceeding. This effect has been observed in the measurement of incident heating rates.

Key words: High enthalpy, Heat transfer, Boundary layer, Experiment, Superorbital expansion tube, Recombination, Damkohler number

1. Introduction

The flat-plate heating in high-enthalpy flow depends on the boundary-layer flow chemistry. There is a large component of chemical enthalpy in the form of dissociated species that is available for heating. The heating rate is evaluated for the two extremes of: (i) a frozen non-catalytic boundary layer, in which none of the chemical energy is returned to the flow and (ii) an equilibrium boundary layer or a fully catalytic surface, in both of which the chemical enthalpy is fully returned to the flow. In the frozen case, the wall and reference conditions are evaluated assuming constant composition across the boundary layer. In the equilibrium and catalytic cases, the composition must be determined from the local gas states. Here the chemical enthalpy which may be liberated through recombination must be included in determining the recovery enthalpy.

For this study it has been assumed that the free stream is in equilibrium. This has been shown to be a reasonable approximation for the flows in this facility (Neely & Morgan 1994). Therefore the tendency towards equilibrium from a dissociated non-equilibrium state should not be a driving force for re-

*Present address: Department of Engineering Science, University of Oxford, Parks Road, Oxford, OX1 3PJ, UK.

combination. If the reference temperature is lower than the temperature at the boundary-layer edge there will be a net tendency toward recombination. The gas-phase Damkohler number is evaluated to determine whether recombination will occur. It is the ratio of the diffusion time to the reaction time. For Damkohler numbers less than unity, the likelihood of the reaction occurring is small. The composition is said to be frozen typically for Damkohler numbers less than the order of 10^{-2} (Roberts 1991).

If the boundary layer remains frozen, there is the possibility of the dissociated species diffusing to the surface and recombining there via catalysis. The surface Damkohler number is the ratio of the characteristic surface reaction rate to the characteristic surface diffusion rate. If the surface Damkohler number is small, then the diffusion rate is much greater than the surface reaction rate and the surface reaction will be frozen and all surfaces will be effectively noncatalytic (East et al. 1980). If it is large then virtually all the dissociated species reaching the wall will be recombined, increasing the heat-transfer rate. Metals are far more likely to be catalytic than non metals. Vitreous materials, such as the quartz substrate and the SiO_2 coatings covering the platinum gauges, should not induce recombination, having low rate constants. However, if the SiO_2 layer was to be eroded, uncovering the platinum, surface catalysis may proceed.

2. Experimental measurement of flat-plate heating rates

The geometries and mounting configurations of the flat plates used in the experiments are shown in Fig. 1.

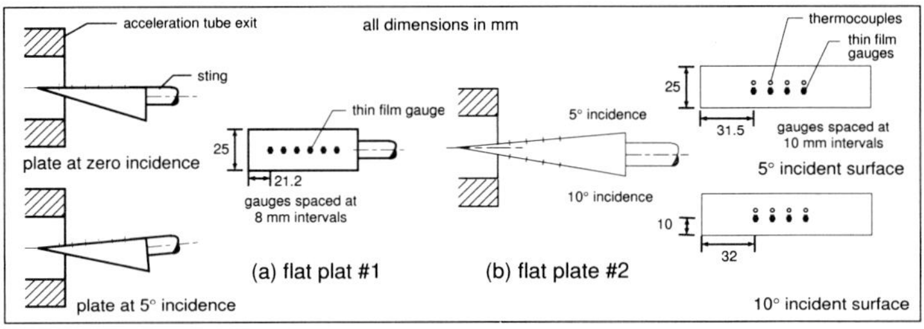

Figure 1. Flat-plate geometries for heat-transfer experiments.

2.1. Discussion of non-incident measurements

The test-flow states in Table 1 were used to calculate conditions at the boundary-layer edge and at the surface using pressure distributions determined by the strong interaction principle for a cold wall. Heating rates calculated from these conditions are compared with measurements for 108 $MJkg^{-1}$ (Fig. 2).

Table 1. Test flows calculated from equilibrium (Neely and Morgan 1994).

Flow condition		104 $MJkg^{-1}$	108 $MJkg^{-1}$
flow velocity	(ms^{-1})	12400	13000
pressure	(kPa)	3.9	5.8
temperature	(K)	6015	5935
density	(10^{-2} kgm^{-3})	0.137	0.215
static enthalpy	($MJkg^{-1}$)	26.5	23.2
total pressure	(MPa)	134	468
dissociation fractions	N, O	0.64, 1.00	0.47, 1.00

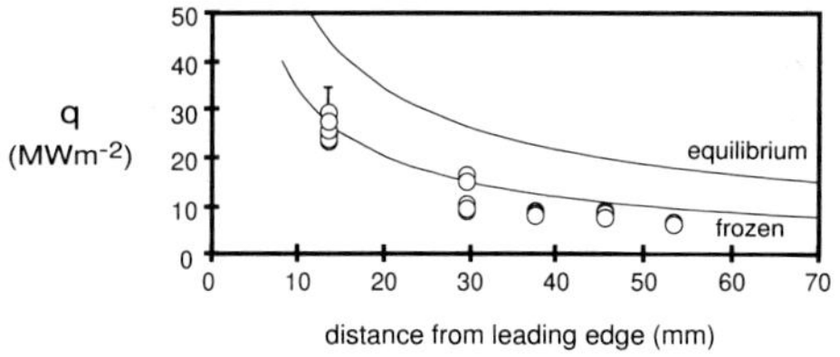

Figure 2. Measured heating rates at zero incidence, 108 $MJkg^{-1}$.

For both recombination reactions the gas-phase Damkohler numbers along the plate are more than two orders of magnitude less than unity, indicating likely chemical freezing in the boundary layer (Fig. 3a). For the length of the

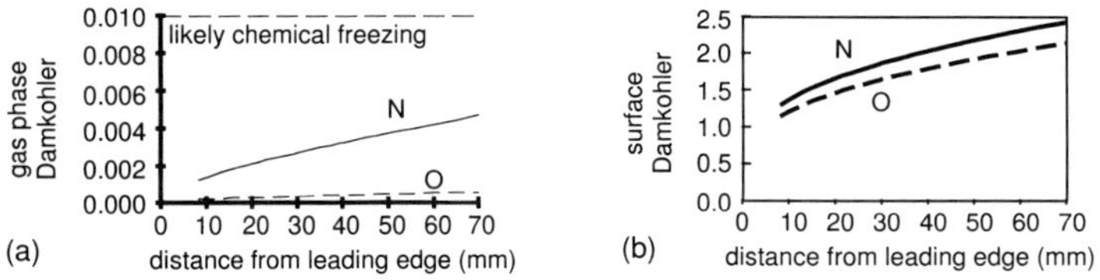

Figure 3. (a) Gas phase and (b) surface Damkohler numbers for recombination along a non-incident surface at 108 $MJkg^{-1}$.

plate, the average temperature within the boundary layer is greater than at the edge despite the cold wall, indicating no net driving potential toward recombination. The net temperature rise within the boundary layer could induce further dissociation of the nitrogen component which, being endothermic, would result in a decrease in the heat-transfer rate, but this is not seen in the equilibrium calculations or the measurements. The surface Damkohler numbers (Fig. 3b) show the diffusion rate and the surface reaction rate to be of similar order, suggesting that recombination will be small.

The measurements agree reasonably with Eckert's (1955) empirical correlation (Fig. 4). Reynolds analogy is found to be less satisfactory. The method of Ott and Anderson (1994), in which Reynolds analogy is corrected to account for nonequilibrium effects, is found to be in better agreement with Eckert's method and the measurements. Van Driest's (1952) numerical calculations are

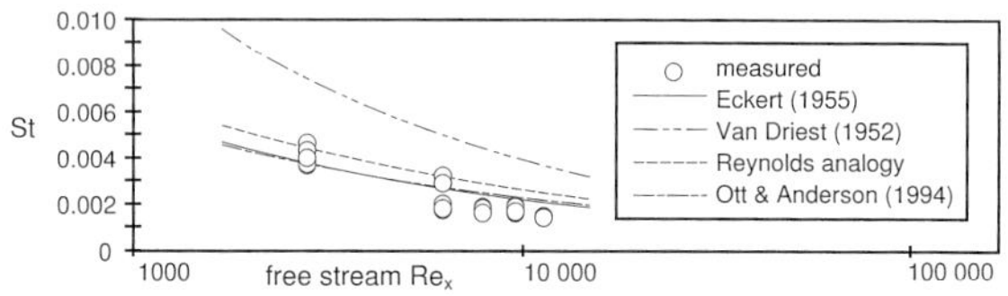

Figure 4. Predictive heating calculations for non-incident 108-MJkg^{-1} flow.

found to significantly over predict the heating rates. This would be expected as these calculations do not account for the possibility of chemical freezing. The experimental results are added to the summary of high-enthalpy heating data presented by Mallinson (1994) (Fig. 5). The measurements at 108 MJkg^{-1} continue the trend observed at lower enthalpies.

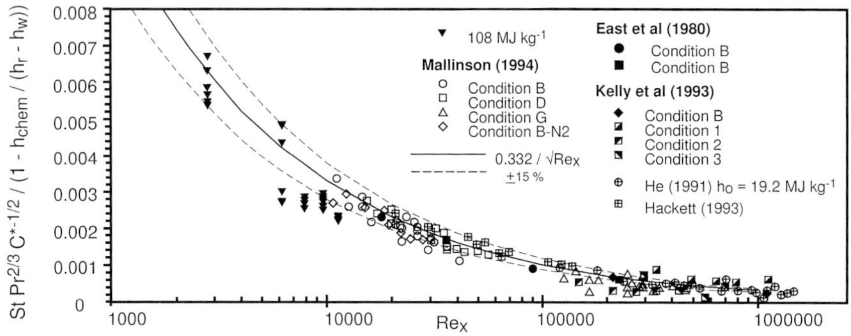

Figure 5. Comparison of non-incident flat-plate heating measurements at high enthalpy (after Fig. 3.5, Mallinson 1994).

2.2. Discussion of incident measurements

The gas-phase Damkohler numbers for the incident flows (Fig. 6a) indicate that nitrogen recombination may no longer be frozen toward the end of the plate at 10° for the 104-MJkg^{-1} flow. At 108 MJkg^{-1} the nitrogen recombination may not be frozen toward the end of the plate at 5°, or along the length of the plate at 10°. The incident data are limited due to deterioration of the gauges via erosion. This may permit recombination catalyzed by the platinum film.

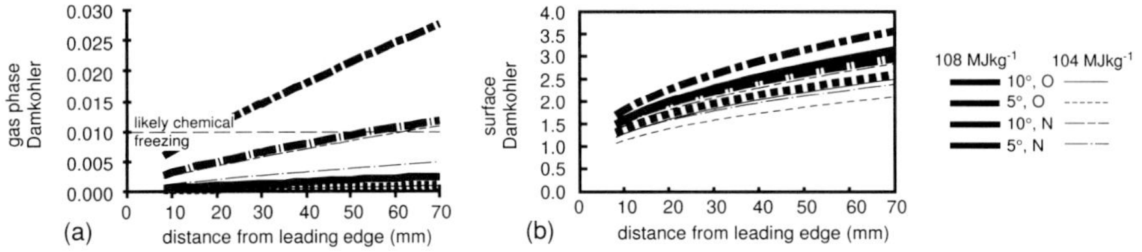

Figure 6. (a) Gas phase and (b) surface Damkohler numbers for recombination along an incident surface.

The surface Damkohler numbers increase with incidence (Fig. 6b), indicating a greater likelihood of recombination. The incident heating rates indicate different behavior (Fig. 7). At the lower incidence, for 108 $MJkg^{-1}$, the measurements follow the frozen predictions as indicated by the gas phase Damkohler numbers. From the limited measurements for 104 $MJkg^{-1}$ at 5°, the agreement is less satisfactory. The leading gauge indicates frozen behaviour as expected but the rearmost gauge indicates recombination.

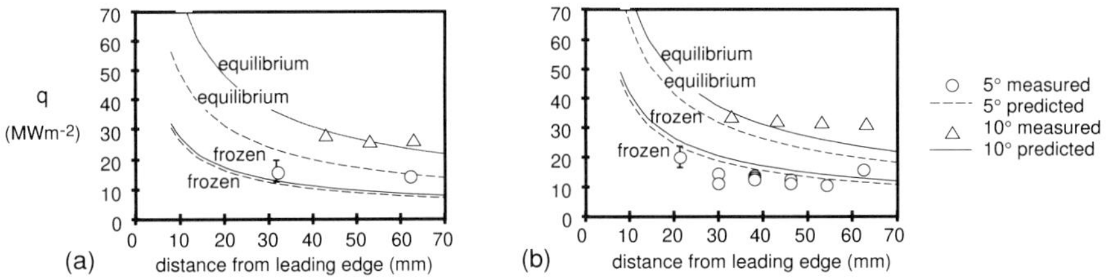

Figure 7. Incident heat-transfer rates for (a) 104-$MJkg^{-1}$ and (b) 108-$MJkg^{-1}$ flows.

At 10°, the measurements for both conditions more closely match the equilibrium predictions, indicating recombination of the dissociated species, in agreement with the Damkohlers. The levels towards the rear of the plate at 10° for 108 $MJkg^{-1}$ are higher than would be expected for even the full contribution of the chemical enthalpy to the heat transfer through recombination. Radiation should not contribute to the heating at the incidences considered.

3. Conclusions

These experiments use the superorbital expansion tube to extend the enthalpy range of heat-transfer measurements in air to beyond 100 $MJkg^{-1}$. At these enthalpies there is total dissociation of the oxygen and significant dissociation of the nitrogen in the free stream. For the geometries considered, the gas-phase and surface Damkohler numbers indicate the likelihood of chemical freezing in the boundary layer on a flat plate at zero incidence in agreement with the measurements. At incidence to the flow, the gas-phase and (to a lesser extent) the surface Damkohler numbers indicate the likelihood of recombination. This is borne out reasonably by the measurements. Hence calculation of the gas-phase and surface Damkohler numbers is found to give a good indication of the chemical behavior in boundary-layer flows at superorbital enthalpies.

Acknowledgement. Financial support for this work was provided by the Australian Research Council.

References

East RA, Stalker RJ, Baird JP (1980) Measurements of heat transfer to a flat plate in dissociated high-enthalpy laminar air flow. J Fluid Mech 97:673.

Eckert ERG (1955) Engineering relations for friction and heat transfer to surfaces in high velocity flow. J Aeronaut Sci 22:585.

Hackett (1993) Aerothermodynamic Heating Due to Shock Wave/Laminar Boundary-Layer Interactions in High-Enthalpy Hypersonic Flow, AIAA 93-3135.

He Y (1991) Transition and Heat Transfer in Compressible Boundary Layer Flow Over a Flat Plate, Ph.D. Thesis, Dept of Mech Eng, The University of Queensland.

Kelly GM (1993) A Study of Reynolds Analogy in a Hypersonic Boundary Layer Using a New Skin Friction Gauge, Ph.D. Thesis, Dept of Mech Eng, The University of Queensland.

Mallinson SG (1994) Shock Wave/Boundary Layer Interaction at a Compression Corner in Hypervelocity Flows, Ph.D. Thesis, Dept of Aerospace and Mech Eng, University College, ADFA, University of New South Wales.

Neely AJ, Morgan RG (1994) The superorbital expansion tube concept, experiment and analysis. Aeronaut J 946:175.

Ott JD, Anderson JD Jr (1994) Effects of nonequilibrium chemistry on the reference temperature method and Reynolds analogy. J Thermophys Heat Trans 8:381.

Roberts GT (1991) Investigation of Surface Recombination Effects on Convective Heat Transfer in Dissociated Air, AASU Memo 91/1, Dept of Aeronautics and Astronautics, University of Southampton.

Van Driest ER (1952) Investigations of Laminar Boundary Layers in Compressible Fluids Using the Crocco Method, NACA TN-2597.

Standard model experiments in high-enthalpy facilities

S. Ueda
National Aerospace Laboratory, Kakuda Research Center
1 Koganesawa, Kimigaya, Kakuda, Miyagi 981-15 Japan

K. Hannemann, G. Eitelberg
DLR SM-SK
Bunsenstrasse 10, 37073 Göttingen, Germany

Abstract: Experiments on the blunt cone model were performed in the HEG high-enthalpy wind-tunnel facility. The model was instrumented for both pressure and heat-flux measurements. It was tested at two diaphragm burst conditions with two respective enthalpies. Holographic interferometry was employed for the flow-field visualization. Test results were compared with numerical calculations. The same model was also tested in the T5 facility at the California Institute of Technology (Caltech). A comparison of the results from these two free-piston driven shock tunnels is also presented.

Key words: Free-piston shock tunnel, Tunnel calibration, Experiment, Heat-flux rate, Wall pressure, Flow visualization

1. Introduction

The development of hypersonic flight vehicles requires accurate knowledge of the flow field around the vehicle. Several facilities have been newly developed for the study of high-enthalpy phenomena. One of these is the HEG in DLR Göttingen Germany (Eitelberg et al. 1992). This facility, a free-piston driven shock tunnel operating in the reflected mode, was established for work in the hypervelocity flow regimen where dissociative effects of air are significant. A standard model experiment was required to evaluate tunnel characteristics. A blunt cone, referred to as ELECTRE, for which flight data exists, was selected as a standard model for reference testing. This model was selected for the following reasons. It exhibits sufficient bluntness to produce dissociation/recombination behind the bow shock in the stagnation region. The expansion from the blunt spherical part to the slender conical section might be rapid enough to display significant non-equilibrium in the expansion fan. It is a simple enough model to be transferred from tunnel to tunnel and to be used to study the tunnel characteristics as they influence the non-equilibrium chemistry as described above. The aim of this study is to provide quantitative data for both numerical code developers as well as facility operators.

2. Experimental

Figure 1 shows a model design 1/6 the size of the ELECTRE flight model. The model came in two halves, the upper one outfitted with pressure transducers and the lower one with thermocouples for heat-flux measurements. A series of 10 shots were conducted on the model in the HEG. Four test conditions, two stagnation enthalpies, and two stagnation pressures were selected based on calibration shots. Table 1 shows test conditions in the HEG. The same model was also tested in the T5 at Caltech (Rousset & Adam 1993). T5 is also a free-piston driven shock tunnel operating in the reflected mode. By comparison, T5 has a smaller nozzle-area ratio than the HEG, and differential interferometry was used instead of holographic interferometry (Kastell et al. 1993). Six test conditions, consisting of two stagnation enthalpies, two stagnation pressures, and three nozzle throat diameters were used.

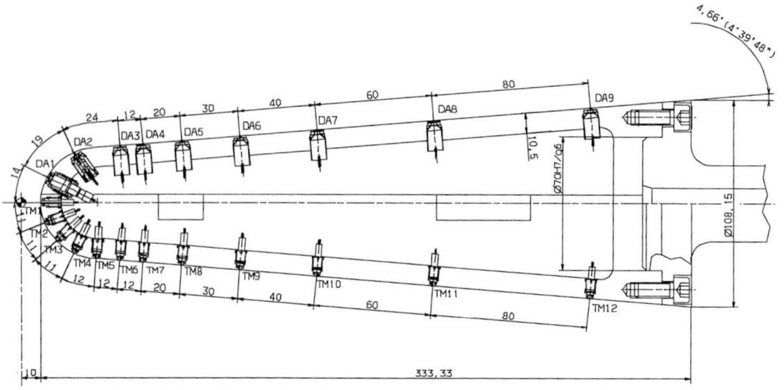

Figure 1. Perspective of the instrumented model.

Table 1. Test conditions in the HEG.

Condition	Reservoir Pressure (MPa)	Enthalpy (MJ/kg)	Nozzle Exit Velocity (km/s)	ρL (g/m^3)
I	38.6	21.1	6.49	0.52
II	90.9	22.3	6.68	1.14
III	45.0	13.2	5.14	0.90
IV	111.0	14.8	5.44	1.95

3. Results and discussion

In free-piston driven shock tunnels like HEG and T5, the mounting of the pressure transducer has to be designed carefully because of the short test time and severe environment. Flush mounting is the best way to improve response time. However, the pressure transducer has to be protected from high temperature and particles in the gas. Pressure-transducer mounting used in the model is shown in Fig. 2a. The pressure is introduced by a hole and goes a long way around a screen to reach the pressure transducer. Several mounting systems were tried to improve response time and prevent damage at the same time. Figure 2b shows the measured pressure history in the conical portion using various transducer mountings. Response time $\tau = \frac{V}{Ac}$, where V is the volume of cavity, A is the area of hole, and c is the speed of sound of the test gas, is required to fill the volume of the mounting system with test gas. Under HEG test conditions, the calculated response times of these mounting are approximately (a) 0.15 ms, (b) 0.07 ms, (c) 0.02 ms at the stagnation region, and (a) 0.4 ms, (b) 0.2 ms, (c) 0.06 ms at the conical portion. These values agree quite well with measured values.

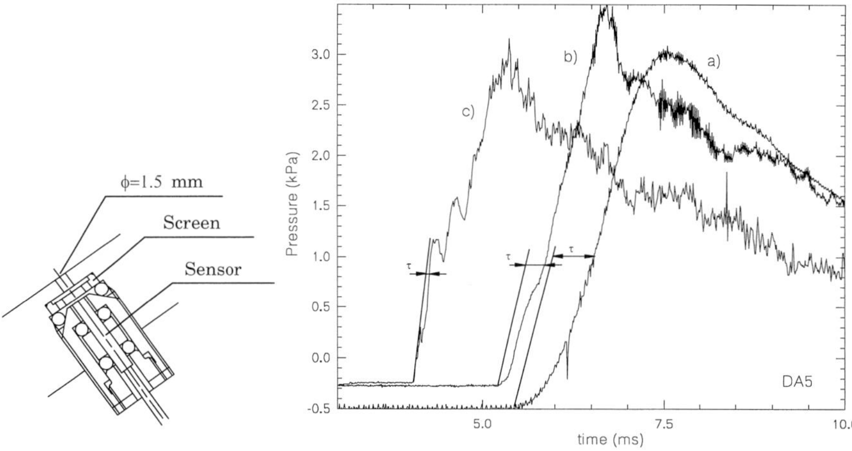

Figure 2. (a) Mounting type b (1.5-mm hole with screen) and (b) response time of various mountings: (a) 1-mm hole with screen, (b) 1.5-mm hole with screen, and (c) 1.5-mm hole without screen.

3.1. Pressure and heat-flux rate

Kulite resistance-type pressure transducers and Medtherm coaxial thermocouples were used for measurement. Measured pressures and heat-flux rates were normalized by calculated free-stream conditions. A non-equilibrium nozzle code was used to calculate these values. Figure 3 shows the pressure coefficient $C_p = \frac{p-p_\infty}{\frac{1}{2}\rho_\infty u_\infty^2}$ and Stanton number $St = \frac{\dot{q}}{\rho_\infty u_\infty (h_s - h_w)}$ distributions along the wall, where ρ, u, p, h, and $\dot{q}$ are respectively density, flow velocity, enthalpy, and heat-flux rate to the wall. Comparison between experimental results and numerical calculations using a non-equilibrium Navier-Stokes code (Brenner et al. 1993) are also presented. Figure 3a shows that slightly higher pressures were measured in the conical portion; this was due to the fact that values of static pressure used to calculate pressure coefficient were insufficiently accurate. Normalization of pressure by pitot pressure is another possible way to compare experimental and numerical results. Figure 3b shows that the Stanton numbers obtained experimentally and numerically are in good agreement. However, the heat-flux rate was not well normalized by Stanton number. The viscous effect has to be considered to obtain better normalization in this regard. Reduced Stanton number by Miller et al. (1983) $St_R = St\frac{(\rho_2/\rho_\infty)^{1/4}}{V^*}$ is one way to consider viscous effect, where V^* and ρ_2 are respectively the parameter of viscous interaction and the density behind a normal shock.

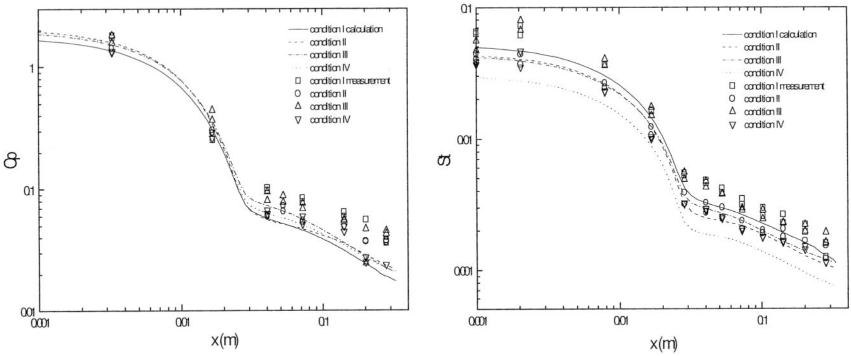

Figure 3. ELECTRE test results obtained in the HEG compared with results obtained by numerical calculation: (a) pressure coefficient and (b) Stanton number.

3.2. Flow visualization

Figure 4a is a finite-fringe interferogram of a HEG shot for the nominal burst pressure condition of 100 MPa at lower enthalpy, condition IV. In the field of view is the spherical part of the model. Figure 4b shows the shock shape obtained by holographic interferometry (HI) and Navier-Stokes calculations. Shock shapes were read from Mach contour graphs and HI photos. The experimentally observed shock shape agrees with the calculated shape downstream of the stagnation region. In the stagnation region, the observed shock standoff distance lies between that predicted by non-equilibrium Navier-Stokes and perfect gas calculation (Van Dyke & Gordon 1959, Billig 1967).

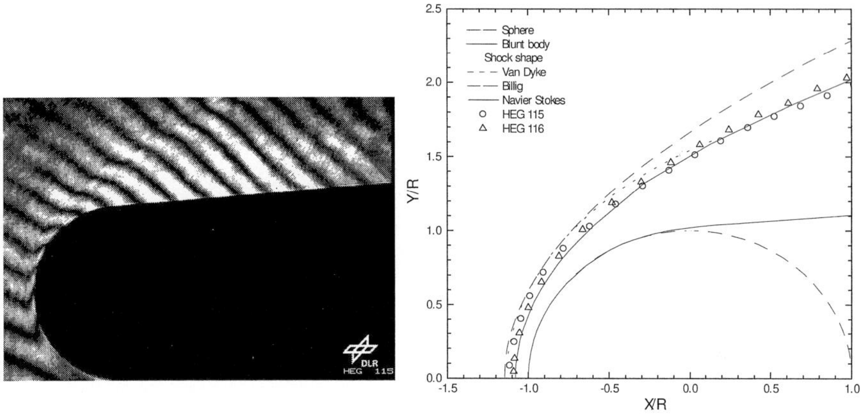

Figure 4. (a) Picture obtained using holographic interferometry. (b) Comparison of shock shape obtained using HI, non-equilibrium Navier-Stokes calculation, numerical calculation according to Van Dyke & Gordon (1959) for an ideal gas around a sphere, and an approximation of Billig (1967).

3.3. Comparison between results obtained the HEG and T5

It is difficult to compare measured results obtained in the HEG and T5 facilities because of the differences in test conditions and nozzle-area ratio. Normalized pressure by pitot pressure and reduced Stanton number by Miller et al. (1983) were used for comparison. Figure 5 shows the results. Test results were averaged for each facility. The vertical line on each set of values shows the standard deviation of test results. The distributions of pressure and the rate of heat flux in the conical portion were flatter at T5 than at HEG. Bellucci et al. (1994) indicated that this is caused by non-uniform density distribution in the T5 nozzle. This is caused by off-design use of contoured nozzle.

4. Conclusions

Standard model experiments were conducted in the HEG and T5 wind-tunnel facilities. Values obtained for the distribution of pressure and heat flux along the wall were normalized and compared. Normalized pressure by pitot pressure and St_R were seen to be good parameters to compare different test conditions and different facilities. Non-uniform density distribution in T5 caused observed differences in the pressure and heat-flux rate distributions. Accurate information on free-stream conditions is required to improve agreement. Shock shape and shock detachment distance were measured and compared with numerical calculations. The observed shock stand-off distance lies between that predicted by non-equilibrium Navier-Stokes and perfect gas calculations.

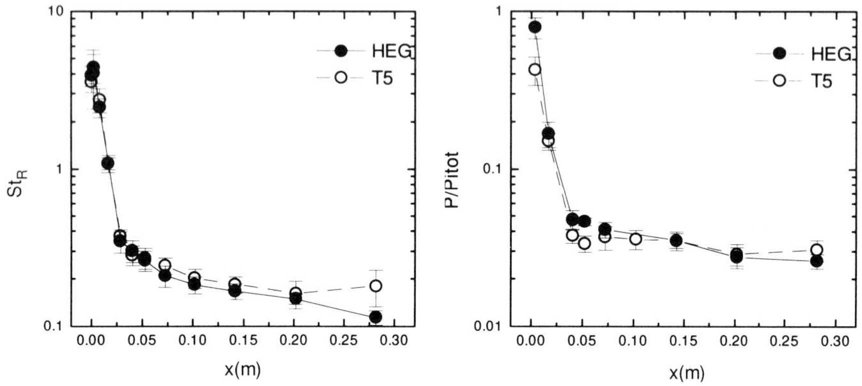

Figure 5. Comparison between HEG and T5 experimental results: (a) pressure normalized by pitot pressure and (b) reduced Stanton number.

Acknowledgement. The authors express their gratitude to J. Muylaert of ESTEC for his support of the model development, D. Kastell for the HI pictures, and the T5 and HEG groups for their active participation in this test series.

References

Bellucci V, Ueda S, Eitelberg G, Muylaert J (1994) Numerical Simulation of Electre in T5 and HEG Hypervelocity Shock Tunnels. EWP-1776.

Billig FS (1967) Shock wave shapes around spherical- and cylindrical-nosed bodies. Journal of Spacecraft and Rockets 4(6):822–823.

Brenner G, Gerhold T, Hannemann K, Rues D (1993) Numerical simulation of shock/shock and shock-wave/boundary-layer interactions in hypersonic flows. Comput Fluids 22(4/5):427–439.

Eitelberg G, McIntyre TJ, Beck WH, Lacey J (1992) The High Enthalpy Shock Tunnel in Göttingen, AIAA Paper 92-3942.

Kastell D, Kortz D, Eitelberg G (1993) In high enthalpy flows studying shock/shock interaction. In: Proceeding of Euromech 309, Optical Flow Diagnostics, Göttingen, Germany.

Miller CG, Gnoffo PA, Micol JR (1983) Heat-transfer distributions for biconics at incidence in hypersonic hypervelocity real-gas flows. In: Proc 14th Int Symp on Shock Tubes and Waves, Sydney, pp 333–341.

Rousset B, Adam P (1993) Electre Experiments in T5. FM 93-2, GALCIT, Caltech, Pasadena, CA.

Van Dyke MD, Gordon HD (1959) Supersonic Flow Past a Family of Blunt Axisymmetric Bodies, NASA TR R-1.

Radiation around a hypersonic ablative flight model

K. Komurasaki, S. Yano, T. Fujiwara
Department of Aerospace Engineering, Nagoya University
Nagoya 464-01 Japan

Abstract: Thermochemical structure around an ablative flight model at hypersonic speed is investigated by using a ballistic range. A polyethylene projectile of 1 cm diameter is launched at 5 km/sec into an observation chamber, and the light emitted from an induced shock layer around the projectile is observed with a spectroscope and a high-response power meter. From the emission spectrograms, ablation products C_2, CN, and CO are found to cause radiation comparable to air molecules N_2 and NO. The emission-power profile along the flight path of the projectile shows a tail layer in which nonequilibriumly excited molecules emit radiation behind the projectile.

Key words: Hypersonic flow, Radiation, Ballistic range, Ablation

1. Introduction

Sample-return missions which return a set of samples from a near-Earth asteroid to Earth have been proposed for scientific interests (Wallace et al. 1985, ISAS 1995). In such missions, the re-entry Mach number is beyond 40 and aerodynamic heating around the capsule will be the order of 10 MW/m^2. Therefore, heat protection becomes a serious problem in developing the capsules (Walberg 1985, Howe 1985, Park 1985). Ablation cooling is an essential technique when the heat flux reaches several MW/m^2 because there exists no material that can endure such an extremely large heat flux.

Reliable data of the heat-transfer rate around the capsule are needed to minimize the ablator weight. However, the prediction is very difficult because at such a hypersonic speed, nonequilibrium radiation plays as important a role as convection. In addition, the interaction between a shock-heated gas and an ablative surface is very complicated (Page and Arnold 1964), and radiation from ablation products, mostly carbon-containing molecules, makes prediction more difficult (Komurasaki et al. 1993). Strong radiation from the carbon-containing molecules enhances radiative heat flux to the vehicle and reduces the effectiveness of ablation cooling.

Available information about chemical reactions and radiative properties of ablation products is very limited. Therefore, measurements on ablative shock layers would be very meaningful for developing a reaction and radiation model of ablation products, and for exactly predicting the radiative-heat-transfer rate over ablation-cooled re-entry vehicles.

The objective of this study is to experimentally examine the thermochemical structure around an ablative projectile to obtain a primary understanding of the phenomena related to ablation, and to develop reaction and radiation models of ablation products.

2. Experimental apparatus

As shown in Fig. 1a, the ballistic range consists of a high-pressure driver, a pump tube, the launch tube, and a test section (vacuum chamber). In the first stage, a high-pressure reservoir and a quick-valve system are utilized, instead of the conventional set of gunpowder and a diaphragm, for the purpose of reproducibility and operational convenience. This gun is capable of launching a projectile of 1 gram weight at the speed of 5 km/sec ($M = 15$) despite its total length of only 6.8 m, which is much shorter than other similar facilities. The projectile flight speed is measured by the laser-cut method. The flight duration of the projectile in the test chamber is enough for the model surface to ablate, due to the extremely large heat flux over the projectile (Yano et al. 1993).

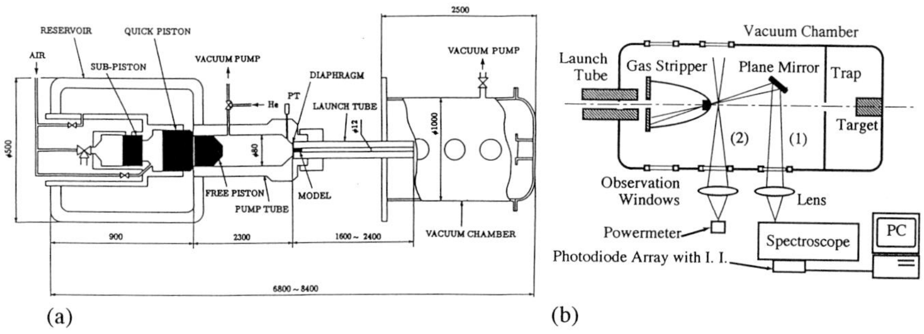

Figure 1. Schematic diagrams of (a) a ballistic range and (b) an optic collection arrangement.

The light emitted from a shock layer is collected from the front of the projectile in an optic system (shown in Fig. 1b). Spectral intensity distributions are obtained using a spectroscope (JAS Co. CT50N). Wide-range spectrograms were taken on instant film (ASA3000), and high-resolution ones are obtained with a multi-channel photo-diode array detector (HAMAMATSU PMA–50). In order to avoid collecting the light induced by the collision between the projectile and a target, a big trap room is assembled around the target. In addition, a gas stripper is set up at the end of the launch tube to prevent the driver gas (helium) from interacting with the shock and tail around the projectile.

The emission-power profile around a projectile is obtained by setting the optical collection path perpendicular to the flight path as shown in Fig. 1b. The temporal variation of the emission power is measured using a Si-PIN photodiode powermeter (YOKOGAWA 3296). The powermeter response time is less than 5 ns, corresponding to a spatial resolution of 0.025 mm. The field of view at the focal point is 7.5 mm high and 1.7 mm wide.

3. Spectral intensity distribution

The spectral intensity distribution of emission for a polyethylene projectile is shown in Fig. 2a. Several band spectra from air molecules like N_2 and NO are identified, while no atomic line spectra are identified.

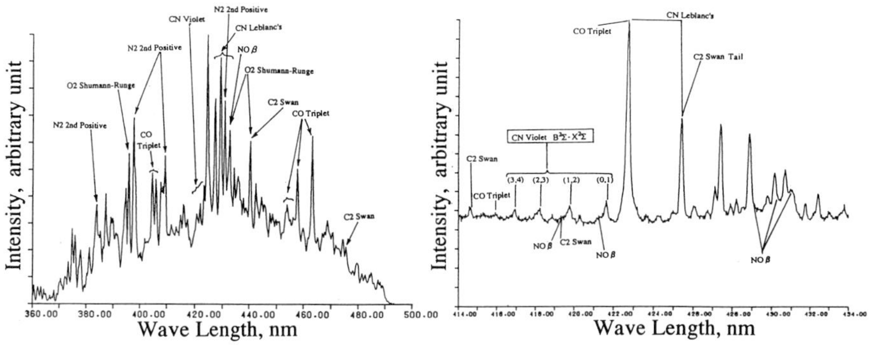

(a) (b)

Figure 2. Spectral intensity distributions of radiation: (a) wide-range spectrogram; (b) high-resolution spectrogram, $V = 4.8$ km/s, $p_\infty = 10$ Pa.

In addition to air spectra, those from several carbon-containing molecules, C_2, CO, and CN, are detected. C_2 is thought to come from polyethylene ablation and subsequent dissociation reactions, while the other molecules, CO and CN, are thought to be produced though the exchange or recombination reactions between dissociated carbon atoms and air molecules.

Figure 2b shows the high-resolution spectral intensity distribution. Several CN violet transition spectra appear on the continuum spectrum. The vibrational temperature of the CN molecules can be estimated from those transitions.

The Boltzmann plot is shown in Fig. 3. Although the CN vibrational temperature is determined to be 3500 ± 700 K from the intensities of the (0-1) and (1-2) transitions, the populations are largely separate from Boltzmann equilibrium. The molecules are populated more in the upper vibrational levels than equilibrium ones. This indicates that CN molecules are in a rising nonequilibrium situation where CN number density increases from zero. The molecules in highly excited vibrational states tend to be in equilibrium with those in free states through radiative de-excitation when collisional de-excitation is not predominant.

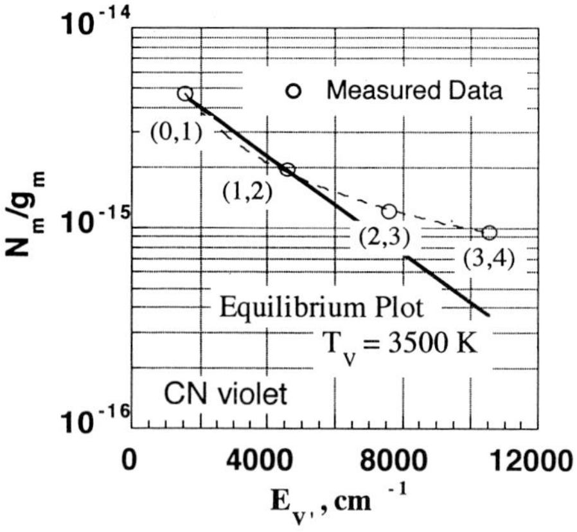

Figure 3. Boltzmann plot for a CN violet band system.

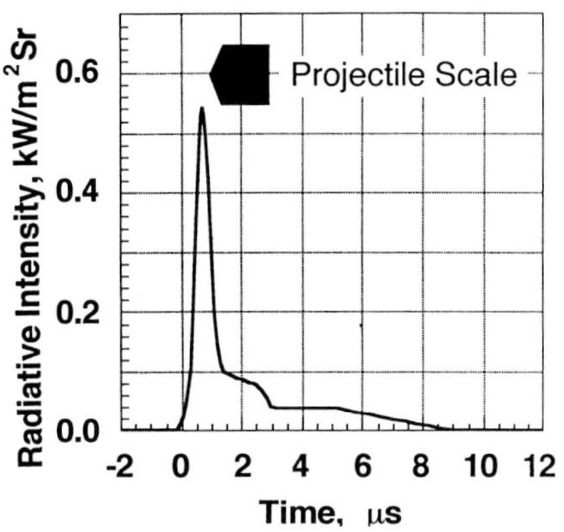

Figure 4. Temporal variation of radiation, $V = 4.5$ km/s, $p_\infty = 1000$ Pa.

4. Emission power distribution

Figure 4 shows the emission-power profile obtained by setting the optical collection path perpendicular to the flight path. Since the flight speed is constant, temporal variation is identical to the spatial distribution in the flight direction. The power increases steeply in the shock layer ahead of the projectile and then decreases. There is still emission in the side region of the projectile, and a weakly illuminated tail follows several centimeters behind the projectile.

The calculated emission power and temperature distributions simulating the experiment are illustrated in Fig. 5. The calculation is based on a two-temperature reaction and nonequilibrium radiation model, not including ablation and carbon reactions (Komurasaki et al. 1994). The calculated emitting region is limited to the portion ahead of the projectile. The discrepancy between measured and calculated profiles might come from carbon reactions. That is, in the side and base regions, the generated carbon-containing molecules (rather than the air molecules) emit radiation.

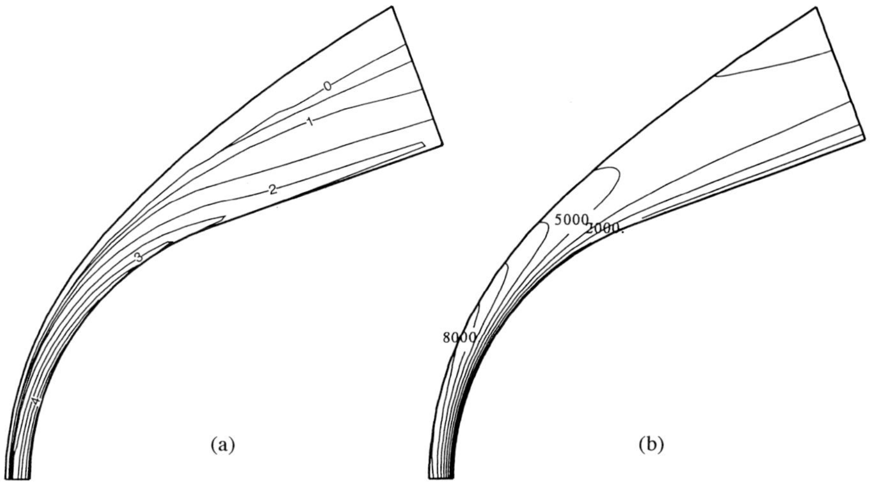

Figure 5. Calculated contour maps of (a) global emission power $\log(J)$, (J, W/m^3) and of (b) translational temperature, T (K); V = 5 km/s, p_∞ = 500 Pa, T_∞ = 300 K.

5. Reactions of ablation products

The ablated carbons are decomposed in the shock layer through the reactions

$$C_3 + M \rightarrow C_2 + C - 174 \text{ kcal/mol} \quad (1)$$

$$C_2 + M \rightarrow C + C - 142 \text{ kcal/mol} \quad (2)$$

Owing to these reactions, heat flux to the projectile surface is considerably reduced. Dissociated carbon can react with oxygen and nitrogen. Among the carbon associated reactions, CN formation reactions are listed as follows:

$$C + N_2 \rightarrow CN + N - 48 \text{ kcal/mol} \quad (3)$$

$$CO + N \rightarrow CN + O - 78 \text{ kcal/mol} \quad (4)$$

$$C + NO \rightarrow CN + O + 28 \text{ kcal/mol} \quad (5)$$

$$C_2 + N \rightarrow CN + C + 36 \text{ kcal/mol} \quad (6)$$

$$C + N + M \rightarrow CN + M + 178 \text{ kcal/mol} \quad (7)$$

In the high-temperature region inside the shock layer, endothermic carbon-associated reactions also contribute to the cooling. However, the exothermic ones would occur in an expanding flow in a tail where the gas is of relatively low temperature.

6. Conclusion

The carbon vapor ablated from polyethylene is dissociated into C_2 and C in a shock layer, absorbing the heat over the front surface of a projectile. The absorbed heat is carried away with the carbon to the tail behind the projectile, where flow is expanding and the gas translational temperature decreases quickly. Therefore, the carbon atoms react with air molecules in the tail. The highly excited carbon-containing molecules release a large part of their vibrational energy through spontaneous emission. The tail, emanating radiation, spreads several centimeters in length behind the projectile. Since carbon-containing molecules are known to radiate strongly even at low temperatures, in an actual case, carbon radiation may increase the heat flux to the vehicle and lower the effectiveness of ablation cooling. In addition, radiation in the tail brings heats to the vehicle base, so that a rear heat shield would be important in vehicle design.

References

Howe JT (1985) Introductory aerothermodynamics of advanced space transportation systems. J Spacecraft and Rockets 22:19–26.

ISAS (ed) (1995) A proposal for exploration of the asteroids, MUSES-C (in Japanese).

Komurasaki K, Fujiwara T (1994) Radiation and shock layer structure around a hypersonic flight model. In: Proc 19th Int Symp Space Tech and Sci, Yokohama, pp 399–404.

Komurasaki K, Kasahara J, Yano S, Fujiwara T (1993) Radiative heat transfer from a shock layer generated around a projectile launched in a ballistic range. In: Shock Waves @ Marseille, 1, pp 127–132.

Page WA, Arnold JO (1964) Shock Layer Radiation of Blunt Bodies at Reentry Velocities, NASA TR R-193.

Park C (1985) Radiation enhancement by nonequilibrium in Earth's atmosphere. J Spacecraft and Rockets 22:27–36.

Walberg GD (1985) A survey of aeroassisted orbit transfer. J Spacecraft and Rockets 22:3–18.

Wallace RA, Blume WH, Hulkower ND, Yen CL (1985) Interplanetary missions for late twentieth century. J Spacecraft 22:316–324.

Yano S, Komurasaki K, Kasahara J, Takaoka H, Fujiwara T (1993) Radiative heat transfer to a hypersonic projectile in ballistic range experiments. In: Proc 19th Int Symp Space Tech and Sci, Yokohama, pp 405–410.

Laser-induced fluorescence measurements in high-enthalpy facilities: application to the study of air plasma/surface interaction

L. Robin, D. Honore, P. Vervisch
CNRS URA 230 – CORIA
University of Rouen
76821 Mont Saint Aignan Cedex, France

Abstract: This study presents experimental results obtained with Laser-Induced Fluorescence (LIF) in continuous high-enthalpy supersonic air flows produced at the Central Research Institute of Machine Building (Tsniimach, Moscow) in ground experimental facilities designed for the study of aerodynamics and heat transfer of supersonic and hypersonic spacecraft. The LIF technique is used to obtain the measurements of NO density, rotational temperature, and velocity evolutions in these flows. Experiments are performed with a narrow-band ArF excimer laser in free jets and in numerous test sections within the boundary and shock layers lying at the vicinity of a Thermal Protection System (TPS) model designed to simulate a misalignment of tiles.

Key words: Laser-induced fluorescence, NO, High-enthalpy air flow

1. Introduction

At the Tsniimach Institute (Moscow), the TT1 arc-jet facility is devoted to the study of local aerodynamics problems arising in supersonic high-temperature flows, while a high-frequency facility has been devoted to characterize the physico-chemistry phenomena encountered in reactive-flow/surface interaction (Anfimov 1992). The main objective of the complete program was to improve reliable data for validation of chemical nonequilibrium boundary-layer and Navier-Stokes codes to compute a new test case of supersonic air-heated flow impinging on a TPS surface. The experimental work presented here is devoted to the study of the aerodynamic flow properties at the vicinity of a misalignment of tiles over a SiO_2 model immersed in a dissociated air-heated flow. An additional aim of the experimental investigation is to point out the great potential of the LIF technique for the study of continuous high-enthalpy air-flow/body interaction.

Fluorescence of NO was induced in the plasma flow by a narrow-band tunable ArF excimer laser via the $D^2\Sigma^+_{(v'=0)} \leftarrow X^2\Pi_{(v''=1)}$ epsilon band. The rotational temperature and NO number density have been determined. The spatial resolution of the LIF technique permitted accurate characterization of boundary and shock layers, as well as shock thickness. Finally, flow velocity is deduced from Doppler-shift measurements of excited rovibrational NO-ϵ lines.

2. Model design and flow conditions

Flight-condition requirements are fulfilled in the TT1 arc-jet facility. The plasma source is a classical DC arc jet. The air flow is accelerated through a supersonic nozzle before its expansion in a low-pressure test chamber at 10^3 Pa. The divergent-shaped nozzle provides an air plasma jet at Mach 5 with a diameter of 300 mm. The stagnation chamber parameters are $P_0 = 1.6 \cdot 10^5$ Pa, $H_0 = 6.7$ MJ/kg, and $T_0 = 4500$ K, with a 4-MW input power supply and an air-flow rate of 0.5 kg/s. The free-stream conditions are $P_{st} = 2 \cdot 10^4$ Pa, $T_A = 1000$ K, and $R_{eu} = 10^6$ m^{-1}. The enthalpy leads to a supersonic, weakly dissociated air-plasma flow (oxygen: 17%). Aerodynamic studies have been performed in the supersonic nonequilibrium flow and at the vicinity of a SiO_2 model with 200 mm length and 100 mm spanwidth, with a step 3 mm high located 100 mm from the water-cooled blunt leading edge. The model is immersed at 15° incidence on the jet axis.

3. Experimental set-up

A tunable ArF excimer laser (Lambda Physik LPX 150/T) is used to induce fluorescence of NO in high-enthalpy air flow via the rotational lines of the ϵ-band $D^2\Sigma^+_{(v'=0)} \leftarrow X^2\Pi_{(v''=1)}$. In narrow-band mode, the $1.8 \cdot 10^{-4}$-nm HWHM pulsed radiation (15 ns) can be tuned with a step increment of 10^{-5} nm over the full spectral range of the broad-band ArF emission centered at 193.40 nm with a 0.5-nm HWHM. To follow through wavelengths the NO-ϵ excitation spectrum, the excitation spectrum of the NO-β band is simultaneously carried out in a home-made static cell (NO, 50 mbar, 300 K).

A schematic of the set-up is depicted in Fig. 1. The probe laser beam is moved perpendicularly to the surface of the model. Furthermore, the model is translated along the jet axis in steps of 10 mm to carry out experiments within different test sections. Fine spatial resolution down to 200 μm is achieved perpendicularly to the surface where the most significant gradients are expected. The NO-ϵ fluorescence is collected at a right angle from the excitation beam pathway on the top of the test chamber through a monochromator. The probe volume is imaged onto the entrance slit of the monochromator used as a tuning 4-Åbandpass filter. The signal is recorded on a 2-GHz numerical oscilloscope (Tektronix DSA602A) allowing its temporal resolution.

4. LIF measurements

ArF broad-band mode experiments provide the signing of species in the plasma and information on the boundary and shock thicknesses. In narrow-band mode, excitation of single rotational levels is devoted to the measurement of the rota-

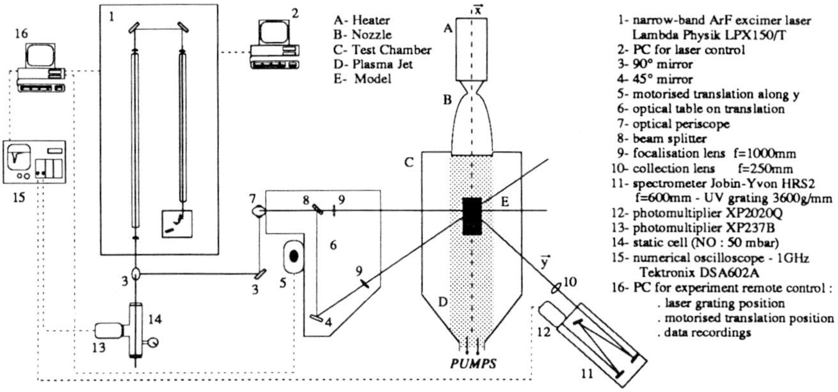

Figure 1. Experimental set-up.

tional temperature and NO density. The highly resolved excitation spectra are used to evaluate flow velocity via the Doppler shift of excitation lines.

The $D^2\Sigma^+_{(v'=0)} \rightarrow X^2\Pi_{(v''=2)}$ NO-ϵ shifted fluorescence is collected after excitation of $D^2\Sigma^+_{(v'=0)} \leftarrow X^2\Pi_{(v''=1)}$ NO-ϵ band via the $R_1(25.5)$, $P_1(38.5)$ and $Q_1(31.5)$ group of lines. These transitions are sufficiently sensitive to temperature. In addition, they are close to the $R_2(30.5)$ NO-β line excited in the NO reference static cell to provide the unshifted reference line for both laser-beam pathways used to measure the Doppler shift. A typical measurement is illustrated Fig. 2 with response of the two incident beam pathways when the excitation wavelength is scanned around the group of ϵ-lines. Temperature is deduced from the intensity ratio of the lines and compared with computed excitation spectra. Density is drawn from the integrated intensity rather than from peak intensity. According to the two excitation pathways separated by a α-degree angle incidence, the shift in frequency $\Delta\lambda$ from the selected transitions in both directions is then proportional to the flow velocity at the beam crossing point ($\frac{\Delta\lambda}{\lambda} = \frac{v}{c}\cos\alpha$). One-step increment of the grating motor fixes the accuracy of the method and corresponds to a shift of $\Delta\lambda = 1.8 \cdot 10^{-4}$ nm and to a flow velocity of 395 m/s for an incidence angle of $\alpha = 45°$.

5. Experimental results

Experiments are performed on both sides of the free jet axis and reveal a sufficient homogeneous region to locate the wide SiO_2 model and assure subsequently a 2D aerodynamic description of the boundary conditions. The unchanged rotational-line distribution and the constant fluorescence intensity point out flat NO number density and temperature (1200 K) profiles in the free-jet section. The Doppler-shifted fluorescence signals provide a flat velocity radial

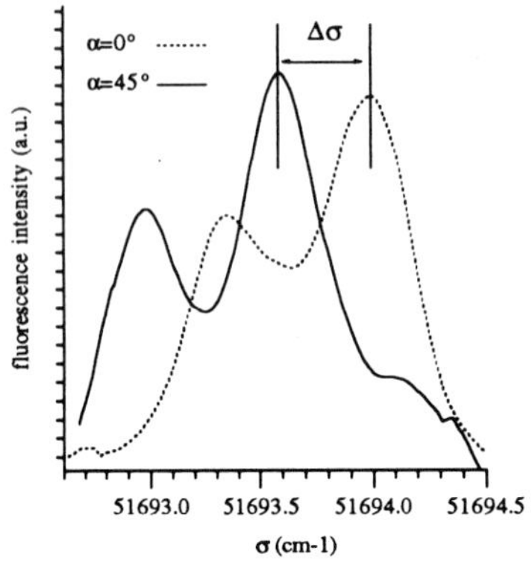

Figure 2. Doppler-shifted NO-ϵ excitation lines observed in the supersonic plasma jet.

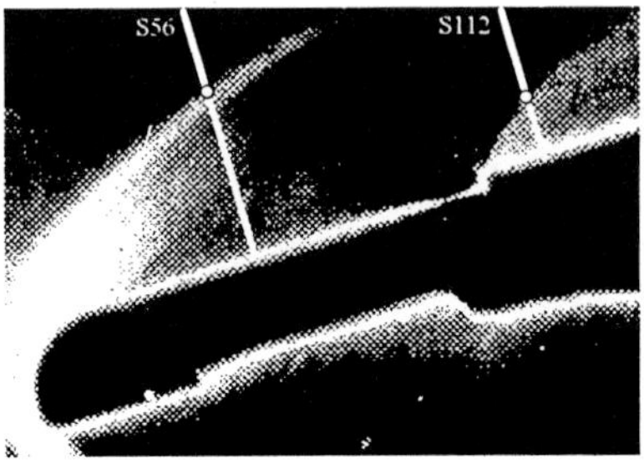

Figure 3. Shadow image of the flow over the model. White lines: sections analyzed by LIF; circle: shock position obtained by LIF.

distribution in the test section with a constant value of 3325 m/s, corresponding to a supersonic flow of Mach 4.9. These results are in good agreement with the calculated parameters: $v = 3300$ m/s and $M = 4.75$, computed with the stagnation-chamber conditions, the nozzle shape, and assumption of perfect-gas flow in equilibrium upstream of the critical convergent throat followed by a frozen expansion downstream in the nozzle divergent region.

The aerodynamic behavior of the supersonic high-enthalpy air flow impinging on the TPS SiO_2 model is shown in Fig. 3. One observes mainly the bow shock which originates from the rounded leading edge and propagates laterally away from the surface, and another less intense shock which rises from the top of the step. Infra-red thermography reveals an important temperature gradient in this region, up to 1500 K at the top of the step and down to 1100 K in the dark region at the foot of the step, where a recirculation zone can be expected. Three sections located at 56, 112, and 156 mm from the leading edge are investigated. The section S_{112} is 12 mm downstream from the step edge.

Figure 4 shows the fluorescence-signal profile from the first section. Measurements are carried out with 500-μm resolution from the free jet ($y > 0$) down to the surface ($y \leq 0$). At the separation region between the free jet and shock layer ($35 \leq y \leq 38$ mm) the fluorescence signal increases exponentially across the shock front because of NO vibrational relaxation, i.e., the higher number density of the NOϵ ($X^2\Pi_{(v''=1)}$)-level and higher vibrational temperature (McMillin et al. 1992). Downstream in the shock layer, the fluorescence intensity evolves slowly down to the model. The number density and rotational temperature do not vary significantly, indicating no significant dissociation of

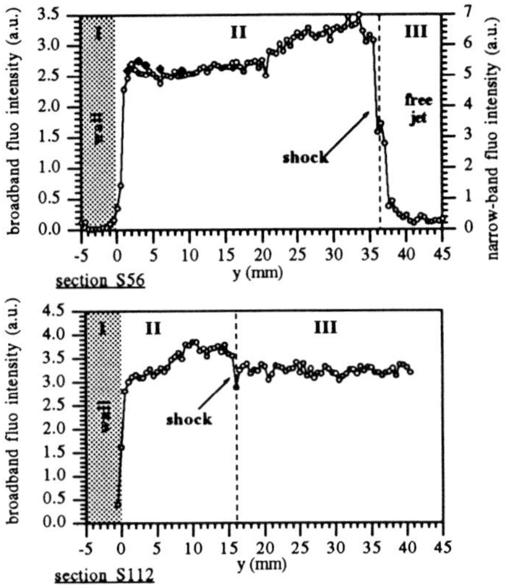

Figure 4. NO-ϵ LIF-signal evolutions. Squares: narrow-band laser excitation; circles: broad-band laser excitation.

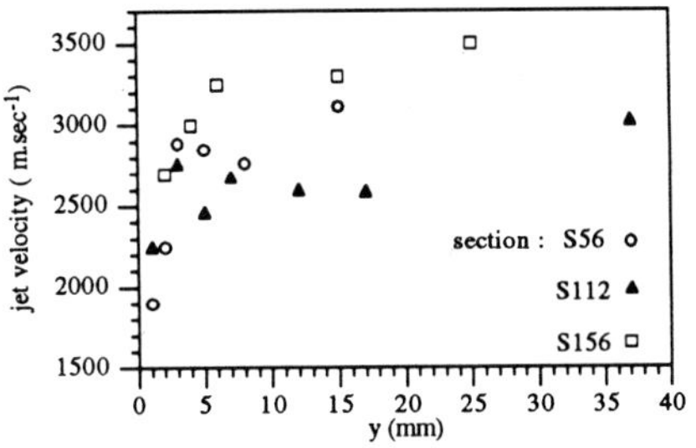

Figure 5. Doppler-shift measurements of the jet velocity over the model.

NO in the shock layer. Near the SiO_2 surface ($0 \leq y \leq 5$ mm), the weak increase of signal does not allow one to locate with accuracy the limit between the boundary and shock layers. In this weakly dissociated jet, the current results do not confirm a predominant effect of wall recombination in comparison to the compression effect of the shock layer near the surface. The equilibrium conditions in the free jet upstream are 1200 K and 10^2 Pa, and the stagnation pressure behind a normal shock is evaluated to $1.9 \cdot 10^3$ Pa, namely, 20-fold greater than the free-jet static pressure. The same order of magnitude is observed on the fluorescence signal jump across the shock propagating laterally away from the model. If one considers first the fluorescence intensity with respect to the distribution of NO $X^2\Pi_{(v''=1)}$ in the absorbent rotational levels and, secondly, the very slight variation of the rotational temperature across the shock, then the NO concentration jump $\simeq$ 22-fold greater in the shock layer than in the free jet, is assumed to be directly proportional to the difference of pressure across the shock. The spatial resolution of 5 points/mm accurately describes the location and intensity of the shock. However, the real thickness of the shock is overestimated because the vibrational relaxation time is lower than the rotational one. The LIF results in this first section are in good agreement with the ombroscopy picture which reveals only the most intense vibrational relaxation zone, without giving information on the intensity on both sides of the shock. The same results are obtained for the location of the second shock ($y \simeq 15$ mm) originating from the step (Fig. 4). Its lower intensity indicates no real change in vibrational relaxation. This is confirmed with the insignificant evolution of the rotational temperature and NO number density.

The axial component of the velocity is affected by the presence of the model. The profiles drawn from the three test sections of interest show a velocity gradient mainly located in the last millimeters close to the surface, in the region where a slight decrease of the fluorescence has been previously observed but not explained (Fig. 5). No discrepancy is observed for the axial component of the velocity across the shock front. Its value is about 3300±200 m/s, which corresponds to the initial free-jet velocity. Upstream of the step, the thickness of the dynamic boundary layer is given at $y = 5$ mm in the first section. Under this limit, the velocity decreases continuously down to $y = 1$ mm to reach its lower value of about 1900 m/s. The thickness of the dynamic boundary layer in front of the step can be expected to be of the same order of magnitude as the step height. One of the main features of the velocity distribution in this second section, 12 mm downstream of the step, lies in the continuous increase in velocity up to the limit of the shock generated at the top of the step. The lower values measured near the wall may be due to a deflection of the flow impinging on the step and a normal component of the velocity, non-negligible in comparison to the lower axial one. The most important velocity gradient is located 5 mm above the surface. For $y > 5$ mm the axial value is approximately constant ($\simeq$ 2500 m/s) up to the limit of the secondary shock, then increases far off to reach the maximum value of the free jet.

6. Conclusion

We have demonstrated the potential of the LIF technique for instrumentation of continuous high-enthalpy facilities. NO number density, temperature, and velocity have been measured in the supersonic air-plasma free jet as well as over the surface of a TPS model, allowing characterization of the aerodynamic behavior of the flow, especially in the vicinity of a misalignment of tiles.

Acknowledgement. This work was sponsored by E.S.A., C.N.E.S. under Dassault Aviation contract. The authors wish to thank I. N. Mourzinov, V. B. Knotko, and V. A. Fadeyev for fruitful discussions and their collaborators for their assistance in operating the test campaign at the Tsniimach Institute.

References

Anfimov N (1992) TSNIIMACH Capabilities for Aerogasdynamical and Thermal Testing of Hypersonic Vehicles. AIAA Paper 92-3962.

McMillin BK, Lee MP, Hanson RK (1992) Planar laser-induced fluorescence imaging of shock-tube flows with vibrational nonequilibrium. AIAA J 30:436.

Use of an expansion tube for imaging studies of hypersonic reactive flows around blunt bodies

M. R. Kamel, C. I. Morris, M. C. Thurber,
S. D. Wehe, R. K. Hanson
High Temperature Gasdynamics Laboratory
Department of Mechanical Engineering
Stanford University, Stanford, CA 94305 USA

Abstract: Initial experiments applying an expansion tube to the study of hypersonic reactive flow around blunt bodies are reported. Two fuel mixtures, $16.7\%CH_4/33.3\%O_2/50\%N_2$ and $5\%C_2H_4/15\%O_2/80\%N_2$, have been successfully accelerated to velocities of 2230 m/s using an expansion tube. The combustion mode observed was that of an adiabatic shock followed by a combustion front, and is consistent with the observations in previous work.

Key words: Expansion tube, Hypersonic flow, Flow imaging, PLIF

1. Introduction

This paper reports on the use of an expansion tube to study hypersonic reactive flows around blunt bodies. Numerous experimental and numerical investigations of such flows have been performed over the past 30 years. Most experimental work has involved firing spheres and blunted cylinders at hypersonic speeds into combustible mixtures, and using schlieren photography to observe the different combustion modes occurring around and behind the body (Lehr 1972). For the cases where the velocity of the body was higher than the Chapman-Jouguet (CJ) velocity of the fuel mixture, a smooth shock-induced combustion front was observed (Chernyi 1968). As the body velocity was decreased, instabilities originating in the induction zone between the bow shock and the combustion region at the front of the body appeared. For body velocities close to the CJ velocities, the instabilities became more pronounced and in some cases led to the formation of cellular detonations.

To demonstrate the potential of the expansion tube for such investigations, a pilot study of hypersonic reactive flow around a spherically blunted cylinder 19 mm in diameter was undertaken in the new Stanford expansion-tube facility. In these initial experiments, methane- and ethylene-based fuel mixtures were accelerated to velocities around 2300 m/s, and OH planar laser-induced fluorescence (PLIF) was used to image the combustion around the body.

A thorough understanding of the characteristics of the hypersonic flows generated by the expansion tube is a prerequisite to such an investigation. Ongoing characterization efforts described by Morris et al. (1995) allowed flow quantities to be determined using the combined information of wall pressure, pitot pressure, IR absorption, and NO PLIF measurements. This foundation in turn allowed selection of the flow condition used in the present study. Following a brief description of the experimental facility, data characterizing this test condition is discussed, with particular attention to flow repeatability and uniformity. Selection of the reactive mixtures is then outlined, with a view towards avoiding premature ignition. Finally, the results of imaging experiments for inert and reactive flows over the blunted cylinder at the chosen test condition are presented and compared to previous studies.

2. Experimental facility

The Stanford expansion-tube facility (Kamel et al. 1995, Morris et al. 1995) consists of four sections: a 180-cm driver with 10-cm inner diameter, followed by a 90-cm buffer, a 275-cm driven section, and a 430-cm expansion section, all with inner diameters of 9 cm. Mounted at 30.5-cm intervals along the downstream end of the driven section and at ports 61.0 cm apart on the expansion section are PCB pressure transducers for shock-speed and wall-pressure measurements. Also located at the second port from the end of the expansion section is an infrared absorption setup used to mark test-gas arrival in characterization runs — a bandpass-filtered InSb IR detector indicates absorption of light from a globar by CO_2 seeded into the test gas.

The expansion section ends in a square viewing chamber in which the test body is mounted, with optical access provided by two opposed pairs of 13×13-cm quartz windows. For characterization runs, the test body can be replaced by a pitot rake equipped with four PCB 113A26 pressure transducers mounted at 19.1-mm intervals across the diameter of the tube.

The quartz ports in the viewing section allow horizontal passage of a laser sheet through the test flow and imaging of fluorescence from above. The sheet is formed from the frequency-doubled output of a Lambda Physik FL3002 dye laser pumped by a Lambda Physik EMG 203 excimer. For PLIF work with nitric oxide (NO) transitions near 226 nm, Coumarin 450 dye was used to achieve laser energies of about 0.3 mJ per pulse. OH PLIF used the somewhat higher pulse energies — roughly 0.7 mJ for this work — achievable near 283 nm with Coumarin 540A dye. In both cases, fluorescence from the flow was imaged onto the 578×384 pixel array of an appropriately filtered Princeton Instruments intensified CCD camera.

Table 1. Freestream characteristics of selected test condition.

Measured Quantities		Calculated Quantities	
Contact Surface Velocity	2230 m/s	Mach Number	7.1
Static Pressure	0.93 psia	Static Temperature	280 K
Pitot Pressure	58 psia		
Steady Flow Test Time	125 μs		

3. Properties of chosen test condition

Previous characterization efforts assisted in the identification of the test condition described in Table 1. On the basis of further control experiments using gas mixtures of pure N_2 and 10% CH_4 + 90% N_2, both deemed sufficiently close gas-dynamically to the reactive mixtures used later, flow core uniformity and flow repeatability during the test time were determined to be good. Figure 1 shows a cross-section of pitot pressure 2 cm downstream of the tube exit. The figure indicates a uniform flow core 3.8 cm in diameter. To investigate the repeatability of the flow, the control mixture of CH_4/N_2 seeded with 0.15% NO was employed, and NO PLIF images of the bow shock were taken for four successive shots, 75 μs into the test time in each case. The shock shapes and standoff distances showed excellent overlap for all shots. Also, pitot traces for different shots agreed to within 5%, and shock speeds in the driven and expansion sections were repeatable to within 1%.

The observed shock standoff distance and shock shapes agree well with correlations from the literature for $M = 7$ flow. Figure 1b compares the shock shapes observed for the two control mixtures to an experimental correlation given by Billig (1967) for $M = 7.1$.

4. Mixture selection

The primary criterion for the reactive mixture was that it not ignite anywhere inside the tube. The most severe limitation comes from the fact that the test gas is subjected to post-reflected-shock conditions for a few microseconds at the driven-expansion section interface due to the finite diaphragm-opening time. Presently, Mylar diaphragms 25 μm thick are used at that interface, but other diaphragms and diaphragm-opening schemes are under investigation.

Premature ignition of the test gas is characterized by one or all of the following effects. A strong reflected shock is recorded by the pressure transducer in the driven section at a location 15.2 cm upstream of the driven-expansion diaphragm. The shock speeds in the expansion section and the pressure jump upon shock passage are considerably higher than in the corresponding inert case.

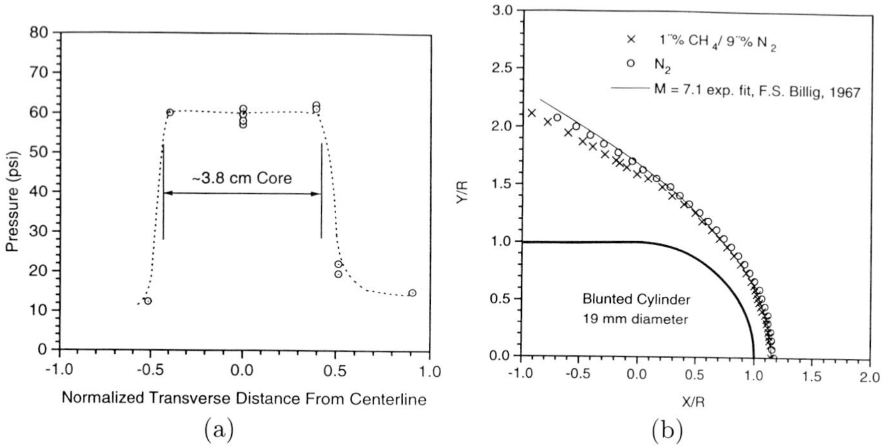

Figure 1. For control experiments performed with the test conditions of Table 1: (a) the profile of pitot pressure across tube diameter 2 cm downstream of exit plane; (b) comparison of the observed bow-shock shape with correlations obtained from the literature for $M = 7.1$ flow for gases at low temperatures and with $\gamma = 1.4$.

In addition, emission from hot CO_2 combustion products is detected by the IR detector as the test gas flows by its viewing port. Finally, the PLIF images in cases of premature ignition show OH fluorescence signals in the freestream, upstream of the bow shock.

Of the several mixtures tested, the following two exhibited no signs of premature ignition in the tube at the chosen test condition, and were therefore used for these experiments:

$$16.7\%CH_4 + 33.3\%O_2 + 50.0\%N_2 \tag{1}$$

$$5\%C_2H_4 + 15\%O_2 + 80\%N_2 \tag{2}$$

5. Results and discussion of reactive experiments

Previous work has shown combustion structure to depend on the ratio of test-gas velocity to CJ velocity. In these experiments, the CJ velocity is 1960 m/s for the methane-based mixture and 1685 m/s for the ethylene mixture. Given the nominal freestream test-gas velocity of 2230 m/s, the ratios of test-gas velocity to CJ velocity for the mixtures are 1.13 and 1.32, respectively. Hence, based on previous work, the combustion mode expected is that of an adiabatic shock followed after an induction length by a combustion front. Figures 2a and 2b are OH PLIF images of the flow of the methane- and ethylene-based

(a) (b)

Figure 2. OH PLIF images for (a) methane-based and (b) ethylene-based mixtures taken 75 μs into the steady test time at the test conditions specified in Table 1.

mixtures, respectively, both taken 75 μs into the steady test time. Both cases show evidence of shock-induced combustion close to the cylinder. The images differ considerably in the thickness of the combustion region and the standoff distance of the combustion front. That difference is consistent with the fact that the ethylene mixture has a shorter induction time than the methane one.

Another feature of the images, more evident in the methane case, is that the boundaries of the OH PLIF signals around the cylinder become almost parallel to the body. This matches previous observations, and is explained by the fact that the bow shock weakens rapidly in the radial direction away from the body's nose, leading to an increase in induction time of the shocked gas. Eventually the induction time approaches infinity, and the reaction front becomes coincident with the flow-field streamlines around the projectile (Alpert and Toong 1972, Lee 1994).

6. Conclusions

The bow-shock shape and standoff distance observed in the control experiments with NO PLIF agreed well with experimental fits from the literature. For the reactive-flow experiments, the features of combustion observed were in accord with the qualitative analysis and observations of previous researchers at corresponding conditions of V_{flow}/V_{CJ}. Thus we conclude that one of the three modes of combustion known to occur when a blunt body is fired into a reactive mixture can be reproduced using an expansion tube. Future experiments will be directed at reproducing the other two combustion modes, namely, the case of an adiabatic shock followed by an oscillating flame front, and the case of a cellular detonation. This will be done by changing the CJ velocity of the mixture and the dimensions of the body. Such experiments will allow future work in the expansion tube aimed at applying PLIF to simulated ram-accelerator flows to be performed with confidence.

Acknowledgement. The authors gratefully acknowledge the contributions of A.F.P. Houwing, M. Di Rosa, F. Levy, and I. Stouklov to this investigation. The work was supported by the U. S. Army Research Office, with Dr. Dave Mann as technical monitor, and the U. S. Air Force Office of Scientific Research, with Dr. Julian Tishkoff as technical monitor.

References

Alpert RL, Toong TY (1972) Periodicity in exothermic hypersonic flows about blunt projectiles. Astro Acta 17:539–560.

Billig FS (1967) Shock-wave shapes around spherical and cylindrical nosed bodies. J Spacecraft and Rockets 4(6):822–823.

Chernyi GG (1968) Supersonic Flow Past Bodies with Formation of Detonation and Combustion Fronts, Problems of Hydrodynamic and Continuum Mechanics, Society for Ind and App Math, Philadelphia, pp 145–169.

Kamel MR, Morris CI, Thurber MC, Wehe SD, Hanson RK (1995) New Expansion Tube Facility for the Investigation of Hypersonic Reactive Flow, AIAA Paper No. 95-0233.

Lee JHS (1994) On the initiation of detonation by a hypervelocity projectile. Presented at the Zeldovich Memorial on Combustion, Voronovo, Russia, Sept 12–17.

Lehr HF (1972) Experiments in shock-induced combustion. Astro Acta 17:589–597.

Morris CI, Kamel MR, Thurber MC, Wehe SD, Hanson RK (1995) Development of an expansion tube for investigation of combustion in supersonic projectile flowfields. Presented at the 2nd International Workshop on Ram Accelerators, Univ of Washington, Seattle, USA, July 17–20.

Spatially and spectrally resolved laser-induced fluorescence measurements in the HEG

M. Wollenhaupt, M. Rosenhauer, T. Müller, W. H. Beck, M. Scheer*
Institute for Fluid Mechanics, German Aerospace Research Establishment
DLR, Bunsenstrasse 10, 37073 Göttingen, Germany

Abstract: NO excitation and fluorescence spectra, excited by an ArF excimer laser at 193 nm and measured in a hot cell (1200 K) in order to find suitable transitions for use in the high-enthalpy shock tunnel HEG, are discussed. NO laser-induced fluorescence measurements at 193 nm were performed in HEG, with the aim of carrying out 2D/2-line thermometry in the free stream and in gas flows around models. Results in HEG at enthalpies up to 21 MJ/kg are given for the space gliders HERMES and HOPE. The difficulties with contaminant emission at these high enthalpies, and means of overcoming them, are discussed.

Key words: NO spectroscopy, Shock tunnel, Laser-induced fluorescence, LIF, Hypersonic flow

1. Introduction

In order to be able to characterize the free-stream flow and examine the flow around test models in the HEG, laser-induced fluorescence (LIF) measurements of NO at 193 nm were carried out in a hot cell and in the HEG. The unambiguous assignment of spectral lines for the five electronic transitions of NO accessible at 193 nm — $\epsilon(0,1)$, $\gamma(3,0)$, $\gamma(4,1)$, $\gamma(5,2)$, and $\beta(7,0)$ — and a study of the behavior of the spectral fluorescence intensity of the chosen NO transitions as a function of temperature were carried out in a specially constructed hot cell.

The difficult environment found in a shock tunnel at enthalpies up to 21 MJ/kg necessitates spectrally resolved measurements of the laser-excited fluorescence in order to ascertain if only the required NO transition is being excited and that no spurious emission is present; this latter point is still one of the severe limitations and handicaps to spectroscopic diagnostic measurements at high enthalpies, and must be taken into account. How this can be done, with reference to emission measurements carried out in HEG, is described.

*Present address: Department of Physics and Astronomy, McMaster University, Hamilton, Ontario, Canada.

Spectrally and 1D spatially resolved LIF images have been recorded in the HEG in the free stream and in the stagnation region of a HERMES model. A quantitative evaluation of these images was partially feasible. Similarly obtained LIF images at the wing leading edge and tip-fin locations of the Japanese HOPE orbital plane could be combined to give 2D temperature field maps (for the present work relative to the free-stream temperature value only).

2. Experimental setup

The experimental setup is described elsewhere (Scheer 1994), and only brief details will be given here. The LIF apparatus consists of two laser sheets counter-propagating through the HEG test section in front of the test model, and two separate image-intensified gated camera systems to capture the fluorescence, either 2D spatially or (by including a monochromator) spectrally/1D spatially resolved. In the latter case, the monochromator slit is placed in the direction of the required spatial resolution. Several checks have to be included in the system to correct for laser spectral and intensity drifts or scatter. The test time in the HEG of ≤ 1 ms necessitates the use of single laser shot measurements. As opposed to single laser measurements, the two-laser/two-camera system, together with an in situ calibration with the heated NO calibration cell carried out in the test section, allows one to combine images so that several apparatus and flow effects then cancel out; this feature showed itself to be very desirable.

3. Results and discussion

3.1. NO spectroscopy

Figure 1 shows NO excitation and fluorescence spectra taken in the heated cell at $p_{NO} = 10$ mbar and $T = 1200$ K. In the 1-nm laser tuning range, over 400 lines could be identified and assigned to their appropriate transitions. Candidate lines suitable for use in the HEG are identified; here the major criteria of single-line excitation and suitably large energy separation of the lower excitation levels ($\Delta E > kT$) are (largely) met. All HEG work was carried out with lines 2 and 8 ($R_{21}(17.5)$ and $R_{22}(27.5)$, respectively) belonging to the NO $\epsilon(0,1)$ band. Tests carried out in the heated cell to determine the behavior of these two lines at different temperatures and pressures revealed that the intensities of the R_{21} satellite branch were stronger compared to the main branches than expected from calculations using published values of Hönl-London factors (Bennett 1970). This is shown in Fig. 2, where it can be seen that the measured satellite branch line intensities relative to their R_{22} neighbors are at least two times stronger than calculated. There is evidence to show that excited electronic-state interactions in NO lead to these unexpected intensity distributions (Amiot and Verges 1982). This topic cannot be further discussed here, but its implications for quantitative LIF measurements must be considered.

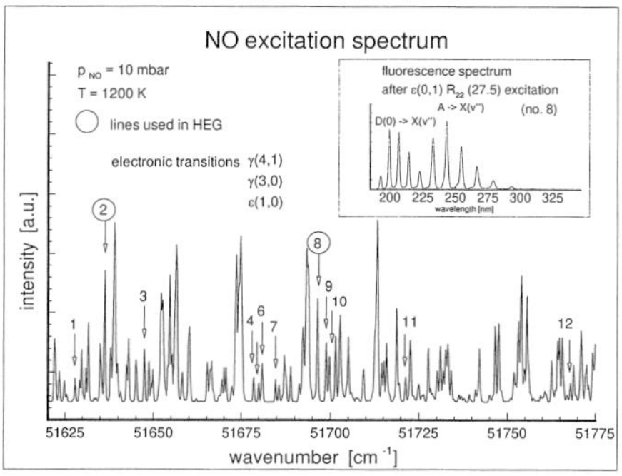

Figure 1. NO excitation spectrum, showing candidate lines (1–12). Fluorescence spectrum for line 8 ($R_{22}(27.5)$) is inset.

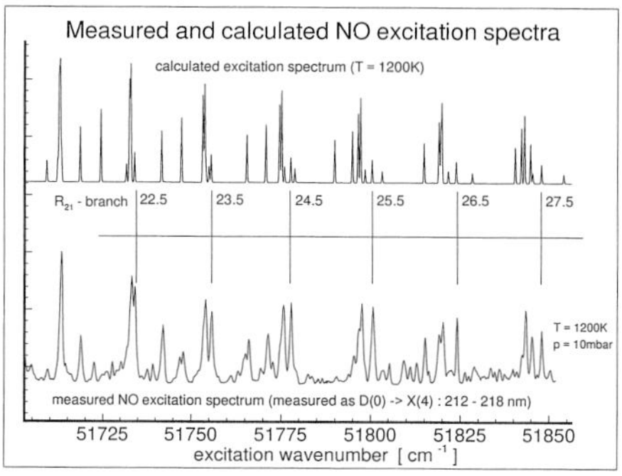

Figure 2. Comparison between measured and theoretical spectra, showing disproportionately large R_{21} branch.

3.2. Spectrally resolved LIF in the HEG

Figure 3 shows one of two images of a spectrally resolved LIF measurement carried out in the stagnation region of a HERMES model placed at an angle of attack of 30° in the HEG at condition I (p_0 = 41 MPa and h_0 = 21 MJ/kg). A good signal-to-noise ratio for the NO fluorescence in the free stream was obtained, as can be seen in the inset spectrum plot. The luminosity from

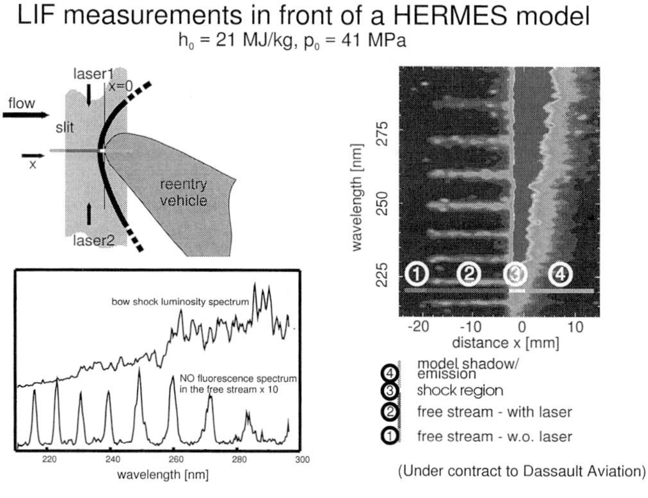

Figure 3. LIF measurement in stagnation region of HERMES. Inset spectrum plot shows characteristic NO lines in the free stream and stray emission behind the shock.

the bow-shock region is considerably stronger than the fluorescence in the free stream, and above about 220 nm completely swamps any fluorescence lines that may be present. However, the two fluorescence lines seen at about 215 and 222 nm in the free stream do not appear behind the shock, even though the luminosity here is not very large. The reasons for this are not yet understood (quenching, absorption, line broadening, etc.).

Evaluation of pairs of such images obtained in the free stream, including corrections for laser shot-to-shot variations, led to values for rotational temperatures which showed scatter that could not be due solely to flow variations. This may be due to the above-mentioned difficulty in carrying out quantitative LIF measurements with the NO lines at 193 nm.

It is clear from the HERMES result that to be able to carry out 2D-resolved LIF measurements under these tunnel conditions, either the spurious emission has to be filtered out, or its nature and source understood and removed. To this end: (i) a special filter was constructed to pass light in the very narrow but tunable region around 200–220 nm where D-X transitions occur; and (ii) emission spectra were recorded in the HEG behind a cylinder bow shock.

3.3. Spurious emission in the HEG

In a series of shots at condition I with test gas N_2, emission spectra were recorded in the stagnation region of a cylinder placed with axis perpendicular to the flow. By suitable normalization, these spectra were combined to give a spectrum over the range 180–850 nm, as shown in Fig. 4. It was possible to identify a large number of the emission lines as originating from Fe, and a simulation of Fe emission spectra, using literature values for line positions and transition probabilities (Fuhr et al. 1981), showed good agreement over major portions of the wavelength range. Nevertheless, several lines were present in the HEG spectrum which were not due to Fe.

Also shown in Fig. 4 are the locations of the NO fluorescence lines (highly magnified). This exemplifies the result shown before — except in a very narrow window around 200–220 nm, the emission completely swamps the LIF signals. Here one can also see why laser excitation at 193 nm has some advantages over the more popular and simpler NO excitation at 226 nm ($\gamma(0,0)$ bands); the latter leads to fluorescence signals in a wavelength range that is overlapped with the spurious emission.

3.4. Spatially resolved LIF in the HEG

Even though the strong emission in the vicinity of the stagnation region makes LIF measurements very difficult, regions further downstream of the model bow (wings, flaps, wakes) are colder and therefore suffer less from this effect. An

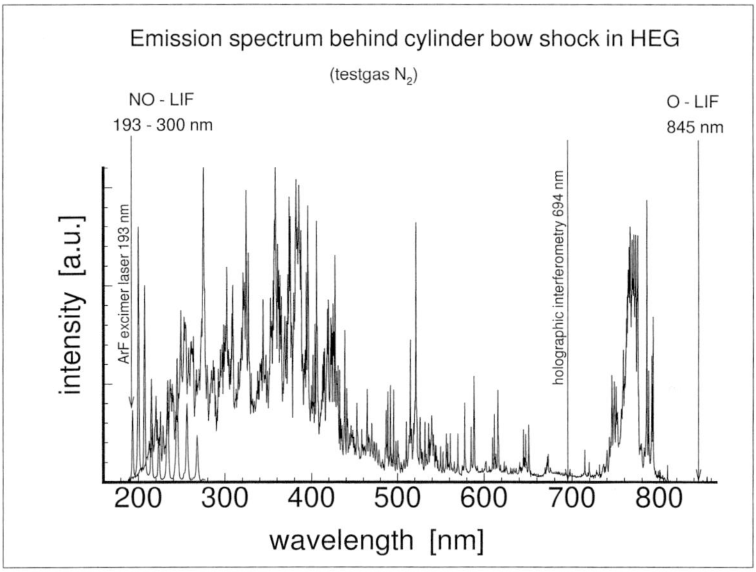

Figure 4. Strong spurious emission behind the HEG bow shock, due largely to Fe lines. A typical NO fluorescence spectrum has been overlaid (not to scale).

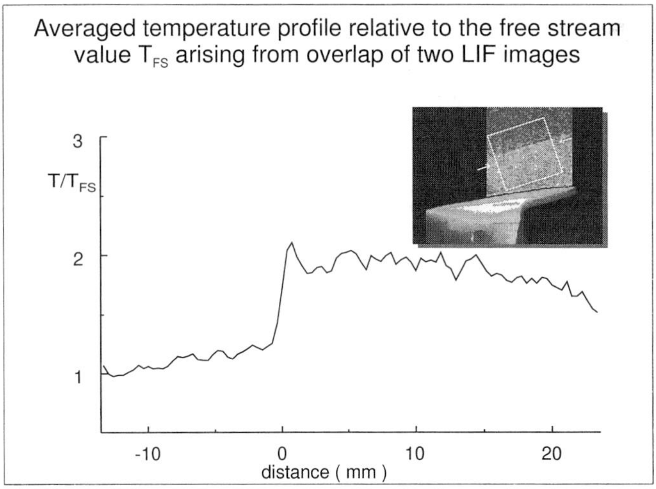

Figure 5. Measured temperature field and plot (relative to calculated free-stream value) at HOPE model wing-tip fin. Model surface is located at $\sim 25 - 29$ mm.

example of this is tests carried out in the HEG on the Japanese space plane HOPE; models with different shapes were placed at an angle of attack of $\alpha = 35°$ and yaw angle $\beta = 0$ and $\pm 3°$ in the HEG such that the two laser beams could probe the regions of interest (wing leading edge and wing-tip fin) on opposite sides of the model. The filters referred to before were used here. The measured LIF images were corrected to provide a visualization of the flow phenomena, as shown in Fig. 5. The LIF image shows clearly the location of the shock relative to the wing. A LIF signal in the free stream can also be seen. By combining the two measured images, it was possible to obtain temperatures (relative to the value in the free stream), as also shown in the figure. The averaged temperature plot shows that temperature varies very little between shock and wing surface.

4. Conclusions

LIF images were obtained in flow studies using two different models in the HEG. As long as one uses laser excitation at 193 nm, the problem of obtaining reliable absolute temperatures (via prior calibrations) still needs to be solved; here further tests are necessary. Steps are underway to suppress the contamination sources in HEG; until then, this still remains a major difficulty. The narrow bandpass filters were essential to carry out 2D LIF work further away from the stagnation region.

Acknowledgement. Our thanks go to J. Lenz and D. Garbe for their support. The financial support of ESA through CNES is gratefully acknowledged. The HERMES and HOPE measurements were carried out under contract to Dassault Aviation (Velizy, France) and Kawasaki Heavy Industries (Gifu, Japan), respectively.

References

Amiot C, Verges J (1982) Fourier transform spectroscopy of the $D^2\Sigma^+ - A^2\Sigma^+$, $E^2\Sigma^+ - D^2\Sigma^+$ and $E^2\Sigma^+ - A^2\Sigma^+$ systems of nitric oxide. Physica Scripta 26:422.

Bennett RJM (1970) Hönl-London factors for doublet transitions in diatomic molecules. Mon Nat R Astr Soc 147:35.

Fuhr JR, Martin GA, Wiese WL, Younger SM (1981) Atomic transition probabilities for iron, cobalt and nickel (a critical data compilation of allowed lines). J Phys Chem Ref Data 10:305.

Scheer M (1994) Temperature and Density Measurements on Hypersonic Wind Tunnels Using Laser Induced Fluorescence of NO, Including a Study of NO Spectroscopy in the Range 193-4 nm, Using a Novel Heated Cell, Diplom thesis, Institute for Fluid Mechanics, DLR, Göttingen.

Bow-shock/disturbance interaction for hypersonic flow over a cylinder

X. Zhong
Mechanical, Aerospace and Nuclear Engineering Department
University of California, Los Angeles
Los Angeles, CA 90095 USA

X. Joubert
UFR de Mécanique
Université Joseph Fourrier, Grenoble, France

Abstract: This paper presents direct numerical simulations of free-stream acoustic and entropy disturbances interacting with the bow shock in front of a cylinder moving at hypersonic speeds. The interaction is an important part of the phenomena of hypersonic boundary-layer receptivity to free-stream disturbances. The back-and-forth reflection and interaction of disturbance waves between the bow shock and body change the disturbance flow field considerably. They create strong entropy and vorticity disturbances behind the shock. It has been found that the smaller the value of the free-stream-disturbance frequency, the closer the entropy and vorticity disturbance spots to the body, which can be important for the boundary-layer receptivity analysis.

Key words: Hypersonic receptivity, Shock-disturbance interaction

1. Introduction

The receptivity of hypersonic boundary layers to free-stream disturbances refers to the processes by which the environmental disturbances initially enter the boundary layers and generate linear instability. Though it is an important aspect of the transition process from laminar to turbulent states, the receptivity phenomena for supersonic and hypersonic boundary layers has not been seriously considered (Reshotko 1990). One difficulty is that the interaction of disturbance waves with the bow shock and other shock waves needs to be considered. Another difficulty is that the real-gas effect becomes important in hypersonic flow.

The presence of the strong bow shock wave in hypersonic flow past a blunt leading edge alters the receptivity phenomena considerably. Figure 1 shows a schematic of a bow-shock/disturbance interaction flow field. When the bow shock is perturbed from its steady configuration, the wave fields behind the bow shock contain three types of disturbance waves: (1) acoustic waves which propagate with the speed of sound relative to the moving fluid and $\delta p \neq 0$, $\delta \rho = \frac{\delta p}{c^2}$,

$\delta u = \delta p/\rho c$, $\delta s = 0$; (2) entropy waves which convect with the moving fluid and $\delta s \neq 0$, $\delta\rho = (\frac{\partial\rho}{\partial S})_p \delta s$, $\delta u = \delta p = 0$; (3) vorticity waves which also convect with fluid velocity and $\delta\omega \neq 0$, $\delta u \neq 0$, $\delta\rho = \delta p = \delta s = 0$. If the disturbances are small, linear analysis can be used to determine the disturbances in flow variables behind the shock and disturbances in shock configurations (Mckenzie and Westphal 1968).

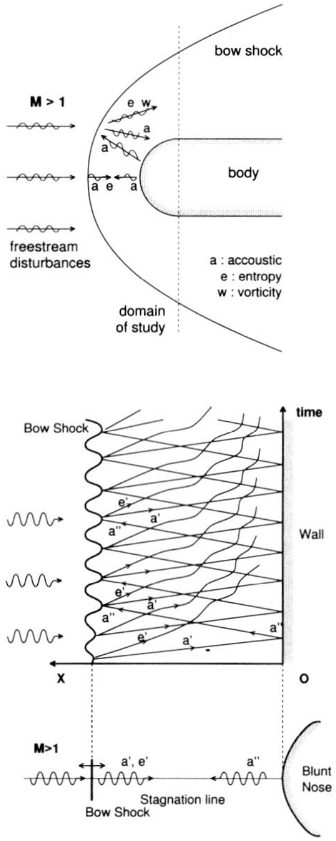

Figure 1. Schematic of bow-shock/disturbance interaction flow field and patterns in the x-t plane.

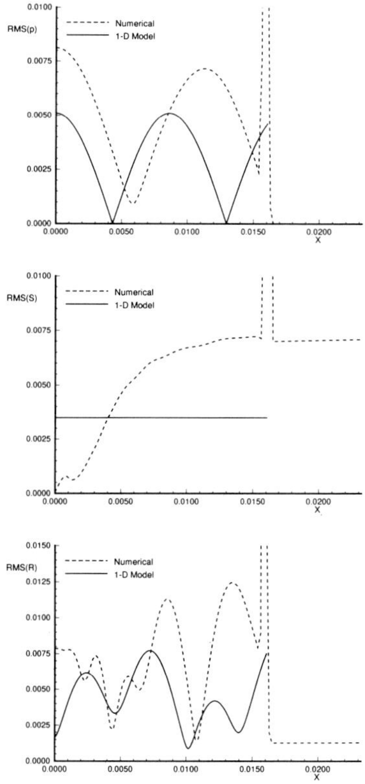

Figure 2. Root-mean-square values of p, s, and ρ perturbations along the stagnation line for Case 2.

The three disturbance waves originating from the bow-shock interaction propagate downstream independently, connected only through the boundary conditions at the shock wave and at the wall. The acoustic waves, which travel along Mach lines, reach the boundary layer and reflect back to the shock upstream. In Morkovin (1960), this reflection and interaction process was assessed using a simplified one-dimensional linear analytical model. A general qualitative knowledge of the process could be obtained. However, this approximate analysis can only be applied to the stagnation line with assumption of uniform flow behind the shock, but the actual flow is nonuniform. It is not clear how the vorticity, a 2D phenomena, is affected by the wave interaction and reflection.

In this paper, we continue our research project (Chiu and Zhong 1995) of studying hypersonic receptivity and stability phenomena involving strong shock waves and real-gas effects. Specifically, we conduct direct numerical simulations of free-stream acoustic and entropy disturbance waves interacting with the bow shock in front of a cylinder moving at hypersonic speeds. The objective is to study the detailed structure of flow disturbances outside the boundary layer which may have strong effects on the stability and transition of the boundary layer.

2. Numerical methods

Time-accurate computations of transient hypersonic flow require high-order spatially and temporally accurate methods with shock-capturing capability. In the present studies, both high-order ENO and TVD schemes solving the Euler equations are used for the numerical simulations. High-order Runge-Kutta schemes are used in temporal discretization. Details of the numerical methods can be found in Chiu amd Zhong (1995).

3. Results and discussions

A Mach 8.03 flow past a circular cylinder is considered as a model of hypersonic flow past a blunt nose. A set of 200×105 body-fitted grids is used. The fluid is assumed to be a perfect gas with $\gamma = 1.4$.

3.1. Case 1: acoustic-wave/bow-shock interaction

We continue our previous studies of free-stream acoustic waves interacting with the bow shock and the blunt body. First, a steady mean flow is obtained numerically. Then, a free-stream acoustic wave is introduced to this numerical mean flow along the free-stream boundary of the computational domain. The amplitude of the acoustic wave is set to be 1% of free-stream unperturbed properties. The following dimensionless wave numbers have been used in the simulations: $\overline{k} = \frac{2\pi}{kd} = 6.260, 3.130, 2.608, 1.956, 1.565$, where k is the wave number in the free stream and d is the standoff distance.

Figure 4 shows the time history of instantaneous pressure, entropy, and density along the stagnation line for the case of free-stream disturbance wave number $\overline{k} = 3.13$. The pressure and entropy contours clearly show the trajectories of independent acoustic and entropy waves behind the bow shock.

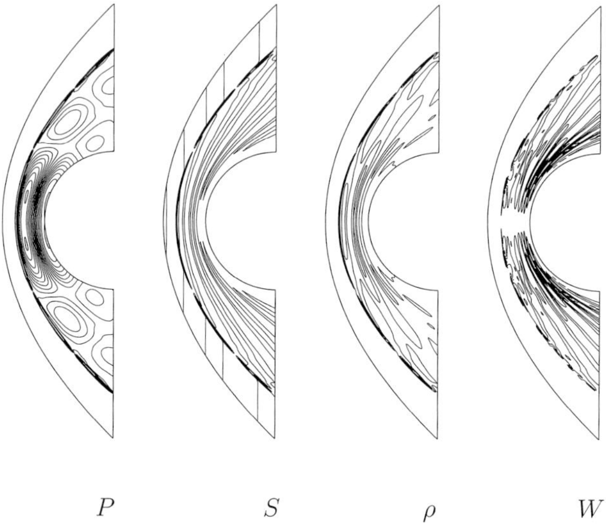

Figure 3. Instantaneous pressure, entropy, density, and vorticity contours, Case 2.

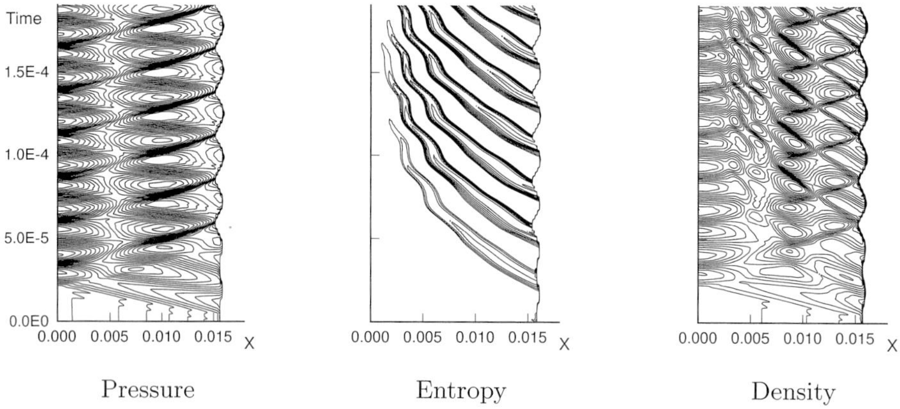

Figure 4. Flow variables along stagnation line as functions of time for Case 1.

The acoustic wave is reflected between the shock and solid body, which results in larger pressure oscillations and bow-shock movements. The oscillation of the bow shock in turn increases the intensity of the entropy and vorticity waves. The figure also shows that the entropy wave is convected independently with the fluid velocity, which decreases to zero at the stagnation point. The nonuniform velocity field behind the bow shock leads to curved entropy-wave lines, which do not reach the wall and oscillate near the wall as velocity does. As a result, the disturbance entropy is very small in a thin layer on the wall. The schematic of the patterns of the acoustic and entropy waves is shown in Fig. 1. On the other hand, the wave pattern of density is more complex than those of pressure and entropy because density disturbances are induced by both acoustic and entropy disturbance waves. The influence of both acoustic and entropy waves can be seen in the density x-t plot. Other results of case 1 can be found in Chiu and Zhong (1995).

3.2. Case 2: entropy-wave/bow-shock interaction

Free-stream entropy waves interacting with the bow shock and the blunt body are computed in this case. The flow conditions are the same as those in Case 1, except 1% entropy disturbance waves are imposed to the free-stream as follows:

$$\delta q = \Delta q e^{ik\{x - u_\infty t\}} \, .$$

Figure 3 shows the instantaneous contours of flow variables. As explained in the previous case, the density contours have complicated patterns because of the combined effects of the entropy and acoustic waves. The disturbance vorticity intensity has a large magnitude near, but not at, the cylinder surface. It is found that the smaller the free-stream disturbances, the closer are the vorticity spots to the wall. Meanwhile, the induced oscillation of the bow shock also creates strong vorticity disturbances which convect near the body surface along stream lines.

Figure 2 shows the root-mean-square (RMS) values of pressure, entropy, and density disturbances along the stagnation line. The RMS values represent the wave strength in the flow field. In the figure, the results of the one-dimensional model of Morkovin (1960) are compared with the numerical solutions. The RMS pressure of the one-dimensional model agrees reasonably well with the numerical solutions. However, the RMS entropy and density results of the model do not agree with numerical solutions, because the model assumes uniform flow behind the shock. The numerical solutions in this figure show that the reflection of the acoustic wave produces the strongest pressure disturbance at the stagnation point, but the maximum density disturbance is not always located at the wall.

Similar to Case 1, the interaction and reflection of the disturbance waves increase the disturbance pressure and shock movement considerably. However, free-stream entropy-wave/bow-shock interaction produces about one order of magnitude weaker disturbances behind the bow shock than the acoustic-wave/bow-shock interaction of Case 1.

4. Conclusions

The interaction of free-stream acoustic and entropy waves with bow shocks has been studied using numerical simulations as a tool. The results and analyses show that the pressure and density disturbances are amplified considerably by the back-and-forth reflection of disturbance waves between the bow shock and the wall. The interaction and reflection also increase the magnitude of oscillation of the bow-shock configuration. The maximum pressure disturbances are always located at the stagnation point, but not the density disturbances because density waves are the combined effects of the entropy and acoustic waves. The entropy and vorticity disturbances have large magnitudes near, but not at, the cylinder surface. These entropy and vorticity disturbance spots get closer to the wall when the frequency of the free-stream disturbances is smaller.

Work is underway to study bow-shock/free-stream disturbance interaction with real-gas effects. Detailed results of the current research can be found though the world wide web at 'http://cfdlab5.seas.ucla.edu/'.

Acknowledgement. This research was supported by AFOSR grant F49620-94-1-0019 with Dr. L. Sakell as the grant monitor. We are grateful to Theodore Lee for reproducing the linear shock interaction results for comparison with current numerical solutions.

References

Chiu CE, Zhong X (1995) Simulation of Transient Hypersonic Flow Using the ENO Schemes, AIAA Paper 95-0469 (to appeared in AIAA J).

Mckenzie JF, Westphal KO (1968) Interaction of linear waves with oblique shock waves. Physics of Fluids 11(11):2350–2362.

Morkovin MV (1960) Note on the assessment of flow disturbances at a blunt body traveling at supersonic speeds owing to flow disturbance in free stream. J of Applied Mechanics 27:223–229.

Reshotko E (1991) Hypersonic stability and transition. In: Hypersonic Flows for Reentry Problems, Springer-Verlag, Vol 1, pp 18–34.

Monte Carlo study of threshold line models for dissociation in hypersonic shock waves

I. D. Boyd
Cornell University
Ithaca, NY 14853 USA

Abstract: A new dissociation model for the Direct Simulation Monte Carlo (DSMC) method is formulated using the threshold line concept. The model is evaluated in a heat bath under equilibrium and nonequilibrium conditions. The model is also employed to calculate one-dimensional hypersonic shock waves of nitrogen and comparison is made with existing experimental data.

Key words: Dissociation modeling, Monte Carlo method, Hypersonic shock waves

1. Introduction

The present study focuses on the well-known mechanism whereby the rate of dissociation is altered under conditions of thermal nonequilibrium behind shock waves. Specifically, the dissociation rate is reduced when the vibrational temperature is lower than the translational value. Several models have been developed to capture this effect, which is important in hypersonic aerospace applications. Of these, two phenomenological approaches are highlighted here. In continuum computational fluid dynamics (CFD), the two-temperature model of Park (1988) is widely used. In this model, an Arrhenius reaction rate coefficient has its temperature dependence governed by: $T_a = {T_t}^s \times {T_v}^{1-s}$ where T_t is the translational temperature, T_v is the vibrational temperature, and s is a modeling parameter that is usually set to 0.5. In the particle-based Direct Simulation Monte Carlo (DSMC) method, the vibrationally favored dissociation (VFD) model examined by Boyd (1992) is considered. In this model, the probability that a particular collision of two particles leads to dissociation is given by:

$$P_d = f(E_c) \times E_v^{\phi}$$

where E_v is the vibrational energy, $f(E_c)$ is a function of the total collision energy, E_c, and ϕ is a modeling parameter. The parameter ϕ is set to 2 for nitrogen.

These dissociation models are unsatisfactory due to their highly phenomenological nature. The purpose of this study is to consider the use of the more detailed threshold line concept for dissociation proposed by Macheret and Rich (1993). Specifically, a form of that model is introduced for use in the DSMC technique.

2. Threshold line model

The threshold line concept is based on the premise that a certain amount of translational energy in a collision is required above the vibrational energy of the molecule to effect the dissociation reaction. In Macheret and Rich (1993), it is proposed that at high vibrational energy levels the energy needed (the threshold) is simply the difference between the activation energy and the molecular vibrational energy (ΔE). By consideration of the collision dynamics, it is proposed that the threshold energy at low vibrational levels is larger than ΔE.

Using these ideas, continuum models expressed in terms of translational, rotational, and vibrational temperatures have been proposed by Macheret et al. (1994) to model nonequilibrium dissociation processes. The first continuum threshold model (Macheret and Rich 1993) was employed by Boyd et al. (1995) to give the first successful predictions of data from a high-altitude, hypersonic flight experiment.

The purpose of this study is to use the threshold concept to develop a nonequilibrium dissociation model that may be employed directly in the DSMC technique. This requires the formulation of a dissociation probability based on the translational collision energy and the rotational and vibrational energies of the dissociating molecule. This has been performed in the work of Macheret et al. (1994). However, the final results of that study are described in terms of continuum temperatures rather than collision energies.

As described in detail by Macheret et al. (1994) the threshold line probability of dissociation for diatom-atom collisions is:

$$P_d \propto \frac{(E_t - F)^{\frac{3}{2}}}{F[(\sqrt{D^*} - \sqrt{\alpha E_v})(\sqrt{D^*} - (2 - \sqrt{\alpha})\sqrt{E_v})]^{\frac{1}{2}}}$$

where $\alpha = 0.25$ for N_2–N_2 and N_2–N. The effective activation energy is

$$D^* = D - E_r + \frac{2E_r^{\frac{3}{2}}}{3\sqrt{6D}}$$

where D is the dissociation energy. For low vibrational levels ($E_v < \alpha D$), the threshold function is

$$F = \frac{(\sqrt{D^*} - \sqrt{\alpha E_v})^2}{1 - \alpha}$$

and for $E_v > \alpha D$, the function is simply

$$F = D - E_v.$$

In the above expressions, E_t is the translational collision energy, and E_r and E_v are the rotational and vibrational energies of the reacting molecule. The above probability model is employed in the present study for both N_2–N_2 and N_2–N dissociation reactions. This approach is adopted due to the requirement for a relatively simple mathematical expression for use in the numerically intensive DSMC technique. The constant of proportionality for P_d is obtained through calibration of the model against rate data under equilibrium conditions as described in the following section.

3. Results

The DSMC threshold line model is used to calculate dissociation rate constants under conditions where the translational, rotational, and vibrational modes are in thermal equilibrium with Boltzmann distributions. An isothermal heat bath is simulated that consists of 50,000 particles. Energy is exchanged between the various modes during collision, but chemical reactions are not processed. Instead, the average dissociation probability is evaluated over all collisions and then converted into a rate constant. Results obtained in this manner are compared in Fig. 1a with the rate coefficient of Byron (1966) for the N_2–N_2 reaction over a wide temperature range. As discussed by Boyd (1992), the coefficient of Byron is employed as it was measured under conditions of thermal equilibrium. It is quite remarkable that the DSMC model provides a temperature dependence that is so close to the Byron rate. It should be noted that the leading constant in the DSMC model is explicitly chosen to give agreement with the Byron rate at $T = 10,000$ K. In Fig. 1b, the dissociation probabilities for various vibrational energy levels are plotted as a function of total collision energy (expressed as a temperature). It is interesting to note that there is a region for all vibrational levels where the dissociation probability exceeds one. The dissociation rates shown in Fig. 1a limit the dissociation probability to a maximum of one.

Comparison is next made of the DSMC threshold line model with the continuum model of Macheret et al. (1994) under nonequilibrium conditions. For simplicity, the translational and rotational modes are equilibrated. Only the effect of vibrational nonequilibrium is considered. Once again, heat-bath simulations are performed. In this case, the translational and rotational modes are given a temperature of 10,000 K. For each of the vibrational levels of the harmonic oscillator that lie below the dissociation energy, a large number of collisions are calculated to determine the state-dependent dissociation rate. These rates are then summed over the Boltzmann populations at different vibrational

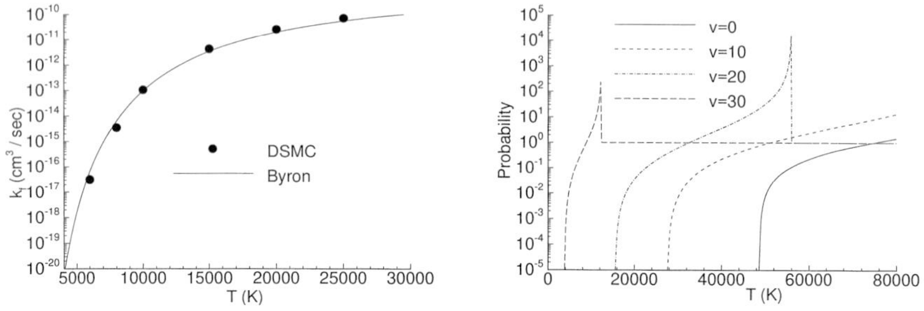

(a) (b)

Figure 1. Properties of the DSMC threshold model: (a) equilibrium dissociation rate constants; and (b) dissociation probabilities as a function of collision energy (temperature) for various vibrational energy levels.

temperatures. The results of this analysis are shown in Fig. 2a for the N_2–N_2 reaction. Comparison is made with the continuum threshold line model. While the two models give good agreement at the lower and upper values of vibrational temperature, the DSMC model predicts a higher dissociation rate

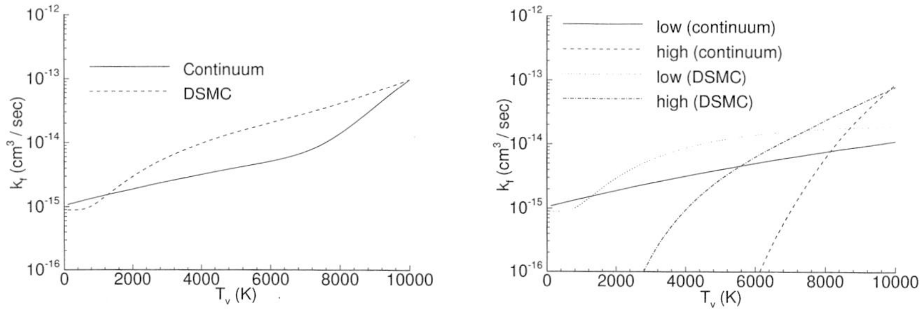

(a) (b)

Figure 2. Comparison of DSMC and continuum threshold models for N_2–N_2 dissociation: (a) nonequilibrium dissociation rate constants; and (b) contribution to k_f of low and high vibrational levels.

for the intermediate values. The reason for this difference is partly explained through consideration of Fig. 2b. This shows the contributions to the overall dissociation rate of the lower ($E_v < \alpha D$) and higher ($E_v > \alpha D$) vibrational levels. The DSMC model generally shows higher dissociation rates from both low and high vibrational levels. It is assumed that these differences arise from the approximations made in obtaining analytical expressions for the continuum model. In particular, integrals are evaluated using the method of steepest descent which can give rise to significant inaccuracies. In addition, the continuum model assumes a continuous distribution of vibrational energy whereas the DSMC approach employs the truncated harmonic oscillator.

Finally, the DSMC threshold model is evaluated through comparison of experimental data in hypersonic shock waves of nitrogen. The data is taken from Kewley and Hornung (1974) who report density profiles obtained from interferograms. Two cases with different upstream velocities are considered. The DSMC simulations of these one-dimensional shocks employ several thousand cells and about 100,000 particles. In the first case, the upstream velocity is 4.8 km/s. This gives a density profile that is dominated by N_2–N_2 dissociation reactions. The DSMC results computed with the threshold line model compare very favorably with the experimental data as indicated in Fig. 3a. The second case considers an upstream velocity of 7.3 km/s. Under these conditions, the density rise across the shock is dominated by N_2–N dissociation reactions. The

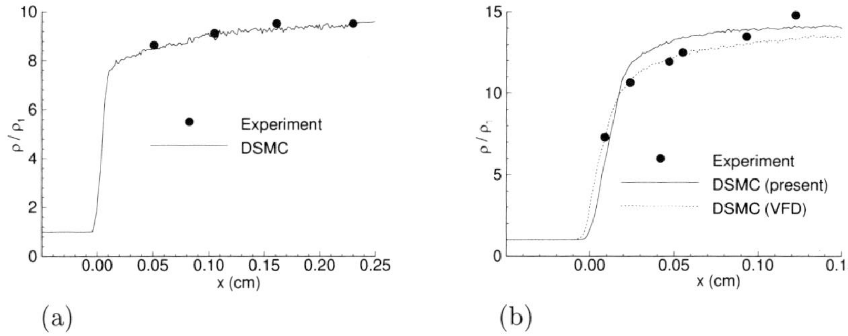

Figure 3. Profiles of density ratio in nitrogen shock waves: (a) U_1=4.8 km/s, (b) U_1=7.3 km/s.

threshold line model is compared with the experimental data in Fig. 3b. Also shown is the result obtained previously for this case using the VFD dissociation model (Boyd 1992). The threshold line model gives reasonable agreement with the experimental data, although it appears that the dissociation rate is slightly overpredicted. By comparison, the VFD model gives excellent agreement with the data. It should be noted that this level of agreement was only obtained with the VFD model by trying several different values of the parameter ϕ.

The level of agreement obtained with the DSMC threshold model and the data for the second shock wave case could be improved through use of a different rate coefficient for the N_2–N reaction. It appears that the data of Byron may be too high. It should be noted that the temperatures in the immediate post-shock region produced by the 7.3-km/s conditions exceed the range examined in the study of Byron. With this in mind, the performance of the new model is considered satisfactory. A more detailed investigation is reported in Boyd (1996).

4. Concluding remarks

A threshold line dissociation model for the DSMC technique has been formulated that determines the probability of reaction based on the energies of each collision. The model reproduced the temperature dependence of measured rate data under equilibrium conditions. It provided good agreement with measurements of density profiles through hypersonic shock waves. The use of the model is recommended over previous DSMC models for vibration-dissociation coupling. The new model is less phenomenological and only requires the determination of a leading-constant multiplication factor for a particular reaction.

Acknowledgement. This work is funded by the Air Force Office of Scientific Research under University Research Initiative F49620-93-1-0373.

References

Boyd ID (1996) A threshold line dissociation model for the direct simulation Monte Carlo method. Phys Fl (in press).

Boyd ID (1992) Analysis of vibration-dissociation-recombination processes behind strong shock waves of nitrogen. Phys Fluids A 4:178–185.

Boyd ID, Candler GV, Levin DA (1995) Dissociation modeling in low density hypersonic flows of air. Phys Fluids 7:1757–1763.

Byron S (1966) Shock tube measurements of the rate of dissociation of nitrogen. J Chem Phys 44:1378–1388.

Kewley DJ, Hornung HG (1974) Free-piston shock tube study of nitrogen dissociation. Chem Phys Let 25:531–536.

Macheret SO, Rich JW (1993) Nonequilibrium dissociation rates behind strong shock waves: classical model. Chem Phys 174:25–43.

Macheret SO, Fridman AA, Adamovich IV, Rich JW, Treanor CE (1994) Mechanisms of Nonequilibrium Dissociation of Diatomic Molecules, AIAA Paper–94 1984.

Park C (1988) Assessment of a two temperature kinetic model for dissociating and weakly ionizing nitrogen. J Thermophysics 2:8–16.

Analytical hypersonic waverider configurations

A. Merlen*, E. Fabre, S. Cieslak
ONERA-IMFL
5 Bvd Painlevé
59000 Lille, France

Abstract: Waverider configurations are deduced from hypersonic flows around axisymmetrical power-law slender bodies at low angles of attack. These inviscid flows have been calculated by analytical techniques and the stream functions are explicitly tabulated. This allows the generation of many waveriders, the aerodynamic coefficients of which are known.

Key words: Hypersonic flows, Waveriders

1. Theory

1.1. Hypotheses

The gas is assumed to be inviscid and thermodynamically perfect, having the specific heat ratio γ. Except in the vicinity of the blunt nose, the slenderness of the body, the high upstream Mach number M_∞, and the small angle of attack α allow the resolution of the problem in the frame of the hypersonic small-disturbance theory. At the limit $M_\infty \to \infty$ and $\alpha = 0$, which leads to the equivalence principle between explosions and hypersonic flows. Well-known self-similar solutions are then available, particularly for power-law bodies Σ', the equation of which is:

$$R' = \frac{\tau L}{2}\left(\frac{x}{L}\right)^n \quad \text{and for the shock: } R_0 = \frac{\tau L}{2\xi_0'}\left(\frac{x}{L}\right)^n$$

where ξ_0' is a positive constant, less than unity. The theoretical results can be summed up roughly by considering that the equivalence principle yields good results for pressure on the obstacle if the power-law exponent n is given a value between 2/3 and 1. Experiments (Kubota 1957) confirm these values. Nevertheless, this theory provides a poor description of the blunting and of the consequences blunting has on the density on the body surface. However, they do show that if the stream function is known, the entropy really acquired at the

*Present address: Laboratoire de Mécanique de Lille, Cité Scientifique, 59655 Villeneuve d'Ascq Cedex, France.

passage through the shock can be reassigned to each streamline to correct the density (Merlen & Andriamanalina).

1.2. Solutions

The definition $M_\infty^2 = \frac{\rho_\infty V_\infty^2}{\gamma p_\infty}$, shows that under the previous hypotheses, the counterpressure p_∞ only appears by way of a linear disturbance in $\frac{1}{M_\infty^2 R_x^2}$. Moreover, α only appears in the term $\frac{\alpha}{R_{0x}}$, R_{0x} being the first approximation of the slope of the shock. This is a typical result of hypersonic small-disturbance theory. The same result holds for all the unknowns of the problem, apart from the fact that they also depend on r, which introduces a dimensionless variable ξ proportional to $\frac{r}{\tau L}\left(\frac{L}{x}\right)^n$. The conditions on the shock can be used to determine explicitly the θ-dependency of the function of subscript 1 (Merlen & Andriamanalina). Finally, by taking as variables: $\eta = \frac{\alpha x}{R_0} = n\frac{\alpha}{R_{0x}} = \frac{2\xi_0'\alpha}{\tau}\left(\frac{x}{L}\right)^{(1-n)}$, and $\zeta = \frac{n^2}{M_\infty^2 R_x^2}$ we write the solution in the form:

$$R = \frac{\tau L}{2\xi_0'}\left(\frac{x}{L}\right)^n(1+\eta\xi_1\cos\theta+\zeta\xi_2) \qquad R' = \frac{\tau L}{2}\left(\frac{x}{L}\right)^n \qquad \xi = \frac{2r\xi_0'}{\tau L}\left(\frac{L}{x}\right)^n$$

$$\begin{aligned} u &= \frac{2nV_\infty}{\gamma+1}\frac{r}{x}(u_0+\eta u_1\cos\theta+\zeta u_2) & v &= \frac{2nV_\infty}{\gamma+1}\frac{r}{x}(\eta v_1\sin\theta) \\ \rho &= \frac{\gamma+1}{\gamma-1}\rho_\infty(\rho_0+\eta\rho_1\cos\theta+\zeta\rho_2) & p &= \frac{2n^2\rho_\infty V_\infty^2}{\gamma+1}\frac{r^2}{x^2}(p_0+\eta p_1\cos\theta+\zeta p_2). \end{aligned} \tag{1}$$

The terms with subscript 0, 1, and 2 are functions of ξ, except for ξ_0', ξ_1, and ξ_2, which are constants, and ξ_0, which is introduced so that the shock equation may be written $\xi = 1$ in the self-similar solution. Small-disturbance hypotheses are expressed by $\eta \ll 1$ and $\zeta \ll 1$. Despite the low values of α considered here, the effect of the incidence is far from negligible. The solution (1) must verify the general equations for any value of θ, η, and ζ. These equations are identically verified for any θ, which validates the form and reduces the problem to three systems of ordinary differential equations in ξ. The first of them gives the self-similar basic flow. The other systems (subscripts 1 and 2) are given in Merlen & Andriamanalina (1992) and give the effect of the incidence and the counterpressure, respectively. We must emphasize that the solutions are calculated once for each pair (n,γ) and provide universal sets of functions and constants. Hence, a particular physical configuration is directly calculated by analytical formulae using the tabulated universal solutions.

2. Comparison with experiments

As an example, we limit this section to the presentation of the results on the lift coefficient. Starting with the definition, the lift is obtained by $F_z = -\int_0^{2\pi}\int_0^L p'R'\cos\theta\,dx\,d\theta$ where only the subscript -1 term is nonzero. Letting $S_{ref} = \frac{\pi\tau^2L^2}{4}$, we obtain: $C_z = \frac{F_z}{\frac{1}{2}\rho_\infty V_\infty^2 S_{ref}} = -\frac{2np_1'\xi_0'}{\gamma+1}\alpha$, which has the same sign as α since p_1 is negative. Experiments realized in the ONERA R3Ch wind tunnel are in good agreement with these analytical results showing less than 5% discrepancy (Fig. 1). The wind tunnel was chosen to fulfill a few necessary conditions to make possible the comparison with this inviscid theory, among them: weak viscous interaction, no real-gas effect (cold hypersonic flow), and strong

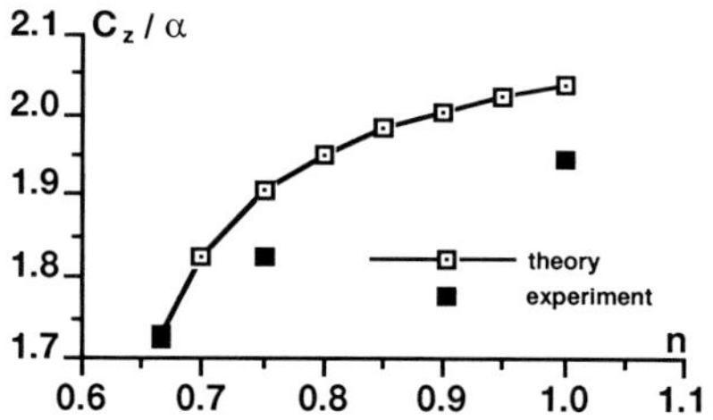

Figure 1. Lift coefficient of the ogives.

shock condition. Of course, this choice led to a compromise on the following physical conditions: $T_i = 1050°\text{K}$, $p_i = 120\ 10^5$ Pa, $M_\infty = 9.95$, $\mathcal{R}_\infty = 2.6\ 10^6$. Three ogives have been tested: $n = 1, 3/4, and 2/3$ with $\tau = 0.4$ and $L = 300$ mm. Because of the slenderness of the bodies, the strong-shock condition is not fulfilled on the rear part of the ogives. However, a higher Mach number would have increased the real-gas effect. Actually, this relatively weak Mach number allowed us to emphasize the influence of second-order terms in the development of the solution, particularly on the shock location and wall pressure. The viscous effects have been studied in the case of the $n = 3/4$ body, for which the self-similarity of the external flow can be extended to the coupled viscous-inviscid flow (Fig. 2) for the axisymmetric case (see Fabre 1995 for the small-incidence effect). As the main result of this comparison between theory and experiment, we can say that the first-order theoretical solution is valid for very high Mach numbers and a cold wall. For lower M_∞, the theory needs to be extended to second-order terms to explain the behavior of the shock shift when $\alpha \neq 0$ and to correct the wall pressure. On the contrary, the aerodynamic coefficients are mainly determined by the first-order terms.

3. Stream functions

The agreement between theory and experiment justifies the use of the theoretical solution to build waverider configurations. Many choices are possible but all are

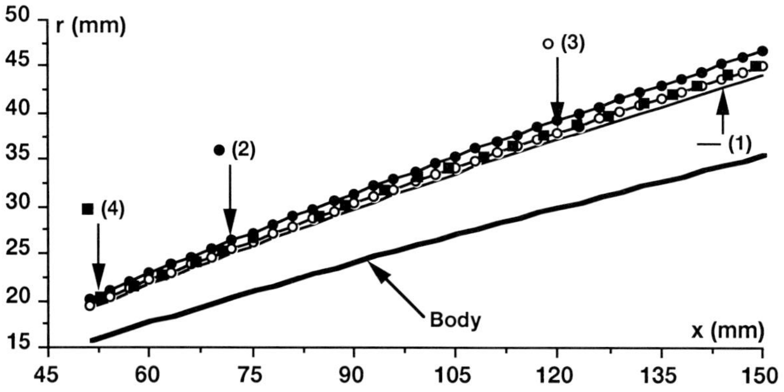

Figure 2. Shock location for $n = 0.75$ and $\alpha = 0.$: (1) shock by inviscid theory; (2) theory for adiabatic wall; (3) theory for cold wall; (4) experiment.

based on a solidification of stream surfaces. In the previous theory, this is made easier by the explicit knowledge of the stream functions which are tabulated in universal dimensionless form. The method for finding the stream function is based on dimensional analysis, but the dimensional considerations are not sufficient if $\alpha \neq 0$ since the flow is not self-similar. Thus, the form of ψ as a function of ζ must be anticipated. From the local equation of the stream functions, it can be shown that, for $\theta \neq 0$, or π:

$$\Psi = \left(\frac{x}{L}\right)^{2n} (|\sin\theta|)^{\frac{2n}{1+n}} \Psi_s \tag{2}$$

where

$$\Psi_s = \left(\frac{x}{L}\right)^{2n} \left\{\rho_0(u_0 - \frac{\gamma+1}{2})\xi^2 + \frac{2n}{1+n}\eta\cos\theta\left[\xi^2\left(\rho_1(u_0 - \frac{\gamma+1}{2}) + \rho_0 u_1\right)\right]\right.$$
$$\left. + n\zeta\left[\xi^2\left(\rho_2(u_0 - \frac{\gamma+1}{2}) + \rho_0 u_2\right)\right]\right\} \tag{3}$$

is the dimensionless form of the stream function in the plane of symmetry. At the zeroth order ($\alpha = 0$), the second stream function is obviously $\Phi = \theta$. We look for the form $\Phi = \theta + \eta\Phi_1(\xi)\sin\theta$. From the local equation, we get a first-order inhomogeneous differential equation for Φ_1. Letting $\Phi_1 = 0$ on the shock, the solution is $\Phi = \theta_i = \theta + \eta\Phi_1 \sin\theta$ with:

$$\Phi_1 = \frac{2}{1-\gamma}\left[\frac{2}{1-\gamma}\rho_0(u_0 - \frac{\gamma+1}{2})\xi^2\right]^{\frac{1-n}{2n}} \int_\xi^1 \frac{\rho_0 v_1 \hat{\xi}}{\left[\frac{2}{1-\gamma}\rho_0(u_0 - \frac{\gamma+1}{2})\hat{\xi}^2\right]^{\frac{1+n}{2n}}} d\hat{\xi}. \quad (4)$$

A more detailed analysis of this stream function, including the Hayes-Probstein entropy correction, is given in Merlen & Andriamanalina (1992).

4. Application to waverider design

As an example, we indicate here a very straightforward principle to build simple hypersonic waveriders. At first we choose a stream surface $\Phi = cst$ as the lower surface (Fig. 3). The upper surface is a cylinder starting from the shock intersection with the lower surface and generated by straight lines parallel to $\overrightarrow{V}_\infty$. Then, the pressure on the upper surface is supposed to be equal to p_∞. Figure 3 shows one of the

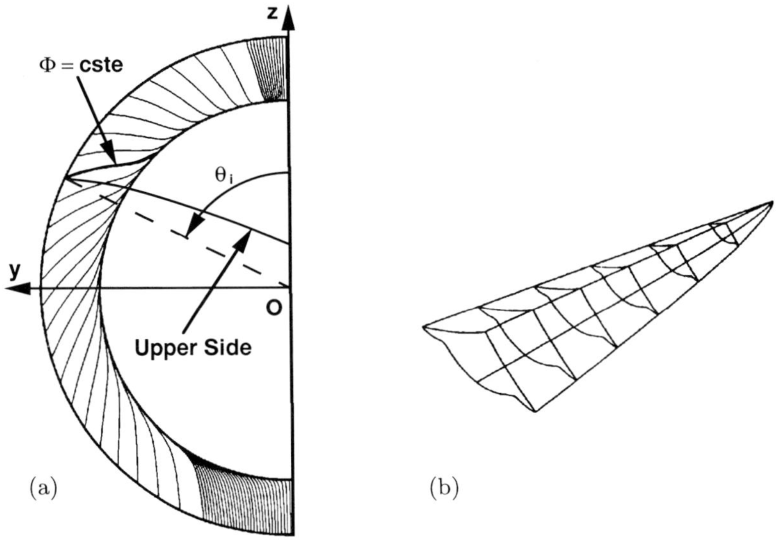

Figure 3. Waverider generation: (a) principle, (b) example.

geometries obtained in this way. Because pressure is known everywhere in the original flow field, we obtain explicitly the aerodynamic coefficients by integration on the surface of the body. It is then possible to evaluate the

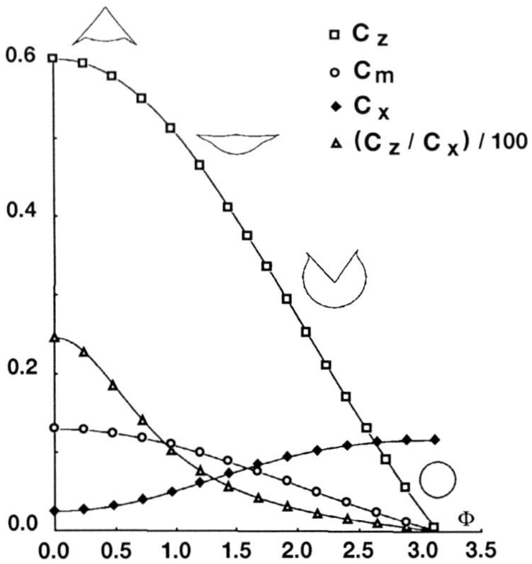

Figure 4. Aerodynamic coefficients versus dihedral angle.

influence of the geometrical parameters on these coefficients. For example, Fig. 4 shows the effect of the dihedral angle on C_z. We must emphasize that the generation of waveriders is reduced to a simple treatment of data files obtained by solving the universal dimensionless problem of Section 1. This allows a very quick evaluation of the characteristics of different configurations. Moreover, optimization is possible taking into account criteria such as a given volume, or a given length, etc.

References

Fabre E (1995) Explosions and Self Similar Hypersonic Viscous and Inviscid flows. Theory-calculation-experiment, Thesis of Université des sciences et technologies de Lille.

Kubota T (1957) Investigation of Flow Around Simple Bodies in Hypersonic Flow, Memorandum No. 40 Guggenheim Aeronautical Laboratory, California Institute of Technology.

Merlen A, Andriamanalina D (1992) Analytical solutions for hypersonic flow past slender power-law bodies at small angle of attack. AIAA Journal 30(11):2683.

A more precise method to determine stagnation-point heat-transfer rates

H. Olivier, H. Grönig
Shock Wave Laboratory, RWTH Aachen
Templergraben 55
52056 Aachen, Germany

Abstract: The theory of Fay and Riddell (1958) shows that the tangential velocity gradient is of great importance for stagnation-point heating. Based on the work of Belotserkovskii (1958) the authors developed an improved method to determine the tangential velocity gradient which, in opposition to the classical Newtonian method, is also valid for high-enthalpy conditions and which, as far as possible, takes into account the real flow properties within the shock layer.

Key words: Hypersonic flow, Stagnation-point heating, Theory of Fay and Riddell, Improved tangential velocity gradient, Integral method

1. Introduction

Most commonly in conjunction with the theory of Fay and Riddell, the Newtonian flow model (Lees 1956) is applied to determine the velocity gradient at the stagnation point of a sphere. It makes use of the one-dimensional momentum equation for a streamline close to the stagnation point. The pressure distribution on the sphere and thus the pressure gradient is computed from Newtonian theory. This finally leads to

$$\left.\frac{du}{dx}\right|_s \frac{D}{u_\infty} = \frac{2}{u_\infty}\sqrt{\frac{2(p_s - p_\infty)}{\rho_s}} \tag{1}$$

where the subscript s indicates the conditions at the boundary-layer edge of the stagnation point and the coordinate x points in the circumferential direction. Since this approach is based on the one-dimensional momentum equation it does not take into account multi-dimensional effects, like the influence of the vorticity produced by the curved bow shock, or the influence of the shock stand-off distance on the shock-layer flow. Especially the shock stand-off distance Δ strongly influences the shock-layer flow and therefore also the tangential velocity gradient. Real-gas effects mainly act on the tangential velocity gradient via the shock stand-off distance.

The aim of the present paper is to derive a method for computing the tangential velocity gradient at a sphere for high-temperature flows, where real-gas effects are important. Vorticity and compressibility effects are also included in the method to be presented. It will be shown that from the conservation equations for the stagnation point an analytical solution can be derived which is valid for perfect as well as real-gas behavior.

2. Improved model of stagnation flow

The conservation equations for mass and and momentum for the radial direction are used in the following form (Hayes and Probstein 1966, Chapter VI, p. 411)

$$\frac{\partial}{\partial \Theta}(\rho u \eta) + \frac{\partial}{\partial r}(\rho v r \eta) = 0 \tag{2}$$

and

$$\frac{u}{r}\frac{\partial v}{\partial \Theta} + v\frac{\partial v}{\partial r} - \frac{u^2}{r} + \frac{1}{\rho}\frac{\partial p}{\partial r} = 0\,, \tag{3}$$

with the velocity component u in the circumferential direction, v in the radial direction, and $\eta = r\sin\Theta$. The conservation equations for energy, for the momentum in Θ direction, and the equation of state are replaced by the integral approaches described below. With the non-dimensional variables

$$\bar{p} = \frac{p}{p_s}, \quad \bar{\rho} = \frac{\rho}{\rho_s}, \quad \bar{u} = \frac{u}{u_{max}}, \quad \bar{r} = \frac{r}{R}, \quad \bar{\Delta} = \frac{\Delta}{R}, \quad u_{max} = \sqrt{2h_0} \tag{4}$$

and

$$Q_v = \bar{\rho}\bar{v}, \quad Q_u = \bar{\rho}\bar{u}, \quad P_v = \bar{\rho}\bar{v}^2 + \frac{p_s}{\rho_s u_{max}^2}\bar{p},$$
$$P_u = \bar{\rho}\bar{u}^2 + \frac{2p_s}{\rho_s u_{max}^2}\bar{p}, \quad P_{uv} = \bar{\rho}\bar{u}\bar{v}, \quad \bar{\eta} = \frac{\eta}{R}$$

and after some algebraic reformulation, similar expressions are achieved as those given by Hayes and Probstein (1966, Chapter VI, p. 412)

$$\frac{\partial}{\partial \bar{r}}(\bar{r}\bar{\eta}Q_v) + \frac{\partial}{\partial \Theta}(\bar{\eta}Q_u) = 0\,, \tag{5}$$

$$\frac{\partial}{\partial \bar{r}}(\bar{r}\bar{\eta}P_v) + \frac{\partial}{\partial \Theta}(\bar{\eta}P_{uv}) - \bar{\eta}P_u = 0\,. \tag{6}$$

The two conservation equations (5) and (6) are integrated along a ray $\Theta = \text{const}$ from the body up to the shock. In order to solve these integral equations it is necessary to make an approximation for the three integrands in these equations. Following the work of Belotserkovskii (1958), the variables $\bar{r}Q_u, \bar{r}P_{uv}$, and $\bar{r}P_u$ are approximated by a linear distribution between the shock and the body. Navier-Stokes computations performed by Wong and Muylaert (1995) show that this approximation is even justified for high-temperature flows with significant real-gas effects (Olivier 1995). The general linear distribution

$$F = F_b + (F_2 - F_b)\frac{\bar{r} - 1}{\bar{\Delta}} \tag{7}$$

and the following boundary conditions for the stagnation-point streamline are introduced into the integral equations

$$\cos\sigma = 1\,,\; \bar{u}_2 = 0\,,\; \frac{d\bar{\Delta}}{d\Theta} = 0\,,\; \bar{\rho}_b = 1\,,\; \bar{p}_b = 1\,,\; \bar{v}_b = 0\,,\; \sin\Theta \approx \Theta\,. \tag{8}$$

The subscript 2 characterizes values immediately behind the shock, whereas subscript b indicates the conditions at the body surface. For the stagnation point the tangential velocity is expressed as

$$\bar{u}_b = \Theta \left.\frac{d\bar{u}_b}{d\Theta}\right|_s . \tag{9}$$

With these boundary conditions the following analytical solution is achieved for the tangential velocity gradient (Olivier 1995)

$$\left.\frac{du}{dx}\right|_s \frac{D}{u_\infty} = \frac{2(1+\bar{\Delta})}{\bar{\Delta}} \frac{p_s - p_2}{\rho_\infty u_\infty^2} \frac{\rho_2}{\rho_s} . \tag{10}$$

It is important to note that, for the derivation of (10), no explicit statement is necessary regarding the thermal and chemical properties of the gas. These properties are part of the basic approximation expressed by (7). Therefore, the equation for the velocity gradient is valid as long as the approximation (7) holds. As expected, according to (10) the tangential velocity gradient at the stagnation point depends on the shock stand-off distance, but it also depends on the density ratio ρ_2/ρ_s, which is a measure of the deviation of the real flow from the constant-density flow model for the stagnation region.

3. Results

3.1. Numerical simulation and theoretical results

According to (10) the velocity gradient has been computed for different flow conditions (see Fig. 1). In this example an equilibrium flow was assumed, where the conditions behind the shock and at the stagnation point were determined by a nine-component air model (Esser et al. 1992).

For the data presented in Figs. 1 and 2 the shock stand-off distance was determined by

$$\frac{\Delta}{R} = \frac{\rho_\infty}{\rho_2} \tag{11}$$

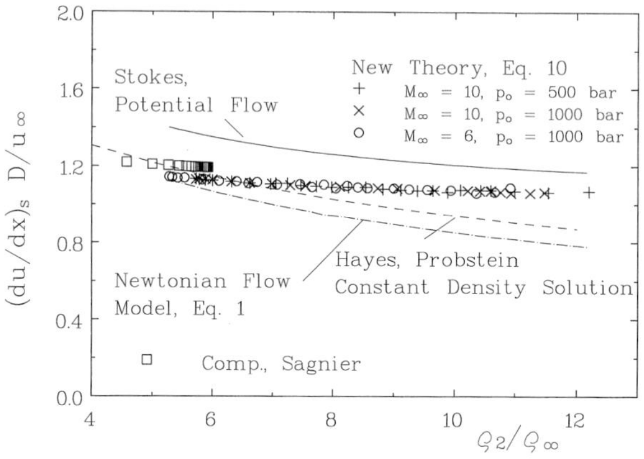

Figure 1. Non-dimensional velocity gradient versus density ratio.

which is a good approximation for flow conditions with high-temperature effects (Olivier 1995). It is worth noting that in Fig. 1, with increasing density ratio the non-dimensional velocity gradient slightly decreases. But, as expected, the dimensional velocity gradient increases, since in this case the shock stand-off distance becomes smaller (for brevity, this graph is not shown in this paper). For the stagnation-point heat flux it was possible to compare the results of non-equilibrium Navier-Stokes calculations with the modified theory of Fay and Riddell which makes use of the new velocity gradient. A collection of numerical results was compiled by Vérant (1994). These computations were carried out at European research institutes, i.e., ONERA-France, ESTEC-Netherlands, and DLR-Germany. Figure 2 shows the collection of these numerical results, where the stagnation-point heat flux is scaled by the square root of the nose radius divided by the stagnation-point pressure. Two curves are also presented according to the theory of Fay and Riddell for the old as well as for the new formulation of the velocity gradient. As expected, for low enthalpies they coincide, but for higher enthalpy flow conditions the formula of Fay and Riddell with the new velocity gradient relation gives higher values which agree very well with the Navier-Stokes computations.

3.2. Experimental results

For high-enthalpy conditions it is very difficult to measure the stagnation-point heat flux with sufficient accuracy in order to prove a theoretical result. Therefore it was decided to determine the derivation of the velocity gradient in the experiment from the Newtonian value (1). With the help of the one-dimensional momentum equation this ratio can be expressed as

$$\frac{du/dx\,|_{s,exp.}}{du/dx\,|_{s,Newt.}} = \frac{D}{2}\sqrt{\frac{-d^2p/dx^2\,|_{s,exp.}}{2(p_s - p_\infty)}}\,. \qquad (12)$$

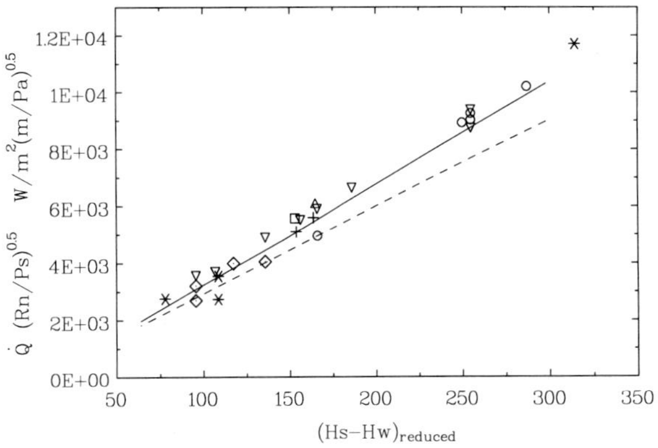

Figure 2. Comparison of Navier-Stokes calculations with the stagnation-point heat flux according to Fay and Riddell and the new velocity gradient: —— Fay and Riddell, Lewis number Le $= 1.4$, viscosity μ (Sutherland, in Hansen 1959), new velocity gradient; – – – – Fay and Riddell, Lewis number Le $= 1.4$, viscosity μ (Sutherland, in Hansen 1959), Newtonian velocity gradient; symbols are from Navier-Stokes calculations, Vérant (1994).

The tests were performed in the Aachen shock tunnel TH2 for different flow conditions. The stagnation temperature was varied between 1500 K and 4700 K and the free-stream Mach number between 6 and 12. In the experiments the pressure distribution in the stagnation region was measured with high spatial resolution. The distribution was approximated by a spline of second order. From this function it was possible to determine the value $d^2p/dx^2 \mid_{s,exp.}$ in (12). In this way the deviation could be determined between the tangential velocity gradient of the experiment and according to the Newtonian theory. These experimental values are shown in Fig. 3 and compared with those according to the new theory (10). Despite the fact that in the experiments only a few flow conditions could be tested, the experimental values clearly show the tendency of the theoretical ones, and the agreement achieved with the values according to the new theory is quite good. Of course, in particular for high stagnation temperatures, the measurements performed in the stagnation region are very sensitive to several experimental parameters. Therefore, the experiments performed can only support the general behavior of the tangential velocity gradient as determined by the theory.

4. Conclusions

The new expression for the velocity gradient represents a significant improvement and a higher accuracy when it is used to compute the stagnation-point heat

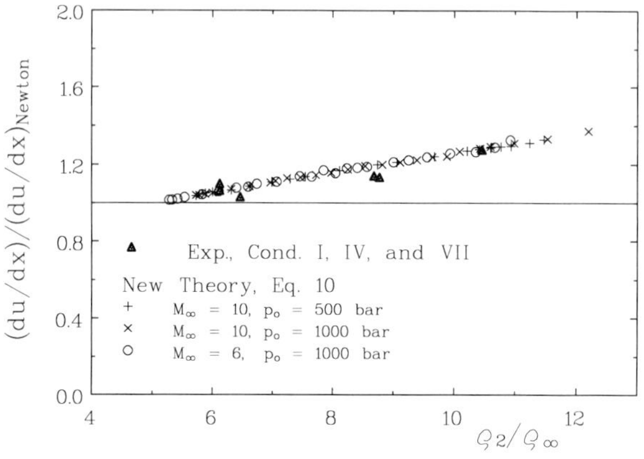

Figure 3. Deviation of the tangential velocity gradient from that according to the Newtonian theory as a function of the density ratio across the normal shock.

flux according to the theory of Fay and Ridell. It is based on a well-developed flow model which is much more realistic than the old one used before. This is confirmed by the good agreement between numerical and experimental results.

References

Belotserkovskii OM (1958) Flow with a detached shock wave about a symmetrical profile. J Math & Mech 22:279–296.

Esser B, Grönig H, Olivier H (1992) High-enthalpy testing in hypersonic shock tunnels. In: Bertin JJ, Periaux J, Ballmann J (eds), Advances in Hypersonics, Defining the Hypersonic Environment, Birkhäuser, Vol. 1, pp 182–258.

Fay JA, Riddell FR (1958) Theory of stagnation point heat transfer in dissociated air. J Aer Sci 25:73–85.

Hansen CF (1959) Approximations for the Thermodynamic and Transport Properties of High Temperature Air, NASA TR R-50.

Hayes WD, Probstein RF (1966) Hypersonic Flow Theory, Inviscid Flow, Academic Press, New York, London.

Lees L (1956) Hypersonic flow. In: Proc. 5th Int. Aeron. Conf., IAS-RAeS, Los Angeles, 1955, pp 241–276.

Olivier H (1995) Influence of the velocity gradient on the stagnation point heating in hypersonic flow. Shock Waves 5:205–216.

Vérant JL (1994) Numerical Enthalpy Rebuilding for Perfect Gas and Nonequilibrium Flows. Application to High Enthalpy Wind Tunnels. Technical report no. 69/6121 SY, ONERA, Chatillon, France.

Wong H, Muylaert J (1995) Private communication, ESTEC, The Netherlands.

Millisecond aerodynamic force measurement using a free-flying side-jet model with built-in gas reservoir

K. W. Naumann, H. Ende, G. Mathieu, A. George
French-German Research Institute Saint-Louis, ISL
5, Rue General-Cassagnou
F-68301 Saint-Louis, France

Abstract: This paper presents the technical realization and first results of a novel development of the ISL millisecond aerodynamic-force measuring technique. It is used to measure the additional force which acts on a model due to interaction between side jets and hypervelocity cross flow. The basic idea is to measure the acceleration of a model, which is set free for the time of test flow and then recaptured again. In addition to prior developments using a free-flying model with built-in pyrotechnical charges, we developed a method to perform such measurements with a free-flying model which carries a gas-filled reservoir. The technical realization includes development of an on-board fast-acting opening device, construction of the model, setup of the gas-supply system, and adaptation of the mounting of the accelerometers. The measurements comprise investigations on the internal flow phenomena, evaluation of the aerodynamic coefficients of the model, and measurements on jet interaction.

Key words: Aerodynamic force measurement, Acceleration measurement, Short-duration flow, Shock tunnel, Jet interaction

1. Introduction

The aim of the work described here is to measure the additional force which acts on a model surface as a result of intense interaction between control side jets and tropospheric hypervelocity cross flow (Naumann et al. 1993a, 1993b).

The experiments to study these effects are carried out in ISL shock tunnel "B" (Fig. 1a), now operating in the reflected mode. It duplicates main sea-level hypervelocity flow conditions up to about 2000 m/s. This significantly facilitates investigations on side-jet interaction, external combustion, or mixing and combustion phenomena in scramjets. The handicap is the short duration of test flow, actually about 3 ms, which complicates aerodynamic force measurements. Our technique uses a fast-acting mechanical support which sets the model free for the time of test flow to measure its acceleration, then recaptures it again. The very short time of free flight allows us to use wires and flexible

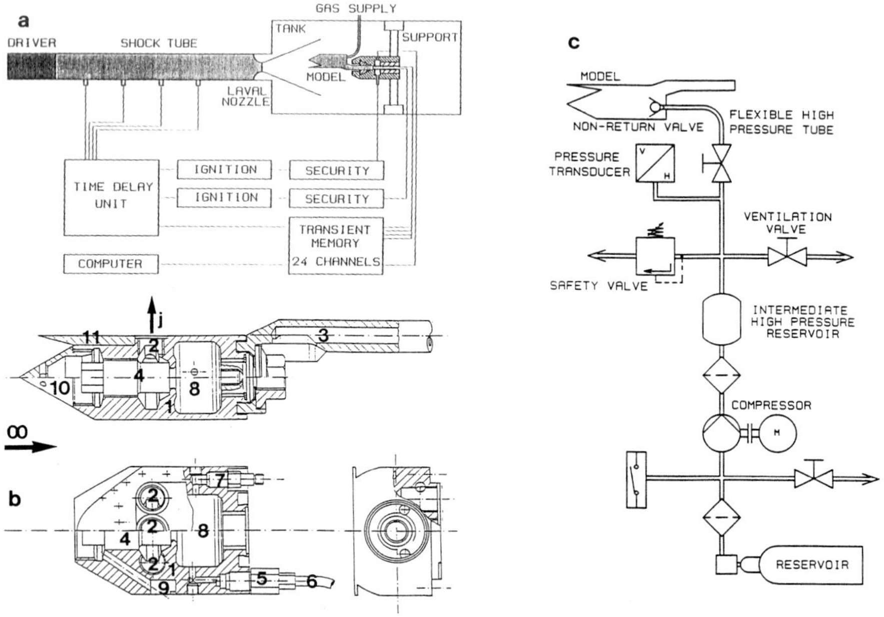

Figure 1. (a) ISL shock tunnel "B" with Laval nozzle and fast-acting mechanical support; (b) model with on-board gas reservoir — (1) body, (2) jet piece, (3) stem, (4) fast-acting opening device, (5) non-return valve, (6) filling tube, (7) pressure gauge, (8) gas reservoir, (9) location for accelerometers, (10) nose cap, (11) wing, (j) jet axis; (c) setup for the gas supply of the model.

pipes without disturbing model motion noticeably. The data-evaluation procedure is straightforward, insensitive to disturbances arising inevitably at the onset of the test flow, and allows us to judge whether the flow field has become quasi-stationary (Naumann et al. 1991a, 1991b). Operation of this technique with free-flying wing-body models equipped with laterally blowing jets supplied by pyrotechnical charges is described in Naumann et al. (1993a, 1993b).

2. Technical realization

The basic problem associated with our measuring task is to supply up to 3 jets of 4-mm throat diameter with at least 10 MPa within about one millisecond, without using a stiff or pressurized tube, in order not to disturb the model's free flight. Of the possible solutions to achieve this (Naumann et al. 1993b), we chose that which uses an on-board reservoir. Figure 1b shows the design of the model; Fig. 1c shows a sketch of the gas-supply system. A compressor fills the reservoir before the test with usually 80.4 MPa. Then the thin and flexible pipe is vented to prevent it from acting upon the model. Immediately before test flow onset the reservoir is opened towards the jets by a fast-acting opening device. It is a development of our own, very compact, and driven by an explosive charge. This accelerates a free piston which weakens the sealing diaphragm of 12-mm diameter so that it ruptures due to the reservoir pressure. Figure 2 shows different aspects of model #3 and its components; Fig. 3 shows the model and the test chamber.

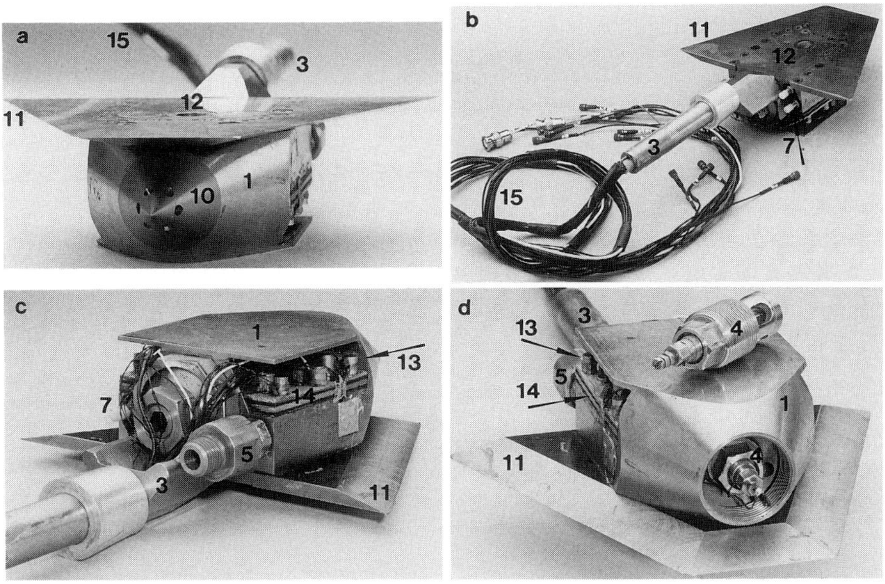

Figure 2. Different aspects of model #3: (1) body, (3) stem, (4) fast-acting opening device, (5) non-return valve, (7) pressure gauges, (10) nose cap, (11) wing, (12) side jet, (13) accelerometers, (14) shock-damping layer, (15) electrical wires.

3. Experiments

For the experiments on internal and nozzle flow we used a nozzle equipped with seven pressure transducers. Figure 4 shows the pressure histories and a sketch of the setup. One millisecond after ignition of the opening device the internal pressure is balanced and provides some milliseconds of quasi-stationary jet flow. The strong pressure drop during the internal equalization of the jet supply gas exhibits significant action of nonideal flow phenomena. These cause a low jet plenum temperature and subsequently condensation effects in the jet flow, which need additional investigation. For our present requirements it is sufficient that the ratio of thrust to plenum pressure is about constant over

Figure 3. Model #3: (a) in the test chamber; (b) retracted and banked for better visibility — (16) model #3, (17) test chamber, (18) mechanical support, (19) Pitot probe with tip aligned to flow direction in the test chamber, (20) flexible gas-supply tube.

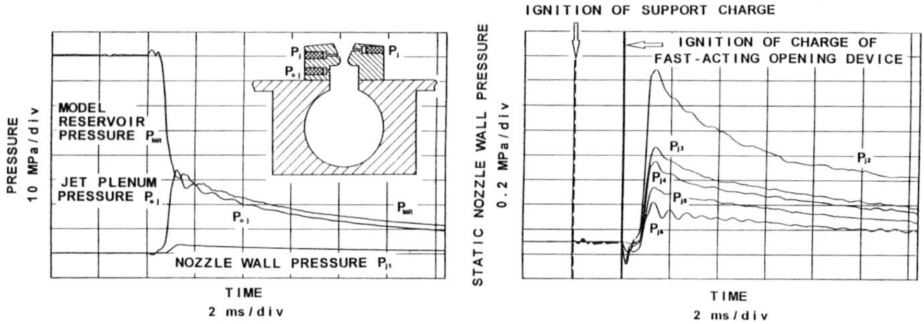

Figure 4. Static pressure histories of model #3 and instrumented nozzle.

time. The aerodynamic coefficients of the model without jet are evaluated as described by Naumann et al. (1993a).

Figure 5a shows the histories of five acceleration components and pitot pressure, Fig. 5b shows lift and drag coefficient histories. The settling times of lift (0.3 ms) and drag coefficient (0.5 ms) are improved compared to prior results (Naumann et al. 1993a) due to improved shock-damping materials for the accelerometers. The evaluation of the additional interaction force acting on the flat wing is carried out as presented in Naumann et al. (1993a) (Fig. 6). The pressure histories yield the relative partial acceleration histories (Fig. 6b). These and the measured total acceleration yield the acceleration component due to the jet/cross-flow interaction (Fig. 6c), and from this and the jet acceleration the interaction coefficient history is calculated (Fig. 6d). In contrast to prior investigations with jets driven by combustion (Naumann et al. 1993a, 1993b), the cold and condensing nitrogen show no quasi-stationary state within 2 ms. Helium

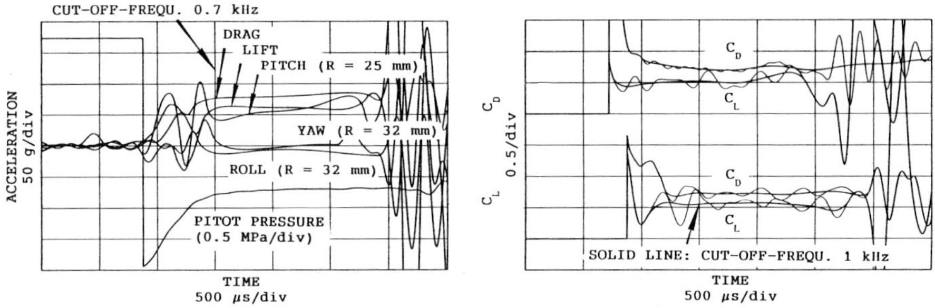

Figure 5. (a) Histories of lift, drag, pitch, roll, and yaw accelerations together with pitot pressure; (b) histories of lift and drag coefficients of both tests with silent jets.

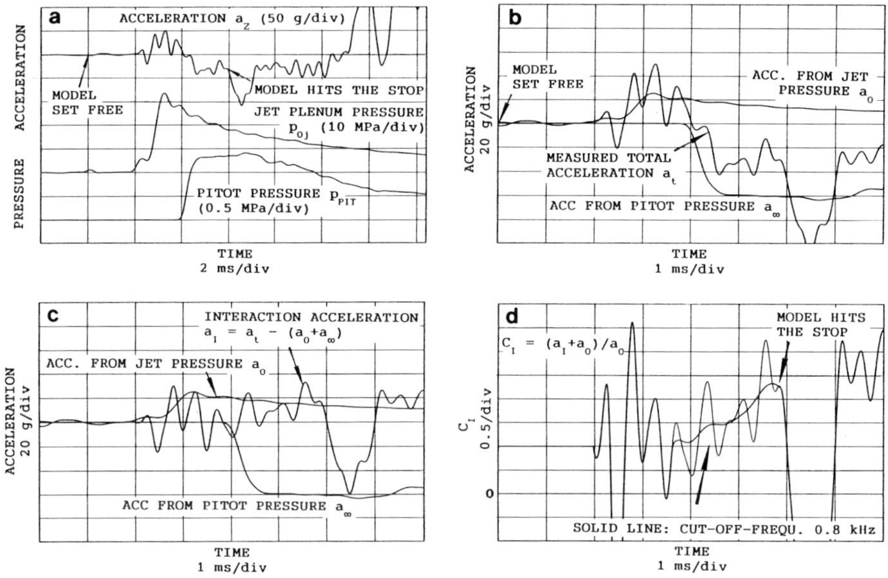

Figure 6. Data histories of an experiment using a nitrogen jet.

(Fig. 7) produced higher model acceleration, less internal pressure loss, and established quasi-stationary flow within 0.5 ms. Hence the nature and state of the gas can significantly influence the time necessary to establish quasi-stationary behavior.

4. Conclusion

The experiments prove that the novel technique to supply jets of free-flying models with stored gas works well. The improved methods to protect the ac-

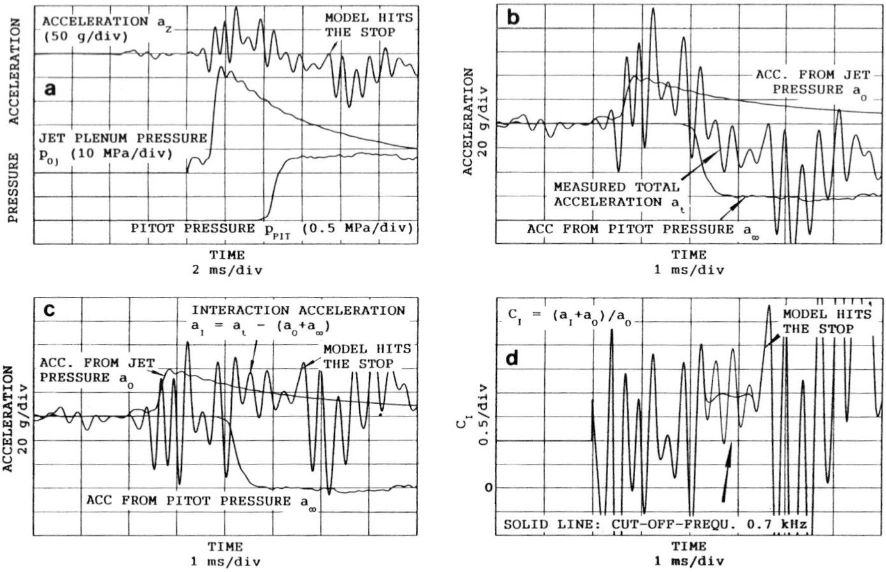

Figure 7. Data histories of an experiment using a helium jet.

celerometers against too heavy shocks originating from the built-in fast-acting opening device brought an improved overall response. Experiments on the internal flow exhibited the existence of strong nonideal and nonisentropic effects. The jet flow shows significant condensation of the nitrogen. The results on interaction force show that with these nitrogen jets, establishing quasi-stationary conditions takes longer than with hot gas from combustibles, or with helium. Further investigations will study these effects.

References

Naumann KW, Ende H, Mathieu G (1991a) Technique for aerodynamic force measurement within milliseconds in shock tunnel. Shock Waves 1:223.

Naumann KW, Ende H, Mathieu G (1991b) Technique for aerodynamic force measurement within milliseconds in a shock tunnel. In: Shock Waves, Proceedings of the 19th ISSW, Sendai, Japan, Vol 2, p 979.

Naumann KW, Ende H, Mathieu G, George A (1993a) Millisecond aerodynamic force measurement with side-jet model in the ISL shock tunnel. AIAA J 31(6):1068.

Naumann KW, Ende H, Mathieu G, George A (1993b) Shock Tunnel Experiments and Approximative Methods on Hypervelocity Side-Jet Control Effectiveness, AIAA Paper 93-1929.

Effects of catalycity and boundary layer reactions on surface heat flux in hypersonic high enthalpy flows

S. L. Gai[†], N. R. Mudford[†], G. T. Roberts[‡], S. G. Mallinson[†]

[†]Department of Aerospace and Mechanical Engineering
University College, University of New South Wales
Australian Defence Force Academy
Canberra ACT 2600, Australia

[‡] Department of Aeronautics and Astronautics
The University, Southampton, United Kingdom

Abstract: The role of chemical reactions in enhancing or inhibiting surface heat flux to bodies exposed to hypersonic, high enthalpy flows is investigated. Attention was directed to both gas phase atomic recombination and catalyzed atomic recombination at the surface. Also examined was the seeming anomaly between the results of an earlier experiment on a flat plate at 0° incidence and the more recent measurements on bluff bodies. In particular, the validity of the hypothesis made in the earlier experiments—that endothermic reactions occurred in the boundary layers, resulting in reduced heat transfer rates beyond what is to be expected when no recombination occurs—has been investigated.

Key words: High enthalpy, Hypervelocity, Catalycity, Surface heat flux

1. Introduction

This study investigates the role of chemical reactions in enhancing or inhibiting surface heat flux to bodies exposed to high enthalpy, hypersonic flows produced in a free-piston shock tunnel. Attention has been given to both gas phase atomic recombination and to surface catalysed atomic recombination at the solid surface. The aim is to test long standing but experimentally unconfirmed theories concerning the extent of the departure of the reactions from chemical equilibrium and the consequent effects on surface heat flux. In particular, attention is directed towards examining the veracity of the hypothesis (see East et al. 1980) that endothermic reactions occur in the boundary layers of sharp leading edge flat plate flows causing reductions in the surface heat flux beyond that which is expected when no recombination occurs via exothermic reactions.

2. Experimental details

The free-piston shock tunnel T3 at the Australian National University in Canberra was used to generate high enthalpy hypersonic flows for the present experiments. The experiments were conducted at two flow conditions, designated B and D, typically representing high and medium enthalpy flows respectively. A conical nozzle of 300 mm exit diameter and two throat sizes 12.7 mm and 25.4 mm diameter were used to generate the free stream. The larger nozzle throat yielded flows with densities about three times higher. Two test gases, air and nitrogen, were used in the experiments. The reservoir and free-stream conditions of these flows are given in Table 1 below.

Table 1. Reservoir and free-stream values for conditions B and D; h = enthalpy, p = pressure, T = temperature, ρ = density, U = velocity, M = Mach number, α_N = nitrogen dissociation fraction, α_O = oxygen dissociation fraction, subcripts 0 and ∞ refer to reservoir and free-stream conditions, respectively.

	Test Gas	Throat mm	h_0 MJ kg^{-1}	p_0 MPa	T_0 K	T_∞ K	ρ_∞ g m^{-3}	U_∞ m s^{-1}	M_∞	α_N	α_O
B	Air	12.7	19.0	22.0	8400	1200	2.5	5500	7.5	10^{-7}	0.82
B	N_2	12.7	19.0	22.0	8800	1250	2.5	5500	7.4	0.08	-
D	Air	12.7	13.0	21.0	7100	930	3.5	4600	7.4	10^{-9}	0.43
D	N_2	12.7	14.0	22.0	8100	1050	3.3	4900	7.4	0.03	-
B	Air	25.4	18.5	21.0	8300	1850	8.8	5200	5.8	10^{-6}	0.72
B	N_2	25.4	18.0	21.0	8800	2100	8.6	5400	5.7	0.07	-
D	Air	25.4	13.5	21.0	7200	1500	11.7	4600	5.8	10^{-7}	0.38

Surface heat flux was measured on a sharp leading edge flat plate at 0° incidence via chromel / alumel surface thermocouples, whose accuracy was ± 15 % (Gai & Joe 1992). In order to assess the effects of varying the degree of catalycity, the model was made such that it was divided lengthwise into two halves, one of which was an aluminium surface and the other was copper plated. The surface thermocouples were located symmetrically with respect to the halves and also staggered such that the fetch of each was unaffected by the other. The effects of a fully non-catalytic surface were investigated in a second series of experiments, by coating the lower half of the model with silicon dioxide SiO_2 which is known to have very low catalycity. A schematic of the model is shown in Fig 1.

3. Theoretical considerations

The Stanton number, S, is defined as

$$S = \dot{q}_w / \rho_e u_e \left(h_r - h_w\right) \tag{1}$$

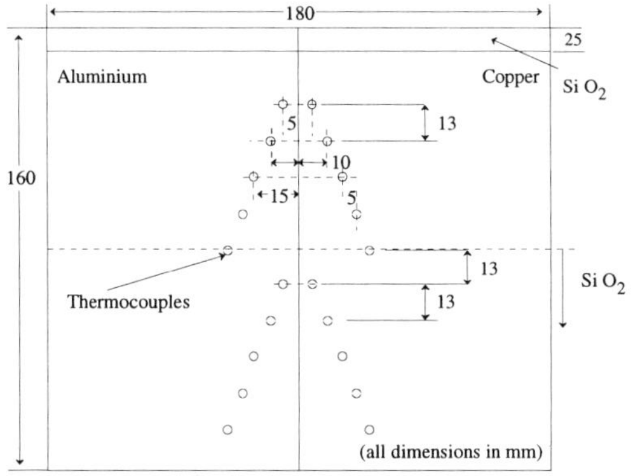

Figure 1. Schematic of the model.

Knowing that $C_f = 0.664\left(C^*/Re_x\right)^{1/2}$, where $C^* = \rho^*\mu^*/\rho_e\mu_e$, C_f is the skin friction coefficient and Re_x is the Reynolds number based on conditions at the edge of the boundary layer and distance x from the leading edge, we have, by Reynolds analogy,

$$S = 0.5C_f\left(\sigma^*\right)^{-2/3} \tag{2}$$

where σ^* is the Prandtl number (0.72 for air and 0.67 for nitrogen). Generalizing this to the frozen flow limit, we obtain (see East et al. 1980)

$$S_f = 0.5C_f\left(\sigma^*\right)^{-2/3}\left\{1 - h_{chem}/\left(h_r - h_w\right)\right\} \tag{3}$$

where h_{chem} is the chemical potential enthalpy of the dissociated species. The value of h_{chem} can either increase due to dissociation and vibrational excitation or decrease due to recombination and vibrational de-excitation. If the flow is in equilibrium, this will correspond to complete recombination either through gas phase boundary layer reactions or an infinitely catalytic wall, so that $h_{chem} = 0$ and we recover the equilibrium expression.

4. Results and discussion

Figure 2 shows results for conditions B and D with the nozzle throat size of 12.7 mm. The heat transfer data are expressed in the terms of the Stanton number against the Reynolds number. Firstly, we note that within the experimental scatter, there are no significant differences between aluminium/copper and SiO_2 coated results. This shows that, for the present experimental conditions, both catalytic and non-catalytic surfaces behave essentially the same indicating negligible surface recombination. Secondly, the results predominantly follow the frozen prediction although over the rear of the plate, there seems to be some tendency to rise towards the equilibrium prediction.

Figure 3 shows results obtained with the larger nozzle throat. We note that, on average, the data seem to lie closer to the equilibrium condition, indicating that with increased density of the flow, greater recombination is taking place, driving the heat transfer towards the equilibrium limit.

Figure 4 shows the air data compared with that of nitrogen at the same flow enthalpies. Using nitrogen eliminates reactions and reaction products involving oxygen. We note that, similar to air, the nitrogen data also fall approximately along the frozen chemistry prediction. In fact, if anything, they are somewhat below the frozen flow prediction on the front part of the plate. Comparing the air and nitrogen results, it is quite evident that the hypothesis of endothermic reactions involving atomic oxygen in the boundary layer, as invoked in East et al. (1980), is not sufficient to explain their lower than frozen boundary layer heat flux results.

Figure 5 shows all the data for condition B in terms of the similitude parameters $M_\infty^3 S$ and $\bar{\chi}$ (Stollery 1970), corrected to take into account the chemical potential enthalpy in a dissociated flow (Mallinson 1994). The data seem to be satisfactorily correlated in terms of these parameters and with the theory (Mallinson, 1994). Also shown in Fig. 5 are the data of Mallinson (1994). Figure 5 summarizes and demonstrates, firstly, that the air and nitrogen data are similar, and, secondly, that the differences between the results for aluminium, copper and glass surfaces are of the same order as the uncertainty in the measurements.

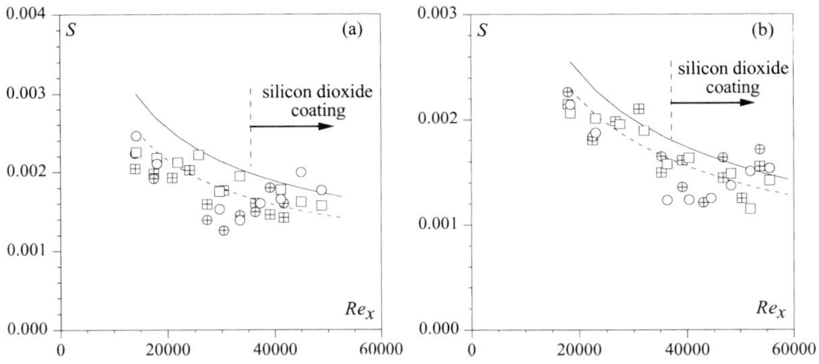

Figure 2. Heat transfer distributions for air with 12.7 mm throat: (a) condition B; (b) condition D. ○ = Al side; □ = Cu side. Open symbols = no SiO_2 coating; crossed symbols = SiO_2 coating on rear of model. Solid line = equilibrium limit; dashed line = frozen limit.

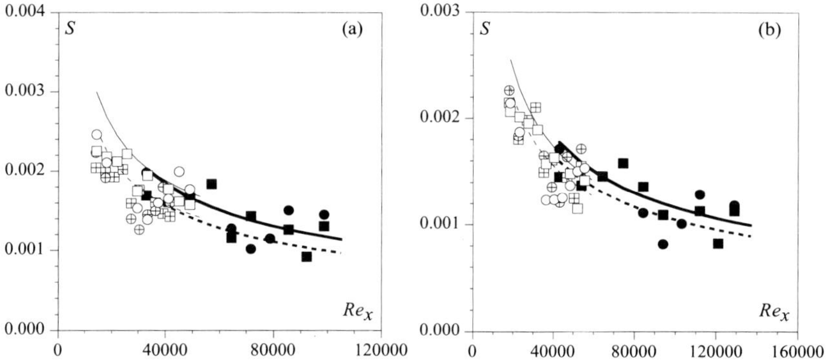

Figure 3. Heat transfer distributions for air with 12.7 mm and 25.4 mm throat: (a) condition B; (b) condition D. ○ = Al side; □ = Cu side. Open symbols = 12.7 mm throat, no SiO_2 coating; crossed symbols = 12.7 mm throat, SiO_2 coating on rear of model; filled symbols = 25.4 mm throat, SiO_2 coating on rear of model. Solid line = equilibrium limit; dashed line = frozen limit; heavier lines pertain to 25.4 mm throat.

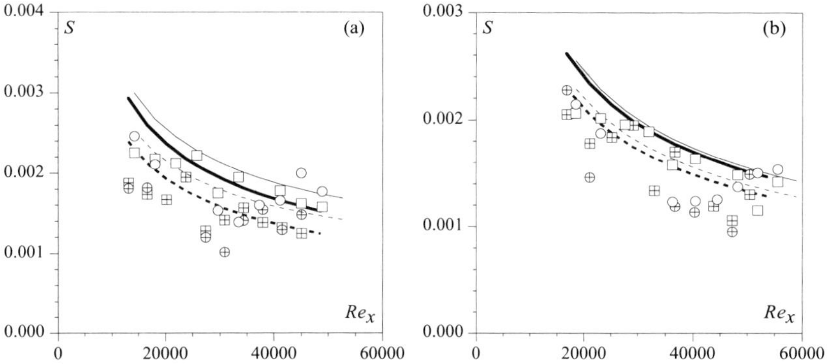

Figure 4. Heat transfer distributions for air and nitrogen with 12.7 mm throat: (a) condition B; (b) condition D. ○ = Al side; □ = Cu side. Open symbols = air , no SiO_2 coating; crossed symbols = nitrogen, no SiO_2 coating. Solid line = equilibrium limit; dashed line = frozen limit; heavier lines pertain to nitrogen.

5. Conclusions

The results show reasonable comparison with the frozen enthalpy theory thus indicating that the boundary layer reactions are frozen and surface reactions are negligible. The nitrogen data are similar but tend to be somewhat below the

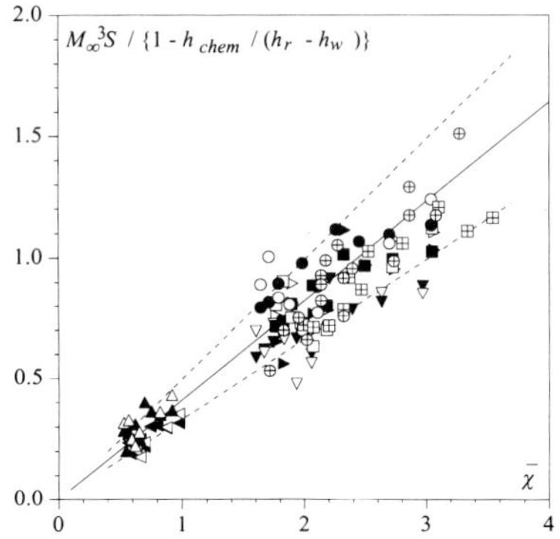

Figure 5. Correlation of flat plate heat transfer for condition B. Open symbols = Al side; filled symbols = Cu side. ○ = condition B-Air, 12.7 mm throat, no SiO_2 coating; □ = condition B-Air, 12.7 mm throat, no SiO_2 coating; △ = condition B-Air, 25.4 mm throat, SiO_2 coating on rear of model; ∇ = condition B-N_2, 12.7 mm throat, no SiO_2 coating; ▷ = condition B-N_2, 12.7 mm throat, SiO_2 coating on rear of model; ◁ = condition B-N_2, 25.4 mm throat, SiO_2 coating on rear of model; crossed circle = condition B-air, Mallinson (1994); crossed square = condition B-N_2, Mallinson (1994). Solid line = frozen theory, Mallinson (1994); dashed lines = $\pm 20\%$.

frozen value, which cannot be explained by taking recourse to the endothermic reactions.

Acknowledgement. We gratefully acknowledge the generous support of the Australian Research Council (Grant A 89131813). We would also like to thank Mr. P.M. Walsh of the Department of Physics and Theoretical Physics, the Faculties, ANU for his assistance with the experiments and Dr. A. Byrne of the RSPhysSE, ANU, for his help in the preparation of the surface coatings.

References

East RA, Stalker RJ, Baird JP (1980) Measurements of heat transfer to a flat plate in a dissociated high-enthalpy laminar flow. J Fluid Mech 97:673.

Gai SL, Joe WS (1992) Laminar heat transfer to blunt cones in high-enthalpy hypervelocity flows. J Thermophysics 6:433.

Mallinson SG (1994) Shock Wave / Boundary Layer Interaction at a Compression Corner in Hypervelocity Flows, PhD Thesis, University of New South Wales, Australia.

Stollery, JL (1970) Hypersonic viscous interaction on curved surfaces. J Fluid Mech 43:497.

Study of experiments sensitive to vibration-dissociation coupling models

J. Olejniczak, G. V. Candler
Aerospace Engineering & Mechanics and Army HPC Research Center
University of Minnesota
Minneapolis, MN 55455 USA

Abstract: A set of experiments to test vibration-dissociation coupling models has been designed for the T5 shock tunnel at Caltech. Using computational fluid dynamics, numerous experimental configurations were analyzed before choosing double-wedge and double-cone geometries. The predicted flow fields for these geometries strongly depend on the choice of the coupling model. This sensitivity is due to a complex shock-wave/boundary-layer interaction that results in oscillations of flow quantities along the body surface. Subsequent measurements of surface pressure and heat-transfer rate, along with interferometry, will provide new validation data for vibration-dissociation coupling models.

Key words: Hypersonic flow, Vibration-dissociation coupling, Shock interaction

1. Introduction

In many hypersonic flows the characteristic times of the fluid motion, chemical dissociation, and vibrational relaxation are of the same order of magnitude. Therefore, these processes must be modeled in a coupled manner. In compressing flows, the vibrational relaxation reduces the dissociation rate, and the dissociation of high-energy oscillators inhibits the vibrational relaxation. The details of this coupling process affect the pressure distribution on a vehicle, and therefore its trim angle of attack.

There are many vibration-dissociation coupling models, however no model has been adequately validated with experiments. This is because there are no experiments specifically designed to test the coupling models. This paper presents a set of experiments designed for the T5 shock tunnel at Caltech that is sensitive to the choice of the vibration-dissociation coupling model. The experiments were designed using a computational fluid dynamics method (Olejniczak and Candler 1995) and two different coupling models. A combination of flow conditions and geometry was found that produces a measurable difference between the flow fields computed using the two models. We found that certain double-cone and double-wedge geometries produce flows that are particularly sensitive to the choice of the model; this sensitivity is discussed below.

We chose the Park (1987) $\sqrt{TT_v}$ and Marrone and Treanor (1963) CVDV models because previous work showed that under T5 conditions, they reproduce the data at either end of the experimental uncertainty (Olejniczak and Candler 1995). Nitrogen was chosen as the test gas to avoid the uncertainties in modeling reacting air. Interferometry will be used to visualize the flow, while surface pressure and heat-transfer rate measurements will provide quantitative data.

2. Design of experiments

To design the experiments, we used computational fluid dynamics to simulate the flow over many different geometries at T5 freestream conditions. Each case was computed with the two different coupling models. The goal was to identify the combination of geometry and freestream conditions that yields the greatest difference in the interferograms and surface quantities between the two computed flow fields.

We found that certain double-wedge and double-cone geometries are attractive because the two coupling models predict dramatically different flow fields. High-enthalpy and high-density freestreams with certain wedge or cone angles are particularly sensitive to the coupling model. Examples are shown in Figs. 1 and 2, where the computed pressure contours are plotted for the flow over double-wedge configurations. The first wedge produces an attached shock which interacts with the detached shock from the second wedge. This produces a complicated flow field with a separation zone near the corner of the wedges. The size of the separation zone depends on the coupling model. There is a range of wedge angles for which the predicted flow field is particularly sensitive to the choice of coupling model; the reason for this sensitivity is discussed below. Similar results are found for double-cone geometries.

The measurable differences between the predictions of the two coupling models are shown in Figs. 3 and 4. Figure 3 shows the computational Mach-Zehnder interferograms corresponding to Fig. 1. The differences in the shock location, fringe number, and fringe shape are easily seen. We can assess the accuracy of these models by comparing the computed interferograms with the experimental interferogram. Figure 4 plots the surface pressure and heat-transfer rate for this case. Note that the pressure and heat-transfer-rate peaks caused by the shock impinging on the second wedge are shifted and that the separation zones are different sizes. These differences will be greater than the uncertainty in the experimental measurements, provided there is sufficient instrument density on the model. Thus, this set of experiments will serve as validation data for vibration-dissociation coupling models.

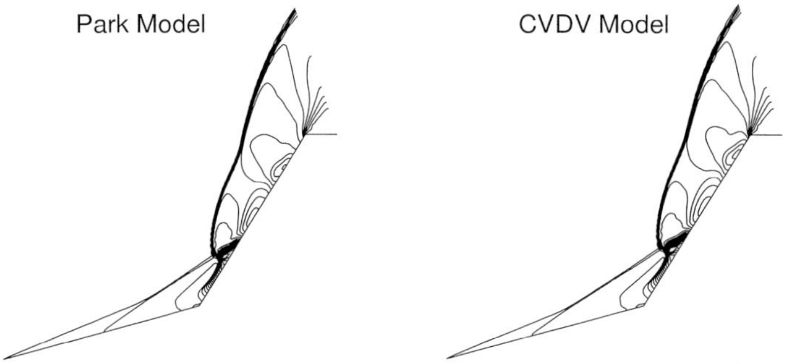

Figure 1. Pressure contours for flow over a double-wedge with a base diameter of 5 cm and angles of 15.5° and 58.5°. The freestream conditions are nitrogen at $u = 6690\,\mathrm{m/s}$, $\rho = 0.01\,\mathrm{kg/m^3}$, and $h_o = 26\,\mathrm{MJ/kg}$.

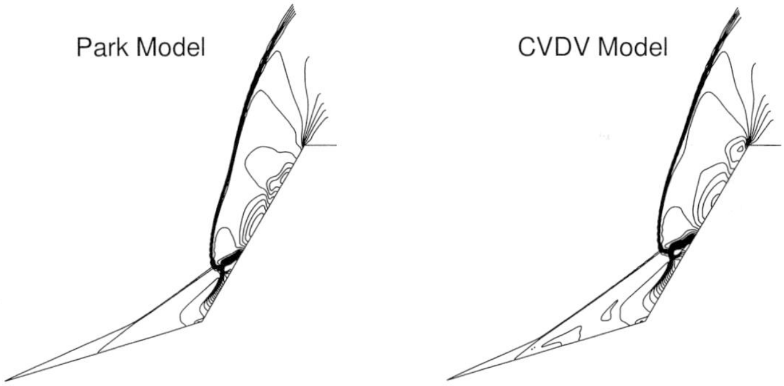

Figure 2. Pressure contours for flow over a double-wedge with a base diameter of 5 cm and angles of 17° and 60°. Freestream conditions as above.

3. Discussion of flow

The flow over the double-wedge and double-cone geometries can be understood by considering a schematic of the flow, shown in Fig. 5. A transmitted shock, formed by the intersection of the oblique shock from the first wedge and the shock from the separation zone, intersects with the bow shock from the second wedge. A transmitted shock from this triple point impinges on the second wedge, causing separation to occur. The gas that is reversed into the separation zone must flow out of the shear layer at the edge of the separation zone. As the second wedge angle is increased the impinging shock becomes more normal,

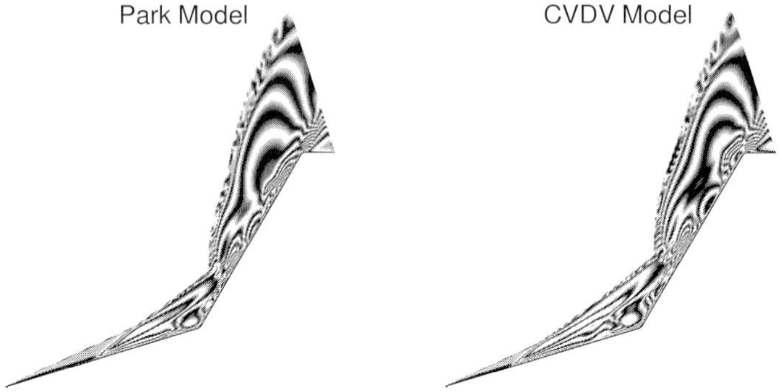

Figure 3. Computed interferograms for the previous figure.

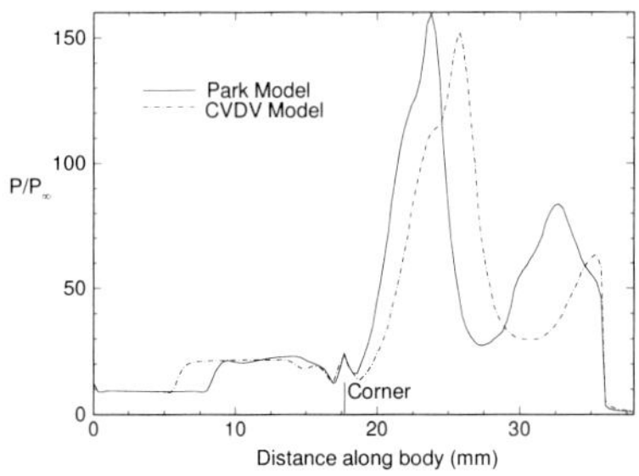

(a)

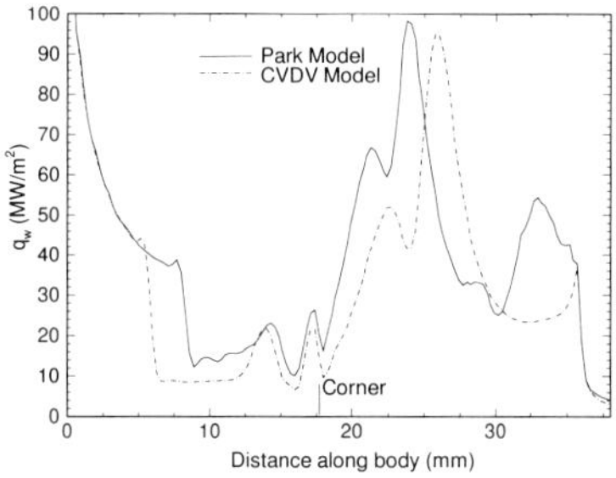

(b)

Figure 4. Computed flow over a double-wedge corresponding to Fig. 2: (a) pressure distribution, (b) heat-transfer-rate distribution.

and more gas is reversed into the separation zone. Eventually a steady shear layer cannot eject enough mass, and the flow becomes unsteady and oscillates. This is the oscillation mechanism identified by Maull (1960) for spiked bodies. This process is highly nonlinear near the point of unsteadiness. Small changes in the bow shock, resulting in changes in the location of the triple point, have a large influence on the size of the separation zone. Thus, different models for the nonequilibrium dissociation rate and the vibrational energy source terms predict widely different flow fields.

Another interesting feature of this flow, which provides a quantitative test of the coupling model, is the underexpanded supersonic jet that is formed along the second wedge surface. This jet is bounded by the shear layer emanating from the triple point and is trapped against the body by the subsonic region behind

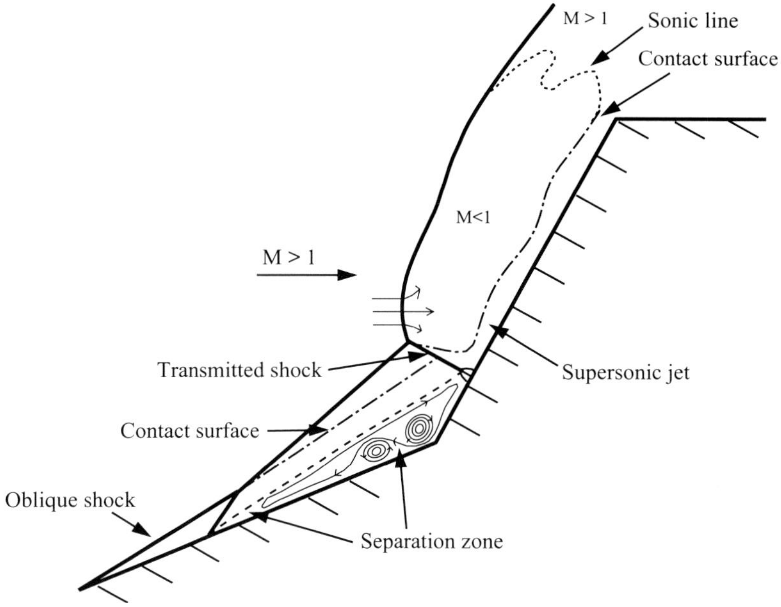

Figure 5. Schematic diagram of the viscous flow over a double-wedge geometry.

the bow shock. This jet produces oscillations in the flow variables along the second wedge surface, as shown in Fig. 6. These oscillations result from the finite rate of transfer between the different energy modes. As in an underexpanded jet, the main mechanism is the reversible transfer between internal and kinetic energy (the $p\,\nabla \cdot \vec{u}$ term in the kinetic-energy equation). The peak height and wavelength of the oscillations are altered by the rate of vibrational-translational energy transfer and vibration-chemistry energy transfer. This latter term differs among the vibration-dissociation coupling models. Figure 6 clearly shows the differences in the predicted oscillations of pressure and heat-transfer rates between the Park and CVDV models. These oscillations are also present in the density, and therefore are seen in the interferograms.

4. Conclusion

This paper presents the design of a set of experiments which will provide new validation data for vibration-dissociation coupling models. The experiments, planned for the T5 shock tunnel, use double-wedge and double-cone geometries, which we have found to be particularly sensitive to vibration-dissociation coupling physics. The mechanism for this sensitivity is a shock-wave/boundary-layer interaction in which the model of the vibration-dissociation coupling has

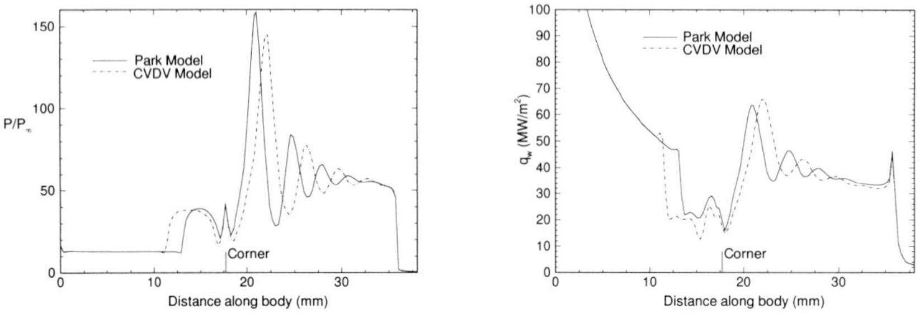

(a) (b)

Figure 6. Computed flow over a double-cone with angles of 25° and 68°. Freestream conditions are the same as the previous figures. (a) Pressure distribution, (b) heat-transfer-rate distribution.

a large influence on the predicted flow field. There are differences in the size of the predicted separation zone and in the oscillation of flow quantities along the body. Therefore, interferometry, along with surface pressure and heat-transfer-rate measurements, will provide data to test vibration-dissociation coupling models.

Acknowledgement. We would like thank Professor H. G. Hornung for providing many useful insights and assisting us in the design of the experiments. This work was supported by the Air Force Office of Scientific Research Grant Number F49620-93-1-0338. Computer time was provided by the University of Minnesota Supercomputer Institute. This work was also sponsored in part by the Army High Performance Computing Research Center under the auspices of the Department of the Army, Army Research Laboratory cooperative agreement number DAAH04-95-2-0003/contract number DAAH04-95-C-0008, the content of which does not necessarily reflect the position or the policy of the government, and no official endorsement should be inferred.

References

Marrone PV, Treanor CE (1963) Chemical relaxation with preferential dissociation from excited vibrational levels. Physics of Fluids 6:1215.

Maull DJ (1960) Hypersonic flow over axially symmetric spiked bodies. J Fluid Mechanics 8:584.

Olejniczak J, Candler GV (1995) Vibrational energy conservation with vibration-dissociation coupling: general theory and numerical studies. Physics of Fluids 7:1764.

Park C (1987) Assessment of Two-Temperature Kinetic Model for Ionizing Air, AIAA Paper No. 87-1574.

Supersonic Flow

Disturbances on the shock front ahead of a concave body due to resonant excitement of vortices

A. O. Serov, L. S. Shtemenko, F. V. Shugaev
Moscow State University, Department of Physics
Chair of Molecular Physics
119899 Moscow, Vorobyovy Gory, Russia

Abstract: A jet is found to arise behind the shock wave reflected from a concave body. The jet gives rise to vortical structures. Experiments were conducted in a shock tube with carbon tetrafluoride used as the test gas. A resonant excitation of the vortical structures arises inside the cavity. Radially unstable cores emit beams of pressure waves which cause disturbances on the shock front.

Key words: Resonant excitement of vortices, Disturbances on shock front

1. Introduction

Sound emission from unstable vortices was investigated by Kopyov et al. (1987) and Vlasov et al. (1978). Milton et al. (1987) showed the formation of vortices behind the shock wave reflected from a cavity. Shock-wave instability was examined in a hypersonic wind tunnel ahead of concave bodies (Johnson 1959). It was suggested that the instability may be the result of the formation of vortices inside the cavity. Disturbances on the shock front were observed in ballistic experiments at high Mach numbers, $M > 8$ (Bedin et al. 1981). However, processes inside the cavity were not investigated.

We studied the formation of vortices and their influence on the detached shock wave in front of concave bodies in supersonic flow.

2. Experimental facility

Experiments were conducted in a single-diaphragm shock tube of a rectangular 34×72 mm^2 cross-sectional area. Carbon tetrafluoride (CF_4) was chosen as the test gas. A model (concave body) was mounted in the test section which had glass windows. The models had the shape of a parallelepiped. The cross-section of some models was a 10-mm square. The flow was three-dimensional. Other models had a width equal to the distance between the glass windows (34 mm), so the flow past the models was close to two-dimensional. The height of the models was $h = 10$ mm. The generatrix of the cavity was either a 10-mm diameter (deep cavity) circle or a 20-mm diameter (shallow cavity) one. Some models had a rectangular cavity. The initial pressure was equal to $p_0 = 1.3 \times 10^3$–2×10^4 Pa. The Mach number of the incident shock wave varied from 2 to 5, the Reynolds number varied from 10^4 to 2.4×10^5 in the flow behind the reflected shock wave, the characteristic length being half the arc of the cavity. A quasi-stationary supersonic flow took place in the test section after the burst of the diaphragm. Its duration was about 150 – 300 μs. We took shadowgraphs and streak-camera records. The pressure on the bottom of the cavity was measured with a Kistler transducer.

3. Results and discussion

The flow pattern inside the cavity was described earlier (Ibrahim et al. 1985; Shtemenko et al. 1993). The characteristic feature of the flow is the formation of the jet. It is laminar at low Reynolds numbers. Vortices are situated at the end of the jet. They also arise near the walls due to the interaction of the transverse shocks with the boundary layer and because of its subsequent separation. In addition, vortices appear as a result of the instability of slipstreams. Figure1 shows successive shadowgraphs. The vortices move toward the plane of

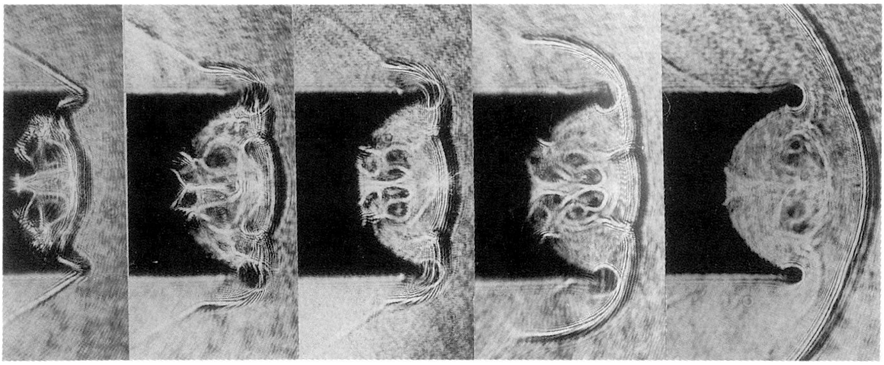

Figure 1. Two-dimensional flow pattern inside the deep cavity: $l/h = 0.5$; $M = 4$; $p_0 = 1.3 \times 10^3$ Pa; $t = 20$ μs, 30 μs, 35 μs, 50 μs, 80 μs; the time is measured from the instant when the incident shock arrives at the bottom of the cavity (l is its depth).

symmetry and then inside the jet toward the reflected shock. Afterwards they stop and turn into two cylindrical vortices (or into a vortex ring when the flow is a three-dimensional one). If the Reynolds number is large enough then the vortical structures become unstable. Waves propagate along the vortex ring i.e., azimuthal instability arises. Distortion of the vortex ring takes place as one can see from Fig. 2. Radial (axisymmetrical) waves propagate along the

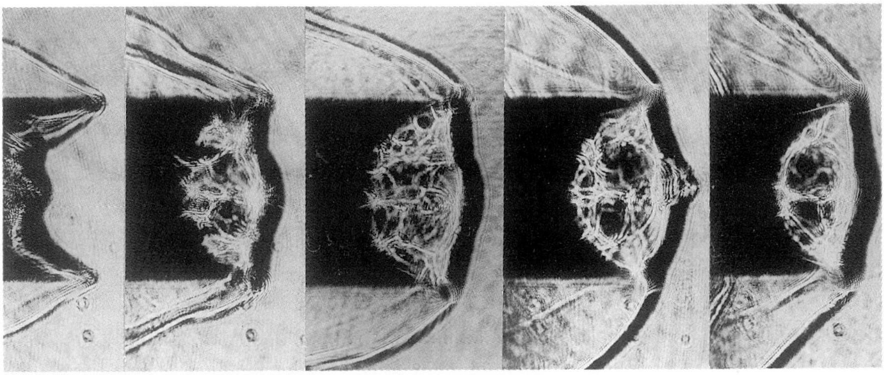

Figure 2. Three-dimensional flow pattern inside the deep cavity: $l/h = 0.5$; $M = 4.5$; $p_0 = 6 \times 10^3$ Pa; $t = 15$ μs; 30 μs, 60 μs, 70 μs, 90 μs.

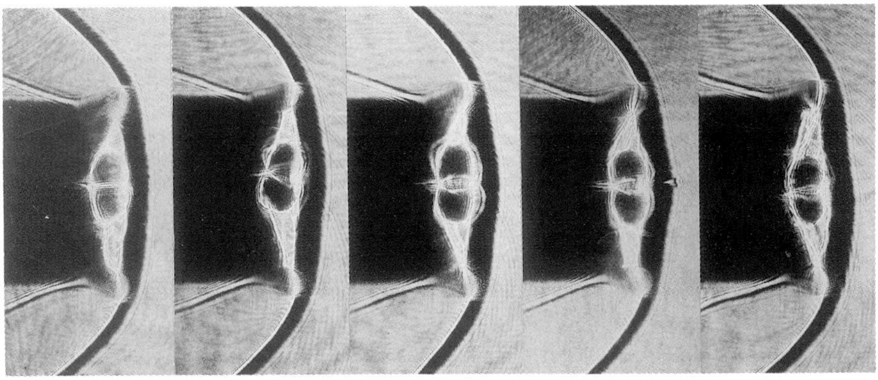

Figure 3. Three-dimensional flow pattern inside the shallow cavity: $l/h = 0.15$; $M = 4.9$; $p_0 = 4 \times 10^3$ Pa; $t = 60$ μs, 80 μs, 90 μs, 100 μs, 130 μs.

vortex ring if the Reynolds number is equal to 2–3×10^5 and the Mach number is equal to 4.5–5. Figure 3 shows the oscillations of the vortex ring.The core of the ring becomes elliptic and disturbances emerge on the shock front. The motion of the shock front is oscillatory. The vortical structures are destroyed periodically, the flow inside the cavity becomes turbulent and the gas flows out of the cavity. Then the vortical structures appear anew and the process repeats. Two oscillations were seen at the streak camera records during 150 μs. Figure 4 shows the flow pattern inside the rectangular cavity. Vortical structures are situated near horizontal walls and not near the plane of symmetry as in the former case. These structures oscillate (the diameter of the vortex core increases and decreases). As a consequence a disturbance appears on the shock front against

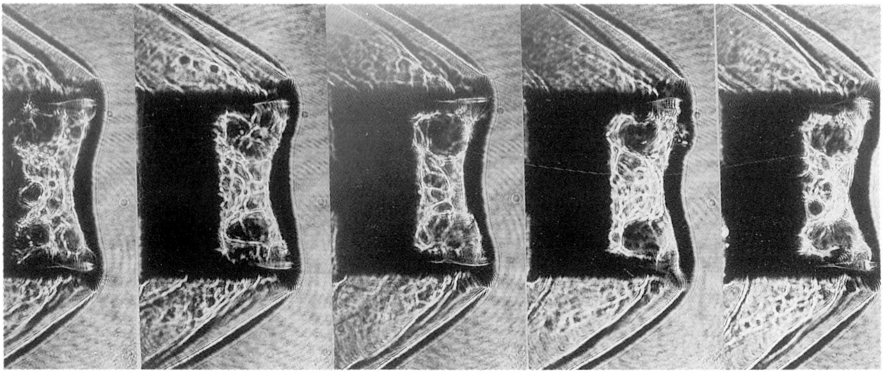

Figure 4. Three-dimensional flow pattern inside the rectangular cavity: $l/h = 0.5$; $M = 4.9$; $p_0 = 4 \times 10^3$ Pa; $t = 50$ μs, 70 μs, 100 μs, 110 μs, 120 μs.

the vortex. The disturbances appear in the two-dimensional flow at the same values of Reynolds number and Mach number.

The amplitude of the pressure oscillations at the bottom of the cavity is equal to 10% of the pressure mean value behind the shock wave in the case of the deep cavity. The period of oscillations coincides with the time during which the vortex moves from the bottom of the cavity to the vortex ring. The pressure waves reflect from the bottom and interact with the vortical structures. At high Reynolds numbers ($Re = 2$–3×10^5, the characteristic length being the diameter of the vortex ring) the vortical structures become unstable and emit acoustic waves that arrive at the shock front.

Disturbances on the shock front appear under appropriate conditions, namely, at $M = 4.5$–5. This fact can be explained as the result of the interaction between acoustic waves and vortices. The interaction may be resonant or not. If there is no resonance then the vortex ring oscillates as a whole, the frequency being equal to that of incident acoustic waves. If the frequency of the acoustic waves is a multiple of the natural frequency of the vortex ring then the ring becomes unstable. Radial waves propagate along its surface. The ring emits acoustic waves and their amplitude may be several times as great as the amplitude of the incident waves.

In our experiments the resonance arises at $M = 4.9$ (shallow cavity). The amplitude of the oscillations grows as the time increases. Radial waves appear at the surface of the core of the vortex ring. One can see this phenomenon from Fig. 3. The frequency of the oscillations of the vortex ring surface is equal to 170–220 kHz as it was found from streak camera records. The frequency of the pressure oscillations at the bottom of the cavity was equal to 150–200 kHz. It is the frequency of the waves that the vortex emits at a resonant excitement.

The frequency of oscillations was equal to 50–70 kHz in the case of deep cavity. In experiments we measured the period of time between two successive disturbances on the shock front. We calculated the amplitude and frequency of the acoustic waves that caused these disturbances. The values coincided with ones recorded with the aid of the pressure transducer. Consequently the disturbances on the shock front are caused by the acoustic waves emitted by unstable vortical structures while being resonantly excited. Figures 2 and 3 show diffraction fringes near unstable vortex rings. We consider them to be a result of the diffraction of light from the grid that is formed by ultrasonic waves in the gas (Shtemenko et al. 1993).

Figure 5 demonstrates the natural frequency of the vortex ring core and the frequency of acoustic waves against Mach number.

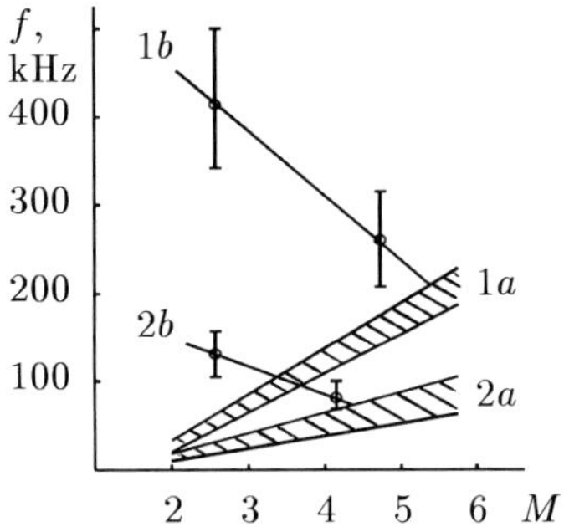

Figure 5. Dependence of the natural frequency of the vortex core and of the frequency of acoustic waves on Mach number: 1 corresponds to shallow cavity, 2 to deep cavity; *a* corresponds to acoustic waves, *b* to vortex ring.

The frequency of acoustic waves may be evaluated if we know the time during which they propagate along the path from the stagnation point of vortices to the bottom of the cavity. The value of the natural frequency of the vortex ring was calculated according to the measured size of the core and the ring velocity. Index 1 relates to the shallow cavity, index 2 relates to the deep cavity, *a* denotes acoustic waves, *b* denotes vortex ring. The size of shaded area denotes the uncertainty in frequency determination. One can see that both values of frequency coincide at $M = 4$–5, i.e., in the range where disturbances on the shock front exist.

4. Conclusion

An experimental investigation of the flow inside the cavity behind the reflected shock wave has been performed. Disturbances on the shock wave in front of a concave body may appear under appropriate conditions. The origin of the disturbances is a resonant interaction between acoustic waves and vortical structures inside the cavity.

References

Bedin AP, Klimov AI, Koblov AI, Mishin GI (1981) Experimental investigation of the flow of relaxing gas past concave bodies. Zhurn Tekhn Fiz 51(10):2119–2122 (in Russian).

Ibrahim MA, Serov AO, Shtemenko LS, Shugaev FV (1985) The reflection of a plane shock wave from a body with groove. Fluid Dynamics (Izv Akad Nauk SSSR MZhG) 20:822–828.

Johnson RH (1959) Instability in hypersonic flow about blunt bodies. Phys Fluids 2:526–532.

Kopyov VF, Leontyev EA (1987) Radiation and scattering of sound from vortex ring. Fluid Dynamics (Izv Akad Nauk SSSR MZhG) 22:398–409.

Milton EE, Takayama K, Archer RD (1987) An evaluation of two-dimensional and axisymmetric focusing of a plane shock wave. In: Proc 16th Int Symp on Shock Waves, Groenig H (ed), Aachen, pp 567–573.

Shtemenko LS, Shugaev FV, Serov AO (1993) Visualization of acoustic waves by means of diffraction. J Flow Vis and Image Processing 1:113–116.

Vlasov EV, Ginevsky AS, Karavasov RK (1978) Investigation of the flow pattern of a jet at various levels of initial turbulence. Uchenye zapiski TSAGI 1:25–32 (in Russian).

An inverse method for calculating nonequilibrium flow behind curved shock waves

S. Mölder
School of Aerospace Engineering, Ryerson Polytechnic University
Toronto, Ontario, M5B 2K3 Canada

Abstract: The theoretical aerodynamic model described here applies to flow of an ideal dissociating gas between a shock and the body producing that shock. The model allows calculation of flow with either axial or planar symmetry and for concave or convex curved shocks which can be either attached or detached at the leading edge of the body. For non-reactive flows over plane wedges we obtain exact results; for flow over circular cones, extremely good agreement is obtained for both weak and strong shocks; much improved agreement over previous methods and with experimental results is obtained for detached shocks on blunt bodies, particularly at the stagnation streamline. For flows in chemical nonequilibrium we get qualitative agreement with the calculations of previous authors.

Key words: Aerodynamics, Hypervelocity, Nonequilibrium

1. Introduction

For supersonic aerodynamics of blunt bodies the *direct problem* is the one where the body shape is specified and the shock shape and the flow between the shock and the body are calculated. For the *inverse problem* the shock is given, and the body shape and the intervening flow are found. Although the direct problem is of usual practical interest, there are design-oriented approaches, where the desired flow is specified, in which case it is convenient to begin by specifying shock-wave shape. Such situations occur in the design of supersonic and hypersonic air inlets. Van Dyke (1958) solved the inverse problem for an ideal gas and found that the solution is poorly convergent. Maslen (1964) based his inverse calculation on the assumption that the streamline curvature behind the shock is equal to the shock curvature and that this curvature remains constant from the shock to the body. This assumption gives good results for moderately inclined body surfaces in hypersonic flow. However it does not correspond to reality for blunt-body flow where shock and streamline curvature have opposite signs. Also, for the flow over a circular cone Maslen's theory assumes a zero streamline curvature behind the shock and therefore a constant pressure in the region between the shock and the body. This is contrary to the exact Taylor-McColl equation solution. In the present inverse method we calculate an exact flow curvature behind the specified shock. The changing flow field is then calculated between the shock and the body assuming the transverse pressure gradient, at a given station, to be a constant as determined by the streamline curvature at the shock at this station.

2. The aerodynamic model

Consider a shock element located a distance y_1 from a reference axis (Fig. 1). In a plane which contains both the upstream and downstream flow vectors, the radius of curvature of the shock is R_a. In a plane which is perpendicular to this plane and to the shock wave the curvature of the shock wave is R_b. It is easy to show that for axisymmetric flow $R_b = y_1/\cos\theta_s$ where θ_s is the shock angle. The corresponding curvatures of the shock are $S_a = -1/R_a$ and $S_b = -1/R_b$. The curvature of the streamline at the shock is calculated using the curved shock relations (Mölder 1971). This curvature is denoted by $(\partial\delta/\partial s)_1$ where δ is the local flow inclination and s is the distance measured along the streamline. In intrinsic streamline coordinates (s, n), the momentum equation across the streamlines is:

$$\frac{\partial p}{\partial n} = \rho V^2 \frac{\partial \delta}{\partial s} \tag{1}$$

In finite-difference form this can be written as:

$$(\Delta n)_1 = \frac{(\Delta p)_1]}{\rho_1 V_1^2 (\frac{\partial \delta}{\partial s})_1} \tag{2}$$

where Δn is measured along a normal to the streamline such that the pressure change is Δp. At this point we know $\Delta p/\Delta n$ but neither Δp nor Δn individually. Now select a small mass flow $\Delta \dot{m}$. This amounts to selecting the calculation step size. It is the mass flow between the streamline passing through (x_1, y_1) and the streamline immediately below, so that: $\Delta \dot{m} = \rho_0 V_0 (\Delta A)_0 = \rho_1 V_1 (\Delta A)_1$.

Using this we can find the height at which the streamline enters the shock, y_2. This allows us to geometrically locate the point (x_3, y_3), at the intersection of the streamline from (x_2, y_2) and the streamline-normal from (x_1, y_1), and to find the pressure $p_3 = p_1 + (\Delta p)_1$. To calculate the other flow quantities at (x_3, y_3), Maslen (1964) assumed that the flow is isentropic from (x_2, y_2) to (x_3, y_3). Thus knowing p_3 and the conditions behind the shock at (x_2, y_2), p_2, and T_2, we can find T_3 from $T_3/T_2 = (p_3/p_2)^{\gamma/(\gamma-1)}$. The energy equation and the equation of state then allow the determination of all flow variables at (x_3, y_3). This process is then repeated for the next lower streamline. The approximate nature of this method comes about from the assumption that the pressure gradient at any given downstream station remains proportional to the streamline curvature at

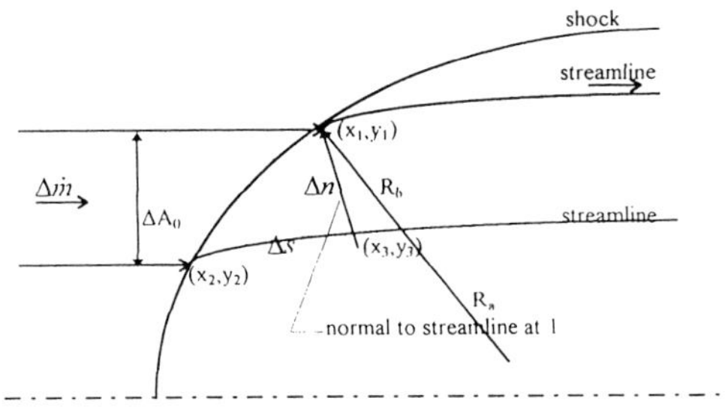

Figure 1. Schematic diagram of curved shock element.

the shock at that station. Thus the method should give results which are exact at the shock and as accurate as the modified Newtonian theory near the body surface.

3. Results

3.1. The aerodynamic model

The present inverse method has been applied to flow behind a plane oblique shock, flow behind a circular cone shock, and flow behind a blunt-body shock consisting of cubic spline segments, in each case for a perfect gas as well as for a chemically reactive gas.

Case A1 — Plane oblique shock. This is the classical oblique shock-on-a-wedge situation for which the streamlines behind the shock are straight and parallel and there is no pressure gradient. Many cases were calculated for Mach numbers ranging from 1.2 to 25 and for shock angles between the Mach angle and 90 degrees. In all cases the calculated wedge angle did not vary from the exact value by more than 0.001 degrees.

Case A2 — Cone flow. For this case the streamline curvature causes the pressure to increase from the shock to the body surface. This flow configuration provides a test of the main feature of this report, namely the assumption of pressure-gradient constancy along the streamline-normals. Figure 2 compares the present results against the classical Taylor-McColl equation solution for flow over a circular cone. For both strong and weak shock waves the difference between the two predictions of cone angle was usually less than one degree, and never larger than 2 degrees.

Case A3 — Blunt body. For blunt-body flow the curved bow shock was specified either as an equation or as a series of points on the shock wave. In the latter case a cubic spline was fitted through the points and the shock

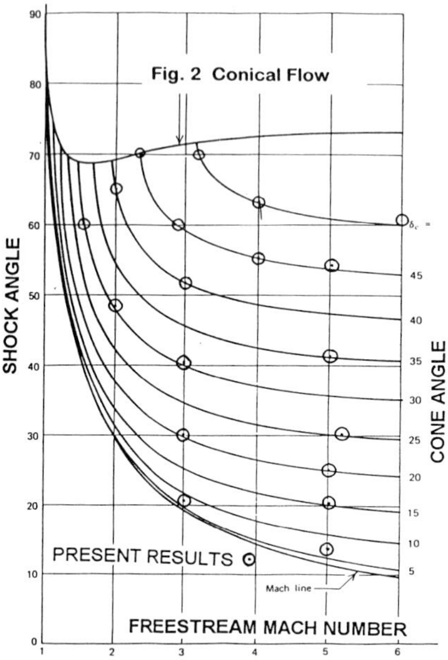

Figure 2. Conical flow.

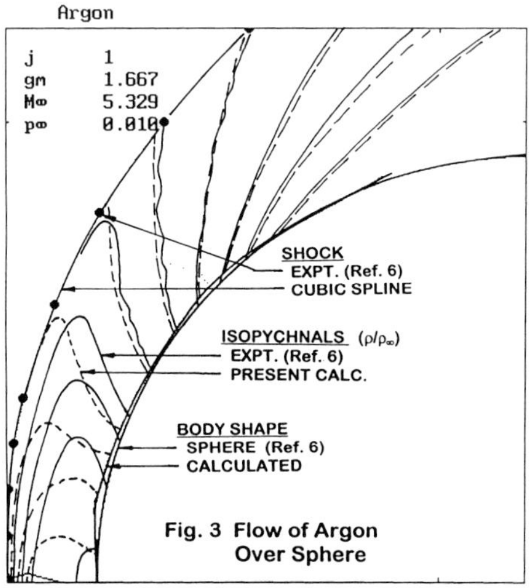

Figure 3. Flow of argon over a sphere.

slope and curvature were then calculated from the spline. In Fig. 3 we compare our calculations with the experimental results of Sedney and Kahl (1961) for Mach number 5.329 flow of argon over a sphere. In this case, of course, no dissociation is present so that we are again testing the aerodynamic aspects of our model. The cubic spline passes through the experimental shock-wave points. Solid contours are lines of measured constant density. Dashed contours are for the corresponding calculations. Good agreement is apparent except in the stagnation region. Lack of agreement is most likely due to the difficulty of specifying the experimental shock shape. Improved agreement in the stagnation region was obtained by perturbing the shock shape within the experimental tolerances. The calculated body is seen to be very close to spherical. Again, small changes in shock shape would lead to even better correspondence. Further refinements were not justified since changes in body shape were of the order of the body boundary-layer thickness, for which no experimental data were available. Aerodynamic features of blunt-body flow are predictable with good accuracy.

3.2. The thermochemical model

The present aerodynamic technique calculates the flow quantities along streamlines, making it particularly suitable for the superposed calculation of chemically reactive flows with time-dependent total differential rate equations. The purpose of this section is to illustrate the application of the simple Lighthill-Freeman nonequilibrium ideal dissociating-gas model (Lighthill 1957, Freeman 1958) where we calculate the dissociation rate (the substantive rate of change of the molecular mass fraction α) along the streamline.

Case C1 — Nonequilibrium flow behind plane shock. Figure 4a shows the nonequilibrium flow of air behind a plane shock at a freestream Mach number of 25. Shown are the lines of constant dissociation rate. As discussed by Stalker (1989), the wedge surface is slightly concave towards the oncoming flow.

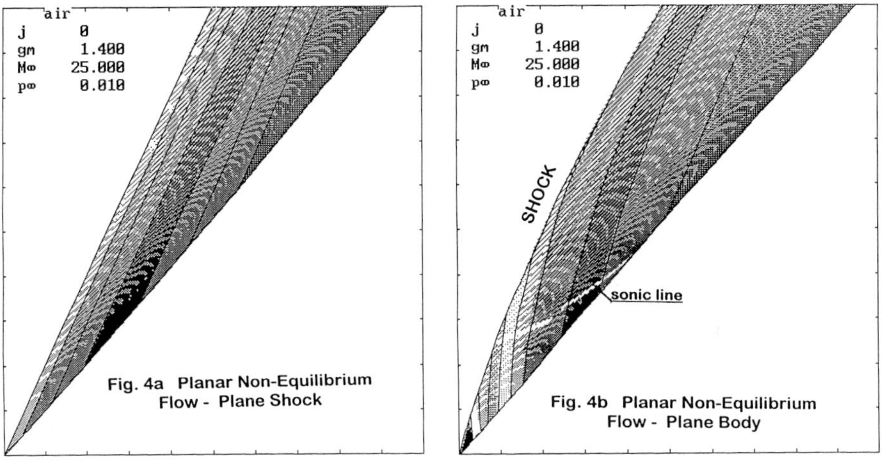

Figure 4. Planar nonequilibrium flow: (a) plane shock; (b) plane body.

Case C2 — Nonequilibrium flow over plane wedge. Figure 4b shows the nonequilibrium flow over a plane wedge. In this case the shock, as well as the lines of constant-dissociation rate, are convex towards the oncoming flow. A sonic line appears, separating the subsonic flow near the leading edge from the supersonic flow further downstream. This clearly illustrates the non-similar nature of nonequilibrium flow over wedge surfaces.

Case C3 — Nonequilibrium flow over bluff body. Figures 5a, b, c are for Mach number 25 flow over a bluff, axi-symmetric, mushroom-shaped body. The three figures are for lines of constant temperature, constant dissociated mass fraction, and constant dissociation rate. The general features of such flows, as described by Hornung (1976), are confirmed.

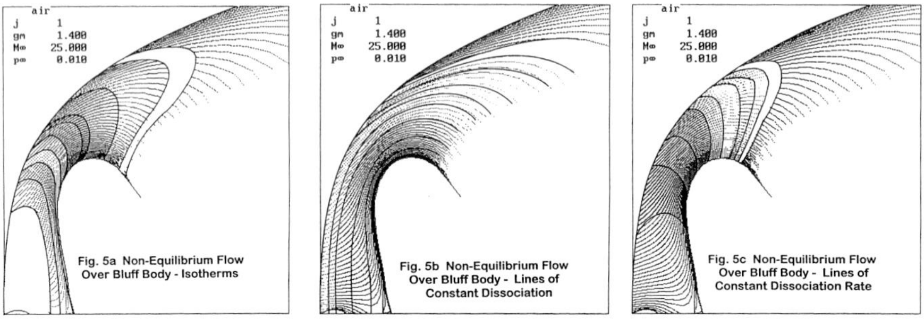

Figure 5. Non-equilibrium flow over bluff body: (a) isotherms; (b) lines of constant dissociation; and (c) lines of constant dissociation rate.

4. Conclusions

An inverse method is presented for the calculation of flow of chemically active gas behind a given shock wave. The method gives very good results for non-reactive flows and is in qualitative agreement with previously calculated ideal dissociated-gas flows.

Acknowledgement. This work was begun under USAF Contract No. F33615-87-C-2748 and The Johns Hopkins University Contracts 602235-0 and 603077-S. Completion was made possible through a sabbatical leave opportunity provided by Dr. G. Eitelberg of DLR Göttingen and Dr. S. Matar of Ryerson Polytechnic University.

References

Freeman NC (1958) Non-equilibrium flow of an ideal dissociating gas. J Fluid Mech 4(4):407.

Hornung HG (1976) Non-equilibrium ideal gas dissociation after a curved shock wave. J Fluid Mech 74:(1)143–150.

Lighthill MJ (1957) Dynamics of a dissociating gas. J Fluid Mech 2(1):1.

Maslen SH (1964) Inviscid hypersonic flow past smooth symmetric bodies. AIAA J 2(6):1055–1061.

Mölder S (1971) Reflection of curved shock waves in steady supersonic flow. CASI Trans 25:44–49.

Sedney R, Kahl GD (1961) Interferometric study of the hypersonic blunt body problem. In: Proc. of the 4th AFBMD/STL Symposium, Pergamon Press.

Stalker RJ (1989) Approximations for nonequilibrium hypervelocity aerodynamics. AIAA J 27(12)1761–1769.

Van Dyke MD (1958)The supersonic blunt-body problem — review and extension. J Aerosp Sci 25:485–495.

Shock-cell structure in a jet at the critical pressure ratio

A. Makhsud, E. Brocher, A. Chambarel, P. Cordonnier
Institut de Mécanique des Fluides
13003 Marseille, France

Abstract: For a jet emanating from a nozzle, the one-dimensional gasdynamics equations lead to the well-known relation for the critical pressure ratio $\pi_c = [(\gamma+1)/2]^{\gamma/\gamma-1}$. At this pressure ratio the jet exit Mach number is equal to unity for one-dimensional flow. In many experimental set-up or practical situations (Crouch et al. 1976), the geometrical configurations of the nozzle do not generate a one-dimensional flow. In particular, experiments on jet noise have often been carried out with internal conical shapes (Ponton & Seiner 1992). The free-surface hydraulic analogy of the two-dimensional flow of a compressible gas is a useful experimental tool to study wave and acoustic phenomena. In this analogy the specific heat ratio γ is equal to 2, the pressure is proportional to the square of the water height h, and the Froude number is equivalent to the Mach number. Therefore, the critical pressure ratio of the "hydraulic" gas is equal to $\pi_c = (3/2)^2$ and the critical height ratio h_{tot}/h_a equal to 3/2, where h_{tot} represents the total height of the flow and h_a the water height on the water table. While running experiments with a converging nozzle (5° half angle) on this table, it was observed that a shock-cell structure exists at the critical height ratio. The existence of a shock-cell structure is important since it may lead to the generation of the so-called "screech" noise. It was then decided to carry out a detailed study of this flow configuration, both experimentally and numerically.

Key words: Shock-cell structure, Critical pressure ratio, Numerical computation

1. Introduction

When a convergent nozzle is operated at subcritical ratio, or when a convergent-divergent nozzle is operated at design pressure ratio, the acoustic spectrum is broad band and smooth, and consists of pure turbulent mixing noise. However, if a convergent nozzle is operated at supercritical pressure ratio (underexpanded), and a convergent-divergent nozzle is operated at off-design Mach number (underexpanded or overexpanded), the resulting acoustic spectrum contains an extra noise contribution, due to the presence of shock structure in the jet flow, in addition to basic turbulent mixing noise (Tanna 1977). The shock-cell structure was investigated by Powell (1953) who observed that the shock-cell spacing is not constant along the jet axis because turbulence in the shear layer produces a momentum exchange at the jet periphery. Tam et al. (1985) have used a multiple-scale model to determine the shock-cell structure of imperfectly expanded supersonic jet in detail. Gutmark et al. (1990) have experimentally studied an underexpanded supersonic rectangular jet over a wide range of pressure ratios.

The hydraulic analogy has been shown to be a valuable tool to simulate the complex interaction process between turbulence, sound, and shock waves. It may be demonstrated that, if the vertical acceleration in the shallow-water flow is smaller than the gravitational acceleration at any point in the flow, the conservation equations for this flow have the same form as the equations for a compressible, two-dimensional flow. Comparison of the momentum equation for the two cases shows that the analogy implies the following correspondence between physical quantities:

Compressible-gas flow	Shallow-water flow
pressure p	$gh^2/2$
density ρ	h
temperature T	h
specific heat ratio $\gamma = Cp/Cv$	2
gas velocity u	water velocity u
speed of sound $c = \sqrt{\gamma RT}$	speed of wavelets $c = \sqrt{gh}$
Mach number $M = u/c$	Froude number $F_r = u/\sqrt{gh}$

where h represents the water depth and g the gravitational constant.

In the analogy, sound waves are simulated by water waves. Despite its limitations (ratio of specific heats = 2, some dispersion of wave speed due to surface tension, simulation restricted to two-dimensional flows) the shallow-water analogy offers several interesting experimental advantages: (i) simple experimental set-up; (ii) geometrical configuration easily modified; (iii) possibility of simulating high Mach-number flows (high Froude number) with small energy expenditure; (iv) visualization of shock-wave/sound-wave propagation by shadowgraph optics; (v) measurement of wave spectrum with non-intrusive method (fiber optics or optical displacement sensors); (vi) possibility of flow-field measurement by PIV; and (vii) frequency of phenomena typically several hundred times smaller than in the gas (some of them visible with naked eye and easily recorded with a video-camera). For amplitudes small compared to the mean value, the linearized relation between pressure and water-level fluctuation is $(\Delta p/p) = 2(\Delta h/h)$. In the present work, a convergent nozzle is operated at critical pressure ratio (critical height ratio) and the shock-cell structure is visualized by using shadowgraphy. Flow visualization showed that a shock-cell structure was present even at the critical height ratio, which cannot be the case for a one-dimensional flow.

2. Experiments

The experimental set-up has previously been described (). The nozzle length is 441 mm. It has walls converging at a half angle of 5° and an exit width D equal to 57 mm. The water level h_a on the table is 27.5 mm so that the total height in the water tank is 41.3 mm at the critical height ratio. The velocities are measured with a hot-film probe Dantec gauge 55 R 42 and the water level with a vertical needle mounted on a micrometer. It was found that the Froude number of the flow at the nozzle exit is not equal to unity as would be the case for a one-dimensional flow at the critical height ratio. The measured Froude number is only 0.83 and the flow is therefore *underexpanded.* After leaving the nozzle, the jet expands and reaches a water level which is even lower than h_a. It is then *overexpanded* and the local Froude number is higher than unity. This is only possible if the streamtubes form converging-diverging channels, the converging part having its origin in the particular nozzle shape. Downstream of this overexpansion, successive compression and expansion waves form the shock-cell structure which has been observed (Fig. 1).

3. Numerical computations

With the usual notations, for a compressible fluid, the Navier-Stokes equations can be written:

$$\begin{cases} \dfrac{\partial}{\partial x_j}(\rho u_j) + \dfrac{\partial \rho}{\partial t} = 0 \\ \rho\left(\dfrac{\partial u_i}{\partial t} + u_j \dfrac{\partial u_i}{\partial x_j}\right) = -\dfrac{\partial p}{\partial x_i} + \dfrac{\partial \tau_{ij}}{\partial x_j} \\ \dfrac{\partial}{\partial t}(\rho e) + \dfrac{\partial}{\partial x_j}\left(u_j(\rho e + p)\right) = \dfrac{\partial}{\partial x_j}\left(u_j \tau_{ij} + \dfrac{\gamma}{P_r}\dfrac{\partial T}{\partial x_j}\right) \end{cases}$$

$$\text{i.e.:} \quad \frac{\partial}{\partial t}(\rho, u_i, T) = \Phi(\rho, u_i, T) \tag{1}$$

$$\text{and} \quad p = (\gamma - 1)\rho\left[e - \tfrac{1}{2}\left(u_j^2\right)\right] = (\gamma - 1)\rho T\ .$$

The form of variational formulation is:

$$\int_\Omega \delta(\rho, u_i, T)\frac{\partial}{\partial t}(\rho, u_i, T)\, d\Omega = \int_\Omega \delta(\rho, u_i, T)\,\Phi(\rho, u_i, T)\, d\Omega$$

and one obtains a differential system discretized by finite elements for the space and Θ-method for the time:

$$([M] + \Theta\Delta t\,[K])\,\Delta\{\rho, u_i, T\}^k = \Delta t\,\{\Phi_n\}^{k+\Theta} \qquad \text{and} \qquad t_{k+1} = t_k + \Delta t\ .$$

An automatic mesh-refinement technique was performed (Sani & Gresho 1992). A parallel computer M.I.M.D. (Ipsc 860) and a parallel algorithm for the resolution are used (Liu 1987). The jet nozzle and the outer space were discretized. A time-dependent flow was calculated before the jet filled a pseudo-cavity.

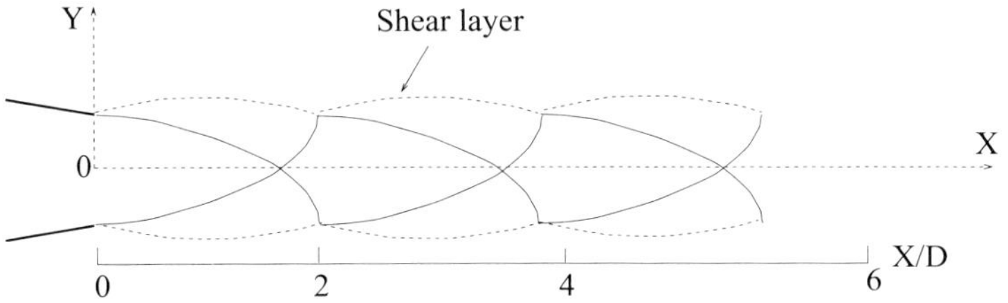

Figure 1. Shock-cell structure.

4. Results and discussion

The Mach number on the jet axis and the velocity profiles have been studied experimentally and computationally. Results for the local Froude-number distribution along the jet axis at the critical height ratio 3/2 are represented in Fig. 2. Experimentally, at the nozzle exit, the flow is subcritical ($Fr = 0.83$), it then accelerates and reaches a supercritical value ($Fr = 1.15$). The jet is then *overexpanded.* This is easily explained by the fact that the flow is conver-

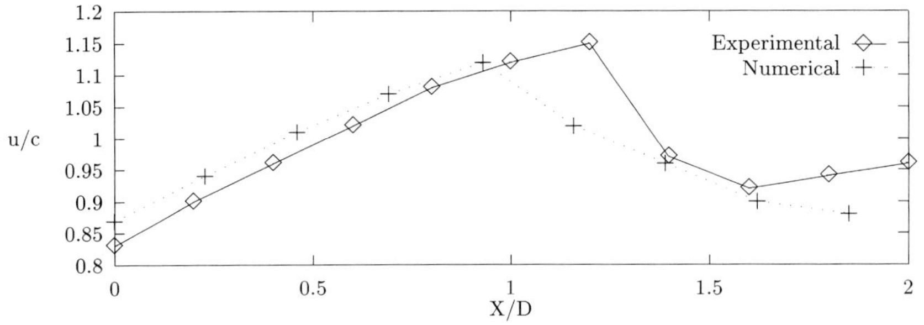

Figure 2. Mach number (Froude number) along the jet axis; X is abscissa, D is nozzle width.

gent at the nozzle exit and has to diverge at some point downstream. At the downstream end of the overexpanded region, a shock wave appears and decelerates the flow to subcritical velocities. The instationary numerical results are in agreement with the experiments on three points:

1. The Mach number increases downstream of the nozzle exit and the maximum Mach number is superior to unity. The difference between the numerical and experimental results at values $X/D \geq 1.75$ is explained by the fact that the numerical domain boundary produces a reflection wave which does not give a stationary flow.

2. Uniformity of the velocity across the jet was tested by measuring the velocity profile at a height Z = 15 mm above the surface of the water table. The comparison of measurements and numerical results is shown in Fig. 3. The agreement is seen to be quite satisfactory, except for the shear-layer velocity profile.

3. The velocity vectors obtained numerically clearly show that the flow downstream of the nozzle exit is at first convergent and then divergent, thereby creating a sonic throat (results to be published).

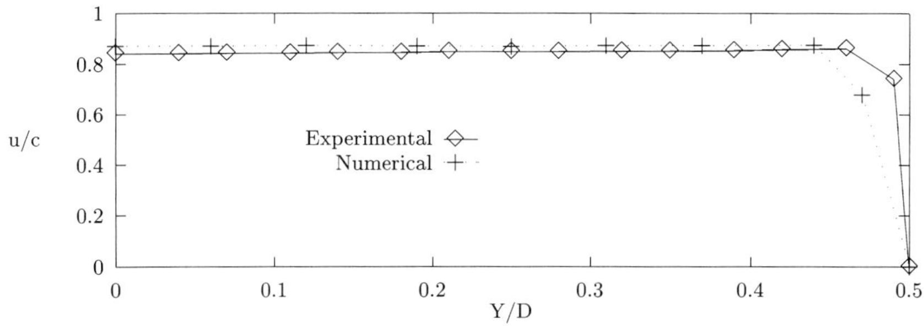

Figure 3. Velocity profiles at the nozzle exit.

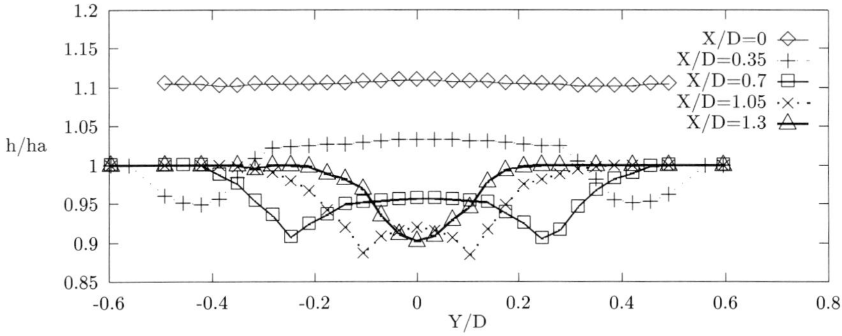

Figure 4. Water-depth profiles downstream of nozzle exit; Y is lateral coordinate.

The local value h of the water depth is interesting for two reasons. First, comparison of this value with the ambient water depth h_a enables one to detect areas of the flow for which either overpressure or underpressure exists, because in the analogy $(h/h_a)^2 = (p/p_a)$. Second, combined with the measurement of the local velocity, it is possible to calculate the local value of the Froude number $Fr = u/\sqrt{gh}$. Figure 4 shows the water-depth profiles at the critical value of $h_{tot}/h_a = 1.5$. It is seen that a fairly strong depression builds up on the jet axis.

5. Conclusions

The hydraulic analogy has shown that a shock-cell structure exists at the critical height ratio (critical pressure ratio) for a convergent nozzle. This process is due to the fact that the streamtubes form converging-diverging channels downstream of the nozzle exit. Good agreement was obtained between the experimental and numerical results, both for the Mach number on the jet axis and for the velocity profiles. Future work with the hydraulic analogy will include detailed measurements of the interaction process of the shear layer with the shock-cell structure by using the PIV technique.

References

Brocher E, Makhsud A (1995) Experimental study on the influence of nozzle inlet geometry on screech noise. In: First Joint CEAS/AIAA Aeroacoustics Conference (16th AIAA Aeroacoustics Conference), CEAS/AIAA Paper 95-067, Munich, Germany.

Crouch RW, Coughlin CL, Paynter GC (1976) Nozzle exit flow profile shaping for jet noise reduction. In: 3rd AIAA Aero-Acoustics Conference, AIAA Paper 76-511, Palo Alto, California.

Gutmark E, Schadow KC, Bicker CJ (1990) Near acoustic field and shock structure of rectangular supersonic jets. AIAA J 28:1163–1170.

Liu WK (1987) Parallel Computation for Mixed-Time Integration. Numerical Method for Transient and Coupled Problems, J Wiley & Sons Ltd.

Ponton MK, Seiner JM (1992) The effects of nozzle exit lip thickness on plume resonance. J of Sound and Vibration 154(2):531–549.

Powell A (1953) On the noise emanating from a two-dimensional jet above the critical pressure. The Aeronautical Quarterly 4:103–122.

Sani RL, Gresho PM (1992) On the finite element modeling of the scalar transport equation. Revue Européenne des Eléments Finis 1(3).

Tam CKW, Jackson JA, Seiner JM (1985) A multiple-scales model of the shock-cell structure of imperfectly expanded supersonic jets. J of Fluid Mechanics 153:123–149.

Tanna HK (1977) An experimental study of jet noise. Part II shock associated noise. J of Sound and Vibration 50(3):429–444.

Mach disc and shock structure in coflowing dual axisymmetric jets

J. Kurian, T. V. Ramana Rao, R. Ramesh Kumar
Department of Aerospace Engineering
Indian Institute of Technology, Madras 600036 India

Abstract: The jet structure of co-axial flows has been investigated experimentally. The effect of variation in the Mach number and semi-divergence angle of the inner nozzle for various combinations of inner and outer chamber pressures were studied. The presence of the outer flow was found to give rise to an exit shock and compress the inner flow, leading to Mach-disc formation. The Mach number and nozzle wall angle of the inner nozzle were found to alter the shock structure of the dual jet. The mass-flow ratio of the outer to inner jet was also seen to modify the flow field.

Key words: Mach disc, Coflowing jets, Shock structure

1. Introduction

Axisymmetric co-axial supersonic jets are found in many high-speed air-breathing propulsion devices. The development and structure of a single freejet has been studied extensively by several investigators. Love et al. (1959) and Crist et al. (1966) studied the factors affecting the jet structure of single underexpanded jets. Love et al. (1959) also studied jets exhausting into supersonic outer streams and observed that an intercepting shock pattern existed at higher pressure ratios before giving way to a Riemann wave, than was the case if the outer stream was subsonic. Buckley (1975), based on his review of earlier studies, concluded that the location of the Mach disc would be unaffected by the coflowing airstream, since the axial Mach-number distribution upstream of the Mach disc is expected to be independent of the external flow conditions. D'Attorre and Harshbarger (1965), however, noted that the distance of the normal shock decreased as the ambient flow velocity was increased. Moreover, Narayanan and Damodaran (1993) observed significant changes in the centerline Mach number distribution upstream of the Mach disc, due to the presence of an outer flow.

The aim of this investigation was to study the influence of the exit Mach number and semi-divergence angle of the inner nozzle, and the chamber pressures of the inner and outer flows on the co-axial jet structure.

2. Experimental facility and test procedure

The experiments were conducted in a two-stream co-axial blowdown freejet facility. Air was the working medium for both streams and was supplied from a common reservoir, but the streams were controlled independently. Suitably designed nozzles could be attached to the inner flow line interchangeably. Nine conical convergent-divergent nozzles with semi-divergence angles (α) of 2.5°, 5°, and 10° designed for exit Mach numbers (M_e) of 1.4, 1.7, and 2.0 were used. The outer line was fitted with co-axial convergent nozzles with semi-convergence angles of 40°. The ratio of the exit area of the outer annular nozzle to the throat area of the inner nozzle (A_O/A_I) was 2. Axial and radial surveys were made using a pitot probe. A cone-static probe (based on the design of Kenworthy and Schetz 1973) was used to make axial surveys. A schlieren system using two concave mirrors in an off-axis arrangement was used for flow visualization. The Mach-disc diameter and location, and the shock cell lengths were measured from the negatives of the schlieren photographs.

3. Results and discussion

The effect of the presence of the outer flow on the jet structure is shown in Fig. 1. Figure 1a is a schlieren photograph of the underexpanded jet ($p_e/p_a = 1.125$) from the $M_e = 1.7$, $\alpha = 2.5°$ nozzle. Alternate expansion (bright) and

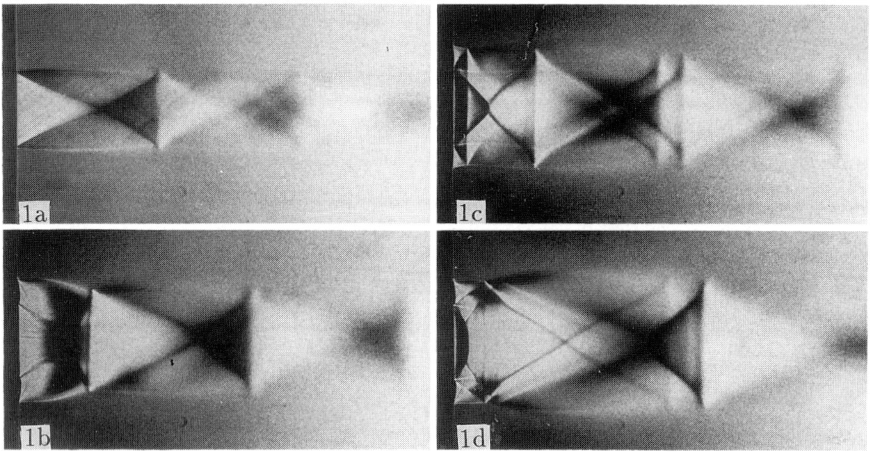

Figure 1. Schlieren photographs: (a) $P_I = 5.56$ atm; (b) $P_0 = 5$ atm; (c) $P_I = 5.56$ atm, $P_0 = 3$ atm; and (d) $P_I = 5.56$ atm, $P_0 = 5$ atm.

compression (dark) regions of the first shock cell are observed and no Mach disc is discernible. Figure 1b is the photograph of the annular jet; the jet is observed to meet the axis and form a shock cell downstream. Figure 1c is the photograph of the dual flow. An annular interaction shock is generated to maintain flow tangency to the inner jet boundary. A similar shock was observed by Love et al. (1959) and called the exit shock — the same terminology will be used here. The intercepting shock formed at the nozzle exit plane meets the jet axis and forms a Mach disc. These shocks proceed further, get reflected at the outer jet boundary to form the second shock cell. As seen in Fig. 1d, increasing the outer-flow chamber pressure moved the first Mach disc closer to the nozzle exit (reduced x_{m_1}) and increased the Mach-disc diameter (d_{m_1}). The greater expanse of the bright region in Fig. 1d relative to Fig. 1b is due to the greater expansion of the outer flow. The variation of d_{m_1} with x_{m_1} for all the nozzles tested is shown in Fig. 2, and is found to be approximately exponential. In contrast, Crist et al. (1966) and Davidor and Penner (1971) observed a linear variation for single sonic freejets. Figures 3a and 3b are the schlieren photographs of the $M_e = 1.7$ and $M_e = 2$, 5° nozzles respectively. The underexpansion ratios of

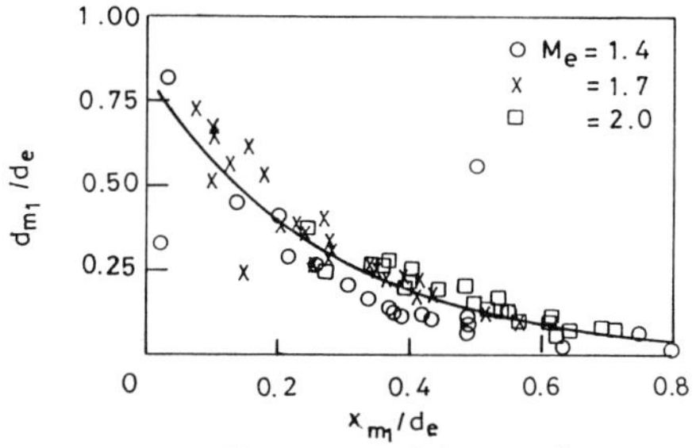

Figure 2. Variation of d_{m_1} with x_{m_1}.

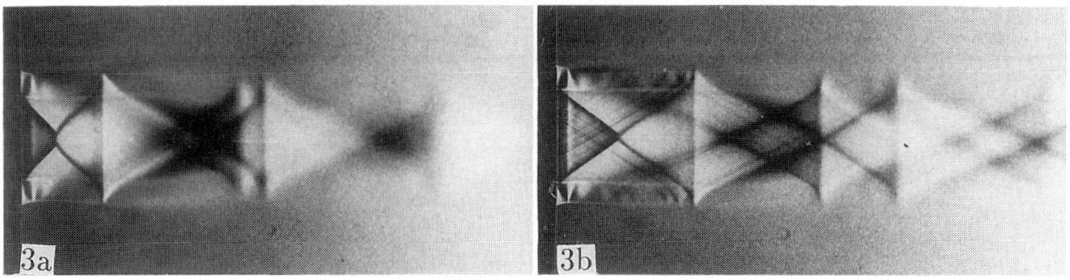

Figure 3. Schlieren photographs: (a) $P_I = 5.56$ atm, $P_0 = 3$ atm; (b) $P_I = 8.82$ atm, $P_0 = 3$ atm.

the inner and outer nozzles are the same; however, considerable difference is observed in the jet structure, indicating a strong Mach-number effect. Increase in the Mach number of the inner jet was found to increase the range of static pressure ratios in which a Mach disc was not formed. A similar trend was

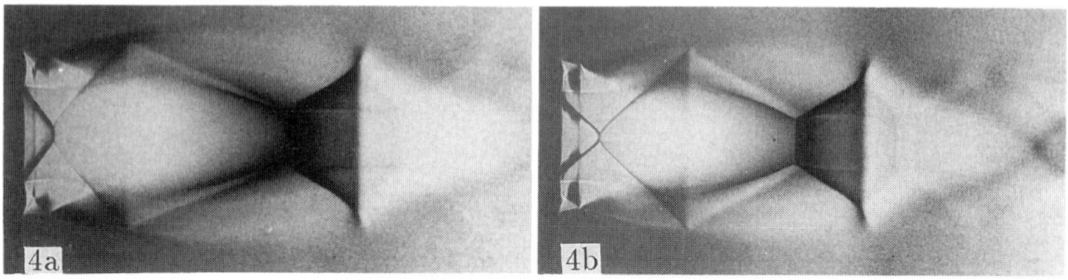

Figure 4. Schlieren photographs, $P_I = 7.8$ atm, $P_0 = 5$ atm, $M_e = 1.4$: (a) $\alpha = 2.5°$; (b) $\alpha = 10°$.

obtained by Love et al. (1959) for freejets exhausting into the ambient. The effect of the semi-divergence angle of the inner nozzle on the jet structure is illustrated in Fig. 4. Increase in the semi-divergence angle is found to increase x_{m_1} and reduce d_{m_1}. However, this trend was not consistently observed for all the nozzles tested. These photographs also show that maximum jet diameter occurs in the second shock cell — a trend followed in almost all the conditions studied. Thus, it is found that the chamber pressures of the inner and outer flows, exit Mach numbers, and semi-divergence angle of the inner nozzle affect the jet structure. Hence, attempts to correlate the jet structure must take all these factors into consideration. The parameter

$$\phi = \frac{(P_I/P_O)(M_e)^{-1/M_e}}{(\alpha/2.5)^{(0.25M_e-0.35)(0.5+1/(M_e)^2)}} \tag{1}$$

was found to reduce the scatter of the data, and reasonably correlate x_{m_1} and the first shock cell length (l_1). The result of the correlation for l_1 is presented in Fig. 5.

Figure 6a gives the variation of x_{m_1} with the chamber pressure of the inner flow, when the chamber pressure of the outer flow is held constant and vice versa. The linear variation observed was partially affected with increase in α. This trend was also not observed for the $M_e = 2$ nozzles. The variation of the axial location of the second Mach disc from the nozzle exit (x_{m_2}) with change in the chamber pressure of the inner and outer flow is shown in Fig. 6b. The linear variation of x_{m_2} with P_I and P_O was not observed to be followed for all the nozzles tested. Plots of d_{m_1} for variation in chamber pressures for the $M_e = 1.7$ nozzles showed that d_{m_1} increased with increase in P_O, and reduced with increase in P_I. Experiments also revealed that the change in mass-flow ratio (due to change in area) of the outer to inner flow affected the dual-jet structure as evident from Figs. 7a and b. Increase in the mass-flow ratio was found to reduce x_{m_1}, increase d_{m_1}, reduce l_1, and increase the second shock

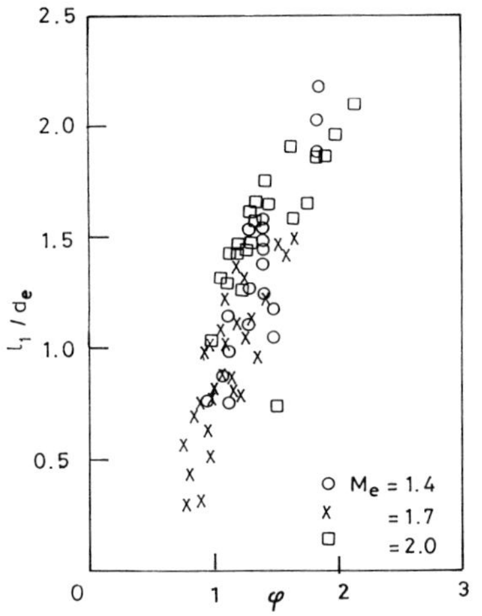

Figure 5. Variation of l_1 with the correlation parameter.

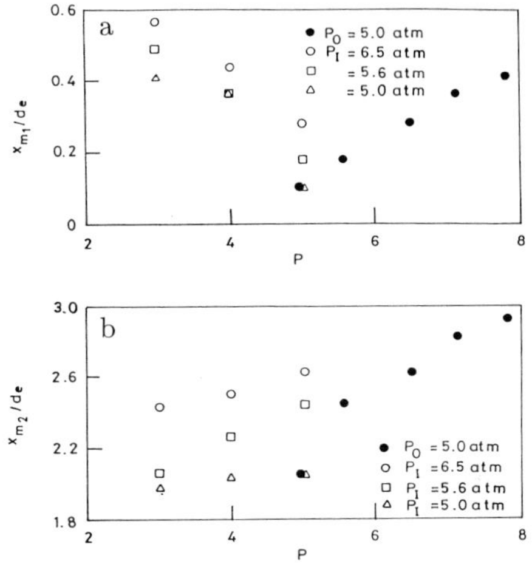

Figure 6. $M_e = 1.7$, $\alpha = 2.5°$: (a) variation of x_{m_1} with chamber pressure; (b) variation of x_{m_2} with chamber pressure.

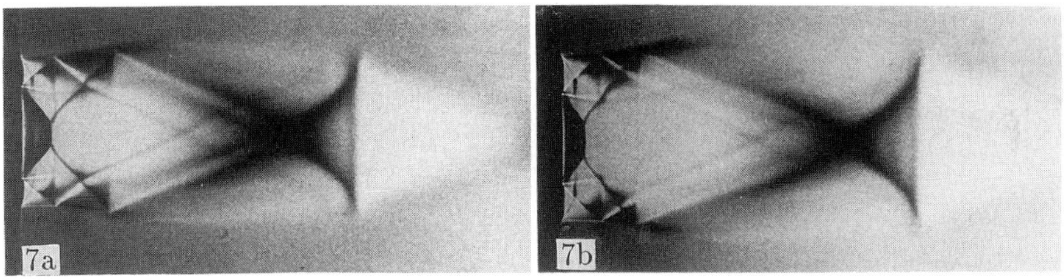

Figure 7. Schlieren photographs, $M_e = 1.7$, $\alpha = 2.5°$, $P_I = 7.12$ atm, $P_O = 5$ atm: (a) $A_O/A_I = 2$; (b) $A_O/A_I = 3$.

cell length. Figure 8 shows the variation in Mach number due to the presence of the outer flow. The Mach-number distribution upstream of the first Mach disc (of the single jet) is altered significantly due to the presence of the outer flow. Lastly, the variation in the shear-layer thickness (b) for the $M_e = 2$ nozzles is shown in Fig. 9. The thickness was defined as the difference in radial locations where the pitot pressure assumed a value of $1 + 0.1 \times (P_{cl} - 1)$ and $1 + 0.9 \times (P_{cl} - 1)$ (cl denotes the centerline value). As seen, the maximum thickness was observed for $\alpha = 5°$ — a similar trend was also observed for the $M_e = 1.7$ nozzles. However, the relative variation at $\alpha = 2.5°$ and $10°$ was not the same at the two Mach numbers.

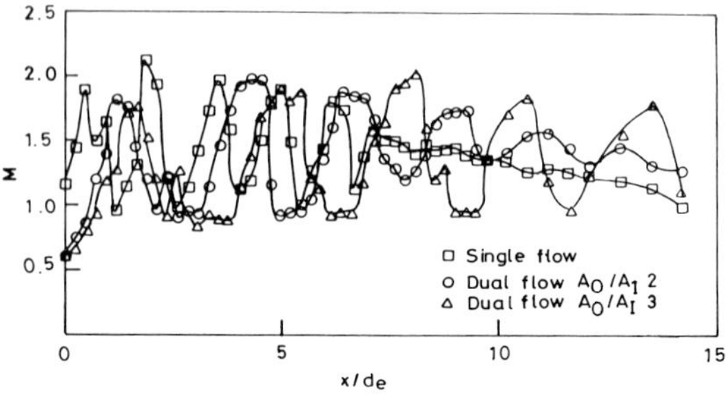

Figure 8. Variation in the centerline Mach-number distribution, $M_e = 1.4$, $\alpha = 10°$.

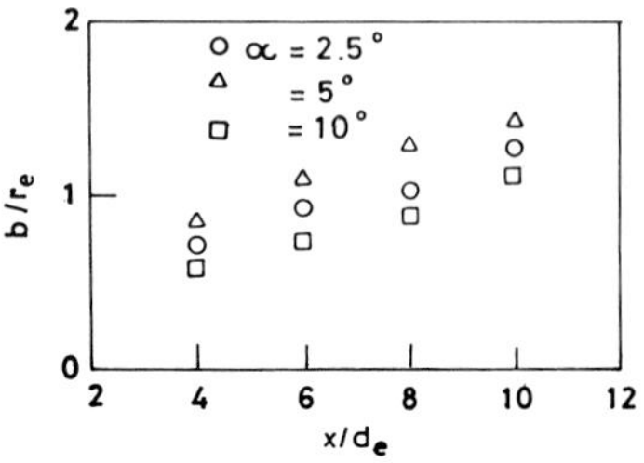

Figure 9. Variation in shear-layer thickness.

4. Conclusions

The presence of the outer flow produced striking changes in the centerline Mach number distribution; and under certain conditions the intercepting shocks formed a Mach disc. The location and size of the Mach disc were found to depend on the chamber pressures of the inner and outer flows. Increase in the Mach number of the inner flow was found to increase the range of the static pressure ratios in which a Mach disc was not formed. While the semi-divergence angle of the inner nozzle was found to modify the shock structure, a discernible trend of variation could not be inferred. A parameter taking into account the variation in the above flow properties was found to reasonably correlate the first shock cell length. The mass-flow ratio of the outer to inner jet was also found to modify the flow field. No discernible trend was observed for the variation in shear-layer thickness with change in the inner-nozzle semi-divergence angle.

References

Buckley FI Jr (1975) Mach disc location in jets in co-flowing airstreams. AIAA J 13:105–106.

Crist S, Sherman PM, Glass DR (1966) Study of the highly underexpanded sonic jet. AIAA J 4:68–71.

D'Attorre L, Harshbarger FC (1965) Parameters affecting the normal shock location in underexpanded gas jets. AIAA J 3:530–531.

Davidor W, Penner SS (1971) Shock standoff distances and Mach disc diameters in underexpanded sonic jets. AIAA J 9:1651–1653.

Kenworthy M, Schetz JA (1973) An Experimental Study of Slot Injection into a Supersonic Stream, NASA CR-2128.

Love ES, Grigsby CE, Lee LP, Woodling MJ (1959) Experimental and Theoretical Studies of Axisymmetric Free Jets, NASA TR R-6.

Narayanan AK, Damodaran KA (1993) Mach disc of dual coaxial axisymmetric jets. AIAA J 31:1343–1345.

A study of three-dimensional supersonic mixing flow phenomena induced by ramp injectors

K. Umii, S. Aso, M. Tannou
Department of Aeronautics and Astronautics, Kyushu University
6-10-1 Hakozaki, Higashi-ku, Fukuoka 812 Japan
Y. Yamane, Y. Ando
Ishikawajima-Harima Heavy Industry Co., Ltd.
M. Fukuda
National Aerospace Laboratory

Abstract: Supersonic mixing flow fields with parallel injection are experimentally investigated. A parallel flow is injected parallel into supersonic flow through a ramp injector. Flow fields are visualized by the schlieren method and oil-flow technique by changing the sweepback angle of the ramp injector. The distributions of surface pressure are also measured in the mixing flow fields. Details of the flow structures of mixing flows due to parallel injection are revealed.

Key words: Stream-wise vortex, Supersonic mixing, SCRAM jet engine

1. Introduction

The development of the SCRAM jet engine has become quite important to the propulsion systems of hypersonic transports and spaceplanes. However, many unsolved problems remain. One of those is the mixing of fuel with supersonic flow in supersonic combustion. For the design of supersonic combustors the mechanism of mixing between free-stream supersonic flow and secondary gaseous flow should be investigated thoroughly. In the case of secondary flow injected normally into supersonic flow, a large total pressure loss is induced. In the present experiments, flow fields with secondary flow injected into supersonic flow with Mach number 3.8 using a ramp injector are investigated.

Flow fields are visualized by the schlieren method and oil-flow technique. Also the distributions of surface pressure are measured. In the experiments the sweepback angle of the ramp injector is varied, and the resulting influence on the geometry of the flow fields is investigated carefully.

2. Experimental apparatus and procedures

A supersonic wind tunnel with nominal Mach number of 4 and with a test section of 150 mm $\times$ 150 mm is used. A flat-plate model with a width of 150 mm is installed in the wind tunnel. Secondary gas is injected into supersonic flow through a ramp injector which is attached onto the flat plate. The flow fields interacting between the streamwise vortices generated by the ramp injector, the turbulent boundary layer developed on the plate, and the shock wave induced by the injection of secondary gas are investigated. Nitrogen gas is used as an injectant. Three types of ramp injector are used in order to investigate the differences of the mixing flow fields due to the configuration of ramp injector. A diagram of the ramp injector models is shown in Fig. 1. A circular nozzle is prepared in back of the ramp. The diameter of the circular nozzle is 5 mm. The

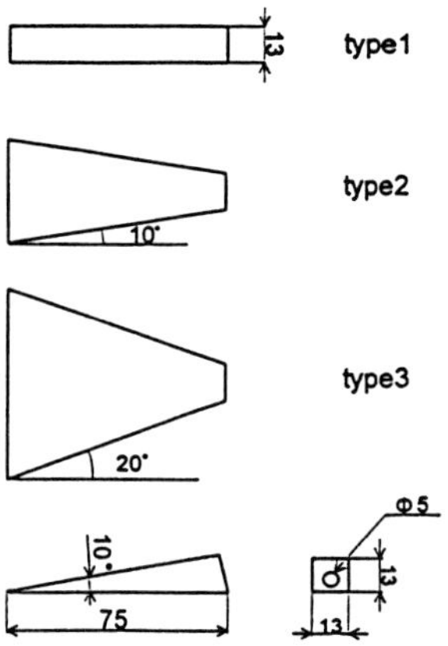

Figure 1. Diagram of ramp injector.

experiments are conducted under the conditions of free-stream Mach number of 3.85–3.90, total pressure of 1.3 MPa, total temperature of 280–290 K and Reynolds number, based on the distance between the leading edge of the flat plate and the forward corner of the ramp, of 2×10^7, with almost adiabatic wall condition.

In the flow fields a primary parameter is the sweepback angle, since the pressure ratio P_c/P_0 (P_c denoting the total pressure of injected secondary flow and P_0 the total pressure of freestream) is fixed as $P_c/P_0 = 0.46$.

3. Experimental results and discussion

Representative schlieren pictures of the flow fields are shown in Fig. 2. All pictures show that a strong shock wave is generated in the forward corner of the ramp, and an expansion wave and shock wave are generated at the apex of the ramp. Downstream of the circular nozzle a barrel shock is generated and a shock wave is seen due to interaction of secondary gas with the flat plate.

(a)

(b)

Figure 2. Representative schlieren pictures: (a) sweepback angle 0° and (b) sweepback angle 20°.

The flow fields are also visualized by the oil-flow technique. Representative oil-flow pictures are shown in Fig. 3. A primary separation line is observed in front of the ramp. On the ramp a streamline that flows away to side of the ramp is observed and on the plate a streamline that flows from the edge to the center is observed. These results suggest that a streamwise vortex near the corner of the ramp is generated, and the vortex becomes larger with the increase in sweepback angle. As the sweepback angle is increased, the region of influence becomes large. However, in the case of a sweepback angle of 20°, the expansion of the jet on the plate conversely becomes more narrow. This is probably because the streamwise vortices generated in the corner of the ramp decrease the expansion of the jet on the plate.

(a)

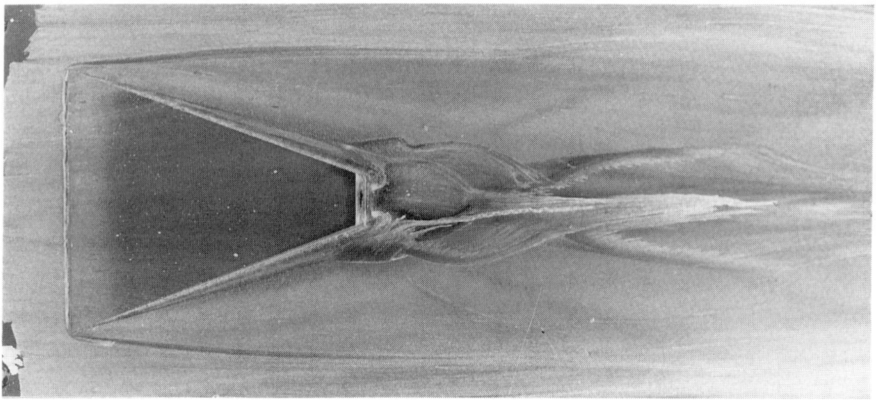

(b)

Figure 3. Representative oil-flow pictures: (a) sweepback angle 0° and (b) sweepback angle 20°.

The distributions of surface pressure are shown in Fig. 4. The pressure is normalized by the static pressure in the undisturbed region. In all pressure distributions a rapid increase of static pressure is observed just behind the ramp, and the higher pressure regions expand parabolically. Further, a rapid drop of static pressure is observed after the increase. This corresponds to the shock wave observed in the schlieren pictures.

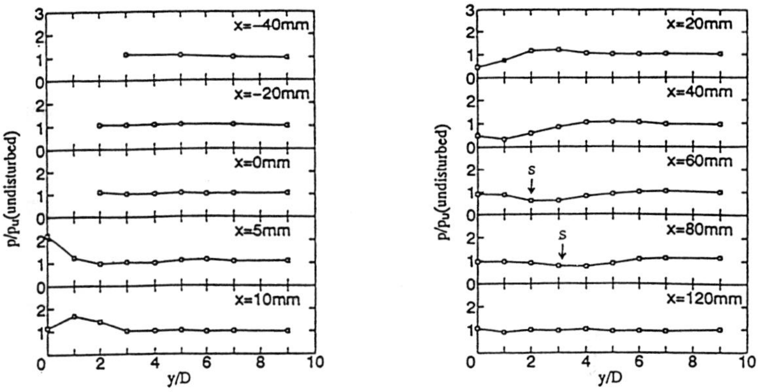

(a)

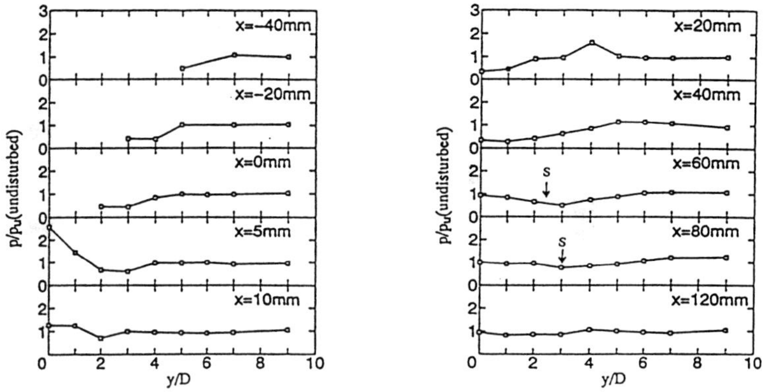

(b)

Figure 4. Representative surface-pressure distributions: (a) sweepback angle 0° and (b) sweepback angle 20°.

On the side of the ramp injector, a lower pressure region is observed. Results point to the existence of streamwise vortices located near the side corner of the ramp; the strength of the vortex increases with the increase of the sweepback angle. In the case of a sweepback angle of 20°, it is observed that these vortices become stronger and induce a region of lower pressure below the vortices, and they decrease the increase of static pressure. These results correspond to those obtained using the oil-flow technique.

4. Conclusions

For all cases, expansion waves and shock waves are generated in the mixing flow fields. It is suggested that streamwise vortices located in the corner of the ramp are generated and these vortices become strong and large with the sweepback angle. Further, it is suggested that the strong streamwise vortices decrease the expansion of the secondary flow on the plate.

References

Anderson GY (1987) An Outlook in Hypersonic Flight, AIAA Paper 87-2074.

Aso S, Tannou M, Okuyama S, Ando Y, Yamane Y, Fukuda M (1994) A Study on Mixing Phenomena in Three-dimensional Supersonic Flow with Circular Injection, AIAA Paper 94-0707.

Drummond JP, Carpenter MH, Riggins DW, Adams MS (1989) Mixing Enhancement in a Supersonic Combustor, AIAA Paper 89-2794.

Waltz IA, Marble FE, Zukoski EE (1991) An Investigation of a Contoured Wall Injector for Hypervelocity Mixing Agmentation, AIAA Paper 91-2265.

Effects of cavity configurations on supersonic internal flow

Y. Watanabe, S. Kudo, F. Higashino
Tokyo University of Agriculture and Technology (TUAT)
2-24-16 Naka-cho, Koganei, Tokyo 184, Japan

Abstract: Effects of cavity configuration and an oblique shock wave impinging over a cavity on supersonic internal flow were investigated experimentally. A shock tunnel with a laval nozzle at the test section was used to generate steady supersonic flows at Mach number $M = 1.83$. The length and the depth of the cavity were varied systematically. The supersonic flow over a cavity was visualized by the schlieren method and by the infinite-fringe method in the form of a Mach-Zehnder interferometer, and the wall static pressure was determined. From the experiment, the effects of the oblique shock wave impinging over a cavity were taken into account. It is found that the incident oblique shock wave does not affect the frequency of the pressure fluctuation inside the cavity, but that the presence of a bow shock wave generated at the rear side of the cavity affects the frequency of the pressure fluctuation downstream of the cavity significantly.

Key words: Supersonic flow, Shock wave, Shear layer, Internal cavity flow

1. Introduction

Supersonic air intake is one of the key components for the development of high-speed propulsion systems, and has been investigated by others (Bogar et al. 1985, Bogar & Sajben 1987). A cavity located at the throat region of the supersonic intake plays a roll in stabilizing the flow. Particularly the normal shock wave, which interacts with the boundary layer and oscillates in the flow direction over the cavity, should be controlled to maintain stable combustion. However, a supersonic flow over a cavity is characterized by a strong oscillation inside the cavity (Zhang & Edwards 1990, Sakamoto et al. 1995), and is very unstable due to the interaction of the shock system with a shear layer or boundary layer. Since the oscillation of the normal shock wave could be affected directly by the pressure fluctuation upstream of the normal shock wave, it is important to investigate the pressure fluctuation downstream of the cavity. However, the mechanism of the pressure propagation downstream from inside the cavity has not been understood. If the flow inside the cavity affects the downstream pressure fluctuation, the flow inside the cavity should be stabilized by some method. In the present study, the effect of cavity configuration on the flow is discussed. The aims of the present study are to observe the change in the vortex structure inside a cavity, and to investigate the frequency and the amplitude of the pressure fluctuation downstream of the cavity.

2. Experimental apparatus and procedures

Figure 1 shows the schematic of the shock tunnel. The high-pressure chamber is 190 mm in inner diameter and 1500 mm in length. The low-pressure section is 77 mm in inner diameter and 3000 mm in length. The test section and the dump tank are located downstream of the low-pressure section. The laval nozzle was installed to obtain a steady supersonic flow at Mach number $M = 1.83$. Figure 2 shows the detail of the cavity configuration. The experiment was performed on two kinds of cavities. One cavity, of length $L = 20$ mm and depth $D_1 = D_2 = 10$ mm, was used to observe the change in the vortex structure in the cavity which corresponds to oscillation of the shear layer. Another cavity, of length $L = 40$ mm, was used to examine the effect of the bow shock wave generated at the rear side of the cavity on the frequency of the pressure fluctuation downstream of the cavity. In this case the depth of the forward side of the cavity D_1 was kept constant at 20 mm, and the depth of the rear side of the cavity D_2 was varied. These changes correspond to a change in cavity dimensions, $D_2/D_1 = 0.5 \sim 1.0$ at $L/D_1 = 2.0$. The location of the incident oblique shock wave was varied from

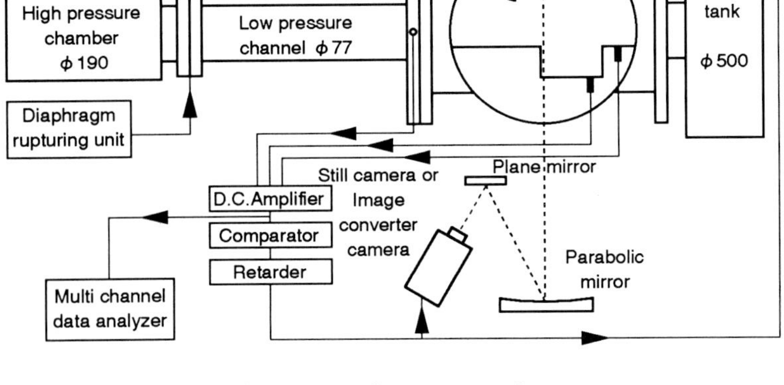

Figure 1. Schematic of experimental apparatus.

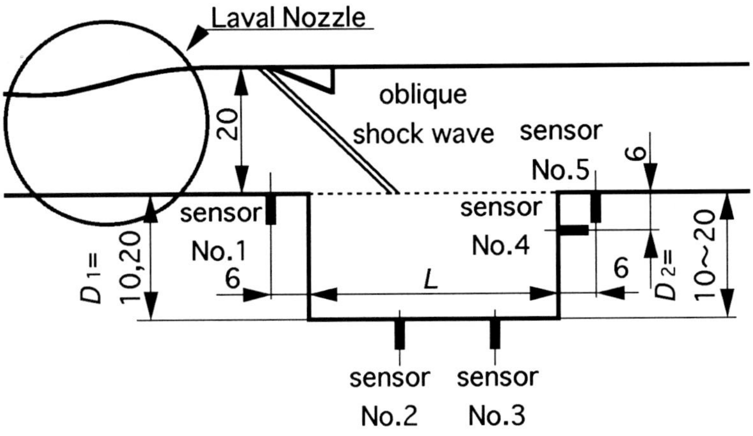

Figure 2. Detail of cavity configuration.

the entrance to the exit of the cavity to examine the effect of impingement of the oblique shock wave on the flow. The flow over the cavity was visualized by the schlieren method and by the infinite-fringe method in the form of a Mach-Zehnder interferometer. An image-converter camera was used to take sequential pictures. The wall static pressure was determined at the points shown in Fig. 2 by piezoelectric transducers.

3. Results and discussion

3.1. The effect of incident oblique shock wave on the flow

Figures 3a and 3b show schlieren photographs which correspond to the case without the incident oblique shock wave and with the oblique shock wave, respectively. The shear layer developed over the cavity and the bow/oblique-shock system were observed by means of these schlieren photographs. The shear layer is bent into the cavity by the incident oblique shock wave. Figure 4a shows the pressure time history determined at the rear side of the cavity (sensor no. 4). Since the shear layer impinges closer to the pressure transducer in the case with the incident oblique shock wave, the amplitude of the pressure fluctuation becomes much larger than in the other case. Figure 4b shows the result of FFT analyses which correspond to the pressure time history shown in Fig. 4a. It is clear from this figure that the frequency of the pressure fluctuation inside the cavity was not affected by the incident oblique shock wave very much. Sequential Mach-Zehnder interferograms were taken by the image-converter camera to clarify the cause of the pressure fluctuation. The frequency of the change in the fringe pattern obtained by sequential interferograms coincides with the

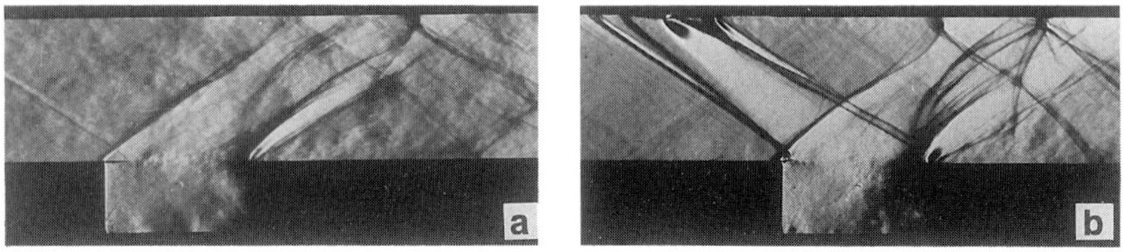

Figure 3. Schlieren photographs: (a) without incident oblique shock wave; (b) with incident oblique shock wave.

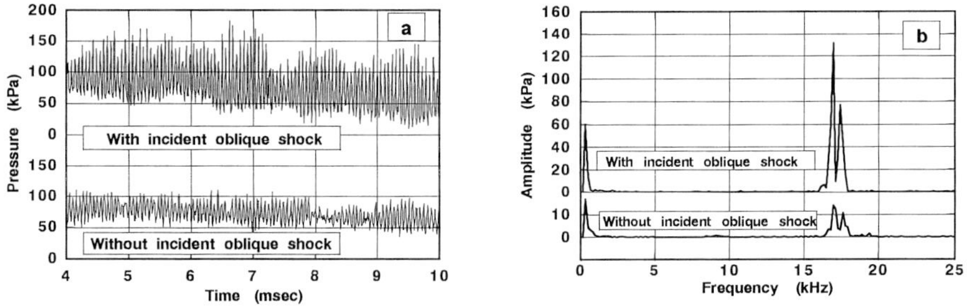

Figure 4. Pressure fluctuation and its frequency inside the cavity: (a) pressure time history; (b) result of FFT analysis.

frequency of the pressure fluctuation. The oscillation frequency of the shear layer also corresponds to the frequency of the change in the fringe pattern.

3.2. The effect of the bow shock wave on the flow

Figures 5a, b, and c show the schlieren photographs. The length-to-depth ratio is $L/D_1 = 2.0$ and the depth of the forward side of the cavity is $D_1 = 20$ mm; these remained the same in all cases. The depth of the rear side wall of the cavity D_2 was changed from 10 mm to 20 mm, which corresponds to the change in cavity dimension D_2/D_1 from 0.5 to 1.0. In the same manner as shown in the schlieren photographs in Fig. 3, the shear layer and the bow/oblique-shock system were observed in Fig. 5a. However, the bow shock wave disappeared in the case of $D_2/D_1 = 0.5$ and 0.75, and instead the attached oblique shock wave was observed. The bow shock wave could be generated by the impingement of the supersonic region in the shear layer on the rear side wall of the cavity. In the case of $D_2/D_1 = 0.5$, the flow passing over the cavity is accelerated abruptly, and deflected downward (with a larger angle compared to the case of $D_2/D_1 = 0.75$) by the expansion fan generated at the corner of the forward side of the cavity. The shear layer impinges on the surface downstream of the cavity with a larger angle in the case of Fig. 5b than in the case of Fig. 5c. This difference of impingement angle causes the difference in the degree of disturbance of the flow in the region downstream of the cavity. Figure 6a shows the pressure time history measured at the point of sensor no. 5 located downstream of the cavity. These measurements correspond to the case shown in Fig. 5. In these waves there are obvious differences in their frequencies and amplitudes. The difference between the case of $D_2/D_1 = 0.75$ and the case of $D_2/D_1 = 0.5$ could be caused by the different angle of impingement of the shear layer. The FFT analysis was

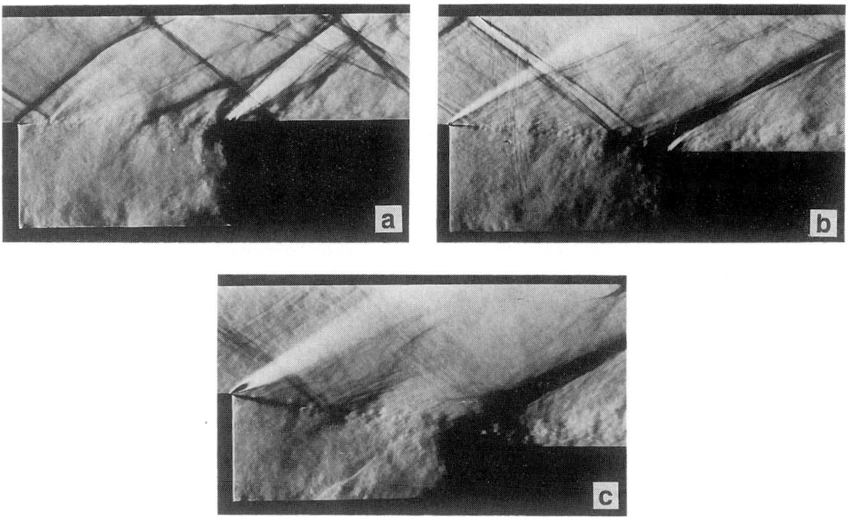

Figure 5. Schlieren photographs: (a) $D_2/D_1 = 1.0$; (b) $D_2/D_1 = 0.75$; (c) $D_2/D_1 = 0.5$.

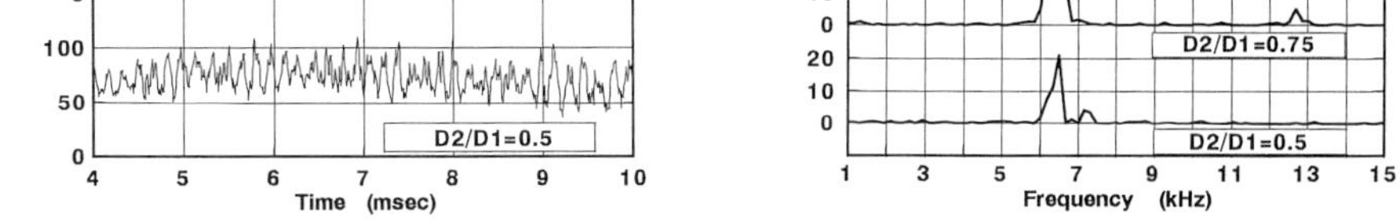

Figure 6. Pressure fluctuation and its frequency downstream of the cavity: (a) pressure time history; (b) result of FFT analysis.

performed on these wave forms to make the difference between the frequency of the pressure fluctuations clear. Figure 6b shows the result of the FFT analysis. The dominant frequency of the cavity with dimension $D_2/D_1 = 0.75$ is almost the same as the value in the case of $D_2/D_1 = 0.5$. Dominant frequencies of the pressure fluctuation inside the cavities are also almost the same value at 6.5 kHz in each case. The only exception is the frequency measured at the point of sensor no. 5 in the case of $D_2/D_1 = 1.0$, which is reduced to a small value by some factor. This might be explained by the fact that as the supersonic region impinges on the rear wall of the cavity, a bow shock wave is generated there only in the case of $D_2/D_1 = 1.0$. The new dominant frequency is thought to be generated by the bow shock wave. These pressure fluctuations are important for the control of the normal shock wave, because the pressure fluctuation could be affected by the oscillation of a normal shock wave significantly. However, it is still unknown how the pressure fluctuation excerts control on the normal shock wave.

4. Conclusions

Effects of the cavity configuration on the supersonic internal flow were investigated experimentally.

The pressure fluctuation inside the cavity corresponds to the change in the fringe structure observed by sequential interferograms, and also corresponds to the oscillation of the shear layer. The dominant frequency of the pressure fluctuation downstream of the cavity was affected significantly by the bow shock wave generated at the corner of the rear side of the cavity.

Acknowledgement. The authors wish to thank to Messrs. H. Kobayashi, T. Osafune, and H. Kakinuma of the Department of Mechanical Systems Engineering, Tokyo University of Agriculture and Technology, for their considerable assistance in conducting the present experiments.

References

Bogar TJ, Sajben M, Kroutil JC (1985) Response of a supersonic inlet to downstream perturbations. J Propulsion 2:118–125.

Bogar TJ, Sajben M (1987) Response of Transonic Diffuser Flows to Abrupt Increases of Back Pressure: Wall Pressure Measurements, AIAA Paper 87–1356.

Sakamoto K, Matsunaga K, Fujii K, Tamura Y (1995) Experimental Investigation of Supersonic Internal Cavity Flows, AIAA Paper 95–2213.

Zhang X, Edwards, JA (1990) An investigation of supersonic oscillatory cavity flows driven by thick shear layers. Aeronautical J December:355–364.

Wave Reflection and Diffraction

Experimental reconsideration of the state of the art of oblique shock-wave reflections in steady flows

A. Chpoun, D. Passerel, J. C. Lengrand
Laboratoire d'Aerothermique du C.N.R.S. Meudon, France
H. Li, G. Ben-Dor
Pearlstone Center for Aeronautical Engineering Studies
Department of Mechanical Engineering
Ben-Gurion University of the Negev, Beer Sheva, Israel

Abstract: The reflection of shock waves over straight surfaces was investigated experimentally. The results for $M_0 = 4.96$ contradict the state of the art regarding the RR $\leftrightarrow$ MR transition. Not only was a hysteresis found to exist, but RR configurations were found to be stable in the dual-solution domain in which theoretically both RR and MR are possible.

Key words: Mach reflection, Transition criterion, Steady flows

1. Introduction

Two shock-wave-reflection configurations are possible in steady flows, namely, regular reflection (RR) and Mach reflection (MR). Two possible transition criteria between these reflections were proposed in the past by von Neumann (1943). They are known as the von Neumann and the detachment criteria. Let the incident shock-wave angles appropriate to these criteria be ω_i^N and ω_i^D, respectively. For $\omega_i > \omega_i^D$, RR is theoretically impossible and for $\omega_i < \omega_i^N$ MR is theoretically impossible. For $\omega_i^N \leq \omega_i \leq \omega_i^D$ both RR and MR are theoretically possible, consequently, the RR $\leftrightarrow$ MR transition could occur at any value ω_i inside this range. Consequently, $\omega_i = \omega_i^N$ and $\omega_i = \omega_i^D$ are the smallest and the largest possible values of ω_i for transition. The dual-solution domain for which both RR and MR are theoretically possible in the (M_0, ω_i)-plane is shown in Fig. 1 for $\gamma = 1.4$. The $\omega_i = \omega_i^D$ and the $\omega_i = \omega_i^N$ criteria divide the (M_0, ω_i)-plane into three domains: a domain in which only RR, a domain in which only MR, and a domain in which both RR and MR are theoretically possible.

The following summarizes the state of the art regarding the RR $\leftrightarrow$ MR transition of planar shock waves in steady flows over straight wedges (see Hornung and Robinson 1982 and Hornung et al. 1979): (1) The RR $\rightarrow$ MR and the MR $\rightarrow$ RR transitions occur at the von Neumann condition. (2) The RR is unstable

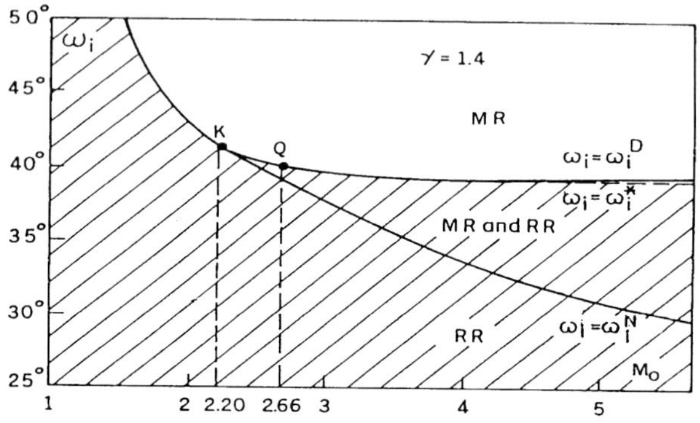

Figure 1. Domains of possible reflection configurations in the (M_0, ω_i)-plane. $\omega_i = \omega_i^D$ and $\omega_i = \omega_i^N$ are the transition lines corresponding to the detachment and von Neumann criteria, respectively. RR is stable for $\omega_i < \omega_i^*$.

in the dual-solution domain. (3) A hysteresis phenomenon does not exist in the RR $\leftrightarrow$ MR transition.

By applying the principle of minimum entropy production, Li and Ben-Dor (1995) showed that the RR is stable in most of the dual-solution domain. The domain $\omega_i \leq \omega_i^*$ in Fig. 1 satisfies the principle of minimum entropy production. This implies that RR wave configurations are physical in that domain. Note that the $\omega_i = \omega_i^*$ line is very close to the $\omega_i = \omega_i^D$ line. Consequently, of the entire $\omega_i^N \leq \omega_i \leq \omega_i^D$ domain for which RR is theoretically possible, only in a very narrow domain, $\omega_i^* < \omega_i \leq \omega_i^D$, is RR aphysical. For this reason it was decided to experimentally investigate the RR $\leftrightarrow$ MR transition with an attempt at establishing a stable RR inside the dual-solution domain.

2. The experimental study

A description of the experimental facility used in the present study is given in Chpoun et al. (1995). Color schlieren photography was employed to record the reflection phenomenon both with still and continuous cameras. When the wind tunnel was started the wedges were placed outside the test section. After a constant flow was established and the nominal conditions were obtained the wedges were brought into the flow one after the other. Then the hysteresis experiments were carried out in two manners: discrete variation of the wedge angle and continuous variation of the wedge angle, with wedge angles changed continuously at a very slow rate of about 0.57 deg/s. A video movie was recorded.

The first set of experiments was aimed at determining the conditions under which the RR $\leftrightarrow$ MR transition occurred. Initially an MR at about $\omega_i \cong 40^\circ$ (in all the experiments $M_0 = 4.96$) was established. Then the angles of attack of the two reflecting wedges were simultaneously and continuously decreased. As a result, the wave angles, ω_i, of the incident shock waves decreased. This resulted in a continuous decrease in the height of the Mach stem until the Mach stem completely vanished. The point where this occurred was defined as the experimental MR $\rightarrow$ RR transition point. Then the direction of the rotation of the wedges was reversed and their angles of attack were simultaneously and continuously increased. While increasing the angles of attack it was observed that the RR was not terminated at the earlier determined MR $\rightarrow$ RR transition point. Instead the RR was maintained until it suddenly changed to an MR. The point where this occurred was defined as the experimental RR $\rightarrow$ MR transition point. The height of the Mach stem of the suddenly formed MR was definitely not zero. The experimental results are shown in Fig. 2 in the (ω_i, ω_r)-plane for $h_{in} = 9$ cm and $w = 6$ cm. MR is marked with triangles and RR is marked with circles. The von Neumann and the detachment transition angles for $M_0 = 4.96$ are $\omega_i^N = 30.88^\circ$ and $\omega_i^D = 39.33^\circ$, respectively. The MR $\rightarrow$ RR transition took place at the von Neumann angle, i.e., $\omega_i^{tr}(\text{MR} \rightarrow \text{RR}) = \omega_i^N = 30.9^\circ$; the RR $\rightarrow$ MR transition occurred at about $\omega_i^{tr}(\text{RR} \rightarrow \text{MR}) = 37.2^\circ$. Thus, it is clear that the hysteresis phenomenon which was hypothesized by Hornung et al. (1979) and up to now has not been confirmed experimentally does exist in the RR $\leftrightarrow$ MR transition in steady flows.

In order to investigate the effect of the geometrical set-up (i.e., length-scale effect) on the transition process, experiments similar to those described above were repeated for three different reflecting wedge surface lengths, w (50, 60, and 70 mm). Each reflecting wedge was tested with three different combinations of inlet cross sections, h_{in} (about 70, 85, and 100 mm). In seven out of the above nine possible combinations of w and h_{in} both the RR $\rightarrow$ MR and the MR $\rightarrow$ RR processes were recorded. The results of these experiments again verified the hysteresis phenomenon in the transition process. All the MR $\rightarrow$ RR experiments showed transition in the vicinity of ω_i^N and all the RR $\rightarrow$ MR experiments indicated transition between ω_i^N and ω_i^D in the vicinity of $\omega_i = 37^\circ$. In addition, the RR $\rightarrow$ MR transition was found to depend on both the inlet cross section and the reflecting-wedge surface length (for details see Chpoun et al. 1995).

The stability of both RR and MR inside the dual solution was also examined. It was found that there is a critical incident shock-wave angle value, say ω_i^{cr}, below which the stable reflection was found to be RR and above which the stable reflection was found to be MR. For the wedge with $h_{in} = 9$ cm and $w = 6$ cm, for which results are shown in Fig. 2, it was found that $\omega_i^{cr} \cong 35.5°$. Consequently, from a stability point of view the dual-solution domain can be divided into two sub-domains. RR is stable in the sub-domain $\omega_i^N \leq \omega_i < \omega_i^{cr}$ and MR is stable in the sub-domain $\omega_i^{cr} < \omega_i \leq \omega_i^{tr}$ (RR $\rightarrow$ MR). The fact that RR was established in spite of the very large disturbance associated with this experimental procedure contradicts the conclusion of Hornung et al. (1979) that inside the dual-solution domain large disturbances should be sufficient to cause the reflection to be MR rather than RR.

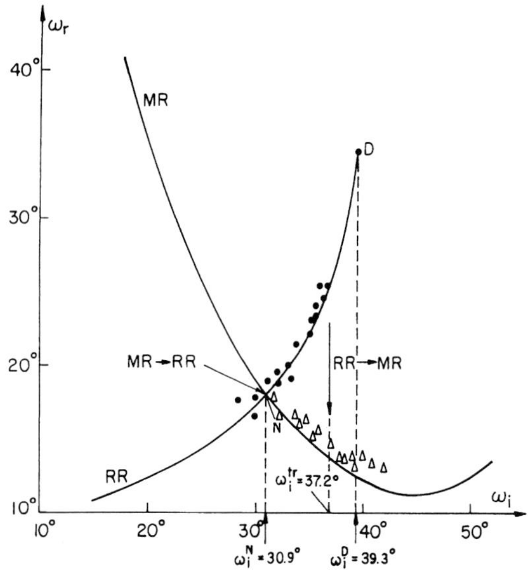

Figure 2. The present experimental results in the (ω_i, ω_r)-plane. $\triangle$ – MR wave configurations, $\bullet$ – RR wave configurations. The hysteresis phenomenon in the RR $\leftrightarrow$ MR transition is clearly evident.

The experimental results also contradict Henderson and Lozzi's (1975) conclusion that the RR $\rightarrow$ MR transition should occur at $\omega_i = \omega_i^N$, "in such a way that mechanical equilibrium of the system is preserved through the process." The sudden transition from RR to MR at $\omega_i > \omega_i^N$ occurred at a point where the mechanical equilibrium requirement was clearly not fulfilled.

3. Conclusions

1. A hysteresis exists in the RR $\leftrightarrow$ MR transition. While the MR $\rightarrow$ RR transition occurs at the von Neumann criterion, i.e., $\omega_i^{tr}(\text{MR} \rightarrow \text{RR}) = \omega_i^N$, the RR $\rightarrow$ MR transition can occur anywhere inside the dual-solution domain $\omega_i^N \leq \omega_i^{tr}(\text{RR} \rightarrow \text{MR}) \leq \omega_i^D$. The exact point at which the RR $\rightarrow$ MR transition takes place depends probably on geometrical parameters associated with the experimental facility and the specific experimental set-up. It is possible that the experimental facilities and set-ups used by Hornung and Robinson (1982) in their experimental investigations were such that the RR $\rightarrow$ MR transition in their facilities occurred at or near the $\omega_i = \omega_i^N$ point. For this reason they failed to obtain a stable RR in the dual-solution domain and to observe the hysteresis associated with the RR $\leftrightarrow$ MR transition.

2. For the geometrical parameters of the wind tunnel and of the wedges and the experimental set-up used in the present study $\omega_i^{tr}(\text{RR} \rightarrow \text{MR}) = 37.2° \pm 1.5°$. This violates Henderson and Lozzi's (1975) argument that the transition should occur at a point where mechanical equilibrium is preserved.

3. There is a critical value, ω_i^{tr}, inside the dual-solution domain below which RR is stable and above which MR is stable.

4. For the geometrical parameters of the wind tunnel, the wedges, and the experimental set-up used in the present study $\omega_i^{cr} = 35.5° \pm 1°$.

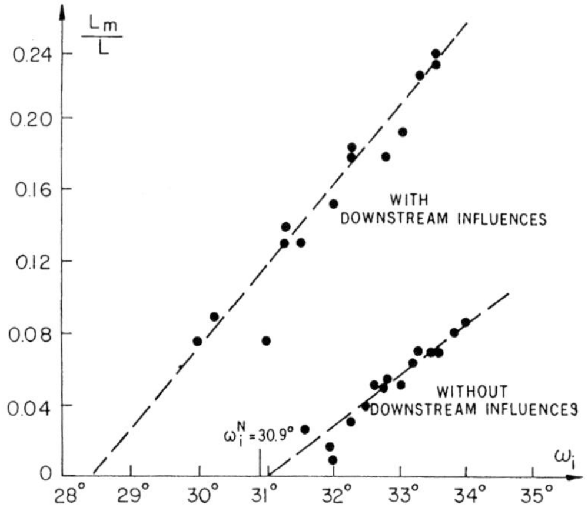

Figure 3. The dependence of the nondimensionalized Mach-stem height l_m/L on the incident shock-wave angle ω_i for two cases with and without downstream influences.

It should be noted here that Azevedo (1989), who obtained Hornung and Robinson's (1982) raw experimental data, showed that the presence of downstream influences at the triple point caused an increase in the height of the Mach stem and, as a result, a shift in the MR $\rightarrow$ RR transition to values of $\omega_i^{tr}(\mathrm{MR} \rightarrow \mathrm{RR}) < \omega_i^N$. This is shown in Fig. 3 in which Hornung and Robinson's unpublished experimental results, as taken from Azevedo's (1989) Ph.D. thesis, for cases with and without downstream influences, are shown. Note that while the results, which were free of downstream influences, indicated that $\omega_i^{tr}(\mathrm{MR} \rightarrow \mathrm{RR}) = 31^\circ \cong \omega_i^N$, the results for which downstream influences were present indicated that $\omega_i^{tr}(\mathrm{MR} \rightarrow \mathrm{RR}) \cong 28.5^\circ < \omega_i^N = 30.9^\circ$. Consequently, it could be concluded that the MR $\rightarrow$ RR transition occurs at the von Neumann point, i.e., $\omega_i^{tr}(\mathrm{MR} \rightarrow \mathrm{RR}) = \omega_i^N$ provided the reflection is free of downstream influences. The fact that Hornung and Robinson (1982) obtained MR at $\omega_i < \omega_i^N$ where MR is theoretically impossible, is yet to be explained.

References

Azevedo DJ (1989) Analytical Prediction of Shock Patterns in a High-Speed Wedge Bounded Duct, Ph.D Thesis, Dept Mech and Aero Eng, State Univ NY Buffalo, NY, USA.

Chpoun A, Passeral D, Li H, Ben-Dor G (1995) Reconsideration of the state-of-the-art of oblique shock wave reflections in steady flows, part I: experimental investigation. J Fluid Mech 301:19-35.

Henderson LF, Lozzi A (1975) Experiments on transition of Mach reflection. J Fluid Mech 68:139–155.

Hornung HG, Oertel H Jr, Sandeman RJ (1979) Transition to Mach reflection of shock waves in steady and pseudo-steady flow with and without relaxation. J Fluid Mech 90:541–560.

Hornung HG, Robinson ML (1982) Transition from regular to Mach reflection of shock waves, part 2: the steady-flow criterion. J Fluid Mech 123:155–164.

Li H, Ben-Dor G (1995) Principle of minimum entropy production in the regular $\leftrightarrow$ Mach reflection transition in steady flow. (Submitted for publication)

Neumann J von, (1943) Oblique reflections of shocks. Explos Res Rep 12, Navy Dept, Bureau of Ordnance, Washington DC, USA.

Experimental and theoretical study of stationary reflection of shock waves

L. G. Gvozdeva, N. B. Sherbak
Gasdynamics Department of High Energy Density Research Center Scientific Association IVTAN Moscow, Russia
A. Chpoun, D. Passerel
Laboratoire d'Aerothermique du C.N.R.S. Meudon, France

Abstract: Results of the joint French-Russian work of a theoretical and experimental investigation of shock-wave reflection phenomenon in steady flows are presented. For experiments, the wind tunnel SH2 in the Laboratoire d'Aerothermique was used. The regular reflection (RR) $\leftrightarrow$ Mach reflection (MR) transition has been studied. It has been found that there exists a phenomenon of hysteresis in the process of transition. The transition from RR to MR occurs near ω_d, the detachment criterion. The transition from MR to RR takes place at $\omega_1 = \omega_N$, the von Neumann criterion. The explanation of the phenomenon is discussed. Numerical simulation was performed. A comparison of numerical-simulation data with experimental data is conducted. The simplest geometry with thin plates is able to provide a qualitative description of hysteresis phenomena.

Key words: Shock wave, Mach reflection, Regular reflection, Hysteresis

1. Introduction

It is well known (see e.g. Bazhenova et al. 1984) that the intersection of two shock waves takes two forms (see Fig. 1). One, which is called regular reflection (RR), takes place at angles of incidence ω_1 close to zero, and the other is Mach reflection (MR), which occurs for angles of incidence close to 90°, near-glancing angles. The transition from one type to the other can occur at the von Neumann criterion, ω_N, or at the detachment criterion, ω_d. In the region between these two boundaries, both systems of reflection are possible. In comparison with the problem of unsteady shock-wave interaction, the steady interaction of shock waves has been studied experimentally and numerically in only a few works. In Pantozopol et al. (1972), it was stated that the transition for air takes place at von Neumann criterion. In Henderson and Lozzi (1979) this observation was confirmed. In the study carried out by Hornung et al. (1979), it was suggested that there should be a hysteresis. In other words, the transition from MR to RR should occur at a different angle than the transition from RR to MR. Nevertheless, the experiments of Hornung and Robinson (1982) showed that there was no hysteresis. The transition occurred at $\omega_1 = \omega_N$.

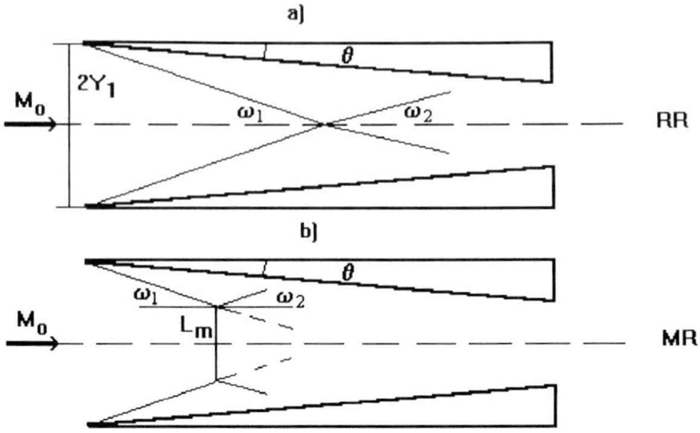

Figure 1. Reflection in stationary case: (a) regular reflection, (b) Mach reflection. M_0 = Mach number of initial flow, θ = wedge angle, ω_1 = angle of incidence, ω_2 = angle of reflection, $2Y_1$ = distance between the plates, L_m = length of Mach stem.

So it was evident that a new series of experiment should be made in order to examine more clearly the conditions of the transition. Gvozdeva suggested that these experiments be undertaken in the Laboratoire d'Aerothermique headed by J. C. Lengrand in Meudon. The work began in 1992; Chpoun and Passerel conducted the experiments and the calculations were done by Sherbak. Passerel's work on this problem will be a major portion of his Doctor's Thesis to be presented in October.

2. Experimental set-up

The wind tunnel SH2 has the following parameters: working gas – air, Mach number of the flow – 4.96, static pressure – 1.683 10^5 N/m^2, static temperature – 76.5 K. A schematic is given in Fig. 2. Two symmetrical wedges allowing a continuous change of the angle between them were placed at the exit of the nozzle. The color schlieren method was used to obtain the photos of the different shock-wave patterns. The stagnation pressure behind the point of intersection was measured. The regular reflection (RR) $\leftrightarrow$ Mach reflection (MR) transition has been studied. The rotation of the plates was around the axis going through the tip of the wedges (see Fig. 2). We believe this experimental design caused the occurrence of the hysteresis phenomenon.

3. Experimental results

3.1. Hysteresis effect

It has been found that a hysteresis can exist through the transition under certain conditions of operation of the special this experimental set-up. The method of

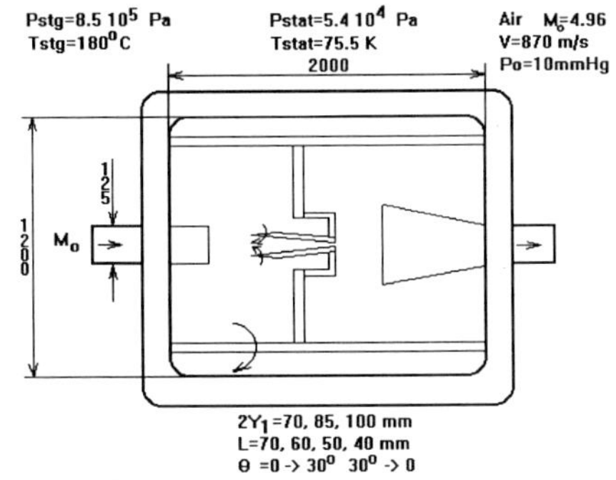

Figure 2. Experimental set-up. Working gas is air, $M_0 = 4.96$, initial pressure in chamber is about 10 torr.

operating is as follows. First, the plates are inserted in the chamber parallel to each other. Then, the flow is started. After reaching stagnation conditions, the plates are rotated smoothly, gradually increasing the angle ω_1 with a velocity $d\omega_1/dt = 0.5°/\text{s}$. The configuration changes from RR to MR. With an opposite rotation, the inverse transition from MR to RR is observed. A number of obtained color schlieren pictures show very clearly that hysteresis takes place. Figure 3 presents the experimental dependence of ω_2 (reflection angle) on ω_1 (angle of incidence). We can clearly see that the transition from MR to RR takes place near the point ω_N and the transition from RR to MR 2° below the ω_d. The length of the Mach stem L_m was measured. The graph of its depen-

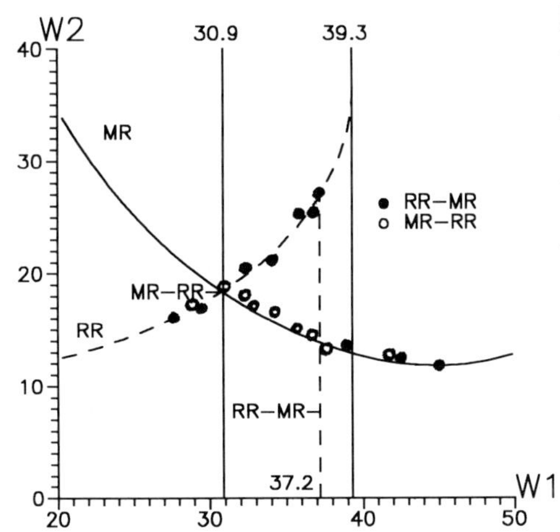

Figure 3. Dependence of reflection angle on incident angle for MR ↔ RR transition.

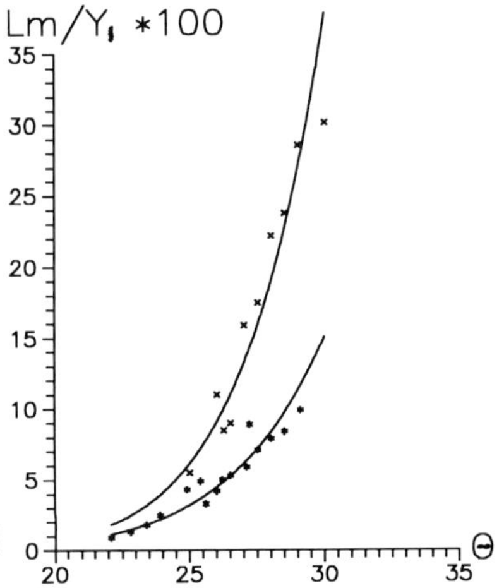

Figure 4. Dependence of Mach-stem height on wedge angle.

dence on θ is given on Fig. 4. It shows the hysteresis effect very clearly. The stagnation pressure was measured with a Pitot tube on the plane of symmetry. The pressure behind the Mach stem was equal to the stagnation pressure of the flow (53.72 KPa). The pressure behind the reflected shock wave in regular reflection was approximately equal to that calculated by the theoretical inviscid values. So it was possible to find what regime was obtained at a given ω_1 not only with schlieren pictures but also with pressure records.

3.2. Dependence of the pattern on geometry

From the theoretical point of view, the transition angle should depend only on local parameters ω, M, γ, P_0, and T_0. It is certainly true for regular reflection for which supersonic flow exists behind the reflected shock wave.

In a Mach reflection, subsonic flow exists behind the Mach stem, so any disturbance can influence the shock-wave pattern. We can expect that the position of the Mach stem depends not only on ω, M, γ, P_0, and T_0, but also on the geometrical parameters of the plates. Furthermore, 3D effects can take place if the width of the plates is too small. Therefore experiments were conducted for different geometrical parameters of the plates. The aim was first to look at the 3D effects and secondly to find the influence of the geometrical parameters. The following set of experiments were made: $2Y_1 = 70$, 80, 100 mm; $L = 70$, 60, 50, 40 mm; plate width = 50, 70 mm.

It was found that there exists a dependence upon geometrical parameters as demonstrated by the fact that the boundary between RR and MR is about $2°$ less than ω_d, while the boundary from MR going to RR lies at ω_N. For the comparison with computed data, we used the experimental data obtained with the larger plate width (70 mm) and a ratio $Y_1/L = 1$.

3.3. Stability of regular and Mach reflection

Next, an attempt was made to verify the stability of the wave pattern. The plates were arranged so that angle ω_1 was more than $40°$. When the flow started, MR occurred. Then the lower plate with its plate holder was removed from the flow and inserted again. This disturbance of the flow was repeated several times. The configuration which appeared at last was determined. With this way of operating it was found that in this region where two types of reflection have been observed to exist, there is in reality no ambiguity. At incident angles greater than $35.5°$, MR occurs, and at lesser angles, RR. So for this mode of generating the flow hysteresis does not take place. The transition angle $\omega_{tr} = 35.5°$.

4. Discussion of experimental results

It seems that the suggestion of Hornung et al. (1979) about the possibility of hysteresis in the case when the change of ω_1 takes place very smoothly is correct. It is possible to get hysteresis with a very slow rotation of the plates. The axis of rotation may also be of importance. At any rate the difference between our experimental set-up and Hornung and Robinson's is mainly in the dimension of the surrounding volume and in the disposition of rotating angle. So the effect of hysteresis depends upon all the details of the set-up.

5. Numerical simulation

The shock-wave reflection problem was investigated by solving the two-dimensional unsteady Euler equations for an ideal (inviscid, non-heat-conducting) gas in Cartesian coordinates. TVD schemes constructed on the basis of Godunov and Steger methods were used for integration. Constructed TVD schemes have a second-order accuracy over space on smooth solutions and formally a third-order on uniform nets. The algebraic interpolation technique was used for grid generation.

Two symmetrical infinitely thin plates are placed at an angle θ inside an incident supersonic flow. At the initial time the parameters of an undisturbed incident supersonic flow with a definite Mach number M are set in all computational regions. Then, the stationary regimes of the flow are determined by the solution of unstationary equations.

In this investigation two parameters were fixed: $\gamma = 1.4$ and $M = 4.96$, in correspondence with experimental values. The ratio of Y_1/L was equal to 1. The angle of the plate θ has been varied over the region 20 to 30 degrees. The number of cells in the characteristic length was 75. Free-stream conditions were imposed on the upper and right boundary of the computational domain and the condition of impermeability was taken at the lower boundary (the symmetry plane) and on the surface of the plate. On the upstream boundary, the flow was supersonic.

The calculations showed that the observed incidence angles of the plane shock waves are rather close to the corresponding theoretical values. Observed parameters behind the incident and the reflected shock waves are in good agreement with the theoretical values.

As was done in experiments, the first series of computations was carried out for angles θ changing during the calculations. This was achieved by the following procedure. First, the calculation was performed for some initial angle $\theta_0 = \theta_d$. After the flow became stationary and a MR was observed, the plate was rotated around the leading edge, so that the new plate position generated a shock wave with the angle smaller than θ_d. When the flow became stationary in this position, the plate was rotated again.

Figure 4 shows the dependence of Mach-stem height normalized by the wedge length on ω_1 for calculations (higher curve) and for experiments (lower curve). The boundary for the transition from RR to MR was 26°, the boundary for the transition from MR to RR was smaller than 23°. The simplest geometry with thin plates therefore gives a qualitative description of the hysteresis phenomenon.

A second series of computations was carried out for a set of fixed plate angles generating incident shock waves with a specified angle θ between θ_n and θ_d. This models the case of a starting flow with a big disturbance before the stationary state. In these latter calculations RR was observed up to $\theta = 26°$. When the angle $\theta > 26°$, MR is observed. The agreement of these calculated angles with experiments seems to be accidental. The whole study shows that further investigations in this region are appropriate and will likely provide needed understanding.

References

Bazhenova TV,Gvozdeva LG, Nettleton MA (1984) Unsteady interaction of shock waves. Progress in Aerospace Sciences 21(4):249–332.

Gilding BH (1988) A numerical grid generation technique. Computers and Fluids 16(1):47–58.

Godunov SK (ed) (1976) Numerical Simulation of Many-Dimensional Problems in Gasdynamics, Nauka, Moscow, p 40 (in Russian).

Henderson LF, Lozzi A (1979) Further experiments of transition to Mach reflection. J Fluid Mech 94:541–560.

Hornung HG, Oertel H Jr, Sandeman RJ (1979) Transition to Mach reflection of shock waves in steady and pseudo-steady flow with and without relaxation. J Fluid Mech 90:541–560.

Hornung HG, Robinson ML (1982) Transition from regular to Mach reflection of shock waves. Part 2. The steady flow criterion. J Fluid Mech 123:155–164.

Pantazopol D, Bellet J-C, Soustre J (1972) Sur les conditions d'apparition de l'effet de Mach dans la'reflexion d'une onde de choc oblique on e'qoulement supersonique. C.R. Acad. Sc., Paris, t.275, Serie A-225.

Steger JL, Warming RF (1981) Flux vector splitting of the inviscid gasdynamic equations with application to finite difference methods. J Comp Phys 40(2):263–293.

Sweby PK (1981) High resolution schemes using flux limiters for hyperbolic conservation laws. SIAM J Numer Anal 25(5):995.

Van Leer B (1981) Flux-vector splitting for the Euler equations. Lecture Notes in Physics 170:507–512.

Mach-stem growth in von Neumann reflection

R. J. Virgona,* F. Higashino
Tokyo-Noko University
2-24-16 Naka-cho, Koganei
Tokyo 184 Japan

Abstract: This research reports on a simple method for predicting the growth of the Mach stem as a function of the wedge angle, θ_w, in von Neumann reflection. In experiments in argon for strong, weak, and very weak shocks the behavior of χ against θ_w could be described by a simple quadratic. Curves for χ vs θ_w derived from the simple theory agree well with experiment and numerical computation results. The theory also predicts that another type of reflection is possible in which the incidence angle of the incident shock, i, remains constant and equal to the glancing incidence angle for very small θ_w which is usually less than 7° for a perfect gas and is also dependent on the Mach number of i.

Key words: von Neumann reflection, Anomalous reflection, Shock tube

1. Introduction

For very weak plane shocks transition from regular reflection at the sonic point $\theta_w = \theta_w^*$ results in von Neumann reflection (VNR) as described by Colella and Henderson (1990). For stronger shocks VNR occurs at $\theta_w = \theta_a$ where the von Neumann three-shock theory (von Neumann 1943) predicts that the reflected shock, r, should lean forward so that β_1, the angle between r and the contact discontinuity, *cd*, is required to be $\beta_1 > 90°$ (Virgona et al. 1992, Virgona 1993). In VNR there appears to be no discontinuity in slope between the incident shock, i, and the Mach stem, s, so the 'triple point,' T, is defined as the point where i begins to curve at the confluence with the leading edge of r. The trajectory path angle of T with respect to the surface of the wedge is taken to be χ. In pseudo-stationary co-ordinates the free-stream angle of incidence, ω_0, of i increases as θ_w becomes smaller until it reaches a maximum at ω_b, which is defined by the glancing incidence theory for which $\chi = \chi_{gl} = 90° - \omega_b$. There can be no reflection for $\omega_0 > \omega_b$, so if $\theta_w > 0°$ when $\omega_0 = \omega_b$ a new type of reflection will occur which has been referred to as anomalous reflection (ANR)

*Present address: Thammasat University, Department of Mechanical Engineering, International Institute of Technology, P.O. Box 22, Thammasat Rangit Post Office, Pathumthani 12121, Thailand.

(Virgona 1993, Fig. 1). For ANR, ω_0 remains fixed as θ_w reduces to zero, and $\theta_w + \chi = \chi_{gl}$, which is referred to as the glancing incidence line.

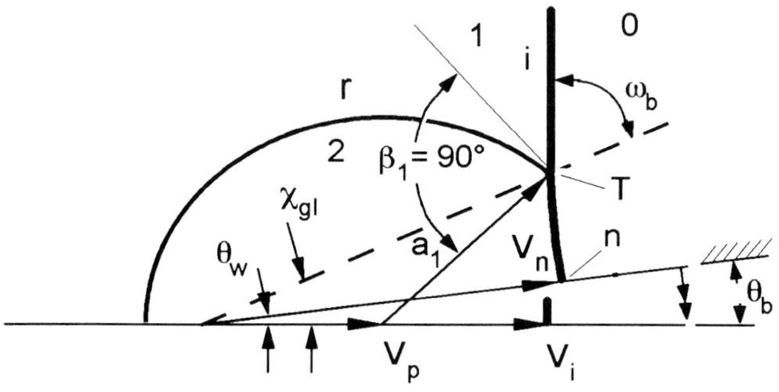

Figure 1. Velocity vector diagram for anomalous reflection, ANR.

2. The theory for χ against θ_w in VNR

In experiments in argon for strong, weak, and very weak shocks the rate of growth of the Mach stem in VNR increased as θ_w was reduced so that the behavior of χ against θ_w could be described by a function of the form

$$\chi = B_2\theta_w^2 + B_1\theta_w + B_0 \tag{1}$$

where the coefficients B_2, B_1, and B_0 are such that (1) is asymptotic to the glancing incidence line at one end (at θ_b, say) and at the other end corresponding to the onset of VNR there is either $d\chi/d\theta_w = 0$ at θ_w^* for very weak reflections (Fig. 2a), or for stronger shocks where transition to VNR takes place from Mach reflection there is a smooth transition in the behavior of χ such that $d\chi/d\theta_w = -Q$ where $-Q$ is taken as $d\chi/d\theta_w$ within the SMR region (Fig. 2b). For very weak shocks the coefficients reduce to

$$\begin{aligned} B_2 &= 1/(4(\theta_w^* - \chi_{gl})), \\ B_1 &= -\theta_w^*/(2(\theta_w^* - \chi_{gl})), \\ B_0 &= \theta_w^{*2}/(4(\theta_w^* - \chi_{gl})), \end{aligned} \tag{2}$$

and $\theta_b = 2\chi_{gl} - \theta_w^*$, while for stronger shocks they become

$$\begin{aligned} B_2 &= \frac{-((\theta_a/2 - Q(\chi_{gl} - Q\theta_a/2 - \chi_a))/(\omega_a - \omega_b) + 1)^2}{4(\chi_{gl} - (Q\theta_a(\chi_{gl} - \chi_a) - (\theta_a/2)^2(Q+1)^2 + \chi_a(\omega_a - \omega_b))/(\omega_a - \omega_b))}, \\ B_1 &= (\theta_a/2 - Q(\chi_{gl} - Q\theta_a/2 - \chi_a))/(\omega_a - \omega_b), \\ B_0 &= (Q\theta_a(\chi_{gl} - \chi_a) - (\theta_a/2)^2(Q+1)^2 + \chi_a(\omega_a - \omega_b))/(\omega_a - \omega_b), \end{aligned} \tag{3}$$

which give $\theta_b = 2(\chi_{gl} - B_0)/(B_1 + 1)$. For $\xi_i \le \xi_v$ it is necessary to know Q, θ_a, and χ_a in order to evaluate (1), so these must be obtained from experiment or

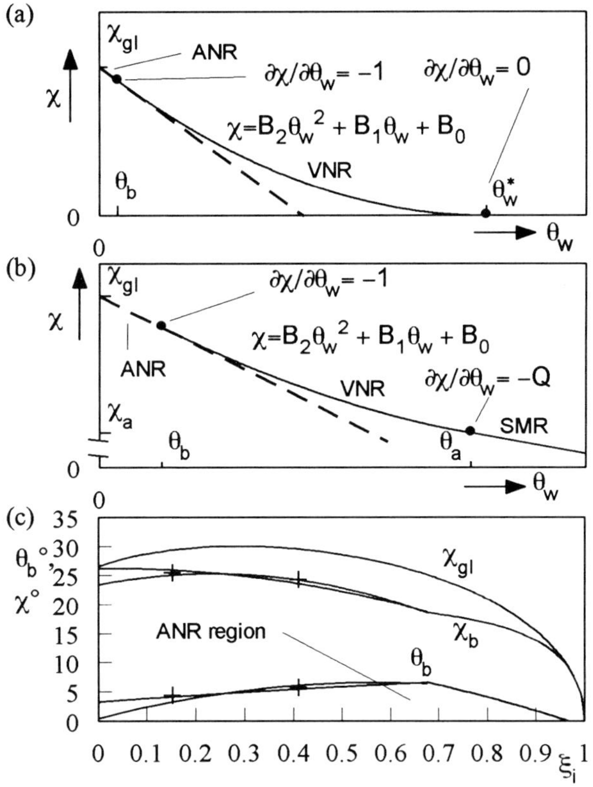

Figure 2. The relationship between χ and θ for VNR and ANR: (a) For $\xi_i \geq \xi_v$; (b) for $\xi_i \leq \xi_v$; (c) existence region for ANR for $\gamma = 5/3$, cross symbols are data derived from Godunov-code numerical computation.

numerical computation. Figure 2c shows the region for θ_b and χ_b derived from the above and using two numerical computation points for the region $\xi_i \leq \xi_v$. There are two curves shown for this region because one curve passes through the $(0, 0)$ point while the other is a smooth fit through the two numerically derived points and the end point of the curve for $\xi_i \geq \xi_v$.

3. Results

In Fig. 3 are shown the curves derived from the above theory against experimental and numerical data for three shock strengths with average $\xi_i = 0.9120$, 0.4106, and 0.1534 for argon. The derived curves for χ vs θ_w agree well with experiment and numerical computation results especially for the case of

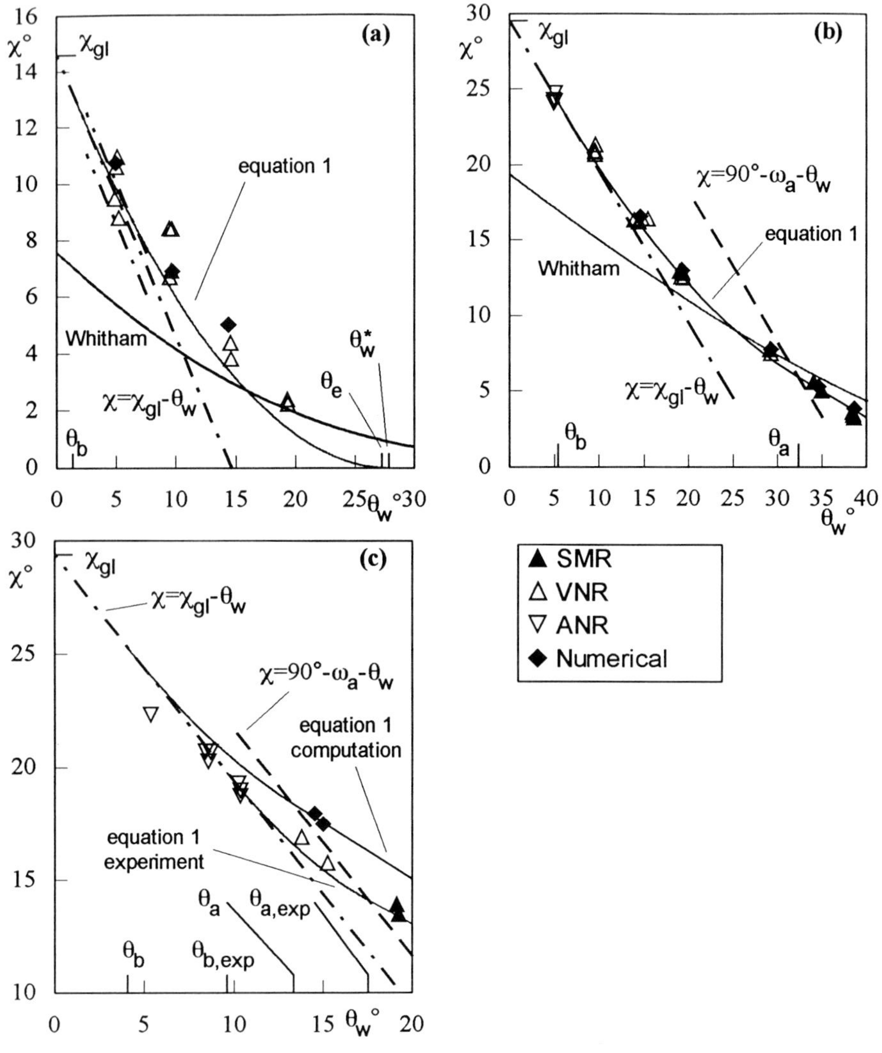

Figure 3. Results for a perfect gas $\gamma = 5/3$: (a) $\xi_i = 0.9120$; (b) $\xi_i = 0.4106$; (c) $\xi_i = 0.1534$. Diamond symbols are numerical computation, other symbols experiment. For Whitham theory, see Henderson (1980).

$\xi_i = 0.4106$. It can be seen that a number of points lie on or close to the ANR line for $\theta_w > 0°$, thus showing that ANR is possible for very small θ_w. Some representative schlieren photographs of the ANR and VNR ranges are shown in Fig. 4, for the values of ξ_i given above.

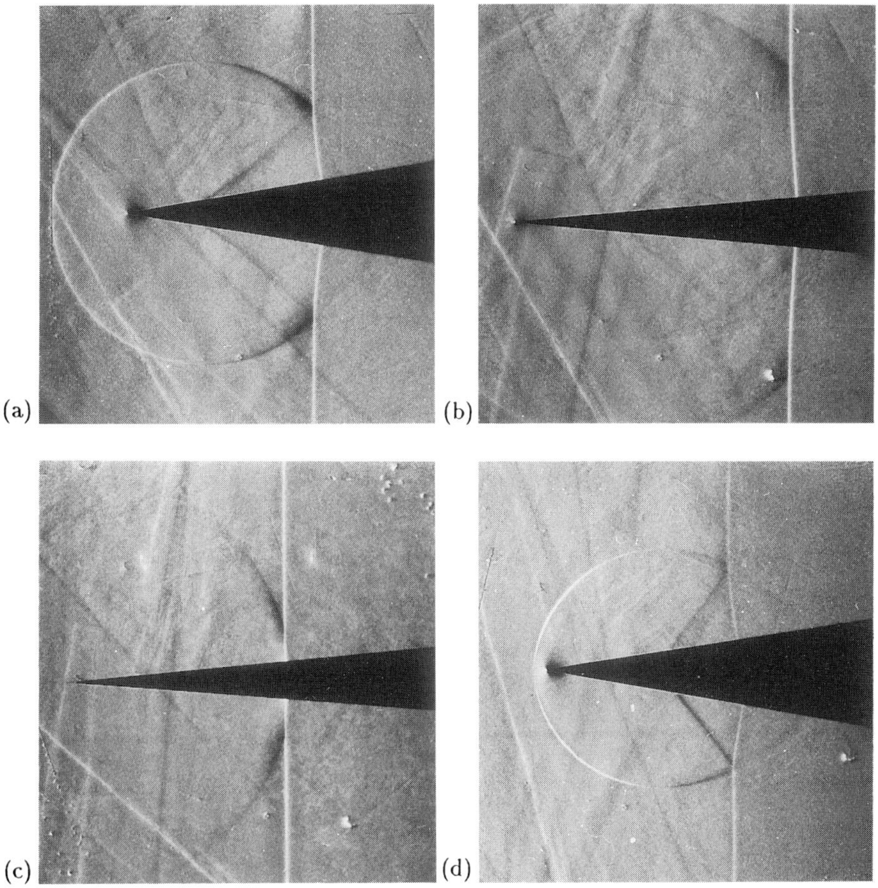

Figure 4. Schlieren photos of VNR and ANR for argon, $\gamma = 5/3$: (a) $\xi_i = 0.4105$, $\theta_{wupper} = 9.53°$, $\theta_{wlower} = 9.47°$, VNR; (b) $\xi_i = 0.4181$, $\theta_{wupper} = 4.91°$, $\theta_{wlower} = 5.09°$, ANR; (c) $\xi_i = 0.9243$, $\theta_{wupper} = 5.16°$, $\theta_{wlower} = 4.84°$, VNR; (d) $\xi_i = 0.1802$, $\theta_{wupper} = 8.08°$, $\theta_{wlower} = 10.88°$, ANR.

References

Colella P, Henderson LF (1990) The von Neumann paradox for the diffraction of weak shock waves. J Fluid Mech 213:71–94.

Henderson LF (1980) On the Whitham theory of shock-wave diffraction at concave corners. J Fluid Mech 99:801–811.

Virgona RJ, Henderson LF, Honma H, Xu DQ (1992) Discrepancies between theory and experiment for shock reflection. In: Takayama K (Ed) Proc 18th Int Symp on Shock Waves, Sendai, Japan, Springer-Verlag, Berlin, pp 189–192.

Virgona RJ (1993) A Study of Shock Wave Diffraction From Regular to Irregular Reflection and the Effects of Gas Viscosity, PhD Thesis, The University of Sydney, Sydney, Australia.

von Neumann J (1943) Oblique reflection of shocks. Explos Res Rep 12, Navy Dept, Bureau of Ordnance, Washington, DC.

Mach-reflection flow at hypersonic speeds

J. Le, H. Ni, G. Hu
China Aerodynamics Research and Development Center
P.O. Box 211, Mianyang, Sichuan, China

Abstract: Experiment and computation on the reflection of a shock wave of $M_s = 7.6$ from a wedge of $\theta_w = 40°$ are investigated. Experiments are carried out in the CARDC shock tube. Computation of both equilibrium and nonequilibrium Mach-reflection flow is based on the laminar full Navier-Stokes (N-S) equation in which the inviscid and source terms are calculated by a second-order upwind TVD scheme and a point-implicit method, respectively. Computational results are in good agreement with those of experiment in all flow fields.

Key words: Mach reflection, Hypersonic flow, TVD scheme

1. Introduction

As judged by comparison with a great number of shock-tube experiments, numerical methods based on the Euler equation with second-order upwind difference schemes such as the Godunov or TVD scheme are sufficiently accurate in the analysis of the perfect inviscid, complex-wave structure of Mach-reflection flow (Glaz et al. 1986). However, for a very strong incident shock wave, the temperature behind the shock is high enough to bring about vibrational excitation and dissociation of the air, resulting in strong effects of the viscosity and nonequilibrium on the experimental flow field. Therefore, further development of the numerical modeling of the viscous nonequlibrium flow field is necessary for the analysis of experimental data and for the validation of computer codes.

To date, there have been a few typical quantitative measurements of density fields widely used for the validation of computer codes for viscous nonequilibrium usteady flow. The purpose of the experiments carried out in CARDC's shock tube at $M_s = 7.6$, $\theta_w = 40°$ is to study the detailed flow structure of the region about the two triple points, including the reflected shock wave, weak slip layer, and Mach stem.

With regard to the numerical simulation, there have been several results only based on the Euler equation (Glaz 1987, Needham 1991, Le and Hu 1992, Clinton and Groth 1993). In the present paper, the laminar full N-S equation is used for numerically simulating both thermal and chemical nonequilibrium Mach-reflection flow. The high-temperature air is assumed to consist of 11 species: O_2, N_2, NO, O, N, O_2^+, N_2^+, NO^+, O^+, N^+ and the reaction-rate coefficients in 32 chemical reactions are taken from the data set compiled by Dunn and Kang (1973). The forward reaction-rate constants K in the chemical source terms are based on Park's two-temperature model of T and Tv: $K = K(\overline{T})$, $\overline{T} = \sqrt{TTv}$. Source terms of three vibrational energy equations with vibration temperature Tv_{O2}, Tv_{N2}, Tv_{ON}, are based on the Landau-Teller method for harmonic oscillators. The relaxation times are obtained from the semi-empirical expressions from Millikan and White. The high-temperature gas-transport coefficients are taken from accurate curve fit (Roop and Gupta 1990).

2. Experiments and results

The experiments were performed in CARDC's shock tube with a test section of 132 mm × 132 mm, which has two pairs of observation windows of 80-mm diameter. The distance between the windows is 400 mm. The model located in the first window is a 2D wedge with a 40° compression corner and height of 88 mm. Double-exposure holographic interferometry was used for quantitatively measuring the density fields and differential interferometry was used for studying the flow structure. Figure 1 is a differential interferogram in which both the first and second slip layers can be distinguished easily. A strong jump in interferomentric fringes against the wall can also be seen. This jump means that values of density begin to increase sharply due to the effect of the cold wall at 298 K. Figure 2 is a holographic interferogram from which the density of the flow field can be evaluated in a very simple way by directly counting the fringes without relying on analytical considerations of three-shock theory near the triple points. The fringe-counting data-reduction technique is obviously inaccurate in chemically reacting flow, as the Gladston-Dale constant K is dependent on the mass concentration of the species. However, in the present

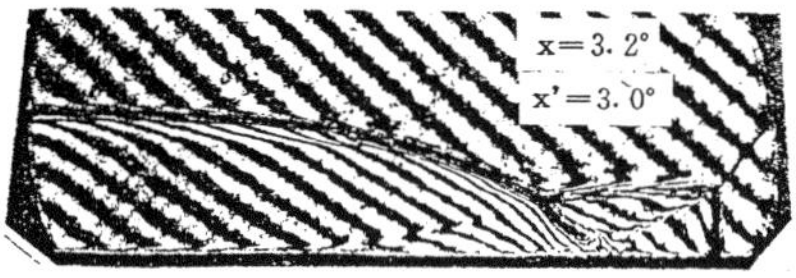

Figure 1. Differential interferogram from CARDC shock-tube experiment. $M_s = 7.6$, $\theta_w = 40°$, $P_0 = 2.63$ KPa, $T_0 = 289$ K.

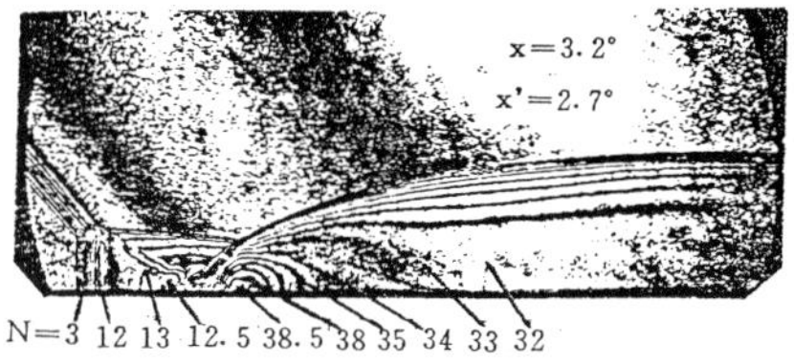

Figure 2. Holographic interferogram from CARDC shock tube. $M_s = 7.6$, $\theta_w = 40°$, $P_0 = 2.63$ KPa, $\rho_0 = 3.21 \times 10^{-5}$ g/cm^3; $\Delta\rho_0 = \lambda/KL$, $\lambda = 6.943 \times 10^{-3}$ cm, $K = 0.225$ cm^3/g, $L = 13.2$ cm; $\rho_N/\rho_0 = 1 + N(\Delta\rho_0/\rho_0) = 1 + N \times 0.729$.

test case, the value of the density some distance behind the incident shock obtained by counting fringes with $K = 0.225 \cdot 10^{-3}$ m^3/kg is very close to the value calculated from the Rankine-Hugoniot conditions for the air equilibrium state. Therefore, the approximate wall density distribution shown in Fig. 3 is evaluated with the same value of K. Strong relaxation effects both behind the incident shock and reflected shock are also indicated in this figure.

3. Numerical calculation and results

Governing equations and difference schemes. The 2D laminar conservative full N-S equations with a gas of N reacting species and M vibrational excited species can be written:

$$\frac{\partial U}{\partial t} + \frac{\partial F}{\partial x} + \frac{\partial G}{\partial y} = \frac{1}{Re}(\frac{\partial Q}{\partial x} + \frac{\partial R}{\partial y}) + S, \tag{1}$$

where $U = (\rho_s, \rho u, \rho v, \rho_m e_{vm}, E)^T$, $F = (\rho_s u, \rho u^2 + p, \rho uv, \rho_m e_{vm} u, \rho u h_0)^T$, $G = (\rho_s v, \rho uv, \rho v^2 + p, \rho_m e_{vm} v, \rho v h_0)^T$, and

$$\begin{aligned} Q = & \left(\rho D_s \frac{\partial x_s}{\partial x}, \tau_{xx}, \tau_{xy}, \rho D_m \frac{\partial x_m}{\partial x} e_{vm} + k_{vm} \frac{\partial T_{vm}}{\partial x},\right. \\ & \left.\sum_{S=1}^{N} \rho D_s \frac{\partial x_s}{\partial x} h_s + k \frac{\partial T}{\partial x} + \sum_{m=1}^{M} k_{vm} \frac{\partial T_{vm}}{\partial x} + u\tau_{xx} + v\tau_{xy}\right)^T \\ R = & \left(\rho D_s \frac{\partial x_s}{\partial y}, \tau_{xy}, \tau_{yy}, \rho D_m \frac{\partial x_m}{\partial y} e_{vm} + k_{vm} \frac{\partial T_{vm}}{\partial y},\right. \\ & \left.\sum_{S=1}^{N} \rho D_s \frac{\partial x_s}{\partial y} h_s + k \frac{\partial T}{\partial y} + \sum_{m=1}^{M} k_{vm} \frac{\partial T_{vm}}{\partial y} + u\tau_{xy} + v\tau_{yy}\right)^T \end{aligned}$$

and $S = (\dot{\omega}_s, 0, 0, \rho_m \dot{e}_{vm} + \dot{\omega}_m e_{vm}, 0)$. The multispecies state equation is $P = \sum_{s=1}^{N} \frac{\rho_s}{M_s} RT$. Introducing the transformation of variables $\tau = t$, $\xi = \xi(x, y)$, $\eta = \eta(x, y)$, Eq. (1) becomes

$$\frac{\partial \widehat{U}}{\partial \tau} + \frac{\partial \widehat{F}}{\partial \xi} + \frac{\partial \widehat{G}}{\partial \eta} = \frac{1}{Re}(\frac{\partial \widehat{Q}}{\partial \xi} + \frac{\partial \widehat{R}}{\partial \eta}) + \widehat{S}. \tag{2}$$

The discrete conservation equation (2) can be written in the following form:

$$\begin{aligned} D_{i,j}^n \Delta \widehat{U}_{i,j} = & -\frac{\Delta \tau}{\Delta \xi}\left(\widetilde{F}_{i+\frac{1}{2},j}^n - \widetilde{F}_{i-\frac{1}{2},j}^n\right) - \frac{\Delta \tau}{\Delta \eta}\left(\widetilde{G}_{i,j+\frac{1}{2}}^n - \widetilde{G}_{j,j-\frac{1}{2}}^n\right) \\ & + \frac{1}{Re}[\frac{\Delta \tau}{\Delta \xi}\left(\widetilde{Q}_{i+\frac{1}{2},j}^n - \widetilde{Q}_{i-\frac{1}{2},j}^n\right) + \frac{\Delta \tau}{\Delta \eta}\left(\widetilde{R}_{i,j+\frac{1}{2}}^n - \widetilde{R}_{j,j-\frac{1}{2}}^n\right)] + \Delta \tau \widehat{S}_{i,j}^n \end{aligned}$$

$$D_{i,j} = \left(I - \Delta \tau \frac{\partial \widehat{S}}{\partial \widehat{U}}\right)_{i,j} \tag{3}$$

The following two-step Range-Kutta method is used so that the conserved variables $\widehat{U}$ can be obtained with second-order accuracy in time and space

$$\overline{U}_{i,j}^{n+1} = \widehat{U}_{i,j}^{n} + \Delta\widehat{U}_{i,j}^{n} \qquad \widehat{U}_{i,j}^{n+1} = \frac{1}{2}(\widehat{U}_{i,j}^{n} + \overline{U}_{i,j}^{n+1}) + \frac{1}{2}\Delta\overline{U}_{i,j}^{n+1}, \quad (4)$$

where numerical flux in ξ-direction is $\widetilde{F}_{i+\frac{1}{2},j} = \frac{1}{2}(\widetilde{F}_{i,j} + \widetilde{F}_{i+1,j} + R_{i+\frac{1}{2},j}\Phi_{i+\frac{1}{2},j})$. The vector $\Phi_{i+\frac{1}{2},j}$ in the upwind TVD scheme can be expressed by $\Phi^{l}_{i+\frac{1}{2},j}$, $\Phi^{l}_{i+\frac{1}{2},j} = \delta(a^{l}_{i+\frac{1}{2},j})(g^{l}_{i+1,j} + g^{l}_{i,j}) - \Psi(a^{l}_{i+\frac{1}{2},j} + \gamma_{i+\frac{1}{2},j})a^{l}_{i+\frac{1}{2}}$; $l = 1, 2, ..., N + M + 3$. The numerical flux $\widetilde{G}$ in the η direction can also be defined similarly.

3.1. Results and discussion

To judge the predictive capabilities of present computer codes, a double Mach-reflection flow of incident $M_s = 8.7$ impinging on a 27° wedge is numerically calculated using N-S and Euler equations. Figure 4 is the schematic diagram

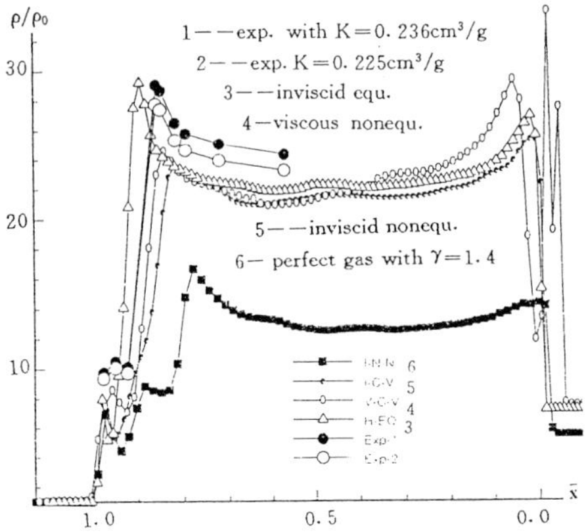

Figure 3. Comparison of density distribution on the wall. $M_s = 7.6$, $P_0 = 2.63$ KPa, $T_0 = 289$ K.

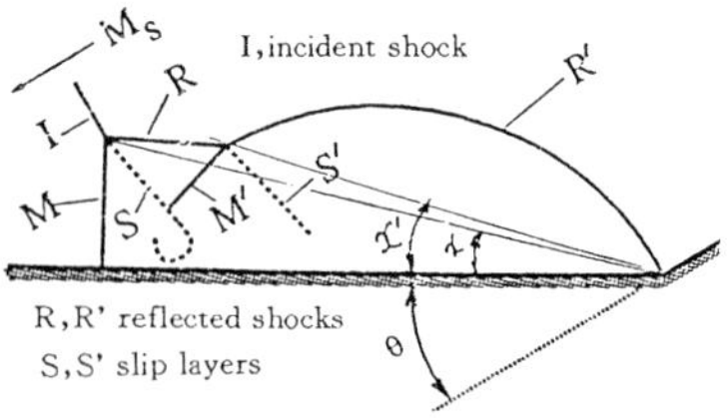

Figure 4. Schematic diagram of double Mach reflection; X, X' first and second triple-point angles.

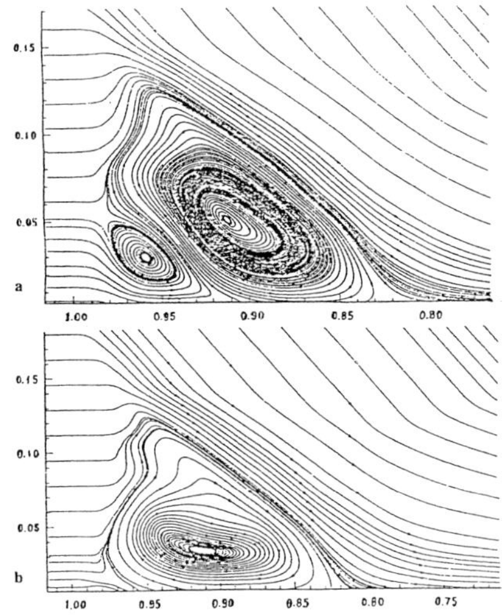

Figure 5. Computed streamlines near Mach-stem region: viscous nonequilibrium (top); inviscid nonequilibrium (bottom). $M_s = 8.7$, $\theta_w = 27°$, $P_0 = 4.1$ KPa, $T_0 = 299.2$ K.

of this flow. Figures 5a and 5b show the computed streamlines of inviscid and viscous nonequilibrium flow near the Mach-stem region. The rollup pattern with two vortices (shown in Fig. 5a) is in substantial agreement with the inter-

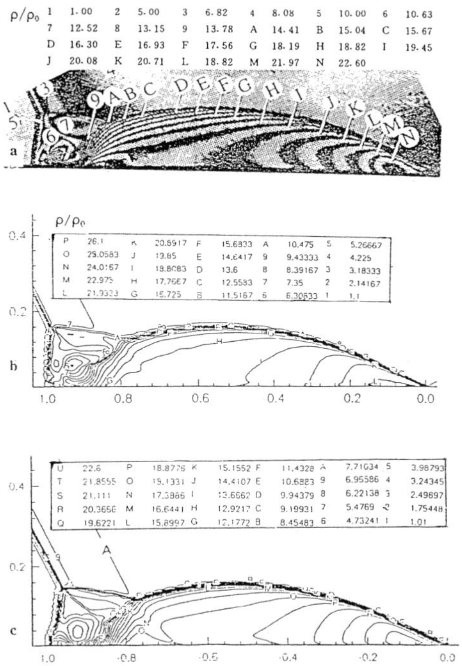

Figure 6. Density contours. $M_s = 8.7$, $\theta_w = 27°$, $P_0 = 4.1$ KPa, $T_0 = 299.2$ K. (a) Exp. from Groth (1993); (b) computed viscous nonequilibrium; (c) computed inviscid nonequilibrium.

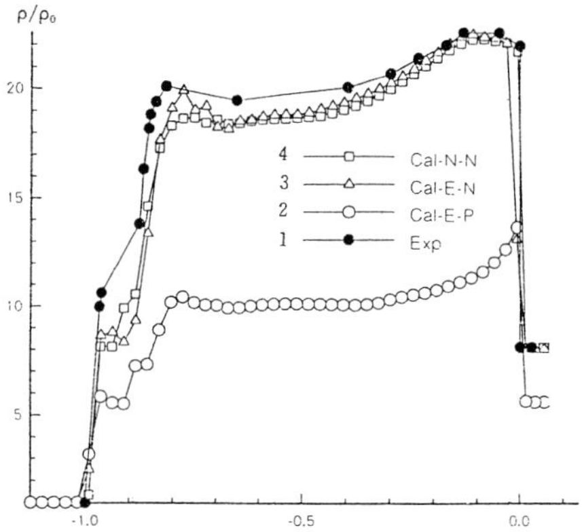

Figure 7. Comparison (cal. and exp.) of density distribution on the wall.

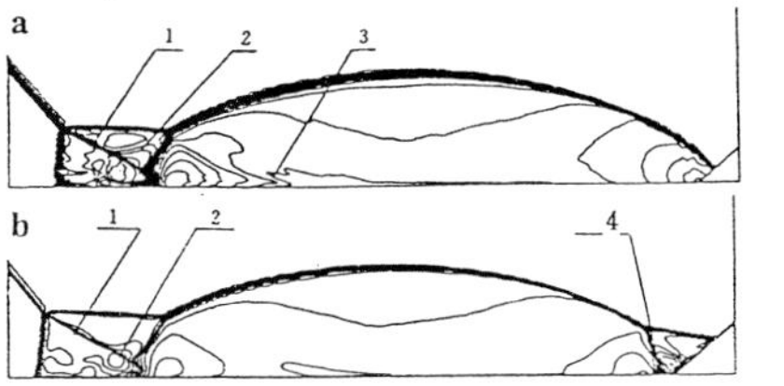

Figure 8. Computed density contours of perfect gas. $M_s = 7.66$, $\theta_w = 40°$, $P_0 = 2.63$ KPa, $T_0 = 289$ K. (a) Inviscid, 1 – first slip layer, 2 – vortex, 3 – second slip layer, $Re = 6.5 \times 10^5$; (b) viscous, 4 – "λ" shock wave.

ferogram from a UTIAS experiment (shown in Fig. 6a). The computed density contours are also displayed in Figs. 6b and 6c. The computed wall density distribution shown in Fig. 7 indicates the good agreement with the measured results of UTIAS tests. A relaxation length of about $0.1L$ can be determined from Fig. 6c, which can be confirmed by the UTIAS interferogram of Fig. 6a. As a second example, the calculation of the double Mach-reflection flow in the present experimental case with $M_s = 7.66$, $\theta_w = 40°$ has also been performed. Figures 8a and 8b are the computed density contours for inviscid and viscous perfect gases, respectively; the first and second slip layer, vortex, and "λ" shock wave in the compression corner are displayed clearly. Comparison of the density distribution on the wall between experiment and computation is illustrated in Fig. 3. The computed results for $\gamma = 1.4$, equilibrium and nonequilibrium gases are reasonable and at the location of $\overline{x} = 0.4$–0.95; densities seem to tend toward equilibrium; the reason for some deviations from experimental results is now being analyzed.

4. Conclusion

The present experiment is successful in attaining quantitative density fields by directly counting fringes without relying on analytical relations of three-shock theory at the triple points. However, in chemically reacting flow, fringe-counting data-reduction techniques provide very rough estimation of the density fields, therefore the reasonable determination of Gladstone-Dale constant K is still a problem to be solved for the accurate comparison of experiment with computation. The present computer codes, validated by experiments, represent a substantial predictive capability for the detailed modeling of the complex nonstationary nonequilibrium and viscous flow with complicated shock structure.

References

Clinton PT, Groth S (1993) TVD Flux-Difference Split Methods for High-Speed Thermochemical Nonequilibrium Flows With Strong Shocks, UTIAS Report No. 350

Dunn MG, Kang SW (1973) Theoretical and Experimental Studies of Reentry Plasmas, NASA CR 2232.

Glaz HM (1987) High Resolution Calculation of Unsteady Two Dimensional Nonequilibrium Gas Dynamics With Experimental Comparisons, AIAA Paper 87-1293.

Glaz HM, Collella P, Glass II (1986) A Detailed Numerical, Graphical, and Experimental Study of Oblique Shock Wave Reflections, UTIAS Report No. 285.

Gupta R, Yos J, Thompson R, Lee K-P (1990) A Review of Reaction Rates and Thermodynamic and Transport Properties for an 11 Species Air Model for Chemical and Thermal Nonequilibrium Calculation to 30000 K, NASA RP 1232.

Le J, Hu G (1992) A high resolution calculation of reflection of oblique shock wave from a compression corner in chemical nonequilibrium gas flow. In: Proceeding of Int Conf on the Methods of Aerophysical Research, Novosibirsk, p 84.

Needham CE (1991) Chemical nonequilibrium effects of the shock reflection. In: 18th Int Symp on Shock Tube and Waves, pp 897–902.

Wave configuration of Mach reflection in steady flows: analytical solution and dependence on downstream influences

H. Li, M. Schotz, G. Ben-Dor
Pearlstone Center for Aeronautical Engineering Studies
Department of Mechanical Engineering
Ben-Gurion University of the Negev, Beer Sheva, Israel

Abstract: Azevedo and Liu's (1993) analytical model for predicting the size of the Mach-reflection wave configuration in steady flows has been improved. Predictions based on the modified analytical model were compared to available experimental results. The agreement was found to be better than that obtained by Azevedo and Liu's (1993) original model. The reason for the better agreement is due to the fact that downstream effects were not neglected in our modified model.

Key words: Mach reflection, Wave configuration, Steady flow

1. Introduction

It was pointed out, both in Ben-Dor (1991) and Ben-Dor and Takayama (1992), that one of the yet unsolved shock-wave reflection problems in steady flows is the mechanism by which the size of a Mach reflection is determined. In such flows, the height of the Mach stem of the Mach-reflection pattern is not uniquely determined. Consider Fig. 1 where the solid lines describe the four discontinuities of a steady Mach-reflection wave configuration, namely: the incident shock wave (i), the reflected shock wave (r), the Mach stem (m), and the slipstream (s). These four discontinuities meet at a single point, the triple point, T. If one selects a new point along the incident shock wave (i) and draws there three lines parallel to the reflected shock (r*), to the Mach stem (m*) and to the slipstream (s*), then one has a new triple point, T*, with the appropriate four discontinuities. The two triple points, T and T*, as well as other triple points which could have been obtained by choosing a different location for T* along the incident shock wave (i), completely satisfy the governing equations of the three-shock theory. However, when experiments with identical initial conditions are repeated in a given facility, only one of the above-mentioned infinite theoretically possible Mach-reflection wave configurations is obtained. Thus, the three-shock theory is incapable of predicting the actual size of the Mach reflection since it is inherently independent of any physical length scale.

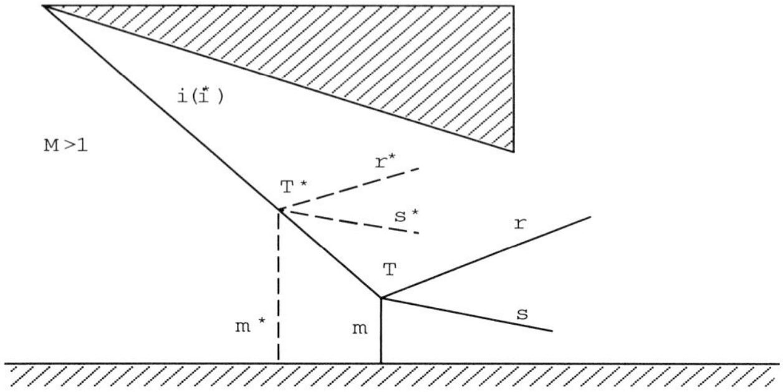

Figure 1. Two theoretically possible Mach-reflection wave configurations for identical initial conditions.

Azevedo and Liu (1993) suggested a physical model for predicting the height of the Mach stem. Consider Fig. 2 where a schematic drawing of a reflecting wedge which generates a Mach reflection is shown. In their model they assumed that the Mach stem, the slipstream, and the bottom surface form an effective one-dimensional converging nozzle, and that the flow in region (3) is isentropic and reaches sonic conditions at the minimum cross-section part of the effective conduit. In addition Azevedo and Liu (1993) assumed that the downstream flow has no influence on the location of the throat and that the throat of this converging nozzle is located at the point where the leading characteristic of the expansion fan, generated at the trailing edge of the reflecting wedge, intersects the slipstream (point E in Fig. 2).

By applying the conservation laws of mass and momentum to the control volume given by ABCDEFA, Azevedo and Liu (1993) developed an analytical model by which the Mach-stem height could be calculated. The predictions of their model are shown in Fig. 3. Although their analytical predictions showed a trend similar to the experimental results obtained for the Mach-stem height dependence on the angle of incidence, the predicted values were far from being satisfactory. This lack of satisfactory agreement is attributed to the above-mentioned oversimplifying assumptions incorporated into their model.

The purpose of the present study is to propose a more realistic model for predicting the Mach-stem height in a Mach reflection observed in steady flows. While adopting Azevedo and Liu's (1993) first two assumptions, the model proposed in the present study takes into account the influence of the downstream flow on the Mach-stem location. Therefore the flow downstream of the Mach stem is analyzed and its effect is incorporated into the analytical model in order to obtain a better model for predicting the Mach-stem location and height.

2. Present study

A physical model describing the interaction, shown in Fig. 2, between the expansion fan (generated at the trailing edge of the reflecting wedge) and the reflected shock wave of the Mach reflection was developed. Unlike Azevedo and Liu (1993) who assumed that the throat of the converging-diverging nozzle is located at the point where the head of the expansion fan intersects the slipstream, in our model the shape of the slipstream was calculated and the throat was located at the minimum cross section. For details see Schotz et al. (1995).

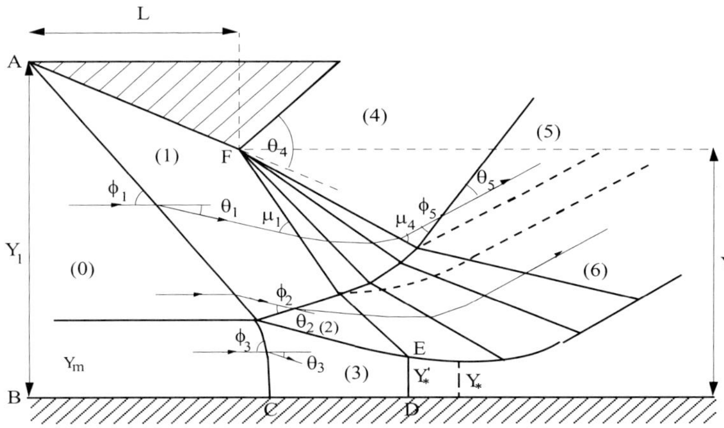

Figure 2. Schematic illustration of Azevedo and Liu's (1993) simplified model. Region (3) is an isentropic converging nozzle.

3. Results and discussion

The normalized Mach stem height y_m/L as a function of the incident shock-wave angle Φ, along with Hornung and Robinson's (1982) experimental results (obtained for $M_0 = 2.84$, 3.49, 3.98, and 4.96) are shown in Figs. 3a–3d, respectively. In addition, Azevedo and Liu's (1993) analytical predictions are also shown in these figures by dashed lines. It is apparent that their predictions consistently underestimate the actual Mach-stem height. Predictions based on the present modified model, for $y_t/L = 0.37$ and a variety of trailing-edge expansion angles θ_4, are shown in Figs. 3a–3d by solid lines. It is evident from these figures that the trailing-edge expansion angle significantly affects the Mach-stem height. Consequently, ignoring its effect, as was done in Azevedo and Liu's (1993) model, is wrong. The results shown in Figs. 3a–3d indicate that the normalized Mach-stem height, y_m/L, increases with increase in the trailing-edge expansion angle. Since Hornung and Robinson (1982) and Azevedo and Liu

(1993) did not specify completely the geometrical shape of the reflecting wedges used in the course of the experiments (no information about θ_4 was given) it is not known whether one or more reflecting wedges were used. However, the present analytical model suggests that:

$$45^\circ > \theta_4 > 20^\circ \text{ for } M_0 = 2.84, \quad 50^\circ > \theta_4 > 38^\circ \text{ for } M_0 = 3.49$$

$$35^\circ > \theta_4 > 25^\circ \text{ for } M_0 = 3.98, \quad 32^\circ > \theta_4 > 25^\circ \text{ for } M_0 = 4.96.$$

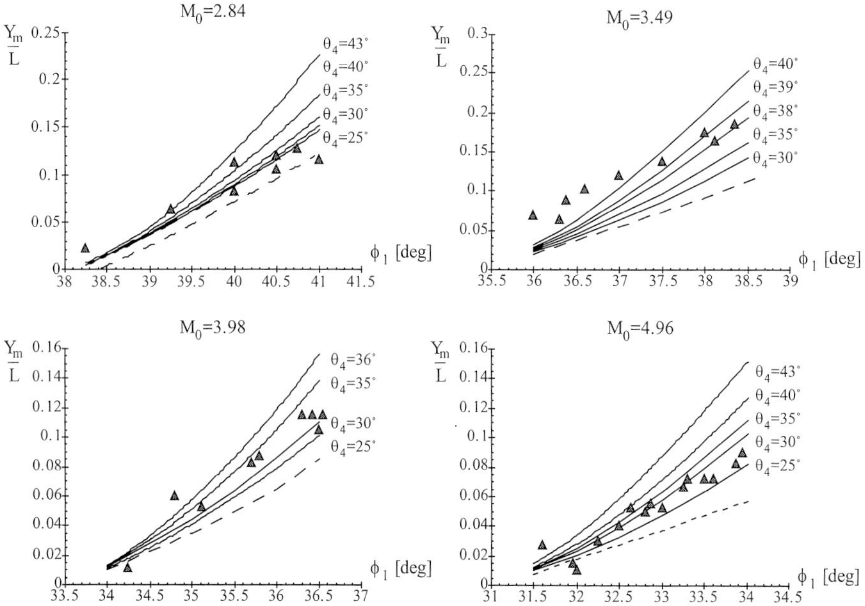

Figure 3. Comparison between experimental results and analytical predictions of the normalized Mach-stem height versus the shock-wave angle of incidence for (a) $M_0 = 2.84$, (b) $M_0 = 3.49$, (c) $M_0 = 3.98$, and (d) $M_0 = 4.96$. Solid triangles – experimental results (Hornung and Robinson 1982), dashed line – Azevedo and Liu's (1993) model, solid line – present model.

Note also that based on Azevedo and Liu's (1993) arguments, the results of Hornung and Robinson (1982) for $M_0 = 3.49$ were most likely influenced by the downstream conditions which caused the Mach stem to be larger in its length than had the reflection been free of downstream influence. Consequently, if the results for $M_0 = 3.49$ are excluded then it is evident that the actual wedge trailing-edge angles in Hornung and Robinson's (1982) experiments were probably within the range $45^\circ > \theta_4 > 25^\circ$.

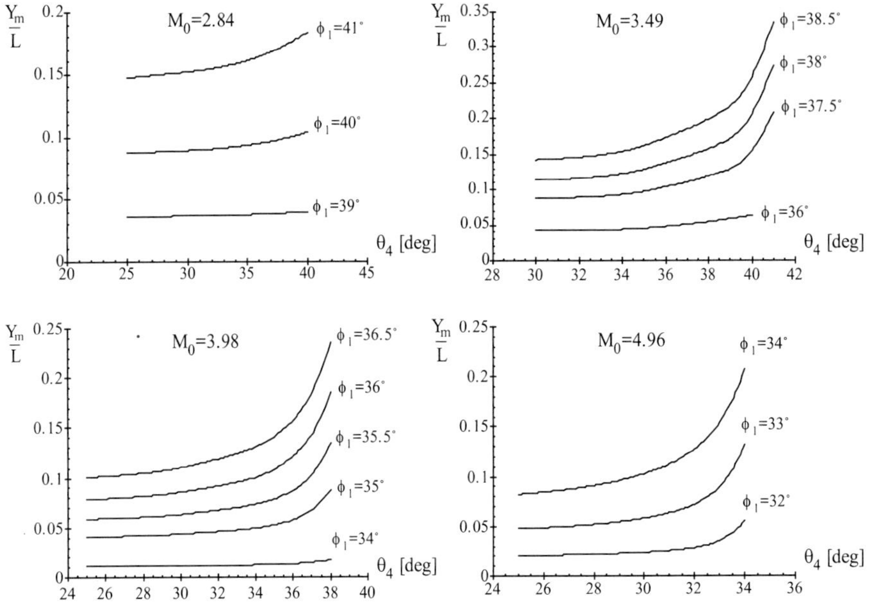

Figure 4. The dependence of the normalized Mach-stem height on the trailing edge angle θ_4 for different shock-wave incidence angles: (a) $M_0 = 2.84$, (b) $M_0 = 3.49$, (c) $M_0 = 3.98$, and (d) $M_0 = 4.96$.

The normalized Mach-stem height, y_m/L, is shown in Figs. 4a–4d as a function of the trailing-edge expansion angle θ_4 (obtained for $M_0 = 2.84$, 3.49, 3.98, and 4.96). The results indicate that the normalized Mach-stem height, y_m/L, decreases as the trailing-edge expansion angle, θ_4, decreases, until it reaches asymptotically a constant value. It is also evident from these figures that as θ_4 increases the rate of increase of the Mach-stem height increases. Figs. 4a–4d also indicate that higher shock-wave angles of incidence result in larger Mach stems for identical wedge trailing-edge angles.

4. Conclusions

The analytical model for predicting the Mach-stem height in steady Mach reflections which was suggested by Azevedo and Liu (1993) has been modified by replacing one of its oversimplified assumptions by a more realistic one. Unlike Azevedo and Liu's (1993) model, which ignored the trailing-edge angle of the reflecting wedge, the present model accounts for this parameter and hence takes into consideration the entire expansion fan and its interaction with both the reflected shock wave and the slipstream whose shapes inside the interaction zone are assumed to be second-order polynomials.

Predictions of the improved model together with those of Azevedo and Liu's (1993) model were compared with the experimental results of Hornung and Robinson (1982). The predictions based on the present model were found to be in better agreement with experimental results than those of Azevedo and Liu (1993).

References

Azevedo DJ, Liu CS (1993) Engineering approach to the prediction of shock patterns in bounded high-speed flows. AIAA J 31(1):83–90.

Ben-Dor G (1991) Shock Wave Reflection Phenomena, Springer, New York.

Ben-Dor G, Takayama K (1992) The phenomena of shock wave reflection—a review of unsolved problems and future research needs. Shock Waves 2:211–223.

Hornung HG, Robinson ML (1982) Transition from regular to Mach reflection of shock waves, part 2: the steady-flow criterion. J Fluid Mech 123:155–164.

Schotz M, Levy A, Ben-Dor G, Igra O (1995) Analytical prediction of the wave configuration size in steady flow Mach reflections. AIAA (to be published).

Non-self-similarity of the initial stages of Mach reflection

J. M. Dewey, A. A. van Netten
University of Victoria
Victoria, B.C., V8W 3P6 Canada

Abstract: Multiple shadowgraph and schlieren photography were used to obtain sequences of photographs showing the initial stages of a Mach reflection progressing along a wedge surface. The wedges used were about 30 cm long. The experiments confirmed the delayed initiation of Mach reflection, and included several cases in which the initiation of Mach reflection occurred more than 20 cm from the tip of the wedge.

The delay in initiation of a Mach reflection was strongly influenced by the ambient gas density and the wedge roughness. After the delayed initiation of Mach reflection, the shock configuration first expanded in a non-self-similar fashion, and in the case of double Mach reflection, the two triple-point trajectories did not extrapolate back to the same point on the wedge. In general, it was observed that the shock configurations eventually became self-similar relative to the tip of the wedge.

Key words: Self-similarity, Shock wave, Mach reflection, Experiment

1. Introduction

Several authors (e.g. Walenta 1980, 1983, 1987; Henderson & Grey 1981; Hornung 1985; Schmidt 1989; and Fuchs & Schmidt 1991), have shown that at low ambient densities Mach reflection does not begin at the tip of the wedge but at a distance of about 100 mean-free-paths from the leading edge. These authors studied the phenomenon using techniques in which a single record was obtained from each experiment. Because of the difficulty of exactly reproducing all the physical conditions from one shock-tube experiment to another, it was difficult to define accurately the triple-point trajectory for a specific incident-shock Mach number, wedge angle, and ambient gas condition. In addition, most of these studies were done using relatively short wedges, less than 10 cm long. These problems have been overcome in the present work by studying the initial stages of Mach reflections using multiple shadowgraph or schlieren photography so that a sequence of photographs was obtained of the development and progression of a Mach reflection in a single experiment on a wedge about 30 cm long.

2. Experimental procedures

2.1. Photographic methods

The experiments were carried out in a 7.5-cm × 25-cm rectangular shock tube, and the wedges were placed in a 25-cm × 25-cm window section which allowed the use of wedges up to 30 cm long. The light source was a ruby laser that could be Q-switched at frequencies up to 25 kHz, and with a pulse duration of about 50 ns. For shadow photography a 6.35-cm × 20.3-cm strip of film was placed directly against the second window, parallel to the surface of the wedge, so as to obtain direct-contact shadowgraphs of the reflection process. For schlieren photography, a system similar to that described by Dewey & Walker (1975) was used. The individual images were recorded with a rotating prism camera, and up to 10 images could be obtained as the shock reflection progressed along the wedge in a single experiment.

2.2. Experimental conditions

Experiments were carried out using plane incident shocks with Mach numbers of approximately 1.3, 2.0, or 2.7. Air at room temperature was the working gas. The reflecting wedges were made of steel or aluminum, and were rigidly mounted with the leading edge about 3 cm above the floor of the shock tube so that the boundary layer from the floor would not be swept up the wedge. The leading edge of the wedge was symmetrical relative to the flow direction, i.e., the angle between the two surfaces was twice the nominal wedge angle, θ_w. To investigate the effect of surface roughness, three types of wedge surface were used: polished (0.05 μm); smooth (1.6 μm); and rough (200 grit sandpaper). The smoothness was measured with a roughness comparator. A range of wedge angles was used so as to form single (SMR), complex (CMR), and double (DMR) Mach reflections.

The experiments in which the incident-shock Mach number was close to 2.7 were carried out at an initial pressure of approximately 11 Torr. Those at Mach 2.0 were done at ambient pressures of 20 or 130 Torr, and those at Mach 1.3 were carried out at either atmospheric pressure or approximately 13 Torr. This was done to investigate the effect of gas density on the initiation of Mach reflection.

The Mach number of the incident shock was measured in each experiment with two piezo-electric time-of-arrival gauges 20 cm apart in the roof of the shock-tube window section. This measurement was confirmed from the photographic record by measuring the shock displacement in the measured time interval between the laser pulses.

2.3. Method of analysis

The shadowgraph or schlieren images were projected at a magnification of 1:1 or 20:1, respectively, on to the flat bed of an x-y digitizer. The surface of the wedge and the positions of all the shocks and other density discontinuities were digitized manually, and the coordinates stored in electronic format. Fiducial markers in the fields of view were also digitized to provide scaling factors and to ensure that there was no significant optical distortion. Figure 1 shows a sequence of digitized shock configurations.

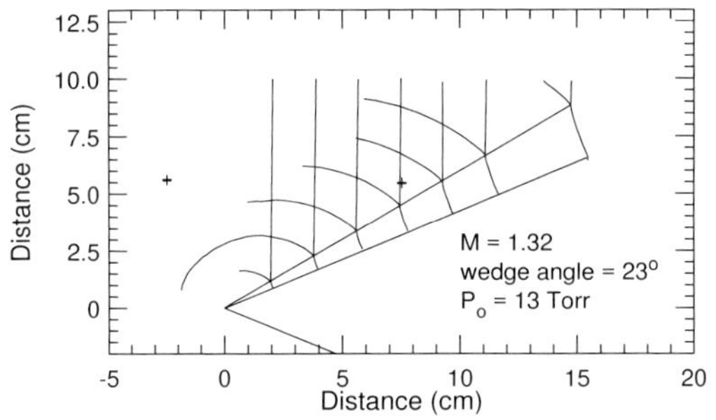

Figure 1. Digitized images of a series of schlieren photographs of a Mach reflection obtained in a single experiment.

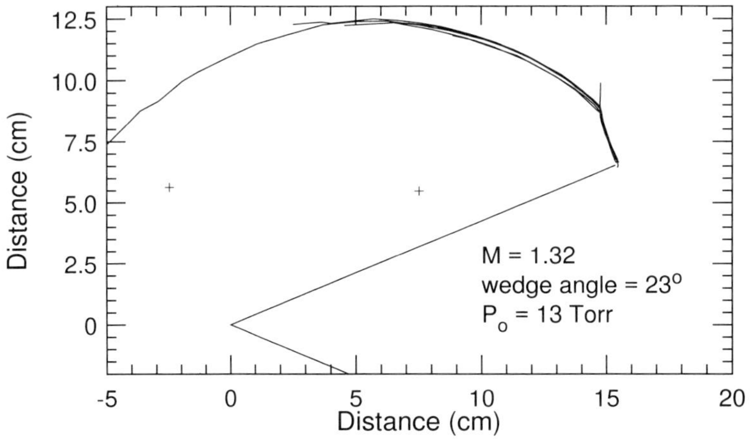

Figure 2. The shock configurations from Fig. 1 replotted after scaling relative to the distance of the incident shock past the leading edge of the wedge. This shows a high order of self-similarity.

The digitized shock configurations from a single experiment were then replotted after being scaled relative to the distance of the incident shock past the leading edge of the wedge. Using this technique, if two or more shock configurations were self-similar, when replotted in this way they would exactly overlay one another. For example, Fig. 2 shows the same series of digitized images as Fig. 1, but replotted after scaling, and showing a high order of self-similarity.

In contrast, Fig. 3 shows four nominal DMRs from a single experiment, replotted after scaling. Here there is a high degree of non-self-similarity.

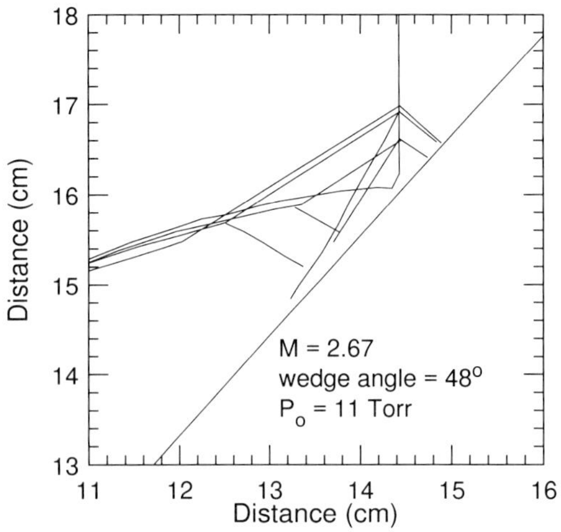

Figure 3. The development of a double Mach reflection after scaling relative to the distance of the incident shock past the leading edge of the wedge. This reflection was clearly non-self-similar, initially appearing as a regular reflection, but gradually developing into a self-similar form.

3. Summary of results

All the experiments with a smooth ($\epsilon = 1.6\mu$m) 23° wedge and incident-shock Mach numbers of 1.3 and 2.7 produced self-similar SMRs at 11 Torr and 761 Torr within 1 cm of the tip of the wedge. The reflections at both ambient pressures were identical, and the scaled shock configurations at the two pressures were exactly similar.

Most of the other experiments were done with wedge angles in the range from 46° to 49°, i.e., close to the angle for transition from regular to Mach reflection. Incident-shock strengths were approximately 2.0 or 2.7 in air at ambient pressures of approximately 11, 20, or 130 Torr. In all cases there was a lack of self-similarity in the initial development of the CMR or DMR. Even on a polished ($\epsilon = 0.05\mu$m) 48° wedge, DMR could not be identified until the incident shock had travelled up to 9 cm along the wedge. Before that, the incident shock appeared to extend to the surface of the wedge. The reflected shock also appeared to extend to the wedge, but to a point a short distance behind the incident shock. There was then a transition period in which the reflection at first appeared to be a CMR, before developing into a DMR. During this phase the two triple-point trajectories did not extrapolate back to the same point on the wedge. Eventually, at a point beyond 20 cm from the tip of the wedge, a self-similar DMR configuration was established. It should be noted that this self-similarity was relative to the tip of the wedge, i.e. the triple-point trajectories extrapolated back to the leading edge, and not to the points on the wedge where they had first appeared.

This pattern of development for Mach reflection extended further up the wedge surface as the ambient-gas density decreased and/or rougher surfaces were used. In the case of very rough surfaces, e.g. 600 grit sandpaper, an apparently self-similar type of regular reflection was maintained for a distance of 30 cm from the tip of the wedge, with wedge angles and shock Mach numbers for which DMR developed on smooth surfaces. Again, in these situations there appeared to be a gap between the the incident and reflected shocks at the wedge surface.

4. Discussion and conclusions

The experiments reported here confirm that there is a delay in the initiation of Mach reflection on a smooth wedge. Where others have suggested that the delay is on the order of 100 mean-free-paths, the present work shows delays many orders of magnitude greater than this. The distance from the leading edge of the wedge at which Mach reflection is initiated increases with decreasing ambient-gas density and increasing surface roughness.

The typical development of DMR is as follows. The initial reflection is similar to a regular reflection, albeit with a gap between the incident and reflected shocks at the wedge surface. At distances which may be more than 20 cm from the tip of a smooth wedge, the Mach-reflection pattern begins to emerge and grows in a non-self-similar manner. The two triple-point trajectories do not extrapolate back to the same point on the wedge. Eventually, the shock configuration becomes self-similar with respect to the tip of the wedge, i.e., the triple-point trajectories are straight and can be extrapolated back to the leading edge.

These findings raise doubts about the validity of the bounds between the various reflection regions determined by experimental observations and reported in such sources as Ben-Dor (1992). It appears that these observations were made using wedges which were not longer than 10 cm and which were machined smooth but not polished, and with ambient pressures of the order of 10 Torr. Under such conditions, the present work indicates that the final reflection state would not have been reached in those experiments; and had longer, smoother wedges been used, Mach rather than regular reflection may have been observed. Such results would have the effect of raising the experimental boundaries between the different reflection regions in the (M_s, θ_w)-plane, thus bringing them closer to the theoretical transition bounds.

Acknowledgement. The authors gratefully acknowledge the assistance of Susan Watkins in the preparation of this paper.

References

Ben-Dor G (1992) Shock Wave Reflection Phenomena, Springer-Verlag, New York.

Dewey JM, Walker DK (1975) A multiply pulsed double-pass laser schlieren system for recording the movement of shocks and particle tracers within a shock tube. J Appl Phys 46(8):3454–3458.

Fuchs J, Schmidt B (1991) Formation of the shock reflection on a wedge. In: Shock Waves, Takayama K (ed), 1, pp 293–298, Springer-Verlag, Berlin/New York.

Henderson LF, Grey PM (1981) Experiments on the diffraction of strong blast waves. Proc Roy Soc Lond A 377:363–378.

Hornung H (1985) The effect of viscosity on the Mach stem length in unsteady strong shock reflection. In: Lecture Notes in Physics, 235, Flow of Real Fluids, pp 82–91, Springer-Verlag, Berlin/New York.

Schmidt B (1989) Structure of incipient triple point at the transition from regular reflection to Mach reflection. In: Rarefied Gas Dynamics: Theoretical and Computational Techniques, Muntz EP et al. (eds), AIAA Inc., Washington DC, 118:597–607.

Walenta ZA (1980) Microscopic structure of the Mach-type reflexion of the shock wave. Arch Mech Warszawa 32(5):819–825.

Walenta ZA (1983) Formation of the Mach type reflection of shock waves. Arch Mech Warszawa 35(2):187–196.

Walenta ZA (1987) Mach reflection of a moving, plane shock wave under rarefied flow conditions. In: Shock Tubes and Waves, Grönig H (ed), VCH Verlagsgesellschaft, Weinheim, pp 535–541.

Numerical confirmation of the hysteresis phenomenon in the regular to Mach reflection transition in steady flows

A. Chpoun
Laboratoire d'Aérothermique, CNRS
4 ter Route des Gardes, 92190 Meudon, France

G. Ben-Dor
Pearlstone Center for Aeronautical Engineering Studies
Department of Mechanical Engineering
Ben-Gurion University of the Negev, Beer Sheva, Israel

Abstract: Numerical calculations based on the Navier-Stokes equations were carried out to investigate the reflection of shock waves over straight reflecting surfaces in steady flows. The results for a flow Mach number of $M_0 = 4.96$ confirmed the recent experimental findings of Chpoun et al. (1995) concerning the transition from regular to Mach reflection. Numerical calculations as well as experimental results showed a hysteresis phenomenon during this transition and the regular reflection was found to be stable in the dual-solution domain in which theoretically both regular- and Mach-reflection wave configurations are possible.

Key words: Hysteresis, Regular reflection, Mach reflection

1. Introduction

In steady flows the reflection of a shock wave over a straight surface leads to two possible wave configurations, namely, regular reflection (RR) and Mach reflection (MR). These wave configurations and the definition of some flow parameters are illustrated in Figs. 1a and 1b, respectively. The regular reflection (RR) consists of two shock waves, namely, the incident shock wave (i) and the reflected shock wave (r). They meet at the reflection point (R), located on the reflecting surface. The flow states are (0) ahead of (i), (1) behind it, and (2) behind (r). The Mach reflection (MR) consists of three shock waves, namely, the incident shock wave, (i), the reflected shock wave (r), and the Mach stem (m). They all meet at a single point known as the triple point (T) which is complemented with a slipstream (s). The Mach stem (m) is usually a curved shock wave which is perpendicular to the reflecting surface at the reflecting point (R).

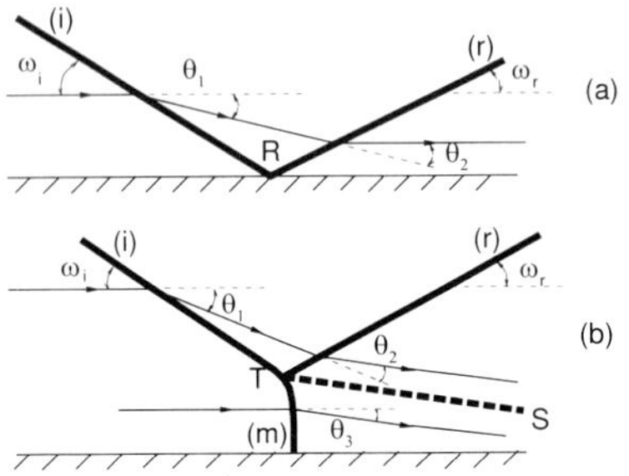

Figure 1. (a) Regular-reflection and (b) Mach-reflection wave configurations.

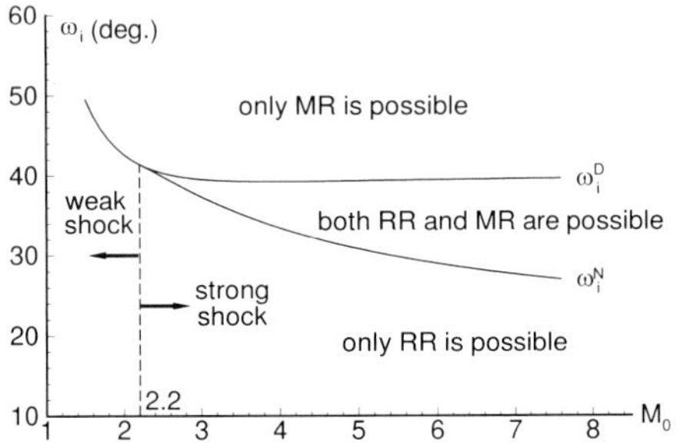

Figure 2. RR $\Longleftrightarrow$ MR transition criteria vs. flow Mach number.

Mainly from theoretical considerations, two criteria were suggested for the RR $\Longleftrightarrow$ MR transition. The first one called the von Neumann (or "mechanical equilibrium") criterion was first suggested by von Neumann in the early 1940s. This criterion corresponds to the situation where the pressure jump across the incident and reflected shock waves (i and r) becomes equal to the pressure jump across the Mach stem (m). The appropriate incident shock angle is noted ω_i^N. The second one, called the detachment criterion, was also suggested by von Neumann and corresponds to the limit where an attached oblique shock can be obtained from the flow at state (1) impinging on the reflecting surface. The corresponding incident shock angle is noted ω_i^D. It should be noted that $\omega_i^N < \omega_i^D$ for strong incident shock waves (Fig. 2). From the analysis of graphical solutions (i and r shock polars), it can be shown (Ben-Dor 1991) easily that for $\omega_i < \omega_i^N$ a Mach-reflection wave configuration is theoretically impossible. On the other hand for $\omega_i > \omega_i^D$, the regular-reflection wave configuration is theoretically impossible. The situation where $\omega_i^N < \omega_i < \omega_i^D$ corresponds to the dual-solution domain where both (RR) and (MR) wave configurations are theoretically possible (Fig. 2).

Since the pioneering work of von Neumann, have been only few expérimental studies related to the RR $\Longleftrightarrow$ MR transition in steady flows. The experiments of Hornung and Robinson (1982) are the more recent ones. Based on their own experiments, as well as those of Henderson and Lozzi (1975, 1979), Hornung and Robinson concluded that in steady flows, the transition from regular to Mach reflection of strong shock waves occurs at the von Neumann condition ($\omega_i = \omega_i^N$) and not at the detachment condition ($\omega_i = \omega_i^D$) as mentioned in previous studies. Based on their length scale concept, Hornung et al. (1979)

hypothesized that a hysteresis should exist in the RR $\Longleftrightarrow$ MR transition. If one starts with an MR at $\omega_i > \omega_i^D$ and then slowly decreases ω_i while M_0 is kept constant, there is a decrease in the Mach-stem height until it vanishes at $\omega_i = \omega_i^N$ where a smooth transition from MR to RR takes place. If the decrease in ω_i continues, then the RR is maintained. If now ω_i is slowly increased, since the flow field is free of significant disturbances, the RR should be maintained until $\omega_i = \omega_i^D$ where a sudden transition from RR to MR should take place. Hornung and Robinson's (1982) experimental attempts to confirm this hypothesis regarding the hysteresis in the RR $\Longleftrightarrow$ MR transition failed and they concluded that "the hysteresis effect predicted by Hornung et al. (1979) could not be confirmed." For this reason they refer to the RR in the dual-solution domain $\omega_i^N < \omega_i < \omega_i^D$ as an unstable regular reflection. Meanwhile, in recent experimental work (Chpoun et al. 1995) which was conducted in the SH2 hypersonic wind tunnel of CNRS Meudon, France, the hysteresis loop hypothesized by Hornung et al. (1979) was observed. The experiments were conducted in a continuous, free jet facility. A few geometrical parameters (plate length, plates cross inlet section, etc.) were varied and the hysteresis was reproduced in all experimental dispositions. Furthermore, unlike previous reports, regular-reflection wave configurations were found to be stable in the dual-solution domain. The hysteresis phenomenon which was observed by Chpoun et al. (1995) is shown in Fig. 3 in the (ω_i, ω_r)-plane. While the MR $\Rightarrow$ RR transition occured at $\omega_i = 30.9°$, which is equal to the von Neumann condition ω_i^N, the RR $\Rightarrow$ MR transition occured at $\omega_i = 37.2°$, which is about 2° smaller than the corresponding detachment condition, $\omega_i^D = 39.3°$. Based on the foregoing

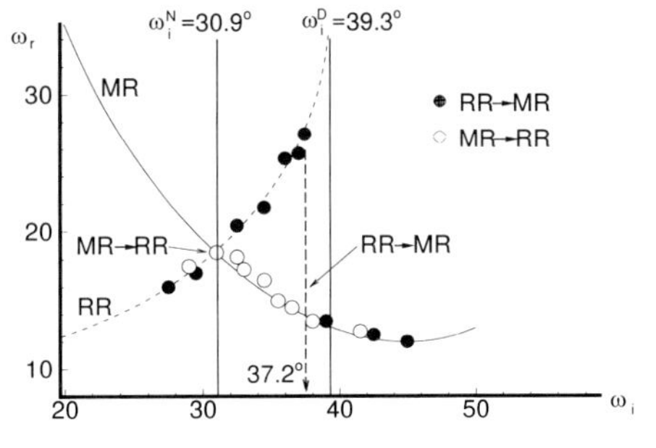

Figure 3. The experimental hysteresis loop in (ω_i, ω_r)-plane.

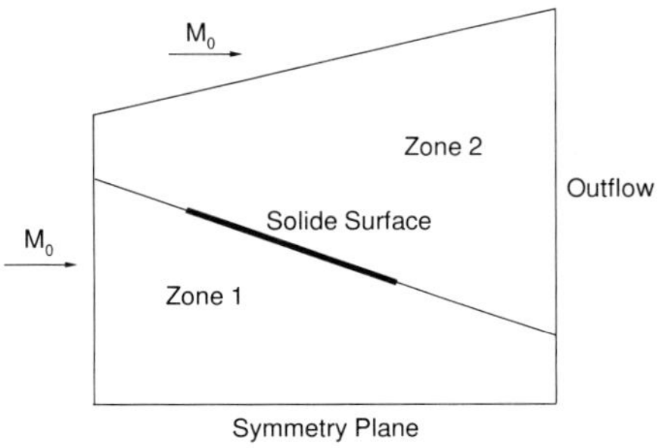

Figure 4. Computational domain and boundary conditions.

introduction and particularly on the recent experimental findings of Chpoun et al. (1995), the following summarizes the state of the art regarding the RR $\Longleftrightarrow$ MR transition in steady flows: (1) the RR $\Rightarrow$ MR transition takes place at nearly the detachment criterion and the MR $\Rightarrow$ RR transition occurs at the von Neumann criterion, therefore, a hysteresis exists in the RR $\Longleftrightarrow$ MR transition; (2) regular-reflection wave configurations are stable in the dual-solution domain. The following sections present numerical work related to shock-wave reflections in steady flows over straight surfaces. The results confirmed the principal experimental findings of Chpoun et al. (1995).

2. Computational work

The numerical results were obtained using a finite-volume zonal code. The code solves the complete Navier-Stokes equations by means of a time-marching method. The evaluation of the inviscid terms are based on the flux-splitting method of Steger and Warming (1981) in combination with upwind-based differencing. The diffusion terms are evaluated using standard central differences. The LU-SGS algorithm is used to solve the system of point Gauss-Seidel relaxation. Viscosities are evaluated using Sutherland's law and thermal conductivities are defined by assuming a constant Prandtl number. The computational domain is divided into two zones shown in Fig. 4. It contains $(2 \times 40 \times 60)$ cells. An infinitely thin solid surface was considered to separate the two zones. The computational domain was large enough for the upstream and upperside boundaries to be specified as freestream conditions, and downstream outflow quantities were extrapolated. To avoid viscous effects, the reflecting straight surface was replaced by a symmetry plane. Although the numerical model deals with an infinitely thin plate model, it reproduces the essential features of the flow which are obtained in the experiments, e.g., the flow compression and expansion at the windward side and the leeward side of the plate, respectively. The sensitivity of numerical results to mesh refinement was analyzed. The study showed that 4800 cells having a size of 0.2 mm near the wall were sufficient to ensure the grid independency of the results. In the same manner, the downstream boundary was moved at different locations and non-significant change was observed in the numerical results. The upstream flow conditions were those of Chpoun et al.'s (1995) experiments. The upstream Mach number was 4.96. The stagnation temperature and pressure were 453 K and 8.5 bar, respectively, leading to a high unit Reynolds number of 1.27×10^7/m which minimized the thickness of the viscous layer along the plate. The plate angles were set to a succession of prescribed values ranging from small values (RR) to large values (MR), then from large to small values with a step of 0.5°. For each angle the numerical results were carried out with initial flow conditions which were obtained from the previous angle's calculations. For each case of calculation, a steady-state flow was achieved. The first calculation was initiated with uniform flow conditions.

3. Results and discussion

Figures 5a–5l (from left to right and from top to bottom) show a sequence of the calculated flow-field density isolines obtained for plate angles ranging from small values (RR) to large values (MR) and then from large values to small values. Figure 5a shows a regular reflection; the incident and reflected shocks as well as

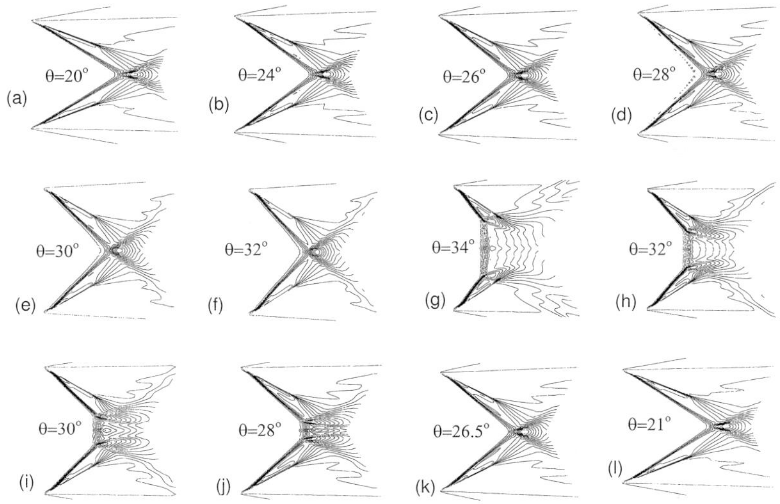

Figure 5. Flow-field numerical density isolines.

leading-edge expansion waves on the leeward side of the plate are evident. When the plate angle is increased step by step the RR is maintained until about 32°, then a sudden transition to Mach reflection with a big Mach stem occurs. The corresponding transition incident shock angle is approximately 44°. Inversely, as the plate angle decreases, a smooth transition to regular reflection is obtained for the plate angle of approximately 27°. The appropriate shock angle is about 36°. In cases with small Mach stems the clear evidence of slipstreams from the density isolines avoids confusion with a regular reflection. It should be noted that, as shown in Fig. 2, the incident shock angles corresponding to the von Neumann and the detachment criteria are, respectively, 30.9° and 39.3° in these flow conditions. The 5° differences between the theoretical and numerical transition angles originate probably from viscosity and the simplified numerical model, i.e., infinitely thin plate rather than a finite-size reflecting wedge.

On the other hand the regular-reflection wave configuration was found to be stable in the dual-solution domain, i.e., the calculations converged systematically to RR wave configurations as the initial flow was uniform or arbitrary. In the same manner, in the experimental part of their work, Chpoun et al. (1995) had shown that in the dual-solution domain an external disturbance changes MR wave configuration to RR. Therefore, the discrete variation of the plate angle shunts the theoretical MR $\Longrightarrow$ RR transition inside the dual-solution domain and the numerical transition angle was found to be larger than the von Neumann criterion. A small angle step probably will lead to a better agreement with the theroretical prediction, i.e., ω_i^N. Nevertheless, a hysteresis loop is clearly evident in the numerical results. In the dual-solution domain, for the same plate angle (28°) both RR and MR solutions were obtained (compare Figs. 5d and 5j).

4. Conclusion

A numerical study has been carried out to investigate the RR $\Longleftrightarrow$ MR transition in steady flows over straight surfaces. The numerical results confirmed the principal experimental findings of Chpoun et al. (1995) and contradicted previous reports on the subject: (1) a hysteresis exists in the RR $\Longleftrightarrow$ MR transition; (2) regular-reflection wave configurations are stable in the dual-solution domain. Furthermore, numerical solutions of flows which contain shock-wave interaction and reflection depend on the initial flow conditions. Therefore, the dependence of numerical solutions on initial flow conditions should be analyzed.

References

Ben-Dor G (1991) Shock Wave Reflection Phenomena, Springer-Verlag, New York.

Chpoun A, Passerel D, Li H, Ben-Dor G (1995) Reconsideration of the state-of-the-art of oblique shock wave reflection in steady flows, part I: experimental investigation. J Fluid Mech (to be published).

Henderson LF, Lozzi (1975) Experiments on transition of Mach reflection. J Fluid Mech 68(1):139–155.

Henderson LF, Lozzi (1979) Further experiments on transition to Mach reflection. J Fluid Mech 94(3):541–559.

Hornung HG, Oertel H, Sandeman RJ (1979) Transition to Mach reflection of shock waves in steady and pseudo-steady flow with and without relaxation. J Fluid Mech 90:541–560.

Hornung HG, Robinson ML (1982) Transition from regular to Mach reflection of shock waves, part 2: the steady-flow criterion. J Fluid Mech 123:155–164.

Hornung HG (1986) Regular and Mach reflection of shock waves. Ann Rev Fluid Mech 18:33–58.

Steger JL, Warming RF (1981) Flux vector splitting of the inviscid gasdynamic equations with application to finite-difference methods. J Comp Phys 40:263–293.

The existence of reflected finite-compression waves in weak pseudo-steady Mach reflections over not-"small" wedge angles

J. J. Liu
Department of Engineering Science, National Cheng-Kung University
Tainan, Taiwan ROC

Abstract: It is well known that the classical three-shock theory fails badly to describe the wave configuration of weak pseudo-steady Mach reflections (MR). The fundamental triple-point trajectory angle (χ) cannot be predicted by the three-shock theory, and the angles defined by the shocks and contact surface near the triple point cannot be determined by the theory even when an experimentally observed χ is used in the three-shock theory. Using existing experimental data for not-"small" reflecting wedge angles, it is shown analytically that the reason for this failure is owing to the existence of reflected finite-compression waves, instead of reflected shock waves, in weak pseudo-steady MR. The predicted χ using the linear pressure-deflection weak oblique shock approximation for the reflected backward-facing finite-compression wave in the otherwise same three-shock theory is found to be in very good agreement with Smith's data (1945) and in reasonably good agreement with Olim and Dewey's data (1992) for incident shock Mach number (M_s) up to 1.51. The "smallness" of a reflecting wedge angle is analytically defined as a function of M_s and the specific heat ratio (γ). Boundaries separating regimes of "small" from not-"small" reflecting wedge angles are determined for M_s up to 2 and $\gamma = 1.4$.

Key words: Mach reflections, Finite-compression waves

1. Introduction

It is a well-known unresolved issue that the von Neumann three-shock theory succeeds in describing regular reflection (RR) and the Mach reflection (MR) of strong incident shocks, but fails badly to describe the MR of weak incident shocks. For weak MR, it fails to predict correctly the angles defined by the shocks and contact surface where they meet at the triple point. This occurs even when experimentally observed triple-point trajectory angles are used in the three-shock theory (Olim and Dewey 1992). In seeking an explanation for the incapability of the three-shock theory to adequately describe weak MR, fundamental assumptions of the theory, such as viscosity, finite curvature of reflected waves, non-self-similarity, and flow non-uniformity, must be reviewed as has been reported by Sternberg (1959), Henderson and Siegenthaler (1980), and Sakurai et al. (1989); but most to not much avail. Recently, Colella and Henderson (1990) numerically simulated weak MR and showed that there is one kind of weak irregular wave reflection which cannot be explained by the conventional three-shock theory, called a von Neumann reflection (vNR). The reflected wave of vNR is not a shock but a smoothly distributed compression wave of finite thickness, and the incident and Mach shocks appear to form a single wave with a continuously turning tangent. Their analysis therefore explains why the conventional theory cannot be applied to vNR. On the other hand, Olim and Dewey (1992) confirmed the observations of other workers that the three-shock theory is unable to describe MR of incident shocks with strengths between Mach numbers 1.1 and 1.5. The resulting three-shock configuration is such that the reflected wave is supersonic and there is a measureable discontinuity in the slopes of the incident and Mach-stem shocks. In this work, we present experimental and analytical evidence to show that the reason for the inapplicability of the three-shock theory to describing weak pseudo-steady MR over not-"small" reflecting wedge angles is owing to the existence of the reflected finite-compression wave, instead of the reflected shock wave. The analysis is then carried out by replacing the reflected shock wave with a linear or a second-order small deflection weak oblique shock approximation for the reflected finite compression wave in an otherwise same three-shock theoretical formulation. The "smallness" of the reflecting wedge angle is analytically defined. For "small" reflecting wedge angles, Liu (1993) showed that the reflected wave in weak MR is a normal Mach wave, when the corresponding three-shock theory requires forward-facing reflected shock solutions.

2. Analysis

A typical pseudo-steady single MR is shown schematically in Fig. 1: there are four discontinuities — the incident shock i, the reflected shock r, the Mach stem m, and the slipstream s — which coincide at the triple point T, making a triple-point trajectory angle χ with the wedge surface having an inclination angle θ_w. Since the analytical formulation of the pseudo-steady von Neumann three-shock theory for a perfect gas is well known, the theory is not discussed here. Subscripts 0, 1, 2, and 3 refer to the flow states as defined in Fig. 2. Following Colella and Henderson's (1990) numerical finding of a band of reflected smoothly distributed compression waves of finite thickness in weak MR, the weak oblique shock approximation linearly relating finite changes between the pressure and flow deflection across a weak oblique shock, as discussed by Liepmann and Roshko (1957) valid for small deflection, is used to describe the reflected finite-compression wave replacing the reflected shock wave in von Neumann type MR.

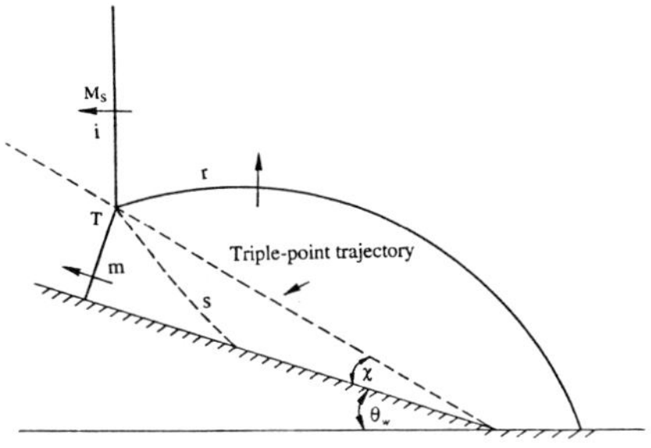

Figure 1. Schematic illustration of a propagating MR over an oblique wall.

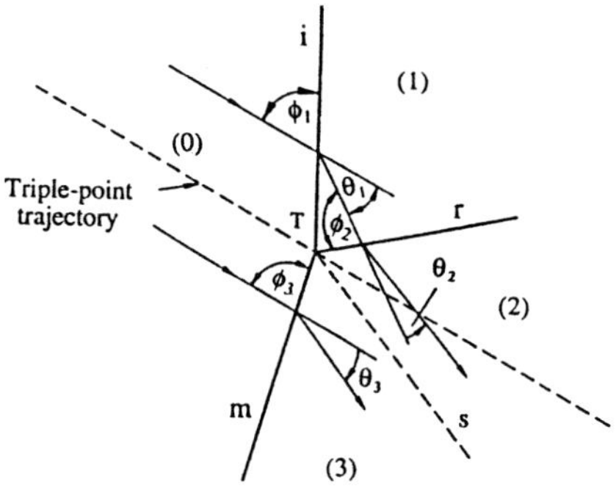

Figure 2. The wave configuration of a MR viewed from the triple point.

The thickness of this reflected compression wave is characterized by a small but finite $\Delta\theta$, so that the weak oblique shock approximation is applicable. Ben-Dor (1992) suggested this simple gas dynamic consideration for reflected compression waves in vNR. However, the pressure-deflection (P-θ) polar solution discussed by him is forward-facing of the triple point. The polar solution should be obtained by drawing a line with the above-calculated gradient, in the direction of decreasing deflection while increasing pressure, on the incident shock polar. This is because Liu (1993) has shown that sound waves downstream of incident shock waves of pseudo-steady MR cannot coalesce into Mach waves which propagate upstream from the triple point. A schematic drawing of a shock polar presentation of a vNR, using MR data reported by Olim and Dewey (1992), is given in Fig. 3. The reflected shock polar is also shown in the figure. It can be seen that the tiny reflected shock polar does not intersect the incident polar except for the Mach-line degeneracy point. The computed χ using the linear weak oblique shock approximation is in close agreement with the experimentally measured value of 12°. The reported uncertainty in the measurement of χ is $\pm 1°$. The γ used is 1.4. The use of a second-order small deflection weak oblique shock approximation for the reflected compression wave gives $\chi = 13.20°$. Figure 4 gives the shock polar solution from the three-shock theory for the same vNR, and one can see the reflected shock wave is forward-facing. The predicted χ (20.47°) differs significantly from the experiment. A graphed (P-θ) solution of vNR over a large wedge angle, where the above-mentioned second-order weak oblique shock approximation is applied, is shown in Fig. 5. The predicted χ (3.95°) agrees closely with the measured value (4°); while the linear weak oblique shock approximation gives $\chi = 2.42°$.

The necessity of classifying reflecting wedge angles of pseudo-steady MR as "small" arises from the existence of physically unrealistic forward-facing

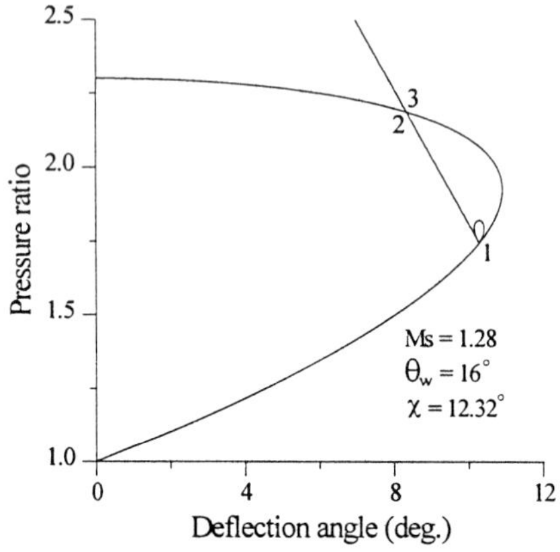

Figure 3. The (P-θ) shock polar solution of a vNR using the linear weak oblique shock approximation.

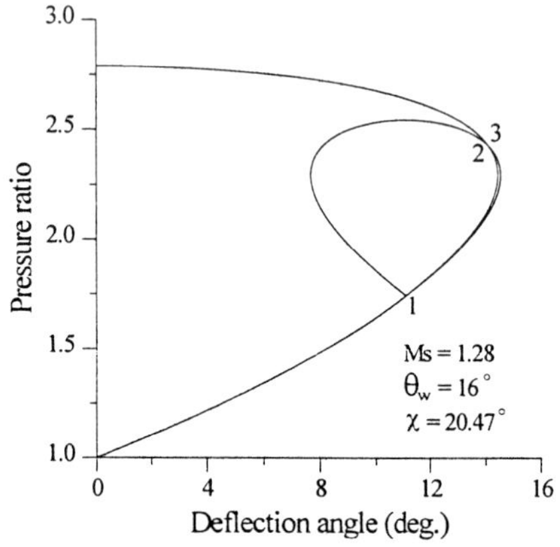

Figure 4. The three-shock theoretical forward-facing reflected shock solution for the same case as shown in Fig. 3.

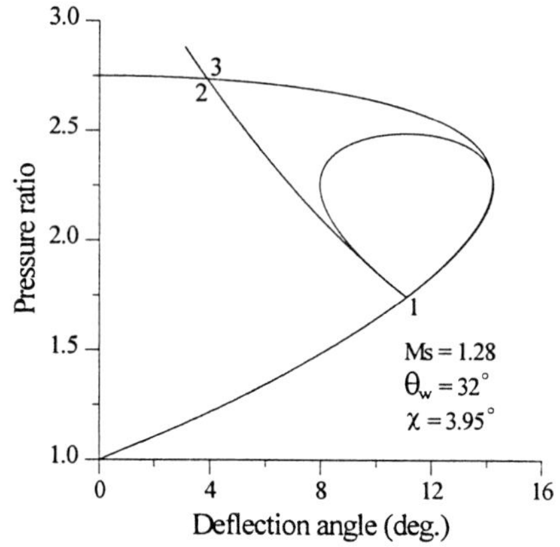

Figure 5. The (P-θ) shock polar solution of a vNR using the second-order weak oblique shock approximation.

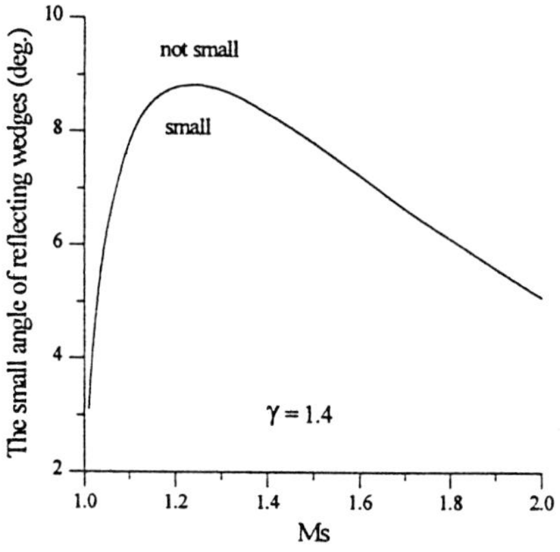

Figure 6. Theoretical limiting small reflecting wedge angles of pseudo-steady MR.

reflected finite-compression solutions when they become sufficiently small for given M_s and γ. A "small" reflecting wedge angle of pseudo-steady MR is thus defined as, for given M_s and γ, a wedge angle below which a backward-facing reflected finite-compression wave solution satisfying the boundary condition assuming a straight Mach stem normal to the wedge surface is theoretically unobtainable. Figure 6 shows the calculated boundary separating "small" from not-"small" angles for $\gamma = 7/5$ and $1.0 < M_s < 2.0$.

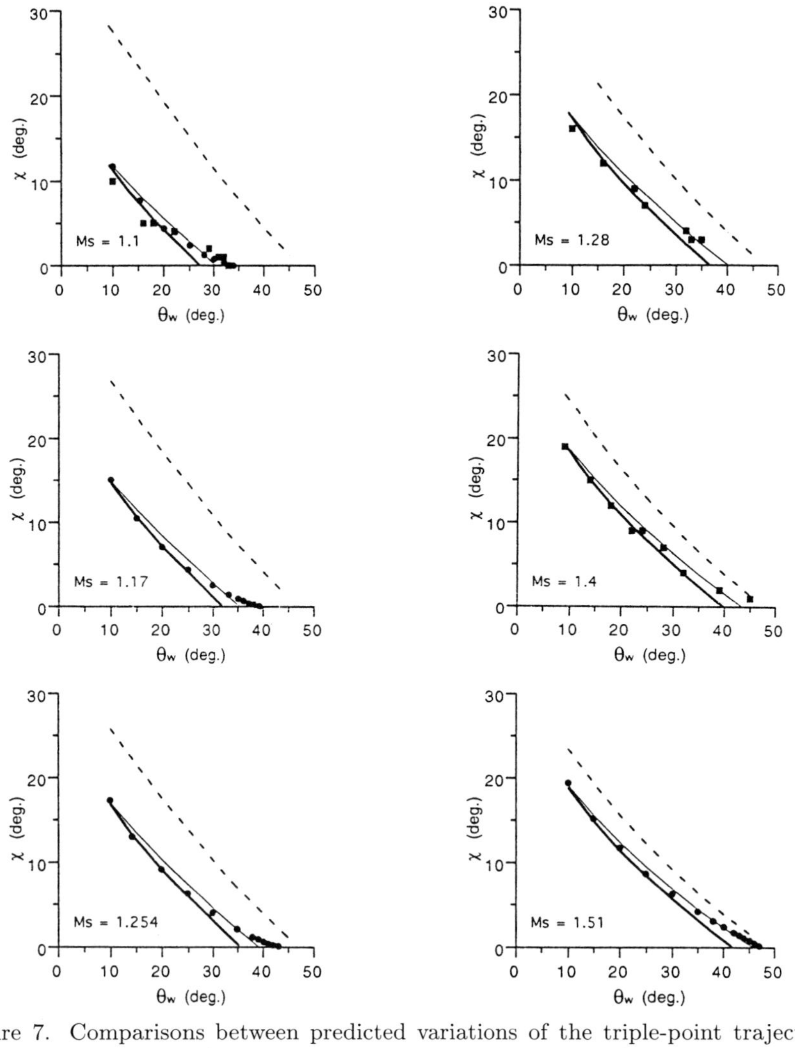

Figure 7. Comparisons between predicted variations of the triple-point trajectory angle with wedge angle for given incident-shock Mach numbers from the three-shock theory (dashed line) and from the first-order (heavy solid line) and second-order (light solid line) weak, reflected oblique shock approximations. Experimental data taken from Smith (1945) (circle) and Olim & Dewey (1992) (square).

3. Discussions and conclusions

In the following, we apply the above-discussed modified three-shock theory to analyze the experimental data of Smith (1945) and Olim and Dewey (1992). Figure 7 shows that predicted triple-point trajectory angles using a linear pressure-deflection weak oblique shock approximation for the reflected backward-facing finite-compression wave in the otherwise same three-shock theory are in very good agreement with Smith's data and in reasonably good agreement with Olim and Dewey's data for M_s up to 1.51. The good agreement holds for flow deflection across reflected finite-compression waves ranging from $\leq 3.3°$ for $M_s = 1.10$ to $\leq 14.8°$ for $M_s = 1.51$, giving the upper limit of the reflecting wedge angle, below which the present theory applies, ranging from 25° for $M_s = 1.10$ to 38° for $M_s = 1.51$. The computed flow Mach numbers downstream of incident shock waves (responsible for the formation of a reflected shock) at these upper limiting wedge angles for $M_s = 1.10$, 1.17, 1.254, 1.28, 1.4, and 1.51 are 1.07, 1.12, 1.20, 1.22, 1.86, and 2.00, respectively. The corresponding reflecting wedge angles are 25°, 29.9°, 34.9°, 35°, 39°, and 38°. One finds that the angles of Mach waves downstream of these incident shock waves are all less than 60°. It can then be argued that the subsequent nonlinear coalescence of downstream Mach waves into a reflected shock wave may take place to render the first-order weak oblique shock approximation for the reflected wave inaccurate. On the other hand, application of the second-order small deflection weak oblique shock approximation gives considerably better agreement with experiment only for wedges within 10° near their transition angles to RR. Figure 7 also shows for a given wedge angle the discrepancy between experiment and three-shock theory, for χ increases rapidly as M_s decreases from 1.51 to 1.1. Therefore, the reflected waves in pseudo-steady weak MR over not-"small" reflecting wedge angles, with inclinations not greater than about 10° less than their transition wedge angles to RR, are backward-facing compression waves of small, finite thickness.

References

Ben-Dor G (1992) Shock Wave Reflection Phenomena, Springer-Verlag.

Colella P, Henderson LF (1990) The von Neumann paradox for the diffraction of weak shock waves. J Fluid Mech 213:71–94.

Henderson LF, Siegenthaler A (1980) Experiments on the diffraction of weak blast waves: the von Neumann paradox. Proc Roy Soc Lond A 369:537–555.

Liepmann HW, Roshko A (1957) Elements of Gasdynamics, Wiley.

Liu JJ (1993) Sound wave structures downstream of incident propagating oblique shock waves. In: Shock Waves @ Marseille, Springer.

Olim M, Dewey JM (1992) A revised three-shock solution for the Mach reflection of weak shocks ($1.1 < M_i < 1.5$). Shock Waves 2:167–176.

Sakurai A, Henderson LF, Takayama K, Walenta Z, Colella P (1989) On the von Neumann paradox of weak Mach reflection. Fl Dyn Res 4:333–345.

Sternberg J (1959) Triple-shock-wave interactions. Phys Fluids 2:179–206.

Smith LG (1945) Photographic Investigation of the Reflection of Plane Shocks in Air, OSRD Rep. 6271, Office Sci. Res. Dev., Washington, DC.

Shock-wave reflections from a cone with attack angles in a symmetric plane

T. Suzuki, Y. Sakamura
Toyama Prefectural University
Kosugi, Toyama 939-03 Japan
X-Z. Yin
University of Science and Technology of China
Heifei, Anhui 230026 China

Abstract: Results from preliminary experiments for three-dimensional shock-wave reflections from a cone are presented. The Mach number of the incident planar shock wave was 1.34. The configurations of shock waves reflected from a cone at a given angle of attack are compared with those from a wedge, which is set at the same angle of attack, and those from other cones with the same angle between the shock-propagating direction and the cone ray.

Key words: Shock-wave reflections, Three-dimensional flows, Cone

1. Introduction

The reflections of oblique shock waves from wedges have been investigated experimentally, analytically, and numerically by many researchers. However, almost all of the previously reported studies belong to two-dimensional (Ben-Dor 1991) or axi-symmetrical (Takayama & Sekiguchi 1977, Milton & Archer 1995) reflections. Few works regarding three-dimensional shock-wave reflections have been reported (Hookham et al. 1995). In three-dimensional shock-wave reflections, interesting and technically important phenomena occur. Shock-wave reflection from a cone positioned at an angle of attack is an example of reflections with a three-dimensional nature. In shock-wave reflection from a cone at a finite angle of attack, the local angle between the incident shock wave and the cone surface varies continuously in the circumferential direction. Under certain conditions, we can simultaneously observe two types of reflections (i.e. a regular reflection (RR) and a Mach reflection (MR) on the cone's lower and upper surfaces, respectively) in a symmetric plane (the vertical meridian plane) as shown in Fig. 1. Hence the *spatial* transition from MR to RR, which cannot be observed in reflections from a wedge, may occur, and there may exist a crescent-shaped Mach shock wave. The phenomena of three-dimensional shock-wave reflections are too complicated to be analyzed theoretically and there is

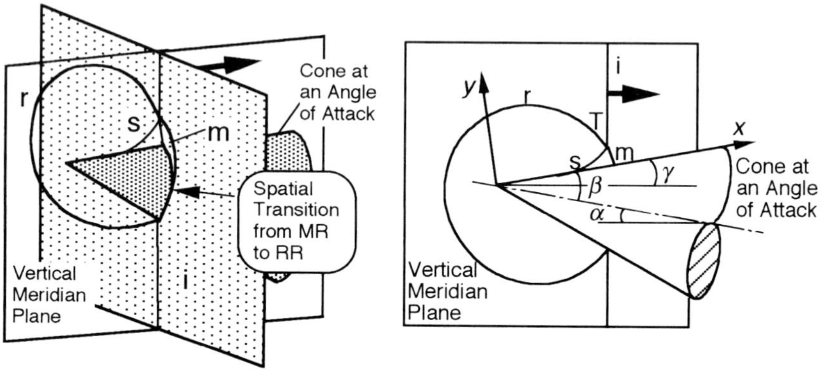

Figure 1. Sketches of shock-wave reflection from a cone at an angle of attack α: i: incident shock; r: reflected shock; m: Mach stem; s: slip surface; T: triple-point; β: semi-apex angle.

little understanding of them. The experimental work reported in this paper is the first step in our plan to study three-dimensional shock-wave reflections from a cone. The configurations of shock waves reflected from a cone at a given angle of attack are compared with those from a wedge, which is set at the same angle of attack, and those from other cones with the same angle between the shock-propagating direction and the cone ray, γ ($\beta - \alpha$ for the upper side of the cone, and $\beta + \alpha$ for the lower, as shown in Fig. 1).

2. Experimental method

The experiments were performed in a shock tube of Toyama Prefectural University. The lengths of the driver and driven sections are 1.45 m and 4.2 m, respectively. The test section of the shock tube has a rectangular cross section of 30 mm × 100 mm and viewing windows of 90 mm × 200 mm. Both the driver and driven gases were air and the Mach number of the produced planar shock wave, M_s, was 1.34. A shadowgraph optical system was used to photograph the wave configuration of the resulting shock-wave reflection.

The cone and wedge models used in the present study are shown in Fig. 2. We prepared cone models with $\beta = 20°$, $30°$, and $40°$, and a wedge model with $\beta = 30°$. These models can be set in the shock-tube test section at arbitrary α ranging from $0°$ to $20°$ as shown in Fig. 2.

In the present experiments, shadowgraphs were taken at four different shock locations for each case in order to visualize the shock-wave configurations. Note that only reflections in the vertical meridian plane from the upper and lower sides of the models at an angle of attack can be visualized by the present

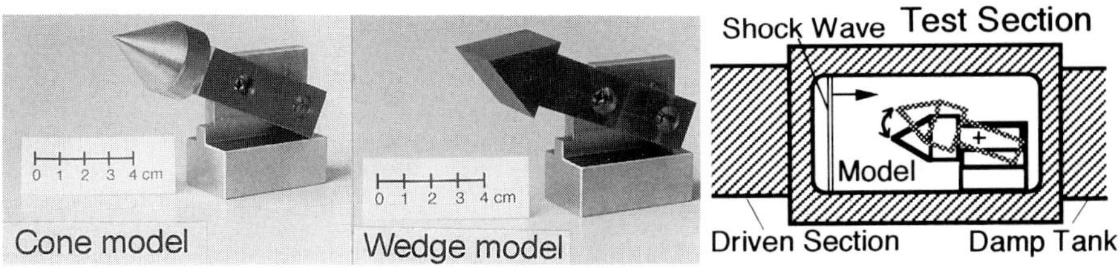

Figure 2. Cone and wedge models.

experimental system. In the case where an MR occurs, the triple-point locations were also measured and the obtained triple-point trajectories were compared. The measurements from photographic negatives (35mm film) were made with the aid of a profile projector of 50 magnifications (Nikon, V-12A).

3. Results and discussion

Figures 3a to 3d show shadowgraphs of shock-wave reflection from a cone model with $\beta = 30°$ at $\alpha = 0°$, $5°$, $10°$, and $20°$, respectively. In the cases of $\alpha = 0°$ and $5°$, both shock-wave reflections from the upper and the lower sides are of the MR type. In all pictures of Fig. 3, the reflected shock waves and slip surfaces are hardly observed because the shadowgraphs only show the information of density change, integrated along the light path. Especially, shock waves reflected from the upper surface of the cone model with high angles of attack ($\alpha > 10°$) were so weak that an accurate determination of the triple point from the shadowgraph pictures was hardly possible. It was found from many additional shadowgraphs taken at smaller intervals of α that the transition from MR to RR, on the lower side, occurred at $\alpha = 10°$ for the present conditions. Thus, the *spatial* transition from a Mach to a regular reflection may occur on the surface of the cone ($\beta = 30°$) at larger attack angles than $10°$. It is smaller than the observed effective wedge angle for a transition in the two-dimensional case, which was $13°$ under the present conditions.

Figure 4 shows the measured locations of the triple points for a cone case and a wedge case, both at zero angle of attack. The x is the distance measured from the apex along the cone ray, and the y is the normal distance measured from the cone surface (see Fig. 1). The symbols show the experimental results. In the case of the Mach reflection on a cone or a wedge placed at $\alpha = 0°$, the triple-point trajectory can be approximately predicted by the shock dynamics theory (Han & Yin 1992). The trajectories calculated using this theory are also plotted for both cone and wedge cases on Fig. 4. In this figure it is seen that

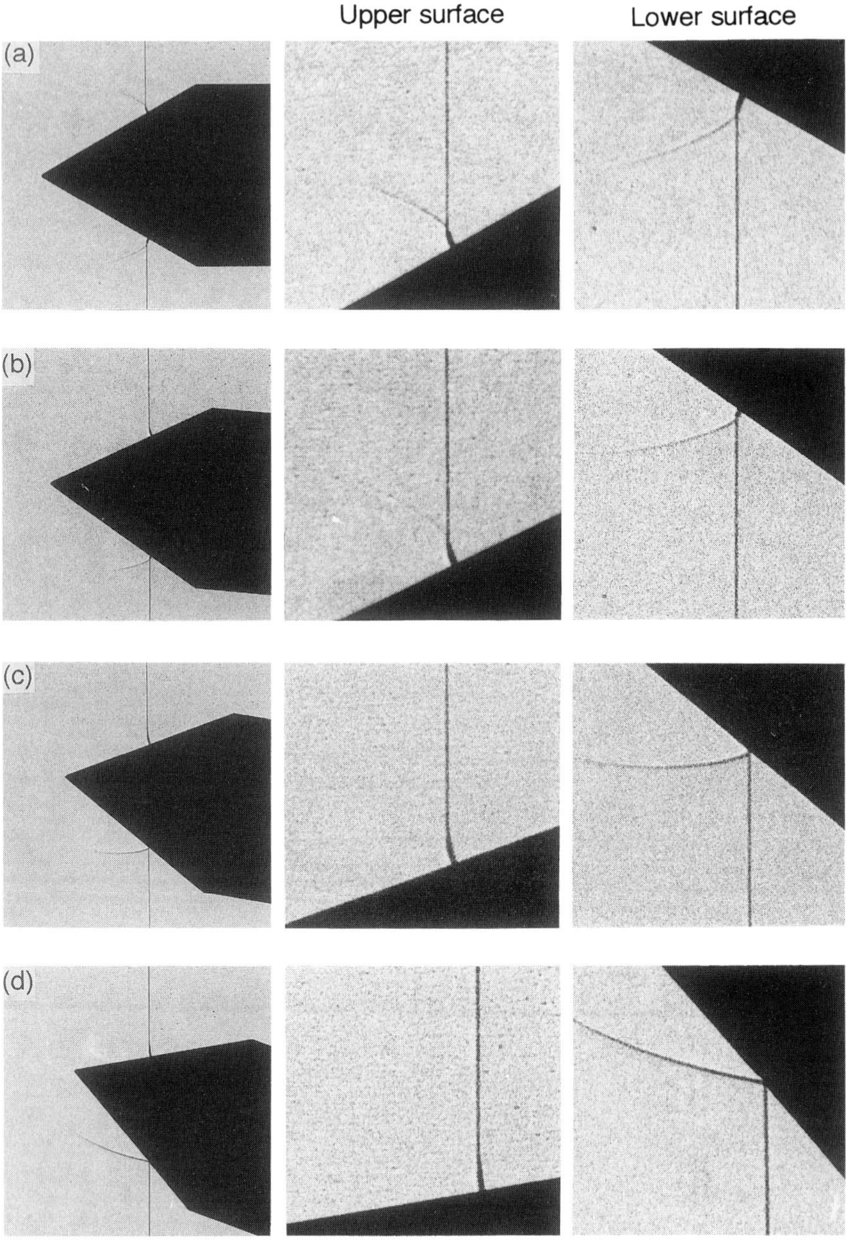

Figure 3. Shadowgraphs of shock-wave reflections from a cone model ($\beta = 30°$): (a) α=0°, (b) α=5°, (c) α=10°, (d) α=20°.

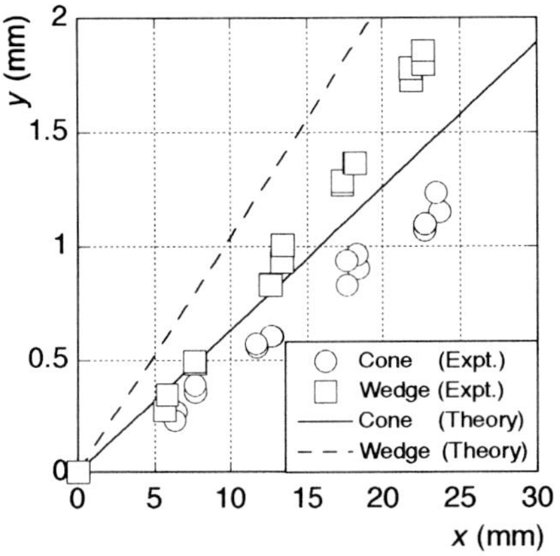

Figure 4. Comparison of the triple-point trajectory for the cone case with that for the wedge case (M_s=1.34, $\alpha = 0°$, $\beta = 30°$).

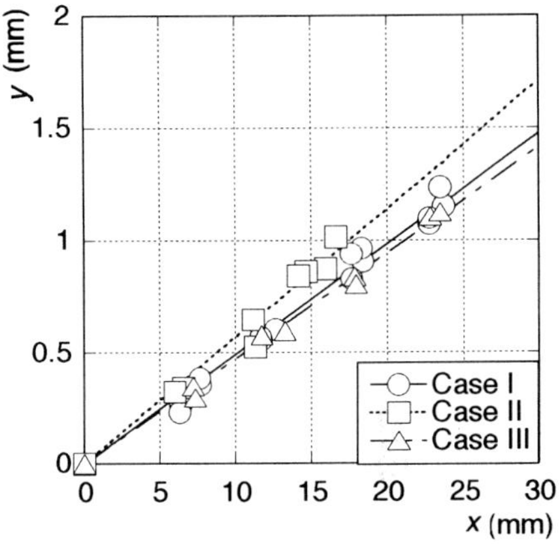

Figure 5. Comparison of the triple-point trajectories for Case I ($\alpha = 0°$, $\beta = 30°$), Case II ($\alpha = 10°$, $\beta = 40°$), Case III ($\alpha = 10°$, $\beta = 20°$) (M_s=1.34).

the triple-point trajectory angle for the cone case is smaller than that for the wedge case although the difference between the two is small. Furthermore it can be said that the trajectories predicted by the shock dynamics theory do not agree well with the experimental ones for both the cone and wedge cases under the present conditions.

In Suzuki & Sakamura (1995), it was shown that the shock-wave reflections from wedge models at a given angle of attack were determined by the effective wedge angle, $\beta - \alpha$ for the upper side of the wedge model and $\beta + \alpha$ for the lower. However, the local reflecting pattern of a shock wave from a cone at a finite angle of attack in a meridian plane may be influenced not only by the local angle between the shock-propagating direction and the cone ray, γ, but also by the local curvature of the cone surface, a cross-flow from windward to leeward, etc. In Fig. 5, the triple-point trajectories on the cone with $\beta = 30°$ at $\alpha = 0°$ (Case I), on the upper side of the cone with $\beta = 40°$ at $\alpha = 10°$ (Case II), and on the lower side of the cone with $\beta = 20°$ at $\alpha = 10°$ (Case III) are plotted. The symbols show the experimental results and the straight lines are given by the least-squares fitting of them. In all three cases, γ is equal to 30°. In this figure, the difference of the triple-point trajectory angle between case I and case III is negligible, but the differences between case II and the others are not. This finding implies that the local reflecting pattern of a shock wave from a cone at a finite angle of attack cannot be determined only by γ. Hence, we should take account of this fact when investigating the three-dimensional shock-wave reflection phenomena.

4. Conclusion

Results from preliminary experiments for a three-dimensional shock-wave reflection from a cone have been presented. The Mach number of the incident planar shock wave was 1.34. The main conclusions from the present study are as follows: (1) the angle of attack that causes the transition from Mach to regular reflection on the lower side is about 10° for a cone with a semi-apex angle of 30°, it is smaller than that obtained for a similar wedge case; (2) the trajectory angle of the triple point for the cone case is smaller than that for the wedge case; (3) the local reflecting pattern of a shock wave from a cone at a finite angle of attack cannot be determined only by the local angle between the shock-propagating direction and the cone ray.

Acknowledgement. The authors wish to express their appreciation to Prof. O. Igra of Ben-Gurion University of the Negev for his kind advice at an early stage of this study. They would also like to acknowledge the assistance of Mr. Y. Kakuda of Toyama Prefectural University for carrying out the experiments. Thanks are also due to Mr. Y. Usui of Toyama Prefectural University for his technical support to prepare the present paper using LaTeX 2_ε.

References

Ben-Dor G (1991) Shock Wave Reflection Phenomena, Springer, New York.

Han Z-Y, Yin X-Z (1992) Shock Dynamics, Kluwer, Dordrecht.

Hookham PA, Rosenblatt M, Takayama K, Watanabe M (1995) Interaction of a planar shock with a cone at an oblique angle: numerical simulation and experiment. In: Shock Waves @ Marseille, Brun R, Dumitrescu LZ (eds), Springer, Berlin, Vol IV, pp 169–174.

Milton BE, Archer RD (1995) Generalized concepts for the internal and external conical Mach reflection of moving shock waves. In: Shock Waves @ Marseille, Brun R, Dumitrescu LZ (eds), Springer, Berlin, pp 157–162.

Suzuki T, Sakamura Y (1995) Shock wave reflections from a cone. Trans Japan Soc Aero Space Sci 38:161–168.

Takayama K, Sekiguchi H (1977) An experiment on shock diffraction by cones. Rep Inst High Speed Mech, Tohoku Univ 36:53–74.

Calculation of focusing of weak shocks using geometrical shock dynamics

J. Cates, B. Sturtevant
Graduate Aeronautical Laboratories
California Institute of Technology
Pasadena, CA 91125 USA

Abstract: The equations of geometrical shock dynamics provide a valuable approximation for many shock-propagation problems. An improved finite difference formulation for geometrical shock dynamics has been developed, and is used to calculate the focusing of plane shock waves reflected from a parabola. The shock pressure jump and the path of the triple point along the shock front were compared to the experimental results of Sturtevant and Kulkarny (1976) for a range of Mach numbers from 1.1 to 1.3. The numerical calculations overestimate pressures near the focus, however, the location of the maximum pressure is accurately predicted for Mach numbers above 1.1. As expected with geometrical shock dynamics, the accuracy deteriorates for weaker shocks.

Key words: Geometrical shock dynamics, Shock focusing

1. Introduction

The equations of geometrical shock dynamics (Whitham 1957) are

$$\nabla \cdot \left(\frac{M}{A(M)} \nabla \alpha \right) = 0, \qquad M = \frac{1}{|\nabla \alpha|}. \tag{1}$$

$\alpha(\mathbf{x}) = a_0 t$ gives the shock position at time t, a_0 being the undisturbed sound speed. $A(M)$ is the Area-Mach-number relation, which gives the relation between the area A of the ray tube and the shock strength M. For a perfect gas, the Area-Mach-number relation is given by

$$A(M) = \exp\left[- \int_{M_0}^{M} \frac{M\lambda(M)}{M^2 - 1} dM \right], \tag{2}$$

where

$$\lambda(M) = \left(1 + \frac{2}{\gamma+1} \frac{1-\mu^2}{\mu}\right)\left(1 + 2\mu + \frac{1}{M^2}\right), \tag{3}$$

$$\mu^2 = \frac{(\gamma-1)M^2 + 2}{2\gamma M^2 - (\gamma - 1)}. \tag{4}$$

These equations form a hyperbolic, second-order partial differential equation describing wave motion of disturbances propagating on the shock front. Discontinuities, called *shock-shocks*, can form along the shock carrying a change in shock angle and Mach number which represent triple points. *Shock-expansions* can also form, such as when the shock diffracts around a sharp corner. At a solid boundary, the boundary condition requires the shock front to be perpendicular to the wall, therefore Mach reflection always occurs and regular reflection is not allowed.

In Cartesian coordinates x, y, Eq. (1) is

$$\frac{\partial(\sigma u)}{\partial x} + \frac{\partial(\sigma v)}{y} = 0, \tag{5}$$

where

$$M = \left[u^2 + v^2\right]^{-1/2}, \qquad \sigma \equiv \frac{M}{A(M)}, \tag{6}$$

and $u = \alpha_x$, $v = \alpha_y$ are the components of the gradient of α, As first noted by Whitham (1959), the equations are analogous to the full potential equation in steady supersonic flow. The extensive numerical methods developed for the supersonic potential equation can be directly adapted to geometrical shock dynamics (Schwendeman 1993).

2. Numerical method

For an arbitrary, non-orthogonal coordinate system defined by $\zeta = \zeta(x, y)$, $\eta = \eta(x, y)$, Eq. (5) can be written in conservative form as:

$$\frac{\partial}{\partial \zeta}\left(\frac{\sigma U}{J}\right) + \frac{\partial}{\partial \eta}\left(\frac{\sigma V}{J}\right) = 0, \tag{7}$$

where U, V are the contravariant velocity components given by

$$U = a_{11}\alpha_\zeta + a_{12}\alpha_\eta, \qquad V = a_{12}\alpha_\zeta + a_{22}\alpha_\eta, \tag{8}$$

and the grid parameters are

$$a_{11} = \zeta_x^2 + \zeta_y{}^2, \qquad a_{12} = \zeta_x\eta_x + \zeta_y\eta_y, \qquad a_{22} = \eta_x^2 + \eta_y^2, \tag{9}$$

$$J = \zeta_x\eta_y - \zeta_y\eta_x. \tag{10}$$

The Mach number is now given by

$$M = [U\alpha_\zeta + V\alpha_\eta]^{-1/2}. \tag{11}$$

In the solution procedure, we consider ζ to be the marching direction and assume all information is known at the ith and all previous levels. The problem is to advance the solution to the level $i + 1$ and obtain α.

At all levels, the quantities α_ζ, α_η are computed from

$$(\alpha_\zeta)_{i,j} = \frac{\alpha_{i,j} - \alpha_{i-1,j}}{\Delta\zeta}, \qquad (\alpha_\eta)_{i,j} = \frac{\alpha_{i,j+1} - \alpha_{i,j-1}}{2\Delta\eta}, \tag{12}$$

where a backward difference is used for the marching direction ζ. For half-points, the velocities are averaged between neighboring points which gives:

$$(\alpha_\zeta)_{i,j+1/2} = \frac{(\alpha_{i,j+1} - \alpha_{i-1,j+1}) + (\alpha_{i,j} - \alpha_{i-1,j})}{2\Delta\zeta}, \tag{13}$$

$$(\alpha_\eta)_{i,j+1/2} = \frac{\alpha_{i,j+1} - \alpha_{i,j}}{\Delta\eta}. \tag{14}$$

Once these quantities are known, U, V follow directly from Eq. (8) and M from Eq. (11).

Since the ζ direction is the marching direction, the ζ derivative term in Eq. (7) will be backward differenced as

$$\frac{\partial}{\partial\zeta}\left(\frac{\sigma U}{J}\right)_{i+1,j} \approx \frac{(a_1 - \theta b_1)\left\{\left(\frac{\sigma U}{J}\right)_{i+1,j} - \left(\frac{\sigma U}{J}\right)_{i,j}\right\} - \theta b_1\left\{\left(\frac{\sigma U}{J}\right)_{i,j} - \left(\frac{\sigma U}{J}\right)_{i-1,j}\right\}}{a_1\Delta\zeta_1 - \theta b_1(\Delta\zeta_1 + \Delta\zeta_2)}, \tag{15}$$

where

$$a_1 = (\Delta\zeta_1 + \Delta\zeta_2)^2, \qquad b_1 = (\Delta\zeta_1)^2. \tag{16}$$

The parameter θ controls the order: $\theta = 0$ gives first-order accuracy and $\theta = 1$ gives second-order accuracy (Shankar 1982).

The η derivative term in Eq. (7) will be written at the level $i+1$ to make the resulting scheme fully implicit:

$$\frac{\partial}{\partial\eta}\left(\frac{\sigma V}{J}\right)_{i+1,j} \approx \frac{1}{\Delta\eta}\left\{\left(\frac{\sigma V}{J}\right)_{i+1,j+1/2} - \left(\frac{\sigma V}{J}\right)_{i+1,j-1/2}\right\}. \tag{17}$$

To simulate viscosity, the analogy of density biasing (or often called artificial compressibility) will be used. The σ in Eq. (17) will be replaced by $\tilde{\sigma}$ defined to be

$$\tilde{\sigma}_{i+1,j+1/2} = (1-\nu)\sigma_{i+1,j+1/2} + \frac{1}{2}\nu\left(\sigma_{i+1,j+2m} + \sigma_{i+1,j+2m-1}\right), \tag{18}$$

with $m = 0$ when $V_{i+1,j+1/2} > 0$ and $m = 1$ when $V_{i+1,j+1/2} < 0$. The artificial velocity employs upwind differencing based on the characteristic signal propagation, which turns off the upwind differencing when the cross flow is elliptic-type (Shankar 1982). In this case, the artificial viscosity coefficient is given by

$$\nu_{i+1,j+1/2} = \mu\left[1 - \frac{a_{22}a^2}{V^2}\right]_{i+1,j+1/2} \tag{19}$$

with

$$\mu = 0 \quad \text{for} \quad \left(a_{22} - \frac{V^2}{a^2}\right)_{i+1,j+1/2} > 0 \quad \text{(elliptic cross flow)}, \tag{20}$$

$$= 1 \quad \text{for} \quad \left(a_{22} - \frac{V^2}{a^2}\right)_{i+1,j+1/2} < 0 \quad \text{(hyperbolic cross flow)}. \tag{21}$$

Given the α values at all previous levels, this formulation results in a set of nonlinear algebraic equations for α_{i+1} which are solved using Newton's method. In general, the artificial viscosity leads to a pentadiagonal system. For most cases, convergence on each step can be achieved while maintaining a tridiagonal system by neglecting derivative terms due to artificial viscosity, i.e., the derivative matrix is calculated as if $\nu = 0$. This often requires more iterations for convergence of each ζ step, but each iteration is significantly faster.

3. Shock-focusing results

The resulting numerical method was used to calculate the focusing of plane shock waves reflected from a parabola. For the initial condition of the calculation, the Mach number of the shock was approximated as the initial Mach number. The computed reflected shock fronts and Mach-number contours for a shock wave with initial Mach numbers $M_0 = 1.3$ and $M_0 = 1.1$ are shown in Fig. 1. The outline of the parabola is shown at the left of each figure, and the focus of the parabola lies on the centerline at $(x, y) = (0, 1)$.

Figure 1a shows an example of the behavior observed for a stronger shock, $M_0 = 1.3$. The initial disturbance from the corner of the parabola travels along the shock front until it reaches the centerline, where a shock-shock discontinuity forms on the shock to enforce the boundary condition. The effect of the discontinuity is clearly shown by the change in the angle of the shock front and the increase in Mach number. The path of the shock-shock propagating outward along the shock front represents the path of the triple point seen experimentally. As the discontinuity propagates outward, it weakens and the Mach number behind the shock-shock decreases.

For weaker shock waves, the shock-shock does not propagate outward, but is swept back toward the centerline as shown in Fig. 1b for $M_0 = 1.1$. Here the shock-shock briefly moves away from the centerline before being pushed back into the centerline. The strength of the shock-shock increases as it is pushed inward, producing a further increase in the centerline Mach number. The strength of the shock-shock continues to increase until the velocity of the shock-shock along the shock front is sufficient for the shock-shock to move outward again. For very weak shocks, the shock-shock is actually pushed to within several grid

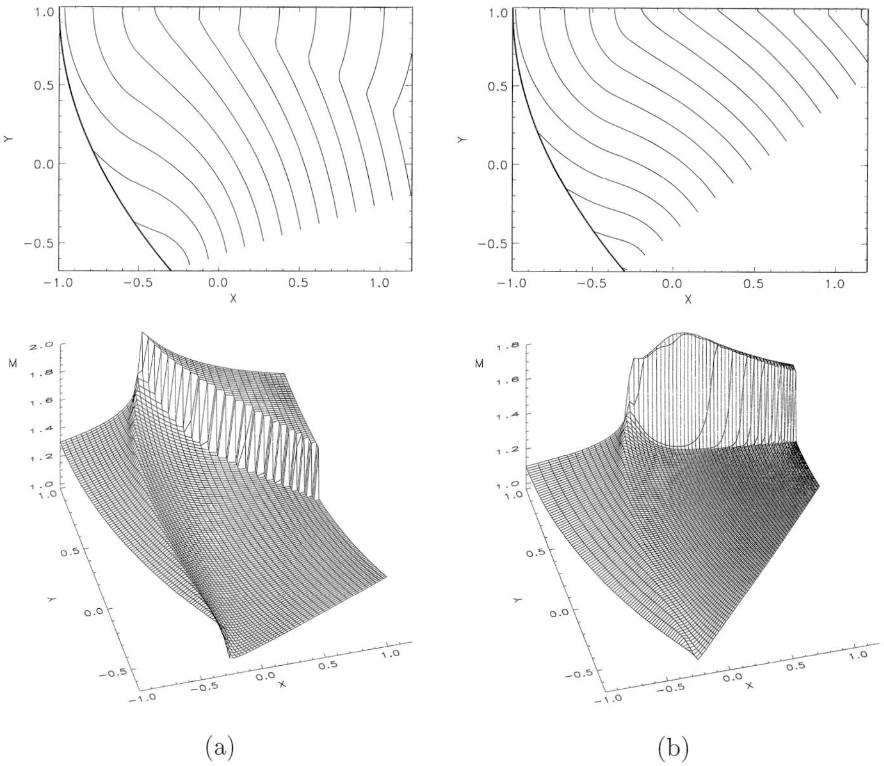

Figure 1. Shock fronts and Mach-number surface for reflection of planar shock waves from a parabola with initial Mach number (a) $M_0 = 1.3$ and (b) $M_0 = 1.1$.

points of the centerline, and the shock-shock path becomes dependent on the grid spacing. From comparison with experiment, one expects these cases to mark where experiment shows a transition to regular reflection and the front becomes crossed and folded.

In Fig. 2, the computed pressure ratio and shock-shock path are compared with experimental results. The experimental pressure ratio is the maximum pressure ratio, which outside the focal region is not the pressure directly at the shock. Within the focal region, the numerical results correctly predict the location of the maximum pressure for the stronger shocks. For the weaker $M_0 = 1.1$ case, the location of the initial peak agrees with the experimental results, but the additional rise in the Mach number due to the requirement in shock dynamics for Mach reflection at the centerline is artificial. The path of the

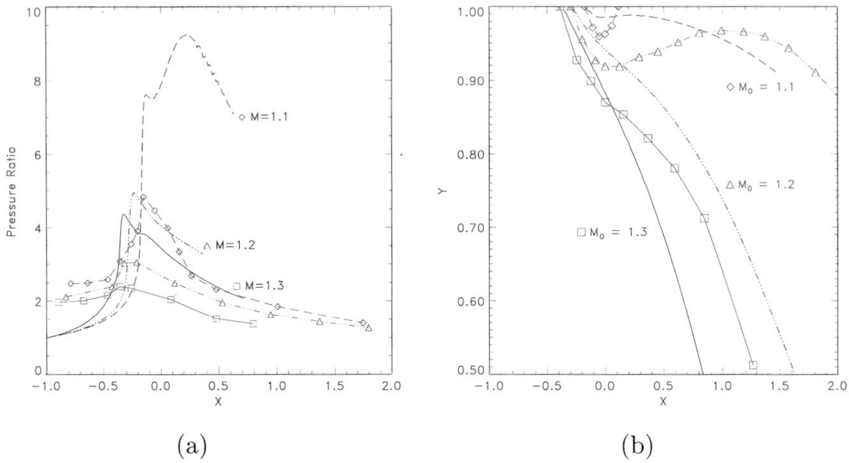

Figure 2. (a) Pressure ratio and (b) shock-shock path computed for reflection of planar shock waves from a parabola compared with experimental results of Sturtevant and Kulkarny (1976). Experimental pressure ratio is the maximum pressure ratio encountered, which outside the focal region is not the pressure directly at the shock.

shock-shock is well predicted for $M_0 = 1.3$, but for weaker shocks the agreement deteriorates. However, the shock dynamics approximation is able to capture the motion of the shock-shock being towards the centerline for weaker shocks.

The finite-difference method developed for geometrical shock dynamics is useful for efficiently computing problems such as shock focusing where strong shock-shock discontinuities occur. The numerical results qualitatively capture the major effects seen experimentally, however quantitative agreement is poor, especially for weaker shocks.

References

Schwendeman DW (1993) A new numerical method for shock wave propagation based on geometrical shock dynamics. Proc R Soc Lond A 441:331–341.

Shankar V (1982) Conservative full potential, implicit marching scheme for supersonic flows. AIAA J 20:1508–1514.

Sturtevant B, Kulkarny VA (1976) The focusing of weak shock waves. J Fluid Mech 73:651–671.

Whitham GB (1957) A new approach to problems of shock dynamics, part I: two-dimensional problems. J Fluid Mech 2:145–171.

Whitham GB (1959) A new approach to problems of shock dynamics, part II: three-dimensional problems. J Fluid Mech 5:369–386.

The shock reflection process over a non-straight surface

S. Kobayashi, T. Adachi
Department of Mechanical Engineering
Saitama Institute of Technology
1690 Fusaiji, Okabe-machi, Ohsato-gun, Saitama, 369-02 Japan
T. Suzuki
Department of Mechanical Systems Engineering
Toyama Prefectural University
5180 Kurokawa, Kosugi-machi, Imizu-gun, Toyama, 939-03 Japan

Abstract: Shock reflection over a non-straight surface has been investigated. The models used in the present experiment are ordinary circular cylindrical concave and convex wedges as well as step-wise simulations of the circular cylindrical wedges. The step height of the step-like wedges is constant and the equivalent radius of curvature is 40 mm. The reflected-wave structure has been photographed with a schlieren apparatus. The transition wedge angle is compared with the analytical formula based on acoustic theory.

Key words: Shock reflection, Wave structure, Unsteady reflection, Transition

1. Introduction

When a planar shock wave is reflected over a non-straight surface, the incidence angle varies as the shock proceeds and the flow field is truly nonstationary. As a result, such problems are less tractable than steady or pseudo-steady problems. In particular, self-similarity does not hold when a characteristic length is present such as the radius of curvature in the case of circular cylindrical wedges. The unsteady reflection over concave and convex circular cylinders has been investigated experimentally and theoretically (Heilig 1969, Ben-Dor et al. 1987) from various viewpoints. The authors, who have studied during the last decade the shock reflection phenomena over multi-guttered wedges and particulate layers (Adachi et al. 1992, Kobayashi et al. 1992), found that the minute reflected waves from such surfaces are useful in understanding the development of the ordinary reflected wave over a smooth surface as well as that of the precursory reflected wave over non-smooth surfaces. In the present paper, step-like wedges are used to simulate circular cylindrical wedges and to investigate the reflection over them.

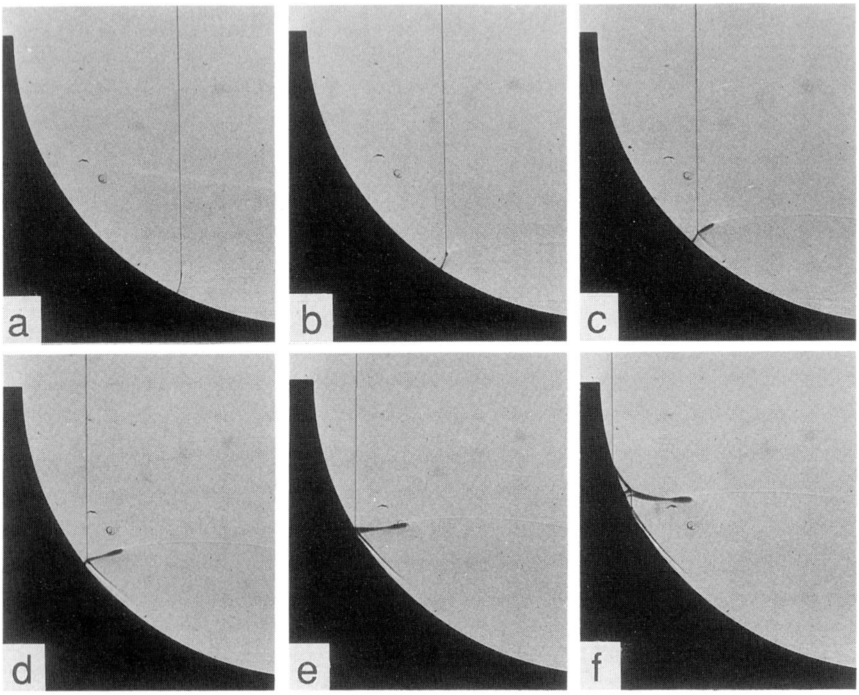

Figure 1. Reflection over a smooth concave wedge (r = 40 mm, M_i = 1.40).

2. Experimental apparatus and method

The experimental setup is the same as that used in our previous paper (Adachi et al. 1992). We used three models with different radii of curvature: r = 20, 30, and 40 mm for the smooth cylindrical wedges. The surface of the step-like wedges is composed of multiple steps with constant height s (1 mm or 2 mm) and an equivalent radius of curvature of 40 mm. The working gas is air and the condition ahead of the incident shock is room temperature and atmospheric pressure. The incident shock Mach number M_i is fixed at about 1.40. The wave structure has been photographed with a schlieren apparatus.

3. Results and discussion

Figure 1 shows shadowgraphs of shock reflection over a cylindrical concave surface. At an early stage, the reflected wave is too weak to be observed clearly. As the incident wave proceeds, a triple point and then a Mach stem appears. The

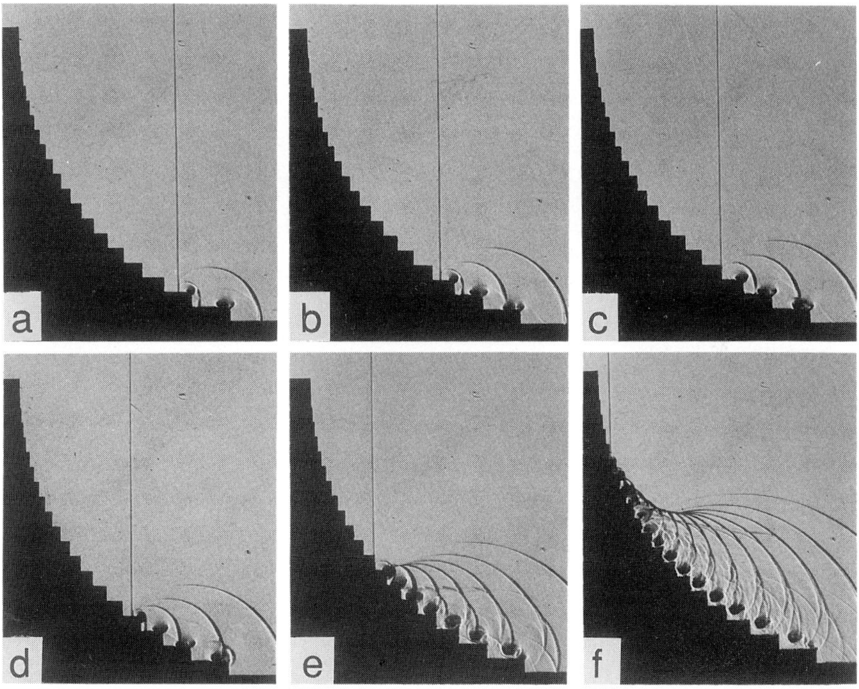

Figure 2. Reflection over a concave step-like model ($r = 40$ mm, $s = 2$ mm, $M_i = 1.41$).

reflected wave is accumulated in the neighborhood of the triple point (Figs. 1c–e). The triple point collides with the surface (Fig. 1e) and the TRR is formed in Fig. 1f. The corresponding reflection over a step-like wedge is shown in Fig. 2. A very weak wave is issued at each step and the "triple point" is barely observable. At an early stage, these weak waves accumulate in the neighborhood of the single triple point, and the triple point can get over the step. However, at a later stage (Figs. 2c–2f), the triple point cannot get over but collides with the step. The resemblance between Figs. 2f and 1f is remarkable. This is more evident in Fig. 3 for $s = 1$ mm. In Fig. 3a, the weak waves accumulate near the foot of the incident shock, which is similar to the accumulated reflected wave in Figs. 1c–e. The difference between the two experiments is that the Mach stem does not grow in the case of the step-like wedge. The reason that the reflected wave over a smooth concave wedge is only restricted near the reflection point is evident in Figs. 2f and 3b. That is to say, the reflected wave over a smooth concave wedge is dispersed away from the reflection point.

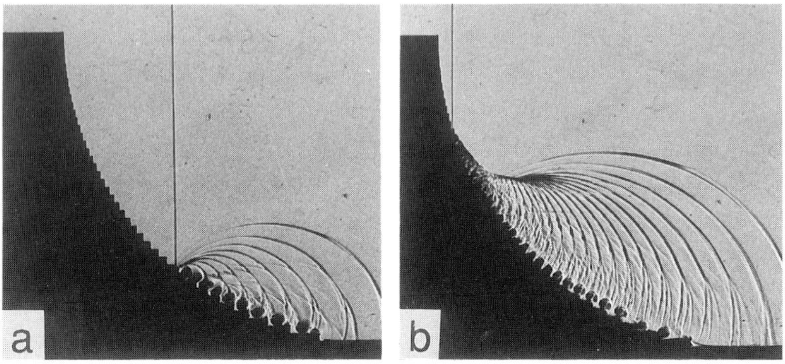

Figure 3. Reflection over a concave step-like model ($r = 40$ mm, $s = 1$ mm, $M_i = 1.41$).

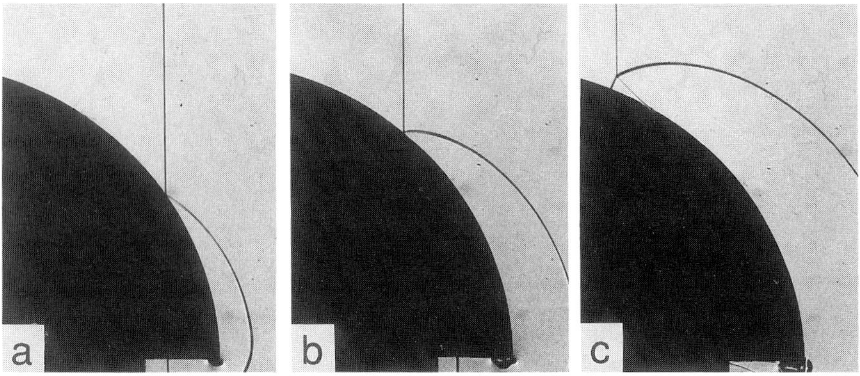

Figure 4. Reflection over a smooth convex wedge ($r = 40$ mm, $M_i = 1.40$).

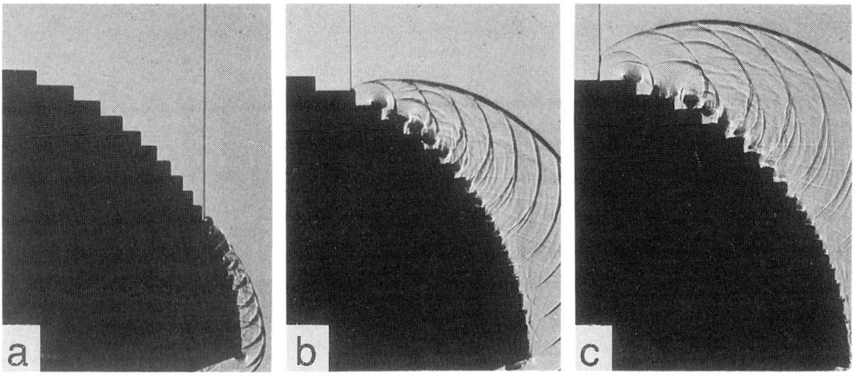

Figure 5. Reflection over a convex step-like model ($r = 40$ mm, $s = 2$ mm, $M_i = 1.41$).

Figure 4 shows the reflection configuration over a cylindrical convex surface. The reflection is RR in Figs. 4a and 4b, and MR in Fig. 4c. As the incident shock propagates (Fig. 4a to Fig. 4b), the angle of reflection increases. The transition from RR to MR has already taken place by Fig. 4c. Figure 5 is the corresponding reflection over a step-like wedge. Contrary to the reflection over the concave step-like wedge, at an early stage, the triple point cannot get over the step until the slope made by the neighboring corners decreases below some value (Fig. 5b). Hereafter, signals issued from each corner catch up with the incident shock near its foot. Consequently, these signals accumulate to form a Mach stem as shown in Fig. 5c. It is noted that the Mach stem is slightly slanted forward at the foot, which is similar to what appears in Fig. 4c.

This experiment suggests that RR cannot persist when the triple point can get over the step. The location of the triple point is the farthest point that the signal generated at the corner can reach. The signal generated at the corner propagates at the sound speed relative to the flow behind the incident shock. Therefore, we have an analytical formula for the transition wedge angle,

$$\theta_{tr} = \tan^{-1}\left\{M_i^{-2}\sqrt{[(\kappa-1)\,M_i^2+2]\,(M_i^2-1)\,(\kappa+1)^{-1}}\right\}, \qquad (1)$$

where κ is the specific heat ratio. When the slope made by the neighboring corners is less than θ_{tr}, the transition from RR to MR occurs. This formula has been derived for the smooth concave wedge as a triple point trajectory angle at glancing incidence (Ben-Dor et al. 1987). However, except for very weak shocks, as shown in Fig. 6 (Takayama et al. 1981), Eq. (1) agrees with the transition wedge angle for convex wedges with surface roughness. For a convex step-like wedge, the transition wedge angle from RR to MR is not determined by a unique value by nature but characterized by upper and lower values defined by two neighboring corners. Thus it is characterized by the angle defined by the corner the triple point can first get over and that by its previous corner. Since the step-like wedge is considered to have surface roughness, it is reasonable the present experimental results over step-like wedges agree with those over rough surfaces.

4. Conclusions

The shock reflection over a non-straight surface has been investigated. Two conclusions may be drawn from this study: (1) We can better understand the development of reflected waves over smooth convex and concave surfaces by considering the reflected wave structure over step-like models. (2) Except for very weak shocks, the analytical formula for the transition wedge angle based on acoustic theory agrees with experiments for convex wedges with surface roughness.

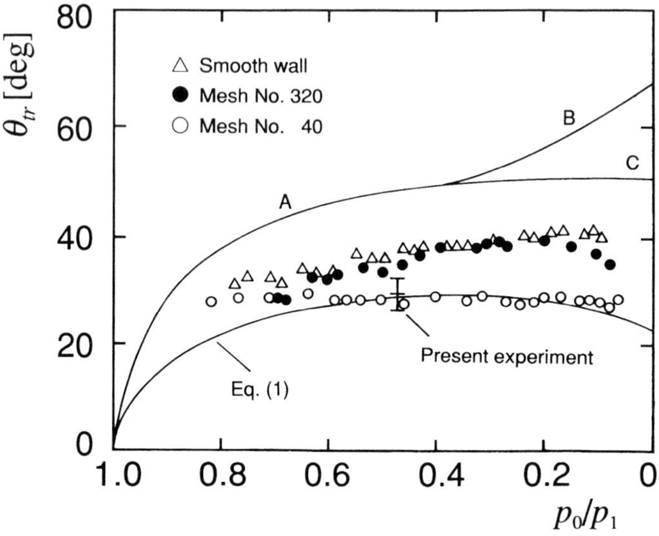

Figure 6. Transition wedge angle vs. inverse pressure ratio across the incident shock. The present experiment: $26° < \theta_{tr} < 32°$ for $s = 2$ mm, $29° < \theta_{tr} < 32°$ for $s = 1$ mm. AB: transition line for steady flow; AC: pseudo-steady flow.

References

Adachi T, Kobayashi S, Suzuki T (1992) An experimental analysis of oblique shock reflection over a two-dimensional multi-guttered wedge. Fluid Dyn Res 9:119–132.

Ben-Dor G, Takayama K, Dewey JM (1987) Further analytical considerations of weak planar shock wave reflections over a concave wedge. Fluid Dyn Res 2:77–85.

Heilig WH (1969) Diffraction of a shock wave by a cylinder. Phys Fluids Suppl 1:154–157.

Kobayashi S, Adachi T, Suzuki T (1992) Unsteady behavior of Mach reflection over a particulate layer. In: Proc 18th Int Symp Shock Waves, pp 239–244.

Takayama K, Ben-Dor G, Gotoh J (1981) Regular to Mach reflection transition in truly nonstationary flows — influence of surface roughness. AIAA J 19:1238–1240.

Oblique shock interactions and the von Neumann paradox

S. Čanić*
Department of Mathematics
Iowa State University, Ames, IA 50011 USA

B. L. Keyfitz†
Department of Mathematics
University of Houston, Houston, TX 77204-3476 USA

Abstract: We present a new theoretical approach to studying weak shock reflection in regimes where shock polar analysis is insufficient. The approach is based on examining the subsonic flow using monotone operator techniques. Our results reveal complicated flow structures in the zone of interaction of waves. Results predicting the region of high velocity gradients in the von Neumann reflection are reported. The precise type and the position of the singularity are described.

Key words: Riemann problem, Shock interactions, von Neumann reflection, Transonic shocks

1. Introduction

When a weak shock wave hits a wedge, different reflection patterns occur depending on the shock strength and the wedge angle. The type of shock interaction that occurs and the transition criteria between different types of reflection have been the subjects of many numerical and experimental studies. The most common theoretical approach is based on shock polar analysis. For certain small Mach numbers and small wedge angles this analysis fails to predict the type of reflection that takes place (the "von Neumann paradox"). The main reason for the inadequacy of this approach is that it assumes that the flow downstream of the reflected wave and near the point of interaction of waves is uniform. However, numerical and experimental studies suggest that in place of a triple-shock configuration dividing the region of uniform flow, as in the case of Mach reflection, for a certain range of parameters, the reflected wave contains a region with high velocity gradients corresponding to a compression wave near the Mach stem (von Neumann reflection).

---

*Supported in part by the Department of Energy under grant DE-FG02ER25220, and by the NSF under grant DMS-9403598.

†Supported in part by the Texas Advanced Research Program under grant 00365-2102-ARP, and by the Department of Energy under grant DE-FG03-94ER25222.

For this reason we have taken a different approach in order to understand the singularities in the equations that govern the flow, and to obtain a bifurcation diagram corresponding to different types of reflection. Besides shock polar analysis we have used monotone operator techniques to study the subsonic flow in the neighborhood of wave interactions. Equations that approximate the flow in the interaction zone have been obtained (Brio & Hunter 1992, Tabak & Rosales 1994) from the compressible gas dynamics equations. The asymptotic equation is the unsteady transonic small disturbance equation, also known as the 2D Burgers equation:

$$\begin{aligned} u_t + uu_x + v_y &= 0, \\ -v_x + u_y &= 0. \end{aligned} \tag{1}$$

This equation was also found (Morawetz 1994) to be an approximation to the full potential equation in the neighborhood of wave interactions. Here u and v denote scaled, normalized velocity components.

Instead of modeling the initial-value problem directly, we study a *two-parameter family of Riemann problems* associated with shock reflection. The initial data consist of two oblique shock waves separating constant states. The two shock waves correspond to the incident shock, $S_1 : x = ay$, and the reflected wave, $S_2 : x = -by$, at time $t = 0$ when the incident shock hits the wedge. The two parameters a and b are related to the angle of the wedge and to the strength of the incident shock. In principle, the solution to the shock-reflection problem could be recovered from this Riemann problem by determining the parameter b from the flow angle a and the conditions at infinity. The uniform states separated by S_1 and S_2 satisfy Rankine-Hugoniot conditions across the shocks, so these are *oblique-shock Riemann data.* In scaling the equations, one finds that the initial states can be given by $U_0 = (0,0)$, $U_1 = (1,-a)$, $U_2 = (1 + a/b, 0)$ (see Brio & Hunter 1992). Propagation of the singularity at the origin at $t = 0$ (the point where the two shocks meet) corresponds to a corner signal which gives rise to a region of subsonic flow.

We study self-similar solutions of this initial-value problem. The system of equations (1) written in self-similar coordinates changes type from hyperbolic to elliptic, corresponding to the regions where the flow is supersonic or subsonic. For $t > 0$, the second shock wave, S_2, either overtakes the incident shock, S_1, or the incident shock travels faster than the reflected wave, depending on the values of a and b. Now, the corner signal, which gives rise to a region where the flow is subsonic and therefore not necessarily uniform, may interfere with the reflected and/or the incident shock wave. In order to describe the flow for all choices of a and b we

(1) study hyperbolic wave interactions,

(2) study interaction of shock waves with the subsonic region, and

(3) obtain a bifurcation diagram in (a,b)-space describing different types of solutions, based on the type of the hyperbolic interaction and on the free-boundary solution.

In Section 2 we summarize results (Canic & Keyfitz 1995) on hyperbolic wave interactions relevant to the study of von Neumann reflection. In Section 3 we present results about the subsonic flow that suggest new, interesting structures in the region of wave interaction. Based on the theory given in Sections 2 and 3, we conjecture the structure of the flow in von Neumann reflection. The bifurcation diagram describing other types of reflection can be found in Canic et al. (1994).

2. Hyperbolic wave interactions

In self-similar coordinates ($\xi = x/t, \eta = y/t$) equation (1) becomes

$$\begin{aligned} (u-\xi)u_\xi - \eta u_\eta + v_\eta &= 0, \\ -v_\xi + u_\eta &= 0. \end{aligned} \tag{2}$$

This system is hyperbolic outside and elliptic inside the parabola P_u: $\xi + \eta^2/4 = u$. This parabola replaces the sonic circle of the Euler equations of compressible gas flow. Finding a self-similar solution of the initial-value problem described above reduces to solving a boundary-value problem in physical space (ξ, η), with data given on a curve sufficiently far from the origin (corresponding to data at infinity at $t = 0$). Because the coefficients of system (2) depend explicitly on ξ and η, this system is not standard for the hyperbolic theory of conservation laws. For example, the characteristics in the hyperbolic region, linearized about a constant state, are straight lines tangent to the sonic parabola.

Hyperbolic wave interactions occur for values of a and b for which either the two shock waves, S_1 and S_2, intersect in the hyperbolic region, or the incident shock meets the wall before interacting with either S_2 or the subsonic region. In these situations, new waves are generated as solutions of a quasi-one-dimensional Riemann problem based at the point of interaction, Ξ_I. In Canic & Keyfitz (1995) we prove the following:

Theorem 1 *Given Riemann data (U_0, U_2) at Ξ_I with hyperbolic states U_0 and U_2, let $D(U_0)$ be the downstream wave locus of U_0. Then there are three regions, R_1, R_2, and R_3, in the complement of $D(U_0)$ such that if U_2 belongs to any of the three regions, there exists a unique Riemann solution between U_0 and U_2, qualitatively different for each region. If U_2 belongs to the complement of $R_1 \cup R_2 \cup R_3$, there may be none, one, or two admissible solutions.*

More details about the structure of quasi-one-dimensional Riemann solutions can be found in Canic & Keyfitz (1995) and Canic et al. (1994).

3. Shock interaction with the subsonic region

In order to understand the interaction of a shock with the subsonic region we need to solve a free-boundary problem for a degenerate elliptic equation (2), where the position of a transonic shock is determined by the Rankine-Hugoniot conditions

$$\begin{aligned} \xi'(\eta) &= -\eta/2 \pm \sqrt{\xi + \eta^2/4 - (u+u_0)/2}, \\ \xi'(\eta) &= (v_0 - v)/(u_0 - u), \end{aligned} \tag{3}$$

with $\xi(\eta_*) = \xi_*$ corresponding the point where the shock meets the sonic curve; $\xi = \xi(\eta)$ is the position of the transonic shock, and u and v are the limits of the velocity components along the transonic shock (see Fig. 1a).

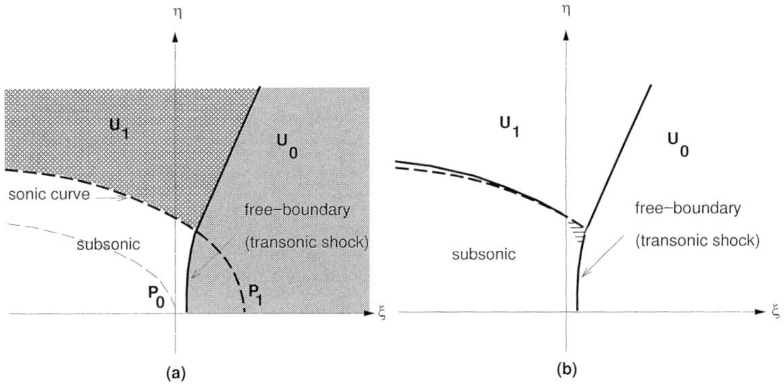

Figure 1. Subsonic boundary-value problems.

In an attempt to capture possible singularities, we first studied a fixed boundary-value problem based on an approximate position of the transonic shock (Canic & Keyfitz 1996a, Canic & Keyfitz 1996b). For this, it is sufficient to solve a second-order, quasi-linear, degenerate elliptic PDE obtained from (2):

$$\left((u-\rho)u_\rho + \frac{u}{2}\right)_\rho + u_{\eta\eta} = 0, \quad \text{where} \quad \rho = \xi + \frac{\eta^2}{4}, \tag{4}$$

and then recover v using the fact that $v = 0$ at the wall. Our results reveal interesting solution structures, as follows.

- The *asymptotic behavior* of a solution is described by the following theorem, proved in Canic & Keyfitz (1996a).

Theorem 2 *If u is constant along the sonic boundary, then there are exactly two possibilities for the asymptotic behavior of a solution at the sonic boundary. Either $u \in C^1$ up to the boundary, and $u_\rho = 1/2$, or $u \in C^\alpha$, where $\alpha \leq 1/2$, and $u = \sqrt{\rho_M - \rho}\, w$, where $w \neq 0$ at the sonic line, $\rho = \rho_M$.*

• *Global existence* of a subsonic solution for an approximate position of the Mach stem is proved in Canic & Keyfitz (1996a) and Canic & Keyfitz (1996b). Using monotone operator techniques adapted for this degenerate elliptic equation we proved that the problem allows two qualitatively different types of solutions, depending on the type of boundary data given along the transonic shock. One solution is *smooth* up to the boundary, another has a *square-root singularity* along the sonic line.

• Another interesting feature of our solution is that the regular subsonic solution (u, v) matches smoothly with the supersonic region, while the singular solution does not (Canic & Keyfitz 1996b). The singular case would require a weak shock wave emerging from the square-root singularity to match the subsonic with the supersonic flow (see Fig. 1b). The existence of a solution of this free-boundary problem has not yet been proved.

• Finally, the *curvature* of the Mach stem is determined by the type of boundary data along the stem. Namely, the first Rankine-Hugoniot condition implies

$$u' = 2\rho'(1 - 2\rho''), \quad \text{where} \quad \rho(\eta) = \xi(\eta) + \eta^2/4. \tag{5}$$

Here u' denotes the derivative of u along the transonic shock $\xi = \xi(\eta)$. Therefore, if the shock steepens monotonically to vertical at the axis, (i.e., $\xi' > 0$), then the shock is convex, (i.e., $\xi'' > 0$ or $\rho'' > 1/2$), if and only if $u' < 0$. Moreover, the higher the gradient of u, the bigger the curvature of the free boundary. Therefore, in order to have a monotonically increasing, convex Mach stem, as observed in the experiments, the data for the approximate free boundary has to be decreasing. By Theorem 2, the solution u must then have a square-root singularity at the point where the Mach stem meets the sonic curve. This argument proves the following:

Theorem 3 *Assume that the incident shock S_1 does not interact with another shock before meeting the sonic curve, and that the Mach stem $\xi = \xi(\eta)$ is an increasing, convex function. Then the subsonic solution must have a square-root singularity at the point where the Mach stem crosses the sonic curve.*

3.1. A conjecture for von Neumann reflection

We conjecture that the following configuration of S_1 and S_2 leads to von Neumann reflection. The incident shock S_1 between the uniform states U_0 and U_1 interacts with the region where U_1 becomes subsonic, giving rise to a free-boundary problem with a convex Mach stem. The flow at the sonic curve has a square-root singularity, which, in turn, gives rise to a weak shock emerging from the singular region. The shock S_2, separating the states U_1 and U_2, interacts with the region where U_2 becomes subsonic, P_2, leading to a free-boundary problem for another transonic shock. The shock generated by the square-root singularity catches up with the transonic shock continuing from S_2 giving rise to the flow pattern shown in Fig. 2.

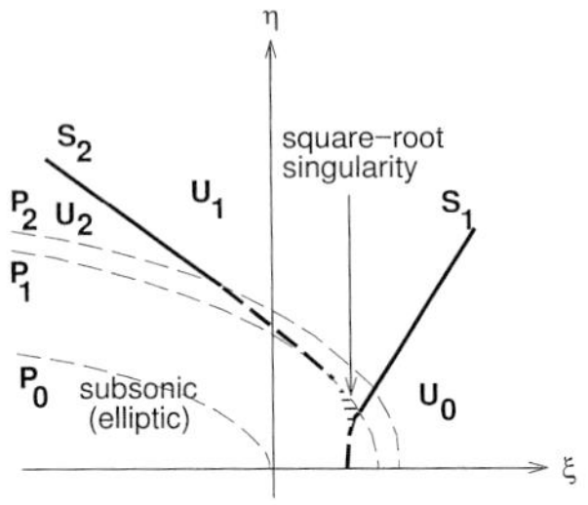

Figure 2. Conjectured structure for von Neumann reflection.

4. Final Remarks

Our results about the square-root behavior of the horizontal velocity component, u, seem to agree with the numerical results described in Tabak & Rosales (1995). Their numerical simulations indicate that the vertical velocity component, v, exhibits a stronger singularity. In the course of our study to prove the existence of a solution of the free-boundary problem we have come across some indications that this might indeed be the case. Nevertheless, this problem is still open.

References

Brio M, Hunter JK (1992) Mach reflection for the two-dimensional Burgers equation. Physica D 60:194–207.

Canic S, Keyfitz BL (1995) Quasi-one-dimensional Riemann problems and their role in self-similar two-dimensional problems. (In preparation)

Canic S, Keyfitz BL (1996a) An elliptic problem arising from the unsteady transonic small disturbance equation. J Diff Equations (to appear).

Canic S, Keyfitz BL (1996b) A smooth solution for a keldysh type equation. Comm PDE 21:319–340.

Canic S, Keyfitz BL, Wagner DH (1994) A bifurcation diagram for oblique shock interactions in the unsteady transonic small disturbance equation. In: Proc 5th International Conference on Hyperbolic Problems, Theory, Numerics and Applications (to appear).

Morawetz CS (1994) Potential theory for regular and Mach reflection of a shock at a wedge. Comm Pure Appl Math XLVII:593–624.

Tabak EG, Rosales RR (1994) Weak shock focusing and the von Neumann paradox. Physics of Fluids A 6(5):1874–1892.

Tabak EG, Rosales RR (1995) Private communication.

Shock-wave interactions at impingement of impulsive supersonic jet upon an obstacle

T. V. Bazhenova, V. V. Golub, A. M. Shulmeister,
T. A. Bormotova, S. B. Bazarov
High Energy Density Research Center
IVTAN, Russian Academy of Sciences
Izhorskaya, 13/19, Moscow, 127412 Russia

Abstract: The unsteady interaction between an impulsive jet and a flat plate was investigated experimentally and numerically. At small distances between the nozzle and plate, the reflected wave plays an important role, bringing about a rise in pressure. At great distances, the head part of the impulsive jet plays the decisive role; on the plate near the axis of the jet, pressure and gas density rise and at the region of the vortex ring the density drops.

Key words: Impulsive jet, Vortex, Reflection

1. Introduction

The primary stage of an impulsive supersonic jet is characterized by the formation of shock waves and vortex rings. A shock-wave system similar to the barrel structure in underexpanded jets develops in the gas jet with a starting shock wave ahead of the outflowing gas (Golub 1994). The present paper is devoted to the experimental and numerical investigation of the unsteady interaction between an impulsive jet and a flat plate.

2. Experimental set-up

A sonic nozzle with diameter of critical cross-section $d_* = 4$ mm was installed at the end of the shock tube, the plate being mounted in front of the nozzle (at the dimensionless distances $\bar{L} = L/d_* = 6$ to 50. The shock tube and vacuum chamber were filled with nitrogen. The initial temperature of the gas was room temperature. The Mach number M_0 of the incident shocks was varied between 2.0 and 4.4. The pressure P_∞ of the ambient gas in the vacuum chamber was between 4 kPa and 50 kPa. The pressure ratio $n = P_a/P_\infty$ was varied from $n = 8$ to 74, where P_a is the pressure at the nozzle exit and P_∞ is the pressure of the ambient gas. The flow was visualized by means of the shadow device and interferometer with a ruby laser. Five pressure transducers and five thin-film temperature gauges were mounted on the plate.

3. Results

The interferogram and sketch of the interaction between the impulsive supersonic jet and the obstacle are shown in Fig. 1. The starting shock wave SW1

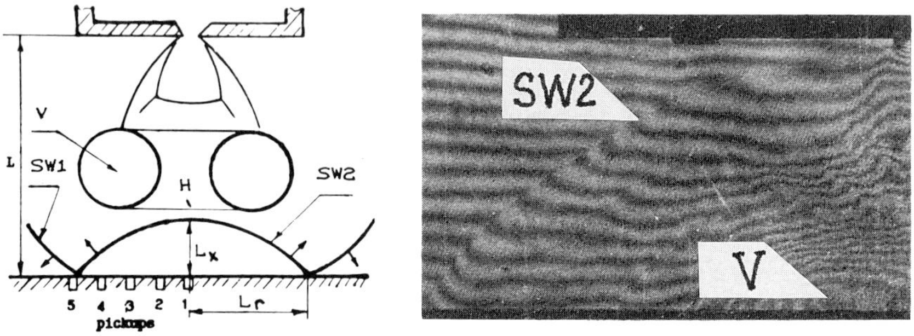

Figure 1. Interaction between the impulsive supersonic jet and the obstacle: sketch (left) and interferogram (right) with $M_0 = 2.75$, $\bar{L} = 10$, and $t = 114$ μs.

(Fig. 1, left) propagating in front of the hot gas and outflowing from the nozzle reaches the plate and is reflected by it, forming shock wave SW2 moving towards the nozzle. The head part of the impulsive jet H and the vortex V are shown in the sketch. The interactions between the high-enthalpy jet flow and vortex V with the reflected shock S2 and with the plate were observed in the later stages of development. In Fig. 1 (right) we present one-half of the axisymmetric interferogram. The vortex ring V touches the plate at this stage, with the reflected shock wave SW2 moving back to the nozzle.

In order to obtain the density distribution, the axisymmetric interferometric fringe distributions were evaluated numerically by solving Abel's integral equation. Since we used finite-width fringe interferometry, there was no need to know the initial values of the density on the shock front. The numerical processing of interferograms is based upon the solution of Abel's integral equation:

$$S(x,y) = 2K/\lambda \int_y^R \frac{(\rho(x,y) - \rho_0)r}{\sqrt{r^2 - y^2}} dr \tag{1}$$

which can be rewritten as an operator formula: $S(x,y) = [A\rho](x,y)$ assuming an axisymmetric jet density field. Here $S(x,y)$ is the fringe shift in the interferogram, K is the Gladstone-Dale constant, λ is the ruby-laser wavelength, x is the coordinate along the jet axis, y is the transverse coordinate, and r is the jet radius at the present cross-section. Abel's equation may be rewritten as:

$$S(y_j) = \frac{RK}{N\lambda} \sum_{i=j}^{N} (\rho_j - \rho_0) a_{ij} \tag{2}$$

where a_{ij} are Pirs's coefficients.

$$\rho(x, y) = [AA^{-1}S](x, y) \tag{3}$$

This formula is precisely the inversion formula of Radon. The density distribution along the symmetry axis of the impulsive jet after impingement was constructed. Figure 2 presents the spatial density distribution for $t = 114\ \mu$s, $L = 40$ mm, and $M_0 = 2.75$. The jet density is normalized by the ambient

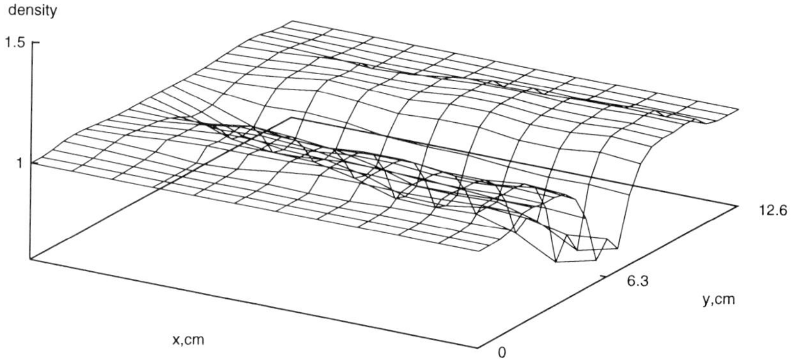

Figure 2. The spatial density distribution for $t = 114\ \mu$s, $L = 4$ cm, $M_0 = 2.75$; $x = 0$ denotes the plate position, with the symmetry axis $y = 6.3$.

gas density ρ_∞ and the coordinates x and y are given in centimeters. At some distance from the axis the density decreases to $0.8\rho_\infty$. Possibly the decrease of the density is related to the interaction of the vortex V with the plate.

The heat flux $q(t)$ into the wall during the impingement process was calculated using the wall temperature history $T(t)$ measured by the thin-film gauges:

$$q(t) = \frac{\lambda}{\sqrt{a\pi}} \int_0^t \frac{dT(t)}{d\tau} \frac{1}{\sqrt{t-\tau}} d\tau \tag{4}$$

where a and λ are coefficients of the thermal conductivity and diffusivity. The pressure and heat-flux histories are shown for two distances between the nozzle exit and the plate, $\bar{L} = 12$ and 38 (Figs. 3a and 3b). The arrival of SW1 at the plate is clearly traced by the initial rise of the pressure recorded by the center pick-ups (Fig. 3a).

After the SW1 reflection process the head part of the impulsive jet H begins to interact with SW2 and after interaction impinges on the plate. The unsteady breaking of the gas jet on the plate leads to the second rise of pressure and heat flux (Fig. 3b). The heat flux rises at the moment of high-enthalpy jet interaction with the plate. The value of the heat flow at $\bar{L} = 38$, $t = 700\ \mu$s rises to $q = 3 \times 10^5$ W/m^2.

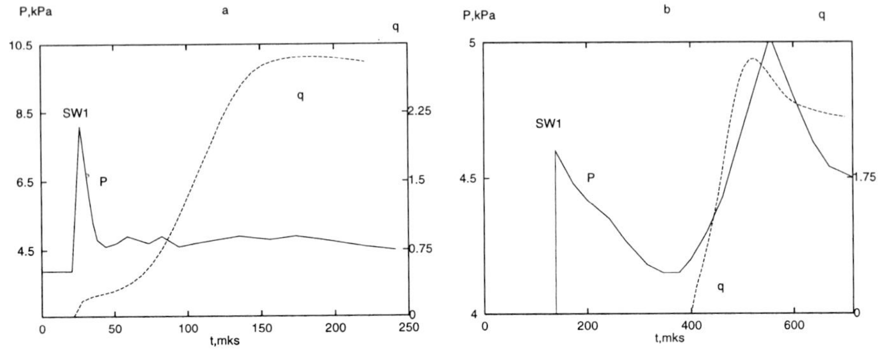

Figure 3. The pressure and heat-flux histories at the obstacle, $M_0 = 4.4$: (a) $\bar{L} = 12$; and (b) $\bar{L} = 38$; q is normalized to 10^{-5} W/m^2.

4. Comparision with the numerical simulation

Numerical simulation was made by solving Euler equations by Godunov's method. The experimental and computed normalized distributions of the pressure and density near the plate are compared. Figure 4a shows the density distribution along the symmetry axis and Fig. 4b shows the distribution along the plate ($M_0 = 2.7$, $\bar{L} = 10$, $t = 114$ μs). As seen in Fig. 4a the calculated density near the plate is in agreement with the experimental data. The density distribution along the plate shows that the main features of the process near the plate are presented both in the calculated data and the experimental data. In the center of the plate the density is slightly increased; at some distance from the center there exists a circular rarefaction zone surrounded by the circular zone of elevated density. This distribution may be interpreted as a result of interaction of the shock wave SW2 and vortex ring V with the plate. The calculated pressure records are in qualitive agreement with the experimental ones.

5. Discussion

The calculated time history of the normalized pressure, density, and gas temperature near the center of the plate is shown in Fig. 5. The first peak on the pressure curve corresponds to SW1 interaction with the plate. It is remarkable that the gas temperature rises slightly at the reflected shock front. Later, the high-enthalpy flow of the jet stagnates near the plate and the gas temperature rises to the value of the stagnation temperature behind the reflected shock wave in the shock tube and then rises higher. Due to unsteady deceleration, the gas temperature near the plate is the higher than the stagnation temperature; this may account for the redistribution of the kinetic energy in nonstationary flows (Stanukovich 1971).

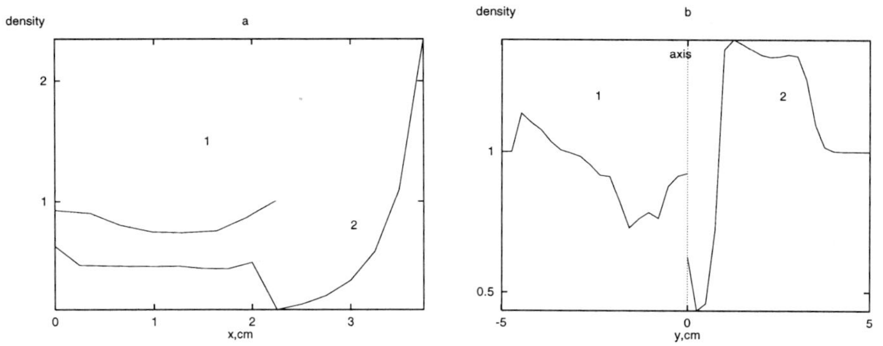

Figure 4. Density distribution along (a) the symmetry axis and (b) the plate; $M_0 = 2.75$, $t = 114\ \mu$s, $L = 4$ cm. (1) corresponds to experiment and (2) to calculation.

Zeldovich and Stanukovich (1947) considered the problem of nonstationary flow interaction with the wall. The flow is generated by the sudden outflowing of gas with stagnation temperature T_{00} through the orifice in the resevoir, mounted opposite the reflecting wall. Using the self-similar solution of the gasdynamic problem for the perfect gas the flow parameters after the interaction with the wall were obtained. The starting shock wave S1 was not taken into account. The one-dimensional case was considered. The nonstationary flow velocity u_0 is equal to $u_0 = 2c_{00}/(\gamma+1)^2$ where c_{00} is the sound velocity at the stagnation temperature, and $\gamma = c_p/c_v$ is the adiabatic index. The temperature T near the reflecting wall after nonstationary flow impingement was found from:

$$c_v T = \frac{u_0^2}{2}\left[\frac{(\gamma^3+\gamma+2)}{(\gamma^2+1)(\gamma+1)}\right]^2 \tag{5}$$

where c_v is the specific heat capacity. The gas-temperature value after impingement related to the stagnation temperature T_{00} is equal to

$$T/T_{00} = [2\gamma/(\gamma-1)]\left[\frac{\gamma^3+\gamma+2}{(\gamma^2+1)(\gamma+1)}\right]^2 \tag{6}$$

In our case $\gamma = 1.4$ which gives a maximum value T/T_{00} equal to 6. Numerical calculation gives $T/T_{00} = 1.25$. The difference is natural because Zeldovich estimation did not take into account the space distribution of the flow energy in the axisymmetric case.

6. Conclusions

(1) At small distances between the nozzle and plate, the reflected wave SW2 plays an important role, bringing about the rise of temperature and pressure

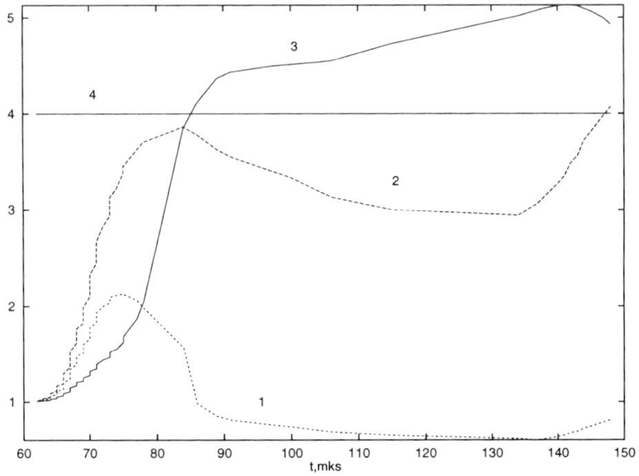

Figure 5. The pressure and temperature histories at the obstacle on the jet axis, $\bar{L} = 10$, $M_0 = 2.75$, (1) corresponds to density, (2) to pressure, (3)to gas temperature, and (4)to stagnation temperature.

(thus, at $\bar{L} = 12$ the pressure in the plate increases 1.5 times). (2) At great distances, the head part of the impulsive jet plays the decisive role. On the plate near the axis of the jet, pressure and temperatures rise and at the region of vortex ring the pressure drops (thus, at $\bar{L} = 38$ the pressure on the plate drops twofold). (3) The gas temperature at the wall at the second stage exceeds stagnation temperature. The increase of the gas temperature at the second stage of interaction leads to an increase in heat flow to the wall.

Acknowledgement. The present research was supported by Grant Nos. 94-01-00218-a and 94-01-01461-a of the Russian Fundamental Research Foundation.

References

Golub VV (1994) Development of shock wave and vortex structures in unsteady jets. Shock Waves 3:279–285.

Stanukovich KP (1971) Unsteady Flows of Continous Media, Nauka Press, Moscow (in Russian).

Zeldovich YB, Stanukovich KP (1947) The reflection of the detonation wave. Doklady Academy Nauk, USSR, 55(7).

Application of the principle of minimum entropy production to the regular-to-Mach-reflection transition in steady flows

H. Li, G. Ben-Dor
Pearlstone Center for Aeronautical Engineering Studies
Department of Mechanical Engineering
Ben-Gurion University of the Negev, Beer Sheva, Israel

Abstract: A new approach based on the principle of minimum entropy production for better understanding the regular-to-Mach-reflection transition in steady flows is presented. To the best of our knowledge the new proposed transition criteria agree with all the available experimental results.

Key words: Minimum entropy production, Shock reflection, Steady flows

1. Introduction

The present understanding regarding the regular $\leftrightarrow$ Mach-reflection transition in steady flows was established by Hornung and Robinson (1982). Since very few experimental works regarding this process were conducted, there were no reasons to discard their conclusions. However, in a recent experimental study which was conducted by Chpoun et al. (1995), results which completely contradict those of Henderson and Lozzi (1975), Hornung and Robinson (1982), and Hornung et al. (1979) were obtained. Hence, there is a need to modify the existing transition criterion to agree with all the available (including the new) experimental results.

Evidently the conservation equations are not sufficient to determine the physical solution which actually takes place. In order to determine the physical solution out of a variety of possible mathematical solutions the irreversibility of the thermodynamic process has to be accounted for. As postulated by Glansdorff and Prigogine (1971) the process must fulfill the principle of minimum entropy production. This principle, which is an a posteriori axiom, has been found to accurately predict the stability of a wide range of phenomena. Since a steady flow field can be regarded as a non-equilibrium steady state of thermodynamics, the principle of minimum entropy production can be adopted to better understand, from a physical point of view, phenomena in which, from a mathematical point of view, multiple solutions are possible.

2. Present study

The principle of minimum entropy production was applied to oblique shock-wave reflections in steady flow. Details of the analysis can be found in Li and Ben-Dor (1996). In the following only the final results are presented. Based on their analysis it was concluded that for given values of γ and M_0 the physical solution must satisfy the condition that $\beta \leq \beta^*$, where, β^* is the angle of incidence between the flow and the deflecting shock wave.

2.1. Application to supersonic-flow deflection

Consider a problem in which a uniform supersonic flow having a Mach number M encounters a wedge having an angle θ. The flow negotiates the wedge by means of a straight attached shock wave. The angle of incidence between the flow and the deflecting shock wave is β. It is well known that the deflecting shock wave is attached provided $\beta \leq \beta_{ext}$. The curves β^* (solid line) and β_{ext} (dashed line) vs. M, for $\gamma = 1.4$, are plotted in the (M,β)-plane shown in Fig. 1. Note that both β^* and β_{ext} approach asymptotic values as $M \to \infty$. $\beta^*(M \to \infty) = 45°$ and β_{ext} $(M \to \infty) = 67.8°$. It is evident from Fig. 1 that the two curves intersect at a point marked by Q. It corresponds to $M_Q = 1.80$ and $\beta_Q = 65.7°$. The shadowed area in Fig. 1, for which $\beta \leq \beta^*$, indicates the domain in the (M,β)-plane in which possible mathematical solutions are physical. An inspection of Fig. 1 indicates the following: (1) for $1 < M < M_Q$ there is a narrow domain in which $\beta_{ext} < \beta < \beta^*$, thus, a "strong" shock solution is physically possible there; (2) for $M > M_Q$ there is a domain in which $\beta^* < \beta < \beta_{ext}$, thus in spite of the fact that this domain lies in the "weak" solution domain, a solution in this domain is aphysical.

It is very important to note here that the facts that the "strong" solution has never been observed in the domain mentioned in (1) and that "weak" solutions have been observed in the domain mentioned in (2) in wind-tunnel experiments should not be used to discard the conclusions mentioned in (1) and (2). This is due to the fact that the analytical results shown in Fig. 1 were obtained under the assumption of "free boundaries" while actual experiments in wind tunnels do not satisfy this constraint.

2.2. Application to regular reflection (RR)

The wave configuration of a steady RR is shown in Fig. 2a. The supersonic flow in state (0) passes obliquely through the incident shock wave i. While passing through i it is deflected by an angle θ_1 and its flow Mach number changes to M_1. The flow in state (1) is then re-deflected by the reflected shock wave r, to become again parallel to the reflecting surface, hence $\theta_2 = \theta_1$. The flow Mach number behind the r is M_2. The angles of incidence are ϕ_1 and ϕ_2.

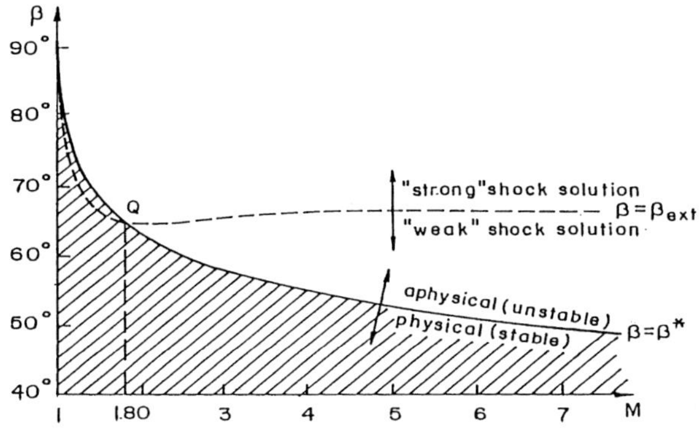

Figure 1. Domains of "strong" and "weak" solutions of the supersonic-flow deflection process and regions of physical and aphysical solution in the (M, β)-plane.

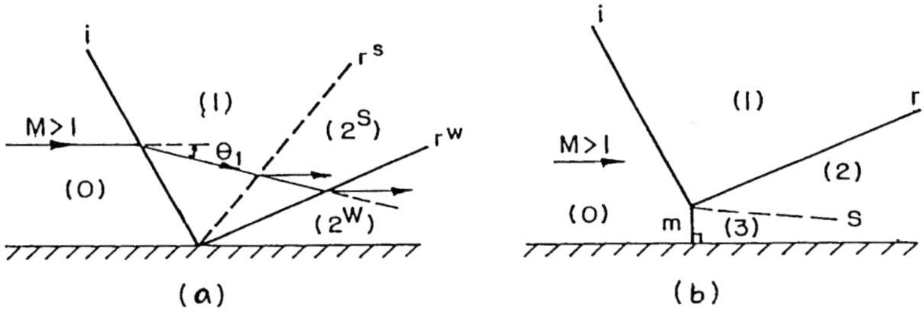

Figure 2. (a) The wave configurations of the "weak" and the "strong" regular reflections (RR). (b) The wave configuration of the Mach reflection (MR).

In order to eliminate the aphysical RR solution out of the two possible mathematical solutions (shown in Figs. 2a and 2b), the principle of minimum entropy production was applied. Owing to the analogy between the RR and the flow-deflection processes, the results shown in Fig. 1 can be used here too. Figure 3 is a reproduction of Fig. 1 for the case of RR. Note the changes in the vertical axis from β to ϕ_2 and in the horizontal axis from M to M_1. Similar to the notation used in Fig. 1, the shadowed area in Fig. 3, which corresponds to $\phi_2 \leq \phi_2^*$, indicates the domain in the (M_1,ϕ_2)-plane in which possible mathematical RR solutions are physical. An inspection of Fig. 3 indicates the following: (1) for $1 < M_1 < M_1^Q$ there is a narrow domain in which $\phi_1^* < \phi_2 < \phi_2^{ext}$, thus a "strong" RR is physically possible there; (2) for $M_1 > M_1^Q$ there is a domain in which $\phi_2^* < \phi_2 < \phi_2^{ext}$, thus in spite of the fact that this domain lies in the "weak" RR domain, an RR solution in this domain is aphysical.

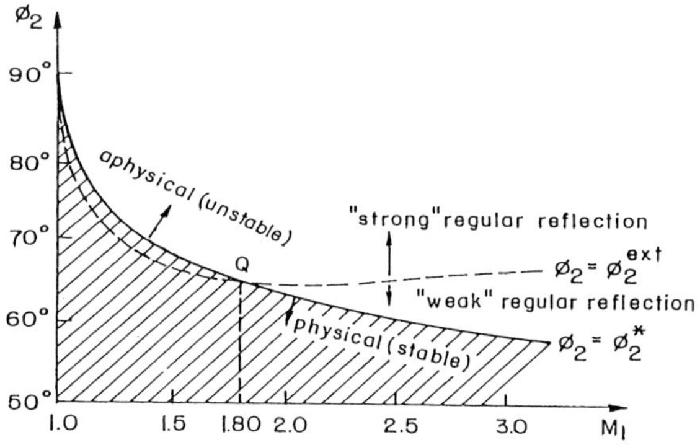

Figure 3. Domains of "strong" and "weak" solutions of a regular reflection and regions of physical and aphysical solutions in the (M_1, ϕ_2)-plane. (Note that M_1 and ϕ_2 are analogous to M and β in Fig. 1.)

The conclusion mentioned in (2) will be discussed following the introduction of the possible RR $\leftrightarrow$ MR transition criteria from in the next section.

2.3. The RR $\leftrightarrow$ MR transition criterion

Figure 4 is a transformed plot of Fig. 3. The curve $\phi_2 = \phi_2^{ext}$ is relabeled as $\phi_1 = \phi_1^{ext}$ in order to stress that it is actually the transition line corresponding to the above-mentioned detachment criterion. In addition to the $\phi_1 = \phi_1^{ext}$ and the $\phi_1 = \phi_1^*$ curves, there is a third curve marked by $\phi_1 = \phi_1^N$ which corresponds to the von Neumann transition criterion. The $\phi_1 = \phi_1^N$ curve originates at point K. For $M_0 < M_0^K$ only the detachment criterion exists and for $M_0 > M_0^K$ both the detachment and the von Neumann criteria are possible. Point K which is a point on $\phi_1 = \phi_1^{det}$, is the origin of $\phi_1 = \phi_1^N$. In addition, $M_0^K = 2.20$ and $M_0^Q = 2.66$. Two more remarks regarding the transformation from the (M_2,ϕ_2)-plane to the (M_0,ϕ_1)-plane should be noted. The first is that the small region in Fig. 3 for which $\phi_2^{ext} < \phi_2 < \phi_2^*$ has vanished owing to the fact that RR is mathematically impossible for $\phi_1 > \phi_1^{det}$. Thus in the range $M_0 < M_0^Q$ the curves $\phi_1 = \phi_1^{det}$ and $\phi_1 = \phi_1^*$ converge smoothly to a single curve. The second remark concerns the range $M_0 > M_0^Q$. Note how the domain $\phi_2^* < \phi_2 < \phi_2^{ext}$ shown in Fig. 3 shrinks down to a very narrow domain for which $\phi_1^* < \phi_1 < \phi_1^{det}$ in Fig. 4. Thus it is practically impossible to distinguish between the $\phi_1 = \phi_1^{det}$

and $\phi_1 = \phi_1^*$. Consequently, the earlier-mentioned conclusion that part of the "weak" RR domain is aphysical is practically meaningless since, as shown in Fig. 4, this domain almost vanishes in the (M_0, ϕ_1)-plane. It should be noted here that based on Fig. 4 the two mathematical solutions are stable since they fulfill the condition that $\phi_1 \leq \phi_1^*$.

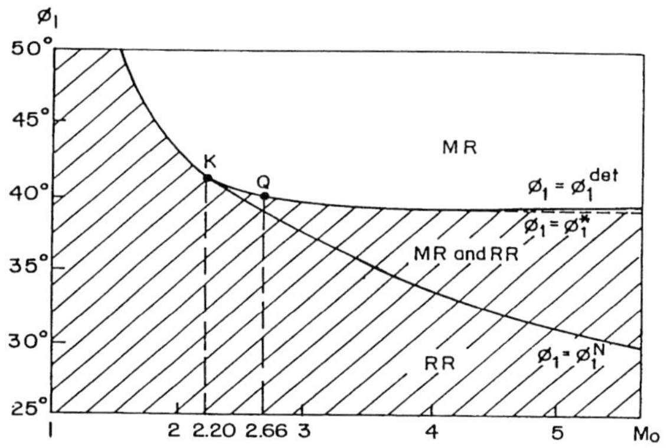

Figure 4. Domains of RR and MR and regions of physical and aphysical solutions.

The suggestion that RR is stable in the $\phi_1^N < \phi_1 < \phi_1^*$ domain disagrees with the state of the art concerning the RR $\leftrightarrow$ MR transition in steady flows which was established by Hornung and Robinson (1982). Their experimental conclusion was that the von Neumann criterion is the correct transition criterion. Furthermore, they noted that Hornung et al.'s (1979) hypothesis of a hysteresis in the RR $\leftrightarrow$ MR transition "could not be confirmed".

However, a recent experimental investigation of the RR $\leftrightarrow$ MR transition in steady flows, which was conducted by Chpoun et al. (1995), clearly validated the existence of the hysteresis phenomenon originally hypothesized by Hornung et al. (1979). Stable RR was recorded in the $\phi_1^N < \phi_1 < \phi_1^* \approx \phi_1^{det}$ domain. Furthermore, their attempts to introduce large disturbances in order to force transition from RR to MR totally failed. The RR in the $\phi_1^N < \phi_1 < \phi_1^*$ domain was perfectly stable. Chpoun et al.'s (1995) results also indicated that the actual angle at which the RR $\to$ MR transition takes place depends on both the geometrical size of the model and the way the experiment is set up. Using different models and experimental setups the values of the RR $\to$ MR transition angles were found to be in the range $35° < \phi_1^{tr}(RR \to MR) \leq 39°$ for $M_0 = 4.96$.

These results, together with the foregoing discussions, led us to propose the following general transition criterion for situations which completely satisfy the "free boundary" constraint under which the principle of minimum entropy production was developed:

For weak shock waves, i.e., $M_0 < M_0^K$:

I. The RR $\leftrightarrow$ MR transition takes place at the detachment criterion.

For strong shock waves, i.e, $M_0 > M_0^K$:

II. The RR $\rightarrow$ MR transition takes place at the detachment criterion.

III. The MR $\rightarrow$ RR transition takes place at the von Neumann criterion.

If the "free boundary" constraint is not satisfied, i.e., if downstream influences can affect the reflection or unsteady effects are present then for strong shock waves, i.e., $M_0 > M_0^K$:

IV. The RR $\leftrightarrow$ MR transition takes place anywhere in the domain $\phi_1^N \leq \phi_1^{tr} \leq \phi_1^* \approx \phi_1^{det}$ and ϕ_1^{tr} depends on the downstream and the unsteady influences.

3. Conclusion

The RR $\leftrightarrow$ MR transition phenomenon was reconsidered with the aid of the principle of minimum entropy production. The reconsideration led to a new set of transition criteria.

References

Chpoun A, Passerel D, Li H, Ben-Dor G (1995) Reconsideration of oblique shock wave reflections in steady flows, part I: experimental investigation. J Fluid Mech 301:19–35.

Glansdorff P, Prigogine A (1971) Thermodynamic Theory of Structure Stability and Fluctuations, John Wiley and Sons, Ltd., London.

Henderson LF, Lozzi A (1975) Experiments on transition of Mach reflection. J Fluid Mech 68:139–155.

Hornung HG, Oertel H Jr, Sandeman RJ (1979) Transition to Mach reflection of shock waves in steady and pseudo-steady flow with and without relaxation. J Fluid Mech 90:541–560.

Hornung HG, Robinson ML (1982) Transition from regular to Mach reflection of shock waves, part 2: the steady-flow criterion. J Fluid Mech 123:155–164.

Li H, Ben-Dor G (1996) Principle of minimum entropy production in the regular Mach reflection transition in steady flows. (Submitted for publication)

Evolution of Courant-Friedrichs structure: self-collimating outflows behind reflected shocks

A. Chernin
Sternberg Astronomical Institute, Moscow University
Moscow, 119899 Russia

P. Voinovich
Advanced Technology Center
P. O. Box 160, St. Petersburg 198103 Russia

Abstract: The evolution of the Courant-Friedrichs structure resulting from the collision of two equal spherical blast waves is simulated numerically on the basis of a locally adaptive unstructured Euler solver. Formation of strong self-collimating axial jets has been demonstrated as a prominent feature of the flow development. The results might supply a possible explanation for some observed astrophysical phenomena.

Key words: Blast waves, Jets, Numerical simulation, Astrophysics

1. Introduction

Almost a half-century ago, Courant and Friedrichs (1948) studied a dynamic structure produced by two identical spherical shock fronts coming into collision. The resulting symmetric structure includes the ring of the Mach front, two surfaces of tangential discontinuity, and two reflected shocks propagating toward the original centers of the blasts. Now we present results of a numerical simulation of the further nonlinear dynamical development of the Courant-Friedrichs structure with the aim of providing a basis to explain the mechanisms that may underlie some observed phenomena.

2. Numerical model

The Sedov-Taylor self-similar point blast solution has been used to set up the initial conditions at a moment in time preceding the collision of shock fronts in a perfect gas with a specific heat ratio of $\gamma = 5/3$. A high-order Godunov-type finite-volume scheme and transiently adaptive unstructured grid approach constitute the principal components of the numerical technique.

Some results are given in the figures in the form of density contour lines. Figures 1, 3, and 5 depict initial conditions for the following images providing

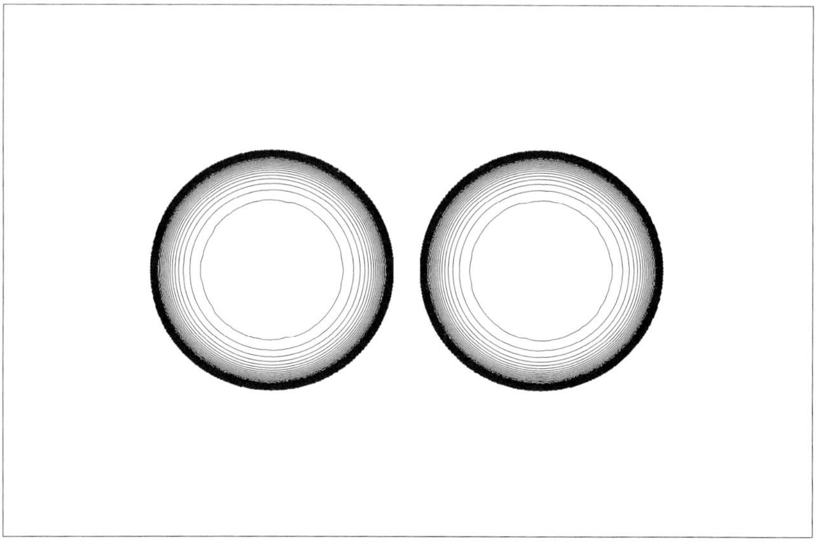

Figure 1. $t = 0.768$.

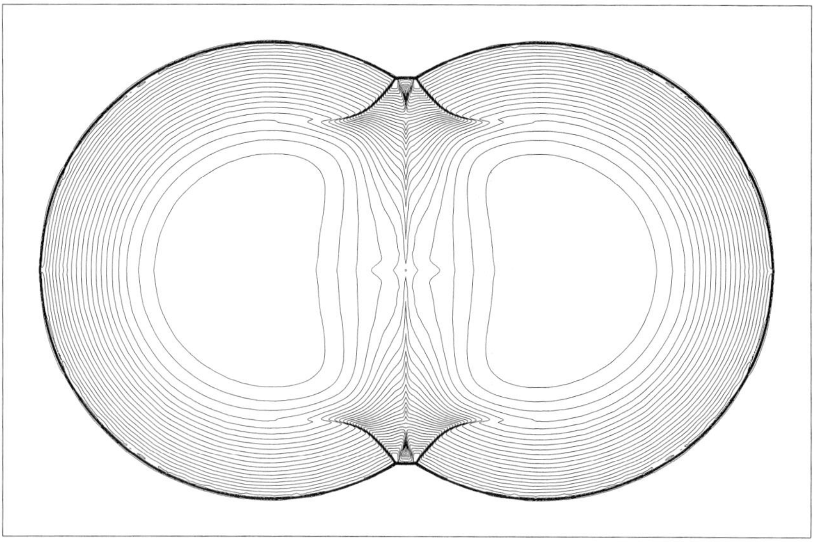

Figure 2. $t = 3.50$.

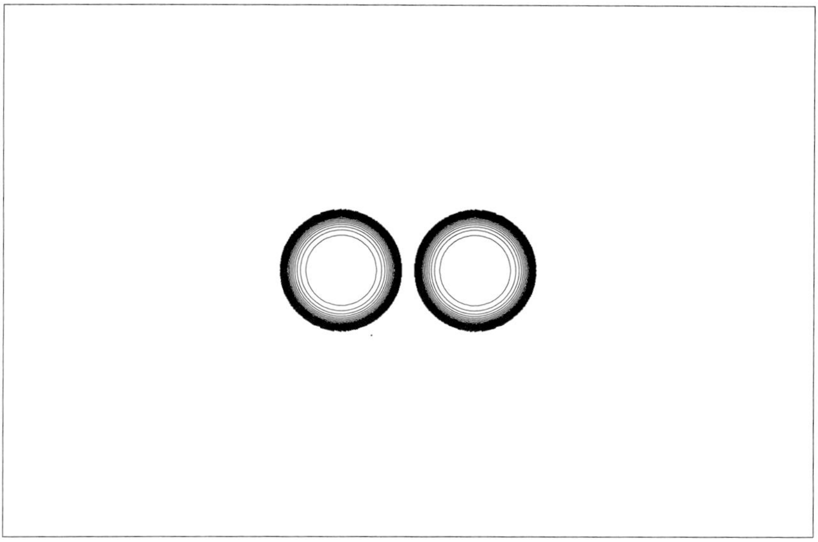

Figure 3. $t = 0.768$.

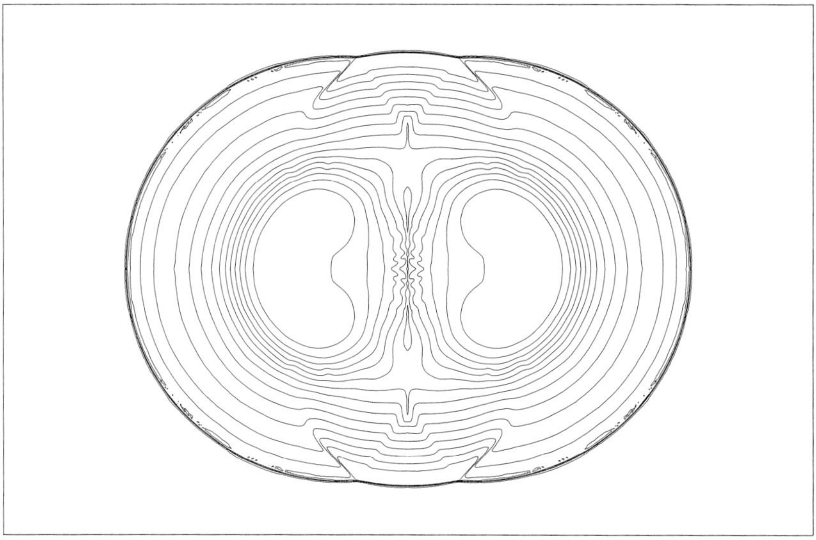

Figure 4. $t = 11.90$.

Figure 5. $t = 0.768$.

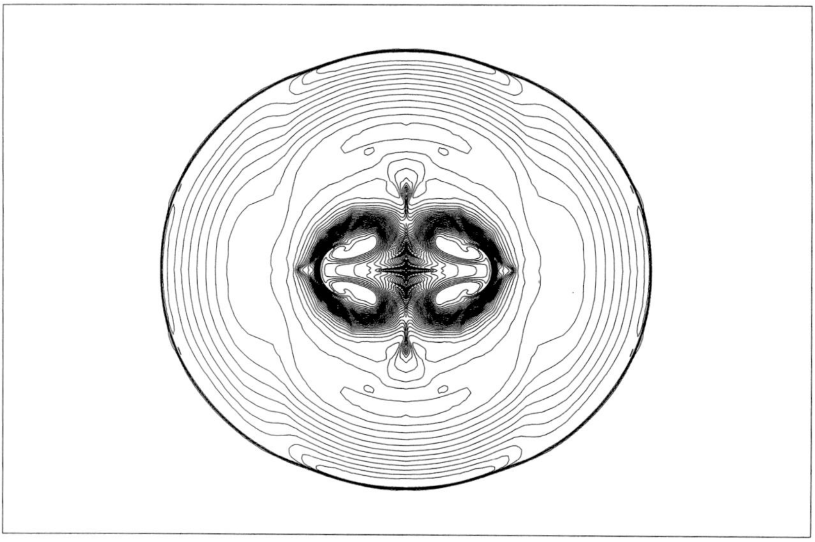

Figure 6. $t = 34.41$.

Figure 7. $t = 865.0$.

a reference scale. The unit time corresponds to the moment of collision of the blast fronts. The contour line spacing is $\Delta\rho = 0.1\rho_0$ for Figs. 1–5, and $\Delta\rho = 0.025\rho_0$ for Figs. 6 and 7, where ρ_0 is the density of ambient gas at rest.

All results were computed on an IBM PC/486 which demonstrates a high performance of the code (for details of numerical technique see Galyukov et al. 1996).

3. Discussion

As can be seen in the figures, the reflected shocks of the Courant-Friedrichs structure are critically affected by the distribution of matter in the area where they propagate. This matter distribution, in turn, is mainly due to the initial shocks that compress the major portion of the gas into two spherical shells behind their fronts. The reflected shocks propagate into the shells and the cavities within shells accelerating in the regions of low gas density.

Another characteristic feature of the flow development is a sort of self-collimation of the outflows behind the fronts of the reflected shocks caused by the rarefaction in the vicinity of the blast centers. As a result, the gas flow behind the reflected shocks tends to transform into directed streams or jets along the symmetry axis. The maximum gas velocity in the jets ranges up to half of the sound speed in the initial gas.

The final stage of the flow development given in Fig. 7 exhibits intensive vorticity generated by shock interactions and passing through a strongly non-uniform density field.

From the point of view of possible astrophysical applications of the process, it can be mentioned that observations provide examples of structures similar to the Courant-Friedrichs configuration; one of them is a system of two giant shells in the Large Magellanic Cloud near 30 Doradus discovered by Meaburn (1980). Collimated binary outflows are also observed in the interstellar medium (Lada 1985).

Acknowledgement. The work was partially supported by the Russian Foundation of Fundamental Investigations (Grant No. 94–01–01456).

The authors express their sincere gratitude to Prof. K. Takayama, Shock Wave Research Center, Institute of Fluid Science, Tohoku University, Sendai, Japan, for valuable support and fruitful discussions.

References

Courant R, Friedrichs KO (1948) Supersonic Flow and Shock Waves, Wiley Interscience, New York.

Galyukov A, Fursenko A, Voinovich P, Saito T, Timofeev E (1996) Efficient 2D and 3D locally adaptive unstructured Euler solvers for unsteady shocked gas flows. In: Shock Waves, Proc of the 20th ISSW, Sturtevant B, Shepherd JE, Hornung HG (eds), World Scientific, Vol 1, pp 691–696.

Lada CJ (1985) Cold outflows, energetic winds, and enigmatic jets around young stellar objects. Ann Rev Astron Astrophys 23:367–318.

Meaburn J (1980) The giant and supergiant shells of the Magellanic Clouds. Mon Not R Astron Soc 192:365–375.

Numerical investigations of the prediction of the Mach-stem height in steady flows

J. Vuillon, D. Zeitoun
I.U.S.T.I. (U.M.R. CNRS 139), Dept. MHEQ, Centre St. Jérome
Université de Provence, 13997 Marseille Cedex 20, France
G. Ben-Dor
Pearlstone Center for Aeronautical Engineering Studies
Department of Mechanical Engineering
Ben-Gurion University of the Negev
Beer Sheva, Israel

Abstract: A two-dimensional supersonic steady flow between a wedge and a symmetry plane was computed by solving the unsteady Euler equations. The numerical method (LCPFCT method) is based on a finite-difference scheme associated with a flux-corrected-transport algorithm. The configuration of the shock-wave reflections is investigated. The numerical results of the Mach-stem height are in good agreement with the experimental data. The results show stable configurations of regular and Mach reflections in the dual-solution domain, and a hysteresis effect in the transition between these two reflections. They also show the influence of the distance h between the wedge trailing edge and the symmetry plane on the RR $\leftrightarrow$ MR transition.

Key words: Steady supersonic flow, Euler equations, Numerical simulation, RR to MR transition

1. Introduction

Generally in a steady flow, as described by Ben-Dor (1991), the wedge-generated shock wave which reflects on a symmetry plane can produce two shock-wave reflection configurations, namely, regular reflection (RR) and Mach reflection (MR). The RR configuration consists of an incident shock wave i and a reflected one r, with a reflection point R which is located on the reflecting surface. The MR configuration contains three shock waves: the incident and reflected ones, as before, the Mach stem m, and in addition, one slipstream s. These four discontinuities interact at a single point called the triple point T, which is located above the reflecting surface. A schematic drawing of these RR and MR wave configurations is shown in Fig. 1. At a fixed freestream Mach number, the transition from one reflection type to another one depends on the angle ω_i between the incident shock wave and the symmetry plane. Two possible tran-

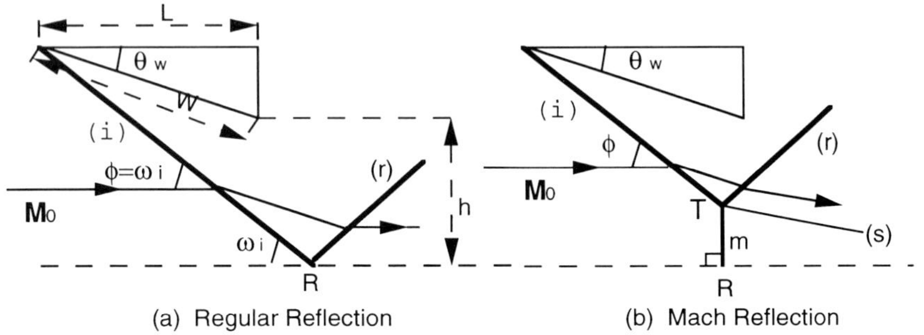

Figure 1. Schematic illustration of the shock-wave configurations between the wedge and the symmetry plane.

sition criteria have been proposed by von Neumann (1963). They are known as the detachment ($\omega_i = {\omega_i}^D$) and von Neumann ($\omega_i = {\omega_i}^N$) criteria which correspond to the values of incident shock-wave angles. The regular reflection does not take place for $\omega_i > {\omega_i}^D$ and the Mach reflection is impossible for $\omega_i < {\omega_i}^N$. In the range of ${\omega_i}^N < \omega_i < {\omega_i}^D$ and for a strong freestream Mach number (i.e., $M_0 > 2.20$), called the dual-solution domain, both reflections are theoretically possible. From their experimental study, Hornung and Robinson (1982) have concluded that: "the RR $\leftrightarrow$ MR transition occurs at the von-Neumann condition and not at the detachment condition." The predicted hysteresis effect in the RR $\leftrightarrow$ MR transition (Hornung et al. 1979) was not confirmed by their results. For this reason, they claimed that the RR configuration in the dual domain was an unstable regular reflection.

Recently, Chpoun et al. (1995) have pointed out this hysteresis effect and have experimentally found a stable RR in the dual domain. Very little numerical work has been devoted to this phenomenon and the aim of the present paper is to numerically study these shock-wave reflections. In order to do that, the unsteady 2D Euler equations have been solved with the LCPFCT method (Oran and Boris 1987). This method is based on a finite-difference scheme associated with a flux-corrected-transport algorithm. The main objectives of this work are firstly, to validate the code with Hornung and Robinson's experimental results; secondly, to put in evidence the hysteresis effect; and finally, to demonstrate that the reflected-wave configurations established inside the dual-solution domain depend on the distance h between the wedge trailing edge and the symmetry plane for given wedge and M_0.

2. Numerical investigations

The numerical description of the problem consists in solving the 2D unsteady Euler equations for describing the supersonic steady flow ($\gamma = 1.4$) between a wedge and a symmetry plane. The LCPFCT code is applied on a cartesian grid, where the ratio between the space steps is equal to the tangent of the wedge angle. The y step is constant and equal to $0.5 \cdot 10^{-2}$ cm in all runs. The computational plane is initialized with the constant upstream conditions and the steady state is supposedly reached when the L2 norm decreases by five orders of magnitude (Vuillon et al. 1995).

2.1. Code validation

For this code validation, the numerical investigations were performed with the values of the nominal freestream Mach number and geometrical data of the wedge used by Hornung and Robinson. In these conditions, the experimental results show MR configurations of the steady flow. Figure 2 represents the evolution of the Mach-stem height versus the incident Mach-number angle for the value of $M_0 = 4.96$. This figure contains Hornung and Robinson's experimental data, Azevedo and Liu's (1993) analytical results, and our computed results. One can notice the good agreement of our results with experimental ones and

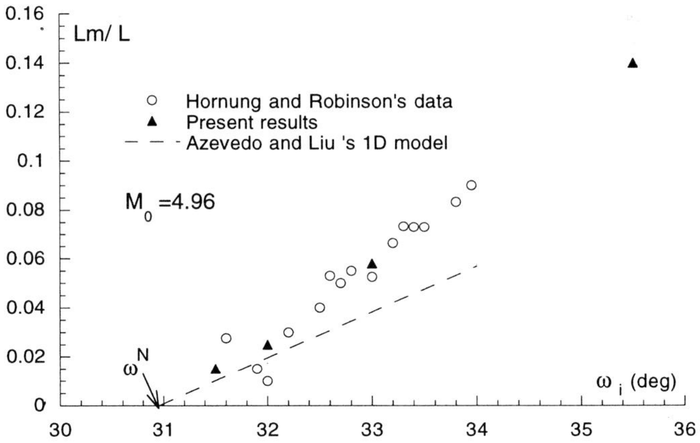

Figure 2. Nondimensional Mach-stem height versus incident shock-wave angle.

the under-prediction given by the 1D analytical model proposed by Azevedo and Liu, certainly due to the strong assumptions listed by those authors. The same agreement has been obtained with lower upstream Mach numbers but are not reported here (see Vuillon et al. 1995).

2.2. Hysteresis effect

In the dual-solution domain (${\omega_i}^N < \omega_i < {\omega_i}^D$), the unsteady computation from an initial constant plane ($M_0 = 4.96$ and $h/w = 0.34$) always leads to a regular reflection. In order to go to an existing stable MR configuration, an over-pressure disturbance is suddenly introduced behind the reflected shock wave (Vuillon et al. 1995). If this is not done, the steady state is always a RR configuration. From this remark, and from an incident Mach-number angle lower than the von Neumann criteria $\omega_i^N = 30.9°$, the wedge angle is increased by 0.5° steps, by a rotation around the wedge trailing edge. Without adding downstream disturbances, stable RR configurations are found until ω_i equals 39°. For $\omega_i = 39.5°$ the RR turns into a stable MR. It is noticed that the detachment criteria is ${\omega_i}^D = 39.3°$. For the value $\omega_i = 40°$, the MR is also obtained, but the reflected shock wave interacts with the wedge trailing edge and the triple point moves upstream until a detached shock wave forms ahead of the wedge (unstarted nozzle flow). Now, from a computed MR configuration as the initial computational plane, it is possible to do the drawback procedure (decreasing the angle from $\omega_i = 39.5°$ by 0.5° steps). Stable MR configurations are obtained until $\omega_i = 31.5°$, with decreasing values of Mach-stem height. The nondimensional Mach-stem height L_m/w is 0.015 for $\omega_i = 31.5°$. For a lower value of ω_i, the Mach stem can not be captured, due to the mesh distribution used, and the steady state is a RR configuration. This behavior is named the hysteresis effect on the RR $\leftrightarrow$ MR transition and is illustrated in Fig. 3, where the Mach-stem height is plotted against the incident shock-wave angle. This effect was also found experimentally by Chpoun et al. (1995).

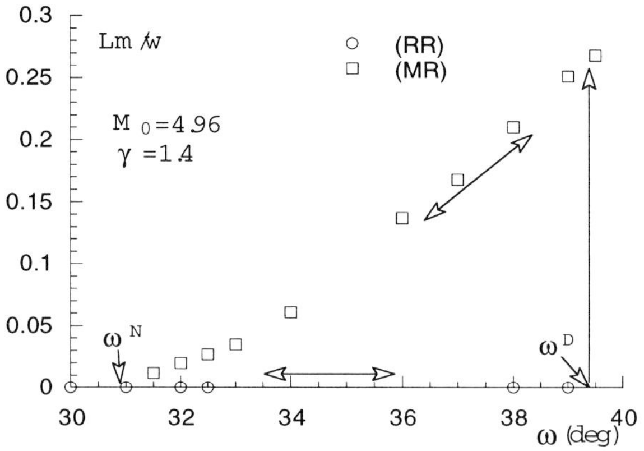

Figure 3. Hysteresis effect in the RR $\leftrightarrow$ MR transition.

2.3. RR ↔ MR transition function of the h distance

For given ω_i and M_0 in the dual-solution domain, the reflected shock-wave configuration strongly depends on the height h between the wedge trailing edge and the symmetry plane (twice this height defines the exit area of the convergent nozzle). Based on theoretical considerations, this distance is bounded by lower and upper limit values (Vuillon et al. 1995). The lower value, h_{min}, corresponds to the case in which the reflected shock wave of the MR configuration grazes the wedge trailing edge. The upper value, h_{max}, is obtained in a stable RR, when the leading characteristic of the trailing-edge expansion interacts with the incident shock wave on the symmetry plane. The transition between RR and MR occurs for a value of h_{tr} such that $h_{min} < h_{tr} < h_{max}$, and it can be found by numerical simulations. In the dual-solution domain, five test cases are chosen and the numerical results are plotted in the (M_0, h_{tr})-plane for different wedge-angle values θ_w in Fig. 4. One can notice, for a fixed value of θ_w, a decrease of h_{tr} with an increase in M_0; and for the same value of M_0 a decrease of h_{tr} with wedge angle.

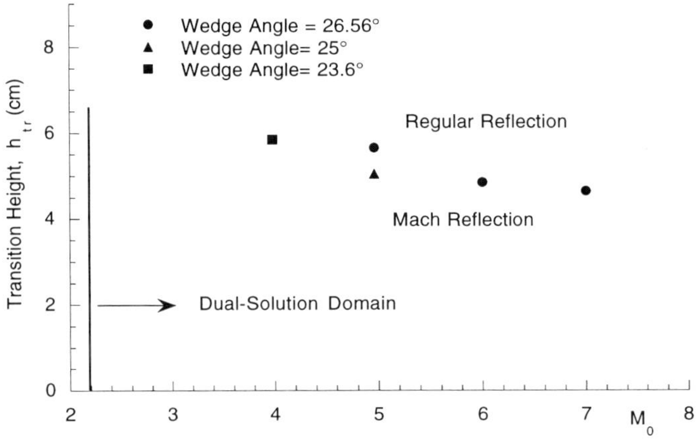

Figure 4. The dependence of h_{tr} versus the incident flow Mach number in the dual-solution domain.

3. Conclusions

The numerical investigations, based on solving the unsteady Euler equations in a cartesian grid, have allowed us to describe the steady shock-wave reflections generated between a wedge and a symmetry plane. The code is validated by the Mach-stem-height data given by Hornung and Robinson. The numerical results have shown firstly, that both regular and Mach reflections are stable in the dual-solution domain; secondly, that the hysteresis effect is produced in the RR $\leftrightarrow$ MR transition; and finally, that the height h between the wedge trailing edge and the symmetry plane influences the RR $\leftrightarrow$ MR transition for given ω_i and M_0. If one takes into account the limiting possibilities of the existing facilities and the strong assumptions of the analytical modeling, the numerical approach is an excellent tool to investigate these shock-wave reflections.

References

Azevedo DJ, Liu SL (1993) Engineering approach to the prediction of shock patterns in bounded high-speed flows. AIAA J 31(1):83–90.

Ben-Dor G (1991) Shock Wave Reflection Phenomena, Springer-Verlag, New York, NY, USA.

Chpoun A, Passerel D, Ben-Dor G (1995) Reconsideration of the state-of-the-art of oblique shock wave reflections in steady flows, part I: experimental investigation. J. Fluid Mech (to appear).

Hornung HG, Robinson ML (1982) Transition from regular to Mach reflection of shock waves, part 2, the steady-flow criterion. J Fluid Mech 123:155–164.

Oran ES, Boris JP (1987) Numerical Simulation of Reactive Flow, Elsevier, New York, USA.

von Neumann J (1963) Collected Works (AH Taub, ed), Vol. 6, Pergamon Press, New York, NY, USA.

Vuillon J, Zeitoun D, Ben-Dor G (1995) Reconsideration of the state-of-the-art of oblique shock wave reflections in steady flows, part II: numerical investigation. J Fluid Mech (to appear).

Interaction of shock waves emitted from a shock tube with a supersonic projectile

Z. Jiang, K. Takayama
Shock Wave Research Center, Institute of Fluid Science, Tohoku University
2-1-1 Katahira, Aoba-ku, Sendai 980-77 Japan

Abstract: This paper reports on shock-wave flows driven by a supersonic projectile released from a shock tube into ambient air. The dispersion control scheme was used for numerical simulations. Two cases were calculated at first and then numerical results were compared with experiments. Three further cases were examined for various Mach numbers of the projectile and different stand-off distances between an initial shock wave and the projectile. It was found that complicated transient phenomena exist in the earlier stages of shock waves, such as interactions between a moving shock wave and a moving body, and induced contact surfaces and moving vortices.

Key words: Supersonic projectile, Moving shock waves, Numerical simulation

1. Introduction

A projectile moving at a supersonic speed in a shock tube drives an initial shock wave in front of it. Once the shock wave is discharged from the open end of the shock tube, shock-wave diffraction and a jet flow develop. Later, when the projectile is released from the shock tube, it interacts with this diffracting shock wave, the jet flow, and also with the initial shock wave. This results in complex moving-shock-wave/moving-body interaction, moving-shock-wave/contact-surface interaction and moving-shock-wave/vortex interaction. Such a transient flow phenomenon shows a significant difference from shock-wave interactions with a body at rest. Therefore, this paper aims at understanding shock-wave/projectile interaction.

A dispersion control scheme based on the dispersion conditions (Jiang 1993, Jiang et al. 1993) was used for solving the Euler equations assuming inviscid and axisymmetric flows. Two cases were calculated first. Then numerical interferograms were constructed from calculated results and compared with experiments to validate the numerical code. Three other cases were carried out with various Mach numbers and different release times of a projectile after an initial shock wave was discharged. Numerical solutions show that interaction among a jet flow, shock-wave diffraction, an initial shock wave, and the bow-shock-wave system around a projectile does produce complicated transient flow phenomena.

2. Governing equations and numerical methods

For axisymmetric problems, a hyperbolic system of conservation laws for perfect gas in cylindrical coordinates can be written as

$$\frac{\partial \mathbf{U}}{\partial t} + \frac{\partial \mathbf{F}}{\partial x} + \frac{\partial \mathbf{G}}{\partial r} + \frac{1}{r}\mathbf{S} = 0, \tag{1}$$

where for the inviscid gasdynamics equations we have

$$\mathbf{U} = \begin{pmatrix} \rho \\ \rho u \\ \rho v \\ e \end{pmatrix}, \quad \mathbf{F} = \begin{pmatrix} \rho u \\ \rho u^2 + p \\ \rho uv \\ (e+p)u \end{pmatrix}, \quad \mathbf{G} = \begin{pmatrix} \rho v \\ \rho uv \\ \rho v^2 + p \\ (e+p)v \end{pmatrix}, \quad \mathbf{S} = \begin{pmatrix} \rho v \\ \rho uv \\ \rho v^2 \\ (e+p)v \end{pmatrix}, \tag{2}$$

where primitive variables in the unknown $\mathbf{U}$ are density ρ, velocity components u and v, and total energy per unit volume e. Fluid pressure is p and the equation of state for perfect gas is given by

$$e = \frac{p}{\gamma - 1} + \frac{1}{2}\rho(u^2 + v^2), \tag{3}$$

where γ, the specific heat ratio, is taken as 1.4. The finite-difference equations of (1) discretized by using the dispersion control scheme are given in the form of half discretion as

$$\left(\frac{\partial \mathbf{U}}{\partial t}\right)^n_{i,j} = -\frac{1}{\Delta x}(\mathbf{H}^n_{i+\frac{1}{2},j} - \mathbf{H}^n_{i-\frac{1}{2},j}) - \frac{1}{\Delta r}(\mathbf{P}^n_{i,j+\frac{1}{2}} - \mathbf{P}^n_{i,j-\frac{1}{2}}) - \mathbf{S}^n_{i,j}, \tag{4}$$

where

$$\mathbf{H}^n_{i+\frac{1}{2},j} = \mathbf{F}^+_{i+\frac{1}{2}L,j} + \mathbf{F}^-_{i+\frac{1}{2}R,j}, \qquad \mathbf{P}^n_{i,j+\frac{1}{2}} = \mathbf{G}^+_{i,j+\frac{1}{2}L} + \mathbf{G}^-_{i,j+\frac{1}{2}R}, \tag{5}$$

$$\begin{cases} \mathbf{F}^+_{i+\frac{1}{2}L,j} = \mathbf{F}^+_{i,j} + \frac{1}{2}\text{minmod}(\Delta\mathbf{F}^+_{i-\frac{1}{2},j}, \Delta\mathbf{F}^+_{i+\frac{1}{2},j}) \\ \mathbf{F}^-_{i+\frac{1}{2}R,j} = \mathbf{F}^-_{i+1,j} - \frac{1}{2}\text{minmod}(\Delta\mathbf{F}^-_{i+\frac{1}{2},j}, \Delta\mathbf{F}^-_{i+\frac{3}{2},j}) \end{cases}, \tag{6}$$

$$\begin{cases} \mathbf{G}^+_{i,j+\frac{1}{2}L} = \mathbf{G}^+_{i,j} + \frac{1}{2}\text{minmod}(\Delta\mathbf{G}^+_{i,j-\frac{1}{2}}, \Delta\mathbf{G}^+_{i,j+\frac{1}{2}}) \\ \mathbf{G}^-_{i,j+\frac{1}{2}R} = \mathbf{G}^-_{i,j+1} - \frac{1}{2}\text{minmod}(\Delta\mathbf{G}^-_{i,j+\frac{1}{2}}, \Delta\mathbf{G}^-_{i,j+\frac{3}{2}}) \end{cases}, \tag{7}$$

$$\begin{cases} \Delta\mathbf{F}^\pm_{i+\frac{1}{2},j} = \mathbf{F}^\pm_{i+1,j} - \mathbf{F}^\pm_{i,j} \\ \Delta\mathbf{G}^\pm_{i,j+\frac{1}{2}} = \mathbf{G}^\pm_{i,j+1} - \mathbf{G}^\pm_{i,j} \end{cases}, \qquad \begin{cases} \mathbf{F}^\pm = \mathbf{A}^\pm\mathbf{U} \\ \mathbf{G}^\pm = \mathbf{B}^\pm\mathbf{U} \end{cases}. \tag{8}$$

Steger and Warming flux vector splitting was used in the equations above. Solutions were evolved in time by the second-order Runge-Kutta method. Reflecting boundary conditions were specified both on solid walls and the centerline of a computational domain. Non-reflecting boundary conditions were applied at the inflow and outflow boundaries. Calculations were carried out on an equally spaced grid with 50 grid points along the radius of a shock tube.

3. Numerical results and discussion

3.1. Verification with experiment

The first case is a shock wave discharged in air from a shock tube with $M_s = 1.6$, and the second is a shock wave caught up by a projectile with a speed of $M_p = 2.2$. Numerical interferograms were constructed using techniques described by Babinsky et al. (1995) to compare with the experimental ones shown in Figs. 1 and 2 (Abe 1990) (experimental results on left, numerical on right). It can be seen that computational results agree well with experiments. Hence the accuracy of the present numerical method is well validated.

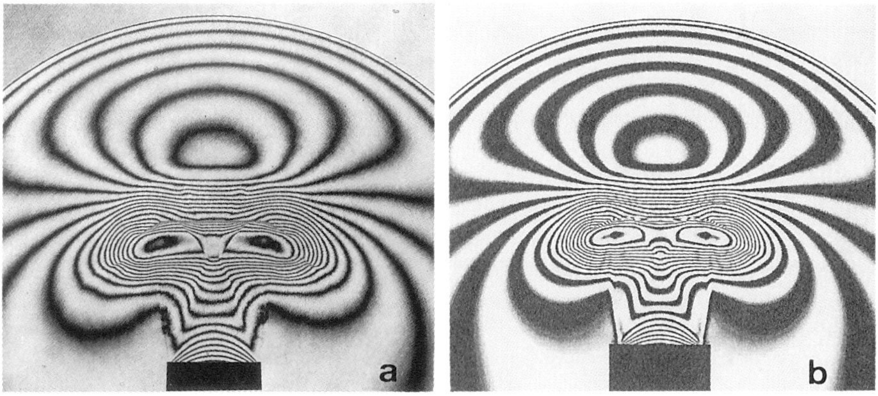

Figure 1. Comparison of numerical and experimental interferograms, $M_s = 1.6$.

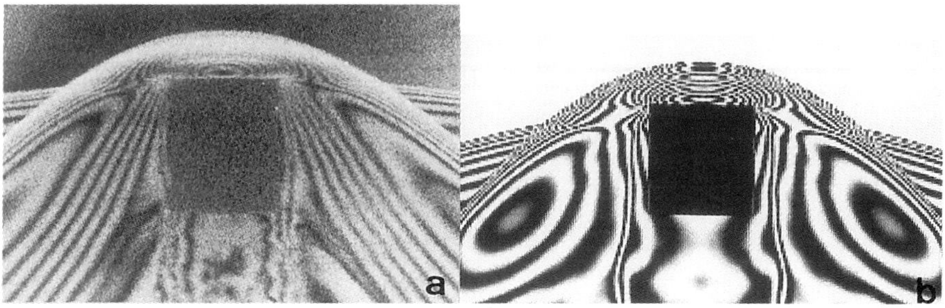

Figure 2. Comparison of numerical and experimental interferograms, $M_p = 2.2$.

3.2. The projectile speed of $M_p = 4.0$

A time sequence of case 3 is shown in Fig. 3 where isobar is plotted in the symmetric lower half and isopycnics in the symmetric upper half (the same display is used in other figures). Figure 3a shows a frame when a projectile is just about to be released and a jet flow develops, in which there are a spherical shock wave with its weak rim sucked into a low-pressure core of a primary vortex ring, and a contact surface between the shock wave and an initial shock wave. Later, as shown in Fig. 3b, the projectile moves through the spherical shock wave and drives out its own bow shock wave. Meanwhile, shock-wave diffraction followed by a contact surface is observed over the rear surface of the projectile due to the body moving through the shock wave that is almost at rest. This diffracting shock wave meets the spherical shock wave in Fig. 3c, distorts the spherical shock wave and changes it into a discontinuous shape consisting of three curved shock waves. The velocity and density discontinuities develop here, and vortices are produced near each discontinuous circle. From the same figure it is found that when the bow shock wave in front of the projectile catches up with the initial shock wave, expansion waves and a contact surface in front of the projectile are produced. Also noticeable is that the contact surface originating from sharp corners of the projectile is going to meet with that of the jet flow. In Fig. 3d, the kink and rolling up of two contact surfaces are observed due to inviscid vortex instability. The similar instability in shear flows appears also near the jet-flow boundary. The contact surface located in front of the projectile is getting thinner and thinner since the fluid flow is leaving the centerline for the side walls of the projectile, and finally vanishes. At this moment, the spherical shock of the jet flow is changed back into a spherical shape again.

3.3. The projectile speed of $M_p = 4.0$ with shorter stand-off distance

The results of the fourth case are shown in Fig. 4 with the same initial conditions as case 3, but a shorter release time of the projectile. From Figs. 3a and 4a, it can be seen that when an initial shock wave arrives at the same position, the projectile in case 4 is going to catch up with its initial shock wave. The contact surface appears in front of the projectile in Fig. 4a where a similar shock-wave diffraction exists over the rear surface of the projectile. Since the contact surface from the projectile meets that of the jet flow earlier, the kink and rolling up of two contact surfaces take on more complicted structures, as shown in Fig. 4b. The flow patterns appear similar to each other in the two cases except for the differences mentioned above.

3.4. The projectile speed of $M_p = 1.25$

Figure 5 shows the shock-wave flow of case 5 with a speed of $M_p = 1.25$. Shock-wave diffraction over the back of a projectile (see Fig. 5a) is found to be similar to the former cases, but the kink of contact surfaces is quite different. Because of the lower Mach number, shock-wave interaction in front of the projectile is invisible in Fig. 5b and the jet flow appears to be the underexpanded one with a repeated expansion-contraction structure.

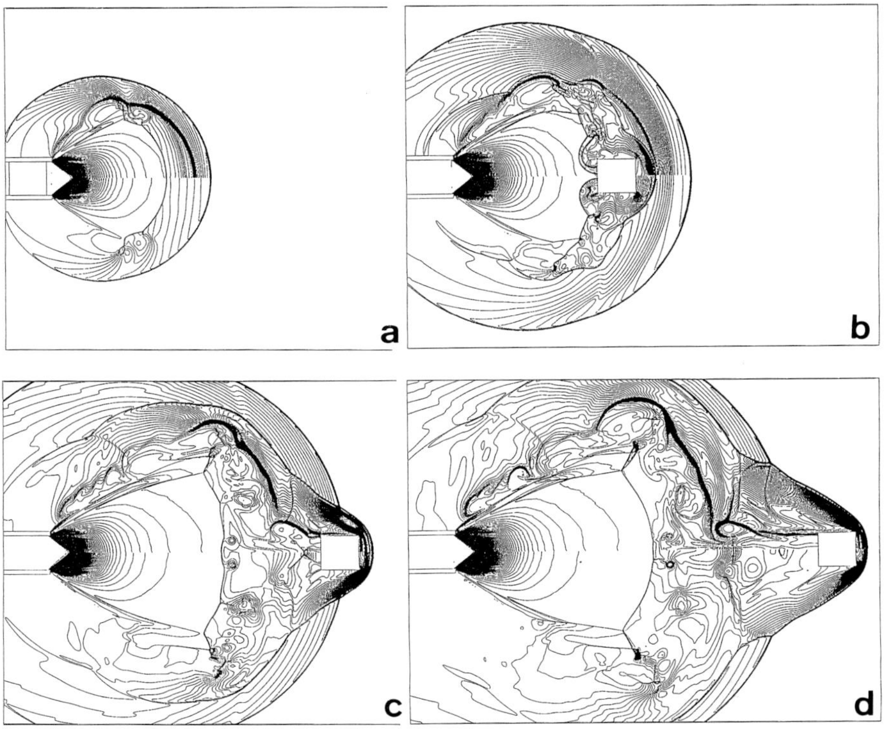

Figure 3. Isobar (lower half) and isopycnics (upper half), $M_p = 4.0$.

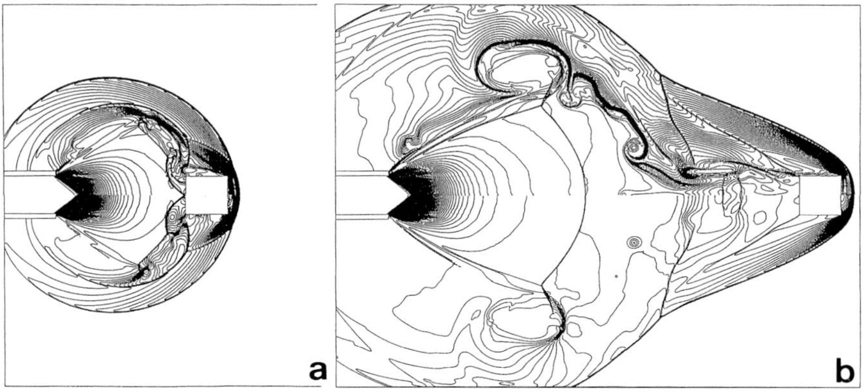

Figure 4. Isobar (lower half) and isopycnics (upper half), $M_p = 4.0$.

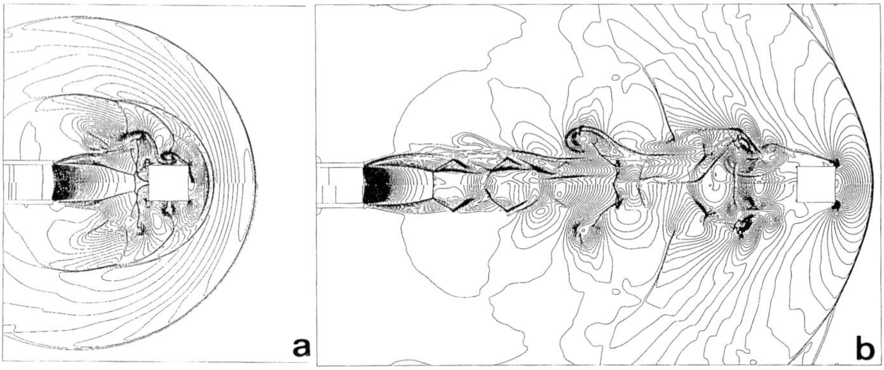

Figure 5. Isobar (lower half) and isopycnics (upper half), $M_p = 1.25$.

4. Conclusions and remarks

Flow patterns change depending mainly on the speed of a projectile because of jet-flow structures varying with the Mach number of the projectile. The stand-off distance between an initial shock wave and a projectile appears to have less effect on flow patterns except for a region behind the projectile. Once they meet, the kink of a contact surface of the jet flow and one produced by a projectile can be observed due to inviscid vortex instability. When the shock-wave system of a projectile catches up with the initial shock wave, the interaction of those two shock waves produces a contact surface and expansion waves behind a new spherical shock wave. In the case of the projectile at a lower speed, this interaction is weaker and invisible.

References

Abe A (1990) Diffraction of Shock Waves Emitted From the Open End of a Shock Tube, Doctoral Thesis, Tohoku University, Sendai, Japan.

Babinsky H, Meguro T, Jiang Z, Takayama K (1995) Numerical visualization of shock wave flow in an expanding tube and comparison with experiment. In: Proc of 4th Symp on Exp and Numerical Flow Visualization.

Jiang ZL (1993) Study on the Finite Difference Theory and Numerical Methods of Week Solution Problems, Doctoral Thesis, Peking University, Beijing, China.

Jiang ZL, Chen YS, Ishii K, Kuwahara K (1993) Numerical dispersion, numerical oscillations, and dispersion control schemes for weak solution probelms. In: Proc of 5th Int Symp on CFD, Sendai, Japan, 1:397.

Numerical investigation of shock-wave reflection problems in steady flows

M. S. Ivanov, S. F. Gimelshein, G. N. Markelov
Institute of Theoretical and Applied Mechanics, Novosibirsk 630090 Russia
A. E. Beylich
Stoßwellenlabor, RWTH Aachen, Aachen 52056 Germany

Abstract: The transition from regular to Mach reflection and back of planar shock waves generated by straight wedges in steady flows was studied numerically with the DSMC method at small Knudsen numbers. The hysteresis effect was observed at increasing and decreasing shock-wave angle. Transformation from one configuration to the other in response to perturbations was shown.

Key words: Mach reflection, Transition, Hysteresis, Stability

1. Introduction

In steady flow, the transition between regular reflection (RR) and Mach reflection (MR) of a wedge-generated shock wave from a symmetry plane can theoretically occur anywhere in the range $\alpha_N < \alpha < \alpha_D$, where α is the angle between the shock and the symmetry plane, α_N is the von Neumann angle, and α_D is the detachment angle. RR is not possible for $\alpha > \alpha_D$ and MR is impossible for $\alpha < \alpha_N$. In the region $\alpha_N < \alpha < \alpha_D$ (dual-solution region), both reflection types are theoretically possible. The dependence of transition angle on the direction of α variation was actively studied from the mid-1970s to the mid-1980s, mostly by experimental methods. The results of such experimental investigations were summarized by Hornung and Robinson (1982) as follows: transition from RR to MR and back in steady flows takes place according to the von Neumann criterion; in the dual-solution domain RR is unstable and a hysteresis effect was not observed. Recent experimental results reported by Ben-Dor et al. (1994), and a numerical study by Ivanov et al. (1995) did, however, exhibit hysteresis. Here we continue the study of Ivanov et al. by considering the influences of flow rarefaction, free-stream, and local perturbations.

2. Method of solution

The Direct Simulation Monte Carlo (DSMC) method for solving the Boltzmann equation is used in the version based on the majorant collision frequency principle (Ivanov and Rogasinsky 1991) in the near-continuum Knudsen-number range, $0.001 \leq \mathrm{Kn} \leq 0.01$. The rectangular computational domain covered one side of the symmetry plane, which was modeled as a specular reflector. Uniform free-stream conditions were imposed on the left and upper boundaries of the computational domain, and specular reflection on the wedge surface.

3. Flow conditions

All results presented here (except for those shown in Fig. 4b) were obtained for the conditions of the argon flow of Hornung et al. (1979), i.e., $\gamma = 5/3$, $M = 16$, $g/w = 0.41$, where g is the distance between the trailing edge of the wedge and the symmetry plane and w is the wedge length. The data in Fig. 4b were obtained for one of the conditions of Hornung and Robinson (1982), i.e., $M = 4.96$, $T_0 = 365$ K, $g/w = 0.34$, $\gamma = 1.4$, nitrogen.

4. Computational results and discussion

4.1. Fixed wedge

Here, the wedge angle was initially fixed such that it generates a shock with α within the range $\alpha_N = 26.16$ and $\alpha_D = 35.37$. A uniform undisturbed flow was assumed in the entire computational domain as the initial condition. Calculations showed that the incident shock flow agreed well with theory. For all α, RR was observed up to $\alpha = \alpha_D$. Figure 1 shows an example of the shock-wave configuration (pressure flow field and streamlines) for the incidence angle $\alpha = 35°$ slightly below α_D. Parameters behind the incident and reflected shock waves are in good agreement with the two-wave solution. When $\alpha > \alpha_D$, MR is observed. An example is shown in Fig. 2 for $\alpha = 36°$ exceeding α_D. The pressure downstream of the Mach stem coincides with the theoretical value. Behind the reflected wave and near the triple point the pressure is up to 10% higher than the pressure downstream of the Mach stem. Thus, RR takes place for all α inside the dual-solution region and MR occurs for $\alpha > \alpha_D$. Similar results were also obtained for nitrogen ($\gamma = 1.4$).

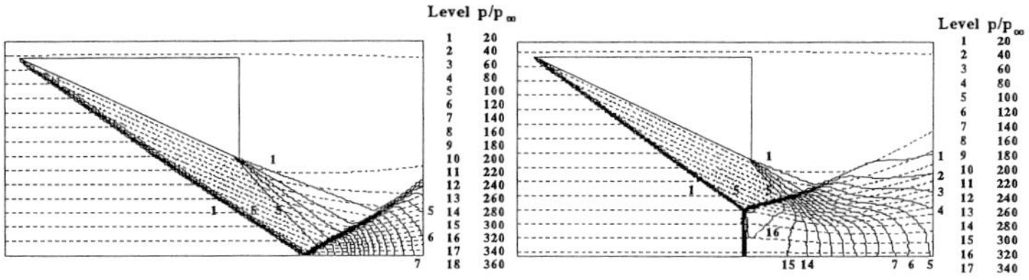

Figures 1 (left) and 2 (right). Pressure field (continuous lines) and streamlines (dashed lines) for uniform flow as initial condition, argon, $M = 16$, Kn $= 0.0025$. Left, $\alpha = 35°$; right, $\alpha = 36°$.

4.2. Moving wedge and the hysteresis effect

Here, initial conditions were chosen to correspond to a calculated MR with $\alpha = \alpha_1 > \alpha_D$. This solution was then used as the initial condition for a calculation with $\alpha = \alpha_2 < \alpha_1$, etc. The angle α was changed by rotating the wedge around its trailing edge. This showed that MR is maintained to values of α well below α_D (see Fig. 3 with $\alpha = 35°$). MR is maintained down to $\alpha = \alpha^* > \alpha_N$. The fact that $\alpha^* > \alpha_N$ may be explained as follows: when $\alpha \to \alpha_N$, the Mach-stem height decreases and becomes comparable with the shock-wave thickness, so MR and RR may no longer be distinguished. Accordingly, α^* decreases with decreasing Kn (see Fig. 4a). DSMC calculations for different flow conditions ($M = 4.96$, $\gamma = 1.4$) also confirmed that the final configuration in the dual-solution domain depends on initial conditions (see Fig. 4b). Thus, numerical solutions exhibit the presence of a hysteresis effect, i.e., of reflection-type dependence on the direction of the incidence angle variation, at α changing through the range $\alpha^* < \alpha < \alpha_D$, both from below and from above (see Fig. 4). The transition from RR to MR takes place at $\alpha = \alpha_D$ when α increases. MR is retained down to $\alpha^* > \alpha_N$ with decreasing α and initial $\alpha > \alpha_D$.

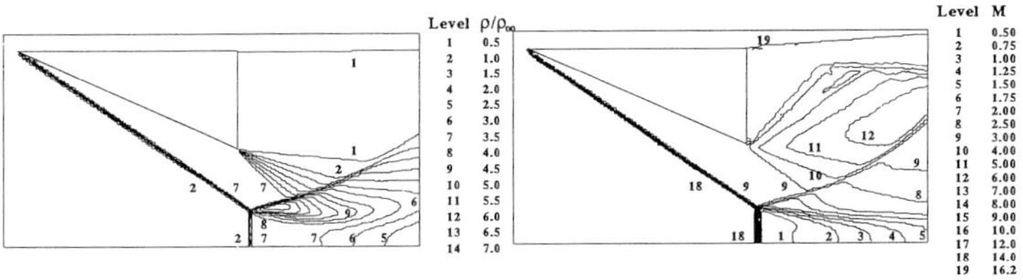

Figure 3. (a) Density flow field and (b) Mach number flow field for the case of Fig. 1.

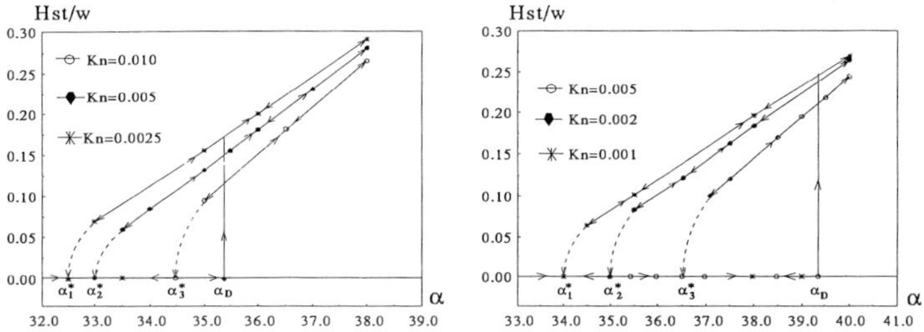

Figure 4. Normalized Mach-stem height vs α: (a) Argon, $M = 16$; (b) Nitrogen, $M = 4.96$.

4.3. MR and RR perturbations

Next, the possibility of MR $\leftrightarrow$ RR transitions was studied in the dual-solution domain and under the effect of external disturbances. MR or RR was prescribed as the initial condition for fixed α and g/w. Then a disturbance was applied during some finite period t_p and the subsequent flow stabilization was observed.

Flow perturbations near the reflection point. RR was prescribed as the initial condition, and a disturbance was introduced in a rectangular region near the reflection point at time $t = 0$, instantaneously stopping all particles present in this region. At $t > 0$, the constraint was removed, and the flow relaxed to a steady state (MR or RR). Figure 5 shows sketches of several perturbation regions for computations with $\alpha = 33.5°$ and $\text{Kn} = 0.005$. "RR" and "MR" indicate the final steady configuration. Figures 5a and b show RR is retained when the disturbance region is of the order of the Mach-stem height, H_{st}, for the MR configuration corresponding to a given α ($H_{st}(33.5°) \approx 0.06w$, see Fig. 4a). When the vertical size of this region is increased up to $2H_{st}$, RR holds. However, a larger horizontal size (up to $2H_{st}$) causes MR formation (Fig. 5d). Figure 6 shows some typical perturbed regions for $\alpha = 35°$; RR is retained when the disturbance region is of the order $0.25H_{st}$ ($H_{st}(35°) \approx 0.132w$, see Fig. 4a). However, when the horizontal size is increased up to $0.5H_{st}$, MR appears (Fig. 6b). If this region is located upstream of the incident shock wave, the perturbations exert a weaker effect and RR holds (Fig. 6c). When the vertical size of this region is increased by a factor of 2 (as in Fig. 5a), MR is formed. Thus, increasing α brings about lower RR stability. RR stability is also appreciably reduced with decreasing Knudsen number. For $\text{Kn} = 0.0025$, RR transforms into MR already for the perturbed region size of about $0.025H_{st}$ (Fig. 6d). The study of MR stability to velocity perturbations showed that this configuration is much more stable than RR. For example, doubling the particle velocities within the regions of order $2H_{st}$ did not destroy MR.

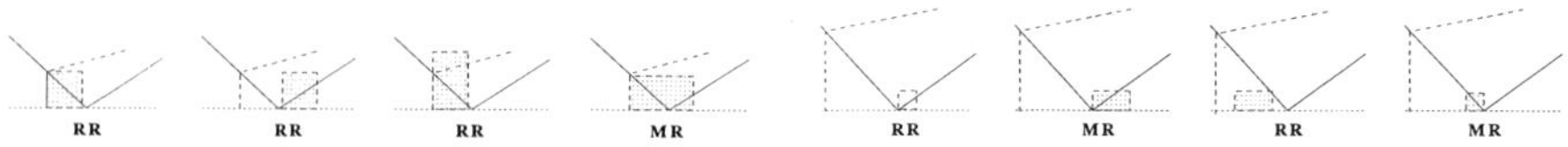

Figures 5 (left) and 6 (right). Shapes and positions of disturbed regions (argon). Solid line – RR configuration, dashed line – MR. Hatched area is disturbed region. On left: $\alpha = 33.5°$, $M = 16$, $\text{Kn} = 0.005$. On right: $\alpha = 35°$.

Free-stream perturbations. The effects of free-stream velocity and density perturbations were studied with MR as initial configuration and $\alpha = 33.5°$. For velocity perturbations, the average velocity of the particles entering the computational domain was changed to $U = xU_\infty$ for a period $t = nt_M$, where $t_M = H_{st}/U_\infty$. As seen in Fig. 7a, sufficiently large n and small x cause MR failure and RR formation, while velocity increases cause MR to be maintained. Likewise, density perturbations of the entering particles ($\rho = x\rho_\infty$) were investigated. Figure 7b shows an appreciable density reduction ($x = 0.5$), as in the case of $u = 0.5u_\infty$, which destroys the MR configuration. Computations with RR as initial configuration at $\alpha = 33.5°$ showed this to be more stable. Velocity perturbations with $x \geq 2$ for $n \geq 1$ were necessary to transform it to MR. Figure 8 illustrates the temporal transformation of RR into MR affected by velocity perturbations. Figures 8a–8c show the unsteady flow pattern for fixed time (the fluctuations in the flow field are related to a strong statistical scattering).

The final steady solution is presented in Fig. 8d. The disturbed region consists of a compression subregion (particles with the velocity $u = 2u_\infty$ overtake the particles with $u = u_\infty$) followed by an expansion subregion (Fig. 8a). Interference of these two subregions with the incident shock wave forms the Mach stem (Fig. 8c). An example of MR→RR transition is shown in Fig. 9. Since $x < 1$, the downstream perturbed region contains the expansion subregion first, with the compression subregion to follow. The former subregion smears the MR, and the latter one forms an RR. Computations for large α showed that at $\alpha \to \alpha_D$ MR stability increases whereas that of RR is reduced.

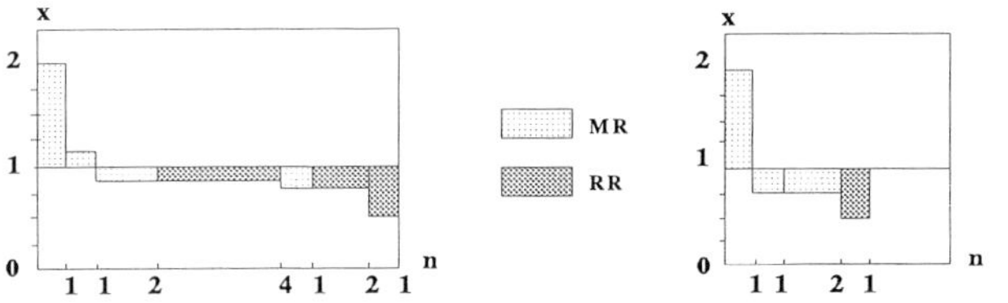

Figure 7. Effect of (a) magnitude and duration of free-stream velocity, and (b) density perturbations on configuration (MR or RR).

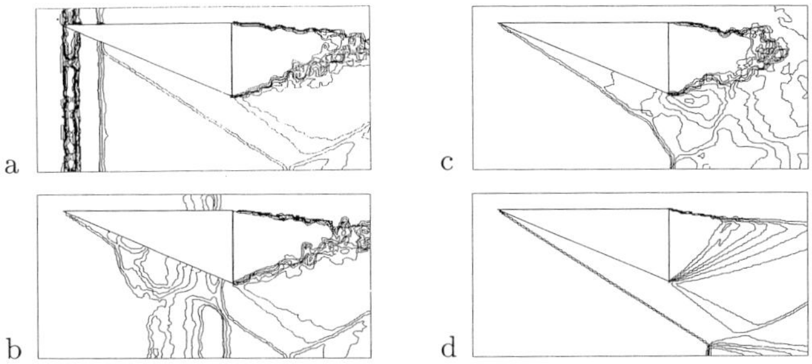

Figure 8. (a) – (d) Free-stream velocity perturbations of RR at $t/t_M = 2$, 8, 18, and final steady solution.

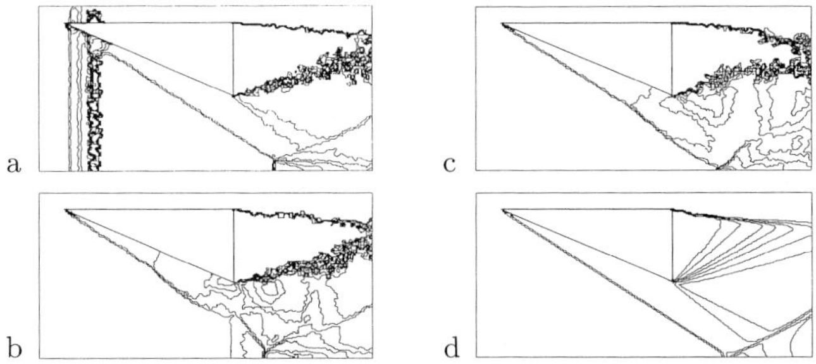

Figure 9. (a) – (d) Free-stream velocity perturbations of MR at $t/t_M = 4$, 20, 28, and final steady solution.

5. Conclusions

The calculations show that: (1) the reflection type depends on the direction of the incidence angle variation, coming from below or from above $\alpha^\star < \alpha < \alpha_D$, i.e., the hysteresis effect exists; (2) the difference $\alpha^\star - \alpha_N$ is reduced when the Knudsen number decreases, and the dependence of H_{st} on α is close to linear within $\alpha^\star < \alpha < \alpha_D$ for various Knudsen numbers; (3) RR is less stable to the flow disturbances near the reflection point than MR. RR stability to this kind of disturbance is appeciably reduced with the flow rarefaction decreasing. The last fact indicates that experiments on the hysteresis effect should be analyzed very carefully, since even small disturbances can transform RR into MR; and (4) free-stream disturbances can transform RR into MR, and vice versa. MR stability increases with α growing ($\alpha^\star < \alpha < \alpha_D$), whereas that of RR is reduced.

Acknowledgement. The authors wish to acknowledge support from the Deutsche Forschungsgemeinschaft and the International Science Foundation (Grant RBW000).

References

Ben-Dor G, Li H, Chpoun A, Passerel D, Vuillon J, Zeitoun D (1994) Analytical, numerical and experimental investigation of the regular $\leftrightarrow$ Mach reflection transition in steady flows. In: 11th Int Mach Reflection Symposium, Univ of Victoria, BC, Canada.

Hornung HG, Oertel H, Sandeman RJ (1979) Transition to Mach reflection of shock waves in steady and pseudosteady flow with and without relaxation. J Fluid Mech 90:541–560.

Hornung HG, Robinson ML (1982) Transition from regular to Mach reflection of shock waves. Part 2. The steady-flow criterion. J Fluid Mech 123:155–164.

Ivanov MS, Gimelshein SF, Beylich AE (1995) Hysteresis effect in stationary reflection of shock waves. Physics of Fluids 7(4):685–687.

Ivanov MS, Rogasinsky SV (1991) Theoretical analysis of traditional and modern schemes of the DSMC method. In: 17th Int Symp on Rarefied Gas Dynamics, Aachen, Germany, pp 629–642.

The interaction of a shock wave discharged from the open end of a shock tube with a flat plate

T. V. Bazhenova, V. V. Golub, A. M. Shulmeister,
T. A. Bormotova, S. B. Bazarov, S. V. Pichugin
High Energy Density Research Center
IVTAN, Russian Academy of Sciences
Izhorskaya, 13/19, Moscow, 127412 Russia

Abstract: This paper is devoted to the investigation of the impingement of a diffracted shock wave on a flat plate and comparison with the numerical simulations.
Key words: Shock-wave diffraction, Vortex ring, Shock-wave impingement on a plate

1. Introduction

During the past four decades, shock waves discharged from the open ends of shock tubes have been studied. The complicated flow field behind a shock wave emitted from the open end of a shock tube was clarified in some details experimentally and numerically (Baird 1987, Abe & Takayama 1989). The purpose of the present paper is to study the impingement of a shock wave discharged from the open end of a shock tube upon a flate plate and the unsteady flow field behind the shock wave. The features of shock-wave reflection after the discharge of a shock of $M_0 = 1.46$ which impinged on the back wall of a square box were reported by Phan & Stollery (1985).

2. Experimental set-up and numerical simulation

A flate plate was mounted opposite to the open end of a shock tube with diameter $d = 20$ mm. The distance L between the open end of the shock tube and the plate was varied from 50 mm to 100 mm. The Mach number of the incident shock wave was varied from 2.0 to 4.7. The low-pressure chamber and vacuum tank were filled with nitrogen to pressures between $P_1 = 4$ and 50 kPa. The flow fields were observed by taking schlieren photographs and interferograms. Computer simulation was based on Euler's equations, which were solved by Godunov's method.

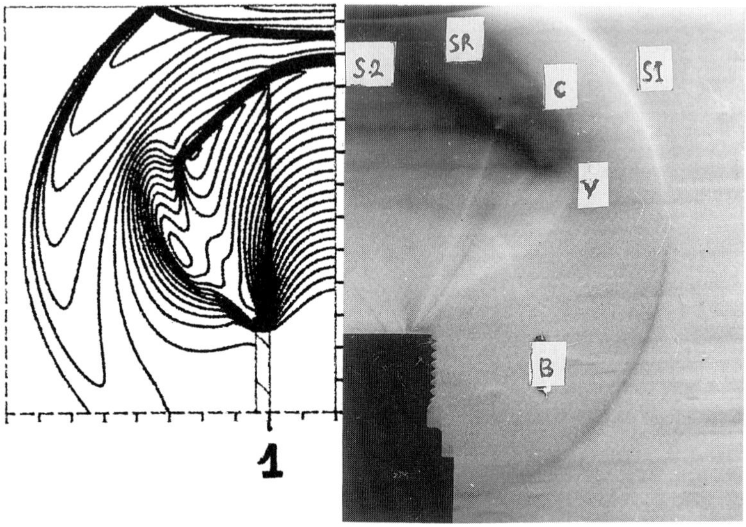

Figure 1. The calculated density contours (left) and schlieren photograph (right) of diffracted shock wave impingement on a flat plate; $M_0 = 3.2$, $P_1 = 4$ kPa, $t = 72$ μs.

3. Results and discussion

The obtained series of shadowgraphs and interferograms have clearly captured the main stages of the process:

- For diffraction into free space of the shock wave discharged from the open end of a shock tube: in the unsteady flow field behind the emitted shock wave there are two kinds of complicated flows depending on the shock-wave Mach number. At low shock-wave Mach numbers, the post-shock flow is subsonic and is dominated by the contact surface and vortex ring with an embedded rearward-facing shock. When the post-shock flow is supersonic, the flow pattern is similar to the barrel structure in the under-expanded stationary jet (Bazarov et al. 1993).

- Figure 1 shows the schlieren photograph and calculated density contours of the impingement of the diffracted shock wave on a flat plate. After the reflection of the spherical shock wave $S1$ from a plate (Fig. 1, right), the reflected shock wave SR arises. The Mach lines of the disturbed flow, the rarefaction fan, and the separation line can be recognized near the open end of the shock tube. The vortex ring V is formed near the head part of the contact surface C of the outflowing gas. The rearward-facing shock $S2$ appears in the outflowing gas. Near the tube exit there occurs a sort of shock B across the boundary of the discharging gas. This shock is seen both in flow discharging in free space and in flow interacting with an obstacle. This shock wave was observed in a previous investigation (e.g. Glass 1974), but the authors don't know where it was discussed. The distance of this shock from the tube exit slightly increases with time.

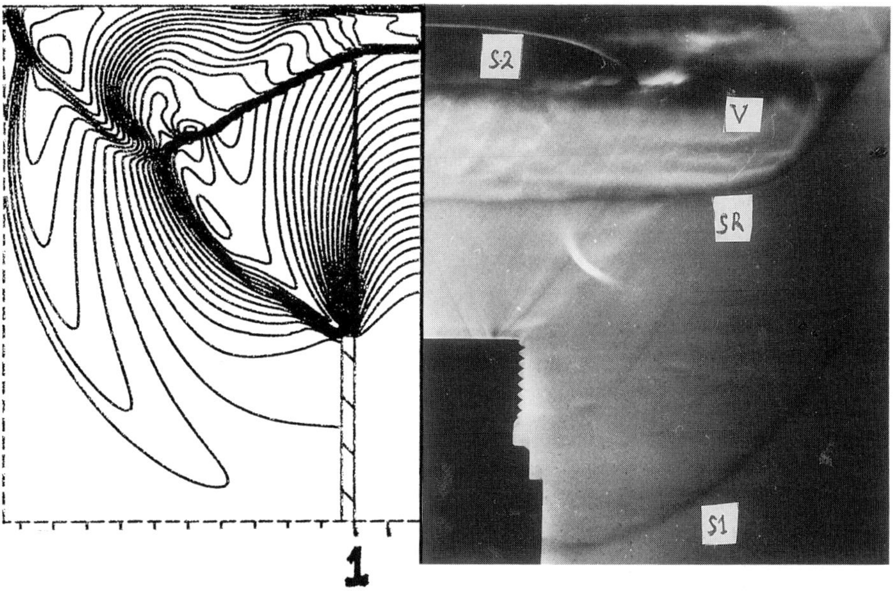

Figure 2. The calculated density contours (left) and schlieren photograph (right) of diffracted shock wave impingement on a flat plate; $M_0 = 4.7$, $P_1 = 4$ kPa, $t = 82$ μs.

As the reflected shock wave SR moves in the upstream direction, it consequently interacts with the contact surface C, secondary shock wave $S2$, and vortex ring V (Figs. 1 and 2). The comparison of calculated density contours and schlieren photographs shows that there is satisfactory agreement between them. Nevertheless, in numerical calculations the bow shock wave B is absent. The schlieren photograph in Fig. 2 shows the reflected shock wave SR in the central part of the outflowing gas, but in the calculated density contours (Fig. 2, left) the SR is seen only in the side part of the outflowing gas. Due to the pressure jump after the shock wave $S1$ reflection near the plate, the velocity of the contact surface moving toward the flate plate is decreased.

The numerical calculation of pressure $\bar{P} = P/P_1$ (curve 1) and density distributions $\bar{\rho} = \rho/\rho_1$ (curve 2) across the flate plate are shown for $t = 72$ μs (Fig. 3a). The pressure and density gain their maxima in the center and are decreased slowly up to the position of the shock wave $S1$ on the plate.

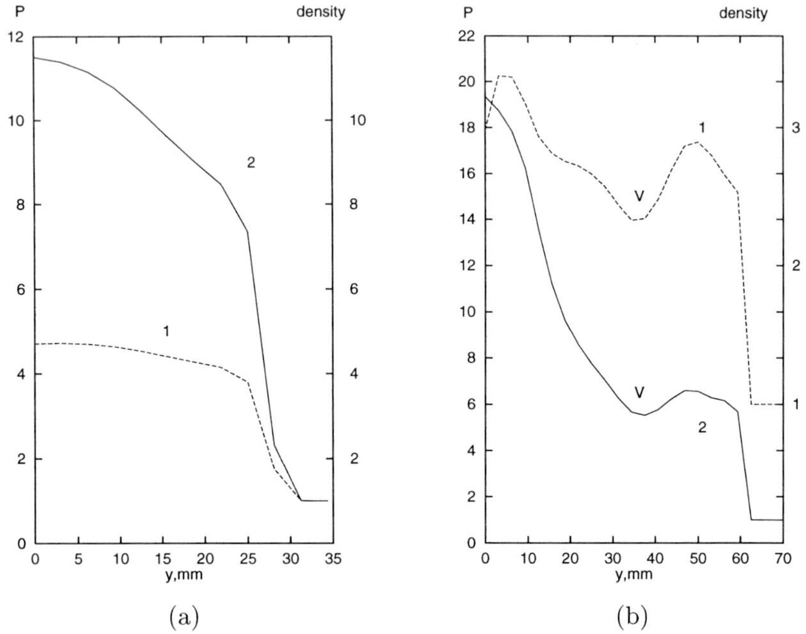

Figure 3. The calculated pressure and density distribution along the plate after impingement of a diffracted shock wave, $P_1 = 4$ kPa, $L = 50$ mm: (a) $M_0 = 3.2$, $t = 72$ μs; (b $M_0 = 4.7$, $t = 82$ μs. Pressure (1) and density (2) are normalized on their initial values.

The shock wave $S2$ after the interaction with the reflected shock wave SR comes close to the plate and occupies a quasi-stationary position. Thus a flow pattern is formed analogous to that when a supersonic under-expanded flow impinges on a plate.

The schlieren photograph in Fig. 2 (right) shows that after the interaction process between SR and the vortex ring V, the shock wave SR passes through it and SR is seen on the back part of the vortex ring V. The vortex ring V moves toward the plate and its size is increased as it comes closer to the plate. The pressure (curve 1) and density (curve 2) distributions show that in the places where the vortex ring is in contact with the plate (Fig. 3b) there are the local minima in pressure and density.

The main stages of the shock-wave/flat-plate interaction are clearly seen on the pressure $\bar{P}$ (curve 1) and temperature $\bar{T} = T/T_1$ (curve 2) histories in the central point of the plate (Figs. 4a and 4b). The initial impingement of the shock wave $S1$ leads to peaks in pressure and temperature. After that there is subsequent decay of pressure and temperature, but when the outflowing gas comes to the plate there occurs a secondary rise of pressure and temperature. Then the quasi-stationary parameters are formed in the central part of the flat plate. When the primary-shock Mach number is varied from 3.2 to 4.7, the qualitative pattern of the interaction process is the same in both cases, but the quantitative influence of the initial Mach number leads to greater pressure and temperature jumps in the initial stage and greater quasi-stationary parameters on the plate.

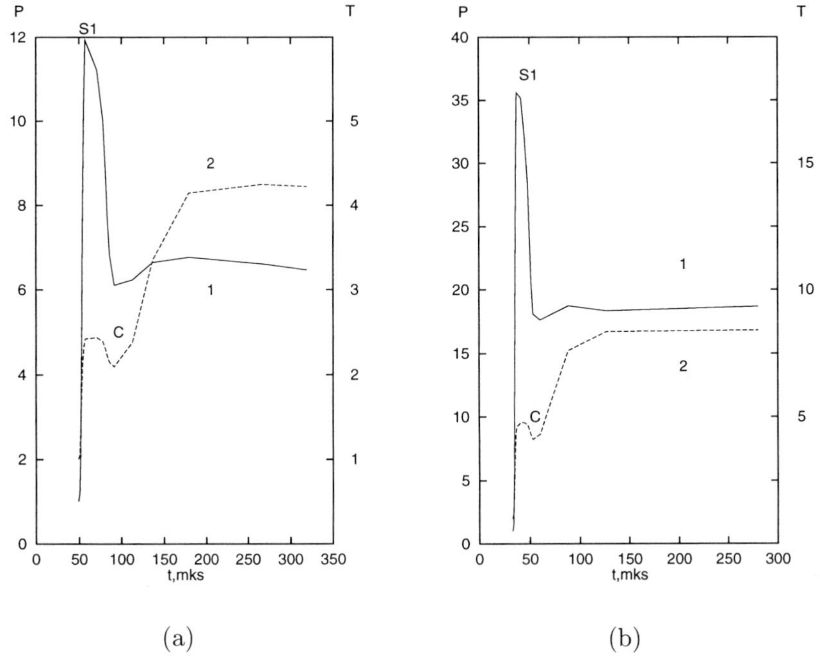

Figure 4. The calculated pressure and temperature time histories after the impingement of a diffracted shock wave on the center point of the plate for $L = 50$ mm: (a) $M_0 = 3.2$; (b) $M_0 = 4.7$. Pressure (1) and temperature (2) are normalized on their initial values.

The quasi-stationary temperature at $M_0 = 4.7$ is near to the stagnation temperature of the flow behind the incident shock wave; at $M_0 = 3.2$ it is less than that.

4. Conclusions

The complex two-dimensional flow with shock waves and vortex ring which arises after the impingement of a shock wave on a plate was studied by experimental and numerical methods. During this interaction, conditions for different kinds of instability are realized. These instabilities and the vortex produced by shock-wave diffraction play a significant role on the resulting surface-pressure and temperature distributions. The following results were obtained:

- When a diffracted shock wave impinges on a plate, regular reflection accounts for the pressure and temperature distributions in the central part of the plate, but on the side part of the plate Mach reflection accounts for such distributions.
- Pressure and density distributions on the plate after shock impingement on the contact surface are determined by the vortex ring.

Acknowledgement. The present research was supported by Grant No. 94-01-00218-a and by Grant No. 94-01-01461-a of the Russian Fundamental Research Foundation.

References

Abe A, Takayama K (1989) A peculiar shape in the transition process of a shock wave discharged from an open end of a shock tube. In: Proc 1989 National Symp on Shock Wave Phenomena, Japan, pp 41–54.

Baird JP (1987) Supersonic vortex rings. Proc R Soc Lond A409:59–65.

Bazarov SB, Bazhenova TV, Golub VV, Shulmeister AM (1993) Three-dimensional shock ejection from a channel. In: Proc 19th Int Symp on Shock Waves, Vol. 4, Sendai, Japan, pp 135–138.

Glass II (1974) Shock Waves and Man, University of Toronto Institute for Aerospace Studies, Toronto, p 169.

Phan KC, Stollery YL (1985) On the effects of shock wave reflection in a confined space. In: Proc 15th Int Symp on Shock Tubes and Waves, pp 135–145.

Analyses of reflection and diffraction of weak shock waves

Y. Takano, N. Iwamoto, T. Goto
Tottori University
Tottori, 680 Japan

Abstract: In the present paper, analyses are made of two-dimensional flow fields generated by reflection and diffraction of weak shock waves on a wedge at an angle of attack. Similarity solutions are obtained for velocity fields in addition to those for density (pressure) fields which had already been obtained by Keller and Blank (1951). Combining these similarity solutions, analytical solutions are constructed that describe reflection processes of weak shock waves on a triple reflector.

Key words: Analysis, Weak shock wave, Reflection, Diffraction

1. Introduction

Gasdynamic phenomena induced by weak shock waves may be modeled using the linearized gasdynamic equations or acoustic equations. Several mathematical theories have been developed for acoustic shock waves based upon the wave equation, which are reviewed by Friedlander (1958). Keller and Blank (1951) presented analytical similarity solutions for pressure fields induced by reflection and diffraction of plane acoustic shock waves from a wedge or a corner. Takano and Karakawa (1993) compared numerical solutions of the two-dimensional Euler equations for pressure fields induced by incident shock waves with Mach number 1.05 with the analytical solutions of Keller and Blank (1951) and obtained fairly good agreement.

In the present paper, we perform two kinds of analyses that extend the analyses of the wave equation of Keller and Blank (1951). First we analyze the acoustic equations and obtain similarity solutions for velocity fields in addition to those for density (pressure) fields induced by the shock waves incident on a wedge at an angle of attack. Secondly we construct analytical solutions for the density fields that describe reflection processes of weak shock waves on a triple reflector by superposing the similarity solutions of the wave equation.

2. Analyses of the flow field around a wedge

The basic equations for the analysis of flow fields generated by reflection of weak incident shock waves around a wedge are acoustic (linearized gasdynamic)

equations. Introducing the new variable $R = r/t$ to the acoustic equations in polar coordinates (r, θ) with the origin at the apex of the wedge, where t is the time elapsed after arrival of the incident shock waves at the apex, we have

$$R\frac{\partial \rho}{\partial R} = \frac{\partial v_r}{\partial R} + \frac{1}{R}\frac{\partial v_\theta}{\partial \theta} + \frac{v_r}{R}, \tag{1}$$

$$R\frac{\partial v_r}{\partial R} = \frac{\partial \rho}{\partial R}, \qquad R\frac{\partial v_\theta}{\partial R} = \frac{1}{R}\frac{\partial \rho}{\partial \theta}. \tag{2}$$

Here ρ is the density normalized by that in the region of undisturbed stationary gas, and v_r and v_θ are the radial and angular components of the velocity, respectively, which are normalized by the acoustic speed of the stationary gas.

Eliminating the velocity components from Eqs. (1) and (2), we obtain

$$\frac{\partial}{\partial R}\left\{(1 - R^2)\frac{\partial \rho}{\partial R}\right\} + \frac{1}{R}\frac{\partial \rho}{\partial R} + \frac{1}{R^2}\frac{\partial^2 \rho}{\partial \theta^2} = 0, \tag{3}$$

which had been derived from the wave equation in two-dimensional space by Keller and Blank (1951). They had presented analytical solutions for the pressure (density) field which are defined within a circular sector of diffraction around the apex. For example, an analytical solution of the variation of density $\rho(R, \theta)$ due to reflection and diffraction around a wedge at an angle of attack, as is illustrated in Fig. 1a, can be written

$$\rho(R, \theta) = f(q, \omega; \alpha_1) + f(q, \omega; \alpha_2), \tag{4}$$

where

$$f(q, \omega; \alpha) = \frac{1}{\pi}\tan^{-1}\frac{(1 - q^2)\sin\alpha}{(1 + q^2)\cos\alpha - 2q\cos\omega}, \tag{5}$$

$$q = \left[\frac{R}{1 + (1 - R^2)^{1/2}}\right]^\lambda, \qquad \omega = \lambda\theta. \tag{6}$$

Here $\alpha_1 = \lambda\psi_1, \alpha_2 = \lambda(\phi_1 - \psi_2)$, $\lambda = \pi/\phi_1$, $\phi_1 = 2\pi - \phi_0$ and $\phi_0 = \psi_1 - \psi_2$, and the angles ψ_1 and ψ_2 $(0 < \psi_2 \leq \psi_1 < \pi/2)$ are indicated in Fig. 1a.

Figure 1b shows a density-contour diagram of $\rho(R, \theta)$ in the case with $\psi_1 = \pi/4$ and $\psi_2 = \pi/12$. It is observed that the incident shock wave coming from right to left diffracts on the upper surface, reflects on the lower surface, and generates a circular sector of diffraction around the apex.

Integrating Eqs. (2) from the boundary of the circular sector ($R = 1$) to R $(0 < R < 1)$, the velocity within the circular sector can be expressed as

$$v_r(R, \theta) = v_r(1, \theta) + \int_1^R \frac{1}{R}\frac{\partial \rho(R, \theta)}{\partial R}dR, \tag{7}$$

(a) (b)

Figure 1. Wave pattern and polar coordinate (a), and density field (b).

$$v_\theta(R,\theta) = v_\theta(1,\theta) + \int_1^R \frac{1}{R^2}\frac{\partial\rho(R,\theta)}{\partial\theta}dR. \tag{8}$$

For this case the boundary conditions for the formulas for velocity components (7) and (8) are expressed as

$$v_r(1,\theta) = \begin{cases} 2\cos\psi_1\cos\theta, & 0 \le \theta < \psi_1 \\ \cos(\psi_1+\theta), & \psi_1 < \theta < \phi_1 - \psi_2 \\ 0 & \phi_1 - \psi_2 < \theta \le \phi_1 \end{cases} \tag{9}$$

$$v_\theta(1,\theta) = \begin{cases} -2\cos\psi_1\sin\theta, & 0 \le \theta < \psi_1 \\ -\sin(\psi_1+\theta), & \psi_1 < \theta < \phi_1 - \psi_2 \; . \\ 0 & \phi_1 - \psi_2 < \theta \le \phi_1 \end{cases} \tag{10}$$

Figure 2 shows the velocity field in the case with $\psi_1 = \pi/4$ and $\psi_2 = \pi/12$. Figure 2a shows vector plots of the flow field combined with the density field and Fig. 2b shows a close-up view near the apex of the wedge. The present analyses predict that the velocity tends to infinity at the apex of the wedge at non-zero angle of attack, although the density converges to a value of $2\pi/\phi_1$ there. It is observed in the close-up view of the velocity field that a flow occurs from the lower surface to the upper surface and a stagnation point exists on the lower surface in the neighborhood of the apex.

3. Analyses of reflection processes from a triple reflector

Figure 3a shows schematic diagrams of reflection processes from a triple reflector consisting of a flat surface and wedges on both sides: (1) an incident shock wave

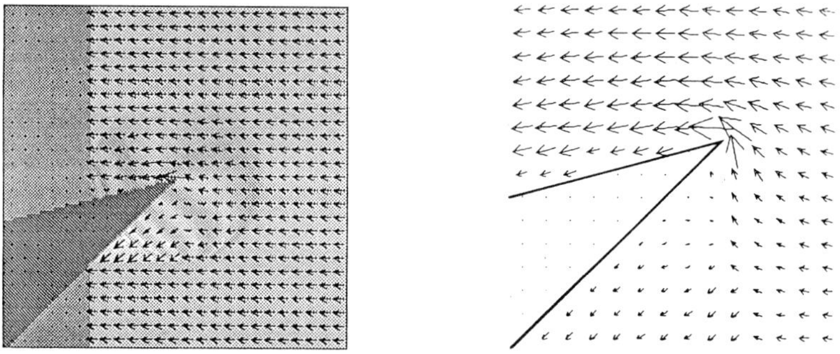

(a) (b)

Figure 2. Velocity field (a) and its close-up view near the apex of the wedge (b).

moving from right to left reflects on the inclined surfaces of the wedges on both sides and circular sectors of diffraction develop from the apexes A and A'; (2) it reflects from the front surface and other circular sectors also develop around corners B and B'. Hence analytical solutions can be constructed for flow fields generated by reflection on a triple reflector by superposing similarity solutions around wedges and corners. As a result of the same kind of analyses as those in the previous section, the pressure (density) field around a corner B between the front surface and the inclined surface at $\phi_0 < \pi/4$ is represented by two combinations of functions $p_1 + p_2$ which are defined as follows

$$p_1(r_B, \theta_B, t_B) = \begin{cases} 0, & r_B/t_B > 1/\cos(\theta_B - \phi_{BS}) \\ 0, & 1 < r_B/t_B < 1/\cos(\theta_B - \phi_{BS}) \text{ and } \phi_{BS} < \theta_B \leq \phi_B \\ 1, & 1 < r_B/t_B < 1/\cos(\theta_B - \phi_{BS}) \text{ and } 0 \leq \theta_B < \phi_{BS} \\ 1 - f(q_B, \omega_B; \alpha_{BS}), & r_B/t_B < 1 \end{cases} \tag{11}$$

$$p_2(r_B, \theta_B, t_B) = \begin{cases} 0, & r_B/t_B > 1/\cos(\theta_B - \phi_{BR}) \\ 1, & 1 < r_B/t_B < 1/\cos(\theta_B - \phi_{BR}) \text{ and } \phi_{BR} < \theta_B \leq \phi_B \\ 0, & 1 < r_B/t_B < 1/\cos(\theta_B - \phi_{BR}) \text{ and } 0 \leq \theta_B < \phi_{BR} \\ f(q_B, \omega_B; \alpha_{BR}), & r_B/t_B < 1 \end{cases} \tag{12}$$

where

$$q_B = \left[\frac{r_B/t_B}{1 + \{1 - (r_B/t_B)^2\}^{1/2}} \right]^{\lambda_B}, \qquad \omega_B = \lambda_B \theta_B. \tag{13}$$

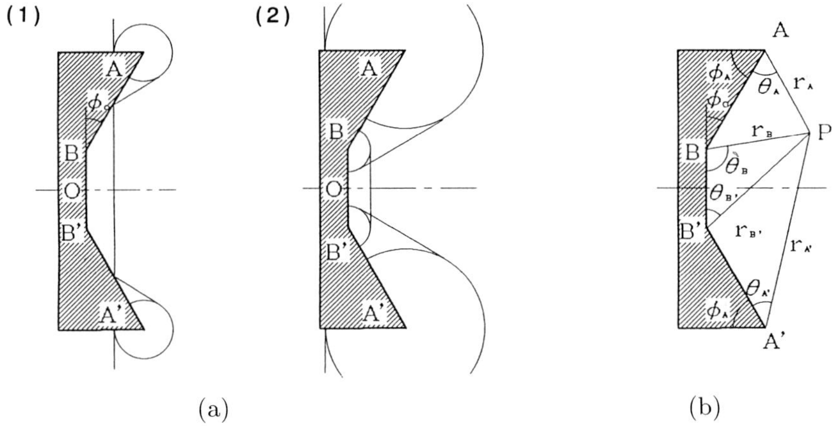

Figure 3. Schematic diagrams of reflection processes from a triple reflector (a) and polar coordinate systems fixed with the reflector (b).

Here a polar coordinate (r_B, θ_B) is defined in Fig. 3b and t_B is the time elapsed after the reflection of the incident shock wave on the front surface, $\phi_{BS} = \pi/2 - 2\phi_0$ and $\phi_B = \pi - \phi_0$ are defined using ϕ_0 indicated in Fig. 3b, $\phi_{BR} = \pi/2$, $\lambda_B = \pi/\phi_B$, $\alpha_{BS} = \lambda_B\phi_{BS}$, and $\alpha_{BR} = \lambda_B\phi_{BR}$. Effects of waves developing around the other corner B' are represented by a function $p_3(r_{B'}, \theta_{B'}, t_B)$ that is a mirror image to $p_1(r_B, \theta_B, t_B)$, and by a function $p_4(r_{B'}, \theta_{B'}, t_B)$ that is the contribution of circular sector around B'. Also, an effect of diffraction from the apex A is expressed as $p_5(r_A, \theta_A, t_A)$. Thus, the pressure field generated by reflection of the incident shock wave on the triple reflector is expressed as

$$p = 1 + p_1(r_B, \theta_B, t_B) + p_2(r_B, \theta_B, t_B) + p_3(r_{B'}, \theta_{B'}, t_B) + p_4(r_{B'}, \theta_{B'}, t_B) + p_5(r_A, \theta_A, t_A). \tag{14}$$

Here a polar coordinate (r_A, θ_A) is defined in Fig. 3b and t_A is the time elapsed after arrival of the incident shock wave at the apex A of the wedge. Figure 4 shows density contour diagrams for the reflection processes on a triple reflector: (a) an incident shock wave moving from right to left reflects on inclined surfaces of the wedges on both sides, and circular sectors of diffraction develop from the apexes; (b) the incident shock just arrives at the front surface; (c) it reflects from the front surface and other circular sectors also develop from both corners; (d) these circular sectors and the normal and oblique reflected shock waves overlap so that complicated wave patterns are observed.

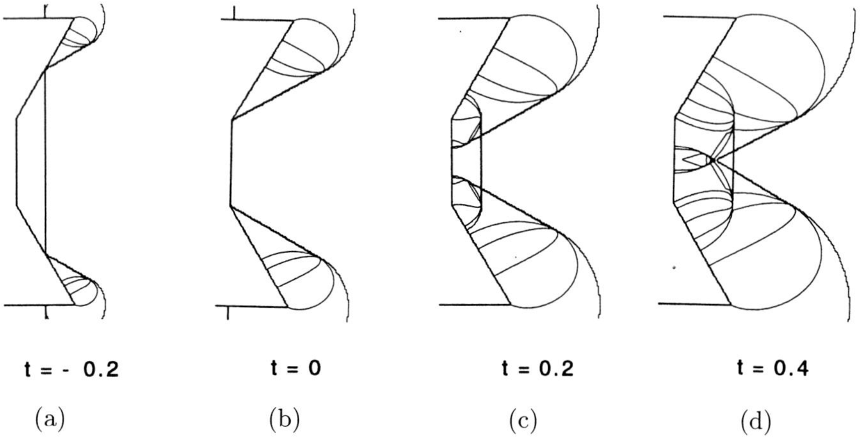

Figure 4. Density contour diagrams of reflection processes: (a) $t = -0.2$, (b) $t = 0$, (c) $t = 0.2$, (d) $t = 0.4$.

4. Concluding remarks

In the present paper, as a result of analyses that use the two-dimensional acoustic (linearized gasdynamic) equations, similarity solutions are obtained for velocity fields in addition to those for the density (pressure) field generated by reflection and diffraction of weak shock waves on a wedge at an angle of attack. Also, it is shown that analytical solutions can be constructed for reflection processes of weak shock waves on a triple reflector by superposing these similarity solutions.

Acknowledgement. The present investigation is partly supported by the Secom Foundation's Research Grant.

References

Friedlander FG (1958) Sound Pulses, Cambridge University Press.

Keller JB, Blank A (1951) Diffraction and reflection of pulses by wedges and corners. Comm Pure Appl Math 4:75–94.

Takano Y, Karakawa K (1993) Comparisons between simulations and analyses of weak shock waves reflected by a wedge at attack angle. In: 5th Int Symp on Computational Fluid Dynamics, Sendai, Vol III, pp 187–192.

Computation and validation of shock-wave interaction with fluid inhomogeneities

J. Falcovitz
Institute of Mathematics, The Hebrew University
Jerusalem 91904, Israel

K. Yamada, H. Nagoya, K. Takayama
Shock Wave Research Center, Institute of Fluid Science
Tohoku University, Katahira 2-1-1, Sendai, Japan

Abstract: We study the interaction between a shock wave propagating in a homogeneous fluid and an enclosed bubble of a different fluid. Two types of such phenomena are considered: the shock-wave breakup of a helium bubble in air and the interaction of a micro-explosion water shock with an air bubble. Numerical simulations are obtained from the 2D GRP scheme extended to treat multi-fluid flows. The scheme employs "volume-of-fluid" information in cells to approximately track interfaces between dissimilar (and immiscible) fluids. Computational results are compared to experimental visualizations and good agreement is demonstrated.

Key words: Shock/bubble interaction, Multi-fluid scheme

1. Introduction

The encounter between a shock wave propagating in a homogeneous fluid and an enclosed bubble of a different fluid gives rise to diverse interaction patterns. Here we specifically consider two types of such phenomena: (i) shock-wave breakup of a helium bubble in air; (ii) interaction of a shock obtained from a micro-explosion in water with an air bubble. In the following we outline the principal features of the numerical scheme. We then present three instances of multi-fluid shock-wave phenomena, comparing experimental observations with numerical simulations.

2. Material Interface Tracking scheme

The numerical computations were performed using the Generalized Riemann Problem (GRP) scheme for compressible flows in an operator-split 2D mode. An extension of GRP into a scheme for the simulation of multi-fluid shock-wave phenomena was developed and named Material Interface Tracking (MIT). In MIT, pressure and velocity continuity at interfaces are obtained by treating the fluids in 'mixed cells' as a homogenized mixture of fluids, having uniform pressure and velocity. The advection of the correct fluid species is obtained by reconstructing the interface piecewise per cell, using solely volume-of-fluid (VOF) information. In situations where an interface cannot be uniquely reconstructed in a particular cell, the scheme reverts to the less accurate antidiffusive measure of preferentially advecting the fluid type matching that of the receiving cell, so as to maintain sharp interfaces.

3. Experimental validation

As preliminary validation of the GRP/MIT scheme, we considered the $M_s = 1.25$ shock-wave breakup of a spherical helium bubble in air (Fig. 1), which compares well with the experimental shadowgraphs obtained by Haas and Sturtevant (1987). The comparison in this case relied mostly on the air/helium interface, which was visible in the shadowgraphs throughout the interaction.

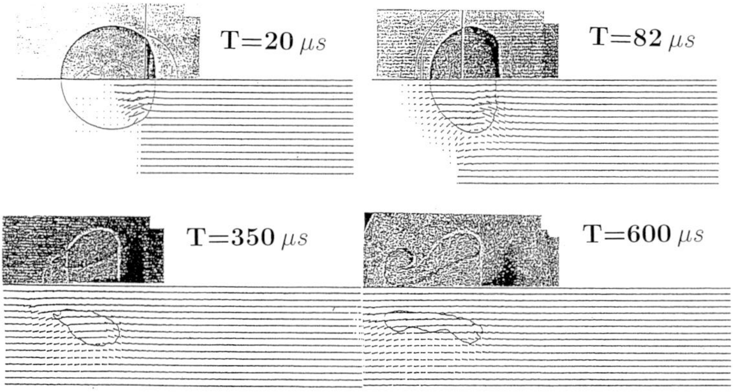

Figure 1. Interaction of $M = 1.25$ air shock with spherical helium bubble (Haas and Sturtevant 1987).

A more detailed observation of the interaction of an $M_s = 1.2$ air shock with a cylindrical He bubble was obtained by recording a series of holographic interferometric pictures at times where the incident shock was passing through the bubble and shortly thereafter (Fig. 2). These interferograms clearly show the following features: shock fronts both in air and within the bubble; the He/air interface; and isopycnics in regions of continuously varying density such as the rarefaction waves arising from the passing of the incident air shock into and around the helium bubble. These features agree well with the GRP/MIT simulation, constituting a physical validation of the method.

The water-shock experiment was conducted as follows. A 10-mg silver azide pellet was immersed in water and was detonated at the moment that an air

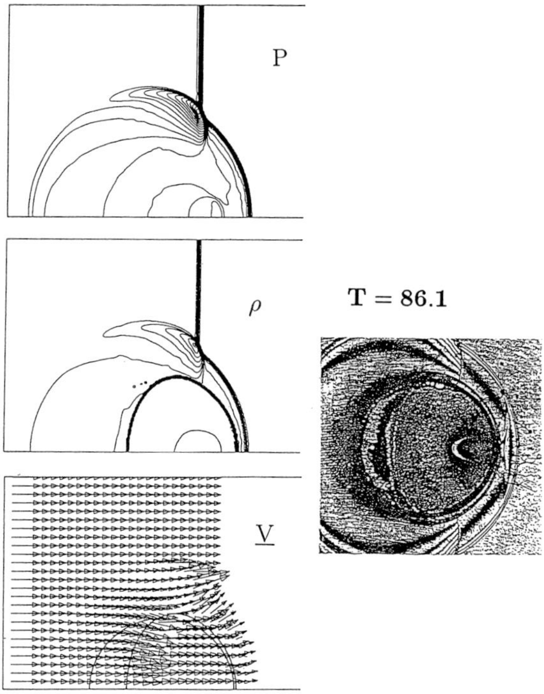

Figure 2. Interaction of $M = 1.20$ air shock with cylindrical helium bubble.

bubble (diameter about 3 mm, initial pressure 0.1 MPa) was located 10 mm away (center-to-center distance). The interaction between the incident shock and the bubble was recorded by a series of shadowgraph pictures taken at 1-μs time intervals (Fig. 3). The shock overpressure at the range of 10 mm was independently measured as about 65 MPa. The following features were observed on the shadowgraphs: the contracting-expanding air bubble; the (nearly spherical) incident wave; a (nearly spherical) rarefaction wave; and a (nearly spherical) rebound shock wave.

The numerical simulation in this case was considerably more involved than in the air/helium case. The (presumably spherical) explosion was computed at a separate 1D computation, starting out from a quiescent compressed gas bubble having the mass and energy densities of burned silver azide, and surrounded by water. At the moment where the water shock overpressure was about 100 MPa, the 1D flow field was recorded and subsequently transformed as initial conditions for the 2D computation. It was found that a simplified scheme where the spherical blast wave was obtained by replacing the silver azide gas by some

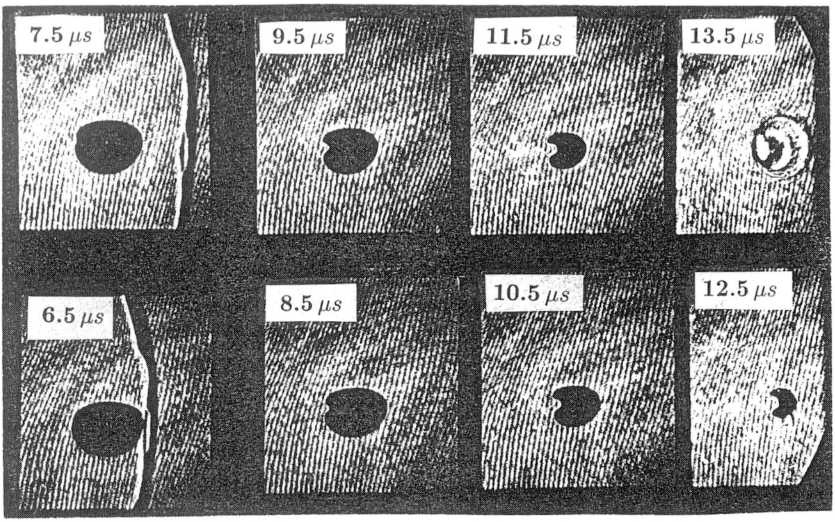

Figure 3. Interaction of shock from 10-mg silver azide explosion in water with air bubble.

equivalent compressed water sphere could be matched to yield the correct shock overpressure. However, the resulting blast wave also contained an incorrect steeper pressure gradient at the shock front. Thus, the silver azide gas bubble had to be maintained in the simulation throughout.

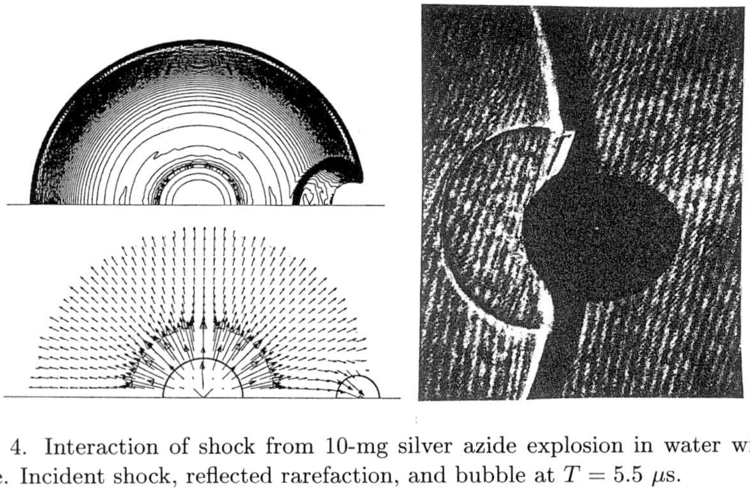

Figure 4. Interaction of shock from 10-mg silver azide explosion in water with air bubble. Incident shock, reflected rarefaction, and bubble at $T = 5.5$ μs.

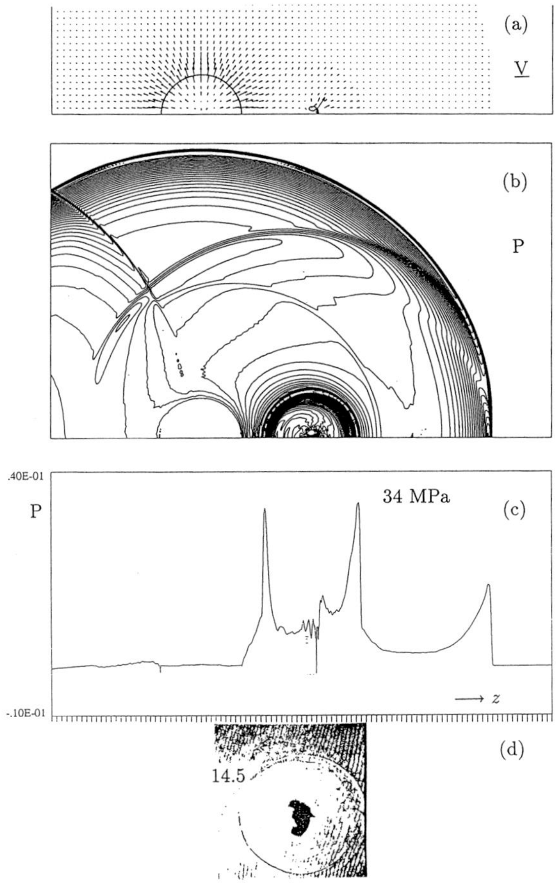

Figure 5. Interaction of shock from 10-mg silver azide explosion in water with air bubble. Rebound shock and bubble at $T = 17.5$ μs: (a) velocity and interface plot; (b) isobars; (c) pressure distribution along axis; (d) shadowgraph (at $T = 14.5$ μs).

Turning to the computation results, all three features previously mentioned were observed in the simulation, with fair — although not complete — agreement with the experimental findings. The rarefaction wave is observed as the incident shock diffracts around the bubble at time 5.5 μs (times are from moment of detonation). A good agreement between the computational and experimental observation is clearly found (Fig. 4). The peak of the computed bubble contraction was observed at about 15 μs. The rebound shock is hence generated from that moment on, and is clearly visible, for example at 17.5 μs (Fig. 5). The shadowgraph shock (Fig. 5) is in qualitative agreement with the computed rebound shock, although it is earlier by about 3 μs. From the recorded time history of average volume and pressure of the bubble, it is found that the peak contraction was at a volume ratio of about 1/300. The corresponding peak bubble pressure was about 320 MPa, whereas by isentropic compression the peak pressure for that contraction ratio would be about 290 MPa. The difference may be attributed to entropy increase during the dynamic compression of the air bubble. A further comparison of computed results with experimental shadowgraphs (the figures are not shown here) indicates the following findings. The first shadowgraph observation of a rebound shock is at about 13.5 μs, which is earlier by about 3 μs than the computation. At about 15.5 μs in the experiment, the rebound shock seems weaker and its front is at a radius of some 6 mm. This front radius is obtained in the computation at about 18.5 μs. Again, the experiment is earlier than the computation by about 3 μs. We do not know the reason for this time discrepancy. But other than that disagreement, the computed and experimental findings agree fairly well.

This study clearly demonstrates the benefit of CFD in enhancing our understanding of the observed phenomena and, furthermore, in estimating effects which cannot be directly measured. In particular, the rebound shock overpressure is computationally estimated as having a peak value of 320 MPa, which is about five times higher than the overpressure of the incident water shock.

Acknowledgement. This study was conducted during the first author's stay as a visiting professor at the Shock Wave Research Center in the Institute of Fluid Science of Tohoku University. The author is grateful to the Center and to its director Prof. K. Takayama for providing a stimulating and pleasant research environment.

References

Ben-Artzi M, Falcovitz J (1984) A second-order Godunov-type scheme for compressible fluid dynamics. J Comp Phys 55:1–32.

Ben-Artzi M, Falcovitz J (1986) An upwind second-order scheme for compressible duct flows. SIAM J Sci Stat Comp. 7:744–768.

Haas JF, Sturtevant B (1987) Interaction of weak shocks with cylindrical and spherical gas inhomogeneities. J Fluid Mech 181:41–76.

The influence of geometric variations on shock focusing in cylindrical cavities

H. Babinsky*, O. Onodera, K. Takayama
Shock Wave Research Center
Tohoku University
Sendai 980 Japan

E. Timofeev, P. Voinovich
Advanced Technology Center
P.O. Box 160, St. Petersburg 198103 Russia

Abstract: A variety of geometric variations to the standard semicircular two-dimensional cavity have been studied experimentally and numerically to investigate the effect on shock-wave focusing at a shock Mach number of 1.4. It was found that introduction of bluntness at the shoulders of the cavity increases the focusing pressures due to significant changes in the shock reflection structure. However, the different flow patterns also change the size and shape of the high-pressure region; thus the suitability of a geometry should be considered with the requirements of a typical application in mind.

Key words: CFD, Unstructured grid, Shock focusing, Interferometry

1. Introduction

During the past few years a number of works have been devoted to the investigation of shock-wave focusing in concave cavities (Takayama 1989). A commonly used geometry is a half-cylindrical cavity with sharp edges pointing towards the incident shock front (see Fig. 1). The aim of the present work is to analyze the influence of some geometric modifications of this shape, seen in Fig. 1, on the focusing process. In all cases the outer edge of the cavity is modified by the introduction of some degree of curvature while the central portion of the cavity remains unchanged. An experimental investigation using holographic interferometry can provide useful information on flow structure and development, however, the most interesting quantity in this problem — the pressure in the focal region — cannot be obtained directly from these results. Nevertheless the experimental data can be used effectively for the validation of a CFD scheme, which in turn offers the whole variety of gasdynamic parameters at any point of interest as well as their development with time.

---

*Present address: Department of Engineering, University of Cambridge, Cambridge, England.

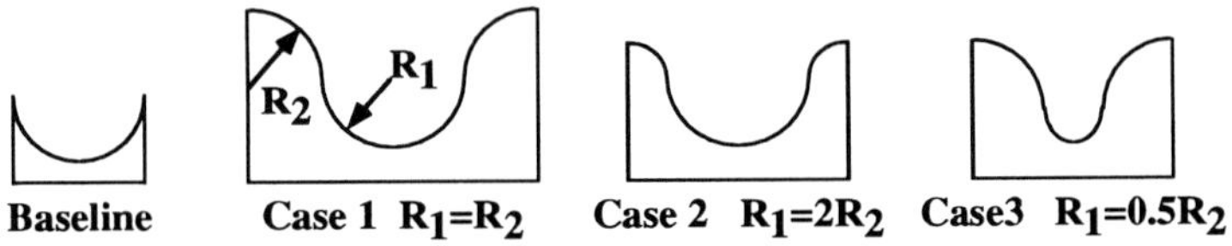

Figure 1. Shock-focusing geometries.

2. Numerical scheme

The problem under investigation involves numerous shocks and slip surfaces of rapidly changing shape combined with a curved model surface. Therefore it is perfectly suited to be simulated by a shock-capturing technique based on locally adaptive unstructured grids. The numerical scheme is described in Galyukov et al. (1996). The grid, consisting of triangles, undergoes a refinement based on the current solution at each time step. The number of total nodes can therefore be kept at a minimum while still providing a solution with very high spatial resolution. The high efficiency of this method and the scheme allow multiple computations on general purpose platforms like workstations and PCs. Some of the results presented here have been obtained in one day of computing on an IBM/486 personal computer. To visualize the results and allow direct comparison with experiments, quasi-experimental images such as interferograms and schlieren photographs have been produced using techniques described in Babinsky et al. (1995). For the purposes of image processing it was required to supply the numerical data on a regular grid. Several approaches have been tried to interpolate the data from the unstructured grid to a rectangular array and it was found that the best results were achieved if the same interpolation routines were used that have also been implemented by the numerical scheme itself, hence the same triangular structure is used for interpolation of data points between nodes. Any other approach leads to unwanted artifacts when visualizing the data. However, the interpolation process is computationally expensive and when many images were created, such as for animations, a Cray T3D supercomputer was used.

3. Results

The experimental results are available as infinite-fringe interferograms, obtained in a shock tube at a shock Mach number of 1.4 in air. The pressure upstream of the shock wave was set to atmospheric conditions and the model width was 60 mm. These parameters have been used in the calculation of numerical interferograms and the two images can be compared directly to evaluate the validity of the numerical scheme. Figure 2 shows such a comparison for case 1, demonstrating the good agreement between CFD and experiment. The maximum disagreement in fringe location was found to be within half a fringe; this is equivalent to a density difference of less than 0.5%. This level is better than the expected experimental accuracy and it can be concluded that the numerical scheme can predict the main flow very satisfactorily. Fine flow structures have been captured well by the present scheme, however, a slight smearing of contact surfaces and slip lines was observed. A similar level of agreement has been found for the other cases investigated.

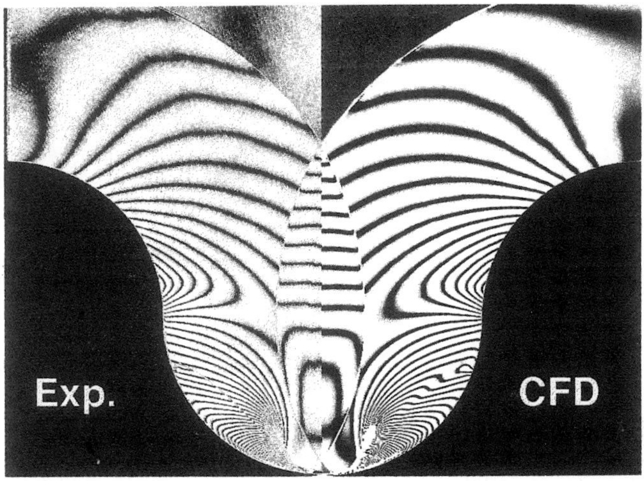

Figure 2. Comparison of experimental (left) and computational (right) interferogram.

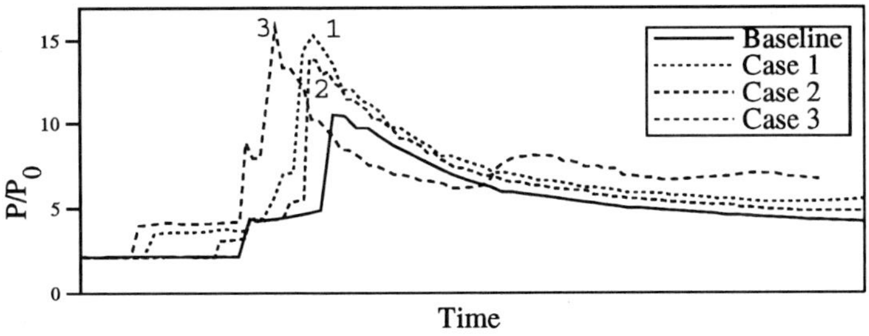

Figure 3. Maximum pressure histories at centerline for all cases.

After establishing confidence in the numerical simulation, the CFD data can be used to compare the various geometries. In Fig. 3 the maximum pressure along the centerline is plotted versus time for all cases. No fixed position was choosen, as the location of the focus on the centerline is initially unknown. It can be seen that the geometrical variations have a significant effect on the achievable pressure levels as well as on the time of focusing. The blunt cases exhibit a higher pressure at the focus and show some additional step-like pressure increases before the actual focusing. A time sequence of shock patterns is given in Figs. 4 and 5, showing shadowgraphs of the flow field together with a schematic of the equivalent location on the pressure curve. In the baseline case, the incoming shock wave undergoes an unsteady shock reflection along the concave side wall of the cavity. Hence, a Mach stem initially appears and then reduces in length. A new shock wave is formed behind the reflection point when the reflection pattern changes from Mach reflection to direct reflection. The focal pressure is achieved when the reflected shock wave and the two new shock waves merge at one point. The blunt cases show a different wave configuration as the incoming shock wave enters the circular part of the cavity. A Mach stem has already formed at this point and the reflected shocks are stronger than in the baseline case. It is this Mach stem which now undergoes inverse Mach reflection and subsequently the wave pattern leading to a focus is different for these cases. Furthermore, the reflected shock waves reach the focal point before focusing takes place, thus also increasing the pressure. The whole history of flow development is significantly different, resulting in a different focusing mechanism and efficiency.

However, the maximum focusing pressure is only one criterion for the selection of a geometry. Other factors include the location of the focus, the sharpness of the pressure peak, and the shape of the high-pressure region. In some applications it may be desirable to have a very sharp focus with low pressures everywhere else in the flow field, whereas other applications may call for a wide area of high pressure. To evaluate the various geometries under these criteria the maximum flow-field pressures, regardless of time, are seen in Fig. 6 for all geometries. It can be seen that the pressure distributions are quite different and the sharpness of the focus strongly depends on the geometry. Of particular importance is the observation that in the blunt cases some very large pressures are observed away from the centerline. These high-pressure regions could easily be missed in an experimental investigation. However, for practical purposes such high-pressure regions in other areas of the flow field are extremely important and may cause the particular geometry to be unsuitable.

4. Conclusions

The combined approach incorporating experimental and CFD methods supplemented by an advanced visualization technique has contributed greatly to the understanding of the presented problem. It was shown that modifications to the geometry of a cavity can have a significant effect on the shock-wave reflection structures observed. In particular these changes have been found to cause large differences in focusing pressure as well as in the shape of the high-

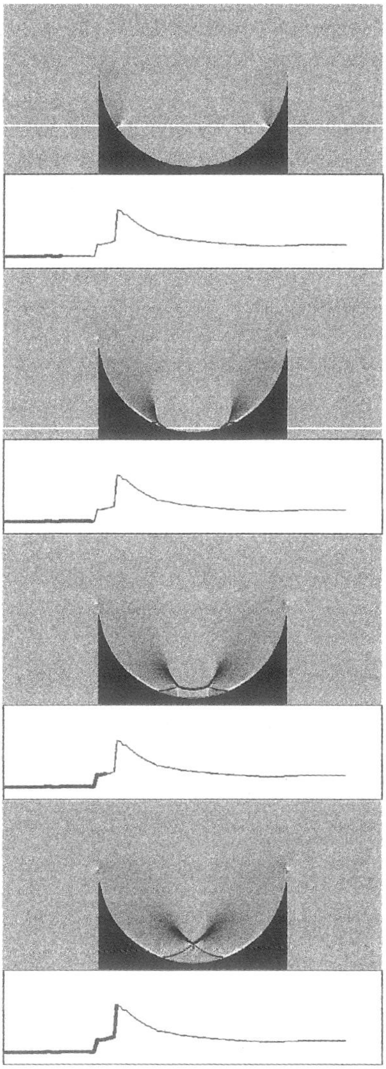

Figure 4. Sequence of numerical schlieren photographs for the CFD solution of the baseline case, together with relevant position on pressure history.

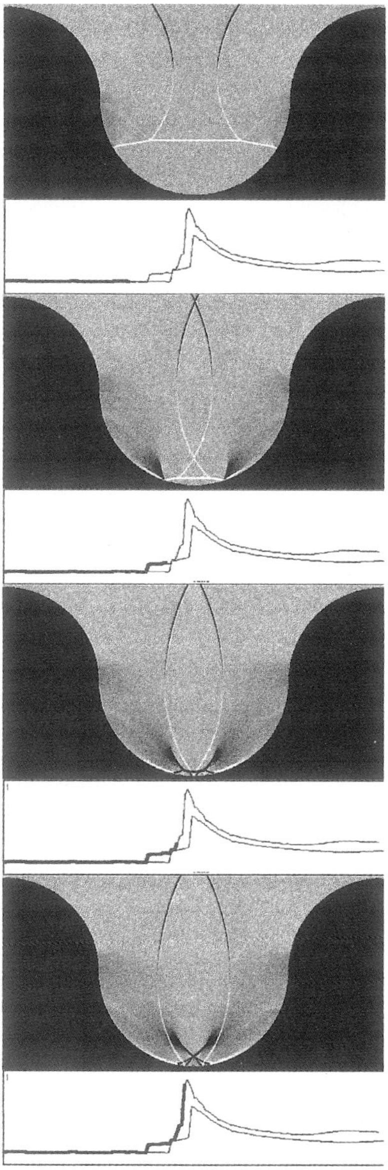

Figure 5. Same as Fig. 4, Case 1.

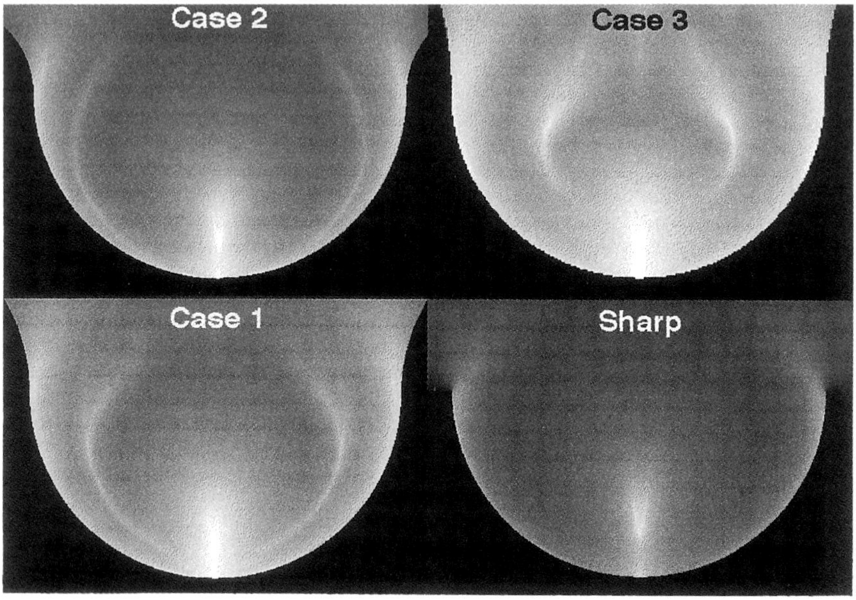

Figure 6. Maximum pressures throughout whole computation time.

pressure region. While blunted cavities can increase the maximum pressure significantly, they also increase the area subject to relatively large pressures and could therefore be unsuitable for some applications. The ideal choice of a focusing geometry cannot be determined generally, but rather depends on the particular requirement of the application.

Acknowledgement. This work was supported in part by the Russian Foundation of Fundamental Investigations (Grant No. 94-01-01456).

References

Babinsky H, Meguro T, Jiang Z, Takayama K (1995) Numerical visualization of shock wave flow in an expanding tube and comparison with experiment. In: 4th ASME/JSME Symp on Exp and Num Flow Vis, Hilton Head, USA, ASME, Vol 218, pp 89–94.

Galyukov A, Fursenko A, Voinovich P, Saito T, Timofeev E (1996) Efficient 2D and 3D locally adaptive unstructured Euler solvers for unsteady shocked gas flows. In: Shock Waves, Proc of the 20th ISSW, Sturtevant B, Shepherd JE, Hornung HG (eds), World Scientific, Vol 1, pp 691–696.

Takayama K (ed) (1989) Proc of Int Workshop on Shock Wave Focusing, Shock Wave Research Center, Inst of Fluid Science, Tohoku University, Japan.

An experimental/numerical approach to shock-wave diffraction

H. Kleine*, Q. Yu, E. Ritzerfeld†, H. Grönig
Shock Wave Laboratory, RWTH Aachen
Templergraben 55
52056 Aachen, Germany

Abstract: Fundamental diffraction processes are investigated by comparing data obtained by numerical simulation with the results of flow visualization. Some small-scale structures and details of the flow pattern that neither experiment nor simulation alone can reveal clearly may be recognized and interpreted by using this combined approach.

Key words: Shock-wave diffraction, Numerical simulation, Visualization techniques

1. Introduction

The process of shock-wave diffraction is usually considered as a fairly well-understood fundamental gasdynamic phenomenon. In fact, for the case of a wave diffracting at a sharp corner most of the essential features were already pointed out in early papers on this issue (Griffith & Brickl 1953, Skews 1967; for further investigations and a review see e.g. Bazhenova et al. 1984 and Hillier 1991). However, even for the apparently most straightforward case of a shock wave diffracting at a sharp 90° corner a number of details had not been discovered in the past. Some of these were discussed in more recent papers (Kleine et al. 1995, Hillier 1995), but a few points still require clarification.

In this contribution, experimental and numerical data are compared to answer some of the open questions concerning the flow patterns that are observed when a shock diffracts at a sharp 90° corner. The same approach is used to describe the shock diffraction around a circular cylinder.

2. Experiments

Most of the experimental data are obtained by applying a number of density-sensitive flow visualization methods, in particular holographic interferometry

*Present address: McGill University, Department of Mechanical Engineering, 817 Sherbrooke Street West, Montreal, PQ, H3A 2K6 Canada.

†Present address: TZN Unterlüß GmbH, Neuensothriether Straße 20, 29345 Unterlüß, Germany.

and color schlieren. For the flows mentioned above and a large number of other gasdynamic processes these techniques provide the almost exclusive means to obtain information about the inner structure of the flow field. The visualization techniques are described in greater detail by Kleine & Grönig (1993) and Kleine (1994). The experiments were conducted in a shock tube with a square cross section of 54.5 mm × 54.5 mm, where shock Mach numbers from $M_S = 1.05$ to $M_S = 3.0$ could be achieved for N_2 or air as test gases.

3. Numerical simulation

Since pressure and temperature behind the incident shock are moderate and since the Reynolds number of the post-shock flow is of the order of 10^5, the problem may be simulated by solving the two-dimensional, time-dependent Euler equations. The case of the 90° corner lends itself directly to using an equally spaced cartesian grid, while for more complex flow fields involving diffraction at curved surfaces, an unstructured, triangular grid with adaptive mesh refinement is chosen. The algorithm for the former case is described by Kleine et al. 1995, while details for the upwind finite-volume scheme used in the latter application are given by Yu & Grönig (1996).

4. Results and discussion

4.1. Shock-wave diffraction at a sharp 90° corner

When a shock diffracts at a sharp 90° corner, the flow field behind it is fully sub- or supersonic, when the shock Mach number is low ($M_S < 1.3$) or sufficiently high, respectively ($M_S \geq M_S^*$; $M_S^* = 2.07$ for $\gamma = 7/5$). The transition regime ($1.3 < M_S < M_S^*$) is characterized by the appearance of secondary shock systems that grow and eventually merge for increasing M_S (Skews 1967). Close to the diffracting corner, the upstream travelling expansion wave accelerates the gas beyond the speed of sound, which is indicated by a number of lambda-shaped shocks on top of the slipstream. The results of numerical simulation, which have been found to be in excellent agreement with experimental data, even suggest that there are supersonic regions embedded in a subsonic flow field where the transition to subsonic flow appears to be shock-free (Kleine et al. 1995). All our subsequent investigations, both experimental and numerical, have led to the same conclusion, although an ultimate proof for the existence of these regions has not been obtained yet.

The existence and the development of the lambda-shaped shock systems on top of the slip stream may be explained by an analogy to the flow patterns observed in stationary transonic flows over solid boundaries (Ackeret et al. 1946). There, for slightly supersonic flow ($1 < M_{flow} \leq 1.2$) the convex curvature of the boundary requires a deceleration of the flow which is established by an

Figure 1. Holographic interferograms of the flow behind a diffracting shock wave in N_2: (a) $M_S = 1.45$, $p_1 = 1.8$ bar; (b) $M_S = 1.58$, $p_1 = 1.25$ bar; (c) $M_S = 2.14$, $p_1 = 0.6$ bar.

almost normal shock. This shock can interact with the boundary layer and induce a local flow separation. For a laminar boundary layer, the separation region can develop in front of the shock thus causing another shock, an oblique shock, the front leg of the lambda-configuration. This front leg shock usually is rather weak and decays with increasing flow Mach number. When the boundary layer becomes turbulent, the separation region induced by the deceleration shock cannot make its presence felt upstream, consequently, the oblique front leg then vanishes. For the quasi-steady flow following the diffracted shock, the visualized flow fields exhibit almost exactly the same phenomena (Fig. 1). Further experiments, in which the Reynolds number of the post-shock flow was varied by changing the initial pressure p_1 while keeping M_S constant, validated qualitatively the analogy to the stationary case. In the numerical results, the main second shock is predicted correctly in all investigated cases. The preceding front leg, however, does not appear in these calculations since its formation mechanism cannot be simulated by the applied Euler code.

In the supersonic regime ($M_S \geq M_S^*$) it was already observed in early investigations (Skews 1967) that a contact surface develops behind the second shock indicating a non-uniformity of the flow field in this region. This can readily be explained by taking into account that this second shock actually consists of two parts: it is formed by the shock on top of the slip stream (see previous paragraph) which merges with a shock that develops out of a Kelvin-Helmholtz-type instability at the right-hand side of the vortex. The contact surface originates at the intersection of these shocks (Fig. 1c).

Numerical simulations (Uchiyama & Inoue 1995 as well as our own calculations) suggest that the slipstream eventually decays into a number of small vortices. Such a pattern cannot be seen on experimental records, since by the time such a decay might be observed the flow field has usually been influenced already by waves that have been reflected from the solid boundaries of the test section. Some flow pictures, however, indicate an onset of vortices along the slip stream. In addition, the observed local density minimum within the slip stream (Kleine et al. 1995) would let a breakup into several vortices appear likely.

4.2. Shock-wave diffraction at a circular cylinder

In the past, the main points of interest concerning the diffraction of a shock wave at a circular cylinder have been to locate the point of transition from regular to Mach reflection and to determine the variation of shock strength along the subsequently diffracted Mach stem (e.g. Heilig 1970). In addition to this, a combined experimental/numerical approach allows one to investigate the rather complicated flow field that is observed after the two diffracted shocks on both sides of the cylinder have crossed. Having established an excellent agreement with respect to the density contours, the numerical data may again be taken to determine other properties of the flow field that are hard or impossible to obtain experimentally, such as local hot spots or locations of maximum pressure loading. Figure 2 indicates what accuracy can be achieved for experimental and numerical density contours.

In this figure, an advanced stage of the diffraction process is shown. The diffracted portions of the shock on the upper and on the lower side of the cylinder have intersected, eventually causing the formation of a Mach stem in the center of the test section. Numerical and experimental density contours only differ by at most 20% of the density increment ($\Delta\rho = 0.054$ kg/m^3). So, with the possible exception of regions where transport processes like viscous forces or heat transfer may be significant (and our calculations therefore not valid), the density at any point of the flow field may be determined with an accuracy of $\pm$ 0.011 kg/m^3. It seems therefore justified to take numerical data for a further analysis of the flow field.

As already indicated by the density contours and further confirmed by the numerical results, the variation of shock strength along the leading shock front is comparatively moderate (61% for $M_S = 2.33$ and 58% for $M_S = 1.35$). In particular, there are no indications that the Mach stem and its associated triple points cause any distinct hot spots in the flow, although along the shock front they create the highest increase of pressure, density and temperature. The actual values of the thermodynamic variables in this region, however, are fairly below the maxima which are found to be in the stagnation region left of the cylinder.

(a)

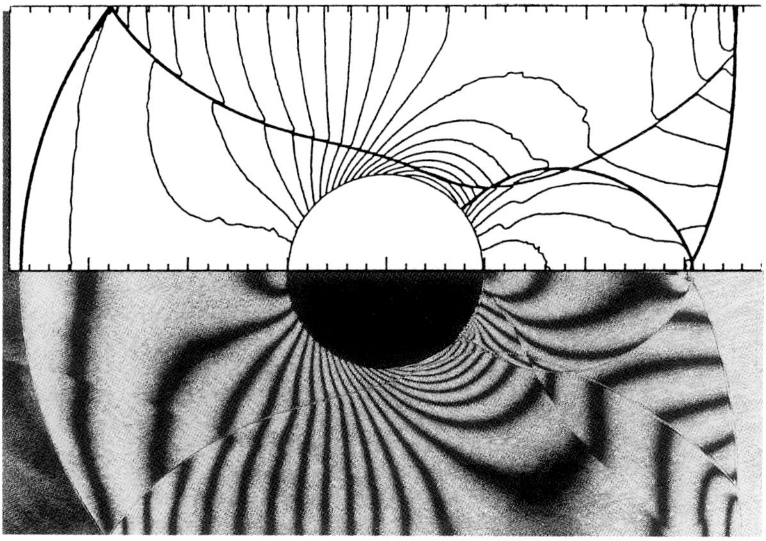

(b)

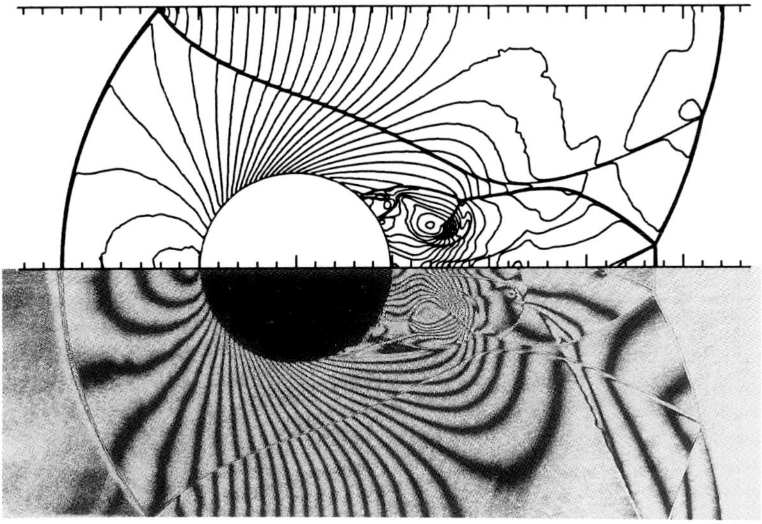

Figure 2. Shock diffraction at a circular cylinder; holographic interferogram and numerically determined density contours: test gas N_2; (a) $M_S = 1.35$, $p_1 = 1$ bar, (b) $M_S = 2.23$, $p_1 = 0.25$ bar.

In the case of $M_S = 2.23$ (Fig. 2b), numerical results indicate that the flow along the cylinder is accelerated to a maximum flow Mach number of $M_f \approx 3.2$ before it is slowed down by the upstream traveling part of the diffracted shock at the cylinder. In the final part (starting at the equator, where $M_f \approx 1.6$), an increase in particle velocity and a decrease in local temperature appear to contribute about equally to this rise of M_f.

References

Ackeret J, Feldmann F, Rott N (1946) Untersuchungen an Verdichtungsstößen und Grenzschichten in schnell bewegten Gasen. Mitt aus dem Inst f Aerodynamik, ETH Zürich.

Bazhenova TV, Gvozdeva LG, Nettleton MA (1984) Unsteady interactions of shock waves. Prog Aerospace Sci 21:249–331.

Griffith WC, Brickl DE (1953) The diffraction of strong shock waves. Phys Rev 89:451–453.

Heilig W (1970) Theoretische und experimentelle Untersuchungen zur Beugung von Stoßwellen an Kugeln und Zylindern. EMI Bericht 2/70, Freiburg.

Hillier R (1991) Computation of shock wave diffraction at a ninety degree convex edge. Shock Waves 1:89–98.

Hillier R (1995) Numerical modelling of shock wave diffraction. In: Brun R, Dumitrescu LZ (eds), Proc 19th Int Symp on Shock Waves, Vol IV, Springer, Heidelberg, Berlin, New York, pp 17–26.

Kleine H, Grönig H (1993) Visualization of transient flow phenomena by means of colour schlieren and shearing interferometry. In: Dewey J, Racca R (eds), Proc 20th Int Congr High Speed Photography & Photonics, Vol 1801, SPIE, Bellingham, pp 400–409.

Kleine H (1994) Verbesserung optischer Methoden für die Gasdynamik, Doctoral thesis, Stoßwellenlabor, RWTH Aachen.

Kleine H, Ritzerfeld E, Grönig H (1995) Shock wave diffraction – new aspects of an old problem. In: Brun R, Dumitrescu LZ (eds), Proc 19th Int Symp on Shock Waves, Vol IV, Springer, Heidelberg, Berlin, New York, pp 117–122.

Skews BW (1967) The perturbed region behind a diffracting shock wave. J Fluid Mech 29:705–719.

Uchiyama N, Inoue O (1995) Adaptive mesh refinement computation of compressible flow. In: Brun R, Dumitrescu LZ (eds), Proc 19th Int Symp on Shock Waves, Vol I, Springer, Heidelberg, Berlin, New York, pp 407–412.

Yu Q, Grönig H (1996) Numerical simulation of the reflection of detonation waves. In: Shock Waves, Proc of the 20th ISSW, Sturtevant B, Shepherd JE, Hornung HG (eds), World Scientific, Vol 2, pp 1143–1148.

Compressibility effects on the propagation and interaction of shock-generated vortices

C. Hébert, M. Brouillette
Département de Génie Mécanique
Université de Sherbrooke
Sherbrooke, Québec, J1K 2R1 Canada

Abstract: This paper examines the vortical flow regimes produced by the impulse of a shock wave emerging from the open end of a shock tube. Spark shadowgraph and schlieren photography were used to characterize the flow field. For shock-wave Mach numbers ranging from M_s = 1.0 to 2.0, three distinct flow regimes were identified. From M_s = 1.0 to 1.43, conventional-looking vortex rings are produced and are characterized by very thin cores and thus high circumferential instability wavenumbers. From M_s = 1.43 to 1.60, a shock wave becomes embedded in the supersonic recirculating region of the vortex ring. From M_s = 1.60 to 2.0, a secondary vortex ring forms ahead of the main vortex ring.

Key words: Vorticity, Compressible turbulence

1. Introduction

Compressible vorticity is important in a wide range of applications (Brouillette et al. 1995, Baird 1987, Minota 1993). Unfortunately, little experimental evidence is available to study the propagation and interaction of vortices when compressibility effects are important (Mandella et al. 1986, Baird 1987, Minota 1993, Brouillette et al. 1995). In the incompressible flow regime, vortex rings can undergo transitions from laminar to turbulent propagation, owing to the development of flow instabilities. Krutzsch (1939) observed the core of a ring on which a circumferential instability appeared spontaneously. Those observations where confirmed experimentally by Maxworthy (1972), and analytically by Widnall and Sullivan (1973).

The present paper examines the various vortical flow regimes produced when an impulse is produced by the refraction of a shock wave from the open end of a shock tube. The effects of compressibility are also identified with respect to the strength of the generating shock wave. Spark shadowgraph and schlieren photography are used to visualize the flow field, and fast-response piezoelectric pressure measurements are used to quantify the characteristics of the flow. More specifically, the effects of Reynolds and Mach numbers on the development of the flow field were examined.

2. Apparatus

A shock-tube facility was built specifically for the study of this flow. The driven section has an inside diameter of 38 mm and a length of 1.85 m, of which the last 165 mm, which constitutes the nozzle of the tube, can be replaced with different sections having inside diameters of 6.4 mm and 12.7 mm respectively. The driver section (inside diameter: 38 mm) has an adjustable length, which allows the production of a variable-width pressure pulse at the nozzle end of the shock tube. Compressed nitrogen is used in the high-pressure section, and the low-pressure section is opened to the atmosphere. The diaphragms are punctured on command with a pneumatically actuated needle plunger. Pressure histories recorded at two locations near the end of the driven section provide us with the features of the generated shock; the Mach number of the shock wave is obtained from time-of-flight measurements between those two locations.

To visualize the emerging shock wave and the ensuing vortical flow field, we used a single-spark shadowgraph/schlieren optical system which is built in a folding Z configuration with two 203-mm spherical mirrors, both with a focal length of 1.52 m; the schlieren system used a horizontal knife edge. The images were recorded with a 35-mm camera on Ilford XP2 black-and-white film (400 ASA), which was then developed using the standard C-41 film development process. The light source, a condenser spark-gap, is triggered after an adjustable delay by a pressure transducer mounted near the nozzle of the driven section.

3. Results

3.1. Experimental conditions

With the adjustable driver length of the shock tube, it was possible to change the shape, amplitude, and width of the pressure pulse produced at the nozzle end. The variation of this parameter gave birth to three flow regimes. The first was produced with a driver length greater than a critical value, for which the pressure pulse comprises a positive step function due to the shock arrival, a constant pressure plateau, and a slowly decaying region due to the arrival of the expansion waves from the driver end. The resulting flow shows evidence of a jet-like structure at the nozzle for an extended period of time and the vortex ring produced by this impulse tends to interact with the jet and break up near the nozzle. The second regime is observed using a driver length for which the aforementionned expansion waves interact with the incident shock wave exactly at the exit of the shock tube. For this critical length, the pressure impulse obtained at the nozzle is maximized for a given driver pressure, but the pressure plateau characteristic of the first regime is reduced. This tended to eliminate the jet flow and thus produce cleaner vortex rings that could propagate over longer distances; the results presented here were all obtained using this so-called critical driver length. The third regime is obtained with a driver length shorter than the critical value; in this case the expansion waves start to interact with the incident shock wave before the latter can travel to the end of the driven section, thus producing a flow field qualitatively similar to the second regime but at a lower pressure than with the critical length.

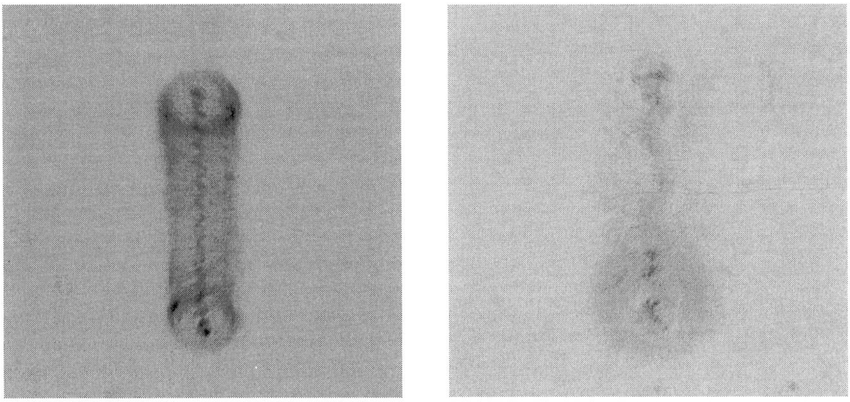

(a) (b)

Figure 1. Spark shadowgraphs of flow field for $M_s = 1.32$: (a) $t = 1.190$ ms; (b) $t = 2.756$ ms.

3.2. Flow regimes

Experiments performed with a generating shock strength between $M_s = 1$ and 2 were able to generate three distinct vortical flow regimes. The first one is illustrated in Fig. 1, showing the flow field produced by a $M_s = 1.32$ shock with a critical driver length. The propagation of the resulting vortex ring is shock-free and, as it procedes downstream, develops a high number of circumferential instabilities around its core region (Fig. 1a). At later times (Fig. 1b), the high wavenumber instabilities seem to have saturated their growth, seemingly increasing the apparent core size of the ring in the process.

Figure 2 shows the second flow regime in which a vortex ring propagates with a rearward-facing shock in the supersonic recirculating region of the ring. The core is no longer observed as a compact structure and is seemingly already highly turbulent. Far downstream of the nozzle (Fig. 2b), the embedded shock seems to disappear as the ring slows down due to the dissipative effects. As the generating shock strength is increased, a secondary vortex ring, with reverse

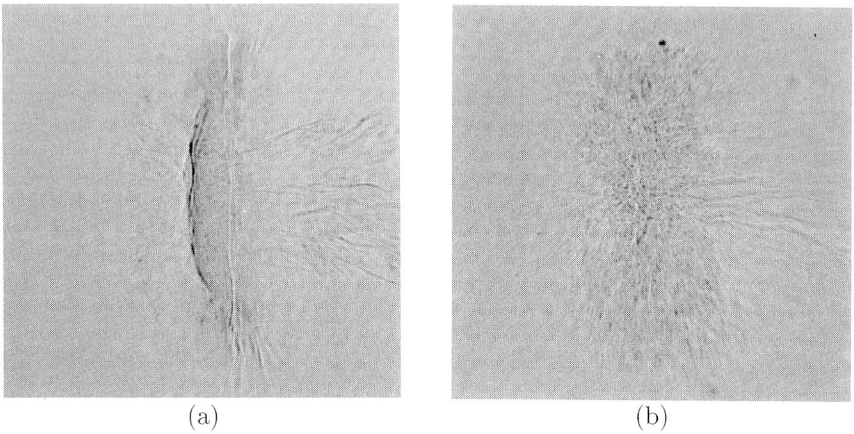

Figure 2. Spark shadowgraphs of flow field for $M_s = 1.51$: (a) $t = 0.693$ ms; (b) $t = 1.768$ ms.

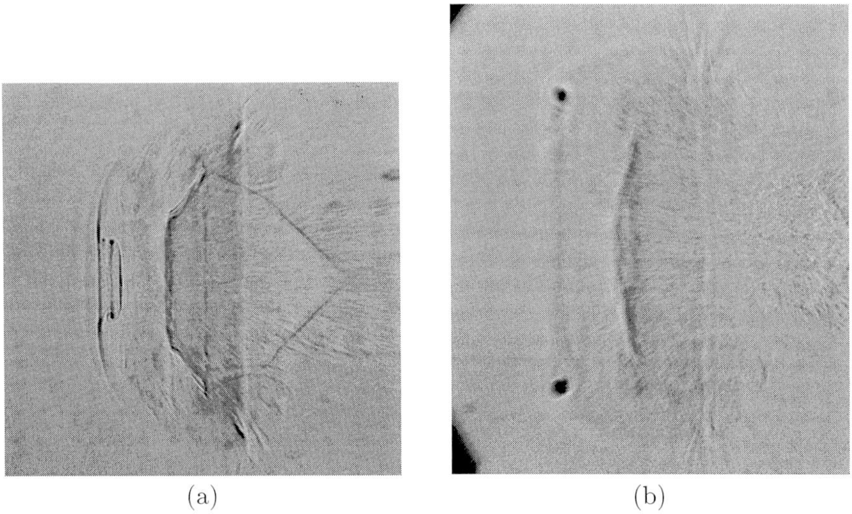

Figure 3. Spark shadowgraphs of flow field for $M_s = 1.65$: (a) $t = 0.469$ ms; (b) $t = 0.738$ ms.

circulation, is formed in front of the main vortex ring (Fig. 3). The mutual induction between the vortices causes the secondary one to increase in size rapidly as it is entrained around the main vortex. The two then seem to merge in a rather poorly defined vortical structure.

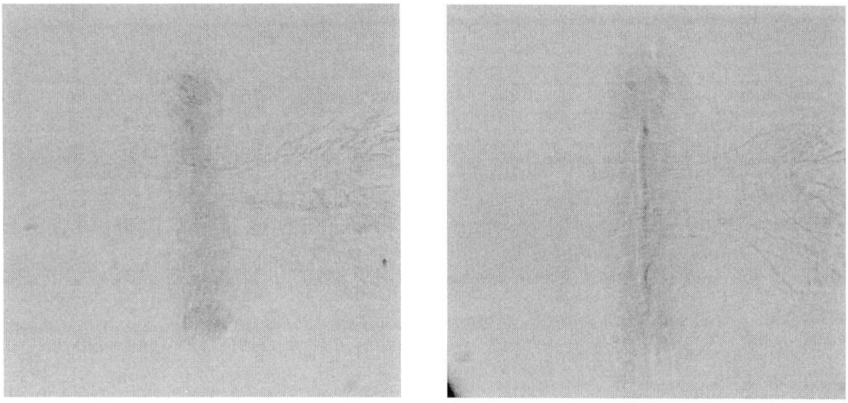

(a) (b)

Figure 4. Spark shadowgraphs of vortex rings at $t = 0.900$ ms: (a) $M_s = 1.42$; (b) $M_s = 1.44$.

3.3. Flow transitions

By varying the strength of the generating shock and using the appropriate critical driver length, it was possible to observe the transition between the three aforementionned flow regimes. Figure 4 shows the first flow transition, between the shock-free compressible regime and the supersonic recirculation regime. This transition is found to be located at $M_s = 1.43$. Previous preliminary studies (Brouillette et al. 1995) had identified this transition at $M_s = 1.34$, but this result was obtained with a driver length larger than the critical length and could thus have been affected by the strong exit-jet characteristic of long drivers.

The onset of the third flow regime takes place when the second counter-rotating vortex forms in the recirculating region of the main vortex ring (Fig. 5b). This secondary counter-rotating vortex results from the displacement of the stagnation point in front of the main vortex because of the total pressure drop across the embedded shock wave. As the shock strength is increased, the formation process is more violent, and the core of the secondary vortex is larger.

4. Conclusions

We have explored the vortical flow field produced by the impulse of a shock wave emerging from the open end of a shock tube. In the $M_s = 1.0$ to $M_s = 2.0$ range, we found that the behavior of the flow can be divided in three regimes: (1) from low supersonic up to $M_s = 1.43$, the vortex ring generated remains shock free, and circumferential instabilities tend to develop around the core region; (2) between $M_s = 1.43$ and $M_s = 1.60$, an embedded shock forms within the

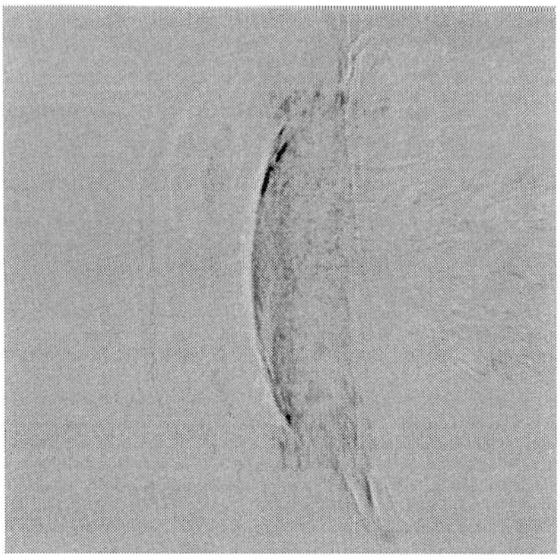

(a)

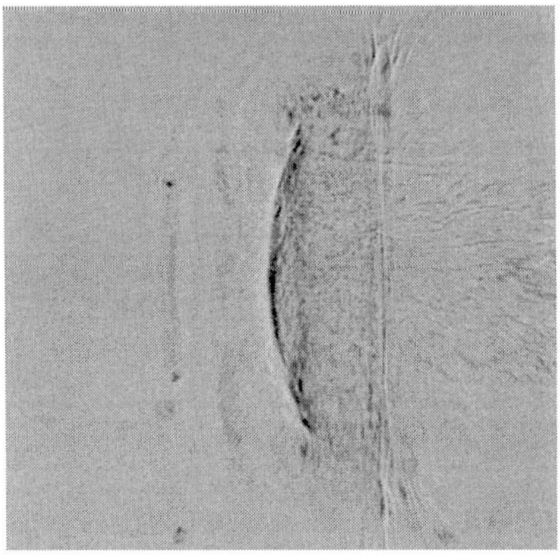

(b)

Figure 5. Spark shadowgraphs of vortex rings at $t = 0.588$ ms: (a) $M_s = 1.58$; (b) $M_s = 1.62$.

recirculating region of the ring, this shock is due to the onset of supersonic flow velocities in the recirculating region; (3) past $M_s = 1.60$, the total pressure drop through the embedded shock causes flow separation which gives birth to a secondary vortex in front of the main vortex.

Acknowledgement. This research is supported by the Natural Sciences and Engineering Research Council of Canada and the Fonds FCAR/Québec.

References

Baird JP (1987) Supersonic vortex rings. Proc R Soc (London) A409:59.

Brouillette M, Tardif J, Gauthier E (1995) Experimental study of shock-generated vortex rings. In: Shock Waves @ Marseille IV, R Brun, LZ Dumitrescu (eds), pp 361–366.

Krutzsch CH (1939) Über eine experimentelle beobachtete Erscheinung an Wirbelringen bei ihrer translatorischen Bewegung in wirklicher Flüssigkeiten. Ann Phys 5:497–523.

Mandella M, Moon YJ, Bershader D (1986) Quantitative study of shock-generated vortex flows. In: Shock Waves and Shock Tubes, Proc 15th Int Symp on Shock Waves, Bershader D, Hanson RK (eds), Berkeley, p 471.

Maxworthy T (1972) The structure and stability of vortex rings. J Fluid Mech 51:15–32.

Minota T (1993) Interaction of a shock wave with a high-speed vortex ring. Fluid Dynamics Research 12:335–342.

Widnall SE, Sullivan JP (1973) On the stability of vortex rings. Proc R Soc Lond A332:335–353.

Shock-wave diffraction by an elastic sphere in water

A. Merlen
Laboratoire de Mécanique de Lille – URA CNRS 1441
Cité Scientifique 59655 Villeneuve d'Ascq, France

P. Pernod, A. Ahyi, A. Kemmou
Institut d'Electronique et Microelectronique du Nord
Cité Scientifique 59655 Villeneuve d'Ascq, France

Abstract: Shock-wave propagation and diffraction on a sphere in water are treated by high-speed cinematography (up to 1 million shadowgraph frames/s), high-power acoustic source and broad-banded hydrophone (from 10 kHz to 10 MHz). The high-power and short-pulse source characteristics are determined, showing the relevance of this kind of source to shock-wave diffraction experiments. Diffraction on an elastic sphere is then investigated emphasizing the easy localization of different phenomena like circumferential (Franz-type) and Rayleigh-type waves. Visualizations show that the flow is far more complicated than the simple diffraction by an infinitely rigid body as it is often supposed in computational fluid dynamics.

Key words: Diffraction, Shocks in liquids, Elastic bodies

1. Introduction

The best way to investigate acoustic diffraction problems is to measure the spatial transient response of the system. The total response of the propagation system may then be simply inferred by convolution of spatial impulse response with the specific signal constituting the source signature. Nevertheless conventional acoustic sources, like piezoelectric transducers, can not be considered as point sources and have generally poor transient responses due to their naturally oscillating behavior. Experimental results can not be easily interpreted because of the great number of diffracting waves and the lack of temporal and spatial resolution. Moreover, the acoustic power radiated by these conventional sources is generally not sufficient to observe the very weak diffusing waves remaining behind the incident wave. We have therefore developed an alternative solution (minisparker), based on underwater electrical discharges, which generates a quasi-spherical impulsive wave front, with a short and simple signature. Moreover, this source, quasi-punctual both in space and time, releases high energy in such a way that a shock wave can be generated.

2. Experimental devices

2.1. Mini-sparker

A coaxial oil-filled capacitor of 18 nF, made for rapid discharges (low inductance of 10 to 40 nH), is charged through resistance and diode protections (Rp, Dp) by a high-voltage power supply adjustable from 1 kV to 10 kV. The discharge is triggered by the use of a 5C22 thyratron tube: a half-period sinusoid of current of 0.4-ms duration and 1000-A maximum is then delivered to the circuit, and a spark is created between the coaxial electrodes tips immersed in water (discharge gap 0.5 mm). As a consequence of this electric arc, a strong acoustic pulse generally followed by an oscillating cavitation bubble is generated (Pernod et al. 1994a, 1994b).

2.2. High-speed shadowgraph system and pressure measurement

Based on the Cranz-Schardin visualization principle, the system consists of 24 light flashes triggered successively after the command with an adjustable time interval (from 100 ns to 1 s). Each flash of 300-ns duration is transformed into a parallel beam by a spherical mirror of 50 cm diameter and a 4-m radius of curvature. The beam then crosses the field of interest where the mini-sparker is placed, and is focused by a second spherical mirror identical to the first one on the 24 lenses of the receiving photographic plate. For its characterization, the pulse emitted by the mini-sparker has been observed by high-speed cinematography and measured by a broad-band hand-made hydrophone (100 kHz to 10 MHz at −7 dB, Nakamura & Otani 1992).

2.3. Characteristics of the mini-sparker emission

The spatial behavior of the emitted wave is shown in the photograph extracted from the 24 photographs (Fig. 1). The pulse presents a well-marked wave front of approximately spherical shape, showing that the source can be considered punctual. The temporal behavior presented in Fig. 2a can be roughly divided into two parts: in the first one, the pressure grows very sharply from 0 to 790 mbar with a rise time of about 100 ns; in the second part, the pressure decreases more slowly in an exponential form, with the total signal duration being about 2 μs. The Fast Fourier Transform of this pulse signature is presented in Fig. 2b. We can see that the frequency spectrum is very large, extending from less than 10 kHz to over than 10 MHz, as can be expected from the steepness of the rising front, and verified by high-pass filtering of the 10-MHz cut-off frequency. Typical maximum frequencies are 2 MHz (−10 dB), 300 kHz (−5 dB), or 200 kHz (−2 dB), the spectrum amplitude being very flat for this last case.

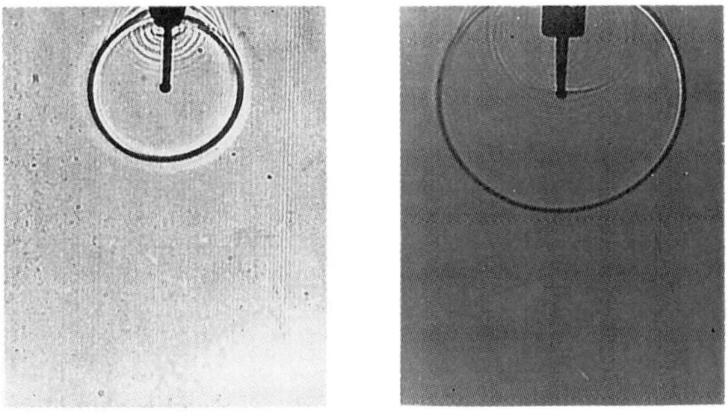

Figure 1. Typical emitted wave.

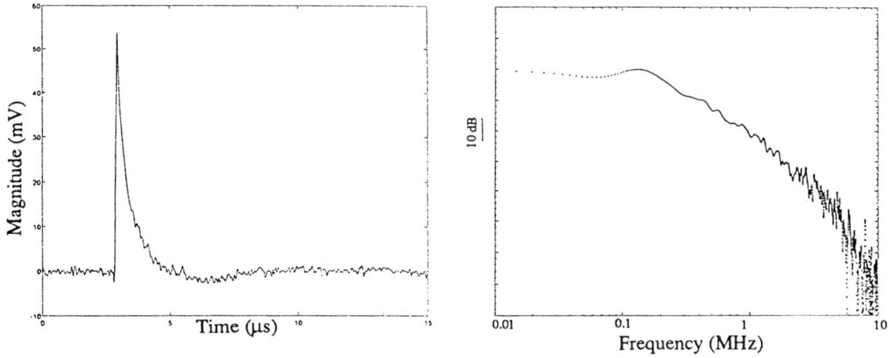

(a) (b)

Figure 2. Pressure measurement and spectrum.

3. Application to diffraction on an elastic sphere

Many papers have been published on investigations of acoustic diffraction of a plane wave by an elastic cylinder or cylindrical shell, which represents a two-dimensional problem. Considering our source properties, we have applied our investigation system to spherical wave diffraction by an elastic sphere, which represents a simple three-dimensional problem owing to symmetry properties.

3.1. Experimental set-up

A 3.5-cm diameter glass sphere is chosen as a diffracting object. It is held by a 0.1-mm thick thread which generates negligible diffraction. The sphere is placed 5 cm above the shock-wave source, as far as possible from the water surface and water-tank sides to avoid echo interferences. Several series of 24 photographs each were used to study the whole diffraction phenomenon with 2-μs intervals.

3.2. Results

Figure 3 shows a series of shadowgraphs of the wave fronts resulting from a pulse coming from below. In our interpretation of these photographs presented in Fig. 4a, we will use the notation (n,m) suggested by Brill and Uberall (1971)

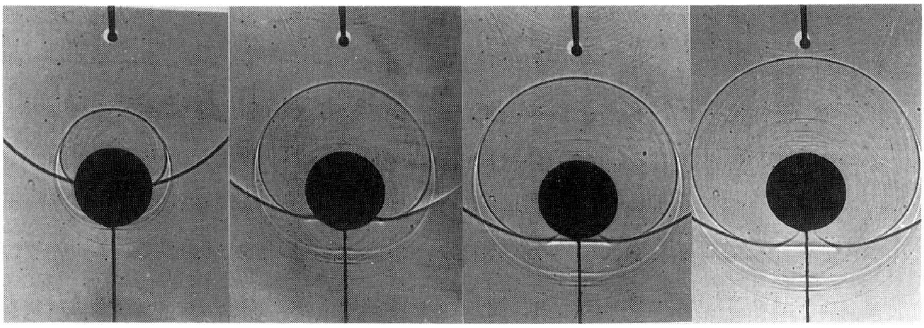

Figure 3. Shock diffraction at different stages. Incident-shock velocity $\approx$ 1480 m/s.

for identification of the different waves emerging from the target: m is the total number of reflections and n is the number of reflections as a shear wave. Five different types of waves are clearly observed: (1) The strong incident wave appears as a spherical wave front centered on the mini-sparker electrode tips and propagates downward. (2) The specular reflection (0,0) corresponds to the backscattered spherical wave front and joins the incident wave at the contact points with the glass sphere. (3) We can observe with a weaker amplitude the direct longitudinally transmitted wave through the sphere (1,0), which has anticipated the incident wave front owing to the greater speed of propagation in the solid medium. A similar transverse wave through the sphere (1,1) should theoretically be present, but is not observed on the photographs probably due to its very weak amplitude and the limited sensitivity of the visualization system. (4) Far more interesting is the very clear observation, in the shadow region of the sphere, of the Franz wave adjoining and prolonging the reflected wave (0,0). (5) We have grouped in the last category all the waves rotating around the sphere which are due to the multiple internal reflections and refractions, including the mode-converted waves. Their number is theoretically infinite; limitations in observations are due to attenuation and the limited sensitivity of the visualization system. All these waves are identified on the transducer signal (Fig. 4b). Franz waves are clearly visualized in Fig. 5a. Apart from all these identified waves, the photographs also show backscattered waves propagating behind the specular reflection which we have not been able to identify. They form a group with different speeds just like the internal reflection radiation waves (Fig. 5b).

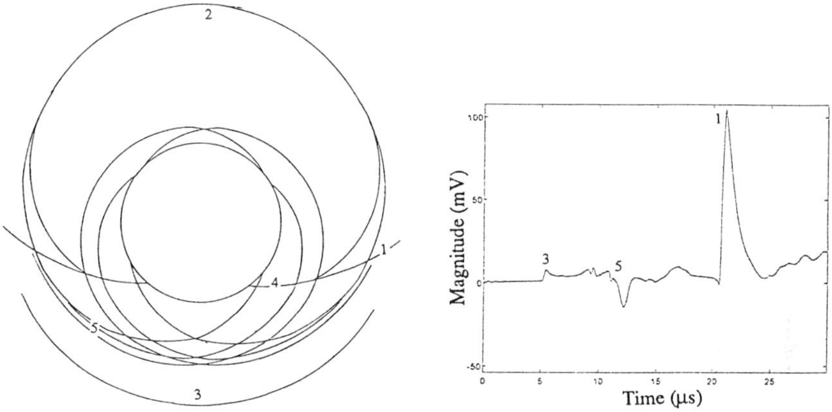

(a) (b)

Figure 4. Wave-front patterns: (a) diagram, (b) signal.

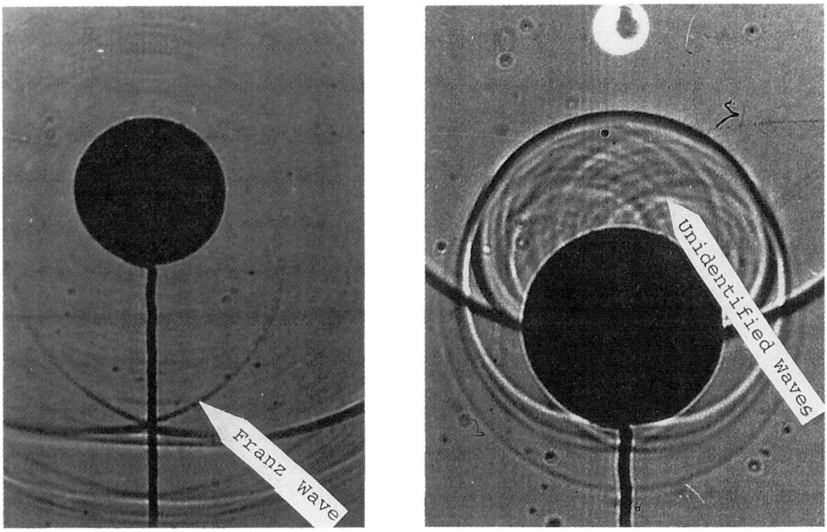

(a) (b)

Figure 5. Visualizations: (a) Franz waves, (b) unidentified waves.

4. Conclusion

A mini-sparker source which generates an optimized spherical wave front with quasi-Dirac-type signature has been developed for impulse-response measurements. Owing to the high acoustic power generated, its association with an ultra-high-speed shadowgraph visualization system is a useful tool for diffraction experiments. Examples with elastic spheres have shown that typical diffused waves reported in the literature (Neubauer & Dragonette 1970, Welton et al. 1980) can be observed with a very high quality of visualization. The same devices can be used for many other wave-structure observations, for example, interactions between shocks and particles, underwater explosions, etc.

References

Brill D, Uberall H (1971) Acoustic waves transmitted through solid elastic cylinders. JASA 50(3):921–939.

Nakamura Y, Otani T (1992) Frequency response of a piezoelectric polymer film hydrophone and an elastic wave induced on the backing surface. Japanese J of Appl Phys 31-1:266–268.

Neubauer WG, Dragonette LR (1970) Observation of waves radiated from circular cylinders caused by an incident pulse. JASA 48(5)2:1135–1149.

Pernod P, Merlen A, Kemmou A, Ahyi A (1994a) Etude par visualisations ultra rapides et mesures de pression d'une source impulsionnelle large bande non directive. Journal de Physique IV, Colloque C5, Supplément au Journal de Physique III, 4:729–732.

Pernod P, Merlen A, Kemmou A, Ahyi A, Delannoy B (1994b) Punctual transient acoustic source for experimental impulse response measurements. In: IMACS, IEEE-SMC, SPRANN 94, IMACS International Symposium on Signal Processing, Robotics and Neural Networks, pp 689–692.

Welton PJ, De Billy M, Hayman A, Quentin G (1980) Backscattering of short ultrasonic pulses by solid elastic cylinders at large k.a. JASA 67(2):470–476.

Physics of a shock-generated freely moving compressible vortex

K. von Ellenrieder, C. T. Kao, D. Bershader*
Stanford University
Stanford, CA 94305 USA

Abstract: This work represents a preliminary study of the physical mechanisms which govern the motion of compressible free vortices. An analysis of results from previous experimental and computational studies suggests that the transport of angular momentum occurring within such vortices can be categorized into three different flow regimes: convective, convective/diffusive, and diffusive. Additionally, the dependence of radial velocity and dissipation in the vortex on unsteady conditions outside of its core are discussed.

Key words: Vortex, Compressibility, Dissipation, Angular momentum

1. Introduction

The two-dimensional motion of an isolated free vortex in an incompressible flow is well understood. Conservation of mass requires that there be no radial velocity in the vortex, and one finds that viscous diffusion is the dominant process governing the fluid's motion. When the compressibility of fluid within the vortex and far from the vortex center are important, an analysis of the flow becomes much more complicated. In studying two-dimensional compressible vortices one must consider the effects of heat conduction, viscous dissipation, compressibility, and convection by radial mass flux on the vortex's motion and structure (Bershader 1995).

This study is based on the vortices examined by Mandella (Mandella 1987 and Mandella & Bershader 1987), which were created by passing a shock of $M \approx 1.8$ over a NACA 0018 airfoil at a 30° angle of attack. The resulting vortex was embedded in a free stream with a Mach number of $M_\infty \approx 0.52$ and a Reynolds number, based on airfoil chord length of 3.7 cm, of $Re_\infty \approx 4.4 \times 10^5$. Using density and pressure measurements of the free vortex, and assuming the radial pressure gradient was balanced by centrifugal forces, Mandella calculated the radial distributions of tangential velocity and angular momentum (Fig. 1). The density distribution was empirically fit to a time-dependent Lorentzian function. During an experimental observation period of several hundred microseconds, the free vortex traveled a distance of more than seven times its

*In memoriam.

own diameter and was convected by the free stream at a remarkably constant velocity of around 1.8×10^2 m/s. This feature, and the rather modest change of vortex structure during the observations, imply that the time scale for associated viscous dissipation effects is substantially longer than experimental times. To see this, consider the classic work by Lamb (1945) concerning incompressible vortices which predicts that vortex decay due to viscosity is exponential:

$$v_\phi = \frac{\Gamma_0}{2\pi r}(1 - e^{-\frac{r^2}{4\nu t}}). \tag{1}$$

Here v_ϕ is tangential velocity, r radial distance from the vortex center, Γ_0 the initial circulation, ν the kinematic viscosity, and t is time. When the exponential term above is the same order as 1.0, or ≥ 0.3, viscous diffusion has significantly affected the vortex's structure. The vortex core size is given by the radius of maximum tangential velocity $r_c \sim 1.0$ mm. When based on r_c and ν the viscous diffusion time is $\sim 1.4 \times 10^{-2}$ s; this is almost two orders of magnitude longer than the experimental observation time ~ 500 μs. Additionally, note that the vortex Reynolds number, based on a circulation strength of $\Gamma \approx 2.5$ m^2/s, is $Re_\Gamma \equiv \Gamma/\nu \approx 1.7 \times 10^5$.

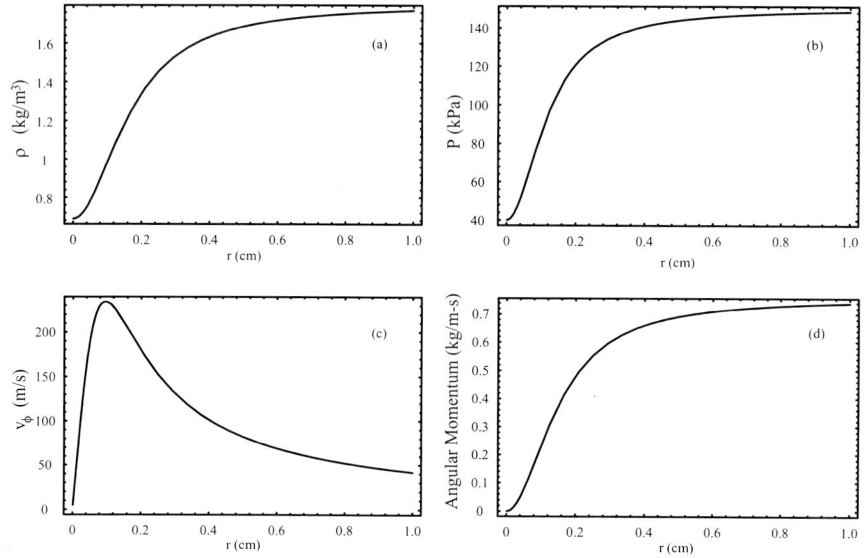

Figure 1. Radial profiles of (a) density, (b) pressure, (c) tangential velocity, and (d) angular momentum within the vortex.

Measurements of pressure and density, denoted as p and ρ respectively, illustrate that the isentropic assumption $p \propto \rho^\gamma$ is not valid within the vortex, and indicate that some sort of dissipation has occurred (Mandella 1987). The key to the observed dissipation may well reside in the radial mass flow associated with unsteady conditions outside of the vortex core.

Colonius et al. (1991) have performed an analytical and computational study in which the pressure, density, and tangential velocity measurements reported by Mandella were used as the initial conditions for a simulation of the free compressible vortex. Flow conditions far from the vortex core were assumed to be constant, and the free vortex was allowed to evolve in time under the action of viscosity. The computational time period used was comparable to the experimental times of Mandella's study, and the resulting radial distributions of pressure, density, and tangential velocity were found to agree with the experimentally measured values. However, the computations predicted much smaller radial velocities than calculated by Mandella.

An analysis of the angular-momentum transport occurring within such compressible vortices suggests that there are distinct flow regimes into which their time evolution can be characterized. Each flow regime is determined by the magnitude of the radial velocity and the core size of the vortex. Additionally, it is shown that small variations in the pressure outside of a vortex's core may influence the radial velocity and dissipation occurring within the vortex.

2. Angular-momentum conservation

For a two-dimensional and cylindrically symmetric fluid flow, the equation governing the conservation of angular momentum, expressed in cylindrical coordinates, is:

$$\rho r^2 \frac{\partial}{\partial t}\left(\frac{v_\phi}{r}\right) + \rho v_r \frac{\partial}{\partial r}(r v_\phi) - \frac{1}{r}\frac{\partial}{\partial r}\left[r^3 \mu \frac{\partial}{\partial r}\left(\frac{v_\phi}{r}\right)\right] = 0, \tag{2}$$

where μ and v_r correspond to viscosity and radial velocity, respectively. The first term in this equation is the local time rate of change of angular momentum, the second its convection by radial mass flux, and the final term is its rate of change due to viscous torque. Using the constants r_c, τ, ρ_∞, μ_0, v_{r0}, and $v_{\phi 0}$, the angular-momentum conservation equation can be cast in the following non-dimensional form:

$$\hat{\rho}\hat{r}^2 \frac{\partial}{\partial \hat{t}}\left(\frac{\hat{v}_\phi}{\hat{r}}\right) + \left\{\frac{\tau v_{r0}}{r_c}\right\}\hat{\rho}\hat{v}_r \frac{\partial}{\partial \hat{r}}(\hat{r}\hat{v}_\phi) - \left\{\frac{\mu_0 \tau}{\rho_\infty {r_c}^2}\right\}\frac{1}{\hat{r}}\frac{\partial}{\partial \hat{r}}\left[\hat{r}^3 \hat{\mu}\frac{\partial}{\partial \hat{r}}\left(\frac{\hat{v}_\phi}{\hat{r}}\right)\right] = 0. \tag{3}$$

As shown in Table 1, quantities characteristic of Mandella's experimental results are chosen for r_c, ρ_∞, μ_0, and $v_{\phi 0}$. Note that $v_{\phi 0}$ cancels from the expression, and that analysis yields the values for the time constant of angular-momentum transport τ, and the radial velocity (at r_c) v_{r0}. An examination of the orders of magnitude of the two resulting dimensionless groups (bracketed terms) suggests that there are three flow regimes into which the transport of angular momentum within a vortex can be categorized: convective, convective/diffusive, and diffusive.

Table 1. Physical constants characteristic of the free vortex.

Constant	Value	Description
r_c	1.0 mm	Vortex core radius
ρ_∞	1.8 kg/m^3	Free-stream density
μ_0	2.0×10^{-5} Ns/m^2	Coefficient of viscosity (air)
$v_{\phi 0}$	2.3×10^2 m/s	Tangential velocity at r_c

2.1. Convective: $\tau v_{r0}/r_c \sim 1.0 \gg \mu_0\tau/(\rho_\infty {r_c}^2)$

In this flow regime convection is the dominant process by which angular momentum is transferred within the vortex. By eliminating τ in the inequality, approximate values may be found for the radial velocity: $v_{r0} \gg 1.1 \times 10^{-2}$ m/s or more roughly $v_{r0} > 3.0 \times 10^{-2}$ m/s. Estimates for the time scale of this motion are given by $\tau \sim r_c/v_{r0}$, which yields a time scale of $\tau \ll 9.1 \times 10^{-2}$ s or $\tau < 3.0 \times 10^{-2}$ s. Note that the radial velocities predicted by Mandella $v_{r0} \sim 0.3$ m/s lie in the convective range, and that for the experimental vortex under investigation, when v_{r0} is negative (inward flow) the angular momentum in the vortex is increasing (Fig. 1d).

2.2. Convective/diffusive: $\tau v_{r0}/r_c \sim 1.0 \sim \mu_0\tau/(\rho_\infty {r_c}^2)$

When the magnitude of each dimensionless group is on of the order of 1.0, both the effects of viscosity and convection must be considered in determining angular-momentum transport within the vortex. An approximate value of v_{r0} in this region is given by: 3.0×10^{-3} m/s $\leq v_{r0} \leq 3.0 \times 10^{-2}$ m/s. The radial velocities reported by Colonius et al., $v_{r0} \sim 6.8 \times 10^{-3}$ m/s, lie in this range.

2.3. Diffusive: $\tau v_{r0}/r_c \ll 1.0 \sim \mu_0\tau/(\rho_\infty {r_c}^2)$

In the diffusive regime viscosity is responsible for the spreading of angular momentum. The time scale for these effects is $\tau >\approx 0.09$ s, which is longer than the time scale predicted above for the incompressible vortex, and is the same as the "slow" time scale of Colonius. Radial velocities in the range $v_{r0} \ll 1.1 \times 10^{-2}$ m/s or $v_{r0} < 3.0 \times 10^{-3}$ m/s are expected for this regime.

2.4. Discussion

In the computational study of Colonius et al. the radial velocity found was two orders of magnitude smaller than the radial velocity experimentally inferred by Mandella. This difference is most likely due to unsteady variations in the experimental ambient-flow conditions. In the computational study, the external pressure, density, and temperature were fixed — the only mechanism for changes to occur in the vortex was viscosity. In the experiment, the vortex forms in a complex, unsteady flow field. The vortex is created in the low-pressure region behind an airfoil, and after being traversed by a shock, is convected away from the airfoil. The radial pressure gradient across the vortex is approximately balanced by its tangential velocity. It is likely that small changes in the external pressure cause the tangential velocity profile to change, and that the radial convection of angular momentum is the mechanism responsible.

Dimensional analysis suggests that unsteady ambient pressure may account for the large radial velocity found by Mandella. When the Reynolds Number is large, and viscous effects are negligible, the radial momentum and the angular-momentum conservation equations, respectively, can be written as:

$$\rho \frac{{v_\phi}^2}{r} = \frac{\partial p}{\partial r}, \qquad \text{and} \qquad \rho \frac{\partial}{\partial t}(r v_\phi) = -\rho v_r \frac{\partial}{\partial r}(r v_\phi). \tag{4}$$

Rearranging these expressions, and using the characteristic values listed in Table 1, an estimate for the radial velocity is found to be:

$$v_{r0} \sim \frac{1}{\tau v_{\phi 0}} \left[\frac{{r_c}^3}{\rho_\infty} \frac{\partial p}{\partial r}\bigg|_c \right]^{1/2}. \tag{5}$$

A 5% variation in the vortex core pressure gradient, $\frac{\partial p}{\partial r}\big|_c \sim 3.4 \times 10^7$ N/m^3 (as estimated from Fig. 1b), occurring over a period of $\tau \sim 500$ μs, may be adequate to produce a radial velocity of $\sim 2.7 \times 10^{-1}$ m/s.

3. Dissipation

With the assumption of cylindrical symmetry, a formal statement of the second law governing the creation of entropy in a fluid system is:

$$\frac{\partial s}{\partial t} + v_r \frac{\partial s}{\partial r} - \frac{1}{\rho T r}\frac{\partial}{\partial r}\left(r\kappa \frac{\partial T}{\partial r}\right) + \frac{\kappa}{\rho T^2}\left(\frac{\partial T}{\partial r}\right)^2 \geq 0 \tag{6}$$

where κ is the thermal conductivity $\sim 2.6 \times 10^{-2}$ J/(msK). The left-hand side of the equation is the total production of entropy per unit mass s, which must be positive. The last two terms containing the temperature T represent the outflow of entropy per unit mass, and the first two terms are the net increase in entropy per unit mass. The radial temperature distribution and the temperature gradient may be determined from Figs. 1a and 1b by use of the ideal gas law. Dimensional analysis reveals that the entropy outflow terms are several orders of magnitude smaller than the unsteady and convective entropy terms. Therefore the role of heat conduction within the vortex may be considered a minor one. The unsteady entropy term may be expressed as a function of density and pressure:

$$\frac{\partial s}{\partial t} = \frac{R}{(\gamma - 1)}\left[\frac{1}{p}\frac{\partial p}{\partial t} - \frac{\gamma}{\rho}\frac{\partial \rho}{\partial t}\right], \tag{7}$$

where R and γ are the gas constant for air and the ratio of specific heats, respectively (Colonius et al. 1991). The rate of change of pressure is found by use of the radial-momentum equation:

$$\frac{\partial p}{\partial t} = \frac{\partial p_\infty}{\partial t} - \int_r^\infty \left[\frac{{v_\phi}^2}{r}\frac{\partial \rho}{\partial t} + \frac{2\rho v_\phi}{r}\frac{\partial v_\phi}{\partial t}\right] dr. \tag{8}$$

Note that for constant ambient conditions, the p_∞ term vanishes. However, small variations of $p_\infty \sim 1.5 \times 10^5$ Pa over time scales on the order of Mandella's experiments may have a significant effect on the dissipation produced. For example, a 10% positive change in p_∞ over a time of $\tau \sim 500\ \mu$s, would produce a value of $\rho T \frac{\partial s}{\partial t} \sim 7.5 \times 10^7$ N/m^{2}s. By comparison, in Mandella's experiments, $\rho T v_r \frac{\partial s}{\partial r} \sim 3 \times 10^8$ N/m^{2}s.

4. Conclusions

The analysis above suggests that the dissipation and radial velocities inside compressible vortices are influenced by unsteady processes occurring outside of their cores. It has also been shown that the nature of the angular-momentum transport mechanisms in such vortices varies according to the vortex radial velocity and core size.

Acknowledgement. The authors are grateful for partial support of this work by NASA Consortium Agreement No. NCC2-5002, and the NASA Graduate Student Researchers Program. Additionally, the first author would like to thank Professor Cantwell for his help and advice.

References

Bershader D (1995) Compressible vortices. In: Fluid Vortices, Sheldon Green (ed), Klewer Academic Publishers.

Colonius T, Lele S, Moin P (1991) The free compressible viscous vortex. J Fluid Mech 230:45–73.

Lamb H (1945) Hydrodynamics, 6th Edition, Dover Publications.

Mandella M (1987) Experimental and Analytical Studies of Compressible Vortices, Ph.D. Thesis, Dept of Applied Physics, Stanford University.

Mandella M, Bershader D (1987) Quantitative Study of the Compressible Vortex: Generation, Structure, and Interaction with Airfoils, AIAA Paper 87-0328.

Study of the propagation of shock waves over bodies by holographic interferometry

J. Le, X. Ye, X. Wu, H. Yang
China Aerodynamics Research and Development Center
P.O. Box 211, Mianyang, Sichuan, China

Abstract: The propagation of shock waves of $M_s = 1.8$ over various bodies such as single cylinders, double cylinders, wedges with back step, or plates with sharp-edged slits in a shock tube or in a channel bend was studied experimentally with a holographic interferometer. Quantitative density fields obtained from interferograms are in good agreement with computations for the case of the wedge with back step.

Key words: Shock wave, Holographic interferometry, Shock tube

1. Introduction

The problem of the propagation and diffraction of shock waves past solid bodies has been investigated extensively in the past both experimentally and computationally. Some excellent results of the quantitative measurement of density fields have been obtained with holographic interferograms, such as: fifteen basic cases of oblique shock-wave reflection (Glaz et al. 1986); and shock-wave diffraction through an aligned baffle system (Reichenbach 1994). However, for some problems with complicated flow structure, including shock waves reflected from a solid wall and the interaction of shock waves with vortices, the quantitative measurements of density fields from time sequences of interferogram frames are still very necessary and important for validating computer codes and for studing complicated flow phenomena. For example, only computational results of the propagation of shock waves through double cylinders in a channel bend have been completed (Fursenko and Sharov 1992), thus investigation of this kind of flow using holographic interferometry is of great value in providing insight into the physical phenomena. The purpose of this study is to quantitatively measure the unsteady flow fields of the propagation of shock waves over some typical bodies using a shock-tube technique in which the incident M_s can be precisely controlled and to compare the results of experiment and computation for one of these typical bodies.

2. Experiment

The experiments were carried out in CARDC's small shock tube with a test section of 32 mm × 32 mm, with two pairs of observation windows of 100 mm × 32 mm. The distance between windows is 400 mm. The driver of the shock tube has a circular section with a diameter of 70 mm. Figure 1 shows the shock tube and the optical system of the holographic interferometer. Usually, a 2D model is located in view of the first window and can be easily exchanged for a channel bend where the double cylinders are placed. Their detailed sizes are shown in Fig. 2. The incident-shock Mach number M_s can be precisely controlled by a specially made plastic diaphragm and by high-accuracy measurements of gas pressure both in the driver and in the driven section. The distances from the transducers to the model are arranged to measure the shock speed and to synchronize the ruby laser. For quantitatively measuring the density fields,

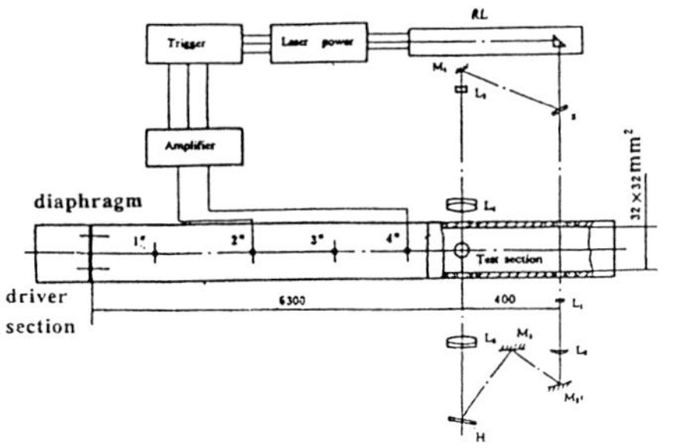

Figure 1. Shock tube and holographic interferometer system.

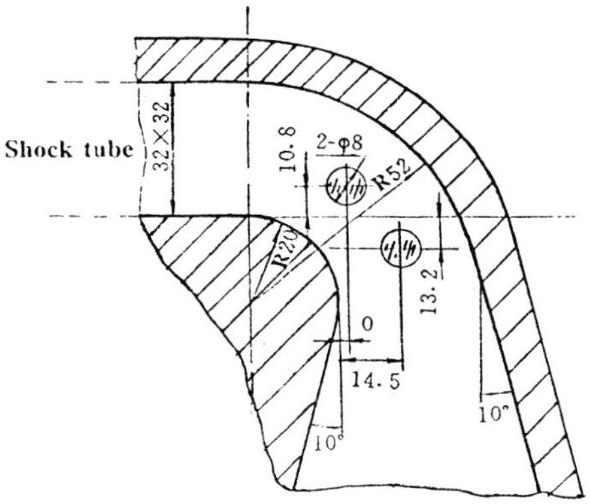

Figure 2. Size of channel bend with cylinders.

double-exposure holographic interferometry was used. The first exposure is performed before the incident shock wave arrives and the second exposure is triggered at the moment at which the shock waves interact with the model. By incrementing triggered delay times (with precison of 1 μs), the frame-by-frome holographic interferograms, separately obtained from a series of tests, can be considered to capture the unsteady process of shock-wave interaction with a body.

3. Results and discussion

Propagation through a slit with sharp edges. The apex angle of the sharp edge is 20° and the distance of two apices is 4 mm. Some of these experimental results are similar to those obtained by shadowgraphs (Amann and Reichenbach 1973). In Fig. 3 are holographic interferograms taken at different instants. After the incident shock wave passes over the model, two intense vortices are formed at the sharp edges but only one diffracting shock wave moves forward. The experimentally determined isopycnics are presented in Fig. 3d. The velocity of the diffracting shock wave along the slit plate can be measured by two instantaneous interferograms (Figs. 3a and b). Then its density ratio (ρ_N/ρ in Fig. 3d) can be calculated by the normal shock-wave relation.

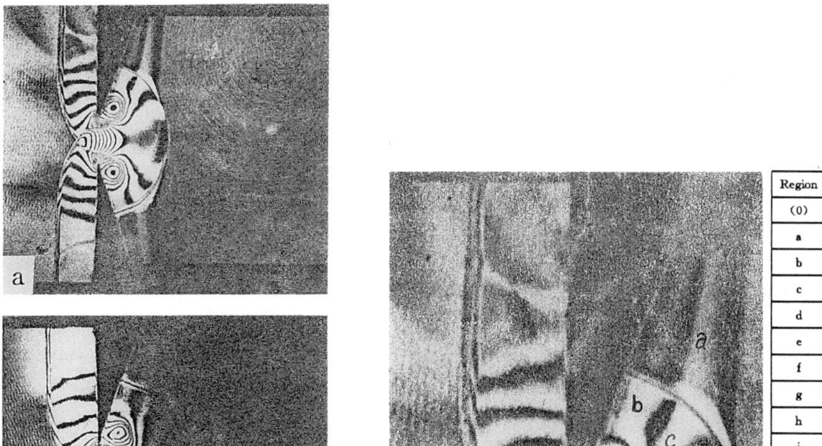

Region	P_N/P_0
(0)	1.00
a	1.26
b	3.29
c	3.55
d	3.81
e	4.07
f	4.33
g	4.59
h	4.85
i	5.11
j	5.37
k	5.63
l	5.89
m	6.15
n	6.41
o	6.67
p	6.93
q	7.19

Figure 3. Holographic interferograms, $M_s = 1.81$, $P_0 = 32$ kPa, $T_0 = 300.5$ K: (a) $t = 0$ μs; (b) $t = 2$ μs; (c) $t = 30$ μs; (d) experimentally determined density contours of case a.

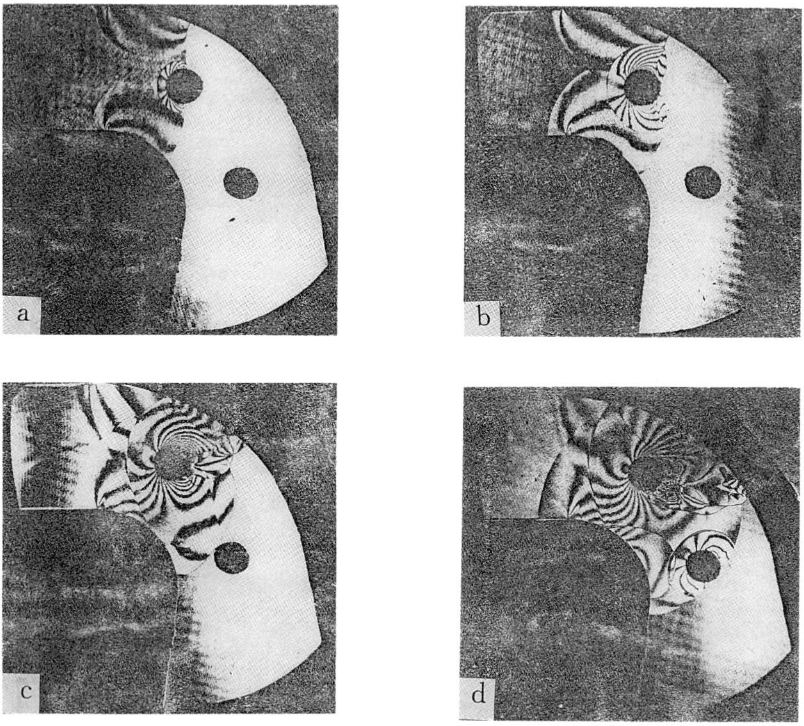

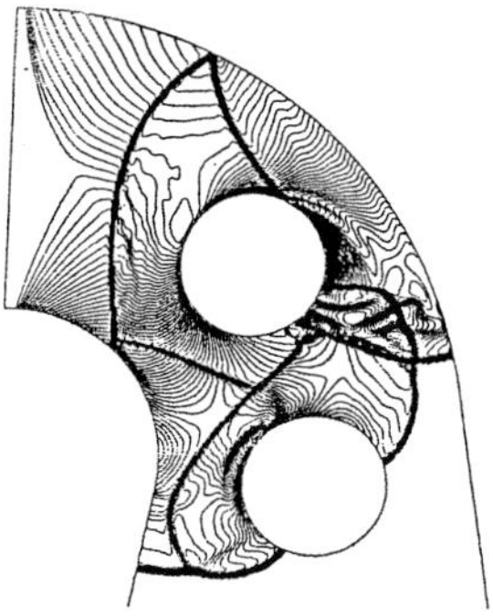

Figure 4. Computed density contours from Fursenko and Sharov (1992); $M_s = 2.1$.

Figure 5. Holographic interferograms from CARDC. $M_s = 1.82$, $P_0 = 32$ kPa, $T_0 = 288$ K: (a) $t = 10$ μs; (c) $t = 24$ μs; (d) $t = 38$ μs; (e) $t = 52$ μs.

Propagation through a channel bend with double cylinders. There is only one paper that computationally treats this unsteady problem based on

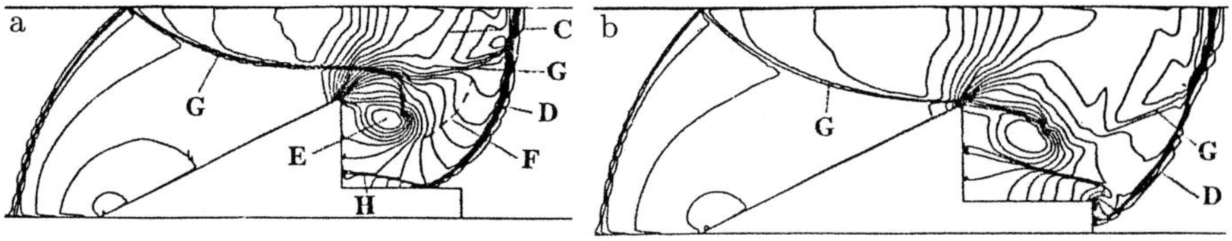

Figure 6. Computed density by ENO scheme. $M_s = 1.80$, $P_0 = 32$ kPa, $T_0 = 285$ K, $Re = 5.3 \times 10^5$, 351×121 grid points: (a) $t = 36.9$ μs; (b) $t = 44.88$ μs.

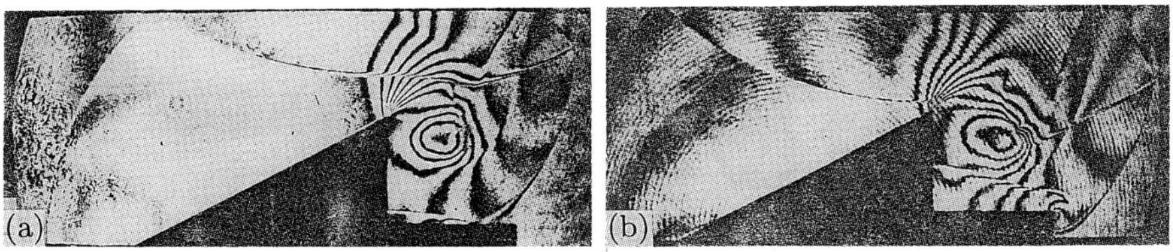

Figure 7. Holographic interferograms. $M_s = 1.80$, $P_0 = 32$ kPa, $T_0 = 285$ K, $\Delta t = 0$ is set at the moment when the Mach stem arrives at the apex of the wedge: (a) $t = 36.9$ μs; (b) $t = 44.88$ μs.

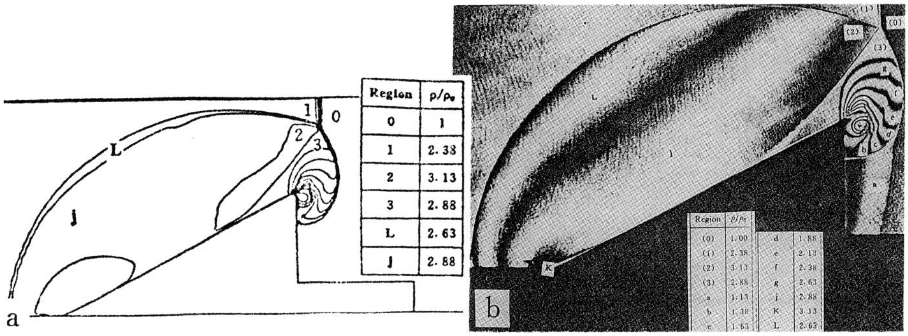

Figure 8. Comparison of density contours from computation and experiement. $M_s = 1.80$, $P_0 = 32$ kPa, $T_0 = 285$ K, $\Delta t = 12$ μs: (a) from computation; (b) from experiment.

the Euler equation with an unstructured finite-volume method (Fursenko and Sharov 1992). Figure 4 shows the result. In contrast with the computation, the experimental results shown in Fig. 5 represent a more complex wave system; in particular, the two vortices formed behind the first cylinder are more extensive. Viscous effects behind so called lambda shock wave are also stronger, thus the flow structure may depend on the Reynolds number.

Propagation past a wedge with back step. Flow fields were computed using the thin-layer approximated N-S equations with second-order ENO schemes (351×121 grid points). The model size and a detailed description of both experiment and computation are given by Le and Li (1994). As shown in Figs. 6 and 7, the vortex (E) has a significant effect on the reflected shock wave (G). The shock wave (G) is distorted a little as it approaches the vortex (shown in Fig. 7a), and then is split into three parts by the vortex (shown in Fig. 7b), its right half is accelerated and its center is completely merged by the vortex. Compared with the experiment, the interaction of the shock wave (G) with the vortex can be more clearly captured by the present computations (shown in Figs. 6a and b). In addition, as shown in Figs. 8a and b, the density contours from experiment and computation are in good agreement.

4. Conclusion

By using the present shock-tube technique, the incident M_s can be precisely controlled and time sequences of holographic interferograms can be taken, making it very ecomomical for us to study quantitatively the problems of unsteady compressible flow. However the accuracy of time measurements is very important and will be further improved.

References

Abe A, Takayama K (1989) Shock wave diffraction from the open end of a shock tube. In: 17th Int Symp on Shock Waves and Shock Tubes, pp 270–275.

Amann HO, Reichenbach H (1973) Unsteady flow phenomena in a shock tube nozzles. In: Proc 9th Int Shock Tube Symp, pp 96–112.

Fursenko AA, Sharov DM (1992) Numerical simulation of shock wave interactions, channel bends and gas nonuniformities. Computers Fluids 21(3):377–396.

Glaz HM, Collella P, Glass II (1986) A Detailed Numerical, Graphical, and Experimental Study of Oblique Shock Wave Reflections, UTIAS Report No. 285.

Le J, Li C (1994) Propagation of shock wave over a wedge. In: Proc 2nd Int Conf of Experimental Fluid Mechanics, pp 786–793.

Reichenbach H (1994) Color interferogram of shock propagation through an aligned baffle system. Shock Waves 4(2).

Accuracy of shock-wave propagation using geometrical shock dynamics

D. W. Schwendeman
Department of Mathematical Sciences
Rensselaer Polytechnic Institute
Troy, NY 12180 USA

Abstract: Shock-wave motion determined by the approximate theory of geometrical shock dynamics is compared with the motion obtained from solutions of the unsteady Euler equations. The problem of two-dimensional shock propagation in a channel with a 90° bend is used to carry out the comparison. Both strong shocks ($M_0 = 2.1$) and weak shocks ($M_0 = 1.1$) are discussed. The best agreement is found in regions where the shock is strengthening. An alternate approach in which the area-Mach-number rule would be replaced by a numerical calculation of the one-dimensional equations for flow in a ray tube is discussed.

Key words: Geometrical shock dynamics, Euler equations, Shock capturing, Shock-wave diffraction

1. Introduction

Geometrical shock dynamics is an approximate theory that determines the motion of the leading shock without an explicit calculation of the flow behind (Whitham 1957, 1959). Numerous investigations have used geometrical shock dynamics to calculate shock-wave motion and in many cases the results have been compared with experiments (e.g. Bryson and Gross 1961 and Henshaw et al. 1986). In general, good accuracy is reported but there is always some error. Geometrical shock dynamics gives an approximation of shock-wave motion as determined by the Euler equations. A detailed study of the accuracy of the approximate theory can be made by comparing numerical calculations of the equations of geometrical shock dynamics with those of the Euler equations. This paper gives some results of this study for the problem of shock-wave diffraction in a channel with a 90° bend.

In geometrical shock dynamics, the shock is moved according to an area-Mach-number rule. The area-Mach-number rule could be obtained from experimental data for a particular problem or set of problems, but in most cases it is taken from an analysis of shock propagation in a channel with gradual area change (Whitham 1958). A study of this problem and the approximations that lead to the area-Mach-number rule give some insight into the accuracy of the geometrical shock-dynamics approximation.

2. Governing equations

The equations of geometrical shock dynamics are

$$\nabla \cdot \left(\frac{M}{A} \nabla \alpha \right) = 0, \qquad M = \frac{1}{|\nabla \alpha|}, \qquad A = A(M), \tag{1}$$

where $\alpha(\mathbf{x}) = a_0 t$ gives the leading shock position $\mathbf{x}$ at a time t and a_0 is the ambient sound speed (Whitham 1959). In (1), $A = A(M)$ is taken to be

$$A = \exp\left(- \int \frac{M\lambda(M)}{M^2 - 1}\, dM \right), \tag{2}$$

where $\lambda(M)$ is a function that varies slowly between 4 when $M = 1$ and 5.0743 when $M \to \infty$ (for $\gamma = 1.4$). The form in (2) was proposed by Whitham following the work of Chester (1954) and Chisnell (1957). The equations in (1) form a hyperbolic system that describes disturbances in M and shock normal $\nabla\alpha/|\nabla\alpha|$ propagating on the leading shock.

The area-Mach-number rule in (2) gives the Mach number $M = U/a_0$ of a shock propagating down a channel with a given cross-sectional area $A(s)$, where s is distance down the channel. It is an approximation of the shock motion as determined by the one-dimensional (hydraulic) equations

$$\begin{aligned} \rho_t + u\rho_s + \rho u_s + \rho u \frac{A'(s)}{A(s)} &= 0, \\ u_t + u u_s + \frac{1}{\rho} p_s &= 0, \\ p_t + u p_s - a^2(\rho_t + u\rho_s) &= 0, \qquad a^2 = \frac{\gamma p}{\rho}, \end{aligned} \tag{3}$$

where ρ, u, p, and a are the density, velocity, pressure, and sound speed in the channel, respectively. The basic assumption that leads to (2) is that the change in $A(s)$ is small, and it can be shown that (2) would be exact if

$$\delta \equiv \left(\frac{1}{U} - \frac{1}{u + a} \right) \left(\frac{p_t + \rho a u_t}{p_t} \right) \tag{4}$$

vanished at the shock. In general, δ is not zero at the shock, and this quantity can be treated as a residual and can be used to study the error in the approximation of (3).

We will consider the motion of the leading shock in two space dimensions $\mathbf{x} = (x, y)$ as determined by a solution of (1). This motion is some approximation of the motion of the shock as determined by a corresponding solution of the unsteady Euler equations

$$\begin{bmatrix} \rho \\ \rho u \\ \rho v \\ E \end{bmatrix}_t + \begin{bmatrix} \rho u \\ \rho u^2 + p \\ \rho v u \\ (E+p)u \end{bmatrix}_x + \begin{bmatrix} \rho v \\ \rho u v \\ \rho v^2 + p \\ (E+p)v \end{bmatrix}_y = 0, \tag{5}$$

where

$$E = \frac{p}{\gamma - 1} + \frac{1}{2}\rho(u^2 + v^2)$$

is the energy. A solution of (5) gives the whole flow field, including the leading shock, and can be used to measure the accuracy of the geometrical shock-dynamics approximation.

3. Numerical methods

To make the desired comparisons, numerical methods are needed to solve (1) and (5). The numerical method used to solve (1) is the same one discussed in Schwendeman (1993). It is a finite-difference formulation of the conservation form in (1). The numerical method used to solve (5) is a second-order Godunov-type finite-difference scheme, similar to the one discussed in Colella (1990).

A numerical scheme is needed also to calculate solutions of (3) for a given $A(s)$, and to calculate the residual (4) at the shock. The desired solution of (3) will always have a shock in the C_+ characteristic, $ds/dt = u + a$, propagating down the channel. For this problem it is convenient to use a numerical scheme based on the characteristic form of (3) for the flow behind the shock and fit the shock in the C_+ characteristic using the normal shock conditions.

4. Results

The problem of two-dimensional shock propagation in a channel with a 90° bend is used to compare the leading shock as determined by a solution of (1) with the motion determined by a solution of (5). This is a good test case because the shock diffraction within the channel produces both regions of shock strengthening and weakening. The gap distance between the inner and outer walls of the channel is normalized to one and the radius of the inner wall is taken to be 0.9368. The channel geometry is the same one discussed in Henshaw et al. (1986) following the experiments in Edwards et al. (1983). The initial shock Mach number is M_0 and the ambient sound speed ahead of the shock a_0 is normalized to one.

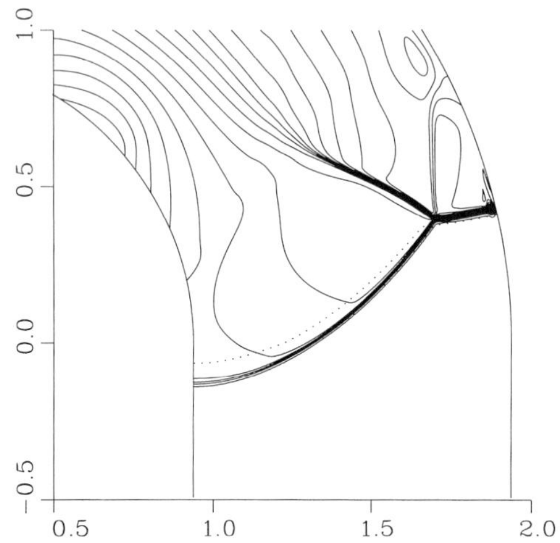

Figure 1. Contours of density (solid curves) and the approximate leading shock position (dotted curve) all at $t = 1$.

Figure 1 shows contours of density in a portion of the channel containing the leading shock at a time $t = 1$. In this example, $M_0 = 2.1$ and $t = 0$ when the shock first meets the 90° bend. The shock weakens near the inner wall and strengthens and forms a Mach stem near the outer wall. The dotted curve in this figure is the leading shock according to geometrical shock dynamics. It is slightly ahead of the Mach stem near the outer wall and shows a significant lag near the inner wall.

The behavior observed in Fig. 1 is consistent with the general belief that geometrical shock dynamics is most accurate in cases when the shock strengthens. This belief is based on a physical intuition of the approximations involved, but is not supported by any mathematical theory. For example, there is no "next correction" in the theory available to assess the error.

One way to make a closer examination of the behavior in Fig. 1 is to calculate $A(s)$ for various rays from the solution of (1) and then determine the motion of the shock along the rays based on solutions of (3). This approach gives an indication of the location of the shock based on a full calculation of (3). The residual in (4) can also be calculated and this quantity gives some measure of the approximation in the area-Mach rule.

Figure 2a shows contours of α in the channel from the numerical solution of (1). These contours give the successive shock positions according to geometrical shock dynamics. The dotted contour curve is $\alpha = t = 1$ ($a_0 = 1$) and this curve

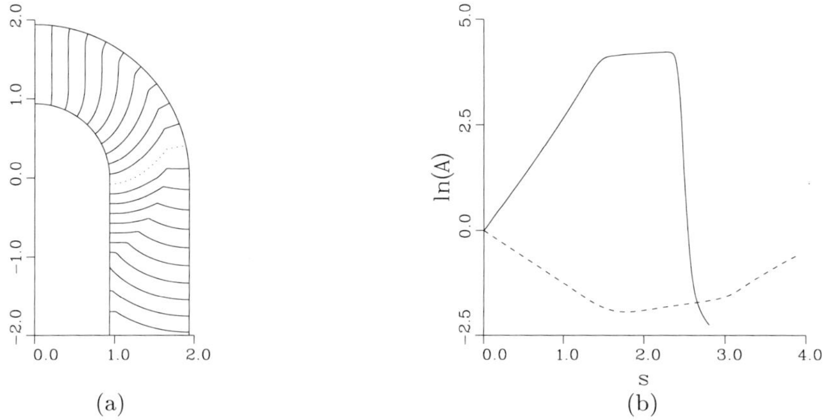

Figure 2. (a) Contours of α and (b) $\ln(A)$ versus s along the channel walls.

is the same one plotted in Fig. 1. From this solution, $A(s)$ can be calculated for various rays and Fig. 2b shows $\ln(A)$ versus s for the ray along the inner wall (solid curve) and the ray along the outer wall (dashed curve).

Numerical calculations of (3) for the two area curves in Fig. 2b give the motion of the shock along the walls without the area-Mach-number rule approximation. The dashed curves in Fig. 3a show this motion and these curves can be compared to the corresponding ones obtained from the solution of (1) which uses the area-Mach-number rule approximation (dotted curves) and the solution of the full Euler equations (solid curves). The set of three curves with the larger values of time t correspond to the inner wall while the other set belongs to the outer wall. Along the outer wall, where the shock strengthens, the agreement is excellent. There is some disagreement in the set of curves belonging to the inner wall and it is interesting to note that the dashed curve given by a calculation of (3) is not closer to the solid curve given by the Euler solution.

Values of δ along the walls are shown in Fig. 3b. The data set with smaller $|\delta|$ belongs to the outer wall and these smaller values are consistent with the better agreement between the dashed and dotted curves in Fig. 3a for the outer wall. The increase in δ near $s = 1.5$ seems to correspond to the change in slope of the area curves in Fig. 2b. It is interesting to note that the values of δ are not particularly small for either wall despite the relatively good agreement between the dashed and dotted for both cases.

A similar study has been carried out for a weak shock case where $M_0 = 1.1$. The general observations noted above for the relatively strong shock still apply

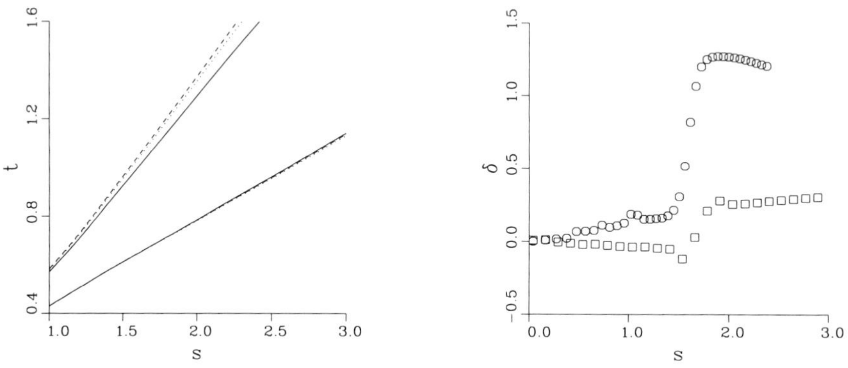

(a) (b)

Figure 3. (a) Shock paths and (b) δ versus s along the channel walls.

and the level of accuracy is about the same in contrast to the general belief that geometrical shock dynamics would be a poor approximation for weak shocks.

Acknowledgement. This work was supported by the National Science Foundation.

References

Bryson AE, Gross RWF (1961) Diffraction of strong shock by cones, cylinders, and spheres. J Fluid Mech 10:1.

Chester W (1954) The quasi-cylindrical shock tube. Phil Mag 45:1293.

Chisnell RF (1957) The motion of a shock wave in a channel, with applications to cylindrical and spherical shock waves. J Fluid Mech 2:286.

Colella P (1990) Multidimensional upwind methods for hyperbolic conservation laws. J Comp Phys 87:171.

Edwards DH, Fearnley P, Nettleton MA (1983) Shock diffraction in channels with a 90° bend. J Fluid Mech 132:257.

Henshaw WD, Smyth NF, Schwendeman DW (1986) Numerical shock propagation using geometrical shock dynamics. J Fluid Mech 171:519.

Schwendeman DW (1993) A new numerical method for shock wave propagation based on geometrical shock dynamics. Proc R Soc Lond A 441:331.

Whitham GB (1957) A new approach to problems of shock dynamics, part I: two-dimensional problems. J Fluid Mech 2:145.

Whitham GB (1958) On the propagation of shock waves through regions of non-uniform area or flow. J Fluid Mech 4:337.

Whitham GB (1959) A new approach to problems of shock dynamics, part II: three-dimensional problems. J Fluid Mech 5:369.

A locally adaptive structured/unstructured Navier-Stokes solver for essentially unsteady shock-obstacle interactions at high Reynolds numbers

D. Ofengeim, E. Timofeev
Ioffe Physical-Technical Institute, Russian Academy of Sciences
Polytekhnicheskaya Str. 26, St. Petersburg 194021 Russia

P. Voinovich
Advanced Technology Center
P.O. Box 160, St. Petersburg 198103 Russia

Abstract: The paper presents a new numerical technique to compute essentially unsteady viscous-inviscid interactions at high Reynolds numbers. Special attention is given to the verification of the method.

Key words: Combined grids, Navier-Stokes equations, Unsteady flow

1. Introduction

The investigation of shock-wave interactions with obstacles is of fundamental and practical interest. The Euler gas model is widely used, providing reasonable results for high Reynolds number flows. Dynamically adaptive unstructured grids and high-resolution schemes contribute greatly to the computational efficiency of the shock-capturing method, as its accuracy is critically affected by resolution at the flow discontinuities. The highly accurate and efficient numerical technique has been developed previously by Fursenko et al. (1993, 1995), however, it cannot accurately predict a flow separation. An extension of the above technique to the Navier-Stokes solver is described in Ofengeim et al. (1995) and Beryozkina et al. (1996). The present paper contains a summary of this approach and focuses on numerical results in comparison with experiments.

2. Numerical model

The numerical model is based on the full Navier-Stokes equations. A thin layer of non-uniform structured orthogonal grid is constructed along the body surface. The rest of the computational domain is covered by an unstructured triangular grid (Fig. 1). In the course of a computation the unstructured grid is adapted to the solution using a hierarchical refinement/derefinement procedure. The structured grid and the adjacent cells of the unstructured grid are fixed.

Nonoverlapping control volumes are established for either grids. An unsplit explicit technique is employed on the unstructured grid. The convective terms are approximated by the second-order Godunov-type scheme (Fursenko et al. 1993), while central differences are applied to the viscous terms. The time-step operator for the structured grid is split according to physical processes. To

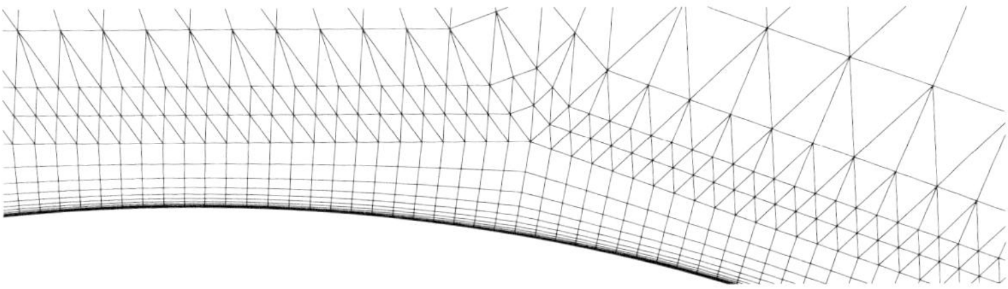

Figure 1. Detail of hybrid structured/unstructured grid.

compensate partially for the severe limitation of the time step, the inviscid operator is also split in space. The respective 1D operators are approximated by a 1D version of the above Godunov-type scheme. A predictor-corrector procedure enhances the temporal accuracy for all operators. The viscous operator on the structured grid is unsplit and contains central differences written implicitly in time for terms with second derivatives along the normal direction to the body surface, and explicitly (a predictor-corrector method) for the remaining viscous terms. The resulting set of linear difference equations with a block tridiagonal matrix of coefficients is solved by the Thomas algorithm.

3. Numerical results and comparison

Verification of the solver presents a problem due to a lack of analytical solutions, reliable computations, and experimental data for unsteady viscous flows with shocks. Consequently, several steady problems have been used as preliminary tests: computations of a boundary layer on a flat plate can be found in Ofengeim et al. (1995); the results for a shock-wave/boundary-layer interaction and for a supersonic flow over a NACA 0012 airfoil are presented below. Shock-wave diffraction over a circular cylinder is also considered as a test problem because of high-quality experimental data by Syshchikova et al. (1990) and Beryozkina et al. (1996) at hand.

All the subsequent computations have been carried out for perfect gases with constant specific heat ratio γ. The Prandtl number is equal to 0.72. The viscosity coefficient is calculated following Sutherland's law.

3.1. Steady tests

The steady-state supersonic $M_\infty = 1.5$, $Re = 10,000$ flow ($\gamma = 1.4$) over an adiabatic NACA 0012 airfoil at zero inclination has been considered as an example without separation. The oblique shock-wave impingement on the laminar flat-plate boundary layer represents another standard benchmark for Navier-Stokes solvers. The flow parameters ensuring separation are: $M_\infty = 2$, $Re = 296,000$, $T_{wall} = 1.7T_\infty$, $\gamma = 1.4$; the primary shock incidence angle is 32.6°.

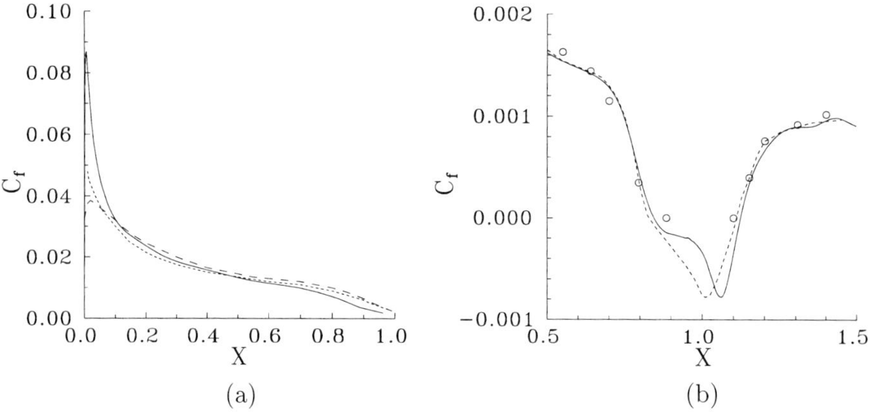

Figure 2. Dependence of the skin-friction coefficient C_f on the longitudinal coordinate X measured from the leading edge: (a) NACA 0012 airfoil; the dotted and dashed lines display computational data by Ivanov et al. (1989) and Schröder & Hänel (1987) respectively; (b) the shock-wave/boundary-layer interaction; the dotted line gives numerical data from Savel'ev (1988), circles are experimental data of Hakkinen et al. (1959). Solid lines represent our calculations.

The solutions are presented in Fig. 2 as skin friction over the surface. The skin friction is known to be very sensitive to both convective and viscous transport, and the good agreement indicates an accurate approximation.

3.2. Shock-wave diffraction over a circular cylinder

The above approach has been applied to the investigation of shock-wave ($M_s = 3.0$) diffraction over a circular cylinder in CO_2 ($\gamma = 1.289$). The Reynolds number based on the cylinder radius r is 80,000; $T_{wall} = 293$ K.

Some results are given in Figs. 3 and 4 for comparison of the shock-tube experiments with Euler computations. The time t_r is measured from the first shock-cylinder collision. The incident shock wave passes a distance of r in one time unit. The number of unstructured-grid nodes varied from 272 for the background grid to 12552 for the final grid at $t_r = 5.8$ (5 refinement levels with

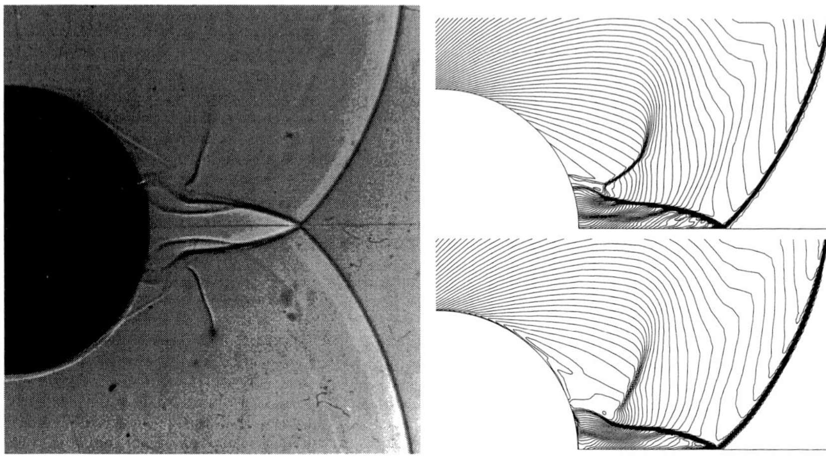

Figure 3. Shock-wave diffraction over a cylinder: the afterbody region for time $t_r = 3.7$. On left, experimental shadowgraph (Syshchikova et al. 1990, Beryozkina et al. 1996); on right, computed density contours: inviscid on top, viscous on bottom.

an initial cell size of $0.4r$). The $0.05r$-thick structured part of the computational grid is equally spaced along the cylinder surface and exponentially stretched in the normal direction with minimal and maximal spacing of $6.8 \times 10^{-5}r$ and $1.6 \times 10^{-2}r$ respectively.

The comparison between experimental data and numerical results shows the good quality of the viscous computations, which describe correctly the wave pattern and the separation location. Pressure histories at fixed points of the cylinder surface also agree with the experiment (see Beryozkina et al. 1996). For $t_r = 3.7$ the Navier-Stokes solver provides correct positions of the rearward-facing stagnation shock wave and the slip surface, and predicts accurately the reflected shock and contact discontinuity near the wall. Additional gasdynamic discontinuities arising near the afterbody, the triple-point configurations at the reflected shock, and the intensity of the latter are reproduced well for $t_r = 5.8$. A fundamental improvement in the solution is obvious compared to the inviscid scheme.

The high efficiency of the proposed adaptive technique made it possible to obtain the presented results using IBM PC/486 personal computers and SUN SPARC workstations while supercomputer power is normally required to achieve the same accuracy with traditional approaches.

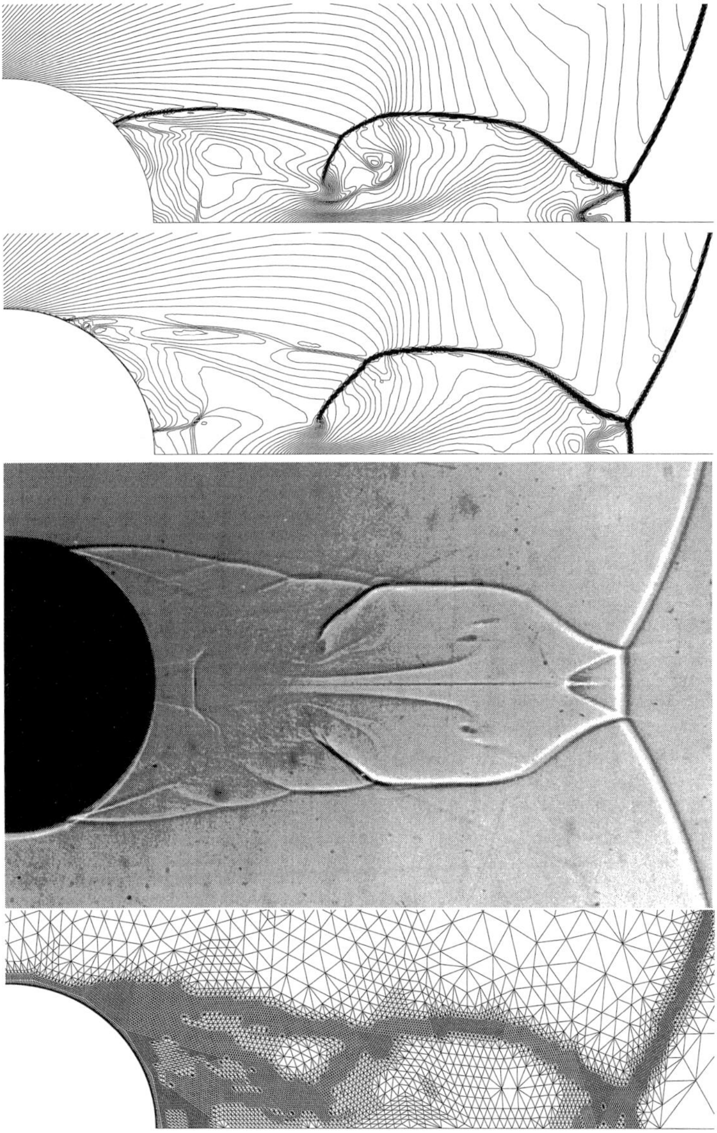

Figure 4. Same as in Fig. 3, $t_r = 5.8$. From top to bottom: density contours for inviscid computation; density contours for viscous computation; experimental shadowgraph (Syshchikova et al. 1990, Beryozkina et al. 1996); grid pattern for the Navier-Stokes solver.

Acknowledgement. Numerous discussions with Drs. M. Syshchikova and M. Beryozkina and their high-quality experimental data are acknowledged sincerely. We thank Prof. Yu. Golovachov and Prof. A. Fursenko for stimulating discussions. E. Timofeev acknowledges the support by Prof. N. Satofuka from Cooperative Research Center, Kyoto Institute of Technology, where the paper was finally prepared for publication. This work was partially supported by the Russian Foundation of Fundamental Investigations (Grants 93-02-14797 and 94-01-01456), Friedrich Alexander University Erlangen-Nürnberg (Contract FAU4001–01IR303) and European Community (COPERNICUS Project COP-94-01490).

References

Beryozkina MK, Ofengeim DM, Syshchikova MP, Timofeev EV, Fursenko AA, Voinovich PA (1996) A combined adaptive structured/unstructured technique for essentially unsteady shock-obstacle interactions at high Reynolds numbers. In: Hyperbolic Problems: Theory, Numerics, Computation, Glimm J, Grove J, Plohr B (eds), World Scientific (to be published).

Fursenko AA, Mende NP, Oshima K, Sharov DM, Timofeev EV, Voinovich PA (1993) Numerical simulation of propagation of shock waves through channel bends. Comp Fluid Dyn J 2:1.

Fursenko AA, Sharov DM, Timofeev EV, Voinovich PA (1995) An efficient unstructured Euler solver for transient shocked flows. In: Shock Waves @ Marseille I, Brun R, Dumitrescu LZ (eds), Springer, pp 371–376.

Hakkinen RJ, Greber I, Trilling L, Abarbanel SS (1959) The interaction of an oblique shock wave with a laminar boundary layer, NASA Memo 2-18-59W.

Ivanov MY, Krupa VG, Nigmatullin RZ (1989) An implicit high-order Godunov scheme for the Navier-Stokes computations. J Numer Math & Math Phys 29:888 (in Russian).

Ofengeim D, Timofeev E, Fursenko A, Voinovich P (1995) A hybrid adaptive structured/unstructured technique for essentially unsteady viscous/inviscid interactions. In: Proc 1st Asian CFD Conf, Hui WH, Kwok Y, Chasnov JR (eds), HKUST, Hong Kong, Vol 1, pp 383–389.

Savel'ev AD (1988) Implementation of Third Order Approximations for Numerical Simulation of Viscous Compressible Gas Flows Near Solid Surfaces, PhD Thesis, Moscow University.

Schröder W, Hänel D (1987) An unfactored implicit scheme with multigrid acceleration for the solution of the Navier-Stokes equations. Comp & Fluids 15:313.

Syshchikova MP, Beryozkina MK, Smirnov IV (1990) The effects upon bodies in the zone of interference of two explosions. In: Unsteady Gas Flows with Shock Waves, Syshchikova MP, Beryozkina MK (eds), Ioffe Physical-Technical Institute, USSR Academy of Sciences, pp 7–29 (in Russian).

Wave Refraction

Vortex-ring impingement on a wall with shock formation

T. Minota
Ariake National College of Technology Omuta, Fukuoka 836 Japan
M. Nishida, M. G. Lee
Kyushu University Fukuoka 812-81 Japan

Abstract: The flow field induced by a high-speed vortex ring approaching a solid plane wall is studied by means of experimental visualization and numerical simulation. As a vortex ring draws close to a wall, an induced boundary layer separates from the wall and forms a wall vortex. A shocklet is generated in the narrow region between the vortex ring core and the wall vortex. Moreover, a new shocklet is formed between the vortex core and the separation point of the boundary layer on the wall. The formation of the former shocklet is by the same mechanism as that in a convergent-divergent nozzle flow. The latter shocklet is caused by the abrupt increase in pressure due to the separation of the flow.

Key words: Vortex ring, Vortex-wall interaction, Formation of shocklet, Flow visualization

1. Introduction

Many studies have been performed to understand flow structure in near-wall turbulence, and have shown that hairpin vortices and bursting phenomena appear in turbulent boundary layers. To understand the physical mechanisms accounting for these phenomena, studies of such fundamental flows as a vortex ring approaching a plane wall have been reported. In an incompressible flow, when a vortex ring approaches a solid wall, secondary and tertiary vortices are generated on the wall owing to viscous effects. The primary vortex is stretched and rebounded by mutual induction with produced vortices (Walker et al. 1987, Orlandi & Verzicco 1993). On the other hand, it is known that in a compressible flow, a turbulent field may be influenced by the shocklets produced and the interactions between them.

The purposes of the present study are to perform optical visualization of a high-speed vortex ring approaching a plane wall and to compare the results with numerical simulations using the axisymmetric Navier-Stokes equations. A shock wave emitted from the open end of a shock tube travels towards wall, hits normal to the wall, and then reflects from it. Immediately after the shock is emitted, the fluid within the shock tube starts to be discharged. A shear layer generated between the emerging and external fluid rolls up, then develops into a strong vortex ring near the open end. Thus the generated vortex ring moves towards the wall at a self-induced velocity, and then interacts with the reflected shock wave traveling towards the open end. After this interaction, the vortex ring approaches the wall, and the boundary layer is induced on the wall, which is followed by a secondary vortex. This wall vortex interacts with the primary vortex ring. The effect of compressibility on this process is studied here.

2. Experimental setup and numerical procedure

A shock wave generated in a shock tube is emitted through a pipe with a 6.0-mm diameter into an atmosphere. The emitted shock wave is followed by a vortex ring as mentioned earlier. A glass plane is placed perpendicularly to the axis of the pipe, at a distance of 16.1 mm from the end of the pipe. Flow visualization was performed by a shadowgraph method with a pulsed dye laser as a light source. The optical system is the same as that described in Minota (1993). Experiments were performed at an incident-shock Mach number of 1.35.

The governing equations for the numerical simulations are the axisymmetric compressible Navier-Stokes equations. They are numerically solved by using the second-order upwind TVD scheme of the Harten-Yee type for convection terms and ordinary central discretization for viscous terms. The computational domain is 21 mm $\times$ 16.5 mm with a grid system of 210 $\times$ 165. No-slip and adiabatic conditions are imposed on the wall.

3. Results and discussion

Figure 1 shows the experimental and computed shadowgraphs at the interaction of the vortex ring and reflected shock, and those after the interaction. In Figs. 1a and b, a slightly bow shock wave connects with the vortex cores and is retarded by the opposing flow due to the vortex. In the computed shadowgraph (Fig. 1b), another shock is embedded in the vortex ring, which is the same rearward-facing shock shown by Baird (1987); but it is not seen clearly in the present experimental shadowgraph (Fig. 1a). A part of the shock wave diffracted in the vortex core travels in the opposite direction and hits the wall. Reflected and intensified shock fronts piled up on the ring surface are seen in Figs. 1c and d. The computation shows that the pressure near the wall at this stage is 1.7 times atmospheric pressure. It may be mentioned that the computed shadowgraphs and experimental ones are in fairly good agreement.

The successive shadowgraphs in Fig. 2 show the development of the wall vortices and the formation of a new shock when the vortex comes close to the wall. The small dark spot shown by "a" on the wall in Figs. 2b–d is a secondary vortex (wall vortex) formed by the concentration of vorticity on the

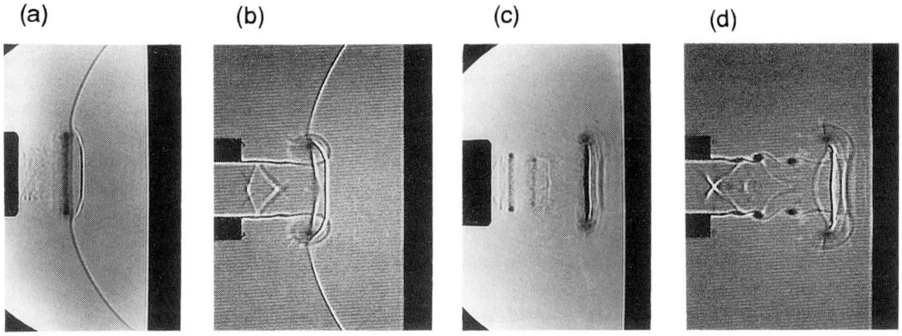

Figure 1. Experimental and computed shadowgraphs: (a) and (c) are experimental shadowgraphs (Mach number of the shock inside the nozzle is 1.35, Reynolds number is 7.1×10^4); (b) and (d) are computed shadowgraphs.

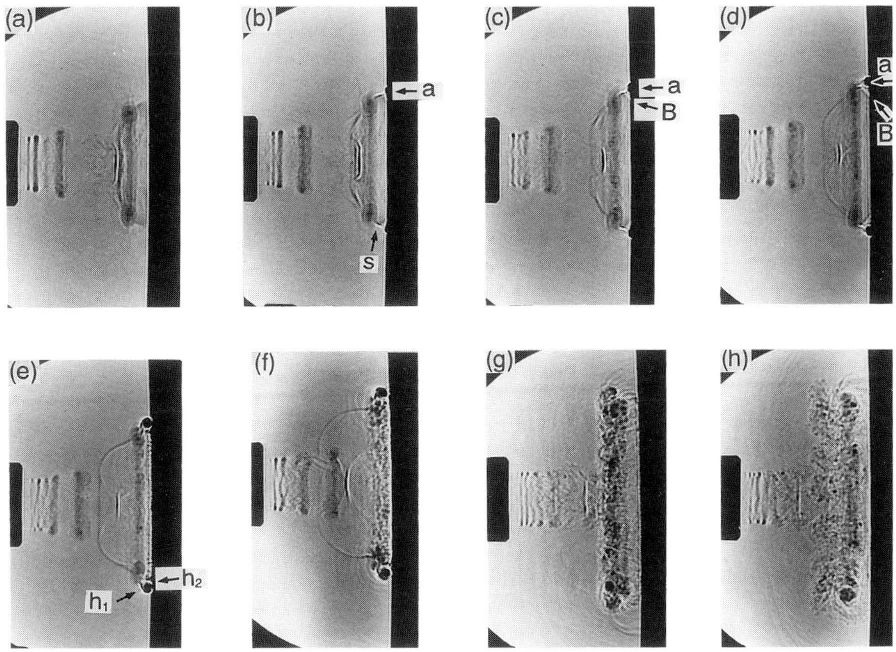

Figure 2. Experimental shadowgraphs showing the development of wall vortices and the formation of shocklets. Times after the stage of Fig. 1a: (a) 66 μs, (b) 68 μs, (c) 70 μs, (d) 74 μs, (e) 82 μs, (f) 89 μs, (g) 114 μs, and (h) 134 μs.

wall. An inward-facing shock shown by "s" is generated between the primary and secondary vortices. In Fig. 2d showing the growth of the secondary vortex, shocklets are formed around the half-circular dark spots (indicated by "a"). In Fig. 2e, these shocklets are observed as two lines "h_1" and "h_2" connecting the primary and the secondary vortex cores. In Figs. 2c and d are seen another new shocklet between the shock "s" and the position B on the wall. This shocklet is also an inward-facing shock. Figures 2f–h illustrate the interaction between the primary and secondary vortex rings. The secondary vortex rises up and winds around the primary vortex. The generated acoustic waves are observed in these figures.

Computational simulations corresponding to Fig. 2 are shown as shadowgraphs, pressure contours, and vorticity contours in Figs. 3 to 5, respectively. The shocklets which are observed in Figs. 2b–d are recognized at almost the same positions in the computed shadowgraphs shown in Figs. 3b–d. As shown in Figs. 4b and c, the shocklet "s" is created from the center of the vortex core towards the wall vortex, and intersects with the shocklet growing from the wall. The vorticity contour line "S" in Figs. 5b and c corresponds to the shocklets "s" in the pressure contours in Figs. 4b and c, respectively.

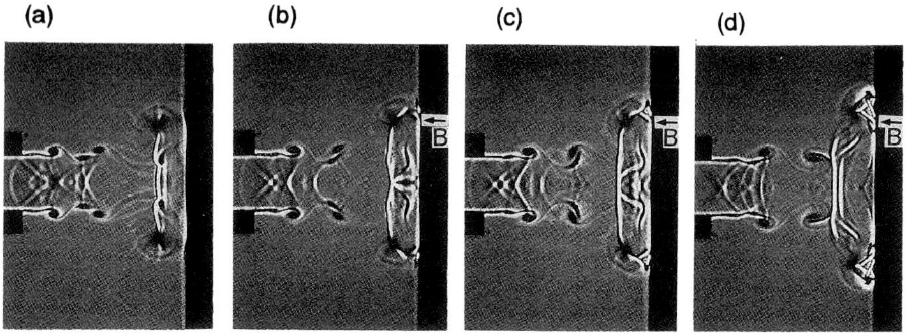

Figure 3. Computed shadowgraph showing the development of the wall vortices and the formation of the shocklets.

The position of the new shocklet on the wall connecting with the shocklet "s" is indicated by an arrow "B" in each figure, which is at $y/R = 0.43$, where y is the radial distance from the central axis and R is the maximum value of y in the computational domain. The point B is the separation point of the boundary layer or the front position of the separation bubble. This can be recognized from the velocity vectors. Figure 6a illustrates the velocity field corresponding to Fig. 3c; the normalized pressure distribution on the wall at the same time is shown in Fig. 6b, where p_1 and γ are the ambient pressure

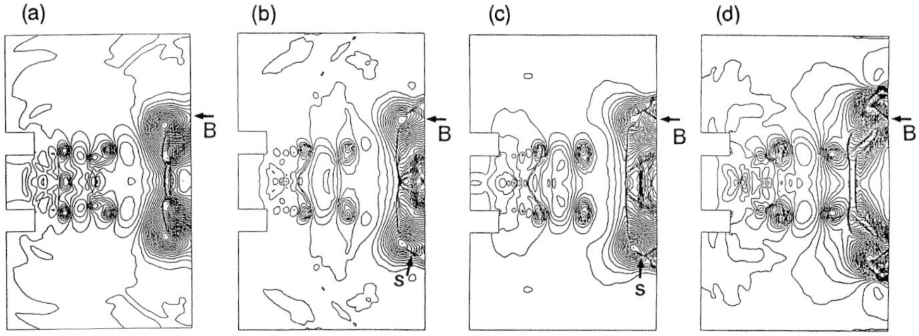

Figure 4. Pressure contours for each step of Fig. 3.

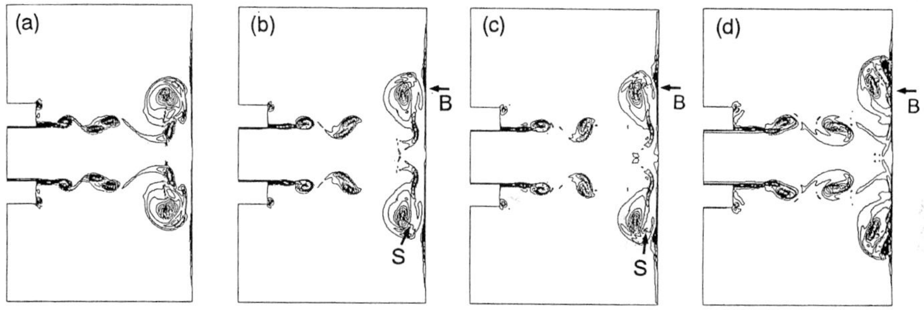

Figure 5. Vorticity contours for each step of Fig. 3.

and the ratio of specific heats, respectively. The fluid impinging on the wall is accelerated while flowing outward along the wall, and the pressure on the wall decreases rapidly. This decrease induces the separation of the boundary layer, then the pressure on the wall rises. Moreover, this abrupt reduction and enlargement cause the formation of the shocklet. The separation bubble generated by such a mechanism becomes a secondary vortex. The arrow "A" in Fig. 6b shows the position of the center of the secondary vortex. On the other hand, the accelerated flow passes through the narrow gap between the primary vortex and the separation bubble or secondary vortex. This is very similar to a convergent-divergent nozzle flow with a normal shock. It has been presumed that the formation of the shocklet "s" between the primary and secondary vortices is owing to the same mechanism as in the convergent-divergent nozzle flow.

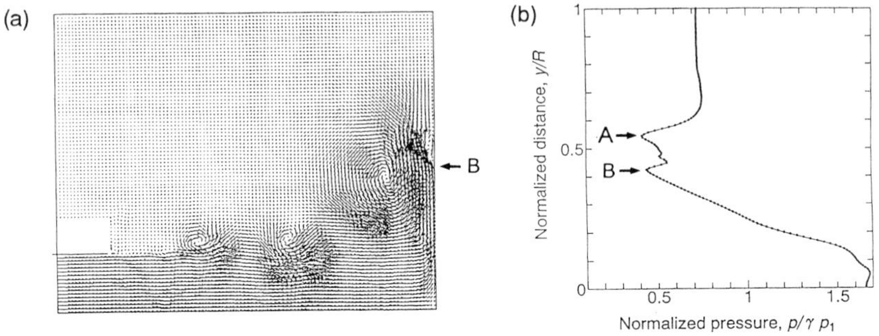

Figure 6. Computed results: (a) velocity field corresponding to Fig. 3c; (b) normalized pressure distribution on the wall.

4. Conclusions

The motion of a high-speed vortex ring which moves towards a plane solid wall has been studied experimentally and numerically. The computed shadowgraphs are in good agreement with the experimental ones. As the vortex ring approaches the wall within a distance comparable to the core diameter, the separation of the boundary layer occurs and a secondary vortex is formed; at the same time a shocklet is generated between both vortices. Moreover new shocklets are generated between the vortex core and the separation point on the wall. The formation of the former shocklet is owing to the same mechanism as that in a convergent-divergent nozzle flow. The latter shocklet is caused by an abrupt increase in pressure due to the separation of the flow. This was confirmed by the simulated results, including pressure contours, vorticity contours, pressure distributions on the wall, and so on.

References

Baird JP (1987) Supersonic vortex rings. Proc R Soc Lond A409:59–65.

Minota T (1993) Interaction of a shock wave with a high-speed vortex ring. Fluid Dynamics Research 12:335–342.

Orlandi P, Verzicco R (1993) Vortex rings impinging on walls: axisymmetric and three-dimensional simulations. J Fluid Mech 256:615–646.

Walker JDA, Smith CR, Cerra AW, Doligalski TL (1987) The impact of a vortex ring on a wall. J Fluid Mech 181:99–140.

Oblique reflection of shock waves in a slow sound-speed fluorocarbon liquid

B. J. P. Mortimer
School of Electrical Engineering, Cape Technikon
PO Box 652, Cape Town, 8000 South Africa

B. W. Skews, L. T. Felthun
School of Mechanical Engineering, University of the Witwatersrand
Private Bag 3 WITS, Johannesburg, 2050 South Africa

Abstract: Very low sound velocities are observed in fluorocarbon liquids. In addition, the nonlinearity parameter B/A for these fluids is exceptionally high. This factor means that the equation of state will be highly nonlinear. This paper presents a fluorocarbon liquid with a sound speed of 655 m/s for use in liquid shock experiments. Analytical, numerical, and experimental results of oblique shock-wave reflection in this medium off of a rigid surface are presented.

Key words: Fluorocarbons, Shock-wave reflection, Nonlinearity parameter

1. Introduction

Research into the propagation and focusing of shock waves in liquids has primarily used water as the medium e.g., Müller (1987), Takayama et al. (1983). Weak shock waves in this medium will travel at close to the acoustic sound speed of 1498 m/s. This value is over four times that of air. The steep rise time of water shocks places several constraints on experimentation. Fluorocarbon liquids have very low sound speeds in comparison to water, being between 470 and 700 m/s at 25°C. The purpose of this paper is to present experimental and numerical results of oblique shock-wave reflection with one of the fluorocarbon liquids, FC-43, as the medium. This liquid has a specific gravity of 1.88, a viscosity of 2.8 centistokes, and a sound speed at 25° of 655 m/s.

The isentropic equation of state for a fluid can be written as:

$$p - {c_o}^2 \rho = \frac{c_o^2}{\rho_o} \frac{B}{2A} \rho^2 \tag{1}$$

where p and ρ are the excess pressure and density respectively and A and B are temperature-dependent quantities (Beyer 1974). The term B/A is called the parameter of nonlinearity and is given by:

$$\frac{B}{A} = 2\rho_o c_o \left(\frac{\partial c}{\partial P}\right)_T + \frac{2c_o \alpha T}{\rho_o C_P} \left(\frac{\partial c}{\partial T}\right)_P \tag{2}$$

where T is the temperature in degrees Kelvin, C_P is the specific heat at constant pressure, and α is the coefficient of thermal expansion. For FC-43, B/A has been measured and reported to be 13.2 (Madigosky et al. 1981). This is the highest value of B/A reported thus far. By comparison, water has a B/A of 5.2.

The equation of state can also be derived in a form known as the Tait equation. This form is widely used for water shock computations.

$$p = B\left(\frac{\rho}{\rho_o}\right)^n - B \tag{3}$$

Here B and n are empirically determined constants. The parameter n has been shown to be related to the nonlinearity parameter B/A (Beyer 1974) by:

$$n = B/A + 1 \tag{4}$$

Thus for FC-43, $n = 14.2$ and the constant $B = \rho_o c_o^2/n = 568$ bar.

2. Oblique reflection of shock waves

A mathematical formulation of the reflection of a shock wave from a rigid wall has been considered by von Neuman (1943) and Courant and Friedrichs (1948) for several different configurations. Two types of reflection were found to be possible; firstly, regular reflection and secondly, irregular or Mach reflection.

Regular reflection was found to occur only under restrictive conditions. The analysis can be extended to include water shock reflection (Müller 1987) if a transformation for fluid pressure is introduced (Holl 1981). These equations yield the angle of reflection β as a function of the incident angle α and the incident ratio of pressure P_{21}:

$$\frac{(1 - P_{12}) \cdot t_\alpha}{1 + \mu \cdot P_{12} + (P_{12} + \mu) \cdot t_\alpha^2} = A \tag{5}$$

$$A^2(1-\mu)^2(t_\alpha - t_\beta) + A[(1-\mu)^2 - (t_\alpha - t_\beta)^2 - (\mu + t_\alpha t_\beta)^2] - (t_\alpha - t_\beta) = 0 \tag{6}$$

where:

$$p_{ik} = \frac{P_i + B}{P_k + B}, \qquad t_i = \tan(i), \qquad \mu = \frac{n-1}{n+1} \tag{7}$$

Using this analysis together with the introduction of a suitable transformation for the equation of state of FC-43 yields a plot of the angle of reflection β as a function of the incident angle α and the incident ratio of pressure P_{21} shown in Fig. 1. From this figure it can be seen that β is always greater than α and that this effect increases with increasing shock strength. The reflection is highly

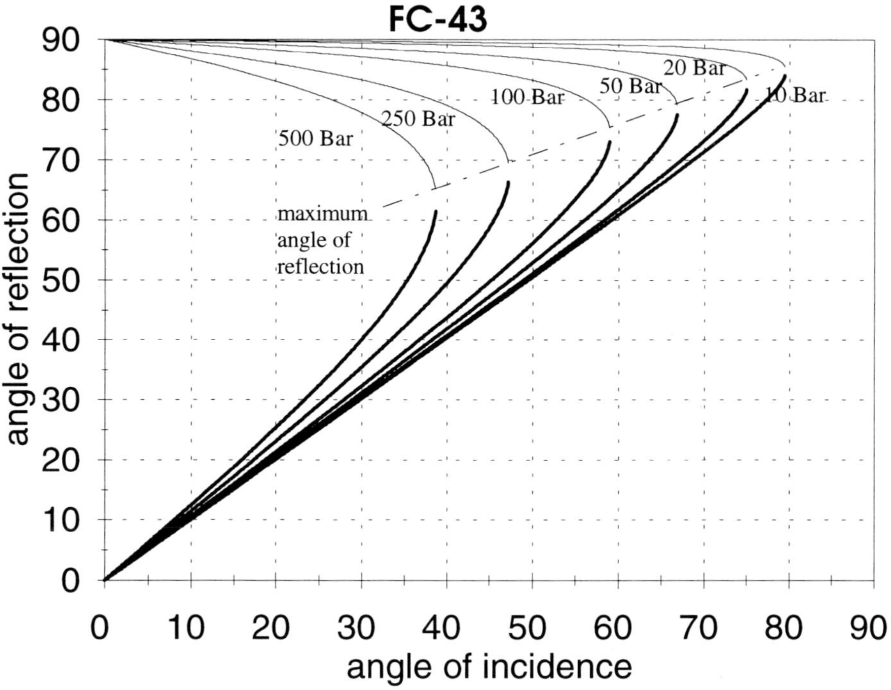

Figure 1. Angle of reflection β as a function of angle of incidence α for different incident shocks.

nonlinear. For a given pressure and incident angle, the deviation in angle of reflection β is greater for FC-43 than for water. For a given pressure, there exists an incident angle at which the reflection will no longer be regular. The reflection then transforms to irregular reflection.

3. Experiments and discussion

The interaction of a shock wave undergoing regular reflection from a rigid surface with FC-43 as the medium has been investigated numerically and experimentally. Wedges were of base angle: 25°, 45°, 75° – 89°, 77° – 87°, and 79° – 85°.

3.1. Computational results

Numerical simulations were conducted using a finite-element code based on the Taylor-Galerkin algorithm (Lohner et al. 1985). This uses a flux-corrected-transport algorithm to maintain high resolution of discontinuities and an adaptive-refinement algorithm to refine the mesh appropriately. In these experiments, the Tait parameters corresponding to FC-43 were used to describe the medium. Shock reflection of various wedges and various incident pressures were simulated. The angle of reflection measured from these experiments is shown in Table 1. Such a simulation is shown in Fig. 2.

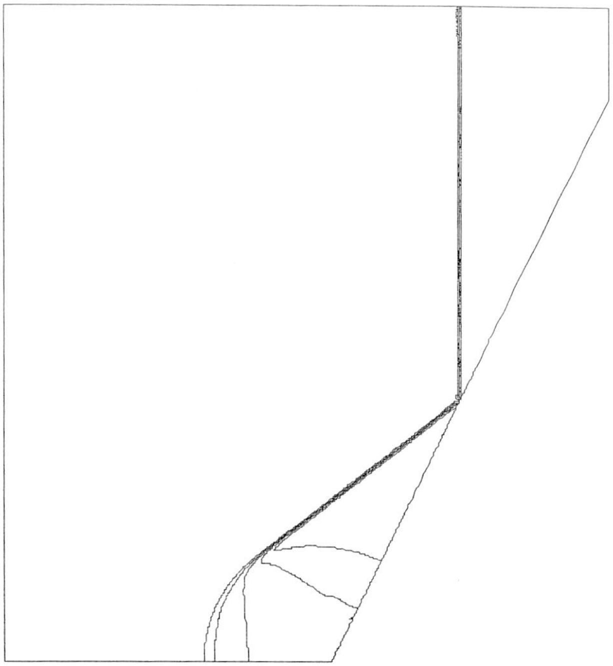

Figure 2. Results from FEM simulation for the reflection of an incident shock of strength 22 bar off a wedge of base angle 25°.

3.2. Shock-tube experiments

Experiments were also conducted using a special vertical shock tube filled with FC-43. The tube is similar to the tube described by Müller (1987), however, the driver section has been modified to facilitate double-diaphragm triggering. A conventional high-pressure air-driver section is pressurized up to 30 MPa. Two pre-scribed metal diaphragms are ruptured and release a shock wave into the air-filled second stage. This shock impinges on a lower test section filled with FC-43. Schlieren photographs were taken of the shock-wave interaction with a wedge mounted at the end of the test section. An example of such an experiment is shown in Fig. 3 — the wave is travelling from top to bottom in the photographs. The light source used in these experiments was a conventional xenon flash of duration of approximately 1 μs.

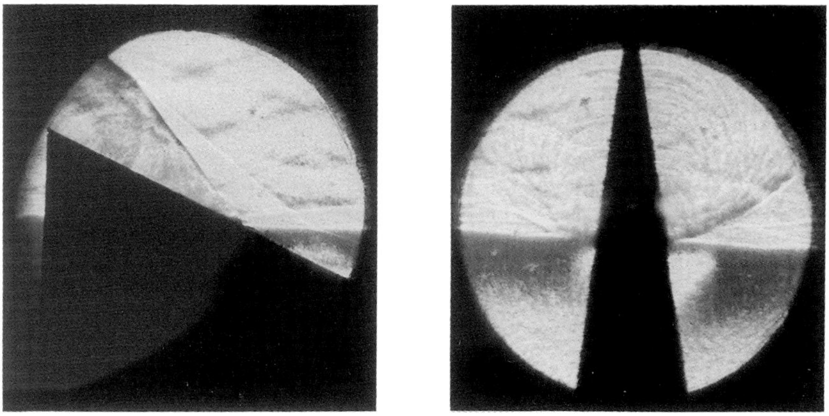

(a) (b)

Figure 3. Schlieren photographs of oblique shock wave reflection off wedges: (a) wedge of 25°, $P_2 = 22$ bar, regular reflection; and (b) wedge of 79° - 85°, $P_2 = 22$ bar, irregular reflection.

For an incident liquid shock of pressure of 22 bar, the Mach number is 1.01. The equivalent gasdynamic inverse pressure ratio is 0.964. These shocks are therefore weak in a gasdynamic sense. The angle of reflection measured from the schlieren photographs for various incident pressures is shown in Table 1.

Table 1. Comparison of the experimental and analytical results.

Wedge Angle	Shock Pressure [bar]	Measured β (degrees) $\beta_{experimental}$	$\beta_{numerical}$	β_{theory}
Wedge: 25°	18	25	25.4	25.33
	22	25.5	25.5	25.41
	28	26	25.6	25.52
Wedge: 45°	18	45.5	45.6	45.8
	22	46	46	45.99
	28	46.5	46.2	46.27
Wedge: 75°	18	80	80	79.95
	22	82	82	irregular
	28	82	82	irregular
Wedge: 79°	10	82	82.5	82.7

The measured angle of incidence for regular reflection with low angles of incidence correspond closely to analytic results. The error in the measurements is less than 0.5° for low angles of incidence. As the corner wave interacts with the reflected shock the waves become curved and it becomes more difficult to measure the angles. Several tests were performed where the reflection is theoretically irregular. However no conventional Mach reflection could be resolved for these weak shock experiments.

4. Conclusions

The use of a slow sound-speed fluorocarbon liquid provides several advantages over water for liquid-shock-wave research. Its slow sound speed places fewer demands on experimentation and its high B/A means that nonlinearities in shocks can be obtained for relatively low peak pressures in comparison to water. An equation of state has been derived and analytical and numerical results of oblique shock-wave reflection given. These compare favorably to low-pressure experimental results validating in part the choice of equation of state parameters.

References

Beyer RT (1974) Nonlinear Acoustics, Naval Ship Systems Command, Washington DC, pp 98–101.

Courant R, Friedrichs RO (1948) Supersonic Flow and Shock Waves, Springer-Verlag, New York, pp 329 (reissued in 1976).

Holl R (1981) Stosswellenfokussierung in Fluiden, Thesis, Doktor-Ingenieur, Technischen Hochshule Aachen.

Lohner R, Morgan K, Zienkiewicz O (1985) An adaptive finite element procedure for compressible high speed flows. Comput Meths Appl Mech Eng 51:441–465.

Madigosky WM, Rosenbaum I, Lucas R (1981) Sound velocities and B/A in fluorocarbon fluids and in several low density solids. J Acoust Soc Am 69(6):1639–1643.

Müller M (1987) Stosswellenfokussierung in Wasser, Thesis, Doktor-Ingenieur, Technischen Hochshule Aachen.

Takayama K, Esashi H, Sanada N (1983) Propagation and focusing of spherical shock waves produced by underwater microexplosions. In: Proceedings of the 14th Int Symp on Shock Tubes and Waves, Archer RD, Milton B (eds), Sydney, pp 553–562.

von Neuman J (1943) Collected Works, Volume VI, (Taub AH, General editor), Pergamon Press, Oxford, London (reissued in 1963).

Refraction of a blast wave propagating parallel to a contact discontinuity: a Free-Lagrange numerical simulation

G. J. Ball, R. A. East
Deptartment of Aeronautics & Astronautics
University of Southampton, Southampton, SO17 1BJ UK

Abstract: A recently developed Free-Lagrange code is used to simulate the refraction of planar blast waves propagating parallel to air/propane and air/freon-114 interfaces. The time evolution of the flow fields is described. The rate of decay of the blast wave in the gas with the slower speed of sound is found to be reduced at the expense of an increased rate of decay in the air.

Key words: Blast wave, Shock refraction, Free-Lagrange method

1. Introduction

Interactions between shock waves and contact discontinuities have been the subject of much recent research, focusing both on underlying physical mechanisms and on their role in practical explosion/blast phenomena. These interactions fall into two broad classes: Richtmyer-Meshkov instabilities and shock refraction events. Here we are concerned only with the latter. In a series of papers (see for example Henderson et al. 1991) Henderson and co-workers have described a variety of wave configurations which may occur as a shock passes obliquely through a contact surface across which the speed of sound changes. The wave structure was found to depend strongly on the shock strength and the shock-contact surface incidence angle, α. Ben-Dor and Rayevsky (1991) have examined the behavior of a shock propagating horizontally ($\alpha = 90°$) in air above a layer of gas of higher density, simulating dust-laden air, for a range of density ratios and shock Mach numbers. They observed similar wave structures to those of Henderson et al., and showed that the leading edge of the higher density layer tended to roll-up into a large vortex, providing a strong mechanism for dust entrainment.

In this paper we consider a related problem, in which a weak planar blast wave is propagated parallel ($\alpha = 90°$) to a 'fast-slow' contact surface between air and propane (sound speed ratio $1:0.73$). The problem may be regarded as an idealization of a flow which might arise following an accidental LPG spillage and explosion. In addition, we examine the more extreme case of refraction at an interface between air and freon 114 (sound speed ratio $1:0.36$). The use of a blast wave, which decays with time, introduces additional physical effects which are absent in the equivalent shock-wave problem, but which may be significant in accidental explosions.

2. Numerical method

The simulation is performed using a recently developed Free-Lagrange code *Vucalm* (Ball 1996), which solves the two-dimensional unsteady Euler equations on an unstructured Lagrangian grid using a Godunov-type method. The working fluid is divided into discrete packets, between which mass exchange is forbidden. Flow variables are stored at a 'particle' at the center of each packet. The approximate locations of the packet boundaries are determined by constructing a Voronoi mesh, in which each particle is enclosed within a N-sided polygonal Voronoi cell, which forms a control volume for the time integration of the Euler equations. The mesh is reconstructed frequently during the simulation to allow the grid connectivity to change naturally under the influence of shear.

The code employs a three-wave approximate Riemann solver, adapted from Toro et al. (1994), which includes an explicit representation of the contact surface in the local Riemann problem. When implemented in the Lagrangian frame, where the cell boundary and the contact surface are coincident by definition, the solver gives zero numerical diffusion at contact discontinuities, which can therefore be sharply resolved without recourse to an interface-tracking routine. A piecewise-linear representation of conserved variables within grid cells is used to obtain nominal second-order spatial accuracy. A slope limiter of the MUSCL type is used to prevent oscillations in the solution near shock waves.

3. Results and discussion

3.1. Problem specification

The starting geometry for our problem is shown in Fig. 1. The flow occurs in a semi-infinite rectangular shock tube, of which only the first 5 meters are represented in the computational domain. The pressure and temperature in

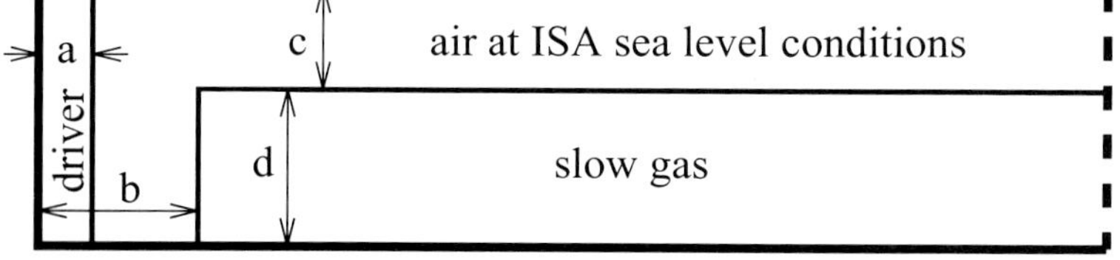

Figure 1. Initial geometry: $a = 0.25\,\text{m}$, $b = 0.75\,\text{m}$, $c = 0.5\,\text{m}$, $d = 0.75\,\text{m}$.

both layers, and in the short driver section, are set to 1.01325 bar and 288.15 K. The particles initially form a square grid, with a pitch of 1/40 m, giving a total of 10^4 cells. To initiate a blast wave, the temperature of the air in the driver section is impulsively increased by a factor of 5. The resulting pressure discontinuity gives rise to a shock propagating to the right and an expansion wave which reflects from the closed end of the tube and then rapidly catches up with the shock to establish a blast wave. In order to characterize the subsequent decay of the blast wave, a calculation was performed using air only. For propagation times between 2 ms and 30 ms, the shock Mach number was found to decay as $M_s = 1 + 0.878t^{-0.475}$ (for t in ms).

3.2. Results for air/propane

Figure 2 shows the flow structure 4 ms after initiation. The contours for this and subsequent plots are of constant $\log_{10}\rho$, with contour interval 0.015. The

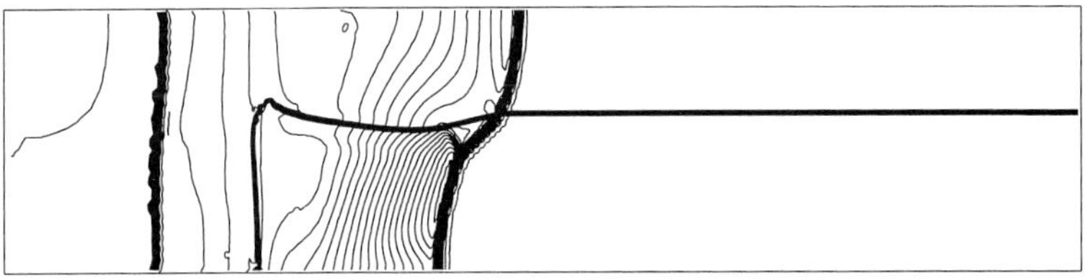

Figure 2. Air/propane solution after 4 ms. Contours are of log ρ.

strong vertical feature to the left is the contact surface marking the edge of the hot driver gas. The shock in the air layer has advanced ahead of that in the propane to form a free precursor shock, and the two waves are linked by an oblique shock, forming a transient Mach reflection in the propane layer. Above the triple point the primary shock in the propane is inclined sharply to the left. Behind the shock structure the pursuing expansion is clearly visible in both layers.

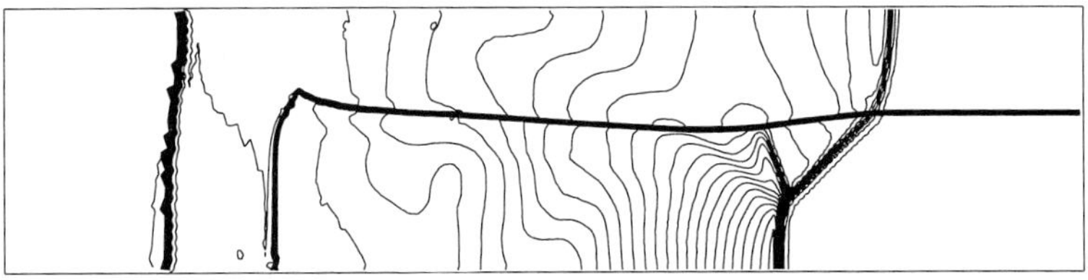

Figure 3. Air/propane solution after 8 ms. Contours are of log ρ.

The subsequent evolution of the flow is shown in Figs. 3 and 4. The triple point moves towards the floor of the tube, while the precursor shock and the

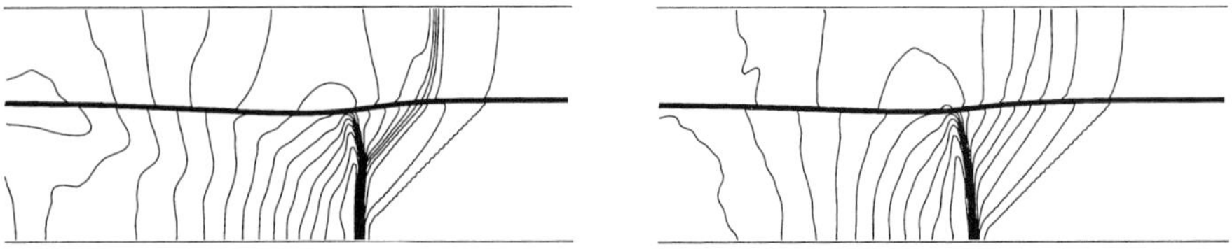

Figure 4. Air/propane solution after 18 ms (left) and 30 ms (right) — log ρ.

linking oblique shock rapidly weaken. An additional weak oblique compression can be seen climbing the rear of the propane shock in Fig. 3. This wave ultimately reinforces the precursor wave (see Fig. 4). Its origin will be discussed in Section 3.3. After 30 ms the precursor wave is little more than a Mach wave, while the propane shock still has a Mach number of about 1.32. At the same stage, the solution for air alone gives $M_s = 1.17$, while for an all-propane case $M_s = 1.18$. It appears that, as a result of the interaction at the contact surface, the blast wave in the gas with the lower sound speed has been sustained at the expense of that in the air, where the rate of decay has been enhanced.

In order to retain the blast wave within the computational domain, the solutions for times beyond 8 ms were obtained in a frame of reference translating at approximately the propagation speed of the wave structure. The domain was truncated to the left at a point where the velocity in the blast had fallen sufficiently to provide a supersonic outflow in the moving frame, thus simplifying the outflow boundary conditions.

3.3. Results for air/freon 114

The initial development of the air/freon flow is qualitatively similar to that of the air/propane case. At 4 ms a transient Mach reflection has been established in the freon (see Fig. 5). The interaction of the blast with the upstream corner of the freon layer has two consequences. Firstly, the difference in post-shock velocities in the two layers causes the freon to begin to roll up into a vortex.

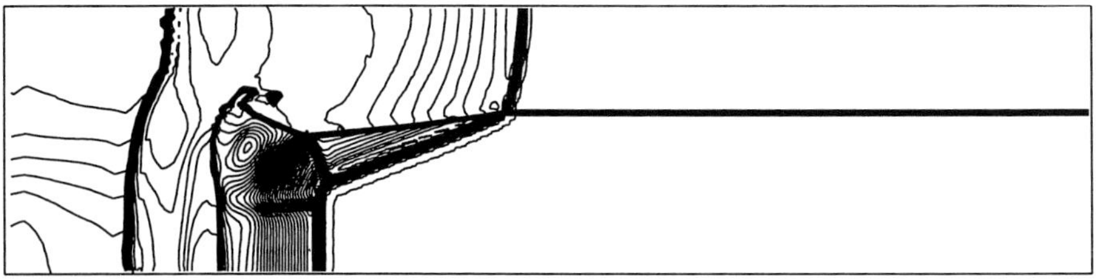

Figure 5. Air/freon 114 solution after 4 ms. Contours are of log ρ.

Secondly, a compression wave appears, aligned almost horizontally and propagating downwards. This latter feature, which is also present, although much weaker, in the air/propane case, seems not to have been observed by other authors.

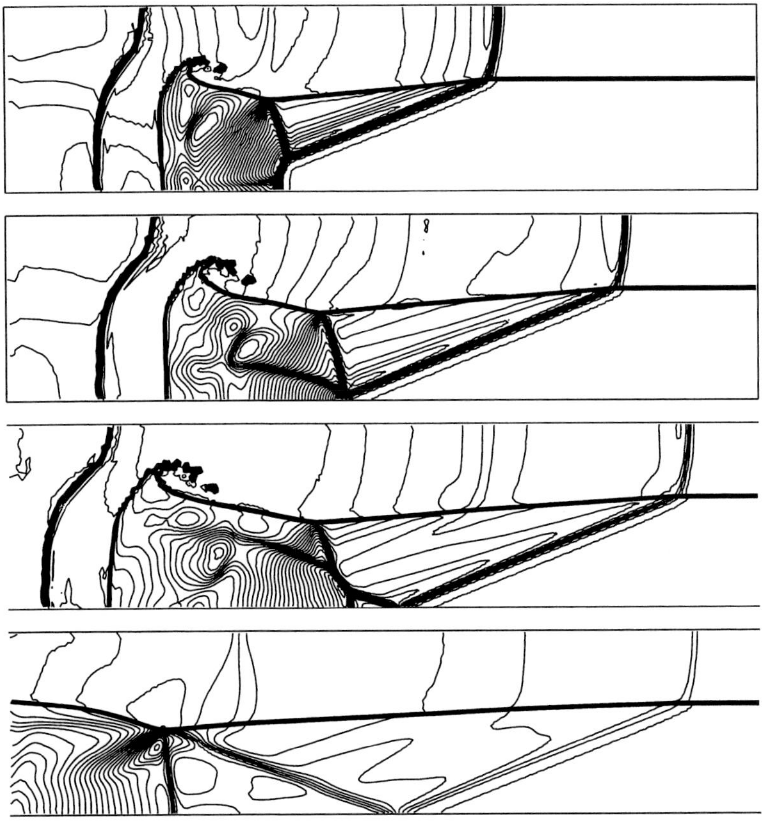

Figure 6. Air/freon-114 solution at (a – d, top – bottom) 6, 8, 10, 16 ms. Reference frame translated at 370 m/s starting $t = 9$ ms. Contours of log ρ.

Subsequent events are shown in Figs. 6a – d. The vorticity at the freon 'corner' is rapidly reduced as the the vortex is traversed by the expansion component of the blast wave, so that the roll-up process is effectively frozen. The freon filament continues to stretch slowly due to the axial velocity gradient. Where the filament thickness is too fine to be resolved on our grid, it breaks into discrete beads. The compression wave reflects from the floor of the tube (Figs. 6a and b), then propagates upwards along the rear of the freon shock (Fig. 6c) and is partially transmitted into the air layer, being refracted in the process so that its propagation direction becomes approximately axial (Fig. 6d). Meanwhile, the precursor shock and the linking oblique shock weaken, the triple point reaches the tube floor, and the transient Mach reflection is replaced by a regular reflection (Fig. 6c). After 16 ms (Fig. 6d) the reflected oblique shock has extended to the contact surface, while the freon shock persists as a nearly normal shock, closely followed by a strong expansion. As in the air/propane case, peak overpressure occurs behind the primary shock in the slower gas, while the pressure excursions in the faster gas rapidly decrease as the wave structure evolves.

4. Conclusions

The flow structures we have observed are in broad accord with those reported by other authors. The use of a decaying blast wave in the present work has two consequences: the vorticity generated at the leading corner of the slow gas layer is rapidly relieved, limiting the extent of entrainment of the slow gas, and the wave structure evolves continuously, exhibiting no self-similar phase, although the rate of evolution falls with time. The rate of decay of the primary blast wave in the slow gas has been shown to be reduced by the interaction, at the expense of an enhanced rate of decay of the precursor shock in the faster gas. An additional compression wave has been observed, originating from the interaction of the blast with the leading corner of the slow gas, which has not apparently been reported elsewhere. It is not clear whether this is particular to blast interactions. The Free-Lagrange code has demonstrated its ability to capture contact surfaces without diffusion, even in the presence of strong pressure waves.

Acknowledgement. The authors are grateful to Mr. D. Webster for his work in developing the graphical user interface for the Vucalm code.

References

Ball GJ (1996) A Free-Lagrange method for unsteady compressible flow: simulation of a confined cylindrical blast wave. Shock Waves 5:311–325.

Ben-Dor G, Rayevsky D (1991) Shock wave interaction with a dense layer. In: Proc International Workshop on Strong Shock Waves, Honma H, Maeno K (eds) Chiba, Japan, pp 145–164.

Henderson LF, Colella P, Puckett EG (1991) On the refraction of shock waves at a slow-fast gas interface. J Fluid Mech 224:1–27.

Toro EF, Spruce M, Speares W (1994) Restoration of the contact surface in the HLL Riemann solver. Shock Waves 4:25–34.

Shock waves generated by the interaction of two vortex rings

T. Minota
Ariake National College of Technology
Omuta, Fukuoka 836 Japan

M. Nishida, M. G. Lee
Department of Aeronautics and Astronautics, Kyushu University
Fukuoka 812-81 Japan

Abstract: This paper describes experimental and numerical studies of the flow field for a head-on collision of two vortex rings and the shock waves generated by it. As the two vortex rings approach, the region between them is compressed, then the pressure becomes high. Moreover, as the rings come close to each other, an inward-facing shock is formed between the two vortex cores. This shock travels together with the stretching of the vortex core line. Computational results of Mach contours and the pressure distribution on the collision plane of the rings show that the formation of this new shock is by the same mechanism as that in a convergent-divergent nozzle flow.

Key words: Vortex ring, Formation of shocklet, Interaction of vortices

1. Introduction

A vortex ring has been used to understand various aspects of vortex motion in a jet flow, vortex interaction, generation of vortex sound, and so on. The motion of interaction between vortex rings exhibits various behaviors depending on initial configuration. Numerous studies of a head-on collision of two vortex rings have been conducted (Shariff 1992). When two vortex rings approach each other, their sizes increase due to mutual induction. In the case of ring collision with shock formation, compressibility plays an important role in the interaction of vortices. However there are not satisfactory studies on the interaction of compressible vortex rings. In a previous work (Minota 1995), the formation and development of a shock connecting with two vortex cores have been observed by means of optical shadowgraphs.

In the present work, the optical visualization and the numerical calculations are carried out under nearly the same conditions. The physical mechanism of the formation of a new shock between the cores is studied by comparison of experimental shadowgraphs with computational results.

2. Experimental setup and numerical calculation

A shock wave generated in a shock tube is separated into two shock waves, and they are emitted at the same time from two nozzles which are placed at a distance of 33.4 mm from each other with a common axis. Two vortex rings formed near the open ends of the nozzles move towards each other at a self-induced velocity. Experiments were performed for an incident-shock Mach number of 1.34. The Reynolds number of the vortex ring in the initial stage is approximately 7.0×10^4. Flow visualization was performed by means of the same shadowgraph method as used in Minota (1995).

The governing equations for the numerical simulations are the axisymmetric compressible Navier-Stokes equations. They are numerically solved by using the second-order upwind TVD scheme of the Harten-Yee type for convection terms and ordinary central discretization for viscous terms. The computational domain is the upper half region, which is 42.9 mm $\times$ 18.5 mm with a grid system of 276 $\times$ 119.

3. Results and discussion

Figure 1 shows the optical shadowgraphs of the flow field of the interaction of shock waves and vortex rings. Computed shadowgraphs corresponding to each moment in Fig. 1 are shown in Fig. 2. A spherical shock wave collides with a vortex ring and is distorted. The bright line (Sd) connecting with the vortex core seen in Fig. 1a shows an impinging shock wave retarded by the opposing flow due to the vortex. In the computed shadowgraphs (Fig. 2a), the shock indicated by "Sd" is the retarded shock mentioned above, and a new rearward-facing shock (Se) is generated. This new shock is not seen clearly in the optical shadowgraph (Fig. 1a). The outer part (Wo) of the impinging shock wave diffracts in the vortex core, and bends as it passes around the core as shown by "Wd" in Fig. 1b. The shock front traveling in the opposite direction interacts with other vortex rings.

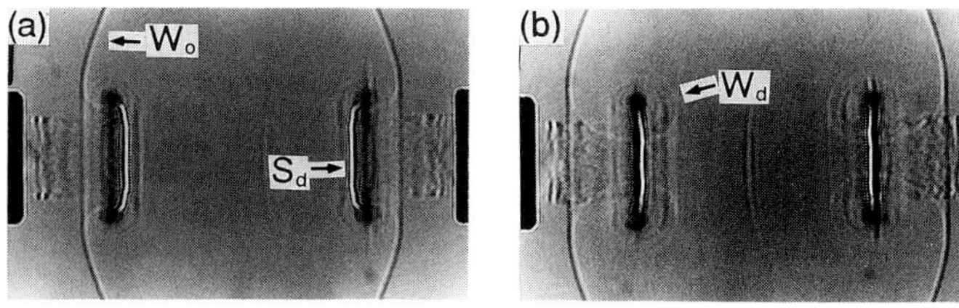

Figure 1. Optical shadowgraphs showing the flow field of the interaction of shock waves and vortex rings. Time: (a) 40 μs; and (b) 50 μs.

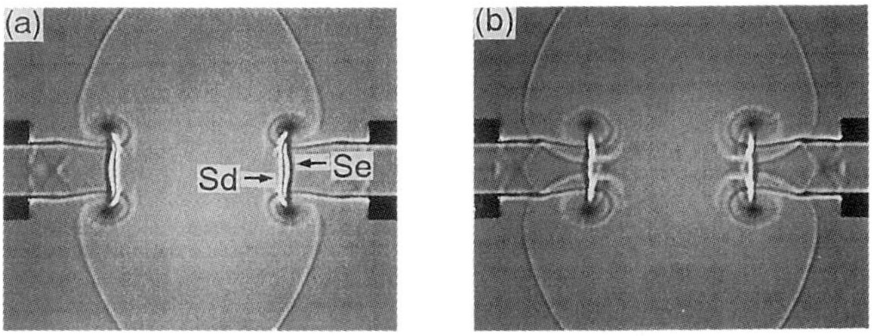

Figure 2. Computed shadowgraphs. Time: (a) 46.5 μs; and (b) 54.2 μs.

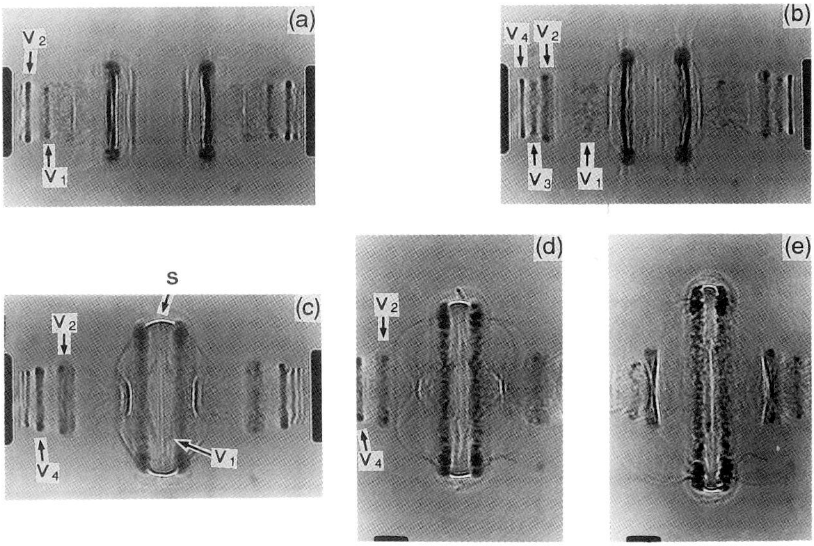

Figure 3. Optical shadowgraphs showing development of the interaction of two vortex rings. Time: (a) 80 μs; (b) 100 μs; (c) 110 μs; (d) 120 μs; and (e) 135 μs.

The successive shadowgraphs in Fig. 3 show the time development of the interaction of two vortex rings. As shown in Fig. 3a, a few shock waves are piled up on the ring surface; these shocks are the waves intensified by the interactions with the vortex. In Fig. 3c, an inward-facing new shock (shown by "s") is seen between the cores of the two vortex rings. The forward motion is blocked and the radial motion is accelerated. Then the vortex core line is stretched and its diameter becomes large. The rate of the diameter development is more than 270 m/s. The inward-facing new shock moves together with the stretching of the vortex core.

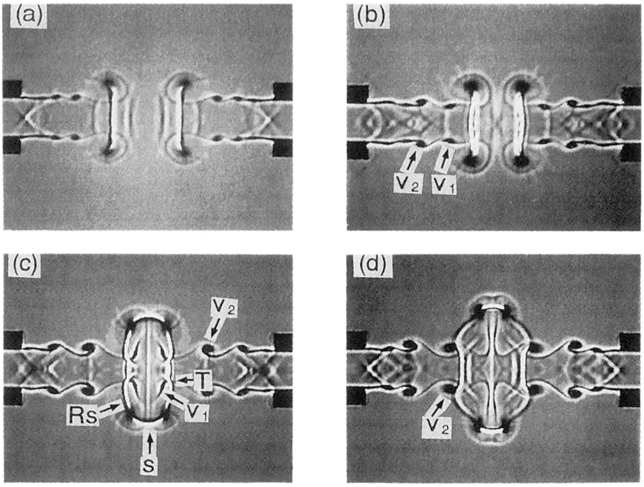

Figure 4. Computed shadowgraphs showing the time development of the interaction of two vortex rings. Time: (a) 80.0 μs; (b) 92.3 μs; (c) 107.6 μs; and (d) 122.0 μs.

A thin shear layer between the jet flow following the vortex ring and the exterior fluid rolls up owing to the Kelvin-Helmholtz instability, and small vortex rings are formed successively. These small vortex rings interact with each other and merge. The small vortex rings "v_1" and "v_2" in Fig. 3a arrive at the positions "v_1" and "v_2" in Fig. 3b, respectively. In Fig. 3b, new rings "v_3" and "v_4" are formed. The ring "v_1" penetrates into the primary vortex, and deforms as indicated by "v_1" in Fig. 3c. The ring "v_3" approaches the ring "v_2", and is merged. Acoustic waves are recognized in Fig. 3e. Computed shadowgraphs corresponding to the optical ones in Fig. 3a–d are shown in Fig. 4. The small vortex "v_1" in Fig. 4b is caught up by the primary vortex in Fig. 4c, and is distorted as shown by "v_1". This is similar to the form of "v_1" in Fig. 3c. Shocks inside the primary ring, indicated by "Rs" in Fig. 4c, have a structure with a triple point shown by "T" through diffraction around the distorted small vortices "v_1". The inward-facing shocks "s" are seen in Fig. 4c–d. These move with the stretching of the vortex ring in a way such as in Fig. 3. The computed shadowgraphs are in fairly good agreement with the optical shadowgraphs concerning the motion of the small vortex rings as well as the process of interaction of the two primary vortex rings and the formation of the new inward-facing shock.

Figure 5 shows the results of computational simulation corresponding to Figs. 4b–d. The top (-A), second (-B), third (-C), and bottom (-D) rows show

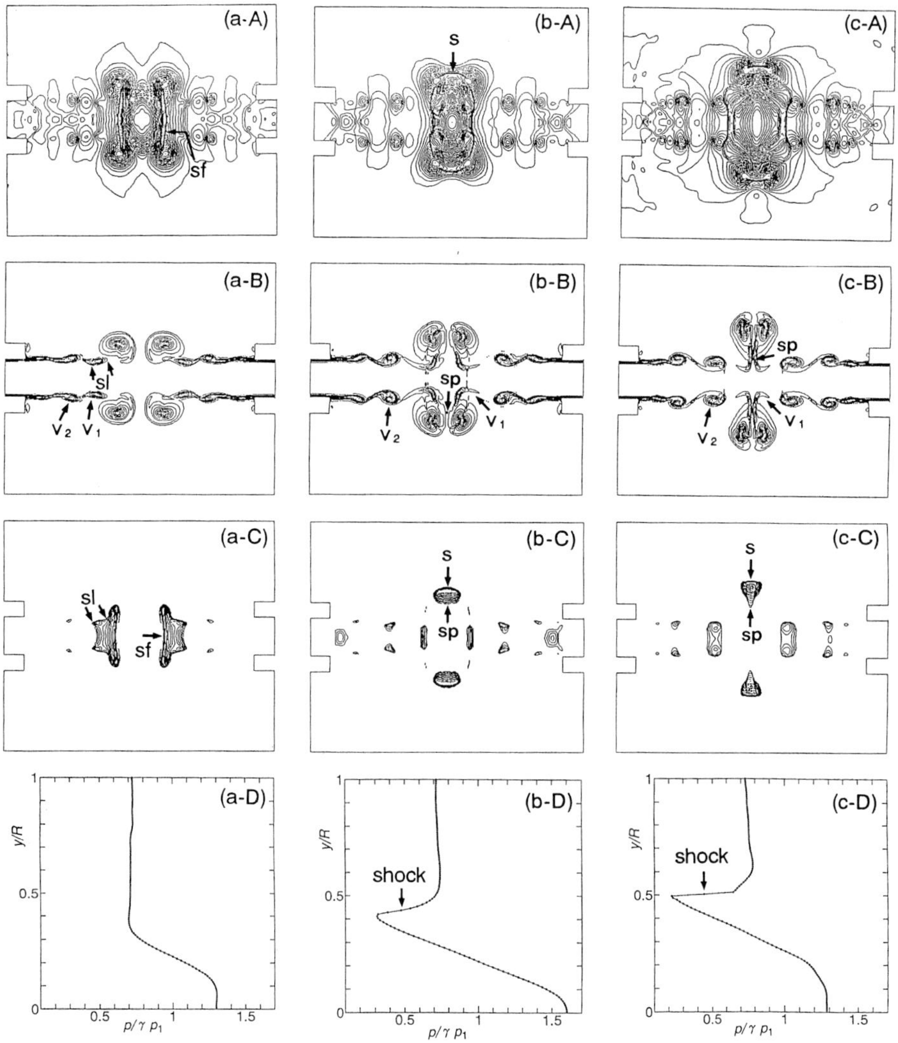

Figure 5. The results of computational simulation corresponding to Figs. 4b–d. Top row shows isobars. Second, third, and bottom rows show vorticity contours, supersonic Mach contours, and the normalized pressure distribution, respectively, where y is the radial distance from the central axis, R the maximum value of y in the computational domain, p_1 the ambient pressure, and γ the ratio of specific heats.

isobars, vorticity contours, supersonic Mach contours, and pressure distribution along the collision plane of the vortices, respectively. The front indicated by "sf" in the supersonic Mach contours of Fig. 5a-C corresponds to outer shock front "sf" in the pressure contours of Fig. 5a-A. The velocity upstream of the shock front "sf" becomes supersonic. The sonic line indicated by "sl" corresponds to the inside segment "sl" of the small vortex "v_1". The penetrating small vortex rings distort as shown in Fig. 5b-B and c-B. These were observed as the vortex "v_1" in the shadowgraphs Fig. 3c and 4c. The inward-facing shock (s) seen in the pressure contours Fig. 5b-A is also recognized as "s" in the Mach-number contours of Fig. 5b-C. The position indicated by "sp" in vorticity contours Fig. 5b-B shows a sonic line "sp" in the Mach contours. The pressure distribution Fig. 5b-D shows that the pressure on the center of the collision plane of the rings is more than 2.2 times the ambient pressure and 1.3 times the initial pressure behind the shock. The pressure of fluid flowing outwards decreases rapidly, and rises suddenly by the formation of a shock. The formation of the inward-facing shock can be explained as follows: the high-pressure fluid on the center of the collision plane is accelerated, and the pressure decreases rapidly. The accelerated fluid passes through the narrow gap surrounded by the two vortices. The flow velocity becomes supersonic downstream of it and a shock is formed in the accelerated flow in order that the low pressure become equal to the ambient pressure, which is higher than the pressure in the accelerated flow. This situation is very similar to a convergent-divergent nozzle flow with a normal shock.

4. Conclusions

The motion of two high-speed vortex rings which approach each other has been studied experimentally and numerically. When the two vortex rings approach each other within a distance comparable to the core diameter, an inward-facing shock is formed between the two vortex rings. This new shock travels together with the stretching of the vortex core line. The formation of the inward-facing shock is by the same mechanism as that in a convergent-divergent nozzle flow. This is confirmed by the computational simulations, i.e., Mach contours and pressure distribution.

References

Minota T (1995) Experiments on shock and vortex interactions. In: Shock Waves @ Marseille, Springer-Verlag, Vol 4, pp 343–348.

Shariff K (1992) Vortex rings. Annual Rev Fluid Mech 24:235–279.

Mechanisms of shock-wave generation resulting from shock-wave/bubble interaction in liquids

K. Yamada, T. Kodama, K. Takayama
Shock Wave Research Center, Institute of Fluid Science
Tohoku University
2-1-1, Katahira, Aoba-ku, Sendai 980 Japan

Abstract: Interactions of a shock wave with a gas bubble were visualized using double-exposure holographic interferometry and high-speed framing photography. The gas bubble was exposed to a spherical shock wave which was generated by detonation of a micro-explosive. The gas bubble contracted by impingement of the shock wave and hence a liquid jet formed; eventually, compression waves or shock waves occurred.

Key words: Flow visualization, Gas bubble, Liquid jet, Shock-bubble interaction

1. Introduction

A flow visualization study was conducted for quantitative observation of interaction between a spherical shock wave and a gas bubble by means of double-exposure holographic interferometry. Processes of shock-bubble interaction from shock impingement to generation of a rebound shock wave were visualized. Four combinations of liquid/gas were examined: silicone-oil/air, water/air, water/helium, and silicone-oil/air-helium (Yamada and Takayama 1995).

2. Experiments

Test liquids were water and dimethylsilicone-oil, and test gases were air and helium. Speed of sound, density, and kinematic viscosity of silicone-oil (Shin-Etsu Chemical Industry Co. 1989; Weissler 1949) were $a_{0s} = 936.5\,\mathrm{m/s}$, $\rho_{0s} = 821\,\mathrm{kg/m^3}$ (293 K), and $\nu_{0s} = 1.0 \pm 0.1 \times 10^{-6}\,\mathrm{m^2/s}$ (298 K), respectively.

An underwater shock wave impinged on a gas bubble released from the bottom of a test tank and rising freely in liquid. A micro-explosive, a 10.0 ± 0.5-mg silver azide (AgN_3) pellet (Chugoku Kayaku Co.), was detonated at a predetermined position in the liquid. The volume of the gas bubble and position of the gas bubble at the moment of initiation of the micro-explosive were specified. The micro-explosive was ignited and subsequently a spherical shock wave interacted with the gas bubble.

3. Results and discussion

Experimental results of silicone-oil/air, water/helium, and silicone-oil/air helium mixed-gas combinations are presented. The shape of the gas bubble before interaction with the shock wave was an oblate spheroidal shape. Typical dimensions of the gas bubbles and overpressures of the incident shock waves are tabulated in Table 1.

Table 1. Typical dimensions of gas bubbles immediately before impingement of an incident shock wave in liquid and overpressure of the incident shock wave. The variables a, b, f, d, and ΔP are major diameter, minor diameter, flattening factor, equivalent diameter of sphere, and overpressure of the incident shock wave, respectively.

Combination of liquid/gas	a (mm)	b (mm)	$f = (a - b)/a$	$d = \sqrt[3]{a^2 b}$ (mm)	ΔP (MPa)
Water/helium	3.9	1.5	0.61	2.8	42.2
Silicone-oil/air	2.0	1.0	0.50	1.6	11.5
Silicone-oil/mixed gas	1.2	0.82	0.29	1.0	26.4

Figures 1a–1d show interferograms for the silicone-oil/air combination. In Fig. 1a an expansion wave reflected from an air bubble and the bubble started to contract. In Fig. 1c, because of enhancement of pressure at the side of the bubble where the shock wave impinged, spherical compression waves were generated. Later, the compression waves coalesced into spherical shock waves (Fig. 1d). When the bubble started to expand, its shape appeared to be a jagged one.

The shock-bubble interaction is not a two-dimensional flow and hence fringes on the interferogram do not correspond to isopycnics. However, by assuming an axisymmetric flow, the density distribution can be obtained by analysis of the interferogram. Figure 2a shows pressure distribution as obtained from Fig. 1a. An axis of axisymmetry is a line through the explosion center and the center of the bubble. In Fig. 2c a maximum pressure was obtained at the side of the bubble where the shock wave first hit.

Figures 3a and 3b show sequential shadowgrams of the shock-bubble interaction for the water/helium combination, in which the incident shock wave propagated from left to right.

In Fig. 3a a dark line in frames 1 and 2 shows the incident spherical shock wave. The incident shock wave interacted with a helium bubble, and the wave

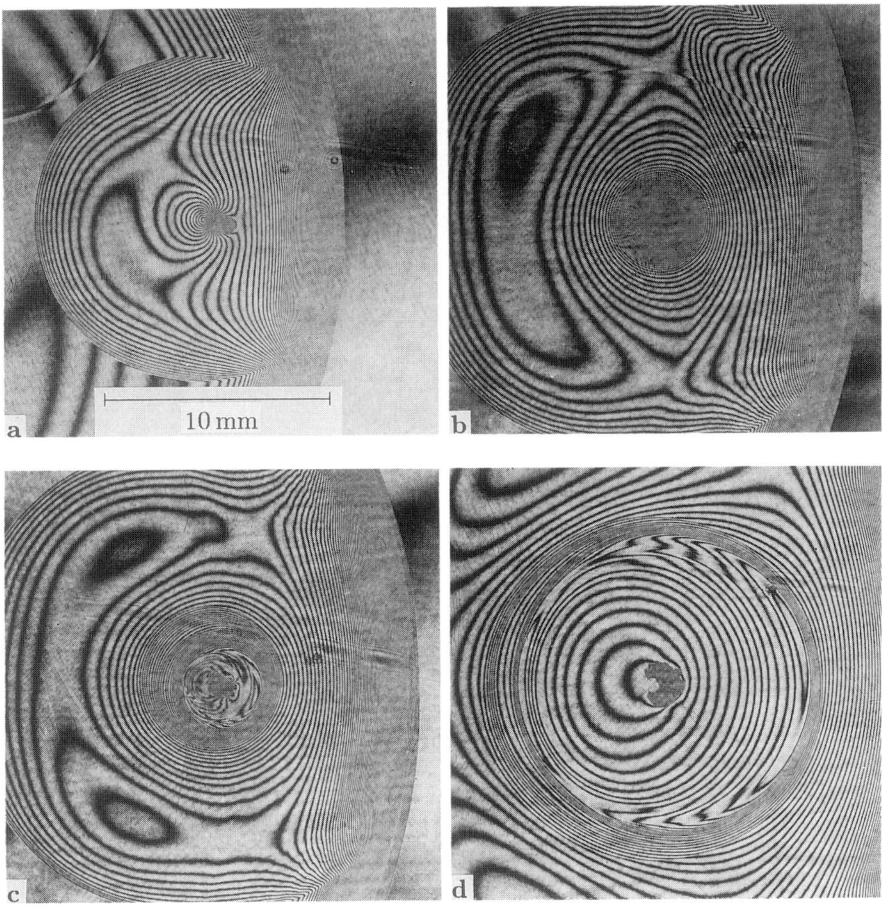

Figure 1. Interferograms of a shock-bubble interaction for the silicone-oil/air combination.

that reflected from the bubble was an expansion wave (frame 2). A liquid jet formed, because liquid particles near the point on the bubble where the shock wave first hit were accelerated due to effect of curvature of the interface and liquid was entrained into the bubble (frames 3–6). The liquid jet penetrated toward the rear surface (frame 6). From these sequential shadowgrams the average speed of the liquid jet was estimated to be 275 ± 10 m/s so that stagnation pressure was thus more than 200 MPa. Spherical shock waves were generated, and at this stage the bubble started to expand (frames 6 and 7).

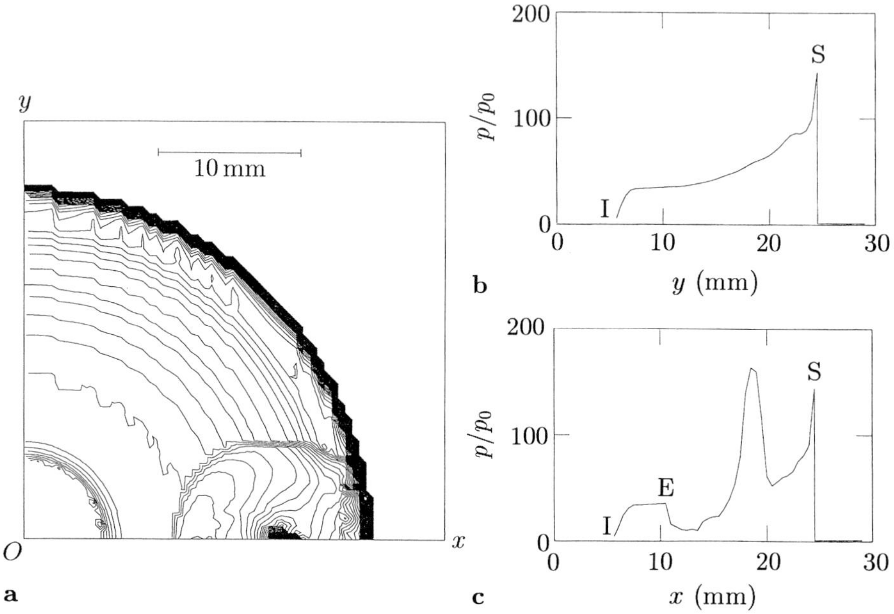

Figure 2. (a) Pressure distribution on the cut-through plane of the axis. The pressure distribution was analyzed from Fig. 1a. Intervals of contours are $5p_0$ where p_0 is atmospheric pressure. (b) Pressure distribution on the line Oy. S – shock front, I – interface of water and explosion product gas. (c) Pressure distribution on the axis Ox. E – wave front of an expansion wave.

In frames 1–3 in Fig. 3b two nearby waves marked A and B were created. In frames 1 and 2, wave A appeares to be stronger at the rear side and weaker at the frontal side. The strength of wave B appeares to be more uniform compared to wave A. The center of wave A is located almost on the rear surface while the center of wave B is in the bubble (frames 1 and 2). Wave A was generated due to water hammer pressure when the liquid jet impinged the rear surface. Wave B is a rebound wave that was generated by expansion of the bubble.

Figure 4 shows shadowgrams of shock-bubble interaction for the silicone-oil/gas consisting air-helium mixture. Concentric spherical waves were generated (frames 3–5). In frame 3 a sonoluminescence can be seen, because temperature was enhanced in the bubble. A counter jet was generated (frames 4–8).

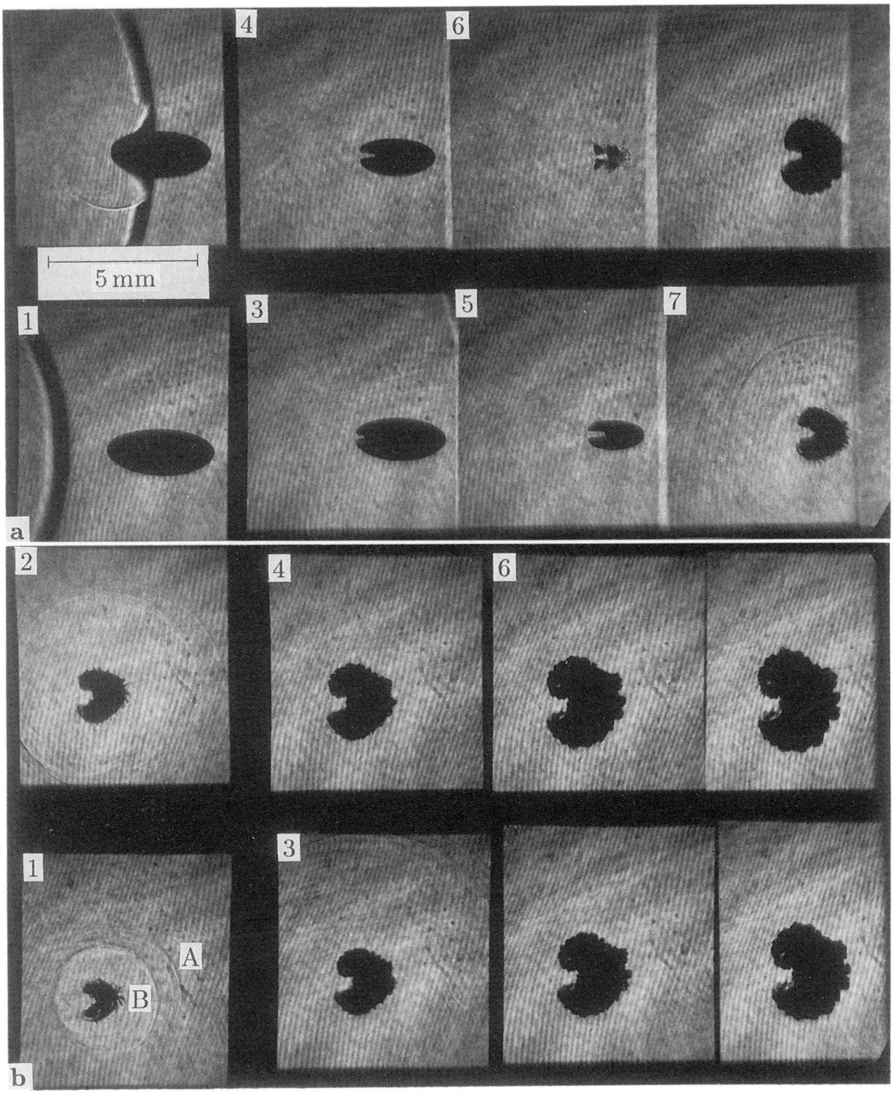

Figure 3. Sequential shadowgrams of a shock-bubble interaction for the water/helium combination. A spherical shock wave propagated from left to right. (a) The interframe time is 2.0 μs. (b) Later stage than (a). The spherical shock wave already passed. The interframe time is 1.0 μs. The exposure time is 1/5 of interframe time for all frames.

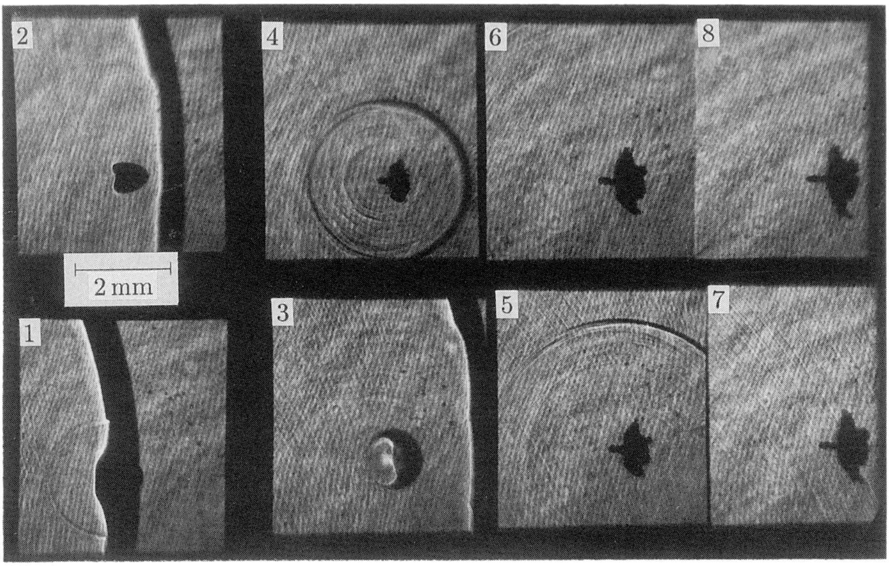

Figure 4. Sequential shadowgrams of a shock-bubble interaction for the silicone-oil/air-helium mixed gas combination. The interframe time is 1.0 μs.

4. Conclusion

Interaction of a shock wave with a gas bubble was investigated experimentally using double-exposure holographic interferometry and high-speed framing photography. Collapse of the gas bubble was well visualized. The pressure profile around the collapsing bubble was analyzed quantitatively.

Acknowledgement. The authors wish to thank to Mr. O. Onodera and Mr. H. Ojima of the Shock Wave Research Center, Institute of Fluid Science, Tohoku University for their assistance in conducting the present experiments.

References

Shin-Etsu Chemical Industry Co. (1989) Explanatory note for KF 96. Shin-Etsu Chemical Industry Co., Japan (in Japanese).

Weissler A (1949) Ultrasonic investigation of molecular properties of liquids. III. Linear polymethylsiloxanes. J Am Chem Soc 71:93.

Yamada K, Takayama K (1995) Interaction of an underwater shock wave with a gas bubble. In: IUTAM Symposium on Waves in Liquid/Gas and Liquid/Vapour Two-Phase Systems, Morioka S, Van Wijngaarden L (eds), Kluwer Academic Publishers, Dordrecht, The Netherlands, pp 161–170.

Passage of two shocks through an interface

S. G. Zaytsev, E. I. Chebotareva, A. N. Aleshin,
E. V. Lazareva, S. N. Titov
The Krzhizhanovsky Power Engineering Institute (ENIN)
Leninsky Pr. 19, Moscow, 117071 Russia

Abstract: We have made an experimental investigation of the evolution of the Richtmyer-Meshkov instability on the passage of incident and reflected shock waves through continuous and discontinuous interfaces. It has been shown that the type of refraction of the reflected shock wave at an interface determines its subsequent evolution.

Key words: Shock wave, Hydrodynamic instability, Turbulence

1. Introduction

We studied the evolution of the mixing zone between gases of different density on the passage of a plane shock wave through the sinusoidal interface modeled by a thin film (Aleshin et al. 1993). It has been shown that the evolution of the interface separating the shock-compressed gases of different density depends upon the type of refraction of the shock. The type of refraction is determined by the following parameters: $a_0 k$, where a_0 and k are the amplitude and the wave number of initial interface, respectively, the Mach number M of the incident shock wave, and the Atwood number $A = (\rho_1 - \rho_2)(\rho_1 + \rho_2)^{-1}$, where ρ_1 and ρ_2 are the densities of the gases separated by the interface. Three typical regimes have been described: soft regular, hard regular, and irregular. For the soft regular regime the growth of the amplitude a_{K12} of the initial interface K_{12} at initial stage of the evolution for $\rho_1 < \rho_2$ is satisfactorily described by the Richtmyer relation:

$$\frac{da_{K12}}{dt} = a^*_{K12} u k A \tag{1}$$

where a^*_{K12} is the amplitude at the end of refraction, u the shock-induced velocity of K_{12}, and A is the Atwood number. For $M = 3$, the relation (1) for the initial stage of K_{12} evolution is applicable at $a_0 k < 2$.

The study of K_{12} evolution for the shock passage through a continuous interface (Zaytsev et al. 1994) has shown that for $a_0 k < 2$ the growth of the amplitude a_{k12} at the initial stage of K_{12} evolution is described by the following relation within 30% error:

$$\frac{da_{K12}}{dt} = \frac{a^*_{K12} ukA}{\Psi} \tag{2}$$

The relation (2) was suggested by Brouillette and Sturtevant (1994). The value Ψ is dependent on the thickness Δ of interface, wave number k, and Atwood number; Ψ was calculated according to Duff et al. (1962). To measure the amplitude a_{k23} of the interface K_{23} formed after the interaction of the reflected shock wave S_{e2b} with K_{12}, the following relation was suggested (Brouillette & Sturtevant 1994):

$$\frac{da_{K23}}{dt} = \frac{da_{K12}}{dt} + \frac{a^*_{K12} ukA}{\Psi(k, \Delta, A)} \tag{3}$$

where a^*_{K23} is the a_{K23} amplitude at the end of refraction of S_{e2b} on K_{12}. Based on experiments, the authors concluded that the relation (3) adequately takes into account the influence of K_{12} evolution on K_{23} evolution (Brouillette & Sturtevant 1994).

2. Experimental data

We performed an experimental investigation of the passage of the reflected shock wave S_{e2b} through the interface K_{23}. The interface K_{12} is formed on the passage of the incident shock wave S_{e1a} through the initial interface K_0. The S_{e1a} propagates in a gas of density ρ_1. The reflected shock wave S_{e2b} propagates in a gas of density ρ_2 ($\rho_2 < \rho_1$). The experiments were performed in shock tubes.

In the first series of experiments performed in a shock tube (Aleshin et al. 1993) the initial discontinuous interface was modeled by a thin film. The mechanical effect of the shock and the heat effect of the shocked flows following the refracted shock wave S_{i1b} resulted in formation of the mixing zone K_{12}. K_{12} of sinusoidal shape and finite thickness Δ contained the initial gases and the products of destruction of the film. The refracted shock wave S_{i1b}, reflecting from the end wall of the shock tube, generated the reflected shock wave S_{e2b} which moved toward K_{12} and refracted on it, thus forming the mixing zone K_{23}.

In the second series of experiments we used a continuous interface obtained with a fast removable plate separating the gases before the experiment. In this case the distribution of concentrations was dependent on molecular diffusion.

The x, t-diagram for run no. 583B is given in Fig. 1; $K^{(1)}_{12}$ and $K^{(2)}_{12}$ are the trajectories of experimental points of the front separating the mixing zone K_{12} from the heavy gas. The distance between $K^{(1)}_{12}$ and $K^{(2)}_{12}$ for fixed time t determines the amplitude $a_{K12}(t)$. S_{e1b} and S_{i2a} are the trajectories of the refracted shock waves generated by the interactions of S_{e1a} with K_0 and of S_{e2b} with K_{12}, respectively.

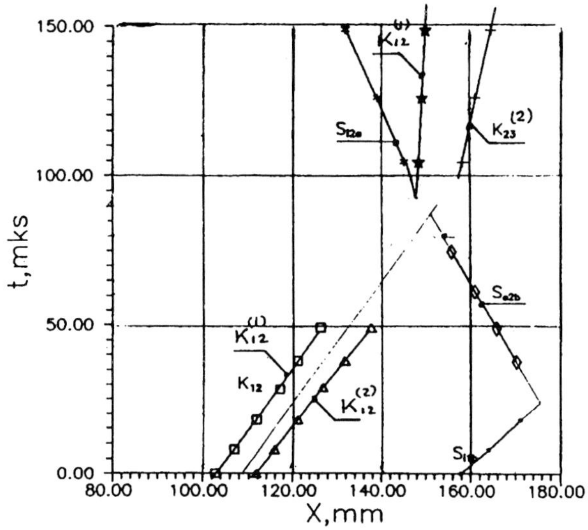

Figure 1. An x, t-diagram for run no. 583B.

The x, t-diagram given in Fig. 2 shows interaction of the incident shock wave S_{e1a} with the continuous interface obtained with a removable plate. The initial conditions and the measured data on K_{12} and K_{23} evolutions are given in Table 1. Numbers of experiments of the first (K_0 – discontinuous interface modeled by a thin film) and the second (K_0 – continuous interface) series are denoted with "B" and "A," respectively. Table 1 gives the experimentally measured values da_{K12}/dt and da_{K23}/dt.

To analyze the evolution of K_{23}, the experimentally measured values da_{K12}/dt are compared to the values calculated from (2) and (3).

$$\chi_1 = \frac{\dfrac{a^*_{K23} u k A}{\Psi(k, \Delta, A)}}{\dfrac{da_{K23}}{dt}} \qquad \chi_2 = \frac{\dfrac{da_{K12}}{dt} + \dfrac{a^*_{K23} u k A}{\Psi(k, \Delta, A)}}{\dfrac{da_{K23}}{dt}}$$

The a^*_{K23} was determined from the trajectories of $K^{(1)}_{12}$, $K^{(2)}_{12}$, $K^{(1)}_{23}$, and $K^{(2)}_{23}$. The $\Psi(k, \Delta, A)$ was calculated according to Duff et al. (1962). The experimentally measured thickness Δ is given in Table 1.

Table 1. Initial conditions and results of experiment.

	583B	**491A**	**560B**	**488A**
A→B→A	Xe→Kr→Xe	Ar→0.5Ar+0.5He→Ar	Ar→He→ Ar	Ar→He→Ar
$\mathbf{P}_0$	0.5	0.5	0.5	0.5
M	3.76	3.22	3.74	3.21
λ, mm	36	11.24	72	4.6
a_0, mm	10		5	
$(da/dt)_1$, mm/μs	0.042	0.017	0.092	0.064
$(da/dt)_2$, mm/μs	0.077	0.067	0.284	0.207
(da/dt), mm/μs	0.114	0.106	1.017	0.632
$a_{K_{12}}(t_R)k$	2.02	1.17	1.49	2.03
A	0.18	0.23	0.76	0.76
regime of refraction	soft	soft	hard	hard
Δ, mm	5.24	3.56	3.08	6.46
χ_1	0.82	0.60	2.90	1.90
χ_2	1.36	0.85	3.30	2.25
Ψ	1.8	2.62	1.2	1.57

3. Discussion

Our previous results (Aleshin et al. 1993) enable us to determine the regime of refraction of S_{e2b} on K_{12} from the experimentally measured interface amplitude $a_{K12}(t_R)$ before interaction of K_{12} with S_{e2b}. Thus, the regime of the refraction of S_{e2b} on K_{12} should be considered as a soft regular for runs 583B and 491A, and as a hard regular regime for runs 560B and 488A. For the runs with a hard regular regime of refraction the amplitude growth rate is expected to follow the relation (3) and hence χ_2 will be 1. This is true, provided that relation (3) adequately describes the influence of K_{12} evolution on K_{23} evolution.

The value χ_1 reflects the ratio between the measured growth rate a_{K23} and the value calculated proceeding on the assumption that K_{23} evolution is independent of K_{12} evolution. Unfortunately, the difference between χ_1 and χ_2, and the differences of each of the values from Table 1 have not allowed us to make a final conclusion whether the relation (3) is adequate or not. However,

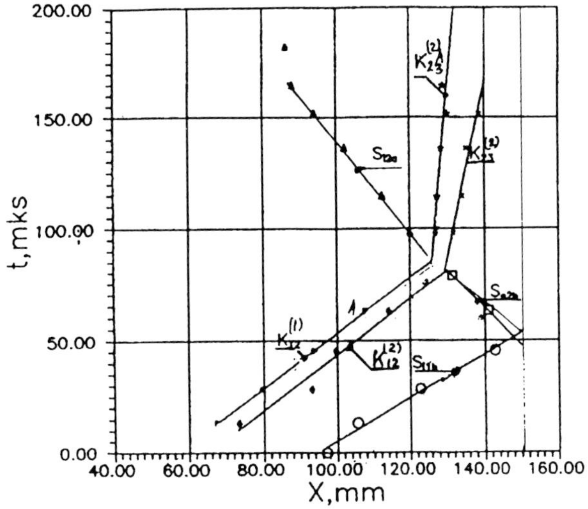

Figure 2. An x, t-diagram for run no. 491A.

the clear excess of χ_1 and χ_2 over 1 confirms the data (Aleshin et al. 1993), indicating that the character of the evolution of the mixing zone is dependent on the type of refraction. Indeed, the amplitude growth rate for hard regular regime is 2–3-fold lower than what is expected from relation (2).

Acknowledgement. The study was partly supported by the Russian Foundation for Fundamental Research (Grant No. 94-01-01144a), by a contract from the French Atomic Energy Commission (CEA/DAM No. 3672), and by the ISTC (Grant No. 94-029).

References

Aleshin AN, Zaytsev SG, Lazareva EV (1993) Experimental and numerical studies on Richtmyer-Meshkov instability. RJCM 1:33–49.

Brouillette M, Sturtevant B (1994) Experiments on the Richtmyer-Meshkov instability: single-scale perturbation on a continuous interface. J Fluid Mech 263:271–292.

Duff RE, Harlow FN, Hirt SW (1962) Effects of diffusion on interface instability between gases. Phys Fluids 5:417–425.

Zaytsev SG, Titov SN, Chebotareva EI (1994) The shock-induced evolution of continuous interface separating gases of different densities. Izv Ross Akad Nauk Mekh Zhidk i Gaza (2):18–26 (in Russian).

Shock-hydrodynamics experiments on the Nova laser

P. L. Miller, T. A. Peyser, P. E. Stry, K. S. Budil
Lawrence Livermore National Laboratory
P.O. Box 808
Livermore, CA 94551 USA

Abstract: We have conducted shock-induced hydrodynamics experiments using the Nova laser at Lawrence Livermore National Laboratory. The laser provides a high-enthalpy source by depositing its energy (about 22 kJ) in a small gold cavity called a *Hohlraum*. The *Hohlraum* serves as a driver section, launching very strong (M $\approx$ 20) shocks into millimeter-scale cylindrical "shock tubes." The flow is imaged radiographically by an electronic framing camera, using a laser-generated X-ray source. Several topics have been addressed with this configuration, including shock-induced mixing at density interfaces (seeded with a variety of perturbations); the development of high-speed, shaped-charge-like jets; the effects of geometry on the planarity of the generated shocks; and shock-shock interactions which develop in the flows. In this paper, we describe the general configuration of our experiments, present an overview of the high-speed jet work, discuss some of our findings, and compare our results with computer simulations we have conducted.

Key words: Jets, Mixing, Richtmyer-Meshkov instability, Nova laser, Compressible flow

1. Introduction

Understanding shock-induced flows, including the Richtmyer-Meshkov instability (Richtmyer 1960, Meshkov 1969) which occurs at shocked density interfaces, is of interest in inertial confinement fusion, astrophysics, and other applications. In this paper, we describe experiments involving high-speed jets generated at a density interface with a hemispherical perturbation. Additional material described in the ISSW oral presentation is not included here, but has been documented in Peyser et al. (1995).

2. Experiment description

The experiments were conducted using the Nova laser facility. Nova has ten separate beam lines, each capable of delivering approximately 2.8 kJ of energy in pulses 1-ns long, at a wavelength of 353 nm. The beams are focused into a 5-m-diameter target chamber, which contains a target positioner holding the target, alignment systems, and various diagnostics.

In our experiments, we use the Nova laser as a high-enthalpy source to drive our miniature shock tubes, and to produce diagnostic X-rays. A small gold cylindrical can, called a *Hohlraum*, serves as the shock-tube driver section (see Fig. 1). The *Hohlraum* dimensions are 3200 μm long by 1600 μm wide, with 1200-μm -diameter holes in both ends and a 45-μm wall thickness. Our miniature shock tube is mounted on a 700-μm hole on the side of the *Hohlraum*. The shock-tube wall is made of beryllium, about 700 μm in internal diameter with 100-μm -thick walls. The working fluid in the shock tube is brominated plastic (2% Br by atomic fraction) in the end closest to the *Hohlraum*, and low-density carbon foam (0.1 g/cc) in the remainder of the tube. The shock-tube length, 2 mm, and external diameter, 0.9 mm, are about the same size as the letter "i" in the word "DIME" on the backside of a U.S. dime.

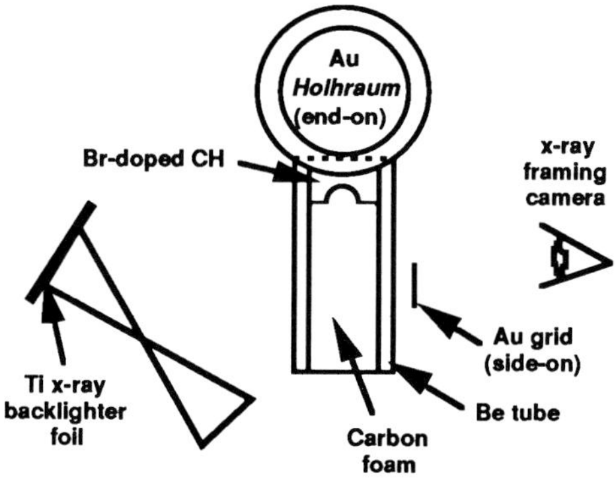

Figure 1. Schematic of the experiment.

The sequence of events in the experiments is as follows. Eight of Nova's beams are focused into the interior of the gold *Hohlraum* can, rapidly heating the interior of the *Hohlraum* to a temperature of about 230 eV (over 2 million K). The plastic-filled end of the shock tube, exposed to the high temperature through the hole in the side of the *Hohlraum* it is mounted over, begins to rapidly ablate, driving a strong shock (tens of Mbars) down the axis of the tube. As the shock propagates down the tube, it crosses the interface between the brominated plastic and the low-density carbon foam. The interface is flat, except for a 150-μm -radius, hemispherical hole in the plastic side (filled with foam), centered on the tube axis. The shock tube and associated parts are made with advanced micro-machining techniques developed at LLNL (Louis et al. 1995).

The primary data collected in these experiments are X-ray images. To provide the X-ray illumination to capture these images, the remaining two of Nova's ten beams, which are not used to heat the *Hohlraum*, are delayed in time and strike a titanium foil located several millimeters behind the shock tube, as viewed from the diagnostic camera. The interaction of the intense laser light with the metal produces X-rays (of about 4.9 keV), providing a backlighting source. The X-ray transmission through the flow is attenuated by a factor which is a combination of path length, type of material, and density. The beryllium tube and unshocked carbon foam are relatively transparent to X-rays of that energy, while the brominated plastic and higher-density materials are more opaque. Therefore, the jet of brominated plastic appears as a dark feature on a lighter background. Images of the resulting flow are captured using an electronic framing camera (Budil et al. 1995), which is basically an X-ray pinhole camera, capturing two-dimensional images with about 100 ps exposures. The data is recorded on film, and later digitized on a densitometer, including wedge corrections.

3. Experimental and numerical results

Two-dimensional, axisymmetric numerical simulations of the jet flow have been conducted using the code CALE. CALE is an arbitrary Lagrangian-Eulerian code that was developed at LLNL. A calculated temporal sequence of the jet behavior, depicting material boundaries, is shown in Fig. 2. Since this is a depiction of material interfaces, densities (and the shock) are not represented. The shock is traveling from top to bottom in the figures. At 0 ns, the initial geometry is evident, and 5 ns after the laser begins to heat the *Hohlraum* the shock is crossing the hemispherical protrusion, causing it to invert. By 10 ns, the hemisphere has fully inverted and is forming a jet, which continues to develop through later times. The sides of the jet, subject to shear, form roll-ups by the Kelvin-Helmholtz instability. Such roll-ups, when viewed in transmission, appear in the data as dark horizontal bands.

Using the known opacities of the materials, the simulation was post-processed to create simulated radiographs, which can then be compared directly with the experimental data. A comparison of data with the results from the simulation is shown in Fig. 3. Cosmetic differences between the two sets of images include the presence of noise, the inclusion of reference grids, and the non-uniformity of the X-ray illumination, all in the experimental data. Features of interest in the experimental data include the jet's bow shock (particularly evident at 10.6 ns), the dark bands of the vortical structure around the head of the jet, swirls of material at 14.8 ns and trailing tendrils from the sides at 19.7 and 21.9 ns, and the growth and thinning of the neck of the jet, particularly at 21.9 ns. Differ-

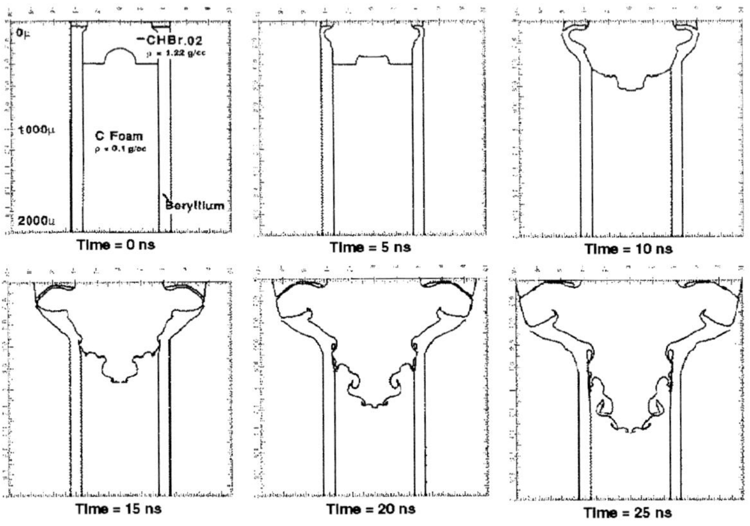

Figure 2. Material boundaries of the high-speed jet experiment, calculated with CALE.

ences between the simulation and the data include faster growth (longer neck) and later development of the swirl features in the simulation.

Using data from a series of separate experiments, four jet features were identified in each picture: the bow-shock location, the tip of the jet material, the center of the dark band from the vortical roll-up, and the "shoulder" or material to the side at the base of the jet. The axial positions of these four features were measured from each data image, and the results are compared in Fig. 4 to the measurements from the simulated X-ray images of the calculation. One data point for the shoulder at 19.5 ns was out of the frame. The time is measured from the beginning of the 1.0-ns laser drive pulse, and the shock crosses the hemispherical protrusion around 5 ns. Axial position is measured from the end of the brominated plastic closest to the *Hohlraum*, with the tip of the hemisphere initially at 250 μm and the flat around it at 400 μm .

The agreement between the simulation and the data is, depending on the context, between excellent and fair. The feature locations from the calculation agree within 5 to 10% of their total travel distance with those from the data, with the exception of the earliest-time data, and the spacing between features at a particular time is also quite similar. The data points at 11.5 ns and 19.5 ns are

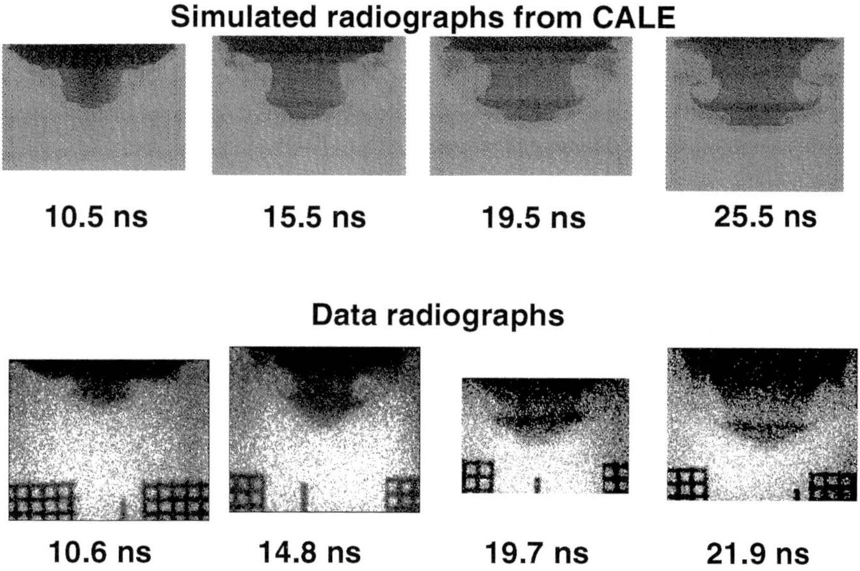

Figure 3. X-ray images from CALE calculation and experimental data.

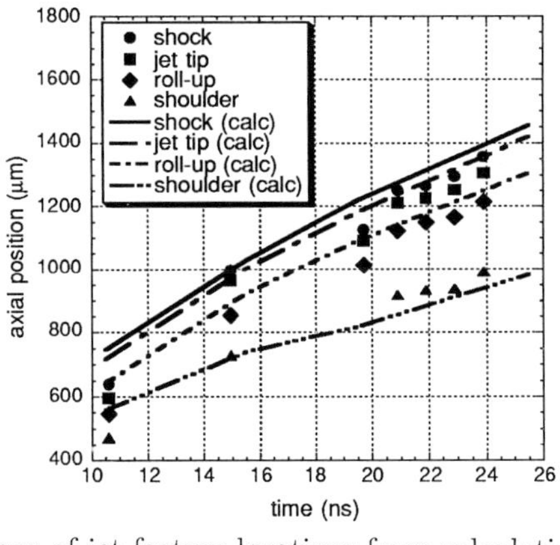

Figure 4. Comparison of jet-feature locations from calculation and experiment.

significantly less advanced than the calculation, however. It should be noted that the drive energy delivered by the laser in each experiment is subject to variations of $\pm 10\%$, or so, and the energies measured for the two lower sets of data were, in fact, lower than the others. One possible approach to correct for this uncertainty in the initial drive would be to assume the shock location is an accurate indicator for the drive condition, and to shift the data points in time to where they agree with the simulated shock location. We do not present such an adjustment here.

4. Conclusions

We have presented results from our shock-induced fluid-dynamics experiments conducted on the Nova laser facility. For this paper, we have focused on our high-speed-jet data and compared it to two-dimensional CALE calculations. The distinct features present in the jets provide good fiducials for comparison with the calculations. The CALE simulations are found to agree in major respects, but differ in some details. We also note that the roll-up observed in the data is possibly the first such direct observation of the Kelvin-Helmholtz instability in this (plasma) regime.

Acknowledgement. This work was performed under the auspices of the U.S. Department of Energy by the Lawrence Livermore National Laboratory under Contract Number W-7405-ENG-48.

References

Budil KS, Perry TS, Bell PM, Hares JD, Miller PL, Peyser TA, Wallace R, Louis H, Smith DE (1995) The flexible x-ray imager. Rev Sci Inst 67(2):1.

Louis H, Demiris A, Budil KS, Miller PL, Peyser TA, Stry PE, Wojtowicz DA, Dimotakis PE (1995) Miniature targets for high-energy density experiments on Nova. Fusion Technology 28(5):1769.

Meshkov EE (1969) Instability of the interface of two gases accelerated by a shock wave. Sov Fluid Dyn 4:101.

Peyser TA, Miller PL, Stry PE, Budil KS, Burke EW, Wojtowicz DA, Griswold DL, Hammel BA, Phillion DW (1995) Measurement of radiation-driven shock-induced mixing from nonlinear initial perturbations. Phys Rev Lett 75:2332.

Richtmyer RD (1960) Taylor instability in shock acceleration of compressible fluids. Commun Pure Appl Math 13:297.

Shock-accelerated heavy-gas layers

J. M. Budzinski, R. F. Benjamin
Los Alamos National Laboratory
Los Alamos, NM 87545 USA

J. W. Jacobs
University of Arizona
Tucson, AZ 85721 USA

Abstract: Shock-accelerated heavy-gas layers are studied experimentally and with front-tracking simulations. Richtmyer-Meshkov instability develops on each interface. Three different patterns form depending on the relative size of the initial perturbations on the upstream and downstream interfaces. The patterns are interpreted in terms of the average vorticity generated on the interfaces.

Key words: Richtmyer-Meshkov instability, Shock acceleration, Shock-induced mixing, Front-tracking

1. Introduction

This study is directed toward a better understanding of the flow of shock-accelerated heavy layers embedded between lower-density fluids. This problem is of fundamental interest because it involves the production, coupling, and transport of vorticity in a system sensitive to initial conditions. Our experimental observations are available in Budzinski et al. (1994), Jacobs et al. (1993), and Jacobs et al. (1995). We report here further analysis of the experimental data and the results of simulations using a front-tracking code.

A fluid interface subjected to a shock acceleration becomes unstable and distorts after the acceleration. This impulsively driven instability is commonly referred to as shock-induced Rayleigh-Talyor instability or Richtmyer-Meshkov (RM) instability. We are studying the impulsive acceleration of two nearby interfaces, and the subsequent growth and coupling of these interfaces. Nearly sinusoidal perturbations are initially imposed on the interfaces. The perturbations grow, distort, and interact, producing distinctive flow patterns. We find that similar initial conditions produce three different flow patterns, identified descriptively as: upstream mushrooms, sinuous, and downstream mushrooms. We report these observations and analysis with simulations using a front-tracking code developed at SUNY Stony Brook (Chern et al. 1986, Grove 1994).

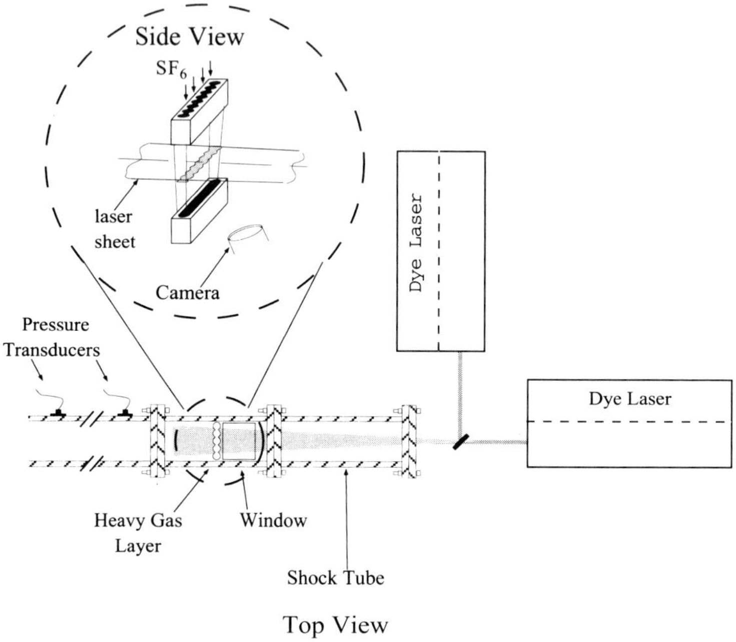

Figure 1. Schematic of shock tube and associated diagnostics. Two images per experiment are acquired: one of the initial conditions of the gas curtain, and the other of the flow pattern at a predetermined time after shock acceleration.

2. Experimental apparatus

Figure 1 shows the experimental apparatus. The layer is generated inside a shock tube by vertically flowing SF_6 through a corrugated nozzle into a test section. In this way, a laminar gas jet (i.e., gas curtain) with a varicose cross section and diffuse boundaries is created with a wavelength of 6 mm and amplitudes up to 1 mm on each side. This flow system has been shown to produce a two-dimensional flow after a shock acceleration (Jacobs et al. 1993, 1995). Small, unsteady, uncontrollable fluctuations in the SF_6 jet cause asymmetries which vary from experiment to experiment. The fluctuations are measured but not controlled. The present experiments use planar laser Rayleigh scattering to capture two frames per experiment, measuring the pre-shock initial conditions and one dynamic condition at varying times after the shock interaction. Two pulsed dye lasers illuminate the test region with coplanar sheets of laser light, and a cooled CCD camera records each pair of images. The images are converted to maps of SF_6 concentration by calibrating the system with scattering from pure air and pure SF_6.

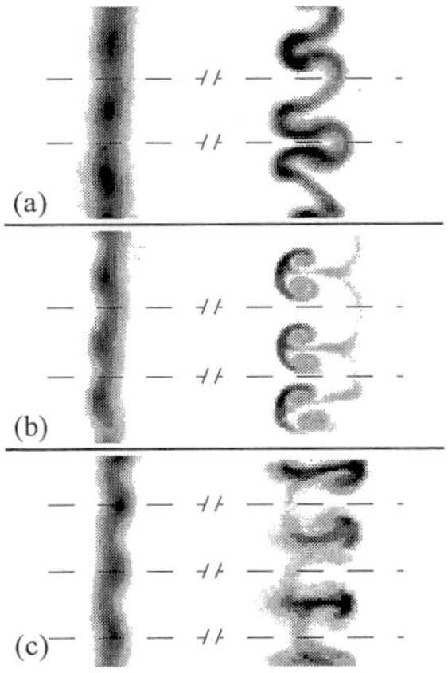

Figure 2. Composite of images from three typical experiments showing the initial and dynamic conditions of each experiment. Each pair of initial and dynamic conditions was recorded on the same event. (a) Upstream mushroom, dynamic image recorded 0.45 ms after the shock interaction. (b) Downstream mushroom, dynamic image at $t = 0.40$ ms. (c) Sinuous, dynamic image at $t = 0.45$ ms.

3. Experimental results

Sample images from the experiments are shown in Fig. 2. We find that asymmetries in the initial conditions correlate with the three post-shock flow patterns in over 90% of the 100+ experiments performed. This correlation is shown in Fig. 3 by plotting the initial upstream amplitude versus the initial downstream amplitude for each experiment, and identifying on the graph which pattern was generated. The post-shock flow patterns are classified visually, and the initial upstream and downstream amplitudes are found from an analysis of the Fourier coefficients of the pre-shock interfaces. On this graph, the experiments which generated similar flow patterns cluster together, with a small amount of overlap between the different clusters. An upstream mushroom pattern develops when the initial perturbations on the upstream side (i.e., on the side first interacting with the shock wave) are larger than the perturbations on the downstream side. A sinuous pattern forms when the initial perturbations on the downstream side are about the same or slightly larger than those on the upstream side. A downstream mushroom develops when the perturbations on the downstream side are much larger than the perturbation on the upstream side.

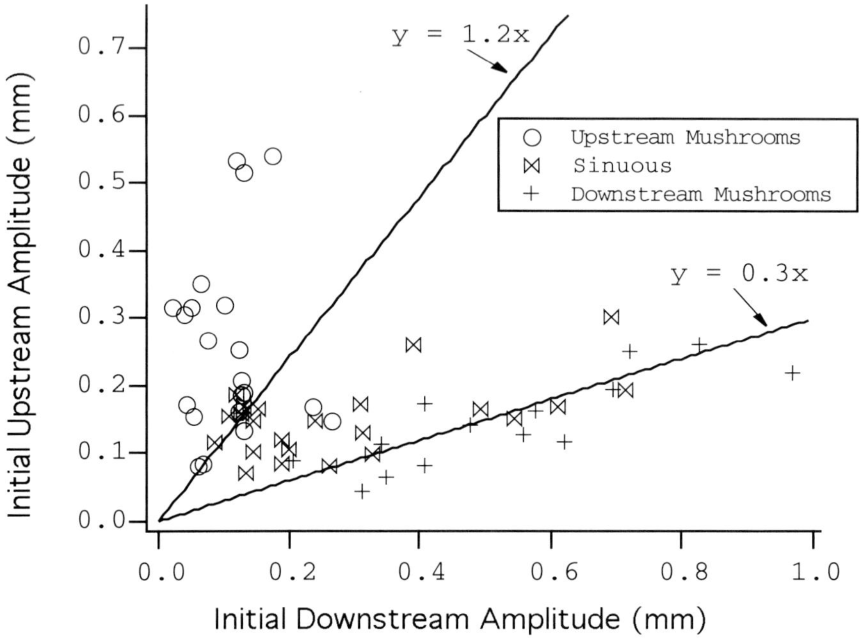

Figure 3. Upstream versus downstream perturbation amplitudes for the experiments show clusters of similar flows.

One way of interpreting the creation of the three patterns is to view the flow evolution within the context of vortex dynamics. The vorticity is generated baroclinically by the shock interaction through the misalignment of pressure and density gradients. The magnitude of vorticity will be greatest where the pressure and density gradients are most misaligned. Assuming the pressure gradients are predominantly from planar waves, most of the vorticity is expected to be in the regions between the peaks and valleys of the initial perturbations. When the perturbation amplitude is greater on one side of the layer, more vorticity is generated on that side. Furthermore, the vorticity is preferentially generated in the lighter fluid. This effect accentuates the offset of the vorticity from the center of the layer. The mushrooms form on the side with the largest initial perturbation amplitude because the heavy fluid is asymmetrically entrained into the side with the most vorticity. The vorticity distribution of a symmetrically perturbed initial layer would be centered on the heavy layer and thus produce a sinuous pattern.

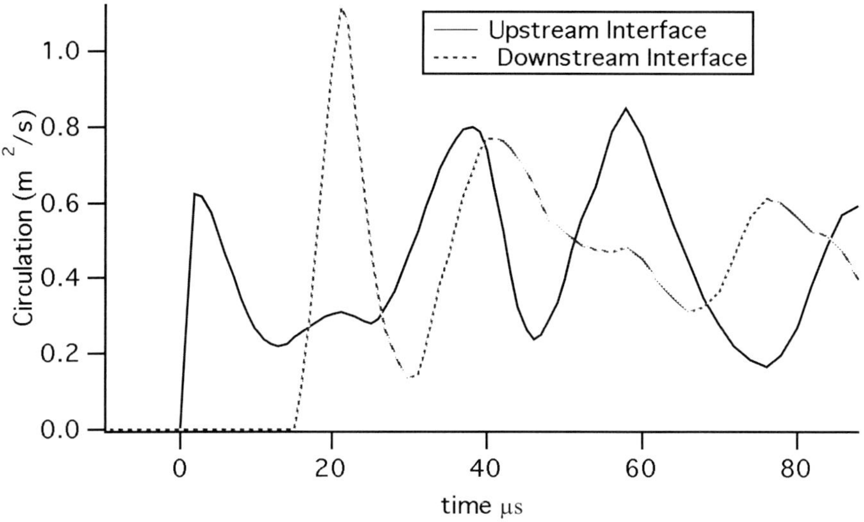

Figure 4. Circulation versus time for a layer with initial symmetric perturbations of 0.33 mm amplitude.

4. Simulations

The simulations, which reproduce all three of the experimental flow patterns, illustrate the complex nature of the wave dynamics and vorticity generation. Every time a pressure wave interacts with a material interface, a transmitted wave and a reflected wave are created. A series of reverberating waves of successively decreasing strengths are generated inside the heavy layer. The simulations show the contribution that each of these waves has on the total circulation generated on the upstream and downstream material interfaces. These contributions are essential to an analysis of the total vorticity generated. The simulations also show the deformation of the shock waves that are reflected and transmitted through the layer. These waves deform to the extent that they break and then create transverse or lateral waves that run along the layer. Interactions of the material interfaces with these lateral waves cause oscillations in the circulation on the upstream and downstream material interfaces. Figure 4 shows the circulation as a function of time for a layer with symmetric perturbations. These oscillations in circulation last for significant periods of time after the first shock interaction. In order to interpret the pattern formation in terms of vorticity on the interfaces, the average of the vorticity over time, or the vorticity after the large oscillations decay, becomes the important parameter.

5. Conclusions

Simulations and experiments are performed on shock interactions with heavy-gas layers. Three flow patterns develop depending on the relative amplitudes of the initial perturbations on the upstream and downstream interfaces. Simulations show large oscillations of the circulation on the interfaces.

Acknowledgement. We are indebted to James Glimm and John Grove for allowing us to use their front-tracking code, and for their expert advice and support. We also would like to recognize the help of Jean-Marc Le Peuvedic in running the front-tracking code. We would like to thank Robert Critchfield for his assistance on the experiments. This work is supported by US DOE, W-7405-ENG-36.

References

Budzinski JM, Benjamin RF, Jacobs JW (1994) Influence of initial conditions on the flow patterns of a shock-accelerated thin fluid layer. Physics of Fluids 6:3510–3512.

Chern IL, Glimm J, McBryan O, Plohr B, Yaniv S (1986) Front tracking for gas dynamics. J Comput Phys 62:83–110.

Grove J (1994) Applications of front-tracking to the simulation of shock refractions and unstable mixing. J Appl Num Math 14:213–237.

Jacobs JW, Klein DL, Jenkins DG, Benjamin RF (1993) Instability growth patterns of a shock-accelerated thin fluid layer. Phys Rev Lett 70:583–586.

Jacobs JW, Klein DL, Jenkins DG, Benjamin RF (1995) Nonlinear growth of shock-accelerated instability in a thin fluid layer. J Fluid Mech 295:23–42.

Experiments on the shock-excited interfacial instability at an air/SF_6 interface

M. Vetter,* B. Sturtevant
Graduate Aeronautical Laboratories
California Institute of Technology
Pasadena, CA 91125 USA

Abstract: Results of flow-visualization experiments of an impulsively accelerated plane interface between air and SF_6 are reported. The shock tube used for the experiments has a larger test section than in previous experiments. The larger extent of uniform test flow relative to nonuniform boundary-layer flow permits unambiguous interpretation of flow-visualization photographs, and the influence of shock-wave/boundary-layer interactions is no longer dominant. The strong wall vortex observed in previous studies is not observed in these experiments. It is found that the thin membrane, which forms the initially plane interface, has a significant influence on the initial growth rate of the interface thickness. However, the measured growth rates after the first reflected shock are independent of membrane configuration and are in good agreement with analytical predictions.

Key words: Richtmyer-Meshkov instability, Flow visualization, Experiment, Shock tube, Turbulent mixing

1. Introduction

The Richtmyer-Meshkov instability occurs when a shock wave impinges upon an interface between two fluids of different densities (Richtmyer 1960, Meshkov 1969). The refraction of the shock wave from spatial perturbations on the interface causes distortion of both the interface and the shock wave. Most of the experiments so far reported in this field have concentrated on the instability of an initially sinusoidal interface with perturbations of wavelength λ and amplitude η_0. Richtmyer (1960) showed that in this case the perturbation amplitude η grows linearly with time. Small-scale perturbations, the subject of this paper, grow rapidly into the nonlinear regime. Perturbations of, for example, $\eta_o/\lambda = 0.1$ and $\lambda = 1$ cm, on an interface which is accelerated impulsively from rest to a velocity of 150 m/s (all values typical for the experiments described here), grow out of the linear regime within 0.7 ms. This finally leads to a tur-

*Presently at: Bayern-Chemie, D-85521 Ottobrunn, Germany.

bulent mixing zone (TMZ) between the two fluids. Mikaelian (1989) adapted results of constant-acceleration experiments of Read (1984) to impulsive acceleration and obtained an analytical expression for the nonlinearly saturated shock-induced thickening of an initially flat interface,

$$\delta = 0.28[u]A't \ , \qquad (1)$$

where δ is the interface thickness, $[u]$ is the change of interface velocity induced by the shock wave, A' is the post-shock Atwood ratio, and t is time. It has recently been suggested (Brouillette & Sturtevant 1989, 1993) that, due to the modest-size shock tubes used for the relatively few experiments on this problem to date (Andronov et al. 1976, Andronov et al. 1982, Brouillette & Sturtevant 1989, 1993), shock-wave/boundary-layer effects have had a large influence on the measured growth rates and on the interpretation of flow-visualization pictures. The goal of this work is to reduce this influence by using a larger test facility.

2. Experimental

The experiments were performed in the GALCIT 17-inch diameter horizontal shock tube (Vetter & Sturtevant 1995). A 27-cm square test section is flanged to the end of this tube, and the transition from circular to square cross section is made with a cookie cutter extending 1.5 m upstream into the shock tube (Fig. 1). A thin (0.5 μm) nitro-cellulose membrane, which forms the initial interface, is placed between the shock tube and the test section. Mach numbers above 1.1 are generated by reducing the initial pressure p_1 below atmospheric.

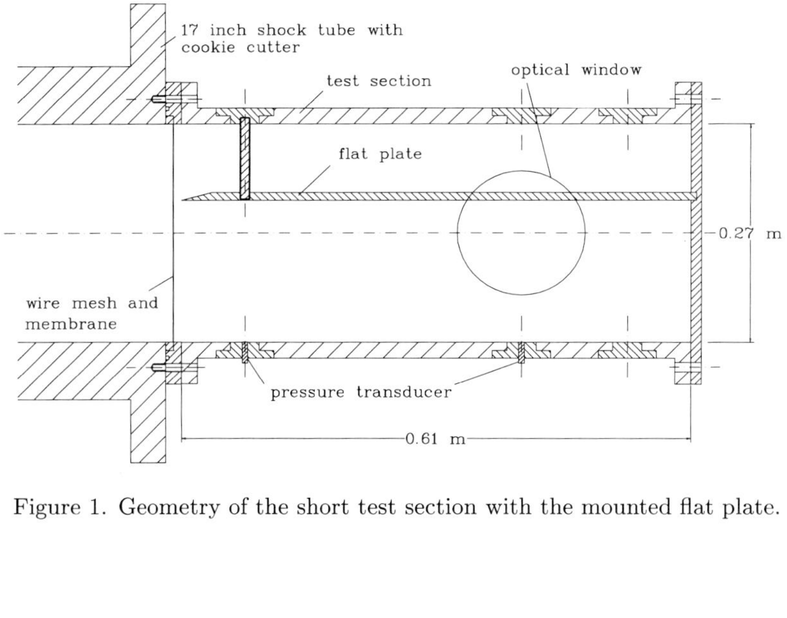

Figure 1. Geometry of the short test section with the mounted flat plate.

For $M_s = 2$, p_1 is as low as 8 kPa. In most of the experiments a wire mesh was installed to support the membrane on the upstream side during the filling procedure. Prior to each run, the membrane was usually pushed slightly into the wire mesh, with an estimated amplitude of about 1 mm. The optical windows for the schlieren visualization system are centered 42 cm downstream of the membrane and are 15 cm in diameter, so the shock-tube wall is not directly visible in the pictures. Therefore, some experiments were performed with a sharp-edged flat plate mounted across the window, extending from the end wall up to a distance of 10 mm downstream of the membrane (see Vetter & Sturtevant 1995). The thickness δ of the interface, defined to be the mean horizontal overall extent of the interface, was measured on both spark-schlieren and high-speed movie photographs. No corrections were made for possible wall effects. The rate of increase of the interface thickness ('growth rate') is obtained by a straight-line least-squares fit to the thickness data.

3. Growth rates

The evolution of the average thickness measured from high-speed motion pictures for four different incident-shock Mach numbers between 1.24 and 1.98 is shown in Fig. 2. The length L of the test section was adjusted in some experiments to insure that the interaction of the reflected shock wave with the interface would be in the field of view. Very small and sometimes no growth

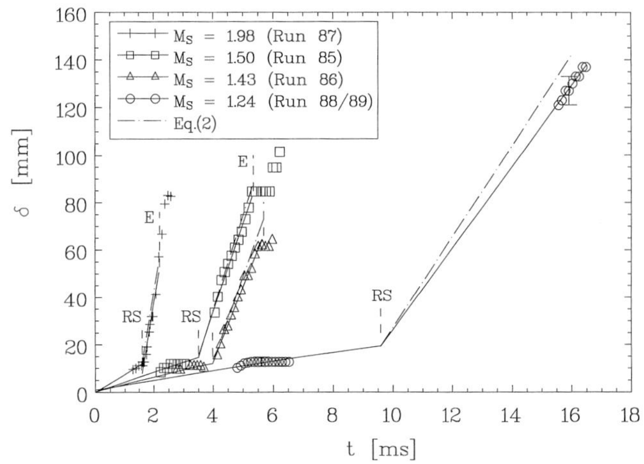

Figure 2. Time evolution of the thickness of the TMZ for the plane discontinuous interface between air and SF_6. Results from high-speed motion pictures (RS: reflected shock; E: expansion).

is observed before the reflected shock interacts with the interface; growth rates are an order of magnitude smaller than predicted by Eq. 1. In experiments in which the wire mesh was placed on the *downstream* side of the membrane to fragment it and enhance mixing, it is observed that the mixing region is up to five times thicker than with the wire mesh on the upstream side. However, this still represents a growth rate only about 2/3 of that predicted by Eq. 1. In this configuration residual wakes of SF_6 from the wire mesh are visible behind the TMZ.

The growth of the interface between the reflected shock and the arrival of the re-reflected expansion wave appears to be linear, and the measured growth rates are in good agreement with Eq. 1 (indicated by the dash-dotted lines in Fig. 2, using the Atwood ratios after the reflected shock), even for the low Mach number with only one datum available. The initial thickness for the post-reflected shock fit is taken from the incident shock fitted result, but nearly the same rate would be obtained if the actual measured pre-reflected shock thickness were used.

Calculated growth rates assuming that the initial thickness is zero are compared with the measurements in Table 1. The index "0" corresponds to the

Table 1. Test conditions and comparison between measured and predicted growth rates of the interface thickness.

	I[a]	II[b]	III[b]	IV[a]	V[a]	VI[b]	VII[b]	
M_s	1.18	1.24	1.43	1.45	1.50	1.50	1.98	–
p_1	55	40	31	29	23	23	8	kPa
L	123	110	62	62	62	62	49	cm
M_T	1.27	1.37	1.67	1.70	1.78	1.78	2.56	–
u_0	56	72	126	126	150	150	287	m/s
$(d\delta/dt)_{0,exp.}$	0.94	2.1	3.0	3.1	4.0	4.2	7.5	m/s
$(d\delta/dt)_{0,Eq.\ 1}$	10.7	14.2	25.5	25.6	30.7	30.7	61.3	m/s
$(d\delta/dt)_{1,exp.}$	19.2	17.0	31.5	35.5	32.6	37.2	74.4	m/s
$(d\delta/dt)_{1,Eq.\ 1}$	20.1	22.2	35.9	37.3	39.8	39.8	75.5	m/s

[a] Results from single spark-schlieren photographs.
[b] Results from high-speed motion pictures.

conditions after the incident shock and the index "1" to those after the reflected shock. The effective initial thickness, due to the starting process, may in fact be finite. Thus the values given for the growth rates after the initial shock are upper bounds. The present results for the growth rate after the reflected shock are summarized in Fig. 3, a plot of growth rate $(d\delta/dt)_1$ normalized by the velocity jump $[u]_1$ induced by the reflected shock vs. the Atwood ratio A_1' after the reflected shock. The measured growth rates fall in a band 1.5 to 23%

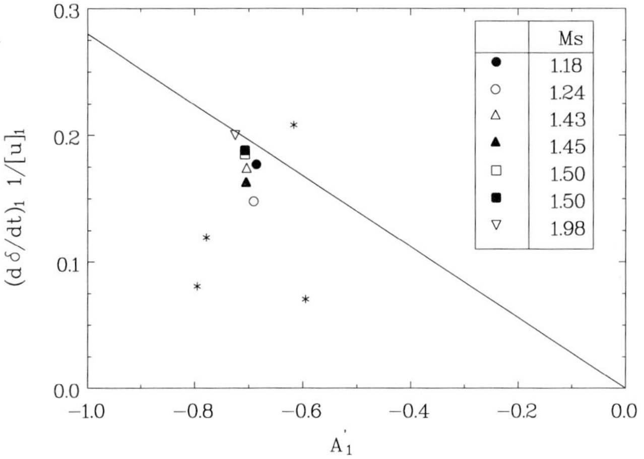

Figure 3. Correlation of reflected-shock growth rates for a discontinuous interface between air and SF_6. Filled symbols: results from single schlieren pictures. Open symbols: results from high-speed motion pictures. —: Mikaelian formula (Eq. 1). $*$: experimental results from Brouillette & Sturtevant (1993).

below the analytical formula of Mikaelian (1989), Eq. 1, with substantially better agreement than in previous investigations. The agreement improves as M_s increases.

4. Discussion and conclusions

Results of growth-rate measurements of impulsively accelerated plane interfaces between air and SF_6 have been presented. Due to the much larger test section than in previous studies, the effects of shock-wave/boundary-layer interaction have been reduced significantly. The resulting increased resolution of schlieren pictures and comparison of pictures taken at the same instant both with horizontal and vertical knife edges permit a better interpretation of the visible flow features. Contrary to previous investigations in smaller facilities, no strong wall vortex was found in these experiments (Vetter & Sturtevant 1995). This suggests that the interactions observed in small tubes may be enhanced by mutual interaction of the wall effects.

The measured growth rate after the incident shock is an order of magnitude less than predicted by Mikaelian (1989), Eq. 1, presumably due to strong damping of mixing by the nitro-cellulose membrane. On the other hand, the growth rate after the reflected shock agrees to within 1.5–23% with Eq. 1, depending on the shock strength, suggesting that the damping effect of the membrane is negated by the reflected shock. The fact that the normalization suggested by Eq. 1 does not remove all of the shock-strength dependence is taken to mean that the growth rate constant 0.28 in Eq. 1 actually depends weakly on initial conditions. Observed growth rates before reshock are not only shock-strength dependent but also apparatus (mesh and membrane) and facility dependent. On the other hand, independently of the pre-reflected-shock conditions, the post-reflected-shock growth rate agrees well with Eq. 1, suggesting that non-linearly saturated Richtmyer-Meshkov mixing is physically similar to saturated Rayleigh-Taylor mixing, from which Eq. 1 is inferred.

Acknowledgement. This research was supported by the U.S. Department of Energy, Lawrence Livermore National Laboratory, under contract agreement DOE W-7405-ENG-48. The first author was also supported in part by Deutsche Forschungsgemeinschaft, Germany.

References

Andronov VA, Bakhrakh SM, Meshkov EE, Mokhov VN, Nikiforov VN, Pevnitskii AV, Tolshmyakhov AI (1976) Turbulent mixing at contact surface accelerated by shock waves. Sov Phys JETP 44:424.

Andronov VA, Bakhrakh SM, Meshkov EE, Nikiforov VV, Pevnitskii AV, Tolshmyakhov AI (1982) An experimental investigation and numerical modeling of turbulent mixing in one-dimensional flows. Sov Phys Dokl 27:393.

Brouillette M, Sturtevant B (1989) Growth induced by multiple shock waves normally incident on plane gaseous interfaces. Physica D 37:248.

Brouillette M, Sturtevant B (1993) Experiments on the Richtmyer-Meshkov instability: small-scale perturbations on a plane interface. Phys Fluids A 5:916.

Meshkov EE (1969) Instability of the interface of two gases accelerated by a shock wave. Sov Fluid Dyn 4:101.

Mikaelian KO (1989) Turbulent mixing generated by Rayleigh-Taylor and Richtmyer-Meshkov instabilities. Physica D 36:343.

Read KI (1984) Experimental investigation of turbulent mixing in Rayleigh-Taylor instability. Physica D 12:45.

Richtmyer RD (1960) Taylor instability in shock acceleration of compressible fluids. Commun Pure Appl Math 13:297.

Vetter M, Sturtevant B (1995) Experiments on the Richtmyer-Meshkov instability of an air/SF_6 interface. Shock Waves 4:247.

Mathematical modeling of gravitational turbulent mixing in non-self-similar problems

V. Y. Neuvazhayev
Russian Federal Nuclear Center – VNIITF
454070, Snezhinsk, Chelyabinsk Region, P.O. 245, Russia

Abstract: Semi-empirical models of different complexity can be applied to describe the turbulent mixing arising as a result of the gas-dynamic instability induced by the action of shock waves and acceleration. In this paper the K and KE models, being the most widely used for scientific applications, are set forth. Considered in detail are four types of problems which can be described, in spite of their complexity, within the framework of the simple models. The significance of the performed investigations lies in fact that all results are obtained in an analytical form which makes it possible to observe general regularities.

Key words: Turbulent mixing, Shock waves, Analytical form

1. The sign-variable acceleration

By compressing the targets in Inertial Confinement Fusion (ICF) problems, the situation in which the acceleration of the boundary changes its sign is typical (Fig. 1). This should be taken into account in turbulent mixing models. The method of separation accounting is based on the addition of the corresponding transport terms (Neuvazhayev 1993a) and is used within the framework of the known K and KE models. In the case of incompressible fluids, the equation for the density profile for the K model is used in the following form:

$$\frac{\partial \rho}{\partial t} = \frac{\partial}{\partial x}\left[0.3L\overline{v}\frac{\partial \rho}{\partial x} - \sqrt{a_s}\sqrt{\frac{2|g|L}{(\rho_1-\rho_2)(\rho_1+\rho_2)}}(\rho_1-\rho)(\rho-\rho_2)\right] \quad (1)$$

$$\text{(a)}\quad \frac{dL}{dt} = 3.06\overline{v} \qquad \text{(b)}\quad \frac{d\overline{v}^2}{dL} + (0.5+0.23A^2)\frac{\overline{v}^2}{L} = 0.11\pi g A \quad (2)$$

Equations (2a) and (2b) for turbulent velocity are deduced for the mean velocity $\overline{v}$ over the mixing zone L, with the width of the mixing zone chosen effectively considering the law of mixed mass conservation. $A = \frac{\rho_1-\rho_2}{\rho_1+\rho_2}$, where ρ_1 and ρ_2 are the initial densities of heavy and light matters, respectively.

The specific coefficient value 0.07 is also presented here: under constant acceleration $L/2 = 0.07 g_o A t_o$; and with attenuation, $L \sim t^{2/7}$ when the acceleration goes to zero.

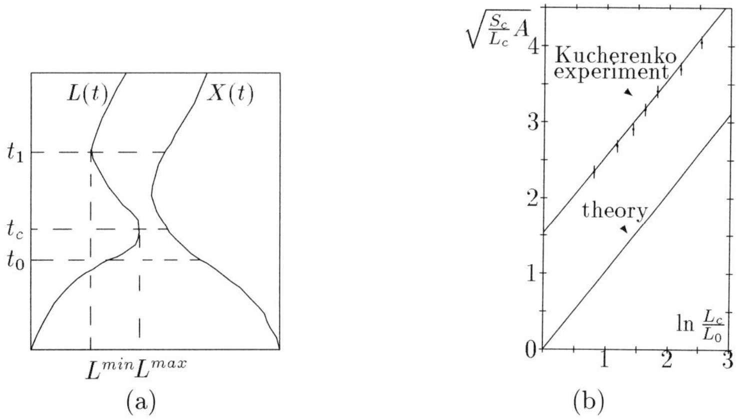

Figure 1. (a) Typical picture of the behavior of the boundary $X(t)$ between light (ρ_2) and heavy (ρ_1) matter under change of acceleration sign: at t_0 and t_1, the acceleration becomes zero; L is the mixing-zone width. (b) Dependence of the value $(S_c/L_c)^{1/2}$ on $\ln(L_c/L_0)$, where L_c is the width of a continuous density profile, L_0 is the initial roughness, and S_c is the displacement of the ampoule if $L = L_c$.

It is essential to note under what conditions the transport term is entered into (1) — the second component on the right side under the differential sign. The transport term is added only in the stable flow stage when the increment $gA < 0$. It is added not at once, but after the period during which the turbulent velocity $\overline{v}$ reduces from $\overline{v}_o$ to some critical value $\overline{v}_c$ according to Eqs. (2a) and (2b): with $t = t_c \overline{v}_c^2 = N \overline{v}_o^2$. By $\overline{v}_o$ we represent velocity at the moment when the acceleration sign changes. Value N is left indeterminate for now. It can be determined by the experiment or calculation.

Equations (2a) and (2b) are used only until critical time t_c, that is, when $t \leq t_c$. When $t \geq t_c$, it would be well to present the new equation for $\overline{v}$ but in so far as it is not clear how to do it, we assume that when $t \geq t_c$, velocity $\overline{v}$ is constant: $\overline{v} = \overline{v}_c$. So, when $t \geq t_c$ in (1), $\overline{v} = \overline{v}_c$ and the mixing-zone width is determined effectively by the density profile. After changing sign, if the acceleration is kept constant as $t \to \infty$, the stationary density profile is elaborated and calculated from the equation

$$\frac{\partial \rho}{\partial x} = \frac{1}{3\overline{v}_o\sqrt{NL}}\sqrt{\frac{2|g|}{(\rho_1-\rho_2)(\rho_1+\rho_2)}}(\rho_1-\rho)(\rho-\rho_2)$$

The limiting value of the mixing-zone width L_∞ is connected with the problem parameters in the following way: $L_\infty = 3.3N|\frac{g_o}{g_1}|L_o$. Notice that setting $N = 0$ means the complete separation $L_\infty = 0$ is taking place (Neuvazhayev 1993b). The experiments of Kucherenko et al. (1994) made it possible to determine one more constant, $\alpha_s = 0.01$.

2. The delay in the turbulent mixing process

Let there be two incompressible fluids with densities ρ_1 and ρ_2. The constant acceleration g is directed perpendicular to the boundary so that the Rayleigh-Taylor instability is taking place. It is known that by the initial continuous boundary there appears a delay in the turbulent mixing process. This problem was studied theoretically by LeLevier et al. (1955), Zababakhin and Ptizin (1988), and others. This problem was studied experimentally by Ardashova et al. (1988), where the relationship between the delay and the value of initial perturbations was set for Atwood number 0.6. The perturbations were preset in the middle of the continuous layer by the particles of matter with intermediate density. The theoretical studies of the delay phenomenon on the basis of the K model of turbulent mixing, conducted by Neuvazhayev (1994) lead to the determination of the analytical relationship between the initial roughness L_o and the mixing delay L_c determined by transference of system S_c at the instant when the turbulent mixing-zone width L is equal to L_c, i.e., to the width of the initial continuous boundary. This formula has the form

$$\sqrt{\frac{L_c}{L_o}} = \exp\left(0.48\sqrt{\frac{S_c A}{L_c}}\right) \tag{3}$$

It results from integrating the equations

$$\frac{\partial \rho}{\partial t} = \frac{\partial}{\partial x} D \frac{\partial \rho}{\partial x} \qquad \frac{d\overline{v}^2}{dL} + (0.5 + 0.23A^2)\frac{\overline{v}^2}{L} = \frac{\pi}{16} L\gamma^2 \tag{4}$$

$$\gamma^2 = \begin{cases} \frac{1.78gA}{L_c} & \text{if} \quad L \le L_c \\ \frac{1.78gA}{L} & \text{if} \quad L \ge L_c \end{cases} \tag{5}$$

Integrating the obtained equations with (2a) leads to the solution for mixing-zone width:

$$\sqrt{L} = \begin{cases} \sqrt{L_o} \exp\left(0.638\sqrt{\frac{gA}{(3.5+0.11A^2)L_c}} t\right) & \text{if} \quad L < L_c \\ \approx \sqrt{L_c} + 0.319\sqrt{\frac{gA}{(3.5+0.11A^2)L_c}}(t - t_c) & \text{if} \quad L > L_c \end{cases} \tag{6}$$

It must be emphasized here that the special feature of the developed solution — the exponential law of mixing in the beginning — is determined by the increment γ which takes the limited value with $L < L_c$. Formula (3) results from (6) if we determine the delay moment t_c by the equality $L = L_c$ and go from time t_c to the transference of the system S_c:

$$S_c = \int_o^t \left(\int g dt \right) dt$$

Then (3) will be true also in case of smoothly changing acceleration g.

Let us discuss the application of formula (3). First we note that when $L_o \to 0$, $S_c \to \infty$, i.e., infinitely small initial roughness corresponds to infinitely large delay in mixing. This result agrees well with experimental data (Ardashova et al. 1988) in which mixing was not observed unless initial perturbations were given. Formula (3) allows one to process experimental data in the plane of variables (x, y): $x = \ln \frac{L_c}{L_o}$; $y = \sqrt{\frac{S_c}{L_c}}$.

In Fig. 2, experimental results are plotted and the slope of the straight line is determined:

$$y = (1.02/\sqrt{A})x + 1.54 \tag{7}$$

The indicated distinction disappeares when setting $L_o = 0.3L_o^{exp}$.

3. The evaporation front

Taking into account turbulent mixing in the evaporation front is a complex problem because it has a number of peculiarities. One of them is the front transference over the matter. This is known to cause stabilization of short wave perturbations. Takabe and Yamamoto (1991) suggested the following formula for increment γ:

$$\gamma = \sqrt{kg} - \beta k v_o \tag{8}$$

where $\beta = 3.5$, Atwood number $A = 1$, v_o is the transference velocity of the eveporation front, and k is the wave number related to the mixing-zone width by the relationship $k = 3/L$. It is obvious that the increment in the form (8) leads to the limitation of the rise of short-wave perturbations. This limitation can be stated in the following way

$$\gamma = \begin{cases} \frac{0.81}{4\beta v_o} g & \text{if} \quad L < L_{cr} \\ 0.9\sqrt{\frac{3}{L} g} - \frac{3}{L}\beta v_O & \text{if} \quad L > L_{cr} \end{cases} \tag{9}$$

In (9) L_{cr} is the critical value of the mixing-zone width when the increment is at its maximum, $L_{cr} = (14.8\beta^2 v_o^2)/g$. Furthermore, much as it was in the previous section, Eqs. (4), (5), and (2a) are integrated and as a result the solution for the mixing- zone width L is obtained. To make the formulas simpler, Eq. (2) in its simplest form was used here without the first term. Then we have the simple expression for the velocity $\overline{v}$: $\overline{v} = \frac{\sqrt{\pi}}{4}\frac{L\gamma}{\sqrt{5+0.23A^2}}$; the equation for mixing-zone width is integrated and the solution has the following form

$$\sqrt{L} = \begin{cases} \sqrt{L_o}\exp\left(\sqrt{\frac{0.07g}{4L_{cr}}}t\right) & \text{if} \quad t \leq t_{cr} \\ \sqrt{0.07g}(t - t_{cr}) + \sqrt{L_{cr}}\left(1 - \ln\sqrt{2\sqrt{\frac{L}{L_{cr}}} - 1}\ \right) & \text{if} \quad t \geq t_{cr} \end{cases} \tag{10}$$

$$t_{cr} = \sqrt{\frac{L_{cr}}{0.07g}}\ln\frac{L_{cr}}{L_o}$$

Classically, in the case without evaporation, $V_o = 0$, and the solution has the following form

$$\sqrt{L} = \sqrt{L_o} + \sqrt{0.07g}\ \ t \tag{11}$$

Comparing the formulas (10) and (11), let us draw attention to the various dependencies on the initial roughness L_o. In the classic case, (11) with zero roughness, the nontrivial solution takes place; but in case (10), when the evaporation front moves over the matter, only the trivial solution corresponds to zero roughness (Neuvazhayev 1994).

4. Thin initially turbulized layer

In many experiments on the boundary, the membrane is present, and the function, especially for gases, is difficult to construct. There are many reasons for the initial turbulization of a thin layer, for example, the removal of a barrier between the gases. The complex targets in the ICF problem have usually thin interlayers.

The problems concerning the atmosphere and the ocean which deal with the determination of the laws of initially turbulized admixture propagation lead to setting up the thin-layer problem.

In accordance with the admixture-density environment-density relationship, several special situations can take place. If the admixture density has an intermediate value, then in accordance with the position of light and heavy media, the intensive mixing by the square law (heavy matter above) or slow mixing with the degree $\frac{2}{7} - \frac{2}{6}$ (light matter above) takes place over the course of time. The admixture spreads out, up, and down according to nearly the same law. If the admixture is heavier than the environment, then in the case when the heavy medium is above, mixing with constant rate directed only to the heavy matter takes place. If the admixture is lighter than the environment, the same situation takes place, and the mixing is directed to the light matter which is below and occurs with the constant rate, i.e., according to the linear law in time.

All these cases are studied on the basis of semi-empirical models. The special case of a thin layer immersed into a heavy one within the framework of the K model was considered by Belenikii and Fradkin (1965). The generalized consideration of this problem can be found in (Neuvazhayev 1983, 1988, 1993a), where the thin layers at the bottom and lid of an accelerating vessel, and the self-simulating problem on spreading of a turbulent mixture layer, were studied.

5. Conclusion

The complexity of the four problems discussed shows that rather simple K and KE models can deal with a fairly wide range of applications quite well.

References

Ardashova RI, Balabin SI, Voloshin NP, Kucherenko YA, Ptizina NV, Chitaikin VI (1988) The experimental investigation of perturbation propagation in gravitationally unstable system with continuous density spreading. Vopr At Nauki Tech Ser Teor Prikl Fiz 1:20–27.

Belenkii SZ, Fradkin ES (1965) Theory of turbulent mixing. In: Proceedings of FIAN, 29:207–237.

Kucherenko YA, Neuvazhayev VE, Pylaev AP (1994) Behavior of gravitational turbulent mixing region under conditions leading to separation. DRAN, 334, 4:445–448.

LeLevier R, Lasher G, Bjorklund (1955) Effect of Density Gradient on Taylor Instability, Report UCRL-4459, Univ. of California Radiation Laboratory.

Neuvazhayev VE (1983) The model properties of the turbulent mixing boundary of accelerating fluids of various densities. Appl Mech & Tech Physics 5:81–88.

Neuvazhayev VE (1988) Blurring of the turbulent mixing layer. Appl Mech & Tech Physics 2:46–51.

Neuvazhyaev VE (1993a) Asymtotic behaviour of a thin interlayer after passing a shock wave. In: Proc 4th Int Workshop on the Physics of Compressible Turbulent Mixing, Linden PE, Youngs PL, Dalziel SB (eds), Cambridge University Press, England, pp 461–477.

Neuvazhyaev VE (1993b) The properties of the turbulent mixing model based on the binary two-velocity model. The separation addition to diffusion models. VNIITF, Snezhinsk Preprint 48.

Neuvazhyaev VE (1994) Dependence of turbulent mixing on the initial roughness in the evaporation front with continuous density profile at the interface. Phys Rev E 50(2):1394–1398.

Takabe H, Yamamoto A (1991) Reduction of turbulent mixing at the ablation front of fusion targets. Phys Rev A44:5142–5152.

Zababakhin EI, Ptizin AR (1988) Convective instability limitation. Vopr At Nauki Tech Ser Teor Prikl Fiz 1:4–6.

Numerical simulation of Richtmyer-Meshkov instability

I. G. Lebo, V. B. Rozanov
Lebedev Physics Institute (FIAN)
Leninsky Pr. 53, Moscow 117924 Russia
V. F. Tishkin, V. V. Nikishin, I. V. Popov
Institute of Mathematical Modelling (IMM), Miusskaya Sq. 4, Moscow 125047
A. P. Favorsky
Keldish Appl. Mathematics Institute (IAM), Miusskaya Sq. 4, Moscow 125047

Abstract: The paper presents results of 2D and 3D Eulerian code "NUT" simulations of Richtmyer-Meshkov instability in shock-tube experiments and laser-driven targets.

Key words: Richtmyer-Meshkov instability, Numerical simulation, Shock wave

1. Introduction and short description of the NUT code

Richtmyer-Meshkov instability (RMI) (Richtmyer 1960, Meshkov 1969) develops near the contact boundary when a shock wave comes through the disturbed interface between two gases. Numerical modeling of RMI is presented for 2D and 3D contact-boundary perturbations. To describe the physical processes in the shock tube, a model of nonviscous, non-heat-conducting ideal gas is used which is presented by a system of Euler equations.

For the approximation of the system of gas-dynamic equations we propose a nonlinear conservative quasi-monotonic difference explicit scheme of a high order of accuracy (analogous schemes for 2D codes have been considered in Vjaznikov et al. 1989). The boundary conditions in this system are given according to the law: $x = 0$, $x = L$, then $u = -u$ (reflection condition). The shock wave moves along the 0Z axis, $z = Z_{boundary}$; the parameters of the shock wave and shocked gas are calculated over the given Mach number and unshocked gas density and pressure. We also solve the equation of thermal conductivity. This effect is very important for the modeling of laser plasma phenomena (see section 3). We have solved it numerically using an implicit scheme.

2. Modeling and comparison with shock-tube experiment

Using the NUT code we performed a series of test simulations modeling the shock-tube experiments made at G. M. Krzhizhanovsky Power Institute, Moscow (Aleshin et al. 1990). First we compared our results with the results of the

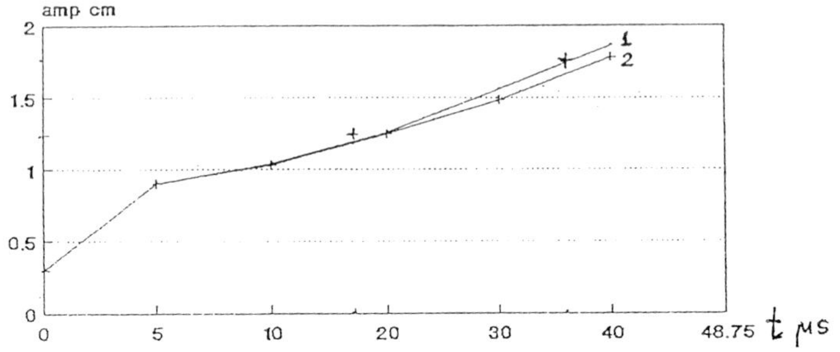

Figure 1. Comparison of perturbation growth velocity: (1) NUT code simulation; (2) results obtained at Moscow Physics and Technology Institute.

group of scientists from the Moscow Institute of Physics and Technology (Belotserkovskii et al. 1994). Figure 1 illustrates the growth of complete amplitudes of perturbations of the contact boundary (a so-called "mushroom structure") between helium and xenon gases (a "complete amplitude" means the distance from the "bottom" to the "top of the spike" of the contact boundary). The initial pressure was 0.5 bar, the Mach number of the initial shock wave was 2.5, the initial disturbance wavelength $\lambda = 8$ mm (these parameters suit the case of a late nonlinear stage, $ka_0 \geq 3$, where k is the wavenumber; $k = 2\pi/\lambda$; a_0 is the initial amplitude of a sinusoidal perturbation; and a_0 is equal to half of the complete initial amplitude). However, the perturbation growth in our case is somewhat higher. We assume that the reason for such a discrepancy lies in the fact that our code is typical of lower approximative viscosity. In the second test, the shock wave goes through the contact boundary between argon and xenon gases. The initial pressure is 0.5 bar, the Mach number, 3.5. Figure 2 shows the calculated growth of the complete amplitude for the perturbation wavelengths $\lambda = 36$ mm, $a_0 = 0.5$ mm. The + symbols denote the perturbation amplitudes obtained in the shock-tube experiments at G. M. Krzhizhanovsky Power Institute (Moscow). The calculations showed good agreement between the complete amplitude growth velocity and the experimental data.

A series of 3D simulations has been carried out. The initial perturbation of the contact boundary (argon-xenon) had the form: $Z_c^{pert} = Z_c^0 + a_0 * \cos(k_x x)\cos(k_y y)$ where $a_0 = 0.5$ mm, $k_x = 2\pi/\lambda_x$, $k_y = 2\pi/\lambda_y$, $\lambda_x = \lambda_y = 36$ mm. An incident shock wave was propagating along the axis OZ, and its velocity corresponded to the Mach number 3.5. As follows from Richtmyer (1960), at a linear stage the perturbations in 2D and 3D geometry have the same velocity provided the condition is fulfilled: $k_{2D} = 2\pi/\lambda_{2D} = k_{3D} = ((2\pi/\lambda_x)^2 + (2\pi/\lambda_y)^2)^{1/2}$ where $\lambda_{3D} = \sqrt{2}\lambda_{2D}$ if $\lambda_x = \lambda_y = \lambda_{2D}$. From this it follows that in 3D geometry the perturbations with $\lambda_{3D} = 36$ mm should be compared in 2D geometry to the perturbations with $\lambda_{2D} \approx 24$ mm. Figure 3

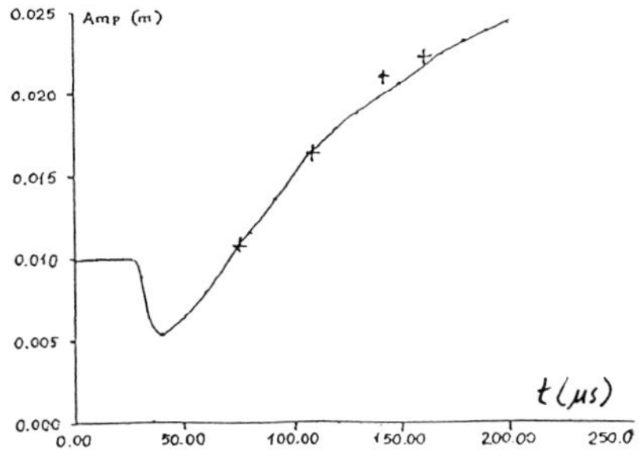

Figure 2. Contact boundary perturbation amplitude growth versus time; + denote perturbation amplitudes obtained in shock-tube experiments.

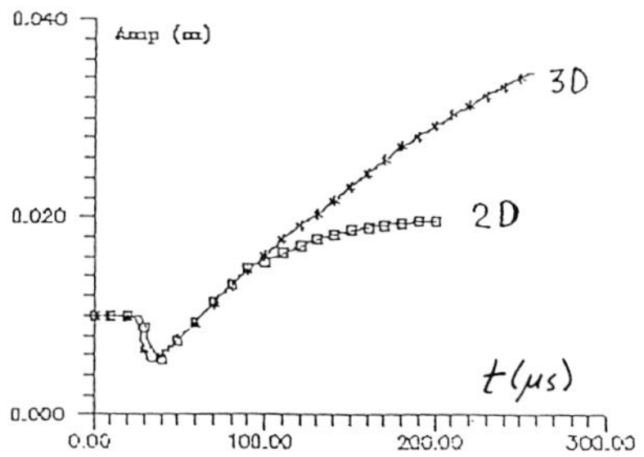

Figure 3. Contact boundary perturbation amplitude growth versus time for 3D ($\lambda = 36$ mm) and 2D ($\lambda = 24$ mm) perturbations, $a_0 = 0.5$ mm.

shows the growth of complete perturbation amplitudes for 3D and 2D geometry. It is seen that at the nonlinear stage ($k^*a > 5.0$, where k is the wavenumber and a the current complete amplitude) the perturbations in 2D geometry grow more slowly. This may be explained by the fact that due to the growth of Kelvin-Helmholtz instability near the top of the spike the processes of vortex-structure formation and "blurring" of such a spike are faster in 2D geometry (see Rozanov et al. 1992).

In order to investigate the influence of the boundary conditions on the development and formation of the vortex structures we have performed a calculation of the Ar-Xe contact boundary perturbation development where: (a) the initial perturbation is a sinusoid with the wavelength $\lambda = 24$ mm, $a_0 = 0.5$ mm; (b) the calculated region of OX axis is twice as high, and the initial perturbation has the following form:

$$Z_c(x,y) = \begin{cases} Z_0 + a_0\cos(k_x x), & \text{if } 0 \le x \le \lambda_x/2; \\ 0, & \text{if } \lambda_x/2 \le x \le \lambda_x \end{cases}$$

that is, "an absolutely elastic wall" (i.e., $u = -u$) is placed at $x = \lambda = 24$ mm; and (c) there is "free flow through a wall" ($du/dx = 0$ at $x = \lambda/2$). The dependence of the perturbation complete amplitude growth is close to that which is observed when the "wall" is located at point $x = \lambda/2$, but there are some differences.

The shape of the perturbation (case b) at $t = 170$ μs is shown in Fig. 4a; Fig. 4b illustrates the amplitude development of the contact boundary corresponding to cases (a), (b), and (c). One can see that a "solitary perturbation" has a somewhat bigger amplitude at $t = 200$ μs. The next series modeled a shock-wave passage through the contact boundary of krypton and xenon gases in 2D geometry. As observed previously, the perturbation of the contact bound-

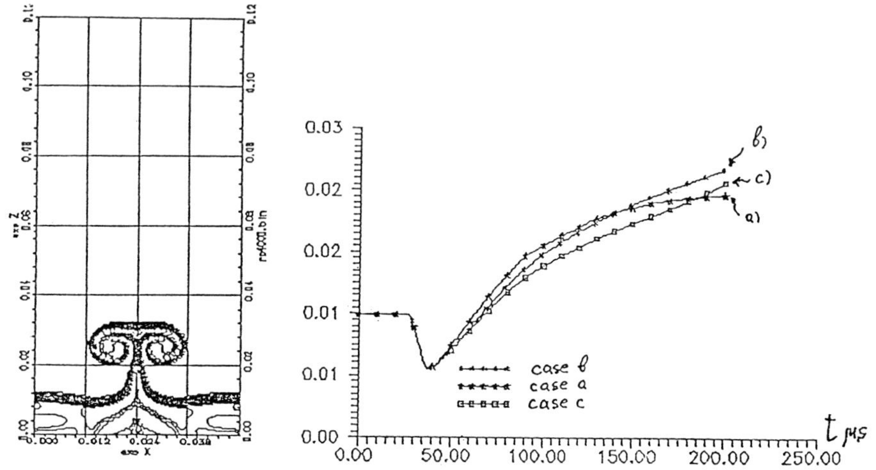

Figure 4. (a) Density isolines near Ar-Xe contact boundary for "solitary perturbation" (case b). (b) Contact boundary perturbation amplitude growth for cases (a), (b), (c).

ary had a sinusoidal shape with wavelength $\lambda = 3.6$ cm and amplitude $a_0 = 0.5$ cm (we compared our results with the results of American scientists from the University of New York at Stony Brook). The initial density in Kr is $\rho = 1.87$ mg/cm^3, and in Xe $\rho = 2.95$ mg/cm^3. In the calculations we set the initial velocity of the unperturbed gas flow $w = -55.5E + 4$ m/s, which allowed us to restrict the calculated region. Below we give the calculation results.

Figure 5 illustrates the amplitude perturbation growth of the contact boundary. A satisfactory agreement is observed between our results and the American results (+ symbols). At the late nonlinear stage of the perturbation growth the amplitude obtained in our calculations is somewhat larger than the American results. We assume that the reason for such a discrepancy is similar to that for test 1. However, this question demands further investigation.

3. Richtmayer-Meshkov instability in a laser target

In contrast to the shock-wave experiments, the development of instabilities in laser targets is essentially influenced by electron (and radiative) heat conductivity. A one-temperature approximation that is true in the dense compressed target (where the electron-ion exchange time is small due to high plasma density) is considered. In order to calculate the development of perturbations in the target, we used as the initial data the 2D Lagrange code "Atlant" data (Lebo et al. 1994). Note that most of the Lagrange codes used in the target compression calculations would not allow us to obtain a "mushroom-like" structure at the contact boundary, due to a strong deformation of the calculation cells.

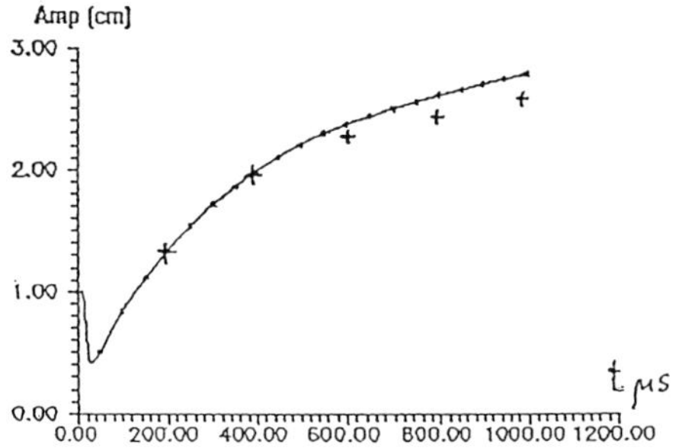

Figure 5. Contact boundary perturbation amplitude growth for Kr-Xe case; + show results of American group from Stony Brook.

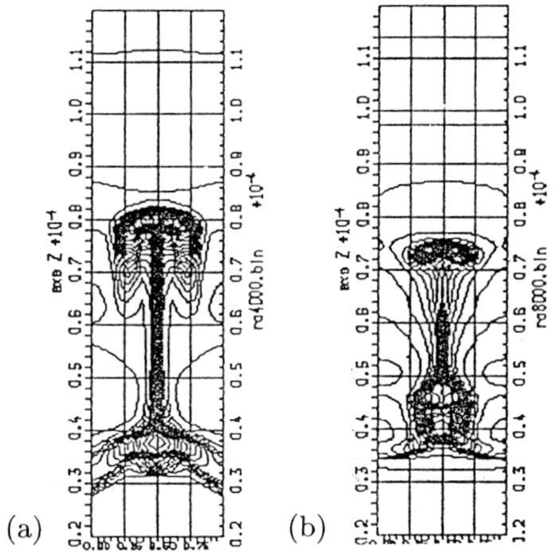

Figure 6. Plasma density isolines near "shell-fuel" and "shell-corona" contact boundaries: (a) without heat conductivity; (b) with heat conductivity.

In order to calculate this problem using the NUT code we have set three subregions: (1) Subregion I with the DT fuel with density $\rho_F = 0.05$ g/cm^3. (2) Subregion II with the nonevaporated part of the shell. The averaged density (ρ_S) and the region thickness (L) have been changed in the calculations so as to obtain $\rho_S * L = 10$ mg/cm^2. (3) Subregion III is the corona. In the calculations the following values of the average corona density were set: $\rho_C = 3$ mg/cm^3 and $\rho_C = 150$ mg/cm^3. The initial temperature in the corona was equal to 1 keV and 0.02 keV. The temperature in subregions I and II was calculated in such a way as to make the pressure at $t = 0$ constant within the whole region. The shock wave moved from the DT fuel to the shell. The gas velocity behind the shock front was $w = 250$ km/s (in one of the variants, 500 km/s; in this case when $\lambda = 5$ μm the "mushroom-like" structure was blurred completely). We have performed 2D calculations of the "shell-fuel" contact boundary perturbation for different wavelengths, $\lambda = 5$, 10, and 20 μm, and initial amplitude $a_0 = 2$ μm. We have performed calculations also with $w = 250$ km/s, but with a higher density of the shell (variant 2: $\rho_S = 0.2$ g/cm^3, variant 3: $\rho_S = 0.8$ g/cm^3, variant 4: $\rho_S = 0.8$ g/cm^3, $\rho_C = 150$ mg/cm^3). In variant 4, the corona density was increased, and the temperature, correspondingly, was decreased. As a result, the heat flow from the corona toward the shell was suppressed. The instability development is strongly affected by the heat conductivity. First, the heat flow from the corona to the shell leads to the formation of another shock wave moving from the corona to DT fuel. As a result, we obtain density and temperature profiles which differ from calculations without heat conductivity (no jump in the density at the contact boundary). Second, the heat conductivity leads to the "blurring" of the vortex structure near the top of a spike. Figure 6a illustrates the isolines of the shell density at $t = 0.237$ ns without heat conductivity; Fig. 6b shows the isolines with allowance for this effect at $t = 0.237$ ns. The vortex structures in the vicinity of the spike top are blurred, and the amplitude of the perturbation is smaller than that in case (a).

4. Conclusions

(1) The NUT code allows one to numerically model the development of such a complicated phenomena as the Richtmyer-Meshkov instability up to its late nonlinear stage, when vortex structures, destruction of juts, and the formation of turbulent mixing zones occur. The comparison of numerical calculations with results from shock-tube experiments has demonstrated good agreement. (2) The 2D perturbations at the nonlinear stage are destroyed faster compared to the 3D perturbations. (3) We started to study the multi-mode perturbation development. (4) The development of instability at the "shell-fuel" contact boundary has some peculiar features in the high-temperature plasma of the laser targets. They are due to the influence of nonlinear heat conductivity. The heat flow at the high temperature of the fuel suppresses the small-scale vortex structures in the vicinity of the spike top, and can reduce the velocity of the perturbation development.

Acknowledgement. The authors are grateful to S. G. Zaytsev and his colleagues for fruitful discussion of the calculation results. This work was supported by the ISTC, Grant No. 29.

References

Aleshin AN, Lazareva EV, Zaytsev SG, Rozanov VB, Gamaly EG, Lebo IG (1990) Dokladi Akademii Nauk SSSR, Fizika 310(5) (in Russian).

Belotserkovskii OM, Demchenko VV, Oparin AM (1994) Doklady Russian Academii of Science 39(2):118 (in Russian).

Lebo IG, Popov IV, Rozanov VB, Tishkin VF (1994) Journal of Russian Laser Research 15(15):136 (in Russian).

Meshkov EE (1969) Izv. Akad Nauk SSSR, Mekhanika Zhidkosti i Gaza (5):151 (in Russian).

Richtmyer RD (1989) Comm Pure and Appl Math 13(2):297.

Rozanov VB, Zmitrenko NV (1992) Preprint FIAN, No. 16 (in Russian).

Vjaznikov KV, Tishkin VF, Favorsky AP (1989) Matematicheskoe modelirovanie 1(5):95 (in Russian).

The effect of refraction regime on the evolution of perturbation for Richtmyer-Meshkov instability

S. G. Zaytsev, A. N. Aleshin, E. V. Lazareva
Krzhizhanovsky Power Engineering Institute (ENIN)
Leninsky Pr. 19, Moscow, 117071 Russia

Abstract: We studied the evolution of an interface after the plane incident shock S_0 passed through it, moving from light to heavy gas. It has been found that the increased $a_0 k$ (where a_0 and k are the amplitude and the wave number of the initial two-dimensional interface K_0, respectively) resulted in reduced depth of penetration of one gas into the other at the final turbulent stage of the interface evolution, and this was due to the change in the type of refraction of S_0 on K_0.

Key words: Shock wave, Hydrodynamic instability, Turbulence

1. Introduction

The passage of the shock S_0 from light (ρ_1) to heavy (ρ_2) gas through a two-dimensional sinusoidal interface K_0 is accompanied by formation of a mixing zone K which separates the shocked flows formed by the refracted (S) and reflected (R) shock waves. At the end of refraction the shapes of S, R, and K are similar to that of K_0 (Fig. 1). Because the generated shock waves S and R are curvilinear, the subsequent shocked flow acquires the component of velocity U_y which is perpendicular to the direction of the motion of S_0 (axis X). Collision of the two flows behind the converging sections of S and R (S_c and R_c, respectively) results in formation of the high-pressure regions P_S^+ and P_R^+ behind these sections, and at the same time, the low-pressure regions P_S^- and P_R^- are formed behind the diverging sections of S and R (S_d and R_d). The properties of the resultant distribution of pressure determine the evolution of the interface K and the shock waves S and R (Richtmyer-Meshkov instability, RMI). These properties are dependent on: (i) the angle α between the normal to S_0 and the tangent to K_0; (ii) the density ratio for the contacting gases; and (iii) the intensity of the incident shock.

After the end of refraction the amplitude a_K of the interface K is increased while the amplitudes a_S and a_R of the shock waves S and R, respectively, are decreased. Visualization and numerical calculations show that when the curvature of K_0 (for unchanged ρ_1, ρ_2, and Mach number M_0 of the incident shock)

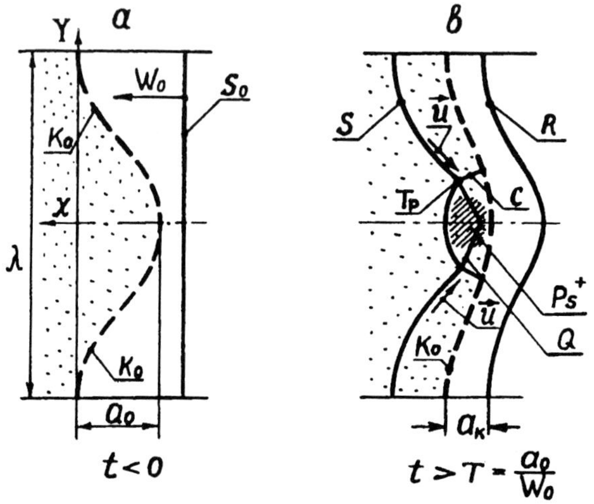

Figure 1. The scheme of excitation of Richtmyer-Meshkov hydrodynamic instability for the hard regime of refraction: (a) dispositions of S_0 and K_0 before the interaction, and (b) dispositions of the refracted (S) and reflected (R) shock waves and the interface K after the end of refraction.

overcomes its critical value, transversal shock waves C appear in the shock-compressed flow behind S, and the interaction of C with S results in a triple point T_p (Fig. 1).

For $\alpha = \pi/2$, the parameters of S, R, and K are determined by one-dimensional equations and denoted by *. For $\alpha_1 < \alpha < \pi/2$ the regime of refraction is regular, i.e., S_0, S, and R meet in one point (Henderson & Macpherson 1968). For $\alpha = Const$, parameters of the flows in the vicinity of the point of intersection of S_0, S, and R remain unchanged during the motion of S_0 through K_0. For $\alpha_1 < \alpha$, the point of intersection of S_0 with R moves aside from K_0, forming a Mach configuration. This is an irregular regime of refraction. These regimes have been studied for $\alpha = Const$ (Henderson & Macpherson 1968).

2. Results of experimental and numerical investigations

Experiments were performed in a shock tube; the internal section of the driven was 72×72 mm^2. Different couples of the inert gases He, Ar, Kr, and Xe were used. The a_0k varied from 0.87 to 7.96, and the initial pressure p_0 was 0.5 atm. The Mach number of the incident shock S_0 varied from 2.5 to 4.0. Numerical calculations were performed using the code of Dr. V. V. Demchenko (Moscow Physico-Technical Institute).

The amplitudes $a_{\rm K}(S)$ of the interface K versus the distance S passed by K are given in Fig. 2. Experimental points 1–4 and 5 are related to regular and irregular refraction, respectively. The experimental and numerical investigations of RMI on a 2D sinusoidal interface K_0 allowed us to find the correlation between the observed changes in the interface evolution and different types of refraction of S_0 on K_0. Thus we can distinguish three regimes of the evolution: soft regular, hard regular, and irregular.

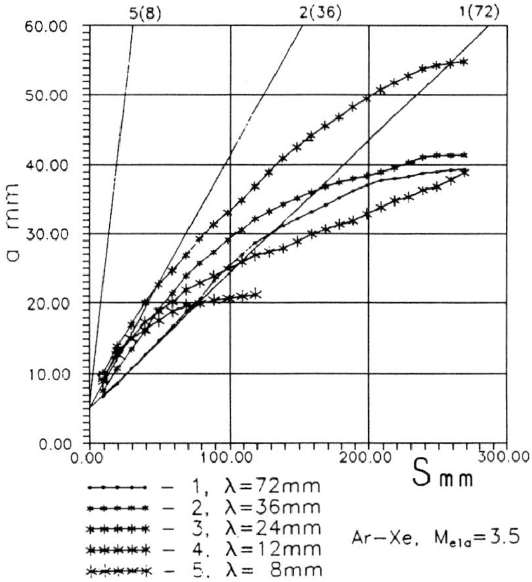

Figure 2. Amplitude $a_{\rm K}(S)$ of the interface K versus the distance S passed by K. Lines 1(72), 2(36), and 5(8) correspond to the values of $a_{\rm K}$ calculated from the Richtmyer relation for K_0 with λ of 72, 36, and 8 mm, respectively.

Soft regular regime. During formation of the high-pressure regions and their subsequent "spreading," the flow between the shock front and interface remains continuous. For this regime the process of K evolution after the refraction of S_0 at K_0 may be divided into three typical stages preceding the formation of a turbulent mixing zone: linear, nonlinear, and transitional. For this regime the maximum of pressure behind the converging section S_C does not exceed $1.2P^*$ (See Fig. 1). For the soft regular regime the K evolution starts from the linear stage which is sufficiently described by the Richtmyer relation:

$$\frac{da_{\rm K}(t)}{dt} = a_{\rm K}^* ukA$$

where $a_{\rm K}^*$ is the amplitude at the end of refraction, u is the velocity given to K by the shock, k is the wave number, and $A = (\rho_2 - \rho_1) \cdot (\rho_1 + \rho_2)^{-1}$.

Hard regular regime. The maximum of pressure behind the converging section S_C does not exceed 2P* (See Fig. 1). For this regime, in the flow between S and K the transversal shock waves C are generated. Interaction of C with S results in a triple point T_p on the S front. During the refraction the interface transforms into the spikes of heavy gas penetrating into the light gas and bubbles of light gas penetrating into the heavy gas. This shape of K is typical for a nonlinear stage. Thus, for the hard regular regime the interface evolution after the end of refraction starts with a nonlinear stage.

Irregular regime. For the experiments corresponding to the points 5 in Fig. 2 the numerical calculations indicate the irregular type of refraction. For this regime, the vortex structures on the heavy gas spikes typical for the transitional stage are formed during the refraction.

The calculated data for the shock passage from He to Xe ($M_0 = 2.5$, $a_0 = 10$ mm, $\lambda = 8$ mm) are given in Fig. 3. At the end of refraction ($t = 4$ μs), a vortex structure (Q zone) appeared on the spike of the heavy gas (Xe) penetrating into the light gas (He). By $t = 9.8$ μs, this vortex structure developed into the typical mushroom-like structure. For this regime of refraction, the maximum pressure in P_S^+ is 3.3P*. The interface evolution after the end of refraction starts with a nonlinear stage.

3. Boundaries for different regimes of refraction

The position of the boundary to distinguish the regular and irregular types of refraction for a sinusoidal interface is approximately determined as follows. The process of refraction is presumably dependent on the character of refraction on the passage of S_0 through the central part of K_0, where the change in the angle α between the direction of S_0 motion and the tangent to K_0 will be relatively small. Here α is close to α_{min} (the vicinity of K_0 inflection point).

The numerically calculated conservation equations for the refraction of a plane shock front on a plane interface arranged at the angle α allow one to determine the wave image of refraction. Thus we determine the boundary between regular and irregular regimes in the (α, M_0)-plane for a fixed A. In Fig. 4 the boundary position is denoted by line 2 in the (a_0k,A)-plane for a fixed M_0. The angle α is counted as α_{min} for a sinusoid with the corresponding a_0k. The points in Fig. 4 correspond to the experiments in which Topler pictures demonstrate a specific network of lines resulting from the interaction of the transversal shock wave C with the reflected shock wave. Experiments performed for the irregular regime recorded the confluence of adjacent mushroom-like structures. Numerical calculations performed for the regimes disposed above line 2 (Fig. 4) demonstrated formation of mushroom-like structures in a heavy gas spike penetrating into light gas. Thus, irregular regimes are disposed above line 2 (Fig. 4).

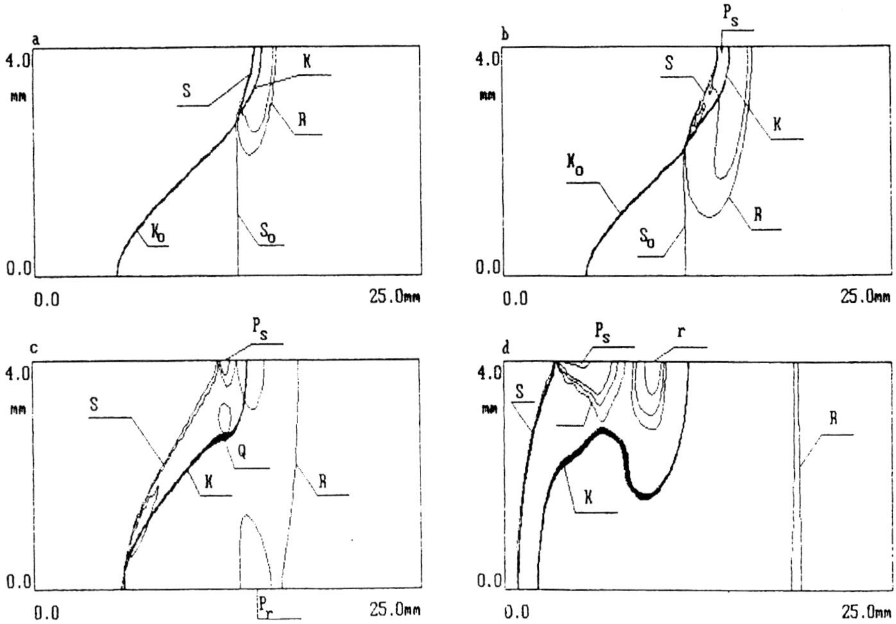

Figure 3. Results of calculation for the passage of the incident shock S_0 from He to Xe through sinusoidal interface K_0 ($\lambda = 8$ mm, $p_0 = 0.5$ atm, $M_0 = 2.5$): (a) $t = 1.0$ μs, (b) $t = 1.6$ μs, (c) $t = 4.0$ μs, and (d) $t = 9.8$ μs.

Line 1 is obtained from numerical calculation using the following empirical condition: $U_y \geq 0.2c$, where c is the local speed of sound (Aleshin et al. 1993). The area between lines 1 and 2 corresponds to the hard regular regime. In the shock-compressed gas behind S, Topler pictures show a specific network of lines of tangential breaks frozen in the flow. The breaks in the S front (triple points) are due to the interaction of transversal shock waves with the S front. The C fronts move along the shock-compressed flow.

For the regimes disposed below line 1, Topler pictures show neither breaks on S front nor a specific network of lines frozen in the flow described above. This area corresponds to soft regimes.

4. Conclusions

The experimental and numerical investigations of the Richtmyer-Meshkov instability for a single-mode two-dimensional initial interface K_0 have shown that when time intervals are large enough for the turbulent mixing zone to be formed, the depth of penetration of one gas into the other decreases with growth of a_0k

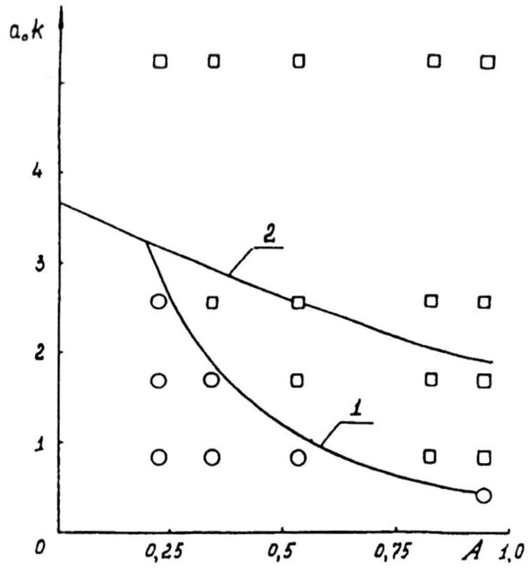

Figure 4. Boundaries of different regimes of refraction in the (a_0k,A)-plane. The irregular regimes of refractions are disposed above line 2. The soft regular regimes are disposed below line 1. The area between lines 1 and 2 corresponds to hard regular regimes.

(a_0 and k are the amplitude and wave number of K_0, respectively). This is caused by the reduced time of transition to the turbulent stage because of the altered evolution of interface which is due to the change in type of refraction. At the turbulent stage the rate of penetration of one gas into the other slightly depends on initial conditions.

Acknowledgement. The study was partly supported by the Russian Foundation for Fundamental Research (Grant No. 94-01-01144a) and by the ISTC (Grant No. 94-029).

References

Aleshin AN, Zaytsev SG, Lazareva EV (1993) Experimental and numerical studies on Richtmyer-Meshkov instability. RJCM 1:33–49.

Henderson LF, Macpherson AK (1968) On the irregular refraction of a plane shock wave at a Mach number interface. J Fluid Mech 32(I):185–202.

A numerical study of shock acceleration of a diffuse helium cylinder

J. A. Greenough, J. B. Bell, P. Colella
Center for Computational Sciences and Engineering
Lawrence Livermore National Laboratory*
Livermore, CA 94550 USA

E. G. Puckett
Department of Mathematics and Institute of Theoretical Dynamics
University of California
Davis, CA 95616 USA

J. W. Jacobs
Department of Aerospace and Mechanical Engineering
University of Arizona
Tucson, AZ 85721 USA

Abstract: The development of a shock-accelerated diffuse helium cylindrical inhomogeneity is investigated using a new numerical method. The new algorithm is a higher-order Godunov implementation of the so-called multi-fluid equations. This system correctly models multiple-component mixtures by accounting for differential compressibility effects. This base integrator is embedded in an implementation of adaptive mesh refinement (AMR) that allows efficient increase in resolution where the computational effort is concentrated when high accuracy, or increased resolution, are required. Qualitative and quantitative comparisons with previous experimental data are excellent. The simulations show that counter-sign vortex blobs are deposited in the jet core by baroclinic generation of the curved shock wave as it traverses the jet. This vorticity deposition occurs over timescales that scale with the shock passage time ($\sim 10\ \mu$s). Three phases of development are identified and characterized. The first is the weak-deformation (WD) phase, where there is weak distortion of the helium jet due to weak vorticity-induced velocity effects. The second phase is the strong-deformation (SD) phase where there is large distortion for the jet and the vortex blobs due to large induced-velocity effects. The last is a relaxation/reorganization (RR) phase where the vorticity field reorganizes into a point-like vortex pair.

*Work performed under the auspices of the US DOE by the Lawrence Livermore National Laboratory under contract W-7405-Eng-48. Additional support under contract W-7405-Eng-48 by the Applied Mathematical Sciences Program and the HPCC Grand Challenge Program of the Office of Scientific Computing at DOE, Defense Nuclear Agency under IACRO 94-831, and the Center for Advanced Fluid Dynamics Applications under contract B295120.

Key words: Shock-accelerated diffuse interfaces, Multi-fluid methods, Adaptive methods

1. Introduction

The evolution of interfaces accelerated by shock waves has been the focus of investigations for many years. This class of problem has applications in such disparate fields as inertial confinement fusion (ICF) and high-speed combustion. The literature is rich in studies on the stability of plane interfaces (Richtmyer 1960, Meshkov 1969, Zaitsev et al. 1985, and Brouillette & Sturtevant 1988). The stability of curved interfaces has received similar attention culminating with the study of Haas & Sturtevant (1987), where cylindrical or spherical shapes, constructed from soap bubbles or nitrocellulose membranes, were filled with either a light or heavy gas and then accelerated by a shock wave. This last study and the previous works relied on such membranes to separate gases. A different class of experiment on acceleration of gaseous cylinders was developed by Jacobs (1992, 1993) that was free from the effect of membranes and used a planar laser-induced fluorescence (PLIF) technique that allowed for excellent visualization of the flow and quantitative information regarding mixing.

The present simulations use a high-resolution adaptive method designed for multi-fluid flows with initial mixed layers. In this paper, the development of the helium mole fraction and vorticity field will be examined. Extensive comparisons with experimental data have also been completed as well as development of a new conceptual model describing this flow, but these cannot be described here due to space limitations.

2. Multi-fluid system and numerical method

The equations used for this study are the so-called multi-fluid equations as developed by Collela et al. (1995) and also described in Puckett & Saltzman (1992). This system is based on a volume-of-fluid approach, where the fraction of a cell volume occupied by a distinguished fluid is followed during the flow evolution. The basic assumptions in this approach are that there is pressure equilibrium within a cell, a single velocity vector for both fluids, and that all changes of state are adiabatic. The equations are given as

$$\frac{\partial f^\alpha}{\partial t} + \nabla \cdot (\boldsymbol{u} f^\alpha) = f^\alpha \frac{\hat{\Gamma}}{\Gamma^\alpha} \nabla \cdot \boldsymbol{u} \tag{1}$$

$$\frac{\partial}{\partial t}(f^\alpha \rho^\alpha) + \nabla \cdot (\boldsymbol{u} f^\alpha \rho^\alpha) = 0 \tag{2}$$

$$\frac{\partial \rho \boldsymbol{u}}{\partial t} + \nabla \cdot (\boldsymbol{u}\boldsymbol{u}\rho) + \nabla p = 0 \tag{3}$$

$$\frac{\partial}{\partial t}\left(f^{\alpha}\rho^{\alpha}E^{\alpha}\right)+\nabla\cdot\left(\boldsymbol{u}f^{\alpha}\rho^{\alpha}E^{\alpha}\right)+pf^{\alpha}\frac{\hat{\Gamma}}{\Gamma^{\alpha}}\nabla\cdot\boldsymbol{u}+f^{\alpha}\frac{\rho^{\alpha}}{\rho}\boldsymbol{u}\cdot\nabla p=0 \tag{4}$$

where f^{α}, ρ^{α}, and E^{α} are the volume fraction, density, and total energy density of fluid component α. The volume fraction is defined as $f^{\alpha}=\Lambda_{\alpha}/\Lambda$ where Λ is the volume of the cell and Λ_{α} is the volume of the cell occupied by fluid α. Γ^{α} is the sound speed gamma for fluid α, and $\hat{\Gamma}=1/\sum_{\alpha}\left(f^{\alpha}/\Gamma^{\alpha}\right)$ represents fraction weighted Γ for the mixture.

The pressure that appears in the above system is defined to be a thermodynamically consistent pressure and defined as $p=\sum_{i}\hat{\Gamma}\left(f^{i}p^{i}/\Gamma^{i}\right)$, where p^{α} is the partial pressure of component α. Note that the formulation is sufficiently general to allow real-gas equation of state (EOS) systems described by pressure given as a function of density and internal energy.

The solution procedure for the system is a higher-order Godunov method following that of Colella (1984) and detailed in Greenough et al. (1995). This base integrator for the multi-fluid system is embedded within an implementation of adaptive mesh refinement (AMR). This methodology is based on the original work of Berger & Oliger (1984) and later Berger & Colella (1989). This is a means for managing a refined grid hierarchy composed of logically rectangular grid patches that allows for efficient increases in resolution by focusing the computational accuracy where errors are deemed large, or in regions where high resolution is required.

3. Problem setup

The initial conditions for this problem are taken from Jacobs (1993). The shock-wave Mach number is 1.094 in air. In Budzinski (1992), a Rayleigh scattering technique is used to measure the actual molar concentrations of helium. It shows the initial profile is Gaussian in shape with a core concentration of approximately 80% helium. The initial half radius is set as $r_{\text{half}}=0.2$, where it is defined as the distance where the helium mole fraction is half the center value. The density ratio between air and helium is $\rho_{\text{helium}}/\rho_{\text{air}}=0.138$.

The computational domain is 9 cm wide by 27 cm long with a base grid of 36 cells wide by 108 cells long. There are two levels of refinement with the first level a factor of 4 more refined than the base grid and the second level a factor of 8 more refined than the first level of refinement. This gives an effective grid resolution of $\Delta x_{\text{fine}}=\Delta y_{\text{fine}}=0.0078125\ cm$. The shock is initialized in air at a distance of $3r_{\text{half}}$ from the jet. All boundaries are inflow/outflow type.

4. Results and discussion

4.1. Flow visualization

Figures 1 and 2 show the evolution of the mole fraction of helium and the development of the vorticity field, respectively. Figure 1a shows the initial conditions and the initial mixed layer. At 88 μs, Fig. 1b, there is a flattening of the jet and a generally weak deformation of its original circular structure. From here out to about 400 μs, spanned in time by Figs. 1c through f, there is strong deformation of the jet as it is inverted and transformed into the vortex pair. After 400 μs, from Fig. 1g on, the flow evolves as a well-defined vortex pair. Comparing these results with those given in Jacobs (1993), the agreement is exceptional.

Based on these observations and for ease and clarity of later exposition, we define three phases of development. The first is the weak-deformation phase (WD) that occurs over short times out to about 90 μs. The second is the strong-deformation phase (SD) that occurs after WD out to about 400 μs. The last is the relaxation/reorganization (RR) phase that occurs after SD.

In the vorticity time sequence, Fig. 2a shows the shock, still essentially planar, leaving positive (in the upper half plane) vorticity behind it. By 66 μs, Fig. 2b, the shock is well out of the jet core and two counter-sign circular vortex blobs are seen located in the jet core. During the WD phase, the vortex blobs remain essentially undistorted. This is due to the fact that vorticity-induced velocity effects are very weak during the WD phase. Also, the blobs are deposited in the jet core over a timescale that scales with the shock passage time ($\sim$ 10 μs). During the SD phase however, there are large induced-velocity effects that distort and stretch the vortex blobs into something like continuous finite-thickness sheets. The underlying vortical field swirls these sheets so that they are reorganized, at later times, into a vortex pair. Note that during the SD phase, there is strong counter-sign vorticity appearing also oriented in sheets, ahead of the distorted blob. Although not shown here, this counter-sign production acts to modulate the overall circulation as it relaxes to a near constant value.

The development of this flow is seen to be determined by the interaction and evolution of the vortex blobs. Once they are present in the flow, they at first interact weakly (during WD), then interact strongly (during SD) as they change from blobs to more point-like vortices (during RR). During the course of the transformation in the vorticity field, the helium jet undergoes a similar transformation as shown. From a modeling point of view, this flow could be accurately depicted by the development of a vortex blob pair in an essentially circular helium jet (slightly flattened on the upstream side in reality). The

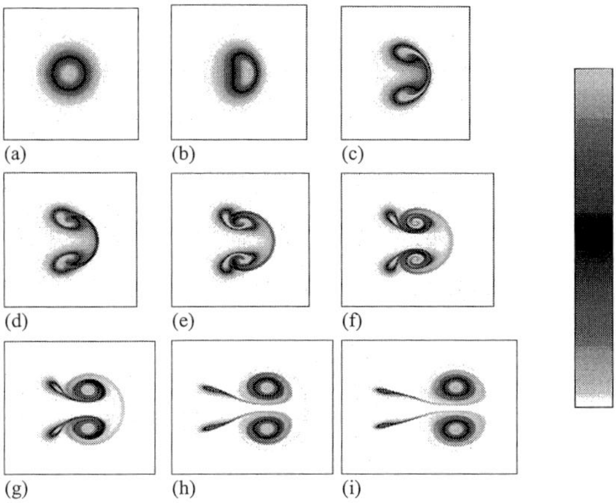

Figure 1. A sequence showing the computed helium mole-fraction evolution. Note that the shock arrival time at the jet is approximately 40 μs and all times reported are relative to the start of the calculation: (a) the initial jet, (b) 88 μs, (c) 202 μs, (d) 259 μs, (e) 318 μs, (f) 436 μs, (g) 561 μs, (h) 943 μs, (i) 1182 μs. The shock moves from left to right. The color bar [sic] to the right varies highest to lowest values from top to bottom.

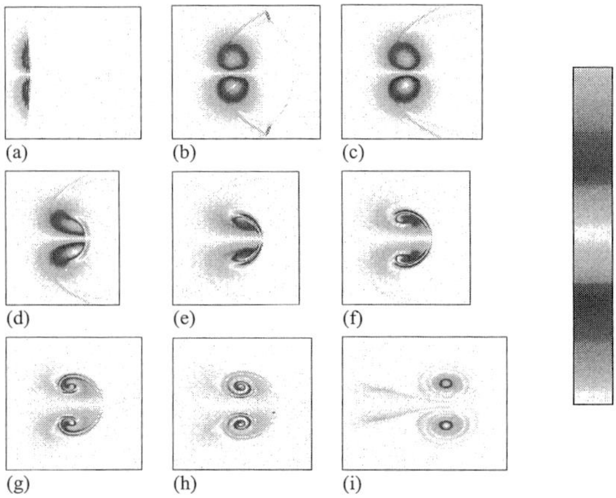

Figure 2. A sequence showing the vorticity evolution. Note that the shock arrival time at the jet is approximately 40 μs and all times reported are relative to the start of the calculation: (a) 44 μs, (b) 66 μs, (c) 88 μs, (d) 141 μs, (e) 202 μs, (f) 259 μs, (g) 318 μs, (h) 436 μs, (i) 1182 μs. The shock moves from the left to right. The color bar [sic] to the right is centered with respect to the data. Values vary highest positive to highest negative from top to bottom.

subsequent development could be predicted by their interaction and reorganization/relaxation to point vortices.

References

Berger MJ, Colella P (1989) Local adaptive mesh refinement for shock hydrodynamics. J Comput Phys 82:64–84.

Berger MJ, Oliger J (1984) Adaptive mesh refinement for hyperbolic partial differential equations. J Comput Phys 53:484–512.

Brouillette M, Sturtevant B (1988) Shock induced Rayleigh-Taylor instability at a continuous interface. Abstract submitted to the Int Workshop on the Physics of Compressible Turbulent Mixing, 24–27 October 1988, Princeton, New Jersey.

Budzinski JM (1992) Planar Rayleigh Scattering Measurements of Shock Enhanced Mixing, Doctoral dissertation, California Institute of Technology.

Colella P (1984) A direct Eulerian muscl scheme for gas dynamics. SIAM J Sci Stat Comput 6:104–117.

Colella P, Ferguson RE, Glaz HM (1995) Multifluid algorithms for Eulerian finite difference methods. In manuscript.

Greenough JA, Bell JB, Colella P (1995) An Adaptive Multifluid Interface-Capturing Method for Compressible Flows in Complex Geometry, AIAA Paper 95-1718.

Haas J-F, Sturtevant B (1987) Interaction of weak shock waves with cylindrical and spherical inhomogeneities. J Fluid Mech 181:42–76.

Jacobs JW (1992) Shock-induced mixing of a light-gas cylinder. J Fluid Mech 234:629–649.

Jacobs JW (1993) The dynamics of shock accelerated light and heavy gas cylinders. Phys Fluids A 5:2239–2247.

Meshkov EE (1969) Instability of the interface of two gases accelerated by a shock wave. Izv Akad Nauk SSSR Mekh Ahidk Gaza 4:1151–157; Russian: Izv Acad Sci USSR Fluid Dyn 4:101–104.

Puckett EG, Saltzman JS (1992) A 3-D adaptive mesh refinement algorithm for multimaterial gas dynamics. Physica D 60:84–104.

Richtmyer RD (1960) Taylor instability in shock acceleration of compressible fluids. Comm Pure Appl Math 23:297–319.

Zaitsev SG, Lazareva EV, Chernukha VV, Belyaev VM (1985) Experimental investigation of the hydrodynamic instability of the interface between media of different density in an acceleration field. Translated from Teplofzika Vysokikh Temperatur 23:535–541.

Hypervelocity Richtmyer-Meshkov instability

R. Samtaney[*], D. I. Meiron
Firestone Flight Sciences Laboratory
D. I. Pullin
Graduate Aeronautical Laboratories
California Institute of Technology
Pasadena, CA, 91125 USA

Abstract: The Richtmyer-Meshkov instability is investigated for shocks sufficiently strong enough to produce active dissociation chemistry. To model the interaction of the flow with chemistry typical of hypervelocity flows, Lighthill's ideal dissociating gas model is employed. While dissociation causes significant changes in density and temperature, the change in growth of the perturbations is small when compared to frozen flow.

Key words: Hypervelocity, Ideal dissociating gas, Growth rate

1. Introduction

Hitherto, most investigations of the Richtmyer-Meshkov (RM) (Richtmyer 1960, Meshkov 1969) instability have been confined to shocks of Mach number $M \leq 4$. In this paper, we present results of *hypervelocity* RM instability. The term "hypervelocity" is used in contrast with the term "hypersonic." Hypervelocity implies not only high Mach number ($M \gg 1$) but also large velocities and therefore high total enthalpy. Thus, in hypervelocity RM instability, the shocks are of strength sufficient to activate dissociation/recombination chemistry and related gas-chemistry interactions. This is the topic of the present paper.

2. Physical parameters, equations and chemistry model, and numerical method

2.1. Physical parameters

The physical domain is a two-dimensional rectangular shock tube of dimensions $[x_l, x_r] \times [0, \lambda/2] \equiv [-0.1, 4.9] \times [0, 0.5]$. A vertical interface separating two gases is perturbed sinusoidally with wavelength $\lambda \equiv 1$, and initial amplitude $A_0 = 0.1$ (see Fig. 1). We use a hydrogen-nitrogen interface, initially in thermal ($T_0 = T_b = 298$ K) and mechanical ($p_0 = p_b = 0.1$ atm) equilibrium.

[*]Also affiliated with the Graduate Aeronautical Laboratories.

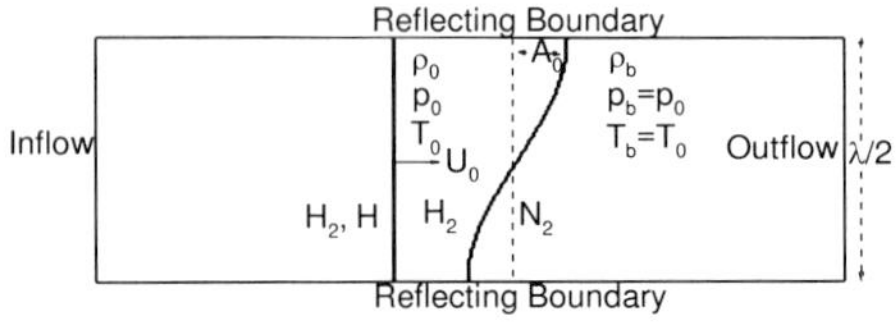

Figure 1. Schematic of the physical domain and parameters.

Although for chemically active flows the Mach number is an ill-defined quantity, for convenience we define a Mach number by $M \equiv U_0/c_0$, where U_0, c_0 are the shock speed and the frozen sound speed in the unshocked hydrogen. In fact, for $T_0 = T_b = 298$ K the frozen and equilibrium sound speeds are identical.

2.2. Equations and chemistry model

Consistent with previous studies of the RM instability, the governing equations of motion are the compressible Euler equations in two dimensions. In this paper we examine two limits: (a) *frozen* (Damkohler number $\Omega = 0$); and (b) *equilibrium* ($\Omega \to \infty$). For $\Omega \to \infty$, we assume that the equilibrium chemical reactions are are adequately modeled by the Lighthill Ideal Dissociating Gas (IDG) model (Lighthill 1957). The law of mass action for the IDG model is

$$\frac{\alpha_k^2}{1-\alpha_k} = \frac{\rho_{d,k}}{\rho_k} \exp\left(\frac{-\theta_{d,k}}{T}\right) \tag{1}$$

In Eq. (1), ρ_k is the density of the kth gas ($k = 1, 2$); α_k is the mass fraction of the monoatomic constituent of the kth gas; $\theta_{d,k}$ is the dissociation temperature of the kth gas, and $\rho_{d,k}$ is the characteristic density. The IDG properties for hydrogen and nitrogen are: $\rho_d = 1800, 130000$ kg/m^3 and $\theta_d = 51900, 113200$ K. Note that in the IDG model the gas is always in a state of vibrational excitation which implies an incorrect limit of specific heat ratio, $\gamma \to 4/3$ at low temperatures.

2.3. Numerical method

The numerical method employed here is a second-order accurate Equilibrium Flux Method (EFM) (Pullin 1980). For equilibrium chemistry computations, we employ operator splitting i.e., the hydrodynamical equations are solved and then the law of mass action (for $\Omega \to \infty$, Eq. 1) for each gas is solved in every computational cell. The physical domain is subdivided into a uniform 2000×200 grid. A message-passing parallel code was implemented on the 512 Intel Paragon at Caltech.

3. Simulation results

In Figs. 2 and 3 density and vorticity contours are shown at "early" and "late" times during the interaction. The incident shock refracts at the interface

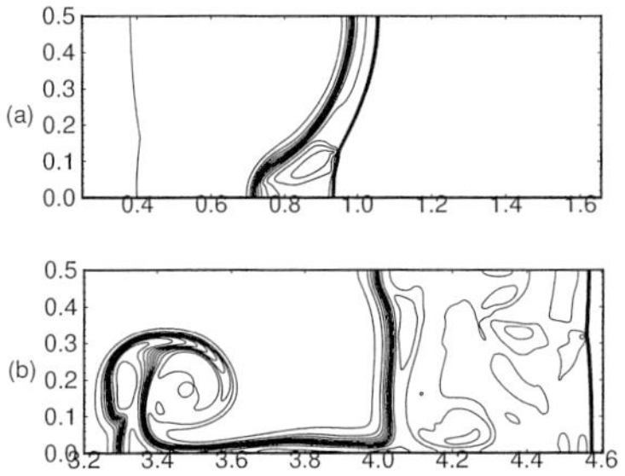

Figure 2. Density contours for an $M = 10$ shock interaction with a H_2-N_2 interface at $t = 1.38 \times 10^{-4}, 6.9 \times 10^{-4}$ for frozen (left column, $\rho = (0.05, 1.0)$) and equilibrium (right column, $\rho = (0.07, 1.8)$) chemistry.

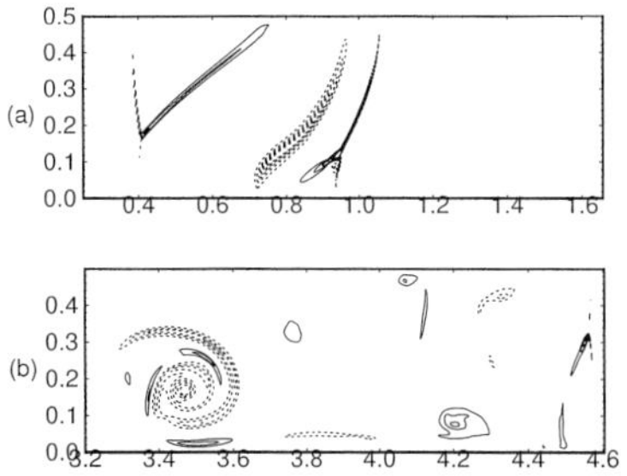

Figure 3. Vorticity contours (dashed negative), $\omega = (-10^5, 10^5)$ for frozen (left column) and equilibrium (right column) chemistry.

and the reflected and transmitted waves are shocks for this low-high acoustic impedance interface. The mean velocities of the reflected and the transmitted shocks, as well as the contact surface are larger for the equilibrium case than the frozen chemistry case. However, the transmitted and the reflected shocks lie closer to the interface for the equilibrium case. During the refraction process, negative vorticity is generated due to *baroclinicity*. As the curved reflected and transmitted shocks straighten, triple points form on the shock fronts due to the nonlinear interactions. Associated with these triple points are shear layers with positive vorticity. As time increases, these triple points traverse the reflected (or transmitted) shock fronts and further reflections from the top and bottom boundaries cause the appearance of shear layers (with alternating signs of vorticity). The contact rolls up and gives rise to the familiar mushroom-shaped morphology of the interface. The bubble rapidly assumes a nearly flat shape. This implies that the growth of the perturbation is mostly due to the relative velocity of the intense spike. The main effect of the endothermic dissociation reaction is to increase the gas density and lower the temperature. The ratio of the peak density (temperature) in the equilibrium chemistry to the frozen chemistry case is approximately 1.7 (0.4). The maximum dissociated fraction of hydrogen (nitrogen) is about 20% (36%). At any instant, the growth rate dA/dt is different for both cases, while the amplitude A of the perturbation shows a very small reduction due to dissociation (see Fig. 4). The growth rate given by the impulse model (Richtmyer 1960) over-predicts the growth rate significantly. In fact, no "linear" period of growth is observed for this case or even for the case with $A/\lambda = 0.01$ (not shown). The first slope change in Γ, Γ_i, (see

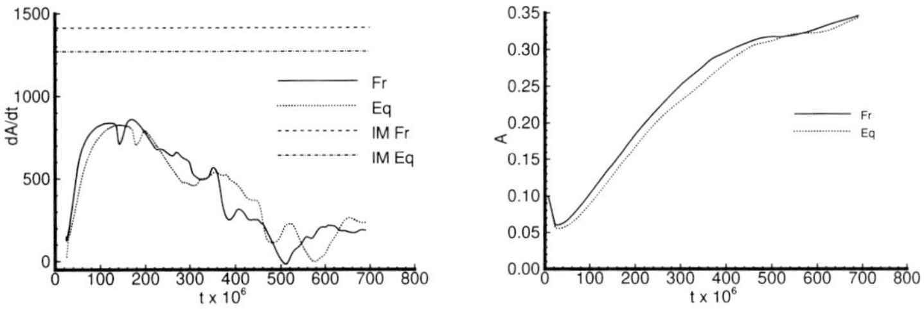

Figure 4. Growth rate (left) and amplitude (right) of the perturbation as a function of time. Eq: Equilibrium, Fr: Frozen chemistry, and IM: Impulse Model.

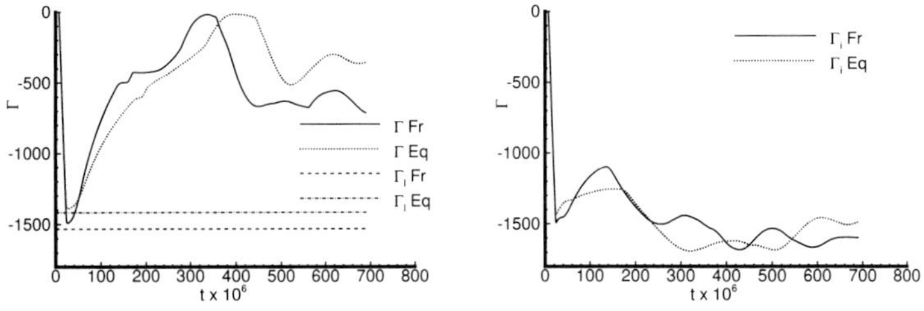

Figure 5. Total circulation (left) and interfacial circulation (right) as a function of time. Γ_l is circulation from a local analysis.

Fig. 5) occurs when the incident shock has completely traversed the interface. The circulation at this point may be determined by means of a local analysis (see Section 4). The circulation Γ shows strong variation with time due to the generation of the shear layers as explained above. The interfacial circulation changes in large part due to the strong compression waves interacting with the interface. At late times, the difference in Γ_i in the two limiting cases is small.

3.1. Mach-number scaling

The hydrodynamic equations are invariant under the following transformation:

$$t \to tM,\ p \to p/M^2,\ u \to u/M,\ E \to E/M^2 \tag{2}$$

In Fig. 6, the amplitude is plotted as a function of tM. While the frozen cases exhibit good scaling behavior, observe the monotonic decrease in A with increase in M for the equilibrium case.

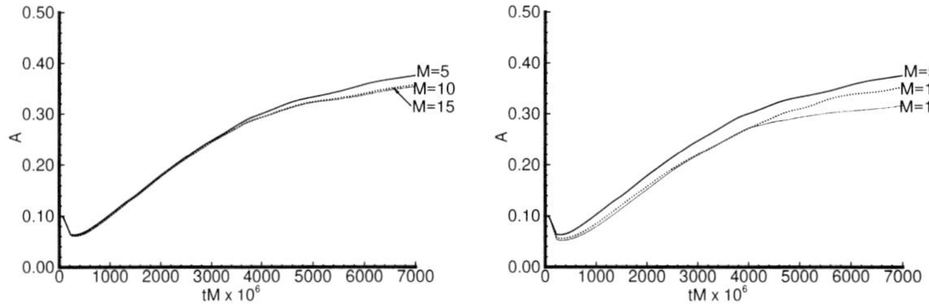

Figure 6. Perturbation amplitude as a function of scaled time, tM, for frozen (left) and equilibrium (right) chemistry.

4. Local analysis

Consider a sawtooth perturbation of a vertical fast-slow interface, where the angle β_0 between the incident shock and the interface is constant. Extending previous work (Samtaney and Zabusky 1994) for $\Omega = \infty$ we obtain the solution for small β_0 in a small neighborhood of the node where all the waves meet. For $\beta_0 = 30°$ and a $H_2{-}N_2$ interface, ρ and T behind the reflected and transmitted shocks, and σ, the vortex sheet density, are plotted in Fig. 7. The differences due to dissociation occur at about $M = 6$. Note that the difference in σ between equilibrium and frozen chemistry is small. For $M > 15$, all the diatomic species have dissociated completely and the resulting gas behaves as a monoatomic perfect gas. The initial baroclinic-circulation generation on a sinusoidal interface is obtained by integrating the σ over the length of the interface. This result was plotted as the horizontal dotted line in the circulation plot (Fig. 5).

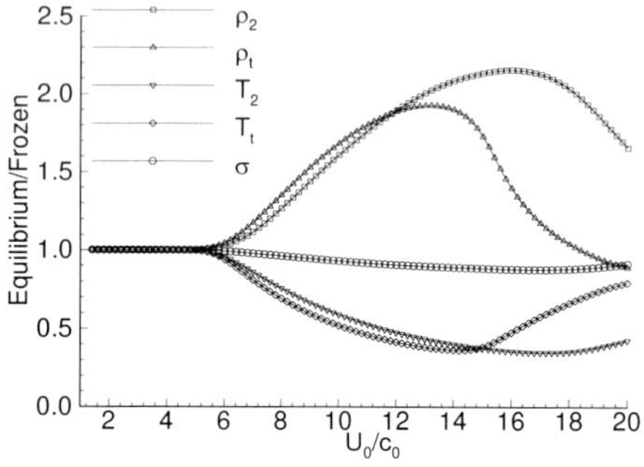

Figure 7. Local solution for equilibrium dissociated flow normalized by the frozen solution for a $\beta_0 = 30°$, H_2-N_2 interface. Subscript '2' ('t') is for values between the reflected (transmitted) shock and contact surface; σ is the vortex sheet strength.

5. Conclusion

In this paper, we examined the effects of dissociation in the interaction of hypervelocity shocks with sinusoidally perturbed density interfaces. As far as the perturbation amplitude is concerned, the chemistry plays a minor role. For frozen cases, Mach-number scaling works well for the entire duration of the simulation. For equilibrium chemistry this scaling is questionable only when dissociation is significant.

Acknowledgement. This work was funded by the Center for Advanced Fluid Dynamics Applications, Lawrence Livermore National Laboratory under grant No. W-7405-ENG-48 Subcontract B295121 and by AFOSR under grant F49620-93-1-0338. RS acknowledges the help of Berna Massingill who provided the mesh archetypes for the Intel Paragon and Thanh Phung who helped optimize the code.

References

Lighthill MJ (1957) Dynamics of a dissociating gas. Part 1. Equilibrium flow. J Fluid Mech 2:1.

Meshkov EE (1969) Instability of the interface of two gases accelerated by a shock wave. Fluid Dynamics 4:101.

Pullin DI (1980) Direct simulation methods for compressible ideal gas flow. J Comput Phys 34:231–244.

Richtmyer RD (1960) Taylor instability in shock acceleration of compressible fluids. Comm Pure and Appl Math XIII:297.

Samtaney R, Zabusky NJ (1994) Circulation deposition on shock-accelerated planar and curved density stratified interfaces: models and scaling laws. J Fluid Mech 269:45–78.

Numerical investigation of Richtmyer-Meshkov instability using front tracking

J. W. Grove, R. L. Holmes*
Department of Applied Mathematics and Statistics
University at Stony Brook, Stony Brook, NY 11794-3600 USA
D. H. Sharp
Complex Systems Group, Theoretical Division
Los Alamos National Laboratory, Los Alamos, NM 87545 USA

Abstract: Front-tracking simulations of Richtmyer-Meshkov instability produce significantly better agreement with experimentally measured growth rates than obtained in previous computations. Careful analysis of the early stages of the shock-acceleration process show that nonlinearity and compressibility play a critical role in the behavior of the shocked interface and are responsible for the deviations from the linear theories.

Key words: Front tracking, Richtmyer-Meshkov instability

1. Introduction

The quantitative understanding of Richtmyer-Meshkov (shock-accelerated) interface instability has challenged scientists for over 30 years. Interest in this instability is motivated by its importance in subjects such as inertial confinement fusion and supernovae dynamics.

The growth of perturbations on a shocked material interface was first studied in Richtmyer's seminal 1960 paper. Richtmyer found that after an initial rapid increase, the growth rate in the solution to a linearization of the Euler equations oscillated about a limiting value. He proposed a formula for this limit, known as the impulsive model, that agreed quite well with the computed limit of the linearized equations. Later shock-tube experiments (Meshkov 1970, Benjamin 1992, Benjamin et al. 1993) produced growth rates that were significantly smaller by as much as a factor of two than the predictions of the impulsive model. In sharp contrast however, many numerical simulations of experiments agreed quite well with the impulsive model (Benjamin et al. 1993, Cloutman & Wehner 1992, Meyer & Blewett 1972). Numerous articles on this instability can be found in the Proceedings of the International Workshops on the Physics of Compressible Turbulent Mixing (Besnard et al. 1991, Dannevik et al. 1992,

*Present address: Courant Institute of Mathematical Sciences, New York University, 251 Mercer St., New York, NY 10012.

Linden et al. 1993), and interesting exceptions are found in Mikaelian (1993) and Yang et al. (1994).

Recent front-tracking simulations of experiments of Benjamin et al. (1993) have substantially improved agreement between experiment, theory, and computation, producing growth rates that are in excellent agreement with those found in Benjamin's experiments. Our computations show that the nonlinear simulation and the linearized theory agree for small initial amplitude perturbations and sufficiently early times, but that there is an abrupt departure of the nonlinear theory from the linear theory at later times caused by a re-acceleration of the interface by shocks produced by nonlinear self-interactions of the reflected and transmitted waves.

2. Experimental configurations

Benjamin's experiments were conducted in a rectangular shock tube where air and sulfur hexafluoride (SF_6) were separated by a thin membrane with an imposed single-wavelength sinusoidal perturbation (Fig. 1 shows a schematic of the geometry). A Mach 1.2 plane shock ruptures the membrane producing a transmitted shock and either a reflected shock or rarefaction wave. The interface perturbation had an amplitude of 0.24 cm and a wavelength of 3.75 cm. The instability consists of the growth of perturbations on the material interface.

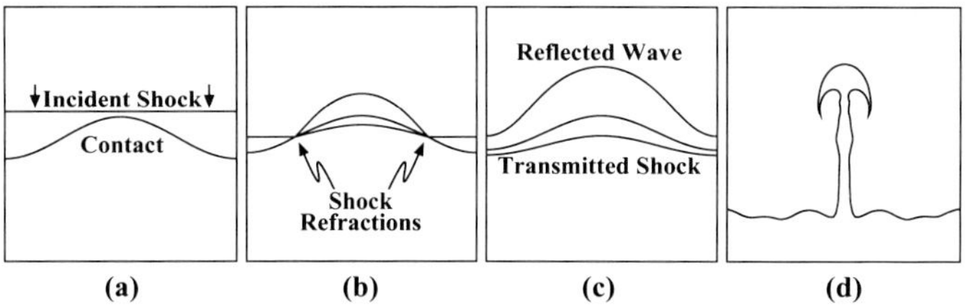

Figure 1. A schematic of the geometry in Richtmyer-Meshkov experiments.

The ambient pressure at the Los Alamos National Laboratory is 0.8 bar and the densities of air and SF_6 are 0.95 g/l and 4.85 g/l, respectively. The gases are assumed to be perfect with $\gamma_{air} = 1.4$ and $\gamma_{SF_6} = 1.09$. The rear wall of the tube was located sufficiently far from the interface that measurements of the growth rate could be made before the arrival of the shock reflected off this rear wall.

We compared the computed amplitude of the perturbations on the interface, $a(t)$, defined to be one-half the total inter-penetration width, and the amplitude growth rate, $\dot{a}(t)$, with Benjamin's measurements. We also measured the flow velocities at the tips of the heavy gas spike and the light gas bubble. Note that $\dot{a}(t) = (v_{spike} - v_{bubble})/2$, where v denotes the velocity at the interface.

3. Comparison of the linear and nonlinear theories

Richtmyer's impulsive model is based on an analogy of the shock acceleration process to an impulsive body force in an incompressible Rayleigh-Taylor instability. If the perturbation is sinusoidal with wave number k, Richtmyer's formula for the perturbation-amplitude growth rate is

$$\dot{a}(t) = kA\Delta u a(0+). \tag{1}$$

Here $A = (\rho_2^* - \rho_1^*)/(\rho_2^* + \rho_1^*)$ is the Atwood ratio using post-shock densities ρ_i^*, Δu is the change in translational velocity of the interface due to the action of the shock, and $a(0+)$ is the shock-compressed amplitude. Equation (1) is assumed hold for $ka(t) \ll 1$. Note that this formula explicitly assumes that compressibility is negligible after the initial shock refraction. Meyer & Blewett (1972) proposed a modification of Eq. 1 for reflected rarefactions in which $a(0+)$ is replaced by the average of the pre- and post-shock amplitudes.

For $ka \ll 1$ the Euler equations can be linearized about the solution of the Riemann problem for a planar shock refracting through a planar material interface. Note that this linearization makes no incompressibility assumption.

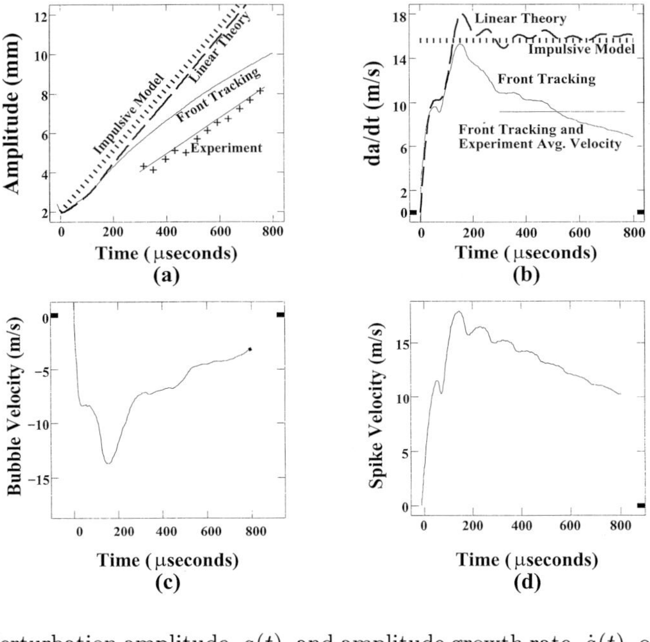

Figure 2. Perturbation amplitude, $a(t)$, and amplitude growth rate, $\dot{a}(t)$, of a shocked air-SF_6 interface.

Figures 2a and b show amplitude and amplitude growth rates from the impulsive model, numerical solution of the linear theory, nonlinear simulations, and experimental results. Also shown is a least-squares fit to the front-tracking amplitude data over the period of experimental observation. We focus on two regions of the growth-rate curves where the linear and nonlinear theories diverge. The first occurs at $t \approx 70$ μs where the growth rate $\dot{a}(t)$ reverses and then begins a second period of rapid increase. Subsequently, at $t \approx 200$ μs, $\dot{a}(t)$ saturates near its maximum value for the linear solution while the nonlinear solution experiences a marked deceleration.

The rapid deceleration of the shocked interface at 200 μs is induced by a series of re-accelerations of the material interface by secondary shocks generated from self-interactions at the reflected and transmitted wave edges. At 33 μs these waves are clearly seen near the edges of the reflected shock (Fig. 3a). As the interaction proceeds, the compression fronts steepen into shocks that collide near the tip of the spike at about 68 μs (Fig. 3b), producing a high-pressure pulse that causes the growth rate to decrease. This deceleration corresponds to the first blip in the growth-rate graph in Fig. 2b. At approximately 207 μs (Fig. 3c) a strong pulse again crosses the interface, leading to the severe (and permanent) deceleration of the bubble. That this decrease is due primarily to action at the bubble side of the interface is clear from Figures 2c and d, which show respectively the separate bubble and spike velocities. By time $t = 344$ μs the strong wave action has moved away from the interface region and the growth-rate curve has smoothed considerably.

Figures 3e and f show the pressure field of the solution to the linearized equations and of the solution to the nonlinear Euler equations at $t = 195$ μs, near the time of the most extreme divergence of the two solutions. The linearized equations of motion do not allow focusing of characteristics and constrain the geometry of the wave fronts to be sinusoidal. These two restrictions prevent the production of the cascading shocks that are present in the nonlinear solution. The linear solution displays a cascading series of acoustic waves, but these do not sharpen into shocks, which greatly reduces their effect on the growth rate of the interface.

4. Comparison to experiments

Comparison of simulations to experiments results in excellent agreement for the growth rates (see Figs. 2a and b, and Grove et al. 1993). However, there is a significant offset in the value of the absolute amplitude of the perturbations. It is believed that this offset is due to the delayed rupture of the membrane used in the experiments to separate the two gases. A detailed discussion of membrane effects is beyond the scope of this paper. We note, however, that a

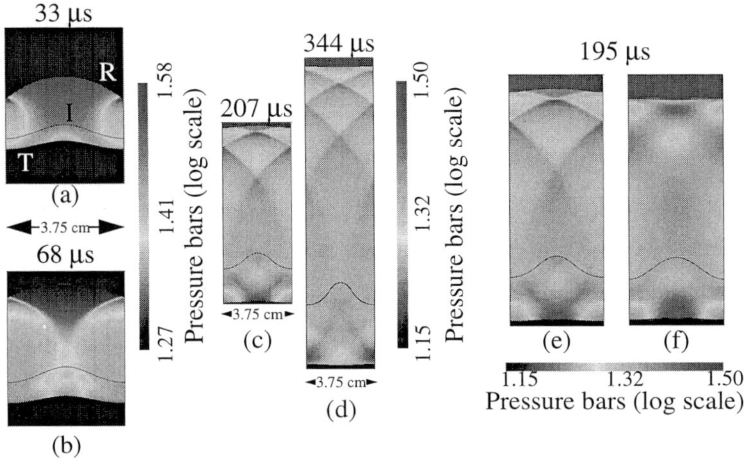

Figure 3. Grey scale plots of the pressure field for the nonlinear simulation and linearized solution of Richtmyer-Meshkov instability. Here T = transmitted shock, I = material interface, R = reflected shock.

computational approach to estimating membrane effects would be complicated. Analyzing the deformation and fragmentation of a thin elasto-plastic membrane would add severe demands to an already demanding computation. There is no established and benchmarked methodology for modeling the fragmentation of such a membrane. Even if one did come up with a model of membrane dynamics, it would be very difficult to know if it was correct, since such a model is likely to require several phenomenological parameters.

5. Conclusion

Simulations of Richtmyer-Meshkov instability show that compressibility and nonlinearity play an important role in the development of Richtmyer-Meshkov instability, up to and including the times measured in experiments. These wave interactions are not captured in the linearized theories commonly used to estimate perturbation growth rates.

Front tracking leads to improved agreement between experiment and simulation perturbation growth rates. It appears worthwhile to study the effect of mass diffusion and membrane effects on the early-time growth of the perturbations to aid in understanding the still unresolved difference in amplitudes.

Acknowledgement. We give our thanks to James Glimm for his encouragement and guidance. This work has been supported by the U.S. Army Research Office and the Mathematical Sciences Institute of Cornell University under subcontract to SUNY Stony Brook, ARO Contract number DAAL03-91-C-0027, the National Science Foundation, Grant number DMS-9057429, the National Science Foundation, Grant number DMS-9201581, the U.S. Army Research Office, Grant number DAAL03-92-G-0185, and the U.S. Department of Energy.

References

Benjamin R (1992) Experimental observations of shock stability and shock induced turbulence. In: Advances in Compressible Turbulent Mixing, WP Dannevik, AC Buckingham, CE Leith (eds), National Technical Information Service, U.S. Dept of Commerce, 5285 Port Royal Rd, Springfield VA 22161, pp 341–348.

Benjamin R, Besnard D, Haas JF (1993) Shock and Reshock of an Unstable Interface, LANL Report LA-UR 92-1185, Los Alamos National Laboratory.

Besnard D, Cavailler C, Dagens L, Figeac P, de Gliniasty M, Haas JF, Holstein PA, Montigny J, Parisot C, Rupert V, Sitt B, Wilke N (1991) Proceedings of Third International Workshop on the Physics of Compressible Turbulent Mixing at Royaumont, France, CEA DAM.

Cloutman LD, Wehner MF (1992) Numerical simulation of Richtmyer-Meshkov instabilities. Phys Fluids A 4:1821–1830.

Dannevik WP, Buckingham AC, Leith CE (1992) Advances in Compressible Turbulent Mixing, National Technical Information Service, U.S. Department of Commerce, 5285 Port Royal Rd, Springfield VA 22161.

Grove JW, Holmes R, Sharp DH, Yang Y, Zhang Q (1993) Quantitative theory of Richtmyer-Meshkov instability. Phys Rev Lett 71(21):3473–3476.

Linden PF, Youngs DL, Dalziel SB (1993) Proceedings of the 4th International Workshop on the Physics of Compressible Turbulent Mixing, Cambridge University Press, Cambridge, England.

Meshkov EE (1970) Instability of a Shock Wave Accelerated Interface Between Two Gases, NASA Tech Trans F-13:074.

Meyer KA, Blewett BJ (1972) Numerical investigation of the stability of a shock-accelerated interface between two fluids. Phys Fluids 15:753–759.

Mikaelian KO (1993) Growth rate of the Richtmyer-Meshkov instability at shocked interfaces. Phys Rev Lett 71(18):2903–2906.

Richtmyer RD (1960) Taylor instability in shock acceleration of compressible fluids. Comm Pure Appl Math 13:297–319.

Yang Y, Zhang Q, Sharp DH (1994) Small amplitude theory of Richtmyer-Meshkov instability. Phys Fluids 6(5):1856–1873.

Richtmyer-Meshkov mixing-zone deformation induced by shock-tube wall boundary layers

G. Jourdan, L. Houas
IUSTI (U.M.R. CNRS 139)
Université de Provence, Marseille, France

Abstract: An experimental investigation of the gaseous mixing zone that originates from Richtmyer-Meshkov instability has been undertaken in a square-cross-section shock tube. Mass concentration fields have been determinated within the mixing zone for the cases when the shock wave passes from a heavy gas to a light one, from one gas to another of close density, and from a light gas to a heavy one before and after being compressed by the reflected shock. The diagnostic method is based on simultaneous three-directional laser absorption of one of the two constituents of the mixing zone. It is shown that the mixing zone is strongly deformed by the wall boundary layer when the latter is turbulent, and becomes more homogeneous when the Atwood number tends to zero.

Key words: Richtmyer-Meshkov instability, Shock/interface interaction, Turbulent mixing zone, Shock-tube experiments, CO_2 laser absorption

1. Introduction

The instability of a shock-accelerated interface separating two fluids of different densities was first studied theoretically by Richtmyer (1960) and experimentally by Meshkov (1970). When the interface is subjected to a shock wave parallel to it, the random small-scale defects due to the making of the interface itself lead to the creation of a mixing zone between the two fluids. Therefore, the density gradients at the interface and the pressure gradient caused by the shock passage are not collinear anywhere and random small-scale vortices are produced.

Experiments in a shock-tube environment are performed in order to understand some of the mechanisms occurring in the implosion of inertial-confinement fusion targets as a consequence of Richtmyer-Meshkov instability. Many experiments for low Mach numbers and atmospheric initial pressure have been undertaken to study this phenomenon. But when the Mach number increases and the initial pressure decreases simultaneously, shock/interface interaction leads to turbulent three-dimensional flow which is difficult to describe. Recently, a multi-directional laser absorption technique has been developed in our laboratory and temperature and species concentration profiles within the mixing zone have been experimentally obtained within nine imaginary identical regions of the square-cross-section shock tube (in the four corners, near the four walls and in the center). Previously, we used several experiments to explore one abscissa with two detectors; now we use three detectors in the same run. The shock-tube wall boundary-layer deformation of the mixing zone has been clearly pointed out with the help of the 3D shape reconstruction of the mixing zone.

2. Experiments

Experiments are performed in a 8.5-cm by 8.5-cm square-cross-section double diaphragm shock tube of about 9 meters total length. The cases when the shock wave passes from the heavy gas to the light one (CO_2/He), from one gas to another of close density (CO_2/Ar), and from the light gas to the heavy one (CO_2/Kr) have been studied for the same initial conditions. The Mach number in the CO_2 is about 3.1 which corresponds to Mach number 2.2, 3, and 3.5 in helium, argon, and krypton respectively, and the initial pressure for the gases is 1500 Pa. CO_2 is used for its spectroscopic properties. The gases are initially separated by a thin membrane (1.5-μm thick mylar) resting over a square grid (of 5×5 wires, diameter of 0.2 mm, and spacing of 17 mm), which allows a regular rupture of the membrane. Figure 1 presents the sketch of the experimental set-up.

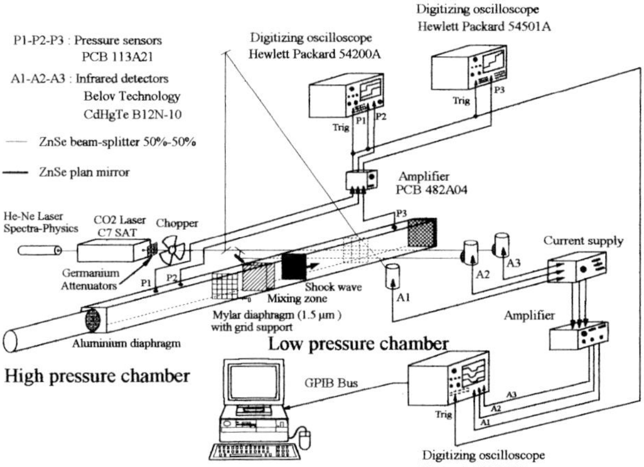

Figure 1. Experimental set-up.

The diagnostic method is a simultaneous three-directional laser absorption technique previously described by Fortes et al. (1994). This method allows the determination of the density and the temperature of one of the mixing constituents (CO_2), and therefore of the concentration. Measurements have been realized at four different abscissa before the return of the first reflected shock, but only the results of the last abscissa are presented in this paper. After compression, the quasi-taylorization of the CO_2/Kr mixing zone does not permit a complete measurement because of the infinite time of the absorption signals, and thus, only CO_2/He and CO_2/Ar results are presented. The accuracy of the concentration and density results is estimated to be about 10%. Concerning the 3D representations, the error bars are evaluated to be about $\pm$ 1 to 4 mm as a function of the errors on the velocity of the mixing zone.

3. Results and discussion

Figures 2, 3, and 4 show the concentration fields and the 3D reconstructions of the three tested mixing zones. Each iso-concentration surface corresponds to a plan cut of the mixing zone perpendicular to the direction of its propagation and is delimited by the shock-tube walls. L is the total thickness of the mixing zone. The simultaneous use of three detectors per run has permitted us to obtain a two-dimensional visualization of the mixing-zone shape because we can estimate the absolute delay in time between the three absorption signals.

These visualizations are given before (Figs. 2a, 3a, and 4) and after (Figs. 2b and 3b) the compression by the reflected shock wave. The first comment that we can make is that each 3D representation of the mixing zone is in good agreement with the corresponding concentration fields. Before the interaction with

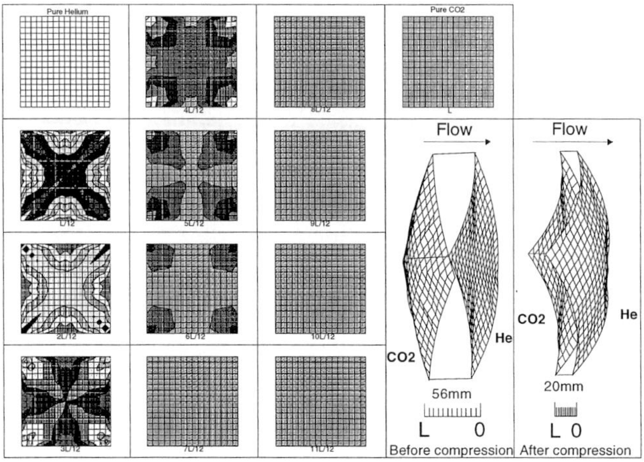

Figure 2. Iso-concentration fields of CO_2/He mixing zone (a) before compression; and 3D reconstruction of the mixing zone (b) before and (c) after compression.

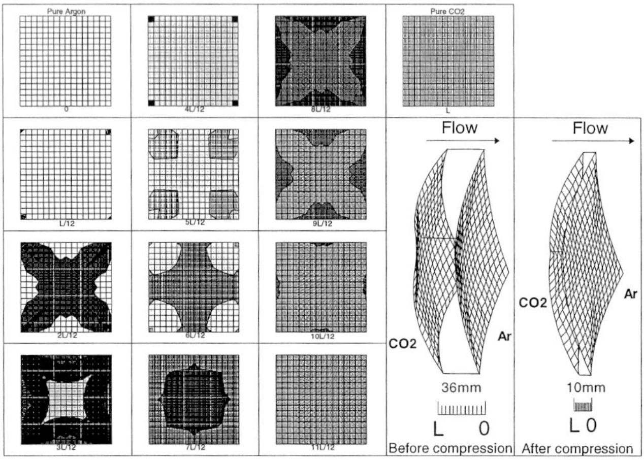

Figure 3. Iso-concentration fields of CO_2/Ar mixing zone (a) before compression; and 3D reconstruction of the mixing zone (b) before and (c) after compression.

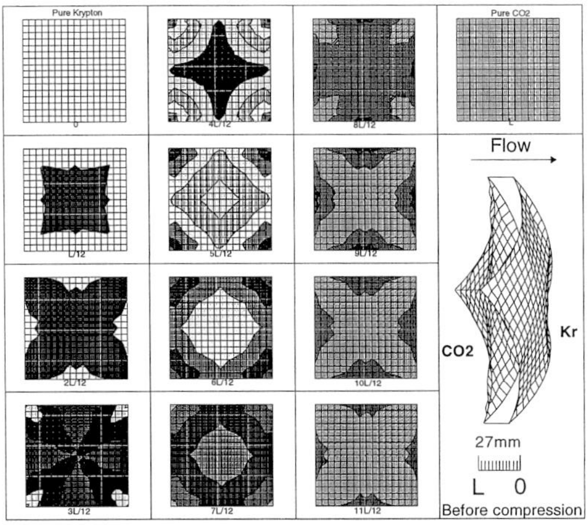

Figure 4. Iso-concentration fields of CO_2/Kr mixing zone (a) before compression; and (b) 3D reconstruction of the mixing zone before compression.

the reflected shock, the CO_2/He mixing zone seems to be the least deformed, since the mixing is present for the first slice $\frac{L}{12}$ in all shock-tube cross sections. Calculations, based on the work of Mirels (1984), have shown that the mixing zone propagates with a boundary layer for CO_2/He which is laminar, transitional for CO_2/Ar, and turbulent for CO_2/Kr. This can explain the different mixing-zone deformations.

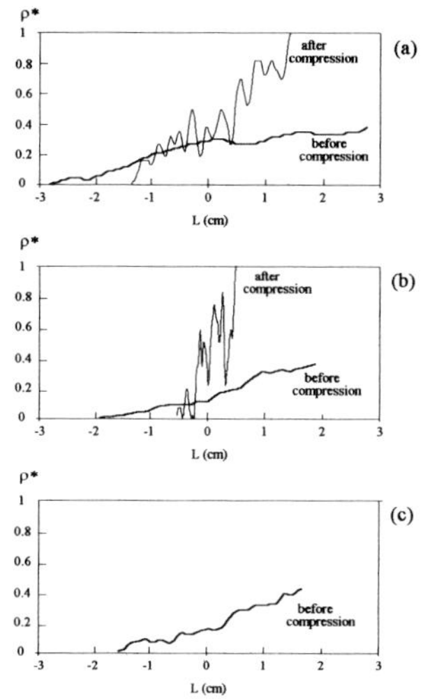

CO_2/He	A	ΔU	$\frac{\rho_{CO2}}{\rho_{He}}$	Time
After incident shock	-0.73	850 m/s	# 10	915 μs
After reflected shock	+0.79	-560 m/s	9.3	60 μs

CO_2/Ar	A	ΔU	$\frac{\rho_{CO2}}{\rho_{He}}$	Time
After incident shock	-0.31	640 m/s	# 1	1215 μs
After reflected shock	+0.5	-585 m/s	3	100 μs

CO_2/Kr	A	ΔU	$\frac{\rho_{CO2}}{\rho_{He}}$	Time
After incident shock	-0.05	535 m/s	# 0.5	1450 μs
After reflected shock	+0.22	-528 m/s	1.5	-

Figure 5. Adimensional density profiles for (a) CO_2/He; (b) CO_2/Ar; and (c) CO_2/Kr mixing zones. A is the Atwood number and ΔU is the mixing-zone velocity jump.

The iso-concentration fields have to be considered as qualitative results. However, some information can be gleaned from their observation. First, the CO_2/He mixing zone (heavy/light with a high Atwood number) is the least homogeneously mixed since the iso-concentration variations reach 70% per slice. The CO_2/Ar and CO_2/Kr mixing zones seem to be more homogeneous and the concentration variation does not exceed 30% per slice. On the other hand, one can say that when the density of the second constituent increases (He, Ar, or Kr), its relative penetration into the CO_2 also increases. As we can observe from Figs. 2a, 3a, and 4, approximately 50% of the rear part of the CO_2/He mixing zone is highly composed of CO_2 (90%). For the CO_2/Ar and CO_2/Kr cases, it is only about 20% of the rear part. Therefore, this is a confirmation that the homogeneity of the mixing increases when the Atwood number tends to zero and also that the mixing zone develops towards the CO_2 in our experimental configuration (for CO_2/He we obtain 60% of the rear part with more than 90% of CO_2 at 164 mm and 335 mm and 50% at 550 mm).

After compression (no result for the CO_2/Kr case) the mixing seems to become more homogeneous (iso-concentration fields are not shown owing to space limitations). The compression seems to be more intense for the CO_2/Ar case and this can be understood since we are nearer taylorization. (Note that each experimental 3D reconstruction of the mixing zone is obtained from a different run; the fact that the CO_2/He and CO_2/Ar mixing zones are bulged in the same direction, before and after the compression by the reflected shock, is coincidental.) This result is quite visible on Fig. 5, where we have plotted the adimensional CO_2 density before and after the compression by the reflected shock for the CO_2/He and CO_2/Ar cases (Figs. 5a and 5b) and only before the compression for CO_2/Kr (Fig. 5c). We also observe that the reflected shock passage has considerably increased the small-scale perturbations which were present within the mixing zone. The density ratio across the shock passage seems to be the same (2.5 for the two cases), but the thickness is decreased by a factor of 2 for CO_2/He and 4 for CO_2/Ar. The difference in the thickness compression can be attributed to the difference in homogeneity of the two mixing zones, or the proximity to the taylorization for the CO_2/Ar case, or both.

Acknowledgement. This study was supported by the C.E.A. Centre d'Etudes de Vaujours-Moronvilliers, on contract No. 304 843/00 D1.

References

Fortes J, Ramdani A, Houas L (1994) CO_2 laser absorption measurements of temperature and density in shock induced Richtmyer-Meshkov mixing zone. Physical Review E 50(4):3041–3049.

Meshkov EE (1970) Instability of a Shock Wave Accelerated Interface Between Two Gases, NASA TT F-13, 074.

Mirels H (1984) Turbulent boundary layer behind constant velocity shock including wall blowing effects. AIAA J 22:1042–1047.

Richtmyer RD (1960) Taylor instability in shock acceleration of compressible fluids. Comm Pure Appl Math 13:297–319.

Vortex methods for early and intermediate time Richtmyer-Meshkov environments

J. Ray, N. J. Zabusky
Department of Mechanical and Aerospace Engineering
P. O. Box 909, Rutgers University
Piscataway, NJ 08855-0909 USA

R. Samtaney
Applied Mathematics
California Institute of Technology
Pasadena, CA 91125 USA

Abstract: The *vortex paradigm* (Hawley and Zabusky 1989) has been used to develop reduced vortex models that predict the time rate of change of the amplitude of harmonics on planar interfaces. Our work represents a substantial improvement over the linearized studies of this environment, e.g. as first conducted by Richtmyer in a pioneering analysis.

The physical parameter domain of our models embraces $1.05 \leq M \leq 4.0$, $\eta = 1.54, 3.0, 5.04$, and $A/\lambda = 0.025, 0.05, 0.1$, where M is the Mach number of the incoming planar vertical shock, η is the density ratio across the interface, and A/λ is the amplitude-to-wavelength ratio of the perturbation to a planar vertical interface. Note, that if the shock moves from a low acoustic impedance to a higher one, $\eta > 1$ and vice versa.

Our models were validated against simulations of the compressible Euler equations.

Key words: Visiometrics, Reduced models, Growth rate

1. Introduction

A linear stability analysis associated with shocks striking a weak, long wavelength harmonic perturbation, ($x = A\cos ky, Ak \ll 1$), on an interface between two gases, was first done by Richtmyer (1960). His "impulse" model states

$$\dot{A} = kA|_{0+}\Delta U \frac{\eta - 1}{\eta + 1} \tag{1}$$

where k the wavenumber, η is the density ratio, $A|_{0+}$ is the amplitude of the harmonic, and ΔU is the velocity of the interface from a 1D interaction, after shock passage. Yang et al. (1994) elucidated Richtmyer's linear analysis.

The vortex paradigm is based on the evolution equation in inviscid flows:

$$\frac{D\omega}{Dt} = \frac{\nabla \rho \times \nabla p}{\rho^2} + \omega \cdot \nabla \vec{u} - \omega \nabla \cdot \vec{u} \tag{2}$$

As the shock traverses the interface, the misalignment of ∇p and $\nabla \rho$ deposits a vortex sheet on the interface which dominates the largely incompressible evolution of the interface, especially for weak shock and low-η cases.

2. Equations, numerics, and parameters

2.1. Numerical method

Consistent with previous numerical studies, we employ 2D compressible Euler equations. We track the interface using a level-set method. The numerical method is a second-order accurate Godunov scheme and is similar to the Eulerian MUSCL scheme.

The physical domain is a rectangular shock tube. The governing parameters are M, η, and Ak. The undisturbed incident gas is initialized with unit density and pressure; the shock is initialized as per the Rankine-Hugoniot conditions.

The test gases are air–CO_2 ($\eta = 1.54, \gamma_0 = 1.4, \gamma_b = 1.297$), air–R22 ($\eta = 3.0, \gamma_0 = 1.4, \gamma_b = 1.172$), and air-$SF_6$ ($\eta = 5.04, \gamma_0 = 1.4, \gamma_b = 1.0935$).

2.2. Convergence of interfacial circulation

We conduct a simulation with $M = 1.5, \eta = 3.0$, and $A/\lambda = 0.05$ on a 1600×80 grid with $\Delta x = \Delta y = 1$. The results are shown in Figs. 1–3. Figure 1a shows excellent convergence of interfacial circulation for the last two grids.

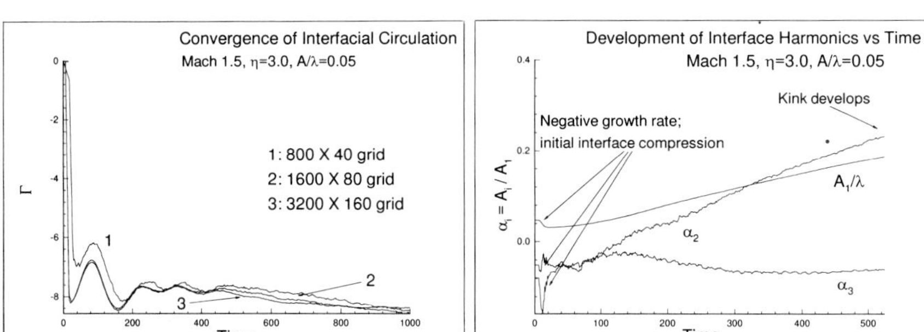

Figure 1. (a) Resolution study; (b) development of interface harmonics.

3. Vortex models

We propose models — analytical and numerical — for the different regimes. The numerical models require minimal inputs from numerical simulations. The growth rate is defined as half the difference between the velocities of the x-extrema of the interface.

3.1. Qualitative aspects

After shock passage, variations in Γ (Fig. 1a) are caused by secondary waves sweeping over the interface. Over time, higher harmonics of the perturbation develop (Fig. 1b) and the vortex sheet rolls-up into a patch (Fig. 2a). Nonlinearity and the rate of roll-up depend on $\Gamma(M, \eta, A/\lambda)$.

The evolution can conveniently be split into four regimes (Fig. 3a): (1)Very Early Time (VET), when the incident shock and dominant compression waves sweep over the interface; (2) Early Time (ET), when the evolution can be modeled by an evolving incompressible vortex sheet; (3) Intermediate Time (IT), when the contact discontinuity becomes a multivalued function of y; and (4) Late Time (LT), when the evolution is nearly incompressible and driven by a turbulent-like collection of vortices. The transition times depend specifically on the details of the run.

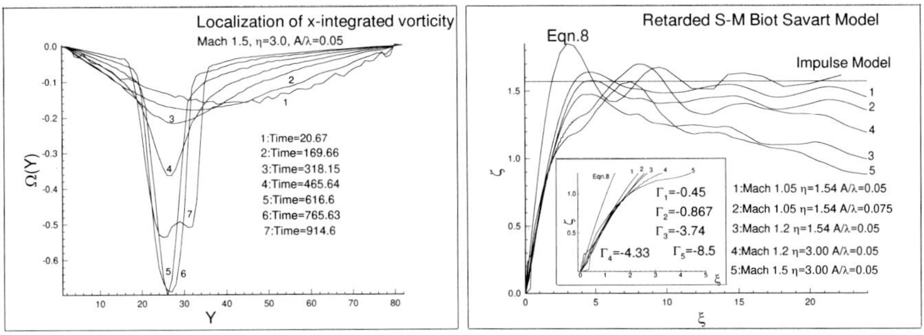

Figure 2. (a) Localization of vorticity; (b) very early time behavior.

The model for VET is corrected by a factor $2\sqrt{\eta}/(\eta + 1)$, which appears in the growth rate of a perturbation on a uniform vortex sheet located on a stratified interface (Chandrasekhar 1961).

3.2. Numerical Biot-Savart model (NBS)

We use this to validate the assumption of incompressibility. We integrate the instantaneous vorticity distribution over the instantaneous interface shape — both extracted from the numerical simulation — to obtain the interface growth rate.

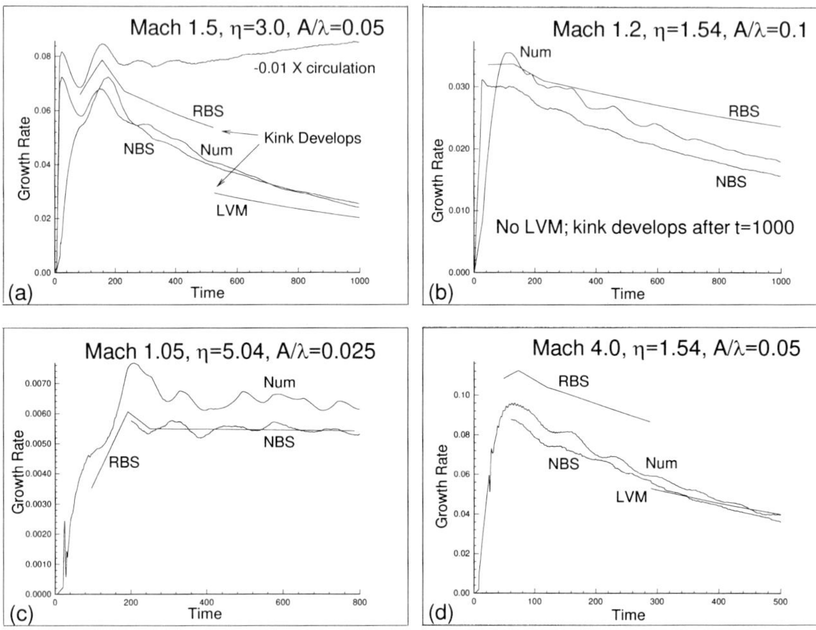

Figure 3. (a)–(d): comparision of models vis-a-vis numerical simulations.

3.3. Reduced Biot-Savart model (RBS)

This Biot-Savart-based model, applicable in ET, assumes incompressibility, a single-mode representation of the perturbation, and a frozen spectral content of the vorticity distribution. We get (Samtaney and Zabusky 1994a)

$$u(x,y,t) = -\frac{1}{4\pi}\int_{\mathcal{S}} \frac{(y-y')\Omega(y')dy'}{|\vec{x}-\vec{x'}|^2}; \quad \dot{A} = \frac{\Gamma(t)k}{4\pi}\int_{-\infty}^{\infty} \frac{\xi \sin(\xi) d\xi}{k^2 A(t)^2 (1-\cos\xi)^2 + \xi^2} \tag{3}$$

where $\Omega(y)$ is the x-integrated vorticity on the interface and $\xi = ky$.

To capture the initial growth rate, $\Gamma(t)$ is obtained by linearly interpolating between $\Gamma(t_1)$ and $\Gamma(t_2)$ for $t_1 \leq t \leq t_2$, $\Gamma(t_2)$ and $\Gamma(t_3)$ for $t_2 \leq t \leq t_3$ and $\Gamma(t) = \Gamma(t_3)$ for $t \geq t_3$, where $\Gamma(t_i)$ are the numerical (simulation) values of interfacial circulation at times of the first three local extrema after the shock has crossed the interface.

3.4. Retarded, Stratification-Modified Biot-Savart model (RSMBS)

We derive a model for VET from SMBS under the assumptions of $M \approx 1$ and $Ak \ll 1$; consequently $\Gamma(t) \approx const.$ We will approximate the circulation on the interface by the first term of the sine expansion (Samtaney and Zabusky 1994b). The effect of compressibility is introduced with a retarded time, as in aeroacoustics (Kambe 1986, Crow 1970). For $M \approx 1$, P/ρ is almost unchanged after shock passage and we take $c_s = \sqrt{\gamma}$. The growth rate is given by

$$\dot{A} = \frac{2\sqrt{\eta}}{\eta+1}\frac{\Gamma k}{4\pi}\int_{-\infty}^{\infty}\frac{(\xi'\sin(\xi'))H(t-r'/c_s)d\xi'}{k^2A(t)^2(1-\cos\xi')^2+(\xi')^2} \tag{4}$$

where $(kr')^2 = k^2A(t)^2(1-\cos\xi')^2+(\xi')^2$. If $Ak \ll 1$, then $r' = c_st \approx y'$ and the expression simplifies to

$$\dot{A} = \frac{2\sqrt{\eta}}{\eta+1}\frac{(\Gamma_1' kA|_{0-})\mathrm{Si}(\xi)}{\pi}, \quad \xi = kc_st = \sqrt{\gamma}kt \tag{5}$$

where $\mathrm{Si}(\xi)$ is the Sine Integral function.

Instead of Γ_1' we use its analytical scaling result (Samtaney and Zabusky 1994b) in Eq. (5) to get

$$\dot{A} = \frac{4\sqrt{\gamma}}{\gamma+1}\left(1+1/M+2/M^2\right)(M-1)\frac{\sqrt{\eta}-1}{\eta+1}\frac{kA|_{0-}}{\pi}\mathrm{Si}(\xi) \tag{6}$$

We plot $\zeta = \dot{A}/(\Gamma_{sc}A|_0k)$ from various numerical simulations and $\mathrm{Si}(\xi)$ versus ξ in Fig. 2b. The departure from the model with increasing interfacial circulation Γ is also evident from Fig. 2(b).

Physically, the oscillations in $\mathrm{Si}(\xi)$ are due to the arrival of signals from vortex images required to enforce the no-flow boundary condition, $\vec{u}\cdot\vec{n} = 0$, at $y = 0$ and $y = \lambda/2$. As $t \to \infty$, $\dot{A} \to \dot{A}_{Richtmyer}$.

3.5. Constant-Symmetric Line Vortex model (LVM)

This incompressible model applies at intermediate times and is analytic. We assume an array of line vortices of constant strength $(-1)^n\Gamma_L$ and location $y = (2n+1)\lambda/4$, $n = \ldots -2, -1, 0, 1, 2, \ldots$, where Γ_L is the first maximum in circulation and can be obtained analytically from Samtaney and Zabusky (1994b). This captures the solenoidal component of the velocity, the compressible component being small for weak shocks and at later times.

Summing the contribution from all the vortices,

$$\dot{A} = -\frac{\beta}{\cosh kA(t)} \quad \text{or} \quad \sinh kA(t) = -\beta kt + \sinh kA(t_s) \tag{7}$$

where $\beta = \Gamma_L k/2\pi$.

For late time and large amplitude ($Ak \gg 1$, $\tanh(Ak) \approx 1$) we get

$$\dot{A} = \frac{1}{kt + C} \qquad C = \frac{\sinh kA|_0}{-\Gamma_L k/2\pi}$$

which is similar in form to Hecht's model for v_{bubble} (Hecht et al. 1994).

4. Conclusion

Our vortex models provide a robust way of estimating the growth of a single-mode perturbation in Richtmyer-Meshkov environments. For $M \approx 1$ and $kA \ll 1$, there exists an analytical model for the growth rate.

Acknowledgement. This work was funded in part by a DOE grant, DE-FG02-93ER25179.A000, monitored by Dr. Daniel Hitchcock. The computations were performed on the CM5 at NCSA, University of Illinois, Urbana-Champaign.

References

Chandrasekhar S (1961) Hydrodynamic and Hydrodynamic Stability, Oxford University Press.

Crow, S (1970) Aerodynamic sound emmision as a singular perturbation problem. Stud Appld Math 49:21–44.

Hawley JF, Zabusky NJ (1989) Vortex paradigms for shock-accelerated density-stratified interfaces. Phys Rev Lett 63:241-1244.

Hecht J, Alon U, Mukamel D, Shvarts D (1994) Simple potential flow models of Rayleigh-Taylor and Richtmyer-Meshkov bubble-fronts. Phys Fluids 6:4019.

Kambe T (1986) Acoustic emissions by vortex motions. J Fluid Mech 173:643–666.

Richtmyer RD (1960) Taylor instability in shock acceleration of compressible fluids. Comm Pure and Appl Math XIII:297.

Samtaney R, Zabusky NJ (1994a) Vorticity Generation and Growth Rate in Richtmyer-Meshkov Instabilities, Caip Tech. Rept. No. 187, CAIP, Rutgers University, Piscataway, NJ 08855-1390 (submitted Phys Flds).

Samtaney R, Zabusky NJ (1994b) Circulation deposition on shock-accelerated planar and curved density stratified interfaces: models and scaling laws. J Fluid Mech 269:45.

Yang Y, Zhang Q, Sharp D (1994) Small amplitude theory of Richtmyer-Meshkov instability. Phys Fluids 6(5):1856.

Visualization of shocked mixing zones using differential interferometry and X-rays

G. Rodriguez, I. Galametz, M.-H. Thorembey, C. Rayer, J.-F. Haas
Commissariat à l'Energie Atomique / Vaujours-Moronvilliers
B.P. 7, F 77181 Courtry, France

Abstract: We compare the behavior of continuous and discontinuous heavy-light interfaces between two gases, sulphur hexafluoride (SF_6)-air or xenon (Xe)-air, subjected to the acceleration by an incident upward propagating shock wave (Mach 1.45 in SF_6 and 1.4 in Xe) and the deceleration by up to three weaker reflected shock waves in a vertical shock tube. Continuous interfaces, obtained after the retraction of a separating diaphragm, have a smooth density profile with an initial thickness of 2.4 and 2.6 cm for SF_6-air and Xe-air respectively. Discontinuous interfaces are initially created by a plastic membrane 0.5 μm thick torn by the passage of the incident shock wave. The images are obtained by differential interferometry and, in the Xe-air case, by X-rays. The visualization illustrates the different development of shock-induced instabilities for the two types of interfaces, and different modes of reflected shock propagation in a polyatomic gas (SF_6) and monatomic gas (Xe). From the X-ray method, profiles of Xe partial density are obtained in the planar region of the mixing zones for the two types of interfaces at all times. In contrast, the differential interferometry provides partial-density profiles for continuous interfaces only and as long as the turbulent scales on the walls do not perturb the signature from the flow in the center. The continuous interfaces do not thicken and never turbulize. The discontinuous interfaces lead to turbulent mixing zones which thicken between reflected shocks. The shapes of average partial-density profiles are similar for the two types of interfaces, for the two gas pairs and, in the continuous case, for the two diagnostics.

Key words: Richtmyer-Meshkov instability, Turbulent mixing, Density measurement, X-rays, Differential interferometry

1. Experimental set-up and diagnostics

The vertical shock tube has been designed for the measurement of the Richtmyer-Meshkov (RM) instability of continuous heavy-light and discontinuous gas interfaces subjected to weak shock waves (Cavailler et al. 1992). It has a square cross section (8 by 8 cm^2) and the distance between the initial interface position and the end plate has been set to 30 cm in the experiments described here. For the shadowgraph or schlieren visualization of the mixing zones arising from the RM instability we normally use a Cordin high-speed framing camera. Images are recorded at times ranging from 1.1 ms, shortly before the arrival of the first reflected shock waves, to 2.7 ms, after passage of the third reflected shock (Fig. 1). For higher quality visualization and measurement of the density averaged along the optical axis, we have adapted to the gas mixing problem (with strong density gradients) the differential interferometry method developed for compressible aerodynamics (Desse & Pegnaux 1993). Figure 2 schematizes the optical set-up for the white-light differential interferometer. As it provides pictures in which the colors correspond to optical path differences proportional to the gradient of optical index, this set-up can be considered a quantitative schlieren system. As we are interested in mostly horizontal shock waves and mixing zones, the two Wollaston biprisms have the same angle of 1 or 0.5° and are rotated so that the vertical density gradient is measured. The single color image recorded per shot is digitalized using a triple CCD camera (for the colors red, green, and blue). The colors are then automatically matched to the optical index and density gradient. By integration from a position of known density, we deduce the partial density and mass concentration in the mixture. The X-ray technique (Rodriguez et al. 1994), for which only Xe significantly absorbs the low-energy ($\leq$ 60 keV) X-ray photons, has been used previously to obtain the Xe partial density. As both techniques integrate the three-dimensional flow field on a two-dimensional image, wall effects are always present and perturb the density measurement at late times.

2. Visualization

The mixing zones arising from the shock interaction with discontinuous (using thin membranes) or continuous (diffusive zone) interfaces for the gas pairs (Xe-air or SF_6-air) have been visualized using the differential interferometer and the X-rays. As X-ray pictures were presented previously (Rodriguez et al. 1995), we show only a few interferometric images for the case SF_6-air. Unfortunately they appear here in black and white.

Figure 3a shows the SF_6-air continuous interface recorded at time denoted 1 in the x-t diagram (Fig. 1). The discontinuous interface at time 1 is shown on Fig. 3b. One sees from top to bottom the downward-propagating reflected shock wave in air and the upward-moving mixing zone with SF_6 on the bottom. The horizontal layers in Fig. 3a indicate that the diffusion zone is still laminar and the grainy pattern below it is the signature of fine-scale mixing in the boundary layer. The contact surface seen on Fig. 3b displays a dark line which is the signature of the ruptured membrane. The turbulent mixing seen a little above and much further below probably lies along the walls because the membrane

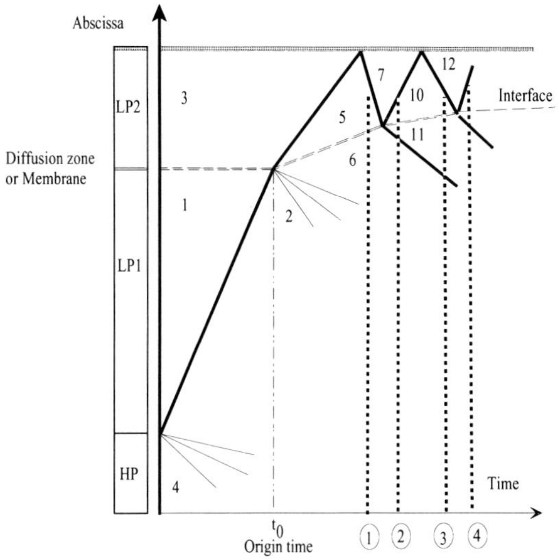

Figure 1. An x-t diagram of the experiment.

has ruptured there at shock passage. Far from the walls the membrane may be still intact and thus prevents gas mixing at this early stage. Figures 4a and 4b recorded at time 2 show the bifurcation of the reflected shock as it refracts into the polyatomic SF_6. A *bubble* can be seen on the walls between the shock and mixing zone which is locally deformed by the interaction. The diffusion zone has been compressed by the reflected shock and the discontinuous interface has been turbulized. Later, the wall bubble seems to turbulize even in the case of the continuous interface. The interferometer photo shown on Fig. 4c was recorded after passage of the second reflected shock on the continuous interface (time 4). The shock appears locally deformed by the wall structures of the mixing zone. It is in fact more disturbed in the discontinuous case (not shown) because the mixing zone is turbulent everywhere. Below this second reflected shock, the turbulent wall layer due to the first reflected shock can be seen. The continuous mixing zone is still laminar but the colored layers tend to be erased by the signature of the small scales on the wall. The measurement of the density profile becomes difficult at this stage.

3. Density profiles

Because the random optical signature of the small scales of turbulent mixing on the walls complicates the data-reduction process, the differential-interferometry method can be used only to obtain concentration and partial-density profiles

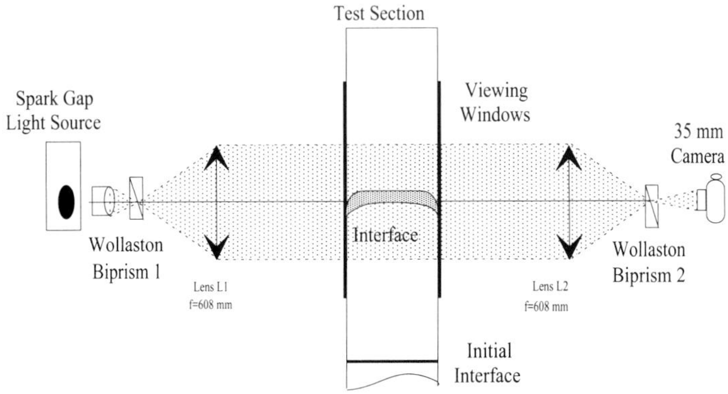

Figure 2. Optical set-up for the differential interferometer.

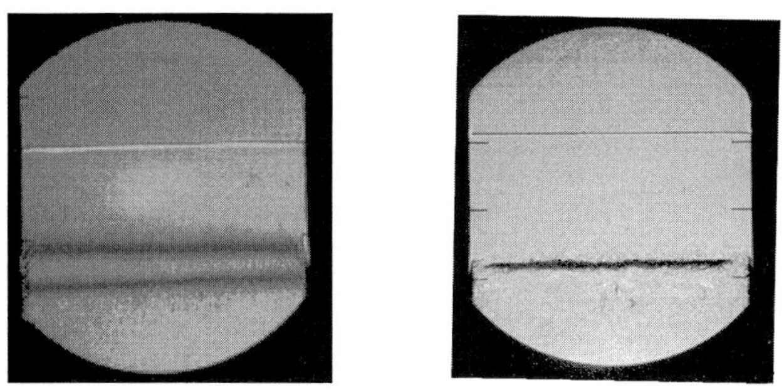

(a) (b)

Figure 3. SF_6-air interface time 1: (a) continuous; (b) discontinuous.

for the continuous mixing zone until time 3 (just before the interaction with the second reflected shock). An additional difficulty in the case of discontinuous interfaces is the pertubating signature of the membrane at early time or its fragments at late time.

Figure 5 shows the partial-density profiles of SF_6 in the SF_6-air mixing zone at times 1, 2, and 3. The profiles are typical of the molecular diffusion process for which the concentrations follow the error function law. Between times 1 and 2, the profiles are steepened due to the hydrodynamic compression by the first reflected shock. There is no widening, however, between times 2 and 3 indicating that no rapid diffusion has happened in this time interval. This is consistent with the absence of turbulent scales in the core of the mixing zone. A comparison of the two diagnostics is possible for the continuous Xe-air mixing zone. The partial densities of Xe (not shown) at times 2 and 3

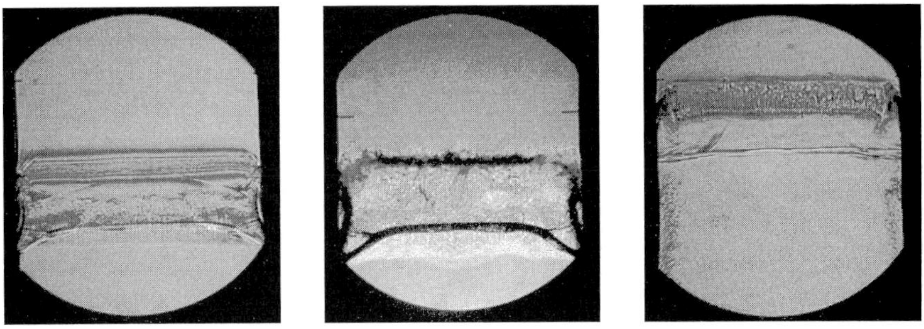

(a) (b) (c)

Figure 4. SF_6-air interface: (a) continuous time 2; (b) discontinuous time 2; (c) continuous time 4.

(between the passage of the first and second reflected shock wave) are similar for the two methods. The X-ray method could be used for the discontinuous Xe-air interfaces, because the membrane (made of low atomic-number elements) is transparent to the X-ray photons and because the absorption process integrates the random scales along the optical axis to yield an optical density on the radiographic film. Figure 6 shows the partial Xe density profiles at times 1, 2, 3, and 4. The profiles widen between times 2 and 3, in accordance with a process of turbulent diffusion.

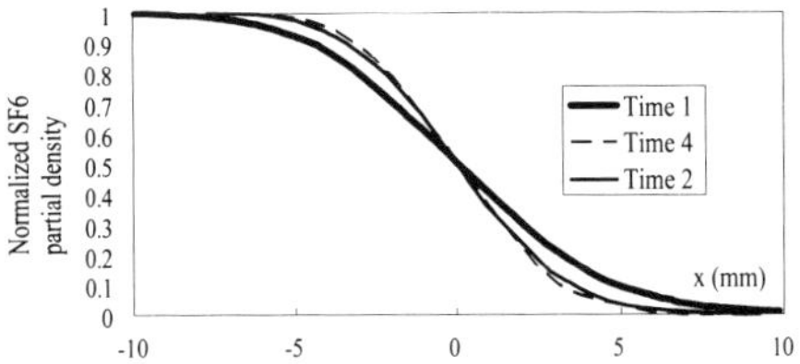

Figure 5. Density profiles for SF6-air continuous interface.

4. Conclusions

With differential interferometry, high-quality visualization of the mixing zones subjected to shock interaction can be obtained, as with conventional schlieren techniques. Our method has provided density measurements for mixing zones which remain essentially controlled by molecular diffusion and as long as turbulent scales on the walls do not perturb the optical signature, i.e. for continuous interfaces until the arrival of the second reflected shock. The results are similar to the radiography and applicable to any gas pairs. The partial-density pro-

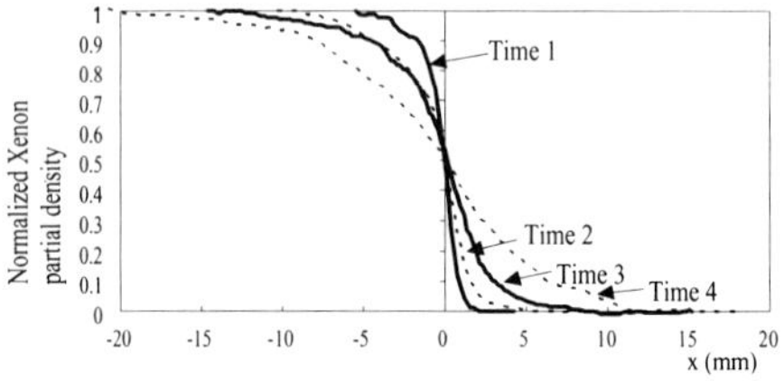

Figure 6. Density profiles for Xe-air discontinuous interface.

files can be used to determine a mixing-zone thickness based on quantitative criteria. In the present experimental conditions with a slow blade retraction for the continuous interface, the mixing zone never turbulizes and its thickness does not increase. The mixing zones created by the shock interaction with a discontinuous interface thicken due to turbulent diffusion and become wider than the continuous interface after the second reshock. There is no significant shape difference between the twice-averaged density profiles of planar diffusive or turbulent mixing zones.

Acknowledgement. We thank G. Delouis for his precious help in the laboratory.

References

Cavailler C, Croso H, Haas JF, Rodriguez G (1992) A new vertical shock shock tube for Rayleigh-Taylor instability studies. In: Proceedings of the 3rd International Workshop on the Physics of Compressible Turbulent Mixing, Royaumont, Wilke N (ed), CEA/DAM, p 27.

Desse JM, Pegneaux JC (1993) Direct measurement of the density field using high speed differential interferometry. Exp in Fluids 15:452.

Rodriguez G, Galametz I, Croso H, Haas JF (1994) Richtmyer-Meshkov instability in a vertical shock tube. In: Proceedings of the 4th International Workshop on the Physics of Compressible Turbulent Mixing, Linden PF, Youngs DL, Dalziel SB (eds), Cambridge University Press, Cambridge, England, p 260.

Rodriguez G, Galametz I, Croso H, Haas JF (1995) Richtmyer-Meshkov instability in a vertical shock tube. In: Shock Waves @ Marseille, Brun R, Dumitrescu L (eds), Proceedings of the 19th International Symposium on Shock Waves, Springer Verlag, Heidelberg, Vol 4, p 275.

Interaction of a turbulent shear flow with a Mach 1.5 normal shock wave

K. Mahesh, S. K. Lele, P. Moin
Department of Mechanical Engineering
Building 500, Stanford University
Stanford, CA 94305 USA

Abstract: Results from direct numerical simulation of the interaction of a Mach 1.5 normal shock with a turbulent shear flow are presented. Interaction with the shock is seen to increase the level of turbulent kinetic energy and thermodynamic fluctuations in the shear flow. The Reynolds shear stress, $\overline{uv}$ is observed to decrease across the shock wave. The observed trends are in agreement with linear analysis in which the influence of upstream anisotropy and entropy fluctuations on the evolution of the turbulence was considered.

Key words: Shock/turbulence interaction, Direct numerical simulation, Linear analysis, Turbulent kinetic energy, Reynolds shear stress, Thermodynamic fluctuations

1. Introduction

Interaction with a shock wave is known (Fernholz and Finley 1981) to alter significantly the mean flow and turbulent fluctuations in a turbulent boundary layer. In recognition of the complex nature of shock wave/boundary layer interaction, recent experiments and computations have studied simpler problems such as the interaction of isotropic turbulence and free shear layers with shock waves. This paper presents results from direct numerical simulation of one such problem: the interaction of a normal shock with an idealized shear flow.

A schematic diagram of the computed flow is shown in Fig. 1. Note that the shock wave is stationary in the mean. The mean velocity at the inflow of the computational domain varies linearly across streamlines while the mean pressure is uniform. The mean density and temperature vary such the mean Mach number is uniform, i.e.

$$\overline{U}_1(y) = U_0 + S(y - y_{\min}), \qquad \overline{P}_1(y) = \frac{1}{\gamma}, \qquad \overline{T}_1(y) = \frac{\overline{U}_1^2(y)}{(\gamma - 1)M_1^2}. \tag{1}$$

The results presented in this paper correspond to a Mach 1.5 shock wave, i.e., $M_1 = 1.5$.

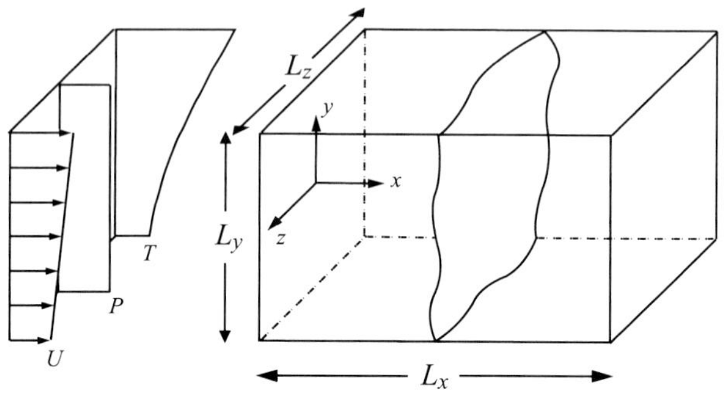

Figure 1. Schematic diagram of the computational setup.

2. Numerical method

The governing equations are the three-dimensional compressible Navier-Stokes equations in conservative form. The computational mesh is uniform in the directions transverse to the shock wave. A non-uniform mesh is used in the streamwise direction to cluster points in the vicinity of the shock. A combination of the sixth-order Padé scheme (Lele 1992) and the sixth-order ENO scheme (Shu and Osher 1989) is used to compute spatial derivatives. In order to reduce the computational cost, the shock-capturing scheme (ENO) is only applied in the shock-normal direction in the vicinity of the shock wave. Time advancement is performed by the third-order Runge Kutta scheme (Wray 1986). Periodic boundary conditions are imposed at the z boundaries while approximately non-reflecting boundary conditions are specified at the y boundaries. Non-reflecting boundary conditions (Poinsot and Lele 1990) are specified at the outflow boundary while turbulent fluctuations are superposed onto the mean flow of Eq. (1) at the inflow boundary.

These turbulent fluctuations are obtained from a separate temporal simulation. The mean flow in the temporal simulation is given by Eq. (1). The initial turbulent field is a superposition of a solenoidal velocity field and entropic thermodynamic field. The initial velocity field is isotropic with $E(k) \sim k^4 \exp(-2k^2/k_0^2)$. The initial pressure fluctuations are set to zero while the initial density fluctuations are set proportional to $-v$; i.e., the initial correlation coefficient between ρ and v equals -1. This ensures that the correlation between u and T will be negative (as in a boundary layer) for a transient period of time. The initial field is advanced in time until the velocity field attains the anisotropy that is typical of a turbulent shear flow. An instantaneous field is then taken and used to generate inflow turbulence in the spatial computation.

3. Results

Table 1 lists the relevant non-dimensional parameters that characterize the upstream turbulence. All quantities are measured at the centerline immediately upstream of the shock. The turbulence upstream of the shock has a low Reynolds number ($R_\lambda = u_{\rm rms}\lambda_1/\nu$), very low fluctuation Mach number ($M_t = \sqrt{\overline{u_i u_i}}/\overline{c}$), and anisotropic Reynolds stresses. The intensity of pressure fluctuations is an order of magnitude lower than those of the density and temperature. Also, the correlation coefficient between the density and temperature fluctuations is -0.99. The upstream turbulence therefore is essentially composed of an anisotropic velocity field along with entropy fluctuations of comparable intensity. The entropy fluctuations are correlated with the velocity field such that the correlation coefficient between u and T is negative.

Table 1. Relevant non-dimensional parameters of the upstream turbulence.

R_λ	M_t	$\overline{u^2}/q^2$	$\overline{v^2}/q^2$	$\overline{uv}/q^2$	$q_{\rm rms}/U_1$	$\rho_{\rm rms}/\overline{\rho_1}$	$T_{\rm rms}/\overline{T_1}$	$p_{\rm rms}/\overline{p_1}$
29.0	0.03	0.53	0.23	-0.22	0.025	0.03	0.03	0.002

Statistics reveal the jump in the mean flow across the shock wave to be very close to that predicted by the Rankine-Hugoniot relations for a laminar shock. For example, the jumps in mean velocity, density, and temperature at the centerline are 0.54, 1.86, and 1.31 respectively, while the corresponding values for a laminar shock ($\gamma = 1.4$) are 0.54, 1.86, and 1.32.

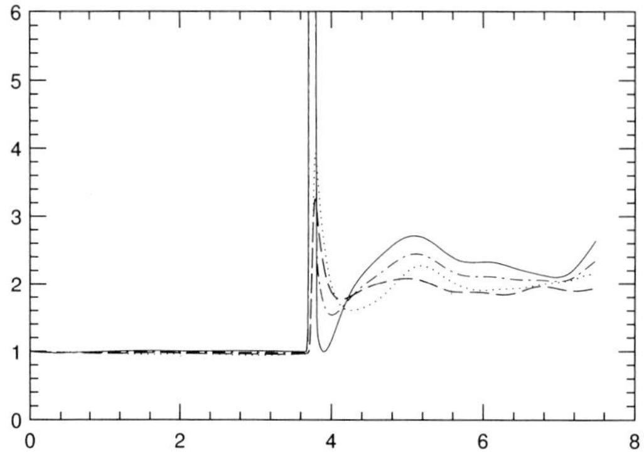

Figure 2. Variation of kinetic energy along the center streamline. All curves are normalized by their values at the inflow; —— ($\overline{u^2}$), ---- ($\overline{v^2}$), ········ ($\overline{w^2}$), —-— (q^2).

3.1. Turbulent kinetic energy

Interaction with the shock wave significantly enhances the kinetic energy of the turbulent fluctuations. The evolution of kinetic energy along the centerline is shown in Fig. 2. Note that λ_1 denotes the streamwise Taylor microscale at the inflow. The high levels in the vicinity of the shock wave are due to intermittency caused by motion of the shock front. Rapid increase in kinetic energy is observed immediately downstream of the shock wave. Similar behavior is observed in the interaction of a shock with isotropic turbulence (Lee et al. 1994). Linear analysis of the isotropic turbulence problem (Mahesh et al. 1995a) reveals rapid conversion of potential energy $(\overline{p^2})$ to turbulent kinetic energy immediately behind the shock, leading to the rapid rise in kinetic energy.

Using downstream values outside the region of rapid variation, $\overline{u^2}, \overline{v^2}$, and $\overline{w^2}$ are amplified by 2.3, 1.9, and 2 respectively, across the shock. These levels of amplification in the shear flow are noticeably higher than that of isotropic turbulence. Linear analysis (Ribner 1954) predicts $\overline{u^2}, \overline{v^2}$, and $\overline{w^2}$ to be amplified by 1.6, 1.38, and 1.38 across a Mach 1.5 shock if the upstream turbulence is isotropic. The higher levels of amplification observed in the shear flow are attributed to two factors: upstream anisotropy and upstream entropy fluctuations with negative $\overline{uT}$.

Incompressible Rapid Distortion Theory was used by Mahesh et al. (1994) to study the influence of upstream anisotropy. The analysis predicted higher levels of kinetic energy amplification across a shock wave in a shear flow than in isotropic turbulence. The relevant parameters responsible for this trend were identified as upstream $E_{11}(k_1)$ and $\overline{u^2}/q^2$ (note that x was assumed to be the shock-normal direction).

Table 1 reveals the presence of significant level of upstream entropy fluctuations with negative correlation between u and T. Linear analysis of the influence of upstream entropy fluctuations (Mahesh et al. 1995b) shows that this would further enhance the amplification of kinetic energy. A preliminary computation of a Mach 1.2 shock wave exhibited a similar strong influence of entropy fluctuations. The sign of $\overline{uT}$ in the Mach 1.2 simulation varied across streamlines upstream of the shock. Kinetic energy on the streamlines with negative upstream $\overline{uT}$ was seen to be amplified by a factor of about 2 while no noticeable change in kinetic energy was observed on the streamlines where $\overline{uT}$ was positive.

3.2. Reynolds shear stress

While kinetic energy is amplified, the Reynolds shear stress, $\overline{uv}$ is seen to *decrease* in magnitude across the shock. The streamwise variation of $\overline{uv}$ and $\overline{uv}/q^2$ on the centerline is shown in Fig. 3. The decrease in magnitude of shear stress

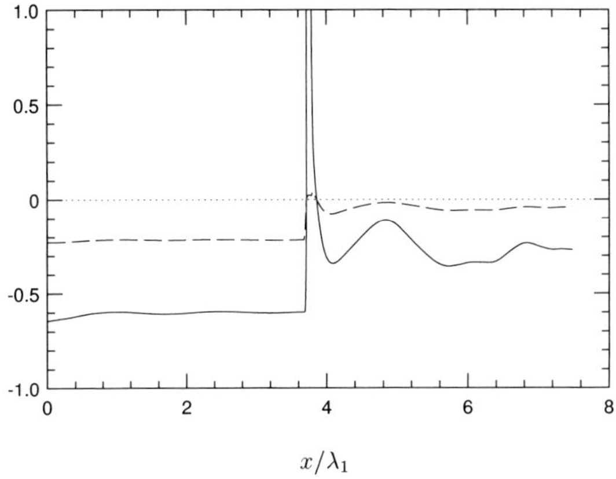

Figure 3. Evolution of shear stress along the center streamline; $\overline{uv}$ is normalized by the value of $u_{\rm rms}v_{\rm rms}$ at the inflow. —— ($\overline{uv}$), ---- ($\overline{uv}/q^2$).

is apparent. This observation is in agreement with the predictions of Mahesh et al. (1994) using Rapid Distortion Theory. Both normal and oblique shocks were considered in this study, which predicted a decrease in shear stress across normal shocks and a slight increase across oblique shocks at sufficiently large angles of obliquity.

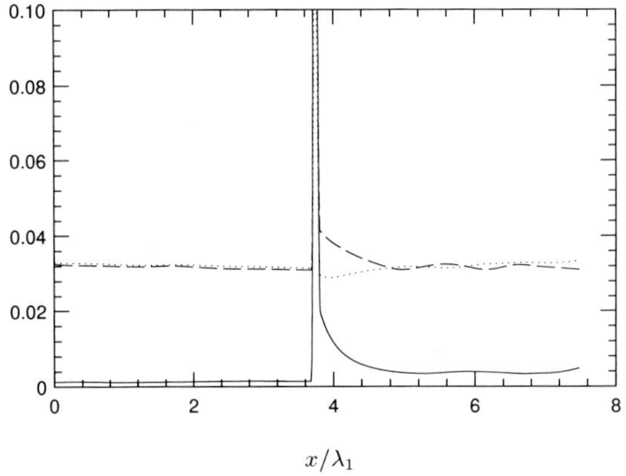

Figure 4. Variation of thermodynamic fluctuations along the centerline; —— ($p_{\rm rms}/\gamma\overline{P}$), ---- ($\rho_{\rm rms}/\overline{\rho}$), ········ ($T_{\rm rms}/\overline{T}$).

3.3. Thermodynamic fluctuations

Thermodynamic fluctuations are seen to be amplified across the shock wave. Figure 4 shows the streamwise evolution of the *intensities* of density, pressure, and temperature. Note that pressure fluctuations are amplified across the shock wave and rapidly decay downstream. This decay is similar to that observed in the interaction of isotropic turbulence with a shock and is inviscid in character. The intensity of pressure fluctuations in the near-field is comparable to those of density and temperature while far-field levels are much smaller. Furthermore, the correlation coefficient between density and temperature in the far-field is nearly -1, i.e., while acoustic fluctuations are significant immediately behind the shock wave, entropy fluctuations dominate in the far-field.

Acknowledgement. This work was supported by the Air Force Office of Scientific Research under Contract F49620-92-J-0128 with Dr. Leonidas Sakell as technical monitor. Computer resources were provided by NASA Ames Research Center.

References

Fernholz FF, Finley PJ (1981) A Further Compilation of Compressible Boundary Layer Data with a Survey of Turbulence Data, AGARDograph 263.

Lee S, Lele SK, Moin P (1994) Interaction of Isotropic Turbulence with a Strong Shock Wave, AIAA Paper 94:0311.

Lele SK (1992) Compact finite difference schemes with spectral-like resolution. J Comput Phys 103:16.

Mahesh K, Lele SK, Moin P (1994) The response of anisotropic turbulence to rapid homogeneous one-dimensional compression. Phys Fluids 6:1052.

Mahesh K, Lee S, Lele SK, Moin P (1995a) The interaction of an isotropic field of acoustic waves with a shock wave. J Fluid Mech 300:383–407.

Mahesh K, Lele SK, Moin P (1995b) The Interaction of a Shock Wave with a Turbulent Shear Flow, report in preparation, Thermosciences Division, Mechanical Engineering Department, Stanford University.

Poinsot TJ, Lele SK (1990) Boundary conditions for direct simulation of compressible viscous reacting flow. J Comput Phys 101:104.

Ribner HS (1954) Shock-turbulence Interaction and the Generation of Noise, NACA TN 3255.

Shu CW, Osher S (1989) Efficient implementation of essentially non-oscillatory shock-capturing schemes, II. J Comput Phys 83:32.

Wray AA (1986) Very Low Storage Time-advancement Schemes, Internal Report, NASA Ames Research Center, Moffett Field, CA.

Waves, General

Converging shock waves: modeling and analysis of structures

H. Wallus, F. Demmig
Universität Hannover
Institut für Plasmaphysik
Callinstraße 38 D-30167 Hannover, Germany

Abstract: We developed a model for the description of converging shock waves, especially for studying the process of fragmentation that occurs during Mach reflection. Nonequilibrium processes like vibrational, dissociative, and ionizational relaxation have been taken into account. In order to give a detailed physical interpretation of specific experimental results, a variety of flow-visualization techniques have been utilizied. A number of numerical methods have been applied to compute converging shock waves.

Key words: Converging shock waves, Time-resolved visualization, Model computation, Pattern recognition, Fragmentation wave system

1. Snapshot model of converging shock waves

Two features due to experimental conditions affect investigations of converging cylindrical shock waves: the shock contour is not perfectly symmetrical and the focus of convergence is not exactly on the symmetry axis of the experimental set up. This can be seen most clearly from time-resolved experiments (Kleine 1986). Our two-dimensional model computations aim to describe shock-wave propagation and reflection, with special attention to the process of fragmentation which occurs as a consequence of the initial shock-front deformation, i.e., its deviation from an ideal circular form. In order to give a detailed interpretation of specific time-resolved experimental results we obtained the initial shock configuration from the first snapshots as initial configuration of the shock entering the test section of the shock tube. The computation then yields snapshots of the converging shock wave, which can be evaluated in the form of isopycnics, shock-front contours, etc., for direct comparison with the corresponding experimental snapshots. As has been explained in more detail in Demmig et al. (1995) the experimental records are digitized, and then are represented by a curve fit. Usually, the first two photographs of the shock just having entered the test section of the tube provide the main information for the formulation of the initial conditions, namely the angular position $r_S(\phi)$ of the shock and the shock Mach number $M_S(\phi)$.

2. Numerical methods

Our preferred method is the UNO scheme (Harten and Osher 1987). In order to reduce computional effort it is desirable to have a fine mesh about the center of convergence and a wide mesh in the rather uniform region behind the initial shock. This meshing can be achieved by the use of polar coordinates quite well, although the coordinate singularity at $r = 0$ has to be treated suitably. For symmetrical shocks, the singularity can be avoided by inserting a thin axis around $r = 0$, which, in this case, is the convergence point. For unsymmetrical shocks the convergence point is offset from the origin, hence the coordinate system has to be modified. Meier (1994) found that the least-complicated solution was to shift the grid by half of a radial step size along a specific diameter, such that the operator functions need not to be evaluated at $r = 0$.

A finite-element method (FEM), which has the advantage of using unstructured grids, is under construction. The matrices of the FEM equation are calculated analytically using bilinear elements with three nodes. A third-order dissipative term is added to suppress oscillations behind the shock front. The λ-scheme of Moretti (1979) might turn out to be a good compromise between difference schemes and methods of bicharacteristics, which are being developed by Meier (1993). A very promising first step toward utilizing the λ-scheme in computations of converging shock waves has been done by vom Endt (1995).

3. Flow visualization

Converging shock waves in molecular gases develop rather complex flow features, especially if nonequilibrium processes are appreciable. This complexity is evident when shock fragmentation occurs resulting in a Mach-reflection configuration and the formation of additional structures after reflection. Without suitable flow visualization aimed at the specific features in question, a physical interpretation is not possible. In addition to the widely used contour maps and 3D plots of relevant physical quantities, we utilized and also developed flow-visualization methods that belong to pattern-recognition techniques. (Demmig and Petersen 1991). The latter allow us to locate and depict the fragmentation wave system that consists of shocks, compression waves, and slip lines. Sometimes it is useful to combine related physical quantities in one plot, as e.g., in Figs. 1c, 2c, and 3c, where the arrows represent the velocity field and their gray tone represents the value of the local energy density.

The process of convergence and reflection of an initially slightly deformed shock wave is illustrated by Fig. 1 through 3 in three large time steps. The isopycnics of Fig. 1a already indicate shock fragmentation. As a consequence of the disturbed three-fold symmetry of the original experimental shock contour, the fragmentation wave system found by the pattern-recognition program looks irregular. In Fig. 1b, full circles denote shock-front points, open circles denote compression waves, and grey points denote slip lines. As can be seen also from Figs. 1a and 2c, the fragmentation is more pronounced in the lower part. Figure 2 shows the shock wave immediately before reflection. Clearly, the convergence 'point' does not coincide with the center of the observation window and the wave system is considerably distorted. The complexity of the flow field behind the reflected shock manifests itself in Fig. 3b. However, in conjunction with Figs. 3a and 3c, a closer interpretation can be given, since behind the reflected shock we observe vortices that originate in the fragmentation wave system during reflection. For comparison, Fig. 4 displays isopycnics from a model computation and an enlarged picture of an experimental snapshot taken at the same instant. Though the agreement of the minute structures cannot be expected to be complete, it is surprisingly good. The model-based computations can obviously facilitate the interpretation of experimental snapshots.

4. The effect of nonequilibrium processes

The propagation of converging shock waves in molecular gases can be affected by nonequilibrium processes like vibrational and dissociative relaxation. The temperature reached in the region near convergence depends on the initial Mach number of the shock wave. Consequently, the fragmentation wave system and, more distinctly, the post-reflection phenomena, will show the influence of nonequilibrium processes. Whether the nonequilibrium effects can be observed depends on the experimental resolution, because the converging shock eventually strengthens and the relaxation time may approach the time characterizing the variation of hydrodynamical quantities. Although vibrational relaxation does not play a decisive role in experiments with Mach number ~ 2 as have been carried out, e.g., by Takayama et al. (1987), an influence of vibrational and dissociation relaxation is already visible for shocks with initial Mach numbers of 2 – 3 (Klingspohn 1994). The formation of the fragmentation wave system will be markedly affected for shocks with initial Mach numbers in the range of 4 – 5. If oxygen is taken as the test gas instead of air these phenomena can be observed already for shocks of Mach 2. The analysis of nonequilibrium effects in real gases can be supported by plots of relaxation variables like the vibrational energy (e.g., in Fig. 5, a gray-shade plot of fractional vibrational energy). The initial shock contour had been chosen to have four-fold symmetry. Figure 5, taken well after reflection, clearly shows regions of increased vibrational energy, which underline the shape of the above-mentioned vortices. The fractional shock front that was determined by pattern-recognition methods is represented by a broken black line. Similar plots can be used in the study of ionizational relaxation, which is neglible) for converging shock waves with an initial Mach number smaller than 6.

5. Conclusion and outlook

Our concept of using the instantaneous shape and velocity of a converging shock wave to predict its future propagation including fragmentation and reflection has been elaborated further through a number of computational and flow-visualization tools. A detailed physical interpretation of features like the fragmentation wave system and structures behind the reflected shock can be given. The model could be refined, if more experimental data were avaible, especially regarding the downstream condition of the incoming shock. We recommend experiments with converging shock waves with initial Mach numbers in the range of 3 – 7, where nonequilibrium processes play a role. These experiments are worthwhile from both basic physical and applied standpoints. A fully established fragmentation wave system is a prerequisite for the study of the physical aspects of this interesting shock-wave flow.

References

Demmig F, Grönig H, Kleine H, Wallus H (1995) Experiments and model computation of cylindrical shock waves with time-resolved deformation and fragmentation. In: Shock Waves @ Marseille, R Brun, L Dumitrescu (eds).

Demmig F, Petersen J (1991) Two-dimensional model computation of converging shock wave propagation, fragmentation, and reflection accounting for nonequlibrium processes. In: Int Workshop Strong Shock Waves, Chiba, pp 221–257.

vom Endt A (1995) Dip. Thesis, Ein Differenzenverfahren mit Stoßfronteinpassung und dessen Anwendung zur numerischen Berechnung konvergenter Stoßwellen in Molekülgasen, U Hannover.

Harten H and Osher S (1987) Uniformly high-order accurate nonoscillatory schemes. SIAM J Numer Anal 24:2:279–309.

Kleine H (1986) Dip. Thesis, Kurzzeitphotographie der Fokussierung zylindrischer Stoßwellen, RWTH Aachen.

Klingspohn M (1994) Dip. Thesis, Ausbreitung, Fragmentation und Reflexion zweidimensionaler konvergenter Stoßwellen in Molekülgasen, U Hannover.

Meier J (1994) Dip. Thesis, Konvergente Stoßwellen im Bereich der Fragmentation und Reflektion, U Hannover.

Meier U (1993) Dip. Thesis, Die Lösung instationärer Stoßwellenprobleme mit zweidimensionalen Charakteristikenverfahren, U Hannover, and unpublished papers.

Moretti G (1979) The λ-scheme. Computers and Fluids 7:191–205.

Takayama K, Kleine H, Grönig H (1987) An experimental investigation of the stability of converging cylindrical shock waves in air. Exp Fluids 5:315–322.

Wallus H (1993) Dip. Thesis, Modellrechnungen zur Ausbreitung konvergenter Stoßwellen auf der Grundlage zeitaufgelößter Experimente, U Hannover.

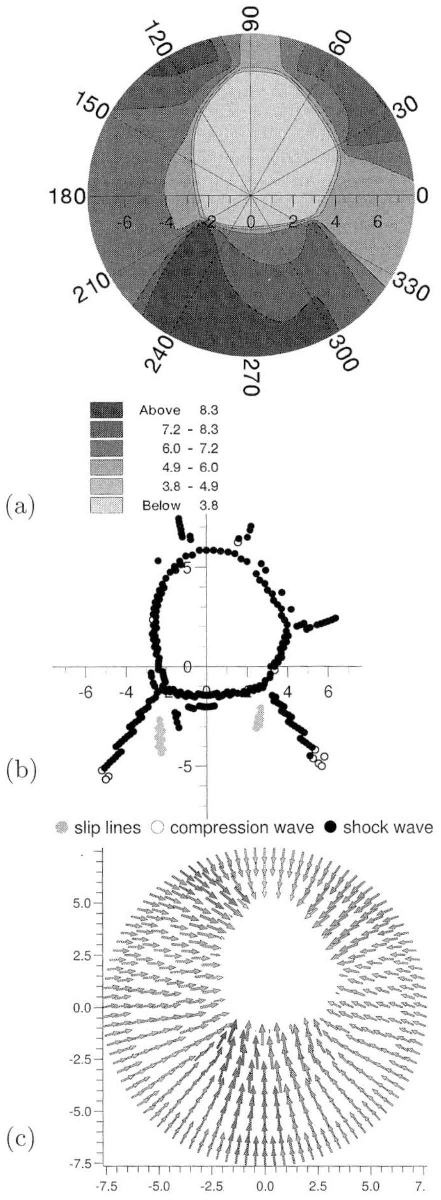

Figure 1. Isopycnics, fragmentation wave system, and velocity field of converging shock before reflection.

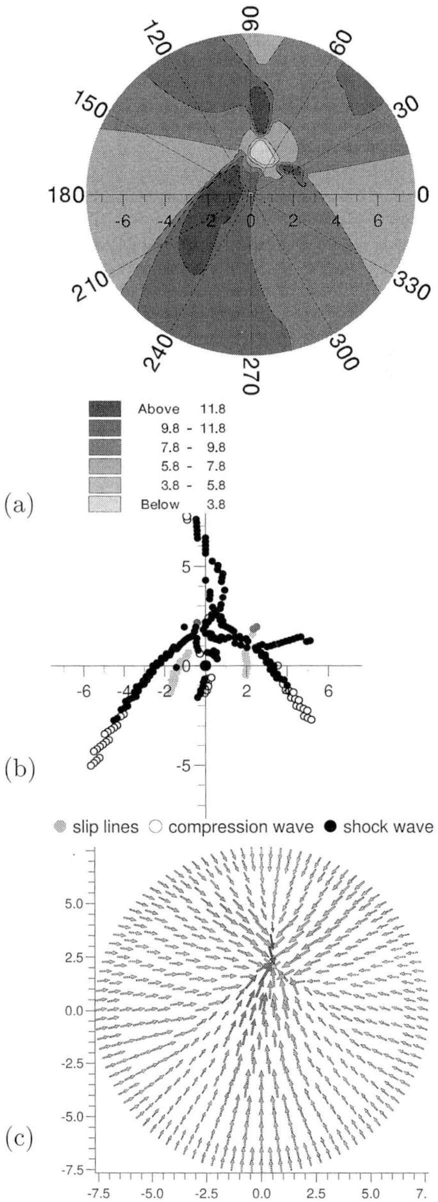

Figure 2. Cf. Fig. 1. Immediately before shock reflection.

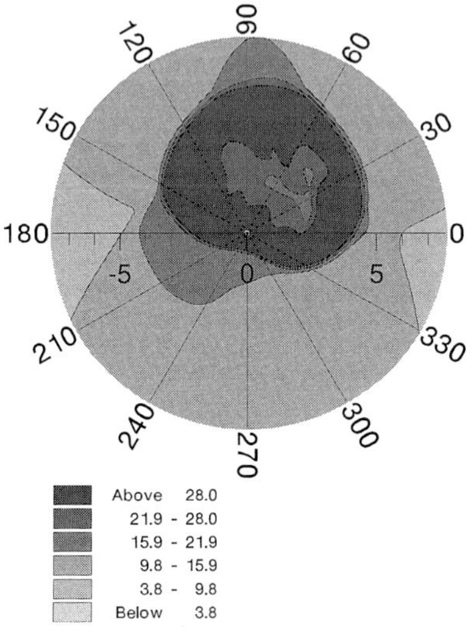

(a)

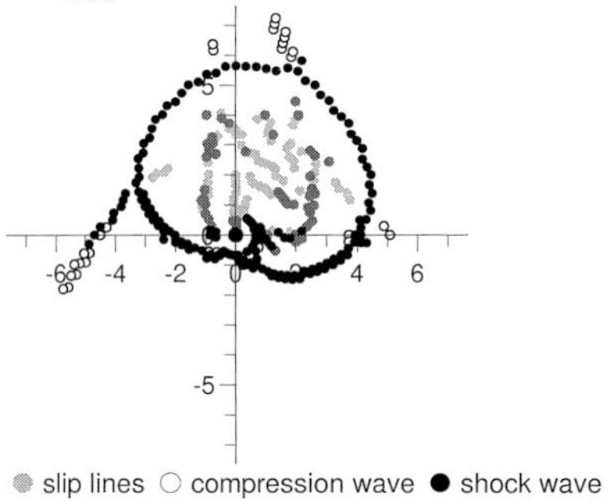

(b)

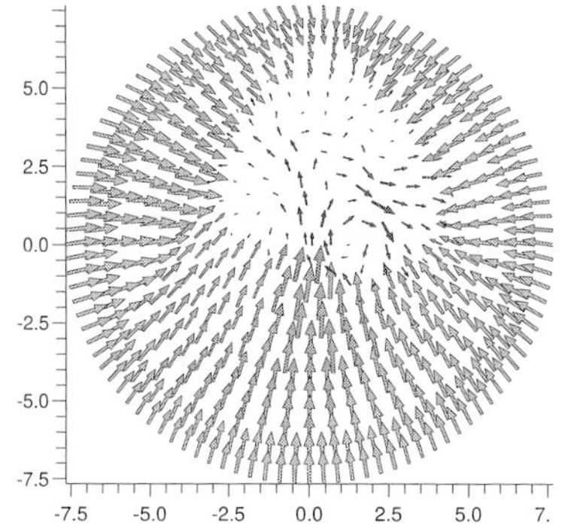

(c)

Figure 3. Cf. Fig. 1. After reflection. Vortices fully established.

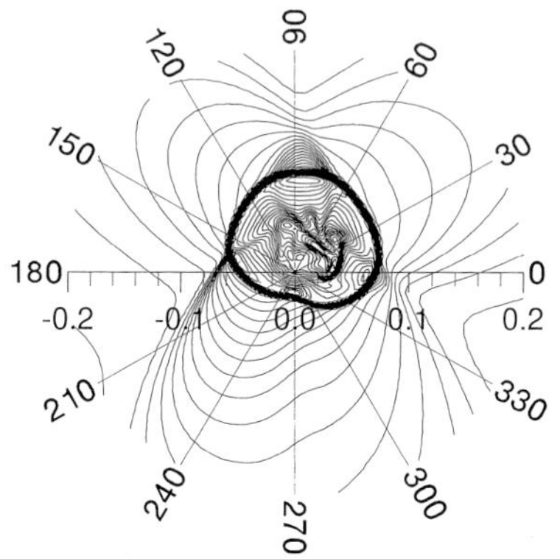

(a)

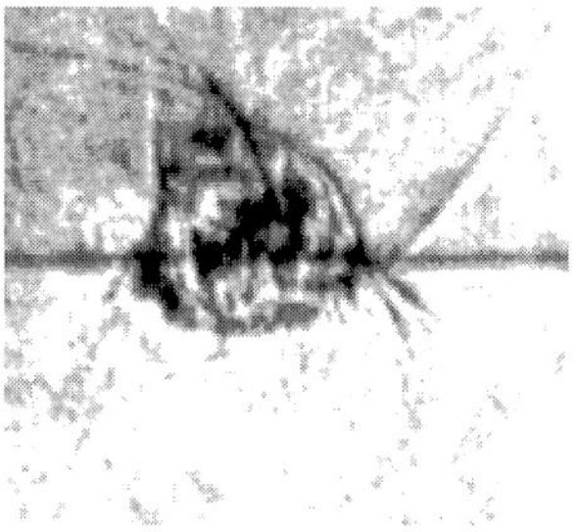

(b)

Figure 4. Isopycnics of model computation and enlarged experimental photo. After reflection.

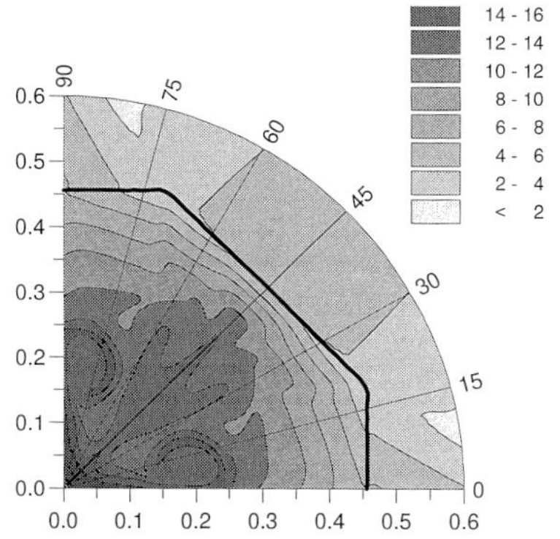

Figure 5. Isoline of vibration energy (in percent). M_S=4. After reflection.

Propagation and stability of strong shocks in gases

M. Mond, I. Rutkevich, E. Toffin
Pearlstone Center for Aeronautical Engineering Studies
Department of Mechanical Engineering
Ben-Gurion University of the Negev, Beer Sheva, Israel

Abstract: The steady state as well as the stability properties of strong shocks propagating in cold monatomic gases are investigated. Excitation of the electronic levels of the atoms as well as ionization processes behind the shock are taken into account in constructing the Hugoniot curve and in reformulating the shock's stability conditions. A parametric study is carried out in order to find domains in parameter space in which shocks become unstable by spontaneously emitting sound- and entropy-vortex waves. Available experimental results agree reasonably well with the theoretical calculations.

Key words: Ionization, Excitation, Shock instability

1. Introduction

Shock waves with high Mach numbers provide a rich source of phenomena and problems of fundamental scientific nature. As the shock's Mach number rises so does the importance of various inelastic processes occurring behind it. In the current work, attention is focused on excitation of the electronic energy levels of the atoms and on ionization. These two processes result in lowering the temperature of the gas behind the gas-dynamical shock and in its subsequent compression beyond the asymptotic classical limit. The steady-state properties of the strong shocks are studied by investigating their Hugoniot curves. It is shown that these latter possess a maximum that represents the maximal degree of compression that is achievable by a shock for given initial conditions.

A deep interest in the problem of shock instability rose during recent years due to some experimental evidence of irregular flows behind strong shocks (Griffiths et al. 1976, Glass and Liu 1977; Glass et al. 1978). The problem of shock instability ranks together with the well-known Kelvin-Helmholtz and Rayleigh-Taylor instabilities as one of the fundamental problems of fluid mechanics. The criteria for shock instability were first obtained by Dyakov (1954) and until recently their straightforward application failed to explain the unstable behavior behind strong shocks. Mond and Rutkevich (1994) have pointed out the importance of the ionization kinetics and, accordingly, modified the criterion for spontaneous emission (SE) of acoustic as well as entropy-vortex waves from shocks. The modified criterion resulted in thresholds for SE that are in reasonable agreement with available experimental data. In the current work, the effects of excitation of the electronic energy levels of the atoms on the steady-state and stability properties of strong shocks are examined. In addition, a parametric study is carried out in order to define the SE domains in the space of parameters like the shock Mach number and the initial pressure of the gas.

2. Steady-state properties of strong shocks in gases

A planar shock wave is considered which propagates with a constant velocity in a cold monatomic gas. It is convenient to describe the shock in a frame of reference that is moving with the shock's normal velocity. As the shock's Mach number rises so does the temperature behind it. Subsequently, excitation of the electronic-energy levels of the atoms as well as their single and multiple ionizations become progressively important as the temperature behind the shock increases to higher and higher values. The classical gas-dynamical shock, with its sharp surface of discontinuity (whose thickness is of the order of a few molecular free-path lengths) gives way to a more complex flow pattern. Thus, the steady-state flow field is divided into four distinct areas: (1) The region ahead of the shock in which the gas is in a cold equilibrium. Due to the low equilibrium temperature in this region it is assumed that all the atoms are in their ground state (region 1). (2) The gas-dynamical shock wave. (3) The region along which the excitation and ionization processes take place, accompanied by a drop in the temperature and a consequent increase of the density. This region is called the relaxation zone (RZ). (4) The region in which excitation and ionization equilibrium is maintained (region 2). The number of free electrons as well as the distribution of excited states reach a constant value in this region. It should be noted that the flow variables in this region are only approximately constants due to the radiation from the hot gas. However, changes caused by radiation occur over a length scale that is much longer than the RZ so the effect of radiation may be neglected.

In the current work, a unified picture is employed in which the gas-dynamical shock together with the RZ constitute an effective surface of discontinuity which is treated as a single unified shock wave. The properties of such shocks and, in particular, their Hugoniot curves are studied by writing the conservation equations that relate both constant-state sides (i.e., regions 1 and 2). The continuity equation and momentum equation have the same form as those for a classical gas-dynamical shock. The energy equation, however, is modified due to the excitation and ionization processes and is

$$e_1 + \frac{p_1}{\rho_1} + \frac{v_1^2}{2} = e_2 + \frac{p_2}{\rho_2} + \frac{v_2^2}{2} + Q_{ion} + Q_{exc} \,, \tag{1}$$

where ρ, v, p, and e are the density, velocity, pressure, and internal energy per unit mass of the gas, respectively, and Q_{ion} and Q_{exc} are the the amounts of energy lost by the gas per unit mass due to ionization and excitation of electronic levels, respectively. The indices 1 and 2 denote the values of the parameters ahead of the shock (region 1) and behind it (region 2), respectively. For a monatomic gas the internal energy per unit mass is

$$e = \frac{p}{(\gamma - 1)\rho}, \tag{2}$$

where γ is the specific heat ratio. The energy-loss terms are

$$Q_{ion} = \frac{NI\alpha}{\mu}, \quad Q_{exc} = \frac{N\bar{E}_a(1-\alpha)}{\mu}, \tag{3}$$

where N is the Avogadro number, μ is the atomic weight of the gas, I is the ionization potential, $\bar{E}_a$ is the average electronic energy level, and $\alpha = n_e/(n_i + n_a)$ is the ionization degree, where n_e, n_i, and n_a are the concentrations of free electrons, ions, and atoms, respectively. The total pressure is determined by the sum of the partial pressures of all the species,

$$p = kT(n_e + n_i + n_a). \tag{4}$$

The average electronic level is calculated according to Boltzmann statistics that yield

$$\bar{E}_a(T) = \frac{1}{D_a} \sum_j g_a^j E_a^j \exp(-E_a^j/kT), \tag{5}$$

where D_a is the atom partition function, and g_a^j and E_a^j are the degeneracy and the value of the jth electronic energy level in the atoms. The partition function is

$$D_a = \sum_j g_a^j \exp(-E_a^j/kT). \tag{6}$$

The ionization degree, α, is calculated through the Saha equation that takes the form

$$\frac{\alpha^2}{1-\alpha} = \frac{G}{n_i + n_a} \left[\frac{m_e k T_2}{2\pi\hbar^2}\right]^{3/2} e^{-I/kT_2}, \tag{7}$$

where $G = 2D_i/D_a$, m_e is the electron's mass, and $\hbar$ is Planck's constant. The quantity D_i is the partition function of the single-charge ions and is calculated in an equivalent manner to Eq. (6). Defining the dimensionless parameters $\theta = T_2/T_1$, $\eta = \rho_2/\rho_1$, and $P = p_2/p_1$, the equations that describe the Hugoniot curve, i.e., $\eta(P)$, may be cast into the implicit parametric form

$$\eta(\theta) = \frac{1 + \Gamma P(\theta)}{\Gamma + P(\theta) - \varsigma\alpha(\eta(\theta),\theta) - \xi(\theta)[1 - \alpha(\eta(\theta),\theta)]}, \tag{8}$$

where $\Gamma = (\gamma + 1)/(\gamma - 1)$, $\varsigma = 2I/kT_1$, $\xi(\theta) = 2\bar{E}_a(T)/kT_1$, and α is

$$\alpha(\eta, \theta) = -\frac{1}{2}\psi + \sqrt{\frac{1}{4}\psi^2 + \psi}, \tag{9}$$

where

$$\psi = b(p_1, T_1)\theta^{3/2}\eta^{-1}e^{-\varsigma/2\theta} \quad \text{and} \quad b(p_1, T_1) = G\left(\frac{m_e}{2\pi\hbar^2}\right)^{3/2}\frac{(kT)^{5/2}}{p_1}. \tag{10}$$

Before turning to the description of the Hugoniot curve, note that the right-hand sides of Eqs. (5) and (6) represent diverging infinite series for free atoms. However, for atoms in a partially ionized plasma only those energy levels should be taken into account that lie within the Debye sphere (Drellishak et al. 1963, Biberman et al. 1986). Thus, the sums in Eqs. (5) and (6) are truncated after $N_{max} = (Ir_D/e^2)^{1/2}$ terms, where $r_D = (kT/4\pi n_e e^2)^{1/2}$ is the Debye radius. A close inspection of Eq. (8) reveals that the Hugoniot curve possesses a maximum for a certain pressure ratio, $P = P_{max}$. This is so because on the one hand, the Hugoniot curve is close to its classical counterpart for low pressure ratios. On the other hand, in the range $P \gg \varsigma$, where the ionization degree is practically 1, the following asymptotic relationship is obtained:

$$\eta \approx \Gamma\left(1 + \frac{\varsigma}{P}\right), \tag{11}$$

which describes a decreasing dependence of η on P. Due to the energy losses during ionization as well as electronic excitation, the degree of maximal compression exceeds its classical asymptotic value. Thus, for example, numerical calculations show that argon at an initial pressure of 5 torr and temperature of 300 K may be compressed by a shock up to 12 times its initial density for shocks with pressure ratio of $P_{max} = 1400$. A universal law for the maximal density ratio behind ionizing shocks was given by Rutkevich and Mond (1994).

3. Stability of strong shocks in gases

The problem of stability of planar shock waves was first investigated by Dyakov (1954). In particular, he and later Kontorovich (1957) formulated the conditions under which acoustic- and entropy-vortex waves are spontaneously emitted from the shock. This phenomenon was termed spontaneous emission (SE) and the criteria for its occurrence are

$$h_c < h < 1 + 2M_2, \quad h_c = \frac{1 - (\eta + 1)M_2^2}{1 + (\eta - 1)M_2^2}, \quad h = -\frac{j^2}{\rho_2^2}\left(\frac{\partial \rho_2}{\partial p_2}\right)_H, \tag{12}$$

where M_2 is the Mach number behind the shock, j is the mass-flow rate across the shock, and the subscript H signifies the derivative along the Hugoniot curve.

A straightforward application of criterion (12) to shocks in ideal gas results in stability of the shock. On the other hand, a series of experiments (Griffiths et al. 1976, Glass and Liu 1977, Glass et al. 1978) revealed irregularities in the flow behind strong shocks in the form of spatial oscillations in the mass density and in the electrons' concentration when the shock's Mach number exceeded a certain threshold.

This apparent discrepancy between theory and experiment has puzzled many researchers over the past twenty years (McIntyre et al. 1991) and was recently resolved by Mond and Rutkevich (1994) who pointed out the crucial role of the electrons' kinetics in the occurrence of SE. It was shown that a frequency range exists for which, on one hand, the perturbations in the electrons' temperature differs from that of the heavy particles while, on the other hand, ionization equilibrium is maintained. This window of thermal non-equilibrium between the electrons and the heavy particles is defined by all wave frequencies ω that satisfy $\nu_\varepsilon \ll \omega \ll 2\nu_{ion}$ where ν_{ion} is the characteristic frequency of ionization by electron impact and ν_ε is the inverse of the relaxation time of the temperature of the heavy particles due to their collisions with the electrons. The latter is proportional to the sum of the elastic-collision frequencies of the electrons with the atoms and the ions. The Coulomb interactions of the electrons with the ions become important as the Mach number rises and the ionization degree approaches unity. Both collision frequencies and ν_{ion} are calculated according to expressions given in Hoffert and Lien (1967). Consequently, the definition of h is replaced by

$$h = -\frac{j^2}{\rho_2^2}\left[\left(\frac{\partial \rho_2}{\partial p_2}\right)_{\rho_1, p_1, \alpha} + \frac{\delta\alpha}{\delta p_2}\left(\frac{\partial \rho_2}{\partial \alpha}\right)_{\rho_1, p_1, p_2}\right]. \tag{13}$$

The ratio $\delta\alpha/\delta p_2$ differs from the derivative $d\alpha/dp_2$ along the Hugoniot curve since the ionization behind the shock is mainly due to electron impact and δT and δT_e are not necessarily equal. In addition, the sound velocity for waves in the non-equilibrium frequency window is also modified in order to take into account the thermal non-equilibrium between the electrons and the heavy particles. A preliminary examination of criterion (12) with the modified h (Eq. 13) and the modified sound velocity for shocks in argon and krypton at initial pressures in the range of 5–10 torr resulted in the occurrence of SE for Mach numbers above 10–15. These results are in reasonable quantitative and qualitative agreement with the experimental results of Glass and Liu (1977) and Glass et al. (1978).

Further numerical investigations show that taking into account the energy losses due to excitation of the electronic energy levels in the atoms affects both the Hugoniot curve and the threshold for SE very little. In addition, an extensive parametric study conducted for argon at room temperature shows that the threshold for SE does not depend on the initial pressure. However, for initial pressures below 11 torr it is impossible to maintain ionization equilibrium without thermal equilibrium in a certain range of the shock Mach numbers. In that regime $\nu_{ion} \ll \nu_\varepsilon$ and hence the classical criterion should be used, resulting in the non-occurrence of SE. These circumstances lead to a significant lowering of the upper limit for SE for low initial pressures. For initial pressures above 11 torr no such restriction exists and SE occurs for Mach numbers above 11.5 (independent of the initial pressure) and below 40–42 (very slight dependence on the initial pressure). It should be noted, however, that for very high Mach numbers such as near the upper limit of the SE, effects like second ionization may play an important role in determining the Hugoniot curve as well as the conditions for SE. For example, the Hugoniot curve is expected to have two maxima and thus to significantly alter the stability condition.

References

Biberman LM, Vorobjev VS, Yakubov IT (1986) Kinetics of Non-Equilibrium Low Temperature Plasmas, Plenum, New York, USA.

Dyakov SP (1954) On the stability of shock waves. Zh Eksp Teor Fiz 27:288–295.

Drellishak KS, Aeschliman DP, Cambel AB (1963) Partition functions and thermodynamic properties of argon plasma. Phys Fluids 6:1280–1284.

Glass II, Liu WS (1977) Effect of hydrogen impurities on shock structure and stability in ionizing monatomic gases: part 1, argon. J Fluid Mech 84:55–77.

Glass II, Liu WS, Tang FC (1978) Effect of hydrogen impurities on shock structure and stability in ionizing monatomic gases: part 2, krypton. Can J Phys 55:1269–1279.

Griffiths RW, Sandeman RJ, Hornung HG (1976) The stability of shock waves in ionizing and dissociating gases. J Appl Phys 9:1681–1691.

Hoffert M, Lien H (1967) Quasi-one-dimensional, nonequilibrium gas dynamics of partially ionized two-temperature argon. Phys Fluids 10(8):1769–1777.

Kontorovich VM (1957) On the stability of shock waves. Zh Exsp Teor Fiz 33:1525–1526.

McIntyre TJ, Houwing AFP, Sandeman RJ, Bachor HA (1991) Relaxation behind shock waves in neon. J Fluid Mech 227:617–640.

Mond M, Rutkevich I (1994) Spontaneous acoustic emission from strong ionizing shocks. J Fluid Mech 275:121–146.

Rutkevich I, Mond M (1994) Similarity relation for maximal gas compression by strong ionizing shocks. J Propulsion and Power 10:906–908.

Blast parameters from unconfined gaseous detonations

S. B. Dorofeev, V. P. Sidorov, A. E. Dvoinishnikov
Applied Chemical Physics Institute
Russian Research Centre "Kurchatov Institute"
Moscow, 123182 Russia

Abstract: Experimental and theoretical data for parameters of the air blast from gaseous detonations are analyzed. Comparison is made on the basis of the main blast parameters themselves and the corresponding TNT-equivalents of the gaseous explosions.

Key words: Gaseous detonations, Blast parameters, TNT-equivalent

1. Introduction

Numerous experimental and theoretical studies of the air blast from gaseous detonations have been conducted. These studies make it possible to proceed with data analysis in order to define the most reliable description of the air blast. Such an analysis has been undertaken earlier (Dvoinishnikov et al. 1993a, 1993b) for the blast waves from HE explosions. One of the important problems in the estimation of explosion effects is the application of the TNT-equivalence concept. This problem has been discussed in different studies (Fishburn 1976; Brossard et al. 1983; Borisov et al. 1985), but many questions are still unresolved. Different values of the TNT-equivalent and different functions of distance have been presented by the different authors. The aim of the second part of this study is to re-evaluate the concept of TNT-equivalent on the basis of selected data for air blasts from high explosives and gaseous detonations.

2. Blast parameters

The comparison of the blast parameters was carried out using Sachs dimensionless variables for unconfined spherical explosions. The data for positive overpressure (P^+) and impulse (I^+) are collected in Fig. 1. A relatively small number of measurements is available for parameters of negative phase (Brossard et al. 1984; Lannoy 1984; Dorofeev et al. 1993). They are not presented here, but these data show good correlation and the comparison was presented by Dorofeev et al. (1993). Figure 1 shows experimental data and the results of calculations. Corresponding experiments have been made with fuel-air and fuel-oxygen mixtures in laboratory and large-scale experiments. Except for the data by Kogarko et al. (1965) and calculations by Fishburn (1976), quite good correlation is observed. Mutual agreement is slightly better for overpressures than for impulses. The comparison shown in Fig. 1 suggests that the data of Brossard et al. (1984) and Dorofeev et al. (1993) are the most reliable. They are based on independent experiments and reveal good mutual agreement. Another important feature of their functions of blast parameters on distance is their refusal to apply $1/R$ behavior for overpressure in the far field. This feature permits reasonable predictions of overpressures at large distances from the explosive cloud to be made. At the same time, the means of representing blast-wave parameters are different in these studies. The only reservation about these data concerns the data for positive impulse that were presented by Brossard et al. (1984) for distances $\overline{R} > 5$. The reason for the behavior shown in the plot needs some explanation or additional validation.

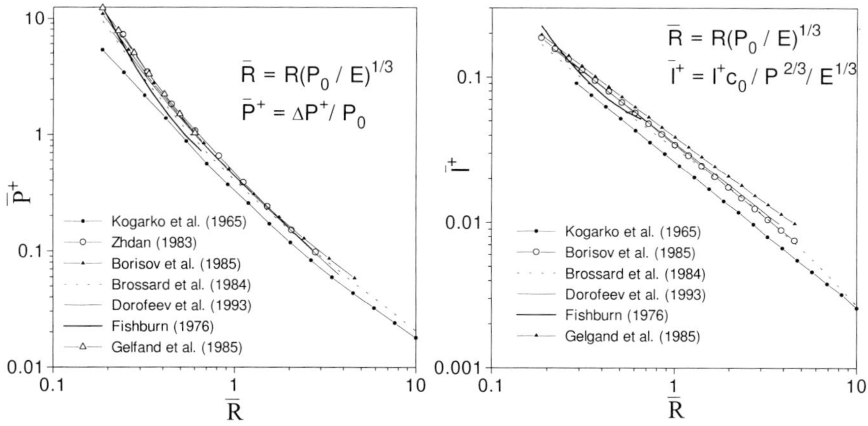

Figure 1. Positive overpressure and positive impulse in air-blast wave versus distance.

3. TNT-equivalents

The concept of TNT-equivalent is supported by the convenience of describing the blast effect of non-ideal explosions by that of some equivalent amount of high explosive. However this convenience does not have a firm basis, since the explosives themselves differ greatly in their intrinsic characteristics. The TNT-equivalent of a gaseous explosion should be defined as the ratio of the energy of the TNT explosion to that of gaseous explosion which has the same value of a given parameter at a fixed distance. In view of this, the TNT-equivalent should be considered (i) as a function of distance and (ii) separately for each parameter of the blast wave. This shows that the usage of some average TNT-equivalent (Brossard et al. 1983; Lannoy 1984) is either incorrect or should be specially validated. In addition, different parameters of air blast are known with different degrees of confidence. Their relative contribution to the blast effect differs itself. Although no need exists to use a TNT-equivalent to describe air-blast parameters from gaseous explosions, an analysis of TNT-equivalents permits a better understanding of the difference between blast waves from HE and gaseous detonation. The following analysis is based mostly on the data selected in the present paper (gaseous detonation) and on the data chosen for high explosives (Dvoinishnikov et al. 1993a, 1993b). Results of calculation of the TNT-equivalents are presented in Figs. 2 and 3. Figure 2 shows that the TNT-equivalent for $\overline{P}^+$ depends strongly on distance. Within the range $\overline{R} = 0.3$ to 2, it increases from 0.3–0.4 to 0.6–0.7 with the $\overline{R}$ increasing for all the data used. Beyond $\overline{R} \approx 2$, which corresponds approximately to overpressures less than 0.1, the analysis is more complicated in view of the lack of data and their limited accuracy. Nevertheless, the behavior of most of the functions indicates the maximum in the range $\overline{R} = 1$ to 3. The choice of data for calculation of TNT-equivalent appears to be very significant. The asymptotic behavior of $\overline{P}^+$ at the far field should be realistic and the same for gaseous and HE explosions. Two examples shown are the data 1 and 3, where the realistic function $1/R^{4/3}$ is used, and the data 6 and 7, where the wrong asymptotic $1/R$ results in an unreliable prediction that the TNT-equivalent approaches 1. While the data for $\overline{P}^+$ show a clear trend, the data for $\overline{P}^-$ do not. The choice of data is very limited and their mutual agreement is poor. The main problem concerns the description for $\overline{R} < 2$. It is not connected with the data for gaseous detonations, which agree well, but with that for HE explosion. In any case, gaseous explosions produce blast waves with relatively stronger negative phase than those of HE with the same positive overpressure. The important question, which remains unclear, is whether it is a factor of 2 or 8.

Figure 3 shows the TNT-equivalent for the impulse. The data used for HE include the mean function obtained by Dvoinishnikov et al. (1993a, 1993b), and the data by Baker (1983). All the values appear to be within the range 0.5–0.6 up to $\overline{R} \approx 4$. The decrease of the equivalent beyond $\overline{R} = 4$ is due to the data for gaseous explosion from Brossard et al. (1984). As was mentioned above, this decrease seems to be unrealistic. Additional analysis is required to clarify this important question. The TNT-equivalent presented by Fishburn (1976), where the comparison was made with ideal point source (Brode 1955), seems also not reliable. The situation with the lack of data for detailed analysis of the TNT-equivalent $\overline{I}^-$ is similar to that for negative overpressure. This was the reason for using also the data for point-source explosions in ideal (Brode 1955) and real gases (Brode 1956). We should notice that all functions grow with distance. The preferable estimate seems to be in between curves 1 and 3 and curves 2 and 3. Again, the negative impulse produced by gaseous detonation

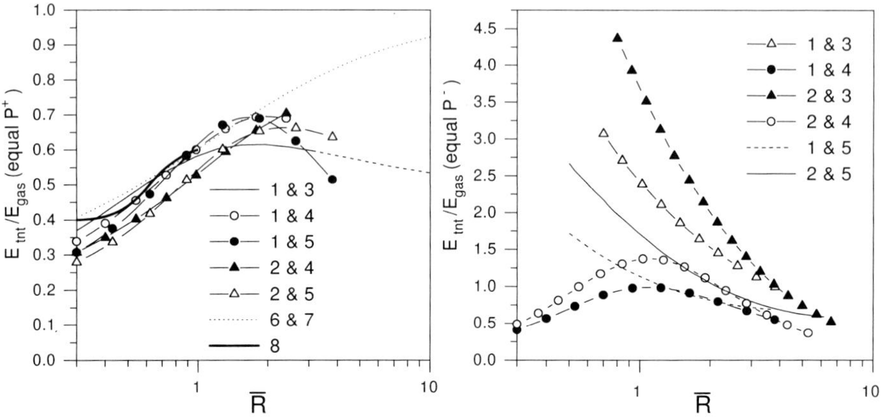

Figure 2. TNT-equivalent for positive and negative overpressure versus dimensionless distance defined by the energy of the gaseous explosion. $\overline{P}^+$: 1 — Dorofeev et al. (1993); 2 — Brossard et al (1984); 3 — Dvoinishnikov et al. (1993b) (spherical charge); 4 — Sadovsky (1952); 5 — Goodman (1960); 6 — Borisov et al. (1984); 7 — Lewis (1981); 8 — Fishburn (1976). $\overline{P}^-$: 1 — Dorofeev et al. (1993); 2 — Brossard et al. (1984); 3 — Dvoinishnikov et al. (1993a) (cylindrical charge); 4 — Lewis (1981); 5 — Dvoinishnikov et al. (1993b) (spherical charge).

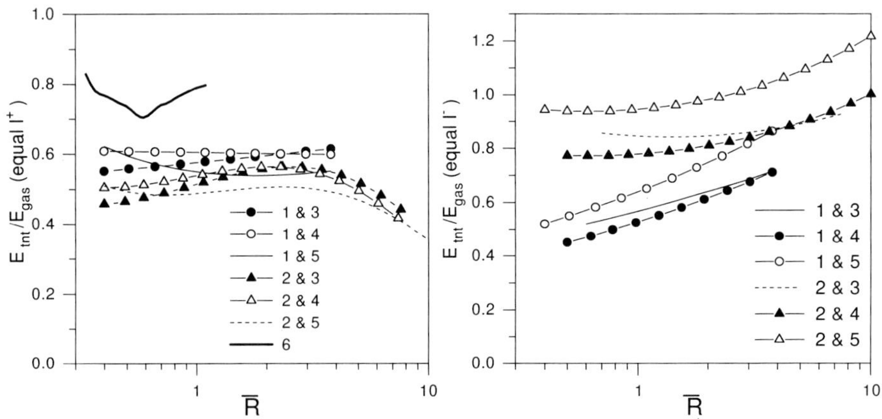

Figure 3. TNT-equivalent for positive and negative impulse. $\overline{I}^+$: 1 — Dorofeev et al. (1993); 2 — Brossard et al. (1984); 3 — Dvoinishnikov et al. (1993a) (mean values); 4 — Dvoinishnikov et al. (1993a) (cylindrical charge); 5 — Baker (1983); 6 — Fishburn (1976). $\overline{I}^-$: 1 — Dorofeev et al. (1993); 2 — Brossard et al. (1984); 3 — Dvoinishnikov et al. (1993a) (cylindrical charges); 4 — Brode (1955); 5 — Brode (1956).

appeared to be greater than that from HE, with the positive zone being the same. This difference, however, is less than that for $\overline{P}^-$.

4. Conclusions

The analysis of the data for the parameters of air blast from unconfined gaseous detonations shows that the description of the air blast from gaseous detonations is at the same level as that for HE explosions. Moreover, the data for the negative phase that is available in the literature seems to be more reliable for the case of gaseous detonation. At the same time, the lack of data for negative phase should be emphasized. In these circumstances the application of the TNT-equivalence concept for estimation of an effect of a gaseous explosion seem to be useless. Nevertheless, the analysis of TNT-equivalents as a function of distance and air-blast parameters can be still useful to clarify the differences in character of air blasts from gaseous detonations. The values of TNT-equivalents still interest many practical engineers. Energetic equivalents are estimated to be within the ranges 0.3–0.7 (P^+), 0.5–0.6 (I^+), 0.5–0.9 (I^-). The mean value of 0.2 ($E_{GAS} = 5E_{TNT}$) presented by Brossard et al. (1983) and Lannoy (1984) is certainly underestimated, probably due to incorrect usage of HE blast parameters. The data for P^- (especially for HE) do not permit the confident evaluation of TNT-equivalent.

References

Baker WE, Cox PA, Westine PS, Kulesz JJ, Strehlow RA (1983) Explosion Hazards and Evaluation, Elsvier, Amsterdam, Oxford, New York.

Borisov AA, Gelfand BE, Tsyganov SA (1985) On modelling of pressure waves forming by detonation and combustion of gas mixtures. J Physics of Combustion and Explosion, Russian Acad of Sci 2:90.

Brode HL (1955) Numerical solutions of spherical blast waves. J Appl Phys 6:766.

Brode HL (1956) Point Source Explosion in Air, Research Memorandum RM-1824-AEC, RAND Corporation, Santa Monica, California.

Brossard J, Leyer JC, Desbordes D, Garnier JL, Hendrickx S, Lannoy A, Perrot J, Saint Clouds JP (1983) Experimental analysis of unconfined explosions of air-hydrocarbon mixtures. characterization of the pressure field. In: 4th Symp Int on Loss Prevention and Safety Promotion in the Process Industries, European Fed of Chem Eng Event No 290 1:D, pp 10–19.

Brossard J, Leyer JC, Desbordes D, Saint Clouds JP, Hendrickx S, Garnier JL, Lannoy A, Perrot J (1984) Air blast unconfined gaseous detonations. Progress in Astron and Aeron 94:556.

Dorofeev SB, Sidorov VP, Dvoinishnikov AE, Alekseev VI, Kuznetsov MS (1993). Experimental study of air blast parameters from fuel rich mixtures detonation. 14th Int Colloquium on Dynamics of Explosions and Reactive Systems, (ICDERS) 2:E1.4.1.

Dvoinishnikov AE, Dorofeev SB, and Gelfand BE (1993a) Analysis of blast wave data from HE explosions. In: Shock Waves @ Marseille, Brun R, Dumitrescu LZ (eds), Vol 4, pp 407–412.

Dvoinishnikov AE, Dorofeev SB, Gelfand BE (1993b) Analysis of the Data on Air Blast Parameters from HE Explosions, Preprint IAE-5605/13, Russian Research Centre "Kurchatov Institute," Moscow.

Fishburn BD (1976) Some aspects of blast from fuel-air explosives. Acta Astronautica 3:1049.

Gelfand BE, Gubin CA, Mikhalkin VN, Shargatov VA (1985) Computation of shock wave parameters by detonation of combustible gaseous mixtures of a variable composition. J Physics of Combustion and Explosion, Russian Acad of Sci 3:92.

Goodman HJ (1960) Compiled Free Air Blast Data on Bare Spherical Pentolite, BRL Report 1092, Aberdeen Proving Ground, Maryland.

Kogarko SM, Adushkin VV, Liamin AG (1965) Investigation of spherical detonation on gas mixtures. Combust Explos Shock Waves USSR 1-2:22.

Lannoy A (1984) Analyse des explosions air-hydrocarbures en milieu libre: etudes deterministe et probabiliste du scenario d'accident. Prevision des effets de surpression. Bulletin Direct Etudes et Recherches EDF, A4.

Lewis DJ (1981) Condensed phase explosions and their blast characteristics. Euromech-139, Aberystwyth.

Sadovsky MA (1952) Mechanical action of blast waves on data of experimental studies. In: Physics of Explosion, USSR Acad of Sci Moscow, Vol 1, pp 70–110.

Zhdan SA (1983) Calculation of gas mixture explosion with regard for shift of chemical equilibrium products. J Physics of Combustion and Explosion, Russian Acad of Sci 1:131.

Numerical simulation of the blast in the World Trade Center

J. D. Baum, H. Luo
Science Applications International Corporation
1710 Goodridge Drive, MS 2-3-1
McLean, VA 22102 USA

R. Löhner
George Mason University
Fairfax, VA 22030 USA

Abstract: This paper describes the application of FEFLO96, a recently developed three-dimensional, adaptive, finite-element, edge-based, ALE shock-capturing methodology on unstructured tetrahedral grids, to the simulation of blast diffraction inside the B-2 level of the World Trade Center garage. The geometry of the B-2 garage level was modeled based on the building blueprints and included all important geometrical features. The shock-diffraction simulation was initiated using the results of a one-dimensional detonation code for a one-thousand pound ANFO charge. The resolution and fidelity of the simulated shock-wave diffraction phenomena, performed via a solution of the transient compressible Euler equations, were enhanced by the application of the classic H-refinement/coarsening grid-adaptation scheme. A large number of shock waves was produced due to shock diffraction around walls, cars, and ramps. Excellent shock adaptation and resolution were obtained at all times. Several phenomena such as three-dimensional shock-diffraction processes, the formation of "hot spots" due to three-dimensional wave focusing at corners, shock reverberation within the parking level, and the formation of three-dimensional Mach stems, were identified and discussed.

Key words: Numerical blast simulation, World Trade Center

1. Introduction

The evaluation of civilian-structure vulnerability to terrorist attacks has become a timely and critical issue after the terrorist attack on the World Trade Center, and the failed attacks on other civilian targets in New York City. Since testing large structures for blast resistance is very expensive, and testing real targets is simply not practical, detailed numerical simulations offer a unique, accurate, and cost-effective approach to assessing vulnerability of these structures to internal explosions.

The numerical simulation of an explosion within a confined environment requires simulating blast initiation and evolution, shock-wave diffraction about many structural components, structural response to the blast, structural failure, and blast transmission through the fragmented structural elements. This list indicates that methodologies from widely disparate fields such as computational fluid dynamics (CFD), computational structural dynamics (CSD), and material modeling must be combined to completely assess structure vulnerability to internal blasts. This coupling has been recently completed. Nevertheless, it was decided that for the first simulation it would be prudent to eliminate the structural coupling and treat all surfaces as non-responding.

Over the past five years we have developed a general CFD methodology capable of efficiently simulating transient shock-shock and shock-structure interactions for realistic, complex, engineering-type, three-dimensional geometries. The unstructured grid approach (Löhner et al. 1987) was chosen as the most suitable for generating surface and volume grids for complex-geometry bodies. Since typical areas of interest require simulation of both steady and transient strong shocks, a high-resolution monotonicity preserving algorithm for unstructured grids, called FEM-FCT was developed. Finally, since the flow fields are typically smooth (except for a few regions where strong gradients appear), efficient adaptive-refinement techniques for transient problems are required (Baum and Löhner 1994).

2. The numerical methodolgy

The primary CFD tools incorporated in the FEFLO methodology include the CAD tool (FECAD), the background-grid generator (PRE-BACK), the advancing-front grid generation (FRGEN3D), the flow solver (FEFLO96), and the post-processing methodology (FEPOST3D). The FEFLO96 flow solver is a new 3D ALE edge-based hydro-solver based on FEM-FCT (Löhner and Baum 1990; Luo et al. 1994). H-refinement is the preferred approach for grid adaptation (Baum et al. 1994, Löhner and Baum 1990). The high-order scheme used is the consistent-mass Taylor-Galerkin algorithm. Combined with a modified second-order Lapidus artificial-viscosity scheme, the resulting scheme is second-order accurate in space, and fourth-order accurate in phase. The critical parameter for the refinement/deletion criteria is density. The recently implemented edge-based data structure reduced memory requirements by about a factor of 2.5 and CPU requirements by about 2.5 (Luo et al. 1994).

3. Numerical results

Solid-body definition. In the absence of CAD data, all wall locations, ramps, exhaust plenum spaces, mechanical rooms, stairways, toll booths, curbs, A/C, and supply and elevator shafts were modeled accurately from the structural and architectural drawings (Fig. 1a). Ten typical car groups were modeled, parked at the correct parking locations.

Blast-wave initiation and diffraction. Blast initiation assumed a vehicle parked near the right curb of the exit lane, leading to the northbound exit to West Street (ramp B). The 1D detonation code modeled a charge of 1000 pounds of ANFO HE placed on the floor, and integrated the solution until the leading shock was only a short distance from the south wall of tower A (the north tower). The solution was then overlayed on a 3D mesh.

The numerical simulation of shock diffraction within the B-2 floor of the World Trade Center is especially suited for an adaptive refinement methodology. The large dimensions of the computational domain (20,249 cm long, 17,743 cm wide, and 580 cm high) are in contrast to the requirement to model the shock as a sharp discontinuity (or at least, minimize shock thickness), and model the smallest physical features of interest (i.e., while modeling the complete domain, also model shock diffraction about a wall that is less than 20 cm thick, a ratio of 1:1000). These contrasting requirements can be fulfilled only by using an adaptive methodology. Still, some compromises on mesh size and total number of nodes had to be made. We initiated the simulation by modeling only the zone fairly close to blast point, while allowing only two levels of refinement. Thus, the initial mesh before refinement had only about 370,000 points and 1.95 million elements. At $t = 25.7$ ms, when the leading shock was close to the boundary of the modeled domain, we remeshed, modeling a larger portion of the garage. We then interpolated the solution from the old mesh to the new mesh, and adapted the solution. By this time we gained access to a Cray YMP/M-92 with 256 MWords static memory, so our mesh size was about 780k points and 4.1M elements. Finally, at $t = 108.9$ ms, we modeled the complete domain. The nonadapted mesh included about 1.54M points and 8.0M elements. After solution adaptation, the final mesh included about 3.52M points and 18.1M elements executing on a new Cray YMP/M-92 with 1 GWords static memory.

The complex shock evolution within the structure is described in great detail in Baum et al. (1994). Due to severe space limitations we will include only two snapshots and demonstrate the complexity of shock evolution in such an environment by analyzing these results.

Examination of the leading shock at $t = 12.51$ ms (Fig. 1b) shows on the right (east) the diffracted wave about the south-east corner room of tower A, and shock diffraction about a column. Closer to the center of the domain we observe a nicely detailed shock diffraction about a column, with almost symmetric rarefaction waves on both sides, and a reflected shock upstream of the column. A similar diffraction process is observed about a column toward the left of the expanding shock. Shock diffraction about the cars situated between these columns is more complex. The reflected pressure amplitude depends on the car profile (i.e., percentage of blocked area from floor to ceiling), car width, and shock incident angle. The impacting shocks were partially transmitted under the cars (wheels were not modeled here).

The dominant shocks on the west (left) side at $t = 95.81$ ms (Fig. 1c) are the shock reflections from the walls of ramps A and B, shock reflection from the second row of cars (reflected in a northwest direction), and shock reflection from the wall of ramp N. The primary shock was partially reflected from the side wall of ramp N, and the row of cars north of ramp T, expanded around the wall surrounding ramp N, and about the vehicles parked west of ramp T. The front propagating south shows, from left to right: a large-amplitude shock and a Mach stem formed near the west wall (between the wall and the adjacent cars); a lower-amplitude shock diffracting about the parked cars; the incident, large-amplitude shock in the traffic lane, followed closely by the reflection from the side wall of ramp N; the broad-side impact of this reflected shock on the row of cars north of ramp T; shock diffraction about the southwest corner of ramp T; and the impact of the shocks reflected from ramps A and B on the central two rows of cars and on the row of cars north to tower N. Naturally, each shock impact produces a reflected shock, plus an assortment of shocks that reverberate between the cars. At the center, shock stagnation at the north wall of ramp T (and the resulting high pressure) explains the significant damage to the cars parked nearby, and the collapse of the wall. Since the cars to the left (west) and at the center of the wall are parked normal to the wall, the shock front is fairly uniform. On the right (east), however, the two vehicles parked diagonally perturbed the flow sufficiently to produce a curved (almost spherical) reflected shock. At the center-right we observe the sharp shock transmitted in the traffic lane between ramp T and the west wall of tower B, the multiple reflections from rooms near the northwest corner of tower B, and from the broad-sided parked cars. At $t = 95.8$ ms, the primary shock diffracts about the south wall of ramp T and the transfer shaft room near the west wall of tower B. The reflected shock from the north wall of ramp T expands at this time near the ramp entrance. In the Secret Service area, we observe the evolution of the primary curved shock and the multiple diffraction processes about the columns. The shock reflected from the two central columns at $t = 89.55$ ms and interacted at $t = 95.81$ ms with the walls of the pipe room attached to the north wall of tower B, to produce a strong reflection, an expansion, and a Mach stem. The weakened incident shock expanded to the north, and hit the PATH wall at $t = 106$ ms. The shock front impacted the presidential limousines at $t = 115.91$ ms. The large-amplitude, spherically reflected shock near the north wall of tower B, with

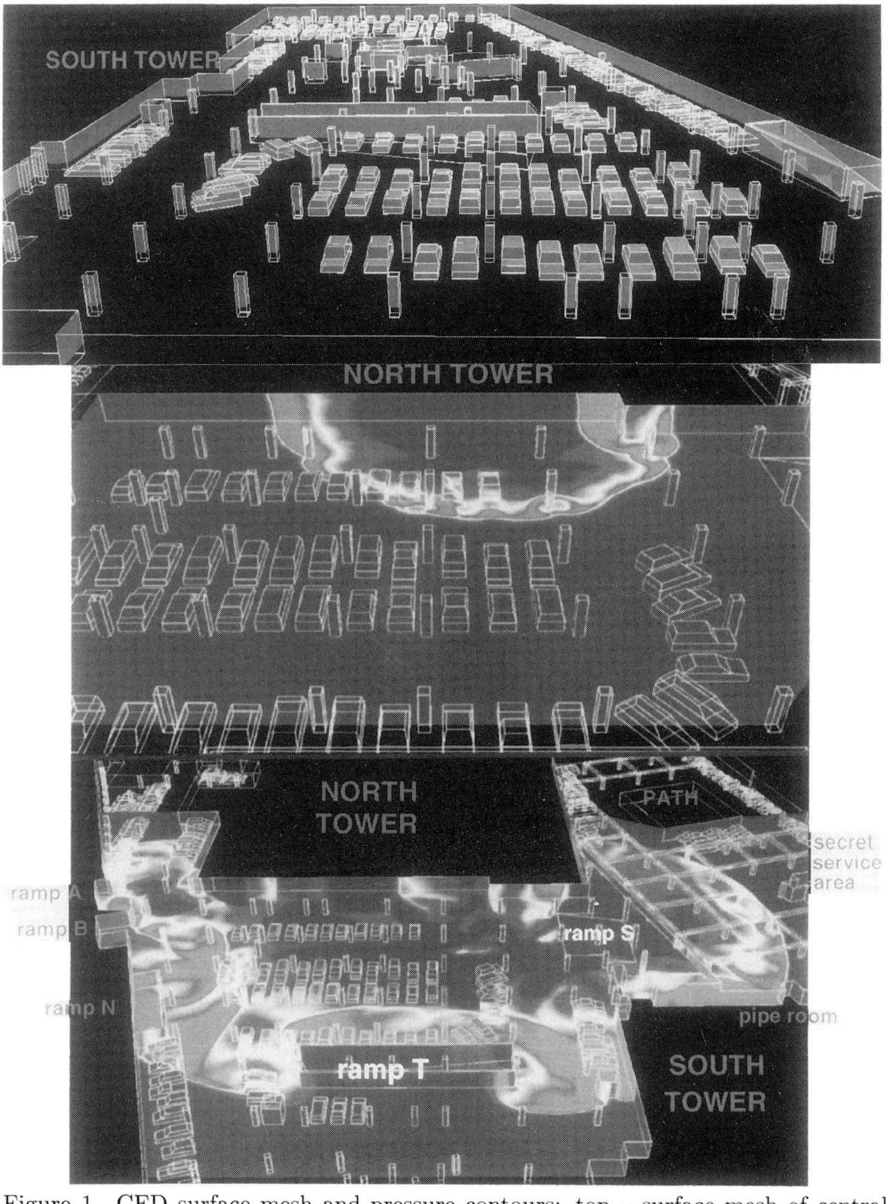

Figure 1. CFD surface mesh and pressure contours: top – surface mesh of central parking area; middle – expanded view of pressure contours at $t = 12.51$ ms; bottom – expanded view of pressure contours at $t = 95.81$ ms.

a circular imprint on the floor, resulted from shock reflection from a corner of the support beams.

The solution was integrated to $t = 287.73$ ms, at which time the shocks have filled the complete computational domain.

4. Summary and conclusions

This paper described the application of a recently developed three-dimensional, adaptive, finite-element, edge-based, ALE shock-capturing methodology on unstructured tetrahedral grids (FEFLO96) to the simulation of blast diffraction inside the B-2 level of the World Trade Center. The evolution of the primary shock and the multitude reflections were studied in great detail. Several shocks were tracked over time. Several interesting 3D shock-diffraction processes were identified and discussed. Among these are the formation of "hot spots" due to three-dimensional wave focusing, shock reverberation between cars, shock diffraction about cars and columns, and the formation of three-dimensional Mach stems. The results demonstrate excellent shock adaptation at all times. All shocks were captured as sharp discontinuities without producing pre- or post-shock oscillations.

Acknowledgement. Funding for this research work was provided by the Defense Nuclear Agency. Drs. Michael E. Giltrud and Douglas A. Sunshine served as the technical program monitors.

References

Baum JD, Löhner R (1994) Numerical simulation of shock-elevated box interaction using an adaptive finite element shock capturing scheme. AIAA J 32(4):682.

Baum JD, Luo H, Löhner R (1994) Numerical Simulation of Blast in The World Trade Center, SAIC Report 1994-23.

Löhner R, Baum JD (1990) Numerical Simulation of Shock Interaction with Complex Geometry Three-Dimensional Structures Using A New Adaptive H-Refinement Scheme on Unstructured Grids, AIAA-90-0700.

Löhner R, Morgan K, Peraire J, Vahdati M (1987) Finite element flux-corrected transport (FEM-FCT) for the Euler and Navier-Stokes equations. Int J Num Meth Fluids 7:1093.

Luo H, Baum JD, Löhner R (1994) Edge-based finite element schemes for the Euler equations. AIAA J 32(6):1183.

Blast-wave interaction with multiple structures

L. H. J. Absil, A. C. van den Berg, J. Weerheijm
TNO Prins Maurits Laboratory
P.O. Box 45, 2280 AA Rijswijk, The Netherlands

Abstract: This paper presents an experimental and numerical study of blast propagation and interaction with multiple urban structures. The main objective of this research is to provide detailed information on blast propagation through urban and industrial areas so as to improve guidelines for the determination of safety distances around potentially hazardous industries.

Key words: Blast waves, Blast interaction with structures, Blast damage assessment, Explosion effects, Safety distances

1. Introduction

Accidental explosions form a threat to communities in the vicinity of hazardous industries. An important aspect of an explosion is the generation of a blast wave, which is responsible for most of the damage at large distances from the explosion source. In the past much effort has been put into developing an understanding of what mechanisms cause explosions. Little attention, however, has been given to assessing the characteristics of the blast wave as it propagates away from the explosion source. Because of this lack of knowledge, current risk analysis is based on the strength of the undisturbed pressure wave: i.e., damage and risk contours are related to the free-field decay of the peak overpressure with distance from the explosion source. Due to the shielding effect of houses on the edge of a city, however, the blast will be attenuated resulting in a reduction of damage to the rest of the city. This modification of the blast wave due to the interaction with surrounding urban structures has not been accounted for in present-day risk analysis. This deficiency has a significant effect on damage-assessment and the determination of safety distances.

On behalf of the Dutch Ministry of Housing, Physical Planning and Environment, a research program was started comprising an experimental and numerical study of blast propagation and interaction with multiple urban structures. The main objective is to generate data that will form the basis for new guidelines for blast damage assessment to urban areas. The first phase of this study was to investigate what parameters play a major role in the shielding effect. Variables that were expected to mainly influence the degree of shielding are the

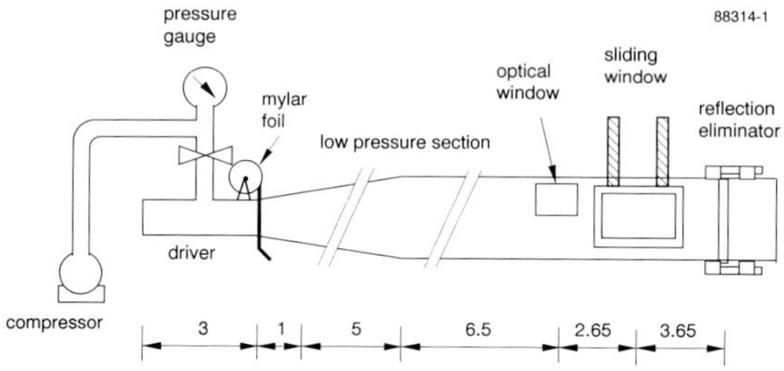

Figure 1. A schematic drawing of the blast simulator (distances given in meters).

peak overpressure and duration of the blast wave, the structure geometry and size, the height of the first structure, the number of structures, and the distance between them.

2. Experimental set-up

The experiments were conducted in a shock tube with a square test section of 40 $\times$ 40 cm^2, see Fig. 1. The 3-m long driver section can be filled with compressed air, up to a maximum of 30 bar, generating blast waves with a peak overpressure of up to 110 kPa and a positive phase duration of 30 ms. The strength of the shock wave can be regulated by varying the initial pressure in the driver section. As scale models, two-dimensional rectangular (5 $\times$ 5 cm^2) cylinders were used and loaded with shock waves of 1 and 10 kPa overpressure: 1 kPa is a typical damage criterion for glass failure, while partial demolition of houses can occur at 10 kPa. These relatively low overpressure levels were selected because the emphasis in the present study is on far-field conditions, where shielding effects can reduce the amount of damage inflicted considerably. The size of urban housing and the mutual distance between houses in a city is typically of the order of 10 meters. Hence, in the experiments the geometry was scaled by a factor of 200. According to the Hopkinson scaling law (Baker et al. 1973) the duration of the blast wave should be scaled proportional to the length scale. The shock-tube driver section was shortened to its smallest size, i.e. 7 cm, to limit the positive phase duration of the blast wave to 4.2 ms, which corresponds to a full-scale duration of about 800 ms. This duration was rather long with regard to the duration of a vapor-cloud explosion blast at far-field conditions, which is typically of the order of 100 ms. Because of the long duration of the blast wave generated in the shock tube, a few experiments were performed with electric detonators to generate the blast wave. With these detonators, blast-wave durations of the order of 0.5 ms could be achieved which correspond to the desired duration of 100 ms at full scale. Highly sensitive piezo-resistive pressure transducers were used to record the pressure-time history at the centers of the front, top, and rear faces of the structures. A Wollaston-prism shearing interferometer (Absil et al. 1992) and a holographic interferometer were used to visualize the shock-wave flow.

3. The BLAST code

For the simulation of a three-dimensional blast-wave/structure interaction, the Computational Fluid Dynamics code BLAST has been developed at TNO-PML (Van den Berg 1990). The code solves the Euler equations which describe inviscid compressible flow. The Flux-Corrected Transport scheme is used for optimum description of shocks and contact discontinuities. The code has the capability to calculate the pressure, density, and temperature distribution around structures for a given infalling blast wave. The BLAST code was used to simulate the flows experimentally studied.

4. Results

A typical example is shown in Fig. 2, which shows the interaction of a blast wave with two rectangular structures. The excellent agreement between the measured interferograms (left) and the isobar pattern as obtained by the numerical simulation (right) indicates that the BLAST code gives a fairly good description of the main features of the shock-wave flow. The first row shows the vertical shock wave striking the structure from the left. After the shock strikes the first structure, part of the shock is reflected backwards, as shown in the second row. The overpressure in the reflected shock is substantially greater, by about a factor of 2, than that of the incident shock. This pressure difference causes a flow from the region of high to low pressure, resulting in the formation of circular expansion waves at the corners of the structure. The reflected frontal overpressure is relieved by the rarefaction wave that travels down the front surface. At the edge of the structure, viscous separation occurs generating a rolled-up vortex featuring a steep density gradient. In the third row, the shock wave has passed the rear face of the first structure, and the compression wave that bridges the pressure difference between the induced air flow above the buildings and the undisturbed flow in between is shown. The compression wave travels downward across the rear face of the first building and subsequently reflects at the earth's surface. The fourth row shows the complicated wave pattern that develops in between the two buildings. The front of the second building will be hit by the weaker diffracted wave, which indicates that it is shielded against the blast wave by the first building.

Next, the infalling wave on the second building is reflected back and the rear face of the first building is hit again, as shown in the last row. This explains why the rear face of a building, which is closely surrounded by other buildings, also shows considerable damage after an accidental explosion.

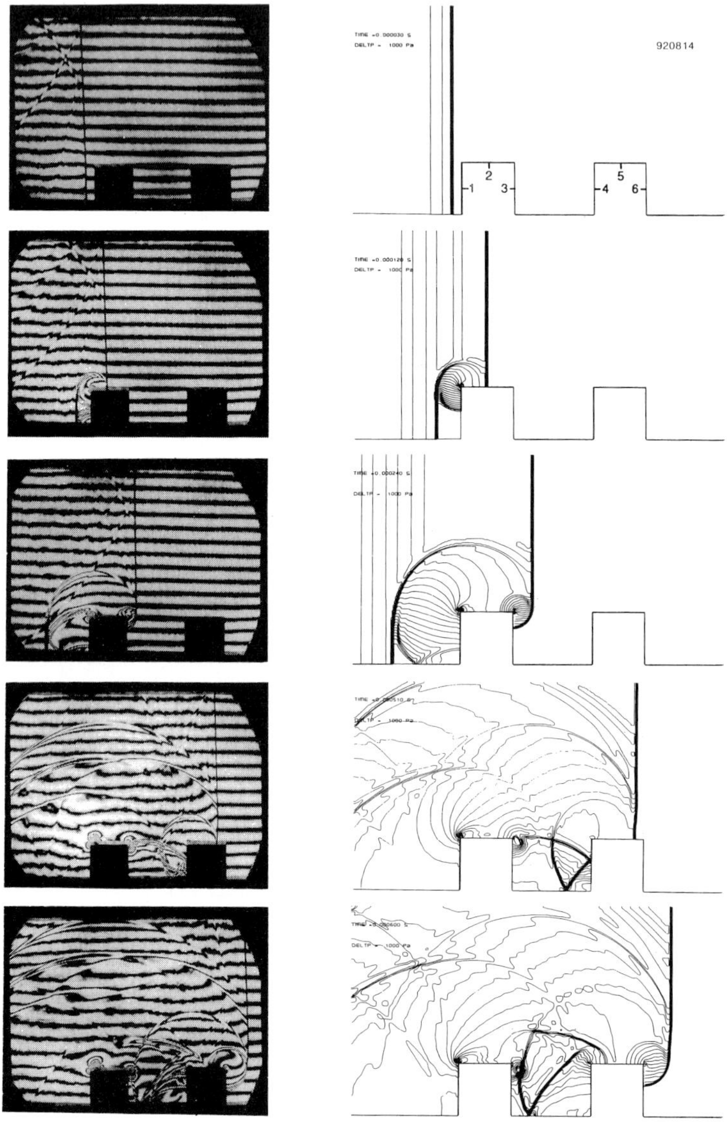

Figure 2. Comparison of interferograms (left) and calculated isobar patterns (right) for a blast wave of 10 kPa and 60-ms duration.

The overpressure versus time records as calculated with BLAST at the midpoints of the front, top, and rear faces of the structures are presented in Fig. 3.

This calculation was made for a blast wave of 10 kPa and 60-ms duration falling in at 10-m high buildings located 15 m apart. In these overpressure signals the complicated wave pattern as seen in Fig. 2 can be readily recognized. The overpressure signal predicted at location 3, at the back of the first building, for instance, shows a sequence of four shock phenomena which can be traced in the flow patterns shown in Fig. 2. The first peak is due to the passage of the infalling blast wave. The gradual pressure build-up is the result of the diffraction of the shock around the first structure. The second pressure jump corresponds with the reflection of the primary wave by the ground. The third pressure increase corresponds with the shock wave reflected directly from the second building. This shock wave is immediately followed by a fourth wave, which is the result of the reflection by the second building and subsequently by the ground. Because the second structure is hit by the weaker diffracted wave, the blast load on the front of the second structure will be lower than that on the front of the first structure. The computation shows that the blast overpressure experienced by the front of the second building is about two times lower than the reflected overpressure of the undisturbed blast wave endured by the front of the first building. A similar reduction in overpressure was found in the experiments using electric detonators. This shows that a considerable shielding effect can occur under realistic conditions.

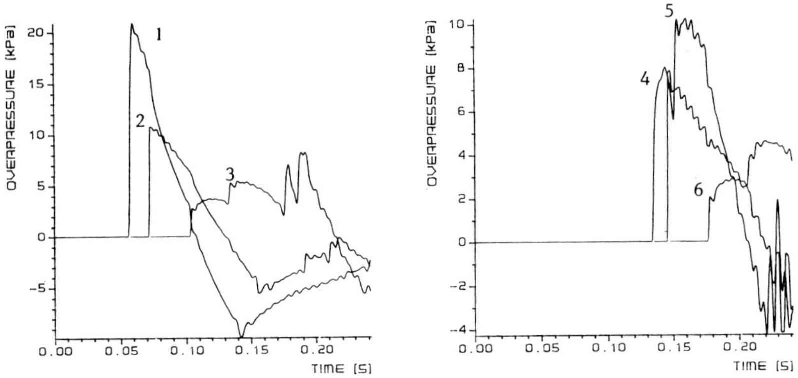

Figure 3. Pressure vs. time signals at stations 1 to 6 (see Fig. 2) as calculated for a blast-wave (10 kPa, 60 ms) interaction with two buildings (10-m high).

The findings of the parametric study can be summarized as follows:

- the shielding effect was found to behave linearly for these rather low overpressures, i.e., nearly the same shielding effect was found for the 1-kPa blast wave as for the 10-kPa blast wave;

• the shielding effect hardly increases after the second structure, i.e., the pressure load on the second, third, and fourth structure is nearly the same;

• the overpressure build-up in between the buildings and the shielding effect were found to be highly dependent on the blast-wave duration and the spacing between the buildings. As the blast wave sustains, multiple wave reflections from the surrounding structures cause a pressure increase on the front and rear faces of the structures;

• the degree of shielding strongly depends on the height of the first structure;

• the geometry of the roof (i.e., flat versus pitched) only had a minor effect on the pressure-load on the structures.

5. Conclusions

The present paper has demonstrated that a significant shielding effect can occur in realistic situations. This shielding effect strongly depends on the duration of the blast wave, the height of the first building, and the distance between the buildings. Good agreement between measurements and numerical simulation was found, indicating that the BLAST code gives a good description of the main features of the diffraction process.

Acknowledgement. The authors are indebted to M. Dirkse for performing the measurements and D. van Leeuwen for performing the calculations. This research was funded by the Dutch Ministry of Housing, Physical Planning and Environment.

References

Absil LHJ, Bakkum EA, van der Meer B, Weerheijm J (1992) Shearing interferometry and diffraction of weak shock waves by scale models. In: Proceedings of the 20th International Congress on High Speed Photography and Photonics, Victoria, Canada.

Baker E, Westine P, Dodge F (1973) Similarity Methods in Engineering Dynamics, Spartan Books, New Jersey.

Van den Berg AC (1990) BLAST 2D — A Code for Numerical Simulation of Two-Dimensional Blast Effects, TNO-PML Internal Report.

Efficient 2D and 3D locally adaptive unstructured Euler solvers for unsteady shocked gas flows

A. Galyukov, A. Fursenko, P. Voinovich
Advanced Technology Center
P. O. Box 160, St. Petersburg 198103 Russia

T. Saito
Cray Research Japan Ltd, Supercomputer Center, Institute of Fluid Science
Tohoku University, 2-1-1 Katahira, Aoba-ku, Sendai 980-77 Japan

E. Timofeev
Ioffe Physical-Technical Institute, Russian Academy of Sciences
Polytekhnicheskaya str. 26, St. Petersburg 194021 Russia

Abstract: This paper presents a set of efficient 2D and 3D Euler solvers for strongly unsteady shocked gas flows. The key points of the numerical methods are: a high-order predictor-corrector Godunov-type finite-volume scheme and local refinement/derefinement procedures supported by the flexible data structures of unstructured triangular (2D) and tetrahedral (3D) grids. The grid generation is performed based on the Watson incremental 2D/3D triangulation providing control over the grid spacing to enhance the resolution in the regions of special interest.

Key words: Unstructured Euler solver, Local grid refinement, Grid generation

1. Introduction

The recent trend toward extensive use of unstructured grids in computational fluid dynamics (CFD) is due to their remarkable adaptability to complex geometries and possible dynamic grid adaptation to the flow pattern. The latter feature appears to be particularly attractive when applied integrally with a shock-capturing scheme, as the accuracy in that case is limited by the resolution at discontinuities, i.e., shock waves, contact surfaces, and slip surfaces. An unstructured grid allows the natural combination of a rather large grid spacing adequate for smooth flow approximation with a very fine grid at discontinuities. When only a small part of the computational domain is covered by the fine grid the local grid refinement leads to high computational efficiency of the method. However, the development of an unstructured code is labor intensive and involves numerous points offering a broad spectrum of alternatives.

2. Grid generation

The principal stage of pre-processing in CFD is the grid generation. Numerous recent papers have been devoted to this subject (see e.g., the review by Barth 1992), but the totally automatic unstructured grid generation for an arbitrary domain remains a challenge.

In the 2D case our grid generation procedure includes the following basic stages: the major boundary nodes are allocated to define the domain geometry; additional nodes are introduced optionally within pre-determined spacing at the boundaries and/or into the domain; the whole set of nodes is triangulated utilizing the technique by Watson (1981) and following the Delaunay criterion; neighboring nodes located prohibitively close are united; the grid can be incrementally refined using the above triangulation technique to meet additional criteria; edge swapping is applied to improve the grid locally; a global smoothing is performed to maintain gradual variation in grid spacing. Some of the steps are invoked iteratively, either individually or in combinations.

The 3D case presents a fundamental problem at the stage of the domain boundary definition; CAD systems are used in many practical engineering applications to create the domain bounding surfaces. For simple cases we utilize a modified version of our 2D grid generator for the surface triangulation. As the visual quality control becomes impractical for 3D unstructured grids, statistical data provided by the grid generation procedures are used extensively.

The basic principles of grid generation in 3D domains are the same as described for two dimensions. A 3D modification of Watson's algorithm performs tetrahedrization of the domain. Nodes positioned too closely to the domain boundary due to grid smoothing are deleted, and the polyhedral cavity surrounding a deletable node is triangulated by the advancing front method (Tanemura et al. 1983).

3. The finite-volume scheme

Constructing the numerical method we proceed from the integral conservation equations applied to node-centered control volumes. In 2D and axisymmetric cases the control volume about a node is confined by the respective parts of medians of the triangular grid elements sharing the vertex. The 3D control volume is bounded by planes passing through the tetrahedron center and the centers of its faces and edges (see e.g., Fursenko 1993).

A Riemann solver is used to compute the mass, momentum, energy, and species fluxes at the control volume interfaces. Linear interpolation within elements (triangles in 2D and tetrahedra in 3D) enhances the spatial accuracy. A limiter is applied when calculating average gradients of primitive gasdynamic variables at each node to prevent spurious oscillations. The second-order accuracy with respect to time is achieved by the two-step predictor-corrector technique. The predictor step of the scheme involves no Riemann solver which contributes to the efficiency of the method. For a detailed description of the scheme see Fursenko (1993).

4. Adaptive grid refinement

To improve the resolution of discontinuities, a grid adaptation procedure is invoked periodically to introduce new grid nodes in the vicinity of shocks, contact surfaces, and slip surfaces and to remove previously inserted nodes where the solution becomes smooth. The classical hierarchical refinement technique (Löhner 1987) is used together with the 2D/axisymmetric solvers. A bisectional refinement procedure based on the method by Rivara & Levin (1992) supplements the solver in the 3D case. A variety of sensors and criteria are used in different applications depending on the specific problem and flow features of interest. The most typical and universal sensor is based on the second derivative of density. The grid nodes are labeled for either refinement or derefinement in two dimensions, while edges are labeled for refinement and nodes for derefinement in three dimensions. Limited gradients at the grid nodes and Riemann solver solutions are optionally used to assign gasdynamic values to the newly inserted nodes. As extra nodes are introduced only at local singularities of the solution, the total amount of grid nodes is normally one to two orders of magnitude lower than for a uniform grid providing the same accuracy. This essentially increases the computational efficiency of the methods.

Some specific features should be emphasized regarding the adaptive grid refinement for transient flow simulations. First, the grid refinement should be reversible; second, the grid rearrangement should be fast compared to the solver; and third, the same data structure should support both the solver and grid refinement/derefinement routines.

5. The data structures

The required memory and efficiency of the grid generators, solvers, and grid adaptation procedures are dramatically affected by the underlying data structures which establish the connectivity of the grid objects. An extremely compact (however suited exclusively for scalar processing) data structure which contains no other objects but nodes has been developed for the 2D unstructured solver. The cross references of nodes are sorted and positioned so that the neighbors of a given node follow each other counter-clockwise. Twelve fixed reference positions are reserved for each node. This is sufficient for all newly inserted nodes and for nodes of the basis grid which originally have 6 or less neighbors. Basis-grid nodes with more neighbors are assigned a doubled reference space.

The data structure for 2D grid generation is also based on nodes, but since the number of neighbors can vary unpredictably references are stored in unidirectional linked lists.

The 3D grid generator, solver, and adaptation procedures utilize the same data structure with tetrahedra as leading elements. A tetrahedron refers to the neighboring tetrahedra, to the edges, and to the vertices (nodes). The edges in turn refer to the nodes they connect. Extra references from nodes to two "parent" nodes are used to support reversibility of the refinement procedure.

The indirect addressing of the grid objects is used in all cases on the basis of dynamic lists of occupied/spare vacancies to avoid expensive procedures for collecting memory units released by the grid derefinement.

6. Vector processing

As mentioned before, the 2D code was developed for efficient operation on scalar single-processor hardware (workstations and PCs). The grid size (in terms of the number of grid objects) in 3D cases normally requires supercomputer resources.

The main loops of the 3D solver which contain recursions are the loops over tetrahedra and over edges. The control volumes, gradients of primitive gasdynamic variables, and increments of conservative variables for the boundary grid nodes are computed in the loops over tetrahedra; the time step limit and increments of conservative variables for internal grid nodes are computed in loops over edges. For efficient vector processing the above loops are split into vectorizable subloops so that within any particular subloop a quantity associated with a specific grid node is updated only once. The tetrahedra (edges) partitioning is performed as follows: zero binary attributes are assigned to all nodes. For each tetrahedron (edge) the attributes of the involved nodes are analyzed and the position of the first common zero bit denotes the subloop number within which the tetrahedron (edge) is to be treated; the respective bit in the attributes is then set to 1.

7. Post-processing

Post-processing for CFD usually implies an appropriate flow visualization technique applied to the computed flow fields. Contour lines are most popular when presenting 2D results; the same method is used in three dimensions, displaying the flow characteristics at the domain bounding surfaces and/or in a set of cross-sections.

For the purpose of quantitative comparison with experimental data, the computed results can be presented in an experiment-like form, e.g., as interferograms, schlieren, and shadowgraphs (see Fursenko et al. 1993a, 1993b).

In some cases extensive data treatment and acquisition is required for multiple time moments, for example, to prepare flow-field animations. For a shock-focusing problem the pressure data reduction was performed with respect to time, i.e., the complete computation for a separate case resulted in a map of maximal pressure values whenever achieved during the strongly transient process of shock interactions (Babinsky et al. 1995). As the respective procedures involved an expensive data conversion into a fixed structured grid independently for every time step, it was accomplished on the massively parallel platform CRAY T3D.

8. Concluding remarks

A high level of automation in pre-processing is achieved and practical experience is gained mostly for 2D and axisymmetric problems. Numerous flows of those classes have been simulated during the past years, including shock-wave reflection, diffraction and focusing, blast-wave interactions, and some engineering applications (see bibliography by the authors given below). The numerical investigation of a shock-wave interaction with the material interface separating two different gases and analysis of the resulting Richtmyer-Meshkov instability are currently in progress.

The described methods and developed software have been shown to be extremely efficient not only in the use of computer resources, but also in terms of the time required to set up a new problem and get reliable data. Preliminary 3D results suggest that respective software can provide comparable labor efficiency on the basis of modern supercomputer architectures in the near future.

9. A viscous extension

The flexible data structures of the unstructured Euler solvers make it possible to combine them with solvers based on alternative grid systems. For the analysis of viscous flows with shocks and transient separations the 2D unstructured solver has been upgraded to approximate the Navier-Stokes equations and coupled with an implicit structured Navier-Stokes solver to compute the flow in the boundary layer at high Reynolds numbers. Results are reported in Ofengeim et al. (1995).

To simulate strong viscous-inviscid interactions over realistic 3D geometries the unstructured solver was combined with a Navier-Stokes solver written for a prismatic grid attached to the body surface.

Acknowledgement. The authors express their sincere gratitude to Profs. F. Durst, Yu. Golovachov, N. Satofuka, M. Schäfer, and K. Takayama, who contributed greatly to the arrangement of this work. We appreciate helpful assistance of Drs. H. Babinsky, Yu. Makarov, and D. Sharov. The sponsorship by the Russian Foundation of Fundamental Investigations (Grants 93–02–14797 and 94–01–01456), Friedrich Alexander University Erlangen–Nürenberg (Contract FAU4001–01IR303), European Community (COPERNICUS Project COP–94–01490), and Ministry of Education, Science and Culture, Japan, is kindly acknowledged.

References

Babinsky H, Onodera O, Takayama K, Timofeev E, Voinovich P (1995) The influence of geometric variations on shock focusing in cylindrical cavities. In: Shock Waves, Proc of the 20th ISSW, Sturtevant B, Shepherd JE, Hornung HG (eds), World Scientific, Vol 1, pp 495–500.

Barth TJ (1992) Aspects of Unstructured Grids and Finite-Volume Solvers for the Euler and Navier-Stokes Equations, AGARD Report 787.

Fursenko AA (1993) Unstructured Euler solvers and underlaying algorithms for strongly unsteady shocked flows. In: Proc 5th Int Symp on CFD, Daiguji H (ed), JSCFD, 1:227–234.

Fursenko AA, Mende NP, Oshima K, Sharov DM, Timofeev EV, Voinovich PA (1993a) Numerical simulation of propagation of shock waves through channel bends. CFD J 2:1.

Fursenko AA, Sharov DM, Timofeev EV, Voinovich PA (1993b) An efficient unstructured Euler solver for transient shocked flows. In: Shock Waves @ Marseille, Brun R, Dumitrescu LZ (eds), I:371–376.

Löhner R (1987) The Efficient Simulation of Strongly Unsteady Flows by the Finite Element Method, AIAA Paper 87–0555.

Ofengeim D, Timofeev E, Voinovich P (1995) A locally adaptive structured/unstructured Navier-Stokes solver for essentially unsteady shock-obstacle interactions at high Reynolds numbers. In: Shock Waves, Proc of the 20th ISSW, Sturtevant B, Shepherd JE, Hornung HG (eds), World Scientific, Vol 1, pp 537–542.

Rivara MC, Levin C (1992) A 3D refinement algorithm suitable for adaptive and multi-grid techniques. Commun in Appl Num Methods 8:281.

Tanemura M, Ogava T, Ogita N (1983) A new algorithm for three-dimensional Voronoi tesselation. J Comput Phys 51:191–207.

Watson DF (1981) Computing the n-dimensional Delaunay tesselation with application to Voronoi polytopes. Computer Journal 24:167.

Viscous Effects

Boundary-layer transition on slender blunt cones at hypersonic speeds

M.A. Zanchetta, R. Hillier
Department of Aeronautics
Imperial College of Science, Technology and Medicine
Prince Consort Road, SW7 2BY, London, UK

Abstract: Experimentally measured heat-transfer data are presented for hypersonic transitional boundary layers. The experiments are performed on a 5° spherically blunted cone in the Imperial College No. 2 gun tunnel, a Mach 9, high Reynolds number facility. The data presented include time-resolved heat-transfer-rate traces, from which information concerning the structure of the transitional hypersonic boundary layer is extracted. The effects of roughness are examined in the reversal regime.

Key words: Transition, Hypersonics, Intermittency

1. Introduction

Kinetic heating poses severe problems for the design of hypersonic vehicles. Regions of high thermal loads include stagnation points, attachment and re-attachment zones, shock and boundary-layer interactions, shock impingement areas, and boundary layer transition regions. Design proposals use nose blunting to reduce the thermal loads experienced in the stagnation regions. The degree of the blunting, via the effect of the resultant entropy layer, strongly influences the location and the extent of the transition region. This important topic is the subject of this paper.

2. Techniques

The test geometry is a 5° cone. Two models are used. The first is a 700-mm chord length model with ten interchangeable spherically blunted nose cones. This model is equipped with 65 thin-film platinum gauges. The second, a 350-mm chord model, has a 25-mm nose radius and is reserved for the purpose of flow visualization. The test facility is a nominal Mach 9 intermittent gun tunnel. Three fully calibrated conditions offer unit Reynolds numbers, Re_∞/m, between 6.5 and 55 $\times$ 10^6, with total temperature in excess of of 1050 K. Run times are typically 20 ms with a 4 ms steady flow window.

3. Results

Thirty different test cases are examined. Each test case is characterized by the nose radius, r_n and the unit Reynolds number, Re_∞/m, or more generally by the nose radius Reynolds number, $Re_n = r_n \times Re_\infty/\text{m}$. Results, shown in Fig. 1, indicate that small amounts of nose blunting, $Re_n < 300,000$, cause a delay in

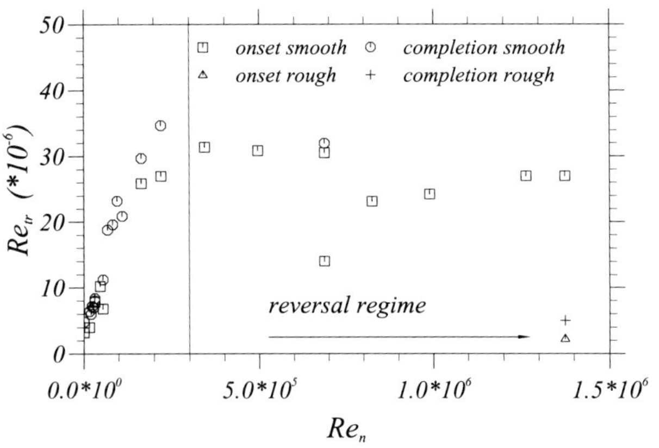

Figure 1. Transition Reynolds number against nose Reynolds number.

transition onset and an increase in the length to transition completion, $Re_{\Delta tr}$. For $Re_n > 300,000$, experiments indicate that the transition behavior is highly sensitive to surface roughness. This regime is denoted the transition reversal regime. A roughness Reynolds number, Re_r, is defined as the product of the roughness height, h, and the unit Reynolds number, Re_∞/m: $Re_r = h \times Re_\infty/\text{m}$. Testing shows that for $Re_r < 500$, the boundary layer is laminar for surface Reynolds numbers in excess of 32×10^6. At $Re_r > 500$, transition completion occurs at 4.2×10^6.

3.1. The structure of the hypersonic transitional boundary layer

In Fig. 2, five heat-transfer-rate time traces are presented. The data are gathered over seven runs with the 3 mm nose at $Re_\infty/\text{m} = 55 \times 10^6$, $Re_n = 165,000$. Due to the nose-blunting entropy-layer effects, there is a significant delay in transition onset, ($Re_{tr} = 25.9 \times 10^6$ cf. $Re_{tr} = 3.2 \times 10^6$ for the sharp nose) and an increase in the length to transition completion ($Re_{\Delta tr} = 3.8 \times 10^6$ cf. $Re_{\Delta tr} = 1.62 \times 10^6$ for the sharp nose). The transition region spans the entire array of gauges. The most upstream gauge, (gauge 02, $s = 0.3505$ m), is wetted by a laminar boundary layer which is characterized by the presence of 'spikes', connected with turbulent or transitional events convecting over the gauge. The

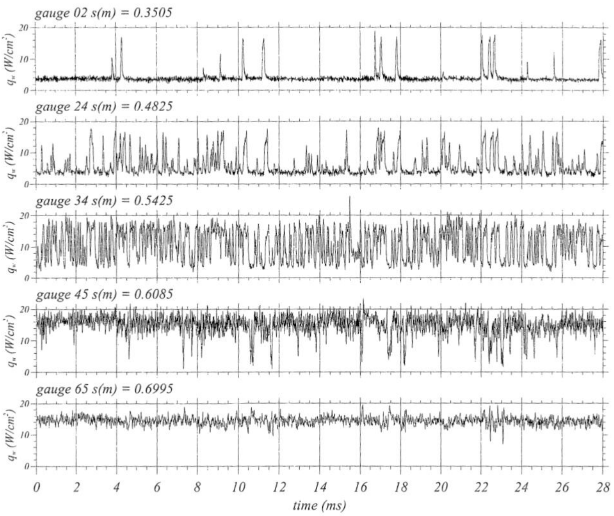

Figure 2. Heat-transfer-rate time histories for five selected gauges; $r_n = 3$ mm, $Re_\infty/\mathrm{m} = 55 \times 10^6$, $Re_n = 165,000$.

duration of the events is 1/10 ms, indicating that they are elongated structures 150δ in length. The structures are tracked as they convect and evolve downstream. Convection speeds calculated between gauge 02 and gauge 24, ($s = 0.4825$ m), indicate that the leading edges and trailing edges of the events travel at 98% and 68% of the local edge velocity, respectively. Further event formation and growth are seen at gauge 34, ($s = 0.5425$ m), where the sensor is witnessing in equal measure laminar and turbulent flow. Due to the speed differential between the leading edge and the trailing edge of the disturbances, successive events merge, leaving only 'stubborn' regions of laminar flow, as seen in the time trace for gauge 45 ($s = 0.6085$ m). Further downstream, the transition process is completed and the boundary layer exhibits purely turbulent behavior, as seen at gauge 65 ($s = 0.6995$ m).

For each gauge the normalized probability density function is determined using 3500 data samples, Fig. 3. The two features visible are the peaks in correspondence to the mean laminar and turbulent heat-transfer-rates, 4 W/cm^2 and 15 W/cm^2 respectively. The distribution at gauge 02 is biased to the 4-W/cm^2, laminar value. Moving downstream, gauges 24, 34, and 45, the effect of boundary-layer transition is to redistribute the PDF from the laminar value

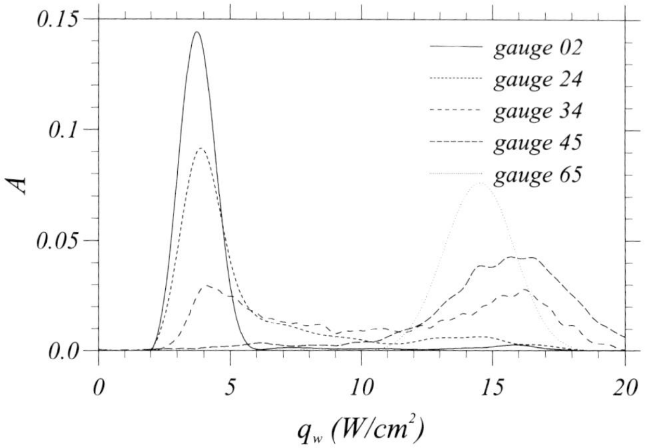

Figure 3. Normalized probability density functions for five selected gauges; $r_n = 3$ mm, $Re_\infty/\mathrm{m} = 55 \times 10^6$, $Re_n = 165,000$.

to the turbulent value. The PDF for gauge 65 is centered purely around the turbulent value of 15 W/cm^2.

It is possible to extract from the time histories the percentage of time each gauge witnesses turbulent flow, represented by the intermittency function, γ. By definition, $\gamma = 0$ for laminar flow and $\gamma = 1$ for turbulent flow. The intermittency distribution against wetted distance for the present case can be found in Fig. 4.

3.2. The reversal regime

The experiments performed in the reversal regime illustrate the boundary-layer sensitivity to roughness in the hemispherical nose region. Liquid-crystal surface-temperature visualization experiments have been performed to study roughness effects. Liquid crystal paints are composed of encapsulated thermochromic particles. When applied to the model surface, they form a covering 10 μm to 20 μm thick. The roughness Reynolds number of the paint, $Re_r < 200$, is insufficient to promote transition and the boundary remains laminar over the entire model length, $Re_{tr} > 22 \times 10^6$, Fig. 5. Controlled application of the paints leads to the formation and deposition of 50 μm particles on the model surface. This enhanced roughness environment, $Re_r = 500$, provides sufficient disturbance to promote transition, Fig. 6. The transition process is characterized by the presence of streak-like structures visible downstream of the sphere–nose junction. The streaks propagate 1.5 nose radii downstream before becoming turbulent. The onset of turbulence is marked by an abrupt change in spreading angle.

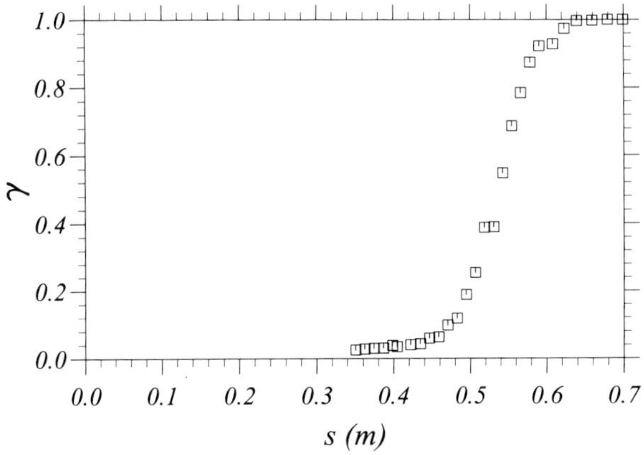

Figure 4. Intermittency function γ, against wetted distance s; $r_n = 3$ mm, Re_∞/m $= 55 \times 10^6$, $Re_n = 165,000$.

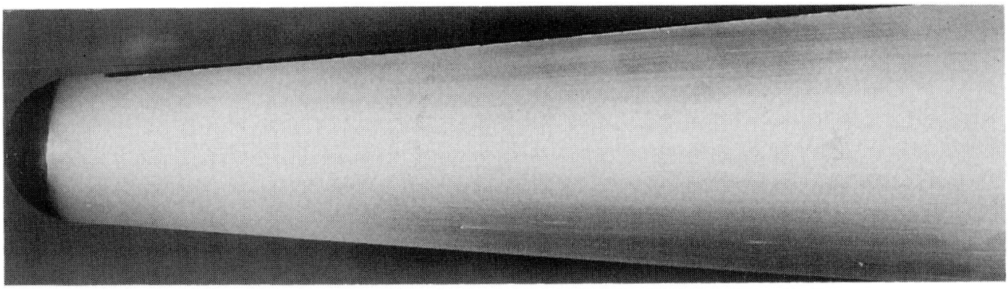

Figure 5. Liquid-crystal surface-temperature visualization, smooth nose configuration; $r_n = 25$ mm, $Re_\infty/\text{m} = 55 \times 10^6$, $Re_n = 1,350,000$.

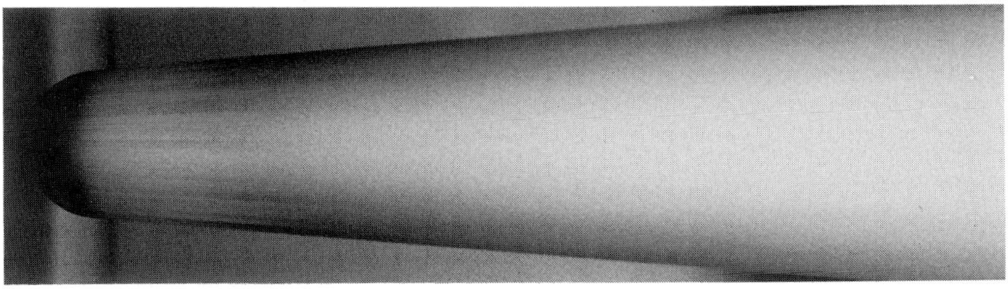

Figure 6. Liquid-crystal surface-temperature visualization, rough nose configuration; $r_n = 25$ mm, $Re_\infty/\text{m} = 55 \times 10^6$, $Re_n = 1,350,000$, $Re_r = 500$.

Due to close circumferential spacing of the streaks, they merge readily once turbulent, giving $Re_{tr} < 5 \times 10^6$.

The sensitivity to isolated roughness elements is explored using 0.13-mm high, 1-mm side length, equilateral triangle trips, located on the sphere–cone junction, ($Re_s = 2.15 \times 10^6$, $Re_r = 1300$.) The effect, Fig. 7, is to cause a highly

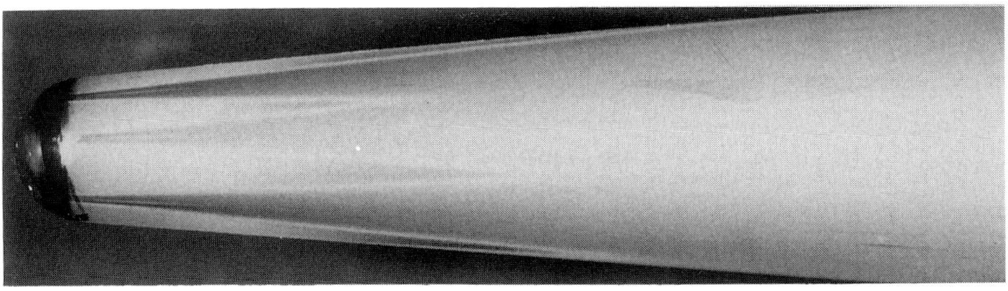

Figure 7. Liquid-crystal surface-temperature visualization, triangular trip elements applied; $r_n = 25$ mm, $Re_\infty/\text{m} = 55 \times 10^6$, $Re_n = 1,350,000$, $Re_r = 1300$.

organized wake, bounded on either side by vortices shed from the rear corners of the trip. The vortices are unstable and undergo transition after travelling two nose radii downstream, giving a transition Reynolds number for the wake of 5×10^6. Transition in other regions of the boundary layer is caused by contamination from the rapidly spreading turbulent wake.

4. Conclusion

The transition location is strongly affected by the presence of the entropy layer associated with nose blunting. The transition location has been experimentally measured for 30 test cases. Transition is delayed by small degrees of nose blunting. Further blunting promotes roughness instability mechanisms. The critical roughness Reynolds number has been determined. The transition process is governed by the formation and growth of turbulent structures. The convection speeds of the disturbances have been calculated and the intermittency distribution has been found.

Acknowledgement. This work is supported by the Defence Research Agency. The authors acknowledge Dr. M. Collyer, Dr. I. Wrisdale, Dr. M. Gilmore, and Dr. T. Caine of DRA Farnborough; and Mr. R. Hutchins and Mr. S. Johnston of Imperial College.

Reflected-shock/side-boundary-layer interaction in a shock tube

M. Nishida, M. G. Lee
Department of Aeronautics and Astronautics, Kyushu University
Fukuoka 812-81 Japan

Abstract: This paper describes the numerical analysis of a reflected-shock/side boundary-layer interaction in a shock tube. The present shock-tube flow is considered as two dimensional and the interaction considered here is such that an initially plane shock reflected from the end wall interacts with the side boundary layer. The computations have been conducted for various Reynolds numbers and incident-shock Mach numbers. Reynolds number of 5×10^5 and incident-shock Mach number of 3.0 were used as typical conditions. Shadowgraphs made from numerical data based on the principle of optical shadowgraph clearly show the bifurcation of a reflected shock wave and the rolling up of a slip line on the shock-tube end wall. Furthermore, the interaction of the shock/boundary layer with suction from the side wall was also investigated to control the interaction.

Key words: Shock/boundary-layer interaction, Shock tube

1. Introduction

It is well known that a reflected shock wave interacts with a side boundary layer in a shock tube which leads to the bifurcation of the reflected shock wave. Concerning the shock/boundary-layer interaction, a number of experimental studies have been conducted. According to these experiments, the flow structure of the shock/boundary-layer interaction is such as shown in Fig. 1 and the simple theory of Mark (1957) can predict that such bifurcation may easily occur for gas with a low ratio of specific heats. On the other hand, numerical simulations of the shock/boundary-layer interaction also have been attempted (Takano 1988, Hataue 1989, Kleine et al. 1992). Takano (1988) did the simulation for nitrogen and argon gases using the thin-layer approximation by means of the flux-corrected-transport method and Beam-Warming scheme as solvers. His result did not show clear shock bifurcation. Although Hataue's simulation using the TVD scheme captured a bifurcated shock, he did not obtain a vortex sheet (Hataue 1989). Kleine et al. (1992) numerically simulated the flow field of this interaction using the MacCormack scheme. They showed the bifurcated reflected shock, but the vortex sheet was not obtained by their numerical simulation.

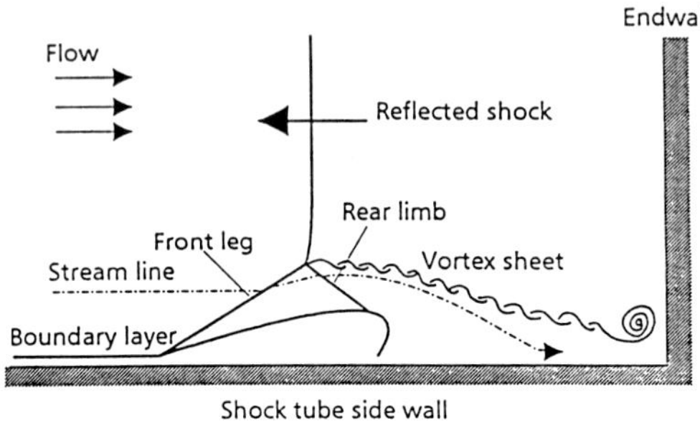

Figure 1. Sketch of a shock/boundary-layer interaction flow field.

In the present work, the numerical calculations of the shock/boundary-layer interaction were carried out for various incident-shock Mach numbers and Reynolds numbers. Furthermore, the interaction of the shock/boundary layer with suction from the side wall was also investigated to control the interaction. The calculations of the shock/boundary-layer interaction with suction were done for various suction speeds. It is the purpose of the present study to investigate the structure of a bifurcated shock wave, the effects of Reynolds number and incident-shock strength on shock/boundary-layer interaction, and the effect of suction on the development of shock bifurcation.

2. Numerical calculation

The two-dimensional flow field treated here is such that an initially plane shock is reflected from the shock-tube end wall and the reflected shock interacts with the side-wall boundary layer. The gas treated is air.

Governing equations for the present problem are two-dimensional Navier-Stokes equations. The computational domain is the lower half region of the duct composed of the left free boundary, lower side wall, upper duct axis, and right duct end wall. The boundary conditions imposed are: no slip and adiabatic wall conditions on the wall, symmetry conditions on the axis, and inflow/outflow condition on the free boundary. Numerical calculations were carried out using the second-order upwind TVD scheme of the Harten-Yee type (Yee 1989).

3. Numerical results

3.1. Structure of bifurcated shock

Computational results for $M_{s1} = 3.0$ and $Re = 5 \times 10^5$ are shown as isopycnics in Fig. 2. The time t' in the figure has been normalized by R/a_1, and $Re = \rho_1 a_1 R/\mu_1$, where R is a half of the duct height, a is the speed of sound,

ρ is the density, μ is the viscosity and the subscript 1 denotes the quantity ahead of an incident shock wave. In this computation, a 400 $\times$ 150 grid system was employed. Like past experimental results, the bifurcation of a reflected shock can be observed in the present numerical result, namely a front-leg shock (indicated by F.L.) and a rear-limb shock (indicated by R.L.) are captured. In addition, it can be clearly seen in the isopycnics that a slip line runs from a triple point towards the end wall and rolls up on the end wall. Strangely, a new discontinuity appears between the front-leg shock and rear-limb shock which has not been shown in any past experiments or computations of reflected shock/side boundary-layer interaction. We have no satisfactory explanation for whether this discontinuity is physical or numerical.

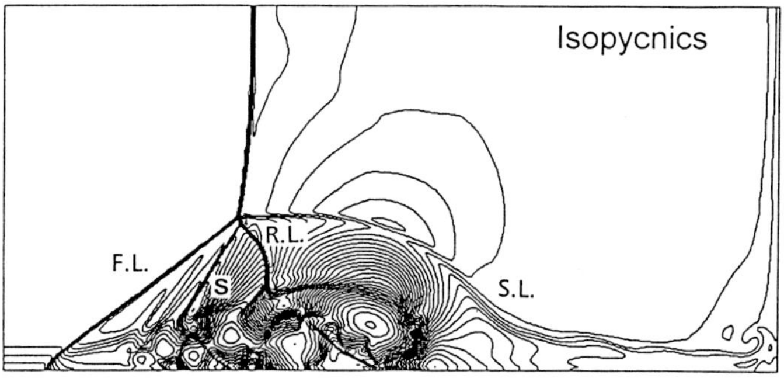

Figure 2. Isopycnics for $M_{s1} = 3.0$ and $Re = 5 \times 10^5$.

3.2. Development of bifurcation in reflected shock

The computed shadowgraph of Fig. 3 illustrates the development of the bifurcation of the reflected shock for $M_{s1} = 3.0$ and $Re = 5 \times 10^5$. This illustration has been developed using the principle of optical shadowgraph. The front-leg shock and rear-limb shock can be clearly identified at $t' = 1.28$. The aforementioned new discontinuity between the front-leg shock and rear-limb shock appears only at $t' = 1.91$ and it is not observed before this time. At $t' = 1.91$, a slip line runs towards the end wall; especially interesting is that this slip line forms a swirl on the end wall. This is very similar to the colored schlieren picture of Kleine et al. (1992). There have been no computational results which succeeded in numerical capture of a vortex sheet.

3.3. Effects of Reynolds number and incident-shock strength

The effect of Reynolds number on the development of a bifurcated shock wave was experimentally investigated by Fokeev and Gvozdeva (1990). Here this

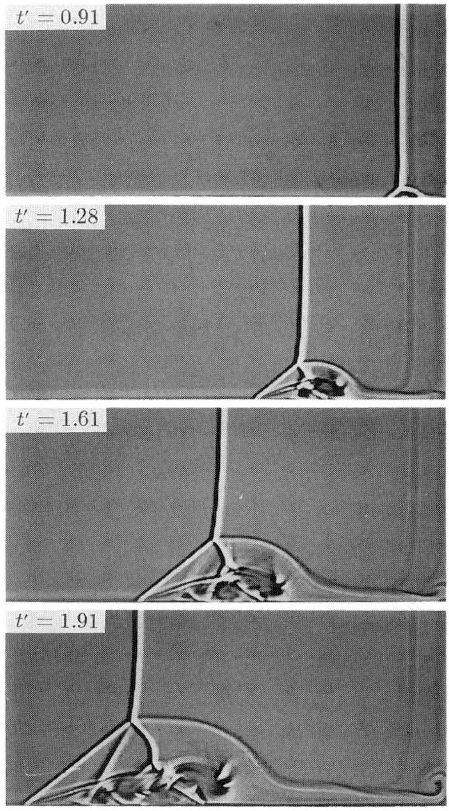

Figure 3. Computed shadowgraphs for M_{s1} = 3.0 and $Re = 5 \times 10^5$.

problem has been numerically studied. Computations were carried out for M_{s1} = 3.0 and $Re = 5.0 \times 10^4$, 1.5×10^5, 5.0×10^5. The height of the bifurcation is almost linearly proportional to the distance from the end wall and slightly increases with a decrease in Reynolds number (see Fig. 4).

The effect of incident-shock strength on bifurcation of a reflected shock wave was also investigated. The calculations were carried out for $Re = 5 \times 10^4$, and M_{s1} = 2.0, 2.5, 3.0, and 3.5. The results are shown in Fig. 4, where the height of a triple point versus distance from the end wall is also illustrated for the four incident-shock Mach numbers. It is seen from the figure that the height of the triple point is increased with incident-shock strength.

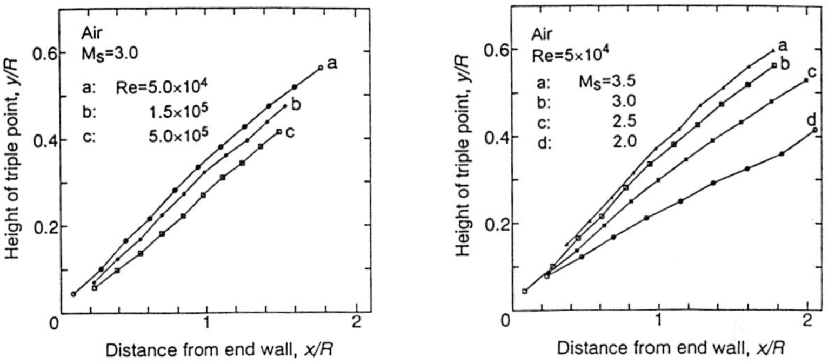

Figure 4. Effects of Reynolds number and incident-shock strength.

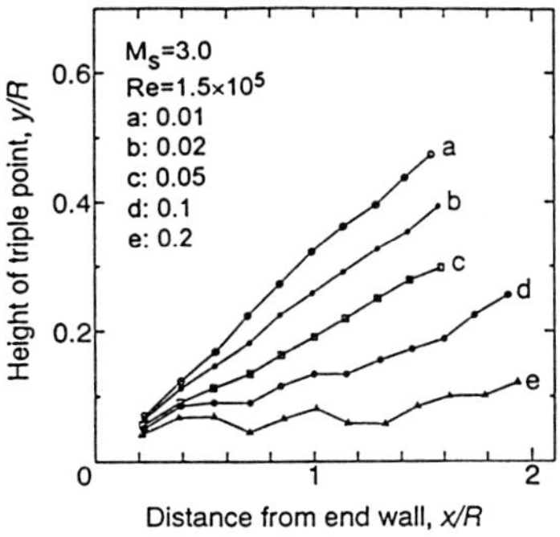

Figure 5. Control of bifurcated shock wave with boundary-layer suction.

3.4. Effect of boundary-layer suction

Active control of the bifurcation of a reflected shock wave is very important for usage of high-pressure and high-temperature gas behind a reflected shock as reservoir gas, because premature contamination is caused by the wall jet in the vicinity of the side wall and as a result, operation time of a shock tunnel is reduced. Xu et al. (1988) experimentally confirmed that the suction of the boundary layer induced by an incident shock wave can control the height of a triple point and reduce the wall jet.

In the present work, the flow field of the bifurcated shock wave with boundary layer suction was numerically analyzed for normalized suction speeds U_{suc}/a_1 of 0.2, 0.1, 0.05, 0.02, and 0.01, where a_1 is the speed of sound ahead of an incident shock wave. The height of the triple point versus a distance from the end wall is shown in Fig. 5, where the values for lines a–e are the ratio U_{suc}/a_1. As expected, the height of the triple point is decreased with an increase in suction speed.

4. Concluding remarks

In the present study, the flow field of reflected shock/side-boundary-layer interaction was numerically analyzed using full Navier-Stokes equations. The present computation captured clear bifurcated shock structure and a slip line rolling up on the end wall; these are very similar to experimental results. Next, in order to reveal effects of Reynolds number and incident-shock strength on the bifurcated shock structure, the height of the triple point was estimated for various Reynolds numbers and various incident-shock Mach numbers. The results show that the height of the triple point is almost linearly increased with distance from the end wall. It is remarkably dependent on incident-shock Mach number, but the effect of Reynolds number is relatively small. Finally, the effect of boundary-layer suction on the bifurcated shock structure was investigated and it was found that the height of the triple point decreased with increasing suction speed.

References

Fokeev VP, Gvozdeva LG (1990) Study of bifurcation of reflected shock waves in channels of various cross-sections. In: Current Topics in Shock Waves, American Institute of Physics, pp 862–866.

Hataue I (1989) Computational study of the shock-wave/boundary-layer interaction in a duct. Fluid Dynamics Research 5:217–234.

Kleine H, Lyakhov VN, Gvozdeva LG, Grönig H (1992) Bifurcation of a reflected shock wave in a shock tube. In: Proc 18th Intl Symp on Shock Waves, pp 261–266.

Mark H (1957) The interaction of a reflected shock wave with the boundary layer in a shock tube. J Aero Sci 24:304–306.

Takano Y (1988) Simulations for effects of side-wall boundary-layer on reflected shock flowfields in shock tubes. In: Proc 16th Intl Symp on Shock Tubes and Waves, pp 645–651.

Xu L, Gourlay CM, Stalker RJ (1988) Reduction of driver gas contamination in a reflected shock tube by boundary layer suction. In: Proc 16th Intl Symp on Shock Tubes and Waves, pp 637–643.

Yee HC (1989) A Class of High Resolution Explicit and Implicit Shock-Capturing Methods, NASA TM–101088.

Effects of shear-layer transition and leading-edge bluntness on control-flap effectiveness

D. Kumar, J. L. Stollery
College of Aeronautics, Cranfield University
Cranfield, Bedfordshire, MK43 0AL UK

Abstract: The effects of shear-layer transition and leading-edge bluntness on flap-control effectiveness have been studied experimentally. A quasi-two-dimensional flat plate equipped with a full-span trailing-edge flap was employed for tests. Leading-edge bluntness effects were simulated by introduction of a hemi-cylindrical leading edge. Flow structure was studied using high-speed schlieren photography and surface-pressure and heat-transfer measurements. Analytical theories have been developed to estimate the flap pressure and laminar and turbulent heat-transfer levels for the sharp and blunt configurations. These are compared against experimental measurements.

Key words: Control flap, Laminar separation, Transition, Bluntness

1. Introduction

Large flap deflections produce extensive regions of separated flow. At low hypersonic Mach numbers, the instability of the separated shear layer promotes transition of the initially laminar boundary layer and leads to either transitional or in severe cases, fully turbulent reattachment. Shear-layer transition produces a substantial increase in heat transfer and hence imposes severe aerothermal loads on the flap. The need to relieve stagnation-point heat transfer at hypersonic speeds dictates the intentional blunting of leading edges. The introduction of bluntness produces a considerable change in the aerodynamic flow structure from that of an equivalent sharp leading edge body. The primary effect is the formation of a bow shock standing off from the leading edge. The shock creates a region of high temperature, low-density flow confined to a region known as the "entropy layer." The increase in leading-edge shock strength together with its highly curved structure give rise to gradients in flow properties (confined to the entropy layer), such as Mach number and unit Reynolds number along and normal to the body surface. The changes in the body flow field due to leading-edge bluntness considerably change the performance of a submerged control flap.

2. Analytical study

Analytical theories employed for estimating the flap heat transfer on both sharp and blunt configurations are deacribed by Kumar & Stollery (1994). In the present work, the leading edge was taken as the origin for the laminar solution whilst the hingeline was used as a virtual origin for the turbulent solutions.

3. Wind-tunnel facility

The Cranfield hypersonic gun tunnel was used; tests were conducted with air as the working gas. An axisymmetric contoured nozzle provided a useful test jet diameter of 15.0 cm at a freestream Mach number of 8.2. The tunnel was equipped with a single-pass schlieren system which used a high-intensity microsecond-duration argon spark source to illuminate density gradients in the flow. Pressure measurements were made using Kulite XCS-190 series pressure transducers. Surface heat-transfer measurements were made using thin-film platinum gauges mounted on a Macor substrate. The temperature history of the gauges was integrated using a low-noise, high-bandwidth analog integrator circuit. The models used are shown in Fig. 1.

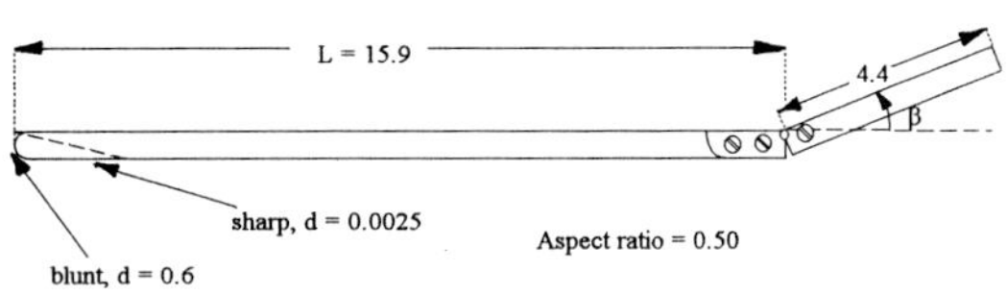

Figure 1. Diagram of control flap model (all dimensions in cm).

4. Results and discussion

In Fig. 2, the single hingeline shock for the $\beta = 5°$ configuration confirms the flow as attached. The compression associated with this shock increases the pressure on the flap. The increase starts at the hingeline and is typical of the pressure distribution for an attached flow configuration. Whilst the flow is attached for the $\beta = 5°$ configuration, the plateau in the pressure distribution for $\beta = 10°$ indicates a well-separated flow structure. This is confirmed by the dual-shock interaction structure in the corresponding schlieren photograph with separation located at $x/L = 0.88$. Further increasing the flap deflection angle to $\beta = 25°$ results in separation occurring at $x/L = 0.6$. There is a small upstream movement in reattachment. The sharp flat-plate heat-transfer measurements of Kumar & Stollery (1994) have shown the flat-plate boundary layer to be laminar. For all the sharp leading-edge configurations shown in Fig. 2, the decrease in the measured heat-transfer distributions below the flat-plate level at the start of their respective interactions is characteristic of laminar interactions. For $\beta = 5°$, the agreement of the measured flap heat-transfer rates with the predictions of laminar reference temperature theory confirms the flap boundary layer as laminar. Thus, transition is not promoted by this flap deflection. For $\beta = 10°$, the increase in the measured flap heat-transfer rates

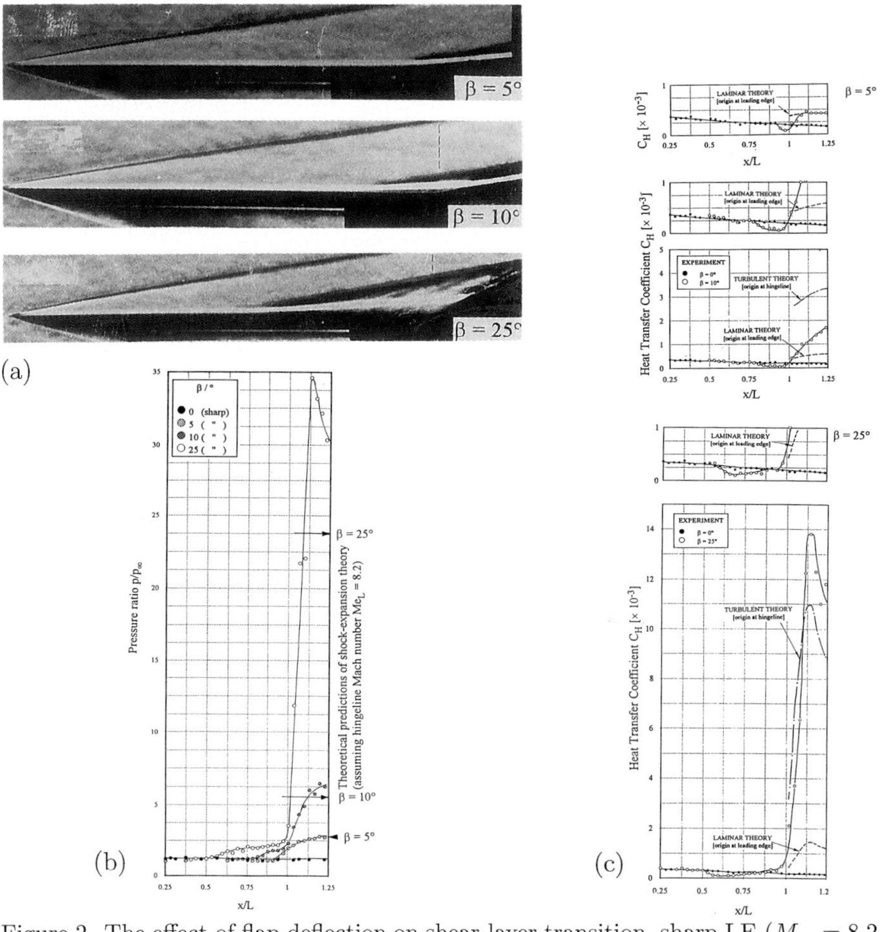

Figure 2. The effect of flap deflection on shear-layer transition, sharp LE ($M_\infty = 8.2$, $Re_\infty/\text{cm} = 9.0 \times 10^4$, $d < 0.0005$ cm, $L = 15.9$ cm): (a) schlieren photographs; (b) pressure distributions; (c) heat-transfer distributions.

above the theoretical laminar level suggests that transition is occurring following the laminar separation. However, as the measured flap heat-transfer remains below the estimated turbulent rate, transition is incomplete at the end of the flap. For $\beta = 25°$, the decrease in heat transfer at separation confirms the local flow as laminar. However, the subsequent increase in heat transfer from within the interaction region suggests rapid transition of the shear layer. This is also evident from the corresponding schlieren photograph which shows shock waves emanating from the edge of the disturbed shear layer. The agreement between the measured flap heat-transfer rates and the predictions of turbulent theory suggest that transition is complete prior to reattachment.

In Fig. 3, the measured flap heat-transfer rates at $x/L = 1.19$ are compared with the predictions of laminar and turbulent reference temperature theory. This clearly shows that flap deflection promotes transition, and in severe cases

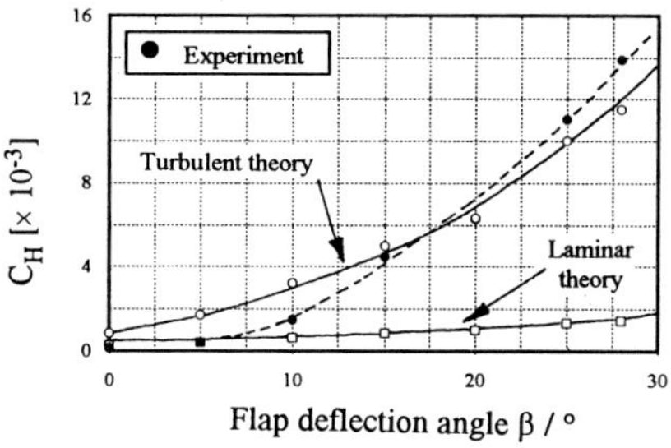

Figure 3. The effect of shear-layer transition on flap heat-transfer rates, sharp LE ($M_\infty = 8.2$, $Re_\infty/\text{cm} = 9.0 \times 10^4$, $x/L = 1.2$ cm).

leads to fully turbulent conditions being realized on the flap. The promotion of separation together with the increased flap adverse pressure gradient are the primary agencies responsible for the promotion of transition with flap deflection. The pressure distribution for the blunt configurations is shown in Fig. 4. The blast effect of the blunt leading edge increases the flat-plate pressure level above that for a sharp leading-edge flat plate. For both $\beta = 5°$ and $\beta = 10°$, the increase in pressure from the hingeline together with the single hingeline shock in the corresponding schlieren photograph confirms the flow on these configurations as attached. The attached flow on the latter configuration contrasts with the well-separated flow structure observed on the corresponding sharp configuration. Thus, for this configuration, bluntness completely surpresses separation. For well-separated flows, ($\beta = 25°$) the delaying of separation from $x/L = 0.6$ on the sharp configuration to $x/L = 0.77$ on the blunt configuration is evident by comparing the pressure distributions and schlieren photographs, shown in Figs. 2 and 5 respectively. The pressure distributions also show that bluntness considerably reduces the pressure recovered on the flap. The calculated flap pressure levels on the blunt configuration, made by assuming the flap to be submerged in an effective supersonic freestream, are shown in Fig. 5. The agreement of these estimates with the measured flap pressure level immediately downstream of reattachment confirms that the loss of stagnation pressure across the leading edge bow shock is responsible for the reduction in flap pressure by bluntness. In a classical theoretical study, Cheng et al. (1961) found that the relative effects of leading-edge bluntness and the viscous displacement effect are measured by the bluntness-viscous interaction parameter $\chi_\epsilon/\kappa_\epsilon^{2/3}$. The sep-

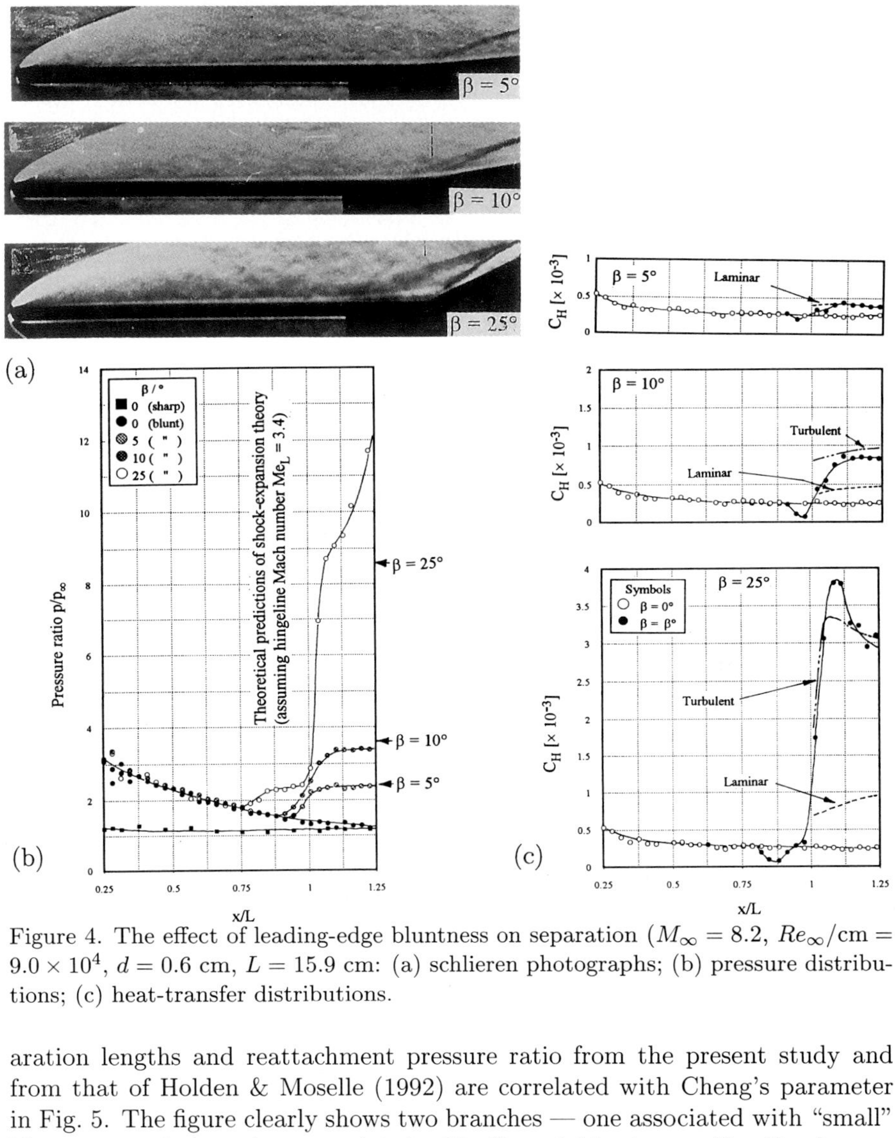

Figure 4. The effect of leading-edge bluntness on separation ($M_\infty = 8.2$, $Re_\infty/\text{cm} = 9.0 \times 10^4$, $d = 0.6$ cm, $L = 15.9$ cm: (a) schlieren photographs; (b) pressure distributions; (c) heat-transfer distributions.

aration lengths and reattachment pressure ratio from the present study and from that of Holden & Moselle (1992) are correlated with Cheng's parameter in Fig. 5. The figure clearly shows two branches — one associated with "small" bluntness and the other associated with "large" bluntness. For the former, changes in the separated flow length occur due to the changes in boundary-layer edge conditions arising within the entropy layer generated by leading-edge bluntness. For "large" bluntness, the correlation shows a substantial decrease in the reattachment pressure ratio with bluntness. This decreases the length of separated flow. The differences between the present study and that of Holden and Moselle's may be due to the transitional nature of the former.

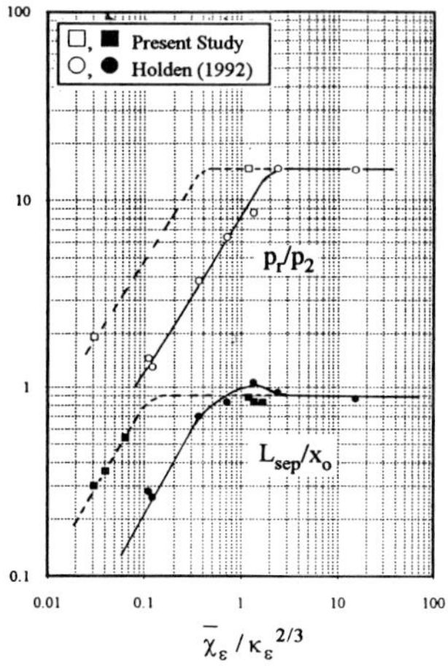

Figure 5. The effect of bluntness on separated flows.

5. Conclusion

Flap deflection promotes separation of laminar boundary layers. The instability of the separated shear layer, together with the adverse pressure gradient promotes transition of an initially laminar boundary layer. Large flap deflections produce extensive regions of separated flows. This results in the onset and sometimes the completion of transition within the separated flow interaction region. Large bluntness substantially delays separation. This is due to the reduction of the pressure recovered on the flap. The reduction in flap pressure by bluntness also alleviates flap heat transfer for both laminar and turbulent flap boundary layers.

References

Ames Research Staff (1953) Equations, Tables and Charts for compressible Flows, NACA-R-1135.

Holden MS, Moselle JR (1992) A Database of Aerothermal Measurements in Hypersonic Flow for CFD Validation, AIAA Paper 92-4023.

Kumar D, Stollery JL (1994) Hypersonic control flap effectiveness. In: Proc of the 19th ICAS Congress, Anaheim, California.

Lukasiewicz J (1961) Hypersonic Flow Blast Analogy, AEDC TN-61-158.

Needham DA (1965) Laminar Separation in Hypersonic Flows, PhD Thesis, University of London (see also Needham DA and Stollery JL, AIAA Paper 66-455, 1966).

Statistical simulation of nonequilibrium hypersonic flows over concave bodies

M. S. Ivanov, S. F. Gimelshein, G. N. Markelov
Institute of Theoretical and Applied Mechanics, Novosibirsk 630090 Russia
Y. E. Gorbachev
International Institute for Interphase Interactions
St. Petersburg 194291 Russia

Abstract: Hypersonic rarefied flows over 2D bodies (a plate, a plate with a flap, an ellipse, an ellipse with deflected rear part) are calculated by the Direct Simulation Monte Carlo (DSMC) method. Various models of vibrational-translational energy exchange and chemical reactions are considered and the influence of the model on the simulated flow is examined.

Key words: Chemically reacting hypersonic rarefied flows, DSMC method

1. Introduction

The study of space-vehicle aerodynamics at the altitudes 80–120 km becomes more and more important. This is primarily caused by the development of a new generation of space transportation systems, for which the mentioned altitudes are determining factors. This requires more comprehensive and detailed understanding of the problems of hypersonic aerodynamics at high flight altitudes. The principal tool for solving the problems of rarefied gas aerodynamics is the Direct Simulation Monte Carlo (DSMC) method. At present, this method is frequently used for studying the flow over space vehicles at altitudes of about 80 km where the real-gas effects are essential, i.e., nonequilibrium excitations of internal (rotational and vibrational) degrees of freedom of molecules and chemical transformations. Many models have been suggested to describe the excitation of internal degrees of freedom and chemical reactions by the DSMC method, varying in their labor intensity and physical validity. The variety of models and algorithms leads to one of the most interesting problems: to what extent does the choice of this or that new model affect the simulated aerodynamics of various bodies. In the present paper this problem is considered.

2. Flow conditions and gas models

The following set of bodies was considered: (a) a flat plate, (b) a plate with a flap, (c) 10% elliptic profile, (d) 10% elliptic profile with deflected rear part. The length of the deflected part (cases b and d) was $0.2L$ where L is the total body length, and the angle of flap deflection was $\delta = 30°$. The flap was deflected towards the free stream. In all cases the angle of attack was 40°. The free-stream velocity and temperature were $U_\infty = 7600$ m/s and $T_\infty = 189$ K, respectively. The wall temperature was assumed constant and equal to $T_w = 1000$ K. The Knudsen number determined from characteristic length L was $Kn = 0.01$.

The calculations were made by the use of the DSMC numerical schemes based on the majorant principle (Ivanov and Rogasinsky 1991). For the intermolecular potential, the Variable Hard Sphere model with an exponent of 0.75 for viscosity-temperature dependence was chosen. The diffuse law with complete accommodation of translational and internal energies was suggested for gas-surface interaction. The Larsen-Borgnakke phenomenological model was employed for simulating the exchange between translational and rotational energy degrees of freedom. Rotation-relaxation collision number Z_r was assumed to be 5. Molecular oxygen was selected as a free-stream gas.

Models used to describe the vibrational-translational (VT) energy exchange:

V1. The Larsen-Borgnakke model with vibration-relaxation collision number $Z_v = 50$. It is the model with the fastest vibrational relaxation.

V2. A model taking into account the quantum nature of vibrations (Ivanov 1993); molecules are considered as anharmonic oscillators (AHO) and only one-step vibrational transitions are involved. It is a model with the slowest vibrational relaxation.

V3. A model for VT exchanges with multi-step transitions (Gimelshein et al. 1994). It employs the energy-dependent form for VT-transition cross sections found using quasiclassical theory for the Morse intermolecular potential.

Models used to describe the gas-phase chemical reactions:

C1. A model with reaction probabilities that depend only on the total energy (i.e. the sum of translational, rotational, and vibrational energies) of a colliding pair. The collision dynamics and algorithm for this case were described in detail in the paper of Gimelshein and Ivanov (1994). This model was used together with model V1 for vibrational excitations.

C2. A model of chemical reactions which takes into account the vibration-dissociation coupling (VDC) (Ivanov 1993). This model was used together with models V2 and V3 for vibrational excitations.

3. Results and discussion

Flat plate. The influence of the gas model on the pressure distribution C_p along the plate is shown in Fig. 1 for various models of nonreacting gas. Here, a TR gas is a gas with translational and rotational degrees of freedom; a TRV gas means the vibrational mode is also considered. Accounting for vibrations leads to a slight C_p reduction along the plate. The choice of vibrational model is not significant for this case (the maximum discrepancy between V1 and V2 models is about 7%).

Figure 2 presents the comparison of C_p distribution along the plate for TRV gas and various models of chemically reacting oxygen. The values of C_p for the V1+C1 model differ appreciably from those for other models on the first half of the plate. This is due to the fact that V2+C2 and V3+C2 models take into account the VDC effect. Since the vibrational mode has no time to be excited in the shock wave near the plate tip, C_p values for these models are almost the same as those for nonreacting gases. Multilevel VT transitions in the V3 model accelerate noticeably the VT relaxation (and hence, chemical reaction rates), compared with the V2 model, and C_p coincides on the lower part of the plate for the V1+C1 and V3+C2 models. Note that the shock wave is attached to the plate and its inclination angle varies depending on the gas model. Inclination-angle reduction causes lower C_p values along the plate.

Plate with a flap. The C_p distribution along the plate with flap for various vibrational models is shown in Fig. 3. The effect of the model in this case is evidently much more significant than for a flat plate. C_p values coincide for various gas models near the body tip but their difference is about 30–50% in the middle of the body and on the flap. The choice of the model for chemical reactions exerts also a considerable effect on C_p, though it is not as significant as the VT model. C_p near the plate tip is slightly lower for V1+C1 than for VDC models, as it was for a flat plate. The maximum difference is observed on the flap.

The influence of vibrational-relaxation and chemical-reaction rates for various models is clearly illustrated by the fields of atomic oxygen mole fraction (Fig. 5). For the V1+C1 model (Fig. 5a) a high concentration of atomic oxygen is observed on the plate tip. The highest increase in concentration for this model occurs in the bow shock wave and is due to rapid growth of translational temperature there. The chemical-reaction rate for VDC models is primarily determined by the vibrational temperature, which increases much slower than translational and rotational temperatures. Therefore, chemical-reaction rates for VDC models are fairly high in the region between the shock and the body. Note that in all above cases, the shock wave was attached to the body.

Elliptic profile. Unlike the case for a plate, the shock wave induced by an ellipse is detached. This is probably the main reason for the smaller influence of the vibrational and chemical-reaction models on the aerodynamic coefficients. As is seen from Figs. 6 and 7, the change of vibrational model hardly influences C_p, and the effect of the chemical-reaction model is 4–5%. The most prominent difference is observed near the body tip.

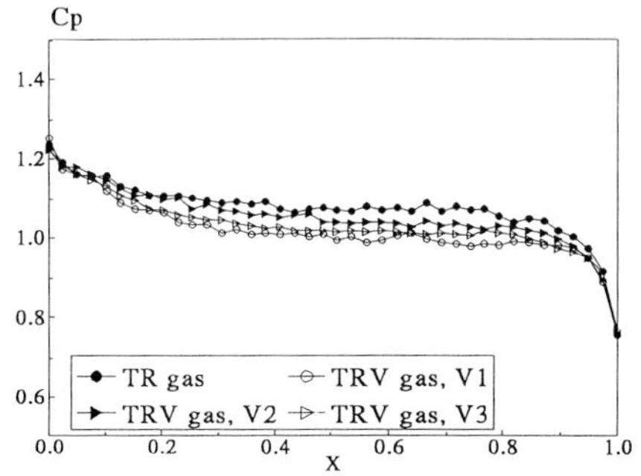

Figure 1. Pressure coefficient over a plate with a flap for various models of molecular vibrations.

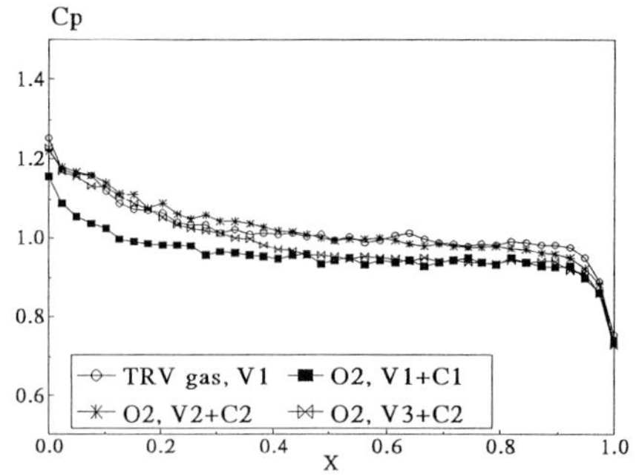

Figure 2. Pressure coefficient over a flat plate for various models of chemical reactions.

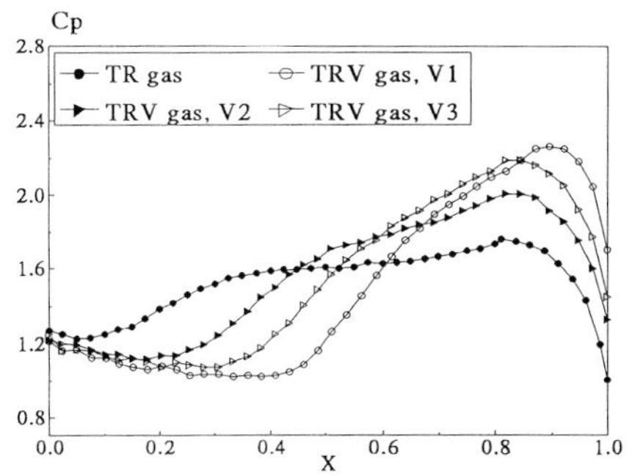

Figure 3. Pressure coefficient over a plate with a flap for various models of molecular vibrations.

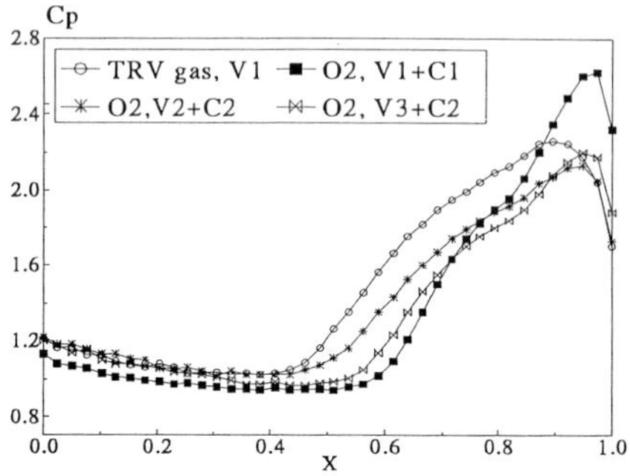

Figure 4. Pressure coefficient over a plate with a flap for various models of chemical reactions.

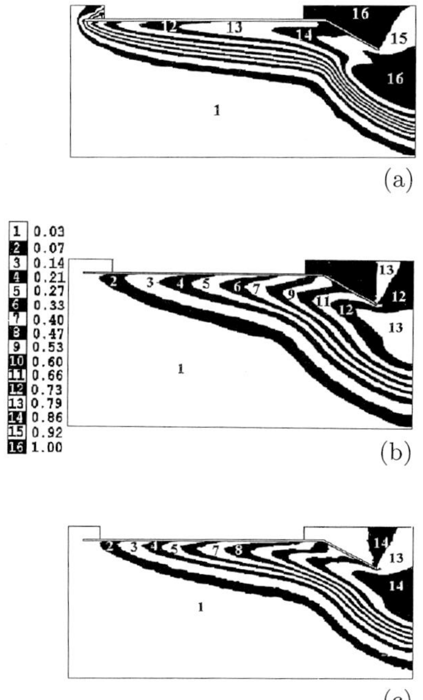

Figure 5. Atomic-oxygen mole fraction about a plate with a flap for model combinations: (a) V1 + C1; (b) V2 + C2; (c) V3 + C2.

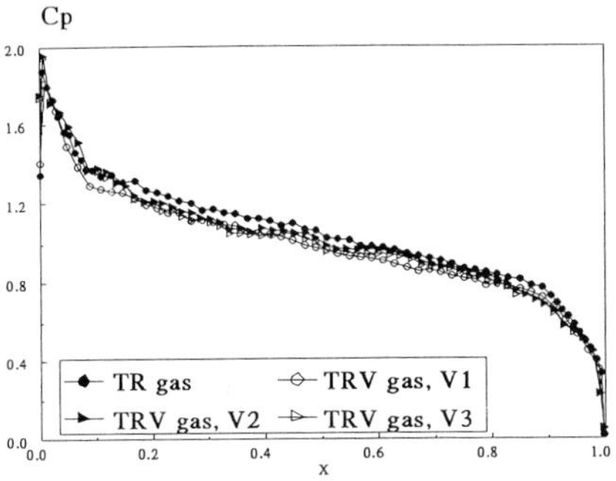

Figure 6. Pressure coefficient over an ellipse for various models of molecular vibrations.

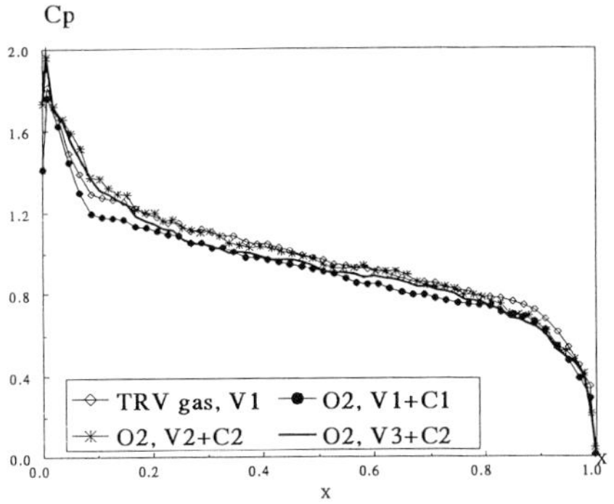

Figure 7. Pressure coefficient over over an ellipse for various models of chemical reactions.

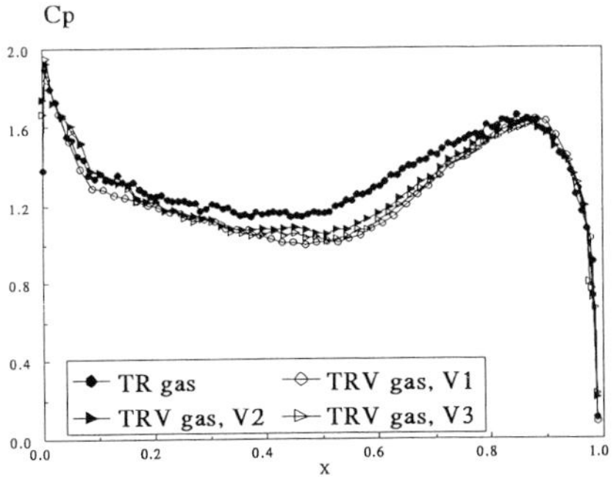

Figure 8. Pressure coefficient over over an ellipse with a flap for various models of molecular vibrations.

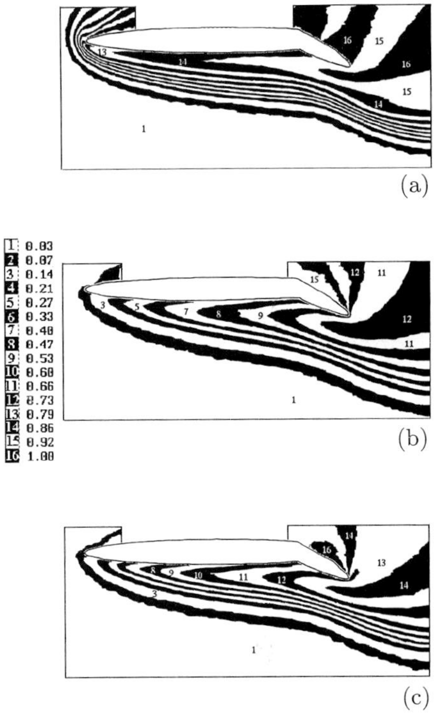

Figure 9. Atomic-oxygen mole fraction about an ellipse with a flap for model combinations: (a) V1 + C1; (b) V2 + C2; (c) V3 + C2.

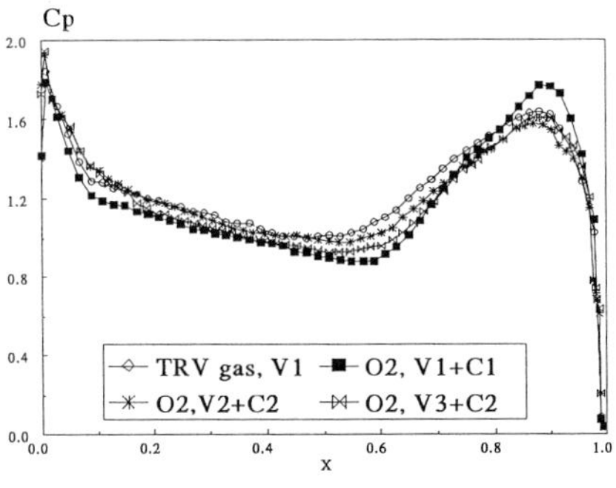

Figure 10. Pressure coefficient over over an ellipse with a flap for various models of chemical reactions.

Elliptic profile with deflected rear part. The deflection of the rear part of the ellipse causes a C_p increase on the second half of the body (Fig. 8). Excitation of the vibrational mode reduces C_p, whereas the choice of vibrational model scarcely affects it, and the maximum discrepancy amounts to 3–5%.

The effect of the chemical model is most conspicuous in the fields of atomic oxygen mole fraction (Fig. 9). However, despite an appreciable difference in the flow structure, the choice of chemical model weakly affects the pressure distribution for the elliptic profile with deflected rear part (Fig. 10).

4. Conclusions

The flow over simple 2D bodies was calculated by the DSMC method for various models for vibrations and chemical reactions. Concave bodies are much more sensible to the gas model variation than convex ones. The choice of the gas model is most significant for a plate with a flap.

For the flow over a plate and a plate with a flap, the choice of the vibrational model makes a larger contribution to aerodynamic coefficients than the chemical model. The reverse situation occurs for elliptic profiles. The determining factor here is whether the bow shock is detached or attached to the body.

Acknowledgement. The authors wish to acknowledge support from the International Science Foundation (Grant No. RBW000).

References

Gimelshein SF, Gorbachev YE, Ivanov MS, Kashkovsky AV (1994) Real gas effects on the aerodynamics of 2D concave bodies in the transitional regime. In: 19th Int Symp on Rarefied Gas Dynamics, Oxford, UK.

Gimelshein SF, Ivanov MS (1994) Simulation of chemically reacting gas flow using majorant frequency scheme of DSMC. In: Rarefied Gas Dynamics: Theory and Simulations (BD Shizgal and DP Weaver, eds), Vol. 159, Progress in Astronautics and Aeronautics, AIAA, Washington, DC, pp 218–233.

Ivanov MS (1993) Transitional regime aerodynamics and real gas effects. In: Proc Int Symp on Aerospace and Fluid Science, Sendai, Japan, pp 77–101.

Ivanov MS, Rogasinsky SV (1991) Theoretical analysis of traditional and modern schemes of the DSMC method. In: 17th Int Symp on Rarefied Gas Dynamics, Aachen, Germany, pp 629–642.

Transition of unsteady boundary layer induced by propagating compression wave

T. Aoki, N. Kondoh, K. Matsuo
Kyushu University
Kasuga, Fukuoka, 816 Japan
S. Mashimo
West Japan Railway Co.
Kita-ku, Osaka, 530 Japan

Abstract: The transition of an unsteady boundary layer induced by a propagating compression wave has been investigated by means of a recently developed laser differential interferometry (LDI) technique using Wollaston prisms. LDI can be made 2 to 3 orders of magnitude more sensitive to optical path change than classical interferometry. The sensitivity was high enough to record the density variations of a compression wave. Measurements were made of time-dependent profiles that gave detailed information on the process of formation of the unsteady boundary layer induced by propagating compression waves in a constant-area duct. It was found that the laminar-to-turbulent transition in the unsteady boundary-layer flow exits describing the density profiles.

Key words: Transition, Unsteady boundary layer, Compression wave, Laser differential interferometry

1. Introduction

The discharged wave from the exit of a tube radiates toward the surrounding area as an impulsive wave and causes an impulsive noise. We can easily understand that this sort of noise due to a discharged wave is similar to the compression wave emerging from the exit portal of a high-speed railway tunnel (Ozawa 1979).

According to the aeroacoustic theory, the magnitude of an impulsive noise from the exit of a tunnel is proportional to the rate of pressure change of the compression wave front arriving at the exit (Ozawa 1979). Thus, we must understand the mechanism of the attenuation and distortion of a compression wave front propagating along a tunnel.

The attenuation and distortion of a compression wave are considered to be related the fact that whether the boundary layer of the flow induced by the compression wave is laminar or turbulent (Matsuo et al. 1991, Aoki et al. 1993). Therefore, in order to estimate the magnitude of an impulsive noise correctly, the transition of the unsteady boundary layer induced by a propagating compression wave in a tunnel is examined and experimental information on the transition is required.

The final objective of our study is to understand the mechanism of the attenuation and distortion of propagating compression waves in a tunnel. In the present paper, experimental investigations are carried out on the transition of the unsteady boundary layer induced by a propagating compression wave in a model tunnel by means of a recently developed laser differential interferometry technique.

2. Experimental apparatus

The open-ended shock tube with a fast-opening gate valve instead of a diaphragm is described in Aoki et al. (1993). Laser differential interferometry (LDI) using Wollaston prisms (Smeets & George 1973) was used for recording the density distribution. A schematic diagram of the optical arrangement is

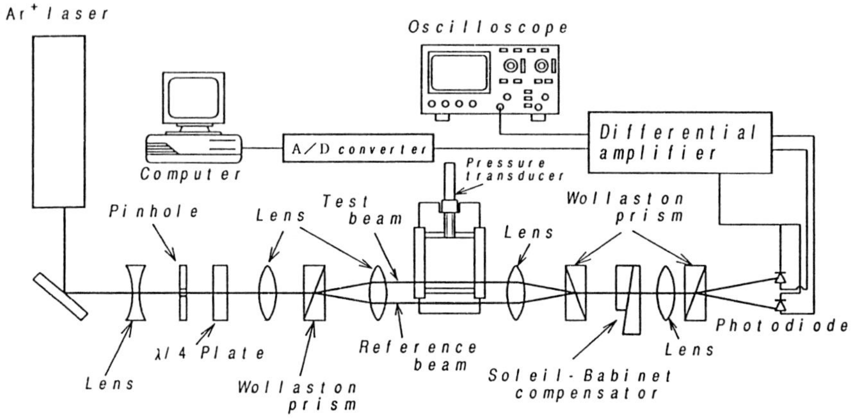

Figure 1. Optical arrangement of laser differential interferometer.

shown in Fig. 1. A continuous argon gas laser is used as a light source and photodiodes are used for detecting the difference of light intensities. The difference of the two detector signals from LDI will therefore be essentially free from vibrations and laser output variations. LDI can be made more sensitive to optical path change by 2 to 3 orders of magnitude than classical interferometry. The sensitivity was high enough to record density variations of a compression wave. This LDI has been used to detect the transition region in the unsteady wall boundary layer induced by a compression wave at $x = 3.15$ m.

Figures 2b and b show schematically two LDI setups, A and B. In the case of setup A, one of the interfering laser beams (test beam) passes through the unsteady boundary layer at a height h. The other beam (reference beam) travels through the main stream of the flow. Therefore the output signal of setup A shows the difference of density between the boundary layer and the main stream. Setup B was used to measure the spatial difference history of density in an unsteady boundary layer at height h.

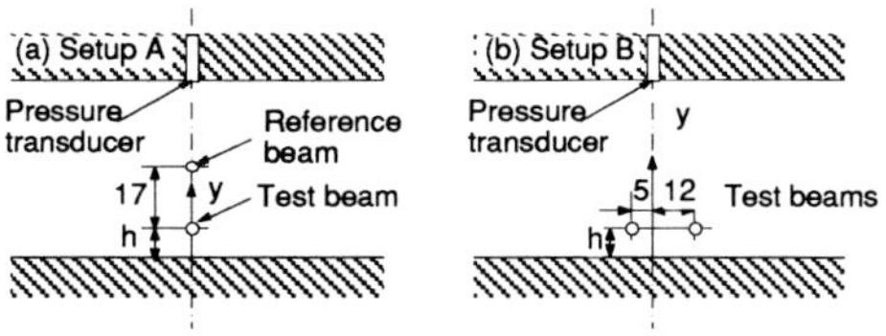

Figure 2. Schematic diagram of optical setup of laser beams.

3. Results and discussion

Typical experimental pressure-time records at $x = 0$ and 3.15 m are shown in Fig. 3a. Let the pressure rise by a compression wave be denoted by $\Delta p_c \equiv p_2 - p_1$, where p_1 and p_2 represent the pressures before and after a compression wave, respectively. The maximum rate of pressure rise $(dp/dt)_{\max}$ at the compression wave front is defined as shown Fig. 3a. Then it is seen that $\Delta p_c = 19$ kPa and $(dp/dt)_{\max} = 6$ kPa/ms. It is clear from the figure that the compression wave form is well controlled by a fast-opening gate valve.

The output signals of the laser differential interferometry at the point $x = 3.15$ m are shown in Fig. 3b (same conditions as Fig. 3a). The figure shows that the both signals for setup A and B detect well at the edge of the boundary layer and the transition. Density distributions of the boundary layer can be calculated by using the signal for the setup A as shown in Fig. 2a. A typical experimental density-time record, the difference of density $\rho - \rho_\infty$ between the boundary layer and the main stream, is shown in Fig. 4, where the compression wave is same as in Fig. 3 and the height of the test beam from the wall is $h = 0.8$ mm. The dotted and dashed lines represent the theoretical results of density profiles in laminar and turbulent boundary layers induced by a shock wave, respectively (Mirels 1956). This experimental data gives detailed information on the process of formation of the unsteady boundary layer induced by a propagating compression wave. The point a denotes the edge of the laminar boundary layer at $h = 0.8$ mm. It can be also concluded from Fig. 4 that the laminar-to-turbulent transition in the unsteady boundary layer flow exits describing the density profiles at point b. It should be noted that the transition time of point b in Fig. 4 coincides with the minus peak on the signal for setup B in Fig. 3b and LDI can be used to locate the unsteady boundary-layer transition induced by a propagating compression wave. The transition time t_r of point b such as

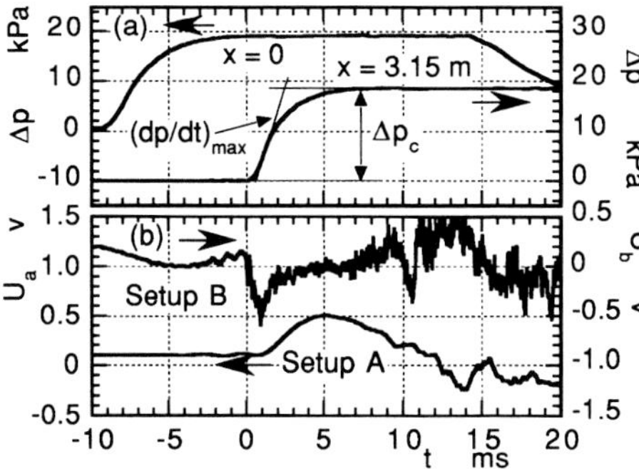

Figure 3. Typical pressure-time record of compression wave and output signals from LDI.

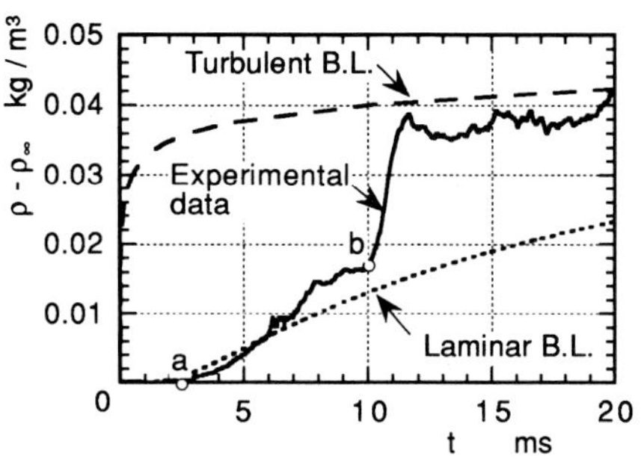

Figure 4. Measured density profiles in unsteady boundary induced by compression wave.

shown in Fig. 4 is determined for every record. The results are plotted in Fig. 5 for the range Δp_c = 15, 17, 19 kPa, $(dp/dt)_{\max}$ = 5, 6 kPa/ms. The value of t_r decreases slightly with Δp_c. To clarify the spatial structure of the unsteady boundary layer, the location of the edge and transition on the boundary layer was measured for $h = 0.4 - 1.6$ mm. Figure 6 shows the experimental results against the length $x_c = at$ (where a is the sonic velocity in front of compression wave) from the wave front for Δp_c = 19 kPa, $(dp/dt)_{\max}$ = 5 kPa/ms. The solid line represents the theoretical results for the edge position in a laminar boundary layer induced by a shock wave (Mirels 1956). Figure X? shows that the solid line agrees well with the experimental point. It looks like the transition propagates from the edge of the boundary layer to the wall. This is a subject for future study.

The transition Reynolds number R_{etr} defined by the wave-fixed coordinate system (Hartunian et al. 1960) is determined for every record. The results are compared with the experimental values for weak shock waves (Asbridge 1959, Hartunian et al. 1960, Thompson & Emrich 1967) as shown in Fig. 7, where the horizontal axis represents the density ratio across the wave front and the shock Mach number. It is seen that the transition Reynolds number of the compression wave agrees well with that of the shock wave. It may be the fact that the characteristic length of transition is relatively longer than that of wave front of the compression wave.

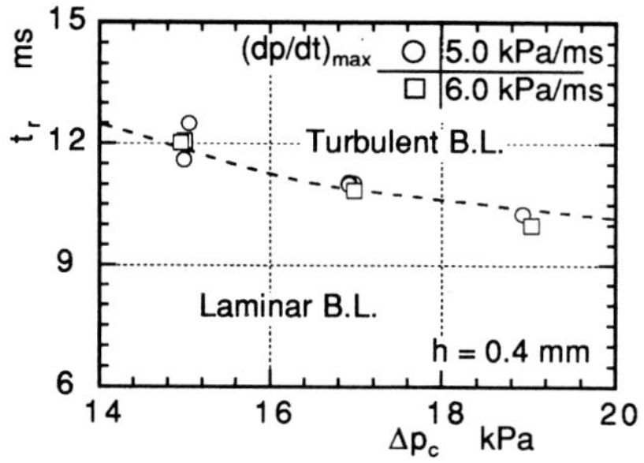

Figure 5. Transition time from laminar to turbulent boundary layer.

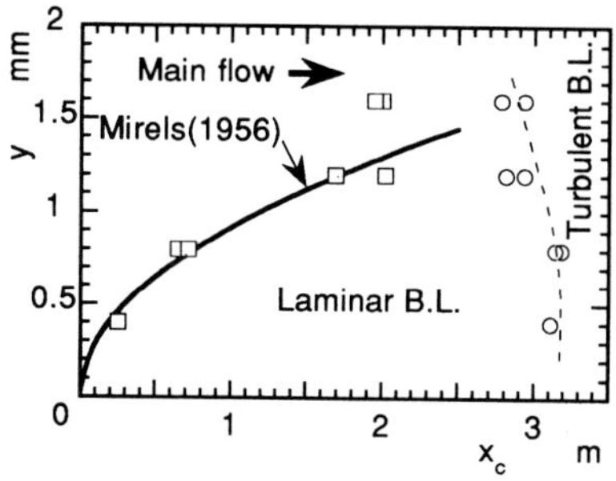

Figure 6. Occurrence of laminar-to-turbulent transition in unsteady boundary layer.

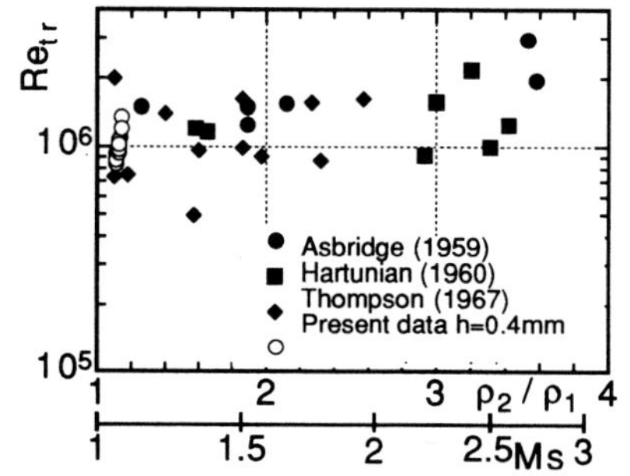

Figure 7. Transition Reynolds number of boundary layer induced by compression wave.

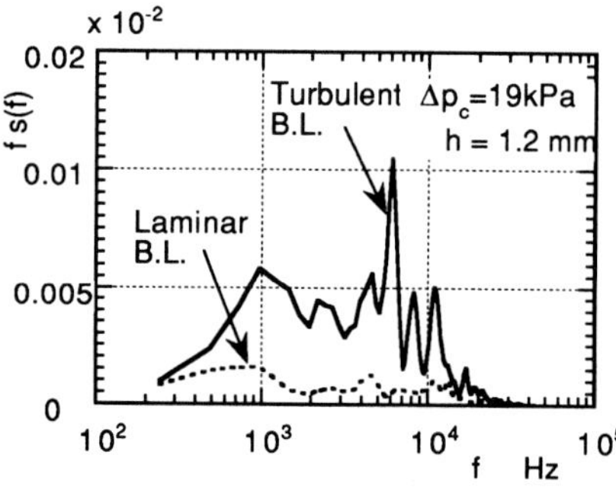

Figure 8. Frequency spectrum of laminar and turbulent boundary layer.

Finally, to obtain information on the boundary-layer structure in the frequency domain, frequency spectrums $fs(f)$ of laminar and turbulent boundary layers measured by the output signal of LDI (setup B) are shown in Fig. 8. Figure 8 indicates that the spectrum curve of the turbulent boundary layer has more peak than that of laminar flow for frequencies between 1 k and 10 kHz. This means that LDI can be used successfully to detect the boundary-layer transition.

4. Concluding remarks

The high-sensitivity interferometry technique, called LDI, was used to record the density variations of compression waves. The measurements were made of time dependent profiles and gave detailed information on the process of formation of the unsteady boundary layer induced by a propagating compression wave. It was found that the laminar-to-turbulent transition in unsteady boundary-layer flow exits describing the density profiles.

References

Aoki T, Matsuo K, Hidaka H, Noguchi Y, Morihara S (1993) Attenuation and distortion of propagating compression wave in high-speed railway model and real tunnels. In: Shock Waves @ Marseille, Vol 3, pp 347–352.

Asbridge JR (1959) An Interferometric Study of Shock Tube Boundary Layer, Lehigh Univ Ins Res Tech 14.

Hartunian RA, Russo AL, Marrone PV (1960) Boundary-layer transition and heat transfer in shock tube. J Aero Space Sci 27(8):587–594.

Matsuo K, Aoki T, Kashimura H, Kawaguchi M, Takeuchi N (1991) Attenuation of compression waves in a high-speed railway tunnel simulator. Proc 7th Intl Symp on Aerodynamic and Ventilation of Vehicle Tunnels, pp 239–252.

Mirels H (1956) Boundary Layer Behind Shock or Thin Expansion Wave Moving into Stationary Fluid. NACA Tech Note 3712, pp 1–53.

Ozawa S (1979) Studies of micro-pressure wave radiated from a tunnel exit. In: Railway Tech Res Rept of Japanese National Railways 1121, pp 1–92 (in Japanese).

Smeets G, George A (1973) Investigation of shock boundary layers with a laser interferometer. In: 9th Int Shock Tube Symposium, pp 431–438.

Thompson WP, Emrich RJ (1967) Turbulent spots and wall roughness effect in shock tube boundary layer transition. Phy Fluids 10(1):17–20.

Structure of multiple shock-wave/ turbulent boundary-layer interactions in a rectangular duct

H. Sugiyama, T. Arai, S. Hongo, T. Kawase, N. Uno
Mechanical Systems Engineering
Muroran Institute of Technology
Muroran, 050 Japan

Abstract: The structure and the turbulence phenomena of multiple shock-wave (called pseudo-shock wave, or PSW)/turbulent boundary-layer interactions in a square duct were investigated using flow-visualization methods and a two-component laser Doppler velocimeter (LDV). First, the PSW was visualized by schlieren photography and laser holographic interferography. Second, the time-mean and fluctuating velocities in the PSW were explored in detail using LDV. Spatial distributions of turbulence intensity, Reynolds shear stress, and turbulence kinetic energy are presented.

Key words: Shock-wave/turbulent boundary-layer interaction, Pseudo-shock wave, Turbulence, LDV, Flow visualization

1. Introduction

When supersonic or hypersonic flow in ducts interacts with downstream blockage devices which could be a flow plug, nozzle, or combustion heat release, a multiple-shock-wave system is produced as the result of shock-wave and duct-wall boundary-layer interactions. A multiple-shock-wave system is called as pseudo-shock wave (PSW) or shock train. The study of PSW has important implications for the design and operation of scramjet/combustor isolators, supersonic gas ejectors, and supersonic wind tunnel diffusers, and so on (Bement et al. 1990).

So far, the characteristics of PSW in ducts have been studied by many researchers — for instance Ikui et al. (1974), Om & Childs (1985), Sugiyama et al. (1988a, 1988b, 1992a, 1992b), and Carroll & Dutton (1990, 1992) — and the characteristics of the PSW were clarified. However, the turbulence phenomena in the PSW have not been fully clarified. This paper describes the turbulence phenomena in the PSW in a square duct that were measred using a two-component laser Doppler velocimeter (LDV).

2. Experimental apparatus and method

A suction-type intermittent supersonic wind tunnel was used in this experiment (Sugiyama et al. 1992a). The test section is a square duct of width $D = 50$ mm and length $L = 1030$ mm. The location of the PSW was fixed in the duct by making small adjustments of the shock location controller. The mean and fluctuating velocities were measured using a fiber-optic two-component LDV system (TSI SYSTEM 9110). The measurement volume diameter was about 0.12 mm and the effective measurement volume length was about 2.4 mm. All of the measurements were made on the mid-plane of the wind tunnel. The two-channel coincidence windows were adjusted to 1 μs. At each measurement location, 2048 velocity samples were taken.

3. Results and discussion

3.1. Overall structure of the PSW

Figure 1 shows the time-mean wall static pressure distributions along the duct measured when the locations of the PSW formation were at $X_f = 140$ mm and 400 mm, respectively. The ordinate is a non-dimensional pressure P/P_a, where P_a is the atmospheric pressure, and X is the distance from the duct entrance. From Fig. 1 it is seen that the pressure rise by the PSW increases as the location of the PSW moves upstream.

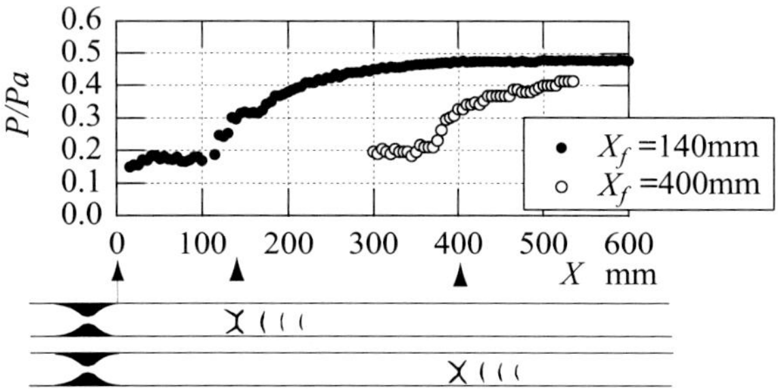

Figure 1. Wall static-pressure distributions along the duct.

Figures 2a and b show a schlieren photograph and laser holographic interferogram of the PSW in a square duct. The location of the PSW was $X_f = 8.0D = 400$ mm. The flow Mach number, boundary-layer thickness and unit Reynolds number just prior to the PSW were $M_\infty = 1.78$, $\delta_\infty = 0.3h$ (where $h = D/2$), and $Re = 13\times10^6$ m^{-1}, respectively. The first shock of the PSW is bifurcated while the following second and third shocks are not. The bifurcated shock consists of a leading oblique shock, a nearly normal shock, and a trailing oblique shock, with the three intersecting at the bifurcation point. A thickening of the boundary layer is observed under the leading oblique shock.

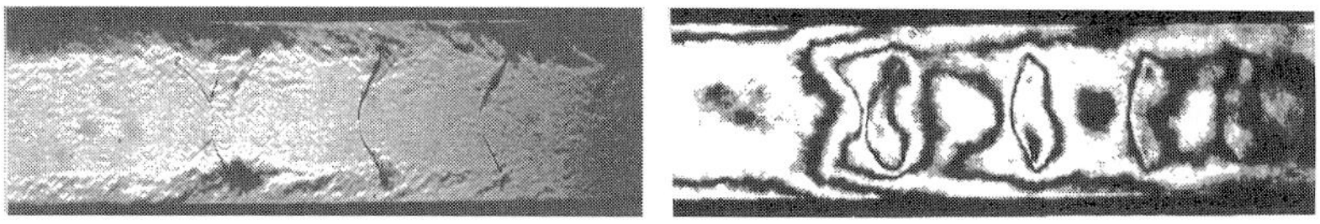

(a) (b)

Figure 2. Visualization of the PSW ($X_f = 8.0D$, $M_\infty = 1.78$, $\delta_\infty/h = 0.3$, $Re = 13\times10^6$ m^{-1}): (a) schlieren photograph, (b) holographic interferogram.

3.2. Mean streamwise velocity profiles in the PSW

Detailed LDV measurements were made of the PSW shown in Fig. 2. Vertical traverses were made at nine stations between the incoming undisturbed region and third shock. LDV measurement stations are shown in Fig. 3. At stations 2 and 3, the leading oblique portion of the first bifurcated shock has reduced the near-wall velocity below that of the incoming boundary layer. At stations 4 and 5, a peak in the velocity occurs at the boundary layer edge (Sugiyama et al. 1992a,b). The core flow reaccelerates following the first shock, and a nearly uniform flow exists outside the boundary layer at station 6.

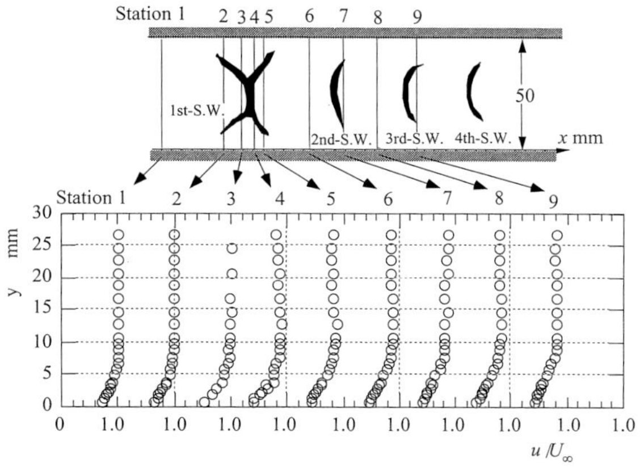

station	X mm	X/D
1	360	7.20
2	387	7.74
3	395	7.90
4	400	8.00
5	405	8.10
6	425	8.50
7	440	8.80
8	455	9.10
9	473	9.46

Figure 3. Mean streamwise velocity profiles in the PSW.

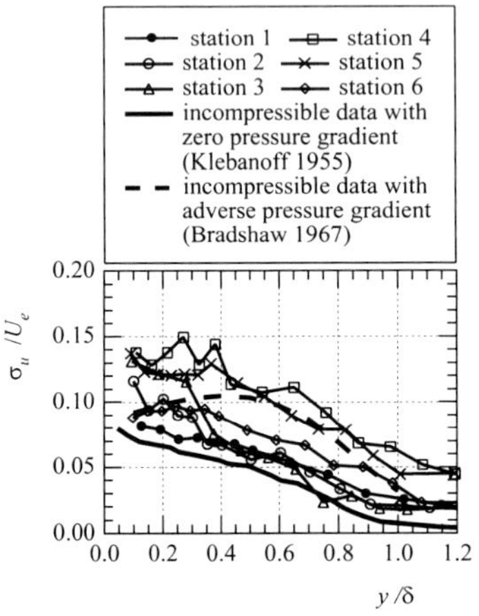

Figure 4. Streamwise turbulence intensity profiles in the PSW.

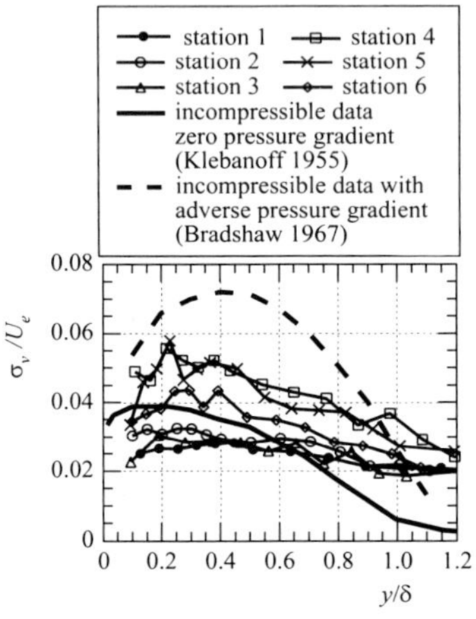

Figure 5. Transverse turbulence intensity profiles in the PSW.

3.3. Turbulence data

Turbulence intensity. Figure 4 shows the standard deviation σ_u of the streamwise velocity u, i.e., turbulence intensity. Here, U_e is the velocity at the outer edge of the boundary layer and δ the boundary layer thickness, which is defined as the distance from the wall for which $u = 0.99U_e$. The incompressible data with zero and adverse pressure gradients over flat plates (Klebanoff 1955, Bradshaw 1967) are also included in Fig. 4. At station 1 in the undisturbed boundary layer, the distribution of turbulence intensity is similar to that of the incompressible flat plate case. At stations 2, 3, and 4, a large increase is caused by the first bifurcated shock to a maximum of $\sigma_u/U_e = 0.15$ for $0.2 < y/\delta < 0.4$. At station 6 in the reacceleration region between the first and second shocks the turbulence intensity decreases. Figure 5 shows the turbulence intensity of transverse velocity, σ_v. At stations 1, 2, and 3, the transverse turbulence intensities are nearly constant across the boundary layer. At stations 4 and 5 under the bifurcated first shock σ_v increases, and at station 6 in the acceleration region σ_v decreases.

Reynolds stress. Figure 6 presents the dimensionless kinetic turbulence shear stress, or Reynolds shear stress, defined as $- < u'v' > /U_e^2$. An increase in the Reynolds stress is caused by the bifurcated first shock. The location of the peak Reynolds stress becomes higher away from the solid wall. At station 6 the Reynolds stress decreases.

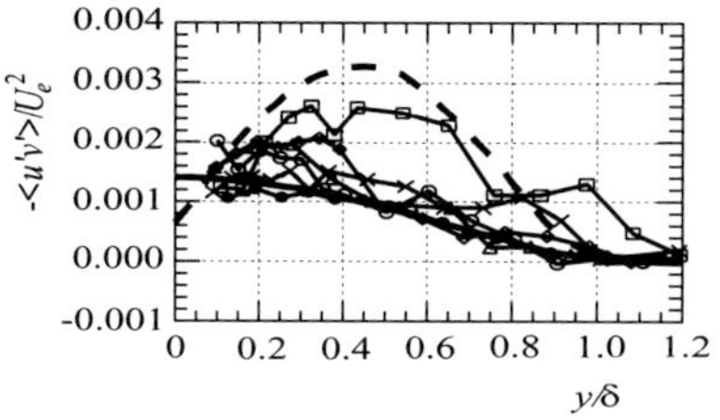

Figure 6. Reynolds shear stress, $-<u'v'>/U_e^2$, in the PSW.

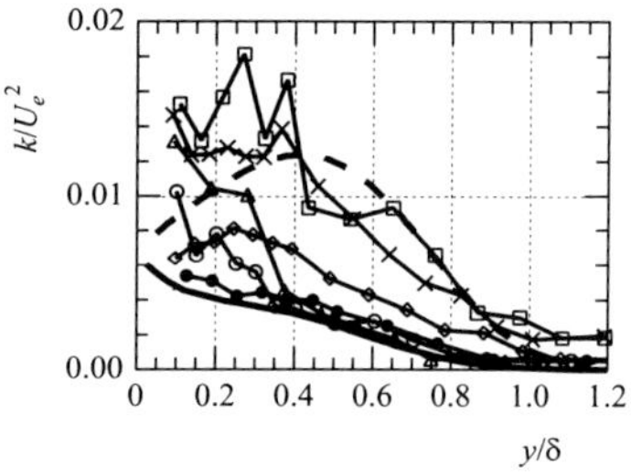

Figure 7. Turbulence kinetic energy, k/U_e^2, in the PSW.

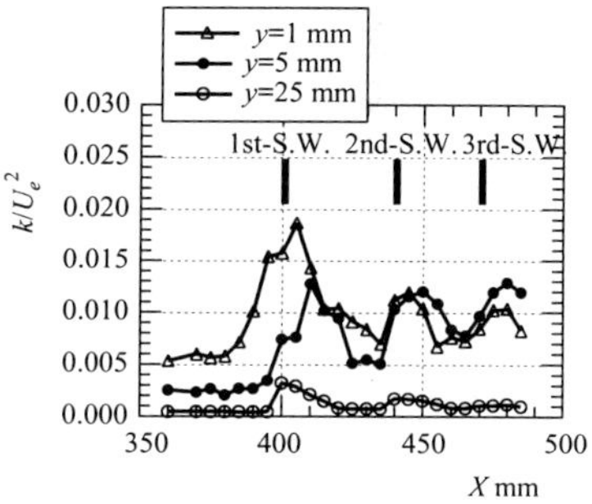

Figure 8. Streamwise distributions of turbulence kinetic energy.

Turbulence kinetic energy. Figure 7 shows the dimensionless turbulence kinetic energy k/U_e^2, where $k = (\sigma_u^2 + \sigma_v^2 + \sigma_w^2)/2$. In this study, the spanwise normal stress σ_w^2 was estimated as $\sigma_w^2 = (\sigma_u^2 + \sigma_v^2)/2$, making $k = 0.75(\sigma_u^2 + \sigma_v^2)$. At station 1 in the incoming, undisturbed boundary layer, the turbulence kinetic energy has almost the same level as in the incompressible case. A large increase in k is caused by the first shock, and the location of the peak moves away from the wall, ranging from $y/\delta = 0.1$ to 0.4. At station 6 in the reacceleration region, the peak level of k decreases. Figure 8 shows the streamwise distribution of k. In the core flow (along the center line of the duct $y = 25$ mm), k is very small and it increases at the first shock because of high frequency oscillation of the multiple shock system. It is clearly seen that in the PSW k increases at each shock, and that k decreases in the reacceleration region between each shock.

4. Conclusions

The internal structure and the turbulence phenomena of multiple shock-wave (PSW)/turbulent boundary-layer interaction in a square duct were investigated experimentally. First, the internal structure of the PSW was visualized by schlieren photography and laser holographic interferography. It was found that the boundary-layer thickness increases at the leading oblique shock of the first bifurcated shock and decreases behind the trailing shock of the first shock, and that there exists a flow reacceleration region between the first and second shocks of the PSW. Next, the time-mean and fluctuating velocities in the PSW were measured in detail using LDV. Spatial distributions of the turbulence intensity, Reynolds shear stress and turbulence kinetic energy were presented. It was found that the turbulence kinetic energy in the boundary layer in the PSW increases under the bifurcated first shock and decreases in the flow reacceleration region between the first and the second shocks.

References

Bement DA, Stevens JR, Thompson MW (1990) Measured operating characteristics of a rectangular combustor/inlet isolator. In: AIAA/SAE/ASME/ASEE 26th Joint Propulsion Conference, AIAA-90-2221, July 16–18, Orlando, FL.

Bradshaw P (1967) The turbulence structure of equilibrium boundary layers. J Fluid Mech 29(4):625–645.

Carroll BF, Dutton JC (1990) Characteristics of multiple shock wave/turbulent boundary-layer interactions in rectangular ducts. J Propulsion and Power 6(2):186–193.

Carroll BF, Dutton JC (1992) Turbulence phenomena in a multiple normal shock wave/turbulent boundary-layer interaction. AIAA J 30(1):43–48.

Ikui T, Matsuo K, Nagai M (1974) The mechanism of pseudo-shock waves. Bulletin of JSME 17(108)731–739.

Klebanoff PS (1955) Characteristics of Turbulence in a Boundary Layer with Zero Pressure Gradient, NACA Report 1247, pp 1135–1153.

Om D, Childs ME (1985) Multiple transonic shock-wave/turbulent boundary-layer interaction in a circular duct. AIAA J 23(10):1506–1511.

Sugiyama H, Takeda H, Zhang J, Abe F (1988a) Multiple shock wave and turbulent boundary layer interaction in a rectangular duct. In: Proc 16th Int Symp on Shock Tubes and Waves, Grönig H (ed), Aachen, pp 185–191.

Sugiyama H, Takeda H, Zhang J, Okuda K, Yamagishi H (1988b) Locations and oscillation phenomena of pseudo-shock waves in a straight rectangular duct. JSME Int J Series II 31(1):9–15.

Sugiyama H, Arai T, Uno N, Takahashi T (1992a) Three-dimensional structure of pseudo-shock waves in a rectangular duct. In: Proc 18th Int Symp on Shock Waves, Takayama K (ed), Sendai, pp 631–636.

Sugiyama H, Arai T, Uno N, Matsumoto T, Homareda M (1992b) Visualization of a pseudo-shock wave in a rectangular duct. In: Proc 6th Int Symp on Flow Visualization, Tanida Y, Miyashiro H (eds), Yokohama, pp 623–626.

Asymptotic defect boundary layer applied to nonequilibrium laminar hypersonic air flows

S. Séror, E. Schall, D. Zeitoun
Lab. I.U.S.T.I. (U.M.R. CNRS 139) Dept. MHEQ, Centre St. Jérôme
Université de Provence, 13397 Marseille Cedex 20, France

J. P. Brazier
ONERA CERT Department of Aerothermodynamics
2, Avenue E. Belin, BP 4025 31055 Toulouse, France

Abstract: The laminar, viscous, nonequilibrium boundary-layer flow around an axisymmetric blunt body has been computed to predict the heat-flux transfer. The extension of the defect formulation to two-dimensional hypersonic flows out of equilibrium including chemical and vibrational nonequilibrium phenomena has been done. This method ensures a smooth matching of the boundary-layer solution with a vortical external flow. Using matched asymptotic expansions, a new set of first-order boundary-layer equations has been derived. The air was considered as a mixture of the five species N_2, O_2, NO, N, O, in which the diatomic species N_2, O_2 are in vibrational nonequilibrium. Comparison between defect calculations and Navier-Stokes calculations (Zeitoun et al. 1994) for the nonequilibrium reacting flow over a sphere-cone shows good agreement. Results of heat-flux predictions are compared to experimental data.

Key words: Boundary layer, Hypersonic flow, Reactive flow, Blunt body, Defect formulation, Asymptotic expansion, Chemical non-equilibrium, Vibrational non-equilibrium

1. Introduction

For hypersonic re-entry flows, the bow shock wave upstream of the spacecraft is the cause of an inviscid vortical flow in the shock layer. Prandtl's equations cannot take into account normal gradients outside of the boundary layer, whereas Van Dyke's second-order matched asymptotic expansions approach (Van Dyke 1962) no longer gives a good matching of the viscous and inviscid solutions when the boundary layer is thick as it is the case for hypersonic re-entry flows. Thus no correct prediction of the influence of the external vorticity on the wall heat flux or the skin friction can be achieved with Van Dyke's approach. To remedy this problem, a defect approach coupled with asymptotic expansions had been proposed (Brazier 1990), but only perfect gas flow had been considered up to now. Moreover, during hypersonic re-entry, an important transfer of the high kinetic energy to the internal energy modes of the gas takes place through this shock wave. This transfer causes a very strong increase in temperature and, therefore, significant modifications of the composition and of the internal energy distribution of the gas. In this work, with regard to the range of Mach numbers investigated, only the chemical and vibrational nonequilibrium processes are taken into account. Energy is thus supplied to the vibration mode and the molecules dissociate. We propose in this paper an extension of the defect approach in the context of hypersonic flows including chemical and vibrational relaxation phenomena. Only steady, two-dimensional planar or axisymmetric laminar compressible reactive flows will be addressed here.

2. Physical modeling

For the considered stagnation conditions, ionization is neglected. The dissociated air can be modeled by a mixture of five species, namely N_2, O_2, NO, N, O. The model assumes seventeen chemical reactions, whose reaction rates are given by Gardiner (1984). The diatomic species N_2 and O_2 are in thermal nonequilibrium and since the vibrational relaxation time for nitrogen monoxide is very small, NO is taken at vibrational equilibrium. The harmonic-oscillator assumption is used and the vibrational exchanges take into account only the translation-vibration (V-T) transfers through the Landau-Teller relaxation theory. A global relaxation time models the exchanges of the molecules with the other species of the mixture. The various relaxation times between species can be found in Druguet et al. (1992).

Only the three major modes, namely translation, rotation, and vibration are accounted for when modeling the partition of internal energy for the various species. Translational and rotational modes are always considered in full equilibrium and thus are characterized by a single temperature T, while vibrational modes may depart from equilibrium. Therefore each of the diatomic species N_2 and O_2 is assumed to have a separate vibrational temperature distinct from the mixture temperature T. Each species of the mixture can be assumed to behave as a perfect gas. The pressure of the mixture is determined from Dalton's law.

The basic equation set is the Navier-Stokes equations together with relaxation equations for the chemical species and the vibration energy. The mass diffusion flux for the I-species are modeled according to the Fick's law using a single diffusion coefficient.

The species viscosity are calculated from the kinetic theory of gases, and Wilke's mixing rule is used to obtain the mixture viscosity. Species translation/rotation thermal conductivities are computed from the Eucken relationship, with mixture values again obtained using Wilke's law. The vibration thermal conductivities of the molecules are expressed in terms of the single diffusion coefficient calculated assuming a constant Lewis number.

3. The defect approach

The proposed strategy is to extend the boundary-layer theory, as developed by Van Dyke (1962) using matched asymptotic expansions, to account for variations of the inviscid flow solution throughout the boundary layer. A way to do it is to use a defect formulation, i.e., not to deal with physical variables in the boundary layer but with the difference between the viscous and the inviscid solutions (Brazier 1990). Therefore applying the method to the Navier-Stokes equations for two-dimensional planar or axisymmetric laminar flows, plus the species conservation equations and the vibrational energy relaxation equations for N_2 and O_2 molecules, new sets of boundary-layer equations have been obtained. The first-order approximation, in the sense of matched asymptotic expansions, includes the usual Prandtl equations plus higher-order terms which ensure a smooth merging of the boundary layer into the inviscid flow.

Thanks to the small-perturbations approach, the calculations of external flow and boundary layer are uncoupled and can be performed separately provided that a specified sequence is respected. First-order external problems must be solved first, then first-order internal. The defect boundary-layer equations are parabolic and can be solved by space marching at a very low cost, like the standard Prandtl equations. The stagnation-point solution is obtained with the help of series expansions with respect to the curvilinear coordinate ξ.

4. Flow past blunt body in the F4 wind tunnel

We have performed a computational study of the high-enthalpy air flow past the blunt body *Electre* defined by the E.S.A for the 1994 workshop. The geometry is essentially a sphere-cone of nose radius $R_0 = 0.035$ m, cone half angle $\alpha = 4.66°$, and overall length $L = 0.4$ m. The wall is assumed to be either fully catalytic or non-catalytic, and at a fixed temperature $T_w = 300$ K. Concerning the boundary conditions relative to the vibrational temperatures, we have considered two extreme cases. First we have assumed that the flow was in thermal equilibrium at the wall, then on the other hand, that there were no vibrational energy exchanges at the wall, i.e., we consider that the wall vibrational heat flux is null. The characteristics of the oncoming flow are given below from the experiment carried out in the F4 hypersonic wind tunnel of ONERA: $\mathcal{M}_\infty = 14.98$, $U_\infty = 5612.4$ m/s, $\rho_\infty = 6.8714\ 10^{-4}$ kg/m^3, $T_\infty = 302$ K, $P_\infty = 71.34$ Pa, $Tv_{N_2\infty} = 5682.5$ K, $Tv_{O_2\infty} = 3618.2$ K, $Y_{N_2\infty} = 0.784812$, $Y_{O_2\infty} = 0.002582$,

$Y_{NO\infty} = 0.002812$, $Y_{N\infty} = 0.003876$, $Y_{O\infty} = 0.205918$. This case corresponds to a Reynolds number based on upstream conditions and nose radius of 900, i.e., typically a low Reynolds-number flow.

5. Discussion of the results

Boundary-layer profiles are displayed in Fig. 1 for a location in the nose region. Longitudinal velocity profiles are plotted in Fig. 1a, and the atomic oxygen mass-fraction profiles are plotted in Fig. 1b. Only the defect approach can match the inviscid solution with a first-order approach. The agreement with the Navier-Stokes solution is fair. Fig. 2a shows the evolutions of the translational and vibrational temperatures along the stagnation streamline in the case of a non-vibrationally catalytic wall. We can notice a strong thermal nonequilibrium in the boundary layer since Tv_{N_2} is about 8000 K at the wall and Tv_{O_2} equals 3300 K. Moreover in the outer flow the vibrational non-equilibrium region of each diatomic species is clearly visible. In the core flow, the relaxation zone of

O_2 is shorter than that of N_2. Figure 2b shows the freezing of the vibrational relaxation process in the expansion zone of the flow. The mass-fraction evolutions along the stagnation streamline are plotted in Fig. 3a. One can notice a strong recombination of atomic species in the boundary layer since the wall is fully catalytic. The evolution of the wall heat flux is given in Fig. 3b for a

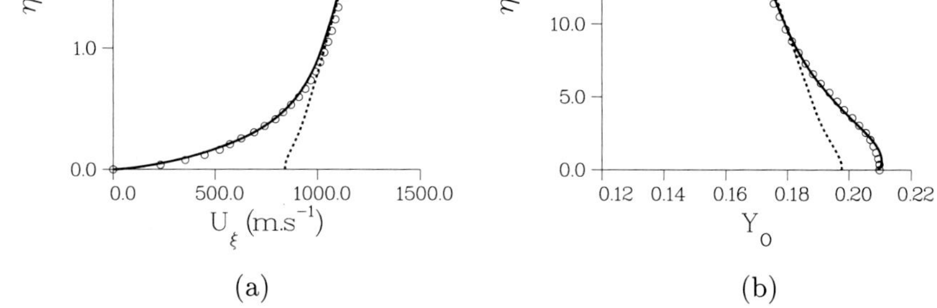

Figure 1. Boundary-layer profiles, $\xi = 2$ cm, non-catalytic wall and wall vibrational heat flux = 0: (a) longitudinal velocity profiles, (b) Y_O mass fraction.

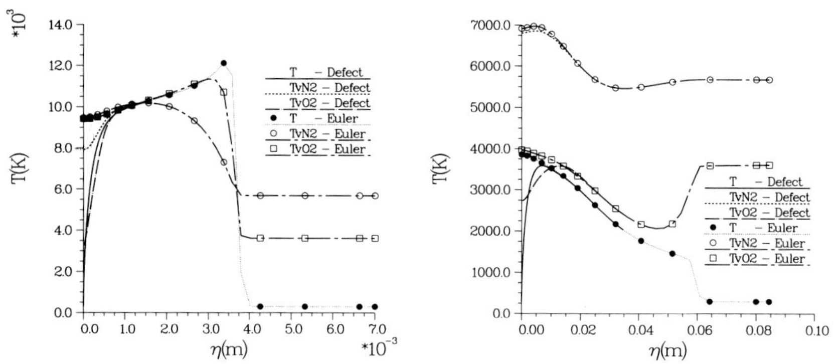

(a) (b)

Figure 2. Temperature profiles for the catalytic wall with a wall vibrational heat flux = 0: (a) $\xi = 0$, (b) $\xi = 20$ cm.

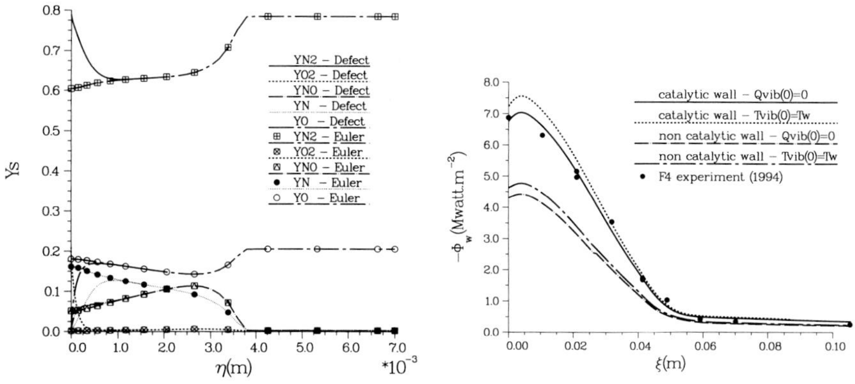

(a) (b)

Figure 3. (a) Mass-fraction profiles along stagnation line for the catalytic wall with a wall vibrational heat flux = 0, (b) wall heat-flux predictions.

fully catalytic wall without vibrational energy exchanges at the wall. This case seems to be closer to the experimental results. The influence of the boundary condition used for the vibration energy equations is clearly visible. For a catalytic wall the assumption of a total deactivation of the vibrational mode of N_2 and O_2 molecules at the wall induces a nose heat flux 13% higher than for a wall without vibrational energy exchanges. For a non-catalytic wall, the deviation in the nose heat flux between these two extreme conditions is 10%. These results confirm the previous one obtained by Tirsky (1993) concerning wall heat-flux predictions for the Buran shuttle.

6. Conclusions

The defect approach has been extended to derive boundary-layer equations from the Navier-Stokes equations for laminar viscous flows out of equilibrium. Together with matched asymptotic expansions methodology, it ensures a correct matching of the inviscid flow with the first-order boundary-layer solution. In this case the agreement with the Navier-Stokes solution is fair. Wall heat-flux predictions for various boundary conditions have shown that the best agreement with the experimental data is achieved for a fully catalytic wall assuming that there are no vibrational energy exchanges at the wall.

Acknowledgement. The authors would like to thank the "Centre de Calcul de l'I.M.T." for allowing access to the Cray YMP-2E computer.

References

Brazier JP (1990) Etude asymptotique des équations de couche limite en formulation déficitaire, Ph.D. Thesis, Ecole Nationale Supérieure de l' Aéronautique et de l' Espace, Toulouse.

Druguet MC, Zeitoun D, Brun R (1992) Inviscid hypersonic nozzle flows in chemical and vibrational nonequilibrium state. In: Computing Methods in Applied Sciences and Engineering, R Glowinski (ed), Nova Science, New York, pp 729–738.

Gardiner WC Jr (1984) Combustion Chemistry, Springer-Verlag.

Tirsky GA (1993) Up to date gasdynamic models of hypersonic aerodynamics and heat transfer with real gas properties. Annual Review of Fluid Mechanics 25:151–181.

Van Dyke M (1962) Second order compressible boundary-layer theory with application to blunt bodies in hypersonic flow. In: Hypersonic Flow Research, FR Ridell (ed), Academic Press, Vol 7, pp 37–76.

Zeitoun D, Schall E, Burtschell Y, Druguet MC (1994) Vibration-dissociation coupling in nonequilibrium hypersonic viscous flows. AIAA Journal 33(1):79–85.

Analysis of nonequilibrium, hypersonic blunt-body flow with the local-similarity method

A. Klomfass, S. Müller, J. Ballmann
Lehr- und Forschungsgebiet für Mechanik, RWTH Aachen
Templergraben 64, 52056 Aachen, Germany

Abstract: A method for the calculation of the stagnation-streamline quantities and the stagnation-point heat transfer in hypersonic flows about spheres or cylinders is suggested which is based on the local similarity of the flow field. Comparisons with two-dimensional Euler and Navier-Stokes calculations and experiments are presented, which include perfect-gas flows and nonequilibrium air flows.

Key words: Nonequilibrium, Stagnation streamline, Local similarity

1. Introduction

A spacecraft entering the atmosphere generates a strong bow shock that leads to the excitation of molecular vibration and dissociation. A nonequilibrium flow prevails if the relaxation times of these processes are of the order of the fluid-dynamic time scale. Such flows are described mathematically by the conservation laws and additional balance equations for the component densities and the vibrational energies. Physical models provide relations for the chemical production rates, the rates of vibrational energy transfer, and the molecular fluxes. Flows about spheres or cylinders with high aspect ratios are suitable for the investigation and validation of these models. The stagnation-streamline quantities and the stagnation-point heat flux are of major interest in this context. We therefore consider a method for the fast calculation of these quantities.

2. The local-similarity method

The stagnation-streamline quantities are usually obtained from an asymptotic steady-state solution of the two-dimensional, unsteady flow problem. However, as two-dimensional hypersonic flows about spheres or cylinders exhibit an approximate local similarity in the stagnation-streamline area, the number of independent spatial variables can be reduced from two to one (Kao 1964). The dimensionally reduced Navier-Stokes equations (DRNSE) are derived from the Navier-Stokes equations in spherical or cylindrical coordinates. In our empirical approach the flow quantities are represented as products of unknown functions of the radial coordinate r and prescribed, assumed functions of the angular coordinate θ, which is here measured from the stagnation streamline. The DRNSE are then obtained by carrying out all differentiations with respect to θ, equating coefficients of like functions of θ, and letting $\theta \to 0$. The velocity components, the temperatures, the mass fractions, and the pressure are here assumed as

$$v_r = \overline{v}_r(r)\cos\theta, \quad v_\theta = \overline{v}_\theta(r)\sin\theta, \quad T^i = \overline{T}^i(r), \quad x_\alpha = \overline{x}_\alpha(r), \tag{1}$$

$$p = p_\infty + \overline{p}(r)\cos^2\theta\,. \tag{2}$$

These relations are motivated by the boundary conditions and Newtonian flow theory, which provides an approximation for the surface pressure. The resulting DRNSE are formally given as

$$\overline{\mathbf{u}}_{,t} + \overline{\mathbf{F}}_{,r} + \overline{\mathbf{G}} = \mathbf{S}\,, \tag{3}$$

with $\overline{\mathbf{u}} = (\rho, \rho\overline{v}_r, \rho\overline{v}_\theta, \rho e^t, \rho_\alpha, \rho_\alpha e^v_\alpha)^T$ and corresponding dimensionally reduced fluxes $\overline{\mathbf{F}}$, and metric terms $\overline{\mathbf{G}}$, which both contain viscous and inviscid contributions. The nonequilibrium source terms $\mathbf{S} = (\mathbf{0}, \omega_\alpha, \Omega^{vib}_\alpha)$ are not affected by the transformation of the equations. The DRNSE are subject to boundary conditions that specify the undisturbed supersonic parallel free stream, a no-slip or slip wall, an isothermal or adiabatic wall, and a fully catalytic or non-catalytic wall.

3. The numerical method

An implicit time-integration method is applied to obtain steady-state solutions of the DRNSE, which are approximated by the finite-difference equations

$$\frac{\overline{\mathbf{u}}_i^{n+1} - \overline{\mathbf{u}}_i^n}{\Delta t} + \frac{\overline{\mathbf{F}}_{i+1}^{n+1} - \overline{\mathbf{F}}_{i-1}^{n+1}}{\Delta r} + \overline{\mathbf{G}}_i^{n+1} = \mathbf{S}_i^{n+1}\,, \tag{4}$$

with $\Delta t = t^{n+1} - t^n$ and $\Delta r = r_{i+1} - r_{i-1}$. The viscous parts of the fluxes and the metric terms still contain radial derivatives of certain flow quantities. These are approximated by second-order accurate central-, left-, or right-sided finite differences as functions of the respective values at points $i-1, i$, and $i+1$. The unknowns $\overline{\mathbf{u}}^{n+1}$, $\overline{\mathbf{F}}^{n+1}$, $\overline{\mathbf{G}}^{n+1}$, and $\overline{\mathbf{S}}^{n+1}$ at time t^{n+1} are extrapolated from their values at t^n by Taylor expansions. For $\mathbf{f}$ being any of these quantities

$$\mathbf{f}^{n+1} = \mathbf{f}^n + \left.\frac{\partial\,\mathbf{f}}{\partial\,\overline{\mathbf{w}}}\right|^n \Delta\overline{\mathbf{w}}^n\,,\ \Delta\overline{\mathbf{w}}^n = \overline{\mathbf{w}}^{n+1} - \overline{\mathbf{w}}^n,\ \overline{\mathbf{w}} = \left(\overline{v}_r, \overline{v}_\theta, \overline{p}, \overline{T}^{tr}, \overline{x}_\alpha, \overline{T}^{vib}_\alpha\right)^T. \tag{5}$$

The introduction of the variables $\overline{\mathbf{w}}$ as calculational quantities simplifies the incorporation of the boundary conditions and the calculation of the Jacobians

$\partial \mathbf{f}/\partial \overline{\mathbf{w}}$ considerably. Introducing (5) into (4) and rearranging terms renders the time-integration scheme

$$\mathbf{a}_i\,\Delta\overline{\mathbf{w}}_{i-1}^n + \mathbf{b}_i\,\Delta\overline{\mathbf{w}}_i^n + \mathbf{c}_i\,\Delta\overline{\mathbf{w}}_{i+1}^n = -\left(\overline{\mathbf{F}}_{i+1}^n - \overline{\mathbf{F}}_{i-1}^n + \Delta r\,\overline{\mathbf{G}}_i^n - \Delta r\,\overline{\mathbf{S}}_i^n\right), \quad (6)$$

where the matrices $\mathbf{a}_i, \mathbf{b}_i$, and $\mathbf{c}_i$ are given in terms of the Jacobians. The system of equations for the interior points $i = 2, \ldots, N-1$ of the computational domain can be written as $\mathbf{A}^n \Delta\overline{\mathbf{w}}^n = \mathbf{RHS}^n$, with a block-tridiagonal matrix $\mathbf{A}$. The products $\mathbf{a}_2\Delta\overline{\mathbf{w}}_1^n$ and $\mathbf{c}_{N-1}\Delta\overline{\mathbf{w}}_N^n$ are included on the right-hand side and the boundary conditions are used to eliminate the unkowns $\Delta\overline{\mathbf{w}}_1$ and $\Delta\overline{\mathbf{w}}_N$. Calculations are initialized with suitable linear distributions of $\overline{\mathbf{w}}$. Steady-state solutions are obtained asymptotically for $t \to \infty$. Due to the implicit formulation, time steps with high CFL numbers can be realized. The Fortran routine NBTRIP of R. Chakravarthy is used to solve $\mathbf{A}^n\Delta\overline{\mathbf{w}}^n = \mathbf{RHS}^n$ for $\Delta\overline{\mathbf{w}}^n$. The approximation of $\overline{\mathbf{F}}_{,r}$ by central differences in (4) is appropriate for flows with thick shock waves. A shock-fitting strategy is used for flows with thin shock waves. In this case only the shock-layer part of the stagnation streamline is the computational domain. The free-stream boundary conditions are then replaced by the post-shock conditions. Since the steady-state shock position is a priori unknown, it is calculated simultaneously with the shock-layer flow. Starting from an initial guess, the shock position at t^{n+1} is determined from the shock speed, which is calculated from the Rankine-Hugoniot relations and a compatibility equation (cf. Klomfass 1995). The computational grid is fitted to the new shock position after each time step.

4. The gas physical model

Air is considered as a mixture of 5 components (N_2, O_2, NO, N, O). The energy is calculated on the basis of independent Boltzmann distributions, where a common temperature $T^{(tr)}$ is assigned to all translational, rotational, and atomic-electronic modes. Individual temperatures $T_{N_2}^{(vib)}, T_{O_2}^{(vib)}$, and $T_{NO}^{(vib)}$ are assumed for the vibrational and electronic modes of the molecular components. The model of a rigid rotor and truncated harmonic-oscillator molecule is used. The vibrational energy exchange in non-reactive molecular collisions consists of VT and VV contributions. The volume specific rates are

$$\Omega_{\alpha\beta}^{(VT)} = \rho_\alpha c_\beta\, k_{\alpha\beta}^{VT}\left(1 - \exp\left(-\theta_\alpha^{(vib)}/T^{(tr)}\right)\right)\left\{e_\alpha^{(vib)}(T^{(tr)}) - e_\alpha^{(vib)}\right\} \quad (7)$$

$$\begin{aligned}\Omega_{\alpha\beta}^{(VV)} = \rho_\alpha c_\beta\, k_{\alpha\beta}^{VV}\,\frac{1}{R_\beta\theta_\beta^{(vib)}}\Bigg\{&\left(e_\alpha^{(vib)} + R_\alpha\theta_\alpha^{(vib)}\right)e_\beta^{(vib)}\exp\left(-\frac{\theta_\alpha^{(vib)} - \theta_\beta^{(vib)}}{T^{(tr)}}\right)\\ &- \left(e_\beta^{(vib)} + R_\beta\theta_\beta^{(vib)}\right)e_\alpha^{(vib)}\Bigg\}. \quad (8)\end{aligned}$$

The fundamental rate constants $k_{\alpha\beta}^{VT}$ are here calculated from curve fits to experimental data, while the $k_{\alpha\beta}^{VV}$ are obtained from the heuristic relation

$$k_{\alpha\beta}^{VV} = \sqrt{k_{\alpha\alpha}^{VT}\, k_{\beta\beta}^{VT}\, \exp\left\{(\theta_\alpha^{(vib)} - \theta_\beta^{(vib)})/T^{(tr)}\right\}}\,, \tag{9}$$

which fits the few available experimental data. The chemical-production rates ω_α are calculated from the usual rate equations, where the 17 major chemical reactions are considered. The preferential dissociation model of Treanor and Marrone is applied for the rate constants $k^{(r')}$ of the endothermic reactions; the rate constants $k^{(r'')}$ of the exothermic reactions are deduced from the principle of detailed balance:

$$k^{(r')} = \Psi\, \tilde{k}^{(r')}(T^{(tr)}) \qquad \text{(endoth. reactions)} \tag{10}$$

$$k^{(r'')} = \tilde{k}^{(r')}(T^{(tr)})\, /\, K_r(T^{(tr)}) \qquad \text{(recombination)} \tag{11}$$

$$k^{(r'')} = \Psi\, \tilde{k}^{(r')}(T^{(tr)})\, /\, K_r(T^{(tr)}) \qquad \text{(exoth. exchange reactions).} \tag{12}$$

Here $\tilde{k}^{(r')}$ denotes the equilibrium rate constants; the preferential dissociation parameter $\Psi = \Psi(T^{(tr)}, T_\alpha^{(vib)})$ accounts for the effect of vibrational nonequilibrium of the respective molecular reactant. The correlations of Park (1990) are used for $\tilde{k}^{(r')}$. The equilibrium constants K_r are calculated from partition functions. The vibrational energy loss in reactive molecular collisions is deduced from the preferential dissociation theory. The volume-specific rates of vibrational energy loss and gain are then

$$\Omega_\alpha^{(vib,diss)} = \omega_\alpha^{(r')}\, e_\alpha^{(vib)}(T_\alpha^*)\,, \quad \frac{1}{T_\alpha^*} = \frac{1}{T_\alpha^{(vib)}} - \frac{1}{T^{(tr)}}\,, \tag{13}$$

$$\Omega_\alpha^{(vib,form)} = \omega_\alpha^{(r'')} \frac{1}{2}\left(e_\alpha^{(0)} - R_\alpha \theta_\alpha^{(vib)}\right)\,, \tag{14}$$

with $e_\alpha^{(0)}$ denoting the (positive) dissociation energy. Equation (14) is derived from (13) and the equilibrium conditions $\Omega_\alpha^{(vib,diss)} = \Omega_\alpha^{(vib,form)}$, $\omega_\alpha^{(r')} = \omega_\alpha^{(r'')}$, and $T_\alpha^{(vib)} = T^{(tr)}$. It states that the formation of a molecule takes place with equal probability for each vibrational level. This process of random recombination is assumed to occur in nonequilibrium, too. Details about the thermochemical model are given in Klomfass (1995). The theories of Newton, Fourier, and Fick are applied to the molecular fluxes. Bulk viscosity, and thermo- and pressure diffusion are as yet neglected. The species viscosities are calculated according to the gas kinetic theory, where the collision integrals are calculated from curve fits (Metz 1995). The mixing rule of Herning and Zipperer is used to approximate the mixture viscosity μ. The conductivities and the diffusion coefficient are approximated by constant Prandtl and Schmidt numbers, where $Pr = Sc = 0.7$ is assumed. Details are given in Klomfass et al. (1993) and Metz (1995).

5. Results and discussion

Several calculations have been carried out to assess the accuracy of the DRNSE. In the following we consider flows about spheres and focus on the shock stand-off distance δ and the stagnation-point heat flux q_{st}. For flows of a perfect gas with $\gamma = 1.4, R = 289\,\mathrm{J/kgK}, \mu = 1.5 \cdot 10^{-5}\,\mathrm{kg/ms}$, and $Pr = 0.7$ the following table shows the comparisons with the results of 2D-Euler calculations and the boundary-layer analysis of Fay and Riddell. These test cases are without

Flow conditions (Wall: 300 K)						DRNSE		2D-Euler/Fay&Rid.	
R_{sph} [m]	V $[\frac{m}{s}]$	p [Pa]	T [K]	Ma	Re	q_{st}	δ/R_{sph}	q_{st} $[\frac{kW}{m^2}]$	δ/R_{sph}
0.152	697	1000	300	2	1.6e5	11.2	0.245	10.7	0.345
0.152	1742	1000	300	5	4.1e5	151	0.139	141	0.156
0.152	3484	1000	300	10	8.1e5	980	0.123	957	0.135
0.152	6967	1000	300	20	16.3e5	6030	0.120	6160	0.129

physical relevance. However, as there is no question concerning the thermochemical equilibration of the boundary layer, the Fay and Riddell results are reliable. As the table above shows, the shock stand-off is significantly underestimated for larger values of δ/R_{sph}. The deviation decreases as δ/R_{sph} becomes smaller. The stagnation-point heat fluxes are acceptably accurate for all cases. Not shown here is that the calculated stagnation pressures and adiabatic stagnation temperatures match the Euler results perfectly. For the chemical and vibrational nonequilibrium flow specified below, Figs. 1 and 2 give a detailed comparison with a full 2D Navier-Stokes calculation.

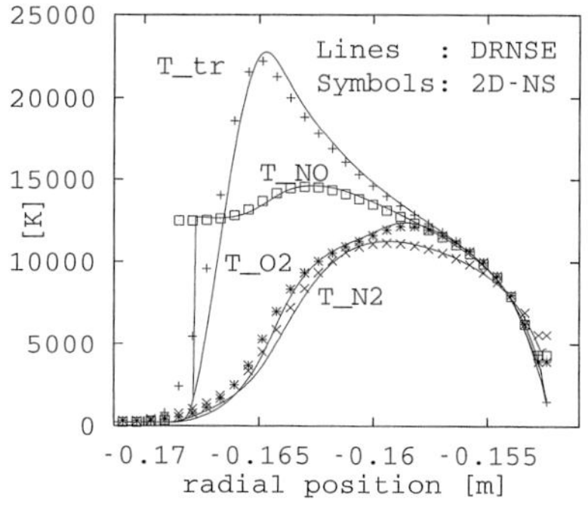

Figure 1. Temperatures.

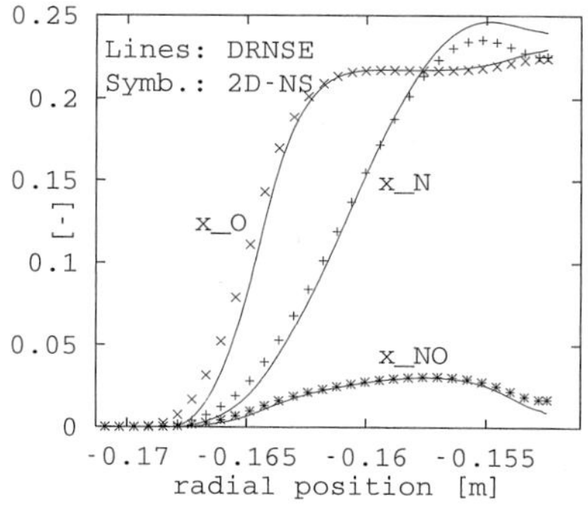

Figure 2. Mass fractions.

Flow conditions (Wall: 1500 K, non-catalytyic)						DRNSE		2D NS	
R_{sph}	V $[\frac{m}{s}]$	p [Pa]	T [K]	Ma	Re	q_{st}	δ/R_{sph}	q_{st} $[\frac{MW}{m^2}]$	δ/R_{sph}
0.152m	7600	4.17	212	26	11400	1.12	≈ 0.081	0.97	≈ 0.085

The deviation of the atomic-nitrogen mass fraction near the wall may be attributed to the different resolution, which is 35 × 160 points for the 2D and 200 points for the DRNSE calculations. The potential of the DRNSE can be assessed from the computing times, which were about 3 h on a Cray Y/MP for the 2D and about 15 min on a standard workstation for the DRNSE calculation. The following examples are flows with higher Reynolds numbers, where thin boundary layers exist. A good agreement with experimentally determined heat fluxes

Flow conditions (Wall: 300 K, non-catalytic)						DRNSE		Exp.	2D-Euler
R_{sph}[m]	V [$\frac{m}{s}$]	p [Pa]	T [K]	Ma	Re	q_{st}	δ/R_{sph}	q_{st} [$\frac{MW}{m^2}$]	δ/R_{sph}
0.0254	3929	11180	1371	5.5	108000	14.8	0.094	15.26	0.096
0.0254	5301	13090	2359	5.3	61000	34.3	0.084	34.51	0.085

(Wen and Hornung 1995) is found. The calculations revealed near-equilibrium boundary layers. In such flows a very high spatial resolution is required to achieve convergence of the surface concentrations and the heat fluxes. About 200 boundary-layer grid points were necessary to resolve the atomic-oxygen recombination near the stagnation point. This resolution could so far only be realized with the DRNSE approach or boundary-layer methods.

6. Conclusions

A useful approximation for the stagnation-streamline quantities in hypersonic blunt-body flows can be obtained from the numerical integration of the dimensionally reduced Navier-Stokes equations. The presented numerical method provides a helpful tool for the investigation of nonequilibrium flows, as it allows a fast, high-resolution calculation of the stagnation-streamline quantities.

Acknowledgement. This work is funded by the Deutsche Forschungsgemeinschaft.

References

Kao HC (1964) Hypersonic viscous flow near the stagnation streamline of a blunt body. AIAA J 2:1892–1897.

Klomfass A (1995) Hyperschallströmungen im Thermodynamischen Nichtgleichgewicht, Dissertation, RWTH Aachen.

Klomfass A, Müller S, Ballmann J (1993) Modelling of Transport Phenomena for the Hypersonic Stagnation Point Heat Transfer Problem, AIAA Paper 93-5047.

Metz S (1995) Numerische Verfahren zur Berechnung des Stoff- und Impulstransports in Wasserstoff-Luft und Methan-Luft Gemischen, Dissertation, RWTH Aachen.

Wen C, Hornung H (1995) Experiments on hypervelocity dissociating flow over spheres. In: Shock Waves @ Marseille, Vol 2, Springer.

Formation of standing shock waves due to evaporation from a spherical condensed phase

Y. Onishi, J. W. Fuchs
Department of Applied Mathematics and Physics
Tottori University, Tottori 680 Japan

Abstract: Transient motions of a vapor due to evaporation processes at a spherical condensed phase placed in an infinite expanse of the vapor have been considered. The Boltzmann equation of BGK type has been solved numerically by a difference scheme subject to the diffusive boundary condition at the interface together with an initial condition associated with sudden change in temperature of the condensed phase. Unlike problems with simple plane geometry, there exists a certain range of parameters in which the evaporation process causes strong expansion near the condensed phase, accelerating the flow to locally supersonic. A standing shock wave is, then, formed in order to connect the supersonic and subsonic regions within the flow field.

Key words: Kinetic theory, Evaporation, Spherical condensed phase, Shock wave, Standing shock wave

1. Introduction

Transient motions of a vapor due to phase-change processes at the condensed phase are of fundamental importance in gas dynamics in the sense that a clear understanding may be obtained as to how the flow fields approach their final states through propagations and continuous interactions of various waves, e.g., shock waves, contact regions, and sometimes expansion waves, produced by these phase-change processes. Problems of this kind, however, are the ones to which the ordinary continuum-based fluid dynamics is not applicable because of the existence of a nonequilibrium region called the Knudsen layer in the vicinity of the condensed phase. It is this nonequilibrium region that is responsible for the phase-change processes and its existence cannot be neglected in any problems, even in the continuum limit, as long as these processes are involved. Therefore, the analysis for flow problems of this kind must necessarily be based on kinetic theory.

The present study is concerned with one of these flow fields, i.e., the transient motions of a gas due to evaporation processes from a spherical condensed phase placed in an infinite expanse of its vapor and the consequent propagation of shock waves through a three-dimensional region. The formation of a standing shock wave may be expected in the flow field during the transition process to its final state. The Boltzmann equation of BGK type (Bhatnager et al. 1954) has been solved numerically for the present study subject to the boundary condition of diffusive reflection at the interface. It may be noted here that the steady state of this problem in the presence of an inert gas has been treated analytically by Onishi (1986) for weak phase-change cases and by Takeuchi & Onishi (1995) for slightly strong cases. Also the numerical study of the steady evaporation from a spherical phase into a vacuum has been given by Sone & Sugimoto (1993).

2. Equations, initial and boundary conditions

Consider a spherical condensed phase with its radius R_1 in an infinite expanse of its vapor at rest. Initially, the gas is in complete equilibrium with the condensed phase at temperature T_∞. The pressure and number density of the gas at this state are P_∞ and N_∞, respectively. At time $t = 0$, the temperature of the condensed phase is suddenly raised to a temperature T_W ($T_W > T_\infty$), which leads to the onset of evaporation from the condensed phase. The Boltzmann equation of BGK type (Bhatnager et al. 1954) in terms of the spherical coordinates $(r,\ \theta,\ \phi)$ may be written for the present problem with spherical symmetry as

$$\frac{\partial f}{\partial t} + \xi\cos\alpha\,\frac{\partial f}{\partial r} - \frac{\xi\sin\alpha}{r}\frac{\partial f}{\partial\alpha} = N\nu_c\,(F_e - f) \tag{1}$$

$$F_e = \frac{N}{(2\pi RT)^{3/2}}\exp\left\{-\frac{\xi^2 - 2U\xi\cos\alpha + U^2}{2RT}\right\} \tag{2}$$

$$\begin{bmatrix} N \\ NU \\ 3NRT \end{bmatrix} = 2\pi\int_0^\pi\int_0^\infty \begin{bmatrix} 1 \\ \xi\cos\alpha \\ \xi^2 - 2U\,\xi\cos\alpha + U^2 \end{bmatrix} \xi^2\sin\alpha\, f\, d\xi\, d\alpha \tag{3}$$

$$P = N\mathrm{k}T \tag{4}$$

$$E = 2\pi\,\frac{m}{2}\int_0^\pi\int_0^\infty \xi^5\cos\alpha\sin\alpha\, f\, d\xi\, d\alpha \tag{5}$$

$$\xi_r = \xi\cos\alpha, \quad (\xi_\theta^2 + \xi_\phi^2)^{1/2} = \xi\sin\alpha \qquad 0\le\xi<\infty \quad 0\le\alpha\le\pi$$

where t is the time; r is the radial coordinate; ξ is the magnitude of the molecular velocity vector $\boldsymbol{\xi}$ ($\xi = |\boldsymbol{\xi}|$) with its components $(\xi_r,\ \xi_\theta,\ \xi_\phi)$; α is an angle between the radial vector $\mathbf{r}$ and the molecular velocity vector $\boldsymbol{\xi}$; f, N, U, T, P, and E are, respectively, the distribution function, the number density, the mean flow velocity in the r direction, the temperature, the pressure, and the energy flux. F_e is the local Maxwellian distribution. $R = \mathrm{k}/m$ is the gas constant per unit mass, k being the Boltzmann constant and m the molecular mass; ν_c is a constant ($N\nu_c$: local collision frequency) related to the transport coefficients of the gas, for example, $\nu_c = P_\infty/(N_\infty \mu)$, μ being the viscosity of gas at the initial equilibrium state.

Now, the initial distribution function is given by

$$f = \frac{N_\infty}{(2\pi R T_\infty)^{3/2}} \exp\left\{-\frac{\xi^2}{2RT_\infty}\right\} \qquad \text{for} \quad 0 \le \alpha \le \pi \tag{6}$$

everywhere in the flow field ($R_1 \le r < \infty$). At time $t = 0$, the temperature of the condensed phase is suddenly changed to a different constant temperature T_W. Since the boundary condition for the distribution function is assumed to be of diffusive type, i.e., molecules leaving the condensed phase after their interactions with it obey a Maxwellian distribution characterized by the velocity (zero here) and the temperature of the condensed phase, the distribution function f in the present case takes for $t \ge 0$

$$f = \frac{N_W}{(2\pi R T_W)^{3/2}} \exp\left\{-\frac{\xi^2}{2RT_W}\right\} \qquad \text{for} \quad 0 \le \alpha \le \frac{\pi}{2} \qquad \text{at} \quad r = R_1 \tag{7}$$

$$f = \frac{N_\infty}{(2\pi R T_\infty)^{3/2}} \exp\left\{-\frac{\xi^2}{2RT_\infty}\right\} \qquad \text{for} \quad \frac{\pi}{2} \le \alpha \le \pi \qquad \text{at} \quad r \to \infty \tag{8}$$

where N_W is the number density of the molecules of gas leaving the interface after their interaction with the condensed phase. N_W is a unique function of T_W and is taken here to be the saturated vapor number density corresponding to T_W. From the Clapeyron-Clausius relation (see e.g., Landau & Lifshitz 1980), it is given by

$$N_W = \frac{P_W}{\mathrm{k}T_W} \qquad \frac{P_W}{P_\infty} = \exp\left\{-\Gamma\left(\frac{T_\infty}{T_W} - 1\right)\right\} \qquad \Gamma = \frac{h_L}{RT_\infty} \tag{9}$$

where P_W is the saturated vapor pressure corresponding to T_W and h_L is the latent heat of vaporization per unit mass. It may be noted that the values of Γ are usually larger than unity, for example, $\Gamma = 19$ at $T_\infty = 25°\text{C}$ and $\Gamma = 13$ at $T_\infty = 100°\text{C}$ for H_2O.

Appropriate nondimensionalization of the present problem gives the following parameters which characterize the motions of the gas:

$$Kn = \frac{l_\infty}{R_1} = \left(\frac{8\gamma}{\pi}\right)^{1/2} \frac{Ma}{Re} \qquad \frac{T_W}{T_\infty} \qquad \Gamma \quad \left(\text{or} \quad \frac{P_W}{P_\infty}\right) \tag{10}$$

where Kn, Re, and Ma are the Knudsen number, Reynolds number, and Mach number, respectively, at the initial state; l_∞ is the molecular mean free path defined by $l_\infty = (8RT_\infty/\pi)^{1/2}/(N_\infty \nu_c) = (\mu/P_\infty)(8RT_\infty/\pi)^{1/2}$. Note that $Ma = 1$ in the present case because the sound speed $c_\infty = (\gamma RT_\infty)^{1/2}$ with $\gamma = 5/3$ has been used here as the reference velocity. Furthermore, a time scale $\tau_\infty = (\gamma/2)^{1/2}(R_1/c_\infty)$ has also been introduced for the present numerical calculations.

3. Results and discussion

The fluid dynamic quantities for various sets of the parameters are obtained as functions of r/R_1 and t/τ_∞. Some of the results are shown in Figs. 1–3. The steady solutions to the Euler equations (with the boundary conditions given by the present numerical analysis) are shown by dashed lines in Fig. 1, whereas in Fig. 2 the steady solutions (Takeuchi & Onishi 1995) to the set of macroscopic equations of Navier-Stokes type and the appropriate boundary conditions derived from the kinetic theory analysis (Onishi 1993) are shown also by dashed lines. Figure 3a shows the weakening process of the initial shock wave and Fig. 3b the local Mach-number distributions at the virtually steady state.

Figure 1 shows the propagation of an initial shock wave followed by a contact region and fairly strong expansion (virtually adiabatic) at an early stage promoted by the three-dimensional expanse. This strong incipient expansion causes a supersonic flow near the condensed phase, which leads to the gradual formation of the standing shock wave at a finite distance from the condensed phase as time goes on. The occurrence of this standing shock wave is necessary in order for the flow field to adjust itself to the subsonic state established behind the initial shock wave propagating toward infinity. When the evaporation is not strong (see Fig. 2), the steady behavior is not of a supersonic-shock-subsonic-type, but simply of a smoothly changing weak subsonic, as can be expected.

Acknowledgement. We acknowledge the finacial support of the Association for Promotion of the Tottori Prefectural Engineering Technology.

References

Bhatnager PL, Gross EP, Krook M (1954) A model for collision processes in gases. Phys Rev 94:511–525.

Landau LD, Lifshitz EM (1980) Statistical Physics, Pergamon Press, New York, pp 251–257.

Onishi Y (1986) The spherical-droplet problem of evaporation and condensation in a vapour-gas mixture. J Fl Mech 163:171–194.

Onishi Y (1993) Kinetic theory based nonlinear treatment for motions of a binary gas mixtures involving slightly strong evaporation and condensation processes II — A system of macroscopic equations, the boundary conditions and the Knudsen-layer corrections. ZAMP 44:827–864.

Sone Y, Sugimoto H (1993) Kinetic theory analysis of steady evaporating flows from a spherical condensed phase into a vacuum. Phys Fluids A 5:1491–1511.

Takeuchi S, Onishi Y (1995) Evaporation and condensation from a spherical condensed phase in a vapor-gas mixture. In: Proc of 33rd Symp of the Chuugoku-Shikoku Branch of Japan Soc Mech Eng (in Japanese), pp 53–54.

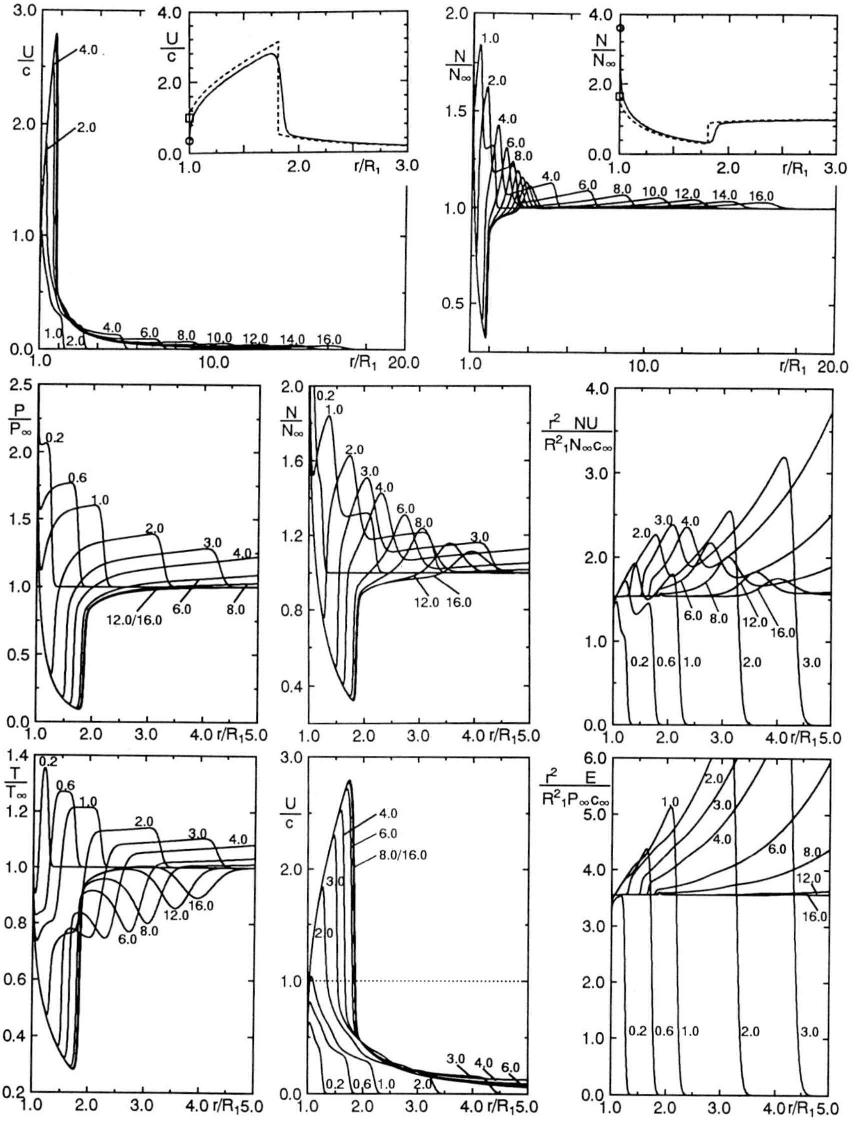

Figure 1. $Kn = 0.01$, $T_W/T_\infty = 1.2$, $\Gamma = 11$ ($P_W/P_\infty = 6.2547$). The numbers in the graphs indicate the values of time t/τ_∞; c is the local sound speed. Dashed lines indicate the steady-state solutions of the Euler equations. The values of $P(1)/P_\infty$ and $T(1)/T_\infty$ at $t/\tau_\infty = 16.0$ are 3.7319 and 1.0375, respectively.

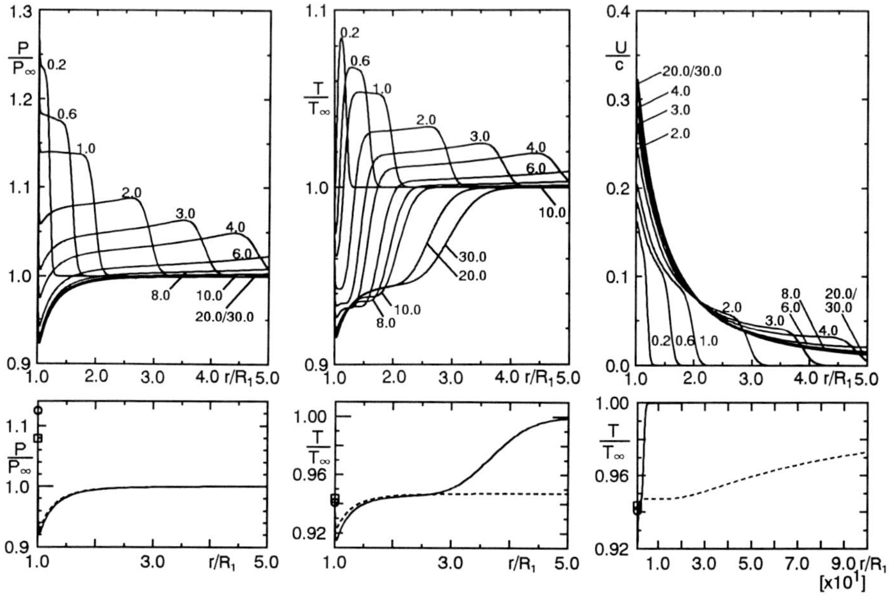

Figure 2. $Kn = 0.01$, $T_W/T_\infty = 1.05$, $\Gamma = 11$, $(P_W/P_\infty = 1.6884)$. The numbers in the graphs indicate the values of time t/τ_∞; c is the local sound speed. Dashed lines indicate the steady solutions (Takeuchi & Onishi 1995) obtained from Navier-Stokes type of equations, to which the present numerical solutions (solid lines) should approach as time goes on.

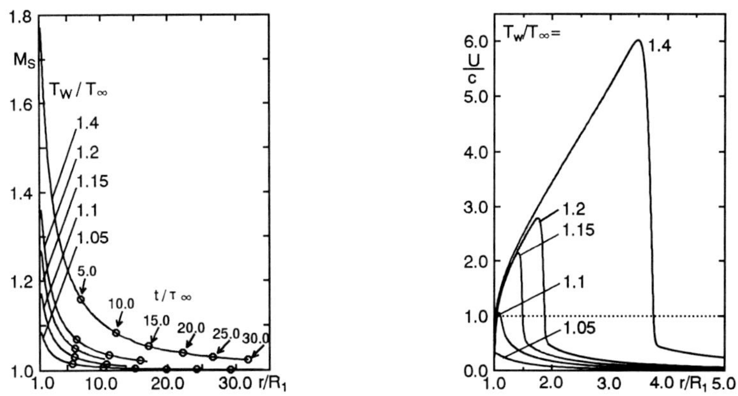

Figure 3. (a) The variations of the initial shock Mach number Ms with the distance from the condensed phase; (b) the local Mach-number distributions at the virtually steady state.

Knudsen-number effects in high-speed microchannel flows

T. X. Nguyen, C. K. Oh, X. N. Saint-Martin-Tillet,
J. V. Piana, E. S. Oran
Naval Research Laboratory, Code 6404
Laboratory for Computational Physics and Fluid Dynamics
Washington, DC 20376-5344 USA

Abstract: The effects of high Knudsen number on the flow in a microchannel are investigated through a series of numerical simulations. Both the Navier-Stokes and Direct Simulation Monte Carlo approaches are used to simulate a high-speed flow of He gas in a two-dimensional microchannel. The results show that the NS leading-edge shock angle is larger than that calculated by the DSMC, causing the two leading-edge shocks to interact sooner. Also, DSMC simulations predict substantial velocity slip and temperature jump at the wall near the leading edges of the channel.

Key words: High Knudsen-number flow, Direct Simulation Monte Carlo, Navier-Stokes, Microchannel flows

1. Introduction

The flow of gas through microdevices, for which the characteristic dimension L is near the order of the molecular mean free path λ, belongs to the transitional regime of fluid dynamics. In this regime, the Knudsen number, $Kn = \lambda/L$, is on the order of 0.1, the value for which the continuum Navier-Stokes (NS) formulation becomes invalid. Molecular approaches must be used to predict accurate slip velocities and temperature jumps at the solid surfaces and their effects on the flow-field structure.

One of the most promising methods applicable to the high Kn regime is the Direct Simulation Monte Carlo (DSMC) technique (Bird 1976), a method that has been used extensively for rarefied gas flows. Recently, DSMC has been used to simulate microchannel flows, in which the relevant channel dimension is on the order of microns and the densities may be relatively high (see, for example, Oh et al. 1995a).

In this work, a series of numerical simulations is used to examine the effects of high Kn on the flow structure in a two-dimensional microchannel. DSMC and NS computations are performed for the same physical system and their results are compared. The DSMC calculations are performed with the DSMC-

MLG algorithm (Cybyk 1994) which was recently optimized on the Thinking Machines CM-5 (Oh et al. 1995b, Nguyen 1995). The second type of calculation is performed with a NS code which has also been optimized on the CM-5 (Weber 1994) and used to study shock/boundary-layer interactions.

2. Numerical methods

2.1. DSMC-MLG approach

The MLG (Boris 1986), a method for managing very large data structures, ensures that particles that are close in physical space are also close in array index space. The MLG achieves this by requiring the spatial coordinates of the particles to satisfy monotonicity constraints.

The combined DSMC-MLG method abandons the spatially fixed grid used in conventional DSMC calculations. Instead, it subdivides the computational domain into templates, which are defined by blocks of nearest-neighbor particles. The locations of the boundary particles define the boundaries of the template. As the fluid particles move during the course of the flow, the particles are resorted into MLG order. The result is a template grid which automatically adapts to the continually changing local number densities of the gas. The combined DSMC-MLG was parallelized on the CM-5, resulting in a very efficient code (80% parallel efficiency achieved) and significant reductions in computational time (Oh et al. 1995b).

2.2. Continuum approach

The NS simulations were performed using a parallelized code based on Flux-Corrected Transport (FCT). FCT (Boris and Book 1976) is a high-order algorithm for integrating the convective transport terms while maintaining conservation, monotonicity, and positivity. The diffusive transport terms are calculated using a central-difference, finite-volume approach, and these terms are combined with the convective transport as source terms in FCT.

3. Results and discussion

The simulations are performed for a 6 μm $\times$ 1.2 μm channel with a Mach 5 inflow of He atoms at 298 K and 1 atm. For this system, $Kn = 0.14$. In the DSMC-MLG simulations, the He atoms are modeled as hard spheres. A total of 110,250 simulated particles are divided into 150 $\times$ 15 time-varying computational templates of 7 $\times$ 7 particles. The solid walls are represented by the fully diffuse boundary conditions and the molecules are reflected according to the Maxwellian distribution based on the wall temperature. In the NS simulations, a rectangular 256 $\times$ 64 grid is used. At the solid walls, no-slip conditions are applied. For both types of calculations, the wall temperature is equal to the free-stream temperature. Figure 1 is a schematic of the flow configuration.

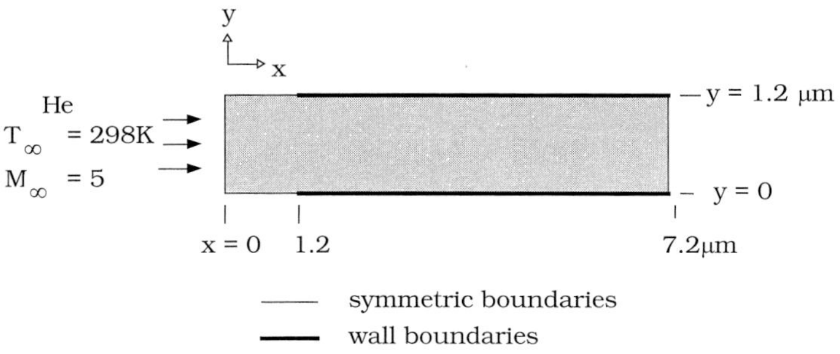

Figure 1. Schematic of the microchannel flow.

Figures 2 and 3 show that NS and DSMC contours of the pressure, p, and viscosity coefficient, μ, differ both qualitatively and quantitatively for the two cases. In the NS calculation, the leading-edge shocks are sharper and more

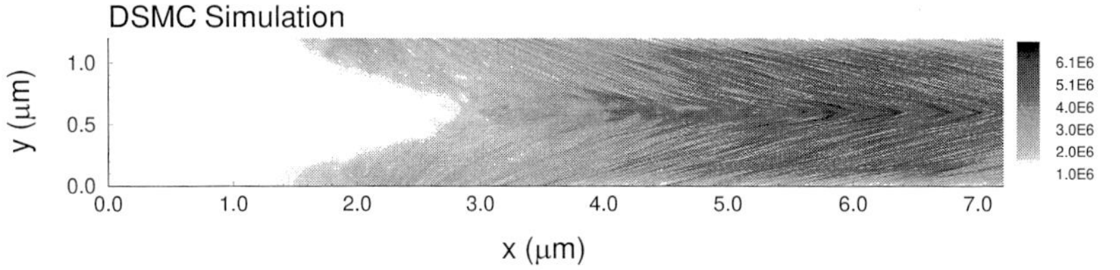

Figure 2. Contours of pressure in the microchannel. The DSMC result represents an instantaneous solution. The streaks in the DSMC result are statistical fluctuations typical of DSMC simulations.

highly curved than the DSMC shocks, indicating the presence of strong viscous interactions between the boundary layer and the outer shock. Furthermore, the

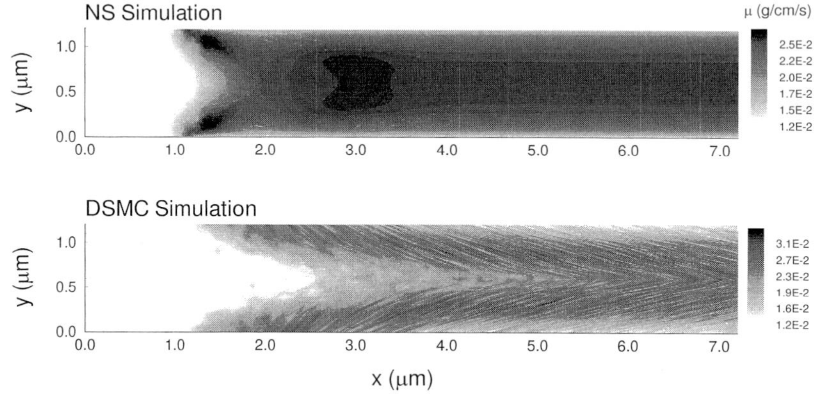

Figure 3. Contours of viscosity coefficient in the microchannel. The DSMC result represents an instantaneous solution.

NS shock angle, on the average, is larger than that calculated by the DSMC method. The shock angle varies between near normal at the wall to $\sim 50°$ near the channel centerline for the continuum case, whereas it is only $\sim 20°$ for the DSMC calculation. The larger shock angle in the NS case means that the two leading-edge shocks interact sooner. The NS shock is also much stronger than the DSMC shock, indicated by the respective pressure ratios of 60 and 7.5 across the shock at the wall.

Figure 4 shows the velocity and temperature profiles at several locations along the length of the microchannel obtained from the DSMC simulation. These velocity profiles show the development of the viscous boundary layer, including the slip velocity generated at the channel walls. The slip velocity is higher near the channel entrance and decreases with distance downstream. The thermal boundary layer, indicated by the temperature profiles, shows a jump in temperature at the walls. The slip velocity and temperature jump along the lower wall of the channel are shown in Fig. 5. The slip velocity is higher near the channel entrance and decreases with distance downstream. Beyond $x = 3.5\ \mu$m, the velocity slip does not change significantly. The temperature jump at the wall initially increases near the channel entrance, reaches a maximum at approximately $x = 2.2\ \mu$m, then decreases further downstream. Thus the DSMC simulations predict the expected velocity slip and temperature jump at the microchannel walls typical of high-Kn flows where the continuum formulation is invalid.

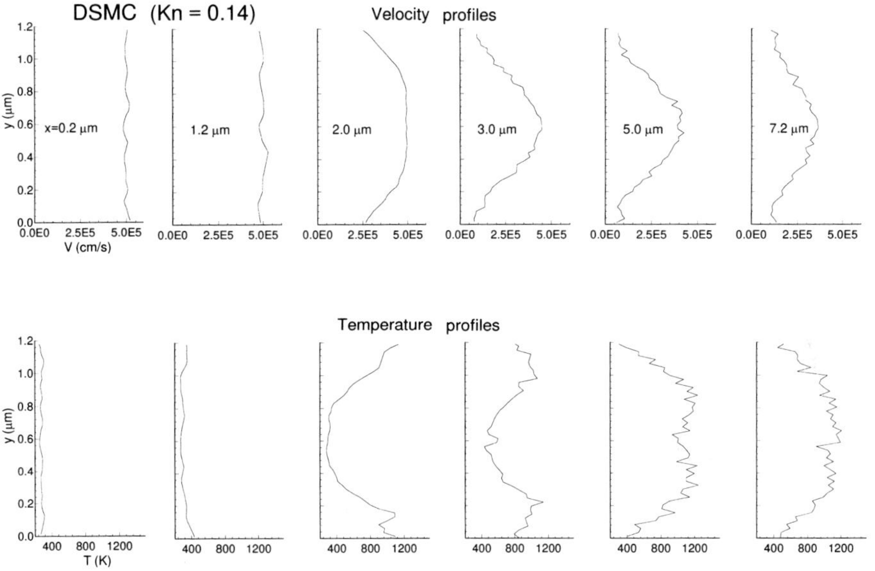

Figure 4. DSMC-computed velocity and temperature profiles at selected locations.

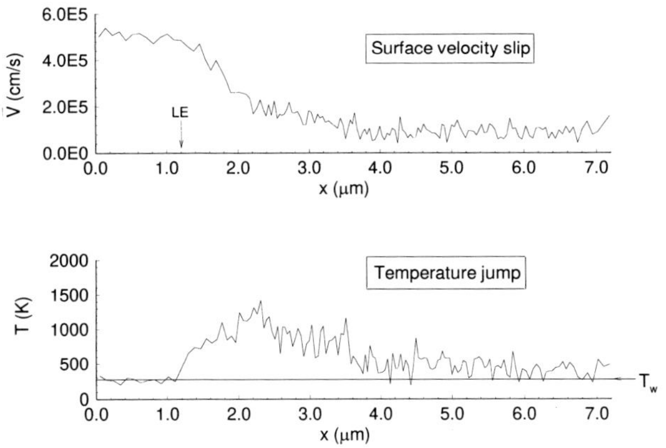

Figure 5. Velocity slip and temperature jump along the lower wall of microchannel.

4. Conclusion

These DSMC results represent one instantaneous solution which requires approximately 7 hours of cpu time on the CM-5 to complete. Even more extensive time-averaging is required to reduce the statistical fluctuations in the solution to an acceptable level, further increasing the computational cost. Each NS simulation requires only $\sim$ 5 minutes of cpu time on the CM-5 for convergence. However, the continuum approach is not valid for $Kn > 0.03$. It therefore becomes important to develop ways of modifying NS calculations to simulate some of the high-Kn effects. These computations will be used as a basis for the derivation of boundary conditions and constitutive relations that extend the validity of the NS simulations.

Acknowledgement. This work was sponsored by ASEE, NRC, ARPA, ONR/NRL, SNECMA, and EM2C-CNRS.

References

Bird GA (1976) Molecular Gas Dynamics, Clarendon Press, Oxford.

Boris JP (1986) A vectorized "near-neighbor" algorithm of order n using a monotonic logical grid. J Comp Phys 66:1–20.

Boris JP, Book DL (1976) Solution of the continuity equation by the method of flux-corrected transport. Methods in Computational Physics 16:85–129.

Cybyk BZ (1994) Combining the Monotonic Lagrangian Grid with Direct Simulation Monte Carlo; a New Approach for Low-Density Flows, PhD Dissertation, University of Maryland.

Nguyen TX (1995) Enhancements of the Parallelized DSMC-MLG Method for Applications to Complex Hypersonic Transitional-Regime Flows, PhD Dissertation, University of Maryland.

Oh CK, Oran ES, Cybyk BZ (1995a) Microchannel Flow Computed with the DSMC-MLG, AIAA-95-2090.

Oh CK, Sinkovits RS, Cybyk BZ, Oran ES, Boris JP (1995b) Massive Parallelization of DSMC Combined with the Monotonic Lagrangian Grid, AIAA-95-0409.

Weber YS (1994) The Numerical Simulation of the Reflected-Shock/Boundary-Layer Interaction in Shock Tubes, PhD Dissertation, University of Maryland.

Studies of separated flow structure and oscillating separation shock waves in shock-wave/turbulent boundary-layer interaction regions induced by a blunt protuberance

S. Maekawa, S. Aso
Department of Aeronautics and Astronautics, Kyushu University
Hakozaki 6-10-1, Higasiku, Fukuoka 812 Japan
K. Karashima, K. Sato
Institute of Space and Astronautical Sciences
Yoshinodai 3-1-1, Sagamihara, Kanagawa 229 Japan

Abstract: The three-dimensional shock-wave/turbulent boundary-layer interaction region induced by a blunt protuberance is examined. In the experiments, oscillating shock waves in the interaction region are also experimentally investigated. Experiments are performed with freestream Mach numbers of 2, 3, and 4 and almost adiabatic wall conditions.

Key words: Oscillating shock wave, Blunt protuberance, Flow visualization

1. Introduction

The shock-wave/turbulent boundary-layer interaction is a very interesting problem and arises in many practical flows. Representative studied were made by Korkegi (1971) and Sedney (1973). One of these phenomena is the oscillating shock-wave motion generated by the shock-induced turbulent boundary-layer separation. This unsteadiness would appear to be an inherent feature of the shock-induced separation, and has been investigated by many researchers (e.g. Sedney and Kitchens 1977 and Dolling and Bogdonoff 1981).

A multi-stage rocket such as H-II is designed to store the maximum amount of fuel, hence pipelines are located outside the rocket body. In such a configuration, the bow shock wave induced by a pipeline fairing interacts with the turbulent boundary layer that is developed on the main body, causing severe aerodynamic heating phenomena and oscillating shock waves in the interaction region.

2. Experimental apparatus and procedures

The experiments are performed in a supersonic wind tunnel of the Institute of Space and Astronautical Sciences with a square test section of 600 × 600 mm and variable Mach number capability over the range of 2–4. A portion of a side

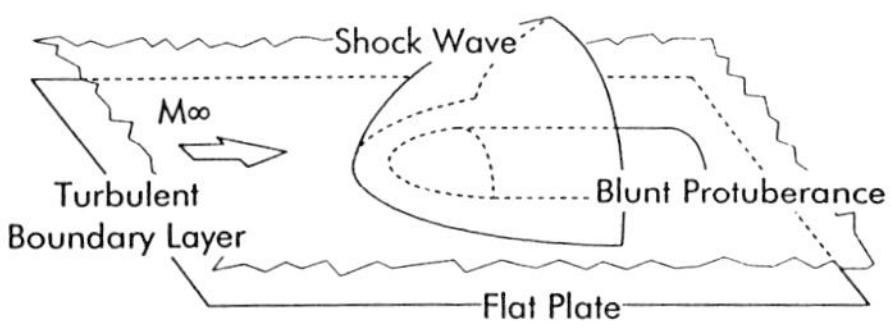

Figure 1. Schematic diagram of the flow field.

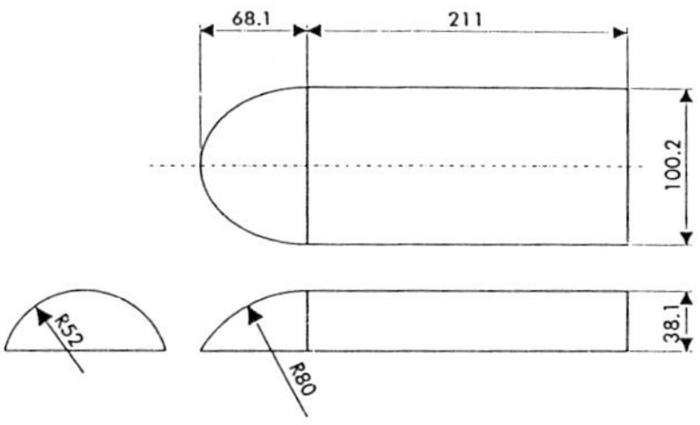

Figure 2. Blunt protuberance model and coordinate system.

wall is an optical window which provides excellent optical access so that flows may be visualized in detail. The boundary-layer thickness is about 50 mm at a freestream Mach number of 4, estimated from schlieren and pitot-pressure measurements (cf. Aso and Maekawa 1993). A schematic of the flow field is shown in Fig. 1.

The configuration of the blunt protuberance is shown in Fig. 2. The x-axis is the main flow direction, the y-axis is lateral, and the z-axis is normal to the side wall. The origin is the tip of the blunt protuberance. The blunt protuberance is located on the side wall for oil-flow visualizations and pressure-distribution measurements or is attached on the flat plate installed in the wind tunnel for visualization at the foot of the body. Experiments are performed under almost adiabatic wall conditions. Total pressure is 0.29 MPa at a freestream Mach number of 2, 0.49 MPa at Mach 3, and 0.59 MPa at Mach 4.

3. Results and discussion

3.1. Visualizations by oil-flow technique

A representative picture and a sketch of an oil-flow pattern are shown in Fig. 3 for Mach number 4. The locations of the primary separation lines, which represent the extent of the interaction, with various Mach numbers are shown in Fig. 4. In Fig. 3, S_1, S_2, and A are the primary separation line, secondary separation line, and attachment line, respectively. The primary separation lines are located at almost the same position irrespective of freestream Mach number. This indicates that the main governing parameter of the interaction regions is not freestream Mach number, but the size of the shock generator, i.e., the blunt

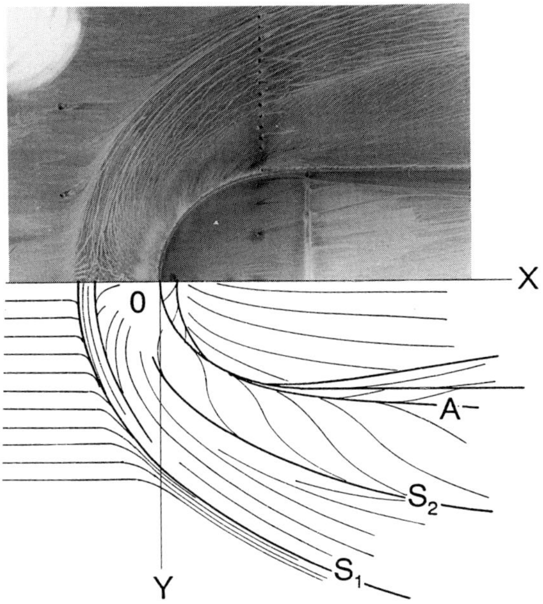

Figure 3. Picture and sketch of an oil-flow pattern for $M_\infty = 4$.

protuberance. Between a secondary separation line and an attachment line, an inverse-S type pattern is observed, which indicates the existence of a separation vortex. Another separation region arises from the shoulder of the blunt protuberance. For the Mach 4 case, just behind the primary separation line, a new separation region exists, suggesting that new vortices in the interaction region are produced.

3.2. Pressure measurements

Pressure measurements on the centerline and $x = 46$ mm are shown in Fig. 5. Pressures are normalized by the value of the undisturbed pressure in front of the interaction region. A pressure plateau is formed close to the secondary separation line due to the separated vortex in the interaction region. The local pressure minimum lies just upstream of the blunt protuberance, due to the small separated vortex located before the protuberance. The maximum pressure is much higher than the pressure plateau. However, the locations at which the pressure starts to rise and the maximum pressures are nearly independent of Mach number. The same tendency is observed for pressure distributions at $x = 46$ mm. Away from the centerline the pressure decreases rapidly, and a

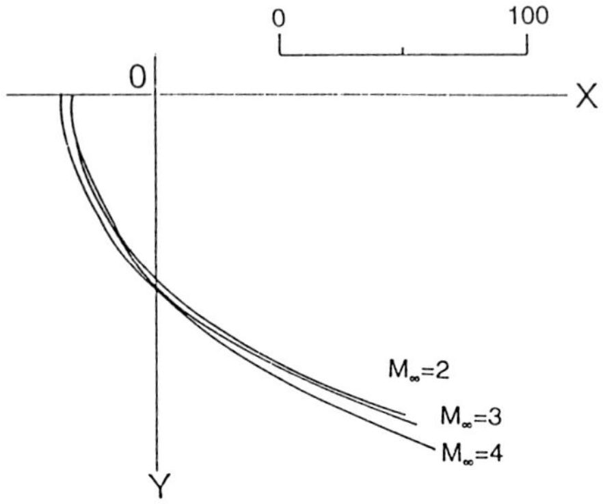

Figure 4. The locations of the primary separation lines.

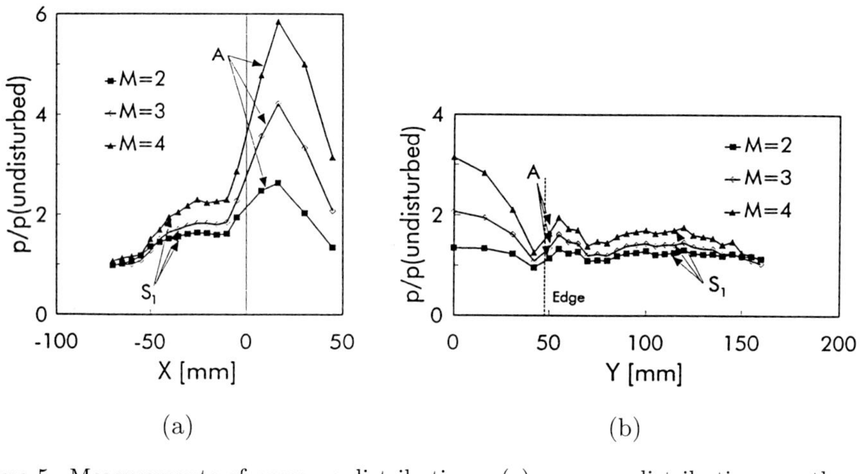

Figure 5. Measurements of pressure distributions: (a) pressure distributions on the centerline, and (b) pressure distributions on $x = 46$ mm.

local pressure minimum is observed upstream of the maximum peak, due to the separated vortex at the corner.

3.3. Visualization by shadowgraph technique

Shadowgraphs are instantaneously taken as randomly shot sequences, shown in Fig. 6. These pictures show the unsteady characteristics of the interaction region. In these pictures, oscillating separated shock waves are observed. Also a slip layer is observed at the intersection point of the separation shock wave and

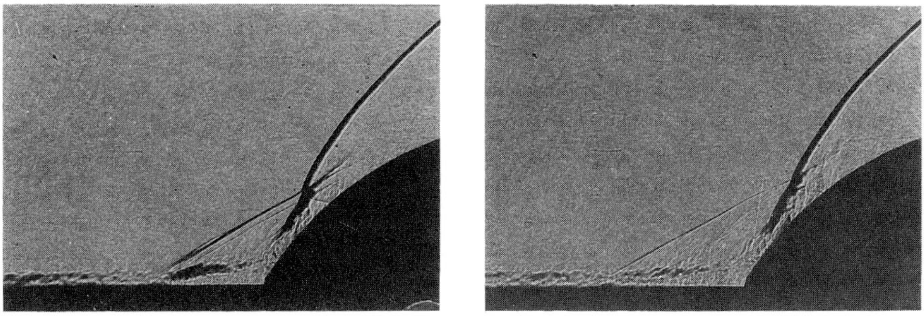

(a) (b)

Figure 6. Instantaneous shadowgraph at $M_\infty = 3$: (a) strong shock separation, and (b) weak shock separation.

the bow shock wave. When a strong-separation shock wave is formed (Fig. 6a), the generating point of the separation shock wave is located slightly further downstream than in the weak-separation shock-wave case (Fig. 6b), and the interaction region decreases. In the strong-separation shock-wave case, a strong slip layer impinges on the blunt protuberance surface.

4. Conclusions

1. The primary separation lines are located at almost the same position irrespective of freestream Mach number. Thus the main governing parameter of these interaction regions is the size of the blunt protuberance.

2. For a freestream Mach number of 4, just behind the primary separation line a new separation region appears, suggesting that new vortices in the interaction region are produced. The position of the primary separation and attachment lines varied very little with Mach number.

3. As the freestream Mach number increases, the maximum pressure becomes higher, but the locations of the start of the pressure rise and the maximum pressures are nearly independent of freestream Mach numbers.

4. The oscillating separated shock waves have been observed by optical flow visualization. The separation phenomenon is highly unsteady. Also there are two significant cases of weak separation and strong separation.

References

Aso S, Maekawa S (1993) Intensive Studies on Unsteady Secondary Separations and Oscillating Shock Waves in Three-Dimensional Shock Waves/Turbulent

Boundary Layer Interaction Region Induced by Sharp/Blunt Fins, AIAA Paper 93-2939.
Dolling DS, Bogdonoff SM (1981) An Experimental Investigation of the Unsteady Behavior of Blunt Fin-Induced Shock Wave Turbulent Boundary Layer Interactions, AIAA Paper 81-1287.
Korkegi RH (1971) Survey of viscous interaction associated with high Mach number flight. AIAA J 9:5:771–784.
Sedney R (1973) A survey of the effects of small protuberances on boundary layer flows. AIAA J 11:6:782–792.
Sedney R, Kitchens CW (1977) Separation ahead of protuberances in supersonic turbulent boundary layers. AIAA J 15:4:546–552.

A numerical study of shock-wave reflections in nonequilibrium chemically reacting flows

S. Aso, T. Mitomo
Department of Aeronautics and Astronautics
Kyushu University, Japan

T. Fujiwara
Department of Aeronautics
Nagoya University, Japan

M. Hayashi
Nishinippon Institute of Technology
Japan

Abstract: In the present study the full two-dimensional Navier-Stokes equations are solved in order to investigate unsteady aerodynamic heating phenomena induced by shock impingement on a ramp surface. The effects of chemical reactions on aerodynamic heating are investigated. The results show that the effects of chemical reactions on shock-induced aerodynamic heating phenomena are quite significant at high temperature.

Key words: Aerodynamic heating, Nonequilibrium flow, Chemical reaction

1. Introduction

Recently, high-speed winged vehicles have been studied. One of the most important problems for designs of such vehicles is the severe aerodynamic heating and pressure rise caused by the impingement of shock waves on vehicle surfaces. In particular, the unsteady aerodynamic heating caused by shock-wave reflections at high shock Mach number has been investigated and the peak heating due to a Mach stem and the second peak heating due to a slip layer have been observed by the present authors (Aso et al. 1989, Ohyama et al. 1993).

In the present study the full two-dimensional Navier-Stokes equations are solved in order to investigate unsteady aerodynamic heating phenomena induced by shock impingement on a ramp surface. When the shock is strong, nonequilibrium chemical reactions occur in the flow. The effects of chemical reactions on aerodynamic heating are investigated. The results are compared with experiments and calculated results without reactions.

2. Numerical methods

In the numerical calculations, nonequilibrium chemically reacting flows are considered. The governing equations are the two-dimensional full Navier-Stokes equations. In conservation form, these are

$$\frac{\partial \boldsymbol{U}}{\partial t} + \frac{\partial \boldsymbol{E}}{\partial x} + \frac{\partial \boldsymbol{F}}{\partial y} = \boldsymbol{J}$$

where each vector is expressed as follows:

$$\boldsymbol{U} = \begin{bmatrix} \rho \\ \rho u \\ \rho v \\ \rho(E + \frac{V^2}{2}) \\ \rho c_i \end{bmatrix}, \qquad \boldsymbol{E} = \begin{bmatrix} \rho u \\ \rho u^2 + p - \tau_{xx} \\ \rho uv - \tau_{xy} \\ \rho(E + \frac{V^2}{2})u + pu - q_x - u\tau_{xx} - v\tau_{xy} \\ \rho c_i u - \rho D_{im}\frac{\partial c_i}{\partial x} \end{bmatrix}$$

$$\boldsymbol{F} = \begin{bmatrix} \rho v \\ \rho uv - \tau_{yx} \\ \rho v^2 + p - \tau_{yy} \\ \rho(E + \frac{V^2}{2})v + pv - q_y - u\tau_{yx} - v\tau_{yy} \\ \rho c_i v - \rho D_{im}\frac{\partial c_i}{\partial y} \end{bmatrix}, \qquad \boldsymbol{J} = \begin{bmatrix} 0 \\ 0 \\ 0 \\ 0 \\ \dot{w}_i \end{bmatrix}.$$

Assuming Stokes' hypothesis of $\lambda = -\frac{2}{3}\mu$, viscous stress tensors are expressed as follows:

$$\tau_{xx} = \frac{2}{3}\mu\left(2\frac{\partial u}{\partial x} - \frac{\partial v}{\partial y}\right), \quad \tau_{yy} = \frac{2}{3}\mu\left(2\frac{\partial v}{\partial y} - \frac{\partial u}{\partial x}\right), \quad \tau_{xy} = \tau_{yx} = \mu\left(\frac{\partial u}{\partial y} + \frac{\partial v}{\partial x}\right).$$

Heat-conduction terms are expressed as follows:

$$q_x = k\frac{\partial T}{\partial x} + \sum_i \rho D_{im}\frac{\partial c_i}{\partial x}h_i, \qquad q_y = k\frac{\partial T}{\partial y} + \sum_i \rho D_{im}\frac{\partial c_i}{\partial y}h_i\ .$$

For chemical reactions of oxygen gas the following elementary reactions are assumed:

$$O_2 + O_2 \rightleftharpoons 2O + O_2, \qquad O_2 + O \rightleftharpoons 2O + O\ .$$

For convective terms a Harten-Yee upwind TVD scheme is used and for viscous terms a conventional central difference is used. For boundary conditions, non-slip conditions are applied for the ramp surface and zero derivatives along the freestream are assumed at the incoming and downstream boundaries. Since in the flows with nonequilibrium chemical reactions the density, temperature,

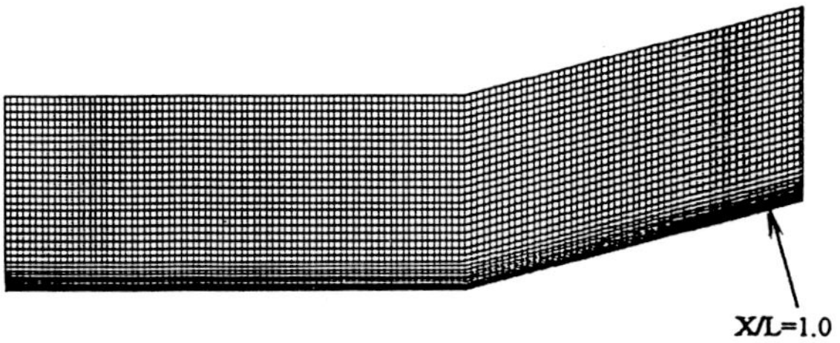

Figure 1. Representative mesh system. Ramp angle is 15 degrees; 515 × 161; every fourth grid shown.

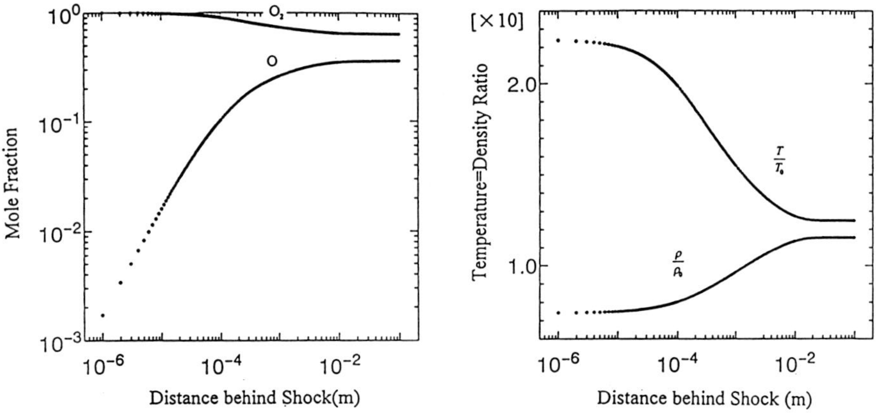

Figure 2. Distributions of chemical species (left) and temperature behind incident shock wave (right); O_2, $M_s = 11.7$, $\theta = 15°$, $p_\infty = 15$ torr, and $T_\infty = 298.0$ K.

and chemical compositions behind the shock change along the distance from the shock front, the one-dimensional, steady Euler equations for nonequilibrium chemically reacting flows are solved first in order to obtain the initial conditions for the incident shock wave. Also, zero physical derivatives normal to incoming flow are imposed for the upper boundary in order to keep the incident shock wave normal. For the energy equation a constant wall temperature condition is assumed. Both the fully catalytic wall case and non-catalytic wall case are considered. Thermal equilibrium is assumed. A representative mesh system is shown in Fig. 1.

3. Numerical results and discussion

Shock-reflection processes in O_2 gas at $M_s = 11.7$, $\theta = 15°$, $p_\infty = 15$ torr, and $T_\infty = 298.0$ K are calculated. The chemical-composition and temperature distributions behind the incident shock wave, which are calculated from the one-dimensional steady Euler equations for nonequilibrium chemically reacting flows, are shown in Fig. 2.

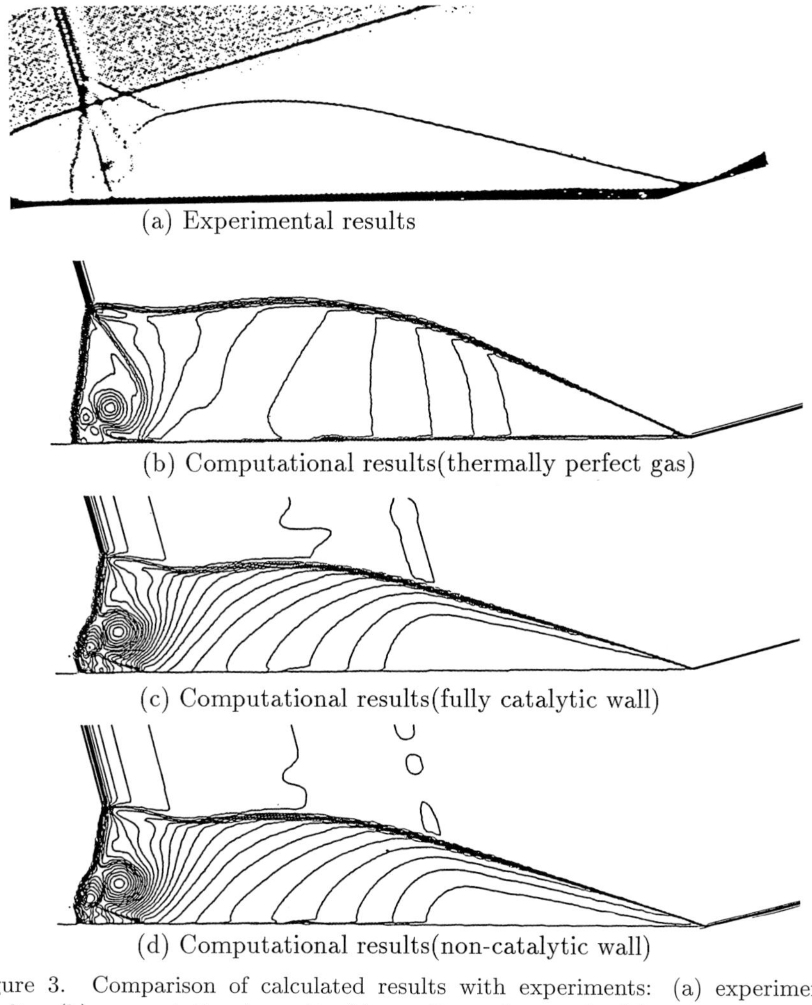

(a) Experimental results

(b) Computational results(thermally perfect gas)

(c) Computational results(fully catalytic wall)

(d) Computational results(non-catalytic wall)

Figure 3. Comparison of calculated results with experiments: (a) experimental results; (b) computational results (thermally perfect gas); (c) computational results (fully catalytic wall); and (d) computational results (non-catalytic wall). O_2, $M_s = 11.7$, $\theta = 15°$, $p_\infty = 15$ torr, and $T_\infty = 298.0$ K.

The calculated results for thermally perfect gases and chemically non-equilibrium gases for fully catalytic and non-catalytic wall conditions are shown in Fig. 3. The calculated shock patterns for chemically non-equilibrium gases with catalytic and non-catalytic wall conditions show quite good agreement with experiments of Weynabts (1968).

The calculated shear-stress and surface heat-flux distributions are shown in Figs. 4 and 5. Though little change in the shear-stress distributions is observed among the three models, quite drastic change in the surface heat-flux distributions is observed. In the fully catalytic wall case, enormous heat flux is seen because all dissociated oxygen recombines at the wall and gives enormous amounts of energy to the wall. These results show that wall catalycity is very important in the estimation of precise aerodynamic heating levels in chemically reacting flows.

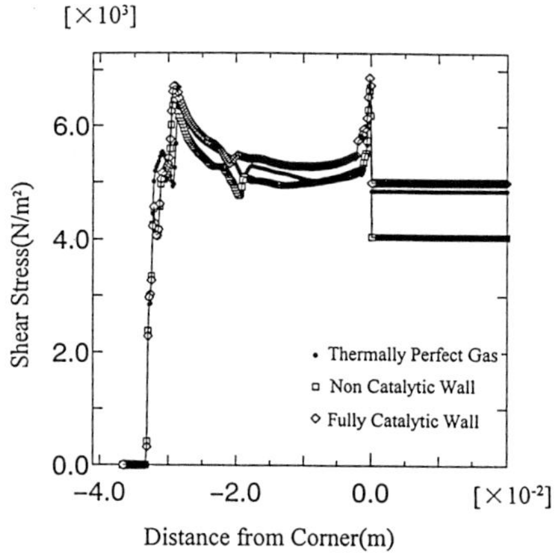

Figure 4. Calculated shear-stress distribution; O_2, $M_s = 11.7$, $\theta = 15°$, $p_\infty = 15$ torr, and $T_\infty = 298.0$ K.

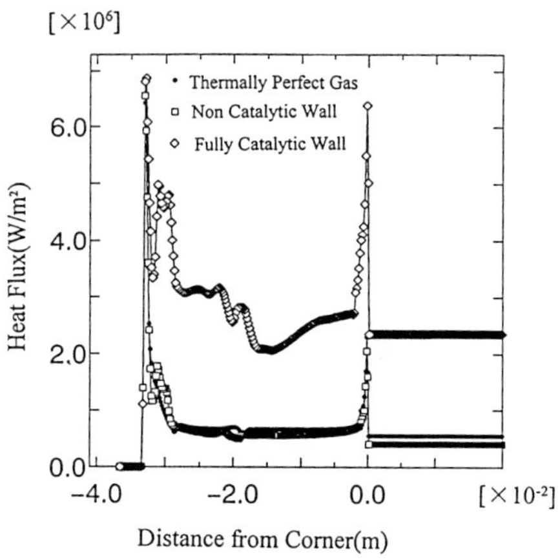

Figure 5. Calculated surface heat-flux distribution; O_2, $M_s = 11.7$, $\theta = 15°$, $p_\infty = 15$ torr, and $T_\infty = 298.0$ K.

4. Conclusions

In the present study the full two-dimensional Navier-Stokes equations are solved in order to investigate unsteady aerodynamic heating phenomena induced by shock impingement on a ramp surface. When the shock is strong, nonequilibrium chemical reactions occur in the flows. The effects of chemical reactions on aerodynamic heating are investigated. The results show the effects of chemical reactions on shock-induced aerodynamic heating phenomena are quite significant at high temperature.

References

Aso S, Hayashi M, Tan A, Tanahashi Y (1989) Surface heat transfer change induced by unsteady shock reflections and numerical simulations on shock reflection. In: Proc 17th ISWST, pp 630–635.

Ohyama K, Aso S, Hayashi M (1993) Numerical simulation of shock induced unsteady aerodynamic heating with highly dense mesh system. In: Shock Waves @ Marseille, R Brun and L Dumitrescu (eds), Springer, pp 395–400.

Weynabts RR (1968) Experimental Investigation of Shock Wave Diffraction over Compression and Expansion Corners, UTIAS TN No.126.

Turbulent spots in boundary layers in a free-piston shock-tunnel flow

D. J. Mee, C. P. Goyne
Department of Mechanical Engineering
The University of Queensland
Brisbane 4072 Australia

Abstract: In this paper, experiments to detect turbulent spots in transitional boundary layers formed on a flat plate in a free-piston shock-tunnel flow are reported. Experiments indicate that thin-film heat-transfer gauges are suitable for identifying turbulent-spot activity and can be used to identify parameters such as the convection rate of spots and the intermittency of turbulence.

Key words: Turbulent spots, Transition, Shock tunnel

1. Introduction

Boundary-layer transition has been observed to take place through formation, growth, and merger of turbulent spots in both incompressible and compressible flows (e.g., Emmons 1951; Clark et al. 1994). In studying transition in high-enthalpy impulse facilities, the frequency of spot production and the growth rate of spots may influence the interpretation of averaged heat-transfer and skin-friction data if an insufficient averaging period is used. Also, recent theoretical analysis by Doorly and Smith (1992) has indicated that there may be differences in spot dynamics at high Mach numbers. Experimental work by Clark et al. (1994) has supported this for Mach numbers up to 1.86.

The aims of the present paper are to demonstrate that turbulent spots can be detected using fast-response heat-transfer instrumentation in a high-enthalpy impulse facility, to obtain an experimental indication of the rate of spot production and spot size, and to investigate the suitability of the present instrumentation for study of spot dynamics in high-speed flows.

2. Detection of turbulent spots

A simplified model of a spot, based on the observations of Schubauer and Klebanoff (1955), as shown in Fig. 1, can be used to determine whether spots might be detectable in hypervelocity flows. The speeds of the leading and trailing edges of the spots are u_t and u_b respectively.

Let the freestream flow speed be u_∞ and the transition length be l_t. Consider a spot initiated at the start of the transition region which passes over a detector

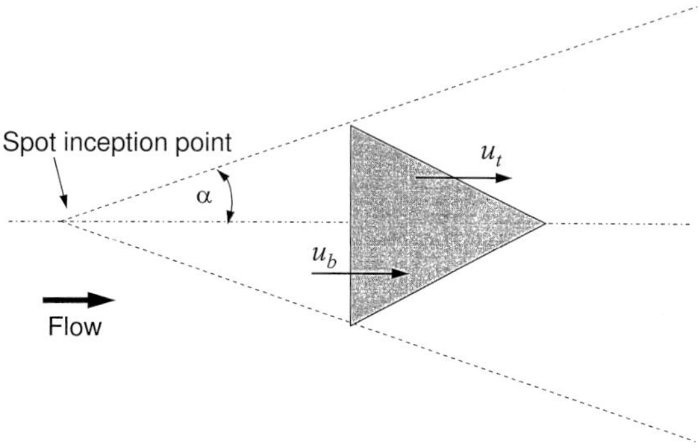

Figure 1. Model planform of a turbulent spot.

located a fraction, f, of the distance l_t from the start of transition. The time between when the apex and when the base of the turbulent spot pass over the detector, t_t, will be given by

$$t_t = \frac{f l_t}{u_\infty} \frac{\frac{u_t}{u_\infty} - \frac{u_b}{u_\infty}}{\frac{u_t}{u_\infty} \frac{u_b}{u_\infty}}. \tag{1}$$

For a typical Mach 5.6 condition in the T4 free-piston shock tunnel at 10-MJ/kg nozzle-supply enthalpy, the transition length on a flat plate with zero pressure gradient is approximately 400 mm and the flow speed is 4000 ms^{-1} (see Section 4). As a first approximation, we take typical low-speed values for $\frac{u_t}{u_\infty}$ and $\frac{u_b}{u_\infty}$ to be 0.88 and 0.50 respectively (Schubauer and Klebanoff 1955). For a gauge at the middle of the transition region, $f = 0.5$ and thus $t_t = 43\ \mu$s. Taking the intermittency to be 0.5 at the middle of the transitional region, the period of laminar flow between the turbulent bursts can be expected to be of a similar duration. Surface-mounted thin-film gauges, such as those used in T4, should be capable of detecting turbulent spots of this size and frequency.

3. Experiments

Experiments were performed in the T4 free-piston shock tunnel at The University of Queensland (Stalker and Morgan 1988). An axisymmetric contoured nozzle was used. The test surface was a flat plate 1500 mm long. This formed one of the inner walls of a duct which captured the flow from an opening 120 mm wide and 60 mm high. Adjustable divergence on the three non-instrumented walls enabled pressure gradients on the test surface to be minimized. Flush-mounted thin-film heat-transfer gauges were used to measure surface heat-transfer rates. These gauges consisted of a platinum film, 1 mm long × 0.3 mm wide × 1 μm thick, painted onto a 2.1-mm-diameter fused quartz cylinder. The frequency response is estimated to be 100 kHz. Locations for gauges were available every 50 mm from a distance of 370 mm to 1470 mm from the sharp leading edge. Not all locations were used. Static pressure measurements were taken simultaneously using PCB type 112A22 piezoelectric pressure transducers located at the same axial location but at a different spanwise location from the corresponding thin-film gauge. The thin-film gauge signals were multiplexed and the signal from each transducer was sampled at 3- or 4-μs intervals. One gauge was sampled at 1-μs intervals for comparison.

4. Results

The present instrumentation was set up to test skin-friction gauges in turbulent boundary layers in the T4 tunnel (Goyne 1995). The experiments reported here were performed at the end of a series of tests for the purpose of seeing if turbulent-spot activity could be detected in T4. The tests were made at a nominal nozzle supply enthalpy of 10 MJ/kg and the flow Reynolds number was varied. The test conditions are given in Table 1.

Table 1. Conditions for experiments.

Shot	Enthalpy MJ/kg	T_∞ K	ρ_∞ kg/m^3	u_∞ m s^{-1}	M_∞	Unit Re m^{-1}
4686	9.9	1260	0.021	3900	5.6	1.6×10^6
4687	9.3	1220	0.031	3850	5.6	2.5×10^6
4688	10.2	1360	0.034	4000	5.6	2.7×10^6

Typical nozzle-supply pressure and test-section Pitot-pressure time histories are shown in Fig. 2a. These results have been processed by a 50-μs moving average filter. The time scales on this and subsequent figures are arbitrary. The flow arrives at the Pitot probe located in the test section 200 μs after the rise in the nozzle-supply pressure. With appropriate time delay, the ratio of Pitot to supply pressure is constant 500 to 600 μs after flow arrival. Previous tests indicate that, subsequently, there is a similar period before driver gas contaminates the test gas (Skinner 1994). During this test time, the pressures remain constant to within about 10%. After the test time, the pressures slowly decrease. The measured static-pressure distributions at the three test conditions, averaged over 500-μs periods during the test time, are plotted in Fig. 2b. The pressures are approximately constant down the test surface for each condition.

Heat-transfer results averaged over a 500-μs period starting at 600 μs after the arrival of flow for the three test conditions are shown in Fig. 3. Shown also are laminar and turbulent levels determined using Eckert's reference temperature method (White 1991) and the theory of Spalding and Chi (1964),

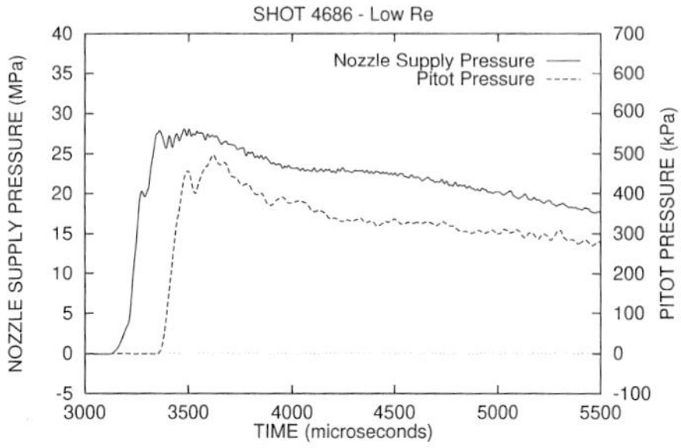

Figure 2(a). Pressure time histories.

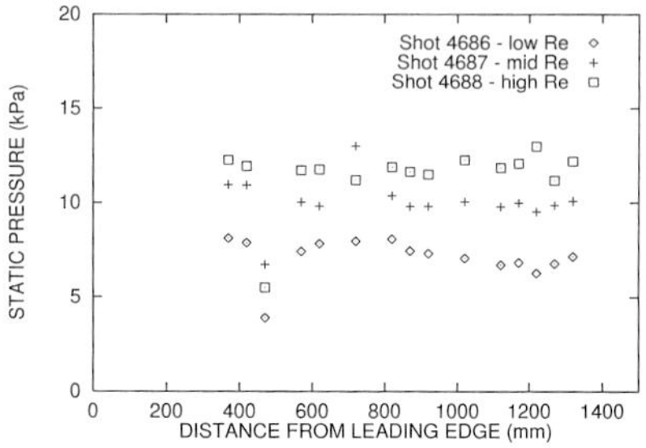

Figure 2(b). Pressure distributions.

respectively. (Sutherland's formula has been used for viscosity and the origin of the turbulent boundary layer has been taken as the leading edge of the plate.) Regions of laminar, transitional, and turbulent flow are evident.

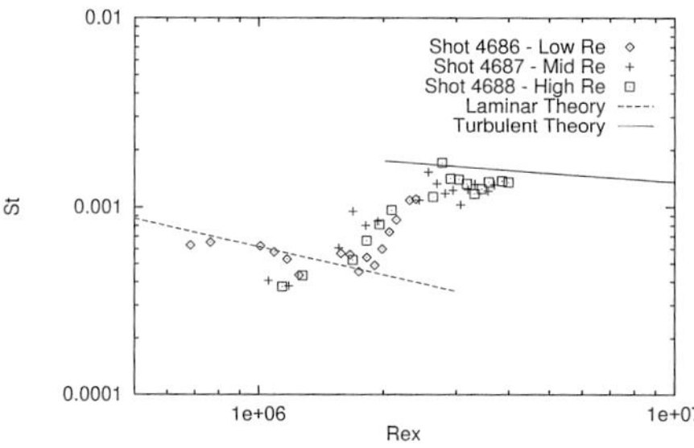

Figure 3. Heat transfer measurements averaged over a 500-μs period.

The time histories of the heat-transfer rates measured using the gauges are plotted for the three test conditions in Fig. 4. The zero level for each trace is plotted in proportion to the distance of the gauge from the leading edge of the test surface. (Thus the plot gives an indication of the development of the flow and enables the propagation of spots down the surface to be identified.) The arrival of flow at the gauge location can be seen from the sudden rise in heat-transfer level, and it can be seen that the flow arrives at later times for gauges progressively further down the plate. Note that results are presented from before the arrival of flow to well beyond the end of the test flow. No filtering or averaging has been applied to the data.

At the low-Re condition, the boundary layer remains laminar to 1120 mm from the leading edge when a turbulent spot appears at time 4300 μs. The spots are characterized by a rapid increase in the heat-transfer rate from the

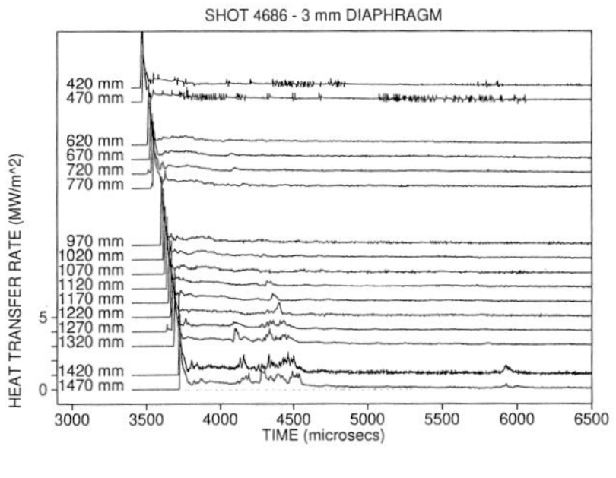

Figure 4(a). Low Re (4686).

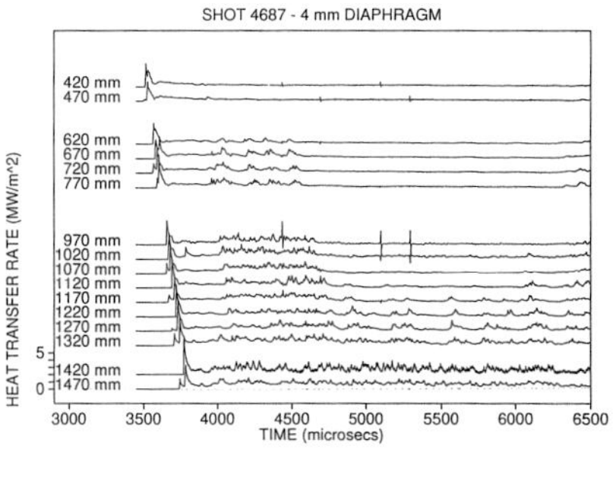

Figure 4(b). Mid Re (4687).

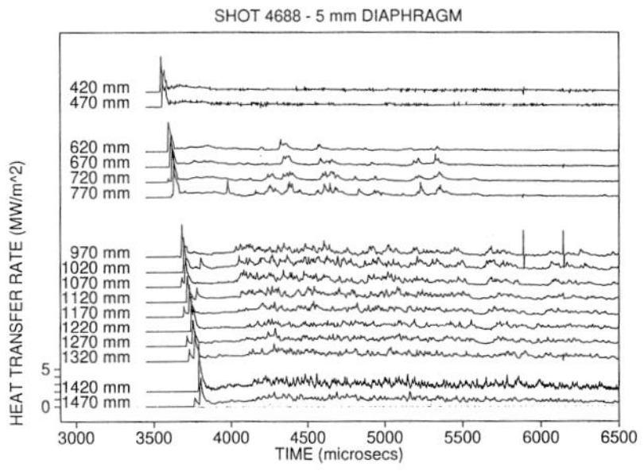

Figure 4(c). High Re (4688).

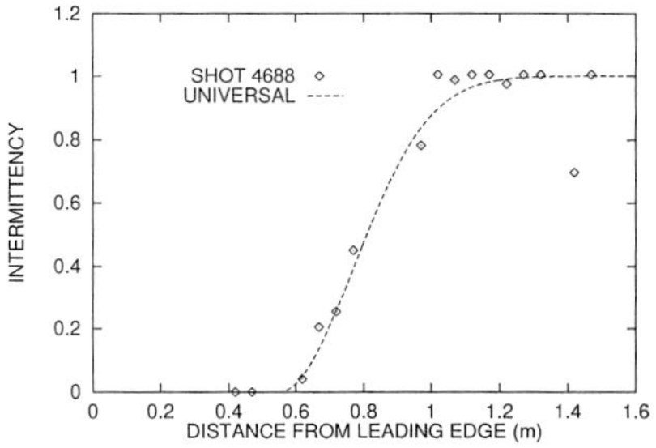

Figure 5. Shot 4688 intermittency.

lower laminar level to the higher turbulent level. This spot grows and others form and the boundary layer is still transitional at the last gauge. The gauge at 1420 mm is the one sampled at the higher rate. The apparent noise on the first two gauges is due to a sampling discretization problem.

At the mid-Re condition, the transition process is clearer. The first spots appear at 620 mm from the leading edge and the growth and merger of spots up to the gauge at 770 mm can be identified. At the gauge at 970 mm, the flow is almost fully turbulent during the test time. After the test time (beyond about 1.5 ms after arrival of flow), the boundary layer towards the front of the plate relaminarizes before transition occurs once again, now further downstream. From this shot, the speed of the trailing edge of the spots has been estimated from the propagation of the last spot before relaminarization to be approximately 60% of the freestream flow speed. This compares with the low-speed flow value of 50% (Schubauer and Klebanoff 1955) and values between 53% and 56% in the work of Clark et al. (1994) for higher speed flows.

The time histories at the high-Re condition show similar trends to the mid-Re condition but with less distinct evidence of the end of the test flow. The durations of the spots are similar to those expected from the simple analysis in Section 2. For the high-Re condition a numerical intermittency meter has been made along the lines of Hedley and Keffer (1974). The distribution of intermittency down the plate for the high-Re condition, based on a 1-ms sample period, are compared with Narasimha's (1957) "universal" intermittency distribution in Fig. 5. Quite good agreement is obtained.

5. Conclusions

The detection of turbulent spots in transitional boundary layers in hypervelocity flows using thin-film heat-transfer gauges has been demonstrated in tests in the T4 shock tunnel. The formation, growth, and merger of spots can be studied under such conditions. It has been demonstrated that factors such as spot-propagation rates and the intermittency of turbulence can be measured. Further tests are planned to study turbulent-spot activity over a range of hypervelocity flow conditions.

References

Clark JP, Jones TV, LaGraff JE (1994) On the propagation of naturally-occurring turbulent spots. J Eng Math 28:1–19.

Doorly DJ, Smith FT (1992) Initial-value problems for spot disturbances in incompressible or compressible boundary layers. J Eng Math 26:87–106.

Emmons HW (1951) The laminar-turbulent transition in a boundary layer. J Aero Sci 18:490–498.

Goyne CP, Paull A, Stalker RJ (1995) A Skin Friction Gauge for Impulsive Flows, AIAA Paper 95-3152.

Hedley TB, Keffer JF (1974) Turbulent/non-turbulent decisions in an intermittent flow. J Fluid Mech 64:645–678.

Narasimha R (1957) On the distribution of intermittency in the transition region of a boundary layer. J Aero Sci 24:711–712.

Schubauer GB, Klebanoff PS (1955) Contributions on the Mechanics of Boundary-Layer Transition, NACA TN 3489.

Skinner K (1994) Mass Spectrometry in Shock Tunnel Experiments of Hypersonic Combustion, PhD Thesis, The University of Queensland, Australia.

Spalding DB, Chi SW (1964) The drag of a compressible turbulent boundary layer on a smooth flat plate with and without heat transfer. J Fluid Mech 18(1):117–143.

Stalker RJ, Morgan RG (1988) Free Piston Shock Tunnel T4 — Initial Operation and Preliminary Calibration, NASA CR-181721.

White FM (1991) Viscous Fluid Flow, McGraw-Hill, New York, p 511.

Influence of reflected-shock/boundary-layer interaction on driver-gas contamination in high-enthalpy shock tunnels

R. S. M. Chue, K. Itoh
National Aerospace Laboratory, Kakuda Research Center
1 Kimigaya, Kakuda, Miyagi, 981-15 Japan

Abstract: A numerical analysis has been carried out to examine the interaction of the reflected shock with a turbulent boundary layer and its role in driver-gas contamination in a high-enthalpy reflected shock tunnel. The results show that wall jetting is responsible for the premature arrival of driver gas in the tailored and over-tailored tunnel conditions, but it does not play a dominant role in the under-tailored case. The effects of nozzle flow at the end plate on driver-gas contamination are also assessed.

Key words: Reflected-shock/boundary-layer interaction, Driver-gas contamination, High enthalpy, Shock tunnel, Test time

1. Introduction

The contamination of the test gas by driver gas is an important problem in high-enthalpy shock-tunnel facilities, as the premature arrival of driver gas severely limits test time. The interaction of the reflected shock with viscous boundary layers has been recognized as a major candidate to cause driver-gas contamination. Mark (1958) established the first theoretical treatment of the resulting bifurcated shock and the transport of boundary-layer gas by wall jetting towards the shock-tube end plate. Based on this mechanism, analytical models were developed to predict the premature arrival of the driver gas (Davies & Wilson 1969, Stalker & Crane 1978). Recently, numerical simulations have provided more detailed information about the contamination process. Wilson et al. (1993) examined the leakage of driver gas for an over-tailored case with laminar boundary layer. However, the tailored condition, which is more relevant to actual tunnel operations, was not examined. Indeed, the effect of the interaction on driver-gas contamination has yet to be quantitatively assessed in order to compare with other mechanisms. The reflected-shock/boundary-layer interaction is therefore analyzed to examine its role in causing driver-gas contamination and the loss of test time for different tunnel operating conditions.

2. The computational model

The present study concentrates on high shock Mach numbers ($M_s \simeq 10$) and high shock-tube pressures that are typical for high-enthalpy facilities. The shock tube investigated has an internal diameter of 30 mm, a length of 4 m, and shock-tube-to-driver-tube area ratio of 0.09. The driver and test gases are helium and air, respectively, which are assumed to be perfect gases. The helium has specific heat ratio of 5/3 and molecular mass of 4.003 kg/kmol. The test gas is assumed to be high-temperature air having a specific heat ratio of 1.25 and molecular mass of 25.9 kg/kmol. The wall boundary layers are assumed turbulent and the algebraic eddy viscosity model of Baldwin & Lomax (1978) is used. Initial conditions are taken at shock reflection with the initial flow field calculated using the boundary-layer theory of Mirels (1964). To isolate the effect of reflected-shock/boundary-layer interaction, the contact region is initially taken to be a plane discontinuous surface free of the effects of diaphragm rupture.

The computational code was originally developed by Takahashi et al. (1994) using the second-order KRC scheme, which is a pointwise nonoscillatory shock capturing method (Itoh et al. 1992). Time integration was done using the second-order Runge-Kutta method. The computational domain covers 26 cm of the end of the shock tube with a 433×121 grid that clusters in the wall regions. The upstream inflow condition and the tube-wall temperature are held fixed. Two sets of computations have been calculated — one for a closed-end shock tube and another for a shock tunnel with a choked nozzle. The shock Mach number is fixed at 10 with initial shock-tube pressure (p_1) at 10 kPa. The total enthalpy behind the reflected shock is 13 MJ/kg.

3. Results and discussion

Figure 1 displays the sequence of temperature and driver-gas mass fraction contours as the reflected shock interacts with the turbulent boundary layer in the closed-end shock tube. The tailored interface condition is enforced in this calculation (i.e., the ratio of transmitted shock pressure to the reflected shock pressure, p_7/p_5, equals 1). The formation of the bifurcated shock and the resulting mechanism of wall jetting of "cold" boundary-layer fluid towards the end plate, which generates a vortex at the corner region, is illustrated in Fig. 1a. The distortion of the contact surface and the leakage of driver gas into the test gas through the bifurcated shock foot can be seen in Fig. 1b. At 50 μs, a stream of driver gas near the edge of the boundary layer and the triple point of the bifurcated shock is driven towards the tube wall and is then partially transported towards the end plate. However, the vortical motion at the contact surface has a counter-clockwise direction which serves to retard further driver-gas leakage. The driver gas transported by wall jetting arrives at the end plate at about 95 μs after shock reflection.

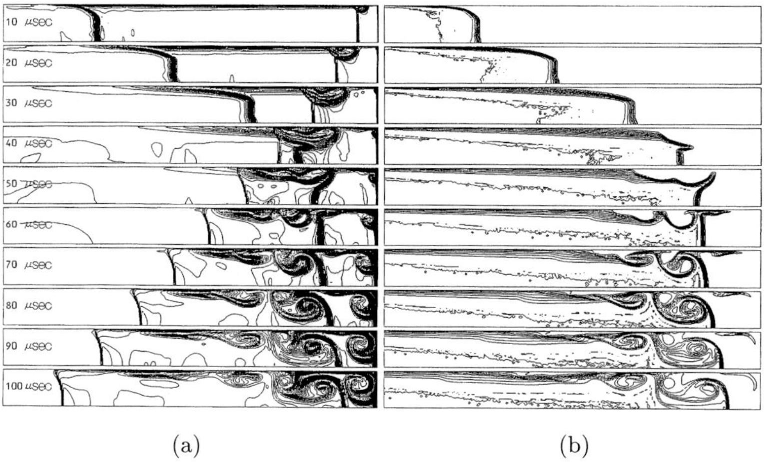

Figure 1. (a) Temperature and (b) driver-gas mass fraction distributions in the shock tube.

The effects of off-tailored tunnel conditions are compared in Fig. 2. The values of p_7/p_5 are, respectively, 0.8 and 1.1 for the under- and over-tailored conditions. The tailored case is redisplayed in Fig. 2b for comparison. Wall jetting is most severe in the over-tailored case and the driver gas arrives at the end plate the earliest, although the overall shape of the contact surface remains largely undistorted. On the other hand, the wall jet in the under-tailored case is too weak to transport the driver gas towards the end plate. However, the contact surface is severely distorted with a large protruded region of driver gas near the center line that flows toward the end plate.

Besides the contamination by "cold" wall fluid and driver gas, Fig. 3 shows that the pressure at the center of the shock-tube end plate becomes quite non-steady with a large pressure peak occurring near 80 μs for both the tailored and off-tailored cases. The quality of the freestream flow would then be affected if the gas is allowed to discharge through a nozzle at the end of the shock tunnel.

An initial effort is made to simulate the effect of nozzle flow at the end of the shock tunnel by assuming the nozzle to be an effective "sink" with choked flow. The nozzle-throat diameter is taken to be 10 mm. The purpose of this simple model is to provide a qualitative picture of how the driver gas may exhaust into the nozzle. Figure 4 shows the temperature distributions and the transport of

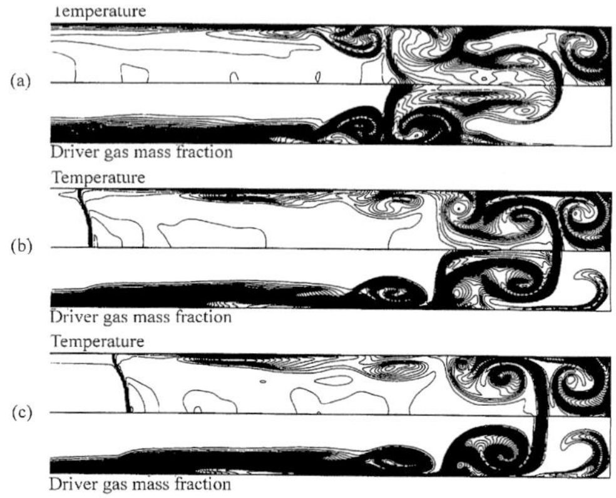

Figure 2. Effect of tunnel operating condition on driver gas contamination at $t = 100\ \mu s$: (a) under-tailored, (b) tailored, and (c) over-tailored condition.

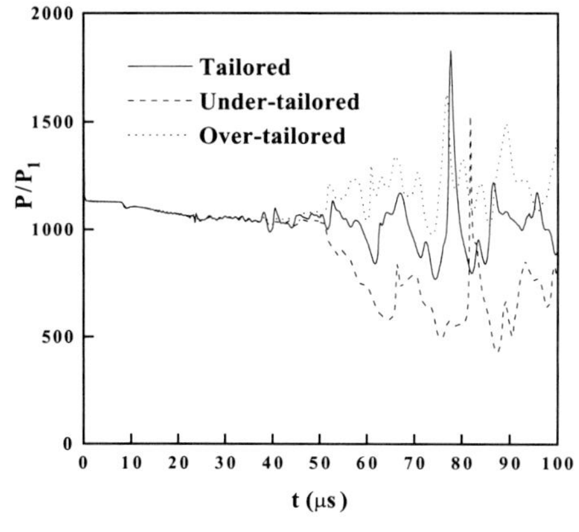

Figure 3. Pressure (p/p_1) history at the center of the shock-tube end plate.

driver gas towards the nozzle for the tailored condition. The early events are quite similar to those without the nozzle, although the corner vortex is stronger with nozzle flow (c.f. Fig. 1). From 80 μs the volume of driver-gas leakage at the end plate appears to be entrained in a vortex close to the end plate with a clockwise motion. As a result, only part of leakage gas is exhausted into the nozzle with the rest being recirculated in the shock tube (100 μs). The time of arrival of driver gas at the nozzle is not significantly different from the closed-end case. A set of off-tailored conditions has also been recalculated with the nozzle flow, and the results are again quite close to those for the closed-end conditions presented in Fig. 2.

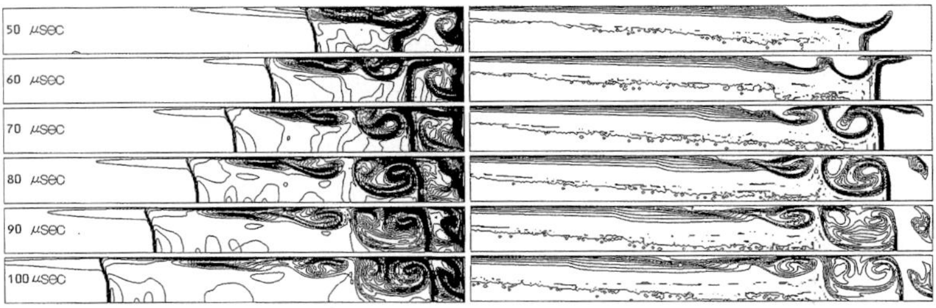

(a) (b)

Figure 4. Effect of nozzle flow on driver gas transport, tailored condition: (a) temperature and (b) driver-gas mass fraction.

Table 1. Time of arrival of driver gas at the end wall.

	Time after shock reflection (μs)
Closed-off end wall	95
With nozzle	90
Analytical approx. by Davies	62
"Ideal" drainage time[a]	226

[a]Where $t_{drainage} = \frac{4M}{A^*}\frac{1}{\rho_5 a_5}\left(\frac{\gamma+1}{2}\right)^{\frac{\gamma+1}{2(\gamma-1)}}$, M = mass of test slug, a = sound speed, and A^* = nozzle throat area. Subscript 5 denotes the condition behind the reflected shock.

The arrival time of the driver gas at the end plate is summarized in Table 1 for the tailored condition. Also included in the table are the values calculated using the analytical approximation of Davies & Wilson (1969), based on the mechanism of wall jetting, and the "ideal" drainage time for reference. The drainage time is the time required to drain the test gas with consideration of mass loss due to boundary-layer effects prior to shock reflection. The approximation of Davies obtained a value on the conservative side and can be a useful tool for predicting test time in actual tunnel operations if the effect of wall jetting is dominant. However, as wall jetting may not always be the cause of driver-gas contamination, as illustrated earlier in the under-tailored condition, the formulation may not always be adequate. In any case, the premature arrival of the driver gas can significantly reduce test time. For the shock tunnel considered in the present study, the arrival time of driver gas is less than 50% of the drainage time.

4. Concluding remarks

The interaction of a reflected shock with a turbulent boundary layer and the contact region has been analyzed. While wall jetting is demonstrated to be an important mechanism to cause contamination of the test gas, it may not necessarily play the dominant role. In particular, the mechanism for contamination is quite dependent on the tunnel operating condition, with under-tailoring appearing to be affected more by the distortion of the contact surface than by wall jetting. This also points out the need to examine other mechanisms, such as non-ideal diaphragm rupture and mixing, before a more precise prediction of driver-gas contamination can be achieved.

Acknowledgement. Special thanks are due to M. Takahashi for his help and discussions, and to G. Eitelberg of DLR-Göttingen for suggesting the calculation with nozzle flow. R. S. M. Chue is supported by the STA Fellowship program of Japan.

References

Baldwin BS, Lomax H (1978) Thin Layer Approximation and Algebraic Model for Separated Turbulent Flows, AIAA–78–257.

Davies L, Wilson JL (1969) Influence of reflected shock and boundary-layer interaction on shock-tube flows. Phys Fluids Supplement I:I-37–43.

Itoh K, Takahashi M, Hiraiwa T (1992) Pointwise non-oscillatory shock capturing scheme. In: Proc of 6th Symp on Computational Fluid Dynamics.

Mark H (1958) The Interaction of a Reflected Shock Wave with the Boundary Layer in a Shock Tube, NACA TM 1418.

Mirels H (1964) Shock tube test time limitation due to turbulent-wall boundary layer. AIAA J 2:38–54.

Stalker RJ, Crane KCA (1978) Driver gas contamination in a high-enthalpy reflected shock tunnel. AIAA J 16:277–279.

Takahashi M, Itoh K, Tani K, Tanno H (1994) Numerical analysis of the driver gas contamination in a reflected shock tunnel. In: Proc of Symp on SHOCK WAVES, JAPAN '94, pp 211–214.

Wilson GJ, Sharma SP, Gillespie WD (1993) Time-Dependent Simulations of Reflected-Shock/Boundary Layer Interaction, AIAA–93–0480.

Suppression of shock-induced separation in dense gases

M. S. Cramer, S. Park
Engineering Science and Mechanics Department
Virginia Polytechnic Institute and State University
Blacksburg, VA 24061 USA

L. T. Watson
Computer Science Department
Virginia Polytechnic Institute and State University
Blacksburg, VA 24061 USA

Abstract: We consider the reflection of oblique compression waves from a two-dimensional, steady, flat-plate laminar boundary layer. The full Navier-Stokes equations are solved numerically through use of a dense-gas version of the Beam-Warming implicit scheme. In the classical perfect-gas theory the compression waves are necessarily shocks. As is well known, the adverse pressure gradient carried by these shocks can lead to boundary-layer separation.

In the present investigation we consider the behavior of Bethe-Zel'dovich-Thompson (BZT) fluids in the context of this classical benchmark problem. BZT fluids are ordinary gases which have specific heats large enough to cause the fundamental derivative of gasdynamics to be negative over a finite range of temperatures and pressures in the single-phase vapor regime. The purpose of the present study is to determine whether the unique dynamics of BZT fluids can suppress boundary-layer separation. We make direct comparisons between the behavior of BZT fluids and more conventional fluids such as steam. The primary result of interest is that this suppression of separation can indeed be achieved. The physical mechanism for the suppression is discussed as are strategies for realizing the suppression in practical configurations.

Key words: Gasdynamics, Shock waves, Shock-induced separation

1. Introduction

An important loss mechanism in both turbomachinery and aerodynamic flows is caused by the shock-induced separation of boundary layers. The physical mechanism behind shock-induced separation is caused by the strong adverse pressure gradients carried by compression shocks which originate either at neighboring blades in the case of turbomachinery, other portions of the fuselage or aerodynamic surfaces in the case of supersonic aircraft, or which form on the wing or fuselage itself in the case of transonic flow. For a given pressure rise, shock waves are more likely to separate the boundary layer because the pressure gradient of the incoming signal is infinite, at least in the context of the inviscid theory. In most flows of practical interest, regions of flow compression are inevitable and in supersonic and transonic flows of perfect gases these compressions inevitably steepen to form compression shocks which in turn carry the strong adverse pressure gradients leading to flow separation.

In recent years it has been shown that such steepening is not inevitable if the class of fluids considered is extended to include high-pressure gases. By high pressure we mean pressures which are so large that the ideal-gas law no longer provides an accurate description of the gas behavior. At such pressure levels more complicated equations of state, such as the van der Waals equation, are required. We will also refer to such high-pressure conditions as dense-gas conditions. It is the principal goal of the present study to demonstrate that shock-induced separation can in fact be suppressed through use of a class of commercially available fluids operating at moderate pressure levels.

The fluids considered here are of the Bethe-Zel'dovich-Thompson (BZT) type. These are ordinary gases which have specific heats which are so large that the fundamental derivative of gasdynamics

$$\Gamma \equiv \frac{a}{\rho} + \left.\frac{\partial a}{\partial \rho}\right|_s = \left.\frac{V^4}{2a}\frac{\partial^2 p}{\partial V^2}\right|_s \tag{1}$$

becomes negative over a finite range of pressures and temperatures in the single-phase vapor regime. Here ρ, p, s, and a are the fluid density, pressure, entropy, and thermodynamic sound speed. The quantity $V \equiv \rho^{-1}$ is referred to as the specific volume. We also refer to the thermodynamic quantity (1) as the intrinsic gasdynamic nonlinearity parameter. For descriptions of the nonclassical dynamics of BZT fluids we refer the reader to the introductions and summaries of Bethe (1942), Zel'dovich (1946), Thompson (1971), Thompson and Lambrakis (1973), Menikoff and Plohr (1989), and Cramer (1991). Here we simply point out that the well-known compression discontinuities of the perfect-gas theory violate the entropy inequality if $\Gamma < 0$ at all points in the neighborhood of the proposed discontinuity. If such an inadmissible discontinuity is introduced into a $\Gamma < 0$ flow, say by a wedge inserted in a supersonic $\Gamma < 0$ flow, it disintegrates into a centered compression reminiscent of the well-known Prandtl-Meyer fans of the perfect-gas theory. This disintegration process was first discussed in the context of steady supersonic flows by Thompson (1971).

The potential advantages of BZT fluids with respect to shock-induced separation should now be obvious. Although the disintegration of the compression discontinuity does not change the polarity of the signal, i.e., it remains a compression, it does in fact significantly reduce the gradient. That is, the incoming compressive signal arriving at the boundary layer is of nonzero length rather than the zero length associated with shock waves. Intuition based on the perfect-gas or even incompressible-flow theory strongly suggests that such a reduction in the adverse pressure gradient experienced by the boundary layer should result in a reduction or even the elimination of shock-induced separation. In the present investigation, we demonstrate that such a suppression of shock-induced separation can indeed occur.

If such a reduction or suppression of separation can be realized in BZT fluids, it is natural to ask whether these nonclassical dynamics can be exploited in technology. One of the most promising applications of BZT fluids is as working fluids in organic Rankine cycles. A general discussion of the nature and the advantages of organic Rankine cycles has been given by Curran (1981) and Angelino and Invernizzi (1993). In order to save space we will restrict our comments to pointing out that BZT fluids tend to be a subclass of retrograde fluids and turn out to be excellent candidates for use as working fluids in such power systems. Not only do BZT fluids have all the advantages of other retrograde fluids, but they also have the potential to reduce or eliminate shock-induced separation in the turbine stage of the cycle.

2. Formulation

To illustrate the suppression of separation we consider the classical benchmark problem of the reflection of an oblique shock from a laminar flat-plate boundary layer. The flow conditions are chosen so that the reflection is regular. As a result the flow is supersonic everywhere in the inviscid part of the flow. The governing equations are taken to be the two-dimensional steady-flow Navier-Stokes equations. These are solved numerically by the well-known Beam-Warming implicit scheme (Beam and Warming 1978), which has been generalized to account for dense-gas effects. A full description of the scheme used and the battery of consistency checks performed have been given by Park (1994).

The equation of state is the state-of-the-art model developed by Martin and Hou (1955). The dense-gas viscosity and thermal-conductivity models chosen were those developed by Chung et al. (1988). Each of these models, the treatment of bulk viscosity, and the physical data for the fluids employed here are described in detail by Park (1994).

Finally, we note that the flows considered here are all single phase. The saturated vapor line correlation of Riedel (1954) was used to ensure that all computed points in the flow field correspond to the single-phase gas regime.

3. Results

In the present study we have compared the behaviors of two fluids. The first is ordinary steam which was chosen because of its widespread use in Rankine power cycles and the fact that is not a BZT fluid, i.e., $\Gamma > 0$ for all pressures and temperatures. The second fluid is the commercially available BZT fluid perfluoro-trihexylamine ($C_{18}F_{39}N$) and will be referred to by its trade name FC-71. As pointed out in Section 2, a full discussion of the physical-property data for each fluid is given by Park (1994) and the references therein.

We have attempted to make direct comparisons between the fluids. For each fluid the freestream pressure, temperature, Mach number, and impingement Reynolds number was fixed at 8.55 atm, 646.15 K, 2.0, and 2.96×10^5, respectively. We note that the freestream pressure of 8.55 atm is only about 4% of the pressure at steam's thermodynamic critical point. As a result the flow of steam can be regarded as that of a perfect gas. For both fluids the incident shock carries a flow deflection angle of 3°. Thus, the incident-shock strength can be regarded as essentially the same for each fluid.

The resultant variation of the plate skin-friction coefficient c_f is depicted in Fig. 1. There x is the coordinate measured parallel to the undisturbed flow direction. The leading edge of the plate was taken to be at $x = 0$ and the shock impingement point is given by $x = L_s$. The circles denote the variation for steam. The separation is seen to be well developed with two local minima in c_f. The + symbols denote the variation of c_f for the BZT fluid FC-71. Here $c_f > 0$ for all x and therefore no separation is observed. A check on the flow details reveals that the values of $\rho\Gamma/a$ immediately upstream and downstream of the incoming compression are -0.04 and -0.16. The theory developed in the references by Bethe, Zel'dovich, Thompson, etc. can be used to show that the initial compression discontinuity in FC-71 is inadmissible and therefore has disintegrated into a centered fan. As a result the adverse pressure gradient is too weak to separate the boundary layer. An examination of a number of shock/boundary-layer interactions and pure boundary-layer flows in dense gases reveals that the reaction of the viscous boundary layer is essentially classical. As discussed by Park (1994), the primary physical mechanism for this suppression is the disintegration of the shock and the resultant weakening of the adverse pressure gradient of the incoming inviscid signal.

Finally, we have also included the computed variation of c_f for FC-71 at the same freestream temperature and Mach number, the same impingement Reynolds number, and the same incoming shock strength as in the previous examples. This variation is given by $\times$ symbols in Fig. 1. The only new feature in this case is that the freestream pressure was taken to be 1 atm instead of 8.55 atm. As a result, the FC-71 behaves as a nearly perfect gas and $\Gamma > 0$

everywhere in this flow. The initial compression discontinuity remains intact as a shock and, as seen in Fig. 1, separates the flow. In fact, the skin-friction variation is very similar to that of steam. The point of this additional calculation is that it supports the idea that the suppression of separation is primarily due to the differences in the incoming inviscid signal and not due to differences in some other detail of the viscous flow. Furthermore, this demonstrates how one can eliminate separation simply by increasing the pressure to bring the state from the ideal-gas conditions to one corresponding to a negative Γ flow.

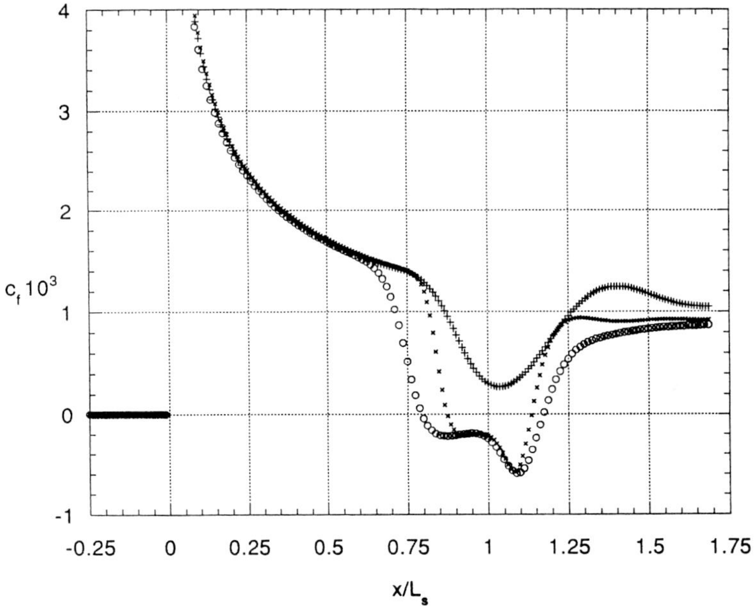

Figure 1. Computed skin-friction coefficients. The open circles denote steam at a freestream pressure of 8.55 atm, the + symbols denote FC-71 at a freestream pressure of 8.55 atm, and the $\times$ symbols denote FC-71 at a freestream pressure of 1 atm.

4. Summary

We have employed a dense-gas version of the Beam-Warming implicit scheme to compare the viscous-inviscid interactions of steam to those for a BZT fluid at the same freestream conditions, the same shock-impingement Reynolds number, and the same incoming-signal strength. It is found that the natural dynamics of BZT fluids in their $\Gamma < 0$ regime leads to a disintegration of the initial compression discontinuity. As a result of this reduction in the adverse pressure gradient, shock-induced separation was suppressed in the BZT fluid. The authors feel that the discovery that the suppression is primarily due to the inviscid dynamics can lead to considerable simplification in attempts to exploit this behavior. One can employ perfect-gas intuition when considering the viscous parts of the flow and restrict any optimizations to the inviscid flow regions.

Acknowledgement. This work was supported by the National Science Foundation under Grant # CTS-8913198. The first two authors would also like to thank Professor S. A. Ragab for many useful discussions and suggestions in the course of this research. We would also like to thank L. T. Watson for making facilities of the Computer Science Department available to S. Park.

References

Angelino G, Invernizzi C (1993) Cyclic methylsiloxanes as working fluids for space power cycles. J Solar Engr 115:130.

Beam RM, Warming RF (1978) An implicit factored scheme for the compressible Navier-Stokes equations. AIAA J 16(4):393.

Bethe HA (1942) The Theory of Shock Waves for an Arbitrary Equation of State, Office Scientific Resesarch & Development Report No. 545.

Chung TH, Ajlan M, Lee LL, Starling KE (1988) Generalized multiparameter correlation for nonpolar and polar fluid transport theories. Ind Eng Chem Res 27:671.

Cramer MS (1991) Nonclassical dynamics of classical gases. In: Nonlinear Waves in Real Fluids, Kluwick A (ed), Springer-Verlag, New York.

Curran HM (1981) Use of organic working fluids in Rankine engines. J Energy 5:218.

Martin JJ, Hou YC (1955) Development of an equation of state for gases. AIChE J 1(2):142.

Menikoff R, Plohr B (1989) Riemann problem for fluid flow of real materials. Rev Mod Phys 61:75.

Park S (1994) Viscous-Inviscid Interactions of Dense Gases, Ph.D. dissertation, Virginia Polytechnic Institute and State University.

Reidel L (1954) Eine neue Universelle Dampfdruckformel-Untersuchungten über eine Erweitung der Theorems der übereinstimmenden Zustände Teil 1. Chem Ing Tech 26:83.

Thompson PA (1971) A fundamental derivative in gas dynamics. Phys Fluids 14:1843.

Thompson PA, Lambrakis K (1973) Negative shock waves. J Fluid Mech 16:187.

Zel'dovich YB (1946) On the possibility of rarefaction shock waves. Zh Eksp Teor 4:363.

Shock waves diffracting over rigid ramps in a viscous, heat-conducting gas (argon)

R. J. Virgona*
Department of Mechanical Systems Engineering
Tokyo Noko University
Naka-cho 2-24-16
Koganei, Tokyo 184 Japan

W. Y. Crutchfield
Center for Computational Sciences and Engineering
Lawrence Livermore National Laboratory, P.O. Box 808
Livermore, California 94550 USA

L. F. Henderson†
I.U.S.T.I.-M.H.E.Q.
Saint-Jérôme Centre
Université de Provence, 13397 Marseille, France

Abstract: Shocks were diffracted over symmetrical steel wedges in argon. The shock Mach number M_i averaged 2.33, so the shocks were strong. The experimental data were simulated by numerically integrating the 2D, compressible, unsteady Navier-Stokes equations for a viscous, heat conducting, polytropic gas. It is found that real-gas effects delay both the onset of Mach reflection (MR), and its evolution to a self-similar state.

Key words: Shock wave, Diffraction, Navier-Stokes, Experiment, CFD

1. Introduction

When a plane incident shock i diffracts over a rigid ramp of apex angle θ either a regular (RR) or an irregular (IR) wave system appears. The criteria which determine when transition from RR to IR (RR $\rightleftharpoons$ IR) takes place have been discussed previously for the perfect gas (von Neumann 1943, Hornung 1986, Henderson 1987, Ben-Dor 1992). Here a study is made of the transition between RR and IR on a ramp for a real gas. A real gas is defined as one which has a finite shear viscosity μ and thermal conductivity k. In a real gas it is found that RR on a ramp generally persists beyond all the perfect-gas transition criteria into regions were it is theoretical impossible (Bleakney & Taub 1949; Kawamura & Saito 1956; Henderson & Lozzi 1975, 1979; Hornung 1986). The

*Present address: Thammasat University, Department of Mechanical Engineering, International Institute of Technology, P.O. Box 22, Thammasat Rangit Post Office, Pathumthani 12121, Thailand.

†Present address: 8 Damour Avenue, East Linfield, Sydney NSW 2070, Australia.

object of this paper is to investigate what effect μ and k have on diffraction and (RR $\rightleftharpoons$ IR).

2. Formulation

The Navier-Stokes equations in conservation form may be written

$$\partial_t U + F(U)_x + G(U)_y = R(U)_x + S(U)_y \tag{1}$$

where

$$U^T = (\rho \;\; \rho u \;\; \rho v \;\; \rho E) \tag{2}$$

$$F^T(U) = \big(\rho u \;\; \rho u^2 + p \;\; \rho uv \;\; u(\rho E + p)\big) \tag{3}$$

$$G^T(U) = \big(\rho v \;\; \rho uv \;\; \rho v^2 + p \;\; v(\rho E + p)\big) \tag{4}$$

$$R^T(U) = \left(0 \;\; (\lambda + 2\mu)u_x + \lambda v_y \;\; \mu(u_y + v_x) \;\; uR_2 + vR_3 + \frac{\mu}{P_r}\gamma e_x\right) \tag{5}$$

$$S^T(U) = \left(0 \;\; \mu(u_y + v_x) \;\; (\lambda + 2\mu)v_y + \lambda u_x \;\; uS_2 + vS_3 + \frac{\mu}{P_r}\gamma e_y\right) \tag{6}$$

$$e = E - (u^2 + v^2)/2 \tag{7}$$

where the superscript T indicates the transpose of a row, i.e. U^T for example is a column vector, $(u,\ v)$ are the $(x,\ y)$ components of velocity, ρ is density, p is pressure, E is specific energy, e is the specific internal energy, and subscripting 2, 3 indicates second and third components of a vector. Argon is well described by the perfect gas law

$$p = \rho RT\ . \tag{8}$$

The viscosity was modeled with

$$\mu = \frac{2.01572 \times 10^{-6}\ T^{1.5}}{171.691 + T}\ \mathrm{kg\ m^{-1}\ sec^{-1}}\ . \tag{9}$$

The coefficient of second viscosity was $\lambda = -2/3\mu$ and the thermal conductivity k was obtained from the Prandtl number $P_r = \frac{\mu C_p}{k} = 0.67$, where C_p is the constant-pressure specific heat. The ratio of specific heats γ was taken as $\gamma = 5/3$.

3. Numerics

The algorithm employs a finite-difference, second-order Godunov scheme with Crank-Nicolson correction (Steinthorsson et al. 1994, 1995). It includes an adaptive mesh refinement to five levels where at each level the refinement was by a factor of 4. The cells were square and at the deepest level of refinement 0.4 microns in length, or about 10 mean-free paths.

4. Results

If the ramp angle θ is less than the detachment criterion for RR $\rightleftharpoons$ IR, that is $\theta < \theta_e$, then by experiment some form of Mach reflection (SMR, CMR, DMR) is obtained. There are resolution difficulties as $\theta \to \theta_e$. In our experiments, as described in Virgona (1993), with $\theta = 52°$ which is close to $\theta_e = 53.776°$, the shock system had the appearance of a regular reflection (RR) except that the i and r shocks did not seem to meet on the ramp surface; there was a small gap g between them (Fig. 1a). The numerics have a finer resolution and can resolve it

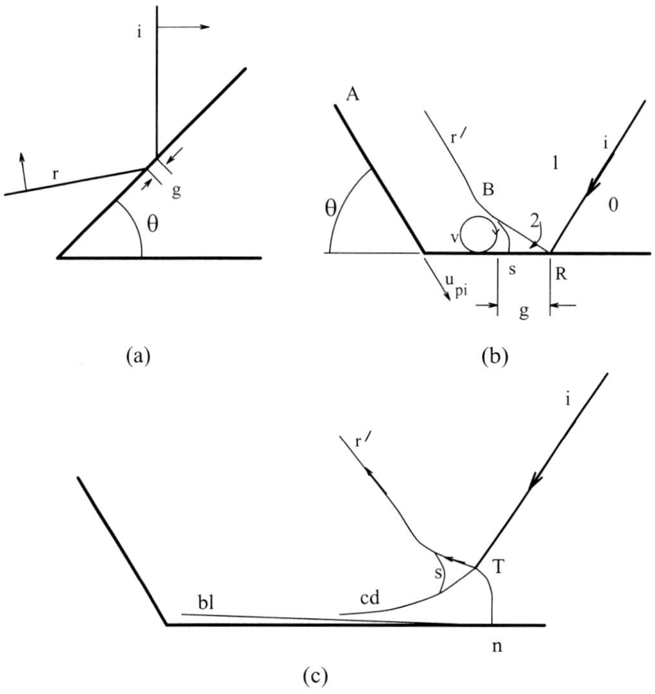

Figure 1. Experimentally observed and numerically resolved diffraction for $\theta = 52$ deg, $\gamma = 5/3$, $M_i = 2.327$: (a) experimental; (b) numerical; (c) after eruption of Mach shock (s is the corner signal and v the vortex).

(Fig. 1b). It comprises a precursor RR with a following and compressive corner signal s. There is a vigorous vortex v downstream of s. The corner signal travels faster than the reflection point and overtakes it, causing the eruption of the Mach shock (Fig. 2). The system then resolves into a self-similar complex Mach reflection (CMR, Fig. 1c).

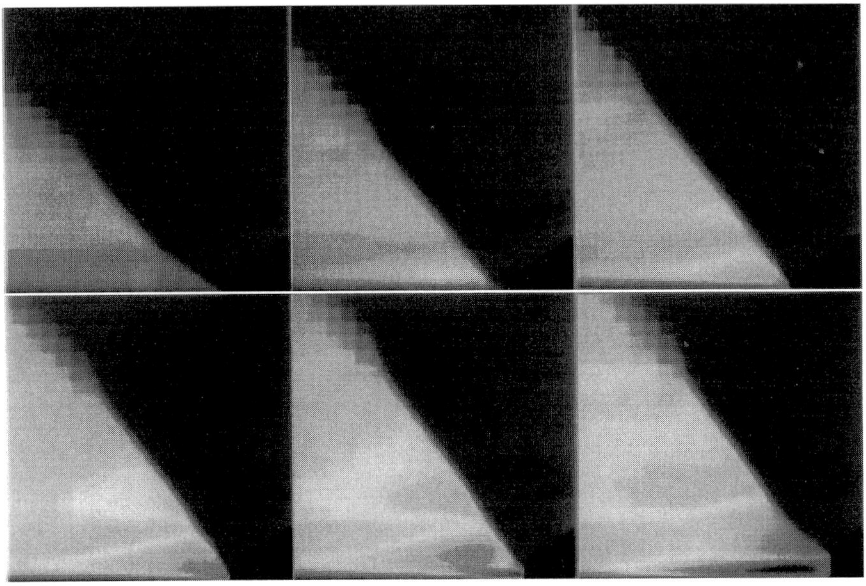

Figure 2. Time sequence of x component of velocity, for $\gamma = 5/3$, $M_i = 2.327$, $\theta = 52°$; ramp surface shown horizontal. Sequence is left to right, top to bottom.

4.1. Shock triple-point trajectory

Some results are presented in Fig. 3a for three values of θ such that $\theta < \theta_e$. Note that the paths are initially curved but become straight with sufficiently large $x(t)$. The straight part corresponds to self-similar CMR, while the curved part corresponds to the non-self-similar evolution. The part which coincides with the x axis ($y = 0$) corresponds to the precursor RR. As shown in Fig. 3b, the initial trajectory leads to a parallax error ϵ in the measurement of the trajectory angle χ, if it is assumed that the trajectory is a straight line passing through the ramp apex, as is sometimes done in experiment. In our case the shocks were photographed after they had traveled about 4 cm along the ramp, so χ depends on a length scale (4 cm). A scale-free trajectory angle ψ is obtained from the slope of the self-similar part of the trajectory. Results for χ are shown in Fig. 4a and for ψ in Fig. 4b. It is remarkable that the χ results support the inviscid θ_e criterion while the ψ results support the von Neumann criterion θ_N. Those results are isothermal with non-slip boundary conditions $T_0 = 273.15$ K and $\mu = 0$ on the ramp surface.

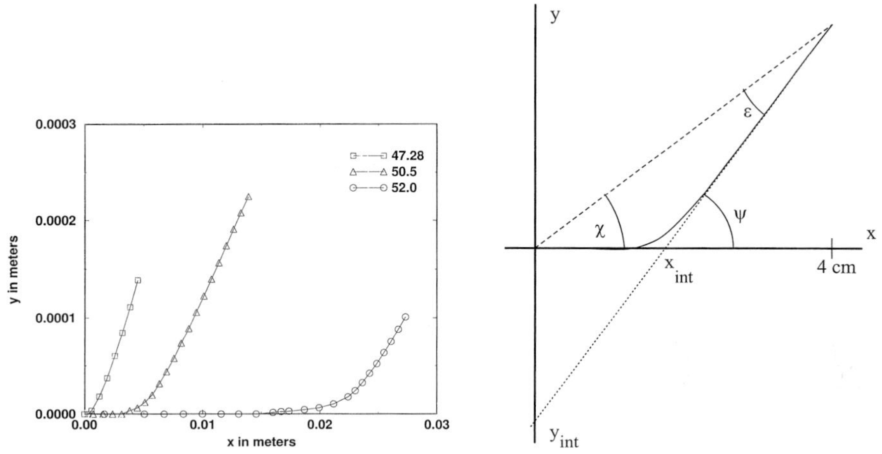

(a) (b)

Figure 3. Shock triple-point trajectories: (a) numerical triple-point trajectories at several wedge angles θ; (b) trajectory angle χ, self-similar trajectory angle ψ, and parallax error angle ϵ.

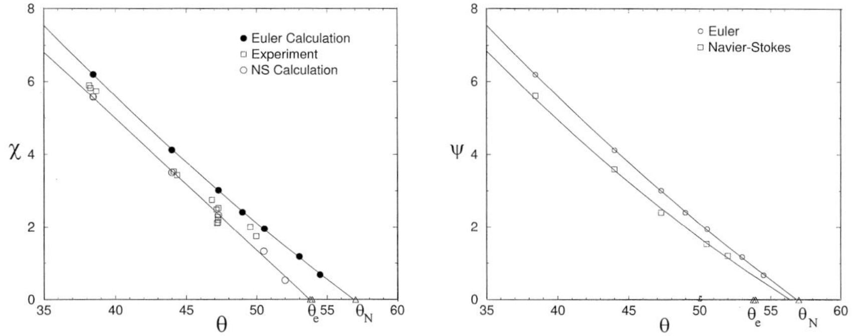

(a) (b)

Figure 4. Trajectory angle versus ramp angle θ for $\gamma = 5/3$, $M_i = 2.327$: (a) χ affected by parallax length scale of 4 cm; (b) scale-free trajectory angle ψ for self-similar flow.

Acknowledgement. This research was supported by the U.S. Department of Energy, Lawrence Livermore National Laboratory, under contract agreement DOE W-7405-ENG-48. We are indebted to Prof. P. Colella for the use of his Euler code data in Fig. 4.

References

Bleakney W, Taub AH (1949) Interaction of shock waves. Rev Mod Phys 21:584–605.

Ben-Dor G (1992) Shock Wave Reflection Phenomena, Springer-Verlag, Berlin.

Henderson LF (1987) Regions and boundaries for reflecting shock wave systems. Z ang Math u Mech (ZAMM) 67:73–86.

Henderson LF, Lozzi A (1975) Experiments on transition of Mach reflexion. J Fluid Mech 68:139–155.

Henderson LF, Lozzi A (1979) Further experiments on transition to Mach reflexion. J Fluid Mech 94:541–560.

Hornung HG (1986) Regular and Mach reflection of shock waves. Ann Rev Fluid Mech 18:33–58.

Kawamura R, Saito H (1956) Reflection of shock waves-1 pseudo-stationary case. J Phys Soc of Japan 11:584–592.

von Neumann J (1943) Oblique reflection of shocks. Explos Res Rep 12, Navy Dept, Bureau of Ordnance, Washington, DC.

Steinthorsson E, Modiano D, Colella P (1994) Computations of Unsteady Viscous Flows Using Solution Adaptive Mesh Refinement in Curvilinear Body-Fitted Grid Systems, AIAA-94-2330; also ICOMP-94-17, NASA TM 106704.

Steinthorsson E, Modiano D, Crutchfield WY, Bell JB, Colella P (1995) An Adaptive Semi-Implicit Scheme for Simulations of Unsteady Viscous Compressible Flows, AIAA-95-1727.

Virgona RJ (1993) A Study of Shock Wave Diffraction from Regular to Irregular Reflection and the Effects of Gas Viscosity, Ph.D. Thesis, The University of Sydney, Sydney, Australia.

Comparison of continuum and molecular approaches for the flow around a perpendicular disk

T. G. Elizarova, I. A. Graur
Institute for Mathematical Modeling, Russian Academy of Sciences
Miusskaya 4a, 125047 Moscow, Russia

A. Chpoun, J. C. Lengrand
Laboratoire d'Aérothermique du CNRS
4 ter, Route des Gardes, 92190 Meudon, France

Abstract: The present work contributes to efforts to define the domain of validity of two continuum approaches, based on Navier-Stokes and quasi-gas-dynamic equations, respectively, by comparing results obtained using each method with those obtained using a direct simulation Monte Carlo method as a reference method. Quasi-gas-dynamic equations are derived from the kinetic equation and include additional diffusion terms. The rarefied flow around a perpendicular disk has been considered for a freestream Mach number varying from 2 to 20.

Key words: Rarefied gas dynamics, Disk, Continuum equations, Monte Carlo simulation

1. Introduction

The interest in defining the domain of validity of continuum approaches for rarefied gas flows is due to the fact that solving continuum equations usually requires less computational resources than methods based on a molecular description. In addition to the usual Navier-Stokes (NS) equations, the quasi-gas-dynamic (QGD) equations were considered. They can be regarded as a new model for the description of viscous gas flows. Results obtained by a direct simulation Monte Carlo (DSMC) method are also presented. The physical modeling is made consistent in all calculations, but DSMC results are free from the near-equilibrium hypothesis that is included in the continuum approaches. Therefore, they can be used as a reference.

The variant of QGD equations used in the present calculations is described in Elizarova et al. (1992, 1994, 1995a, and 1995b). QGD equations may be obtained by averaging the kinetic equation with additional diffusion terms on the left-hand side. They differ from NS equations by the structure of the dissipative terms but they were proved to reduce to NS equations in the limiting case of small Knudsen numbers. In Cartesian coordinates, the QGD system is written as:

$$
\begin{aligned}
\rho_t + \nabla_i \rho u^i &= \nabla_i \frac{\mu}{p}(\nabla_j \rho u^i u^j + \nabla^i p), \\
(\rho u^k)_t + \nabla_i \rho u^i u^k + \nabla^k p &= \\
& \nabla_i \frac{\mu}{p} \nabla_j \rho u^i u^j u^k + \nabla_i \frac{\mu}{p} \nabla^i p u^k + \nabla_i \frac{\mu}{p} \nabla^k p u^i + \nabla^k \frac{\mu}{p} \nabla_i p u^i, \\
E_t + \nabla_i (E+p) u^i &= \\
& \nabla_i \frac{\mu}{p} \nabla_j (E+2p) u^i u^j + \nabla_i \frac{\mu}{p} \nabla^i \frac{p u^k u_k}{2} + \frac{\gamma}{\gamma-1} \nabla_i \frac{\mu}{\rho} \nabla^i p + \nabla_i \kappa \nabla^i T.
\end{aligned}
$$

Considering the perfect-gas equations, one obtains a closed system of equations which describes the space-time evolution of the macroscopic parameters of the gas, where $\mathbf{u}$ = velocity, ρ = density, p = pressure, E = total energy, T = temperature, μ and κ denote viscosity and heat-conductivity coefficients, respectively, and γ denotes the specific heat ratio.

In previous work (Elizarova et al. 1994, 1995a, and 1995b), the flow along a flat plate was considered and QGD equations were found to bring some improvement with respect to NS equations. In contrast with this simple configuration we now consider an infinitely thin circular disk perpendicular to a supersonic freestream. The corresponding axisymmetric flow features shock waves, expansion waves, and possibly recirculation zones. It is expected to be more sensitive to the choice of the equations or methods used for modeling. The problem is also more difficult from a computational point of view due to strong gradients and large differences in the density levels encountered.

2. Computational work

The comparison requires that an identical physical modeling of the gas properties be used in all calculations. This is obtained by considering a monoatomic hard-sphere gas in DSMC, whose transport properties are represented in the continuum approaches by a $\mu \propto T^{1/2}$ viscosity law and a constant Prandtl number ($Pr = 2/3$). Temperature T identifies with the translational temperature and γ is equal to 5/3. The wall temperature T_w is taken to be equal to the stagnation temperature T_0. Molecules impinging onto the disk surface are assumed to be reflected diffusively with full accommodation at wall temperature T_w. The freestream Mach number Ma_∞ was successively taken equal to 2, 5, 10, and 20. Most calculations were carried out for a freestream Knudsen number $Kn_\infty = \lambda_\infty / D$ equal to 0.1, where λ_∞ is the mean free path and D denotes the diameter of the disk. The corresponding values of the Reynolds numbers $Re = \rho_\infty V_\infty D/\mu_\infty$, $Re_2 = \rho_2 V_2 D/\mu_2 = \rho_\infty V_\infty D/\mu_2$, and $Re_w = \rho_\infty V_\infty D/\mu_w$

Table 1. Flow parameters (Reynolds numbers given for $Kn_\infty = 0.1$).

Ma_∞	Re	Re_2	Re_w	p_{i2}/p_∞	p_{fm}/p_∞	p_2/p_∞	T_2/T_∞	ρ_2/ρ_∞
2	33.0	22.9	21.6	6.34	12.61	4.75	2.078	2.286
5	82.4	28.0	27.0	37.17	67.38	31.0	8.68	3.571
10	164.8	29.1	28.1	147.3	262.5	124.75	32.12	3.883
20	329.6	29.4	28.4	588.0	1043	499.8	125.9	3.970

are given in Table 1; V denotes the velocity,and subscripts $()_\infty$ and $()_2$ refer to conditions in the freestream and behind a Rankine-Hugoniot shock wave, respectively. In addition, the dimensionless pressure on the forward-facing side of the disk is given in Table 1 as estimated by the modified Newtonian approximation (p_{i2}) and by a free-molecule calculation (p_{fm}).

NS calculations were carried out using a commercial code (INCA) based on a finite-volume upwind implicit method with flux vector splitting.

QGD calculations used a code developed at the Institute for Mathematical Modeling (IMM) based on second-order finite-volume space approximation in explicit form and adapted to an axisymmetric geometry.

DSMC calculations were carried out using the code DISIRAF developed at the Laboratoire d'Aérothermique du CNRS based on Bird's ideas (Bird 1976).

For NS, QGD, and DSMC calculations, the grid was refined in the radial ($r-$) direction near the symmetry axis (NS, QGD) and/or near the disk edge (QGD, DSMC). The grid was also refined in the axial ($x-$) direction near the disk plane (NS, QGD, DSMC). A typical mesh consisted of 4000–5000 grid points. The smallest grid intervals were $\delta x \approx \delta r \approx 0.15\,\lambda_\infty$.

Both NS and QGD calculations assumed the gas to be perfect and used the same formulation for velocity slip and temperature jump at the wall.

3. Results and discussion

Due to space limitations, only part of the results will be presented here. Based on grid-sensitivity studies, the results are free from discretization error except for NS results near the edge of the disk. They are given in dimensionless form based on freestream values. Lengths are reduced by the disk radius R. More details can be found in Elizarova et al. (1994) regarding the numerical methods and in Lengrand et al. (1995) for the present calculations.

The distributions of reduced velocity and temperature along the stagnation streamline are given in Fig. 1 for $Kn_\infty = 0.1$ and $Ma_\infty = 2$ and 5. The disk plane was taken as the origin for x. QGD temperature profiles depart

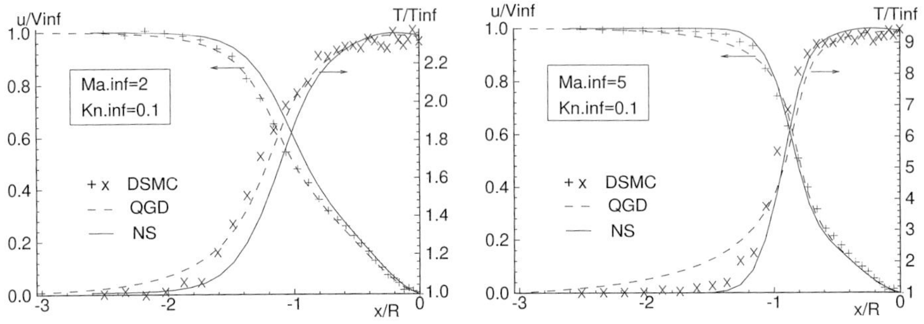

Figure 1. Velocity and temperature profiles ahead of the disk.

from the freestream value earlier than NS and DSMC profiles. Compared with NS results, DSMC and QGD results exhibit differences in the same direction. Thus, additional terms of QGD equations act in the correct direction but they occasionally over-correct the deficiencies of NS equations. The density profiles (not shown) are less sensitive to the model.

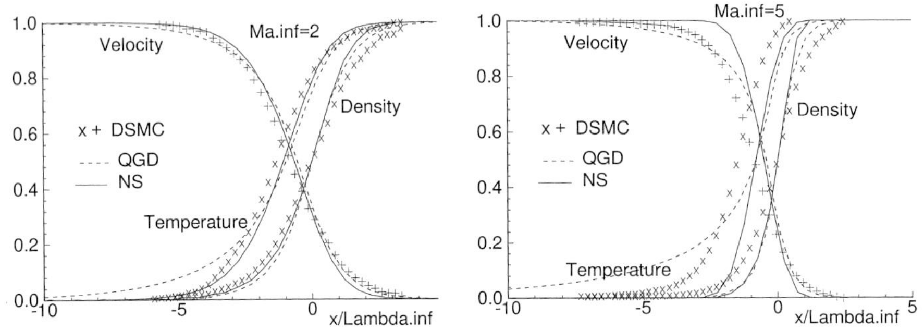

Figure 2. One-dimensional shock-wave structure.

One-dimensional shock-wave structures were calculated for the same Mach numbers by the same three methods. However, NS calculations were carried out with a code developed at IMM rather than with the INCA code. Flow parameters are presented (Fig. 2) in the usual dimensionless form based on Rankine-Hugoniot parameter ratios. Abscissas x are divided by λ_∞. The point where $\rho = (\rho_\infty + \rho_2)/2$ is taken as the origin for x. The same observations can be made for the disk stagnation line and for the 1D shock structure.

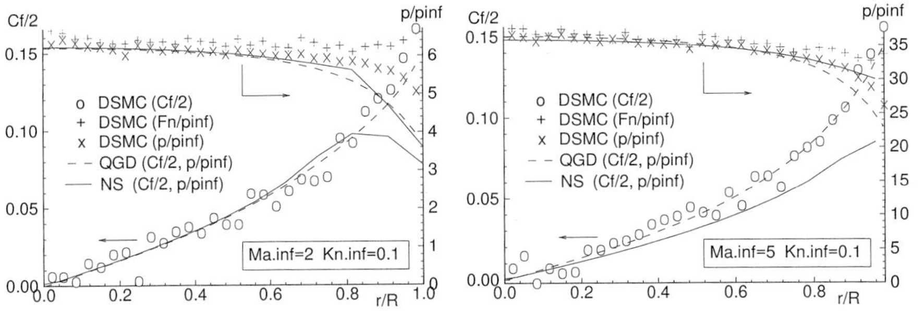

Figure 3. Wall quantities.

Wall quantities (pressure and shear stress F_t) are of interest for practical applications. These quantities are plotted in Fig. 3 for the same flow conditions as in Fig. 1. DSMC makes a difference between normal stress F_n due to molecular normal momentum exchange and pressure p deduced from the equation of state near the wall. Except for near the disk edge, where results are affected by discretization, all pressures are close to one another and close to the Newtonian estimation p_{i2}. However the skin-friction coefficient $C_f/2 = F_t/(\rho_\infty V_\infty^2)$ is much better predicted by QGD than by NS equations.

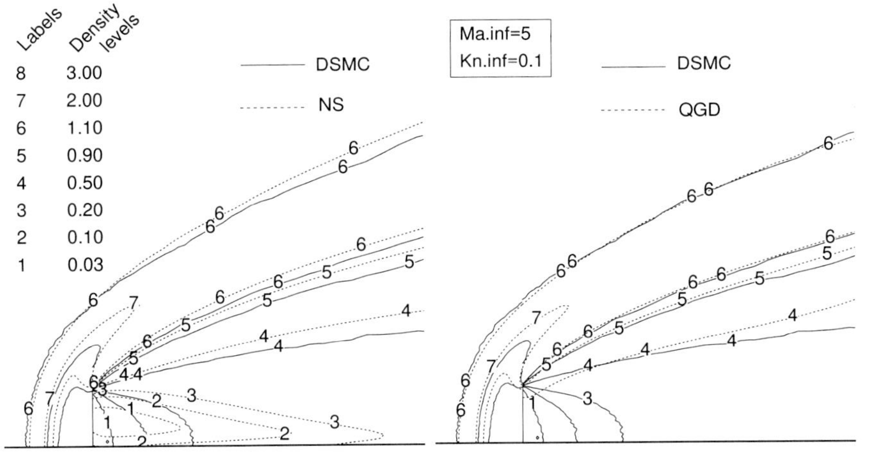

Figure 4. Two-dimensional flow field.

All the above results were obtained from calculations that included also the wake of the disk. Examples of 2D flow fields are given in Fig. 4. The main differences between the results are observed in the near wake, where continuum approaches result in much higher densities and temperatures. This is attributed to a very large local mean free path (10–100 λ_∞) which makes the continuum formulations unrealistic and very sensitive to the model. However the far wake tends to coincide for all three calculations.

Other QGD and NS calculations were carried out for smaller Knudsen numbers. For $Kn_\infty = 0.01$, results coincide except in the near wake for large Mach numbers. For $Kn_\infty = 0.005$, they coincide in the whole flow field.

4. Conclusion

The flow around a disk placed perpendicular to a supersonic rarefied flow was calculated using NS and QGD equations. DSMC calculations were used as a reference. Additional terms in QGD equations act in the right direction to correct (and occasionally over-correct) the differences between NS and DSMC results.

QGD equations are clearly superior to NS ones for calculating the distribution of skin friction along the forward-facing side of the disk under near adiabatic conditions. Subject to later confirmation, one can expect from Reynolds' analogy that the same conclusion will hold for heat transfer in a cold-wall situation. At low Knudsen numbers, QGD equations tend to coincide with NS ones.

More work is needed on other flow configurations (e.g., jet expansion) and other conditions (e.g., cold wall) to generalize the conclusions deduced from the present work and from a previous one where a flat plate was considered.

References

Bird GA (1976) Molecular Gas Dynamics, Clarendon Press, Oxford.

Elizarova TG, Chetverushkin BN, Sheretov YV (1992) Quasi-gas-dynamic equations and computer simulation of viscous gas flows. In: Lecture Notes in Physics 414, pp 421–425.

Elizarova TG, Graur IA, Lengrand JC, Chpoun A (1994) Rarefied Gas Flow Simulation Based on Quasi Gas-Dynamic Equations, Lab. d'Aérothermique du CNRS, RC 94-4.

Elizarova TG, Graur IA, Chpoun A, Lengrand JC (1995a) Comparison of continuum and molecular approaches for rarefied gas flows. In: Proceedings of the 19th Int Symp on Rarefied Gas Dynamics, Oxford (GB), July 25–29, 1994, Oxford Univ Press, Vol 2, p 780.

Elizarova TG, Graur IA, Lengrand JC, Chpoun A (1995b) Rarefied gas flow simulation based on quasi gas-dynamic equations. AIAA J 33(12):2316–2324.

Lengrand JC, Chpoun A, Graur IA, Elizarova TG (1995) Supersonic Rarefied Flow Around a Perpendicular Disk, Lab. d'Aérothermique du CNRS Report R-95-6.

The numerical simulation of unsteady shock-wave/boundary-layer interactions in reacting gas mixtures

Y. S. Weber, Wright Laboratory
Wright-Patterson AFB, OH 45433-7913 USA

E. S. Oran, J. P. Boris, Naval Research Laboratory
Washington DC 20376-5344 USA

J. D. Anderson, Jr., University of Maryland
College Park, MD 20742 USA

Abstract: The reflected-shock/boundary-layer interaction occurring near the end wall in shock tubes has been simulated for both a nonreacting and reacting gas. Specifically, we compare computations in calorically perfect oxygen to dissociating oxygen to assess the effects of endothermic chemical reactions. The results indicate a complex structure consisting of an unstable shear layer in the bifurcation region with embedded vortices. The reduction in specific heat ratio leads to a decrease in the size of the bifurcation angle and reflected-shock velocity. For the conditions examined here, embedded lambda shocks were evident beneath the main bifurcated structure.

Key words: Bifurcation, Reactive, Reflected-shock/boundary-layer interaction

1. Introduction

Numerous analytic and experimental studies on bifurcations of reflected shocks in shock tubes have been performed (Mark 1959, Strehlow et al. 1959, Honda et al. 1975, Brossard et al. 1985). However, due to the high resolution requirements, there have been very few computational simulations. Wilson (1993) simulated reflected-shock interactions for an axisymmetric configuration and assessed the flow quality and impact on test time. Y. Weber (1994a) examined the detailed structure of the separated zone and the impact of the growth of the bifurcation region on the outer inviscid flow for two-dimensional configurations. Details of the separated-flow region show that it is composed of an unstable shear layer with embedded vorticity near the wall. Similar structures were found in simulations of inviscid flows in which reflected shocks interacted with stratified density gradients (Glowacki et al. 1985). In the present work, numerical simulations assess the effect of the chemical dissociation of oxygen on the reflected-shock/boundary-layer interaction. The simulation captures the properties of vortical structures down to the size of a few grid cells.

2. Governing equations and numerical method

Numerical simulations were performed using a reactive Euler solver (Weber J 1994) that was modified to include the viscous transport terms of the Navier-Stokes equations (Weber Y 1994b). For nonreacting O_2, a calorically perfect gas assumption was used to close the equations. For the reacting gas, thermodynamic equilibrium was assumed and the sensible enthalpy was obtained from data in the JANAF tables (Stull 1971). Rate constants for the chemical source terms were obtained from the Dunn and Kang (1973) reaction mechanism. The viscosity and thermal conductivity were computed using Sutherland's law for the nonreactive gas and a mole-weighted form of the Chapman-Enskog equation (Bird et al. 1960) for the reactive gas. Since the objective of these studies is to assess the impact of energy absorption caused by dissociation on the reflected-shock/boundary-layer interaction, diffusion has been neglected.

The convective terms were integrated using the Flux-Corrected Transport (FCT) algorithm (Boris et al. 1993). FCT is a high-order, monotone, positivity preserving algorithm for solving conservation equations. The viscous terms were treated as source terms in the momentum and energy equations. The chemistry integration is time-split from the fluid-dynamic integration. FCT is used to convect the species number densities in the absence of source terms. The resulting species number densities are then used as initial values for the chemical integration which is handled using the Selected Asymptotic Integration Method (SAIM) (Young et al. 1977), which is a robust, efficient second-order ordinary differential-equation solver.

3. Simulation parameters

The effect of chemical dissociation on the reflected-shock/boundary-layer interaction is studied by comparing shock bifurcations in nonreacting O_2 (Case I) and dissociating O_2 (Case II). The initial shock-pressure ratio and temperature are 23.7 and 700 K, respectively. The unit Reynolds number is 2.349×10^3 and is referenced to the conditions behind the incident shock. For Case II, these parameters are based on the initial, frozen state of the gas in Region 2. Based on differences in the specific heat ratio, induced velocity and shock speed are 2% higher in Case II. There were 1024 cells in the x-direction (spaced uniformly) and 256 in the y-direction (clustered near the wall).

4. Results

Comparisons of temperature fields in Fig. 1 show that the reactive-oxygen flow

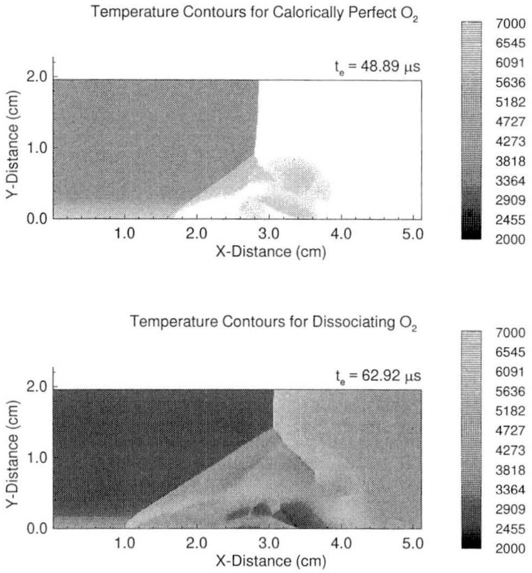

Figure 1. Temperature contours for calorically perfect and reactive O_2.

field is significantly cooler. The maximum temperature is 4800 K compared to the nonreacting value of 8000 K. The trend is opposite for the density field. Fig. 1 also shows the unstable shear layer with embedded vorticity pockets.

The pressure contours of Fig. 2 show qualitatively similar bifurcation structures. The bifurcation angle for Case I is 37° and for Case II is 33°. In both cases, there is a weak shock wave between the bifurcated foot and tail shock, due to the irregular displacement surface over which the inviscid flow travels. As the flow changes direction to remain parallel to the displacement surface, weak shock and expansion waves are generated. The shock located between the bifurcated foot and tail interacts with the tail shock and deforms it. The pressure contours for both cases also show the low-pressure center in the region where the shear layer is reversed. For Case II, the reflected-shock pressure is higher, as shown in the centerline pressure distribution in Fig. 3. This may be attributed to the decrease in specific heat ratio due to the chemical reactions occurring in the flow field.

Small shocklets located between the main bifurcated structure and the lower wall appear only in Case II. They are located near the strong expansion in the bifurcation region. This strong expansion (shown as a white region near the shocklets) expands the flow and causes small lambda shocklets to form. Near the center of the low-pressure field, the specific heat ratio and temperature are low. The lower specific heat leads to a strong expansion which eventually leads

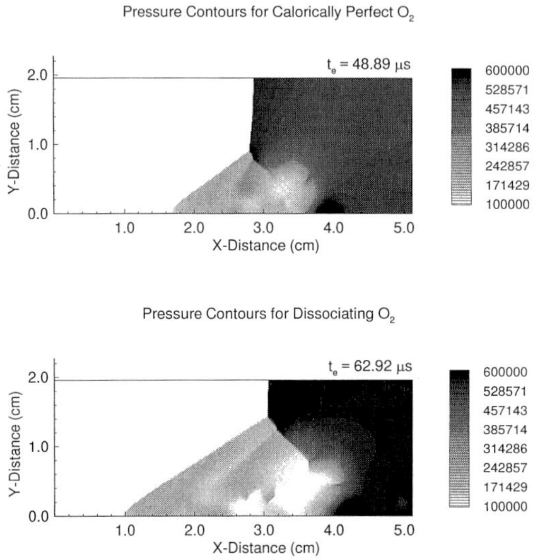

Figure 2. Pressure contours between calorically perfect and reactive O_2.

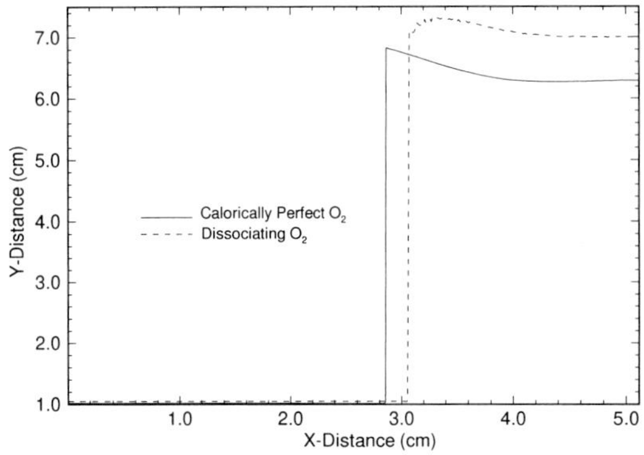

Figure 3. Comparison of centerline pressure profiles for calorically perfect and reactive O_2.

to a shock, or series of shocklets in the case presented here.

Figure 4, the mass-fraction distribution for both O_2 and O in the reactive flow field, shows that there is very little dissociation behind the incident shock, but significant dissociation behind the reflected shock. The reaction zone appears as a gradient in either of the mass fraction contours. The low-energy

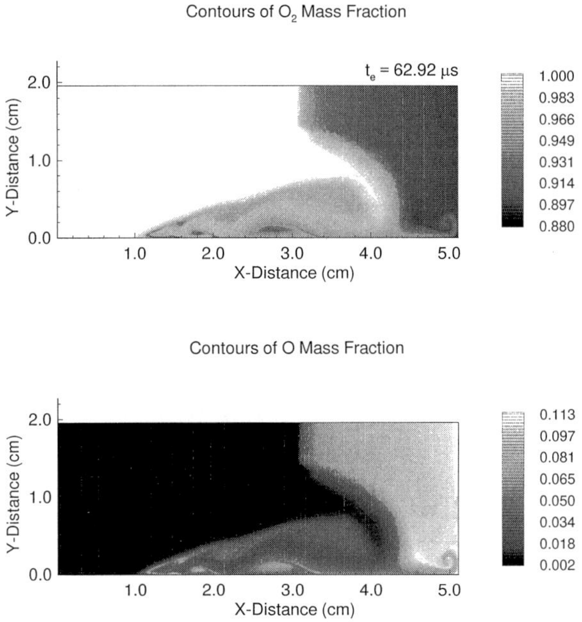

Figure 4. Contours of O and O_2 mass fraction for the reactive flow field.

pockets of vorticity contain monatomic oxygen which has been trapped in there by the fluid mechanics of the interaction.

In the reactive-flow calculation, there is a decrease in shock velocity. This can be attributed to energy absorption by the vibrational energy modes in the gas, as well as an increase in the zero-point energy of the dissociated mixture over the calorically perfect gas.

In conclusion, the chemical reactions lead to a decrease in the specific heat ratio, the reflected-shock velocity, and the bifurcation angle. In addition, the current reactive flow simulation yields embedded shocklets which were not present in previous studies.

Acknowledgement. Funding was provided in part by NRL through ONR and the ARPA Applied and Computational Mathematics Program and the PALACE KNIGHT program of the US Air Force. Computational support was provided by the Army HPC Research Center, the National Center for Supercomputing Applications, the University of Maryland Institute for Advanced Computer Studies, and NRL.

References

Bird RB, Stewart WE, Lightfoot EN (1960) In: Transport Phenomena, John Wiley and Sons, Inc.

Boris JP, Landsberg AM, Oran ES, Gardner JH (1993) LCPFCT — A Flux-Corrected Transport Algorithm for Solving Generalized Continuity Equations, NRL Memorandum Report 6410-93-7192.

Brossard J, Charpentier N (1985) Experimental study of shock wave reflection in a narrow channel. In: 15th Int Symp on Shock Tubes and Shock Waves, D Bershader, R Hanson (eds), Berkeley, CA.

Dunn MG, Kang S (1973) Theoretical and Experimental Studies of Reentry Plasmas, NASA CR-2232.

Glowacki WJ, Kuhl AL, Glaz HM, Ferguson RE (1985) Shock wave interaction with high sound speed layers. In: 15th Int Symp on Shock Tubes and Shock Waves, D Bershader, R Hanson (eds), Berkeley, CA.

Honda M, Takayama K, Onodera O, Kohama Y (1975) Motion of reflected shock waves in shock tube. In: 10th Int Symp on Shock Waves and Shock Tubes, G Kamimoto (ed), Japan.

Mark H (1959) The Interaction of a Reflected Shock Wave with the Boundary Layer in a Shock Tube, NACA TM-1418.

Strehlow RA, Cohen A (1959) Limitations of the reflected shock technique for studying fast chemical reactions and its application to the observation of relaxation in nitrogen and oxygen. J Chem Phys 30:257.

Stull DR (1971) JANAF Thermochemical Tables, National Bureau of Standards NSRDS-NBS 37.

Weber JW (1994) Physical and Numerical Aspects of Two-Dimensional Detonation Simulations Including Detailed Chemical Kinetics on a Massively Parallel Connection Machine, PhD Dissertation, University of Maryland.

Weber YS (1994a) The numerical simulation of shock bifurcation near the end wall in a shock tube. In: 25th AIAA Fluid Dynamics Meeting, AIAA 94-2307.

Weber YS (1994b) The Numerical Simulation of the Reflected-Shock/Boundary-Layer Interaction in Shock Tubes, PhD Dissertation, University of Maryland.

Wilson GJ, Sharma SP, Gillespie WD (1993) Time-dependent simulations of reflected-shock/boundary layer interaction. In: 31st AIAA Aerospace Sciences Meeting, AIAA 93-0480.

Young TR, Boris JP (1977) A numerical technique for solving stiff ordinary differential equations associated with the chemical kinetics of reactive-flow problems. J Appl Phys 81:2424.

Author Index

Abdul-Baki, I., 1113
Abgrall, R., 1545
Absil, L.H.J., 685
Achasov, O.V., 1131
Adachi, T., 429
Adam, S., 1225
Adomeit, G., 929, 935, 1027
Ahrens, T.J., 1401, 1449, 1455, 1461
Ahyi, A., 513
Aizik, F., 147
Akamatsu, T., 1649
Aleshin, A.N., 575, 611
Anderson, J., 801
Ando, Y., 355
Aoki, T., 723
Aoyama, T., 1641
Arai, T., 729
Archer, R.D., 1267
Aslam, T., 97
Aso, S., 355, 759, 765

Bélanger, J., 1119
Babinsky, H., 141, 495, 1599
Baillion, M., 869
Bakker, P.G., 1587
Ball, G.J., 173, 557
Ballmann, J., 741
Bauer, H.-J., 947
Baum, J.D., 679
Bazarov, S.B., 441, 477
Bazhenova, T.V., 441, 477, 1169
Bdzil, J., 97
Beck, W.H., 185, 281
Becker, M., 1231
Behnia, M., 1267
Bell, J.B., 617
Belov, V.Y., 1425
Ben-Dor, G., 147, 369, 393, 405, 447, 459, 1303, 1333, 1357
Benjamin, R.F., 587
Bershader, D., 165, 519, 911, 1605
Beylich, A.E., 471
Billiotte, M., 869
Blaha, J., 1249
Blumenthal, R., 935
Bonazza, R., 1189
Boris, J., 801
Bormotova, T.A., 441, 477
Boulton, D., 1033
Boyd, I., 293
Brück, S., 215
Braun, S., 1327
Braun-Unkhoff, M., 857
Brazier, J.P., 735
Breitung, W., 1101
Brennen, C.E., 1213
Britan, A., 1285
Brocher, E., 343
Brouillette, M., 507, 1089
Brown, M.S., 1611
Bruckner, A.P., 1063
Brun, R., 33, 227
Budil, K., 581
Budzinski, J.M., 587
Burcat, A., 1045
Burtschell, Y., 227, 1545, 1569

Cadman, P., 863, 1039
Candler, G.V., 323
Čanić, S., 435
Cardoso, M., 1545
Casey, R.T., 1267
Cates, J., 423
Causon, D.M., 1009, 1195
Chadwick, K., 179
Chaix, A., 227
Chambarel, A., 343
Chauvin, A.H., 983

Chebotareva, E.I., 575
Chen, H., 833
Chernin, A., 453
Chpoun, A., 369, 375, 405, 795
Christiansen, W.H., 135
Chue, R.S.M., 777
Cieslak, S., 299
Clifford, L.J., 1033
Colella, P., 617
Cordonnier, P., 343
Cramer, M.S., 783, 1163, 1327
Crutchfield, W.Y., 789
Cullis, I.G., 997
Cummings, E.B., 1611

Damion, J.P., 1527
Daru, V., 1527
Das, H.K., 1279
Davidson, D.F., 923, 941
Davies, G.A.O, 1207
Davis, J., 191, 203
Deguchi, Y., 1629
Delale, C.F., 1237
Demmig, F., 661
Dewey, J.M., 399
Di Rosa, M.D., 923
Diaconescu, G., 1351
Dihu, R., 1137
Dorofeev, S.B., 77, 673, 1101
Druguet, M.C., 227, 1545
Dumitrescu, L.Z., 227, 1569
Dumitrescu, M.-P., 227, 1487, 1581
Dupré, G., 893
Dvinyaninov, M., 1045
Dvoinishnikov, A.E., 673

East, R.A., 173, 557
Efimenko, A.A., 1101
Eitelberg, G., 185, 221, 257, 1575
El-Beshbeeshy, M., 1189
Elizarova, T.G., 795
Elperin, I., 1303
Elperin, T., 147, 1285, 1309
Emelyanov, A.V., 1169
Ende, H., 311
Eremin, A.V., 881, 917, 1169
Exberger, R., 911

Fabre, E., 299
Falcovitz, J., 489, 1309
Favorsky, A.P., 605
Felthun, L., 551
Fieweger, K., 929, 935, 1027
Figura, E.S., 997
Flanagan, O., 1207
Fletcher, D., 1605
Fortov, V.E., 1443
Frank, P., 857
Frohn, A., 905, 977
Frolov, S.M., 1015
Frost, D.L., 121, 1113, 1387
Fu, Y., 1243
Fuchs, J.W., 747
Fujikawa, S., 1219
Fujita, M., 1431
Fujiwara, T., 263, 765, 1051, 1071, 1157, 1175
Fukuda, M., 355
Fursenko, A., 691
Fuyuto, T., 1071

Gai, S.L., 233, 239, 245, 317
Galametz, I., 647
Gallagher, K.G., 1401
Gallis, M.A., 179
Galyukov, A., 691, 1593
Gao, D.Z., 1515
Garen, W., 1231
Gelfand, B.E., 935, 1027, 1321
George, A., 311
Gilmore, J., 911, 1605
Gimelshein, S.F., 471, 717
Goel, B., 1443
Golub, V.V., 441, 477

Gong, M., 1243
Gorbachev, Y.E., 717
Goto, T., 483
Gou, G.X., 1515
Goyne, C.P., 771, 1623
Grönig, H., 147, 153, 305, 501, 1143
Graur, I.A., 795
Greenough, J.A., 617
Griffin, D., 1497
Grove, J.W., 629
Gu, J.-H., 1509
Gvozdeva, L.G., 375, 1345, 1393

Hänel, D., 1077
Hébert, C., 507
Haas, J.-F., 647
Halat-Augier, C., 893
Han, Z., 1243
Hannemann, K., 221, 257, 1575
Hanson, R.K., 275, 923, 941
Harvey, J.K., 179
Hayashi, A.K., 1051, 1071, 1157
Hayashi, M., 765
He, Y., 953
Heberlein, D.C., 1137
Henderson, L.F., 789
Hertzberg, A., 135
Herzler, J., 839
Hidaka, Y., 899
Higashino, F., 361, 381, 1201
Higgins, A.J., 1063
Hill, L., 1413
Hillier, R., 699, 1207
Hindelang, F.J., 827
Hironde, N., 203
Hitchings, D., 1207
Hoeijmakers, H., 1363
Holden, M.S., 179
Holmes, R.L., 629
Hongo, S., 729
Honma, H., 159, 1181, 1255, 1641
Honore, D., 269
Hornung, H.G., 1611
Hornung, K., 1261
Horvath, T.J., 221
Houas, L., 635, 869
Houwing, A.F.P., 991, 1617
Hu, G., 387
Huq, N., 863

Igra, O., 147, 1285, 1303, 1309, 1339
Ikeda, K., 1655
Ingram, D.M., 1009
Inomata, T., 959
Inoue, A., 959
Itoh, K., 43, 777, 1503
Itoh, S., 1431
Ivanov, M.S., 471, 717
Iwamoto, N., 483
Iwamoto, R., 899

Jacobs, J.W., 587, 617
Jacobs, P.A., 1539
Jacobs, P.G., 1491
Jiang, B., 1009
Jiang, J.P., 1285, 1309
Jiang, Z.L., 465
Joubert, X., 287
Jourdan, G., 635

Kambe, T., 1021
Kameda, M., 1201
Kamel, M.R., 275
Kaneko, R., 1481
Kaneshige, M., 1119
Kao, C.T., 165, 519
Kapila, A., 1413
Karashima, K., 759
Kastell, D., 221
Kawase, T., 729
Kedrinskii, V.K., 107
Kellerer, H., 947
Kemmou, A., 513
Kendall, M.A., 1491

Kern, R.D., 833
Keyfitz, B.L., 435
Khomik, S.V., 1321
Kleine, H., 501, 1351
Klenk, W., 905
Klomfass, A., 741
Kluwick, A., 1327
Knowlen, C., 1063
Knowles, J.K., 1449
Kobayashi, S., 429
Kochurko, A.S., 1101
Kodama, T., 569
Koffi-Kpante, K., 869
Koike, T., 899
Komp, K.H., 935
Komurasaki, K., 263
Komuro, T., 1503
Kondoh, N., 723
Koppenwallner, G., 203
Koreeda, J., 1181
Kotani, M., 1219
Krek, R., 185, 1575
Kruse, T., 851, 875, 887
Kubota, S., 1431
Kudo, S., 361
Kumar, D., 711
Kurian, J., 349, 1279
Kurokui, Y., 1641

Löhner, R., 679
Labracherie, L., 869
Labuda, S.A., 1131
Lackner, K.S., 87
Lagutov, Y.P., 1345, 1393
Laskin, A., 971
Lazareva, E.V., 575, 611
Le, J., 387, 525
Leblanc, J.E., 1175
Lebo, I.G., 605
Lee, J.H.S., 1113, 1351, 1387
Lee, J.J., 1089
Lee, M.G., 545, 563, 705
Legoete, T., 1375
Lele, S.K., 653
Lenartz, M., 153
Lengrand, J.C., 369, 795
Levy, A., 1333, 1357
Li, H., 369, 393, 447, 1333
Li, K.Y., 1083
Li, Z.-F., 1509
Lifshitz, A., 971, 1045
Liu, J.J., 411
Lord, S.J., 1207
Luo, H., 679

Mölder, S., 337
Müller, S., 741
Müller, T., 281
Méolans, J.G., 983
Maekawa, S., 759
Maeno, K., 159, 1181, 1255, 1641
Mahesh, K., 653
Makhsud, A., 343
Makris, A., 1113, 1387
Mallinson, S.G., 233, 317
Manion, J.A., 839
Maršík, F., 1237, 1249
Markelov, G.N., 471, 717
Marsh, S.P., 1419
Martin, R., 227
Mashimo, S., 723
Mathieu, G., 311
Matsui, H., 965
Matsui, Y., 1481
Matsumoto, M., 1219
Matsumoto, Y., 1201
Matsuo, K., 723
Maul, J., 197
Maus, J.R., 1533
McIntyre, T.J., 991, 1617
McKenzie, N.R., 1533
Medvedev, S.P., 1321
Mee, D.J., 771
Meiron, D.I., 623

Menikoff, R., 87
Merlen, A., 299, 513
Messitt, D.G., 1551
Meyer, S., 911
Mick, H.-J., 875, 881
Miller, G.H., 1467
Miller, P., 581
Milne, A.M., 1033
Milthorpe, J., 239
Milton, B.E., 1267
Minota, T., 545, 563
Minucci, M.A.S., 1551
Mitomo, T., 765
Miura, Y., 1473
Moin, P., 653
Mond, M., 147, 667, 1437
Morgan, R.G., 251, 1491, 1563
Mori, S., 1051
Morris, C.I., 275
Mortimer, B., 551
Mouti, M., 983
Mudford, N.R., 173, 233, 245, 317
Muelder, S.A., 53
Muylaert, J., 191, 203, 1587
Myrabo, L.N., 1551

Nagamatsu, H.T., 1551
Nagano, S., 1431
Nagayama, T., 1629
Nagnibeda, E.A., 983
Nagoya, H., 489
Nakatani, A., 1655
Naumann, C., 857
Naumann, K.W., 311
Neely, A., 251
Nerenberg, J., 1387
Neuvazhayev, V.Y., 599
Nguyen, T., 753
Ni, H., 387
Nikishin, V.V., 605
Nishida, M., 545, 563, 705
Nishimori, Y., 1255
Norikan, K., 1137
Noskov, M.A., 1015
Nozawa, H., 1255

O'Keefe, J.D., 1455
Odagawa, M., 159
Ofengeim, D., 537
Ogawa, H., 1157
Oh, C., 753
Ohama, Y., 1181
Okamoto, M., 1473
Olejniczak, J., 323
Olivier, H., 305
Onishi, Y., 747
Onodera, O., 495, 1481, 1521
Oran, E., 753, 801
Oya, M., 965

Paál, G., 1231
Page, N., 1497
Paillard, C.E., 893
Palma, P.C., 991, 1617
Park, S., 783
Passerel, D., 369, 375
Patel, D.L., 1137
Patz, G., 1057
Paull, A., 1557, 1623
Penyazkov, O.G., 1131
Pernod, P., 513
Petersen, E.L., 923, 941
Petrie, P.J., 1539
Peyser, T., 581
Pfahl, U., 929, 1027
Piana, J., 753
Pichugin, S.V., 477
Polenov, A.N., 1321
Popov, I.V., 605
Pruiksma, J., 1369
Pu, Y.K., 1083
Puckett, E.G., 617, 1467
Pullin, D.I., 623

Quenett, C., 1003

Quirk, J.J., 1125

Röhrig, M., 923, 941
Rabbath, P.A.B., 991
Ragozin, D.S., 1131
Ramana Rao, T.V., 349
Ramesh Kumar, R., 349
Ray, J., 641
Rayer, C., 647
Redlinger, R., 1101
Reichenbach, H., 17
Resnyansky, A.D., 1425
Ritzerfeld, E., 501
Roberts, G.T., 173, 317
Robin, L., 269
Rodriguez, G., 647
Rom, J., 1149
Romensky, E.I., 1425
Rosenband, V., 1407
Rosenhauer, M., 281
Roth, P., 1, 845, 851, 875, 881, 887, 1077, 1273, 1315
Rozanov, V.B., 605
Rutkevich, I., 667, 1437
Ruyten, W., 1635

Séror, S., 735
Saint-Martin-Tillet, X., 753
Saito, T., 141, 691
Sakamura, Y., 417
Sakurai, A., 1021
Samtaney, R., 623, 641
Sanderson, S.R., 209
Sasoh, A., 1063, 1481
Sato, H., 159, 1219
Sato, K., 759, 899, 1503
Saunders, R., 1009
Schading, G.N., 1273
Schall, E., 735
Scheer, M., 281
Schnerr, G.H., 1225, 1237
Schotz, M., 393
Schuricht, P.H., 173
Schwendeman, D., 531
Seiler, F., 1057
Seitz, M., 1381
Sekine, N., 1521
Sellschop, R., 1375
Serov, A.O., 331
Shamoun, B., 1189
Sharma, S., 911, 1605
Sharov, Y.L., 1345, 1393
Sharp, D.H., 629
Shaw, L.L., 53
Shepherd, J.E., 1119
Sherbak, N.B., 375, 1393
Shida, H., 1629
Shimaura, N., 1201
Short, M., 1125
Shtemenko, L.S., 331
Shugaev, F.V., 331
Shulmeister, A.M., 441, 477
Shumova, V.V., 881, 917, 1169
Sidorov, V.P., 673, 1101
Simeonides, G., 203, 1587
Sislian, J.P., 141
Skews, B., 551, 1357, 1375, 1381
Smeets, G., 1003, 1057, 1363
Smeulders, D., 1363, 1369
Smith, M., 1635
Sommer, T., 887
Sopuch, P., 1249
Sorek, S., 1357
Srulijes, J., 1057
Stalker, R., 1497, 1623
Stewart, D., 97
Stollery, J.L., 711
Strecker, J.J.F., 1315
Stry, P., 581
Stuhler, H., 977
Sturtevant, B., 191, 209, 423, 593
Sudnitsin, O., 1563
Sugiyama, H., 729

Sumida, W.K., 1137
Sun, X.W., 159
Suzuki, T., 417, 429

Takahashi, K., 959
Takano, Y., 483
Takasugi, N., 1219
Takata, T., 1455
Takayama, F., 1021
Takayama, K., 141, 465, 489, 495, 569, 1481, 1521, 1599, 1655
Taki, T., 1021
Tamagawa, M., 1649
Tan, T.-H., 1419
Tanno, H., 1503
Tannou, M., 355
Tarkenton, G.M., 1163
Tatschl, R., 1015
Teodorczyk, A., 1095
Thibault, P. A., 1107
Thoma, M.L., 827
Thomas, G.O., 1095
Thorembey, M.-H., 647
Thurber, M.C., 275
Timnat, Y.M., 1407
Timofeev, E., 495, 537, 691, 1593
Tishkin, V.F., 605
Titov, S.N., 575
Toffin, E., 667
Tomita, K., 1655
Tsang, W., 839
Tsuchiya, K., 965
Tulis, A.J., 1137
Tumino, G., 203

Ueda, S., 257
Umii, K., 355
Uno, N., 729
Uphoff, U., 1077

van den Berg, A.C., 685
van Dongen, M., 1363, 1369
van Gool, C., 1369
van Netten A.A., 399
Vasiliev, E., 1303
Vervisch, P., 269
Vetter, M., 593
Virgona, R.J., 381, 789
Voinovich, P., 453, 495, 537, 691, 1593
von Ellenrieder, K., 165, 519
Vorobiev, O.Y., 1443
Vuillon, J., 459

Wallus, H., 661
Walpot, L.M.G., 1587
Walpot, W., 203
Wang, B., 153
Wang, B.-Y., 1291, 1297
Wang, D.-Z., 1297
Wang, R., 863
Wang, Y.-C., 1213
Ward, S.M., 1095
Watanabe, N., 1641
Watanabe, Y., 361
Watson, L.T., 783
Weber, Y., 801
Weerheijm, J., 685
Wehe, S.D., 275
Whiting, E., 911
Widdecke, N., 905
Williams, G.H., 1425
Williams, W., 1635
Wilson, G.J., 67
Winfree, N.A., 1449
Wintergerst, K., 857
Wittig, S., 947
Woiki, D., 845
Wolanski, P., 1015
Wollenhaupt, M., 281
Wu, Q.-S., 1291, 1297
Wu, X., 525

Xu, Y.-H., 1291, 1297

Xu, Y.-Q., 1291

Yamada, K., 489, 569
Yamane, Y., 355
Yang, H., 525
Yang, J.M., 141
Yang, W., 1461
Yano, S., 263
Yao, J., 833
Ye, X., 525
Yin, X.-.Z, 417
Yoshida, K., 1649
Yoshida, T., 1051
Yoshinaka, A., 1113
Yu, H.-R., 1509
Yu, Q., 501, 1143

Zabusky, N.J., 641
Zanchetta, M.A., 699
Zaretsky, E., 1339, 1437
Zaslonko, I.S., 917
Zaytsev, S.G., 575, 611
Zeitoun, D., 227, 459, 735, 1545, 1569
Zerf, G., 1261
Zhang, F., 1107
Zhang, X.-Y., 1509
Zhao, W., 1509
Zhong, X., 287
Ziborov, V.S., 881

Keyword Index

Ablation, 263
Absorption, 911
Absorption spectra, 923
Absorption spectroscopy, 857, 887
Acceleration measurement, 311
Acetone pyrolysis, 899
Acetylene, 851
Adaptive methods, 617
Advanced ceramics, 1407
Aerodynamic force measurement, 311
Aerodynamic heating, 765
Aerodynamics, 337
α-methylnaphthalene, 1027
Aluminum, 1137
Aluminum dust-air mixture, 1083
Angular momentum, 519
Animation, 141
ARAS, 881, 959, 965
Astrophysics, 453
Asymptotic expansion, 735
Atomic oxygen, 911
Atomization, 1027, 1267
Attenuation law, 147
Automobile engine, 1521

Backwards step, 239
Baffle system, 1291
Ballistic range, 263
Base flow, 179
Bifurcation, 801, 1225
Bimodal distribution, 1315
Blast damage assessment, 685
Blast interaction with structures, 685
Blast parameters, 673
Blast wave, 77, 453, 557, 685, 1321, 1387
Blast-wave propagation, 1351
Blunt body, 735
Blunt cone, 221, 239
Blunt protuberance, 759
Bluntness, 711
Bone, 1655
Boundary layer, 67, 245, 251, 735, 1497, 1533, 1581, 1623
Boundary-layer blockage, 1563
Boundary-layer formation, 1563
Break-up, 1027, 1261
Break-up time, 1027
Bubble, 1201
Bubble dynamics, 159

C_2, 887
Catalyticity, 317
Cavitation, 159, 1207
Cavitation noise, 1213
Cavitation structure, 107
Ceramic armor, 1449
CFD, 495, 789, 1143, 1225, 1569
CFD validation, 1635
Chemical in-tube accelerators, 1149
Chemical kinetics, 845, 851, 857, 875, 929
Chemical nonequilibrium, 735, 977, 1545
Chemical reaction, 765
Chemical reactions, finite-rate, 1063
Chemical separation, 1473
Chemically reacting hypersonic rarefied flows, 717
Cloud cavitation, 1213
Cloud of particles, 1279
CN emission, 869
CO_2 laser absorption, 635
Coflowing jets, 349
Collision broadening, 923
Combustible gas mixture, 1057
Combustion, 1003, 1057, 1551
Combustion synthesis, 1407

Combustion torch, 1051
Comet Shoemaker-Levy 9, 1455
Communication blackout, 1175
Compaction, 1413
Compaction waves, 1357
Comparative rate, 953
Comparative rate measurements, 839
Compositional change, 1473
Composites, fibrous and laminated, 1425
Compressibility, 519
Compressible flow, 581
Compressible foam, 1345, 1393
Compressible turbulence, 507
Compressible turbulent boundary layer, 1015
Compression corner, 233
Compression wave, 723, 1339
Computational model, 1533
Condensation, 1219, 1231, 1243, 1255
Condensation coefficient, 1219
Condensation shock, 1249
Cone, 417
Continuum equations, 795
Control flap, 711
Control jets, 173
Converging shock waves, 661
Corundum, 1401
Critical pressure ratio, 343
Cryogenic liquid, 159
Curvature effect, 87, 97
Cylinder flow, 215

Damkohler number, 251
Decay, 1339
Decomposition, 827, 953, 971
Defect formulation, 735
Deflagration, 77, 929, 1107
Deflagration-to-detonation transition (DDT), 929, 1015, 1107, 1413
Deformation, 1261
Density measurement, 647
Detection, 1557
Detonation, 77, 97, 153, 929, 1071, 1077, 1101, 1107, 1119, 1125, 1143, 1149
Detonation driver, 153, 1509
Detonation initiation, 107, 1131
Detonation, over driving, 1503
Detonation propagation limits, 1113
Deuterated ethylene, 833
DFWM, 1611
Diaphragm rupture, 1539
Diaphragmless shock tube, 141
Diesel engine, 929
Differential interferometry, 647
Diffraction, 141, 483, 513, 789
Direct numerical simulation, 653
Direct Simulation Monte Carlo (DSMC), 717, 753
Disk, 795
Dispersed waves, 1641
Dissipation, 519
Dissociation, 33, 245
Dissociation modeling, 293
Dissociation-recombination, 917
Disturbances on shock front, 331
Drag, 1261
Drag coefficient, 1285
Driver, 1491, 1497
Driver gas, 1557
Driver-gas contamination, 777, 1581
Droplet, 1027, 1267
Dry-powder dispersion, 1315
Dust deposit, 1321
Dust-gas suspensions, 147
Dust lofting, 1309
Dusty gas, 1297, 1303, 1309
Dusty shock, 1291

Edge interactions, 97
Effective drag coefficient, 1279
Effective stress, 1363
Ejecta, 1461

Elastic bodies, 513
Electro-optic shutter, 53
Embedded shock formation, 1237
Emission-absorption spectroscopy, 1169
Engine silencer, 1521
Entrainment, 1267
Entropy, 1491
Environmental, 947
Equation of state, 1437, 1443
Equilibrium gas dynamics, 1369
Ethylene combustion, 1033
Euler equations, 459, 531
Evaporation, 747, 1027, 1219
Excitation, 667
Exhaust gas-pipe system, 1521
Expansion tube, 67, 275, 1539, 1563
Explosions, 77, 685
Explosive compaction, 1407
Explosive, granular, 1413
Extracorporeal shock wave lithotripsy (ESWL), 107, 1655

Fay and Riddell theory, 305
Finite-compression waves, 411
Finite-volume code, 1539
Flame acceleration, 1083
Flame, spontaneous, 1015
Flow establishment times, 179
Flow imaging, 275, 991
Flow visualization, 165, 257, 545, 569, 593, 729, 759, 1189, 1225, 1617, 1635
Fluorocarbons, 551
Foam, 1375, 1381
Foam, flexible, 1363, 1369
Foam, permeable, 1351
Fourier transform, 1599
Fracture, 1655
Fragmentation wave system, 661
Free-Lagrange method, 557
Free piston, 1491, 1497
Free-piston driver, 43
Free-piston shock tunnel, 43, 227, 257, 1503, 1515
Friction, 1413
Fringe analysis, 1599
Front tracking, 587, 629, 1195
Fuel-coolant interaction, 121
Fullerene formation, 863
Furonitrile, 971

Gas bubble, 569
Gas phase, 851, 875
Gas-wall interaction, 197
Gaseous detonation, 673, 1113
Geometrical shock dynamics, 423, 531
Global simulation method, 1009
Godunov methods, 1467
Griceite, 1401
Grid, adaptive, 1143
Grid generation, 691
Grid, local refinement, 691
Grids, combined, 537
Grid, unstructured, 495
Growth rate, 623, 641

H_2S, 965
Halis, 191
Halocarbons, 1045
Halomethanes, 959
Heat engine, 135
Heat flux, 245, 257
Heat transfer, 251, 1491
HEG shock tunnel, 185, 221, 1575, 1587
Heterogeneous detonation, 1089, 1137
HFC-134a, 1255
HHK, 203
High enthalpy, 191, 251, 317, 777, 1587
High-enthalpy effects, 179
High-enthalpy flow, 269, 1515, 1635
High-enthalpy nozzle flow, 1575
High Knudsen-number flow, 753

High pressure, 923, 935, 941
High-speed photography, 53
High temperature, 833
Historical report, 17
Hollow projectile, 1063
Holographic interferometry, 141, 525, 1599
Homogeneous condensation, 1237
Hood, 1481
Hugoniot jump conditions, 87
Hydraulic shock, 1207
Hydrazine, 893
Hydrocarbons, 929
Hydrodynamic code, 997, 1009
Hydrodynamic instability, 575, 611
Hydrogen, 1101
Hydrogen-air, 935
Hydrogen combustion, 1009
Hyperboloid flare, 185
Hypersonic facility, 227, 1487, 1581
Hypersonic flow, 173, 179, 191, 221, 245, 263, 275, 281, 293, 299, 305, 323, 387, 699, 735, 1563, 1635
Hypersonic nozzle, 1569
Hypersonic receptivity, 287
Hypervelocity, 185, 191, 233, 239, 317, 337, 623, 1119, 1149
Hysteresis, 375, 405, 471

Ideal dissociating gas, 623
Ignition, 1027, 1057, 1413
Ignition-delay time, 929, 1027
Ignition, spontaneous, 1021
Ignition times, 941
Image-converter camera, 53
Image processing, 1641
Imaging spectroscopy, 1181
Impact, 1449, 1455, 1461
Improved tangential velocity gradient, 305
Impulsive jet, 441
Induction zone, 997
Integral method, 305
Interaction of vortices, 563
Interference heating, 173
Interferometry, 495, 647
Interferometry, finite-fringe, 1641
Intermittency, 699
Internal cavity flow, 361
Intersecting characteristics, 1237
Ionization, 667
IR diode laser absorption spectroscopy, 905
IR-laser absorption, 899
IR-spectroscopy, 1273
Iron, 1401
Isentropic compression, 1497

Jet interaction, 311, 1131
Jets, 453, 581
Jet, liquid, 569
Jetting, 1461, 1467
Jupiter, 1455

Kármán integral equations, 1563
Kinetic theory, 747
Kinetics, 893, 997
Kinetics, shock-tube, 1

Laminar separation, 711
Large-scale experiments, 1101
Laser absorption, 1635
Laser diagnostics, 923, 941, 1051
Laser differential interferometry, 723
Laser-flash photolysis, 881
Laser holography, 1261
Laser-illuminated photography, 53
Laser-induced fluorescence (LIF), 269, 281, 1629
Laser photolysis, 965
LDV, 729
Leading edge bluntness, 233
Light extinction, 1189

Linear analysis, 653
Liner implosion, 1419
Lineshape, 911
Liquid-spray combustion, 863
LITA, 1611
Living time, 1027
Living-tissue cell, 1649
Local similarity, 741
Low temperature, 935, 1255

Mach disc, 349
Mach reflection, 97, 369, 375, 387, 393, 399, 405, 411, 471, 1255
Magnetohydrodynamics, 1157
Mantle-core structure, 1473
Material interfaces, 1195
Measurements, 191
Melting, 1413
Methane ignition, 1045
Methyl halides, 959
MHD shocks, 1163
Micro capsule, 1649
Microchannel flows, 753
Microinhomogeneities, 107
Minimum entropy production, 447, 1249
Mixing, 581
Model computation, 661
Modeling, 857, 893
Molecular dynamics, 1219
Monte Carlo method, 293, 795
Moving shock waves, 465
Multi-component flows, 1195
Multi-fluid methods, 489, 617
Multi-pulsed ruby laser, 53
Multipass technique, 827
Multiphase flow, 1107

n-decane, 1027
Navier-Stokes equations, 537, 753, 789
Nd-B-Fe powder, 1431
NH_4Cl/Ar aerosols, 1273
Nitric oxide (NO), 269, 827, 905
Nitromethane, 1089
NO spectroscopy, 281
NO_x, 857
Non-deuterated ethylene, 833
Nonequilibrium, 33, 221, 337, 741, 765, 1587
Nonequilibrium condensation, 1225
Nonequilibrium Navier-Stokes, 203
Nonequilibrium radiation, 1169, 1181
Nonlinearity parameter, 551
Non-uniform distribution, 1201
Nova laser, 581
Nozzle starting process, 1545
Nozzles, 1297, 1303, 1563, 1587
Numerical blast simulation, 679
Numerical computation, 221, 343, 453, 459, 465, 501, 605, 1033, 1063, 1071, 1125, 1297, 1363, 1425, 1521, 1527, 1575, 1593

Off-axis loading, 1425
OH, 923, 941
Optical-flow visualization, 1521
Optical method, 197, 1345
Origin of the Moon, 1473
Oscillating shock wave, 759
Oxidation, 965
Oxy-hydrogen detonation, 1509
Ozone, 827

Particle coagulation, 1315
Pattern recognition, 661
Percolation, 1393
Perfluoropropylene, 953
Permeability, 1363
Phase transformation, 1449
Phase transition, 1249
Photolysis, 845
Piston-stopping device, 1487
Planar laser-induced fluorescence, 275, 991, 1617, 1635

Plasmas, 1163
Plasma, anisotropic, 1175
Plasma, nonideal, 1443
Plasma, waves in, 1175
PLIF, 275, 991, 1617, 1635
Polycyclic aromatic hydrocarbons, 863
Porous material, 1321, 1333, 1357, 1375, 1381
Porous medium, 1113, 1387
Porous wall, 1481
Precursor shock, 1369
Pressure, 947
Pressure amplification, 1375, 1381
Pressure fluctuations, 1527
Pressure measurement, 1057
Pressure-wave refrigerator, 1593
Projectile, 1119
Projectile acceleration, 1057
Pseudo-shock wave, 729
Pulsating instabilities, 1125
Pulse laser, 159
Pyrolysis, 845, 959

Quantum fluid, 1327

Radiation, 263
Ram accelerator, 1057, 1063
Raman scattering, 1605
Rankine-Hugoniot relations, 1369
Rapid phase transformation, 121
Rarefaction focusing, 107
Rarefaction waves, 1243
Rarefied gas dynamics, 795
Rate coefficients, 959
Rate constant, 953
RDX, 1137
Reaction mechanism, 899
Reaction rate, 899
Reactive flow, 735, 801, 1033
Real gas, 191
Real-gas effects, 203
Recombination, 251, 1169
Red blood cell, 1649
Reduced mechanism, 1009
Reduced models, 641
Reentry vehicle, 1175
Reflected-shock/boundary-layer interaction, 777, 801
Reflection, 141, 417, 429, 441, 447, 483, 551, 1291, 1375, 1381
Reflection, anomalous, 381
Reflection, regular, 375, 405, 1157, 1333
Relaxation models, 977
Retrograde gases, 1231
Reynolds shear stress, 653
Richtmyer-Meshkov instability, 581, 587, 593, 605, 629, 635, 647
Riemann problem, 435
Rotational-vibrational excitation, 905
RR to MR transition, 459

Safety distances, 685
Scaling, 77
Scavenger, 953
SCRAM jet engine, 355, 1003
Self-ignition, 929, 935
Self-similarity, 399
Separation, 185, 191
70-deg cone, 179
Shaped-charge jet, 53
Shear layer, 221, 361
Shock acceleration, 587
Shock acceleration, diffuse interfaces, 617
Shock attenuation, 1351
Shock capturing, 87, 531
Shock-cell structure, 343
Shock compression, 1419
Shock consolidation, 1431
Shock diffraction, 477, 501, 531, 1291
Shock diffraction, reactive, 1009
Shock focusing, 423, 495, 1131
Shock heating, 1419

Shock impingement, 209, 477
Shock-induced mixing, 587
Shock-induced separation, 783
Shock instability, 667
Shock interaction, 87, 323, 435
Shock interaction, boundary-layer, 191, 203, 705, 729
Shock interaction, bubble 489, 569
Shock interaction, disturbance, 287
Shock interaction, dust 1309
Shock interaction, foam, 1387
Shock interaction, interface, 635
Shock interaction, sphere, 1285
Shock interaction, turbulence, 653
Shock interaction, turbulent boundary-layer, 729
Shock interference, 209
Shock loading, 107
Shock polar, 1157
Shock reflection, 141, 417, 429, 441, 447, 483, 551, 1291, 1375, 1381
Shock refraction, 557
Shock-on-shock interaction, 209, 215
Shock structure, 349
Shock tunnel, 153, 191, 203, 281, 311, 771, 777, 991, 1533, 1539, 1551, 1605, 1617, 1623
Shock-vortex interaction, 165, 1021
Shock-wave diffraction, 477, 501, 531, 1291
Shock-wave impingement, 209, 477
Shock-wave propagation, 1195, 1297, 1425
Shock-wave reflection, 141, 417, 429, 441, 447, 483, 551, 1291, 1375, 1381
Shock wave, standing, 747
Shock wave, strong, 1181
Shocklet, formation of, 545, 563
Shocks in liquids, 513
Short-duration flow, 311
Silicon, 875
Single-pulse shock tube, 839
Skin friction, 1623
Soot, 863, 947
Spherical condensed phase, 747
Spontaneous acoustic emission, 1437
Stability, 471
Stability of shocks, 1437
Stagnation-point heating, 305
Stagnation streamline, 741
Starch, 1137
Steady flow, 369, 393, 477
Steady supersonic flow, 459
Stream-wise vortex, 355
Stress relaxation time, 1339
Sulfur, 845
Superfluid, 1327
Superorbital expansion tube, 251
Supersonic combustion, 991
Supersonic expansion, 1169
Supersonic flow, 361
Supersonic mixing, 355
Supersonic projectile, 465
Surface heat flux, 317

T5, 191, 203
Temperature measurements, 869
Temperature probe, 1497
Temperature, shock, 1401
Terrestrial planets, 1473
Test time, 777
Tetra-ethoxysilane decomposition, 839
Thermal conductivity, 1401
Thermal decomposition, 833, 887, 893, 959, 1273
Thermal grating, 1611
Thermal nonequilibrium, 977
Thermodynamic fluctuations, 653
Three-dimensional flows, 417
Three-dimensional simulation, 1101
Time-of-flight mass spectrometry, 833
Time-resolved visualization, 661

Titan, 869
TNT, 1137
TNT-equivalent, 673
Transition, 179, 429, 471, 699, 711, 723, 771
Transition criterion, 369
Transition to detonation, 1083
Transonic flow, 1225
Transonic shocks, 435
Transport terms, 33
Tuned operation, 43, 1503
Tunnel calibration, 257
Tunnel sonic boom, 1481
Turbulence, 575, 611, 729
Turbulent kinetic energy, 653
Turbulent mixing, 593, 599, 647
Turbulent mixing zone, 635
Turbulent spots, 771
TVD method, 215, 387, 1357, 1545
Two-phase analysis, 1357
Two-phase flow, 1071, 1077, 1201, 1225

Underwater shock wave, 1431
Unsteady boundary layer, 723
Unsteady flow, 67, 537, 1279
Unsteady reflection, 429
Unstructured Euler solver, 691

Vacuum ultraviolet, 911
Vapor explosion, 107, 121, 1189
Velocimetry, 1611
Vibration-dissociation coupling, 323, 983
Vibrational nonequilibrium, 735, 881
Vibrational populations, 33, 983
Vibrational relaxation, 917, 1605
Viscous nonequilibrium flows, 215
Visiometrics, 641
Visualization techniques, 17, 501
Void fraction, 1189
von Neumann reflection, 381, 435
Vortex, 441, 519
Vortex, resonant excitement of, 331
Vortex ring, 477, 545, 563, 1021
Vortex-wall interaction, 545
Vorticity, 507

Wake flow, 221
Wall pressure, 257
Water droplet, 1261
Wave configuration, 393
Wave machinery, 135
Wave processes, 1593
Wave splitting, 1369
Wave structure, 429
Wave transmission, 1089
Waveriders, 299
Waves in plasma, 1175
Weak shock wave, 483, 1231, 1521
World Trade Center, 679

X-rays, 647

Proceedings of the 20th International Symposium on

SHOCK WAVES

Proceedings of the 20th International Symposium on

SHOCK WAVES

Volume II

Pasadena, California, USA   July 1995

Editors

B Sturtevant

J E Shepherd

H G Hornung

Graduate Aeronautical Laboratories
California Institute of Technology

World Scientific
Singapore • New Jersey • London • Hong Kong

Published by

World Scientific Publishing Co. Pte. Ltd.
P O Box 128, Farrer Road, Singapore 912805
USA office: Suite 1B, 1060 Main Street, River Edge, NJ 07661
UK office: 57 Shelton Street, Covent Garden, London WC2H 9HE

This work relates to Department of Navy Grant N00014-95-1-0301 issued by the Office of Naval Research. The U.S. Government has a royalty-free license throughout the world in all copyrightable material contained herein.

This work is sponsored in part by the Defense Nuclear Agency under Grant DNA001-95-1-0007. This manuscript is submitted for publication with the understanding that the U.S. Government is authorized to reproduce and distribute reprints for Government purposes notwithstanding any copyright notions thereon.

British Library Cataloguing-in-Publication Data
A catalogue record for this book is available from the British Library.

PROCEEDINGS OF THE 20TH INTERNATIONAL SYMPOSIUM ON SHOCK WAVES (Vol. II)

ISBN 981-02-2593-8 (Set)
ISBN 981-02-2957- 7 (Vol. I)
ISBN 981-02-2958-5 (Vol. II)

Printed in the United States of America.

Contents

Volume I

Preface v

Paul Vielle Lecture

Shock-tube kinetics in homogeneous and heterogeneous reaction systems
Roth, P. 1

Otto Laporte Lecture

120 years of shock-wave visualization — a fascinating activity
Reichenbach, H. 17

Plenary Lectures

Transport and relaxation behind shock waves: old and new problems
Brun, R. 33

Tuned operation of a free-piston shock tunnel
Itoh, K. 43

High-resolution imaging of hypervelocity metal jets using advanced high-speed photographic techniques
Shaw, L.L., Muelder, S.A. 53

Numerical simulation of high-enthalpy impulse facilities
Wilson, G.J. 67

Blast effects of confined and unconfined explosions
Dorofeev, S.B. 77

Anomalous physical effects from artificial numerical length scales
Menikoff, R., Lackner, K.S. 87

Curved detonation fronts in solid explosives: collisions and boundary interactions
Bdzil, J.B., Aslam, T.D., Stewart, D.S. 97

Bubbly cavitation in intense rarefaction waves and its effects
Kedrinskii, V.K. 107

Wave propagation in vapor explosions
Frost, D.L. 121

Irving Glass Memorial Session

Wave machinery for chemical processing and high-efficiency heat engines
Christiansen, W.H., Hertzberg, A. 135

Holographic interferometric animation of shock diffraction and reflection using a highly repeatable shock tube
Yang, J.M., Babinsky, H., Takayama, K., Saito, T., Sislian, J.P. 141

Universal attenuation law of planar shock waves propagating in dust-gas suspensions
Aizik, F., Ben-Dor, G., Elperin, T., Igra, O., Mond, M., Grönig, H. 147

Development of a detonation driver for a shock tunnel
Lenartz, M., Wang, B., Grönig, H. 153

Behavior of laser-induced bubbles and pressure waves in liquid nitrogen
Sato, H., Odagawa, M., Sun, X.W., Maeno, K., Honma, H. 159

The interaction of a two-dimensional strong compressible vortex with a shock wave
Kao, C.T, von Ellenrieder, K., Bershader, D. 165

Hypersonic Flow

Interference heating effects caused by a 3D transverse jet in hypersonic flow
Mudford, N.R., Roberts, G.T., Schuricht, P.H., Ball, G.J., East, R.A. 173

Comparison between shock-tunnel measurements on a planetary-probe configuration and DSMC predictions
Holden, M.S., Chadwick, K.M., Gallis, M.A., Harvey, J.K. 179

Hyperboloid flare experiments in the HEG
Krek, R.M., Eitelberg, G., Beck, W.H. 185

Experiments on the Halis axisymmetric configuration in the T5 shock tunnel
Davis, J., Sturtevant, B., Muylaert, J. 191

Shock-tube measurements for aero-thermodynamic gas-wall interaction
Maul, J. 197

Computational analysis of experiments in T5 and HHK for the study of flap-induced separation on an axisymmetric configuration
Muylaert, J., Walpot, L., Hironde, N., Tumino, G., Simeonides, G., Koppenwallner, G., Davis, J. 203

Shock-interference heating in hypervelocity flow
Sanderson, S.R., Sturtevant, B. 209

Investigation of shock-shock interactions in hypersonic reentry flows
Brück, S. 215

Recompression of nonequilibrium flow in the wake of a blunted cone
Kastell, D., Hannemann, K., Eitelberg, G., Horvath, T.J. 221

Flow in the TCM2-Marseille hypersonic facility: some new experimental results
Brun, R., Burtschell, Y., Chaix, A., Druguet, M.-C., Dumitrescu, L.Z., Dumitrescu, M.-P., Martin, R., Zeitoun, D. 227

Leading edge bluntness effects in hypervelocity compression corner flow
Mallinson, S.G., Gai, S.L., Mudford, N.R. 233

Hypersonic high-enthalpy flow over a blunt-stepped cone
Gai, S.L., Milthorpe, J.F. 239

Surface heat flux to the Viking aeroshell in hypersonic reacting flow
Mudford, N.R., Gai, S.L. 245

Heat transfer in a superorbital chemically frozen boundary layer
Neely, A., Morgan, R. 251

Standard model experiments in high-enthalpy facilities
Ueda, S., Hannemann, K., Eitelberg, G. 257

Radiation around a hypersonic ablative flight model
Komurasaki, K., Yano, S., Fujiwara, T. 263

Laser-induced fluorescence measurements in high-enthalpy facilities: application to the study of air plasma/surface interaction
Robin, L., Honore, D., Vervisch, P. 269

Use of an expansion tube for imaging studies of hypersonic reactive flows around blunt bodies
Kamel, M.R., Morris, C.I., Thurber, M.C., Wehe, S.D., Hanson, R.K. 275

Spatially and spectrally resolved laser-induced fluorescence measurements in the HEG
Wollenhaupt, M., Rosenhauer, M., Müller, T., Beck, W.H., Scheer, M. 281

Bow-shock/disturbance interaction for hypersonic flow over a cylinder
Zhong, X., Joubert, X. 287

Monte Carlo study of threshold line models for dissociation in hypersonic shock waves
Boyd, I.D. 293

Analytical hypersonic waverider configurations
Merlen, A., Fabre, E., Cieslak, S. 299

A more precise method to determine stagnation-point heat-transfer rates
Olivier, H., Grönig, H. 305

Millisecond aerodynamic force measurement using a free-flying side-jet model with built-in gas reservoir
Naumann, K.W., Ende, H., Mathieu, G., George, A. 311

Effects of catalycity and boundary layer reactions on surface heat flux in hypersonic high enthalpy flows
Gai, S.L., Mudford, N.R., Roberts, G.T., Mallinson, S.G. 317

Study of experiments sensitive to vibration-dissociation coupling models
Olejniczak, J., Candler, G.V. 323

Supersonic Flow

Disturbances on the shock front ahead of a concave body due to resonant excitement of vortices
Serov, A.O., Shtemenko, L.S., Shugaev, F.V. 331

An inverse method for calculating nonequilibrium flow behind curved shock waves
Mölder, S. 337

Shock-cell structure in a jet at the critical pressure ratio
Makhsud, A., Brocher, E., Chambarel, A., Cordonnier, P. 343

Mach disc and shock structure in coflowing dual axisymmetric jets
Kurian, J., Ramana Rao, T.V., Ramesh Kumar, R. 349

A study of three-dimensional supersonic mixing flow phenomena induced by ramp injectors
Umii, K., Aso, S., Tannou, M., Yamane, Y., Ando, Y., Fukuda, M. 355

Effects of cavity configurations on supersonic internal flow
Watanabe, Y., Kudo, S., Higashino, F. 361

Wave Reflection and Diffraction

Experimental reconsideration of the state of the art of oblique shock-wave reflections in steady flows
Chpoun, A., Passerel, D., Lengrand, J.C., Li, H., Ben-Dor, G. 369

Experimental and theoretical study of stationary reflection of shock waves
Gvozdeva, L.G., Sherbak, N.B., Chpoun, A., Passerel, D. 375

Mach-stem growth in von Neumann reflection
Virgona, R.J., Higashino, F. 381

Mach-reflection flow at hypersonic speeds
Le, J., Ni, H., Hu, G. 387

Wave configuration of Mach reflection in steady flows: analytical solution and dependence on downstream influences
Li, H., Schotz, M., Ben-Dor, G. 393

Non-self-similarity of the initial stages of Mach reflection
Dewey, J.M., van Netten A.A. 399

Numerical confirmation of the hysteresis phenomenon in the regular to Mach reflection transition in steady flows
Chpoun, A., Ben-Dor, G. 405

The existence of reflected finite-compression waves in weak pseudo-steady Mach reflections over not-"small" wedge angles
Liu, J.J. 411

Shock-wave reflections from a cone with attack angles in a symmetric plane
Suzuki, T., Sakamura, Y., Yin, X.-.Z 417

Calculation of focusing of weak shocks using geometrical shock dynamics
Cates, J., Sturtevant, B. 423

The shock reflection process over a non-straight surface
Kobayashi, S., Adachi, T., Suzuki, T. 429

Oblique shock interactions and the von Neumann paradox
Čanić, S., Keyfitz, B.L. 435

Shock-wave interactions at impingement of impulsive supersonic jet upon an obstacle
Bazhenova, T.V., Golub, V.V., Shulmeister, A.M., Bormotova, T.A., Bazarov, S.B. 441

Application of the principle of minimum entropy production to the regular-to-Mach-reflection transition in steady flows
Li, H., Ben-Dor, G. 447

Evolution of Courant-Friedrichs structure: self-collimating outflows behind reflected shocks
Chernin, A., Voinovich, P. 453

Numerical investigations of the prediction of the Mach-stem height in steady flows
Vuillon, J., Zeitoun, D., Ben-Dor, G. 459

Interaction of shock waves emitted from a shock tube with a supersonic projectile
Jiang, Z., Takayama, K. 465

Numerical investigation of shock-wave reflection problems in steady flows
Ivanov, M.S., Gimelshein, S.F., Markelov, G.N., Beylich, A.E. 471

The interaction of a shock wave discharged from the open end of a shock tube with a flat plate
Bazhenova, T.V., Golub, V.V., Shulmeister, A.M., Bormotova, T.A., Bazarov, S.B., Pichugin, S.V. 477

Analyses of reflection and diffraction of weak shock waves
Takano, Y., Iwamoto, N., Goto, T. 483

Computation and validation of shock-wave interaction with fluid inhomogeneities
Falcovitz, J., Yamada, K., Nagoya, H., Takayama, K. 489

The influence of geometric variations on shock focusing in cylindrical cavities
Babinsky, H., Onodera, O., Takayama, K., Timofeev, E., Voinovich, P. 495

An experimental/numerical approach to shock-wave diffraction
Kleine, H., Yu, Q., Ritzerfeld, E., Grönig, H. 501

Compressibility effects on the propagation and interaction of shock-generated vortices
Hébert, C., Brouillette, M. 507

Shock-wave diffraction by an elastic sphere in water
Merlen, A., Pernod, P., Ahyi, A., Kemmou, A. 513

Physics of a shock-generated freely moving compressible vortex
von Ellenrieder, K., Kao, C.T., Bershader, D. 519

Study of the propagation of shock waves over bodies by holographic interferometry
Le, J., Ye, X., Wu, X., Yang, H. 525

Accuracy of shock-wave propagation using geometrical shock dynamics
Schwendeman, D.W. 531

A locally adaptive structured/unstructured Navier-Stokes solver for essentially unsteady shock-obstacle interactions at high Reynolds numbers
Ofengeim, D., Timofeev, E., Voinovich, P. 537

Wave Refraction

Vortex-ring impingement on a wall with shock formation
Minota, T., Nishida, M., Lee, M.G. 545

Oblique reflection of shock waves in a slow sound-speed fluorocarbon liquid
Mortimer, B.J.P., Skews, B.W., Felthun, L.T. 551

Refraction of a blast wave propagating parallel to a contact discontinuity: a Free-Lagrange numerical simulation
Ball, G.J., East, R.A. 557

Shock waves generated by the interaction of two vortex rings
Minota, T., Nishida, M., Lee, M.G. 563

Mechanisms of shock-wave generation resulting from shock-wave/bubble interaction in liquids
Yamada, K., Kodama, T., Takayama, K. 569

Passage of two shocks through an interface
Zaytsev, S.G., Chebotareva, E.I., Aleshin, A.N., Lazareva, E.V., Titov, S.N. 575

Shock-hydrodynamics experiments on the Nova laser
Miller, P.L., Peyser, T.A., Stry, P.E., Budil, K.S. 581

Shock-accelerated heavy-gas layers
Budzinski, J.M., Benjamin, R.F., Jacobs, J.W. 587

Experiments on the shock-excited interfacial instability at an air/SF_6 interface
Vetter, M., Sturtevant, B. 593

Mathematical modeling of gravitational turbulent mixing in non-self-similar problems
Neuvazhayev, V.Y. 599

Numerical simulation of Richtmyer-Meshkov instability
Lebo, I.G., Rozanov, V.B., Tishkin, V.F., Nikishin, V.V., Popov, I.V., Favorsky, A.P. 605

The effect of refraction regime on the evolution of perturbation for Richtmyer-Meshkov instability
Zaytsev, S.G., Aleshin, A.N., Lazareva, E.V. 611

A numerical study of shock acceleration of a diffuse helium cylinder
Greenough, J.A., Bell, J.B., Colella, P., Puckett, E.G., Jacobs, J.W. 617

Hypervelocity Richtmyer-Meshkov instability
Samtaney, R., Meiron, D.I., Pullin, D.I. 623

Numerical investigation of Richtmyer-Meshkov instability using front tracking
Grove, J.W., Holmes, R.L., Sharp, D.H. 629

Richtmyer-Meshkov mixing-zone deformation induced by shock-tube wall boundary layers
Jourdan, G., Houas, L. 635

Vortex methods for early and intermediate time Richtmyer-Meshkov environments
Ray, J., Zabusky, N.J., Samtaney, R. 641

Visualization of shocked mixing zones using differential interferometry and X-rays
Rodriguez, G., Galametz, I., Thorembey, M.-H., Rayer, C., Haas, J.-F. 647

Interaction of a turbulent shear flow with a Mach 1.5 normal shock wave
Mahesh, K., Lele, S.K., Moin, P. 653

Waves, General

Converging shock waves: modeling and analysis of structures
Wallus, H., Demmig, F. 661

Propagation and stability of strong shocks in gases
Mond, M., Rutkevich, I., Toffin, E. 667

Blast parameters from unconfined gaseous detonations
Dorofeev, S.B., Sidorov, V.P., Dvoinishnikov, A.E. 673

Numerical simulation of the blast in the World Trade Center
Baum, J.D., Luo, H., Löhner, R. 679

Blast-wave interaction with multiple structures
Absil, L.H.J., van den Berg, A.C., Weerheijm, J. 685

Efficient 2D and 3D locally adaptive unstructured Euler solvers for unsteady shocked gas flows
Galyukov, A., Fursenko, A., Voinovich, P., Saito, T., Timofeev, E. 691

Viscous Effects

Boundary-layer transition on slender blunt cones at hypersonic speeds
Zanchetta, M.A., Hillier, R. 699

Reflected-shock/side-boundary-layer interaction in a shock tube
Nishida, M., Lee, M.G. 705

Effects of shear-layer transition and leading-edge bluntness on control-flap effectiveness
Kumar, D., Stollery, J.L. 711

Statistical simulation of nonequilibrium hypersonic flows over concave bodies
Ivanov, M.S., Gimelshein, S.F., Markelov, G.N., Gorbachev, Y.E. 717

Transition of unsteady boundary layer induced by propagating compression wave
Aoki, T., Kondoh, N., Matsuo, K., Mashimo, S. 723

Structure of multiple shock-wave/turbulent boundary-layer interactions in a rectangular duct
Sugiyama, H., Arai, T., Hongo, S., Kawase, T., Uno, N. 729

Asymptotic defect boundary layer applied to nonequilibrium laminar hypersonic air flows
Séror, S., Schall, E., Zeitoun, D., Brazier, J.P. 735

Analysis of nonequilibrium, hypersonic blunt-body flow with the local-similarity method
Klomfass, A., Müller, S., Ballmann, J. 741

Formation of standing shock waves due to evaporation from a spherical condensed phase
Onishi, Y., Fuchs, J.W. 747

Knudsen-number effects in high-speed microchannel flows
Nguyen, T.X., Oh, C.K., Saint-Martin-Tillet, X.N., Piana, J.V., Oran, E.S. 753

Studies of separated flow structure and oscillating separation shock waves in shock-wave/turbulent boundary-layer interaction regions induced by a blunt protuberance
Maekawa, S., Aso, S., Karashima, K., Sato, K. 759

A numerical study of shock-wave reflections in nonequilibrium chemically reacting flows
Aso, S., Mitomo, T., Fujiwara, T., Hayashi, M. 765

Turbulent spots in boundary layers in a free-piston shock-tunnel flow
Mee, D.J., Goyne, C.P. 771

Influence of reflected-shock/boundary-layer interaction on driver-gas contamination in high-enthalpy shock tunnels
Chue, R.S.M., Itoh, K. 777

Suppression of shock-induced separation in dense gases
Cramer, M.S., Park, S., Watson, L.T. 783

Shock waves diffracting over rigid ramps in a viscous, heat-conducting gas (argon)
Virgona, R.J., Crutchfield, W.Y., Henderson, L.F. 789

Comparison of continuum and molecular approaches for the flow around a perpendicular disk
Elizarova, T.G., Graur, I.A., Chpoun, A., Lengrand, J.C. 795

The numerical simulation of unsteady shock-wave/boundary-layer interactions in reacting gas mixtures
Weber, Y.S., Oran, E.S., Boris, J.P., Anderson, J.D. 801

Author Index 807

Keyword Index 815

Volume II

Chemical Physics

UV- and IR-absorption measurements in a shock tube using multiple reflections
Thoma, M.L., Hindelang, F.J. 827

High-temperature thermolysis of ethylene-d_4
Kern, R.D., Chen, H., Yao, J. 833

Single-pulse shock-tube study of tetraethoxysilane decomposition
Herzler, J., Manion, J.A., Tsang, W. 839

Bimolecular S-atom reactions studied in pyrolysis and photolysis shock-wave experiments
Woiki, D., Roth, P. 845

Laser-flash-photolysis study of the decomposition of vibrationally excited CO_2
Eremin, A.V., Shumova, V.V., Ziborov, V.S., Mick, H.-J., Roth, P. 881

Formation of C_2 during high-temperature pyrolysis of fullerene C_{60} in shock waves
Sommer, T., Kruse, T., Roth, P. 887

Thermal decomposition of gaseous hydrazine behind a reflected shock wave
Halat-Augier, C., Dupré, G., Paillard, C.E. 893

Shock-tube study of acetone pyrolysis
Hidaka, Y., Sato, K., Iwamoto, R., Koike, T. 899

Measurements of rotational-vibrational excitation behind incident shock waves by tunable diode laser absorption spectrometry
Klenk, W., Widdecke, N., Frohn, A. 905

Absorption lineshape measurement of atomic oxygen at 130 nm
Meyer, S., Bershader, D., Gilmore, J., Sharma, S., Exberger, R., Whiting, E. 911

Spectroscopic detection of highly excited molecular-state transformations behind shock waves and master-equation implementation
Zaslonko, I.S., Eremin, A.V., Shumova, V.V. 917

Measurement of the absorption spectrum of the OH A-X (0,0) 306-nm band-head at high pressures
Davidson, D.F., Röhrig, M., Petersen, E.L., Di Rosa, M.D., Hanson, R.K. 923

Self-ignition of n-heptane/iso-octane mixtures and n-decane with air under engine conditions
Fieweger, K., Pfahl, U., Adomeit, G. 929

Self-ignition of H_2-air mixtures at high pressure and low temperature
Blumenthal, R., Fieweger, K., Komp, K.H., Adomeit, G., Gelfand, B.E. 935

High-pressure shock-tube measurements of ignition times in stoichiometric $H_2/O_2/Ar$ mixtures
Petersen, E.L., Davidson, D.F., Röhrig, M., Hanson, R.K. 941

Soot formation from rich hydrocarbon oxidation under elevated pressure conditions
Kellerer, H., Bauer, H.-J., Wittig, S. 947

Single-pulse shock-tube study on the decomposition of C_3F_6
He, Y. 953

Direct measurements of rate coefficients for thermal decomposition of methyl halides using shock-tube ARAS technique
Takahashi, K., Inoue, A., Inomata, T. 959

Examination of the oxidation mechanism of H_2S
Tsuchiya, K., Oya, M., Matsui, H. 965

Thermal decomposition of furonitrile: studies with a single-pulse shock tube
Laskin, A., Lifshitz, A. 971

Comparison of relaxation models with shock-tube data
Stuhler, H., Frohn, A. 977

Various models of coupling of vibrational-dissociational kinetics behind shock waves
Nagnibeda, E.A., Méolans, J.G., Mouti, M., Chauvin, A.H. 983

Combustion

Optical diagnostic measurements in a supersonic-combustion ramjet
McIntyre, T.J., Houwing, A.F.P., Rabbath, P.A.B., Palma, P.C. 991

Numerical investigation of induction-zone development in hydrogen-oxygen mixtures
Figura, E.S., Cullis, I.G. 997

Shock-tube investigations of H_2 combustion in a high-temperature supersonic air flow
Smeets, G., Quenett, C. 1003

Numerical simulation of reactive shock diffraction around an object containing a cavity
Jiang, B., Ingram, D.M., Causon, D.M., Saunders, R. 1009

Modeling of explosion in explosion behind a reactive shock wave by the joint velocity-composition PDF method
Frolov, S.M., Noskov, M.A., Tatschl, R., Wolanski, P. 1015

Spontaneous ignition of a reactive gas mixture by shock-vortex interaction
Takayama, F., Kambe, T., Taki, T., Sakurai, A. 1021

Shock-tube investigations of atomization, evaporation, and ignition of n-decane and α-methylnaphthalene droplets
Pfahl, U., Fieweger, K., Adomeit, G., Gelfand, B.E. 1027

Numerical modeling of chemistry and gas dynamics during shock-induced ethylene combustion
Clifford, L., Milne, A., Boulton, D. 1033

Shock-tube combustion of liquid sprays of ethanol: effect of additives on ignition delays and combustion characteristics
Cadman, P. 1039

The effect of halocarbons on methane ignition
Burcat, A., Dvinyaninov, M., Lifshitz, A. 1045

LIPF measurements and modeling of jet ignition through an orifice between two separated rooms
Mori, S., Yoshida, T., Hayashi, A.K., Fujiwara, T. 1051

Gasdynamic limits of ignition and combustion of a gas mixture in ISL's RAMAC 30 scram accelerator
Seiler, F., Patz, G., Smeets, G., Srulijes, J. 1057

Ram-accelerator operation with simplified geometries
Sasoh, A., Knowlen, C., Higgins, A.J., Bruckner, A.P. 1063

Detonation

Triple-shock structure in dusty gas detonations
Hayashi, A.K., Fuyuto, T., Fujiwara, T. 1071

Detailed modeling of detonation in heterogeneous flows
Uphoff, U., Hänel, D., Roth, P. 1077

Detonation-tube studies on aluminum dust-air mixtures under weak ignition conditions
Pu, Y.K., Li, K.Y. 1083

Detonation propagation in a one-dimensional heterogeneous condensed-phase explosive
Lee, J.J., Brouillette, M. 1089

Transmission of a detonation across an air gap
Teodorczyk, A., Thomas, G.O., Ward, S.M. 1095

Large-scale confined hydrogen-air detonation experiments and their numerical simulation
Breitung, W., Dorofeev, S.B., Efimenko, A.A., Kochurko, A.S., Redlinger, R., Sidorov, V.P. 1101

Transition from deflagration to detonation with a dense aluminum-dust slug in a combustible gas
Zhang, F., Thibault, P. A. 1107

Effect of confinement on the propagation of a gaseous detonation in a porous medium
Makris, A., Frost, D.L., Abdul-Baki, I., Yoshinaka, A., Lee, J.H.S. 1113

Detonation initiation by hypervelocity projectiles
Bélanger, J., Kaneshige, M., Shepherd, J.E. 1119

Numerical investigation of pulsating detonations
Quirk, J.J., Short, M. 1125

Initiation of detonation by gasdynamic methods in a half-limited space
Achasov, O.V., Labuda, S.A., Penyazkov, O.G., Ragozin, D.S. 1131

Characterization of heterogeneous detonations of powders dispersed in nitrogen
Tulis, A.J., Sumida, W.K., Norikan, K., Dihu, R., Patel, D.L., Heberlein, D.C. 1137

Numerical simulation of the reflection of detonation waves
Yu, Q., Grönig, H. 1143

Analysis of detonation on a hypervelocity projectile in the in-tube chemical accelerator
Rom, J. 1149

Plasma and Radiation

Regular-shock-reflection phenomena in magnetohydrodynamics
Ogawa, H., Fujiwara, T., Hayashi, A.K. 1157

Nonlinear magnetohydrodynamic waves
Tarkenton, G.M., Cramer, M.S. 1163

Mechanisms of the formation of highly excited ArI and ArII level populations in shock-heated jets
Bazhenova, T.V., Eremin, A.V., Emelyanov, A.V., Shumova, V.V. 1169

Interaction between shock-generated plasma and electromagnetic field in a reentry vehicle
Leblanc, J.E., Fujiwara, T. 1175

Time-resolved imaging spectroscopy of nonequilibrium shock-front radiation in air
Koreeda, J., Ohama, Y., Maeno, K., Honma, H. 1181

Multi-phase Flows

Measurements of void fraction in a vapor explosion by a light-extinction technique
El-Beshbeeshy, M., Bonazza, R., Shamoun, B. 1189

A hybrid front-tracking scheme for shock wave propagation in gas/water media
Causon, D.M. 1195

Shock waves in dilute/non-dilute bubbly flows
Kameda, M., Shimaura, N., Higashino, F., Matsumoto, Y. 1201

Hydraulic shock loading due to supercavitating projectiles
Davies, G.A.O, Flanagan, O., Hillier, R., Hitchings, D., Lord, S.J. 1207

Shock waves and noise in the collapse of a cloud of cavitation bubbles
Wang, Y.-C., Brennen, C.E. 1213

Shock-tube-deduced condensation coefficient
Fujikawa, S., Kotani, M., Sato, H., Takasugi, N., Matsumoto, M. 1219

Unsteady shock formation in transonic flow with nonequilibrium phase transition
Schnerr, G.H., Adam, S. 1225

Condensation phenomena behind incident and reflected shock waves in heavy vapors at low Mach numbers
Garen, W., Becker, M., Paál, G. 1231

On the formation of embedded shock waves in rarefaction waves by homogeneous condensation
Delale, C.F., Maršík, F., Schnerr, G.H. 1237

Condensation induced by rarefaction waves and reflected rarefaction waves
Fu, Y., Han, Z., Gong, M. 1243

Condensation shock wave as the consequence of the principle of minimum entropy production
Maršík, F., Sopuch, P., Blaha, J. 1249

Shock-wave reflection around a two-dimensional wedge in low-temperature gases with phase change
Maeno, K., Nozawa, H., Nishimori, Y., Honma, H. 1255

Complex droplet deformation and breakup modes at intermediate Weber numbers from laser holography
Zerf, G., Hornung, K. 1261

Entrainment of liquid droplets into shock-induced airstreams
Archer, R.D., Milton, B.E., Behnia, M., Casey, R.T. 1267

Combined sublimation and thermal decomposition of NH_4Cl/Ar aerosols in shock waves
Schading, G.N., Roth, P. 1273

Effective sphere drag coefficient for unsteady gas-particle flows
Kurian, J., Das, H.K. 1279

Acceleration of a sphere behind planar shock waves
Britan, A., Elperin, T., Igra, O., Jiang, J.P. 1285

Dusty shock reflection and diffraction in aligned baffle systems
Wang, B.-Y., Xu, Y.-H., Xu, Y.-Q., Wu, Q.-S. 1291

Numerical analysis of wave patterns of shock propagation in a dusty gas along convergent-divergent nozzles
Wu, Q.-S., Wang, D.-Z., Xu, Y.-H., Wang, B.-Y. 1297

Dusty gas flow in a nozzle
Elperin, I., Igra, O., Ben-Dor, G., Vasiliev, E. 1303

Head-on collision of a planar shock wave with a dusty gas layer
Jiang, J.P., Elperin, T., Igra, O., Falcovitz, J. 1309

An experimental and theoretical study of shock-wave induced particle coagulation
Strecker, J.J.F., Roth, P. 1315

Experimental verification of the concept of amplification of shock loading by means of compressible covering as applied to the propagation of blast waves over a dust deposit
Medvedev, S.P., Polenov, A.N., Khomik, S.V., Gelfand, B.E. 1321

Effect of thermal expansion on the shock speeds of first, second, and fourth sound in He II
Braun, S., Kluwick, A., Cramer, M.S. 1327

Analytical prediction of regular reflection over porous surfaces
Levy, A., Li, H., Ben-Dor, G. 1333

Experimental observation of shock-wave evolution in an elastomeric foam
Zaretsky, E., Igra, O. 1339

Experimental investigation of interaction of shock wave with closed cellular compressible foam
Lagutov, Y.P., Sharov, Y.L., Gvozdeva, L.G. 1345

Blast-wave propagation in foam
Kleine, H., Diaconescu, G., Lee, J.H.S. 1351

Experimental and numerical investigation of the head-on collision of planar shock waves with rigid porous materials
Levy, A., Sorek, S., Ben-Dor, G., Skews, B.W. 1357

Shock-induced wave phenomena in open-cell permeable and flexible foams
van Dongen, M.E.H., Smeets, G.V.R., Hoeijmakers, H.W.M., Smeulders, D.M.J. 1363

Equilibrium gas dynamics and shock relations for flexible foams
van Dongen, M.E.H., van Gool, C.J., Pruiksma, J.P., Smeulders, D.M.J. 1369

Axisymmetric tests of shock-wave impact on PUR foam
Skews, B.W., Sellschop, R.M., Legoete, T. 1375

Shock impact on porous plugs with a fixed gap between the plug and a wall
Seitz, M.W., Skews, B.W. 1381

Attenuation of a blast wave with a cellular material
Makris, A., Frost, D.L., Nerenberg, J., Lee, J. H.S. 1387

Structure of deformation waves in porous compressible materials — influence of percolation
Gvozdeva, L.G., Lagutov, Y.P., Sharov, Y.L., Sherbak, N.B. 1393

Condensed Matter

Ultra-high-pressure thermal-conductivity measurements of griceite and corundum
Gallagher, K.G., Ahrens, T.J. 1401

Combustion synthesis of advanced ceramics assisted by explosive compaction
Timnat, Y.M., Rosenband, V. 1407

Dynamic compaction of granular materials in a tube with wall friction, applied to deflagration-to-detonation transition
Hill, L.G., Kapila, A.K. 1413

Investigation of HE-driven cylindrical liner implosions
Tan, T.-H., Marsh, S.P. 1419

Modeling shock-wave propagation in obliquely loaded unidirectional composites
Resnyansky, A.D., Romensky, E.I., Belov, V.Y., Williams, G.H. 1425

Shock consolidation of powders by utilizing a converged underwater shock-wave assembly
Kubota, S., Nagano, S., Fujita, M., Itoh, S. 1431

Thermodynamic relations, Hugoniot adiabatics, and stability of strong shocks in solids
Rutkevich, I., Zaretsky, E., Mond, M. 1437

Numerical simulation of shock-wave generation with KALIF
Vorobiev, O.Y., Fortov, V.E., Goel, B. 1443

Effects of solid-solid phase changes on impact mechanics
Knowles, J.K., Winfree, N.A., Ahrens, T.J. 1449

Comet Shoemaker-Levy-9 impact on Jupiter
Ahrens, T.J., Takata, T., O'Keefe, J.D. 1455

Impact jetting of geological materials
Yang, W., Ahrens, T.J. 1461

Numerical computation of jetting impacts
Puckett, E.G., Miller, G.H. 1467

Compositional changes of the Moon and Earth by impacts
Miura, Y., Okamoto, M. 1473

Technology and Diagnostics

Experimental investigation of tunnel sonic-boom suppression
Sasoh, A., Onodera, O., Takayama, K., Kaneko, R., Matsui, Y. 1481

Discarding the tuned-piston conditions for Stalker tubes: possible gains and drawbacks
Dumitrescu, M.-P. 1487

Compact entropy-raising driver for impulse facilities
Kendall, M.A., Morgan, R.G., Jacobs, P.A. 1491

Temperature distribution in free-piston compressors
Griffin, D.K, Stalker, R.J., Page, N.W. 1497

Soft-landing operation for free-piston driver by adjusting piston-stopper height
Tanno, H., Itoh, K., Komuro, T., Sato, K. 1503

Preliminary experimental results of the detonation-driven shock tube
Yu, H.-R., Zhao, W., Li, Z.-F., Gu, J.-H., Zhang, X.-Y. 1509

Operational characteristics of a free-piston shock tunnel with a light-gas gun as the shock tube
Gou, G.X., Gao, D.Z. 1515

Characteristics of shock waves in exhaust systems and exhaust gas flows of automobile engines
Sekine, N., Onodera, O., Takayama, K. 1521

Influence of the curvature of the diaphragm on flow quality in shock tubes
Daru, V., Damion, J.P. 1527

Modifications to the AEDC shock-tunnel code and comparisons with data
Maus, J.R., McKenzie, N.R. 1533

Numerical simulation of diaphragm rupture
Petrie, P.J., Jacobs, P.A. 1539

Starting-phase numerical simulation of the TCM2 wind-tunnel nozzle
Cardoso, M., Burtschell, Y., Druguet, M.C., Zeitoun, D., Abgrall, R. 1545

Combustion driver with interface compression and computer code for hypersonic shock tunnel to produce stagnation temperatures to 20,000 K and Mach numbers of 35
Nagamatsu, H.T., Messitt, D.G., Myrabo, L.N., Minucci, M.A.S. 1551

A simple shock-tunnel driver-gas detector
Paull, A. 1557

Boundary-layer blockage in expansion-tube nozzles
Sudnitsin, O., Morgan, R.G. 1563

Throat design for hypersonic nozzles
Dumitrescu, L.Z., Burtschell, Y., Zeitoun, D. 1569

Latest calibration results of the HEG contoured nozzle
Hannemann, K., Krek, R., Eitelberg, G. 1575

Trapping the boundary layer: a method to diminish flow contamination in shock tunnels
Dumitrescu, M.-P. 1581

High-enthalpy nozzle flow sensitivity study and effects on heat transfer
Walpot, L.M.G., Muylaert, J., Bakker, P.G., Simeonides, G. 1587

Wave processes in a pressure-wave refrigerator
Galyukov, A., Voinovich, P., Timofeev, E. 1593

Quantitative holographic interferometry of shock-wave flows using Fourier transform fringe analysis
Babinsky, H., Takayama, K. 1599

Single-pulse spontaneous Raman-scattering measurements in an expanding nitrogen/oxygen admixture
Gilmore, J.O., Sharma, S., Fletcher, D., Bershader, D. 1605

Laser-induced thermal acoustics (LITA) measurement of gas properties
Cummings, E.B., Hornung, H.G., Brown, M.S. 1611

PLIF imaging of nitric oxide in high enthalpy flows
Palma, P.C., McIntyre, T.J., Houwing, A.F.P. 1617

A skin-friction gauge for impulsive facilities
Goyne, C.P., Paull, A., Stalker, R.J. 1623

Imaging and quantitative measurement of transverse injection in supersonic flow using LIF
Shida, H., Nagayama, T., Deguchi, Y. 1629

On the role of laser absorption in planar laser-induced fluorescence imaging of high-enthalpy flows
Ruyten, W.M., Smith, M.S., Williams, W.D. 1635

High-speed image processing of interferograms in experiments on weak shock waves
Kurokui, Y., Aoyama, T., Watanabe, N., Maeno, K., Honma, H. 1641

Medical Applications

Experimental and mathematical model of living-tissue cell damage induced by plane shock waves
Tamagawa, M., Akamatsu, T., Yoshida, K. 1649

Application of extracorporeal shock wave on bone
Ikeda, K., Nakatani, A., Tomita, K., Takayama, K. 1655

Author Index 1661

Keyword Index 1669

Volume II

Chemical Physics

UV- and IR-absorption measurements in a shock tube using multiple reflections

M. L. Thoma, F. J. Hindelang
Universität der Bundeswehr München
D-85577 Neubiberg, Germany

Abstract: Various experiments concerning the thermal decomposition of atmospheric trace gases (O_3, NO) in a shock tube have been carried out using a multiple-reflection cell (MRC) previously designed and described by the authors (Thoma et al. 1994). In the experiments either 8 or 16 inner beam reflections were used in order to increase the absorption length initially given by the tube diameter up to a factor of 17.3.

Key words: Ozone, Nitric oxide, Decomposition, Multipass technique

1. UV-laser absorption experiments with ozone

The thermal decomposition of ozone has recently been studied by the authors using a frequency-doubled ion laser at $\lambda = 257.25\,\mathrm{nm}$ (Thoma and Hindelang 1993). In numerous subsequent experiments based on the same experimental equipment and the use of the multiple-reflection cell (MRC), the characteristics of this MRC were tested in detail. Due to the increased absorption length (and the additional use of a reference beam in order to minimize laser fluctuations) the detection limit could be improved considerably.

1.1. Two-beam experiments

As shown in Fig. 1, two laser beams (one represents the direct transit, the other is reflected 8 times) have the same longitudinal shock-tube coordinate while being different in absorption length by a factor of 8.63. They are used to simultaneously measure the thermal decomposition of ozone behind an incident shock wave; see Thoma et al. (1994) for experimental details. A direct comparison of two simultaneously detected absorption signals leads to the following results: The signal with 8 reflections shows significantly less noise (see Fig. 2). The measured initial ozone concentrations are the same within an uncertainty of 1–2%. The experiments are in very good agreement with numerical calculations. Due to the use of fixed flat metal mirrors[1] the MRC shows an excellent mechanical ruggedness and the results are reproducible.

[1]These mirrors were constructed at the mechanical workshop of the authors' facility.

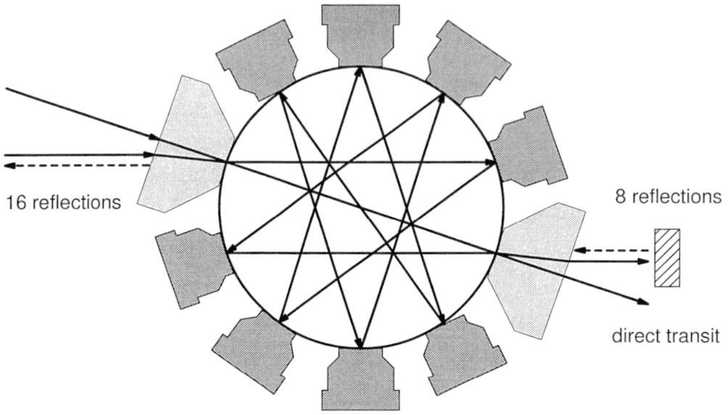

Figure 1. Different optical beam paths (10 cm, 86.3 cm, 172.6 cm) depending on the number of reflections inside the MRC. In order to obtain 16 reflections, the transmitted beam is reflected by a concave mirror outside the tube.

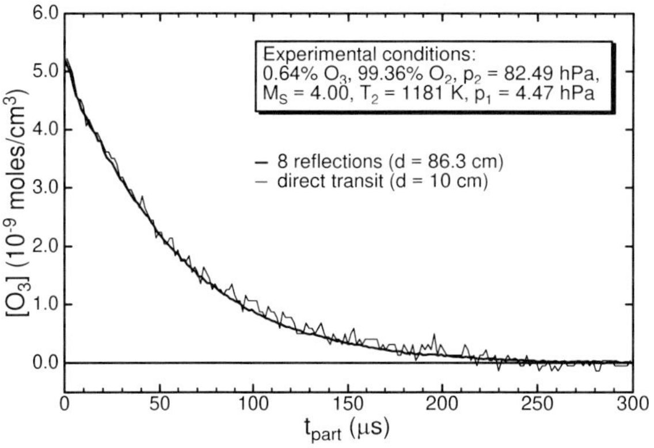

Figure 2. Thermal decomposition of O_3 behind an incident shock wave obtained by two simultaneously detected absorption signals having different absorption lengths d.

1.2. Reference-beam experiments

In order to maximize the signal-to-noise ratio several measurements have been carried out using a reference beam while the main beam has been reflected 8 or 16 times inside the MRC.

Figure 3 shows two typical results demonstrating the lowest detectable ozone concentration: about 0.01 Pa ozone partial pressure at room temperature (wavelength 257.25 nm).

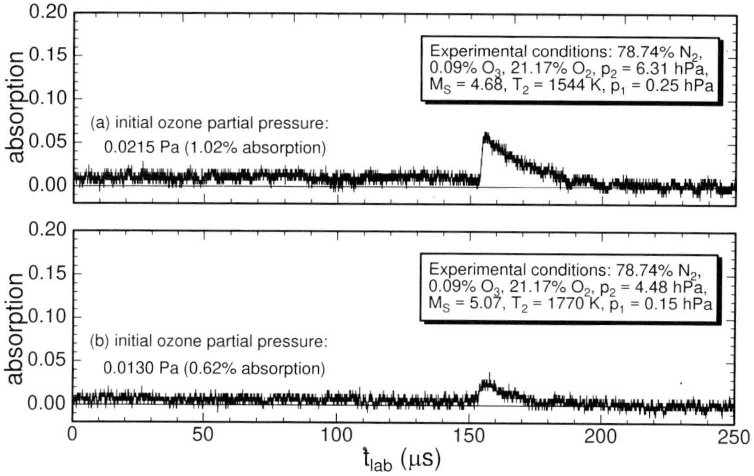

Figure 3. Signals of two different reference-beam measurements in a shock tube (inner diameter 10 cm) using 16 reflections inside the MRC (absorption length 172.6 cm).

1.3. Wall adsorption – boundary-layer transition

In more than 30 of our experiments a second absorption peak could be recorded — at a distance t_t with respect to the arrival of the shock front — after having observed a complete decomposition of the ozone concentration (see Fig. 4). This signal peak obviously is due to a certain amount of ozone which has initially been adsorbed at the cold shock-tube walls (maybe also at the mirrors' surfaces) and "evaporates" when the boundary layer becomes turbulent; it can be obtained by both beams of a typical two-beam experiment as shown in Fig. 4.

According to Mirels (1958), t_t can be used to define a transition Reynolds number Re_t, which describes a boundary-layer transition (laminar $\rightarrow$ turbulent):

$$Re_t = \frac{\rho_2}{\mu(T_2)} \cdot u_2^2 \cdot \frac{v_S}{v_S - u_2} \cdot t_t \tag{1}$$

where ρ_2, $\mu(T_2)$, and u_2 denote the density, viscosity, and the flow velocity relative to the wall (immediately behind the shock); v_S is the shock velocity in laboratory-fixed coordinates. In a Mach-number range from 3.9 to 6.3 a characteristic pressure dependence $Re_t(p_1)$ can be obtained. In Fig. 5 it is compared to experimental results of Fujii et al. (1979) obtained in a 5-cm inner diameter shock tube with test gases mainly consisting of argon.

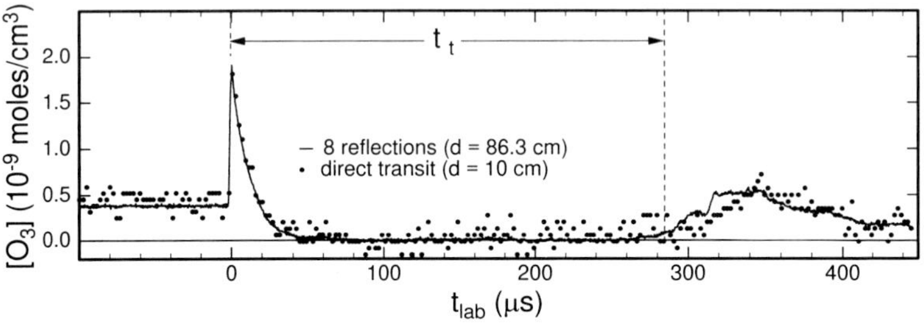

Figure 4. Boundary-layer transition in a two-beam experiment. After the transition at t_t: initially wall-adsorbed ozone molecules occur in the paths of the two laser beams. Experimental conditions: 2.3% O_3, 97.7% O_2, $p_2 = 10.6\,\mathrm{hPa}$, $M_S = 4.77$, $T_2 = 1564\,\mathrm{K}$, and $p_1 = 0.40\,\mathrm{hPa}$.

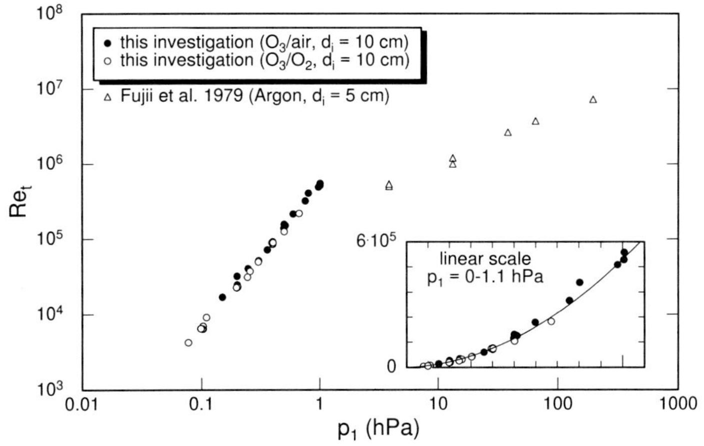

Figure 5. Transition Reynolds number Re_t as a function of the initial pressure p_1. Values from this investigation (○, •) can be well described by the parabolic fit function: $Re_t\,(p_1) = 3.34 \cdot 10^4 \cdot p_1 + 4.82 \cdot 10^5 \cdot p_1^2$; d_i is the inner diameter of the tube.

2. IR-laser absorption experiments with nitric oxide

An infrared tunable diode laser has been used in order to to scan vibration-rotation transitions at the fundamental bands of NO ($\lambda \approx 5.2\,\mu\mathrm{m}$). A reference signal is produced by means of a multi-coated CaF_2 beam splitter having a reflection of about 10%. The signal beam itself passes the MRC via 8 gold-coated mirrors (reflectivity > 99.3%, substrates: stainless steel) and two CaF_2 windows, then traverses a Czerny-Turner monochromator which ensures single-mode operation and minimizes the detection of background emission caused by the hot flow in the shock tube. Finally, the beam is focused onto a liquid-nitrogen-cooled InSb detector. The reference beam also passes a monochromator before being focused onto another detector of the same type. In Fig. 6 a

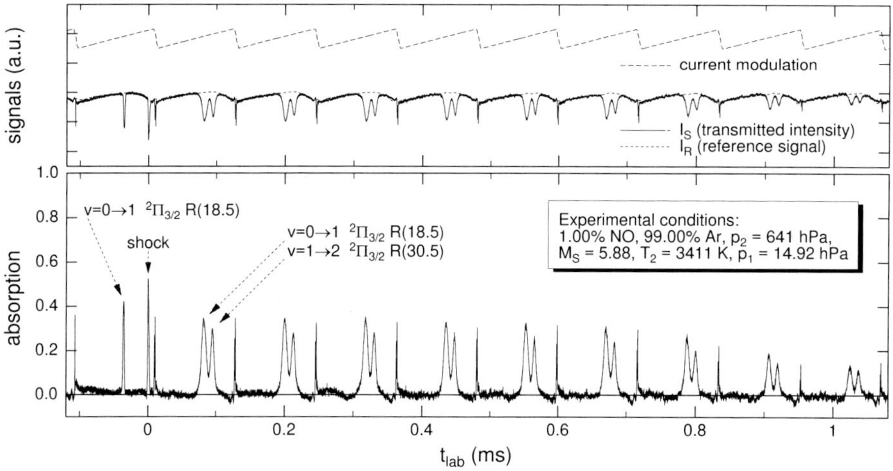

Figure 6. Thermal decomposition of NO behind an incident shock wave, detected with a tunable diode laser using 8 reflections inside of the MRC (optical path length 86.3 cm). After the arrival of the shock front (schlieren peak), a second vibration-rotation transition can be detected

typical result is presented, demonstrating the suitability of the MRC for IR-absorption measurements behind incident shock waves[2]: the reference signal I_R is smoothed numerically, then the absorption can be determined via the relation $A = 1 - (I_S - I_{S0})/(I_R - I_{S0})$ where I_{S0} is the signal intensity's zero level. The spectral positions of the two vibration-rotation transitions agree with those given by Amiot et al. (1978)[3]. Rate constants of the two reactions

$$NO + NO \rightarrow N_2O + O \tag{2}$$

$$NO + O \rightarrow O_2 + N \tag{3}$$

which both contribute with comparable sensitivity to the thermal NO decomposition have been investigated within the temperature range 3300 – 4200 K. Most of the results can be described using the rate constants given by Thielen and Roth (1984) for reaction 2 and those of Hanson and Salimian (1984) for reaction 3. In both cases the rate constants of this investigation scatter between these values and values about 30% higher.

[2]Diode laser parameters: temperature 50 K, current saw-tooth frequency 8.5 kHz.

[3]There, $v = 0 \rightarrow 1\,^2\Pi_{3/2}$R(18.5): 1935.4952 cm^{-1}, $v = 1 \rightarrow 2\,^2\Pi_{3/2}$R(30.5): 1935.5408 cm^{-1}.

3. Summary

The optical detection limit in a shock tube ($d_i = 10\,\mathrm{cm}$) was successfully improved by means of multiple-beam reflections:

(1) Experiments with initial ozone partial pressures even down to ~0.01 Pa (100 nbar) could be performed; as the maximum partial pressure in the ozone layer at ~ 20 – 25-km altitude is about 0.02 Pa (200 nbar), UV-absorption shock-tube experiments with atmospheric O_3/air composition are now possible. The rate constants obtained ($O_3 + M \rightarrow O + O_2 + M$, temperature range 1100 – 2700 K) agree with previously determined values (Thoma and Hindelang 1993).

(2) The optical IR arrangement described above (two monochromators, multiple reflections) allowed us to improve the quality and the detection limit of tunable diode-laser diagnostics in shock tubes. Recommended values for the rate constants of the two reactions (2) and (3) could be verified.

Acknowledgement. The authors thank W. C. Gardiner Jr., whose "beamaverage" subroutine has been very useful to check the time resolutions of the different laser beams used in the experiments.

References

Amiot C, Bacis R, Guelachvili G (1978) Infrared study of the $X^2\Pi$ $v = 0$, 1, 2 levels of $^{14}N^{16}O$, preliminary results on the $v = 0$, 1 levels of $^{14}N^{17}O$, $^{14}N^{18}O$, and $^{15}N^{16}O$. Canadian Journal of Physics 56:251–265.

Fujii N, Konshi M, Ando H, Asaba T (1979) Evaluation of boundary-layer effects in shock-tube studies of chemical kinetics. Int J Chem Kin 11:285–304.

Hanson RK, Salimian S (1984) Survey of rate constants in the N/H/O system. In: Combustion Chemistry (ed. W.C. Gardiner Jr.), Springer, New York.

Mirels H (1958) The wall boundary layer behind a moving shock wave. In: Proc of the Symposium on Boundary Layer Research (ed. H. Görtler), Freiburg, Germany, Springer, Berlin.

Thielen K, Roth P (1984) Resonance absorption measurements of N and O atoms in high temperature NO dissociation and formation kinetics. In: 20th Symposium (International) on Combustion, pp 685–693.

Thoma ML, Hindelang FJ (1993) Experiments on shock-heated ozone dissociation in oxygen/air using UV laser spectroscopy. In: Sock Waves @ Marseille (eds. Brun R, Dumitrescu L), Springer, Berlin.

Thoma ML, Kaschow R, Hindelang FJ (1994) A multiple-reflection cell suited for absorption measurements in shock tubes. Shock Waves 4:51–53.

High-temperature thermolysis of ethylene-d_4

R. D. Kern, H. Chen, J. Yao
Department of Chemistry
University of New Orleans
New Orleans, Louisiana 70148 USA

Abstract: A total of nine mixtures of various compositions of ethylene in a neon-1% argon diluent were investigated behind reflected shock waves over the temperature and total pressure ranges of 2020–2358 K and 0.26–0.47 atm, respectively. Both fully deuterated and non-deuterated ethylene were employed as the reactant. Reactant and product profiles were obtained by dynamic analysis of the reflected shock zone by time-of-flight (TOF) mass spectrometry. The product profiles for acetylene and diacetylene were modeled using a 21-step mechanism. Attempts to acquire TOF data for the H_2 profiles were not successful. Profiles for C_2D_4, C_2D_2, C_4D_2, and D_2 were modeled using the same mechanism adjusted only for the different molecular weights and thermochemical values of the major species. The overall isotope effect for the thermolysis of ethylene was determined to be minimal.

Key words: Thermal decomposition, Deuterated ethylene, Non-deuterated ethylene, High temperature, Shock tube, Time-of-flight mass spectrometry

1. Introduction

One of the goals of experimental chemical kinetics is to record simultaneously the concentrations of all species from H atoms to high-molecular-weight products that appear during the observation time of the reaction. A shock tube coupled to a time-of-flight (TOF) mass spectrometer has the potential to achieve this goal. However, there are difficulties involved in obtaining comparable mass spectral sensitivities for low- and high-mass species. In order to assess the TOF technique with regard to the analysis of "low-" and "high-" mass compounds, a relatively well-known system, the thermal decomposition of ethylene, was selected. Rate constants for the decomposition over the temperature and total pressure range suitable for TOF study are available from a previous shock-tube investigation (Kiefer et al. 1983). The products of the decomposition are known to be C_2H_2, H_2, and C_4H_2; the latter two define the "low" and "high" m/e products for this particular reaction system. In addition, the thermolysis of C_2D_4 provides the opportunity to examine any overall isotope effect and to assess the product range of m/e 4 to m/e 52.

2. Experimental

Nine reacting mixtures were prepared. Their compositions and experimental conditions are listed in Table 1. The gases used are: C_2H_4 (Matheson, 99.5%), C_2D_4 (Cambridge Isotope Laboratory, 98%), 99% Ne-1% Ar (Air Products, 99.999%). In addition, two calibration mixtures were prepared: 2% C_2H_2 (Matheson, 99.6%) and 2% D_2 (Aldrich Chemicals, 99.8%). All mixtures contained a diluent consisting of 99% Ne-1% Ar. The initial pressure P_1 was 5.6 torr for mixtures A–H and 3.7 torr for Mixture I. The mass spectral sensitivity factors for C_2H_4, C_2D_4, C_2H_2, and D_2 were determined by shocking the appropriate mixtures at non-reacting temperatures. The factor for deuterium was derived by measuring the temporal-peak-height ratios m/e 20 : m/e 4 over the temperature range 1600–2200 K. The factor is reasonably constant at these elevated temperatures but remarkably different from the value measured at room temperature. Attempts to determine a mass spectral factor for H_2 were frustrated by the relatively low sensitivity of the TOF to m/e 2. The mass spectral factor for C_2D_2 is assumed to be the same as that determined for C_2H_2. This information is used to construct the various concentration-reaction time profiles for the key species.

Table 1. Summary of experimental conditions.

Mixture	Compositions	T_5 (K)	P_5 (atm)
A	0.75% C_2H_4	2020 – 2300	0.33 – 0.38
B	1.0% C_2H_4	2030 – 2300	0.34 – 0.40
C	1.5% C_2H_4	2100 – 2350	0.35 – 0.47
D	2.0% C_2H_4	2100 – 2350	0.39 – 0.47
E	0.75% C_2D_4	2160 – 2358	0.37 – 0.42
F	1.0% C_2D_4	2020 – 2300	0.34 – 0.40
G	1.5% C_2D_4	2070 – 2300	0.36 – 0.42
H	2.0% C_2D_4	2110 – 2450	0.39 – 0.47
I	2.0% C_2D_4	2116 – 2355	0.26 – 0.30

The apparatus and experimental procedures have been discussed previously (Kern and Xie 1991) as has the electronic digitization of the mass spectral peak heights (Kern et al. 1995). A recent improvement to the data-processing procedure is the development of software written in Turbo-Pascal. The program provides several options for displaying interesting parts of the mass spectra and allows the automatic reading of the m/e values and their respective amplitudes for all mass peaks of interest appearing during a given experiment. Computer files are then created to input the concentration-reaction time data into the modeling programs (Kee et al. 1989).

3. Results and discussion

Representative reaction profiles are displayed in Fig. 1. Solid lines represent modeling calculations based on the ethylene mechanism (Kiefer et al. 1983) and an abbreviated acetylene mechanism suitable for the temperature and pressure range herein derived from a comprehensive set of rate constants for acetylene pyrolysis (Kiefer et al. 1992). Ethylene pyrolysis has been studied by a variety of shock-tube techniques including single-pulse shock-tube analysis, TOF mass spectrometry, IR emission (product analysis), atomic resonance absorption spectroscopy, and laser schlieren densitometry. One of the initiating steps, reaction (1), has undergone extensive theoretical examination. References to these efforts are found in a recent review (Kern and Xie 1991). The two initiation reactions are pressure dependent in the range investigated herein.

$$C_2H_4 + M \rightarrow C_2H_2 + H_2 + M \tag{1}$$

$$C_2H_4 + M \rightarrow C_2H_3 + H + M \tag{2}$$

Rate constant k_1 is 233 times greater (at 1700 K) and 40 times greater (at 2200 K) than k_2. The decomposition occurs primarily via 1,1 elimination to singlet vinylidene followed by rapid isomerization to acetylene.

The thermal data for C_2D_4, C_2D_2, and D_2 are from a recent tabulation (Burcat and McBride 1994). The thermal properties of C_4D_2 were calculated using the group-additivity method. The major products of the reaction are acetylene and molecular hydrogen; diacetylene appears only at higher temperatures. Modeling reveals that the concentrations of H (or D) atoms are below the TOF detectability limit ($< 10^{-9}$ mol/cm^3). Attempts to record reliable profiles for H_2 were not successful despite extensive adjustment of the TOF ion source and drift-tube magnets. It was possible to obtain profiles for D_2 using C_2D_4 as the initial reactant. However, the most successful strategy is to employ different magnet settings for high and low masses.

A pM analysis (Gardiner 1977) for ethylene and acetylene is shown in Fig. 2 for experiments at selected low, average, and high temperatures for the mixtures

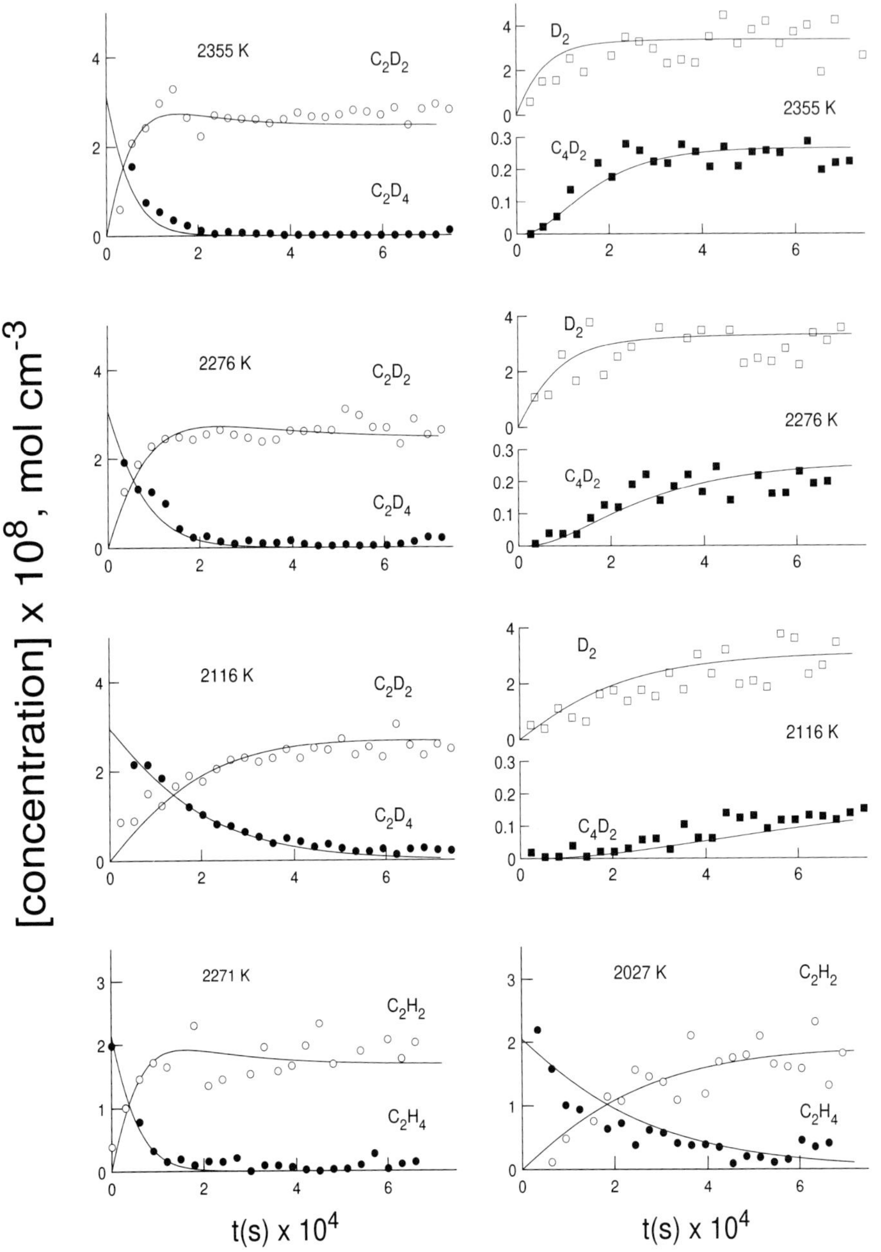

Figure 1. Reaction profiles of key species for ethylene decomposition. Lowermost pair at 2271 K and 2027 K represent runs from mixture B; all other profiles from I.

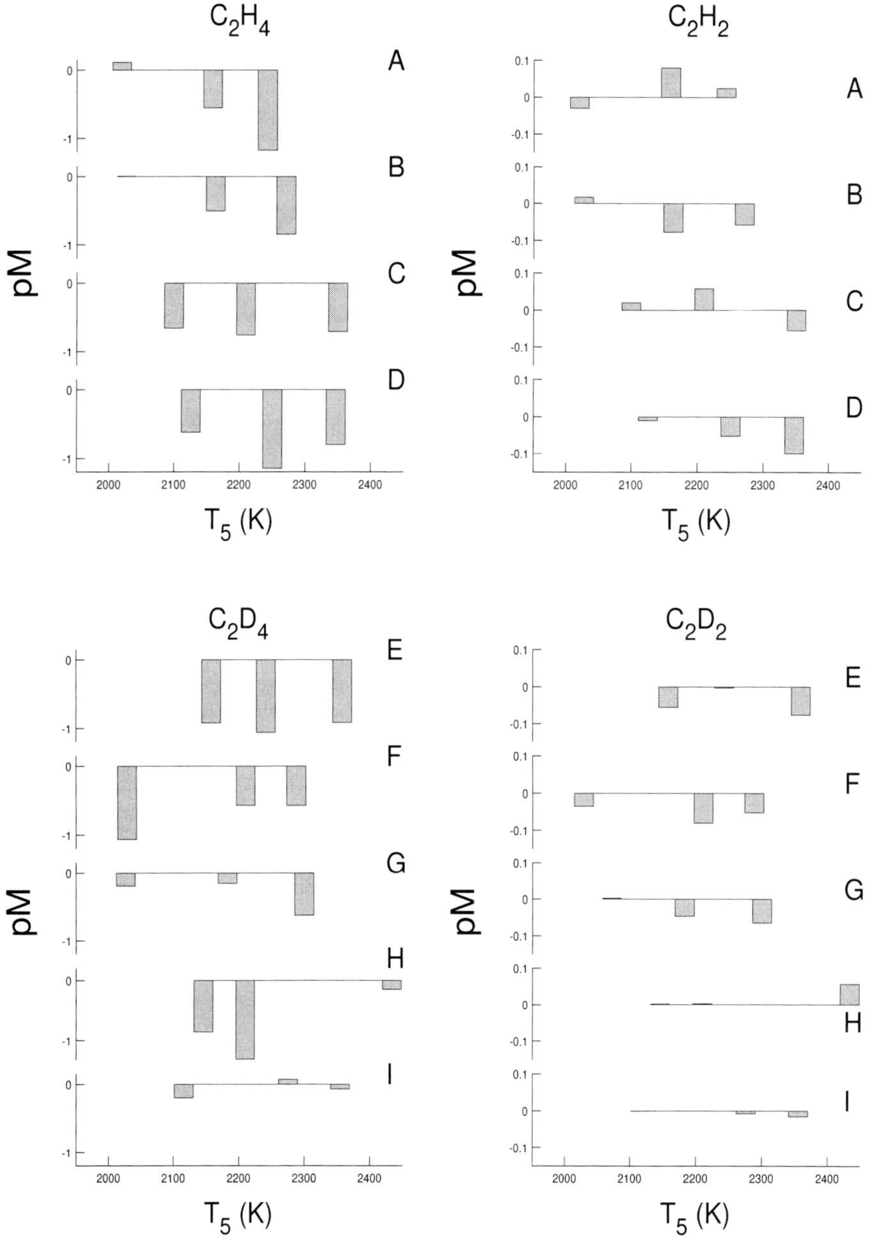

Figure 2. The pM analysis results for ethylene and acetylene in mixtures A–I.

listed in Table 1. The results are indicative of reasonable agreement between the TOF data and the modeling predictions. Preliminary analysis of the C_2D_4 data at lower temperatures revealed that the C_2D_2 profiles constructed by C_2D_4 disappearance, $[C_2D_2]_t = [C_2D_4]_0 - [C_2D_4]_t$, were consistent with C_2D_2 profiles constructed with the TOF data employing the same sensitivity factor for C_2D_2 as for C_2H_2.

The carbon-atom balance achieved in all experiments is close to 100% which indicates low soot yields for ethylene pyrolysis under these reaction conditions. The fact that reasonable fits are obtained for the C_2D_4 experiments using rate constants derived from C_2H_4 decomposition indicates that the overall isotope effect is minimal. We conclude that it is possible but not routine to record reaction profiles for m/e 4. However, reliable profiles for m/e 1 and m/e 2 are not yet attainable.

Acknowledgement. Support of this work is provided by the Division of Chemical Sciences, Office of Basic Energy Sciences, Office of Energy Research, U.S. Department of Energy under contract DE-FG0585ER/13400.

References

Burcat A, McBride B (1994) Ideal Gas Thermodynamic Data for Combustion and Air-Pollution Use, Technion-Israel Inst Tech Report TAE 697.

Gardiner WC (1977) The pC, pR, pP, pM and pS method of formulating the results of computer modeling studies of chemical reactions. J Phys Chem 81:2367–2371.

Kee RJ, Rupley FM, Miller JA (1989) Chemkin-II: A Fortran Chemical Kinetics Package for the Analysis of Gas-Phase Chemical Kinetics, Sandia National Laboratory Report No. SAND 89-8009.

Kern RD, Xie K (1991) Shock tube studies of gas phase reactions preceding the soot formation process. Prog Ener Combust Sci 17:191–210.

Kern RD, Chen H, Kiefer JH, Mudipalli PS (1995) Thermal decomposition of propargyl bromide and the subsequent formation of benzene. Combust Flame 100:177–184.

Kiefer JH, Kapsalis SA, Al-Alami MZ, Budach KA (1983) The very high temperature pyrolysis of ethylene and the subsequent reaction of the product acetylene. Combust Flame 51:79-93.

Kiefer JH, Sidhu SS, Kern RD, Xie K, Chen H, Harding LB (1992) The homogeneous pyrolysis of acetylene II: the high temperature radical chain mechanism. Combust Sci and Tech 82:101-130.

Single-pulse shock-tube study of tetraethoxysilane decomposition

J. Herzler, J. A. Manion, W. Tsang
National Institute of Standards and Technology
Gaithersburg, MD 20899 USA

Abstract: Tetraethoxysilane has been decomposed in single-pulse shock-tube experiments. The rate expression for its decomposition process is 6.7E16 $\times$ $\exp(-38970/T)$ s^{-1} over the temperature range of 1160 – 1285 K and pressures of about 1.5 bar. The main products are ethene and ethanol. By comparison with decomposition experiments with tetrapropoxysilane, dimethyldiethoxysilane, and trimethylethoxysilane, we could identify the initial reaction steps. The main reaction channel is the elimination of ethene, followed by ethanol elimination of the formed silanol.

Key words: Single-pulse shock tube, Comparative rate measurements, Tetraethoxysilane decomposition

1. Introduction

Thermal chemical vapor deposition (CVD) of TEOS (tetraethoxysilane) is widely used in the semiconductor industry for silicone-dioxide thin-film preparation in integrated-circuit fabrication (Takeuchi et al. 1994). As device dimensions shrink into the submicron region, material requirements and deposition conditions become more stringent and CVD processes must be understood at an increasingly detailed mechanistic level. Because of the close coupling of gas-phase and surface reactions under normal atmospheric CVD conditions, the kinetics and the mechanisms of the gas-phase reactions must be fully understood. The single-pulse shock-tube method used in this study is a very suitable technique for determining the mechanisms and rate constants for the decomposition of large organic molecules like TEOS. This is due to the short reaction time and through gas chromatography the possibility of working at very low concentrations of reactants. Surface and chain reactions can be completely eliminated. True thermal unimolecular rate constants are being measured. Use of an internal standard to determine reaction temperature allows determination of extremely accurate rate expressions.

2. Experimental section

Our experiments were carried out in a heated single-pulse shock tube designed to permit us to study relatively nonvolatile compounds. The shock tube and the associated gas handling and preparation system were all maintained at 108°C. Analysis of the reactants and products was by gas chromatography using a wide-bore, 30-m carbowax capillary column in the programmed temperature mode for the determination of the heavier components and a 2-m × 1-mm i.d. column packed with Hayesep S[1] combined with a 1-m × 1-mm i.d. column packed with Hayesep N for the lighter species. Table 1 contains a listing of the mixtures used in the present study.

Mesitylene is used as the chemical inhibitor. It inhibits chain processes through the reactions:

$$\begin{aligned} H^* + (CH_3)_3C_6H_3 &\Longrightarrow CH_3^* + (CH_3)_2C_6H_4 \\ &\Longrightarrow H_2 + (CH_3)_2C_6H_3CH_2^* \\ CH_3^* + (CH_3)_3C_6H_3 &\Longrightarrow CH_4 + (CH_3)_2C_6H_3CH_2^* \end{aligned}$$

Table 1. Mixtures used in the present study: 4-m.c.h. is 4-methylcyclohexene; 1,3,5-t.m.b. is mesitylene (1,3,5-trimethylbenzene); TMOS is tetramethoxysilane; tetralin is 1,2-dihydronaphtaline; TPOS is tetrapropoxysilane; TMEOS is trimethylethoxysilane; DMDEOS is dimethyldiethoxysilane; and TMMEOS is trimethylmethoxysilane.

140 ppm TEOS	140 ppm 4-m.c.h.	1% 1,3,5-t.m.b.	
270 ppm TEOS	140 ppm 4-m.c.h.	2% 1,3,5-t.m.b.	
420 ppm TEOS	140 ppm 4-m.c.h.	1% 1,3,5-t.m.b.	0.5% TMOS
420 ppm TEOS	140 ppm 4-m.c.h.	1% 1,3,5-t.m.b.	
210 ppm TEOS	70 ppm 4-m.c.h.	0.5% 1,3,5-t.m.b.	
420 ppm TEOS	140 ppm t-butanol	1% 1,3,5-t.m.b.	
140 ppm TEOS	140 ppm 4-m.c.h.	1% 1,3,5-t.m.b.	700 ppm TMOS
200 ppm TMOS	140 ppm tetralin	1% 1,3,5-t.m.b.	
140 ppm TPOS	140 ppm 4-m.c.h.	1% 1,3,5-t.m.b.	
420 ppm TMEOS	140 ppm t-butanol	1% 1,3,5-t.m.b.	
380 ppm DMDEOS	140 ppm t-butanol	1% 1,3,5-t.m.b.	
420 ppm TMMEOS	140 ppm t-butanol	1% 1,3,5-t.m.b.	

We used the well-known decomposition of 4-methylcyclohexene to propene and butadiene as the internal temperature standard, $k = 2{\cdot}10^{15}{\times}\exp(-33500/T)$

[1]Certain commercial materials and equipment are identified in this paper in order to specify adequately the experimental procedure. In no case does such identification imply recommendation or endorsement by the National Institute of Standards and Technology, nor does it imply that the material or equipment is necessarily the best available for the purpose.

s^{-1} (Tsang 1981). To get an accurate temperature determination at temperatures above 1225 K, t-butanol was used as the temperature standard. For its decomposition to isobutene and water, the rate coefficient of Newman et al. (1979) corrected by a factor of k (cyclohexene decomposition [Tsang 1981]) / k (cyclohexene decomposition [Newman et al. 1979]) was used. The temperatures were calculated considering the second decomposition channel of t-butanol (t-butanol $\Longrightarrow CH_3^* + (CH_3)_2C^*OH \Longrightarrow CH_3^* + H^*$ + acetone).

From the measured t-butanol, isobutene, and acetone concentrations, the average reaction temperature can then be calculated by the relation

$$\ln\left(\frac{[\text{t-butanol}]_i}{[\text{t-butanol}]_f}\right) = 6.0 \times 10^{14} \times \exp(-33640/T) \times t \times \left(\frac{1 + [\text{acetone}]_f}{[\text{isobutene}]_f}\right)$$

where the subscripts i and f refer to final and initial concentrations; t is the residence time of 500 μs, and T is the reaction temperature.

The TEOS decomposition in the presence of tetramethoxysilane (TMOS) and the decomposition of tetrapropoxysilane (TPOS), trimethylethoxysilane (TMEOS), and dimethyldiethoxysilane (DMDEOS) were studied to get a better insight into the mechanism of TEOS decomposition.

The t-butanol, mesitylene, and TEOS were purchased from Aldrich Chemicals, 4-methylcyclohexene from Wiley Organics, while TMOS, TPOS, TMEOS, trimethylmethoxysilane, and DMDEOS were from United Chemical Technologies. Except for vigorous degassing and removing of ethanol from the ethoxysilanes by a 0.2-nm molecular sieve, they have been used without further purification. The argon from Matheson was of Ultrapure Grade.

3. Results

We found the following rate coefficients and products for the decomposition of the ethoxysilanes and of tetrapropoxysilane (Table 2, Fig. 1).

Table 2. Rate coefficients and products of the decomposition of some ethoxysilanes and of TPOS.

Substance	A [s^{-1}]	E_a/T	Main Products
TEOS	6.7E16	38970	ethene, ethanol
DMDEOS	3.5E16	38980	ethene, ethanol
TMEOS	2.8E17	42440	ethene
TPOS	8.2E16	38910	ethene, propene, propanol

TEOS decomposition: The ratios of ethene and ethanol relative to consumed TEOS increase with rising temperature. The product distribution and the rate coefficient are independent of the initial TEOS concentration. *TMOS decomposition:* We found very good agreement with results of Chu et al. (1995) that TMOS decomposes about a factor of 100 slower than TEOS. *TEOS decomposition in the presence of TMOS:* The presence of a surplus of TMOS influences neither the rate coefficient nor the product distribution of TEOS decomposition. The ratio of formed ethene to consumed TEOS and the ratio of formed ethanol to consumed TEOS remain the same. *TPOS decomposition:* The ratio of ethene, propene, and propanol relative to consumed TPOS increases with increasing temperature. The ratio of propene to ethene is about 3:1. *TMEOS decomposition:* The ratio of ethene relative to consumed TMEOS slightly decreases with increasing temperature (0.9 → 0.8). The ethanol production is negligible.

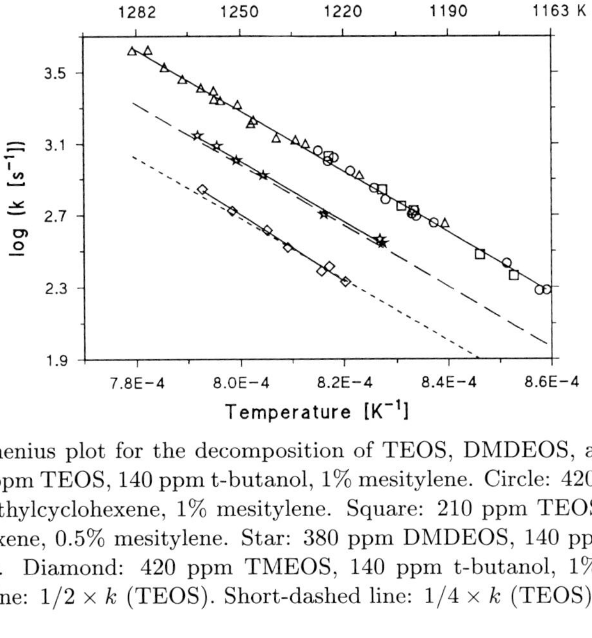

Figure 1. Arrhenius plot for the decomposition of TEOS, DMDEOS, and TMEOS. Triangle: 420 ppm TEOS, 140 ppm t-butanol, 1% mesitylene. Circle: 420 ppm TEOS, 140 ppm 4-methylcyclohexene, 1% mesitylene. Square: 210 ppm TEOS, 70 ppm 4-methylcyclohexene, 0.5% mesitylene. Star: 380 ppm DMDEOS, 140 ppm t-butanol, 1% mesitylene. Diamond: 420 ppm TMEOS, 140 ppm t-butanol, 1% mesitylene. Long-dashed line: $1/2 \times k$ (TEOS). Short-dashed line: $1/4 \times k$ (TEOS).

4. Discussion

The initial step of TEOS decomposition can be a simple bond fission R1, a four-center elimination R2, or a six-center elimination R3:

$$\begin{array}{lll} \text{R1} & \text{TEOS} \Longrightarrow & (C_2H_5O)_3SiOCH_2^* + CH_3^* \\ \text{R2} & \text{TEOS} \Longrightarrow & (C_3H_5O)_3SiOH + C_2H_4 \\ \text{R3} & \text{TEOS} \Longrightarrow & (C_2H_5O)_2SiO + C_2H_4 + C_2H_5OH\ . \end{array}$$

A subsequent reaction of R2 could be

$$\text{R2a} \quad (C_2H_5O)_3SiOH + TEOS \implies (C_2H_5O)_3SiOSi(C_2H_5O)_3 + C_2H_5OH \ .$$

In the case of TMEOS decomposition, the ratio of ethene production to TMEOS consumption of 80 – 90% shows that the four-center elimination is the favored pathway. Therefore we assume that for TEOS, R2 is also the major pathway compared to R1. This assumption is supported by the results of the TPOS decomposition where we could determine the products of both reactions, ethene for the bond fission (Rp1a, Rp1b) and propene for the elimination (Rp2):

$$\begin{array}{lll} \text{Rp1a} & \text{TPOS} & \implies (C_3H_7O)_3SiOCH_2^* + C_2H_5^* \\ \text{Rp1a}_1 & C_2H_5^* & \implies C_2H_4 + H \\ \text{Rp1b} & \text{TPOS} & \implies (C_3H_7O)_3SiOC_2H_4^* + CH_3^* \\ \text{Rp1b}_1 & (C_3H_7O)_3SiOC_2H_4^* & \implies (C_3H_7O)_3SiO^* + C_2H_4 \\ \text{Rp2} & \text{TPOS} & \implies (C_3H_7O)_3SiOH + C_3H_6 \ . \end{array}$$

The ratio of propene to ethene of 3:1 shows that the elimination is also the major pathway.

The ratio of the rate coefficients of TEOS, DMDEOS, and TMEOS decomposition

$$k(\text{TEOS})/k(\text{DMDEOS})/k(\text{TMEOS}) \approx 4/2/1$$

implies that R3 is negligible because otherwise k(TMEOS) should be slower because the TMEOS decomposition can only occur by R1 and R2 not by R3. Experiments with trimethylmethoxysilane showed that this substance is clearly more stable than TMEOS in the temperature range of our measurements so that we can assume that Si–C bond breaking is a minor DMDEOS and TMEOS decomposition pathway in our experiments. Only in the upper temperature range of the TMEOS experiments could influence of the Si–C bond breaking be observed causing a slight upward deviation from the $1/4 \times k$ (TEOS) line.

Ethanol production: Ethanol production by R2a as proposed by Takeuchi et al. (1994) and Chu et al. (1991) can be excluded as an important pathway because (i) with an excess of TMOS the ratio of ethanol to consumed TEOS remains the same compared to mixtures without TMOS and no significant amount of methanol could be observed (the reaction of TMOS + $Si(OC_2H_5)_3OH$ should be similarly fast as R2a so that in the presence of excess TMOS mostly methanol instead of ethanol should be produced and the ethanol production should be reduced if R2a is a relevant reaction of ethanol production); (ii) the ratio of ethanol to consumed TEOS is independent of the TEOS concentration (if R2a was an important reaction the ethanol production and the rate coefficient of TEOS consumption should increase with higher TEOS concentration because

R2a is a bimolecular reaction); and (iii) the amount of ethanol prduced in the TMEOS decomposition experiments is only very small (the silanol formed by ethene elimination from TMEOS would react with remaining TMEOS forming $(CH_3)_3SiOSi(CH_3)_3$ + ethanol if R2a was so fast that it would be an important source of ethanol).

The above discussion shows that ethanol must be formed by a unimolecular reaction. Because we excluded R3 as a major pathway and because ethanol is also produced by DMDEOS we have to assume that ethanol is produced by

$$\text{R4} \qquad (C_2H_5O)_3SiOH \Longrightarrow (C_2H_5O)_2SiO + C_2H_5OH \; .$$

Our measured activation energy of 77 kcal/mol shows that the C–O bond fission ($\Delta_{H,r}^{298} \approx 101$ kcal/mol, Ho & Melius 1994) which was proposed by Desu (1989) and Ivanov et al. (1985) as an important reaction channel, is negligible compared to the ethene elimination and the C–C bond fission. The six-center elimination leading to ethane and acetaldehyde can also be neglected, because only small amounts of these substances could be detected.

References

Chu JCS, Breslin J, Wang NS, Lin MC (1991) Relative stabilities of tetramethyl orthosilicate and tetraethyl orthosilicate in the gas phase. Materials Letters 12:179–184.

Chu JCS, Soller R, Lin MC, Melius CF (1995) Thermal decomposition of tetramethyl orthosilicate in the gas phase: an experimental and theoretical study of the initiation process. J Phys Chem 99:663–672.

Desu SB (1989) Decomposition chemistry of tetraethoxysilane. J Am Ceram Soc 72:1615–1621.

Ho P, Melius CF (1995) Theoretical study of the thermochemistry of molecules in the Si-O-H-C system. J Phys Chem 99:2166–2176.

Ivanov MY, Kupryashkina TN, Kompaniets VZ, Kopteva MA, Konoplev AA (1985) Kinetics and mechanism of pyrolysis of tetraethoxysilane (TEOS). J Appl Chem USSR 58:1691–1695.

Newman CG, O'Neal HE, Ring MA, Leska F, Shipley N (1979) Kinetics and mechanism of the silane decomposition. Int J Chem Kin 11:1167–1182.

Takeuchi H, Izumi H, Kawasaki A (1994) A decomposition study of TEOS in thermal CVD. In: Mat Res Soc Symp Proc 334, pp 45–50.

Tsang W (1981) Comparative-rate single pulse shock tube studies on the thermal stability of polyatomic molecules. In: Shock Waves in Chemistry, Lifshitz A (ed), Marcel Dekker, New York.

Bimolecular S-atom reactions studied in pyrolysis and photolysis shock-wave experiments

D. Woiki, P. Roth
Institut für Verbrennung und Gasdynamik
Gerhard Mercator Universität Duisburg
47048 Duisburg, Germany

Abstract: The reactions of S atoms with O_2, H_2, and NO have been studied in pyrolysis and photolysis reaction systems behind reflected shock waves. S atoms were generated either by the high-temperature pyrolysis of COS or by the laser-flash photolysis of CS_2, respectively. Atomic resonance absorption spectroscopy has been applied to follow both the decay of the S atoms and the formation of the respective atomic reaction product O, H, or N. The reaction systems used have a large excess of the molecular reactants thus providing pseudo-first-order conditions for the consumption of S.

Key words: Chemical kinetics, Sulfur, Pyrolysis, Photolysis

1. Introduction

Reliable high-temperature data of the bimolecular S-atom reactions:

$$S + O_2 \overset{k_1}{\rightleftharpoons} SO + O \qquad \Delta h^0_{298} = -23.3 \text{ kJ/mol} \tag{R1}$$

$$S + H_2 \overset{k_2}{\rightleftharpoons} SH + H \qquad \Delta h^0_{298} = +80.3 \text{ kJ/mol} \tag{R2}$$

$$S + NO \overset{k_{3a}}{\rightleftharpoons} SO + N \qquad \Delta h^0_{298} = +109.4 \text{ kJ/mol} \tag{R3a}$$

$$\overset{k_{3b}}{\rightleftharpoons} SN + O \qquad \Delta h^0_{298} = +144.5 \text{ kJ/mol} \tag{R3b}$$

are limited. Reaction (R1) was investigated by Saito et al. (1986), who measured O-atom concentrations in shock-heated $COS/O_2/Ar$ mixtures. The only available high-temperature rate coefficient k_2 is included in the reaction mechanism used by Fifer et al. (1973) to interpret their measurements in $SO_2/H_2/Ar$ systems. Reaction (R3) was investigated by Oya et al. (1991), who monitored NO and S_2 emissions in shock-heated COS/NO/Ar systems.

In the present investigation S-atom reactions were studied in highly diluted systems behind reflected shock waves using atomic resonance absorption spectroscopy (ARAS) for sensitive measurements of S, O, N, and H atoms. S atoms were generated either by pyrolysis of a few ppm COS or by laser-flash photolysis of CS_2.

2. Experimental

The experiments were carried out behind reflected shock waves in two different pressure-driven diaphragm shock tubes (I) and (II), both of which can be pumped down to ultra-high vacuum conditions to minimize impurity effects. The optical diagnostic for ARAS consists of a microwave-excited discharge lamp, the absorption path in the shock tube separated by LiF windows, a vacuum uv monochromator, and a solar-blind photomultiplier. It was calibrated by reaction systems with well-defined concentrations of the respective atomic species. For photolysis experiments an ArF-excimer laser pulse at 193 nm was coupled into the measurement plane of shock tube (II) through a quartz-glass end plate. For a detailed description of the experimental setup and the calibration procedures, see Woiki et al. (1993), Woiki and Roth (1992), and Thielen and Roth (1984).

3. Results and discussion

Reaction (R1) was studied in the temperature range 1140 K $\leq T \leq$ 3310 K at pressures around 1.0 bar by monitoring the SI resonance absorption at 147.4 nm. The relative initial concentrations used in the pyrolysis experiments were 5 to 10

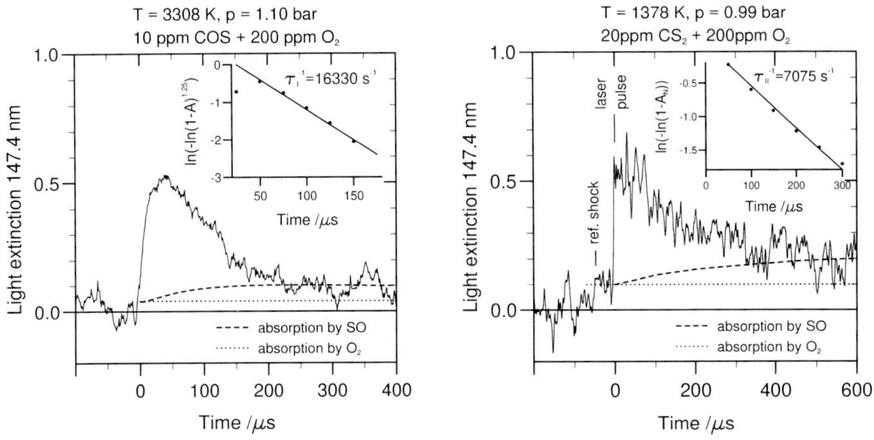

Figure 1. Examples of measured light extinction of SI resonance radiation in $COS/O_2/Ar$ pyrolysis (left) and $CS_2/O_2/Ar$ photolysis (right) behind shock waves.

ppm COS with 50 to 200 ppm O_2 and 20 or 25 ppm CS_2 with 200 to 300 ppm O_2 in the photolysis experiments. Typical examples of measured profiles are shown in Fig. 1. All absorption profiles were transferred into first-order plots after subtracting the interference absorptions of O_2 and SO and the reciprocal time constants τ_I^{-1} or τ_{II}^{-1} were determined assuming pseudo-first-order conditions for reaction (R1); see insets of Fig. 1. They can be related to the rate coefficient k_1

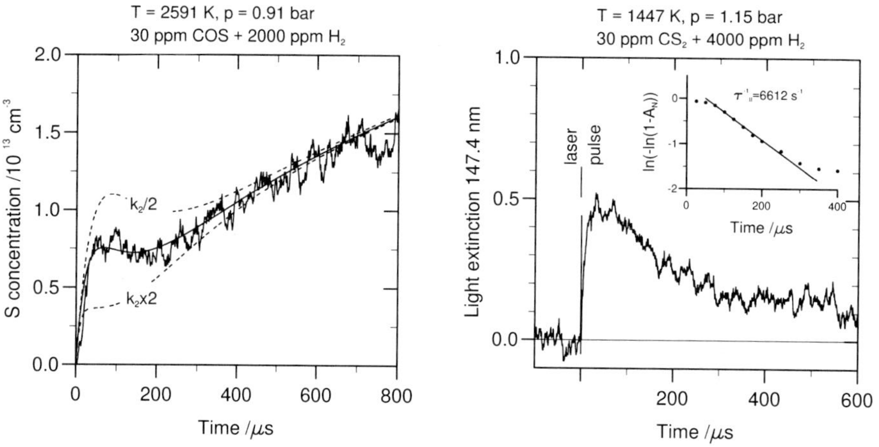

Figure 2. Examples of measured and calculated S-concentration profiles behind shock waves in $COS/H_2/Ar$ pyrolysis system (left) and light extinction of SI resonance radiation in $CS_2/H_2/Ar$ photolysis system (right).

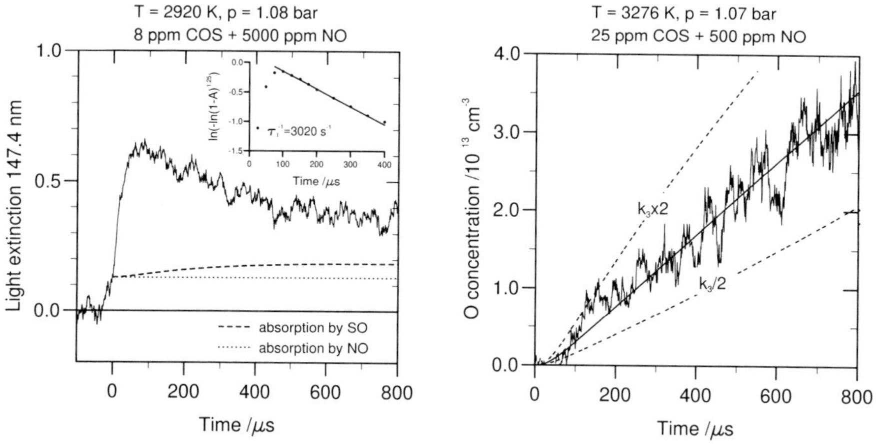

Figure 3. Examples of measured SI light extinction (left) and O concentration (right) behind shock waves in COS/NO/Ar systems.

by: $\tau_I^{-1} = (k_1 \times [O_2]_0)$ or $\tau_{II}^{-1} = (k_1 \times [O_2]_{t=0} + k_{CS_2} \times [CS_2]_{t=0})$. The expression $(k_{CS_2} \times [CS_2]_{t=0})$ takes into account the background reaction between S and the remaining CS_2 after the photolysis. The k_1 values obtained were controlled by transferring the absorption signals into S-concentration profiles, which were compared with computer simulation based on a reaction mechanism given by Woiki and Roth (1995a). For the experiments with the lowest excess of O_2, where the first-order conditions were not sufficiently fulfilled, small variations of the first-order rate coefficients were necessary to obtain best fits. All individual k_1 values are summarized in the respective Arrhenius diagrams of Fig. 4. O-atom concentration profiles, which were additionally measured in $COS/O_2/Ar$ pyrolysis experiments, were highly sensitive to the rate coefficient k_1; typical examples are given in Woiki and Roth (1995a). Their interpretation, based on computer simulation, results in k_1 values which fit perfectly to the previous ones. The comparison of the present result with that of Saito et al. (1986) is also given in Fig. 4. The deviation of about a factor of 2 can be explained by differences in the rate coefficients for the COS pyrolysis used in both studies.

Reaction (R2) was studied in $COS/H_2/Ar$ pyrolysis and $CS_2/H_2/Ar$ photolysis experiments at temperatures between 1260 K and 2660 K. The relative initial concentrations of the S precursors were 10, 20, or 30 ppm COS and 25 or 30 ppm CS_2. The H_2 concentrations in the pyrolysis experiments were between 200 and 2000 ppm and in the photolysis experiments between 1000 and

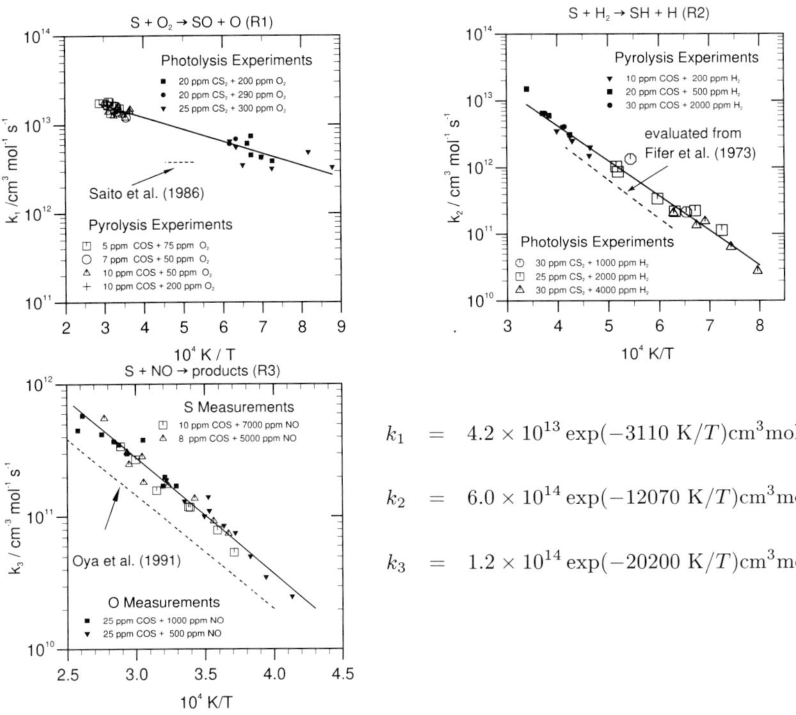

$$k_1 = 4.2 \times 10^{13} \exp(-3110\ \mathrm{K}/T)\mathrm{cm}^3\mathrm{mol}^{-1}\mathrm{s}^{-1}$$

$$k_2 = 6.0 \times 10^{14} \exp(-12070\ \mathrm{K}/T)\mathrm{cm}^3\mathrm{mol}^{-1}\mathrm{s}^{-1}$$

$$k_3 = 1.2 \times 10^{14} \exp(-20200\ \mathrm{K}/T)\mathrm{cm}^3\mathrm{mol}^{-1}\mathrm{s}^{-1}$$

Figure 4. Arrhenius diagrams of measured rate coefficients k_1, k_2, and k_3 in comparison with literature values.

4000 ppm. Figure 2 shows typical examples of S-concentration and S-absorption profiles measured in pyrolysis (left side) and photolysis (right side) experiments, respectively. A first-order analysis was possible only for the profiles obtained from the photolysis experiments. The individual values of the rate coefficient k_2 were determined from the reciprocal time constants τ_{II}^{-1} of the S decay. In

Table 1. Simplified reaction mechanism for high-temperature COS/NO/Ar reaction systems, $k_j = A \times \exp(-T_A/T)$.

	Reaction			k_j /cm^3 mol^{-1} s^{-1}		Reference
				A	T_A	
1	S + O_2	⇌	SO + O	4.3×10^{13}	3100	this study
(3)	S + NO	⇌	products	1.2×10^{14}	20200	this study
3a	S + NO	⇌	SO + N	$0.69 - 0.95 \times k_3$		
3b	S + NO	⇌	SN + O	$k_3 - k_{3a}$		
4	COS + Ar	⇌	CO + S + Ar	2.9×10^{14}	32000	Woiki and Roth (1992)
5	COS + S	⇌	CO + S_2	6.0×10^{13}	4560	Woiki et al. (1993)
6	SO_2 + Ar	⇌	SO + O + Ar	2.9×10^{16}	58620	Mallard et al. (1994)
7	SO + Ar	⇌	S + O + Ar	4.0×10^{14}	53890	Mallard et al. (1994)
8	SN + O	⇌	SO + N	5.0×10^{13}	0	estimate
9	SO + NO	⇌	SO_2 + N	3.0×10^{11}	0	estimate
10	S_2 + NO	⇌	SO + SN	1.0×10^{11}	0	estimate
11	NO + Ar	⇌	N + O + Ar	9.6×10^{14}	74700	Mallard et al. (1994)
12	NO + N	⇌	N_2 + O	2.0×10^{13}	0	Mallard et al. (1994)
13	NO + O	⇌	N + O_2	2.5×10^{13}	23200	Mallard et al. (1994)
14	NO + NO	⇌	N_2O + O	9.0×10^{12}	35000	Mallard et al. (1994)

the pyrolysis experiments, the dissociation of H_2 and secondary reactions of SH radicals with H_2 are of importance at higher temperatures and restrict the application of the first-order analysis. Therefore the rate coefficient k_2 was determined by a fitting procedure of the measured S-concentration profiles using a reaction mechanism given by Woiki and Roth (1995b). The high sensitivity of the S profiles to the rate coefficient k_2 allowed a comparably direct determination of this value; see the variations $k_2 \times 2$ and $k_2 \times 0.5$ in Fig. 2. All obtained rate coefficients k_2 are summarized in Fig. 4.

Reaction (R3) was studied by S- and O-atom measurements in pyrolysis experiments in the temperature range between 2626 K and 3874 K with initial mixtures containing 8 to 25 ppm COS and 500 to 7000 ppm NO. Photolysis experiments on reaction (R3) were frustrated by the too-slow S consumption at lower temperatures. Typical S-absorption and O-concentration profiles are shown in Fig. 3. The S profiles were analyzed by applying the first-order method after subtracting the interfering absorptions of NO and SO. The O profiles were analyzed by computer simulation based on the reaction mechanism of Table 1. It includes either a direct O formation via reaction channel (R3b) or the interaction of N formation via reaction (R3a) followed by a fast conversion of N into O via the reaction N + NO. The simulation was found to be independent of the branching ratio of reaction (R3) but was highly sensitive to the overall rate coefficient k_3 as shown by the variations of k_3 in Fig. 3. All obtained k_3 values are given in the Arrhenius diagram of Fig. 4. S and O measurements agree perfectly. Information on the branching ratio of reaction (R3) was provided by N-atom profiles measured in reaction systems with 100 ppm COS and 100 ppm NO, where the N formation via reaction (R3a) is sufficiently high, but the N consumption is low because of the small excess of NO. All measured N profiles could be fitted by computer simulation using branching ratios k_{3a}/k_3 between 0.69 and 0.95. With higher temperatures, reaction (R3b) becomes more important.

Acknowledgement. The financial support of the Deutsche Forschungsgemeinschaft is gratefully acknowledged.

References

Fifer RA, Moreau R, Bauer SH (1973) The high temperature reduction of sulfur dioxide by hydrogen. In: Proc 9th Int Shock Tube Symp, pp 731–741.

Mallard WG, Westely F, Herron JT, Hampson RF (1994) NIST chemical kinetics database, Version 6.0 NIST standard reference data, Gaithersburg MD.

Oya M, Tsuchiya K, Asaba T (1991) The rate constant of the elementary reaction, NO + S $\rightleftharpoons$ N + SO, in shock waves. In: Proc 18th Int Shock Waves Symp, pp 799–804.

Saito K, Ueda Y, Ito R, Kakumoto T, Imamura A (1986) Measurements of the bimolecular rate constants for S + O_2 $\rightarrow$ SO + O and CS_2 + O_2 $\rightarrow$ CS + SO_2 at high temperatures. Int J Chem Kinet 18:871.

Thielen K Roth P (1984) Resonance absorption measurements of N and O atoms in high temperature NO dissociation and formation kinetics. In: 20th Symp (Int) on Combustion, The Combustion Institute, Pittsburgh, PA, pp 685–693.

Woiki D, Roth P (1992) Shock tube measurements on the thermal decomposition of COS. Ber Bunsenges Phys Chem 96:1347–1352.

Woiki D, Markus MW, Roth P (1993) A shock tube-laser flash photolysis study of the reaction COS + S $\rightleftharpoons$ CO + S_2. J Phys Chem 97(38):9682–9685.

Woiki D, Roth P (1995a) Oxidation of S and SO by O_2 in high-temperature pyrolysis and photolysis reaction systems. Int J Chem Kinet 27:59–71.

Woiki D, Roth P (1995b) A shock tube study of the reaction S + H_2 $\rightleftharpoons$ SH + H in pyrolysis and photolysis systems. Int J Chem Kinet 27:547–553.

Kinetics of C_2-radical reactions during shock-induced pyrolysis of acetylene

T. Kruse, P. Roth
Institut für Verbrennung und Gasdynamik
Gerhard-Mercator-Universität Duisburg
47048 Duisburg, Germany

Abstract: Reactions involving C_2 radicals were studied during pyrolysis of acetylene behind reflected shock waves. The experiments with mixtures of 5 to 50 ppm C_2H_2 diluted in argon were performed in the temperature range 2900 K $\leq T \leq$ 5500 K at pressures between 1.5 and 2.5 bar. C_2 radicals were detected by Ring Dye Laser Absorption Spectroscopy (RDLAS). Additionally, C atoms and C_3 radicals were simultaneously monitored either by Atomic Resonance Absorption Spectroscopy (ARAS) or by using a CCD camera coupled to a spectrograph. By computer fittings of the experimental C_2 profiles, rate coefficients for the reactions:

$$C_2H + M \underset{}{\overset{k_2}{\rightleftharpoons}} C_2 + H + M \qquad \text{(R2)}$$

$$C_2 + C_2 \underset{}{\overset{k_3}{\rightleftharpoons}} C + C_3 \qquad \text{(R3)}$$

$$C_2 + M \underset{}{\overset{k_4}{\rightleftharpoons}} C + C + M \qquad \text{(R4)}$$

were obtained. Furthermore, our experiments indicate that a modification of the heat of formation of C_2 by about 6.5 kcal/mol seems to be necessary.

Key words: Chemical kinetics, Gas phase, Acetylene

1. Introduction

Elementary reactions involving C_2 radicals are not only important in combustion kinetics but also in diamond synthesis from gas-phase reaction of hydrocarbons or during formation and decomposition of fullerenes. In the present study, C_2 radicals were quantitatively measured with high sensitivity by Ring Dye Laser Absorption Spectroscopy (RDLAS) during the pyrolysis of acetylene highly diluted in argon. To confirm a proposed reaction mechanism, some additional experiments were made by recording C and C_3 concentrations by Atomic Resonance Absorption Spectroscopy (ARAS) and emission spectroscopy, respectively.

2. Experimental

The experiments were performed behind reflected shock waves in a pressure-driven diaphragm shock tube of 80 mm diameter. The shock tube and the test-gas mixing vessel were operated under ultra-high vacuum conditions. For a more detailed description of the shock tube, see Thielen and Roth (1987).

The narrow linewidth ring dye laser system used to measure the C_2 concentration was described earlier by Markus and Roth (1992). The laser beam was coupled into the measurement plane of the shock tube via an optical fiber and split into probe and reference beam. Silicon photodetectors were used to monitor the incident and transmitted laser light. The time resolution of about 3 μs was limited by the beam diameter and the shock speed. The frequency of the dye laser was fixed at $\lambda = 467.440$ nm corresponding to the center of the overlapping $R_1(37)$, $R_2(36)$, and $R_3(35)$ rotational lines in the (1-0) band of the C_2 ($d^3\Pi_g \leftarrow a^3\Pi_u$) transition, which was found to be most suitable for a quantitative diagnostic in the studied temperature range. With the knowledge of the oscillator strengths of the C_2 transitions from Naulin et al. (1988), the absolute C_2 concentration was determined from the measured absorption by applying the Lambert-Beers law.

Carbon atoms were detected at 156.1 nm by ARAS using a microwave discharge lamp which was operated at 6 mbar in a mixture of He containing 1% CO. A McPherson 0.5-m vacuum monochromator with a 150-μm slit width coupled to a photoelectron multiplier was used to select the respective C-atom line. The relation between measured absorption and C-atom concentration was determined by calibration experiments based on thermal decomposition of CH_4 at temperatures $T \geq 3800$ K. The detailed description of the ARAS setup for C-atom measurements was given by Thielen and Roth (1987).

The time-resolved emission during C_2H_2 thermal decomposition was recorded in the wavelength range from 330 to 600 nm by a combination of a grid spectrograph and an intensified CCD camera. The spectral resolution was 2 nm and the time resolution 20 μs for a measurement time of about 1 ms. A detailed description of the above system was given by von Gersum et al. (1994).

3. Results and discussion

An example of an emission spectrum obtained during thermal decomposition of acetylene at 3450 K is shown in Fig. 1. It was recorded with the CCD camera/spectrograph system, with a reaction time of 200 μs. Two main features are of importance: strong emission of the $C_2(d^3\Pi_g \leftarrow a^3\Pi_u)$ Swan bands at $\lambda > 430$ nm and weak emission of the $C_3(A^1\Pi_u \leftarrow X^1\Sigma_g^+)$ bands in the spectral range 360 nm $< \lambda <$ 420 nm. The C_3 concentration can be estimated quantita-

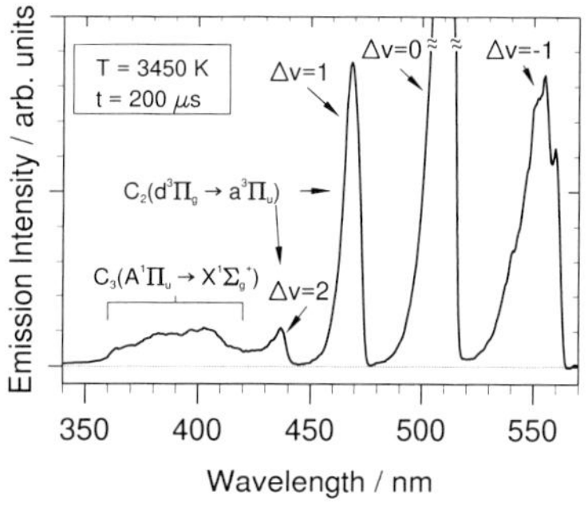

Figure 1. Emission spectrum of C_2 and C_3 during acetylene pyrolysis at a reaction time of 200 μs.

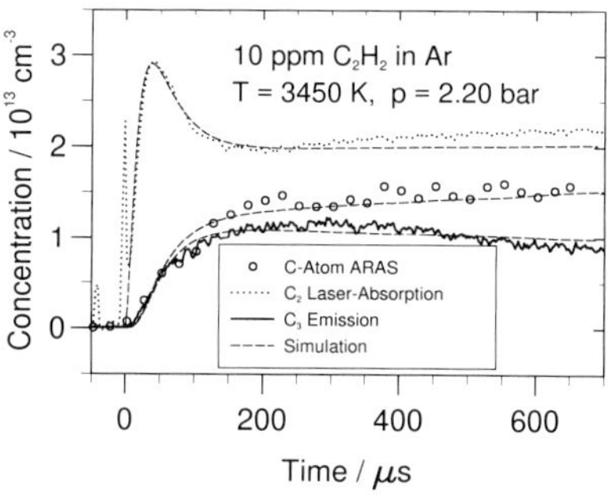

Figure 2. Comparison of measured and calculated C, C_2, and C_3 concentrations during acetylene pyrolysis.

tively from the ratio of the total C_3 and C_2 emissions. With the knowledge of the absolute C_2 concentration together with literature values of both oscillator strengths given by Naulin et al. (1988) and Becker and Tatarczyck (1979), the C_3 concentration can be determined.

Typical measured concentration profiles of all three carbon radicals obtained from a shock-tube experiment at $T = 3450$ K and $p = 2.2$ bar are shown in Fig. 2. After the shock arrival, the C_2 concentration shows a fast rise during the first 30 μs followed by a rapid decrease to a nearly constant value. The C_2 decrease is accompanied by an increase of C and C_3 concentration. Similar results were obtained in other C_2H_2-decomposition experiments.

The observed species behavior can be described by the following simplified reaction mechanism, which is part of a more extended scheme of about 80 rections given by Kiefer et al. (1992). Because of the low reactant concentration only a few reactions are of importance:

$$C_2H_2 + M \overset{k_1}{\rightleftharpoons} C_2H + H + M \quad \text{(R1)}$$

$$C_2H + M \overset{k_2}{\rightleftharpoons} C_2 + H + M \quad \text{(R2)}$$

$$C_2 + C_2 \overset{k_3}{\rightleftharpoons} C + C_3 \quad \text{(R3)}$$

$$C_2 + M \overset{k_4}{\rightleftharpoons} C + C + M \quad \text{(R4)}$$

The fast C_2 formation proceeds via reactions (R1) and (R2), whereas the rapid decrease of C_2 to a quasi-stationary level and the formation of C and C_3 can be explained by reaction (R3). At temperatures above 3900 K, the C_2 dissociation (R4) must also be taken into account.

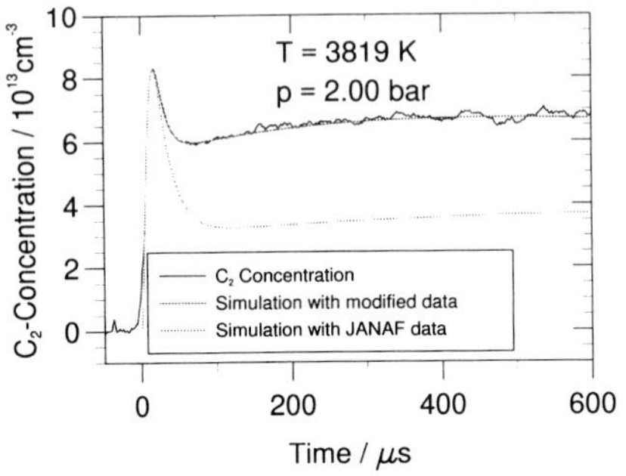

Figure 3. Simulation with original and modified thermodynamical data for an experiment with 50 ppm C_2H_2.

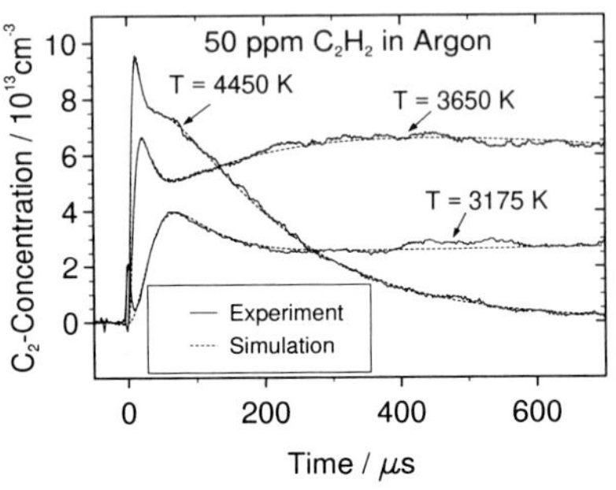

Figure 4. Measured and calculated C_2 concentration profiles at three different temperatures.

A comparison of calculated and measured concentration profiles of C, C_2, and C_3 for an individual shock-tube experiment is also shown in Fig. 2. The experimentally observed nearly stationary concentration of C, C_2, and C_3 for times $t \geq 300$ μs could only be verified by computer simulation based on the above reaction mechanism by choosing the equilibrium constant of reaction (R3) to be equal to $K_{c,3} = 1.4$. The thermodynamical data for C, C_2, and C_3, taken from the JANAF thermochemical tables, Chase et al. (1985) result in an equilibrium constant of $K_{c,3} = 9.2$ for the present experimental conditions, which is inconsistent with the experimental findings. Figure 3 demonstrates as a further example the sensitivity of the computer simulation with respect to the equilibrium constant $K_{c,3}$ for an experiment with 50 ppm acetylene in argon at 3810 K. The simulation based on the JANAF data, represented by the dotted line, yields a clearly too low C_2 concentration, while the simulation with the modified $K_{c,3}$ value, represented by the dashed line, agrees nearly perfectly with the experimental profiles. The change in $K_{c,3}$ can be achieved by increasing the Gibbs free energies of C or C_3 by about 13 kcal/mol or by decreasing the Gibbs free energy of C_2 by 6.5 kcal/mol. As will be discussed later, a reduction of the heat of formation of C_2 by about 6.5 kcal/mol is most likely.

Figure 4 shows further examples of C_2-concentration profiles measured in mixtures containing 50 ppm C_2H_2, which were heated to three different temperatures, in comparison with computer simulations. The initial C_2 increase is sensitive to reactions (R1) and (R2). The subsequent peak is caused by the bimolecular reaction (R3), while reaction (R4) is responsible for the significant C_2 decay at higher temperatures (see experiment at 4450 K), from which the rate coefficient k_4 can be determined. The rate coefficient k_1 is well known in the temperature range 1850 K $\leq T \leq$ 3000 K from earlier H-atom studies of Frank and Just (1980). For higher temperatures up to 4000 K, rate-coefficient values can be obtained from a RRKM calculation, based on parameters selected to fit the low-temperature values of Frank and Just (1980). With the knowledge of k_1, the rate coefficient k_2 can be evaluated from fittings of the initial C_2 increase. Results for k_2 obtained in the described way are shown in the Arrhenius diagram of Fig. 5. The solid line in Fig. 5 represents a RRKM calculation using the most recent values for the excitation energies and the heat of formation of C_2H from Kiefer et al. (1992). The best agreement with the experimental points was achieved with $\Delta h_f^0 = 191.5$ kcal/mol for C_2, which is equivalent to a reduction of 6.5 kcal/mol compared to the JANAF value. The rate expression resulting from the present experiments can be represented by:

$$k_2 = 5.2 \times 10^{18} \times (T/1000)^{-3.59} \times \exp(-106.6 \text{ kcal mol}^{-1}/RT)\text{cm}^3\text{mol}^{-1}\text{s}^{-1}$$

The expression of k_2 is considerably higher than recent estimates by Kiefer et al. (1992).

The decrease of 6.5 kcal in the heat of formation corresponds exactly to the change in Gibbs free energy required to change the equilibrium constant $K_{c,3}$. Thus, by decreasing the Δh_f^0 JANAF value of C_2 from 198.5 kcal/mol to 191.5 kcal/mol, both the measured steady-state value of C_2 and the low activation energy for the H-atom abstraction by reaction (R2) can be explained. A similar low value for the heat of formation of C_2 was recently measured by Urdahl et al. (1991).

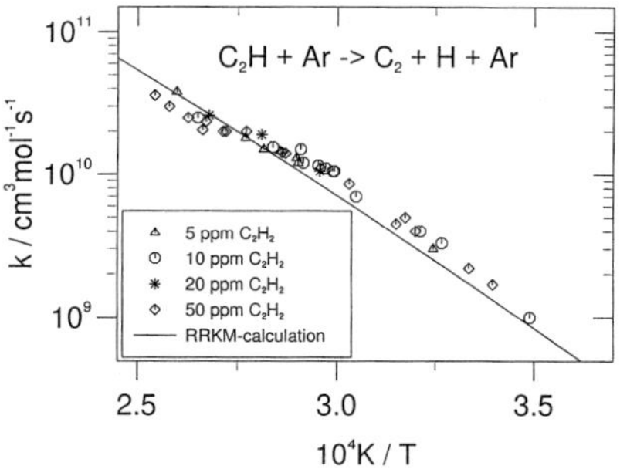

Figure 5. Rate coefficient of the reaction $C_2H + M \rightleftharpoons C_2 + H + M$. The solid line corresponds to a RRKM calculation.

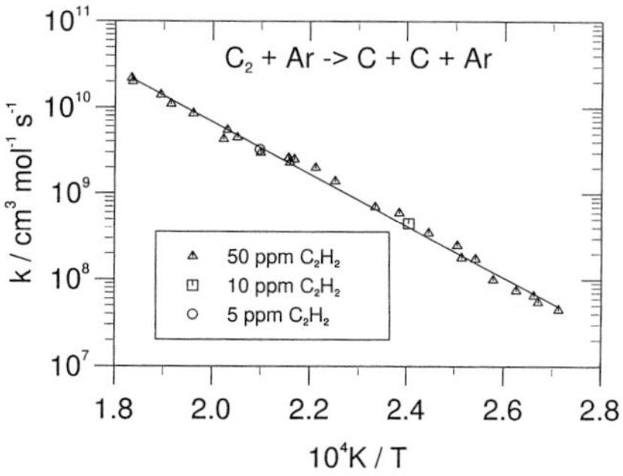

Figure 6. Arrhenius diagram of the reaction $C_2 + Ar \rightleftharpoons C + C + M$.

The rate coefficient for the bimolecular reaction (R3) evaluated at different reaction conditions was found to be independent of temperature. It can be approximated by a mean value of $k_3 = (2.0 + / - 0.4) \times 10^{14}$ cm^3 mol^{-1} s^{-1}. The rate coefficients for the C_2 dissociation (R4) evaluated from the signal decrease at higher temperatures are shown in Fig. 6. The data points can be represented by the Arrhenius expression $k_4 = 7.4 \times 10^{15} \times \exp(-138$ kcal mol$^{-1}/RT)$ cm^3 mol^{-1} s^{-1}. The activation energy of 138 kcal mol^{-1} is close to the bond dissociation energy of $D_0 = 144$ kcal mol^{-1} (see Gaydon 1968).

Acknowledgement. The authors thank Mr. L. Jerig for his help in conducting the experiments and Dr. V. Smirnov for performing the RRKM calculations.

References

Becker KH, Tatarczyk T (1979) Lifetime measurements of electronically excited $C_3(^1\Pi_g)$ radicals in different vibrational states. Chem Phys Lett 60:502–506.

Chase MW Jr, Davies CA, Downey JR Jr, Frurip DJ, McDonald RA, Syverud AN (1985) JANAF thermochemical tables. J Phys Chem Ref 14(1).

Frank P, Just T (1980) High temperature decomposition of acetylene and diacetylene at low relative concentrations. Comb and Flame 38:231–248.

Gaydon AG (1968) Dissociation Energies and Spectra of Diatomic Molecules, Chapman and Hall, London 1968.

Kiefer JH, Sidhu SS, Kern RD, Xie K, Chen H, Harding LB (1992) The homogeneous pyrolysis of acetylene II: the high temperature radical chain mechanism. Combust Sci and Tech 82:101–130.

Markus MW, Roth P (1992) Ring dye laser-based absorption measurements of CH radicals behind shock waves. In: Proc 18th Int Symp Shock Waves, Springer Verlag, Sendai, Vol 2, pp 1011–1018.

Naulin C, Costes M, Dorthe G (1988) C_2 radicals in a supersonic molecular beam. Radiative lifetime of the $d^3\Pi_g$ state measured by laser-induced fluorescence. Chem Phys Lett 143:496–500.

Thielen K, Roth P (1987) Resonance absorption measurements of N, O, and H atoms in shock heated $HCN/O_2/Ar$ mixtures. Comb and Flame 69:141–154.

Urdahl RS, Bao Y, Jackson WM (1991) An experimental determination of the heat of formation of C_2 and the C-H bond dissociation energy in C_2H. Chem Phys Lett 178:425–428.

von Gersum S, Kruse T, Roth P (1994) Spectral emission during high temperature pyrolysis of fullerene C_{60} in shock waves. Ber Bunsenges Phys Chem 98:979–982.

A shock-tube study on the reaction of methyl with nitric oxide

M. Braun-Unkhoff, C. Naumann, K. Wintergerst*, P. Frank
DLR, Institut für Physikalische Chemie der Verbrennung
70569 Stuttgart, Germany

Abstract: The reaction of $CH_3 + NO$ has been studied behind reflected shock waves at temperatures between 1400 and 2000 K and at total pressures between 1.5 and 3.5 bar. Reaction progress was observed by measuring OH radicals by applying narrow-linewidth laser absorption; additionally, H and I atoms were measured using the well-known ARAS technique. From modeling the H- and OH-absorption profiles it was concluded that the reaction of CH_3 with NO may proceed via three reaction channels.

Key words: Chemical kinetics, NO_x, Absorption spectroscopy, Modeling

1. Introduction

Nitric oxide is often formed in flames as an unwanted by-product from either atmospheric or fuel-bound nitrogen (see for example Miller and Bowman 1989). Due to the complexity of technical combustion, extensive modeling seems to be the most promising tool in developing strategies like the well-known reburn process to reduce NO_x emissions in stationary sources. Therefore, a chemical-reaction mechanism is required which would contain reliable data on both the kinetics and product distributions of the relevant elementary reactions. The reaction between methyl radicals and nitric oxide may play an important role in the complex NO_x cycle leading either to toxic NO, NH_3, and HCN or to molecular nitrogen.

For theoretical and thermodynamical reasons, the reaction of methyl with NO should proceed at least via two channels:

$$CH_3 + NO \overset{k_1}{\rightleftharpoons} HCN + H_2O \qquad \Delta H^\circ_{298,reac} = -\ 79\ \text{kcal mol}^{-1} \qquad \text{(R1)}$$
$$CH_3 + NO \overset{k_{2a}}{\rightleftharpoons} H_2CN + OH \qquad \Delta H^\circ_{298,reac} = +\ 12\ \text{kcal mol}^{-1} \qquad \text{(R2a)}$$
$$H_2CN + M \overset{k_{2b}}{\rightleftharpoons} HCN + H + M \qquad \text{(R2b)}$$

*Present address: Institut für Physikalische Chemie und Elektrochemie III, Universität Karlsruhe, 76128 Karlsruhe, Germany.

Reaction R1 is assumed to be the main channel regarding the values for the heat of reaction and the reaction barrier (see for example Lifshitz et al. 1993). Reaction R2a is rate limiting for the H-atom production because the primarily formed H_2CN decomposes very fast under the conditions of the present work.

2. Experimental

Details of the shock-tube technique as well as of the optical setup have been described in more detail elsewhere (Frank and Just 1985). The measurements were performed behind the reflected shock front close to the end flange (5 mm). Atomic resonance absorption spectroscopy (ARAS) was used to monitor simultaneously time-dependent H- and I-absorption profiles by using an oxygen spectral filter (for H atoms at $\lambda = 121.6$ nm) and a 1-m-vacuum-UV monochromator (for I atoms at $\lambda = 145.9$ nm). Microwave-excited discharge lamps containing 1% H_2 or CH_3I diluted with He served as light sources.

Narrow-linewidth laser absorption of OH radicals at $\lambda = 308.6091$ nm ($Q_1(5)$ line of the $A^2\Sigma^+ \leftarrow X^2\Pi$ system) was applied to monitor time-dependent OH concentrations. An Ar-ion laser was used to pump a single-frequency ring dye laser which is actively stabilized. The laser light is intracavity frequency-doubled using an angle-tuned $LiIO_3$ crystal. In order to increase the detection limit of OH, (1) the optical pathway was doubled by means of a mirror and (2) a two-beam technique was applied to cancel laser fluctuations. The laser mode structure was observed by passing a portion of the fundamental beam into a scanning interferometer (2-GHz free spectral range). The wavelength of the fundamental was monitored by a Burleigh WA 20 wavemeter. Additionally, a methane-air flame was used to ensure that the laser wavelength was tuned exactly to the OH-absorption peak.

The H- and I-concentration profiles are based on calibration measurements (Braun-Unkhoff et al. 1990, Wintergerst and Frank 1993). The OH-absorption profiles were transferred into concentration profiles by calculating the value for the absorption cross-section σ from spectroscopic constants according to established procedures (Hanson 1993). The validation of the calculated values was checked in a separate set of experiments. No deviation between observed and calculated absorption profiles could be seen (Braun-Unkhoff et al. 1993).

Because of the high sensitivity of the optical measurement technique, very small initial concentrations could be used. Due to the very fast bimolecular reaction of methyl radicals, mixtures of 2000–5000 ppm NO and 2–5 ppm CH_3I diluted with argon were used for the H-absorption measurements, whereas for the OH-absorption measurements mixtures of 3% NO and 32–50 ppm $(CH_3)_2N_2$ were used. Methyliodide and azomethane served as methyl precursors. The thermal dissociation of $(CH_3)_2N_2$ ($T \geq 1250$ K) or CH_3I ($T \geq 1350$ K) is sufficiently fast to produce CH_3 radicals nearly instantaneously (Braun-Unkhoff et al. 1990). By measuring the I-absorption profiles the initial methyl concentration could be determined in situ very precisely. The initial concentrations of azomethane were permanently controlled by means of a gas chromatograph equipped with a flame ionization detector (FID), the relative error being about $\Delta c/c \leq 0.05$.

It should be pointed out that it was very important (1) to use freshly distilled NO and (2) to look for possible impurities like N_2O and NO_2. The background absorption of NO was measured in separate experiments for 1200 K $\leq T \leq$ 2000 K and found to be: $\sigma_{121.6} = 4.76 \times 10^{-22} \times T + 1.98 \times 10^{-18}$ cm^2 and $\sigma_{145.9} = (5.4 \pm 0.6) \times 10^{-18}$ cm^2.

3. Results

The measured H and OH absorption profiles were modeled by applying a reaction scheme of about 30 reactions and 42 species. For the main reaction channel R1, the Arrhenius expression derived by Lifshitz et al. (1993) for 1100 K $\leq T$ $\leq$ 1330 K was used, which is in good agreement with the one evaluated by Wolff and Wagner (1988) for 1800 K $\leq T \leq$ 2150 K.

A typical experimental profile of the total absorption measured at $\lambda = 121.6$ nm is shown in Fig. 1. The signals correspond to the superposition of the NO background absorption (within the observation time interval practically constant) and the increasing H ARAS signal. Sensitivity analysis has shown that the H profiles are very sensitive with respect to reaction R2a (Fig. 2, straight line). As reaction progress proceeds, the influence of the main channel R1 becomes also important (Fig. 2, long-dashed line). Bimolecular reactions of methyl radicals and of methyl + H (see Frank and Braun-Unkhoff 1987) have only a minor influence on the H profiles compared to reactions R1/R2a (Fig. 2, dotted line) because of the very low initial concentrations of the radical precursor CH_3I. The pyrolysis of methyl provides for additional production of H atoms at the

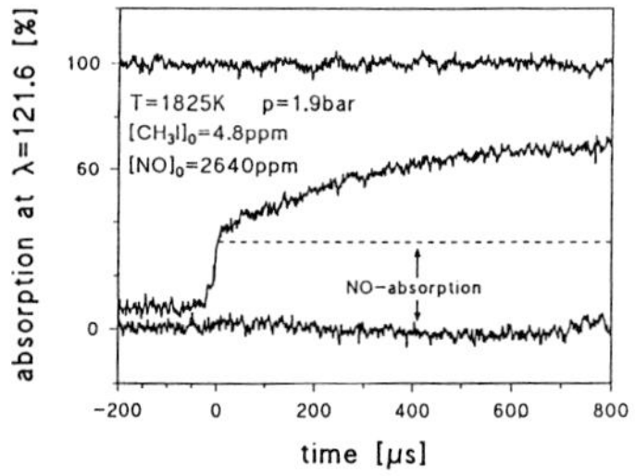

Figure 1. Absorption profile at 121.6 nm.

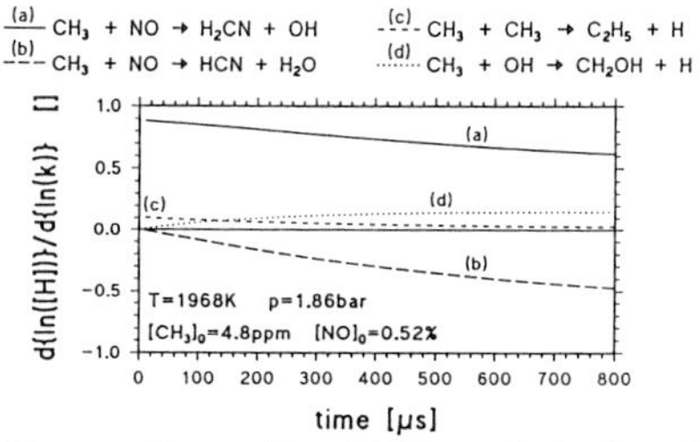

Figure 2. Sensitivity plot for H-absorption measurements.

upper limit of the investigated temperature range. As an example, Fig. 3 shows the comparison between measured and calculated profiles. It also indicates the sensitivity of the profile with respect of varying k_{2a} by $\pm 50\%$.

A typical experimental absorption profile measured at $\lambda = 308.6091$ nm is seen in Fig. 4. The absorption rises immediately without any induction period indicating that the OH species must be produced directly by an initiation reaction. Sensitivity analysis revealed that the two NO-removal reactions R1/R2a

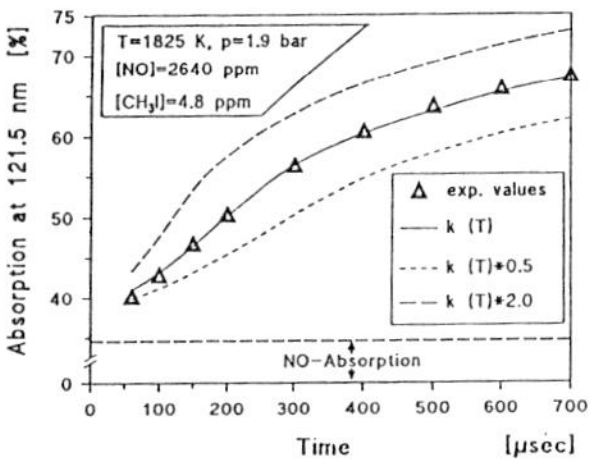

Figure 3. The k_{2a} sensitivity of the measured total absorption.

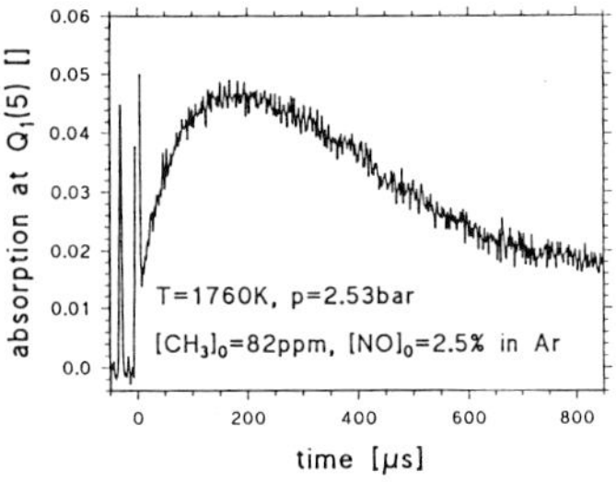

Figure 4. OH-absorption profile.

dominate at the beginning of the observation time interval (Fig. 5, straight and long-dashed curves). Again, the influence of secondary reaction steps increases as reaction progress proceeds: bimolecular reactions of OH with CH_3 and with OH itself (see Braun-Unkhoff et al. 1993) are — besides the bimolecular methyl reactions — very important for reproducing the experimental OH profiles (Fig. 5, dotted and dash-dotted lines). Good agreement between observed and calculated OH-absorption profiles is achieved (Fig. 6, straight line). The measured OH profiles are very sensitive to variations of the k value of the reaction R2a (Fig. 6, dotted and dashed lines).

The temperature dependence of the rate coefficient of the reaction R2a is displayed separately for the H-atom profiles (Fig. 7, circles) and for the OH profiles (Fig. 7, stars). The following rate expressions for the reaction R2a, $CH_3 + NO \rightarrow H_2CN + OH$, were derived (uncertainties one standard deviation):

$$\text{H profiles:} \quad k_{2a} = (7.1 \pm 3) \times 10^{12} \exp(-(12140 \pm 610\ \text{K}/T))\ \text{cm}^3\text{mol}^{-1}\text{s}^{-1}$$

$$\text{OH profiles:} \quad k_{2a} = (1.1 \pm 2) \times 10^{12} \exp(-(11240 \pm 570\ \text{K}/T))\ \text{cm}^3\text{mol}^{-1}\text{s}^{-1}.$$

4. Discussion

The Arrhenius expressions given above for reaction R2a derived from modeling either the H or the OH profiles, differ significantly by a factor of 3–4. This

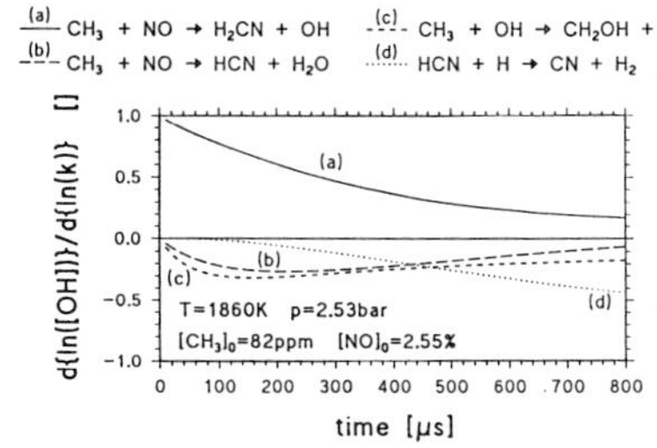

Figure 5. Sensitivity plot for OH-absorption measurements.

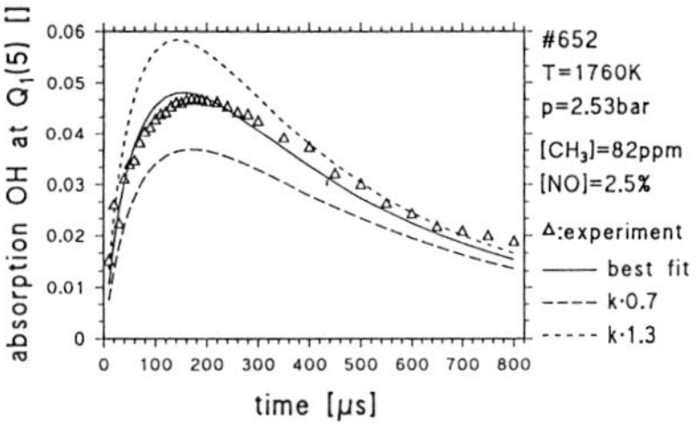

Figure 6. The k_{2a} sensitivity of the measured OH absorption.

experimental finding may be explained by a third reaction channel which would produce additional H atoms even at the lower end of the investigated temperature range. Two possible pathways were found to be compatible with our measured OH- and H-concentration profiles:

$$CH_3 + NO \overset{k_3}{\rightleftharpoons} NH + H_2CO \qquad \Delta H^\circ_{298,reac} = +\ 1\ \text{kcal mol}^{-1} \qquad \text{(R3)}$$

$$CH_3 + NO \overset{k_4}{\rightleftharpoons} NH_2 + HCO \qquad \Delta H^\circ_{298,reac} = -\ 1\ \text{kcal mol}^{-1} \qquad \text{(R4)}$$

The subsequent reactions of NH and NH_2, which had to be considered because of the high NO level, are multichannel reactions. For the NH + NO reaction, the dominant products are considered to be $N_2O + H$ providing the needed H atoms; for the $NH_2 + NO$ reaction, the dominant products are believed to be $N_2 + H_2O$ — in this case the needed additional H atoms are produced from the very fast decay of HCO (Baulch et al. 1992).

Modeling the H-absorption profiles by using a third NO-removal channel (either R3 or R4) reproduces the H profiles (see Fig. 8). Modeling the OH-absorption profiles favors reaction R4 as the third reaction channel because the shape and size of OH profiles are changed only marginally, whereas the use of reaction R3 would reproduce the measured OH profiles in a less sufficient way.

Under the assumptions mentioned above, an Arrhenius expression for the third NO-removal channel R4 is suggested:

$$k_4 \quad = \quad 3 \times 10^{14} \exp(-(20000\ \text{K}/T))\ \text{cm}^3\text{mol}^{-1}\text{s}^{-1}.$$

It should be pointed out that up to now there are no proofs for the existence of this reaction step. Further measurements are required: if NH or NH_2 would be identified as products this would support the suggestion that the reaction of methyl with NO proceeds via at least three pathways.

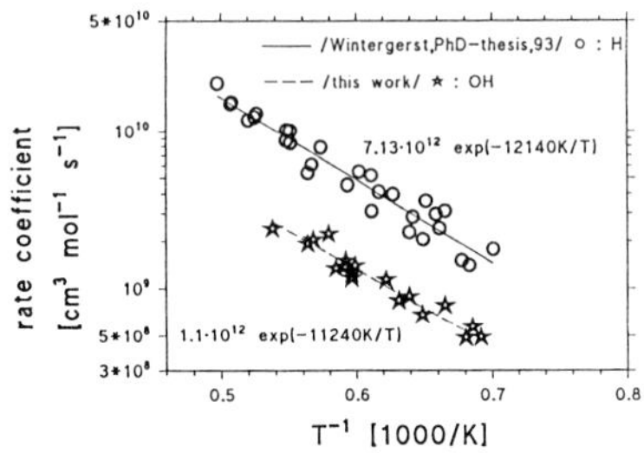

Figure 7. Arrhenius plot of the reaction R2a: $CH_3 + NO \rightarrow H_2CN + OH$.

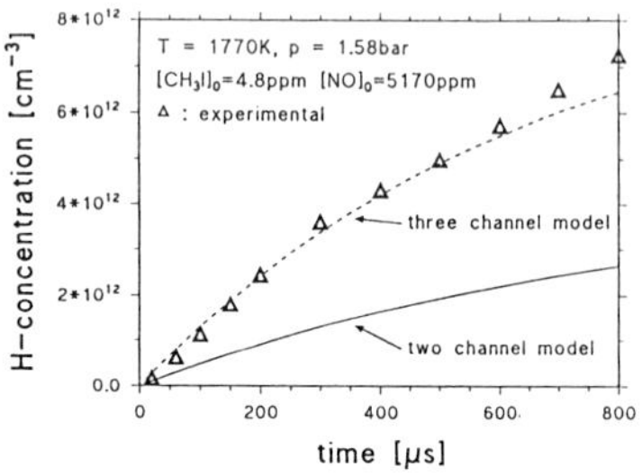

Figure 8. Effect of NO-removal reactions on measured H profiles.

Acknowledgement. The authors wish to thank Prof. T. Just for his interest in this study and for helpful discussions. We would also like to thank N. Ackermann and M. Kallfass for technical assistance. We are also very grateful to Dr. C. Wahl and M. Kapernaum for preparing and analyzing azomethane and methyliodide.

References

Baulch DL, Cobos CJ, Cox RA, Esser C, Frank P, Just T, Kerr JA, Pilling MJ, Troe J, Walker RW, Warnatz J (1992) Evaluated kinetic data for combustion modelling. J Phys Chem Ref Data 21(3)411–737.

Braun-Unkhoff M, Frank P, Just, T (1990) High temperature reactions of benzyl radicals. Ber Bunsenges Phys Chem 94:1417.

Braun-Unkhoff M, Naumann C, Frank P (1993) A shock tube study of the reaction $CH_3 + O_2$. In: Shock Waves @ Marseille, Springer, pp 203–208.

Frank P, Just T (1985) High-temperature reaction rate for $H + O_2 \rightarrow OH + O$ and $OH + O_2 \rightarrow H_2O + H$. Ber Bunsenges Phys Chem 89:181.

Frank P, Braun-Unkhoff M (1987) A shock tube study on the reaction $CH_3 + CH_3 \rightarrow C_2H_5 + H$. In: Proceedings of the 16th Int Shock Tube Symp, pp 379–389.

Hanson RK (1993) The role of lasers in shock tube studies of chemical kinetics. In: Shock Waves @ Marseille, Springer, pp 7–14.

Lifshitz A, Tamburu C, Frank P, Just T (1993) The reaction $CH_3 + NO \rightarrow HCN + H_2O$, experimental and modeling study. J Phys Chem 97(16)4085–4090.

Miller JA, Bowman CT (1989) Mechanism and modeling of nitrogen chemistry in combustion. Prog Energy Combust Sci 15(4)287–338.

Wintergerst K and Frank P (1993) Direct measurement of the reaction $H + CO_2$ at elevated temperatures. In: Shock Waves @ Marseille, Springer, pp 77–82.

Wolff T and Wagner HG (1988) An investigation of the reaction between CH_3 radicals and NO at high temperatures. Ber Bunsenges Phys Chem 92:678.

Formation of soot, fullerene (C_{60}), and PAHs from liquid-spray combustion of hydrocarbons in a shock tube

R. Wang, P. Cadman, N. Huq
Combustion Physics Group, University of Wales
Wales, Aberystwyth, SY23 3BZ UK

Abstract: The liquid-spray combustion/pyrolysis of heptane, toluene, and other hydrocarbons has been studied in a shock tube at temperatures between 1000 and 3000 K. The soot and polycyclic aromatic hydrocarbons (PAHs) have been monitored and more than 20 of the latter identified. Fullerene (C_{60}) has been found in the products of benzene oxidation and possibly in small quantities in other fuels. Benzophenone was found as an important product of toluene combustion. The amount of PAH increased with the amount of soot produced. The total soot yield formed 'bell'-shaped curves in these liquid-spray combustion experiments as has been found in gaseous mixtures.

Key words: Soot, Polycyclic aromatic hydrocarbons, Fullerene formation, Liquid-spray combustion

1. Introduction

The process of liquid-spray combustion is used extensively throughout modern society in compression ignition engines (diesel), burners, and gas turbines. Compared to gaseous premixed combustion, liquid-spray combustion has been found to produce larger amounts of, by an order of magnitude, soot (carbon) particulates together with condensed hydrocarbons (PAHs). Concerns about the emission of these particulate pollutants has led to stricter controls on the amounts that can be emitted from engines.

Liquid-spray combustion has been studied in engines, flames or burners, rapid compression machines, flow systems, and shock tubes (Cadman 1995). Although practical devices such as engines, turbines, etc. are the final means to test theories and parameters, they often do not lend themselves well to investigations of the fundamental processes which occur because of the complexity of the net combustion and non-uniform local conditions. Other techniques are often preferable (Kamimoto & Kobayashi 1991); while not simulating a practical device completely, they do simulate some of the conditions occurring but in a more uniform and controlled manner so that the effect of changing single parameters can be identified more easily.

The combustion of heptane has been studied by Adomeit and co-workers (Cierzki & Adomeit 1989, 1993; Fieweger et al. 1995) and Burcat and Dyinyaninov (1993) at pressures up to 49 bar. Adomeit and co-workers (Cierzki & Adomeit 1989, 1993; Fieweger et al. 1995) have also investigated the soot production from gaseous heptane in shock waves online using laser-absorption techniques (Fieweger et al. 1995). They found the presence of cool flames in heptane combustion and the occurrence of very sharp pressure profiles which they called strong ignition (Cierzki & Adomeit 1989, 1993). Frenklach and Yuan (1987) have investigated the shock-wave combustion of gaseous benzene/alcohol mixtures and found a reduction in the soot which they ascribed to the oxidation of soot by OH and the reactions of hydrogen atoms. Williams and Alexiou (1994) have investigated the shock-tube combustion of gaseous mixtures of toluene and the production of soot from them. They also found a reduction in soot which they attributed to the lower mole fraction of the aromatic in the reaction mixtures. In this work we have investigated the production of soot and PAHs from liquid-spray fuels such as toluene, heptane, and benzene by post-reaction analysis and online monitoring.

2. Experimental

A 11-m long, 63.5-mm internal diameter stainless steel shock tube (5–6-m driver and 6-m driven section) equipped with optical windows and pressure transducers was used for this work (Cadman 1995). An injection system constructed in this laboratory was fitted onto the end plate of the tube. Melinex diaphragms were used (175-μm thickness) to separate the two sections. The shock was initiated by a 12-V/4-A pulse through a hot wire between the two layers of diaphragm material. This was found to give reproducible bursting of the diaphragm.

The fuel injector was controlled by a 12-V pulse from the same control unit for synchronization with the diaphragm bursting. PCB pressure transducers placed at the end plate, main observation window, and along the tube were used to measure the incident- and reflected-shock velocities which were utilized to calculate the reflected-shock temperature and pressures using CHEMKIN and a program written at Aberystwyth. The main observation window was situated 10 cm from the end plate where the spray disappearance and rate of soot appearance could be monitored using a small 20-mW 488-nm argon-ion laser, photodiode, and interference filter.

Emission measurements were made at the same point via other windows using optical fibers, beam splitters, and photomultipliers with spectral isolation using narrow band-pass interference filters (454 and 546 nm). A second window system 20 cm from the end plate had a He-Ne laser-beam/photodiode system for detection of soot further down the tube. Most reaction of the hydrocarbon species occurred between the 10-cm port and the end plate. Eight transient signals could be captured with a 20-MHz transient recorder (32K per channel) and stored in a microcomputer for subsequent analysis. Filters were placed in the exhaust lines of the evacuation system so that immediately after the combustion the shock tube could be evacuated and the soot and PAHs trapped for further analysis.

The filter papers (Whatman 3) which had been previously weighed before use were then re-weighed to obtain soot yields (with 90% confidence error limits to ± 0.5 mg); the filtrate was extracted from the carbon by soxhlet extraction using dichloromethane as a solvent. The extracted material was then analyzed by high-pressure liquid chromatography on a Chromspher 3PAH column for PAHs and an Anachem 8μm cyano 250-mm column for fullerene. The analyzed material was then detected with a Waters diode-array detector. This had an advantage that the uv-visible absorption spectrum could be measured for each unknown material detected as an aid to identification and quantitative calibration. Some samples were also analyzed via GC mass spectrometry on a Kratos MS25 mass spectrometer equipped with a GLC and electron-impact detector. The online detection of soot production was also measured via laser absorption using the argon-ion and He-Ne laser systems described above.

3. Results

The spray-combustion/pyrolysis of heptane, toluene, and a few experiments with benzene were investigated over the temperature range 1000–3000 K and at pressures between 2 and 10 bar in mixtures containing 1% and 10% oxygen/argon or argon only. The evaporation of the droplets took place within 200 μs in the reflected shock, depending on the fuel. Some fuels evaporated within a few microseconds.

The collected soot plus its condensed hydrocarbons were separated by soxhlet extraction and analyzed by HPLC when over 50 different PAHs were detected and about 20 identified (Fig. 1). This identification was confirmed by GC mass spectrometry (we thank Mr. J. Heald for these analyses).

Figure 1 shows the PAHs so far identified which have up to seven rings fused together. There were additional larger PAHs found by HPLC, but these are not volatile enough to confirm their structure by GC mass spectrometry. The total number and amount of PAHs estimated from these analyses increases with the amount of soot in a fairly linear manner and this is shown in Fig. 2.

The fraction of soot which is PAH is about 1.4×10^{-4} and seems to be independent of the conditions used in its production. The yields were found to give a bell-shaped curve for both heptane and toluene as shown in Fig. 3.

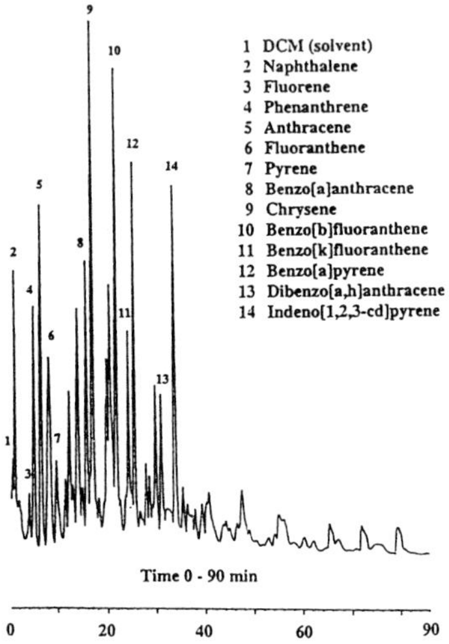

Figure 1. Chromatogram of soot extract from toluene combustion.

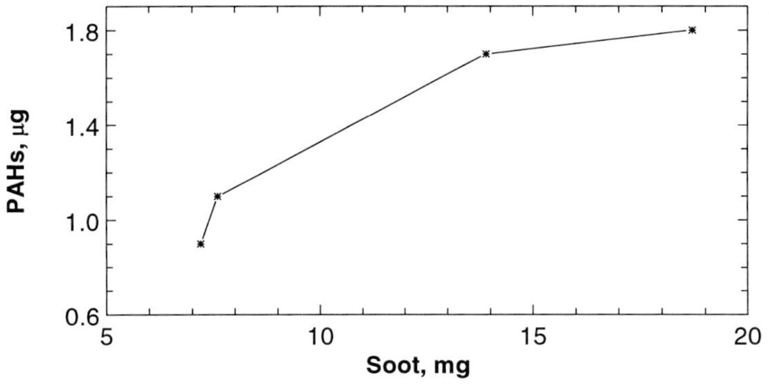

Figure 2. PAH yields against soot yields from heptane combustion at different temperatures.

Maximum production of soot appeared at about 1700 K for heptane pyrolysis and for the 1% and 10% oxygen mixtures. The yield was lower for both 10% oxygen and the pure argon case than for 1% oxygen. For toluene the results were similarly within the experimental error for both 1% oxygen and the argon-only case, except that the peak yield may be shifted to slightly higher temperatures. The total yield was much higher in the case of toluene than for heptane, especially if the fuel/oxygen ratio is taken into account. The influence of pressure

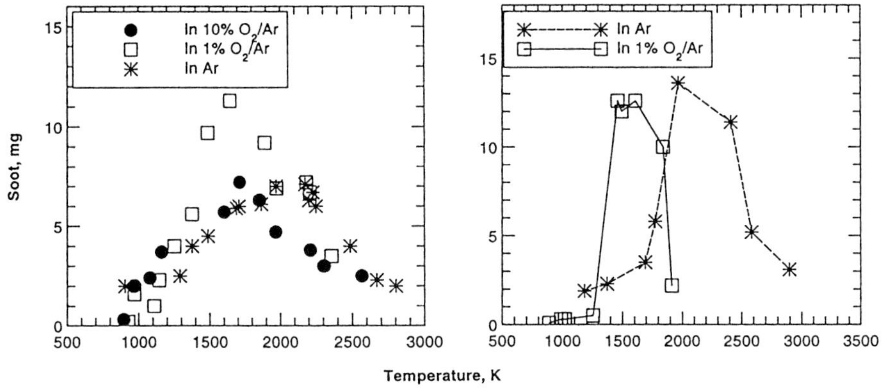

Figure 3. Soot yield versus temperature from heptane (left) and toluene (right).

on soot yield could not be determined with any accuracy for the experiments but any effect seemed minimal within this range. The main influences on soot production were chemical type, fuel/oxygen ratio, and temperature.

The main PAHs so far identified from the large number detected are polycyclic aromatics having a 2- to 7-ring structure. Some found in the soot extract are carcinogenic e.g., benzo[a]pyrene. The results of analysis indicate that similar PAHs are formed along with the carbon particulates regardless of temperature, pressure, oxygen concentration, and fuel used. However, the total amount produced and the amounts of individual products were dependent on the combustion conditions and fuel used.

The presence of fullerene (C_{60}) was detected in the soot extract from toluene combustion but not heptane, using a special column for its separation from PAH (although the amount was small). The uv-absorption spectrum and retention time on the column coincided with that of pure samples of fullerene (Mer Corp., Tucson, AZ, USA). Its detection was confirmed by mass-spectrometric analysis.

In other experiments using benzene sprays larger amounts of fullerene were found (Wang & Cadman 1995). Among the PAHs present, fluoranthrene and pyrene were there in the largest amounts. They may play a pivotal role in the production of other larger PAHs and soot. The mass-spectrometric analysis showed that the molar mass increased by 24 each time (C_2), which is expected for the formation of an additional fused ring. The growth of the larger PAHs therefore may occur by a series of benzene-ring additions to the existing smaller PAHs (Fig. 4), and eventually form soot or fullerene. Other reactions must also occur as single-ring compounds such as biphenyl, terphenyl, and benzophenone were found. The latter was only detected in the combustion of toluene.

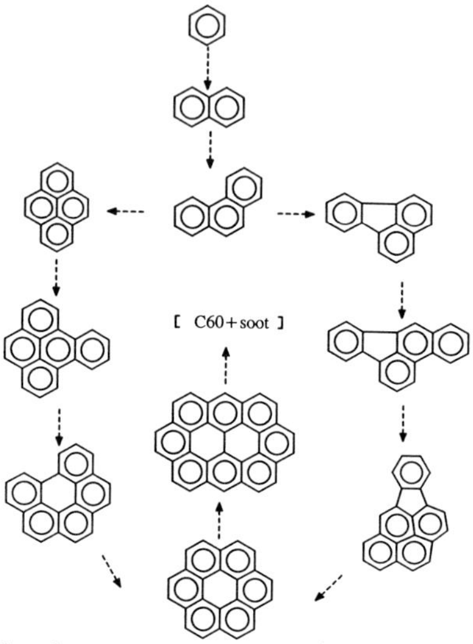

Figure 4. Molecular mass progress to form soot and fullerene.

References

Burcat A, Dyinyaninov M (1993) Ignition delays of n-pentane. In: Shock Waves @ Marseille, R Brun (ed), Springer-Verlag, p 197.

Cadman P (1995) Combustion of liquid hydrocarbon sprays. In: Shock Waves @ Marseille, R Brun (ed), Springer-Verlag, p 179.

Cierzki H, Adomeit G (1989) Ignition delays in n-heptane air mixtures. In: 17th International Symposium on Shock Waves, Y Kim (ed), p 707.

Cierzki H, Adomeit G (1993) Shock tube investigation of self ignition in n-heptane. Comb Flame 93:421.

Fieweger K, Cierzki H, Adomeit G (1995) Ignition characteristics of fuel-air mixtures. In: Shock Waves @ Marseille, R Brun (ed), Springer-Verlag, pp 160–167.

Frenklach M, Yuan T (1987) Effect of alcohol addition on soot formation from benzene. In: 16th International Symposium on Shock Waves, p 487.

Kamimoto T, Kobayashi H (1991) Diesel combustion. Prog Ener Comb Sci 17:163.

Wang R, Cadman P (1995) C_{60} detection in benzene combustion. Fullerene Sci and Tech 3:553.

Williams A, Alexiou A (1994) Soot formation in pyridine and tolulene-pyridine mixtures. Fuel 73:1280.

Determination of nonequilibrium temperatures of CN by simultaneous wavelength measurements on the $B^2\Sigma^+ \rightarrow X^2\Sigma^+$ violet system

L. Labracherie, M. Billiotte, K. Koffi-Kpante, L. Houas
IUSTI (UMR CNRS 139)
Marseille, France

M. Baillion
Aerospatiale, Division Espace et Défense
Les Mureaux, France

Abstract: The free-piston facility of the Marseille laboratory is used to perform radiation measurements downstream of a 5.7-km/sec normal shock wave propagating in a N_2-CH_4-Ar mixture at an initial pressure of approximately 2 mbar. The intensity of spontaneous emission for the $B^2\Sigma^+ \rightarrow X^2\Sigma^+$ electronic transition of CN molecules is recorded at three different wavelengths during the same run, and the rotational and vibrational temperatures are determined by matching these three spectral lines with synthetic spectrum calculations.

Key words: CN emission, Temperature measurements, Titan

1. Introduction

In the hypervelocity regime considered for the Huygens entry (about 6 km/sec), many complex processes take place behind the shock wave such as dissociation, vibrational and electronic excitation of molecules, as well as weak ionization. Furthermore, these physical and chemical processes have the same time scale as the flow itself, so that the forebody shock layer is in thermochemical nonequilibrium. The main components of the Titan atmosphere are N_2, CH_4, and possibly Ar. Then, because of the rapid dissociation of methane in the shock layer, a large part of the radiation is due to spontaneous emission for the $B^2\Sigma^+ \rightarrow X^2\Sigma^+$ electronic transition of CN molecules (Nelson et al. 1989). Baillion and Taquin (Baillion et al. 1993) have computed that the upstream chemical composition which leads to the maximum total radiative heat flux level was a 3%-CH_4, 20%-Ar and 77-% N_2 (in mole fraction) mixture. Experimental evidence of the choice of both the worst chemical upstream composition and the level of radiative heat flux was then to be performed in ground testing facilities.

2. Experiments

The Marseille free-piston facility (Labracherie et al. 1993) is used to perform spectroscopic experiments downstream of a 5.7-km/sec normal shock wave at an initial pressure of approximately 2 mbar. The intensity of radiation for the $B^2\Sigma^+ \rightarrow X^2\Sigma^+$ electronic transition of CN molecules is recorded behind the shock at the test-section passage, 3.5 m downstream of the main diaphragm. The emitted light behind the shock is focused onto a rectangular fiber optic such that the solid angle of light collection is less than 10^{-3} sr and the spatial broadening at the center of the shock tube is about 0.8 mm. The scheme of the optical apparatus for emission measurements is shown in Fig. 1. Two emission-

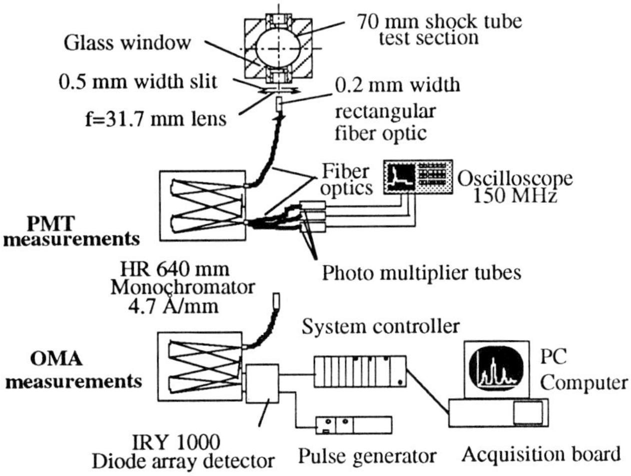

Figure 1. Optical apparatus for emission recording.

measurement devices are available: a time-evolution emission-intensity measurement with photo multiplier tubes (PMT) and a single-spectrum acquisition with an optical multi-channel analyzer (OMA). The time-dependent intensity of radiation is simultaneously recorded at three wavelengths with a rectangular bundle of 200-μm fiber optics at the monochromator exit. The horizontal spacing between the slits was previously fixed to fit in with the three spectral wavelengths of interest. This device is particularly attractive in reducing the number of shots needed, especially because the three scanned wavelengths correspond to identical experimental conditions. For single-shot OMA acquisition, the delay and the duration of the gate pulse can be tuned if the spectrum is to be measured in the equilibrium or in the nonequilibrium region. In the present study, only spectra in the equilibrium region have been measured because for nonequilibrium acquisitions, the gated time should be less than 200 ns, and in this case the signal-to-noise ratio is too low.

3. Results

3.1. Effects of argon addition on CN emission

In order to evaluate the effect of argon addition in a N_2-CH_4 upstream mixture on the nonequilibrium intensity of spontaneous emission, four different upstream gas mixtures were tested. The argon initial composition varies between 0% and 20%, while the methane molar fraction is 3% whatever the mixture. The shock-speed measurement is achieved by recording the time-of-flight between the three platinum heat-flux gauges located at the test-chamber wall. The error on shock-velocity and initial-pressure measurements is about 200 m/s and 0.01 mbar, respectively. The reproducibility of test conditions, which was evaluated by comparing the radiation profiles under the same initial test conditions, was found to be less than 3%. The comparison of radiation intensity profiles, recorded at 3882 Å for each of the four upstream mixtures, is shown in Fig. 2.

The peak intensity ranges, in arbitrary units, from 0.56 for the 5%-Ar mixture, to 0.64 for the 20%-Ar mixture, which is a maximum relative variation of up to 10%. Unfortunately, the presence of oscillations on the top part of the signal could lead to an underestimate of the value of the peak intensity. The characteristic excitation time is about 250 ns and no evidence of time increasing with argon addition has been observed from the four different emission records. However, the response time of the apparatus should be reduced to under 10 ns for a correct measurement of the emission excitation time. The total decay time of radiation varies from 10 μs for the 0%-Ar test mixture to 6 μs for the 20%-Ar test mixture, that is, a distance from 5.7 cm to 3.4 cm for present test conditions. Under the conditions of interest in the present experiment, translational, rotational, and vibrational relaxation are fast compared to the rates of chemical reactions in the gas. Then, the radiation decay time represents the chemical adjustment of the gas since all degrees of freedom of the molecules are in thermodynamical equilibrium. Therefore, if more heavy argon atoms are present in the shock layer, faster dissociation of CN molecules by means of collisional processes is to be expected.

3.2. Determination of rotational and vibrational temperatures

Figure 3 shows a comparison of the CN violet spectrum measured in the equilibrium region by OMA with the calculation by a line-by-line code (Arnold et al. 1969).

The temperature is determined from a comparison between the computed and the measured spectra: the temperature is iterated until the synthetic spectrum best fits the experimental one. The value of the equilibrium temperature is found to be 7800 K $\pm$ 600 K and is in good agreement with theoretical calculations. Theoretically, the ratio of two intensities in the same vibrational band is only a function of the rotational temperature, and the ratio of two intensities in different vibrational bands depends only on the rotational and vibrational temperatures. Thus, rotational and vibrational temperatures can be determined by measuring three intensities: two in the (0,0)

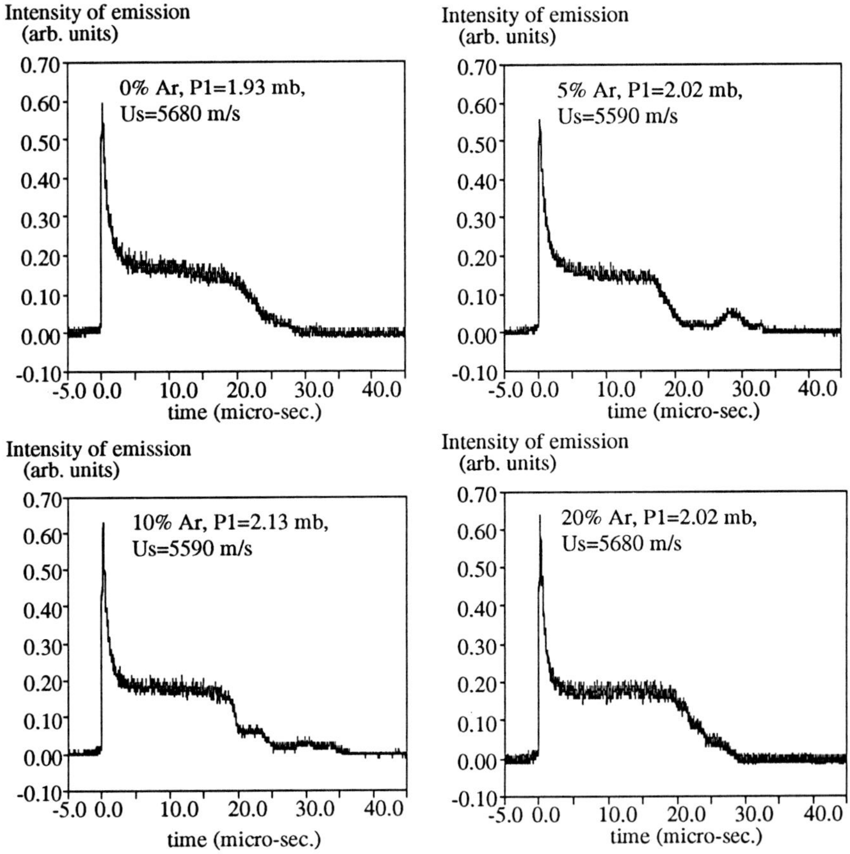

Figure 2. Comparison of CN emission profiles for various argon compositions in the upstream mixture.

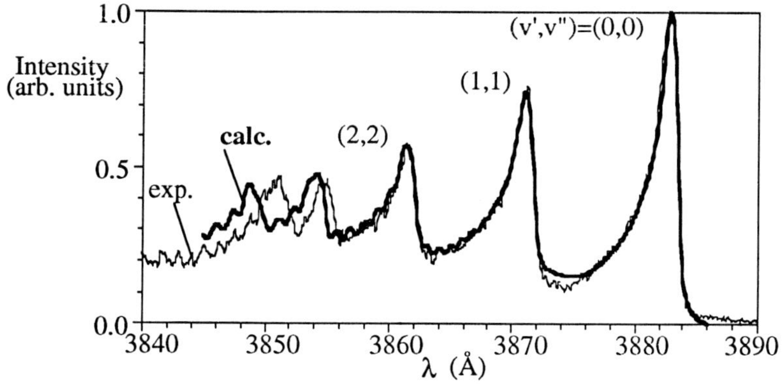

Figure 3. Comparison of experimental and computed CN spectra.

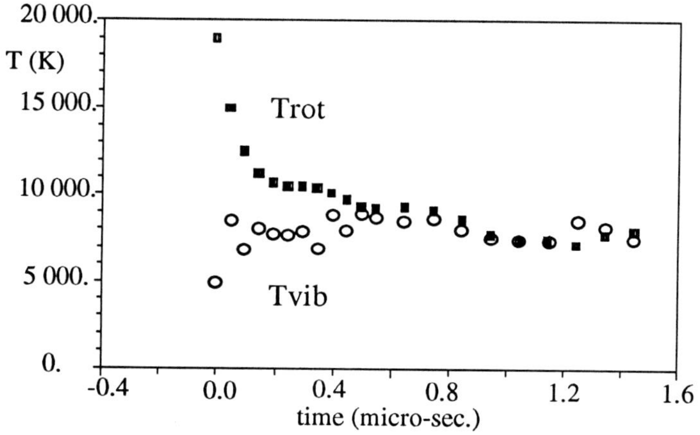

Figure 4. Experimental evolution of rotational and vibrational temperatures within the shock layer.

band and a third one in the (1,1) band. However, the overlapping of the different bands and the slit function of the instrument prevent one from obtaining a discrete relationship between the ratios of measured intensities and temperatures (Blackwell & Scott 1989, Labracherie et al. 1995). The temperature profiles calculated from time-dependent intensity measurements (3869.6 Å, 3875.8 Å, and 3879.7 Å) are shown in Fig. 4. The error on the rotational temperature determination is at least 20% and may be up to 50% for the vibrational temperature. The measured vibrational relaxation time is about 500 ns, which is twice the value calculated with the theoretical Millikan and White formula at a temperature of 10000 K. The spatial broadening of the radiation measurement at the center of the test section may account for this discrepancy. However, the temperature profiles are in good agreement with numerical calculation (Baillion et al. 1993, Park & Bershader 1991).

4. Conclusion

Rotational and vibrational temperatures of CN have been determined from a comparison between measured and computed intensities in the CN violet spectrum. However, the use of high-frequency amplifiers (typical rise time in the range of 10 ns) would allow one to obtain further information in the nonequilibrium region. The present experiments confirm the choice of the 20%-Ar, 3%-CH_4, 77%-N_2 mixture as the most radiative atmosphere for the Huygens probe.

References

Arnold JO, Whiting EE, Lyle GC (1969) Line by line calculation of spectra from diatomic molecules and atoms assuming a Voigt line profile. J Quant Spectrosc Radiat Transf 9:775–798.

Baillion M, Taquin G, Soler J (1993) Huygens radiative probe environment. In: Shock Waves @ Marseille, Brun R and Dumitrescu L (eds), Springer-Verlag, II:339–346.

Blackwell HE, Scott CD (1989) Nonequilibrium Shock Layer Temperature Profiles from Arc Jet Radiation Measurements, AIAA Paper 89-1679.

Labracherie L, Dumitrescu MP, Burtschell Y, Houas L (1993) On the compression process in a free-piston shock tunnel. Shock Waves 3(1):19–23.

Labracherie L, Billiotte M, Houas L (1995) Nonequilibrium determination of temperature profiles by emission spectroscopy. J Quant Spectrosc Radiat Transf 54(3):573–579.

Nelson HF, Park C, Whiting EE (1989) Titan Atmospheric Composition by Hypervelocity Shock Layer Analysis, AIAA Paper 89-1770.

Park CS, Bershader D (1991) Studies of radiative emission from the simulated shock layer of the Huygens probe. In: Shock Waves, Takayama K (ed), Springer-Verlag, I:671–676.

A shock-tube study of the reaction $Si + H_2 \rightarrow SiH + H$

H.-J. Mick, T. Kruse, P. Roth
Institut für Verbrennung und Gasdynamik
Gerhard Mercator Universität Duisburg
47048 Duisburg, Germany

Abstract: In the present investigation the very fast thermal dissociation of silane was used as a Si-atom source to study its reaction with H_2. Initial mixtures of 1 to 10 ppm SiH_4 and 500 to 1000 ppm H_2 diluted in Ar were used and both SiH and Si absorptions were measured in the reacting shock-heated gas. The experiments were performed in the temperature range 2030 K $\leq T \leq$ 3170 K at pressures between 1.1 and 1.8 bar. The formation of SiH radicals is caused by the reaction $Si + H_2 \rightleftharpoons SiH + H$ (R3), and its consumption is mainly due to the reaction $SiH + H_2 \rightleftharpoons SiH_2 + H$ (R4). From fittings of the calculated SiH profiles to the measured ones, the rate coefficients k_3 and k_4 were determined to be: $k_3 = 2.0 \times 10^{15} \exp(-16\,000\,\mathrm{K}/T)$ $\mathrm{cm^3\,mol^{-1}\,s^{-1}}$ ($\pm$35%) and $k_4 = 4.8 \times 10^{14} \exp(-11\,900\,\mathrm{K}/T)$ $\mathrm{cm^3\,mol^{-1}\,s^{-1}}$ ($\pm$35%). In contrast, measured Si profiles were not suited for distinguishing between the above reactions, because both are in a similar way sensitive to Si atoms. Therefore the measured profiles were fitted by small variations of k_3 and k_4 obtained from SiH profiles. The Si profiles observed were well fitted using this method, indicating that SiH and Si measurements are consistent, resulting in identical rate coefficients k_3 and k_4.

Key words: Chemical kinetics, Gas phase, Silicon

1. Introduction

In spite of the major role of silane chemistry in chemical-vapour-deposition (cvd) processes for semiconductor manufacturing and production of new ceramic materials, the elementary reactions of silane and its fragments are rarely studied. Particularly, high-temperature kinetic data of bimolecular reactions including silicon atoms are not available in the literature. Only one room-temperature study was performed by Husain and Norris (1978). They measured Si-atom absorptions in numerous reaction systems and determined rate coefficients of elementary reactions of Si atoms with other species. In the case of $Si + H_2 + He$, Husain and Norris (1978) were able to give an upper limit for the rate coefficient of $k \leq 3.6 \times 10^{14}$ $\mathrm{cm^6\,mol^{-2}\,s^{-1}}$. To our knowledge no further investigations concerning this reaction were performed.

In the present investigation initial $SiH_4/H_2/Ar$ mixtures were used to study the reaction of Si atoms with H_2 by detecting Si and SiH concentration profiles. Previous shock-tube experiments on the dissociation of silane revealed that highly diluted silane is a very effective source for silicon atoms at temperatures above 1800 K (see Mick et al. 1993). Under these conditions SiH_4 decomposes very rapidly to form Si atoms via SiH_2 as the intermediate species.

2. Experimental

The experiments were conducted behind reflected shock waves in a stainless steel shock tube of 79 mm inner diameter. The tube is constructed as a heatable ultra-high vacuum apparatus and can be evacuated through a special end-plate valve using a forepump, a turbo-molecular pump, and a liquid-nitrogen-cooled titanium sublimation pump, down to pressures of about 1×10^{-8} mbar. Typical leak-plus-outgassing rates were 5×10^{-6} mbar/min. The gases used were carefully mixed in a stainless steel cylinder by the partial-pressure method. The gas-mixing setup can be heated and evacuated like the shock tube. Due to the high sensitivity of the ARAS detection technique the test gases were of ultra-high purity: Ar (99.9999%), SiH_4 (99.995%), H_2 (99.9999%). Prior to each run the tube was evacuated to pressures of 5×10^{-7} mbar or lower. Diaphragms made of aluminum, 30 or 50 μm thick, were ruptured by increasing the driver-gas (H_2) pressure. Temperatures and pressures behind the reflected shock waves were calculated from the incident shock speed, which was measured by four thin-film gauges placed along the test section at known intervals. A more detailed description of the apparatus is given by Roth and Just (1977).

The ARAS technique was used to observe Si-atom absorptions at a wavelength $\lambda = 251.6$ nm. The optical arrangement consisted of a pulsed hollow-cathode lamp, the shock-tube absorption zone, a 0.25-m Jarrel-Ash monochromator, and a photomultiplier. Since the spectral shapes of the lines emitted by the hollow-cathode lamp are not known in detail a simple relation between the directly measured absorption and the atom concentration does not exist. Therefore a series of carefully controlled shock-wave calibrations has been done to relate the measured absorption to the corresponding Si concentration. The Si-atom calibration was carried out by Mick et al. (1993).

The ring dye laser technique was used to measure SiH concentrations in the Q-branch of the $A^2\Delta - X^2\Pi(0,0)$ band near 413.53 nm. The laser light was introduced into the measurement plane of the shock tube via an optical fiber. The intensities of the transmitted radiation were monitored by silicon detectors. The line center positions of the $Q_1(10)$ doublet used for the present absorption measurements are well known from the literature (see Klynning et al. 1979). Nevertheless, missing data on the pressure broadening of the absorption coefficient requires its experimental determination. Therefore the characteristic line shape, which is strongly dependent on pressure, of the partly resolved Λ-type doublet was analyzed. A more detailed description is given by Markus and Roth (1995).

3. Results and discussion

3.1. SiH-radical measurements

SiH absorption was measured in $SiH_4/H_2/Ar$ reaction systems in the temperature range 2030 K $\leq T \leq$ 3170 K at pressures between 1.1 and 1.8 bar. Initial mixtures of 5 to 10 ppm SiH_4 and 500 to 1000 ppm H_2 diluted in Ar were used. The absorption profiles measured were transformed into SiH concentrations using the spectroscopic data given above. A typical example is illustrated in Fig. 1 showing the time-dependent SiH concentration for an experiment with 10 ppm SiH_4 and 500 ppm H_2. The time zero indicates the arrival of the reflected shock

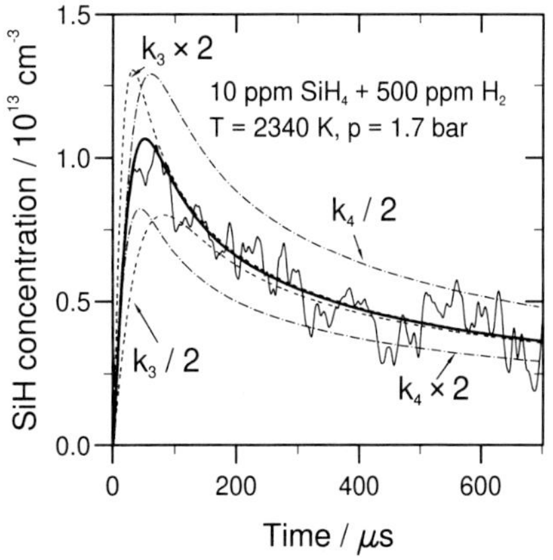

Figure 1. Measured and calculated SiH concentration profiles. The dashed profiles correspond to simulations with varied values of k_3 and k_4, respectively.

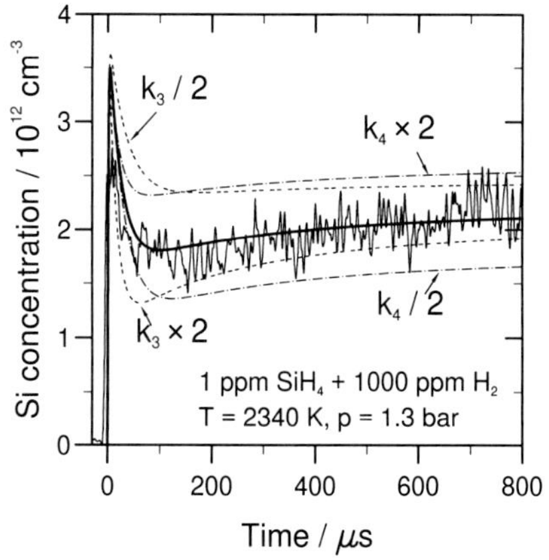

Figure 2. Measured and calculated Si concentration profiles. The dashed profiles correspond to simulations with varied values of k_3 and k_4, respectively.

wave in the measurement plane. The signal increases during the first 80 μs followed by a continous removal for the remaining reaction time. Experiments performed under the same experimental conditions, but without H_2, show no measurable absorption by SiH. Therefore, in the present case the formation of SiH radicals is obviously caused by the reaction: $Si + H_2 \rightleftharpoons SiH + H$ (R3). A sensitivity analysis based on a larger reaction scheme including silicon-hydrides identifies the reaction $SiH + H_2 \rightleftharpoons SiH_2 + H$ (R4) to be responsible for the consumption of SiH radicals. The data evaluation was done by computer simulations based on the simplified reaction scheme of Table 1. In that case the

Table 1. Simplified reaction mechanism of the SiH_4/H_2 system highly diluted in Ar; $k_i = A \times \exp(-E_a/T)$ $cm^3\,mol^{-1}\,s^{-1}$.

No	Reaction	Rate Coefficient		Reference
		A	E_A	
1	$SiH_4 + Ar \rightleftharpoons SiH_2 + H_2 + Ar$	9.9×10^{15}	24 000	Mick et al. (1993)
2	$SiH_2 + Ar \rightleftharpoons Si + H_2 + Ar$	9.1×10^{13}	15 100	Mick et al. (1993)
3	$Si + H_2 \rightleftharpoons SiH + H$	2.0×10^{15}	16 000	see text
4	$SiH + H_2 \rightleftharpoons SiH_2 + H$	4.8×10^{14}	11 900	see text
5	$H_2 + Ar \rightleftharpoons H + H + Ar$	2.2×10^{14}	48 300	Baulch et al. (1972)

calculated SiH profiles were fitted to the measured ones with the rate coefficients k_3 and k_4 as variable parameters. An individual example of this procedure is illustrated as a smooth solid line in Fig. 1. Comparable nearly perfect fits were obtained for all other experiments of this series. The influence of variations of rate coefficients k_3 and k_4 on the calculated SiH profiles is demonstrated by the dashed lines in Fig. 1. It can be seen that reaction (R3) influences the SiH profile at early reaction times during the formation of SiH radicals. In contrast, reaction (R4) dominates the system during the consumption of SiH radicals, i.e. at reaction times $t \geq 100$ μs.

The individual rate coefficients k_3 obtained from this data reduction are summarized as solid symbols in the Arrhenius diagram of Fig. 3. The data points can be represented by $k_3 = 2.0 \times 10^{15} \exp(-16\,000\ K/T)$ cm^3 mol^{-1} s^{-1}, which is shown as a solid line in Fig. 3. The combined uncertainty from the SiH diagnostic and the experimental scatter is about $\pm 35\%$. The preexponential factor of the Arrhenius expression given above is close to the collision frequency and seems to be too high. The activation energy obtained from this data evaluation of $E_a = 133$ kJ/mol is about 10 kJ/mol smaller than the reaction enthalpy calculated with the thermochemical data of Kee et al. (1994) and Chase et al. (1985). This discrepancy is within the uncertainty of the reaction enthalpy. In the literature, reaction (R3) was discussed by Husain and Norris (1978) based on Si-atom absorption measurements in a static system. They determined an upper limit for the three-body reaction $Si + H_2 + He$ of $k \leq 3.6 \times 10^{14}\,cm^6\,mol^{-2}\,s^{-1}$. The linear extrapolation of the present result to a three-body reaction and to room temperature results in a much smaller value indicating non-Arrhenius behavior.

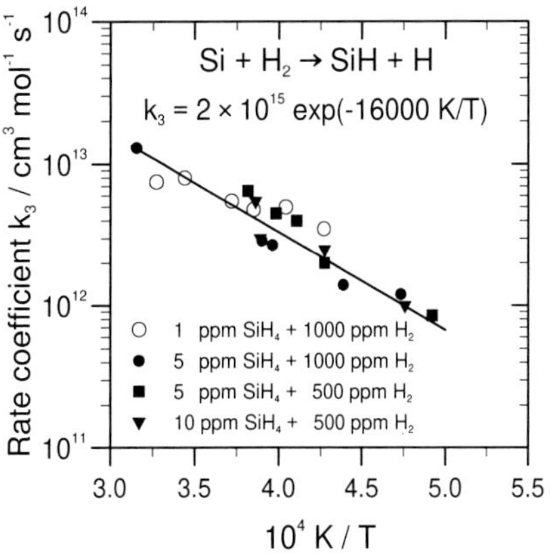

Figure 3. Arrhenius diagram of the reaction Si + $H_2 \rightleftharpoons$ SiH + H including rate coefficients obtained from SiH profiles (solid symbols) and from Si profiles (open circles).

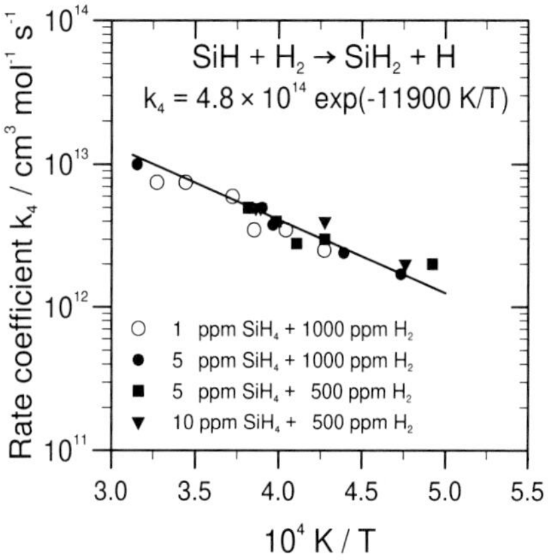

Figure 4. Arrhenius diagram of the reaction SiH + $H_2 \rightleftharpoons SiH_2$ + H including rate coefficients obtained from SiH profiles (solid symbols) and from Si profiles (open circles).

Individual rate coefficients k_4 obtained from fitting the SiH profiles at late reaction times are summarized as solid symbols in the Arrhenius diagram of Fig. 4. The data points can be represented by $k_4 = 4.8 \times 10^{14} \exp(-11\,900 \text{ K}/T)$ cm^3 mol^{-1} s^{-1}, which is shown as a solid line in Fig. 4. The combined uncertainty due to experimental scatter and the uncertainty of the SiH diagnostic is about $\pm 35\%$. The activation energy of $E_a = 99$ kJ/mol is about 13 kJ/mol smaller than the calculated reaction enthalpy and therefore within the uncertainty limit of the thermochemical data of Kee et al. (1994) and Chase et al. (1985).

3.2. Si-atom measurements

Si-atom absorption measurements were performed in the temperature range 2110 K $\leq T \leq$ 3060 K at pressures between 1.2 and 1.4 bar. Initial mixtures of 1 ppm SiH_4 and 1000 ppm H_2 were used. The measured light extinction signals at wavelength $\lambda = 251.6$ nm are equal to absorption by Si atoms because no interference absorption of other species was observed. The transformation to Si concentrations was done by applying the respective calibration curve. A typical Si concentration profile is illustrated in Fig. 2. The signal shows a very rapid increase at reaction time zero followed by a fast consumption to about 50% of the maximum value within a time period of 100 μs. For later reaction times the Si concentration increases slightly. This behavior can well be described by the simplified reaction mechanism given in Table 1. The high sensitivity of the rate coefficients k_3 and k_4 with respect to Si concentration is demonstrated by the dashed profiles in Fig. 2, which were calculated from variations of k_3 and k_4. It can be seen that both rate coefficients are in a similar way sensitive to the calculated Si profiles. Thus Si-atom measurements are not suited for distinguishing between the rate coefficients k_3 and k_4. Because of this problem, the calculated Si profiles were fitted to the measured ones by small variations of k_3 and k_4 obtained from the SiH profiles. The smooth solid line in Fig. 2 demonstrates the result of such a fitting procedure. Comparable good fits were obtained for all other experiments of this series. The individual rate coefficients k_3 and k_4 obtained from this data reduction are summarized as open circles in Fig. 3 and Fig. 4, respectively. The values are within the uncertainty limit of the rate coefficients obtained from SiH measurements. Both sets of kinetic data are consistent.

Acknowledgement. The authors thank Mrs. C. Kmiecik for her help in conducting the experiments. The financial support of the Deutsche Forschungsgemeinschaft is gratefully acknowledged.

References

Baulch DL, Drysdale DD, Horne DG, Lloyd AC (1972) Homogeneous Gas Phase Reactions of the H_2-O_2 System, Vol. 1, Butterworth, London.

Chase MW Jr, Davies CA, Downey JR Jr, Frurip DJ, McDonald RA, Syverud AN (1985) JANAF thermochemical tables. J Phys Chem Ref 14(1).

Husain D and Norris PE (1978) Kinetic study of reactions of ground state silicon atoms, $Si[3p^2(^3P_j)]$, by atomic absorption spectroscopy. J Chem Soc Faraday II 74:106.

Kee RJ, Rupley FM, Miller JA (1994) The Chemkin Thermodynamic Data Base, Sandia Report SAND87-8215B, Sandia National Laboratories.

Klynning L, Lindgren B, Sassenberg U (1979) On the Λ-type doubling in the ground state of SiH. Physica Scripta 20:617–619.

Markus MW, Roth P (1995) Development of a quantitative ring dye laser absorption diagnostic for free SiH radicals. J Quant Spectrosc Radiat Transfer (submitted).

Mick HJ, Smirnov VN, Roth P (1993) ARAS measurements on the thermal decomposition of silane. Ber Bunsenges Phys Chem 97:793–798.

Roth P, Just T (1977) Atomabsorptionsmessungen zur kinetik der reaktion $CH_4+O \rightarrow CH_3+OH$ im temperaturbereich 1500 K$\leq T \leq$ 2250 K. Ber Bunsenges Phys Chem 81(6):572.

Laser-flash-photolysis study of the decomposition of vibrationally excited CO_2

A. V. Eremin, V. V. Shumova, V. S. Ziborov
High Energy Density Research Center
Russian Academy of Sciences, 127412 Moscow, Russia

H.-J. Mick, P. Roth
Institut für Verbrennung und Gasdynamik
Gerhard-Mercator-Universität Duisburg
D-47048 Duisburg, Germany

Abstract: The photodissociation of CO_2 induced by the pulse of an excimer Ar-F laser ($\lambda = 193$ nm) was studied in CO_2/Ar, CO_2/He, and CO_2/N_2/Ar mixtures, which were first heated up by a reflected shock wave to temperatures 860 K $\leq T \leq$ 2800 K. ARAS measurements of $O(^3P)$ atoms carried out during photolysis of CO_2 under equilibrium conditions could be approximated by

$$\frac{[O(^3P)]_{phot}}{[CO_2]_0 \cdot j} = \frac{\sigma}{Q_{CO_2}} \exp(-10630 \text{ K}/T) \ .$$

where j is the photon flux, Q_{CO_2} the vibrational partition function, and $\sigma = 1.5 \times 10^{-16}$ cm^2 the cross-section of laser quanta absorption. The temperature dependence of the present result indicates that CO_2 molecules of energy level $E^*/k = 10630$ K are mostly responsible for the absorption of the laser energy. The interpretation of results, obtained under vibrational nonequilibrium conditions in relaxing CO_2/N_2/Ar mixtures revealed that the eleventh level of the bending mode of CO_2 is mostly responsible for the absorption of laser radiation. The correlation of this experimentally observed behavior with the geometry of the lowest electronic states of CO_2 indicates that the only possibility for laser-induced dissociation is to go via the spin-forbidden channel to $CO_2(^3B_2)$, which decomposes to form $CO(^1\Sigma) + O(^3P)$.

Key words: ARAS, Laser-flash photolysis, Vibrational nonequilibrium

1. Introduction

Laser photolysis of shock-heated gases provides essential progress in chemical kinetics of elementary reactions and gives an insight into the role of molecular excited states during reaction, see Hanson (1993) and Matsui et al. (1994). Moreover, the flash photolysis of polyatomic molecules in vibrational nonequilibrium makes it possible to learn more about the role of different vibrational modes during light absorption. These problems of reacting species in nonequilibrium gases with highly excited vibrational levels were the subjects of numerous discussions, e.g., Koshi et al. (1991) and Matsumi et al. (1991). Nevertheless, further investigations are necessary to understand these phenomena in detail.

A notable example is the photodissociation of carbon dioxide under the influence of uv absorption by the red-shifted wing of the ($X^1\Sigma - {}^1B_2$) band of CO_2. Recently, Matsumi et al. (1991) found that photolysis of CO_2 at 131 nm and 147 nm occurs mostly via the spin-allowed channel to form $O(^1D)$ atoms, whereas during photolysis at 157 nm, 6% of the oxygen atoms formed were observed to be in the $O(^3P)$ state. On the other hand, Koshi et al. (1991) found in shock-tube photolysis experiments at 193 nm that most of the O atoms formed were in the $O(^3P)$ state. The goal of the present work is to study the detailed dynamics of carbon-dioxide photodissociation at various temperatures, mixtures, and different vibrational excitation levels.

2. Experimental

The experiments were performed in an ultra-high-vacuum shock tube of the Institut fuer Verbrennung und Gasdynamik. The emission of an Ar-F excimer laser (193 nm) was coupled into the measurement plane through an end plate made of quartz glass. The energy of the laser emission was varied in the range 10 mJ $\leq E \leq$ 120 mJ. The laser pulse with a duration of about 13 ns was synchronized with the shock wave and triggered with a variable delay time of 0 to 1 ms with respect to the arrival of the reflected shock wave in the measurement plane. The shock tube was equipped with an ARAS diagnostic for measuring O atoms at wavelength $\lambda = 130.5$ nm during photolysis of CO_2. The transformation of measured absorption into concentration requires a calibration, which was performed by Mick et al. (1993). Interference absorption of the OI radiation by CO_2 was taken into account. The absorption cross-section determined from the present experiments in the temperature range 1200 to 2800 K can be represented by $\sigma_{OI}(CO_2) = (2.1 \pm 1.0) \times 10^{-17} \exp(2090 \pm 230\mathrm{K}/T)$ cm^2. The absorption profiles measured were evaluated considering this interference absorption. A more detailed description of the experimental setup is given by Woiki et al. (1993).

Additional experiments on CO_2 emission at $\lambda = 2.7$ μm and $\lambda = 4.3$ μm in mixtures of 1% CO_2 diluted in 99% N_2 were carried out at the High Energy Density Research Center. A detailed description of the shock tube used is given by Eremin and Ziborov (1993). From these experimental results it was possible to determine the vibrational energy relaxation of the asymmetric mode v_3 or the respective temperature T_3 as well as the energy relaxation of the bending mode v_2 expressed by T_2.

3. Results and discussion

The first series of experiments verified that the absorption by CO_2 was not saturated by the laser pulse. Saturation can happen especially at low temperatures i.e., low level of vibrational excitation. For this purpose the $O(^3P)$-atom yields were measured under nearly constant experimental conditions only varying the energy of the laser pulse. It was found that the yield of photolysis was a linear function of laser energy, i.e., the effect of saturation was negligible up to photon densities of about $j = 2 \times 10^{16}$ phot cm^{-2}. The photolysis yields obtained were therefore normalized with the photon density to become independent of the slightly varying laser pulse energies.

The next group of experiments included measurements of the temperature dependence of $O(^3P)$ atom yields during photolysis under thermal equilibrium conditions. Experiments were carried out in mixtures containing 1000 to 5000 ppm of CO_2 diluted in Ar or He. The temperature and pressure behind the reflected shock waves were varied in the range 800 K $\leq T \leq$ 2800 K and 0.2 bar $\leq p \leq$ 2.5 bar. In all of the experiments vibrational equilibrium was established after 1 μs. The experiments showed a strong temperature dependence of the O-atom yield after photolysis of CO_2 while no dependence on pressure was found. All of the results obtained are summarized in the Arrhenius diagram of Fig. 1. The measured O-atom yields were, in the present case, normalized with the initial CO_2 concentration and the photon flux j. The data points scatter around a straight line, which can be approximated by

$$\frac{[O(^3P)]_{\rm phot}}{[CO_2]_0 \cdot j} = 6.2 \times 10^{-19} \exp(-5480 \text{ K}/T) \text{ cm}^2 \text{ phot}^{-1} \; . \qquad (1)$$

An equivalent description of the experimental results is possible by introducing the CO_2 vibrational partition function, which is a weak function of temperature

$$\frac{[O(^3P)]_{\rm phot}}{[CO_2]_0 \cdot j} = \frac{1.5 \times 10^{-16}}{Q_{CO_2}} \exp(-10630 \text{ K}/T) \text{ cm}^2 \text{ phot}^{-1} \; . \qquad (2)$$

These experimental results can be compared with the photolysis efficiency of the unknown excited state of CO_2, which absorbs the laser energy resulting in the decomposition product $O(^3P)$:

$$[O(^3P)]_{\rm phot} = q \cdot j \cdot \sigma \, [CO_2]^* \; , \qquad (3)$$

where q, j, and σ are the quantum yield of photolysis, the photon flux, and the absorption cross-section, respectively. The property $[CO_2]^*$ is the concentration of CO_2 molecules in the energy state E^* which are able to absorb the offered photon energy. For equilibrium conditions their population can be determined

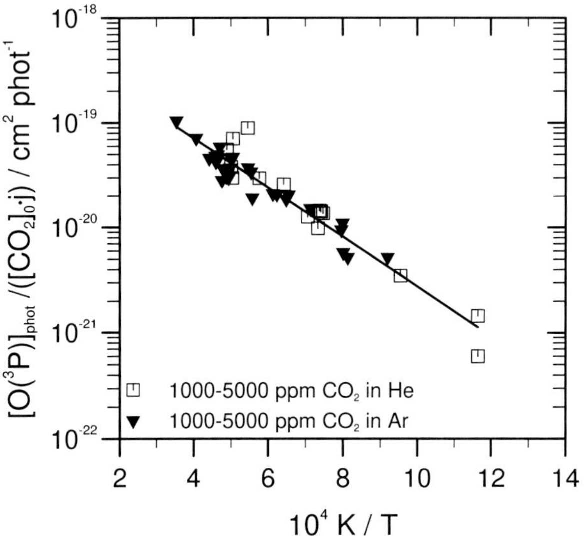

Figure 1. Temperature dependence of $O(^3P)$-atom yields in excimer laser photolysis of CO_2 behind reflected shock waves at thermal equilibrium conditions.

from the Boltzmann distribution:

$$\frac{[CO_2]^*}{[CO_2]_0} = \frac{\exp(-E^*/k\,T)}{Q_{CO_2}} \quad . \tag{4}$$

The combination of Eqs. (3) and (4) yields

$$\frac{[O(^3P)]_{phot}}{[CO_2]_0 \cdot j} = \frac{q \cdot \sigma}{Q_{CO_2}} \exp\left(-\frac{T^*}{T}\right) \ , \tag{5}$$

where $T^* \equiv E^*/k$ has been substituted. Comparison of the experimental result of Eq. 2 with the theoretical relation of Eq. 5 allows identification of the energy state and the cross-section of the absorbing CO_2 species. For an assumed quantum yield of photolysis of $q = 1$ (see Koshi et al. 1991), the following properties can easily be determined:

$$\begin{aligned} \sigma &= 1.5 \times 10^{-16}\ \mathrm{cm}^2 \\ T^* &= E^*/k = 10630 \pm 1000\mathrm{K} \ , \end{aligned} \tag{6}$$

which characterize the light-absorbing CO_2 state.

The remaining problem is to identify the vibrational mode and level of the CO_2 molecules absorbing the energy of the photolysis laser. The measured temperature dependence of the O-atom yield during photolysis under vibrational equilibrium conditions indicates that either the 11th rotational level of the bending mode $E_b(11)/k = 10550$ K or the 3rd level of the asymmetric mode $E_a(3)/k = 10140$ K must be considered. A distinction between the two is possible by making use of the different relaxation behavior of the two vibrational modes. It is known from relaxation measurements in CO_2/N_2 mixtures that the asymmetric mode of CO_2, which is strongly coupled to the slowly relaxing N_2 vibration, remains "cool," whereas the relaxation of the bending mode is fast.

We have therefore performed a further series of experiments, in which photolysis of CO_2 was studied under vibrational nonequilibrium conditions in mixtures of 1000 ppm/2000 ppm of CO_2 plus 10% N_2 in Ar. The photolyzing laser was fired into the relaxing mixtures with different delay times compared to the time of shock arrival. A typical result is illustrated in Fig. 2. The measured O-atom

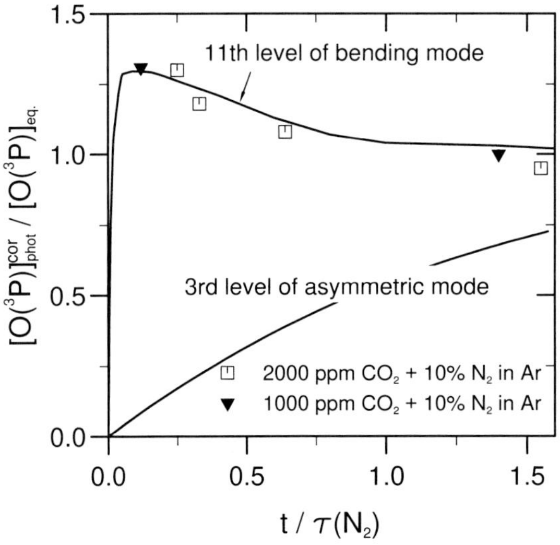

Figure 2. Time dependence of the photolysis yield at vibrational nonequilibrium in comparison with the calculated populations of different vibrational levels of CO_2.

yield $[O(^3P)]_{phot}$ was corrected because of the different quenching influence of N_2 in the mixtures and normalized with the measured O-atom concentrations at equilibrium. The time axis in Fig. 2 was made dimensionless by the vibrational relaxation time of N_2. Figure 2 also contains the calculated relaxing population densities of the 11th level of the bending mode and the 3rd level of the asymmetric mode. From the good agreement between measured $O(^3P)$ concentration during photolysis and the population density of the 11th level of the bending mode of CO_2, we conclude that the dynamics of the CO_2 photolysis goes in the following way: the energy of the 193-nm radiation is first absorbed by the 11th level of the v_2 mode of ground state $CO_2(X^1\Sigma)$ resulting in a vibrationally excited $CO_2(^1B_2)$ state with an energy of about 7.3 eV. This energy is below the limit of the spin-allowed decomposition of CO_2. It is therefore most probable that CO_2 undergoes a transition to the triplet $CO_2(^3B_2)$ state, which in turn decomposes to form the observed $O(^3P)$ and $CO(^1\Sigma)$.

Acknowledgement. The authors would like to thank Mr. D. Woiki for the helpful discussions, Mr. L. Jerig for the excellent engineering support and Mr. T. Kruse for the help in conducting the experiments. The financial support of the Russian Fundamental Research Foundation (Gr.No.94-03-08881a) and Deutsche Forschungsgemeinschaft is gratefully acknowledged.

References

Eremin AV, Ziborov VS (1993) Non-equilibrium radiation from the CO_2 ($^1B_2 - X^1\Sigma_{1g}$) in shock-heated flows. Shock Waves 3:11.

Hanson RK (1993) The role of lasers in shock tube studies of chemical kinetics. In: Shock Waves @ Marseille, R Brun, L Dumitrescu (eds), Vol 2, p 7.

Koshi M, Yoshimira M, Matsui H (1991) Photodissociation of O_2 and CO_2 from vibrationally excited states at high temperatures. Chem Phys Lett 176(6):519.

Matsumi Y, Shafer N, Tonokura K, Kawasaki M, Huang Y-L, Gordon R (1991) Doppler profiles and fine-structure branching ratio of $O(^3P_j)$ from photodissociation of carbon dioxide at 157 nm. J Chem Phys 95(10):7311.

Matsui H, Koshi M, Oya M, Tsuchiya K (1994) Improvement of chemical kinetics data at high temperatures by piston-actuated shock tube, excimer laser photolysis, and atomic resonance absorption spectroscopy. Shock Waves 3:287.

Mick H-J, Burmeister M, Roth P (1993) Atomic resonance absorption spectroscopy measurements on high-temperature CO dissociation kinetics. AIAA J 31(4):671.

Woiki D, Markus MW, Roth P (1993) A shock tube-laser flash photolysis study of the reaction $COS+S \rightleftharpoons CO+S_2$. J Phys Chem 97(38):9682.

Formation of C_2 during high-temperature pyrolysis of fullerene C_{60} in shock waves

T. Sommer, T. Kruse, P. Roth
Institut für Verbrennung und Gasdynamik
Gerhard-Mercator-Universität Duisburg
D-47048 Duisburg, Germany

Abstract: The formation of C_2 during thermal decomposition of fullerene C_{60} was studied behind shock waves by time-dependent absorption and emission measurements. The experiments were performed in the temperature range 2450 K $\leq T \leq$ 3450 K at pressures 1.1 bar $\leq p \leq$ 2.0 bar. C_2-concentration profiles were measured with a ring dye laser spectrometer at $\nu = 21393.12\ \mathrm{cm}^{-1}$. Simultaneously, the time behavior of spectrally resolved light emitted from the shock-heated aerosol was recorded by an intensified CCD camera. A simplified reaction mechanism with both C_2-formation and consumption reactions was proposed and the results were discussed in terms of kinetic parameters.

Key words: Thermal decomposition, C_2, Absorption spectroscopy

1. Introduction

The existence of fullerenes as a new carbon modification has been established during the last few years. The C_{60} molecule discovered to be stable by Kroto et al. (1985) was proposed to be a soccer-ball shaped cage where the carbon atoms are placed at the edges of 12 pentagons and 20 hexagons. The proposed structure was confirmed in an immense number of experimental studies. In contrast to many investigations concerning the stability of C_{60} molecules, only a few papers dealt with its chemical kinetics during decomposition. O'Brien et al. (1988) found that the decisive step in fragmentation during photolysis of fullerene is the loss of C_2 followed by a rearrangement of the remaining fragments.

The aim of the present work was to study the formation of C_2 during pyrolysis of fullerene C_{60} at high temperatures behind reflected shock waves and to describe the experimental results by a kinetic model. Two different diagnostic tools were applied to follow the shock-induced reactions in C_{60}/Ar mixtures. A ring dye laser spectrometer was used to detect time-dependent C_2 formation by absorption measurements in the (1–0) band of the $\mathrm{d}\,^3\Pi_g - \mathrm{a}\,^3\Pi_u$ transition. Simultaneously the light emitted from the shock-heated C_{60}/Ar mixture was dispersed by a spectrograph and the time behavior of the spectrally resolved emission was recorded by an intensified CCD camera.

2. Experimental

The experimental setup used to study fullerene behavior at high temperatures consists of an aerosol generator for dispersing the fullerene powder, the main shock tube for heating the C_{60}-containing aerosol to high temperatures, a ring dye laser spectrometer for the measurement of C_2-absorption signals, a spectrograph with intensified CCD camera for the detection of spectral- and time-resolved emission, and an IR-diode laser system for the determination of the initial concentration of C_{60} via measurements of CO and CO_2 after a complete oxidation of C_{60} in $C_{60}/O_2/Ar$ mixtures. A detailed technical description of the experiment is given by Sommer and Roth (1995).

3. Results

The measurements of both the spectral emisson of high-temperature C_{60}/Ar mixtures and the absorption of C_2 were performed behind reflected shock waves at temperatures between 2450 K and 3450 K and at pressures between 1.1 bar and 2.0 bar.

Detectable C_2-absorption signals were recorded only for temperatures above 2650 K. Representative time-resolved absorption profiles converted to C_2 concentrations via Lambert-Beer's law are shown in Fig. 1. It is obvious that both the formation rate of C_2 after shock arrival and the C_2 peak concentration depend on temperature. Higher temperatures lead to a higher yield of C_2 and also the maximum concentration is reached at shorter times. In the temperature range of detectable C_2 concentrations this time of increase and the peak concentration varied from 600 μs and 1 ppm C_2 at 2650 K up to 10 μs and 20 ppm C_2 at 3450 K. At temperatures $T > 2800$ K the measured C_2-absorption profiles show a characteristic decrease after reaching maximum concentration.

The emission of high-temperature C_{60}/Ar mixtures was recorded in the wavelength range between 360 nm and 600 nm. Typical examples obtained from two different shock-tube experiments at time $t = 150$ μs after shock arrival are illustrated in Fig. 2. The characteristic band structures with peaks at $\lambda \approx 467$ nm, $\lambda \approx 516$ nm, and $\lambda \approx 560$ nm were identified to belong to different progressions ($\Delta v = -1, 0, +1$) of the $d\,^3\Pi_g - a\,^3\Pi_u$ transition of C_2. The measured signals in the range between 360 nm and 420 nm correspond to the $A\,^1\Pi_u - X\,^1\Sigma_g^+$ transition of C_3. Similar spectra were obtained for all experiments at temperatures $T > 2650$ K, whereas no characteristic emission of either C_2 or C_3 was observed at lower temperatures. It was found that the time behavior of the emission signals was in excellent agreement with the C_2-absorption measurements.

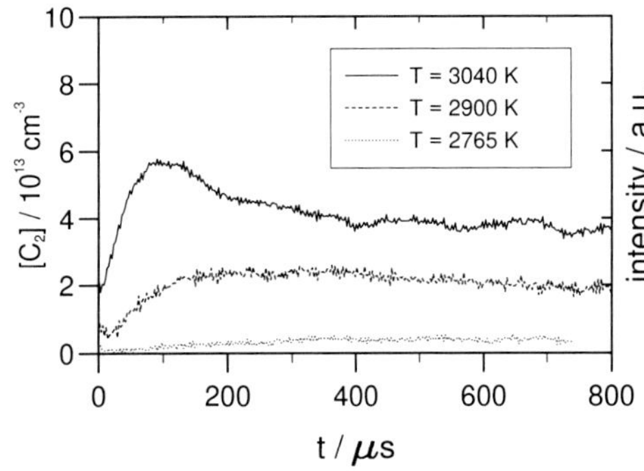

Figure 1. Examples of C_2 concentrations measured during pyrolysis of C_{60} at different post-shock temperatures.

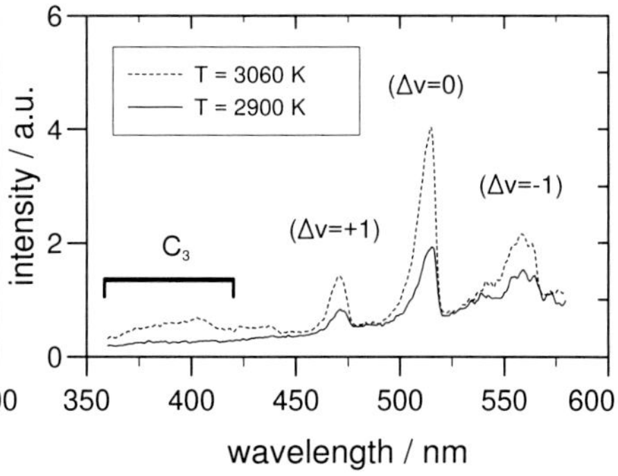

Figure 2. Examples of spectral emission of shock-heated C_{60}/Ar mixtures at two post-shock temperatures, $t = 150$ μs after shock arrival.

From infrared-absorption measurements of CO and CO_2 during complete fullerene oxidation an initial concentration of about 40 ppm C_{60} was obtained for all experiments. Shock-tube experiments with both pure Ar or O_2/Ar mixtures without any fullerenes showed no absorption of C_2, CO, and CO_2. Also in C_{60}-pyrolysis experiments, no absorption of CO and CO_2 could be detected.

4. Discussion

It can be assumed that several subsequent processes occur in the shock-heated C_{60}/Ar aerosol. Due to the relatively high vapor pressure of C_{60}, the solid agglomerated carbonaceous particles evaporate rapidly at high temperatures. This process starts behind the incident shock wave, where temperatures are in the range 1300 K $\leq T \leq$ 1650 K, and continues behind the reflected wave much more rapidly because of the higher temperature level. An upper limit for the fullerene particle evaporation of about 30 μs for conditions used during the present study can be calculated based on thermodynamic data for fullerene C_{60}. This value was experimentally confirmed by laser-light scattering and extinction measurements behind shock waves (von Gersum et al. 1994). We therefore can assume for all our experiments that the initial reactant C_{60} is in the vapor phase for nearly the complete observation time and that the C_{60} pyrolysis is dominated by homogeneous gas-phase kinetics.

A first step towards an interpretation of the absorption measurements is to determine the C_2 slope at early reaction time and to relate it to the concentration of the initial reactants. In this way an apparent rate coefficient for the appearance of C_2 during high-temperature pyrolysis of C_{60} can be defined:

$$k_{app} = \left. \frac{d\,[C_2]\,/\,dt}{[C_{60}]\,[Ar]} \right|_{t\approx 0} \tag{1}$$

All individual values for apparent rate coefficients obtained in this way are summarized in the Arrhenius diagram of Fig. 3 (see open squares). A least-square fit to the data points represented by the line in this diagram leads to an activation energy of about 495 kJ mol^{-1} and can be expressed by:

$$k_{app} = 2.7 \times 10^{17} \exp(-495\ \mathrm{kJ/mol}/RT)\ \mathrm{cm^3\ mol^{-1}\ s^{-1}} \tag{2}$$

It is clear from the C_2-concentration measurements of Fig. 1 that for a more detailed kinetic interpretation at least one C_2-formation and one C_2-consumption reaction must be considered. It is near at hand to assume a direct C_2 abstraction from C_{60} and a removal of the C_2 concentration by a bimolecular $C_2 + C_2$ reaction, which was recently observed during high-temperature pyrolysis of acetylene highly diluted in argon (Kruse and Roth 1996).

$$C_{60} + Ar \longrightarrow C_2 + C_{58} + Ar \qquad \text{(R1)}$$
$$C_2 + C_2 \rightleftharpoons C_3 + C \qquad \text{(R2)}$$

The remaining question is what happens to the fullerene fragments. With respect to our measurements we can consider two extreme cases: a further fragmentation <u>*without*</u> any C_2 formation (decomposition model I),

$$C_{58} + Ar \xrightarrow{x\neq 2} C_x + C_{58-x} + Ar$$
$$\left.\begin{array}{l} C_{58-x} + Ar \longrightarrow \\ C_x + Ar \longrightarrow \end{array}\right\} \sum_{y\neq 2} C_y + Ar$$

or a further breakdown of the fullerene fragments <u>*with*</u> C_2 formation during any partial reaction step (decomposition model II),

$$C_{58} + Ar \longrightarrow C_2 + C_{56} + Ar$$
$$C_{58-x} + Ar \longrightarrow C_2 + C_{56-x} + Ar \qquad x = 2, 4, 6 \ldots$$

Computer simulations using the SENKIN program package were performed based on the two proposed decomposition models.

Due to the fact that the thermodynamic equilibrium between C, C_2, and C_3 is not reached within the detection time of 800 μs at temperatures lower than 3500 K, reaction (R2) is assumed to be quasi-equilibrated. The rate coefficient of the reverse reaction was calculated based on the equilibrium constant with a slightly modified value of the heat of formation of C_2 (see Urdahl et al. 1991).

This variation of the thermodynamic data was confirmed by recent measurements in different reaction systems (Kruse and Roth 1996). The rate coefficient of reaction (R2) is taken from Kruse and Roth (1996):

$$k_2 = 2.0 \times 10^{15} \exp(-75 \text{ kJ/mol}/RT) \text{ cm}^3 \text{ mol}^{-1} \text{ s}^{-1} \quad (3)$$

Rate coefficient k_1 was varied during computer simulations to achieve a good fitting of the measured C_2 profiles. In the case of decomposition model II, as a lower limit, all rate coefficients of subsequent C_2-forming reactions were assumed to be equal to k_1, although fullerene fragments are expected to be less stable than C_{60}. Typical examples of calculated C_2 profiles in comparison with experimental traces are illustrated in Fig. 4. In the case of decomposition model II, it was not possible to fit the measured C_2 profiles by varying the rate coefficient k_1 (see dotted line). If more realistic, higher rate coefficients compared to k_1 were considered for the decomposition of fragments, the deviation between measured concentration profiles and simulations based on decomposition model II would be even greater.

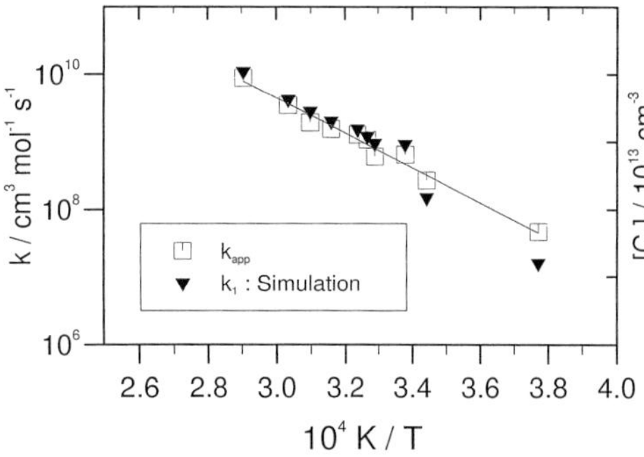

Figure 3. Arrhenius diagram of the apparent rate coefficient and of rate coefficient k_1 determined from simulations based on the decomposition model I.

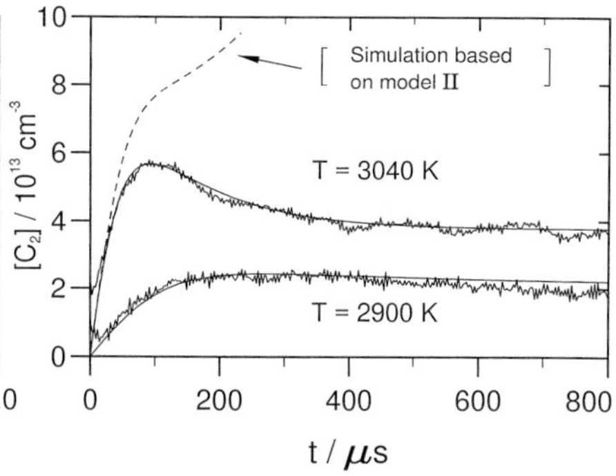

Figure 4. Comparison between measured (noisy lines) and calculated C_2 concentrations: decomposition model I (solid lines) and decomposition model II (dashed lines).

For the decomposition model I nearly complete agreement between calculated and measured C_2 profiles for the two examples in Fig. 4 as well as for all other experiments was obtained. The individual rate coefficients k_1 used as best-fit parameters are summarized in the Arrhenius plot of Fig. 3 (filled symbols). The deviations from the earlier proposed apparent rate coefficient k_{app} are comparatively small. Although this fact cannot be regarded as a verification of the proposed decomposition model I, it nevertheless must be interpreted as a hint for the correctness of the mechanism. The differences in the k_1 and k_{app}

data values of Fig. 3 at the low-temperature end are clearly caused by uncertainties in the determination of $d[C_2]/dt|_{t=0}$ due to the relatively low signal-to-noise ratio. Because of the minor differences in the two sets of data points in Fig. 3, we assume the rate coefficient k_1 for the ejection of C_2 from C_{60} is equal to the apparent rate coefficient k_{app} expressed by Eq. (2). The activation energy of 495 kJ mol^{-1} or 5.1 eV determined in the present study is in between values of 4.0 eV and 7.8 eV reported by Eckhoff et al. (1993) and Kolodney et al. (1994), respectively. Theoretical calculations often based on Hartree-Fock methods predict much higher values between 11 and 14 eV.

5. Conclusion

Time-resolved C_2-concentration measurements during high-temperature pyrolysis of fullerene C_{60} were performed behind shock waves under well-defined conditions using a ring dye laser spectrometer and an intensified CCD camera. C_{60} seems to be stable at temperatures below 2650 K and only at higher temperatures an increasing amount of C_2 could be observed. A simplified reaction mechamism based on C_2-formation and consumption reactions was proposed. Computer simulations were compared with the experimental C_2 traces and rate coefficients for the abstraction of C_2 from C_{60} were determined.

References

Eckhoff WC, Scuseria GE (1993) A theoretical study of the C_2 fragmentation energy of C_{60} and C_{70}. Chem Phys Lett 216:399.

Kolodney E, Tsipinyuk B, Budrevich A (1994) The thermal stability and fragmentation of C_{60} molecule up to 2000 K on the millisecond time scale. J Chem Phys 100(11):8542.

Kroto HW, Heath JR, O'Brien SC, Curl RF, Smalley RE (1985) C_{60}: Buckminsterfullerene. Nature 318:162.

Kruse T, Roth P (1996) Kinetics of C_2-radical reactions during shock-induced pyrolysis of acetylene. In: Shock Waves, Proc of the 20th ISSW, Sturtevant B, Shepherd JE, Hornung HG (eds), World Scientific, Vol 2, pp 851–856.

O'Brien SC, Heath JR, Curl RF, Smalley RE (1988) Photophysics of buckminsterfullerene and other carbon cluster ions. J Chem Phys 88(1):220.

Sommer T, Roth P (1995) C_2 formation during high-temperature pyrolysis of fullerene C_{60} in shock waves. J Phys Chem (to be published).

Urdahl RS, Bao Y, Jackson WM (1991) An experimental determination of the heat of formation of C_2 and the C–H bond dissociation energy in C_2H. Chem Phys Lett 178(4):425.

von Gersum S, Kruse T, Roth P (1994) Spectral emission during high-temperature pyrolysis of fullerene C_{60} in shock waves. Ber Bunsenges Phys Chem 98:979.

Thermal decomposition of gaseous hydrazine behind a reflected shock wave

C. Halat-Augier, G. Dupré, C. E. Paillard
Laboratoire de Combustion et Systèmes Réactifs (C.N.R.S) and
University of Orléans
45071 Orléans Cedex 2, France

Abstract: The thermal decomposition of hydrazine highly diluted with N_2 or N_2/He ($\geq$ 98 mol%) has been studied behind a reflected shock wave between 200 and 1300 kPa, 1100 and 1400 K. For $P \geq 500$ kPa, the overall reaction order is equal to unity. A partial helium substitution for nitrogen in N_2H_4/N_2 mixtures has shown that the overall rate constant and the fall-off depend on the nature of diluent. From the experimental data found for the high-pressure range, a kinetic model taking into account the diluent influence on the rate constant has been built.

Key words: Hydrazine, Thermal decomposition, Kinetics, Shock tube, Modeling

1. Introduction

The thermal decomposition of hydrazine has been studied by several researchers in a shock tube and with argon as a diluent. Halat-Augier et al. (1994) have shown that a large discrepancy exists between the rate constants found by different authors. However, there is agreement that the mechanism of hydrazine decomposition depends on temperature and pressure. The thermal decomposition studies of hydrazine carried out by Eberstein and Glassman (1965) in a flow reactor at low temperature and pressure confirm the complexity of the reaction mechanism. The aim of the present study is to obtain a large amount of experimental data in order to build a detailed kinetic model of hydrazine thermal decomposition. For this purpose, a shock tube and a uv absorption technique have been used to clarify the thermal decomposition of hydrazine diluted with N_2 or with N_2/He over a large temperature and pressure range.

2. Experimental set-up

The shock tube used for this study, made of stainless steel, 80 mm i.d., has been already described by Halat-Augier et al. (1994). Near its downstream end, it is coupled to a uv absorption system able to follow the time-absorption signal of hydrazine at 250 nm, from which the hydrazine concentration is deduced.

The thermal-decomposition kinetics of hydrazine highly diluted with N_2 or N_2/He ($\geq$ 98 mol%) has been studied behind a reflected shock wave over a large temperature and pressure range: 1100–1400 K and 200–1300 kPa.

3. Results

3.1. Experimental results

An incubation period preceding the thermal decomposition of hydrazine has been observed by Michel and Wagner (1965). For $T < 1150$ K and $P < 1$ MPa, or for $T \leq 1200$ K and $200 \leq P$ (kPa) ≤ 300, the present study has also permitted observation of an incubation period prior to decomposition. This incubation period which is both dependent on pressure and temperature shows that the mechanism of hydrazine thermal decomposition is complex. However, the rapid reaction which follows the incubation period has a reaction order towards hydrazine equal to unity. For $P < 500$ kPa, the rate constant varies with pressure. This variation could be due to the existence of a fall-off range (Fig. 1a). When $P \geq 500$ kPa, the overall reaction order is equal to one and from the Arrhenius plot (Fig. 1b) the expression of the high-pressure rate constant is deduced: $k_{\infty N_2}(\mathrm{s}^{-1}) = 1.9\times\ 10^{12}\exp(-22900/T)$. The average deviation between the experimental half-life times and the values calculated with the above expression is equal to 8%.

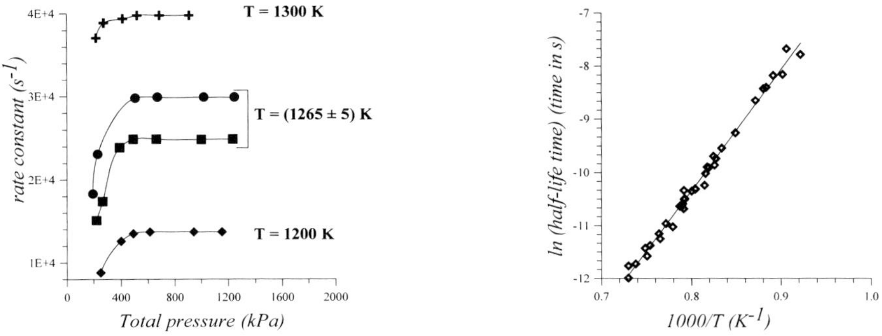

(a) (b)

Figure 1. (a) Variation of the rate constant of N_2H_4 thermal decomposition versus total shock pressure at different shock temperatures for (1% N_2H_4, 99% N_2); (b) logarithmic variation of half-life time for N_2H_4 decomposition versus inverse shock temperature for dilutions higher than 98 mol% N_2 and for $P \geq 500$ kPa.

The thermal-decomposition study of hydrazine diluted either with N_2 or with N_2/He has shown that the fall-off, as well as the overall rate constant, depends on the nature of the diluent. For nitrogen, the overall rate constant reaches the value of the high-pressure rate constant for $P \geq 500$ kPa, whereas, for nitrogen/helium, this value is not reached at 900 kPa. The comparison between $k_{\infty N_2}$ and the rate constants found at 900 kPa and for different compositions of N_2/He confirms this observation: $k_{(20-40)\%\mathrm{He}}(\mathrm{s}^{-1}) = 3.5\times\ 10^{11}\exp(-21100/T)$, $k_{60\%\mathrm{He}}(\mathrm{s}^{-1}) = 2\times\ 10^{11}\exp(-20500/T)$. The half-life time values calculated with these expressions deviate from the experimental one by about 6%. At 1200 K, the deviation between $k_{\infty N_2}$, $k_{(20-40)\%\mathrm{He}}$ and $k_{60\%\mathrm{He}}$ is equal to 11% and 30%, respectively. The above expressions show that the addition of helium in a N_2H_4/N_2 mixture slows down the reaction rate. This effect is a function of the amount of helium added to the initial mixture. Moreover, the incubation period observed with helium is longer than that observed with nitrogen alone. The diluent probably acts as a collision partner in the initiation step of the mechanism.

The different results given above can be explained if one considers that the mechanism of hydrazine thermal decomposition is mostly governed by the initiation step: $N_2H_4 + M \rightleftharpoons N_2H_4^* + M$, $N_2H_4^* \rightleftharpoons (N_2H_4)^{\#}$ and $(N_2H_4)^{\#} \rightarrow 2NH_2$. The formalism of Rice-Ramsperger-Kassel-Markus (RRKM) theory allows one to explain the pressure effect on the reaction rate and the observation of an incubation period which might correspond to the time required to reach the equilibrium between the formation of the activated complex and the reverse reaction. The diluent effect can be also clarified: the theoretical expression for the rate constant given by the RRKM theory takes into account the diluent concentration, the diluent collision efficiency, and frequency. If these parameters are modified because the diluent is changed, the corresponding rate constant is modified. The collision efficiency of He being lower than that of N_2 means the rate constant with He as a diluent is smaller (as well as the overall reaction rate) than with N_2. Moreover, in the case of N_2/He, where the collision efficiency is lower, the equilibrium between the bimolecular activation and deactivation processes is reached more slowly, leading to a longer incubation period.

3.2. Kinetic modeling

The results of the kinetic modeling concern the experimental data found for $P \geq 500$ kPa. The kinetic model which has been built consists of 48 elementary reactions and 11 chemical species. It is shown in Tables 1 and 2.

This model is based on the compilation of Hanson and Salimian (1984) which has been modified by adding several reactions (13, 44–48), revising the rate constants of some of them (12, 16, 30, 42–43), and changing the initiation step

Table 1. Kinetic model of hydrazine thermal decomposition; $k = A\ T^b \exp(-T_a/T)$ with A in $s^{-1}K^{-b}$ or $cm^3mol^{-1}s^{-1}K^{-b}$.

	Reactions	A	b	T_a (K)
1	$N_2H_4 \rightleftharpoons 2NH_2$	1.04×10^{12}	0	23300
2	$N_2H_3 + M \rightleftharpoons N_2H_2 + H + M$	7×10^{15}	0	28000
3	$N_2H_2 + M \rightleftharpoons NH_2 + N + M$	5×10^{16}	0	20000
4	$N_2H_2 + M \rightleftharpoons N_2H + H + M$	10^{16}	0	25000
5	$N_2H_2 + M \rightleftharpoons NH + NH + M$	3.16×10^{16}	0	50000
6	$N_2H + M \rightleftharpoons N_2 + H + M$	1.99×10^{13}	0	10000
7	$N_2H + M \rightleftharpoons NH + N + M$	10^{15}	0	35000
8	$NH_3 + M\ NH_2 + H + M$	2.2×10^{16}	0	47036
9	$NH_3 + M \rightleftharpoons NH + H_2 + M$	6.31×10^{14}	0	47000
10	$NH_2 + M \rightleftharpoons NH + H + M$	3.16×10^{23}	−2	46000
11	$NH + M \rightleftharpoons N + H + M$	3.16×10^{21}	−2	42000
12	$H_2 + M \rightarrow 2H + M$	7.61×10^{5}	−1.4	52530
13	$NH_2 + H_2 \rightleftharpoons H + NH_3$	3.24×10^{13}	0	6492
14	$N_2H_4 + H \rightleftharpoons N_2H_3 + H_2$	1.3×10^{13}	0	1260
15	$N_2H_4 + H \rightleftharpoons NH_2 + NH_3$	4.46×10^{9}	0	1560
16	$N_2H_3 + H \rightleftharpoons N_2H_2 + H_2$	10^{13}	2	800
17	$N_2H_3 + H \rightleftharpoons NH_2 + NH_2$	1.58×10^{12}	0	0
18	$N_2H_3 + H \rightleftharpoons NH + NH_3$	10^{11}	0	0
19	$N_2H_2 + H \rightleftharpoons N_2H + H_2$	10^{13}	0	500
20	$N_2H + H \rightleftharpoons N_2 + H_2$	3.98×10^{13}	0	1500
21	$NH_3 + H \rightarrow NH_2 + H_2$	5.42×10^{5}	2.5	4990
22	$NH_2 + H \rightleftharpoons NH + H_2$	1.905×10^{13}	0	0
23	$NH + H \rightleftharpoons N + H_2$	5.01×10^{13}	0	1000
24	$N_2H_4 + NH \rightleftharpoons NH_2 + N_2H_3$	10^{12}	0.5	1000
25	$N_2H_2 + NH \rightleftharpoons N_2H + NH_2$	10^{13}	0	1000
26	$N_2H + NH \rightleftharpoons N_2 + NH_2$	1.99×10^{11}	0.5	1000

considered as a bimolecular one by Hanson and Salimian. In order to take into account the experimental results found for $P \geq 500$ kPa, a unimolecular initiation step ($N_2H_4 \rightleftharpoons 2NH_2$) has been chosen, the rate constant of which depends on the nature of the diluent. The initiation rate constants for N_2- and N_2/He-diluted mixtures were obtained from the adjustment of the experimental overall rate constant found at high pressure in a N_2 and a N_2/He bath, respectively. This adjustment has been made between 500 and 1340 kPa, in order to find a good agreement between the calculated concentrations of hydrazine and the experimental data in the temperature range studied. As the initiation step (reaction 1) governs the mechanism, it has been assumed that the overall rate constant is close to that of the initiation step.

This can also justify that the initiation rate constant can be deduced from the adjustement of the experimental one. The initiation rate constant, given in

Table 2. Kinetic model of hydrazine thermal decomposition; $k = AT^b \exp(-T_a/T)$ with A in $s^{-1}K^{-b}$ or $cm^3mol^{-1}s^{-1}K^{-b}$.

	Reactions	A	b	T_a (K)
27	$NH + NH \rightleftharpoons NH_2 + N$	1.99×10^{11}	0.5	1000
28	$NH + NH \rightleftharpoons N_2H + H$	7.94×10^{12}	0.5	500
29	$N_2H_4 + NH_2 \rightleftharpoons N_2H_3 + NH_3$	3.98×10^{11}	0.5	1000
30	$N_2H_3 + NH_2 \rightleftharpoons N_2H_2 + NH_3$	2×10^{10}	0	0
31	$N_2H_2 + NH_2 \rightleftharpoons N_2H + NH_3$	10^{13}	0	2000
32	$N_2H_2 + NH_2 \rightleftharpoons NH + N_2H_3$	10^{11}	0.5	17000
33	$N_2H + NH_2 \rightleftharpoons N_2 + NH_3$	10^{13}	0	0
34	$NH_3 + NH_2 \rightleftharpoons N_2H_3 + H_2$	7.94×10^{11}	0.5	1080
35	$NH_2 + NH_2 \rightleftharpoons NH + NH_3$	6.31×10^{12}	0	5000
36	$NH_2 + NH_2 \rightleftharpoons N_2H_2 + H_2$	3.98×10^{13}	0	6000
37	$NH + NH_2 \rightleftharpoons N_2H_2 + H$	3.16×10^{13}	0	500
38	$N_2H_4 + N_2H_2 \rightleftharpoons 2N_2H_3$	2.51×10^{10}	0.5	5000
39	$N_2H_3 + N_2H_2 \rightleftharpoons N_2H_4 + N_2H$	10^{13}	0	5000
40	$N_2H_2 + N_2H_2 \rightleftharpoons N_2H + N_2H_3$	10^{13}	0	5000
41	$N_2H + N_2H \rightleftharpoons N_2H_2 + N_2$	10^{13}	0	5000
42	$NH + N \rightleftharpoons N_2 + H$	3×10^{13}	0	0
43	$N_2H + N \rightleftharpoons NH + N_2$	3.16×10^{13}	0	1000
44	$NH + NH \rightleftharpoons N_2 + 2H$	3.16×10^{13}	0	1000
45	$N + N + M \rightarrow N_2 + M$	2.8×10^{17}	−0.75	0
46	$N_2 + M \rightarrow 2N$	2.3×10^{29}	−3.5	113240
47	$N + H_2 \rightleftharpoons NH + H$	2.37×10^{9}	0	16606
48	$H + H + M \rightarrow N_2 + M$	5.44×10^{18}	−1.3	0

Table 1, is strictly valid for $P \geq 500$ kPa and N_2H_4/N_2 mixtures. The half-life times calculated by the model differ by about 20% with those found experimentally. Whereas the induction period has not been successfully modeled, the hydrazine concentration profiles show a good agreement with the experimental data (Fig. 2). The kinetic modeling predicts the same effect as that observed experimentally in the case of a helium substitution for nitrogen. It confirms that the diluent only interferes in the mechanism as a collision partner for the bimolecular activation and deactivation processes. Finally, for $P \geq 500$ kPa, the kinetic model of hydrazine thermal decomposition has been validated towards hydrazine on the entire temperature, dilution, and composition range studied.

4. Conclusion

The kinetic study of hydrazine thermal decomposition carried out between 1100 and 1400 K, 200 and 1300 kPa, and for a dilution higher than 98 mol% N_2 or N_2/He has shown a pressure dependence on the mechanism until the shock pressure reaches 500 kPa. For $P \geq 500$ kPa, the kinetic model built according to the experimental data allows one to predict fairly well the hydrazine behavior as

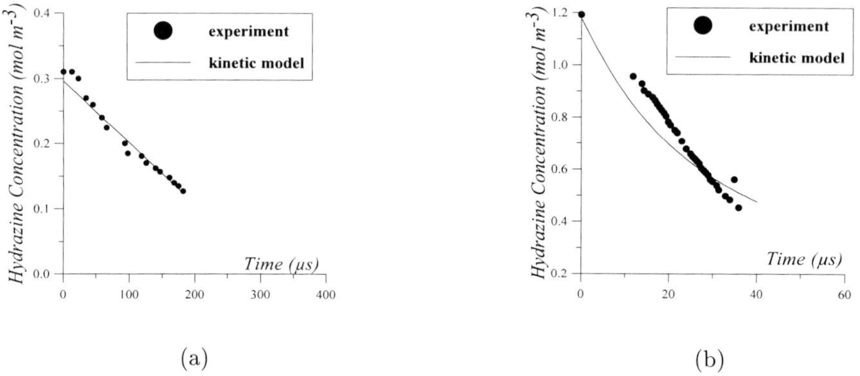

Figure 2. Comparison between experimental and computed hydrazine concentrations: (a) 1155 K, 600 kPa, 99.5 % N_2; (b) 1262 K, 1235 kPa, 99 % N_2.

a function of the nature of the diluent, of mixture composition, and temperature.

Acknowledgement. C. Halat-Augier wishes to thank the Centre National d'Etudes Spatiales (CNES) and MATRA for financial support.

References

Eberstein IJ, Glassman I (1965) The gas-phase decomposition of hydrazine and its methyl derivatives. In: 10th Symp (Int) on Comb, The Combustion Institute, Pittsburgh, pp 365–374.

Halat-Augier C, Dupré G, Paillard CE (1994) Explosion hazard due to an adiabatic compression of gaseous hydrazine. In: International Conference on Spacecraft Propulsion, Toulouse, France.

Hanson RK, Salimian S (1984) Survey of rate constants in the N/H/O systems. In: Combustion Chemistry, Gardiner WC (ed), Springer Verlag.

Michel KW, Wagner HG (1965) The pyrolysis and oxydation of hydrazine behind shock waves. In: 10th Symp (Int) on Comb, The Combustion Institute, Pittsburgh, pp 353–364.

Shock-tube study of acetone pyrolysis

Y. Hidaka, K. Sato, R. Iwamoto
Department of Chemistry, Faculty of Science
Ehime University, Bunkyo-cho, Matsuyama 790 Japan
T. Koike
Department of Chemistry, National Defense Academy
Hashirimizu, Yokosuka 239 Japan

Abstract: The pyrolysis of acetone was studied behind reflected shock waves in the temperature range 1050 – 1650 K at total pressures between 1.5 and 3.5 atm. The study was carried out by using time-resolved IR-laser absorption and single-pulse techniques. From a computer-simulation study, mechanisms and rate-constant expressions that could explain all our data are discussed. A 54-reaction mechanism satisfactorily modeled our data. The rate constants of reactions (1) – (4) were evaluated as $k_1 = 1.64 \times 10^{16}\exp(-79.9\ \mathrm{kcal}/RT)\ \mathrm{s}^{-1}$, $k_2 = 3.20 \times 10^{14}\exp(-8.39\ \mathrm{kcal}/RT)\ \mathrm{cm^3mol^{-1}s^{-1}}$, $k_3 = 5.31 \times 10^{11}\exp(-9.74\ \mathrm{kcal}/RT)\ \mathrm{cm^3mol^{-1}s^{-1}}$, and $k_4 = 1.25 \times 10^{12}\exp(-28.0\ \mathrm{kcal}/RT)\ \mathrm{s}^{-1}$.

$$(CH_3)_2CO \rightarrow CH_3CO + CH_3 \qquad (1)$$
$$(CH_3)_2CO + H \rightarrow CH_2CH_3CO + H_2 \qquad (2)$$
$$(CH_3)_2CO + CH_3 \rightarrow CH_2CH_3CO + CH_4 \qquad (3)$$
$$CH_2CH_3CO \rightarrow CH_2CO + CH_3 \qquad (4)$$

Key words: Reaction rate, Acetone pyrolysis, Reaction mechanism, Shock-tube experiment, IR-laser absorption

1. Introduction

The pyrolysis of acetone has been studied by many groups. Many investigations have been carried out, from which the decomposition products and the Arrhenius parameters were determined especially at temperatures below 1100 K. At temperatures above 750 K, it has been reported that the reaction is a chain process. It seems to be well established that reactions (1) – (4) are the chain initiation and propagation steps of the decomposition (Berces 1972).

$$(CH_3)_2CO \rightarrow CH_3CO + CH_3 \qquad \Delta H_0 = 79.9\ \mathrm{kcal/mol} \qquad (1)$$
$$(CH_3)_2CO + CH_3 \rightarrow CH_2CH_3CO + CH_4 \qquad \Delta H_0 = -6.8\ \mathrm{kcal/mol} \qquad (3)$$
$$CH_2CH_3CO \rightarrow CH_2CO + CH_3 \qquad \Delta H_0 = 28.2\ \mathrm{kcal/mol} \qquad (4)$$
$$CH_3CO \rightarrow CH_3 + CO \qquad (5)$$

On the other hand, at temperatures above 1100 K, only two reports have been made. Capelin et al. (1974) studied this chemistry in the temperature range 573 to 1473 K by flash vaporization into a heated reaction chamber made of silica tubing. From a gas-chromatograph analysis of the products, they proposed the following pyrolysis mechanism:

$$(CH_3)_2CO \rightarrow CH_2CH_3CO + H \qquad \Delta H_0 = 96.5\ \text{kcal/mol} \qquad (1a)$$
$$(CH_3)_2CO + H \rightarrow CH_2CH_3CO + H_2 \qquad (2)$$
$$(CH_3)_2CO + CH_3 \rightarrow CH_2CH_3CO + CH_4 \qquad (3)$$
$$CH_2CH_3CO \rightarrow CH_2CO + CH_3 \qquad (4)$$
$$CH_3 + CH_3 \rightarrow C_2H_6 \qquad (6)$$
$$CH_3 + H \rightarrow CH_4$$

Ernst et al. (1976) studied acetone pyrolysis in the temperature range 1350 – 1650 K using a shock-tube technique. From time-resolved uv-absorption data at 290 nm, they proposed initiation reaction (1) with the Arrhenius expression at the high-pressure limit $k_1{}^{\infty} = 2.7 \times 10^{16}\exp(-81.7 \pm 2.9\ \text{kcal}/RT)\ \text{s}^{-1}$. They assumed a mechanism with reactions (1), (3), (4), (5), and (6).

In the present paper, we carried out the study of acetone pyrolysis at higher temperatures using a shock-tube technique. We proposed a detailed reaction mechanism to reproduce our experimental data.

2. Experimental

Two shock tubes of 4.1 cm i.d. were used in this study. One was a standard shock tube connected to an IR-laser absorption apparatus (Hidaka et al. 1993a). In this apparatus, profiles of IR-laser absorption in the acetone pyrolysis were measured at 3.39 μm. The other was a new magic-hole type shock tube on which both single-pulse and time-resolved spectroscopy was performed (Hidaka et al. 1993b). The reacted mixtures were analyzed on three gas chromatographs (Shimazu GC-3BT1, Shimazu GC-3BT2, Shimazu GC-8AT), which had thermal-conductivity detectors. The gas chromatographs were serially connected. An effective heating time was determined using a method that was described previously (Hidaka et al. 1985). The effective heating times at 1200, 1300, 1400, and 1500 K were determined to be 1.97, 1.89, 1.80, and 1.71 ms, respectively.

The gas compositions that were used are (a) 1% $(CH_3)_2CO$, 99% Ar, and (b) 4% $(CH_3)_2CO$, 96% Ar. The initial pressure used in this study was 50 torr. The computer calculations used in this study were essentially the same as described previously (Hidaka et al. 1994a, 1994b, and 1985). The computer routine was Gear-type integration of the set of differential equations describing the chemical kinetics under constant density conditions for reflected shock waves. Reverse reactions were automatically included in the computer program through equilibrium constants computed from polynomial fits to standard thermochemical data. The basic thermochemical data source was the JANAF table (Stull and Prophet 1971), the data of Benson (1976), and the table of coefficient sets for NASA polynomials (Burcat 1984) with the following alterations. The heats of formation used for the species C_2H, C_2H_3, C_3H_3, and C_4H_5 were 135.3 kcal/mol, 71.7 kcal/mol, 83.6 kcal/mol, and 72.3 kcal/mol, respectively (Ervin et al. 1990, Lias et al. 1988).

3. Results and discussion

The 4% acetone diluted with Ar was heated in the temperature range 1050 – 1650 K and the IR-laser absorption of heated mixtures was observed. At temperatures below 1100 K, the absorption intensities after a schlieren-spike remained constant. At temperatures between 1100 K and 1450 K, the absorption intensities increased with time. Above 1450 K, the absorption intensities increased with time, arrived at a maximum value, and then decreased with time. Time-profiles of the total absorption data A_t were determined. The A_t was defined by the following equation:

$$A_t = \log(I_f/I_t)/\log(I_f/I_0)$$

where I_f was the signal voltage corresponding to the full intensity and I_0 and I_t were the signal voltages corresponding to the absorption intensities at the reflected shock front ($t = 0$) and at time t, respectively. Three typical time-profiles of A_t for the 4% $(CH_3)_2CO$, 96% Ar, mixture are shown in Fig. 1. In order to know the calculated A_t of the reactant and each product at 3.39

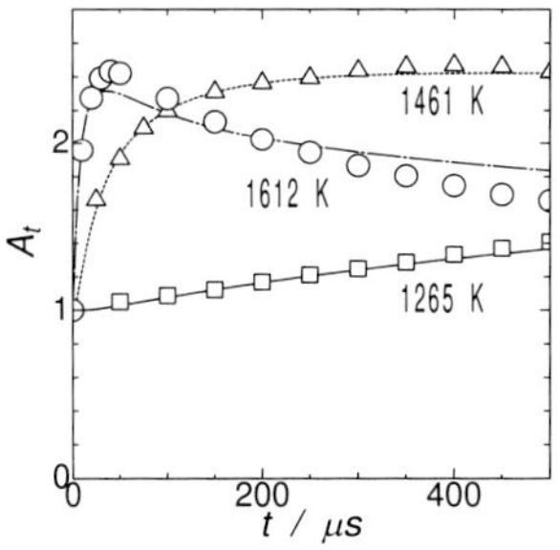

Figure 1. Comparison of the observed time-profiles A_t (symbols) with the calculated ones using 4% $(CH_3)_2CO$, 96% Ar, mixture. Curves are the results calculated by using the 54-reaction mechanism.

μm, their extinction coefficients were measured. The equations for extinction coefficients obtained for main species were $\epsilon((CH_3)_2CO) = 9.93 \times 10^3 - 2.73 \times$

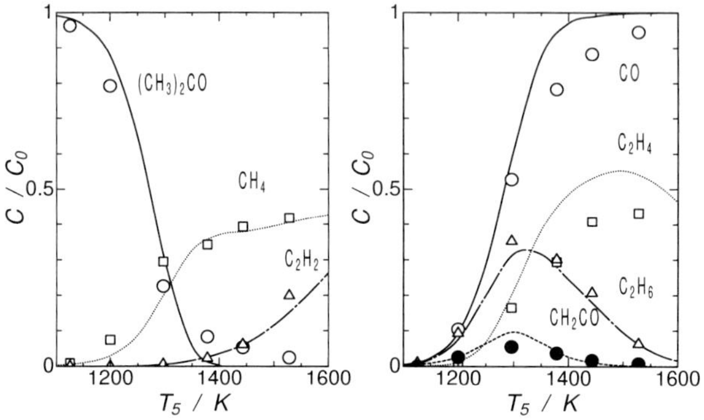

Figure 2. Comparison of the observed product distributions (symbols) with the calculated distributions using 1% $(CH_3)_2CO$, 96% Ar, mixture, where C_0 is the initial concentration of acetone and C is the concentration of each species. Curves are the calculated results.

T cm^{2}mol^{-1}, $\epsilon(C_2H_6) = 5.43 \times 10^4 - 17.5 \times T$ cm^{2}mol^{-1}, $\epsilon(C_2H_4) = 3.38 \times T$ cm^{2}mol^{-1}, $\epsilon(C_2H_2) = -1.44 \times 10^3 + 1.03 \times T$ cm^{2}mol^{-1}, $\epsilon(CH_4) = 3.46 \times 10^4 - 16.7 \times T$ cm^{2}mol^{-1}, $\epsilon(CH_2CO) = 1.15 \times 10^3 + 0.875 \times T$ cm^{2}mol^{-1} for $(CH_3)_2CO$, C_2H_6, C_2H_4, C_2H_2, CH_4, and CH_2CO, respectively. The extinction coefficients were used for simulations.

The mixture of 1% $(CH_3)_2CO$, 99% Ar, was heated in the temperature range 1100 – 1530 K using the single-pulse shock tube. The reacted mixture was analyzed by the gas chromatographs. The product distribution is shown in Fig. 2. Under our experimental conditions, the main products that were detected were CO, CH_4, C_2H_6, C_2H_4, C_2H_2, and CH_2CO (ketene). The species p-C_3H_4 (propyne), a-C_3H_4 (allene), C_3H_6, C_3H_8, and 1,3-C_4H_6 were also detected in smaller quantities.

To interpret our data, a computer modeling study was performed. By trial and error methods, we tried to determine a mechanism and rate constant expressions that could predict all our results. A 54-reaction mechanism including reactions (1) – (4) was able to predict all our data, including the small quantities of p-C_3H_4, a-C_3H_4, C_3H_6, C_3H_8, and 1,3-C_4H_6. The reactions and the rate constant expressions except for reactions (1) – (4) used for this mechanism have been shown in the other reports (Hidaka et al. 1994, 1993a, 1993b, 1992, and 1989). The initial step in the thermal decomposition of acetone was studied in detail by Ernst et al. (1976). From the absorption results at 290 nm, they showed that reaction (1) was the initiation step and played an important role in the acetone pyrolysis. We also adopted reaction (1) as the initiation step. Reaction (1) was very sensitive to the consumption of acetone. To explain all our experimental data consistently, this reaction with $k_1 = 1.64 \times 10^{16}\exp(-79.9$ kcal$/RT)$ s^{-1} was indispensable. The values of k_1 at 1400 K and 1500 K were 5.52×10^3 s^{-1} and 3.74×10^4 s^{-1}, respectively. The densities we used were 1.7 $\times$ 10^{-5} – 2.5×10^{-5} mol/cm^3. The k_1 values evaluated by Ernst et al. (1976) were 7.2×10^2 s^{-1} (1400 K) and 3.2×10^3 s^{-1} (1500 K) at the density of 2 $\times$ 10^{-5} mol/cm^3. The values we evaluated for reaction (1) were hihger than those of Ernst et al. (1976) by a factor of about 8 – 12, but were consistent with those in the high pressure limit of Ernst et al. (1976). The evaluated activation energy of 79.9 kcal/mol was the same as the ΔH_0 value of reaction (1). From these data, reaction (1) is inferred to be close to the first-order limit under our experimental conditions. Capelin et al. (1974) proposed reaction (1a) as the initiation reaction. However, when reaction (1a) is assumed to be the initiation reaction and corresponding rate-constant expressions for reactions (2) – (4) and (1a) were assumed, the A_t curves shown in Fig. 1 were not predicted. Furthermore, the use of reaction (1a) gave a large amounts of ketene and a small amount of methane compared to observed levels. The product yields shown in Fig. 2 were never predicted when reaction (1a) was assumed as the initiation reaction. Hence, we concluded that reaction (1a) was not important as the initiation reaction under our experimental conditions.

Reactions (2) and (3) were used for H-atom abstraction from $(CH_3)_2CO$. Kondratiev (1972) estimated the rate constant $k_2 = 4.57 \times 10^{13}\exp(-8.39\text{kcal}/RT)$ cm^{3}mol^{-1}s^{-1} in the temperature range 298 – 873 K, which was calculated on the basis of two studies. As long as we used the extrapolated value of Kondratiev for reaction (2), computed A_t values were smaller than the observed A_t at a given time and the yield of ketene was always lower than the observed one. One way to match the computed concentrations to the observed ones was to increase the k_2 value by a factor of about 7. The best fits for our data were obtained by using $k_2 = 3.20 \times 10^{14}\exp(-8.39\text{kcal}/RT)$ cm^{3}mol^{-1}s^{-1}. The rate-constant expression evaluated for reaction (3) was $k_3 = 5.31 \times 10^{11}\exp(-9.74\text{kcal}/RT)$ cm^{3}mol^{-1}s^{-1}. This expression is consistent with that of Kondratiev(1972), $k_3 = 3.89 \times 10^{11}\exp(-9.87\text{kcal}/RT)$ cm^{3}mol^{-1}s^{-1} (over range 350 – 710 K), obtained on the basis of the data of various authors taken at lower temperatures.

It is known that at low temperatures the CH_2CH_3CO radicals produced by reactions (2) and (3) decompose by reaction (4) and produce ketene (Berces 1972). Under our experimental conditions, reaction (4) also played an important role in the ketene formation. The best fits for our data were obtained by using $k_4 = 1.25 \times 10^{12}\exp(-28.0\text{kcal}/RT)$ s^{-1}.

References

Benson SW (1976) Thermochemical Kinetics, 2nd edition, Wiley, New York.

Berces T (1972) The decomposition of aldehydes and ketones. In: Comprehensive Chemical Kinetics, Vol 5, Decomposition and isomerization of organic compounds, Chapter 3, Bamford CH, Tipper CF (eds) Elsevier, Amsterdam, ppp 258–265.

Burcat A (1984) Thermochemical data for combustion calculations. In: Combustion Chemistry, Gardiner WC Jr (ed), Springer-Verlag, New York, p 456.

Capelin BC, Ingram G, Kokolis J (1974) The pyrolytic identification of organic molecules, II A quantitative evaluation. Microchemical Journal 19:229.

Ernst J, Spindler K, Wagner HG (1976) Untersuchungen zum thermischen zerfall von acetaldehyde und aceton. Ber Bunsenges Phys Chem 80:645.

Ervin KM, Gronert S, Barlow SE, Gilles MK, Harrison AG, Bierbaum VM, Depuy CH, Lineberger WC, Ellison GB (1990) Bond Strengths of ethylene and acetylene. J Am Chem Soc 112:5750.

Hidaka Y, Kimura K, Kawano H (1994) High-temperature pyrolysis of ketene in shock waves. Combust Flame 99:18.

Hidaka Y, Taniguchi T, Tanaka H, Kamesawa T, Inami K, Kawano H (1993a) Shock-tube study of CH_2O pyrolysis and oxidation. Combust Flame 92:365.

Hidaka Y, Higashihara T, Ninomiya N, Oshita H, Kawano H (1993b) Thermal isomerization and decomposition of 2-butyne in shock waves. J Phys Chem 97:10977.

Hidaka Y, Nakamura T, Tanaka H, Jinno A, Kawano H, Higashihara T (1992) Shock tube and modeling study of propene pyrolysis. Int J Chem Kinet 24:761.

Hidaka Y, Nakamura T, Miyauchi A, Shiraishi T, Kawano H (1989) Thermal decomposition of propyne and allene in shock waves. Int J Chem Kinet 21:643.

Hidaka Y, Shiba S, Takuma H, Suga M (1985) Thermal decomposition of ethane in shock waves. Int J Chem Kinet 17:441.

Kondratiev VN (1972) Rate Constants of Gas Phase Reactions, translated by Holtschlag LJ, Springfield.

Lias SG, Bartmess JE, Liebman JF, Holmes JL, Levin RD, Mallard WG (1988) Gas-phase ion and neutral thermochemistry. J Phys Chem Ref Data 17:109.

Stull DR, Prophet H (1971) JANAF Thermochemical Tables, 2nd Edition, NSRDS-NBS 37.

Measurements of rotational-vibrational excitation behind incident shock waves by tunable diode laser absorption spectrometry

W. Klenk, N. Widdecke, A. Frohn
ITLR, Institut für Thermodynamik der Luft- und Raumfahrt
Universität Stuttgart
Pfaffenwaldring 31, 70550 Stuttgart, Germany

Abstract: An experimental technique suited for the investigation of rotational-vibrational excitation processes of diatomic molecules is described. The experiments have been performed in an ultra-high-vacuum shock tube using mixtures of air, nitric oxide, and argon as test gases. The tunable diode laser absorption spectrometer (TDLAS) is equipped with two laser diodes which can be tuned simultaneously in different wavelength ranges near 5 μm for state-selective measurements of population densities in the infrared absorption bands of the NO molecule. Scanning of complete absorption profiles with TDLAS have been performed with repetition rates up to 1 MHz. The "line by line–layer by layer" technique used to calculate the absorption spectrum of NO has been extended. In the new model different rotational and vibrational temperatures can be used.

Key words: Rotational-vibrational excitation, Nitric oxide, Infrared diode laser absorption spectroscopy

1. Introduction

High temperatures behind shock waves cause excitation of rotational and vibrational degrees of freedom as well as dissociation and ionization of the molecules. The thermodynamic variables and species concentrations for such conditions are of fundamental interest for hypersonic aerodynamics. Complex numerical codes which include models for the excitation of internal energy states have been developed by different authors for the simulation of high-enthalpy nonequilibrium flow fields in air. For these models more state-selective physico-chemical reaction data are needed (Park 1990).

Tunable diode laser absorption spectroscopy is a non-intrusive diagnostic technique used for state-selective measurements. In recent years this technique has been used for studying reaction rates at high temperatures behind shock waves (Brandt and Roth 1987, Coerdt and Gronig 1989, Hanson 1995, Moser and Hindelang 1992).

2. Description of the experimental facility

Shock tubes are well suited for studying relaxation processes behind shock waves. Since these processes depend strongly on the purity of the test gas the experiments have been performed in a pressure-driven ultra-high-vacuum shock tube in order to avoid contaminations. To meet very high purity requirements, the complete facility can be baked at temperatures of 600 K before an experiment and may be evacuated to pressures below 10^{-7} mbar. The driven section with a length of 12 m has a circular cross section with an inner diameter of 72 mm. The driver section is equipped with a double-diaphragm system. The bursting pressure can be adjusted very precisely in a wide range of driver pressures between 30 bar and 100 bar. Mixtures of helium, hydrogen, and nitrogen were used as driver gases and mixtures of synthetic air, argon, and nitric oxide as test gases. In the first part of the test section the pressure was measured by two Kistler pressure transducers type 603 B. From the signals of these gauges the pressure jump across the shock wave and the shock-front velocity have been determined. The variation of the ion concentration behind the incident shock wave as well as the velocity of the contact surface are determined with two Langmuir probes. The second part of the test section is equipped with a tunable infrared diode laser absorption spectrometer.

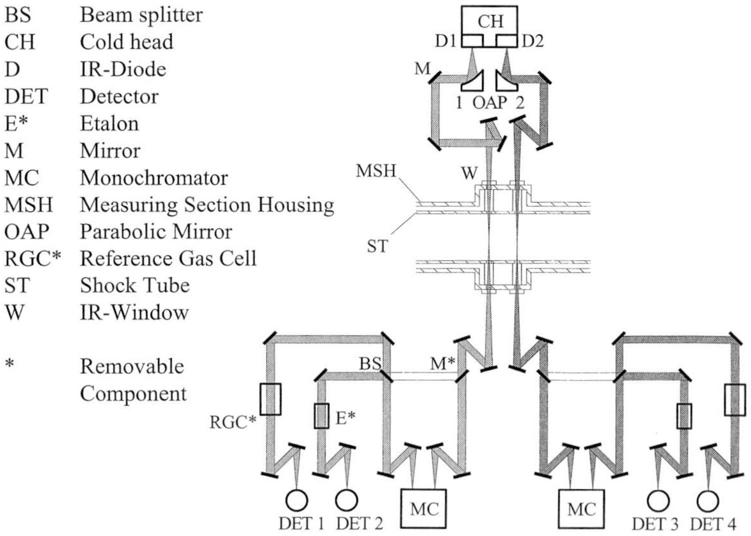

Figure 1. Schematic diagram of optical arrangement of TDLAS.

2.1. The tunable diode laser absorption spectrometer (TDLAS)

To investigate the excitation processes behind incident shock waves a two-channel fast-tuning infrared diode laser absorption spectrometer has been developed. Figure 1 shows a schematic diagram of the optical arrangement. In order to get a highly resolved spectral signal with high time resolution, the TDLAS is tuned with scan frequencies up to 1 MHz. The two laser diodes are tuned independently of each other at wavelengths near 5 μm, which corresponds to wave numbers of approximately 2000 cm^{-1}. The signals of the fast HgCeTe detectors are recorded with a 200-MHz transient recorder. The detection limit of TDLAS depends strongly on the emission characteristics of the laser diode that was used and on the line strength of the choosen transition. Planning shock-tube experiments with TDLAS therefore requires a careful selection of suitable transitions.

3. Absorption spectroscopy in shock-tube flows

Applying Beer-Lambert's absorption law, the conditions behind the shock wave can be related to the transmission I/I_0 of the laser beam by the well-known equation

$$I/I_0 = e^{-\alpha_\nu \, l \, c} \, . \tag{1}$$

Here l is the length of the optical path. The concentration of the absorbing species c varies with time. The integration of the absorption coefficient α_ν over the wave number leads to the line strength

$$S = \int_{Line} \alpha_\nu \, \mathrm{d}\nu \, . \tag{2}$$

The specific absorption coefficient α_ν is a function of temperature T and of wave number ν. Broadening effects like Doppler-broadening or pressure-broadening have been eliminated by integrating α_ν over the complete line shape. For this reason the line strength $S(T)$ does not depend on the shape of the absorption line. The line strengths of transitions of nitric oxide can be calculated using the "line by line–layer by layer" technique (Goldman and Schmidt 1974). This technique using only one temperature has been extended. In the new model different translational, rotational, and vibrational temperatures can be used. A computer program has been developed to calculate the absorption spectrum of the $\Delta v = 1$ bands of $N^{14}O^{16}$ at elevated temperatures with regard to the $^2\Pi_{1/2}$ state and the $^2\Pi_{3/2}$ state, the Λ-doubling, and the Λ-splitting (Widdecke et al. 1995). Figure 2 shows three calculated absorption profiles of several transitions near 1906 cm^{-1}. The first profile, calculated at the initial conditions, is composed of two transitions originating from the vibrational ground

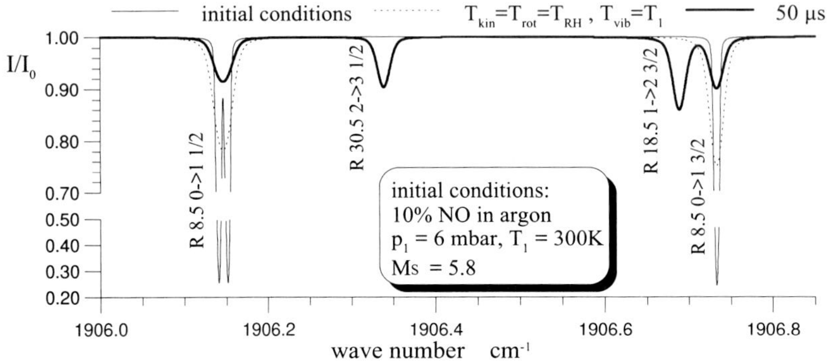

Figure 2. Calculated absorption profiles, at initial conditions, immediately behind and 50 μs after passage of the incident shock wave.

state. Immediately after passage of the shock wave, the translational temperature corresponds to the temperature T_{RH} given by the Rankine-Hugoniot equations. The full rotational excitation needs only a few collisions and the change in temperature due to this process can be neglected. The absorption spectrum for this case ($T_{kin} = T_{rot} = T_{RH}$, $T_{vib} = T_1$, no vibrational excitation and no chemical reactions) consists of the same two transitions as the spectrum described before. Using the results of nonequilibrium relaxation calculations of Widdecke the spectrum expected 50 μs (laboratory time) after passage of the shock wave has been calculated. The population densities of the vibrational excited states have increased at the expense of the population densities of the vibrational ground states.

4. Comparison of experimental results and calculations

Absorption measurements have been performed using mixtures of nitric oxide and argon as test gases. These measurements have been used to calibrate TDLAS. The temperature and the pressure behind incident shock waves at Mach numbers near $M_s = 6$, in a test gas consisting of a mixture of 90% argon and 10% nitric oxide, are almost constant. The decomposition of NO is very slow compared with the rotational and vibrational excitation of the NO molecules. The measurements presented in Fig. 3 demonstrate the excitation process. The diode-laser emission was modulated with a periodic sawtooth current at 500 kHz over three different wave number sectors. The transmission signal taken before the arrival of the shock wave (timestep $= -1$ μs) corresponds well with the calculated transmission signal shown in Fig. 2. The transitions originating from the vibrational ground state (R-branch, $j = 8.5$) are the only ones recorded

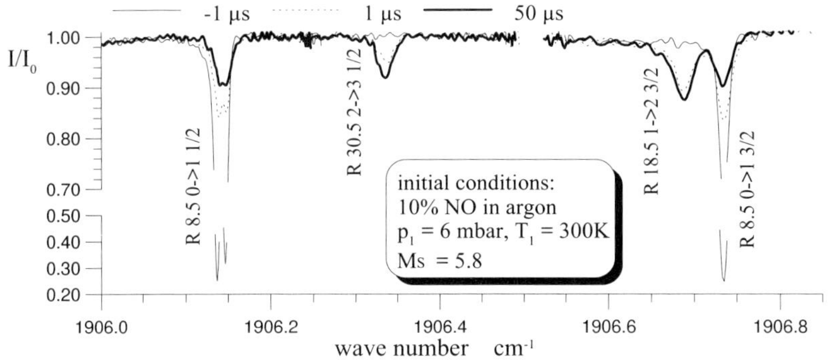

Figure 3. Measured absorption profiles behind incident shock wave.

before the arrival of the shock wave. The compression of the test gas due to the shock wave and the decrease of the population density of the rotational level $j = 8.5$ due to the temperature rise leads to a drop in number density of the absorbing molecules. The difference between the calculated transmission signal immediately behind the shock wave ($T_{kin} = T_{rot} = T_{RH}$) and the first measured profile 1 μs after passage of the shock wave is due to the vibrational excitation of the NO molecules. The measured transmission signals consist of two additional transitions. The population densities of these transitions, originating from the first and from the second excited vibrational level, increase with time, while the population densities of the transitions originating from the vibrational ground state decrease. The transmission signals recorded 50 μs after passage of the shock wave when vibrational excitation is complete are in good agreement with the calculated profiles.

An example of investigations of population densities of excited states of nitric oxide produced behind an incident shock wave in air is shown in Fig. 4. The population densities of all transitions increase due to the growth of the NO mole fraction. The difference between measured and calculated profiles can be explained by slightly different kinetic temperatures and pressures. The ratio of the line-intensities of different transitions, which is characteristic for the excitation state, corresponds well with the theoretical results.

5. Conclusions

It has been shown that the tunable infrared diode laser absorption spectrometer that was described can be used to investigate excitation processes behind incident shock waves. Measurements over complete absorption profiles with repetition rates up to 1 MHz are possible. The measurements correspond well

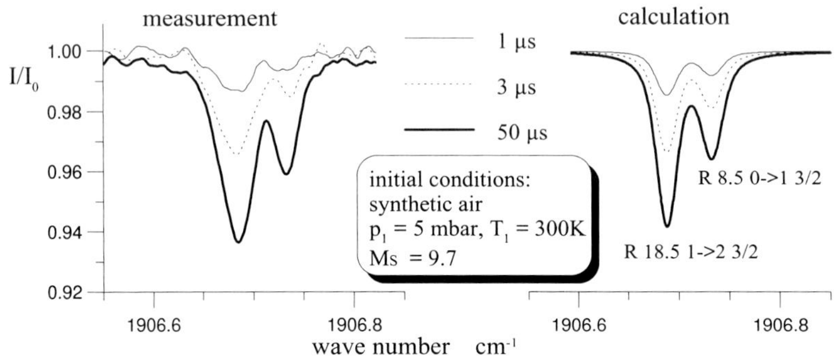

Figure 4. Comparison of measured and calculated absorption profiles.

with calculations performed with a computer program developed at ITLR. The basis of this program is a three-temperature absorption line model linked with a nonequilibrium relaxation code. In the future TDLAS will be used to investigate the time history of different excited levels of nitric oxide.

Acknowledgement. The authors are grateful for financial support by Deutsche Forschungsgemeinschaft.

References

Brandt O, Roth P (1987) Temperature measurements behind shock waves using a rapid scanning IR-diode laser. Phys Fluids 30:1294–1298.

Coerdt R, Grönig H (1989) Determination of pressure and temperature of gases by parameter fitting the FFT of an absorption line. Applied Optics 28:3021–3023.

Goldman A, Schmidt SC (1974) Infrared spectral line parameters and absorptance calculations of NO at atmospheric and elevated temperatures for the $\Delta v = 1$ bands region. J Quant Spectrosc Radiat Transfer 15:127–138.

Hanson RK (1995) The role of lasers in shock tube studies of chemical kinetics. In: Shock Waves @ Marseille, Brun R, Dumitrescu L (eds), Springer, Vol 2.

Moser L K, Hindelang F J (1992) Boundary layer effect on thermal NO decomposition behind incident shock waves. Shock Waves 2:129–132.

Park J (1990) Nonequilibrium Hypersonic Aerothermodynamics, A Wiley-Interscience Publication.

Widdecke N, Klenk W, Frohn A (1995) Experimental techniques for the investigation of relaxation processes behind strong shock waves, ZFW 19:3:213–218.

Absorption lineshape measurement of atomic oxygen at 130 nm

S. Meyer*, D. Bershader†, J. Gilmore
Stanford University, Stanford, CA 94305 USA

S. Sharma, R. Exberger
MS 230-2, NASA-Ames Research Center
Moffett Field, California 94035 USA

E. Whiting
Thermosciences Institute
Moffett Field, California 94035 USA

Abstract: Absorption of vacuum ultraviolet light by atomic oxygen has been measured in the EAST Facility at NASA-Ames Research Center. Tunable VUV is generated with a Raman-shifted ArF excimer laser, and absorption is measured over a range of wavelengths in the region of the 130.49 nm line. The lineshape due to collision broadening by atomic oxygen is measured at various wavelengths up to 1200 Lorentzian linewidths away from the line center. The empirical width is three times that given by the theory in NEQAIR5. A possible explanation for this difference is discussed. This experiment demonstrates the viability of high-order Raman shifters as light sources in single-shot measurements.

Key words: Atomic oxygen, Lineshape, Absorption, Vacuum ultraviolet

1. Introduction

Shapes of spectral lines can play an important role in radiative transfer by strong lines, such as the prominent oxygen triplet near 130 nm. Atomic absorption is a candidate for ground-state concentration measurements, but the lineshape must be known to evaluate such measurements. The lineshape for this transition is also of general interest because of its importance for radiative transport.

Lineshape predictions are uncertain because of limited experimental data on collisional broadening, especially in the line wings. Experiments with air species are primarily limited to Stark broadening, with little emphasis on collisions with neutrals. This experiment was conducted to measure self-broadening of the O line at 130.49 nm.

---

*Present address: Molecular Physics Laboratory, SRI International, 333 Ravenswood, Menlo Park, California 94025 USA; smeyer@mplvax.sri.com.

†In memoriam.

Relevant energy levels are shown in Fig. 1. The ^3P ground state is split into three closely spaced j levels (j = 2, 1 and 0) with relative populations of 5.3:3.1:1 at the nominal test temperature of 6060 K.

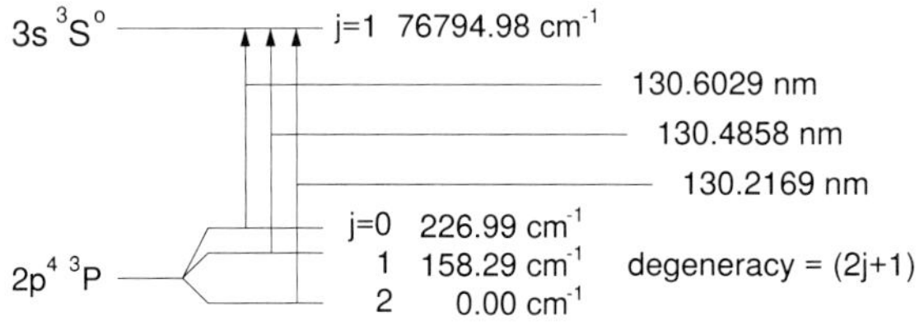

Figure 1. Energy levels of the atomic oxygen triplet at 130 nm (Moore 1993).

2. Experimental description

Transmission through the test gas is related to atomic density, path length, absorption cross section, and lineshape by Beer's Law (Lochte-Holtgreven 1968). The lineshape is determined empirically by measuring transmission over a range of wavelengths where the test conditions are known. Molecular oxygen is dissociated in the Electric Arc-driven Shock Tube (EAST) Facility at NASA-Ames Research Center (Sharma et al. 1988). The shock tube provides an atomic oxygen sample in thermodynamic equilibrium with a minimum of interfering species, especially molecular oxygen and free electrons. The test condition selected is 6.8 km/sec into 1 torr oxygen because it provides nearly complete dissociation with only slight ionization.

Composition of the test gas is determined from shock speed assuming equilibrium. Measured shock speed is corrected for attenuation with an approximate model in which laboratory time is converted into particle time to determine the upstream location where the gas was shocked. Effective shock speed, at that location, is determined from measured shock attenuation.

Establishment of chemical equilibrium is evaluated with emission measurements. Time-resolved traces show an initial spike due to the nonequilibrium overshoot, which is consistent with nonequilibrium calculations performed with NONEQ (Sharma & Whiting 1994). Absorption is measured after this nonequilibrium region passes, 6.4 μsec after shock arrival. Emission spectra recorded with a gatable OMA at various delay times characterize the test gas (Fig. 2). Electronic temperature deduced from the absolute intensity of O lines is consistent with equilibrium calculations based on shock speed with a typical uncertainty of 0.05 km/sec. Thus, temperature and composition are known for the VUV measurements.

Figure 3 is a schematic of the instrumentation. Light from a tunable, pulsed ArF laser is focused into a Raman shifter to produce VUV. The Raman shifter is filled with hydrogen at 350 torr and 77 K, and generates linearly polarized beams at numerous wavelengths. One of these overlaps the O line at 130.49 nm (Döbele et al. 1986). This beam is separated from the others by a pair of 30° MgF_2 prisms. Incident and transmitted intensities are measured by KBr photomultipliers which are insensitive to residual light at 193 nm. Several O_2 Schumann-Runge absorption lines overlap the ArF tuning range, and these lines are used for wavelength calibration (Yoshino et al. 1984).

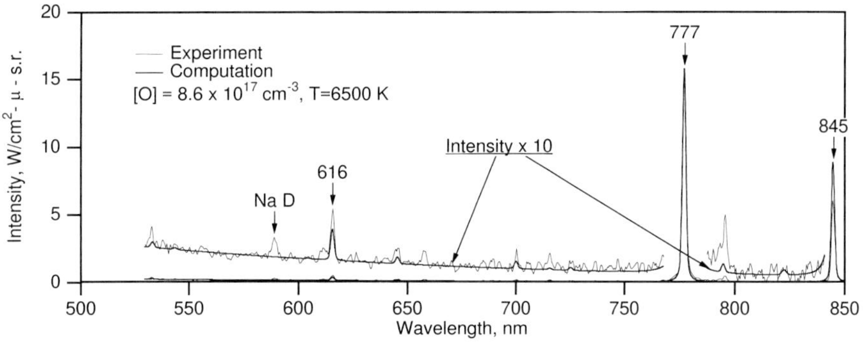

Figure 2. Equilibrium emission spectra used to verify test condition.

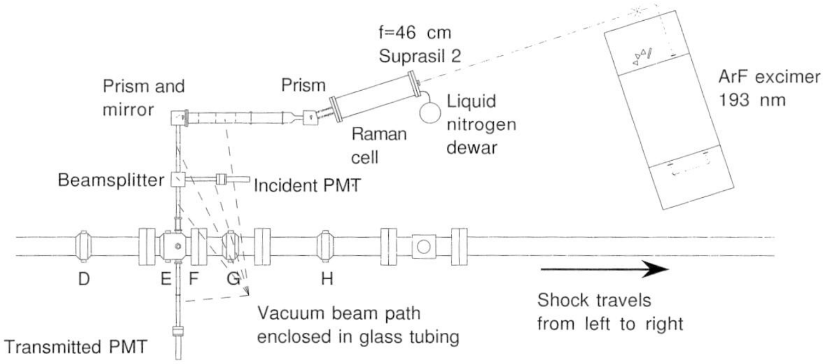

Figure 3. Schematic of shock tube and VUV instrumentation.

3. Results and discussion

Absorption data were collected on 11 runs, with a mean effective shock speed of 6.83 km/sec and a standard deviation of 0.16 km/sec (Table 1). Results of the absorption measurements are shown as a function of wavelength in Fig. 4. Shaded areas show the tuning ranges of the laser. The gap in the tuning region is due to absorption of the ArF pump laser by atmospheric O_2 between the laser and the Raman cell. In this spectral region, the laser energy entering the Raman cell is not high enough for Raman shifting to 130 nm.

Table 1. Test conditions.

Shock Speed	$6.83 - 0.16$ km/sec	6.83 km/sec	$6.83 + 0.16$ km/sec
Temperature	5400 K	6100 K	6800 K
Number density of O	9.6×10^{17}cm^{-3}	9.0×10^{17}cm^{-3}	8.4×10^{17}cm^{-3}
Number density of O_2	6×10^{15}cm^{-3}	2×10^{15}cm^{-3}	5×10^{14}cm^{-3}
Number density of e^-	2×10^{13}cm^{-3}	7×10^{13}cm^{-3}	3×10^{14}cm^{-3}

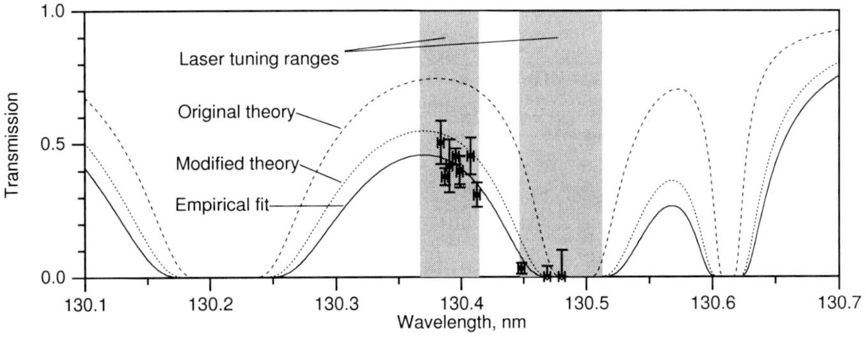

Figure 4. Comparison of measured transmission to theoretical models.

Measurements show complete attenuation of the beam near the line center at 130.49 nm, with increasing transmission as the laser is tuned away from the line center. Vertical bars are due to uncertainties in the reference intensity, and horizontal bars are due to the finite laser bandwidth and uncertainties in the laser frequency calibration. The absence of any measurable signal at the line center demonstrates the absence of stray or scattered light in the experiment. Far in the wings, contributions from the line at 130.22 nm become important. An empirical curve is fitted by adjusting the Lorentzian linewidth to match the data. Adjacent components of the triplet are included, and the widths of the three components are set equal for reasons described below. The Lorentzian linewidth deduced from the data is 100×10^{-6} nm.

Theoretical absorption profiles are calculated using NEQAIR5 for a gas in equilibrium behind a 6.83-km/sec shock. The calculated linewidth is three times smaller than the empirical value. Agreement is improved by an alternative implementation of the theory described by Ali & Griem (1965). In the wings, the lineshape is determined by collisions between oxygen atoms, or resonance broadening, which is proportional to the density of colliders resonant with the atom of interest. The linewidth calculation in NEQAIR5 only includes collision partners in the same j level as the absorber (i.e., j = 1 for 130.49 nm). However, the $^3S^\circ$ excited state is connected by dipole-allowed transitions to all three j levels of the lower state. Since the resonance interaction between two atoms can be described by exchange of virtual photons, it is independent of the external radiation field. Therefore, collisions with oxygen atoms in all three j levels should be equally effective in broadening the line. Most atoms are in ^3P, so virtually all O atoms contribute to the resonance broadening. This interpretation is supported by the measurements.

The most likely source of error in the measured lineshape is the presence of other absorbing species in the gas and broadening caused by collisions with species other than oxygen atoms. Residual O_2 density, at the nominal test condition, is $2 \times 10^{15} cm^{-3}$, corresponding to a background absorption estimated at 5–10 percent. The presence of scattering particles displaced from the walls during shock passage is checked by tests in pure argon. No attenuation of the beam is observed in these tests. Electrons and oxygen ions, O^+, are collision partners which compete with atomic oxygen. However, at the highest expected number density of electrons, the linewidth for electron broadening predicted by NEQAIR5 is only 4 percent of the measured linewidth. The broadening due to O^+ is not known, but the concentration of O^+ is essentially the same as that of free electrons. Thus, even if the ion-atom collisions are effective, they do not make a significant broadening contribution due to the low ion concentration. Finally, uncertainties in the linewidth of the adjacent line at 130.22 nm introduce an additional uncertainty of about 10 percent. Further details are given by Meyer (1995).

The experimental technique demonstrated here provides a direct measurement of the atomic lineshape. Unlike curve-of-growth and effective linewidth methods, which integrate over the entire line, this technique measures absorption at individual positions on the line. The instrumentation developed here can be used to measure the number density of atomic oxygen at conditions where the lineshape in the wings has been determined, or to make further lineshape measurements under suitably defined conditions.

4. Conclusions

The lineshape of the $^3P_1$–$^3S^\circ$ transition of atomic oxygen at 130.49 nm has been measured in the far wing region. The empirical linewidth in this spectral region is larger than that predicted by the NEQAIR5 code. Agreement between theory and experiment is improved significantly by including collisions with all j-levels of the ^{3}P ground state in the linewidth calculation. Finally, we have demonstrated that a Raman-shifted excimer laser is an effective, tunable VUV light source for single-shot experiments where signal averaging is not practical.

Acknowledgement. Support for S. Meyer provided by NASA-Ames Research Center through Interchange Number NCA2-723 to Stanford University and for E. Whiting through NAS2-14031 to Eloret Institute is gratefully acknowledged. The authors thank M. Dyer and D. Huestis of SRI International for assistance with the Raman shifter, D. Fletcher, C. Park, and W. Huo for valuable discussions, and W. Warren for operation of the EAST Facility.

References

Ali AW, Griem HR (1965) Theory of resonance broadening of spectral lines by atom-atom impacts. Phys Rev 140:A1044–1049.

Döbele HF, Hörl M, Röwekamp M, Reimann B (1986) Detection of atomic oxygen by laser-induced fluorescence spectroscopy at 130 nm. Appl Phys B 39:91.

Lochte-Holtgreven W (ed) (1968) Plasma Diagnostics, North-Holland Publishing Co., Amsterdam, pp 66–134.

Meyer SA (1995) Vacuum Ultraviolet Absorption Measurements of Atomic Oxygen in a Shock Tube, PhD Thesis, Stanford University.

Moore CE (1993) Tables of Spectra of Hydrogen, Carbon, Nitrogen, and Oxygen Atoms, CRC Press, Inc.

Sharma SP, Park C, Dannenberg RE (1988) Operating Characteristics of a 60 cm and a 10 cm Electric Arc-driven Shock Tube, AIAA 88-0142.

Sharma SP, Whiting EE (1994) Modeling of Nonequilibrium Radiation Phenomena: An Assessment, AIAA 94-0253.

Yoshino K, Freeman DE, Parkinson WH (1984) Atlas of the Schumann-Runge absorption bands of O_2 in the wavelength region 175–205 nm. J Phys Chem Ref Data 13:207.

Spectroscopic detection of highly excited molecular-state transformations behind shock waves and master-equation implementation

I. S. Zaslonko
Semenov Institute of Chemical Physics
Russian Academy of Sciences, 117334 Moscow, Russia

A. V. Eremin, V. V. Shumova
High Energy Density Research Center "IVTAN" Association
Russian Academy of Sciences, 127412 Moscow, Russia

Abstract: This work is devoted to the detailed description of kinetic processes occurring during the dissociation-recombination of small polyatomic molecules at high temperatures. The results of interpretation of spectroscopic data concerning the decay of CO_2 and SO_2 in shock waves at T = 2500 to 4000 K, P = 10 to 30 atm are presented.

Key words: Dissociation-recombination, Vibrational relaxation

1. Introduction

New theoretical approaches to the description of the essential deviations from the Arrhenius kinetics of the dissociation-recombination of small polyatomic molecules required the last several years to develop. The experiments testify to the existence of significant depletion of molecular states lying below the dissociation barrier, as well as population of over-barrier states at early stages of decay (Zaslonko et al. 1990b; Eremin and Ziborov 1993). These facts were not described hitherto, and the question of kinetics of over-barrier state populations depending on the decay and recombination processes was not solved.

This work is devoted to the solution of this problem, that is, to analysis of deep stages of dissociation, description of the whole energetic spectrum of electronically and vibrationally excited states of small polyatomic molecules during the transition to equilibrium given by the balance of thermal decay and recombination.

2. Kinetic model

The common approach to such problems consists of solution of a system master equation (ME) considering the whole molecular energetic spectrum. Since kinetic rate constants of specific level reactions are unknown in most cases, a promising way is to solve the ME in terms of a smooth energetic approximation.

This idea is realized in the so-called "step-ladder" model of ME solution by Zaslonko et al. (1990a), which considers the molecular energetic spectrum as a system of equidistant levels with energy E_i and population

$$A_i = \frac{A_0 \rho(E_i) \exp(-\frac{E_i}{T}) dE}{Q_v \sum_i f(E_i) dE} \tag{1}$$

where $\rho(E_i)$ is the state density in the Whitten-Rabinovich approximation, $f(E_i)$ is their distribution function (DF), E_0 is dissociation threshold energy, Q_v is the vibrational partition function, and A_0 is full concentration of reagent. The dissociation is considered as a limit of rising movement over the ladder.

The previous studies consider only the beginning of dissociation of highly diluted gas mixtures, which are far from equilibrium and the recombinative processes are negligible. But these processes have a vast influence on the population of over-barrier states and on the whole process while the system approaches equilibrium. In this work the step-ladder model of unimolecular decay is expanded to describe the late stages of dissociation along with the quasi-stationary one in the framework of a united approach.

The population E_i of any microscopic state under consideration is given by the solution of the ME:

$$\frac{dA_i}{dt} = -W_+ + W_- + W_f(i) + W_{bm}(i) + W_{ev}(i) + W_r(i) \tag{2}$$

that is, the kinetic mechanism includes the vibrational relaxation within both ground and electronically excited states (process rates are W_+ and W_-), energy exchange between them (process rates W_{ev}), spontaneous decay of over-barrier states (processes rates W_f), bimolecular reactions (W_{bm}), and recombination (W_r). The collisional transitions between only adjacent levels are taken into account. The model considered assumes the quasi-equilibrium between over-barrier active excited states of parent molecules and dissociation products due to rapid establishment of partial equilibrium between spontaneous decay and recombination.

The "step" of ladder $\Delta E = E_{i+1} - E_i$ is a model parameter; in our calculations it was varied from 0.5 to 2.0 kT.

To represent the DF transformation in the course of dissociation-recombination processes the three-dimensional presentation of DF in normalized form was used. The normalizing function (F_{eq}) is suggested to be:

$$F_{eq}(E,t)|_{E<E_0} = \frac{\rho(E_i) \exp(-\frac{E_i}{T})}{Q_v} \tag{3}$$

$$F_{eq}(E,t)|_{E>E_0} = \frac{\rho(E_i)\exp(-\frac{E_i}{T})}{Q_v} \times \frac{B_{eq}(t)}{B(t)} \times \frac{k_{diss}}{k_f(E_i)}, \tag{4}$$

where $B_{eq}(t)/B(t)$ is the ratio of equilibrium and actual concentrations, k_{diss} is the apparent decay rate coefficient, and $k_f(E_i)$ is the microscopic dissociation rate constant of level i.

In such terms the state of abruptly heated gas (for example, behind a shock wave) will be given by a δ-function at the beginning coordinate; the steady state of the system will be given by a straight line parallel to the (E,t)-plane; the transformation of the δ-function into the line will represent the vibrational relaxation in the system (see Fig. 1). The time delay of DF formation at higher levels attains ≈ 2 μs.

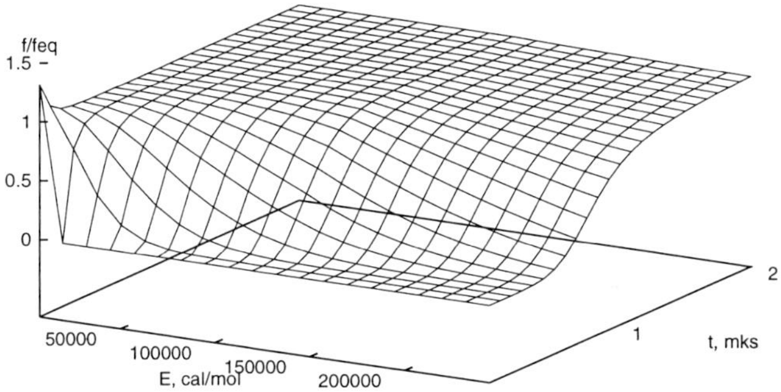

Figure 1. Vibrational relaxation into the ground state of CO_2 at $T = 2500$ K, $P = 1$ atm in the absence of dissociation. F/F_{eq} is normalized to equilibrium distribution function; E is energy.

3. Results

The approach developed permitted us to simulate all the observable kinetic characteristics of shock-heated CO_2 and SO_2: vibrational relaxation within ladder structure from bottom to top; establishment of steady state of highly excited state distribution including electronically excited states; quasi-stationary dissociation; and transition to thermodynamic equilibrium. In Fig. 2 the transformation of $CO_2(X^1\Sigma)$ DF during the dissociation of CO_2 is represented. One can see the quasi-stationary stage of dissociation (1) and the nonstationary one (2), which is linked with the increasing influence of the recombination at late stages of the process.

Based on the "step-ladder" model, the apparent rate constant of CO_2 dissociation is given by the equation

$$k_{diss} = \frac{\sum_i k_f(i)A(i)}{A}, \tag{5}$$

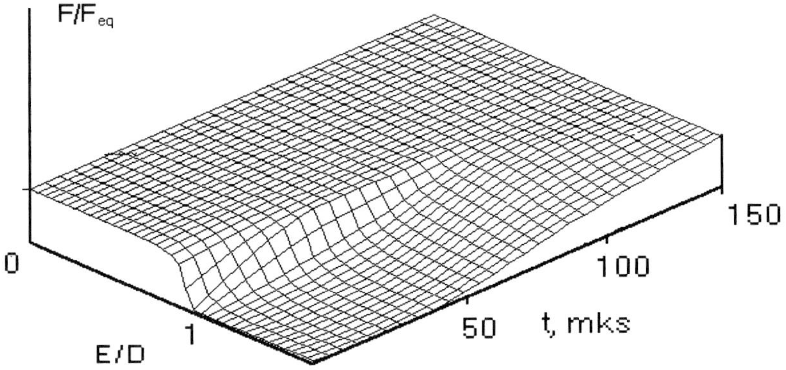

Figure 2. Transformation of CO_2 highly excited state distribution during thermal decomposition at T = 3920 K, P = 5 atm in 1% CO_2 in Ar. E_0 is dissociation barrier.

as a combination of dissociation rate constants $k_f(i)$ of specific over-barrier levels and was calculated to be $k_{diss} = 1.0310^{14} \exp(-48600\text{K}/T)$ cm^3 mol^{-1} s^{-1}, which is close to the NIST database value ($k_{diss} = 3.06 \cdot 10^{14} \cdot \exp(-51400\text{K}/T)$ cm^3 mol^{-1} s^{-1}). But the macroscopic parameter of dissociation process doesn't give any reliable information about the mechanism of the process.

To evaluate the mean portion of energy transferred per collision, ΔE, which is the principal parameter of the process, and to get insight into the process mechanism, it was necessary to simulate the behavior of specific molecular levels. In Figs. 3a and 3b the time profiles of population of $CO_2(^1B_2)$ over-barrier states with E = 6.0 eV measured by emission-absorption technique in two regimes and their "step-ladder"-model simulations with various values of ΔE are represented ($E_0(CO_2)$ = 5.45 eV). The strong depletion of this over-barrier state (Fig. 3a) during the quasi-stationary decay shows that the population mechanism is mostly a recombinative one. At higher pressure (Fig. 3b) the quasi-stationary population of this level is higher due to increase of activating collisions. One can see that the best fit with experiments uses calculation with ΔE = 0.8 kT.

To derive some quantitative conclusions about the CO_2 decay mechanism, the branching ratio of spin-forbidden and spin-allowed channels of dissociation was analyzed. Table 1 shows the calculated ratio $\alpha = W_{forb}/W_{all}$, where W is decay rate through the corresponding channel for a given ΔE. At high temperatures the role of an allowed channel is shown to increase and attain $\approx$ 5% at 4500 K. At low temperatures the increasing role of forbidden conversion $^1B_2 \rightarrow {}^3B_2$ state and decreasing of effective energy of dissociation was

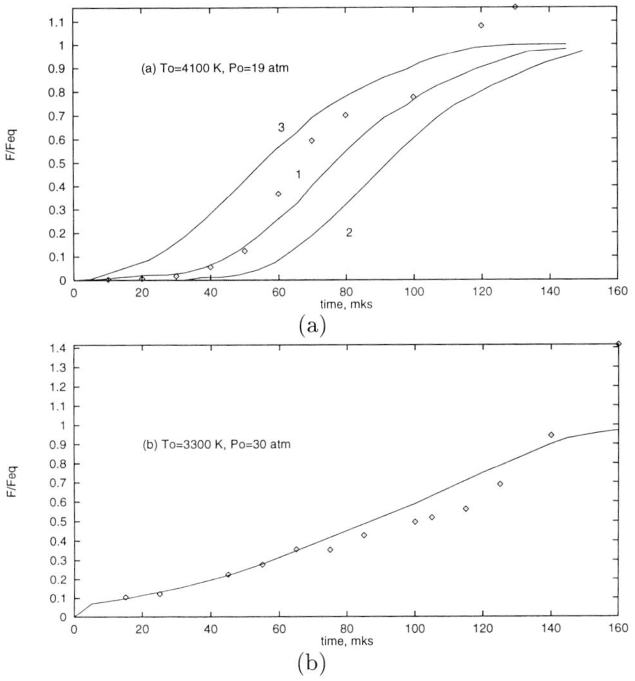

Figure 3. Time evolution of $CO_2(^1B_2)$ state ($E = 6.0$ eV) DF. Symbols correspond to emission-absorption measurements (Eremin and Ziborov 1993); lines correspond to "step-ladder" simulations at $\Delta E = 0.8$ kT (1), 0.4 kT (2), and 1.6 kT (3). (a) $T = 4100$ K, $P = 20$ bar; (b) $T = 3300$ K, $P = 30$ bar.

revealed. This result is in good agreement with experimental observations by Eremin and Ziborov (1993).

Table 1. Branching ratio of spin-forbidden and spin-allowed channels of CO_2 dissociation for various temperatures and pressures.

Temp, K	2500	3000	3500	4000	4500
$P = 2 \cdot 10^6$ Torr	83.30	42.92	28.65	21.68	17.34
$P = 3 \cdot 10^6$ Torr	66.21	35.84	23.70	17.69	14.01

Opposite to CO_2, which has a practically stable fragment CO in the temperature range considered, the SO_2 fragment SO dissociates at the same temperatures as the parent molecule with the comparable rate constant. So the recombinative effects are so slight that the SO_2 dissociation quasi-stationary stage lasts very long and DF remains depleted during the whole process.

Comparison of the temperature dependence of the $SO_2(^1B_2)$ state ($E = 5.43$ eV) depletion, as measured by Zaslonko et al. (1990b) in "hot bands," and by Bazhenova et al. (1993) with emission-absorption technique and "step-ladder"-model simulations, was carried out. The calculations proved that the low-lying electronic SO_2 states with the high density of vibrational levels near the barrier play a main role in the decay, according to conclusions made by Plach and Troe (1984) and Hippler et al. (1988). The depletion of under-barrier states was shown to characterize the relation between the processes of intramolecular energy transfer.

In Fig. 4 the results of "hot-band" measurements of $SO_2(^1B_1)$ DF at 3500 K and 4500 K and step-ladder simulations are presented. Comparison of experimental findings with theoretical treatment allowed evaluation of the mean portion of energy transferred per collision for SO_2, which was found to be close to results for CO_2.

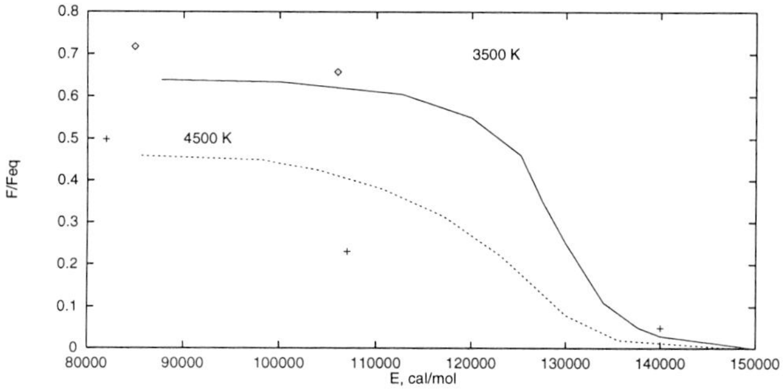

Figure 4. Distribution function of $SO_2(^1B_1)$ at 3500 K (◇) and 4500 K (+) by "hot-band" measurements by Zaslonko et al. (1990b); and step-ladder simulations (lines).

Acknowledgement. The authors are grateful to the Russian Fundamental Research Foundation (Grant No. 94-03-08881a) for financial support of this work.

References

Bazhenova TV, Eremin AV, Kubatin VM, Ziborov VS, Roth P, Woiki D (1993) 14 Int Colloq Dynam Expl and React Systems, ICODERS, Coimbro, Portugal, 1–6 Aug 1993, Vol 1, p 251.

Eremin AV, Ziborov VS (1993) Shock Waves 8(4):475.

Hippler H, Nahr D, Plach HJ, Troe J (1988) J Phys Chem 92(19):5503–5506.

Plach HJ, Troe J (1984) Int J Chem Kinetics 16:1531–1542.

Zaslonko IS, Haiduchenya LV, Mukoseev YK, Tereza AM (1990a) Sov J Chemical Physics 9(4):443.

Zaslonko IS, Haiduchenya LV, Mukoseev YK (1990b) Sov J Chemical Physics 9(5):589.

Measurement of the absorption spectrum of the OH A-X (0,0) 306-nm bandhead at high pressures

D. F. Davidson, M. Röhrig, E. L. Petersen,
M. D. Di Rosa, R. K. Hanson
High Temperature Gasdynamics Laboratory
Department of Mechanical Engineering
Stanford University
Stanford, CA 94305 USA

Abstract: We report detailed absorption spectrum measurements for a section of the OH A-X (0,0) bandhead near 306 nm at 60 atm and 1735 K. This spectra was taken using narrowline laser absorption in a new high-pressure shock-tube facility. The variation of the OH absorption coefficient with wavenumber is presented, as well as best-fit collision widths and collision shifts.

Key words: Absorption spectra, OH radical, High pressure, Laser diagnostics, Collision broadening

1. Introduction

The requirement of higher output power in combustion devices is met in many cases by increasing the reaction-chamber pressure. Working-chamber pressures for devices such as advanced gas turbines, rocket engines, and ram accelerators may reach several tens and even hundreds of atmospheres. In these high-pressure devices, the chemistry of combustion may differ in fundamental ways from that found in atmospheric combustors. Because of the demand for better knowledge of such chemistry, a new high-pressure, high-purity shock tube has been built at Stanford. We are developing a program which investigates detailed chemical kinetics and ignition times of fuel-oxidizer systems at elevated pressures and temperatures.

In the past, to study chemical processes in a shock tube, we have used narrow linewidth laser absorption of various combustion species. In particular, we have found that absorption by the transient OH radical is sensitive, selective, and non-intrusive, and yields valuable information about the combustion process (Rea et al. 1987). Accurate measurements of OH-concentration time histories of use in chemical kinetic studies were easily determined from absorption measurements because the behavior of the OH lineshape and the temperature and pressure dependence of the OH absorption coefficient is well established. To extend this method to higher pressures requires a knowledge of the variation of the OH collision-broadening and collision-shift parameters with pressure.

We have chosen to study the collision broadening in the spectral region between 32622 to 32632 cm^{-1} which includes two peak features, near 32625 and 32630 cm^{-1}, which are strongly sensitive to the collision-width parameter. In this study we report the OH absorption coefficient and collision broadening and shift parameters which should permit the use of narrow linewidth absorption in a shock tube to measure OH-concentration histories at high pressures.

2. Experimental method

The experimental discussion is divided into four parts: OH generation in the shock tube, prediction of OH concentrations, laser and optical setup, and the determination of absorption coefficient.

OH radical concentrations were generated by the ignition of very dilute stoichiometric $H_2/O_2/Ar$ gas mixtures at high pressures behind reflected shock waves. These high pressures were attained using a new high-purity, pressure-driven shock-tube facility designed to attain reflected shock pressures as high as 1000 atm (see Petersen et al. 1996). The driven section is 5 m long, 5 cm diameter, and the driver section is 3 m long 7.6 cm diameter with a converging double-diaphragm, breech-loading interface. Shock speeds and pressures are monitored with 6 piezo-electric transducers. Ultimate pressures of $< 10^{-6}$ torr and leak/outgassing rates of $< 10^{-5}$ torr/min are regularly achievable. Gas mixtures of 1000 ppm H_2/500 ppm O_2 or 5000 ppm H_2/2500 ppm O_2 in argon were used. The majority of the reflected shock conditions centered near 60 atm and 1735 K.

OH concentrations were calculated using the 25 reaction O/H subset of the GRIMECH v1.2 Natural Gas Combustion mechanism and thermodynamic data (Frenklach et al. 1994). Rate coefficients for two reactions (1) $H + O_2 + Ar = HO_2 + Ar$ and (2) $H + OH + Ar = H_2O + Ar$ were reduced by half to effect slightly better fits in the ignition times and the long time decay, respectively. The sensitivity analysis shown in Fig. 1a indicates that the peak OH concentration is most sensitive to the rate coefficients for the reactions (3) $H + O_2 = OH + O$ and (4) $O + H_2 = H + OH$. The rate coefficients for both these reactions are well established, and do not contribute significant uncertainty to the absorption coefficient. At the conditions of the present experiments, 1700 K and 60 atm, the calculated peak OH concentration increased by 5.8% when the rate coefficient for reaction (1) was reduced by half; this is indicative of the expected systematic uncertainty from the chemical kinetics. The OH concentration is not significantly sensitive to reactions in the mechanism with large uncertainties at 1700 K, as would be the case at 1200 K.

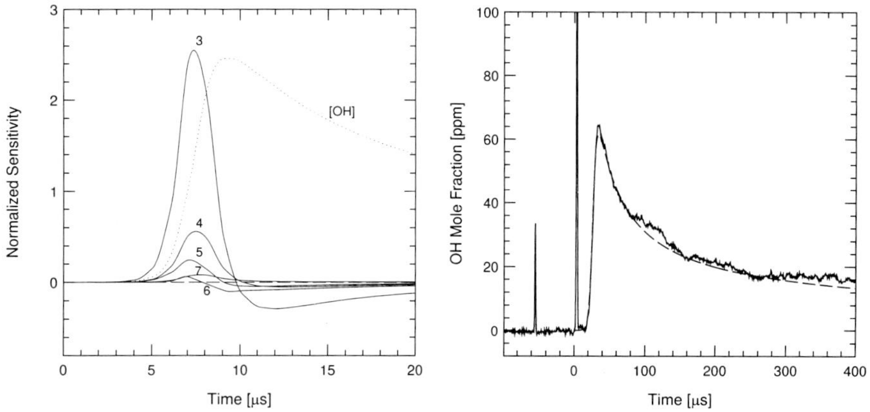

(a) (b)

Figure 1. (a) Normalized sensitivity, S, of OH mole fraction to kinetic rate parameters. $S = (d[\mathrm{OH}]/dk_i)(k_i/[\mathrm{OH}]_{max})$. Conditions: 1700 K, 60 atm, 5000 ppm H_2/2500 ppm O_2/Argon. k_3: $H + O_2 = O + OH$; k_4: $O + H_2 = H + OH$; k_5: $H + HO_2 = O_2 + H_2$; k_6: $OH + H_2 = H + H_2O$; k_7: $H + HO_2 = OH + OH$. Dotted line: relative OH concentration. (b) Example absorption data and calculation. Conditions: 1830 K, 58.40 atm, 1000 ppm H_2/500 ppm O_2/Argon, 32629.72 cm^{-1}, $k_\lambda = 20.5\,atm^{-1}cm^{-1}$. Dashed line: GRIMECH v1.2 calculation.

The OH-absorption measurements were made using narrow linewidth laser absorption in a simple Beer's Law experiment. The CW narrow linewidth laser radiation was generated using an Ar^+-pumped ring-dye laser and temperature-tuned intracavity doubling with an AD*A crystal. This laser radiation was separated into two beams: a reference beam I_0, and a transmitted beam, I, which passed through the shock tube. The OH-absorption time history was recorded in each of a series of 22 shocks. After each shock the tuning of the laser was changed, eventually covering the range from 32622 to 32632 cm^{-1} (306.542 to 306.448 nm). This wavelength region was selected as it has the strongest high-pressure OH-absorption feature accessible with our Spectra-Physics™ temperature-tuned frequency-doubling system. It includes two peaks, near 32625 and 32630 cm^{-1}, which are strongly sensitive to the collision-width parameter. Wavenumber measurements accurate to $\pm 0.03\,cm^{-1}$ were made using a Burleigh WA-20 wavemeter.

The fractional transmission is related to the temperature-, pressure- and wavelength-dependent absorption coefficient k_λ through the Bouguer-Lambert relation

$$I/I_0 = \exp(-k_\lambda P_{\mathrm{total}} X_{\mathrm{OH}} L) \qquad (1)$$

The transmission data were first converted to absorbance (i.e., $\ln I/I_0$), and then a best-fit absorption coefficient for each data trace was determined by bringing the measured absorbance into coincidence with the product of the calculated OH mole fraction X_{OH}, the known reflected shock pressure P_{total}, the path length $L = 5\,\mathrm{cm}$, and varying k_λ. An example data trace is shown in Fig. 1b. The data show good signal-to-noise ratio and no absorption in the absence of OH concentration.

3. Results and discussions

The spectral shape of the region 32622 to 32632 cm^{-1} contains significant contributions from the broadened rotational lines R_1(7-11) and the $Q_{2,1}$(7-11) which are all summed. The calculated absorption spectra requires the OH line positions (Stark et al. 1994), the Einstein A coefficients and J′ assignments (Goldman & Gillis 1981), and a numerically simple method of calculating the Voigt function. The Voigt lineshape requires the collision width $\Delta\nu_{coll}$, a collision shift-to-width ratio Δ, and the Doppler broadening. For argon, all lines were considered to have the same collision shift and width. As a first estimate to the collision broadening and shift, the collision width suggested by Rea et al. (1987), and a fixed shift-to-collision width ratio $\Delta = -0.3$ was used. This collision width can be expressed as

$$\Delta\nu_{coll} = 0.141\ P\ (300/T)^{0.78}\ [\mathrm{cm}^{-1}] \tag{2}$$

Small corrections were made to the measured absorption coefficient to bring shock-to-shock variation in T_5 and P_5 to the normalized values of 1735 K and 60 atm.

The calculated absorption spectrum is compared to the experimental absorption shape in Fig. 2a. The collision width suggested by Rea et al. (1987) gives a fair fit to the present data. The collision broadening of the majority of the data falls within ±10% of that given by Eq. (2).

It is possible to improve the fit to the data by using a collision width 10% larger that that of Eq. (2) and a larger shift-to-collision width ratio $\Delta = -0.35$. This fit is shown in Fig. 2b. Using the temperature dependence of Eq. (2), the collision-width measurements of Shirinzedeh et al. (1985) at 300 K, Engleman (1969) at 293 K, and Rea et al. (1987) at 1400–4100 K can be related to the nominal temperature of 1735 K. In these cases, the collision width is linear from an effective pressure of 0.35 to 60 atm. The larger collision shift-to-width ratio differs from an earlier low-pressure measurement (Shirinzadeh et al. 1985) of 0.22 in argon and 0.083 in air for the 306.84 nm OH P1(2) line. Kessler et al. (1993) found a collision shift-to-width ratio of 0.1 in a low-pressure hydrogen-air flame confirming Shirinzadeh's shift in air. The increase in shift-to-collision-

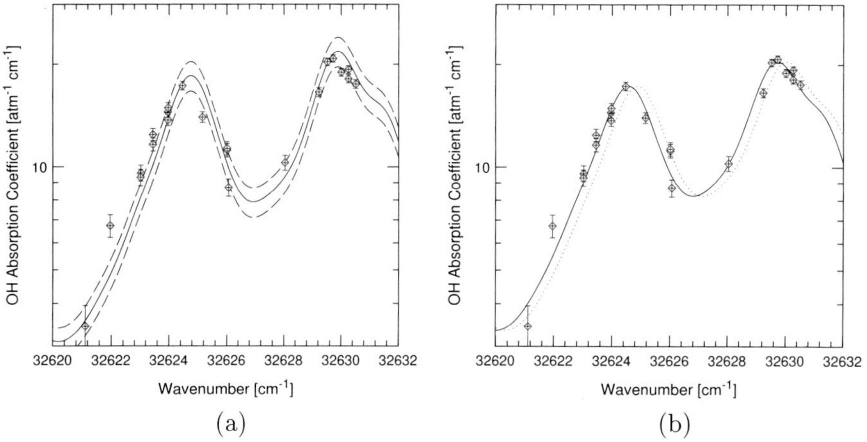

Figure 2. (a) High pressure OH A-X (0,0) absorption spectra. Conditions: 1735 K, 60 atm, argon broadening. Solid line: fit using Rea et al. (1987) collision width $\Delta\nu_{\text{coll}} = 2.145\ \text{cm}^{-1}$, shift-to-collision-width ratio $\Delta = -0.3$. Dashed lines: $\pm 10\%$. (b) Solid line: best-fit collision width $\Delta\nu_{\text{coll}} = 2.39\ \text{cm}^{-1}$, collision shift-to-width ratio $\Delta = -0.35$. Dashed lines: $\Delta = -0.22$.

width ratio with pressure is one explanation of a non-linear variation of this ratio with pressure. It has been suggested (Royer 1980) by statistical lineshape theory that the asymmetry (undetectable in the present experiments) and the shift vary quadratically with the number density, while the width varies only linearly. In the present experiments, the maximum number densities we have achieved are only 10 times that found at NTP so a large quadratic component in the pressure dependence of the shift is not expected. A preferred explanation is that the temperature dependence of the shift is different than that of the collision width. These differences have been seen in NO (Chang et al. 1992).

4. Conclusions

We have measured the absorption spectra of the OH A-X (0,0) band head from 32622 to 32632 cm^{-1}. The calculated values of OH concentration from the GRIMECH v1.2 mechanism agree with the measured OH concentration at 1735 K, 60 atm if a collision width of $2.39\,\text{cm}^{-1}$ and a collision shift of -0.83 cm^{-1} are used. These measurements confirm that there is a linear dependence of the collision width on pressure from below 1 atm to 60 atm. The collision shift appears to vary differently than the width.

Acknowledgement. This work was supported by the Aerospace Sciences Directorate of AFOSR with Dr. Julian Tishkoff as contract monitor.

References

Chang AY, Di Rosa MD, Hanson RK (1992) Temperature dependence of collision broadening and shift in the NO A←X (0,0) band in the presence of argon and nitrogen. J Quant Spectrosc Radiat Transfer 47:375.

Engleman R Jr (1969) Collision broadening of transient absorption spectra. J Quant Spectrosc Radiat Transfer 9:391.

Frenklach M, Wang H, Bowman CT, Hanson RK, Smith GP, Golden DM, Gardiner WC, Lissianski V (1994) An optimized kinetics model for natural gas combustion. In: 25th Symp (Int) on Combustion, The Combustion Institute, Pittsburgh, Work-in-Progress Abstracts, p 212.

Goldman A, Gillis JR (1981) Specral line parameters for the A-X (0,0) band of OH for atmospheric and high temperatures. J Quant Spectrosc Radiat Transfer 25:111.

Kessler WJ, Allen MG, Davis SJ (1993), Rotational level-dependent collisional broadening and line shift of the A-X (1,0) band of OH in hydrogen-air combustion gases. J Quant Spectrosc Radiat Transfer 49:107.

Petersen EL, Davidson DF, Rohrig M, Hanson RK (1996) High-pressure shock-tube measurements of ignition times in stoichiometric $H_2/O_2/Ar$ mixtures. In: Shock Waves, Proc of the 20th ISSW, Sturtevant B, Shepherd JE, Hornung HG (eds), World Scientific, Vol 2, pp 941–946.

Rea EC Jr, Chang AY, Hanson RK (1987) Shock-tube study of the pressure broadening of the A-X (0,0) band of OH by Ar and N_2. J Quant Spectrosc Radiat Transfer 37:117.

Royer A (1980) Shift, width and asymmetry of pressure-broadened spectral lines at intermediate densities. Phys Rev A 22:1625.

Shirinzadeh B, Bakalyar DM, Wang CC (1985) Measurement of collision-induced shift and broadening of the ultraviolet transitions of OH. J Chem Phys 82:2877.

Stark G, Brault JW, Abrams MC (1994) Fourier-transform spectra of the A-X (0,0) bands of OH and OD. J Opt Soc B 11:3.

Self-ignition of n-heptane/iso-octane mixtures and n-decane with air under engine conditions

K. Fieweger, U. Pfahl, G. Adomeit
Institut für Allgemeine Mechanik, RWTH Aachen
D-52064 Aachen, Germany

Abstract: Various stoichiometric mixtures of iso-octane and n-heptane with air have been investigated using the reflected-shock technique at an initial pressure of 40 bar and an initial temperature between 700 and 1200 K. Because of the inhomogeneity of the self-ignition process leading to a deflagrative first phase of combustion, two different ignition-delay times were defined. The differences of the two reference fuels iso-octane and n-heptane and their mixtures with respect to their knocking properties are illustrated by the results of this study. On the one hand there are differences in ignition-delay times of more than one order of magnitude; on the other hand, n-heptane, when included as a small part in mixtures, effects a fast transition from the first deflagrative phase of the self-ignition into a detonation-like process. Furthermore, stoichiometric mixtures of n-decane with air were investigated using a heatable, high-pressure shock tube.

Key words: Self-ignition, Ignition-delay time, Diesel engine, Chemical kinetics, Hydrocarbons, Deflagration, Detonation, DDT

1. Introduction

Self-ignition occurs in internal combustion engines in two different ways: in the diesel engine, the inflammation of the injected and vaporized fuel starts the combustion process. In spark-ignited (SI) engines the auto-ignition of the unburned gas with subsequent strong pressure waves — knock — is undesirable because of the high thermal and mechanical load on the engine. The results of shock-tube investigations of the self-ignition process of undiluted fuel/air mixtures are of relevance for both processes. Ignition-delay times are used to adjust and validate full and reduced kinetics (e.g., Chevalier et al. 1990), which are used in computational models of the diesel-engine process. The different self-ignition modes investigated in this study are closely related to different knocking modes observed in engine experiments (e.g., König et al. 1990) and represent — besides the ignition-delay time — a second important feature characterizing a fuel and its knocking properties.

In this study, stoichiometric mixtures of iso-octane with varying n-heptane fraction and octane number (ON) (0% = ON 100, 10% = ON 90, 20% = ON 80, 40% = ON 60) have been investigated as two-component SI-model fuels. Furthermore, n-decane was chosen as a diesel-model fuel because its physical and chemical properties are closer to a real diesel fuel than n-heptane.

2. Experimental design

The most important technical data of the three different shock tubes used in this study are summarized in Table 1. The first tube has been described in detail

Table 1. Technical data for the shock tubes.

No.	Cross-section	Max time	Max static pressure	Operating temperature	Length high-pressure part	Length low-pressure part
1	○ 87 mm	10 ms	500 bar	293 K	5.3 m	6.3 m
2	○ 90 mm	10 ms	500 bar	up to 423 K	6.1 m	6.4 m
3	□ 54 mm	5 ms	20 bar	293 K	3.1 m	5.1 m

in recent publications, e.g., Ciezki and Adomeit (1993), Fieweger et al. (1993). The second tube is very similar to the first with respect to geometry and operating pressure, but can be heated up to a temperature of 423 K in order to avoid condensation of vaporized fuel in the test gas. The third shock tube consists of a test section with a length of 200 mm and optical access to the whole cross-section through Plexiglas windows. This enables the visualization of the ignition process by means of a Cranz-Schardin camera (shadow photographs) and holography. All devices are equipped with pressure gauges and quartz windows with photomultipliers located 15 mm from the end wall of the driven section in order to record histories of pressure and emission at 431 nm (CH band).

3. Results and discussion

3.1. Gasdynamic aspects of the self-ignition process

The auto-ignition process of hydrocarbons in practical systems like engines, but also in experimental devices such as shock tubes and rapid compression machines, proceeds in different modes depending on temperature, pressure, fuel, geometry, impurities, etc. A fundamental division was made by Oppenheim and coworkers (Oppenheim 1985, Vermeer et al. 1972). At high temperatures strong ignition occurs, which is an explosion of the reactive mixture connected with the instantaneous development of a shock that forms a detonation. At lower temperatures the ignition process is mild, starting initially as a constant-pressure flame. Due to inevitable inhomogeneities of temperature, radical concentration, or particles (hot spots, exothermic centers) in the test gas, this deflagrative process starts at one or several kernels. In the later phase the mild self-ignition process may lead to a secondary explosion and a transition to detonation (DDT). The first deflagrative phase of the self-ignition is connected with a smooth pressure rise. In case of transition, this is followed by strong pressure peaks with steep gradients resulting from shock waves arising from a secondary *explosion in the explosion*. The fuel n-heptane always shows a fast transition with — at high pressures — a negligible short deflagrative phase, whereas fuels with high octane number (e.g., iso-octane, benzene; see Fieweger et al. 1993) do not show a DDT at sufficiently low temperatures and pressures.

In order to interpret the results obtained with our high-pressure shock tubes and to find a criterion for the determination of ignition-delay times for experiments with several types of self-ignition, we performed a set of experiments using the third shock tube described above. The self-ignition process was visualized using a Cranz-Schardin camera. A comprehensive survey of the results of this study can be found in a second publication (Blumenthal et al. 1995). Figure 1 shows a series of photographs (frames 1, 6, 11: $\Delta t = 200\ \mu s$, 12–14:

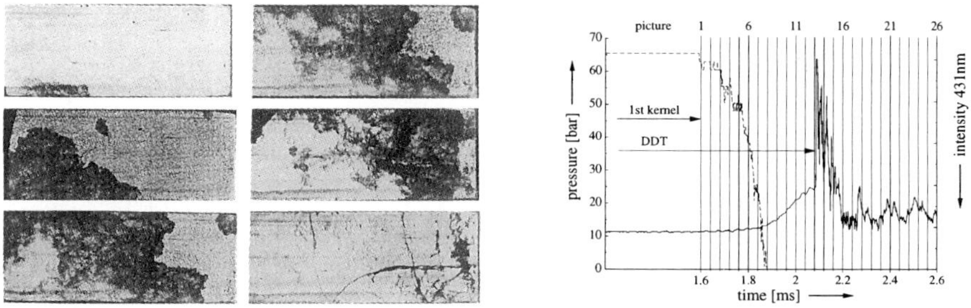

Figure 1. Correlation between the first flame kernel and the first increase of CH-band emission at 431 nm, iso-octane/air, 13.5 bar, 1028 K. Frames 1, 6, 11, 12–14.

$\Delta t = 200\ \mu s$) and the corresponding pressure and emission histories at 15 mm in front of the end wall (right-hand side of the photographs). The vertical lines mark the instants where pictures were taken during this run. They are numbered at the top of the diagram. The formation of the first kernel and its growth can be seen very clearly. The first pressure rise begins at picture 6, but the first increase of the emission signal is recorded already at picture 1. The deflagrative combustion phase lasts until picture 14, when an auto-explosion near the end wall arises. The ensuing shock waves, reflected at the end wall, can be seen both in the picture and in the pressure record.

3.2. Ignition-delay times

Whereas the validation of kinetic schemes is normally done zero-dimensionally by modeling a homogeneous reactor, such an ideal, homogeneous self-ignition process of undiluted fuel/air mixtures is approached in shock tubes only at high temperatures; these temperatures are thus of minor interest for engine applications. As a consequence, we decided to evaluate two different ignition-delay times τ_i from our experiments. The first one, denoted in the following Arrhenius diagrams as *1st kernel*, is defined by the occurrence of the first flame kernels, which are indicated by the first measurable emission at 431 nm. The second time results from the transition to a detonation-like process connected with the formation of shock waves and is marked with DDT (see Fig. 1). In the case of a two-step ignition process the preignition is marked with τ_1. These three times are marked in the right half of Fig. 3, showing histories of pressure and emission at 431 nm of a two-step ignition process of 60% iso-octane and 40% n-heptane at 40 bar.

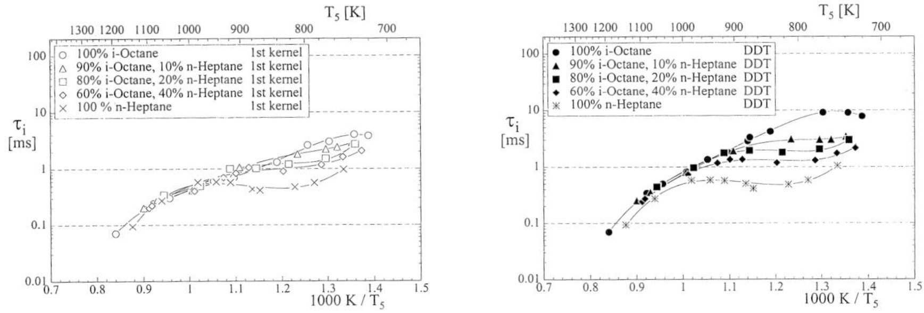

Figure 2. Ignition-delay times $\tau_{1st\ kernel}$ and τ_{DDT} at 40±2 bar, $\Phi = 1.0$.

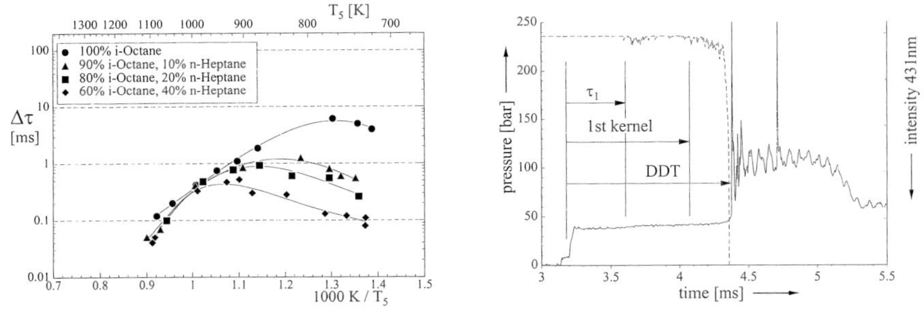

Figure 3. Left: transition time $\Delta\tau$ from first flame kernel to secondary explosion. Right: typical histories, 60% iso-octane/40% n-heptane with air, 39 bar, 831 K.

Figure 2 shows the ignition-delay times $\tau_{i-1st\ kernel}$ and τ_{i-DDT} obtained in the manner described above. As could be expected, the data of the mixtures are located within the range limited by the results of the pure fuels. For temperatures above 1000 K there is no significant difference between the ignition-delay times of the mixtures and pure iso-octane. This is caused by the fact that in the high-temperature region iso-octane consumes all radicals produced by the n-heptane decay (Pitsch and Peters 1995). For temperatures below about 900 K a more or less s-shaped dependence can be observed, which is very pronounced for n-heptane, but not for iso-octane in this temperature range. This negative temperature coefficient (NTC) is connected with a two-step ignition process (see also Fig. 3), which was observed also for mixtures containing only 10% n-heptane, and can be explained by alkylperoxy radical reactions (see e.g., Chevalier et al. 1990). With lower n-heptane fraction in the mixtures the s-shaped dependence becomes less pronounced. Nevertheless, the self-ignition process at lower temperatures ($T < 900$ K) is mainly influenced by the low-temperature kinetics of n-heptane. This can also be seen clearly in the left part of Fig. 3 where the difference in ignition-delay times $\Delta\tau = \tau_{DDT} - \tau_{1st\ kernel}$ from the occurrence of the first kernels to the DDT is plotted. The transition times of pure iso-octane and of the mixtures are identical for temperatures above 1000 K. At lower temperatures the influence of the n-heptane fraction in the mixtures leads to much faster transition.

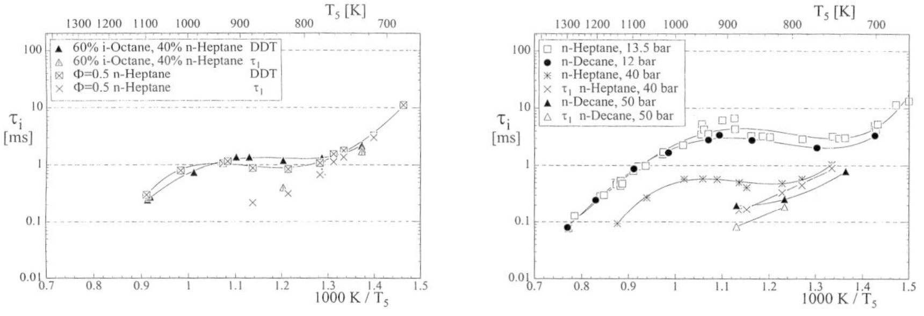

Figure 4. Left: ignition-delay times of a mixture 60% iso-octane/40% n-heptane, Φ=1.0, compared with results of n-heptane, Φ=0.5, both 40 bar. Right: ignition-delay times of n-decane, compared with results of n-heptane, $\Phi = 1.0$, all DDT.

Considering as a first approach iso-octane to be inert in the first phase of self-ignition, the corresponding equivalence ratio Φ for n-heptane in a stoichiometric mixture of 40% n-heptane and 60% iso-octane with air is 0.37. The ignition-delay times τ_i and the preignition times τ_1 of this mixture compared to n-heptane data with $\Phi = 0.5$ (Ciezki and Adomeit 1993) are in good agreement (left part of Fig. 4).

First results for n-decane obtained with our heated high-pressure shock tube are presented in the right half of Fig. 4. The results of stoichiometric n-decane/air mixtures at 12 and 50 bar are compared to n-heptane data at similar pressures. The self-ignition process of n-decane is very similar to n-heptane showing a NTC in the temperature range between 700 and 900 K connected with a two-step ignition process and a very fast DDT. The results at lower pressures differ only for temperatures below 1000 K. This is in agreement with the different cetane numbers (CN) of the two fuels (n-decane: CN = 77, n-heptane: CN = 56). Also the longer chain of n-decane will accelerate the low-temperature kinetics (Pitsch and Peters 1995). Similar behavior was found at high pressures for the experiments performed up to now. This work is in progress and will be extended in the future to a two-component diesel-model fuel containing 70% n-decane and 30% α-methylnaphtalene.

Acknowledgement. This work was supported by the Deutsche Forschungsgemeinschaft as a project of the Sonderforschungsbereich 224 "Motorische Verbrennung".

References

Blumenthal R, Fieweger K, Komp HK, Adomeit G (1995) Gas-dynamic features of self-ignition of non-diluted fuel/air mixtures at high pressure. In: 15th Int Colloquium on the Dynamics of Explosions and Reactive Systems, Boulder (to be published).

Chevalier C, Louessard P, Mueller UC, Warnatz J (1990) A detailed low-temperature reaction mechanism of n-heptane auto-ignition. In: Int Symp COMODIA 90, Kyoto, p 93.

Ciezki HK, Adomeit G (1993) Shock-tube investigation of self-ignition of n-heptane-air mixtures under engine relevant conditions. Comb Flame 93:421–433.

Fieweger K, Ciezki HK, Adomeit G (1993) Comparison of shock-tube ignition characteristics of various fuel-air mixtures at high pressures. In: Shock Waves @ Marseille, R Brun, LZ Dumitrescu (eds), Springer-Verlag, p 161.

König G, Maly RR, Bradley D, Lau AKC, Sheppard CGW (1990) Role of Exothermic Centers on Knock Initiation and Knock Damage, SAE Paper No. 902136.

Oppenheim AK (1985) Dynamic features of combustion. Phil Trans R Soc Lond A 315:471–508.

Pitsch H, Peters N (1995) Techn Hochschule Aachen, personal communication.

Vermeer DJ, Meyer JW, Oppenheim AK (1972) Auto-ignition of hydrocarbons behind reflected shock waves. Comb Flame 18:327–336.

Self-ignition of H_2-air mixtures at high pressure and low temperature

R. Blumenthal, K. Fieweger, K. H. Komp, G. Adomeit
Institut für Allgemeine Mechanik
RWTH Aachen, Templergraben 64, 52056 Aachen, Germany

B. E. Gelfand
Institute of Chemical Physics
Russian Academy of Sciences, 117977 ul. Kosygina, 4 Moscow, Russia

Abstract: In this work the reflected-shock technique has been used to determine ignition-delay times τ_i of mixtures of 15% H_2 in air. These ignition-delay times have been compared to calculated τ_i using a recently published reaction mechanism. Pressure and temperature have been varied in the range 3 bar $\leq p \leq$ 50 bar and 700 K $\leq T \leq$ 1200 K. At high temperatures mild and strong ignition originating from the end wall have been observed in contrast to randomly distributed nonhomogeneous mild ignition at lower temperatures. The difference in calculated and measured ignition-delay time strongly increases with decreasing temperature. Both inhomogeneities in the mixture — including particles — and a partially inaccurate modeling of the reaction mechanism should be regarded as responsible for these discrepancies.

Key words: Self-ignition, Hydrogen-air, High pressure, Low temperature

1. Introduction

The determination of ignition-delay times of premixed homogeneous combustible mixtures is important for practical applications (knocking in internal combustion engines [ICE], safety in nuclear power plants) as well as for the proof of chemical reaction mechanisms (Chevalier et al. 1990, Mueller et al. 1992, Maas & Warnatz 1988).

The H_2-air system has been investigated (e.g., Bhaskaran et al. 1973), but ignition-delay times for pressures $p > 4$ bar have not been found in the literature. In the low-pressure, high-temperature region, experiments and calculations fit together quite well (see our results and those of Maas & Warnatz 1988), whereas at elevated pressures and reduced temperatures a disagreement has been observed (Voevodsky & Soloukhin 1965). The data of Skinner & Ringrose (1965) (8% H_2 + 2% O_2 + 90% Ar, $p \approx 5$ bar, $T \approx 960 - 1080$ K) have been used by Maas & Warnatz (1988) to validate their reaction mechanism. Deviations of one order of magnitude can be observed but have not been mentioned by the authors. In our work we will show results of mixtures of 15% H_2 in air both in the mild and the strong ignition regimes at elevated pressures $p \leq 50$ bar yielding large differences in τ_i between experiment and calculation. We will discuss the reasons for the differences in calculated and experimentally observed τ_i in the different ignition regimes.

2. Experimental set-up and reaction mechanism

The experimental results have been obtained using three shock tubes with differing geometries and sizes. The most important technical data are summarized in Table 1. A more detailed description of shock tubes 1 and 3 can be found

Table 1. Technical data for the shock tubes.

No.	cross section	max time	max static pressure	length high-pressure part	length low-pressure part	location
1	□ 34 mm × 50 mm	1 ms	100 bar	0.5 m	1.2 m	Moscow
2	□ 54 mm × 54 mm	5 ms	20 bar	3.1 m	5.1 m	Aachen
3	○ 87 mm	10 ms	500 bar	5.3 m	6.3 m	Aachen

in recent publications of Gelfand et al. (1993) and Ciezki & Adomeit (1993), respectively. The second shock tube consists of a test section with a length of 200 mm and optical access to the whole cross-section through Plexiglas windows. These windows are necessary to visualize the ignition process by means of a Cranz-Schardin camera (shadow photographs) and holography. Four pressure gauges are positioned perpendicular to the Plexiglas windows. The distances between the gauges are 50 mm; one gauge is located 15 mm from the end wall. Opposite to the pressure gauges, four quartz glass cylinders are placed which allow the measurement of OH-band emission at 307 nm.

The experimental results are compared with the most recent reaction mechanism published by Baulch et al. (1994). Zero-dimensional constant volume calculations for the ignition-delay times have been performed using SENKIN and CHEMKIN (Kee et al. 1993, Lutz et al. 1992).

3. Results

In order to describe the experimentally observed ignition process as exactly as possible, different ignition-delay times have been introduced. For the inhomogeneous mild ignition with transition to detonation (DDT), two ignition-delay times seem to be useful: one that indicates the first recognition of any flame kernel and one that indicates the instant the secondary explosion occurs leading to the transition to detonation. Whenever possible the occurrence of the first flame kernel has been identified by the shadow photographs. However, if this has not been possible the first OH-band emission or the first smooth pressure rise has been used. The transition (DDT) has been identified by the first sharp pressure peak. There are two reasons for using two ignition-delay times. First, the occurrence of a first flame kernel may be caused by some disturbances (e.g., temperature, particles) and, as a consequence, may result in ignition-delay times that are too short. Second, the process of deflagration to detonation transition is also important for real applications. The importance of the time lag between combustion initiation and transition to detonation is obvious. As a result the two ignition-delay times could be regarded as the lower boundary of the ignition-delay time of a homogeneous reactor. The ignition at the end wall and the mild ignition without DDT are characterized by only one ignition delay time corresponding to the first pressure peak and the first recognition of a flame, respectively. At high temperatures the development of a detonation after a mild ignition initiated at the end wall takes place very quickly. It would be hard to see these differences in τ_i in the Arrhenius plot. Thus, we plotted only one τ_i (first pressure peak) and used the same mark as in the regime of strong ignition, because both symbolize the very rapid occurrence of a detonation.

Figure 1 shows ignition-delay times for pressures $p = 3$ to 4 bar and $p = 5$ to 8 bar. As one can see, good agreement between measurement and calculation is achieved at $p = 3$ to 4 bar for temperatures $T \geq 1040$ K (ignition at the end wall and DDT). For lower temperatures, in the regime of inhomogeneous mild ignition, a discrepancy is observed leading to calculated ignition-delay times that are 10 times longer compared with the measured τ_i. By increasing the pressure ($p = 5$ to 8 bar, Fig. 1) agreement between experiment and theory is found for $T \geq 1065$ K. For $p = 9$ to 11 bar (Fig. 2) the disagreement between experimentally observed τ_i and calculated ignition-delay times becomes remarkable even at high temperatures. Decreasing the temperature to $T \approx 825$ K one obtains $\tau_{i,calc} \approx 800$ ms compared with measured $\tau_{i,exp} \approx 5$ ms. Experiments at still higher pressures ($p = 12$ to 18 bar) show that this disagreement increases, particularly with decreasing temperature (see Fig. 2). For example, the experimentally observed ignition-delay time for $p = 12$ to 18 bar at $T = 730$ K yields $\tau_{i,exp} \approx 10$ ms in contrast to calculated ignition-delay time value of $\tau_{i,calc} \approx 6$ s. There seems to be no considerable change in τ_i when the pressure is increased further. Figure 3 shows our results for $p = 35$ to 47 bar. Even for high temperatures, differences between experiment and theory are observed. The pressure dependence of τ_i is represented in Fig. 3 for different temperatures $T = 800$ K, 900 K, and 1100 K. For the experiments in the transition regime, the instant of transition to detonation has been plotted. The disagreement of about two orders of magnitude in the whole pressure range at $T = 900$ K is obvious. By

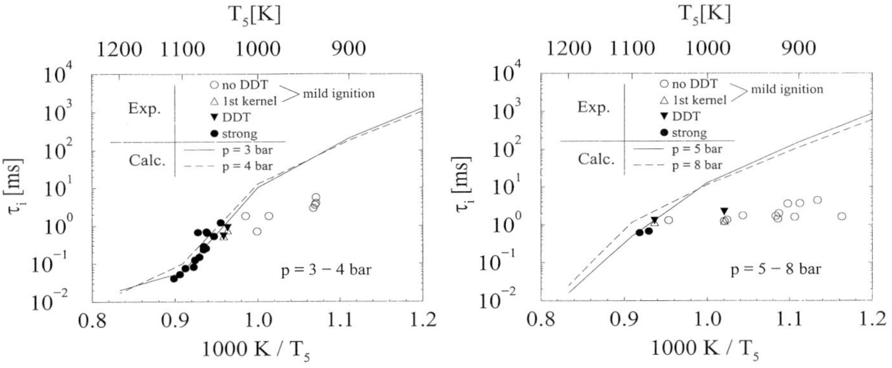

Figure 1. Ignition-delay times for p = 3 to 4 bar and p = 5 to 8 bar. Symbols = experiments; lines = calculations; open symbols = mild ignition.

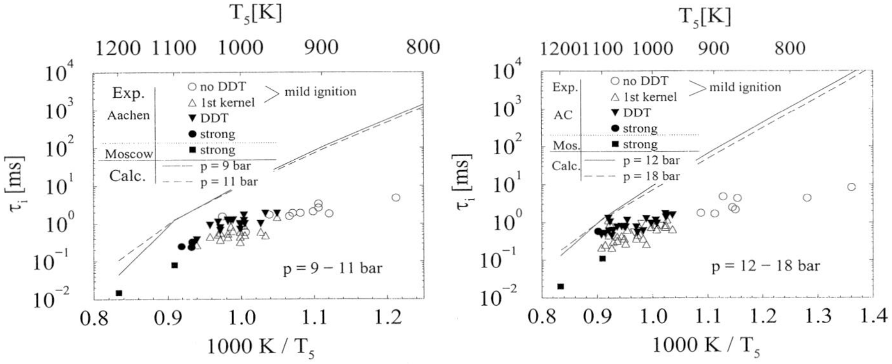

Figure 2. Ignition-delay times for p = 9 to 11 bar and p = 12 to 18 bar. Symbols = experiments; lines = calculations; open symbols = mild ignition.

decreasing the temperature to T = 800 K this disagreement becomes even more evident leading to an offset in τ_i with $\tau_{i,calc} \approx 1000 \times \tau_{i,exp}$.

4. Discussion and conclusions

In order to interpret the results we will show in the following that — depending on the temperature — different effects seem to be responsible for the observed ignition behavior. It can be seen that the largest disagreement has been observed at low temperatures when self-ignition takes place inhomogeneously in the mixture. Many reasons for this behavior can be found, e.g., stochastic phenomena as explained by Terao (1977), sensitivity of the combustible mixture

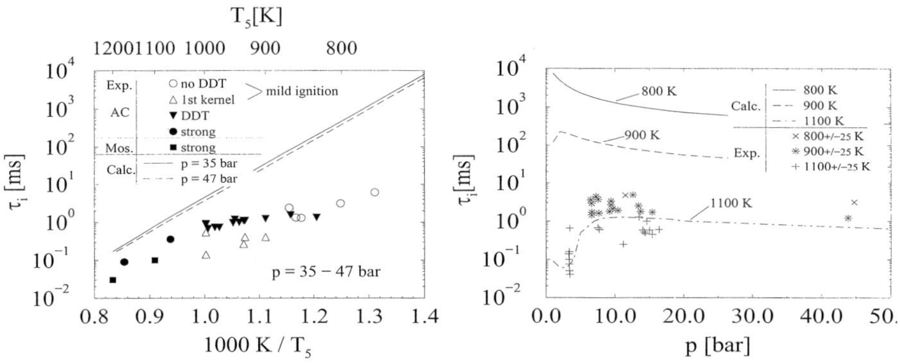

Figure 3. Ignition-delay times for $p = 35$ to 47 bar and $T = 800$, 900, 1100 K. Symbols = experiments; lines = calculations; open symbols = mild ignition (left graph).

($(\partial\tau_i/\partial T)_p$, Oppenheim 1985) or the influence of particles as noted by Elsworth et al. (1969). Temperature disturbances alone are not the reason for the large deviations, particularly at low temperatures, because in order to explain the difference of τ_i at $T \approx 750$ K, disturbances of $\Delta T > 200$ K are necessary. Disturbances of this magnitude may be caused by geometric effects in the shock tube (e.g., crevices), but due to the fact that all three shock tubes show the same results this seems to be unlikely. This is also confirmed by further experiments (holography) showing that inhomogeneous mild ignition starts with and without contact to the walls at various locations, which vary from experiment to experiment. Hence, the influence of particles appears to be important (Elsworth et al. 1969). Nevertheless, the quantitative results obtained in our study are of interest because of two facts. First, considerable disagreement has also been found at high temperatures and high pressures when combustion starts from the end wall (Fig. 3). Thus, the reaction mechanism appears to need correction in the high-pressure range. Second, in order to gather important information for applications (ICE, nuclear power plant) studies of the ignition behavior under practically relevant conditions are required. Because inhomogeneities, in particular particles, cannot be avoided in real applications, the ignition-delay times obtained in our study are the actual times one would observe in applications. In this low-temperature regime more detailed calculations including the modeling of the influence of particles and other inhomogeneities on the self-ignition are necessary. Nevertheless, zero-dimensional calculations are directly comparable to experimental results in the high-temperature regime.

These observations indicate that at low temperatures the influence of particles as well as the present reaction mechanism are responsible for the large differences in τ_i. Temperature disturbances alone cannot be the cause of these discrepancies. The deviations at high pressures and high temperatures seem to be caused by the reaction mechanism that has to be adjusted in this range.

References

Baulch DL, Cobos CJ, Cox RA, Frank P, Hayman G, Just T, Kerr JA, Murrells T, Pilling MJ, Troe J, Walker RW, Warnatz J (1994) Summary table of evaluated kinetic data for combustion modeling: supplement 1. Comb Flame 98:59–79.

Bhaskaran KA, Gupta MC, Just T (1973) Shock tube study of the effect of unsymmetric dimethyl hydrazine on the ignition characteristics of hydrogen-air mixture. Comb Flame 21:45–48.

Chevalier C, Louessard P, Mueller UC, Warnatz J (1990) A detailed low-temperature reaction mechanism of n-heptane auto-ignition. In: Int Symp COMODIA, pp 93–97.

Ciezki HK, Adomeit G (1993) Shock-tube investigation of self-ignition of n-heptane-air mixtures under engine relevant conditions. Comb Flame 93:421–433.

Elsworth JE, Haskell WW, Read IA (1969) Non-uniform ignition processes in rapid-compression machines. Comb Flame 13:437–438.

Gelfand BE, Popov OE, Medvedev SP, Khomik SV, Kusharin AY, Agafanov GL (1993) The preliminary investigation of self ignition phenomena in H_2-air mixtures at high pressure. Dokl RAS.

Kee RJ, Rupley FM, Miller JA (1993) CHEMKIN II: A FORTRAN Chemical Kinetics Package for the Analysis of Gas Phase Chemical Kinetics, Technical Report SAND89-8009B, UC-706, Sandia Nat Lab.

Lutz AE, Kee RJ, Miller JA (1992) SENKIN: A FORTRAN Program for Predicting Homogenous Gas Phase Chemical Kinetics with Sensitivity Analysis, Technical Report SAND87-8248, UC-401, Sandia Nat Lab.

Maas U, Warnatz J (1988) Ignition processes in hydrogen-oxygen mixtures. Comb Flame 74:53–69.

Mueller UC, Peters N, Linan A (1992) Global kinetics for n-heptane ignition at high pressures. In: 24th Symp (Int) on Comb, pp 777–784.

Oppenheim AK (1985) Dynamic features of combustion. Phil Trans R Soc Lond A 315:471–508.

Skinner JB, Ringrose JM (1965) Ignition delays of H_2-O_2-air mixtures at relatively low temperatures. J Chem Phys.

Terao K (1977) A stochastic theory of shock-heated ignitions and its applications. In: 11th Int Symp on Shock Tubes and Waves, pp 390–397.

Voevodsky VV, Soloukhin RI (1965) On the mechanism and explosion limits of hydrogen-oxygen chain self-ignition in shock waves. In: 10th Symp (Int) on Comb, pp 279–283.

High-pressure shock-tube measurements of ignition times in stoichiometric $H_2/O_2/Ar$ mixtures

E. L. Petersen, D. F. Davidson, M. Röhrig, R. K. Hanson
High Temperature Gasdynamics Laboratory
Department of Mechanical Engineering
Stanford University, Stanford, CA 94305 USA

Abstract: Ignition-delay times of stoichiometric H_2-O_2 mixtures highly dilute in argon were measured using a new high-pressure shock tube. Experiments were conducted in reflected shock waves for temperatures between 1175 and 1880 K, pressures up to 87 atm, and hydrogen mole fractions of 2% by volume or less. The preliminary results indicate a clear distinction between chain-branching effects at high temperatures and low pressures and chain-termination effects at low temperatures and high pressures. No effect of argon dilution on measured ignition times is evident.

Key words: Ignition times, Shock tube, High pressure, Laser diagnostics, OH

1. Introduction

Shock tubes have been widely used to measure the ignition-delay time of hydrogen-oxygen mixtures. For example, Schott and Kinsey (1958), Asaba et al. (1965), and Jachminowski and Houghton (1971) used shock-tube OH-absorption measurements to ascertain H_2-O_2 ignition times. OH emission was used by Skinner and Ringrose (1965) and Belles and Lauver (1965). Ignition times were derived from pressure measurements in a shock tube by Fujimoto (1963), Craig (1966), and Cheng and Oppenheim (1984), and either interferometric or schlieren techniques were employed by White and Moore (1965) and Meyer and coworkers (1971, 1973). While these experiments covered a wide temperature range (900–2600 K), none were conducted at pressures above 5 atm.

To extend the study of fundamental combustion chemistry to higher pressures, a new shock tube has been built at Stanford University. Some of the first experiments conducted in this facility have involved measurement of ignition-delay times of hydrogen-oxygen mixtures at elevated pressures. This paper presents an overview of the H_2-O_2 ignition-time study and gives critical details of the new shock tube.

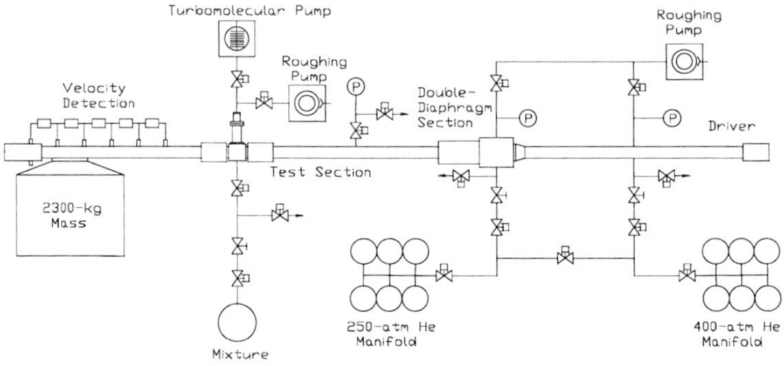

Figure 1. Stanford high-pressure shock tube. The 3-m, 7.6-cm-diameter driver is helium driven; the 5-m test section has a 5-cm inner diameter. Six piezoelectric pressure transducers and 5 time-interval counters are for velocity detection. The 2300-kg mass is to absorb end-wall vibrations.

2. Experimental method

All experiments were conducted using reflected shock waves. Figure 1 presents a schematic of the shock-tube facility, which is capable of attaining reflected-shock pressures as high as 1000 atm. The 3-m, 7.6-cm-diameter driver is helium-driven from a 400-atm helium manifold. Either double or single diaphragms can be used, depending on the range of conditions required; the diaphragms were made from either lexan or pre-scored aluminum for the test conditions presented herein. The 5-m, 5.0-cm-diameter driven section can be turbomolecular pumped to ultimate pressures below 10^{-6} torr with a 10^{-5}-torr/min combined leak and outgassing rate. Six 1-μs-response PCB113A pressure transducers and five 0.1-μs Philips PM6666 time-interval counters are used to obtain the incident-shock velocity. As shown in Fig. 1, a 2300-kg mass is attached to the driven section to minimize anticipated end-wall vibrations. Optical-access ports are available at three different axial locations (2 incident, 1 reflected).

Reaction progress of the H_2-O_2-Ar mixtures was monitored using narrow-linewidth laser absorption of OH. A Spectra-Physics 380 ring-dye laser operating on Rhodamine 6G dye was pumped by a Spectra-Physics 164 argon-ion laser; through intracavity frequency doubling by a temperature-tuned AD*A crystal, uv light near 306 nm was produced. This wavelength corresponds to the OH A-X (0,0) bandhead. The uv beam was passed through the reflected port of the shock tube located 20 mm from the end wall via 12.7-mm-thick uv-silica windows with a 6-mm optical diameter.

The mole fraction of OH produced in the shock tube was determined from the measured incident and transmitted laser light through Beer's Law. For the experiments herein, the target wavenumber ranged from 32622 to 32632 cm^{-1}. The corresponding high-pressure values of the OH absorption coefficient were obtained from a parallel study conducted in our laboratory (Davidson et al. 1996).

Three different stoichiometric mixtures of H_2-O_2-Ar were used: 2% H_2/1% O_2/97% Ar, 1% H_2/0.5% O_2/98.5% Ar, and 0.1% H_2/0.05% O_2/99.85% Ar. Each was a premixed Ultra High Purity Linde mixture.

Reflected-shock conditions were determined using the incident-shock velocity and standard 1D relations invoking the Sandia thermochemical database. Attenuation of the incident shock wave was typically 3%/m. The accuracy of the velocity measurement coupled with boundary-layer effects limited knowledge of the reflected-wave temperature to ± 20 K.

3. Results and discussion

A total of 94 hydrogen-oxygen ignition-time measurements were taken. Temperatures varied from 1175 to 1880 K, and pressures ranged from 33 to 87 atm. The resulting ignition-delay times (τ) were between 5 and 350 μs. Figure 2 shows a sample pressure trace and a typical OH profile. The arrival of the incident and

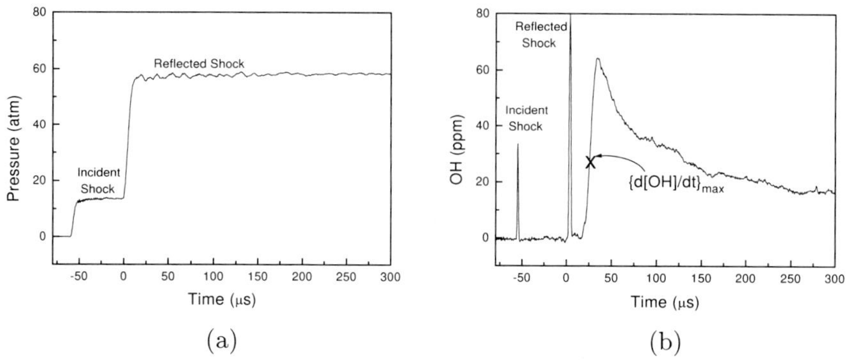

Figure 2. Sample results for 0.1% H_2/0.05% O_2/99.85% Ar; T = 1830 K, P = 58.4 atm. (a) Pressure trace, (b) OH profile from laser-based absorption diagnostic; ignition time is defined as maximum OH formation rate.

reflected shock waves are evident in the OH trace by the schlieren spikes due to laser beam steering, and $t = 0$ is defined at the beginning of the reflected-shock schlieren spike. Following a finite induction time, combustion occurs, leading to an exponential growth of OH followed by a slower decay. Ignition time is defined herein as the time when the maximum slope in the OH formation is reached. Note that ignition is not evident in the Fig. 2a pressure trace because of the high argon dilution used. This result was seen in most of the data, so the OH diagnostic was utilized exclusively to deduce τ. Previous investigators were able to use pressure as an induction time diagnostic because their mixtures were either undiluted (Craig 1966) or contained 90% argon or less (Fujimoto 1963; Cheng and Oppenheim 1984).

Figure 3 presents the results of the present study along with representative findings from the literature. Two trends are clearly seen in Fig. 3, depending on the temperature and pressure levels. At high temperatures and relatively low pressures, the data tend to follow the shallower line with an 8050-K (16.0-kcal/mol) slope. This trend holds for the present data for pressures near 90 atm as well as for the 1- and 2-atm correlations obtained by Schott and Kinsey (1958), White and Moore (1965) and Cheng and Oppenheim (1984). The other major trend is seen at lower temperatures and higher pressures. These data follow a steeper line. Note that the transition between the two regimes is pressure dependent, i.e., the transition occurs at higher pressures for increasing temperature.

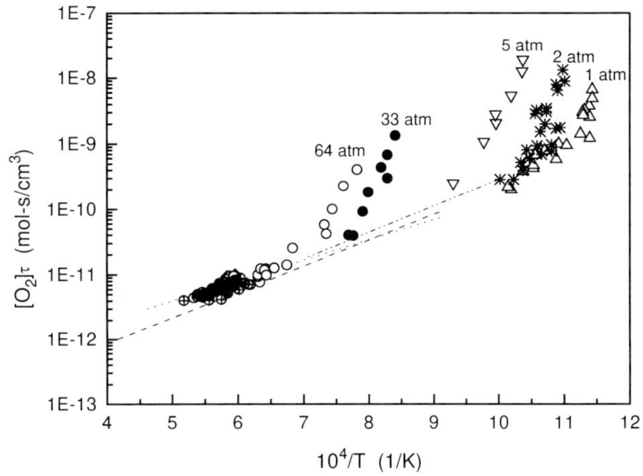

Figure 3. H_2-O_2 ignition time results. Current study: • 33 atm; ⊕ 57 atm; ∘ 64 atm; ⋄ 87 atm. Previous studies: △ Craig (1966), 1 atm; ∗ Craig (1966), 2 atm; ▽ Skinner and Ringrose (1965), 5 atm; - - - Schott and Kinsey (1958), 1 atm; · · · White and Moore (1965), 1 atm; - ·· - Cheng and Oppenheim (1984), 2 atm.

The results seen in Fig. 3 can be explained by chemical kinetics. At high temperatures and relatively low pressures, detailed kinetic modeling indicates that the $H + O_2 \rightarrow OH + O$ chain-branching reaction is dominant. From the latest GRI-Mech 1.2 mechanism (Frenklach et al. 1994), the activation energy for this reaction is 14.4 kcal/mol; this is close to the 16.0-kcal/mol slope obtained from the ignition time plot. At lower temperatures and higher pressures, the dominant kinetic reaction can be shown to be the chain-terminating $H + O_2 + M \rightarrow HO_2 + M$. The transition from chain branching to chain termination accounts for the slope change evident on the $\tau[O_2]$ plot. Further kinetic details on the high-pressure H_2-O_2 oxidation data will be presented in a later paper.

As mentioned above, the level of argon dilution in our experiments ranged from 97 to 99.85%. The study by Schott and Kinsey (1958) ranged from approximately 99% argon dilution to less than 80%. Skinner and Ringrose (1965) and Cheng and Oppenheim (1984) used mixtures containing 90% argon. White and Moore's correlation covers pure H_2-O_2 and H_2-air mixtures, while Craig (1966) studied H_2-air ignition times. Therefore, as Fig. 3 indicates, our highly diluted data are well within the typical ignition-time scatter seen in less-diluted studies. A natural conclusion of this work is that the ignition times in H_2-O_2 mixtures are insensitive to inert-diluent levels to within the accuracy of the measurements.

Also evident in the Fig. 3 comparison with other investigators is the good agreement independent of the varying definitions of ignition-delay time. For example, both Craig (1966) and Cheng and Oppenheim (1984) used the sharp increase in their pressure signal; Schott and Kinsey (1958) specified ignition when the OH concentration reached 1×10^{-6} mol/l; Skinner and Ringrose (1965) defined τ at the maximum OH signal; and White and Moore (1965) used the sharp density change from their interferogram as the ignition time. Thus the agreement between the present $d[OH]/dt_{max}$ definition and earlier ones is an indication that the rapid production of OH in the H_2-O_2 reaction makes the measured value of τ less definition sensitive.

4. Conclusions

A new high-pressure shock tube was used to obtain ignition-delay times in stoichiometric H_2-O_2-Ar mixtures for pressures between 33 and 87 atm and temperatures from 1175 to 1880 K. When combined with previous, lower-pressure studies, the transition from chain-branching dominance due to $H + O_2 \rightarrow OH + O$ at high temperatures to the chain-terminating $H + O_2 + M \rightarrow HO_2 + M$ reaction at low temperatures and high pressures is evident. Comparisons with previous investigations also indicate that the measured H_2-O_2 ignition times are insensitive to inert-gas dilution for 0–99.85% diluent levels.

Acknowledgement. This work was supported by the Office of Naval Research, with Dr. R. Miller as contract monitor. Thanks are due to Chris Morris for LaTeXassistance.

References

Asaba T, Gardiner WC Jr, Stubbeman RF (1965) Shock-tube study of the hydrogen-oxygen reaction. In: 10th Sym (Int) on Comb, The Combustion Institute, Pittsburgh, pp 295–302.

Belles FE, Lauver MR (1965) Effects of concentration and of vibrational relaxation on the induction period of the H_2-O_2 reaction. In: 10th Sym (Int) on Comb, The Combustion Institute, Pittsburgh, pp 285–293.

Cheng RK, Oppenheim AK (1984) Autoignition in methane-hydrogen mixtures. Comb and Flame 58:125–139.

Craig RR (1966) A Shock Tube Study of the Ignition Delay of Hydrogen-Air Mixtures Near the Second Explosion Limit, AFAPL-TR-66-74, AF Aeropropulsion Lab, Wright-Patterson.

Davidson DF, Röhrig M, Petersen EL, Di Rosa MD, Hanson RK (1995) Measurement of the absorption spectrum of the OH A-X (0,0) 306-nm bandhead at high pressures. In: Shock Waves, Proc of the 20th ISSW, Sturtevant B, Shepherd JE, Hornung HG (eds), World Scientific, Vol 2, pp 923–928.

Frenklach M, Wang H, Bowman CT, Hanson RK, Smith GP, Golden DM, Gardiner WC Jr, Lissianski V (1994) An optimized kinetics model for natural gas combustion. In: 25th Sym (Int) on Comb, The Combustion Institute, Pittsburgh, Work-in-Progress Abstracts, p 212.

Fujimoto S (1963) Chemical reaction in a shock wave. I. The ignition delay of a hydrogen-oxygen mixture in a shock tube. Bull Chem Soc Japan 36:1233–1236.

Jachimowski CJ, Houghton WM (1971) Shock-tube study of the initiation process in the hydrogen-oxygen reaction. Comb and Flame 17:25–30.

Meyer JW, Oppenheim AK (1971) On the shock-induced ignition of explosive gases. In: 13th Sym (Int) on Comb, The Combustion Institute, Pittsburgh, pp 1153–1164.

Meyer JW, Cohen LM, Oppenheim AK (1973) Study of exothermic processes in shock ignited gases by the use of laser shear interferometry. Comb Sci and Tech 8:185–197.

Schott GL, Kinsey JL (1958) Kinetic studies of hydroxyl radicals in shock waves. II. Induction times in the hydrogen-oxygen reaction. J Chem Phys 29:1177–1182.

Skinner GB, Ringrose GH (1965) Ignition delays of a hydrogen-oxygen-argon mixture at relatively low temperatures. J Chem Phys 42:2190–2192.

White DR, Moore GE (1965) Structure of gaseous detonation. IV. Induction zone studies in H_2-O_2 and CO-O_2 mixtures. In: 10th Sym (Int) on Comb, The Combustion Institute, Pittsburgh, pp 785–795.

Soot formation from rich hydrocarbon oxidation under elevated pressure conditions

H. Kellerer*, H.-J. Bauer, S. Wittig
Institut für Thermische Strömungsmaschinen
Universität Karlsruhe
Kaiserstrasse 12, 76128 Karlsruhe, Germany

Abstract: Two different methods for determination of particle diameter and number density were applied to high-pressure soot formation from n-heptane behind a reflected shock in a shock tube. Disturbance by PAH and uncertainties concerning the wavelength dependence of the reflective index for soot particles restrict multi-wavelength-extinction measurements to qualitative results during early stages of soot growth. However, the scattering-extinction technique proved very suitable for the time-resolved determination of particle diameters in the Rayleigh regime. Soot formation at high pressures was found to be accompanied by high particle number densities in the range of $N \approx 10^{12} - 10^{13} \cdot 1/\text{cm}^3$, which is considerably higher than at atmospheric conditions.

Key words: Soot, Pressure, Environmental

1. Introduction

One of the limits of high-pressure combustion in combustion chambers for engine applications is the problem of high radiative heat transfer as a result of the soot loading of the flame. The pressure in these devices ranges from 15 to 45 bar for advanced aero engines and even higher for Diesel engines. Therefore, understanding mechanisms of soot formation at elevated pressures is of growing importance in the design of advanced combustion chambers.

To date, experimental studies of sooting processes have focused primarily on low-pressure or atmospheric conditions. As only few experiments at high-pressure conditions were carried out, soot formation at high pressures is still an insufficiently understood subject. Fundamental investigations on the influence of pressure on soot formation have been carried out in shock tubes for rich oxidation by Müller and Wittig (1991). They could demonstrate that methane soot yield rises proportionally to the fuel partial pressure, whereas total pressure has a strong promoting influence at 30 bar that diminishes towards higher pres-

*Corresponding author.

sures. Extensive measurements of shock-tube pyrolysis by Bauerle et al. (1994) indicated that the effect of pressure on soot formation is strongly dependent on the hydrocarbon fuel used. For understanding high-pressure soot formation, information on particle diameter and number density are of major interest. High-pressure shock-tube measurements so far exploited two-wavelength-extinction or extinction-emission techniques. However, these methods do not allow a clear assignment of particle diameters and the results did not correspond with theoretical modeling. Therefore, soot formation under rich burning conditions ($\phi = 5$) has been studied in a high-pressure shock tube at the *Institut für Thermische Strömungsmaschinen.* A three-wavelength-extinction technique (Dittmann et al. 1992) together with an extinction-scattering method have been applied in our measurements to obtain particle diameters and number densities.

2. Experimental

A conventional shock tube was used which was essentially that described by Müller and Wittig (1991). Optical investigations were carried out behind the reflected shock 10 mm in front of the end wall. The applied pressure ranged from 30 to 100 bar, temperature from 1600 to 2100 K, and carbon concentration was varied between 2.0 and $8.0 \cdot 10^{18}/cm^3$.

3. Optical setup

Two different optical investigation techniques were used to determine soot particle diameters and number densities (Fig. 1): three-wavelength-extinction method and a light-scattering-extinction method first applied to shock tubes by Graham and Homer (1973). The extinction method is based on the Lambert-

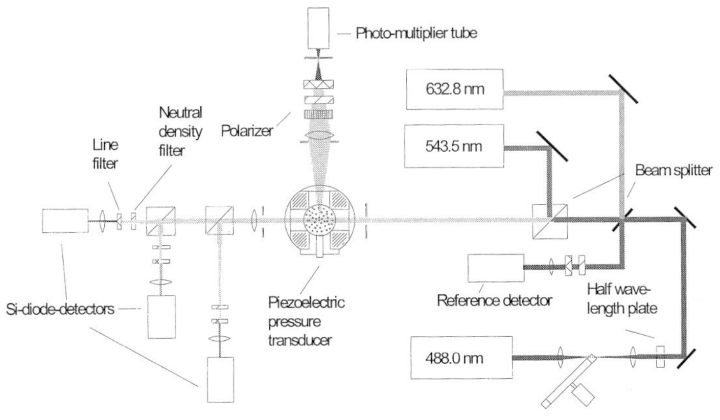

Figure 1. Optical system.

Beer Law that describes the extinction of a laser beam by a particle cloud:

$$\frac{I}{I_0} = \exp\left(-l \cdot N \cdot \int_0^\infty \frac{d^2\pi}{4} \cdot Q_{ext}(m, \alpha, \sigma_g)\, dd\right) \quad (1)$$

Here, l is the extinction length, N the particle number density, d the volumetric particle diameter, and Q_{ext} the extinction coefficient. Q_{ext} can be calculated applying Mie theory as a function of the refractive index m, the dimensionless particle size $\alpha = \frac{d\pi}{\lambda}$, and the geometric standard deviation of the assumed particle size distribution σ_g. The chosen distribution was a log-normal distribution with $\sigma_g = 0.2$, based on results achieved by transmission electron microscopy (TEM) of primary particles. This is in accordance with studies by Bauerle et al. (1994). The wavelength-dependent values for the real and imaginary parts of the refractive index were calculated using a dispersion model with constants suggested by Lee and Tien (1981). Exploiting extinction at two different wavelengths, the two unknowns, particle diameter and number density, can be derived. This is performed by evaluating the measured extinction ratio in terms of Mie calculations. The extinction ratio is defined as:

$$ER_{12} = \frac{\ln(\frac{I}{I_0})_{\lambda 1}}{\ln(\frac{I}{I_0})_{\lambda 2}} = \frac{\int_0^\infty p(d)\, \frac{d^2\pi}{4}\, Q_{ext}(m, \alpha_1, \sigma_g)\, dd}{\int_0^\infty p(d)\, \frac{d^2\pi}{4}\, Q_{ext}(m, \alpha_2, \sigma_g)\, dd} \quad (2)$$

and therefore only dependent on particle diameter. However, for clear assignment, a third wavelength is necessary. In our experiment, laser light at 488, 543.5, and 632.8 nm was sent through the shock tube and after passing interference filters separately measured by diode detectors.

To determine particle diameters by the scattering-extinction method, in addition to attenuation at a single wavelength, the angular scattering coefficient Q_{vv} has to be measured. The scattering coefficient can be calculated by:

$$Q_{vv} = N \frac{\lambda^2}{4\pi^2} \int_0^\infty p(d)\, i_\perp(\Theta, d, \lambda, m)\, dd \quad (3)$$

where $i_\perp$ denotes the scattering function. For the scattering measurements a 3-W Ar^+-laser at 488 nm was used as a light source. Scattered light was detected at a scattering angle of 90°. To discriminate scattering against emission, a motor-powered mechanical beam chopper (40 kHz) was used. Laser light polarization was rotated by a $\lambda/2$-plate to allow vv scattering. The scattered light was filtered by a polarizer and a small-banded interference filter (488±0.45 nm) before being focused on a pinhole in front of a photo-multiplier tube. The scattering-extinction system was calibrated by scattering from argon atoms of known scattering cross section (Rudder and Bach 1968).

4. Experimental results and discussion

A typical example for extinction ratios in shock-tube soot formation is shown in Fig. 2. Start of measurement time is triggered by the arrival of the incident shock in the measurement plane. For one pair of wavelengths the curve is characterized by a maximum that is reached shortly after start of soot formation indicated by the beginning of laser extinction. The maximum is followed by

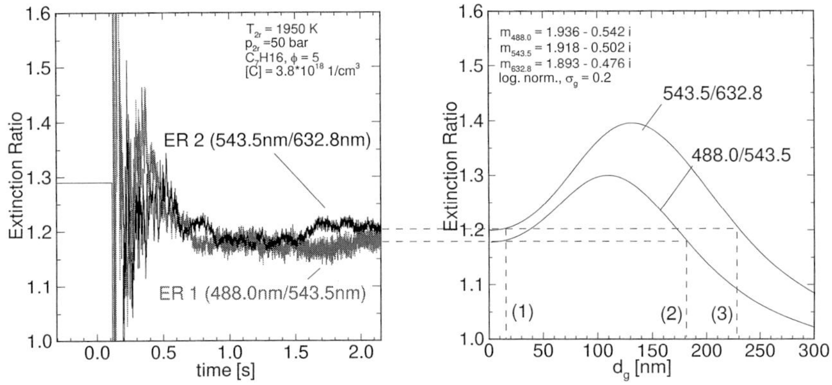

Figure 2. Time-resolved measurements of extinction ratios (left) and interpretation in terms of geometric mean particle diameter (right).

a smooth slope that tends towards a constant value. This behavior has often been interpreted in a way such that the final values have to be determined on the right side of the theoretical ER curve (Fig. 2 (2),(3)). Obviously, this leads to large particles which have to be considered as agglomerates due to their small primary particles shown by TEM. The assumption of a homogeneous and solid sphere, required for Mie calculations, has then to be questioned. The introduction of a third wavelength allows the problem to be solved. It could be demonstrated that the diameters (2) and (3) are wrong assignments. The measured soot particle diameters are located in the Rayleigh regime (1). This conclusion is supported by theoretical work on absorption behavior of agglomerates indicating that absorption is predominantly determined by the size of the primary particles (Dittmann 1995). Therefore, the ER-value peak right after the start of soot formation cannot be attributed to particle growth. In agreement with absorption measurements carried out in a premixed flat flame by Weiner and Harris (1989) it can be concluded that large polycyclic hydrocarbons (PAH) with absorption spectra ranging up to the near infrared are responsible for this phenomenon. Also, changing C-H ratio during early particle growth may significantly influence the wavelength dependence of the refractive index making the determination of particle diameter even more difficult.

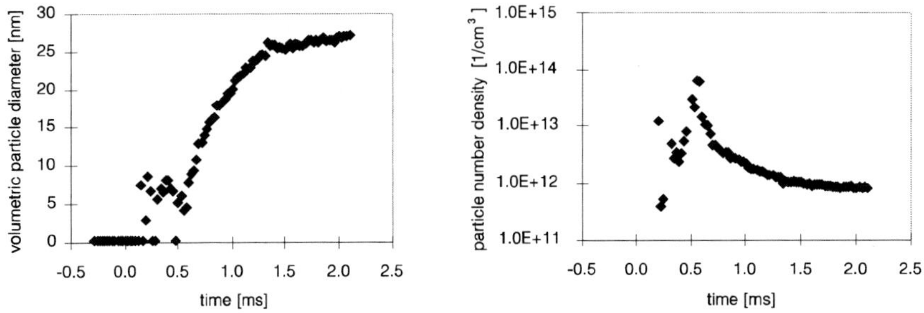

Figure 3. Measurement of soot particle diameter and number density: C_7H_{16}, $\phi = 5$, Ar = 98%, T = 1750 K, p_{2r} = 25 bar.

Whereas extinction measurements do not give accurate results for particle diameters and number densities in early stages of soot formation, particle growth can be excellently followed by scattering-extinction measurements. This is demonstrated in Fig. 3. After a rapid growth that follows particle inception, the rise in particle diameter decreases with time to yield a volumetric diameter of 27 nm at the end of the observation time. The fast growth is the result of particle coagulation which reduces the number density quickly. The measured final particle number density ($\approx 10^{12}/cm^3$) was found to be significantly higher than at atmospheric flame conditions. Enhanced soot production at high pressures (Kellerer et al. 1995) is therefore not a result of larger particles but accompanied by an increase in number density. This result can be attributed to reduced coagulation coefficients in the transition regime between free-molecular and continuum flow.

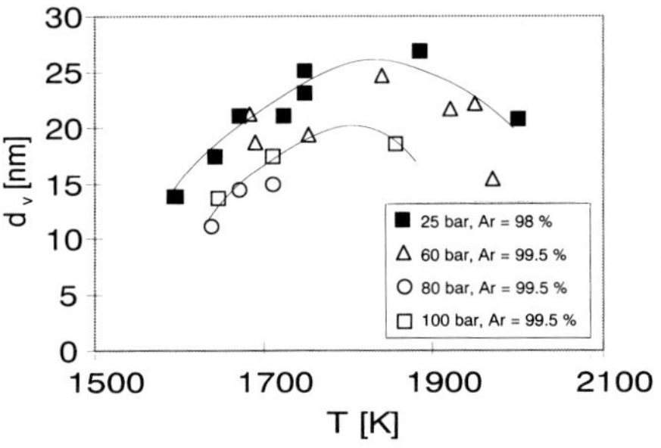

Figure 4. Soot particle sizes at different pressures ($C_7H_{16}, \phi = 5$: Ar = 98%, 15 and 25 bar; Ar = 99%, 60, 80, and 100 bar).

Measured particle sizes taken 1.5 ms after reflected shock arrival are shown for different pressure conditions in Fig. 4. The largest particles are detected at temperatures around 1850 K with diameters decreasing toward lower and higher temperatures. The comparison at constant carbon concentration (25 bar, Ar = 98% to 100 bar, Ar = 99.5%) indicates that an increase of pressure at constant carbon concentration leads to smaller particle sizes. Particle number densities up to $> 10^{13}$ could be determined at 100 bar.

5. Conclusion

Two different methods for the determination of particle diameter and number density in the early stages of high-pressure soot formation have been investigated. It could be shown that the three-wavelength-extinction technique can only be used for soot-particle characterization at later stages of soot formation. Applications are suitable for flame and exhaust measurements where no absorption due to PAH is to be expected. Combined scattering-extinction allows one to follow particle growth with good time resolution. Determined particle diameters were in the range of 15 – 30 nm. Increased total pressure resulted in smaller particle diameters. Therefore, enhanced soot formation at high pressures is accompanied by high particle number densities in the range of $N \approx 10^{12} - 10^{13}/\mathrm{cm}^3$.

Acknowledgement. The authors gratefully acknowledge support by the Deutsche Forschungsgemeinschaft within the *Graduiertenkolleg Energie- und Umwelttechnik.*

References

Bauerle S, Karasevich Y, Slavov S, Tanke D, Tappe M, Thienel T, Wagner HG (1994) Soot formation at elevated pressures and carbon concentrations in hydrocarbon pyrolysis. In: 25th Symposium (Int.) on Combustion.

Dittmann R (1995) The optical cross sections of soot agglomerates: the model of the virtual refractive index and first-order multiple scattering. In: 4th Int. Congress Optical Particle Sizing, Preprint, pp 59–68.

Dittmann R, Feld H-J, Samenfink W, Wittig S (1992) Multiple wavelength extinction technique for particle characterization in dense particle clouds. In: 5. European Symposium on Particle Characterization, Partec 5. Nürnberg, March 1992, Preprints 2, p 507.

Graham SC, Homer JB (1973) Light scattering measurements on aerosols in a shock tube. In: 9th Symposium (Int.) on Shock Tubes, pp 712–719.

Kellerer H, Bauer H-J, Wittig S (1995) Soot formation in a shock tube under elevated pressure conditions. In: 15th Int. Coll. on the Dynamics of Explosions and Reactive Systems (to be published).

Lee SC, Tien CL (1981) Optical constants of soot in hydrocarbon flames. In: 18th Symposium (Int.) on Combustion, pp 1159–1166.

Müller A, Wittig S (1991) Influence of temperature and pressure on soot formation in a shock tube under high pressure conditions. In: 18th Symposium (Int.) on Shock Waves, pp 759–764.

Rudder RR, Bach DR (1968) Rayleigh scattering of ruby-laserlight by neutral gases. Journal of the Optical Society of America 58:9.

Weiner AM, Harris SJ (1989) Optical detection of large soot precursors. Combustion and Flame 77:261.

Single-pulse shock-tube study on the decomposition of C_3F_6

Y. He
Institute of Mechanics
Chinese Academy of Sciences
Beijing 100080, Peoples Republic of China

Abstract: Perfluoropropylene C_3F_6 has been decomposed in the presence of scavenger H_2 in a single-pulse shock tube. The products CH_4, C_2F_4, CF_3H, and C_2F_3H are obtained and used as a measure of the rate of the cleavage of the vinyl-methyl C–C bond in C_3F_6.

The rate expression for this reaction $C_3F_6 \rightarrow CF_3 + C_2F_3$ has been found to be

$$k(C_3F_6 \rightarrow CF_3 + C_2F_3) = 10^{17.1\pm0.2} \exp\left(-\frac{85000 \pm 2000}{RT}\right)\ \mathrm{s}^{-1}$$

$$(1090\ \mathrm{K} < T < 1190\ \mathrm{K})$$

The results are consistent with a bond dissociation energy for the primary vinyl C–CF_3 bond in perfluoropropylene of 85 kcal/mol.

Key words: Decomposition, Perfluoropropylene, Scavenger, Rate constant, Comparative rate

1. Introduction

Perfluoropropylene C_3F_6 has been decomposed in comparative-rate single-pulse shock-tube studies, in the presence of H_2 as the scavenger. The decyclization of 4-methyl-1-cyclohexene (4-MCH) is used as the internal standard. The experiment is carried out at a pressure of 0.1 MPa, a temperature range of 1090 to 1190 K, and residence time of 1×10^{-3} s.

Dalby (1964) studied the flash photolysis of C_3F_6 and his analysis gave the reaction key way as

$$C_3F_6 + \mathrm{hv} \rightarrow CF_2 + CFCF_3 \tag{1}$$

with final products C_2F_4 and C_4F_8. There have been no previous studies on C_3F_6 decomposition, but there have been many papers on hydrocarbon C_3H_6 decomposition. It is well known that the decomposition reaction of C_3H_6 involves auto-acceleration and gives the reaction products CH_4, C_2H_4, C_3H_8, C_4

alkene, and a small amount of higher hydrocarbons. Laidler and Loucks (1972) consider the initiative reaction to be

$$C_3H_6 \rightarrow CH_3 + CHCH_2 \tag{2}$$

Because of the negative charactor of fluorine atom, the reaction activity of the molecule C_3F_6 should be quite different from C_3H_6. The purpose of this paper is to investigate the stability of the molecule C_3F_6, particularly concentrating on studies of the primary decomposition channel as the molecule loses its stability at high temperature.

2. Experiment

The experiments are carried out in a heated single-pulse shock tube maintained at 80°C (Cui et al. 1992). Gas chromatography with a flame ionization detector and a 2-m GDX-103 packed column are used for analysis. We have obtained a good separation of the constituents with relative retention time being on the order $CH_4 < C_2F_4 < C_2F_2H_2 < CF_3H < C_2F_3H < C_3F_6 < C_3H_6 < 1,3C_4H_6$, in which the latter two products are the products of decyclization 4-MCH.

The shock-tube initial pressure P_1 is 8000 Pa, and the corresponding reflected shock density is 0.3 NPT.

Two reaction mixtures were prepared in steel tanks and studied:

A: 1040 ppm $C_3F_6 + 1.24\%\ H_2 + 200$ ppm (4-MCH) + Ar.

B: 1400 ppm $C_3F_6 + 2.0\%\ H_2 + 200$ ppm (4-MCH) + Ar.

3. Result

According to the distribution of the products obtained from the gas chromatography, a reaction scheme has been put forward to connect the products with the possible primary decomposition channel of C_3F_6 as follows.

The primary step in the pyrolysis of C_3F_6 is the cleavage of the perfluoromethyl-perfluorovinyl bond

$$C_3F_6 \rightarrow CF_3\cdot + CFCF_2\cdot \tag{3}$$

The trifluoromethyl radical then abstracts hydrogen to form fluoroform CF_3H and a hydrogen atom H,

$$CF_3\cdot + H_2 \rightarrow CF_3H + H\cdot \tag{4}$$

Then a series of elimination, insertion, and abstraction reactions occur. The following reaction scheme is proposed

$$CF_3H \rightarrow CF_2: + HF \tag{5.1}$$

$$CF_2 + H_2 \rightarrow CF_2H_2^* \rightarrow CFH : +HF \tag{5.2}$$

$$CFH : +H_2 \rightarrow CFH_3^* \rightarrow CH_2 : +HF \tag{5.3}$$

$$CH_2 : +H_2 \rightarrow CH_3 \cdot +H \cdot \tag{5.4}$$

$$CH_3 \cdot +H_2 \rightarrow CH_4 + H \cdot \tag{5.5}$$

where the insertion products $CF_2H_2^*$ and CFH_3^* are energized intermediate. The possible channel of the recombination of CF_2: could be

$$CF_2 : +CF_2 :\rightarrow C_2F_4 \tag{5.6}$$

since the product C_2F_4 was found in our experiment. Inspection of the scheme above shows that all the obtained stable products, CH_4, CF_3H, and C_2F_4, are indicators of radical CF_3. Based on the analysis above, the carbon mass balance of $CF_3\cdot$ can be described as

$$[CF_3\cdot] = [CH_4] + [CF_3H] + 2[C_2F_4] \tag{6}$$

The extent of pyrolysis of C_3F_6 can be determined from the measurements of concentrations of the final stable products $[CH_4]$, $[C_2F_4]$, and $[CF_3H]$. The other part of the intermediate radical yielded by (3) is the perfluorovinyl $C_2F_3\cdot$. It abstracts hydrogen H_2 to form C_2F_3H,

$$C_2F_3\cdot + H_2 \rightarrow C_2F_3H + H \cdot \tag{7}$$

Then we have the carbon mass balance

$$[C_2F_3\cdot] = [C_2F_3H] \tag{8}$$

The rate constant of the decyclization of 4-MCH, which we used as a comparative rate, is well established (Tsang 1981) as

$$k_c(CH_3C_6H_9 \rightarrow C_3H_6 + 1,3C_4H_6) = 10^{15.3} \exp\left(-\frac{33400}{T}\right)\ s^{-1} \tag{9}$$

Both (6) and (8) are used to measure the extent of reaction (3). By using and measuring the concentrations of products CH_4, CF_3H, and C_2F_4, the comparative rate result is shown in Fig. 1, and the quantitative expression is

$$\begin{aligned}\log k(C_3F_6 \rightarrow CF_3\cdot + C_2F_3\cdot)\\ = 1.25 \log k_c(4MCH \rightarrow C_3H_6 + 1,3C_4H_6) - 1.85\end{aligned} \tag{10}$$

Substituting (9) into (10), we have

$$k(C_3F_6 \rightarrow CF_3\cdot + C_2F_3\cdot) = 10^{17.3} \exp\left(-\frac{83000}{RT}\right)\ s^{-1} \tag{11}$$

On the other hand, using (7) and measuring the extent of reaction (3) by concentration of product C_2F_3H, the result is shown in Fig. 2 and the comparative rate expression is

$$\log k(C_3F_6 \rightarrow CF_3\cdot + C_2F_3\cdot) = 1.31 \log k_c(4MCH \rightarrow C_3H_6 + 1,3C_4H_6) - 2.52$$

This in turn leads to the rate expression

$$k(C_3F_6 \rightarrow CF_3\cdot + C_2F_3\cdot) = 10^{17.5} \exp\left(-\frac{87000}{RT}\right)\ \mathrm{s}^{-1} \tag{12}$$

Taking the average of the values for the active energy and the preexponential at the factor given by (11) and (12), we arrive at the most likely expression of the rate constant for the pyrolysis of perfluoropropylene

$$k(C_3F_6 \rightarrow CF_3\cdot + C_2F_3\cdot) = 10^{17.4\pm0.2} \exp\left(-\frac{85000\pm2000}{RT}\right)\ \mathrm{s}^{-1} \tag{13}$$

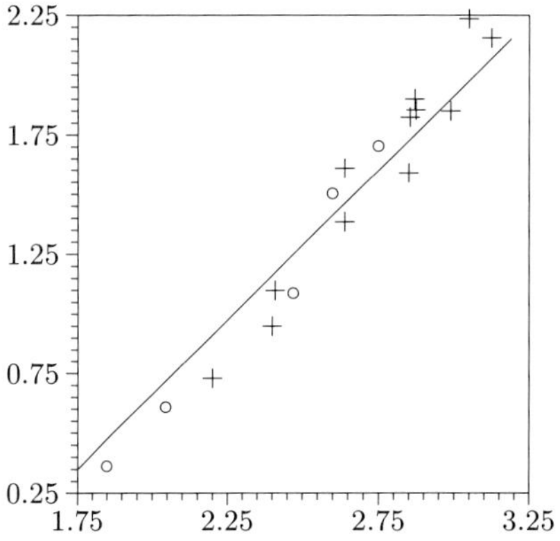

Figure 1. Comparative rate studies on the decomposition of C_3F_6 with 4-MCH as internal standard. Data are based on the carbon balance of $CF_3\cdot$.

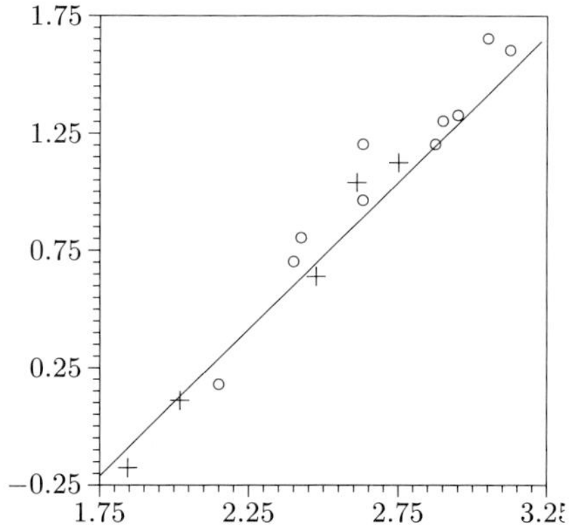

Figure 2. Comparative rate studies on the decomposition of C_3F_6 with 4-MCH as internal standard. Data are based on the carbon balance of $C_2F_3\cdot$.

The horizontal axis is $\log k(4MCH \rightarrow C_3H_6 + 1,3C_4H_6)/s$

The vertical axis is $\log k(C_3F_6 \rightarrow CF_3 + C_2F_3)/s$

(+) 1040 ppm C_3F_6, 1.24% H_2, 200 ppm 4-MCH in argon

(○) 1400 ppm C_3F_6, 2.0% H_2, 200 ppm 4-MCH in argon

4. Discussion

From the experimental results we find the primary step in the pyrolysis of C_3F_6 as the breaking of carbon bond CF_3–C_2F_3; it has a bond dissociation energy

BDE(CF_3–C_2F_3) = 85 kcal/mol. This is consistent with the radical formed by this bond cleavage derived by analyzing the distribution of the final stable products mentioned earlier.

We will now discuss the validity of this argument from thermochemical considerations. We have known that the cleavage of the C–F bond can not possibly be a factor in these studies. The C–F bond strength is 126 kcal/mol for C_2F_4 and 126.8 kcal/mol for C_2F_6. These values should not change much for C_3F_6. The energy of the C–F bonds both in the vinylic group and those in the methylic group of C_3F_6 should be much greater than the determined reaction activation energy of 85 kcal/mol. The possible channels that have to be considered are the breaking of either the vinylic C=C bond or the vinylic C–C bond. The energy of the vinyl C=C bond in C_2F_4 is 76 kcal/mol. This value is much smaller than the value of the vinylic bond in C_2H_4. BDE($CH_2 = CH_2$) = 172 kcal/mol (Cox and Pilcher 1970). This effect gives to the C–F bond some double bond characteristics and strength; meanwhile, the C=C bond should be weakened. The dissociation energy for the reaction of the primary vinyl C–F bond in fluorinated ethylenes

$$C_2H_{4-n}F_n \rightarrow C_2H_{4-n}F_{n-1} + F \tag{14}$$

can be expressed as

$$\begin{aligned} &\mathrm{BDE}(C_2H_{4-n}F_n \rightarrow C_2H_{4-n}F_{n-1} + F) \\ &\quad = \Delta H_f^\circ(C_2H_{4-n}F_{n-1}) + \Delta H_f^\circ(F) - \Delta H_f^\circ(C_2H_{4-n}F_n) \\ &\quad\quad n = 1, 2, 3, 4 \end{aligned} \tag{15}$$

where ΔH_f° is the formation enthalpy for the substance expressed in the braket. The values of the formation enthalpies for the fluorinated ethylene are adopted from Lide (1992–1993): $\Delta H_f^\circ(C_2F_4) = -169$ kcal/mol, $\Delta H_f^\circ(C_2HF_3) = -117$ kcal/mol, $\Delta H_f^\circ(C_2H_3F) = -33.2$ kcal/mol, $\Delta H_f^\circ(C_2H_4) = 12.5$ kcal/mol, $\Delta H_f^\circ(F) = 18.4$ kcal/mol, and $\Delta H_f^\circ(H) = 52.0$ kcal/mol. Combining (15) for n = 4 and n = 3, we have BDE($C_2F_3H \rightarrow C_2F_3 + H$) − BDE($C_2F_4 \rightarrow C_2F_3 + F$) = $\Delta H_f^\circ(C_2F_4) - \Delta H_f^\circ(C_2F_3H) + \Delta H_f^\circ(H) - \Delta H_f^\circ(F)$. Substituting the values of these formation enthalpies, the calculated difference of the BDE for these two processes is 18.4 kcal/mol. It was known that the dissociation energy BDE($C_2F_3H \rightarrow C_2F_3 + H$) is 108 kcal/mol. Then the BDE($C_2F_4 \rightarrow C_2F_3 + F$) is 126 kcal/mol. Combining (13) for n = 0 and n = 1, we have BDE($C_2H_3F \rightarrow C_2H_3 + F$)–BDE($C_2H_4 \rightarrow C_2H_3 + H$) = 11.9 kcal/mol. The BDE($C_2H_4 \rightarrow C_2H_3 + H$) is not quite well known now. If we adopt the value for the vinyl-H bond in propene of 110 kcal/mol, then the value BDE($C_2FH_3 \rightarrow C_2H_3 + F$) is 122 kcal/mol.

As the number of the fluorine atoms varies from one to four, the double-bond character of the C–F bonds in fluoronated ethylenes remains and there is no change. It is expected that on average there is a 24-kcal/mol decrease of the C=C bond energy in fluoronated ethylenes for each H atom substituted by a fluorine atom. This leads to the C=C bond strength in trifluoroethylene of 100 kcal/mol. By using the data of the formation enthalpies of free radicals $\Delta H_f^\circ(CF_2) = -46$ kcal/mol, $\Delta H_f^\circ(CFH) = 26$ kcal/mol, we have the value of the BDE $(C_2F_3H \rightarrow CF_2 + CFH) = 97$ kcal/mol.

The determined value of 85 kcal/mole for the activation energy of the pyrolysis of perfluoropropylene is consistent with this predication. Based on this value and the formation enthalpies of C_2F_3 and CF_3 mentioned above, the formation enthalpy of C_3F_6, $\Delta H_f^\circ(C_3F_6) = -258$ kcal/mol, is obtained.

Acknowledgement. The author would like to acknowledge the assistance of Fan Bencheng and Cui Jiping for operating the shock-tube during experiments and for helpful discussion. The project was supported by the National Natural Science Foundation of China.

References

Cox JD, Pilcher G (1970) Thermochemistry of Organic and Organometallic Compounds, Academic Press, New York.

Cui JP, Fan BC, He YZ(1992) The application of single pulse shock tube in chemical kinetics studies. J Chinese Chem Phys 5:374.

Dalby FW (1964) Flash photolysis measurement of the kinetics of CF_2 reactions. J Chem Phys 41:2297.

Laidler KJ, Loucks LF (1972) The decomposition and isomerization of hydrocarbons. In: Comprehensive Chemical Kinetics (Bamford CH and Tipper CFH, eds) Elsevier, Vol. 5, p 1.

Lide DR (1992–1993) (ed) Handbook of Chemistry and Physics, Section 5, CRC Press.

Tsang W (1981) Comparative-rate, single-pulse shock tube studies on the thermal stability of polyatomic molecules. In: Shock Waves in Chemistry (Lifshitz A, ed), Marcel Dekker, New York, Basel, p 59.

Direct measurements of rate coefficients for thermal decomposition of methyl halides using shock-tube ARAS technique

K. Takahashi, A. Inoue*, T. Inomata
Department of Chemistry, Sophia University
7-1 Kioi-cho, Chiyoda-ku, Tokyo 102, Japan

Abstract: The rate coefficients for the thermal decomposition reaction of methyl iodide and methyl bromide, $CH_3X(+M) \rightarrow CH_3 + X(+M)$, were determined behind reflected shock waves over the temperature range of 1180–1600 K and the total concentration range of 8.6×10^{-6}–4.0×10^{-5} mol·cm^{-3}. Due to the high sensitivities of I- and Br-ARAS, CH_3X greatly diluted with argon or nitrogen was used in the experiments. Thus the rate coefficients could be directly measured without the influence of secondary reactions. The low-pressure limit rate coefficients (in cm^{3}mol^{-1}s^{-1}) were:

$$k_1^{2nd}(CH_3I\text{-}Ar) = (2.28 \pm 0.26) \times 10^{15} \exp[-(165 \pm 6)\ \mathrm{kJ \cdot mol^{-1}}/RT]$$
$$k_1^{2nd}(CH_3Br\text{-}Ar) = (4.04 \pm 0.49) \times 10^{16} \exp[-(227 \pm 6)\ \mathrm{kJ \cdot mol^{-1}}/RT]$$
$$k_1^{2nd}(CH_3I\text{-}N_2) = (1.14 \pm 0.12) \times 10^{16} \exp[-(172 \pm 7)\ \mathrm{kJ \cdot mol^{-1}}/RT]$$

Key words: Methyl halides, Halomethanes, Thermal decomposition, Pyrolysis, Rate coefficients, Shock tube, ARAS

1. Introduction

In high-temperature kinetic studies using a shock tube, complicated secondary reactions often occur together with the focus reaction. Due to its high sensitivity, atomic resonance absorption spectroscopy (ARAS) enables us to use very dilute test-gas mixtures, resulting in the suppression of secondary reactions. Although numerous rate coefficients for high-temperature reactions have been determined by combining H-, O-, N-, and S-ARAS with a shock tube, few works using halogen-ARAS have been reported. In this work, the sensitivities of I- and Br-ARAS were preliminarily examined behind reflected shock waves. As an application of halogen-ARAS to kinetic studies, we investigated the pyrolysis of methyl iodide and methyl bromide. The decompositions of both methyl halides are initiated through the simple bond-fission reaction of the carbon-halogen bonds, $CH_3X(+M) \rightarrow CH_3 + X(+M)$ (X: I or Br) (R1). In

*Present address: Tokyo Gas Co. Ltd., 1-16-25 Shibaura, Minato-ku, Tokyo 102, Japan.

high-temperature ranges above 1000 K, there are no previous data on these rate coefficients except that published by Saito et al. (1980). They monitored the reactant concentration with UV absorption spectroscopy. In the present study, we tried to perform the halogen-ARAS measurements at low concentrations of CH_3X, and to determine the rate coefficients for reaction (R1) without the influence of secondary reactions.

2. Experimental design

The experiments were carried out behind reflected shock waves in a 6.2-cm i.d. stainless-steel shock tube. Driver and test sections of the shock tube were 1.7-m and 3.8-m long, respectively, so that a dwell time above 1 ms could be obtained in all of the experiments. The associated vacuum line and the apparatus for shock-velocity measurements have been described previously (Takahashi et al. 1989). A microwave discharge lamp, in which helium containing 0.03% I_2 or 0.1% CF_3Br was flowing, was used as a light source for I- or Br-ARAS. The thermal decomposition of CH_3I and CH_3Br was studied over the temperature range of 1180–1600 K and the total concentration range of 8.6×10^{-6}–4.0×10^{-5} mol·cm^{-3}, using 0.5–20 ppm CH_3X diluted with argon or nitrogen. Liquid CH_3I and CH_3Br were purified by trap-to-trap distillation. Argon and nitrogen (purities were 99.9995% and 99.9998%, respectively) were used without further purification.

3. Results and discussion

3.1. Sensitivities of I- and Br-ARAS

Calibration experiments for I- and Br-ARAS were performed by dissociating CH_3I and CH_3Br behind reflected shock waves. The observed equilibrium absorbance was related to the calculated equilibrium concentration of halogen

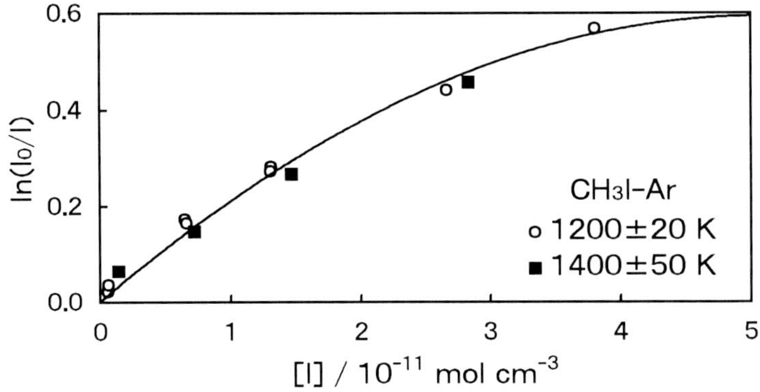

Figure 1. Calibration curve for I($^2P_{3/2}$)-ARAS at 183.0 nm.

Table 1. Apparent absorption coefficients in the low-absorption range.

Atom	Transition	λ / nm	ϵ / cm^2mol^{-1}	Reference
I	$^4P_{5/2}$-$^2P_{3/2}$	183.0	2.9×10^9	This work
Br	$^4P_{5/2}$-$^2P_{3/2}$	157.7	7.8×10^8	This work
O	$^3S_1$-^3P_J	130.2–130.6	9.9×10^9	This work
OH	$A^2\Sigma^+$-$^2\Pi_i$ (0,0)	306.4	1.0×10^7	Takahashi (1988)

Absorption coefficient ϵ in base e.

atoms. Figure 1 shows the calibration curve for $I(^2P_{3/2})$-ARAS at 183.0 nm. The curve is independent of temperature. The apparent absorption coefficient ϵ (base e), which means the tangential slope of the calibration curve, somewhat decreases with the iodine-atom concentration due to self-absorption or self-reversal. In the low-absorption range, however, ϵ can be regarded as approximately constant. Table 1 shows ϵ in the low-absorption range, examined by using this method. The ϵ values of both I and Br atoms are smaller than that of O atoms measured with the same optical system, but are much larger than those of diatomic molecules such as the OH radical.

3.2. Rate coefficients for thermal decomposition of CH_3X

Reaction (R1) might be composed of two channels producing $X(^2P_{3/2})$ and $X(^2P_{1/2})$. $X(^2P_{3/2})$ is the ground electronic state, and $X(^2P_{1/2})$ lies 7603 cm^{-1} (X=I) and 3685 cm^{-1} (X=Br) above ground (Dagdigian & Campbell 1987).

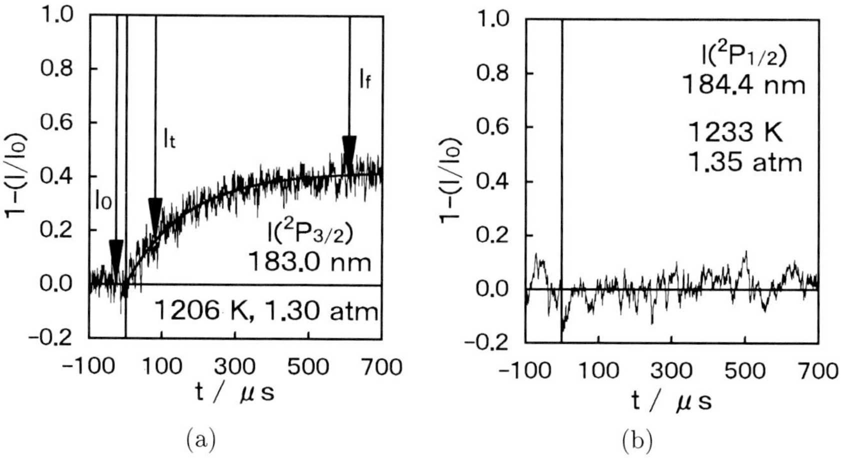

Figure 2. Typical absorption profiles for: (a) $I(^2P_{3/2})$ at 183.0 nm and (b) $I(^2P_{1/2})$ at 184.4 nm in 8 ppm CH_3I-Ar system.

In this case, we must monitor the two electronic states of halogen atoms to determine the overall rate coefficients for reaction (R1). Figure 2 shows typical absorption profiles for $I(^2P_{3/2})$ at 183.0 nm and $I(^2P_{1/2})$ at 184.4 nm in the CH_3I-Ar system. $I(^2P_{3/2})$ increases with the progress of CH_3I pyrolysis and finally reaches an equilibrium value, but $I(^2P_{1/2})$ is not detected during the reaction time. A similar tendency was found for $Br(^2P_{3/2})$ and $Br(^2P_{1/2})$ in the CH_3Br-Ar system. No detection of $X(^2P_{1/2})$ indicates that the channel producing $X(^2P_{1/2})$ directly does not occur or that the quenching reaction of $X(^2P_{1/2})$ to $X(^2P_{3/2})$ is much faster than the channel producing $X(^2P_{1/2})$. In either case, as the rate of producing of $X(^2P_{3/2})$ becomes equal to the decomposing rate of CH_3X, the overall rate coefficients for reaction (R1) are obtained by monitoring $X(^2P_{3/2})$. As mentioned in the last section, Lambert-Beer's law approximately holds in the low-absorption range. Therefore, the absolute concentration of halogen atoms is not required for the determination of the rate coefficients when low-concentration CH_3X is used. If we assume that CH_3X is decomposed only through reaction (R1), the first-order rate coefficient k_1^{1st} can be directly derived from $\ln[\ln(I_t/I_f)/\ln(I_0/I_f)] \equiv A_t = -k_1^{1st}t$, where I_0, I_t, and I_f are the transmission light intensities at time $t = 0$, t, and ∞, respectively, as shown in Fig. 2(a). From the linear relationship between A_t and reaction time, k_1^{1st} was obtained. To check the validity of this procedure, we tried to convert absorbance into absolute concentration for some observed profiles, and

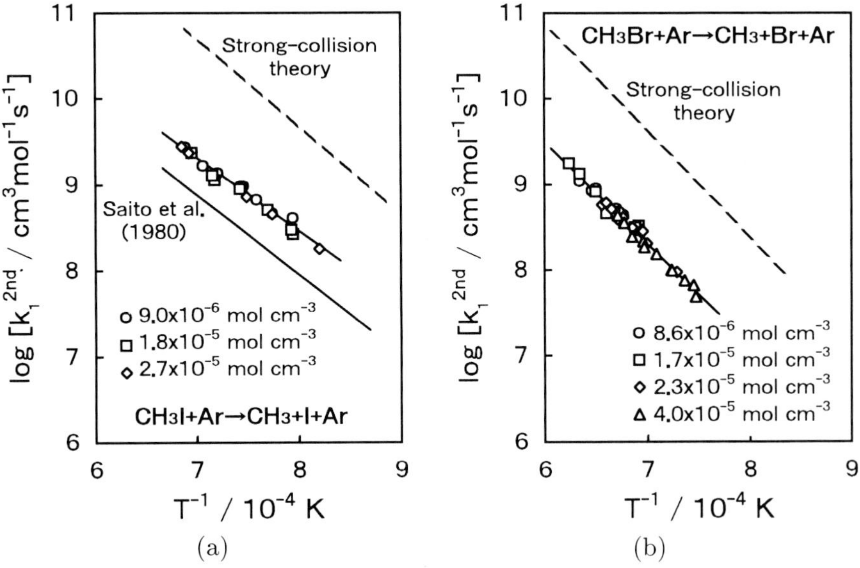

Figure 3. Arrhenius plots of k_1^{2nd} in: (a) CH_3I-Ar and (b) CH_3Br-Ar systems.

to fit the concentration profile to the form of $[I]_t = [CH_3X]_0 \cdot [1 - \exp(-k_1^{1st} t)]$ by least-squares method. No difference was found between the k_1^{1st} values obtained by the two methods. In CH_3I-Ar and CH_3Br-Ar systems, because Arrhenius plots of k_1^{1st} depended on the total concentration [M], the second-order rate coefficients k_1^{2nd} were calculated from the relation $k_1^{2nd} = k_1^{1st}/[M]$. In each system, the Arrhenius plots of k_1^{2nd} obtained under some total concentrations follow a single straight line, as shown in Fig. 3. This fact indicates that reaction (R1) for CH_3I and CH_3Br proceeds in the low-pressure region under the present experimental conditions. Arrhenius expressions for the low-pressure limit rate coefficients (in $cm^3 mol^{-1} s^{-1}$) are:

$$k_1^{2nd}(CH_3I\text{-}Ar) = (2.28 \pm 0.26) \times 10^{15} \exp[-(165 \pm 6)\ kJ \cdot mol^{-1}/RT] \quad (1)$$

$$k_1^{2nd}(CH_3Br\text{-}Ar) = (4.04 \pm 0.49) \times 10^{16} \exp[-(227 \pm 6)\ kJ \cdot mol^{-1}/RT] \quad (2)$$

Similarly, the low-pressure limit rate coefficient for thermal decomposition reaction of CH_3I diluted with N_2 is:

$$k_1^{2nd}(CH_3I\text{-}N_2) = (1.14 \pm 0.12) \times 10^{16} \exp[-(172 \pm 7)\ kJ \cdot mol^{-1}/RT] \quad (3)$$

These rate coefficients were determined assuming that secondary reactions were negligible. To confirm this assumption, sensitivity analyses were carried out. In addition to reaction (R1), the reaction scheme composed of $CH_3X+X=CH_3+X$ (R2), $X+X+M=X_2+M$ (R3), and $CH_3+CH_3=C_2H_6$ (R4) was used in the calculation. The rate coefficients for reactions (R2)$-$(R4) were quoted from the reports of Westbrook (1982) and Westbrook et al. (1983). The results for CH_3I-Ar and CH_3Br-Ar systems are shown in Fig. 4. In both systems, the contribution of secondry reactions (R2)$-$(R4) to halogen-atom concentration is much smaller

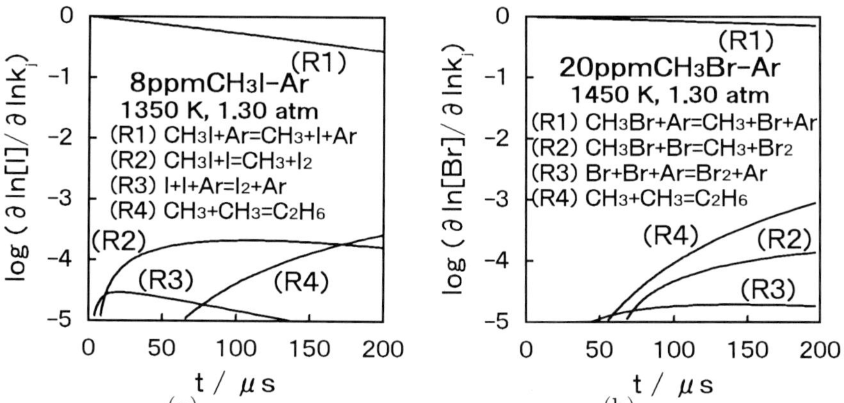

(a) (b)

Figure 4. Sensitivity analyses for: (a) CH_3I-Ar and (b) CH_3Br-Ar systems.

than that of the focus reaction (R1). Furthermore, the concentration profile of halogen atoms calculated from reactions (R1)–(R4) perfectly agreed with that calculated from reaction (R1) only. We can conclude from these results that the rate coefficients for reaction (R1) were determined without the influence of secondary reactions. As found in Fig. 3(a), the k_1^{2nd}(CH_3I-Ar) obtained in this work is about double the value reported by Saito et al. (1980). They have measured the rate coefficient in an initial concentration of CH_3I four powers of ten higher than that of the authors, using the UV absorption of the CH_3I molecule. There is thus some possibility that their results are influenced by secondary reactions. The theoretical calculation of the low-pressure limit rate coefficients ${k_1}^{SC}$ was performed by using the appropriate RRKM strong-collision formulation proposed by Troe (1977a, 1977b). In the CH_3I-Ar and CH_3Br-Ar systems, the calculated ${k_1}^{SC}$ is shown with a dashed line in Fig. 3. In both systems, ${k_1}^{SC}$ is larger than the observed k_1^{2nd}, which indicates a weak-collision process in reaction (R1). The collision efficiency factor β_c calculated from the relation $\beta_c = k_1^{2nd}/{k_1}^{SC}$ is $\approx$0.05 in CH_3I-Ar and CH_3Br-Ar systems at 1400 K, while $\approx$0.10 in the CH_3I-N_2 system. These values are reasonable because the energy-transfer efficiency depends on the bath gas rather than the reactant gas.

References

Dagdigian PJ, Campbell ML (1987) Spin-orbit effects in gas-phase chemical reactions. Chem Rev 87:1.

Saito K, Tahara H, Kondo O, Yokubo T, Higashiyama T, Murakami I (1980) The thermal gas-phase decomposition of methyl iodide. Bull Chem Soc Jpn 53:1335.

Takahashi K (1988) The Addition Effects of Halogen-Containing Compounds on Hydrocarbon Ignition Using Shock Tube Technique, Thesis, Sophia University.

Takahashi K, Inomata T, Moriwaki T, Okazaki S (1989) The addition effects of methyl halides on ethane ignition behind reflected shock waves. Bull Chem Soc Jpn 62:2138.

Troe J (1977a) Theory of thermal unimolecular reactions at low pressures. I. Solutions of the master equation. J Chem Phys 66:4745.

Troe J (1977b) Theory of thermal unimolecular reactions at low pressures. II. Strong collision rate constants. Applications. J Chem Phys 66:4758.

Westbrook CK (1982) Inhibition of hydrocarbon oxidation in laminar flames and detonations by halogenated compounds. In: 19th Symposium (Int) on Combustion, The Combustion Institute, Pittsburgh, p 127.

Westbrook CK, Dryer FL, Schug KP (1983) Numerical modeling of ethylene oxidation in laminar flames. Combust Flame 52:299.

Examination of the oxidation mechanism of H_2S

K. Tsuchiya, M. Oya
National Institute for Resources and Environment
16-3 Onogawa, Tsukuba, Ibaraki 305 Japan
H. Matsui
Department of Mechanical Engineering, The University of Tokyo
7-3-1 Hongo, Bunkyo-ku, Tokyo 113 Japan

Abstract: The oxidation mechanism of hydrogen sulfide at high temperatures has been investigated by the laser-photolysis shock-tube atomic-resonance absorption method. Rate constants for reaction $S + O_2 \rightarrow O + SO$ and $SO + O_2 \rightarrow O + SO_2$ were determined directly by conducting KrF laser photolysis in $COS/O_2/Ar$ and $SO_2/O_2/Ar$ mixtures behind reflected shock waves. The reaction of $HS + O_2$ was examined by preparing HS radical in ArF photolysis of $H_2S/O_2/Ar$ mixtures. Kinetic modeling for the H_2S oxidation process based on the recently accumulated experimental data for the key elementary reaction has been conducted. The induction times of SO_2 observed in the shock-tube study by Bradley and Dobson (1967) are consistent with the present mechanism.

Key words: H_2S, Oxidation, Shock tube, Laser photolysis, ARAS

1. Introduction

The oxidation mechanism of hydrogen sulfide is important for understanding the atmospheric pollution that has been a serious problem for a long period, and some kinetic studies have been conducted (Bradley and Dobson 1967, Frenklach et al. 1981). Recently, rate constants and the detailed mechanisms of the elementary reactions, such as $H + H_2S$ and $O + H_2S$, important in H_2S oxidation at high temperatures, were studied. There still remain problems in the consecutive reactions of the H_2S oxidation process. In this study, first the rate constants for the following two O-atom producing reactions of the H_2S oxidation process were directly measured by the laser-photolysis shock-tube technique:

$$S + O_2 \rightarrow SO + O, \quad (1)$$

$$SO + O_2 \rightarrow SO_2 + O, \quad (2)$$

and then the reaction

$$HS + O_2 \rightarrow \text{products} \quad (3)$$

was examined by preparing HS using an ArF excimer laser photolysis of H_2S in $H_2S/O_2/Ar$ mixtures. Based upon the experimental results of this work and previous ones, a kinetic model for an H_2S oxidation mechanism at high temperatures was constructed. The validity of the present model has been examined by comparing numerical simulation for the induction times of SO_2 with those of the shock-tube study by Bradley and Dobson (1967).

2. Experimental

The present experiments were conducted in a piston-actuated shock tube 5 cm diameter and 4 m long. The details of the laser-photolysis shock-tube atomic-resonance absorption spectrometry system were described previously (Matsui et al. 1994). Experiments to obtain the rate constants of reaction (1) were carried out on O_2/COS mixtures diluted in Ar (O_2/COS = 302/124, 625/62 ppm). COS was used as a precursor of $S(^3P)$ atoms, i.e., $S(^1D)$ atoms produced in the photolysis were very rapidly converted to $S(^3P)$ in the present experimental conditions. The initial concentration of S atoms was always kept low enough so that a pseudo-first-order kinetic condition for the decay of S atoms was maintained. The rate constants of reaction (2) were determined from the formation rate of O atoms in the KrF photolysis of O_2/SO_2 mixtures in Ar (O_2/SO_2 = 20200/282 ppm). ArF photolysis of H_2S/O_2 mixtures diluted in Ar were conducted to examine reaction (3) and the time dependencies of O and H atoms were monitored at temperatures ~1400, ~1580, and ~1800 K.

3. Results and discussion

3.1. The rate constants of $S(^3P) + O_2 \rightarrow O(^3P) + SO$

Absorption profiles of S and O atoms were measured over the temperature range 980 – 1610 K. The absorption of S or O atoms was obtained by subtracting the background signal caused by the absorption of COS and O_2, and was converted to concentration using each calibration curve. Sulfur atoms produced in the photolysis of COS showed exponential decay; and O atoms showed increase initially due to (1) then followed a slow decay. The first-order decay rate for S atoms was determined by a least-squares fit. The net decay rate of S atoms due to the reaction with O_2 was obtained by subtracting the contribution of the S + COS reaction (typically 4 – 18% of the whole rate). Oxygen atoms were produced by the reaction of S atoms with O_2 and decreased due to reaction with COS. The numerical simulation based upon the rates of S + O_2, S + COS, and O + COS reproduced well the concentration profiles of O atoms observed in the present experiment. The decay rate for S atoms was measured in the initial concentration ratio of $(O_2)_0/(S)_0$ = 130 – 540 and $(COS)_0/(S)_0$ = 40 – 120, but dependence of the decay rate on the initial concentration was not observed: thus the effect of consecutive reactions on the decay rate of S atoms is unimportant. A least-squares fit of the experimental data gives the following Arrhenius expression for the rate constant for (1) in the temperature range of 980 – 1610 K: $k_1 = (2.5\pm0.6)\times10^{-11}\exp[-(15.3\pm2.5)$ kJ mol$^{-1}/RT]$ cm^{3}molecule^{-1}s^{-1}.

3.2. The rate constants of $SO + O_2 \rightarrow O + SO_2$

The KrF laser photolysis in the mixtures of 2.02% O_2 and 282 ppm SO_2 diluted in Ar was conducted and absorption profiles at 130.5 nm were measured in the temperature range of 1130 – 1640 K. The net absorption of O atoms was obtained by correcting the background signal due to SO_2 and vibrationally excited O_2. The amount of O atoms produced by the photolysis of SO_2 remained constant in the mixture containing SO_2 only, while in the mixture containing both SO_2 and O_2, the amount showed a gradual increase and approached a steady level of approximately twice as much as the initial amount of O atoms produced by the photolysis of SO_2. This suggests that the decay of O atoms due to the reactions of $O + SO_2 \rightarrow SO + O_2$ and $O + SO + Ar \rightarrow SO_2 + Ar$ is sufficiently slow in the present experimental conditions. The concentration of SO radicals produced by the photolysis of SO_2 was determined from the initial amount of O atoms by subtracting a small amout of O atoms originating from the photolysis of O_2. Thus, the rate constant of reaction (2) was obtained by a simple fitting method for the concentration profiles of O atoms. A least-squares analysis gave the following Arrhenius expression for the rate constant for (2) in the temperature range of 1130 – 1640 K: $k_2 = (1.7\pm0.9)\times10^{-12}\exp[-(34\pm11)$ kJ mol$^{-1}/RT]$ cm^{3}molecule^{-1}s^{-1}.

3.3. Evaluation of the contribution of the $HS + O_2$ reaction

Typical examples of the concentration profiles of O and H atoms in a $H_2S/O_2/Ar$ mixture are shown in Fig. 1. The time dependence of O atoms is most sensitive to $O + H_2S$ and $H + O_2$ reactions, especially in the initial stage of the reaction course, but as a whole, without taking into account the $HS + O_2$ reaction (3), it is impossible to get an excellent agreement between the calculated profiles and the observed ones. Thus, reaction (3) is suggested to be sufficiently sensitive to the profile of O atoms in the 193-nm photolysis of H_2S/O_2 mixtures. Several reaction channels may be possible for $HS + O_2$, i.e.,

$$\begin{aligned} HS + O_2 &\rightarrow HSO + O, && (3a)\\ &\rightarrow SO_2 + H, && (3b)\\ &\rightarrow SO + OH, && (3c)\\ &\rightarrow HO_2 + S. && (3d) \end{aligned}$$

Kamiya (1995) has recently performed ab initio calculations of the potential energy surface of the $HS + O_2$ system to examine the product channels of reaction

(3). According to his calculations, the most probable product channel of the reaction is O + HSO (3a). Thus, in this study, analysis by kinetic simulation was performed by assuming that channel (3a) dominated over all the other reaction channels. The magnitude of k_3a was determined by fitting the calculated profiles of O atoms to observed ones. Reaction mechanisms and rate parameters used in the simulation are listed in Table 1. The best agreement between the kinetic simulation and the experiment for the time dependence of O atoms is attained by letting $k_3a = 3.1\times10^{-11}\exp(-75\ \mathrm{kJ\ mol^{-1}}/RT)\ \mathrm{cm^3 molecule^{-1} s^{-1}}$ for (3a). The results of the numerical simulation with this rate constant are shown as solid curves in Fig. 1.

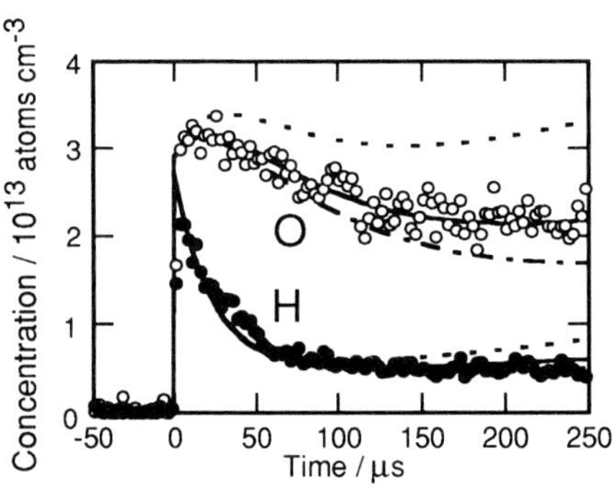

Figure 1. An example of the concentration profiles of O and H atoms after the ArF laser photolysis behind reflected shock waves ($T = 1398 \pm 5$ K, 51 ppm H_2S + 3050 ppm O_2). Solid curve: calculated profiles in the reaction mechanism of Table 1. Dotted curve: sensitivity check of k_3a by a factor of 5.

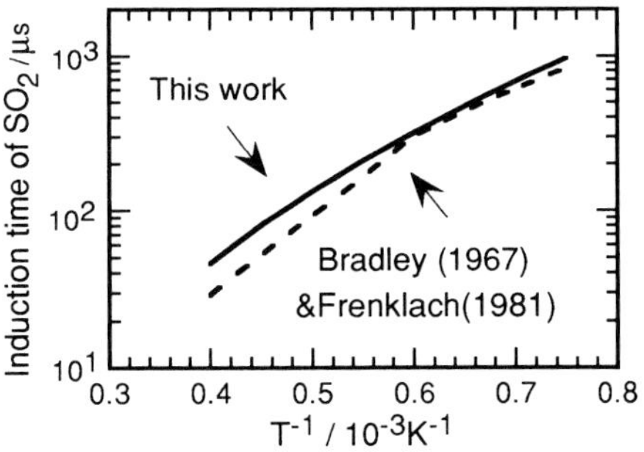

Figure 2. Comparison of the induction time for SO_2 production obtained by numerical simulation (based upon the reaction mechanisms of Table 1) with those of experiment obtained by Bradley and Dobson (1967) and Frenklach et al. (1981) (P = 0.4 atm, 4% H_2S + 6% O_2 diluted in Ar).

3.4. Induction times of SO_2

Bradley and Dobson (1967) measured induction times behind incident shock waves over the temperature range of 1350 – 2450 K in 2 – 6% H_2S and 3 – 9% O_2 mixtures diluted in Ar. They reported that induction times for appearance of OH and SO_2 were identical at temperatures below 1600 K, but the former became longer than the latter at temperatures above 1700 K. Frenklach et al. (1981) conducted regression analysis for experimental data for the induction times of SO_2 obtained by Bradley and Dobson to construct their model for H_2S oxidation. In order to confirm the validity of the reaction scheme constructed in this study, numerical simulation of the induction times of SO_2 was performed and compared with the experimental results of Bradley and Dobson subjected to

Table 1. Reaction mechanisms of H_2S/O_2 system; $k = AT^B \exp[-E(\text{kJ mol}^{-1})/RT]$ in molecule-cm-s units.

Elementary reaction	A	B	E	Reference
$H_2S + Ar = H_2 + S + Ar$	3.63E-11	0.00	268.4	Shiina (1996)
$H + H_2S = H_2 + HS$	3.20E-10	0.00	20.7	Yoshimura (1992)
$S + H_2S = HS + HS$	1.38E-10	0.00	30.8	Shiina (1996)
$S + H_2 = H + HS$	2.63E-10	0.00	82.5	Shiina (1996)
$S + HS = H + S_2$	4.47E-11	0.00	0.0	Mihelcic (1970)
$HS + HO_2 = H_2S + O_2$	1.66E-12	0.00	0.0	Frenklach (1981)
$H_2S + O = OH + HS$	2.02E-10	0.00	32.0	Tsuchiya (1994)
$H_2S + O = H + HSO$	0.50E-10	0.00	32.0	Tsuchiya (1994)
$H_2S + OH = HS + H_2O$	5.25E-12	0.00	0.0	Perry (1976)
$HS + O = SO + H$	1.26E-10	0.00	0.0	Tsuchiya (1994)
$HS + O_2 = HSO + O$	3.05E-11	0.00	75.3	This work
$S + O_2 = SO + O$	2.53E-11	0.00	15.3	This work
$SO + O_2 = SO_2 + O$	1.70E-12	0.00	34.0	This work
$SO + OH = SO_2 + H$	8.40E-11	0.00	0.0	Jourdain (1979)
$HSO + Ar = H + SO + Ar$	1.40E-08	0.00	245.0	guess
$HSO + O_2 = HO_2 + SO$	1.66E-12	0.00	0.0	Frenklach (1981)
$S_2 + O = SO + S$	1.66E-11	0.00	0.0	Singleton (1988)
$H + O_2 = OH + O$	3.30E-10	0.00	70.3	Baulch (1992)
$O + H_2 = OH + H$	8.49E-20	2.67	26.3	Baulch (1992)
$OH + H_2 = H_2O + H$	1.70E-16	1.60	13.8	Baulch (1992)
$OH + OH = H_2O + O$	2.50E-15	1.14	0.4	Baulch (1992)
$H + O_2 + Ar = HO_2 + Ar$	1.70E-30	-0.80	0.0	Baulch (1992)
$H + HO_2 = OH + OH$	2.80E-10	0.00	3.7	Baulch (1992)
$H + HO_2 = H_2 + O_2$	7.10E-11	0.00	5.9	Baulch (1992)
$H + HO_2 = H_2O + O$	5.00E-11	0.00	7.2	Baulch (1992)
$O + HO_2 = OH + O_2$	5.40E-11	0.00	0.0	Baulch (1992)
$HO_2 + OH = H_2O + O_2$	4.80E-11	0.00	-2.1	Baulch (1992)
$O_2 + Ar = O + O + Ar$	3.01E-06	-1.00	494.1	Tsang (1986)
$H_2 + Ar = H + H + Ar$	9.88E-06	-1.10	436.8	Cohen (1983)
$H + OH + Ar = H_2O + Ar$	2.30E-26	-2.00	0.0	Baulch (1992)

the regression analysis. The computed induction time is defined as attainment of 1/10 of the SO_2 final molar fraction. The calculated ignition delay times for SO_2 agree excellently with the experimental results, as shown in Fig. 2. Thus, it may be reasonable to conclude that the present reaction scheme for the oxidation of H_2S is satisfactory. According to the reaction mechanism of this work, the most important reaction intermediate appears to be HSO, which is produced in the reactions of $HS + O_2$ as well as $O + H_2S$. It decomposes immediately to H + SO at high temperature to supply SO, which plays a role of the second chain carrier i.e., SO is oxidized to final products SO_2 via the reaction with O_2 and reproduces O atoms.

References

Baulch DL, Cobos CJ, Cox RA, Esser C, Frank P, Just T, Kerr JA, Pilling MJ, Troe J, Walker RW, Warnatz J (1992) Evaluated kinetic data for combustion modelling. J Phys Chem Ref Data 21:411.

Bradley JN, Dobson DC (1967) Oxidation of hydrogen sulfide in shock waves. I. Absorption studies of OH and SO_2 in H_2S-O_2-Ar mixtures. J Chem Phys 46:2865.

Cohen N, Westberg KR (1983) Chemical kinetic data sheets for high-temperature chemical reactions. J Phys Chem Ref Data 12:531.

Frenklach M, Lee JH, White JN, Gardiner Jr WC (1981) Oxidation of hydrogen sulfide. Combust Flame 41:1.

Jourdain JL, LeBras G, Combourieu J (1979) Kinetic study of some elementary reactions of sulfur compounds including reactions of S and SO with OH radicals. Int J Chem Kinet 11:569.

Kamiya K (1995) Private communication.

Matsui H, Koshi M, Oya M, Tsuchiya K (1994) Improvement of chemical kinetic data at high temperatures by piston-actuated shock tube, excimer laser photolysis, and atomic resonance absorption spectrometry. Shock Waves 3:287.

Mihelcic VD, Shindler RN (1970) ESR-spektroskopische untersuchung der reaktion von atomarem wasserstoff mit H_2S. Ber Bunsenges Phys Chem 74:1280.

Perry RA, Atkinson R, Pitts Jr JN (1976) Rate constants for the reactions OH + $H_2S \rightarrow H_2O$ + SH and OH + $NH_3 \rightarrow H_2O + NH_2$ over the temperature range 297–427K. J Chem Phys 64:3237.

Shiina H, Oya M, Yamashita K, Miyoshi A, Matsui H (1996) Kinetic studies on the pyrolysis of H_2S. J Phys Chem (in press).

Singleton DL, Cvetanovic RJ (1988) Evaluated chemical kinetic data for the reactions of atomic oxygen (^3P) with sulfur containing compounds. J Phys Chem Ref Data 17:1377.

Tsang W, Hampson RF (1986) Chemical kinetic data base for combustion chemistry. Part 1. Methane and related compounds. J Phys Chem Ref Data 15:1087.

Tsuchiya K, Yokoyama K, Matsui H, Oya M, Dupre G (1994) Reaction mechanism of atomic oxygen with hydrogen sulfide at high temperature. J Phys Chem 98:8419.

Yoshimura M, Koshi M, Matsui H (1992) Non-Arrhenius temperature dependence of the rate constant for the H + H_2 S reaction. Chem Phys Lett 189:199.

Thermal decomposition of furonitrile: studies with a single-pulse shock tube

A. Laskin*, A. Lifshitz
Department of Physical Chemistry
The Hebrew University, Jerusalem 91904 Israel

Abstract: The decomposition of furonitrile highly diluted in argon was studied behind reflected shocks in a heated, pressurized driver single-pulse shock tube over the temperature range 1135–1380 K and overall densities of $\sim 2.5 \times 10^{-5}$ mol/cm^3. Furonitrile undergoes unimolecular decomposition via three parallel channels:

furonitrile → $CH_3C{\equiv}C{-}CN$ + CO, $k_1 = 10^{15.95} \exp(-83.0 \times 10^3/RT)\mathrm{s}^{-1}$
furonitrile → $CH{\equiv}C{-}CN$ + CH_2CO, $k_2 = 10^{14.20} \exp(-78.5 \times 10^3/RT)\mathrm{s}^{-1}$
furonitrile → C_2H_2 +CH(CN)CO, $k_3 = 10^{15.50} \exp(-85.0 \times 10^3/RT)\mathrm{s}^{-1}$

The first channel is the major decomposition channel, the second and third have approximately the same rates but differ in their Arrhenius parameters. Additional products such as CH_4, C_2H_4, C_2H_6, $CH_2{=}C{=}CH_2$, $CH_3{-}C{\equiv}CH$, HCN, and $CH_2{=}CHCN$ are found in the post-shock mixtures in small quantities and are produced by free-radical processes. The initiation of free radicals proceeds by an ejection of a hydrogen atom from the allylic position in the major reaction product: $CH_3C{\equiv}C{-}CN \rightarrow CH_2^{\bullet}C{\equiv}C{-}CN + H^{\bullet}$ and by cleavage of the furonitrile ring without an H atom migration: furonitrile $\rightarrow CH_2^{\bullet}C{\equiv}C{-}CN + HCO^{\bullet}$.

Key words: Furonitrile, Decomposition

1. Experimental

1.1. Apparatus

The thermal reactions of furonitrile were studied behind reflected shocks in a pressurized driver, 52-mm i.d. heated single-pulse shock tube. The tube and the gas-handling system were maintained with the aid of a computer-controlled thermostating system at 100(±1)°C in order to insure high enough vapor pressure of furonitrile (boiling point 148°C) for mixture preparation and shock-tube experiments. The tube had a 4-m long driven section divided in the middle by a 52-mm i.d. ball valve. The driver had a variable length up to a maximum of 2.7 m, and could be varied in small steps in order to obtain the best cooling

*In partial fulfillment of the requirements for a Ph.D. degree to be submitted by A. Laskin to the Senate of The Hebrew University.

conditions. Cooling rates were approximately 5×10^5 K/s. A 36-L dump tank was connected to the driven section at 45° angle near the diaphragm holder in order to prevent reflection of transmitted shocks and in order to reduce the final pressure in the tube. After running an experiment, gas samples were transferred from the tube through a heated injection system to a Hewlett-Packard Model 5890A gas chromatograph (GC) operating with flame ionization and nitrogen-phosphor detectors.

Reflected-shock temperatures were calculated from the extent of decomposition of 1,1,1-trifluoroethane ($CH_3CF_3 \rightarrow CH_2{=}CF_2 + HF$) which was added in small quantities to the reaction mixture to serve as an internal standard. The temperatures were calculated from the relation:

$$T = -(E/R) \ln \left\{ -\frac{1}{At} \ln(1 - \chi) \right\} \tag{1}$$

where t is the reaction dwell time, E is the activation energy of the standard reaction (72.6 kcal/mol), A is the pre-exponential factor ($10^{14.51}$ s^{-1}), and χ is the extent of decomposition of CH_3CF_3.

1.2. Materials and analysis

Reaction mixtures containing 0.3% furonitrile and 0.1% 1,1,1-trifluoroethane in argon were prepared manometrically and stored at high pressure in 12-L glass bulbs. Both the bulbs and the line were pumped down to better than 10^{-5} torr before the preparation of the mixtures. Gas-chromatographic analyses of the post-shock mixtures were performed on two 2-m Porapak N columns connected to flame-ionization and nitrogen-phosphor detectors.

Two of the reaction products, CH_2CO and CH(CN)CO, could not be determined quantitatively owing to their reaction with water absorbed on the injection system and on tubings leading to the GC. In several tests, gas samples were collected in bulbs containing small amounts of methyl alcohol. These compounds react with the alcohol to form the corresponding methyl esters which were clearly identified in the GC.

The concentrations of the reaction products $C_5(pr)$ were calculated from their GC-peak areas using the following relations:

$$C_5(pr)_i = A(pr_i)t/S(pr_i) \times \{C_5(\text{furonitrile})_0/A(\text{furonitrile})_0\} \tag{2}$$

$$C_5(\text{furonitrile})_0 = \{p_1 \times \%(\text{furonitrile}) \times \rho_5/\rho_1\}/100RT_1 \tag{3}$$

$$A(\text{furonitrile})_0 = A(\text{furonitrile})_t + 1/5 \sum N(pr_i) \times A(pr_i)_t/S(pr_i) \tag{4}$$

In these relations $C_5(\text{furonitrile})_0$ is the concentration of furonitrile behind the reflected shock prior to decomposition and $A(\text{furonitrile})_0$ is the calculated GC-peak area of furonitrile prior to decomposition (Eq. 4) where: $A(pr_i)_t$ is the peak area of a product i in the shocked sample, $S(pr_i)$ is the product's sensitivity relative to furonitrile, and $N(pr_i)$ is the product's number of carbon atoms; ρ_5/ρ_1 is the compression behind the reflected shock and T_1 is the initial temperature.

2. Results

Figure 1 shows an Arrhenius plot of the rate constants for the overall decomposition of furonitrile. They were calculated as first-order rate constants from the relation:

$$k_{total} = -\ln\{[\text{furonitrile}]_t/[\text{furonitrile}]_0\}/t \tag{5}$$

The value obtained is: $k_{total} = 10^{16.18}\exp(-83.1 \times 10^3/RT)\text{s}^{-1}$ where R is expressed in units of cal/(K mol). Figures 2–4 show first-order rate constants

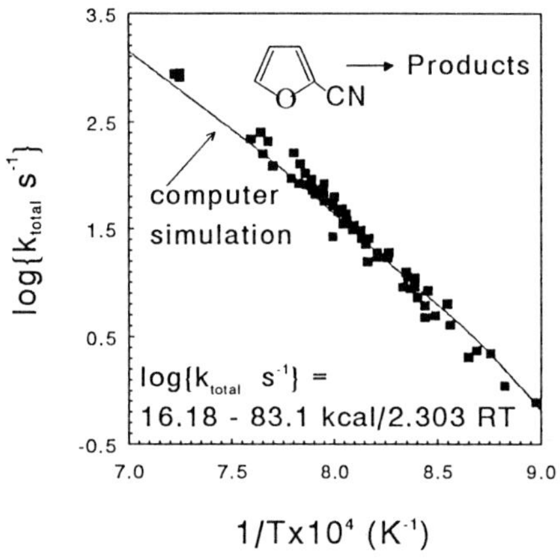

Figure 1. Total decomposition of furonitrile.

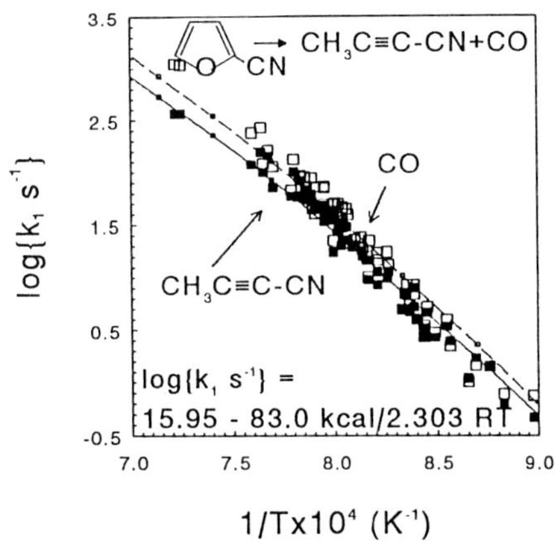

Figure 2. Unimolecular production of $CH_3C{\equiv}C{-}CN$ and CO.

for the production of three major reaction products plotted against reciprocal temperature. They were calculated from the relation:

$$k_{product} = \frac{[\text{product}]_t}{[\text{furonitrile}]_0 - [\text{furonitrile}]_t} \times k_{total} \tag{6}$$

The three products shown in these plots are formed by direct unimolecular cleavage of the ring. The rate constants for CO, which is believed to be formed simultaneously with $CH_3C{\equiv}C{-}CN$, are plotted together with the latter on one figure (Fig. 2). The lines in the figures are results of the computer simulation as will be discussed later.

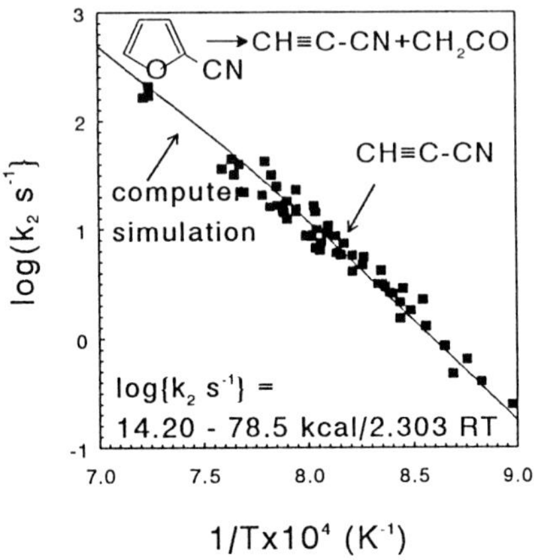

Figure 3. Unimolecular production of CH≡C–CN.

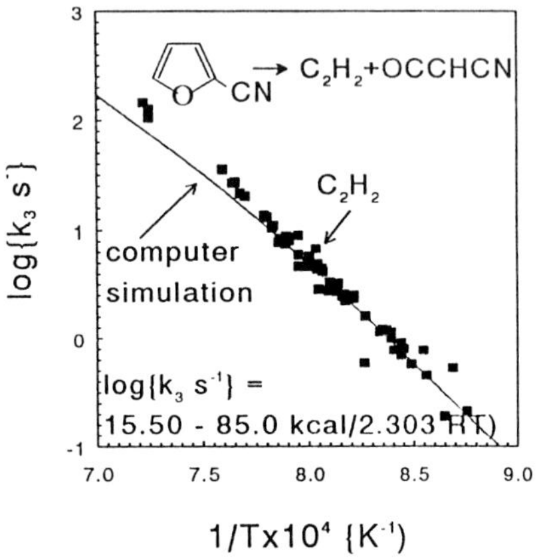

Figure 4. Unimolecular production of C_2H_2.

3. Discussion

3.1. Reaction mechanism

The three unimolecular reaction channels in furonitrile, i.e.,

$$\begin{aligned}
&\text{furonitrile} \to CH_3C{\equiv}C\text{–}CN + CO, && k_1 = 10^{15.95}\exp(-83.0\times10^3/RT)\text{s}^{-1}\\
&\text{furonitrile} \to CH{\equiv}C\text{–}CN + CH_2CO, && k_2 = 10^{14.20}\exp(-78.5\times10^3/RT)\text{s}^{-1}\\
&\text{furonitrile} \to C_2H_2 + CH(CN)CO, && k_3 = 10^{15.50}\exp(-85.0\times10^3/RT)\text{s}^{-1}
\end{aligned}$$

are very similar to the ones observed in the past for the dissociation of furan, with approximately the same rates (Lifshitz et al. 1986). The channel that produces CH_2CO and C_2H_2 in furan appears in furonitrile as two different channels owing to decreased symmetry in the latter.

In furan decomposition, the major reaction channel produces $CH_3C{\equiv}CH$ and CO. It was unequivocally shown (Lifshitz et al. 1986) that C_2H_2, which appeared in much smaller concentrations, was formed not as a secondary decomposition product from $CH_3C{\equiv}CH$ but directly by a unimolecular cleavage of the ring. The compounds CH≡C–CN and C_2H_2 in furonitrile are believed to be produced by the same type of ring cleavage as in furan.

As has been mentioned before, several additional products such as CH_4, C_2H_6, $CH_3C{\equiv}CH$, $CH_2{=}C{=}H_2$, HCN, and others, were found in the post-shock mixtures, although at lower concentrations. All these products are formed by free-radical reactions. It is suggested that similar to findings in other molecules with the same ring structure, cleavage of the ring which takes place in reaction (1), but without H-atom migration from position (5) to position (4), takes place here too and is responsible for the initiation of free radicals: furonitrile → $HCO^{\bullet}$ + $CH_2^{\bullet}$-C≡C–CN. At higher temperatures reaction (9) (see Table 1) becomes the main source for free radicals.

Table 1. Reaction scheme for furonitrile decomposition.

	Reaction	A	E	ΔS_r^0	ΔH_r^0
1.	C_4H_3O–CN → CO+CH_3C≡C–CN	9.50×10^{15}	83.0	41.05	28.96^i
2.	C_4H_3O–CN → CH_2CO+CH≡C–CN	2.58×10^{14}	78.5	39.88	47.80^i
3.	C_4H_3O–CN → C_2H_2+CH(CN)CO	3.16×10^{15}	85.0	42.79	54.10^i
4.	C_4H_3O–CN → $HCO^{\bullet}$+$CH^{\bullet}$=CHCN	5.00×10^{15}	86.0	35.50	81.46^e
5.	Ar+$HCO^{\bullet}$ → $H^{\bullet}$+CO+Ar	1.18×10^{14}	16.1	25.22	17.65^n
6.	$CH^{\bullet}$=CHCN → $H^{\bullet}$+CH≡C–CN	1.00×10^{13}	38.0	23.94	42.54^e
7.	C_4H_3O–CN+$H^{\bullet}$ → $C_4H_3O^{\bullet}$+HCN	1.50×10^{14}	24.0	12.12	23.42^e
8.	CH_3C≡C–CN+$H^{\bullet}$ → $C_3H_3^{\bullet}$+HCN	8.00×10^{13}	7.0	8.04	-20.19^e
9.	CH_3C≡C–CN → $CH_2^{\bullet}C$≡C–CN+$H^{\bullet}$	1.00×10^{15}	90.0	36.48	92.67^e
10.	$CH_2^{\bullet}C$≡C–CN+C_4H_3O–CN → CH_3C≡C–CN+C_4H_2O–$CN^{\bullet}$	1.00×10^{11}	19.0	0.61	21.94^e
11.	$C_4H_3O^{\bullet}$ → CO+$C_3H_3^{\bullet}$	1.00×10^{16}	74.0	36.97	-14.65^e
12.	$C_4H_3O^{\bullet}$ → C_2H_2+$CH^{\bullet}$=CO	2.58×10^{14}	75.0	35.66	21.59^e
13.	$C_4H_3O^{\bullet}$ → CH_2CO+$C_2H^{\bullet}$	3.16×10^{15}	70.0	39.08	47.73^e
14.	$C_2H^{\bullet}$+$C_2H^{\bullet}$ → C_4H_2	5.00×10^{12}	0	−41.04	-154.49^e
15.	Ar+a-C_3H_4 → $C_3H_3^{\bullet}$+$H^{\bullet}$+Ar	1.00×10^{17}	70.0	34.01	91.63^n
16.	Ar+p-C_3H_4 → $C_3H_3^{\bullet}$+$H^{\bullet}$+Ar	1.00×10^{17}	70.0	32.98	93.09^n
17.	a-C_3H_4 → pC_3H_4	1.48×10^{13}	60.4	1.03	-1.46^n
18.	CH_3C≡C–CN+$H^{\bullet}$ → CH≡C–CN+$CH_3^{\bullet}$	1.12×10^{14}	78.0	3.29	-14.98^n
19.	p-C_3H_4+$H^{\bullet}$ → C_2H_2+$CH_3^{\bullet}$	1.12×10^{14}	7.8	5.59	-8.53^n
20.	CH_3CN+$H^{\bullet}$ → $CH_3^{\bullet}$+HCN	1.12×10^{14}	7.0	7.14	-7.31^n
21.	CH_4 → $CH_3^{\bullet}$+$H^{\bullet}$	7.80×10^{14}	103.8	34.22	107.77^n
22.	Ar+CH_2CO → $CH_2^{\bullet\bullet}$+CO+Ar	3.67×10^{15}	59.3	35.58	78.46^n
23.	C_2H_2+Ar → $C_2H^{\bullet}$+$H^{\bullet}$+Ar	1.80×10^{16}	107.1	33.96	130.19^n
24.	$CH^{\bullet}$=CO+$H^{\bullet}$+Ar → CH_2CO+Ar	1.00×10^{15}	0	−30.53	-104.05^e
25.	$CH_3^{\bullet}$+C_4H_3O–CN → CH_3CN+$C_4H_3O^{\bullet}$	5.00×10^{12}	31.0	4.98	30.73^e
26.	$H^{\bullet}$+C_4H_3O–CN → H_2+C_4H_2O–$CN^{\bullet}$	1.00×10^{14}	15.0	9.45	7.72^e
27.	$CH_3^{\bullet}$+C_4H_3O–CN → CH_4+C_4H_2O–$CN^{\bullet}$	1.00×10^{12}	15.0	2.87	6.83^e
28.	C_4H_2O–$CN^{\bullet}$ → CO+$CH_2^{\bullet}C$≡C–CN	1.00×10^{16}	68.0	40.43	7.02^e
29.	C_4H_2O–$CN^{\bullet}$ → $CH^{\bullet}$=CO+CH≡C–CN	2.50×10^{14}	63.5	33.32	37.24^e
30.	C_4H_2O–$CN^{\bullet}$ → $C_2H^{\bullet}$+CH(CN)CO	3.20×10^{15}	70.0	39.65	69.65^e

^{i} Results of this investigation, ^{n} NIST-Chemical Kinetics Data Base, ^{e} estimated
C_4H_3O–CN is furonitrile, C_4H_2O–$CN^{\bullet}$ is furonitrile radical, $C_4H_3O^{\bullet}$ is furan radical

3.2. Reaction scheme and computer modeling

In order to account for the observed product distribution we have constructed a reaction scheme containing 28 species and 30 elementary reactions. The scheme is shown in Table 1. The third and fourth columns give the two parameters A and E for the forward rate constants, k_f, in units of cm^3, mol, sec, and kcal/mol. Columns 5 and 6 give the standard entropy change, ΔS_r^0, and the standard enthalpy change, ΔH_r^0, at 1250 K. Rate parameters for reactions 1–

3 were based on the experimental results (see Figs. 2–4). The others were either estimated or taken from NIST Chemical Kinetics Data Base (Westly et al. 1994). The thermodynamic functions of the various species were either taken from various literature sources or estimated using the NIST Structure and Properties program (Stein et al. 1994). The rate constants of the back reactions were calculated from the forward rate constants and the equilibrium constants of the reactions.

Figures 5–6 show results of the computer simulation for four products which are formed by free-radical reactions. The agreement between the calculated and the experimantal results is very good.

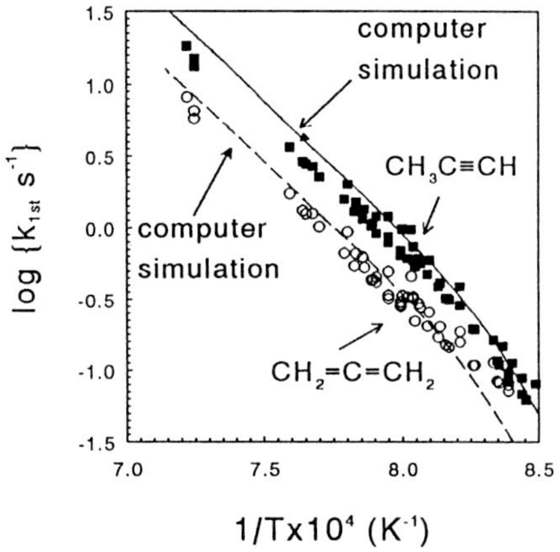

Figure 5. Production of $CH_3C{\equiv}CH$ and $CH_2{=}C{=}CH_2$.

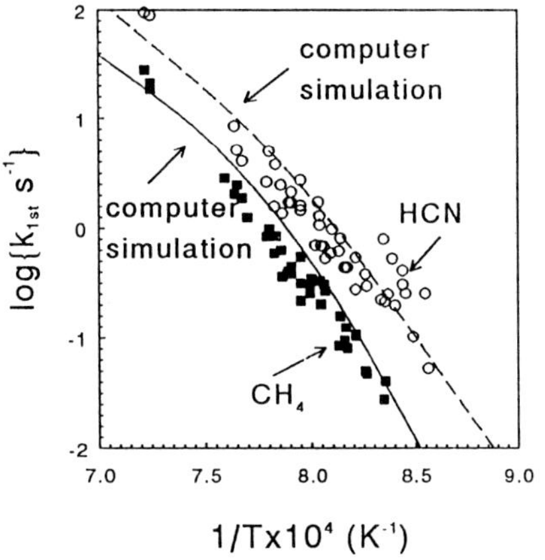

Figure 6. Production of methane and hydrogen cyanide.

Acknowledgement. This work was supported by a grant from the G.I.F., the German-Israeli Foundation for Scientific Research and Development. The authors thank Dr. P. Frank, who served as the cooperative investigator in this research, for his advice and encouragement.

References

Lifshitz A, Bidani M, Bidani S (1986) Pyrolysis of furan behind reflected shocks. J Phys Chem 90:5373–5377.

Stein SE, Rukkers JM, Brown RL (1993) NIST Standard Reference Data 25.

Westly F, Herron JT, Cvetanovic RJ, Hampson RF, Mallard WG (1994) NIST Chemical Kinetic Standard Reference Database (Version 5).

Comparison of relaxation models with shock-tube data

H. Stuhler, A. Frohn
Institut für Thermodynamik der Luft- und Raumfahrt
Universität Stuttgart
Pfaffenwaldring 31, 70550 Stuttgart, Germany

Abstract: The effects of thermal nonequilibrium on relaxation processes behind strong shock waves in O_2-N_2-Ar mixtures are studied. Rate constants are modeled with dependence on vibrational and electronic excitation. The mean internal energies of the reacting particles in all reactions are considered in the electronic and in the vibrational energy equations. Theoretical results of different approaches are compared with experimental data.

Key words: Thermal nonequilibrium, Chemical nonequilibrium, Shock-tube experiments, Relaxation models

1. Introduction

Chemical and physical aspects of relaxing air behind shock waves have been studied in the last few years by many authors. In the developed relaxation models the values of the reaction rates and the excitation rates of the internal degrees of freedom at high temperatures are not well known. Also, the interaction between vibrational and electronic excitation and chemical reaction rates is still the subject of research. Some models which are able to explain the influence of vibrational excitation on relaxation processes are the CVDV model (Marrone & Treanor 1963) and the T-T_{vib} model of Park (1990). Recently, new approaches for reaction rates as functions of the vibrational and electronic temperatures have been developed (Hansen 1993, Macheret et al. 1994). It is, in general, not possible to determine which of these theoretical approaches is most reliable. Only for the frequently used semi-empirical T-T_{vib} model has it been shown that good agreement with several experimental investigations exists (Park 1989). Because this model has some physical shortcomings, a physically more satisfying model has been developed (Stuhler & Frohn 1995). One of the most important improvements is that all rate constants are determined with dependence on vibrational and electronic excitation. The average internal energies of reactants and products are considered in the electronic energy equation and in the different vibrational energy equations. These equations are based on an 'almost Boltzmann' distribution within the internal degrees of freedom. The present relaxation models are based on several parameters or approaches which may be varied without offending basic physical relations. Therefore, the effects of these approaches on the relaxation processes are evaluated.

2. Results

Several calculations have been made and compared with experimental data. A typical temperature profile showing the principal behavior of the internal degrees of freedom is shown in Fig. 1. The translational temperature immediately behind the shock is given by the Rankine-Hugoniot relations. Behind the shock front the relaxation of internal degrees of freedom and chemical reactions takes place. The vibrational temperature of oxygen relaxes quickly to its maximum of about 8500 K. At this point the energies of the dissociated, highly excited O_2 molecules and the energy-exchange mechanisms compensate each other. The relaxation of the vibrational temperature of nitrogen is slower. Collisions of neutral particles are not very effective in exciting the electronic degrees of freedom. Futhermore, the few excited species react quickly and produce mainly nonexcited species in the dominating endothermic reactions. Therefore the electronic temperature is low at the beginning of the relaxation process. As soon as sufficient electrons are available the fast energy-exchange mechanisms with the vibrational modes of N_2 (Lee 1992) cause a rapid rise of the electronic excitation. This behavior might explain the experimental results of Sharma (1991). The first-produced nitric oxide molecules have an excitation temperature of about 12500 K. This temperature is mainly dependent on the mean energies of the reacting particles in exchange reactions (Stuhler & Frohn 1995).

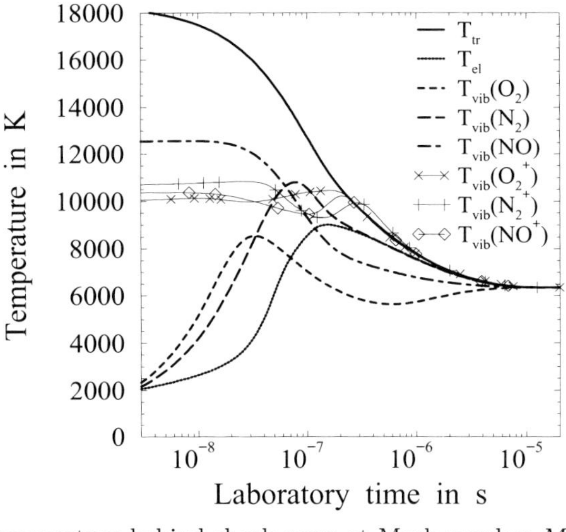

Figure 1. Temperature behind shock wave at Mach number $M = 18$. Initial conditions are $\psi_{O_2} = 0.21$, $\psi_{N_2} = 0.79$, $T_1 = 287$ K, and $p_1 = 100$ Pa.

Because the dissociation rates are dependent on the relaxation of the internal degrees of freedom, an incubation time Δt_p is observed. Experimental results for incubation times of oxygen diluted in Ar have been presented in the paper of Wray (1962), where the O_2 concentration was monitored by its uv-absorption at 1470 Å. Absorption coefficients were found to be almost independent of the vibrational temperature for $T_{vib} \geq 6000$ K. Evaluating the transmission of radiation at T_{tr} and at T_{vib} results in nearly the same dissociation rates. Incubation times as plotted in Fig. 2 were calculated using up to 16 electronic exited states of oxygen. Effects of vibrational nonequilibrium were modeled according to the CVD and CVDV approaches. The values calculated in this manner are within the range of the experimental scatter. Therefore it is not possible to determine which model is the best. It should be noted that for temperatures $T \leq 6500$ K, some measured incubation times were negative before a correction for boundary-layer effects and shock-front curvature was applied.

The dissociation of nitrogen is compared with the experiments in which uv-absorption at 1176 Å was used to detect the disappearance of N_2 (Appleton et al. 1968). Measurements were obtained for the temperature range from 8000 K to 15000 K. The agreement with theoretical results is quite good also for the experiments not shown. Experimental results were evaluated under the assumption that the nitrogen molecules are vibrationally relaxed before the slower dissociation begins. This is not in agreement with the present theoretical investigations which show that the dissociation of highly excited N_2 molecules

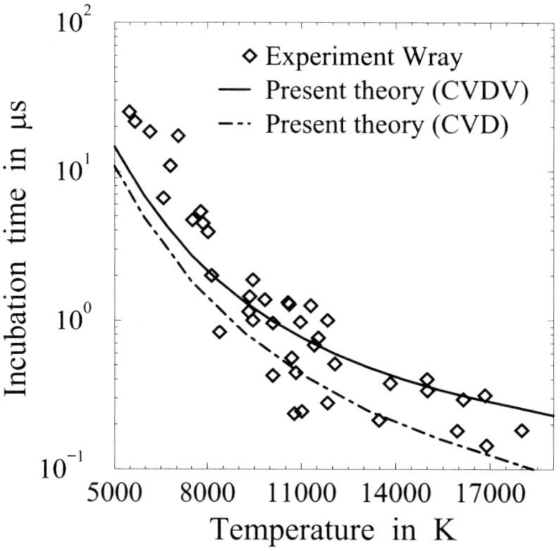

Figure 2. Incubation time Δt_p for oxygen dissociation reaction. Effects of vibrational excitation are modeled according to the CVD and CVDV models.

will lower the vibrational temperature about some thousand Kelvin. Values of the absorption coefficients, and therefore the measured rate constants, will change. Since Appleton used absorption coefficients obtained at the end of the initial relaxation, no corrections are made in the presented evaluation. Further studies of the dissociation rate of N_2 applying atomic resonance absorption spectroscopy (Roth & Thielen 1985) are compared with the present theoretical investigations. Good agreement was found for the temperature range between $T = 4500$ K and $T = 6500$ K. But the measured temperature exponent of $\alpha = -3.33$ is not in agreement with these calculations. The temperature exponents of dissociation reactions here computed are in the range from $\alpha = 0$ to $\alpha = -2.5$. The value $\alpha = 0$, which is obtained when the electronic energy of excited species is high $\mathcal{E}_{el,i} \geq \mathcal{D}_i - 1.5\mathcal{R}T$, becomes especially important for the dissociation of oxygen at high temperatures, because the O_2 molecules have low energy levels with small dissociation barriers. According to the available energy theory, the temperature exponent has the value $\alpha = -1.5$, but the corrections for thermal nonequilibrium also have to be taken into account. These corrections are defined with the coupling factor $V = \vec{k}(T, T_{el}, T_{vib})/\vec{k}(T)$. Because V is dependent on the temperatures, its value will vary during the relaxation process. When evaluating V at the local maximum of the vibrational temperature, an additional temperature dependence, which is proportional to $T^{-0.8}$, is obtained at 6000 K. For temperatures below 6000 K these maxima are difficult to determine in the considered Ar-N_2 mixture.

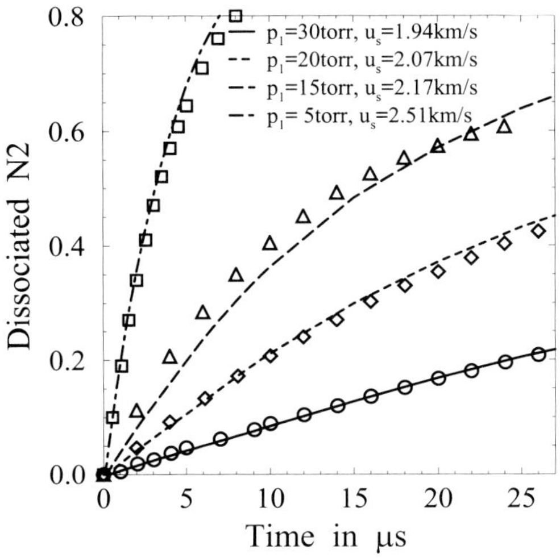

Figure 3. Dissociation of nitrogen diluted in argon. Experimental results are taken from Appleton et al. (1968).

Ionization relaxation times are used to study the theoretical predictions of the electron density in air. These characteristic times are defined as the ratio of the maximum electron concentration to its maximum slope. Experimental results have been reported for Mach numbers between $M = 7.2$ and $M = 22$ which were obtained with electrostatic probes or microwave-reflection measurements (Frohn & De Boer 1967, Schäfer & Frohn 1972, Lin & Teare 1963). Figure 4 shows that the computed results are in acceptable agreement with experimental data. To determine the sensitivity of the models, different parameters were used. For example, the variation of the characteristic temperature of the CVDV approach leads to a difference for the ionization relaxation time of about 50% at Mach number $M = 20$. For low Mach numbers the ionization relaxation times were not affected by this variation.

3. Summary

Immediately behind the shock front one has large deviations from thermal equilibrium. The calculated effective reaction rates range from negligible to small depending on the model used. The incubation times would provide a nice test case, if more precise experimental data were available. After the initial relaxation of internal degrees of freedom, the coupling factors are typically in the range between $V = 0.1$ and $V = 0.8$ and rise monotonically during the relaxation process. Since all considered models show this behavior, it seems to be simple to fit any model to one single experiment. The ionization-relaxation times are, for example, in good agreement with the calculations of Schäfer & Frohn (1972) using a CVD model and with the results of the T-T_{vib} model (Park

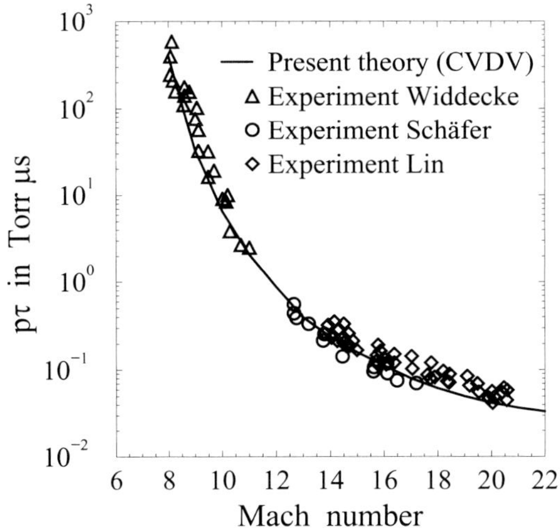

Figure 4. Ionization relaxation times in air for Mach numbers $22 > M > 8$.

1990) although completely different approaches and rate constants were used. But for a reliable prediction of the relaxation processes behind shock waves, it is necessary to obtain agreement with all available experiments.

Acknowledgement. Financial support of the Deutsche Forschungsgemeinschaft is gratefully acknowledged.

References

Appelton JP, Steinberg M, Liquornik DJ (1968) Shock-tube study of nitrogen dissociation using vacuum-ultraviolet light absorption. J Chem Phys 48(2):599–608.

Hansen CF (1993) Vibrational nonequilibrium effects on diatomic dissociation rates. AIAA J 31(11):2047–2051.

Frohn A, De Boer PCT (1967) Ion density profiles behind shock waves in air. AIAA J 5(2):261–264.

Lin S, Teare J (1963) Rate of ionization behind shock waves in air. Phys Fluids 6(3):355–375.

Lee JH (1992) Electron-Impact Vibrational Relaxation in High Temperature Nitrogen, AIAA 92-0807.

Levine RD, Bernstein RB (1991) Molekulare Reaktionsdynamik, Teubner.

Marrone PV, Treanor CE (1963) Chemical relaxation with preferential dissociation from excited vibrational levels. Phys Fluids 6(9):1215–1221.

Macheret SO, Fridman AA, Adamovich IV, Rich JW, Treanor CE (1994) Mechanisms of Nonequilibrium Dissociation of Diatomic Molecules, AIAA 94-1984.

Park C (1988) Two-Temperature Interpretation of Dissociation Data for N_2 and O_2, AIAA 88-0458.

Park C (1990) Nonequilibrum Hypersonic Aerothermodynamics, John Wiley.

Roth P, Thielen E (1985) Measurements of N-atom concentrations in dissociation of N_2 by shock waves. In: Proc 15th ISSW, pp 245–252.

Sharma SP (1991) Vibrational and rotational temperature measurements in a shock tube. In: Proc 18th ISSW, pp 683–690.

Schäfer JH, Frohn A (1972) Ionization behind shock waves in nitrogen-oxygen mixtures. AIAA J 10(8):985–987.

Stuhler H, Frohn A (1995) Rate Constants and Internal Energies of Reactants and Products in Thermal Nonequilibrium, AIAA 95-2058.

Wray KL (1962) Shock-tube study of the coupling of the O_2-Ar rates of dissociation and vibrational relaxation. J Chem Phys 37(6):1254–1263.

Widdecke N, Klenk W, Frohn A (1995) Experimental techniques for the investigation of relaxation processes behind strong shock waves. ZFW 3:492–497.

Various models of coupling of vibrational-dissociational kinetics behind shock waves

E. A. Nagnibeda
St. Petersburg University
198904 St. Petersburg, Russia

J. G. Méolans, M. Mouti, A. H. Chauvin
Université de Provence
13397 Marseille, Cedex 20, France

Abstract: The kinetics of vibration-dissociation coupling behind shock waves in diatomic nitrogen is considered on the basis of nonlinear differential equations for level populations and conservation equations. The populations of each vibrational level are presented here. Molecular energy spectrum is simulated as an anharmonic Morse oscillator. The analysis of the relaxation zone behind the shock wave is given using different models of vibration-dissociation coupling and with taking into account vibrational-vibrational (VV) and translational-vibrational (TV and VT) energy exchanges.

Key words: Vibrational populations, Vibration-dissociation coupling

1. Introduction

A complete treatment of nonequilibrium kinetics remains a very hard task and phenomenological treatments may contribute to the knowledge and understanding of these phenomena. In particular the values of the vibrational populations and those of the dissociation rate constants of diatomic gases may considerably depend on the degree of vibrational nonequilibrium. Two methods of treating such relaxing and dissociating media behind a plane shock are presented here. Both of them use the anharmonicity of vibrations (Morse oscillator). The first model is mainly applied to equilibrium free streams where Mach numbers are high ($M = 10$ to $M = 16$) while the second model is applied to free streams where a strong nonequilibrium exists. To describe the bound-bound transitions the first model uses a vibrational cross section taking into account the influence of the intermolecular potential angle dependence; then for the bound-free transitions we use a calculation based on the Marrone-Treanor model (Marrone & Treanor 1963): the effects of all the levels are taken into account. According to the second model, dissociation is permitted only from the upper level and 50 excitation levels in molecular nitrogen are considered. On the basis of the numerical solution of the nonlinear equations for level populations together with conservation equations, level populations in molecular nitrogen are presented in the relaxation zone behind a shock wave.

2. Governing equations and models of coupling

Equations for level populations n_{v_1} of energy level v_1 have the form:

$$\frac{\partial n_{v1}}{\partial t} + \frac{\partial n_{v1}\, u}{\partial x} = \frac{Z(N_2-N_2)}{n}\left[\sum_{v_2,v_1',v_2'} -n_{v_1} n_{v_2} P_{v1\to v1'}^{v2\to v_2'} + n_{v_1'} n_{v_2'} P_{v_1'\to v1}^{v_2'\to v_2'} \right.$$
$$\left. + \sum_{v_2,v_2'} -n_{v_1} n_{v_2} P_{v1\to d}^{v2\to v_2'} + n_N^2\, n_{v_2'} P_{d\to v1}^{v_2'\to v2}\right] \tag{1}$$
$$+Z(N_2-N)\left\{\sum_{v_1'} -n_{v_1} P_{v1\to v_1'} + n_{v_1'} P_{v_1'\to v1} + [-n_{v_1} P_{v1\to d} + n_N^2 P_{d\to v1}]\right\}$$

where u is the macroscopic velocity, $Z(N_2-N_2)$ is the number of collisions per molecule per second, $P_{v1\to v_1'}^{v2\to v_2'}$ is the transition probability of a molecule from the v_1 to the v_1' vibrational level and from the v_2 to the v_2' level, $Z(N_2-N)\ P_{v_1\to v_1'}$ is the inelastic frequency of molecule-atom collisions, n_N is the number density of atoms N, and n is the total molecular number density. $P_{v1\to d}^{v2\to v_2'}$ and $P_{v_1\to d}$ represent the dissociation probabilities respectively for molecule-molecule and molecule-atom collisions: in fact according to the model given below, these probabilities do not depend on the vibrational level of the partner. These equations are also supplemented with the equation governing the atomic population evolution. In the relaxation equations the various inelastic frequencies ZP derive from inelastic vibrational cross sections by the usual relationships. For the first model the calculation of the cross section used here has been previously given (Méolans & Chauvin 1991). Following Marrone & Treanor (1963), preferential dissociation from the high vibrational states is assumed. The dissociation process is governed for each level as follows:

$$\left\{\frac{\partial [N_2]_{v_1}}{\partial t} + \frac{\partial [N_2]_{v_1} u}{\partial x}\right\}_{Diss} = \sum_{M=N_2,N} -k_{fv_1}(N_2-M)[N_2]_{v_1}[M] + \text{backward terms} \tag{2}$$

where $[N_2]$ and $[M]$ are respectively the N_2 and the M molar concentrations. The forward rate constant $k_{f_{v_1}}(N_2-M)$ is related to $P_{v_1\to d}$ by comparing Eq. (1) to Eq. (2). Using the Marrone-Treanor model under the assumption of preferential dissociation from high levels and using the equations governing the kinetics, one obtains:

$$k_{fv_1}(N_2-M) = k_{feq}(N_2-M)\,(Q(T)/Q(-U))\ \exp\ (\varepsilon_{v_1}[1/kU + 1/kT]) \tag{3}$$

where Q is the partition function, U is an additional parameter of the model, and ε_{v_1} is the vibrational energy of the v_1^{th} level. The result does not require the harmonic oscillator assumption. The backward probability $P_{d\to v_1}$ calculation is deduced by applying the detailed balancing principle. Furthermore we have used empirical values of U obtained in the following way: from Eq. (3), we calculate the expression of the mean dissociation rate k_f which according to Marrone & Treanor (1963) depends on U. Then this expression is fitted to the semi-empirical k_f expressions (Park 1993). The model was simplified in the case when only single quantum energy transfers take place, and in the second model dissociation occurs only from the upper vibrational level and recombination may be neglected. Moreover it is suggested that all molecules reaching the rth level dissociate immediately and that the dissociated molecule number is defined as a molecule stream to the rth level through VV and TV transitions. Equations taking into account these simplifications are given by Nagnibeda (1995). The transition probabilities are taken from Billing's calculations (Billing & Fisher 1979) and from the SSH theory modified for anharmonic oscillators (Capitelli 1986). We consider both equilibrium and nonequilibrium Boltzmann vibrational distribution in free stream with vibrational temperature T_V: $T_V = T$ for equilibrium free stream and $T_V > T$ for the nonequilibrium case. Nonequilibrium vibrational distribution before the shock front may take place for example either after the expansion of a gas in a nozzle or may be provided by some kind of vibrational excitation. In previous works (Kovalev & Nagnibeda 1991, Nagnibeda 1995) it is shown that the relaxation zone and anharmonic effects after the shock front depend strongly on conditions in the free stream and have different characters in the cases of strong and weak nonequilibrium in the free stream.

3. Results and discussion

In the case of a shock-heated gas the shock wave is propagating essentially in a free stream characterized by a vibrational population equilibrium ($T_{V_0} = T_0$). The results obtained using the first model are given for various conditions at the distance $x = 7.05 \times 10^{-1}$ mm from the shock front (Fig. 1). Up to the eighteenth level (which is the upper population calculated) the populations are monotonously decreasing. The curves corresponding to $M = 10$ and $M = 12$ are calculated for the same conditions of pressure and temperature in the free stream ($P_0 = 390$ Pa, $T_0 = 271$ K), and also for the curve corresponding to $(M = 10)_N$, but in this case the shock wave propagates in a nonequilibrium free stream and the initial vibrational temperature is $T_{V_0} = 620$ K while the corresponding temperature is $T_0 = 271$ K. We have only a very small difference between the case $M = 10$ and the case $(M = 10)_N$. Moreover the results corresponding to Mach numbers $M = 14$ and $M = 16$ are obtained with equilibrium conditions in the free stream and with $T_0 = 200$ K and $P_0 = 300$ Pa. Generally speaking we can observe that the populations increase when the Mach number increases. For the first eighteen levels the population profile decreases when v increases. Moreover the results obtained with an equilibrium free stream are close to those obtained with a free stream in moderate nonequilibrium conditions: the discrepancy existing immediatly behind the shock vanishes quickly along the relaxation zone (Fig. 2). Figures 3 and 4 correspond to strong nonequilibrium conditions in the free stream ($T_{V_0} > T_0$). This condition becomes of important either in the case of vibrational energy pumping or after expansion of the flow. Fifty excitation levels in molecular nitrogen are considered. In Fig. 3 level populations for molecular nitrogen are presented at the distance $x = 4 \times 10^{-2}$ mm from the shock front. The curves correspond to the following conditions: $M = 5$, $T_0 = 5500$ K, $T_0 = 500$ K, $P_0 = 1600$ Pa; and $M = 3, 6$, $T_{V_0} = 4500$ K, $T_0 = 800$ K. Figure 4 depicts level populations at the distance $x = 7 \times 10^{-1}$ mm from the shock front for the same conditions in the free stream as in Fig. 3. One can see the increase of level populations with increasing Mach numbers, as was shown above. At the same time there exists noticeable increase of level populations at the middle levels in the relaxation zone, which takes place because of the very intensive stream of vibrational quanta to the middle levels in the case $T_{V_0} > T_0$. Comparison of these results with those shown in Figs. 1 and 2 shows the increasing of level populations after the shock front in the case of strong vibrational nonequilibrium in the free stream and also the increase of the length of the relaxation zone.

4. Conclusions

The results show the influence of the conditions in the free stream on level populations after the shock. Three cases are considered: (1) Equilibrium free stream ($T_0 = T_{V_0}$). If we denote $T^{(1)}$ and $T_V^{(1)}$ gas temperature and vibrational temperature immediately after the shock we may write in this case $T^{(1)} = T_{V_0} < T^{(1)}$. As usual TV vibrational excitation takes place after the shock front and after a short excitation zone the molecular populations decrease. (2) Nonequilibrium free stream $T_0 < T_{V_0}$. The structure of relaxation zone in this case may be different in dependence on translational and vibrational temperatures $T^{(1)}$ and $T_V^{(1)}$ just after the shock front. In the case $T_0 < T_{V_0} = T_V^{(1)} < T^{(1)}$ the character of relaxation is similar to the first case: actually vibrational relaxation and dissociation also go through translational-vibrational energy transfers with $T_V(x) < T(x)$. (3) Nonequilibrium free stream with strong vibrational excitation and not very intensive shock wave $T_0 < T^{(1)} < T_{V_0} = T_V^{(1)}$. The numerical results show a quite different character of vibrational relaxation in this case. The explanation is that vibrational energy supply provides intensive VV exchanges and a current of vibrational quanta to the middle and upper levels and after that VT exchanges with $T_V(x) > T(x)$ take place. Strong VV exchanges may appreciably enhance the dissociation rate in this case.

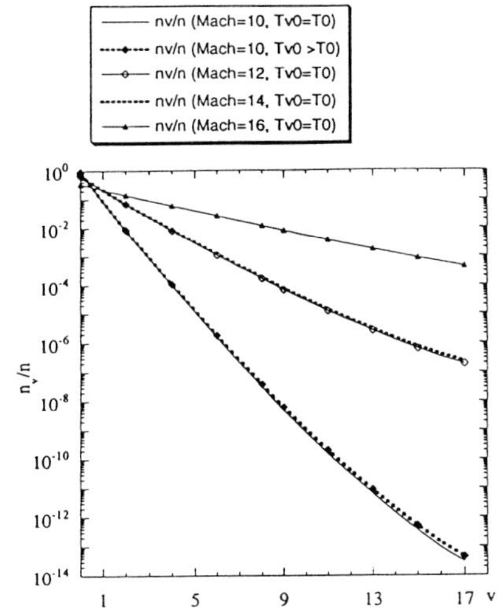

Figure 1. Levels populations $n_v|n$.

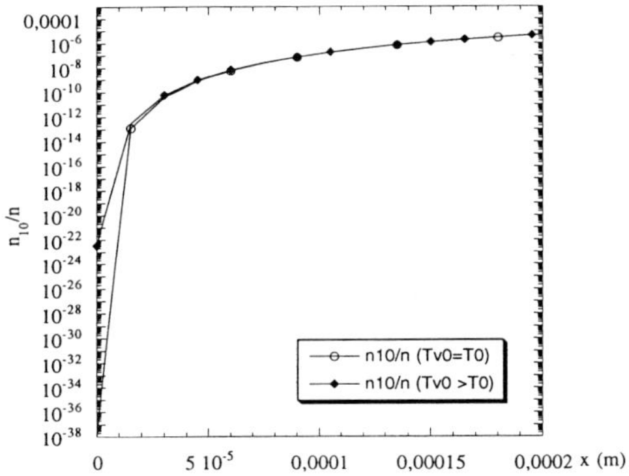

Figure 2. Comparison of level populations at $x = 7 \times 10^{-1}$ mm from the shock for equilibrium and weak nonequilibrium free stream.

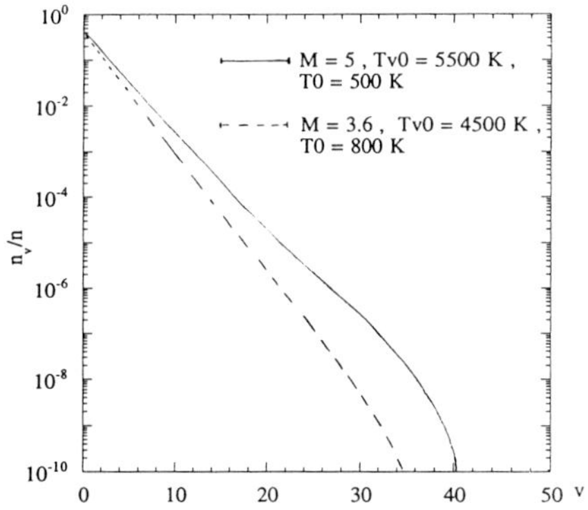

Figure 3. Level populations $n_v|n$ versus v at $x = 4 \times 10^{-2}$ mm.

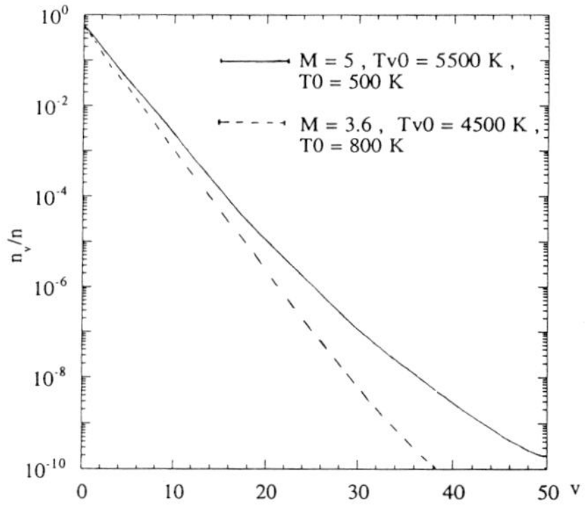

Figure 4. Level populations $n_v|n$ versus v at $x = 7 \times 10^{-1}$ mm.

References

Billing GD, Fisher ER (1979) VV and VT rate coefficients in N_2 by a quantum-classical model. J Chem Phys 43:395.

Capitelli M (ed) (1986) Nonequilibrium Vibrational Kinetics, Springer, Berlin, p 64.

Kovalev II, Nagnibeda EA (1991) Numerical investigation of the dissociation rate in vibrational nonequilibrium gas. In: Proc of the 10th Conference on RGD, Moscow (in Russian).

Marrone PV, Treanor CE (1963) Chemical relaxation with preferential dissociation from excited vibrational level. Phys Fluids 6:1215–1221.

Méolans JC, Chauvin AH (1991) Scattering cross sections for homonuclear diatomic molecule collisions: influence of the angular potential dependence. AIAA–91–1340, 26th Thermophysics Conf, Honolulu, Hawaii.

Nagnibeda EA (1995) The structure of the relaxation zone behind shock waves in the reacting gas flows. In: Proc of the Second European Symp on Aerothermodynamics for Space Vehicles, ESTEC, Noordwijk, pp 299–303.

Park C (1993) Review of chemical-kinetic problems of future NASA mission I: Earth entries. J of Therm and Heat Transf 7:385–398.

Combustion

Optical diagnostic measurements in a supersonic-combustion ramjet

T. J. McIntyre*, A. F. P. Houwing, P. A. B. Rabbath, P. C. Palma
Aerophysics and Laser-based Diagnostics Research Laboratories
Department of Physics, Faculty of Science
Australian National University, Canberra, 0200 Australia

Abstract: Experimental measurements are presented of combustion in a supersonic flow environment. Streamwise injection of hydrogen into a heated supersonic air stream has been studied in a free-piston driven shock tunnel, using surface-pressure measurements, shadowgraphy, and planar laser-induced fluorescence.

Key words: Supersonic combustion, Shock tunnel, Flow imaging, Planar laser-induced fluorescence

1. Introduction

The development of a supersonic-combustion ramjet (SCRAMJET) is foreseen for the next generation of hypersonic vehicles. Current research has been focusing on examining ways to increase fuel-air mixing and combustion efficiency (see Paull et al. 1995). Pressure measurements have yielded information about the flow field in such experiments but provide little information on the extent of mixing which occurs in the flow. Optical techniques, however, can provide non-intrusive two-dimensional qualitative and quantitative measurements. Among the simplest are schlieren and shadowgraph techniques while one of the more advanced techniques currently used for combustion studies is planar laser-induced fluorescence (PLIF) as demonstrated by Palmer et al. (1995) and McMillin et al. (1995). This technique can provide two-dimensional detection of the OH molecule which is naturally present in combustion regions.

2. Experimental arrangement

2.1. SCRAMJET model

The model used for these experiments was a rectangular, constant-cross-section duct with central injection as shown in Fig. 1. The duct has a 25×50 mm cross-section and is about 400 mm in length. The injector is 5-mm thick, the upstream section of which forms the splitter plate that extends far enough into

*Present address: Dept. of Physics, University of Queensland, Brisbane, 4072 Australia.

the flow so that shock waves generated at the tip pass outside the duct. Optical access is obtained via windows mounted in the side and top plates of the duct and test section. Because of the design of the model, images could only be recorded starting about 30 mm downstream from the injector and extending for a further 180 mm. Kulite pressure transducers were distributed along the floor of the duct also beginning some 30 mm downstream from the injector as shown in Fig. 1. Free-stream conditions were generated using the free-piston

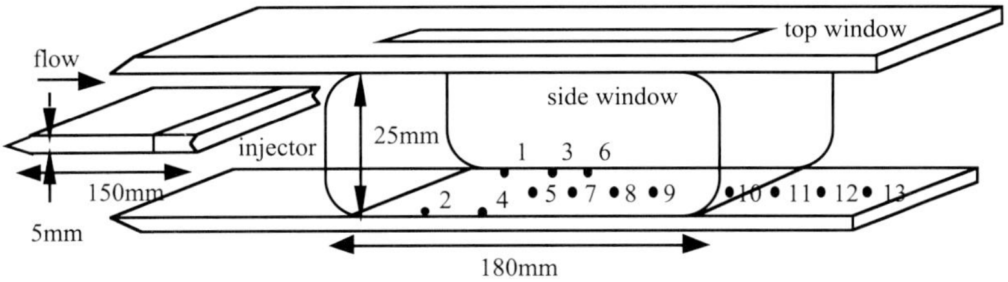

Figure 1. Schematic of the SCRAMJET model.

driven shock tunnel T3 at the Australian National University. Operation of this tunnel provides a range of conditions with a constant-pressure test time of approximately 1–2 ms. The flow is terminated by the arrival of the driver gas which gradually mixes into the test gas. Injection of the hydrogen into the flow was achieved by the use of a Ludwieg tube and a fast-acting valve. The valve was triggered by detection of the recoil of the tunnel and set so that injection began several milliseconds before the arrival of the test gas and lasted well beyond the test time. The design of the injector is such that the fuel enters the flow at sonic conditions. The operating conditions for the tunnel and SCRAMJET are given in Table 1. The initial shock-tube pressure, shock velocity, and nozzle reservoir pressure are measured, from which the nozzle reservoir temperature and nozzle exit conditions are calculated using equilibrium and non-equilibrium codes. For comparison, the conditions at the inlet of the duct are measured by a PCB pressure transducer mounted in the roof of the duct just upstream of the injector. The pressure and duration of the hydrogen flow are measured by a PCB pressure transducer mounted just prior to the injector.

Table 1. Operating conditions.

	Nozzle Reservoir	Free-stream	Injector
Pressure	15 MPa	110 kPa	687 kPa
Temperature	3200 K	1100 K	250 K
Velocity	–	2390 m/s	1215 m/s
	Enthalpy 4.0 MJ/kg	Mach Number 3.7	Equiv. ratio 0.92

2.2. Shadowgraph system

A shadowgraph system described by McIntyre et al. (1995) was employed to provide qualitative flow visualization to identify regions for closer examination by other techniques. It has the advantage of being relatively easy to implement and provides a wide field of view. Triggering was achieved using the nozzle reservoir PCB transducer and employing a variable delay unit to allow observations at different stages of the test time. The pulse length of the laser is ~500 ns which is sufficiently short to produce an instantaneous image of the flow.

2.3. Planar laser-induced fluorescence arrangement

The PLIF measurements presented here focus on detection of the OH molecule indicating regions in the duct where combustion is occurring. The experimental arrangement is shown in Fig. 2. The light source for the system was a Nd:YAG pumped dye laser using Rhodamine 6G dye. The output was frequency doubled using a WEX system providing up to several millijoules of radiation at about 283 nm. The output is tunable, allowing selection of various rovibrational transitions in the (1,0) band of the $A^2\Sigma \leftarrow X^2\Pi$ electronic transition. The light was formed into a 25-mm wide sheet and passed into the test section through fused-silica windows mounted in the roof of the test section and SCRAMJET. The fluorescence from this sheet was captured by an intensified CCD (ICCD) camera through Schott Glass UG5 and WG305 filters, as well as a spectrally selective mirror. These filters and mirror discriminate against flow luminosity and laser scattering but allow passage of most of the fluorescence in the (1,1) band and the collisionally populated (0,0) band of the OH $A \leftarrow X$ transition. Selection of the excitation transition and laser power were such to give optimal signal to noise for flow visualization purposes. Triggering was again from the nozzle reservoir transducer and a camera gate of some 400 ns was used to discriminate against flow luminosity. The 8.5 ns pulsewidth of the pump laser ensured that good temporal resolution was achieved.

3. Results

3.1. Pressure distributions

The existence of combustion in the flow was first investigated using the Kulite pressure transducers mounted along the floor of the duct. To distinguish between effects resulting from combustion and those caused by other flow features, experiments were performed with: an air test gas and hydrogen injection; a nitrogen test gas with hydrogen injection; and lastly an air test gas with no injection. Little difference was observed between the last two cases, so results are only presented for the nitrogen test gas or 'non-combusting' case and the air test gas with injection, or 'combusting' case. All measurements presented

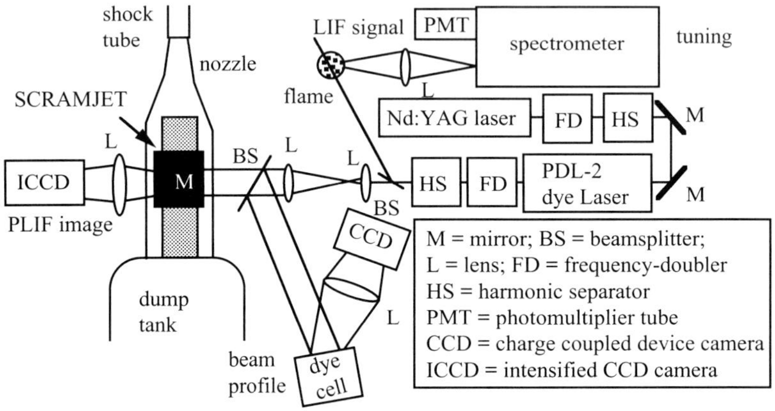

Figure 2. Experimental arrangement of the PLIF system.

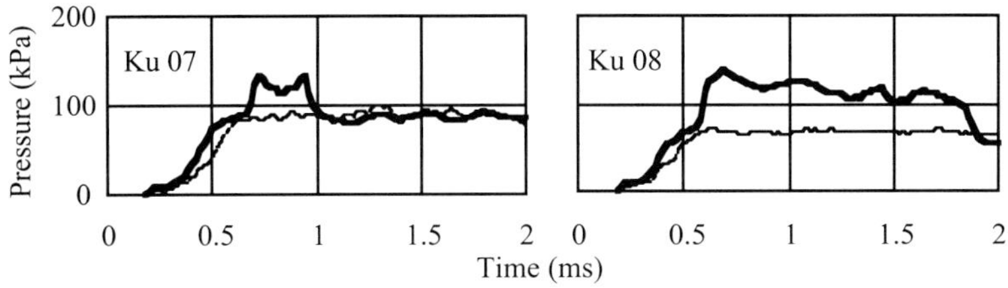

Figure 3. Pressure traces along the floor of the duct. Heavy lines: air case; thin lines: nitrogen case. Approximate positions are shown in Fig. 1.

here are for a nozzle reservoir enthalpy of 4 MJ/kg. Figure 3 presents pressure traces at two locations along the floor of the duct for the both test-gas conditions. Here one sees a significant difference in pressure histories. Sudden rises are observed on both transducers, which appear to provide evidence of shock waves moving along the duct as a result of the combustion.

3.2. Shadowgraph imaging

Shadowgraph images of the flow are shown in Fig. 4. In each case, flow is from left to right with injection some 30 mm upstream of the image. The hydrogen jet can be observed entering the left of the picture as well as oblique shock waves probably generated at the entrance to the duct where the boundary layers develop. For the nitrogen test-gas case, the hydrogen jet is discernible for only about half the length of the duct before apparently mixing with the test gas. In contrast, the air test-gas case shows what appears to be a clearly

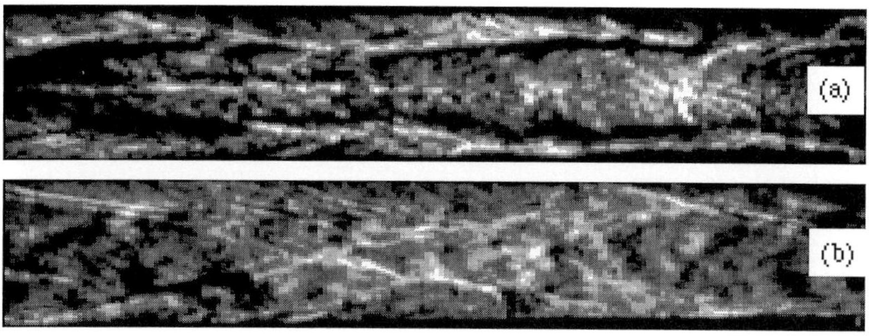

Figure 4. Shadowgraph images. Flow is from left to right: (a) air, (b) nitrogen. Each image is 25 mm high corresponding to the duct height.

defined hydrogen jet that penetrates along the length of the duct. A mixing layer develops between the two streams, evident as a dark region presumably where combustion is occurring. The oblique shocks are observed to both reflect off the shear layer and also to pass through into the hydrogen where a near normal shock is observed.

3.3. Planar laser-induced fluorescence

Evidence to show that combustion was indeed only occurring in the mixing layer was sought using PLIF detection of OH. Initial PLIF measurements were unsuccessful due to the high luminosity of the flow (see McIntyre et al. (1995)). To overcome these problems, further filtering in the form of a spectrally selective mirror was incorporated in the detection optics. This mirror reflects above 80% of the incident light in the wavelength range 310–350 nm while transmitting all other wavelengths. With this arrangement, the image shown in Fig. 5 was obtained. Clearly visible in the image is the presence of a laser-induced fluorescence signal corresponding to the location of the darker regions observed in the shadowgraph images. This result confirms that the dark region observed in the shadowgraph images is a mixing layer and it is only here that combustion is occurring. Confirmation that the fluorescence was laser-induced was achieved by blocking part of the laser sheet in some of the experimental runs.

4. Conclusions and further work

The existence of supersonic combustion in a two-dimensional model tested in a free-piston driven facility has been demonstrated. Internal surface measurements showed significant pressure rises for the case in which hydrogen was injected into an air test gas over that observed for a nitrogen test gas. Shadowgraph images showed that only a relatively narrow mixing layer existed between the co-flowing air and hydrogen streams. PLIF measurements confirmed that

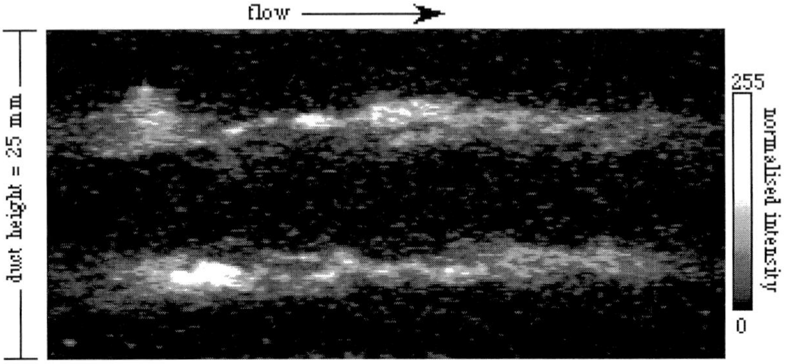

Figure 5. Planar laser-induced image for air test gas. Flow is from left to right. Excitation is through the $(1,0)R_1(10)$ transition. Image is compressed to 50% in the vertical direction because of frame interlacing.

this layer was the only region within which combustion was occurring, and highlights the importance of enhancing fuel-air mixing at supersonic conditions to improve SCRAMJET performance.

Acknowledgement. The authors would like to thank Paul Walsh for his technical assistance with the tunnel and SCRAMJET. Inputs from Sudhir Gai, Neil Mudford, and Sam Mallinson from the Australian Defence Force Academy, and from Allan Paull, Richard Morgan, Maria Pulsonetti, and Ray Stalker from the University of Queensland are gratefully acknowledged. This research has been financially supported by the Australian Research Council and NASA Grant NAGW 1467.

References

McIntyre TJ, Rabbath PAB, Houwing AFP (1995) Imaging of combustion processes in a supersonic combustion ramjet. In: Proceedings XII International Symposium on Air-Breathing Engines (ISABE), Melboune, Australia.

McMillin BK, Seitzman JM, Palmer JL, Hanson RK (1995) Dual-laser PLIF imaging techniques for shock tube studies of mixing and combustion. In: Shock Waves @ Marseille, Springer Verlag, pp 321–326.

Palmer JL, Hanson RK (1995) Single-Shot OH PLIF Thermometry in a Reacting, Supersonic Free Jet, AIAA-95-0517.

Paull A, Stalker RJ, Mee DJ (1995) SCRAMJET thrust measurement in a shock tunnel. Aero J 99:161–163.

Numerical investigation of induction-zone development in hydrogen-oxygen mixtures

E. S. Figura
Fluid Gravity, Chiltlee Manor, Liphook, Hampshire, GU30 7AZ, UK
I. G. Cullis
Warheads Department, DRA Fort Halstead, Sevenoaks, Kent, TN14 7BP, UK

Abstract: Experimental studies have been performed to investigate induction times in detonable gaseous mixtures. In this paper, we investigate the development of the induction zone in a hydrogen-oxygen mixture, using a hydro-kinetic computational code.

Key words: Induction zone, Hydrodynamic code, Kinetics

1. Introduction

The propagation and transition to detonation of reactive shock waves is a highly complex phenomenon involving chemical and gas-dynamical processes with widely differing time scales. Any attempt to model such systems numerically requires large computing resources in order to resolve the chemical kinetics and the interaction with the gas dynamics. This is a problem even for the simplest systems (e.g. hydrogen-oxygen) and quickly becomes very severe for more complicated mixtures. One approach to alleviate the problem is to replace the full chemical kinetics model with a much simpler, two-stage model which can reproduce all essential features of the flow. However, such *induction parameter models* require accurate knowledge of the induction time and energy-release rates for the mixture and such information, if available, is limited and often has high uncertainty associated with it.

In this paper, we investigate the development of the induction zone in hydrogen-oxygen mixtures. Our simulations support experimental work at the University College of Wales, Aberystwyth, which is attempting to resolve this process. The experiments involve sending a shock through diluted hydrogen-oxygen mixtures which contain an obstacle. As the shock passes over the obstacle, a wave is reflected off the front face. Reactions will start to develop behind this wave and the delay can be measured. Once ignited, the reaction front quickly overtakes the reflected shock. At the leading edge of the obstacle, the mixture will be highly compressed and starts reacting immediately, leading to a rather complicated wave structure.

2. Experimental details

The experimental apparatus consisted of a shock tube with a windowed test section at one end in which the obstacle (a block of dural, height 3.8 cm, width 2.5 cm) was placed (see Fig. 1).

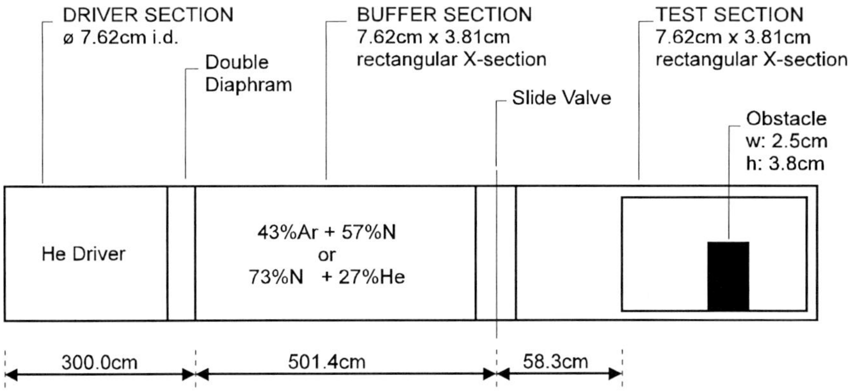

Figure 1. Schematic of the experimental apparatus.

The shock tube had a helium driver section coupled to an acoustically matched buffer section. The length of the test section (~30 cm) allowed the obstacle to be placed at such a distance that the contact surface between the buffer and test gases did not interact with the wave system resulting from the shock-obstacle interaction during the period of observation. A spark-schlieren camera system was used to capture the shock with an interframe spacing of either 5±1 μs or 10±1 μs. Pressure gauges (not shown in Fig. 1) placed along the shock tube monitored the passage of the shock. In addition, a photomultiplier tube recorded (via an optical fiber) the emission from around the back face of the obstacle — thus giving a measure of the induction time. The gas in the test section was at an initial pressure of 76 torr. In the experiments, two mixtures were studied: $2H_2/O_2/7N_2$ and $2H_2/O_2/7Ar$. This provided data on the effect of different diluents on the development of the induction zone.

3. Numerical calculations

The numerical calculations were performed using GRIM, a 2D multi-material hydrodynamic code which includes a finite-rate, non-equilibrium chemistry model. The hydrocode solves the Euler equations for conservation of mass, momentum, and energy. These equations are solved using a two-stage process. The initial phase is Lagrangian during which the grid conceptually distorts with the fluid. The velocities determined by this step are used in a high-order accurate advection stage to return the mesh back to the initial Eulerian cell boundaries.

The chemistry model is introduced as the first phase of the computation, prior to the Lagrangian phase. We solve the species conservation equations by calculating the chemical production rate for each species. The computation of the evolution of temperature and species mass fractions is performed under the assumption that each cell is a constant-pressure reactor. The computed energy release is fed into the flow field as a source term in the energy equation during the Lagrangian phase.

3.1. Geometry and initial conditions

A computational mesh of 300 × 76 cells was used (30.0 cm × 7.6 cm), the geometry being 2D Cartesian. The rectangular obstacle was located at a distance of 13.9 cm. Since the buffer test-gas contact surface did not interact with the shock-wave system, we did not need to model the buffer section. The computing time was therefore minimized by starting the calculation at time = 0 with the shock front in the test gas being located at the back edge of the obstacle. The necessary post-shock jump conditions were calculated using a shock-tube code (Mitchell and Kee 1992) which simply calculates the shock jump conditions in the presence of chemical energy release. The input to this code included shock velocity, which was deduced from the pressure-gauge records. For the present gas mixtures, the post-shock conditions are shown in Table 1.

Table 1. Post-shock conditions for each gas mixture.

	$2H_2/O_2/7N_2$	$2H_2/O_2/7Ar$
v_{shock} (cm/s)	1.169×10^5	8.540×10^4
ρ (g/cm^3)	3.8581×10^{-4}	3.8254×10^{-4}
ϵ (erg/g)	7.2498×10^9	3.4668×10^9
u (cm/s)	8.7645×10^4	5.6088×10^4
γ	1.36276	1.53806
p (dyn/cm^2)	1.09056×10^6	7.30239×10^5
T (K)	789.20	724.79
M	3.05	2.46
$Y[H_2]$	0.017369	0.012773
$Y[O_2]$	0.137852	0.101369
$Y[N_2$ (Ar)]	0.844779	0.885858

$Y[...]$ = mass fraction

Inflow boundary conditions were implemented at the left-hand edge of the mesh in order to correctly simulate the incident shock. The chemical reactions used for the computations can be found in Figura et al. (1992).

3.2. Case 1: $2H_2/O_2/7N_2$

Schlieren photographs from the experiment for this mixture are shown in Fig. 2.

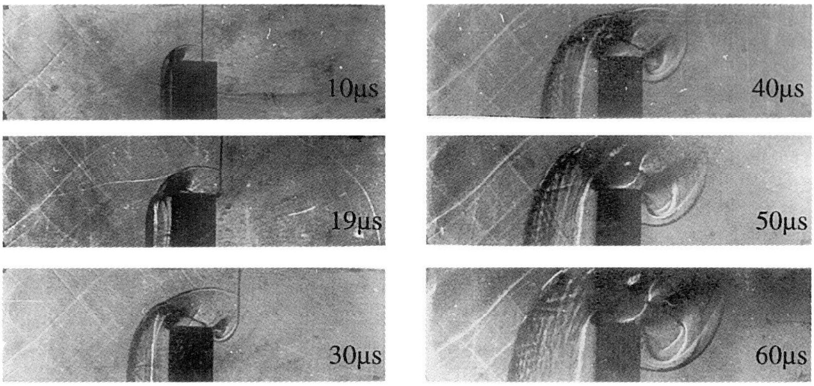

Figure 2. Schlieren photographs showing time history of shock system.

The incident shock is traveling from left to right and is not sufficiently strong to produce a detonation. In this particular case, time = 0 coincides with the shock having already traveled ~5 mm along the top of the obstacle. The second frame (10 μs) shows that the mixture has ignited at the obstacle face and has driven a wave which has almost reached the reflected shock front. By the third frame (19 μs), the reaction wave has overtaken the reflected shock; meanwhile, the incident shock is starting to diffract around the front obstacle corner. There are also signs of a vortex forming around the top left-hand corner of the obstacle. In the next two frames, expansion is cooling the reactive wave as it turns the obstacle front corner. The wave structure proceeds towards the upper wall and at 50 μs it hits the upper wall. Further structure is seen around the diffracting incident wave.

Density contours from the numerical simulation are shown in Fig. 3. In this case, time = 0 corresponds to the incident shock having reached the obstacle. At 10 μs, we see that ignition has started at the front face of the obstacle and is driving a wave outwards. This wave reaches the reflected shock at 20 μs. The appearance of the vortex forming around the top left-hand corner of the obstacle can also be seen. In the next two frames, expansion is cooling the reactive wave as it turns the obstacle front corner. At 50 μs the reactive wave structure hits the upper wall. This leads to a Mach wave configuration propagating towards the left of the test section, which is to be clearly seen at later times (60 – 100 μs). At the same time, another Mach stem is forming at the top wall, closer to the incident shock front. This eventually catches the incident shock and finally overtakes it (60 – 70 μs). Ultimately, this wave system reaches the bottom wall and forms another Mach-type reflection (100 μs).

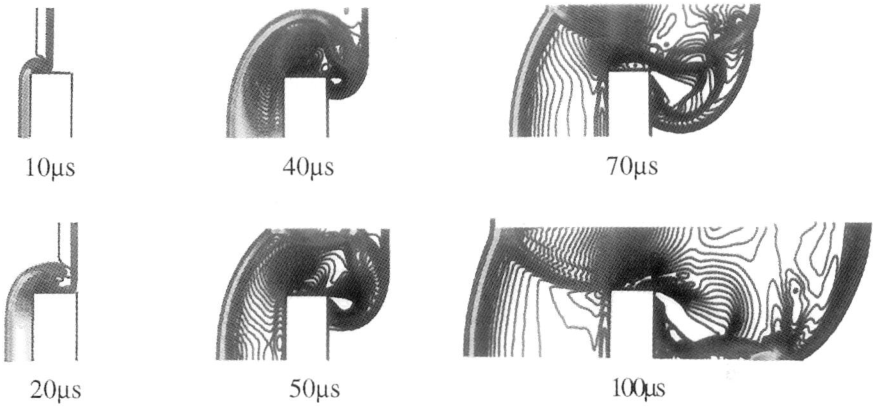

Figure 3. Density contours.

3.3. Case 2: $2H_2/O_2/7Ar$

Schlieren photographs from the experiment are shown in Fig. 4. The interframe spacing is nominally 5 μs. In the frame labeled 14 μs we see that a reaction wave is already starting to appear at the obstacle face. This builds in intensity

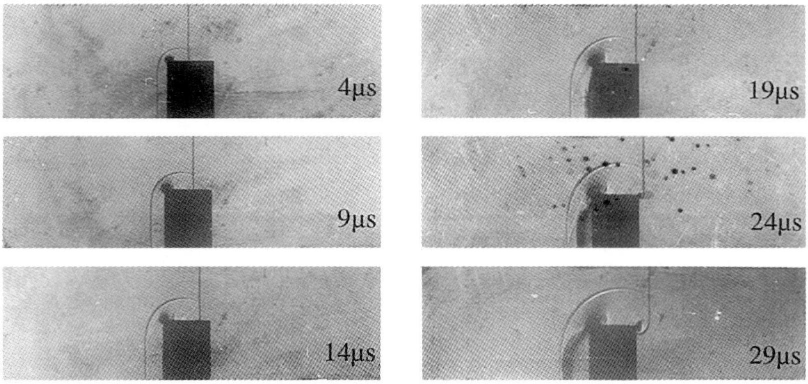

Figure 4. Schlieren photographs showing time history of shock system.

in the next frame. By 24 μs it is clearly advancing towards the reflected shock and at 29 μs it has reached the wave. The same process is observed in the numerical simulation, Fig. 5.

However, in the simulation the reactive wave is seen to reach the reflected shock quicker than is observed in the experiment. There are two possible reasons for this. The first is that the cell size is too large which leads to some 'smearing'

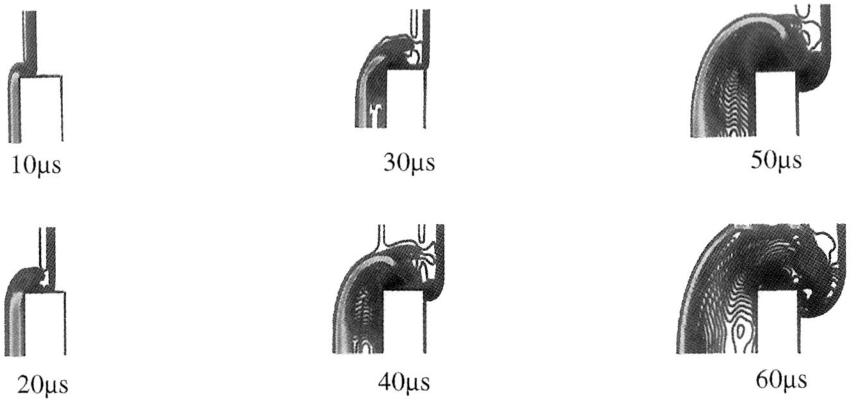

Figure 5. Density contours.

of temperature around the reflected shock front. Since we are close to reaction threshold, this could shorten the induction time and so account for this observed difference. The second possibility is that the basic reaction set is not sufficiently complete and that some of the missing exotic species (e.g. O_3) actually play an important role. In addition, dissociation reactions are not present; these act as a 'temperature sink' and so delay the onset of the reactive wave.

4. Conclusions

The results from this study show that the numerical simulations predict very well the chemistry and flow field produced by the shock-wave/obstacle system. The nitrogen-diluted mixture requires a higher Mach-number shock than the argon-diluted mixture in order to initiate reactions. This is expected since the diatomic nitrogen molecule can absorb more excess energy before dissociating. In the case of the argon-diluted mixture, we need to investigate both the grid resolution and the completeness of the reaction set to see if better agreement with the experimental data can be achieved.

Acknowledgement. We would like to thank Mr. R. Williams of the Shock Waves Group at U.C.W. Aberystwyth for the experimental schlieren photographs.

References

Figura ES, Thomas GO, Milne AM (1992) Studies of reactive triple shock collision. Shock Waves 2:203–209.

Mitchell RE, Kee RJ (1992) A General Purpose Computer Code for Predicting Chemical Behavior Behind Incident and Reflected Shocks, Sandia Report SAND82-8205, UC-401.

Shock-tube investigations of H_2 combustion in a high-temperature supersonic air flow

G. Smeets, C. Quenett
French-German Research Institute of Saint-Louis
F-68301 Saint-Louis, France

Abstract: Experiments are performed in a constant-area duct having a cross section of 60 × 60 mm. The shock-tube-generated flow has a stagnation enthalpy of about 5 MJ/kg (4200 K stagnation temperature) corresponding to the real flight of a hypersonic vehicle with Mach number 9 and ~ 40-km flight altitude. The parameters of the flow at the entrance of the duct are approximately: flow velocity, 2500 m/s; static temperature, 1850 K; and static pressure, 1.4 bar. By simultaneous monitoring of the evolution the static pressures and velocity profiles, a quantitative determination of the combustion efficiencies in the duct has been possible. In case of a special injection mode, a rather effective mixing could be obtained leading to about 70% combustion efficiency 30 cm (5 times the hight of the duct) downstream of the injection location.

Key words: Shock tube, Experiment, Supersonic, Combustion, Scramjet

1. Introduction

Attempts made in the past suggest that the greatest challenge in realizing an operational scramjet motor is finding the right way for rapidly mixing and efficiently combusting the injected hydrogen without introducing substantial losses of momentum of the supersonic flow. In the following is given a summary of C. Quenett's thesis (Quenett 1995). For shock-tube-generated flow with flight Mach number 9, first, by means of flow visualizations, detailed pictures of the very complicated combusting flow pattern in a test chamber of constant cross section are presented. Then, quantitative data on simultaneous evolution of wall pressure and flow velocity along the duct are recorded. Using the two simultaneously recorded independent quantities, together with a balance of energy and momentum of the flow obtained by the conservation equations, the combustion efficiency can be estimated.

2. Shock tube and experimental arrangements

The experiments were performed in a shock tube of 10 cm constant inner diameter, having driver and driven tube sections of 3 m and 19 m length, respectively. The driver gas was cold hydrogen at about 120 bar and the driven section was filled with air or nitrogen of 0.2 bar initial pressure. The test time, terminated by the arrival of the first driver gas, was close to 1 ms. A schematic of the experimental arrangement mounted to the end of the driven section is shown in Fig. 1. The shock-heated air arriving at the end of the tube is first expanded in a divergent tube in order to increase the flow Mach number to about 3. The flow parameters at the entrance of the following constant-area duct having a cross section of 60 mm by 60 mm are approximately: $u = 2500$ m/s, $T = 1850$ K, $p = 1.4$ bar. They correspond to the expected state of the flow at the combustor entrance (after inlet compression) during the real flight of a hypersonic vehicle with Mach number 9 and at about 40 km flight altitude.

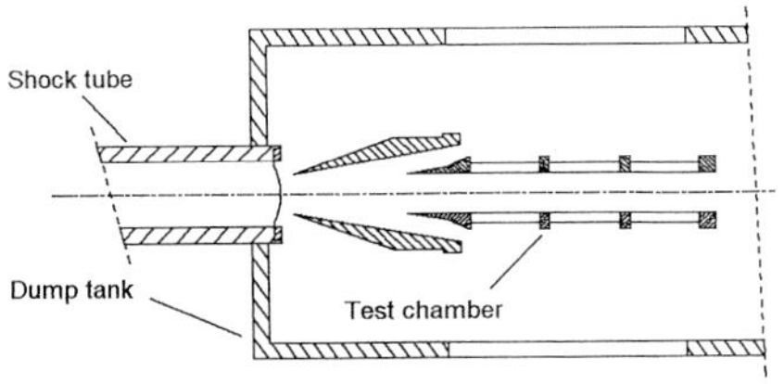

Figure 1. Schematic of the experimental arrangement downstream of the shock tube.

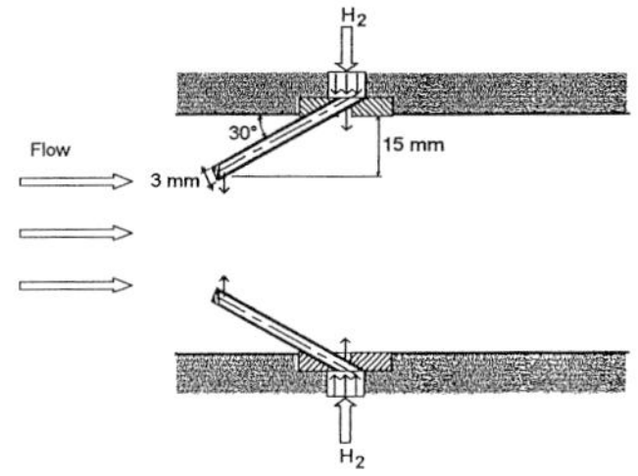

Figure 2. Combined injection: from holes in the side wall and additionally from tubes immersed in the flow.

Two different ways of injection are compared: (a) normal to the flow from arrays of holes in two opposite sidewalls arranged perpendicularly to the main stream direction, and (b) only partly from holes in the sidewall and additionally from the top of small cylindrical tubes immersed in the flow at a 30° angle as shown in Fig. 2. In both cases, the total rate of injected hydrogen is approximately 49 g/s corresponding to 75% stoichiometric amount for the mass rate of the passing air flow of about 2.3 kg/s. The combustion of the injected hydrogen can be observed along the duct up to about 30 cm downstream of the injection location. At the high static temperature, the chemical reactions are rather fast and the combustion will be mainly mixing controlled.

3. Diagnostic techniques

For flow visualizations, a differential interferometer based on Wollaston prisms is applied. As described by Smeets (1974 and 1990), schlieren-like pictures are the result. In addition to the interferograms, pictures of the near-uv radiation from chemiluminescent OH radicals are recorded by means of a CCD camera with image intensifier. Thus, the locations of significant OH formation characterizing the combustion reaction can be identified.

Wall pressures are recorded with conventional quartz piezo-transducers mounted flush to the wall of the duct. The same transducers serve for pitot-pressure measurements of the upstream air flow at the entrance of the constant-area test chamber. Heat fluxes to the wall are also recorded in a conventional manner, using thermocouples implanted flush to the wall.

For precise and highly time-resolved records of the flow velocities and their changes resulting from combustion, a Doppler technique is used based on a spectrometric analysis of the Doppler shift of monochromatic laser light scattered from particles embedded in the flow. Polydisperse TiO_2 particles, tracer particles capable of surviving at the high flow temperatures, are used having an average size of 0.32 μm. After being added to the air in a special disperser, the particles are put into the driven tube together with the air. The velocity records shown were achieved using a Michelson interferometer having a large optical path difference for sensitive wavelength detection and autostabilized by means of a Pockels cell. This technique was specially developed for shock-tube applications (see Smeets and George 1978 and Smeets 1990).

4. Visualizations of flow pattern

The size of the flow sector in Fig. 3 is 35 mm by 90 mm. It ranges from the lower wall to slightly above the middle of the duct and shows the region immediately downstream of the injection location. In the lower left corner, the jets of hydrogen can be seen ejecting from an array of holes in the side wall perpendicularly to the flow. A strong oblique shock is formed by the injection. A second (weak) shock is created by a small step in the wall caused of a mismatch of the small plate containing the injection holes. A layer of hydrogen at the wall having about 1-cm thickness can be clearly seen. In Fig. 4, showing a sector of the duct 24 cm further downstream, the visible layer at the wall has substantially expanded. As such expansion is not to be seen on comparative pictures made with nitrogen instead of air, it is to be attributed to combustion processes within this turbulent layer.

Figure 3. Hydrogen jets from the side wall and the flow immediately downstream of the injection location.

Figure 4. Picture showing a sector of the duct 24 cm further downstream.

The pictures become more complex in case of the alternative injection mode (partly from holes in the sidewall and additionally from the top of small cylindrical tubes immersed in the flow as shown in Fig. 2). The flow immediately downstream of the injection location (Fig. 5) shows two additional elements: another oblique shock wave (caused by the immersed tubes from the opposite side) and a second hydrogen layer at about 2 cm distance from the wall. The interaction of the oblique shock waves with the hydrogen layers is expected to strongly increase turbulent mixing and, by this, lead to more complete combustion during the transit time of the flow. The picture in Fig. 6, again taken 24 cm further downstream in the duct, shows over the entire duct a completely turbulent flow having lost all its specific structures like shock waves or layers. To make sure that the flow has not been thermally choked, a wedge was mounted into the duct. The attached shock waves clearly seen give proof that the flow is still supersonic in the center of the duct.

Figure 5. Flow immediately downstream of the injection location in the case of combined injection.

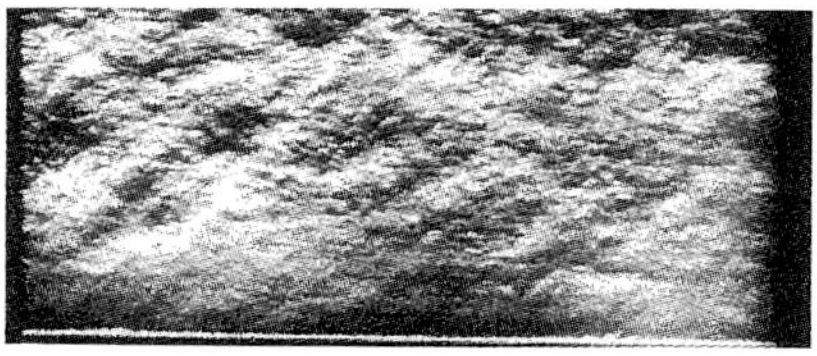

Figure 6. Picture taken 24 cm further downstream.

Figure 7 shows the near-UV radiation from chemiluminescent OH radicals recorded by means of a CCD camera with image intensifier. Most parts of the picture show some OH emission. The locations of maximum OH production and hence maximum combustion activity are in front of the hydrogen jets and where the two oblique shock waves intersect the injected hydrogen layer.

5. Quantitative determination of the flow parameters

The parameters of the air flow entering the duct (as indicated above) are determined as precisely as possible on the basis of the following measured quantities: (a) velocities of the incident shock along the driven tube using a number of shock detectors in the tube wall; (b) static pressure histories at the end of the driven tube by pressure transducers in the tube wall; and (c) profiles of pitot pressures and flow velocities in both sections, at the end of the tube as well as after the isentropic expansion in the divergent tube using a pitot rake and the Doppler technique.

For assessing the combustion efficiencies achieved, both the static pressures and velocity profiles of the flow with and without hydrogen injection are mea-

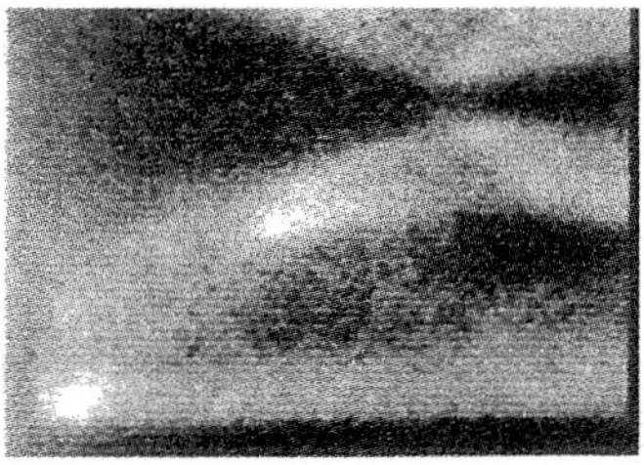

Figure 7. UV radiation from chemiluminescent OH radicals in the case of combined injection.

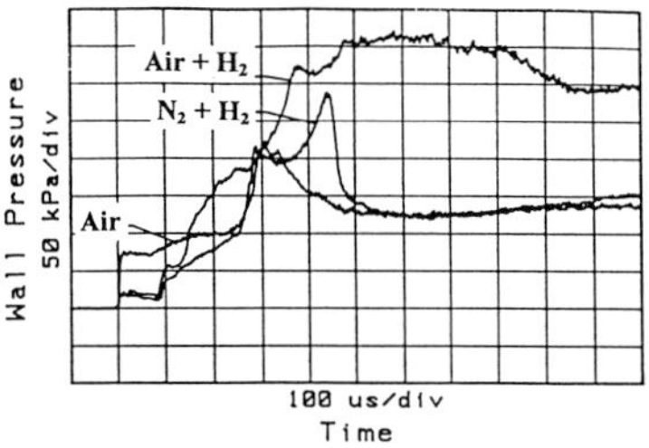

Figure 8. Static pressure signals (combined injection).

sured at the end of the duct 30 cm downstream of the injection location. As an example, Fig. 8 shows static pressure signals for the case of combined injection from the wall and from the small tubes. The three traces correspond to comparative experiments with air and hydrogen injection, with nitrogen and injection of hydrogen, and with air without any injection. Corresponding velocity signals received at 20 mm distance to the wall are given in Fig. 9. For further data reduction, average values of the recorded pressures and velocities over the period of relative constant flow condition between 0.6 ms (termination of the unsteady starting processes in the duct) and 1.0 ms (arrival of the first polluting driver gas) are used.

6. Estimation of combustion efficiencies

The theoretical model used for defining combustion efficiency is based on the simplifying assumption that at some distance downstream of the injection location the injected hydrogen has been (at least on a large scale) uniformly mixed with the air and that there exist two types of stream tubes: one containing combustion products in thermal equilibrium and the other still containing the non-reacted mixture. A reaction rate can then be defined as the ratio of the mass flow of the reacted gas to the total mass flow. Using the 1D conservation equations for the two streamtubes individually and taking into account that the static pressures have to adjust themselves, both a static pressure increase as well as a corresponding velocity decrease can be calculated as functions of the reaction rate. To allow for losses of momentum and energy in the boundary layers and which also may be caused by the tubes immersed in the flow, additional terms can be added to balance equations leading to a slight shifting of the calculated curves.

By fitting the experimental data (average values of all measured static pressures and velocities) to the calculated curves, the reaction rate can be deter-

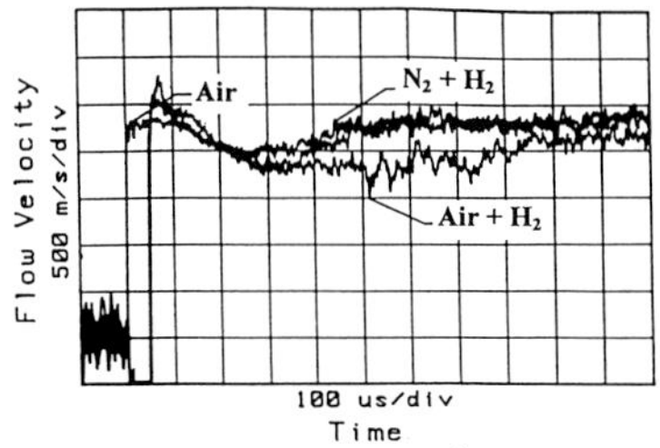

Figure 9. Velocity signals (constant flow condition).

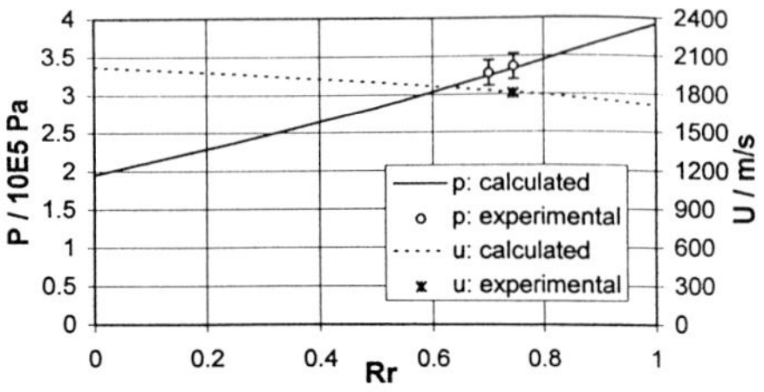

Figure 10. Best fit determination of reaction rate.

mined. For the case of combined injection, Fig. 10 shows a best fit achieved for a reaction rate of 70% to 75% and with additional terms accounting for 4% losses of both momentum and energy. The best fit for the case of only side-wall injection is achieved for about 25% reaction rate and for losses of 2% momentum and 3% energy. A net thrust is estimated from the increase of momentum of the flow if both an idealized inlet and an idealized thrust nozzle are branched to the combustor tube. For the case of the combined injection mode, a specific impulse was calculated close to 20,000 m/s.

7. Conclusions

By simultaneous monitoring of the evolution of both the static pressures and velocity profiles, a quantitative determination of the combustion efficiencies in the duct has been possible. Together with heat-flux measurements, the losses of momentum and energy could be quantified. In the case of a combined injection mode, partly from holes in the sidewall and additionally from the top of small cylindrical tubes immersed in the flow under a 30° angle, a rather effective mixing could be obtained leading to about 70% combustion efficiency 30 cm (5 times the hight of the duct) downstream of the injection location. The momentum and energy losses were determined to be about 4% each.

References

Smeets G (1974) Observational techniques related to differential interferometry. In: Proc 11th Int Congr on High-Speed Photography, London, pp 283–88.

Smeets G, George A (1978) Instantaneous laser Doppler velocimeter using a fast wavelength tracking Michelson interferometer. Rev Sci Instum 49(11):1549–1556.

Smeets G (1990) Interferometry and Velocity Interferometry, Lecture Series 1990–05 on Measurement Techniques for Hypersonic Flows, VKI.

Quenett C (1995) Stossrohruntersuchungen zur H_2-Verbrennung in einer heissen Ueberschallstroemung, Thesis, University of Essen, Germany.

Numerical simulation of reactive shock diffraction around an object containing a cavity

B. Jiang, D. M. Ingram, D. M. Causon, R. Saunders
Centre for Mathematical Modelling and Flow Analysis
The Manchester Metropolitan University
Chester Street, Manchester M1 5GD, UK

Abstract: A reduced reaction mechanism for hydrogen combustion has been derived using the global simulation method. The mechanism has been incorporated into a flow hydrocode and used to study the interaction between a plane detonation wave an object containing a cavity. In the first part of the paper the global simulation method is discussed and a reduced mechanism is derived. The second part of the paper describes the validation of the method and discusses the results relevant to the cavity problem.

Key words: Hydrogen combustion, Reduced mechanism, Hydrocode, Reactive shock diffraction, Global simulation method

1. Introduction

The object of this study is to investigate the diffraction of a detonation wave around a simple object containing a cavity by means of numerical simulation. Reactive shock-on-shock interactions and shock-wave diffraction around obstacles often occur in accidental explosions in and around chemical plants, storage tanks, and pipe lines. It is therefore essential to have available methods capable of predicting the blast loading to which these structures are subjected. For many years, enduring effort has been expended to understand detonation processes but until now, there is lacking a comprehensive theory to describe their complicated structure (Nettleton 1987). The reason for this is that the change in gas composition and liberation of chemical energy add considerably to the complexity of the mathematical analysis.

The chemistry generates a self-sustaining reactive shock via the energy released from the reactions behind the flame. These reactions are strongly temperature dependent. As a result, any simplification of the detailed chemistry must be treated with extreme caution. At high temperatures, a substantial amount of intermediates will exist in equilibrium thus reducing the energy released. If the reaction mechanism is oversimplified, this may lead to over prediction of the released energy. In order to derive an appropriate simplification, the detailed reaction mechanism must be studied. Here, important species which exist under high temperatures have been isolated and included in the reduced mechanism. The resulting reduced mechanism links the species through only a few of the most important elementary reactions, for which rate constants are evaluated, so that the reduced mechanism will preserve the macroscopic properties of the detailed model (Jiang et al. 1995).

2. Deriving reduced reaction mechanisms

In the following section we describe a global simulation method for obtaining reduced reaction mechanisms from the detailed rate kinetic mechanisms published in the literature. Simplified reaction mechanisms are attractive because, in addition to being extremely stiff, detailed mechanisms contain many hundreds of reversible chemical reactions in tens of chemical species. Simplified reaction mechanisms of the type obtained in this work have previously been derived by workers such as Paczko et al. (1986) and Peters & Kee (1987). However, these mechanisms use algebraic combinations of rate coefficients obtained from the detailed mechanism and are thus computationally expensive. The global simulation approach permits us to derive a reduced reaction mechanism and then obtain new rate coefficients through simulation.

As with most reduction processes, it is necessary to begin by seeking a model of the chemical processes which contains not only the species in which we are primarily interested but sufficient detail to give a good approximation to the detailed mechanism. Such model mechanisms are obtained using the steady state and partial equilibrium hypotheses (Williams 1985).

2.1. The global simulation method

Suppose that an appropriate reduced mechanism has been obtained which has the following form

$$\begin{aligned} \text{Reactants} + \text{Intermediates} &\rightleftharpoons \text{Intermediates} \\ \text{Intermediates} &\rightleftharpoons \text{Products} \end{aligned}$$

and where the reaction rates are obtained in the normal way

$$\frac{dc_i}{dt} = \sum_{k=1}^{M} (\nu''_{i,k} - \nu'_{i,k}) \times \left(\widehat{k}_{f,k} \prod_{j=1}^{N} c_j^{\nu'_{j,k}} - \widehat{k}_{b,k} \prod_{j=1}^{N} c_j^{\nu''_{j,k}} \right) \quad i = 1, 2, ..., N - N_e, \quad (1)$$

where N is the number of species and N_e is the number of atomic species; ν' and ν'' are stoichiometric factors for reactants and products respectively; c is the mole concentration; $\widehat{k}_f$ and $\widehat{k}_b$ are forward and backward rate constants for the reduced scheme.

Table 1. Coefficients for the reduced hydrogen scheme (mole cm^{-3} s^{-1}).

reac-	forward				backward			
tion	*a*	*b*	*c*	*d*	*a*	*b*	*c*	*d*
i	8.07E1	-2.06E1	-1.29E4	2.42E-3	-1.10E1	8.90E0	2.51E3	-6.35E-4
ii	-8.17E0	7.79E0	1.58E3	-3.67E-4	7.59E0	2.96E0	2.42E3	-4.49E-5
iii	-2.81E1	1.40E1	5.36E3	-1.03E-3	1.67E2	-4.64E1	-2.41E4	4.77E-3
iv	-2.64E1	1.34E1	5.57E3	-1.04E-3	8.24E1	-2.09E1	-1.51E4	2.32E-3

We now rewrite (1) in terms of the production (+) and consumption (−) rates,

$$\frac{dc_i}{dt} = \left(\frac{dc_i}{dt}\right)^+ - \left(\frac{dc_i}{dt}\right)^- \tag{2}$$

and equate this to the reaction rates from the detailed mechanism

$$\left(\frac{dc_i}{dt}\right)^{\pm}_{reduced} = \left(\frac{dc_i}{dt}\right)^{\pm}_{detailed}. \tag{3}$$

Substituting (1) into (3) leads to a system of linear simultaneous equations whose solution defines the rate constants for the reduced reaction scheme for a given set of initial conditions.

By repeatedly solving for the rate constants over a range of temperatures we can determine, by a least-squares method, the coefficients for a form of Arrhenius reaction rate law.

2.2. Reduced mechanism for hydrogen combustion

A reduced combustion mechanism for hydrogen has been obtained using the global simulation method to reduce the detailed mechanism proposed by Lui & Rogg (1991). The derived mechanism contains five chemical species and four reactions,

$$\begin{aligned}
&\text{i.} & H_2 + M_1 &\rightleftharpoons 2H + M_1\\
&\text{ii.} & O_2 + H + M_2 &\rightleftharpoons OH + O + M_2\\
&\text{iii.} & H + O + M_3 &\rightleftharpoons OH + M_3\\
&\text{iv.} & H + OH + M_4 &\rightleftharpoons H_2O + M_4
\end{aligned}$$

The coefficients for the reaction rate in the form of a modified Arrhenius expression, $\log \widehat{k} = \log a + b \cdot \log T + c/T + dT$, are shown in Table 1, while the third body efficiency for the mechanism are: $M(H_2)=M(O_2)=0.75$ and $M(H_2O)=15.2$.

2.3. Validation

The reduced mechanism described has been coupled to an existing flow hydrocode (Ingram et al. 1995) which uses the LSODE package (Hindmarsh 1983) to integrate the system of ODEs resulting from the reaction mechanism. The flow hydrocode solves the time-dependent compressible Navier-Stokes equations for laminar flow with ordinary diffusion terms for the chemical species. In order to validate this approach we have applied the hydrocode to two problems (i) shock-induced detonation (Takano 1993) and (ii) reignition of a detonation wave (Jones et al. 1995). Figure 1 shows the results obtained with the present method which compare extremely favorably with those obtained in the literature.

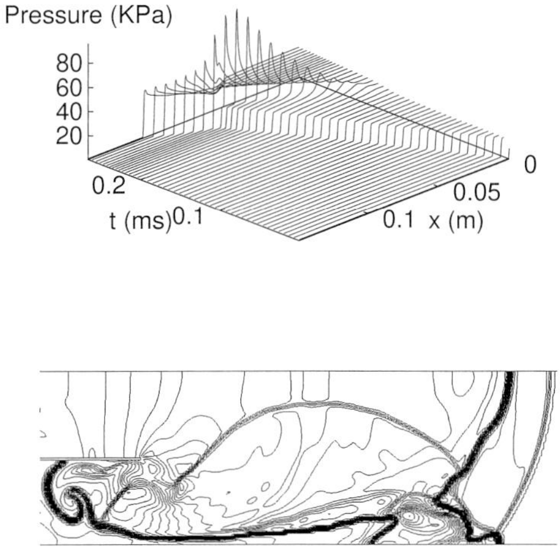

Figure 1. Shock induced detonation (top) and reignition of a detonation wave (bottom).

3. Results

The authors have simulated the interaction of a planar reactive shock wave with a simple ⊏ shaped object. The object was placed in a shock tube filled with an H_2-O_2-Ar (2:1:7 mole) mixture. The initial detonation wave was generated using a 1D calculation, which ignited an undiluted mixture of H_2 and O_2, and tracked the resulting wave into the diluted mixture. This solution was evolved until a CJ detonation wave was observed. The two-dimensional calculation was started by positioning the wave just in front of the obstacle.

Because of the high dilution in the shock tube, the combustion wave and pressure wave disassociate as the blast wave diffracts around the back of the obstacle and a pocket of unburnt gas forms around the front tips of the "c" (Fig. 2). After being reflected at the centerline, the shock wave passes through the flame front forming a high-temperature pocket. This pocket grows and causes the diffracted flame front to bend. The high temperatures and pressures inside this pocket accelerate the combustion processes and accelerate the flame front.

Figure 2. Detonation wave interaction, $t = 25$, 30, and 38 μs.

The marked growth of this pocket can be clearly observed by $t = 43\,\mu$s, and between this time and $t = 45\,\mu$s transition to detonation occurs along the centerline of the geometry (Fig. 3). By $t = 47\,\mu$s the shock wave has begun to reflect from the back of the cavity and the spread of the detonation zone can be clearly observed.

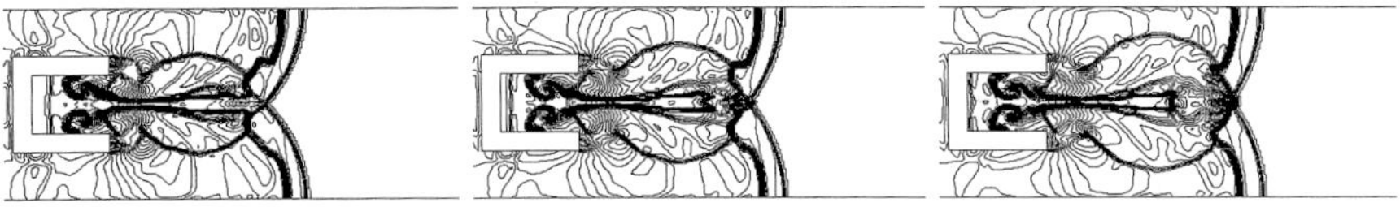

Figure 3. Detonation wave interaction, $t = 43$, 45, and 47 μs.

Figure 4 shows the growth of the detonation front until $t = 61\,\mu$s when the detonation wave has completely filled the shock tube. Behind the detonation wave a weak shock wave can be observed as can the collapse of the region of uncombusted gas along the centerline of the shock tube. Figure 4 also shows the reflection of the diffracted shock wave from the end wall of the obstacle and its passage through the combustion zone inside the cavity.

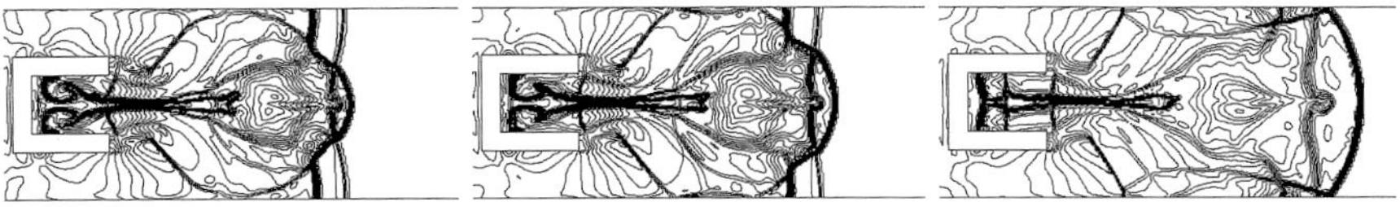

Figure 4. Detonation wave interaction, $t = 52$ 55, and 61 μs.

The calculations were performed on a 500×100 grid on a Silicon Graphics Power-Challenge computer using one processor and took about 18 hours.

4. Conclusions

A reduced reaction mechanism for hydrogen combustion has been derived using the global simulation method and incorporated into a hydrocode. The resulting hydrocode has been validated against several test problems before being applied to simulate the diffraction of a detonation wave around a simple obstacle with a cavity. The results obtained show a highly complex regime which includes disassociation of the reaction zone and the shock wave and the subsequent return to detonation. The calculations demonstrate the utility of the present method in simulating complex reacting flow fields numerically.

References

Hindmarsh A (1983) ODEPACK, a systematised collection of ode solvers. In: Scientific Computing, (R Stepleman, ed), North-Holland, Amsterdam, pp 55–64.

Ingram DM, Batten P, Lambert C, Causon DM (1995) Hydrodynamic code calculations of a blast on a tank farm. In: Shock Waves @ Marseille: Proceedings of the 19th International Symposium on Shock Waves, Marseille, France 1993, (R Brun and L Dumitrescu, eds), Vol. IV, pp 419–424, Springer-Verlag.

Jiang B, Ingram DM, Causon DM, Saunders R (1995) A global simulation method for obtaining reduced reaction mechanisms for use in reactive blast wave flows. Shock Waves 5(1/2):81–87.

Jones D, Sichel M, Oran E (1995) Reignition of detonations by reflected shocks. Shock Waves 5(1/2):47–57.

Lui Y, Rogg B (1991) Thermal radiation in non-premixed flames. In: Combustion and Reaction Kinetics International Jahrestaung-Fraunhofer-Institute fuer Treib-und Explosvsstoff, Fraunhofer-Inst fuer Treib-und Explosivsstoffe, 22nd International Annual Conference on ICT 1991, pp 105.1–105.12, Pfinztal-Berghausen, Germany.

Nettleton M (1987) Gaseous Detonations — Their Nature, Effects and Control, Chapman and Hall, New York.

Paczko G, Lefal P, Peters N (1986) Reduced reaction schemes for methane, methanol and propane flames. In: 21st International Symposium on Combustion, pp 739–748.

Peters N, Kee R (1987) The computation of stretched laminar methane-air diffusion flames using a reduced four-step mechanism. Comb and Flame 68:17–26.

Takano Y (1993) Numerical simulation for shock-tube experiments of reflected-shock waves in combustible gas. JSME International Journal, Series II, 36(2):300–306.

Williams FA (1985) Combustion Theory, 2nd edition, Benjamin/Cummings, Menlo Park, California, USA.

Modeling of explosion in explosion behind a reactive shock wave by the joint velocity-composition PDF method

S. M. Frolov, M. A. Noskov
N. N. Semenov Institute of Chemical Physics
Kosigin Str., 4, Moscow, 117977 Russia

R. Tatschl
AVL LIST GmbH, Kleiststrasse 48, A-8020 Graz, Austria

P. Wolanski
Institute of Heat Engineering, Warsaw Technical University
Nowowiejska 25, Warsaw, 00-665 Poland

Abstract: A possible explanation for explosion in explosion is proposed; that it is due to enhancement of local chemical activity by dissipation of turbulence in a compressible turbulent boundary layer behind shock waves. Temperature fluctuations result in the significant shortening and nonuniform spatial distribution of auto-ignition delay time. Estimations based on the Zel'dovich mechanism show a possibility of coupling between spontaneous flames and pressure waves.

Key words: Deflagration-to-detonation transition, Compressible turbulent boundary layer, Spontaneous flame

1. Introduction

The phenomenon of Deflagration-to-Detonation Transition (DDT) has remained the subject of detailed studies for a long period of time. One of the most important features of DDT is the 'explosion in explosion' process discovered by Meyer et al. (1970). Experimental observations revealed that detonation arises due to a localized explosion between the flame front and shock waves generated by flame, most frequently in the near-wall boundary layer. One of the possible explanations for the 'explosion in explosion' is in the existence of mechanisms for energy dissipation in boundary layer, namely viscous dissipation of the flow kinetic energy and the dissipation of turbulence. However, regarding the DDT phenomenon, an increase in mean temperature $\overline{T}$ due to viscous dissipation fails to explain the shortening of auto-ignition delay τ_i as described by Meyer et al. (1970). It is known (Cheng et al. 1994) that the supersonic shear layer in a flow of Mach number 2 exhibits temperature fluctuations of magnitude $\sim 0.2\overline{T}$. Such fluctuations can significantly affect τ_i. If the auto-ignition event arises first at a certain point within a boundary layer, then, depending on the spatial distribution of τ_i, in the vicinity of this point, the following condition can be met:

$$u_{sp}/a_s \approx O(1) \tag{1}$$

where $u_{sp} = |\partial\tau_i/\partial l|^{-1}$ is the spontaneous flame velocity (Zel'dovich et al. 1988), a_s is the local sound speed, and $\partial/\partial l$ is the derivative along the path. When Eq. (1) holds, the Zel'dovich coupling mechanism comes into effect, and the interaction of spontaneous flame with the accompanying pressure waves can result in 'explosion in explosion'. This paper deals with the effect of turbulent temperature fluctuations on τ_i in a boundary layer behind a shock wave.

2. Formulation and results of calculations

Consider a planar shock wave of constant Mach number. The shock wave propagates in an inert gas along a smooth rigid wall. At $t = 0$ it enters the reactive-mixture cloud. The aim of this study is to follow the motion of some fixed mixture particles which enter the boundary layer and to accumulate the information on auto-ignition sites and times to obtain the mathematical expectation of auto-ignition delay and its spatial distribution. We use the system of coordinates attached to a shock wave. The turbulent nature of the flow in the boundary layer results in a variety of instantaneous spatial distributions of flow variables and scalars. Therefore, we use the joint velocity-composition probability-density-function (jvc-pdf) method (Pope 1985). The advantage of this method is that it provides the solution without additional closure assumptions for mean rates of chemical reactions by applying detailed reaction mechanisms of fuel oxidation. The jvc-pdf is obtained by the Monte Carlo method. The solution of the problem is significantly simplified by the fact that we are interested in the pre-ignition period of the reactive-mixture evolution. This allows one to neglect the effect of pre-ignition chemical transformation on the mean flow, keeping the latter fixed.

Two cases were considered, namely with a cold (I, $T_w = 320$ K) and a hot (II, $T_w = 850$ K) tube wall. Initial conditions are as follows: $U_\infty = 1\,300$ m/s, $D = 2\,000$ m/s; $T_\infty = 1\,000$ K, $\nu_\infty = 10^{-5}$ m^2/s; $P_\infty = 1.515$ MPa, $\gamma = 1.4$. A stoichiometric CH_4–O_2 mixture was considered. The oxidation mechanism was taken from Frolov et al. (1995). For case I, the mathematical expectation $\overline{\tau}_{i1} = 1.43 \cdot 10^{-3}$ s; for case II $\overline{\tau}_{i2} = 2.64 \cdot 10^{-4}$ s. In the free stream $\tau_{i\infty} = 2.6 \cdot 10^{-3}$ s. Figure 1 shows particle trajectories in case II. Figure 2 shows the probability-density function of τ_i for cases I, II depending on the distance to shock wave. Curves in Fig. 2 represent the mathematical expectation of τ_i versus distance. It follows from Fig. 2a (case I) that there exists a certain distance from the shock wave where the probability of auto-ignition is zero. For

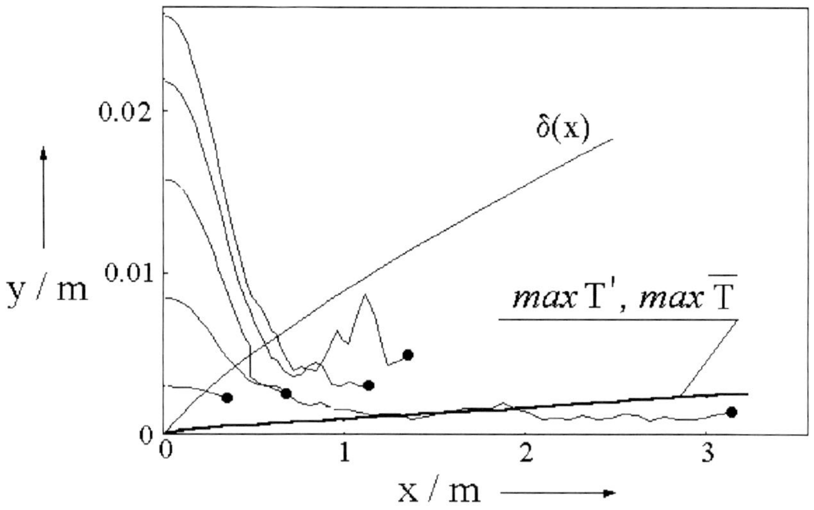

Figure 1. Particle trajectories behind a shock wave in case II (hot wall). Curves show the loci of maximum mean temperature and RMS temperature fluctuations. Solid points indicate the locations of particle auto-ignition; δ is the boundary-layer thickness.

case II (Fig. 2b) this distance is small. Based on the curves in Fig. 2 one can determine the spontaneous flame velocity as $u_{sp} \approx |\partial \overline{\tau}_i / \partial x|^{-1}$. Clearly, Eq. (1) is satisfied. Figure 3 shows the apparent trajectory of the spontaneous flame for a single flow realization (case II). Point *1* denotes the first auto-ignition event within the boundary layer, point *2* the second event, etc. Several auto-ignition events can occur simultaneously (points *4*, *4'*, *4''*; *7*, *7'*; *9*, *9'*). In spite of the probabilistic character of distinct auto-ignition events, the spontaneous flame propagates downstream the shock wave.

3. Conclusion

The results of Monte Carlo modeling of mixture auto-ignition behind a shock wave indicate that turbulent temperature fluctuations in the boundary layer can play a significant role in shortening the overall ignition delay, which is observed experimentally (Meyer et al. 1970). A possibility of coupling between the spontaneous flame and pressure disturbances has been proved by applying the Zel'dovich criterion Eq. (1). This implies that the temperature fluctuations in the boundary layer, arising as a result of turbulent energy dissipation, can become a reason for localized chemical energy release with pressure build-up

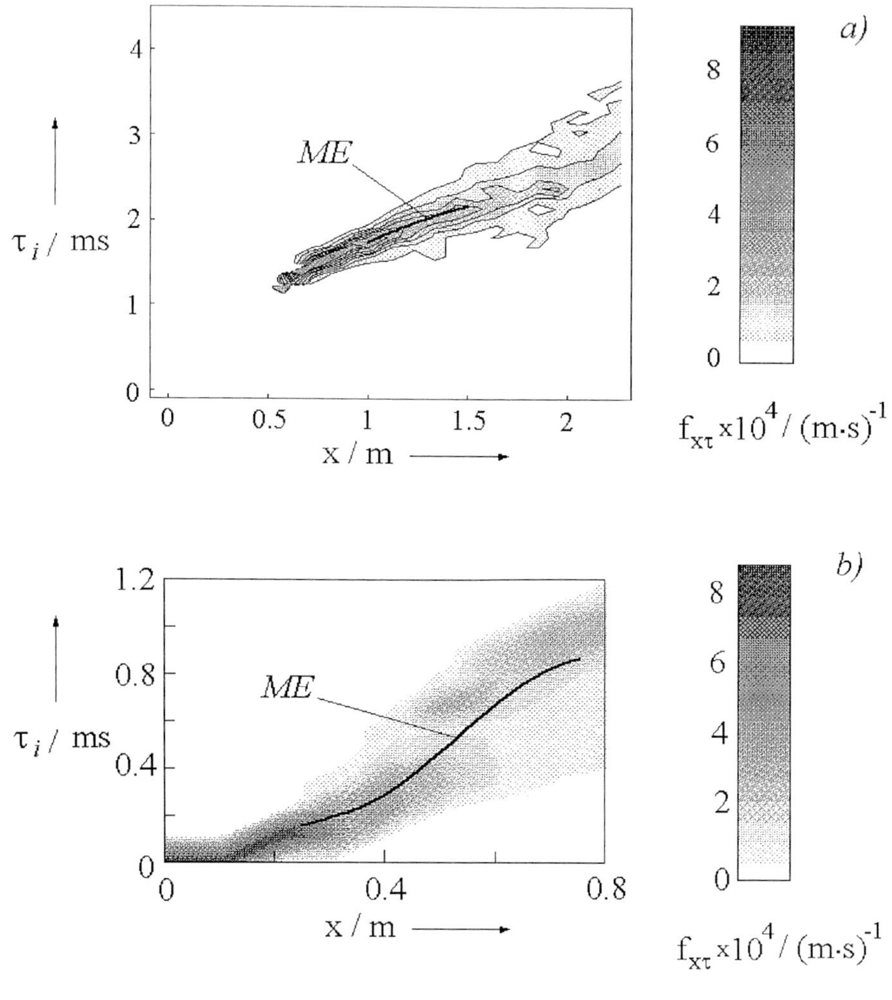

Figure 2. Probability-density function of auto-ignition delay time for cases with cold (a) and hot (b) tube wall. *ME* is the mathematical expectation of τ_i at distance x from shock wave.

and formation of a secondary shock wave. In general, the processes under study can be relevant to DDT. However, this study is only a small step forward toward deeper understanding the phenomenon.

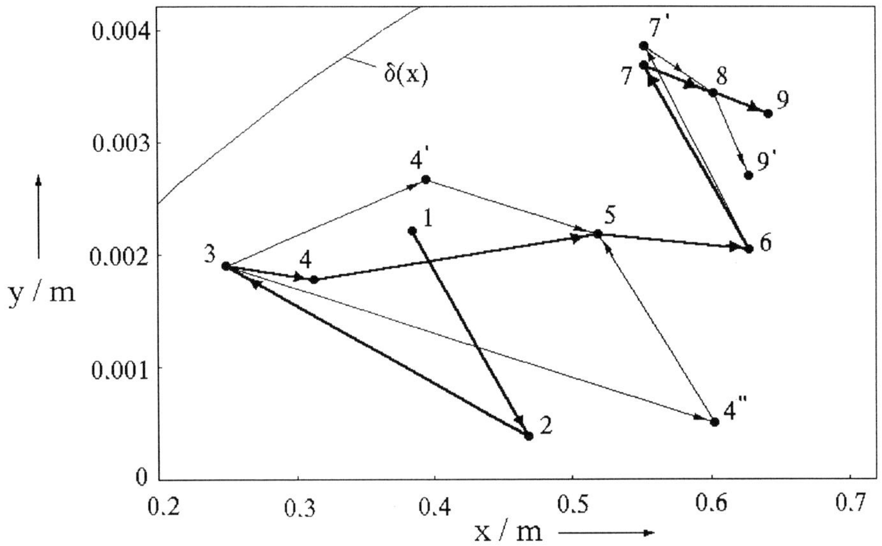

Figure 3. Single realization of spontaneous flame inside the boundary layer behind a shock wave (hot wall). Point *1* – first auto-ignition event, *2* – second event, etc; δ is the boundary layer thickness.

Acknowledgement. This research was supported by Grant #MEN/D0E-94-182, the Joint US-Polish Research Foundation.

References

Cheng TS, Wehrmeyer JA, Pitz RW, Jarett O, Northan JR, Northan GB (1994) Raman measurement of mixing and finite-rate chemistry in a supersonic hydrogen–air diffusion flame. Comb Flame 99:157.

Frolov SM, Basevich VY, Brandstaetter W, Tatschl R (1995) Numerical simulation of turbulent flames. In: Proc FIRE User Meeting, Graz, Austria.

Meyer JV, Urtiew PA, Oppenheim A (1970) On the inadequacy of gasdynamic processes for triggering the transition to detonation. Comb Flame 14:13.

Pope SB (1985) PDF methods for turbulent reactive flows. Prog Energy Combust Sci 11:119.

Zel'dovich YB, Gelfand BE, Tsyganov SA, Frolov SM, Polenov AN (1988) Concentration and temperature nonuniformities (CTN) of combustible mixtures as a reason of pressure waves generation. Progr Astr Aeron, Dynamics of Explosions 114:99.

Spontaneous ignition of a reactive gas mixture by shock-vortex interaction

F. Takayama*, T. Kambe†, S. Taki‡, A. Sakurai§

Abstract: A computational investigation has been made concerning the interaction between a vortex ring and a plane shock wave in a reactive gas mixture ($H_2 + O_2$). It is shown that a self-intensification of the shock wave takes place due to a mechanism of wave convergence similar to a case of non-reactive gas. This mechanism increases the temperature at a localized region of the wave convergence and, as a result, acts as a trigger of spontaneous ignition. Ignition (delay) time due to shock-vortex interaction is reduced about 0.3 times compared with that of a single vortex ring.

Key words: Shock wave, Vortex ring, Shock vortex interaction, Spontaneous ignition

1. Introduction

Phenomena resulting from the interaction between a vortex ring and a plane shock wave in an axisymmetric arrangement have been studied numerically and experimentally (Takayama et al. 1993, Minota 1993). These investigations clarified, among other things, self-intensification of the shock wave due to a mechanism of wave convergence, in which the maximum temperature attains about 2.5 times the temperature behind the shock wave in the localized region.

The present study investigates the effect of spontaneous ignition given by the shock-vortex interaction in a reactive gas mixture ($O_2 + H_2$) using a numerical simulation. In a computation, thirteen elementary chemical reactions for hydrogen and oxygen are considered, and the numerical schemes are TVD (Yee 1987) and a multi-level grid refinement method (Taki 1993).

2. Basic equations

As a basic governing system of equations, axisymmetric compressible full Navier-Stokes equations and continuity equations of chemical species are considered

*Department of Electronic Engineering, Iwaki Meisei University, Chuohdai 5-5-1, Iwaki, Fukushima, 970, Japan.

†Department of Physics, University of Tokyo, Hongo 7-3-1, Bunkyo-ku, Tokyo, 113, Japan.

‡Department of Mechanical Engineering, Hiroshima University, Kagamiyama 1-4-1, Higashi-Hiroshima, Hiroshima, 724, Japan.

§Tokyo Denki University, Nishiki-cho 2-2, Kanda, Chiyoda-ku, Tokyo, 101, Japan.

together with the equation of state of a perfect gas. As for the thermodynamic properties, the constant pressure specific heat, enthalpy, and free energy are approximated by functions of the temperature because the system takes a wide range of temperature. We consider only the molecular transport properties. The diffusion coefficient of each species is expressed using the mean binary diffusion coefficient between each species and the mole fraction of each species. The transport coefficients, such as the mean binary diffusion coefficient between each species, and viscosity and heat conductivity coefficients, are expressed using the Chapman-Enskog formula.

3. Chemical reactions

The chemical reactions of hydrogen and oxygen are modeled in terms of thirteen elementary reactions as shown in Table 1, each including forward and backward reactions. These reactions consist of these seven species: H_2, O_2, H_2O, OH, HO_2, O, and H. An induction time for this model of chemical reaction compares well with the results of the shock-tube experiment (Voevodsky and Soloukhin 1965), where the induction time is a time from initial start to the instant of maximum gradient for production of the OH radical.

Table 1. Hydrogen-oxygen reaction model.

Reaction step	Forward Rate			
	A	E(K)	n	Factor
$H_2 + O_2 \rightleftharpoons 2OH$	1.7×10^{10}	24,200		
$H + O_2 \rightleftharpoons OH + O$	2.2×10^{11}	8,455		
$O + H_2 \rightleftharpoons OH + H$	1.8×10^{7}	4,480	1	
$H + O_2 + M \rightleftharpoons HO_2 + M$	1.5×10^{9}	−500		17 for H_2O
$H + HO_2 \rightleftharpoons 2OH$	2.5×10^{11}	950		
$OH + H_2 \rightleftharpoons H + H_2O$	2.2×10^{10}	2,590		
$OH + OH \rightleftharpoons O + H_2O$	5.5×10^{10}	3,520		
$H + OH + M \rightleftharpoons H_2O + M$	1.6×10^{16}	0	−2	3 for H_2O
$H_2 + HO_2 \rightleftharpoons OH + H_2O$	2.0×10^{8}	12,000		
$OH + HO_2 \rightleftharpoons H_2O + O_2$	1.6×10^{10}	0		
$O + HO_2 \rightleftharpoons OH + O_2$	4.8×10^{10}	500		
$H + H + M \rightleftharpoons H_2 + M$	1.0×10^{11}	0	−1	
$O + O + M \rightleftharpoons O_2 + M$	1.9×10^{7}	−900		2 for O

Forward reaction rate $R = AT^n \exp(-E/T)$ in kmol-sec-m^3-K units.

4. Numerical method

The basic system of differential equations are discretized by the Harten-Yee type TVD scheme (Yee 1987) for the convection terms, by the central difference for the transport terms, and by the first-order Lomax-Bailey implicit scheme for the chemical-reaction terms, respectively. The finite-difference scheme of the basic equation for the variable $f = [\rho, \rho u, \rho v, \rho e, \rho Y_j]^T$, where ρ is the density, u and v the velocities in the axial and radial directions, e the specific total energy per unit mass, Y_j the mass fraction of chemical species j, is given as follows:

$$f^{n+2} = L_z(\Delta t)L_r(\Delta t)L_c(2\Delta t)L_r(\Delta t)L_z(\Delta t)f^n \tag{1}$$

where $L_r(\Delta t)$ and $L_z(\Delta t)$ are the TVD central-difference operator with respect to the r and z directions, and $L_c(\Delta t)$ is the Lomax-Bailey implicit-difference operator for a time step Δt, respectively.

Computation grids for the numerical scheme are refined by an adaptive multi-level grid refinement method (Taki 1993) because the field includes compact objects such as a shock wave, a vortex, and regions of chemical reactions. the grid-refinement method is able to set the grid size from first rough level to fifth fine level and to change the grid level responding to the gradient of prefixed physical quantities.

5. Boundary and initial conditions

The upper half of a cylindrical tube is taken as the computing domain, shown in Fig. 1. A shock wave (S_V) of Mach number M_v is ejected out through a hole on the left end wall of the cylindrical tube at $t = 0$, which serves to generate a

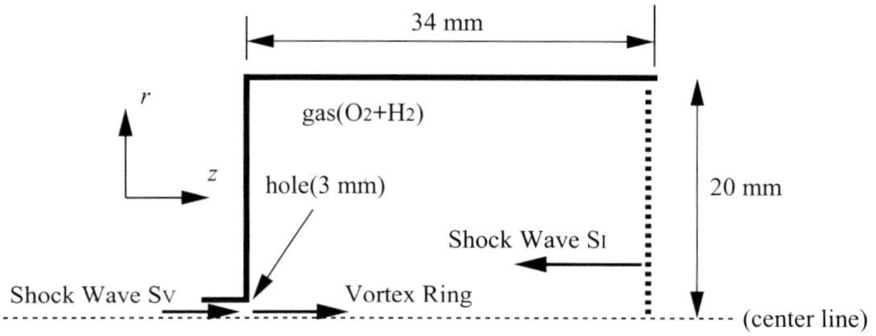

Figure 1. Computational domain and boundary condition.

vortex ring. The right end of the tube is kept open, where a plane shock wave (S_I) of Mach number M_I propagates to the left direction at $t = t_I$ and impinges on the vortex. Mach numbers of both shock waves are set as $M_v = 1.4$ and $M_I = 1.225$. The gas mixture in the cylindrical tube consists of hydrogen and oxygen ($H_2 + O_2$), at rest initially, with the pressure p_0 and temperature T_0, which are prescribed such that the pressure and temperature behind the shock wave S_I take the values: $p_I = 10\,\mathrm{kPa}$ and $T_I = 900\,\mathrm{K}$. The computation grid is set to three levels of equal space, in which the first level is 0.3 mm, the second 0.1 mm, and the third level 0.033 mm. The temperature and the OH are employed as the monitoring physical quantities for the grid-refinement method.

6. Results

The results from the computation clearly exhibit both the convergence of the shock wave S_I and a spontaneous ignition. Figure 2 shows the computed density black-and-white pictures of the time series. The axial and radial coordinates are denoted by z and r respectively, and the lower half of the computer diagram is the mirror inversion of the upper half and denoted by negative r values. It is

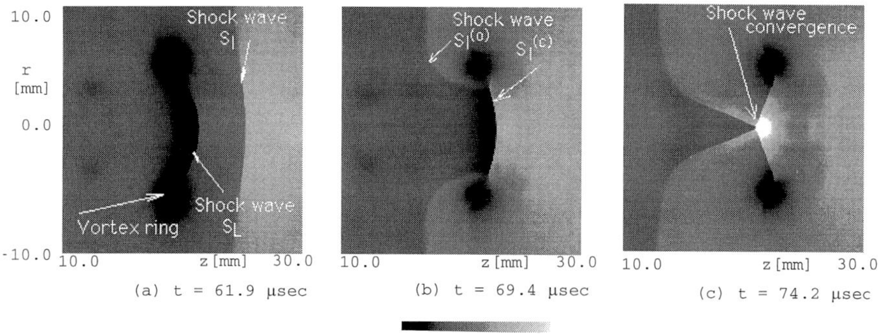

Figure 2. Computed density black-and-white pictures at three sequential times for $M_v = 1.4$ and $M_I = 1.225$.

seen that the shock wave S_L is formed at the central region of the ring due to the Laval nozzle mechanism of the vortex ring. The impinging plane shock wave S_I is deformed by the interaction with the vortex ring (Fig. 2a), the outer deformed shock wave $S_I^{(o)}$ is diffracted to the centerline direction by the rotational motion of the vortex ring, and the inner wave $S_I^{(c)}$ is retarded by the opposing flow in the ring (Fig. 2b). The shock waves $S_I^{(o)}$ and $S_I^{(c)}$ finally converge on the centerline region (Fig. 2c).

Figure 3 shows contours of (a) the density, (b) the temperature, and the mass fraction of (c) the H_2O and (d) the OH in an axisymmetric cross-section, including the symmetry axis at the time $t = 74.37\,\mu s$ immediately after the spontaneous ignition. The central dark region indicates the shock-wave convergence as a result of interaction with the vortex ring (Fig. 3a). The temperature in this region attains about 1900 K, intensified by the wave convergence and the chemical reaction (Fig. 3b). Molecules of H_2O and OH in the localized region of high temperature are produced increasingly by the wave intensification which accelerates the rate of the chemical reactions (Fig. 3c and d). The ignition, defined at the instant of the maximum gradient of production of OH, has already taken place at the time of Fig. 3.

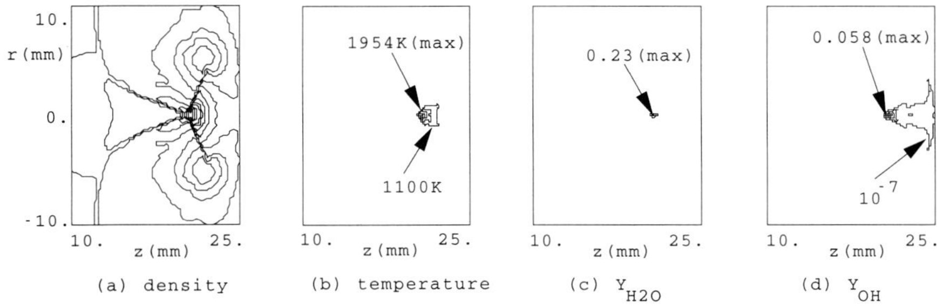

Figure 3. (a) Density, (b) temperature, (c) mass fraction of H_2O, and (d) mass fraction of OH, contours at $t = 74.37\,\mu$s for $M_V = 1.4$ and $M_I = 1.225$.

The ignition-delay time is investigated for three cases (shown in Table 2) with varying time history of maximum temperature and maximum mass fraction of OH, Y_{OH}. Case 1 is for $M_V = 1.4$ and $M_I = 1.225$; and case 2 is for the same condition but without the interacting shock wave; in case 3 a vortex moves into the gas mixture with higher values of temperature ($T_0 = 900$ K) and pressure ($p_0 = 10$ kPa), which is the same as the condition behind the shock wave S_I with $M_I = 1.225$ in case 1. Results for case 1 and case 2 are shown in Fig. 4a, and for case 3 in Fig. 4b. The ignition-delay time of case 1 is measured from the time of deviation of Y_{OH} for the case of no ignition, to the time of the maximum gradient of Y_{OH} and is found to be about 53 μs, whereas in case 2 the time histories of the maximum temperature and the maximum mass fraction of the OH are kept flat. The ignition delay for case 3 is counted from time 0 because of the higher initial values of T_0 and p_0. Figure 4 shows that the spontaneous ignition takes place in case 1 and case 3, and that the ignition-delay time in for each case is $t_d = 20\,\mu$s and 67 μs, respectively. The ignition-delay time for case 1 is about 0.3 times less than for case 3. This result shows that the self-intensification by the shock-vortex interaction accelerates the spontaneous ignition, and also suggests that the shock-vortex interaction in a turbulent reactive gas mixture, where various large- and small-scale vortex rings and various strength shock waves exist, stimulates chemical reactions and is expected to trigger spontaneous ignition.

Table 2. Initial and boundary conditions for the three cases.

Case	Shock Wave S_V	Shock Wave S_I	Initial Condition of Tube
1	$M_V = 1.4$	$M_I = 1.4$ ($T_I = 900$ K, $p_I = 101$ kPa)	$T_0 = 798$ K, $p_0 = 64.1$ kPa
2	$M_V = 1.4$	-	$T_0 = 798$ K, $p_0 = 64.1$ kPa
3	$M_V = 1.4$	-	$T_0 = 900$ K, $p_0 = 101$ kPa

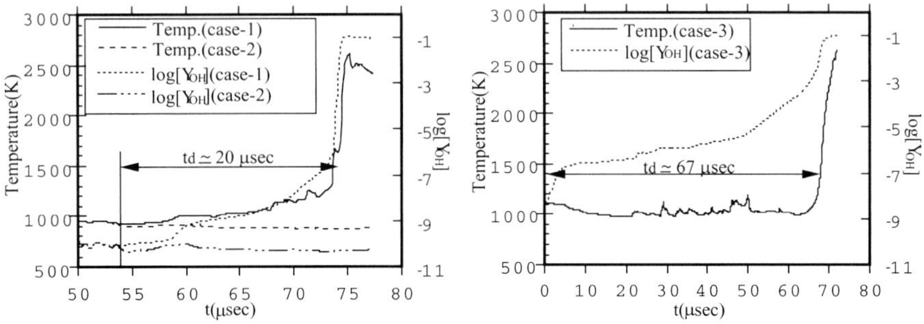

(a) Maximum temperature and maximum mass fraction of OH for case 1 and case 2.

(b) Maximum temperature and maximum mass fraction of OH for case 3.

Figure 4. Time history of the maximum temperature and maximum mass fraction of OH (Y_{OH}) for three cases: (a) results for case 1 and case 2, and (b) for case 3.

7. Conclusions

A computer study has been carried out to simulate the shock-vortex interaction in a reactive gas mixture. The ignition-delay time measured from the start of interaction between the shock wave and the vortex is about 0.3 times less than for the case of a single vortex with conditions as behind the prescribed interacting shock wave. This fact shows that the self-intensification of the shock-vortex interaction accelerates the spontaneous ignition in the reactive gas mixture and suggests that shock-vortex interaction in turbulent reactive gas mixtures is expected to trigger spontaneous ignition.

References

Minota T (1993) Interaction of shock wave with a vortex ring. Fluid Dyn Res 12:335.

Takayama F, Ishii Y, Sakurai A, Kambe T (1993) Self intensification in shock-vortex interaction. Fluid Dyn Res 12:343.

Taki S (1993) Numerical simulations of interactions of flamelets with shock waves in the premixed gas. Progress in Astronautics and Aeronautics 151:274.

Voevodsky VV, Soloukhin RI (1965) On the mechanism and explosion limits of hydrogen-oxygen chain self-ignition in shock wave. In: 10th Symp on Combustion, p 279.

Yee HC (1987) Upwind and Symmetric Shock-Capturing Schemes, NASA TM-89464.

Shock-tube investigations of atomization, evaporation, and ignition of n-decane and α-methylnaphthalene droplets

U. Pfahl, K. Fieweger, G. Adomeit
Institut für Allgemeine Mechanik
RWTH Aachen, Templergraben 64, 52056 Aachen, Germany

B. E. Gelfand
Institute of Chemical Physics
Russian Academy of Sciences, 117977 ul. Kosygina, 4 Moscow, Russia

Abstract: In this work the shock-tube technique is used to investigate the ignition process of single fuel droplets. This phenomenon consists of different physical and chemical steps. Shadow photographs are taken to observe the break-up of the parent drop behind the incident shock as well as the atomization and evaporation of the tiny drops behind the reflected shock. The pressure history and the CH-band emission are recorded to determine the ignition-delay time. Furthermore, characteristic times describing the break-up process and the evaporation behind the reflected shock are calculated. The investigated fuels are n-decane and α-methylnaphthalene with parent drop size of $d_0 = 2$ mm. The break-up of α-methylnaphthalene leads to longer times and to larger diameters of tiny drops compared to n-decane because of the higher surface tension and higher dynamic viscosity. Furthermore, the living times and the ignition-delay times are longer for α-methylnaphthalene.

Key words: Droplet, Break-up, Atomization, Evaporation, Ignition, Break-up time, Living time, Ignition-delay time, n-decane, α-methylnaphthalene

1. Introduction

The self-ignition behavior of atomized fuel in a gaseous oxidizer is a fundamental problem in diesel engines and combustion chambers. This phenomenon consists of different physical and chemical processes. First, tiny droplets originate from the injected fuel. Evaporation of these droplets is accompanied by mixing of fuel vapor with oxidizer. The self-ignition starts after a chemical ignition-delay time (Yoshizawa et al. 1981). These processes can be studied in detail using the shock-tube technique. Measurements of chemical ignition-delay times with n-decane (Fieweger et al. 1996) and previous investigations of the break-up and evaporation processes behind shock waves (e.g., Ranger and Nicholls 1969) give a sound basis for these studies. The break-up process is observed by means of shadow photography. Characteristic break-up times and diameters of secondary drops as well as living times behind the reflected shock are calculated, and the ignition-delay times are measured.

2. Experimental set-up

The experimental results have been obtained using two different shock tubes. The first shock tube has a total length of 8.2 m and is constructed for a maximum initial pressure behind the reflected shock of 20 bar. It has a square cross-section of 54 mm. The test section with a length of 200 mm is equipped with Plexiglas windows to visualize the break-up and evaporation processes by taking shadow photographs with a Cranz-Schardin camera. A second shock tube with a total length of 12.5 m and a circular cross-section of 90 mm is used to measure ignition-delay times at initial pressures up to 50 bar.

3. Results and discussion

3.1. Break-up and atomization

Figure 1 shows the break-up process of an n-decane droplet (upper line) at 12.2 bar and 1255 K and of an α-methylnaphthalene droplet (lower line) at 11.9 bar and 1243 K. Although the dynamic viscosity and the surface tension of the two fuels are very different, the break-up process looks very similar. After the incident shock passes the drop, the break-up process does not start immediately. This delay is characterized by the break-up time τ_B (Ranger and Nicholls 1969; see Eq. 1). The shattering of the parent drop starts after about $0.2\tau_B$ (Gelfand 1993). This characteristic break-up time differs only weakly for the investigated fuels (Fig. 2). However, with an increase of the pressure p_5, the characteristic break-up time decreases because of the higher density behind the incident shock wave (Fig. 2). The diameter of the tiny drops d_{sec} produced by stripping was calculated by Ingebo and Foster (1957) and Gelfand (1993).

$$\tau_B = \frac{d_0}{u_2} \cdot \sqrt{\frac{\rho_d}{\rho_2}} \qquad d_{sec} = 3.9\, d_0 \cdot \frac{\left(\frac{\nu_d}{\nu_2}\right)^{0.25}}{(\mathrm{We}\cdot\mathrm{Re})^{0.25}}\,, \tag{1}$$

where d_0 denotes the size of parent drop, ρ_d the density of fuel, u_2 and ρ_2 the gas velocity and density behind incident shock wave, and ν_d and ν_2 the kinematic viscosity of drop and gas.

The results for d_{sec} are shown in Fig. 2 for n-decane, α-methylnaphthalene, and two different mixtures of both at $p_5 = 13$ and 50 bar. The Re and We numbers for these investigations vary in the range of $(5-30)\cdot 10^4$ and $(3-18)\cdot 10^4$. These Re and We numbers lead to a so-called "explosion" mode of drop atomization (Gelfand 1993).

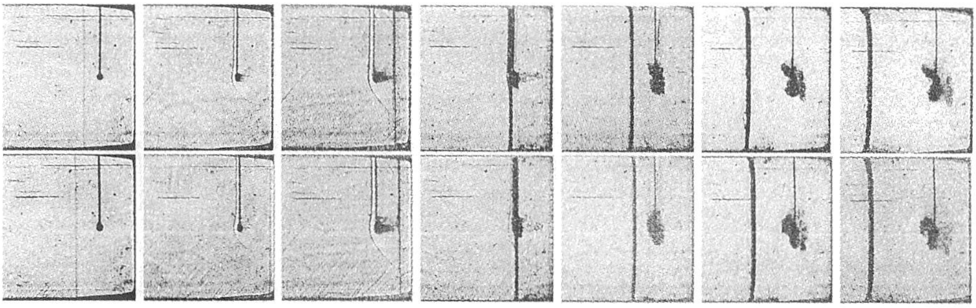

Figure 1. N_2 atmosphere, $\Delta t = 20\ \mu s$ between frames; upper line: n-decane, $p_5 = 12.2$ bar, $T_5 = 1255$ K; lower line: α-methylnaphthalene, $p_5 = 11.9$ bar, $T_5 = 1243$ K.

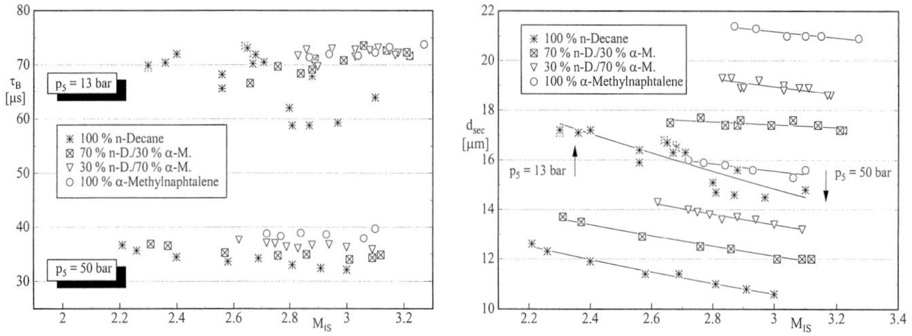

Figure 2. Characteristic break-up time τ_B and secondary drop diameter d_{sec} vs. incident Mach number for the same conditions as ignition-delay time measurements.

3.2. Evaporation

The break-up process generates a cloud of tiny drops behind the incident shock wave. These tiny drops are accelerated and heated up by the gas stream with velocity u_2 until the reflected shock passes the drops again. The living time τ_L of the secondary drops is the time which is required for complete evaporation behind the reflected shock. These times are calculated using a program by Klingsporn and Renz (1994). The same conditions were chosen as for the ignition-delay time measurements (Fig. 6). The results are shown in Fig. 3. Each calculation was done twice with two different initial temperatures of the secondary drops. The first initial temperature was 293 K, the initial temperature of the parent drop before break-up. The second initial temperature was the equilibrium temperature T_{Gl} of the tiny drops at condition 2. This is the maximum which the temperature can reach after heating up behind the incident shock. This leads to a range of living times (Fig. 3). The results show that the living times increase with increasing fraction of α-methylnaphthalene. For higher pressures p_5 the living times decrease because of the higher evaporation rate at higher pressures and because of the smaller diameters of tiny drops (Fig. 2). Figure 4 shows a comparison of n-decane and α-methylnaphthalene at similar initial conditions. The different evaporation rate behind the reflected shock wave is obvious.

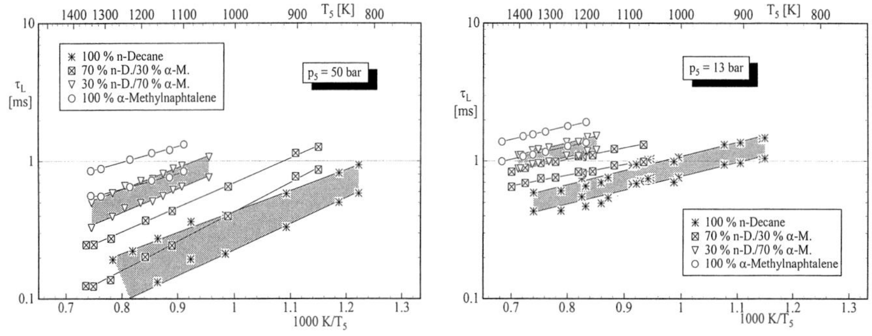

Figure 3. Living time τ_L vs. 1000 K/T_5 for the same conditions as ignition-delay time measurements.

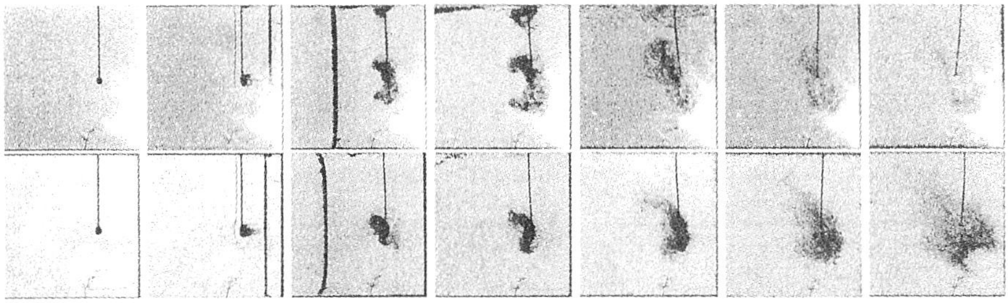

Figure 4. Both lines: frames 1–4: $\Delta t = 80$ μs and frames 4–7: $\Delta t = 480$ μs; upper line: n-decane, $p_5 = 15.8$ bar, $T_5 = 925$ K; lower line: α-methylnaphthalene, $p_5 = 12.0$ bar, $T_5 = 1179$ K.

3.3. Ignition

The evaporation of the liquid drops leads to the formation of an ignitable fuel/air mixture. The ignition process itself is strongly influenced by the fuel, pressure, temperature, and equivalence ratio ϕ (Fieweger et al. 1996). As an example, Fig. 5 shows the break-up process of an n-heptane droplet followed by the formation of the fuel/air mixture. In frame 6, upper line, most of the fuel is already

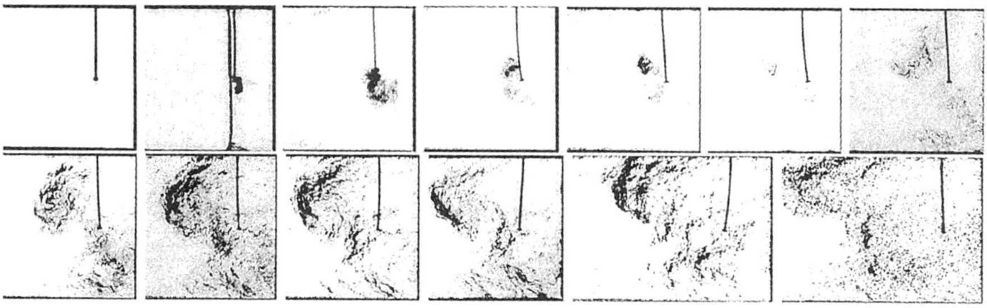

Figure 5. N-heptane, $p_5 = 16.0$ bar, $T_5 = 1214$ K, $\tau_i \approx 0.6$ ms; between frames 1–11 $\Delta t = 100$ μs, between frames 11–13 $\Delta t = 500$ μs.

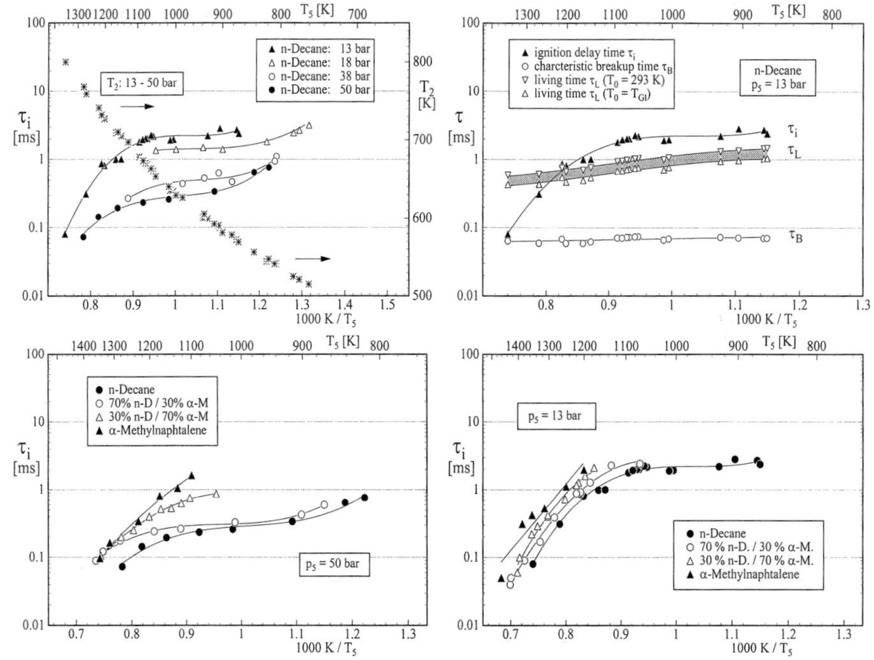

Figure 6. Measured ignition-delay times vs. 1000 K/T_5; upper left: additonal T_2; upper right: additional calculated break-up time τ_B and living time τ_L.

evaporated. Frame 7 shows the beginning of the ignition. The frames in the lower line show the entire subsequent combustion process which is finished when the fuel is consumed. The ignition-delay time is the interval measured between the reflected shock passing the cloud of secondary drops and the first CH-band emission (431 nm); it is plotted in Fig. 6 vs. 1000 K/T_5 for different pressures. A comparison of droplet ignition-delay times with ignition-delay times for premixed stoichiometric fuel/air mixtures (Fieweger et al. 1996) shows in general longer times for the droplets. However, in the range of the s-shaped dependence, the droplet ignition-delay times may be shorter because of the strong dependence of the chemical ignition-delay time on the equivalence ratio. Also plotted in Fig. 6, upper left, is the temperature behind the incident shock T_2. With an increase of T_2, the atomization and evaporation behind the incident shock increase, which shortens the ignition-delay times behind the reflected shock. Figure 6, upper right, shows the comparison between the characteristic break-up time τ_B, the living time region τ_L, and the ignition-delay time τ_i for n-decane at 13 bar. For temperatures above 1200 K, the living times are much longer than the ignition-delay times. In this region the condition behind the incident shock has more influence on the formation of ignitable fuel/air mixtures than for lower temperatures. The break-up times do not depend on the temperature. Generally speaking, the ignition-delay times are shorter for n-decane, which is the more ignitable fuel, compared to α-methylnaphthalene. Indeed, at 13 bar the ignition-delay times increase continuously with an increasing fraction of α-methylnaphthalene. This behavior, however, is different at 50 bar as can be seen in Fig. 6, lower left. For temperatures below 1100 K, the ignition-delay times for a 30%/70% n-decane/α-methylnaphthalene mixture hardly differ from pure n-decane, whereas above 1250 K, the n-decane/α-methylnaphthalene mixtures behave almost as pure α-methylnaphthalene.

References

Fieweger K, Pfahl U, Adomeit G (1996) Self-ignition of n-heptane/iso-octane mixtures and n-decane with air under engine conditions. In: Shock Waves, Proc of the 20th ISSW, Sturtevant B, Shepherd JE, Hornung HG (eds), World Scientific, Vol 2, pp 929–934.

Gelfand BE (1993) Droplet Shattering Phenomena, Institute of Chemical Physics, RAS, Moscow.

Ingebo R, Foster H (1957) Drops Size Distribution for Cross Current Breakup of Liquid Jet in Air Streams, NASA TR-4087.

Klingsporn M, Renz U (1994) Vaporization of a binary unsteady spray at high temperature and high pressure. Int J Heat Mass Transfer 37(suppl 1):265–272.

Ranger AA, Nicholls JA (1969) Aerodynamic shattering of liquid drops. AIAA Journal 7(2):285–290.

Yoshizawa Y, Tomito M, Kawada M (1981) A study on the ignition of a fuel droplet in high temperatures stagnant gas. In: 13th Int Symp on Shock Waves, Niagara Falls, USA.

Numerical modeling of chemistry and gas dynamics during shock-induced ethylene combustion

L. Clifford, A. Milne
Department of Mathematical Sciences
St. Andrews University
Scotland

D. Boulton
Fluid Gravity Engineering
5 Alexandra Place
St. Andrews, Scotland

Abstract: We present the results of a numerical study of shock-induced ethylene combustion. We use an induction-parameter model (IPM) which provides a simplified approach to the inclusion of chemistry into reacting-flow calculations. We describe the derivation of two functions that model respectively the induction times and the subsequent energy release. These functions are found by curve fitting to data calculated using a chemical-kinetics reaction data set. We show some results of coupled gas-dynamics and reactive-flow calculations using the induction-parameter and chemical-kinetics models. The results from the two approaches are in good agreement with experimental results, but using the induction-parameter model leads to savings of more than 95% in the computational time taken.

Key words: Numerical modeling, Ethylene combustion, Shock tube, Reactive flow

1. An induction parameter model for ethylene combustion

The induction-parameter model (IPM) (Taki and Fujiwara 1981, Oran et al. 1980) provides a simplified approach for the inclusion of chemistry into reacting-flow calculations. The primary benefit is the reduction in computation time. For any fuel system it is formulated by determining, for a wide range of initial pressures and temperatures: (i) an induction delay time, which corresponds to the time for significant increase in radical concentrations, and (ii) a reaction time, which corresponds to the time for fuel consumption and energy release. These parameters are determined from calculations using a chemical-reaction-rate equation-integration package and an appropriate chemical-reaction scheme. For hydrocarbons, the latter can be quite complex with over one hundred reaction steps. It should be noted that these reaction schemes can be significantly different from those which are applicable for propagating low-velocity flames where other physical parameters dominate.

Clifford et al. (1995) derived an IPM model for the shock-induced combustion of ethylene in an argon atmosphere. We outline here the basis of this model. In an IPM the detailed reaction mechanism is replaced by the following reactions:

(1) Non-exothermic Induction Reaction.

The rate of reaction is given by

$$\omega_\alpha \equiv \frac{d\alpha}{dt} = -\frac{1}{\tau_i} \tag{1}$$

where τ_i is the induction time.

The induction time represents the delay between the fluid first experiencing a change in conditions (such as passing through a shock) and the subsequent energy release (as a result of chemical reactions). During the induction period, no energy is released so long as $\alpha > 0$; α represents the fraction of induction time remaining during this period. The chemical composition of the fluid is not assumed to change during the induction period.

(2) Exothermic Reaction.

The energy-release phase starts as soon as $\alpha < 0$. The rate of reaction is given by:

$$\omega_\beta \equiv \frac{d\beta}{dt} = -kT^2\beta^2 \exp\left(-\frac{E}{RT}\right) \tag{2}$$

where β is a measure of the chemical energy yet to be released (so that when β reaches 0 all the chemical energy has been released and the reaction is complete); k is the reaction rate and E is the activation energy, which are found by plot-fitting of data obtained using a detailed reaction mechanism. We note here that in the IPM model the changes in the chemical species present as the reaction proceeds are not modeled.

The continuity equation for α (similarly for β) is written as:

$$\frac{\partial(\rho\alpha)}{\partial t} + \nabla \cdot (\rho\alpha\nu) = \omega_\alpha, \tag{3}$$

the solution of which is split into two stages:

(a) Lagrangian phase

$$\frac{\partial(\rho\alpha)}{\partial t} = \omega_\alpha \tag{4}$$

followed by

(b) Van Leer advection phase

$$\frac{\partial(\rho\alpha)}{\partial t} + \nabla \cdot (\rho\alpha\nu) = 0 \tag{5}$$

and similarly for β.

The energy released (ΔE_{ipm}) at each time step is computed as:

$$\Delta E_{ipm} = Q\Delta\beta, \tag{6}$$

where $\Delta\beta$ is the change in β over the time step, and Q is a constant which can be regarded as the maximum amount of energy that can be released by a reaction.

2. Shock-tube simulations

To model a shock tube we use a hydrocode called EDEN. This package incorporates finite-rate nonequilibrium chemistry into a two-dimensional Fluid-In-Cell (FLIC) hydrocode. The complete time cycle is divided into different phases and the changes in the fluid variables due to only some of the terms in the hydrodynamic equations are then calculated in each phase. The changes in the variables over a whole time step are then found by summing the changes in each phase. The chemistry solver is thus written in a self-contained package and added to the other routines.

The hydrodynamic equations that we solve are:

Conservation of mass	$\frac{\partial\rho}{\partial t} = -\nabla \cdot (\rho\mathbf{v})$
Conservation of species	$\frac{\partial n_i}{\partial t} = -\nabla \cdot (n_i\mathbf{v}) + Q_i - L_i n_i$
Conservation of momentum	$\frac{\partial(\rho\mathbf{v})}{\partial t} = -\nabla \cdot (\rho\mathbf{v}\mathbf{v}) - \nabla \cdot P$
Conservation of energy	$\frac{\partial(\rho\epsilon)}{\partial t} = -\nabla \cdot (\rho\epsilon\mathbf{v}) - P\nabla\mathbf{v} - \nabla \cdot (q) + \Delta E$

where ρ is the mass density, $\mathbf{v}$ is the flow velocity, P is the pressure scalar, ϵ is the specific internal energy, q is the heat flux, and ΔE is the chemical energy released by the reactions.

In the chemistry solution routines, at each time step, the change in mass fractions of the species and the energy released are calculated. This is the first stage of the calculation. The initial chemistry time step is taken as the hydro time step determined from Courant conditions and is reduced if the energy release during this phase is too large. This revised chemistry time step then becomes the hydro time step. The energy released is then used as a source term in the second, Lagrangian phase, in which the velocity and energy are calculated based on acceleration effects due to pressure gradients. The final phase is a Van Leer advection phase, into which the species mass densities from the chemical phase are input and the momentum and energy are transported.

2.1. Experimental details

The shock tube used consists of three horizontal sections: driver, buffer, and test sections. The driver section is filled with helium up to 20 bar. The buffer section is filled with an inert gas mixture and allows the incident shock to stabilize before it enters the reactive test gas. The test section containing the reactive gas mixture is 76 mm high. The test and buffer gases are matched acoustically so that the incident shock propagates from the buffer section into the test section without any changes in the transmitted pressure and velocity. The test and buffer sections have the same initial pressure and are separated by a slide valve, the valve being opened at the same time as the shock is initiated in the drive gas by the bursting of a diaphragm that separates the driver and buffer gases.

A rectangular obstacle of height 38 mm (half the tube height) is placed in the test section, which produces a wide range of gas-dynamic conditions as part of the shock is reflected from the obstacle and part is diffracted around it.

For the purposes of the numerical modeling it is only necessary to model the test-gas region as the shock entering it from the buffer zone can be modeled by the appropriate inflow boundary conditions. The initial conditions have the incident-shock conditions being present to within a few cells of the leading edge of the obstacle. The spatial resolution is 1 mm per cell. The obstacle and walls of the tube are modeled by suitable perfectly reflecting boundaries.

2.2. Numerical results

We present here some representative results for a shock with an incident Mach number in the test section of 3.65. The described behavior of the system is generally the same for any Mach number for which the conditions in the incident shock wave do not lead to combustion, but where the reaction proceeds behind the reflected shock wave. The major differences are in induction times and the strength of the subsequent blast wave.

When the shock hits the obstacle the unrestricted upper part passes over the top of the obstacle then expands into the region behind it. The lower part of the incident shock wave is reflected by the obstacle. Conditions behind the reflected shock produce a reaction along the leading edge of the obstacle with a short induction time (a few microseconds). This sharp increase in pressure produced by the reaction at the wall results in a blast wave rapidly propagating outward from the wall, which quickly catches up to the reflected shock and rapidly steepens to form a detonation wave.

After a time the reactive wave hits the upper part of the shock tube and reflects. Figures 1 and 2 show density contours of the wave structure after 80

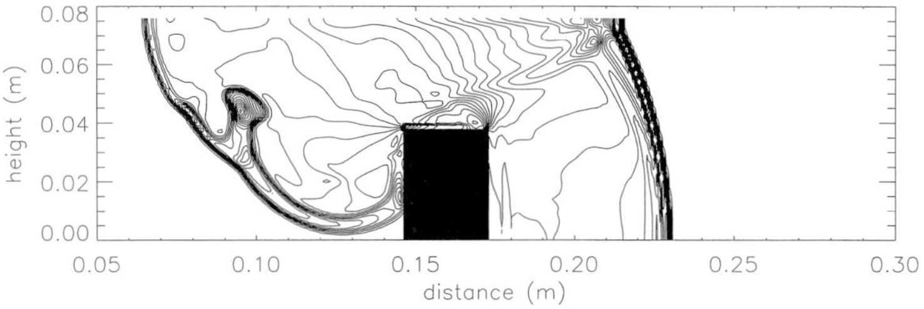

Figure 1. Density contours using the IPM model at Mach 3.65 after 80 μs.

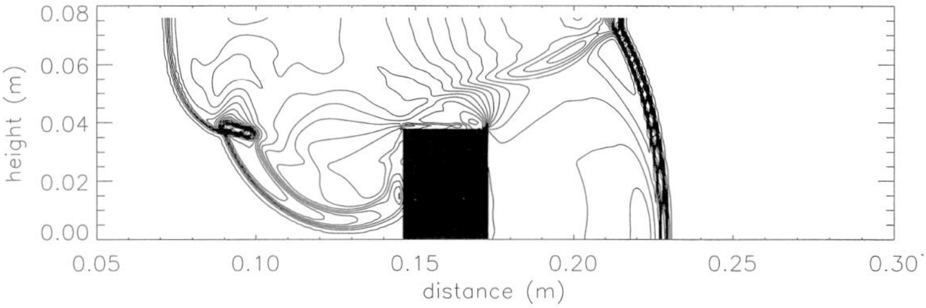

Figure 2. Density contours for the chemical-kinetics simulation at Mach 3.65 after 80 μs.

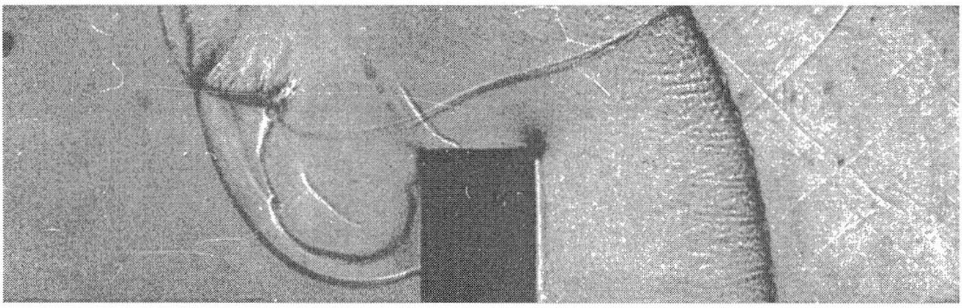

Figure 3. Schlieren photograph at Mach 3.65 after approximately 80 μs.

μs has elapsed. We see that there is good agreement between the two models. The small differences in the speeds of the waves are because the IPM model does not incorporate changes in the chemical species as the reaction proceeds, and so the values of C_v and γ in the two models are not quite the same and the detonation waves proceed at slightly different speeds.

Comparisons between model and experiments are discussed in Thomas et al. 1995. There is good agreement between them as shown by Fig. 3.

Finally, as the motivation for constructing an IPM model is to decrease the computational time needed to simulate complex reactive-flow problems, we compare the computer time taken to produce the data shown in Figs. 1 and 2. Using an ESCOM Pentium-based personal computer the IPM simulation needed less than 5% of the computational time of the chemical-kinetics simulation.

3. Conclusion

We derived an induction-parameter model for reactions involving the oxidation of ethylene in an argon atmosphere. We have found the model to be acceptably accurate in terms of induction-time and energy-release predictions, and to provide large savings in computer time (because of the reduction of the set of equations to be solved to two and the number of species in the system to just three). We used the model in a 2D simulation of complex behavior of a shock passing over a rectangular obstacle. This is a good test of the model since a wide range of temperature and pressure environments are present in the problem. There is good agreement between experiment and model with the model reproducing all the main features of the experiment. We believe we have shown that induction-parameter models can be derived for hydrocarbon oxidation that are sufficiently accurate to be used with some confidence in reactive-flow models and that the increase in speed of these models means that they could be useful for problems that are currently impractical in terms of computational facilities.

References

Clifford LJ, Milne AM, Murray BA (1995) Numerical modelling of chemistry and gas dynamics during shock induced ethylene combustion. Combustion and Flame (submitted).

Oran ES, Boris JP, Young TR, Flanigan M, Picone M (1980) Simulations of Gas Phase Detonations: Introduction of an Induction Parameter Model, NRL Memorandum Report 4255.

Taki S, Fujiwara T (1981) Numerical simulation of triple shock behaviour of gaseous detonation. In: 18th Symp (Int) Combust, Pittsburgh, PA, The Comb Inst, pp 1671–1681.

Thomas GO, Sands CJ, Milne AM, Clifford LJ (1995) Studies of shock inititated combustion. Combustion and Flame (submitted).

Shock-tube combustion of liquid sprays of ethanol: effect of additives on ignition delays and combustion characteristics

P. Cadman
Combustion Physics Group, University of Wales
Wales, Aberystwyth, SY23 3BZ UK

Abstract: The shock-tube combustion of liquid sprays of ethanol containing small quantities of additives such as isopropyl nitrate (IPN) and/or water has been investigated at pressures around 12 bar and temperatures between 1000–1800 K. Two types of combustion were found, type 1 where the pressure profile increased smoothly with time, and type 2, characterized by a very sharp pressure profile and more noise. The latter had characteristics of an unsustained detonation. IPN was found to reduce the ignition delay but water lengthened it and changed type 2 to type 1. IPN shortened the delay of fuel sprays containing water but did not give type-2 combustion except perhaps at very high temperatures. The soot yield was very small in all the combustions of this fuel that were investigated.

1. Introduction

The shock tube has been widely used to measure the combustion characteristics and ignition delays of homogeneous mixtures of fuel/gaseous oxidants (Cadman 1993; Cadman & Denning 1990, 1995; Cadman et al. 1996). At Aberystwyth we have tried to extend measurements to the combustion of two-phase systems, including the oxidation of soot particulates (Cadman et al. 1996, Frenklach & Yuan 1987) and the liquid-spray combustion of toluene and tetradecane (Cadman 1993). In previous work (Cadman 1993) the combustion of tetradecane and toluene sprays was found to give two types of combustion, one where the pressure increase with time was smooth and one where there was a very sharp pressure rise which was followed by smaller ringing-type signals.

The velocity of the ignition throughout the system was over 10^7cm s^{-1}. It was suggested that this type of combustion could be caused by a detonation-type of ignition in the droplets/vapor pockets which exist in the high-temperature and pressure conditions. Fuels often contain additives to modify the combustion in some way, e.g., diesel fuels can contain small amounts of organic nitrates (such as isopropyl or octyl nitrate) which are thought to reduce the ignition delay. In previous work we have looked at the effect of isopropyl nitrate (IPN) and t-butyl peroxide on the ignition delays of tetradecane and toluene. Water and alcohols are important in modifying the production of soot during combustion of larger hydrocarbons. While being immiscible with normal fuels, water is completely miscible with ethanol and so its effect can be investigated easily. In this study we have investigated the effect of water and IPN on the ignition delays of ethanol.

2. Experimental

An 11-m long 63.5-mm internal diameter stainless steel shock tube (5–6-m driver and 6-m driven section) equipped with optical windows and pressure transducers was used for this work (Cadman 1993). An injection system constructed in this laboratory was fitted onto the end plate of the tube. Melinex diaphragms were used (175-μm thickness) to separate the two sections. The shock was initiated by a 12-V/4-A pulse through a hot wire between the two layers of diaphragm material. This was found to give reproducible bursting of the diaphragm. The fuel injector was controlled by a 12-V pulse from the same control unit for synchronization with the diaphragm bursting. PCB pressure transducers placed at the end plate, main observation window, and along the tube were used to measure the incident- and reflected-shock velocities which were utilized to calculate the reflected-shock temperature and pressures using CHEMKIN and a program written at Aberystwyth.

The main observation window was situated 10 cm from the end plate where the spray disappearance and rate of soot appearance could be monitored using a small 20-mW 488-nm argon-ion laser, photodiode, and interference filter. Emission measurements were made at the same point via other windows using optical fibers, beam splitters, and photomultipliers with spectral isolation using narrow band-pass interference filters (454 and 546 nm). A second window system 20 cm from the end plate had a He-Ne laser-beam/photodiode system for detection of soot further down the tube. Most reaction of the hydrocarbon species occurred between the 10-cm port and the end plate. Eight transient signals could be captured with a 20-MHz transient recorder and stored in a microcomputer for subsequent analysis. The online detection of soot production was also measured via laser absorption using the argon-ion and He-Ne laser systems described above.

3. Results

The combustion of ethanol sprays was investigated as described previously (Cadman 1993); 40-mg aliquots of fuel were injected immediately before the shock wave. The ignition delay was measured from the shock arrival to the first temperature increase as measured by the ratio of light emissions at the two wavelengths visible (454 and 546 nm). The shape of the pressure profile after combustion indicated which type of combustion occurred.

Ethanol was injected into 10% oxygen/argon mixtures and shock heated to 900–1700 K and 11–12 bar reflected pressure, where both the pressure profile and ignition delay were measured. For ethanol/oxygen alone, laser-absorption measurements showed that the droplets had disappeared on passage of the reflected shock. This was different from previous fuels (Cadman 1993) where droplets lasted for up to 200 μs. Type-2 ignition was found to be present, particularly at the higher temperatures (Fig. 1). The pressure profiles showed ringing as the waves reflected across the tube at about 80–100-μs intervals. The

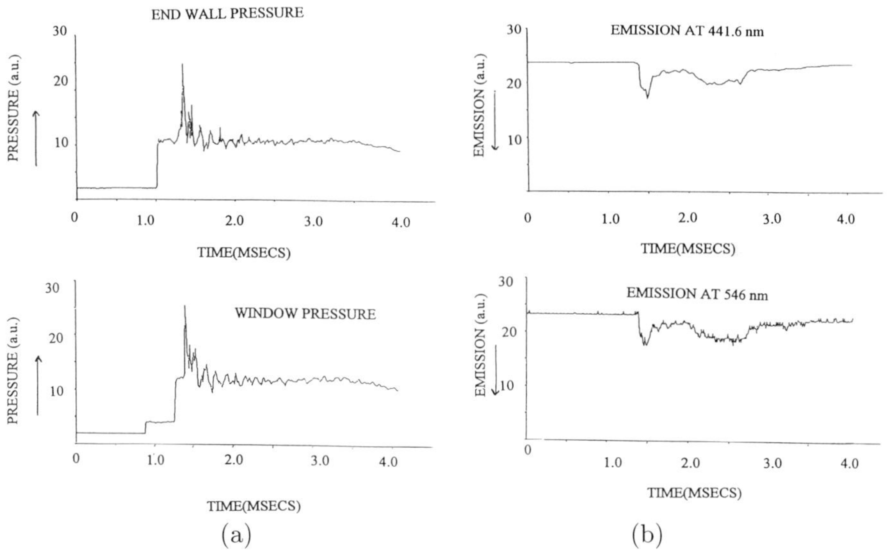

Figure 1. (a) Pressure and (b) emission profiles for ethanol/10% O_2 in argon; T = 1288 K, P = 10.6 bar.

emission profiles showed that the light output occurred in a similar periodic fashion but often with a higher frequency than the pressure profiles, suggesting that small pockets of combustion were occurring but were not always strong enough to affect the pressure output. This is in keeping with the idea of small micro-detonations occurring but not being transmitted throughout the whole reaction volume.

The light output from ethanol combustion is much less than that from either toluene or tetradecane. Ethanol containing 10% water was then investigated under similar conditions. There was a large difference in the type of ignition and only type 1 was found except for a couple of experiments at the highest temperatures, and even these combustions were borderline type 2. Figure 2

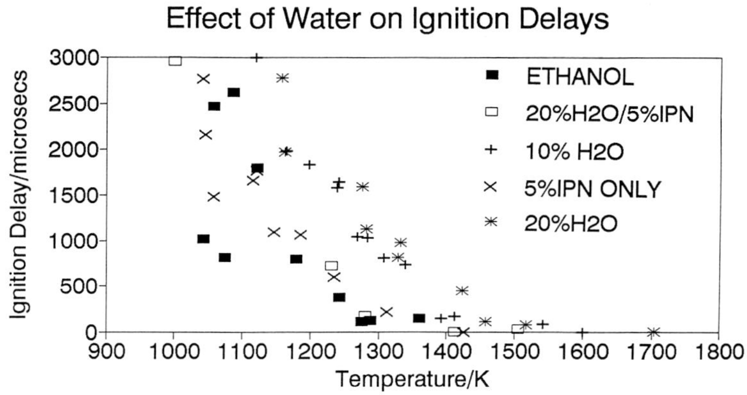

Figure 2. Effect of water on ignition delay — ethanol/5% IPN/10% O_2.

shows the ignition delays for oxygen/ethanol and oxygen/ethanol/water mixtures. The presence of water increases the ignition delays over those without water. Twenty-percent water/ethanol results are also shown in Fig. 2. The ignition delays are similar to those with 10% water, i.e., no further lengthening seemed to occur and, in addition, no type-2 combustion was found.

The combustion of ethanol containing 5% IPN (an 'ignition improver') was investigated similarly. The ignition delays for these mixtures are shown in Fig. 2, where little difference was found compared to ethanol only. The presence of 5% IPN has little effect upon the ignition delays for ethanol in contrast to its effect on toluene and tetradecane. Type-1 combustion occurred except for two higher-temperature experiments (at $>$ 1300 K) where very weak type-2 combustion was found (Fig. 3). Mixtures of 75% ethanol, 20% water, and 5% IPN were investigated. The ignition delays displayed in Fig. 2 show that the ignition delays are shorter than for those mixtures without IPN but containing water, and similar to those with 100% ethanol or with IPN only. Emission profiles indicated little light emission, even compared to ethanol alone and showed that this system was on the borderline between ignition and non-ignition, resulting in erratic ignition/combustion. In the higher-temperature range the pressure profiles indicated that when ignition did take place, despite the weak combustion, type-2 ignition could still occur.

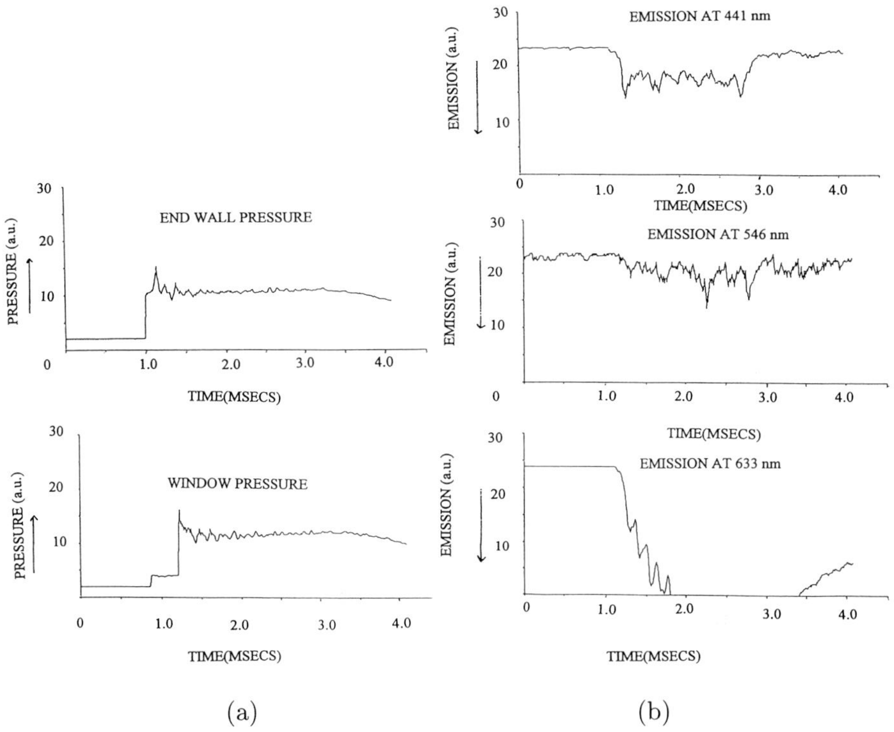

Figure 3. (a) Pressure and (b) emission profiles for ethanol/5% IPN/10% O_2. T = 1426 K, $P = 11.8$ bar.

The effect of adding water to ethanol seems to be to lengthen the ignition delay and change type-2 into type-1 combustion, i.e., it changes the essential nature of the combustion, an effect which can be partially counteracted by also adding IPN to the system. If the linear regression lines are plotted (Fig. 4) the effect of the additives on the ignition delay is shown more clearly. Ethanol/water mixtures have longer delays mainly because of the temperature-independent factor. The addition of IPN (with and without water) causes similar delays but both are slightly longer than ethanol alone. Online laser-absorption monitoring of the soot volume fraction showed that these systems are very clean with little soot production. The small soot production by alcohol fuels has been confirmed by investigations with alcohol/hydrocarbon/water combinations (Cadman et al. 1996, Frenklach & Yuan 1987, Alexiou & Williams 1991).

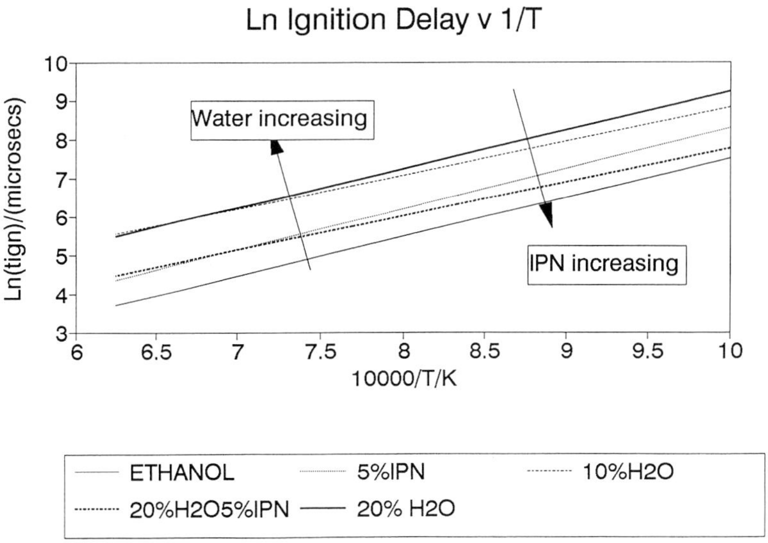

Figure 4. Comparison of best-fit lines.

Acknowledgement. We thank the EPSRC for financial support.

References

Alexiou A, Williams A (1991) Soot formation in methanol/toluene mixtures. Fuel 73:1280.

Cadman P (1993) Shock tube combustion of liquid hydrocarbon sprays at high temperatures. In: Shock Waves @ Marseilles, R Brun, LZ Dumitrescu (eds), Springer-Verlag, p 179.

Cadman P, Denning RJ (1990) Oxidation of soot particulates in shock waves. In: 17th International Symposium on Shock Waves, Y Kim (ed).

Cadman P, Denning RJ (1995) Shock tube study of high temperature oxidation of soot by nitric oxide. In: First Symposium on Clean Air, published as Energies for Clean Technology, M da Graca Carvalho (ed), Gordon Breach.

Cadman P, Wang R, Kunz A (1995) Soot formation in liquid sprays of propanol/tetradecane/water in reflected shock waves (to be published).

Frenklach M, Yuan T (1987) Effect of alcohol addition on shock initiated formulation of soot from benzene. In: 16th International Symposium on Shock Waves, Aachen.

The effect of halocarbons on methane ignition

A. Burcat, M. Dvinyaninov
Faculty of Aerospace Engineering
Technion – Israel Institute of Technology, Haifa 32000 Israel
A. Lifshitz
Department of Physical Chemistry
The Hebrew University, Jerusalem 91904 Israel

Abstract: Ignition-delay times in mixtures containing methane, oxygen, and various halocarbons, diluted in argon, were measured behind reflected shocks in a single-pulse shock tube. The temperature range covered was 1500–2000 K at pressures p_5 of ~10 atm. Five different halocarbon additives: $CBrF_3$, $CHCl_2F$, $C_2H_2F_2$, C_2F_6, and CF_4 were tested in order to examine their effect on the ignition delay of methane. All the additives had either an enhancing effect on the oxidation of methane or no effect at all. The attempts to increase the ignition delay of methane by adding compounds, which are known to reduce flame velocities, were unsuccessful.

Key words: Methane ignition, Halocarbons

1. Introduction

It has been shown in numerous studies (Burcat et al. 1971) that methane has the longest ignition-delay time among the simple aliphatic hydrocarbons. It was also shown (Lifshitz et al. 1971) that by adding small amounts of ethane or propane to methane, it was possible to shorten and control its ignition-delay time. These phenomena were rationalized by the fact that the C–H bond in methane is considerably stronger than the C–C bond in larger hydrocarbons. The initiation due to $M + CH_4 \rightarrow M + CH_3^\bullet + H^\bullet$ in methane is therefore much slower than the equivalent one in other hydrocarbons which proceeds according to the reaction: $R_1{-}R_2 \rightarrow R_1^\bullet + R_2^\bullet$. Whereas it is rather simple to shorten the ignition delay of methane, we are not aware of any attempt to increase its delay.

From a chemical-kinetic view point, when halocarbons are added as additives, there is competition between the higher initiation rates in compounds such as CH_3Cl, CH_3Br, and others (relative to CH_4) and the tendency of fluorine, chlorine, and bromine atoms to combine with hydrogen atoms to produce the inert HF, HCl, and HBr molecules.

This article describes an attempt to increase the ignition delay of methane by addition of halocarbons which are used as fire extinguishers and are known to decrease flame velocities of hydrocarbons (Westbrook 1982). The ability to increase and control the ignition delay of methane is of great relevance to the operation of ram accelerators (Higgins et al. 1993) since the precise timing of the sudden energy release (ignition) of the high-pressure fuel-oxygen mixture, relative to the position of the projectile, is a crucial factor.

As will be shown, the high initiation rates of halocarbons override the inhibiting effect caused by the production of the unreactive HX molecules. Thus, the addition of halocarbons to methane-oxygen mixtures not only does not increase its induction time, it sometimes shortens it.

2. Experimental

The ignition-delay times were measured behind reflected shocks in a pressurized driver single-pulse shock tube made of 54-mm i.d. stainless steel tubing. The driven section was separated from the driver by 'Mylar' polyester film of various thicknesses depending upon the desired shock strength. Reaction mixtures for running the experiments were prepared manometrically, and were stored at high pressure in stainless steel cylinders.

Reflected-shock parameters were calculated from the measured incident-shock velocities using the three conservation equations and the ideal-gas equation of state. The incident-shock velocities were measured with two miniature, high-frequency pressure transducers (PCB) placed 200 mm apart in the downstream end of the driven section. The signals generated by the shock waves passing over the transducers were connected in parallel and were fed through a piezo amplifier to one beam of a Nicolet Model 3091 dual-beam digital oscilloscope. Time intervals between the two signals shown on the oscilloscope (approximately 250 μs) were obtained digitally with an accuracy of 2 μs corresponding to 15–20 K depending upon the shock temperature.

The duration of the induction period was determined from pressure observations with the aid of a third transducer located at the center of the end plate of the driven section. It was taken as the time interval between the reflection of the shock wave from the end plate and the sudden increase in pressure caused by the onset of ignition. Reaction dwell times varied between 2.0 to 2.2 ms, which set an upper limit for the delays that could be measured. The induction times were read from the second beam of the oscilloscope with an accuracy of $\pm 20\%$ at low induction times (around 100 μs) and $\pm 10\%$ at high induction times in which the pressure jumps were sharper. A typical pressure record showing the shock heating and the ignition phenomenon is shown in Fig. 1.

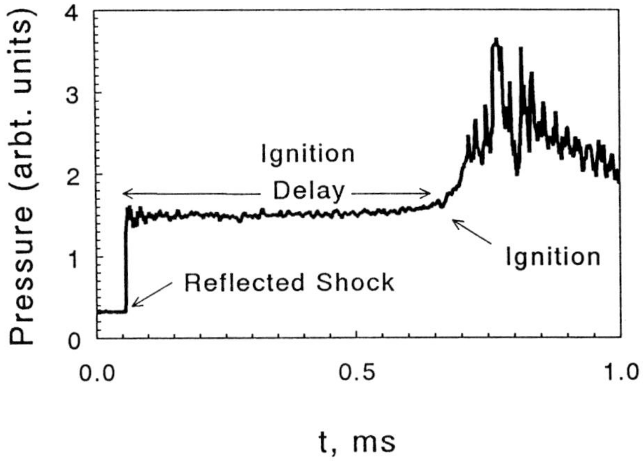

Figure 1. A pressure record showing the shock heating and the ignition process.

3. Results

We have measured ignition-delay times in six mixtures containing 3.5% methane and 7% oxygen diluted in argon; to five of these mixtures 0.3% (~10% of the methane concentration) of various halocarbons were added. The halocarbons tested were: $CBrF_3$, $CHCl_2F$, $C_2H_2F_2$, C_2F_6, and CF_4. Several experiments without additives were also performed for comparison. The experimental conditions and the results of the experiments are given in Table 1 and are shown graphically as plots of $\log \tau$ vs. $1/T$ in Fig. 2. The correlation which relates the ignition-delay times of pure methane to the temperature and the concentrations is given by (Lifshitz et al. 1971):

$$t_{ign} = 3.62 \times 10^{-14} \exp(46.5 \times 10^3/RT) \times [\mathrm{Ar}]^0 [\mathrm{CH_4}]^{0.33} [\mathrm{O_2}]^{-1.03}\ \mathrm{s} \qquad (1)$$

and is represented in the figure by a straight line. It confirms the results obtained in several experiments performed without additives.

As can be seen none of the halocarbons which were added to the reaction mixtures caused an increase in the ignition delay of methane. All the data points scatter slightly below the points of pure methane, indicating that no inhibiting effect which delays the ignition is obtained.

4. Discussion

Lifshitz et al. (1990) have recently demonstrated the existence of a correlation between the temperature required to obtain a given ignition-delay time ($T_{(t_{ign})}$) in mixtures of various fuels and oxygen, and the bond dissociation energy of the fuel. The correlation is shown in Fig. 3 where $1/T_{(t_{ign})}$ is plotted vs. D(C–X)

Table 1. Experimental conditions and ignition delay of methane in mixtures of 3.5% methane and 7% oxygen in the presence of 0.3% additives.

No	additive	p_5	T_5	t_{ign}
7	none	10.47	1728	107
8		10.52	1770	83
9		8.68	1540	403
10		12.46	1924	27
11	$CBrF_3$	9.99	1675	97
12		10.11	1688	89
13		9.50	1623	142
14		9.47	1599	225
15	$CHCl_2F$	8.85	1537	435
16		10.42	1710	113
17		10.81	1737	75
18		12.57	1919	25
19	CF_4	8.91	1558	250
20		10.57	1735	87
21		12.07	1868	46
22		12.46	1900	33
23		10.80	1735	111
24		9.49	1630	232
25		10.26	1681	243
26		9.52	1605	296
27	C_2F_6	9.14	1565	188
28		8.44	1487	465
29		10.63	1717	83
30		12.23	1863	27
31		8.48	1487	556
32		10.21	1664	120
33		11.85	1832	42
34		11.25	1773	51
35	$C_2H_2F_2$	9.66	1651	119
36		8.68	1542	405
37		8.92	1542	403
38		12.37	1894	29
39		14.11	2087	14
40		12.40	1911	17
41		10.13	1677	126
42		10.24	1677	140

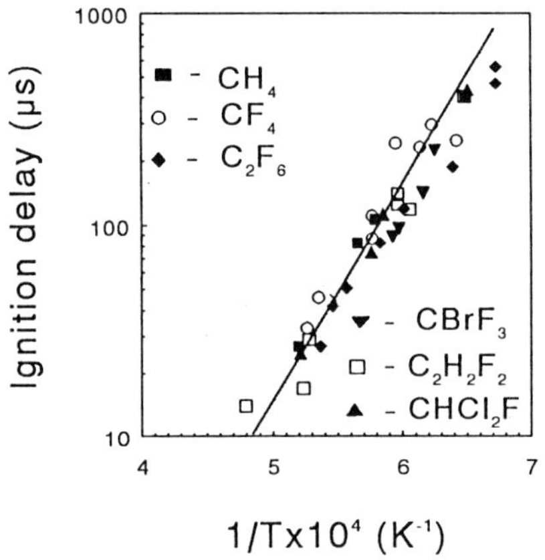

Figure 2. Ignition delay in methane-oxygen-argon mixtures, with and without additives.

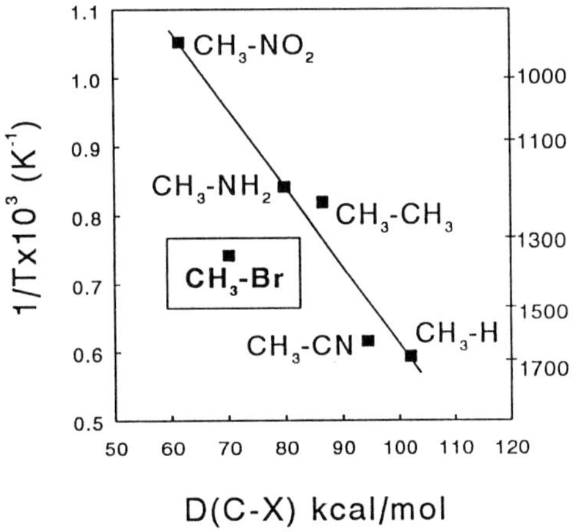

Figure 3. The temperature for which the ignition delay is 318 μs as a function of the bond-dissociation energy.

for mixtures of the same equivalence ratio, where D(C–X) is the bond strength of the weakest bond in the fuel. For the purpose of comparison, $T_{(t_{ign})}$ corresponding to 318 μs was chosen. The molecules used in this correlation start with CH_3NO_2 with a bond energy of ~60 kcal/mol, CH_3NH_2, CH_3Br, C_2H_6, CH_3CN up to CH_4 with a bond strength of ~104 kcal/mol.

Whereas the data points of five molecules can be practically connected with a straight line, as can be seen in Fig. 3 the data point for CH_3Br is considerably below the line. Its position on the line corresponds to a temperature $T_{(t_{ign}=318\mu s)}$ of approximately 1040 K whereas the actual point is positioned at ~1350 K. This behavior reflects a strong inhibition of the combustion process by the bromine atoms which are generated during the course of methyl-bromide consumption.

Figure 3 shows, however, that an ignition delay of $t_{ign} = 318$ μs in methane requires a temperature of ~1670 K. This temperature is still considerably higher than the one required for CH_3Br to show a delay of 318 μs namely, 1350 K. These observations imply that the inhibition in halocarbons caused by recombination of Br or Cl with H is significant, but not strong enough to override the high initiation rates due to the weak C–Cl or C–Br bonds in the molecules. The inability to increase the ignition delay of methane by addition of halocarbons is a direct consequence of this unfavorable competition.

Acknowledgement. M. Dvinyaninov thanks The Israeli Ministry of Absorption and The V.P.R. Struck Fund for the Advancement of Science for their financial support.

References

Burcat A, Scheller K, Lifshitz A (1971) Shock tube investigation of comparative ignition delay times for C_1, C_5 hydrocarbons. Comb and Flame 16:29–33.

Higgins AJ, Knowlen C, Bruckner AP (1993) An Investigation of Ram Accelerator Gas Dynamic Limit, AIAA-93-2181, pp 1–12.

Lifshitz A, Scheller K, Burcat A, Skinner GB (1971) Shock tube investigation of ignition in methane-oxygen-argon mixtures. Comb and Flame 16:311–321.

Lifshitz A, Bidani M, Carroll HF (1990) The role of initiation vs. chain branching in ignition processes. In: Current Topics in Shock Waves, Proceedings of the 17th International Symposium on Shock Waves & Shock Tubes, American Institute of Physics Conference Proceedings, Kim YW (ed), pp 702–706.

Westbrook CK (1982) Inhibition of hydrocarbon oxidation in laminar flames and detonations by halogenated hydrocarbons. In: 19th Symp (Int) on Combustion, The Combustion Institute, pp 127–141.

LIPF measurements and modeling of jet ignition through an orifice between two separated rooms

S. Mori, T. Yoshida, A. K. Hayashi*, T. Fujiwara
Department of Aerospace Engineering, Nagoya University
Nagoya 464-01, Japan

Abstract: The ignition and combustion processes of combustion products of a jet of methane-air ejected through an orifice were studied by the LIPF (Laser-Induced Predissociation Fluorescence) method in a combustion chamber separated by a wall with an orifice. Upon visualizing OH and hot-O_2 species, it was seen that the burned gas jet near the orifice in the driven cell was mixed with the fresh gas to ignite the unburned mixture. We clarified the features of jet ignition through an orifice between two rooms and the effect of the orifice size on the total number of OH molecules.

Key words: Laser diagnostic, Combustion torch

1. Introduction

A number of studies on the mechanism of jet ignition in a divided chamber have been carried out in the few past decades. For example, Yamaguchi et al. 1985 predicted that the most favorable for lean burning is composite ignition and well-dispersed burning followed by wrinkled laminar burning, and Cattolica and Vosen (1985) showed that the fluid dynamics of the combustion torch has a significant influence on its chemical structure and development of the subsequent ignition process in the main combustion chamer, which was studied by using OH-LIF. In the present study the jet ignition mechanism is observed through the ignition driver radicals of OH and ignition supporter of hot O_2 by using the LIPF (Laser-Induced Predissociation Fluorescence) method. The fluorescence of OH LIPF is normally very weak, hence OH- and hot-O_2 (above 600 K) fluorescence profiles are measured by a high-power tunable KrF and ArF excimer laser (200 – 400 mJ). A cylindrical combustion chamber is separated by a wall with an orifice, and both rooms are filled with the fuel-lean methane-air mixture.

Numerical analysis of jet ignition is performed for geometry similar to the experimental setup to study quantitative as well as qualitative properties by

*Present address: Department of Mechanical Engineering, Aoyama Gakuin University, Tokyo 157, Japan.

modeling the OH emission and flame propagation. A single-step reaction model for the methane-air reaction mechanism is used to simulate the jet-ignition phenomena in the present case.

2. Experimental apparatus and measurements

A methane-air mixture with an equivalence ratio of 0.717 was used for the present experiments at an initial temperature of 293 K and an initial pressure of 1 atm. Figure 1 shows a schematic of the inner view of the cylindrical combustion chamber. The chamber is equipped with three quartz glass windows on the side walls. Two of the windows are used for the laser path and the third one for the photographic images of fluorescence signals. The partition wall with orifices of 5-, 10-, and 20-mm diameter is set at the center of the chamber. The whole cylidrical chamber is set vertically and the electric spark ignition is placed at the bottom of the cell. The pressure is measured 90 mm from both the upper and lower ends of the chamber.

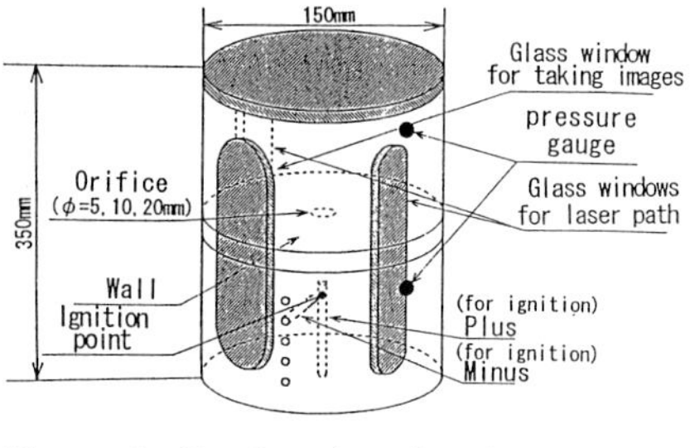

Figure 1. Combustion chamber.

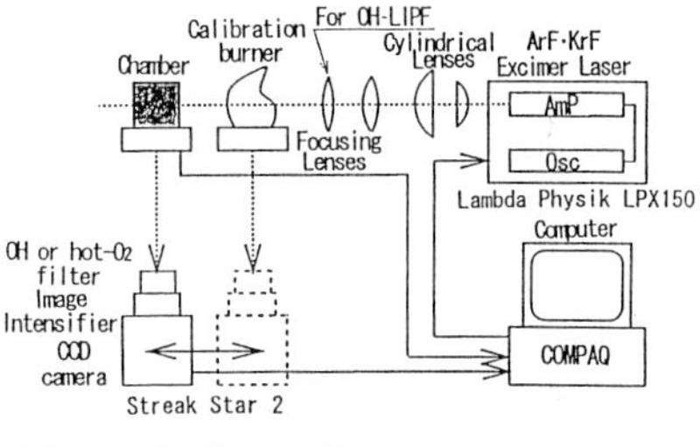

Figure 2. Laser diagnostic system.

Figure 2 shows the whole system for LIPF measurement. The ultraviolet laser light is pulsed from the excimer laser system of KrF and ArF (LaVision). The laser light of OH LIPF (KrF 248 nm) is focused to a sheet of about 40 mm $\times$ 1 mm, whereas that of hot-O_2 LIPF (ArF 193 nm) to a sheet of about 90 mm $\times$ 1 mm after passing through the cylindrical and convex lenses. For the LIPF measurements the fluorescence images of molecules are taken by a CCD camera with filters for OH and hot-O_2 fluorescence signals.

3. Experimental results

3.1. OH and hot-O_2 concentrations by LIPF measurements

Since the lean methane-air mixture is used in the main chamber, O_2 molecules exist before and after combustion: the quantitative values of hot O_2 can not be compared before and after combustions due to the absorption by hot-H_2O product. OH and hot-O_2 concentration images shown in Figs. 3–5 are taken in the regions from the orifice exit to the indicated location in the upper cell (the driven cell).

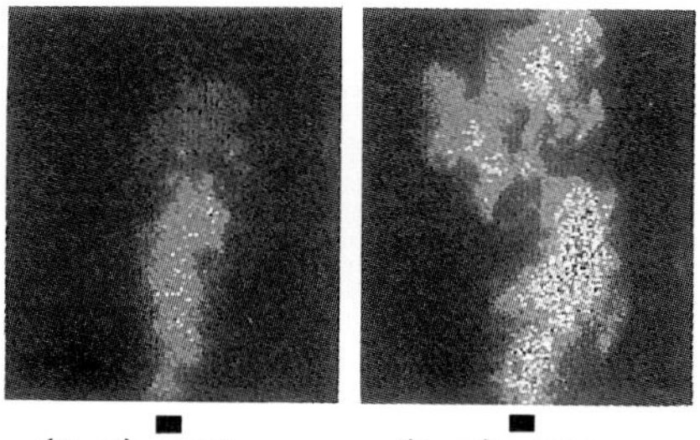

(3.1) 128ms (3.2) 134ms

Figure 3. Fluorescence profiles of jet flame, hot-O_2 LIPF ($\phi = 5$ mm).

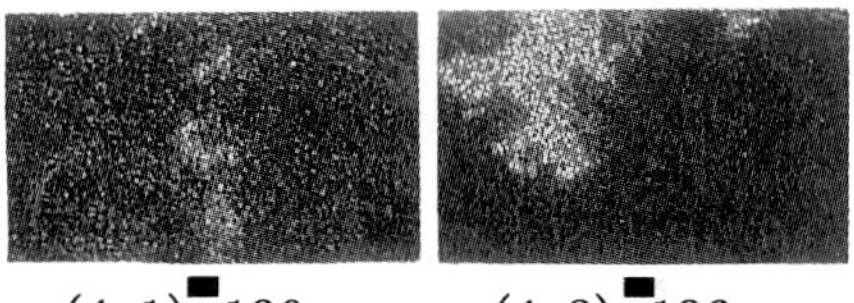

(4.1) 130ms (4.2) 136ms

Figure 4. Fluorescence profiles of jet flame, OH LIPF ($\phi = 5$ mm).

Figure 3 shows the profiles of hot-O_2 concentrations. The combustion jet with hot O_2 starts to flow into the upper cell (the driven cell) about 106 ms after the spark ignition at the bottom cell (the drive cell). The jet keeps its shape while it becomes larger until 130 ms. At 128 ms (Fig. 3.1), the jet length is about 5 cm. But eventually the cold surrounding mixture cools off the hot jet to stop increasing its size. At 134 ms (Fig. 3.2), there appears a high-temperature zone near the top of the jet (this zone is hard to see from this black and white picture). After that, the jet stops its quick expansion, but several high-temperature spots appear at the top of the picture. Previously we thought the jet reignites at its head, but this result indicates that the reignition starts at the upper part of the jet. Figure 4 shows the profiles of OH concentrations. From 118 ms to 130 ms, several separated OH islands appear in the jet, which is different from the results of hot-O_2-LIPF profiles, since the unburned mixture among the several separated OH islands in the jet is above 600 K.

The combustion jet speed in the upper chamber, when the separation wall has an orifice of 10 mm diameter, is faster than that when the wall with the 5-mm orifice is used, but the ignition and combustion characteristics are similar. Figure 5 shows the profiles of hot-O_2 concentrations in the case of using the 20-mm-diameter orifice. At 77 ms (Fig. 5.1) up to the position 3 cm from the orifice, the jet expands and convolutes the unburned mixture further above this position. At 86 ms (Fig. 5.2) the jet stops glowing and the high-temperature zones exist. Then reignition starts with at the upper part of the region of inverse triangle shape. The results of OH LIPF are similar to that of hot-O_2 LIPF as far as the features of the jet and the reignition are concerned.

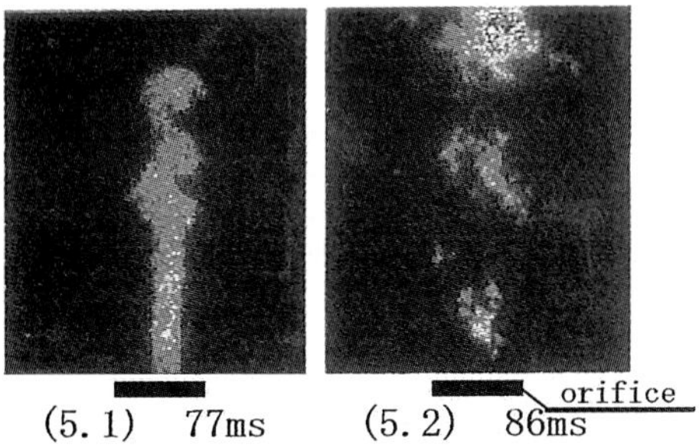

Figure 5. Fluorescence profiles of jet flame, hot-O_2 LIPF ($\phi = 20$ mm).

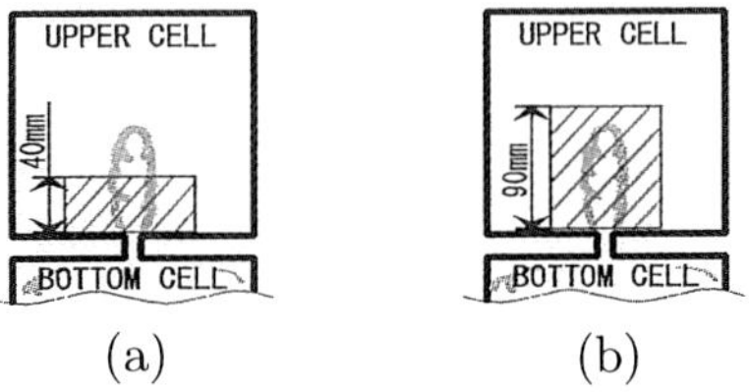

Figure 6. Position of LIPF frame shown by hatched region for (a) Fig. 4 and (b) Figs. 3 and 5.

3.2. Probable numbers of OH molecules for reignition

We calculated a probable total number of OH molecules for the reignition in the jet. At first we measured the fluorescence intensity of OH molecules (the number of OH molecules) in the jet just downstream the orifice, without considering any influence of quenching in the unburned mixture. Secondly, we assumed that the average jet velocity is taken into account during the jet expansion and the density in the jet around the orifice is homogeneous. The calculation shows that the jet speed is inversely proportional to the square of the orifice size from the continuity equation. Figure 7 shows the calculated relative total number of OH molecules in the jet, which implies that the total number of OH

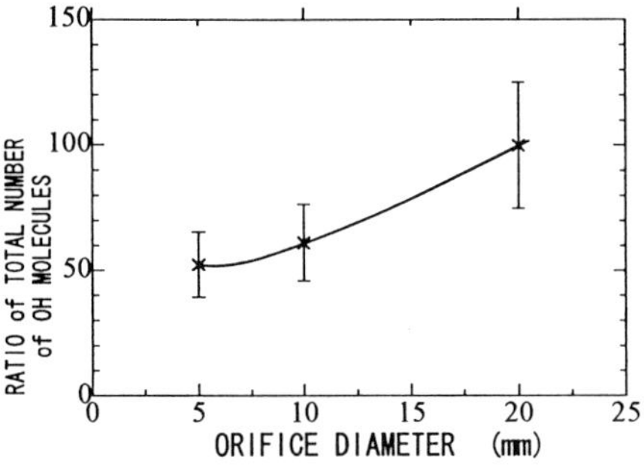

Figure 7. Number of OH molecules passing through the orifice.

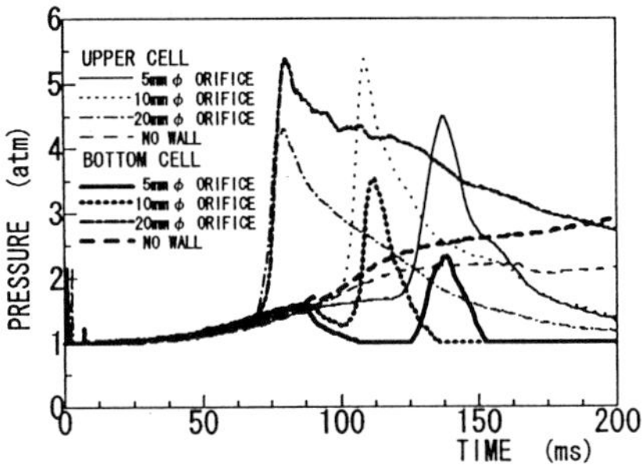

Figure 8. Effect of the orifice diameter on the pressure profiles.

molecules passing through the orifice increases as the orifice diameter increases. This result is consistant with and supports the following pressure profiles in both chambers.

3.3. Effect of orifice diameter on pressure profiles

Figure 8 shows the pressure changes measured in both the upper and lower cells. The first pulses near 0 ms show the noises of the electric spark. The solid lines are the pressure profiles when the orifice diameter of 5 mm is used. The pressure change obtained at the upper cell shown by the thin solid line has one peak, whereas that at the bottom cell shown by the thick solid line has two peaks. The mixture in the upper cell is compressed by the combustion in the bottom cell, hence the pressure in the upper cell increases slowly at first and suddenly increases faster due to the jet ignition. This phenomena is confirmed by LIPF measurement. The dotted lines show the pressure profiles using the orifice diameter of 10 mm; which indicates jet ignition occurs earlier than the case using the 5-mm orifice. The chain lines show the pressure profiles using the 20-mm orifice, detecting much earlier ignition. The broken lines are the pressure change without the separation wall, in which case the laminar flame propagates in the chamber of 35 cm in length.

4. Numerical simulation: modeling and computations

Numerical analysis is performed for almost the same conditions as the experiments having the wall with an orifice of 10 mm in diameter. The full Navier-Stokes equations are used to model the jet ignition with a Harten-Yee, non-MUSCL modified-flux TVD scheme for convective terms, a central-reference scheme for viscous terms, and a Crank-Nicholson semi-implicit scheme for species-production terms. In the present calculation we used a single-step reaction model. The boundary conditions on the walls are given by the isothermal, non-slip, and non-catalytic conditions, and those on the center line are given by the mirror reflection rule.

Figure 9 (left) shows the numerical results of temperature distribution at 24.69 ms. The disconnected regions of high temperature in the jet are recognized as similar to the hot-O_2 profile. Figure 9 (right) shows the case with the ignition point different from the case in Figure 9 (left). Full combustion occurs when the jet hits the top wall, showing different features from the experimental results.

5. Discussion and conclusion

Jet ignition was studied experimentally using the OH LIPF and hot-O_2 LIPF measurement techniques and numerically using a one-step model. The combustion jet ejected from the orifice was cooled for some interval by its expansion and the cold unburnt mixture and did not increase its size. We think that the temperature of the OH radicals and the hot-O_2 molecules in the jet is above 600 K. At the head of the jet, the burned mixture in the jet is mixed with the unburned mixture and becomes easy to ignite. After the induction period,

Figure 9. Numerical results of temperature distribution while the flame is near the orifice (left) and with a different ignition point (right).

reignition occurs and full combustion starts. The smaller orifice diameter allows less total OH molecules to pass through the orifice for the reignition. This gives the possibility of the minimum number of OH molecules for reignition using this chamber. In the future, the OH- and hot-O_2-molecule profiles at higher positions in the upper cell and more accurate combustion mechanisms will be studied.

The numerical results share a few features with the experimental results, however, numerical calculations with a one-step reaction mechanism do not predict well the overall or detailed experimental results; for example, the time sequence of the phenomena. We may know the reason for the difference in flame-propagation speed between the experiment and modeling, which shows the necessary improvement on the methane-air reaction mechanism.

Acknowledgement. We thank Dr. H. Karasawa of Hitachi Ltd., Dr. N. Yoshikawa, Dr. M. Hishida, A. Saitoh, Dr. D. Weide of LaVision Ltd., and S. Mizoguchi of Sogo Yozai Ltd. for cooperation with us throughout this study.

References

Cattolica R, Vosen S (1985) Combustion-torch ignition: fluorescence imaging of OH concentration. Comb Flame 68:267.

Hishida M, Hayashi AK (1992) Numerical simulation of the interaction between a flowfield and exothermic reaction on pulsed jet systems. Preprint of 14th ICDERS, C9.3.1.

Lavision 2D-Messtechnik GmbH. LI(P)F-Techniques. LDS Instruction Manual 5:5–8.

Yamaguchi S, Ohiwa N, Hasegawa T (1985) Ignition and burning process in a divided chamber. Bomb Comb Flame 59:177.

Gasdynamic limits of ignition and combustion of a gas mixture in ISL's RAMAC 30 scram accelerator

F. Seiler, G. Patz, G. Smeets, J. Srulijes
French-German Research Institute of Saint-Louis
F-68301 Saint-Louis, France

Abstract: ISL has performed experiments in a 30-mm caliber ram accelerator, called RAMAC 30, at superdetonative flight speeds, applying the new concept of accelerating a sub-caliber projectile flying in a tube at supersonic speed by self-synchronized ignition and combustion of a pre-filled combustible gas mixture. The gasdynamic limits of ignition and combustion of this gas mixture are discussed by means of rail-tube experiments.

Key words: Ram accelerator, Projectile acceleration, Combustible gas mixture, Ignition, Combustion, Pressure measurement

1. Introduction

In 1988, ISL decided to build two ram accelerators: a 30-mm tube, called RAMAC 30, and a 90-mm one, RAMAC 90 (Giraud et al. 1993). The RAMAC 30 facility is used for basic research, mainly in the superdetonative flight regime, to investigate the establishment of stable ignition and combustion around the projectile's midbody in a supersonic, superdetonative ram-combustion mode. Figure 1 explains the principle of the scram accelerator process. A sub-caliber vehicle consisting of a cylindrical midbody with a conical forebody and an afterbody moves through a combustible gas mixture in a cylindrical tube. On the conical forebody of the ram projectile an attached bow shock is formed in the high-Mach-number upstream flow, which undergoes one or more reflections between the tube wall and the midbody surface. The mixture is expected to ignite and combust in this toroidal section: the combustor of constant diameter.

2. Rail-tube concept

The RAMAC 30 design follows the prototype ram accelerator of Hertzberg et al. (1986) in which fin-stabilized bodies are accelerated in a circular tube. In the first RAMAC 30 design, however, the alternative concept of a ram tube with inner rails in combination with cylindrical and finless projectiles (see Seiler et al. 1995) was chosen. The forebody cone angle is 14 – 16 degrees

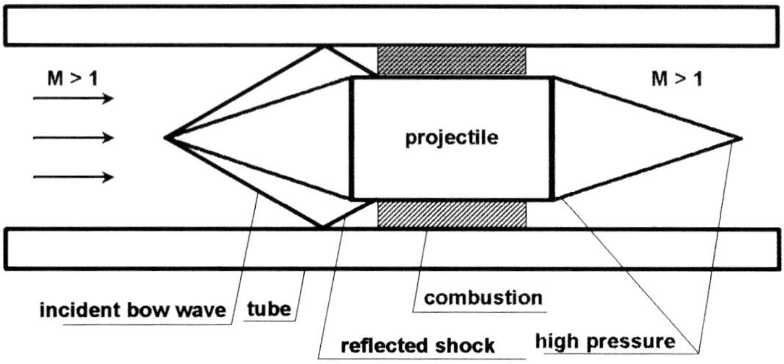

Figure 1. Principle of scram accelerator process.

and the back angle is similar. The combustor zone has a length of 45 – 60 mm and the projectile mass is 125 – 135 g. The whole facility concept is described by Seiler et al. (1995) with a powder gun as pre-accelerator; the models are accelerated to an initial velocity of about 1800 m/s.

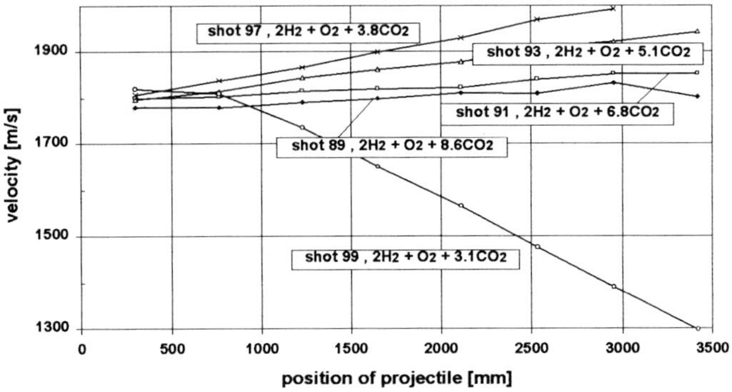

Figure 2. Flight velocities obtained with hydrogen-based gas mixtures.

The rail-tube experiments were made with aluminum projectiles and reactive gas mixtures, based on hydrogen (H_2), oxygen (O_2), and carbon dioxide (CO_2) as diluent. Best ignition result and avoidance of unstart effect was obtained with the gas mixture $2H_2 + O_2 + 3.8CO_2$ and 27-bar fill pressure, having a Chapman-Jouguet detonation velocity of about 1450 m/s (see Fig. 2, shot 97). The velocities were determined using the signals from electromagnetic coil sensors placed in the tube wall identifying the passage of a magnet inside the projectile.

3. Limits of ignition and combustion

The experiments show that in the superdetonative mode there exist stability limits. One limit is set by thermal choking followed by an unstart which means that the combustion wave moves in front of the projectile. Another failure limit is found to be the decoupling of the ignition and combustion from the projectile.

Figure 3 shows a set of typical signals received with the electromagnetic sensors and the wall pressure gauges for three measuring locations M16, M18, and M19 of experiment 97 along the tube. The pressure signals are correlated in time directly to the position of the projectile. They show a very strong pressure increase by a front shock placed at the beginning of the combustor zone with a peak pressure of more than 1000 bar. This first wave is probably produced by the reflection of the incident shock wave at the tube wall forming the first reflected shock. In the expansion generated at the corner of the projectile shoulder the pressure strongly decreases. A second wave follows initiating a high-pressure region. This second jump is perhaps located where the third reflected shock is formed from the second one inducing combustion, which looks similar to the well-known detonation phenomena, described by Chapman and Jouguet (Zierep 1990). Increasing the CO_2 content makes the combustible gas

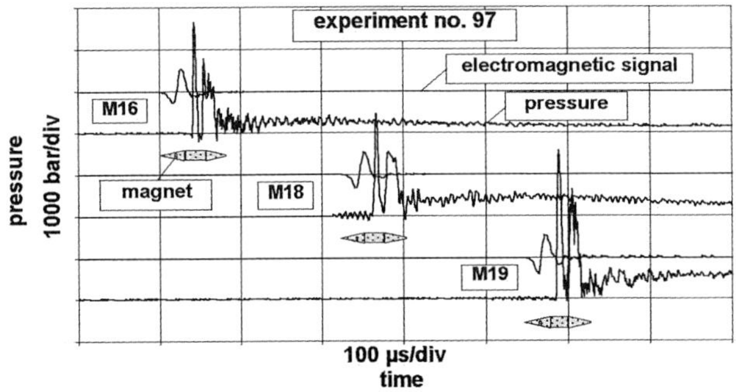

Figure 3. Electromagnetic and pressure signals received for shot 97.

mixture more and more phlegmatic and it needs a third wave to be ignited, as seen in Fig. 4 for shot 89 at measuring station M19. Then the gas reaction occurs behind the projectile and the pressure gain is lost. Decoupling of the combustion (or no combustion) depends on the mole parts of the inert component of the gas mixture, i.e., the CO_2 content in the $H_2/O_2/CO_2$ mixture used herein. This outcome defines an upper dilution limit of the combustible gas mixture for getting the mixture ignited in the combustor before expansion in the divergent back region. The temporal position of the second combustion-inducing shock inside of the constant-diameter combustor channel relative to the position of the front shock is depicted in Fig. 5 versus projectile position along the tube. Its location relative to the end cross-section of the combustor is also marked. The position of the first wave inside the combustor slit is approximately the same for all experiments. For shot 97, as described above, the ignition is well localized inside of the combustor. Increasing the content of the CO_2 diluent shows that the combustion-inducing second shock moves backwards (shots 93, 91, 89). When its position passes the end cross-section of the combustor, the projectile acceleration vanishes and a third wave appears.

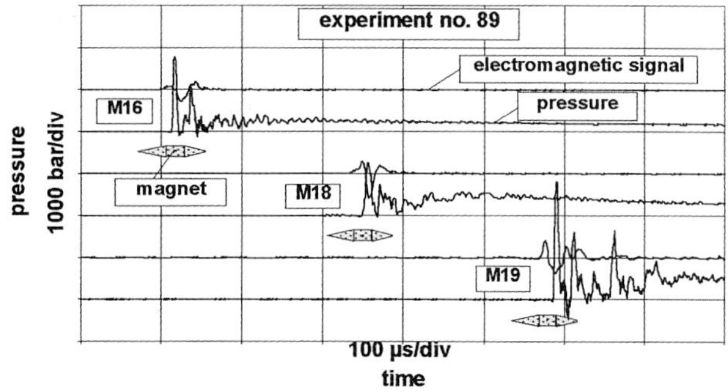

Figure 4. Electromagnetic and pressure signals received for shot 89.

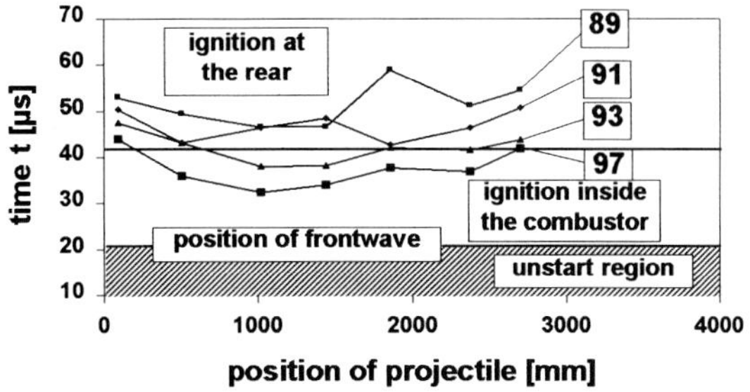

Figure 5. Location of the combustion-inducing second shock wave.

The time t_r between the front wave and onset of combustion at the second or third shock can be considered as a sort of relaxation (or induction) time; this is shown in Fig. 6 averaged along the projectile path versus the mole parts x of the CO_2 diluent contained in the stoichiometric H_2/O_2 mixture. As expected, it rises with increasing mole number x of the CO_2 content and drops with decreasing x, i.e., increasing heat release. If the heat release is too high, the experiment shows that the second wave moves towards the front shock and a so-called unstart begins. To avoid thermal choking followed by an unstart,

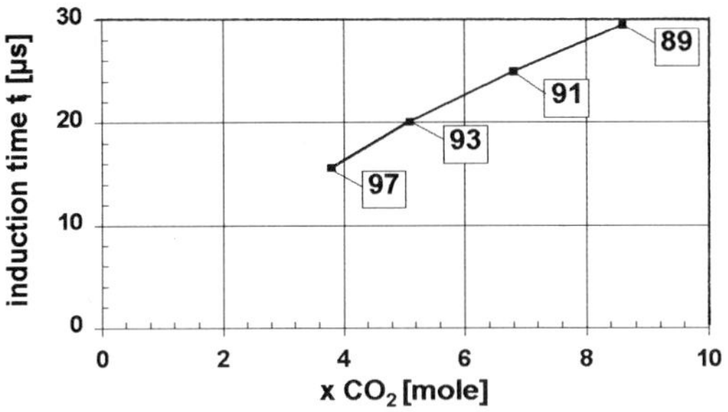

Figure 6. Relaxation time versus x mole CO_2.

the heat release must be adapted to the flow Mach number M in front of the combustion. For a one-dimensional flow, as present in the channel between the projectile and wall, Zierep (1990) gives a relation describing the maximal heat input q as:

$$\left(\frac{q}{c_p T}\right)_{\max} = \frac{\left(M^2 - 1\right)^2}{2\,(\chi + 1)\,M^2} \,. \tag{1}$$

In this relation $\chi = c_p/c_v$, with c_p the specific heat at constant pressure and c_v that at constant volume. T is the gas temperature in front of combustion. Figure 7 shows the distribution of the normalized heat release as a function of the flow Mach number M. We see two regions: a zone in which no stable

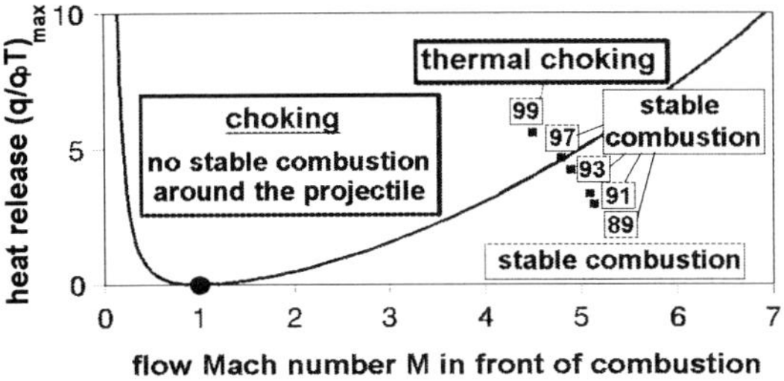

Figure 7. Maximum heat release inside the combustor channel.

combustion occurs and the flow is choked thermally and a zone with stable combustion by limiting the heat input. Five points are inserted representing the firings 89, 91, 93, 97, and 99. Looking at Fig. 2 for shot 99, we have a projectile unstart predicted by the above-given relation. The other four shots are at the border (97) and inside (89, 91, 93) the allowed region, giving a projectile acceleration as shown in Fig. 2.

4. Conclusions

The experiments performed proved that there are two limits for operating a ram accelerator in superdetonative ram-combustion mode: one for ignition and one for combustion of the gas mixture used. The following limits for the RAMAC 30 of ISL have been found for the actual flow conditions in front of the combustion zone, calculated with the code of Smeets et al. (1992) giving the flow conditions $p \approx 85$ bar, $T \approx 385$ K, and axial flow velocity $u \approx 1750$ m/s:

1. Ignition: There exists an upper limit for the parts of the diluent contained in the fuel/oxidizer gas mixture. With a stoichiometric hydrogen/oxygen mixture diluted with CO_2, a maximum of about $8CO_2$ mole parts are allowed for having ignition within the combustor of 45 – 60-mm length.

2. Combustion: There exists an upper limit of heat release to avoid thermal choking followed by an unstart, i.e., the detonation wave moves in front of the projectile. With hydrogen-based stoichiometric combustible gas mixtures the lower limit for the CO_2 is about 4 mole parts.

3. The following stoichiometric H_2/O_2 mixture diluted with x mole parts CO_2 must be used for stable operation of RAMAC 30: $2H_2 + O_2 + xCO_2$ with $4 < x < 8$.

References

Giraud M, Legendre JF, Simon G (1993) RAM accelerator in 90 mm caliber or RAMAC 90: experimental results concerning the transdetonative combustion mode. In: 14th Int. Symp. on Ballistics, Québec, Canada.

Hertzberg A, Bruckner AP, Bogdanoff DW (1986) The ram accelerator: a new chemical method of achieving ultra-high velocities. In: 37th ARA Meeting, Québec, Canada.

Seiler F, Patz G, Smeets G, Srulijes J (1995) The rail tube in ram acceleration: feasibility study with ISL's RAMAC 30. In: Second Int. Workshop on Ram Accelerator, RAMAC II, University of Washington, Seattle, USA.

Smeets G, Gatau F, Srulijes J (1992) Rechenprogramm für Abschätzungen zur Ram-Rohrbeschleunigung, ISL Report RT 507/92.

Zierep J (1990) Strömungen mit Energiezufuhr, G. Braun Verlag, Germany.

Ram-accelerator operation with simplified geometries

A. Sasoh*, C. Knowlen, A.J. Higgins, A.P. Bruckner
University of Washington
Aerospace and Energetics Research Program, AERB FL-10
Seattle, WA 98195 USA

Abstract: Investigations on ram-accelerator operation with simplified geometries are made with axisymmetric, hollow projectiles in experiment and with a quasi-one-dimensional, quasi-steady numerical flow analysis which includes finite-rate chemistry. In the experiment, positive thrusts were obtained. The numerical analysis deals with subdetonative and transdetonative operation modes by taking a generalized choking condition into account. When an effective induction length becomes shorter than the projectile characteristic length, the choking occurs on the projectile (transdetonative mode), yielding a thrust which is larger than that calculated by the thermally choked Hugoniot theory.

Key words: Ram accelerator, Hollow projectile, Finite-rate chemical reactions, Numerical analysis

1. Introduction

A ram accelerator is characterized by its potential of launching a heavy payload at high speeds (Hertzberg et al. 1988, Heiser and Pratt 1994). This potential exists basically owing to the fact that pressure is always highest immediately behind a projectile, and encourages one to apply this technology to space launchers, hypersonic ground test facilities, and hypervelocity impact experiment, etc. The concept of a ram accelerator was developed and experimentally validated at University of Washington, being experimentally followed by ARL (Kruczynski 1993) and ISL (Giraud et al. 1992, Seiler et al. 1993), who showed that the scaling-up of a ram accelerator was feasible. However, in order to realize the above applications, the performance of the apparatus needs to be further improved.

So far neither experimental diagnostics nor computational fluid dynamics has proved successful for thoroughly understanding the flow field in a ram accelerator. If possible, simplification of the flow field will promote understanding to a large extent. One way to obtain axisymmetric flow passage is to use a hol-

---

*Present address: Shock Wave Research Center, Institute of Fluid Science, Tohoku University, 2-1-1 Katahira, Aoba, Sendai 980-77, Japan.

low projectile; if the projectile is supported by its cylindrical body, it does not need fins. In this study, ram-accelerator operation with hollow projectiles was experimentally tested. In order to analyze the ram-accelerator thrust characteristics more in detail, quasi-one-dimensional numerical analyses with finite-rate chemistry were also conducted.

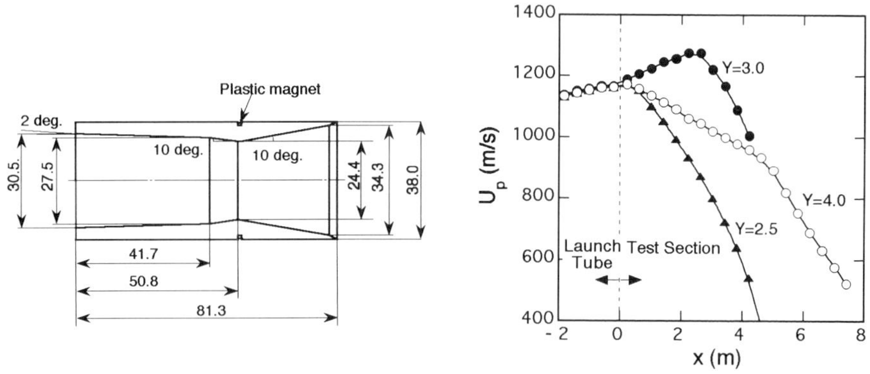

Figure 1. Hollow projectile experiment: geometry (left); projectile velocity histories (right), $2.8CH_4 + 2O_2 + YCO_2$.

2. Hollow-projectile operation

The hollow projectile shown in Fig. 1a was used in this study. It is made of 7075T651 aluminum alloy. The outer diameter of the projectiles is 38.0 mm, which is 0.1 mm smaller than the tube diameter. The projectile has a 10-degree inverse cone section upstream. In order to avoid damage from rupturing diaphragms, the apex angle of the inverse cone at the lip is made large: 30 degrees. The cross-sectional-area ratio of the throat to the tube is 0.41, equaling that for a conventional center-body projectile (Knowlen et al. 1994). The downstream section is a diverging inverse cone which is truncated by a base. The cross-sectional-area ratio of the flow passage at the base of the tube also equals that for a conventional projectile with fins being taken into account. A magnetized plastic ring is installed into the groove on the outside of the projectile. The projectile weighs 0.105 kg. The experiments were conducted using the ram accelerator at Aerospace and Energetics Research Program, University of Washington (Hertzberg et al. 1988, Bruckner et al 1991).

Figure 1 shows examples of projectile velocity histories. The entry speeds were about 1170 m/s. For $Y = 3.0$, when the projectile enters the ram-acceleration section, the mixture is ignited behind the projectile, producing high pressure of the order of 30 MPa. The projectile is accelerated up to $x = 2.6$

m. A velocity gain ΔU_p of 108 m/s is obtained. At $x = 3.0$ and higher, the pressure histories indicate the occurrence of unstart — a shock wave disgorges over the projectile, which is decelerated by the high pressure behind the shock wave.

For $Y = 4.0$, the mixture is not ignited. After entering the ram acceleration section, the supersonic diffuser operation is made — the projectile coasts with supersonic flow passing through. The drag over the projectile is relatively small. At about 900 m/s, a normal shock appears in front of the projectile. Behind the shock wave the flow becomes subsonic, resulting in failure of the supersonic diffuser operation. For $Y = 2.5$, due to rapid heat release by combustion, unstart occurs immediately after entering the ram-acceleration section.

For $Y = 3$, from the thermally choked Hugoniot theory (Knowlen and Bruckner 1992), the non-dimensional thrusts $F/(p_1 A_t)$ are estimated to be 3.5 and 2.7 for $U_{p0} = 1170$ m/s and 1270 m/s, respectively. However, the mean value of the measured non-dimensional thrust is 2.0 — much smaller than the theoretical value. A simple calculation shows that if a pressure of 30 MPa is exerted only on the inside of the projectile the amount of the expansion of the projectile outer diameter becomes comparable with the clearance between the projectile and the tube wall. Therefore, under such high pressure, mechanical friction force against the tube wall due to this expansion is not negligible in thrust production. Also, in the clearance the effect of viscosity is so large that the static temperature is increased and the mixture can be ignited there. In order to solve these problems associated with the clearance, it is preferable to increase the clearance. However, increasing the clearance makes the projectile attitude unstable. Further study should be focused on solving these problems.

3. One-dimensional analysis

Numerical analyses of quasi-steady-state, quasi-one-dimensional channel flows with finite-rate chemical reactions are conducted in the way that is described in Sasoh et al. (1995). The flow passage made between the projectile and the tube wall is assumed to have a converging-diverging cross-sectional variation. The cross-sectional area is assumed to smoothly vary all through the flow field.

In the present calculation, an initial tube pressure is assumed to be 50 atm. The center-body projectile is composed of a cone of 10-degree half-apex angle on the upstream part and another cone in the opposite direction of 4.5-degree half-apex angle downstream. In the standard projectile, the distance between the throat and the exit, L_0, is 183 mm. The cross-sectional-area ratio of the throat to the tube is 0.41.

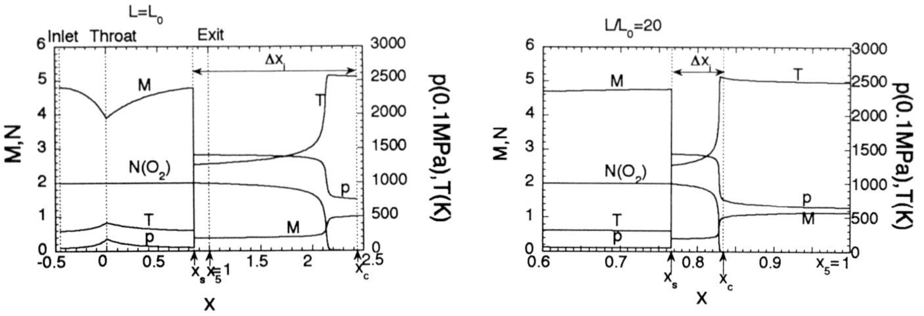

Figure 2. Example of calculated flow field: in subdetonative mode (left); in transdetonative mode (right), $2.8CH_4 + 2O_2 + 5.7N_2$, $M = 4.8$, single-step reaction model.

A generalized choking condition is expressed by (Sasoh et al. 1995)

$$\frac{dq}{c_pT} - \frac{dA}{A} - \frac{dW}{W} = 0 \tag{1}$$

where A, c_p, dq, T, and W are cross-sectional area, specific heat at constant pressure, heat release, temperature, and molecular mass, respectively.

Examples of variations of flow-field properties and Mach number in the subdetonative mode are shown in Fig. 2. The axial distance from the entrance is normalized by the distance between the throat and the exit of the projectile region, L, and denoted by x. Let one define an effective induction length of the energy release Δx_i by the distance $x_c - x_s$, where x_c is choking location and x_s is shock location. In this figure, Δx_i is longer than $x_5 - x_s$ (x_5 corresponds to the exit of projectile region). For large L, it is possible that Δx_i is smaller than $x_5 - x_s$ and the choking point moves onto the projectile. In the example of the transdetonative mode, the ratio of Δx_i to $x_5 - x_s$ is about 0.29, being much smaller than unity. Downstream of the choking point, M continues to increase through supersonic expansion; M at the exit becomes 1.15.

Figure 3 shows relations between a calculated nondimensional thrust and a projectile Mach number for different projectile dimensions or for different reaction models. For $L/L_0 = 40$ the nondimensional thrust in the transdetonative mode first increases and then decreases with increasing M_1. This calculated thrust characteristic well reproduces the experimentally observed one (Higgins et al. 1993). The data in the presently employed chemical-reaction model (almost the same as the one is employed in Weirs and Candler 1994), based on Westbrook and Dryer (1981) and Coffee (1981), were determined mainly from results of low-pressure systems and are not necessarily accurate for the present experimental conditions. Usually, effective reaction rates increase with increasing pressure. Moreover, effects of turbulence shorten the induction length. For these reasons, the calculated induction length may not quantitatively agree with the experimentally observed value. However, thrust characteristics in the transdetonative mode can be well reproduced by giving an effective value of L/L_0.

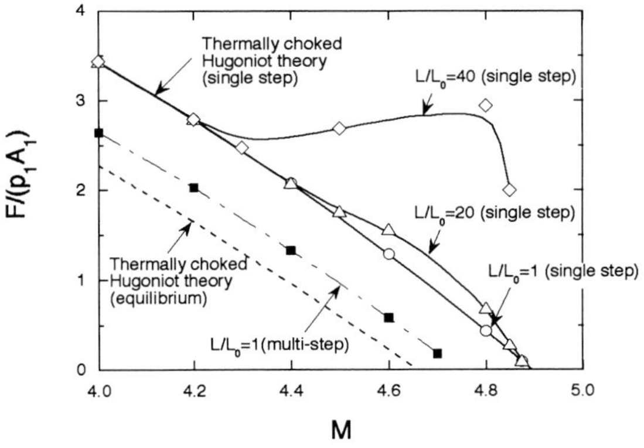

Figure 3. Calculated nondimensional thrust versus Mach number, $2.8CH_4 + 2O_2$ $+ 5.7N_2$.

4. Summary

With the hollow projectile, a positive thrust is experimentally obtained. In order to improve the performance, one needs to solve the aerodynamic and mechanical problems associated with the clearance between the hollow projectile and the tube wall.

Ram-accelerator thrust characteristics are well described by the present quasi-one-dimensional analysis with finite-rate chemistry. When the induction length becomes smaller than the projectile characteristic length, choking occurs on the projectile, being accompanied with supersonic expansion downstream. In this case, the normalized thrust becomes larger than that calculated by the thermally choked Hugoniot theory. The detailed thrust characteristics in the transdetonative mode warrant further investigation of quantitative estimation of the effective induction length of the high-pressure, turbulent chemical reactions.

Acknowledgement. The authors appreciate the valuable suggestions and encouragements made by Professor Abraham Hertzberg. Also the authors thank Gilbert Chew for his considerable help in designing the hollow projectiles through CFD computation, and Messrs. Tom Imrich and Josh Elvander for their help with the experiment. This work was conducted by a U.S.-Japan cooperative science program funded through Japan Society for Promotion of Science.

References

Bruckner AP, Knowlen C, Hertzberg A, Bogdanoff DW (1991) Operational characteristics of the thermally choked ram accelerator. J Propulsion Power 7:828–836.

Coffee TP (1981) On simplified reaction mechanisms by oxidation of hydrocarbon fuels in flames by C. K. Westbrook and F. T. Dryer. Combustion Science and Technology 43:333–339.

Giraud M, Legendre JF, Simon G, Catoire L (1992) Ram accelerator in 90 mm caliber, first results concerning the scale effects in the thermally choked propulsion mode. In: 13th Int Symp on Ballistics, Stockholm, also ISL Rep CO 210/92.

Heiser WH, Pratt DT (1994) Hypersonic Airbreathing Propulsion, AIAA Educational Series, pp 546–551.

Hertzberg A, Bruckner AP, Bogdanoff DW (1988) Ram accelerator: a new chemical method for accelerating projectiles to ultrahigh velocities. AIAA J 26:195–203.

Higgins AJ, Knowlen C, Bruckner AP (1993) An Investigation of Ram Accelerator Gas Dynamic Limits, AIAA Paper 93–2181.

Knowlen C, Bruckner AP (1992) Hugoniot analysis of the ram accelerator. In: Shock Waves, Proc 18th Int Symp on Shock Waves, Vol. I, Springer-Verlag, Heidelberg, pp 617–622.

Knowlen C, Higgins AJ, Bruckner AP (1994) Investigation of Operational Limits to the Ram Accelerator, AIAA Paper 94-2967.

Kruczynski DL (1993) New Experiments in a 120-mm Ram Accelerator at High Pressures, AIAA Paper 93–2589.

Sasoh A, Knowlen C, Higgins AJ, Bruckner AP (1995) Effect of Finite Rate Chemical Reaction on Ram Accelerator Thrust Characteristics, AIAA 95–2492.

Seiler F, Patz G, Smeets G, Srulijes J (1993) Status of ISL's RAMAC 30 with rail stabilized projectiles. In: Proc First Ram Accelerator Workshop, Saint-Louis, France, September.

Weirs VG, Candler GV (1994) Ram-Accelerator Simulation Using Elemental Conservation Equations, AIAA Paper 94–2966.

Westbrook CK, Dryer FL (1981) Simplified Reaction Mechanisms for the Oxidation of Hydrocarbon Fuels in Flames. Combustion Science and Technology 27:31–43.

Detonation

Triple-shock structure in dusty gas detonations

A. K. Hayashi
Aoyama Gakuin University, Setagaya, Tokyo 157 Japan
T. Fuyuto
Toyota Central R & D Labs., Inc. Nagakute Aichi 480-11 Japan
T. Fujiwara
Nagoya University, Nagoya 464-01 Japan

Abstract: Two-dimensional triple-shock structures in a cornstarch-dust/oxygen mixture are numerically calculated using unsteady two-phase Euler equations. The 'continuous description' is used for the solid phase. The governing equations are integrated using a second-order explicit Harten-Yee non-MUSCL modified-flux-type TVD-upwind scheme and a second-order central-difference scheme. The present results show the detailed dynamic structure of cornstarch-oxygen two-phase detonation with equivalence ratio of 0.45 in a 100-mm wide channel. The calculated smoked-foil record has wider and longer etching profiles than that of hydrogen/oxygen detonation near the region of triple-point collisions. The solid phase has the velocity and temperature lags which provide a stronger Mach shock wave in the triple shock. The length of the diamond structure in the present case is about 44 cm.

Key words: Two-phase flow, Detonation, Numerical simulation

1. Introduction

The detonation structure of a two-phase mixture is quite different from that of a single-phase mixture, mainly due to the lags between the two phases. The detonations in a dust-gas mixture have been studied experimentally by Kauffman et al. (1979), Wolanski et al. (1984), Fan and Sichel (1988), Ohyagi et al. (1993), Zhang et al. (1993), and Li et al. (1993). The triple-shock structures of single-phase and two-phase detonations are known to differ: in two-phase mixtures the detonation cell structure is larger than in single-phase mixtures. Particle-gas interaction plays a crucial role in the ignition and combustion processes and detonation structure. Spinning detonation is observed more frequently in the two-phase media than in the single-phase media.

This paper reports, for the first time, on computational modeling of the two-dimensional two-phase detonation structure in a cornstarch-dust/oxygen mixture. The two-phase Euler equations are integrated by the second-order explicit Harten-Yee non-MUSCL modified-flux-type TVD-upwind scheme for convection terms and the second-order central-difference method for diffusion terms. The dynamic behavior of triple shocks is studied.

2. Numerical simulation

The two-dimensional governing equations for two-phase detonation are described by two-phase Euler equations which are set up for both gas and solid-particle phases with the assumption that both phases can be taken as continuous media. The gaseous mixture consists of O_2, H_2O, CO_2, and the gaseous volatiles of cornstarch. The solid particles are treated as rigid spheres with constant specific heat and uniform internal temperature, which do not collide each other. In order to use a TVD scheme, the Euler equations have a conservative form and therefore, their solid-phase stress term is introduced (Krier and Gokhale 1978) in the momentum and energy equations. External forces are neglected. The constant-pressure specific heat of each species is provided by a polynominal function of temperature, whose coefficients are obtained from the JANAF table. The molecular viscosity, thermal conductivity, diffusion, drag force, and heat conduction between the two phases are considered. In the high-temperature atmosphere, cornstarch starts to liquify, then gasify, followed by a gas-phase reaction to combustion. The rates to model the present two-step reaction mechanism are obtained from the ignition-delay data of Ohyagi et al. (1993) and Kobayashi et al. (1989), which are presented in detail by Hayashi et al. (1994).

The Euler equations are transformed from a physical coordinate system to a computational one in order to deal with the complex geometry and non-uniform grid system. Then they are integrated as described above.

Initially, the cornstarch-oxygen mixture (with equivalence ratio of 1.0, temperature of 293 K, pressure of 0.1 MPa, and solid-particle diameter of 14.5 μm), flows into a stationary detonation wave which is transplanted from the calculated one-dimensional detonation results and is in equilibrium between the two phases both upstream and downstream. The perturbed regions with the high temperature of 2930 K and high pressure of 5.0 MPa are provided to a couple of computational regions. Initially there is a lag between the two phases. The particle mass per unit volume of the mixture is 750 kg/m^3, heat capacity is 1000 J/kg, enthalpy of combustion is -677 kcal/mol at 0.1 MPa, and temperature is 293 K. The numerical boundary conditions are set up with the constant upstream values while the downstream values are determined using the second-order extrapolation. The numerical domain of interest is divided into 51×181 grid points each of which is equivalent to a physical value of 1×1 mm^2.

3. Numerical results and discussion

The previous study (Hayashi et al. 1993) showed that the triple shock wave was not formed steadily and its computation did not converge. However, the termperature profiles of both phases indicated the effect of temperature lag between the two phases.

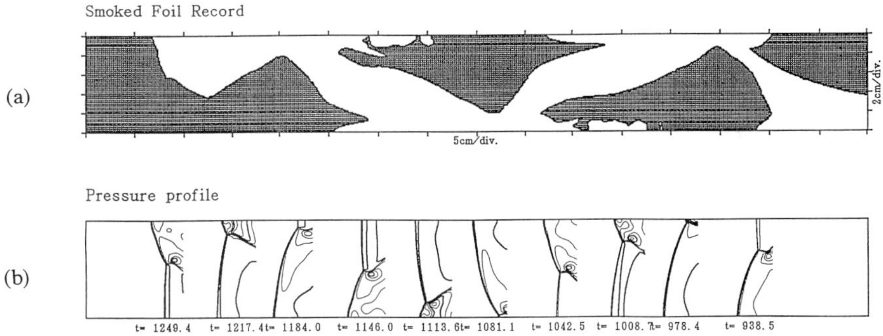

Figure 1. Numerical results of the single triple-shock cornstarch-oxygen two-phase detonation: (a) the computed smoked-foil record and (b) the pressure profiles.

Figure 1 shows the numerical simulation for a 100-mm tube width when the perturbation occurs at a single location on the wall; Fig. 1a is the computed smoked-foil record and Fig. 1b shows the pressure profiles. The results indicate clearly the case of the single triple-shock propagation. The right end of the tube in the figure is about 170 cm downstream the point where the perturbation is given in the wave coordinate system. A typical diamond cell size of the smoked-foil record is about 44 cm. In the present case, with the numerical pressure condition of 6.0 MPa, the etched area is large and the cell shape is distorted compared with the single-phase case. Due to the two-phase flow, wider high-pressure (and high-temperature) regions near and behind the triple point are seen compared to the single-phase flow.

Figure 2 shows the sequential shocks and slip boundary profiles of Fig. 1. Figure 2a is the state that the triple point almost collides to the wall ($t = 1081$ μs); Fig. 2b is the state that the Mach shock wave moves upstream after colliding with the wall, and the solid-phase and combustion regions behind the incident shock wave, flow into the wall due to the slip jet stream and are separated upstream and downstream ($t = 1113$ μs). Figure 2c is the state that the triple point moves upstream and away from the wall and the upstream directed flow becomes a clockwise circulation ($t = 1145$ μs).

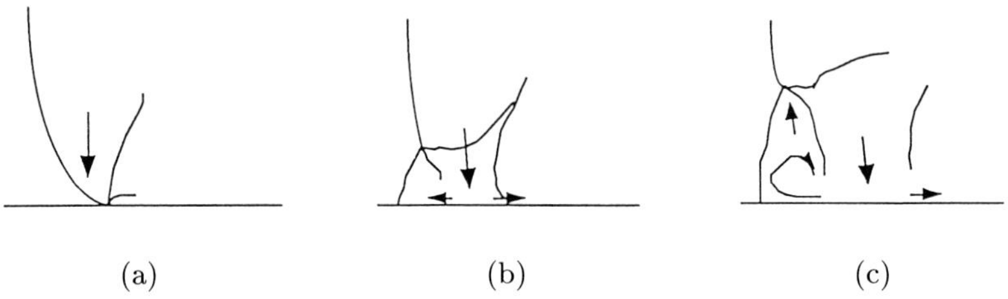

Figure 2. Sequences of the dynamics of the shock waves and slip boundary near a triple point in the two-phase detonation: (a) triple-point collision; (b) triple-point reflection and separation of flame upstream and downstream; and (c) generation of vortex.

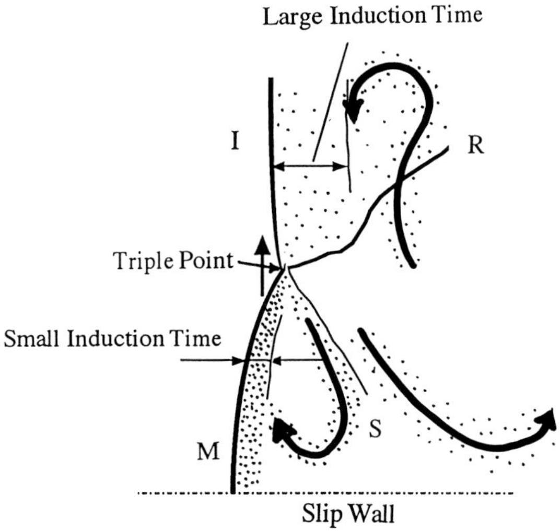

Figure 3. Close-up schematic picture of the shock waves and slip boundary configuration in the two-phase detonation.

Figure 3 is a close-up schematic of the shock and slip boundary profiles at the time 1145 μs, which consists of Mach shock (M), incident shock (I), reflected shock (R), and slip line (S). It is recognized from the solid velocity-vector profiles and solid number-density profiles that there are large lags between solid and gas phases behind the Mach shock, which provide the wide high pressure and temperature region behind it. This proves the appearance of the wide etched region in the smoked foil record in Fig. 1a. Contrary to the situation behind the Mach shock, the state behind the incident shock has the weaker velocity

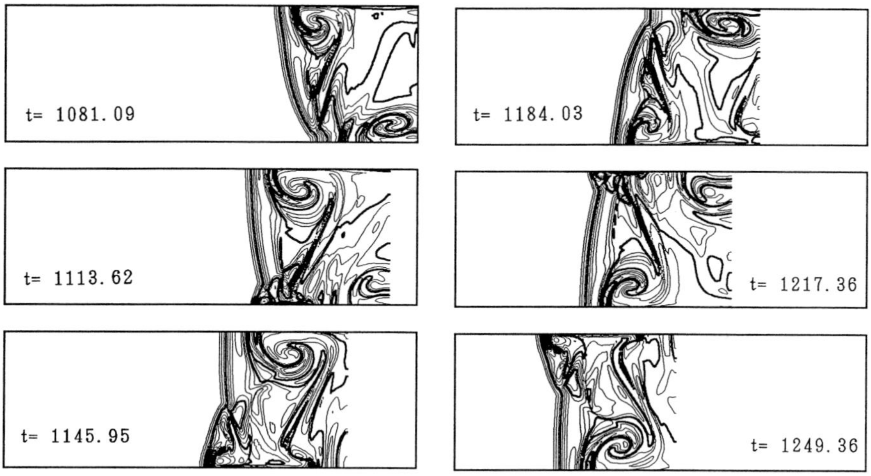

Figure 4. Numerical results of the single triple-shock cornstarch-oxygen two-phase detonation: the cornstarch particle-number-density profiles.

and temperature lags, which provides a longer induction period for ignition and combustion to the two-phase mixture. However the slip line, which exists between the reflected shock and Mach shock, has a strong jet stream due to a large velocity lag between both sides. The two-phase jet stream of the Mach shock side becomes a clockwise circulation which pushes up the Mach shock which transverses the tube. The two-phase stream of the reflected shock side along the slip line becomes another counter-clockwise circulation. The third counter-clockwise circulation is the one which appeared at the previous collision with the other side of the wall. This circulation crosses the reflected shock. Similar phenomena happen in the single-phase detonation case.

Figure 4 shows the cornstarch-particle number-density profiles in the single triple-shock two-phase detonation. The figure clearly shows that the clockwise circulation is growing, and the solid particles are entrained into the slip jet flow to burn.

4. Conclusions

The cornstarch/oxygen two-dimensional two-phase detonation is simulated numerically to find the two-phase flow dynamics near the triple point for the single triple shock mode. It is understood that the two-phase lag influences much of the triple-shock structure in two-phase flows. Due to this lag in velocity and temperature, the diamond cell configuration of the smoked-foil record becomes longer in the two-phase case than that in the single-phase case.

References

Fan B-C, Sichel M (1988) A comprehensive model for the structure of dust detonations. In: Proceedings of the 22nd Symp. (Int.) on Combustion, The Combustion Institute, pp 1741–1750.

Hayashi AK, Fuyuto T, Fujiwara T (1994) Cornstarch-air two-phase detonation phenomena. Archivum Combustionis 14(1–2):73.

Hayashi AK, Fuyuto T, Fujiwara T (1993) Numerical simulation of shock waves through combustible two-phase media. In: Shock Waves @ Marseilles, R Brun (ed), Vol. II, pp 191–196.

Kauffman CW, Wolanski P, Ural E, Nicholls JA, Van Dyke R (1979) Initiated combustion of grain dust. In: Proceedings Int. Symp. on Grain Dust, pp 164–190.

Kobayashi S, Adachi T, Suzuki, T (1989) Study on Ignition of Fine Particles by Shock Waves, Japanese Government Scientific Research Report, No. 62460101, p 98.

Krier H, Gokhale SS (1978) Modeling of convective mode combustion through granulated propellant to predict detonation transition. AIAA J 16(2):177–183.

Li Y-C, Alexander CG, Wolanski P, Kauffman CW, Sichel M (1993) Experiemntal investigations of accelerating flames and transition to detonation in layered grain dust. In: Progress in Astronautics and Aeronautics, Vol. 154, Dynamic Aspects of Explosion Phenomena (AL Kuhl, J-C Leyer, AA Borisov, WA Sirignano, eds), pp 170–184.

Ohyagi S, Tanaka T, Yoshihashi T, Suzuki S (1993) Initiation process of dust layer detonations. In: Proc. Japanese Shock Wave Symp., Sendai, pp 539–541.

Wolanski P, Lee D, Sichel M, Kauffman CW, Nicholls JA (1984) The Structure of Dust Detonations. Progress in Astronautics and Aeronautics, Vol. 94, Dynamics of Shock Waves, Explosions, and Detonations, pp 241–263.

Zhang F, Greilich P, Ven AVD, Gronig H (1993) Two-headed detonation structure in cornstarch-oxygen mixtures. Progress in Astronautics and Aeronautics, Vol. 154, Dynamic Aspects of Explosion Phenomena (AL Kuhl, J-C Leyer, AA Borisov, WA Sirignano, eds), pp 216–227.

Detailed modeling of detonation in heterogeneous flows

U. Uphoff, D. Hänel, P. Roth
Institut für Verbrennung und Gasdynamik
Universität Duisburg
D-47048 Duisburg, Germany

Abstract: A one-dimensional model is proposed describing the propagation of detonation waves in H_2/O_2 mixtures enriched with solid carbon particles, assuming both phases to be reactive. The chemical reactions of the gas phase are described by a detailed reaction mechanism; the surface reactions between gas and particles are modeled using three global reactions. The system of equations is solved using state-of-the-art flux-difference-splitting schemes for both phases coupled with an adaptive moving grid to resolve discontinuities in an accurate way. Results obtained indicate the formation of a weak double-head structure.

Key words: Detonation, Two-phase flow

1. Introduction

Numerical simulations of unsteady detonation phenomena in homogeneous fluids with detailed kinetics have been performed by various authors (Bourlioux et al. 1991, Weber et al. 1992), indicating the need of special numerical treatment to resolve the structure of the reaction zone correctly. As a step towards the understanding of detonations in heterogeneous gas-particle mixtures, different kinds of simplifications have been assumed to reduce the computational effort and to isolate the effects of the phase interactions. Ignition processes and detonation structure in dusty media have been studied by Gelfand et al. (1989) using only the gas-phase governing equations with source terms for the particle processes. Detonation propagation has been investigated by Khasainov and Veyssière (1988) assuming a steady behavior of the wave complex.

In the present approach a model based on the time-dependent governing equations using the Eulerian approach for both phases is taken as a basic concept for the construction of a numerical method. The aim was to improve both the flexibility in the description of phase interactions and the modeling of the combustion process. The use of high-resolution flux-difference-splitting schemes for both gas and particle phase ensures accuracy for a wide Mach-number range.

2. Model and governing equations

In the model used, the gas phase was assumed to be an ideal mixture consisting of n species; the solid phase was assumed to consist of spherically shaped, non-interacting, and electrically neutral particles. The surface of the particles was assumed to be smooth and porosity effects were neglected. The governing equations for two interacting continua consisting of gas (index G) and particles (index P) can be written in the following short form:

$$\mathbf{U}_t + \mathbf{F}_x = \mathbf{Q} \ ,$$

$$\mathbf{U} = \begin{pmatrix} \mathbf{U}_G \\ \mathbf{U}_P \end{pmatrix}, \quad \mathbf{F} = \begin{pmatrix} \mathbf{F}_G \\ \mathbf{F}_P \end{pmatrix}, \quad \mathbf{Q} = \begin{pmatrix} \mathbf{Q}_G \\ \mathbf{Q}_P \end{pmatrix} . \tag{1}$$

The vector $\mathbf{U}$ includes the quantities to be conserved, i.e., mass, momentum, energy, and species for each phase, respectively. $\mathbf{F}$ consists of the corresponding inviscid and viscous fluxes. Interactions between the phases were represented by source terms for the transfer of mass, momentum, and heat, which are collected in the vector $\mathbf{Q}$. It also includes the chemical source terms.

The H_2/O_2 gas-phase combustion process was modeled in the well-known way based on a system of elementary reactions between the gaseous species. The solid material was assumed to be be oxidizable by the species O_2, O, and OH, where the number of reactive collisions between the particle surface and the gas molecules was quantified by reaction probabilities α_j which were obtained experimentally. The particle surface reactions resulted in a decrease in the mass of the suspended particles and caused an increase of the gaseous species formed during heterogeneous reactions.

2.1. Numerical method

The numerical method used was based on the semi-discrete written form of (1):

$$\frac{\partial\left(\Delta x_i \mathbf{U}_i\right)}{\partial t} = \left(\mathbf{F}^{(rel)}_{i+1/2} - \mathbf{F}^{(rel)}_{i-1/2}\right) + \Delta x_i \mathbf{Q}_i = \mathbf{Res}_i \quad i = 1 \ldots N \ . \tag{2}$$

The cell boundaries were allowed to move, therefore variable grid spacing Δx_i and numerical fluxes $\mathbf{F}^{(rel)}_{i\pm1/2}$ relative to the moving boundaries were considered. The numerical flux of the gas-phase equations was modeled using the Harten-Yee TVD method. The corresponding flux of the solid phase was obtained using the HLL approximative Riemann solver, extended to second-order accuracy by a MUSCL extrapolation. Essential for calculation of embedded detonation waves is a very fine resolution of the front. This was achieved by a dynamically moving, adaptive grid. Grid points were clustered in regions of interest fulfilling the demand of an equidistribution of the arc-length of the solution. The adaption process was controlled by monitor functions, which consist of weighted gradients of those variables whose gradients are important for the solution. This idea can be illustrated as follows:

$$(\Delta x)_i \cdot M_i = const., \qquad M_i = f\left(c_1 \frac{\partial X_1}{\partial x} \ldots c_k \frac{\partial X_j}{\partial x} \ldots\right). \tag{3}$$

Details can be found in Weber et al. (1992). The system of equations (2) and (3), governed by very different scales in length and time was integrated in time by an implicit solver for stiff ODE's (DASSL).

3. Results

Calculations were performed assuming a vessel with adiabatic walls containing a stoichiometric H_2/O_2 mixture enriched with solid particles. A well-tested reaction mechanism consisting of 9 species and 18 elementary reactions (He 1991) was used to describe the gas-phase reactions. The heterogeneous particle oxidation was modeled by assuming three global reactions of O, OH, and O_2 with the carbonaceous surface. Reaction probabilities shown in Table 1 were taken from Roth et al. (1990) and from von Gersum and Roth (1992).

Table 1. Reaction probabilities for heterogeneous reactions, $\alpha_j = A_j \exp(-B_j/T)$.

Reaction	A_j	B_j
$C(s) + O \rightarrow CO$	0.23	0
$C(s) + O_2 \rightarrow 2\ CO$	0.1679	10060
$C(s) + 2\ OH \rightarrow 2\ CO + H_2$	0.25	0

Particles and gas were assumed to be initially at rest having the same initial temperature. Two different test cases were studied: the ignition behavior and detonation formation in a suspension with a linear temperature gradient at the left boundary, and the detonation propagation in a suspension initiated by a pressure and temperature discontinuity. In the first example the initial conditions were chosen as follows: temperature at the left boundary was 1600 K changing to 650 K on a length of 2 cm, the initial pressure was 2 atm. Pressure and CO mass-fraction profiles are shown in Fig. 1. The time step between the curves is 1.4 μs. A detonation of increasing intensity is developing. An impression of the grid-adaption process can be obtained from Fig. 2 showing traces of the gridpoints during the calculation. The initial grid consisted of 100 equally spaced points and the adaption process was controlled by monitoring temperature- and mass-fraction gradients. Figure 4 shows profiles for particle temperature and particle velocity. The velocity relaxation of the solid phase is comparable with the development of the detonation structure. The second example considers the propagation of a detonation into the suspension for

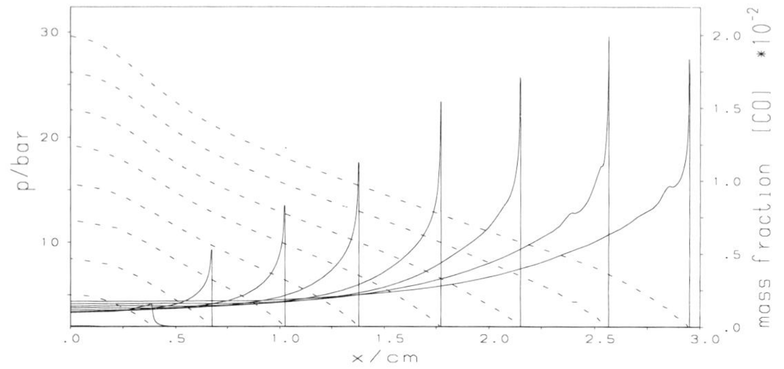

Figure 1. Detonation formation in a stoichiometric H_2/O_2 mixture with $n_P = 0.75 \times 10^6$ particles cm^{-3} and $d_{p0} = 4\ \mu m$.

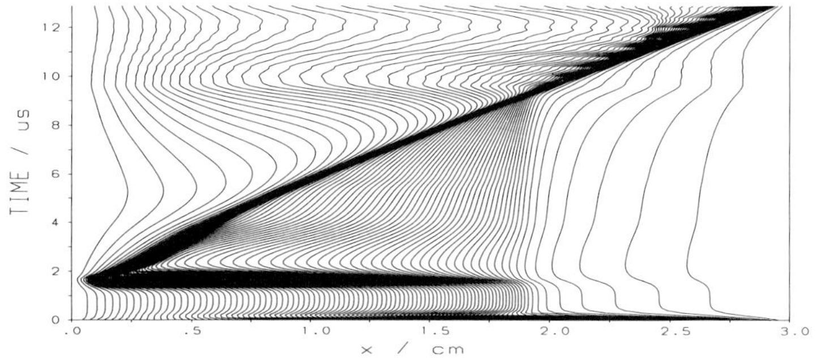

Figure 2. Grid-point movement versus time.

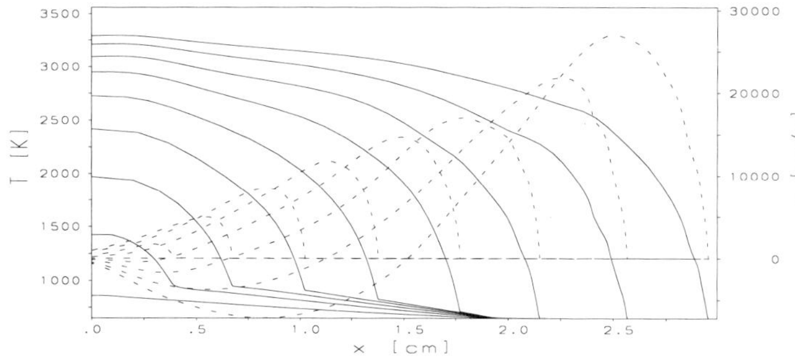

Figure 3. Particle temperature (solid lines) and particle velocity (dashed lines).

longer time. The initial mixture has a temperature of 300 K and a pressure of 1 atm. Combustion was started with a pressure and temperature discontinuity having the properties of the von Neumann spike of the corresponding gas-phase ZND detonation (Zhang and Lee 1994). The particle concentration and diameter were chosen to be $n_p = 0.75 \times 10^6$ cm^{-3} and $d_p = 8$ μm. The calculated pressure profiles of Fig. 4 show behind the front shock the formation of a pressure wave, indicating a supersonic region behind the CJ point, where the heat release of the heterogeneous reaction takes place. From the mass fraction of CO the big difference in the time scale of the particle combustion compared to the homogeneous reactions becomes evident. An increase in the particle number density causes a decrease in the speed of the leading front and lowers temperature and pressure behind the front. An increase in the reactivity of the heterogeneous reaction by a factor of two results in a faster combustion of the solid material and increases the pressure hump behind the leading shock of the following wave, while the detonation velocity remains unaffected.

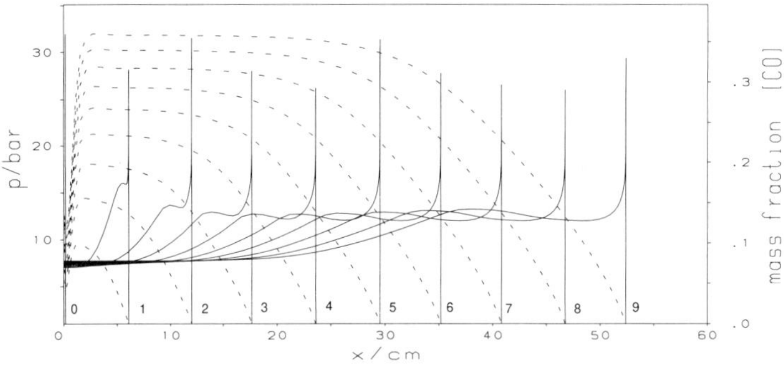

Figure 4. Detonation propagation in a stoichiometric H_2/O_2 mixture enriched with carbonaceous particles, $n_P = 0.75 \times 10^6$ particles cm^{-3} and $d_{p0} = 8$ μm. Profiles are plotted every 21 μs.

4. Conclusion

The present numerical investigation on detonation phenomena in gas/particle mixtures indicates the strong influence of the different time scales of the relaxation and combustion processes of both phases on the flow situation. The ignition and formation of the gas-phase detonation is dominated by the homogeneous reactions. For the initial conditions of the mixture the gas-phase detonation shows an oscillatory behavior which is not influenced by the presence of solid particles. An increase in the particle load results in a delayed ignition and wave formation due to the exchange of heat and momentum between the phases.

Regarding longer time scales, the presence of reactive particles and their reactions with the gas phase control the region behind the CJ point of the gas-phase detonation. A pressure wave caused by the additional heat release due to particle combustion seems to be a characteristic feature of heterogeneous detonation in the range of parameters used in this study.

As a closing comment, it must be remarked that the computational effort for test cases like those presented here is high. Although an adaptive grid strategy was employed, the fine resolution of the induction and the reaction zone of the gas phase is very time consuming. The steep gradients in the radical concentrations requires the evaluation of many jacobians $\partial \mathbf{Res}/\partial \mathbf{U}$.

Acknowledgement. The authors whish to thank the German Science Foundation (DFG) for their financial support.

References

Bourlioux A, Majda AJ, Roytburd V (1991) Theoretical and numerical structure for unstable one-dimensional detonations. SIAM J Appl Math 51(2):303–343.

Gelfand BE, Frolov SM, Bartenev AM, Tsyganov SA (1989) On direct initiation of detonation in a dusty medium. Arch Combust 9 No. 1/4.

von Gersum S, Roth P (1992) Soot oxidation in high temperature N_2O/Ar and NO/Ar mixtures. In: 24th Combustion Symposium, The Combustion Institute pp 999–1006.

He L (1991) Etude des phenomenes d'allumage, Applications aux melanges H_2/O_2, Dissertation L'Universite de Provence.

Khasainov BA, Veyssière B (1988) Steady, plane double-front detonations in gaseous detonable mixtures containing a suspension of aluminium particles. AIAA Prog in Astronautics and Aeronautics 114:248–298.

Roth P, Brandt O, von Gersum S (1990) High temperature oxidation of suspended soot particles verified by CO and CO2 measurements. In: 23rd Combustion Symposium Orléans.

Weber H-J, Mack A, Roth P (1992) Numerical simulation of ignition processes and combustion wave propagation in H_2O_2 reaction systems. Prog in Astronautics and Aeronautics AIAA 151.

Zhang F, Lee JHS (1994) Friction-induced oscillatory behaviour of one-dimensional detonations. Proc R Soc 446:87–105.

Detonation-tube studies on aluminum dust-air mixtures under weak ignition conditions

Y. K. Pu, K. Y. Li
Institute of Mechanics, Chinese Academy of Sciences
Beijing 100080, People's Republic of China

Abstract: A new pneumatic dispersion system for obtaining a good-quality uniform dust suspension in a horizontal test tube was developed. The initial intensity of dispersion-induced turbulence and the quality of dust-cloud suspension in the test tube was examined. A detonation tube for studies of flame acceleration in fine aluminum dust-air mixtures and its transition to detonation under weak ignition conditions was constructed in the course of the present study. The detonation tube consists of a 12-m long test tube which is equipped with ten pairs of well-developed dispersion systems. Aluminum dust, spherical in shape (6-μm diameter), was used for tests. Experimental results demonstrated that a dust flame keeps accelerating in aluminum dust-air mixtures almost linearly through the tube length, 9 m from the ignition plate. The maximum values of flame-propagation velocity approach 2 km/s at the distance of 9 m from the ignition plate, then the flame keeps the maximum propagation velocity constantly between the distance of 9 m and 12 m from the ignition plate.

Key words: Flame acceleration, Transition to detonation, Aluminum dust-air mixture

1. Introdution

Many kinds of fine dust clouds can support the propagation of detonation at a very high rate of energy release, causing catastrophic damage to human life and property in industries. Aluminum powder is one of the strongest. The experimental data of Pu (1990) in the form of explosion characteristics demonstrated that fine aluminum dust is much more violent than organic dust (cornstarch, etc.). With regard to dust-explosion alleviation and dust-detonation prevention for industries, the investigation of the mechanism of fine aluminum-dust flame acceleration and transition processes which occur in a confined tube under industrial explosion conditions becomes necessary. Detonation propagation in premixed substances is one of the basic aspects of combustion. Dust mixtures are insensitive to detonation, greatly complicating the investigation of such detonations. It has not yet been made clear whether a transition to stable, self-sustained detonation can be realized for an organic or a metallic dust-air mixture or can be initiated by a weak ignition energy. In previous research, some attempts were made to study the transition to detonation of cornstarch dust flame in a horizontal tube (Zhang 1991). The above experimental results demonstrate that generally a precursor shock having a critical Mach number develops and divides the transition process from deflegration to detonation into a slow and fast transition stage. The slow transition stage strongly depends on the initial ignition energy and the initial dispersion-induced turbulence level. It is believed that to predict how the slow transition stage varies with the initial conditions, one needs the support of good experimental data. Part of the difficulty, intrinsic to the above-mentioned investigation, stems from the dearth of good experimental data, which prevents one from making realistic assumptions, discarding terms of secondary importance, or developing relations between key parameters. Hence, gathering of reliable experimental data from dust flame acceleration remains a priority in the present study. The investigation of flame acceleration in transition to detonation in a fine aluminum-dust/air mixture was carried out in a detonation tube under the initial conditions of ignition and dispersion, which were thought to meet the requirements for obtaining smooth dust flame propagation and acceleration from ignition to transition, and finally to detonation. A uniform dust-cloud suspension in the test tube was provided by the developed dispersion systems with a knowledge of dispersion-induced turbulence intensity. The weak ignition condition, having a certain variable energy range, was produced by igniters which contain a certain amount of black powder.

2. Experimental details

The horizontal dust-detonation tube for the present studies is 14 m long, with a 0.14-m inner diameter and a 10-mm wall thickness. It consists of a 12-m test section and a 2-m exhaust section with a 10-m^3 dump tank. The test section is equipped with 20 individual well-developed dispersion systems, shown in Fig. 1. The pneumatic dispersion system developed in the present study is essentially a hollow spherical version of a circular perforated pipe which is used in the 20-liter sphere and 1-m^3 vessel. Twenty dispersion spheres are equally spaced on the wall of the tube, in the order shown in Fig. 1. In each system, air flows from a pressurized cylinder (0.23 liter) and forces the dust in the dust chamber (0.01 liter) through a perforated hollow sphere into the test tube. There are 82 dispersion orifices drilled on each hemispherical surface with diameters of 1 to 1.8 mm. The dispersion process is intrinsically transient during the 500 ms, and the turbulence intensity decays during this process. The initial overpressure in the air reservoir is 0.6 MPa. The turbulence measurements accomplished by hot

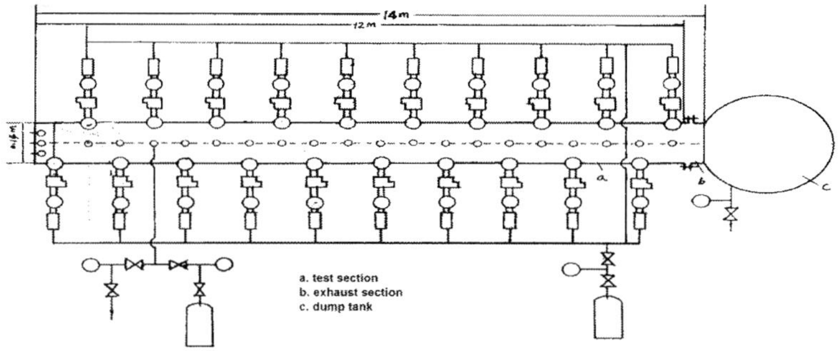

Figure 1. Diagram of the horizontal combustion-tube facility.

probes and anemometers are carried out in air. The voltage signals coming from anemometers for the instantaneous flow velocity of each experiment are recorded and stored in a personal computer memory (floppy disks), then processed on the basis of the ensemble-averaging method (Pu 1989) by a computer. In the measurements the hot-wire probe is fixed at the center of the cross-section located in the middle position between two adjacent dispersion spheres.

The initial concentration distribution of dispersed dust along the axis of the 12-m section is roughly examined by observation of the dust-layer thickness at the bottom of the tube after dispersion. The qualitative evaluation of the duration of dispersed dust-particle suspension is monitored by transmission measurements through the 0.14-m inner diameter of the test tube at two cross-sections located at the one-third position between two adjacent spheres.

To achieve planar flame propagation, six igniters are mounted uniformly on the closed end of the test section. Each igniter contains 0.1 to 1 g of black powder inside a paper capsule and is ingited by a small fuse head. The total ignition energy produced by six igniters is about 277 to 2770 J (6 to 60 J/liter), the rate of pressure rise in the closed tube being about 0.2 to 2 MPa/s. The flame is detected by sixteen photodiodes which are located on the tube wall in orifices spaced uniformly along the axis of the tube. The signals are recorded by a 16-channel recorder with a sampling rate of 0.1 μs. Eight piezoelectric transducers having a sensitivty of 0.001 to 0.002 MPa/mV are used to indicate the pressure profile in the detonation process. The transducers are located on the wall at one cross-section of the 12-m distance of the tube from the igintion plate. The signals are recorded by a 8-channel recorder with a sampling rate of 1 μs. Atomized aluminum-dust/air mixtures with a dust concentration of 103 to 820 g/m^3 are used for the tests. Scanning electron micrographs indicate that the aluminum particles are spherical in shape. Mean particle diameters of 6 and 16 μm are used for the testing.

3. Experimental results and discussion

The examination of dispersion, formation, and suspension of dust clouds in the test tube. The developed dust-dispersion systems installed in the test tube create a homogeneous dust-cloud suspension in the tube. This was proved by observing that the thickness of the dust layer along the tube is practically the same for some time after dispersion.

The measured results of dispersion-induced turbulence parameters (Pu 1995), as a function of time, at the center of the cross-section located in the middle position between two adjacent dispersion spheres showed that the dispersion process in the test tube is intrinsically transient for a period of 200 ms. The peak values of mean flow and RMS turbulence velocity in the transient dispersion process both occur about 20 ms after activation of the solenoid valve and are 8 m/s and 3 m/s, respectively. Starting from 20 ms to 100 ms after activation of the solenoid valve, both velocities decrease rapidly to 2.5 m/s and 1.2 m/s, respectively, then from 100 ms to 200 ms decrease slowly to almost zero. In the present study the shortest ignition-delay time employed is 60 ms. This means that during flame propagation from 60 ms to 200 ms after the solenoid valve is activated, the combustion process goes on with decaying transient turbulence.

The measurement results of light transmission (Pu 1995), as a function of time, in two cross-sections located at the one-third position between two adjacent dispersion spheres demonstrated that in both sections the suspension of dispersed-dust clouds start from 20–30 ms after activation of the solenoid valve, then continue until 500 ms (for the first section) and 300 ms (for the second section). Combining the experimental results obtained from the above two measurements of dispersion-induced turbulence and from light transmission, it is recognized that the dispersion systems developed here have a duration of dust-cloud suspension in the test tube between 20 ms to 300–500 ms.

The influence of initial turbulence intensity at ignition on the flame acceleration. There are five different ignition-delay times (varied from 60 to 300 ms) employed for testing. Figure 2 shows the effect of ignition-delay time on the aluminum-dust flame acceleration. Combining the experimental results obtained from Fig. 2 and the measured results of the dispersion-induced turbulence parameters (Pu 1995), the influence of initial turbulence intensity at ignition on flame acceleration can be figured out. It is believed that the influence of initial turbulence on the flame acceleration process is mainly from the initial conditions of the dust-air mixture formation.

The influence of dust concentration and particle diameter on dust flame acceleration. Figure 3 presents the influence of concentration of the alumunum-dust/air mixture (varied between 103 and 820 g/m^3, 6-μm particle

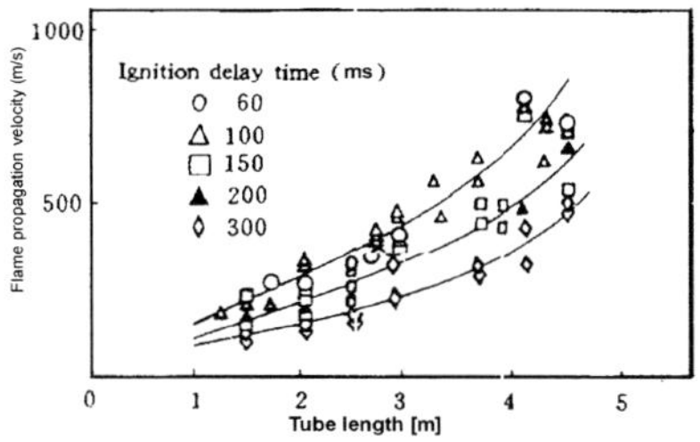

Figure 2. The influence of dispersion-induced turbulence intensity on flame acceleration.

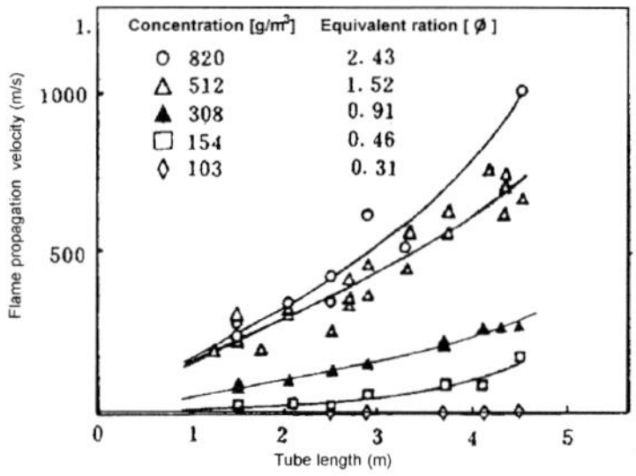

Figure 3. The influence of dust concentration on flame acceleration.

diameter) on flame acceleration. The experimental results demonstrated that when the dust concentration is larger than 500 g/m^3, the flame acceleration becomes significant. Two aluminum-dust particle diameters are used for testing. The experimental results show the significant effect of particle diameter on flame acceleration, since the flame propagation in the aluminum-dust/air mixtures has a much faster acceleration rate for the 6-μm particle diameter than for the 16-μm particles.

Flame acceleration in a fine aluminum-dust/air mixture. Atomized aluminum-dust flame propagation was carried out in the contructed combustion tube as explained above. Figure 4 shows the propagation velocity along the axis of the tube. These results demonstrate that the fine aluminum-dust flame keeps accelerating through 9 m of the tube length from the ignition plate, almost linearly, at a rate of about 250 m/s per 1 m of tube length. The maximum flame propagation velocity approaches almost 2000 m/s at the distance of 9 m from the ignition plate. Then the flame keeps the above maximum value of propagation velocity constantly between the distance of 9 to 12 m of tube length from the ignition plate.

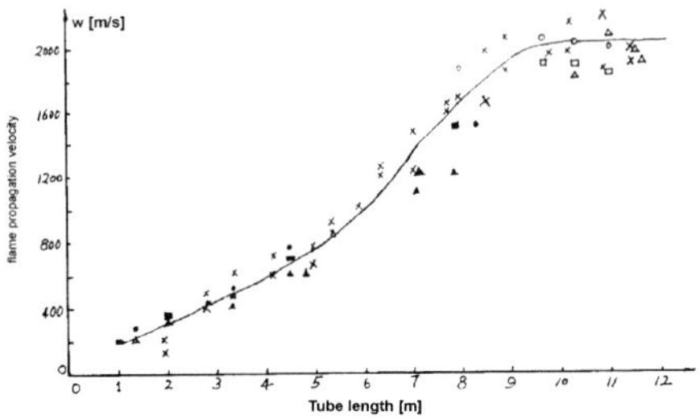

Figure 4. Dust-flame propagation velocity as a function of tube length.

4. Summary

(1) A combustion tube has been constructed to study acceleration of dust flame and its transition to detonation under conditions of weak ignition, similar to what could happen in industries producing aluminum powder. A local accidental blast producing a good suspension of dust and followed by weak ignition can happen under real powder industry conditions.

(2) The developed test tube successfully creates a uniform dust-cloud suspension in the tube, produced by a well-constructed dispersion system. Therefore, it becomes possible to examine the effect of initial ignition energy and initial dispersion-induced turbulence intensity on the transition processes of dust flames.

(3) Accelerating flame propagation obtained in the fine atomized aluminum-dust/air mixture with the present combustion conditions is characterized as steep, smooth, and monotonous. The maximum value of flame-propagation velocity approached is around 2000 m/s at a distance of 9 m from the ignition plate, then keeps constant from 9 to 12 m from the ignition plate.

References

Pu YK, Yuan SX, Ding DY, Wang BL, Tang MJ (1990) The investigation of explosion characteristics in fine aluminum dust-air mixtures. In: Proceedings of the Fourth International Colloquium on Dust Explosion, Porabka-Kozubnik, Poland, pp 159.

Zhang F, Grönig H (1991) Transition to detonation in corn starch dust-oxygen and air mixtures. Combustion and Flame 86:21–32.

Pu YK, Li YC, Kauffman CW, Bernal LP (1989) Determination of turbulence parameters in closed explosion vessels. In: AIAA Progress in Astronautics and Aeronautics, Vol 132, p 107.

Pu YK, Hu S, Jarosinski J,(1995) Combustion tube studies of dust flame acceleration. In: Shock Waves 5:103-108.

Detonation propagation in a one-dimensional heterogeneous condensed-phase explosive

J. J. Lee, M. Brouillette
Département de Génie Mécanique
Université de Sherbrooke
Sherbrooke, Québec, J1K 2R1 Canada

Abstract: The mechanisms of detonation in a solid-liquid heterogeneous explosive mixture are investigated using a one-dimensional configuration. The simplified arrangement used consists of a long rectangular channel divided into a donor section and an acceptor section by a slab of solid inert material, then filled with a liquid explosive. Using nitromethane (NM) with 15% diethylenetriamine (DETA) by weight as a liquid explosive and glass or steel for the solid slabs, the propagation of a detonation across one or more solid slabs of inert material is investigated. Detonation is initiated in the donor section and reinitiation of a detonation in the acceptor section is found to depend on the material properties of the solid slab and its thickness. The critical slab thickness above which detonation cannot reinitiate is measured and is found to be much smaller for steel than for glass. Arrangements using two and more solid slabs were also tested. For the two-slab arrangements, the thickness of the slabs and the gap between them was varied. The critical gap width below which reinitiation no longer occurred in the acceptor section was found to depend weakly on the thickness of the slabs.

Key words: Nitromethane, Heterogeneous detonation, Wave transmission

1. Introduction

Previous experiments (Lee et al. 1995a, 1995b) with a condensed-phase explosive consisting of nitromethane (NM) in a packed bed of solid glass beads have suggested the presence of two distinct modes of detonation propagation. It was put forth that in one mode, the inert glass beads do not participate significantly in the detonation wave process, with the wave propagating only in the liquid explosive occupying the voids between the beads. This mode is analogous to gaseous detonations in porous media (Makris et al. 1995) where the inert material constituting the porous media acts only as a solid boundary to the local detonation process. In the other mode, wave propagation through the inert media becomes an important mechanism in the overall detonation process. The possibility of a global detonation propagation supported by wave transmission through the solid inert material in the heterogeneous explosive mixture arises due to the comparable shock impedances of the liquid and the solid, resulting in an appreciable fraction of wave energy transmitted through the solid.

In the present work, we investigate these transmission mechanisms of detonation in heterogeneous condensed-phase explosives using a one-dimensional configuration of liquid explosive and solid inert material. The arrangement of the explosive consists of a donor and an acceptor section separated by an inert layer. This configuration closely resembles typical gap-test experiments for condensed explosives and is analogous to gas-phase experiments performed by Edwards et al. (1981) and Bjerkedvedt et al. (1986) where air gaps were introduced in a gaseous explosive. The critical slab thickness below which a detonation can successfully reinitiate in the acceptor section is measured, and is found to depend mostly on the shock-impedance mismatch between the liquid explosive and the inert solid. Additional experiments were performed using two or more slabs to more closely simulate a continuous liquid-solid heterogeneous mixture. Detonation reinitiation in the acceptor section is measured for two identical inert slabs separated by a NM-filled gap of a certain width. Finally, an arrangement of alternating layers of glass slabs and NM was tested and detonation velocity was measured.

2. Experimental details

The explosive charges consist of rectangular channels 280 mm in length and approximately 15 mm in width and height, constructed with 1-mm thick cardboard soaked in paraffin (Fig. 1).

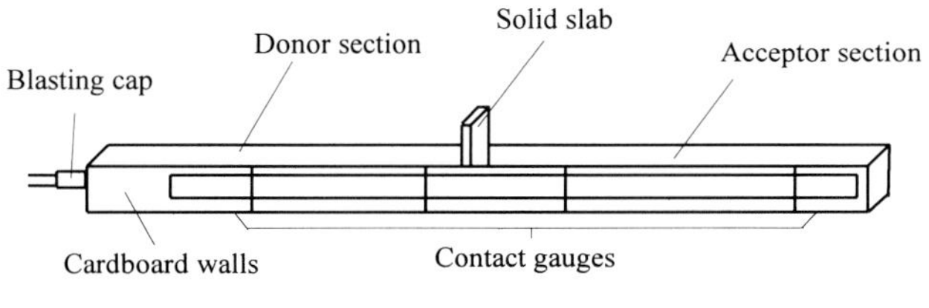

Figure 1. Schematic of the explosive charge.

The liquid explosive used in this study is the same as that used in previous experiments with liquid-bead mixtures (Lee et al. 1995b), i.e., NM with 15% diethylenetriamine by weight. Detonation-wave-front curvature effects are minimized since the charge width is 15 times larger than the unconfined critical diameter (Lee et al. 1995a). The detonation was initiated in the donor section

Table 1. Trials for the single-glass-slab tests.

Thickness (mm)	Detonation (glass)	Failure (glass)	Detonation (steel)	Failure (steel)
0.10			1	
0.20			1	
0.25			1	
0.30			2	
0.38			1	3
0.70				2
1.0	2			
1.26	1			
1.7	4	1		
1.9	1	2		
2.0		3		
2.32		1		

by a number-8 electric blasting cap. The inert solid slabs consisted of Pyrex glass or cold-rolled C1010 steel plates of varying thickness. The donor-section length was over seven times the charge width in order to minimize initiation transients and finite charge-length effects. The critical slab thickness for detonation reinitiation in the acceptor section was measured by gradually increasing the thickness of the slab until detonation did not occur. Experiments were also performed with many slabs in succession using only glass for the slab material. In these tests, slabs of equal thickness separated by a small NM-filled gap were placed in the channel. The critical gap width was measured by reducing the gap width until reinitiation no longer occurred in the acceptor section. Tests were performed with two slabs and 38 slabs. In the 38-slab arrangement, 1-mm thick slabs were used with 0.66-mm NM-filled gaps between each slab. Velocity measurements were taken along this 63-mm array of alternating inert material and liquid explosive. Velocity measurements were performed in all tests using time-of-arrival contact gauges placed against the channel wall. These gauges provided detonation-velocity measurements as well as confirmation as to whether the detonation had reinitiated or not in the acceptor section.

3. Results and discussion

3.1. Single-slab experiments

In the single-slab experiments, measurements of the critical thickness were performed with glass and steel, which exhibit very different material properties, in particular, densities and shock impedances. The results of simple go/no-go observations for glass and steel are shown in Table 1. The values of the critical thickness are found to be 1.8 mm $\pm$ 0.07 mm for glass and 0.34 mm $\pm$ 0.04 mm for steel. The velocities upstream and downstream of the slab were consistently found to be 5.88 km/s, which is close to 5.83 km/s, the Chapman-Jouguet (CJ) detonation velocity for the NM-DETA mixture as calculated by the CHEETAH thermodynamic equilibrium code (Fried 1994). In the case of failure, no velocity-gauge signals were detected downstream of the slab.

Detonation reinitiation in the acceptor section results from transmission of the detonation pressure from the donor section through the slab and into the explosive in the acceptor section. The shock pressure transmitted into the acceptor section is attenuated due to three main effects: the impedance mismatch between the inert solid and the liquid explosive, axial expansion waves behind the leading shock of the detonation (Taylor waves), and lateral expansion waves propagating inwards from the sides of the charge causing wave-front curvature. The main cause of shock-pressure reduction is the impedance mismatch between the liquid and the solid, which can be calculated using the impedance-matching method and the shock-Hugoniot data of the substances. The Taylor wave, however, causes an attenuation of the shock pressure in the solid due to the axial expansion waves which catch up with the leading shock, resulting in a decay of the pressure pulse. The thicker the solid slab, the larger the decay, and above a certain critical thickness, the transmitted shock has decayed too much to reinitiate a detonation in the acceptor section. Shock decay due to curvature effects is considered small in the present experiments since the slab is thin (typically 7 to 50 times smaller than the charge width), hence the process is close to one dimensional. As long as the transmitted shock pressure is high enough and its duration is long enough, detonation reinitiation can occur. Shock-initiation criteria are discussed by Hardesty (1976), Campbell et al. (1961), and Walker (1979) for NM; and by Lee et al. (1995a) and Engelke (1980) for DETA-sensitized NM.

Using shock-Hugoniot data for the substances involved (Marsh 1980, Lysne and Hardesty 1973), and CJ detonation parameters calculated from the CHEETAH code, the transmitted shock strength in the acceptor was estimated by the impedance-matching method. With a detonation pressure P_{CJ} of 11.9 GPa, the transmitted wave in the acceptor is found to be 10.1 GPa for glass and 3.9 GPa for steel. The higher shock impedance of steel compared to glass results in a smaller transmitted shock in the acceptor section. This is consistent with the experimental results which show that the critical thickness is much smaller for steel than for glass. Since the shock is less attenuated by the glass, it can propagate much further before decaying below the pressure necessary for reinitiation. These critical-thickness results imply that a large impedance mismatch between the explosive and the inert material may have a strong inhibiting effect in explosives dependent on transmission mechanisms. This effect would be especially apparent in a condensed explosive with a high fraction of inert material such as that used by Lee et al. (1995a, 1995b).

3.2. Multiple-slab experiments

Additional experiments using multiple-slab configurations were done to more closely represent a continuous 1D heterogeneous explosive, i.e., a very large number of alternating layers of explosive and inert material. For the two-slab tests, two slab thicknesses were used: 1 mm and 1.7 mm. The critical gap width was found to be 0.35 mm $\pm$ 0.07 mm for both 1-mm and 1.7-mm slabs. This shows that the reinitiation process does not depend strongly on the slab thickness. This weak dependence may be a consequence of the complex wave-interaction process for the two-slab configuration, which may lead to a strong coupling between the slabs causing the detonation-reinitiation process to depend on some of the many secondary reflected and transmitted waves. Additional tests were performed with 0.15-mm thick slabs, but no critical thickness was found as the detonation transmitted through the two-slab assembly for any gap thickness.

The 38-slab arrangement showed interesting results. A fairly stable detonation was found to propagated readily through the assembly of 1-mm glass slabs with 0.66-mm layers of NM between them. The detonation velocity was found to be 4.03 km/s. This value is identical within experimental error to the detonation velocity in heterogeneous mixtures of DETA-sensitized NM in packed beds of glass beads (Lee et al. 1995a). It should be noted that the gap width in the glass-slab arrangement was chosen so that the volume fraction of NM in the explosive was the same as in the glass-bead explosive, i.e. 40%.

4. Conclusions

The critical thickness for which a detonation in DETA-sensitized NM can successfully transmit from a donor to an acceptor section is measured for two materials with different shock impedances. The critical thickness is 1.8 mm $\pm$ 0.07 mm for glass and 0.34 mm $\pm$ 0.04 mm for steel. The results are consistent with calculations that show that the transmitted pressure in the acceptor section is much lower for steel than for glass due to the more severe impedance mismatch between the former and the liquid explosive (NM). Impedance mismatch between the explosive and the inert material in a heterogeneous explosive may therefore strongly inhibit detonation propagation if the mismatch is severe. For a two-slab configuration, the critical gap width of NM necessary for reinitiation was found to be 0.35 mm $\pm$ 0.07 mm for both 1-mm and 1.7-mm slabs. The dependence of the critical gap width on slab thickness is thus weak. For a multiple-slab arrangement of alternating glass slabs 1-mm thick and NM layers 0.66-mm thick, a stable detonation was found to propagate at 4.03 km/s. This value is the same as that in a NM-glass-bead mixture with the same proportion of liquid explosive to solid inert material. Through these experiments, we hope to further elucidate the effects of the material properties of the inert material on heterogeneous explosives.

Acknowledgement. The authors would like to thank Jean-Philippe Dionne for his assistance with the CHEETAH code calculations. This research is supported by ICI Explosives Canada and by the National Sciences and Engineering Research Council of Canada.

References

Bjerketvedt D, Sonju OK, Moen IO (1986) The influence of experimental condition on the reinitiation of detonation across an inert region. Progress in Astronautics and Aeronautics 106:109–130.

Campbell AW, Davis WC, Travis JR (1961) Shock initiation of detonation in liquid explosives. Physics of Fluids 4:498–510.

Edwards DH, Thomas GO, Williams TL (1981) Initiation of detonation by steady planar incident shock waves. Combustion and Flame 43:187–198.

Engelke R (1980) Effect of a chemical inhomogeneity on steady-state detonation velocity. Physics of Fluids 23(5):875–880.

Fried L (1994) CHEETAH, 1.22 User's Manual, UCRL-MA-117541.1.

Hardesty DR (1976) An investigation of the shock initiation of liquid nitromethane. Combustion and Flame 27:229–251.

Lee JJ, Brouillette M, Frost DL, Lee JHS (1995a) Effect of diethylenetriamine sensitization on detonation of nitromethane in porous media. Combustion and Flame 100:292–300.

Lee JJ, Frost DL, Lee JHS, Dremin AN (1995b) Propagation of nitromethane detonation in porous media. Shock Waves 5:115–119.

Lysne PC, Hardesty DR (1973) Fundamental equation of state of liquid nitromethane to 100 kbar. J Chem Phys 59(12):6512–6523.

Makris A, Shafique H, Lee JHS, Knystautas R (1995) Influence of mixture sensitivity and pore size on detonation velocities in porous media. Shock Waves 5:89–95.

Marsh SP (ed) (1980) LASL Shock Hugoniot Data, University of California Press, Berkeley, California.

Walker FE (1979) Initiation and detonation studies in sensitized nitromethane. Acta Astronautica 6:807–813.

Transmission of a detonation across an air gap

A. Teodorczyk,* G. O. Thomas,† S. M. Ward
Department of Physics
University College of Wales
Aberystwyth, SY23 3BZ UK

1. Introduction

The transmission of a detonation across an air gap has been postulated as one means whereby a transition from a deflagration to detonation (DDT) can be engineered in a controlled manner. Such DDT events are of both fundamental and practical interest. From a practical viewpoint it is of interest to identify the exact conditions under which a transition to detonation may occur whilst the process by which DDT occurs is of continuing fundamental interest. The use of an attenuating inert air-gap configuration provides one potential means for generating the conditions where transition to detonation occurs in a controlled and reproducible manner. The approach is attractive to those wishing to investigate the transition process in greater detail as it avoids the initial stochastic flame-acceleration phase usually associated with DDT events. The studies themselves grew from a practical need to generate controlled transition to detonation as a reproducible means of testing certain aspects of detonation arrester performance.

In the present paper, results are presented from an experimental program to investigate the behavior of incident detonation in a range of gaseous mixtures as they propagate across inert gaps. The experimental results are also compared with numerical predictions of the wave propagation in the inert and acceptor sections.

The specific points of interest in the present studies were the influence of gap length and the potential for re-initiation as a function of mixture reactivity. The later we investigated by varying the fuel.

2. Methodology

It has already been demonstrated (Thomas et al. 1991) that if a steady detonation is incident on an inert pocket of gas the detonation fails, instead giving

*Permanent address: University of Technology, Warsaw, Poland.
†Corresponding author.

a transmitted shock followed by expanding detonation products. If the shock transmitted into the reactive mixture beyond the inert region is sufficiently strong, and the mixture in this region is sufficiently reactive, transition to detonation may occur. The sequence of events envisaged is as follows (see Fig. 1). An energetic booster section initiates a steady detonation in the main donor section. This steady incident detonation wave is then rapidly attenuated by an appropriate means, in the present case by an inert air gap. Subsequently, as this decoupled wave propagates into the reactive acceptor section, flame and shock acceleration are observed with the possibility of transition to detonation. Very large pressures can be generated locally in the vicinity of the transition-to-detonation point.

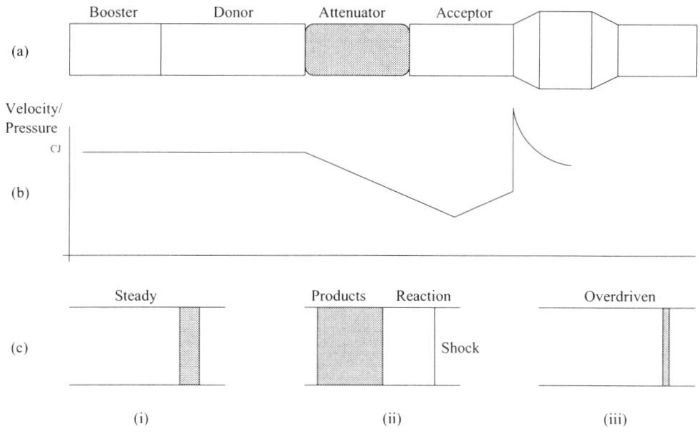

Figure 1. Schematic showing (a) the different sections used; (b) anticipated velocity history; and (c) summary of wave-front characteristics at different positions — (i) steady detonation, (ii) decoupled shock and reaction zones, and (iii) overdriven wave following transition to detonation.

3. Experimental details

The experimental apparatus, shown schematically in Fig. 2, consisted of a 1-m long booster section, a 3-m long donor section, an air gap, and a 1-m long acceptor section. All sections had an internal diameter of 50 mm. A detonation arrester was also located downstream of the acceptor section followed by a 0.5-m long end section.

The air gap was separated from the combustible mixture by two slide valves formed from a 1-mm stainless steel plate mounted between two flanges. The plates were extracted rapidly by a pneumatic piston arrangement actuated just prior to ignition of the booster mixture.

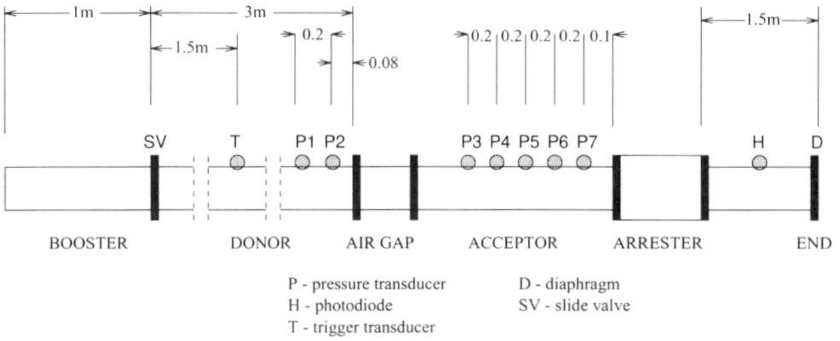

Figure 2. Schematic of general experimental arrangement.

In some tests, re-initiation downstream of the attenuating section was stimulated by introducing an obstacle section comprised of five annular disks, 2 mm thick with 40% blockage ratio. The longitudinal spacing between the disks was 0.1 m.

All experiments were conducted at ambient atmospheric pressure and temperature, using hydrogen-, ethylene-, and propane-air mixtures. The booster used to initiate detonation in the initial donor section was a stoichiometric oxy-acetylene mixture, again at atmospheric pressure and ignited by a low-energy spark. A number of pressure and photo-diode gauges were used to monitor shock and flame propagation and their locations are also given in Fig. 2.

4. Results

The least reactive mixture tested was that of propane-air. With both 25-cm and 10-cm air gaps, fast deflagration waves were re-established in the acceptor section, but no transition to detonation was obtained. Typical pressure histories with the 10-cm long gap are shown in Fig. 3a. In this case, for propane-air, a shock/flame complex propagated in the acceptor tube with an almost steady velocity of about 1100 ms^{-1} with some evidence of small local explosions in the reaction zone which did not lead to transition.

To aid the transition process, the short obstacle section described above was then placed in the acceptor section just after the attenuating section. The transmitted shock/flame velocity was now slightly lower (about 950 ms^{-1}) and despite stronger local pressure impulses due to interaction with the obstacles, the combustion wave was not accelerated further. This indicates the strong destructive role of wall obstacles on the flame behavior in less reactive mixtures.

For hydrogen-air, the most reactive mixture tested, a detonation wave was re-initiated in the acceptor tube in almost all cases, for both 10- and 25-cm long air gaps. Figure 3b shows a series of pressure histories obtained with hydrogen for a 25-cm long gap. Transient pressure pulses resulting from a transition to detonation can be seen clearly on gauge PT5.

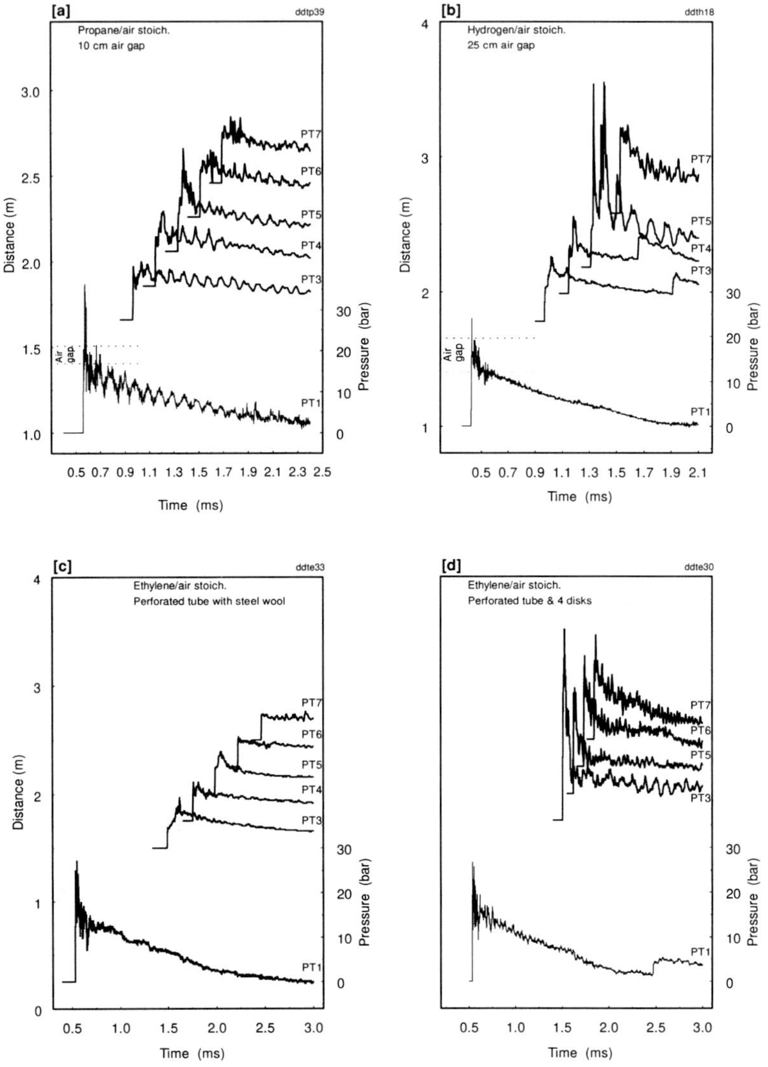

Figure 3. Pressure histories: (a) for propane-air detonation incident on a 10-cm long air gap; (b) showing the attenuation and subsequent re-initiation of detonation in hydrogen-air as it traverses a 25-cm long air gap; (c) for an incident ethylene-air detonation propagating across a section lined with acoustically absorbing material; and (d) showing re-initiation of detonation in ethylene-air when obstacles were placed after a section lined with acoustically absorbing material.

For the ethylene-air mixture, the shock/flame complex had the steady-state velocity of the transmitted wave, which was again of the order of 1100 ms^{-1} at the end of acceptor tube. With the stronger local explosions in the flame zone the combustion wave was steadily accelerated.

For some of the tests, the effectiveness of a different attenuating mechanism was investigated. Figures 3c and 3d show the pressure histories obtained with ethylene-air when a 50-cm long section filled with an acoustically absorbing material was used. To achieve this, an initially enlarged attenuating section was lined with a 2-cm thick layer of steel wool retained by a coarsely perforated mesh. Thus an absorbing wall could be formed, whilst retaining the internal diameter at 50 mm.

In Fig. 3c the incident detonation is clearly attenuated efficiently as the transverse waves associated with the incident detonation are attenuated. The transmitted wave is now not sufficiently strong to re-initiated detonation in the acceptor section. However, when the obstacle section was placed immediately downstream of the absorbing section, transition to detonation was readily obtained, as can be seen from the pressure records shown in Fig. 3d.

To investigate the re-initiation process further, a series of numerical simulations have been initiated to model interaction of an established detonation with the air gaps. Non-reactive calculations have been performed using a computer code to implement the Random Choice Method. This cannot model the transition process, but it does allow the propagation and subsequent decay of the blast wave in the attenuator and acceptor sections to be modeled. This is possible by treating the detonation front as an instantaneous gasdynamic discontinuity.

The results obtained for the blast decay were in reasonable agreement with the experimental profiles in most cases. In general, the code overpredicted the strength of the transmitted wave. One probable cause for this is that the initial detonation-wave pressure profile is based on an ideal tube, with no wall losses.

5. Conclusions

The present work has demonstrated that the use of an attenuating inert air-gap section allows the subsequent re-initiation of detonation to be achieved in a reproducible manner in many instances. Whilst direct onset of DDT may not be possible with all mixtures (none was observed with propane-air with the present configurations), the use of an obstacle region can promote rapid onset of DDT. This later onset of transition is due to local increases in temperature arising from the reflection and diffraction of the transmitted blast wave.

Using this approach, DDT could be engineered to occur over a limited spatial range, some 40 cm in extent. This was achieved using an apparatus less than 5 m in overall length and has been shown to be possible for both hydrogen-air and ethylene-air mixtures. Other studies by Thomas and Teodorczyk (1995) have shown that through the use of another attenuating mechanism, DDT can be obtained in essentially the same geometry for propane-air.

Both 10-cm and 25-cm air gaps were tested. On balance, it was concluded that an air gap of 25 cm was too long; and the use of obstacles caused strong instabilities and pressure oscillations which resulted in a more random history of the process.

References

Thomas GO, Sutton P, Edwards DH (1991) The behaviour of detonation waves at concentration gradients. Comb and Flame 84:312–322.

Thomas GO, Teodorczyk A (1995) Testing procedures for explosion arresters: some problems and potential solutions. In: Proceedings Second European Conference on Major Hazards OnShore/Offshoer, UMIST, Manchester 24–26 October 1995, Institution of Chemical Engineers, UK (to appear).

Large-scale confined hydrogen-air detonation experiments and their numerical simulation

W. Breitung[1], S. B. Dorofeev[2], A. A. Efimenko[2]
A. S. Kochurko[2], R. Redlinger[1], V. P. Sidorov[2]

[1]Forschungszentrum Karlsruhe, Institut für Neutronenphysik und Reaktortechnik (INR), P.O. Box 3640, 76021 Karlsruhe, Germany

[2]Russian Research Center Kurchatov Institute
Applied Chemical Physics Institute, 123182 Moscow, Russia

Abstract: Large-scale (263 m^3) H_2-air (16 to 25% volume) detonation experiments in complex 3D geometry are described. The effect of confined gaseous detonations was simulated numerically in 3D calculations. The simulation was shown to be a reliable tool for predicting the blast effect of confined detonations in complex 3D geometry for sufficiently reactive mixtures and to simulate the main features of unstable propagation with corresponding pressure field.

Key words: Large-scale experiments, Detonation, Three-dimensional simulation, Hydrogen

1. Introduction

The understanding of the conditions required for detonation onset and propagation is both of fundamental and practical interest. Different modes of combustion processes can be expected during industrial accidents. The numerical tools for the description of possible consequences of these events need to be developed. Presently, the level of understanding and computer capabilities allow one to build appropriate models for the description of pressure fields resulting from particular combustion modes. The description of the entire sequence of events, starting from slow burning to fully developed detonation, is hardly possible in the framework of one model. The objective of the present work is to study the pressure field generated by confined 3D detonations experimentally and numerically. The second goal is evaluation of the 3D code by means of the experimental data.

2. Experiments

The experiments were performed in the RUT facility near Moscow (Dorofeev et al. 1993, Breitung et al. 1994). The geometry of the enclosure is shown in Fig. 1. The facility was initially filled with air at ambient pressure and temperature

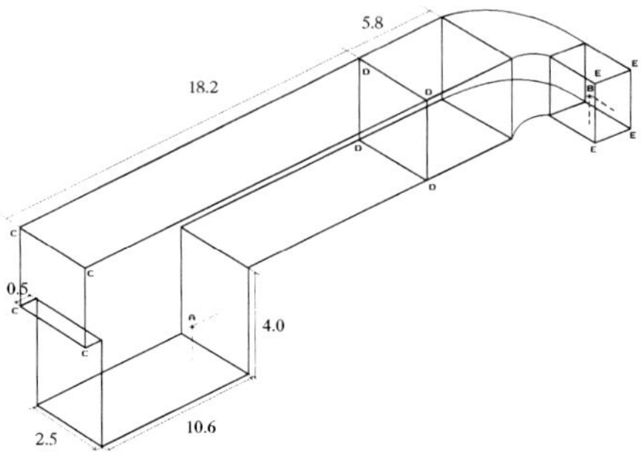

Figure 1. Geometry of RUT-facility test section. Dimensions are in meters. Points of ignition are designated by A and B.

(1 bar, 283 K). Hydrogen was added slowly at the bottom of the lower room and continiously mixed with air by two fans. The observed nonuniformity of the gas mixture during filling was less than $\pm 0.5\%$ of hydrogen. The H_2-air mixture was ignited with high-explosive (HE) charges. Ten H_2-air detonation experiments were carried out in the RUT facility. The first five tests investigated the effect of location of the detonation ignition and the reproducibility of the obtained data. Tests 6 to 10 studied the effect of hydrogen concentration on detonation loads for a given initiation point. Two characteristic detonation propagation regimes were observed in the tests: stable propagation throughout the 3D geometry for 20% and 25% hydrogen; and near-limit, unstable propagation with quenching and re-ignition for 16% hydrogen. In the first regime the characteristic size of the RUT channel (2.5×2.3 m) is much larger than the critical tube diameter $d_c = 10 \ldots 13\lambda$. The detonation cell widths λ for 20% and 25% hydrogen are about 6 cm and 2 cm, respectively (Guirao et al. 1989). In test 10 with 25% H_2 the front pressure remained above Chapman-Jouguet (CJ) values for all gauges, showing a stable propagation throughout the whole volume. The detonation cell width for 16% H_2 is about 25 cm (Guirao et al. 1989). In this case, the size of the RUT channel is less than the critical tube diameter ($2.5 \ldots 3.25$ m) and a decaying detonation wave could be expected at a cross-section enlargement. This was indeed observed at the plane where the RUT channel connects to the lower room. Measured pressure histories of test 6 (16% H_2) are shown in Fig. 2 (left). In this figure consecutive pressure histories are shown. The pressure in the wave front was near CJ values initially in the channel, but decayed to adiabatic isochoric complete combustion (AICC) values later when entering the volume with the lower room. The time resolution of the pressure recordings was sufficient to resolve spatial wave structure above several centimeters characteristic size. The

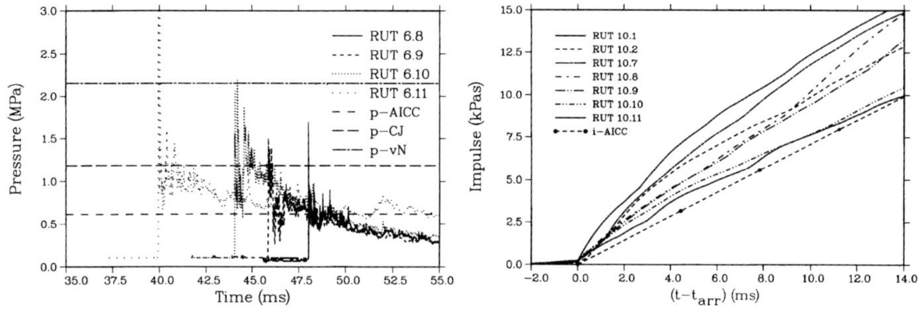

Figure 2. Decay of detonation wave in test 6 (16% H_2) (left). Effect of gauge location on measured impulse (right).

cellular structure of the detonation front could be resolved very well in the case of the lean H_2-air mixture. The space interval between the pressure spikes was about 15 $\div$ 30 cm, which is close to the transverse wave spacing (cell width) in the mixture. The measured peak pressure depends on the mutual orientation of the incoming wave front and the loaded surface. In the case of a perpendicular orientation, "side-on" pressures up to the theoretical von-Neumann values were observed. In the case of a normally reflected detonation, significantly higher "end-on" pressure is observed. The peak pressure ratio between end-on and side-on pressures is approximately 3.5 MPa/1.5 MPa = 2.25. This value agrees with earlier calculations, which predicted a ratio of 2.2. The ignition location influences only the wave/surface orientation. With the changing of mixture composition, the measured peak pressure scales roughly with the AICC pressure of the mixture ($P_{max}/P_{AICC} \simeq 2$). The impulse of the detonation wave $i = \int P(t)dt$ governs the maximum deformation of structural elements. The relevant integration time is the characteristic time of the detonation peak during which $P(t) > P_{AICC}$. At later times the impulse basically increases linearly with time due to the quasi-stationary pressure. The impulse to a wall depends on its orientation relative to the propagating wave and its distance from the point of ignition (curves 10.11, 10.10, 10.9, 10.8, see Fig. 2, right). The impulse increases with increasing distance from the point of ignition. Increased impulses are also measured for side-on position near reflecting surfaces (curves 10.7, 10.2) with the highest value obtained for the normally reflected position (curve 10.1). The late impulses increase basically linearly in time and with a slope close to P_{AICC} (curve i-AICC). The long-term pressure history is determined by shocks which reverberate in the confined geometry. Measured pressures oscillated with decreasing amplitude. The time average is close to the AICC pressure of the

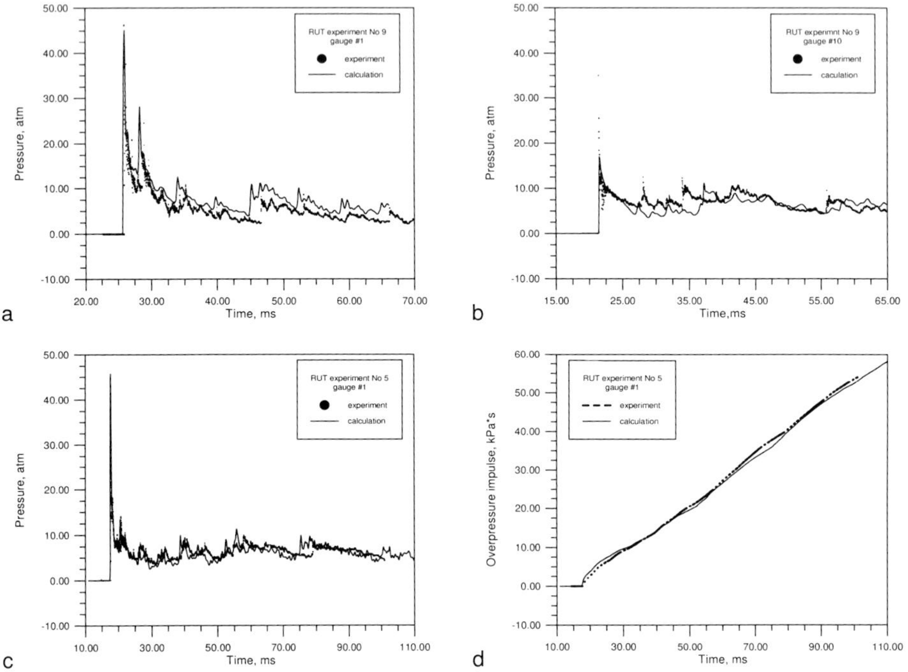

Figure 3. Comparison of experimental and calculated data. Pressure results for 25% H_2: (a) normal reflection, ignition point B; (b) side-on gauge position, ignition point B; (c) normal reflection, ignition point A; (d) overpressure impulse, corresponding pressure is shown in Fig. 3c.

mixture and is the same for all wall locations. The long-term pressures measured for different mixtures correlate with calculated AICC pressures, but are slightly below the corresponding values.

3. Numerical simulation

Code 3b02 (D3D) was developed for numerical simulation of gas detonation in complex 3D geometries. The code is based on the following models: conservation equations for momentum, energy, and species masses of a two-component reacting gas; one-step reaction kinetics with Arrhenius-type reaction rate; constant specific-heat equations of state; 3D geometry approximated by cubic numerical cell; and arbitrary room shapes, combinations, and flow obstacles. The RUT tests with fully developed detonation ($20 \div 25\%H_2$) were simulated with a numerical cell size of $(10\ \text{cm})^3$. Three pressure-time and one overpressure impulse comparisons are presented; as an example, Fig. 3a compares the data for experiment 9 with 25% H_2. Very good agreement can be observed for the peak pressure. Since the model used has no loss mechanisms like radiation or friction, the calculated pressure and sound speed at late stages are slightly above the measured values. All major peaks of the measurements can be easily identified in the calculations. The code models the complex wave system in 3D geometry quite well. Figure 3b compares pressure data for side-on gauge position. The

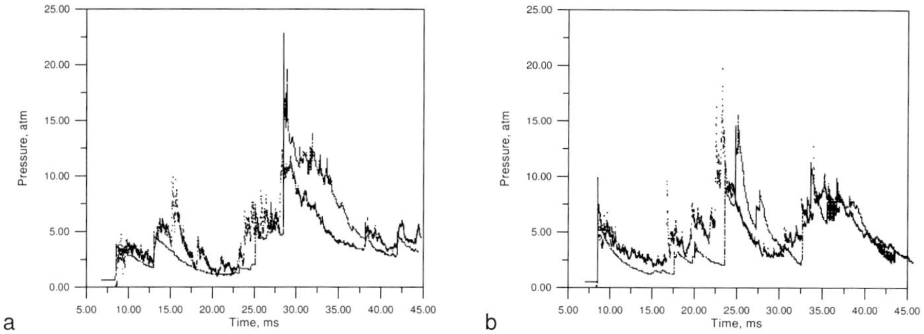

Figure 4. Comparison of experimental and calculated pressure histories in test 6: gauge # 5 (a); gauge # 8 (b). Dots – experiment; solid line – calculation.

calculated detonation peak pressure is less then in a normally reflected wave by a factor of 2.3, which agrees with experimental observations. Also the later, much more smooth pressure history is reproduced satisfactorily. To check the effect of ignition location, Fig. 3c compares pressure data with ignition point in the lower room. Ignition at this point produced many shock reverberations in the lower room. The code is also able to model this case with acceptable accuracy. Measured and calculated overpressure impulses $i = \int (P(t) - P_0)dt$ are compared in Fig. 3d. The conclusion from a systematic comparison of measured and calculated impulses is that the predicted values are slightly conservative by up to 10%. Attempts to model the near-limit detonations at 16% H_2 with the one-step kinetic model (as for $20 \div 25\%$ H_2) failed; the reason being the large effective activation energy of 16% H_2 mixture. Extensive analysis (Efimenko et al. 1994) showed that a one-step kinetic mechanism, based upon adiabatic ignition behind the leading shock, is not able to drive stable detonation propagation for large activation energies. For the simulation of detonation properties for mixtures with large effective activation energies, a model with two parallel reactions was proposed. The proposed model was implemented in simulations of RUT experiments with a 16% H_2 mixture (tests 6 and 8). A mesh size of 1 cm was used, since simulations require resolution of chemical scale. Due to that requirement, only 2D simulation is currently possible. Simulations show that two different regimes of re-ignition occur: in every reflection and only after multiple reflections. The second regime is exhibited in the place where the mixture was highly pre-pressed, and the explosion was characterized by extremely high pressures (over 50 MPa). Both experiments and simulation have indicated locally more severe consequences of detonations in insensitive mixtures than in near-stoichiometric mixtures. Good agreement is observed for transducers located near the exit from the channel (Fig. 4). In the lower far corner from the exit from the channel the simulated detonation wave decoupled somewhat earlier than in the experiment. Because of this, the relative height of spikes was distorted, but qualitative correlation between calculation and experiment is satisfactory. The results of these calculations show that application of the two-parallel-reactions kinetic scheme to detonation modeling appears to be fruitful.

4. Conclusion

The experiment provided the first data on H_2-air detonation loads at large scale (263 m^3) and in complex 3D geometry. Reproducible pressure data were obtained. The data include all important characteristics of 3D detonations, such as curved and oblique fronts, side-on and reflected end-on pressures, cellular structure, detonation decay, and re-ignition. The measured pressure fields were used to evaluate the three-dimensional code 3b02 (D3D). Very good agreement was observed for the initial detonation peak and for all tests with fully developed detonation (20 and 25% H_2). Near-limit detonations at 16% H_2 were simulated with the proposed two-parallel-reactions model in 2D calculations. Simulated results showed good agreement with the main features of near-limit detonation propagation. Satisfactory agreement between calculated and experimental pressure histories was observed.

References

Breitung W, Dorofeev SB, Efimenko AA, Kochurko AS, Redlinger R, Sidorov VP (1994) Large scale experiments on hydrogen-air detonation loads and their numerical simulation. In: Proceedings of ARS'94 Int Topical Meeting of Advanced Reactor Safety, Pittsburg, PA, Vol 2, pp 733–745.

Dorofeev SB, Sidorov VP, Velmakin SM, Zhernov AV (1993) Large Scale Hydrogen-Air Detonation Experiments, Report RRC Kurchatov Institute, RRCKI-80-05/59, Moscow, Russia.

Efimenko AA, Dorofeev SB, Shepherd JE (1994) Chemical reaction kinetics and detonation modelling. Presented at the Twenty-Fifth Symposium (Int) on Combustion, University of California, Irvine.

Guirao CM, Knystautas R, Lee JH (1989) A Summary of Hydrogen-Air Detonation Experiments, McGill University, Sandia Natl Labs, Rep NUREG/CR-4961, SAND 87-7128.

Transition from deflagration to detonation with a dense aluminum-dust slug in a combustible gas

F. Zhang, P. A. Thibault
Combustion Dynamics Ltd.
132 4th Avenue SE, Medicine Hat, AB, Canada T1A 8B5

Abstract: This paper investigates a DDT scenario in which a slug of dense reactive dust suspension is inserted near the end wall of a tube filled with a combustible gas. A theoretical analysis of this scenario indicates that the insertion of the dust suspension can result in a dramatic enhancement of the pressures generated by the DDT process. Experiments were performed in an 8-cm internal diameter tube filled with a lean mixture of 6.75% acetylene in air. A 100-mm cloud of aluminum dust was dispersed at the end of the tube. The experiments indicate that pressures between 300 and 700 atm can be achieved at the end wall with a dust concentration of 5 – 20 kg/m^3.

Key words: DDT, Deflagration, Detonation, Multiphase flow

1. Introduction

Deflagration to detonation transition (DDT) can result in significantly higher pressures than that of the Chapman-Jouguet (CJ) detonation in the initial mixture. Among the various scenarios proposed, the one described by Craven and Greig (1968) produces the largest peak pressure at the end wall. In that scenario, the precursor shock of the deflagration wave is reflected off an end wall. Detonation onset occurs in the shock-reflected region between the flame front and the wall. The detonation wave then propagates towards the wall and its reflected peak pressure can be of the order of 10 times that of the reflected CJ detonation in the initial mixture. Experimental evidence of such high pressure was first reported by Kogarko (1958).

Pressures generated in the Craven and Greig scenario can be further increased by enhancing the initial incident and reflected pressures at the end wall. Gelfand et al. (1975) first observed that the peak pressure behind a compressible foam on the back wall is considerably larger than for a normal rigid-wall reflection. Here, we consider the effect of adding a dense cloud of combustible particles to a lean reactive gas mixture. Enhancement of pressures generated by DDT depends on the momentum and energy exchange between the gas and the particles, and on the rate of heat release due to particle burning.

2. Theoretical considerations

When a deflagration wave achieves a critical flame velocity in a confined tube, the onset of detonation occurs via localized auto-explosions between the flame front and the precursor shock. If the tube length is reduced appropriately, the flame velocity does not reach its critical value. In this case, the precursor shock will be reflected off the end wall of the tube, and the localized auto-explosions can occur in the vicinity of the flame front. The resulting detonation propagates towards the wall in the reflected shock state, and its reflection at the wall creates a high peak pressure. Figure 1 shows the theoretical pressures behind the flame front, precursor shock, reflected shock, CJ detonation, and reflected detonation. These are calculated for a mixture of 6.75% acetylene in air with $p_0 = 1$ atm and $T_0 = 300$ K. The states of the flame front and the precursor shock are obtained from the deflagration Hugoniot, and the reflected-shock state is based on the thermally and calorically perfect gas law. The detonation and its reflected-wave states are calculated using an equilibrium code.

If we insert a slug of dense dust suspension at the end of the tube, the DDT scenario can be illustrated by the wave diagram and the pressure profiles near the wall. These are qualitatively depicted in Fig. 2. The precursor shock S_1 impacts on the dense multiphase slug, creating a transmitted shock S_{1a} propagating into the slug and a reflected shock S_2 backwards into the pre-compressed gas. The transmitted shock is then reflected off the end wall as S_{3a} and propagates back into the multiphase slug. Because of the transfer of the momentum and heat between the two phases, the pressure in the slug behind the reflected shock S_{3a} can be several times higher than that for a normal rigid-wall reflection. The temperature, however, can remain below the auto-ignition temperature. As the shock S_{3a} arrives at the interface between the gas and the multiphase slug, the refraction there creates an expansion wave R_{4a} moving back into the slug, and a shock S_3 transmitted into the gas. If the temperature behind the shock S_3 is close to the gas auto-ignition temperature and yet the flame front is on the left of the interface, localized auto-explosions can occur ahead of the flame front. In this case, onset of detonation takes place and the detonation wave propagates into the multiphase slug to create a high detonation pressure. Its reflection on the end wall can result in a very large peak pressure. Note that detonation D_{5a} may fail due to the reflected rarefaction wave R_{5a}, but can re-initiate upon wave reflection at the end wall.

The pressure enhancement produced by the presence of the dust cloud is shown in Fig. 3, which displays the pressure increase, or ratio, of the reflected shock, the CJ detonation, and the reflected detonation. The solid particles with a density $\rho_p = 2700$ kg/m^3 and a particle diameter $d_p = 5$ μm, are dispersed with a concentration $\sigma_p = 10$ kg/m^3. For the system with the multiphase slug,

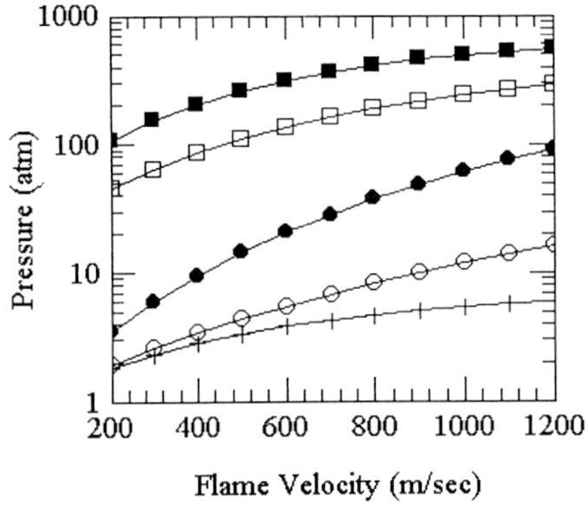

Figure 1. Pressure behind different fronts for 6.75% C_2H_2 in atmospheric air. Flame (+); precursor shock (open circles); reflected shock (closed circles); CJ detonation on reflected shock (closed squares); reflected CJ detonation (open squares).

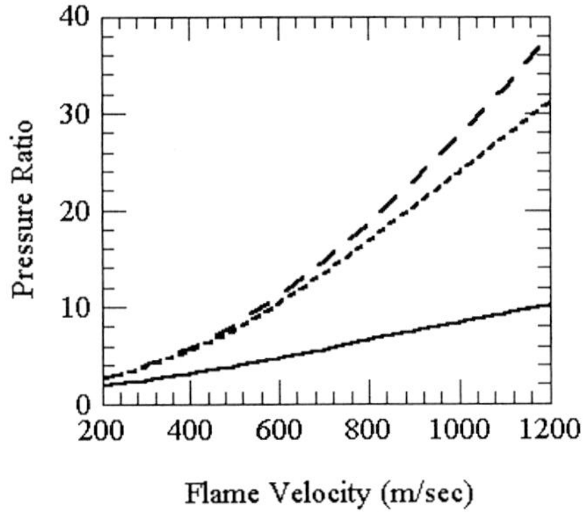

Figure 3. Equilibrated pressures for 6.75% C_2H_2 in atmospheric air with a slug of Al-dust suspension $\sigma_p = 10$ kg/m^3 at the end wall. The values are normalized by those from Fig. 1. Reflected shock (solid line); CJ detonation (dotted line); reflected CJ detonation (dashed line).

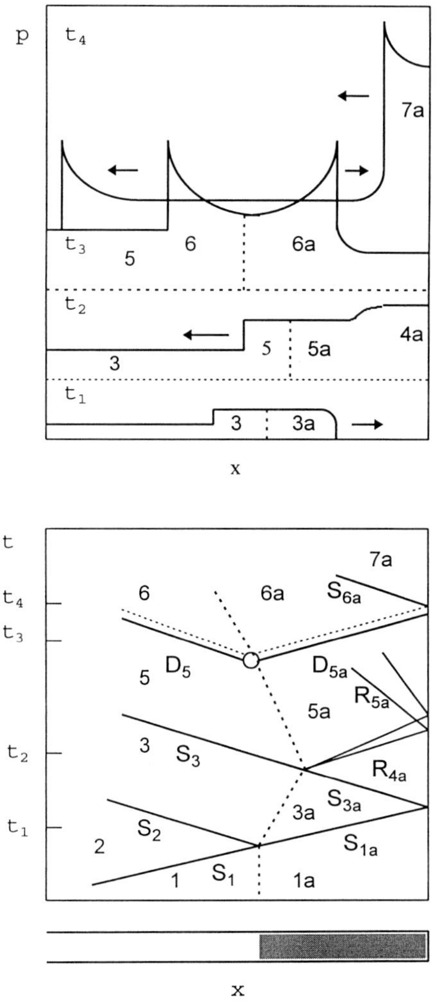

Figure 2. Schematic wave diagram and pressure profiles for a DDT scenario with a slug of dense dust suspension at the end wall.

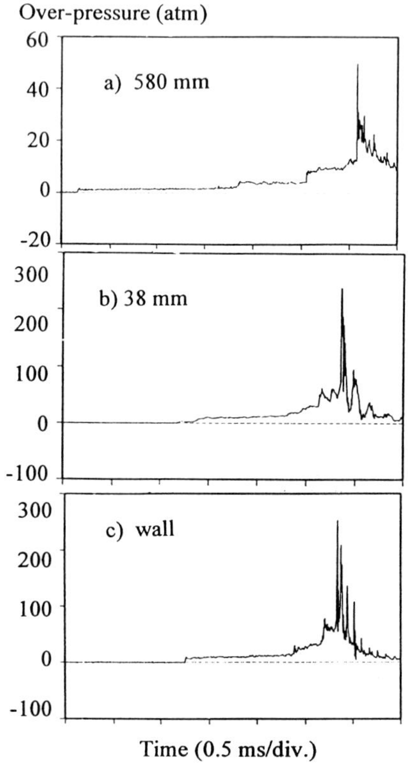

Figure 4. Gaseous DDT overpressures. Time ranges from 120.5 – 124 ms.

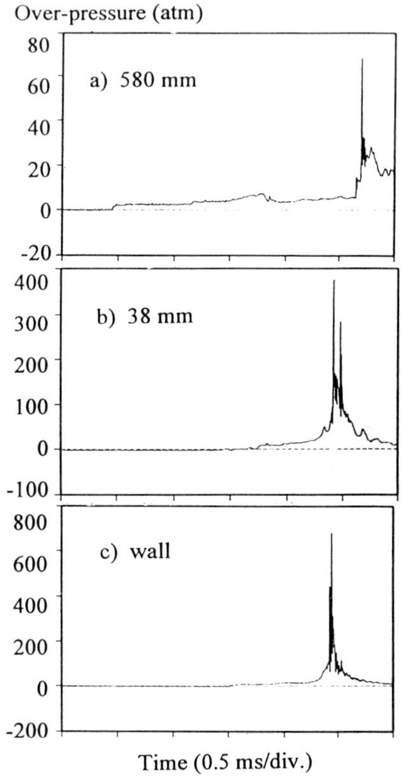

Figure 5. DDT with a 100-mm Al-dust slug of $\sigma_p = 10$ kg/m^3. Time ranges from 120 – 123 ms.

the pressures for the reflected waves are obtained assuming a fully equilibrated relaxation. The CJ detonation pressures are calculated on the basis of the phase-frozen assumption. The pressure values for the system without the multiphase slug are obtained from Fig. 1. Due to the assumptions made, the calculations for the multiphase slug are approximate estimates of the achievable pressures. For a flame velocity between 500 and 600 m/s, the pressure ratio of the reflected shocks can be as high as 3.9 to 4.7. The pressure ratio of the reflected detonation ranges between 8.0 to 11, which results in pressures that are 115 to 195 times larger than the CJ detonation pressure for the initial mixture.

If the dense dust is inert and sufficiently fine, the rate of energy transfer from the gas to the dust competes with the heat-release rate of the gas. As a result, DDT may not occur or the formed detonation wave can fail. If the dust

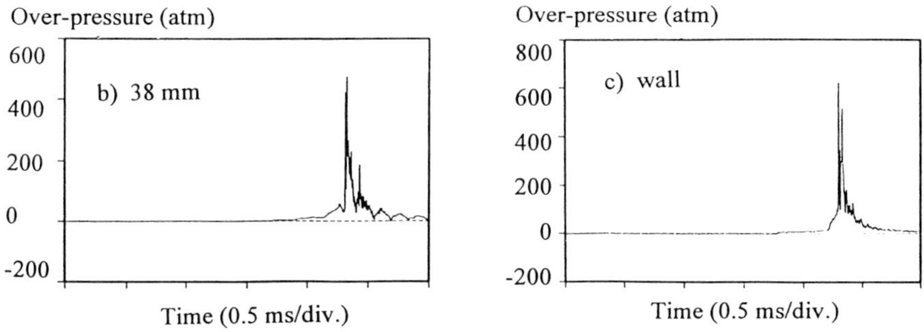

Figure 6. DDT with a 100-mm Al-dust slug of $\sigma_p = 5$ kg/m^3 for 120 to 123 ms.

is reactive, but not too dense, it can react with the gas-detonation products and the remaining oxygen to amplify the gas detonation.

3. Experimental results and conclusions

In order to examine the amplification process, experiments were carried out in a tube 8 cm in diameter and 4.5 m in length. A gas mixture of 6.75% acetylene in air with $p_0 = 1$ atm and $T_0 = 300$ K was used and ignited at one end by an electrical spark of about 200 mJ. A dense dust suspension slug was created at the end of the tube using a dust dispersion system that is similar to that described by Zhang and Groenig (1991). Aluminum dust with a mean particle diameter of 5 μm served as combustible dust, and glass spheres 3.2 μm in mean diameter were chosen for chemically inert dust. One Kulite piezo-resistive pressure transducer HKS375 was mounted at the end wall, and two on the side wall, 580 mm and 38 mm away from the end wall. The time resolution of data-acquisition system was 4 μs.

As a reference, Fig. 4 provides the DDT overpressure records for the gas mixture without the multiphase slug. After the precursor shock (shock pressure = 3.2 atm) is reflected off the end wall (reflected shock pressure = 11 atm), localized auto-explosion occurs between the flame front and the end wall. This can be recognized by the third pressure jump in Fig. 4a and by the pressure jumps around $t = 123.1$ ms in Figs. 4b and 4c. The blast wave then propagates towards the end wall and reflects off it, creating an overpressure of 255 atm. Note that the theoretical reflected peak pressure of the CJ detonation has a value of 260 atm for the flame velocity of 500 m/s, which corresponds to a reflected shock pressure of 14 atm and temperature of 726 K (see Fig. 1).

Figure 5 gives the DDT overpressure records with a slug of aluminum-dust suspension 100 mm in thickness, and a dust concentration $\sigma_p = 10$ kg/m^3. Evolution of a localized auto-explosion can be recognized by the pressure jump at about $t = 122.3$ ms in Fig. 5b. Consequently a detonation wave is formed close to the end wall, which can be seen in Figs. 5b and 5c. The maximum pressure at the end wall reaches a value of 700 atm. From the overpressure records it is unclear whether the first auto-explosion occurs in the multiphase slug or in the gas. Using the same dust and slug thickness, but reducing the dust concentration to 5 kg/m^3, a similar DDT scenario can be obtained. This case is displayed in Fig. 6, where the maximum pressure is 635 atm. Again, it is difficult to determine if the first auto-explosion, observed at about $t = 122.25$ ms in Fig. 6b, takes place in the multiphase slug or in the gas.

When increasing the dust concentration to 20 kg/m^3, no DDT has been observed for the three tests performed. Changing the dust slug to 300 mm while keeping the concentration of 10 kg/m^3, detonation transition is observed to occur on the first reflection from the interface. Finally, with a 100-mm slug of glass-dust suspension ($\sigma_p = 10$ kg/m^3), DDT has not been observed for the three tests performed. We conclude that under the proper choice of parameters, the effect of the dense combustible multiphase slug is to enhance the pressures generated by the DDT process in the end gas. The experiments performed in this study have achieved peak pressures between 300 and 700 atm. Based on theoretical calculations, it is expected that pressures exceeding 1000 atm are possible.

Acknowledgement. We would like to thank D. Wilmut and C. Murray who performed the experimental tests.

References

Craven AD, Greig TR (1968) The development of detonation over-pressures in pipelines. Chem E Symposium Series 25:41–50.

Gelfand BE, Gubin SA, Kogarko SM Popov OE (1975) Investigation of propagation and reflection of pressure waves in porous media. Zh Prikl Mekh Tekh Fiz 16:74–77.

Kogarko SM (1958) Investigation of the pressure at the end of a tube in connection with rapid nonstationary combustion. Soviet Physics – Technical Physics (ZhTF) 28:2041.

Zhang F, Groenig H (1991) Detonation structure of corn starch particle-oxygen mixtures. In: Dynamics of Detonation and Explosions: Detonation, Kuhl AL et al. (eds), AIAA Progr Astronaut Aeronaut, AIAA Inc., Washington 133:342–355.

Effect of confinement on the propagation of a gaseous detonation in a porous medium

A. Makris,* D. L. Frost, I. Abdul-Baki, A. Yoshinaka, J. H. S. Lee
McGill University, Mechanical Engineering Department
817 Sherbrooke St. W., Montreal
Quebec, Canada H3A 2K6

Abstract: Experiments have been carried out to study the effect of confinement on the velocity and lean propagation limit of a gaseous detonation propagating in a porous medium consisting of a packed bed of spherical beads. If the confinement of the porous medium is changed from a rigid tube to a weak paper tube, there is no effect on the results for a gas mixture with a highly irregular detonation cellular structure (ethylene-oxygen) down to a tube-diameter-to-bead-diameter ratio of 3. However, for the same ratio, detonations in a gas mixture with a regular cellular structure (argon-diluted acetylene-oxygen) fail more readily with weak confinement. The results are consistent with the existence of local and global failure mechanisms proposed by Lee (1995).

Key words: Gaseous detonation, Porous medium, Detonation propagation limits

1. Introduction

The propagation limits for a gaseous detonation wave, in general, are a function of the characteristic detonation length scale (i.e., the ZND reaction-zone thickness or cell size λ) and a physical dimension that characterizes the boundary condition of the system (e.g., the tube diameter d). For example, the detonation limit in a rigid tube can be estimated from the ratio of the cell size to tube diameter (e.g., $\lambda/d \approx \pi$). For a wide range of gas mixtures, the critical tube diameter, d_c, corresponding to the transmission of a detonation from a tube to an unconfined region, is proportional to the cell size, i.e., $d_c = 13\lambda$ (e.g., Mitrofanov & Soloukhin 1964; Matsui & Lee 1979). Although there are exceptions to these correlations, in general they do provide a guideline for successful self-sustained propagation.

In a heterogeneous system such as a porous medium comprised of inert solid particles permeated with an explosive gas, additional length scales are

---

*Present address: MedEng Systems, Inc., 2470 Don Reid Dr., Ottawa, Ontario, Canada K1H 9P5.

introduced into the problem, i.e., the particle diameter and the void size. For spherical particles of the same size, the pore size is approximately one-third of the particle diameter. In this case, the criterion for self-sustained propagation in a tube or successful transmission through an orifice must involve this third length scale. In a previous study, Makris et al. (1993) found that the pore size of the medium, being the smaller physical dimension of the system, actually controls the detonation propagation. They found a correlation for the propagation velocity and limits in a porous medium based on the ratio of the void size to d_c of the mixture. For critical transmission of gaseous detonation into a porous medium through an orifice, the characteristic pore size is found to be the governing length scale for most of the mixtures considered, suggesting that the limits for successful transmission are governed by a local phenomenon (Makris et al. 1995b).

In general, when the confinement provided by the tube walls is reduced, the tube diameter may influence the propagation limits as well as the detonation velocity. For the propagation of detonations in liquids, due to the high pressures generated, the reaction products are effectively unconfined. For the case of the propagation of a detonation in a liquid explosive in a porous medium, it is found that the tube diameter plays a major role in the propagation limits (Lee et al. 1995). The failure of the outer confinement causes expansion waves to propagate towards the center of the charge resulting in a global curvature of the wave. This in turn causes the effective reaction-zone length scale to increase and the overall detonation velocity to decrease relative to the ideal Chapman-Jouguet (CJ) detonation value. Thus for condensed explosives in porous media, the charge diameter plays an important role.

For a gaseous explosive in a porous medium, the effects of the degree of confinement on the propagation limits and detonation velocity are not clear. Propagation velocities for gaseous detonations in porous media within a rigid tube range from V_{CJ} down to as low as $0.25V_{CJ}$ near the lean or rich mixture propagation limits (e.g., Lyamin et al. 1991; Makris 1993). For gaseous detonation in a weakly confined porous medium, the lateral expansion will influence the local processes within a void, at least near the outer boundary of the porous medium. If the strength of the wall material is minimized, it is not known if the characteristic pore size will still dominate the phenomenon, or whether the tube diameter will influence the propagation limits. This paper presents results of a recent study of the failure of gaseous detonations in a porous medium weakly confined within a paper tube. The lean propagation limit and the average propagation velocity have been determined for ethylene-oxygen and argon-diluted acetylene-oxygen mixtures.

2. Experimental

A detonation is initiated at the top end of a vertical detonation tube (146-mm inner diameter) comprised of two sections and propagates 2.6 m in a premixed gas mixture at 1 atm before it transmits into the porous medium which is contained within a 76-cm long paper tube. The paper tube itself is wax-coated on both sides and is suspended from an orifice plate located just upstream of the porous medium. The porous medium is comprised of mono-sized ceramic spheres with diameters ranging from 2–13 mm. Prior to initiation of the detonation, the outer shell surrounding the paper tube is flushed with argon to provide an inert atmosphere. Pressure transducers within the upper section of the tube are used to ensure that a fully developed CJ detonation wave has been established prior to entry into the porous medium. Four ionization probes inserted into the paper tube at regular intervals are used to monitor the transmitted wave velocity within the porous medium. For unsuccessful detonation transmission, the wave fails in less than 20 cm of travel in the porous medium, except very near the propagation limit where the failure distance may be longer, and large segments of the paper tube are found intact after the test. By varying the mixture composition for each paper tube and sphere diameter it is possible to identify the critical condition for transmission.

3. Results and discussion

Figure 1 shows a comparison between the failure limits for detonation propagation in ethylene-oxygen mixtures confined within a rigid tube as opposed to the weak confinement provided by a paper tube with a diameter of 38 mm. If the porous medium consists of 2-mm beads (see Fig. 1a), the propagation velocities and failure limits for both the rigid and paper tubes are the same. Near the limit, the critical tube diameter in the absence of a porous medium is about 40 mm (Makris et al. 1995a), several orders of magnitude larger than the average void dimension of the porous medium. Thus the transmission of the detonation is strongly dependent on the local interaction of the detonation wave with the porous medium. The critical equivalence ratio for failure ϕ_c for this case is about 0.25. For the larger 13-mm beads, the results again are largely identical except near the limit, where there is slightly more scatter in the paper-tube results. Even for a ratio of the paper-tube diameter to bead diameter of about 3, the propagation velocities and limits are governed primarily by the characteristic scale of the pores indicating that local processes control the wave propagation. These results are consistent with the previous findings for mixtures with irregular cellular structures for the transmission of a detonation to a porous medium through an orifice (Makris et al. 1995b).

Figure 2 shows similar conditions as in Fig. 1b, except that the ethylene-oxygen mixture has been replaced by acetylene-oxygen mixtures diluted with 75% argon. High argon-diluted mixtures tend to increase the shocked temperature and render the wave more stable, i.e., the propagation mechanism relies less on the instability associated with the cellular structure. It can be observed in Fig. 2 that in this case the paper tube has a relatively large effect on the lean propagation limit, shifting it from about 0.3 for the rigid tube to slightly over 0.6 for the paper tube.

The results shown above support the idea of "local" and "global" failure mechanisms proposed by Lee (1995) which depend on the characteristics of the gas mixture. Gases with high values of the stability parameter E/RT (where E is the activation energy), such as ethylene-oxygen, exhibit a highly irregular cellular pattern associated with the detonation propagation and small perturbations can cause failure of the wave or subsequent re-transition to detonation. A porous medium artificially generates "hot-spots" due to shock-wave/particle interactions which assist the re-transition to a quasi-detonation. Since the propagation is dominated by the local processes in the individual pores, the overall curvature of wave for weak confinement has little effect on the velocity or propagation limits. Gases with low values of E/RT, such as highly argon-diluted acetylene-oxygen mixtures, have a more stable structure and the propagation mechanism is more dependent on auto-ignition by the shock than on instability processes. For such gases, the $d_c = 13\lambda$ critical tube-diameter correlation has been known to break down and global length scales of the mixture influence

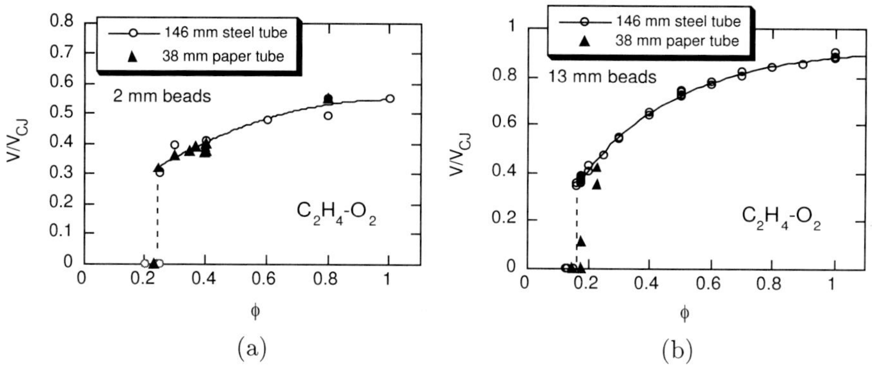

Figure 1. Detonation velocity in porous media as a function of equivalence ratio ϕ showing a comparison between the propagation limits for rigid and weak confinement in ethylene-oxygen mixtures. Propagation limit for rigid tube shown by dashed lines. Bead diameters of 2 mm (a) and 13 mm (b) were used.

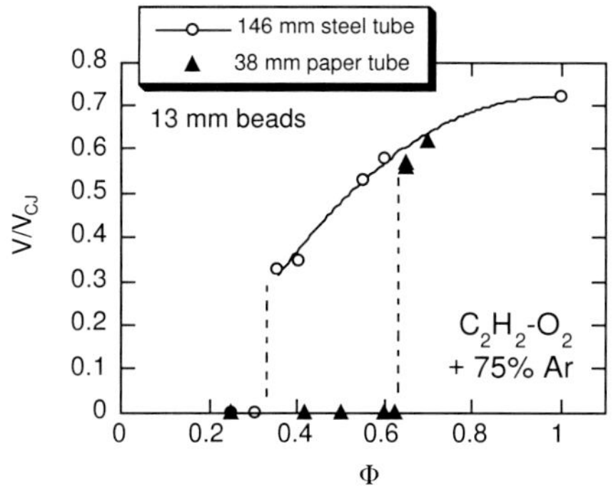

Figure 2. Effect of confinement on velocity and lean propagation limit for argon-diluted acetylene-oxygen mixtures.

the propagation limits. Makris et al. (1995b) found that for 75% argon-diluted acetylene-oxygen mixtures, the lean propagation limit was shifted closer to a stoichiometric equivalence ratio when a 10-mm orifice was placed just upstream of the porous medium with a typical void size of 3 mm (in contrast with gases with regular cellular structure for which no effect was observed). This is consistent with the present findings that weakening the confinement causes the wave to fail more readily since the tube diameter, which represents the global length-scale of the mixture, influences the propagation velocity and limits.

4. Conclusions

The present results indicate that the degree of confinement has a weak influence on the failure of a normal unstable detonation in a porous medium. Although the previous studies (Makris et al. 1993, 1995a,b) indicate that the propagation mechanism is local, and governed by the events in a void (since the correlation for the velocity and limits is based on the ratio of the cell size to void size), the expansion waves from the failure of the outer confinement can influence the local processes inside a void if the tube diameter is of the same order of magnitude as the particle size. For gas mixtures with a high stability parameter E/RT, if the tube diameter is more than 3 times the bead size, the weak confinement provided by a paper tube has little influence on the velocities and lean failure limit. This is in sharp contrast with more stable mixtures (with low values of E/RT), where for a tube/bead ratio of 3, the propagation is strongly influenced by the global effects. The lateral expansion results in a global curvature of the wave which in turn causes the effective reaction-zone length scale to increase. Thus although the local phenomenon still controls the self-sustained propagation of the detonation, the global curvature changes the overall reaction-zone length and thus the failure is governed by both the paper-tube diameter as well as the void size. However, the dependence is not a strong one because the maximum increase in the reaction-zone length scale due to curvature is at most a factor of two according to Schelkin's criterion prior to complete failure due to global curvature alone. Hence a weak dependence is observed.

Acknowledgement. The authors acknowledge the assistance of I. Kaderi in the experiments. Support was provided by a joint program with the Natural Sciences and Engineering Research Council of Canada and ICI Explosives, Canada.

References

Lee JHS (1995) On the critical diameter problem. In: Combustion Science and Technology Book Series (in press).

Lee JJ, Brouillette M, Frost DL, Lee JHS (1995) Effect of diethylenetriamine sensitization on detonation of nitromethane in porous media. Combustion and Flame 100:292–300.

Lyamin GA, Mitrofanov VV, Pinaev AV, Subbotin VA (1991) Propagation of gas explosion in channels with uneven walls and in porous media. In: Dynamic Structure of Detonation in Gaseous and Dispersed Media, Borisov A (ed), p 51.

Makris A, Papyrin A, Kamel M, Kilambi G, Lee JHS, Knystautas R (1993) Mechanisms of detonation propagation in a porous medium. Prog in Astro and Aero AIAA 153:363–380.

Makris A, Shafique H, Lee JHS, Knystautas R (1995a) Influence of mixture sensitivity and pore size on detonation velocity deficits in porous media. Shock Waves (in press).

Makris A, Oh TJ, Lee JHS, Knystautas R (1995b) Critical diameter for the transmission of a detonation wave into a porous medium. In: Proc 25th Symposium (International) on Combustion, The Combustion Institute, pp 65–72.

Makris A (1993) The Propagation of Gaseous Detonations in Porous Media, Ph.D. Thesis, Dept of Mech Eng, McGill University.

Matsui H, Lee JHS (1979) On the measure of the relative detonation hazards of gaseous fuel-oxygen and air mixtures. In: Proc 17th Symposium (International) on Combustion, The Combustion Institute, pp 1269–1280.

Mitrofanov VV, Soloukhin RI (1964) The diffraction of multi-front detonation waves. Soviet Phys Dokl 9(12):1055–1058.

Detonation initiation by hypervelocity projectiles

J. Bélanger*, M. Kaneshige, J. E. Shepherd
California Institute of Technology
Pasadena, CA 91125 USA

Abstract: We report experimental studies on the initiation of detonations by hypervelocity projectiles in hydrogen-oxygen-nitrogen mixtures. The experimental facility T5 at Caltech was modified to launch 25-mm diameter projectiles at speeds around 3000 m/s. Flow visualization and pressure measurements are reported for several mixture conditions. A clear transition between shock-induced combustion and detonation is observed. Transition to detonation is observed with an overtaking wave occurring in the shocked but unreacted gas.

Key words: Detonation, Projectile, Hypervelocity

1. Introduction

Experiments on detonations initiated by hypervelocity projectiles ($U > U_{CJ}$) have been carried out in the T5 shock tunnel. Our goal is to understand the conditions needed to produce stabilized detonation waves. Effects of initial pressure have been explored in this preliminary study. There are two principal configurations (Fig. 1) that we anticipate observing for hypervelocity projectile

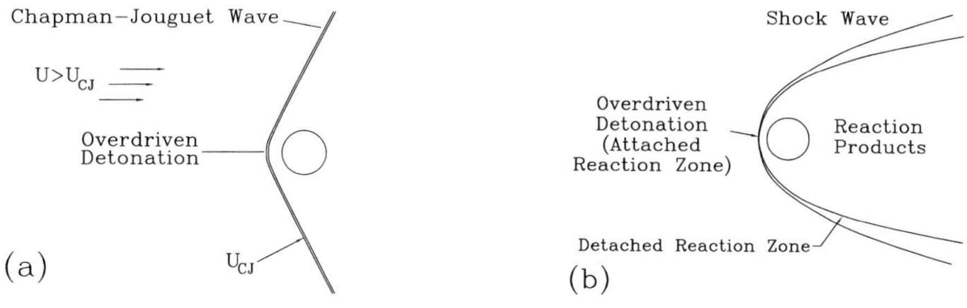

Figure 1. Hypothetical wave configurations: (a) attached reaction zone, (b) detached reaction zone.

flight in detonable mixtures. To date, only the second configuration (Fig. 1b), shock-initiated combustion of a limited extent, has been observed.

*University of Minnesota, Minneapolis-St. Paul, MN USA.

Key factors in the initiation and stabilization of a detonation wave on a projectile are the ratio of the detonation cell size to projectile diameter and the ratio of the projectile speed to the mixture Chapman-Jouguet (CJ) speed (Shepherd 1994). For the mixture used in these experiments, CJ velocities and estimated cell sizes (30 times the ZND reaction zone length) are given in Table 1 for a range of wave speeds.

In previous experiments using projectiles (for example, Lehr 1971), stabilized detonation waves have not been observed because the projectiles were either too slow or too small. In the present tests, we used a very sensitive mixture (stoichiometric hydrogen-oxygen) diluted with a small amount of nitrogen to reduce the Chapman-Jouguet wave speed (2400 m/s) to significantly less than the projectile velocity. The estimated detonation cell width of this mixture for an initial pressure of 1 bar is approximately 2 mm, a factor of 10 less than the projectile diameter.

Table 1. Detonation parameters for $2H_2 + O_2 + N_2$.

P_0(bar)	1.0	0.50	0.25	0.10
U_{CJ}(m/s)	2400	2370	2340	2300
U (m/s)	λ (mm)	λ (mm)	λ (mm)	λ (mm)
2400	1.9	3.3	6.4	16
2500	1.3	2.4	4.7	11
2600	0.94	1.8	3.5	8.8
2700	0.71	1.4	2.7	6.7
2800	0.55	1.1	2.2	5.3
2900	0.44	0.87	1.7	4.2
3000	0.35	0.70	1.4	3.3

2. Facility

T5 has been modified by the addition of a launch tube, extending from the nozzle throat, and a test section/target section assembly, which is mounted on the downstream door of the dump tank. A schematic of the apparatus is shown in Fig. 2.

The high-enthalpy gas generated by T5 accelerates a 25-mm diameter nylon sphere (about 10 grams) through the 3-m long launch tube. Passing through the T5 dump tank, the sphere breaks a timing wire, ruptures a mylar diaphragm, and enters a test section mounted on the downstream door of the dump tank. This test section has a 150-mm square inside cross-section and is 760 mm long. It is equipped with three evenly spaced pressure transducers, two optical ports (for projectile timing by laser triggers), and optical windows for flow visualization centered 550 mm from the entrance. The projectile impacts an assembly of aluminum honeycomb and plates in the target section. Using a shock speed of 5000 m/s and helium gas in the shock tube section of T5, the projectile muzzle velocity is approximately 3000 m/s.

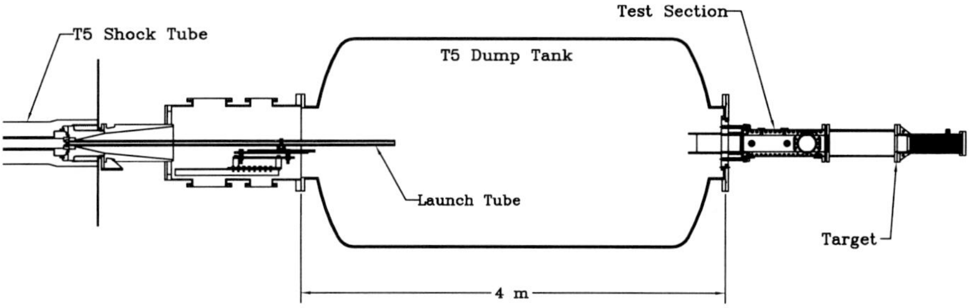

Figure 2. Schematic of T5 with gas gun modification and detonation test section.

3. Results

A summary of results from several experiments is given in Table 2.

Table 2. Summary of selected T5 detonation experiments.

T5 Shot	Projectile Velocity (m/s)	Wave Speed 1 (m/s)	Wave Speed 2 (m/s)	Mixture	Initial Pressure (bar)
861	2940	2860	2860	N_2	0.25
862	2820	3060	3020	$2H_2+O_2+N_2$	1.00
863	2930	3060	3110	$2H_2+O_2+N_2$	1.00
865	2690	2380	3200	$2H_2+O_2+N_2$	0.10
1007	2740	2780	2710	N_2	0.25
1009	2420	3460	2980	$2H_2+O_2+N_2$	0.50
1010	2560	3460	2860	$2H_2+O_2+N_2$	0.50
1011	2640	3250	3060	$2H_2+O_2+N_2$	0.25
1012	2790	2580	3200	$2H_2+O_2+N_2$	0.25

Differential interferograms and pressure-transducer traces from five experiments are shown in Figs. 3 through 6. The optical windows are 165 mm in diameter, which allows a clear view of portions of the top and bottom walls. The projectile is 25 mm in diameter and is shown moving from right to left. The images were made with a differential interferometer using a Q-switched

Nd:YAG laser. In Figs. 4a and b, the round shadows at the top of the window are caused by chips in the optical glass windows. Small pieces of debris from the diaphragms are occasionally visible in the flow as conical shocks, sometimes in front of the projectile (Fig. 4a) and somtimes in the wake (Fig. 3a).

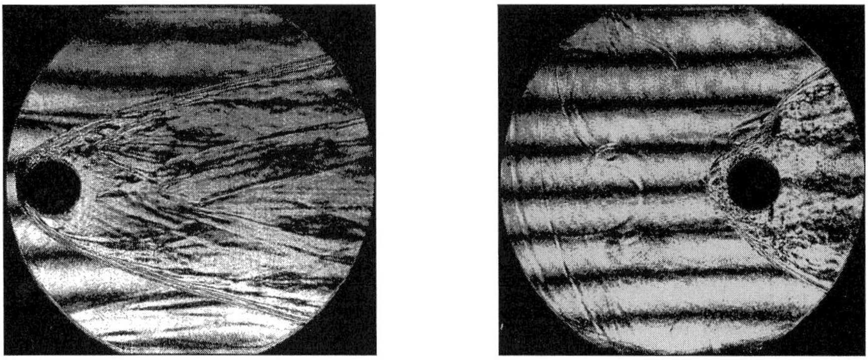

Figure 3. (a) T5 Shot 861 (b) T5 Shot 865

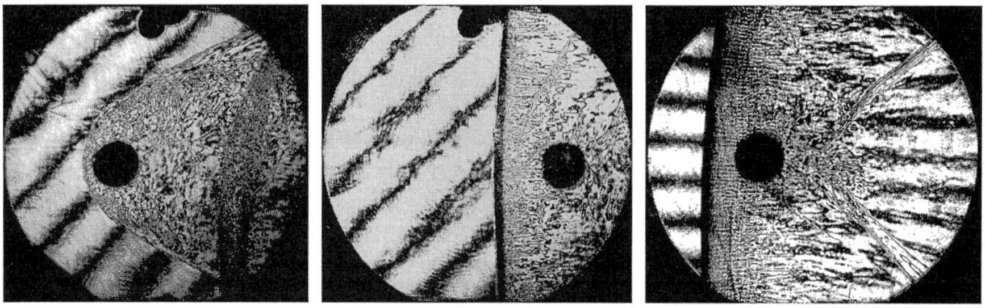

Figure 4. (a) T5 Shot 1012 (b) T5 Shot 1010 (c) T5 Shot 863

The pressures were recorded with PCB 113A24 transducers sampled at 1 MHz. The plots show the data normalized by the initial pressures. In Fig. 6, the clipping visible in T5 shot 1010 was caused by a fault in the data-acquisition system. The dashed lines show the path of the projectile, as determined from the laser triggers and photographs.

As a reference case, a non-reacting experiment was performed (Figs. 3a and 5). The pressure signals and wave shapes compare favorably with the known results for hypersonic shock waves on spheres. Experiments involving reacting mixtures with increasing initial pressures (and consequently decreasing cell sizes) are shown in Figs. 3b through 4c and 5 through 6.

Although only four initial pressures were examined, it is apparent from the data that a critical transition occurs near 0.25 bar. At 0.10 bar, the bow wave is similar to the non-reactive shock, although the photograph and pressure data show the effects of combustion. Figure 4a, at 0.25 bar, shows a bow wave similar to that at 0.10 bar, but more interesting phenomena are also visible. The reaction zone detaches from the shock wave at about 1.5 diameters behind the front of the sphere. Similar detachments have been observed in previous studies (Lehr 1971). This indicates that the initial reaction process is being quenched due to the expansion behind the curved bow shock wave.

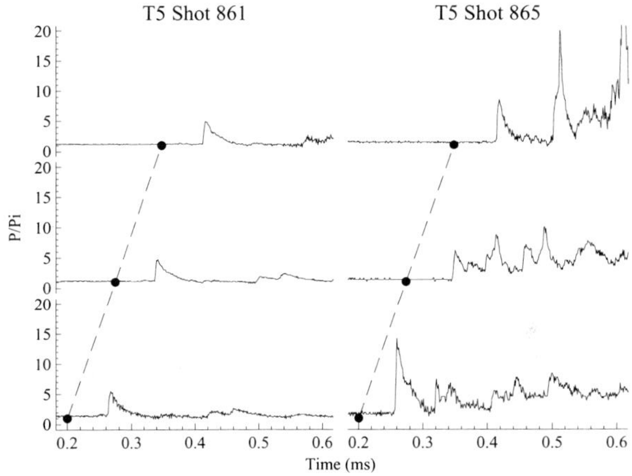

Figure 5. Pressure transducer plots for T5 shots 861 and 865.

More significantly, a second wave is seen to be overtaking the projectile from behind. This wave appears to be a detonation propagating in shocked but unreacted gas. The origin of this wave is not certain, but may be a result of processes occurring when the projectile enters the test section. The pressure data and the photograph together indicate that this was a Deflagration to Detonation Transition (DDT) event.

Figures 4b and c show clear detonation events, with cellular structure visible behind the leading waves. Sooted foil records confirmed the photographic observations. The nearly normal angle of the waves was unexpected. As shown in Fig. 1, the overdriven leading wave is expected in steady flow to decay to a CJ wave away from the body and the steady flow wave angle is predicted to be about 63 degrees. Laser trigger data show that the projectile velocities were constant ($\pm$ 50 m/s) during the flight, while the pressure data indicate that the apparent leading wave velocities were not constant. These variations are evident in the data in Table 2. An acceleration of the leading wave in T5 shot 1012 is consistent with the second shock wave overtaking the first. A deceleration of the leading waves in T5 shots 1010 and 863 suggest that the waves were overdriven but weakened as they propagated. We conclude that the waves observed in shots 863 and 1010 are transient configurations.

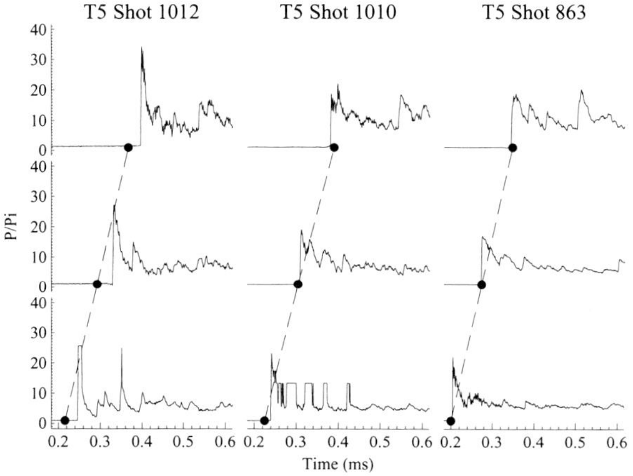

Figure 6. Pressure transducer plots for T5 shots 1012, 1010, and 863.

4. Conclusion

These results demonstrate new aspects of hypersonic-projectile behavior in detonable gases. There are clearly very significant implications for combustion processes in ram accelerators and other hypersonic propulsion schemes. Further exploration in this ongoing program will examine the effect of body size, velocity, and the unsteadiness of the waves.

Acknowledgement. We are indebted to H. G. Hornung for his avid interest, advice, and generosity with the T5 facility. Bahram Valiferdowsi expertly operated T5 and the staff of the GALCIT shops made many timely contributions.

References

Shepherd JE (1994) Detonation waves and propulsion. In: Combustion in High Speed Flows, J Buckmaster, TL Jackson, A Kumar (eds), Kluwer Academic Publishers, pp 373–420.

Lehr HF (1971) Experiments on shock-induced combustion. In: Astronautica Acta, Pergamon Press, pp 589–597.

Numerical investigation of pulsating detonations

J. J. Quirk*
ICASE Mail Stop 132 C
NASA Langley Research Center
Hampton, VA 23681 USA

M. Short
School of Mathematics
University of Bristol
Bristol, BS8 1TW UK

Abstract: We present an overview of a numerical investigation into the behavior of a one-dimensional detonation wave driven by an idealized chain-branching reaction, and highlight the dangers of drawing conclusions from under-resolved simulations.

Key words: Detonations, Pulsating instabilities, Numerical simulations

1. Introduction

This study forms part of a concerted theoretical-computational effort to fathom the range in behavior of one-dimensional, pulsating detonation waves. It aims to bridge the gap between studies which strive for realism in the reaction modeling (e.g., Sussman 1994) and studies which make do with one-step Arrhenius kinetics in the interests of obtaining a better theoretical understanding of the chemical-compressible interactions which underpin this type of instability (e.g., Bourlioux et al. 1991).

Here, following standard arguments, we assume that the essential dynamics of the complex chain branching reactions which drive detonations in gaseous mixtures can be represented by three idealized stages: initiation, chain-branching, and chain-termination with rates k_I, k_B, and k_C, respectively. Thus we employ the following idealized chain-branching reaction model involving fuel (F), an intermediate radical (Y), and products (P) with the assumption that all the heat release (Q) occurs in the chain-termination stage.

*Present address: Graduate Aeronautical Laboratories, Mail Code 205-45, California Institute of Technology, Pasadena CA 91125 USA.

$$
\begin{array}{llcll}
\text{Initiation :} & F & \rightarrow & Y & ; \; k_I = \exp\left[\frac{1}{\epsilon_I}\left(\frac{1}{T_I} - \frac{1}{T}\right)\right] \\
\text{Chain-branching :} & F+Y & \rightarrow & 2Y & ; \; k_B = \exp\left[\frac{1}{\epsilon_B}\left(\frac{1}{T_B} - \frac{1}{T}\right)\right] \\
\text{Chain-termination :} & Y & \rightarrow & P & ; \; k_C = 1.
\end{array} \tag{1}
$$

The initiation and chain-branching rates k_I and k_B are taken to have Arrhenius temperature-dependent forms with inverse activation energies ϵ_I and ϵ_B and cross-over temperatures T_I and T_B, respectively. The termination reaction is assumed to be independent of temperature (T), and so is modeled as having a constant rate, which is scaled to unity because of our adopted nondimensionalization. This reaction scheme together with the reactive Euler equations constitute our flow model of a detonation (see Short and Quirk 1995).

Short and Dold (1994) have used a normal-modes approach with the shooting technique of Lee and Stewart (1990) to analyze the behavior of small-amplitude disturbances from the plane, steady detonation waves admitted by the model above. They introduced the chain-branching cross-over temperature (the temperature at which the chain-branching and chain-termination rates are equal, here denoted by T_B) as a bifurcation parameter. In essence, a change in this parameter represents a change in the chemical composition of the air-fuel mixture used to propagate the detonation wave and so can be expected to lead to changes in the wave's behavior. Their analysis provides certain theoretical results which can be used to lend credibility to the present numerical investigation.

2. Numerical method

If a direct numerical simulation is to produce reliable results, it must be able to resolve all of the physical scales pertinent to the problem under investigation. Here this was done using a sophisticated mesh-refinement algorithm (Quirk 1994) with an integration scheme based on Roe's linearized Riemann solver (see Short and Quirk 1995). The simulations were started with the exact traveling-wave solutions admitted by the flow model and were run at several mesh resolutions (80, 100, 160, and 320 mesh points in the half-reaction length) to determine the grid sensitivity of the numerical results. Furthermore, the traveling waves were propagated upwards of 1000 half-reaction lengths to be sure that their dynamic behavior was fully settled. Consequently, although one-dimensional, the present simulations are not cheap to run and would be prohibitively expensive without mesh refinement (in the worst case, the equivalent uniform-mesh computation would require 400,000 iterations on a mesh containing 512,000 cells).

3. Results

We considered a bifurcation study where T_B was increased from 0.8 with the following parameters fixed: $Q = 3$, $\epsilon_I = 1/20$, $\epsilon_B = 1/8$, $T_I = 3$, ratio of specific heats $\gamma = 1.2$, and detonation overdrive equal to 1.2. Figure 1 shows the variation in the steady detonations admitted by the flow model across this range. Observe the increase in the induction length of the wave and the decrease in the height of the radical spike as T_B is increased. These variations hint at dramatic changes in the dynamical behavior of the detonations.

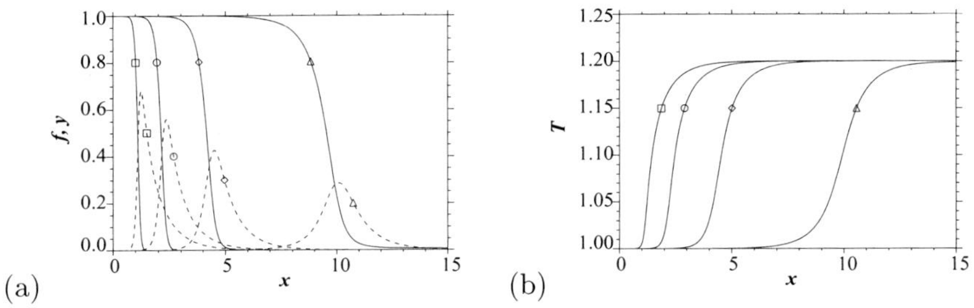

Figure 1. Steady detonation structure as T_B varies from 0.8 to 0.95 in steps of 0.05: (a) mass fractions of fuel (f, solid line) and radical (y, dashed line) and (b) normalized temperature.

Figure 2 summarizes the results from the corresponding normal-modes analysis of Short and Dold (1994). The analysis predicts that the boundary between stable and unstable detonations lies at $T_B = 0.802$ and indicates that as T_B is increased beyond this value the detonation becomes increasingly unstable, picking up progressively higher frequency modes. For example, at $T_B = 0.95$ there are nine unstable roots. The analysis also predicts that the most dangerous mode i.e., the one with the largest growth rate is always the lowest-frequency root. Figure 3 shows that the present simulations have no difficulty in predicting the stability boundary. Furthermore, for $T_B = 0.82$ the numerics correspond well to the analysis in that they both predict a single-mode oscillation with a period $T \approx 35$. A grid-convergence study for this case with 80, 100, 160, and 320 points in the half-reaction shows that there is only a slight decrease in the maximum shock pressure as the grid resolution is increased (see Fig. 4).

As T_B is increased further the mechanisms underlying the pulsations become more complex and involve such large-amplitude waves that the normal-mode analysis is no longer pertinent to the long-time behavior of the detonation. In fact the mechanisms become so involved that direct simulations also prove to be unreliable unless an extremely fine grid is used. For example, when $T_B = 0.85$, a

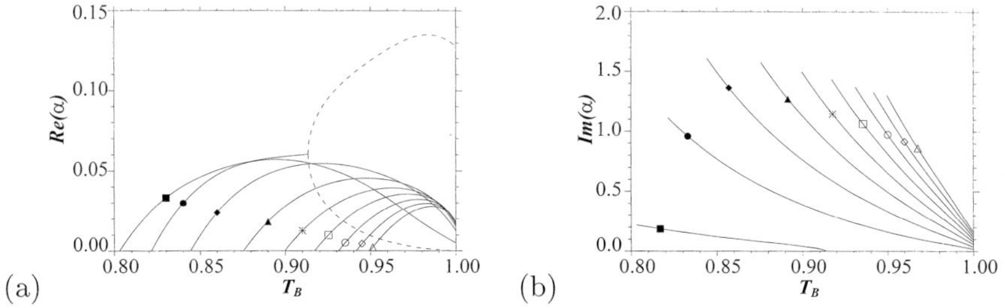

Figure 2. Normal-mode results (Short and Dold 1994): (a) growth rate $\mathrm{Re}(\alpha)$ and (b) frequency $\mathrm{Im}(\alpha)$ as T_B is varied. The dashed line in (a) corresponds to purely real roots.

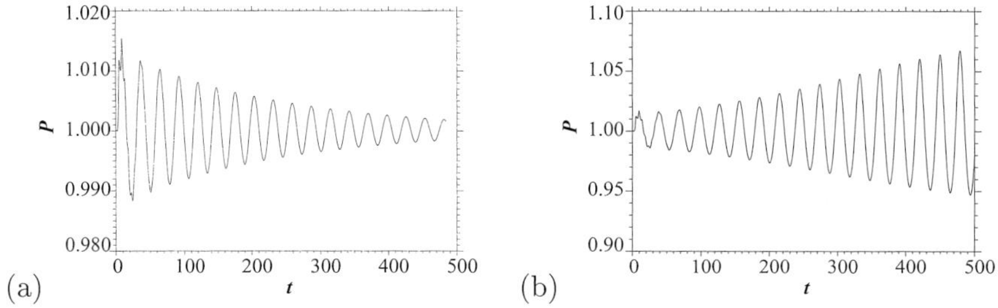

Figure 3. Shock-pressure histories near the stability boundary $T_B = 0.802$: (a) $T_B = 0.800$ and (b) $T_B = 0.805$.

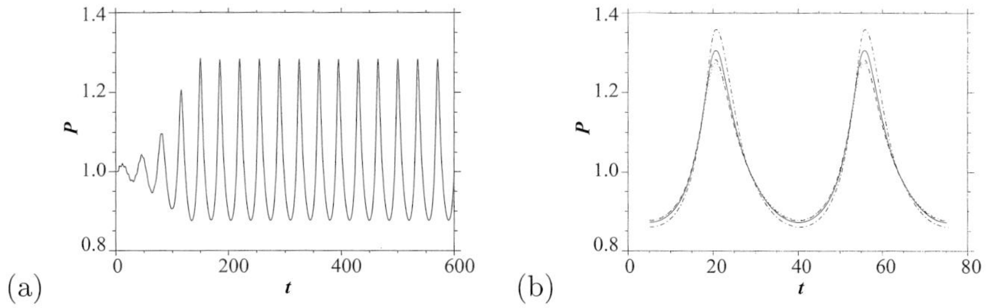

Figure 4. Shock-pressure history for $T_B = 0.82$: (a) 160 points/half-reaction length (b) grid-convergence study.

simulation with a grid resolution of 160 points/half-reaction erroneously predicts a two-mode oscillation (see Fig. 5a), because when the resolution is doubled the simulation predicts an attracting single-mode limit cycle which appears to be the correct dynamical behavior, as further increases in grid resolution have no effect (see Fig. 5b).

Similarly, for $T_B = 0.86$ an under-resolved computation predicts wildly chaotic pulsations instead of the two-mode pulsation shown in Fig. 6a achieved with 320 points/half-reaction. The dynamics behind this mode of pulsation can be gleaned from Fig. 6b, which shows several pressure snapshots centered on the larger of the two spikes in the pressure-history signature. The circles indicate the position where the radical concentration is greatest. Observe that a secondary detonation wave forms behind the lead shock, thus the pulsation is far from being a perturbation about a mean flow. Also note that there is a fifteen-fold variation in the induction length over the course of the pulsation, and so computational methods which rely on a fixed width of refinement (e.g., Bourlioux et al. 1991 or Sussman 1994) are ill-suited to this type of flow.

With further increases in T_B the pulsations become more severe until a point is reached where the shock temperature actually drops below the chain-branching cross-over temperature, causing the detonation to quench. The ability of the present reaction model to reproduce quenching represents an important improvement in realism over the classical one-step Arrhenius reaction model and is discussed in depth by Short and Quirk (1995).

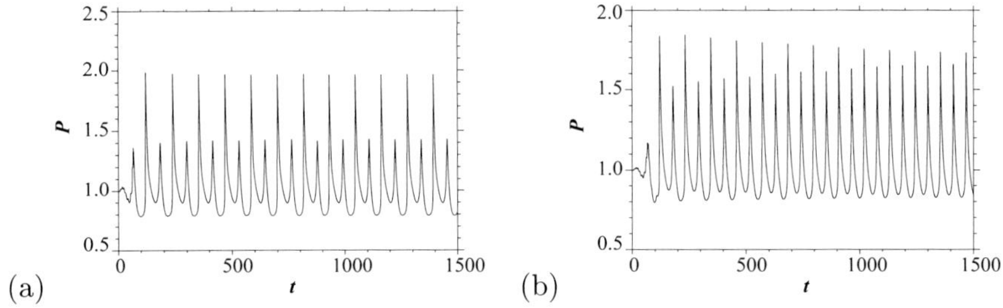

Figure 5. Shock-pressure history for $T_B = 0.85$: (a) 160 points/half-reaction length and (b) 320 points/half-reaction length.

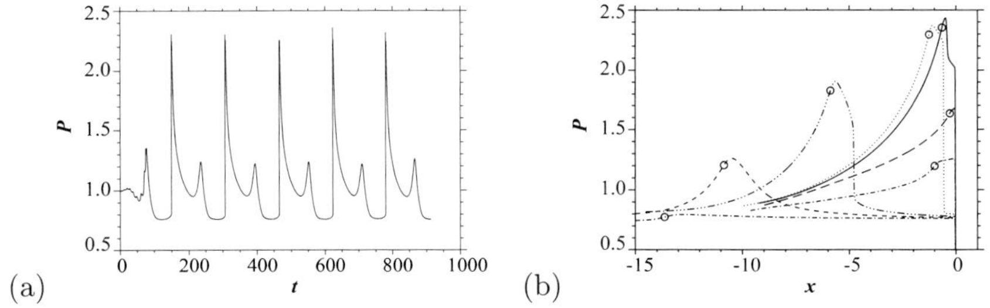

Figure 6. $T_B = 0.86$ case: (a) shock pressure history and (b) pressure snapshots centered on the larger of the two pressure spikes in the history trace.

4. Conclusions

Here we have only scratched the surface regarding the behavior of pulsating detonations and the interested reader is directed to Short and Quirk (1995) for a more comprehensive treatment. Nevertheless we have shown why great care needs to be taken with simulations of reactive phenomena. Thus, while the present grid resolutions might appear excessive compared to previous numerical studies of pulsating detonations (e.g., Bourlioux et al. 1991), they are a prerequisite for accurate results. Moreover, Figs. 5 and 6 show that algorithmic arguments such as whether to track or capture shocks are irrelevant if the grid is too coarse to resolve the smooth but small-scale structures of the detonation wave. With the present reaction model it is essential to resolve the radical spike accurately, because it directly affects the release of chemical energy and so ultimately dictates the grid resolution required for a reliable simulation.

Currently, two-dimensional detonation simulations are limited to around 20 points/half-reaction (Quirk 1994), as opposed to the 320 points/half-reaction needed here to reach an accurate solution for a one-dimensional flow. Therefore, conclusions drawn from multi-dimensional detonation simulations with chain-branching kinetics should be viewed with a healthy degree of scepticism, at least until the computations concerned are shown to be grid converged.

Acknowledgement. This research was supported by the National Aeronautics and Space Administration under NASA Contract No. NAS1-19480 while the authors were in residence at the Institute for Computer Applications in Science and Engineering (ICASE), NASA Langley Research Center, Hampton, VA 23681.

References

Bourlioux A, Majda AJ, Roytburd V (1991) Theoretical and numerical structure for unstable one-dimensional detonations. SIAM J Appl Math 51:303–343.

Lee HI, Stewart DS (1990) Calculation of linear instability: one-dimensional instability of plane detonation. J Fluid Mech 216:103–132.

Quirk JJ (1994) A parallel adaptive grid algorithm for computational shock hydrodynamics. Appl Num Math (to appear).

Short M, Dold JW (1994) Linear stability of a one-dimensional detonation wave for a model 3-step chain branching reaction. Mathematical and Computer Modeling (to appear).

Short M and Quirk JJ (1995) On the non-linear stability and detonability limit of a one-dimensional detonation wave for a model 3-step chain-branching reaction. J Fluid Mech (submitted).

Sussman MA (1994) A Computational Study of Unsteady Shock Induced Combustion of Hydrogen-Air Mixtures, AIAA Paper 94-3101.

Initiation of detonation by gasdynamic methods in a half-limited space

O. V. Achasov, S. A. Labuda, O. G. Penyazkov, D. S. Ragozin
Heat and Mass Transfer Institute
15, P. Brovki Str., 220072 Minsk, Belarus

Abstract: The shock-wave and jet methods allowing initiation of detonation in gaseous mixtures are studied. It is shown that these methods may be used in the working cycle of the pulse detonation engine. The conditions under which they can be practically realized are determined.

Key words: Detonation initiation, Shock focusing, Jet interaction

1. Introduction

With the recent energetic problems, a need arises for a propulsion concept that is less expensive but offers higher performance than propulsion systems currently in use. The Pulse Detonation Engine (PDE) offers a possible solution. PDEs employ an air-breathing, constant-volume process, marked by high pressure and temperatures. These are conditions with higher power densities than conventional engines (Oppenheim 1970). Therefore, investigation of the different stages of the PDE working cycle, and the detonation-initiation process in particular, is of great importance. The gasdynamic initiation is very attractive because it does not require additional energy sources. It is well known that when a plane shock wave collides with a circular concave wall, the reflected shock wave from the concave wall forms a focus or a caustic at which the pressure and the temperature can be enhanced (Takayama 1990). It has been observed that the shock-wave energy required for initiation of detonation can be substantially decreased when focusing phenomenon is used (Borisov et al. 1990, Chan et al. 1990). It was demonstrated that fast turbulent mixing of hot combustion products with reactants can lead to the onset of a detonation (Carnasciali et al. 1991). The interaction of supersonic jets may enhance pressure and temperature and, consequently, cause detonation initiation. Moreover, to prevent the detonation in conveyances, the separate injection of the fuel and oxidizer with mixing following may be applied.

Thus, in the present work, the following gasdynamic methods for the initiation of gaseous detonation in a semiclosed channel with a concave wall are studied: (a) plane shock-wave focusing due to the reflection from concave axisymmetric or two-dimensional surfaces; (b) interaction of supersonic jets of the combustible mixture; (c) separate injection of acetylene and oxygen jets into the combustion chamber with the supersonic mixing following; and (d) fast turbulent mixing of the hot combustion products with the detonative mixture (prechamber initiation).

2. Description of experiments

The experiments were carried out with the use of both shock-tube technique and detonation-chamber models. A two-diaphragm shock tube with unsteady flow expansion was used to perform axisymmetric model experiments. The pressure variation in different cross-sections was recorded by piezoelectric pressure gauges with a 1-mm spatial resolution. Experimental results were recorded and processed by an automatic data acquisition system including digital oscilloscopes and a central computer.

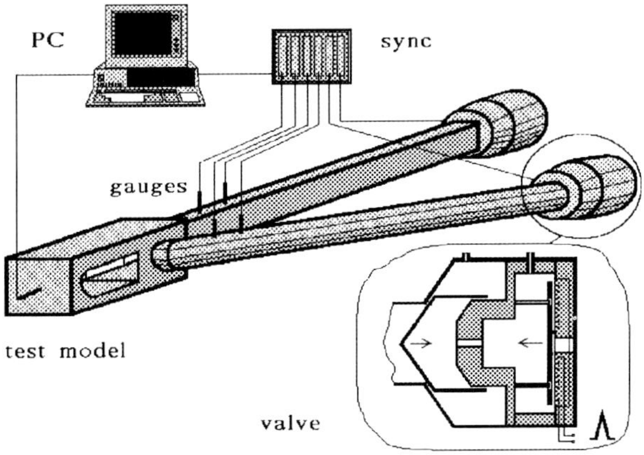

Figure 1. Schematic of the experimental set-up.

Experiments with a two-dimensional concave reflecting surface and with interacting jets were carried out using a set of two coupled diaphragmless shock tubes (Fig. 1). The test section was equipped with quartz windows allowing direct photographic observation of the ignition process. The schlieren device optically aligned with a high-speed (1.3 μs/frame) photographic camera was used to visualize the process. To fix the instant at which the luminosity of the reacting gas mixture commences, the test volume was focused on a photomultiplier cathode. To eliminate light from easily excited impurities, a monochromatic interference filter was used to pass only the desired portion of the emission spectrum of the mixture. The high-pressure valve with a forced electropneumatic start was used to synchronize the shock-tube start with the high-speed camera operation and also to eliminate the influence of diaphragm breaks.

3. Results and discussion

3.1. Detonation initiation by the reflected shock-wave focusing

Our experiments determined the lowest shock-wave intensity required to initiate gaseous detonation. It is well known that detonation conditions are specified not only by the energy contributed to a reactive gas, but also by the power (Lee et al. 1975). Therefore, we took into account these two characteristics. We have found that the reflecting circular surface reduces considerably both the energy and the power thresholds of detonation. The observed reduction of the detonation threshold with the pressure increase corresponds to the results of Chan et al. (1990). Also, we have found that the detonation threshold decreases as the oxygen concentration in a hydrogen-oxygen mixture increases, which is in agreement with the results of Westbrook et al. (1984). Comparative analysis of the detonation-initiation conditions in the stoichiometric mixtures of acetylene, hydrogen, and carbon monoxide with oxygen have shown that the highest and the lowest detonation-initiation thresholds correspond to the third and the first mixtures, respectively.

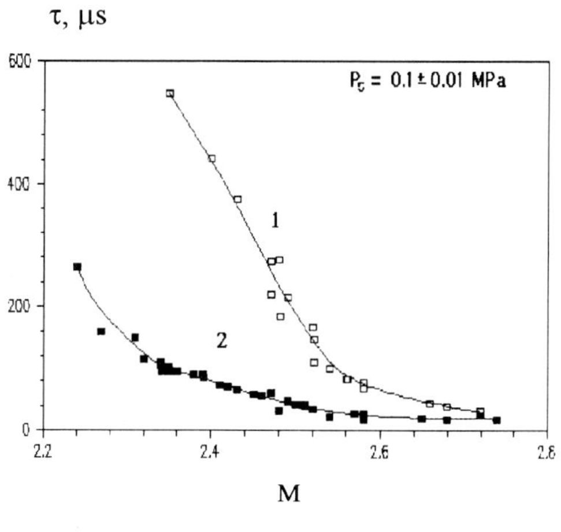

Figure 2. Ignition delay in stoichiometric hydrogen-oxygen mixture vs. incident shock-wave Mach number: 1, normal reflection; 2, reflection from the concave cavity.

As follows from the plot of ignition delay (a time interval between the first and the second pressure peaks recorded at the cavity bottom) as a function of the incident shock-wave Mach number (Fig. 2), the reflecting circular surface reduces ignition delay significantly.

3.2. Prechamber self-initiation of gaseous detonations

It was found previously (Zeldovich et al. 1956) that detonation may be initiated by means of a quasiplanar wave flowing out from a tube of a diameter d. In this case, the tube plays the part of a prechamber, being a "detonator" of the main volume. As shown by Bikovskii and Mitrofanov (1980), the stable detonation wave may be obtained in a cylindrical tubular channel under certain conditions. Based on the above results, we have made an attempt to initiate the detonation in a model monocyclic detonative engine with a circular prechamber. We have studied two types of detonation initiation, namely, direct and inverse types. Under the inverse scheme, ignition takes place at the exit of the detonation chamber and the detonation wave propagates towards the bottom. When the direct type is used, the ignition takes place at the chamber bottom and the detonation wave propagates towards the exit of the chamber. We have found that all power characteristics corresponding to the inverse scheme are higher compared with those of the direct scheme (Fig. 3). The maximum difference has been observed at times of 50–150 μs.

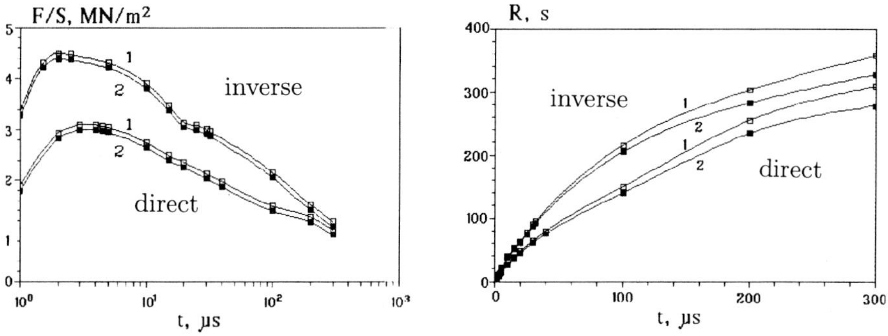

Figure 3. Experimental results of the thrust and specific impulse measurements in vacuum (1) and at sea level (2) for the stoichiometric hydrogen-oxygen mixture. Initial pressure is 0.1 MPa.

3.3. Detonation initiation by supersonic-jet mixing

We have found that the detonative regime occurs when mixing supersonic jets of fuel (acetylene) and oxidizer (oxygen). Transition to detonation takes place at the minimum distance downstream from the screen nozzle when thermal decomposition of acetylene occurs in the prechamber. At the fixed fuel stagnation temperature, the detonation regime exists at the equivalent ratio $\alpha < 0.9$ only. Preliminary thermal decomposition of acetylene in the prechamber leads to the detonation at $\alpha < 1.1$. Also, we have observed that the use of an obstacle (diaphragm) with a transparency of 30% placed at the supersonic part of the combustion chamber allows detonation at a lower stagnation temperature.

3.4. Detonation initiation by supersonic-jet interaction

First we studied the process of the interaction of nonreactive supersonic air jets in a semiclosed concave cavity and found that the interaction of two opposite gas jets injected from the lateral channel walls results in a complex gasdynamic structure (Fig. 4). A weak shock is generated and radiates out, followed by a transient starting jet exiting from the opening itself. In the case of two jets, they can intersect each other as a result of head-on collisions. At first, two fronts undergo, along the line of centers, a normal interaction producing a reflected shock. At the next instant, quadruple shock intersections are formed on both sides of the line of centers. When the intersection angle acquires a certain critical value, triple-shock intersections set in. A turbulent vortex "bubble" heads the transient jet flow and most of the entrainment and mixing occurs in this vortex bubble. When this complex gasdynamic structure collides with a circular concave wall, the reflection from the concave wall forms a focus at which the pressure and the temperature can be enhanced. We have found that the focusing degree is higher when jets inclined to the concave wall are used.

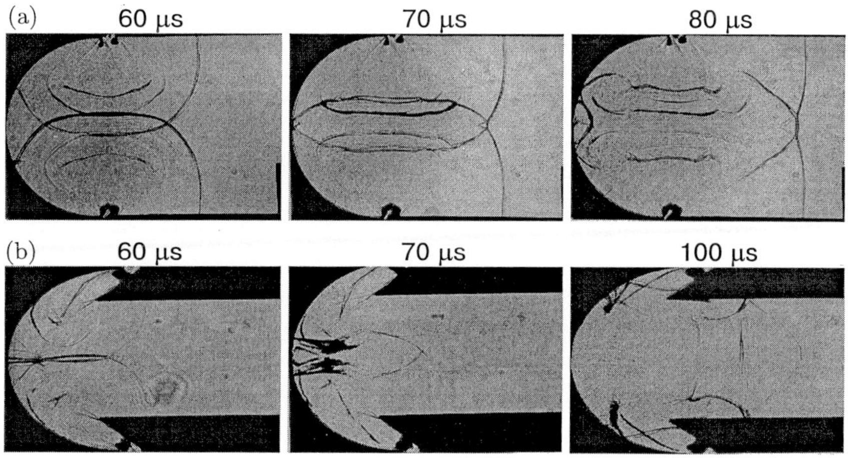

Figure 4. Shadowgraphs of the normal (a) and inclined (b) interaction of air jets. Initial pressure is 30 kPa, shock wave Mach number is 2.37.

Experiments have been carried out on the direct initiation of detonation by means of such jets. A stoichiometric mixture of hydrogen with oxygen was used with initial pressure of 20 kPa. Normal jets produced by the reflection of the shock wave with Mach number $M \geq 2.0$ led to the onset of detonation only in the jet reservoir, indicating that pressure and temperature enhancement in the focus region is not high enough. In contrast, inclined-jet interaction led to the onset of detonation in the concave cavity at $M \leq 1.95$, indicating higher focusing degree.

4. Conclusion

Thus, we have studied several different methods of detonation initiation, namely, plane shock-wave focusing due to the reflection from concave axisymmetric or two-dimensional surfaces; interaction of supersonic jets of combustible mixtures; separate injection of acetylene and oxygen jets into the combustion chamber with supersonic mixing following; fast turbulent mixing of hot combustion products with the detonative mixture (prechamber initiation). We have determined the conditions under which these methods can be practically realized.

References

Bikovscii FA, Mitrofanov VV (1980) Detonative combustion of the gas mixture in the cylindrical chamber. Fizika goreniya i vzriva 16(5):107–117 (in Russian).

Borisov AA, Zamanscii VM, Kosenkov VV, Lisyanskii VV, Skachkov GI, Troshin KY, Gelfand BE (1990) Ignition of gaseous combustible mixtures in focused shock waves. In: Current Topics in Shock Waves, Kim YW (ed), AIP Conference Proceedings, American Inst of Physics, New York, pp 696–701.

Carnasciali F, Lee JHS, Knystautas R (1991) Turbulent jet initiation of detonation. Comb and flame 84:170–180.

Chan CK, Lau D, Thibault PA, Penrose JD (1990) Ignition and detonation initiation by shock focusing. In: Current Topics in Shock Waves, Kim YW (ed), AIP Conference Proceedings, American Inst of Physics, New York, pp 161–166.

Lee JH, Knystantas R, Guirao CM (1975) Critical power density for direct initiation of unconfined gaseous detonation. In: Proc 15th Int Symp on Combustion, The Combustion Institute, Pittsburg, pp 53–67.

Oppenheim AK (1970) Introduction to Gasdynamics of Explosions, Springer, Wien, New York.

Takayama K (ed) (1990) Proc Int Workshop on Shock Wave Focusing, Tohoku Print, Sendai, Japan.

Westbrook CK, Pitz WJ, Urtiew PA (1984) Chemical kinetics of propane oxidation in gaseous detonation. In: Progress in Astronautics and Aeronautics, Bowen JR, Manson N, Oppenheim AK, Soloukhin RI (eds), Inst of Aeronautics and Astronautics Inc, New York, pp 151–174.

Zeldovich YB, Kogarko SM, Simonov NN (1956) Experimental investigation of the spherical detonation in gases. Tekhn Fizika 26:8–12 (in Russian).

Characterization of heterogeneous detonations of powders dispersed in nitrogen

A. J. Tulis, W. K. Sumida, K. Norikane, R. Dihu
IIT Research Institute, Chicago, Illinois 60616 USA
D. L. Patel, D. C. Heberlein
U.S. Army CECOM, Fort Belvoir, Virginia 22060 USA

Abstract: A 7.32-m long 152-mm i.d. horizontal detonation tube incorporating a 1.83-m long highly instrumented section was used to investigate and characterize the heterogeneous detonation of energetic powder mixtures dispersed in gaseous nitrogen, and to assess phenomena such as spinning and/or multiple-front detonations. Hybrid powder of various β-lactose (CHO), flaked (4 m^2/g:S/M) aluminum (Al), 7.5-μm ammonium perchlorate (AP), and 15-μm RDX powder compositions were evaluated. The detonations that were achieved with CHO/Al/AP and CHO/Al/AP/RDX provided no evidence of spinning detonation, but did provide good confirmation of multiple-front detonations, especially when RDX was present in large quantities. The Al appears to be more sensitive to initiation of detonation, under appropriate conditions, than the RDX; in detonations both in air and in nitrogen, presence of Al resulted in lower induction times than for RDX alone. Small particle size of the Al is a prime requisite.

Key words: Heterogeneous detonation, RDX, TNT, Starch, Aluminum

1. Introduction

The achievement and phenomenology of heterogeneous two-phase solid particulate/gas detonations have become subjects of considerable interest and investigation in recent years, involving fuel, energetic material, and molecular explosive powders dispersed in air (Tulis 1995). Detonations of fuel powders dispersed in air were documented over a decade ago, e.g., detonation of aluminum (Al) powder (Tulis and Selman 1982) and grain dust (Kauffman 1980) achieved in one-dimensional detonation/shock tubes. Furthermore, about two decades ago, it was discovered that molecular explosive powders such as PETN, RDX, and TNT could be dispersed in air and detonated (Tulis et al. 1983) much more effectively and independently of the severe stoichiometry and homogeneity conditions requisite in the case of liquid or solid fuels. Results of these studies demonstrated that the detonation phenomenology of fuels such as Al, and monopropellants such as RDX, are considerably different when dispersed in air. Whereas the solid-powder fuels are critically dependent upon stoichiometry, homogeneity, and particle size, the monopropellant powders, in which stoichiometry is intrinsic, are generally independent of these same criteria, except for minimum concentrations required to sustain propagation of detonation. The monopropellant powders detonate readily and stably, yielding detonation characteristics that correlate well with analytical thermochemical Chapman-Jouguet (CJ) computations. On the other hand, fuel powders are very difficult to detonate, even in one-dimensional detonation-tube studies, in which case they generally manifest spinning detonations. Further studies have demonstrated that the severe criteria for achieving stable detonation in fuel particles dispersed in air could be considerably relaxed by sensitization with small amounts of molecular explosive powders such as RDX (Tulis et al. 1993). Nevertheless, the phenomenological aspects of these homegeneous detonations are still not well understood nor have they been adequately investigated.

PETN and RDX detonate individually as discrete particles when dispersed in a gaseous medium such as air, and it has been recently demonstrated that RDX dispersed in nitrogen will readily detonate (Tulis et al. 1994). Since the addition of Al powder to RDX dispersions in air provides considerably improved detonation characteristics, it is evident that the Al is reacting within the detonation zone such that the overall detonation and blast output are enhanced, e.g., the Al participates in the detonation mechanism. Organic high explosives such as TNT and RDX are metastable molecular CHNO compounds. Carbohydrates (CHO) such as starch do not contain nitrogen. Nevertheless, the supply of oxygen in CHO compounds is as great if not greater than that in molecular explosives, particularly fuel-rich explosives such as TNT.

A major factor in the energetics of chemical reactions is, of course, the heat of formation of the reactants. However, the heat of formation of the products is of the greatest importance, i.e., the formation of Al_2O_3 yields much greater energy than CO_2, which in turn releases much greater energy than H_2O. Hence, if an appropriate amount of Al is added to CHO leading to production of Al_2O_3, the energy release is about 1.83 kcal/g, which is greater than that of most explosives.

The CHO compound used in this work was disaccharide β-lactose anhydride (CHO hereafter). It has a very small particle size, about 1 μm, but because of its great tendency to agglomerate requires treatment with an appropriate de-agglomerating agent. However, it was determined that in the case of admixing flaked Al with CHO, the flaked Al itself serves quite well to mitigate agglomeration.

2. Analytical

Figures 1 through 3 illustrate the analytical TIGER Code CJ results for detonation temperatures, pressures, and velocities for a particular composition experimentally evaluated, i.e., 70 percent of 80:20 CHO:Al and 30 percent AP, over a concentration range from 0.1 to 100 kg/m^3, demonstrating the influence of the RDX additive of 0-, 10-, 25-, and 50-weight percent. It is interesting to note that the influence of the RDX, even at 50 percent, is manifested mainly as increased CJ temperature, with only a slight increase in CJ velocity, and an essentially negligible influence on CJ pressure (Dihu et al. 1995).

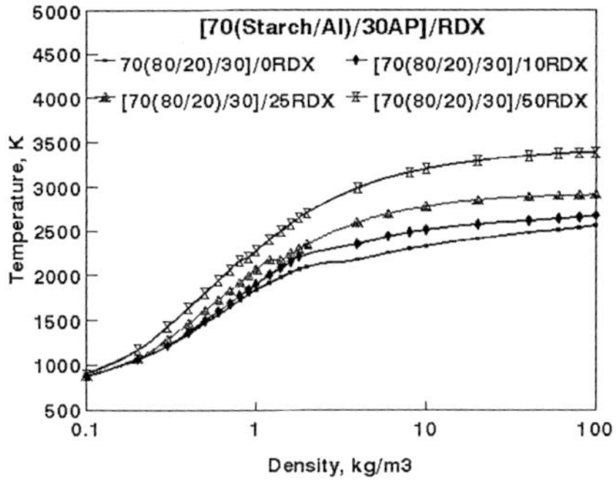

Figure 1. CJ temperature versus density for CHO/Al/AP/RDX system in N_2.

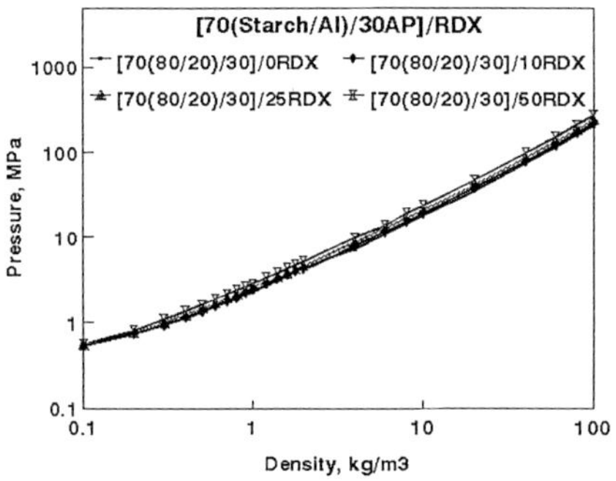

Figure 2. CJ pressure versus density for CHO/Al/AP/RDX system in N_2.

3. Experimental procedure

A 7.32-m long 152-mm internal diameter horizontal detonation tube incorporating a 1.83-m long highly instrumented section was used to investigate and characterize the heterogeneous detonation of energetic powder mixtures dispersed in gaseous nitrogen. The instrumented section is constructed of stainless steel and contains one set each of diametrically opposed piezoelectric pressure transducers and fiber-optic light-detector probes at each of the eleven stations. However, to better assess phenomena such as spinning and/or multiple-front detonations, five sensor locations at four of the eleven stations were altered so that two stations utilized five piezoelectric pressure transducers at 60° intervals and two stations utilized five fiber-optic light-detector probes. The experimental procedures in these studies were essentially identical to those previously published (see Egghart et al. 1992). Two lasers and two video cameras were used for monitoring concentration of the powders as well as establishing the timing. Initiation of detonation was achieved using a 20-g explosive charge of C4 in addition to the standard detonator and A4 pellet. The CHO was micron sized, the Al was flaked type 40XD with a S/M of 3–4 m^2/g (less than one micron if spherical), the AP was 7.5 μm, and the RDX was 15 μm.

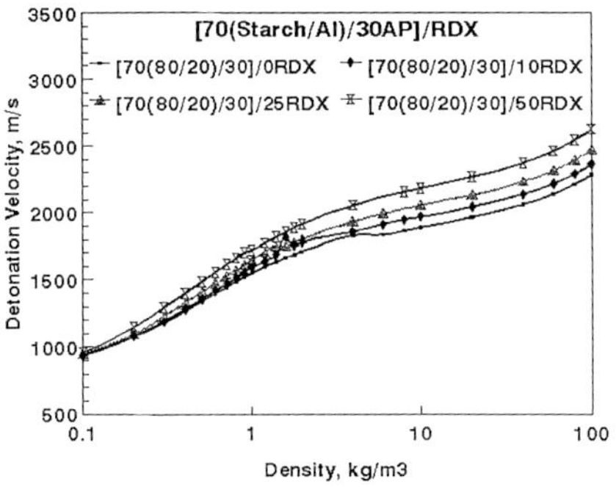

Figure 3. CJ velocity versus density for CHO/Al/AP/RDX system in N_2.

This investigation involved various hybrid CHO, Al, AP, and RDX powder compositions. A dry-nitrogen gas-pressurized vessel and appropriate nozzles dispersed the candidate powders into a flowing stream of dry nitrogen gas to achieve relatively homogeneous concentrations. Based on analytical results that involved a comprehensive parametric investigation of CHO/Al/AP and CHO/Al/AP/RDX in nitrogen (Dihu et al. 1995), the two compositions [70(90/10)/30]/RDX and [70(80/20)/30]/RDX of [(CHO/Al)/AP]/RDX (where RDX variations were 0, 10, 25, and 50) were comprehensively evaluated experimentally at a nominal concentration of 3.0 kg/m^3. Only data from detonations that resulted in stable and consistent CJ characteristics over the entire instrumented section of the detonation tube were used. The data were then averaged over the eleven instrumented stations and again averaged for three replicated experiments. In addition, experiments were conducted with four CHO/Al weight ratios, i.e., 90:10, 80:20, 70:30, and 60:40, all with 30 percent by weight AP, and no RDX, to assess the influence of the amount of Al. Finally, experiments were also conducted on the three binary compositions CHO/AP 50:50, Al/AP 40:60, and Al/CHO 50:50. Calibration experiments were always conducted in order to obtain concentration information and timing requirements.

4. Experimental results

Experimental results conducted at a nominal concentration of 3.0 kg/m^3 are presented in Table 1, in which corresponding analytical values are included for comparison. The nominal 3.0 kg/m^3 concentration was about the highest that could be experimentally homogeneously achieved in the detonation tube.

Table 1. Analytical CJ and experimental detonation results for some quaternary, ternary, and binary powder compositions at a nominal concentration of 3 kg/m^3 dispersed in nitrogen.

	Analytical		Experimental	
Composition	Pressure MPa	Velocity m/s	Pressure MPa	Velocity m/s
[(CHO/Al)/AP]/RDX				
90[70(90/10)/30]/10	4.84	1613	5.93	1814
75[70(90/10)/30]/25	5.69	1662	8.19	1878
50[70(90/10)/30]/50	6.90	1914	7.00	1893
[(CHO/Al)/AP]/RDX				
90[70(80/20)/30]/10	6.16	1817	6.01	1700
75[70(80/20)/30]/25	6.60	1865	6.56	1936
50[70(80/20)/30]/50	7.47	1983	7.61	2070
[(CHO/Al)/AP] 0 RDX				
[70(90/10)/30]	4.32	1520	2.27	1240
[70(80/20)/30]	5.77	1760	6.04	1790
[70(70/30)/30]	6.08	1771	3.42	1600
[70(60/40)/30]	6.30	1793	4.97	1868
Binary Composition				
CHO/AP 50:50	5.04	1640	7.65	1777
Al/AP 40:60	7.25	1900	4.82	2019
Al/CHO 50:50	4.81	1550	fail	fail

5. Conclusions

The CHO/Al/AP compositions are detonable in nitrogen, which means that direct heterogeneous particle–particle mechanisms are involved, evidently requiring a primary decomposition/detonation of the AP to release its oxygen, and possibly also involving decomposition of the CHO molecule. However, since detonation of CHO/Al powers dispersed in nitrogen was not achieved, under the test conditions imposed, it appears that the particles of CHO and Al would have to be in intimate contact (as in the bulk state in which this composition did detonate). Since both CHO/AP and Al/AP powder dispersions in nitrogen did detonate, decomposition/detonation of the AP would appear to be a requisite first step. With RDX present, it is concluded that the detonation of the RDX particles was the primary event.

The detonations that were achieved with CHO/Al/AP and CHO/Al/AP/RDX provided no evidence of spinning detonation, but did provide good confirmation of multiple-front detonations, especially when RDX was present in large quantities. The Al appears to be more sensitive to initiation of detonation, under appropriate conditions, than the RDX; in detonations both in air and in nitrogen, presence of Al resulted in lower induction times than for RDX alone. Small particle size of the Al is a prime requisite.

Acknowledgement. The theoretical and experimental research described in this publication was conducted under U.S. Army Contract DAAK-70-93-0011, "Insensitive Dispersed Explosives" in support of the Army Mine Neutralization Program.

References

Dihu R, Tulis AJ, Sumida WK, Heberlein DC, Patel DL (1995) Analytical TIGER code assessment of Chapman-Jouguet detonation characteristics of starch/aluminum/ammonium perchlorate powders dispersed in nitrogen. In: Proc 4th Symp (Int) on Behavior of Dense Media Under High Dynamic Pressures, Tours, France, pp 221–226.

Egghart H, Heberlein DC, Tulis AJ, Sumida WK (1992) Detonation tube studies of heterogeneous solid powder-gas fuel/oxidizer detonation. In: Proc 18th (Int) Pyrotechnics Sem, IIT Research Institute, pp 1083–1107.

Kauffman CW (1980) Analysis of Recent Explosions, Grain Elevator and Proc Soc Tech Conf, Portland, Oregon.

Tulis AJ, Selman JR (1982) Detonation tube studies of aluminum particles dispersed in air. In: Proc 19th (Int) Symp on Comb, The Combustion Institute, p 655.

Tulis AJ, Austing JL, Heberlein DC (1983) Detonation tube studies of reactive particles in two-phase detonations. In: Proc 14th (Int) Symp on Shock Tubes and Waves, New South Wales Univ Press, pp 447–454.

Tulis AJ, Sumida WK, Heberlein DC, Patel DL, Egghart H (1993) Detonation tube studies of particle size and RDX sensitization with regard to spinning and/or multiple-front detonations. Archivum Combustionis 13(3–4):253–260.

Tulis AJ, Sumida WK, Joyce RP, Heberlein DC, Patel DL (1994) Phenomenological aspects in explosive powder/gas two-phase detonations. In: Proc 25th (Int) Symp on Comb, The Combustion Institute, pp 79–87.

Tulis AJ (1995) Recent advances in two-phase hybrid solid particles/gas heterogeneous detonations. In: Proc 21st (Int) Pyrotechnics Sem, Moscow, Russia (in press).

Numerical simulation of the reflection of detonation waves

Q. Yu, H. Grönig
Shock Wave Laboratory, RWTH Aachen
Templergraben 55, 52056 Aachen
Germany

Abstract: The reflection of an unsteady, two-dimensional gaseous detonation wave propagating over a wedge has been numerically investigated. In order to calculate the unsteady detonation waves around complex geometries, a two-dimensional code solving the inviscid compressible reacting Euler equations was developed by using a high-order extension of Godunov's method, an ideal-gas law, and a two-step reversible reaction model combined with an adaptive, unstructured mesh refinement algorithm. The cellular configuration, the overdrive degree of the Mach stem, and the transition from Mach to regular reflection of detonation waves were investigated by changing the wedge angle. The numerical results obtained for the stoichiometric hydrogen-oxygen mixture are in qualitatively good agreement with the experimental data.

Key words: CFD, Detonation, Adaptive grid

1. Introduction

The reflection of detonation waves is one of the typical phenomena under transient conditions when the detonation waves encounter a change in composition, pressure, temperature, or confinement. The transient features of the detonative process have been revealed by many investigators (Edwards et al. 1984, Gavrilenko & Prokhorov 1982, Meltzer et al. 1991) during the last few decades. Most experimental investigations had the objective to determine the shape of the reflected wave generated during Mach or regular reflection, the critical wedge angle for transition from Mach to regular reflection, and the change in cell size as a function of overdrive by using schlieren, streak photography, and smoke foils. The two- and three-shock theory were also extended to calculate the critical and trajectory angles for transition from Mach to regular reflection. The detailed results for the stoichiometric hydrogen-oxygen mixtures diluted by argon are summarized in Table 1. It was shown that the calculations did not compare well with the experiments. A recent study (Li et al. 1995) has shown that better results were obtained by using modified models based on the two- and three-shock theory.

In this paper, a numerical method was used to calculate the reflection of detonation waves. The purpose of the present work is not only to calculate the critical and trajectory angles for transition from Mach to regular reflection but also to analyze the cellular configuration and the overdrive degree of the Mach stem by changing the wedge angle.

Table 1. Critical wedge angle: stoichiometric hydrogen-oxygen-argon.

Comments	Initial Pressure (bar)	Dilution (%)	Theoretical Data (deg.)	Experimental Data (deg.)
Gavrilenko et al.	0.05–1.0	0	34 ± 0.4	39–41
		70	37 ± 0.4	
Edwards et al.	0.2	10	65	46–50
Meltzer et al.	0.2	0	34	40–45
		50	34	
		70	34	

2. Physical model and numerical schemes

When a detonation wave propagates over a wedge, a reflected wave is generated by the wedge surface. The flow field behind the incident detonation front is different from the one behind the non-reactive shock front. The incident detonation front consists of many triple points which produce a cellular configuration. The propagation process depends on the nature of the fuel and the composition and initial pressure of the mixture. The detonation considered in this study is defined to have an equilibrium configuration which is a regularly repeatable structure with a steady time-averaged velocity.

The detonation propagation in rectangular tubes is considered as a three-dimensional rectangular mode. Experimental results (Strehlow et al. 1970) have shown that a rectangular mode detonation in reality contains two independent, orthogonal, two-dimensional planar modes which are observed in narrow rectangular tubes. The strength of the transverse waves and the cell size observed for the planar mode detonation equal that observed for the rectangular mode detonation under the same initial conditions. According to this opinion, it is possible that the two-dimensional numerical predictions can be compared directly with experimental data for detonations propagating in rectangular tubes.

The flow of detonation waves is assumed to be inviscid. Thus, viscosity effects are not included in this paper. The two-dimensional compressible reacting Euler equations combined with an ideal gas law and a two-step chemical reaction model (Korobeinikov et al. 1972) are given by

$$\rho_t + \nabla(\rho\mathbf{v}) = 0, \tag{1}$$
$$(\rho\mathbf{v})_t + \nabla(\rho\mathbf{v}\mathbf{v}) + \nabla P = 0, \tag{2}$$
$$(\rho E)_t + \nabla(\rho\mathbf{v}E + \mathbf{v}P) = 0, \tag{3}$$
$$(\rho\alpha)_t + \nabla(\rho\mathbf{v}\alpha) = \rho\omega_\alpha, \quad \text{and} \tag{4}$$
$$(\rho\beta)_t + \nabla(\rho\mathbf{v}\beta) = \rho\omega_\beta, \tag{5}$$

with

$$E = e + Q\beta + \frac{v^2}{2}, \tag{6}$$
$$P = (\gamma - 1)\rho e, \tag{7}$$
$$\omega_\alpha = -k_1\rho exp\left(-\frac{E_1^+}{T}\right), \quad \text{and} \tag{8}$$
$$\omega_\beta = -k_2P^2\left[\beta^2 exp\left(-\frac{E_2^+}{T}\right) - (1-\beta)^2 exp\left(-\frac{E_2^+ + Q}{T}\right)\right]. \tag{9}$$

In the above equations P, ρ, T, $\mathbf{v}$, and E denote the pressure, density, temperature, velocity, and total specific energy. These variables are nondimensionalized by the constant preshock state. The reaction progress parameters α and β are $1 > \alpha > 0$ and $\beta = 1$ in the induction period, and $\alpha = 0$ and $1 > \beta > 0$ in the exothermic period. The variables of the CJ state point can be obtained at $\beta = \beta_{eq}$.

The governing equations are solved by using a fractional step scheme, in which the global procedure involves two processes per time step. In the first fractional step, the non-reacting hydrodynamic part is solved by using an upwind finite-volume scheme (Lottati 1990) for solving multi-dimensional hyperbolic systems of conservation laws with an adaptive mesh refinement on triangular, unstructured grids (Löhner 1989, Holmes & Lamson 1986). In the second fractional step, the constant volume combustion is considered and the reaction progress parameters α and β are advanced explicitly by integrating the ordinary differential equations. The boundary conditions are set by using the slip conditions.

3. Results and discussion

According to the experimental results of Edwards et al. (1984) and Meltzer et al. (1991), the cell width behind the incident detonation front is 6 to 8 mm for the stoichiometric hydrogen-oxygen mixture diluted by argon at an initial pressure of 0.2 bar. The width of the computational domain is selected so that a two-dimensional cellular structure of detonation waves with the mode number 12 can be established with a cell width of 6 mm. The unit length scale is selected to be 5×10^{-3}m. The finest mesh with the induction length scale is obtained by using the mesh refinement. The nondimensionalized parameters in Eqs. (8) and (9) (Korobeinikov et al. 1972) are $Q = 28.0$, $\gamma = 1.3$, $E_1^+ = 34.64$, $E_2^+ = 6.929$, $k_1 = 3.16 \times 10^4$, and $k_2 = 2.58 \times 10^{-2}$. The Mach number of the CJ detonation wave is 5.1, which is the same as the experimental data (Edwards et al. 1984, Meltzer et al. 1991). In order to obtain the two-dimensional detonation front, the detonation is initiated by putting six pockets of unburned material behind the initial detonation front with the ZND profile. The numerical results indicated that a stable, two-dimensional cellular structure can be rapidly established before the detonation wave propagates over the wedge.

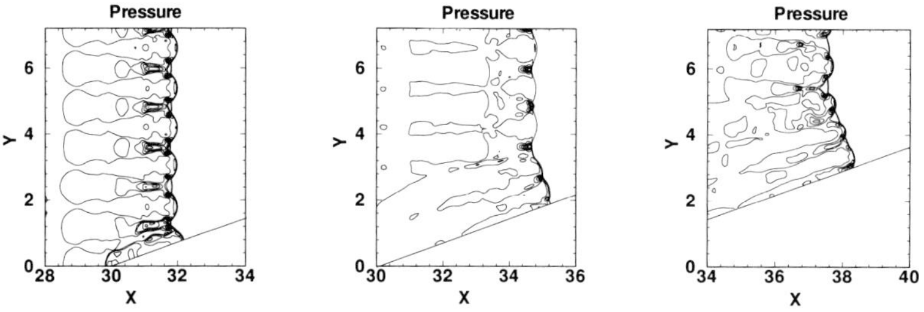

(a) $t = 3.0$ (b) $t = 3.5$ (c) $t = 4.0$

Figure 1. The computational isobars of detonation waves in $2H_2 + O_2$ mixture with an initial pressure of 0.2 bar and a wedge angle of 20° at different times. $P_{MIN} = 6.0$, $P_{MAX} = 70.0$, $\Delta P = 6.4$.

Figure 1 shows the isobars of the detonation wave entering a convergent section at different times. The angle of the wedge is 20°. As shown in Fig. 1, once the detonation front encounters the wedge, the reflected wave and new triple point are formed. The new reignition sources behind the Mach stem are formed due to the collision of the transverse waves with the reflected wave, collisions between the transverse waves, and the reflection of the transverse waves on the wedge surface. A two-dimensional cellular structure behind the Mach stem is established when the local explosions occur continuously. The process of forming the two-dimensional cellular structure of the Mach stem without any artificial disturbance is also shown in Fig. 2. The Mach stem appears between the new triple point and the wall of the wedge and is vertical to the wall. The reflected wave is kinked and the contact surface is not formed due to the existence of the reaction zones behind the incident detonation front and the Mach stem. This phenomenon was observed in schlieren photographs from the experimental study of Meltzer et al. (1991).

Figure 3a gives a comparison of calculated trajectory angles with the experimental ones for detonation waves at different wedge angles. The critical wedge angle for the transition from Mach to regular reflection in the computations is 50°. This angle was determined experimentally to be 45°.

Figure 3b shows the influence of increasing the overdrive degree of the Mach stem on the cell width behind the Mach stem. The cell width behind the Mach stem is larger than that obtained by the measurements (Edwards et al. 1984, Meltzer et al. 1991) in the case of strong overdriven detonations. The numerical results indicate that the computations need to consider the three-dimensional transverse wave structure behind strong overdriven detonations and the boundary layer effects on the spacing of triple points.

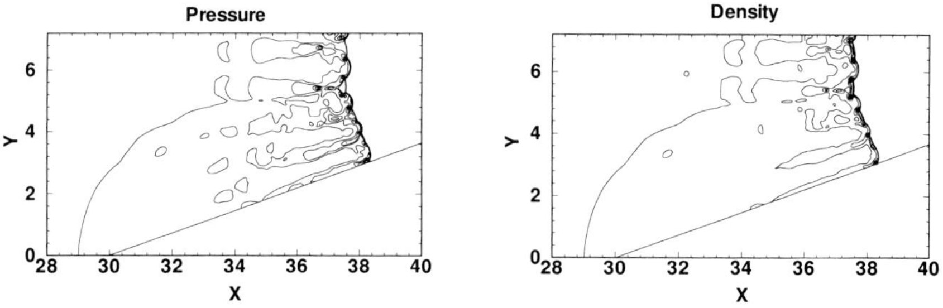

Figure 2. The computational isobars and isopycnics of detonation waves in $2H_2 + O_2$ mixture with an initial pressure of 0.2 bar and a wedge angle of 20°. $P_{MIN} = 6.0$, $P_{MAX} = 66.0$, $\Delta P = 6.0$; $\rho_{MIN} = 2.2$, $\rho_{MAX} = 12.0$, $\Delta\rho = 0.98$.

4. Conclusion

In this approach, the basic shapes of the incident detonation front, Mach stem, and reflection wave were numerically investigated by use of a two-step reversible reaction model for the stoichiometric hydrogen-oxygen mixture. The numerical results are in qualitatively good agreement with experimental data. The following three aspects about the reflection of detonation waves will be investigated by further numerical simulations. (1) Describing accurately the chemistry dependent reflection of detonation waves by using a more detailed chemical model. (2) The influence of the incident-wave cell size on the wall-induced triple-point trajectory and critical angle by changing the initial pressure or the dilution of argon in stoichiometric mixtures of hydrogen-oxygen. (3) The boundary-layer

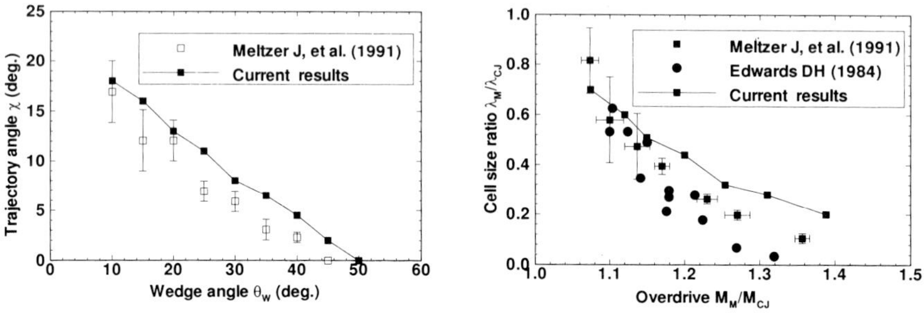

(a) (b)

Figure 3. (a) Comparison of calculated trajectory angles for detonation waves with the measured results; (b) comparison of calculated overdriven cell size ratio with the measured results.

effects on the formation of the cellular structure behind the Mach stem and the shape of the trajectory of the wall-induced triple point.

References

Edwards DH, Walker JR, Nettleton MA (1984) On the propagation of detonation waves along wedges. Archivum Combustionis 4(3):197–209.

Gavrilenko TP, Prokhorov ES (1982) Compressed detonation wave in a real gas. Combustion, Explosions and Shock Waves 17(6):689.

Holmes DG, Lamson SH (1986) Adaptive triangular meshes for compressible flow solutions. In: Proceedings of First International Conference on Numerical Grid Generation in CFD, Landshut FRG, pp 413–424.

Korobeinikov VP, Levin VA, Markov VV, Chernyi GG (1972) Propagation of blast waves in a combustible gas. Astronautica Acta 17:529–537.

Li H, Ben-Dor G, Grönig H (1995) Analytical reconsideration of the oblique reflection of detonation waves and comparison with experiments (to be published).

Löhner R (1989) Adaptive H-Refinement on 3-D Unstructured Grids for Transient Problems, AIAA-89-0365.

Lottati I, Eidelman S, Drobot A (1990) A Fast Unstructured Grid Second Order GODUNOV Solver (FUGGS), AIAA-90-0699.

Meltzer J, Shepherd JE, Akbar R, Sabet A (1991) Mach reflection of detonation waves. In: Kuhl AL, Leyer J-C, Borisov AA, Sirignano WA (eds) Shock Waves, Explosions and Detonations, Progress in Astronautics and Aeronautics, AIAA, Washington DC, 153:78–94.

Strehlow RA (1970) Multi-dimensional detonation wave structure. Astronautica Acta 15:345–357.

Analysis of detonation on a hyper-velocity projectile in the in-tube chemical accelerator

J. Rom
Technion-Israel Institute of Technology
Haifa 32000, Israel

Abstract: The initiation of shock-induced combustion and the subsequent acceleration of a projectile in the in-tube chemical launchers for accelerating projectiles to hypervelocity are investigated using an energy-balance analysis. This analysis can be applied to the two types of chemical launchers, the Ram Accelerator (RA) and the External Propulsion Accelerator (EPA). It is shown that in the EPA, where the shock-induced combustion is initiated by the hypersonic flow of a detonable mixture flowing into a forward-facing step, the drag of this blunt step is more than sufficient to initiate shock-induced combustion at a projectile velocity well below the Chapman-Jouguet detonation velocity. The energy-balance analysis is also applied to evaluate the maximum velocity which can be achieved in the in-tube accelerators when the available chemical-reaction energy is utilized. It is shown that in the RA, due to the high drag (caused by the thermal and shock waves) which chokes the flow between the projectile and the tube wall as the projectile Mach number is increased, the maximum projectile velocity is limited to about 1.2 – 1.3 times the Chapman-Jouguet detonation velocity. It is observed that the experimental measurements of the maximum velocity achieved in the RA are at most at 1.3 times the Chapman-Jouguet detonation velocity, which is in agreement with the results of this analysis. In the EPA, the total drag coefficient of the projectile with the step is about 0.28 for the range of hypersonic Mach numbers, while the useful chemical-reaction energy increases with increasing flight Mach number. Therefore, the maximum projectile velocity in the EPA can reach a level of six times the Chapman-Jouguet detonation velocity, which is well above the escape velocity into space.

Key words: Detonations, Chemical in-tube accelerators, Hypervelocity

1. Introduction

The in-tube chemical launchers for accelerating projectiles to hypervelocity utilize, in the superdetonative mode of operation, the possibility of generating continuous thrust by initiating shock-induced combustion in the premixed fuel/oxidizer mixture. The first method proposed for an in-tube chemical launcher was the Ram Accelerator (RA), originated and developed by A. Hertzberg and his colleagues at the University of Washington (Hertzberg et al. 1988). The concept of the RA, operating in the superdetonative mode, is based on utilization of the scramjet cycle, where the projectile acts as a free centerbody and the tube as an extended cowling. The sharp-nosed-projectile diameter is slightly less than the tube diameter (typically 70 to 80%), therefore the nose shock wave is reflected from the tube wall onto the projectile center-body. Under proper conditions this reflected shock wave initiates a combustion process so that when the chemical-reaction products are expanded on the rear part of the projectile, thrust is generated.

Another method for operating the chemical in-tube accelerator is based on the utilization of the external propulsion cycle, proposed by Rom (Rom 1990, Rom and Avital 1992). In the External Propulsion Accelerator (EPA) the projectile diameter is much smaller than the tube diameter so that there is no interaction between the flow over the projectile and the tube wall over the complete length of the projectile. The shock-induced combustion is established by aerodynamic means on the projectile, by a forward-facing step on the projectile shoulder, or by the blunt leading edge of a ring wing positioned on the rear part of the projectile.

For high enough projectile velocity, the heating and compression of the combustible mixture by this shock wave can result in initiation of a combustion front in the region between the shock wave and the face of the step. Due to the interaction between the shock wave from the projectile nose with the detached shock-detonation wave established in front of the step, the high-pressure, high-temperature chemical-reaction products are confined by the contact surface generated at the wave intersection point. Therefore, the aerodynamically confined region filled with the chemical-reaction products is in effect an "external combustion chamber" which extends further on the rear part and in the base region of the projectile and is used to produce thrust on the projectile.

2. Initiation of shock-induced combustion on a blunt body

Discussions on the initiation of detonation on a blunt-nosed projectile are presented by Lee (1994, 1984). He estimated the energy required for initiation of detonation by a hypervelocity projectile by the use of the blast-wave analogy, thus equating the blast energy required for the initiation of detonation with the work done by the drag force of the traveling projectile.

When we consider the initiation of detonation ahead of a blunt body in hypersonic flow, it was pointed out by Zeldovich (1992) that detonation will occur when the induction time is shorter than the flow time from the detached shock wave to the face of the body. The induction time is determined by the chemical-reaction kinetics of the mixture at the conditions behind the detached shock wave while the flow time is a function of the detachment distance, which varies with the radius of curvature of the blunt nose of the body.

Analyses of the hypersonic flow over blunt bodies with chemical reactions were presented by Sichel and Galloway (1967), Galloway and Sichel (1969), and Chernyi (1968). It was shown in these studies that the establishment of combustion in the stagnation region depends on the ratio of the ignition-delay distance to the shock-standoff distance. Sichel and Galloway (1967) define a parameter relating the ignition-delay time, the body radius of curvature, and the flow velocity. When this ignition-delay parameter is less than one, then combustion will occur between the detached shock wave and the body surface.

The forward-facing step has an infinite radius of curvature at the stagnation region; it is expected that above a minimum projectile velocity, the temperature behind the detached shock wave will be higher than the auto-ignition temperature and a combustion process will be initiated. Above the step, the shock wave will curve, reducing the radius of curvature as the distance above the step increases. As discussed by Lee (1994), when the shock-wave curvature is reduced below a critical value, and hence the flow divergence is too severe, then the reactions may be quenched even when the initial shock temperature in the stagnation point exceeds the auto-ignition limit. Therefore, in order to insure stabilization of the reaction front behind the detached shock wave, a minimum step height is needed.

This qualitative argument is in agreement with the results of the analysis of Tivanov and Rom (1993) where the minimum-step-height concept was introduced and evaluated by an approximate analysis. The minimum step height evaluated there and this analysis are in agreement with the phenomenological analysis of Sichel and Galloway (1967).

2.1. The minimum Mach number for initiation of detonation

The minimum Mach number for the initiation of shock-induced combustion on a projectile with a forward-facing step in the EPA can be evaluated by equating the critical energy required for direct initiation of cylindrical detonation with the work done by the drag force of the blunt step on the hypervelocity projectile. Due to the high drag of the blunt step, the minimum projectile velocity that produces energy to balance the critical initiation energy for detonation is then much below the detonation velocity. Since the projectile in the EPA is expected to fly well above the detonation speed, there is no problem securing the energy required to initiate detonation.

The energy balance is applied also to the projectile in the RA. In this case, the drag coefficient is equated with that of the drag of the projectile including the effects of the interaction between the projectile and the tube wall. As the projectile speed increases, the drag force on the projectile increases so as to provide the energy for initiation of the detonation. At the lower initial speeds, the required drag for initiation of detonation is provided by the perforated-piston obturator which is used in the RA for initiation of the process in the subdetonative mode.

2.2. The maximum velocity by the energy balance in the EPA

The energy balance requires that the work done by the drag force acting on the hypervelocity projectile be balanced by the energy released by the combustion of the gas mixture. Following the discussion by Lee (1994), it is possible to evaluate the maximum velocity that can be reached by the projectile when all the combustion energy of the reacting mixture is utilized. The maximum velocity is found to be directly proportional to the ratio of the outside diameter of the external combustion zone to the projectile diameter and inversely proportional to the drag coefficient. In the EPA the thickness of the external combustion layer increases as the flow Mach number increases. It was found from the numerical calculations, which are reported in part in Rom et al. (1995), that for the 32-mm diameter projectile with a 1-mm step the combustion-layer thickness varies from 3 step heights at Mach number 5 to 5.5 step heights at Mach number 6 and to 17 step heights at Mach number 10. The value of the drag coefficient for the projectile with the cone with the 1-mm step is about 0.27 for Mach numbers above 5 (for cone-cylinder without the step the drag coefficient is about 0.1). Thus the maximum projectile velocity for the 32-mm projectile in the 120-mm tube is about six times the detonation velocity, which is well above escape velocity. Of course as the projectile velocity increases, the problems of heating and ablation of the projectile surfaces and the aerodynamic and acceleration loads on the projectile structure become more critical.

2.3. The maximum velocity by the energy balance in the RA

It is of interest to evaluate the maximum velocity of the projectile in the RA by this energy-balance analysis, particularly since there are some experimental results on the operational limits in the RA. Using the energy-balance analysis it is possible to evaluate the maximum velocity of the projectile in the RA. Such an evaluation was first done by Lee (1994).

The projectiles used in the RA of the University of Washington and at the Army Research Laboratory (ARL) in Aberdeen, as presented in Hertzberg et al. (1988) and Nusca (1994), have a tube-diameter-to-projectile-diameter ratio of 1.3 and a drag coefficient of about 0.25 at Mach numbers above 5 in free flight without the tube-wall interference. However, the multiple shock-wave reflections from the tube wall to the projectile center section including effects of shock-wave–boundary-layer interactions and the combustion/detonation which they initiate cause drag and thermal choking at increasing Mach numbers. The effective drag coefficient increases to values of one and even higher. This is similar to the choking that occurs in intakes of ramjet engines. As the projectile velocity increases, the thermal-choking effects due to the shock losses and heating increase. Then the total drag loss consumes an increasing portion of the available reaction energy until the energy required to overcome the drag force will be equal to the available energy from the chemical reaction. At this condition the projectile will coast at constant velocity since the drag is equal to the thrust. However, conditions of "unstart" occur when the structure of the projectile fails due to thermal and ablation losses at lower velocity than the maximum velocity. This result is supported by many experimental measurements of projectile velocity in the RA facilities of the University of Washington (Knowlen et al. 1994, Hinkey et al. 1993), the ARL, and the French-German Institute of Saint Louis.

3. Summary and conclusions

It is shown by the energy-balance analysis that projectiles in in-tube chemical accelerators can initiate shock-induced combustion on the rear part of the projectile for flight velocities below the Chapman-Jouguet detonation velocity. It is shown that using the energy-balance analysis it is possible to evaluate the maximum velocity that can be achieved in the in-tube chemical accelerators by balancing the drag work of the flying projectile with the total available chemical reaction energy of the fuel/oxidizer mixture.

In the Ram Accelerator case, the high drag experienced by the projectile due to the excessive heat release which results in choking of the flow at increasing Mach number limits the projectile velocity to about 1.3 times the Chapman-Jouguet detonation velocity. This result is supported by the measurements obtained in the RA at the University of Washington (Knowlen et al. 1994, Hinkey et al. 1993) and in other RA facilities.

Using this analysis, it is shown that the maximum projectile velocity in the External Propulsion Accelerator can reach six times the Chapman-Jouguet detonation velocity, which is well above the escape velocity into space.

Therefore, it is suggested by the results of the present analysis that the appropriate speed ranges for applications of the RA and the EPA are as follows: (a) for speeds up to 2.5 km/sec, the regular RA is sufficient; (b) speed ranges from 2.5 km/sec to about 3.5 km/sec can be achieved by fin-guided or rail-guided subcaliber projectiles in hybrid external propulsion and ram accelerator facilities; (c) the EPA is appropriate for applications above 3.5 km/sec and up to escape velocity into space.

Acknowledgement. This research was supported in part by the European Research Office of the U.S. Army under contract N68171-94-C-9065.

References

Chernyi G (1968) Supersonic flow past a sphere of a hot gas mixture with consideration of ignition delay time. Akad Nauk SSSR, Mekh Zhidkosti Gaza 1:20–32.

Galloway AJ, Sichel M (1969) Hypersonic blunt body flow of $H_2 + O_2$ mixtures. Astro Acta 15:89–105.

Hertzberg A, Bruckner AP, Bogdanoff DW (1988) Ram accelerator: a new chemical method for accelerating projectiles to ultrahigh velocities. AIAA Journal 26(2):195–203.

Hinkey JB, Burnham EA, Bruckner AP (1993) Investigation of Ram Accelerator Flow Fields Induced by Canted Projectiles, AIAA Paper 93-2186.

Knowlen C, Higgins A, Bruckner A (1994) Investigation of Operational Limits to the Ram Accelerator, AIAA Paper 94-2967.

Lee JHS (1984) Dynamic parameters of gaseous detonations. Ann Rev Fluid Mech 16:311–316.

Lee JHS (1994) On the initiation of detonation by a hypervelocity projectile. Paper presented at the Zeldovich Memorial Conference on Combustion, September 12–17, 1994, Voronovo, Russia.

Nusca MJ (1994) Reacting Flow Simulations for Large Scale Ram Accelerator, AIAA Paper 94-2963.

Rom J (1990) Method and Apparatus for Launching a Projectile at Hypersonic Velocity, U.S. Patent 4,932,306, June 12, 1990.

Rom J, Avital G (1992) The External Propulsion Accelerator: Scramjet Thrust Without Interaction With Accelerator Barrel, AIAA Paper 92-3717.

Rom J, Nusca MJ, Kruzcynski D, Lewis M, Gupta AK, Sabean J (1995) Investigations of the Combustion Induced by a Step on a Projectile Flying at Hypersonic Speeds in the External Propulsion Accelerator, AIAA Paper 95-0259.

Sichel M, Galloway AJ (1967) Regimes of exothermic blunt body flow. Astro Acta 13:137–145.

Tivanov G, Rom J (1993) Investigation of Hypersonic Flow of a Detonable Gas Mixture Ahead of a Forward Facing Step, AIAA Paper 93-0611.

Yungster S, Eberhardt S, Bruckner AP (1991) Numerical simulation of shock induced combustion generated by high speed projectiles in detonable gas mixtures. AIAA Journal 29(2):187–199.

Zeldovich YB (1992) Selected Works of Ya. B. Zeldovich, Volume 1 (Ostriker JP, editor), Princeton University Press, Princeton, New Jersey, pp 441–451.

Plasma and Radiation

Regular-shock-reflection phenomena in magnetohydrodynamics

H. Ogawa, T. Fujiwara, A. K. Hayashi
Department of Aerospace Engineering
Nagoya University
Nagoya 464-01 Japan

Abstract: The "total pressure-deflection angle" shock polars are obtained to apply to magnetohydrodynamic (MHD) shock-reflection problems. They are categorized into four kinds, based on the type of shocks. Among several possible reflections, the regular-reflection phenomena are studied here using shock polars, revealing that only four incidence-reflection combinations are possible for regular reflection.

Key words: Magnetohydrodynamics, Regular reflection, Shock polar

1. Introduction

Oblique shock reflections in magnetohydrodynamic (MHD) fluids may be observed: (i) in interplanetary space where a propagating shock wave interacts with the Earth's ionosphere or magnetosphere; (ii) in an electromagnetic shock tube; and (iii) in a plasma accelerator. Since the magnetic field plays an important role in such flow fields, these phenomena must be analyzed by the use of the magnetohydrodynamic equations. In the case of purely hydrodynamic shocks, researchers have studied extensively oblique shock-wave reflection phenomena (Ben-Dor 1992), revealing that there exist 10 different wave configurations associated with the reflection of shock waves over an oblique surface. Recently, the authors (Ogawa et al. 1995a) numerically studied MHD shock-reflection phenomena occurring when an MHD plane shock collides with a two-dimensional perfectly conducting wedge which is initially surrounded by a stationary ideal MHD fluid. We found four different wave configurations, depending on the initial strength of the magnetic field, for the gasdynamic Mach number 2 and the wedge angle 30 degrees. The calculated wave configurations contained fast, slow, and intermediate shocks, and their mutual interactions.

It is well known that the MHD Rankine-Hugoniot relations yield six *types* (type is used for incident shock) of solution; i.e., fast shock, slow shock, and four types of intermediate shocks (Kennel et al. 1989). Although intermediate shocks had been ruled out by the so-called MHD evolutionary conditions (Jeffrey and Taniuti 1964), Wu (1990) recently carried out with high resolution a series of numerical analyses on one-dimensional magnetohydrodynamic Navier-Stokes equations, which revealed the persistency of intermediate shocks. Along this line, we analyze shock polars involving intermediate shocks. Further discussions are presented in Ogawa et al. (1995b) along with examples of shock polars and regular-reflection solutions.

2. MHD Rankine-Hugoniot relations

When MHD shocks propagate with an angle to the upstream magnetic field, a coplanar plane can be defined by the upstream velocity vector and magnetic field (Kennel et al. 1989). Consider Fig. 1 and attach a frame of reference to the shock. The x-axis is normal to the shock pointing from upstream (1) to downstream (2). The y- and z- axes are on the shock surface and at right angles to each other. The velocity and magnetic field do not have z-components, since the coplanar plane is selected as the x-y-plane.

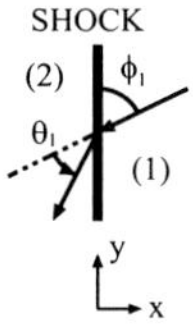

Figure 1. Reference frame attached to shock.

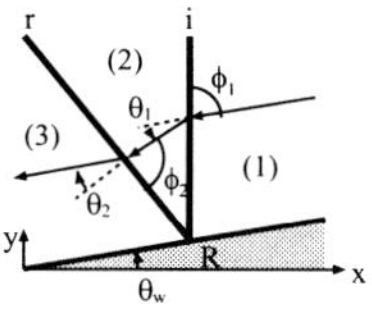

Figure 2. Configuration of regular reflection.

Denote the density, pressure, magnetic field, and velocity vectors in state i by ρ_i, p_i, $\boldsymbol{B}_i = (B_{xi}, B_{yi})^T$, $\boldsymbol{V}_i = (V_{xi}, V_{yi})^T$, respectively. Then the Rankine-Hugoniot relations for ideal MHD can be reduced to the following single polynomial of order three (Kennel et al. 1989):

$$R(X) = 0, \tag{1}$$

where

$$R(X) = X^2 H(X) - \frac{\sin^2(90^\circ - \phi_1)}{M_{A1}^2} J(X), \tag{2}$$

$$H(X) = \frac{\gamma+1}{\gamma-1} X + \frac{V_{c1}^2 - V_{x1}^2}{V_{x1}^2}, \quad J(X) = \frac{\gamma}{\gamma-1} X^2 - X\left(Y - \frac{1}{\gamma-1}\right) + \frac{Y(1-Y)}{\gamma-1}, \tag{3}$$

$$V_{c1}^2 = \frac{\gamma+1}{\gamma-1} C_{I1}^2 - \frac{2C_{S1}^2}{\gamma-1}, \quad X = \frac{V_{x2}}{V_{x1}} - \frac{1}{M_{I1}^2}, \quad Y = 1 - \frac{1}{M_{I1}^2}, \tag{4}$$

$$M_{A1}^2 = \frac{V_{x1}^2}{C_{A1}^2} = \frac{\mu_0 \rho_1 V_{x1}^2}{B_1^2}, \quad M_{I1}^2 = \frac{V_{x1}^2}{C_{I1}^2} = \frac{\mu_0 \rho_1 V_{x1}^2}{B_x^2}, \quad M_{S1}^2 = \frac{V_{x1}^2}{C_{S1}^2} = \frac{\rho_1 V_{x1}^2}{\gamma p_1}, \tag{5}$$

where γ and μ_0 are the specific-heat ratio and vacuum magnetic permeability. Here, $90^\circ - \phi_1$ is the shock normal angle, which is defined as an angle between the x-axis and the upstream magnetic field (Fig. 1). Once the upstream conditions are given, Eq. (1) can be solved for V_{x2}; the downstream conditions are obtained from the magnetic induction equation and conservation equations of mass, momentum, and energy. Since a polynomial of order three yields three roots, at most three different types of shock can exist for a given set of upstream conditions (Kennel et al. 1989). It can be found from Eqs. (1)–(5) that the shock solutions depend upon ϕ_1, M_{S1}, and $\beta_1 \equiv \beta_1 = 2\mu_0 p_1/B_1^2$.

3. Shock polar

Shock-interaction phenomena are often studied using shock polars in ordinary hydrodynamics. Shock polars might work well also in understanding MHD shock-wave interaction phenomena. In MHD, the total pressures are continuous across entropy discontinuities, so that it may be appropriate to use a "total pressure-deflection angle" shock polar. The total pressure p_T is defined as the sum of the flow static and magnetic pressures, $p_T = p + B^2/(2\mu_0)$.

The upstream tangential velocity, V_{y1}, has completely disappeared from Eqs. (1)–(5), which reflects translational invariance in the tangential direction. Thus, we work on a frame of reference where the upstream flow velocity and magnetic field are parallel, making the problem simple. In this frame of reference, the downstream magnetic field and velocity vectors are parallel as well.

As shown in Fig. 1, we denote the deflection angle of the magnetic field across the shock by θ_1. The "total pressure-deflection angle" shock polar is the graphical representation of the relationship between the downstream total pressure p_{T2} and the deflection angle θ_1, for a fixed value of the upstream 'flow' gasdynamic Mach number $M_1 = V_1/C_{S1}$. The parameter specifying a point on the shock polar is ϕ_1. The shock polar shape can be determined by the two parameters, M_1 and β_1.

It can be found that the shock polar can generally be classified into four kinds: (i) a shock polar containing slow shock only — SO, (ii) that of intermediate shock only — IO, (iii) that of fast shock only — FO, and (iv) that consisting of fast and intermediate shocks — FI. There exist no shock polars consisting of (a) fast and slow shocks — FS, and (b) intermediate and slow shocks — IS, as shown below.

The propagation speed of a slow shock is lower than the upstream intermediate speed, $V_{x1} < C_{I1}$. The inequality gives

$$M_1 < \left(\frac{\gamma\beta_1}{2}\right)^{-1/2}. \tag{6}$$

On the other hand, the propagation speeds of fast and intermediate shocks are higher than C_{I1}:

$$\left(\frac{\gamma\beta_1}{2}\right)^{-1/2} < M_1. \tag{7}$$

Since no ϕ_1 satisfies both Eqs. (7) and (8), shock polars of the kinds FS and IS do not exist.

4. Application to the regular-reflection problem

Shock polars in MHD discussed in the previous section are applied to a two-dimensional, coplanar shock-reflection problem. Our shock polars are based upon the assumption that the upstream magnetic field and velocity vectors are parallel. Such a situation is realized in the case of regular reflection (RR) on a perfectly conducting surface (Ogawa et al. 1995a) which is a coplanar plane.

4.1. Two-shock theory in MHD

The two-shock theory is an analytical model to describe the flow field in the vicinity of the RR reflection point. The wave configuration of RR is shown schematically in Fig. 2. It consists of two shocks: the incident shock i and the reflected shock r. These two shocks intersect at the reflection point R located on the reflecting surface. It is assumed that the flow field is two dimensional, and there is no magnetic field component normal to the two-dimensional plane. Then, the magnetic field and flow velocity vector ahead of the incident shock are parallel to the wedge surface, which is the case in which we can apply shock polar analyses. We choose the two-dimensional plane as the x-y-plane where the x-axis is normal to the incident shock. The wedge angle, θ_w, is the angle between the x-axis and the reflecting surface.

We attach a frame of reference to the reflection point R. The flow is deflected by θ_1 after passing through i, and changes to state (2). The flow in state (2) enters r with the incident angle ϕ_2, and is deflected by the angle θ_2, and changes to state (3). If $\theta_1 = \theta_2$ is possible, enabling the streamlines and magnetic field in state (3) to be parallel to the wedge surface, there exists a RR solution for a given state (1). Otherwise, another wave configuration prevails.

4.2. Graphical solution of regular reflection

Figure 3 represents a shock polar solution of the RR flow field. The deflection angle θ is measured with respect to the wedge surface. State (1) is the point where $p_T = p_{T1}$ and $\theta = 0°$. A shock polar (I-polar) is drawn for M_1 and β_1. The I-polar gives all the states behind an incident oblique shock. State (2) is on the I-polar. Once the incident angle ϕ_1 and shock type are given, M_2, β_2, and other physical parameters in state (2) are obtained. A shock polar (R-polar) is drawn for M_2 and β_2; note that θ is measured with respect to the wedge surface. All the states behind a reflected shock are represented by the R-polar. State (3) should be on the R-polar if it exists. Since the condition for RR to exist is $\theta_1 - \theta_2 = 0°$, state (3) is determined by the point where R-polar intersects with the $\theta = 0°$ axis. The intersection gives ϕ_2, from which the condition in state (3) is obtained.

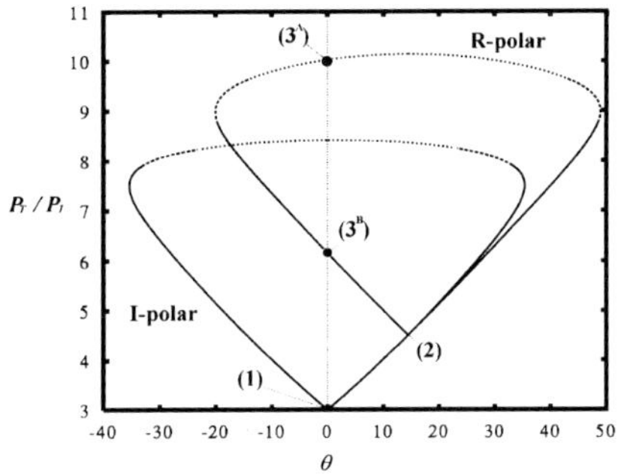

Figure 3. Shock-polar solution of regular reflection for $M_{S1} = 2.0$, $\beta_1 = 0.5$, and $\theta_w = 30°$.

4.3. Possible combinations of shocks

Out of three types of MHD shock (fast, slow, and intermediate), the number of their possible combinations are nine, five of which can be discarded for field-aligned flow fields.

Since the magnetic field and flow velocity vector are assumed parallel,

$$B_x V_{y2} - B_{y2} V_{x2} = 0. \tag{8}$$

When the incident shock is a fast shock,

$$V_2^2 = V_{x2}^2 + V_{y2}^2 > C_{I2}^2 + V_{y2}^2 = C_{A2}^2 + \frac{B_{y2}^2}{\rho_2}\left(\frac{V_{x2}^2}{C_{I2}^2} - 1\right) > C_{A2}^2, \tag{9}$$

because the shock-normal component of the flow velocity behind the incident shock is greater than the intermediate characteristic speed in state (2), $C_{I2} < V_{x2}$. This indicates that shock polars of kinds IO, FO, and FI are possible in state (2), as we have seen. Hence, intermediate and fast shocks are allowed as the reflected shock. If the incident shock is an intermediate shock or a slow shock due to the condition $V_{x2} < C_{I2}$, we have

$$V_2^2 = V_{x2}^2 + V_{y2}^2 < C_{I2}^2 + V_{y2}^2 = C_{A2}^2 + \frac{B_{y2}^2}{\rho_2}\left(\frac{V_{x2}^2}{C_{I2}^2} - 1\right) < C_{A2}^2. \tag{10}$$

Thus, only shock polars of kind SO can exist in state (2), so that a slow shock is permitted as the reflected shock in this case. Consequently, there are four possible combinations of incident and reflected shocks, as shown in Table 1.

Table 1. Possible combinations of incident and reflected shock types.

	Incident shock	Reflected shock		Incident shock	Reflected shock
1	fast	fast	3	intermediate	slow
2	fast	intermediate	4	slow	slow

5. Conclusions and related discussions

(1) The "total pressure-deflection angle" shock polars are obtained for field-aligned cases. Based on the MHD shock type included in the shock polar, shock polars can generally be classified into four kinds; FO, FI, IO, and SO.

(2) The "total pressure-deflection angle" shock polars are applied to a regular-reflection problem. Simple analyses prove that four possible combinations of incident and reflected shocks exist, as shown in Table 1. It should be noted that the field-aligned condition is assumed in the proof, which is valid for regular reflection on a perfectly conducting surface.

Acknowledgement. This research was made possible partly by support from the Science Research Fund of the Japanese Ministry of Education given to the first author, who is currently a Research Fellow of JSPS (Japan Society for Promotion of Science).

References

Ben-Dor G (1992) Shock Wave Reflection Phenomena, Springer-Verlag, New York, pp 2–5.

Jeffrey A, Taniuti T (1964) Nonlinear Wave Propagation, Academic Press, New York, pp 222–223.

Kennel CF, Blandford RD, Coppi P (1989) MHD intermediate shock discontinuities. part 1. Rankine-Hugoniot conditions. J Plasma Phys 42:299.

Ogawa H, Fujiwara T, Hayashi AK (1995a) Novel features of magnetohydrodynamic shock wave reflection from perfectly-conducting wedge: unsteady reflection and β dependence (to be published in Trans Jpn Soc Aeronaut and Space Sci).

Ogawa H, Fujiwara T, Hayashi AK (1995b) Shock polar analyses of regular reflections in magnetohydrodynamics. Phys Plasmas 2:3282.

Wu CC (1990) Formation, structure, and stability of MHD intermediate shocks. J Geophys Res 95:8149.

Nonlinear magnetohydrodynamic waves

G. M. Tarkenton
Institute for Advanced Physics
8406 Red Willow Drive
Austin, TX 78736 USA
and
Physics Department
University of Texas at Austin
Austin, TX 78712 USA

M. S. Cramer
Engineering Science and Mechanics Department
Virginia Polytechnic Institute and State University
Blacksburg, VA 24061 USA
and
Institute for Advanced Physics
8406 Red Willow Drive
Austin, TX 78736 USA

Abstract: We present a calculation of the quadratic nonlinearity parameter for small-amplitude modes of ideal magnetohydrodynamics (MHD). In so doing, we find that most of the modes have positive nonlinearity, but in one case the slow magnetosonic wave is expected to steepen backwards. We plot the nonlinearity parameter as a function of the angle between the wave vector and the ambient magnetic field, as well as the wave-normal plots showing the propagation speed as a function of angle. This work ties several earlier efforts together to present a unified treatment of these waves within the ideal MHD model.

Key words: MHD shocks, Plasmas

1. Introduction

We consider waves propagating in an ideal magnetohydrodynamic fluid and analyze the nonlinear wave speed as a function of the inclination angle between the magnetic field $\vec{B}$ and the wave vector $\vec{k}$. This subject has appeared in analyzing the earth's bow shock and its interaction with the solar wind; see Spangler and Plapp (1992). Others have considered similar systems: Kakutani and Ono (1969) analyzed weakly nonlinear waves in a cold collisionless plasma, deriving a Korteweg-deVries equation governing the dispersive wave evolution. They discovered some interesting behavior as the angle between the ambient magnetic field and the propagation direction changed. The key physical prediction was the existence of rarefaction waves in addition to compression waves. Similar phenomena were observed in laser-plasma coronae by Bezzerides et al. (1978), demonstrating experimental evidence for expansion shocks. Work in this area continued with Gedalin (1993) who extended the previous work to a hot, two-fluid plasma. He demonstrated explicit solitary wave forms in a number of special cases, generalizing the original cold plasma work of Kakutani and Ono (1969). In the present study, we treat the plasma in the magnetohydrodyan-mics (MHD) limit and derive explicit formulas for the speed of weak, obliquely propagating shock waves. We make no restrictions on the angle between the propagation direction and the background magnetic field, and we recover the shock speeds for uncoupled acoustic and Alfven waves in the appropriate limiting case and MHD waves in another.

2. Sketch of calculation

We begin with the equations of single-fluid magnetohydrodynamics and consider a static background with uniform density ρ_0 and a uniform magnetic field $\vec{B}_0$. We ignore both viscous and resistive dissipation. We establish a coordinate system such that the waves propagate in the x-direction; then the ambient magnetic field will make angle θ with respect to the x-axis. The disturbances are simple, right-running waves of small amplitude. Under these conditions, we applied the reductive perturbation technique of Taniuti and Wei (1968) to derive the nonlinear corrections to the shock speed. In the process, we obtained the well-known (c.f. Jeffrey and Tanuiti 1964) expressions for the linear wave speeds:

$$\lambda_0^2 = \frac{1}{2}\left[\left(a_0^2 + v_A^2\right) \pm \sqrt{\left(a_0^2 - v_A^2\right)^2 + 4a_0^2 v_A^2 \sin^2\theta}\right], \tag{1}$$

where a_0 is the sound speed and $v_A = \sqrt{B_0^2/4\pi\rho_0}$ is the linearized Alfven speed, B_0 being the magnitude of the background field. The positive sign in (1) corresponds to the fast wave, while the negative sign goes with the slow wave. Applying the reductive perturbation approach, we then calculated the quadratic nonlinearity parameter Γ; this gives us the first nonlinear correction to the shock speed:

$$\lambda = \lambda_0 + \Gamma\frac{u_+ + u_-}{2},$$

where u_+ and u_- are the appropriately chosen upstream and downstream (nondimensional) disturbance amplitudes, respectively. For our system, we find

$$\frac{\Gamma}{\lambda_0} = \left(\frac{v_A^2 \sin\theta}{\lambda_0^2 - a_0^2}\right) \frac{\frac{v_A^2\sin^2\theta}{\lambda_0^2 - a_0^2}\left(\frac{a_0^2}{\lambda_0^2 - a_0^2}\Gamma_0 + \frac{a_0^2}{2\lambda_0^2} - 1\right) + 2 + \frac{v_A^2}{2\lambda_0^2}}{1 + \frac{a_0^2 v_A^2 \sin^2\theta}{(\lambda_0^2 - a_0^2)^2}},$$

where Γ_0 is the nondimensional gas-dynamic steepening parameter, which for a polytrope of index γ is $\Gamma_0 = (\gamma + 1)/2$. The leading term in parentheses arises from employing the y-component of the perturbing magnetic field as the measure of the wave's amplitude. When considering the limiting behavior of certain modes, we will need to change the measure of the wave's amplitude to some other parameter, such as the density disturbance $\rho - \rho_0$; this will alter the leading factor due to the relationship among the components of the right eigenvector of the system.

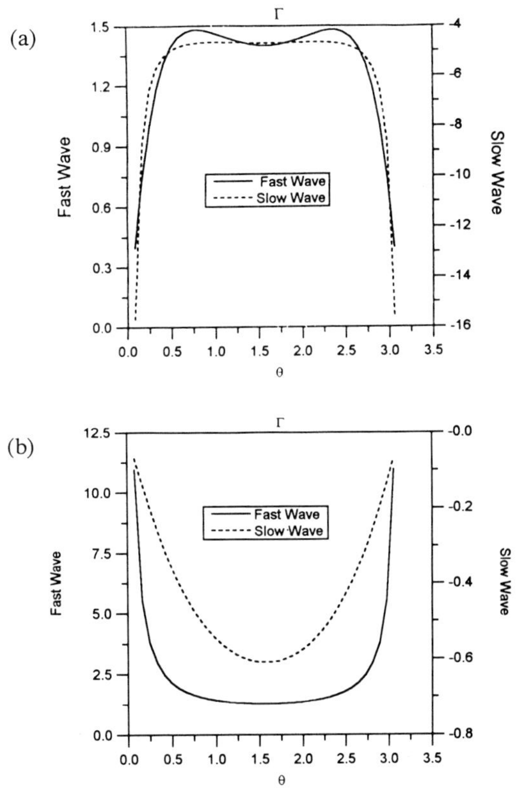

Figure 1. Scaled quadratic nonlinearity parameter Γ for the fast (solid lines) and slow (dashed lines) magnetosonic waves: (a) $a_0/v_A = 0.7$; (b) $a_0/v_A = 2.0$.

3. Discussion

In Fig. 1, we show the angular dependence of the steepening parameter for the fast and slow waves in the cases where the sound speed is less than the Alfven speed and the sound speed is greater than the Alfven speed. For the first case, the steepening parameter is positive for all angles; this means that a wave with a positive amplitude will steepen in the direction of propagation, while a negative

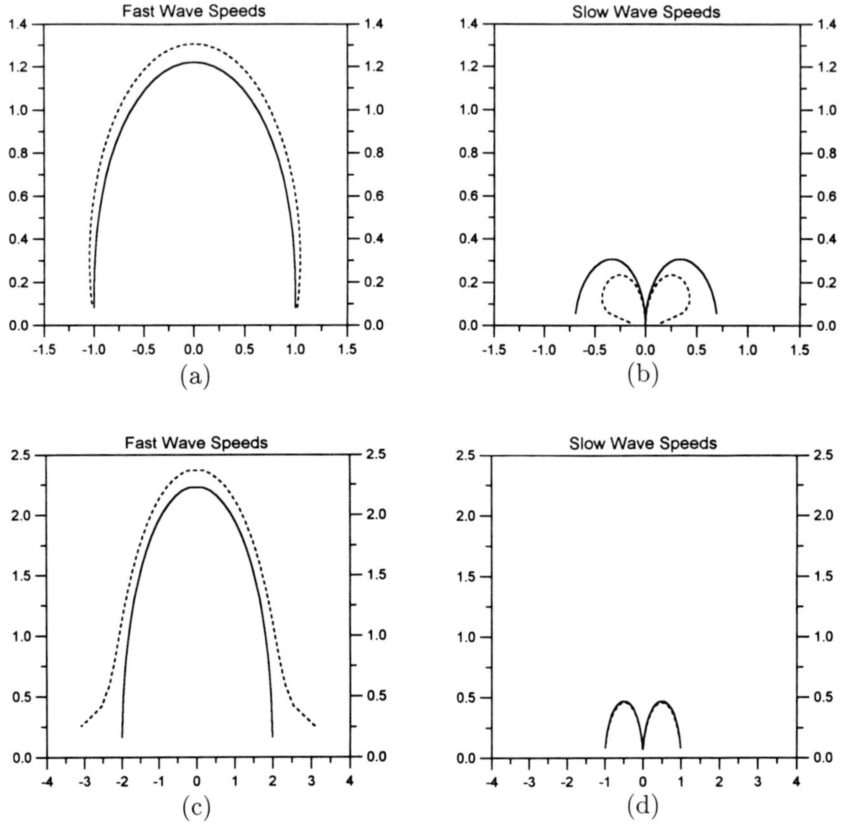

Figure 2. Wave-normal plots for the two cases shown in Fig. 1: (a) and (b) show the fast and slow waves for $a_0/v_A = 0.7$, respectively; (c) and (d) show the fast and slow waves for $a_0/v_A = 2.0$, respectively. The solid lines are the linear wave speeds and the dashed lines are the shock speeds.

amplitude will break up into a centered fan. A singularity is observed for the slow-wave mode of the first case, i.e., $a_o = 0.7v_A$. This singularity has its origin in our use of the magnetic field perturbation as a measure of the magnitude of the disturbance; in the limit of small angles, the slow wave is essentially acoustic and therefore has no magnetic amplitude. In the second case, the fast wave has essentially the acoustic nonlinearity, whereas the slow wave has a negative nonlinearity. This will steepen negative-amplitude waves into shocks, while smoothing out positive-amplitude disturbances. This is the origin of the rarefaction waves discovered by Kakutani and Ono (1969).

The next result appears in Fig. 2. We show the shock-wave normal plots (c.f. Anderson 1963) for the two cases of Fig. 1 for both the linear and nonlinear cases. In each case we used a positive shock with $u_+ = 0.1$ and $u_- = 0.0$. In the usual way these are polar graphs such that the radial distance between the origin and the curve is the wave speed scaled with the Alfven speed. The angle made by this radial line is then the angle θ between the wave vector and $\vec{B}_o$. We see that for $a_0 < v_A$, the nonlinear wave moves faster than its linear counterpart for the fast waves; this reflects the fact that $\Gamma > 0$ for these modes. For the slow mode (Fig. 2b), the values of Γ are negative and the shock speed is computed to be less than the linear wave speed. For the other case where $a_0 > v_A$, the signs of Γ for the fast and slow modes are opposite each other and therefore the shock and linear wave speeds have the same relation to each other as in the cases depicted in Fig. 2a and Fig. 2b. In the slow-wave cases, the actual shock does not satisfy the usual Lax causality (speed-ordering) conditions and therefore will disintegrate into a centered fan. On the other hand, shocks of the opposite polarity ($u^+ < 0$) are admissible and will propagate with the speed given by (2) and (3).

4. Conclusions

We have derived the first nonlinear correction to the shock speeds for the fast and slow magnetosonic waves of ideal MHD. The nonlinearity parameter we found was positive in all cases except for the slow wave with $a_0 > v_A$. This has augmented the analysis carried out by previous authors who used different models for the plasma and included dispersive effects. We have also provided a clean theoretical description of when negative nonlinearity should appear, and have shown that the first-order correction is relatively large for all obliquity angles except the small ones. In the latter case additional nonlinear effects might arise.

References

Anderson JE (1963) Magnetohydrodynamic Shock Waves, MIT Press, Cambridge, MA.

Bezzerides B, Forslund DW, Lindman EL (1978) Existence of rarefaction shocks in a laser-plasma corona. Phys Fluids 21(12):2179.

Gedalin M (1993) Nonlinear waves in two-fluid hydrodynamics. Phys Fluids B 5(7):2062.

Jeffrey A, Taniuti T (1964) Non-Linear Wave Propagation, Academic Press, New York.

Kakutani T, Ono H (1969) Weak non-linear hydromagnetic waves in a cold collision-free plasma. J Phys Soc Jpn 26(5):1305.

Spangler S, Plapp B (1992) Characteristics of obliquely propagating, nonlinear

Alfven waves. Phys Fluids B 4(10):3356.
Taniuti T, Wei CC (1968) Reductive perturbation method in nonlinear propagation. J Phys Soc Jpn 24:941.

Mechanisms of the formation of highly excited ArI and ArII level populations in shock-heated jets

T. V. Bazhenova, A. V. Eremin, A. V. Emelyanov, V. V. Shumova
High Energy Density Research Center "IVTAN" Association
Russian Academy of Sciences, 127412 Moscow, Russia

Abstract: The experimental and numerical investigation of the mechanisms responsible for the formation of highly excited ArI and ArII level populations in ionized supersonic jets is carried out. The jet parameters chosen provided nearly frozen recombination process at the jet flow. However negligible for the flow parameters, recombinative collisions lead to essential overpopulations of highly excited atomic levels and hence to the nonequilibrium radiation of the electronically excited atoms. On the other hand, such intensive radiation resulted in the depletion of low-lying optically active states and provided the appearance of inverse populations at a number of transitions ($\lambda = 8520$Å and $\lambda = 4880$Å).

Key words: Supersonic expansion, Nonequilibrium radiation, Recombination, Emission-absorption spectroscopy

1. Introduction

Expanding flows of shock-heated gases are extensively used for investigations of the kinetic processes leading to the nonequilibrium populations of atomic and molecular state formation. Of particular interest is the possibility of creating an inverse population of electronically excited states and laser generation in such flows.

The general features of recombination in expanding gas flows are commonly known. The nonequilibrium character of processes in such flows leads to the nonequilibrium population of some atomic levels and results in the appearaence of intensive radiation. It is proved that during supersonic plasma expansion, the approximately full freezing of recombination occurs. But the slight recombination accompanied by the rapid radiative cooling of some levels leads to the formation of inverse populations. Hitherto no direct observations of kinetics of recombining atoms specific levels were carried out. The inverse population of ArI levels was observed in ionized argon jets only in the near-infrared range ($\lambda = 2 \div 3$ μm) (Losev et al. 1981).

This work is devoted to the experimental and numerical investigation of the mechanisms responsible for the formation of highly excited ArI and ArII level populations and conditions of inverse-population formation in ionized supersonic quasi-stationary jets. The direct spectroscopic observation of different atomic- and ion-level concentrations and the detailed kinetic description of them is carried out.

2. Experimental set-up

The experiments were carried out in two-dimensional jets of shock-heated Ar. The experimental setup consisted of 50-mm diameter shock tube linked with a two-dimensional vacuum chamber equipped with multichannel emission-absorption spectroscopy. The direct measurements of the separate excited level populations were made at the wavelengths 8520 ± 45Å, 3520 ± 35Å, and 4880 ± 15Å at the distance 4.5 mm from the exit of slit sound nozzle with half-width $h^* = 1$ mm. Besides that, the experiments on the resonant absorption of argon-ion laser light at the wavelength 4880Å were carried out. The jet stagnation parameters were $T_0 = (7 \div 13) \times 10^3$ K, $P_0 = 1 \div 20$ atm. The impurity content of argon was less then $5 \times 10^{-3}\%$.

3. Results

Experiments performed revealed an inverse population on the $4p'[3/2] - 4s'[1/2]^0$ ArI transition for the whole range of temperatures investigated. For these conditions, the amplification coefficient k_{852} and its dependence on T_0 were for the first time measured. The k_{852} value is found to be $(0.1 \div 7) \times 10^{-2}$ cm^{-1} (Fig. 1). At the wavelength 4880Å, amplification only for temperatures 9000 $\leq T_0 \leq$ 10000 K was observed (Fig. 2). At the wavelength 3520Å, no amplification for $T_0 = 7000 \div 13000$ K was noticed.

The experimental data treatment allowed us to determine the population temperatures of specific atomic and ion levels:

$$T_i^* = -\frac{E_i}{k} \ln \left(\frac{g_0 N_i}{g_i N_0} \right), \tag{1}$$

where E_i is level energy, N_0 is total atoms or ions concentration, $g_{i,0}$ are the partial weights of levels i and 0. The values N_i are determined from the measured amplification coefficient κ:

$$\kappa = \frac{1}{l} \ln \left(\frac{I_a - I_e}{I_0} \right) = -k_\lambda P = \sigma_{eff} \cdot \left(N_2 \frac{g_1}{g_2} - N_1 \right), \tag{2}$$

where N_1, N_2 are 4p and 4s populations, and k_λ is the absorption coefficient.

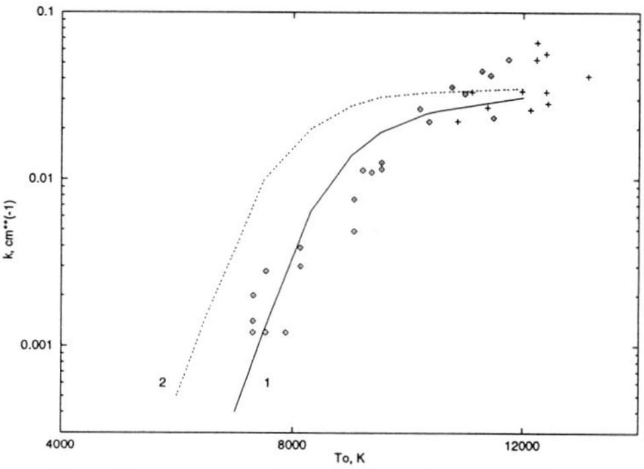

Figure 1. Experimental and computed amplification coefficient at $\lambda = 8520Å$ at the point $x = 4.5h^*$ on the jet axis over the stagnation temperature T_0. ◇, + correspond to measurements at pressures $P_0 = 5 \div 20$ atm and $P_0 = 1 \div 3$ atm respectively; 1, 2 correspond to calculations at $P_0 = 1$ and 20 atm respectively.

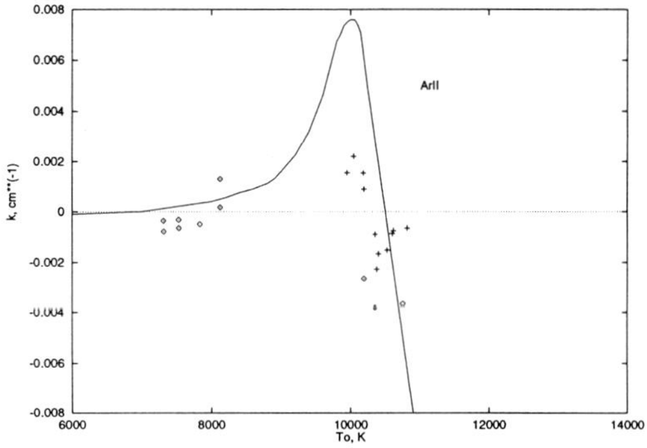

Figure 2. Experimental and computed amplification coefficient at $\lambda = 4800Å$ at the point $x = 4.5h^*$ on the jet axis over the stagnation temperature T_0. ◇ corresponds to emission-absorption, + corresponds to laser resonant measurements, — corresponds to calculation at $P_0 = 20$ atm.

In the absorption channel the signal $I_a = I_s + I_e$ was registered, where I_s is the transmitted part of the light-source signal and I_e is the emission signal. I_s is a function of the absorption coefficient k_λ:

$$I_s = I_0 \int_{line} \exp(-k_\lambda Pl) d\lambda \tag{3}$$

where I_0 is the total light-source amplitude, P is the pressure, and l is the optical path length. The emission signal I_e according to the Kirchgoff law is:

$$I_e = \int_{line} D \left(\frac{2hc^2}{\lambda^5} \right) \cdot \exp\left(-\frac{hc}{\lambda kT} \right) \cdot [1 - \exp(-k_\lambda Pl)] \tag{4}$$

The light-source intensity according to Wine's law is:

$$I_0(T_0) = D\left(\frac{2hc^2}{\lambda^5}\right)\cdot \exp\left(-\frac{hc}{\lambda k T_0}\right)\Delta\lambda_f \tag{5}$$

where T_0 is the source temperature and D is the apparatus coefficient. Using Eqs. (1)–(5) the level-population temperatures were determined and analyzed assuming that only collisions with free electrons in the jet are essential. In Fig. 3 the temperatures of 4p,4s (ArI) and 4p (ArII) levels as functions of calculated electron concentration at a fixed point of the jet are represented.

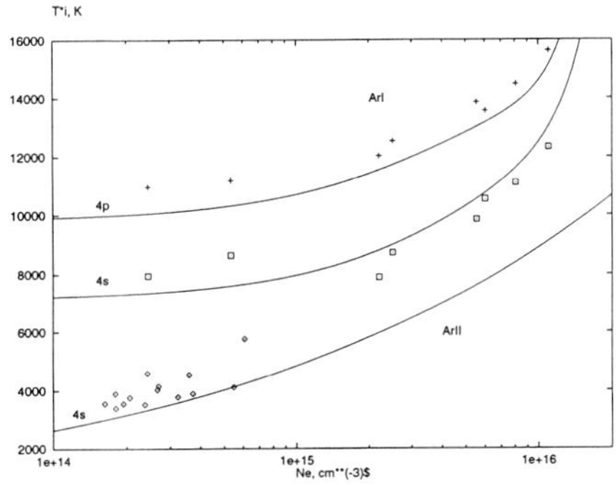

Figure 3. Temperatures of 4p,4s (ArI), and 4p (ArII) levels at the fixed point $x = 4.5h^*$ on the axis of jet over the electron concentration. +, □ correspond to emission-absorption, ◇ corresponds to laser resonant measurements, — correspond to calculations.

4. Kinetic description

The analysis of data obtained was made on the basis of a detailed kinetic description of the process. The flow parameters of a two-dimensional jet are determined by Euler's equations for the iso-entropic rarefaction of partialy ionized perfect gas.

The electron concentration in the jet was shown to be nearly frozen; the recombination was found to have only slight influence on the gasdynamical jet parameters and the specific heat coefficient γ was calculated to be constant.

The electron temperature T_e was determined from the equation of electron energy balance:

$$\frac{dT_e}{dt} = \frac{2}{5R\alpha\rho}\left(\frac{dP_e}{dt} + (Q_e)_{el} + (Q_e)_{inel} - (Q_e)_{rad}\right), \tag{6}$$

where $(Q_e)_{el}$, $(Q_e)_{inel}$, and $(Q_e)_{rad}$ are the energy lost in elastic and inelastic collisions and due to radiation. The detailed kinetic description of seven lower ArI excited states and of five ArII lower states was performed. The collisional energy exchange and radiative processes in atoms and ions as well as ionization-recombination processes were considered. The reaction $Ar^{++} + e + e \rightarrow Ar^{+} + e$ was shown to be negligible at the temperatures considered, since the Ar^{++} concentration was very low. The rate coefficients were taken from Bergman & Vainstein (1963) and Gordietz et al. (1971). It was found that the best agreement with our experimental results was the numerical simulation with the rate coefficient $k_{(3p-4s)}$ equal to

$$k_{(3p-4s)} = 1.08 \cdot 10^{-12} (\delta\varepsilon)^{-3/2} \exp(-\beta) G(\beta) \quad \mathrm{cm}^3\ \mathrm{s}^{-1} \tag{7}$$

where $\delta\epsilon$ is the energy of transition given in Rydberg units, $\beta = \Delta E/kT$, and $G(\beta)$ is the tabulated function. The value $k_{(3p-4s)}$ obtained from the extrapolation of data from Kitaeva et al. (1969) was less by a factor of 1.5 and led to shifting of calculations from the experimental data at $\approx$ 1000 K over the T_0 axis.

5. Discussion

The results of simulations, compared with the experimental data, allow us to understand the mechanism of the inversion appearance in ArI and ArII.

In Fig. 4 the computed profiles of Ar 4p′[3/2] and Ar 4s′$[1/2]^0$ concentrations along the jet coordinate x for the stagnation parameters T_0 = 11000 K, P_0 = 1.3 atm are shown. One can see that near the nozzle exit where the density of the gas is still high, the populations of both levels are increasing due to some energy gain from the recombination. At small distances from the nozzle the population of the lower level 4s is higher than that of 4p. At the distance $x/h_* = 1.2$ the collision probability is decreasing and the radiative process leads to the depletion of the excited level population. The competition of the two processes gives a different profile of the depletion of these two levels, that is, the 4s level depletion goes faster and the inverse population occurs.

The influence of the secondary ion A^{++} recombination process on the formation of ArII level populations is negligible. The most intensive depleting process in ArII is the radiative cooling of specific levels, so the only cause of the formation of inverse populations is the difference in the levels' radiative life times. The life time τ of the excited state 4s is less than that of 4p (τ(4p-4s) = 9.09 10^{-9} s, τ(4s-3p) = 3.85 10^{-10} s), which is why the inversion becomes possible.

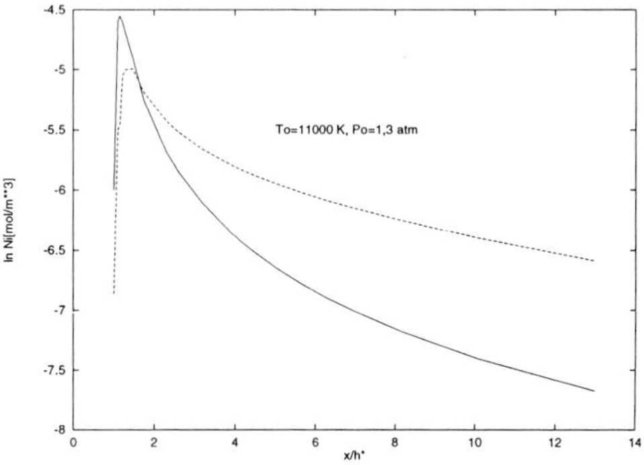

Figure 4. Computed profiles of (ArI) 4p and (ArI) 4s concentrations at the jet coordinate.

6. Conclusions

The detailed analysis of electron-excited level population kinetics, taking into account the recombinative, radiative, and energy transfer processes with the participation of free electrons, allowed the simulation of the amplification coefficient k_{852} appearance and variation with the stream coordinate. The analysis of experimentally measured ArI nonequilibrium populations revealed that this effect is mainly determined by the recombination rates and leads in particular to the inversion appearance at the 4p′[3/2]-4s′[1/2] ArI levels.

The nonequilibrium character of ArII level populations is shown to be independent of recombination and to be due to the difference of ArII excited levels' radiative life times and collisional transition rates between them. The close values of lower (4p and 4s) ArII excited-level concentrations in the supersonic jet determine the rather narrow range of jet stagnation parameters resulting in the slight inversion at this transition. The nonequilibrium population of higher ArII excited levels (4d-4p) does not result in the inversion at this transition, due to rapid 4d-level radiative cooling.

References

Losev SA, Smekhov GD, Fotiev VA (1981) Rus J Quantum Electronics 8(1):168–169.

Bergman IL, Vainstein LA (1963) Rus J Proceedings of Rus Academy of Sciences, Ser. Physical 27(8):1018–1021.

Gordietz BF, D'imova IA, Shelepin AA (1971) Russ J Applied Spectroscopy 15(2):204.

Kitaeva VF, Odintsov AI, Sobolev NN (1969) Russ J Successes of Phys Sciences 99(3):361.

Interaction between shock-generated plasma and electromagnetic field in a reentry vehicle

J. E. Leblanc, T. Fujiwara
Department of Aerospace Engineering
Nagoya University, Nagoya, 464-01 Japan

Abstract: The communication blackout experienced by reentry vehicles is a significant, still unsolved engineering problem. A simulation of shock-generated plasmas is performed, including searching for a communication window. A steady-state flow field in chemical nonequilibrium is generated with a full chemistry model of air at reentry conditions. Then a ray-tracing program searches for a window. The electromagnetic spectrum is limited by atmospheric attenuation and by the spacecraft plasma, leaving a narrow usable band. These frequencies are too low; a window has to be forced by external means. A magnetostatic field imposed on a plasma allows certain polarization modes to pass at frequencies lower than the plasma frequency. Results determine that a window at radar X-band might be possible downstream of the stagnation point with magnetic fields around 0.4 tesla.

Key words: Communication blackout, Reentry vehicle, Waves in plasma, Anisotropic plasma

1. Introduction

During reentry, the bow shock wave will generate electron densities of the order of 10^{20} e^-/m^3, enough to disrupt communications. This is known as *blackout*, a problem since the beginning of space travel and still unsolved. Much research was conducted during the 1960s, but recent literature reflects the lack of ongoing research. The February 1994 launch of the H-II rocket by the Japanese NASDA, with the Orbital Reentry Experimental Vehicle (OREX), has brought attention to the blackout problem. Steady and reliable communication is critical for the future unmanned Japanese shuttle. Our purpose (Leblanc and Fujiwara 1995) is to identify possible windows or methods to open one. The main cause of blackout is reflection or absorption of electromagnetic energy at frequencies (ω) lower than the plasma frequency (ω_p). For $\omega < \omega_p$ a plasma behaves like a conductor, while for $\omega > \omega_p$ the plasma is practically transparent. Secondary losses are collisions between electrons and ions. Some of the solutions previously proposed (Bachynski and Cloutier 1963; Rotman et al. 1964) included: (1) placing the antenna on the lee side and communicating by satellite relay; (2) seeding the flow field with electron-absorbing chemicals; (3) increasing the frequency of transmission signal; (4) using extremely low frequencies with long skin depth; and (5) applying a magnetostatic field to the plasma, allowing certain polarized modes to be transmitted. The latter method will create an anisotropic condition where certain polarization modes will travel through the plasma aided by the gyromagnetic effect. Objections are weight and power requirements, but maturing superconductor technology might provide strong magnetic fields with relatively compact systems. Liquid-hydrogen propulsion systems are likely to be used, providing the cryogenics for a superconductor magnet. We investigate this possible solution by coupling a Computerized Fluid Dynamic (CFD) program with a ray-tracing algorithm to predict the wave path through the flow field. The electron density and temperature are the inputs to the plasma simulation. Ions are not used.

2. Physics of waves in plasmas

The analysis of small-amplitude waves in plasma was applied in great detail to the communication blackout problem by Bachynski, Rotman, Sims, Papa, Hodara, and many others (Rotman et al. 1964). The following is a brief summary of the results obtained by Graf and Bachynski (1962).

Assuming a wave uncoupled from the plasma density or dielectric constants ($\nabla K = 0$), a flat and infinite boundary in the y-z plane, with x-z as the plane of incidence ($\partial/\partial y = 0$), and the z-component of the phase velocity inside the plasma the same as at the bow shock wave interface ($\partial/\partial y = -jk\sin\theta$), the wave inside the plasma is

$$\mathbf{E}(x, z, t) = \mathbf{E}_n \exp[(\alpha + j\beta)x - jkz\sin\theta + j\omega t] \tag{1}$$

where $k = 2\pi/\lambda$ and $\mathbf{E}_n$ is the vector at the interface. The attenuation coefficient $\alpha(k, \theta, K_o)$ and phase coefficient $\beta(k, \theta, K_o)$ are strong functions of the complex dielectric constant $K_o = 1 - \omega_p^2/(\omega + j\nu_c)\omega = K_r - jK_i$.

For a plasma sharply bound and with a magnetostatic field applied ($B_o \neq 0$) perpendicular to the bow shock wave (x-axis, i.e., incidence parallel to the magnetic field), the plasma dielectric characteristic is given by

$$K_{\parallel\pm} = K_o + \left(\frac{1 - K_o}{1 + f^2}\right)\left\{1 - \frac{\sin^2\theta}{2K_o} \pm \sqrt{\frac{\sin^4\theta}{4K_o^2} + \frac{f^2\sin^2\theta}{K_o} - f^2}\right\} \tag{2}$$

where $f = (\nu_c + j\omega)/\omega_b$. The magnetic field B_o alters the plasma characteristics by inducing cyclotron gyrations about the field lines of force. Two waves are identified and named Right- and Left-handed elliptically polarized modes

(R-wave and L-wave). The $(K_{\|+})$ solution can be identified as the R-wave. Similarly, $(K_{\|-})$ is the L-wave. The anisotropic plasma is birefringent; two modes will penetrate the interface, be refracted at different angles, and in general, be elliptically polarized, even if the incident wave is plane polarized. The polarization of the reflected and refracted waves can be calculated from the transmission and refraction coefficients and is a well-known diagnostic technique.

Solutions indicate the R-wave will have a pass band for frequencies lower than the first cutoff at ω_p. Since the relation among ω, ω_p, ν_c, ω_b, and θ is a complex one, a simplified diagram given by Bachynski and Cloutier (1963) for a lossless ($\nu_c = 0$), boundless plasma is recreated in Fig. 1. There is a threshold for B_o relative to ω_p. When $\omega < \omega_p$ no propagation is possible unless $\omega_b > \omega_p$, then the electrons are dominated by the magnetic field and the wave passes with little attenuation. The minimum field needed for a given frequency ν is

$$B_{min} = \frac{\omega m_e}{e} \approx 3.57 \times 10^{-11} \nu. \tag{3}$$

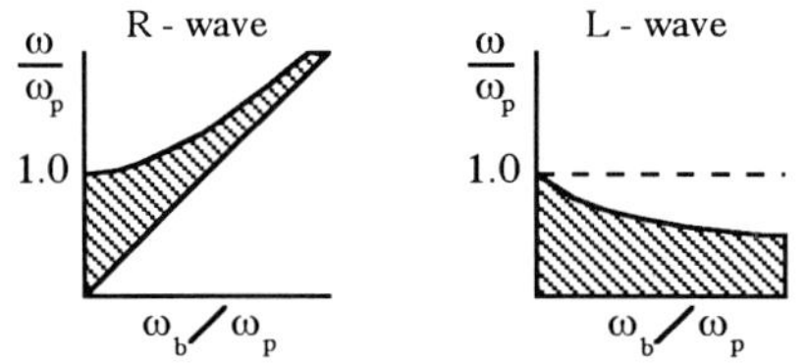

Figure 1. Pass bands and stop bands (shaded) in an anisotropic, lossless, boundless plasma.

3. Flow-field simulation

The CFD code was developed by Reddy and Fujiwara (1989). It is a Navier-Stokes code with an eleven-species, fifteen-elementary-reaction model. We generated the solution for an axisymmetric hemisphere-cone of nose radius 1.0 m and cone half angle 20° for standard air at 70 km altitude and Mach numbers $M = 35$, zero angle of attack. Figure 2 shows the radial profiles of electron density at three locations downstream from the stagnation point.

4. Blackout simulation

A ray-tracing algorithm calculates the path of the electromagnetic wave in a sharply-bound, anisotropic, heterogeneous plasma. Our hypothesis depends on the effectiveness of the gyromagnetic effect, so the plasma must be fully immersed in the magnetostatic field. The equations assume a magnetic field

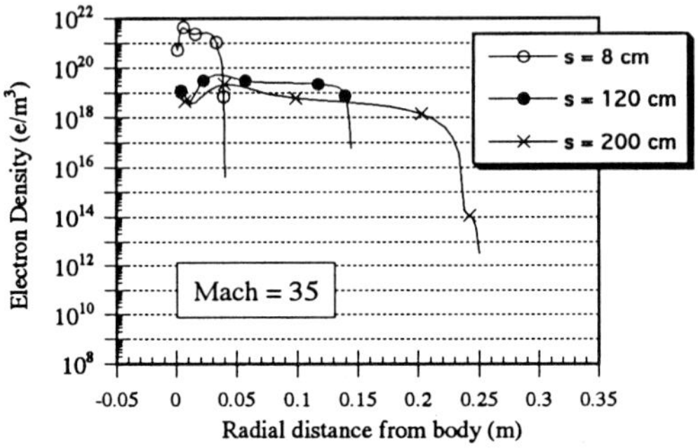

Figure 2. Radial profiles of electron density at $s = 1$ cm, 120 cm, and 200 cm downstream from the stagnation point.

orientation fixed to the interface; the physical equivalent would be to rotate the antenna and magnet together for maximum reception. The dielectric characteristics are calculated from Eq. (2).

The interface (shock wave) is assumed locally flat and discontinuous as evidenced in Fig. 2. At the boundary: (1) the wave will be partly reflected and partly transmitted, and (2) the transmitted part will be refracted in either one of two directions depending on the relative polarization. The coefficients of reflection (ρ) and transmission (τ) are functions of both angle of incidence and polarization. The initial refraction is calculated using Snell's law ($\eta_o \sin\theta = \eta_1 \sin\phi$). The plasma discretization is based on the existing CFD grid. Starting from any layer j, the ray is extended in a straight line to the intersection with the next layer $j+1$; the density gives the index of refraction η_{j+1} and the gradient gives the normal vector $\hat{\mathbf{n}} = -\nabla n_e(\xi, \eta, \zeta)$. The angle of incidence is $\theta = \cos^{-1}(\hat{\mathbf{n}} \cdot \vec{\mathbf{r}}_i)$ and Snell's law gives the refraction ϕ. The path is extended to the new layer $j+2$ and the procedure repeated. The criteria for the signal to be considered lost (maximum attenuation) is a reduction of -50 dB. The individual rays are traced until they either become too attenuated (more than -50 dB), suffer severe refraction (sideways), are reflected (cutoff), or reach the body.

5. Results and analysis

To overcome the blackout, very high frequencies are desirable, but atmospheric attenuation becomes severe beyond 40 GHz. Low frequencies are not aided by the gyromagnetic effect, so a window is not likely below VHF. Therefore, the possible frequencies are X and K_a bands. We scanned the shock at 10, 20, and 30 GHz. The magnetostatic field was oriented perpendicular to the shock, with its magnitude varied from $B_o = 0.3$ through 1.0 T, as needed to force a window open. A representative scan is shown in Fig. 3 for the 10-GHz case at Mach $= 35$; the angle varies from $\pm 40°$ in $10°$ increments. Of the three possible orientations, the only one resulting in a window for $B < 1.0$ T was the B_{ox} field. Table 1 summarizes the result for the window between 160 and 220 cm downstream, as shown in Fig. 3, for the B_x field. Once opened, all the waves were similarly attenuated, so the standard deviation for each average is small. To open a window for $\omega < \omega_p$, the field has to be larger than the critical value given by Eq. (3). As a consequence, higher frequencies demand stronger fields, which require larger and heavier magnets.

Table 1. Result of scans over the $M = 35$ shock for the window shown in Fig. 3.

	0.3 T	0.4 T	0.5 T	0.8 T	1.0 T
10 GHz	closed	−7 dB	−2 dB	—	—
20 GHz	closed	closed	closed	−4 dB	−2 dB
30 GHz	closed	closed	closed	closed	closed

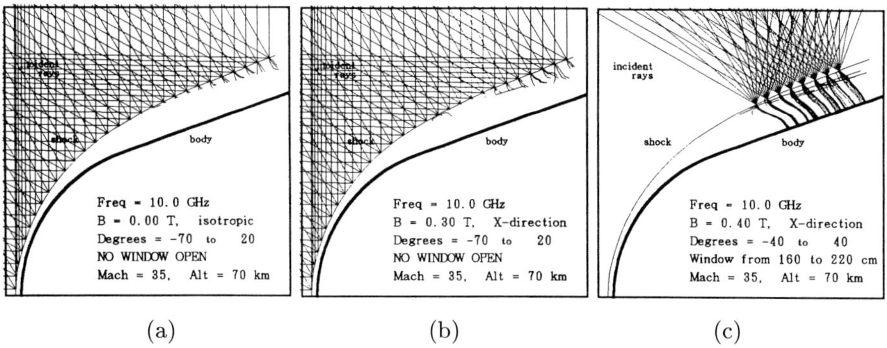

Figure 3. Result for $M = 35$, $\nu = 10$ GHz: (a) isotropic ($B = 0$); (b) weak ($B < B_{min}$); (c) open ($B > B_{min}$).

Windows have the following characteristics: (1) The angle of incidence is always within $\pm 40°$ of the normal. Angles greater than $40°$ were either reflected or severely refracted. (2) All rays follow similar paths inside the shock layer. After refraction, a ray has approximately the same direction as any other, so it encounters the same gradients. (3) Average attenuation was less than 10 dB; few rays reached the body with -40 or -30 dB.

6. Conclusions

The analysis indicates only a narrow portion of the radio spectrum is applicable to the communication blackout problem. Some of the limitations are beyond the control of technology. Useful frequencies are radar X and K_a bands. The windward plasma is too dense for those frequencies, but a window could be opened if a magnetostatic field is imposed between the antenna and the bow shock wave. Fields of about 0.4 T (4,000 gauss) are strong enough to open a window for X band (10 GHz), with very little attenuation (less than −10 dB) downstream from the nose. A minimum field strength ($B > 0.03$ T/GHz) and polarization mode (right-hand circular) are required to take advantage of electron cyclotron effects induced by the magnetostatic field. Practical limitations on the implementation of a strong magnetic field such as weight, energy requirements, and effect on surrounding systems including the antenna, still need to be addressed.

Acknowledgement. The authors wish to thank Mr. Toshichika Kawai for his generous Kombulu Scholarship.

References

Bachynski MP, Cloutier GG (1963) Communications in the presence of plasma media. In: Dynamics of Manned Lifting Planetary Entry, John Wiley & Sons, Inc., pp 206–298.

Graf KA, Bachynski MP (1962) Electromagnetic waves in a bounded, anisotropic plasma. Can J Phys 40:887–905.

Leblanc JE, Fujiwara T (1995) Comprehensive analysis of communication with a reentry vehicle during blackout phase. Trans Jpn Soc Aero Space Sci (submitted).

Reddy KV, Fujiwara T (1989) HYREFS code user's manual. In: Memoirs of the Faculty of Engineering, Nagoya University, Vol. 40, No. 1.

Rotman W, Moore HK, Papa R (eds) (1964) Electromagnetic Aspects of Hypersonic Flight: Second Symposium on the Plasma Sheath, Its Effect upon Reentry Communication and Detection, Spartan Books, Baltimore, Maryland.

Time-resolved imaging spectroscopy of nonequilibrium shock-front radiation in air

J. Koreeda,* Y. Ohama,† K. Maeno, H. Honma
Department of Mechanical Engineering, Chiba University
1-33 Yayoi-cho, Inage-ku, Chiba 263, Japan

Abstract: This paper describes the imaging spectroscopic technique applied to the experimental study of radiation emanating from strong shock waves in air. Shock velocity ranges from 9.7 to 11.6 km/s and initial pressure from 13.3 to 41.6 Pa in air. The spectral diagnostic system consists of an imaging spectrograph, a streak camera, a gated image-intensified CCD camera, and a personal computer for data acquisition/processing. This spectral diagnostic system is capable of simultaneous wavelength-, intensity-, and time-resolved spectroscopic measurements on the nanosecond order. Wavelength range is chosen primarily to investigate the first negative band of N_2^+. Spectra with enhanced spectral resolution are obtained. The emissions corresponding to N_2^+ narrowband profiles are found to be oscillatory in time.

Key words: Imaging spectroscopy, Nonequilibrium radiation, Strong shock waves

1. Introduction

Currently, there is a world-wide interest in understanding and generating high-enthalpy flows. This is due to aerospace projects related to transatmospheric vehicles, maneuvering reentry vehicles, aeroassisted orbital transfer vehicles (AOTV), and advanced space transportation-system concepts (Park 1993). High-enthalpy flows are generated around these vehicles. Because of the high temperature and low pressure, nonequilibrium high-temperature effects take place, e.g., vibrational and electronic excitation and relaxation, radiation, dissociation, ionization, etc. Great effort is being made to perform numerical computations of high-temperature flows, since it is difficult to simulate them experimentally around full-configured models. Although an experiment in a shock tube can not reproduce all the complex flow around a hypersonic vehi-

---

*Present address: INPE/DMC, Av. dos Astronautas 1758, Cx. Postal 515, S. J. Campos, SP, CEP 12201-970, Brazil.

†Present address: Kawasaki Heavy Industries Co., 3-1-1, Higashi-Kawasaki, Chuo-ku, Kobe 650, Hyogo, Japan.

cle, it is necessary for validation of numerical simulations. Furthermore, strong shock-wave experiments can provide physical constants and the basis for more accurate thermochemical models, both having direct applications in numerical simulations.

In the 1960s, many experimental and theoretical works were done to understand the natural radiation emission behind shock waves in both air and nitrogen, e.g., Avco-Everett Research Laboratory's investigations (Keck et al. 1959; Allen et al. 1961). In Avco's experiments, spectrographs, photomultipliers, and rotating-drum cameras were employed. Recently, great efforts have been made at NASA Ames to complement and to obtain more accurate experimental data in relation to Avco's results (Sharma 1991, 1993). NASA's analyses are based on time-frozen spectra obtained in a spectrograph/linear-diode-array combination. The present investigation shows results obtained by the imaging spectroscopic technique, in which two-dimensional data give simultaneous wavelength-intensity-time information. More details about the imaging spectroscopic technique and its application to shock-wave radiation measurements can be found in Koreeda et al. (1995).

2. The imaging spectroscopic technique

2.1. The spectral diagnostic system

A free-piston, double-diaphragm shock tube is used to generate strong shock waves in air (Honma 1993). Figure 1 shows the layout used to obtain the spectra of the spontaneous radiation emanating from shocks waves. The spectral diagnostic system consists of an imaging spectrograph, a streak camera, a gated image-intensified CCD camera, two focusing lenses, and personal computer/workstation for data acquisition and processing (Koreeda et al. 1995).

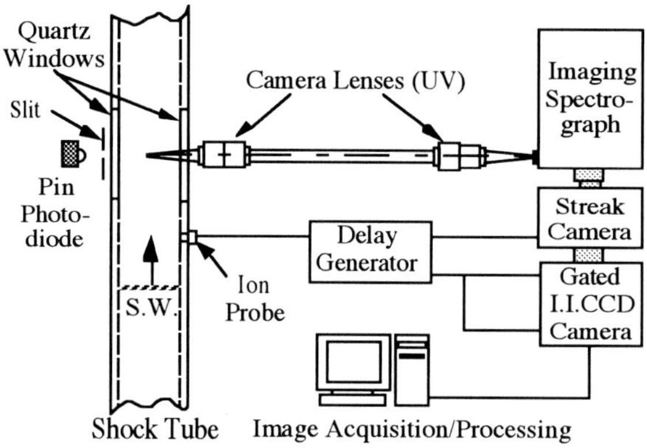

Figure 1. Layout of the spectral diagnostic system.

2.2. Experimental procedure

Depending on the spectral-diagnostic-system configuration, different experiments can be performed. By removing the streak camera and connecting the intensified CCD camera directly to the imaging spectrograph, and varying the delay and gating times, time-frozen measurements are performed. With the streak camera, as shown in Fig. 1, simultaneous wavelength-intensity-time measurements are made.

2.3. Spectral, spatial, and temporal resolutions

The grating plates of the spectrograph used in the present experiments are of 600 and 1200 grooves/mm. The spectral resolutions are 0.043 and 0.021 nm/pixel (pixel of the CCD camera), respectively. The spatial resolution in the direction parallel to the shock tube is determined by the spectrograph entrance slit width (100 μm). The temporal resolution of the streak camera depends on the horizontal slit width in its entrance optics and the sweep speed. Setting the slit narrower for a given sweep speed results in an improved time resolution. However, the signal/noise ratio deteriorates. For a given slit width, the temporal resolution improves if faster sweep speeds are used, and as in the first case, the signal/noise ratio deteriorates. In the present experiments, several combinations are considered, leading to temporal resolutions from 67 to 533 ns.

3. Time-resolved imaging spectroscopy

By using the streak camera, a two-dimensional image containing wavelength, intensity, and time information is obtained: the streak image. Imaging spectroscopic analysis can be performed in different ways. For example, the contour plot of a typical streak image is shown in Fig. 2. The shock velocity is $u_s = 11.2$ km/s with corresponding Mach number $M_s = 32.4$, at initial pressure $p_1 = 14.7$ Pa in air. The spectral and sweep-time resolution are 0.043 nm/pixel and 0.025 μs/pixel, respectively. Due to the streak-camera slit width, in this case 400 μm, there is a temporal intensity spreading over approximately 500 ns. For the same sweep-time resolution but with a 100-μm streak slit, the temporal intensity spreading falls to 130 ns, though the image signal/noise ratio deteriorates. The 339.9-nm – 366.3-nm wavelength range is chosen to investigate the second positive band of N_2, including $N_2(2+)(0,1)$ and $N_2(2+)(1,2)$, and the first negative band of N_2^+, including $N_2^+(1-)(1,0)$, $N_2^+(1-)(2,1)$, and $N_2^+(1-)(3,2)$. From the contour plot, we note a larger relaxing time of $N_2^+(1-)(1,0)$ at 358.2 nm, compared to other spectral lines. Three-dimensional plots from the same spectral data of Fig. 2 are shown in Fig. 3. In these plots the time origin is chosen to correspond to the peak of radiation emission integrated in the 339.9-nm – 366.3-nm range. Each spectrum is the result of a 100-ns integration time to simulate time-frozen spectra. Streak images corresponding to molecular elec-

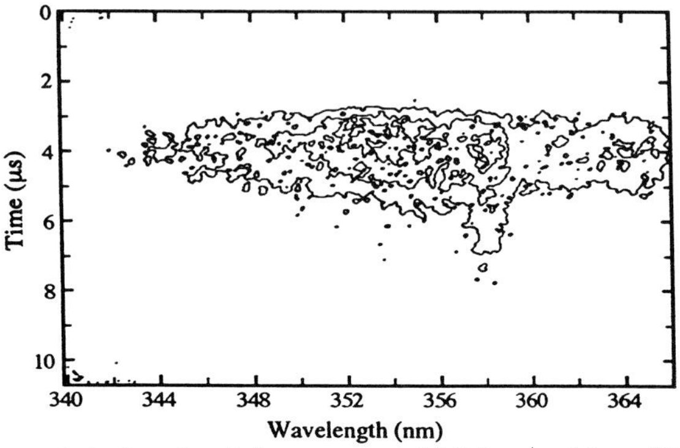

Figure 2. Contour plot of a streak image; $u_s = 11.2$ km/s, $M_s = 32.4$, $p_1 = 14.7$ Pa in air. Sweep time = 0.025 μs/pixel, spectrograph slit = 100 μm, streak camera slit = 400 μm, spectral resolution = 0.043 nm/pixel.

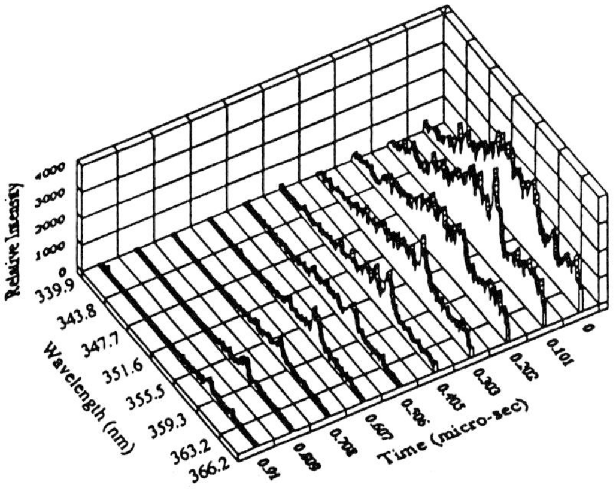

Figure 3. Spectral behavior behind the peak of radiation. Same experimental parameters as in Fig. 2.

tronic transitions with temporal profiles larger than 1 μs are observed. We call them "large" streak images. Figures 2 and 3 correspond to a "large" streak image. However, "slim" streak images are also obtained, in which the temporal profiles are within 1 μs. The spectrum corresponding to a "slim" streak image is shown in Fig. 4. Here, measurements of $N_2^+(1-)$ are obtained with enhanced resolution. Other similar time-resolved measurements are not available, but the results can be compared to time-frozen measurements. In Fig. 4, the spectrum corresponding to the peak of radiation is time-integrated in 200 ns for comparison with Sharma's experiment (1991). A sharper spectrum is obtained in the present work. Furthermore, the present spectra are time resolved.

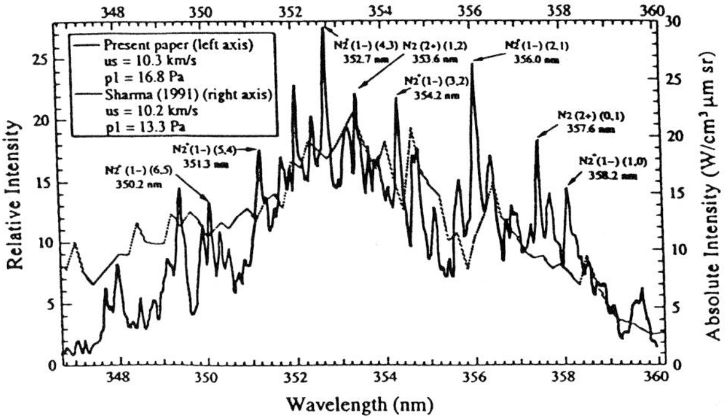

Figure 4. Nonequilibrium radiation spectrum corresponding to the total emission peak, processed from a streak image; $u_s = 10.3$ km/s, $p_1 = 16.8$ Pa in air. Integration time = 200 ns, spectral resolution = 0.021 nm/pixel, and temporal broadening = 133 ns.

4. Time-oscillatory behavior of molecular transitions

The narrowband profiles (monochromator-like) corresponding to transitions of $N_2^+(1-)$ (1,0), (2,1), (3,2), (4,3), and (5,4), with respective wavelengths 358.2 nm, 356.0 nm, 354.2 nm, 352.7 nm, 351.3 nm, are shown in Fig. 5. The profiles are obtained by integrating the emission intensities over a 0.4-nm wavelength interval. An oscillating behavior is observed in these emissions. A possible explanation of this phenomenon can be attributed to temporal oscillations of the vibrational populations. In Fig. 5, it can be also observed that when the transitions (1,0) and (3,2) have local peak values, at the same time, the transitions (2,1) and (4,3) approximately have local minimum values and vice-versa.

5. Concluding remarks

A spectral diagnostic system is set up to investigate the radiation emission of strong shock waves in air. The imaging spectroscopic technique allows, for the first time, simultaneous measurements of wavelength, intensity, and time of the nonequilibrium shock-front radiation. "Large" and "slim" streak image types are observed. In the "large" streak images, greater contribution from the first negative band of N_2^+ is observed. A time-oscillating behavior of $N_2^+(1-)$ (1,0), (2,1), (3,2), (4,3), and (5,4) is observed. This phenomenon is for the first time observed in strong shock-front measurements.

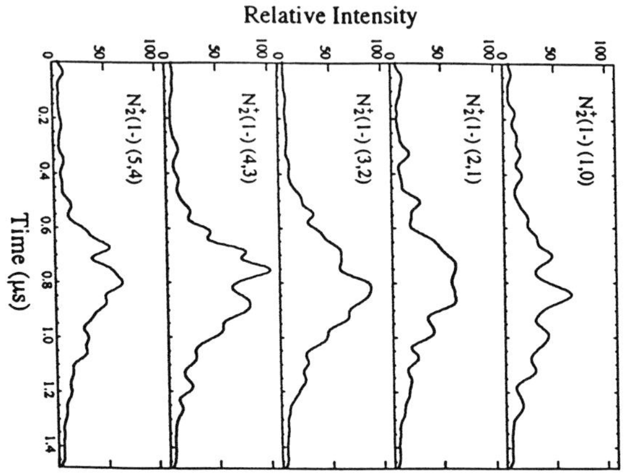

Figure 5. Oscillating behavior of $N_2^+(1-)$ emissions; u_s = 10.3 km/s, p_1 = 16.8 Pa in air, integration time = 200 ns, integrated spectral bandwidth of each profile = 0.4 nm, spectrograph slit = 100 μm, spectral resolution = 0.021 nm/pixel, temporal broadening = 133 ns.

Acknowledgement. The financial assistance received from the Japan Society for the Promotion of Science and the Ministry of Education, Science and Culture of Japan under Grant-in-Aids for Scientific Research 05402052 and 05302051 is acknowledged with thanks.

References

Allen RA, Camm JC, Keck JC (1961) Radiation From Hot Nitrogen, Avco Everett AFBSD-TR-61-2, Research Rep 102.

Honma H (1993) Shock tube studies of strong shock waves. In: Proc Intl Symp on Aerosp and Fluid Science, Kamiyama S (ed), Sendai, Japan, pp 128–148.

Keck JC, Camm JC, Kivel B, Wentink T Jr (1959) Radiation from hot air. Part II: Shock tube study of absolute intensities. Annals of Physics 7:1–38.

Koreeda J, Ohama Y, Honma H (1995) Imaging spectroscopy of the nonequilibrium shock front radiation in air. Shock Waves (to be published).

Park C (1993) Review of chemical-kinetic problems of future NASA missions, I: Earth entries. J Thermophysics and Heat Transfer 7(3):385–398.

Sharma SP (1991) Research activity at the shock tube facility at NASA Ames. In: Proc Intl Workshop on Strong Shock Waves, Honma H, Maeno K (eds), Chiba, Japan, pp 1–21.

Sharma SP (1993) Research on nonequilibrium phenomena at Ames EAST facility. In: Shock Waves @ Marseille, Brun R, Dumitrescu L (eds), Marseille, France, pp 241–250.

Multi-phase Flows

Measurements of void fraction in a vapor explosion by a light-extinction technique

M. El-Beshbeeshy*, R. Bonazza, B. Shamoun
Department of Nuclear Engineering & Engineering Physics
University of Wisconsin
Madison, WI 53706 USA

Abstract: Experiments consisting of the pouring of hot water onto a pool of liquid refrigerant R12 are performed. Vapor explosions are observed when pouring boiling water onto a pool of refrigerant R12, both without and with a flow of air in the refrigerant to increase its initial void fraction. A light-extinction technique is used to measure the void fraction just before the pouring of the water.

Key words: Vapor explosions, Void fraction, Light extinction, Flow visualization

1. Introduction

Direct contact between a hot and a cold, highly volatile liquid causes the cold liquid to undergo very fast evaporation; if the evaporation time scale is much shorter than the time it takes the pressure build-up to propagate through the media, a pressure discontinuity can actually form. The risk of explosive interaction is important in the development of a technique, called catalytic extraction process (CEP), for the disposal of toxic chemical waste. The process consists of the injection of a room-temperature aqueous solution of toxic, organic compounds into a pool of molten iron; the large temperature rise and the presence of the iron induce a chemical cracking of the toxic solutes into their harmless components. The need to assess the danger of an explosive fuel-coolant interaction (FCI) and the possibility of suppressing the tendency to spontaneous explosions by increasing the void fraction of the mixture is the motivation behind the present work.

The present experiments are a cryogenic simulation of the real process, with water and liquid refrigerant R12 (freon) in place of the molten metal and the room-temperature aqueous solution, respectively. In these experiments a stratified, rather than injection, geometry is used, with the water being poured on top of the freon. The purpose of the project is to measure the void fraction

*Present address: Department of Mechanical Engineering, University of Alexandria, Egypt.

prior, during, and after the FCI by means of a light-extinction technique and to investigate the existence of a void-fraction threshold for no explosion.

2. Experimental apparatus

A schematic of the test section is presented in Fig. 1. The test chamber has a rectangular cross section, 205 × 25 mm. The bottom and the side walls of the chamber are machined out of phenolic resin, the front and back walls are machined out of Lexan. Liquid refrigerant (R12 at 243 K) is collected at the bottom of the chamber. Two nozzles with orifices flush with the bottom of the test section allow for the injection of air into the cold liquid in order to vary the initial value of the void fraction. In the right wall is a slot, closed by a sliding gate controlled by an air cylinder; on the other side of the slot is a tank, also made of phenolic resin, where water at 330 K is collected just prior to a run. When the slide gate is lifted water pours over the cold liquid, thus creating a stratified geometry. The cold liquid starts to evaporate and a vapor explosion occurs causing a shock to propagate through the vapor film layer between the two liquids. Three piezoelectric transducers located in the bottom of the chamber (indicated by p_1, p_2, p_3 in Fig. 1) measure the pressure upon passage of a shock. A schematic of the optical setup is presented in Fig. 2. A

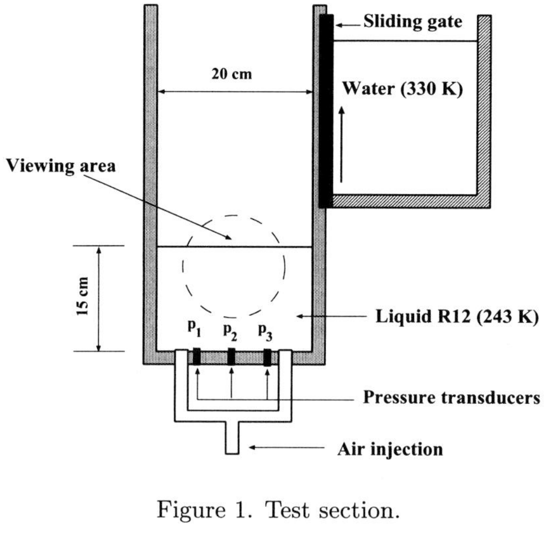

Figure 1. Test section.

2-W, CW argon ion laser ($\lambda = 514$ nm) produces a beam of 2.1-mm diameter; a plano-concave lens of focal length 5 mm amplifies the beam diameter to 115 mm. A plano-convex lens of focal length of 500 mm and clear aperture of 115 mm collimates the beam which then traverses the test section through the back

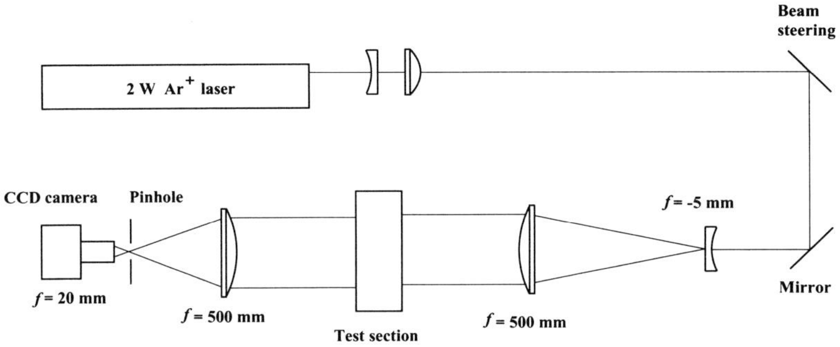

Figure 2. Optical setup.

and front walls. A second plano-convex lens, identical to the first one, focuses the beam; a 1-mm pinhole is located at the focal point of the second lens. A 20-mm, $f = 2.8$ lens images the beam onto the 256×256-pixel array of a CCD camera. The circular beam is imaged so as to fit entirely on the sensor array, thus each pixel images a square of 445 μm per side.

The output of the sensor array is digitized at a speed of 200 fps, with an 8-bit precision. Sequences of 31 images are acquired, digitized, and stored.

3. Experimental technique and image processing

The present technique is based on the theory of light extinction which describes the attenuation of the intensity of a beam of monochromatic light traversing a volume of the two-phase fluid (Ohba & Itoh 1978). The scattering of a plane, monochromatic wave of wavelength λ incident on a single, homogeneous sphere of diameter D of arbitrary material is governed by the Mie equations (Born & Wolf 1980). Since in the present case the scatterers have diameters ranging from a few hundreds of microns to few millimeters and the light wavelength is 514 nm, the Mie scattering cross section, σ_s, of the bubbles can be evaluated in the limit $(\pi\lambda/D) \ll 1$. When a light beam traverses a volume containing a dispersion of scatterers of constant diameter D (monodisperse particles) the ratio of the transmitted to incident light intensity (transmittance) follows $\tau = \exp(-\sigma_e nl)$ where σ_e is the extinction cross section of one scatterer, n the number density of the scatterers, and l the length of the beam's optical path through the volume. In the present case $\sigma_e = \sigma_s = \sigma$ because of negligible absorption. The cross section σ can be related to an extinction coefficient Q through $\sigma = (Q\pi/4)D^2$. For a polydisperse bubble distribution, the average extinction cross section, $\overline{\sigma}$, accounts for the fact that the beam encounters scatterers of different diameter (and therefore different cross section) along its path. The average extinction cross section can be calculated from

$$\overline{\sigma} n l = \sum_{i=1}^{N_D} n_i \sigma_i \tag{1}$$

where the index i scans the diameter ranges. In this case the transmittance is given by

$$\tau = \frac{I}{I_0} = \exp(-\overline{\sigma} n l) , \tag{2}$$

so that

$$\ln \tau = -\sum_{i=1}^{N_D} n_i Q_i \frac{\pi}{4} D_i^2 l . \tag{3}$$

Equation (3) needs to be modified by the introduction of a correction coefficient R_i (Gumprecht & Sliepcevich 1953) that accounts for scattered light reaching the CCD sensor. The corrected expression for the transmittance is then

$$\ln \tau = -\sum_{i=1}^{N_D} n_i R_i Q_i \frac{\pi}{4} D_i^2 l . \tag{4}$$

Introducing $\overline{R}\,\overline{Q} = \sum_{i=1}^{N_D} n_i R_i Q_i D_i^2 / \sum_{i=1}^{N_D} n_i D_i^2$ and $\overline{D}_2 = \left(\sum_{i=1}^{N_D} n_i D_i^2 / \sum_{i=1}^{N_D} n_i \right)^{1/2}$ one can write

$$\ln \tau = -\overline{R}\,\overline{Q}\,\overline{D}_2^2 n \frac{\pi}{4} l . \tag{5}$$

The product $n\pi\overline{D}_2^2$ has units of $1/L$ and represents the interfacial area distribution, here indicated with γ. From Eq. (5),

$$\gamma = \frac{-4 \ln \tau}{\overline{R}\,\overline{Q}\, l} ; \tag{6}$$

γ can therefore be measured by measuring τ and by calculating the product $\overline{R}\,\overline{Q}$. For the present diameter values $Q_i \approx 2$ and $R_i \approx 1$ for all diameters.

The other quantity of interest is the void fraction given by

$$\alpha = \sum_{i=1}^{N_D} n_i \frac{\pi}{6} D_i^3 . \tag{7}$$

In the present work only γ can be measured directly and an empirical $\gamma - \alpha$ correlation (Veteau 1981) is used, *viz.*

$$\gamma = 10\alpha^{2/3} \tag{8}$$

where γ is in cm^{-1}.

Since the CCD output is linear in light intensity, the transmittance can be evaluated as

$$\tau(x, y) = \frac{p(x, y) - p_d}{\overline{p}_0(x, y) - p_d} ; \tag{9}$$

where $p(x, y)$ and $p_0(x, y)$ are the image and background pixel value distributions and $p_D = 11$ is the CCD's dark noise.

4. Results

Figure 3 shows the histograms of the void-fraction distributions of R12 refrigerant injected with air at the rate of 630 cc/s and with no air injection. An explosive interaction was observed in both cases; the corresponding pressure traces are shown in Fig. 4. The peak pressure in the case of freon without air injection is about twice as high as that for the case of air-injected freon. Conversely, the time integral of the pressure traces (essentially the impulse consequent to the interaction) from the middle transducer is 5.1 MPa·s for air-injected freon and 2.6 MPa·s for the case with no air injection. The pressure

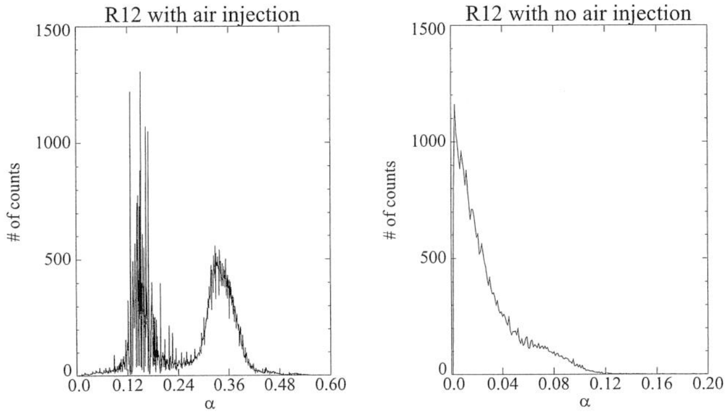

Figure 3. Histograms of the void-fraction distribution.

traces also show multiple wave reverberations in the case of no air injection and a constant pressure decrease, right after the peak, in the case of air injection prior to the fuel-coolant interaction; this is expected from the intense diffraction and attenuation the shock wave undergoes in a porous medium like the air-injected freon.

Measurements of the void fraction after direct contact between the cold liquids and the water were hindered by the formation of a thick fog of extremely small and numerous bubbles and droplets: the intensity of the light transmitted through the fog was always below the sensitivity of the CCD's sensor.

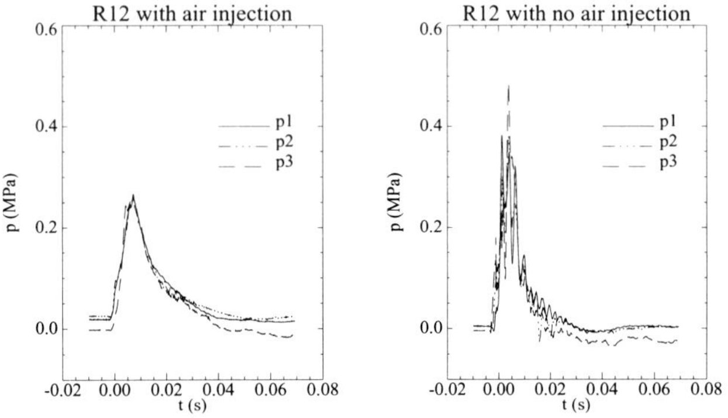

Figure 4. Pressure traces.

5. Conclusions

A new experimental technique has been used to measure the void fraction in a cold liquid prior to interaction with a hot fuel and to determine its effect on the likelihood of a spontaneous vapor explosion. Results show that, even with a pre-interaction average void fraction of about 30%, a vapor explosion will occur when water is poured onto liquid R12. The peak pressure decreases while the time integral of the pressure increases when the pre-interaction void fraction is increased by air injection. More work is needed to successfully apply the technique to measure void fraction during and just after the fuel-coolant interaction.

Acknowledgement. The authors gratefully acknowledge Molten Metal Technology of Waltham, MA, for the support for this program.

References

Born M, Wolf E (1980) Modern Optics, Pergamon Press, Oxford, p 633.

Gumprecht RO, Sliepcevich CM (1953) Scattering of light by large spherical particles. J Phys Chem 57:90.

Ohba K, Itoh T (1978) Light Attenuation Technique for Void Fraction Measurement in Two-Phase Bubbly Flow; Parts I, II, III, Tech Rep Osaka Univ 28:487.

Veteau (1981) Contribution à l'étude des techniques de mesure de l'aire interfaciale dans les écoulements à bulles, Ph.D. Thesis, Institut National Polythechnique, Grenoble, France.

A hybrid front-tracking scheme for shock wave propagation in gas/water media

D. M. Causon
Centre for Mathematical Modeling and Flow Analysis
The Manchester Metropolitan University
Manchester M1 5GD UK

Abstract: A hybrid front-tracking algorithm is presented for applications to shock-wave propagation in gas/gas and gas/water media. The hybrid scheme, which combines a Godunov-type scheme with the random-choice method, resolves shock waves and rarefactions accurately and maintains zero-width material fronts. The method is being used to study shock-wave interaction with a gas bubble in water.

Key words: Front-tracking, Shock-wave propagation, Multi-component flows, Material interfaces

1. Introduction

Euler calculations using conservative variable schemes exhibit strong oscillations around material interfaces. These are not the well-known Gibbs oscillations commonly associated with classical shock capturing schemes; they are due to non-physical pressure fluctuations which would occur even with a monotone scheme. The problem does not arise in single-component flow calculations.

One modeling approach for multi-component flows is to adopt a 'multi-phase' formulation with individual species equations and Dalton's law of partial pressures with appropriate equations of state. Alternatively, one can convect an initially piecewise-constant profile of the ratio of specific heats, γ, at the local flow velocity. The dissipated profile provides an 'effective' intermediate value of γ at each point on the material interface. These approaches will be discussed in Section 2 and in Section 4 it will be shown that neither approach yields oscillation-free profiles, particularly in the more demanding gas/water cases. Recently, Karni (1994) proposed a new primitive variable algorithm. For gas/gas problems in one spatial dimension, this algorithm yields oscillation-free solution profiles but incurs conservation errors which limit it to shock waves of weak to moderate strength. Here, we describe a hybrid conservative/primitive variable Euler method which retains conservation and is applicable to both gas/gas and gas/water media.

2. Models for multi-component flows

For clarity of exposition, we restrict attention to two-component shock-tube flows in one space dimension on the interval $0 \leq x \leq 1.0$, assuming a diaphragm located at $x = 0.5$ separating two different ideal gases.

We solve the Euler equations,

$$\frac{\partial \mathbf{U}}{\partial t} + \frac{\partial \mathbf{F}(\mathbf{U})}{\partial x} = 0 \text{ with } \mathbf{U} = \begin{bmatrix} \rho \\ \rho u \\ e \end{bmatrix}, \quad \mathbf{F}(\mathbf{U}) = \begin{bmatrix} \rho u \\ \rho u^2 + p \\ (e+p)u \end{bmatrix}, \tag{1}$$

in conservative variable form, where ρ is the density, u is the velocity, and e the total energy per unit volume respectively. The equations are closed with the Tammann equation of state (Tammann 1912)

$$p = \left(e - \frac{\rho u^2}{2} - P_c\right)(\gamma - 1) - P_c \tag{2}$$

where for air $\gamma = 1.4$, $P_c = 0$, and for water $\gamma = 7$, $P_c = 3000$ atm. We now consider three simple conservative-variable models for multi-component flow calculations.

Model I (multi-phase approach)

This model treats a multi-component flow problem using a multi-phase approach with individual species equations for the mass conservation of each component. The full equation set is

$$\begin{bmatrix} \rho_1 \\ \rho_2 \\ \rho u \\ e \end{bmatrix}_t + \begin{bmatrix} \rho_1 u \\ \rho_2 u \\ \rho u^2 + p \\ (e+p)u \end{bmatrix}_x = 0, \tag{3}$$

where $\rho = \rho_1 + \rho_2$ and Dalton's law of partial pressures is used to calculate the pressure at each point (after numerical mixing occurs).

Model II (advected γ, P_c profiles)

Here, we obtain an 'effective γ and P_c' at each point by including an advective transport equation for γ and P_c in the equation set

$$\begin{bmatrix} \rho \\ \rho u \\ e \\ \rho\gamma \\ \rho P_c \end{bmatrix}_t + \begin{bmatrix} \rho u \\ \rho u^2 + p \\ (e+p)u \\ \rho u \gamma \\ \rho u P_c \end{bmatrix}_x = 0, \tag{4}$$

where the initial profiles for γ and P_c are assumed to be piecewise constant.

Model III (front-tracking only)

The third model uses an advection equation for a progress variable ϕ which propagates at the local flow velocity and tracks the material interface. A suitable initial distribution for the progress variable ϕ on the range $0 \leq x \leq 1.0$ (with the diaphragm located at $x = 0.5$) is $\phi(x, 0) = x - 0.5$. The position of the material interface at any later time corresponds to the point at which ϕ crosses the x-axis. In contrast to the scheme described next, the above models make no special provision for the accurate resolution of the material interface.

3. A hybrid front-tracking method

The author's hybrid scheme combines a high-resolution shock-capturing finite-volume scheme with a multi-component random-choice method (RCM). The position of the material interface is obtained by tracking the front in the same way as in model III of Section 2. The idea is to capitalize on the ability of the RCM to maintain zero-width material interfaces while retaining the good shock capturing properties of a Godunov-type scheme. Accordingly, the two schemes are as follows:

(i) A two-step MUSCL Hancock scheme (Van Leer 1984) the first stage of which is

$$\mathbf{U}_i^{n+\frac{1}{2}} = \mathbf{U}_i^n - \frac{\Delta t}{2\Delta x}(F(\mathbf{U}_{i+\frac{1}{2}}^n) - F(\mathbf{U}_{i-\frac{1}{2}}^n)), \tag{5}$$

where the flux vector functions are evaluated at the mid-points of cell faces using a linear reconstruction of the flow solution within each mesh cell, i.e.

$$\mathbf{U}_{i\pm\frac{1}{2}}^n = \mathbf{U}_i^n \pm \frac{1}{2}\delta\mathbf{U}_i^n, \text{ where } \delta\mathbf{U}_i^n = B(\mathbf{U}_{i+1}^n - \mathbf{U}_i^n, \mathbf{U}_i^n - \mathbf{U}_{i-1}^n),$$

$$\text{and } B(x, y) = \begin{cases} \max[0, \min(kx, y), \min(x, ky)] & \text{if } x > 0 \\ \min[0, \max(kx, y), \max(x, ky)] & \text{otherwise} \end{cases}$$

is a slope limiter such that $k = 1$ corresponds to the well-known minmod limiter and $k = 2$ is the superbee limiter. We use $k = 1.5$. The second step is

$$\mathbf{U}_i^{n+1} = \mathbf{U}_i^n - \frac{\Delta t}{\Delta x}\left(F(\mathbf{U}_{i+\frac{1}{2}}^*) - F(\mathbf{U}_{i-\frac{1}{2}}^*)\right) \tag{6}$$

where the interface fluxes $F(\mathbf{U}_{i+\frac{1}{2}}^*) = F(\mathbf{U}_i^{n+\frac{1}{2}} + \frac{1}{2}\delta\mathbf{U}_i^n, \mathbf{U}_{i+1}^{n+\frac{1}{2}} - \frac{1}{2}\delta\mathbf{U}_{i+1}^n)$ are found by solving a local Riemann problem $R(\mathbf{U}_L, \mathbf{U}_R)$ with left/right states $\mathbf{U}_L, \mathbf{U}_R$ respectively. We use an approximate Riemann solver due to Harten et al. (1983). The HLL interface flux can be written as

$$F(U^*) = \begin{cases} F_L & \text{if } S_L > 0 \\ F_{LR} & S_L \leq 0 \leq S_R \\ F_R & \text{if } S_R < 0 \end{cases}, \tag{7}$$

where S_L, S_R are wave speed estimates, $F_L = F(U_L), F_R = F(U_R)$ and

$$F_{LR} = \frac{S_R F_L - S_L F_R + S_L S_R (U_R - U_L)}{S_R - S_L}. \tag{8}$$

Our wave speed estimates S_L, S_R are

$$S_L = \min(u_L, u_R) - c_{max},\ S_R = \max(u_L, u_R) + c_{max},$$

where $c_{max} = \max(c_L,\ \bar{c},\ c_R)$ and $c_{L,R} = \sqrt{\gamma(p_{L,R} + P_C)/\rho_{L,R}}$, $\bar{c} = \sqrt{\gamma\bar{p}/\bar{\rho}}$, $\bar{\rho} = \frac{1}{2}(\rho_L + \rho_R)$, $\bar{p} = (\bar{e} - P_C)(\gamma - 1) - P_C$ and $\bar{e} = \frac{1}{2}(e_L + e_R)$.

(ii) A multi-component random-choice method based on Glimm's method (Glimm 1955) in which a sequence of Riemann problems is solved exactly, i.e.,

$$\overline{\mathbf{U}}_i^{n+1} = \tilde{\mathbf{U}}_i(t_n + \Delta t,\ x_i + \theta_n \Delta x) \tag{9}$$

where $\overline{\mathbf{U}} = [\rho, u, p]^T$ and $\tilde{\mathbf{U}}_i$ is the solution of two Riemann problems $R(\mathbf{U}_{i-1}^n, \mathbf{U}_i^n)$ and $R(\mathbf{U}_i^n, \mathbf{U}_{i+1}^n)$ at $t = t_n + \Delta t, x_{i-\frac{1}{2}} \leq x \leq x_{i+\frac{1}{2}}$ within cell i. Here, θ_n is a random number on the interval $[0,\ 1]$ obtained via van der Corput sampling. The MUSCL-Hancock scheme is applied throughout the flow field except at cells bounding the known location of the material interface; these are updated using the random-choice method.

4. Results

Numerical experiments involving single-γ gas, two-γ gas, gas/water shock-tube problems and shock collision problems in 1D and 2D show that the method is robust and maintains zero-width material interfaces. Example calculations presented in Figures 1 – 3 relate to a two-γ gas version of the well-known Sod shock-tube problem with $\gamma_L = 1.667$, $\gamma_R = 1.2$ and two gas/water problems. The calculations were performed with a Courant number $v = \Delta x/|u + c| = 1.0$ and 160 cells. The strong oscillations found in the conservative variable solutions are eliminated. The method is currently being used to study shock-wave focusing in water with applications to extracorporeal shock-wave lithotripsy. Here, cavitation bubbles are known to appear in the focal region of the lithotripter (Takayama 1993). The bubbles collapse asymmetrically giving rise to a high-speed liquid microjet which can cause local hemorrhages. It is hoped to use the hybrid method to study the interaction between a shock wave and an air bubble in water. Initial results are encouraging (Figs. 4 and 5).

References

Glimm J (1955) Solution in the large for non-linear hyperbolic systems of equations. Comm Pure and Appl Math 697–715.

Harten A, Lax PD, Van Leer B (1983) On upstream differencing and Godunov-type schemes for hyperbolic conservation laws. SIAM Review 25(1):35–61.
Karni S (1994) Multicomponent flow calculations by a consistent primitive algorithm. J Comp Phys 112:31–43.
Takayama K (1993) Application of underwater shock wave focusing to the development of extracorporeal shock wave lithotripsy. Jpn J App Phys 32(5B):2192–2198.
Tammann G (1912) Uber Zustandsglecchungen im Gebiete Kleiner Volumen. Annalen der Physik 4(36):975–1013.
Van Leer B (1984) On the relation between the upwind differencing schemes of Godunov, Engquist-Osher and Roe. SIAM J Sci Stat Comput 5:1–20.

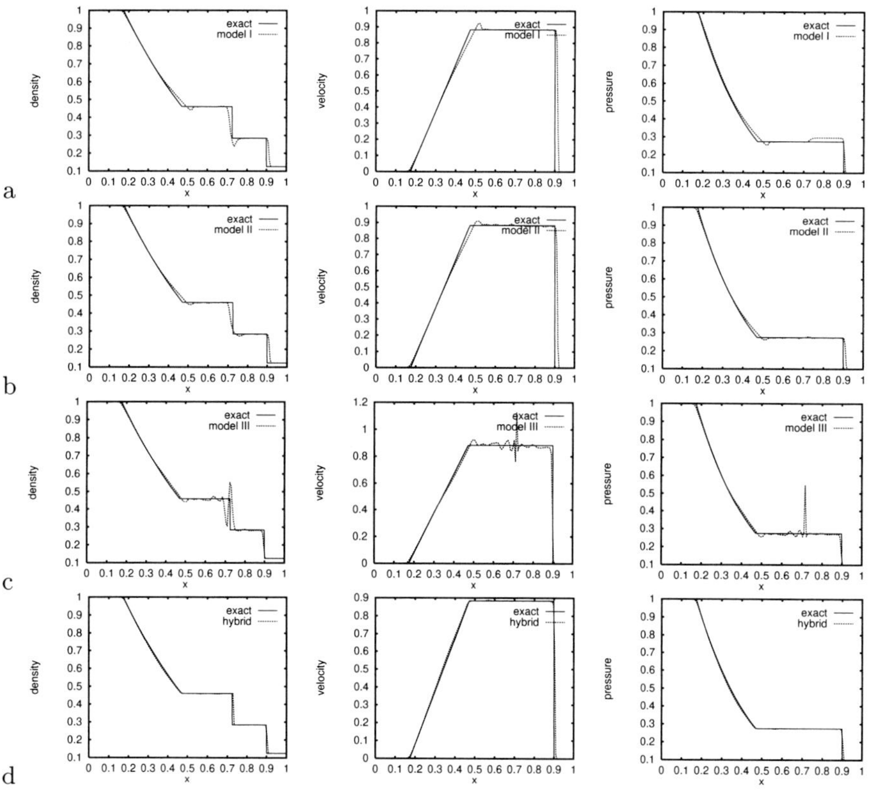

Figure 1. Conservative variable vs front tracking for a gas/gas problem: (a) Model I, (b) Model II, (c) Model III, and (d) hybrid algorithm.

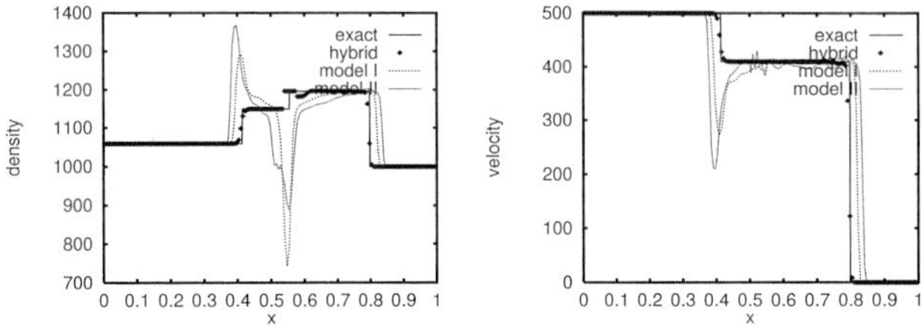

Figure 2. Conservative variable vs front tracking for gas/water problem 1.

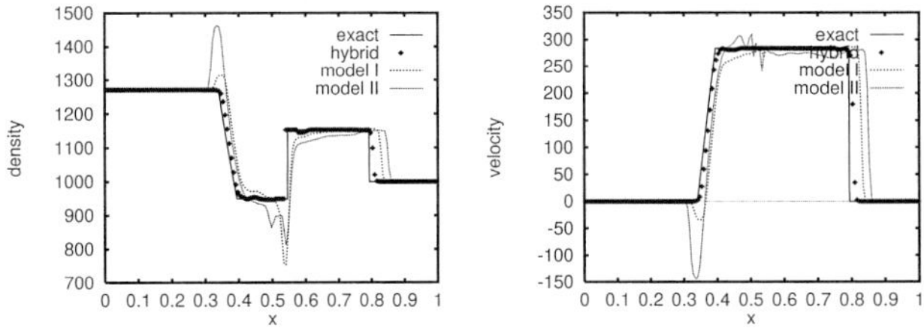

Figure 3. Conservative variable vs front tracking for gas/water problem 2.

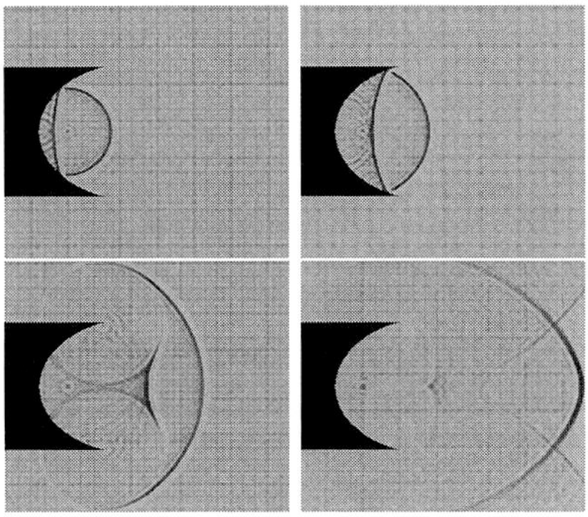

Figure 4. Shock-wave focusing in water.

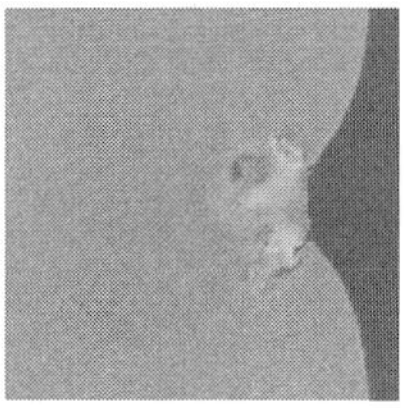

Figure 5. Interaction between a planar shock wave in helium and a neon gas bubble.

Shock waves in dilute/non-dilute bubbly flows

M. Kameda, N. Shimaura, F. Higashino
Tokyo University of Agriculture and Technology (TUAT)
2–24–16, Naka-cho, Koganei, Tokyo 184 Japan

Y. Matsumoto
The University of Tokyo
7–3–1, Hongo, Bunkyo-ku, Tokyo 113 Japan

Abstract: The effects of a non-uniform spatial distribution of gas bubbles in a liquid on unsteady shock-wave phenomena in the liquid containing the bubbles are investigated experimentally and numerically. Results show that the non-uniformity of the bubble distribution substantially alters the profile of the shock waves, even though the gas volume fraction is $O(10^{-3})$. The wavelength of relaxation oscillation at the back of the shock is an effective length scale to estimate the effect of the non-uniformity on the pressure profiles. When the scale of the non-uniformity is sufficiently small compared to the wavelength, the profiles are close to a profile obtained numerically by using a uniform distribution of bubbles.

Key words: Two-phase flow, Shock wave, Bubble, Non-uniform distribution

1. Introduction

Wave propagation in a liquid containing gas bubbles is studied theoretically by many researchers. For a small gas-volume fraction region, van Wijngaarden (1968) proposed a well-known mathematical model for the wave propagation. The macroscopic state of the bubbly mixture is described by its gas volume fraction α, density $\rho = (1-\alpha)\rho_l$, momentum $\rho \boldsymbol{U} = (1-\alpha)\rho_l \boldsymbol{U}_l$, and pressure P. The equations are written as

$$\alpha = \sum_{k=1}^{K} v_k / \Delta V, \qquad v_k = \frac{4}{3}\pi R_k^3, \tag{1}$$

$$\frac{\partial}{\partial t}[(1-\alpha)\rho_l] + \nabla \cdot [(1-\alpha)\rho_l \boldsymbol{U}_l] = 0, \tag{2}$$

$$\frac{\partial}{\partial t}[(1-\alpha)\rho_l \boldsymbol{U}_l] + \nabla \cdot [(1-\alpha)\rho_l \boldsymbol{U}_l \boldsymbol{U}_l] = -\nabla P + (1-\alpha)\rho_l \boldsymbol{g}, \tag{3}$$

$$R_k \frac{d^2 R_k}{dt^2} + \frac{3}{2}\left(\frac{dR_k}{dt}\right)^2 = \frac{1}{\rho_l}\left\{p_{gwk} - \frac{2\sigma}{R_k} - \frac{4\mu_l}{R_k}\frac{dR_k}{dt} + \frac{\rho_l |\boldsymbol{U}_{gk} - \boldsymbol{U}_l|^2}{4} - P\right\}, \tag{4}$$

where ΔV is the volume of a cell, K the total number of bubbles in the cell, $\boldsymbol{g}$ gravitational acceleration, p_{gw} gas pressure at the bubble wall, σ surface tension, μ viscous coefficient, and subscripts l and g liquid and gas phase, respectively. Although in van Wijngaarden's original formulation the bubble radius R is a continuum variable like ρ, the bubble motions are tracked individually here. The gas pressure p_g should be estimated by solving the conservation equations for mass, momentum, and energy of the bubble interior directly, because the internal processes cannot be described by a simple relation, such as the polytropic approximation $p_g = p_{g0}(R_0/R)^{3n}$.

On the other hand, for a not small gas-volume fraction region, in which the distance between neighboring bubbles is $O(R)$, it is not clear whether van Wijngaarden's theory is valid. The bubble–bubble interaction may require some corrections of the Rayleigh-Plesset equation (4), because it is a equation for a bubble in an infinite liquid. Beylich & Gülhan (1990) stated that (4) should be corrected even in the small gas-volume fraction region ($\alpha \sim 0.3\%$). Watanabe & Prosperetti (1994) and Kameda et al. (1995) reported a large difference between experiment and van Wijngaarden's theory. Recently, we found that a key to this puzzle may lie in the non-uniform spatial distribution of bubbles (Kameda & Matsumoto 1995, 1996). The spatial distribution of bubbles is assumed to be uniform in almost all theoretical approaches, although not all the experimental data were obtained under those conditions. In this paper, we show that the non-uniformity substantially alters the structure of the shock waves; when the non-uniformity is taken into account, there is a good agreement between numerical results obtained by using Eqs. (1) – (4) and experimental results in the range of $O(10^{-3})$ – $O(10^{-2})$ of α. We also attempt to clarify the effect of the non-uniformity.

2. Experiments

Figure 1 shows a schematic diagram of the experimental apparatus. Experiments were done with a vertical glass shock tube with an inner diameter D of 18 mm and a total length of 5.4 m. The tube is separated into two parts: a 1.6-m long high-pressure gas chamber and a 3.8-m long low-pressure (atmospheric) gas section. The liquid is introduced in the low-pressure section up to a certain level, about 2 m. A shock wave is generated by rupturing a thin plastic membrane (material: Lumirror) separating the high-pressure chamber from the low-pressure gas section. Four piezoelectric pressure transducers are flush-mounted on the wall of the tube to measure the shock waves. A number of glass capillary tubes of 0.1 mm in inner diameter and 80 mm in length are mounted at the bottom of the tube to serve as bubble generators. A photograph is taken by a 35-mm framing camera each time just before the membrane is ruptured. The initial bubble radius and local gas volume fraction at the pho-

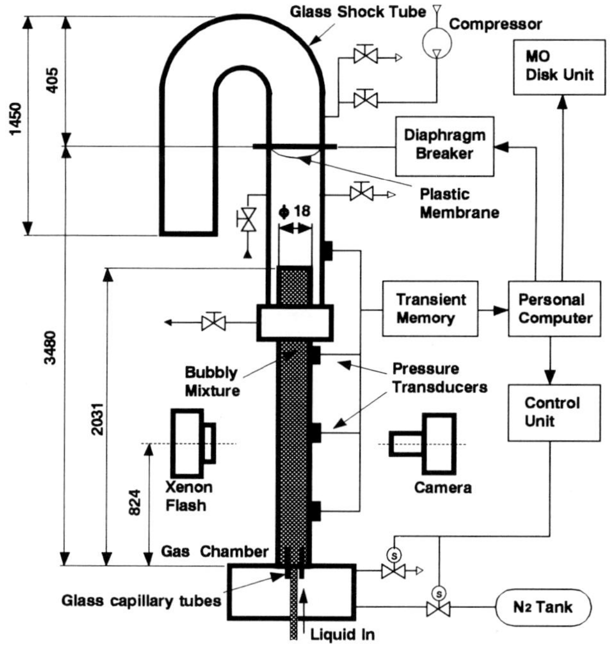

Figure 1. Experimental setup.

tographic observation section are measured from the photograph. The initial bubble radius and local gas volume fraction at other points are obtained by accounting for the hydrostatic effect.

In the present experiments, the liquid used is silicone oil (dimethyl siloxane polymer, Shinetsu Chemical Co. KF96-50); the density, kinematic viscosity, and surface tension are 960 kg/m^3, 50×10^{-6} m^2/s, and 20.8×10^{-3} N/m, respectively, at 298.15 K. The gas inside the bubbles is nitrogen.

3. Results and discussion

When bubbles are generated from the bottom of the shock tube into a still liquid, two features of non-uniform distribution of the bubbles are observed in the tube: one is the random distribution in the axial direction of the tube, another is the parabolic distribution in the lateral direction. Figure 2 shows three instantaneous photographs of bubbles in the tube. Although the bubbles rise regularly (like a straight chain) just after their generation, their path trembles so that the distribution becomes non-uniform. When the distribution becomes non-uniform, the bubbles are relatively concentrated along the axis of the shock tube. The gas volume fraction near the axis is higher than that near the side wall of the tube.

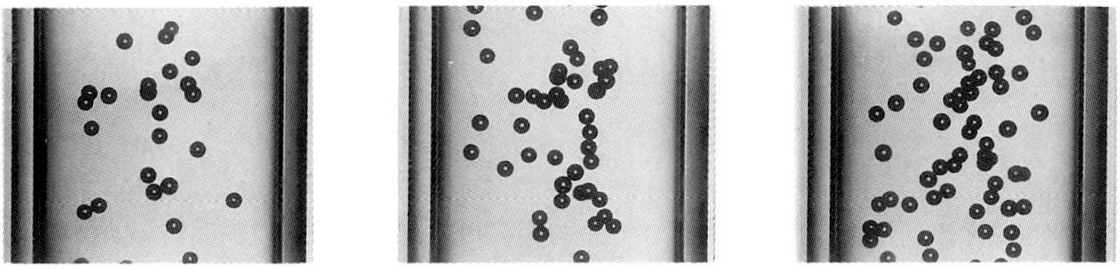

Figure 2. Instantaneous distribution of bubbles in the shock tube, $\alpha_{L0} = 1.07\%$, $R_{L0} = 0.62$ mm.

The random distribution of bubbles suppresses the relaxation oscillation behind the shock front. On the other hand, the period of the relaxation oscillation increases due to the parabolic distribution. An experimentally obtained pressure profile of a shock wave is compared with two numerical profiles in Figs. 3a and 3b. The two numerical results are obtained by: (1) An axisymmetric 2D calculation. The initial arrangement of bubbles is given randomly in the axial direction, and parabolically in the lateral direction. (2) A 1D calculation with uniform distribution of bubbles in the axial direction. These calculations are carried out using Eqs. (1) – (4) by a finite-difference scheme (Kameda & Matsumoto 1995, 1996). For low gas volume fraction ($\alpha_{L0} = 0.15\%$) (Fig. 3a), the amplitude of pressure oscillation behind the shock front of the 1D uniform numerical result is larger than the others, although the period of the pressure oscillation is not obviously different. For higher gas volume fraction ($\alpha_{L0} = 1.07\%$) (Fig. 3b), the period of the oscillation of the 1D uniform numerical result is shorter than the others.

Figure 3c shows an experimental result where the spatial distribution of the bubbles in the axial direction is fairly uniform. The upwardly directed bubbly flow is made in the shock tube in order to obtain the uniform distribution of bubbles. The average flow rate of the liquid is 7×10^3 mm^3/s, the average flow velocity is 28 mm/s, and the mean gas volume fraction $\alpha_{L0} = 0.16\%$. The Reynolds number, which is calculated by using the average flow velocity, the inner diameter of shock tube, and the kinematic viscosity of the liquid, is 11. One can see that the experimental result agrees well with the 1D uniform numerical one.

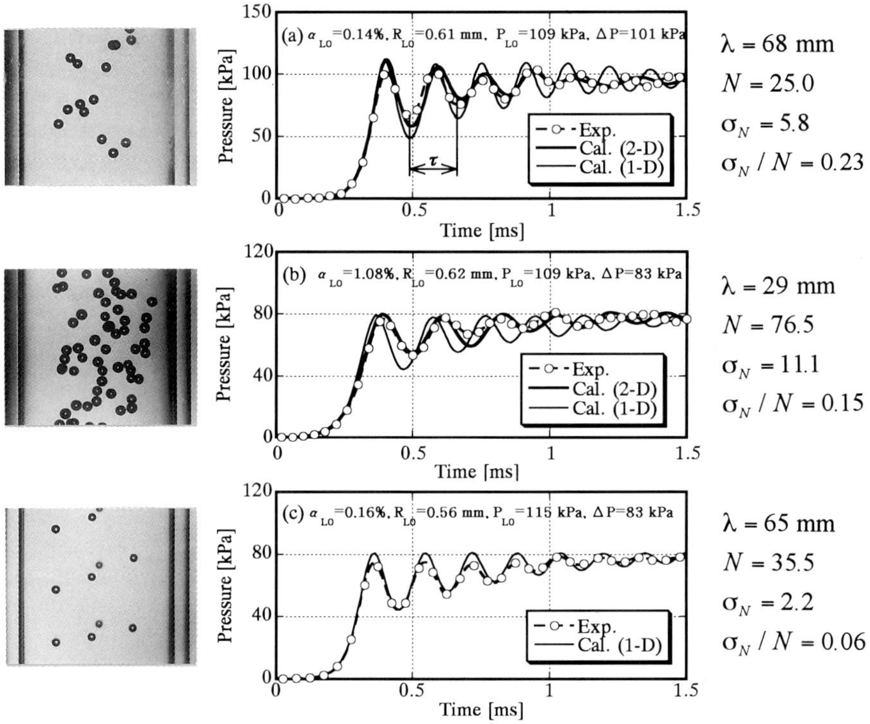

Figure 3. Comparisons of experimental and numerical results. The photographs show instantaneous bubble distributions.

The effect of the non-uniform distribution of bubbles is correlated well with a wavelength behind the shock front λ, which is defined as

$$\lambda = C_S \tau, \qquad C_S = \sqrt{\frac{P_{L0} + \Delta P}{\alpha_{L0}(1 - \alpha_{L0})\rho_l}}, \tag{5}$$

where τ is the period defined in Fig. 3a, and C_S is the propagation velocity of the shock wave in steady state.

The effect of the random distribution is estimated as follows. Using λ, we can define a cell whose volume is $\lambda \times \pi D^2/4$. The averaged total number of bubbles in the cell N and its variance σ_N are listed in Fig. 3. The suppressed oscillation shown in Fig. 3a occurs when the ratio σ_N/N becomes large. If σ_N/N is sufficiently small, the pressure profiles of the shock wave are close to a profile obtained by using a uniform distribution of bubbles shown in Fig. 3c.

The effect of the parabolic distribution of bubbles is correlated with parameter D/λ, because the inner diameter of the shock tube D is the characteristic size of the concentrated bubble formation in the cross section of the shock tube. When the parameter D/λ increases, the period of the relaxation oscillation becomes longer than that with a uniform distribution of bubbles. If one fixes all the experimental conditions except for α_{L0}, λ is approximately proportional to $\alpha_{L0}^{-1/2}$ with a small value of α_{L0}. Thus, the effect of the parabolic distribution becomes significant when the initial gas volume fraction is large.

4. Conclusions

Non-uniform spatial distribution of bubbles has a considerable influence on the shock-wave structure in a bubbly mixture. A characteristic wavelength of the relaxation oscillation behind the shock front is one of the most effective reference scales to estimate the effect of the non-uniformity. When the characteristic length of the non-uniformity is not small compared to the wavelength, the pressure profile of the shock wave alters from the profile obtained by using a uniform distribution of bubbles.

Acknowledgement. This study was supported by a Grant-in-Aid for Scientific Research Program of the Japanese Ministry of Education, Science and Culture under Grant Nos. 05003524 and 07750181.

References

Beylich AE, Gülhan A (1990) On the structure of nonlinear waves in liquids with gas bubbles. Phys Fluids A 2:1412–1428.

Kameda M, Matsumoto Y (1995) Structure of shock waves in a liquid containing gas bubbles. In: IUTAM Symposium on Waves in Liquid/Gas and Liquid/Vapour Two-Phase Systems, Morioka S, van Wijngaarden L (eds), Kluwer, Dordrecht, pp 117–126.

Kameda M, Matsumoto Y (1996) Shock waves in a liquid containing small gas bubbles. Phys Fluids 8 (in press).

Kameda M, Matsumoto Y, Haneda Y (1995) Propagation of shock waves in dilute bubbly liquids. In: Shock Waves @ Marseille, Brun R, Dumitrescu LZ (eds), Springer, Berlin, Vol 3, pp 141–146.

van Wijngaarden L (1968) On equations of motion for mixtures of liquid and gas bubbles. J Fluid Mech 33:465–474.

Watanabe M, Prosperetti A (1994) Shock waves in dilute bubbly liquids. J Fluid Mech 274:349–391.

Hydraulic shock loading due to supercavitating projectiles

G. A. O. Davies, O. Flanagan, R. Hillier, D. Hitchings, S. J. Lord
Department of Aeronautics
Imperial College of Science, Technology and Medicine
Prince Consort Road, London, SW7 2BY UK

Abstract: This paper presents the results of using a second-order Godunov-type scheme adapted to the problem of modeling shock waves and cavitation zones in liquids. Various one-dimensional validation tests are presented. A two-dimensional simulation of a high-speed projectile in water is included, together with data produced in a parallel experimental study.

Key words: Hydraulic shock, Cavitation

1. Introduction

The problem of pressure-wave propagation in liquids is important in many practical engineering problems. The present study was stimulated by the very large overpressures which can be generated in aircraft fuel tanks due to penetration by high-speed debris or projectiles. The severity of the overpressures is an immediate consequence of the very low compressibility of liquids; if a projectile at 200 m/s enters a water-filled tank at one bar the instantaneous peak stagnation point pressure is of order 3800 bar.

The second important phenomenon associated with the high-speed projectile, and a critical requirement in the modeling, is the substantial zone of cavitation formed downstream of it, generally with a volume many hundreds or thousands that of the projectile itself. This arises from the tendency for the liquid to produce substantial rarefaction in the base region of the projectile. Since the liquid is able to support little or no tensile stress, a cavitation void must open in the wake. The rapidity of the process almost certainly means that there is insufficient time for liquid to evaporate to achieve an equilibrium vapor pressure in the cavity. Essentially we have a supercavitating flow; either a vacuum forms, or the cavity is partially occupied by air or gas entrained from external to the tank through the entry hole (see the schematic in Fig. 1).

2. The experiments

The experimental work was conducted in a very rigid tank, 25 cm long in the projectile flight direction, and 1 m by 1 m in the transverse directions. The

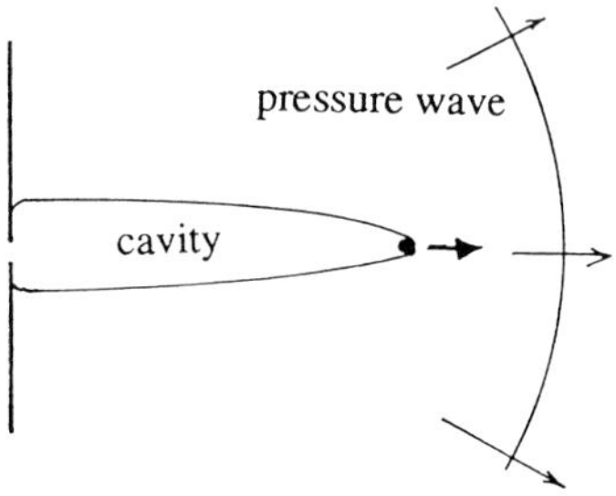

Figure 1. Schematic of supercavitating projectile.

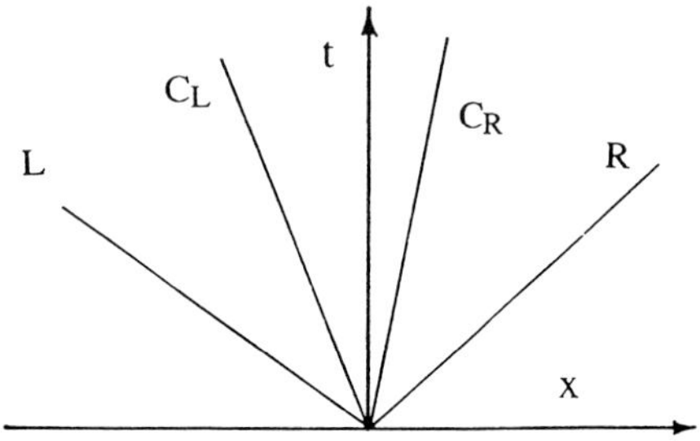

Figure 2. Riemann problem for cavitating fluid.

tank was water filled, with a free surface. The useful duration of the experiment was the time up to the return of reflected waves from the free surface and the side walls to the measuring instrumentation, during which phase the flow can be regarded as axisymmetric. Spherical projectiles were fired from a gas gun, with a helium gas driver giving maximum projectile velocities of 670 m/s. The projectiles were held in a large-diameter sabot, which was trapped at the barrel exit so that the projectile alone penetrated a melinex-covered hole on the entry wall of the tank. Experiments involved high-speed filming and pressure histories using blast gauges.

3. Computations

The computations used a second-order Godunov-type scheme to solve the two-dimensional axisymmetric Euler equations on a mesh of rectangular cells in the x-r plane. The basic solver involved a piecewise linear reconstruction of the flow variables within a cell and the solution of the associated Riemann problem at the cell interface. Generally a (one-dimensional) Riemann solver will decompose the interface flow field into left-and right-facing pressure waves separated by a contact surface across which the pressure and normal velocity are continuous, but the density and transverse velocity may be discontinuous. For the hydraulic shock problem we have to anticipate the possibility of cavitation, so that the Riemann solver includes the option of two contact surfaces, with a vacuum void between them as indicated in Fig. 2.

The second problem faced by the solver was the actual description of the appropriate equation of state. If a cell is fully occupied by water, at pressures greater than or equal to zero bar, we have used the Tait equation of state given by

$$p = B[(\rho/\rho_r)^n - 1] \tag{1}$$

Here ρ_r is the critical lower density at which the water pressure is zero bar (and the liquid is on the verge of cavitation) and B and n are constants with values of 2995 bar and 7.44 in the present simulations. If the cell average density is less than the critical value the cell contains both vacuum voids and also liquid water at zero bar; in our modeling we treat this as a homogeneous mix for the cell and a Riemann solver has been formulated for both this and the fully liquid state.

3.1. One-dimensional demonstration problems

To illustrate the generation of shock waves and cavitation zones the code was run for a range of one-dimensional simulations. In these the water is initialized at one bar, with a solid left-hand boundary to the tube and a through-flow condition at the right-hand end. If the velocity is negative, directed therefore towards the solid end boundary, then a reflected compression or shock wave will occur. Figure 3 shows such a case, with an initial velocity of 200 m/s. The axes are labeled as pressure in bar, versus a normalized axial distance x/at, where x is the distance from the wall, a is the ambient speed of sound for water, and t is the elapsed time since the reflection commenced. The simulation shows capture of the wave over two cells, and the pressure rise and wave location are in close agreement with the exact theoretical values according to the Tait equation.

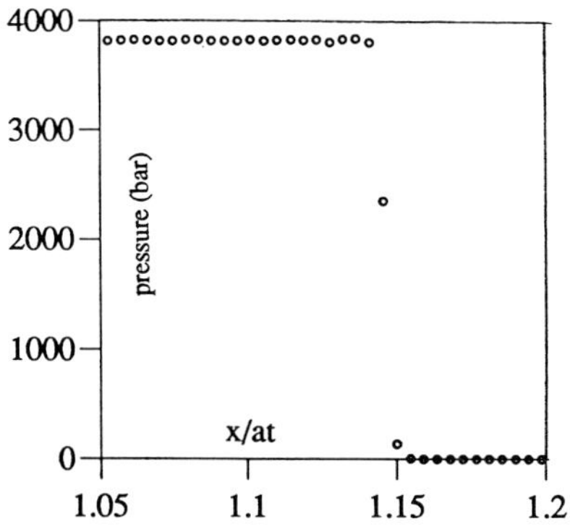

Figure 3. Shock simulation.

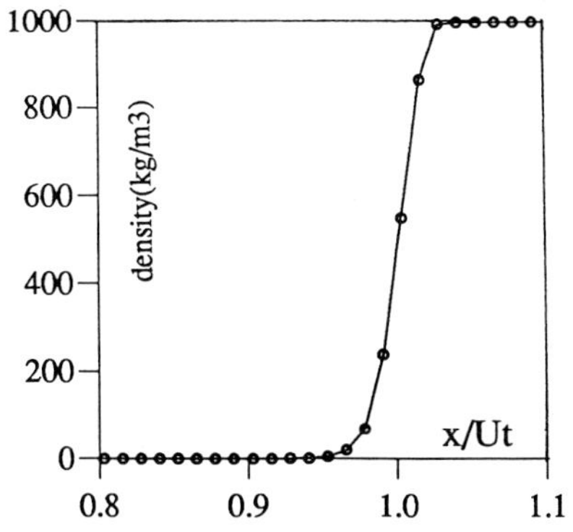

Figure 4. Cavitation simulation.

Next we consider cavitation, which is established easily by initializing with a positive velocity. We assume that cavitation occurs when the pressure falls to zero bar, which is achieved with a velocity of 0.07 m/s or above. The problem with simulating such a low-velocity case is that since the cavitation front moves at less than one ten-thousandth of the acoustic wave speed in the liquid, and the latter controls the stable time step for the simulation, it takes very many time steps for the cavitation zone to grow beyond even the first cell. We have therefore again used a 200-m/s case, so that the cavitation front formed will propagate away from the end wall at approximately 199.93 m/s. Figure 4 shows the computed density distribution, normalizing x now by the elapsed time and the initial fluid velocity. The front is smeared over several cells and falls to a density one ten-thousandth that of water; a limit is set within the code to prevent the density falling below this value.

The third one-dimensional simulation represents the response of a cavitation zone to an incident shock wave. Here we initialize a 200-m/s velocity left to right, with the tube now sealed at both ends. The resultant shock wave formed at the right-hand boundary propagates towards the cavitation zone at the left-hand side, first meeting at a time ta/L equals 0.78, where a is the initial speed of sound and L is the tube length. After the shock/cavitation interaction commences an expansion wave is sent back into the shock. The most critical feature is the behavior of the cavitation zone which is propelled back towards the left-hand wall. The time sequence of Fig. 5 shows a well-behaved contraction of the bubble.

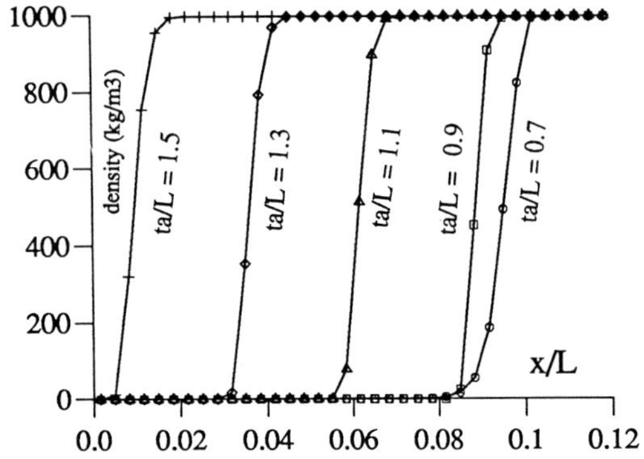

Figure 5. Density profile through cavitation front, after interaction with a shock.

4. Two-dimensional modeling of the projectile in water

Projectile simulation requires a very large range of mesh sizes to represent a small projectile adequately as well as to incorporate the large scale of the test tank. We have taken the approach of simulating the projectile by a source which inputs momentum and energy into the flow field at the projectile position according to the projectile drag. In turn, a detailed CFD simulation was performed to provide a drag time history for the supercavitating projectile and these values were also calibrated against projectile deceleration histories determined from the experiments. The simulation here is performed for an initial projectile entry speed of 670 m/s, giving a flight distance of 0.25 m before impact with the far face of the tank. The projectile speed falls in the simulation

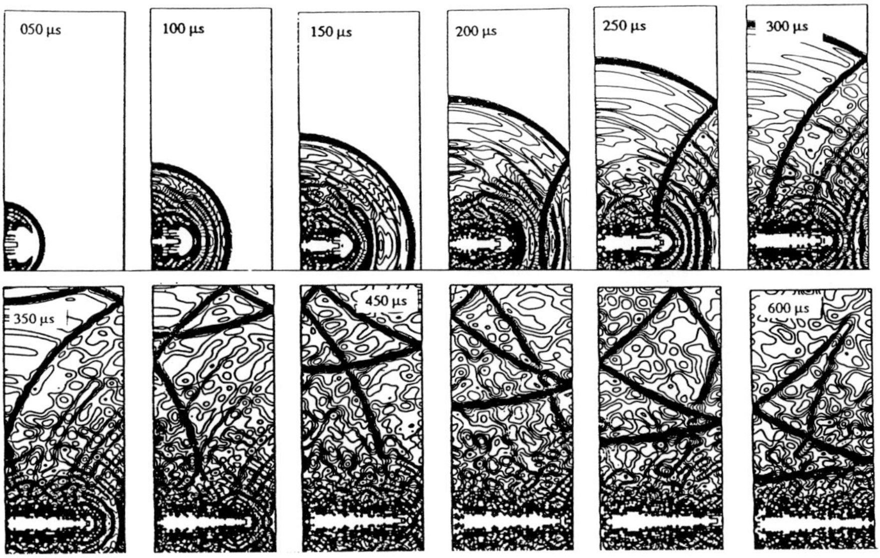

Figure 6. Isobars at various elapsed times, for the entry of a 670 m/s projectile into a tank 0.25m wide.

in response to the projectile drag law. Figure 6 shows a set of isobar contours, at different elapsed times, where the propagation and subsequent reflection of waves is clear, as well as the development of a substantial cavity zone. The background noise occurs principally because of the step-like manner in which the projectile source is moved through the tank. Pressure time histories were experimentally recorded at various locations in the tank, and can be found documented in Flanagan (1993). One such case is shown in Fig. 7 for a location half-way across the tank, and offset radially by 100 mm from the projectile flight path. Three experimental traces are shown which indicate the level at which the experiments can be reproduced, differences reflecting factors such as run-to-run variations in projectile speed and exact flight path, blast-gauge resonances, and inclusion of extra debris into the tank. Nonetheless, levels of pressure and the associated time scales are reproduced satisfactorily in the CFD.

5. Concluding remarks

A liquid equation of state, including cavitation, has been incorporated into the Riemann solver for an Euler code. One-dimensional validation tests and two-dimensional comparisons with experiment indicate satisfactory performance. A specific area of value would be to use local mesh adaptation, or a front-tracking

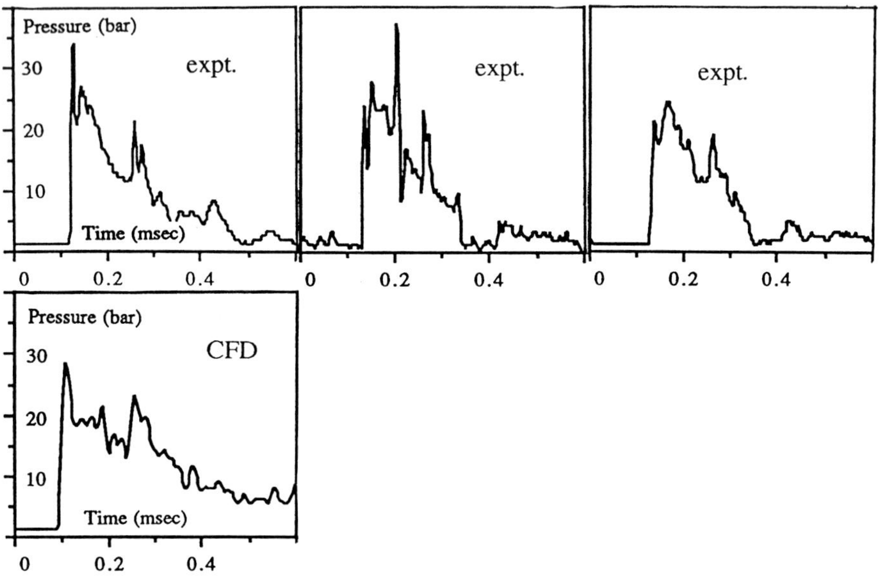

Figure 7. Experiment and CFD. Pressure time traces.

routine, to follow the cavitation interface. This is to address the fact that in many simulations the cavitation front may take tens or hundreds of time steps to move one cell, which can potentially lead to smearing or resolution problems with the interface.

Acknowledgement. This work was supported by the Ministry of Defence and monitored by staff of DRA Farnborough; we are grateful for many helpful observations from Dr. Rodger Munt and Dr. Sarah Bishop.

References

Flanagan TO (1993) Structural Damage by Hydraulic Shock, Ph.D. Thesis, London University.

Shock waves and noise in the collapse of a cloud of cavitation bubbles

Y.-C. Wang, C. E. Brennen
California Institute of Technology
Pasadena, CA 91125 USA

Abstract: Calculations of the collapse dynamics of a cloud of cavitation bubbles confirm the speculations of Mørch and his co-workers and demonstrate that collapse occurs as a result of the inward propagation of a shock wave which grows rapidly in magnitude. Results are presented showing the evolving dynamics of the cloud and the resulting far-field acoustic noise.

Key words: Cloud cavitation, Shock wave, Cavitation noise

1. Introduction

In many flows of practical interest one observes the periodic formation and collapse of a "cloud" of cavitation bubbles (see, for example, Bark and Berlekom 1978, Shen and Peterson 1978). Such a structure, which has a foamy appearance and consists of numerous small bubbles, is termed "cloud cavitation." Clouds of cavitation bubbles are often shed from a cavitating foil and the coherent dynamics of these clouds can result in a collapse process which has much greater potential for noise production and damage than the individual bubbles, acting independently, would have (see, for example, Soyama et al. 1992).

Most previous theoretical studies of the dynamics of cavitating clouds have been linear or weakly nonlinear analyses but have not, as yet, shown how a cloud would behave during the massively nonlinear response in a cavitating flow. The linearized behavior of a collapsing layer of bubbly fluid next to a solid wall was first modeled by van Wijngaarden in 1964. Later d'Agostino and Brennen (1983) investigated the linearized dynamics of a spherical cloud (see also d'Agostino and Brennen 1989) and showed that the interaction between bubbles leads to a coherent dynamics of the cloud, including natural frequencies that can be much smaller than the natural frequencies of individual bubbles. Other studies of cloud dynamics are summarized by Brennen (1995).

Several years ago, Mørch and his co-workers (Mørch 1980, 1981; Hanson et al. 1981) speculated that the collapse of a cloud of bubbles involves the formation and inward propagation of a shock wave and that the geometric focusing of this shock at the center of the cloud creates the enhancement of the noise and damage potential associated with cloud collapse. The present paper extends a previous computational investigation of this flow (Wang and Brennen 1994) and examines the formation, development, and acoustic consequences of the shock wave.

2. Basic equations

Consider a spherical cloud of bubbles surrounded by an unbounded pure liquid at rest at infinity, as shown in Fig. 1. The pure liquid is assumed incompressible, with a density ρ_L. It is assumed that the population of bubbles per unit liquid volume, η, within the cloud, is uniform initially and that there is no coalescence or break-up of bubbles. Since relative motion between two phases and the mass of liquid vaporized or condensed are both neglected it follows that η remains both constant and uniform within the cloud. The basic equations (see

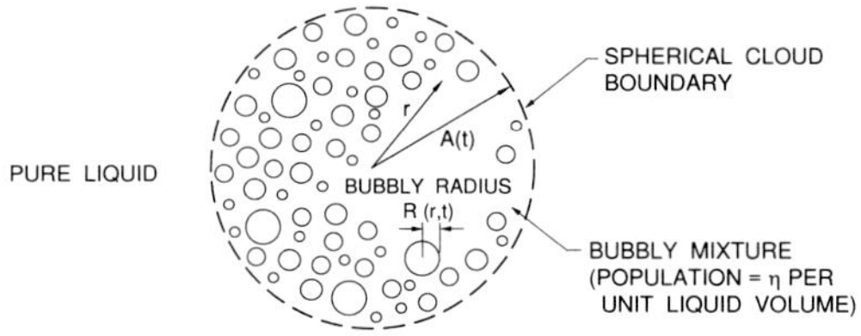

Figure 1. Schematic of a spherical cloud of bubbles.

d'Agostino and Brennen 1989) consist of the following dimensionless forms of the continuity and momentum equations for the spherical bubbly flow:

$$\frac{1}{r^2}\frac{\partial(r^2u)}{\partial r} = \frac{12\pi\eta R^2}{3+4\pi\eta R^3}\frac{DR}{Dt} \quad ; \quad r \le A(t)\ , \tag{1}$$

$$\frac{Du}{Dt} = -\frac{1}{6}(3+4\pi\eta R^3)\frac{\partial C_P}{\partial r} \quad ; \quad r \le A(t)\ , \tag{2}$$

where $\frac{D}{Dt} \equiv \frac{\partial}{\partial t} + u\frac{\partial}{\partial r}$ is the Lagrangian derivative, $u(r,t)$ is the mixture velocity, $R(r,t)$ is the individual bubble radius, $C_P(r,t) = (p(r,t) - p_0)/(\frac{1}{2}\rho_L U^2)$ is the pressure coefficient in the cloud, ρ_L is the liquid density, $p(r,t)$ is the mixture pressure, p_0 is the initial equilibrium pressure in the mixture, and $A(t)$ is the radius of the cloud. The bubble population per unit liquid volume, η, is related to the void fraction, α, by $(\frac{4}{3}\pi R^3)\eta = \alpha/(1-\alpha)$. The variables and equations are non-dimensionalized using the initial bubble radius, R_0, a reference flow velocity, U, the time scale, R_0/U, and the dynamic pressure, $\frac{1}{2}\rho_L U^2$. To incorporate the bubble/bubble interactive effects, the dynamics of the bubbles are modeled using the Rayleigh-Plesset equation which connects the local mixture pressure coefficient, C_P, to the bubble radius, R:

$$R\frac{D^2R}{Dt^2}+\frac{3}{2}\left(\frac{DR}{Dt}\right)^2+\frac{\sigma}{2}[1-R^{-3k}]+\frac{4}{Re}\frac{1}{R}\frac{DR}{Dt}+\frac{2}{We}[R^{-1}-R^{-3k}]+\frac{1}{2}C_P = 0\ , \tag{3}$$

where $\sigma = (p_0 - p_v)/(\frac{1}{2}\rho_L U^2)$ is the cavitation number, p_v is the vapor pressure, $We = \rho_L U^2 R_0/S$ is the Weber number, S is the surface tension of liquid, $Re = U R_0 \rho_L/\mu_E$ is the Reynolds number, μ_E is the effective viscosity of liquid which incorporates the various bubble-damping mechanisms described by Chapman and Plesset (1971), and k is the effective polytropic index of the gas inside the bubble.

The boundary condition on the surface of the cloud, $r = A(t)$, can be derived from the solution of incompressible liquid flow outside the cloud (see Wang and Brennen 1995):

$$C_P(A(t),t) = C_{P\infty}(t) + \frac{2}{A(t)}\frac{d[A^2(t)u(A(t),t)]}{dt} - u^2(A(t),t) \tag{4}$$

where $C_{P\infty}(t)$ is the known and imposed driving pressure at infinity and is chosen as a simple sinusoidal form:

$$C_{P\infty}(t) = \begin{cases} \frac{1}{2}C_{PMIN}\{\cos(\frac{2\pi}{t_G}t) - 1\} & ; \quad 0 < t < t_G \\ 0 & ; \quad t < 0 \quad \text{and} \quad t > t_G \end{cases} \tag{5}$$

where C_{PMIN} is the minimum pressure coefficient at infinity and t_G is the dimensionless duration of the low-pressure perturbation. Consequently, for a cloud flowing with velocity U past a body of size D, the order of magnitude of t_G will be D/R_0, and C_{PMIN} will be the minimum pressure coefficient of the flow. At the center of the cloud, the symmetry of the problem requires $u(0,t) = 0$. Also, at time $t \leq 0$, it is assumed that the whole flow field is in equilibrium. It is also assumed, for simplicity, that all the bubbles have the same initial radius R_0. Thus we have the following initial conditions: $R(r,0) = 1$, $\frac{DR}{Dt}(r,0) = 0, u(r,0) = 0, C_P(r,0) = 0$. The above equations (1), (2), and (3) with the above boundary conditions can, in theory, be solved to find the unknowns $C_P(r,t)$, $u(r,t)$, and $R(r,t)$ for any bubbly cavitating flow with spherical symmetry. However, the nonlinearities in the Rayleigh-Plesset equation and in the Lagrangian derivative, D/Dt, present considerable computational impediments. An integral method which uses the Lagrangian coordinate system, (r_0, t), has been developed (see Wang and Brennen 1995) in which r_0 is the non-dimensional initial radial position at time $t = 0$.

3. Results and discussion

We have chosen the following typical flow for the computational results presented here. A cloud of nuclei, composed of air bubbles of initial radius $R_0 = 100$ μm in water at 20°C, is convected through a low-pressure region with velocity $U = 10$ m/sec. Other parameters are: initial void fraction, $\alpha_0 = 0.5\%$; cavitation number, $\sigma = 0.45$; the minimum pressure coefficient, $C_{PMIN} = -0.75$; the

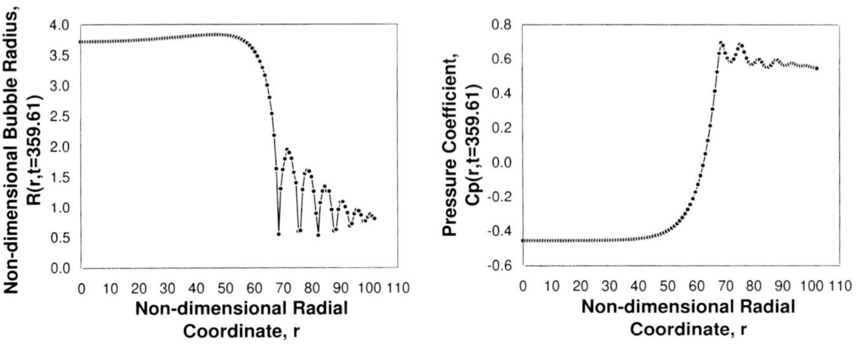

Figure 2. The structure of a shock wave accompanying the collapse of the cloud of cavitation bubbles. Left: The dimensionless bubble-size distribution in the cloud as a function of the dimensionless radial coordinate at the dimensionless time, $t = 359.61$. Right: The pressure-coefficient distribution in the cloud as a function of the dimensionless radial coordinate at the same time as on the left.

non-dimensional cloud radius, $A_0/R_0 = 100$; the non-dimensional duration of the low-pressure perturbation, $t_G = 250$, which corresponds to the ratio of the length scale of the low-pressure perturbation to the initial radius of the cloud, $D/A_0 = 2.5$. The Reynolds number, $Re = UR_0\rho_L/\mu_E$, is taken as 0.05.

The structure of the developing shock wave is illustrated by the results presented in Fig. 2. This structure is similar to that of the steady planar shocks analyzed by Noordij and van Wijngaarden (1974). The shock wave strengthens not only because of the geometric effect of an inwardly propagating spherical shock but also because of the coupling of the single-bubble dynamics with the global dynamics of the flow through the pressure and velocity fields. When the shock reaches the center of the cloud, very high pressures are produced and cause a volumetric rebound of the cloud. The time history of the radius of the cloud is shown in Fig. 3. Note that, unlike single bubbles, the cloud radius, $A(t)$, only decreases to a size marginally smaller than its equilibrium size during the collapse process. But each volumetric rebound causes an acoustic peak in the far field, as shown in Fig. 3. The normalized measure, p_a, of the acoustic noise, p_a^*, can be calculated from the volumetric acceleration of the cloud and is given by

$$p_a(t) = \frac{p_a^* r}{\frac{1}{2}\rho_L U^2 D} = \frac{2R_0}{D}\left[A^2(t)\frac{d^2A(t)}{dt^2} + 2A(t)(\frac{dA(t)}{dt})^2\right] . \tag{6}$$

After several collapse cycles, the cloud begins to oscillate in a regular way as shown in Fig. 3 (right). This regular oscillation contributes large peaks in the noise spectrum at the natural frequency of the cloud and at higher harmonics,

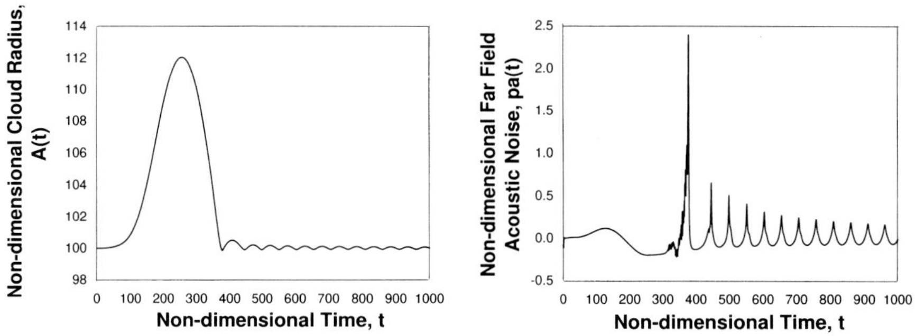

Figure 3. Left: The time history of the dimensionless cloud radius. Right: The time history of the far-field acoustic noise calculated from the data shown in the left figure.

as shown in Fig. 4. In the present example, the first natural frequency of the cloud is about 0.02 (2 kHz) and the natural frequency of bubbles in the cloud is 0.158 (15.8 kHz).

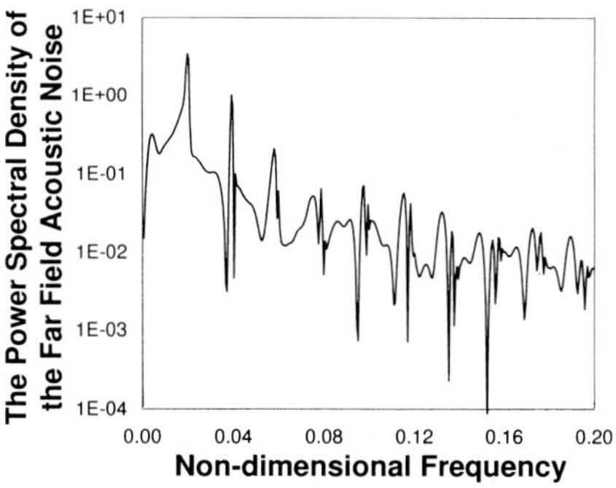

Figure 4. The non-dimensional power spectral density of the far-field acoustic noise in Fig. 3 as a function of dimensionless frequency. The lowest cloud natural frequency is about 0.02. The natural frequency of bubbles in the cloud is 0.158.

4. Conclusions

We have shown how a shock wave develops as part of the nonlinear collapse of a spherical cloud of cavitation bubbles. The focusing of the shock produces very high pressures at the center of the cloud and then causes rebound of the cloud. The volumetric acceleration induces a large pulse in the far-field noise. Moreover, there are further but weaker shocks which arrive at the center and thus produce a train of acoustic impulses which, eventually, lead to a regular oscillation of the cloud at the first cloud natural frequency. These results suggest that shock focusing may be one of the major mechanisms for the enhanced noise and damage potential associated with cloud cavitation.

Acknowledgement. The authors are very grateful for the support for this research provided by the Office of Naval Research under Contract N00014-91-J-1295.

References

Bark G, Berlekom WB (1978) Experimental investigations of cavitation noise. In: Proc 12th ONR Symp on Naval Hydrodynamics, pp 470–493.

Brennen CE (1995) Cavitation and Bubble Dynamics, Oxford University Press.

Chapman RB, Plesset MS (1971) Thermal effects in the free oscillation of gas bubbles. ASME J Basic Eng 93:373–376.

d'Agostino L, Brennen CE (1983) On the acoustical dynamics of bubble clouds. In: ASME Cavitation and Multiphase Flow Forum, pp 72–75.

d'Agostino L, Brennen CE (1989) Linearized dynamics of spherical bubble clouds. J Fluid Mech 199:155–176.

Hanson I, Kedrinskii VK, Mørch KA (1981) On the dynamics of cavity clusters. J Appl Phys 15:1725–1734.

Mørch KA (1980) On the collapse of cavity cluster in flow cavitation. In: Proc First Int Conf on Cavitation and Inhomogenieties in Underwater Acoustics, Springer Series in Electrophysics 4, pp 95–100.

Mørch KA (1981) Cavity cluster dynamics and cavitation erosion. In: Proc ASME Cavitation and Polyphase Flow Forum, pp 1–10.

Noordij L, van Wijngaarden L (1974) Relaxation effects, caused by relative motion, on shock waves in gas-bubble/liquid mixtures. J Fluid Mech 66:115–143.

Shen Y, Peterson FB (1978) Unsteady cavitation on an oscillating hydrofoil. In: Proc 12th ONR Symp on Naval Hydrodynamics, pp 362–384.

Soyama H, Kato H, Oba R (1992) Cavitation observations of severely erosive vortex cavitation arising in a centrifugal pump. In: Proc Third IMechE Int Conf on Cavitation, pp 103–110.

van Wijngaarden L (1964) On the collective collapse of a large number of gas bubbles in water. In: Proc 11th Int Conf Appl Mech, Springer-Verlag, Berlin, pp 854–861.

Wang Y-C, Brennen CE (1994) Shock wave development on the collapse of a cloud of bubbles. In: Cavitation and Multiphase Flow Forum, FED 194:15–19.

Wang Y-C, Brennen CE (1995) The noise generated by the collapse of a cloud of cavitation bubbles. In: ASME/JSME Symp on Cavitation and Gas-Liquid Flows in Fluid Machinery and Devices, FED 226:17–29.

Shock-tube-deduced condensation coefficient

S. Fujikawa, M. Kotani, H. Sato, N. Takasugi
Toyama Prefectural University
Toyama 939-03
Japan

M. Matsumoto
Nagoya University
Nagoya 464-01
Japan

Abstract: A theory is proposed for the deduction of the condensation coefficient from the measurement of the liquid-film growth on the shock-tube end wall in a shock-wave reflection region. The method is proved theoretically and demonstrated experimentally to be suitable for the deduction of the condensation coefficient. Molecular dynamics simulation is performed to visualize dynamic processes of evaporation and condensation and to compute the condensation coefficient.

Key words: Shock tube, Molecular dynamics, Condensation, Evaporation, Condensation coefficient

1. Introduction

It is challenging to measure the phase-transition rate from a vapor to liquid phase at a vapor-liquid interface, and to determine the condensation coefficient of vapor molecules at the interface, because this work is a gateway to the connection between macroscopic and microscopic phase-transition phenomena. Measurements of the phase-transition rate and the condensation coefficient have been made for many kinds of vapors for more than a half century, but the subject remains open. This can be attributed to the lack of appropriate experimental methods and to the lack of basic understanding of condensation phenomena.

A schematic diagram showing changes in thermodynamic states of a vapor exposed to a shock wave is indicated on the pressure-volume plane in Fig. 1. The initial state of the vapor prior to the arrival of the incident shock wave is denoted by O. The state of the main stream changes from state O to state 2 through two stages of adiabatic compression; first by the incident shock wave and, subsequently, by the reflected shock wave. However, the state at the inner surface of the end wall changes instantaneously from state O to state 2W keeping the pressure behind the reflected shock wave constant; the state is a supersaturated one under which the phase transition from vapor to liquid phase takes place, resulting in a liquid film which is formed uniformly on the end-wall surface.

Based on the above principle of phase transition, the authors have previously published an experimental determination of the condensation coefficient (Fujikawa et al. 1995) and a corresponding theoretical analysis (Fujikawa 1995). However, questions as to why the condensation coefficient can be reasonablly determined by the present method still remain. In the present paper, a theoretical basis for the deduction of the condensation coefficient of methanol vapor is given, and the deduced condensation coefficient is presented. Results of molecular dynamics simulation are presented on dynamic processes of evaporation and condensation at the interface.

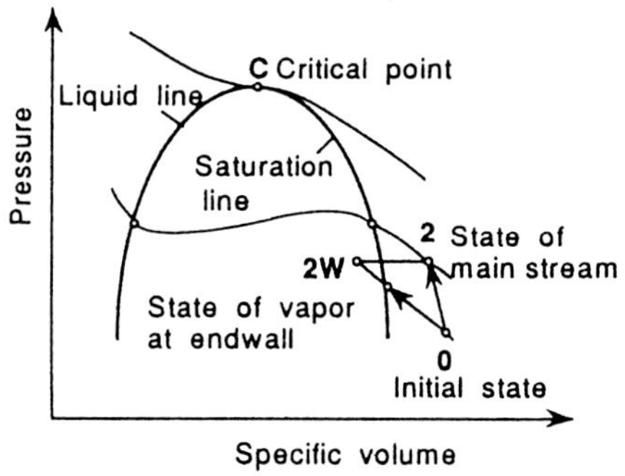

Figure 1. Change of thermodynamic states by shock wave.

2. Theory of liquid-film growth

The theory of the liquid-film growth has been derived from one-dimensional gasdynamics equations with the kinetic boundary conditions for the vapor and thermal equations for both the liquid film and the shock-tube end wall (Fujikawa 1995). The growth of the liquid film is described by

$$d(t) = \frac{\beta_1}{{\beta_2}^2}\left\{\exp({\beta_2}^2 t)\cdot \operatorname{erfc}(\beta_2 t^{1/2}) + \frac{2}{\pi^{1/2}}\cdot \beta_2 t^{1/2} - 1\right\} + d_0 \tag{1}$$

in which

$$\beta_1 = \frac{\sigma_c(p_\infty - p_0^*)}{\rho_l T_0}\cdot\left(\frac{2T_\infty}{R}\right)^{1/2}$$

$$\beta_2 = \frac{\sigma_c L}{\lambda_s {T_0}^2}\cdot\left(p_\infty - p_0^* + \frac{p_0^* S T_0}{T_\infty}\right)\cdot\left(\frac{2D_s T_\infty}{R}\right)^{1/2}$$

$$S = \frac{2.3026BT_\infty}{C^2}$$

where B and C denote the parameters in Antoine's equation for the saturated vapor pressure (Fujikawa 1995); d denotes the thickness of the liquid film; d_0 the initial thickness of the liquid film; D_s the thermal diffusivity of the end wall; L the latent heat for evaporation and condensation; p_0^* the initial saturated vapor pressure; p_∞ the vapor pressure behind the reflected shock wave; R the gas constant; t the time; T_0 the initial temperature of the vapor, the liquid film, and the end wall; T_∞ the vapor temperature behind the reflected shock wave; λ_s the thermal conductivity of the end wall; ρ_l the density of the liquid film; and σ_c the condensation parameter. There is a relation between the condensation parameter σ_c and the condensation coefficient α as follows:

$$\alpha = \frac{2\pi^{1/2}\sigma_c}{1 + 0.798\pi^{1/2}\sigma_c} \tag{2}$$

The growth behavior is drastically influenced by the condensation parameter. In Eq. (1), for $\beta_2 t^{1/2} \ll 1$, the growth of the liquid film is in proportion to the condensation parameter, while for $\beta_2 t^{1/2} \gg 1$ it becomes independent of the condensation parameter as time elapses. That is, in the range of $\beta_2 t^{1/2} \cong 1$, there exists a transition phenomenon in the liquid-film growth. Defining a transition time as τ in Eq. (1), we get

$$\tau = \frac{\lambda_s^2 p_\infty {T_0}^4}{2\rho_\infty {\sigma_c}^2 D_s L^2 \{(p_\infty - p_0^*)T_\infty + p_0^* S T_0\}^2} \tag{3}$$

where ρ_∞ denotes the vapor density behind the reflected shock wave. Equation (1) is valid for the condition of

$$\frac{t_\infty}{\tau} \ll 1 \tag{4}$$

where t_∞ is

$$t_\infty = \frac{D_\infty}{\{a_\infty \sigma_c (p_\infty - p_0^*)/p_\infty\}^2}$$

in which a_∞ denotes the sound speed of the vapor behind the reflected shock wave and D_∞ the thermal conductivity of the vapor behind the reflected shock wave.

3. Mechanism of transition phenomenon

The growth behavior of a liquid film is strongly dependent on the change of the saturated vapor pressure which is the function of the temperature of the liquid-film surface.

For $\beta_2 t^{1/2} \ll 1$, the temperature of the liquid-film surface is given by (Fujikawa 1995)

$$T_l(t, x=0) \cong T_0 + \frac{a_\infty L T_\infty \rho_\infty \sigma_c}{\lambda_s T_0} \cdot \left(1 - \frac{p_0^*}{p_\infty}\right) \cdot \left(\frac{8D_s}{\gamma\pi}\right)^{1/2} \cdot t^{1/2} \tag{5}$$

where x denotes the distance from the liquid-film surface toward the vapor and γ the specific heat. The temperature changes according to the square root of the time and its gradient is dependent on the condensation parameter. For $\beta_2 t^{1/2} \gg 1$, on the other hand, the temperature of the liquid-film surface is given by (Fujikawa 1995)

$$T_l(t, x=0) \cong T_0 + \frac{\beta_1 \rho_l D_s^{1/2} L}{\beta_2 \lambda_s} \tag{6}$$

That is, the temperature approaches asymptotically the constant one over time. The final temperature is here independent of the condensation coefficient.

Now, let us consider the balance of heat flux at the vapor-liquid interface as follows:

$$\rho_l L \frac{dd(t)}{dt} \cong \lambda_l \left(\frac{\partial T_l}{\partial x}\right)_{x=0} \cong \lambda_s \left(\frac{\partial T_s}{\partial x}\right)_{x=-d} \tag{7}$$

$$\propto \frac{T_s(t, x=-d) - T_0}{t^{1/2}} \cong \frac{T_l(t, x=0) - T_0}{t^{1/2}} \tag{8}$$

where T_l denotes the temperature of the liquid film, T_s the temperature of the end wall, and λ_l the thermal conductivity of the liquid film. For $\beta_2 t^{1/2} \ll 1$, from Eqs. (5) and (7), we get $dd(t)/dt = \text{const}$, i.e., $d(t) \propto t$. It can therefore be found that the process before the transition time is the one in which the temperature of the liquid-film surface increases and the temperature also diffuses into the interior of the end wall. The growth behavior of the liquid film is thus dependent on the condensation coefficient because it is strongly dependent on the temperature of the liquid-film surface. After the transition time, on the other hand, the temperature of the liquid-film surface approaches the asymptote given by Eq. (6); in consequence, Eq. (7) becomes $dd(t)/dt \propto t^{-1/2}$. That is, we obtain $d(t) \propto t^{1/2}$. At this time stage, the temperature of the liquid-film surface becomes constant, and the temperature of the interior diffuses into it. The growth behavior of the liquid film becomes independent of the condensation coefficient.

4. Measurements of condensation coefficient

Figure 2 (circle symbols) shows the time variation of the liquid-film thickness, $\Delta d = d - d_0$.

Theoretical curves for $\sigma_c = 0.040$, 0.045, and 0.050, i.e., in turn $\alpha = 0.13$, 0.15, and 0.16 are also shown. It is found that the most probable values of the condensation parameter and the coefficient are 0.045 and 0.15, respectively, in the present experimental conditions. The magnitude of experimental error associated with the present estimation of the condensation parameter and the coefficient can be evaluated by the following parameter $Z = (p_{\rm exp} - p_0^*)/(p_\infty - p_0^*)$, in which $p_{\rm exp}$ denotes the actual vapor pressure behind the reflected shock wave. For the present experiment, we obtain $Z = 0.98$, which means that the above most probable value of the condensation parameter is smaller than the true value by 2%. Thus, we can obtain the corrected value of the condensation parameter, 0.046 (the corrected condensation coefficient = 0.15).

In Fig. 2, results (square symbols) for conditions different from those above are shown. In this case, Z is determined to be 1.08. The corrected condensation parameter and the coefficient are thus 0.049 and 0.16, respectively.

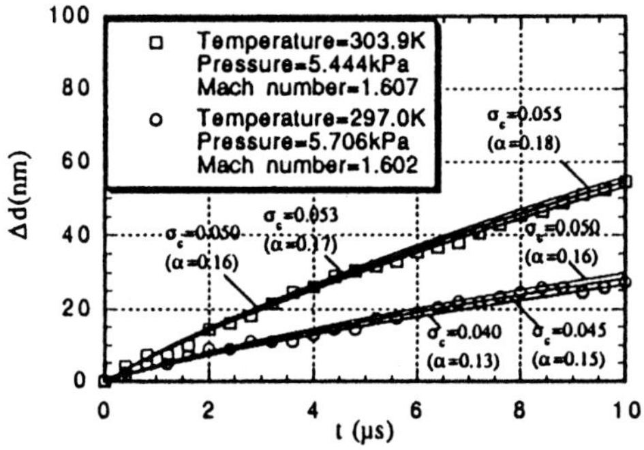

Figure 2. Time variation of liquid-film thickness.

5. Molecular dynamics simulation

In order to understand the dynamic processes of evaporation and condensation, a simulation of molecular dynamics is performed. In most studies to date on the condensation coefficient α, the condensation process has been considered as a unimolecular chemical reaction, which means that the evaluation of α is equivalent to estimating how many vapor molecules are reflected after colliding with the liquid surface. To avoid confusion, let $\alpha_{\rm self}$ be the condensation coefficient from this point of view. To estimate $\alpha_{\rm self}$ quantitatively, we have developed an autocorrelation function method. The result is $\alpha_{\rm self} \cong 0.9$, which is very different from the experimental values in the preceding sections.

In visualizing the molecular configurations, we sometimes observed that vapor molecules colliding with the liquid surface apparently drove other molecules out of the liquid. The significance of this molecular exchange can be estimated through the correlation of two fluxes on the vapor side of the interfacial region, $J_{\rm coll}$ for the flux of vapor molecule collision, and $J_{\rm ent}$ for the flux entering the vapor region. With a memory function $\kappa(t)$, the relation is expressed as

$$J_{\rm ent}(t) = J_{\rm evap}(t) + \int_0^\infty \kappa(t')J_{\rm coll}(t-t')dt' \tag{9}$$

where $J_{\rm evap}$ is the entering flux which is independent of $J_{\rm coll}$. Taking the ensemble average for each term in Eq. (9), we obtain the following expression for the total condensation coefficient:

$$\alpha = \frac{\langle J_{\rm evap}\rangle}{\langle J_{\rm coll}\rangle} = 1 - \int_0^\infty \kappa(t')dt' \tag{10}$$

The kernel $\kappa(t)$ is evaluated from the simulation data as

$$\langle \Delta J_{\rm coll}(t_1)\Delta J_{\rm ent}(t_2)\rangle = \begin{cases} 0 & t_1 > t_2 \\ C\kappa(t_2 - t_1) & t_1 \le t_2 \end{cases} \tag{11}$$

where Δ denotes the deviation from its average and C a constant representing the mean square variation of $\Delta J_{\rm coll}$. The estimated value of α is $0.1 \sim 0.3$ for methanol at 300 K.

6. Concluding remarks

The theoretical basis for the shock-tube deduction of the condensation coefficient has been presented on the basis of gasdynamic theory with kinetic boundary conditions. It has been elucidated that the condensation coefficient must be deduced within the transition time during which the liquid film begins to grow in proportion to the time. The condensation coefficient of methanol vapor has been found to be $0.15 \sim 0.16$. The molecular dynamics simulation has demonstrated that the above value of the condensation coefficient is a quite reasonable one and that the difference between α and $\alpha_{\rm self}$ is caused by molecular exchange at the interface.

References

Fujikawa S, Matsumoto M, Kotani M, Sato H (1995) Molecular study of evaporation and condensation of an associating liquid (shock-tube experiment and molecular dynamics simulation). Trans JSME 61:215–222.

Fujikawa S (1995) Theory of film condensation on shock-tube endwall behind reflected shock wave. Trans JSME 61:635–643.

Unsteady shock formation in transonic flow with nonequilibrium phase transition

G. H. Schnerr, S. Adam
Institut für Strömungslehre und Strömungsmaschinen
Universität Karlsruhe (TH), Kaiserstrasse 12
D-76128 Karlsruhe, Germany

Abstract: Here we present new modes of periodic shock formation in two-phase Laval nozzle flows with nonequilibrium condensation. In circular-arc nozzles different symmetric oscillation modes can be distinguished depending on the reservoir conditions and the nozzle geometry. In slender quasi-1D nozzles a sharp frequency minimum exists and at least one minimum was found for strongly curved nozzles. In Laval nozzles with parallel outflow we detected a new unsymmetric mode of oscillation with complex moving oblique shock systems above a threshold value of the water-vapor pressure in the supply. The frequency dependence on the vapor pressure ratio depicts the characteristic structure of a bifurcation problem. Experiments and numerical simulations are in excellent agreement. From our knowledge this is the first example of bifurcation in inviscid unsteady transonic flow.

Key words: Two-phase flow, Transonic flow, Nonequilibrium condensation, Bifurcation, Flow visualization, CFD

1. Introduction

Strong heat addition after nonequilibrium phase transition in expansion flows of water-vapor/carrier-gas mixtures causes oscillations with periodically moving shocks and high-frequency pressure fluctuations. Self-excitation develops because the upstream moving shock reduces the maximum nucleation rate and therefore the heat release to the flow for a certain time. Then, momentarily subcritical heat addition accelerates the flow again which increases the nucleation rate and leads to the formation of a new shock which again moves upstream. This phenomenon is controlled by the cooling rate of the nozzle $(-dT/dt)$, the reservoir conditions, and 2D wall-curvature effects. The periodic interaction of the flow with the homogeneous condensation process repeats with typical frequencies of 200 Hz up to 1200 Hz and is important to many technical applications, e.g., to steam turbines where it reduces the efficiency and excites flutter of the blades.

2. Physical model and numerical scheme

The process of homogeneous condensation is modeled with the classical nucleation theory of Volmer, Frenkel, and Zel'dovich and the Hertz-Knudsen droplet growth law. Surface tension and mass accommodation coefficient are determined corresponding to steady experiments. Because viscosity effects are not important, we solve the time-dependent 2D Euler equations in conservation form coupled to four additional conservation equations to calculate the liquid phase. Air and water vapor are treated as perfect gases, whereas the properties of the mixture depend on the condensate mass fraction g. The hyperbolic system of 8 conservation equations is solved by a MUSCL-type finite-volume method on a body-fitted grid. The source terms are treated by applying a fractional-step method with explicit, second-order-accurate time integration. Flux vector splitting of van Leer in the version for real gases is used to calculate the fluxes at the cell interfaces based on the frozen speed of sound. With the fully vectorized code a CPU time in the range of $5 - 10$ min is required to establish periodic solutions (measured performance 1.8 GFLOPS on SNI S600/20). Details of the numerical scheme may be found in Schnerr et al. (1994a,b).

3. Symmetric oscillations in nozzles S1, S2 — frequency minima

In our previous work we investigated instabilities in different circular-arc nozzles with emphasis on the frequency dependence on the water-vapor content in the mixture as well as on the effect of 2D wave phenomena (Schnerr et al. 1994b). For this type of nozzle geometry the transition from steady to unsteady flow occurs in a narrow range of the water-vapor pressure ratio Φ_0. We distinguish three different modes of oscillation with one minimum of the frequency for slender nozzles whereas two minima exist for flows in strongly curved nozzles (Fig. 1). The critical temperature gradient of the nozzles S2 and S1 is the same (1D theory), $(-dT/dx)^* = 8.1$ K/cm, which means that the different frequency distribution for identical reservoir conditions is caused by 2D wave effects. Stronger wall curvature tends to stabilize the flow near the walls and reduces the frequency of the oscillation near the limit of stability. In slender circular-arc nozzles the shocks always are symmetric with respect to the nozzle axis and almost normal, whereas complex systems of curved, but still symmetric shocks develop for higher nozzle wall curvatures (Schnerr et al. 1994b).

4. Unsymmetric oscillations in nozzle A1 — bifurcation

In the following we describe numerical and experimental investigations of the periodic shock formation in Laval nozzles with parallel outflow (exit Mach number $M_e = 1.2 - 1.4$, isentropic flow), and compare them with instabili-

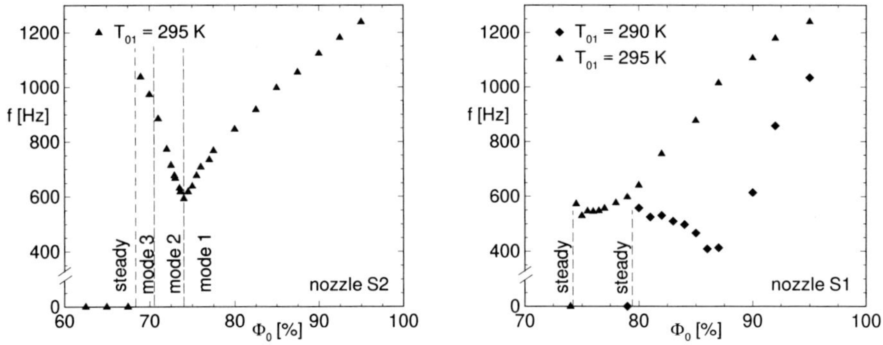

Figure 1. *Frequency minima* of self-excited flow oscillation in a slender circular-arc nozzle S2 (left), total throat height $2y^* = 30$ mm, radius of wall curvature $R^* = 400$ mm, and in a strongly curved nozzle S1 (right), $2y^* = 120$ mm, $R^* = 100$ mm. Numerical results of the frequency dependence on the water-vapor pressure ratio $\Phi_0 = p_v(T_{01})/p_{v,saturation}(T_{01}) \cdot 100$ [%], T_{01} reservoir temperature.

ties in circular-arc nozzles. The important difference between these two nozzle shapes with respect to the condensation process is the variation of the cooling rate $(-dT/dt)$, which is nearly constant in circular-arc nozzles, but decreases strongly nonlinearly in nozzles with parallel outflow where it vanishes at the nozzle exit. As a consequence the steady-state flow regime is limited to very low values of $\Phi_0 < 40\%$. However, below $\Phi_0 \approx 70\%$ the oscillations are still symmetric. Above this characteristic threshold we detected a new unsymmetric mode of oscillation with moving oblique shock systems, although the nozzle was perfectly symmetric with respect to the centerline.

All details of the experimentally determined flow field are well confirmed by numerical solutions of the Euler equations. Figure 2 shows one period of the unsymmetric oscillation. Downstream of the nozzle throat the flow depicts a system of moving oblique shocks decreasing in strength toward the nozzle exit. Near the nozzle throat an unsteady Mach reflection occurs ahead of the position where the very upstream shock becomes more and more curved and separates from the remaining shock system. It weakens and disappears in the oncoming subsonic flow. The maximum static-pressure increase with respect to the adiabatic flow is 27% of the reservoir pressure. Maximum nucleation rates grow in near-wall regions around the throat up to $J \approx 10^{20}$ m^{-3}s^{-1}. Obviously, and in contrast to circular-arc nozzles the condensation process concentrates near the upper and lower nozzle walls, where alternating intensive droplet formation develops, depicted by the condensate mass fraction contours g/g_{max}. By convection the condensate moves in the main flow direction toward the nozzle

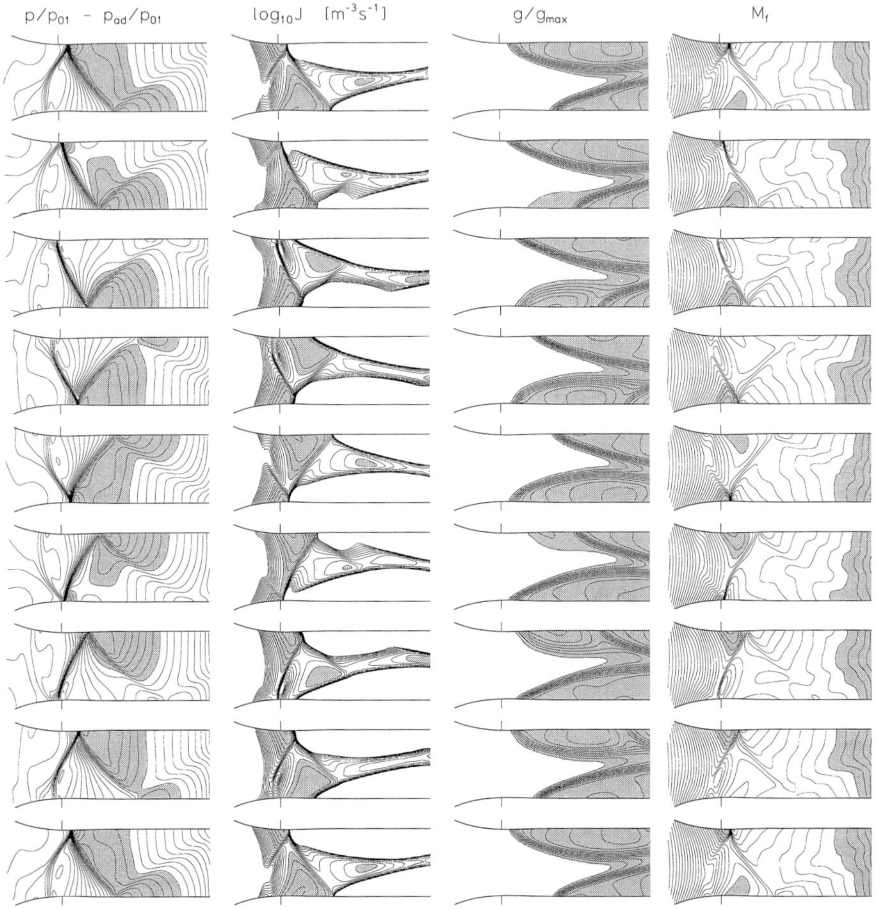

Figure 2. Numerical simulation of one period of *unsymmetric* oscillation in nozzle A1 — pressure disturbance with respect to the adiabatic flow $(p-p_{ad})/p_{01}$, increment 1%, shaded area $\Delta p/p_{01} > 20\%$; nucleation rate $\log_{10} J[\mathrm{m}^{-3}\mathrm{s}^{-1}]$, increment 1, shaded area $\log_{10} J > 15$; normalized condensate mass fraction g/g_{max}, increment 0.05, shaded area $g/g_{max} > 0.05$; frozen Mach number M_f, increment 0.02, shaded area $M_f > 0.90$; frequency $f = 1075$ Hz, reservoir conditions $T_{01} = 295$ K, $p_{01} = 1.0$ bar, $\Phi_0 = 90\%$, $x = 15.0\, \mathrm{g}_{H_2O}/\mathrm{kg}_{dry\,air}$. Flow from left, time increases from top.

exit. Simultaneously the shock system moves upstream. The frozen Mach-number contours M_f show that the entire flow field remains subsonic during the whole period whereas in cicular-arc nozzles the flow monotonically accelerates to $M_f > 1$.

The agreement between experiment and inviscid numerical simulation proves that this unsymmetric oscillation mode is not controlled by viscous effects but by the interaction of transonic flow and instantaneous heat addition caused by homogeneous condensation. Above the threshold value of Φ_0 our numerical simulation yields both the symmetric and the unsymmetric mode, whereas only the unsymmetric mode was observed in experiments. Generation of the unsymmetric flow oscillation requires instantaneous superposition of a small unsymmetric disturbance on the pre-existing symmetric solution. This indicates that the transition is a stability problem.

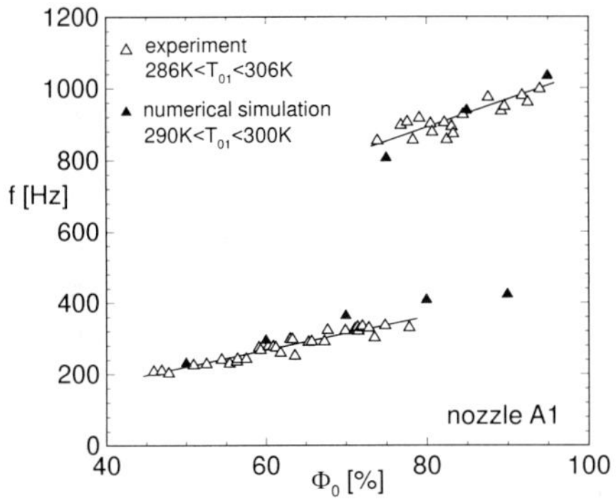

Figure 3. *Bifurcation* — Frequency dependence on the vapor-pressure ratio Φ_0 in nozzle A1 — parallel outflow $M_{exit} = 1.2$ (isentropic flow).

The experimental results shown in Fig. 3 depict the sudden jump of the frequency from a lower branch which represents the symmetric mode to the upper branch with unsymmetric oscillations. The frequency typically increases from 200–450 Hz to 800–1050 Hz. Obviously the frequency is less affected by variations of the reservoir temperature T_{01}. Comparison with numerical results shows convincing agreement with experiments.

Decreasing Φ_0 in the unsymmetric flow state yields flow patterns as shown in Figs. 4a–d. The oblique shocks steepen and the frequency, i.e. the velocity, of the shocks relative to the flow decreases. For $T_{01} = 295$ K, $p_{01} = 1.0$ bar and $\Phi_0 = 68.5\%$ the flow is still unsymmetric whereas $\Phi_0 = 68.0\%$ yields the symmetric solution.

For nozzles with higher exit Mach numbers up to 1.4 (isentropic flow) the unsymmetric structure of the flow field still remains, but now the flow accelerates to supersonic Mach numbers in the divergent nozzle section. Therefore the formation of unsymmetric oscillations is independent of the boundary condition at the nozzle exit (subsonic or supersonic outflow).

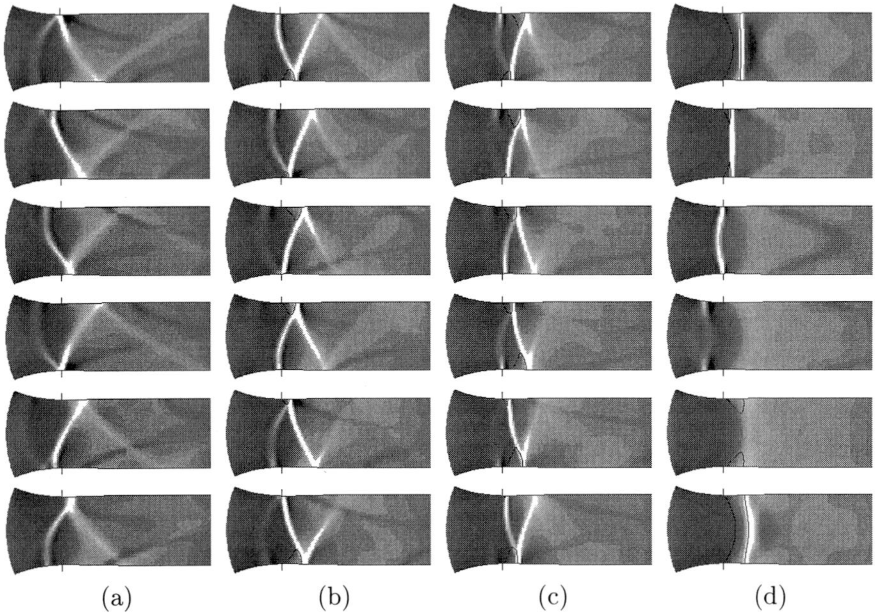

Figure 4. Numerical schlieren pictures of self-excited flows in nozzle A1: (a) $\Phi_0 = 95.0\%$, $f = 1106$ Hz; (b) $\Phi_0 = 75.0\%$, $f = 892$ Hz; (c) $\Phi_0 = 68.5\%$, $f = 759$ Hz; (d) $\Phi_0 = 68.0\%$, $f = 382$ Hz; ($T_{01} = 295$ K, $p_{01} = 1.0$ bar). The black lines indicate sonic lines, flow from left, time increases from top.

Acknowledgement. The authors would like to express their gratitude to the Deutsche Forschungsgemeinschaft (DFG) for supporting these experimental and theoretical/numerical research on condensing flows (contract Schn 352/13–3).

References

Schnerr GH, Adam S, Mundinger G (1994a) Frequency control of shock oscillations in high speed two-phase flow. In: Proc Fourth Triennial Int Symp on Fluid Control, Measurement and Visualization (FLUCOME'94), CERT-ONERA, Aug 29–Sept 2, 1994, P Hebrard (ed), ENSAE Toulouse France, Vol 2, pp 957–962.

Schnerr GH, Adam S, Mundinger G (1994b) New modes of periodic shock formation in compressible two-phase flows. In: Proc IUTAM Symposium Waves in Liquid/Gas and Liquid/Vapor Two-Phase Systems, Kyoto, May 9–13, S Morioka, L van Wijngaarden (eds), Kluwer Academic Publishers (1995), pp 377–386.

Condensation phenomena behind incident and reflected shock waves in heavy vapors at low Mach numbers

W. Garen, M. Becker, G. Paál
Department of Natural Science Technology
Fachhochschule Ostfriesland
Constantiaplatz 4, 26723 Emden, Germany

Abstract: Spontaneous condensation on the walls and in the core flow of a shock tube producing very low Mach number shocks has been investigated. The condensation behavior of regular and retrograde vapors has been compared. The rate of droplet growth has been described by existing theories. The applicability of new optical methods to monitor the growth of liquid film on the walls and to deduce droplet growth in the core flow has been demonstrated.

Key words: Condensation, Weak shock waves, Retrograde gases

1. Introduction

Condensation phenomena are usually investigated in expanding flows because of the rapid temperature drop leading to the supersaturation of the vapors (Wegener 1969, Sislian et al. 1976). One of the commonly occurring flow configurations is a rarefaction wave in a shock-wave tube. In our investigations, however, it was possible to detect condensation in the compressed flow behind incident and reflected shock waves. This uncommon phenomenon occurs only under the simultaneous presence of two factors: (a) the usage of *retrograde* fluids whose thermodynamic characteristics are different from the majority of fluids (called *regular* fluids), and (b) a shock wave with a very low Mach number.

Film condensation phenomena have been investigated by many authors (Fujikawa et al. 1987, Garen et al. 1989, Peters et al. 1989, Goldstein 1964, Teske et al. 1991) and one of the commonly used experimental techniques to measure film thickness is the intensity reflection method (Vasicek 1960) which uses the interference of two beams reflected from the air-liquid and the liquid-wall interfaces, respectively. An extended version of this technique was used in this work and its advantages will be elucidated later. Experiments have been performed for regular fluids (methanol, water) and for a typical retrograde fluid (octane) and the possibility of core condensation was investigated.

2. Theoretical background

Vapors of retrograde fluids will, contrary to regular ones, condense through adiabatic compression under certain conditions (Thompson et al. 1973). Retrograde fluids have heavy molecules and many internal degrees of freedom, and thus a large heat capacity. Their saturation boundary curve in the T-S diagram is "tongue-shaped," i.e. it has a positive slope on the vapor side (Fig. 1a). Let us consider an isentropic compression which is well approximated by a very weak shock wave. It is easy to see (Fig. 1b) that the slope of the insentropic compression curve (1-2) is larger than that of the vapor pressure curve, contrary to the case of regular fluids. Condensation is thus possible through isentropic compression.

Due to the small deviations from the equilibrium state, ideal gas behavior will be assumed. The slope of the vapor-pressure curve (obtained from the Clausius-Clapeyron equation) and that of the adiabatic compression curve are, respectively:

$$\left.\frac{dp}{dT}\right|_{Cl-Cl} = \frac{r(T)}{T^2}\frac{p(T)}{R_s} \qquad \text{and} \qquad \left.\frac{dp}{dT}\right|_{s} = \frac{\kappa}{\kappa-1}\frac{p(T)}{T} \tag{1}$$

where $r(T)$ is the heat of evaporation, R_s is the the specific gas constant, and κ is the adiabatic exponent. We can introduce the parameter B defined as:

$$B = \frac{\left.\frac{dp}{dT}\right|_{s}}{\left.\frac{dp}{dT}\right|_{Cl-Cl}} = \frac{c_p(T)\,T}{r(T)} = \frac{h(T)}{r(T)} \tag{2}$$

where c_p is the specific heat capacity at constant pressure and h is the specific enthalpy. Hence, if $B < 1$, the fluid is regular, if $B > 1$, the fluid is retrograde. For stronger shocks under ideal shock conditions similar criteria can be set up using the Rankine-Hugoniot relations in place of the isentropic compression.

In Fig. 2a, two pressures, p_i and p_{iS} (pressure behind the shock and the saturated vapor pressure, respectively; $i = 2$ for incident and $i = 5$ for reflected shocks) are compared as a function of the Mach number, both nondimensionalized by p_1, the initial pressure ($P_{2S} = p_{2S}(T_2)/p_1$; $P_2 = p_2/p_1$; $P_{5S} = p_{5S}(T_2)/p_1$; $P_5 = p_5/p_1$). Figure 2a shows both for the incident and the reflected shock waves that the vapor pressure falls below p_2 only within a region of Mach numbers whose boundaries depend on the initial conditions p_1, T_1 and the fluid properties. This region, called the retrograde region, shrinks with decreasing p_1 and T_1 and below a certain initial pressure it completely disappears. Condensation through compression is possible only within this region. Within the retrograde region the degree of supersaturation ($S_2 = p_2/p_{2S}$, $S_5 = p_5/p_{5S}$) varies with Mach number as seen in Fig. 2b. In our experiments iso-octane

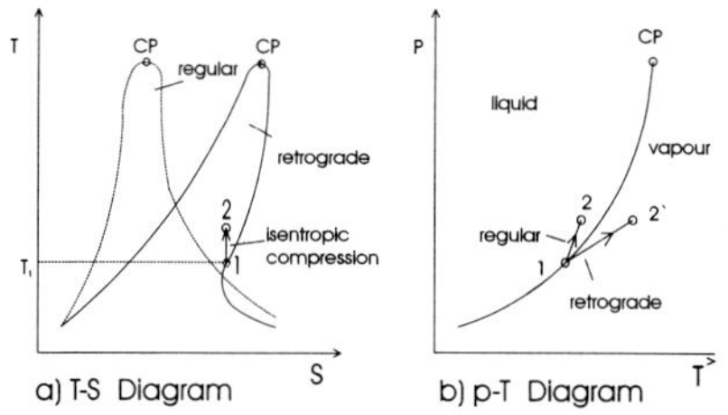

Figure 1. (a) T-S diagram, (b) p-T diagram.

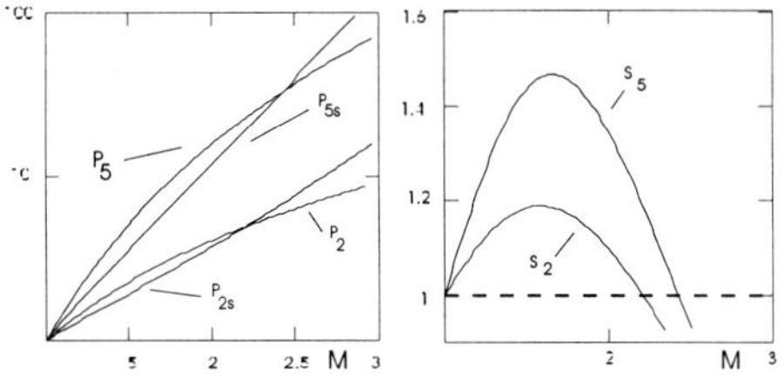

Figure 2. Pressure ratios (left) and saturation ratios (right).

was used and it exhibited retrograde behavior only within a relatively narrow Mach-number range both for incident and reflected shock waves.

3. Experimental setup and instrumentation

First, the shock-wave apparatus will be briefly presented here (Fig. 3); for a detailed description see (Garen et al. 1974). The shock tube is made of aluminum with a rectangular cross section of 40 mm × 40 mm and a length of 2600 mm. The usual shock-tube diaphragm was replaced by a special valve which opens the high-pressure section independently of the pressure difference between the high- and low-pressure sections. This configuration has several advantages: it is capable of producing highly repeatable shocks from very low up to medium Mach numbers and since no fragments of the bursting diaphragm are to be removed, the shock tube remains clean inside.

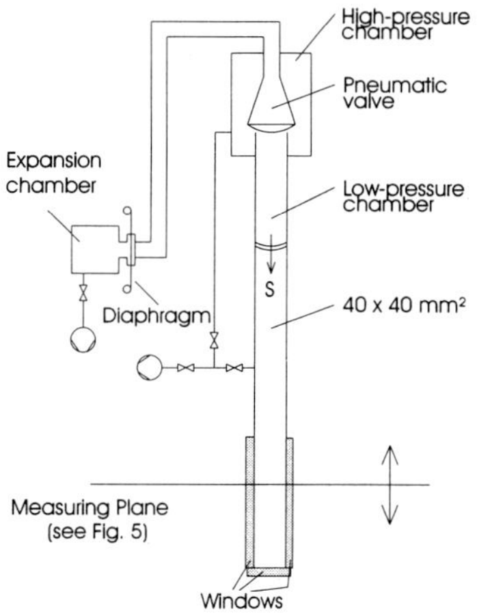

Figure 3. Experimental setup.

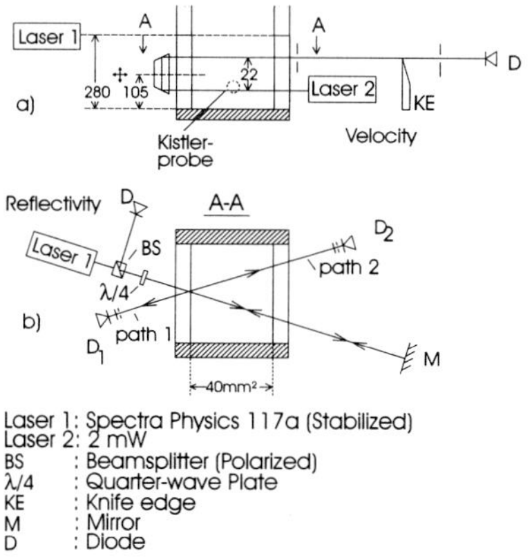

Figure 4. Measuring arrangement.

The thermodynamic properties of the test fluids are summarized in Table 1. In order to test our apparatus and instrumentation, the first measurements were taken using methanol which allowed the comparison of our results with those from the literature (Teske 1994). Next, iso-octane was used as a typical retrograde test fluid and distilled water as a regular test fluid.

Table 1. Thermodynamic properties of the fluids used at 22°C.

	M	c_p	Δh_{evap}	γ_{theor}	$B = \frac{c_p T_1}{r(T_1)}$
	$\frac{\text{kg}}{\text{kmol}}$	$\frac{\text{kJ}}{\text{kgK}}$	$\frac{\text{kJ}}{\text{kg}}$		
Methanol	32.04	1.407	1192	1.126	0.348
Iso-octane	114.22	1.64	307.74	1.046	1.572

4. Optical instrumentation

As shown in Fig. 4a, the velocities of the incident and reflected shocks are measured by a double schlieren setup. The distance of the two beams as well as the position of the whole setup along the test section can be varied. The pressure in the tube is monitored using a Kistler probe (Type 603B). In Fig. 4b the optical technique is shown. The standard reflection setup to measure film growth on the wall, consisting of a frequency stabilized He-Ne laser with 5 mW power and photodiode D1 (path 1) was supplemented by a second photodiode D2 which measures the thickness at the same point using the beam after three reflections (path 2). This configuration has a double advantage: the arrival of the shock can be detected by D2 through the schlieren effect; and by building the ratio of the signals D2 and D1, scattering effects due to the possible appearance of small droplets in the core flow can be detected. In order to avoid reflection back to the laser cavity, a polarizing beam splitter combined with a quarter wave plate was placed into the beam path.

5. Results and discussion

As already mentioned, the first test measurements were taken with methanol as a regular test fluid. The time histories and the intensity at diodes D2 and D2 are presented in Fig. 5. The decreasing frequency of the intensity variation corresponds to the well-known growth rate of the liquid film thickness proportional to $t^{\frac{1}{2}}$ (Teske 1994). The arrival of the incident and the reflected shock can be seen on the lower curve in the form of spikes.

In Fig. 6 signals from the two photodiodes for the retrograde fluid iso-octane are shown. The initial pressure p_1 is very nearly the saturated vapor pressure at room temperature ($p_1 = 49$ mbar, $p_s = 50$ mbar). The aforementioned law applies only after the reflected shock whereas between the incident and reflected shocks the the rate of growth of the film thickness follows a completely different law — the increasing frequency indicates an increasing growth rate.

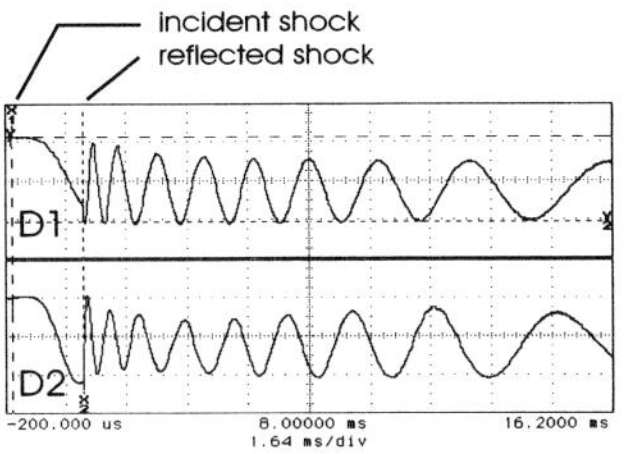

Figure 5. Time evolution of film thickness for methanol.

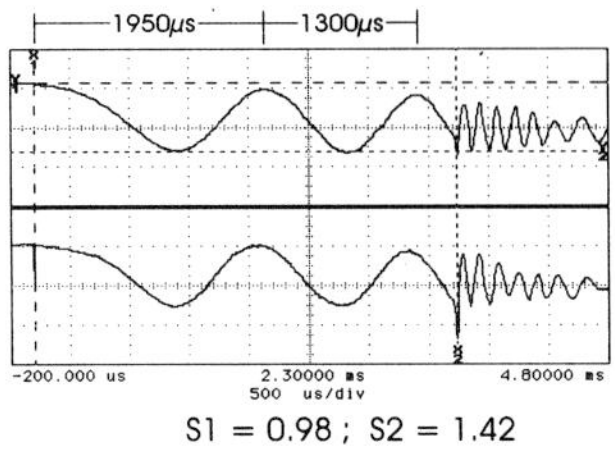

Figure 6. Time evolution of film thickness for iso-octane.

In Fig. 7 the initial pressure is clearly below the saturation pressure (38 mbar). After the passage of the incident shock the vapor will get just above the saturation point ($S_2 = 1.08$). As a result, the film condensation starts with a time lag. In this figure the benefit of employing a second diode becomes particularly clear — the difference between the two signals carries important information. The signal of D1 indicates that during the first 1.2 ms after the shock arrival no film condensation takes place. The signal of D2, however, shows a slight but monotonic intensity drop, which is attributed to the appearance of tiny droplets in the core flow. This hypotheses is further confirmed by the ruffled look of the D2 signal indicating Mie scattering from the droplets.

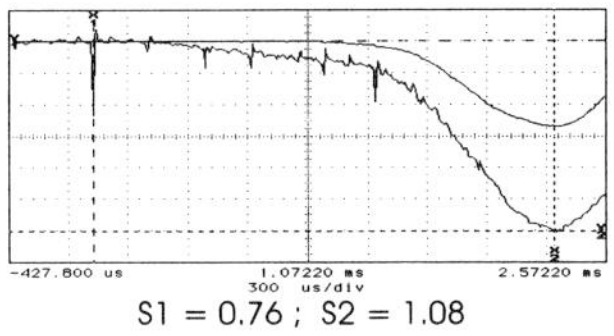

Figure 7. Time evolution of film thickness for iso-octane.

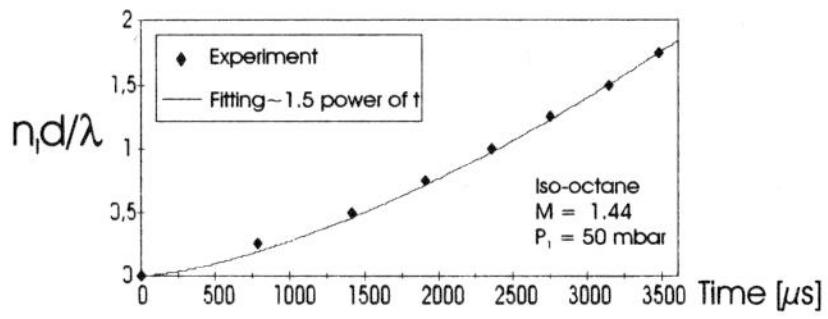

Figure 8. Normalized film thickness as a function of time.

In Fig. 8 the time history of the film thickness, calculated from signals like those in Fig. 6, is presented. It is compared with the theory which describes droplet growth in a supersaturated vapor under the assumptions that the transport coefficients are constant during the shock passage and that the mean free path is much smaller than the droplet radius (Gyarmathy et al. 1965). The agreement is surprisingly good, which may indicate that contrary to regular fluids, the condensation starts away from the wall in the form of rapidly growing droplets that may deposit on the wall, thus building the film. (Due to the lack of space the results obtained for water vapor cannot be presented here — they will be published elsewhere.)

6. Conclusions

(1) The results above have shown that for the retrograde fluid iso-octane, under sufficiently weak shocks the time history of the film condensation follows a different law than for regular fluids. There is evidence that droplets are formed in the core flow that may initiate film formation on the walls. (2) Our shock-wave apparatus proved to be ideally suited for these studies, which require a narrow range of small Mach numbers. (3) Our film-thickness measuring technique using two photodiodes proved fruitful both in obtaining the exact time of the shock-wave arrival and in indicating droplet formation in the core flow.

References

Fujikawa S, Okuda M, Akamatsu T, Goto T (1987) Non-equilibrium vapor condensation on a shock tube endwall behind a reflected shock wave. J Fluid Mech 183:293–324.

Garen W, Synofzik R, Frohn A (1974) Shock tube for generating weak shock waves. AIAA Journal 12(8):1132–1134.

Garen W, Diestelhorst-Becker N, Becker M (1989) Reflection and transmission measurements in shocked water vapors. In: Proc. of the 17th Int. Symp. on Shock Waves, Leigh University, Bethlehem, Pennsylvania, pp 796–801.

Goldstein R (1964) Study of water vapor condensation on shock tube walls. J Chem Phys 40:2793–2799.

Gyarmathy G, Meyer H (1965) Spontane Condensation, VDI Forschungsheft 508, VDI-Verlag, Düsseldorf.

Peters F, Paikert B (1989) Experimental results on the rate of nucleation in supersaturated n-propanol, ethanol and methanol vapors. J Chem Phys 91:5672–5678.

Sislian JP, Glass II (1976) Condensation of water vapor in rarefaction waves in homogeneous nucleation. AIAA Journal 14:1731.

Teske T, Obermaier F (1991) Comparative measurements of film condensation on the side and end walls of a shock tube. In: 18th Int. Symposium on Shock Waves (K Takayama, ed), Sendai, pp 485–490.

Teske T (1994) Parametrische Untersuchung der Filmkondensation reiner Gase und von Gemischen mit Inertgas im Stossrohr, PhD Thesis, Univ. Göttingen.

Thompson PA, Lambrakis KC (1973) Negative shock waves. J Fluid Mech 60:187–208.

Vasicek A (1960) Optics of Thin Films, Norbe Holland, Amsterdam.

Wegener PP (1969) Gas dynamics of expansion flows with condensation and homogeneous nucleation of water vapor. In: Nonequilibrium Flows 1, (PP Wegener, ed), Marcel Dekker, New York.

On the formation of embedded shock waves in rarefaction waves by homogeneous condensation

C. F. Delale*, F. Maršík
Institute of Thermomechanics
Academy of Sciences of the Czech Republic
Dolejškova 5
182 00 Prague 8, Czech Republic
G. H. Schnerr
Institut für Strömungslehre und Strömungsmaschinen
Universität Karlsruhe (TH), Kaiserstrasse 12
D-76128 Karlsruhe, Germany

Abstract: A theory of embedded shock formation by homogeneous condensation in the centered rarefaction wave of a shock tube is presented. The necessary and sufficient conditions for the existence of such embedded shock waves are exhibited in the parabolic approximation of the family of intersecting characteristics. In particular, the embedded shock origin is precisely located. Predictions of the theory, including an estimate for the average embedded shock speed, are compared with the results of experiments with visualized shocks for the expansion of water vapor in nitrogen (or air).

Key words: Homogeneous condensation, Intersecting characteristics, Embedded shock formation

1. Formation of condensation shocks

We consider a mixture of a condensable vapor and a carrier gas, which for time $t' < 0$ is at rest, in the driver section (region 4 in Figs. 1a and 1b) of a shock tube with initial relative humidity φ_4, initial specific humidity ω_4, and initial temperature T'_4 (here all primed quantities denote the actual dimensional variables whereas the corresponding unprimed variables are the dimensionless variables normalized as in Delale et al. 1995a). At $t' = 0^+$, just after the diaphragm at $x' = 0$ is instantly removed, a contact discontinuity moves off along path OC, a shock wave travels to the right along path OD, and a rarefaction wave centered

*Present address: Department of Mechanical Engineering, İstanbul University, 34850 Avcılar, İstanbul, and Department of Mathematics, TÜBİTAK Marmara Research Center, 41470 Gebze, Kocaeli, Turkey.

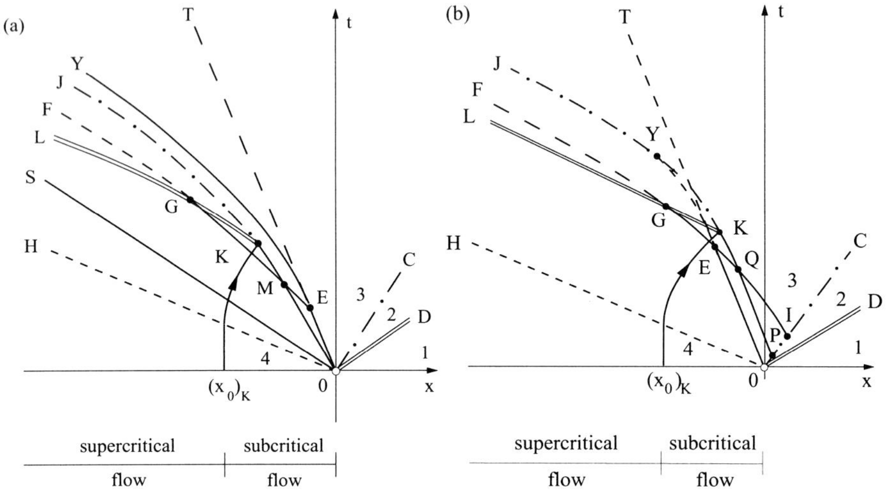

Figure 1. Wave diagram for the formation of an embedded condensation shock KL with origin K falling (a) in the interior of rarefaction wave, (b) in region 3 beyond the rarefaction wave tail EY.

at O travels to the left. The lines OH and OT are respectively the rarefaction wave head and the virtual isentropic rarefaction wave tail. The line OS represents the constant thermodynamic state along which saturation is reached. The condensable vapor is then cooled along pathlines in metastable states reaching relatively high supersaturations until condensation becomes visible (onset of condensation) due to the significant number of condensation nuclei formed by thermal fluctuations (homogeneous nucleation). The curve EF in Figs. 1a and 1b shows a typical wave front for the onset of condensation corresponding to maximum nucleation rates attained along pathlines. The characteristics or waves emanating either from the center O or the contact discontinuity OC remain almost straight as long as the effect of homogeneous condensation on the flow variables is negligibly small. However, as these characteristics cross the condensation wave front EF of Fig. 1a (or IEF of Fig. 1b), they begin to curve appreciably towards the head of the rarefaction wave due to substantial heat addition. Depending on the amount of heat released by condensation, the flow field may be continuous (subcritical flow) or it may include an embedded shock wave formed by excessive heat addition (supercritical flow). The complete asymptotic solution for subcritical flows is already discussed in Delale et al. (1995a) using asymptotic theory. In what follows we discuss the conditions under which embedded shock waves appear. The location of the embedded shock origin is then given explicitly.

We first need to parameterize both families of intersecting characteristics emanating from either the center O or from the contact discontinuity OC. The characteristics emanating from the center O are straight up to the condensation wave front EF prior to shock formation (see Fig. 1a) and are given by

$$x = -\alpha t \quad , \quad 0 < \alpha_E \le \alpha < \alpha_S \tag{1}$$

where α_S and α_E are the corresponding slopes of the saturation line OS and isentropic rarefaction wave tail OE. Denoting the condensation wave front by $t = t_\ell(\alpha)$, which is presumably to be obtained from the subcritical asymptotic solution of Delale et al. (1995a), (1) for the left-running characteristics emanating from the center O is only valid for $t \le t_\ell(\alpha)$. For $t > t_\ell(\alpha)$ this family of characteristics will be curved due to substantial heat addition as shown in Fig. 1a (a typical characteristic of the family is shown by the curve OMK). In the first approximation the equations of the characteristics can be assumed to take the parabolic form

$$t = A(\alpha)\,[\, x + \alpha\, t_\ell(\alpha)\,]^2 - \frac{x}{\alpha} \tag{2}$$

for $0 < \alpha_E \le \alpha < \alpha_s$ and $t \ge t_\ell(\alpha)$ where the single coefficient $A(\alpha)$, which will presumably be determined from the semi-analytical construction of the subcritical algorithm of Delale et al. (1995a), characterizes a measure of the curvature of the characteristics due to heat addition. In particular as $A(\alpha) \to 0$ (negligible heat addition), the characteristics become straight, i.e., $t \to -x/\alpha$. Since the characteristics in the x-t plane are curved concave downward, it follows that $A(\alpha) < 0$ for $0 < \alpha_E \le \alpha < \alpha_S$ and $t \ge t_\ell(\alpha)$. Now the position of the subcritical condensation wave front EF, given by $t_\ell(\alpha)$, together with the coefficient $A(\alpha)$ determine whether or not the family of characteristics will intersect in the rarefaction wave (existence of embedded shock waves in the rarefaction wave). It can be shown (details are given in Delale et al. 1995b) that two neighboring characteristics with initial slopes α and $\alpha + d\alpha$ intersect if

$$\Delta(\alpha) \equiv e^2(\alpha) + 4\,\alpha^3\, t_\ell(\alpha)\,\frac{dA}{d\alpha} \ge 0 \tag{3}$$

where $e(\alpha)$ is given by

$$e(\alpha) = 1 + 2\,\alpha^2 A(\alpha) \left[\, t_\ell(\alpha) + \alpha\,\frac{dt_\ell}{d\alpha} \right] \;. \tag{4}$$

In some cases embedded shock waves can also be formed due to intersecting characteristics emanating from the contact discontinuity OC (see Fig. 1b). Parameterizing the family of characteristics emanating from the contact discontinuity OC by β, chosen as the time of origin of a particular characteristic (e.g., curve PQK in Fig. 1b) emanating from the contact discontinuity, we obtain the equation of the characteristics

$$t = \bar{A}(\beta)\,[\,x - (V_c + \alpha_E)\,\beta + \alpha_E\,\bar{t}_\ell(\beta)\,]^2 + \left(1 + \frac{V_c}{\alpha_E}\right)\beta - \frac{x}{\alpha_E} \tag{5}$$

for this family in the parabolic approximation where β_I denotes the time when the condensation wave front meets the contact discontinuity, V_c is the normalized velocity of the contact discontinuity OC, and $t = \bar{t}_\ell(\beta)$ is the condensation wave front IE in this region. The single coefficient $\bar{A}(\beta)$, presumably to be obtained from the asymptotic solution in region 3, characterizes a measure of the curvature of characteristics in region 3. In particular $\bar{A}(\beta) = 0$ for $t < \bar{t}_\ell(\beta)$. It can easily be deduced that $\bar{A}(\beta) < 0$ for $0 \le \beta < \beta_I$ and $t > \bar{t}_\ell(\beta)$. It is also important to notice that as $\beta \to 0$, we are on the rarefaction wave tail OEY (see Figs. 1a and 1b) and thus $\alpha \to \alpha_E$. Consequently we have $\bar{t}_\ell(0) = t_\ell(\alpha_E)$ and $\bar{A}(0) = A(\alpha_E)$. The necessary and sufficient condition that two characteristics with parameters β and $\beta + d\beta$ intersect in the heat-addition zones can then be shown (details are given in Delale et al. 1995b) to be

$$\bar{\Delta}(\beta) \equiv \bar{e}^2(\beta) - 4\left(1 + \frac{V_c}{\alpha_E}\right)\frac{d\bar{A}}{d\beta} \ge 0 \tag{6}$$

where

$$\bar{e}(\beta) \equiv -2\,\bar{A}(\beta)\left[V_c + \alpha_E - \alpha_E\,\frac{d\bar{t}_\ell}{d\beta}\right] \tag{7}$$

(existence of embedded shock waves in region 3).

Depending on the amount of heat released by condensation, the embedded shock origin K may lie in the interior of the rarefaction wave (Fig. 1a), or beyond the rarefaction wave tail in region 3 (Fig. 1b), or, in the limit, on the rarefaction wave tail, with $(x_0)_K$ denoting the initial position of the particular pathline K lies on. It can be shown that K lies in the interior of the rarefaction wave if $\Delta(\alpha_E) < 0$, beyond the rarefaction wave tail in region 3 if $\Delta(\alpha_E) > 0$, and on the rarefaction wave tail if $\Delta(\alpha_E) = 0$. When K lies in the interior of the rarefaction wave, it is identified as the point where the characteristic with initial slope α_K, given by $\Delta(\alpha_K) = 0$, meets the envelope of intersecting characteristics given by (3) tangentially. For the case where the embedded shock origin lies beyond the rarefaction wave tail in region 3, it is identified as the point where the characteristic emanating from the contact discontinuity OC (Fig. 1b) with parameter β_K given by $\bar{\Delta}(\beta_K) = 0$ meets the envelope of the family of intersecting characteristics given by (6) tangentially. The explicit expressions for the coordinates of the embedded shock origin K in each case are given in Delale et al. (1995b). In the limit when $\alpha_K \to \alpha_E$ (or equivalently $\beta_K \to 0$), the embedded shock origin K lies on the rarefaction wave tail.

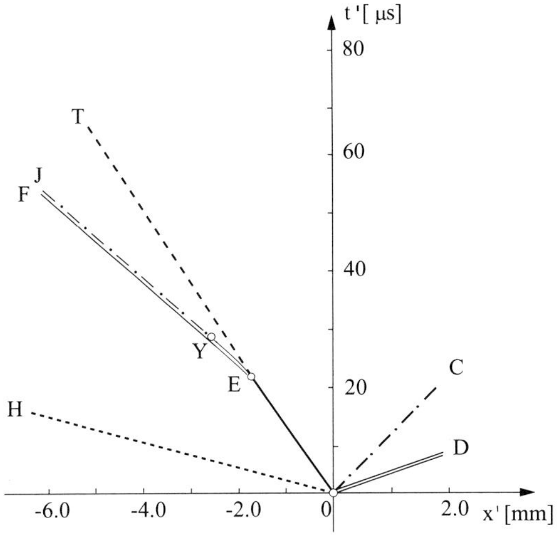

Figure 2. The condensation wave front EF and the envelope YJ of the family of intersecting characteristics for the expansion of water vapor in nitrogen with initial driver conditions $T_4' = 295.3$ K, $\varphi_4 = 0.9727$, and $\omega_4 = 0.01776$.

2. Results

Predictions of the embedded shock formation theory of the preceding section for the experiments of Glass et al. (1977) for the expansion of water vapor in nitrogen are obtained using the classical nucleation theory together with the Hertz-Knudsen droplet growth law as the condensation model. For these experiments the existence conditions for the formation of embedded condensation shock are satisfied, as visualized by the schlieren photographs as well. Identifying $A(\alpha)$ and $t_\ell(\alpha)$ for these experiments from the subcritical asymptotic solution of Delale et al. (1995a), we show that the embedded shock origin K, for the example studied, lies beyond the rarefaction wave tail (for details see Delale et al. 1995a). The results are shown in the x-t diagram in Fig. 2 where EF is the condensation wave front and YJ represents the envelope of the family of intersecting characteristics emanating from O with OEY denoting the rarefaction wave tail. The maximum nucleation rate along a pathline passing through Y is of the order of 10^{26} m^{-3} s^{-1} and the value of the condensation mass fraction at Y is of the order of 10^{-3}. Since the embedded shock origin K lies close to Y beyond the rarefaction wave tail (the location of point K by (12) and (13) requires the subcritical asymptotic solution of region 3, which is not readily available), these values can also be taken as typical values at point K. Using the experimental onset point of Glass et al. (1977), we find an average shock speed of 183.5 m/s which compares well with the visualized value of 190.5 m/s under similar conditions. The example studied here supports the validity of the assumptions underlying the presented shock formation theory. Of course, a full discussion of the asymptotic theory of shock-tube flows with an embedded condensation shock (supercritical flows) requires a shock-fitting technique for the condensation shock.

Acknowledgement. This work is supported in part by the Deutsche Forschungsgemeinschaft (DFG), in part by the Grant Agency of the Czech Republic, in part by the Scientific and Technical Research Council of Turkey (TÜBİTAK), and in part by the Turkish Academy of Sciences (TÜBA).

References

Delale CF, Schnerr GH, Zierep J (1995a) Asymptotic solution of shock tube flows with homogeneous condensation. J Fluid Mech 287:93–118.

Delale CF, Schnerr GH, Maršík F (1995b) Theory of embedded shock formation in rarefaction waves by homogeneous condensation (revised manuscript submitted to Eur J Mech B/Fluids).

Glass II, Kalra SP, Sislian J (1977) Condensation of water vapor in rarefaction waves. III. Experimental results. AIAA J 15:686–693.

Condensation induced by rarefaction waves and reflected rarefaction waves

Y. Fu, Z. Han, M. Gong
University of Science & Technology of China, Dept. of Modern Mechanics
Hefei, Anhui 230026, People's Republic of China

Abstract: Homogeneous condensation induced by unsteady rarefaction waves and reflected rarefaction waves in a vapor-gas mixture was investigated experimentally. It is shown that the temperature of condensation onset during very fast unsteady expansion in a vapor-gas mixture is much lower than that during equilibrium processes in the atmosphere. It is of interest to note that the size of droplets approximates a constant, but the number density and the mass density of droplets change rapidly in the region of static flow.

Key words: Shock tube, Rarefaction waves, Condensation

1. Introduction

Usually, the processes of the occurrence and disappearance of clouds and fog in the atmosphere can be described by equilibrium thermodynamics. In the area of weather modification, such as rain stimulation and hail mitigation by artificial means, the process of very fast unsteady changes in gas, water vapor, and droplet mixtures is important. Unfortunately, we do not exactly know the inherent characters of this process. Because fast unsteady condensation and evaporation occur in supersonic tubes, turbines, and combustion engines etc., many studies have been made of these phenomena. By using expansion and diffusion cloud chambers, and supersonic and expansion tubes, Kotake & Glass (1981) and Peters & Paikert (1989) investigated the processes of unsteady nucleation and condensation in vapor-gas mixtures. Marble (1969) and Nettleton (1977) reviewed the phenomena of evaporation. Goossens et al. (1986, 1988) and Smolders et al. (1989, 1992) studied the structure of a shock wave in fog and the evaporation caused by passage of a shock wave.

We report on homogeneous condensation in a vapor-gas mixture induced by unsteady rarefaction waves and reflected rarefaction waves in a shock tube. Our purpose is to investigate the process of phase transformation during very fast unsteady condensation in a vapor-gas mixture. A two-wavelength light-extinction method was used to calculate droplet size and concentration. The homogeneous condensation process, including changes in droplet number density and mass density, was observed.

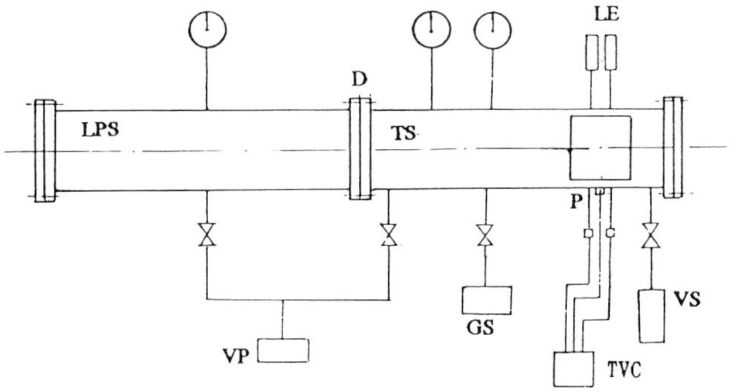

Figure 1. Schematic of experimental facilities. TS: test section; LPS: lower-pressure section; D: diaphragm; VP: vacuum pump; VS: vapor supply; GS: gas supply, LE: two-wavelength light-extinction facility; TVC: transient wave converter; P: pressure transducer.

2. Experimental method and facility

The experiments were conducted in a conventional shock tube with a square cross-section of 94×94 mm^2 as shown in Fig. 1. The driver section is 2.35 m long and the driven section is 2.0 m long. There is a Mylar diaphragm between driver and driven sections. In order to create the interaction between rarefaction waves and reflected rarefaction waves with a vapor-gas mixture, the driver section is used as a test section. A viewing window of 100×120 mm^2 is located 15 mm away from the end of the test section. Two semiconductor lasers ($\lambda_1 = 670$ nm, $\lambda_2 = 780$ nm) are used together with two photodiodes as a light detector for measuring droplet size and droplet number density. The pressure in the interaction region of rarefaction waves and reflected rarefaction waves is measured by a piezoelectric pressure transducer at a point 10 mm from the end of the test section. Pressure and light signals are simultaneously recorded on a transient wave converter and a microcomputer.

The pressure in the test section is brought to a certain value, then the section is filled with a vapor-gas mixture. The Mylar diaphragm is ruptured using electrical heating of a wire which is attached to its surface.

The vapor-gas mixture is expanded adiabatically through the rarefaction waves and reflected rarefaction waves before condensation occurs. The onset of condensation and the growth of the droplets are detected by means of the two-wavelength light-extinction method, called the DQ method, described by Goossens et al. (1988). The DQ method is based on the principle that the extinction of a plane light wave in a cloud of droplets is a function of wavelength. With this method, the mean droplet radius and droplet number density can be calculated and the width of the droplet-size distribution function can be estimated. According to the Lambert-Beer Law, the intensity of a light beam passing through a scattering medium over a distance L is described as follows:

$$I = I_0 \exp(-\beta L) \ . \tag{1}$$

Considering that the scattering medium consists of droplets with a certain size distribution, and a number density N, and that the refractive indices of the media are given, the extinction coefficient β can be written as

$$\beta = N\pi\langle r^2 Q\rangle \tag{2}$$

where Q is the extinction efficiency, r is droplet radius, and $\langle r^2 Q\rangle$ denotes the average extinction cross-section. The extinction efficiency Q is a known function of r/λ and can be calculated for spherical particles using Mie theory.

The dispersion quotient DQ is defined by Goossens et al. (1988) as a ratio of the extinction coefficients β_1 to β_2 measured by two different wavelengths. It can be written as

$$DQ = \frac{\beta_1}{\beta_2} = \frac{\langle r^2 Q_1\rangle}{\langle r^2 Q_2\rangle} \ . \tag{3}$$

Usually, the droplet-size distribution $F(r)$ is proposed to satisfy a 0th-order log-normal distribution with a modal droplet radius r_m and a width parameter ε:

$$F(r) = \frac{1}{\sqrt{2\pi}\varepsilon r_m} \exp\left[-\left(\frac{\ln(r/r_m)}{\sqrt{2}\varepsilon}\right)^2 - \frac{\varepsilon^2}{2}\right] \ . \tag{4}$$

For a constant ε, the value of DQ first strongly decreases with increasing particle size and goes to an absolute minimum. The value of the minimum of DQ depends on ε. When the droplets formed by condensation become sufficiently large, the minimum DQ value can be accurately measured.

In the present experiments, we assume that the condensed droplets have equal size at the same time (Hastings & Hodgson 1979). When two different wavelength light-extinction measurements are made, the droplet radius r and the number density N can be calculated from Eq. (1) and Eq. (2). The mass of droplets per unit volume ρ_p can be expressed by the droplet number density N and the density of water material ρ_w by:

$$\rho_p = \frac{4}{3}\pi\langle r^3\rangle \rho_w N \tag{5}$$

where $\frac{4}{3}\pi\langle r^3\rangle$ is the average volume of droplets.

3. Experimental result and discussion

Typical experimental results of homogeneous condensation due to unsteady expansion in a vapor-gas mixture are shown in Fig. 2. The states before and after the rupture of the diaphragm are given in Table 1, where p_1 is the initial pressure in the driven section of the shock tube; p_4 and T_4 are the initial pressure and temperature in test section of the shock tube; p_{41} is the ratio of the pressure p_4 to the pressure p_1; and p_5 and T_5 are pressure and temperature in the region behind the end of reflected rarefaction waves. Note that the parameters in Table 1 present the condition of isentropic flow in which no condensation occurs.

Table 1. State before and after diaphragm rupture.

p_1 (atm)	p_4 (atm)	p_{41}	T_4 (K)	p_5 (atm)	T_5 (K)
0.1	1.7	17.0	298.0	0.0455	105.96

The signals of pressure and two-wavelength light extinction are illustrated in Fig. 2a in which the condensation of the vapor-gas mixture takes place under the effect of unsteady rarefaction waves and reflected rarefaction waves. Then, reflected compression waves cause the water droplets to evaporate. The expansion of this gas mixture is regarded as isentropic before the onset of condensation. So the temperature of the onset can be calculated from the isentropic relation

$$\left(\frac{T_c}{T_4}\right)^{1/2} = \left(\frac{\rho_c}{\rho_4}\right)^{\frac{\gamma-1}{2}} = \left(\frac{p_c}{p_4}\right)^{\frac{\gamma-1}{2\gamma}} \tag{6}$$

where subscript c represents the state of condensation. As soon as condensation occurs, the state of the gas-droplet mixture no longer obeys the isentropic relation because latent heat is released.

The temperature T_c (111 K) is calculated with p_4, T_4, and p_c. The pressure p_c (0.05 atm) is observed the onset of condensation. The droplet radius r, number density N, and mass density ρ_p are calculated and shown in Figs. 2b, 2c, and 2d. It follows from Fig. 2b that droplet size grows so quickly that it takes almost no time for droplet radius to reach a constant 510 nm at the beginning of condensation. On the other hand, it takes about 2.0 ms for both the number and the mass density of droplets to change from zero to the first peak. In the region between the end of reflected rarefaction waves and the beginning of the reflected compression waves, in which the gas-droplet mixture is at rest, the droplet radius r almost remains constant, but the number density and mass density of droplets decrease obviously after the first peak, then both of them rapidly increase to the second peak at 13.7 ms. It takes about 7.7 ms for this process. This phenomenon is interesting and represents the non-equilibrium condensation process of a vapor-gas mixture.

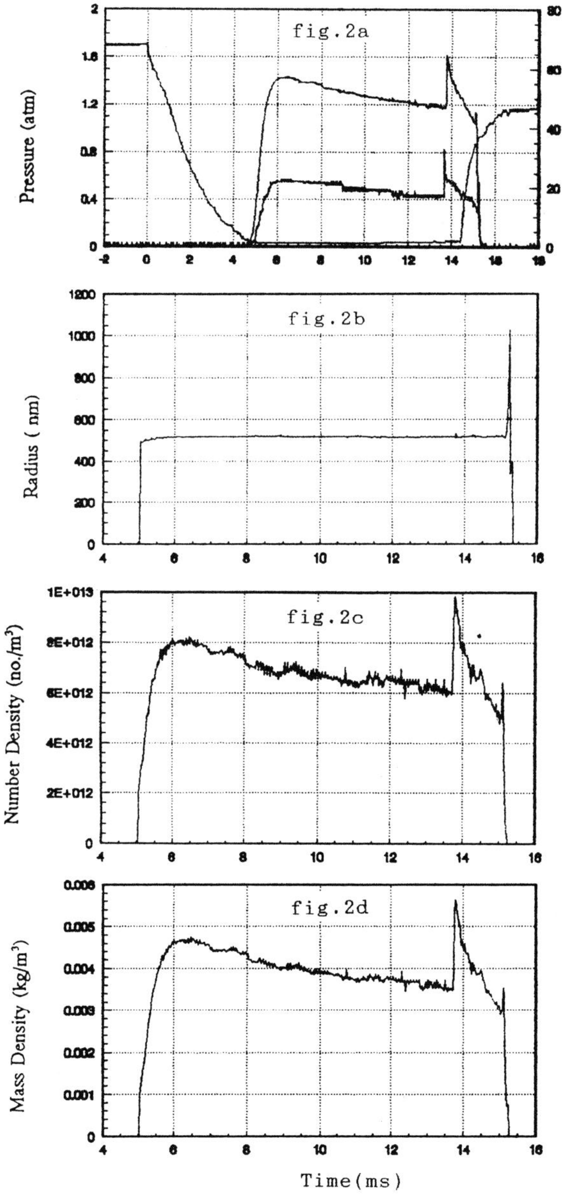

Figure 2. Typical recordings of induced condensation with initial conditions $P_1 = 0.1$ atm, $P_4 = 1.7$ atm, and $T_4 = 298$ K: (a) pressure and light extinction; (b) droplet radius; (c) number density; and (d) mass density.

From the experimental results above, it is evident that the unsteady condensation of a vapor-gas mixture is different from the condensation that occurs in the natural atmospheric environment. In the natural atmospheric environment, water vapor and supercooled droplets usually become ice crystals at about −40°C. But in the process of very fast unsteady expansion with rapid decrease of temperature, the temperature of condensation onset is much lower than in the natural atmospheric environment. This delay process may result from an additional resistance of the pressure decrease against the condensation of water vapor. But it is difficult to explain another interesting phenomenon: that the number density and mass density of droplets decrease from 6.0 ms to 13.7 ms, and both of them rapidly increase to the second peak at 13.7 ms. This phenomenon happens in the region of static flow where the pressure remains almost constant. We try to explain this phenomenon from the viewpoint of non-equilibrium thermodynamics.

References

Goossens HWJ, Berkelmans MJCM, van Dongen MEH (1986) Experimental investigation of weak shock wave propagating in a fog. In: Proceedings of the International Symposium on Shock Waves and Shock Tubes, Stanford, USA, pp 721–727.

Goossens, HWJ, Cleijne JW, Smolders HJ, van Dongen MEH (1988) Shock wave induced evaporation of water droplets in a gas-droplet mixture. Exp Fluids 6:561–568.

Hastings DL, Hodgson JP (1979) The formation of an aqueous fog in a shock tube. J Phys D, Appl Phys 12:2111–2122.

Kotake S, Glass II (1981) Flows with nucleation and condensation. Prog Aeros Sci 19:129–196.

Marble FE (1969) Some gasdynamic problems in the flow of condensing vapors. Astronaut Acta 14:585–614.

Nettleton MA (1977) Shock-wave chemistry in dusty gases and fogs: a review. Comb and Flame 28:3–16.

Peters F, Paikert B (1989) Nucleation and growth rates of condensing water vapor in argon from shock tube experiments. Exp Fluids 7:521–530.

Smolders HJ, Willems JFH, de Lernge HC, van Dongen MEH (1989) Wave induced growth and evaporation of droplet in a vapor-gas mixture. In: Proceedings of the 17th International Symposium on Shock Waves and Shock Tubes, Bethelehem, USA, pp 802–807.

Smolders HJ, de Kok CJGM, van Dongen MEH (1992) Condensation and evaporation relation induced by non-linear waves. In: Proceedings of the 18th International Symposium on Shock Waves and Shock Tubes, Sendai, Japan, pp 475–480.

Condensation shock wave as the consequence of the principle of minimum entropy production

F. Maršík, P. Sopuch, J. Blaha
Institute of Thermomechanics
Academy of Sciences of the Czech Republic
Dolejškova 5, 182 00 Prague 8, Czech Republic

Abstract: The principle of minimum entropy production appears as a new unifying concept for the description of condensation and evaporation. The Chapman-Jouguet condition for these phenomena is equivalent to this principle and gives new relations for the well-measurable quantities: pressure, density, and amount of condensate. The onset of condensation demonstrated by the condensation shock is experimentally studied for water, ethanol, and pentanol (with argon as a carrier gas) in the expansion fan in a shock tube. To validate this principle the pressure jump of the condensation shock and its velocity are measured.

Key words: Phase transition, Condensation shock, Minimum of entropy production

1. Introduction

The homogeneous nucleation of water vapor in a carrier gas was first studied experimentally in the expansion part of a shock tube using light scattering by Barschdorff (1975) and the numerical solution was added by Sislian and Glass (1976). The nucleation and growth rate of droplets in the argon was measured by Peters and Paikert (1989) using the Mie-scattering method.

The temperature gradient in the expansion part is given by

$$\frac{\mathrm{d}T_4}{T_L \mathrm{d}x} = \frac{\mathrm{d}T_4}{a_L T_L \mathrm{d}t} = -\frac{2(\kappa_4 - 1)}{(\kappa_4 + 1)L_{TS}} \tag{1}$$

and can be adjusted by changing the distance of the test section from the diaphragm (L_{TS}). The speed of sound in the driver section is a_L, then by (1) for $L_{TS} = 219$ mm (see Fig. 1) and $\kappa_4 = 1.63$, the temperature gradient drop equals 0.2 K/μs.

The general validity of the principle of the minimum entropy production as a new unifying concept for the description of condensation accompanying a shock has to be proved. The application on the phase transition was first mentioned by Maršík et al. (1994). This principle gives new relations for well-measurable quantities such as pressure, density, and mass fraction of condensate. The correspondence of the principle with the Chapman-Jouguet condition for chemical reactions (combustion) was demonstrated by Barrère and Fabri (1971).

2. Balance laws on singular surfaces

Any disturbances that propagate through a continuum with a finite speed can be regarded as a wave. A singular surface is a wave which has normal and discontinuous derivatives of the trajectory $\mathbf{x} = \mathbf{x}(\mathbf{X}, t)$ of a material point across the surface. The most important singular surface (SS) is the shock wave, i.e., the derivative $\dot{\mathbf{x}} = \mathbf{v}$ (the velocity field) is discontinuous. Let $+$ denote the quantities before and $-$ the quantities behind the SS (see Fig. 1). We assume the gradients of all quantities, e.g., pressure p, temperature T, velocity v, enthalpy h, entropy s etc., are negligibly small in the vicinity of the SS. Balance laws for the one-dimensional (shock tube) case in nondimensional forms are (Landau and Lifschic 1986) for mass:

$$\tilde{j} = \frac{\rho^+ U^+}{(p^+ \rho^+)^{\frac{1}{2}}} = \frac{\rho^- U^-}{(p^+ \rho^+)^{\frac{1}{2}}}, \quad \text{for} \quad U^+ + v^+ = U^- + v^- \tag{2}$$

where U^+ and U^- are the velocities of the SS with respect to the fluid before and behind, respectively. For momentum (Rayleigh line):

$$\tilde{p} = -\tilde{j}^2 \tilde{\nu}, \quad \text{for} \quad \tilde{p} = \frac{p^- - p^+}{p^+}, \quad \tilde{\nu} = \frac{\nu^- - \nu^+}{\nu^+} = \frac{\rho^+ - \rho^-}{\rho^-}. \tag{3}$$

For energy (Rankine-Hugoniot relation):

$$\tilde{h} = \frac{h^- - h^+}{p^+ \nu^+} = (1 + \frac{\tilde{\nu}}{2})\tilde{p}. \tag{4}$$

For entropy:

$$\frac{T^- ds^-}{p^+ \nu^+} = d\tilde{h} - (1 + \tilde{\nu}) d\tilde{p} = \frac{\tilde{\nu}^2}{2} d\tilde{j}^2. \tag{5}$$

3. Minimum entropy production and the Chapman Jouguet condition

From these conservation equations, one finds the entropy produced in the fluid going through the SS is

$$\rho^+ U^+ (s^- - s^+) = \sqrt{p^+ \rho^+}\, \tilde{j} (s^- - s^+) \quad \left[\frac{\mathrm{J}}{\mathrm{m^2 s K}}\right]. \tag{6}$$

The extreme condition of the entropy production (6) calculated from Eqs. (3–5) (for $s^+ = const$ and $s^- = s^-(\tilde{p}, \tilde{\nu})$) is

$$\frac{T^- ds^-}{p^+\nu^+} = \frac{\tilde{p}}{2} d\tilde{\nu} - \frac{\tilde{\nu}}{2} d\tilde{p} = \frac{1}{2}\Big(\tilde{p} - \tilde{\nu}\frac{d\tilde{p}}{d\tilde{\nu}}\Big) d\tilde{\nu} = 0 \ . \tag{7}$$

The local minimum condition of the entropy production (6) behind the SS is determined by its second differential $\tilde{\nu} \mathrm{d}^2\tilde{p}/(\mathrm{d}\tilde{\nu})^2 < 0$. The Chapman-Jouguet condition can thus be restated (Maršík et al. 1994) as: *The entropy production is minimum at the stable Chapman-Jouguet point.*

3.1. Condensation shock wave

The densities, pressures, and enthalpies in the balance laws (2), (3), (4) for the mixture of carrier gas (Ar) and condensing vapor have the form

$$\rho = \rho_c + \rho_v + \rho_l, \quad p = p_c + p_v, \quad h = h_v + c_c(h_c - h_v) - c_l h_{lv}, \tag{8}$$

where $c_c = \rho_c/\rho$, $c_l = \rho_l/\rho$, and $h_{lv} = h_{lv0} - k_v \hat{R}_v T$ is the heat of evaporation. Specific heats of the gaseous mixture are

$$\overline{c}_p = c_{pv} + c_c(c_{pc} - c_{pv}), \quad \overline{c}_v = c_{vv} + c_c(c_{vc} - c_{vv}) = \overline{c}_p - \hat{c}_0 \hat{R}_v \ . \tag{9}$$

The state equation is taken in the ideal gas mixture form

$$p = p_c + p_v = \rho(\hat{c}_0 - c_l)\hat{R}_v T, \quad \hat{c}_0 = \frac{c_c(M_v - M_c) + M_c}{M_c}, \tag{10}$$

where M_c and M_v are the molecular mass of carrier gas and vapor, respectively. Substituting the temperature from enthalpy in the energy equation (4) by equation (10) we obtain the general form of Rankine-Hugoniot (R-H) adiabat.

The nondimensional liquid fraction behind the condensation wave (CW), the evaporation heat, and the adiabatic exponent of gaseous mixture are

$$c_l^- = c_l^+ + \tilde{c}_l(\hat{c}_0 - c_l^+), \quad \tilde{q}_l = \frac{h_{lv0}}{\hat{R}_v T^+}, \quad \overline{\kappa} = \frac{\overline{c}_p}{\overline{c}_v},$$

respectively. The condensate change c_l^-, (or $\tilde{c}_l$ on the singular surface) is called a condensation or evaporation shock and its magnitude is unknown. We determine it by principle of the minimum entropy production (7).

The entropy for a system with a phase transition is defined by the formula

$$T ds = dh - \nu dp - (\mu_l - \mu_v) d\tilde{c}_l \ . \tag{11}$$

In the region (+) before the CW, the so-called nucleation region, we consider $d\tilde{c}_l = 0$. In the region (–) behind the CW we assume equilibrium two-phase flow where the chemical potential of the liquid, μ_l, and vapor μ_v, are equal. At this condition the droplets have critical size. The transition between these two regions takes place on the condensation (evaporation) wave.

We apply the extremum condition (7) to the entropy rise (11) while requiring conditions which are balance of momentum (3) and balance of energy (14). The method of Lagrange multipliers gives the necessary conditions for the extremum of the entropy production (Barrère and Fabri 1971)

$$\tilde{\nu} = \frac{\tilde{p}}{\overline{\kappa} + (\overline{\kappa}+1)\tilde{p}} \, . \tag{12}$$

Introducing (12) into the *R-H* adiabat (4) under the typical experimental condition $c_l^+ = 0$, $c_l^- \ll \hat{c}_0 = 0.45$ and $k_v = 0$ reduces the *R-H* adiabat to the relations between Mach number M_{CW} of the *CW* and heat release $\tilde{Q}\tilde{c}_l$ through the condensation discontinuity

$$M_{CW} = \frac{\sqrt{2(\overline{\kappa}+1)\tilde{Q}\tilde{c}_l} + \sqrt{2(\overline{\kappa}+1)\tilde{Q}\tilde{c}_l + 4}}{2} \, , \tag{13}$$

for

$$\tilde{Q} = \frac{\overline{\kappa}-1}{\overline{\kappa}}\tilde{q}_l - 1 \begin{cases} > 0 & (\doteq 3.1 \quad \text{for water}) \quad \ldots \text{regular fluids} \\ < 0 & (\doteq -0.75 \quad \text{for } C_6F_{14}) \quad \ldots \text{retrograde fluids.} \end{cases} \tag{14}$$

The quantity $\tilde{Q}\tilde{c}_l > 0$ represents the heat release corresponding to condensation.

The Mach number of the condensation shock as a function of pressure jump is

$$M_{CW} = \sqrt{1 + \frac{\hat{p}^2}{4(1+\hat{p})}} + \frac{\hat{p}}{2}\sqrt{\frac{1}{(1+\hat{p})}} \, , \qquad \text{for} \;\; \hat{p} = \frac{(\overline{\kappa}+1)\tilde{p}}{\overline{\kappa}} \, . \tag{15}$$

Here $\tilde{p} = (p^- - p^+)/p^+$ (where + and − denote the quantities in front of and behind the shock wave, respectively), and $\overline{\kappa} = \bar{c}_p/\bar{c}_v$ is the specific heat ratio of the whole mixture, see Eq. (9).

4. Experimental method

The driver part was evacuated and filled with the desired vapor to pressure p_v (see Fig. 1). Argon was added to the desired pressure and the acetate diaphragm was ruptured by a needle. The *CW* Mach number, M_{CW}, was measured from the time shift of the pressure jump recorded by two pressure transducers at different places ($L_{PT} = 320$ mm) (see Fig. 2a). The pressure-time course was recorded by a two-channel oscilloscope (the upper trace is the transducer closer to the diaphragm) and evaluated from a photograph of the oscilloscope trace.

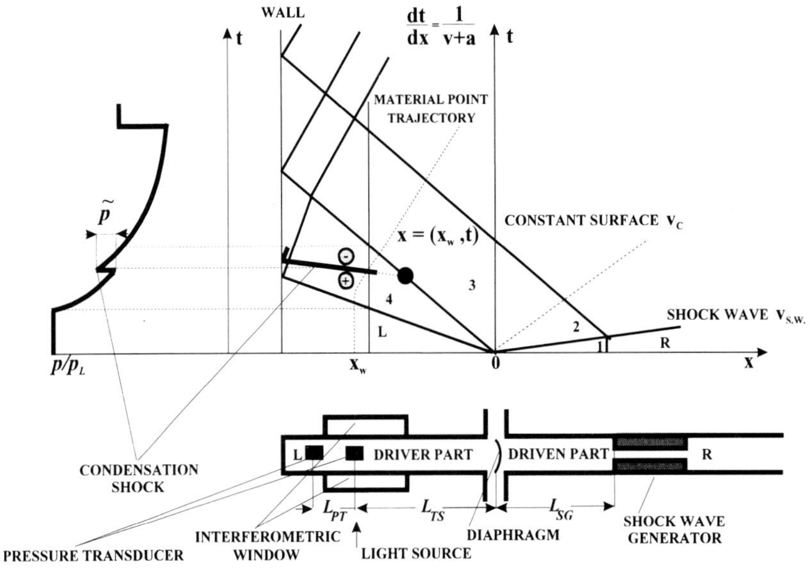

Figure 1. The principle of the condensation shock measurement in the shock tube.

5. Results and discussion

The influence of the condensation shock on the pressure and density jump is evident from Fig. 2a. Its intensity depends on the amount of the condensed liquid, $\tilde{c}_l$, and on the equilibrium properties of the vapor and the liquid. The quantity $\tilde{Q} = (\overline{\kappa} - 1)/(\overline{\kappa}\tilde{q}_l) - 1$ seems to be very important, where $\tilde{q}_l$ is nondimensional evaporation heat and $\overline{\kappa}$ is specific heat ratio of the expanding mixture of gases.

The position of the condensation shock in the expansion depends on the nucleation model, particularly on the surface tension, σ. The onset of condensation and the surface tension are here coupled with the amount of condensate c_l. Hence, precise pressure and M_{CW} measurements give additional information about the wetness, so that the nucleation rate can be determined more accurately. Further, it follows from equations (13) and (15) that the chemical composition and molecular masses play also a role. A comparison of predictions of the principle of the minimum of entropy production (15) with experiments performed for water, ethanol, and pentanol at different pressures is shown in Fig. 2b.

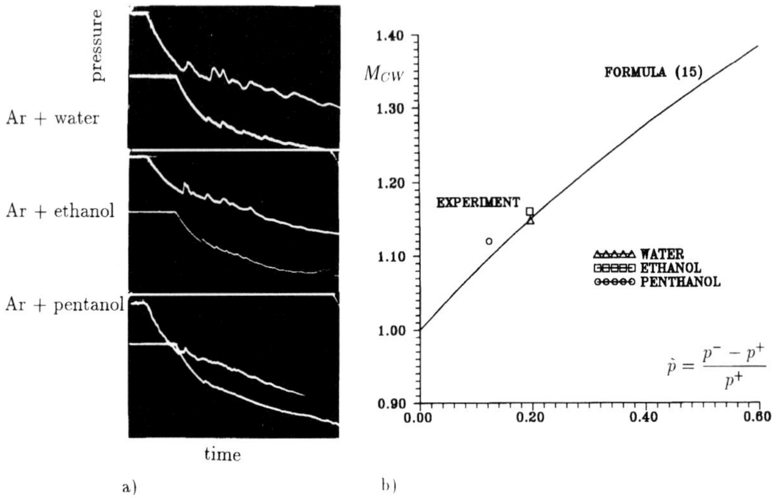

Figure 2. (a) The pressure time course at two different places (see Fig. 1); (b) condensation shock-wave Mach number versus pressure jump $\tilde{p}$.

Acknowledgement. This work was supported by the Grant Agency of the Academy of the Czech Republic, No. 276403 and by the Grant Agency of the Czech Republic, No. 101/93/0445.

References

Barrère M, Fabri J (1971) Ondes de choc avec combustion. In: Chocs et ondes de choc (Tom I, ed), A. L. Jaumotte, Mason & Cie, Paris, pp 192–223.

Barschdorff D (1975) Carrier gas effects on homogeneous nucleation of vapor in shock tube. Phys Fluids 18(5):529–535.

Landau LD, Lifschic EM (1986) Hydrodynamics, Nauka, Moskva (in Russian).

Maršík F, Sopuch P, Blaha J (1994) New nucleation theories and principle of minimum entropy production. In: Proceedings of the 12th International Conference on the Properties of Water and Steam, September 11–16, 1994, Orlando, Florida, USA.

Peters F, Paikert B (1989) Nucleation and growth rate of homogeneously condensing water vapor in argon from shock tube experiments. Experiments in Fluids 7:521–530.

Sislian JP, Glass II (1976) Condensation of water vapor in rarefaction waves I. Homogeneous nucleation. AIAA J 15(2):1731–1737.

Shock-wave reflection around a two-dimensional wedge in low-temperature gases with phase change

K. Maeno, H. Nozawa, Y. Nishimori, H. Honma
Department of Mechanical Engineering, Chiba University
1-33 Yayoi, Inage-ku, Chiba 263 Japan

Abstract: Shock-wave propagation and reflection around a two-dimensional wedge in low-temperature conditions have been studied. The problem also includes phase-change phenomena at a wall surface under appropriate test conditions. In order to clarify the interaction of shock-wave propagation with the phase change experimentally, we have used a diaphragmless shock tube with a cold test section that is cooled down to about 150 K by liquid nitrogen. Flow visualization is carried out using the shadowgraph and schlieren methods. The triple-point trajectories of Mach reflection are observed, and it is found that trajectories in the near-saturated conditions of cold HFC-134a indicate that development is a little slower than normal. Numerical simulation of shock-wave reflections on the two-dimensional wedge surface has been also carried out with an Euler equation system by using Yee's symmetric TVD scheme, and condensation is simulated by distributing the 'sink' at the wedge surface.

Key words: Low temperature, Condensation, Mach reflection, HFC-134a

1. Introduction

The shock-wave reflection and flow field around a two-dimensional wedge in low-temperature gases are of great interest. These cryogenic flows sometimes include phase-change phenomena between vapor and liquid. Cryogenic high-speed flow with phase change is one of the important problems for developing LACE (Liquefied Air Cycle Engine) or cryogenic wind tunnels (Schnerr 1989). Shock-wave propagation and/or reflection in cryogenic fluids were first investigated using helium (Liepmann et al. 1973). The effects of phase change on shock-wave propagation and reflection, however, have been reported in a limited number of papers. Behind an incident shock wave in near-saturated conditions, the vapor is adiabatically compressed and the temperature increases, but the inside walls of the tube that have large heat capacity remain at low temperature, so that the heated vapor close to the wall is cooled rapidly and condenses on the wall surfaces and in the thermal boundary layer. This phenomenon makes the flow field behind the shock wave very complex.

This paper reports on an investigation of the flow field and the triple-point trajectories of SMR (Single Mach Reflection, Ben-Dor 1992) of shock waves on a two-dimensional wedge in low-temperature gases.

2. Experiment

The experiments were carried out using the Chiba University diaphragmless cryogenic shock tube (Maeno & Orikasa 1986). In Fig. 1 (left), the double-piston mechanism works similarly to diaphragm breaking, and a 20-mm × 20-mm square test section is mounted vertically and is cooled down to about 150 K by liquid nitrogen. This shock tube has a great advantage for experiments performed low-temperature conditions because it does not have a diaphragm and thus is free from contamination because there is no need for diaphragm changing. This advantage allows high-frequency operation in cryogenic conditions.

The test section about 500 mm long is in a vacuum chamber so as to be insulated from room temperature. A section adapter from circular to square is about 450 mm upstream from the end wall of the test section. Initial temperature is measured by a CA thermocouple and pressure histories are measured by two piezo-electric pressure transducers. The thermocouple and transducers are mounted on the wall at 226 mm, 126 mm, and 276 mm upstream from the end wall, respectively. A two-dimensional wedge (removable) is set at the bottom. Two optical windows (32-mm diameter) are set near the end wall on both side walls for flow visualization and optical measuring. Flow visualizations are performed using shadowgraph and schlieren methods using a xenon-pulsed flash lamp as shown in Fig. 1 (right). Liquid nitrogen is introduced into the cooling section on the upper part of the test section.

Argon and HFC–134a are employed as test gases. HFC–134a ($CH_2F - CF_3$) is a kind of environmentally acceptable fluorocarbon (JAR & JFGA 1990) and its general properties are shown in Table 1. Argon is not condensed under our test conditions (~150 K). On the other hand, HFC–134a can be condensed under our conditions. In our experiments using HFC–134a, a shock wave propagates into its near-saturated or superheated vapor.

Table 1. General properties of HFC-134a.

Chemical Name	1,2,2,2–Tetrafluoroethane	Chemical Formula	$CH_2F - CF_3$
Molecular Weight	102.031	Heat of Vaporization	177.99 kJ/kg[a]
Boiling Point	246.97 K[b]	Freezing Point	172 K[b]
Isobaric Specific Heat	0.08518 kJ/(kg·K)[c]	Specific Heat Ratio	1.1186[c]

[a]at 298.15 K, [b]at 101.325 kPa, [c]vapor, at 298.15 K, 101.325 kPa

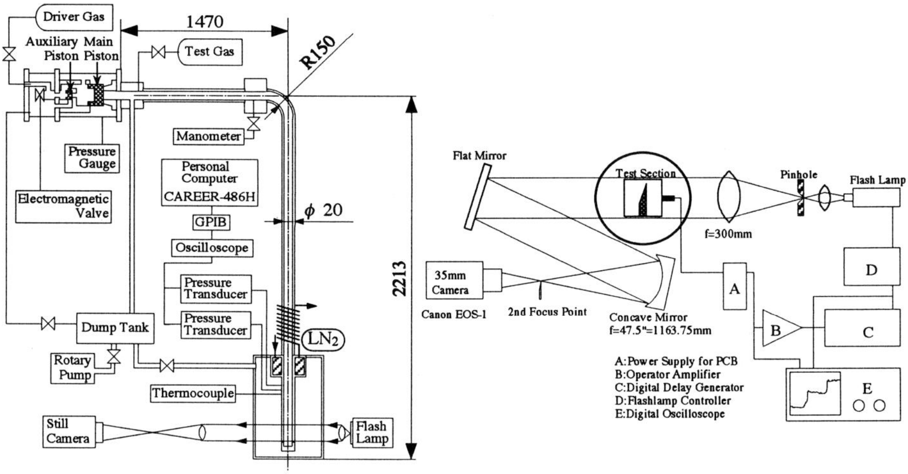

Figure 1. Experimental apparatus: (left) schematic overview, (right) flow-visualization settings.

3. Numerical calculation

Numerical simulations are also performed for the inviscid and compressible gases, i.e., solution of an Euler equation system by the finite-difference method. Yee's symmetric TVD scheme (Yee 1986) is used, and a grid system with 300 $\times$ 80 nodes is used for the calculation. The two-dimensional Euler equations for the generalized curvilinear coordinate system (ξ, η) are given as follows:

$$\frac{\partial \hat{\boldsymbol{Q}}}{\partial t} + \frac{\partial \hat{\boldsymbol{E}}}{\partial \xi} + \frac{\partial \hat{\boldsymbol{F}}}{\partial \eta} = 0, \tag{1}$$

where $\hat{\boldsymbol{Q}}$ is the conservation vector and $\hat{\boldsymbol{E}}, \hat{\boldsymbol{F}}$ are the Euler fluxes for the ξ, η directions, respectively. The state equation is used for ideal gas as follows:

$$p = (\gamma - 1)\left[e - \frac{1}{2}\rho(u^2 + v^2)\right]. \tag{2}$$

In our calculation, modified boundary conditions are used to simulate the condensation at the wall. Instead of mirror boundary conditions, 'sink' boundary conditions are introduced. Figure 2 shows a schematic drawing of a calculated grid system and sink boundaries. In Fig. 2b, a shock wave propagates into a saturated or superheated vapor. Since there is no condensation on the wall in the initial state, mirror boundary conditions are used before the shock propagation. On the other hand, after the shock passes, condensation occurs on the wall and wedge surfaces if the conditions are satisfied behind the shock wave. Therefore sink boundary conditions are applied to the shock-passed wall region. The sink boundary conditions simply simulate the volume decrease due to the condensation at the wall and wedge surfaces. Volume change per unit area has the dimension of speed. The sink boundary conditions are distributions of sink velocity normal to the wall added to the mirror conditions. The magnitude of the sink velocity v_s is defined as follows:

$$v_s = \alpha_s u_2, \tag{3}$$

where α_s and u_2 are a sink coefficient and the stream velocity behind the shock wave, respectively. In our numerical results, comparison with experimental data is performed mainly on the triple-point trajectories.

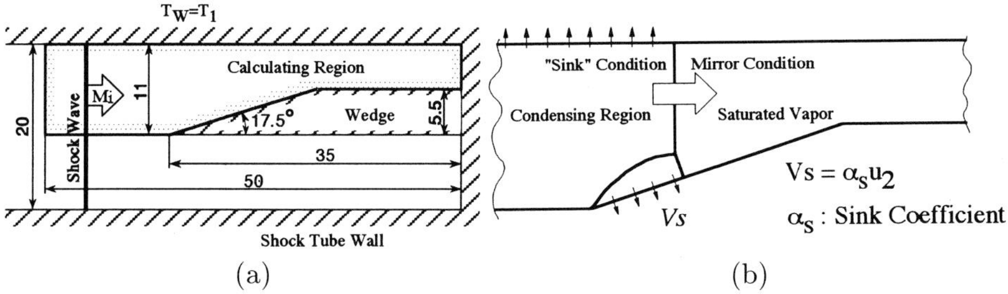

Figure 2. Calculating layout: (a) schematic drawing, (b) boundary conditions.

4. Results and discussion

Figure 3 shows three typical pictures taken under different initial temperatures: (a) normal temperature, (b) low temperature in saturated condition, and (c) low temperature in superheated condition. SMR can be found in pictures (a) and (b), and then the position of the triple point can be measured from these pictures. Figure 3c shows strange shock-wave behavior, namely, a curved incident shock wave and disappearance of the slipline in Mach reflection. These disturbances are observed at an initial temperature less than about 200 K. These phenomena seem to originate from the high rate of condensation at the wall and wedge surfaces, and from volume decrease due to the condensation. Mach waves from the wall and wedge surface are clearly visible in pictures (b) and (c). These Mach waves come from the film and/or drop-wise condensation on the wall and wedge surfaces.

Figure 4a shows the triple-point trajectories of Mach reflection in HFC-134a vapor. The experimental data are indicated by points, and lines are calculated results. Experimental results show that the triple-point trajectories of

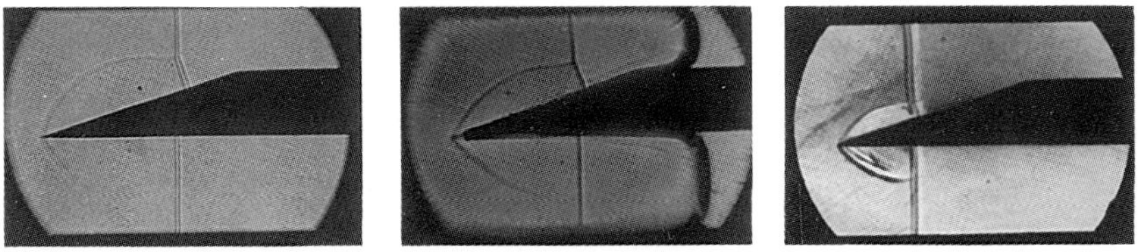

(a) (b) (c)

Figure 3. Pictures of flow visualization results, test gas is HFC-134a: (a) shadowgraph, $T_1 = 293$ K, $M_i = 3.82$, $p_1 = 4.00$ kPa; (b) shadowgraph, $T_1 = 194$ K, $M_i = 2.76$, $p_1 = 7.71$ kPa; (c) schlieren, $T_1 = 198$ K, $M_i = 2.98$, $p_1 = 4.27$ kPa.

low-temperature condensing conditions are less developed than normal ones. Calculation results show that triple-point trajectories develop at a slower rate if the sink boundary conditions are applied than if they are not. The results suggest that in accordance with the experimental tendency, one may simulate the condensation on the wall and wedge surfaces.

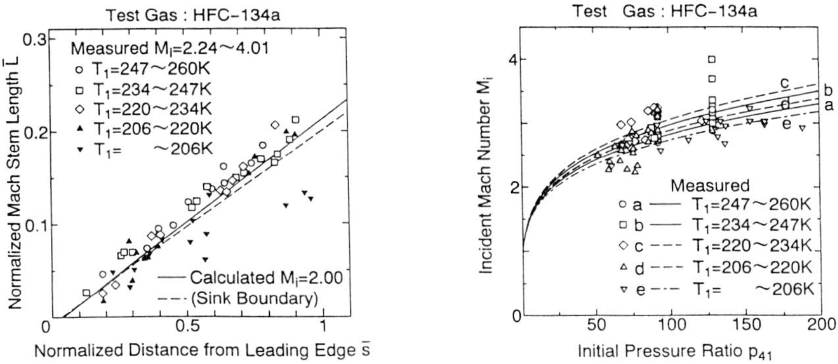

(a) (b)

Figure 4. (a) Triple-point trajectories in HFC-134a, (b) incident shock Mach number versus initial pressure ratio of HFC-134a at various initial temperatures.

Figure 4b shows the relation of the incident-shock Mach number at the test section with the initial ratio of the high pressure in the driver chamber to the low pressure in the driven tube. Curves are plotted by least-mean-square fitting with measured data. These results show that in the range of initial temperatures greater than 220 K, incident-shock Mach number increases with decreasing initial temperature. If the temperature is less than 220 K, on the other hand, it decreases as the initial temperature decreases. In both cases, the tube performances are thought to be affected. This is due to the decrease in sound speed in the highest range of temperature, and mainly to the condensation at temperatures lower than 220 K.

5. Conclusion

Shock-wave propagation and/or reflection in low-temperature test gases with phase change have been investigated experimentally and numerically. Present work is mainly focused on the effects on the flow field of condensation behind the shock wave.

The shock wave and its reflection are seen to be disturbed in low-temperature condensing conditions. Triple-point trajectories indicate the difference between non-condensable conditions and condensable conditions. In the case of condensable conditions, triple-point trajectories develop at a slower rate than those of non-condensable conditions. Decreasing the initial temperature, the Mach number of the incident shock wave increases at temperatures greater than 220 K, and the Mach number decreases in the range below 220 K. The tendency is due to the condensation of HFC–134a vapor to the wall surface.

In our calculation, sink boundary conditions are introduced. By these boundary conditions, simulated triple-point trajectories also develop at a slower rate. Present calculations, however, neglect the energy transfer, multiphase flow, and condensed film growth. Further improvements of calculations of these phenomena should be made. Our results give basic knowledge of shock-wave propagation with phase changes. Further experiments and calculations may be needed to clarify the problems.

References

Ben-Dor G (1992) Shock Wave Reflection Phenomena, Springer-Verlag.

JAR & JFGA (1990) Thermophysical Properties of Environmentally Accepted Fluorocarbons HFC-134a and HCFC-123, published by Japan Assoc. of Refrigeration and Japan Freon Gas Assoc.

Liepmann HW, Cummings JC, Rupert VC (1973) Cryogenic shock tube. Phys of Fluids 16–2:332.

Maeno K, Orikasa S (1986) Study on shock waves in low temperature gas by means of a non-diaphragm shock tube. In: Bershader D, Hanson R (eds) Proc 15th Int Symp Shock Waves and Shock Tubes, pp 563.

Schnerr G (1989) 2-D transonic flow with energy supply by homogeneous condensation: onset condition and 2-D structure of steady Laval nozzle flow. Exp in Fluids 7:145.

Yee HC (1986) Numerical Experiments with a Symmetric High-Resolution Shock-Capturing Scheme, NASA TM 88325.

Complex droplet deformation and breakup modes at intermediate Weber numbers from laser holography

G. Zerf, K. Hornung
Universität der Bundeswehr München
D-85577 Neubiberg, Germany

Abstract: Displacement data and deformation shapes of water droplets from 50 to 330 μm in diameter have been obtained in the acceleration phase behind incident shock waves through pure water vapor by means of in-line double-pulse holography. For low Weber numbers ($3.5 \leq We \leq 14$) similar deformation and fragmentation behavior as in the literature has been observed. However, for $14 \leq We \leq 22$ fine surface instabilities and for $22 \leq We \leq 45$ the formation of a thick trunk in the downstream direction occur. The drag coefficient drastically differs from the standard drag curve due to deformation effects.

Key words: Water droplet, Deformation, Breakup, Drag, Laser holography

1. Introduction

Deformation and fragmentation form an issue of continuing interest since pulse-laser techniques give access to smaller particles and higher velocities as compared to existing classical studies. The present contribution reports on experiments with water droplets in their own vapor. In this case mass exchange (condensation and evaporation) may have an influence on the deformation and drag behavior together with forces of various kinds like the surface tension σ, windforces, and shearing forces. The Weber number $We = \rho u^2 d_p / \sigma$ is one important characteristic quantity especially in the critical situation (We_c), when the applied aerodynamic pressure forces begin to exceed the restoring forces due to surface tension.

2. Experimental setup and procedure

A horizontal shock tube (see Fig. 1) of 10 cm diameter is used, filled with water vapor at saturation conditions at room temperature in its test section. The shock strengths have been varied from $M = 1.3$ to 2.0 in order to simulate different flow conditions around the droplets. A vibrating aerosol generator (a modified model TSI 3450) produces a stream of slightly nonuniformly spaced droplets of nonidentical sizes which is introduced into the shock tube perpendicularly to the flow axis, previously filled with water vapor. The nonuniformity of the stream is achieved by running the droplet generator out of the range of $3.5D_{ori} < \lambda_{ori} < 7D_{ori}$ with D_{ori} and λ_{ori} being the diameter and the oscillation wavelength of the orifice respectively, as determined experimentally by Schneider and Hendricks (1964). Water vapor is produced in an external vessel

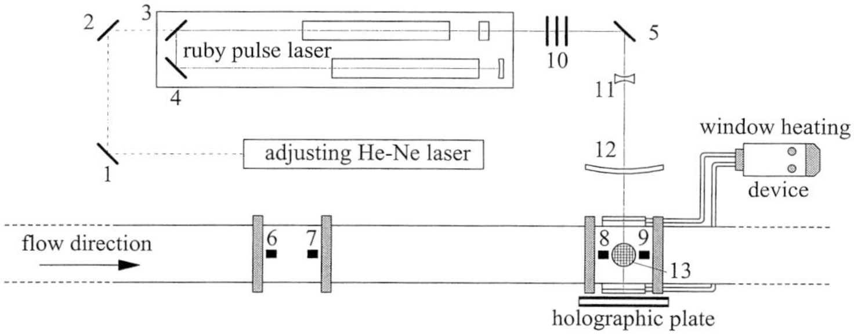

Figure 1. Shock tube and holographic camera: 1–4 mirrors, 5 gold coated mirror, 6–9 piezo probes, 10 grey filter, 11–12 optical lenses, 13 droplet generator (on top).

which is heated 2 K above room temperature to encourage the water-vapor filling procedure. Controlled slight heating of the observation windows excludes condensation during the filling procedure.

In-line holography is then used to monitor the droplets in the acceleration phase behind shock waves of various strengths. The flash lights and the Pockels cell of a 20-ns ruby laser are independently triggered by suitably situated upstream piezo probes. Two holograms in the undeformed and deformed state of the droplet stream are placed onto the same holographic plate (Agfa-Geveart 8E75 HD). For the present investigation holography is a powerful tool, since it allows the simultaneous measurement of many particles regardless of their position along the line of sight.

Careful reconstruction of the holograms uses a three-dimensional micropositioning device connected to an image-processing system. More details of the experimental setup are described in a previous paper (Zerf and Hornung 1994).

3. Experimental results

The interpretation of the deformation and fragmentation shapes as well as of the drag measurements take into account the time dependence of the gas velocity in the post-shock region due to boundary-layer effects caused by wall condensation as first described by Luger (1990) and Zerf and Hornung (1994). The theoretical description is analogous to that of Mirels (1966).

3.1. Deformation and fragmentation shapes

The results of our studies cover a range of Weber numbers from the deformation threshold up to fragmentation ($3.5 \leq We \leq 45$). Although in our investigation the droplets are surrounded by their own vapor and are much smaller in size than in the studies of Simpkins (1971), Reinecke and Waldman (1975), and Häßler (1970), we observed a similar deformation and fragmentation behavior for low Weber numbers ($3.5 \leq We \leq 14$). The classical vibrational and bag breakup modes as described by Reinecke and Waldman (1975) were observed and we found approximately the same critical Weber number We_c as compared to the study of Simpkins (1971). This means that the deformation and fragmentation shapes in this Weber-number range seem to be well characterized by the droplet's diameter, as well as by the Reynolds number and do not strongly depend on the nature of the surrounding carrier gas.

However, for intermediate Weber numbers ($14 \leq We \leq 45$), new interesting deformation modes occur, different from what one knows from the parachute-like shapes of droplets in air. They are accompanied by the formation of jets. In a previous work (Zerf and Hornung 1994) we have discussed first results of this behavior. After shock passage the fluid is not in phase equilibrium. The external fluid motion as well as the interface are affected by condensation of water vapor onto the droplet's surface. This may contribute to the different deformation and fragmentation behavior as compared to previous studies. As

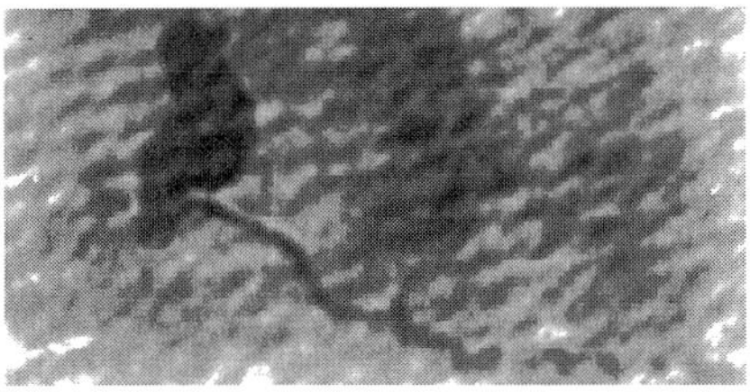

Figure 2. Jet instability 157.5 μs after shock passage at $We = 19.9$, $d_p = 246$ μm.

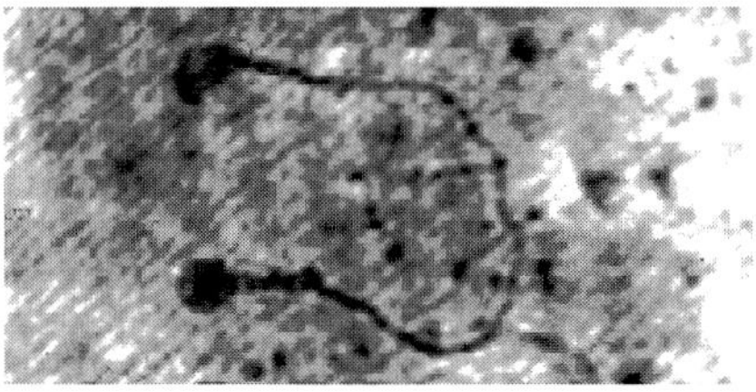

Figure 3. Connected whiskers of one droplet $d_p = 165$ μm at $We = 14$, $t_p = 239$ μs.

a result of an extended series of experiments one can clearly distinguish two cases:

(1) For smaller droplets, $d_p \leq 250$ μm, and Weber numbers between $We = 14$ and $We = 22$, the formation of very thin jets is predominant (see Fig. 2 and Fig. 3, $M = 1.64$). These fine surface instabilities do not have a certain propagation direction and can occur anywhere at the downstream hemisphere of the droplet. Different initial internal motions, induced during the

droplet-generation process, may contribute to this arbitrariness. Figure 4 shows the time development ($t_p = 185$ μs and $t_p = 235$ μs) of such a behavior by two successive holograms of a 250-μm droplet at $M = 1.53$ and $We = 15$. Some other interesting fragmentation shapes connected with these surface instabilities have been shown (Zerf and Hornung 1994).

(2) For Weber numbers between $We = 22$ and $We = 45$ the formation of a relatively thick trunk from the droplet's center in the downstream direction is observed (see Fig. 5, $M = 1.78$ and $d_p = 320$ μm). The end of this trunk again disintegrates into smaller whiskers. This fragmentation behavior occurs when the droplets are larger.

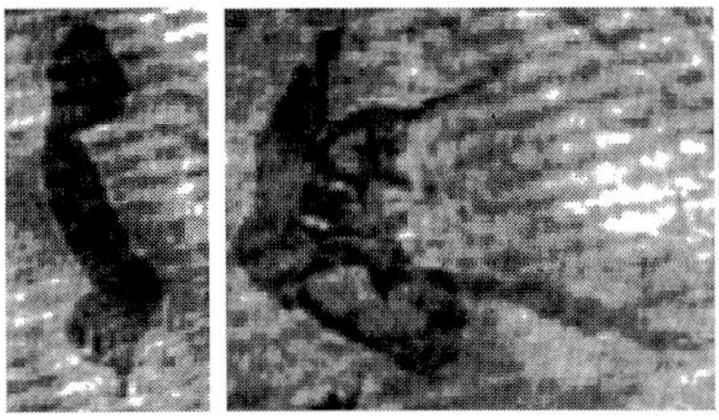

Figure 4. First and second deformation state of the same droplet ($\Delta t = 50$ μs).

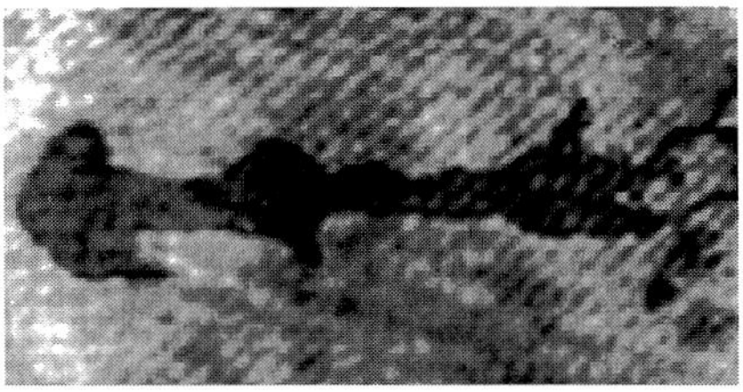

Figure 5. Formation of a trunk in the wake region at $We = 41.5$, $t_p = 212$ μs.

By inspection of a large number of holograms, it became clear that in the intermediate Weber number range ($14 \leq We \leq 45$) the deformation shapes can not be described only by the magnitude of the Weber number. Other characteristic values like the Mach number or alternatively the droplet diameter which are correlated by the Hugoniot equations are also needed to define this deformation and fragmentation behavior.

3.2. Drag measurements

The droplet's center-of-mass displacement gives information about the wind forces acting on the droplet, the trajectory, and the drag coefficient. In many studies the Weber number is kept low to exclude deformation effects. In contrast, our investigation is dedicated to the regime of strong deformation. In this case the deformation effects are so large that corrections like a special acceleration term (see Temkin and Kim 1980) add only a small contribution. Lateral influence between the droplets (see Mulholland et al. 1988) of the stream is small in our case, which follows from a separate numerical study we made.

The balance of forces on a single droplet leads to the following horizontal component of its equation of motion:

$$\frac{du_p}{dt} = c_d \frac{3}{4} \frac{\rho_2}{\rho_l d_p} \left| \mathbf{u_2} - \mathbf{u_p} \right| (u_2 - u_p) \qquad \text{and} \qquad \frac{dx_p}{dt} = u_p \tag{1}$$

where ρ_l, ρ_2, $\mathbf{u_2}$, and c_d are the density of distilled liquid water, the density and the velocity vector in the post-shock region, and the drag coefficient, respectively. For our calculations we describe c_d to be proportional to the standard drag coefficient given by:

$$c_{ds} = \frac{24}{Re}\left(1 + \frac{1}{6}Re^{\frac{2}{3}}\right) \tag{2}$$

The data reductions carefully take into account the outlet velocity of the vibrating orifice as well as the variation of the gas state behind the shock wave.

Figure 6 gives an overview of our displacement-time values in nondimensional units (index s indicates state immediately behind shock). Because the droplets slightly vary in size, a large number of data points could be derived from one experiment. For small displacements we observed approximate agree-

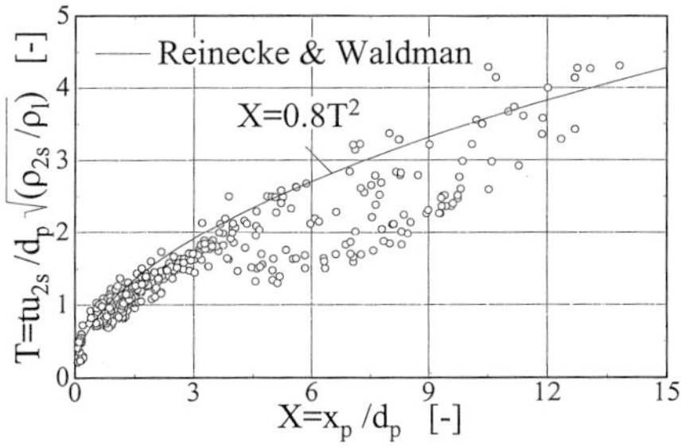

Figure 6. Displacement versus time in dimensionless units ($1.3 \leq M \leq 2.0$).

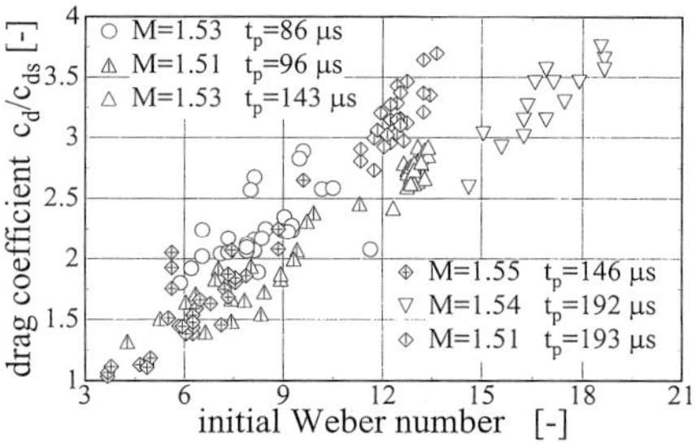

Figure 7. Deviations from the standard drag curve ($M \approx 1.53$).

ment with $X = 0.8T^2$ of Reinecke and Waldman (1975), but then the values are lower and scatter considerably. However, this is the range of violent and complex deformations, which have a large influence on the momentum exchange.

Examples for resulting deviations of the drag coefficient from the standard drag curve are shown in Fig. 7. Here the Mach numbers are chosen to be very close ($M = 1.51 \ldots 1.55$) in order to be able to discuss more clearly the Weber-number dependence. Figure 7 contains no systematic trend of the acceleration time t_p with the droplet diameter ($\propto We$, since $M \approx$ const.). One may take this as an indication that c_d/c_{ds} does not itself depend on Reynolds number. The scatter in Weber number within one experiment is due to the nonuniformity of droplet diameter. An increase of the drag coefficient caused by deformation is observed, but its values show a large scatter for the same Weber number. The reason certainly is the large shape variation within one shock-tube experiment. Even droplets having the same diameter and belonging to the same experiment do not have exactly the same deformation shape. Such violent and unstable deformations are difficult to predict and describe in detail. However, in practical applications such situations frequently occur and a mean drag coefficient, even with an appreciable scatter, forms a quantity of technical importance.

4. Conclusions

We have extended the knowledge on deformation to the "droplet in vapor" situation, especially to Weber numbers up to the onset of fragmentation. We confirmed known structures for $3.5 \leq We \leq 14$ and found new ones for $14 \leq We \leq 45$. Moreover, it could be shown that an approximate drag law can still be fitted to displacement data even in the case of large and chaotic droplet deformations.

Acknowledgement. The authors would like to thank F. Hindelang for continuous support.

References

Häßler G (1970) Breakup of large water drops under the influence of aerodynamic forces in a steady stream of steam at subsonic velocities. In: Proc of Third Conf (Int) on Rain Erosion and Related Phenomena, Hampshire, England.

Luger P (1990) Stoßwellen in einkomponentigen Zweiphasengemischen, Ph.D. Thesis, Universität der Bundeswehr, München.

Mirels H (1966) Flow nonuniformity in shock tubes operating at maximum test times. Physics of Fluids 9(10):1907–1912.

Mulholland JA, Srivastava RK, Wendt JOL (1988) Influence of droplet spacing on drag coefficient in nonevaporating, monodisperse streams. AIAA J 26(10):1231–1237.

Reinecke WG, Waldman GD (1975) Shock layer shattering of cloud drops in reentry flight. In: AIAA 13th Aerosp Meeting, Pasadena, CA, Pap. 75–152.

Schneider JM, Hendricks CD (1964) Source of uniform-sized liquid droplets. Review of Scientific Instruments 35(10):1349–1350.

Simpkins PG (1971) Non-linear response of deforming drops. Nature Physical Science 233:31–33.

Temkin S, Kim SS (1980) Droplet motion induced by weak shock waves. J Fluid Mech 96:133–157.

Zerf G, Hornung K (1994) Laser holographic studies on deformation and fragmentation of droplets in their own vapour. In: Proc of 7th Symp (Int) on Appl of Laser Techn to Fluid Mech, Lisboa 2:34.4.1–34.4.8.

Entrainment of liquid droplets into shock-induced airstreams

R. D. Archer, B. E. Milton, M. Behnia, R. T. Casey
School of Mechanical and Manufacturing Engineering
The University of New South Wales
Sydney 2052, Australia

Abstract: Experiments on the direct entrainment of liquid drops from the wall of a duct into a shock-driven airstream have been performed. The entrainment process ranges, with time, from no effect to a droplet-distortion process generating vortices, spray, and eventually atomization. These regimes have been classified using Weber number and dimensionless time. The droplet-distortion and entrainment processes appear to be different from those for freely suspended droplets. The effect of the gas-stream duct-wall boundary layer is significant, with a paired vortex system drawing off liquid in a cloud where it is atomized.

Key words: Droplet, Entrainment, Atomization, Experiment

1. Introduction

The two-phase process involved in transport of a liquid (secondary fluid) into a gas stream (primary fluid) has received much attention. Examples of its practical relevance are in fuel preparation for combustion, refrigerant systems, pipeline flows, and heat exchangers. Each case is usually examined individually on a phenomenological basis by both experimental and numerical studies. Both the introduction of the liquid into the gas stream and its subsequent transport downstream have received significant attention. The latter encompasses the mixed-gas/vapor phase driving the liquid phase either as gas-stream-borne droplets or as a separated liquid wall film. The current research program is related to the second case where the formation, movement, and re-entrainment of the liquid deposited on the wall are examined.

Liquid may be deposited onto walls either by condensation or direct impact. For both mechanisms, data are currently available. The reverse process of liquid transfer from wall films to the gas stream either by evaporation or direct entrainment needs more attention. Most work has concentrated on evaporation and there has been little research into the processes involved in direct liquid entrainment to an airstream. Previous work includes both numerical simulation (Morioka and Matsui 1975) and shock-tube experiments directed to the mixing of streams of the same phase. Borisov et al. (1981), Frolov et al. (1984) and

Milton et al. (1991) used shock waves for creating appropriate air flows over stationary liquid films. These experiments showed that substantial delay occurred before entrainment commenced and that the entrainment itself resulted from an interfacial wave on the liquid surface with the leading edge playing a primary role. The fluid properties have also been highlighted as being of importance in the type of surface wave structure developed. Here, basic processes are examined in individual droplets.

2. Shock-wave interaction with liquid droplets

Similarities might be expected between the entrainment atomization of droplets affixed to walls and the shattering processes associated with free droplets embedded in a gas stream. Free droplets have been extensively studied. The distortion and shattering processes have been described and categorized. Most of the work relates to droplets suspended in the airstream free of the duct-wall boundary layer. Some experiments used free-falling droplets (Merrington and Richardson 1947, Hinze 1948), while others considered suspended droplets either in wind tunnels (Lenard 1904, Lane 1951) or shock tubes (Priem 1957, Engel 1958, Ranger and Nicholls 1972). Wierzba and Takayama (1988) classify breakup as in Table 1.

Table 1. Classification of breakup mechanisms for freely suspended droplets.

Breakup Type	Approximate We Range	Breakup Type	Approximate We Range
Vibrational	8 to 15	Chaotic	30 to 65
Bag	11 to 30	Stripping	60 to 400,000
Bag and Stamen	19 to 35	Catastrophic	$> 1{,}000$

Table 2. Experimental conditions.

Shock Wave		Droplet Size			
Mach Number, M_s	Flow Velocity, m/s	Volume, μL	Equatorial Dia., mm	Height, mm	Equivalent Dia., mm
1.054	30.4	25	5.0	2.0	4.69
1.093	51.5	50	6.4	2.5	5.91
1.112	61.5	100	9.2	2.6	7.44
1.146	79.1				

These limits vary widely with experiment type, droplet size, and fluid used. The demarcation between chaotic, stripping, and catastrophic modes is obscure, requiring more precise criteria. Dimensionless breakup times are similarly poorly defined and variation with Weber number (We) needs to be taken into account.

We concentrate on the entrainment of individual liquid droplets which are originally affixed to the wall and develop local interfacial instability. In addition, the interaction of the wall boundary layer with the droplet becomes apparent. Water droplets were used in three sizes of volumes: 25, 50, and 100 μL. Shock Mach numbers were selected to simulate the flow velocity in automobile engine manifolds and ports (Table 2).

3. Droplet-entrainment processes

The test gas was air, initially at room conditions. Droplets were placed on the lower walls of the shock tube and transverse shadowgraphs taken of the interaction with a weak incident shock wave and the subsequent flow. Tests used either individual droplets or droplets placed along the shock-tube test-section floor in increasing size downstream and spaced to prevent interaction. Any downstream motion from the initial drop location could be discerned, the flow in all cases being from right to left (see Fig. 1).

The interaction processes can be described as depicted in Fig. 1. First, the shock wave traverses the droplet without any discernible effect, consistent with the findings on freely suspended droplets. With time, a deformation of the droplet becomes apparent with shortening length and increasing vertical height. A small spur consistently appears at the upper, trailing-edge corner. Shortly afterwards, a wave pattern appears on the fairly flat upper surface of the drop. The leading edge then becomes distorted, building into a peak (or tower) which, although lacking the preceding trough, is in fact somewhat similar to the standing wave in film experiments (Milton et al. 1991). The appearance of a vortex at the upper, trailing-edge corner shows the start of liquid removal which then, if the airstream velocity is sufficient, develops into a spray. Finally, shattering or atomization of the droplet occurs and it is noted that the bulk of the liquid has by this time been displaced downstream.

The interaction of the wake and the boundary layer differs from that in freely suspended droplets. A paired vortex structure appears behind the droplet with rotation towards the wall in the region close to the wall (anti-clockwise in side view) and rotation away from the wall in the region above the droplet in the flow field (clockwise). Both vortices draw off fluid from the droplet in a cloud-like pattern. With further progression, instabilities appear at the surface of this cloud and discrete, atomized micro-droplets of the order of 100 μm in diameter are shed in distinct streams emanating from the perturbations which may be interpreted as wavelet crests. Most of the micro-droplets are carried upwards

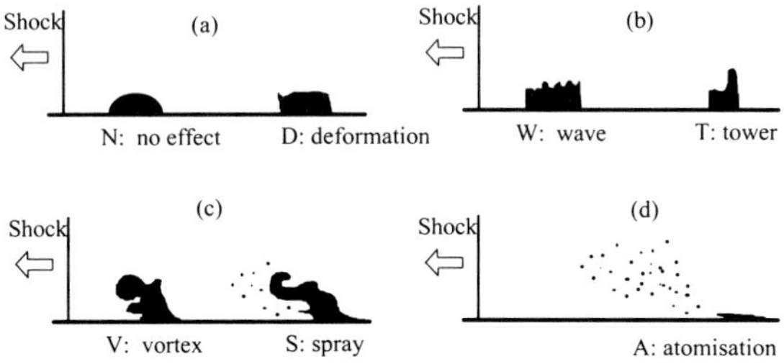

Figure 1. Sketch of the droplet entrainment sequence (time increasing from left to right, top to bottom).

into the flow by the upper vortex whilst a small proportion is directed onto the wall as a thin film. For shock-induced airstream speeds greater than 60 m/s, vortices are not clearly defined before breakup.

4. Classification of droplet entrainment

The droplet entrainment process has been categorized according to the shape changes observed with time, Fig. 1. These are, in order: no effect, N; initial deformation, D; upper surface wave appearance, W; tower formation, T; initial vortex development, V; commencement of spray from the vortex, S; and finally atomization, A. They have been correlated here in terms of Weber number (We) against dimensionless time, t^* in Fig. 2, which shows the results of 139 measurements. The Weber number is $\mathrm{We} = \rho v^2 d/\sigma$ where ρ and v are the airstream density and velocity respectively, d is the equivalent droplet diameter, and σ is the liquid/air surface tension; dimensionless time is $t^* = tv/d$. Another correlation, $\mathrm{We}/\mathrm{Re}^{0.5}$, which is sometimes used in freely suspended droplet shattering and in jet breakup to a spray, does not provide significant improvement.

Figure 3 approximately delineates the regimes of Fig. 2. Note the following: (i) The T (tower) regime overlaps the W (wave) section which forms earlier. At high We, the tower regime tends to disappear. (ii) The V (vortex) and S (spray) regimes substantially overlap. The vortex always tends to produce some spray from its outer periphery, the amount progressively increasing with the delay time. At longer times, a vortex is not always distinguishable and a spray forms directly, more noticeable at higher We. (iii) The D (deformation) and W (wave) regions overlap substantially at low We and the former may then extend partially into the T (tower) regime. (iv) Full atomization only occurs at high values of t^* exceeding about 100.

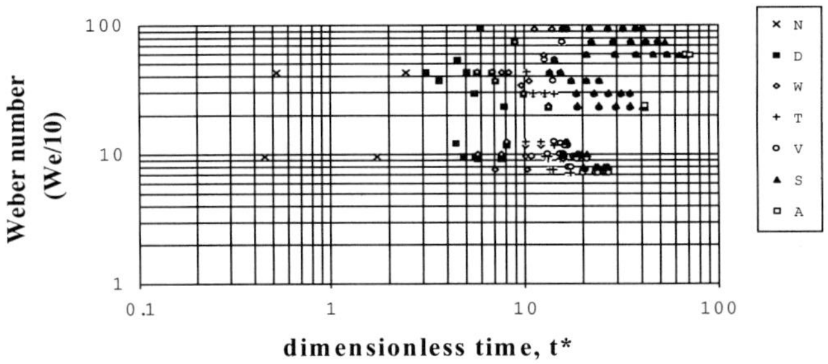

Figure 2. Classification of the entrainment sequence.

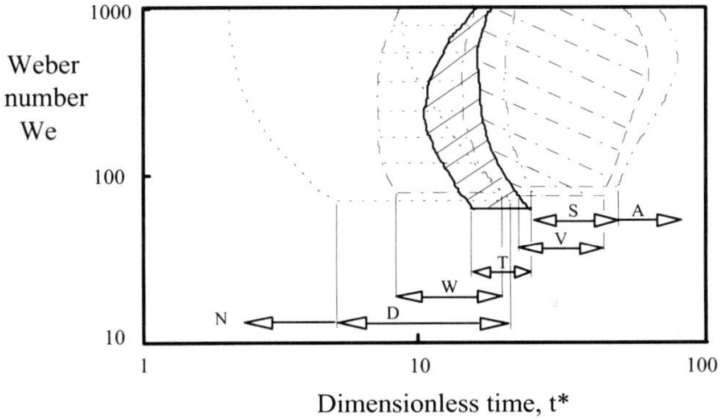

Figure 3. Approximate entrainment regimes.

5. Comparison with freely suspended droplets

The droplets due to gravitation and surface tension are not initially hemispherical but are of a flattened shape, more extreme with increased droplet volume. Their equatorial diameter, height, and equivalent hemispherical diameter are shown in Table 2. We and Re have been calculated for all three (using double the height for diameter). We values range from 58 to 671 and Re values from 7.6×10^3 to 4.1×10^4. For comparison with freely suspended droplets, the equivalent diameter has been used and values of We then range from 66 to 529.

The breakup times are much longer than those for freely suspended droplets. The duct-wall boundary layer reduces the effective velocity relative to the core flow. In most tests, although entrainment of some type occurred, atomization (catastrophic shattering) did not until values of t^* were 2 to 10 times those for freely suspended droplets.

6. Conclusions

Experiments on the direct entrainment of water droplets from a duct wall into an airstream have been carried out using a shock tube. The entrainment process, after a delay, begins with distortion of the droplet and the formation of surface waves which often develop into a tower-like wave between the droplet center and its leading edge. A paired vortex system develops behind the droplet and fluid is drawn into this system in the form of large clouds. Instabilities of the surfaces of these clouds atomizes the liquid and distributes it predominantly into the flow field in spiral streams of micro-droplets of the order of 100 μm in diameter. The breakup process has been classified into 7 regimes which have been plotted on a Weber-number/dimensionless-time basis. The boundaries of these regions overlap considerably. Comparison with freely suspended droplets indicates that the wall-affixed droplets have more consistent processes and require higher values of We and longer times for atomization.

References

Borisov A, Gelfand B, Sheparchev S, Timofeev E (1981) Mechanism of mixture formation behind a shock wave sliding over a liquid surface. Phys of Comb and Explosion 5, Akad Nauk, Siberia Sect. Novosibirsk, 85–94.

Engel OG (1958) Fragmentation of waterdrops in the zone behind an air shock. J Nat Bur of Standards 60(3):245–280.

Frolov S, Gelfand B, Timofeev E (1984) Interaction of a liquid film with high speed gaseous flow behind a shock wave. Phys of Comb and Explosion 5, Akad Nauk, Siberia Sect. Novosibirsk, 107–114.

Hinze JO (1948) Critical speeds and sizes of liquid globules. J App Sci Res A1:273–288.

Lane WR (1951) Shatter of drops in streams of air. Ind & Eng Chem 43(6):1312–1317.

Lenard P (1904) Uber regen. Meteorologische Zeitschrift 21:249.

Merrington AC, Richardson EG (1947) The break-up of liquid jets. Proc Phys Soc 59(1):1–13.

Milton BE, Behnia M, Takayama K (1991) Interaction of liquid films with shock induced airflows. Proc 18th Int Symp Shock Waves, Sendai, 2:1265–1270.

Morioka S, Matsui G (1975) Pressure-wave propagation through a separated gas-liquid layer in a duct. J Fl Mech 70(4):721–731.

Priem RJ (1957) Breakup of water drops and sprays with a shock wave. Jet Prop 27(10):1084–1087.

Ranger AA, Nicholls JA (1972) Atomization of liquid droplets in a convective gas stream. Int J Heat & Mass Transf 15(6):1203–1211.

Wierzba A, Takayama K (1988) Experimental investigations of the aerodynamic breakup of liquid drops. AIAA J 26(11):1329–1335.

Combined sublimation and thermal decomposition of NH_4Cl/Ar aerosols in shock waves

G. N. Schading, P. Roth
Institut für Verbrennung und Gasdynamik
Gerhard-Mercator-Universität Duisburg
D-47048 Duisburg, Germany

Abstract: The sublimation and thermal decomposition of NH_4Cl particles suspended in argon was investigated behind incident shock waves under high-temperature conditions. The decrease in particle size during sublimation was observed by laser-light scattering under three fixed angles. The detected light intensities were evaluated in terms of time-dependent particle size. Simultaneously, the expected decomposition product HCl was followed by IR-absorption measurements. A tunable IR-diode laser was rapidly tuned over the (0-1) P6 vibrational-rotational line of HCl at $\nu = 2752.03491$ cm^{-1}. The measured absorption profiles were evaluated in terms of time-dependent HCl concentrations.

Key words: Thermal decomposition, IR-spectroscopy, NH_4Cl/Ar aerosols

1. Introduction

In recent years, several experiments were performed to study the evaporation of solid, suspended particles under high-temperature conditions; see Timmler and Roth (1989, 1991) and Schading and Roth (1992). The optically measured decrease in particle size was used to determine the temperature-dependent evaporation coefficient. A comparison with theoretical calculations showed in some cases a rather great difference between experimental findings and theoretically expected behavior. These discrepancies motivated the present work, which is based on simultaneous measurements of gas-phase and particle properties.

In the experiments reported here, the sublimation and thermal decomposition of NH_4Cl particles were investigated under high-temperature conditions behind incident shock waves. The reduction in particle size was measured by laser-light scattering together with Mie theory. Simultaneously, the appearance of the decomposition product HCl was followed by Infrared Diode Laser Spectroscopy (IDLS). A specially designed aerosol generator was used to generate a nearly monodisperse aerosol with a log-normal size distribution. The results obtained were compared with theoretical considerations.

2. Experimental

The experiments were performed in a conventional diaphragm-type shock tube, which was equipped with an optical light-scattering detection system for particle diagnostics and an Infrared Diode Laser Spectrometer (IDLS) for gas-phase concentration measurements. The shock tube was used as a high-temperature isothermal-isobaric reactor, in which the initial aerosol was nearly instantaneously heated up by a shock wave to a constant temperature and pressure level. The gas properties behind the shock waves were calculated from the incident shock velocity by applying one-dimensional gasdynamic equations.

The aerosols were generated by heterogeneous condensation. The particles are nearly spherical and have a log-normal size distribution. The initial size and the size distribution were determined by an optical particle counter. The aerosol was made to flow through the running section of the shock tube for a while in order to get it homogeneously dispersed along the tube. The laser-light scattering system used to measure the reduction in particle size consisted of three light fibers placed at fixed angles of 25°, 45°, and 75° relative to the outcoming laser beam. The infrared diode laser was used to monitor the time-dependent HCl concentration by absorption measurements at the (0-1) P6 transition, $\nu = 2752.03491\ \text{cm}^{-1}$. The optical setup consisted of a liquid-N_2 cooled diode laser, the shock-tube absorption path, a 0.5-m Czerny-Turner monochromator, and an InSb detector. The wavelength of the laser beam was calibrated by inserting a Fabry-Perot etalon with a spectral range of $0.05\ \text{cm}^{-1}$ into the optical path. More details about the experimental setup are given by Timmler and Roth (1991) and Schading and Roth (1994).

3. Results and discussion

In 32 experiments the sublimation and thermal decomposition of NH_4Cl particles were studied behind incident shock waves in the temperature range $530\ \text{K} \leq T \leq 630\ \text{K}$ at pressures $1.0\ \text{bar} \leq p \leq 1.75\ \text{bar}$. The initial mean diameter of the particles varied between 0.48 μm and 0.54 μm. An example of measured scattered light intensities during particle sublimation obtained under three different angles is shown in Fig. 1. In front of the shock, the scattered light signals are on a constant level resulting from the initial particle properties. Immediately behind the shock, the signals rise rapidly due to increase of the particle number density, small change in the refractive index, and changes in the density of the particle material. The relaxation of particle temperature and particle velocity is finished at the time when the maxima of scattered light signals appear. The following decrease of the signals results from the particle sublimation. The evaluation of the measured signals was done by comparing them with intensities calculated from Mie theory assuming spherical particles with a log-normal size distribution. An extensive description of the data-reduction procedure can be found in Timmler and Roth (1991) and Roth and Fischer (1985). As an example, the evaluation of the experiment of Fig. 1 is shown in Fig. 2, where the square of the particle radius is plotted versus time. The agreement between the

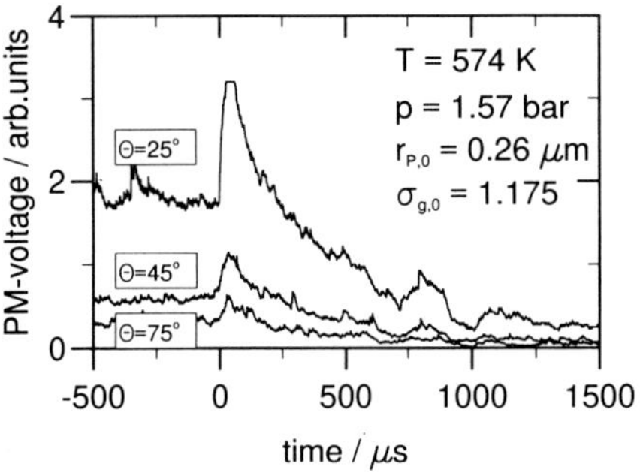

Figure 1. Example of scattered-light signals obtained from a particle ensemble under three different angles.

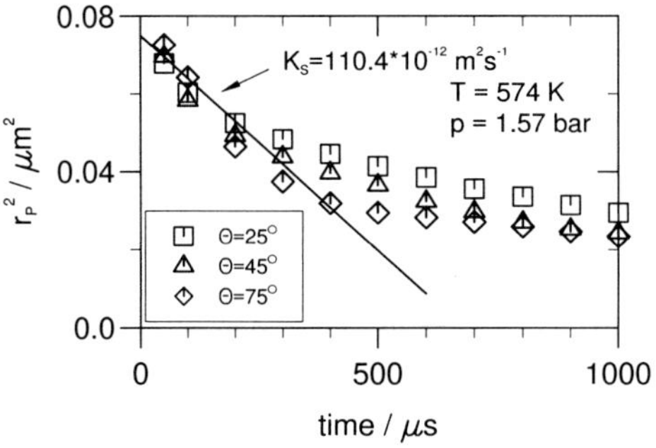

Figure 2. Evaluation of the individual shock-tube experiment of Fig. 1 showing the time-dependent size of NH_4Cl particles.

three signals evaluated is quite good. A sublimation rate K_S can be introduced which is equal to the rate of particle surface-area reduction at $t = 0$

$$K_S = -\left.\frac{d\,r_P^2}{dt}\right|_{t=0} \tag{1}$$

The sublimation rate deviates from the expected linear behavior for later reaction times due to Knudsen effects, which are significant for small particles. A summary of all K_S values obtained is given in Fig. 5; for a more detailed discussion, see remarks below.

In the present case, it is useful to express the observed decrease in the particle size by an equivalent property, and that is the concentration of solid NH_4Cl (mole of solid NH_4Cl per unit volume of suspension). The application of the mass conservation equation for the dispersed phase results in:

$$K_{NH_4Cl} \equiv \left.\frac{d\,[NH_4Cl]_{solid}}{dt}\right|_{t=0} = \frac{n_P}{M_{NH_4Cl}}\rho_P^0 \cdot 2\pi r_P \cdot K_S \tag{2}$$

In this equation r_P, n_P, M_{NH_4Cl}, and ρ_P^0 are the particle radius, the particle number concentration, the molecular weight of NH_4Cl, and the density of the particle material, respectively. The disappearance rate of the solid material expressed in concentration units obviously depends on the particle radius. The results of the present light-scattering measurements represented by K_{NH_4Cl} are summarized in the Arrhenius diagram of Fig. 6, see squares. The scattering of the data is relatively high.

A typical shock-tube experiment showing a sequence of laser absorption scans is shown in Fig. 3. In the scans behind the shock front the HCl absorption lines can clearly be identified. The time-dependent increase in absorption indicates an increase in the concentration of the absorbing HCl. Data reduction from measured absorption to HCl concentration was done based on the the Lambert-Beer law, see Schading and Roth (1994). Only the line center values together with the line strength of the vibrational-rotational line were used. The example of Fig. 4 illustrates the time-dependent HCl concentration evaluated

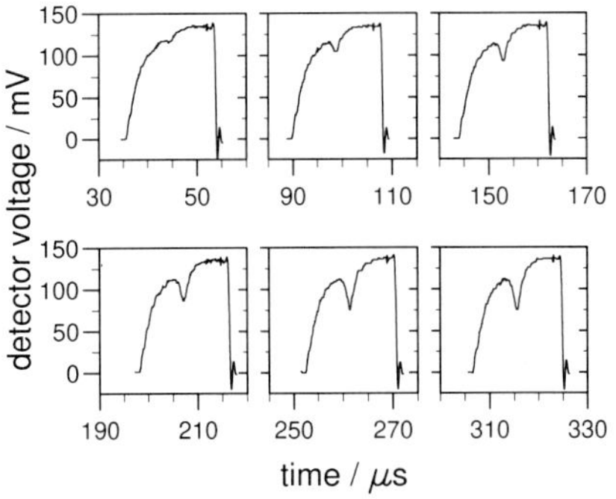

Figure 3. Sequence of IR-diode laser scans with HCl absorption lines at $\nu = 2752.03491$ cm^{-1} obtained from a shock-heated NH_4Cl/Ar aerosol.

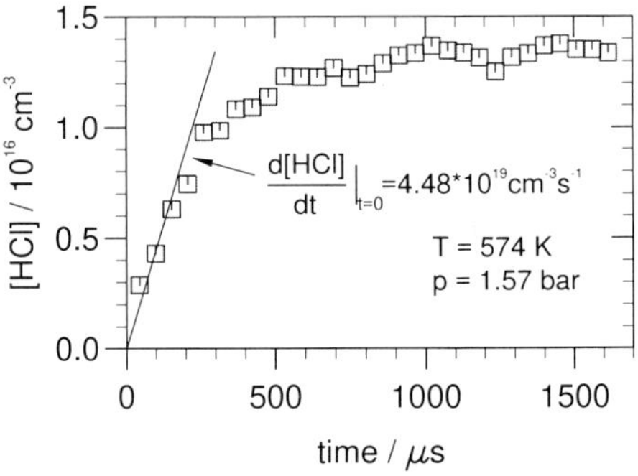

Figure 4. Evaluation of the laser scans shown in Fig. 3 together with the initial HCl formation rate.

from the sequence of absorption spectra shown earlier. A first data reduction is possible by introducing a HCl formation rate at $t = 0$, see slope in Fig. 4.

$$K_{HCl} = \frac{d\,[HCl]}{dt}\bigg|_{t=0} \tag{3}$$

The results of all experiments are also included in Fig. 6, see triangles.

The comparison of the two independent data sets in Fig. 6, namely the disappearance rate K_{NH_4Cl} of the solid dispersed material and the formation rate of the dissociation product HCl, seem to agree quite well within the uncertainty limits of the experiments. It must be concluded that the dissociation products HCl and NH_3 are directly ejected from the solid phase.

$$(NH_4Cl)_{solid} \longrightarrow HCl + NH_3$$

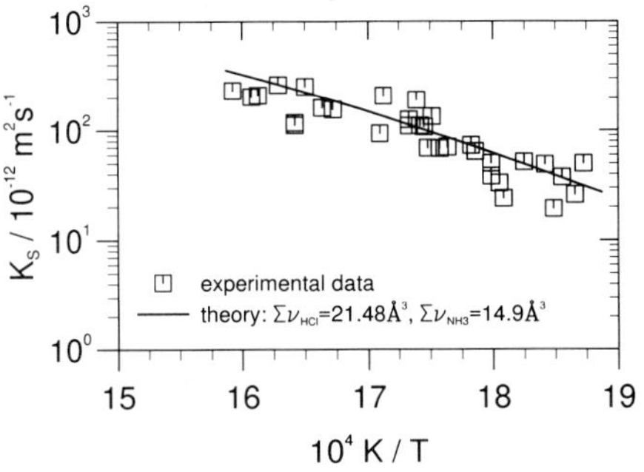

Figure 5. Sublimation rates of solid NH_4Cl compared with theoretical calculations.

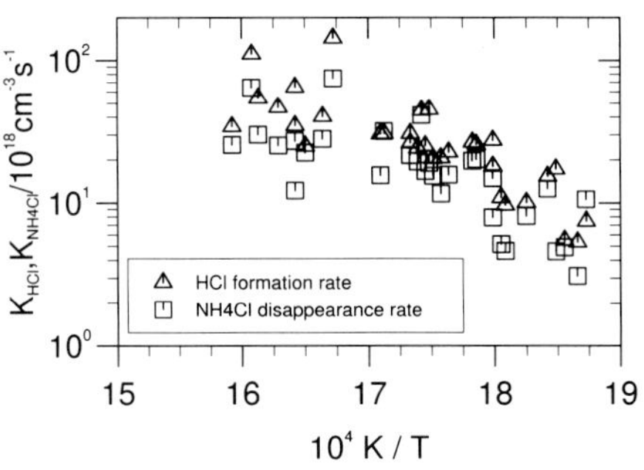

Figure 6. Comparison of measured disappearance rate of solid NH_4Cl and formation rate of HCl.

Another explanation could be that the rate of dissociation of gaseous NH_4Cl into HCl and NH_3 is very fast compared to the rate of formation of gaseous NH_4Cl by sublimation.

A further argument, which confirms the present observation, comes from a comparison between measured and calculated NH_4Cl sublimation rates. Under the assumption of complete dissociation of solid NH_4Cl into gaseous HCl and NH_3 at the particle surface, the decrease in particle size is caused by the diffusive mass transport of the dissociation products into the ambient gas phase. According to Timmler and Roth (1991) the measured K_S values can be expressed by:

$$K_S = -\frac{d\,r_P^2}{dt} = (D_{\mathrm{HCl},G} M_{\mathrm{HCl}} + D_{\mathrm{NH_3},G} M_{\mathrm{NH_3}}) \frac{p_D}{\rho_P^0\, R\, T_P} \tag{4}$$

The sublimation rate K_S is mainly determined by the temperature-dependent dissociation pressure of the particle material p_D in the above equation. The diffusion coefficients $D_{\mathrm{HCl},G}$ and $D_{\mathrm{NH_3},G}$ can be calculated according to Fuller et al. (1966). The solid line in Fig. 5 shows the calculated sublimation coefficient using the following values for the diffusion volumes: $\Sigma\nu_{\mathrm{Ar}} = 16.1\text{Å}^3$ and $\Sigma\nu_{\mathrm{NH_3}} = 14.9\text{Å}^3$. The diffusion volume for HCl was estimated according to Fuller et al. (1966) by the sum of the atomic diffusion volumes of H and Cl to be $\Sigma\nu_{\mathrm{HCl}} = 21.48\text{Å}^3$. Obviously, the agreement between the experimental data and the calculated values is very good, confirming the rapid decomposition of solid NH_4Cl into the gaseous products HCl and NH_3.

4. Conclusion

Laser-light scattering and IR-diode laser absorption were successfully applied to study the combined sublimation and thermal decomposition of NH_4Cl particles behind shock waves. From the agreement between the rates of solid NH_4Cl sublimation and gaseous HCl formation, it must be concluded that either HCl and NH_3 are directly formed at the solid surface or that the dissociation of NH_4Cl vapor primarily formed is very fast compared to its formation by particle sublimation.

Acknowledgement. This work originated in the Sonderforschungsbereich 209 of the Universität Duisburg. The financial support of the Deutsche Forschungsgemeinschaft is gratefully acknowledged.

References

Fuller EN, Schettler PD, Giddings JC (1966) A new method for prediction of binary gas-phase diffusion coefficients. Ind Eng Chem 58:19–47.

Roth P, Fischer R (1985) An experimental shock wave study of aerosol droplet evaporation in the transition regime. Phys Fluids 28(6):1665.

Schading GN, Roth P (1992) Evaporation of solid cyanuric acid particles suspended in argon. J Aerosol Sci 23:157.

Schading GN, Roth P (1994) Thermal decomposition of HCl measured by ARAS and IR diode laser spectroscopy. Comb Flame 99:467–474.

Timmler J, Roth P (1989) High temperature evaporation of solid particles. J Aerosol Sci 20:1011.

Timmler J, Roth P (1991) High-temperature evaporation rates of solid KBr and NH_4Cl aerosol particles. Part & Part Syst Charact 8:79.

Effective sphere drag coefficient for unsteady gas-particle flows

J. Kurian, H. K. Das
Department of Aerospace Engineering
Indian Institute of Technology, Madras 600036 India

Abstract: A horizontal shock tube has been used to measure the sphere drag coefficient applicable to a cloud of particles for unsteady gas particulate flow. Measurement of cloud concentration with time behind the propagating shock front was done using a laser light-extinction technique. First derivative of the observed intensity versus time signals gives a velocity-time history of the cloud from which the sphere drag coefficient could be determined. The drag coefficient was determined in the range of Reynolds numbers between 70 and 1100.

Key words: Effective drag coefficient, Cloud of particles, Unsteady flow

1. Introduction

Considerable research has been conducted by various groups to measure the steady drag coefficients of spherical particles for a wide range of Reynolds numbers. These measurements were done using different techniques, e.g., Arnold (1911), Liebster (1927), and Schmiedel (1928) derived drag coefficients for low Reynolds numbers from the determination of the terminal velocity of spheres falling through various liquids. Roos and Willmarth (1971) obtained values of sphere drag by towing a sting-mounted sphere through water. Crowe et al. (1971) measured drag coefficients at low Reynolds numbers by studying the motion of small spheres in a microballistic range. Bailey (1974) made an extensive series of measurements of sphere drag coefficients for Reynolds numbers up to 10^6 in an aeroballistic range. Heinrich et al. (1963) obtained values in a wind tunnel for Reynolds numbers between 2×10^3 and 2×10^4. Based on the results of such experiments, standard drag coefficients for spheres in steady-flow situations is well known for a wide range of Reynolds numbers.

Though the values of drag coefficient for spheres are well known for steady flows, their applicability is questionable for unsteady-flow situations, which are in practice more significant. Measurements of sphere drag coefficients for unsteady cases have also been done, and some of the important results in the last three decades are found in Selberg and Nicholls (1968), Buckley (1968), Rudinger (1970), Karanfilian and Kotas (1978), Outa et al. (1981), Sommerfeld (1985), and Igra and Takayama (1991, 1993). Selberg and Nicholls (1968) used a shock tube to accelerate single particles and photographed their displacement to determine the drag coefficient. Buckley (1968) experimentally investigated the drag of particles accelerating within particulate clouds at transonic slip-flow conditions using a light-extinction technique. Similar technique was also used by Rudinger (1970) who derived effective drag coefficients for flows of suspensions of spherical glass particles in air. Outa et al. (1981) calculated particle drag from streak recordings of dispersed glass spheres in air inside a shock tube. Sommerfeld (1985) used laser-doppler velocimetry to find a drag law describing particle acceleration behind a shock wave inside a shock tube filled with a uniform gas-particle mixture. Most of these experiments were limited to Reynolds number values up to 1000. Values of unsteady sphere drag coefficient for Re up to 6000 are available in Karanfilian and Kotas (1978), who measured the drag on a sphere which was subjected to a simple harmonic motion. One of the latest reported unsteady sphere-drag measurements is in Igra and Takayama (1993). They also used post-shock-wave flow (and double-exposure holography) to accelerate spheres kept on the floor of a shock tube. Values of unsteady drag coefficient for $6000 < \mathrm{Re} < 101000$ were reported. The recent shock-tube study by Rodriguez et al. (1993) reports on unsteady sphere drag coefficient for a similar Reynolds-number range. One conclusion that can be made from all these studies is that values of the unsteady drag coefficient for spheres is considerably different from those in steady-flow cases. Also, unsteady sphere drag coefficients measured by different methods do not agree with each other in overlapping Reynolds-number ranges. Further, the applicability of a steady drag coefficient for single spheres in calculations for gas-particle flows is questionable because of the interaction among particles.

The purpose of the present paper is to experimentally obtain values for the unsteady sphere drag coefficient applicable to a cloud of particles in a wide Reynolds-number range. The method used in the present experiments is based on the light-extinction technique.

2. Details of experiments

2.1. Shock tube and instrumentation

The experiments were conducted with a stainless steel shock tube 50 mm in diameter and 8.5 m long. The driven section is 7 m long and aluminum-foil diaphragms between 0.05 and 0.25 mm thick were used. Rupture of diaphragm was accomplished with high pressure on the driver side. A pressure transducer mounted just downstream of the diaphragm closes a solenoid valve which stops the supply of high-pressure air to the driver section immediately after rupture. The downstream end of the driven section was connected to a settling tank with a filter to prevent particles from entering the vacuum pump.

A particle-injection system was designed to produce a uniform flow rate of particles into the driven section immediately downstream of the diaphragm before each shock-tube run. The injection system consisted of an inlet system, a particle reservoir, a convergent-divergent (C-D) nozzle, a mixing chamber, and an exhaust system. The exit of the particle reservoir was placed in the diverging portion of the C-D nozzle causing particles to get sucked by the low pressure created in the air stream when a sufficiently high-pressure air was fed into the inlet system. Particles were mixed with air in the mixing chamber and the gas-particle mixture was fed to the shock tube from the exhaust system. The pressure level in the driven section was continuously monitored as injection proceeded and the initial condition was taken as that value just prior to diaphragm rupture.

The particles used in the experiments were Ballotini spheriglass and precipitated calcium-carbonate powder. Particle material densities were 2500 kg/m^3 and 2790 kg/m^3, respectively. Average particle sizes were determined using a Malvern laser-diffraction particle sizer. Two types of glass particles had medium diameters 6.1 μm and 34.6 μm. Medium diameter for calcium-carbonate powder was 19.26 μm. The loading ratio of the gas-particle mixture was calculated using the particle concentration obtained from laser-extinction measurements at two stations along the length of the driver section. A beam from a 15-mW He-Ne laser was used as the light source and a photodiode was used as a receiver, output of which was captured by a digital storage oscilloscope.

2.2. Laser light-extinction technique

The principle of operation of the laser light-extinction method is as follows: a beam of light from a continuous-wave, He-Ne laser is made to traverse the shock-tube test section and to fall on a photo-detector. The intensity of the laser beam was measured before the start of the gas-particle injection as I_0. With the gas-particle flow in the test section, the intensity is reduced to I which is related to the initial intensity by the Lambert-Beer equation as

$$\frac{I}{I_0} = \exp(-k\rho_p) \tag{1}$$

where k is a constant which includes the extinction coefficient and ρ_p is the cloud density. In the present experiments, a pin photodiode (PIN SPOT 8D, United Detector Technology) is employed as a photo-detector. The output voltage E of the photodiode is directly proportional to the intensity of light incident on it. Hence, the Lambert-Beer law can be written as the ratio of the voltage outputs from the photo-diode with and without particles in the flow as

$$\frac{E}{E_0} = \exp(-k\rho_p) \ . \tag{2}$$

Determination of k applicable to our experimental situation was done making use of Oswalds' equipment in which a measured mass of particles was dispersed in a known volume by feeding high-pressure air through a solenoid valve connected to the equipment. The path length of laser light in the Oswalds' equipment is the same as that in the shock-tube test section. The extinction of light under these conditions is calibrated in the calibration volume against the known concentration, hence known loading ratio. The calibration curve obtained in our arrangement is given in Fig. 1.

3. Calculation of drag coefficient

An equation for the drag coefficient in terms of quantities measured in the shock tube was obtained (Buckley 1968) combining the equations relating the optical properties of a cloud of particles with those describing its motion. Using continuity and momentum equations separately for both particles and gas in a gas-particle mixture in a shock-fixed coordinate system and Eq. (2) for the light-extinction signal, the expression for the drag coefficient of a cloud of particles was obtained as

$$C_d = \frac{4D_p\rho_s\frac{d}{dt}(\frac{E}{E_0})}{3\rho_g V_s\frac{E}{E_0}\ln(\frac{E}{E_o})(1-\frac{u_g}{u_p})^2} \tag{3}$$

where D_p = average particle diameter, ρ_s = density of particle material, V_s = shock velocity, u_g = gas velocity, and u_p = particle velocity.

But when measurements are done immediately behind the shock front, the equation is simplified as

$$C_d = \frac{4D_p\rho_s(\rho r)\frac{d}{dt}(\frac{E}{E_0})}{3\rho_g V_s\frac{E}{E_o}\ln(\frac{E}{E_o})(\rho r-1)^2} \tag{4}$$

where ρr is the density ratio across the shock front and is found from the normal shock relation. Hence, together with a knowledge of the initial pressure and temperature in the shock tube, the only measurements required to evaluate the drag coefficient are those of E as a function of time and V_s.

4. Results and discussion

Prior to each run, the interior surface of the driven section of the shock tube was cleaned. For each type of particle, runs were made with five different incident Mach numbers, e.g., 1.2, 1.4, 1.5, 1.55, and 1.6.The initial velocity of the mixture in the shock tube was neglected and the shock velocity was computed from the reading of a microsecond counter using time-distance method as well as from the pressure increase after the shock. Loading ratio η is found from the calibration curve (Fig. 1). The Reynolds number was calculated from the

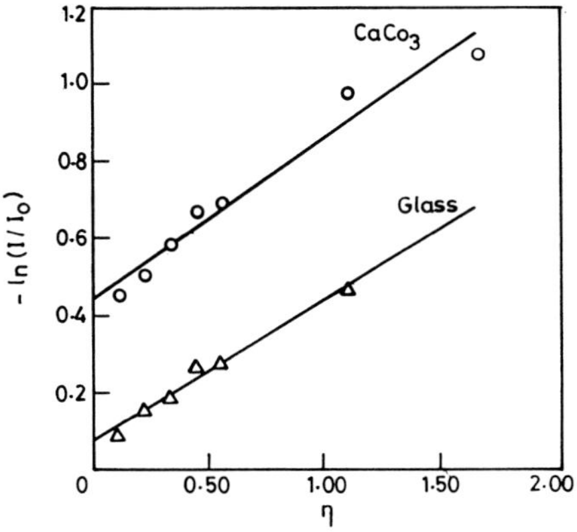

Figure 1. Loading-ratio determination using light-extinction method.

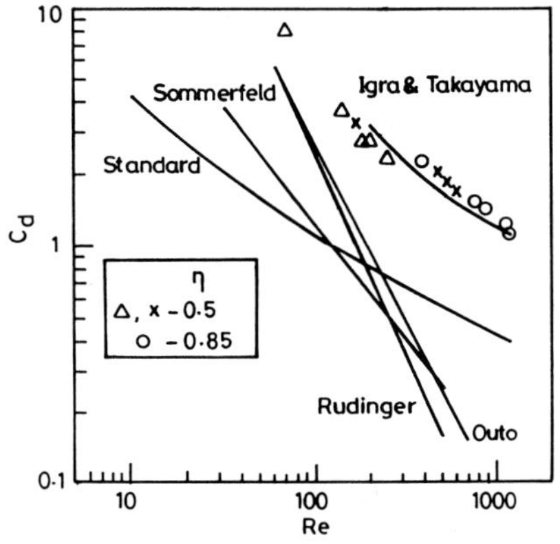

Figure 2. Correlation between drag coefficient and Reynolds number and experimental results.

slip velocity between gas and particle and the particle diameter. The range of Reynolds numbers obtained was between 70 and 1100. The experimental points are shown in Fig. 2 where values of drag coefficients are plotted against the particle Reynolds number. For comparison, the steady-flow standard drag curve and some of the previous correlations for unsteady drag are also shown in the figure. The trend seen in three of the experimental correlations is that the unsteady drag coefficient is higher than standard drag values up to a Reynolds number of about 200. Thereafter the C_d values are smaller than the steady-flow values. Present experimental points do not show this trend, but they are about 100% more than the steady-flow values and compare well with the unsteady values for a single sphere obtained by Igra and Takayama (1991) for Re $>$ 200. Igra and Takayama justified the increased drag as a result of the unsteady nature of the flow around the sphere. One interesting point is that our results, which are applicable for a cloud of particles, agree with the sphere values of Igra and Takayama. The particle-particle interaction does not change the C_d values considerably.

5. Conclusions

The present experimental results indicate that the unsteady drag coefficient for a sphere applicable to a cloud of particles is different from values obtained in previous works. Unsteady drag coefficients are much greater than standard drag coefficients for the whole range of experiments. To find the effect of particle-particle interactions, a systematic study employing much higher loading ratios is necessary.

References

Arnold HD (1911) Limitations imposed by slip and inertia terms upon Stokes law for the motion of spheres through liquids. Phil Mag 22 (6):755–775.

Bailey AB (1974) Sphere drag coefficient for supersonic speeds in continuum and free molecular flows. J Fluid Mech 65:401–410.

Buckley FT Jr (1968) An Experimental Investigation of the Drag of Particles Accelerating Within Particulate Clouds at Transonic Slip Flow Conditions by a Light Extinction Technique, Ph.D. Dissertation, University of Maryland.

Crowe CT, Babcock WR, Willoughby PG (1971) Drag coefficients for particles in rarefied low Mach number flows. In: Int Symp on Two Phase Systems, Technion, Haifa, Israel, Paper 3-3.

Heinrich HG, Nicuum RJ, Mark EL (1963) The Drag Coefficient of a Sphere Corresponding to a One Meter Robin Sphere Descending from 260000 Ft, University of Minnesota, Contract Rep. A. F. 19 (604)-8-34.

Igra O, Takayama K (1991) Shock tube study of the drag coefficient of a sphere in a nonstationary flow. In: Shock Waves, Proceeding of the 18th ISSW, Sendai, Japan, pp 491–497.

Igra O, Takayama K (1993) Shock tube study of drag coefficient in a nonstationary flow. Proc R Soc London A 442:231–247.

Karafilian SK, Kotas TJ (1978) Drag on a sphere in unsteady motion in a liquid at rest. J Fluid Mech 87:1185–1196.

Liebster M (1927) Ber den Wiederstand von Kugeln. Ann d Physik 83:541–562.

Outa E, Tajima K, Suzuki S (1981) Cross sectional concentration of particles during shock process propagating through a gas-particle mixture in a shock tube. In: Proc of the 13th Int Symp on Shock Waves.

Rodriguez G, Grandebouef P, Khelifi M, Hass JF (1993) Drag coefficient measurement of spheres in a vertical shock tube and numerical simulation. In: Shock Waves @ Marseilles, Vol III, Springer-Verlag, pp 43–48.

Roos FW, Willmarth WW (1971) Experimental results on sphere and disc drag. AIAA J 9:285–291.

Rudinger G (1970) Effective drag coefficient for gas particle flow in shock tubes. JSME J Basic Eng Ser D92:165–172.

Schmiedel J (1928) Experimentelle Untersuchungen ber die Fallbewegung von Kugeln und Scheiben in reibenden Fl ssigkeiten. Phys Z 29:593–610.

Selberg BP, Nicholls JA (1968) Drag coefficient of small spherical particles. AIAA J 6:401–407.

Sommerfeld M (1985) The unsteadiness of shock waves propagating through gas particle mixtures. Exp in Fluid 3:197–206.

Acceleration of a sphere behind planar shock waves

A. Britan, T. Elperin, O. Igra, J. P. Jiang
The Pearlstone Center for Aeronautical Engineering Studies
Department of Mechanical Engineering
Ben-Gurion University of the Negev
Beer Sheva, Israel

Abstract: The trajectory and velocity of different-diameter spheres were measured by using different optical methods and performing experiments in two different shock tubes. Based upon the obtained results and using a simple one-dimensional flow model, the effect on the sphere motion of the shock-wave interactions with the sphere was studied.

Key words: Shock-wave/sphere interaction, Drag coefficient

1. Introduction

To predict the final flow which results from the sphere's interaction with a shock wave in a shock tube we may consider the sphere's presence in the channel as a simple constriction in the channel cross-section area (F_c). For a given gas, the flow field on both sides of the constriction depends solely upon the incident shock-wave Mach number, M_s, and the area ratio between F_c and the sphere's cross section (F_p), i.e., $\bar{F} = F_c/F_p$. Two-dimensional, non-steady flow calculations (see, e.g., Britan et al. 1986) predict for the case under investigation that the flow pattern consists of two shock waves (transmitted and reflected) and a contact surface. The present study centers on the effects that changes in the shading parameters $\bar{F}$ have on the flow pattern and on the sphere motion behind the incident shock-wave front. Acceleration of a single 38-mm diameter sphere (ρ_p = 89.44 kg/m^3) was studied in an 8-m long and 0.09-m diameter shock tube having a 0.08-m × 0.08-m square, 3-m long channel. The motions of wave fronts, the sphere, and the interfaces were recorded using a laser-shadowgraph optical system through Plexiglas windows. For recording the shadowgraphs a high-speed camera and Cu-vapor stroboscopic laser were used; about 20 flow pictures with a 100-μs time interval between neighboring frames were taken during each test. Acceleration of light plastic spheres (D_p = 12, 15, and 20 mm, ρ_p = 43.18 kg/m^3) were tested in a small-size shock tube with a 1.5-m long channel and a 0.8-m long driver section. Both the driver and channel were made from a 31-mm × 31-mm square steel tube. A simple optical system with a He-Ne laser and a photo detector were used to measure the duration of the sphere's intersection (blocking) of the laser beam. Pressure transducers were used for triggering the system and for measuring the shock-wave velocity and trajectory.

2. Experimental results and discussion

From the set of photographs shown in Fig. 1 all the main details associated with shock diffractions on the sphere can be seen. The reflected shock wave in front of the sphere has a bow shape which interacts with transmitted parts of the incident shock wave; and two triple points of the Mach configuration are clearly seen in Fig. 1b. The area-ratio parameter $\bar{F}$ does not affect the process at this initial stage since at the considered time the shock waves have not as yet reached the channel walls. The characteristic parameter governing the flow field near the sphere is the speed of sound behind the incident shock wave (Britan et al. 1986). This first stage of the flow-formation process is terminated by reflection of triple points from the channel side walls. The new transmitted planar shock wave and two reflected cross waves of the original Mach configuration are clearly seen in Fig. 1c. The flow is disturbed by the sphere's support which protrudes from the channel floor. The presence of this support generates turbulence in the main stream. After the reflection of crossing waves from the channel center plane, the flow approaches a steady state.

In Fig. 1d choked flow with a standing shock wave near the flow throat, similar to an over-expanded flow, can be seen. The standing shock eventually breaks into several weak shock waves and disappears due to the sphere's motion; see Fig. 1g. Since the sphere's transverse displacement is very small in comparison with its lateral displacement we analyzed only the displacement along the x-direction. Results of these measurements are marked in Fig. 2a by cross points. A sixth-order polynomial was fitted to the experimental results. The time interval during which the sphere's motion can be studied is limited by the appearance of a reflected shock wave from the channel end wall. Therefore, only measurements taken during a time period of 1.6 ms are shown in Fig. 2a; where time $t = 0$ and $x = 0$ correspond to Fig. 1b. A visible sphere displacement begins only after all transient disturbances have subsided. At this stage of the process we assume that the one-dimensional flow field has not been changed by the sphere's presence and we have used the Rankine-Hugoniot shock relations to determine the flow density (ρ) and velocity (v_g) behind the incident shock wave.

The expression for the force acting on the sphere by the post-shock-wave flow is:

$$f = \frac{\pi}{4} D_p^2 C_D \frac{1}{2} \rho \left| v_g - v_p \right| (v_g - v_p) - \frac{1}{2} \frac{\pi}{6} D_p^3 \rho \dot{v}_p \tag{1}$$

where $\dot{v}_p$ is the sphere acceleration. When the density of the sphere $\rho_p \gg \rho$ the

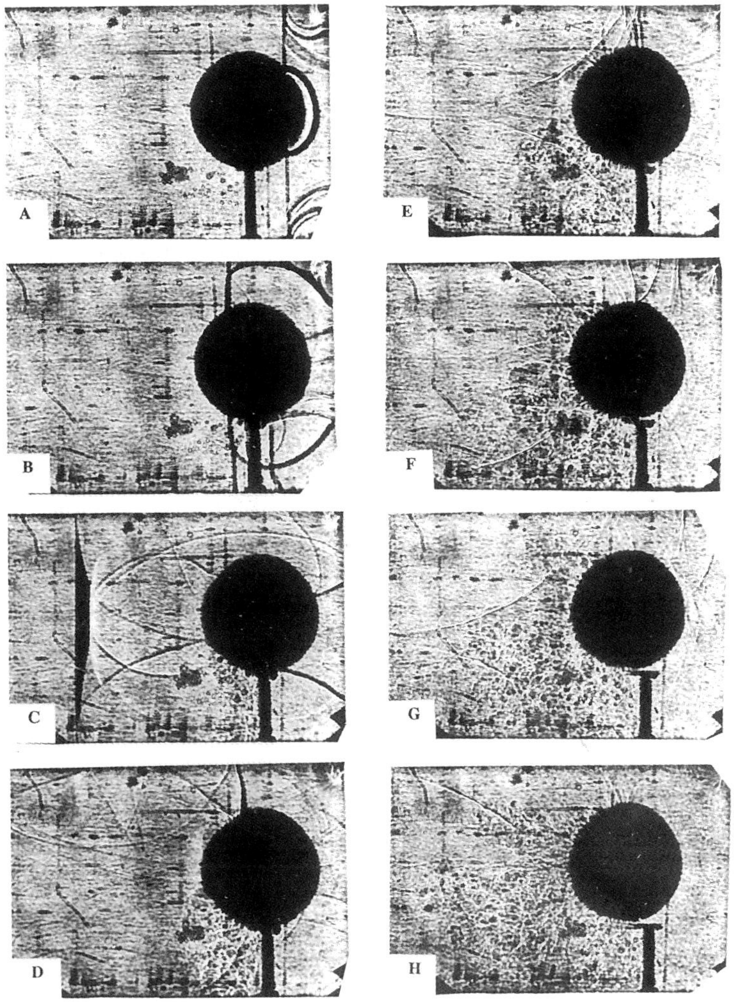

Figure 1. Photograph of interaction between a planar shock wave and a sphere ($D_p = 38$ mm, $M_s = 1.5$, distortion in the sphere shape is caused by imperfections in the optical system).

sphere's acceleration is:

$$\frac{dv_p}{dt} = \frac{3}{4D_p}\frac{\rho}{\rho_p}C_D \left| v_g - v_p \right| (v_g - v_p) \tag{2}$$

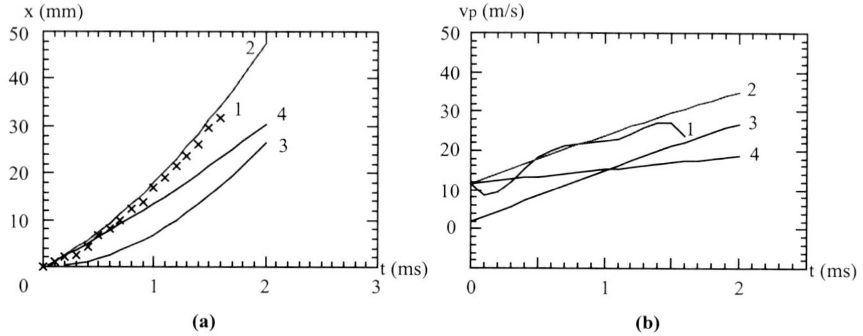

Figure 2. (a) Measured and caculated trajectories of a sphere ($D_p = 38$ mm, $M_s = 1.5$). 1 – experiment; 2 – calculation for $v_i = 11$ m/s; 3 – calculation for $v_i = 0$ m/s; 4 – calculation for $v_i = 11$ m/s and C_D as in Igra and Takayama (1993); (b) variations in the sphere velocity deduced from its trajectory ($D_p = 38$ mm, $M_s = 1.5$). 1 – based on curve 1 in Fig. 2a; 2 – calculation for $v_i = 11$ m/s; 3 – calculation for $v_i = 0$ m/s; 4 – calculation for $v_i = 11$ m/s and C_D as in Igra and Takayama (1993).

Following Igra and Takayama (1993) we define an acceleration modulus, $A_c \equiv \frac{dv_p}{dt} D_p (v_g - v_p)^{-2}$, as a parameter emphasizing the difference between a non-steady and a steady drag force. Since in all our measurements $A_c \leq 0.015$ we can safely use the steady drag-coefficient correlation of Igra and Ben-Dor (1988).

We suppose, for simplicity, that just after the incident shock wave has passed the sphere the sphere acquires an initial velocity v_i associated with the collision process between the incident shock wave and the sphere. It can be seen from Fig. 2b that $v_i = 11$ m/s and at the initial stage of the sphere's motion its velocity may drop below this value due to its interaction with shock waves reflected from the channel side walls (see Figs. 1b and c). After some time the sphere accelerates and begins to move with an almost constant acceleration.

To analyze the effect of different formulations proposed for the drag coefficient, in addition to the steady correlation proposed by Igra and Ben-Dor (1988), we also used the unsteady correlation proposed by Igra and Takayama (1993). At early times (see Figs. 2a and b) the results based upon this unsteady correlation for C_D agree fairly well with experimental findings; however, at a later time better agreement is obtained when the steady correlation for C_D is used (curve 2 in Figs. 2a and b). This should be expected since at early times the flow around the sphere is definitely unsteady; it approaches a steady flow at a later time. The initial velocity, v_i, plays an important role in determining the sphere's motion. When $v_i = 0$ the calculated trajectory is very far from the experimental results; see curve 3 in Figs. 2a and b. The above-mentioned non-steadiness in the flow is more pronounced for the sphere velocity $v_p(t)$ than for its trajectory $x_p(t)$. However, the non-steady nature of the sphere motion is not so strong and curve 2 in Figs. 2a and b predicts correctly the experimental results because the initial perturbations due to the shading effect have practically no influence on the flow.

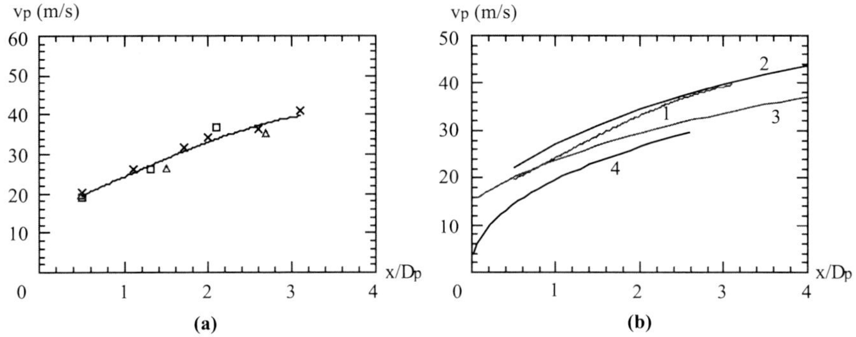

Figure 3. (a) Sphere velocity vs. distance (x/D_p), $M_s = 1.25$, $D_p = 20$ mm (cross points), 15 mm (square points), 12 mm (triangle points). The curve fit for experimental points is shown by line; (b) comparison between experimental and calculation for $v = f(x/D_p)$. 1 – curve fit to experimental points; 2 – computational results for $M_s = 1.3$, $v_i = 15$ m/s; 3 – computational results for $M_s = 1.25$, $v_i = 15$ m/s; 4 – computational results for $M_s = 1.25$, $v_i = 1.5$ m/s.

To study whether the quasi-one-dimensional approach will hold for low $\bar{F}$, i.e., with stronger shading effect, the value of $\bar{F}$ must be reduced, at least by half. This will require using a sphere with diameter $D_p \geq 50$ mm. Acceleration of such a large and heavy sphere would be slow and could be a problem because of limitations on the shock-tube test time. We therefore employed an alternative method to reduce the parameter $\bar{F}$ by using a small-size shock tube and a set of small spheres having diameters of $D_p = 12$, 15, or 20 mm. For these spheres the appropriate values of $\bar{F}$ were: $\bar{F} = 8.5$, 5.4, and 3, respectively. Using two laser beams positioned at a distance $\Delta x > D_p$ allows measuring the sphere's velocity at each beam position and the time passed while the sphere travels between the two beams. A third-order polynomial fit which provides a good fit to the obtained experimental results is shown by the line in Fig. 3a. Extrapolation of this fitting curve (in Fig. 3a) to $x/D_p = 0$ results in $v_i = 15$ m/s, which is close to $v_i = 11$ m/s used for the large sphere ($D_p = 38$ mm). The main reason for the observed scatter in the experimental findings shown in Fig. 3a is the inaccuracy in reading the initial sphere position, with respect to the laser beam, at different runs; it is only within ±10% . At the same time, it should be noted that there is no systematic discrepancy between different points on the plot shown in Fig. 3a in spite of the large variations in the parameter $\bar{F}$. This means that similar to the previous case, the shading effect on the sphere acceleration is marginal in the small-size shock tube.

On the basis of pressure measurements (time of arrival of shock waves) and calculation of the contact surface trajectory we constructed an x-t diagram for the small-size shock tube. It was found that during a significant part of the test time the sphere moves close to the contact surface. Flow parameters in this region may exceed, by up to 10 – 15%, the values reached immediately behind the shock-wave front due to mass losses through the wall boundary layer (Mirels 1966). For estimating the dependence of the sphere's trajectories on the flow non-uniformities, a set of calculations was made for the post-shock-wave flow appropriate to $M_s = 1.25$ and a second set was evaluated for $M_s = 1.30$. Because of the difference in the M_s used (about 4%) the expected difference in the post-shock-wave gas pressure is 12%; with an 8% difference in density and a 13% difference in the flow velocity. One can see from Fig. 3b that at early stages of the sphere motion ($x/D_p \leq 1$) experimental results are closer to the calculated values obtained for $M_s = 1.25$. However at later times (for $x/D_p > 2.5$) they approach the results obtained when using $M_s = 1.30$. This can be explained by changes in the flow properties that occur along the post-shock gas flow. The discrepancy between experimental and calculated results for $v_i = 1.5$ m/s (curve 4) is large and shows the effect that the velocity, v_i , has on the obtained results.

References

Britan A, Rudnitskii A, Starik A (1986) Numerical investigation of the process of shock reflection from a wall with a slot. Zh Prikl Mech Tekh Fiz 6:903–909.

Igra O, Ben-Dor G (1988) Dusty shock waves. Applied Mechanics Reviews 41:381–437.

Igra O, Takayama K (1993) Shock tube study of the drag coefficient of a sphere in a non-stationary flow. Proc R Soc Lond A442:231–247.

Mirels H (1966) Flow nonuniformity in shock tubes operating at maximum test times. Phys Fluids 9:1907–1912.

Dusty shock reflection and diffraction in aligned baffle systems

B.-Y. Wang
Institute of Mechanics, Chinese Academy of Sciences
Beijing 100080 China

Y.-H. Xu, Y.-Q. Xu, Q.-S. Wu
Department of Mechanics, University of Science and Technology of China
Hefei 230026 China

Abstract: This paper describes shock propagation through dilute gas-particle suspensions in baffle systems. The governing equations are solved by using high-resolution schemes and the computational results show the reflection and diffraction patterns of a planar incident shock wave in an aligned baffle system.

Key words: Dusty shock, Shock reflection, Shock diffraction, Baffle system

1. Introduction

Shock-induced flows have inspired avid interest for more than 100 years of research (Hillier 1993). Most of the existing work mainly concerns shock phenomena in pure gases. Recently, investigations have been extended to treat the problem of shock reflection and diffraction in various gas-particle suspensions (i.e. dusty gases) since many applications (dust explosion, ablation, and erosion, etc.) involve such dusty gas. For instance, shock propagation in baffle systems filled with a dilute dusty gas and the post-shock flow in the two-phase medium are important to understand for explosion safety analyses.

In this work, numerical simulation of the two-phase flow induced by a planar incident shock wave in an aligned two-dimensional baffle system is performed. Some experimental and computational studies have been done for the pure gas case (Reichenbach and Kuhl 1993; Luo et al. 1993) but none for the dusty gas case until now. To formulate the unsteady compressible two-phase flow in these systems with complex geometry, two-dimensional and time-dependent Euler equations are presented based on the two-continuum model, where the interaction between the gas and particles is taken into account. The governing equations are solved by using the operator-split GRP scheme (Falcovitz et al. 1993) and NND-4 scheme (Zhang et al. 1988), which are adopted respectively to the gas and particle phases. The dusty shock reflection and diffraction patterns, and the unsteady two-phase flow structures behind the shock front are obtained for a two-dimensional baffled tube. These computational results reveal the fundamental features of the post-shock flow field: turbulent mixing and particle relaxation are the principal dissipation mechanisms of the incident shock.

2. Governing equations

As mentioned above, the problem under consideration is the reflection and diffraction of a planar incident shock wave propagating in an aligned 2D baffle system filled with a dilute dusty gas. Here, based on the two-continuum model, we suppose that: (1) The gas is compressible and perfect. Its viscosity and thermal conductivity are ignored except for the interaction between the gas and particles. (2) The particles are rigid and inert. They are uniform solid spheres without Brownian motion and internal temperature gradient. (3) As a dilute gas-particle system, the partial pressure due to particle contribution can be neglected. (4) The interaction mechanisms involve viscous friction and heat transfer between the gas and particles.

Under these assumptions, the Euler equations governing time-dependent, two-dimensional, and two-phase flows can be written in the quasi-conservative form as (in the follwing expressions, the subscript p refers to the particles)

$$\frac{\partial U}{\partial t} + \frac{\partial F(U)}{\partial x} + \frac{\partial G(U)}{\partial y} = H(U) \tag{1}$$

where

$$U = \begin{pmatrix} \rho \\ \rho u \\ \rho v \\ \rho E \\ \rho_p \\ \rho_p u_p \\ \rho_p v_p \\ \rho_p E_p \end{pmatrix} \quad F = \begin{pmatrix} \rho u \\ \rho u^2 + p \\ \rho uv \\ (\rho E + p)u \\ \rho_p u_p \\ \rho_p u_p^2 \\ \rho_p u_p v_p \\ \rho_p E_p u_p \end{pmatrix} \quad G = \begin{pmatrix} \rho v \\ \rho uv \\ \rho v^2 + p \\ (\rho E + p)v \\ \rho_p v_p \\ \rho_p u_p v_p \\ \rho_p v_p^2 \\ \rho_p E_p v_p \end{pmatrix} \quad H = \begin{pmatrix} 0 \\ -f_x \\ -f_y \\ -Q \\ 0 \\ f_x \\ f_y \\ Q \end{pmatrix}$$

Here u and v are velocity components in x and y directions. $E = CT + (u^2+v^2)/2$ is the total specific energy (C stands for gas specific heat at constant volume C_v or particle heat capacity C_m). The variables p, ρ, and T are respectively pressure, density, and temperature. The three gas parameters can be associated by the state equation:

$$p = \rho RT \tag{2}$$

where R is the gas constant. The appearance of some interphase terms is the main difference between our Eq. (1) for the dusty gas and the corresponding equation for pure gas. For the present problem, the interaction parameters are introduced as follows: f_x, f_y, and q are respectively the drag force components and heat transfer to be exterted on particles per unit volume; $Q = u_p f_x + v_p f_y + q$ is dissipation work due to the interphase effects.

3. Numerical method

In constructing a two-dimensional scheme for time-dependent compressible flows, well-known techniques of operator-splitting for temporal and spatial operators (Sod 1978; Strang 1968) are employed and then the single 2D conservation system (1) is decomposed into a set of three 1D systems:

$$\frac{\partial U}{\partial t} + \frac{\partial F(U)}{\partial x} = 0; \frac{\partial U}{\partial t} + \frac{\partial G(U)}{\partial y} = 0; \frac{dU}{dt} = H(U) \tag{3}$$

In this way, there is no coupling between the flows of the gas and particle phases for the first two systems. The high-resolution schemes GRP and NND-4 are applied respectively to solve the gas and particle flows. Here, the GRP method is an "analytic" second-order extension of the classical Godunov scheme. Due to its ability to capture shocks with high resolution without spurious oscillations, the GRP scheme is developed for predicting the complex shock-wave pattern involving transmission, reflection, and diffraction in the baffled tube. The third system is an ordinary differential equation which contains the interaction terms between the gas and particles. It is solved by a second-order predictor-corrector method.

Let Lx, Ly, and Lr to be the difference operators, the following combined five-step scheme can be constructed for each time step:

$$U_{ij}^{n+1} = Lr\left(\frac{\Delta t_n}{2}\right) Lx\left(\frac{\Delta t_n}{2}\right) Ly\left(\Delta t_n\right) Lx\left(\frac{\Delta t_n}{2}\right) Lr\left(\frac{\Delta t_n}{2}\right) U_{ij}^{n} \tag{4}$$

where i, j, and n are space and time nodes; Δt_n is the time interval determined by the CFL condition.

4. Computational results

With the above algorithm, computations were performed to simulate the dusty shock reflection and diffraction process in the aligned 4-baffle system with the following parameters for the particle phase: diameter $d = 10\ \mu$m, material density $\sigma_p = 2500$ kg/m^3, mass loading ratio $\alpha = 1.0$, heat capacity ratio $\beta = 1.0$; and with the following expressions for the interphase interactions: drag coefficient $C_D = 0.48 + 28Re^{-0.85}$, Nusselt number $Nu = 2.0 + 0.6Pr^{1/3}Re^{1/2}$ (*Re* is the slip Reynolds number, *Pr* is the Prandtl number). Corresponding calculations for the pure gas were also carried out in order to illustrate particle effects on the induced flows. The computation was first made under the same conditions as for the pure gas interferogram test by Reichenbach and Kuhl (1993), where the incident shock Mach number M_S is 1.23. In this weak-shock case, the obtained isopycnics for the pure and dusty gases are plotted respectively in Fig. 1a and 1b and the interferogram pictures are shown in Fig. 2.

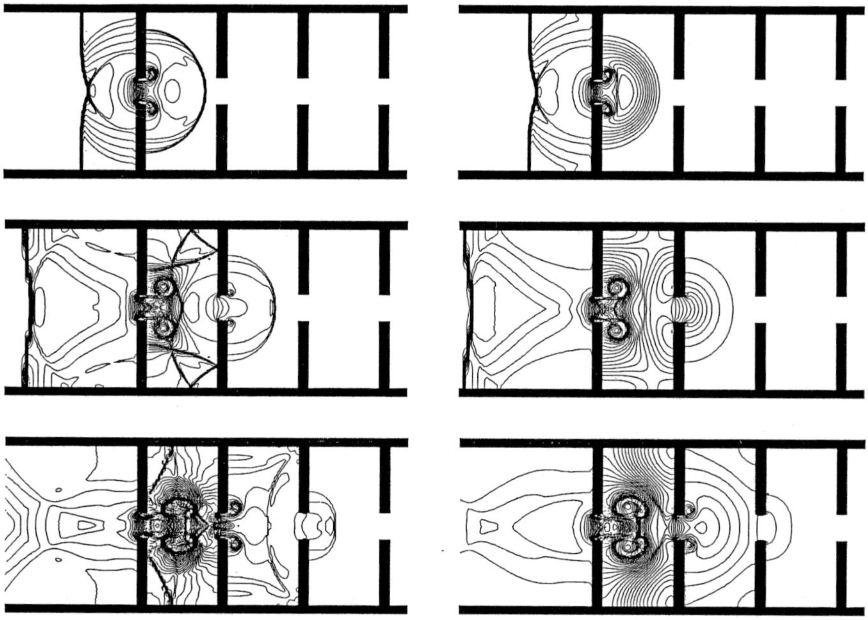

(a) (b)

Figure 1. Density contours of gas phase ($M_S = 1.23$): (a) pure gas and (b) dusty gas.

Figure 2. Shock propagation interferograms in pure gas ($M_S = 1.23$).

The time sequence of the flow field in Fig. 1a takes the same instants as the flow visualization test: they are successively at $t = 96.04\ \mu$s, $186.08\ \mu$s, and 276.39 μs. The comparison of numerical and experimental results indicates the excellent agreement between the computation and experiment. It verifies the accuracy and effectiveness of the numerical method. Figure 1b presents the numerical results of the shock-induced flow in the dusty gas case, where the frozen shock fronts have the same x-position as for the pure gas. However, the arrival time takes different values as $t = 108.94$ μs, 220.61 μs, and 315.42 μs. The decrease in the strength and speed of the transmitted shock wave can be attributed to particle-dissipation mechanisms.

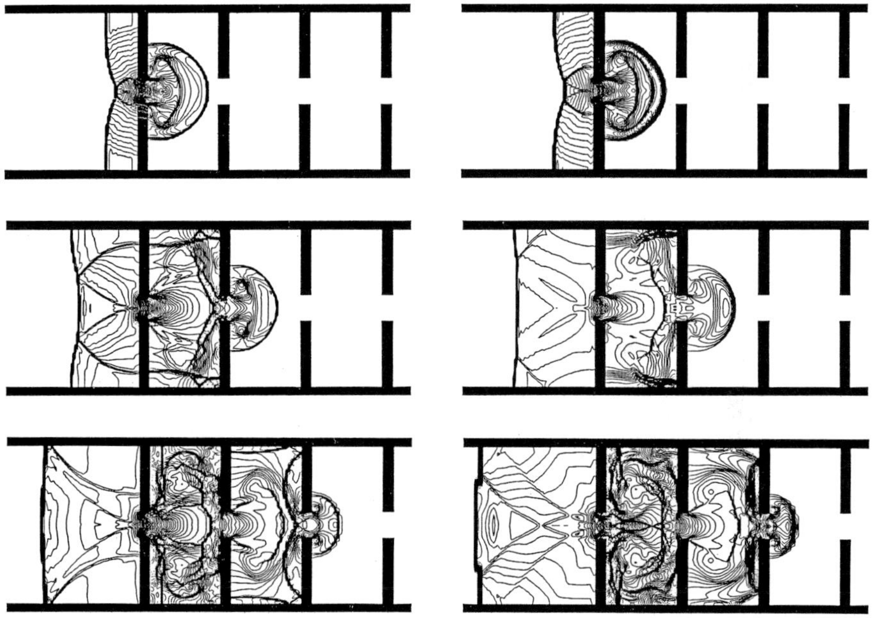

(a) (b)

Figure 3. Density contours of gas phase ($M_S = 2.50$): (a) pure gas and (b) dusty gas.

For the case of a stronger incident shock wave ($M_S = 2.5$), Fig. 3 shows the gas density contours in the post-shock flow. The time instants in sequence are respectively $t = 48.65$ μs, 104.71 μs, 160.11 μs for the pure-gas case and $t = 61.43$ μs, 132.21 μs, 198.95 μs for the dusty-gas case. Similarly, at these moments the shock front arrives at the same position as in the case of $M_S = 1.23$. These numerical results reveal the fundamental features of the dusty shock reflection and diffraction patterns. When an incident shock wave propagates through a dusty gas in a baffled tube, reflection and diffraction occur and the strength of the transmitted shock is greatly reduced because of double mechanisms: baffle blocking and particle dissipation. For the dusty gases, the reflection and diffraction patterns in baffle systems become more complicated compared with those in corner systems (Wang et al. 1993). It is seen that at each sharp corner of the orifices in the baffle plate, a shear layer is generated and then rapidly rolls up in a spiral vortex. Moreover, each pair of the corners results in a mushroom-shaped region, where turbulent mixing is a dominant process. Therefore turbulent mixing, particle relaxation, and their interactions consist of complex fluid-dynamic effects.

5. Conclusion

Numerical simulations of the dusty shock induced flow in the aligned two-dimensional baffle system indicate that the reflection and diffraction patterns become quite complex due to the presence of baffle plates and particles. Turbulent mixing and particle relaxation are the main dissipation mechanisms.

Acknowledgement. The authors are indebted the National Natural Science Foundation of China for its financial support.

References

Falcovitz J, Alfandary G, Ben-Dor G (1993) Numerical and experimental study of a head-on reflection of a regular reflection. International Journal for Numerical Method in Fluids 17:1055–1077.

Hillier R (1993) CFD and unsteady shock diffraction. In: 19th Int Symp on Shock Waves, Book of Abstracts, Vol 1, Marseille, pp 14–15.

Luo H, Baum JD, Lohner R, Cabello J (1993) Adaptive Edge-Based Finite Element Schemes for the Euler and Navier-Stokes Equations on Unstructured Grids, AIAA-93-0336.

Reichenbach H, Kuhl AL (1993) Shock-induced turbulent flow in baffle systems. FhG/EMI E12/93.

Sod GA (1978) A survey of several finite difference methods for systems of nonlinear hyperbolic conservation laws. J Comp Phys 27:1–31.

Strang G (1968) On the construction and comparison of difference schemes. SIAM J Num Anal 5:506–517.

Wang BY, Wu SQ, Zhu H, Xu YH (1993) Numerical simulation of dusty shock diffraction around a 90 degree corner. In: 19th Int Symp on Shock Waves, Book of Abstracts, Vol 1, Marseille, pp 422–423.

Zhang HX (1988) Non-oscillation and non-free-parameter dissipation difference scheme. Acta Aerodynamica Sinica 6:143–165.

Numerical analysis of wave patterns of shock propagation in a dusty gas along convergent-divergent nozzles

Q.-S. Wu, D.-Z. Wang, Y.-H. Xu
Department of Mechanics, University of Science and Technology of China
Hefei 230026 China
B.-Y. Wang
Institute of Mechanics, Chinese Academy of Sciences
Beijing 100080 China

Abstract: Based on a simplified model of quasi-one-dimensional unsteady flow, the wave patterns of shock propagation in the quiescent or moving dusty gases ahead of the shock front along convergent-divergent nozzles are analyzed by using high-resolution numerical methods — the GRP scheme and the NND-4 scheme. These results are compared with those for the pure gas.

Key words: Shock propagation, Nozzles, Dusty gas, Numerical calculation

1. Introduction

Investigation of the behavior of shock propagation in convergent-divergent nozzles is both interesting and important in scientific research and engineering problems such as the starting process of single- or double-driver shock tunnels. For the former, the gas in the nozzle ahead of the shock wave is quiescent while for the latter, it is moving. Since the late 1960s many researchers have experimentally and numerically studied the features of shock-wave propagation in nozzles (Smith 1966; Amann 1969; Kashimura et al. 1986; Prodromou and Hillier 1991). However, only the simple case, in which the working medium is a pure gas and it is quiescent ahead of the shock front, was considered. For the quiescent dusty gas ahead of the shock front and for the moving pure or dusty gas ahead of the shock front, studies remain to be carried out. In this paper, on the basis of the simplified model of quasi-one-dimensional unsteady flow, the wave patterns of shock propagation in the quiescent or moving dusty gases ahead of the shock front along convergent-divergent nozzles are analyzed by using high-resolution numerical method — the GRP scheme (Ben-Artzi and Falcovitz 1986) and the NND-4 scheme (Zhang and Li 1992). The results obtained are compared with those for the pure gas.

2. Physical model and numerical method

Suppose that the dusty gas is a dilute suspension and the general assumptions suitable for this two-phase flow system are adapted (Crow 1982). In addition, suppose that, before the incident shock wave arrives at the entrance, the two-phase flow behind the shock front has approached an equilibrium state except for the relaxation zone, and there is a quiescent equilibrium gas-particle mixture or a steady equilibrium two-phase flow ahead of the shock front.

The shape of the nozzles is given by

$$A = \begin{cases} A_{in} & x_{min} \le x \le x_{in} \\ A_* + (A_{in} - A_*)\left(\dfrac{x - x_*}{x_{in} - x_*}\right)^2 & x_{in} < x \le x_* \\ A_* + (A_{ex} - A_*)\left(\dfrac{x - x_*}{x_{in} - x_*}\right)^2 & x_* < x \le x_{ex} \\ A_{ex} & x_{ex} < x \le x_{max} \end{cases}$$

Where A_{in}, A_*, and A_{ex} are the entrance, throat, and exit areas of the nozzles. The shape parameters of the nozzle are given as follows: $A_{in}/A_* = 4.0$, 1.2, 1.2, and $A_{ex}/A_* = 8.0$, 8.0, 2.0 respectively for nozzles No. 1, 2, 3.

Employing the two-continuum model, the governing equations are written in the quasi-conservation type of time-dependent 1-D compressible flows for the dusty gas (Wang and Wu 1991). The operator-splitting technique is applied to deal with the coupling effects between the gas and particles. The split, homogeneous and partial differential equations of the gas phase and the solid phase are numerically solved by the GRP scheme and the NND-4 scheme respectively. The ordinary differential equations with interphase coupling effects are solved by the predictor-corrector method.

3. Numerical results and discussion

For comparison, the following six cases are under computational consideration: cases 1 to 3 correspond to the pure gas and cases 4 to 6 to the dusty gas; particle diameter $d = 10\mu m$, the loading ratio $\alpha = \rho_p/\rho_g = 1.0$, the ratio of specific heats $\beta = C_m/C_v = 1.0$, the material density of particles $\sigma_p = 2500$ kg/m^3. Concerning the flow condition ahead of the incident shock, cases 1 and 4 are at a quiescent state, cases 2 and 5 at a subsonic state, and cases 3 and 6 at a subsonic-supersonic state. The possible wave patterns obtained by changing the incident-shock Mach number and the nozzle-shape parameters for cases 1 to 6 are schematically shown in Figs. 1 to 6, where S_i, S_1, S_2, S_r, C, and R refer to the incident shock, the transmitted (primary) shock, the secondary shock, the reflected shock, the compressible wave, and the rarefaction wave, respectively. For cases 1 – 6, it can be seen that the number of possible wave patterns obtained are 7, 8, 5, 5, 3, and 3 respectively.

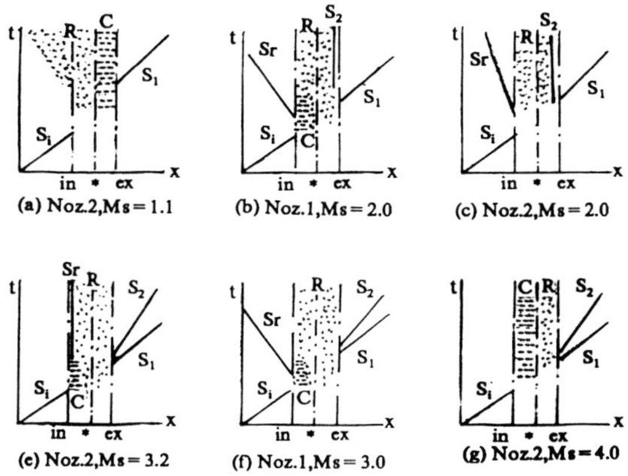

Figure 1. Wave patterns for a quiescent pure gas ahead of a shock.

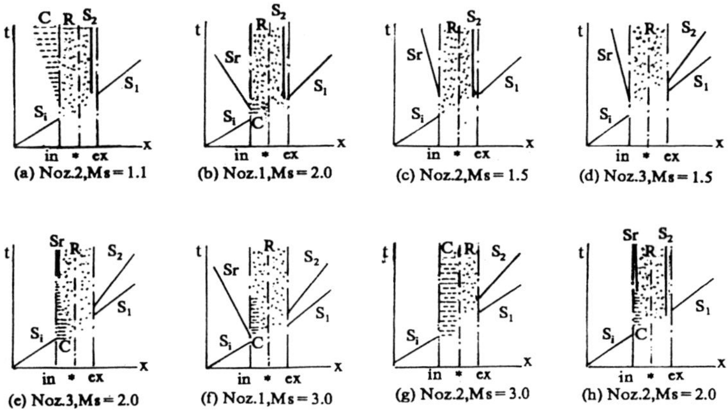

Figure 2. Wave patterns for a pure gas with a subsonic flow ahead of a shock ($M_{1*} = 0.99$).

Comparing Fig. 1 for pure gas with Fig. 4 for dusty gas under quiescent conditions, we can determine: (1) As long as the nozzle-throat flow induced by the shock wave is sonic or supersonic (the flow is expansible after the throat), the secondary shock will be produced after the primary shock (see all the cases of Fig. 1 except for Fig. 1a). Depending on the shock strength and the convergent-divergent ratio of nozzles, the secondary shock wave either stops in the divergent zone or propagates downstream. However, the incident-frozen-shock Mach number M_{sf} in the dusty gas is less than the incident-shock Mach number M_s in the pure gas for the same nozzle and the same wave pattern structures (see Fig. 1f, g and Fig. 4c, d) because the wave-pattern structures depend on the drive-pressure ratio for the same nozzle. Under the same drive-pressure ratio, M_{sf} is less than M_s due to the relaxation of two-phase interaction behind the shock wave. (2) When the incident shock wave enters the nozzle, whether

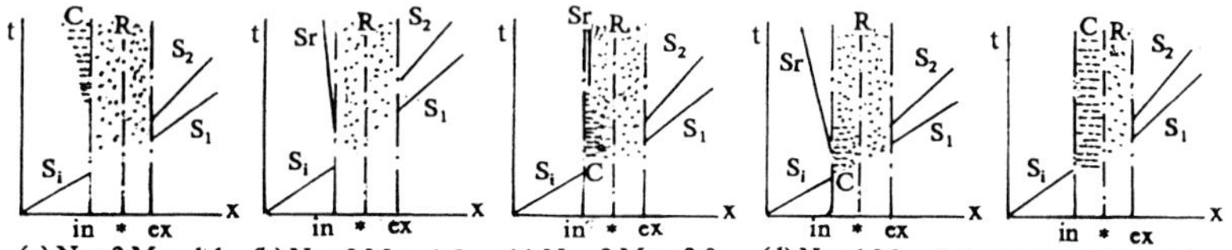

(a) Noz.2,Ms = 1.1 (b) Noz.2,Ms = 1.5 (c) Noz.2,Ms = 2.0 (d) Noz.1,Ms = 2.0 (e) Noz.2,Ms = 3.0

Figure 3. Wave patterns for a pure gas with a subsonic-supersonic flow ahead of a shock ($M_{1*} = 1.0$).

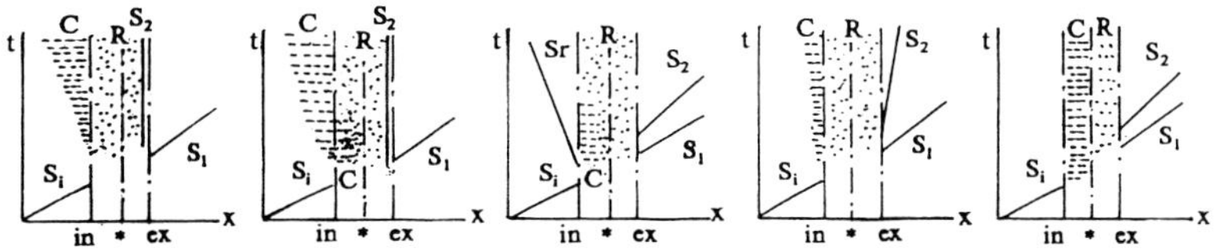

(a) Noz.2,Msf = 1.1 (b) Noz.1,Msf = 1.1 (c) Noz.1,Msf = 2.0 (d) Noz.2,Msf = 1.5 (e) Noz.2,Msf = 3.0

Figure 4. Wave patterns for a quiescent dusty gas ahead of a shock.

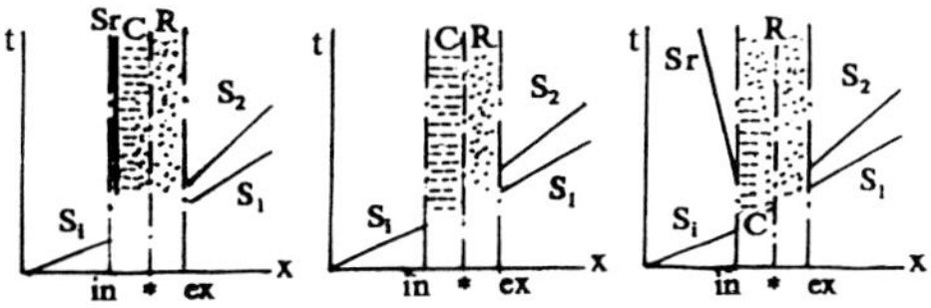

(a) Noz.3,Msf = 1.1 (b) Noz.2,Msf = 2.0 (c) Noz.1,Msf = 1.5

Figure 5. Wave patterns for a dusty gas with an equilibrium subsonic flow ahead of a shock ($M_{1f*} = 0.99$).

there is reflection (and of what type) depends upon the nozzle shape and the incident-shock strength. For the pure gas, in general, the shock wave will be reflected except for the cases of a very weak or a very strong incident shock. Under these conditions there will be either rarefaction wave or no wave reflection. However, for the dusty gas, the cases of reflecting compressible wave are more than those of reflecting rarefaction wave because the equilibrium drive pressure behind the relaxation zone in the dusty gas is higher than that in the pure gas and thus it is more difficult to form the reflected shock wave. (3) The pure- and dusty- gas flow features in these nozzles are basically similar: provided that the induced flow behind the nozzle throat is supersonic, the flow ahead of the secondary shock wave in the nozzle divergent zone is expansive. In the nozzle convergent zone, the flow is all expansible for the small convergent ratio or the weak incident shock; for the large convergent ratio or the strong incident shock, the flow is firstly compressible and then gradually changes to expansible; for the very strong incident shock, the flow is all compressible. However, as mentioned earlier, M_{sf} is less than M_s for the same wave patterns.

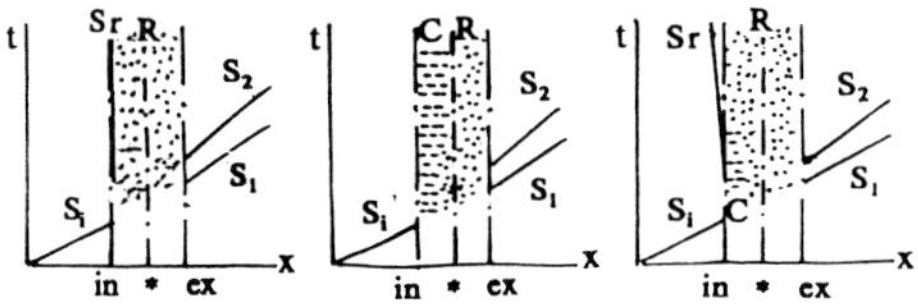

(a) Noz.2,Msf= 1.1 (b) Noz.2,Msf=2.0(c) Noz.1,Msf=1.5

Figure 6. Wave patterns for a dusty gas with an equilibrium subsonic-supersonic flow ahead of a shock ($M_{1f*} = 1.0$).

For a moving medium ahead of the incident shock wave, the wave-pattern structures and the features of flow-parameter distributions for the pure gas are about the same in the quiescent state and the subsonic flow. But for the same wave pattern, the corresponding incident-shock Mach number M_s is less in the subsonic flow ahead of the shock (see Figs. 1c, 1d, 2c, and 2d). When the steady flow ahead of the incident shock is subsonic-supersonic, there is no case in which the secondary shock wave stops in the divergent zone; the total number of wave patterns is less than that in the quiescent state and the subsonic flow ahead of the incident shock. For the dusty gas, the wave patterns are approximately the same in the subsonic and the subsonic-supersonic flow ahead of the incident shock. The secondary shock may not stop in the nozzle divergent zone, but the reflected shock wave may. Based the computational pressure distribution, this secondary shock is very weak and nearly dispersed. All of these different characteristics are produced by the interactions between the incident shock wave and the steady equilibrium two-phase flow, and the relaxation effects between the two phases behind shock waves.

4. Concluding remarks

The numerical results indicate that the characteristics of the shock propagation in a dusty gas along convergent-divergent nozzles are different from those in a pure gas: for the same nozzle and the same incident shock strength ($M_{sf} = M_s$) there are different wave-pattern structures; for the same induced wave patterns, the necessary incident-shock strength and the nozzle parameter are different. Also, these differences change with the quiescent or moving state of the medium ahead of the incident shock.

Acknowledgement. The support from the National Natural Science Foundation of China is gratefully acknowledged.

References

Amann HO (1969) Experimental study of starting process in a reflection nozzle. Phys of Fluids 12:150–153.

Ben-Artzi M, Falcovitz J (1986) An upwind second-order scheme for compressible duct flows. SIAM J Sci Stat Comput 7:744–768.

Crowe CT (1982) Review—numerical models for dilute gas-particle flows. J of Fluids Engineering 104:297–303.

Kashimura H, Iwata N, Nishida M (1986) Numerical analysis of the wave propagation in a duct with an area change by random choice method. Bulletin of JSME 29:1440–1445.

Prodromou P, Hillier R (1991) Computation of unsteady nozzle flows. In : Takayama K (ed) Proc 18th Int Symp on Shock Waves, Vol. 2, Sendai, pp 1113–1118.

Smith CE (1966) The starting process in a hypersonic nozzle. J Fluid Mech 24:625–640.

Wang BY, Wu QS (1991) Numerical investigation of dusty gas shock wave propagation along a variable cross-section channel. In: Takayama K (ed) Proc 18th Int Symp on Shock Waves, Vol. 1, Sendai, pp 521–526.

Zhang HX, Li ZW (1992) Numerical simulation of hypersonic laminar wake flow. Acta Mechanica Sinica 24:389–399.

Dusty gas flow in a nozzle

I. Elperin, O. Igra, G. Ben-Dor
The Pearlstone Center for Aeronautical Engineering Studies
Department of Mechanical Engineering, Ben-Gurion University of the Negev
Beer-Sheva, Israel
E. Vasiliev
Department of Computational Mechanics, Faculty of Mathematics
Volgograd State University, Volgograd, Russia

Abstract: The flow field developed in a dust-laden gas flow through a converging-diverging nozzle is studied. The conservation equations governing such two-phase flows are solved numerically using second-order accuracy modification of Godunov's scheme for the gaseous and solid phases. The numerical results reveal the effect of various parameters of the gas-solid particle suspension on the flow field developed inside and outside of the converging-diverging nozzle.

Key words: Shock waves, Dusty flows, Nozzles

1. Introduction

Analysis of two-phase flows has been a subject of intense research for a long time. The motivation for this interest is the fact that such flows appear in many engineering applications; for example, in coal-mine explosions, in nozzle flows of solid-propellant rockets, and in flows associated with volcanic eruptions. Since such flows are difficult to simulate in laboratory conditions, most of the published research in this field consists of numerical simulations and most of them treat the two-phase flow as being quasi-one-dimensional flow. This, of course, is an over simplification which might lead to wrong conclusions.

The object of the present study is to provide an accurate numerical analysis of gas-solid particles, two-phase nozzle flow, and its expansion outside of the nozzle. The study will center on the effects that solid particles have on the two-phase flow. The numerical method used in the present work allows the analysis of both the starting-flow processes in the nozzle and the eventually reached steady-state flow.

2. Theoretical background

A two-phase flow, composed of gas-solid particles in suspension, through an axisymmetrical converging-diverging nozzle is considered. The governing equations for the suspension flow are developed for each phase, treating each as a continuous medium. The solid particles, considered to be a pseudo-gas, are of equal size, uniformly distributed, and assumed to be rigid spherical particles. A more detailed description of the proposed physical model can be found in Ishii et al. (1987).

For this model, the conservation equations are written as follows for the gaseous phase:

$$\frac{\partial \rho}{\partial t} + \frac{\partial}{\partial x}\rho u + \frac{\partial}{\partial r}\rho v = -\frac{\rho v}{r} \tag{1}$$

$$\frac{\partial \rho u}{\partial t} + \frac{\partial}{\partial x}(\rho u^2 + p) + \frac{\partial}{\partial r}(\rho u v) = -\frac{\rho u v}{r} - \rho_p A(u - u_p) \tag{2}$$

$$\frac{\partial \rho v}{\partial t} + \frac{\partial}{\partial x}(\rho u v) + \frac{\partial}{\partial r}(\rho v^2 + p) = -\frac{\rho v^2}{r} - \rho_p A(v - v_p) \tag{3}$$

$$\frac{\partial e}{\partial t} + \frac{\partial}{\partial x}(e + p)u + \frac{\partial}{\partial r}(e + p)v$$

$$= -\frac{(e + p)v}{r} - \rho_p A(u_p(u - u_p) + v_p(v - v_p)) - \rho_p c B(T - T_p) \tag{4}$$

and for the solid phase:

$$\frac{\partial \rho_p}{\partial t} + \frac{\partial}{\partial x}\rho_p u_p + \frac{\partial}{\partial r}\rho_p v_p = -\frac{\rho_p v_p}{r} \tag{5}$$

$$\frac{\partial u_p}{\partial t} + u_p\frac{\partial u_p}{\partial x} + v_p\frac{\partial u_p}{\partial r} = A(u - u_p) \tag{6}$$

$$\frac{\partial v_p}{\partial t} + u_p\frac{\partial v_p}{\partial x} + v_p\frac{\partial v_p}{\partial r} = A(v - v_p) \tag{7}$$

$$\frac{\partial T_p}{\partial t} + u_p\frac{\partial T_p}{\partial x} + v_p\frac{\partial T_p}{\partial r} = B(T - T_p) \tag{8}$$

Standard gasdynamics notations have been used in Eqs. (1)–(8).

This set of nonlinear partial differential equations, (1)–(8), was solved numerically using a second-order modification of the Godunov scheme described in Vasiliev (1995). Essentially, this modification, which was prepared for solving a system of two-dimensional conservation laws, employs the evaluation of corrections at the two-dimensional stencil. Corrections are constant over a cell in the computational domain and are added to the main parameters before calculating fluxes through the cell boundaries. The same algorithm is used for calculating corrections for the solid and the gaseous phases. As mentioned, the modification of the Godunov scheme yields a second-order accuracy in space and in time. The scheme provides a high-quality, efficient method for calculation of perfect-gas flows with internal discontinuities.

3. Results and discussion

Results discussed below were obtained for the nozzle geometry described in Ishii et al. (1987). Prior to flow initiation a diaphragm is located at the nozzle's minimum cross-sectional area. It separates the converging and the diverging parts of the nozzle. In the converging part a quiescent gas-solid particle suspension is present in a state of equilibrium at the appropriate reservoir conditions (index "0"). To the right of the diaphragm, in the diverging part of the nozzle, the gas is set at ambient conditions (index "∞"). The reservoir and surrounding (ambient) conditions, as well as the physical properties of the suspension and relevant constants are: for the gaseous phase $\gamma = 1.4$; $c_p = 1034$ J/(kgK) at $T_0 = 3200$ K; $\mu = 7.974 \times 10^{-5}$ kg/(ms) at $T_0 = 3200$ K; $Pr = 0.71$. For the solid particles $c = 1686$ J/(kgK); $\rho_m = 3990$ kg/m^3; $d_p = 0.5, 1, 2, 4, 6, 8, 10$, and 50 μm. The reservoir conditions are as follows: $p_0 = 945$ kPa; $T_0 = 3200$ K; $\rho_0 = 1.0$ kg/m^3; $\rho_{p0} = 0, 0.3, 0.5$ and 0.7 kg/m^3. The surrounding conditions (ambient) are: $p_\infty = 102$ kPa; $T_\infty = 300$ K.

At $t = 0$ the diaphragm is instantaneously removed and flow through the nozzle starts. Since the reservoir and the ambient conditions are kept constant a steady flow will eventually be reached. The computational domain includes the space inside the nozzle and part of the surroundings close to the right of the nozzle exit. In the present analysis it is assumed that particles impinging on the nozzle wall are absorbed by the wall.

In Fig. 1 distributions of the solid-phase density for different particle sizes are shown for a steady flow in the nozzle. As would be expected, an increase in the

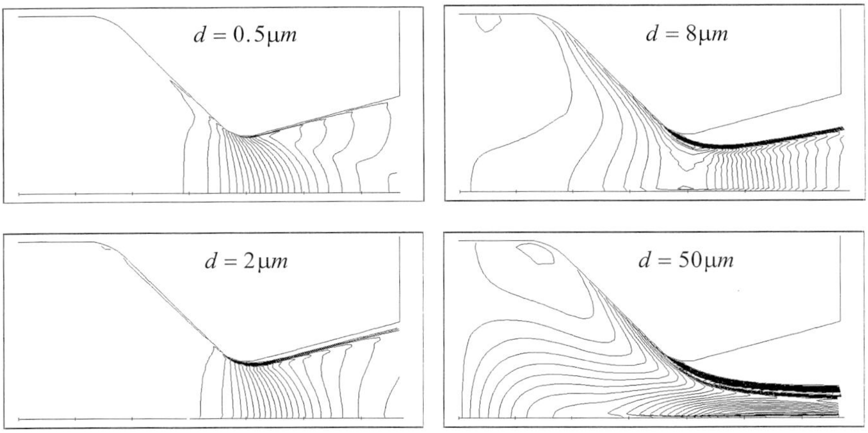

Figure 1. Isopycnics in the solid phase for $\rho_{p0} = 0.5$ kg/m^3.

particle size results in an increase in the particle free zone. The boundary of the solid-phase flow detaches from the wall in proximity to the nozzle throat. This is shown more clearly in Fig. 2, in which only the limiting particle streamlines are depicted for the same conditions as those presented in Fig. 1.

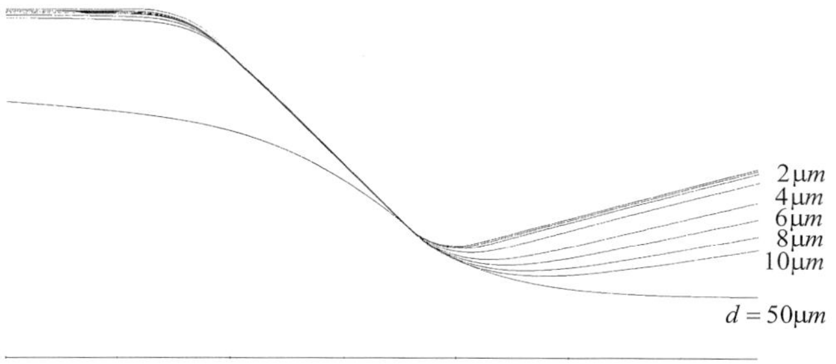

Figure 2. Limiting streamlines in the solid phase of the suspension for $\rho_{p0} = 0.5$ kg/m^3.

In Fig. 3 isopycnics in the solid phase, for particle sizes of $d = 0.2$ μm and $d = 4$ μm, are shown for two different values of the dust-loading ratio ($\eta = 0.3$ and 0.7). In all cases shown in Fig. 3 the difference between two neighboring isopycnics is equal to one tenth of the initial particle spatial density. As is apparent from Fig. 3 the density distribution in the solid phase, for a given particle size, is practically independent of the dust-loading ratio, η. Increase in η causes proportional increase in the spatial particle density throughout the flow field. However, changes in the dust-loading ratio mildly affect the locations of the limiting particle streamlines.

The effect of the dust presence on the gaseous phase in a steady flow through the nozzle is shown in Fig. 4. Near the wall, in the dust-free zone, the gas isopycnics demonstrate the expected flow expansion in the converging part of the nozzle. This expansion is changed abruptly when the isopycnics penetrate the dust region. The flow field of the gaseous phase, in the case of suspension with very small particles ($d = 0.5$ μm), is similar to a single-phase (gas) flow having a ratio of specific heats of $\gamma = 1.1868$ (see Fig. 5). This value of γ is the equivalent γ of gas-solid-particle suspension in an equilibrium flow condition. On the other hand, the flow field of the gaseous phase of the suspension with large particles ($d = 50$ μm) is similar to a frozen flow whose isopycnics are also shown in Fig. 5 ($\gamma = 1.4$). This could be expected since very small particles adhere very quickly to the gaseous flow conditions.

In Fig. 6 density distributions in the gaseous phase (isopycnics), for an overexpanded jet, just outside the nozzle, are shown. Isopycnics are shown for the case when the flow field is steady inside the nozzle and almost steady in

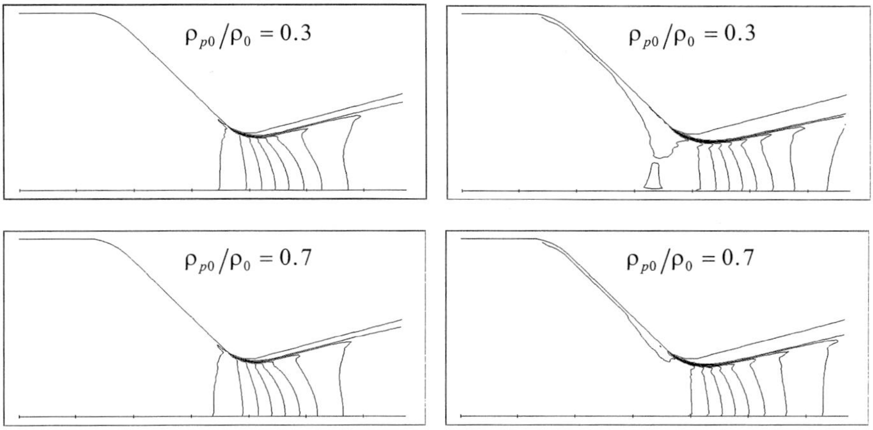

Figure 3. Isopycnics in the solid phase for $d = 2$ μm (left) and $d = 4$ μm (right).

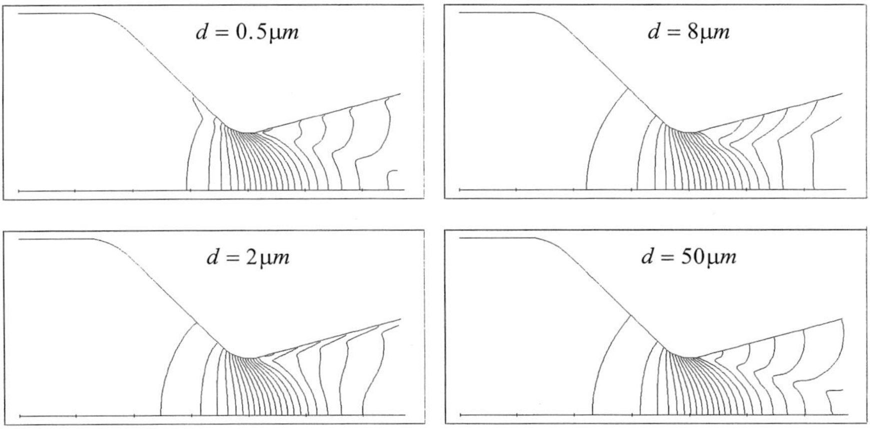

Figure 4. Isopycnics in the gaseous phase for $\rho_{p0} = 0.5$ kg/m^3.

Figure 5. Isopycnics in a pure gaseous flow.

the free jet outside of the nozzle. It can be seen that changes in the particle diameter affect the wave pattern near the triple point. Increase in the particle diameter, from $d = 0.5$ μm to $d = 2$ μm, causes smearing of the shock structure due to relaxation processes.

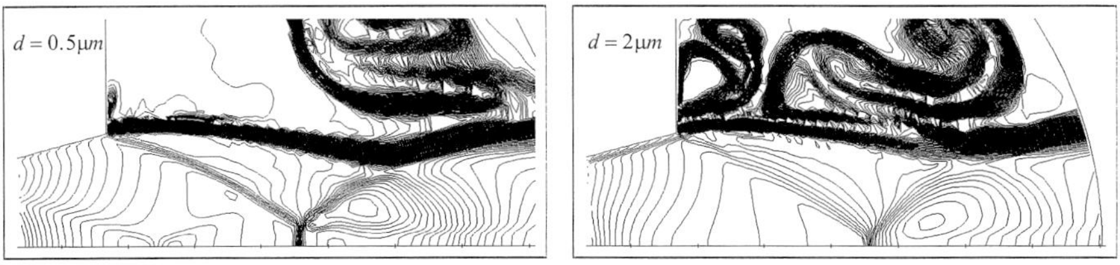

Figure 6. Isopycnics in the gaseous phase for $\rho_{p0} = 0.5$ kg/m^3.

4. Conclusion

Two-phase nozzle and overexpanded jet flows were simulated numerically using a new modification to Godunov's scheme. The effect of the solid-particle size and loading ratio on the flow field was analyzed. The two-dimensional character of the converging-diverging nozzle geometry results in a nonhomogeneous solid-particle distribution across the nozzle section and limiting particle streamlines were obtained. It was also noticed that the presence of solid particles affects the structure of the triple shock waves in the overexpanded jet, outside the nozzle.

References

Ishii R, Umeda Y, Kawasaki K (1987) Nozzle flows of gas-particle mixtures. Phys Fluids 30(3):752–760.

Vasiliev E (1995) W-modification of Godunov's scheme and its application for two-dimensional nonstationary flows of the dusty gas. Nonstationary two-phase flow. Sov J Comp Math (to be published, in Russian).

Head-on collision of a planar shock wave with a dusty gas layer

J. P. Jiang, T. Elperin, O. Igra
The Pearlstone Center for Aeronautical Engineering Studies
Department of Mechanical Engineering
Ben-Gurion University of the Negev
Beer Sheva, Israel

J. Falcovitz
Institute of Mathematics
The Hebrew University
Jerusalem, Israel

Abstract: The head-on collision of a planar shock wave with a dusty gas layer placed near a rigid wall was studied numerically. It was demonstrated that the air and solid particles in the dusty gas flow away from the rigid wall due to the formation of a rarefaction wave after the interaction between the wave reflected from the rigid wall and the dusty-gas/pure-air interface. Notably, not only a shock wave but also a compression wave can cause this phenomenon. This flow problem was motivated by the more complex configuration of dust lifting and entrainment behind sliding shock waves.

Key words: Dust lofting, Shock-wave/dust interaction, Dusty gas

1. Introduction and theoretical background

Shock-wave propagation in dusty gases was studied intensively during the past four decades due to its appearance in many engineering cases; for example, in the nozzle flow of solid-propellant rockets, in chemical or nuclear explosion-generated blast waves over a dusty terrain, and in coal-mine explosions. The present work studies numerically the interaction between a planar shock wave and a dusty gas layer placed near the end wall of a conduit. In the considered conduit the dusty gas layer (suspension) occupies the region $0 \leq x \leq \ell$; the remaining part is filled with pure air. The flow in the conduit is treated as a one-dimensional dusty gas problem by the following conservation equations of mass, momentum, and energy (see, e.g., Miura and Glass 1981) for the gaseous phase:

$$\frac{\partial \rho}{\partial t} + \frac{\partial \rho u}{\partial x} = 0; \tag{1}$$

$$\frac{\partial \rho u}{\partial t} + \frac{\partial \rho u^2}{\partial x} + \frac{\partial p}{\partial x} = -\frac{\sigma}{m} D; \tag{2}$$

$$\frac{\partial}{\partial t}\rho\left(c_v T + 0.5u^2\right) + \frac{\partial}{\partial x}\rho u\left(c_p T + 0.5u^2\right) = -\frac{\sigma}{m}\left(vD + Q\right); \tag{3}$$

and for the solid phase:

$$\frac{\partial \sigma}{\partial t} + \frac{\partial \sigma v}{\partial x} = 0; \tag{4}$$

$$\frac{\partial \sigma v}{\partial t} + \frac{\partial \sigma v^2}{\partial x} = \frac{\sigma}{m}D; \tag{5}$$

$$\frac{\partial}{\partial t}\sigma\left(c_m\theta + 0.5v^2\right) + \frac{\partial}{\partial x}\sigma v\left(c_m\theta + 0.5v^2\right) = \frac{\sigma}{m}\left(vD + Q\right). \tag{6}$$

In order to make this set of equations solvable we used the equation of state for a perfect gas $P = \rho RT$; for the drag force, D, and heat transfer rate, Q, we used the correlations proposed by Miura and Glass (1981). In the numerical computations a reference distance, L, is defined as $L = \frac{4}{3}\frac{\rho_p}{\rho_1}d$ (index p denotes solid particles); d is the dust particle diameter. Standard gasdynamic notations are used in Eqs. (1)–(6). This set was solved numerically using a second-order explicit TVD (total variation disminishing) scheme for the gaseous phase and a MacCormack scheme for the solid phase with a second-order central-difference approximation for the interaction term between the two phases (see, e.g., Hirsch 1990, Chapters 17 and 21). At the conduit inlet a non-reflecting boundary condition was prescribed. On the end wall the boundary condition was a zero velocity for both the gas and the solid particles.

2. Partially dispersed shock wave

Consider an incident shock wave with Mach number $M = 2.11$ which hits, at $t = 0$, the dusty gas layer interface. The dusty gas layer occupies originally the region $0 < x < 60L$. The calculation domain was $-20L \leq x \leq 60L$ and it was divided into 800 cells. The solid phase of the suspension was composed of glass beads of $d = 10$ μm and material density $\rho_p = 2.5$ g/cm^3. The initial conditions of the carrying gas, air ($\gamma = 1.4$) were: $\rho_1 = 1.23 \times 10^{-3}$ g/cm^3, $T_1 = 300$ K, and $P_1 = 1$ atm while the dust-mass-loading ratio was set at $\alpha = 1$, and $c_v = c_m = 717$ J/(kg·K). Ahead of the incident shock wave the suspension was in a state of quiescent equilibrium. Results are shown in Fig. 1, where the dusty gas extends from $x = 60L$ down to the location where the particle velocity changes abruptly from zero to a maximum (in Figs. 1a and 1b) or minimum (in Figs. 1c and 1d). Figure 1a shows that 2.78 ms after the incident shock wave has arrived at the dusty gas layer interface, the following wave pattern is observed. A transmitted, partially dispersed shock wave propagates into the dusty gas layer. This shock wave, at the considered time, is close to the dusty gas layer rear end. A reflected shock wave is located at $x/L = -2$. In Fig. 1b ($t = 3.56$ ms), the transmitted shock wave reached the conduit's end wall and was reflected back into the dusty gas layer. The front of the reflected shock wave is located at $x/L = 55.5$. Behind the reflected shock wave the velocity of both phases is reduced to zero. The reflected shock wave propagates inside the dusty gas layer which, at the considered time, extends from $x/L = 60$ to $x/L = 51.5$.

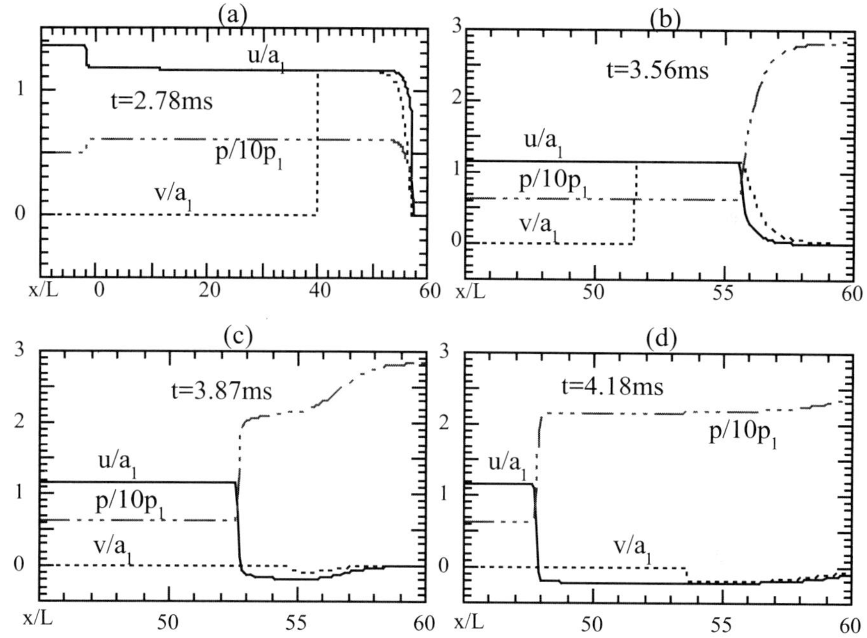

Figure 1. Flow parameters at various times for $M = 2.11$ and $\alpha = 1$: u is gas velocity, v is solid-phase velocity, p is gas pressure; $a_1 = 350$ m/s, $p_1 = 1$ atm, $L = 2.72$ cm.

In Fig. 1c ($t = 3.87$ ms) the reflected shock wave propagated to $x/L = 52.5$ while the dusty gas layer interface is placed at $x/L = 54.5$ where the particle velocity begins to change from zero to a minimum value. Now the reflected shock wave emerges out of the dusty gas layer. It is apparent from Fig. 1c that in the dusty gas layer, near the interface, the velocity of both phases is negative, i.e., it is directed opposite to the original flow direction. The pressure in this area is also lower than that encountered behind the reflected shock wave from the conduit end wall. The reason for this change is the formation of a rarefaction wave, which propagates into the dusty gas layer when the reflected shock wave reaches the dusty gas interface and is transmitted into the pure gas. The negative dust velocity can attain values as high as $-0.2a_1 = -70$ m/s, see Fig. 1d (for $t = 4.18$ ms).

3. Compression wave

In this case the region -30 cm $< x < 0$ is filled with pure air while the region $0 \le x \le 5$ cm contains dusty gas at $t = 0$. An incident air shock wave with a Mach number $M = 1.27$ arrives in $x = -29$ cm at $t = 0$. The dusty gas is composed of air and polyester beads of $d = 10$ μm and $\rho_p = 0.7$ g/cm^3. The pre-shock air parameters are: $\rho_1 = 1.0 \times 10^{-3}$ g/cm^3, $T_1 = 303$ K, and $P_1 = 0.896 \times 10^5$ Pa, while $\alpha = 35$, $c_v = 717$ J/(kg·K), and $c_m = 1200$ J/(kg·K) (Gelfand et al. 1983). The incident shock wave propagates into a quiescent suspension. The computation domain covers the region -30 cm $\le x \le 5$ cm and it was divided into 700 cells. In the present computations we multiply C_D, the steady drag coefficient, used in Miura and Glass (1981) by 1.5, in order to account for the unsteady nature of the flow (Igra and Takayama 1993).

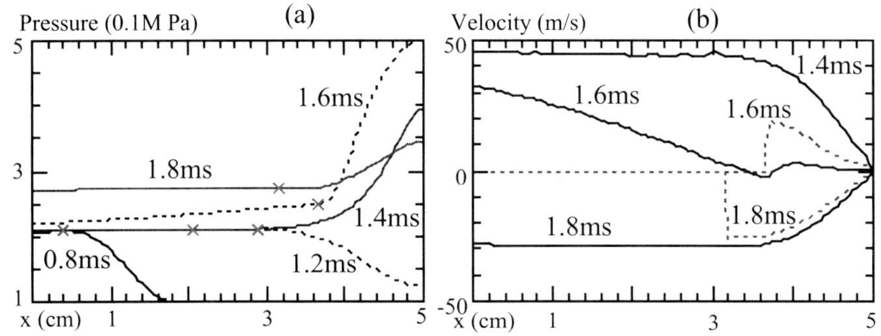

Figure 2. Flow parameters at various times for $M = 1.27$ and $\alpha = 35$: (a) gas pressure; (b) velocity of gas (solid line) and solid phase (dashed line).

Some of the obtained results are shown in Fig. 2. The location of the leading edge of the dusty gas layer (the interface) at different times, obtained from computations, is marked by crosses in Fig. 2a. In this figure the pressure profile obtained for $t = 0.8$ ms shows that after the head-on collision of the incident shock wave with the dusty gas layer interface the pressure behind the reflected shock wave, propagating back into the gas, is only 0.85 of what would have been encountered in a similar reflection from a rigid wall. The transmitted shock wave in the dusty gas layer is transformed into a fully dispersed shock wave, or a compression wave, due to its interaction with the solid particles in the suspension. The head of this dispersed wave has not reached the end wall at time $t = 0.8$ ms, (see Fig. 2a).

At $t = 1.4$ ms the dispersed transmitted shock wave has reflected from the end wall as a compression wave, not a shock wave as was the case in the results shown in Fig. 1b, so there is no sharp jump in the pressure profile at this time. Up to time $t = 1.4$ ms the pressure is practically constant within the region between the reflected shock wave into the gas and the leading edge of the dusty gas layer. Thus the reflected wave from the conduit's end wall does not emerge out of the dusty gas layer until $t = 1.4$ ms.

In Fig. 2b spatial velocity variations are shown. The dusty gas occupies the region from the end wall ($x = 5$ cm) up to the location where the solid phase velocity changes sharply from zero to a maximum (at 1.6 ms) or minimum (at 1.8 ms). The analysis of the parameters in Figs. 2a and 2b at time $t = 1.6$ ms reveals that the reflected compression wave resulted in a compression wave being transmitted into the pure air ($x < 3.65$ cm) and a rarefaction wave being reflected back into the dusty gas ($x > 3.65$ cm). It should be noted that at $t = 1.6$ ms, in the vicinity of the dusty gas layer interface, the gaseous-phase velocity becomes negative due to the rarefaction wave (see Fig. 2b at the position $x = 3.65$ cm). Finally, the rarefaction wave causes not only the gas, but also the dust particles to flow away from the end wall, as indicated by the negative velocity of both phases for $t = 1.8$ ms (see Fig. 2b).

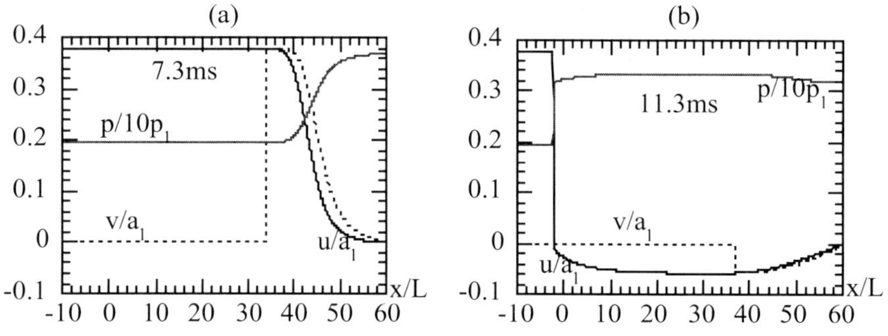

Figure 3. Flow parameters at various times for $M = 1.3$ and $\alpha = 1$: u is gas velocity, v is solid-phase velocity, p is gas pressure; $a_1 = 350$ m/s, $p_1 = 1$ atm, $L = 2.72$ cm.

In another case analyzed, we employed the same initial conditions as in the computations presented in Section 2, except for changing the incident shock-wave Mach number to $M = 1.3$. This was done in order to obtain a fully dispersed shock wave in the dusty gas. Now a compression wave reflects from the end wall as is shown in the pressure curve of Fig. 3a. Negative velocities (for both phases) are clearly observed in the results shown in Fig. 3b where the reflected compression wave has emerged from the dusty gas and converged into a shock wave in the pure gas region (see Fig. 3b at $x/L = -2.5$).

This flow behavior can be illustrated by an x-t diagram. Such diagrams are shown in Fig. 4a (for the case discussed in Section 2) and Fig. 4b (for the case discussed in Section 3). In Fig. 4a, the transmitted wave and the reflected wave from the conduit's end wall are partially dispersed shock waves with a sharp front followed by a compression wave. It is this sharp front that plays the dominant role in the process. In Fig. 4b both the transmitted and reflected waves are compression waves (fans) without a sharp front.

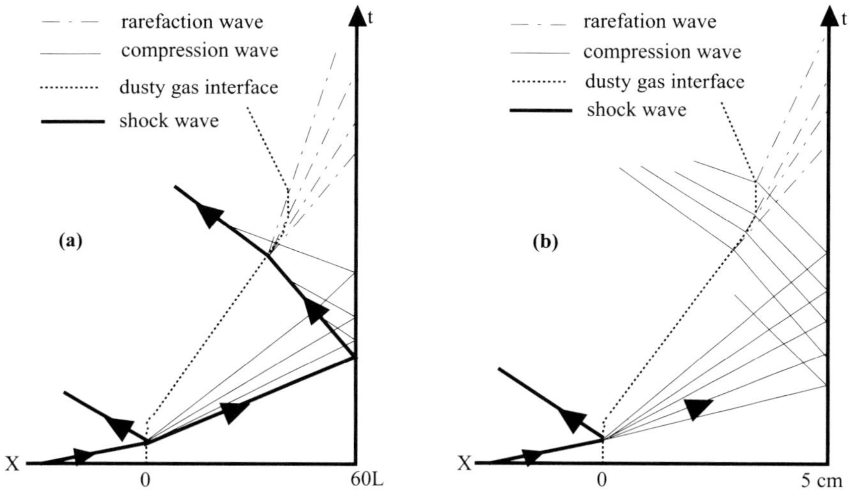

Figure 4. Wave pattern in x-t plan for (a) a partially dispersed shock wave; and (b) a compression wave in dusty gas.

References

Gelfand BE, Gubanov AV, Timofeev EJ (1983) Interaction of shock waves in air with a porous screen. Izvestiya Akademii Nauk SSSR Mekhanika Zhidkostii Gaza 4:78–84.

Hirsch C (1990) Numerical Computation of Internal and External Flows, John Wiley & Sons, England.

Igra O, Takayama K (1993) Shock tube study of the drag coefficient of a sphere in a non-stationary flow. Proc R Soc Lond 442:231–247.

Miura H, Glass I (1981) On a Dusty-Gas Shock Tube, UTIAS Report 250, CN ISSN 0082–5255.

An experimental and theoretical study of shock-wave induced particle coagulation

J. J. F. Strecker, P. Roth
Institut für Verbrennung und Gasdynamik
Gerhard-Mercator-Universität Duisburg
47048 Duisburg, Germany

Abstract: In the present study dry powder containing primary particles of two different sizes was dispersed in an expansion-wave-driven aerosol generator. The resulting bimodal distributed aerosol was allowed to fill the shock tube and was accelerated by a shock wave. During the shock-induced velocity relaxation of the particles, relative motion of the two size classes occurred. The resulting coagulation was monitored by in-situ light scattering of individual particles using a detection system with two intersecting measurement volumes modified for fast flow conditions. From the particle-size distributions measured both in front of and behind the incident shock, the degree of coagulation was determined employing the third moment of the count diameter of a discrete size distribution. The shock-induced particle coagulation was also simulated based on a one-dimensional model assuming two interacting continua for the gas and particle phase. The interphase flux terms were restricted to the exchange of momentum and energy both caused by drag forces. The particle-number concentration was simulated by the General Dynamic Equation using the flow-field-induced coagulation as the only relevant source term.

Key words: Dry-powder dispersion, Bimodal distribution, Particle coagulation

1. Introduction

Brownian motion of aerosol particles is the normal source for particle coagulation. Additionally, particle collisions can occur due to relative particle motion induced by gravity or flow fields. Examples are coagulation in sedimentation chambers, in turbulent flows, or in acoustic waves. Shock waves and the associated shock-induced fluid flow can also cause particle coagulation due to particle relaxation. The lighter particles are accelerated faster than the heavier ones, thus causing particle collision. In a stationary shock situation, the gas-phase properties change very rapidly in the shock front while the behavior of the particle cloud is quite different and more complex. The particles are not directly influenced by the quasi-discontinuity. They move with their initial velocity through the shock zone and are decelerated by drag forces to the lower gas velocity. The processes of aerosol compression and particle coagulation occur always simultaneously. The compression increases the particle-number concentration, whereas the coagulation decreases it. A quantitative prediction of both interaction processes needs to consider the detailed processes.

2. Experiments and results

The experiments were conducted in a stainless-steel shock tube of 50 mm inner diameter. The low- and high-pressure sections were 5.5 and 2.5 m in length, respectively. The end of the low-pressure section was connected to a vacuum tank of 80 l volume separated by a thin diaphragm allowing absorption of the shock wave. The driven section could be evacuated below 0.05 mbar and was for each run first filled with argon and then connected to the aerosol generator. The gas/particle mixture flew slowly through the low-pressure section initiated by a vacuum pump. During this flow, the pressure was reduced from about 800 mbar to its final value of 400 mbar. By this procedure a very homogeneous particle-concentration distribution along the full length of the low-pressure tube was achieved. The test section was located at the end of the low-pressure part, equiped with an in-situ xenon light-scattering system. The detection device consisted of two intersecting measurement volumes employed for individual particle sizing.

The aerosol generator used consisted of a small conventional shock tube of 50 mm inner diameter connected to a glass vessel of 20 l volume used as an aerosol chamber. Two powders containing primary particles of different sizes were deposited on a flat plate and inserted near the diaphragm. For each run the low-pressure tube and the glass vessel were evacuated and the high-pressure section was filled with the dispersion fluid. After bursting of the diaphragm, the generated expansion wave dispersed the powder. More details and results of this shock- and expansion-wave-driven powder disperser have been published earlier; see Rajathurai et al. (1990), Strecker and Roth (1990).

The aerosols used consisted of argon containing primary particles of two different size classes. Several combinations were used: Latex with primary particles $d_p^* = 1.27\ \mu\text{m}/2.5\ \mu\text{m}$ and $d_p^* = 1.23\ \mu\text{m}/2.83\ \mu\text{m}$, respectively. In the bimodal distributed aerosols, the coagulation was studied by varying the shock velocity from 372 to 775 ms^{-1}, resulting in Mach numbers between 1.16 and 2.40. An individual shock-tube experiment showing optically measured particle-size distributions both in front of and behind the shock wave is shown in Fig. 1. The pre-shock aerosol (left part) contains primary particles of mean diameters $d_p^* = 1.27\ \mu\text{m}/2.5\ \mu\text{m}$ and agglomerates. The aerosol behind the shock wave

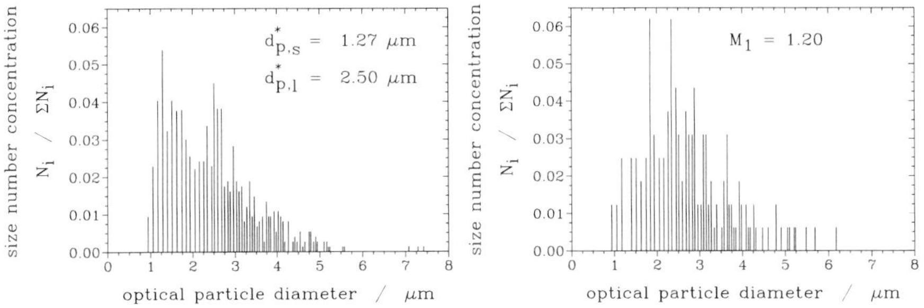

Figure 1. Optically measured particle-size distribution in front of (left) and behind (right) a shock wave, indicating particle coagulation.

(right part of Fig. 1) shows a very different size-distribution behavior. The number concentration of the smaller primary particles is reduced whereas the agglomerates are increased, indicating particle coagulation. Similar results were obtained from other shock-wave experiments.

In order to describe the coagulation behavior it is most convenient to employ the third moment of the count diameter of a discrete size distribution, which is:

$$M_{3,0} = \frac{\sum_i (n_{p,i} \cdot d_{p,i}^3)}{\sum_i n_{p,i}} = d_m^3. \tag{1}$$

In this equation $n_{p,i}$ and $d_{p,i}$ are the particle-number concentration and particle diameter of size class i, respectively. $M_{3,0}$ is equal to the cube of the mass average diameter d_m^3. This definition can be applied to an aerosol containing both primary particles and agglomerates ($M_{3,0}$) as well as to the primary particles of the powder ($M_{3,0}^*$). The ratio of these moments characterizes the degree of agglomeration in the aerosol:

$$\alpha_{3,0} = \frac{M_{3,0}}{M_{3,0}^*}. \tag{2}$$

For aerosols containing only primary particles, $M_{3,0}$ is equal to $M_{3,0}^*$ resulting in $\alpha_{3,0} = 1$. If agglomerates are present in the aerosol, $\alpha_{3,0}$ is always greater than 1. From the optical size measurements of particles both in front of $(\alpha_{3,0})_1$ and behind $(\alpha_{3,0})_2$ the shock wave, the degree of agglomeration was determined. From both these values, the efficiency η of shock-induced coagulation was calculated:

$$\eta = \frac{(\alpha_{3,0})_2/(\alpha_{3,0})_1 - 1}{2/(\alpha_{3,0})_1 - 1} \cdot 100\%. \tag{3}$$

If the initial aerosol remains unaffected by the shock wave, η is equal to 0. On the other hand, collection of all particles of the small-size fraction by the large ones results in $(\alpha_{3,0})_2 = 2$ or $\eta = 100\%$.

Results of all shock-tube experiments performed in the present study are summarized in Fig. 2. The efficiency η is shown as a function of the Mach number. The results clearly show that particles of the bimodal distributed aerosols coagulate due to shock-induced particle collisions. The strong decrease

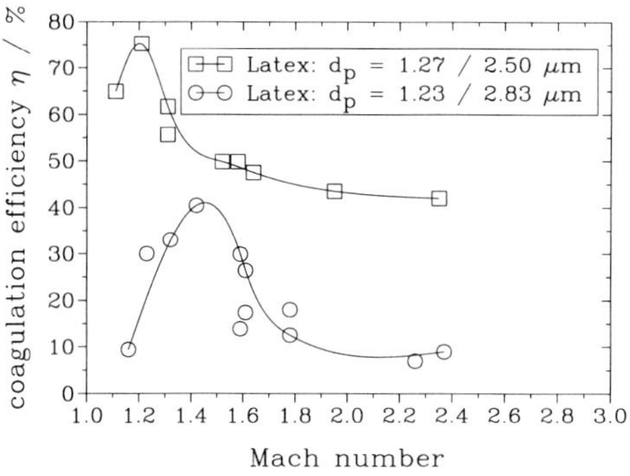

Figure 2. Results of shock-tube experiments on particle coagulation.

of η with increasing Mach number indicates that a further process e.g., particle breakup, must also be considered in the relaxation zone. In the first set of experiments (open squares) the number concentrations of the small and the large particles were roughly the same whereas in the second set of experiments (open circles) the ratio was about 2.5.

3. Coagulation model and computer simulation

The shock-induced particle coagulation was simulated based on a one-dimensional model assuming two interacting continua for the gas and particle phases. The interphase flux terms were restricted to the exchange of momentum and energy both caused by drag forces. The particle-number concentration was simulated by the General Dynamic Equation (Friedlander 1977) using flow-field-induced coagulation as the only relevant source term:

$$n_{p,i} \cdot \frac{dv_{p,i}}{dx} + v_{p,i} \cdot \frac{dn_{p,i}}{dx} = \left[\frac{dn_{p,i}}{dt} \right]_{coag.,s} \qquad i = 1...6. \tag{4}$$

For an aerosol containing spherical particles with a bimodal distribution, collisions between particles of different size classes including the resulting agglomerates were taken into account. The coagulation source term is the product of the collision efficiency, the collision cross-section of the particles, the number concentrations, and the relative velocity. For particle Stokes numbers $Stk > 0.1$, the collision efficiency is given by Hinds (1982):

$$\eta_{coll.} = \left(\frac{Stk}{Stk + 0.25} \right). \tag{5}$$

The velocity of an agglomerate formed was assumed to be equal to the velocity of the large particle before collision. The post-shock initial conditions of the particles were assumed to be equal to the pre-shock conditions, whereas the initial gas properties were calculated from the post-shock equations of ideal gases. Figure 3 shows examples of calculated particle-number concentrations

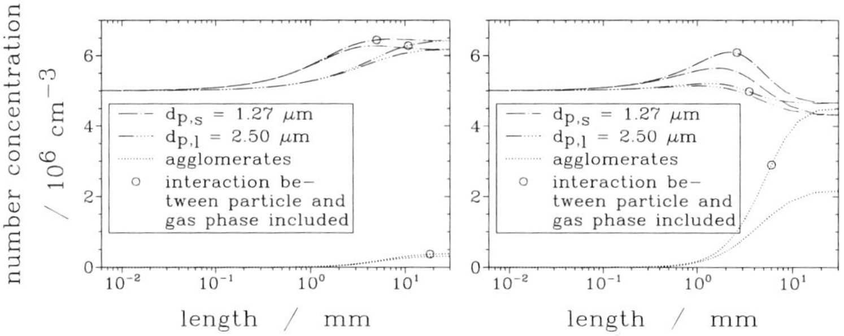

Figure 3. Number concentrations calculated for a Mach number of $M = 1.2$ and initial concentrations of $n_{p,small} = n_{p,large} = 5 \cdot 10^6$ cm^{-3} (left) and $n_{p,small} = n_{p,large} = 5 \cdot 10^7$ cm^{-3} (right).

in the relaxation zone behind a stationary shock front for bimodal distributed aerosols of low (left) and high (right) initial number concentration. The primary Latex particles have diameters of $d_p^* = 1.27$ μm and 2.5 μm. The agglomerates formed are indicated by the dotted lines. Curves marked with open circles were obtained taking the full gas-particle interaction terms into account. The other curves were calculated neglecting the particle effects on the gas-phase properties. It is evident that for particle-number concentrations $n_{p,i} < 10^7$ cm^{-3} (left part of Fig. 3) the particles do not significantly affect the gas flow. The difference in the coagulation efficiencies is below 1%. For higher particle-number concentrations (right part of Fig. 3), the particles decelerate the gas flow resulting in a further increase of the particle-number concentration. This causes an increasing formation of agglomerates resulting in a 16% higher coagulation efficiency. In Fig. 4 calculated coagulation efficiencies are plotted versus Mach number. The initial data of the results with open symbols were taken from the experiments of Fig. 2. It is obvious that the calculated coagulation efficiencies strongly deviate from the measured ones.

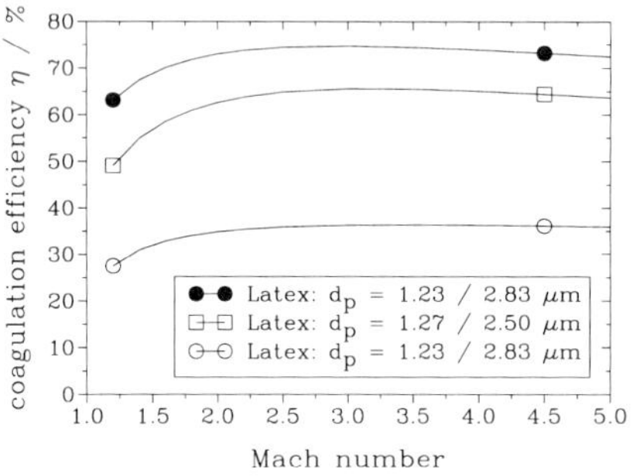

Figure 4. Coagulation efficiencies: open symbols – calculated with data base used in the experiments; closed circles – calculated with $n_{p,small} = n_{p,large}$.

4. Discussion

The discrepancy between experimental data and results obtained from computer simulation indicate that the model describing shock-induced coagulation must be improved. Further processes like agglomerate breakup or improved collision modeling must be taken into account. Results obtained from experiments with agglomerated aerosols (Strecker and Roth 1994) showed that even weak shock waves can break up particles due to intensive drag forces in the relaxation zone thus increasing the number concentration. The modeling of agglomerate disintegration is quite difficult due to a lack of knowledge of the mechanical process during particle breakup.

Acknowledgement. The authors thank Mrs. B. Nelius for her help in conducting the experiments. This work originated in the special research program "Aerosols" of the University of Duisburg. The financial support of the Deutsche Forschungsgemeinschaft is gratefully acknowledged.

References

Friedlander SK (1977) Smoke, Dust and Haze, New York, John Wiley & Sons.

Hinds WC (1982) Aerosol Technology Properties, Behavior, and Measurement of Airborne Particles, New York, John Wiley & Sons.

Rajathurai AM, Roth P, Fißan H (1990) A shock and expansion wave-driven powder disperser. J of Aerosol Sci Technol 12:613–619.

Strecker JJF, Roth P (1990) Generation of monodisperse and of bimodal aerosols by the expansion wave-driven generator. J Aerosol Sci 21:681–684.

Strecker JJF, Roth P (1994) Particle breakup in shock waves studied by single particle light scattering. J of Part & Part Syst Charact II:222–226.

Experimental verification of the concept of amplification of shock loading by means of compressible covering as applied to the propagation of blast waves over a dust deposit

S. P. Medvedev, A. N. Polenov, S. V. Khomik, B. E. Gelfand
Institute of Chemical Physics
Russian Academy of Sciences
Moscow, Russia

Abstract: This investigation of the interaction between blast waves and dust deposits was performed using specially developed shock-tube techniques. The experiments revealed the strong dependence of the value of pressure transmitted through a dust deposit on the duration of the compression phase of the primary air blast wave. Additional experiments on the propagation of shock (blast) waves over a layer of polyurethane foam were conducted to demonstrate that the main features of the transmission shock loading through a dust deposit and porous compressible covering are similar in appearance. A simplified mechanistic simulation was proposed to describe the results of the tests both on shock-dust and shock-foam interactions.

Key words: Dust deposit, Blast waves, Porous material

1. Introduction

Adachi et al. (1984) were first to report evidence of unusual trends in pressure profiles under a shock-loaded dust deposit. These data and a more systematic study performed by Gelfand et al. (1989) show that the propagation of a shock wave over a dust deposit gives rise to the significant overloading of a rigid wall under the deposit. All the mentioned experiments were performed using conventional shock tubes. Thus, the phenomenon of our interest has been examined in the case of planar stepwise shock waves. Real blast waves, arising in the environment as a result of an accident, are characterized by triangular (blastwise) pressure profiles. The few illustrative examples (Gelfand et al. 1989) which concern the interaction between blastwise pressure waves and dust deposits are not a sufficient basis for drawing ultimate conclusions on the practical danger of the phenomenon of amplification of shock loading. Thus, additional experimental research is in order. The present work reports the results of a more comprehensive study of the interaction between the blastwise shock waves and rigid wall covered with dusty material.

2. Experimental

2.1. Transmission of blast loading through the dust layer

It should be noted that the problem of the laboratory simulation of blastwise shock waves is worthy of special attention. As is known (see e.g., Gelfand et al. 1989), the shock tube in its simplest form seems to be quite suitable for producing shock waves of triangular pressure profile. This can be arranged by using a sufficiently short high-pressure section (HPS). These experiments have been carried out in a horizontal shock tube of square inner cross-section (40 $\times$ 40 mm) and of 1-m length. An assortment of HPSs of lengths 0.5, 0.22, 0.07, and 0.01 m was available. Nitrogen or helium at the pressure 0.2–1 MPa was used as the driver gas. The low-pressure section (LPS) was filled with air at 0.1 MPa in all runs. The layer of dust under investigation was located in the LPS at the distance 0.35–0.45 m from the bursting membrane. The parameters of blast waves both above and below the layer were measured by means of oppositely mounted piezoelectric gauges.

The variation of the HPS length together with that of the driver-gas pressure provides the significant change in the duration of the shock compression phase. This could be varied in the range $T = 0.4$ –6 ms. It should be emphasized that the change in the shock-wave shape and duration is our principle interest. Thus, everything presented below concerns experiments with a dust deposit of a fixed depth of 20 mm. Plexiglass dust with particle diameters of 5–10 μm was chosen because it is precisely the material which exhibits clearly defined "amplification" properties (Gelfand et al. 1989).

The basic idea of the tests was to change the duration of the shock-wave interaction with the dust deposit while keeping the shock-front overpressure at a fixed value. Figure 1 presents the selected sequence of the pressure records. The pressure profiles in Fig. 1a are related to the loading by a stepwise shock wave. Obviously, the responses of the bottom pressure gauge depend heavily on the shock duration. There exists a critical value $T_c \approx 0.6$ ms at which the loading transmitted through the dust deposit is equal to the air shock-front overpressure. If $T > T_c$ (Figs. 1a, b, c) the bottom pressure will exceed the surface one. If $T < T_c$ (Fig. 1d) the bottom pressure will be lower than the surface one.

2.2. Blast loading of the layer of polyurethane foam

As mentioned previously (Gelfand et al. 1989) the process of shock loading of a surface covered with a dust layer is quite similar to interaction of shock waves

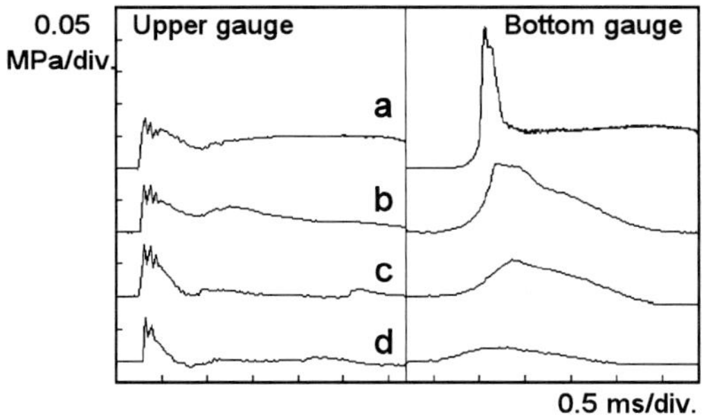

Figure 1. Pressure records over (left side) and under (right side) the deposit of 20 mm in depth. High-pressure section length m: a – 0.5; b – 0.22; c – 0.07; d – 0.01.

with walls protected by a porous compressible material, like polyurethane foam. Contrary to dust system, the properties of polyurethane foam are known with some precision. Furthermore, in this application, the use of polyurethane foam instead of dust allows us to exclude the effect of dust dispersion behind the shock wave. The use of the foam provides high-fidelity visualization of the dynamics of the layer. It is noteworthy that a lot of experimental results related to normal shock reflection from a rigid wall covered with compressible foam are available (see, e.g., survey given by Skews 1995). Nevertheless, there is a gap in the studies of shock (blast) wave propagation over a foam layer.

The experiments with the polyurethane foam layer were performed in a horizontal shock tube of 1 m length. The sidewalls of the LPS (rectangular inner cross-section 47 × 83 mm) were equipped with windows providing the possibility of visualizing the process by the shadowgraph technique. Figure 2 presents results of experiments with a polyurethane foam layer of 20 mm thickness. The upper trace in Fig. 2a represents the gas pressure over the layer and the bottom one is the pressure under the layer at the same cross-section. As evident from the figure, this is the case of stepwise loading. Figure 2b presents results of experiments with a shock wave with a triangular pressure profile of the same amplitude as in Fig. 2a. When the pressure profiles presented in Fig. 2 are compared with that of Fig. 1, it is apparent that, in spite of the obvious differences between dust and foam materials, the mechanism of the transmission of the shock loading is the same in both cases. A sequence of the shadowgraphs illustrate shock and layer dynamics. Not considering all details, we merely note here that the time of maximum pressure under the layer is consistent with the instant of time at which the maximum compression of the foam layer is achieved. It should be emphasized that the maximum displacement of the foam surface is approximately the same under both stepwise and blastwise pressure profiles.

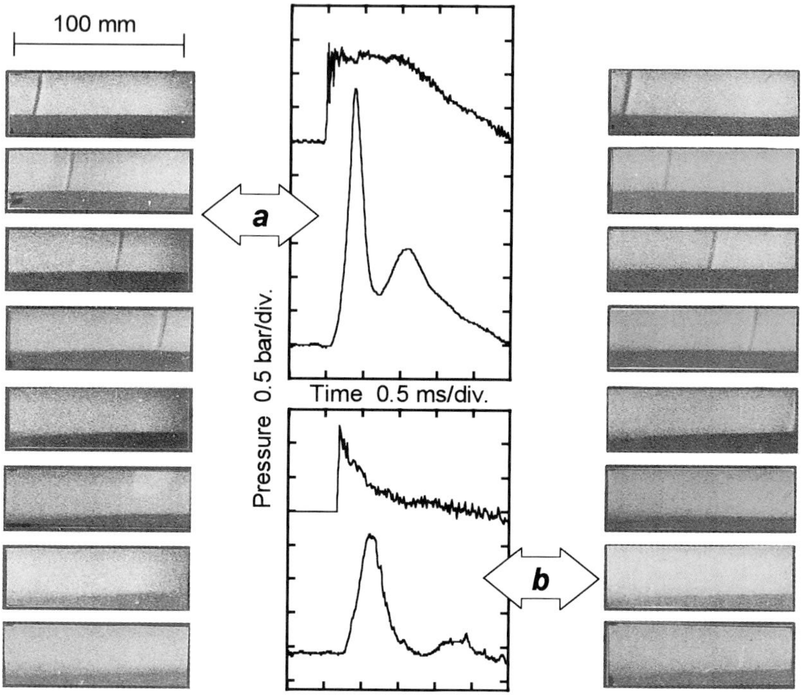

Figure 2. Pressure records and shadow pictures in the interaction between shock waves and a polyurethane foam layer: a – stepwise shock wave; b – blast wave.

3. Simulation and discussion

Phenomenological consideration of the experimental findings both for the dust and foam layer can be based on the simplified mechanistic model of a compressible layer which has been developed by Gelfand et al. (1987, 1989). It should be recalled here, that in the framework of this model we assume that a column of mass m with a unit cross-sectional area "cut" in a layer of a given thickness can be replaced by a mechanical system composed of a point mass m and the combination of an ideally plastic Coulomb element with a zero restoring force and an elastic element with elasticity coefficient k and damping coefficient c. For the polyurethane foam, this simulation is readily apparent from consideration of stress-strain curves. The sample of foam can be easily compressed by more than 50% even under negligible overpressure. Thus, during the first stage of shock loading, the foam is freely accelerated to some velocity v. Then the elasticity properties of the foam skeleton begin to play a crucial role in the interaction with the support. For the dust layer we should take into account that the powder particles are not initially closely packed. The overpressure of the shock wave displaces particles to a new position, thus compacting the layer. To a first approximation (Gelfand et al. 1989), the particle acceleration is not affected by the external forces (except overpressure) until the dust is more or less closely packed. Then, owing to the elastic properties of the particle material, the loading is transmitted to the bottom rigid wall. Hence, both for the dust and foam layer, the motion of mass m is governed by the equation $mx''(t) + cx'(t) + kx = p(t)$ with the initial conditions $x(0) = 0$ and $x'(0) = v$.

The velocity v before the packing exerts primary control over the maximum transmitted overpressure. In spite of a number of simplifications the described model seems to be attractive for the interpretation of the available experimental evidence. As can be readily seen, the decrease of pressure immediately behind the shock front (as in blastwise waves) inevitably gives rise to the lowest value v in comparison with the stepwise loading. Besides, by the time of the beginning of the elasticity interaction the current overpressure $p(t)$ will be much lower than that at the front of the blast wave. Both of these factors lead to the decrease of the transmitted loading in the case of blast waves in comparison with the case of a stepwise shock wave.

Figure 3 shows an example of the comparison between measured and calculated pressure profiles under the layers for various stepwise and blastwise shock waves (middle record in Fig. 4b is for a deposit of 10 mm depth, all the others are for 20-mm deposits). The calculations were performed under two assumptions: (1) the elasticity coefficient k is proportional to shock overpressure; (2) the relative compression of the layer does not depend on the pressure profile (Fig. 2). As is seen, the proposed simulation satisfactorily predicts the dynamics and maximum value of transmitted loading. The discrepancies between the measured and predicted pressure profiles are caused by the significant simplifications that were adopted in the model. For the sake of further improvement of the mechanistic model, in part, it will be reasonable to take into account the non-zero restoring force at the initial stage of the layer compaction. Another supplement was proposed recently by Lee et al. (1995), who suggested taking into account the dependence of elasticity on the deformation over the whole region of the stress-strain curve.

4. Concluding remarks

The experimental procedure based on specially designed conventional shock tubes seemed to be efficient for the study of the interaction between blast-wise shock waves and dust and foam deposits. It was found that the loading transmit-

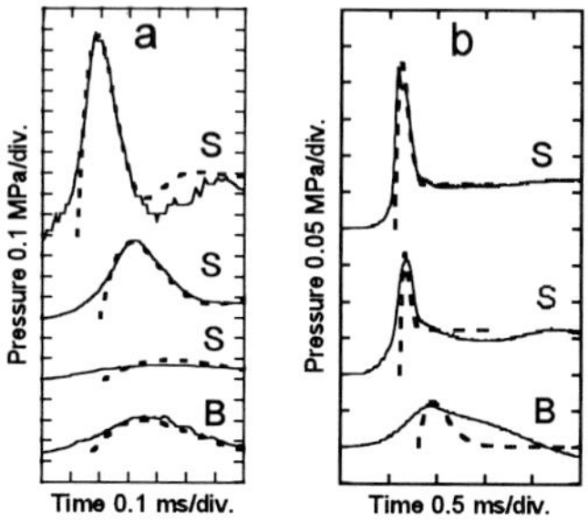

Figure 3. The comparison between measured and calculated pressure profiles: a – polyurethane foam layer; b – Plexiglass dust deposit; S – stepwise loading; B – blastwise loading.

ted through a given compressible deposit is strongly influenced by the duration and shape of the air shock wave. There exists a range of rather short shock durations at which the loading of the rigid wall covered with porous material is lower than that in the case of an uncovered wall. Thus, the compressible layers can play both protective and destructive roles.

To summarize, the present study demonstrates that the concept of "blast-wave amplification by means of compressible covering" is ambiguous. In every case it is necessary to take into account the real structure of the air shock wave. One way of doing this is to use the experimental data on the interaction between the stepwise shock waves and dust deposits.

References

Adachi T, Suzuki T, Tanabe K (1984) The pressure waves in a dust layer over which a shock wave is propagating. In: Theoretical and Applied Mechanics, Vol 32, Tokyo, Univ of Tokyo Press, pp 211–219.

Gelfand BE, Medvedev SP, Polenov AN, Frolov SM (1987) On the influence of porous compressible covering on the behavior of shock loading of construction. Zh Tekh Fiz 57:831–833.

Gelfand BE, Medvedev SP, Borisov AA, Polenov AN, Frolov SM, Tsyganov SA (1989) Shock loading of stratified dusty systems. Arch Combust 9:153–165.

Lee JJ, Frost DL, Lee JHS (1995) Transmission of a blast wave through a deformable layer. In: Shock Waves @ Marseille. Brun R, Dumitrescu LZ (eds), Springer, Berlin, Vol 3, pp 181–186.

Skews B (1995) Shock wave impact on porous material. In: Shock Waves @ Marseille. Brun R, Dumitrescu LZ (eds), Springer, Berlin, Vol 3, pp 11–20.

Effect of thermal expansion on the shock speeds of first, second, and fourth sound in He II

S. Braun, A. Kluwick
Institut für Strömungslehre
Technische Universität Wien
Vienna, Austria

M. S. Cramer
Engineering Science and Mechanics Department
Virginia Polytechnic Institute and State University
Blacksburg, VA 24061 USA

Abstract: We consider one-dimensional, weakly nonlinear first, second, and fourth sound waves in He II. The first correction (in amplitude) to the shock speed is computed for each sound mode. The expressions obtained are exact with respect to the coefficient of thermal expansion β. It is shown that the commonly made assumption of negligible β can lead to significant error in the shock speeds for first and fourth sound away from the lambda line. This contrasts with the calculation of the linear sound speeds and the shock speed for second sound where the $\beta = 0$ approximation yields accurate results. Near the lambda line, the exact expressions for all three modes are seen to contain fundamentally different singularities than those found in the commonly employed $\beta = 0$ theory.

Key words: Quantum fluid, Superfluid, Shock waves

1. Introduction

At ordinary pressures the transition from ordinary liquid helium (He I) to superfluid helium (He II) occurs at approximately 2.17 K. The remarkable properties of He II include the existence of quantized vorticity, the ability to flow through microscopic passages with virtually no resistance, and the existence of multiple sound modes. The primary focus of the present investigation is the influence of the coefficient of thermal expansion

$$\beta \equiv -\frac{1}{\rho}\frac{\partial \rho}{\partial T}\bigg|_{p,w^2} \tag{1}$$

on the first, second, and fourth sound modes. Here ρ is the mass density of He II, T is the absolute temperature, p is the pressure, and w is the counterflow or slip velocity. First sound is the direct analog of ordinary sound in any liquid or gas in that it carries nearly isentropic perturbations in the density and particle velocity. Second sound has no analog in ordinary fluids and carries perturbations in the temperature, entropy, and counterflow velocity; in second sound both the pressure and bulk velocity of the fluid are nearly constant. Fourth sound carries pressure and temperature perturbations in clamped He II, i.e., that where the velocity of the normal component is rendered stationary in flows through densely packed powders or narrow capillaries.

A commonly made simplification in the study both of sound and shock waves in superfluid helium is the assumption that the thermal expansion coefficient (1) is negligibly small. As pointed out by Putterman (1974), such an assumption would seem reasonable because the size of β is on the order of 10^{-3} to 10^{-2} K^{-1} at pressures and temperatures of practical interest. In fact, if one simply neglects β everywhere it appears, the resultant linearized sound speeds agree well with the exact, i.e., $\beta \neq 0$, versions; further discussion of this point can be found in Putterman's (1974) text.

As one might expect, the accuracy of the $\beta = 0$ assumption is seen to depend on the size of the terms premultiplying β. It turns out that β tends to appear in the nondimensional combinations

$$\beta T \quad \text{and} \quad G \equiv \frac{\beta a^2}{c_p}, \tag{2}$$

where G is the well-known Grüneisen parameter, a is the $\beta = 0$ version of the linearized speed of first sound, and c_p is the specific heat at constant pressure. The quantity s is the fluid entropy. The first quantity in (2) is seen to be relatively small at most temperatures and pressures. However, because the size of a and c_p are of the order of 200 m/s to 300 m/s and 10^3 $m^2/(s^2\text{-}K)$, the Grüneisen parameter G can be of order one at many pressures and temperatures. For example, if we use the data of Maynard (1976), we find that $\beta T \approx -0.03$ and $G \approx -1.05$ at a temperature of 1.6 K and a pressure of 20 bar. Thus, at this temperature and pressure, the assumption of negligible β is by no means obvious.

The purpose of the present report is therefore to provide an overview of the accuracy of the $\beta = 0$ approximation in the estimation of shock speeds for first, second, and fourth sound. For the most part, the results for fourth sound will be a review of the prior work of Cramer and Kluwick (1993). However, our presentation provides corrections to the estimates valid near the lambda line. The results for first and second sound are new.

The overall approach presented here will be to compute the shock speeds for each mode without the use of the $\beta = 0$ approximation. We then compare our results, which are exact with respect to the thermal expansion coefficient (1), to those obtained in the full $\beta \to 0$ limit.

2. Outline of analysis

The governing equations of the superfluid are taken to be the well-known Landau two-fluid equations recorded in most texts on superfluid helium, e.g., Putterman (1974). The particular form chosen is that presented by Cramer and Sen (1990). The equations used in the analysis of fourth sound are the well-known clamped equations. Again, these are reviewed in the references by Putterman (1974) and Cramer and Kluwick (1993). Throughout, the effects of dissipation will be ignored and all shocks will therefore be regarded as discontinuities.

In the present investigation we restrict our attention to small disturbances in the form of simple right-running waves. The weak nonlinear effects will be taken into account through use of the well-known multiple-scale approach of Tanuiti and Wei (1968).

To complete our evaluation of the $\beta = 0$ approximation, we employ the physical data of Maynard (1976). Asymptotic results for the neighborhood of the lambda line were taken from the work of Ahlers (1973, 1982) and Goldner and Ahlers (1992).

3. Results

As is well known, the multiple-scale approach of Tanuiti and Wei (1968) yields an inviscid Burgers equation. The wave and shock speeds corresponding to this equation can be written

$$\lambda\left(1 + \frac{\Gamma}{\lambda}u\right) \qquad \text{and} \qquad \lambda\left(1 + \frac{\Gamma}{\lambda}\frac{u^+ + u^-}{2}\right), \tag{3}$$

where λ is the linear wave speed of the mode in question and u is an appropriate nondimensional measure of the disturbance form and amplitude, again for the particular mode in question. The superscripts $\pm$ denote values of the disturbance immediately ahead of and behind the shock. The second material constant Γ is the quadratic nonlinearity or steepening coefficient. The factor Γ clearly yields the first nonlinear correction to the shock speed and its estimation is the principal focus of the present investigation.

We begin by reviewing the results for fourth sound as presented by Cramer and Kluwick (1993). The exact expression for Γ/λ for the fourth-sound mode was given in equations (31) and (35) of Cramer and Kluwick's article and in graphical form in their Fig. 3. The result of their work clearly demonstrated that the indiscriminate neglect of our equation (1) leads to significant error in the nonlinear wave speed, shock speeds, and even the Doppler shifts.

We now consider the propagation of shock waves in first sound. As pointed out in most standard references on superfluids, e.g., Putterman (1974), the $\beta = 0$ expression for the linear sound speed is an excellent approximation to the exact result. On the other hand, the expression for the quadratic steepening parameter Γ is seen to differ from the $\beta = 0$ version. These two approximations for Γ have been plotted and are compared in Fig. 1. The solid lines denote the exact variation of Γ/λ for first sound whereas the dashed lines represent the $\beta = 0$ approximation. Inspection of Fig. 1 reveals that there is less error

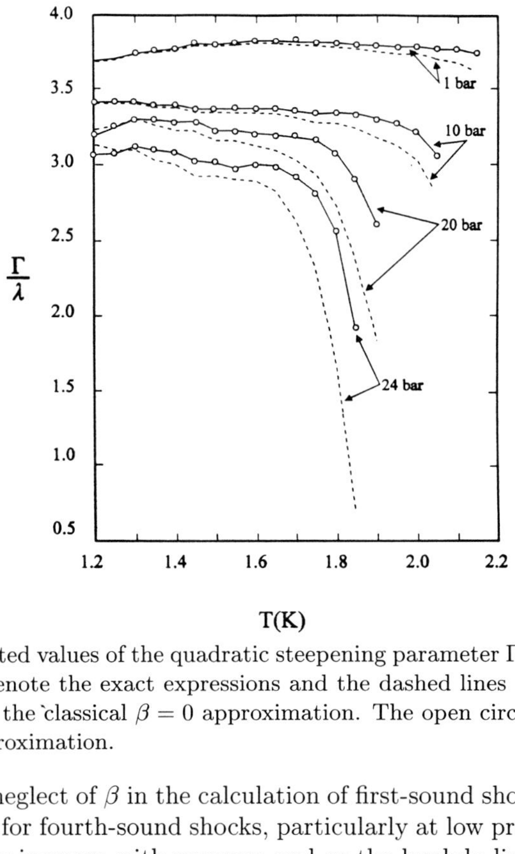

Figure 1. Computed values of the quadratic steepening parameter Γ/λ for first sound. The solid lines denote the exact expressions and the dashed lines denote the results corresponding to the classical $\beta = 0$ approximation. The open circles denote a modified small β approximation.

involved in the neglect of β in the calculation of first-sound shock waves than in the calculations for fourth-sound shocks, particularly at low pressure. However, the discrepancies increase with pressure and as the lambda line is approached. The influence of the thermal-expansion coefficient is clearly non-negligible at pressures on the order of 20 bar.

As in the case of fourth-sound shock waves, a modified small β approximation was developed which retains terms involving G but neglects terms involving

βT. The variations of Γ/λ based on this approximation are denoted by the open circles in Fig. 1. This approximation is seen to be in excellent agreement with the exact theory which serves to reinforce the idea that it is the neglect of the Grüneisen parameter which leads to the error in the classical $\beta = 0$ theory.

Finally, we consider the effect of the thermal expansion on second-sound shock waves. In this case, we found that the $\beta = 0$ theory and the exact theory are in excellent agreement over most temperatures and pressures. This observation is consistent with the experimental evidence of Dessler and Fairbank (1956) which generally agrees well with the classical $\beta = 0$ theory. In order to save space we will present more detailed comparisons at a later date.

We close this section by describing the asymptotic behaviors of Γ/λ in the neighborhood the lambda line. The idea here is to evaluate the limiting values of Γ/λ as given by the exact theory and the conventional $\beta = 0$ theory. To accomplish this, the first two authors have developed a self-consistent approximation based on the singularity laws of Ahlers (1973, 1982) and Goldner and Ahlers (1992). As a result of their analyses, it was found that both the exact and $\beta = 0$ theories yield $\Gamma/\lambda \to -\infty$ as $T \to T_\lambda(p)$ for each mode. The function $T_\lambda(p)$ yields the He I $\to$ He II transition temperature at the pressure p. The values of Γ/λ for first, second, and fourth sound were found to be of order

$$\frac{1}{\epsilon \ln^3 \epsilon}, \quad \frac{1}{\epsilon^{\frac{1}{3}} \ln \epsilon}, \quad \text{and} \quad \frac{1}{\epsilon \ln \epsilon} \tag{4}$$

when no $\beta = 0$ approximation was made, where $\epsilon \equiv 1 - T/T_\lambda \ll 1$ in the neighborhood of the lambda line. However, when the classical $\beta = 0$ approximation is made and the $\epsilon \to 0$ limits are then taken, it was found that Γ/λ is of order

$$\frac{1}{\epsilon \ln^2 \epsilon}, \quad \frac{1}{\epsilon^{\frac{1}{3}}}, \quad \text{and} \quad \frac{1}{\epsilon} \quad \text{as} \quad \epsilon \to 0 \tag{5}$$

for first, second, and fourth sound, respectively. Upon comparison of (4) and (5) it is seen that the $\beta = 0$ assumption produces asymptotic results which are in error by a factor of $\ln \epsilon$. A quick explanation for this error is that $|\beta| \to \infty$ as $\epsilon \to 0$. The assumption that $\beta = 0$ is obviously asymptotically incorrect which, according to our computations, affects all three sound modes. Complete details on the asymptotic behavior as well as the behavior away from the lambda line will be presented at a later date.

4. Summary

We have considered one-dimensional, right-running, weakly nonlinear shock waves in superfluid helium. It has been shown that the classical $\beta = 0$ assumption yields inaccurate estimates of the shock speed in first and fourth sound at many pressures and temperatures. Simultaneously, our work has demonstrated that the shock speeds of second sound are accurately predicted by the $\beta = 0$ theory except very near the lambda line. In first and second sound the importance of the thermal expansivity (1) increases as the pressure increases and as the lambda line is approached.

An obvious conclusion of these studies is that it is not enough to examine the numerical value of β or even βT to determine whether the $\beta = 0$ approximation can be made. Even so, the validity of the $\beta = 0$ approximation also depends on the precise sound mode of interest. Clearly, no simple rules exist. In experiments or applications involving moderate pressures or temperatures near the lambda line, the best procedure is to include the thermal expansivity from the outset.

References

Ahlers G (1973) Thermodynamics and experimental tests of static scaling and universality near the superfluid transition in He^4 under pressure. Phys Rev A 8:530.

Ahlers G (1982) Static and dynamic critical phenomena near the superfluid transition in ^4He. In: Phase Transitions, Levy M, Le Guillou JC, Zinn-Justin J (eds), Plenum, New York.

Cramer MS, Sen R (1990) Mixed nonlinearity and double shocks in superfluid helium. J Fluid Mech 221:233.

Cramer MS, Kluwick A (1993) The effect of thermal expansion on nonlinear fourth sound. Phys Fluids A 5(5):1239.

Dessler AJ, Fairbank WM, (1956) Amplitude dependence of the velocity of second sound. Phys Rev 104(1):6.

Goldner LS, Ahlers G (1992) Superfluid fraction of ^4He very close to T_λ. Phys Rev B 45(22):13129.

Maynard J (1976) Determination of the thermodynamics of He II from sound-velocity data. Phys Rev B 14(9):3836.

Putterman SJ (1974) Superfluid Hydrodynamics, North-Holland/American Elsevier, New York.

Tanuiti T, Wei CC (1968) Reductive perturbation method in nonlinear wave propogation. J Phys Soc Jpn 24:941.

Analytical prediction of regular reflection over porous surfaces

A. Levy, H. Li, G. Ben-Dor
Pearlstone Center for Aeronautical Engineering Studies
Department of Mechanical Engineering
Ben-Gurion University of the Negev, Beer Sheva, Israel

Abstract: An analytical model for solving the flow field associated with regular reflections of planar shock waves over porous layers was developed. The analytical predictions of the proposed model were compared to experimental results. Very good to excellent agreement was evident.

Key words: Porous materials, Regular reflection

1. Introduction

The reflection of shock waves over different wedges (see Ben-Dor 1991) and the interaction of shock (compaction) waves with porous media are two phenomena which have gained much attention in the past two decades due to their application to many engineering fields.

A comprehensive study concerning the propagation of planar shock/compaction waves in rigid porous media was initiated a few years ago in our department. Based on our experience in investigating these phenomena, we felt confident enough to combine them and investigate the regular reflection of oblique shock waves over porous surfaces. Another motivation was a paper presented by Kobayashi et al. (1993) at the 19th ISSW. In that paper, they proposed two different analytical models for describing the phenomenon. Then, they presented and solved the analytical model of their simpler case. Unfortunately, however, that case was the less realistic model. Consequently, we decided to try to solve their second case, the more realistic model.

2. The physical models

The two physical models proposed by Kobayashi et al. (1993) are shown schematically in Figs. 1a and 1b. Their first physical model, shown in Fig. 1a, was termed by them as the simple sink model. The incoming flow [state (0)] parallel to the surface of the porous layer is deflected clockwise by an angle θ_1 as it passes across the incident shock wave, i. Then, as it passes across the reflected shock wave, r, it is deflected counterclockwise by an angle θ_2. Since the flow in state

(2) can penetrate into the porous layer, the boundary condition there is

$$\theta_1 - \theta_2 = \delta \tag{1}$$

where δ is the overall deflection angle induced by the "sink" effect which is introduced by the porous layer. In summary, Kobayashi et al.'s (1993) first physical model accounts only for the flow of the gas behind the reflected shock wave into the porous surface and completely ignores the fact that a shock wave should be transmitted into the porous layer as shown schematically in Fig. 1b, in which their second physical model is shown. This physical model was correctly called by them the realistic model. Following the presentation of their realistic model, Kobayashi et al. (1993) claimed that "realistic as it is, it is not easy to solve the whole flow field since one must deal with the coupled problem between the pure gas phase and the porous phase." Their solution of their first physical model was empirical as they were not able to analytically relate δ to M_0, ϕ_1, and the porous-layer properties. Instead, for each experiment they fitted an appropriate value of δ by which the angle between the incident and reflected shock waves, ω_{ir}, as obtained by solving the well-known two-shock theory with the boundary condition given by equation (1), agreed with the experimental one.

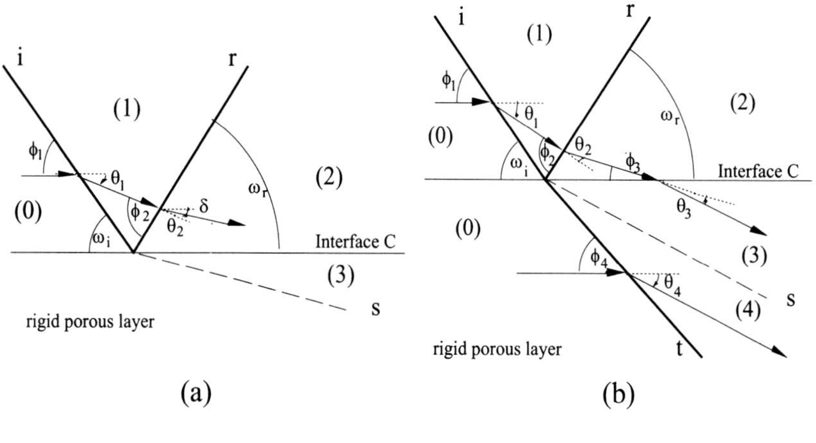

Figure 1. The two physical models proposed by Kobayashi et al. (1993): (a) the simple sink model and (b) the realistic model.

3. The present analytical model

The aim of the present study is to formulate the governing equations of the physical model shown in Fig. 1b. While the manner of describing the gaseous-phase flow through the various shock waves outside the porous layer in states

(0), (1), and (2) is and well known, the manner of treating the flow inside the porous material is much more complicated and far less known.

The macroscopic balance equations for a two-phase saturated porous medium, as obtained by Levy et al. (1995), were adopted in the present study. The macroscopic conservation equations were obtained by an averaging process over a representative elementary volume of the microscopic balance equations which were originally presented by Bear and Bachmat (1990). In addition, the porous material was assumed to be rigid and the gaseous phase to be an inviscid, non-conductive perfect gas.

The classical assumptions of the well-known two-shock theory (see Ben-Dor 1991), that the flow field is pseudo-steady and all the discontinuities are straight, were adopted. The justification of these assumptions lies in the recently obtained experimental results of Skews (1994) where measurements of the triple-point trajectory indicated that it is "a straight line within the accuracy of the measurement." Consequently, if Mach reflection is self-similar, it is expected that the regular reflection, which is simpler, will also be self-similar. In addition, Skews (1994) noted that in his experiments a number of broad waves were evident in the external flow when the incident shock wave reflected as a Mach reflection. When the reflection was regular "the larger scale wave structures were no longer apparent" and the flow behind the reflected shock wave near the reflection point, where our model is applied, approached uniform conditions.

As shown in Li et al. (1995), the properties of the gaseous phase which flows inside the porous layer could be transformed to obtain properties which result in governing equations similar to those of a pure gas. Using this novel approach the overall treatment of the phenomenon could turn out to be much more convenient than the treatment of the original equations, since the gas inside the pores could be treated using models originally developed for pure gas. Due to lack of space the governing equations are not presented in the following. The full derivation is given in Li et al. (1995).

4. Results and discussion

In the following, comparisons between predictions of the proposed analytical model with the experimental results of Skews (1992) and Kobayashi et al. (1993) are given.

Skews (1992) experimentally investigated the oblique reflection of shock waves from porous materials. The porous material in his case was silicone carbide (SiC) matrix with a mean pore size of about 2.5 mm. Based on water-displacement measurements, Skews (1992) estimated the porosity to be 0.8. Levy et al. (1993), who used the same material in a different experimental study, calculated the porosity from $\varphi = 1 - \rho_c/\rho_s$, where ρ_c is the porous material density and ρ_s is the density of the solid material of which the skeleton is made. The porosity value as obtained by Levy et al. (1993) was 0.73. The tortuosity, τ, of the investigated porous material had to be determined. This was done by using the procedure outlined by Li et al. (1995). The tortuosity was $\tau = 0.7$.

Kobayashi et al. (1993) conducted experiments with a rubber foam having a porosity of 0.98 and a dusty layer whose porosity was 0.44. Both Skews (1992) and Kobayashi et al. (1993) measured the angle, ω_r, between the reflected shock wave, r, and the reflecting porous surface. The values of ω_r as a function of M_i for two wedge angles, $\theta_w = 40°$ and $50°$, as measured by Skews (1992), are shown in Fig. 2 as squares and circles, respectively. The dashed lines describe the dependence of ω_r on M_i as calculated using the two-shock theory for a solid reflecting surface. The solid lines were calculated using the proposed analytical model for $\varphi = 0.8$ which was reported by Skews (1992) and $\varphi = 0.73$ which was reported by Levy et al. (1993). Figure 2 indicates that the proposed analytical model is capable of very well reproducing the experimental results. Better agreement is obtained with the smaller porosity $\varphi = 0.73$. Up to $M_i = 1.4$ the agreement with both wedges is excellent. It is also evident from Fig. 2 that for $M_i > 1.4$ much better agreement is obtained with the larger-wedge ($\theta_w = 50°$) experiments. The reason for the discrepancy between the experimental results and the analytical predictions for $\theta_w = 40°$ and $M_i > 1.4$ is not known.

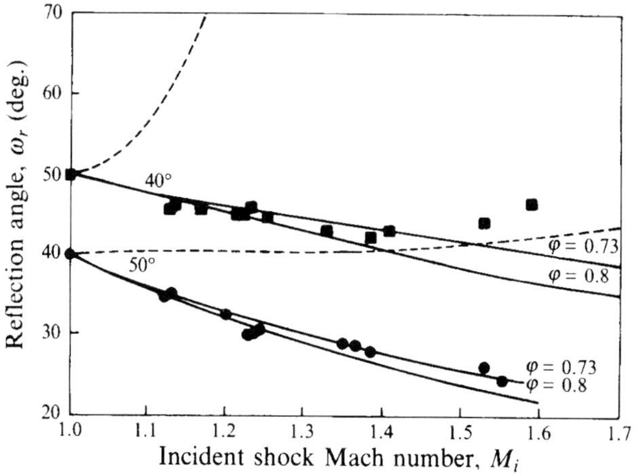

Figure 2. The reflected shock-wave angle, ω_r as a function of M_i for two different wedge angles, $\theta_w = 40°$ and $50°$.

Kobayashi et al. (1993) presented experimentally measured values of reflection angle, ω_r, as a function of incident angle, ω_i, for two incident shock-wave Mach numbers, $M_i = 1.2$ and 1.41. Their results are reproduced in Figs. 3a and 3b. The dashed line in these figures describes the relation between ω_r and ω_i as obtained from the two-shock theory for reflection over solid walls. Note that such a wall could be regarded as a porous material for which $\varphi = 0$. The

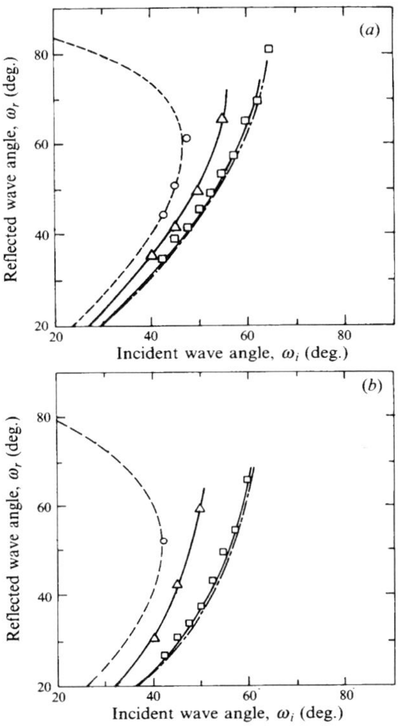

Figure 3. The reflection angle, ω_r, as a function of the incident angle, ω_i, for two different incident shock-wave Mach numbers, (a) $M_i = 1.2$ and (b) 1.41.

dashed-dotted line in these figures describes the other limiting case of $\varphi = 1$. Naturally, results over any actual porous material for which $0 < \varphi < 1$ should lie in between these two limiting lines. The experimental results marked by open circles were obtained over a solid reflected surface ($\varphi = 0$); the triangles were obtained over a dusty layer with $\varphi = 0.44$; and the squares were obtained for a rubber foam with $\varphi = 0.98$. The predictions of the proposed analytical model for $\varphi = 0.44$ and 0.98 are shown in Figs. 3a and 3b by solid lines. Excellent agreement is evident with both porosities and both incident shock waves.

In order to calculate the analytical results shown in Figs. 3a and 3b, the tortuosities of the rubber foam and dust layer were required. Based on the fact that the tortuosity was found to be slightly smaller than the porosity in the case of a silicon carbide (0.7 as compared with 0.73), the tortuosity of the dust layer was set at 0.4 ($\varphi = 0.44$) and that of the rubber foam was set at 0.9 ($\varphi = 0.98$). Due to this way of choosing the tortuosity values, the sensitivity of the analytical results for the reflection angel, ω_r, with respect to the tortuosity was checked. The calculations for $\tau = 0.4$ and 0.5 resulted in $|\omega_r(\tau = 0.5) - \omega_r(\tau = 0.4)|/\omega_r(\tau = 0.4) < 0.01$. Thus it was clear that the predicted values of ω_r did not depend strongly on τ and that our analysis was valid.

5. Conclusions

An analytical model for describing regular reflection over porous surfaces was developed. The very good to excellent agreement which was evident when the analytical predictions were compared with the experimental results validates the various assumptions used in the course of developing the model and confirms the validity of the entire physical model.

References

Bear J, Bachmat Y (1990) Introduction to Modeling of Transport Phenomena in Porous Media, Kluwer Academic Press, Dordrect, Germany.

Ben-Dor G (1991) Shock Wave Reflection Phenomena, Springer-Verlag, New York, NY, USA.

Kobayashi S, Adachi T, Suzuku T (1993) Regular reflection of a shock wave over a porous layer: theory and experiment. In: Shock Waves @ Marseille, Springer-Verlag.

Levy A, Ben-Dor G, Skews BW, Sorek S (1993) Head-on collision of normal shock waves with rigid porous materials. Exp Fluids 15:183–190.

Levy A, Sorek S, Ben-Dor G, Bear J (1995) Evolution of the balance equations in saturated thermoelastic porous media following abrupt simultaneous changes in pressure and temperature. Transport in Porous Media (to be published).

Li H, Levy A, Ben-Dor G (1995) Analytical prediction of regular reflection over rigid porous surfaces. J Fluid Mech 282:219–232.

Skews BW (1992) Oblique reflection of shock waves from rigid porous materials. In: 10th Mach Reflection Symposium, University of Denver, Denver, Colorado, USA.

Skews BW (1994) Oblique reflection of shock waves from rigid porous materials. Shock Waves 4:145–154.

Experimental observation of shock-wave evolution in an elastomeric foam

E. Zaretsky, O. Igra
Pearlstone Center for Aeronautical Engineering Studies
Department of Mechanical Engineering
Ben-Gurion University of the Negev, Beer Sheva, Israel

Abstract: Evolution of a gas shock of low Mach number ($M = 1.03$ to 1.15) propagating through flexible polyurethane foam was studied experimentally in a shock tube. The gas compression wave is responsible for forming a complex wave structure in the foam. The propagation velocity of the gas compression wave is close to the velocity of sound in air. The amplitude of this wave decays with increasing propagation distance as $p^* = p^*(0) \exp(-x/0.017)$. The stress relaxation time in the gas-foam system was found to be equal to $\tau = 25$ μsec.

Key words: Compression wave, Decay, Stress relaxation time

1. Introduction

It is known that for a material whose compression curve (pressure vs. volume) has a non-convex character, shock-wave propagation in the material may have a complex structure. The compression curves of elastomeric foams with varying porosity are non-convex (Gibson and Ashby 1988). Skews et al. (1993) showed that a shock propagating into foam from the foam-gas interface contains at least two waves: the gas wave and the foam compression wave. In shock-tube experiments (Skews et al. 1993, Gelfand et al. 1983, Gvozdeva et al. 1986) it was shown that far from the gas-foam interface a stationary pressure pulse propagates into the foam with a constant velocity. Thus a complex wave pattern should be expected to exist in proximity of the interface region of the foam.

The present work is an attempt to trace the evolution of a shock- generated pressure pulse in the foam from its initiation as a shock-like state up to its final form as a compression wave.

2. Experimental

Experiments were conducted in an 80-mm diameter, 4-m long shock tube designed specifically for experiments with polyurethane foams. The incident shock-wave Mach number was deduced from pressure pick-ups placed along the tube's wall. The incident shock-wave Mach number was within the range $M = 1.03$ to 1.15. A Kistler 603 pressure transducer, T_1, was embedded at the center of the shock-tube end wall. An additional Kistler 603 pressure transducer, T_2, was positioned at the middle of the end-wall radius; it stuck out from the end wall up to the air-foam interface to record the gas shock-wave arrival. Polyurethane foam samples composed of 1, 2, 3, and 4 layers (80 mm in diameter and about 10 mm in thickness) were placed adjacent to the shock-tube end wall. The gauge readings were fed, for processing, into a Tektronix TDS 540 digital oscilloscope.

In order to improve the temporal resolution of measurements, the Kistler 603 pressure transducer was employed without a charge preamplifier. Thus, later parts of pressure readings were accompanied by an exponential decrement of the transducer signal. The true transducer readings were derived from our pressure readings using the following procedure. The rear (the decayed) part of the pressure signal is relatively smooth. During this time interval, the pressure value should be equal to the pressure p_s which the foam experiences after several reflections of the travelling waves from the front and rear ends of the foam sample. The rear part of the signal was exponentially fitted and extrapolated upstream to the shock front. The reconstructed pressure values p/p_s were obtained by dividing the "live" pressure by the extrapolated one.

The reproducibility of the incident shock-wave Mach number in experiments with the samples of different thickness was about 0.5%. The inaccuracy in deducing propagation velocities of the pressure disturbances did not exceed 6%.

3. Results

Parameters of the shock-tube experiments are summarized in Table 1. The propagation velocity of the zero-pressure disturbances in the foam, C_o, was obtained by dividing the initial foam sample thickness (the thickness of the n-layer sample is $n \times 9.8$ mm) by the time elapsed between pressure pulses from the transducers T_2 and T_1.

Table 1. Shock Mach number M_1 and velocity C_o.

M_1	1.03±0.005	1.08±0.005	1.10±0.005	1.15±0.005
C_o m/sec	337±3	332±5	336±5	338±3

C_o are equal, independent of the shock strength, and very close to the sound velocity in air at normal conditions, i.e., 340 m/sec.

Figure 1 shows the front part of time-pressure readings recorded at the end wall during the collision of incident shock waves of different strengths with the foam samples. The evolution of a pressure front is apparent in this figure. The presence of two different compression waves, with different sets of propagation characteristics, may be recognized in successive pressure profiles obtained for the same Mach number. The waves are separated at the point $p*$ by the kink in the pressure profile. In comparison with the highly spread, high-pressure compression wave, the low-pressure wave may be considered as an unspread pressure jump with amplitude $p*$, propagating into the undisturbed foam with a velocity almost equal to C_o. On the t-h-plane (time-Lagrangian coordinates), propagation of this pressure jump (precursor) may be described by a single characteristic having a slope C_o, while the high-pressure wave should be described by a fan of characteristics (see Fig. 2a). Propagation of this precursor through the foam is accompanied by a strong decay. This is shown schematically in Fig. 2b.

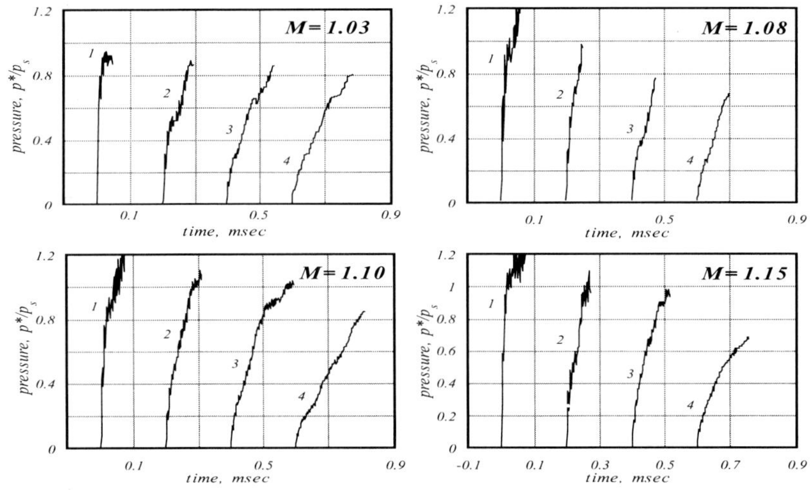

Figure 1. Wave-front evolution with increasing number of foam sample layers.

4. Discussion

As was shown by Zaretsky and Ben-Dor (1994), the propagation velocity of the zero-pressure disturbance in a foam is equal to

$$C_o^{(f)} = \sqrt{\frac{E'_f}{\rho_f}} = \sqrt{\frac{(1-\nu)}{(1+\nu)(1-2\nu)} \frac{E'_f}{\rho_f} \left(\frac{\rho_f}{\rho_s}\right)^2} \quad (1)$$

where E' is the material's constraint modulus (after its collision with shock, the foam material is characterized by 1D strain state), E is the material's Young

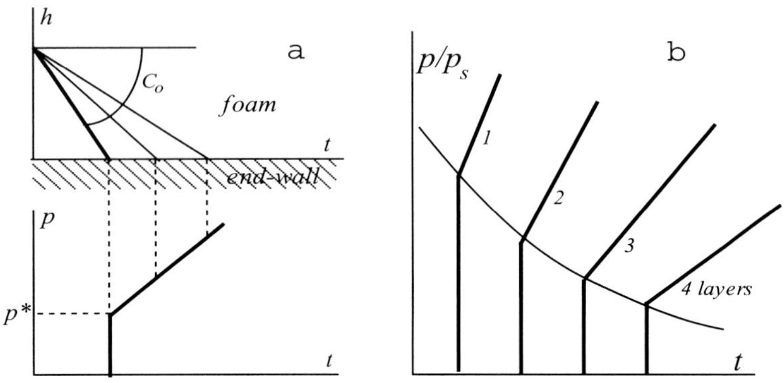

Figure 2. (a) steep gas and spread foam wave fronts, and (b) gas precursor decay.

modulus, ρ is its density, and ν its Poisson ratio. Indexes f and s stand for foam and the foam skeleton material, respectively. Calculation of $C_o^{(f)}$ via (1) for the polyurethane foam used (density $\rho_f = 30$ kg/m^3 $\nu = 1/3$, $E_s = 45$ MPa, and $\rho_s = 1200$ kg/m^3) yields $C_o^{(f)} = 38$ m/sec. Recall that the actual value of the precursor velocity in the present experiments was $C_o = 340$ m/sec, the sound velocity in air. These results are evidence that the precursor descends from the gas shock, penetrates into the foam, and decays during its propagation.

The propagation equation of the compressive stress P_x in the media, having initial density ρ_o along a characteristic C, may be described in Lagrangian form, (Duwall 1963) as:

$$dp_x + \rho C du = -F(p_x, \rho)\, dt \quad \text{on} \quad \frac{dh}{dt} = \frac{\rho}{\rho_o} C \tag{2}$$

where x is the direction of wave propagation, u is particle velocity, and h is Lagrangian coordinate. The dissipation function $F(p_x, \rho)$ is related to the strain rate $\partial\varepsilon_x/\partial t$ generated in the media by stress p_x,

$$\frac{\partial \varepsilon_x}{\partial t} = \frac{\tau}{\eta} F(p_x, \rho) \tag{3}$$

where τ is the stress relaxation time and η is the coefficient of viscosity. Now (2) can be rewritten as:

$$dp^* + \rho C_o du^* = -F(p^*, \rho)\, dt \quad \text{on} \quad \frac{dh}{dt} = \frac{\rho}{\rho_o} C_o \tag{4}$$

From conservation of mass and momentum across the wave front, $p^* =$

$\rho_o C_o u^*$, one obtains

$$du^* = \frac{dp^*}{\rho_o C_o} \tag{5}$$

Combining (4) and (5) yields

$$\frac{dp^*}{dt} = -\frac{F(p^*, \rho)}{2} \quad \text{on} \quad x = C_o t \tag{6}$$

or

$$\frac{dp^*}{dx} = -\frac{F(p^*, \rho)}{2C_o} \quad \text{along characteristic} \quad C_o \tag{7}$$

Since the air compression wave propagates into the undisturbed foam and the propagation velocity of this wave is almost equal the undisturbed sound velocity in air, the equality $\rho = \rho_o$ may be accepted. Assuming that the precursor decay is caused by the viscous friction between the foam and the air compression wave which propagates through it, and assuming a linear character for the dissipation function, one obtains, $F(p^*, \rho_o) = \frac{1}{\tau} p^*$. Thus, (7) yields

$$\frac{dp^*}{dx} = -\frac{1}{2\tau C_o} p^* \tag{8}$$

The solution of (8) is

$$p^*(x) = p^*(0) \exp\left(-\frac{x}{2\tau C_o}\right) \tag{9}$$

where $p*(0)$ is the pressure (compression stress) amplitude generated by a shock at the leading edge of the foam.

The parameters $p*(0)$ and τ in (8) were determined from the experimentally measured values of amplitude of the precursor pressure $p*/\ p_s$ in foam samples of different thickness (Table 2). Within the accuracy of the experimental measurements, the stress relaxation time τ is independent of the shock strength while the initial pressure $p*$ demonstrates an almost linear dependence on M_1 with $p* \to p_s$ when $M_1 \to 1$. The latter agrees with doubling of the pressure amplitude for an acoustic wave reflection from a rigid wall (Courant and Friedrichs 1976).

Table 2. Parameters of gas wave decay in polyurethane foam.

M_1	1.03	1.08	1.10	1.15
$p*(0)/p_s$	1.2±0.1	1.4±0.1	1.5±0.1	1.7±0.1
τ, μsec	25±2	25±2	25±2	25±2

5. Conclusion

The evolution of a gas shock wave of low Mach number ($M = 1.03$ to 1.15) which propagates through flexible polyurethane foam was studied experimentally in a shock tube. The gas compression wave is responsible for forming a complex wave structure in the foam. The propagation velocity of the gas compression wave is close to the velocity of sound in air. The amplitude of this wave decays with distance as $p^* = p^*(0)\exp(-x/0.017)$.

The stress relaxation time in the gas-foam system was found to be equal to $\tau = 25\ \mu$sec.

References

Courant R, Fredrichs KO (1976) Supersonic Flow and Shock Waves, Springer-Verlag, New York, Heidelberg, Berlin.

Duwall GE (1963) Propagation of plane shock waves in stress-relaxing medium. In: Stress Waves in Elastic Solids, Kolsky H and Prager W (eds), Springer-Verlag, Berlin, pp 21–32.

Gelfand BE, Gubanov AV, Timofeev EJ (1983) Interactions of shock waves in air with a porous screen. Izv Akademii Nauk SSSR Mechanika Zhidkosti i Gaza 4:78–84.

Gibson LG, Ashby MF (1988) Cellular Solids; Structure and Properties, Pergamon Press, Oxford, New York.

Gvozdeva LG, Faresov YM, Brossard J, Charpentier N (1986) Normal shock wave reflection on porous compressible material. In: Dynamics of Explosions, Bowen JR, Leyer JC, Soloukhin RI (eds), Progress in Astronautics and Aeronautics 106:155–165.

Skews BW, Atkins MD, Seitz MW (1993) The impact of a shock wave on porous compressible foam. J Fluid Mech 253:245–265.

Zaretsky E, Ben-Dor G (1994) Compressive stress-strain relations and shock Hugoniot curves of flexible foams. J Eng Mater Tech 117:278–284.

Experimental investigation of interaction of shock wave with closed cellular compressible foam

Y. P. Lagutov, Y. L. Sharov, L. G. Gvozdeva
Russian Academy of Science
High Energy Density Research Center Scientific Association IVTAN
Izhorskaya 13/19, Moscow 127412 Russia

Abstract: The interaction of a shock wave with closed cellular porous compressible polyethylene was investigated. It was experimentally shown that the significant increase in pressure at a rigid surface covered by a layer of compressible porous material under shock-wave action is due to impact of the compressed porous material on the surface. Disturbances in gas and foam were visualized simultaneously. The velocities of deformation waves inside the foam and the densities of the foam behind the fronts of these waves were determined. The pressure profiles on the side and back walls of the shock tube were recorded.

Key words: Shock wave, Compressible foam, Optical method

1. Introduction

As has been found (Gelfand et al. 1975), the pressure rise in shock-wave reflection from the closed end of the shock tube can be greater than the normal reflection pressure when a layer of compressible foam is present. The great interest in this problem has been caused by the work of Gvozdeva and Faresov (1984). Skews (1993) made a review of investigations. The experimental method of simultaneous visualization of the propagation of disturbances in transparent gaseous phase outside the sample and in nontransparent porous compressible material, the pressure measurements at three points on the end wall, and the analysis of the results of one of the experiments were all presented by Lagutov and Sharov (1995).

2. Experimental conditions and experimental method

Experiments were made in a shock tube with square cross-section 72×72 mm^2. Both air and carbon dioxide were used as the working gas. Experiments were conducted with 72×72×54 mm^3 sample made of elastic polyethylene foam with a closed cellular structure, initial density ρ_0 = 33.2 kg/m^3 and porosity α = 0.97. The characteristic dimension of the cells was approximately 0.5 mm. A cubic sample was placed into the closed experimental chamber with optical

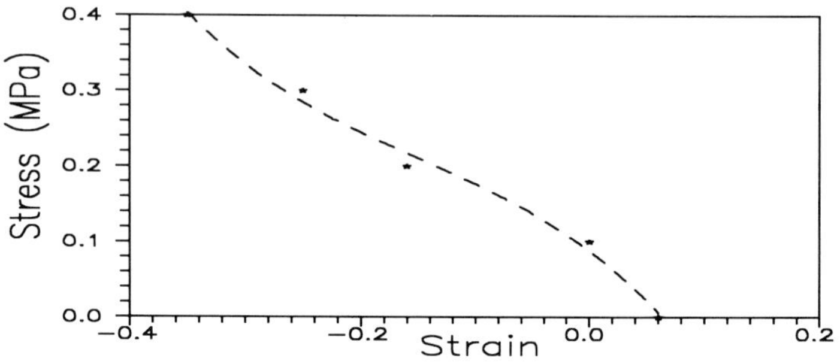

Figure 1. Stress-strain curve for polyethylene foam.

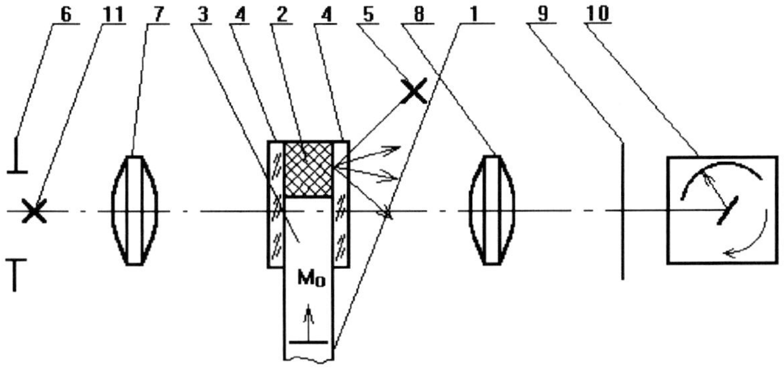

Figure 2. Optical scheme of experiments.

windows to investigate the stress-strain dependence for the polyethylene foam in the region from vacuum ($\sim$ 10 Pa) to 400 KPa. The change in dimensions of the sample under the action of slowly changing external gas pressure was measured. Only along one axis the dimension of the sample changed significantly (Fig. 1). Then in all experiments this axis was parallel to the axis of the shock tube (see Fig. 2, 1) where the sample of foam (2) was installed near a closed end inside the experimental chamber (3) supplied with optical glasses (4).

A photo flash lamp (5), placed before the reception part of the schlieren device but out of its field of view has been used for sample illumination. A Toepler-system schlieren device consisting of the entrance diaphragm (6), collimating (7) and reception (8) lenses, and visualization diaphragm (9), as well the streak camera (10) were used for visualization and registration of the process. The streak camera was placed so that the image of its slit at the plane of object was parallel to the direction of propagation of the incident shock wave and cut out the middle part of the sample. On the side of the sample a system of equidistant dark lines was drawn perpendicular to the tube axis with a spacing of 10 mm for visualization of propagation of compression waves inside the porous material. The slit and thread were applied as entrance and visualization diaphragms.

The photo flash lamp was equipped with five light fibers (not shown in Fig. 2). Four of them were used to supply the uniform and more powerful illumination of the side part of the sample with drawing lines. The fifth light fiber was used for illumination of entrance diaphragm (6). In the other case the separate pulsed lamp (11) was used for the purpose. The start of the lamps was synchronized with the process. Pressures at the side and back walls of the shock tube were measured.

3. Analysis of experimental data

Figure 3 shows a photo record and an x-t diagram of the interaction of a shock wave in air ($P_0 = 100$ KPa, $M_0 = 1.55$) with the sample. It is seen that the incident shock wave, propagating with the velocity $U_0 = 535$ m/s, at the time $t = 0$ acts on a free surface of the sample F. At this moment the reflected shock wave R_1 appears and propagates up the stream of gas with the velocity $U_1 = 270$ m/s. The surface F moves toward the closed end of the tube with constant velocity $V_F = 77$ m/s, and then oscillates with a period of $T \simeq 675$ mks.

As a result of the shock-wave impact a compression wave front S_2 appears in the sample and propagates toward the closed end of the tube with velocity $V_2 = 121$ m/s. The breaks of lines on the photo record show that the front of a weak disturbance S_1 moves before the front S_2 with the velocity $V_1 = 224$ m/s. At the time t_1 the front of the weak disturbance S_1 reaches the closed end of the tube and reflects from it. However, it is difficult to see the reflected front S_1, and therefore, it was not marked in Fig. 3.

At the time t_2 the front of compression wave S_2 reflects from closed end of the tube and propagates in the opposite direction through the compressed material in the form of a compression wave front S_3 with the velocity $V_3 = 56$ m/s. Then at the time t_2 after impact of the wave S_2 on the end wall a further movement of the sample toward the end wall by inertia and by the pressure of the gas behind the reflected shock wave has occurred and a further compression of the foam has resulted.

The maximum pressure at the end wall was reached during the time interval 500 to 700 mks from the beginning of the process. This is seen from oscillograms on Fig. 3. The beginning of the interaction on the oscillograms was determined from the measured velocity V_1 (or V_2). From Figs. 3 and 4 it is seen that the back wall of the sample does not move during the entire compression time.

The compressed sample, without changing its dimension, moves with constant velocity down the flow and impacts on the end wall forming a significant increase in pressure (see oscillograms for gauges G1, G2, G3 in Fig. 3). Thus the impact of the compressed foam on the rigid wall is a reason for the significant increase in the pressure.

At the time t_3 (Fig. 3) the front of reflected compression wave S_3 arrives at the free surface of the sample F. A second reflected shock wave R_2 (Figs. 3 and 4) moves up the flow. At the same time t_3 the rarefaction wave S_4 is formed in the foam. This front S_4 propagates with velocity $V_4 = 30$ m/s from F to the end wall. At the time t_4 the front of the compression wave S_5 is formed and propagates from the wall with velocity $V_5 = 38$ m/s. Then the velocities of the rarefaction wave (S_6) and weak compression wave (S_7) are approximately equal to the same value 38 ± 1 m/s. From the slope of the side lines of the sample it is seen that the parameters of the foam are uniform within each region (0–7) in Fig. 3.

From Fig. 3 the following densities of foam were determined: $\rho_1 = 1.06\rho_0$; $\rho_2 = 2.70\rho_0$; $\rho_3 = 5.60\rho_0$; $\rho_4 = 3.57\rho_0$; $\rho_5 = 4.35\rho_0$; $\rho_6 = 3.70\rho_0$; and $\rho_7 = 4.35\rho_0$ in the corresponding regions behind the fronts $S_1 - S_7$ (see also Lagutov and Sharov 1995). The time dependencies of the thicknesses of the foam layers behind the fronts $S_1 - S_7$ are shown on the x-t diagram. Further, it is seen that at the time t_1 the front of the weak disturbance S_1 reaches the closed end and on the oscillograms of the gauges G1, G2, G3 the first increase in the pressure appears. The next increase in the pressure at time t_2 is due to the compression wave front S_2 whose reflection S_3 at the closed end of the tube serves to bring the moving sample to rest. During the time interval $t_2 - t_3$ the maximum pressure on the closed end of the tube is achieved. Then the pressure decreases and achieves its minimum value at the time t_4.

Figures 3 and 4 show that the propagation of disturbances inside porous compressible polyethylene in shock action can be considered as one dimensional.

From the oscillogram of the gauge G4 (Fig. 3) it is seen that the arrival of fronts S_1 and S_2 together causes the first increase in the pressure. The following increase is due to the gas pressure behind the reflected wave R_1. The next increase in the pressure is due to the reflected wave R_2. After that the pressure decreases and oscillates at a constant level.

The gauge (G5) nearest to the back wall (Fig. 3) gives the arrival of the front S_1 (first pressure increase), the reflected front S_1 (second increase), the front S_2 (third increase), the reflected front S_2 (fourth maximum pressure increase). After that the pressure decreases and oscillates at a constant level.

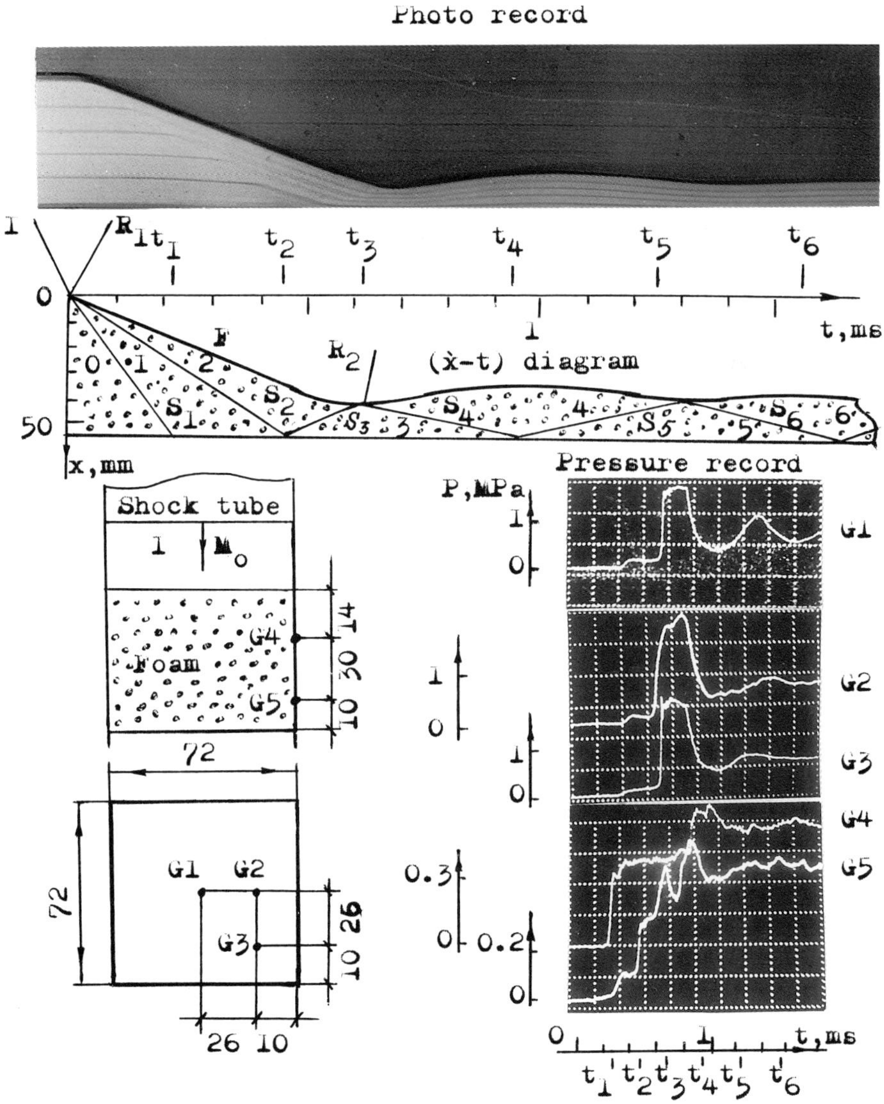

Figure 3. Interaction of shock wave with foam. $M_0 = 1.55$, $P_0 = 100$ KPa in air, I, R_1, and R_2 are incident, first, and second reflected shock waves; S_1 is weak leading deformation wave, S_2 – S_7 are incident or reflected deformation waves, G1 – G5 are pressure gauges.

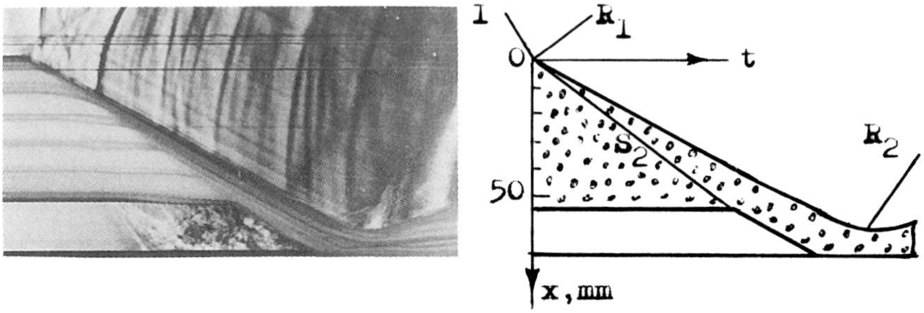

Figure 4. Photo records and x-t diagrams: interaction of shock wave with foam placed at the distance of 20 mm from the rigid wall, $M_0 = 2.9$, $P_0 = 100$ KPa in carbon dioxide.

4. Conclusion

A visualization technique has been demonstrated for propagation of the disturbances both inside and outside nontransparent closed cellular compressible porous materials under shock-wave impact. The x-t diagrams and the velocities for propagation of disturbances inside and outside the porous material, and the densities of the porous material behind the disturbance fronts, have been determined. It has been shown that the propagation of disturbances inside the sample can be considered as one dimensional. It was shown experimentally that the significant increase in the pressure at a rigid surface covered by a layer of compressible material under shock-wave action in comparison with the case of normal reflection without such a layer is due to the impact of the compressed foam.

Acknowledgement. This work is supported by Russian Foundation of Fundamental Science (94-02-05304a) and the International Science Foundation (M94000, M94300).

References

Gelfand BE, Gubin SA, Kogarko SU, Popov DE (1975) Study of special features of propagation and reflection of pressure in porous medium. Soviet Physics-Applied Mathematics and Technical Physics 6:74–77.

Gvozdeva LG, Faresov YM (1984) Interaction of shock waves in air with a wall covered with a layer of porous compressible materials. Letters to Journal of Technical Physics 19(14):1153–1156.

Lagutov YP, Sharov YL (1995) Propagation of disturbances in porous compressible polyethylene with closed pores under shock wave impact. Letters to Journal of Technical Physics 21(2):64–69.

Skews BW (1993) Shock wave impact on porous materials. In: Shock Waves @ Marseille, Brun R, Dumitrescu LZ (eds), Springer, Vol 3.

Blast-wave propagation in foam

H. Kleine*, G. Diaconescu, J. H. S. Lee
McGill University, Department of Mechanical Engineering
817 Sherbrooke Street West, Montreal, Quebec, H3A 2K6 Canada

Abstract: The decay of a spherical blast wave in polyurethane foam is investigated experimentally. Up to a certain distance from the charge, the propagation of the wave is almost identical to that of an air blast. From that point on, however, the wave is strongly attenuated, which is shown by a significant drop of its velocity and the pressure jump across it. In contrast to previous studies on this subject, any influence of confining walls on the wave propagation can be ruled out.

Key words: Blast-wave propagation, Shock attenuation, Permeable foam

1. Introduction

In recent years, the propagation of a shock wave in foam material has been the subject of several investigations. A comprehensive review paper on this topic was given by Skews (1995). Although some models have been developed to describe the wave phenomena in the foam (e.g. van Dongen et al. 1995), the primary concern in these studies has been the transmission of a shock wave through a (mostly deformable) porous medium and the final impact on a rigid surface covered with this material. Less effort has been spent on the actual propagation process of the wave inside the medium. In the configuration used in almost all experimental works a block of foam is placed snugly in a shock tube adjacent to its end wall. Side-wall pressure records may be taken for monitoring the refraction of the incident wave at the front surface of the foam block (Henderson et al. 1989, Skews 1991) and for recording the wave that is being transmitted into the foam. These records may be used to estimate whether or to what extent the wave is attenuated and spread while propagating through the foam. However, considerable friction forces can be generated when the foam is compacted during this process, which may significantly influence the phenomenon. As a result, the pressure-transducer readings may not represent the true propagation process. Furthermore, multiple wave reflections and transmissions occur at the gas/foam interface, which result in complex wave-interaction processes and make interpretation of obtained results even more difficult.

*On leave from Stoßwellenlabor (Shock Wave Laboratory) RWTH Aachen, Templergraben 55, 52056 Aachen, Germany.

Most of the shortcomings mentioned above do not occur if a spherical blast wave is generated inside the foam material. In particular, wall effects are eliminated. By monitoring the wave, useful data on foam properties and on the propagation process may be obtained.

2. Experiments

In the present study, the decay of blast waves in polyurethane foam of three different densities has been investigated (ρ_{foam} = 24, 29, 31 kg/m^3). For all kinds of foam the volume fraction of foam material was found to be $\epsilon = 0.05 \pm 0.01$ (or, equivalently, the porosity of the material, defined as the volume fraction of the gas, was $\alpha = 1 - \epsilon = 0.95 \pm 0.01$). A charge of 8 g Detaprime was placed in foam blocks of dimensions 432 mm × 305 mm × 305 mm (see Fig. 1). The minimum distance of the charge with respect to the surfaces of the foam block was 150 mm.

The propagation of the wave that is established after the charge has been detonated was monitored by means of several contact gauges that were placed along a radial line with distances ranging from 10 to 200 mm with respect to the charge. The gauges may be arranged in this way because they do not present a major obstacle to the wave that would alter its propagation: their overall thickness is less than 0.3 mm and the wave passes them side-on. In each experiment, seven gauges were placed inside the foam block as indicated in Fig. 1. The signals were recorded with two LeCroy Digital Oscilloscopes that provided a sampling rate of 2 Msample/s.

In order to get additional data about the wave propagation (other than just the times of arrival from the contact gauges) a pressure transducer (PCB type 113A20) was located at selected positions inside the foam block. In order to avoid spurious signals due to an acceleration of the transducer, it was mounted rigidly on the bottom wall of the blast chamber. Because of its size and the necessary protective mounting, the presence of this transducer may be expected to influence the propagation of the wave. For this reason, only one transducer was used per experiment. In these cases, the contact gauges were placed only between the charge and the transducer. The pressure records were also meant to verify up to what distance from the charge the contact gauges could be expected to yield reasonable signals. In order to provide some reference data, air-blast measurements were also performed. In these experiments, the gauges were placed at the same distance and orientation with respect to the charge as in the foam blocks.

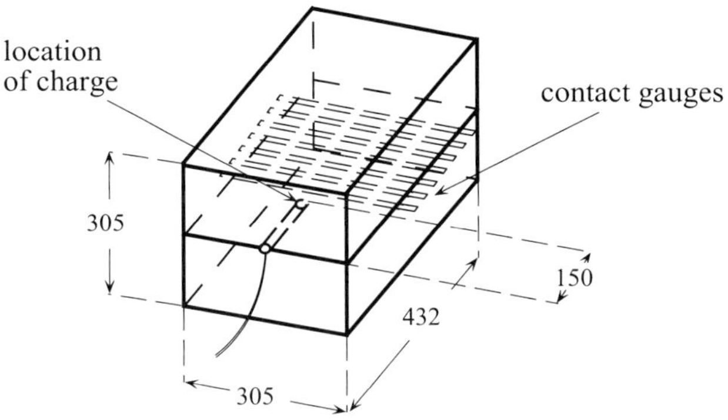

Figure 1. Dimensions (in mm) of foam blocks and arrangement of gauges.

3. Results and discussion

The experimental data are summarized in Figs. 2 and 3. Error bars associated with the individual measurements have approximately the size of the symbols used. The time-of-arrival curve of the air blasts is found to be in almost perfect

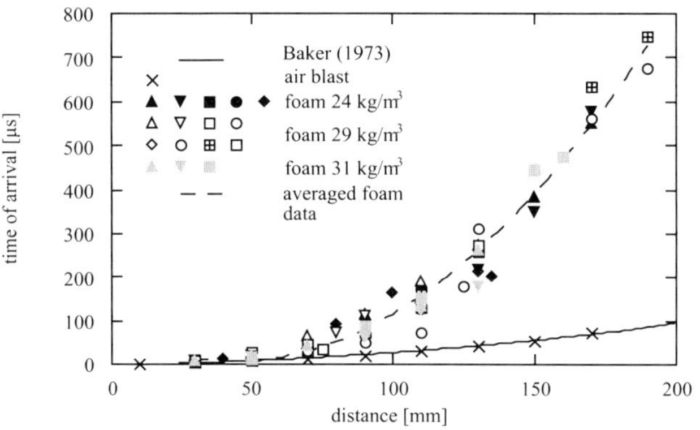

Figure 2. An x-t diagram of the propagation of a blast wave in foam.

agreement with the data given by Baker (1973). From the data presented in Fig. 2 the following main conclusions may be drawn:

- For a certain distance from the charge, the shock trajectory in foam is almost indiscernible from that of an air blast. In our case, this distance amounts to about 5 cm for an 8-g charge.

- Beyond this distance, the blast waves in foam start to decelerate rapidly with respect to a corresponding wave in air.

- The scattering of the data appears to increase with increasing distance from the charge — however, for the range of foam density tested here, no visible trend can be detected that would indicate an influence of ρ_{foam} on the shock trajectory. It seems therefore justified for the range of ρ_{foam} investigated here to describe the attenuation of the blast wave inside the foam by a curve for averaged data (Fig. 2).

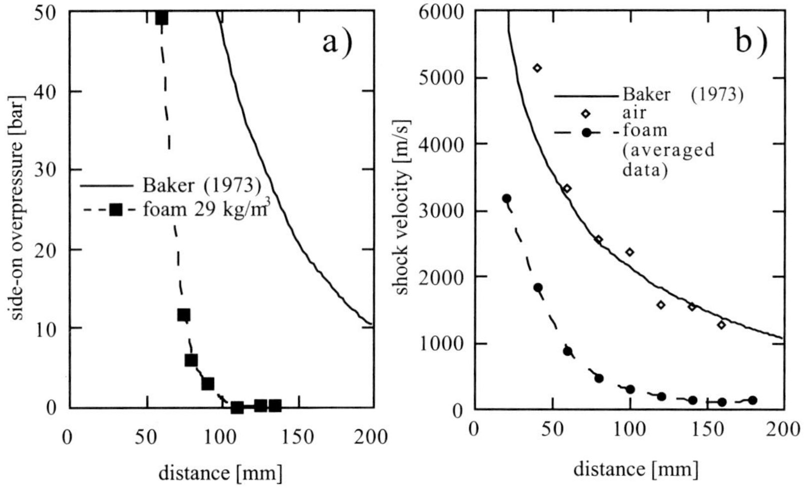

Figure 3. Side-on overpressure (a) and shock velocity (b) vs. distance from charge.

At a distance of about 19 cm from the charge, the velocity of the blast wave in the foam has decayed to about 110 m/s as opposed to 1100 m/s for an air blast at the same location, indicating a severe loss in kinetic energy (Fig. 3b).

Pressure records confirm the attenuation process indicated by the measurement of the shock trajectory (Fig. 3a): while the side-on overpressure initially corresponds to the values obtained for air blasts, it drops significantly once the wave has traveled a certain distance (in this case 4 – 5 cm as already observed for the shock trajectories in Fig. 2). The pressure jump rapidly decays to values of less than 100 mbar. The pressure readings indicate, however, that the decaying initial shock is then followed by an extended compression wave. This wave may eventually reach an amplitude that exceeds by more than an order of magnitude the pressure jump caused by the incident blast wave (Fig. 4).

The formation of the compression wave has also been observed in a number of previous studies (e.g. Henderson et al. 1989, Skews 1991).

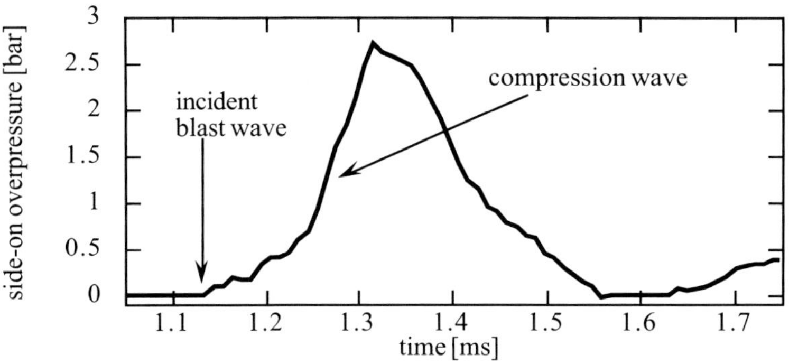

Figure 4. Pressure record obtained in a foam block ($\rho_{foam} = 29$ kg/m^3) at 12.5 cm from the charge.

For the time being, no model exists for the observed rapid onset of shock attenuation within the foam. The pseudo-gas model suggested by Gel'fand (Gel'fand et al. 1983, Faresov 1988) cannot be expected to describe these processes since it does not take into account the internal structure of the foam. In spite of its apparent limitations and thermodynamic inconsistencies that come along with its simplicity, it has been shown to give results that agree fairly well with some experimental data (e.g. the pressure behind the wave reflected from the gas/foam interface, Skews 1991).

In a self-similar solution for the propagation of a spherical blast wave, for which the strong shock limit is taken, the application of the pseudo-gas model yields that for a blast wave in foam the decay law has the same form as for an air blast, i.e.

$$R_S \sim t^{2/5}, \tag{1}$$

where R_S is the radius of the wave. This result seems to be confirmed by our experiments since the decay curves for blast waves in air and foam coincide in the vicinity of the charge where the strong shock limit may be applicable. This agreement, however, is fortuitous and can solely be attributed to the fact that in the pseudo-gas model the caloric equation for the foam medium is identical to the one for a perfect gas, with the exception of a constant factor α, the porosity ($\alpha = 1$ corresponds to the case of a perfect gas).

It is, on the other hand, not surprising that the trajectory for a blast wave in foam is almost identical to the one in air: the small amount of solid material (here 5%) may be considered to have an almost negligible influence on the propagation of a strong shock. At a later stage, however, once the wave has decayed, the internal structure of the foam begins to play a significant role for the further propagation of the wave. A model describing these processes has yet to be developed.

Acknowledgement. This work was started in 1993 by one of us (JHS Lee) while on sabbatical leave at the Shock Wave Research Center of Tohoku University, Sendai. The support of Professor K. Takayama is gratefully acknowledged. The authors would also like to express their thanks to I. Abdul-Baki, L. Alakkuttussery, and J. F. Viau for their assistance in preparing the manuscript.

References

Baker WE (1973) Explosions in Air, University of Texas Press, Austin, London, pp 150–163.

Faresov YM (1988) Interaction of shock waves with a porous compressible medium. In: Unsteady Interaction of Shock and Detonation Waves in Gases, Korobeinikov VP, Urtiew PA (eds), Hemisphere Publishing Corp, New York, Washington, Philadelphia, London, pp 48–62.

Gel'fand BE, Gubanov AV, Timofeyev EI (1983) Interaction of shock waves in air with a porous barrier. Izv Akad Nauk SSSR Mekh Zhidk Gaza 4:79–84.

Henderson LF, Virgona RJ, Di J, Gvozdeva LG (1989) Refraction of a normal shock wave from nitrogen into poly-urethane foam. In: Proc 17th Int Symp on Shock Waves and Shock Tubes, Kim YW (ed), AIP Conference Proc 208, pp 814–818.

Skews BW (1991) The reflected pressure field in the interaction of weak shock waves with a compressible foam. Shock Waves 1:205–211.

Skews BW (1995) Shock wave impact on porous media. In: Shock Waves @ Marseille III, Brun R, Dumitrescu LZ (eds), Springer, Heidelberg, Berlin, New York, pp 11–20.

van Dongen MEH, Smeulders DMJ, Kitamura T, Takayama K (1995) On the modelling of wave phenomena in permeable foam. In: Shock Waves @ Marseille III, Brun R, Dumitrescu LZ (eds), Springer, Heidelberg, Berlin, New York, pp 163–168.

Experimental and numerical investigation of the head-on collision of planar shock waves with rigid porous materials

A. Levy, S. Sorek*, G. Ben-Dor
Pearlstone Center for Aeronautical Engineering Studies
Department of Mechanical Engineering
Ben-Gurion University of the Negev, Beer Sheva, Israel
B. W. Skews
School of Mechanical Engineering
University of Witwatersrand, Johannesburg, South Africa

Abstract: The head-on collision of planar shock waves with rigid porous materials was investigated experimentally in a 75-mm × 75-mm shock tube. The rigid porous samples were made of silicon carbide (10 and 20 ppi) and alumina (30 and 40 ppi). The length of the samples was in the range 40–100 mm. The shock-wave Mach-number range was 1.2–1.7. The pressure histories at various locations along the side wall and at the end wall were recorded in each experiment. The problem was also investigated numerically. A TVD-based computer code capable of solving the governing equations was developed. The agreement between the experimental and numerical results was very good. Based on this agreement a parametric investigation using the computer code was carried out.

Key words: Porous materials, Two-phase analysis, Compaction waves, TVD

1. Introduction

If a detailed analysis of the flow field which develops inside a porous medium which is struck head-on by a shock wave is required, then the process should be analyzed using the multi-phase approach. In this approach, the porous medium is considered as multi-phase in which the various phases interact with each other. Unlike the mixing theory approach of Mazor et al. (1994) which leads to a single-phase model, a two-phase modeling approach was carried out. The governing equations of the considered problem were developed. The macroscopic-mass, linear-momentum, and energy-balance equations for a two-phase saturated porous medium, as obtained by the dimensional analysis which Levy et al. (1995a) conducted on the macroscopic-balance equations of Bear and Sorek (1990) and Sorek et al. (1992) were adopted in the present study.

*Also affiliated with the Water Resource Research Center, J. Blaustein Desert Research Institute, Sede Boker Campus, Israel.

The modeling was based on conceptualizing the porous medium as a continuum composed of interacting compressible solid and multiple fluid phases. Macroscopic physical laws expressing mass, momentum, and energy balances for the fluid and the solid matrix were formulated on the basis of Representative Elementary Volume (REV) concepts, presented by Bear and Bachmat (1990). The solid phase was assumed to be rigid and thermoelastic and the fluid phase was considered as a Newtonian air. A detailed derivation of the three-dimensional macroscopic governing equations describing the flow field in the porous media was presented by Levy et al. (1995a). Since the obtained set of equations was too complex to solve analytically, it was decided to solve numerically the one-dimensional set as was developed and presented by Levy et al. (1995b).

2. Present study

In the present study, the one-dimensional version of the governing equations was solved. The set of governing equations consisted of six partial differential equations and six unknowns. Due to the complexity of the equations, an analytical solution of the set was impossible. Instead, a numerical solution was performed. Details of the numerical method are given in the following section.

3. The numerical method

An upwind TVD shock-capturing scheme, originally developed by Harten (1983), was extended to solve the problem of two-phase flow which described wave propagation and interaction in saturated porous media. Details of the scheme for solving the governing equations can be found in Levy et al. (1995b).

The boundary condition on the shock-tube end wall was simulated by using the image-point method. The computational grid system was composed of 500 nodes. The computations were performed with a DecStation 5000/260 and an Indy R4400/150. The performance of the presently developed TVD scheme was checked by simulating the well-known 1D shock-tube problem. Figure 1 illustrates the analytical solution (solid lines) and the numerical predictions (open circles) for the fluid's density, velocity, and pressure distributions. The comparison between the analytical solutions and the numerical predictions of the present simulation clearly indicates that the developed numerical code reproduced the 1D shock-tube problem excellently.

4. Experimental study

The head-on collision of planar shock waves with rigid porous materials was investigated experimentally in order to validate the predictions of the physical model and the numerical code. The experiments were conducted in the 75-mm

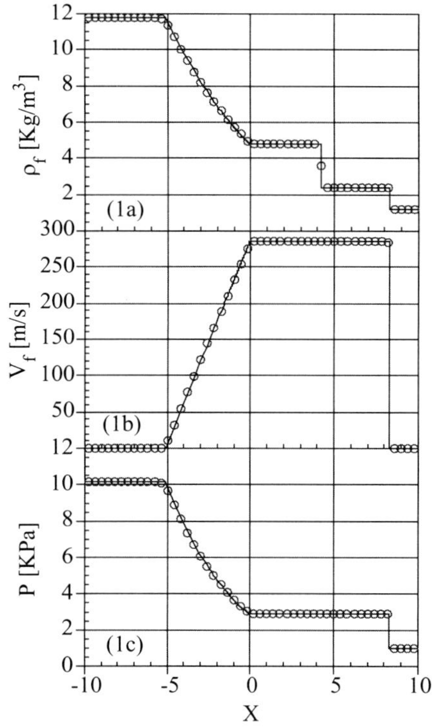

Figure 1. Comparison between the analytical solutions (solid lines) and the numerical predictions (open circles) of the classical 1D shock-tube problem: (a) density, (b) velocity, and (c) pressure.

× 75-mm shock tube of the School of Mechanical Engineering of the University of Witwatersrand in Johannesburg, South Africa. The incident shock-wave Mach-number range was $1.2 \leq M_i \leq 1.7$, and the initial pressures and temperatures throughout the study were about 830 mbar and about 288 K.

The rigid porous samples were made of silicon carbide (SiC) and alumina (Al_2O_3). The SiC manufactured porous material had either 10 or 20 pores per inch and the Al_2O_3 manufactured porous material had either 30 or 40 pores per inch. Twelve experiments were conducted with each sample. The samples were mounted at the end of the driven section of the shock tube in such a way that their back edges were supported by the shock-tube end wall (for more details see Levy et al. 1993).

5. Comparison between the numerical predictions and the experimental results and discussion

In order to validate the physical model and the numerical code, the governing equations were solved numerically for different samples and a variety of initial conditions and then compared to experimental results. In order to solve the governing equations and compare the solution to the experimental results, the various parameters which appear in the physical model were estimated for each sample (for more details see Levy et al. 1995a).

Figures 2a, 2b, and 2c represent typical experimental results and their numerical simulations. The pressure history of the pure gas just ahead of the front edge of the porous material is shown in Fig. 2a. The pressure histories of the gas occupying the pores of the porous material along the shock-tube side wall and at its end wall are shown in Figs. 2b and 2c, respectively. P_1 is the pressure ahead of the incident and the transmitted shock waves, P_2 is the theoretical pressure reached behind the incident shock wave, and P_5 is the theoretical pressure which would have been reached had the incident shock wave collided head-on with a solid end wall. The symbols represent the experimental results and the solid lines are the numerically predicted values. In general, the agreement between the experimental and the numerical results is seen to be very good. Although Figs. 2a to 2c describe the results of only one experiment, similar agreements were obtained in the comparisons with all the experiments which were conducted in the course of the present study. Details of all these comparisons are given in Levy (1995).

6. Conclusions

The general macroscopic-balance equations for a saturated rigid porous medium were developed. Their one-dimensional version was solved numerically using a TVD-based numerical code which was developed during the course of this study. The numerical predictions were compared to experimental results and good to excellent agreement was evident for different porous materials and for a wide range of initial conditions. To the best of the authors' knowledge, this is the first time that compaction waves are successfully simulated inside a porous medium.

Dedication This study which is a part of Dr. Avi Levy's Ph.D. research, is dedicated to Professor Ami Harten. It was decided in the early stages of this study to develop a TVD-based computer code for simulating the phenomenon. For this reason Professor Ami Harten was asked to act as an external examiner. Ami was quite skeptical about the possibility of developing the code. However, as reported in this study a TVD-based code was finally developed and the agreement between its numerical predictions and the experimental results was

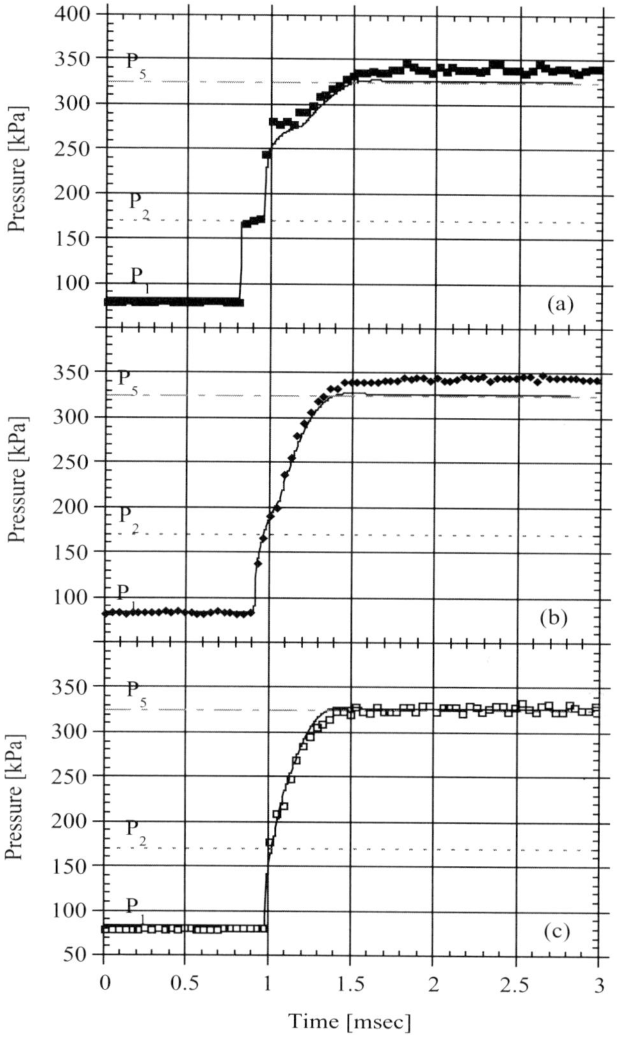

Figure 2. Comparison between the numerical predictions (solid lines) and the experimental results (squares and diamonds) of the pressure histories of the gas at various locations: (a) just ahead of the front edge of the porous material, (b) along the side wall, and (c) at the end wall.

found to be very good to excellent. Unfortunately, Ami did not live to see the results.

References

Bear J, Bachmat Y (1990) Introduction to Modeling of Transport Phenomena in Porous Media, Kluwer Academic Publishers, Dordrecht, The Netherlands.

Bear J, Sorek S (1990) Evolution of governing mass and momentum balance equations following an abrupt pressure impact in a porous medium. Transport in Porous Media 5:169–185.

Harten A (1983) High resolution schemes for hyperbolic conservation laws. J Comput Phys 49:357–393.

Levy A (1995) Wave Propagation in a Saturated Porous Media, Ph.D. Thesis, Department of Mechanical Engineering, Ben-Gurion University of the Negev, Beer Sheva, Israel (in Hebrew).

Levy A, Ben-Dor G, Skews BW, Sorek S (1993) Head-on collision of normal shock waves with rigid porous materials. Experiments in Fluids 15:183–190.

Levy A, Sorek S, Ben-Dor G, Bear J (1995a) Evolution of the balance equations in saturated thermoelastic porous media following abrupt simultaneous changes in pressure and temperature. Transport in Porous Media (to be published).

Levy A, Ben-Dor G, Sorek S (1995b) Numerical investigation of the propagation of shock waves in rigid porous materials: part I — development of the computer code and comparison with experimental results. (Submitted for publication).

Mazor G, Ben-Dor G, Igra O, Sorek S (1994) Shock wave interaction with cellular materials. part I: analytical investigation and governing equations. Shock Waves 3:159–165.

Sorek S, Bear J, Ben-Dor G, Mazor, G (1992) Shock waves in saturated thermoelastic porous media. Transport in Porous Media 9:3–13.

Shock-induced wave phenomena in open-cell permeable and flexible foams

M. E. H. van Dongen, G. V. R. Smeets, H. W. M. Hoeijmakers
Eindhoven University of Technology
PO Box 513, 5600 MB, Eindhoven, The Netherlands
D. M. J. Smeulders
Delft University of Technology
PO Box 5028, 2600 GA Delft, The Netherlands

Abstract: The stress-strain relation and permeability values for the foam Recticel BPS-60 are specified. It is shown that permeability strongly depends on foam compression. A method is presented to measure both gas pressure and effective foam stress at the end wall of a shock tube. Foam collapse and densification induced by shock-wave reflection are shown. Finally, a numerical model is proposed considering foam and adjacent gas as one inhomogeneous medium.

Key words: Flexible foam, Permeability, Effective stress, Experiment, Shock tube, Numerical model

1. Introduction

Wave propagation in permeable and flexible foams is very interesting from a gasdynamical point of view because of its explicitly non-linear two-phase character. A survey of previous research on wave propagation through foams is given by Skews (1995). Important experimental, theoretical, and numerical contributions were given by, among others, Monti (1970), Gel'fand et al. (1983), Bashenova and Gvozdeva (1988), Skews et al. (1993), Baer (1992), and van Dongen et al. (1995a). In this paper several aspects of wave propagation in permeable foams are discussed: (i) the strong dependence of permeability on compression ratio, (ii) the possibility of measuring the effective stress and gas pressure independently during wave experiments, and (iii) a proposal for a numerical scheme in which foam and adjacent gas are considered as one inhomogeneous medium.

2. Governing equations

To describe the propagation of shock waves through a porous foam, the conservation laws of mass and momentum are used. In the one-dimensional situation (van Dongen et al. 1995a):

$$\frac{\partial \mathbf{U}}{\partial t} + \frac{\partial \mathbf{F}}{\partial x} = \mathbf{Q}, \tag{1}$$

where

$$\mathbf{U}^T = [\phi_f \rho_f, \phi_p, \phi_f \rho_f v_f, \phi_p v_p], \tag{2}$$

$$\mathbf{F}^T = [\phi_f \rho_f v_f, \phi_p v_p, \phi_f \rho_f v_f^2 + \phi_f p_f, \phi_p v_p^2 + \phi_p p_p / \rho_p], \tag{3}$$

$$\mathbf{Q}^T = [0, 0, m_f, m_p / \rho_p], \tag{4}$$

and where the index p stands for *p*orous material (i.e., the solid phase) and f for *f*luid (i.e., the gas phase). In the equations, ϕ_f stands for the volume fraction of the gas, ρ_f for the density, p_f for the pressure, and v_f for the velocity. The same terms are defined for the porous material. The term m_f $(= -m_p)$ represents the interaction force between the phases:

$$m_f = p_f \frac{\partial \phi_f}{\partial x} + \frac{\phi_f \eta}{k}(v_p - v_f) + \phi_f^2 b_f \rho_f (v_p - v_f)|v_p - v_f|. \tag{5}$$

In these equations, η is the viscosity of the gas, k the permeability, and b_f the second Forchheimer coefficient. Both coefficients depend on the foam deformation. Due to a large contact surface between the two phases and a very large heat capacity of the porous material it is assumed that the propagation of waves in the foam is an isothermal process. The compressive stress in the solid phase, p_p, consists of two components:

$$p_p = p_f + \sigma / \phi_p, \tag{6}$$

where σ represents the effective stress. This is the force per unit (total) area adding up to the pore pressure p_f. For an incompressible solid, σ is uniquely related to the porosity ϕ_f. Closure of the model is obtained by the ideal gas equation.

3. Physical properties of foam

When a foam is subjected to a mechanical load, it will display a very peculiar mechanical behavior. In Fig. 1a, this effective stress, σ, is depicted versus the strain, ε, for foam Recticel BPS-60, which is used both in the experiments and in the numerical simulations. The strain, which indicates the relative change in length, is related to the porosity by $\varepsilon = \frac{\phi_{f0} - \phi_f}{1 - \phi_f}$. Here, ϕ_{f0} is the porosity when the foam is uncompressed. Fig. 1a gives both the measurements with the foam constrained in a tube as well as with the foam unconstrained. Three regimes are visible: an elastic regime, a collapse regime, and a densification regime. In the elastic regime, a linear relation exists between the stress and the strain, characterized by the elasticity modulus E if unconstrained and K_p if constrained.

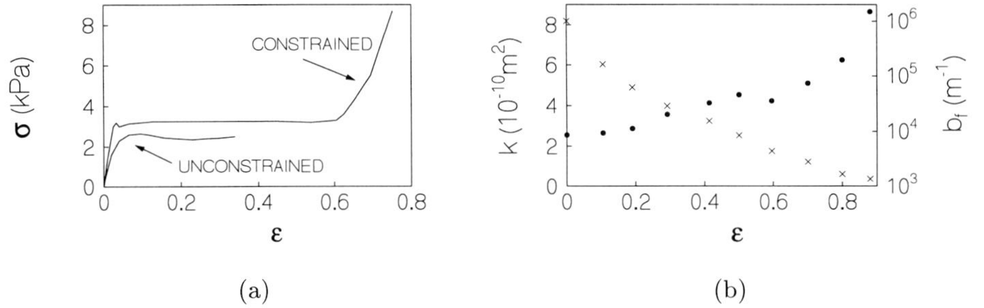

Figure 1. (a) Stress-strain measurements of foam BPS-60; (b) permeability parameters k ($\times$) and b_f ($\bullet$) as a function of the compressive strain.

This is followed by a collapse regime in which the stress remains constant at a value σ^*. Then the stress increases rapidly upon further compression. A second important physical property of interest is the permeability, characterized by the parameters k and b_f. Both have been experimentally determined for a wide range of ε values. The results are given in Fig. 1b. Permeability is very much affected by the rate of compression. The parameter b_f varies two decades for ε varying between 0.3 and unity. The physical properties are given in Table 1. In

Table 1. The physical properties of foam BPS-60.

ϕ_{f0}	0.983±0.002
K_p (10^5 Pa)	2.12±0.23
E (10^5 Pa)	1.02±0.16
σ_c^* (kPa)	3.53±0.10
σ_u^* (kPa)	2.83±0.15
k_0 (10^{-10} m^2)	8.19±0.08
b_{f0} (10^3 m^{-1})	8.66±0.08
ρ_p (kgm^{-3})	1682±10

this table, k_0 and b_{f0} are the permeability and second Forchheimer coefficient in the uncompressed situation. Furthermore, σ_c^* is the constrained value of the collapse stress and σ_u^* the corresponding unconstrained value.

4. Experiments

Experiments were performed in a shock tube whose end wall is depicted in Fig. 2. In this figure, the positions are given of the pressure transducers and the initial position of the foam material. Transducer T1 is mounted 21.5 cm in front of the end wall, so it measures only the gas pressure. Transducers T2 and T3 are mounted in the end wall of the tube. T2 is in contact with the foam material and T3 measures only the gas pressure (van Dongen et al. 1995b). This implies that the difference of T2 and T3 is the effective stress σ. Figure 3 presents for

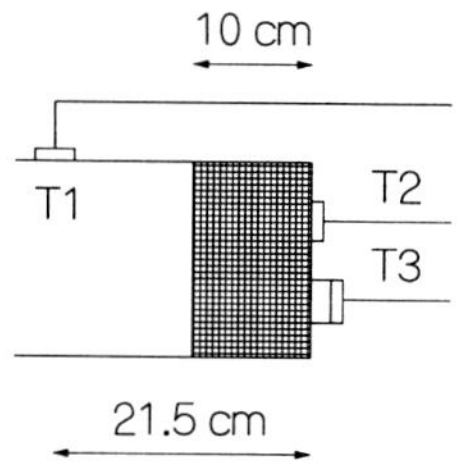

Figure 2. Part of the test section with the different pressure transducers and the initial position of the foam material.

two different shock Mach numbers the time histories of the pressures at the end wall for transducers T2 and T3 and the difference of the signals, which is the effective stress σ. With increasing Mach number the pressure peak increases. After this peak, the pressure relaxes towards its value as if no foam were present. Also, It is clearly visible that the wave travels in a dispersive way. For a shock Mach number of 1.15, the stress σ shows a fast increase upon the arrival of the wave front and is then fixed at its dynamic collapse value $\sigma^* = 9.5 \pm 0.5$ kPa. Apparently the value of the collapse stress is a function of the strain rate of the foam and is therefore different from its static value of 3.25 ± 0.10 kPa (see Table 1). After about 2.5 ms, σ starts to decrease again: the foam expands. When the Mach number is increased, we notice that the first jump towards the collapse stress is followed by a further increase. Comparison with the static effective-stress measurements shows that the foam must have entered its densification regime.

5. Numerical model

The conservation laws (1) are solved for the foam and the gas in front of the foam as if one material were present with varying properties in space (e.g., $\phi_p = 0$ for gas and $\phi_p > 0$ for foam). Since the energy equation is not accounted for in the present isothermal model, this affects the relation between the density and the pressure in the gas.

The method used to solve (1) is the operator-splitting technique. The homogeneous part is solved using a finite-volume method employing a second-order Runge-Kutta time integration. The inhomogeneous part is solved using a Newton method. An example of the results of the method is depicted in Fig. 4. Here

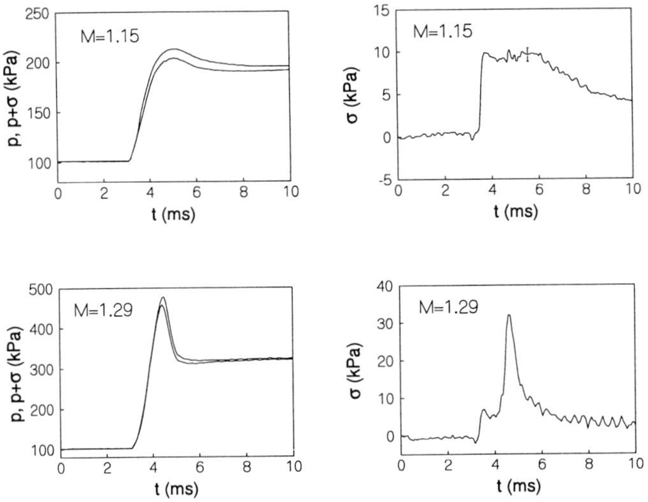

Figure 3. The pressure measured by T3 and the sum of pressure and effective stress measured by T2 for two different strengths of the incident shock wave at the end wall (left). The difference of the two signals, σ, is depicted on the right. Test gas: air. Initial pressure: 1.0 bar.

a contour plot of the volume fraction of the solid ϕ_p is shown. For the given shock Mach number, ϕ_p reaches a value at the end wall of 0.094, corresponding to a value of the strain ε of 0.82. From Fig. 1a it becomes clear that the foam has entered its densification regime indeed (as is observed in the experiments, see Fig. 3). The contour plot also shows a large foam compression, approximately a factor 5, in agreement with experimental results (Gvozdeva 1995).

6. Conclusions

The physical properties of foam BPS-60 were measured. We have shown that there is a clear dependence of the permeability parameters on the foam compression ratio. The experimental shock-tube results show that foam Recticel BPS-60 already enters its densification regime at a shock Mach number of 1.29. Also, the value of the collapse stress apparently depends on the rate of foam compression. The numerical model presented describes quantitatively the interaction of a shock wave with a porous foam. Since in the model gas and foam are considered as one inhomogeneous medium, wave reflections automatically emerge. This model needs to be developed further, incorporating the energy equation and the dynamic value of the effective stress.

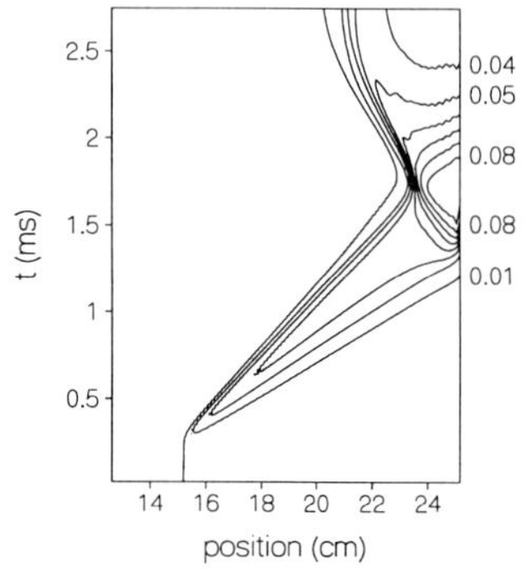

Figure 4. A contour plot of the volume fraction of the solid ϕ_p. The shock Mach number was 1.33. Test gas: isothermal. Initial pressure: 1.0 bar.

References

Baer MR (1992) A numerical study of shock wave reflections on low density foam. Int J Shock Waves 2:121.

Bazhenova TV and Gvozdeva LG (1988) Unsteady interactions of shock waves. In: Shock Tubes and Waves, Grönig H (ed), VCH Publ, Weinheim/New York, pp 41–49.

Gel'fand BE, Gubanov AV, Timofeev EI (1983) Interaction of shock waves in air with a porous screen. Izv Akad Nauk SSSR, Mekh Zhidkostii Gaza 4:79.

Gvozdeva LG (1995) Private Communication.

Monti R (1970) Normal shock wave reflection on deformable solid walls. Meccanica 5:285.

Skews BW (1995) Shock wave impact on porous materials. In: Shock Waves @ Marseille III, Brun R, Dumitrescu LZ (eds), Springer Verlag, Berlin, pp 11–20.

Skews BW, Atkins MD, Seitz MW (1993) The impact of a shock wave on porous compressible foams. J Fluid Mech 253:245.

van Dongen MEH, Smeulders DMJ, Kitamura T, Takayama T (1995a) On wave phenomena in permeable foam. ACUSTICA 81:63.

van Dongen MEH, Smeulders DMJ, Kitamura T, Takayama K (1995b) On the modelling of wave phenomena in permeable foam. In: Shock Waves @ Marseille III, Brun R, Dumitrescu LZ (eds), Springer Verlag, Berlin, pp 163–168.

Equilibrium gas dynamics and shock relations for flexible foams

M. E. H. van Dongen, C. J. van Gool, J. P. Pruiksma
Eindhoven University of Technology
P.O. Box 513, 5600 MB Eindhoven, The Netherlands
D. M. J. Smeulders
Delft University of Technology
P.O. Box 5028, 2600 GA Delft, The Netherlands

Abstract: Rankine-Hugoniot relations are formulated for shock waves in flexible foams, taking into account the contribution of the solid material to the total stress. It is shown that depending on foam properties, wave splitting may occur, which is related to a decrease of the speed of sound when the foam collapses. Precursor waves, reported in literature, are explained.

Key words: Rankine-Hugoniot relations, Flexible foam, Equilibrium gas dynamics, Precursor shock, Wave splitting

1. Introduction

Wave propagation in a flexible and permeable foam is very much affected by the peculiar nonlinear elastic behavior of the solid polyurethane material (Skews et al. 1993). This paper deals with the "equilibrium" gas dynamics of wave propagation in foams and with "equilibrium" shock relations. Here, "equilibrium" means that there is no velocity slip between solid and gas and that there is also thermal equilibrium. For open-cell permeable foam, this holds for waves with large time constants and gradient lengths. The mechanical properties of the foam will be taken into account so that the treatment can be considered as an extension or modification of the pseudo-gas model by Gel'fand et al. (1983) and of the model by Rayevski et al. (1990). The effective stress σ is defined as the normal compressive stress acting in a cross-section of the foam material in excess of the pore pressure. In open-cell foams, the relation between the effective stress and the foam compressive strain $\varepsilon = (L_0 - L)/L_0$, with L_0 and L the initial and actual length of the foam, can easily be obtained by compressing the foam at constant pore pressure. An important and very useful assumption for the description of wave propagation in foams is that the foam material itself is incompressible so that the deformation of the foam is fully attributed to the bending and twisting of the slender polyurethane rods that constitute the structure of the material (Gibson and Ashby 1988). With constant solid density ρ_p the compressive strain is proportional to the solid volume fraction

ϕ_p: $\varepsilon = 1 - \phi_{p0}/\phi_p$, with the subscript 0 referring to the undisturbed stress-free state. It can be shown that in this case the effective stress σ is a unique function of ϕ_p, for a given ϕ_{p0}. A schematic example of the relation between σ and the compressive strain is shown in Fig. 1. In the figure, we can discern

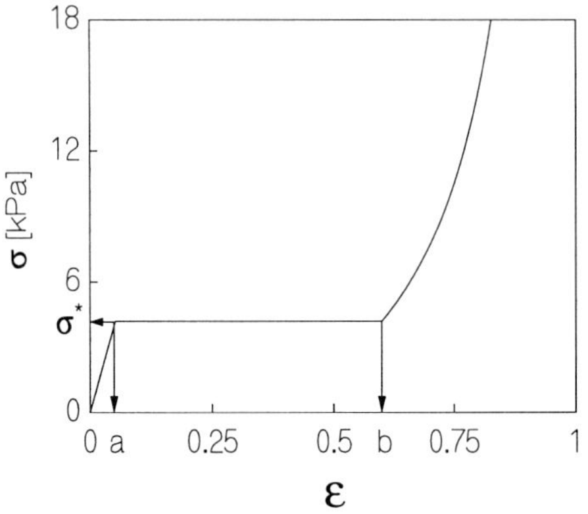

Figure 1. Schematic stress-strain relation for polyurethane foam.

three regimes: (i) for $\varepsilon < a$ the foam shows conventional elastic behavior, (ii) for $a \leq \varepsilon \leq b$ the foam collapses and the effective stress has a constant value σ^*, and (iii) for ε in between b and its maximum value, the foam is in its so-called densification regime where σ increases rapidly. As a simple model, we will take:

$$\sigma(\varepsilon) = \begin{cases} K_p\varepsilon & \text{for } 0 \leq \varepsilon < a \\ \sigma^* & \text{for } a \leq \varepsilon < b \\ \sigma^* \left(\frac{1}{1-\varepsilon} - \frac{b}{1-b}\right) & \text{for } b \leq \varepsilon < \phi_{f0}, \end{cases} \tag{1}$$

where K_p is the constrained modulus of the foam.

2. Equilibrium shock waves in a compressible foam

The pores of the foam are filled with a gas of density ρ_f and volume fraction $\phi_f = 1 - \phi_p$. We shall assume that the gas satisfies the ideal gas law $p = \rho_f RT$, with a constant specific heat c_{vf}, and that the solid has a constant heat capacity c_p. Since there is no relative motion between solid and gas, the relative mass fractions of the solid and the gas are not changed by passing a shock front. Defining the total density $\rho = \phi_p\rho_p + \phi_f\rho_f$ and the shock compression ratio $r = \hat{\rho}/\rho$, useful relations expressing mass conservation are:

$$\hat{\phi}_p = r\phi_p, \qquad \frac{\hat{\rho}_f}{\rho_f} = r\frac{1-\phi_p}{1-r\phi_p}, \qquad \frac{\hat{\rho}_f}{\hat{\rho}} = \frac{\rho_f}{\rho}\frac{1-\phi_p}{1-r\phi_p} \tag{2}$$

For the sake of definiteness we assume an arbitrary pre-shock state; the superscript ˆ refers to the post-shock state. The conservation laws of mass, momentum, and energy now take their usual form for a coordinate system fixed to the wave front:

$$\rho u = \hat{\rho}\hat{u} \tag{3}$$
$$p_t + \rho u^2 = \hat{p}_t + \hat{\rho}\hat{u}^2 \tag{4}$$
$$\frac{1}{2}u^2 + i = \frac{1}{2}\hat{u}^2 + \hat{\imath}, \tag{5}$$

with u the velocity of the foam with respect to the wave front, $p_t = p + \sigma$ the total stress in a cross-section, and i the enthalpy of the gas-solid system:

$$i = c_{vt}T + \frac{p_t}{\rho}, \tag{6}$$

with $\rho c_{vt} = \phi_f \rho_f c_{vf} + \phi_p \rho_p c_p$. The velocities u and $\hat{u}$ can be eliminated to obtain the Rankine-Hugoniot equation:

$$\hat{p}_t - p_t = 2\rho\frac{\hat{\rho}}{\rho + \hat{\rho}}(\hat{\imath} - i). \tag{7}$$

Combination of (6) and (7), and using the definitions for r and the auxiliary relations (2) yields a relation between the temperature jump and the compression ratio r:

$$\frac{\hat{T}}{T} = \frac{1 + \delta\frac{r-1}{r}\left(1 + \frac{\hat{\sigma}+\sigma}{p}\right)}{1 - \delta(r-1)\frac{1-\phi_p}{1-r\phi_p}}, \tag{8}$$

with $\delta = \frac{1}{2}\frac{\rho_f}{\rho}\frac{R}{c_{vt}}$. This parameter δ is of the order 10^{-2}, so that for moderate compression ratios the isothermal approximation is quite reasonable. Isothermal shocks will be discussed in the next section. From (8) other shock relations can be derived such as $\frac{\hat{p}}{p} = r\frac{\hat{T}}{T}$. The relations between u, $\hat{u}$, and r are found by combining (3), (4), and (8). After some rearrangements we obtain:

$$u = r\hat{u} = \left[\frac{r}{r-1}\frac{p}{\rho}\left\{r\frac{1-\phi_p}{1-r\phi_p}\frac{\hat{T}}{T} + \frac{\hat{\sigma}-\sigma}{p} - 1\right\}\right]^{1/2}. \tag{9}$$

To simplify the algebra we will proceed by considering the limiting case $\delta \to 0$, so that all changes are assumed isothermal.

3. Isothermal shocks

For the isothermal case, (9) can be simplified to:

$$u = r\hat{u} = \left[\frac{p}{\rho}\left\{\frac{r}{1-r\phi_p} + \frac{r}{r-1}\frac{\hat{\sigma}-\sigma}{p}\right\}\right]^{1/2}. \tag{10}$$

The equilibrium sound speed c follows from (10) by taking the limit $r \to 1$ and using the definition $\varepsilon = 1 - \phi_{p0}/\hat{\phi} = 1 - \phi_{p0}/r\phi_p$. We then obtain with L' Hôpital's rule from (10):

$$c = \lim_{r \to 1} u = \left[\frac{1}{\rho} \left(\frac{p}{\phi_f} + \frac{\phi_{p0}}{\phi_p} \frac{d\sigma}{d\varepsilon} \right) \right]^{1/2}. \tag{11}$$

The shock Mach number $M_0 = u/c$ follows from (10) and (11). In Fig. 2 we have plotted M_0 versus the compression ratio r for a foam with realistic properties: $\phi_{p0} = \phi_p = 0.044$, $\sigma = 0$, $\sigma^* = 4.2$ kPa, $(d\sigma/d\varepsilon)_0 = K_p = 50$ kPa, $b = 0.66$. It appears that compression shocks are possible in the elastic regime $r < r_c$ and also for sufficiently high compression ratios. In between there is a range of compression ratios for which the shock Mach number would be smaller than unity, so that a shock solution does not exist. In order to analyze the physical background of this behavior and to see what wave pattern to expect in the intermediate regime of compression ratios, we have to consider the equilibrium gas dynamics of isothermal compressible foams.

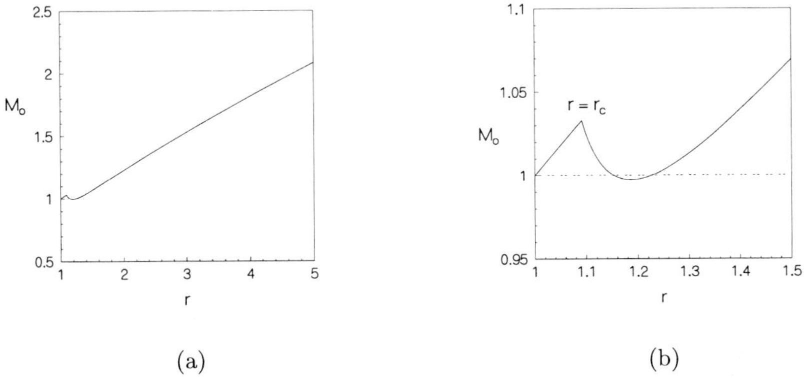

Figure 2. (a) Shock Mach number versus isothermal compression ratio. Isothermal compression ratios close to unity are plotted separately in (b); r_c is the critical compression ratio. A physically realistic solution demands $M \geq 1$.

4. Equilibrium gas dynamics of isothermal foams

The one-dimensional conservation laws for mass and momentum for isothermal foam can be written in their characteristic form:

$$\left(\frac{\partial}{\partial t} + [u \pm c] \frac{\partial}{\partial x} \right) (u \pm \Gamma) = 0, \tag{12}$$

where $c^2 = \frac{d(p+\sigma)}{d\rho}$ and the gasdynamic variable $\Gamma = \int \frac{c}{\rho} d\rho$. The sound speed c was already given in (11). In evaluating Γ and c as a function of ρ, relations similar to (2), expressing the invariance of mass fractions, should be used. An example of $c(\rho)$ for a foam that satisfies the stress-strain relation (1) is shown in Fig. 3a. For values of ρ/ρ_0 close to unity, the sound speed decreases with

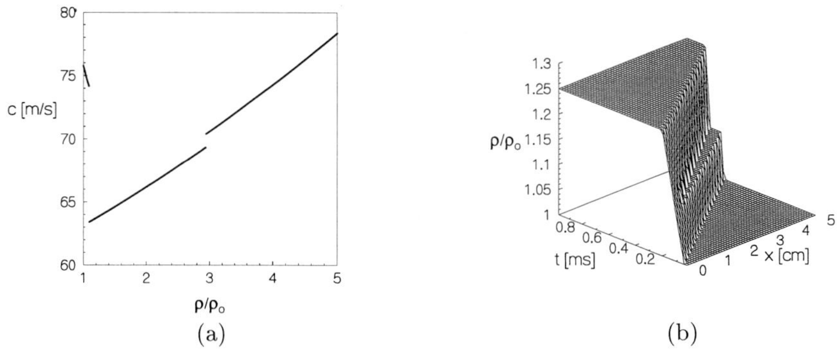

Figure 3. (a) Isothermal sound speed versus compression ratio; (b) the development in space of a ramp-shaped compression wave in time. Two waves are formed because of the negative jump in the sound speed.

increasing ρ because of the last term in (11). The solid volume fraction changes rapidly with ρ. When the foam collapses, $d\sigma/d\varepsilon$ vanishes and the speed of sound jumps to a lower value. From then on, c gradually increases and shows a small secondary jump when $d\sigma/d\varepsilon$ is no longer zero. The negative jump in c will cause wave splitting (Zel'dovich and Raizer 1967). This can be explained as follows, assuming that we start with an initial ramp-shaped disturbance in time as is indicated in Fig. 3b. Because we are dealing with a right-running simple wave, we know from the method of characteristics (Thompson 1984) that the state of the medium is constant along the characteristic path with constant speed $(u + c)$. Since u varies continuously and c has a negative jump at $\rho = \rho_c$, the wave will split up into two waves. From an analysis of $\frac{dc}{d\rho} + \frac{c}{\rho}$, it follows that both compression waves will steepen up, so that two different shock waves are generated: a "fast" weak shock wave over which σ jumps to σ^* and a secondary shock. This behavior can be observed in the space-time diagram of Fig. 3b. This also clarifies why there is no physical shock solution in the intermediate regime of compression ratios. This corresponds to the situation that the second shock wave is slower than the first one, so that no steady-state solution exists. When the compression ratio r increases, the velocity of the second shock increases, whereas the precursor shock is not influenced. When r is so large that the speed of the second shock is equal to or greater than the speed of the first one, a steady-state shock solution does exist.

5. Discussion and conclusions

We have found that the nonlinear mechanical properties of foam may cause wave splitting, depending on the foam parameters and on the strength of the wave. Such wave splitting does not always occur. The supersonic condition $u \geq c$, combined with $\hat{\sigma} = \sigma^*$ and $\sigma = 0$, leads to a quadratic equation for r, with a discriminant which is only positive in a restricted domain of parameters. A second observation is that if wave splitting occurs for shocks of moderate strength, the first wave becomes a precursor wave. Such precursor waves have been observed experimentally (Gvozdeva et al. 1995). If the total compression is much larger than that of the precursor shock, the major shock wave is the second one, starting from the level $\sigma = \sigma^*$. In this case the relevant equilibrium sound speed has to be evaluated at the collapse condition with $d\sigma/d\varepsilon = 0$, and $\rho = \rho_c$.

Acknowledgement. The authors are much indebted to Professor L. G. Gvozdeva for her stimulating comments.

References

Gel'fand BE, Gubin SA, Timofeev AI (1983) Interaction of shock waves in air with a porous screen. Izv Akad Nauk SSSR Mekh Zhidk Gaza 4:79.

Gibson LJ, Ashby M (1988) Cellular Solids — Structure and Properties, Pergamom Press.

Gvozdeva LG, Faresov YM, Sharov YL (1995) Structure of deformation waves in porous compressible materials under the influence of shock waves. In: Shock Waves @ Marseille, Brun R and Dumitrescu LZ (eds), Vol IV, p 175.

Rayevski DK, Gvozdeva LG, Faresov YM, Bressard J, Bailly P (1990) Reflection of shock and explosion waves from surfaces covered with layers of polyurethane foam. In: Proc 11th Int Coll Dynamics of Explosions and Reactive Systems, AIAA, p 171.

Skews BW, Atkins MD, Seitz MW (1993) The impact of a shock wave on porous compressible foams. J Fluid Mech 253:245.

Thompson PA (1984) Compressible-Fluid Dynamics, Rensselaer Polytechnic Institute.

Zel'dovich YB, Raizer YP (1967) Physics of Shock Waves and High-Temperature Hydrodynamic Phenomena, Academic Press.

Axisymmetric tests of shock-wave impact on PUR foam

B. W. Skews, R. M. Sellschop, T. Legoete
School of Mechanical Engineering
University of the Witwatersrand
PO WITS, 2050, South Africa

Abstract: A series of tests have been conducted on the shock-wave impact on conical polyurethane foam samples. Such tests are necessary to remove the wall-friction influences present with one-dimensional tests in a shock tube. Cones with different apex angles made of foams with different densities were used. It is found that pressure amplification is present on the mounting wall and that this is a function of radial position and foam type. It is inferred that the wave within the material is inclined to the axis. Substantial deformation of the cone is evident. The reflected wave at the surface is difficult to detect and appears to be an expansion wave.

Key words: Foam, Porous, Shock-wave reflection, Pressure amplification

1. Introduction

There has been considerable interest over the past decade in studies of the impact of shock waves on compressible open-cell foams. This has largely been due to the pressure amplification which occurs when a layer of compressible material is placed against a rigid wall, i.e., the peak pressure measured at a wall with foam present is greater than that without. The original investigations were done by Gelfand et al. (1975) and Gvozdeva et al. (1985). This work was continued by Skews et al. (1993) who showed that there were separate waves in the gas and the skeleton moving at different velocities, and that there were small gas flows, both in and out of the foam, during the compression process. All the quantitative tests to date have been conducted on a plug of foam that fills the cross-section of a shock tube. However, there is significant drag between the specimen and the walls of the tube which results in curvature of the foam face (Seitz and Skews 1991), and some pulling away of the material from the wall, resulting in edge jets (Skews et al. 1993). These extraneous effects result in the flow not being strictly one dimensional as is generally assumed in attempts to numerically predict the behavior. This paper describes preliminary studies to overcome these effects. Use is made of conical foam specimens with the base of the cone attached to the rear wall of the shock tube. The process is then two-dimensional axi-symmetric, rather than two-dimensional plane.

The principal features of the one-dimensional studies are that a compression wave passes through the specimen before detectable foam motion takes place, followed by the compaction front which is the start of foam particle motion. Based on this behavior, a simplified wave pattern in the material would be as indicated in Fig. 1. Only the leading edges of the waves are indicated and it is clear that the actual phenomena will be more complicated because of interactions on the axis of symmetry.

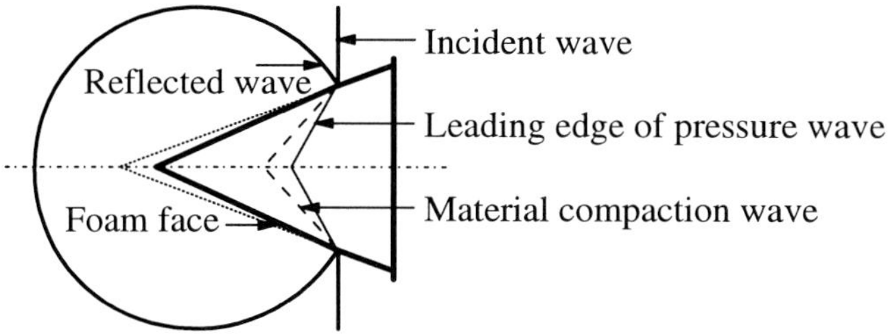

Figure 1. Schematic diagram of wave processes in a compressible foam cone.

2. Experiments

The shock tube should have a transverse dimension at least twice that of the diameter of the cone base in order to ensure that the flow remains axi-symmetrical for as long as reasonably possible. An inexpensive tube was manufactured from commercially available square tubing with internal dimensions of 260×260 mm, enabling models with 150-mm bases to be used. The tube was externally stiffened with ribs in order to reduce vibrations in the walls. A 50-mm thick back wall on flexible mounts was utilized to prevent tube vibrations from affecting the transducers used for measuring back-wall pressures. All tests reported were run at a nominal Mach number of 1.2 into atmospheric air (83 kPa). The conical foam samples were manufactured by gluing a polyurethane foam block to a circular hardboard disk which was then placed in a lathe. The foam was machined while frozen with a stream of liquid nitrogen. Most tests were conducted on three cones with base diameters of 150 mm and heights of 180 mm (apex angle of 45°), having bulk densities of 16, 28, and 32 kg/m^3. Water-displacement tests, done with care to remove all interstitial air, gave void fractions of 0.973, 0.960, and 0.960 respectively. Two additional cones with 60° and 90° apex angles and base diameters of 121 and 120 mm, respectively, made from the 16-kg/m^3 material were also tested. The cones were attached to the rear wall of the shock tube by small strips of double-sided tape. At the end of each test it was found that the cone had separated from the wall, presumably due to rebound in the recovery phase of the foam after compression.

3. Results and discussion

Back-wall pressures on the axis of symmetry are shown in Fig. 2 for the 16-kg/m^3 foam and different cone-apex angles and an incident-shock Mach number of 1.215. Pressures are non-dimensionalized with respect to the pressure that would be encountered on the back wall with no foam present. It is noted that the pressure trace is very similar in shape to that obtained in the one-dimensional tests (Skews 1991). Peak pressures are higher than without foam, and the amplification for the 45° cone is somewhat larger than for the physically shorter models. On the basis of the presumed behavior shown in Fig. 1, both the gas and the material compaction wave will reflect obliquely off the back wall. The waves are conical and the effect of the reflection angle and interaction between the gas motion and that of the cell walls is not immediately clear. Numerical techniques, coupled with pressure surveys, would be required to elucidate the internal dynamics. After the main pressure surge due to wave reflection there is subsequent oscillation of the pressure trace ascribed to wave oscillation within the body of the foam. The wave patterns in the present situation are clearly more complex than for one-dimensional tests. The period of oscillation increases as the specimen gets longer.

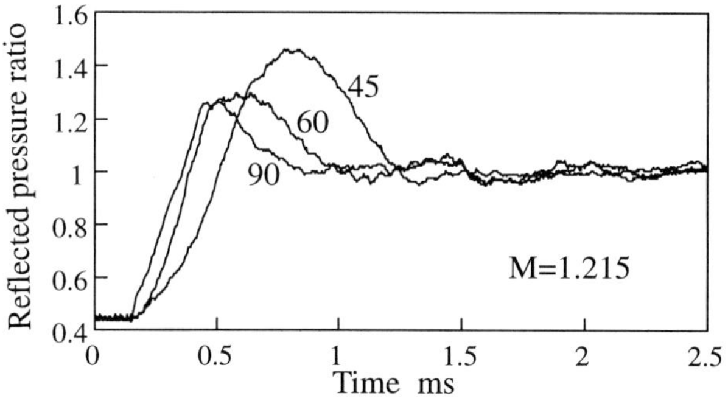

Figure 2. Centerline back-wall pressures for different apex angles.

Some idea of the behavior of the internal reflected wave can be obtained from the differences in the pressure traces taken at different radial positions behind the foam. Figure 3 shows the results for two of the 45°-apex-angle specimens. For the 16-kg/m^3 specimen the transducer just inside the outer rim of the specimen shows a trace essentially of that of the outside gas, i.e., a step function at the instant of reflection followed by nearly steady conditions thereafter. Further into the material the wave spreads out into a compression, as in the one-dimensional study. At the same time, the peak pressure increases. The slower velocity of the wave in the material is clear from the later wave-arrival time at the transducers closer to the cone axis. The results for the 28-kg/m^3 material are generally similar except that the amplifications are much larger. Even the edge transducer shows amplification and the multiple oscillation of the trace is more marked. As shown later this behavior correlates with the distortion. On the other hand, the more dense material (32 kg/m^3) shows behavior similar to that of the lightest specimen. The trend in peak pressure is given in Fig. 4. The 16- and 32-kg/m^3 materials give almost identical results, very different from that of the intermediate material. Attempts have been made to derive some information from the arrival-time data, but the accuracy of measurements is insufficient. The indications from the analysis that has been done, however, are that the transmitted wave in the material is curved concave forward. Also the time between the head of the wave and peak pressure condition increases towards the axis of the specimen as would be expected due to the spreading of the wave.

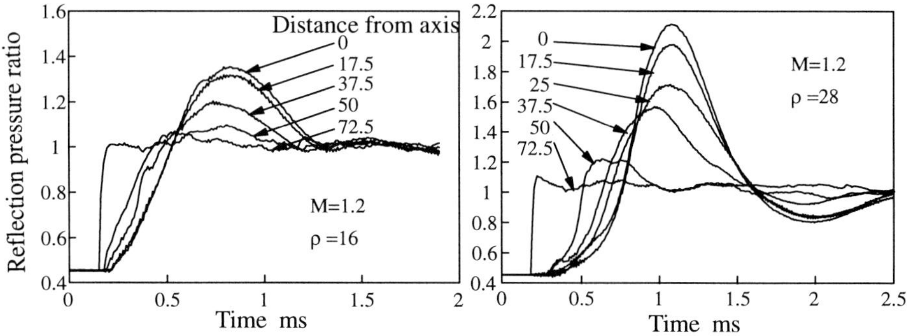

Figure 3. Radial pressure histories on the back wall for two foams.

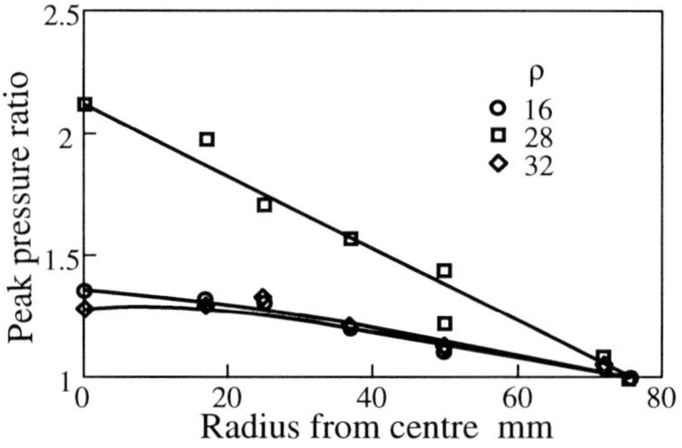

Figure 4. Radial variation of peak pressure.

Schlieren photography was undertaken on the three 45° cones. Considerable difficulty was encountered in trying to get an image of the reflected wave. It is clearly very weak and this is made worse for visualization by its being axisymmetric. The situation appears to be similar to that on a perforated wall in a transonic wind tunnel where wave cancellation occurs. A number of pictures taken using high schlieren sensitivity showed some shading behind the wave. The reflected wave appears to be an expansion wave. More sensitive methods, such as interferometry, will be needed to evaluate the wave properly. After reflection off the back wall of the shock tube the shock moves back down the cone with the shock taking on a typical diffraction profile. In some images there is a faint indication of the variable pressure within the body emerging into the external flow and disturbances at the foot of the shock, but these are too indistinct for meaningful conclusions to be drawn.

Quantitative data of the distortion of the foam was obtained from tracings of the schlieren photographs. General trends are established from this data. For more exact data, improved experimentation will be needed, particularly in the positioning of the specimen at exactly the same position every time and the use of more accurate devices to make measurements. The current results are shown in Fig. 5. Measurements of the diameter are taken at positions one-quarter and three-quarters of the cone height from the apex and are non-dimensionalized by the original diameter at the same position. Time is measured from the shock arrival at the apex. At the 75% position, distortions of about 20% are reached by the time the shock that is reflected off the back wall has passed the apex of the cone. At this stage the transverse collapse is still accelerating with the least-dense foam moving the fastest (at about 35 m/s). At the front measuring station more interesting effects occur. Firstly all three specimens show a contraction followed by a leveling off, and then in two cases, an expansion. The result of these motions is that the edge of the foam does not remain straight. The reason for this behavior is that the flow into the thinner section near the apex has ceased, so there is no longer a pressure gradient and the pressure in this section of the foam is uniform at the surrounding gas value. Further towards the back, equilibrium has not been achieved, both because of the thicker section and because of the later imposition of the pressure. Particularly noticeable from the data is that the collapse of the 28-kg/m^3 foam is much larger than the other two. This is probably the reason that much higher peak pressures are reached, as noted earlier. The foam property causing this higher velocity needs to be established. It does not appear to be related either to the density or the porosity, and thus skeleton elasticity and permeability considerations have to be taken into account.

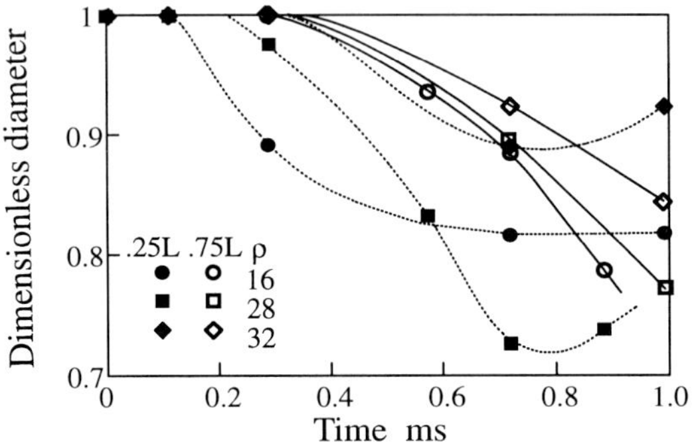

Figure 5. Diametral distortion of the specimens.

4. Conclusions

The general behavior of the collapse of conical foam bodies attached to a wall has shown that pressure amplification occurs associated with significant collapse of the foam material. Reflected waves at the surface appear to be expansion waves and the main wave within the material spreads out in time. Density and void-fraction effects are not sufficient to account for the results. Because of the complex nature of the coupling between the phases, the geometry that was tested also presents an excellent test case for computational modeling.

References

Gelfand BE, Gubin SA, Kogarko SM, Popov OE (1975) Investigation of the special characteristics of propagation and reflection of pressure waves in a porous medium. Fh Prikl Mekh Tekl Fiz 6:74–77.

Gvozdeva LG, Faresov YM, Fokeev VP (1985) Interaction of air shock waves with porous compressible materials. J Appl Mech & Tech Phys, 3:401–405.

Seitz MW, Skews BW (1991) Three-dimensional effects in the study of shock wave loading of compressible porous foams. In: Shock Waves, Proceedings of the 18th International Symposium on Shock Waves, Takayama K (ed), Springer Verlag.

Skews BW (1991) The reflected pressure field in the interaction of weak shock waves with a compressible foam. Shock Waves 1:205–211.

Skews BW, Atkins MD, Seitz MW (1993) The impact of a shock wave on porous compressible foams, J Fluid Mech 253:245–265.

Shock impact on porous plugs with a fixed gap between the plug and a wall

M. W. Seitz, B. W. Skews
School of Mechanical Engineering
University of the Witwatersrand
PO WITS, 2050, South Africa

Abstract: The pressure in a fixed-volume gap behind a porous compressible foam plug is measured following the interaction with a plane shock wave. The effect of the drag on the gas moving within the uncollapsed foam lattice is shown to have a significant effect on the magnitude of the peak pressure when a gap is present, with the density of the foam only being a second-order effect. For no back-wall gap, the rate of pressure rise behind the foam is also shown to be a function of the drag within the foam, while the magnitude of the rise is related to the density of the foam. It is proposed that the peak-pressure magnitudes attainable with increases in foam length are linked to the internal flow properties of foams. Pressure histories, schlieren photographs, and high-speed oblique photographs are used to study the phenomena.

Key words: Foam, Porous, Shock-wave reflection, Pressure amplification

1. Introduction

The physical behavior of a compressible porous foam plug interacting with a moving shock wave, and the resulting flow features in the foam and surrounding gas, have previously been extensively researched (see review by Skews 1993). Most characteristic of this interaction process is the large pressure amplification experienced behind the foam plug when the plug is placed against a rigid back wall. The magnitude of this pressure amplification is dependent on the shock-wave strength, the properties and geometry of the foam, and the three-dimensional boundary effects. To explain the phenomenon, various physical and numerical models have been developed that give close approximations to experimental observations (Skews et al. 1993, Baer 1992, Olim et al. 1993). Experimental comparisons are, however, often misleading due to the difficulty in extrapolating external measurements to the internal gas flow and solid-phase behavior. Such limitations have resulted in various novel experiments aimed at confirming assumptions on the internal processes occurring during the compression of the foam (Seitz and Skews 1991).

2. Experimental procedures

To experimentally characterize the flow of gas within a foam matrix during shock-wave interaction, a foam plug has been allowed to collapse normally resting against the back wall of a shock tube and then with a small fixed gap behind the full back face area of the foam. This gap enables any gas flowing from the back of the foam plug during the compression phase to be captured in a control volume behind the foam. By investigating the changes in pressures in this region and at the side walls adjacent to the foam plug, a qualitative indication of the flow resistance of uncollapsed foam matrix subjected to large pressure gradients can be established.

The shock-wave tests were conducted in a fully automated shock tube. The test section is 180 mm high by 76.2 mm wide, with a schlieren view field of 180×180 mm square. Along with schlieren photographs, oblique high-speed photographs were taken to examine the three-dimensional behavior of the foam plug during compression. The shock-wave Mach number selected for the tests was Mach 1.51, with the initial expansion chamber pressure of 0.83 bar. Of interest for this discussion is a PCB high-speed pressure transducer located in the center of the back wall of the test section (transducer 1), and a pressure transducer located 18 mm from the back wall on the side wall of the test section (transducer 2). The pressure histories from these pressure transducers were digitized and recorded using an ADAM 2-MHz analog-to-digital converter integrated into the automated shock-tube facility.

Four commercially available polyurethane foam types were selected, although only results pertaining to foams 1 and 4 will be discussed here. All the foams are permeable (foam type 1 being the most permeable, foam 4 the least), with a homogeneous dispersion of open and partially open foam cells, with no sealed cells (bubbles). Flow resistance tests of foams 1 and 4 give pressure-drop results over a 50-mm thick sample of material of 0.35 and 2.06 kPa for a superficial velocity of 1.61 m/s. The densities are 19.7 and 37.2 kg/m^3 respectively. Pore densities are approximately 530 pores/cm^2 for both foams although the lighter foam has a larger average pore size. The foam plugs used were 180×76 mm in cross-section, with uncompressed lengths of nominally 75 and 35 mm. The pressure behind the reflected shock wave is about 3.3 bar for the lighter foam and 3.7 for the heavier one, giving gas velocities in the direction of the foam face of about 100 and 75 m/s respectively. The fixed volume gap behind the foam plugs was created using honey-comb spacers placed between the back wall of the shock tube and the foam plug. Two honey-comb thicknesses of 6 mm and 12 mm were used, with a cell size of 5 mm. The extent to which the foam extruded into the honey comb during compression was established to be between 1 and 2 mm. To ensure that no mechanical forces were applied to the pressure transducer by the honey-comb material, the back-wall transducer was recessed 2 mm from the back surface of the honey comb.

3. Observations, results, and discussion

A piston analogy is the simplest physical model describing the origin of the high back-wall pressures experienced behind a foam during a shock-wave interaction process. The 'piston' in this model is the portion of the foam plug that collapses during the progressive compression of the foam (Fig. 1). This foam 'piston' is 'leaky' with an increasing mass as it is accelerated by the pressure difference into a fixed volume, being finally brought to rest by the pressure rise in the ever-reducing volume ahead of the piston. This compressed foam region (piston) has a significantly reduced volume when compared to uncompressed foam ($\sim 80\%$ less) as the gas in the pores is forced out ahead of the advancing collapsed foam front. The permeability of the fully collapsed foam is very low, thus it has been observed experimentally and numerically that little gas passes through the 'piston' during the compression phase.

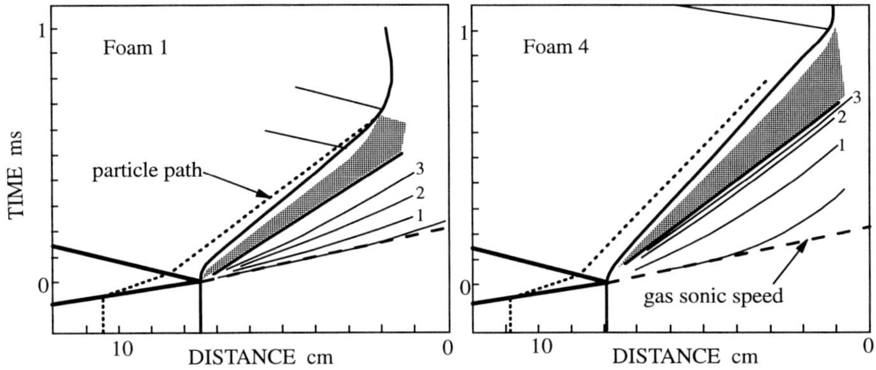

Figure 1. Schematic wave diagrams for the two foams.

The relative velocity of the compressed foam front into the uncompressed foam is illustrated in Fig. 1 for foams 1 and 4. This velocity is higher than the calculated gas velocity in the flow field external to the foam as the advancing compressed 'piston' engulfs the foam. This feature is evident in the figures by the increasing width of the 'piston'. As the front face of the foam moves, a layer of material relaxes back to its original geometry due to pressure equilibrium being achieved.

The physical gas movement ahead of the 'piston' is strongly influenced by the drag characteristics of the uncompressed open-pore structure. For high-drag foams with many partially closed cells and a small pore structure, the flow of gas is retarded, resulting in higher static pressures within the foam lattice. The pressure contours in Fig. 1 illustrate the compression fans within the foam, with the stagnation pressure gradient in the low-drag foam 1 being gradual, while the gradient developed in foam 4 is steeper close to the foam 'piston'. This is consistent with the observation of differing rates of pressure rise against the back wall, with the denser foams exhibiting the steeper pressure profiles (Fig. 2).

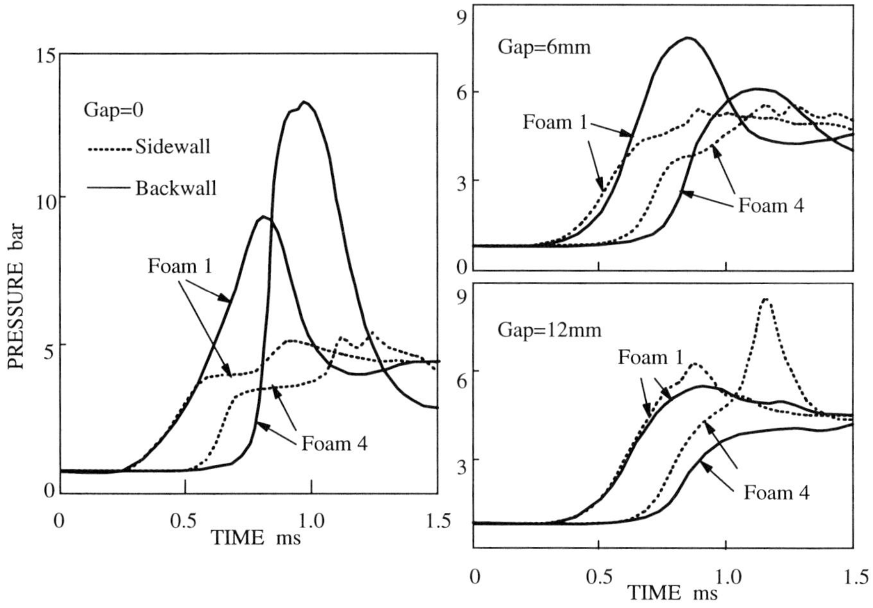

Figure 2. Pressure traces with no gap, a 6-mm gap, and a gap of 12 mm.

A further consequence of the flow retardation within the foam is the non-linear increase in the maximum peak pressure attainable with a change in foam-plug length (Gvozdeva and Faresov 1985). This non-linearity is now thought to be due to the reduction in 'piston' compaction as the gas within the collapsing pores is not being fully forced out due to the poor drainage of the gas already ahead of the foam 'piston', i.e., high static pressure within the foam lattice. This physical behavior is observed for foams 1 and 4 and further general tests conducted on 100-mm long specimens of these foams.

With the introduction of a small volume gap behind the foam plug, a theoretical change in peak back-wall pressure for a change in the gap volume may be calculated using simple adiabatic compression relationships. Experimentally, however, when the gap behind the foam is increased from 6 to 12 mm, the changes in peak pressures are not comparable to these values. The peak back-wall pressure for foam 4 is 13.8 bar (absolute pressure) for no gap, while for the 6- and 12-mm gaps the back-wall pressures are 6.1 and 4 bar respectively (Fig. 2). The pressure at the front face of the foam (behind the reflected shock wave) during the initial compression of the foam is 3.5 bar, with a rise to 4.6 bar as the foam is brought to rest fully collapsed (not a function of gap size). For the 12-mm back-wall gap it may thus be argued that there is a deviation from the theoretical values due to a net gas flow inward toward the back of the foam plug due to side-wall leakage. This leakage flow does not, however, account for the unexpected findings of the peak back-wall pressures experienced for foam 1 of 7.7 and 5.5 bar for gaps of 6 and 12 mm, respectively. These higher back-wall pressures would result in a net outflow of gas from the back-wall region into the front foam region if edge leakage occurs, which is not experimentally observed. From Fig. 2 therefore, a clear relationship between foam density and peak back-wall pressure appears to exist for no back-wall gap, but this trend is not observed when gaps are introduced behind the foam plug, with foam 1 exhibiting the highest peak pressures for foams with back-wall gaps.

As discussed previously, during the collapse of a foam plug the gas within the foam is compressed and driven forward into the low-pressure open pores of the uncollapsed foam. For the case where the foam is against the back wall, a very small volume of the compressed gas is finally expelled through the back face, with the majority remaining compressed within the pores of the foam. The back-wall pressure transducers thus closely follow the pressure within the foam lattice. When, however, a gap is introduced behind the foam plug, a larger volume of gas must be expelled through the back face of the foam to fill the control volume gap. As foam 1 has the most open structure, gas is easily and rapidly driven from the foam, allowing a degree of pressure equilibrium to be reached between the gas within the foam lattice and the gas in the control-volume gap. For foam 4 however, the high-pressure gas flow is restricted and equilibrium is not reached prior to the relaxation of the foam following peak compression. This phenomena thus accounts for the lower than expected pressures measured in the gap for foam 4 compared to foam 1. The back-wall pressure, with a fixed back-wall gap, is thus dependent on the permeability of the foam, rather than just the density of the foam material. The side-wall transducers adjacent to the fully compressed foam plug support this hypothesis as the side-wall pressures measured are significantly higher than the back-wall pressures with the gap introduced (Fig. 2).

4. Conclusions

The flow resistance of a high-pressure gas flowing from within a porous compressible medium can be established using a fixed gap behind a foam plug. This feature is important as it indicates the flow characteristics within a foam lattice subjected to transient high pressure gradients. These flow characteristics have a significant influence on the features of the high back-wall pressure.

References

Baer MR (1992) A numerical study of shock wave reflections on low density foam. Shock Waves 2:121–124.

Gvozdeva LG, Faresov YM (1985) Calculating the parameters of steady shock waves in porous compressible media. Zh Tekh Fiz 55:773–775.

Olim M, van Dongen MEH, Kitamura T, Takayama K (1993) On numerical simulation of propagation of shock waves in porous foams. In: Japan Symp Shock Waves, Sendai, pp 261–264.

Seitz MW, Skews BW (1991) Three-dimensional effects in the study of shock wave loading of porous compressible foams. In: Shock Waves, 18th International Symposium on Shock Waves, Takayama K (ed), Springer-Verlag.

Skews BW, Atkins MD, Seitz MW (1993) The impact of a shock wave on porous compressible foams. J Fluid Mech 253:245–265.

Skews BW (1993) Shock wave impact on porous materials. In: 19th International Symposium on Shock Waves, Universite de Provence, Marseille.

Attenuation of a blast wave with a cellular material

A. Makris*, D. L. Frost, J. Nerenberg, J. H. S. Lee
McGill University
Mechanical Engineering Department
817 Sherbrooke St. W., Montreal
Quebec, Canada H3A 2K6

Abstract: The attenuation of a blast wave with closed-cell polyethylene foams has been studied experimentally as a function of foam thickness and density for blast wave strengths ranging from M = 1.2 to 2.7. For weak blasts, the transmitted pressure beneath the foam is amplified whereas if the blast strength overcomes a threshold related to the stress-strain behavior of the material, the pressure may be attenuated by up to 70%. If the thickness of the foam is insufficient to absorb the blast energy, densification of the foam can lead to a large amplification of the reflected pressure compared to the reference pressure recorded with no foam present.

Key words: Blast wave, Porous medium, Shock-foam interaction

1. Introduction

The ability to attenuate the peak pressure and impulse of a blast wave in air has important practical applications such as in minimizing structural damage due to accidental explosions and in the design of protective apparel used in explosive disposal operations. In order to mitigate the effects associated with blast waves, multi-layered laminations which include rigid components, as well as cellular materials for blast energy absorption, have been considered. For a proper design, it is necessary to understand the interaction between a blast wave and a compressible cellular material. This interaction and the role of the physical and dynamic properties of the material on the characteristics of the transmitted wave are poorly understood at present.

A number of investigators since the early 1970s have studied the reflection and transmission of a shock wave through a foam layer in a shock tube (see the recent review by Skews 1995). A common interesting feature of the studies is that the pressure recorded on the wall behind the foam is higher than the reflected pressure recorded with no foam present. This effect has been attributed

*Present address: MedEng Systems, Inc., 2470 Don Reid Dr., Ottawa, Ontario, Canada K1H 8P5.

to the momentum imparted to the foam by the shock being transferred to the wall as the foam comes to rest. The majority of past work was performed in shock tubes using low-density open cell foams, where the gas can flow relative to the foam material itself. Mazor et al. (1994) demonstrated analytically that cellular materials, such as foams, cannot be used to reduce the pressure loading acting on an end-wall of a shock tube. This appears to be consistent with previous experimental work, however, it does not appear to be true for blast waves which have a triangular pressure profile with a finite duration. Rayevsky et al. (1989) found that it was possible to attenuate weak blast waves (with Mach numbers up to 1.2), whereas stronger blast waves were amplified. This was later confirmed by Lee et al. (1995), who reported that the amplification/attenuation of the blast pressure was found to be a function of the strength of the blast wave, the type of foam (open- or closed-cell) and its characteristic nonlinear stress-strain relation. The present paper describes a systematic experimental study of the effect of blast wave strength, foam density and foam thickness on the attenuation of a blast wave. A closed-cell polyethylene-based foam has been used in order to avoid the added complexity associated with the blast-induced flow through an open-cell foam skeleton.

2. Experimental details

The blast waves were generated by detonating a quantity of C4 plastic explosive suspended 1 m above a large aluminum witness plate. Square foam samples (23 cm long and 19 mm thickness per layer) were placed at the center of the plate with a pressure transducer flush mounted below the sample. A square steel frame was placed around the foam to minimize shock interactions at the foam edges. The Mach number of the blast wave just before it impacted the foam ranged from 1.2 to 2.7 for charge masses of 18 and 770 g, respectively. To determine the arrival time of the blast wave at the top surface of the foam, a contact probe consisting of two pieces of aluminum foil separated by a small gap and connected by a capacitive discharge circuit was used. Two different elastomeric closed-cell polyethylene foams were used with densities of 45 and 96 kg/m^3. The elastic response of the foams as characterized by the stress-strain curve, obtained under quasi-static conditions (0.2 mm/s), is shown in Fig. 1. The curves consist of an initial approximately linear elastic region, followed by a plateau, where elastic buckling of the cell walls takes place and most of the energy absorption occurs (at a stress of 0.2–0.3 MPa and 0.5 MPa for the low- and high-density foams, respectively). It is well known that the plateau stress level increases with strain rate as a result of the visco-elastic behavior of the material. The plateau region is followed by the densification regime, where the stress increases sharply with further strain as a result of the total collapse of the skeleton walls.

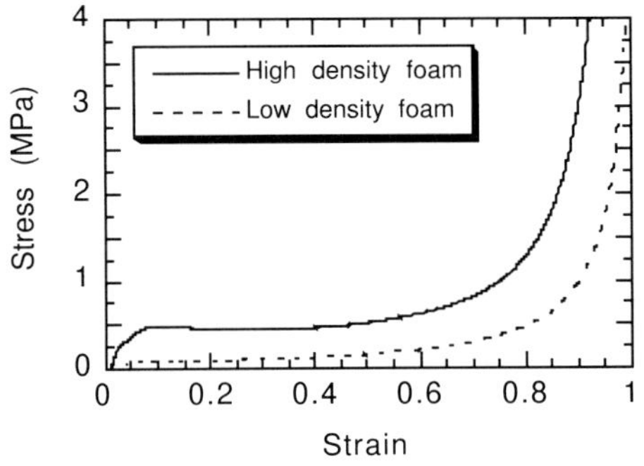

Figure 1. Quasi-static stress-strain curves for high and low density closed-cell polyethylene foams.

3. Results and discussion

For relatively weak blasts (25 g C-4), the peak pressure recorded under the low-density foam is amplified by a factor of 1.5 to 2, compared to the reference with no foam present, as shown in Fig. 2a. This is consistent with previous findings for shock propagation through foams placed at the end wall of a shock tube. As the thickness of the foam increases, the pressure amplification is not changed, but the duration (and impulse) of the transmitted wave increases, similar to what was observed by Gvozdeva et al. (1985). It appears that the increased total momentum acquired by the foam due to its larger mass affects the duration of the pulse. Fig. 2a also shows that the wavefront steepens with propagation distance in the foam. According to Mazor et al. (1994), if the first and second derivatives of the stress with respect to the strain are positive, as in the elastic buckling region, the transmitted compression waves can transform into a shock provided the foam is thick enough. However, when the foam deformation is within the linear elastic domain, it exhibits a retrograde behavior and transmitted compression waves spread out instead of steepening and transforming into a shock. This is indicated by the slow initial rise of the wave in Fig. 2 for portions of the front where pressure levels are clearly within the linear elastic region (i.e., about 0.1 MPa) and the steepening of the wave profile for higher pressures. The velocity of the shock wave in the foam varied from 350 to 460 m/s, in comparison with the foam sound speed of about 100 m/s. By using multiple foam layers, it was found that the shock velocity in the foam decreases as it propagates in the foam, consistent with what was observed by Ben-Dor et al. (1994).

The energy absorption capacity of a foam is determined by the area under the stress-strain curve. If the foam length is insufficient to attenuate the wave impulse, the cellular material will collapse. This is clearly depicted in Fig. 2b for a relatively strong blast (170 g C4; M $\approx$ 1.7) in the low-density foam. For a single layer of foam the initial pressure wave amplitude is dramatically reduced, i.e., by 60-70%, until the foam densifies and the pressure rapidly amplifies. By increasing the foam thickness, densification of the foam can be avoided and attenuation of the blast wave can be achieved, although the duration of the transmitted pressure pulse increases. When 5 layers of the low density foam are used, the pressure wave is clearly maintained below the threshold value for foam densification. This finding is in sharp contrast to what has been observed in shock tube studies involving cellular materials, where the presence of the foam serves to amplify the pressure for rigid wall reflection. Figure 2b also indicates that when the blast energy is dissipated, the initial rate of wave pressure rise is reduced with propagation distance in the foam.

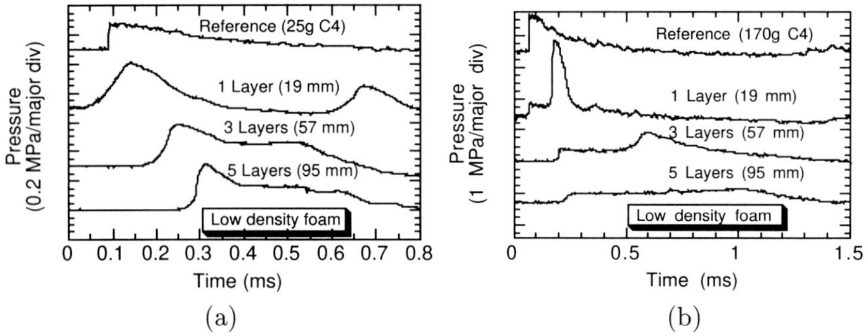

Figure 2. Effect of foam layer thickness on reflected pressure recorded behind *low*-density polyethylene foam for: (a) 25 g C4 charge, and (b) 170 g C4 charge.

A similar effect is observed for the higher density foam, as shown in Fig. 3a. The critical charge mass m_{crit} at which attenuation is first observed is increased due to the larger dynamic plateau stress (i.e., > 0.5 MPa) and a stronger blast is required to achieve densification. As the blast wave propagates within the foam and energy is being absorbed by buckling of the cellular material, the wave exhibits a decaying front profile with distance of propagation, similar to the low-density foam in Fig. 2b. If the incident shock strength is increased for a single layer of high density foam, a steepening of the compression wavefront occurs, as illustrated in Fig. 3b. For the higher charges, the pressure trace exhibits an initial plateau region of 0.6–0.7 MPa, which is slightly higher than the quasi-static plateau stress (i.e., 0.5 MPa). A single layer of high-density foam appears to be sufficient to attenuate the blast from 170 g C4, compared to approximately five layers of low-density foam. This is a clear indication of the differences in energy absorbing capacity of the two foams, as illustrated by their stress-strain curves.

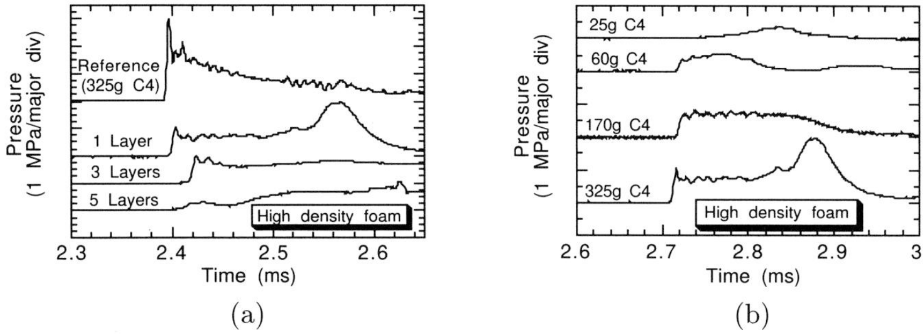

Figure 3. Reflected pressure recorded below *high*-density polyethylene foam. (a) Effect of foam layer thickness for 325 g C4 charge, (b) Effect of charge mass for single foam layer (19 mm).

The results are summarized in Figs. 4a,b which show the ratio of the pressure recorded under the foam to the reference pressure without any foam, as a function of charge mass. The graph indicates whenever two distinct peaks were observed, or equivalently, a plateau pressure followed by a densification peak. The behavior is qualitatively similar for the two types of foam, although the cross-over from amplification to attenuation (indicated by the dashed line) occurs at a different blast strength. For the low-density foam $m_{crit} \approx 50$ g, whereas for the high-density foam $m_{crit} \approx 100$ g due to the difference in plateau stress for the materials.

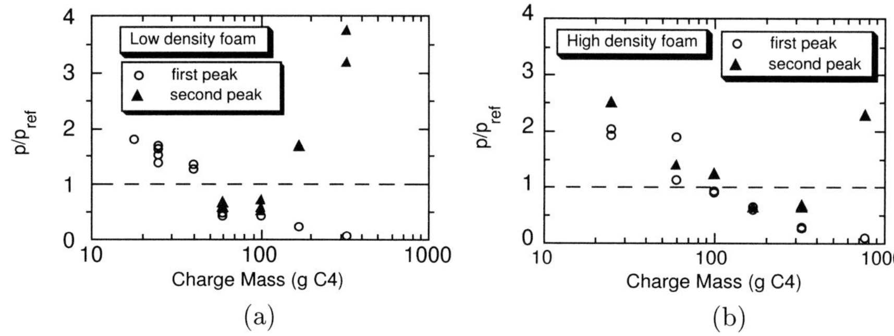

Figure 4. Ratio of pressure recorded under a single foam layer to reference reflected pressure in absence of foam as a function of charge mass for: (a) low-density foam, and (b) high-density foam. Points located below the dashed line correspond to attenuation of the reference reflected blast wave in the absence of foam.

4. Conclusions

It has been conclusively demonstrated that a cellular material can be used to successfully attenuate a blast wave. Pressure reductions of up to 70% were achieved with the foams tested. For effective energy absorption to take place, the blast strength must overcome the threshold, or plateau, stress value of the material. For weaker shocks, the pressure measured behind the material may actually be amplified (as has commonly been observed in previous studies). On the other hand, if the blast energy absorbing capacity is inadequate, then densification will occur with the potential for large amplification of the reflected pressure that would be measured with no foam present. The current findings demonstrate the necessity of carefully considering the anticipated blast strength and (dynamic) stress-strain characteristics of a cellular material for effective blast wave attenuation.

Acknowledgement. The authors would like to thank Don Cox and Ron Vanderbeek from Energy, Mines and Resources Canadian blast chamber facilities, Chris Graham and Allen Stanish from Med-Eng Systems for technical assistance, and Ron Purvis from the RCMP Bomb Data Center for ensuring that the required explosives were always available.

References

Ben-Dor G, Mazor G, Igra O, Sorek S, Onodera H (1994) Shock wave interaction with cellular materials. Shock Waves 3:167.

Gvozdeva LG, Faresov YuM, Fokeev VP (1985) Interaction between air shock waves and porous compressible materials. Sov Phys Appl Math and Tech Phys 3:111.

Lee JJ, Frost DL, Lee JHS (1995) Transmission of a blast wave through a deformable layer. In: Brun R, Dumitrescu LZ (eds), Shock Waves @ Marseille, III:181.

Mazor G, Ben-Dor G, Igra O, Sorek S (1994) Shock wave interaction with cellular materials. Shock Waves 3:159–165.

Rayevsky DK, Gvozdeva LG, Faresov Y, Brossard J, Bailly P (1989) Reflection of shock and explosion waves from surfaces covered with layers of polyurethane foam. AIAA Prog Astr and Aero 134:171.

Skews BW (1995) Shock wave impact on porous materials. In: Brun R, Dumitrescu LZ (eds) Shock Waves @ Marseille III:11.

Structure of deformation waves in porous compressible materials — influence of percolation

L. G. Gvozdeva, Y. P. Lagutov, Y. L. Sharov, N. B. Sherbak
High Energy Density Research Center, Scientific Association IVTAN
Russian Academy of Science, Izhorskaya 13/19, Moscow 127412 Russia

Abstract: The results of experimental investigations of the interaction of a shock wave with compressible polyurethane foam are presented. It is shown that the trailing edge of the compressed zone in a sample coincides with the free surface F of the sample (a plane surface of the sample was normally impacted by the shock wave) if the surface F is nonpenetrable for a gas behind the shock wave. The thickness of compressed zone linearly increases with time. For a sample with a penetrable surface F, the back of the compressed zone does not coincide with F, but lies inside the sample. The pressure, density, and temperature abruptly increase at the forward edge of the compressed zone (the shock front) and have a constant value through the compressed zone. Across the trailing edge of the compressed zone the density and the temperature abruptly decrease, while the pressure remains the same. Pressures at the side and back walls of the shock tube were measured.

Key words: Shock wave, Compressible foam, Percolation

1. Introduction

The present paper continues the study of Gvozdeva et al. (1993) of the structure of deformation waves inside porous compressible foams struck by a shock wave. Gas dynamic and physical behavior of the transmitted compression wave have been established, and extended qualitative characteristics of the process are given.

Reflection of a shock wave from a sample of porous compressible foam exhibits many interesting features. Gelfand et al. (1975) and Gvozdeva and Faresov (1984) found that end-wall pressures could exceed those obtained from reflection from a rigid wall. Usual theoretical models of the process suppose that refraction of the shock wave at the front face of the foam takes place. Assumptions are made that there is no gas flow across the front face of the foam and that the material of the foam can be treated as a pseudo-gas. However experimental data of Skews et al. (1993) and Gvozdeva et al. (1993) have shown that the usual assumption that the foam face is itself a gas contact surface is not correct. Gvozdeva et al. (1993) have discovered that in a sample of polyurethane foam the compression wave is practically a shock wave, with a relatively fast rise in density, followed by a region of constant density, and then a rarefaction wave near the foam surface. In the present study experiments have been done to investigate this phenomena, and special attention has been paid to the effect of gas flows on the reflection process.

2. Experiments

Experiments were done in a shock tube of square cross-section 72 × 72 mm^2. Air, nitrogen, carbon dioxide, argon, and helium at initial gas pressure $P_0 = 10$ – 100 KPa were used as the working gases. Mach numbers of incident shocks M_0 were 1.5 – 2.2. Experiments were done with polyurethane foam with porosity $\alpha = 0.97$. The thickness of the sample was 40 mm. The sample of foam was installed in the shock tube near its closed end so that all sides of it except one are in contact with the surface of the shock tube. The surface of the sample F turned to the diaphragm was free. The experimental section had optical glass windows. A system of equidistant vertical lines was drawn on the side surface of the foam with a spacing of 10 mm. A high-speed streak camera simultaneously recorded both a schlieren pattern of the gas in front of the free surface of the sample (registered in transmitted light) and the side surface of the sample using reflected light from an external light source (Lagutov and Sharov 1995). Pressures at the side and back walls of the shock tube were measured.

3. Results

In the experiments both photographic records of the propagation of the disturbances in the transparent gaseous phase and in the nontransparent compressible foam and pressure records were obtained. In order to study the influence of percolation on the process, some experiments were made with the sample covered with a thin layer of mylar to prevent gas leakage through the surface of the foam.

Figure 1 shows photo records of the propagation of the disturbances in the transparent gaseous phase outside the sample and also in the nontransparent porous compressible polyurethane. Figure 1a is for an exposed open cell surface. Figure 1b has this exposed surface F covered by a thin film of mylar of thickness about 20 μm. The initial pressure was $P_0 = 100$ KPa, and the Mach number of the incident shock wave was $M_0 = 1.6$.

It is seen from Fig. 1a that the distance between the first and second lines along x-axis on the side surface of the sample along the direction of the propagation of the shock wave is approximately constant, but slightly different, before and after the interaction of the shock wave with the sample. Therefore, the part

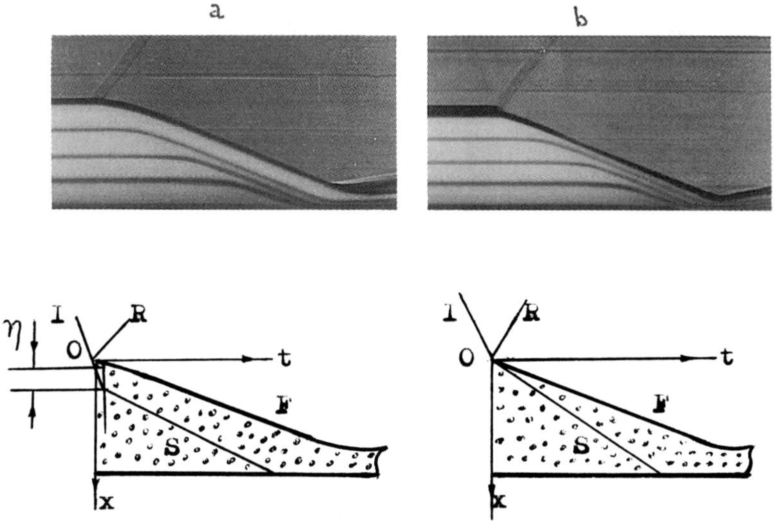

Figure 1. Gas percolation through foam surface. Top: photo records; bottom: x-t diagrams. $M_0 = 1.6$, $P_0 = 100$ KPa in air; (a) porous surface, (b) surface sealed with mylar film. I, R, S are incident, reflected, and deformation waves; F is free foam surface, η is shocked-gas penetration depth.

of the sample between the first and second lines is not highly compressed and is relatively penetrable for the shock wave and for the gas behind it. Then the walls of cells collapse and the compression wave in the porous sample formed. It should be noted that in this experiment the cell's walls collapse somewhere between the first and second lines. The distance between the lines along the x-axis on the side surface of the sample is decreased and becomes approximately equal to the distance after formation of the compression wave.

For the case when the free surface of the sample F is not penetrable for a gas behind the shock wave (Fig. 1b) the distances between the lines along the x-axis on the side surface of the sample are uniformly decreased for the whole thickness of the sample. Thus it is possible to form several conclusions about the structure of the compression wave.

For the sample with a nonpenetrable free surface, the thickness of the zone of compressed porous material with constant parameters between the trailing edge of the compressed zone and the leading edge of the compression wave increases linearly with time. The trailing edge of the compressed zone coincides with the free surface of the sample.

For the sample with a penetrable free surface, the trailing edge of the compressed zone does not coincide with F and is inside the sample. The pressures are the same on both sides of the trailing edge and the density abruptly changes across it.

The pressure, density and temperature abruptly increase across the leading edge of the compression wave and then keep a constant value through the compressed zone. Across the trailing edge the density and temperature abruptly decrease, and the pressure remains the same.

Figure 1 also shows x-t diagrams for the propagation process of the incident I and reflected R shock waves in the gas in front of the sample and x-t diagrams for the propagation of the compression wave S inside the sample for a penetrable free surface F (Fig. 1a) and for a nonpenetrable surface F (Fig. 1b). If we assume (1) that the velocity of incident shock wave I is not changed inside a porous foam layer of thickness η until collapse of the cell's walls and (2) the cell's walls collapse instantaneously, then the shocked-gas penetration depth η (the coordinate for the initial formation of the compression wave) can be determined (Fig. 1a). Some results of determining η for different gases, for Mach numbers M_0, and for initial pressures P_0 are presented in Table 1.

Table 1. Determination of η.

Gas	CO_2	N_2	Ar	N_2	N_2	Air	Air
P_0, KPa	100	100	100	100	100	14	66
M_0	1.67	1.59	1.58	2.19	1.62	1.71	1.64
η, mm	6	6	11	5	5	11	9

Figure 2a shows pressure records for the gauges G1 – G5 corresponding to the experiment of Fig. 1a. The pressure records for the experiment of Fig. 1b are presented on Fig. 2b. As seen from photo records for the gauges G1, G2, and G3, the propagation of disturbances inside porous compressible polyethylene from shock wave impact can be considered as one dimensional.

Oscillograms for the gauge G4 show the pressure behind the first reflection wave (the first pressure increase) and behind the second reflection wave (the second pressure increase). In oscillograms for the gauge G5 are seen three successive pressure increases: the first is the arrival of the weak compression wave, the second signals the arrival of the front of the compression wave, and the third represents the arrival of the reflected compression wave. The oscillograms of Fig. 2 show that the pressure level for the experiment with a penetrable free surface F is lower then the pressure level for the experiment with a nonpenetrable surface.

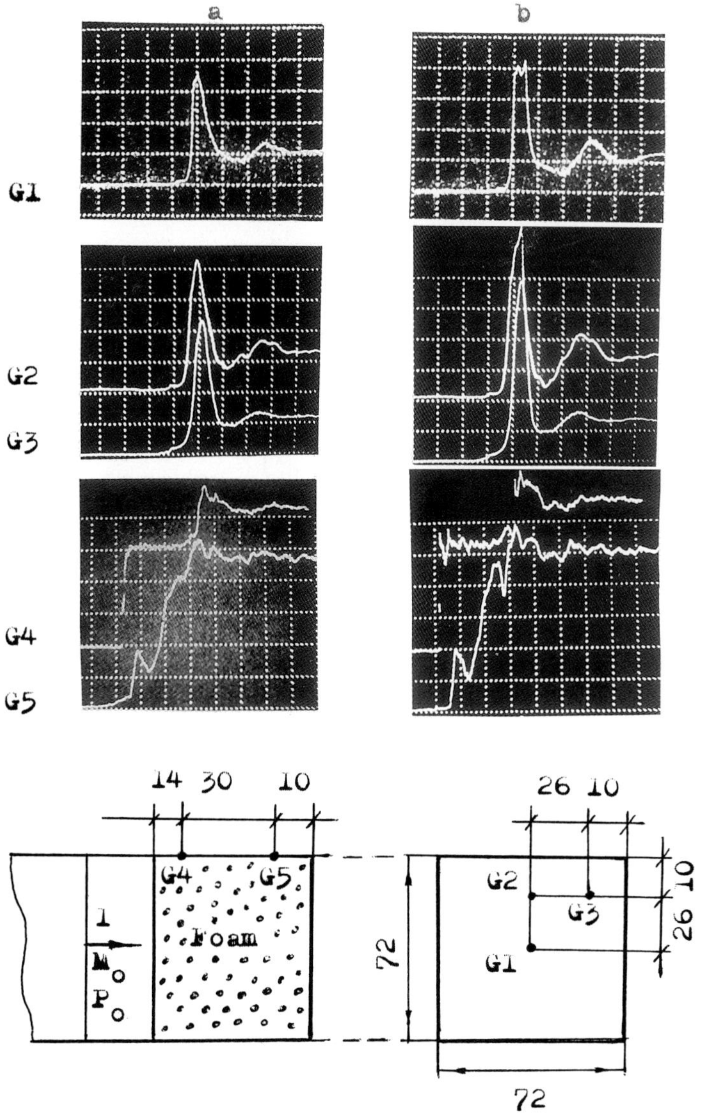

Figure 2. Pressure records for gauges G1 – G5: (a) case for Fig. 1a, (b) case for Fig. 1b.

4. Conclusion

We have investigated in the shock tube the penetration of shocked gas into a porous compressible material with open cells during the initial stage of shock-wave interaction with a foam material. It is shown that at distances greater than h from the free surface of the sample, the cell's walls collapse. The penetration of the gas behind the incident shock wave into the porous material prevents the cells from collapsing for distances less than η. A zone of compressed material in the foam is formed where the cells have collapsed. The trailing edge of this zone is fixed relative to the free surface of the sample. The forward front propagates in the direction of the incident shock wave.

The pressure is constant across the trailing edge and is equal to the pressure behind the first reflection wave. Thus across the leading edge of the compression wave the density and pressure increase. On the trailing edge which does not coincide with free surface F, only a density jump is found.

Acknowledgement. This work is supported by Russian Foundation of Fundamental Science (94-02-05304a) and the International Science Foundation (M94000, M94300).

References

Gelfand BE, Gubin SA, Kogarko SU, Popov DE (1975) Study of special features of propagation and reflection of pressure in porous medium. Soviet Physics-Applied Mathematics and Technical Physics 6:74–77.

Gvozdeva LG, Faresov YM (1984) Interaction of shock waves in air with a wall covered with a layer of porous compressible materials. Letters to Journal of Technical Physics 19(14):1153–1156.

Gvozdeva LG, Faresov YM, Sharov YL (1993) Structure of deformation wave in porous compressible materials under the Influence of shock waves. In: Shock Waves @ Marseille III, Shock Waves in Condensed Matter and Heterogeneous Media, Brun R and Dumitrescu LZ (eds) Springer, pp 175–180.

Lagutov YP, Sharov YL (1995) Propagation of disturbances in porous compressible polyethylene with closed pores under shock wave impact. Letters to Journal of Technical Physics 21(2):64–69.

Skews BW, Atkins MD, Seitz MW (1993) The impact of a shock wave on porous compressible foams. J Fluid Mech 253:245–265.

Condensed Matter

Ultra-high-pressure thermal-conductivity measurements of griceite and corundum

K. G. Gallagher, T. J. Ahrens
Lindhurst Laboratory of Experimental Geophysics
Seismological Laboratory, California Institute of Technology
Pasadena, CA, 91125 USA

Abstract: Time-dependent interface temperatures were measured for 500Å to 1-μm iron and stainless steel films encapsulated within LiF and Al_2O_3 transparent anvils and shocked to between 220 and 318 GPa (Gallagher et al. 1994). The corresponding anvil shock pressures are 140–170 GPa in LiF and 112–242 GPa in Al_2O_3. Thermal conductivity, k, is related to measured diffusivity, $\kappa = k/\rho C_p$ where ρ and C_P are the density and specific heat of the shocked state. Existing models predict that on transformation from STP to high T and P, k would increase by a factor of ~ 10 for LiF and decrease by a factor of ~ 5 for Al_2O_3. Experimentally determined values of k are a factor of $\sim 10^3$ smaller than the STP values for Al_2O_3 and are a factor of $\sim 10^2$ smaller than the STP values for LiF. A possible explanation for the difference between theory and experiment is that the temperature dependence of conductivity may be sensitive to defect concentration, making existing theories inappropriate.

Key words: Griceite, Corundum, Thermal conductivity, Shock temperature, Iron

1. Introduction

The experimental determination of thermal conductivities of dielectrics such as griceite (LiF) and corundum (Al_2O_3) at ultra-high pressures is important because of the use of these materials in experiments to obtain shock temperatures for metals. Those experiments observe the greybody radiation emitted from the surface of the metal through a transparent anvil. The shocked metal's Planck radiation peaks are in the visible wavelengths because the samples are 10^4 K, and metals are opaque at those wavelengths so their interiors cannot be optically observed. In order to keep the surface of the metal at high pressure after the shock wave has reached it, the metal is put in contact with a transparent window such as griceite or corundum. The internal temperature of the shocked metal is then obtained from the measured metal-window interface temperature and the contrast in thermal conductivities of the two materials.

Measurements of conductivities of dielectrics at high pressures and comparison with theory are also of intrinsic interest. In this paper, we summarize the theory of the effects of temperature and pressure on k for dielectrics and describe our initial experimental results.

2. Theoretical

2.1. Thermal conductivity

Roufosse & Jeanloz (1983) produced a theory which Bass et al. (1987) used to calculate the thermal conductivity of dielectric windows used in shock-temperature experiments on iron samples. They assumed a temperature dependence and density dependence of k as follows:

$$k_{T,0} = A + B/T \tag{1}$$

$$k_{T,P} = k_{T,0} \left(\frac{\rho_P}{\rho_0} \right)^{2\gamma + 5/3} \tag{2}$$

where $A_{\mathrm{Al_2O_3}} = -2.6\,\mathrm{W/mK}$, $B_{\mathrm{Al_2O_3}} = 1.2 \times 10^4\,\mathrm{W/m}$, $A_{\mathrm{LiF}} = -0.2\,\mathrm{W/mK}$, and $B_{\mathrm{LiF}} = 3.7 \times 10^3\,\mathrm{W/m}$ are measured at ambient pressure and high temperature and γ, ρ_0, and ρ_P are the Grüneisen parameter and the initial and compressed densities of the material. Tang (1994) uses a model for conductivity with a theoretical dependence on T and ρ_0, originally developed by Leibfried

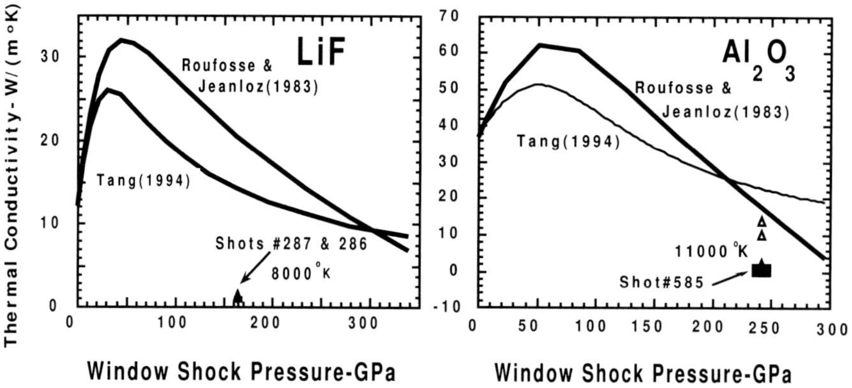

Figure 1. Comparison of thermal conductivity of griceite and corundum calculated along their Hugoniot with the Roufosse & Jeanloz (1983) and Tang (1994) methods. For griceite, both theories predict a much higher conductivity than is measured, and for corundum the Tang theory is generally closer to the present data.

& Schlömann (1954). Their equation is:

$$\frac{k}{k_0} = \left(\frac{\rho}{\rho_0}\right)^{5/3} \exp\left[3\gamma_0\left(1-\frac{\rho_0}{\rho}\right)\right]\frac{T_0}{T} \tag{3}$$

Equation (3) agrees with the Eqs. (1) and (2) in both slope and amplitude at low pressures, but at higher pressures the the conductivities are lower (Fig. 1).

2.2. Radiative transfer

During the conductivity experiments, the temperature of the metal that is in contact with our dielectric window can be as high as 10^4 K. Thus, it is unclear how much of the observed decay in temperature results from radiative heat transfer as opposed to thermal conduction. However, if the total radiative energy that can be lost from the metal via thermal radiation, E_{rad}, during the 250 ns of the experiment is compared to the total energy loss implied by the decrease in temperature in our sample, ΔE_{total}, we can determine if the energy from radiative transfer is significant in our experiment. For example, for shot 286, the peak temperature is 10,200 K, so the total greybody radiation energy is at most 170 J/m^2 (Eq. 4). The temperature decrease of the sample is 2000 K in 250 ns, so the energy is 1841 J/m^2 (Eq. 5). Thus the ratio $E_{rad}/\Delta E_{total}$ is 1/11, so to first order, radiative heat transfer may be ignored.

$$E_{rad} = \int_0^{250\,ns} \sigma T_t^4 dt \tag{4}$$

$$\Delta E_{total} = \rho \ell C_v \Delta T \tag{5}$$

where σ is 5.670×10^{-8} W/m^{2}K^4, T_t is the measured time-dependent temperature of the radiating surface, ΔT is the change in temperature of the sample with density ρ, thickness ℓ, and heat capacity C_v.

2.3. Heat flow

To measure thermal conductivity of the window material, a sample configuration in which a thin metal film is sandwiched between two dielectric anvils was used. Symmetric heat flow from the metal film into the anvils occurs upon shock compression of the metal while the shock wave is traveling through the second anvil. Previous work on the shock temperatures of metals (Bass et al. 1987, Ahrens et al. 1990) used a thick metal film with an anvil on one side and a metal driver plate on the other. Thus the heat-flow configuration was approximated as an infinite half-space. Unlike the half-space configuration, the thin-film configuration does not have a simple analytic solution to the coupled thermal conduction equations for the anvil and film, thus finite-difference calculations were performed using the 'ConMan' code (King et al. 1989) (Fig. 2). The

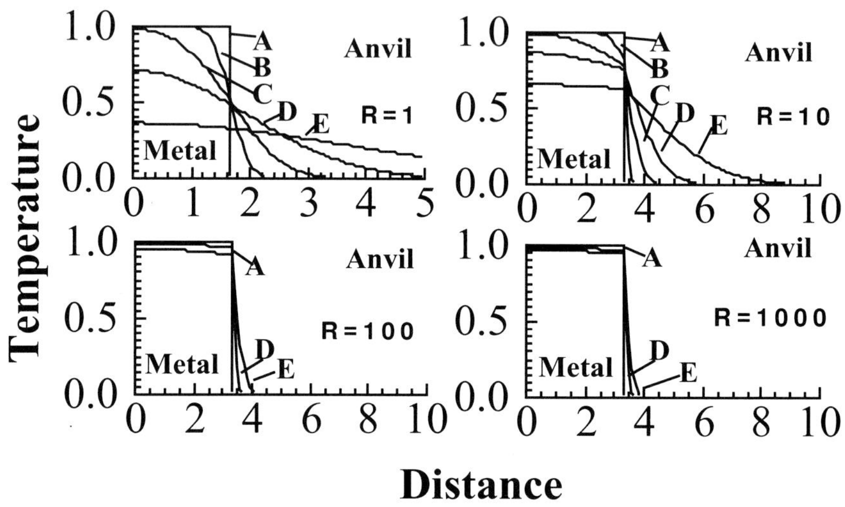

Figure 2. Symmetric heat flow-models for 'sandwich' configuration of shock-temperature experiments. Temperature is normalized by the difference between the Hugoniot temperatures of metal and the anvil. Distance is normalized by 1/2 of the thickness of the metal film, 2ℓ, and normalized time is $t' = t\kappa_2/\ell^2$ where t is real time. A: $t' = 0$; B: $t' = .125$; C: $t' = 1$; D: $t' = 5$; E: $t' = 25$.

solution is symmetric across the center plane of the thin film. Shown are four plots for differing values of the ratio of thermal diffusivities, $R = \kappa_{metal}/\kappa_{anvil}$. For higher values of R, the interface temperature is closer to the interior temperature of the film, and it changes more slowly with time.

3. Experimental

Time-resolved radiance from the samples was measured at six wavelengths (e.g., Fig. 3). This is compared to predictions of interface temperature from the model shown in Fig. 2 for various values of R. We conclude that the best-fit value of $R \approx 100$. The k_{metal}, calculated via the Weidemann-Franz law, was assumed to be valid, whereas the k_{anvil}, was considered to be the unknown quantity. The measurement of R together with the separately measured complete equations of state of the metal and anvil materials can be used to calculate a value for k_{anvil} in the shocked state. The thermal conductivities determined from these data are shown in Table 1. The corundum anvils did not behave consistently when shocked above 229 GPa. However, the radiative signals from experiments with Al_2O_3 are usable up to at least 242 GPa.

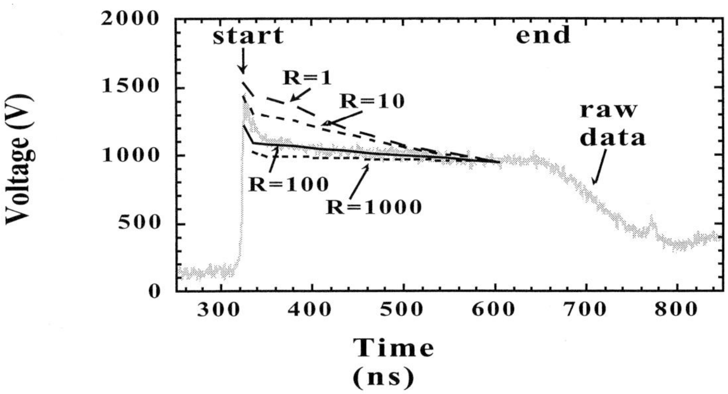

Figure 3. Comparison of raw data for shot 286 and calculated curves for varying R, the ratio of thermal diffusivities of the Fe film and the LiF window. The best fit is $R \approx 100$ and an emissivity for Fe of 0.6.

4. Results

The thermal conductivities measured are reported in Table 1. For LiF the Roufosse and Jeanloz theory overpredicts the conductivity by a factor of 10^3 and the Tang theory overpredicts it by a factor of 10^2. For Al_2O_3 the Roufosse and Jeanloz theory overpredicts experiment by a factor of 10^3, as does the

Table 1. Experimental and theoretical high temperature and pressure thermal conductivities of corundum (Al_2O_3) and griceite (LiF).

shot #	Anvil	P_h (GPa)	T_h (K)	k_{STP} (W/mK)	$k_{T,0}$ (W/mK) calc.	$k_{T,P}$ (W/mK) calc.	$k_{T,P}$ (W/mK) meas.
285	Al_2O_3	242	2753	46	1.7	18	~0.02
286	LiF	266	2100	3	1.5	24.2	0.02
287	LiF	165	2087	3	1.6	24.1	0.01–0.002

Tang theory. If we use the metal film's peak temperature instead of its average temperature we get a lower conductivity, and in fact in the case of corundum the conductivity predicted by Tang agrees with experiment. However this is less physically meaningful because the temperature is very high for a short time and most of the thermal decay occurs while the sample is at a lower temperature.

5. Discussion and conclusions

We conclude that the thermal conductivities of dielectrics shocked to 150–250 GPa are much lower than has been previously calculated. These low conductivities can be used in data analysis for shock-temperature experiments on metals. A low window conductivity implies a smaller correction from the metal's observed interface temperature to its internal temperature. The reason why theory overpredicts the measured values for thermal conductivity is poorly understood. It may be that the defects in the shocked state inhibit phonon transport. An example of this is that k is much lower for amorphous solids that for crystalline solids of the same composition. More research on this question is required.

Acknowledgement. This research was supported by NSF. California Institute of Technology, Division of Geological and Planetary Sciences, contribution number 5583.

References

Ahrens TJ, Bass JD, Abelson JR (1990) Shock temperatures in metals. In: Shock Compression of Condensed matter – 1989, Proc APS Topical Conf held in Albuqerque, New Mexico, Schmitt, Johnson and Davison (eds), New York, Elsevier, pp 851–857.

Bass JR, Svendsen B, Ahrens TJ (1987) The temperature of shock compressed iron. In: High Pressure Research in Mineral Physics, Manghnani and Syono (eds), Tokyo, Terra, pp 393–402.

Gallagher KG, Bass JD, Ahrens TJ, Fitzner M, Abelson JR (1994) Shock temperature of stainless steel and a high pressure - high temperature constraint on thermal diffusivity of Al_2O_3. In: High-Pressure Science and Technology 1993, AIP Conference Proceedings 309, Schmidt, Shaner, Samara and Ross (eds), New York, AIP Press, pp 963–966.

King SD, Raefsky A, Hager BH (1989) ConMan: vectorizing a finite element code for incompressible two-dimensional convection in the earth's mantle. Phys Earth Plan Int 59:195–207.

Leibfried G, Schlömann E (1954) Wärmeleitung in elektrisch isolierenden kristallen. Nachr Akad Wiss Göttingen, Math Phys Klasse I A 4:71.

Roufosse MC, Jeanloz R (1983) Thermal conductivity of minerals at high pressure: the effect of phase transitions. J Geophys Res 88(B9):7399–7409.

Tang W (1994) The pressure and temperature dependence of thermal conductivity for nonmetal crystals. Chinese J High Press Phys 8(2):125–132.

Combustion synthesis of advanced ceramics assisted by explosive compaction

Y. M. Timnat, V. Rosenband
Faculty of Aerospace Engineering
Technion-Israel Institute of Technology
Haifa 32000 Israel

Abstract: Novel techniques for obtaining advanced ceramics by explosive compaction are discussed, showing their advantages and the areas for future study.

Key words: Combustion synthesis, Advanced ceramics, Explosive compaction

1. Introduction

Combustion synthesis, which is based on utilization of internal chemical energy of initial reactants, presents an attractive alternative to conventional methods of materials preparation owing to its demonstrated advantages (Munir 1988, Varma & Lebrat 1992). Gases entrapped in the matrix can lead to a relatively porous product (Rice & McDonough 1985); thus consolidation must be accomplished during or immediately following the reaction to eliminate porosity.

A novel technique for final-products consolidation is explosive compaction; it needs no expensive and bulky installation, since the force-inducing part of it is replaced by an inexpensive explosive. It allows one to create in the initial powder mixture very high pressures of the order of 50 GPa. Explosion provides conditions of equal loading over the entire surface of intricate articles, an unobtainable condition in presses.

The high-pressure action on the hot reaction products may improve the properties of the final products, since the common action of high temperatures and pressures behind the combustion wave front essentially influences the material structure. The shock-wave velocities reach a value of about 10^3 m/s, several orders of magnitude higher than the combustion-wave velocities (usually below 0.1 m/s). So the time of interaction is very short, providing the possibility of selective action of high pressure on any stage of the synthesis.

If the characteristic width of the post-effect zone at the combustion wave is larger than the specimen dimension, then by varying the detonation initiation time, one may obtain from the same initial composition relatively homogeneous final specimens with different structures and properties. Otherwise, the shock-wave loading will result in a specimen with a gradient of structure and physico-mechanical parameters.

Different facilities for the concomitant occurrence of combustion synthesis and explosive compaction may be used. The simplest experimental setup is shown in Fig. 1 (Adadurov et al. 1992). In the first case (Fig. 1a), explosion products behind the gliding detonation wave compress the thin-walled cylindrical container with the specimen under investigation. In the second case (Fig. 1b), one uses the normally incident detonation wave to accelerate a metal piston compressing the reaction products in a massive metallic container. In both cases, reaction is initiated by a heated helix at the bottom of the container. Detonation was initiated at some moment after combustion-wave propagation.

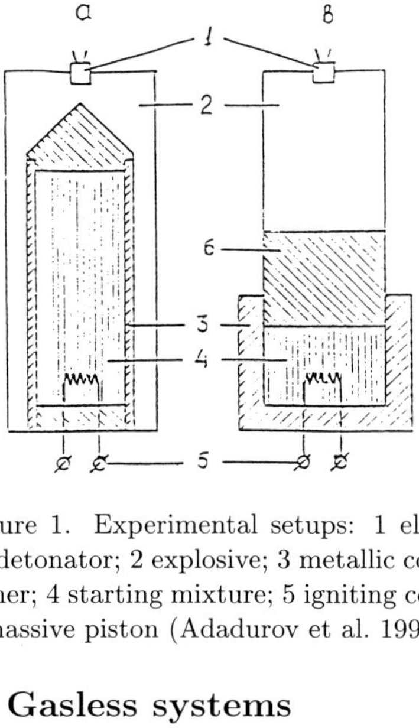

Figure 1. Experimental setups: 1 electrodetonator; 2 explosive; 3 metallic container; 4 starting mixture; 5 igniting coil; 6 massive piston (Adadurov et al. 1992).

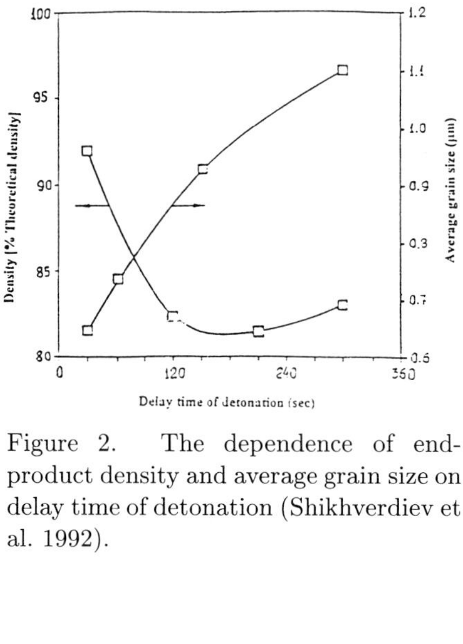

Figure 2. The dependence of end-product density and average grain size on delay time of detonation (Shikhverdiev et al. 1992).

2. Gasless systems

The concomitant occurrence of combustion synthesis and dynamic loading is more convenient to use for mixtures, whose combustion is practically gasless. Shock compression (Adadurov et al. 1992) was applied at the moment when the heating zone reached the opposite end of a cylindrical specimen. The shock compression was assumed to influence all the stages of the combustion synthesis. Maximal pressures achieved were in the range of 7–50 GPa. The products of afterburning and structuring were densely packed rods of titanium carbide or boride. No new phases were found.

These data may be interpreted as follows. Shock-compression heat is added to the heat released in chemical reaction. So that specimens may be heated to temperatures much above 2000 K, the specimen is cooled at atmospheric pressure resulting in annealing of new or pressure-induced phases. Further attempts to apply dynamic pressures on the last stages of combustion synthesis to obtain densely packed articles included experiments on shock compression of IV–VI group metal carbides and borides preheated up to 1500–3000 K. Rods, tubes, and disks were obtained with up to 97% density of that theoretically predicted, showing a finely dispersed structure with grain size of 1–3 μm.

Experiments on combustion synthesis with explosive impacting of Ti + C mixtures (Shikhverdiev et al. 1992) have shown the possibility of influencing the final carbide structure by a delay between the time when the combustion wave reaches the end of the specimen and detonation initiation. The dependence of final specimen density on detonation delay is shown in Fig. 2. Density decreases as delay increases because the material is not ductile at longer delays, as the ductile-brittle transition temperature for TiC lies between 800 and 1200 °C. Results indicate the possibility of obtaining high-density ultrafine-grained ceramics (grain size less < 1 μm) formed by recrystallization.

Kecskes et al. (1990a,b) found that microstructures of samples obtained by combustion synthesis with explosive compacting had significantly different grain sizes and morphologies than equivalent hot-pressed materials. To produce fully dense TiC and TiB_2 samples, C/M values (explosive-charge-mass-to-compression-plate-mass ratio) of 0.44 and 0.22 were used. They found that the density of TiC is maximum for C/Ti of 1.0 and drops off for lower ratios; however, at lower C/Ti ratios, the samples are closer to their full density. Densification with dynamic consolidation becomes more effective with decreasing C/Ti ratios. Probably, as the C/Ti ratio decreases, the increasing number of carbon vacancies results in less restrictive motion of the dislocations and thus easier deformation of the TiC grains. The TiC and TiB_2 samples have been produced at 98% of theoretical density and micro-hardness values were equal to or greater than commercially available hot-pressed materials.

In studies by Grebe & Advani (1992) on the synthesis of titanium carbide, the explosive loading conditions — the densification pressure controlled by the ratio of explosive-charge mass to drive-plate mass (C/M) and the delay (t_d) between the full passing of the combustion wave and explosive detonation — critically affected the density and microstructure of the final compacted products. Two C/M ratios (0.2, 0.4) were used corresponding to peak pressures of about 1.5 and 2.0 GPa in the hot reaction product (with $0 < t_d \leq 15$ s).

The density of the combustion-synthesized and explosively densified compacts increased with increasing C/M ratio. In general, higher final compact densities and smaller grain sizes were observed with increasing t_d between reaction completion and densification. At C/M = 0.4 the density of the compacts reached a maximum of 95% of theoretical for a $t_d = 8$ s. For lower t_d, outgassing of impurities from the combustion-synthesized products is still incomplete and results in delamination of the explosively densified compact. With $t_d > 10$ s, the TiC product is at temperatures well below its melting point and is becoming less plastic, making compaction more difficult. Thus, there is an optimum time delay which assures maximum densification.

The value of t_d strongly affects the grain size of the final compacts. At C/M = 0.4, the grain size decreases from 163 to 11 μm when t_d increases from 0 to 8 s. With further increase in t_d, grain size increases to 82 μm. This indicates a complex interplay of cooling effects subsequent to reaction completion and effects of dynamic loading during explosive densification.

The structural and mechanical properties of titanium carbides and diborides prepared by combustion synthesis assisted by explosive compaction were studied by Wang et al. (1994). Densities higher than 99% and 97% of the theoretical densities were achieved for TiB_2 and TiC, respectively. The micro-hardness, elastic modules, and compressive strength of TiB_2 and TiC are comparable or in excess of those reported for hot-pressed TiB_2 and TiC. The micro-hardness of TiC and TiB_2 prepared by combustion synthesis and explosive compaction exceed those reported for hot-pressed TiB_2 and TiC. The micro-hardness did not reflect the hardness of the compacts as bulk materials. The measured hardness of TiB_2 and TiC was essentially lower, indicating weak grain boundary bonding (Fig. 3), which might be inherent to materials prepared using this technique. This process is very rapid, so there may not have been sufficient time for the atoms to rearrange to produce atomic-level bonding between the grains.

3. Gas-evolving and gas-solid systems

Schemes of shock-wave compression were also suggested for gas-evolving systems (Rabin et al. 1990). Densely packed specimens of TiC-Al_2O_3 were prepared from mixtures of titanium oxides, aluminum, and soot. Such dispersed-phase composites containing alumina (TiC-Al_2O_3) reinforced with titanium carbide exhibit good electrical conductivity, high hardness, strength, and fracture toughness. A possible experimental assembly, shown in Fig. 4, includes special vent holes to allow hot gases to escape during the exothermic reaction. Samples reacted without consolidation exhibited swelling and resulted in products containing over 50% porosity. Application of the dynamic compression pulse to the hot sample resulted in densification. The dynamically compressed samples ranged from 85% to over 95% of theoretical density. The highest density was achieved by allowing the hot sample to degas and equilibrate for about 5 s after the reaction and using a stress amplitude of about 1 GPa. All samples appeared to be well-bonded and exhibited a fine interconnected composite microstructure with micro-hardness about 20 GPa, a value consistent with typically reported micro-hardness for TiC-Al_2O_3 materials processed conventionally.

Figure 3. Electron micrograph of sintering necks between TiC particles in combustion synthesized/dynamically densified TiC-Ni material (LaSalvia et al. 1994).

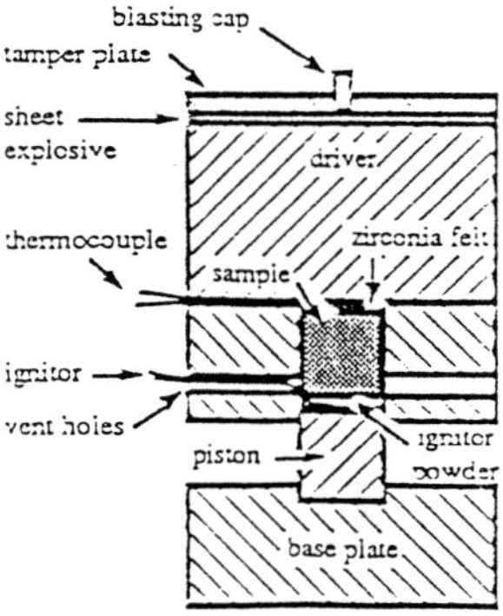

Figure 4. Arrangement used to conduct combustion synthesis and subsequent dynamic consolidation (Rabin et al. 1990).

Special apparatus necessary for combustion synthesis with explosive compaction of gas-solid systems was used for a Ti-N-O combustion system (Molokov & Mukasjan 1992). This approach promises to resolve some technological problems: fabricating pore-free materials in the gas-solid system, layered materials, etc. Here Ti reacts with air in a two-stage process, first a titanium-nitrogen interaction, then reaction with oxygen in the post-combustion period. Explosive treatment can alter the heat-release conditions rapidly, bringing the sample into contact with a cooler environment (the liner), and can create a marked increase in thermal conductivity. The cooling rate at the front on quenching by explosion is at least 10^4 °C/s. The interaction of the nitride phase with oxygen was limited and the final product, $TiN_{0.8}$, compacts with 90 – 97% of theoretical density and finely dispersed structure (grain size of 1 μm).

4. Conclusions

The concomitant occurrence of combustion synthesis and shock-wave treatment of synthesis products may be considered as a new efficient method of preparing materials. The data show that varying the explosive compacting parameters of the reactive mixture can affect the physical characteristics of the compacted ceramic and control the structure of the synthesized material. Additional concerns regarding the mechanical properties of densified materials (cracks, fracture toughness, macro-hardness, etc.) and process scale-up have yet to be addressed.

References

Adadurov GA, Borovinskaja I, Gordopolov Y, Merzanov AG (1992) The technological backgrounds of shock wave SHS compacting. J Eng Phys 63:538–546.

Gordopolov YA, Fedorov VM, Molokov IV (1989) Explosive working of SHS products. In: Proc 10th Int Conference on High-Energy Effects on Materials, Ljublyana, Yugoslavia, pp 144–153.

Gordopolov YA, Merzanov AG (1991) The use of shock waves in the SHS research. In: Proc 13th Int Colloq on Dynamics of Explosions and Reactive Systems, Nagoya, Japan.

Grebe HA, Advani A (1992) Combustion synthesis and subsequent explosive densification of titanium carbide ceramics. Metall Trans 23A:2365–2372.

Kecskes LJ, Benck RF, Netherwood PH (1990a) Dynamic compaction of combustion synthesized hafnium carbide. J Am Ceram Soc 73:383–387.

Kecskes LJ, Kottke T, Niiler A (1990b) Microstructural properties of combustion synthesized and dynamically consolided titanium boride and titanium carbide. J Am Ceram Soc 73:1274–1282.

LaSalvia JC, Meyers MA, Kim DK (1994) Combustion synthesis/dynamic densification of TiC-Ni cermets. Materials Synthesis and Processing 2:255–274.

Merzanov AG (1992) Self-propagating high-temperature synthesis and powder metallurgy: unity of goals and competition of principles. Advances in Powder Metallurgy and Particulate Materials 9:341–368.

Molokov IV, Mukasjan AS (1992) Explosive treatment of SHS gas-solid systems. Int J of Self-Propagating High-Temperature Synthesis 1:155–159.

Munir ZA (1988) Synthesis of high temperature materials by self-propagating combustion methods. Ceramic Bulletin 2(2):342–349.

Murr LE, Monson T (1988) Shock-induced microstructures in explosively fabricated super-conductors. J of Metals 40:19–23.

Rabin BH, Korth GE, Williamson RL (1990) Fabrication of titanium carbide-alumina composites by combustion synthesis and subsequent dynamic consolidation. J Am Ceram Soc 73:2156–2157.

Rice RW, Mc Donough WJ (1985) Intrinsic volume changes of self-propagating synthesis. J Am Ceram Soc. 68:232–239.

Shikhverdiev RM, Kudashev RR, Efimov OY, Zaripov NG, Gordopolov YA (1992) Shock wave effects in the Ti-C SHS system. Int J of Self-Propagating High-Temperature Synthesis 1:64–66.

Varma A, Lebrat JP (1992) Combustion synthesis of advanced materials. Chem Eng Sci 47:2179–2194.

Wang L, Wixom MR, Thompson LT (1994) Structural and mechanical properties of TiB_2 and TiC prepared by self-propagating high-temperature synthesis/dynamic compaction. Materials Science 29:534–543.

Dynamic compaction of granular materials in a tube with wall friction, applied to deflagration-to-detonation transition

L. G. Hill
Los Alamos National Laboratory
Los Alamos, NM 87545 USA

A. K. Kapila
Rensselaer Polytechnic Institute
Troy, NY 12180 USA

Abstract: A theoretical problem is considered in which a granular material is pushed through a tube of arbitrary cross-section by a constant-velocity piston against the resistance of compaction work and wall friction. The crushing of the material is dictated by a simple yet physically reasonable compaction law. By considering two special cases — the limit of vanishing friction and the quasi-static limit — we identify the two basic compaction wave structures. We then consider the general case in which the two waves interact. Estimates suggest that for typical deflagration-to-detonation tests, explosive at the wall melts on time scales that are short compared to the experiment.

Key words: Compaction, DDT, Friction, Granular explosive, Ignition, Melting

1. Introduction

Many investigators have examined deflagration-to-detonation transition (DDT) of granular explosives using shock-tube-like experiments (e.g., McAfee et al. 1989). Often a piston is driven explosively into the bed, and the resulting compaction work is thought to initiate combustion which ultimately leads to detonation. We examine an aspect of the DDT tube problem that has received little attention, namely the resistance and resulting energy dissipation due to wall friction. Wall friction has a large effect in ram-pressed charges (Elban & Chiarito 1986) — essentially the quasi-static limit of the DDT tube test — so that one expects even greater effects for the longer aspect-ratio DDT tubes. The resistance mechanism is that axial stress applied by the piston is transmitted to the walls, giving rise to a proportional frictional drag. Since particles are interlocked, resistance at the wall is transmitted throughout the interior. The entire bed resists motion in proportion to how hard it is pushed upon — much like the ubiquitous "chinese finger" but acting in compression.

2. Rheology

Because a confined granular material exhibits solid-like properties it is appropriate to adopt the generalized definition of pressure used in elasticity, i.e., $p = -(\sigma_x + \sigma_y + \sigma_z)/3 = -(\sigma_r + \sigma_\theta + \sigma_z)/3$, where all stresses are force per unit *total* area, normal stress components are positive in tension, and pressure is positive in compression. We define an ideal granular material whose loading state is dictated by a *compaction law* (Herrmann 1969) of the form $\phi = \rho/\rho_s = f(p)$, where ϕ is the solid volume fraction, ρ is the mixture density, and the subscript s means "solid". To determine analytic solutions we choose a simple but physically realistic form for $f(p)$, where S_0 and ϕ_0 are the zero-pressure slope and volume fraction, respectively:

$$\phi = 1 - \frac{(1-\phi_0)^2}{S_0\,p + (1-\phi_0)} \,. \tag{1}$$

We consider Class-A granular HMX explosive as an example, and infer its compaction law from pressing data (Elban & Chiarito 1986) using the fact that the experimental geometry enforced nearly uniaxial strain. Both this task and the formulation of the equations of motion in Section 3 are greatly simplified by the inference that Poisson's ratio, ν, remains constant and equal to its solid value during crushing. This follows from pressing data (e.g., Campbell et al. 1988) for which the ratio of normal to axial stress, which depends only on ν (specifically, $\sigma_n/\sigma_z = \nu/(1-\nu)$), remains constant. The inferred compaction law is shown in Fig. 1 together with a least-squares fit to Eq. (1).

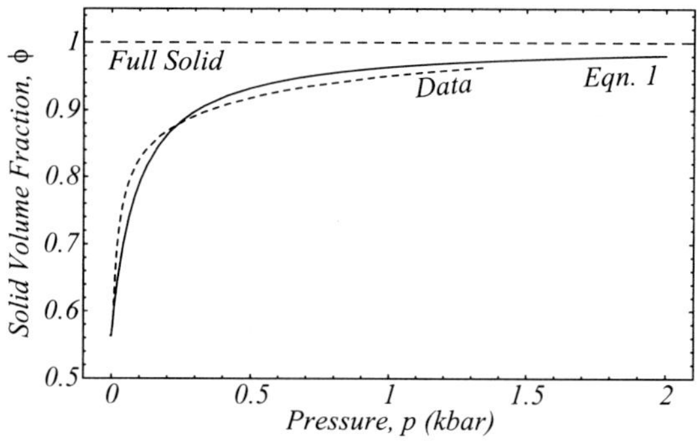

Figure 1. Compaction law for Class A HMX, $\phi_0 = 0.563$: data vs. fit to Eq. (1).

The wall drag is assumed to obey a standard friction law except that, because the solid area fraction in contact with the wall increases as the material is crushed, the friction coefficient is not constant. Intuitively one expects that the coefficient should be weighted by the solid area ratio, $\mu/\mu_s = A_s/A = \phi$. Equating the solid area and volume fractions is a common practice following from the statistical argument that for many randomly arranged particles, each plane within a volume element intersects the same solid area. This behavior for μ can be more rigorously justified, but here we merely note that the limits $\phi \to 0$ (no material) and $\phi \to 1$ (solid material) are both sensible.

3. Formulation

Consider a cylindrical semi-infinite tube of arbitrary cross-section, filled with powder and sealed at the origin by a piston, as shown in Fig. 2. The initial

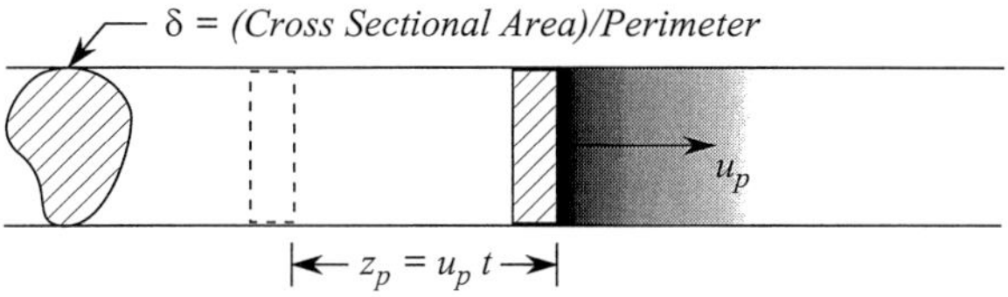

Figure 2. Schematic diagram of the problem.

conditions are uniform mixture density and zero pressure. At $t = 0^+$ the piston is moved impulsively to the right at a constant velocity u_p. The resulting material motion is assumed to be always and everywhere one dimensional. To simplify the equations we transform to a Lagrangian coordinate system, whereby the spatial coordinate z is replaced by the mass-weighted spatial coordinate h. Then, incorporating the above results and assumptions, the dimensionless equations for mass and momentum conservation become:

$$\frac{\partial \phi}{\partial \tilde{t}} + \phi^2 \frac{\partial M}{\partial \tilde{h}} = 0\,, \tag{2}$$

$$\frac{\partial M}{\partial \tilde{t}} = -\frac{\partial \tilde{p}}{\partial \tilde{h}} - \tilde{p}\,. \tag{3}$$

Here, $M = u/a_0$ is the Mach number, and the "tilded" quantities have been scaled as follows: $\tilde{p} = S_0\, p$, $\tilde{h} = h/l_f$, and $\tilde{t} = a_0 t/l_f$. The reference quantities l_f (the "friction" length) and a_0 (the longitudinal sound speed) are given by:

$$l_f = \frac{1-\nu}{\nu}\frac{\delta}{\mu_s}, \quad a_0 = \sqrt{\frac{3(1-\nu)}{1+\nu}\frac{1}{\rho_s S_0}}\,, \tag{4}$$

with δ the area-to-perimeter ratio of the cross section.

4. Results

4.1. Zero friction limit

In the absence of friction the second term on the right-hand side of Eq. (3) does not appear. The piston drives a steady compaction shock (sw) into the powder, whose speed M_{sw} and amplitude $\tilde{p}_{sw}$ are given by:

$$M_{sw} = \frac{M_p}{2(1-\phi_0)}\left(1+\sqrt{1+\frac{4(1-\phi_0)^2}{M_p^2}}\right), \qquad \tilde{p}_{sw} = \phi_0\, M_p\, M_{sw}\,, \tag{5}$$

where $M_p = u_p/a_0$ is the piston Mach number. In reality the shock has a finite thickness which depends on its amplitude (e.g., McAfee et al. 1989).

4.2. Quasi-static limit

This case corresponds to a vanishingly small piston speed. Over correspondingly long time intervals Eq. (2) survives intact but the left-hand side of Eq. (3) is negligible. Subject to the requirements that the initial piston pressure is zero and that the powder far upstream is undisturbed, an analytic solution emerges:

$$\tilde{p}(\tilde{t},\tilde{h}) = \tilde{p}(\tilde{t},0)e^{-\tilde{h}}, \quad \phi(\tilde{t},\tilde{h}) = 1 - \frac{(1-\phi_0)^2}{\tilde{p}(\tilde{t},\tilde{h})+(1-\phi_0)}, \tag{6}$$

$$\tilde{p}(\tilde{t},0) = \phi_0(1-\phi_0)\left[\exp\left(\frac{\phi_0}{1-\phi_0}M_p\tilde{t}\right)-1\right]. \tag{7}$$

The pressure and compaction fields, mapped back to the Eulerian frame, are shown in Fig. 3. The wave is not truly steady as the pressure is always rising. But, due to the compaction law shape, the compaction field assumes a steady profile $\approx 5\,l_f$ wide following a start-up transient. The downstream state of the developed frictional wave (fw) is fully compacted, so its (Eulerian) speed is:

$$M_{fw} = \left(\frac{1}{1-\phi_0}\right)M_p\,. \tag{8}$$

4.3. General case

One expects the general solution, corresponding to nonzero friction and order-unity piston speeds, to involve a friction-attenuated shock. One may further surmise that to a good approximation the problem can be divided into two distinct regions: the leading shock for which inertial forces dominate, and the downstream flow in which frictional forces dominate. A perturbation solution valid for short times, or equivalently, small friction coefficient (not presented

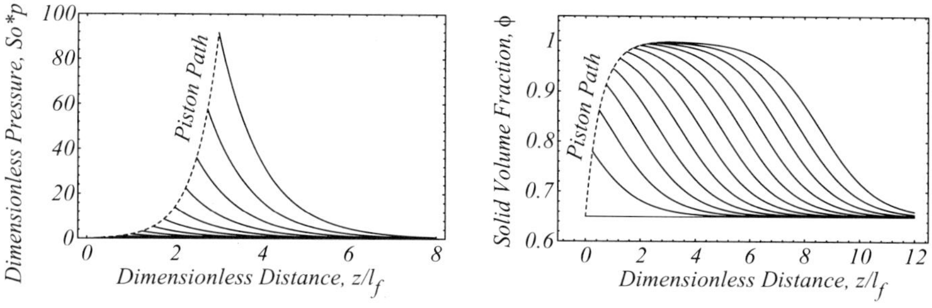

Figure 3. Pressure, compaction fields in quasi-static limit; $\phi_0 = 0.65$, $\Delta\tilde{t} = 0.25$.

here), finds the correction to the particle speed between the piston and the shock to be time independent, which supports this notion. In this inertialess approximation Eq. (3) formally reduces to that for the quasi-static limit, but the initial piston pressure is the zero-friction shock pressure corresponding to the prescribed piston speed, and the solution extends only to the shock location $\tilde{h}_{sw}(\tilde{t})$, at which point it is matched to shock jump conditions. There are now two coupled first-order ODEs that must be solved numerically. A phase-plane analysis of the two equations, also not presented here, shows that below a critical value of M_p, $M_{pcrit} = 1 - \phi_0$, the strength of the leading shock decays to zero asymptotically. Above M_{pcrit} its amplitude decays to a finite value.

Figure 4 compares the results of this inertialess approximation (again mapped to the Eulerian frame) to the results of a full numerical solution. Their agreement is remarkable, with respect to both pressures and shock location.

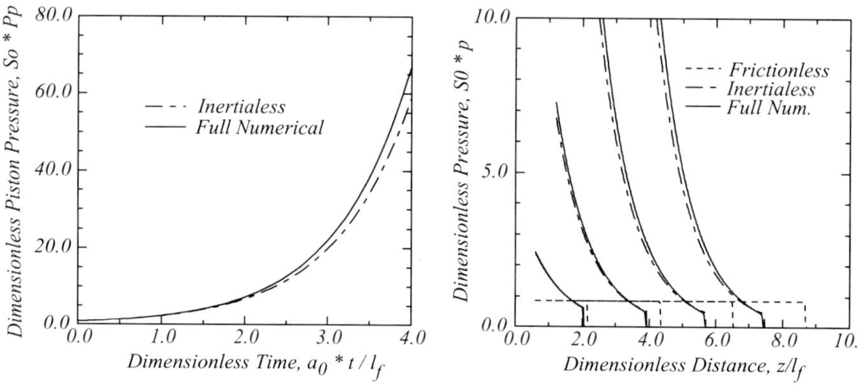

Figure 4. Pressure for full problem; $\phi_0 = 0.65$, $\Delta\tilde{t} = 1$, $M_p = 0.6$, $M_{pcrit} = 0.35$.

5. Wall temperature

Heat generation at the wall can be equated to the frictional work there: $\dot{q_w} = \dot{w_f} = \mu_s \, \phi \, (-\sigma_n) \, u$. The problem is simplified by noting that most of the heat flux flows into the (metal) wall, and that the temperature distribution is, for short times, confined to a thin boundary layer. The wall temperature is then:

$$T_w(z,t) = T_{w0} + \left(\frac{3\nu}{1+\nu}\right)\left(\frac{\mu_s}{\sqrt{\pi} b_w}\right) \int_0^t \frac{\phi(z,\tau)\, p(z,\tau)\, u(z,\tau)\, \mathrm{d}\tau}{\sqrt{t-\tau}}, \qquad (9)$$

where b_w is the wall heat-penetration coefficient. After the leading shock passes one may briefly assume a constant state. Then for a steel wall and $u_p \approx 100$ m/s, $\mathrm{d}T_w/\mathrm{d}t$ is of order 100 C/μs. The melting temperature (247 C) is reached in order 1 μs — the same time scale as the leading shock rise. One expects a melt layer to lubricate, decreasing further frictional resistance and energy dissipation. The question of wall ignition may therefore depend strongly on the difference between the melting and critical temperatures of the explosive.

6. Conclusions

Without wall friction, an impulsively started piston drives a narrow compaction shock through the tube. In the quasi-static limit with wall friction a different kind of compaction wave occurs whose width is proportional to the tube diameter and inversely proportional to the friction coefficient. In general, both wave types are present, and their interaction is such that the amplitude of the leading shock is attenuated. Neglecting inertia behind the shock yields an excellent approximation to the wave structure. In a typical DDT test the rate of work at the wall is sufficient to cause rapid melting and, possibly, ignition.

Acknowledgement. This work was supported by the U.S. D.O.E. and by the N.S.F. We thank the "DDT team" for their interaction. AKK also thanks Don Schwendeman for sharing his numerical expertise and Arwen Warlock Dixon for helpful discussions.

References

Campbell RL, Elban WL, Coyne PJ (1988) Side-wall pressure measurements in quasi-static compaction of porous beds of HMX powders and ABL 2523 casting powder. JANNAF Propulsion Sys Hazards Mtg Los Angeles, CA.

Elban WL, Chiarito MA (1986) Quasi-static compaction study of coarse HMX explosive. Powder Technology 46:181.

Herrmann W (1969) Constitutive equation for the dynamic compaction of ductile porous materials. J Appl Phys 40:2490.

McAfee JM, Asay BW, Campbell AW (1989) Deflagration to detonation transition in granular HMX. In: 9th Symp (Int) on Det Portland, OR, p 265.

Investigation of HE-driven cylindrical liner implosions

T.-H. Tan, S. P. Marsh
Los Alamos National Laboratory
Los Alamos, NM 87545 USA

Abstract: A 2D hydrocode calculation shows that when a hollow core HE cylinder is initiated to drive a thin stainless steel tubing, the convergence velocity can exceed 1 cm/μs to produce 15 Mbar pressure at impact as the density increases to 19.5 g/cm^3. The temperature from shock heating rises rapidly above 8 eV, and the result is a combination of radiation and plasma emissions. We have carried out several experiments with a wide array of diagnostics to investigate the implosion dynamics and final state interaction phenomena, and the results are found to be in reasonable agreement with the code predictions.

Key words: Liner implosion, Shock compression, Shock heating

1. Introduction

Several experiments have been performed to investigate the complex dynamics inside the cavity formed by the phased implosion of an HE-driven cylindrical liner. They are motivated by the results of hydrocode calculations which show that extremely high pressure and temperature can be reached when a liner is imploded at very high velocity. For example, when a hollow-core HE cylinder is initiated to drive a thin stainless steel tubing, the convergence velocity is predicted to reach above 1 cm/μs, thus producing a 15-Mbar pressure as the density is shock compressed to 19.5 g/cm^3. The temperature from the heating rises rapidly above 8 eV and is followed by a combination of radiation and plasma emissions. The process continues for many microseconds as the phased implosion proceeds along the axis, and the evolving dynamics can be observed from the open end of the cylinder. The presence of structures in the expansion front and the maintenance of high temperature during the expansion are predicted, too. Simulation also shows that the plasma flow can be perturbed and hydrodynamically redirected to converge into a smaller region. These are intriguing code predictions that if proven valid will provide us with a means to investigate a variety of topics in interesting dynamic regimes. In these exploratory experiments, the HE initiation simultaneity and liner integrity are closely monitored to ensure that the implosion is symmetrical and free from evidence of material failure prompted by instabilities. Radiographs of implosions are compared with those from numerical simulations. Finally, multiple diagnostics that include an extensive array of time- and spatially resolved instruments are employed to measure the history and profiles of radiation emission and plasma expansion. In this paper we include samples from code calculations and a discussion of the device and experimental setup, diagnostic instrumentation, and data and code comparisons.

2. Sample calculations

The MESA Eulerian hydrocode calculation is the tool that provides the critical guidance in the design and the execution of the experiments. A few examples from the implosion of a stainless steel liner are provided here to facilitate the discussion. Similar calculations have been extended to other materials for comparison but will not be presented. The equation of state (EOS) input and zone size play a critical role in the outcome of the simulations. These EOS values are often not totally derived from measurement, but rather from calculations based on a complicated set of physical models to cover the entire domain from solid to ionized gas. An end cap is incorporated to seal the cavity from gas leaks under intense dynamic loading.

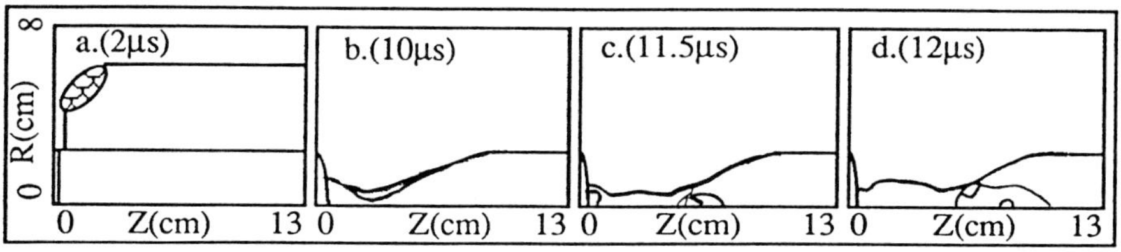

Figure 1. Samples of 2D hydrocode calculation at different times.

Snapshots from the calculations are shown in Fig. 1. Time starts at zero when the corner of the HE cylinder is initiated. To count from the firing unit discharge, an additional 15.2-μs burn time must be added. Figure 1a shows the device setup and the progression of the HE detonation wave at 2 μs after the corner initiation. Figure 1b illustrates the liner implosion at 10.5 μs, right after the liner collapse on axis. The maximum compressed density is calculated to be 19.5 g/cm^3, while maximum pressure and temperature are 15 Mbar and 8 eV, respectively. At this temperature the generated hot gas will be partially ionized. The simulated event at 11.5 μs is shown in Fig. 1c, where the plasma expansion envelope is moving at about 6 cm/μs, and closer examination reveals a wavy structure, suggesting the presence of instability. The dynamic profile at 12 μs is illustrated in Fig. 1d for comparison with the radiograph. At later times, the calculation shows the hot gas to continue expanding and persisting with strong sign of instability at the expansion front. The temperature remains high but may be a numerical artifact since the code does not account for heat loss. The bulk of the hot gas continues to expand at very high velocity with the fastest component exceeding 14 cm/μs. A cluster of much higher density liner debris traveling at 1.8 cm/μs eventually impacts the port to generate a 25-eV temperature.

3. Experiments

3.1. Device description and a typical experimental setup

The device description and arrangement of a typical experiment are illustrated in Fig. 2. The device, which consists of a simple hollow core PBX-9501 HE cylinder (13.5 cm o.d $\times$ 5.08 cm i.d. $\times$ 13.5 cm long), is corner initiated to implode a 5.08-cm o.d., 0.081-cm thick, seamless 304 stainless steel tubing. The ring initiation is accomplished with a single detonator by means of a thin-wall conical structure fabricated from C2 explosives. A 0.48-cm thick stainless steel sealing disk is welded to one end of the liner. The open end is glued to an expansion chamber to facilitate the fielding of diagnostics. The chamber is made either from plastic or aluminum and may be evacuated or backfilled to any level of pressure. To measure the plasma expansion and propagation of uv and higher frequency radiation the chamber is evacuated. A plastic chamber is employed when a vacuum above a few millitorrs is adequate and it enables cameras to view sideways through the chamber. A metal chamber is used when higher vacuum is needed.

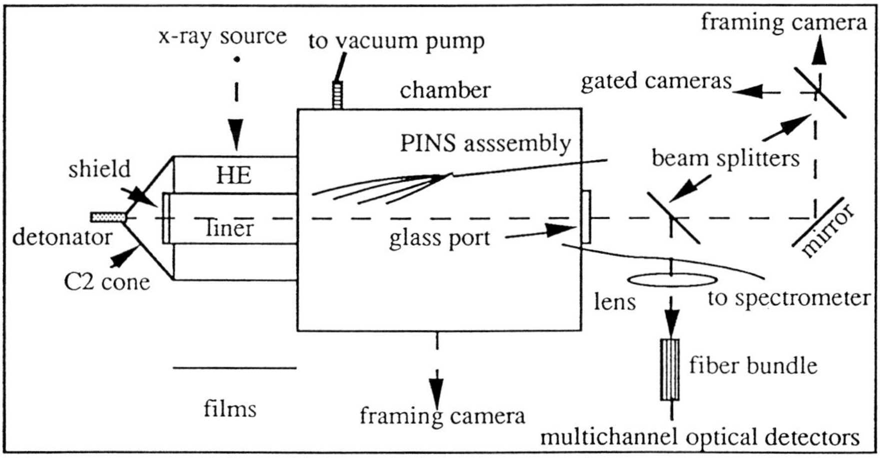

Figure 2. Schematic of device and experimental setup.

3.2. Itemized diagnostics and functions

An extensive array of diagnostic instrumentation is fielded in each shot, and designed to assess the performance and to reveal temporal and spatial information that can be correlated with the simulations. Some instruments are not shown in Fig. 2; all instruments are not fielded in every shot. Often, eight electrical pins are placed in a collar around the HE cylinder to measure the initiation simultaneity. A radiograph is taken to assess the implosion symmetry and over-all dynamic behavior. A time- and spectrally resolved multichannel optical-fiber detector system is employed to record light-emission history inside the imploding cavity. These detectors with color or neutral density (ND) filters are calibrated and arranged to cover four orders of magnitude in intensity. Sharing the view along the axis with the aid of a beam splitter are several time-resolved cameras. A gated 1024-channel spectrometer is fielded via an optical fiber to grab an interval of light emission for absolute spectrum determination, and it provides an in-situ calibration for the fiber detectors. An IMACON framing camera that views sideways through the plastic chamber is fielded to record the expansion of plasma and liner debris. Optical pins were introduced once to measure the plasma expansion velocity; but they are intrusive to the experiment. Photoelectric diodes (XRD) are employed to measure the uv, vuv, and soft x-ray emissions.

3.3. Results and comparisons with code predictions

In all the experiments the electrical-pin data have consistently shown that the initiation around the corner of the HE cylinder is simultaneous to within 25.5 ns and starting at 15.2 μs after the firing discharge. All the experimental times are measured from the discharge signal. A radiograph is shown in Fig. 3. Aside from showing implosion symmetry, it also provides a wealth of dynamic information after the initial axial collapse, which includes the size and shape of the imploding liner over its entire length, signatures of plasma and jet emissions, shock reflection into the expanding HE gas, and behavior of the sealing cap. The profile from the calculation overlaps remarkably well with the observed features.

The signal of a typical fiber optic channel is shown in Fig. 4a. The start at 25.5 μs corresponds exactly to the calculated time of collapse. It rises steeply to reach a peak in about 1 μs and may be explained by plasma expansion into the imploding wall. The temperature between 26.25 to 27.25 μs is estimated to be 8 eV from intensity calibration, which is consistent with the fit to a blackbody-like spectrum from the spectrometer and in agreement with the prediction. Continuous intensity increase before the second peak may be attributed to emissions from the expanding plasma surface, hot spots generated by instability, and con-

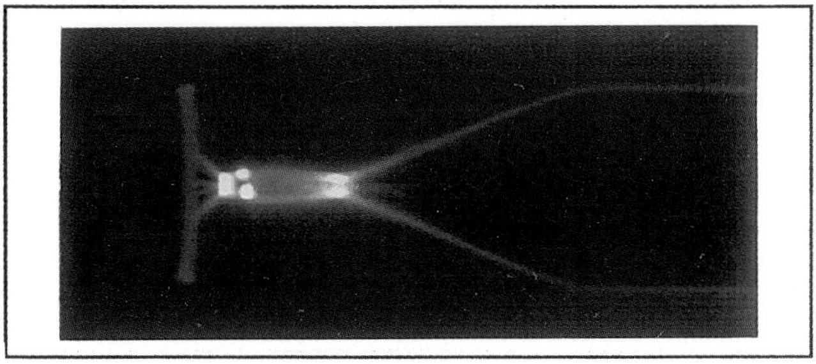

Figure 3. A 200-ns radiographic exposure taken at 26.9 μs.

tinuous implosion. The calculation predicts maximum temperature to remain between 8 and 10 eV during the expansion. The location of the second peak at 32 μs coincides with the calculated arrival of higher density plasma at the glass port. The third peak at 36.6 μs correlates well with the predicted final impact by the massive cluster. This is supported by the observations from the side viewing IMACON and the axially imaging rotating mirror camera. The absence of light earlier than 25.51 μs suggests that the vacuum has no leak.

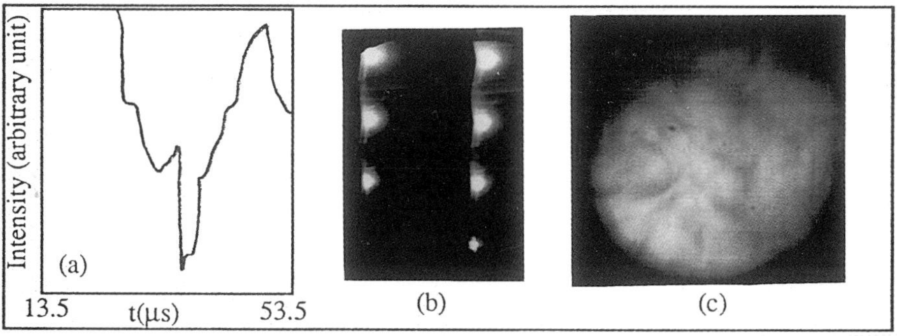

Figure 4. (a) Light emission inside the imploding cavity; (b) expanding debris in 0.5-μs interval; and (c) axial view of expanding plasma with 200-ns exposure.

IMACON images (with camera viewing sideways through the plastic chamber) wall were recorded at 0.5-μs intervals. Using a 1-ND filter, the early unsaturated records reveal the impact at the window by the transparent plasma. The fastest plasma velocity is thus estimated to be between 12–17 cm/μs, consistent with the calculation and supported by the optical pin measurements. The set of IMACON images, shown in Fig. 4b, was taken with 2-ND filters. The sequence is from left to right and bottom to top. The self-luminous debris enters the chamber at 29.5 μs and advances at an estimated 1.9 cm/μs, again consistent with the prediction and is responsible for producing the third peak in Fig. 4a.

The usefulness of ultra-fast framing photography of the imploding liner cavity is constrained by its inability to accommodate the rapid variation in the object location and light emission intensity. In one setting the IMACON images were saturated because of insufficient attenuation. In another setting, a rotating mirror camera focusing on the expansion chamber yielded no information on the collapsing cavity. However, the images of a cloud-like expansion over several microseconds corresponds closely to the debris observed in the IMACON. The punch-through at the port is clearly visible. Finally, a well-focused snapshot taken along the axis by a fast-gated electronic camera at 1.1 μs after the liner collapsed is shown in Fig. 4c. The appearance of structures in the plasma expansion implies nonuniform density and/or temperature distribution, strongly indicating the presence of turbulence and hot spots as suggested by the simulation. Further study of this structure may shed light toward understanding the nature of plasma behaviors in a phased implosion process. In the one experiment where soft x-ray detectors are fielded, the unfiltered channel with a threshold of 16 eV sees a reasonable amount of vuv and soft x-rays emitted from the expanding plasma. Only a small signal is detected in the high-energy carbon filtered channel. This data is consistent with the presence of hot spots in the expanding plasma surface.

4. Conclusions

We have demonstrated that a very thin stainless steel cylindrical liner can be driven by HE to implode symmetrically and intact to very high velocity. Extreme conditions are created as a result. We are pleasantly surprised to find that the data are in reasonable agreement with many of the predictions from the numerical simulation. It appears that we may have found a powerful tool to study interesting ultra-high pressure and temperature phenomena in the regimes that are normally investigated only with electromagnetically driven, pulsed power devices.

Acknowledgement. We are indebted to C. Findley, N. Gray, H. Oona, P. Rodriquez, S. Sterbenz, and the supporting crews at the PHERMEX and Ector experimental facilities for their contributions. Research supported by DOE.

Modeling shock-wave propagation in obliquely loaded unidirectional composites

A. D. Resnyansky*
Lavrentyev Institute of Hydrodynamics
Novosibirsk 630090 Russia

E. I. Romensky
Sobolev Institute of Mathematics
Novosibirsk 630090 Russia

V. Y. Belov, G. H. Williams
University of Wollongong
Wollongong, NSW 2522 Australia

Abstract: This paper addresses the calculation of shock-wave propagation for obliquely loaded unidirectional composites. The behavior of Glass Fiber-Reinforced Plastics (GFRP) and Lexan/Aluminum laminated composites have been computed for loading by waves propagated both along the symmetry axis of the material and in off-axis directions. Additional wave splitting occurs in the latter case. It is shown that microstructural peculiarities underlie this behavior. Good agreement with the available experimental data has been established.

Key words: Fibrous and laminated composites, Shock-wave propagation, Numerical simulation, Off-axis loading

1. Introduction

Composite materials play a very significant role in modern industry. Their behavior is well described in static conditions by the anisotropic models of elasto-plasticity. However, the behavior of constructions for times comparable to the reaction time of the material must be described by kinetic models.

Further, complex loading requires a consideration of the interphase peculiarities connected with the off-axis action of the deformation. A large amount of experimental data is available for unidirectional fibrous composites. However, only a few of them are in dynamic conditions (Walsh & Harding 1985, Chiem & Lee 1994). Sets of data dealing with shock-wave propagation are even fewer (Lundergan & Drumheller 1972, Sve & Okubo 1974). Unfortunately, they are concerned with laminated structures. We will analyze the data (Sve

*Present address: Mathematics Department, University of Wollongong, Wollongong, NSW 2522, Australia.

& Okubo 1974) for off-axis shock-wave loading. We use models (Resnyansky & Romensky 1992, Resnyansky & Romensky 1993a) obtained with the help of the phenomenological homogenization method (Resnyansky & Romensky 1993b). The analysis of unidirectional structures by the models developed used simple statements of thin-rod stretching (Resnyansky & Romensky 1993b) or tube tests (Resnyansky et al. 1995). Shock-wave propagation has been computed only for structures associated with the direction of the wave propagation (Bordzilovsky et al. 1994). In this paper we apply our model to the shock-wave propagation problem for the off-axis case.

2. Fibrous composites

A general model for the case of finite deformations (described, e.g., in Resnyansky et al. 1995) is reduced for the fibrous composite in the one-dimensional case to the following:

$$
\begin{gathered}
\frac{d\varphi}{dt} = \frac{a_{21}a_{22}a_{33}(\psi_{11}-\psi_{22}) + a_{33}((a_{22})^2-(a_{21})^2)\psi_{12}}{(\rho/\rho_*)} \, , \quad \frac{d\tilde{\Delta}_{11}}{dt} = -\tilde{\lambda}_{11} \, , \\
\frac{da_{11}}{dt} + a_{11}\frac{\partial u}{\partial x} = -a_{11}\psi_{11} - a_{12}\psi_{21} \, , \quad \frac{da_{21}}{dt} + a_{21}\frac{\partial u}{\partial x} = -a_{21}\psi_{11} - a_{22}\psi_{21} \, , \\
\frac{da_{12}}{dt} = -a_{11}\psi_{12} - a_{12}\psi_{22} \, , \quad \frac{da_{22}}{dt} = -a_{21}\psi_{12} - a_{22}\psi_{22} \, , \quad \frac{da_{33}}{dt} = -a_{33}\psi_{33} \, , \qquad (1) \\
\frac{d\rho_*}{dt} = \rho_*(\psi_{11}+\psi_{22}+\psi_{33}) \, , \quad \frac{dE}{dt} - \frac{\sigma_{11}}{\rho}\frac{\partial u}{\partial x} = 0 \, , \quad \rho\frac{du}{dt} - \frac{\partial \sigma_{11}}{\partial x} = 0 \, , \\
\sigma = -2\rho G\frac{\partial E}{\partial G} \, , \quad G = A^T A \, , \quad E = E(G, S, \tilde{\Delta}) \, , \quad \tilde{\Delta} = B^{-1}\Delta B \, ,
\end{gathered}
$$

where φ denotes the angle between the fiber and the x_1-axis and we suppose the fiber lies in the $(x_1,\ x_2)$-plane. Here $t,\ x,\ u,\ \rho,\ \sigma,\ T$ are the standard quantities. The tensor of structural parameters, Δ, is responsible for microstresses resulting from the difference in elastic deformations of constituents, and B is the matrix of rotation around the x_3-axis on the angle φ. Λ with components λ_{ij} and Ψ with components ψ_{ij} are given tensors which describe the irreversible behavior of the composite. ρ_* is the material density in the unloaded state. The internal energy E is a given function of a self-adjoint metric tensor of elastic strains G, the entropy S, and Δ (an equivalent of the equation of state). A with components a_{ij} is the tensor of elastic distortion. Tensors Λ and Ψ, as well as the potential E, are chosen, as a first approximation, from the linear consideration of the case of small elastic deformations (Resnyansky & Romensky 1992).

Assuming uniform deformation ($\partial u/\partial x = \dot{\varepsilon}$) for the case of a thin rod (in this case the constitutive equations for a_{ij} at $i \neq j$ are replaced by the conditions

$\sigma_{ij} = 0$), we can obtain a set of $(\sigma - \varepsilon)$ curves for a given $\dot{\varepsilon}$. We are interested in a set of curves depending on the fiber orientation. It should be noted that in our model, φ corresponds to the fiber location in the reference (unloaded) state. Therefore, the real fiber position should be corrected from φ by a possible adiabatic loading. For unidirectional GFRP with the same properties as in (Bordzilovsky et al. 1994) the resultant $(\sigma - \varepsilon)$ curves 1, 2, 3, at initial fiber position $\varphi_0 = 0,\ 0.2,\ 0.7$ rad, respectively, are presented in Fig. 1. The nonlinear behavior at $\varphi_0 = 0.2$ rad is caused by the change in orientation during the rod stretching. The mode at $\varphi = 0$ corresponds to stretching along the fibers, the mode at $\varphi = 1.57$ corresponds to the case of a load applied to the stretched rod in the transverse direction. In the mode at $\varphi = 0.7$, which is close to the case $\varphi = 1.57$ qualitatively, the composite's yield limit falls down to the yield limit of the matrix. During the deformation at $\varphi = 0.2$ the transition from the last regime to the first one occurs when the fiber orientation aligns with the tension axis. Thus, an intermediate range of angles is of special interest and

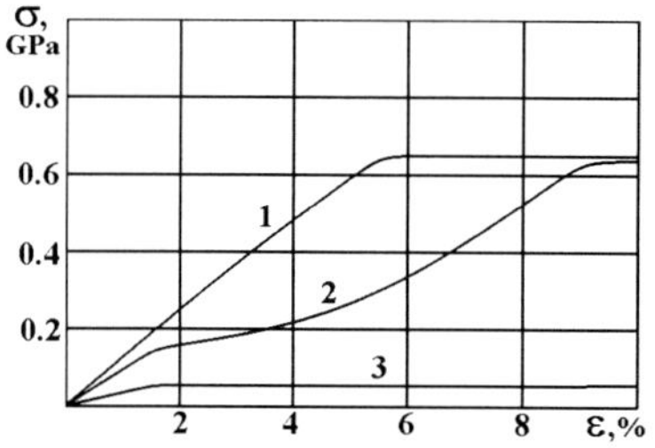

Figure 1. The stress-strain curves 1, 2, 3 for GFRP at $\varphi_0 = 0,\ 0.2,\ 0.7$, respectively.

important because, firstly, nonlinear behavior occurs and secondly, peculiarities of the phase interaction arise from the interphase sliding, debonding, etc. The latter is not considered by the model except for the material sliding. However, a number of these effects could be modeled.

For $\varphi_0 = 0.2,\ 0.7,\ 1.2$, profiles of the shock wave, induced by a plane impulse of velocity $v_0 = 0.8$ km/sec, are plotted in Fig. 2 at the moment when the first wave reaches the right boundary of the composite sample. Here v_0 is the velocity behind the last front. In cases (a) and (c) two-wave splitting occurs. However, the nature of the first waves in these cases is rather different. In the case (a) the first "precursor" is responsible for plastic deformation in the glass and elastic deformation in the epoxy matrix because in the latter phase the elastic strain limit is higher. The elastic precursor is overlapped by this wave whose intensity is rather high. On the other hand the first wave in the case (c) is the real elastic precursor; microstresses are unchanged behind it. As can be seen from Fig. 1, the yield stress decreases as the off-axis angle increases. Hence, with increasing

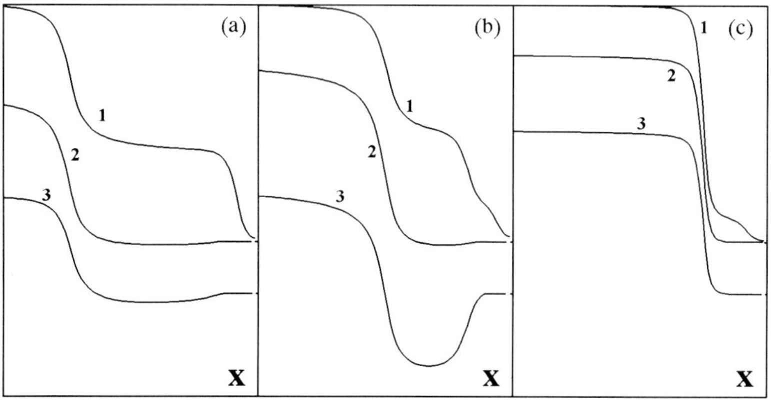

Figure 2. The shock-wave profiles: (a) $\varphi_0 = 0.2$; (b) 0.7; (c) 1.2. Curve 1 – velocity; 2 – the value of structural parameter $\tilde{\Delta}_{11}$; 3 – the value of $\varphi_0 - \varphi - 0.01$.

φ_0, the intensity of the first wave decreases as well and the intensity of the elastic precursor, whose speed is determined by the elastic moduli of the composites, rises. On the other hand, when the speed of the first wave is decreased, the resultant plastic wave overtakes the first wave and the wave structure reduces to the two-wave profile again (Fig. 2a).

This phenomenon needs to be proved experimentally. Unfortunately, shock-wave data are not available for fibrous composites. Nevertheless, similar data are available for laminated structures. Therefore, in the following section we will analyze the behavior of these structures.

3. Laminated composites

In the case of laminated composites, the general system of equations reduces to an analogous system (1) with specific right-hand parts which are of the form used in Resnyansky & Romensky (1993a). We assume that there is no motion in the directions transverse to the wave-propagation axis. Then, the structural parameter $\tilde{\Delta}_{22} = \tilde{\Delta}_{33} = \tilde{\Delta}_+$ is the only one involved in the computed system. We will consider the experimental data for the loading of the composition Lexan/Aluminum by a wave generated in air (Sve & Okubo 1974). Fortunately, mechanical characteristics of the constituents are such that wave splitting occurs. Results and analyses of numerical simulation at $v_0 = 0.025$ km/sec (Fig. 3), corresponding to the present case, are analogous to those in the previous section. We have computed the loading with the following initial angles between the normal to the lamina and the direction of the shock-wave propagation: (a) $\varphi_0 = 47°$ and (b) $\varphi_0 = 90°$ (see Sve & Okubo 1974). These

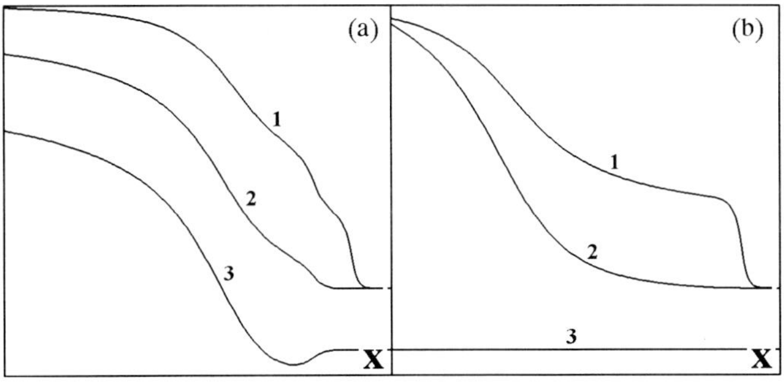

Figure 3. The shock-wave profiles at (a) $\varphi_0 = 47°$; (b) $90°$. Curve 1 – velocity; 2 – the value of structural parameter $\tilde{\Delta}_+$; 3 – the value of $\varphi_0 - \varphi - 0.01$.

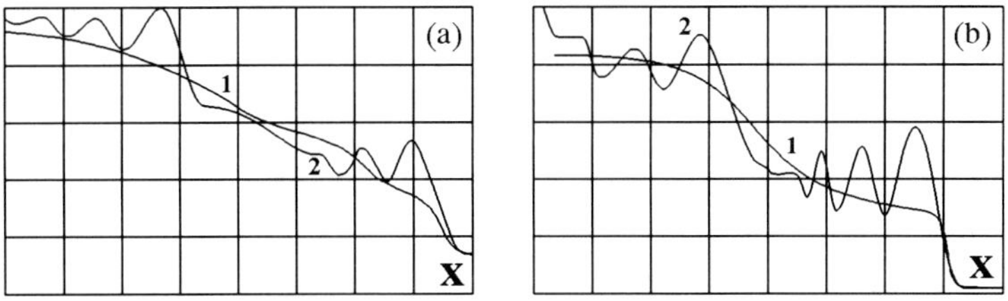

Figure 4. The shock-wave profiles of velocity at (a) $\varphi_0 = 47°$; (b) $\varphi_0 = 90°$; 1 – calculation, 2 – the experiment.

cases are structurally analogous to the cases of loading along fibers and the off-axis loading in the previous section (Fig. 2b and Fig. 2a, respectively). Explanations of the microstructural behavior during the shock-wave propagation are practically the same. Exceptions are in the microstress behavior behind the first plastic wave. This results from the difference in respective mechanical properties of Lexan and Aluminum in comparison with the pair Glass – Epoxy. In this case the constituent of the composite irreversibly deformed in the first wave is stretched whereas in the previous case the corresponding phase was compressed. A comparison of the calculations with the experiment (Sve & Okubo 1974) (curve 2 in Fig. 4) in reduced coordinates shows good agreement. It should be noted however that the properties of the constituents for the materials computed may be quite different from the materials used in Sve & Okubo (1974).

4. Conclusion

A phenomenological explanation of shock-wave splitting observed in experiments has been proposed. The microstresses arising during deformations are taken into account. The many interphase peculiarities occurring in the off-axis tests (debonding, splitting, fracture) are not considered but the present consideration was sufficient for successful modeling. The phenomenological homogenization method was designed with a careful consideration of the mechanical processes across the structural elements. Some strain-rate peculiarities with the loading along the interphase surfaces still should be considered. Unfortunately, there is only a small number of suitable dynamic experiments available at present concerning this subject.

Acknowledgement. The first two authors were supported by the International Science Foundation, Grant No. RCC000. The second author was also supported in part by the Russian Foundation for Fundamental Researches, Grant No. 95-01-00942.

References

Bordzilovsky SA, Karakhanov SM, Merzhievsky LA, Resnyansky AD (1994) Shock-wave propagation in unidirectional composites. DYMAT Journal 1(4):333.

Chiem CY, Lee WS (1994) Dynamic behaviour of tungsten-base composite under shear loading. DYMAT Journal 1(1):15

Lundergan CD, Drumheller DS (1972) The propagation of transient stress pulses in an obliquely laminated composite. In: Dynamics of Composite Materials, (Lee EH, ed), ASME, New York, pp 35–47.

Resnyansky AD, Romensky EI (1992) Dynamic strain model for a fibrous thermoviscoelastic composite. Fizika Gorenia i Vzryva 28(4):120.

Resnyansky AD, Romensky EI (1993a) Model of dynamic deformation of a laminated thermoviscoelastic composite. Fizika Gorenia i Vzryva 29(4):123.

Resnyansky AD, Romensky EI (1993b) Model of viscoelastic unidirectional composite with coupled thermal and elastic processes. In: Shock Waves @ Marseille, (Brun R, Dumitrescu LZ, eds), Springer-Verlag, Vol. 3, pp 273–278.

Resnyansky AD, Romensky EI, Belov VY, Williams G (1995) Plasticity phenomena at the high-rate loading of the fibrous composite. In: Proc 5th Symp (Int) on Plasticity and Its Current Appl, Sakai, Japan (to appear).

Sve C, Okubo S (1974) Experiments on pulse propagation in an obliquely laminated composite. J Appl Mech 39:1052.

Walsh LM, Harding J (1985) Effect of strain rate on the tensile failure of woven reinforced polyester resin composites. J de Physique 46(C5):405.

Shock consolidation of powders by utilizing a converged underwater shock-wave assembly

S. Kubota, S. Nagano, M. Fujita, S. Itoh
Department of Mechanical Engineering
Kumamoto University 2-39-1 Kurokami, Kumamoto, Japan

Abstract: We tried to consolidate Nd-B-Fe magnet powder to get a high-performance permanent magnet. A rapid plastic deformation during the shock consolidation of the powder may lead to superior magnetic properties. In this paper, the processes of the consolidation of the magnet powder were investigated by numerical calculations.

Key words: Shock consolidation, Nd-B-Fe powder, Underwater shock wave

1. Introduction

Shock consolidation is one of the typical techniques used to produce bulk material from powder. It is a well-known technique to use cylinders for these consolidations. In this technique, for the most part, procedure explosives have been used in direct contact with the tubular powder container. While this technique gives good consolidation for soft powders, such as pure copper and mild steel, it is not successful with hard powder (Gourdin 1986).

The two most severe limitations of this process are cracking and holing in the center of the bulk. The cracks may be caused by the effects of rarefaction waves after shock reflection which occurs at the free end of the cylinder. This rarefaction wave makes the temperature of the bulk go down rapidly as it propagates into the bulk. This excessive temperature change causes thermal stress and leads to many cracks. A central hole is generated due to the effect of Mach reflection of the shock waves (Meyers and Wang 1988).

We have developed a consolidation technique which can be applied to difficult-to-consolidate powders, using a converging underwater shock wave. It was confirmed that this technique has a higher and longer pressure duration on the powder than that of the cylinder technique.

The consolidation of magnetic powder is very interesting for the manufacture of high-performance permanent magnets (Petrie and Page 1992). They pointed out that a rapid plastic deformation during the shock consolidation of the powder leads to superior magnetic properties. In this paper, we present the

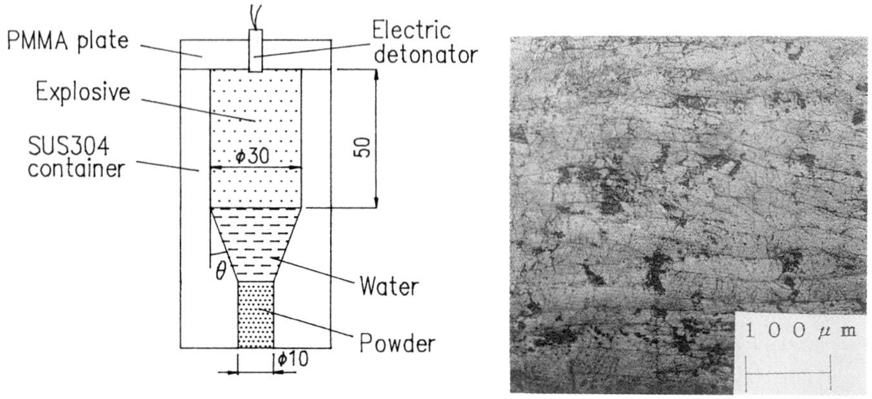

(a) (b)

Figure 1. (a) Schematic illustration of shock-compaction assembly; (b) optical micrograph of compacted Nd-B-Fe powder.

results of the consolidation of Nd-B-Fe powder by experiment. The processes of the consolidation of the magnetic powder were investigated by numerical calculation. Finally, the converging process of the underwater shock wave is also verified.

2. Numerical calculation

Figure 1a illustrates the underwater-shock-consolidation assembly. This assembly consists of three parts: an explosive container, a water tank, and a powder container. A high explosive called SEP, manufactured by ASAHI Chemical Industry Co. Ltd., was used in the experiments. The detonation velocity of SEP was 6970 m/s. The initial density was 1310 kg/m^3.

Figure 1b shows an optical microgragh of compacted Nd-B-Fe powder. It is well compacted. When explosives detonate using a suitable electric detonator, the detonating wave originates and propagates from top to bottom of the explosive chamber. This detonation wave immediately causes an underwater shock wave after impinging on the water in the tank. The underwater shock wave travels into the water tank, where it interacts with the tank wall to converge under increased shock pressure. This increased shock pressure and shock duration facilitates the consolidation of the powder. The pressure level and shock duration of the consolidating powder can be easily controlled by adjusting the mass and the detonation-wave velocity of the explosive. The explosive chamber has a height of 50 mm and an inner diameter of 30 mm. The powder chamber has a 10-mm diameter and a length of 20 mm. We used SUS304 steel for the water chamber. Numerical calculations were carried out using the developed Arbitrary Lagrangian Eulerian method (ALE method) (Amesden et al. 1980). The governing equations are two-dimensional inviscid equations of motion. For details of this procedure refer to the original paper. The equations of state for all materials concerned are necessary in the calculations. We used the Jones-Wilkins-Lee (JWL) equation of state for the explosive (Lee et al. 1973). JWL parameters were obtained by previous expansion-tube tests (Itoh et al. 1994).

For numerical calculation of the powder-compaction process it is important to have a reliable model for the shock response of the powdered materials. Petrie and Page (1991) proposed the Hugoniot model of the powder by utilizing cold-pressure models. The cold pressure is the pressure that would be measured in the absence of thermal effects. In this case the only form of energy storage in the compacts is in the elastic deformation of the solid-phase material in the compact (Petrie and Page 1991). We used this Hugoniot model for the equation of state of the powder. Parameters for Petrie-Page shock Hugoniot for Nd-B-Fe powder are listed in Table 1.

Table 1. Material properties for Nd-B-Fe.

ρ_{00} (kg/m^3)	a	b (Pa^{-1})	ρ_{s0}(kg/m^3)	K (Pa)	Γ
3770	0.5658	2.41×10^{-8}	7494	1.67×10^{11}	1.69

3. Results and discussion

The pressure contour maps obtained at a wall angle of $\theta = 20°$ are shown in Fig. 2a. The time described in the figure was measured from the instant when the explosive was detonated by an electric detonator. The condensed region of the pressure contour maps corresponds to the shock wave. The shock wave in SUS304 travels faster than in the powder. The pressure maps obtained in the case of a wall angle of $\theta = 30°$ are indicated in Fig. 2b. At the entrance of the powder chamber, the underwater shock wave has been deformed as shown in this figure. This deformed underwater shock wave often caused the blow up of the powders in the powder chamber (Hokamoto et al. 1994).

Figure 3 indicates the pressure histories obtained at the center of this assembly at a wall angle of $\theta = 20°$. At 13.1 μs, the pressure across the underwater shock wave reaches about 28 GPa due to the effects of the converging from the wall of the assembly. When the underwater shock wave impinges on the powders, the reflected shock wave returns back to the water region; the strength of the reflected underwater shock wave is less than that of the incident shock wave due to the effects of the expansion wave coming from the explosive.

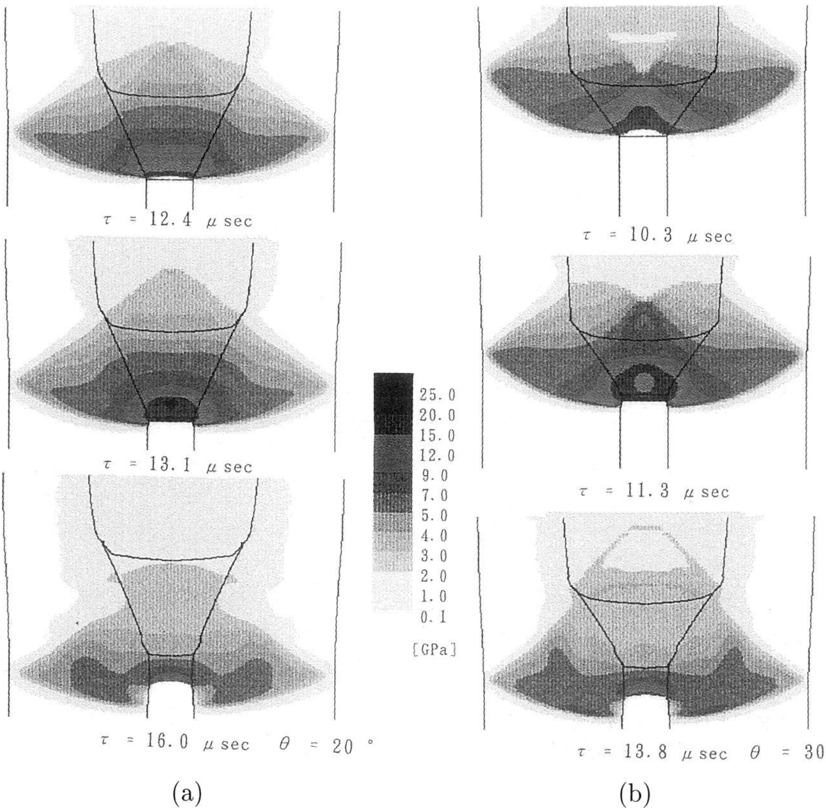

Figure 2. Pressure contour maps: (a) in the case of a wall angle of $\theta = 20°$; (b) in the case of a wall angle of $\theta = 30°$.

The pressure of the water which acts on the powders is shown in Fig. 4a. As we can see from this figure, an effective consolidation time, defined by the time when the pressure is greater than 10 GPa, is 1.4 μs. This long duration of high pressure is the reason why this shock compaction assembly consolidates powders quite well. The peak pressure was about 28 GPa, three times the pressure when the explosive initially impinged on the water. This result was caused by the effects of converging underwater shock waves in this assembly.

Figure 4b shows the pressure and the density changes on both sides of the boundary between the water and powder regions. At the instant when the underwater shock wave hits the powder, the powder is immediately compressed up to its solid-state density ρ_s. As the shock wave travels into the powder, the

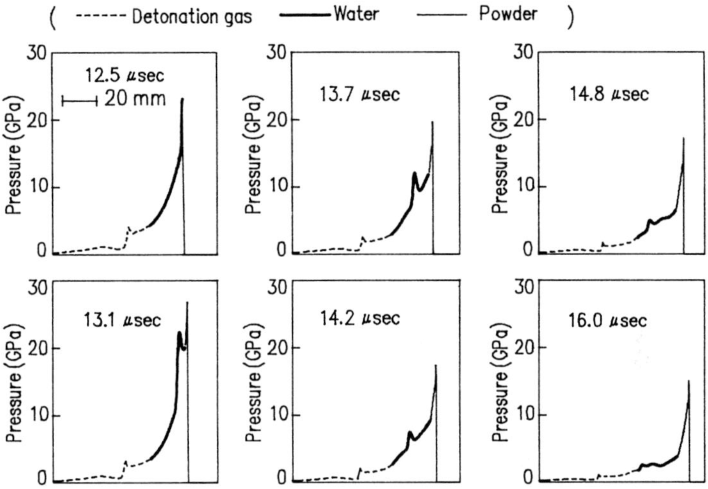

Figure 3. Pressure histories obtained along the center line of the shock compaction assembly when $\theta = 20°$.

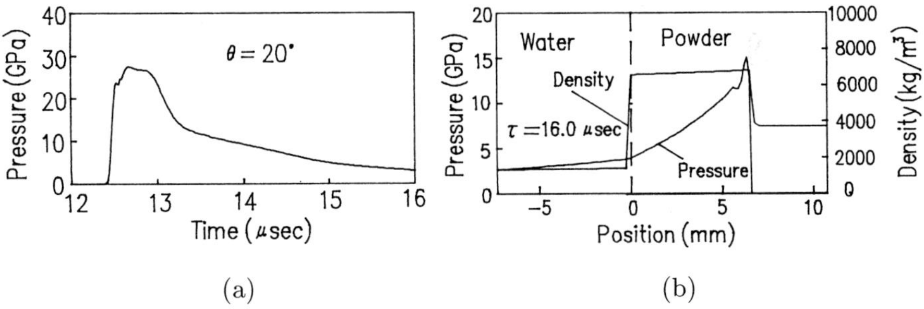

Figure 4. (a) Pressure of water which acts on the powder; (b) pressure and density of the water and bulked material.

region with solid density ρ_s becomes larger. These results were obtained 16 μs after the explosive was detonated.

4. Conclusion

Shock compaction of Nd-B-Fe powder was investigated by numerical calculations. The shock compaction of this powder was carried out using a compaction assembly to produce difficult-to-consolidate powder. This assembly used converging underwater shock waves. When the converged angle of $\theta = 20°$ was used, the peak pressure reached about 28 GPa at the instant when the underwater shock wave impinged on the powder. It was about three times the pressure obtained when the explosive first impinged on the water. This high pressure was caused by the effects of converging underwater shock waves and reflection from the compressed powder. It was confirmed that a long duration (up to 1.4 μs) of high pressure (over 10 GPa) was created in this assembly. This long duration is expected to be useful in various metal-processing procedurs such as the transformation from hBN to wBN or cBN, and so on.

Acknowledgement. We would like to thank Miss Kim Shephard, a student of the University of Newcastle, Australia, for her assistance in revising this paper for publication in English.

References

Amesden AA, Ruppel HM, Hirt CW (1980) SALE: A Simplified ALE Computer Program for Fluid Flow at All Speeds, Los Alamos Scientific Laboratory Report, LA-8095,UC-32.

Gourdin WH (1986) Dynamic consolidation of metal powders. Progress in Materials Science 30:39–80.

Hokamoto K, Itoh S, Fujita M, Chiba A (1994) Explosive shock consolidation of high speed steel powders. Iron and Steel 80(2):125–130.

Itoh S, Kubota S, Kira A, Nagano S, Fujita M (1994) On underwater shock wave generated by underwater explosion of high explosives (I). J of the Japan Explosives Society 55(5):202–208.

Lee M, Finger M, Collins W (1973) JWL Equation of State Coefficients for High Explosive, UCID-16189, Lawrence Livemore Laboratory.

Meyers MA, Wang SL (1988) An improved method for shock consolidation of powders. Acta Metall 36(4):925–936.

Petrie MW, Page NW (1991) An equation of state for shock loaded powders. J Appl Phys 69(6):3517–3524.

Petrie MW, Page NW (1992) Shock consolidation of rapidly solidified Neodymium based earth magnet powders. Acta Metall 40(11):3195–3203.

Thermodynamic relations, Hugoniot adiabatics, and stability of strong shocks in solids

I. Rutkevich, E. Zaretsky, M. Mond
Pearlstone Center for Aeronautical Engineering Studies
Department of Mechanical Engineering
Ben-Gurion University of the Negev, Beer Sheva, Israel

Abstract: A theoretical study of the caloric equation of state and of the thermodynamic functions is carried out for shock-compressed materials. The obtained equation of state is consistent with a given Hugoniot adiabatic $D = D(U)$, where D is the shock's velocity and U is the flow velocity behind it. For the linear law $D(U)$ the Gruneisen parameter and the sound velocity along the shock adiabatic and along the initial isentrope are calculated for aluminum and for copper. For these metals the criterion of stability of shocks under spontaneous emission of sound from shock waves is investigated and it is shown that spontaneous emission may occur in an experimentally available range of parameters.

Key words: Equation of state, Stability of shocks, Spontaneous acoustic emission

1. Introduction

Modeling of the equation of state and the thermodynamic functions of shock-compressed materials represents an important and unavoidable part of the general theory of shock waves in solids. The thermodynamic properties that influence the creation, existence, and stability of strong shock waves are still not attainable from direct experiments. Therefore, a straightforward theory of shock waves based on a known equation of state and known expressions for the specific internal energy ϵ and the specific entropy s, which is well developed in fluid mechanics (Landau and Lifshitz 1987), cannot be applied to shocks in solids. On the other hand, the existing experimental Hugoniot data in the (U, D) plane, where D is the shock velocity and U is the particle velocity behind the shock, can be used in theoretical constructions of the equation of state (Zel'dovich and Raizer 1966; Al'tshuler et al. 1987). The purpose of the present work is to construct the equation of state consistent with a given Hugoniot adiabatic in the (U, D) plane, to calculate the Gruneisen parameter and the sound velocity, and to check the criterion of Kontorovich (1957) for spontaneous emission of sound from shocks in metals.

2. Caloric equation of state

The problem of determination of the caloric equation of state (CES) representing the connection between the pressure p, the density ρ, and the temperature T, may have a continuum of solutions consistent with a given dependence $D = D(U)$. The most economical way to reach a closed thermodynamic description of the material is to assume the form of the CES that contains only one unknown function of one of the thermodynamic variables. The following structure of the CES is employed:

$$p = \rho_1 c_1^2[\bar{\rho} - 1 + \alpha(T - T_1)]\ \Phi(\bar{\rho}) \qquad (\bar{\rho} = \rho_2/\rho_1) \tag{1}$$

where ρ_1, c_1, and T_1 are the density, the isothermal sound velocity, and the temperature of the unperturbed material ahead of the shock, and α is the coefficient of thermal expansion. The function $\Phi(\bar{\rho})$ should be determined in terms of a given Hugoniot adiabatic $D(U)$. Assuming that the thermal-expansion coefficient α and the specific heat C_V are constant, we obtain the following expressions for the thermodynamic functions ϵ and s in terms of $\Phi(\bar{\rho})$:

$$\varepsilon(T,\rho) = \varepsilon(T_1,\rho_1) + C_V(T - T_1) + c_1^2 \int_1^{\bar{\rho}} \left[\bar{\rho}^{-1} - \bar{\rho}^{-2}(1+\alpha T_1)\right]\ \Phi(\bar{\rho})d\bar{\rho} \tag{2}$$

$$s(T,\rho) = s(T_1,\rho_1) + C_V \left[\ln(T/T_1) - \Gamma_1 \int_1^{\bar{\rho}} \bar{\rho}^{-1}\Phi(\bar{\rho})d\bar{\rho}\right] \tag{3}$$

Here $\Gamma_1 = \alpha c_1^2/C_V$ is the Gruneisen parameter of the unperturbed material. Using Eqs. (1) and (2) and the thermodynamic definition of the Gruneisen parameter, $\Gamma = \rho^{-1}(\partial p/\partial \varepsilon)_\rho$, results in the following simple relationship between Γ and Φ:

$$\Gamma = \Gamma(\bar{\rho}) = \Gamma_1 \Phi(\bar{\rho})/\bar{\rho} \tag{4}$$

Therefore, within the framework of the CES (Eq. 1) the problems of determination of $\Phi(\bar{\rho})$ and $\Gamma(\bar{\rho})$ are equivalent. As is well known, the Hugoniot relations across the shock that represent the conservation laws for mass and momentum totally determine the shock adiabatic in the plane (p,ρ) if it is known in the plane (U,D). Thus, for the popular linear model

$$D = c_0 + SU \tag{5}$$

the equation of the shock adiabatic in the plane (p,ρ) has the following form (see, for example, Zel'dovich and Raizer 1966)

$$\frac{p_H(\rho)}{\rho_1 c_0^2} \equiv P(\bar{\rho}) = \frac{(\eta-1)^2\bar{\rho}(\bar{\rho}-1)}{(\eta-\bar{\rho})^2} \qquad \eta = max\frac{\rho_2}{\rho_1} = \frac{S}{S-1} \tag{6}$$

The internal energy along the shock adiabatic ε_H is a known function of $\bar{\rho}$ given by the Hugoniot relation that expresses the conservation of energy across the shock: $\varepsilon_H = \varepsilon_1 + p_H(\bar{\rho})[1-\bar{\rho}^{-1}]/2\rho_1$. Employing this formula, Eqs. (1)–(3), and the thermodynamic relationship $p/\rho^2 + T\,(\partial s/\partial\rho)_T = (\partial\varepsilon/\partial\rho)_T$ written along the shock adiabatic results in a first-order, nonlinear differential equation for the function $\Phi(\bar{\rho})$:

$$\frac{d\Phi}{d\bar{\rho}} = -\frac{[\Gamma_1(1-\bar{\rho}^{-1})\Phi^2 - 2\Phi]dP/d\bar{\rho} + \Gamma_1\bar{\rho}^{-2}P(\bar{\rho})\Phi^2 + a\Phi^2 + b\Phi^3}{2P(\bar{\rho})} \tag{7}$$

where

$$a = 2(1+\alpha T_1\Gamma_1)^{-1} \qquad b = a\Gamma_1(1+\alpha T_1 - \bar{\rho})\bar{\rho}^{-2}$$

Here $P(\bar{\rho})$ is the normalized pressure along the Hugoniot adiabatic. The initial condition for Eq. (7) is $\Phi(1) = 1$. In the phase plane $(\bar{\rho}, \Phi)$ the differential equation (7) has two singular points, $A \equiv (\bar{\rho} = 1,\ \Phi = 1)$ and $B \equiv (\bar{\rho} = \bar{\rho}_m,\ \Phi = 2\bar{\rho}_m/\Gamma_1[\bar{\rho}_m - 1])$. Here $\bar{\rho}_m$ is the maximal value of the compression degree by the shock. In the case of the linear dependence $D(U)$, the maximal compression degree is $\bar{\rho}_m = \eta = S/(S-1)$. It can be shown that A is the saddle point and B is the nodal point of Eq. (7). There is only one physically appropriate solution, whose image in the phase plane is the separatrix that outgoes from the the point A and enters into the point B. For the linear dependence $D(U)$ when $P(\bar{\rho})$ is given by Eq. (6), the relevant solution of Eq. (7) was obtained numerically.

It should be noted that the linear law for $D(U)$ and the dependence $P(\bar{\rho})$ given by Eq. (6) are invalid in some small vicinity of the initial point $\bar{\rho} = 1$ within the framework of the single-shock theory. It is known that the section of the experimental straight line (5), for which $c_0 < D_* < D < c_{1\ell}$, where $c_{1\ell}$ is the longitudinal sound velocity in an unperturbed elastic material and D_* is the value of D when the plastic flow is first initiated behind the elastic shock, corresponds to the situation with two shock waves. The first represents an elastic precursor, while the second one propagates with the velocity $D < c_{1\ell}$ in a medium where the tangential stresses are relaxed and the sound velocity is less than $c_{1\ell}$. Although in the literature the term "plastic wave" is often used for the second shock wave, here it is the elastic front that remains after the absorption of the shear strain energy (Gilman 1992). The shock splitting disappears when D becomes larger than $c_{1\ell}$. Strictly speaking, the single-shock theory is correct only in this case. However, even for $D < c_{1\ell}$ such a theory represents a reasonable approach to calculating the second shock wave if the pressure p_H behind the second shock is much larger than the dynamic limit of elasticity p_* reached behind the precursor. According to experimental data for Al and Cu given by Mitchell and Nellis (1981), the experimental points belong to the range $\bar{\rho} > 1.09$ and the condition $p_H \gg p_*$ is satisfied for both metals. This allows one to employ the single-shock theory in the available experimental range.

3. The Gruneisen parameter and the sound velocity

Numerical calculations of the Gruneisen parameter normalized to its initial value at $T_1 = 300$ K were carried out for aluminum ($S = 1.35$, $\Gamma_1 = 2.09$) and for copper ($S = 1.5$, $\Gamma_1 = 1.98$). For both metals there is an initial interval of the values of $\bar{\rho}$ where $\Gamma(\bar{\rho})$ is a weakly increasing function. The initial increase of $\Gamma(\bar{\rho})$ follows from the inequality $(d\bar{\Gamma}/d\bar{\rho})_{\bar{\rho}=1} > 0$ which has been proven for the linear law $D(U)$. The normalized sound velocity $\bar{c}_H = c_H/c_0$ was calculated along the shock adiabatic as a function of the compression degree $\bar{\rho}$ for Al and for Cu. The quantity c_H was determined from the relation $c_H^2 = (\partial p/\partial \rho)_s$ with $s = s_H(\bar{\rho})$. The calculated values of c_H for aluminum are in reasonable agreement with experimental data which exist in the range $\bar{\rho} < 1.8$ (Al'tshuler 1965). The normalized sound velocity $\bar{c}_H = c_H/c_0$ was calculated along the initial isentrope $s = s_1$ for Al and for Cu. The quantity c_s was determined from the relation $c_s^2 = (\partial p/\partial \rho)_s$ with $s = s_1$. Unlike the monotonically increasing dependence $c_H(\bar{\rho})$, the calculated dependence $c_s(\bar{\rho})$ has a maximum and becomes a decreasing function for high values of $\bar{\rho}$ that are still not attainable in experiments. In addition, the function $R(\bar{\rho}) = \bar{\rho} c_s(\bar{\rho})$ was calculated. The sign of the derivative $dR/d\bar{\rho}$ is responsible for the creation of shock waves from continuous compression waves due to the well-known Riemann mechanism of wave breaking. This mechanism works when $dR/d\bar{\rho} > 0$. In spite of the non-monotonic behavior of $c_s(\bar{\rho})$, the function $R(\bar{\rho})$ increases monotonically for both Al and Cu. Therefore, the shocks in these materials are genuine, i.e., they can be described by discontinuous solutions of the corresponding Euler equations. Testing the Riemann criterion $dR/d\bar{\rho} > 0$ is necessary when the interaction of shocks with small perturbations is investigated.

4. Conditions for spontaneous emission of sound

The spontaneous emission (SE) of sound from a shock represents a special kind of instability of the shock wave: it occurs when, in the region behind the shock, the reflection coefficient of an acoustic wave impinging on the shock becomes infinite for some value of the angle of incidence. The SE can be manifested in the form of two- and three-dimensional perturbations that appear in the flow behind the shock. The latter becomes non-planar and its front suffers a corrugation. This phenomenon was first considered by Dyakov (1954) and the criterion of SE was obtained by Kontorovich (1957) in the following form

$$h_c \equiv \frac{1 - M_2^2(\bar{\rho}+1)}{1 + M_2^2(\bar{\rho}-1)} < h \equiv -V_2^2 \left(\frac{d\rho_2}{dp_2}\right)_H < 1 + 2M_2 \tag{8}$$

Here $M_2 = V_2/c_H < 1$ is the Mach number of the flow behind the shock in the frame of reference where the shock is at rest, $V_2 = D - U$ is the flow velocity behind the shock in this frame. The results of numerical calculations of both functions $h(\bar{\rho})$ and $h_c(\bar{\rho})$ for Al and for Cu show the existence of a certain critical degree of compression $\bar{\rho}_c$, beyond which SE may occur. The existence of SE for sufficiently high values of $\bar{\rho}$ was shown theoretically by Rutkevich et al. (1994) on the basis of a CES of the type (1), in which $\Phi(\bar{\rho})$ was assumed a power function. This assumption led to a weakly nonlinear dependence $D(U)$, whose asymptotic behavior was linear for large values of U. The finite curvature of the obtained Hugoniot shock adiabatic $D(U)$ provided a higher threshold for SE than in the case of the strictly linear model. Since for Al and for Cu the linear dependence $D(U)$ is in the best correspondence with existing experiments (Mitchell and Nellis 1981), the lowering of the threshold of SE provided by this model seems to be a reliable prediction.

5. Conclusions

In the present work, a new approach to construction of a CES has been developed. The closed description of the thermodynamic properties of the shock-compressed medium that is consistent with a given law $D(U)$ has been obtained. The numerical calculations of the Gruneisen parameter Γ, the sound velocity along the shock adiabatic c_H, and along the initial isentrope c_s were carried out of the linear law $D(U)$ for aluminum and for copper. Some new qualitative properties such as non-monotonic behavior of Γ and c_s as functions of the compression degree $\bar{\rho}$ were found. It was shown that the Riemann condition for the creation of shock waves is satisfied in aluminum and copper in the whole range of $\bar{\rho}$ values. The criterion for SE was checked and it was found that SE may occur behind strong shocks.

References

Al'tshuler LV (1965) Use of shock waves in high-pressure physics. Soviet Physics Uspekhi 8:52–91.

Al'tshuler LV, Brusnikin SE, Kuz'menkov EA (1987) Isotherms and Gruneisen functions for 25 metals. J Appl Mech Tech Phys (1):129–141.

Dyakov SP (1954) On the stability of shock waves. Zh Eksp Teor Fiz 27:288–295.

Gilman JJ (1992) The plastic wave myth. In: Shock Compression of Condensed Matter 1991, Schmidt SC, Dick RD, Forbes JW, Tasker DG (eds), Elsevier Science Publishers BV, pp 387–389.

Kontorovich VM (1957) On the stability of shock waves. Zh Eksp Teor Fiz 33:1525–1526.

Landau LD, Lifshitz EM (1987) Fluid Mechanics, 2nd Edition, Pergamon Press, Oxford.

Mitchell AC, Nellis WJ (1981) Compression of aluminum, copper and tantalum. J Appl Phys 52:3363–3374.

Rutkevich I, Zaretsky E, Mond M (1994) Resonant reflection of two-dimensional acoustic waves from shocks in solids. J de Physique IV, Colloque C8:723–728.

Zel'dovich IB, Raizer YP (1966) Physics of Shock Waves and High Temperature Hydrodynamic Phenomena, Academic Press, New York.

Numerical simulation of shock-wave generation with KALIF

O. Y. Vorobiev, V. E. Fortov
High Energy Density Research Center
142432 Moscow, Russia

B. Goel
Forschungszentrum Kalrsruhe, INR
D-76021 Karlsruhe, Germany

Abstract: The present paper contains an investigation of the dynamics of thin-foil acceleration by light-ion beams and an analysis of various experiments performed recently (Bauming et al. 1993). The results of simulations depend mainly on two factors: power-density profile and the equation of state for the strongly coupled dense, hot plasma. The peak beam power is estimated to be 0.15 ± 0.04 TW/cm^2. The available equations of state such as SESAME and Bushman do not reproduce experimental data. After checking carefully the input-beam parameters we decided to use a new equation of state. The equation of state used in calculations has been presented in a simple analytical form containing fitting parameters. These parameters have been adjusted to describe the experimental data adequately. This analytical form of the equation of state is used to facilitate numerical stability of the code and to provide handy parameters to study the effect of their variation on the obtained numerical results and on the equation of state determined by different authors.

Key words: Shock waves, Nonideal plasma, Equations of state

1. Experiments

Experiments analyzed here have been described extensively in Bauming et al. (1993). Thin aluminum foils of different thicknesses were irradiated with a KALIF beam. The velocity of rear surface was measured by the ORVIS system. Two types of experiments are analyzed. (1) Foil acceleration experiments: in these experiments, thin subrange (i.e., the thickness of the target was larger than the range of protons) aluminum foils of various thicknesses were irradiated by the KALIF beam. The back-surface velocity was recorded with ORVIS. The thickness of the targets varied between 22 and 75 μm. (2) Pressure measurements: In these experiments the aluminum foils, whose thicknesses slightly exceeded the range of protons, were covered with LiF windows to avoid shock reflections. The impedance of LiF is approximately the same as that of aluminum. The velocity at the Al-LiF interface was recorded as a pressure signal.

2. Mathematical model

The Euler equations of motion are used to describe the dynamics of matter under the impact of a beam. They express conservation laws for mass, momentum, and energy. The caloric equation of state is used to close the equations of motion. The radiative energy transfer is treated in the radiation heat-conduction approximation. In this case, the energy flow is proportional to the temperature gradient, the proportionality coefficient being expressed by the Rosseland mean free path. Thus the conductive energy flow consists of two terms: electron conductivity and radiation conductivity. The electron conductivity coefficient λ_e (Polyshchuk et al. 1991) and the Rosseland mean free path ℓ (Tsakiris and Eidman 1987) are functions of density and temperature and were determined by wide-range semi-empirical formulas. The maximum target temperature reached in present KALIF experiments does not exceed a few tens of eV. Therefore, the energy of the radiation field is not very important in the calculations reported here and the simple approximation of radiation heat conduction is deemed to be adequate to calculate the radiative heat transfer. The absorption of the beam energy by the target is calculated in a single-particle approximation. Stopping power is calculated by semi-empirical formulas developed in Polyshchuk et al. (1991) and Ebeling et al. (1991). These formulas describe the enhancement of the stopping power as a function of heating and the expansion of matter. The hydrodynamical calculations were performed using the Godunov method on the moving grid. The nodes of the grid moved with contact velocities calculated by solving the Riemann problems between the neighboring cells.

3. Wide-range semi-empirical equation of state

For the numerical simulation of foil acceleration by a KALIF beam it is necessary to use an equation of state (EOS) describing, with reasonable accuracy, the shock compression of matter and the plasma thermodynamics in the energy deposition region. In the following calculations, a semi-empirical equation of state in the form of generalized Mie-Gruneisen equation of state is developed which describes the experimental data on shock compression of the solid samples in the pressure range up to several Mbar. Following the usual procedure, the internal energy of matter is divided into two parts: the cold energy, E_c, is only a function of density of matter, while the energy of thermal motion, E_t, is a function both of temperature and density. As our simple equation-of-state model does not contain the temperature, we shall use (p, ρ) variables. For calculations we need functions $P(\rho, E)$ and $C_s(\rho, E)$. To describe Hugoniot data for solid matter we choose the cold pressure in traditional form

$$P_c(\rho) = \sum_{i=1}^{5} a_i \delta^{(i/3+1)}, \qquad \delta \geq 1 \tag{1}$$

$$P_c(\rho) = \frac{\rho_0 C_{s0}^2}{(N-M)}(\delta^N - \delta^M), \qquad \delta \leq 1 \tag{2}$$

where $\delta = \frac{\rho}{\rho_0}$ and ρ_0 and C_{s_0} are normal density and sound velocity respectively. At given binding energy E_b one of the coefficients in (2), for example M, can be expressed through another one as

$$M = 1 + \frac{C_{s0}^2}{E_b(N-1)} \tag{3}$$

Coefficients a_i are found by solving the system of linear equations to ensure the properties of material under normal conditions (normal sound velocity, Grüneisen coefficient, zero pressure) and by using the Thomas-Fermi-Dirac model at high compressions. The cold component of the total pressure dominates strongly over the thermal component in shocked states of solid matter up to several Mbars. The thermal pressure is related to the thermal energy through the Grüneisen parameter γ, which is represented as a function of energy and density

$$P_t = \gamma(E, \rho) E_t \rho \tag{4}$$

We use the following dependence of Grüneisen parameter on the internal energy

$$\gamma = \gamma(\rho)_\infty + \frac{\gamma(\rho)_0 - \gamma(\rho)_\infty}{1 + (\frac{E}{E_0})^\alpha}, \tag{5}$$

where E_0 and α are fitting parameters. According to (6), $\gamma(\rho)_0$ tends to $\gamma(\rho)_\infty$ for hot matter. To describe the asymptotic behavior of γ in the limit of high compressions (Fermi gas) and expansions (ideal gas) we choose both functions γ_0 and γ_∞ having asymptotic value 2/3 for $\rho \to \infty$ and $\rho \to 0$.

$$\gamma(\rho)_0 = \frac{2}{3} + \left(\gamma_s - \frac{2}{3}\right) \frac{2}{(\frac{\rho_0}{\rho} + \frac{\rho}{\rho_0})} \tag{6}$$

$$\gamma(\rho)_\infty = \frac{2}{3} + \left(\gamma_{min} - \frac{2}{3}\right) \left(\frac{2}{x + 1/x}\right)^\beta, \qquad x = \frac{\rho_0}{\rho}\xi, \tag{7}$$

where γ_s is Grüneisen coefficient at normal conditions, ρ_0 is the normal density, γ_{min} is some minimum value of γ in dense plasma, and β and ξ are fitting parameters. Parameter ξ gives the density of plasma where γ_∞ reaches its minimum value. This density approximately corresponds to the critical point and is about an order of magnitude smaller than the normal density. Parameter β determines the rate of increase of γ from its plasma value to the asymptotic value 2/3 of ideal gas. This parameter together with α are the most important parameters to fit simulation results to experimental data. As for parameter E_0, it can be matched by fitting shock-wave experiments with porous samples. To describe compression of porous samples the thermal component of pressure in the EOS must be correct for densities close to the normal density. Results of Hugoniot adiabate calculations for solid and porous samples ($E_0 = 30$ kJ/g) are shown in Fig. 1 together with experimental data. We find that our simple analytical EOS allows us to describe available experiments on isentropic expansion of shocked material. Results of these experiments as well as simulation results of the shock compression and expansion of aluminum are shown in Fig. 2.

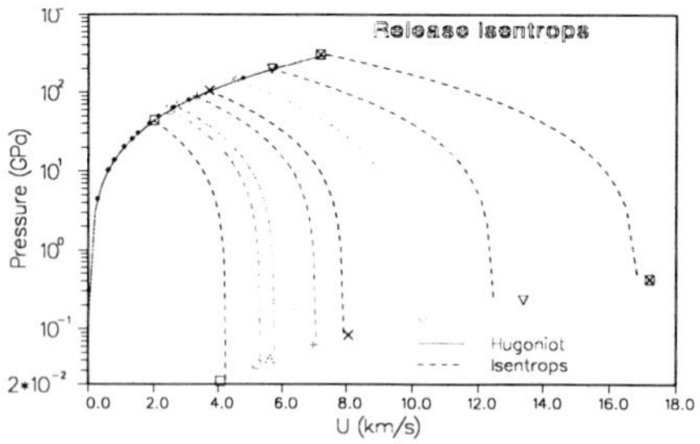

Figure 1. Calculated Hugoniot curve and release isentropes from different pressures together with experimental data. Pressures < 2 Mbar – experiments (Bakanova et al. 1983), pressures > 2 Mbar – experiments (Glushak and Zharkov 1989), lines – calculations.

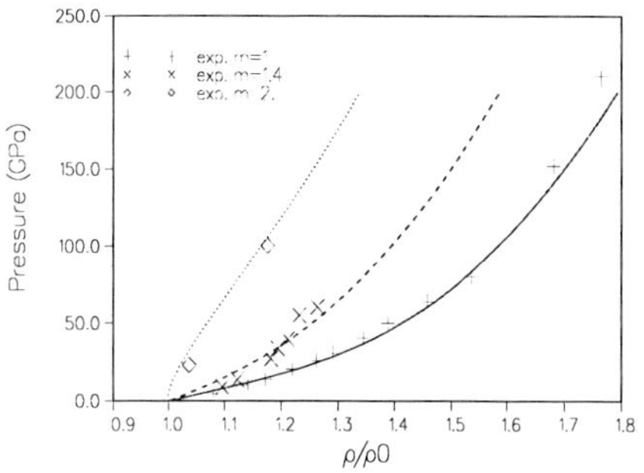

Figure 2. Calculated Hugoniot curves of porous samples and experimental data (Lawrence Livermore 1982). ◇ – experiments (Kormer et al. 1962) (porosity $m = 2$), × – experiments (Marsh 1980), lines – calculations.

4. Simulation results and experimental data

The purpose of the simulations presented below is to give an adequate description of experiments on aluminum-foil acceleration performed with KALIF using reasonable parameters for the equation of state. The time evolution of the power density was determined recently using different diagnostic methods.

To begin our analysis we used SESAME and Bushman equations of state and performed calculations with the experimentally observed time evolution of KALIF beam and assumed thereby peak power densities of 0.15 TW/cm^2 for 75- and 50-μm foils. Calculated velocity profiles give inadequate description of experimental data for the time interval between 10 and 20 ns, when the density falls to an order of magnitude less value and temperature increases up to 10 eV (for example, see Fig. 3). These states of matter correspond to the region of strongly coupled plasma. Note that uncertainty in equation of state models is larger and experimental information is lacking in this region.

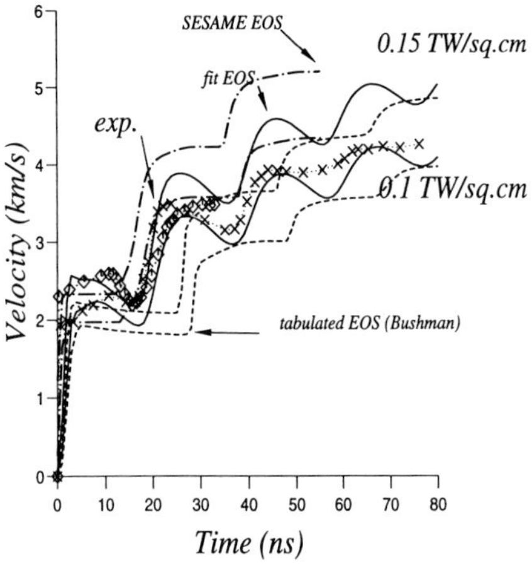

Figure 3. Experimental data and simulation results for 75-μm foil.

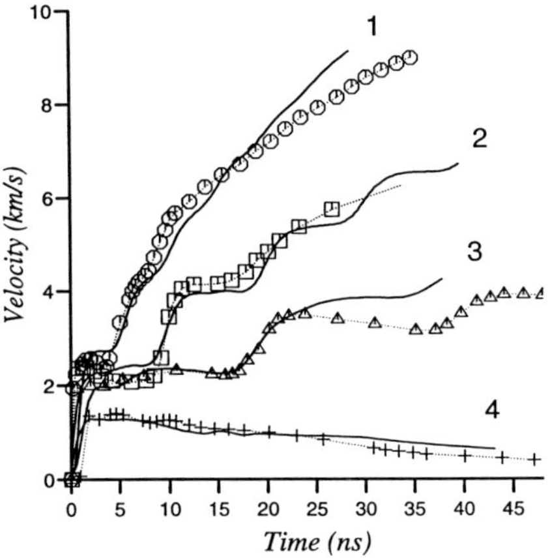

Figure 4. Experimental data and simulation results. Curves are for various foil thickness: (1) 33 μm, (2) 50 μm, (3) 75 μm, (4) 33 μm Al + LiF.

To rectify this situation we used our semiempirical EOS model which contains several fitting parameters. By varing these parameters we could descibe acceleration experiments.

Using experimentally measured profiles different experiments have been analysed. In this way we can determine the dependence of $\gamma(\rho, E)$ by matching the coefficients β, α and ξ in formulas (5) – (7) and by fitting the calculated results to the experiments. Thus determined coefficients as well as the parameters of the cold curve are given in the table below.

Table 1. Coefficient for EOS.

α	β	E_0	γ_{min}	γ_s	ξ	
1.1	0.02	30	0.2	2.1	0.1	
a_1	a_2	a_3	a_4	a_5	N	M
−15009.612	51732.56	−66509.87	37631.99	−7845.07	3.5	1.93633

Results of numerical simulations performed using analytical EOS with parameters presented in Table 1 are shown in Fig. 4 together with experimental data for foils of different thickness.

5. Conclusions

We have calculated the foil acceleration experiments using the newly determined temporal evolution of power and particle energy for B_θ-diode at KALIF.

This analysis provides information on the equation of state in the region of strongly coupled non ideal plasma. In this region experimental information is very scarce thus the present data can aid theoreticians who are developing equations of state for strongly coupled non ideal plasma. We have assumed in our analysis that the beam is completely reproducible. This is only true to a certain extent. The accuracy of measured beam parameters is between 10-20%, except for peak power value. Normalisation of peak power dates back many years and uncertainty may be large. These fact limits the accuracy of the present analysis. Nevertheless, the parameters used in our calculations are within experimental errors and improving accuracy of beam parameters will also improve the accuracy of measured equation of state data.

References

Bakanova AA, Dudoladov IP, Zhernokletov M (1983) Prikladnaja Matematika i Technicheskaja Phisika (in Russian) 2:76–81.

Baumung K, Bachmann H, Kanel GI, Karow HU, Licht V, Rusch D, Singer J, Stoltz O (1992) Hydrodynamic target experiments with proton beams at KALIF. In: Proc Ninth International Conference on High-Power Particle Beams, D Mosher, G Cooperstein (eds), Washington DC, USA, Vol I, p 68.

Baumung K, Kanel GI, Utkin A, Razorenov S (1993) Light-ion beam-target interaction experiments on KALIF. Nuovo Cimento 106 A:1771.

Bushman AV, Fortov VE, Lomonosov IV (1989) Proc Enrico Fermi School, Elsevier, p 249.

Ebeling W, Förster A, Fortov VE, Gryaznov VK, Polishchuk AY (1991) Thermophysical properties of hot dense plasmas. In: Teubner-Texte zur Physik, Bd 25, Teubner Verlag, Stuttgart-Leipzig.

Glushak BL, Zharkov AP, Zhernokletov M, Ternovoy V, Filimonov A, Fortov V (1989) Russian JETP 96:1301.

Goel B, Vorobjev OY, Ni AL (1992) Generation of ultra high pressure with light ion beams. In: Proc Ninth International Conference on High-Power Particle Beams, D Mosher, G Cooperstein (eds), Washington DC, Vol II, p 957.

Kormer S, Funtikov A, Urlin V, Kolesnikova A (1962) JETP (in Russian) 42:686–701.

Lawrence Livermore National Laboratory (1982) Report UCID-118574-82-2.

Marsh SP (ed) (1980) LASL Shock Hugoniot Data, U of CA Press, Berkeley.

Polyshchuk AY, Fortov VE, Khloponin VS (1991) Calculation of the effective collision lengths of ions and protons in a plasma using the local electron-density approximation. Sov J Plasma Phys 17:523.

Tsakiris GD, Eidman K (1987) An approximate method for calculation Planck and Rosseland mean opacities in hot, dense plasmas. JQSRT 38:353.

Effects of solid-solid phase changes on impact mechanics

J. K. Knowles*, N. A. Winfree*, T. J. Ahrens†
*Division of Engineering and Applied Science
†Lindhurst Laboratory of Experimental Geophysics
California Institute of Technology
Pasadena, CA 91125 USA

Abstract: This paper describes an idealized problem that models the response of an elastic solid undergoing a shear-induced phase transformation due to dynamic loading.

Key words: Phase transformation, Impact, Solids, Ceramic armor

1. Introduction

Elastic materials whose strain-energy potentials are non-convex functions of strain can be used to model the macroscopic effects of solid-solid phase changes (Abeyaratne and Knowles 1990, 1991). In the presence of such potentials, the boundary initial-value problems of finite elasticity fail to possess unique solutions, despite enforcement of the second law of thermodynamics. It has been shown that uniqueness is restored if a *nucleation criterion* governing the initiation of the phase transition and a *kinetic relation* governing its evolution are added to the field equations and jump conditions of elastodynamics (Abeyaratne and Knowles 1991).

We describe a model problem involving dynamic loading of an infinite elastic medium subject to a concentrated shear force whose magnitude increases linearly in time at a given rate. Because the problem involves a concentrated load generating singularly large strains and stresses, nucleation always occurs, but imposition of a kinetic relation remains necessary.

We compare the distribution of energy densities in the medium in the presence of a phase change with the corresponding distribution in a material incapable of phase change.

2. The problem

Consider an infinite elastic medium, initially undeformed and at rest in a configuration that is taken as reference. Let r, ϕ, z be cylindrical coordinates in this configuration (Fig. 1). At time $t = 0$, an axially directed shear traction is applied as a line load along the z axis, its magnitude at time t being

λt, where the loading rate λ is given. The deformation is taken to be finite anti-plane shear. The only nonvanishing component of displacement is the axial component, $u_z \doteq u(r,t)$, and the corresponding shear stress of interest is $\sigma_{rz} = \sigma_{zr} \doteq \sigma(r,t)$. The material we consider is an incompressible neo-Hookean solid (a material which can sustain finite anti-plane shear) which posesses the shear stress-shear strain relationship of Fig. 2. This curve corresponds to an elastic material whose strain energy potential is a non-convex function of strain. The strains and stresses engendered by the concentrated loading at hand will be singular at $r = 0$, and hence will always be sufficient to nucleate a phase change from phase 1 to phase 3 there, the unstable phase being absent (Abeyaratne and Knowles 1991). The interface between the low-strain phase and the high-strain phase, the *phase boundary*, is located at $r = \dot{s}t$ at time t, where $\dot{s}$ is the phase-boundary velocity, assumed constant. We assume the phase boundary propagates subsonically, so that $0 < \dot{s} < c_3$. The expanding phase boundary is preceded by a shear wave; the expected wave pattern at time t is as shown in Fig. 1.

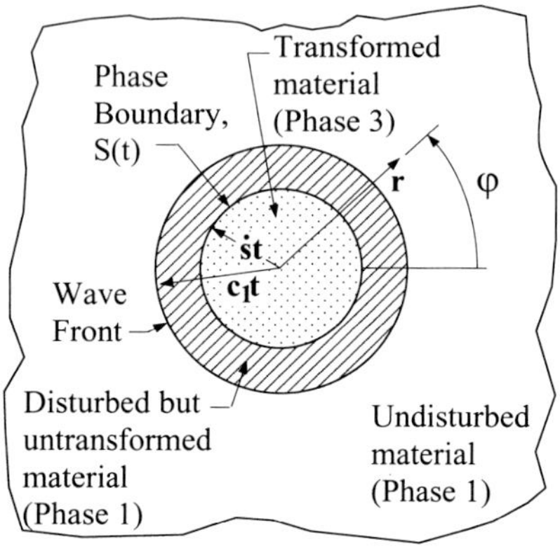

Figure 1. The problem: a shear traction is applied along the z axis (out of page) and ramped from 0 at rate λ. A shear wave propagates radially, possibly followed by a phase boundary.

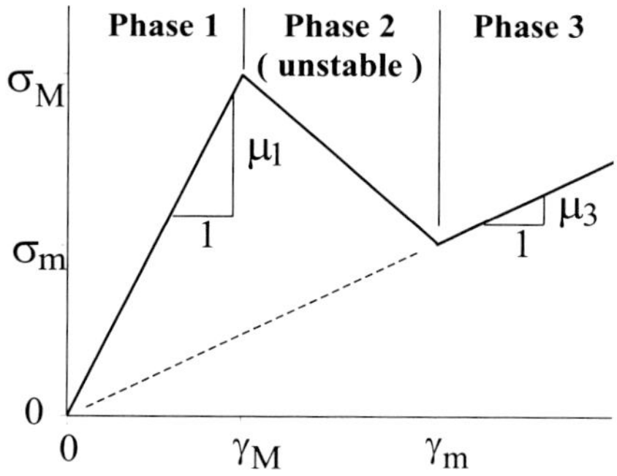

Figure 2. Stress-strain curve of the material.

3. The boundary initial-value problem

Away from strain discontinuities, the classical field equations of finite elastodynamics (Abeyaratne and Knowles 1990) reduce to:

$$c_3^2 \nabla^2 u(r,t) - \ddot{u}(r,t) = 0 \,, \qquad \text{for } 0 < r < \dot{s}t \,, \tag{1}$$

$$c_1^2\nabla^2 u(r,t) - \ddot{u}(r,t) = 0\ , \qquad \text{for } \dot{s}t < r < \infty\ , \tag{2}$$

where $c_1 = \sqrt{\mu_1/\rho}$ and $c_3 = \sqrt{\mu_3/\rho}$ are shear wave speeds in the low- and high-strain phases, respectively, and ρ the referential mass density. The initial and boundary conditions are:

$$u(r,0+) = 0\ , \qquad \dot{u}(r,0+) = 0\ , \tag{3}$$

$$\lim_{r\to 0} 2\pi r\sigma(r,t) = \lambda t\ . \tag{4}$$

Across the strain discontinuities at the phase boundary, $r = \dot{s}t$, and at the wave front, $r = c_1 t$; the jump conditions of elastodynamics specialize to

$$u(c_1t-,t) = u(c_1t+,t) = 0\ , \qquad u(\dot{s}t-,t) = u(\dot{s}t+,t)\ , \tag{5}$$

$$\sigma(\dot{s}t+,t) - \sigma(\dot{s}t-,t) + \rho\dot{s}[\dot{u}(\dot{s}t+,t) - \dot{u}(\dot{s}t-,t)] = 0. \tag{6}$$

4. The solution for constant $\dot{s}$

As posed, the problem has a similarity solution, in which

$$u(r,t) = \begin{cases} c_1 t\phi_3(r/(c_1t),\dot{s})\ , & \text{for } 0 < r < \dot{s}t, \\ c_1 t\phi_1(r/(c_1t),\dot{s})\ , & \text{for } \dot{s}t < r < c_1t, \\ 0\ , & \text{for } c_1t < r, \end{cases} \tag{7}$$

where ϕ_1 and ϕ_3 are fully and explicitly determined functions of $r/(c_1t)$ and $\dot{s}$, the formulas for which are omitted here for brevity; a more detailed account of this problem will appear elsewhere. Thus, Eq. (7) provides a one-parameter family (parameter $\dot{s}$) of solutions of our boundary initial-value problem. To determine $\dot{s}$, a kinetic relation must be invoked, as sketched below.

The solution (Eq. 7) must be such that the associated strains γ outside and inside the moving phase boundary lie in the intervals corresponding respectively to phases 1 and 3 (Fig. 2). This requirement ultimately leads to the restriction that the point $(\dot{s},\lambda)$ lie in the shaded region of Fig. 3.

Not every member of the family of solutions (Eq. 7) conforming to Fig. 3 satisfies the second law of thermodynamics. When this restriction is imposed in the form described in Abeyaratne and Knowles (1990), it leads to the requirement that $(\dot{s},\lambda)$ lie in the shaded region shown in Fig. 4. Thus, prior to the imposition of a kinetic relation, only those members of the one-parameter family of solutions (Eq. 7) corresponding to the shaded region in Fig. 4 satisfy the field equations, the jump conditions at the phase boundary and at the shear wave, the phase-segregation conditions, and the entropy inequality.

The curves forming the boundaries of the shaded regions in Figs. 3 and 4 can be fully and explicitly determined from the similarity solution (Eq. 7).

5. The kinetic relation

As shown in Abeyaratne and Knowles (1990, 1991), an appropriate way to describe the kinetics of the phase transition in the present setting is to require that the driving force at the phase boundary be a specified function, determined by the material, of the phase boundary velocity $\dot{s}$. By driving force, we mean the generalization to dynamics of the jump across the phase boundary of the Gibbs free energy of quasi-statics; see the discussions in Abeyaratne and Knowles (1990, 1991). When this notion is properly implemented in the present problem, the set of points $(\dot{s}, \lambda)$ admitted by the kinetic relation consists of a curve within the shaded region in Fig. 4. This curve allows one to uniquely determine a phase-boundary velocity $\dot{s}$ for each given load rate λ in a certain range, and hence to single out a unique member of the one-parameter family of solutions (Eq. 7) to the underlying boundary initial-value problem.

6. Energy distribution

The stored elastic energy W per unit reference volume for our elastic material at a strain γ is given by the area under the stress-strain curve between 0 and γ (Fig. 2). The *total* energy per unit volume is given by $E = W + (1/2)\rho\dot{u}^2$; here, E depends on position r and time t. For each solution among the one-parameter family (Eq. 7), E can be determined as a function of r, t, *and the phase-boundary velocity* $\dot{s}$. The value of $\dot{s}$ must be determined by the kinetic relation. Here, however, we assume $\dot{s}$ to be given, and in Fig. 5 we plot E versus r for a given t and *for a given* $\dot{s}$ for one of the phase-transition solutions (Eq. 7).

For comparison, we consider a second elastic material in which the stress-strain relation replacing that of Fig. 2 is $\sigma = \mu_1\gamma$ *for all strains* γ. Thus our

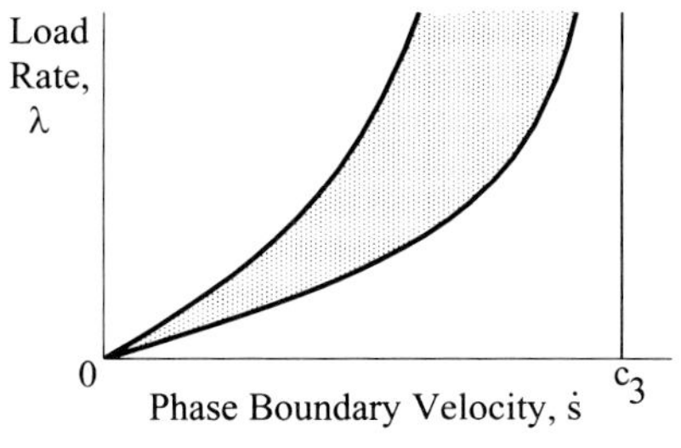

Figure 3. Shaded region indicates valid solutions after enforcing phase-segregation requirements.

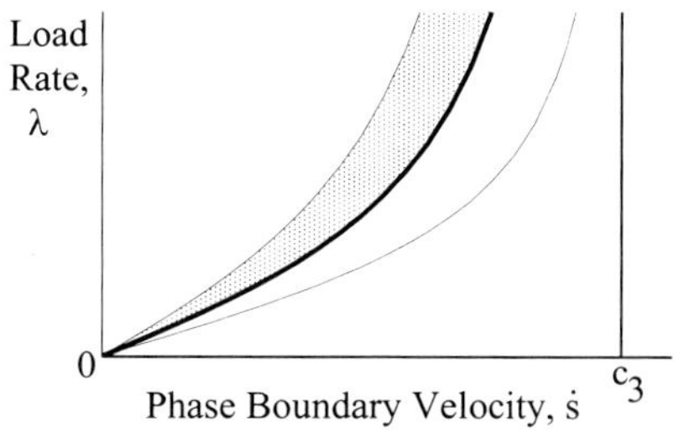

Figure 4. The second law eliminates more solutions.

second material is capable of existing in a single phase only, and its properties in this phase are identical with those of our two-phase material *in its low-strain phase.* The solution of the present boundary initial-value problem for this new material is obtained by standard methods, and from this solution, one may compute the distribution of total energy density as a function of r for each t for purposes of comparison with the energy density distribution for the two-phase material described above (Fig. 5).

Motivated by applications to ceramic armor, we consider the total energy density in the untransformed material ahead of the phase boundary and compare it to the energy density of the same particles in the single-phase material. E_1 and E_2 are the respective one-phase and two-phase total energy densities in this region; we plot the ratio E_2/E_1 in Fig. 6. This quantity is a function of only the ratio of shear wave speeds of the two phases, c_3/c_1, and of the ratio of the phase-boundary velocity to the shear wave speed in the original phase, $\dot{s}/c_1$.

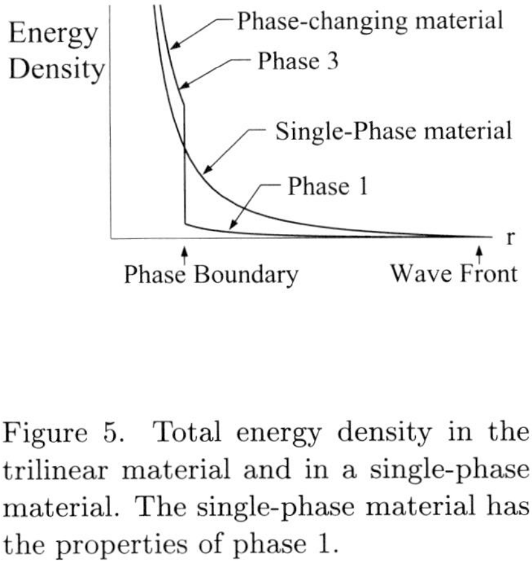

Figure 5. Total energy density in the trilinear material and in a single-phase material. The single-phase material has the properties of phase 1.

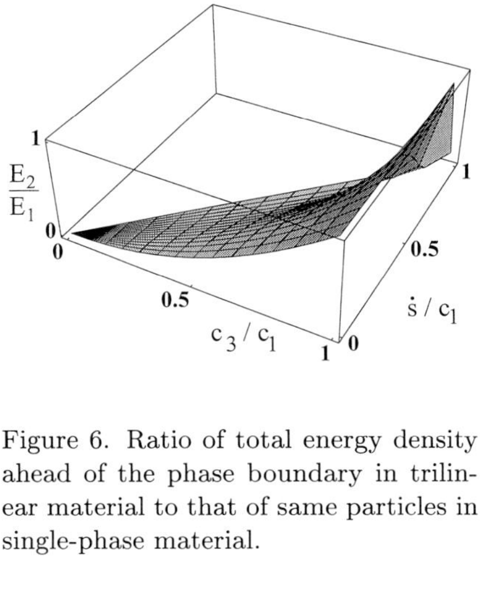

Figure 6. Ratio of total energy density ahead of the phase boundary in trilinear material to that of same particles in single-phase material.

7. Concluding remarks

A phase transformation in the present idealized model decreases the total energy to which the untransformed material is exposed. This "shielding" effect is enhanced with an increasing shear modulus in the original phase and with a decreasing shear modulus in the new phase. A faster phase-boundary velocity also increases the shielding effect. In this model, the phase-boundary velocity cannot exceed the shear-wave speed of the new phase.

Acknowledgement. Research supported by the Army Research Office. Contribution number 5593, Division of Geological and Planetary Sciences, California Institute of Technology.

References

Abeyaratne R, Knowles JK (1990) On the driving traction acting on a surface of strain discontinuity in a continuum. J Mech Phys Solids 38:345–360.

Abeyaratne R, Knowles JK (1991) Kinetic relations and the propagation of phase boundaries in solids. Arch Rational Mech Anal 114:119–154.

Comet Shoemaker-Levy-9 impact on Jupiter

T. J. Ahrens*, T. Takata†, J. D. O'Keefe
Lindhurst Laboratory of Experimental Geophysics
California Institute of Technology, Pasadena, California 91125 USA

Abstract: Calculations describing the passage and stopping of 0.4- to 2-km diameter fragments of Comet Shoemaker-Levy 9 (SL9) in the atmosphere of Jupiter predict a luminous $\sim 10^1$-second-duration shock wave. An expansion plume (analogous to a nuclear fireball) induces bright ($\sim 10^3$ K), $\sim 10^2$-km diameter plumes that rise to $\sim$ 3000 km. A SL9 progenitor diameter of 5 ± 2 km and a total impact energy of 1.2×10^{30} erg is inferred from comparing calculations and observations.

Key words: Comet Shoemaker-Levy 9, Impact, Jupiter

1. Introduction

The most spectacular and energetic planetary event ever witnessed by humankind was the impact at 60 km/s of fragments of Comet Shoemaker-Levy 9 (SL9) on Jupiter, 16 to 21 July 1994. Although the velocity of the fragments was well determined from knowledge of the orbit, fragment sizes and density were uncertain. Numerical models indicated that the parent body was < 2 km in diameter and the total expected impact energy was 8×10^{28} erg. Photometric data yielded an estimated (maximum) 7.4-km diameter parent body, or a total impact energy (solid ice) of 4×10^{30} erg. Thus there was an uncertainty of a factor of ~ 50 for the impact energy. Even taking into account the long human history of cometary observations, the discovery of such a large number (~ 22) of well-defined fragments is unique.

In this report we summarize our calculational methods and predictions (Ahrens et al. 1994a, Ahrens et al. 1994b, Takata et al. 1994) and compare these to the observations of the SL9 impacts (Takata et al. 1995).

2. Approach

We modeled the interaction of the SL-9 fragments with Jupiter's atmosphere in three stages: (1) the passage and transfer of the kinetic energy of the comet fragment to the atmosphere of Jupiter, (2) the expansion and rise of the at-

---

*Correspondent.
†Present address: Geological Institute, U. of Tokyo, Hongo, Bunkyo-ku, Tokyo, Japan.

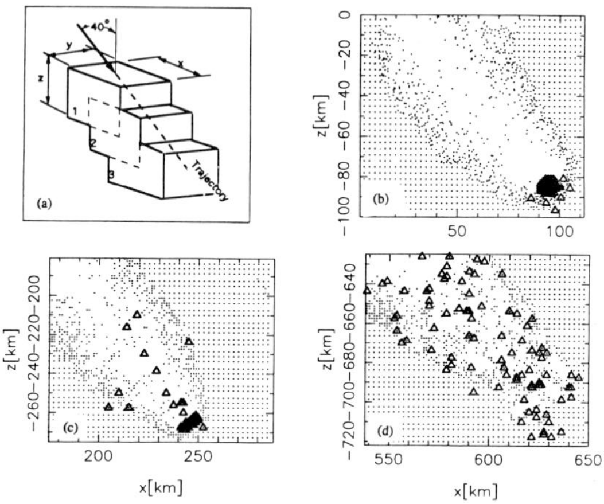

Figure 1. (a) Calculational grids employed to conduct SPH calculations for a 10-km fragment. Jupiter's atmosphere (dots) and bolide fragments (triangles) for times (b) 2.0, (c) 5.0, and (d) 21 seconds.

mosphere heated by the stopping of the comet, and (3) the radiation from the heated plume.

We used the Smoothed Particle Hydrodynamics (SPH) method to calculate the full three-dimensional flow and thermodynamic fields in the comet fragment and the atmosphere of Jupiter upon passage and stopping of 2- and 10-km diameter comet fragments impacting Jupiter at an angle of 40° from the zenith at 60 km/s. We scaled these "entry" results to a 0.4-km diameter fragment and then performed additional plume calculations. We calculated the flow that is driven upward by expansion effects. Finally, we calculated the greybody thermal radiation spectrum to predict spectral emission.

We described passage and stopping of the projectile in the atmosphere (Fig. 1). Because the density in the upper atmosphere is much less than that of cometary nucleii, the impactor will traverse many diameters before being stopped. For a 2-km bolide, we employed a series of 37, ~50,000-particle computational regions.

Two equation-of-state (EOS) models were used for the comet and atmosphere. A fluid, no strength, water-ice EOS (O'Keefe and Ahrens 1982) was used for the 280 cometary ice particles. For Jupiter's atmosphere, we assumed an ideal-gas EOS with a polytropic exponent of 1.4 and specific heat of 4×10^7 erg/(g°K) appropriate for 89% H_2 and 11% He.

3. Results

The penetration of large projectiles ($>$ 0.4-km diameter) into the atmosphere of Jupiter can be described in terms of four regimes: (1) initial, prior to distortion, (2) crushing and distortion, (3) disintegration and stopping, and (4) buoyancy driven plume.

Initial, Prior to Distortion. When the bolide enters Jupiter's atmosphere it produces a high-temperature shock wave in the atmosphere in front of the bolide. The high-speed flow of the atmosphere tangential to the melted material of the bolide (Reynolds' number $\sim 10^{13}$) drives surface (Kelvin-Holmholtz) instabilities that grow and shed bolide material, as shown by Keith and Banks (1990).

Crushing and Distortion. The dynamic pressure at the front of the bolide exceeds the strength of the impactor. We assume a crushing strength of 3×10^{-3} bar. The bolide will crush at a height of $\sim$ 360 km above the 1-bar level. The SL-9 fragments will be reduced to rubble before stopping in Jupiter's atmosphere, in 5–10-s time scales (Fig. 1). Besides crushing, distortion of the bolide occurs. Rayleigh-Taylor instabilities with wavelengths comparable to the bolide diameter grow on a time scale of $\sim$ 0.4 s. Evidence of this is seen in a few particles that are in front of the impactor in Fig. 1. Crawford et al. (1994a) observed similar phenomena in their SL-9 calculation.

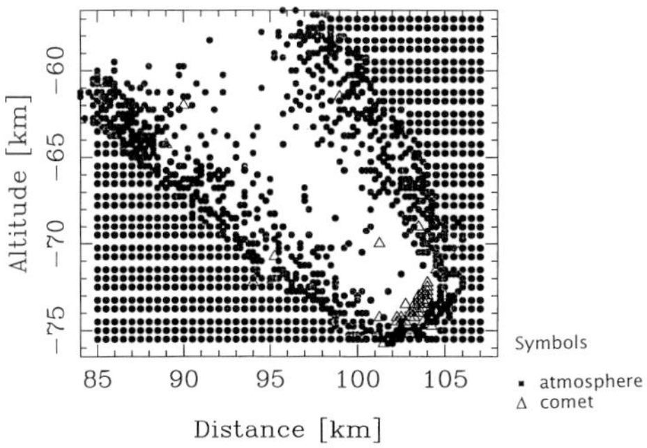

Figure 2. SPH calculation of particle distribution for a region along mid-plane of the trajectory within the incident plane for 2-km SL-9 fragment 2.6 s after impact. Circles and triangles represent Jupiter and comet particles, respectively.

The relative diameter increases from unity to maximum of $\sim$ 3 for a 2-km impactor (Fig. 2). After the disintegrated comet has been "stopped" by the atmosphere, most of the projectile material lines the conical hole in Jupiter's atmosphere. The subsequent expansion-driven flow then propels comet fragments and vapor upward in Jupiter's atmosphere. The resultant energy distribution (for the first SPH entry calculations) yielded deposition per unit length of trajectory, both for the 2- and 10-km calculations (Takata et al. 1994).

To describe the expansion-driven flow fields (plumes) which result from the energy deposition, we constructed a second SPH grid of particles and assigned excess (relative to ambient) energies to particles in a linear array lying at 40° to the zenith. The plume from the 2-km impactor used 250,000 atmospheric particles in an array with spatial dimensions in a cube, 1000 km/side.

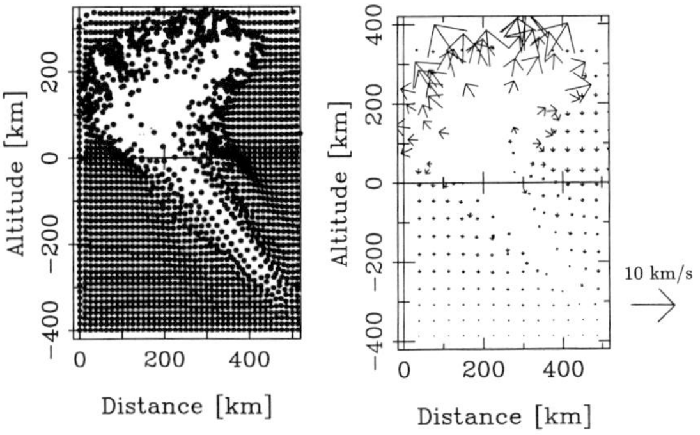

Figure 3. Ejecta plume 75.2 seconds after 2-km diameter cometary fragment impacts Jupiter: (a) mid-plane particle distribution, (b) mid-plane particle velocity distribution. Circles and triangles have some meaning as in Fig. 2.

Upward ejecta velocities (Fig. 3b) do not exceed ~ 2 and 10 km/s for 0.4- and 2-km diameter cometary fragments, and hence, *neither* cometary or Jovian atmosphere escape the planet.

4. Plume height and impact energy

The height achieved by the SL9 fragment shock-heated gas "plume" is dependent on impact energy and to a degree, studied in part by Crawford et al. (1994b) and Boslough et al. (1995), on the effective diameter of the impacting fragment.

The plumes observed by Hammel et al. (1995) for fragments A, E, G1, and W reached maximum heights of 2966 ± 370, 2916 ± 170 (lower bound), 3346 ± 170, and 2666 ± 170 km (lower bound), respectively. The nearly constant height of these plumes (Fig. 4) is surprising as much larger variations in fragment sizes are inferred both from the pre-impact photometric data and the areas of opaque material deposited on the cloud tops. Here 0 km is the 1-bar datum. Interpolation of maximum plume heights between our SPH simulations for a 0.4- and 2-km diameter bolide yields peak heights of $\sim$ 500 and $\sim$ 3000 km, respectively (Fig. 4).

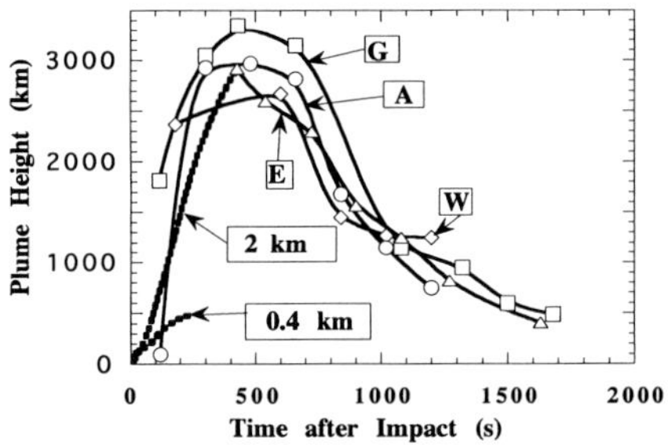

Figure 4. SPH calculation for $D = 0.4$- and 2-km plume heights and HST (Hammel et al. 1995) observations.

Applying SPH results to observed plume heights gives $D_{SPH} = 2.0 \pm 0.1$, 2.0 ± 0.05, 2.1 ± 0.04, and 1.9 ± 0.5 km for the A, E, G1, and W fragments. Upper limits for the diameter of each fragment are available from Weaver et al.'s (1995) optical data. We assume that the squares of these upper limits are proportional to the observed brightness of each fragment (Table 1).

Table 1. Calculated diameter of 5 of the 22 SL9 fragments. Fragments A, G1, E, and W diameters were obtained from plume heights of SPH simulations and the HST observations (Hammel et al. 1995). Lower- and upper-limit values result from variance in scaling.

Fragment ID	Upper Limit diameter (km) from Weaver et al. (1995)	**Best-estimate diameter (km) present work**	Lower Limit diameter (km) present work	Upper Limit diameter (km) present work
A∗	1.4	**2.01**	1.91	2.11
B†	1.68	**1.12**	0.36	1.63
E	2.8	**2.00**	1.95	2.04
G1	4.06	**2.11**	2.07	2.15
W	2.38	**1.93**	1.88	1.98
Total 22 fragments progenitor diameter (km)	7.43	**4.98**	3.20	6.79

∗Bold face: Plume heights from Hammel et al. (1995). †Typical value for fragment B observed by Weaver et al. (1995), whose plume heights were *not* measured.

We compare our SPH calculations to the Hubble Space Telescope (HST) data for the A, E, G1, and W plume heights (Fig. 1) to determine their diameters, and then scale Weaver et al.'s (1995) catalog (Table 1). This yields 5.0 ± 1.8 km for an (ice) comet progenitor diameter. The corresponding equivalent energy of a solid-ice projectile at 60 km/s is 1.2 (1.8, -0.8) $\times 10^{30}$ erg.

Acknowledgement. Research supported by NASA. Contribution No. 5581. Division of Geological and Planetary Sciences, California Institute of Technology.

References

Ahrens TJ, Takata T, O'Keefe JD, Orton GS (1994a) Impact of Comet Shoemaker-Levy 9. Geophys Res Lett 21:1087–1090.

Ahrens TJ, Takata T, O'Keefe JD, Orton GS (1994b) Radiative signatures from impact of Comet Shoemaker-Levy 9 on Jupiter. Geophys Res Lett 21:1551–1553.

Boslough MB, Crawford DA, Trucano TG, Robinson AC (1995) Numerical modeling of Shoemaker-Levy 9 impacts as a framework for interpreting observations. Geophys Res Lett 22:1821–1824.

Crawford DA, Boslough MB, Trucano TG, Robinson AC (1994a) The impact of comet Shoemaker-Levy 9 on Jupiter. Shock Waves 4:47–50.

Crawford DA, Trucano TG, Boslough MB, Kipp ME, McGlaun JM (1994b) The influence of body shape on the deformation and break-up of Comet Shoemaker-Levy 9 fragments as they enter the Jovian atmosphere. Lunar & Planet Sci (abstract) XXV:295–296.

Hammel HB, Beebe RF, Ingersoll AP, Orton GS, Mills JR et al. (1995) Hubble Space Telescope imaging of Jupiter: atmospheric phenomena created by the impact of Comet Shoemaker-Levy-9. Science 267:1288–1296.

Keith JC, Banks NE (1990) Hypervelocity impact and aerodynamic breakup of liquids. Int J Impact Engng 10:309–322.

O'Keefe JD, Ahrens TJ (1982) Cometary and meteorite swarm impact on planetary surfaces. J Geophys Res 87:6668–6680.

Takata T, Ahrens TJ, Harris AW (1995) Comet Shoemaker-Levy 9: fragment and progenitor impact energy. Geophys Res Lett 22:2433–2436.

Takata T, O'Keefe JD, Ahrens TJ, Orton GS (1994) Comet Shoemaker-Levy-9: impact on Jupiter and plume evolution. Icarus 109:3–19.

Weaver HA et al. (1995) The Hubble Space Telescope (HST) observing campaign on Comet Shoemaker-Levy 9. Science 267:1282–1288.

Impact jetting of geological materials

W. Yang, T. J. Ahrens
Lindhurst Laboratory of Experimental Geophysics, Seismological Laboratory
California Institute of Technology, Pasadena, California 91125 USA

Abstract: To understand jetting of earth materials, gabbro slabs were accelerated to 1.5 – 2 km/s and impacted gabbro, novaculite, and porous sandstone targets at inclination angles of 30° – 60°. The ejecta were collected using a catcher box filled with styrofoam and the particles extracted using chloroform. Jetting angles and ejecta masses are determined. The mass of the ejected particles per unit area (~ 50 mg/cm^2) of the impactor remains almost independent of the impact velocity, inclination angle, thickness of the target, sample mineralogy, and density. Hydrodynamic models are used to calculate the jetting mass, angle, and velocity. The metallic jetting model predicts ~ 6 times more ejecta than experimentally measured as the inclination angle increases. X-ray diffraction of the recovered ejecta shows that it is still in crystalline form, which agrees with thermodynamic calculations. Because the theoretical jetting model for thin metal plates provides a poor description of the experiments, the application of metal plate theory to planet-sized objects appears to be questionable.

Key words: Impact, Jetting, Ejecta

1. Introduction

When two objects collide over a range of angles and impact velocities, high-velocity material is released from the contact zone between the two objects. This phenomena is called jetting. Jetting has been extensively studied in the case of symmetric and asymmetric collisions of thin metal plates both theoretically and experimentally (Birkhoff et al. 1948, Wang 1989, Pack and Curtis 1990). Kieffer (1977) applied the symmetric collision theory to the impact of geologic materials and predicted that molten silicate jets would be produced upon impact at velocities as low as 1 – 2 km/s. Melosh and Sonett (1984), applying the theory of symmetric collisions to the collision of two unequal spheres, suggested that a mega-impact event on the proto-Earth may have ejected enough mass in a solid, liquid, and vapor plume to form the Moon. More recently, Vickery (1993), using Melosh and Sonett's model, calculated the jetting velocity, ejecta mass, and composition in the case of an asteroid impact on Earth. She concluded that for such events, the initially molten, now glassy objects that are recognized as tektites, which in several cases have strewn fields of 10^3 – 10^4 km, did not originate from the jetting process. As most of the tektites strewn fields can either be related to specific craters, or a terrestrial origin (Cohen 1963, Taylor and Epstein 1962, Blum et al. 1992), it appears that tektites are associated with the high-speed ejecta, which is lofted at large angles $> 45°$ to the local horizontal. Although there have been several studies and applications of the jetting theory to geologic materials, virtually no experimental constraints are available for the jetting sources involving geological materials. This work summarizes our laboratory results and compares these to theoretical results.

2. Experiments

Experiments were conducted using a 40-mm propellant gun to accelerate 5-mm thick gabbro discs to 1.5 – 2 km/s. Upon impact with inclined (30° – 60°) rectangular target slabs (5- to 12-mm thick gabbro, novaculite, and porous sandstone), ejecta were collected in a catch box filled with styrofoam (32 kg/m^3, 0.1-m thick). Figure 1 and Table 1 give the experimental configuration and conditions. As a result of the impacts, a series of elliptically shaped craters, ~ 5 cm long and ~ 2 cm wide, were formed. By careful measurement of the relative

Table 1. Experimental parameters.

Shot #	Flyer Plate		Target			Impact	Impact v
	ρ (g/cm^3)	t (mm)	m	ρ (g/cm^3)	t (mm)	Angle	(km/s)
846	2.89	5.04	gabbro	2.89	4.98	30°	1.51
852	2.90	4.99	gabbro	2.90	4.79	30°	1.52
855	2.90	4.99	gabbro	2.90	4.87	30°	1.51
905	2.88	5.16	novaculite	2.24	10.2	30°	1.95
906	2.87	5.13	nocaculite	2.24	10.2	45°	1.95
907	2.88	5.16	novaculite	2.24	10.0	30°	1.52
927	2.89	5.35	sss	1.80	11.9	45°	1.49
928	2.90	4.57	sss	1.83	11.8	45°	1.83
940	2.90	4.89	gabbro	2.98	10.2	60°	1.52
941	2.90	5.42	gabbro	2.88	10.1	60°	2.02

Here v = velocity; m = material; t = thickness; ρ = density; gabbro = San Marcos Gabbro, Escondido, CA; novaculite = Norton, Black Hard Arkansas sharpening stone, AK; sss = Saltwash South Sandstone, UT.

positions of the target and the crater created by the ejecta, the jetting angle was determined. Then the styrofoam which contained the ejecta was dissolved in chloroform and the ejecta were recovered. Table 2 summarizes the results. The design of the present planar experiment has several advantages over the traditional sphere impacting a half-space target. The ejecta are totally separated from other debris, the jetting angle is determined easily by geometrical measurement, and the experimental results can be compared with theoretical predictions. By changing the impact angle, we can simulate the different regimes of a sphere impacting a half-space target. Figure 2 shows typical X-ray diffraction patterns of recovered ejecta. The results show that no amorphous phase

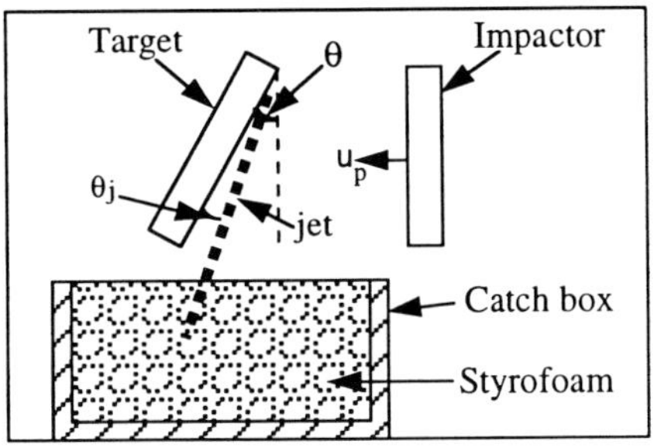

Figure 1. Configuration of oblique impact experiments: u_p – impact velocity, θ – impact angle, and θ_j – jetting angle.

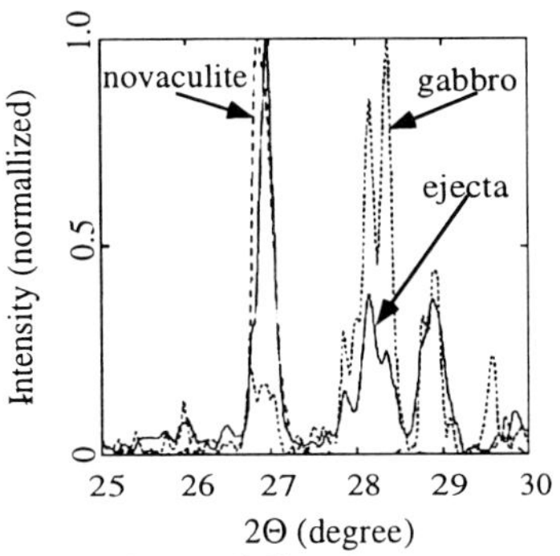

Figure 2. X-ray diffraction patterns (35 kV, Cu anode) of the ejecta from shot 906 (Table 1).

is present, thus the ejecta are still in crystalline form. Comparison of X-ray diffraction patterns of ejecta and ground-up unshocked materials (Fig. 2) shows that their patterns match. Thus the ejecta are, unlike tektites, composed of both the target and impactor materials. Tektites are largely glass which have major element compositions similar only to target materials. They contain, in some cases, only minute traces of the projectile materials. Scanning electron microscope (SEM) images of the ejecta from the present experiments show that the particles are less than a few microns in diameter despite the fact that grain sizes of the impactor and target are as large as 3 mm.

Table 2. Theoretical results and experimental values (in parentheses).

Shot No.	846	852	855	905	906	907	927	928	940	941
Jet mass (mg/cm^2)	53 (66)	52 (53)	52 (-)	64 (30)	153 (62)	64 (30)	155 (42)	140 (54)	299 (45)	323 (52)
Jetting angle (degree)	6.7	6.8	6.8 (6.5 ± 0.5)	5.1 (4.7 ± 0.5)	6.1 (5.5(± 1)	5.2 (8.1 ± 0.5)	6.6 (9.0 ± 1)	5.8 (10 ± 1.5)	3.8 (3.2 ± 0.6)	4.3 (10 ± 1.5)
Jet velocity (km/s)	5.78	5.82	5.86	7.43	4.92	5.79	3.80	4.61	2.76	3.70

3. Theoretical model

Analytical asymmetric collision models are described by Wang (1989) and Pack and Curtis (1990). In our treatment we take the collision point o as the origin of a moving coordinate system (referenced to the laboratory coordinate system). The flow pattern of the impactor and target plates is shown in Fig. 3. We assume that the flow depicted in Fig. 3 is incompressible and the surface between the impactor and the target remains planar. Then Bernoulli's theorem yields equal flow velocities of the jet, u_j, slug, u_s, and incoming flow, u_1, and u_2. If we further assume that the jet and slug move along one straight line, using the conservation of mass and momentum and geometric relations, we get the jet mass m_j, the average jet velocity in the moving coordinates u_j, the jet velocity v_j, and jetting angle θ_j in laboratory coordinates as:

$$\theta_2 = \arctan(m_1 u_p/(m_1 + m_2)/\tan\theta) \tag{1}$$

$$m_j = m_{1j} + m_{2j} = [m_1(1 - \cos(\theta - \theta_2)) + m_2(1 - \cos\theta_2)]/2 \tag{2}$$

$$u_j = (m_{1j} u_p/\tan\theta + m_{2j} u_p/\sin\theta)/m_j \tag{3}$$

$$v_j^2 = u_j^2 + (u_p/\sin\theta)^2 + 2u_j u_p \cos\theta_2/\sin\theta \tag{4}$$

$$\theta_j = \arcsin(\sin(\theta_2) u_j/v_j) \tag{5}$$

Calculated values for m_j, θ_j, and v_j are given in Table 2. In order to estimate the pressure and temperature that the materials have experienced, Bernoulli's equation is used again and the flow is assumed now to be compressible

$$e + pv + u^2/2 = e_o + p_o v_o + u_o^2/2 \tag{6}$$

Substituting e and p with the Hugoniot relations $e = p(v_o - v)/2$ and $p = c_o^2(v_o - v)/[(1-s)v_o + sv]^2$, we can solve (6) for p and v at the stagnation point. With other thermodynamic parameters, the shock temperature and post-shock temperature the materials have experienced can be calculated (Yang et al. 1992). The parameters used are listed in Table 3.

Table 3. Parameters used in pressure and temperature calculations.

	γ_0	θ_0 (K)	c_0 (km/s)	s	P (GPa)	$T_{\rm hmax}$ (K)	$T_{\rm pmax}$ (K)
Gabbro	5.2[a]	1022[b]	5.6	0.15	18	815	304
sss	0.7[c]	999[c]	2.90	0.72	7	454	364
Novaculite	0.7[c]	999[c]	2.35	1.1	21	1196	411

Here γ_0 – Grüneisen constant; θ_0 – Debye temperature; a – calculated from the formula $\gamma_0 = \alpha c_0^2/c_p$, where α – thermal expansion coefficient, c_p – heat capacity (Touloukian et al. 1989); b – obtained by fitting c_p to Debay's formula; c_0 and s are obtained by fitting the shock-wave data (Marsh 1980); $T_{\rm hmax}$ and $T_{\rm pmax}$ are maximum shock and post-shock temperatures, respectively; sss – Saltwash South sandstone; d – from Swegle (1990).

4. Discussion and conclusions

From Table 2 we can see that the calculated masses of ejecta are significantly greater than those of the experiments, except for shot 846 and 852 (Table 1). By plotting the ejecta masses versus inclination angle on Fig. 4, we immediately see that while the theory predicts an increase of ejecta mass as the inclination

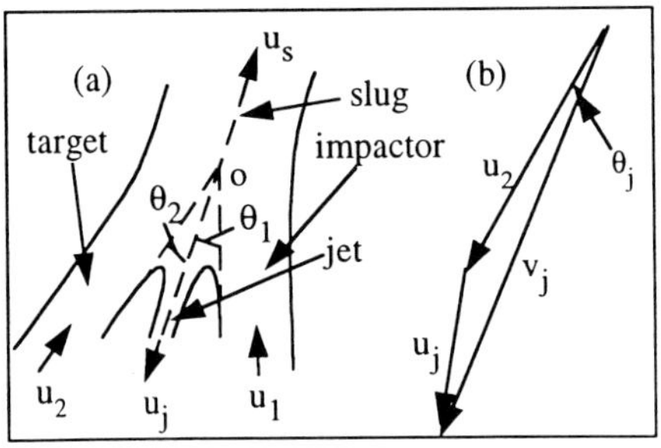

Figure 3. (a) Flow pattern of the impactor and target in a moving frame, o – collision point; (b) velocity vectors are given in laboratory coordinates; v_j – jet velocity, θ_j – jetting angle.

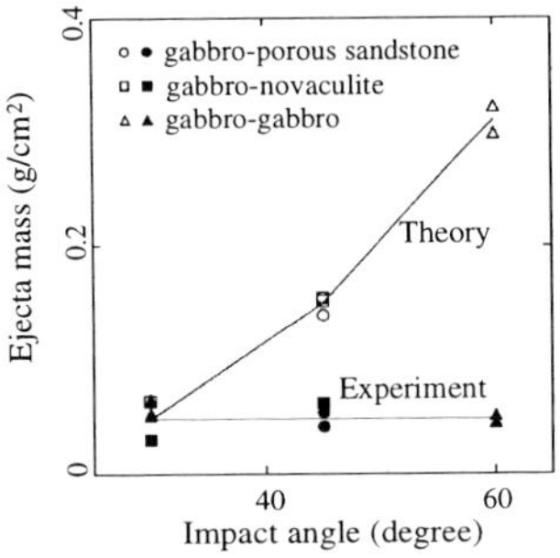

Figure 4. Ejecta mass per unit area versus impact angle.

angle and plate thickness increase, the experiments show an approximately constant amount of ejecta mass ($\sim$ 50 mg/cm^2) despite the fact that the impact velocity, material, and plate thickness were varied considerably. Since increasing the thickness of the target does not appear to increase the mass of ejecta, we conclude that only part of the total thickness of the impactor and target contribute to the formation of a jet. This thickness is about 5 mm (shots 846 and 852) with our experimental conditions. In Melosh and Sonett's model, and also in Vickery's calculations, the thickness which is assumed to contribute to jetting is chosen as the radius of the impacting sphere. In light of the present experiments the above assumption appears to be unsupported. Obviously different jet ejecta masses can be inferred by assuming different thicknesses. It seems that the present jetting theory does not obviously apply to geological materials. Although our experiments were conducted at relatively low velocities, the impactor and target materials both experienced shock pressures far in excess of their strengths, hence they should behave like fluids. In addition, the theoretical model predicts little ejecta mass change when impact velocity changes. Thus we believe that the discrepancy between experiments and theory will not disappear at high impact velocities (e.g., 10 km/s). From the X-ray diffraction patterns (Fig. 2), we find that the ejecta contain significant amounts of impactor and target materials. Since most tektites do not contain significant amounts of an extraterrestrial component (Koeberl and Shirey 1993), by analogy we do not believe they are produced by the jetting process upon asteroid impact. However, the tektites may be produced by other processes after jetting occurs during impact events. The jetting angles (Table 2) are all less than 10°. This suggests that oblique (or asymmetric) impact is very different from symmetric impact. Thus applying symmetric impact theory (as was used by Melosh and Sonett, and Vickery), does not appear to be appropriate for asymmetric impacts. The calculated shock and post-shock temperatures are all below the melting points of the materials used. This agrees closely with the X-ray diffraction and SEM results in which no melting was detected. We also conclude, in contrast to Kieffer (1977), that impact velocities greater than 2 km/s are required to produce molten jets for the silicate materials used in our experiments.

Acknowledgement. Research is supported by NASA under NAGW 1941. Division of Geological and Planetary Sciences. Contribution No. 5582.

References

Birkhoff G, Macdougall DP, Pugh EM, Sir Taylor G (1948) Explosives with lined cavities. J Appl Phys 19:536.

Blum JD, Papanastassiou DA, Koeberl C, Wasserburg GJ (1992) Geochimica et Cosmochimica Acta 56:483.

Cohen AJ (1963) In: Tektites, O'Keefe JA (ed), Univ of Chicago Press, Chicago, pp 189–386.

Kieffer SW (1977) Impact conditions required for formation of melt by jetting in silicates. In: Impact and Explosion Cratering, Roddy DJ, Pepin RO, Merrill RB (eds), Pergamon Press, New York, pp 751–769.

Koeberl C, Steven SB (1993) Detection of a meteoritic component in Ivory-Coast tektites with rhenium-osmium isotopes. Science 261:595.

Marsh SP (1980) LASL Shock Hugoniot Data, Univ of California, Berkeley, CA.

Melosh HJ, Sonett CP (1984) In: Origin of the Moon, Hartmann WK, Philips RJ, Taylor GJ (eds), Lunar and Planetary Institute, Houston, Texas, pp 621–624.

Pack DC, Curtis JP (1990) On the effect of asymmetries on the jet from a linear shaped charge. J Appl Phys 67:6701.

Swegle JW (1990) J Appl Phys 68:1563.

Taylor MP, Epstein S (1962) Oxygen isotope studies of Ivory Coast tektites and impactite glass from the Bosumtwi crater, Ghana. Science 153:173.

Touloukian YS, Judd WR, Roy RF (1989) Physical Properties of Rocks and Minerals, Hemisphere Publishing Corporation, New York.

Vickery AM (1993) The theory of jetting — application to the origin of tektites Icarus 105:441.

Wang J (1989) Refraction of shock waves and impact of free surfaces in condensed matter. In: Proceedings of the International Conference on Applied Mechanics in Beijing, Zheng Z (ed), Pergamon Press, Beijing, pp 512–517.

Yang W, Bond GM, Tan H, Ahrens TJ, Liu G (1992) Dynamic consolidation of superhard materials. J Mater Res 7:1501.

Numerical computation of jetting impacts

E. G. Puckett
University of California, Davis
Davis, CA 95616 USA

G. H. Miller
University of Chicago
Chicago, IL 60637 USA

Abstract: Numerical computations based on a high-order Godunov method for multiple phases are presented. Computations of asymmetric oblique impacts reveal an instability that may reduce the jet velocity below predicted values. At small impact angles the peak pressure experienced by jetted materials is substantially less than that given by the standard model. Calculated temperatures and the fraction of melt and vapor created by a jetting collision must be reduced commensurately.

Key words: Godunov methods, Shock waves, Jetting

1. Introduction

We have developed an Eulerian high-order Godunov method for shock-capturing computations with multiple condensed phases. The thermodynamic model is based on a Mie-Grüneisen equation of state and a linear Hugoniot. This numerical method includes a volume-of-fluid algorithm for tracking solid-solid and solid-vacuum interfaces and an adaptive mesh refinement algorithm for locally resolving important features of the flow field in space and time. Details are presented in Miller & Puckett (1996). We have used this computational model to study shock refraction in experimental assemblies (Miller & Puckett 1994). Here, we present new computational results aimed at understanding the fluid-dynamic regime in asymmetric collisions that lead to jetting.

2. Jetting

The theory that provides jet mass and momentum is based on the fully developed flow resulting from the collision of thin plates (Birkhoff et al. 1948, Walsh et al. 1953, Harlow and Pracht 1966). In that theory, a length scale is provided by the plate thicknesses. In problems involving the collision of spheres (e.g., Melosh & Sonett 1986, Vickery 1993) the relevant length scale is not at all obvious, and in fact fully developed flow of the sort described in the thin-plate theory never occurs. Likewise the relevant scale governing the short-time behavior of plate impacts (or equivalently the impact of plates that are thick compared to their lengths) is not known. There is some experimental work addressing these issues. Yang et al. (1992) have shown good agreement between theory and jet properties in experiments with plates that are thin compared to their lengths, while Yang & Ahrens (1995) found a striking disagreement between thick-plate experimental results and the steady-state thin-plate theory. To reconcile the theory with experiment it is necessary to understand how the jet develops in the self-similar regime, and its subsequent transition to steady state. Preliminary work aimed at addressing that question is described here.

The inset to Fig. 1a shows the initial geometry of our computation — an aluminum (Al) plate strikes a 50° Al ramp at $V_{\text{imp}} = 2$ km/s. The bottom boundary is a reflecting wall. The problem domain is 10 cm by 5 cm. Figure 1a shows the configuration of the material interfaces 900 ns after impact. A jet has formed, and the slip plane joining the impacted materials has undergone a Kelvin-Helmholtz instability.

Figure 1b shows the pressure contours at 900 ns. It is apparent from this figure that the material flowing from the colliding materials into the jet is not compressed by a single shock or collection of shocks, but rather by a complex and dispersed compressional fan.

The vortices we observe are similar to those observed in shock welding in the irregular regime (e.g.,Bahrani et al. 1967, Godunov et al. 1971). Theories for the origin of these features derive the vortex length scale from the plate thicknesses (e.g., Hunt 1968, Godunov et al. 1970, Bazdenkov et al. 1985, Gupta & Kainth 1990). It seems likely that the vortex formation evident in our computations is responsible for the wave formation observed in shock welding. If so, we believe the theories for wave formation that draw on analogy to von Karman vortex streets are incorrect since we find only a single vortex sheet in our results. Godunov et al. (1971) reached this same conclusion using an argument based on the Reynolds and Strouhal numbers.

3. Theoretical considerations

The model (Kieffer 1977) that is commonly used to calculate the pressure at the stagnation point is incorrect, and hence in many cases will give inaccurate values for the stagnation pressure. The calculation is based on transforming the problem to a reference frame in which the point of collision is stationary and then using Bernoulli's law

$$I + |\vec{U}|^2/2 = \text{const.} \qquad \text{(along streamlines in stationary, isentropic flow)} \quad (1)$$

to calculate the enthalpy $I = E + PV$ at the stagnation point. The model assumes the streamline connecting the stagnation point $\vec{x}_{sp}$ to some upstream point $\vec{x}_0$ crosses a single shock. Consequently, (1) together with a form of (1)

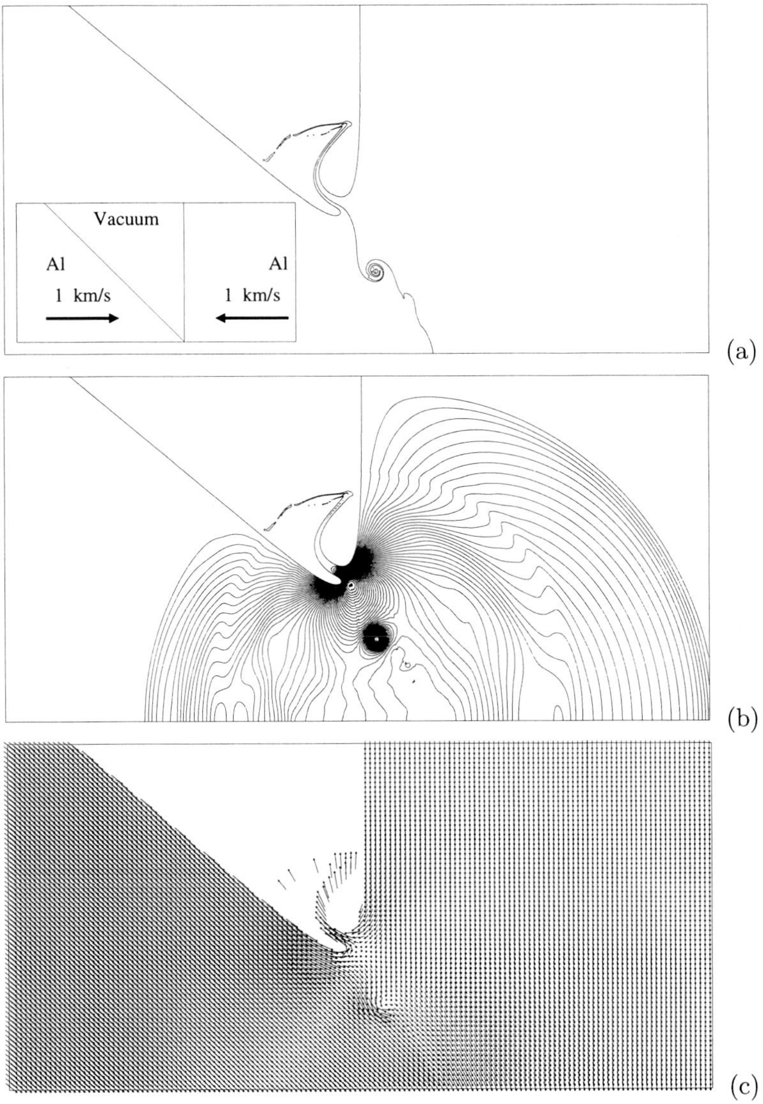

Figure 1. (a) Material interfaces at 900 ns after impact and at 0 ns (inset); (b) pressure contours from 0.2 to 10.2 GPa in steps of 0.2 GPa, and superimposed solid-vacuum interface boundary; (c) velocity vectors relative to the stagnation point.

that holds across shock waves can be used to obtain the value of the enthalpy at the stagnation point: $I_{sp} = I_0 + |\vec{U}_0|^2/2$. In order to obtain the stagnation pressure P_{sp} from I_{sp} one needs two additional equations relating the variables P, V, and E. One of these is the equation of state $P(E, V)$. In the standard model one closes the system of equations with the Rankine-Hugoniot energy jump condition $E = E_0 + (P + P_0)(V_0 - V)/2$. However, this assumes that the energy in the post-shock state, say E_1, is identical to the energy E_{sp} at the stagnation point. This cannot be true, for in applying Bernoulli's law to connect the post-shock state (P_1, V_1, E_1) to the stagnation-point state (P_{sp}, V_{sp}, E_{sp}) one must assume that the entropies of these points are equal: $S_1 = S_{sp}$. If $E_1 = E_{sp}$ and $S_1 = S_{sp}$, then the two thermodynamic states must be equal, $(P_1, V_1, E_1) = (P_{sp}, V_{sp}, E_{sp})$ and consequently by conservation of energy $\vec{U}_1 = \vec{U}_{sp} = 0$. This implies zero mass flux $\rho_1 U_1$ across the shock, which is impossible.

Since $\partial P/\partial V|_{\text{Hugoniot}} < \partial P/\partial V|_S < \partial P/\partial V|_I < 0$, for a given stagnation enthalpy I_{sp}, closure of the problem using the energy-jump equation will underestimate P_{sp} as shown in Fig. 2. The lower bound (b) is what Kieffer calculated, but this point is not physically attainable as discussed above. For a single-shock process, the true pressure must lie on an isentrope ($S_1 > S_0$) connected to the Hugoniot $\mathcal{H}$ at pressure lower than P_b. The intersection of this isentrope S_1 with the isenthalp $I = {U_0}^2/2$ will lie between (a) and (b).

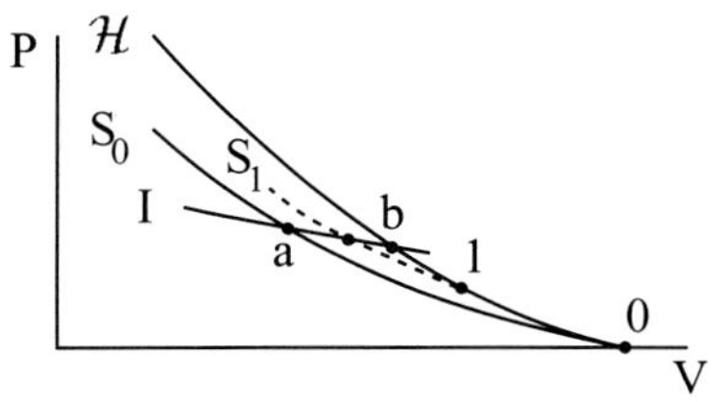

Figure 2. Bernoulli's law specifies the enthalpy $I = I_{sp}$ at the stagnation point. If the path to the stagnation point is isentropic S_0, the pressure P_a will be a maximum. If dissipation is maximized, which we think implies a single shock $\mathcal{H}$, the pressure P_b will be a minimum.

We also note that this pressure calculation is not well posed for asymmetric oblique collisions (e.g., Fig. 1) since in the stationary reference frame the initial kinetic energy of the plate and wedge are different, and so the stagnation pressures calculated from them using Bernoulli's law differ.

The computations shown in Fig. 1 are nearly self-similar (Jones et al. 1951). Under these circumstances the flow is not stationary and (1) must be modified. Instead,

$$I_{sp} + \frac{1}{2}|x_{sp}/t|^2 = I_0 + \frac{1}{2}|\vec{U}_0'|^2 - \int_{s_0}^{s_{sp}} |\vec{U}'|^2 ds \tag{2}$$

where $\vec{U}' = \vec{U} - \vec{x}/t$, and ∇' is the gradient operator with respect to $\vec{\xi} = \vec{x}/t$. The integral is taken along the integral curve defined by $d\vec{\xi}/ds = \vec{U}'(\vec{\xi}(s))$ from some initial point s_0 to the stagnation point s_{sp} (Fig. 1c). At small angles α the stagnation-point enthalpy computed from (2) is generally less than that given by (1), principally because the upstream velocities in the self-similar regime are 0 and V_{imp} whereas they are $V_{\mathrm{imp}} \cot\alpha$ and $V_{\mathrm{imp}} \csc\alpha$ in the steady-state regime.

The steady-state theory applied to the self-similar computations (Fig. 1) gives a stagnation pressure of 4.0 GPa or 10.0 GPa depending on whether the calculation is done for the plate or the wedge. The peak pressure in Fig. 1 is 10.2 GPa. However, in a similar computation with a 20° angle the theory gives 48.4 and 55.4 GPa whereas the numerical computation gives 27.5 GPa. At 15° the disagreement is still larger: theory gives 94.1 or 101.4 and computation gives 35.9. For the cases described the discrepancy between theory and computations is largest near the onset of jetting (13.2° with the present geometry and 2 km/s horizontal velocity) where the upstream velocity U_0 differs most from the steady-state value. Geological and planetary applications that seek to explain tektites, chondrules, and the formation of the Moon (e.g., Kieffer 1975, Melosh & Sonett 1986, McKinnon 1989, Vickery 1993) are more nearly self-similar than steady-state, and it is under such conditions that our computations suggest the commonly used theory fails most egregiously.

4. Conclusions

Oblique, asymmetric collisions lead to vortex sheets that in turn lead to Kelvin-Helmholtz instabilities. Jetting models based on the idealized symmetric geometry presented in Birkhoff et al. (1948), i.e., assuming locally smooth, linear, and stationary material interfaces, are wrong in detail and possibly lead to inaccurate conclusions.

A conclusion that is independent of our computations is that the model commonly used to compute the stagnation-point pressure is incorrect. Our computations support this result by revealing more complex wave interactions than commonly supposed, and by demonstrating poor agreement between the theoretical and the computed peak pressures. Because of this we suggest that calculations of the thermodynamic state of jetted materials by the commonly used model require reevaluation. Our numerical results suggest the discrepancy is largest at small jetting angles, with the commonly used model overestimating the stagnation pressure by as much as a factor of three in the examples given.

Acknowledgement. This work was supported by NSF grants DMS-9104472 and DMS-9316529 to EGP and EAR-9304263 to GHM. Computational support was provided by the National Energy Research Supercomputing Center at the Lawrence Livermore National Laboratory and the Pittsburgh Supercomputing Center.

References

Bahrani AS, Black TJ, Crossland B (1967) The mechanics of wave formation in explosive welding. Proc Roy Soc London A 296:123.

Birkhoff G, MacDougall DP, Pugh EM, Taylor G (1948) Explosives with lined cavities. J Appl Phys 19:563.

Bazdenkov SV, Demichev VF, Morozov DKh, Pogutse OP (1985) Possible mechanism of wave formation in explosive welding. Comb Expl Shock Waves 21:124.

Godunov SK, Deribas AA, Zabrodin AV, Kozin NS (1970) Hydrodynamic effects in colliding solids. J Comput Phys 5:517.

Godunov SK, Deribas AA, Kozin NS (1971) Wave formation in explosive welding. J Appl Mech Tech Phys 7:398.

Gupta RC, Kainth GS (1990) Swinging wake mechanism for interface wave generation in explosive welding of metals. J Appl Mech 57:514.

Harlow FH, Pracht WE (1966) Formation and penetration of high-speed collapse jets. Phys Fluids 9:1951.

Hunt JN (1968) Wave formation in explosive welding. Phil Mag 17:669.

Jones DM, Moira P, Martin E, Thornhill KC (1951) A note on the pseudo-stationary flow behind a strong shock diffracted or reflected at a corner. Proc Roy Soc London A 209:238.

Kieffer SW (1975) Droplet chondrules. Science 189:333.

Kieffer SW (1977) Impact conditions required for formation of melt by jetting in silicates. In: Impact and Explosion Cratering, Roddy DJ, Pepin RO, Merrill RB (eds), Pergamon, New York, p 751.

McKinnon WB (1989) Impact jetting of water ice, with applications to the accretion of icy planetesimals and Pluto. Geophys Res Lett 16:1237.

Melosh HJ, Sonett CP (1986) When worlds collide: Jetted vapor plumes and the Moon's origin. In: Origin of the Moon, Hartmann WK, Phillips RJ, Taylor GJ (eds), Lunar & Planetary Institute, Houston, p 621.

Miller GH, Puckett EG (1994) Edge effects in molybdenum-encapsulated molten silicate shock wave targets. J Appl Phys 75:1426.

Miller GH, Puckett EG (1996) A high-order Godunov method for multiple condensed phases. J Comput Phys (submitted).

Vickery AM (1993) The theory of jetting: application to the origin of tektites. Icarus 105:441.

Walsh JM, Shreffler RG, Willig FJ (1953) Limiting conditions for jet formation in high velocity collisions. J Appl Phys 24:349.

Yang W, Ahrens TJ, Miller GH, Petach MB (1992) Jet ejecta mass upon oblique impact. In: Shock Compression of Condensed Matter 1991, Schmidt SC, Dick RD, Forbes JW, Tasker DG (eds), Elsevier, New York, p 1011.

Yang W, Ahrens TJ (1995) Impact jetting of geological materials. Icarus 116:269.

Compositional changes of the Moon and Earth by impacts

Y. Miura
Faculty of Science, Yamaguchi University, Yoshida, Yamaguchi 753, Japan
M. Okamoto
Faculty of Science, Okayama University, Tsushima-naka, Okayama 780, Japan

Abstract: Chemical change can be observed in impact mixtures formed by solid-melt-gas reactions with short impact times. Chemical change of plagioclases in the lunar brecciated and basaltic rocks (especially Na loss) followed by progressive impacts can be found to make the Na-rich gas atmosphere on the Moon. Impact experiments on gabbroic anorthosite of similar lunar highland crust rocks reveal chemical changes between coarse- and fine-grained fragments with significant chemical loss. A new theory of the origin of the Moon (i.e. multiple impact evolution theory) is proposed by applying compositional change after impact and anomalous distribution of overall lunar impact craters, which can explain unsolved problems.

Key words: Chemical separation, Compositional change, Earth planetary bodies, Heavy and light parts by impact, Mantle-core structure, Origin of the Moon

1. Introduction

The effect of shock waves on Earth and planetary materials (including artificial impact materials) has been studied mainly by solid-state reaction (French 1968). The detailed study of impact processes within seconds (Gault et al. 1968; Melosh 1989; Miura et al. 1992a, 1992b), shows overall reactions of gas-, melt- and/or solid-gas-melt-solid states which can be observed by indirect and post-shock evidence of anomalous composition, structure, and texture of quenched solid materials (Miura 1991; Miura et al. 1995a, 1995b). Therefore, compositional analyses of fine-grained impact materials are the clue to the impact reactions in artificial and Earth-Moon planetary materials (Miura 1991; Miura et al. 1992a, 1992b, 1995a, 1995b). The main purpose of the present study is to elucidate compositional changes of artificial and Earth-Moon planetary bodies by impact experiments and impact-crater analysis.

2. Experimental

The bulk chemical composition of impact fragments was analyzed by analytical scanning electron microscopy (ASEM) at Yamaguchi University; this is the same method and samples used by Miura (1991) and Miura et al. (1992b, 1995a, 1995b). Composition of Apollo lunar samples offered by the Johnson Space Center (JSC), NASA, USA was determined by using an electron microprobe analyzer (EPMA) of the same laboratory. For supporting data, the 174 lunar impact craters reported by Wilhelm (1984) are used for anomalous distribution analysis of the craters and previous problems of lunar origin.

3. Results and discussions

3.1. Chemical dynamics of Na on lunar samples

Composition analyses of Apollo lunar samples reveal normal and anomalous chemical composition of plagioclase feldspars forming the lunar crust (Miura 1988; Ryder 1993). In fact, plagioclases — (Ca, Na, K)$Al_2Si_2O_8$ — of breccia and basalt show anomalous composition (i.e. deficient in Na and Al) as listed in Table 1. The anomalous composition indicates that K, Na, and Al elements with low melting and boiling points are deficient under high temperature after impact. The main supply of Na gas in the dynamic and faint atmosphere (i.e. 30 torus Na gas) is evaporated elemental Na from lunar surface materials by progressive impacts (Miura and Kato 1992; Miura and Graham 1994).

Table 1. Anomalous compositions of lunar plagioclases by impact.

Samples Oxides	12002 [12]*	14066 [7]	15455 [12]	67435 [6]	76535 [12]
Al_2O_3 (wt%)†	−1.0	−0.5	−1.0	−1.2	0.0
Na_2O (wt%)†	−1.2	−0.5	−1.0	−0.5	−0.5
An (mol.%)	91.1	81.6	94.3	94.3	96.3
Age (Ga)	3.24	3.93	3.92	4.35	4.23
Rock	Basalt	Breccia	Breccia	Troctolite	Breccia

* The number of point analyses by EPMA.
† Excess amounts are calculated from the terrestrial standard plagioclases (normalized by Si; cf. Miura 1988).

3.2. Core-mantle structure formation by impacts

Mantle-core concentric structure (Miura et al. 1995a) can be formed by artificial impact experiments which indicate that large metallic spherule (Fe and Co oxides, CM of ca. 84 wt%) becomes the core of the concentric texture and that lithophile elements of Si, Al, Mg, Ca, K, Na (ca. 56 wt%) cover the core of the mantle of the concentric texture (cf. Miura et al. 1995a), though small metallic spherule (Fe-core or Cu-grain) makes a small metallic region in the mantle structure. The starting composition of a metallic projectile or target granite was completely changed to a metal-rich or silicate-rich mixture. The metallic iron-rich part of the Moon is estimated as 2.3 (wt%), which suggests that the Moon, of which the major parts are crust and mantle structure without a large clear core, is formed of light aggregates by the impact collision process.

3.3. Chemical change experiments by impact on lunar crustal rock

In order to prove the chemical separation of impact fragments, an impact experiment (velocity of 2 km/s) using an epoxy projectile (without iron) was done on a target rock of gabbroic anorthosite which is similar to most lunar highland rock. The final bulk composition of two fragments was obtained from the whole excavated volume of the crater (89.8 cm^3, 251.4 g), coarse-grained fragments (86.2% of the excavated crater, 216.7 g), and fine-grained ejecta (9.5% of the crater, 23.9 g). The loss amounts of 4.3% of the crater (i.e. 100 − 86.2 − 9.5) are presumed to be the evaporated portion of the present impact experiment. This chemical separation provides some of the best evidence for compositional change by impact process.

3.4. Giant impact theory of Moon formation from lunar rock research

Much lunar rock research gives the oxygen isotopic ratio of O^{16}/O^{18} the same value on the Moon as on the Earth (Wanke and Dreibus 1986). This means that the Moon composition is different from meteorites from the asteroid belt and SNC (Mars-origin?) meteorites, and thus suggests that the theory of simple capture (Gerstenkorn 1955) from other solar space regions is not correct; but points instead to aggregate origin near the Earth. Bulk composition of the Moon and the Earth are not the same, because the Moon reveals itself to be volatile-poor, water-poor, refractory-rich (Al, Ca, Th, and REE), with a high FeO/MgO value (i.e. 0.33 of the Moon, but only 0.23 of the Earth), which indicates that the Moon is not formed by simple accretion (Schmidt 1959) or precipitation (Ringwood 1970).

From the anomalous angular momentum of the Moon compared with that of the Earth, different impacts are required for two bodies, which indicates that it is unlikely that the Moon was formed by a single impact (Baldwin 1963; Wilhelm 1984), fission (Darwin 1879), accretion, or simple giant impact (Hartman and Davis 1975; Cameron and Ward 1976).

Giant impact theory, in which a planetesimal body (Mars-sized body, MPB) hit the proto-Earth to form the Moon can explain (a) water-poor and refractory-rich elements (by heating of impact), (b) the same oxygen isotopes (formed within the same area near impact), (c) no big definite iron core (formed by silicate-rich portion by impact), (d) different ratio of FeO/MgO (formed during ejection by impact), (e) different angular momenta of the Earth and the Moon by different impact materials and time, (f) normal event to form other planets of the solar system, and (g) huge energy source to form magma ocean by impact.

However, the giant impact theory has several problems, including different inclination, large impact craters on the near side of the Moon, little consideration of orbital direction of impact ejection, and the anomalously large size of the Moon (Miura et al. 1994). Therefore, another impact model (e.g. multiple impacts and evolution on the Moon, Miura et al. 1994) is required to explain the above problems by using lunar impact-crater data.

3.5. Multiple impacts and accretion model of the Moon formation

There are 131 (75%) lunar impact craters on the near side and 43 (25%) on the rear side of the Moon. Area percentages of those on the near and rear sides are 61% and 39%, respectively. This suggests that large impacts are mainly on the near side.

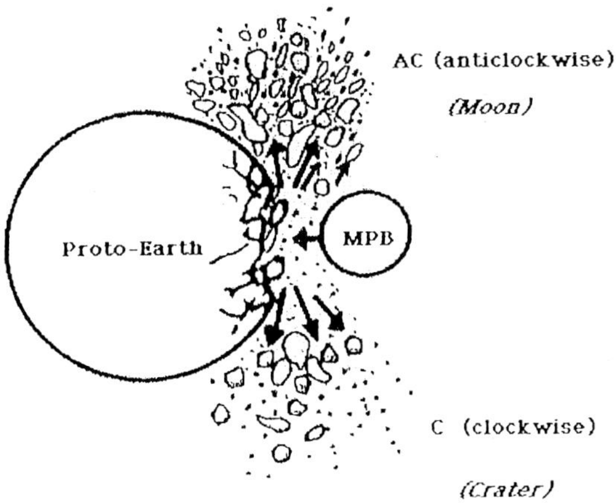

Figure 1. Ejection direction of first giant impact between the MPB and the proto-Earth which creates two orbits of anticlockwise (AC) and clockwise (C) materials. AC impact fragments form the proto-Moon by accretion, whereas C impact fragments collide head-on with the Earth or proto-Moon to make impact craters.

Among the 48 large craters (> 300 km in diameter, both sides of the Moon), more large fragments with a clockwise (C) orbit (e.g. collided to the western side, 70 area %) are found than with an anticlockwise (AC) orbit (30 area %). Regarding the 130 small impact craters (< 250 km in diameter), compared to other areas, more impacts with large fragments on the southern near and northern rear sides (66 area % and 47% by number vs. 34 area % and 53% by number) incline slightly toward the lunar rotation plane ($\sim 6°$ to orbit, compared with large inclination of Earth, $\sim 23°$), though the number of craters is similar in the two large areas, resulting in the slight inclination of the Moon.

In order to explain the compositional difference between the Moon and Earth and the proposed Earth fragmental ring (consisting of two C and AC orbits) that leads to lunar craters (cf. Fig. 1), the dynamic model of multiple impact evolution of the Moon is proposed as follows (Miura et al. 1994): (a) proto-Earth collides with an evolved planet (Mars-sized body, MPB) by giant impact; (b) ejecta fragments with both AC (mainly for the Moon formation) and C (mainly for lunar large impact craters) orbits make the Earth's ring. The anomalous size of the proto-Moon can be explained by the great accretion process of the impact fragments from both the ring and ejecta of lunar craters. The slight inclination of the Moon due to multiple impacts is different than the inclination of the Earth (due to a single giant impact). The proto-Moon, which was pulled by the Earth's gravity, was hit periodically by large inner C orbital fragments on the near side and thus moved away from the Earth. Thus the present model of the Moon is a multiple-impact model of "giant impact, multiple impacts, accretion, and evolution" which accommodates some ideas from previous individual theories.

4. Summary

Our results are summarized as follows: (1) Chemical change of plagioclases in lunar brecciated and basaltic rocks (especially Na loss) followed by progressive impacts can be found to make the faint Na-gas atmosphere on the Moon. (2) Impact experiments on gabbroic anorthosite of similar lunar highland crust rocks show chemical changes between coarse and fine fragments with some chemical loss. (3) A multiple-impact evolution theory of the Moon is proposed using compositional difference and distribution of lunar impact craters. The new model can explain unsolved problems on the origin of the Moon.

Acknowledgement. The authors thank Professor Kazuyoshi Takayama for helpful discussions. Senior author acknowledges sample support from JSC, NASA, USA for the analysis of the Apollo lunar samples. This study is supported by Monbusho Grants, 1990FY-1993FY and Yamaguchi-prefectural Grant Award of Science and Technology, 1992FY.

References

Baldwin RB (1963) The Measure of the Moon, Univ of Chicago, p 488.

Cameron AGW, Ward W (1976) The origin of the Moon. Lunar Planet Sci VII, NASA-LPI, USA, Vol 7, p 120.

Darwin G (1879) On the procession of a viscous spheroid and the remote history of the Earth. Phil Trans R Soc London 170:447.

French B (1968) Shock metamorphism as a geological process. In: Shock Metamorphism of Natural Materials, French BM, Short NM (eds), Mono, Baltimore, pp 1–17.

Gault DE, Quaide WL, Oberbeck VR (1968) Impact cratering mechanics and structures. Shock Metamorphism of Natural Materials, French BM, Short

NM (eds), Mono, Baltimore, pp 87–99.
Gerstenkorn H (1955) Uber Gezeitenreibung beim Zweikorper-problem. Zeit Astrophys 36:245.
Hartmann WK, Davis DR (1975) Satellite-sized planetesimals and lunar origin. Icarus 24:504.
Melosh HJ (1989) Impact Cratering, Oxford University Press, p 245.
Miura Y (1988) Normal and anomalous compositions of lunar feldspar — I. Lunar plagioclases. In: Lunar Planet Sci XIX, NASA-LPI, USA, pp 794–795.
Miura Y (1991) Evidence for shock wave effect of meteoritic impact. Shock Waves 1:35.
Miura Y, Kato T (1992) Shock metamorphism on the Moon. Celestial Mechanics and Dynamical Astronomy 54:245.
Miura Y, Kato T, Imai M (1992a) Evidence of earth catastrophe by anomalous shocked quartz at the K/T boundary. Celestial Mechanics and Dynamical Astronomy 54:249.
Miura Y, Takayama K, Kato T, Kawashima N, Yamori A (1992b) Shock metamorphism in artificial impact craters. In: Proc 18th Int Symp on Shock Waves, Takayama K (ed), Sendai, pp 403–408.
Miura Y, Graham RA (1994) Shock metamorphism on the surface of the Moon. Shock Waves 3:293.
Miura Y, Okamoto M, Iancu G, Arayama A (1994) Origin of the Moon and its multi-impact model. In: Proc 27th ISAS Lunar Planet Symp, Sagamihara, Japan, pp 15–18.
Miura Y, Takayama K, Kawashima N, Kato T (1995a) Shock metamorphic evolution of meteorites and impact craters. In: Shock Waves @ Marseille, Brun R, Dumitrescu LZ (eds), Springer, pp 405–410.
Miura Y, Takayama K, Kato T, Kawashima N, Imai M, Iancu G, Okamoto M (1995b) Shocked quartz, silica and carbon materials in meteorites and impact craters. In: Shock Waves @ Marseille, Brun R, Dumitrescu LZ (eds), Springer, pp 399–404.
Ringwood AE (1970) Petrogenesis of Apollo 11 basalts and implications for lunar origin. J Geophys Res 75:6453.
Ryder G (1993) Catalog of Apollo 17 Rocks, Report JSC-NASA (USA) No 26088, Vol 1, p 411.
Schmidt OY (1959) A Theory of the Origin of the Earth, Lawrence & Wishart, London, p 139.
Wanke H, Dreibus G (1986) Geochemical evidence for the formation of the Moon by impact-induced fission of the proto-Earth. In: Origin of the Moon, Hartmann WK, Phillips RJ, Taylor GJ (eds), Hawaii, LPI, pp 649–672.
Wilhelms DE (1984) The Moon. In: The Geology of the Terrestrial Planets, Carr MH (ed), NASA SP-469, pp 107–205.

Technology and Diagnostics

Experimental investigation of tunnel sonic-boom suppression

A. Sasoh, O. Onodera, K. Takayama
Shock Wave Research Center
Institute of Fluid Science, Tohoku University
2-1-1 Katahira, Aoba-ku
Sendai 980 Japan

R. Kaneko, Y. Matsui
East Japan Railway Co.
Tokyo Kotsu Kaikan 7th Floor, 2-10-1 Yuraku-cho, Chiyoda-ku
Tokyo 100 Japan

Abstract: In order to investigate the generation of weak shock waves in train tunnels and the emission of the resulting sonic booms from train-tunnel exits and to search for methods for the reduction of these sonic booms, experimental studies were conducted using the scaled train-tunnel simulator. The installation of a hood at the entrance of the tunnel was effective for low piston speeds. The installation of a partially perforated wall at the exit of the tunnel simulator was significant for higher piston speeds. Throughout the series of train-tunnel simulator experiments, the combination of entrance hood and perforated wall significantly reduced shock overpressures for piston speeds ranging from 60 to 100 m/s.

Key words: Tunnel sonic boom, Hood, Porous wall

1. Introduction

The success of the high-speed train transportation system, Shinkansen, in Japan is closely related to the size of the country and also the population distribution. Japan is a mountainous country so that, in order to have relatively straight railway tracks, many short and long train tunnels have been constructed. The process of weak shock-wave generation by the train should be studied by shock-wave researchers as a leaky piston problem. In a previous paper (Sasoh et al. 1995), the construction of a train-tunnel simulator and the preliminary experiments were reported. Using this simulator, extended experiments were carried out for the reduction of tunnel sonic booms by installing hoods at the entrance of the test section and/or by installing perforated walls along the exit of the test section. The details of this study are reported in Takayama et al. (1995).

2. Evaluation of tunnel sonic-boom strength

The experiments were conducted using the train-tunnel simulator (details are described in Sasoh et al. 1994). The diameter and the length of the test section are 40 mm and 25 m, respectively.

When a simple compression wave is emitted from the exit of the test section, an impulsive pressure wave is generated (Matsuo and Aoki 1992). If the pressure gradient of a simple wave is finite, the overpressure of the impulsive wave is proportional to the time derivative of the pressure variation at the exit of the tube (Rudinger 1957; Stollery et al. 1981; Kashimura et al. 1994). However, this relation does not hold in the case of shock waves. In the present experiment, the strength of the sonic boom was defined by an overpressure measured at a point shown in Fig. 1 by using a capacitor-type microphone (Brüel & Kjær 4135, its gain performance is ± 0.3 dB under 5 kHz and ± 2.0 dB under 100 kHz). The evaluation of peak overpressures of the impulsive sound wave is related to its energy flux.

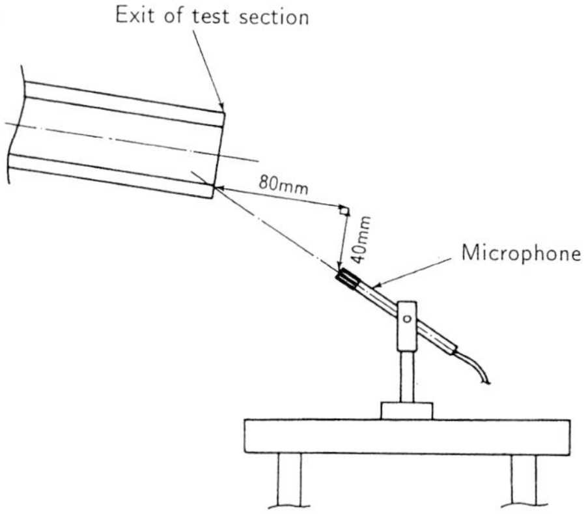

Figure 1. Arrangement of overpressure measurement by using a microphone at the exit of the test section.

3. Effect of hood at entrance

Ozawa (1979) confirmed the effect of installation of a hood at the entrance of a tunnel on the reduction of the overpressure of the impulsive sound wave. In the case of lower overpressures, compression waves will never coalesce into a shock wave. In this particular case, the installation of an entrance hood is effective. Iida (1993) confirmed this effect by a numerical simulation. In practice, in some of the Shinkansen tunnels, entrance hoods have been often used for suppression of tunnel sonic booms.

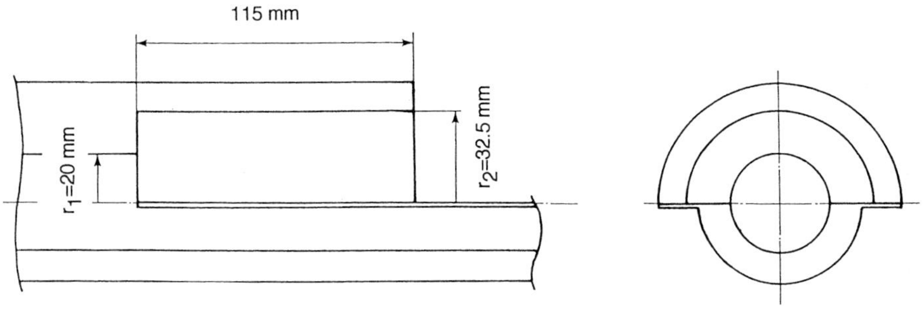

Figure 2. Configuration of hood installed at the entrance of the test section.

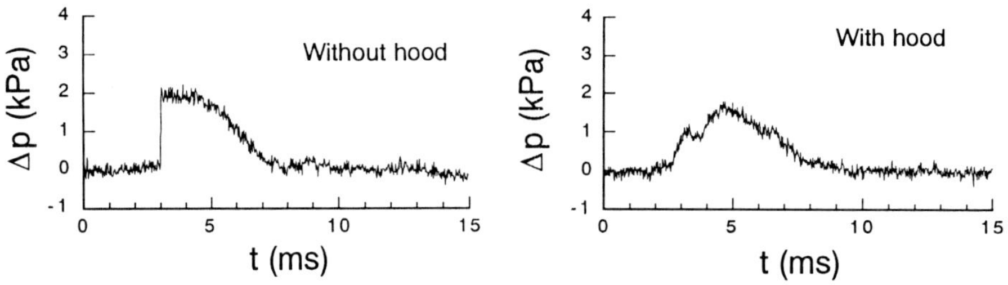

Figure 3. Time variations of overpressure for piston speed of 70 m/s.

The test hood, as is shown in Fig. 3, is a 65-mm diameter, 115-mm long co-axial half-cylinder. Figure 3 shows pressure histories with and without the hood. With the hood, when a piston enters into the test section that has a sudden two-step area contraction at its entrance, compression waves are built up at the steps. Hence, the rise of compression-wave heads became more gradual and the profile of the compression waves had two maxima. The coalescence of compression waves into a shock wave was significantly delayed.

4. Effect of porous wall near exit

It is known that if a shock wave passes through a porous or perforated wall, the mass suction through the wall subsequently induces a loss of momentum and energy fluxes and thereby decreases the shock-wave strength (Wu and Ostrowski

1971; Szumowski 1971; Deckker and Koyama 1983; Onodera and Takayama 1990, 1992). This phenomenon was applied to suppress tunnel sonic booms.

The porous material used in the experiment is made of aluminum and has a density of from 200 to 300 kg/m^3 and porosity of about 0.9. The thickness of the perforated aluminum wall was 5 mm. The coverage of the porous wall over the circumference of the test section was 1.0 and 0.16 in Figs. 4 and 5, respectively. The length of the porous wall section was 1.0 m and 0.5 m, in Figs. 4 and 5, respectively.

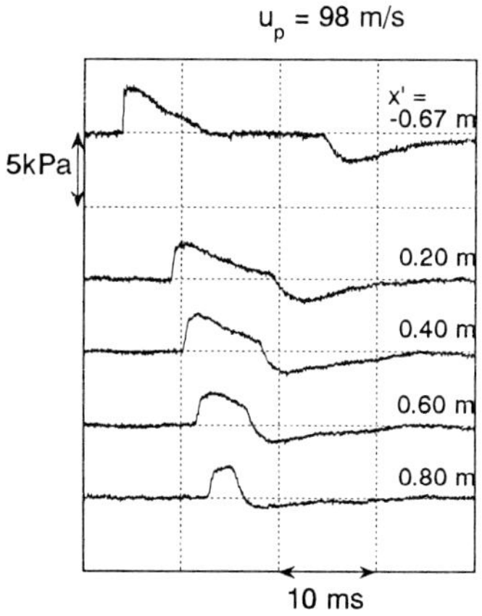

Figure 4. Time variations of overpressures along porous wall.

Figure 4 shows overpressure histories which were measured for $u_{\rm p} = 98\,{\rm m/s}$ along the porous wall installed completely around the circumference of the test section near the exit (the length is 0.5 m). The distance x' was measured from the entrance of the porous part of the test section. A shock wave was already generated at $x' = -0.67\,{\rm m}$ before entering the perforated section. When the shock wave passed through the perforated section, due to mass outflow through the perforations, expansion waves were generated at individual holes of the perforated wall. These expansion waves hence overtook the shock wave from behind and attenuated it (Onodera et al. 1992). As a result, the initial shock wave overpressure was lowered and, with further propagation, even the initial steep rise of overpressures became smeared. This trend is more pronounced for high piston speeds.

5. Hybrid usage

As seen in Fig. 5, the combination of the entrance hood and the porous wall effectively suppresses shock-wave overpressures for a wide range of piston speeds from $u_{\rm p} = 55\,{\rm m/s}$ to $105\,{\rm m/s}$. As mentioned above, the porous wall is found to effectively attenuate shock waves especially for high piston speeds. For low piston speeds, the pressure gradient of the head of the compression wave remains finite just by installing the entrance hood. Combining simultaneously the entrance hood and the porous wall, the peak overpressure was decreased by 50 to 80% of that obtained in the absence of the entrance hood and the perforated wall.

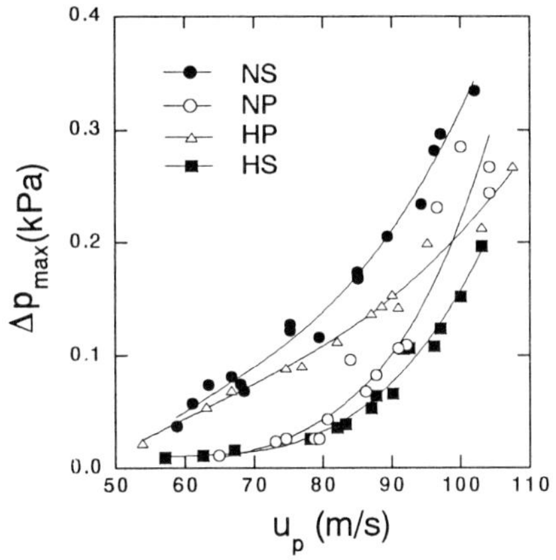

Figure 5. Effects of the entrance hood and/or perforated wall on overpressure reduction.

6. Conclusion

In this study, methods of suppression of tunnel sonic booms were experimentally examined. The installation of an entrance hood at the test section indeed delayed the formation of shock waves. This effect is significant for relatively slow piston speeds. By installing a porous wall at the exit of the test section, the gradient of the shock-wave overpressure became less steep. This effect is particularly pronounced for higher piston speeds. The combination of the entrance hood and porous wall can effectively decrease the shock overpressures for a wide range of piston speeds, $u_p = 60 \sim 100\,\mathrm{m/s}$.

References

Deckker BEL, Koyama H (1983) Motion of a weak shock wave in a porous tube. In: Proc 14th Int Symp Shock Tubes Waves, Archer, Milton (eds), Sydney, pp 239–246.

Iida M (1993) Numerical studies of compression waves generated by train entering tunnels, Ph.D. Thesis, University of Tokyo.

Kasimura H, Yasunobu T, Aoki T, Matsuo K (1994) Emission of a propagating compression wave from an open end of a tube (2nd Report, relation between incident compression wave and impulsive wave). Tran Jpn Soc Mech Engineers (in Japanese) B60:71.

Matsuo K, Aoki T (1992) Wave problems in high-speed railway tunnels. In: Shock Waves, Proc of the 18th Int Symp on Shock Waves, K Takayama (ed), Springer-Verlag, Heidelberg, pp 95–102.

Onodera H, Takayama K (1990) Shock wave reflection over wedges with suction. Tran Jpn Soc Mech Engineers (in Japanese) B56:112.

Onodera H, Takayama K (1992) An analysis of shock wave propagation over perforated wall and its discharge coefficient. Tran Jpn Soc Mech Engineers (in Japanese) B58:1408.

Ozawa S (1979) Studies of micro-pressure wave radiated from a tunnel exit. Railway Tech Res Rep, The Railway Tech Res Inst, Japan Nat Railways 1121.

Rudinger G (1957) The reflection of pressure waves of finite amplitude from an open end of a duct. J Fluid Mech 3:48.

Sasoh A, Onodera O and Takayama K (1994) Scaled train tunnel simulator for weak shock wave generation experiment. Rev Sci Instrum 65:3000–3004.

Sasoh A, Saito T, Funabashi S, Onodera O, Takayama K, Kaneko R, Matsui H (1995) A numerical and experimental study of sonic booms generated in high-speed train tunnels. In: Shock Waves @ Marseille, Brun and Dumitrescu (eds), III, pp 353–358.

Stollery JL, Phạn KC, Garry KP (1981) Simulation of blast waves by hydraulic analogy. In: Shock Tube Research, Proc 13th Int Symp, CE Treanor, JG Hall (eds), State Univ New York Press, pp 781–789.

Szumowski AP (1971) Attenuation of a shock wave along a perforated tube. In: Shock Tube Research, Proc 8th Int Symp Shock Tubes, JL Stollery, AG Gaydon, PR Wen (eds), Chapman & Hall, paper 14.

Takayama K, Sasoh A, Onodera O, Kaneko R and Matsu Y (1995) Experimental investigation on tunnel sonic boom. Shock Waves 5:127–138.

Wu JHT, Ostrowski PP (1971) Shock attenuation in a perforated duct. In: Shock Tube Research, JL Stollery, AG Gaydon, PR Wen (eds) Chapman & Hall, paper 15.

Discarding the tuned-piston conditions for Stalker tubes: possible gains and drawbacks

M.-P. Dumitrescu
Lab. I.U.S.T.I. (U.M.R CNRS 139) Dept. MHEQ
Université de Provence
13397 Marseille, Cedex 20, France

Abstract: All free-piston facilities have to face one major operating constraint, which is for the piston to land softly at the end of the compression tube. This is usually done by using an adequate compression ratio, ensuring that there is enough pressure of the driver gas at the end of the useful stroke of the piston to stop it. A stopping device has been designed that has proven its efficiency in the TCM2 free-piston shock tunnel, thus enabling the use of higher compression ratios.

Key words: Hypersonic facility, Piston-stopping device

1. The elementary theory approach

The main feature of the Stalker tube is that the high initial parameters of the driver gas are obtained by means of a dynamic compression using a free piston. The piston motion being determinant for all subsequent phases, a careful investigation is needed in order to ensure that the facility has a correct — and safe — operating cycle. In other words, this is to say that the "tuned-piston" set of conditions (Hornung 1988) has to be respected. These conditions state that:

(a) when the diaphragm bursts, the piston should have the necessary speed u for compensating the pressure decay induced by the flow of the driver gas towards the shock tube;

(b) under the effect of the driver-gas pressure, the piston should come to rest at the very end of the compression tube.

Denoting by λ the compression ratio (Fig. 1), the first condition is written:

$$\frac{u}{a_{\mathrm{He}_0}} = 0.5625\,\lambda^{1/3}\,\frac{d^2}{D^2} \tag{1}$$

Since the final aim is the successful operation of the facility, this implies that it should work very similarly to a classical shock tube having a driver section filled with gas at the conditions obtained at the end of the compression stroke of the piston, i.e., at diaphragm bursting time. Assuming this similarity, the pressure of the driver gas should stay constant for a time t^{**}, enough for the run

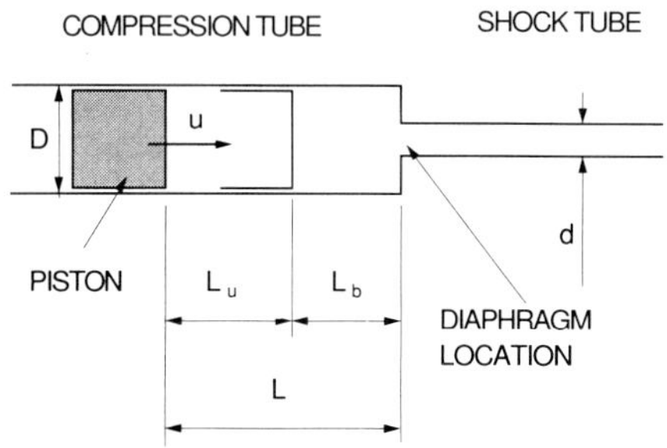

Figure 1. Piston stroke (useful and braking).

to be completed. The value of t^{**} may be determined approximately as follows: given the geometry of the facility (mainly the area ratio between the shock tube and the compression tube) and the incident-shock Mach number, the expansion wave generated by a decay in driver-gas pressure must not reach the end of the shock tube before at least t^*, which is the desired test time (Brun et al. 1996). For the TCM2 facility, assuming an M_S of about 10, the ratio t^{**}/t^* is about 1.6. Coupled with the speed requirement, this leads to a supplementary condition:

(c) the speed of the piston should remain constant during t^{**}.

Of course, this requirement cannot be exactly fulfilled, but we may assume that a speed decay of 20 percent from the initial value is acceptable. Therefore, during t^{**}, the piston will travel a distance L_u:

$$L_u = 0.9 \cdot u \cdot t^{**} \tag{2}$$

During t^{**}, the piston deceleration must not exceed the value:

$$d_p = \frac{0.2}{t^{**}} u = \frac{\pi D^2 p_B}{4 M_p} \tag{3}$$

which leads to the maximum permissible ratio p_B/M_p as a function of t^{**}:

$$\frac{p_B}{M_p} = \frac{0.2 \cdot 0.5625 \cdot \lambda^{1/3} \cdot \left(\frac{d}{D}\right)^2 \cdot a_{\mathrm{He}_0}}{\frac{\pi D^2}{4} \cdot t^{**}} \tag{4}$$

where M_p is the piston mass and p_B the driver-gas pressure. Assuming now that the braking stroke L_b could be neglected, the maximum usable compression ratio

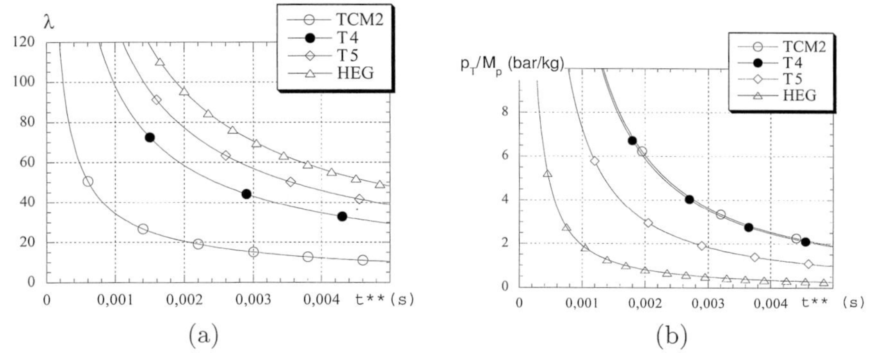

Figure 2. (a) Theoretical λ_{max} vs. t^{**} (no braking), and (b) max pressure-piston mass ratio vs. t^{**}.

would be:

$$\lambda = \left(\frac{L}{0.9 \cdot 0.5625 \cdot \left(\frac{d}{D}\right)^2 \cdot a_{\mathrm{He}_0} \cdot t^{**}} \right)^{3/4} \tag{5}$$

which is plotted in Fig. 2a as a function of t^{**}. This may be substituted in the expression for the ratio p_B/M_p, which is plotted in Fig. 2b. If now one tries to satisfy both conditions (b) and (c), it is clear that this leads to very low usable compression ratios, since the piston is still traveling fast at the end of the useful stroke L_u, and thus the remaining length L_b must be sufficient to allow for the piston slow-down (with an even lower deceleration than at the beginning, due to the pressure decay produced by the flow of the driver gas into the shock tube). For TCM2 the maximum usable compression ratio in these conditions is limited to an approximate value $\lambda = 8$, which is roughly the third of the theoretical value for $t^{**} = 1.6$ ms.

2. The stopping device

To overcome this compression-ratio limitation, a stopping device has been designed and tested in the TCM2 facility (Fig. 3). The basic idea is that, while still counting on the pressure forces acting on the front face of the piston for stopping it, a short braking length L_b may be achieved only by raising the pressure in front of the piston, starting from the moment when the useful stroke L_u has been traveled. To this purpose, a plug has been mounted in front of the piston on a collapsible rod of a suitable length. When the piston reaches the end of the useful stroke, the plug comes to shut off the shock-tube entrance, thus blocking the outflow of driver gas. The sharp pressure rise that is produced stops the piston over a much shorter distance; but care must be taken not to exceed the ultimate design pressure of the facility. A series of small orifices has been drilled into the plug, whose total free cross-section was chosen so as to prevent excessive pressure build-up. The device has since been successfully used in more than 50 runs, stopping the piston from speeds up to 200 m/s over a length L_b of 20 cm.

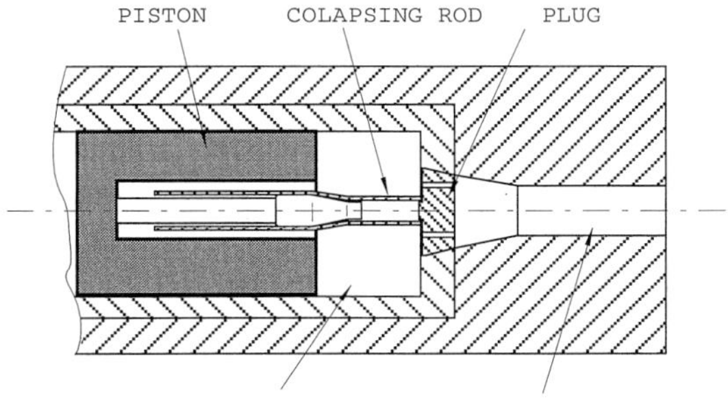

Figure 3. The piston-stopping device.

3. Concluding remarks

The use of the piston-stopping device described above resulted in the capability of the TCM2 facility to be operated at higher compression ratios — typically in the 12 – 16 range — than were attainable in the previous classical configuration. If the recovery factor is defined as being the ratio between the bursting pressure of the diaphragm, p_B, and the reflected shock pressure measured at the end plate of the shock tube, p_5, then, for the measured values obtained in TCM2, the maximum ratio was 1.52. The reason for such a high recovery ratio is believed to be the compliance to condition (c) as defined above. These remarks tend to confirm the guess that if a suitable stopping device is available, condition (b) may be discarded, thus gaining a higher attainable compression ratio and a larger safety margin (in case of premature burst of the diaphragm).

References

Brun R, Burtschell Y, Chaix A, Druguet M-C, Dumitrescu LZ, Dumitrescu M-P, Martin R, Zeitoun D (1996) Flow in the TCM2-Marseille hypersonic facility: some new experimental results. In: Shock Waves, Proc of the 20th ISSW, Sturtevant B, Shepherd JE, Hornung HG (eds), World Scientific, Vol 1, pp 227–232.

Hornung H (1988) The Piston Motion in a Free-Piston Driver for Shock Tubes and Tunnels, Graduate Aeronautical Laboratories California Institute of Technology Report FM88-1.

Compact entropy-raising driver for impulse facilities

M. A. Kendall, R. G. Morgan, P. A. Jacobs
Department of Mechanical Engineering
The University of Queensland, Brisbane 4072
Australia

Abstract: A free-piston driver that employs entropy-raising shock processes with diaphragm rupture has been constructed. Results from a range of conditions with helium and argon driver gases are reported. Significant performance gains were achieved with argon while losses cancelled out any improvements in helium. Heat transfer is identified as the dominant loss form. Measurements compare well with predictions from a quasi-one-dimensional numerical code.

Key words: Free piston, Driver, Entropy, Heat transfer

1. Introduction

A well-proven driver technique for shock tubes is the free-piston compressor (Stalker 1972). In these compressors, driver-gas pressure and temperature are limited by the volumetric compression ratio. Hence, when applied to high-enthalpy applications, quasi-isentropic drivers can become very large and expensive. To achieve comparable performance with a smaller compressor, an alternative driver technique is explored. In particular, we will study a compression process which involves entropy-raising mechanisms. Such a process will produce a given temperature rise with a smaller volumetric compression ratio than an isentropic compression alone. Entropy-raising drivers using shocks and expansions (Knöös 1969), throttling processes (Anfimov 1992), and shocks alone (Terao and Tomonobu 1994) have been developed with some success. A concept that cascades entropy-raising expansions with quasi-isentropic free-piston compressions appears to be the most simple and attractive in theory (Bogdanoff 1990).

Preliminary experiments with a limited entropy change achieved slight increases in driver performance that correlate well with analytical and numerical predictions (Kendall and Morgan 1993). In these experiments an extra diaphragm was placed within the driver at the front face of an extended buffer. However, to make the concept practical, larger entropy changes are required. One way of achieving this is to allow the piston to pass over the diaphragm station and to continue compression in the adjacent downstream tube.

A pilot facility has been constructed to demonstrate this effect and investigate the practical operational problems. Experimental results are presented with a brief description of the mechanisms involved. Heat-transfer losses are shown to be an important factor in reducing driver performance. A sample of pressure and heat-transfer data is compared with predictions from a quasi-one-dimensional Lagrangian numerical code.

2. Apparatus

Figure 1 outlines features of the pilot entropy-raising arrangement. Note that the driver is split into two sections, separated by a diaphragm. During operation, the piston is released, compressing the gas in V_1 until the point of diaphragm D_1 rupture. Upon rupture a shock is generated in V_2 that reflects off the diaphragm D_2, and again off the piston face. These entropy-producing shocks are still processing the gas while the piston recompresses the driver gas, until it comes to rest at the buffer at the point of diaphragm D_2 rupture. Locations of the pressure transducer (P_1) and four thermocouples are shown. Driver dimensions and gauge locations relative to diaphragm D_1 are summarized in Table 1. The piston, with a mass of 1.08 kg, traversed a tube of a 38-mm bore. A distinguishing feature of the piston is it passes the ruptured diaphragm D_1 at speeds of up to 70 m/s. To ensure that the piston passes the diaphragm safely, the diaphragm fitting has been flared out to a larger bore. An area change at this point introduces an alignment and gas-seal problem that was overcome with three wear rings arranged to maintain piston orientation through the location of two rings at all times.

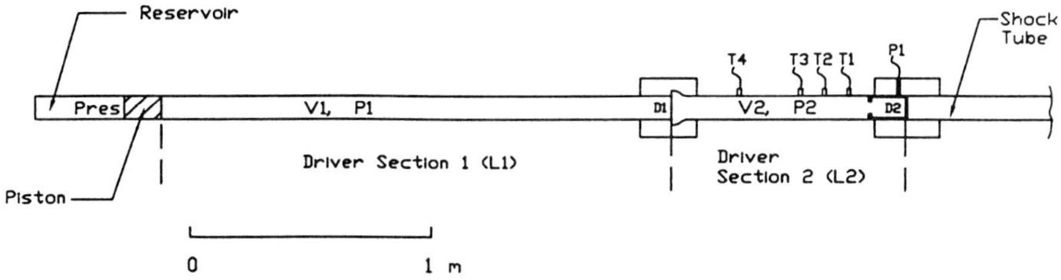

Figure 1. A schematic of the pilot entropy-raising driver.

Driver performance was gauged by the final gas temperature achieved from a fixed volumetric compression ratio of 27. Mean driver temperature at the point of final diaphragm (D_2) rupture was determined from the gas mass, volume (known through buffer impact), and pressure. Heat transfer was measured with thermocouples. The data were compared with simulations produced by a quasi-one-dimensional Lagrangian computer code (Jacobs 1994). Within this code, viscous effects and heat transfer are incorporated to provide an accurate

Table 1. Specifications of the pilot driver.

L_1 (mm)	L_2 (mm)	λ	D (mm)	V_{res} (m^3)	T_4 (mm)	T_3 (mm)	T_2 (mm)	T_1 (mm)	P_1 (mm)
2340	1020	27	38.1	0.018	340	590	690	790	998

description of the processes, however, the mixing between different gas slugs is not modeled.

3. Conditions

Quasi-isentropic reference conditions were established through the removal of diaphragm D_1 and operation of the driver as a standard free-piston compressor. The diaphragm consistently used in the D_2 location was 1.0-mm thick aluminum sheet, which ruptured at approximately 15.5 MPa. Performance of entropy-raising conditions was determined through comparison with the standard driver.

A range of operating conditions was tested. The piston passed through Mylar, aluminum, and steel diaphragms allowing different pressure ratios to generate a range of theoretical entropy changes. Driver section 2 was either evacuated (below 0.5 mm Hg), or set at the pressure needed for tailoring the primary shock generated by D_1 rupture. According to analytical theory (Kendall and Morgan 1994), an evacuated cavity would induce the greatest entropy change upon the rupture of diaphragm D_1. Numerical predictions indicate that high shock speeds induced within the driver induce large heat losses to the driver walls. Tailored conditions offer the most uniform driver mixture predictions and reduced heat transfer at the expense of theoretical entropy change. Helium and argon driver gases were also used to explore their suitability.

4. Results and analysis

Experimental results are summarized in Table 2, with the terminology outlined in Fig. 1. Final conditions are denoted by subscript "f". Conditions 1–4 are characterized by a helium driver gas, with the conventional quasi-isentropic configuration being tested as condition 1. The "entropy-raising" helium conditions did not achieve any real temperature increases over the reference isentropic condition. For example, in condition 3B, a final temperature of 1525 K was achieved in comparison to 1610 K for the isentropic reference. However, significant performance increases of between 38% and 55% were achieved with argon. The highest temperature, 2410 K, was achieved in condition 6A.

The disparity between the entropy-raising helium and argon results is highlighted in an energy analysis. Temperature (T), the gas constant ($\overline{R} = 8.3144$

Table 2. Summary of experimental results.

Condition	Description	D_1 (MPa)	Initial Conditions P_{res} (MPa)	P_1 (kPa)	P_2 (kPa)	Final Conditions P_f (MPa)	T_f (K)
Helium							
1	Isentropic	0	1.1	101.5	101.5	15.0	1610
2	Evacuated	1.6	1.3	160.5	0	15.45	1525
3A	Evacuated	4.7	1.6	136.5	0	14.95	1730
3B	Tailored	4.7	1.46	91	145.5	14.85	1520
4	Evacuated	8.03	1.85	151.5	0	15.2	1590
Argon							
5	Isentropic	0	1.14	102.5	102.5	15.0	1570
6A	Evacuated	4.7	1.4	96.5	0	14.7	2410
6B	Tailored	4.7	1.25	54.3	104.5	15.0	2290
7A	Evacuated	8.03	1.65	96.5	0	14.3	2340
7B	Tailored	8.03	1.45	50.1	126.5	15.1	2175
8	Evacuated	15.75	1.65	71.8	2.8	10.6	2290

J/g-mol/K), work done on the driver gas (w), and driver losses (L) are linked by

$$\Delta T = \frac{\gamma - 1}{\overline{R}} (w - L) \tag{1}$$

where losses (L) are the sum of heat transfer to the driver walls, piston friction, and diaphragm work. In the experiments heat transfer was the dominant loss term, accounting for 65–85% of the overall losses. Note that in the ideal situation of no losses, temperature change is a function of work and γ only. Therefore, performance in all argon and helium conditions would be identical.

Figure 2 shows that when viscous and heat-transfer losses are included in the analyses, however, differences between the gases arise. The line through the origin represents the no-losses case (100% efficiency). Each point is labeled by its experimental condition. Closest agreement between helium and argon is seen in reference conditions 1 and 5 respectively, indicating that losses in an isentropic compressor are almost identical. This is not surprising, considering that temperature, friction factor, and velocity are nearly matched throughout compression. Furthermore, losses are minimized through low time scales at elevated temperatures.

In the other experiments, however, the driver gas is compressed quasi-isentropically, and then processed by a series of shocks. Theoretically, shocks raise the gas entropy and hence performance. In practice, however, the shock-heated gas convects to the driver walls, hence reducing performance. The overall performance of the driver is sensitive to the tradeoff between these parameters.

For the helium conditions, all final temperatures are within 10% of the reference, while energy entering the system has been increased by up to 60%. This indicates that for the configurations tested, any gains in entropy were offset by heat transfer.

In argon, the tradeoff was modified. Shocks driven by argon were slower than helium, which resulted in lower heat transfer. The performance benefit can be seen in condition 6B, which had a higher final temperature than condition 3B. Further performance gains expected from stronger shocks, however, are once again offset by losses. For example, work done in condition 8 is nearly double 6B while final temperature is nearly constant. This indicates that an argon limit has been reached with this experimental geometry. Further evidence of a limit lies in the similarity between evacuated and tailored conditions while the difference between expected overall entropy change is large.

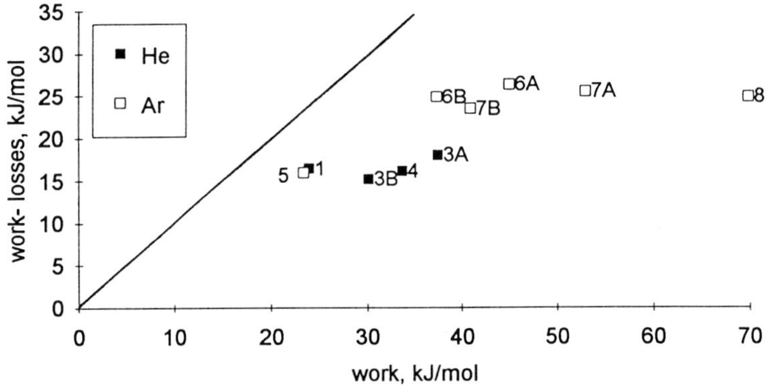

Figure 2. Performance and efficiency as a function of driver gas.

Figure 3 contains a sample of measured and predicted pressure and heat-transfer histories for condition 7B. The history with noise is measured data while the smoother, thicker line is the numerical approximation. Time $t = 0$ corresponds to D_1 rupture. The first peak corresponds to the primary shock, followed by subsequent reflections. Agreement between the predictions and experiment is good with both the amplitude and frequency of the shocks being modeled well. Similar comparisons were also achieved with the references and other tailored conditions. It was difficult to model the evacuated conditions numerically due to large pressure ratios over diaphragm D_1.

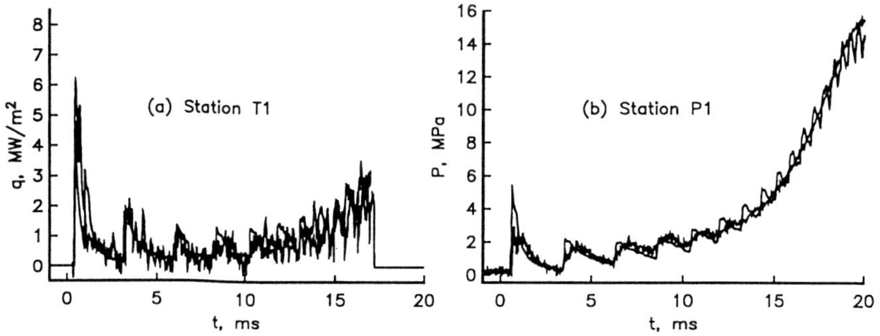

Figure 3. Measured and numerically estimated (a) heat-transfer history at T_1; and (b) pressure history at P_1.

5. Conclusions

It has been demonstrated that the performance of a conventional free-piston driver can be enhanced through the inclusion of entropy-raising shock processes. The extent of these gains is governed by a tradeoff between entropy-producing shocks and heat losses of the processed gas. Temperature gains of up to 55% were achieved with argon. Any entropy changes in helium, however, were offset by greater losses. It was also shown that a quasi-one-dimensional computational code with viscous effects is suitable in the modeling of tailored entropy-raising driver conditions.

References

Anfimiv (1992) TSNIIMASH capabilities for aerogasdynamical and thermal testing for hypersonic vehicles. In: AIAA Aero Ground Test Conf 92.

Bogdanoff D (1990) Improvements of pump tubes for gas guns and shock tube drivers. AIAA J 28:483.

Jacobs P (1994) Quasi-one-dimensional modelling of a free-piston shock tunnel. AIAA J 32:1.

Kendall M, Morgan R (1993) Non isentropic shock tunnel driver concepts. In: Proceedings Eighth National Space Symp, Brisbane 8:257.

Kendall M, Morgan (1994) A high performance, compact free piston driver for shock tube: practical considerations. In: Proceedings Ninth National Space Symp, Sydney 9:187.

Knöös S (1969) Study of a bypass piston shock tube. In: Proceedings Seventh International Shock Tube Symposium, Toronto, 28:213.

Stalker R (1972) Development of a hypervelocity wind tunnel. Aero 3:34.

Terao K, Tomonobu F (1994) Shock tube using free piston driven by detonation waves. Jpn J Appl Phys 33:2811.

Temperature distribution in free-piston compressors

D. K. Griffin, R. J. Stalker
University of Queensland
Department of Mechanical Engineering
St. Lucia, Queensland, 4072 Australia

N. W. Page
University of Newcastle
Department of Mechanical Engineering
Newcastle, 2308, New South Wales

Abstract: Results of an experimental investigation into the thermodynamic conditions generated by a $\phi = 108.25$ mm free-piston driver are presented. The driver studied is the first stage of a free-piston shock tunnel. The first set of experiments involved the measurement of the bulk gas temperature of helium for compression ratios ranging from $\lambda = 9.2$ to $\lambda = 60$. The percentage reduction in the bulk temperature from the isentropic temperature ranged from 11.4% at $\lambda = 9$ to 32.6% at $\lambda = 60$. The second set of experiments involved measuring the radial temperature profile. This was done to confirm the hypothesis that a jet of cooler gas forms in front of the piston under certain conditions in free-piston compression tubes. This necessitated the development of a temperature probe to measure the relative temperature of the driver gas. The development and operation of the probe is discussed. The experiments confirmed the presence of a weak jet at low compression ratios. At higher compression ratios, the jet was not observed and was assumed to have degenerated into a turbulent zone.

Key words: Free piston, Isentropic compression, Drivers, Boundary layer, Temperature probe

1. Introduction

Free-piston compression tubes are commonly used to generate high-pressure high-temperature supplies of gas for driving high-energy impulse facilities such as shock tunnels, expansion tubes, and gas guns (Stalker 1967). It has been identified that the ideal case of isentropic compression without heat loss is not practically met (Knoos 1971, Page & Stalker 1983). It is important to both the designer and the experimenter that these heat losses are quantified. This enables the thermodynamic state of the compression-tube gas after diaphragm rupture to be determined, and hence, the conditions generated in the driven tube.

Studies on the incompressible flow of water inside cylinders (Khatir 1977) experimentally demonstrated that the flow field inside a piston-cylinder arrangement is perturbed from one-dimensional flow by the presence of the wall boundary layer. As the piston moves along the cylinder, the wall boundary layer is scraped up. At the start of the piston stroke, the boundary layer rolls up off the piston face and forms a ring vortex ahead of the piston. This vortex grows with piston travel and, at larger distances, a second phenomena has been observed. An axial jet forms, which injects entrained liquid from the vortex region ahead of the piston. As the piston travels further, this jet was seen to break up.

It was the aim of this investigation to confirm the presence and determine the structure of the non-ideal jet/vortex effects in the compressible regime of a free-piston compression tube. The presence of a cold jet in the driver gas has been suggested as a possible contributing mechanism to the reported loss in plateau pressure in free-piston shock tunnels (Page & Stalker 1983).

2. Bulk temperature measurement

To determine the bulk or average temperature generated in the compression tube, a series of experiments was conducted with the end of the tube sealed off. The experiments were conducted in a $\phi = 108.25$ mm compression tube. The volume of the buffer was 12.4 l and the volume and length of the compression tube was 24.85 l and 2700 mm respectively. The buffer gas was nitrogen and the driver gas was helium while the mass of the piston was 2.7 kg. The various conditions tested are summarized in Table 1.

Table 1. Summary of the experimental conditions tested.

Condition identifier	Compression ratio (λ)	Compression tube fill (kPa)	Reservoir fill (MPa)	Isentropic temp. (K)	Bulk temp. (K)
A	9.2	230	3.0	1316	1166
B	14.9	155	3.0	1818	1661
C	31.0	95	3.0	2963	2279
D	46.6	75	3.0	3882	2832
E	56.3	65	3.0	4404	3228
F	60.0	37	1.6	4598	3101

The bulk temperature was measured just as the piston rebounded according to the following method. The peak pressure P_c was measured with a piezoelectric pressure transducer. The volume occupied by the driver gas was determined by measuring the clearance volume V_c from which the peak compression ratio λ_c can be calculated. The perfect-gas equation can then be used to compute the average temperature of the driver gas.

Table 1 shows the calculated bulk temperatures for each of the conditions tested with the isentropic temperature, the isentropic temperature being the temperature achieved if there was not heat transfer from the gas to the wall. The temperature losses range from 11.4% at $\lambda = 9$ to 32.6% at $\lambda = 60$. These results are used to normalize the data from the temperature probes.

3. Temperature-probe development

To test the hypothesis that a jet of cooler gas forms ahead of the piston, a gas-temperature probe was developed. The probe sampled a small fraction of the driver gas during the shot and expanded it through a choked nozzle over a heat-flux sensing element. Data from pressure measurements in the gas were then combined with heat-flux measurements to calculate the local gas temperature. The heat-flux measurements were made through the monitoring of the temperature indicated by a chromel-alumel thermocouple pair spot welded onto the probe body.

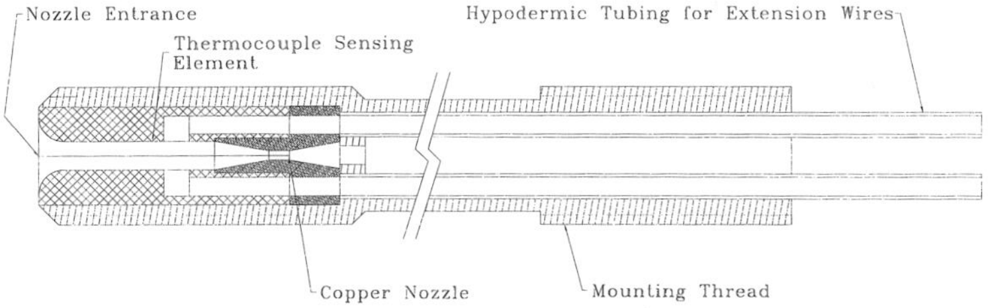

Figure 1. Assembly diagram of stagnation-temperature probe.

3.1. Probe design

The thermal and physical conditions the probe was subjected to were extremely arduous. A schematic representation of the probe is shown in Fig. 1. The sensing-element body was made of mild steel with an interference fit in the outer stainless steel casing. To prevent melting of the nozzle throat during the test, the throat needed to be made of copper. One of the most difficult aspects of the design was to find a castable material to protect the thermocouple extension wires. This material was subjected to the full gas flow. It was found the most successful combination was a diamond-epoxy composite. The diamond, having a very high thermal conductivity, limits the peak surface temperature by quickly conducting the heat from the surface. The epoxy resin provided the mechanical strength to bond the diamond together.

Each half of the chromel-alumel thermocouple junction consists of 25-μm thick foil spot welded to the wall of the steel nozzle body. This was the most successful method of forming the thermocouple junction evaluated in a series of probe prototypes.

3.2. Probe data reduction

The primary quantities measured during a shot are the thermocouple voltage from each probe and the pressure transducer voltage. The thermocouple signal is used to calculate the heat transfer from the gas to the wall. Due to the unique method of forming the heat-flux sensing element, a two-stage scheme was needed to calculate the surface heat transfer. In the first stage, the heat-flux flowing from the thermocouple material into the steel nozzle was calculated. This took into account the variable thermal properties of the steel substrate and the two-dimensional flow of heat. The second stage took this result and calculated the heat conducted through the thermocouple element from the gas. The second stage represents a nonlinear inverse heat conduction problem and was solved using the method of Carasso (1992).

The heat flux from the driver gas to the inner wall of the nozzle can be modeled with the flat-plate, laminar boundary-layer equation,

$$St^* = \frac{0.6}{\sqrt{Re^*}} \tag{1}$$

with the dimensionless groups being evaluated at the Eckert reference temperature.

The pressure ratio across the probe is maintained above the critical sonic pressure ratio during the shot, ensuring choked flow in the throat of the probe. Neglecting viscous effects, this fixes the axial Mach-number distribution in the probe for any driver pressure and temperature. Knowing the stagnation pressure and density, the velocity of the gas flowing over the heat-flux gauge can then be calculated. Using this information and the boundary-layer equation, the heat transfer to the wall of the probe for any stagnation pressure and temperature can be calculated. Given that the pressure and heat transfer are measured during a shot, the stagnation-temperature history at the sensing element can be easily calculated by using the scheme in reverse.

The data sought are the local gas temperatures at various radial locations in the compression tube. The data-analysis scheme is mathematically ill-posed and greatly amplifies noise from the raw signals. The data was digitally filtered. A data-averaging technique was also used to improve the signal quality. The traces from a series of identical shots with different probes in each location were ensemble averaged and normalized against each other.

Once the traces were generated, they were compensated for the calibration factor. The bulk temperature measured for each condition was used to normalize the data so that an absolute measure of temperature could be calculated.

4. Temperature measurements

The radial temperature profile for three different test conditions was measured. They were for conditions A, B, and C outlined in Table 1. The probe length for Condition A was 135 mm, 75 mm for Condition B, and 42 mm for Condition C. Three probes were inserted into the driver gas. One was stationed on the tube centerline, one at $r/r_0 = 0.41$, and the final probe near the wall at $r/r_0 = 0.80$. The resultant traces are shown in Fig. 2.

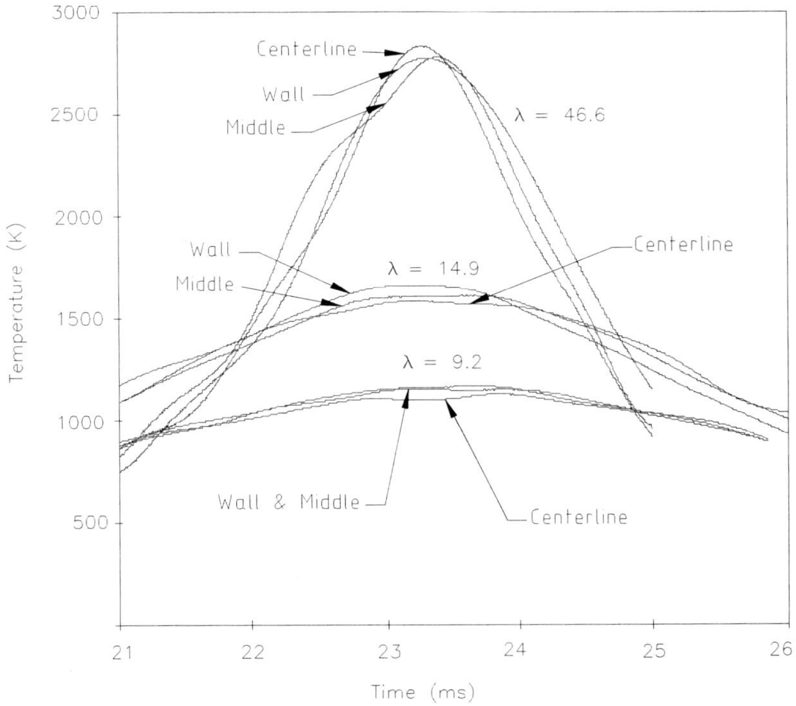

Figure 2. Temperature profiles for $\lambda = 9.2$, $\lambda = 14.9$, and $\lambda = 46.6$. Centerline, Middle, and Wall refer to the radial positions of the probes being $r/r_0 = 0.00$, $r/r_0 = 0.41$, and $r/r_0 = 0.80$, respectively.

5. Discussion and conclusions

For the $\lambda = 9.2$ condition, a small temperature deficit can be seen between the probe trace on the centerline and the ones at the middle and at the wall. This suggests that there exists a weak jet of slightly cooler gas ahead of the piston. For the $\lambda = 14.9$ condition, the wall temperature is hotter than the middle temperature which in turn is hotter than the centerline trace. In this condition, the jet has grown in size but remains relatively weak. The $\lambda = 46.6$ shot indicates that the jet has disappeared or degenerated into a turbulent zone.

The growth of the cross-sectional area of the jet and the transition from a distinct jet to a turbulent zone was noted by Khatir (1977). The results of this investigation indicate that the jet structure is qualitatively similar in the compressible regime and the incompressible regime previously tested.

The main conclusion is that although the temperature of the driver gas is less than the isentropic value, there exists a high degree of thermal homogeneity. Procedures that predict the performace of a device using a homogenous driver temperature will yield results very close to ones that take into account two or three-dimensional compression-tube thermal gradients.

Acknowledgement. This research was supported by the Australian Government, Defence Science and Technology Organisation.

References

Carasso AS (1992) Space marching difference schemes in the nonlinear inverse heat conduction problem. In: Inverse Problems, Vol 8, pp 25–43.

Jacobs PA (1994) Quasi-one-dimensional modeling of a free-piston shock tunnel. AIAA J 32(1):137–145.

Khatir FA (1977) An Investigation into the Turbulent Fluid Zone Produced by a Moving Piston Inside a Cylinder, BSc. Honours Project, University of Southamton, Department of Aeronautics and Astronautics.

Knoos S (1971) Theoretical and experimental study of piston gas-heating with laminar energy losses. AIAA J 9(11):2119–2127.

Page NW, Stalker RJ (1983) Pressure losses in free piston driven shock tubes. In: Shock Tubes and Waves, Proc of the 14th International Symposium on Shock Tubes and Waves, pp 118–125.

Stalker RJ (1967) A study of the free-piston shock tunnel. AIAA J 5(12):2160–2165.

Soft-landing operation for free-piston driver by adjusting piston-stopper height

H. Tanno, K. Itoh, T. Komuro, K. Sato
National Aerospace Laboratory
Kakuda Research Center
1-Koganesawa, Kimigaya, Kakuda, Miyagi 981-15 Japan

Abstract: An experimental study of soft-landing operation for a free-piston driver was carried out. By adjusting piston stopper height according to a new tuning principle, piston collision speeds under 10 m/s were accomplished. This is much lower than the minimum piston collision speed without a piston stopper (25 m/s). By comparison with experimental data, the present CFD method was evaluated as to its usefulness for setting operation conditions. It was concluded that soft-landing operation by adjusting piston stopper height can be realized without any devices such as piston brakes, even in large-size facilities.

Key words: Free-piston shock tunnel, Over driving, Tuned operation

1. Introduction

In a free-piston shock tunnel (FPST), it is essential to improve the effective duration in which the driver-gas condition is held constant. An over-driving operation, which compensates driver-gas drainage into shock tube by the use of high piston speeds, is an useful method for this. However, it increases the risk of piston collision with the end of the compression tube, since it needs high piston speed at diaphragm rupture. Thus, soft-landing operation is required especially for large-sized facilities which have massive pistons.

In this study, in the scope of operation of the large-sized FPST (NAL-HIEST), a new concept for soft-landing operation was incorporated into the tuning principle (NAL formula) and examined with the small free-piston test driver (NAL-CTA).

2. New concept for tuned operation

The over-driving condition is satisfied if the normalized piston speed $\beta \geq 1$ (Stalker 1967). Under certain operating conditions, given values of β and compression ratio λ, piston motion has a point at which piston speed (u_p) and piston acceleration (du_p/dt) become zero simultaneously (Fig. 1). The present concept/method of soft-landing operation is to catch the piston at this point by the use of a piston stopper of the proper height. Therefore, the tuning problem for soft-landing operation becomes an eigenvalue problem to solve for the natural frequency, ω, which satisfies $u_p = 0$ and $du_p/dt = 0$ for given β and λ,

$$\beta \equiv u_r/U_R, \quad \omega \equiv \alpha \frac{P_0 V_0}{W_p (A^*/A)^2 a_r^2} \tag{1}$$

where u_r is piston speed at diaphragm burst, U_R is reference speed (see Hornung et al. 1990), P_0 and V_0 are driver-gas initial pressure and volume, W_p is piston mass, a_r is driver-gas sonic speed at initial condition, A^*/A is area ratio of shock tube and compression tube; and

$$\alpha \equiv 2(\gamma - 1)(\frac{\gamma + 1}{2})^{\frac{\gamma+1}{\gamma-1}} .$$

Itoh (1994) formulated a β-ω relation for FPST soft-landing operation, which is called the NAL formula. On the basis of the NAL formula, the operating condition can be tuned for certain β and ω so that the piston comes to rest at the point of inflection. In Fig. 2, the solid line shows the NAL formula β-ω relation. On this line, piston motion has a point which satisfies $u = 0$, $du/dt = 0$, so that soft-landing operation will be accomplished.

3. Experiment and CFD

In order to demonstrate the new concept for soft-landing operation, experiments were carried out with NAL-CTA (shown in Fig. 3) free-piston driver which has a 15-cm transparent (acrylic) test window in the end section of the compression tube, through which piston motion can be observed directly by a high-speed video camera at 6000 frames/s. From this photo image, piston position was measured with an accuracy of ± 3 m/s. To measure diaphragm burst timing precisely, a high-sensitivity photo diode and pressure transducer were placed after the diaphragm, and used to trigger the measurement system.

Since the total length of the compression tube is 2.2 m, whereas the transparent test window is 15 cm, the compression ratio was set at 16.0–20.0 in order to estimate the piston velocities accurately from the observed images. An area ratio A^*/A between between the exit orifice and compression tube was chosen to be 0.09. Considering the strength of the transparent window, the maximum diaphragm burst pressure was chosen to be 10 MPa, thus the piston mass was determined to be 120 g by the present principle for tuning. Variable parameters were pressure for piston driver and initial pressure of compression-tube gas.

In this study, CFD was also done to decide the operating condition. The code used in this study was already reported by Tani et al. (1993).

4. Results and discussion

4.1. Result without piston stopper

Before the soft-landing operation experiment, a preliminary experiment without the piston stopper was carried out and the piston collision speed (U_c) was measured. As seen in the example of the high-speed camera image sequence shown in Fig. 1, piston motions could be measured directly. From these images, the piston velocity profile was estimated using a cubic spline smoothing function. Figures 4 and 5 show examples of the piston-velocity profile and compression-tube pressure history, respectively.

In the β-ω map shown in Fig. 2, the absolute values of U_c are indicated. In spite of the variation of ω and β, all of values of U_c were higher than 25 m/s even for experiments carried out around the predicted soft-landing curve because of piston re-acceleration.

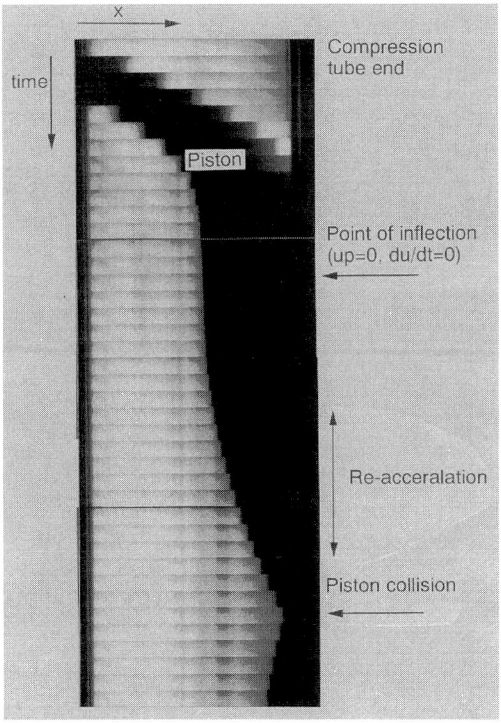

Figure 1. High-speed camera image of piston motion with point of inflection.

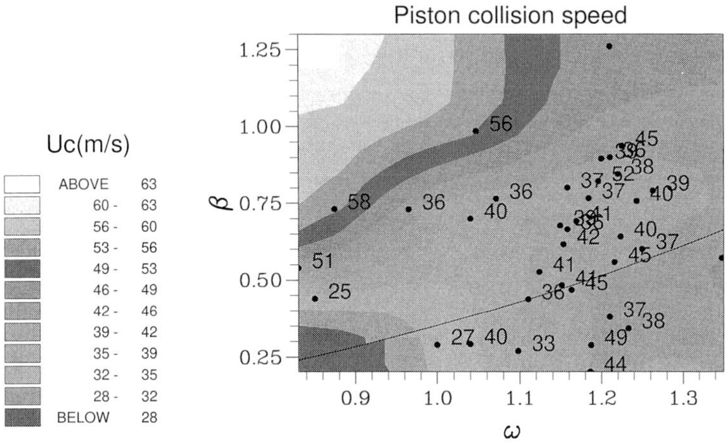

Figure 2. Piston collision speed on β-ω map.

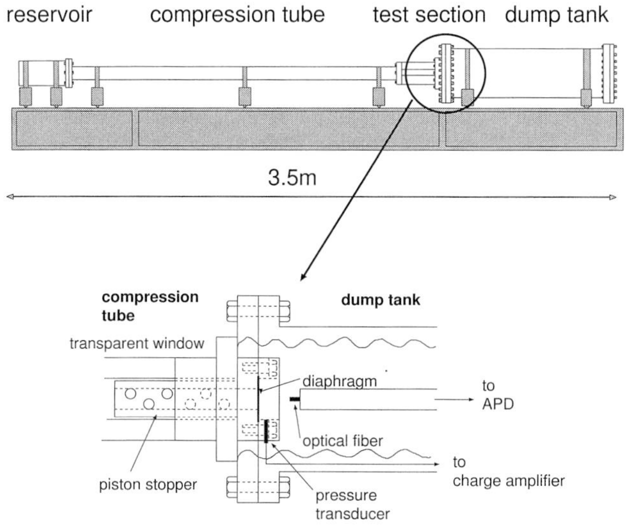

Figure 3. Schematic of NAL-CTA.

Preliminary experiment also showed that measured compression ratio ($\lambda_m = V_0/V_m$) at diaphragm rupture became $\sim$ 1.3–1.5 times larger than that of an

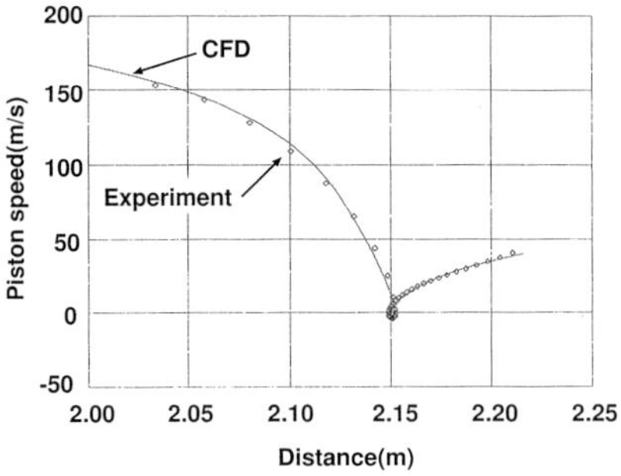

Figure 4. Piston-velocity profile (without piston stopper).

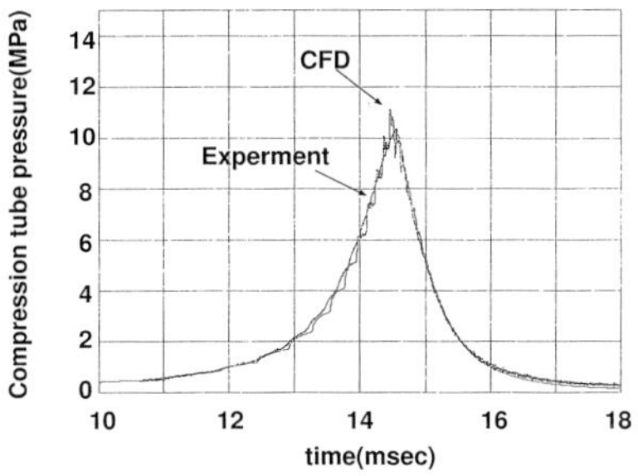

Figure 5. Compression-tube pressure history (without piston stopper).

adiabatic compression process. This phenomena is suspected to be caused by driver-gas leakage through the piston ring. Taking the driver-gas leakage into account, ω was compensated by the following equation

$$\omega^2 = \omega_0^2 \frac{V_m}{V_r} \tag{2}$$

where V_m is measured volume, V_r is volume assumed as adiabatic compression, and ω_0 is value without leakage.

In Fig. 2 the compensated value of ω is shown. To predict piston motion more accurately, the CFD code was improved by intoducing a simple leakage model, and the improved results are shown in Figs. 4 and 5. As is seen in the figures, the improved code predicted the experiments well.

4.2. Results with piston stopper

With the NAL formula and the CFD code, a soft-landing operation condition with a certain piston-stopper height was chosen. Figure 6 shows experimental results of piston collision speed for various piston-stopper heights. Piston-stopper height (l_{ps}) was varied from 70 mm through 90 mm with 5-mm pitch. In this series of experiments ($\beta = 0.45$, $\omega = 1.05$), the minimum piston collision speed was 7 m/s which was in the case of an 80-mm piston stopper. Piston speed increases with deviation of the height of the piston stopper from 80 mm. This fact is consistent with CFD prediction. The present soft-landing operation method was also examined under different conditions ($\beta = 0.53$, $\omega = 1.24$) and a soft landing with a 6-m/s piston collision speed was achieved.

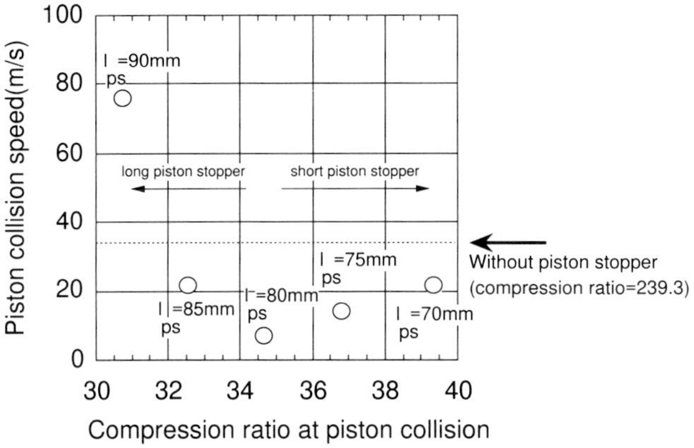

Figure 6. Piston collision speed with variation of piston-stopper length.

5. Conclusion

Comparing CFD and experimental results, it is concluded that the present soft-landing operation method of adjusting the piston-stopper height is quite effective for improvement of the free-piston driver. With the present method, soft-landing operation can be realized without any devices such as piston brakes.

References

Itoh K (1994) Exact and approximate formula for a tuned operation of a free piston shock tunnel. In: 4th International Workshop Shock Tube Technology, Brisbane.

Stalker RJ (1967) A study of the free piston shock tunnel. AIAA J 5(12):2160–2165.

Tani K, Itoh K, Takahasi M, Tanno H (1994) Numerical study of free piston shock tunnel performance. Shock Wave Int J 3:313–319.

Preliminary experimental results of the detonation-driven shock tube

H.-R. Yu, W. Zhao, Z.-F. Li, J.-H. Gu, X.-Y. Zhang
LHD, Institute for Mechanics, Chinese Academy of Sciences
Beijing 100080 China

Abstract: The wave diagram of flow in a detonation-driven shock tube is presented. Four types of igniter were investigated and the effects of the damping tube clarified. Driving capability was verified by experiments.

Key words: Shock tube, Detonation driver, Oxy-hydrogen detonation

1. Introduction

It is predicted that an oxy-hydrogen detonation-driven shock tunnel is capable of generating high-enthalpy flow at high density (Yu et al. 1992). In order to verify the prediction and to explore some related technical issues, a detonation-driven shock tube has been constructed at the Institute of Mechanics, Chinese Academy of Sciences. The purpose of this study is to develop a powerful igniter by which mixtures of rich hydrogen and oxygen can be detonated dependably, and to clarify the effects of the damping tube. Driving capabilities of an oxy-hydrogen driver at initial pressure of 1 MPa were measured.

2. Basic shock-tube cycle

The detonation-driven shock tube (see Fig. 1 schematic) consists of the conventional parts of a shock tube, i.e., driver and driven section, additional damping section which is attached to the end of the driver, and a detonation-initiation tube near the main diaphragm. The driver and driven section have the same circular cross-section diameter of 100 mm and equal length of 5.6 m. The diameter and length of the damping tube are 190 mm and 1.8 m respectively.

Before operation, hydrogen and oxygen are fed simultaneously into the driver and the initiation tube by means of individual critical nozzles. In the initiation tube the gaseous mixture is ignited by a spark igniter at the end. When the burnt gas jets into the driver, a detonation wave is formed at once and propagates leftwards. Behind the detonation front, a uniform static state (region 4) exists which corresponds to the initial state of a conventional shock tube. The bursting of the main diaphragm causes a central rarefaction wave which originates at the main diaphragm. The detonation driver is superior to a conventional combustion driver because of the constant temperature and pressure

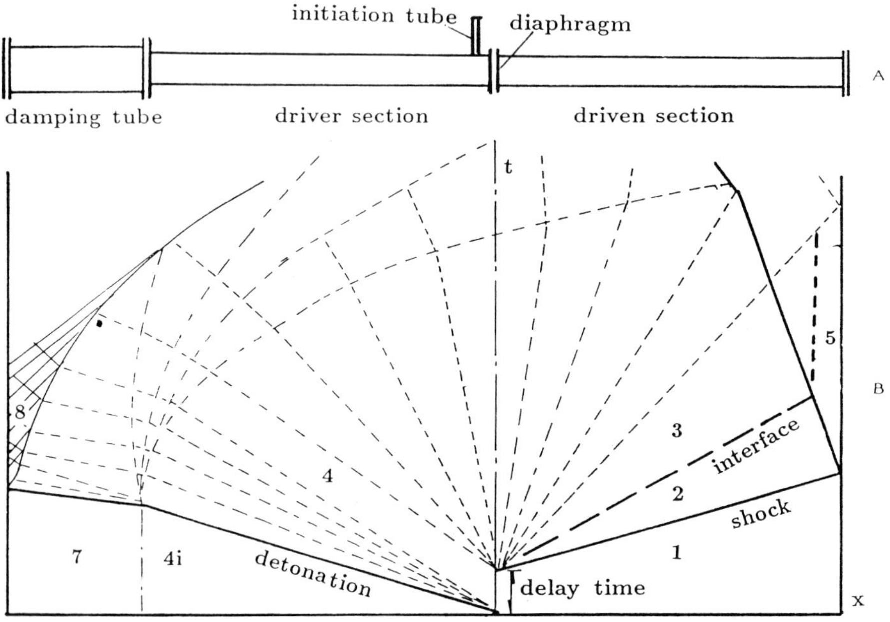

Figure 1. Wave diagram of flow in a detonation-driven shock tube.

of the driver gas at the head of the central rarefaction wave and lower heat loss. When the detonation front arrives at the end of the driver, the thin diaphragm used to separate the driver and damping tube is broken. The detonation wave propagates directly into the damping tube and high reflected pressure is averted.

3. Direct initiation of a detonation

The enthalpy of the test gas in a shock tunnel is entirely dependent on the incident shock Mach number M_s. M_s is increased by raising both ratios of pressure and sound speed across diaphragm. The maxima of both pressure and temperature behind the detonation front are reached at stoichiometric composition. Because hydrogen is lighter than oxy-hydrogen burnt gas, addition of hydrogen increases the sound speed of the mixture. For a given driven condition and initial pressure of driver gas, the maximum incident shock Mach number is attained at over-stoichiometric composition.

The ratio of initial pressure ratios in driver and in driven section were calculated on the assumption of equilibrium everywhere, and are shown in Table 1. This table clearly shows that the optimum ratio is about 3 for $M_s = 3$ and about 5 when M_s exceeds 9.

Table 1. Value of initial pressure ratio p_{4i}/p_1.

M_s	H_2/O_2					
3	2.8	2.4	2.5	2.6	2.7	3.1
6	28	21	20	21	22	25
9	177	107	97	96	99	112
12	812	438	372	352	362	407

The parameter that characterizes the phenomenon of direct initiation detonation is the ignition energy. If the ignition energy exceeds the critical threshold value, the blast wave generated by the igniter and reaction front is always coupled in the form of a detonation wave. However, quantitative determination of critical energy for given initial conditions of the mixture have not been possible. Experimental results show that even for the same type of igniter, the critical energies for same mixture can differ by three orders of magnitude (Lee 1977). The general behavior of the critical energy versus mixture composition at a given pressure is in the form of a U-shaped curve with the critical energy approaching infinity at the lean and rich limits. Since the optimum mixture ratios for attaining strong shock waves are close to the rich limit, a powerful igniter is needed for the operation of a detonation driver. In order to choose a reliable igniter, by which a detonation can be initiated directly in an oxy-hydrogen mixture rich in hydrogen, four types of igniter were tried.

In all our experiments the initial pressure was 1 MPa. The pressure-time curves measured are used to diagnose whether a detonation is initiated directly or not. The detonation is initiated directly if the measured pressure curves are as in Fig. 2a. If they are as in Fig. 2b, the detonation has not been initiated.

Explosive igniter. A hydrogen/oxygen mixture ratio of up to 4 can be detonated immediately by 20 mg of Tetryl. Further improvement of the explosive igniter has been abandoned due to contamination of the burnt product. *Turbojet igniter.* This igniter can generate a long torch, but the ignition intensity is rather weak. *Electric explosive wire igniter.* Experimental results show that this igniter is capable of initiating detonation of an oxy-hydrogen mixture with a ratio of hydrogen to oxygen up to 4.5. *Hot-gas jet igniter.* The burnt gas in the initiation tube jets into the detonation driver. It was found that mixtures with ratios $H_2/O_2 \leq 5.5$ can directly initiate a detonation, because the hot flow carries a large amount of energy.

The explosive wire and the hot-gas jet satisfy all the requirements of an igniter to be used in a detonation driver, but the latter is the most clean and efficient method. A sketch of the initiation tube is shown in Fig. 3. The diameter of the jet is only 20 mm. Such a small orifice is not difficult to construct on the driver wall, even at high pressure conditions. There is no need for a diaphragm

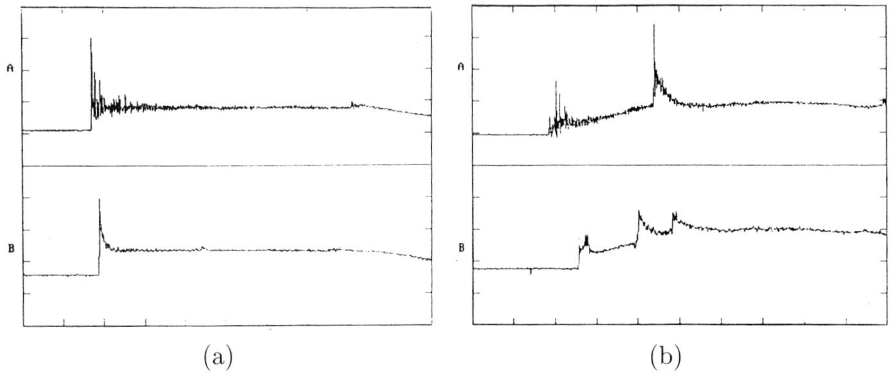

Figure 2. Pressure-time curves.

between the initiation tube and driver, so the operation is convenient. For a fixed composition, the critical energy for direct initiation of detonation decreases as initial pressure increases. In practical operation of a detonation-driven shock tunnel, general initial pressure in the driver is higher than that in the above experiments. Therefore the initiation tube developed is expected to be used successfully.

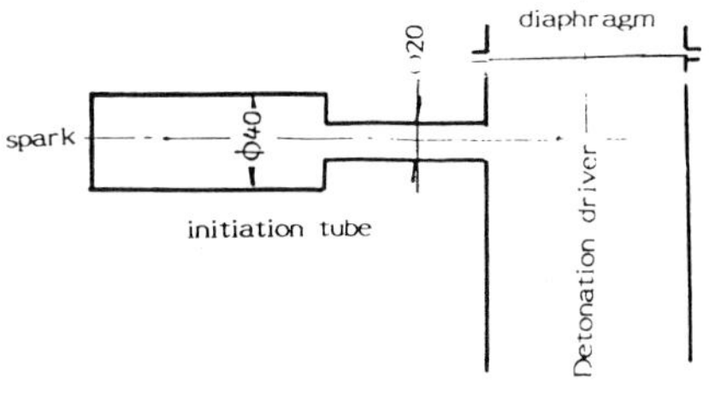

Figure 3. Sketch of the initiation tube and experimental setup.

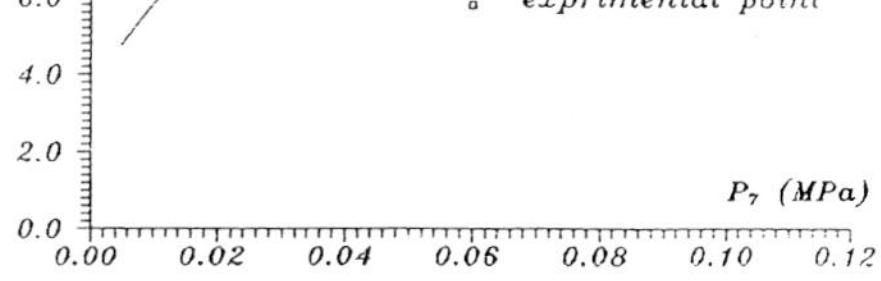

Figure 4. Reflected peak vs. initial pressure in damping tube.

4. Effect of the damping tube

The damping tube is capable of reducing the load on the structure and increasing the holding time. The holding time means here the duration of a constant state at the main diaphragm.

A detonation wave initiated at the main diaphragm propagates leftwards to the end of the driver. When it arrives at the end closure, a reflected shock wave is formed. The von Neumann peak in the reflected region is higher than 200 times the initial pressure. Such high pressure may seriously jeopardize the structure of the driver in the case of higher initial pressure. If a damping tube is attached to the end of the driver and a thin diaphragm is employed instead of the end closure, the detonation wave arrives the end of driver, breaks the thin diaphragm, and propagates directly into the damping tube.

This flow is equivalent to the flow in the detonation-driven shock tube when the detonation is initiated at the end of the driver. In this case, the reflected shock wave disappears at the end of the driver but appears on the end wall of the damping tube. Figure 4 shows the relation between the reflected peaks on the end wall of the damping tube and the initial pressure in the damping tube. It can be seen that the reflected peaks decrease with initial pressure decrease. The minimum of reflected pressure on the end wall of the damping tube is reached at vacuum.

The second advantage of the damping tube lies in the increase of actual driving duration. In case of the driver without a damping tube, the detonation front must be reflected on the closure of the driver and a reflected shock propagates backwards. If a damping tube is attached to the driver end and the initial pressure is low enough, the detonation wave propagates into the damping tube and a rarefaction wave is reflected towards the main diaphragm from the end of the driver. Because the speed of the shock is always higher than that of the rarefaction, the rarefaction arrives at the position of the main diaphragm later than the reflected shock. Experimental results show that the holding time of a detonation driver with a damping tube is about 130% of that without a damping tube.

5. Shock Mach-number measurements

The incident-shock Mach number M_s has been measured at the initial pressure of 1 MPa in the driver and at a ratio of $H_2/O_2 = 3$. The data, shown in Fig. 5, are approximately in agreement with the numerical results.

6. Conclusions

The results of the preliminary experiments in the detonation-driven shock tube verify that the incident-shock Mach numbers are approximately in agreement with numerical predictions given the assumption of equilibrium. The extreme high peak of reflected pressure on the end of the driver can be eliminated by attaching a damping tube to the end of the driver. If the initial pressure in the damping tube is low enough, the reflected pressure on the end wall of the

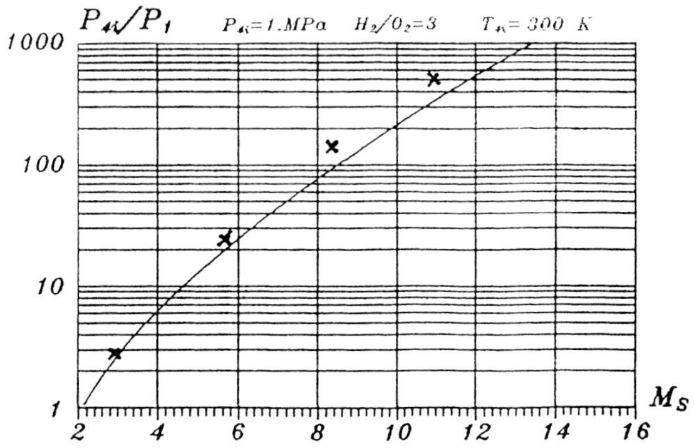

Figure 5. Relations between M_s and p_{4i}/p_i at $H_2/O_2 = 3$.

damping tube can be decreased lower than 5 times the initial pressure in the driver. An initiation tube with contraction cross section has been developed which satisfies the requirements of the detonation driver. It was found that the recoil motion of the detonation-driven shock tube is very light.

Acknowledgement. This work was supported by the NSF of China and sponsored by the Chinese Academy of Sciences.

References

Edwards DH, Williams GT, Breeze JC (1959) Pressure and velocity measurements on detonation waves in hydrogen-oxygen mixtures. J Fluid Mech 6:497–516.

Lee JHS (1977) Initiation of gaseous detonation. Ann Rev Phys Chem 28:75–104.

Yu H-R, Esser B, Lenartz M, Groenig H (1992) Gaseous detonation driver for a shock tunnel. Shock Waves 2:245–254.

Operational characteristics of a free-piston shock tunnel with a light-gas gun as the shock tube

G. X. Gou, D. Z. Gao
China Aerodynamics Research & Development Center
P. O. Box 211 Mianyang, Sichuan, China

Abstract: This paper describes preliminary calibration results in the CARDC free-piston shock tunnel. The facility is operated with a two-stage light-gas gun with a 200-m ballistic range as the shock tube. The experimental results have demonstrated its capability to produce high-velocity flow for a short duration. When using a 50-mm i.d. shock tube and air as the test gas, shock Mach numbers are 8–11.9, stagnation temperatures are 6000–10000 K, pressures are 50–140 MPa, and durations of the reservoir pressure plateau are about 1.5–2.5 ms. The compression process and flow phenomena in the compression tube and shock tube must be further studied and analyzed.

Key words: Shock tube, Free-piston shock tunnel, High-enthalpy flow

1. Introduction

The free-piston shock tunnel has been confirmed to be capable of producing the high-velocity air flow necessary for studying real-gas effects on hypersonic flight. This type of facility was pioneered by R. J. Stalker (Stalker 1989) and now has been significantly developed in many countries (Morrison et al. 1989, Eitelberg 1993, Maus et al. 1993, Labracherie et al. 1993). Since 1990, an effort to obtain high-enthalpy flow has been made by making use of a light-gas gun as the shock tube in the Ballistic Range Laboratory of the Hypervelocity Aerodynamics Institute, CARDC. The shake down was made in spring of 1992, and further calibration tests were completed in June 1994. Principal results and discussion of the tests are presented in this paper, together with the problems of extreme heat loads to the facility.

2. Description of the facility

Figure 1 shows the principal components of the facility, for which the powder chamber, compression tube, and test section/blast chamber are common for both the free-piston shock tunnel and the ballistic range. The shock tube consists of the launch tube of the range and a measurement section with a 50-mm i.d. (used in 1994) or 37.6-mm i.d. (used in 1992). The measurement section

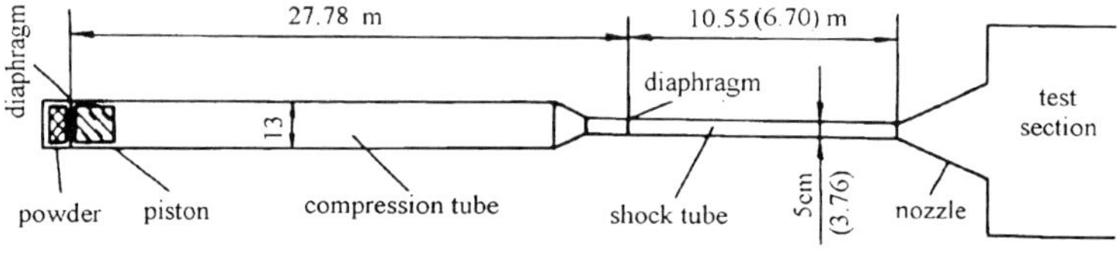

Figure 1. Schematic diagram of CARDC free-piston shock tunnel.

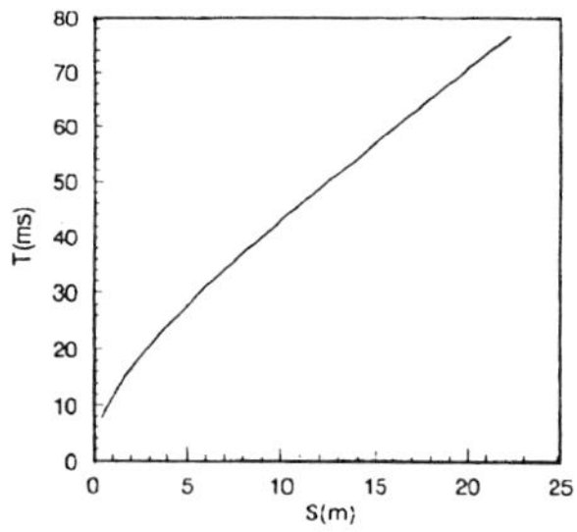

Figure 2. Piston trajectory (No. 11).

is 0.55 m long with probes on it used for measuring shock speeds and reflected shock pressure near the end of the shock tube. Hydrogen was used as the driver gas and air or nitrogen as the driven gas. The driver pressure can be measured just upstream of the main diaphragm or about 3.2 m upstream. The piston speed is measured by a microwave instrument and because of the limitation of the probe position, the measurement can be made only in the first 23.6 m of the compression tube. A 10-deg half-angle conical nozzle with a replaceable throat section has been adopted, and has an exit diameter of 242 mm. Kistler 6005 and 6211 piezoelectric gauges are used to measure reflected and driver pressure, and STPP-3 transducers made by the shock-tunnel laboratory of our institute are used to measure pitot pressure.

3. Calibration results

3.1. Compression process

The piston motion is influenced by many factors, for example, powder mass, piston mass, the charge pressure of the compression tube, and the friction between the piston and tube. The piston is made of rigid plastic and weighs 33.4 kg. The charge pressure of the compression tube is 1 MPa, and powder mass varies from 2.2 kg to 3.2 kg. Moreover, the characteristics of diaphragm burst affect the driver pressure. All of the diaphragms employed for these shots are flat ones and made of stainless steel with a four-petal design. They burst very well except for the 3.2-kg powder mass shot.

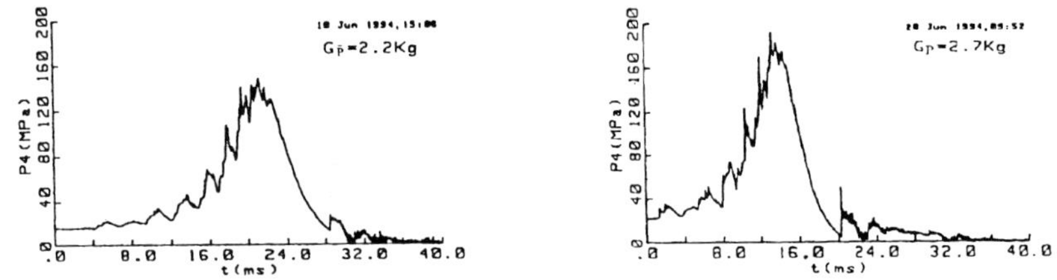

Figure 3. Driver pressure near the diaphragm: No. 4 (left) and No. 16 (right).

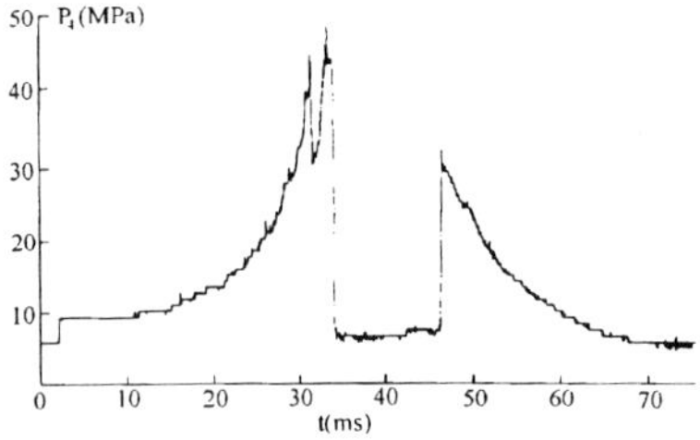

Figure 4. Driver pressure 2.7 m upstream of the diaphragm (No. 11).

Figure 2 shows a typical piston trajectory. The maximum speeds are from 360 m/s to 410 m/s. Driver-pressure histories measured just upstream of the diaphragm are presented in Fig. 3, which clearly shows that there are strong pressure oscillations while pressure increases. The other history shown in Fig. 4 is measured before the cross-section contraction at the end of the compression tube (i.e. 3.2 m upstream of the diaphragm). Although the peak pressure is not captured because of the gauge position, the pressure history before the diaphragm burst has been obtained. It is obvious that with increasing pressure there are no remarkable oscillations. This phenomenon suggests that the oscillations occur in the tube between the cross-section contraction and diaphragm. For the influence of the geometry on driver pressure, Labracherie et al. (1993) made experiments and similar results were demonstrated. In addition, good shot-to-shot repeatability is demonstrated in our tests.

3.2. Calibration results in shock tube

Figure 5 shows typical reservoir-pressure histories. These look like the usual overtailored ones, but according to evaluation, the corresponding tailored shock Mach numbers are about 13 and 14 respectively. Furthermore, following the arrival of the primary shock at the end of the shock tube is a quite strong secondary shock by which the reservoir gas is compressed again and pressure increases up to high stagnation conditions. In all probability, the aforemen-

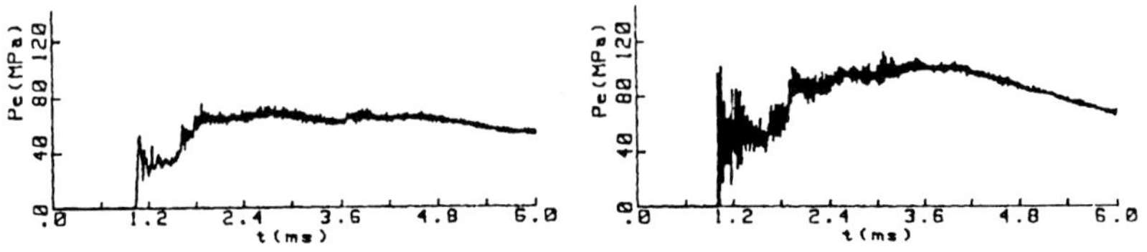

(a) (b)

Figure 5. Reservoir pressures: (a) $M_s = 11.9$ (No. 4), and (b) $M_s = 10.2$ (No. 16).

tioned phenomena are related to the compression characteristics of the facility. Further investigation will be made. Principal test results obtained in 1994 are given in Table 1. In the table, P_{4i}, P_1, are charge pressures in the compression tube and the shock tube respectively, G_p is the powder mass, V_p is the piston velocity, M_s is the shock Mach number, and P_0, T_0 are the reservoir pressure and temperature respectively. It is unfortunate that the diaphragm bursting pressures were not measured exactly. Moreover, for most shots summarized in the table, the durations of the reservoir pressure plateau were about 1.5–2.5 ms.

Table 1. CARDC free-piston shock-tunnel data (1994).

Run No.	P_{4i} MPa	P_1 MPa	G_p kg	V_p m/s	M_s	P_o MPa	T_o K
2	1.0	0.020	2.2	–	11.56	70.7	10020
3	1.0	0.040	3.2	–	11.8	146.1	10445
4	1.0	0.013	2.2	–	11.9	65.3	10147
5	1.0	0.013	2.2	–	11.8	58.6	10383
6	1.0	0.027	2.2	360	9.17	50.8	6884
7	1.0	0.040	2.2	366	9.22	58.5	6175
11	1.0	0.027	2.2	360	9.26	51.4	6836
16	1.0	0.027	2.7	410	10.2	95.5	8230
17	1.0	0.040	2.7	407	9.61	91.2	7565
18	1.0	0.067	2.7	401	8.45	85.5	6090
22	1.0	0.080	2.7	414	8.92	108.2	6644
23	1.0	0.020	2.7	407	10.23	97.8	7800
24	1.0	0.30	2.7	407	8.44	228.3	5900

3.3. Pitot-pressure survey

During tests in 1994, a pitot-pressure survey was made. Because the thermal protection problems of gauges on the rake have not been prefectly solved, the data taken from the rake were often not complete and only taken at the nozzle exit position. Figures 6 and 7 show a typical history and the radial distribution of pitot pressure.

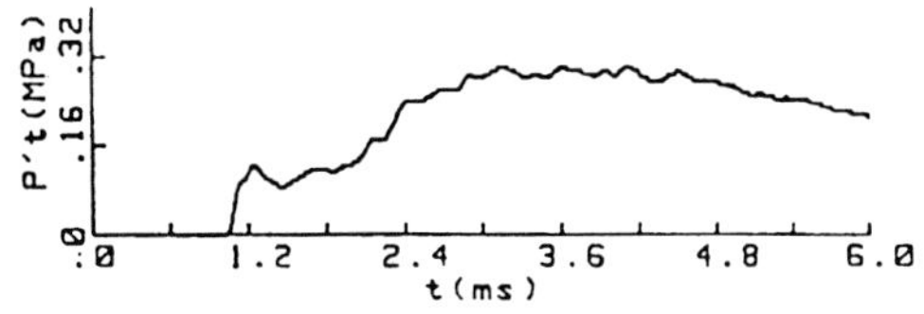

Figure 6. Pitot-pressure history (No. 19).

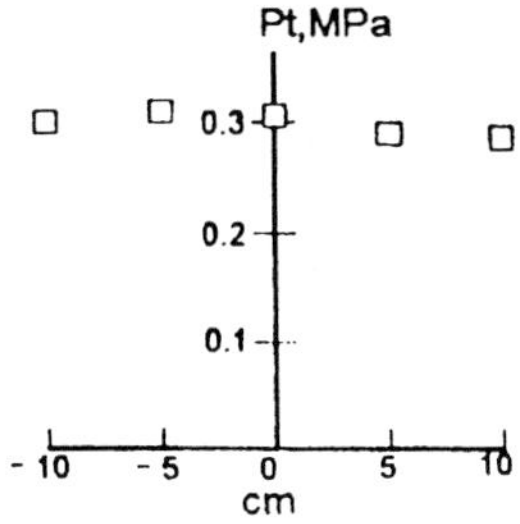

Figure 7. Pitot-pressure distribution in nozzle exit plane (No. 15).

In addition, severe ablations were encountered at the nozzle and measurement sections of the shock tube. These problems were encountered in other facilities too, and some good experience for solving these problems has been presented (Hornung 1992, Maus et al. 1993). It is clear that repeated tests must be made, and designs for gauges and thermal protection of the nozzle and shock tube are presently being improved for our facility.

4. Conclusion

According to the tests, the following conclusions have been reached:

1. The piston-compression process is influenced by many factors and it is important that these factors are effectively controlled for improving the performance of the facility.

2. The driver-pressure histories have shown that there are strong pressure oscillations during the increase of pressure which are due to the geometry of the compression tube between the cross-section contraction and diaphragm.

3. The reservoir-pressure histories suggest that some phenomena in the shock tube seem to be different from what is expected by classical theories, and

that the real flow pattern may be found by analysis of the compression process in the compression tube.

4. Severe ablations take place at the nozzle and shock tube. It is necessary to adopt new materials and re-design the corresponding parts.

References

Eitelberg G (1993) Calibration of the HEG and Its Use for Verification of Real Gas Effects in High Enthalpy Flows, AIAA 93-5170.

Hornung HG (1992) Performance Data of the New Free-Piston Shock Tunnel at GALCIT, AIAA 92-3943.

Labracherie L, Dumitrescu MP, Burtschell Y, Houas L (1993) On the compression process in a free-piston shock tunnel. Shock Waves 3:19–23.

Maus J, Blanks J, Dewitt J (1993) Calibration Tests of a New Free-Piston Shock Tunnel, AIAA-93-5003.

Morrison WRB, Stalker RJ, Duffin J (1989) New generation of free-piston shock tunnels. In: Proc 17th ISSWST, Bethlehem, p 582.

Stalker RJ (1989) Recent developments with free piston drivers. In: Proc 17th ISSWST, Bethlehem, p 96.

Characteristics of shock waves in exhaust systems and exhaust gas flows of automobile engines

N. Sekine
Fuji Heavy Industries Ltd., 3-9-6, Osawa, Mitaka, Tokyo 180 Japan
O. Onodera, K. Takayama
Shock Wave Research Center, Institute of Fluid Science Tohoku University
2-1-1, Katahira, Aoba, Sendai 980-77 Japan

Abstract: In conjunction with the reduction of exhaust noise of gasoline engines, shock-tube experiments were made on the initiation and propagation of shock waves of $Ms = 1.1$ in a branched tube which has a silencer. Holographic interferometric flow visualization was intensively carried out in a branched section of the exhaust gas pipe. Pressures were measured along the exhaust gas-pipe system in the catalyst. The silencer placed at the rear end of the exhaust gas-pipe system in the catalyst is effective in reducing shock-induced noise. The shape of the most effective silencer was found to be a double-chamber-type with a pipe inserted in it combined with a sound-absorption lining at the rear end of it.

Key words: Automobile engine, Engine silencer, Exhaust gas-pipe system, Numerical simulation, Optical-flow visualization, Shock tube, Weak shock wave

1. Introduction

The non-linear transition of compression waves discharged from the exhaust ports of automobile engines into shock waves often generates unpleasant exhaust noises. This has been one of the problems for automobile engine manufacturers to resolve urgently (Sekine et al. 1988; Sekine et al. 1989). To clarify the characteristics of exhaust gas-pipe systems and to determine the shape of a silencer, a shock tube having a branch which is analogous to an automobile exhaust gas-pipe system having a catalyst (hereinafter, Exhaust Pipe Front: EPF) was used for a shock wave of approximately $Ms = 1.1$ in air. In the previous work (Sekine et al. 1993) the insertion loss of silencers was discussed by referring to pressure measurements along a shock tube. It was then found that the most favorable configuration suited for exhaust-pipe silencers was a double-chamber-type having a pipe inserted in it. Its first-stage chamber can attenuate the shock waves and the second-stage chamber serves as a silencer suppressing sound waves. Measured exhaust noises in prototype engines adopting the double-chamber silencer confirmed that these experimental results are useful.

2. Shock waves in exhaust gas-pipe systems

The experiment was carried out using a water-cooled 2.0-liter 4-cylinder, 4-cycle gasoline engine. The engine was connected to an electrical dynamometer through a transmission. The measurements were conducted in firing condition under full load at an engine rotation speed of 2000 rpm. Initial exhaust-gas temperature was at the end of the silencer approximately 500 K.

Pressure transducers (Kistler Model 603B) were set at stations Nos. 1–7 along the exhaust gas pipe as shown in Fig. 1. The exhaust noise was measured simultaneously at the station No. 8.

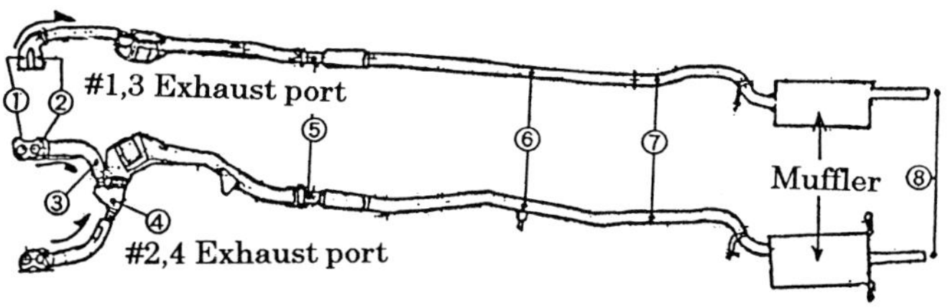

Figure 1. Pressure measurement of the exhaust gas-pipe system.

Two low-pressure channel shock tubes, with a 30 mm $\times$ 30 mm cross section and 800 mm long, were connected to each high-pressure chamber (550 mm long) and were connected with each other, resulting in a branched shock tube. These high-pressure chambers had a volume nearly equal to the automobile cylinder volume.

The two-low pressure channels responded to inlet pipes forming a manifold with a 1000-mm long outlet pipe, which is analogous to an automobile exhaust gas-pipe system in a catalyst, EPF. The side walls of this setup were made from transparent PMMA plates through which holographic interferometric flow visualization was carried out. The upper and lower walls were made from steel along which pressure transducers were installed (Sekine et al. 1988; Sekine et al. 1989). Figure 2 shows the experimental setup.

High-pressure driver gas was nitrogen at approximately 71 kPa and at room temperature, and the diaphragm was a 9-μm thick Mylar film which separated the low-pressure channels at ambient condition and the high-pressure chambers. The individual diaphragms were successively ruptured at time intervals of 30 ms. The time interval was controlled by successive detonations of 10-mg silver azide pellets which were pasted on the tip of a glass fiber and were attached

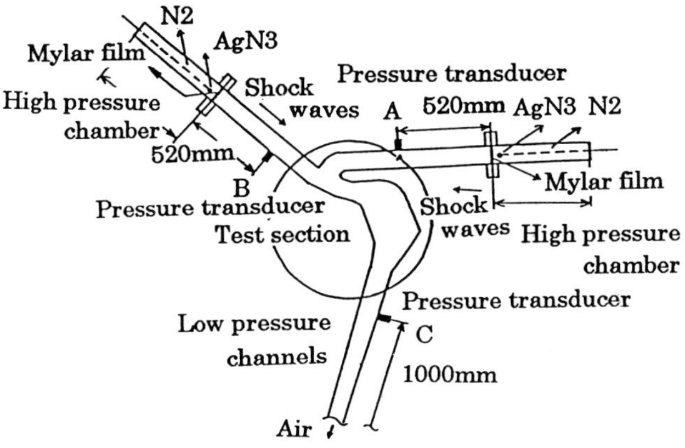

Figure 2. Experimental setup of shock tube.

to the diaphragms. The pellets was detonated by radiating them with a Q-switched YAG lasers. This time interval corresponded to that of an exhaust valve at an engine speed of 2000 rpm. The shock-wave Mach number thus created was $Ms = 1.1 \pm 1.0\%$ at the entrance of the EPF. Measured pressure signals were stored on a transient recorder with sampling rate of 5 μs/word.

Figure 3 shows pressure variations measured along the exhaust gas pipe. Compression waves discharged from the exhaust-gas port eventually became weak shock waves propagating along the EPF. In particular, pressure variation measured at position Nos. 3 and 4 showed a sharp peak pressure, which indicates that its variation can create pronounced high-frequency components of exhaust noises. The shock wave attenuated with propagation along the EPF.

However, shock waves reappear propagating along the Rear Exhaust Pipe (EPR), which followed the rear end of the EPF as shown at Nos. 6 and 7 in Fig. 3. This measurement clearly illuminates the propagation characteristics of shock waves appearing in automobile engine exhaust gas-pipe systems.

Figure 4a shows sequential interferograms of shock-wave propagation over an inlet pipe which represents the engine exhaust ports 1 and 3. The shock wave diffracts at the corner where the two inlet pipes merge and a reflected shock wave is generated. Figure 4b shows sequential interferograms of shock-wave propagation along the second inlet pipe with a delay time of 30 ms from the arrival of the shock wave shown in Fig. 4a. The shock wave propagating along the second inlet pipe interacts with the reflected shock waves of the first shock wave which remained after the delay time of 30 ms.

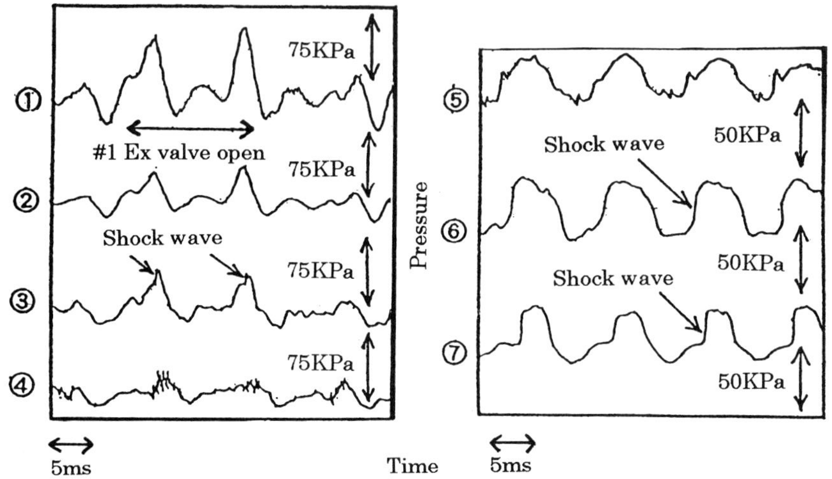

Figure 3. Pressure variations along exhaust gas pipe, 2000 rpm, full load.

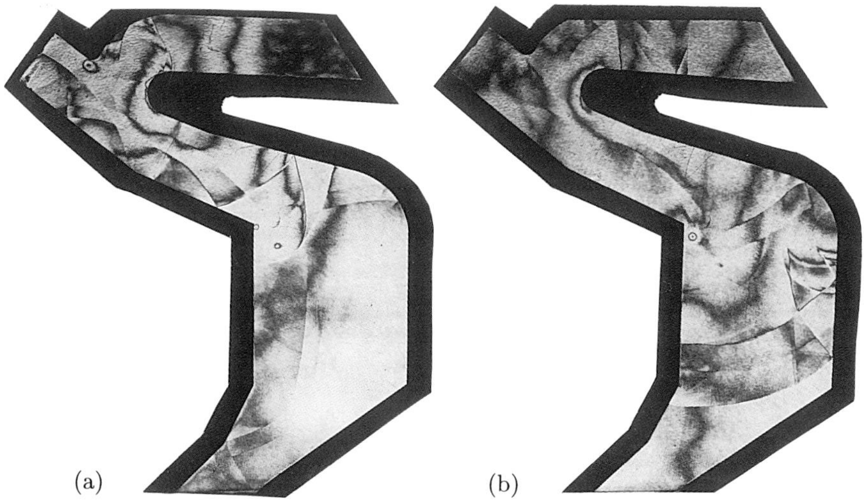

Figure 4. Sequential interferograms of shock-wave propagation in (a) the inlet pipe and (b) along the second inlet pipe.

3. Effects of shock waves on silencer characteristics

A shock tube was again used to study the attenuation characteristics of overpressure on silencers (Sekine et al. 1993). Frequency characteristics of overpressures were analyzed from results of pressure measurements along a shock tube. These frequency characteristic values represent insertion loss derived by acoustic theory. Figure 5 shows the insertion loss of various configurations of the silencers. High-frequency components over 1 kHz of overpressure were well attenuated when inlet and outlet pipes were inserted along the center line, whereas a double-chamber-type silencer can more effectively attenuate the high-frequency components in the range from approximately 1 kHz to 3 kHz than any other types of silencer. However, overpressures in the frequency range over 3 kHz were increased. A resonator having many holes and a sound-absorption lining could suppress overpressures of the frequency range over 3 kHz. The frequency range of unpleasant exhaust noises caused by shock waves was from approximately 1 kHz to 4 kHz (Sekine et al. 1993). Based on these results, a double-chamber silencer having a resonator with holes and sound-absorption lining at the rear end of the double chamber was examined.

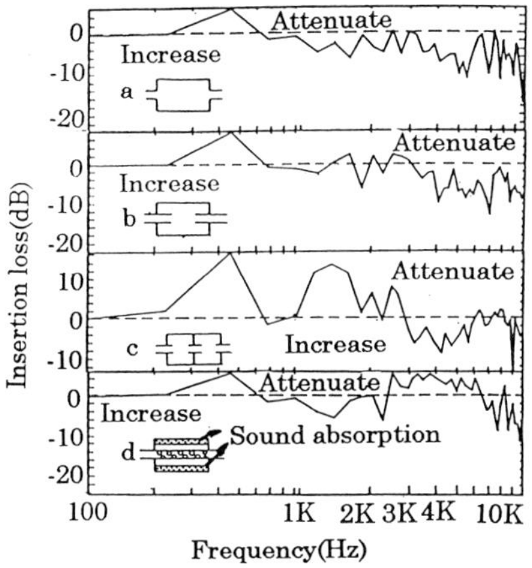

Figure 5. Insertion loss of various silencers.

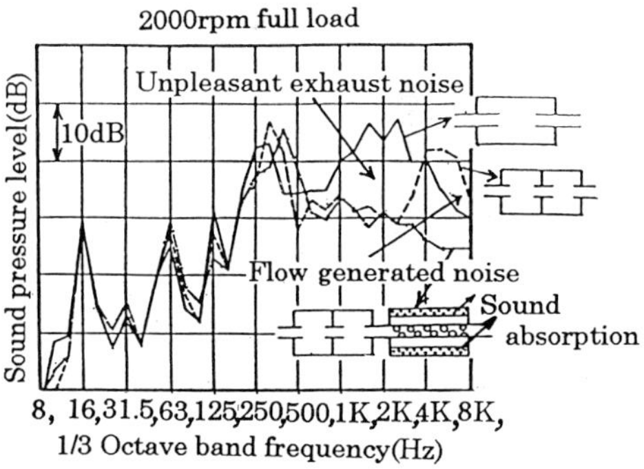

Figure 6. Spectra of sound pressure levels.

4. Measurement of sound pressures

Experiments were carried out with the engine described in Section 2; and the method and condition of the experiments were the same as those described in Section 2.

Figure 6 shows spectra of sound pressure levels measured at the outlet of the silencer. The frequency range of the noise was from approximately 1/3 octave band 1 kHz to 4 kHz. The sound intensity over 1/3 octave band of 4-kHz range in the double-chamber silencer is higher than that in a single-chamber silencer. This is the so-called flow-generated noise (Sekine et al. 1993). A double-chamber silencer combined with a sound-absorption lining can reduce both shock-induced noises and flow generated noises. This type of silencer is found to be more effective than any other type of silencer. The present findings obtained by the shock-tube experiments agreed well with previous experiments. These results have been successfully adopted in the design of practical exhaust gas-pipe systems. Subaru automobiles (model LEGACY produced in 1993) have been already equipped with an EPF and a double-chamber silencer having a resonator with multi-hole and sound-absorption lining.

5. Conclusions

The results obtained are summarized as follows: (1) Simple shock tubes can simulate shock-wave propagation in automobile exhaust gas-pipe EPF. Complex wave interactions were well visualized in the EPF. (2) By using a double-chamber silencer, weak shock waves can be attenuated more effectively than in single-chamber silencers. However, the double-chamber silencer still contains pronounced high-frequency noises which are mainly generated by reflected shock waves. (3) A resonator having holes and sound-absorption lining can more effectively suppress high-frequency noises than any other type of silencer. (4) Consequently, a double-chamber silencer combined with a sound-absorption lining at the rear end of the double chamber was successfully used.

References

Sekine N, Onodera O, Takayama K (1988) Shock Wave Development and Propagation in Automobile Exhaust Systems, SAE Paper 880082.

Sekine N, Onodera O, Takayama K (1989) Generation and propagation of shock waves in the exhaust pipe of a 4 cycle automobile engine. In: Proc of the 17th ISSW, p 208.

Sekine N, Onodera O, Takayama K (1993) Effects of shock waves on silencer characteristics in exhaust gas flows of automobile engines. In: Shock Waves @ Marseille, R Brun and L Dumitrescu (eds), Springer.

Influence of the curvature of the diaphragm on flow quality in shock tubes

V. Daru, J. P. Damion
Laboratoire SINUMEF and Laboratoire de Métrologie Dynamique
ENSAM, 151 Boulevard de l'Hopital, 75013 Paris, France

Abstract: The initial curvature of the diaphragm in shock tubes has been shown in a previous paper to originate periodic disturbances of the flow which should theoretically be uniform between the waves. With the help of numerical simulations, we here extend this work to the different context of high pressure ratios (which are of concern for example in shock tunnels and expansion tubes).

Key words: Shock tube, Pressure fluctuations, Numerical simulation

1. Introduction

In a previous paper (Daru & Damion 1995), we have shown that the initial curvature of the diaphragm in a shock tube, due to the pressure difference between the driver and the driven gas, gives rise to perturbations of the flow. Numerical simulations, which were in very good agreement with the experimental data, have revealed that the perturbations are periodic, originated by the multiple reflections of a transverse wave which is created because the initial discontinuity is not normal to the wall of the tube.

This work was done in the context of the characterization of pressure transducers, which implies low pressure ratios (1.3 to 2). In fact, this type of perturbation is widely observed in the very different context of shock tunnels and expansion tubes. Of course, different causes can originate noise in the flow, as for example the opening time of the diaphragm. In the case where the diaphragm is thin (cellophane), the opening time is generally very low compared to the sound speed, and this then cannot explain the observed disturbances. Cellophane is used for example in expansion tubes for the secondary diaphragm. In this type of facility, capable of producing high-enthalpy flows for aerothermodynamic investigations, the run times available are reduced by large disturbances which disrupt the flow. Paull and Stalker (1991) have modeled these disturbances by lateral acoustic waves, and have used the acoustic theory to show that the gas interface can act as a filter for transverse acoustic waves propagating from the driver to the driven gas.

In expansion tubes, not only the primary diaphragm but also the secondary diaphragm can originate the production of noise. Roberts et al. (1995) have studied experimentally the effect of the secondary diaphragm on the flow quality. More precisely, they have studied how disturbances can be created by multiple shock reflections occurring between the diaphragm and the driver-test-gas interface, before the diaphragm bursts. In fact, since the secondary diaphragm is curved when it bursts, transverse waves as described above are probably one of the sources of the disturbances observed in the flow.

In this paper, we extend our study to the context of high pressure ratios. We show, with the help of numerical experiments, that the predictions given by the acoustic theory are well verified in many cases, but the reasons for this are not so clear. In particular, the disturbances originate not only in the driver gas, but in both gases.

2. Numerical simulation

The numerical simulations rely on the solution of Euler equations in cylindrical coordinates, with the assumption of axisymmetrical flow. Although viscosity and real-gas effects should be taken into account in the high-pressure and temperature range, the Euler equations constitute an adequate model for wave propagation and interaction phenomena. Of course, the attenuation of waves due to the viscous effects is not represented.

We will consider three configurations for the repartition of gases as driver/driven gas: Air/Air, Ar/He, He/Ar. For each case, the specific heat ratio γ is constant, equal to 1.4 for air, and to 1.658 for Ar or He. We limit ourselves to cases where γ is the same for the driver and driven gas, so as to avoid numerical difficulties encountered in multicomponent flow calculations (typically numerical oscillations, Karni 1994). An additional equation for the concentration of gas should be solved if one needs to know the temperature field, but here knowledge of the pressure and density fields is sufficient for our purpose.

Each gas is characterized by the value of γ and the gas constant R (R is equal to 287 for air, 208 for Ar, 2078 for He). The perfect gas law of state $p = R\rho T$, where p is the pressure, ρ the density, and T the temperature, is used to set the values of density at the initial state.

The numerical scheme is a centered second-order accurate predictor-corrector scheme, with a corrective term added to make it TVD and eliminate numerical oscillations (for more details see Daru & Damion 1995). We use a Cartesian uniform mesh made of 320$\times$60 cells. At the initial state, the discontinuity is supposed to be spherical and a mean value is computed in each cell. The computations are done for a 2-m long tube, with a diameter of 15 cm.

3. Results and interpretation

3.1. Origin of the disturbances

As was the case in the low-pressure-ratio range, the fact that the initial discontinuity is not normal to the wall gives rise to a transverse shock wave which moves towards the center of the tube, where it reflects as a wave expanding towards the wall. Multiple reflections then occur between the center and the wall, producing periodic disturbances whose period, in a frame of reference moving with the fluid velocity, is proportional to the diameter of the tube and inversely proportional to the local sound speed at which the reflections propagate. A Doppler shift should be applied in a fixed frame.

3.2. Analysis of the results

The essential difference with the low-initial-pressure-ratio case is that the ratio of the sound speeds a_2 (related to the shock-heated test gas) and a_3 (related to the expanded driver gas) is no longer close to 1. The large disturbances that are created at the center of the tube, where the transverse wave coalesces, first occur in the gas with the largest sound speed, the wave moving at the local sound speed in each gas. This is illustrated in Fig. 1, where we can see the density field at the beginning of the computation for two cases where $a_2 > a_3$ (a) or $a_2 < a_3$ (b). We then observe a decoupling of the wave across the interface

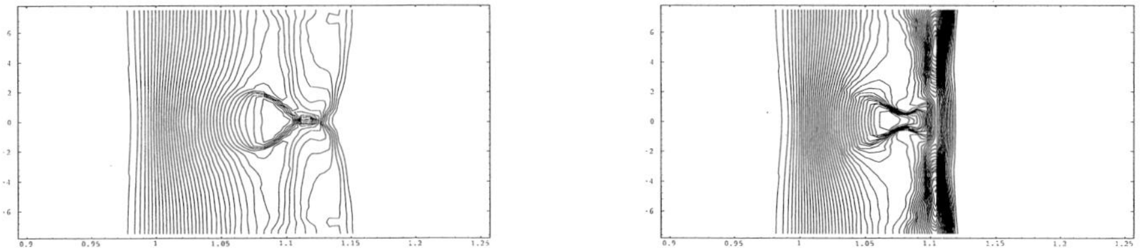

(a) (b)

Figure 1. Iso-density lines. Initial state: (a) $a_2/a_3 = 1.44$, He, $T = 293°\text{K}$ (HP), Ar, $T = 293°\text{K}$ (LP), $p_4/p_1 = 300$. (b) $a_2/a_3 = 0.68$, Air, $T = 3000°\text{K}$ (HP), Air, $T = 293°\text{K}$ (LP), $p_4/p_1 = 50$.

(Fig. 2). The amplitude of the waves in the driven-gas region is very small, but we can see the occurrence of the first peak at the center of the tube in each gas (right-end figures) and notice that the period of the phenomenon is approximately twice as big in the driven gas. A (z, t) diagram of the density (Fig. 3) is useful to see the evolution of the process as time goes on. The amplitude of the disturbances is indeed much larger in the driver gas, and that seems to confirm the predictions of the acoustic theory. However, the theory is based on the assumption that the disturbances, originating in the driver gas, are transmitted through the interface, and then decay if $a_2 > a_3$. We see here that there are disturbances created in both gases, and that transmission occurs from the driven gas to the driver gas only. This is more visible in Fig. 4, which represents the density history at the center of the tube, for a ratio $a_2/a_3 = 4.31$ (Ar/He case). In this figure, we observe very clearly that the first disturbance created in the driven gas propagates across the interface into the driver gas, where the flow is still quiet, and then interacts with the disturbance specific to the driver gas. In fact, it seems in these results that the initial local amplitude of the transverse wave, rather than transmission or reflection of the wave across the interface, plays a very important role in the subsequent decay with time, as seen in Fig. 5, where the initial amplitude of the transverse wave is smaller and the decay more rapid in the driven-gas region.

In fact, we have observed a decay with time in all cases, which can be more or less rapid. This could be explained by energy conservation reasons, the wave being "split" at each period due to the reflection on the wall. There also exists an axial decay, the reason for it probably lying in the spherical nature of the wave. Indeed, dissipative effects of the numerical scheme are also of influence.

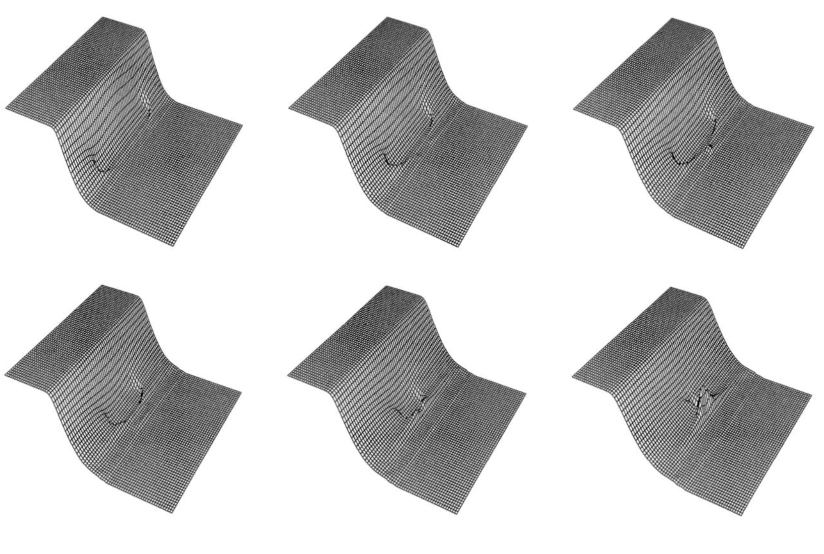

Figure 2. Three-dimensional representation of the density field. Initial state: Air, 293°K (HP and LP), pressure ratio 100, $a_2/a_3 = 2.1$.

The case where we employ He for the driver gas and Ar for the driven gas is another illustration of the fact that mechanisms other than the filter effect of the interface are involved. If we set the initial temperature to 293°K in both

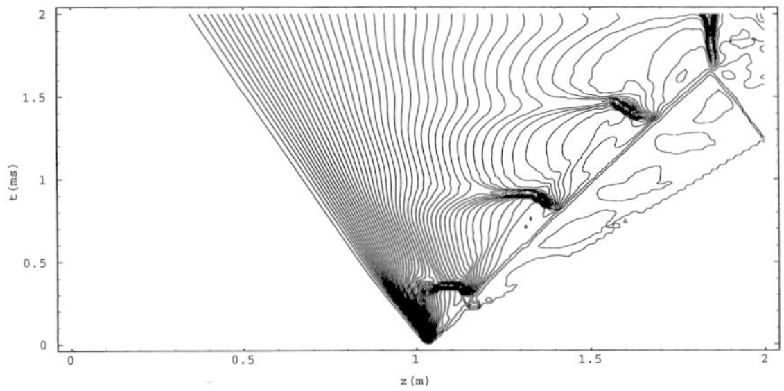

Figure 3. Iso-density lines in the (z, t) plane at the center of the tube, same conditions as in Fig. 2.

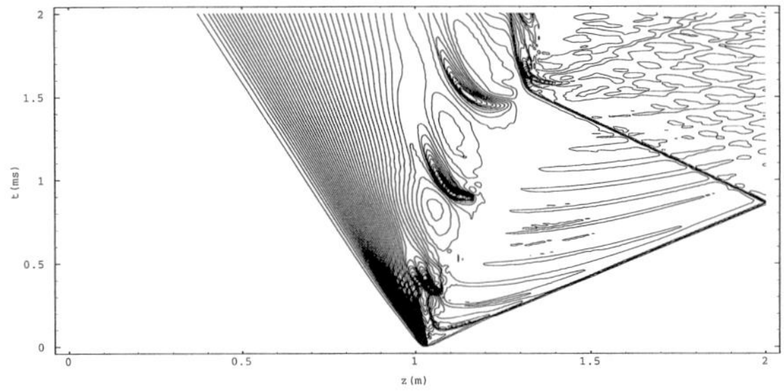

Figure 4. Pressure history at the center of the tube. Initial state: Ar, 293°K (HP), He, 293°K (LP), pressure ratio 5, $a_2/a_3 = 4.31$.

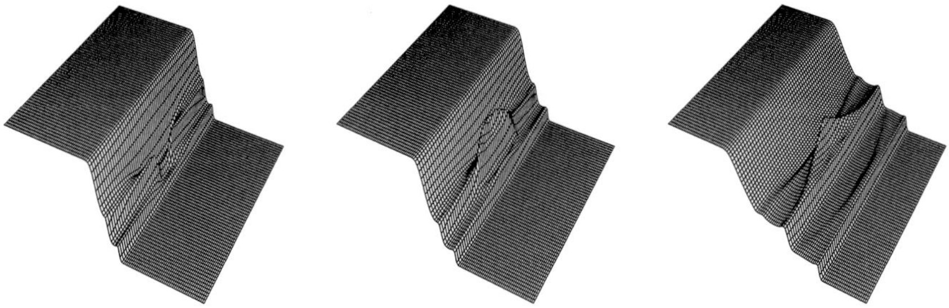

Figure 5. Three-dimensional representation of the density field. Initial state: Air, 293°K (HP and LP), pressure ratio 5, $a_2/a_3 = 1.27$.

gases and let the initial pressure ratio increase, the ratio a_2/a_3 stays below 1 until $p_4/p_1 \approx 88$ when it is equal to 1, and then it is greater than 1. Figure 6 represents the history of the pressure at the center of the tube for the three cases $a_2/a_3 < 1$ (a), $a_2/a_3 \approx 1$ (b), $a_2/a_3 > 1$ (c). We see that decay occurs in the driven gas in both cases (a) and (c), which is in contradiction with the acoustic theory.

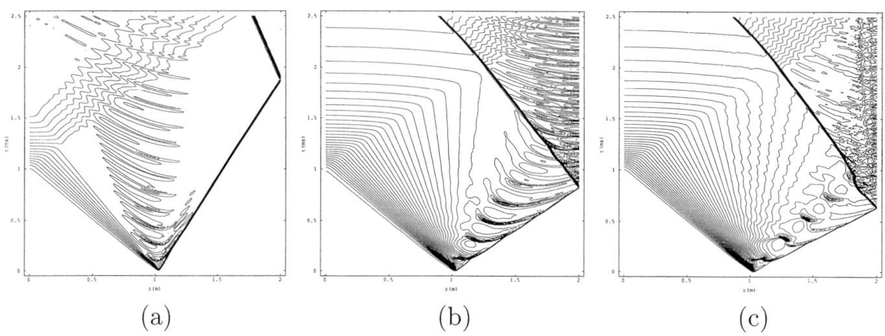

Figure 6. Pressure history at the center of the tube, He, 293°K (HP), Ar, 293°K (LP): (a) $a_2/a_3 = 0.45$, $p_4/p_1 = 5$; (b) $a_2/a_3 \approx 1$, $p_4/p_1 = 88$; (c) $a_2/a_3 = 1.44$, $p_4/p_1 = 300$.

4. Conclusion

The initial curvature of the diaphragm in shock tubes, when the opening time is negligible, can originate disturbances which seriously corrupt the quality of the flow. Complex interactions of waves occur. The acoustic theory can explain the role of the gas interface, but does not predict the local behavior of the disturbances. Numerical simulation can then provide a complementary and more precise tool to predict such phenomena.

References

Daru V, Damion JP (1995) Analysis of the flow perturbations in a shock tube due to the curvature of the diaphragm. In: Shock Waves @ Marseille, Brun R, Dumitrescu LZ (eds), Springer Verlag, Vol 1, pp 451–456.

Karni S (1994) Multi-component flow calculations by a consistent primitive algorithm. J Computational Physics 112(1):31–43.

Paull A, Stalker RJ (1991) Acoustic waves in shock tunnels and expansion tubes. In: Proc 18th ISSW, Takayama K (ed), Sendai, pp 697–704.

Roberts GT, Morgan RG, Stalker RJ (1995) Influence of secondary diaphragm on flow quality in expansion tubes. In: Shock Waves @ Marseille, Brun R, Dumitrescu LZ (eds), Springer Verlag, Vol 1, pp 203–208.

Modifications to the AEDC shock-tunnel code and comparisons with data

J. R. Maus, Micro Craft Technology, Inc.
Lt. N. R. McKenzie, USAF
Arnold Engineering Development Center*
Arnold AFB, TN 37398 USA

Abstract: Modifications to a mathematical model of the free-piston shock tunnel have been made to more accurately describe the boundary-layer flow on the shock-tube wall. These modifications are designed to yield improved predictions of facility performance and better estimates of available test time. Results of computations with the revised code and comparisons with data are presented.

Key words: Shock tunnel, Boundary layer, Computational model

1. Introduction

The limited productivity and the large number of operational parameters inherent with free-piston shock tunnels make a facility computational model an important tool for planning tests and selecting initial run conditions. A computer code to solve the one-dimensional gas dynamic equations, based on a scheme developed by von Neumann and Richtmyer (1950), has long been used at AEDC to describe the G-Range light-gas-gun launch cycle. This code was adapted to describe the flow in shock tubes, shock tunnels, and particularly free-piston shock tunnels. As with all one-dimensional formulations, the effects of viscous shear and heat convection phenomena, involving lateral gradients of velocity and temperature, can only be approximated. In the code used at AEDC, empirically determined coefficents based on one-dimensional steady flow considerations are employed to model the effects due to friction and heat transfer. In the present work the modeling of the shock-tube boundary layer is improved to simulate the effective mass loss of test gas due to boundary-layer growth. This mass loss influences the incident-shock speed and the available test time for a reflected shock tunnel.

---

*The research reported herein was performed by the Arnold Engineering Development Center (AEDC), Air Force Materiel Command. Work and analysis for this research were done by personnel of Micro Craft Technology, Inc./AEDC Operations, technical services contractor for the AEDC aerospace flight dynamics facilities. Further reproduction is authorized to satisfy the needs of the U. S. Government.

2. Description of code

In the computational model, the flow domain is divided into several regions, each possibly with different initial conditions. Each of these regions is assumed to have its own equation of state and is subdivided into a number of cells. The system of mass points that describes the initial mass distribution in the duct is labeled and tracked as the flow field evolves in time. Each mass point contains one-half the mass of the adjacent cells and is located at the interface of the two cells. The dependent thermodynamic variables, pressure, specific volume, and internal energy, are computed in the interior of the cell between mass points J and $J+1$. This is illustrated schematically in Fig. 1. To advance the solution one step, dt, in time, Newton's second law is used to compute the velocity and displacement of each mass point. The new axial locations of the mass points

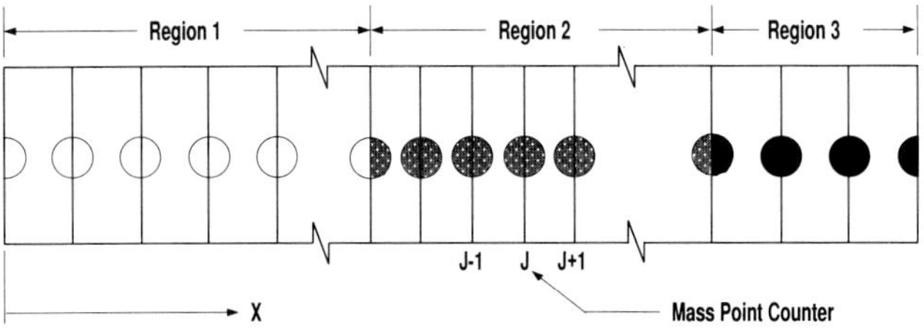

Figure 1. Mass points, cells, and regions.

allow the volume of each cell to be determined. Since the mass in each cell is known, the density or specific volume of material in the cell can be computed. The energy equation and the equation of state can be solved to determine the internal energy, E, and the pressure, P, in each cell at $t + dt$. Friction and heat-loss coefficients, C_f and C_h, based on one-dimensional, steady-flow considerations, are included in the formulation of the momentum and energy equations.

3. Shock-tube boundary layer

In a series of reports and papers in the 1950s and 60s, Mirels (1956, 1964) investigated the effect of the shock-tube boundary layer on incident-shock speed and maximum test time. The modification to the code described in this paper leans heavily on Mirels' work. The development of the boundary layer between the initial shock and the contact surface is shown schematically in Fig. 2.

In a laboratory-fixed coordinate system (X^*, Y^*, t^*), shown in the upper part of the figure, the flow is unsteady, depending on space coordinates and time. Transformation to a coordinate system (X, Y, t) that moves with the shock results in a steady flow between the shock and the interface. In this

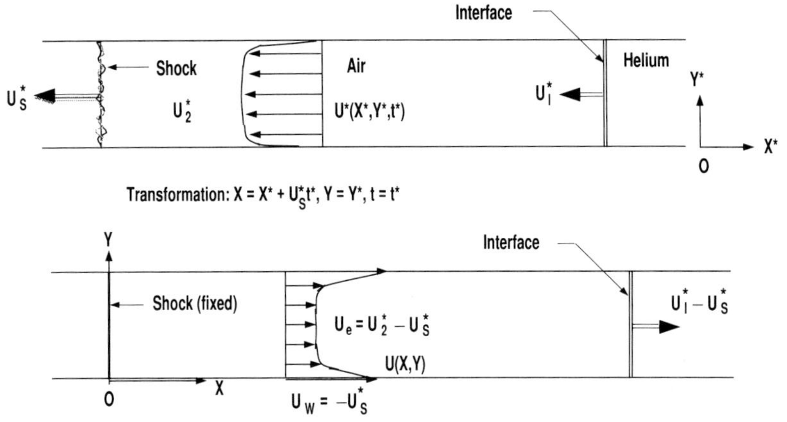

Figure 2. Shock-tube boundary-layer transformation.

coordinate system, the steady-flow boundary-layer equations apply in the test gas. Moreover, to a first approximation, the pressure in the test gas is constant, so the boundary-layer equations take the form for a flat plate. The interface is still moving in the transformed coordinates, so the driver gas flow is unsteady and the analysis does not extend easily into this region.

The displacement thickness is negative in the transformed coordinates, meaning that the boundary layer causes an effective outflow from the shock tube. In the code, this effect is modeled by reducing the mass of the individual elements at each time step, in accordance with the computed local mass outflow. Either a laminar or a turbulent boundary layer can be specified in the code. The displacement thickness variation, for either case, can be obtained from a solution of the von Karman momentum integral equation.

4. Results and comparison with data

4.1. Conventional shock-tube results

To examine the applicability of the code to high-velocity shock-tube flows where non-ideal gas effects become important, but without the complications of a free-piston driver, an example from AEDC Tunnel J was used. Tunnel J was a reflected shock tunnel with an electrically heated driver that operated at AEDC in the 1960s and early 1970s. The data shown in Fig. 3 (Pate et al. 1972) were obtained with heated hydrogen as the driver gas and air as the driven gas. Computations were carried out, treating hydrogen as an ideal diatomic gas, for perfect air with no boundary-layer effects, for equilibrium air, and for equilibrium air with boundary-layer effects (mass loss and friction). The

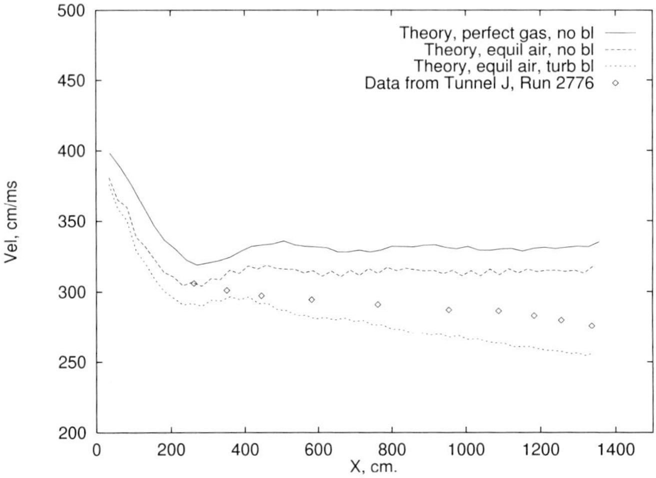

Figure 3. Comparison of computed shock speed with data from Tunnel J.

equilibrium-air computation shows a reduction in the shock speed due primarily to the higher density in the shock-processed test gas. The addition of boundary-layer effects causes a further reduction in shock speed with a rate of decay somewhat greater than that exhibited by the data. The oscillations in the computed shock speed arise from the difficulty in determining the location of a computationally captured shock. This feature is seen in all computed shock-speed results.

4.2. Free-piston shock-tunnel results

The device toward which this code is directed is the free-piston shock tunnel at AEDC known at the Impulse Facility, which was described in some detail by Blanks & Dewitt (1994). For the purposes of this model, the facility is divided into five regions as illustrated in Fig. 4. These are: (1) the propellant gas (burning gunpowder), (2) the piston, (3) the driver gas (helium), (4) the test gas (air), and (5) the initial low-pressure gas in the test section and dump tank. The results of two computations with the code are shown in Figs. 5 and 6. Both computations were carried out for an equilibrium-air test gas and with $C_f = 0.009$ and $C_h = 0.0002$.

Computational results in Fig. 5 show the shock and air-helium interface trajectories in X-t space with and without boundary-layer mass loss. There is no major difference in the computed shock speed for the two cases. No attempt has been made to track the reflected shock in these results, so the trace for the incident shock simply stops at the end of the shock tube. The

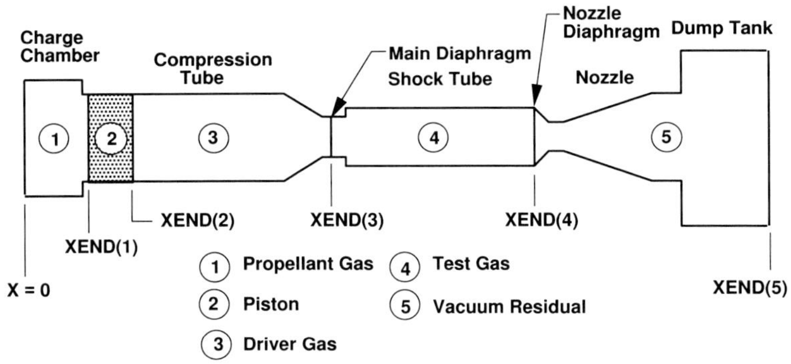

Figure 4. Initial region boundaries for model of Impulse Facility.

interface, on the other hand, continues to be tracked and passes on through the nozzle throat. The calculations show that near the end of the shock tube, with mass loss, the trajectories of the shock and interface become nearly parallel, indicating no benefit can be derived from further lengthening of the tube. This finding is consistent with the conclusion of Mirels (1964). Although no data on helium arrival are available, it is thought that the 3-ms test time calculated with boundary-layer mass loss is much more realistic than over 6 ms obtained without loss.

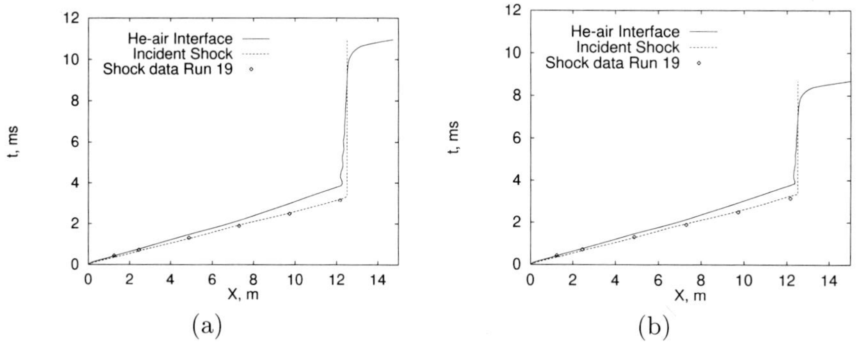

Figure 5. Shock and interface tracectories for Run 19 (a) without boundary-layer mass loss, (b) with boundary-layer mass loss.

A comparison of computational results for nozzle stagnation pressure, P_5, with experimental data is shown in Fig. 6. The computed results with and without mass loss differ only slightly and agree reasonably well with the data.

Although the pressure oscillation shown in this figure is undesirable, it is encouraging that the mathematical model predicts this behavior fairly well.

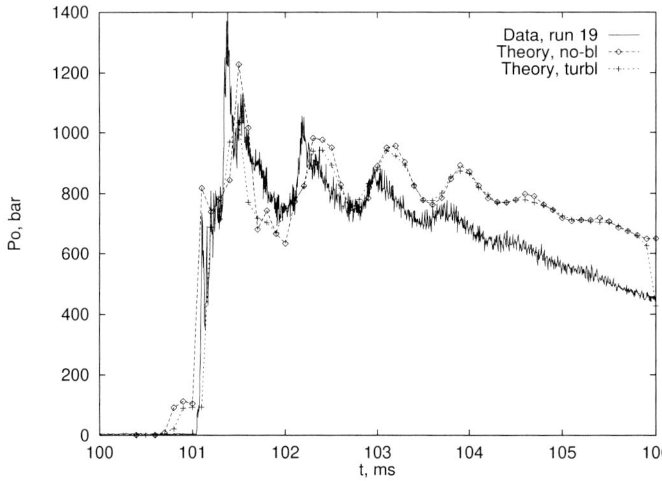

Figure 6. Comparison of computed stagnation pressure with data from Run 19.

5. Summary

The mathematical model for the AEDC free-piston shock tunnel has been modified to include the effective test-gas loss due to the growth of the boundary layer in the shock tube. The modification has a minor influence on the computed shock speed and stagnation pressure but causes a substantial reduction in the computed time at which helium first arrives in the facility test section.

References

Blanks JR, Dewitt JR (1994) Calibration Tests of the AEDC Free-Piston Shock Tunnel, AIAA 94-2526.

Mirels H (1956) Attenuation in a Shock Tube Due to Unsteady-Boundary-Layer Action, NACA TR 1333.

Mirels H (1964) Shock tube test time limitations due to turbulent-wall boundary layer. AIAA Journal 2:84–93.

Pate SR, Siler LG, Stallings DW, Wagner DA (1972) Development of the AEDC-VKF Tunnel J — a Real Gas High Density, True Velocity, Hypersonic, Test Facility, AIAA 72-993.

von Neumann J, Richtmyer RD (1950) A method for the numerical calculation of hydrodynamic shocks. J Applied Physics 21:232–237.

Numerical simulation of diaphragm rupture

P. J. Petrie, P. A. Jacobs
Mechanical Engineering Department
University of Queensland
Australia

Abstract: We have performed numerical simulations of diaphragm rupture for the cases of (1) the relatively slow rupture of the primary diaphragm in a shock tunnel or expansion tube and (2) the fast rupture of the light secondary diaphragm in an expansion tube. The code was based on an unstructured-grid finite-volume formulation of the Euler equations for two-dimensional planar or axisymmetric flow. The flow field that develops for each type of diaphragm rupture is illustrated and the flow mechanisms behind previously reported experimental observations are examined.

Key words: Diaphragm rupture, Finite-volume code, Shock tunnel, Expansion tube

1. Introduction

Shock tunnels and expansion tubes are used to generate high-energy, hypersonic flows for the ground testing of hypervelocity vehicles. Flow in each facility is usually initiated by the rupture of a diaphragm that separates high- and low-pressure gas. In a shock tunnel, the test gas is first processed by the *incident* shock, brought to rest through a *reflected* shock, and then expanded (in a quasi-steady process) through a converging-diverging nozzle. The expansion-tube facility differs in that the shock-processed test gas is expanded (in an unsteady process) to the test-section conditions. In the ideal situation the test gas is not stagnated after being processed by the the incident shock.

In this paper two different types of diaphragm rupture are examined. The first is the relatively slow rupture of the *primary* diaphragm initially separating the driver gas from the driven (test) gas. The second is the fast rupture of the light *secondary* diaphragm in an expansion tube when it is struck by the shock wave used to process the test gas.

The simulations were performed with a finite-volume code solving the Euler equations for compressible, inviscid flow. The flow domains were discretized with unstructured grids with triangular elements, and a solution-adaptive remeshing scheme was used to focus the computational effort in regions where the flow-field gradients were high. This allowed the capture of shock waves and shear layers with greater resolution than possible with fixed-grid simulations at the same (or similar) computational expense.

2. Opening of a primary diaphragm

Consider the speed of the shock developed after rupturing of the primary diaphragm. Miller and Jones (1975) measured shock speeds within the NASA Langley six-inch expansion tube. Air, argon, carbon dioxide, and helium were used as test gases. When the pressure ratio across the diaphragm was high, ($> 10^3$) it was observed that the measured incident-shock speed exceeded the theoretical shock speed determined by assuming one-dimensional flow and instantaneous diaphragm opening. Other researchers (cited by Miller & Jones 1975) have also noted higher than expected shock speeds. White (1958) and Ikui and co-workers (1969) showed that the maximum one-dimensional theoretical shock speed is higher if a finite opening time is assumed. The diameter of the driver section of the NASA Langley six-inch expansion tube was greater than the diameter of the driven section. Alpher and White (1958) developed a theory for predicting shock strength when the cross-sectional area of the driver and driven tubes are not equal. The shock speeds measured by Miller were significantly greater than speeds predicted by the above-mentioned one-dimensional methods.

Satofuka (1970) showed that multi-dimensional simulations with gradual diaphragm opening can produce higher shock speeds than one-dimensional theories of White (1958) and Ikui and co-workers (1969). Numerical simulations of the flow through the gradually opening primary diaphragm are presented here. The geometry of the domain is based on the primary diaphragm station of the NASA Langley expansion tube (Miller & Jones 1975). The simulations are at a higher grid resolution and at higher initial pressure ratio than those performed by Satofuka (1970). The speed of the shock was computed and compared with the experimentally measured shock speed. The opening diaphragm was modelled as a dilating iris (Cambier and co-workers 1992, Vasil'ev & Danil'chuk 1994). The opening time of the diaphragm was 200 μs. The diameter of the driver section was 165.1 mm and the length was 2.44 m. The diameter of the driven tube was 152.4 mm. The transition from driver diameter to driven diameter occurred after the diaphragm location and extended over a length of 190.5 mm. Both the driver and driven gases were helium. The driver gas conditions (pressure and temperature) were $p = 35 \times 10^6$ Pa, $T = 342.0°$K, and the conditions in the driven section were $p = 3450$ Pa, $T = 297.6°$K. The driver-gas fill conditions were at a pressure high enough to cause significant deviation from ideal gas behavior. The Redlich-Kwong equation of state for helium (Aungier 1995) was used.

Figure 1 compares the computed shock speed with the experimentally measured shock speed from Miller and Jones (1975) and the maximum shock speed estimated by a one-dimensional flow model which includes the effects of gradual diaphragm opening (Ikui and co-workers, 1969) and a larger diameter for the driver section (Alpher and White, 1958). The maximum shock speed computed by the current multi-dimensional simulation is higher than the shock speed predicted by the one-dimensional theory and is closer to the experimental measurements. Note that the computed shock speed trace does not show the same rate of decay observed in the experimental data after the maximum speed has been reached, presumably because viscous losses are not included.

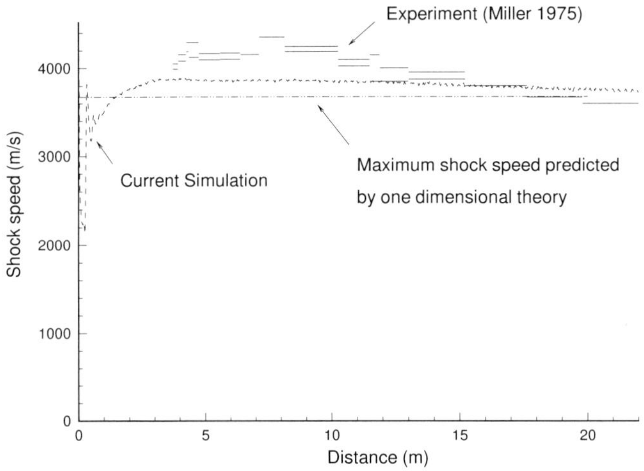

Figure 1. Comparison of computed shock speeds with experimental data.

The development of the flow during and after diaphragm rupture for a constant area tube is displayed as a series of snap-shots of the density contours (Fig. 2). The opening time for this simulation was 320 μs. The shock front becomes planar within a relatively short distance. The interface initially has a convex shape due to the finite opening time of the diaphragm but, at a later time, it can be seen that the shape of interface is concave. At approximately 80 μs the interaction of the interface with the tube wall causes an oblique upstream-facing shock to develop, causing the density and the pressure of the flow behind the interface to be higher at the wall than at the center. This region of higher pressure near the wall appears to push the interface into the concave shape. The evolution of the contact surface was also observed by Vasil'ev and Danil'chuk (1994).

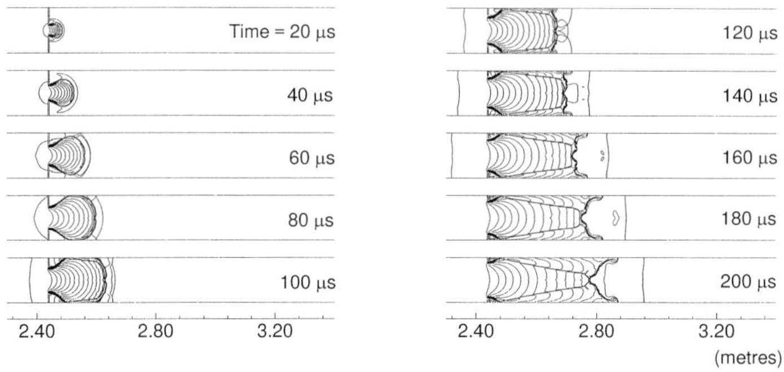

Figure 2. Time history of density contours for primary diaphragm opening.

3. Secondary diaphragm rupture

Consider the fate of a light, secondary diaphragm within an expansion tube. The secondary diaphragm has a much smaller mass than a typical primary diaphragm and is ruptured when struck by the incident shock. Bakos and Morgan (1994) hypothesized that, when the diaphragm is ruptured, it breaks free of the tube wall and is carried along by the expanding flow, much like a solid piston. What happens to the diaphragm fragments as they are convected along the tube is of interest as it appears that the fragments interfere with the expanding flow for a significant time. One of the consequences of the diaphragm acting like an accelerating piston is the production of a reflected shock which propagates upstream through the (already shock-processed) test gas. Although this reflected shock is initially strong, it weakens as the diaphragm fragments accelerate along the tube.

Experimental work by Roberts and co-workers (1993) suggests that the strength of the reflected shock can be reduced by using a pre-deformed diaphragm. Numerical simulations have been performed to examine a possible mechanism causing the reduction in the strength of the reflected shock when pre-deformed diaphragms are used instead of planar diaphragms. In the simulations, the flows were assumed to be two dimensional and the initial deflection of the center of the pre-deformed diaphragm was one quarter of the tube diameter. On arrival of the incident shock, the diaphragm was assumed to break into seven fragments with the subsequent motion of the fragments determined by the pressure acting on their surfaces. The fragments had three degrees of freedom: two in translation and one in rotation. The planar diaphragm was modeled as one piece.

Figure 3 shows the flow and pre-deformed diaphragm evolution over a period of 90 μs. When struck by the incident shock, the fragments close to the tube

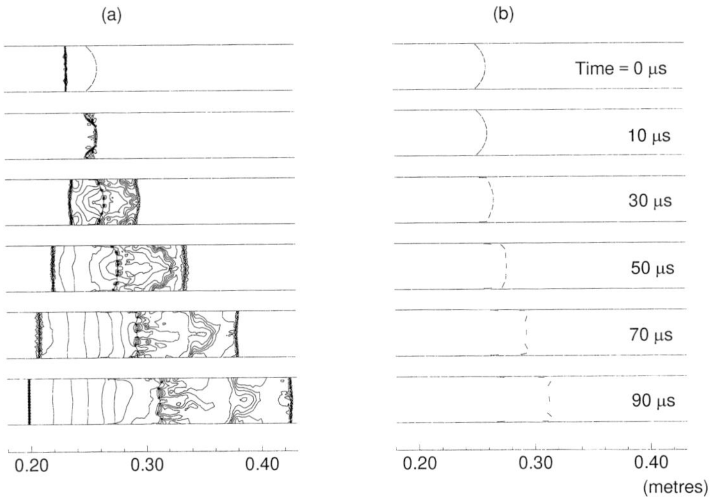

Figure 3. Simulation of secondary diaphragm rupture. Time history of (a) density contours, and (b) diaphragm fragments.

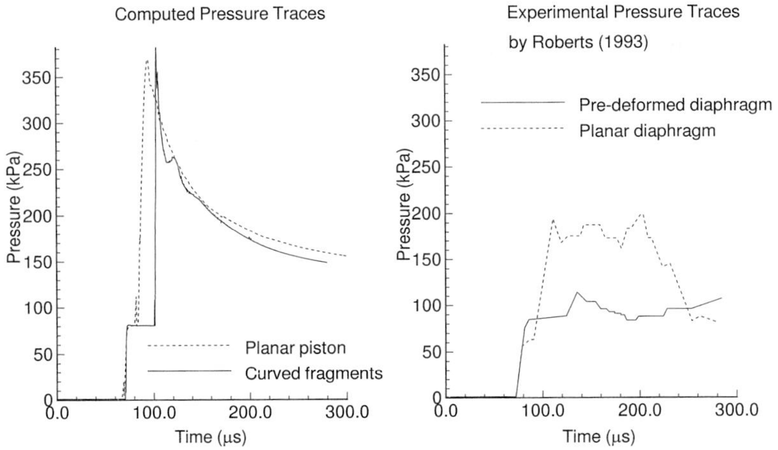

Figure 4. Pressure traces 15 mm upstream from diaphragm location.

wall are pushed towards the tube wall. Figure 4 compares computed and experimental pressure traces 15 mm upstream from the original diaphragm location for the pre-deformed diaphragm. The comparison is not close, but we expect that the simulations will improve when ionization and three-dimensional effects are considered.

4. Concluding remarks

An axisymmetric finite-volume Euler solver was used to simulate the flow through a slowly-opening diaphragm. The computed speed of the developing shock wave is higher than that estimated by previous one dimensional models and is closer to the experimental mesurements.

A two-dimensional version of the code was also used to examine the interaction between the incident shock and a light secondary diaphragm in an expansion tube. Although the current simulations did not compare well with the available experimental data, we expect that the inclusion of three-dimensional and real-gas effects will greatly improve the simulation.

References

Alpher RA, White DR (1958) Flow in shock tubes with area change at the diaphragm section. Journal Fluid Mechanics 3:457–470.

Aungier RH (1995) A fast, accurate real gas equation of state for fluid dynamic analysis applications. Journal of Fluids Engineering 117:277–281.

Bakos RJ, Morgan RG (1994) Chemical recombination in an expansion tube. AIAA Journal 32(6):1316–1319.

Cambier JL, Tokarcik S, Prabhu DK (1992) Numerical simulations of unsteady flow in a hypersonic shock tunnel facility. In: AIAA 17th Aerospace Ground Testing Conference, Nashville, TN, AIAA Paper 92-4029.

Ikui T, Matsuo K, Nagai M (1969) Investigations of the aerodynamic characteristics of the shock tubes. Bulletin of JSME 12:783–792.

Miller CG, Jones JJ (1975) Incident Shock-Waves Characteristics in Air, Argon, Carbon Dioxide, and Helium in a Shock Tube with Unheated Helium Driver, Technical Report TN D-8099, NASA Langley Research Center.

Roberts GT, Morgan RG, Stalker RJ (1993) Influence of secondary diaphragm on flow quality in expansion tubes. In: Shock Waves @ Marseille, R Brun and L Dumitrescu (eds), Springer.

Satofuka N (1970) A Numerical Study of Shock Formation in Cylinderical and Two Dimensional Shock Tubes, Technical Report 451, ISAS, University of Tokyo.

Vasil'ev EI, Danil'chuk EV (1994) Numerical solution of the flow formation problem for a shock tube with transverse diaphragm removal. Izvestiya AN SSSR, Mekhanika Zhidkosti i Gaza n2 Mar–Apr:147–154.

White DR (1958) Influence of diaphragm opening time on the shock-tube flows. Journal Fluid Mechanics 4:585–599.

Starting-phase numerical simulation of the TCM2 wind-tunnel nozzle

M. Cardoso, Y. Burtschell, M. C. Druguet, D. Zeitoun
Lab. I.U.S.T.I. (UMR 139 CNRS) Dept. MHEQ, Universite de Provence
Marseille, France

R. Abgrall
I.N.R.I.A. Sophia Antipolis, Project SINUS
06902 Sophia BP 93, France

Abstract: A numerical method adapted to the computation of hypersonic flows is presented and applied to the simulation of the transient and steady flows in the TCM2 facility, particularly between the nozzle and the end of the shock tube. Obviously, a complex wave system has been observed in the nozzle part as well as vortices which deform the He/air interface after interaction. Moreover, taking into account the real-gas effects (chemical nonequilibrium, vibrational equilibrium) it has been possible to evaluate the maximum nozzle throat temperature. Lastly, the duration of the starting phase and of the steady one (He contamination) have been determined.

Key words: Shock wave, Chemical nonequilibrium flow, Nozzle starting process, TVD scheme

1. Introduction

The development of a numerical facility to predict unsteady flow in the nozzle of a hypersonic wind tunnel plays a very important role in the design process. Indeed, since 1992, the free-piston shock tunnel TCM2 has existed in Marseille. Until now, this wind tunnel has been tested only as a shock-tube facility (Brun et al. 1992). But to perform experiments in the whole tunnel, it is mandatory to understand correctly the complex phenomena occurring during the nozzle starting phase. In order to improve the use of the TCM2 wind tunnel (especially the use of the nozzle), we have simulated the starting phase and the evolution to the steady phase of the flow inside the shock tube (end) until the nozzle exit. Such numerical simulations allow a better understanding of the aerodynamical processes and the physico-chemical phenomena. More precisely we want to evaluate the starting phase time and then the hot (usable) steady phase. The end of the hot flow is marked by the arrival of the interface (driver-gas contamination). Concerning the physico-chemical phenomena, it is mandatory to evaluate the concentrations of the different species both at the end of the shock tube and about the nozzle throat since the flow temperature depends on them.

And it is important to evaluate the temperature at the nozzle throat and how long it remains high in order to see how the throat should be protected against the high temperature and what kind of materials must be used to overcome the heat.

2. Governing equations

The governing equations are the axisymmetric Euler equations formulated for an air mixture in chemical nonequilibrium with a mass evolution equation for each species. Thus, the conservative vector is:

$$U = (\rho_s(s = 1, \ldots, 5), \rho_{He}, \rho u, \rho v, \rho e) \ . \tag{1}$$

Finally, the vibrational energy is taken at thermal equilibrium.

3. Chemical model

In this five-species air mixture, each species is considered a perfect gas. The chemical source terms are derived from the reactions that occur between these species, given by the following 17-reaction scheme:

$$\begin{aligned} N_2 + M &= 2N + M \\ O_2 + M &= 2O + M \\ NO + M &= N + O + M \\ N_2 + O &= NO + O \\ NO + O &= N + O_2 \end{aligned} \tag{2}$$

with M = N, O, NO, N_2, O_2. The forward and backward reaction rates are calculated with the translational temperature.

4. Numerical algorithm

We use a finite-volume method on a structured grid with an axisymmetric formulation. For the discretization of the Euler equations on each cell we use a MUSCL method where the TVD properties are respected by using the Van-Leer limiter. The flux at each cell interface is approximated by the HLLEMR solver (Flandrin et al. 1994) in order to prevent Roe solver errors (Quirk 1992). To obtain a high precision the chemical source terms are implicited.

5. Mesh, initial and boundary conditions

The mesh is a structured grid of 600×34 points whose geometry corresponds to the TCM2 design (Dumitrescu et al. 1996). In order to get easily the initial conditions of the simulation, we make some assumptions (see Fig. 4, x-t diagram): first, we consider the instantaneous rupture of the diaphragm without incident-shock reflection on it; second, to prevent expansion waves the shock tube is infinitely supplied by helium 35 cm before the end of the shock tube which is filled by air. In the chemical nonequilibrium case, the air is taken at equilibrium. For the boundary conditions, we have used a supersonic inlet to simulate the arrival of the helium, and a reflecting condition for the wall. The creation of two virtual points for all the boundaries allows us to treat every cell interface with second-order accuracy.

6. Results and discussion

6.1. Qualitative and quantitative points of view

After the instantaneous rupture of the diaphragm at the incident shock-wave arrival (Bird 1957), we can observe the creation of an interface and an expansion fan while the shock reflection on the convergent part takes several DMR configurations (Ben-Dor 1992). Then, all the discontinuities seem to follow the schematic evolution of the nozzle flow which has been presented by Tokarcik-Polsky & Cambier (1993) (Fig. 1). Both frozen and chemically nonequilibrium computations seem to have the same behavior for the shock waves, but we can note near the symmetry axis the appearance of a curvature of the primary shock

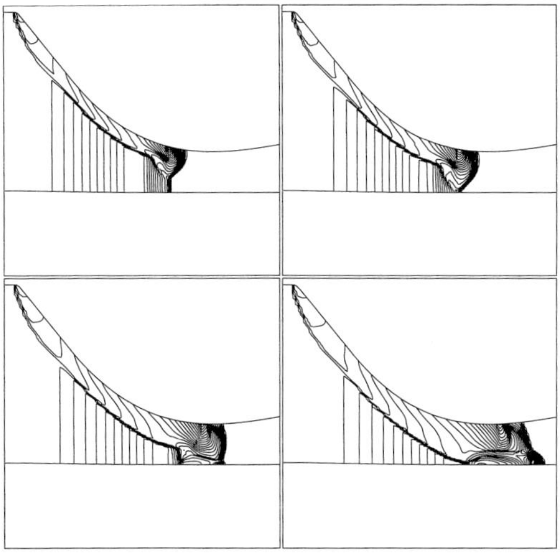

Figure 1. Starting phase at times (in seconds) $5.83e^{-6}$, $6.22e^{-6}$, $7.11e^{-6}$, $8.8e^{-6}$.

in the chemical case that doesn't exist in the frozen case (Fig. 2). Moreover, the gradient due to the relaxation zone behind the primary shock is well visible. Finally, the appearance of a vortex near the axis following the reflection shock

(Tokarcik-Polsky & Cambier 1993) is observed (Fig. 3). It seems that other vortices are also created, in particular at the corner between shock tube and nozzle. We can see the plane reflected shock wave and the interface distorted by the main vortex which seems to take an increasing importance.

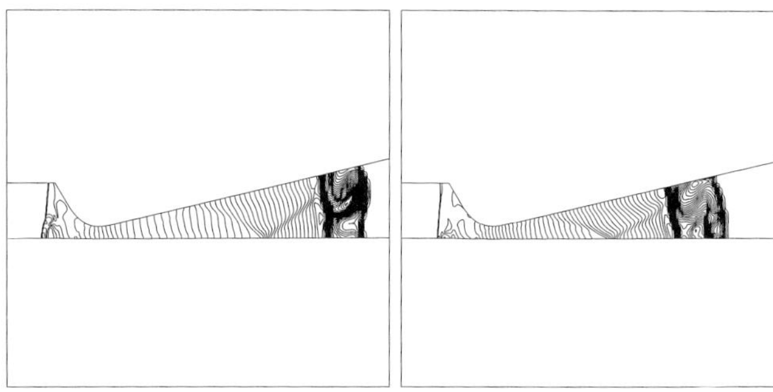

Figure 2. Primary shock comparison between the frozen case (left) and the nonequilibrium case (right).

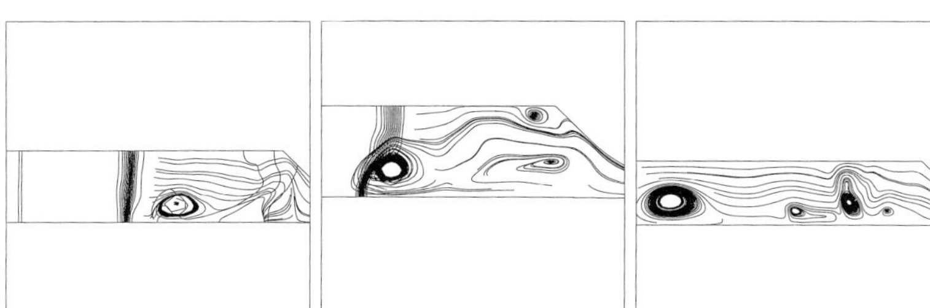

Figure 3. Interaction between the vortex and the contact discontinuity.

6.2. Time history

While taking into account the real-gas effects, we remark that during the starting phase, the peak temperature at the nozzle throat is four times less in the nonequilibrium case (11,000 K) than in the frozen one (44,000 K) as we can see in Fig. 5. In the frozen case, which is the longest computation, we can observe than the steady state is obtained at the throat after 0.25 ms (0.3 ms at the exit station) whilst the helium arrival occurs at 0.85 ms. Thus the steady state lasts about 0.6 ms. Regarding Fig. 6a, we remark that the Mach number becomes equal to unity at the beginning of steady flow. Lastly, on Fig. 6b we can see the time evolution of species concentrations respectively at the throat station and 25 cm after it where the effects of the multiple shocks and expansions are well

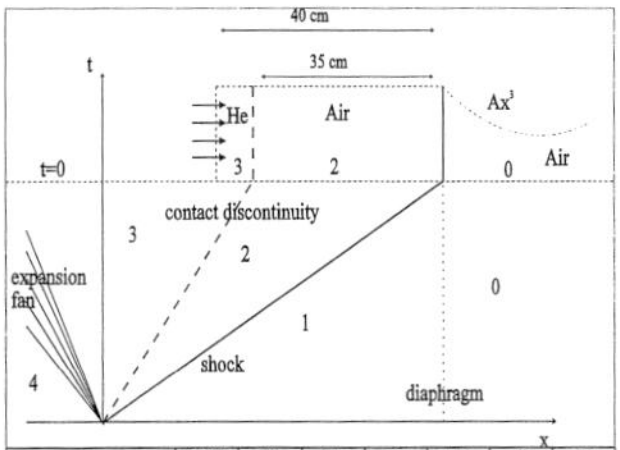

	MACH	P4	T4	P1	T1	P0	T0
FROZEN	9.43	163	1631	0.15	293	0.001	293
CHEMICAL NONEQUIL	9	163	1631	0.15	293	0.001	293

Figure 4. The x-t diagram.

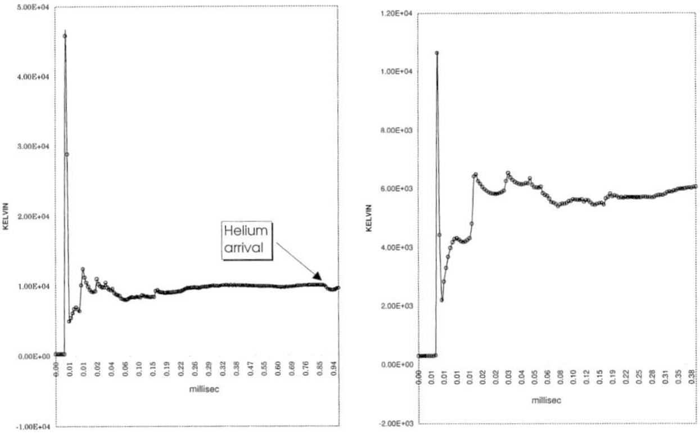

Figure 5. Temperature comparison at nozzle throat between the frozen case (left) and the nonequilibrium case (right).

visible. However, the last picture gives the concentrations at the exit nozzle since the kinetic is then frozen (Fig. 6c).

7. Concluding remarks

This simulation has allowed better understanding of the flow behavior of the TCM2 facility. Indeed, it has been possible to give an order-of-magnitude increase of the steady and unsteady phase duration, and of the temperature supported by the nozzle throat. We have also noted the importance of the vortex evolution and its influence on the contact discontinuity. The latter is a problem of numerical treatment that S. Karni has examined and that R. Abgrall has managed to solve. Finally, as the helium contamination is a real problem, improved treatment of the contact discontinuity and simulation of the boundary layer are required to improve the agreement with reality.

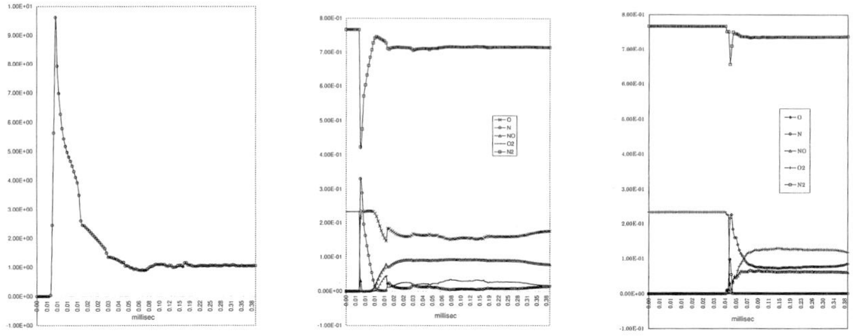

(a) (b) (c)

Figure 6. (a) Time evolution of the throat Mach number; (b) time evolution of the throat species concentrations; and (c) time evolution of the species concentrations 25 cm after the throat.

Acknowledgement. The authors would like to thank C. Sedes for allowing them access to her computer.

References

Abgrall R (1991) Private communication.

Abgrall R (1992) An extension of Roe's upwind scheme to algebraic equilibrium real gas models. Computer and Fluids 19(2):171–182.

Ben-Dor G (1992) Shock Wave Reflection Phenomena, Springer.

Bird GA (1957) A Note on Multiple Diaphragm Shock Tube, British Crown Copyright, Britannic Majesty's, London.

Brun R, Burtschell Y, Dumitrescu L, Dumitrescu M, Houas L, Labracherie L, Zeitoun D (1992) Computation and experimentation of free piston shock tunnels: example of the Marseille facility. Prepared for: NATO Advanced Research Workshop, ONERA, France.

Dumitrescu LZ, Burtschell Y, Zeitoun D (1996) Throat design for hypersonic nozzles. In: Shock Waves, Proc of the 20th ISSW, Sturtevant B, Shepherd JE, Hornung HG (eds), World Scientific, Vol 2, pp 1569–1574.

Flandrin L, Charrier P, Dubroca B (1994) A robust finite volume method for computations on two-dimensional unstructured hybrid meshes. In: Computational Fluid Dynamics, ECCOMAS.

Karni S (1994) Multicomponent flow calculations by a consistent primitive algorithm. J Computational Physics 112:31–43.

Quirk JJ (1992) A Contribution to the Great Riemann Solver Debate, NASA Contractor Report 191409, ICASE Report No. 92-64.

Tokarcik-Polsky S, Cambier J-L (1993) Numerical Study of the Transient Flow in the Nozzle Section of a Shock Tunnel, AIAA Paper 93-2018.

Combustion driver with interface compression and computer code for hypersonic shock tunnel to produce stagnation temperatures to 20,000 K and Mach numbers of 35

H. T. Nagamatsu, D. G. Messitt, L. N. Myrabo
Department of Mechanical and Aeronautical Engineering
Rensselaer Polytechnic Institute
Troy, NY 12180-3590 USA

M. A. S. Minucci
Instituto de Estudos Avançados, CTA
São José dos Campos, SP, Brazil

Abstract: Designers of hypersonic air-breathing vehicles will rely excessively on expensive flight experiments because the ability to conduct ground tests at high flight Mach number and enthalpy is limited. An increase in the capability of existing ground test facilities is needed to reduce the number of required flight tests. Hypersonic shock tunnels can achieve higher reservoir enthalpy by operating in the equilibrium interface mode. By loading a heavy gas, such as SF_6 or argon, into a small buffer region between the shock tube driver and driven sections, the contact-surface mixing is decreased. This results in a longer, more uniform test time and a quicker rise to the equilibrium condition. Computer predictions of a Combustion Driven Hypersonic Shock Tunnel, operating in the equilibrium interface mode, indicate that maximum reservoir temperatures of 20,000 K can be reached with a corresponding (equilibrium) freestream flight Mach number of 35.

Key words: Combustion, Shock tube, Shock tunnel

1. Introduction

High-enthalpy shock tunnels operate by generating a reservoir for a hypersonic nozzle by heating and compressing the test gas by a strong incident shock wave. The high pressure and temperature in the driver may be generated by compression with a piston, or by combustion of stoichiometric hydrogen and oxygen with a mixture of helium. A combustion-driven hypersonic shock tunnel was operated at the General Electric Research and Development Center (cf. Nagamatsu et al. 1959, Nagamatsu and Sheer 1985). This shock tunnel has been donated to RPI and is currently in storage.

The Combustion Driven Hypersonic Shock Tunnel (CDHST) has a 20-foot long driver designed for a maximum pressure of 100,000 psi. The facility was operated in the reflected mode to obtain hypersonic gasdynamic data for the development of ICBM nose cones ($M_\infty = 20$ and $T_0 = 6,000$ K) (cf. Nagamatsu et al. 1959). Magnetohydrodynamic experiments were conducted at the end of the 100-ft long driven tube with an incident-shock Mach number of 30 and temperature of 10,000 K where the air was completely dissociated and ionized.

The test envelope of shock tunnels, such as the CDHST, may be expanded by operating in the equilibrium-interface mode (cf. Copper 1962). The equilibrium-interface technique uses successive shock-wave/contact-surface interactions to further compress and heat the test gas. This technique is currently used by researchers at RPI, Boeing, and NASA (cf. Minucci et al. 1993a) to extend the high-enthalpy capabilities of their facilities.

The test time of a shock tunnel can be limited by contamination of the test gas by the driver gas, which typically has a lower molecular weight and higher thermal velocity. This contamination is caused by diffusion through the contact surface, where there is a large thermal gradient. When operating in the equilibrium-interface mode, additional contamination can be caused by the interaction of the reflected shock waves with the shock-tube boundary layer. A new technique has been shown to reduce the spreading of the contact surface and enhance compression of the test gas. This technique involves using a small buffer region between the driver and driven sections containing a gas heavier than the driver gas, such as argon, SF_6, or xenon (cf. Minucci et al. 1993b).

1.1. Equilibrium interface with buffer gas

The equilibrium-interface method is used in the room-temperature helium-driven RPI Hypersonic Shock Tunnel (HST) to achieve high-enthalpy reservoir conditions of 840 psia and 4,100 K with the driver loaded to 2,000 psi. The RPI HST uses a double diaphragm section to initiate the diaphragm bursting. The quality of the reservoir conditions and the duration of the test time when operating in the equilibrium interface mode have been observed to depend upon the gas with which this chamber is filled (cf. Minucci et al. 1993a). Helium performs the worst, with heavier buffer gases generating a more rapid rise and a longer duration of the measured equilibrium pressure. The heavier gases also generate a slightly higher pressure level (Minucci et al. 1993b). Measurements of the heat-transfer rate on the shock-tube wall for different buffer gases indicate that the contact-surface spreading is reduced.

2. Combustion-driven shock tunnel

The General Electric CDHST (cf. Nagamatsu et al. 1959), has been donated to RPI and is currently in storage. The CDHST has a 6-inch-diameter driver section, which is loaded with stoichiometric amounts of hydrogen and oxygen, mixed with helium. The gases are injected at several locations through orifices angled 45° to produce a swirling motion, enhancing mixing of the reactants. The mixture is ignited by a simultaneous discharge across 18 modified spark plugs. The spark plugs are mounted flush with the inner surface of the driver tube, and are arranged in a spiral pattern. Extremely reliable and reproducible driver-combustion results have been obtained with this method (cf. Nagamatsu et al. 1959).

The driver section was tested prior to initial operation in the CDHST to characterize the combustion process (Nagamatsu and Martin 1959). Combustion results were obtained for different mixtures of varying percent helium at total initial pressures ranging from 25 to 500 psia. The measured ratio of peak to initial pressure in the driver is presented in Fig. 1 for different driver helium content. Also included in Fig. 1 are the calculated post-combustion pressure and temperature (cf. Lewis and von Elbe 1961). This calculation assumed complete combustion and no dissociation of the combustion products. Detonation was experimentally observed in the driver tube for the higher-pressure mixtures containing 65% helium, with the peak pressure up to 50 times the initial pressure.

The driver section is separated from the driven section by a single stainless steel diaphragm located at the beginning of the conical transition from the driver diameter of 6 inches to the 4-inch diameter of the driven tube. This section will be modified to take advantage of the buffer gas when operating in the equilibrium interface regime. A double diaphragm section will replace the single diaphragm upstream of the transition section. The double diaphragm section can be loaded with the buffer gas (argon or xenon) at high pressure prior to loading the driver. The test will be initiated when the combustion in the driver causes the pressure to increase beyond the bursting pressure of the first diaphragm. This represents an improvement over the buffer gas technique as currently used in the RPI helium-driven shock tunnel, where an unknown amount of the buffer gas is vented in order to initiate diaphragm bursting (cf. Minucci 1991).

2.1. Analysis of equilibrium interface for the combustion driver

The equilibrium-interface technique can be used to raise the reservoir pressure and temperature by operating the shock tunnel at an over-tailored condition. The successive reflections of the shock wave from the contact surface and the shock-tube end wall heat and compress the reservoir gas. The shock-

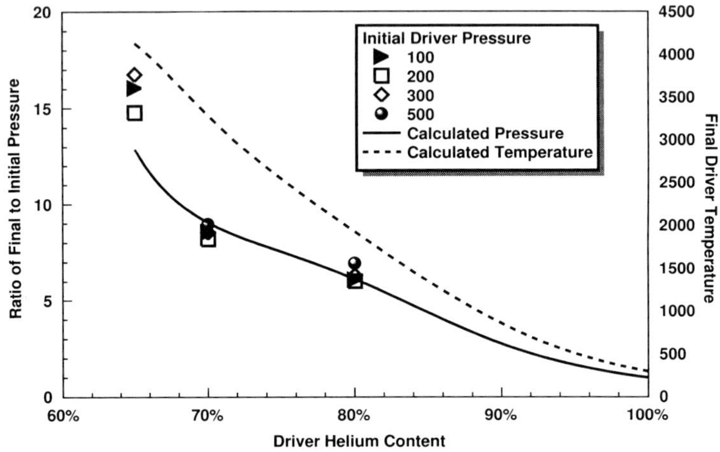

Figure 1. Measured and calculated ratio of post-combustion pressure to initial pressure, and calculated temperature.

wave/contact-surface interactions were modeled for equilibrium air by calculating discrete jumps due to each reflected shock wave (Minucci et al. 1993a). These computations allow the equilibrium pressure and temperature to be predicted given the incident shock-wave Mach number. Figure 2 presents the equilibrium pressures and temperatures for a combustion driver with various post-combustion temperatures. These calculations are for the CDHST with 30,000 psi of helium driving equilibrium air. Similar curves may be generated for other driver pressures. These curves can be used in conjunction with Fig. 1 and the shock tunnel P_4/P_1 relation (Minucci et al. 1993a) to select driver initial conditions for a specified reservoir condition.

The calculations using this method were compared with actual equilibrium pressures measured in the RPI HST. The computed pressures were appreciably less than the measured pressures. The calculated pressure and temperature in the reservoir increased by discrete jumps as each reflected shock wave passed. However, the experimentally observed pressure in the reservoir increased much more smoothly, indicating the flow is better represented by a series of compression waves instead of shock waves. Because of viscous effects and contact-surface spreading in the actual shock tube, the shock-wave/contact-surface interaction results in a gradual reflection of the shock wave. The discrete wave-tracking computer code is unable to correctly simulate this interaction.

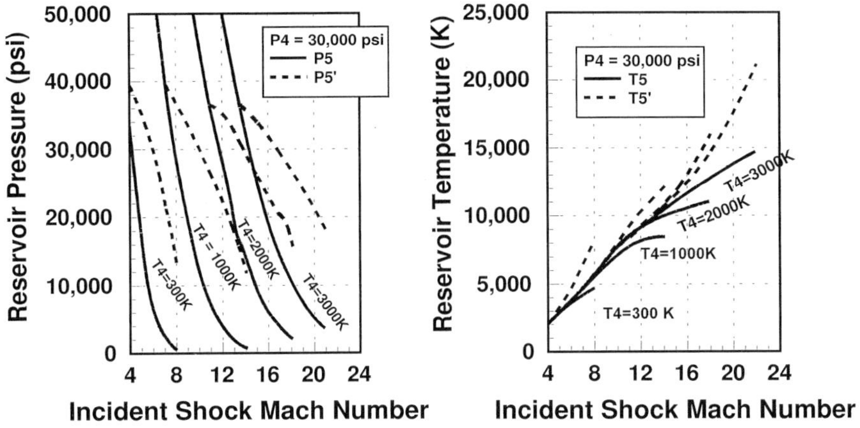

Figure 2. Variation of the equilibrium pressure and temperature with the the incident shock-wave Mach number for various driver temperatures.

Due to this deficiency, another computer code has been developed to simulate the shock-tube flow. This code solves the Euler equations for the entire shock tube, and is not restricted to discrete shock-wave interactions and sharp contact discontinuities. The computer program implements a high-order accurate Roe solver (Roe 1986) modified for quasi-one-dimensional flow of equilibrium air with additional source terms to correct for viscous, heat-transfer, and diaphragm loss effects. In addition, the artificial compression method, (cf. Harten 1978), allows for control over the sharpness of the captured contact surface.

The nature of the multiple corrections makes it necessary to perform an extensive calibration of the code to get useful data. This process is continuing, with detailed comparisons to data on shock-wave speed and position, and pressure and temperature measurements from the CDHST, the RPI HST, and a shock tunnel being operated in Brazil (Minucci et al. 1993b). The code allows construction of wave diagrams and analysis of test conditions.

3. Conclusion

The test envelope of hypersonic shock tunnels, such as the CDHST, may be extended by operating in the equilibrium-interface mode. A disadvantage of this technique is contamination of the test gas by mixing at the contact interface due to the spreading of the interface and the subsequent interactions with the reflected shock waves. A small buffer-gas region located between the driver and driven sections of the shock tube has been demonstrated to reduce the interface spreading, resulting in longer test times and a more uniform equilibrium region. Computer simulations of the CDHST have shown that maximum reservoir temperatures of 20,000 K can be achieved.

Acknowledgement. The authors would like to thank M. Nascimento and Dr. C. Andrade for their contributions to the experiments and analysis, respectively.

References

Copper JA (1962) Experimental investigation of the equilibrium interface technique. Phys of Fluids 5:844–849.

Harten A (1978) The artificial compression method for computation of shocks and contact discontinuities: III. Self-adjusting hybrid schemes. Mathematics of Computation 32:363–389.

Lewis B, von Elbe G (1961) Combustion, Flames and Explosions of Gases, 2nd Edition, Academic Press, New York.

Minucci MAS (1991) An Experimental Investigation of a 2-D Scramjet Inlet at Flow Mach Numbers of 8 to 25 and Stagnation Temperatures of 800 to 4,100 K, Ph.D. Dissertation, Rensselaer Polytechnic Institute, Troy, NY.

Minucci MAS, Nagamatsu HT (1993a) Hypersonic shock-tunnel testing at an equilibrium interface condition of 4100 K. J Thermophysics and Heat Transfer 7:251–260.

Minucci MAS, Nascimento MAC, Nagamatsu HT (1993b) Improving the Performance of the Equilibrium Interface Technique in Shock Tube Flows, AIAA Paper 93-5007.

Nagamatsu HT, Geiger RE, Sheer RE Jr (1959) Hypersonic shock tunnel. ARS J 29:332–340.

Nagamatsu HT, Martin ED (1959) Combustion investigation in the hypersonic shock tunnel driver section. J Applied Physics 30:1018–1021.

Nagamatsu HT, Sheer RE Jr (1985) Hypersonic Gas Dynamics, AIAA Paper 85-0999.

Roe PL (1986) Characteristic-based schemes for the Euler equations. Annual Review of Fluid Mechanics 18:337–365.

A simple shock-tunnel driver-gas detector

A. Paull
Department of Mechanical Engineering
The University of Queensland, 4072 Australia

Abstract: A device has been developed to detect the arrival of the driver gas in a shock tunnel. The detector is small enough to be used in conjunction with other experiments. It works by choking a duct when the specific-heat ratio is increased past a critical value. Times are given for the onset of a 7.5-percent contamination level in flows with freestream enthalpies ranging from 3–9 MJ/kg. These results are compared with and are shown to be in agreement with measurements made with a mass spectrometer. Results displaying the rate at which the test gas is contaminated are also given.

Key words: Driver gas, Detection

1. Introduction

The time available for testing in reflected shock tunnels is limited by the time at which the test gas becomes contaminated by the driver gas. Prior to the results of Skinner (1994), who used a mass spectrometer to determine the time of arrival of helium in the test gas, it had been assumed that the interface between the driver and test gases is one which is sharp. However, this was found to be incorrect. It was found that no definite interface exists, but rather, after a period during which no helium is detected, steadily increasing concentrations of helium are observed. There are existing estimates of the times during which there are acceptable contamination levels (the test time), however, Skinner (1994) indicates that in some instances these estimates may be inaccurate. Broadly, these estimates over-estimated the test time at higher enthalpies (greater than 12 MJ/kg), whereas at lower enthalpies the estimates are, in general, correct or an underestimate of the available test time. To further complicate the issue, Skinner (1994) has also observed that there is significant variation between test times at the same conditions but during different tests.

It is clear from these observations that if the maximum test time is to be sought for each test, a driver-gas detector which can be used routinely in conjunction with experiments and which can quantify the available test time is required. Although one of the many virtues of a mass spectrometer is its ability to accurately detect the presence of helium in an air test gas, the mass spectrometer used by Skinner (1994) is large, filling most of the test section. Hence, it is not possible to incorporate it with experiments which are not specifically designed to use it. This has lead to the development of a much smaller device specifically designed to detect the driver gas.

The device described here uses the change in specific-heat ratio to detect the presence of the driver gas and will work only if the driver gas has a specific-heat ratio different from that of the test gas. In general, this is the case, as reflected shock tunnels use a helium-argon mixture for the driver gas and a predominantly diatomic gas such as air for the test gas. Hence, the driver-gas specific heat is greater than that of the test gas.

2. Driver-gas detector — description

The driver-gas detector which is described here works by exploiting the difference in shock angles produced by a wedge in supersonic flows of gases with different specific-heat ratios. It can be shown, for a thermally perfect gas, that if the Mach number and wedge angle remain unchanged, the shock angle will increase as the specific-heat ratio of the gas increases. Furthermore, the maximum angle of the wedge for which the shock is attached to the leading edge of the wedge decreases as the specific-heat ratio increases.

It should be understood that these remarks are only true if the Mach number of the flow remains unchanged. If the Mach number increases and the specific-heat ratio is unchanged, the shock angle will decrease. Furthermore, the maximum wedge angle for which the shock is attached will increase.

This is an important point to consider, because in reflected shock tunnels, the test flow is produced by expanding the gas from a stagnated region through a nozzle. Hence, as the specific-heat ratio of the gas increases then so does the Mach number of the flow exiting the nozzle. Therefore, the increase in shock angle on a wedge expected from an increase in specific-heat ratio is offset to some degree by the increase in Mach number of the flow. The overall effect still results in an increase in shock angle and a decrease in the maximum wedge angle as the specific-heat ratio increases; however, this change is significantly smaller than that which would be expected had the change in Mach number not been considered.

Figure 1 displays the change in the maximum wedge angle for which the shock is attached as the contamination level is increased. The gas is assumed to be thermally perfect. The contamination level is expressed as a volumetric percentage. The corresponding uncontaminated-air total enthalpy is 3.2 MJ/kg. Figure 1 also displays the Mach number of the flow as a function of the contamination level for the nozzle under consideration. It can be seen that as the contamination level increases (so that the specific-heat ratio increases) the Mach number increases and the maximum wedge angle to which the shock is attached decreases. Thus, for example, if a wedge of 46 degrees was used it would have an attached shock in the uncontaminated flow which would remain attached until the contamination level reached approximately 2.5 percent. However, above this value the shock will be detached.

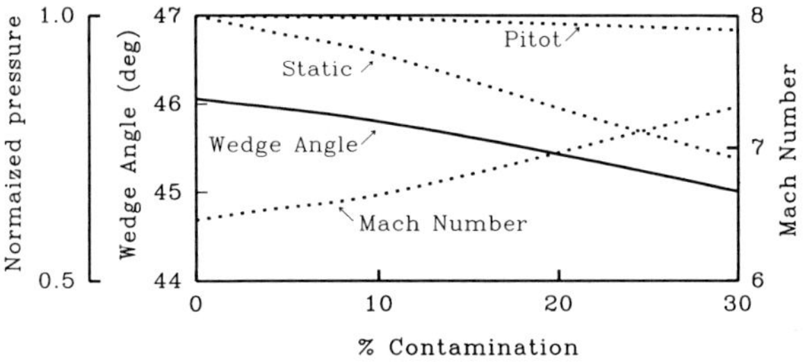

Figure 1. Pitot and static pressures, freestream Mach number, and maximum wedge angle to which a shock is attached as a function of percent contamination by volume. Pressures normalized by uncontaminated levels.

In the device described here the detachment of the shock is used to choke a duct. Figure 2 is a scaled drawing of the driver-gas detector used here. It can be seen that the detector consists of a square duct (15 × 15 mm) with a wedge at the exit of the duct and a pressure transducer mounted in the wall of the duct. The inclination of the wedge is adjustable. The angle of the wedge is adjusted so that (i) uncontaminated test gas produces a shock attached to the wedge which propagates downstream out of the duct without striking the opposite wall and (ii) when a predetermined contamination level is reached the shock becomes detached and moves forward allowing the shock to reflect off the opposite wall. This then produces a flow which is choked in the duct. The subsequent pressure rise created by the choked flow in the duct is detected by the pressure transducer.

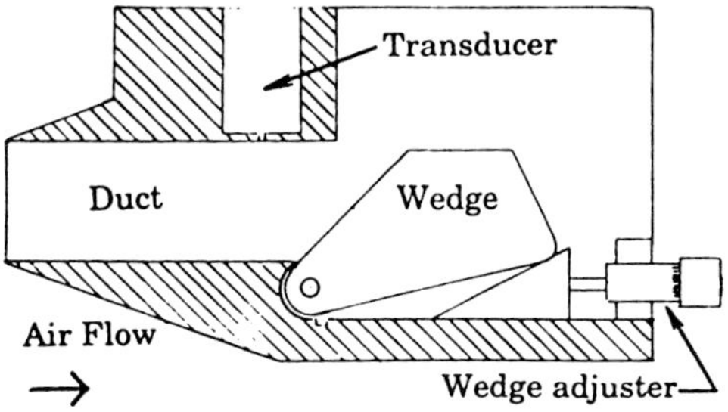

Figure 2. Schematic of driver-gas detector. Scale is 1:1:2.

The level of contamination which is detected is determined by the angle of the wedge. As the angle of the wedge is increased the duct will choke at a lower level of contamination. It can be seen from Fig. 1 that to obtain an accurate measure of the contamination level the wedge angle has to have a reasonably fine adjustment. As can be seen in Fig. 2, the wedge angle was adjusted using a vernier thread and could be adjusted to within 0.01 degrees, which is approximately an order of magnitude better than required.

A variation of this technique was used by Paull and King (1995). A much smaller angled wedge (10 degrees) was used so that the shock was always attached to the nose. However, the wedge was sufficiently long so that multiple reflections of the shock off the wedge and opposite wall would occur. As the specific-heat ratio increases, the shock angle increases, therefore more reflections are possible. Furthermore, if there are a sufficient number of reflections, choking will occur. The location of the wedge with respect to the exit of the duct was adjusted so that at a prescribed level of contamination there was a sufficient number of shock reflections to choke the device.

3. Results

The tests were done in the free-piston shock tunnel T4. A range of test conditions were targeted to determine the available test time. To directly compare with the results obtained by Skinner (1994), the angle of the wedge was set so that if the level of contamination exceeded 7.5 percent by volume, the flow in the duct would choke.

This angle was determined by deliberately seeding the air test gas with 5 percent and 10 percent by volume of driver gas. It is argued that immediately after the start of flow the test gas has not been contaminated and therefore the test gas has the level of driver gas in it equal to the level of deliberately introduced seeding. Hence, if the test gas has been seeded with 5 percent of driver gas and it does not choke immediately after the start flow, and yet when seeded with 10 percent of the driver gas it does choke immediately, the the detector has been set to detect levels of contamination between 5 percent and 10 percent. It will be said that this detection level is 7.5 percent.

As the Mach number of the flow changes with total enthalpy, and as real-gas effects will also influence the calibration as the flow enthalpy is increased, it is clear that the calibration process is required for each enthalpy tested and for each level of detection. This has been done in the results presented below.

Figure 3 displays the time histories of the pressure inside the duct for the three different tests when the air test gas was (i) unseeded, (ii) 5-percent seeded, and (iii) 10-percent seeded with driver gas. In addition, the pitot pressure measured outside the duct is displayed. After the flow chokes it can be seen that the pressure in the duct is approximately equal to that of the pitot pressure,

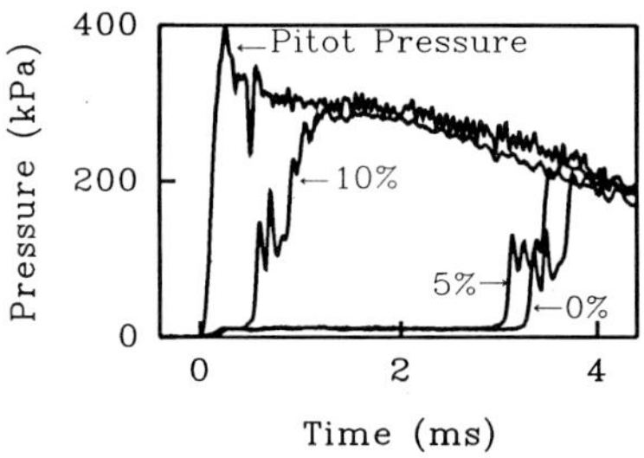

Figure 3. Time histories of pressure measurements in the duct.

as it should be. It can also be seen that the process to reach fully choked flow in the duct takes approximately 300 ms. It will be assumed that choking is initiated when the first shock is observed, that is, when the pressure first rises to approximately half the pitot pressure. The time at which choking is initiated is more important than the time at which a fully choked flow is achieved, as the initiation time more accurately indicates the time at which the specific-heat ratio increased sufficiently to detach the shock from the wedge. Although fully choked flow is not achieved at this time, for brevity, this will be referred to as the time of choking.

With 10-percent seeding of the test gas, the flow choked at approximately 0.4 ms, that is, immediately after the starting flow. However, with 5-percent seeding, the flow chokes at approximately 3 ms. At this much later time the level of contamination has increased so that the level of seeding, in addition to the level of contamination, is sufficient to choke the duct. From these results it is concluded that the detector has been set to choke at a contamination level between 5 percent and 10 percent, that is nominally 7.5 percent. Hence, it can be concluded from the unseeded case that 7.5-percent contamination occurs 3.2 ms after the arrival of the starting flow. Furthermore, from the 5-percent seeded case, it can be deduced that the level of contamination at 3 ms after the arrival of the starting flow is 2.5 percent.

By setting the detector to detect different contamination levels, a time history of the contamination level can be obtained through the compilation of a number of tests. These results are displayed in Fig. 4. Results from Skinner's (1994) mass-spectrometer tests are also displayed. It can be seen that reasonable agreement occurs, although there is significantly more scatter in the mass-spectrometer results.

Finally, the time after arrival of the starting flow for 7.5-percent contamination to occur at different total enthalpies is displayed in Fig. 5. In addition, the results from Skinner (1994) and Paull and King (1995) are displayed for

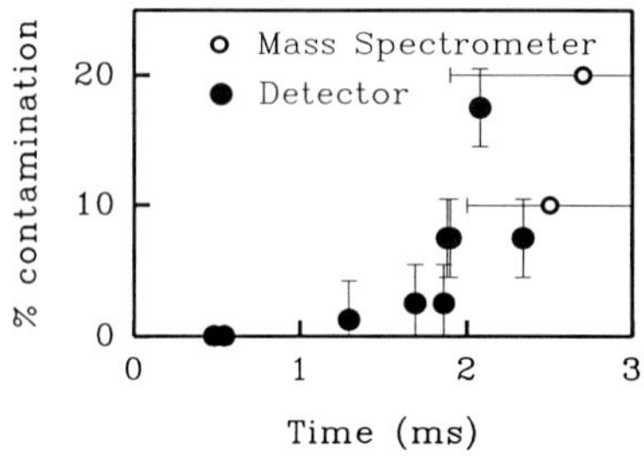

Figure 4. Contamination levels as a function of time. Total enthalpy = 8.6 MJ/kg.

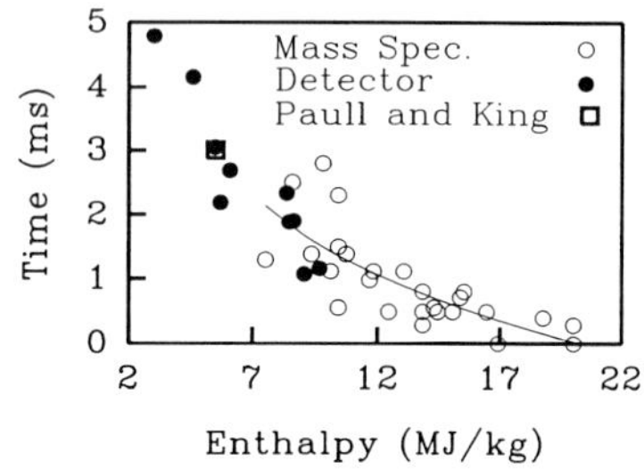

Figure 5. Time after arrival of the starting flow for $7.5 \pm 2.5\%$ contamination as a function of total enthalpy.

respectfully, 10- and 13.75-percent contamination levels. The line is a best fit to the mass-spectrometer results. It can be seen that in the region of overlap, the mass-spectrometer results and the results from the device described here are in good agreement. Furthermore, the trends in contamination time outside the region of overlap from both devices are also in good agreement.

These results show that as the total enthalpy increases from 3.2 MJ/kg to 9.5 MJ/kg, the time after arrival of the start flow to a 7.5-percent contamination level decreases from approximately 5 ms to 1 ms. From the results of the 10-percent seeded tests, the starting flow constitutes approximately 0.3–0.5 ms of these times, so that the test times range between approximately 4.5 ms at 3.2 MJ/kg to 0.5 ms at 9.5 MJ/kg.

4. Conclusions

A device that is small enough to be used in conjunction with many experiments and which will detect a predetermined level of driver-gas contamination has been developed.

Acknowledgement. This work was completed under grants from the Australian Research Council, the Australian Space Office, and NASA Langley Research Center.

References

Paull A, King MD (1995) A driver gas detection device for shock tunnels. Shock Waves 4(5):289–291.

Skinner KA (1994) Mass Spectrometry in Shock Tunnel Experiments of Hypersonic Combustion, Ph.D. Thesis, University of Queensland, Australia.

Boundary-layer blockage in expansion-tube nozzles

O. Sudnitsin, R. G. Morgan
University of Queensland
Brisbane, QLD, 4072 Australia

Abstract: The results of a first-order perfect-gas correction for the effects of the boundary-layer formation within expansion tubes with nozzles are presented. The analytical model developed to describe the boundary-layer formation within the expansion tube and an expansion nozzle located at the end of the acceleration tube is based on the Kármán integral equations. The results of this analytical model are compared with experimental data from an expansion diffuser. The model provides a useful tool for the preliminary design of nozzles for such facilities.

Key words: Boundary-layer formation, Boundary-layer blockage, Nozzle flow, Expansion tubes, Hypersonic flow, Kármán integral equations

1. Introduction

Scale modeling of hypersonic flows cannot be achieved with complete matching of all non-dimensional scaling parameters. Laboratory testing generally involves only partial similarity and can be justified if the phenomenon of interest is controlled primarily by matchable parameters. For example, the binary scaling parameter allows accurate modeling of binary finite-rate dissociation processes, simultaneously reproducing viscous effects.

However, many processes of interest in hypersonic flow do not follow binary scaling, and exact simulation requires full-size models. Combustion, recombination, and gas radiation are examples of such processes. The size of the test section, therefore, limits the size of flight vehicle which can be tested in this way. Expansion of laboratory test flows enables larger models to be tested, but the associated drop in pressure limits the range of flight conditions which can be reproduced. Because of their high total-pressure simulation capability, expansion tubes can potentially provide improved performance over other existing facilities.

In the superorbital expansion tube, it is of interest to model rarefied flow phenomena, for which it is necessary to reduce gas density by means of a nozzle. The limited test size of an expansion tube may be increased by means of an expansion nozzle located at an appropriate section of the tube (Sudnitsin and Morgan 1994). The starting process associated with nozzle flows reduces the steady test time available (Leyva 1994), and may provide an additional limitation on model size.

A nozzle was tested successfully by Miller and Jones (1983) on an expansion tube. However, for their purpose, the unmodified expansion tube was found to be better and the use of the nozzle was discontinued.

Despite the reduction of the binary scaling parameter associated with the use of nozzles, direct simulation over a useful range of flight conditions may still be obtained.

However, certain problems need to be addressed if such nozzles are to be used. Firstly, reservoir pressure must be sufficient to reproduce real-flight conditions. A simple ideal-gas analysis (Sudnitsin and Morgan 1994) provides a quick assessment of operational conditions in terms of the important non-dimensional parameters. Using this approach, the characteristics of simulated flow in terms of total pressure, driver-noise attenuation (Paull and Stalker 1992) and test-gas temperature may be found for three different configurations of an expansion tube with the divergent nozzle placed at A — the end of the driver section, B — the end of the shock tube, and C — the end of the acceleration tube. Configurations and conditions which may give superior performance and for which further investigation is justified can thus be easily identified.

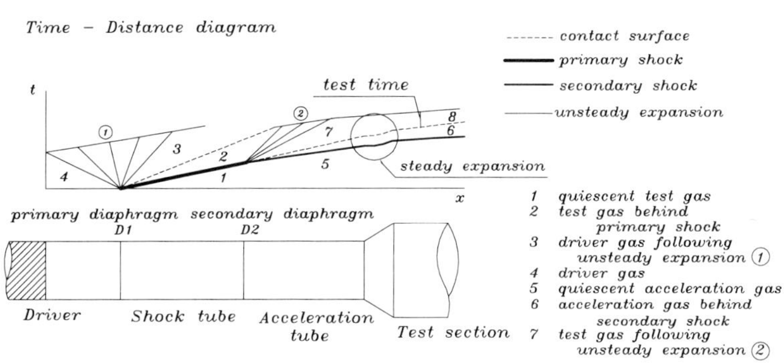

Figure 1. Configuration C.

According to the results of the performance comparison in terms of nose-to-tail pressure ratio, configuration C has been chosen for initial investigation (Fig. 1). This configuration is advantageous because it is relatively simple to modify the existing facility and it can be shown that its performance at some operating conditions is comparable to configurations A and B (Sudnitsin and Morgan 1994).

One aspect of nozzle design concerns the influence of boundary layer on the effective test-flow area achieved in the core flow. This effect is small for low-Mach-number flows, and often not corrected for in the design stage because direct calibration under operating conditions can be used to determine precisely the expanded-flow parameters. However, in superorbital expansion flows, significant boundary-layer blockage may arise, and the nozzle geometric area ratio may be quite different to that seen by the core flow. Consequently, an analysis is presented which couples the boundary-layer displacement thickness to the expansion process, giving an improved indication of the state of the core flow.

The present analysis has been done for a small-scale expansion tube at the University of Queensland (X1), and the nozzle is currently under construction. Experimental data for a diffuser on an expansion tube obtained by General Applied Science Laboratories, Inc. (GASL) (Bakos et al. 1992, Bakos 1994) were used to validate the analytical results prior to designing the nozzle for X1 at the University of Queensland.

2. Boundary-layer analysis

The present analysis investigates the growth of the boundary layer on the walls of the expansion tube which will reduce the size of the available test core from A to A' (Fig. 3) and change the area ratio of the nozzle, resulting in a decrease in the pressure ratio associated with it. Flow is assumed to be compressible, and the formation process has been divided into two stages, formation within the tube and growth within the divergent nozzle.

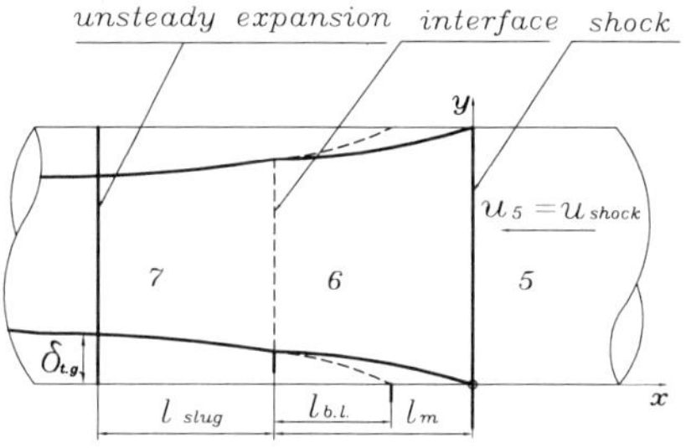

Figure 2. Boundary-layer formation in the acceleration tube.

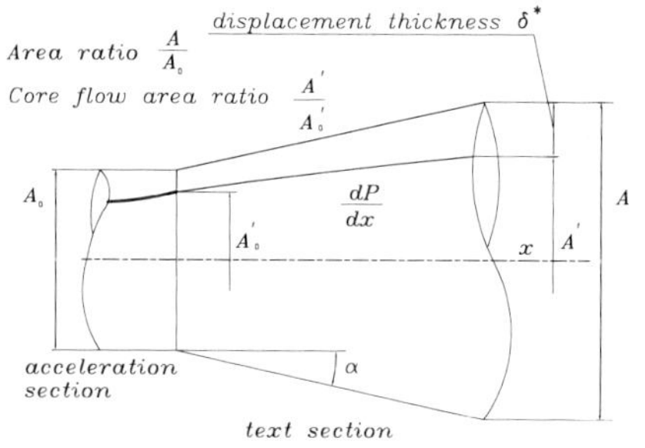

Figure 3. Boundary-layer formation in the nozzle.

2.1. Boundary-layer formation within the acceleration tube

A boundary layer grows between the head of the rarefaction wave and the shock. Figure 2 schematically demonstrates the different stages of the boundary-layer formation and flow for each region.

Boundary layers within the shock tube act as an aerodynamic sink for the acceleration gas in region 6 between the shock and interface (Fig. 2). Test gas is also lost by this process in region 7, behind the interface. The boundary layer at the exit plane of the acceleration tube grows with time. The worst boundary-layer thickness occurs with the arrival of the secondary unsteady expansion head (Fig. 2). Useful gas flow in this region is completed at this time.

For the present calculations a worst-case approach was adopted, assuming the origin of the test-gas boundary layer coincided with the shock location. Therefore, region 6 (Fig. 3) is considered infinitesimally small and $l_{b.l.} = l_m$, which is a good approximation at high shock speeds. Thus, the test-gas boundary layer develops under the influence of conditions behind the interface (region 7 in Fig. 3).

Table 1. Sample predicted condition for X1, targeting supersonic combustion.

Regions Fig. 1	Pressure, Pa	Temperature, K	Velocity, m/s	Speed of sound, m/s
4	5E+07	2000	0	1827[a]
7	6E+5	3082	3727	1112[b]
8	25E+3	1250	4200	709

Primary shock speed $U_{sh1} = 3000$ m/s; secondary shock speed $U_{sh2} = 4500$ m/s.
Speed-of-sound ratio across the driver gas-test gas interface $a_3/a_2 = 0.7$.
[a] Driver gas is a mixture of He and Ar with $\gamma = 1.67$ and $\mathcal{R} = 1000$ J/(kg K).
[b] Test and acceleration gas are air with $\gamma = 1.4$ and $\mathcal{R} = 287$ J/(kg K).

In the sample calculation for X1 (Table 1), it was assumed that the boundary layer behind the interface in the acceleration tube is turbulent. The characteristic thicknesses δ — the boundary layer thickness, δ^* — the displacement thickness, and θ — the momentum thickness, which define integral deficits, were calculated from Hayes and Probstein (1959) and act as starting conditions for the nozzle boundary-layer calculation.

2.2. Boundary-layer formation within the expansion nozzle

The modeling of the nozzle boundary layer assumed that the surface can be represented by a flat plate with a favorable pressure gradient, that the flow is compressible, and that the Kármán-based method of Walz (1969) may be used to quantify the development of δ^* and θ with x.

The approach that was used allows the bulk boundary-layer properties to be calculated without solving for the internal profiles. These properties can then be used in a one-dimensional approach to calculate nozzle-exit conditions.

The physical information required to complete the calculations was incorporated into the differential equation for the velocity, density, and area-ratio variations along the nozzle. After the initial values of the variable parameters (Walz 1969) were estimated, final bulk properties of the boundary layer were obtained by solving the system of differential equations. A conical nozzle with an area ratio of 9 was used to generate the axial-pressure distribution that was used for the computations. When the effect of the boundary-layer displacement thickness was added to the contour, the geometric area ratio that was needed to expand the core flow to the correct pressure was found to be 20. This illustrates the importance of boundary-layer blockage for these flow conditions. The increase of displacement thickness is due to the entrainment of new fluid in the nozzle and also to the expansion of the boundary-layer gas. In this example, 66% of the downstream displacement thickness is due to entrainment in the nozzle, illustrating the importance of minimizing nozzle length. The experimental validation of this result is yet to be made.

3. Comparison with experimental data

In order to validate the analysis above, it was applied to data from a diffuser placed at the end of the expansion tube at GASL (Bakos et al. 1992). The 0.85-m long diffuser with the initial conditions M17 (Bakos 1994) produced a boundary layer with a thickness $\delta_{exit} \approx 6.25$ mm (from an exit Mach number profile from Bakos et al. 1992). The nozzle-starting condition, M17, and the history of the boundary layer were deduced from Bakos at al. (1992) where the boundary-layer thickness was measured to be $\delta_0 \approx 25$ mm at the acceleration-tube exit.

The diffuser was designed to produce an increase in static pressure of 11.9 (perfect-gas approximation) relative to the incoming flow. After correction of the contour to account for the displacement thickness, the pressure increase that is expected is 8. The actual experimental pressure ratio (Table 2) (Bakos et al. 1992) was measured to be 8.8. Therefore it can be seen that the inviscid calculation of a diffuser contour somewhat overestimates the pressure ratio, whereas the boundary-layer-corrected contour can predict the pressure ratio much more realistically. The fact that the viscously corrected and experimental pressure ratios are not exactly the same may be partly due to the fact that the boundary-layer thickness is not precisely known from the data that is available. Further validation will be provided by the planned experiments.

4. Conclusions

The present work indicates the importance of viscous effects in expansion tubes with nozzles. It demonstrates the significant effect of the displacement-thickness correction on the contour and the predicted pressure ratio. As the velocities and Mach numbers of the phenomena of interest increase, it will be important to correct for viscous effects from the early stages of nozzle design, rather than by calibrating for viscous effects after construction. A simple correction technique is provided to assist with this type of problem.

Acknowledgement. This work was partially supported by the NASA Langley Research Center (Grant NAGW 674, contract monitor R.C. Rogers).

References

Bakos R (1994) An Investigation of Test Flow Nonequilibruim Effects on Scramjet Combustion, Ph.D. Thesis, University of Queensland.

Bakos RJ, Tamago J, Rizkalla O, Pulsonetti MV, Chinitz W, Erdos JI (1992) Hypersonic mixing and combustion studies in the hypulse facility. J Propulsion and Power 8:900–906.

Hayes WD, Probstein RF (1959) Hypersonic Flow Theory, Academic Press, New York.

Leyva IA (1994) Study of the addition of a divergent nozzle to an expansion tube for increasing test time. In: 18th AIAA Aerospace Ground Testing Conference, Colorado Springs, CO, Paper No. AIAA 94–2533.

Miller CG, Jones JJ (1983) Development and performance of the NASA Langley Research Center expansion tube/tunnel, a hypersonic-hypervelocity real-gas facility. In: AIAA 14th International Symposium on Shock Tubes and Waves, Sydney, Australia.

Paull A, Stalker R (1992) Test flow disturbances in an expansion tube. J Fluid Mech 245:493–521.

Sudnitsin O, Morgan RG (1994) Expansion tube area change optimisation. In: 4th International Workshop on Shock Tube Technology, Brisbane, Australia.

Walz A (1969) Boundary Layers of Flow and Temperature, The M.I.T. Press.

Throat design for hypersonic nozzles

L. Z. Dumitrescu, Y. Burtschell, D. Zeitoun
Lab. I.U.S.T.I. (U.R.A. CNRS 1168) Dept. MHEQ
Université de Provence, 13397 Marseille, France

Abstract: The design of a conical nozzle for the TCM2 free-piston shock tunnel is discussed. Based on Navier-Stokes computations (with full vibrational and chemical nonequilibrium), a throat contour is developed, featuring a smooth curvature distribution, blending smoothly into the straight conical expansion. By comparison with simpler contours, it is shown that a nearly disturbance-free flow is predicted in the nozzle. Other points are also taken up, such as the design of the subsonic region and the determination of the maximum throat area.

Key words: Hypersonic nozzle, CFD

1. Introduction: scope

The design of a hypersonic nozzle for the TCM2 facility, a high-enthalpy, free-piston shock tunnel, will be discussed. Consistent with the main intended application of that facility, i.e., basic studies in hypersonics, particular attention has been paid towards ensuring the high quality of the nozzle flow itself. The design is based on full Navier-Stokes flow computations, awaiting planned experimental confirmations.

2. Conical vs. contoured nozzles?

In the quest for ideal free-stream conditions, contoured hypersonic nozzles were, until recently, the rule; however, especially in off-design conditions, strong disturbances are experimentally found, mostly focused on the centerline (Hannemann et al. 1996). But it is now recognized that full duplication of free-flight conditions is unattainable: therefore, since a combination of CFD and testing is anyway necessary for hypersonic studies, the slight flow divergence of the flow in a conical nozzle could be accommodated. On the other hand, conical nozzles have certain definite advantages: insensitivity to off-design reservoir conditions, shorter length, leading to shorter flow settling time (an important feature for impulse facilities), less tendency to boundary-layer transition (due to a continuously expanding flow); and, as will be shown, a quite remarkable degree of exit-flow uniformity is actually obtained.

Based on these considerations, and on our own former experience with conical and contoured nozzles (Dumitrescu 1969, Brun & Guibergia 1967), it was decided to develop a conical nozzle of 8° half-angle, and an expansion ratio of 165, giving a Mach number of 5.9 in nominal reservoir conditions (about 200 bar, 6000 K).

3. The throat geometry

All computations show that most of the variations of the flow parameters, including chemical and/or vibrational freezing, occur along the first few inches downstream the throat, and that the origin of disturbing wavelets is also there; for this reason, the design of the throat geometry has been given particular attention.

It is known that a transonic flow over an airfoil is very sensitive to the local curvature radius, and moreover to any curvature discontinuity; bearing that in mind, a systematic study was undertaken on the effects of varying the wall curvature at the throat of the nozzle, where the flow changes from subsonic to supersonic velocity. A sketch of the throat geometry is shown in Fig. 1; all dimensions are referred to the throat circular cross-section radius R_T. Navier-Stokes computations of the flow in a short length of the nozzle have been carried out. Details of the procedure are given in the following section; here the results will be presented. First, the standard throat design was considered, i.e., two

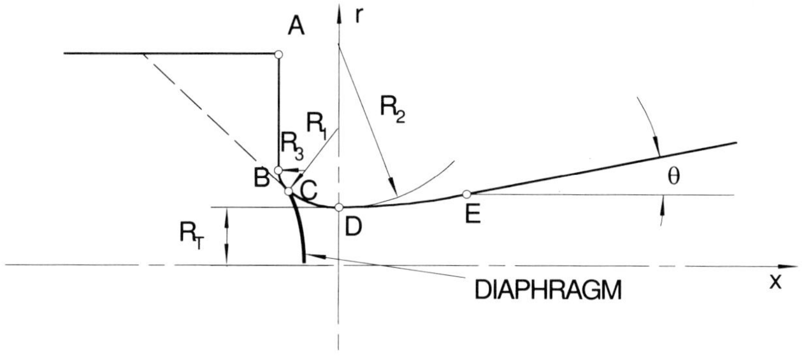

Figure 1. Throat contour geometry.

circular arcs (with one radius, R_1, in the subsonic part and another, R_2 beyond the throat itself), continued by the conical expansion. As can be gathered from the iso-Mach contours displayed in the three left-side diagrams of Fig. 2, rather strong perturbation waves are generated at the junctions between the circular-arc segments, and at the transition to the conical expansion; taking equal radii on both sides of the throat (and increasing the curvature radius) is beneficial, but still insufficient. Then a new type of contour was investigated, in the form

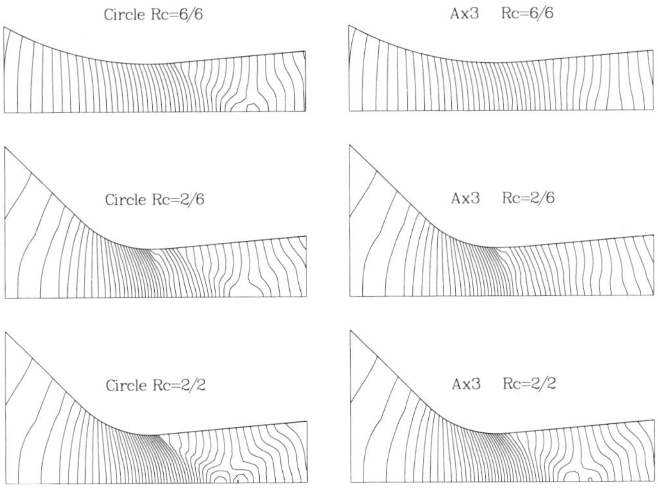

Figure 2. Iso-Mach lines for various throat geometries.

of a cubic parabola: $r = Ax^3 + Bx^2$ (Fig. 1), which has the property that its radius of curvature increases smoothly, from a given value at the throat, to infinity at the point of tangency to the conical expansion line. If, in addition, the curvature radii R_1 and R_2 are chosen equal, the nozzle contour becomes free of any curvature discontinuity. The right-side diagrams of Fig. 2 show that all sharp disturbances are practically eliminated with the choice $R_1 = R_2 = 6 \times R_T$ and a cubic parabola contour. In Fig. 3 the boundary-layer displacement thickness is shown as it develops along the throat walls: again, the smoother evolution is for the contour being adopted, thus hopefully reducing the propensity to premature transition.

It remains now to design the subsonic part of the nozzle: numerical experiments have shown that any changes of the wall shape in the region where the local Mach number is well below unity do not affect perceptibly the uniformity of the supersonic flow. Therefore, our throat design has been completed in the subsonic part by an arbitrary transition, with a radius R_3, to the vertical line representing the shock-tube end wall (AB in Fig. 1). This flat shape of the end-wall was chosen in order to avoid the development of a complicated wave pattern at the reflection of the primary shock, which would increase the duration of the

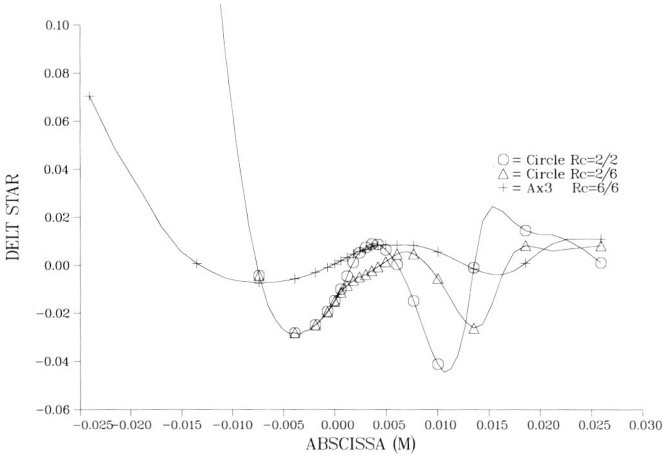

Figure 3. Boundary-layer thickness evolution for various throat geometries.

starting process. For convenience, in the computations, the subsonic part has been replaced by a 45° straight line; results are insensitive to the change.

4. Flow modeling and computational procedure

The axisymmetric Navier-Stokes equations are solved for air in thermal and chemical nonequilibrium, using Park's 17-reaction scheme and Landau-Teller's vibrational-relaxation model. The numerical scheme is second-order implicit with Gauss-Seidel relaxation; details are given elsewhere (Zeitoun et al. 1994). For the throat region, a non-uniform mesh of 101×71 points was used; for the full-length nozzle (see next section), the grid consisted of 171×71 points.

5. Computational results for the full nozzle

In view of planned experimental comparisons, the flow in the full nozzle has also been computed, and detailed distributions of the flow parameters, chemical composition, and translational and rotational temperatures have been produced for several operating conditions of the tunnel. In Fig. 4, as an example, iso-Mach lines are drawn for two throat geometries: it may be seen that, while the chosen throat geometry ensures an almost perfectly smooth flow (on paper), non-compliance with the requirement of smooth curvature generates perturbation waves which are reflected several times along the nozzle. Figure 5 presents

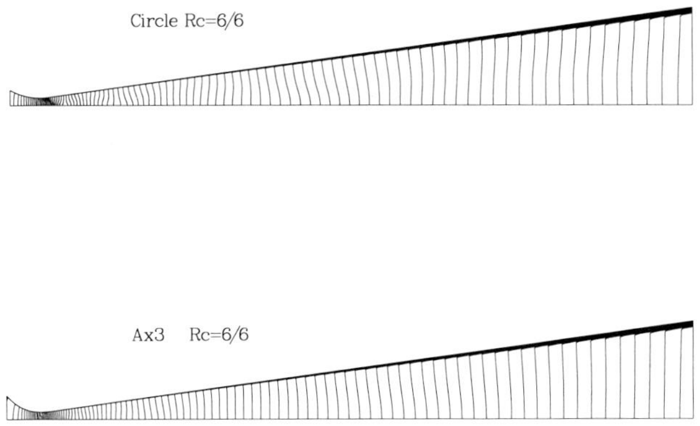

Figure 4. Iso-Mach lines for the full nozzle, with two throat geometries.

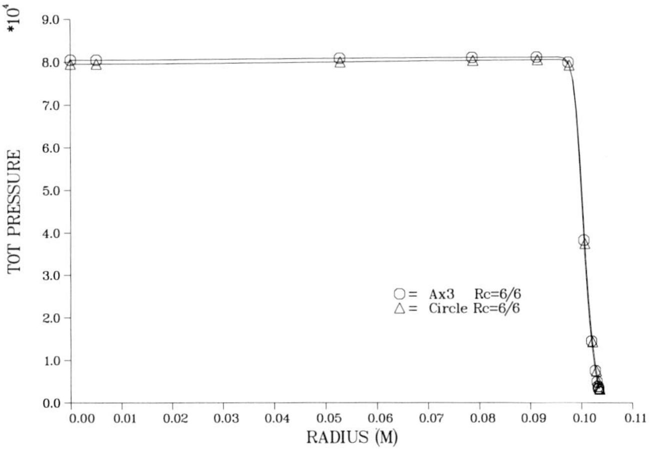

Figure 5. Total pressure distribution across the nozzle exit plane.

the total pressure distribution across the nozzle exit plane: it compares advantageously with computed (and experimental) distributions in contoured nozzles (Krek et al. 1996).

6. Other design points

To complete the nozzle design, two more points have to be considered: the first concerns the placement of the nozzle diaphragm. This problem has to be tackled empirically: we have chosen the position shown in Fig. 1, at the entrance of the designed throat contour. It is hoped that, since the possible break in the wall smoothness occurs in the subsonic region, this will allow the boundary layer to heal itself, thus avoiding premature transition.

A second point, strangely not often discussed in the literature, concerns the throat cross-sectional area: as the mass flow-rate per unit area is known, given the reservoir conditions (behind the reflected shock), the maximum throat area is clearly defined by the condition that all the gas found initially in the shock tube should pass through the nozzle in the desired test time (in our case, one millisecond). The actual design area should give some allowance to shock attenuation and mixing at the contact surface (say, a factor of 0.6 – 0.7).

7. Concluding remarks

A conical nozzle has been manufactured according to the design discussed herein; it is provided with a number of transducer ports along its walls and is presently being subjected to a detailed flow investigation. Results will be made public as soon as available.

References

Brun R, Guibergia J-P (1967) Influence de la sphéricité d'un écoulement libre hypersonique sur la distribution de la pression et du flux de chaleur à la surface de corps de révolution élancés. Journal de Mécanique, Paris 6(1):79–88.

Dumitrescu LZ (1969) Studies in Shock Tubes, (in Romanian), Editura Academiei, Bucharest.

Hannemann K, Krek R, Eitelberg G (1996) Latest calibration results of the HEG contoured nozzle. In: Shock Waves, Proc of the 20th ISSW, Sturtevant B, Shepherd JE, Hornung HG (eds), World Scientific, Vol 2, pp 1575–1580.

Zeitoun D, Boccaccio E, Druguet M-C, Imbert M (1994) Reactive and viscous flow in hypersonic nozzles. AIAA J 32(2):333–339.

Latest calibration results of the HEG contoured nozzle

K. Hannemann, R. Krek, G. Eitelberg
DLR - Institute for Fluid Mechanics, Bunsenstraße 10
D-37073 Göttingen, Germany

Abstract: Results obtained in calibrating the nozzle exit flow of the HEG facility at four conditions are discussed. A rake was used to measure pitot-pressure profiles at two axial locations in the test section. For a finer resolution of the boundary layer, a smaller rake was positioned at the nozzle exit. The experimental data are compared with results obtained from nozzle-flow calculations using an upwind TVD Thin-Layer Navier-Stokes code, which takes into account thermal and chemical nonequilibrium.

Key words: HEG, High-enthalpy nozzle flow, Numerical simulation

1. Introduction

A space vehicle re-entering earth's atmosphere from low-earth orbit is subjected to flows that place the vehicle through extreme physical conditions. The re-entry velocity of low-earth orbit is approximately 7 – 8 km/s. This leads to the development of a strong bow shock in front of the vehicle. The shock dramatically alters the gas properties behind it, with temperatures reaching over 10000 K, and the air experiences dissociation of the oxygen and nitrogen molecules. One type of ground-based facility that can simulate the relaxation phenomena in re-entry flow conditions is the free-piston shock tunnel, of which the High Enthalpy Shock Tunnel Göttingen (HEG) is currently the largest.

2. HEG facility and instrumentation

The free-piston-driven shock tunnel HEG is a reflected type shock tunnel. It consists of three main sections: the driver, consisting of an air buffer and a compression tube, the shock or driven tube, and the nozzle/test section (Eitelberg 1994). The reservoir for the nozzle is formed between the end wall of the driven tube and the reflected shock. The gas expands through a convergent-divergent nozzle into the test section. The 3.75-m long contoured nozzle has a throat diameter of 0.022 m and an exit diameter of 0.88 m, resulting in an area ratio of 1600. Details of the design procedure are given in Hannemann (1990). The reservoir conditions are determined by measuring the incident-shock speed, the pressure in the tunnel end wall, and calculations with the equilibrium code ESTC (McIntosh 1968, Eitelberg 1994). A permanent probe is installed within the test section to measure the pitot pressure and stagnation-point heat-transfer rate on a sphere of radius 10 mm. A rake was utilized to measure the uniformity of the flow exiting the HEG nozzle. The rake consists of 16 pitot-pressure probes on two perpendicular arms. An extra pitot-pressure measurement was made by a sensor located at the center of the rake. On two of the arms, the innermost sensor locations are ±50 mm from the center, while on the perpendicular arms, the innermost sensor locations are ±100 mm. The sensors are then equispaced at 100 mm along the arms. In order to obtain a finer resolution of the boundary-layer pitot-pressure profile, a small boundary-layer rake was also used, which consists of five pitot-pressure sensors, equispaced at 30 mm (Krek and Eitelberg 1994). The tests were performed at four conditions as described in Table 1.

Table 1. Reservoir pressure and total enthalpy of HEG test conditions.

Condition		I	II	III	IV
p_0	[MPa]	39	91	45	110
h_0	[MJ/kg]	21	22	13	15

3. Experimental results

3.1. Repeatability

The shot-to-shot repeatability is presented in Figs. 1 and 2, where the normalized pitot pressure and stagnation-point heat transfer of the permanent probes are plotted for 10 different shots in one test campaign. The traces were measured for condition I, which was used in more than 50% of all HEG shots performed. Both the pitot pressure as well as the heat transfer show a deviation of ±10%, which, for a shock tunnel, is a good repeatability.

3.2. Selection of test window

The pressure histories for the central pressure probe and the probe located at $r = 250$ mm from the center on the main rake are shown in Fig. 3 for condition I. The pressure signal of the central probe reveals a strong transient overshoot for the first 1 – 1.5 ms from signal rise. This overshoot is also measured on pitot probes up to a radial distance of approximately 100 mm. Further out from the nozzle axis, the overshoot becomes less pronounced, until it is no longer measurable (see Fig. 3). Two pressure readings obtained from the boundary-layer rake are shown in Fig. 4. The pressure signals resulting from the transducers located closest to the wall also reveal a large overshoot. This effect is reduced when moving closer to the nozzle centerline and can therefore be explained with the establishment of the boundary layer. The time scales for the boundary-layer

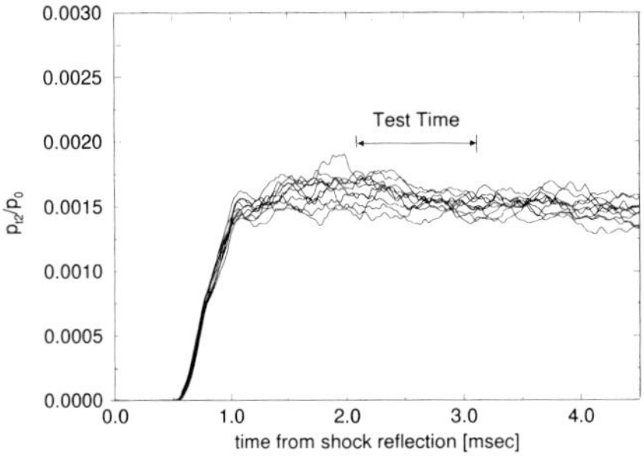

Figure 1. Pitot-pressure signals from permanent probe for ten different shots.

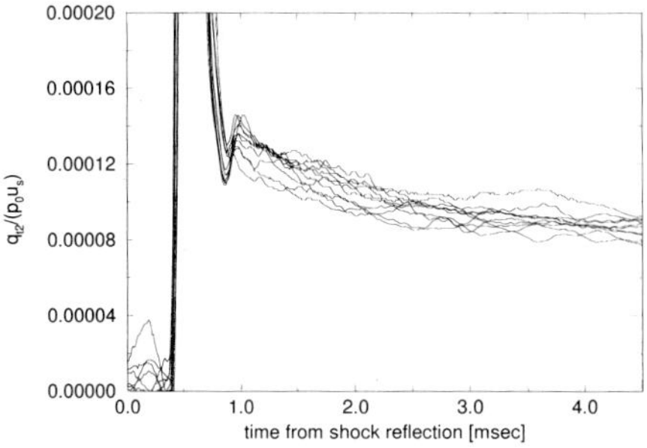

Figure 2. Heat-transfer signals from permanent probe for ten different shots.

growth and the pressure overshoot in the center of the nozzle flow are similar. It is thought that as the boundary layer develops, compression waves are generated and propagate downstream. These compression waves coalesce in the central region of the nozzle exit. For further evaluation, an approximate 1-ms test-time window, starting after the decay of the pressure overshoot, is utilized.

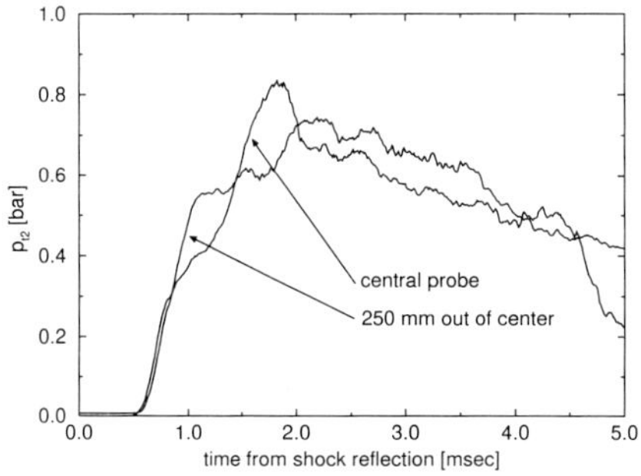

Figure 3. Pitot-pressure signals from the main rake.

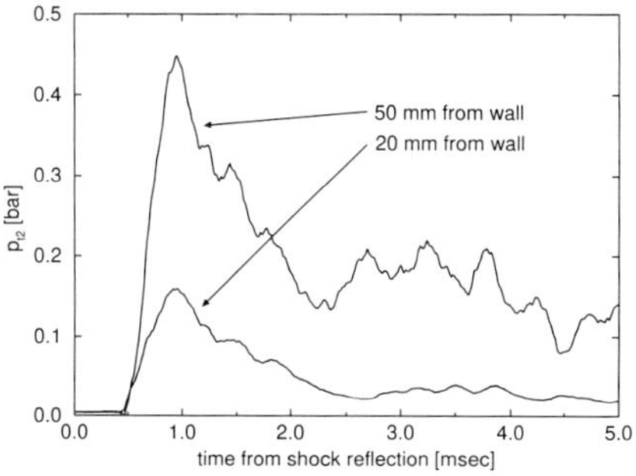

Figure 4. Pitot-pressure signals from the boundary-layer rake.

3.3. Flow uniformity

The uniformity of the flow is a major issue when it comes to testing various models. It is important to know the inflow quality to realistically evaluate the results obtained from the models with a high degree of certainty. The pitot-pressure profiles for all four conditions are plotted in Fig. 5. To obtain these profiles, the main rake was tested at two axial locations. After repeatability experiments were conducted with the rake in its downstream position it was moved upstream to measure the axial uniformity of the flow. No axial variation of the flow can be detected. From Fig. 5 it can be seen that there is a systematic variation in some of the probe measurements, e.g., the probe at $r = 250$ mm is consistently high. To establish if this is a flow effect or a systematic error, the main rake was rotated through 180° and one shot was conducted at condition II. It can be seen that these variations have now been mirrored about the nozzle axis, which can also be shown by numerically re-rotating the results (see Fig. 5b). This suggests a gauge effect. Therefore, it can be assumed that the uniformity of the nozzle flow is better than measured. The probes are calibrated statically before and after the test campaign and the variation to each other and the manufacturer's calibration factors is less than 2%. There is a need for a dynamic calibration.

4. Numerical code

Numerical simulations of the flow in the HEG nozzle have been performed using the implicit, upwind TVD Euler/Thin-Layer Navier-Stokes code NSHYP (Brenner et al. 1993). The three-dimensional, time-dependent Thin-Layer Navier-Stokes equations in conservation form for a multicomponent mixture of perfect gases are considered. For the simulations discussed in this paper the fluid is assumed to be air. The relevant chemical reactions were taken from Park (1989). The determination of the vibrational energy of the molecular species is based on the assumption of a harmonic oscillator. The transition of energy between the vibrational modes and the translational modes is approximated using the Landau-Teller formulation. The vibrational relaxation times were obtained from correlations of Millikan and White and Park (see Park 1989). The calculation of the transport properties is based on the hypothesis of Stokes, Fourier, and Fick. The viscosity of the single species is obtained from the Blottner curve fits, and the mixture viscosity from Wilke's mixture rule. The thermal conductivity is obtained using the Eucken correction. A constant Lewis number of $Le = 1.4$ is used to calculate the diffusion coefficient. At the subsonic inflow boundary of the nozzle both velocity components are extrapolated from the integration domain. Due to the low velocities and high densities it is assumed that the flow here is in chemical and thermal equilibrium. Therefore, the full state of the gas entering the nozzle can be calculated by using the total enthalpy and the entropy of the nozzle reservoir. At the outflow boundary all conservative variables are extrapolated. At the nozzle wall no-slip conditions apply. It is assumed that the wall is isothermal ($T_{wall} = 300$ K) and fully catalytic. Calculations of laminar as well as turbulent nozzle flow, where the transition was assumed to take place immediately downstream of the nozzle throat, were performed. The latter has been carried out using the algebraic turbulence model of Baldwin and Lomax (1978).

5. Comparison of experimental and numerical results

In Fig. 5 the computed pitot-pressure profiles ($0.92\rho u^2$) resulting from turbulent and laminar calculations are plotted together with the measured pitot-pressure profiles. The numerical profiles are only shown for x = 111 mm (conditions I, III) and x = 150 mm (conditions II, IV) downstream of the nozzle exit. In

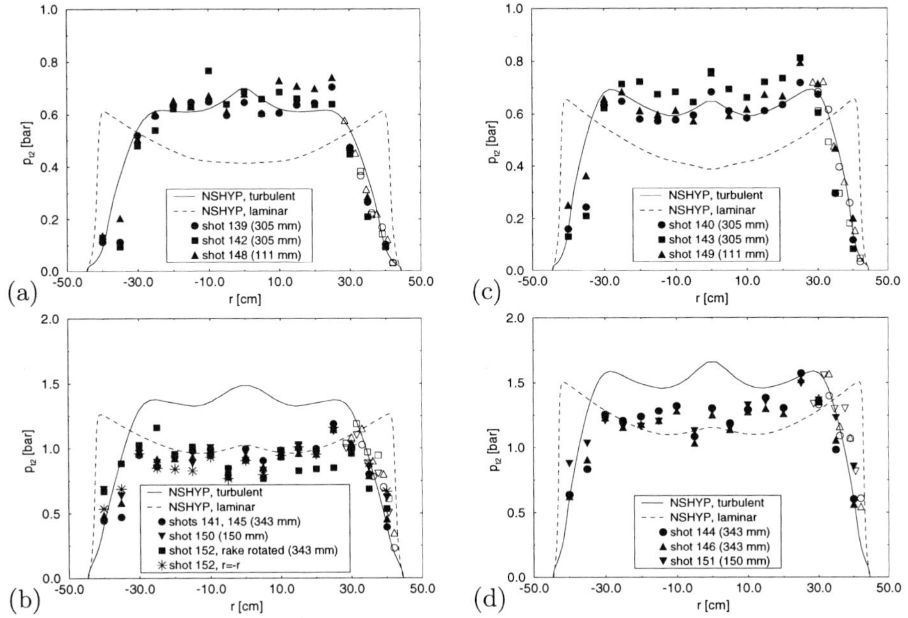

Figure 5. Calculated and measured pitot-pressure profiles (open symbols represent data from boundary-layer rake at nozzle exit): (a) condition I; (b) condition II; (c) condition III; (d) condition IV.

order to obtain the data at these locations, the nozzle has been extended in the calculations using a conical contour. For conditions I and III, the calculated pressure profiles resulting from the turbulent simulations reproduce the measured pressure profiles, also for the boundary layer. For condition III, the compression on the centerline is more evident in the measured profiles than in the calculated profile. For condition I the opposite effect can be observed. For conditions II and IV, the measured data in the boundary-layer follow the numerical profile resulting from the turbulent calculations. In the core region, however, the measured pressure profiles are closer to the one obtained with the laminar calculation. This behavior, which is currently not understood, appears to be strange insofar as a higher pressure implies a higher Reynolds number and therefore an even "more turbulent" flow. No information about the boundary-layer state (e.g. point of transition) is currently available.

6. Conclusions

The HEG facility has been calibrated at four test conditions. Calibration results indicate that there is a transient overshoot measured by the central pressure probes, with steady flow established after this overshoot. The pitot-rake measurements indicate that the four conditions are repeatable from shot to shot and there is little or no axial variation of the flow exiting the nozzle. Further, the comparison of measured and calculated turbulent pitot-pressure profiles shows good agreement for the low-pressure conditions I and III. For the high-pressure conditions II and IV, the turbulent calculations and the measurements agree only in the boundary layer. The measured pitot-pressure levels in the core flow could not be matched. These conditions are not yet completely understood.

Acknowledgement. The financial support of ESA/CNES is gratefully acknowledged. The authors are also thankful to Juergen Lenz and Dirk Garbe for the tunnel operation and the rake instrumentation.

References

Baldwin BS, Lomax H (1978) Thin Layer Approximation and Algebraic Model for Separated Turbulent Flows, AIAA Paper 78-257.

Brenner G, Gerhold T, Hannemann K, Rues D (1993) Numerical simulation of shock/shock and shock/boundary-layer interactions in hypersonic flows. Comput Fluids 22:427–439.

Eitelberg G (1994) First Results of Calibration and Use of the HEG, AIAA Paper 94-2525.

Hannemann K (1990) Design of an Axisymmetric, Contoured Nozzle for the HEG, DLR-FB 90-04.

Krek R, Eitelberg G (1994) Classical Characterization of HEG (Nozzle, Freestream Flow Field), DLR-IB 223-94 A50.

McIntosh MK (1968) Computer Program for the Numerical Calculation of Frozen and Equilibrium Conditions in Shock Tunnels, Department of Physics, School of General Studies, Australian National University, Canberra ACT.

Park C (1989) Nonequilibrium Hypersonic Aerothermodynamics, John Wiley & Sons, New York.

Trapping the boundary layer: a method to diminish flow contamination in shock tunnels

M.-P. Dumitrescu
Lab. I.U.S.T.I. (U.M.R CNRS 139) Dept. MHEQ
Université de Provence, 13397 Marseille, Cedex 20, France

Abstract: In many reflected-shock tunnels, the test gas is prematurely contaminated by the driver gas (usually He) creeping through the foot of the bifurcated reflected shock. The available test time is thus severely limited. In an attempt to overcome this limitation, a boundary-layer suction technique is proposed. The shock-tube endwall is provided with a slit at its periphery, through which part of the boundary layer is absorbed. Proof-of-concept tests demonstrate the correct operation of the device.

Key words: Hypersonic facility, Boundary layer, Driver-gas contamination

1. The problem

As has been reported earlier (Slade et al. 1993), when a shock tunnel is operated in the reflected-shock mode, the test gas is contaminated by the driver gas much earlier than the arrival of the contact region in the vicinity of the end plate of the shock tube. The reason seems to be the presence of a lambda pattern at the foot of the reflected shock, which induces a small transverse component for the velocity of the boundary layer (in shock-fixed coordinates) and, when reaching the contact region, creates a region through which the driver gas starts to creep rapidly towards the end plate (Fig. 1a).

The boundary layer thus formed generates a vortex at the edge of the end plate (Kleine et al. 1991), which accelerates further the flow of cold He towards the nozzle entry, thus bringing about premature contamination of the test gas (Fig. 1b). As has also been reported (Dumitrescu & Brun 1969), the mixing between the cold boundary layer and the initial hot-gas slug induces a decay in the total enthalpy, whereas the pressure stays constant, thus making the cooling effect undiscernible on the pressure traces.

2. Preliminary tests

To get the order of magnitude of the characteristic speed of the propagating turbulent front, a first series of tests at moderate Mach numbers (5.5 . . . 6.5) and 0.8 bar initial pressure, has been conducted in the TCM2-Marseille free-

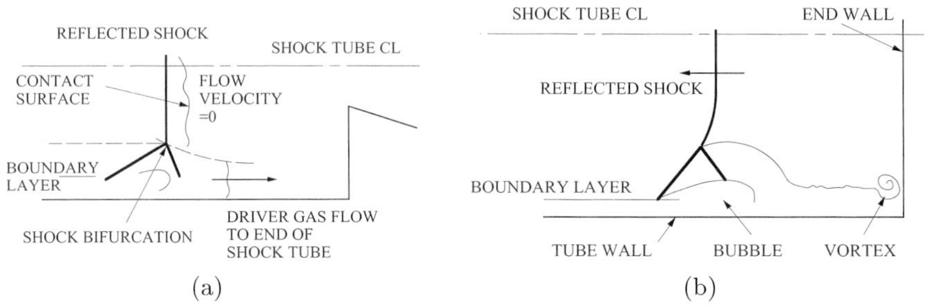

Figure 1. Contamination mechanism: (a) after Slade et al. (1993); (b) after Kleine et al. (1991).

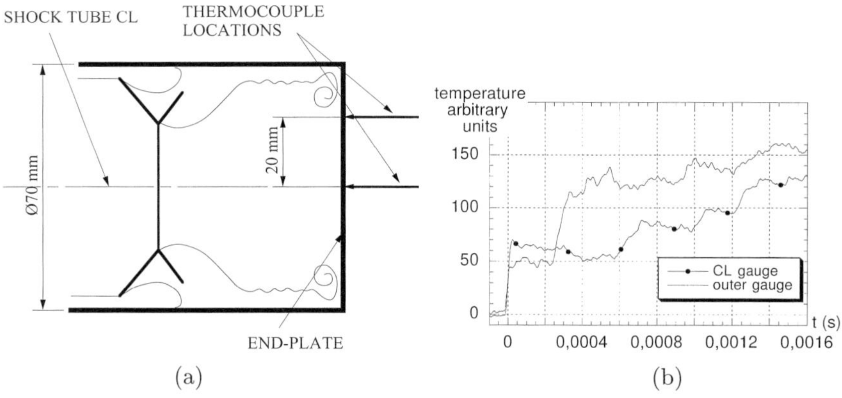

Figure 2. Preliminary investigation: (a) experimental setup; (b) temperature recordings.

piston shock tunnel. The setup chosen is shown in Fig. 2a: the end plate was fitted with two thermocouples mounted on the centerline and at 20 mm radial distance. The temperature recordings (Fig. 2b) show, as expected, flat plateaus of almost the same value for both gauges but, after a short time, a sharp rise appears at the eccentric station. The same rise occurs somewhat later at the centerline, thus indicating that the vortex-induced turbulent flow reached the gauge. The average inwards-propagating front speed may therefore be assumed to be around 110 m/s. It is clear that the arrival of the turbulence does not necessarily imply that He is already present in that region, but, since most of the impurities (and even ablated particles from the shock tube's walls) are initially contained within the boundary layer, their presence is most likely to occur in that region, and it seems to be worth trying to eliminate them as much as possible. Also, one would like to avoid mixing of the clean flow through the nozzle with gas coming from the tube boundary layer, which is turbulent and of uncertain thermodynamic state.

3. Device design

Since the TCM2 facility has a relatively short shock tube (6 m), all the unfavorable effects listed above are present and thus have to be avoided in order to get a decent test time. A closer look reveals that most of the shortfalls would be, if not removed, at least diminished if the boundary layer could be discarded in the vicinity of the end plate. The design requirements are, among others, a reasonable cost and a certain flexibility of the device, in order to accommodate several operating regimes of the facility. The ease of operation, cleaning, and repair were, of course, taken into account to get an acceptable compromise.

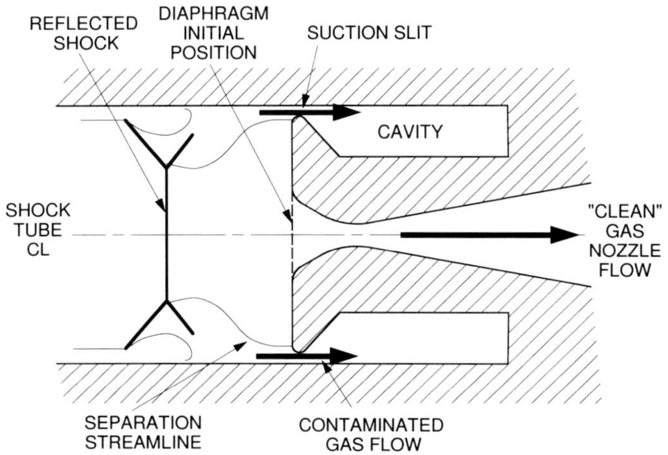

Figure 3. New end-plate design.

These considerations led to the design shown in Fig. 3. The idea consists of creating an absorbing slit around the perimeter of the end plate, followed by an enlarged cavity, thus absorbing part of the boundary layer across the sonic "throat" thus formed. There is no need for a secondary diaphragm or any other pumping device.

The size of the slit has to be quite small, in order to preserve the plane form of the reflected shock, but the mass flow across the sonic throat has to be large enough to bleed the boundary layer during the useful test time. This has to be correlated with the size of the cavity, since the pressure has to be kept below the maximum value for which sonic speed is reached and maintained in the slit during the test time. Initially, the pressure differential is null since the pressure inside the cavity is equal to that in the shock tube (p_1), but, immediately after the reflection of the incident shock, the pressure ratio is very high (p_{51}), ensuring that sonic speed is reached. The aim is to split the "reservoir gas" (left behind the reflected shock) in two regions: the contaminated boundary layer and a "clean" test-gas slug, which would then be drawn separately inside the trapping cavity and inside the nozzle, the impact point of the separating streamline lying somewhere in the vicinity of the slit lip.

4. Proof-of-concept tests

In an attempt to validate the concept, an experimental setup has been designed as shown in Fig. 4. The wall temperature is measured at the same locations as in the previous tests (on the centerline and at 20 mm radial distance) and pressure pickups are fitted on the end-plate centerline (behind a hole in the thermocouple) and inside the cavity.

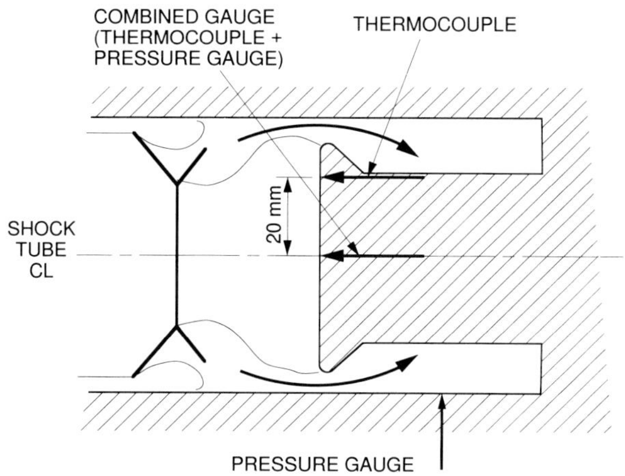

Figure 4. Proof-of-concept experimental setup.

The same initial conditions have been used as in the previous test: the results are shown in Figs. 5a and 5b. As may be seen, the temperature reaches a flat plateau (of the same value as in the previous tests) immediately after shock reflection, but the sharp rise indicating turbulence is delayed beyond the target test time (1 ms) at both locations. No nozzle has been fitted, since there are no means of ascertaining if the flow at the thermocouple locations is induced by the trapping slit or by the flow through the nozzle itself. The pressure build-up inside the cavity is fairly linear during the test time, and well below one half of p_5, thus indicating that sonic speed is maintained inside the slit.

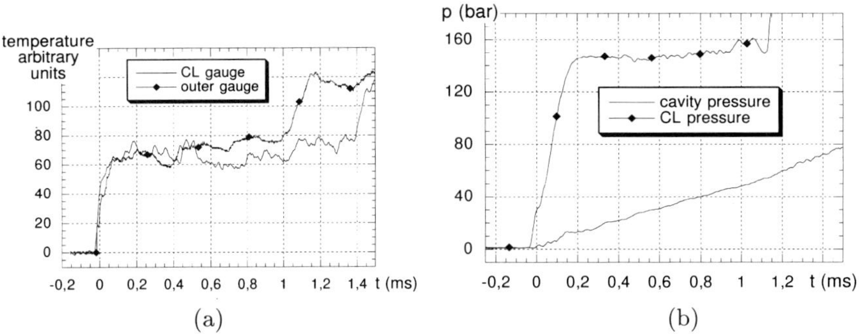

Figure 5. Proof-of-concept test results: (a) temperature recordings; (b) pressure recordings.

5. Concluding remarks

Although only proof-of-concept tests have been carried out, the device seems to work satisfactorily. There are, however, some questions which still have to be answered: (a) what is the right size of the slit; (b) what is the minimum acceptable volume for the cavity; (c) what diagnostic device should be used to ascertain that the device is efficient. The right sizes for the slit and for the cavity are dependent on the test conditions (in fact on the boundary-layer thickness). The slit may easily be made adjustable by use of a replaceable lip (which may also prove useful in case of wear), and the minimum volume of the cavity should be estimated for the case of the lowest operational value of the initial pressure p_1 in the shock tube. Further investigations and developments are now under study.

Acknowledgement. The author wishes to gratefully acknowledge the help of Ph.D. student A. Chaix in running the tests and of Ph.D. student M. Zechner from SWL Aachen who designed and built the special thermocouple used in combination with the pressure gauge.

References

Dumitrescu LZ, Brun R (1969) Sur les conditions réservoir d'une soufflerie à choc réfléchi. In: Comptes Rendus de l'Académie des Sciences de Paris, A 268, pp 827–830.

Kleine H, Lyakhov VN, Gvozdeva LG, Grönig H (1991) Bifurcation of a reflected shock wave in a shock tube. In: Shock Waves, Proc 18th Int Symp on Shock Waves, Takayama K (ed), Sendai, Japan, Springer-Verlag, Heidelberg, pp 261–266.

Slade JC, Crane KC, Stalker R (1993) Driver gas detection by quadrupole mass spectrometry in shock tunnels. In: Shock Waves @ Marseille, Proc 19th Int Symp on Shock Waves, Brun R, Dumitrescu LZ (eds), Marseille, France, Springer-Verlag, Heidelberg, pp 293–298.

High-enthalpy nozzle flow sensitivity study and effects on heat transfer

L. M. G. Walpot[1,2], J. Muylaert[1], P. G. Bakker[2], G. Simeonides[1]
[1]Aerothermodynamics Section YPA, ESA-ESTEC
P.O. Box 299, 2200 AG, Noordwijk, The Netherlands
[2]Department of Aerospace Engineering, Delft University of Technology
P.O. Box 5058, 2600 GB Delft, The Netherlands

1. Introduction

The characterization of hypersonic short-duration wind-tunnel flow fields is typically based on the measurement and/or theoretical determination of the total pressure and total enthalpy in the tunnel reservoir and the measurement of the pitot pressure in the test section. As the nozzle expansion is complicated in high-enthalpy facilities by thermochemical nonequilibrium or non-isentropic effects, the free stream is obtained by the computational modeling of the nozzle expansion process, at least as long as detailed test-section flow-field measurements are not available. The subsequent discussion focuses on the sensitivity of computed test-section flow field properties upon a number of assumptions made in the calculation of the nozzle expansion and on the adequacy of test-section and nozzle-wall measurements for the validation of computational models. Three high-enthalpy wind tunnels are considered: the ONERA F4 arc jet facility, and the DLR HEG, and the Caltech T5 piston-driven shock tunnel.

2. Nozzle flow

High-density real-gas effects may become important in the first stages of nozzle expansion in hypersonic wind tunnels operating at high reservoir pressure, as for example in the VKI Longshot facility discussed by Simeonides (1992). The operation of high-enthalpy facilities, however, is for the moment restricted to relatively low reservoir pressures primarily to avoid the erosion of internal walls of the wind tunnel structure observed at high-enthalpy/high-pressure operation. Consequently, the operating conditions considered here are free of high-density real-gas effects, the presence of which would call for increased computational complexity, as illustrated in Fig. 1 where the maximum density encountered in the nozzle reservoir is seen to be almost an order of magnitude lower than the level where intermolecular forces become important.

The expansion along nozzle #2 of the ONERA F4 facility has been computed for a reservoir pressure of 430 bar and reduced total enthalpy of 260 by the LORE1D code of Walpot (1994) incorporating various thermal and chemical nonequilibrium models, thermal equilibrium, full equilibrium, and vibrational melting (sudden relaxation, proposed by Boudreau 1993) at various streamwise stations. Relevant flow-field parameters at the nozzle exit are summarized in Table 1 and compared with the centerline laminar predictions of the 2D nonequilibrium Navier-Stokes solver TINA (Netterfield 1992). The Mach-number evolution along the nozzle is shown for the various models in Fig. 2. The use of 4- rather than 2-temperature models is seen to have little influence on the free-stream conditions, and the same goes the V-V coupling; it is likely that it will be more significant above 2000 K. The Treanor-Marrone coupling inverts the relative magnitudes of T_{n2} and T_{O2} for this test case. A significant difference is noted between the Park and Dunn & Kang chemical reaction rates where the former give a somewhat lower Mach number. Thermal and full equilibrium computations yield lower Mach-number distributions by virtue of the increasingly higher temperatures. Lastly, the vibrational melting model yields temperatures that are significantly higher than the thermal equilibrium case due to the irreversibility of the process and the associated entropy rise.

Table 1. Comparison of nozzle exit flow conditions for various methods.

chem. coupl.	DK 2T 2D NS	2T DK	2TPk	4T DK	4T DK TM	4T DK TMVV	4T Pk VV	DK equil	DK th eq	DK 2T melted
Ma	12.37	13.22	12.67	13.13	13.44	13.42	12.58	7.20	11.16	9.12
$\frac{Re}{10^5 m}$	1.517	1.604	1.478	1.567	1.657	1.653	1.452	0.601	1.266	1.030
u	5822	5785.5	5802.2	5819.7	5797.1	5798.9	5836.3	6102.0	5862.1	5847.6
$10^3 \rho$	.6535	.63357	.6250	.6267	.6352	.6350	.6203	.56896	.63514	.66549
p	99.71	83.89	90.82	85.179	81.64	81.86	92.42	292.12	121.23	189.00
T	456.	399.8	441.9	410.12	384.6	385.9	452.6	1770.4	577.02	855.26
T_{n2}	4094.	3911.1	3871.4	4248.5	4235.7	3978.0	4153.3	1770.4	577.02	855.26
T_{o2}	4094.	3911.1	3871.4	2408.3	5005.2	4979.7	2560.2	1770.4	577.02	855.26
T_{no}	4094.	3911.1	3871.4	3229.8	3346.8	4209.9	3917.1	1770.4	577.02	855.26
c_{n2}	.79	.78806	.78746	.7879	.7908	.7906	.7874	.79736	.78741	.78874
c_{o2}	.01572	.02549	.03408	.02471	.01752	.01788	.03272	.19469	.02635	.02297
c_{no}	.01920	.02557	.02685	.02570	.01962	.01995	.02680	.00564	.02696	.02410
c_n	.0	.0	.0	.0	.0	.0	.0	.0	.0	.0
c_o	.1740	.1608	.15159	.1615	.17201	.17147	.15298	.00229	.15926	.16417
p_{t2}	19937	19391	19251	19410	19510	19517	19333	19625	19975	20852

The transverse pitot-pressure distributions, measured and computed by the TINA solver in the test section of the F4 nozzle, are shown in Fig. 3. Reasonable agreement is found between the measurements and the 2D nonequilibrium computations assuming a laminar nozzle-wall boundary layer. The corresponding 1D results with various thermochemical models are also shown to be in reasonable agreement with the measured mean pitot pressure. Closer examination of Table 1, however, shows that the pitot pressure is almost totally insensitive to the thermochemical behavior or non-isentropicity of the flow expansion along the nozzle. Consequently, the measurement of pitot pressure alone in the test section is not adequate to characterize the test-section flow field in high-enthalpy facilities (in contrast to perfect gas/equilibrium facilities). Similar comments

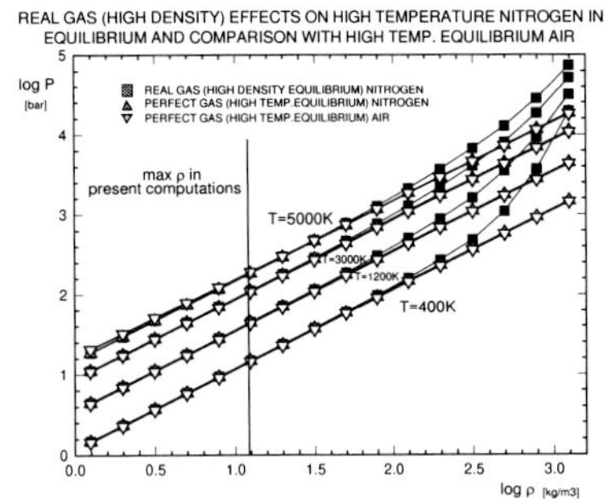

Figure 1. High-density real-gas effects.

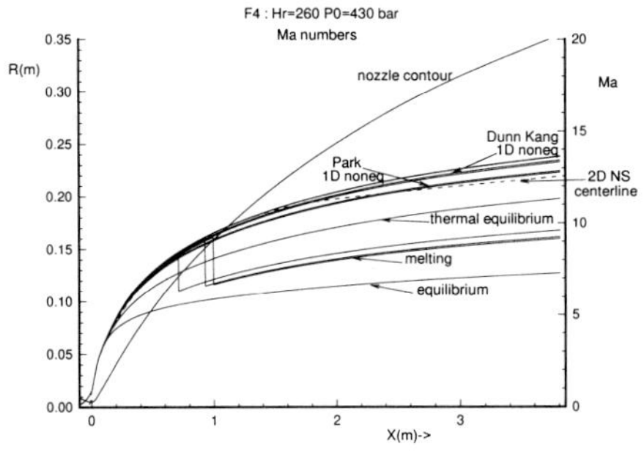

Figure 2. Mach-number distributions in F4.

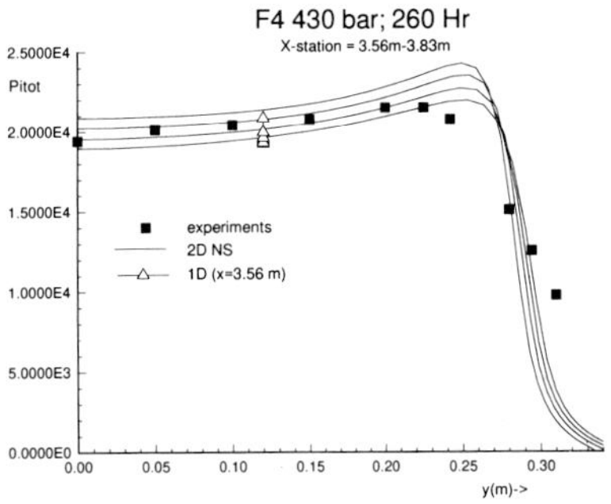

Figure 3. Comparison of measured and computed pitot pressure in F4 test section.

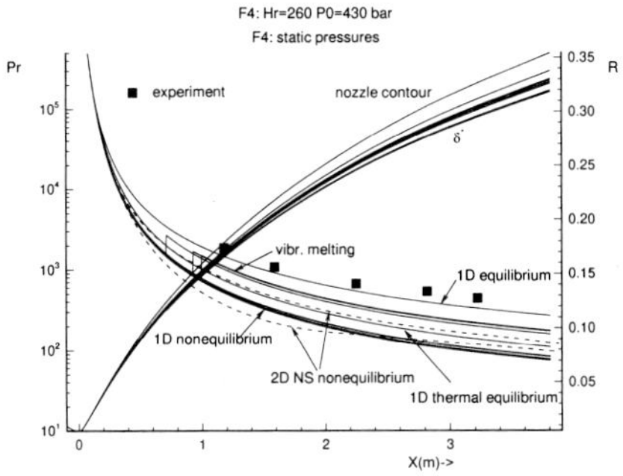

Figure 4. Pressure distributions in F4.

apply to the free-stream velocity and density, whereas a significative sensitivity is observed in the free-stream pressure and temperature.

In order to improve the level of characterization of the nozzle expansion and, thus, of the test-section free-stream, surface-pressure measurements were performed along the F4 nozzle wall. The measured data are compared with the TINA 2D nonequilibrium computations and the LORE1D results with various thermochemical models in Fig. 4. Although it appears that the best prediction of the measured wall-pressure distribution is provided by the 1D full thermochemical equilibrium assumption along the nozzle (see also Verant 1994), it is noted that the 1D computations tend to provide a transverse mean of the actual pressure field which is of the order of 0.7 times the respective 2D predictions along the nozzle wall (see comparison between the nonequilibrium LORE1D and TINA wall computed pressures in Fig. 4). Therefore, if the 1D results are extrapolated to the nozzle wall by such a factor, the model assuming a sudden vibrational relaxation will be the closest to the measurements, whereas the equilibrium results will be higher. An assumption of sudden vibrational relaxation, possibly caused by the presence of contaminants eroded from the tunnel structure, or by water vapor as proposed by Boudreau at lower enthalpies, appears to the authors to be more plausible than a full thermochemical equilibrium assumption, at least for this high-enthalpy test case. Upcoming non-intrusive measurements are expected to shed more light on nozzle-expansion assumptions.

3. Standard model Electre

Having noted that pitot-pressure measurements are not sufficiently sensitive to give a complete characterization of the flow field, the Electre standard model was designed to provide further insight into the flow exiting the nozzle of high-enthalpy facilities. With reference to Fig. 5, it is noted that the best match to the pressure distribution measured over Electre in the F4 tunnel is provided by the full equilibrium nozzle-flow assumption, although non-equilibrium flow is assumed around the Electre configuration. With regard to the heat-transfer comparison, however, the experiments exhibit a significantly higher reduction in heat flux around the shoulder of the model than any of the computations and/or assumptions for the nozzle expansion are able to capture. Notably, the computational results are mesh independent with typically 50 or more points distributed within the attached boundary layer. The effect of thermal equilibrium/non-equilibrium over Electre was checked for the melting vibration nozzle-flow assumption and found to be negligible. In the HEG case, depicted in Fig. 5, the scatter in the measured data covers effectively the entire range of the computed results. The mean computed curves, however, which closely correlate the mean measured data, are represented by the computations assuming vibrational melting along the nozzle. This would be consistent with the argument developed above that the melting vibration nozzle situation represents the F4 nozzle wall pressure measurements if 1D results are extrapolated to the wall.

With reference to Fig. 6, increasing the Lewis number from 1.0 to 1.4 is seen to cause an increase in stagnation-point heating of about 8%. Measured and predicted stagnation heating values are given in Fig. 7 for the present F4 and HEG Electre cases, and the HAC configuration tests in the Caltech T5 tunnel (Muylaert et al. 1996). The Fay-Riddell estimates (for an equilibrium boundary layer and Lewis number of 1.0), assuming a Prandtl number of 0.6, are in close agreement with the Verant (1994) estimates based on the correlation of high-enthalpy computational results. However, with the exception of the F4 case (which exhibits the lowest pitot pressure in Fig. 7), the experimental data are found to be in better agreement with the Fay-Riddell formulation assuming a higher Prandtl number. The same applies to the TINA predictions for the HEG and the low-pressure T5 cases, whereas the high-pressure T5 TINA results correlate better with the low-Prandtl-number Fay-Riddell and Verant

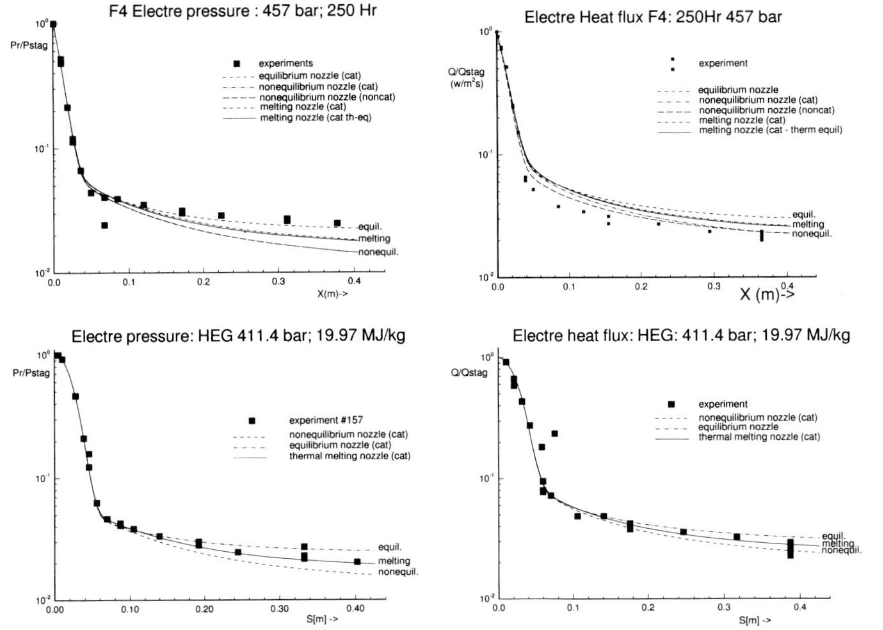

Figure 5. F4 and HEG Electre normalized pressure and heat-transfer distributions.

estimates. All in all, these comparisons illustrate the relevance of transport and thermodynamic properties, especially viscosity and Prandtl number; note in particular their strong dependence upon both pressure and temperature at high-enthalpy conditions (Gupta et al. 1991).

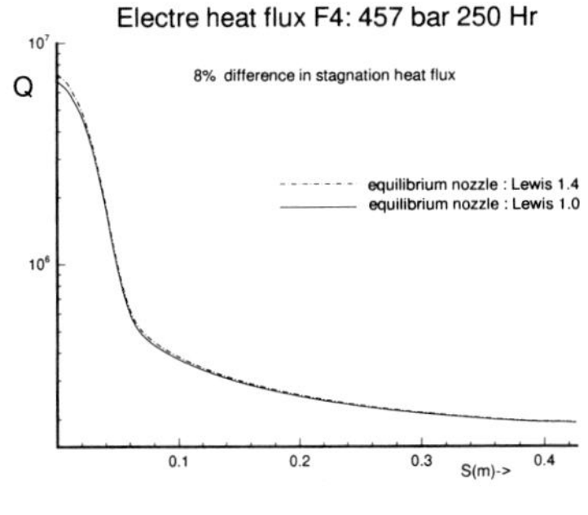

Figure 6. Lewis-number effect on heat-transfer distribution over Electre.

Case	Conditions pitot [bar] - enthalpy [MJ / kg]	TINA predictions	Measured values	Fay-Riddell values			Verant values
				Pr=0.60	Pr=0.72	Pr=0.90	
F4 equil. nozzle F4 noneq. nozzle F4 melting nozzle	0.2 - 20	6.65 / 7.20[1] 6.41 / 5.04[2] 6.55	6.9	6.4	5.7	5.0	6.7
				(35mm nose radius of curvature)			
HEG equil. nozzle HEG noneq. nozzle HEG melt. nozzle	0.6 - 20	9.95 9.77 9.73	9.4	11.6	10.4	9.1	11.6
				(35mm nose radius of curvature)			
T5 HAC RUN#662	7.5 - 11	76	62+/-1.6	76	69	60	72
T5 HAC RUN#730	7.9 - 19	130	89+/-0.6	135	121	106	137
T5 HAC RUN#723	2.9 - 11	39	38+/-1.8	45	41	36	45
T5 HAC RUN#719	3.3 - 17	64	64+/-1.3	82	72	64	80
				(3.0mm nose radius of curvature)			

[1] increasing Lewis number from 1.0 to 1.4
[2] catalytic / non-catalytic

Figure 7. Measured and predicted stagnation-point heating, MW/m^2

4. Conclusion

One-dimensional nozzle-flow computations have shown, through comparison with 2D Navier-Stokes solutions, to be an efficient means for performance of parametric sensitivity studies. Examination of various thermochemical non-equilibrium models has shown that the most important differences are found between the Park and Dunn & Kang reaction rate models. Static pressure, temperature, and Mach number have been identified as the most sensitive to thermochemical modeling and, particularly, to the assumptions of non-equilibrium, thermal or full equilibrium, and sudden vibrational relaxation along the nozzle, as opposed to density, velocity, and pitot pressure. The F4 high-enthalpy results show evidence that nozzle-flow expansion is far from the originally assumed non-equilibrium case, an explanatory hypothesis being vibrational melting triggered by flow contaminants. Similar indications are provided by the Electre results in F4 and HEG high-enthalpy conditions without, however, being conclusive. Reduction of flow contamination and non-intrusive flow-field measurements will provide further insight. Concerning the Fay-Riddell formulation for stagnation heating, a strong sensitivity to transport properties and Prandtl number has been seen with uncertainties in Prandtl number being as significant as uncertainties in the stagnation-region velocity gradient.

Acknowledgement. The authors wish to acknowledge M. Netterfield for his contribution towards the TINA computations and D. Devezeaux for his useful comments.

References

Boudreau AH (1993) Characterization of Flow Fields in Hypersonic Ground Test Facilities, VKI/AEDC Course Methodology of Hypersonic Testing.

Gupta RN, Lee KP, Thompson RA, Yos JM (1991) Calculations and Curve Fits of Thermodynamic and Transport Properties for Equilibrium Air to 30000 K, NASA Reference Publication 1260.

Muylaert J, Walpot L, Hironde N, Tumino G, Simeonides G, Koppenwallner G, Davis J (1996) Computational analysis of experiments in T5 and HHK for the study of the flap-induced separation on an axisymmetric configuration. In: Shock Waves, Proc of the 20th ISSW, Sturtevant B, Shepherd JE, Hornung HG (eds), World Scientific, Vol 1, pp 203–208.

Netterfield M (1992) Validation of a Navier Stokes Code for Thermochemical Non-equilibrium Flows, AIAA 92-2878.

Simeonides G (1992) Hypersonic Shock Wave Boundary Layer Interactions over Compression Corners, Ph.D. Thesis, U. Bristol, von Karman Institute.

Verant J (1994) Numerical Enthalpy Rebuilding for Perfect Gas and Nonequilibrium flows — Application to High Enthalpy Wind Tunnels, ONERA RT 69/ 6121/SY, May 1994.

Walpot L (1994) Development of an Efficient Nonequilibrium Navier-Stokes Solver, Technical LR-Report, TU Delft, Dept. Aerospace Engineering.

Wave processes in a pressure-wave refrigerator

A. Galyukov, P. Voinovich
Advanced Technology Center
P. O. Box 160, St. Petersburg 198103 Russia

E. Timofeev
Ioffe Physical-Technical Institute, Russian Academy of Sciences
Polytekhnicheskaya Str. 26, St. Petersburg 194021 Russia

Abstract: The paper provides some results concerning the numerical study of strongly transient gasdynamic processes in a pressure-wave refrigerator (PWR). A hierarchical set of numerical models from the simplest 1D to a fully 3D formulation is introduced. The computations show that 1D solutions give a reasonable foundation for the understanding of PWR operational principles but cannot predict satisfactorily the refrigeration efficiency. Good agreement with experiment is achieved by considering 2D and 3D effects of gas mixing by overlap of rotating nozzles and expansion tubes.

Key words: Pressure-wave refrigerator, Numerical simulation, Wave processes

1. Introduction

The pressure-wave refrigerator (PWR), called also "thermal separator," represents a simple facility (see Fig. 1) for gas cooling by decompression which has many applications in energy transformation and chemical processes. A rotating gas distributor contains nozzles from which the high-pressure gas enters radially mounted expansion tubes. The opposite ends of the tubes are closed or, alternatively, are connected to tanks for shock damping. The gas motion in the tubes and nozzles, as their openings coincide with each other, is similar to that in a shock tube. After a nozzle has passed a tube opening, the expanded gas comes out of the tube into the receiver in which the distributor rotates. The process repeats periodically in each nozzle and tube.

Some experimental results concerning the PWR can be found in Marchal et al. (1985), Yu et al. (1989) and Fang et al. (1991). Prior to the present study, the authors could find no reports of numerical study of a PWR in the literature. Our numerical approach and results were first published in Galyukov et al. (1994). This paper includes new data.

Progressive overlapping of high- and low-pressure channels is a universal feature of pressure-wave machines. The phenomenon influences essentially the

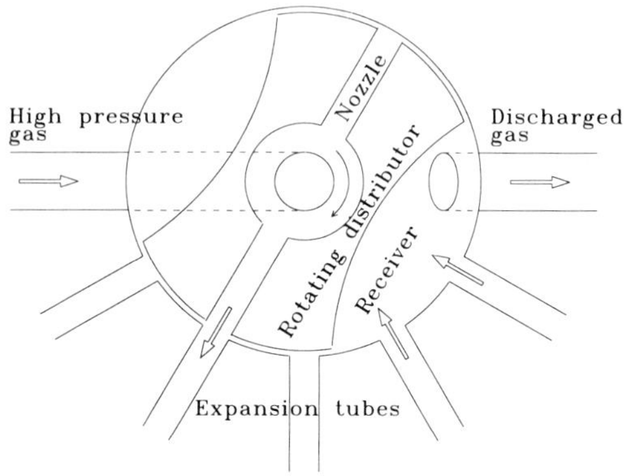

Figure 1. A conceptual sketch of the pressure-wave refrigerator.

wave processes and efficiency of the devices and requires 2D (in the rotation plane) or 3D modelling to simulate adequately the processes of gas mixing. The respective codes are typically CPU-time and memory consuming. Therefore, investigation of an actual PWR with long expansion tubes over a number of cycles is practicable today only on the basis of 1D models. The alternative employed in the present paper is to analyze 1D–3D results in combination.

2. Numerical models

It has been demonstrated that the role of viscous effects on the gas motion in the pressure-wave supercharger (PWS) (which is conceptually close to the PWR) is negligibly small (Zhang & So 1990). Unlike PWS channels, the expansion tubes of the PWR have a rather high length-to-diameter ratio which might lead to attenuation of shock waves. Based on the experiments by Fang et al. (1991), the computations reported here were carried out for ratios for which the above phenomenon could be neglected. Centrifugal and Coriolis forces were also disregarded because the respective terms are negligible at rotational speeds typical for PWRs. Consequently, we proceeded from the unsteady 1D–3D Euler equations written in an integral form and supplied with appropriate initial and boundary conditions.

Several shock-capturing finite-volume numerical techniques based on both structured and unstructured grids were employed in the computations (Fursenko et al. 1993). High-resolution explicit predictor-corrector Godunov-type schemes and local adaptation of the grid in the unstructured grid case constitute the basis of the methodology. The structured approach implies Cartesian grid (non-mapped grid for round cross-sections) and dimensional splitting.

As the feeding nozzles of the PWR rotate, the boundary conditions at a tube's open end repeat periodically in time and the flow in each tube develops cyclically. The expansion tubes are all identical as are the nozzles, so it is sufficient to consider the gas motion in a single nozzle-tube pair with the appropriate periodical conditions. The 1D computations were performed for many periods starting from an initial step function until the solution became periodical with prescribed accuracy. Since the gas in the tubes is continuously being heated up by shocks resulting from the high-pressure gas input and those reflected from the tubes' ends, a heat-sink term has been introduced into the energy equation for the 1D model to maintain the average temperature and provide a convergent cyclic solution.

The 2D and 3D computations were performed with the intention of studying only the early development of the contact interface separating the gas in the tube from a new portion of expanded gas: no periodical solutions were achieved.

3. Results and discussion

All the results below were obtained for air, which is here treated as a perfect gas with a ratio of specific heats of $\gamma = 1.4$ and a uniform initial temperature of $T = 293$ K. As the 1D solver is extremely fast, massive computations were performed to analyze the influence of the main design parameters on PWR efficiency. A good qualitative agreement between 1D numerics and experimental results is observed. However, the values of the efficiency ratio are considerably overestimated (see below). Figure 2 shows the density-field development as wave diagrams which clearly indicate the principal wave processes in the expansion tube. The presented converged solutions were computed using the 1D model for parameters taken from Fang et al. (1991). The top-left image in Fig. 2 corresponds to the minimum efficiency ratio, because the cooled gas is heated up by the strong reflected shock wave. The highest efficiency ratio is attained provided the reflected shock approaches the tube end immediately after the cooled gas emission has finished and the opening closes (Fig. 2, top right). It is essential that the maximum efficiency ratio correspond to the minimum flow rate so that in practice a compromise might be beneficial as shown in Fig. 2, bottom left. Small tanks attached to the tubes at their closed ends reduce the negative influence of reflected waves (Fig. 2, bottom right) and therefore considerably improve the PWR efficiency.

Two-dimensional and three-dimensional simulations of the flow which develops in the nozzle and in the respective part of an expansion tube were performed to analyze the effect of progressive overlapping of the tube and nozzle openings on the gas stirring in the tubes. The 2D and 3D results are qualitatively analogous. A 2D example is given in Fig. 3 which illustrates the flow-pattern

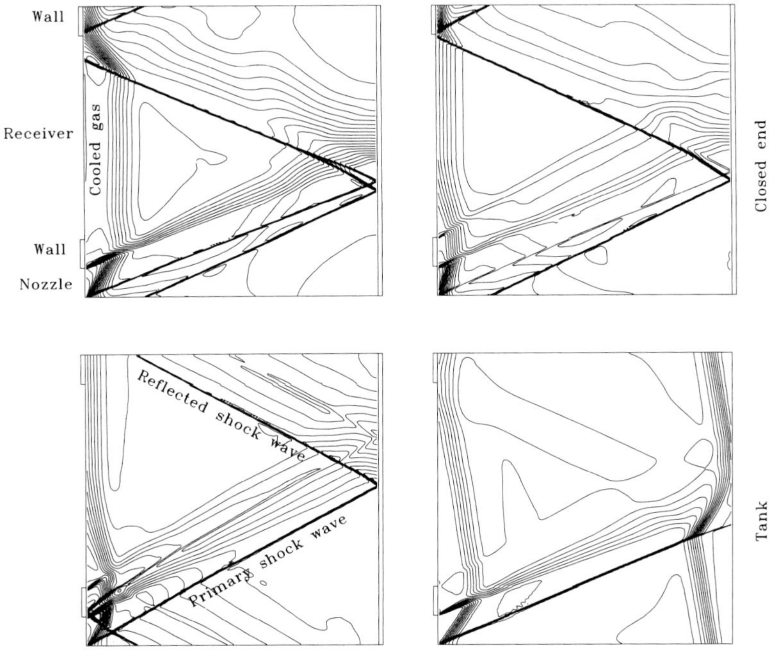

Figure 2. Wave diagrams for an expansion tube (1.8 m long). Distributor rotor speed: 3000 rpm – top left, bottom right; 3500 rpm – top right; 4000 rpm – bottom left.

development at early stages of the PWR cycle. Two plane-parallel rectangular channels, each 20 mm wide, were considered at an initial pressure ratio of 5 atm to 1 atm and a linear overlapping velocity of 60 m/s. One can clearly see that the gasdynamic processes in the nozzle rotation plane significantly affect the general flow pattern (see Galyukov et al. 1994 for a detailed analysis).

Figure 4 provides numerical results (curves) for the PWR efficiency ratio η versus distributor rotor speed in comparison with the experiment (Fang et al. 1991, crosses in Fig. 4a) and, simultaneously, illustrates our combined approach to the PWR study. The design parameters and initial conditions have been taken from Fang et al. (1991): initial pressure ratio, 5.5 atm to 1.8 atm; distributor diameter, 120 mm; diameter of nozzles and tubes, 10 mm; length of expansion tubes, 2.5 m; expansion tubes without tanks. The refrigeration efficiency ratio is given relative to the ideal adiabatic expansion:

$$\eta = \frac{1 - T_g/T_{HP}}{1 - (P_{LP}/P_{HP})^{\frac{\gamma-1}{\gamma}}} \quad , \tag{1}$$

where T_{HP} and P_{HP} are temperature and pressure in the high-pressure channel

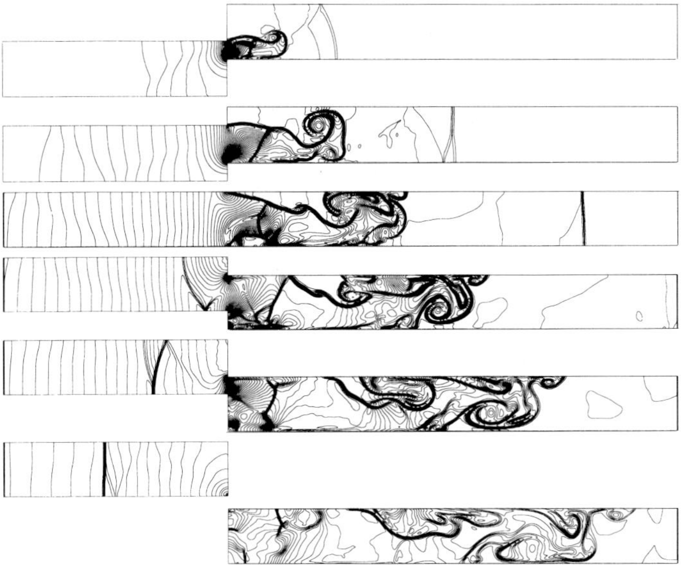

Figure 3. Density contours for the nozzle (left) – tube (right) overlapping process.

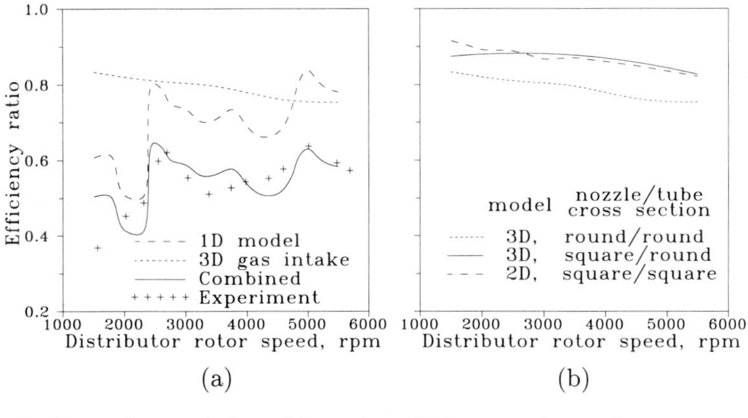

Figure 4. Dependence of the refrigeration efficiency ratio on the rotor speed: (a) models and experiment; (b) individual efficiency of gas intake.

(nozzle); P_{LP} is the pressure in the low-pressure channel (receiver); and T_g is the average temperature of cooled gas. The dashed curve (η_1) in Fig. 4a corresponds to the one-dimensional model and characterizes the negative influence of the reflected shock waves. It is clearly seen that the PWR efficiency is considerably overestimated. Effective 3D gas mixing, caused by the jet arising as soon as the round openings start overlapping, reduces the PWR efficiency. The dotted curve in Fig. 4a represents the reduction in refrigeration efficiency η_2 exclusively by gas mixing due to the overlapping process. A good quantitative agreement with the experiment is achieved by integrating both the above effects. The combined efficiency ratio $\eta_\Sigma = \eta_1\eta_2$ is displayed in Fig. 4a by the solid line.

Figure 4b depicts the efficiency reduction due to gas mixing for different types of nozzle/tube cross-sections. The diameter of round cross-sections and the sides of the square are equal. For a given rotor speed the efficiency ratio is greater for the nozzle/tube pair which provides a higher value of the integral overlapping area over the overlap time. The result is closely related to the conclusion by Galyukov et al. (1994) that an increase of the nozzle/tube diameter ratio (which in turn raises the above integral) enhances PWR efficiency.

Acknowledgement. This work was supported in part (the methodology) by the Russian Foundation of Fundamental Investigations (Grant 94–01–01456). The authors would like to thank Prof. A. Fursenko and Prof. Yu. Golovachov who contributed greatly to the arrangement of this research. We wish to express our thanks to Prof. Ya. Berman, Dr. O. Man'kovskiy and Prof. K. Takayama for numerous stimulating discussions. The third author wishes to thank Prof. N. Satofuka for his kind help during the author's stay at the Cooperative Research Center, Kyoto Institute of Technology, where the paper was finally prepared for publication.

References

Fang YQ, Zheng J, Liu RJ, Zhu C, Fan J, Hu DP (1991) Experimental study of gas wave refrigeration. In: Proc 18th Int Symp on Shock Waves, II: 1335–1338.

Fursenko AA, Mende NP, Oshima K, Sharov DM, Timofeev EV, Voinovich PA (1993) Numerical simulation of propagation of shock waves through channel bends. Comp Fluid Dyn J 2:1.

Galyukov A, Timofeev E, Voinovich P (1994) Numerical analysis of transient gas flow in a pressure wave refrigerator. In: Computational Fluid Dynamics 94, S Wagner, EH Hirschel, J Periaux, R Piva (eds), John Wiley & Sons, Chichester, New York, Brisbane, Toronto, Singapore, pp 678–684.

Marchal P, Malek S, Viltard JC (1985) Skin-mounted rotating thermal separator efficiently recovers NGL from associated gas. Oil and Gas Journal 62:120.

Yu HR, Fang YQ, Yu W (1989) The thermal separator: a new approach for refigeration pressurized gases. In: Current topics in shock waves, YW Kim (ed), American Institute of Physics, New York, II, pp 659–664.

Zhang HS, So RMC (1990) Calculation of the material interface in a pressure-wave supercharger. J of Power and Energy 204:151.

Quantitative holographic interferometry of shock-wave flows using Fourier transform fringe analysis

H. Babinsky*, K. Takayama
Shock Wave Research Center
Tohoku University
Sendai 980 Japan

Abstract: This paper demonstrates that Fourier transform fringe analysis can be used for quantitative evaluation of interferograms in shock-wave research. This technique is capable of providing whole-field density information from finite-fringe interferograms. The main steps of this technique are presented, examples are given, and problems specific to shock-wave research are discussed.

Key words: Holographic interferometry, Fringe analysis, Fourier transform

1. Introduction

One of the most widely used tools in experimental research of shock-wave phenomena is holographic interferometry (Takayama 1983). However, due to the difficulties involved with the quantification of interferograms it is often limited to a flow-visualization technique with the added advantage that some quantitative information can be extracted by simple fringe counting. For both basic research and the validation of CFD schemes, an automatic interferogram analysis technique that can provide high-quality data at high spatial resolution is desirable. Fourier transform fringe analysis, originally introduced by Takeda et al. (1981) has recently been used in transonic aerodynamic testing by Parker (1993) and is in principle well suited for the investigation of shock-wave flows. It is the purpose of this paper to determine the usefulness of this technique for shock-wave research and demonstrate its implementation.

2. Fourier transform fringe analysis

Interferometry reveals the phase change experienced by coherent light passing through a test section filled with gas of variable density. By determining the distribution of the optical phase change, the density field inside the working section can be calculated easily, provided the flow is two dimensional. Relevant formulas and descriptions can be found in Vest (1979). Finite-fringe interferom-

*Present address: Dept. of Engineering, University of Cambridge, Cambridge, England.

etry superimposes on the phase-change distribution an additional linear phase variation to introduce uniformly distributed parallel reference fringes in the interferogram. Phase variations due to density changes are visible as displacement of the reference fringes with the direction of displacement revealing the sign of the phase change.

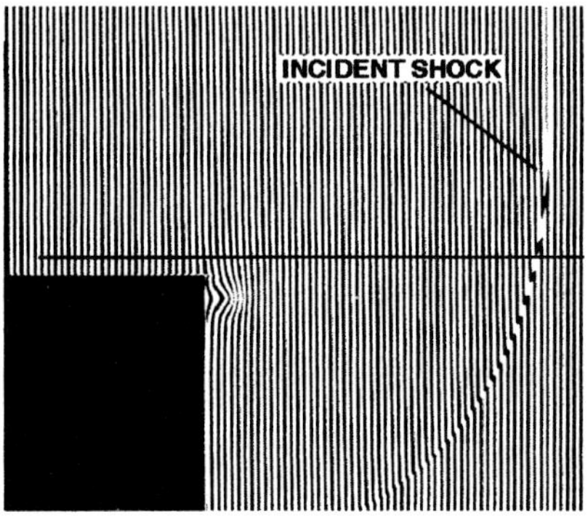

Figure 1. Computer-generated finite-fringe interferogram.

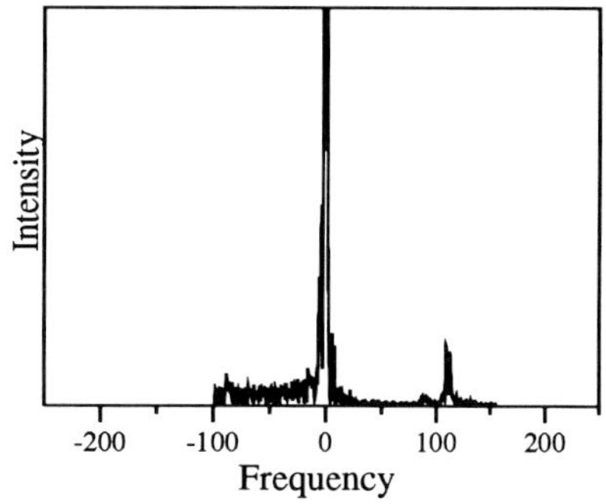

Figure 2. FFT spectrum of greyscale signal along the line indicated in Fig. 1.

Fourier transform fringe analysis makes use of finite-fringe interferograms. Such an image is given in Fig. 1, created from a CFD simulation using techniques described in Babinsky et al. (1995). Using numerical data to create the image allows a critical evaluation of the performance of the analysis because the density distribution obtained from the interferogram can be compared directly to the original data. The original analysis, proposed by Takeda et al. (1981), followed one-dimensional arguments, therefore a slice of image data shall be considered (see Fig. 1). Undisturbed fringes can be interpreted as sinusoidal variations of constant frequency whereas regions of variable density cause the fringe spacing to vary, thus introducing frequency modulations. By shifting the image information in frequency space by $-f_0$, where f_0 is the frequency of the undisturbed finite fringes, the phase modulation can be recovered — this is equivalent to removing the carrier frequency f_0. Figure 2 shows the frequency spectrum of the brightness data taken along the line. The finite-fringe frequency can be determined from the location of the first-order peak. The spectrum after shifting by $-f_0$ is shown in Fig. 3. Note that the negative portion has been discarded and that the range of the new spectrum is limited to approximately $\pm f_0$. It is advisable to smooth the edges of the domain with a suitable windowing function to avoid high-frequency artifacts. The modified spectrum is treated with an inverse Fourier transform giving a complex result where the phase angle is equivalent to the desired phase variation. The absolute signal is a measure of the light intensity at each point. Due to the use of the arctan function in the phase calculation, the data is 'wrapped' into a range of $\pm\pi$ and must by 'unwrapped' by a suitable algorithm. In principle this is straightforward, a multiple of 2π is simply added to the data wherever a phase jump of 2π is found. The result after performing this operation and subsequent calculation of the density is given in Fig. 4 and compared with the original density distribution. The agreement is very good, however, some high-frequency noise can be observed. Because of the spectral shifting the spatial resolution of this analysis is limited by $\pm f_0$, which is equivalent to the spacing between undisturbed finite fringes. In practice it is therefore advisable to apply a low-pass filter at this frequency to reduce the noise. For optimum spatial resolution, a large number of finite fringes (high f_0) should be used, however, this number is often limited by the image resolution after digitization.

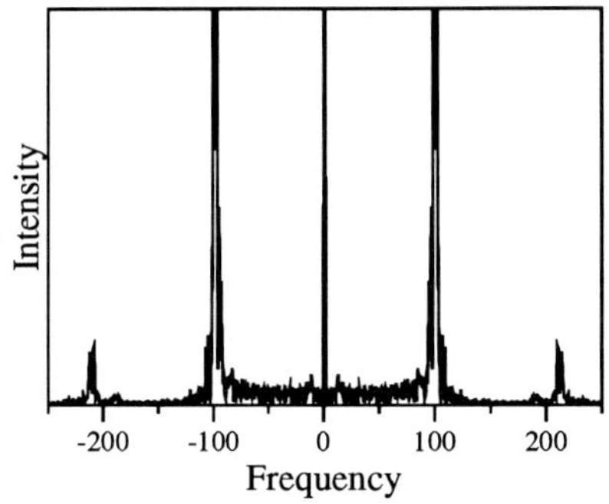

Figure 3. FFT spectrum after shifting by $-f_0$.

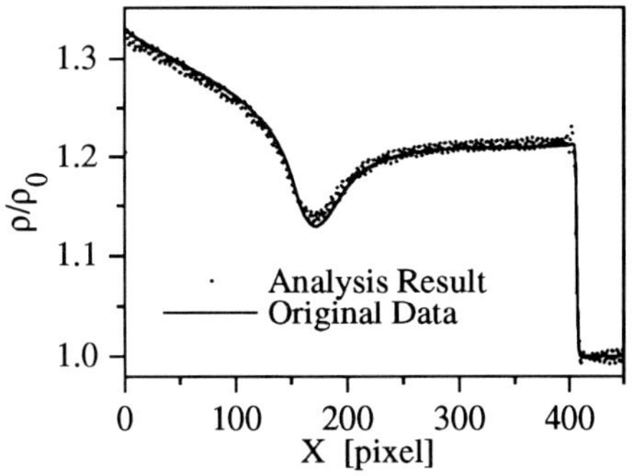

Figure 4. Density distribution after unwrapping, compared with original data.

3. Application to shock-wave research

Bone et al. (1986) improved this technique by introducing a two-dimensional Fourier transformation. This has several advantages over the one-dimensional approach: the spectrum can be shifted in frequency space in the x- and y-directions, thus allowing fringe directions different from the vertical. Furthermore it is possible to retain spectral information beyond f_0 in the direction perpendicular to the shift. Figure 5 shows a shock reflection on a wedge installed in a 60-mm $\times$ 150-mm shock tube ($M_s = 1.2$ in air; 15° wedge angle; partial view). The image was digitized with a scanner to a resolution of 1024 $\times$ 1024 pixels. Before any shifting in frequency space, the data centered around the origin of the spectrum (zeroth-order) must be removed. Otherwise this peak would come to rest at $-f_0$ where it would introduce unwanted noise. Parker (1993) suggested removing the central order by subtracting the spectrum of an image containing only the intensity distribution, but no interference fringes (i.e., a shadowgraph). In principle this technique is highly successful. However, it depends on the existence of a high-quality shadowgraph of the flow field. In shock-wave research it is usually not feasible to record such an image simultaneously. Although it is possible, in principle, to obtain a similar picture during the reconstruction of the interferogram (see Parker 1993), this has not been successful here, thus simple high-pass filtering was used. Another problem is caused by nonlinearities associated with the process of developing the film and subsequent scanning of the image. This leads to distorted fringes which do not display the expected sinusoidal greyscale variation and therefore cause a significant second-order peak in the frequency spectrum. It was found that this phenomenon can be reduced significantly if a nonlinear scaling function is applied to the original image data. The shape of this function is chosen to result in approximately sinusoidal fringes. In addition a small region of the spectrum centered around the second-order peak is filtered before performing the inverse FFT. Although the spatial resolution is theoretically limited to $\pm f_0$ it was found that some useful information can be preserved beyond this range, particularly in directions parallel to the fringe orientation. Therefore a larger cut-off frequency can be chosen for the final low-pass filtering if the zeroth-order and second-order peaks are carefully removed. The wrapped phase distribution that was finally obtained after analysis is seen in Fig. 6. The unwrapping of such data is complicated by a number of difficulties.

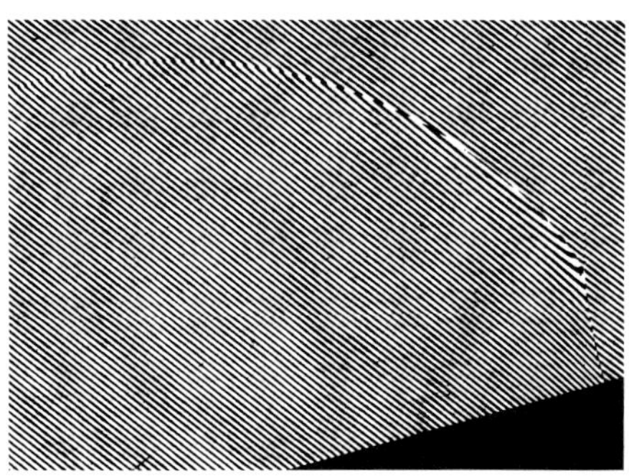

Figure 5. Finite-fringe interferogram of a shock-wave reflection on a 15° wedge.

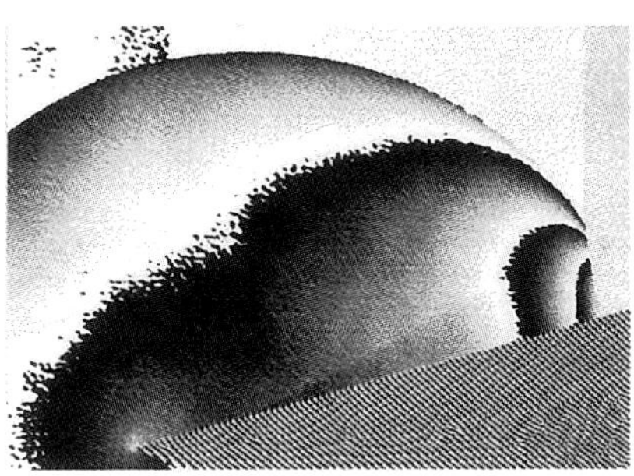

Figure 6. Wrapped phase distribution obtained from Fig. 5 after analysis.

The foremost problem is that shock-wave flow fields typically exhibit jumps in phase that are not caused by phase wraps but are due to density jumps associated with shock waves. If the density change through a shock wave is large (causing a phase change $> 2\pi$) then the unwrapping algorithm may confuse this signal with a phase wrap and attempt to correct it. A further problem is caused by image deficiencies such as dust particles or optical distortions. These can also introduce jumps in the wrapped phase distribution which should be ignored by an unwrapping algorithm. The most successful approach to deal with these problems is to generate a mask that covers all areas that may cause erroneous unwrapping and apply a suitable unwrapping algorithm to each connected, problem-free region. Many strategies have been suggested to find such regions automatically and it was found that a combination of several approaches is most successful.

Vandenhouten and Grebe (1995) suggested excluding all points where the magnitude of real and imaginary signal is very low. This technique has the added advantage that solid objects are also masked. However, in the vicinity of shock waves, the signal intensity is not necessarily reduced. Bone (1991) suggested a criterion based on second derivatives which can mask shock waves automatically. This criterion was also implemented here, together with an automatic exclusion of locally inconsistent points. After choosing suitable cut-off values for each criterion, a mask is created automatically, as shown in Fig. 7. It can be seen that, despite a combination of several masking criteria, the final result is still insufficient to guarantee successful unwrapping. In particular it can be seen that the incoming shock wave is not detected at all. This is due to the fact that the shock strength is almost an exact multiple of an equivalent phase jump of 2π. In addition, this shock wave is straight, which causes the criterion based on second derivatives to fail as well. Nevertheless any unwrapping across this shock must be prohibited and the density values on either side have to be prescribed manually from other measurements or theoretical considerations.

Figure 7. Automatically generated mask.

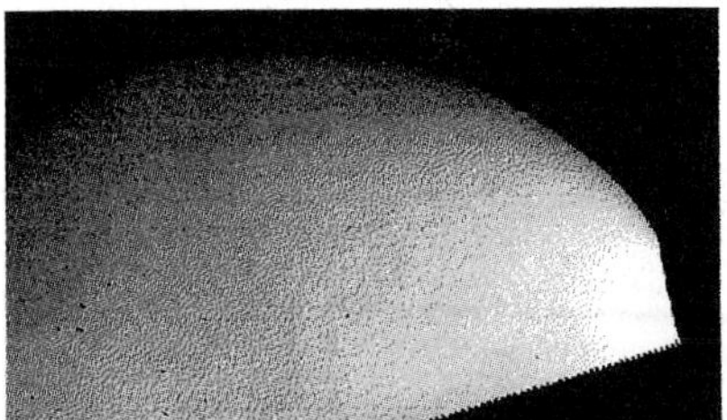

Figure 8. Final result of analysis, density is represented as greyscale map.

In contrast, the reflected shock wave has been partially detected. This is due to the fact that it is curved and of variable strength. Far away from the reflection point this wave is weak enough to allow unwrapping and should therefore not be masked. However, gaps in the mask can still be found in regions where unwrapping should not be permitted. It is therefore suggested the automatically created mask be used as a basis for further manual intervention. For example, the incident shock wave should be added by hand and the gaps in the mask of the reflected wave should be closed. A certain degree of knowledge about the flow field is necessary in this step. Once the mask has been edited the starting points for unwrapping must be marked (locations of known density — one per interconnected region) and unwrapping can be performed. A non-recursive flood-fill algorithm (Bone 1991) is used for this purpose. The final result can be seen in Fig. 8, giving the flow density throughout the flow field. The quality of such information is far beyond the data that could be obtained by fringe tracking in an infinite-fringe interferogram and as such it is invaluable for further post-processing or comparison with CFD.

4. Conclusions

A technique has been suggested for analysis of finite-fringe interferograms of shock-wave flows which can provide whole-field density information. It was shown how to treat problems associated with nonlinearities due to development and digitization. The difficulties caused by shock waves and poor image quality have been discussed and methods are suggested to overcome these. A test case based on numerical data as well as a typical shock-wave flow field have been analyzed with good results. The quality of the final data is very encouraging and it is particularly suited for further post-processing techniques such as tomography, or comparison with numerical data.

References

Babinsky H, Yang JM, Takayama K (1995) Animated visualization of shock wave flowfields for dynamic comparison between experiment and computational prediction. In: Proc SPIE 2410:101.

Bone DJ, Bacher HA, Sandemann RJ (1986) Fringe-pattern analysis using a 2-D Fourier transform. Appl Opt 25:10:1653.

Bone DJ (1991) Fourier fringe analysis: the two-dimensional phase unwrapping problem. Appl Opt 30:25:3627.

Parker SC (1993) The Quantitative Analysis of Transonic Flows by Holographic Interferometry, PhD Thesis, University of Warwick, UK.

Takayama K (1983) Application of holographic interferometry to shock wave research. In: Proc SPIE 398:174.

Takeda M, Ina H, Kobayashi S (1981) Fourier transform method of fringe-pattern analysis for computer-based topography and interferometry. J Opt Soc Am 24:5:824.

Vandenhouten R, Grebe R (1995) Phase reconstruction and unwrapping from holographic interferograms of partially absorbent phase objects. Appl Opt 34:8:1401.

Vest CM (1979) Holographic Interferometry, John Wiley and Sons.

Single-pulse spontaneous Raman-scattering measurements in an expanding nitrogen/oxygen admixture

J. O. Gilmore*
Stanford University, Stanford, CA 94305 USA
S. Sharma, D. Fletcher
NASA-Ames Research Center, Moffett Field, CA 94035 USA
D. Bershader†
Stanford University, Stanford, CA 94305 USA

Abstract: The results of single-pulse spontaneous Raman-scattering measurements in an expanding nitrogen/oxygen admixture are reported. The relative proportions of N_2 and O_2 approximated air. The expansions were from 2500 K, 12 kg/m^3. Raman scattering of KrF excimer laser radiation by N_2 and O_2 was recorded, and vibrational temperatures were determined by fitting computed band shapes to the data. The results were consistent over varying spectral resolution, laser focal volume, and laser frequency (near/exact O_2 resonance). The data indicate a stronger coupling of the N_2 and O_2 vibrational energies than reported by White in shock compressions of a similar admixture.

Key words: Raman scattering, Vibrational relaxation, Shock tunnel

1. Introduction

The distribution of vibrational energy in flows of common gases has been of great interest for more than fifty years. Investigations of both compressing and expanding flows have been conducted. Despite the activity, there is a continuing interest in the expanding flow problem since the body of data collected from nozzle expansions is less consistent than that from shock compressions.

The presently reported experiments employed Raman spectroscopy to directly measure vibrational-state populations in expansions of an admixture of N_2 with O_2 that approximated air. The conditions of the experiments were such that the mean thermal energy was comparable to or less than the vibrational quanta. Accordingly, vibrational-vibrational (V-V) energy exchange between the modes of the components was significant.

*Present address: Stanford University, Department of Mechanical Engineering, Stanford, CA 94305 USA, email:john_gilmore@qmgate.arc.nasa.gov.
†Posthumous.

2. Diagnostic technique

N_2 and O_2 are inherently nonpolar due to their charge symmetry, and thus have no infrared spectra from which to determine vibrational-state populations. However, these molecules are Raman active, and the strengths of spectral lines due to Raman scattering are proportional to state populations. Gillespie (1993) demonstrated a single-pulse Raman diagnostic technique for near-atmospheric density flows of pure N_2. The technique used a pulsed, KrF excimer laser and an Intensified Charge Coupled Device (ICCD) detector to overcome the extreme feebleness of the Raman effect.

When applying this diagnostic technique to a high-temperature admixture of N_2 with O_2, there are several new considerations. The most important is the potential for absorption of the laser energy by O_2 through its Schumann-Runge system. The strongest bands are $0 \leftarrow 6$ and $2 \leftarrow 7$. Once absorbed, subsequent emission could interfere with observation of the Raman effect. In the present experiments, it was possible to eliminate most of the fluorescence by tuning the laser. When present by purposeful excitation, or as residual, the O_2 fluorescence was restricted to a spectral range that did not interfere substantially with the lower vibrational transitions of the O_2 Raman Q branch.

3. Experiment

The present experiments were conducted in the Electric Arc-driven Shock Tube (EAST) at NASA Ames Research Center. The facility was operated as a reflected-shock tunnel with a two-dimensional nozzle placed near the end of the tube. Details of the facility and nozzle are described by Gillespie (1993). The test gas was a 3.76/1 (molar) admixture of N_2 with O_2. Impurities totaling less than 25 ppm were comprised mostly of Ar and Kr, with less than 1 ppm of H_2O.

Figure 1 illustrates the experiment apparatus. The optical measurements began after the establishment of a quasi-steady nozzle flow. A tilted lens focused a KrF excimer laser pulse into the nozzle along its centerline. Light collected from the meridional focus was dispersed by a spectrograph, and detected by an ICCD array. The laser's wavelength was calibrated by exciting OH and O_2 fluorescence in a nearby flame. Further details are given by Gilmore et al. (1995).

Data were collected for the cases of pure N_2, pure O_2, as well as for the admixture. As shown in Table 1, all cases had similar reservoir temperatures and densities. The pure N_2 data allowed a check with Gillespie (1993). The pure O_2 data verified that the spectral region of the N_2 Raman band shape was free of O_2 fluorescence when the laser was tuned to exciting lines. The different vibrational relaxation rates of the pure gases also allowed evaluation of the data analysis techniques over the range of excitation expected in the admixture. Figure 2a shows a typical Raman band shape for the case of pure N_2.

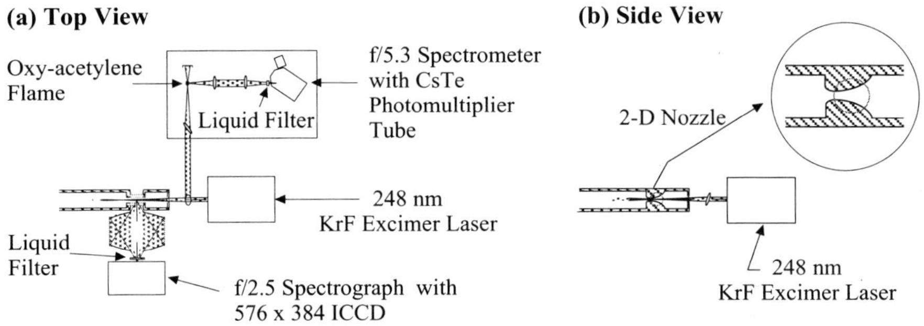

Figure 1. Experiment apparatus.

Table 1. Nominal thermodynamic states of the test gases in the nozzle reservoir.

Test Gas	Density [kg/m^3]	Temperature [K]	N_2 mole fraction	O_2 mole fraction	NO mole fraction	O mole fraction
N_2/O_2	12	2500	0.7780	0.1978	0.0235	7×10^{-4}
N_2	12	2600	1	0	0	0
O_2	15	2500	0	0.9986	0	1.4×10^{-3}

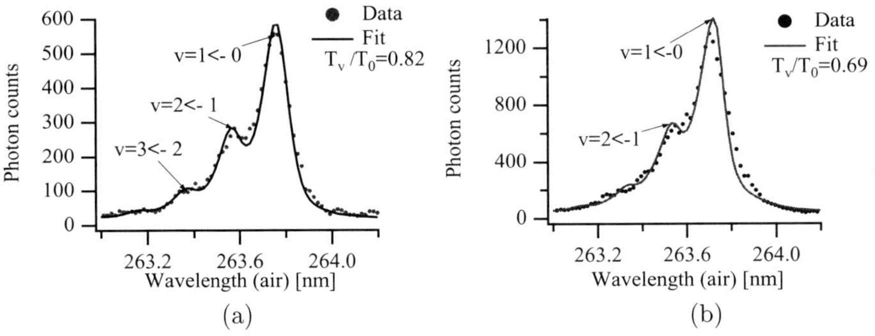

Figure 2. N_2 fundamental Stokes-Raman band shapes. The first several vibrational transitions are indicated: (a) pure N_2 4.1 cm from the throat, (b) admixture 2.9 cm from the throat.

Two sets of data were collected for the case of the admixture. The nominal reservoir conditions for both sets were the same, but differences were introduced in the details of the measurement technique.

The first set of data from the admixture was acquired with the laser tuned to minimize fluorescence. The laser was tuned to 40,257 cm^{-1} (vacuum), which was removed from O_2 absorption features. The optics were adjusted to obtain a nominal spectral resolution of 0.1 nm. The spatial resolution was 9 mm. Figure 2b presents a typical N_2 Raman band shape from this data set along with a synthetic band shape from the data analysis. The vibrational bands of the Q branches of both N_2 and O_2 were barely resolved at these thermodynamic conditions. No Raman scattering from NO was observed.

Another set of data was acquired with the laser tuned to excite fluorescence. The laser was tuned to 40,251 cm^{-1} which excited the 2←7 R(11) and 0←6 P(13) lines of O_2. The beam was also defocused by a factor of two relative to the latter data set. This resulted in a proportional decrease in Raman signal under room conditions. The optics were adjusted to obtain a nominal spectral resolution of 0.17 nm. An O_2 Raman/fluorescence band shape from this data set is in Fig. 3.

The spectrum in Fig. 3 shows the maximum fluorescence observed. The P(13) and R(11) lines of the 2→8 band were evident. Lines from the 0→7 band were not evident since the $\Delta v'' = 1$ band near 257 nm is amongst the strongest for the 2←7 pumped transitions, and amongst the weakest for the 0←6 pumped transitions. Overlaid on the $\Delta v'' = 1$ fluorescence/fundamental Raman band shape is the $\Delta v'' = 2$ band shape observed near 267 nm at the same nozzle location, but in a separate experiment. These data illustrate only the O_2 fluorescence since the Raman overtone was beneath the detection limit of the experiment. As evident from Fig. 3, the fluorescence did not interfere significantly with the lower vibrational transitions of the O_2 Raman Q branch. These transitions determined the vibrational and rotational temperatures in the data analysis. The small interference that approached the amount of photon statistical noise in the Raman Q branch was accounted for by including the measured $\Delta v'' = 2$ band shape in the data fitting procedures.

4. Analysis and discussion

The N_2 and O_2 spectra were analyzed by iteratively fitting computed band shapes to the data. Further details are given by Gilmore et al. (1995).

Figure 4 shows the results from the band fits for the cases of pure N_2 and pure O_2. Also included in the figure for comparison are the results of calculations assuming a Landau-Teller/Millikan-White formulation of the vibrational relax-

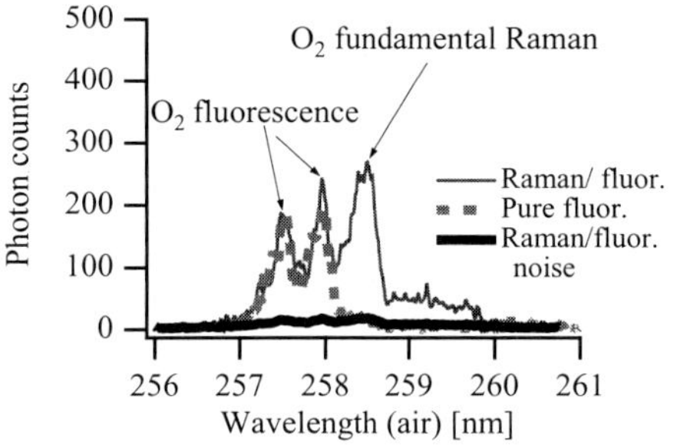

Figure 3. O_2 fundamental Raman and $\Delta v'' = 1$ fluorescence; $\Delta v'' = 2$ fluorescence and photon noise are overlaid.

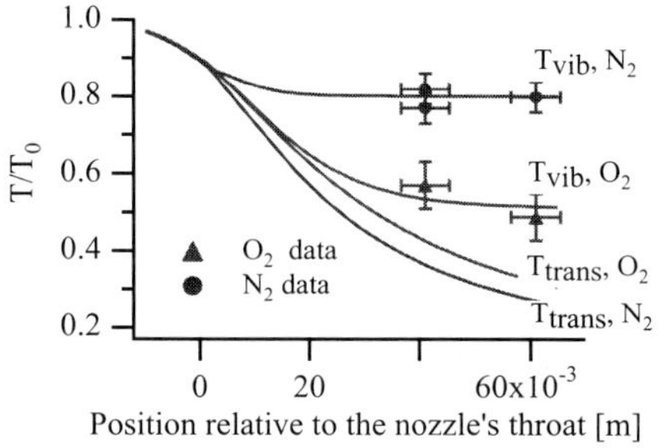

Figure 4. Results from the band fits to the pure-gas data. The spatial resolution is indicated.

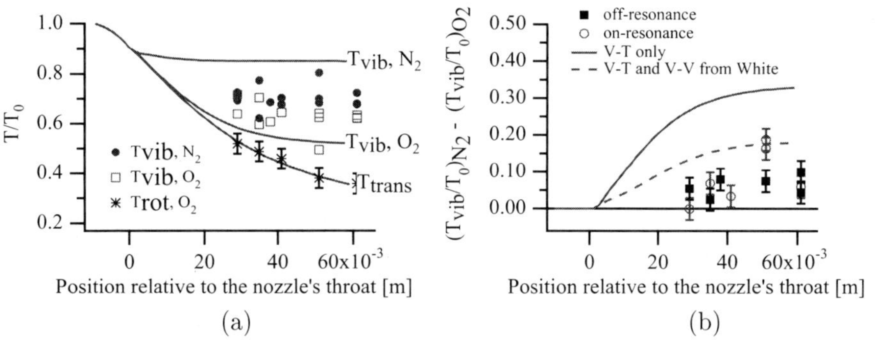

Figure 5. Results from the band fits to the admixture data: (a) calculation shows V-T behavior; (b) calculations show V-T behavior and V-T/V-V behavior from White.

ation. This model incorporates the effects of vibrational-translational (V-T) interactions. For the case of pure N_2, the present results agree well with those of Gillespie (1993). The pure O_2 results also compare well to the calculation. Thus, V-T interactions accounted for the relaxation of the pure gases.

Figure 5 shows the admixture results from both data sets. Also included are calculations for this case with the relaxation model that described the behaviors of the pure gases. Figure 5 shows that the data acquired with differing spectral resolution, laser focal volume, and laser frequency yielded similar results. When compared to the calculation, the data show that N_2 relaxed more rapidly, and O_2 less rapidly than V-T interactions could account for. This indicates that another energy transfer mechanism, the V-V mechanism, was active.

White and Millikan (1964) and White (1968) studied a similar admixture in a shock tube. The present results agree with White's conclusion that N_2 relaxed more rapidly than could be accounted for by V-T interactions. However, the present results also indicate that O_2 relaxed less rapidly than could be accounted for by V-T interactions. White did not observe this.

Figure 5b shows the normalized difference between the N_2 and O_2 vibrational temperatures determined from the present experiments and by two calculations. One calculation represents behavior due to V-T interactions (as in Fig. 5a). Another incorporates the equivalent N_2 relaxation time determined by White and Millikan (1964). The present data show that the O_2 vibrational temperatures were less than those of N_2, but the differences approached the experimental uncertainties. Comparison with the calculation incorporating White's results indicate a stronger coupling of the vibrational energies in the expansion.

Qualitatively, the stronger coupling in the case of an expansion as compared to a shock compression is likely due to the different environments for the relaxation in the two cases. The amount of coexistent vibrational nonequilibria in an expansion may increase the significance of non-linear cross terms in the species vibrational-energy transport equations. Also, the presence of additional constituents, such as NO, in the expansion may promote coupling by lowering the energy barrier to V-V exchange. Further comments are in Gilmore et al. (1995). Quantitative analysis of this coupling will be included in our future research.

5. Conclusions

Recent direct measurements of N_2 and O_2 vibrational populations in an expanding admixture have been reported. The results were consistent over a range of spectral resolution, laser focal volume, and laser frequency. N_2 relaxed more rapidly and O_2 less rapidly than V-T interactions could account for. The data indicate a stronger coupling of the N_2 and O_2 vibrational energies than reported by White in shock compressions of a similar admixture.

References

Gillespie W (1993) Raman Scattering Measurements of Vibrational Relaxation in Expanding Nitrogen, Ph.D. thesis, SUDAAR 636, Department of Aeronautics and Astronautics, Stanford Univ, Stanford, CA USA.

Gilmore J, Sharma S, Fletcher D, Bershader D (1995) Single-Pulse Raman Scattering Measurements in an Expanding Nitrogen/Oxygen Admixture, AIAA Paper 95-2125, Am Inst of Aeronautics and Astronautics, Washington, DC.

White D, Millikan R (1964) Vibrational relaxation in air. AIAA J 2:1844–1846.

White D (1968) Shock-tube study of vibrational exchange in N_2-O_2 mixtures. J Chem Phys 49:5472–5476.

Laser-induced thermal acoustics (LITA) measurement of gas properties

E. B. Cummings, H. G. Hornung
Graduate Aeronautical Laboratories
California Institute of Technology
Pasadena, CA 91125 USA

M. S. Brown
Metrolaser
Irvine, CA 92714-6428 USA

Abstract: Laser-induced thermal acoustics (LITA), an optical four-wave mixing technique, has been applied to measure sound speed and thermal diffusivity in gases. Signals from experimental studies of LITA are accurately predicted and interpreted by the use of an analytical theoretical model.

Key words: LITA, DFWM, Thermal grating, Velocimetry

1. Introduction

Laser-induced thermal acoustics (LITA) is a nonlinear optical technique for remotely measuring the sound-speed and transport properties of a fluid with accuracy comparable to the best conventional intrusive methods (Cummings 1995). LITA excels in high-pressure environments, where many other diagnostics fail. The development of the technique has addressed issues including the need to make single-pulse measurements in challenging environments by the use of affordable apparatus. We describe the physics of LITA and outline a derivation of an analytical expression for the LITA signal. We provide experimental validation of the expression and present results of a exploratory LITA experiments.

The accuracy of LITA measurements (typically ~0.1% for sound speed, ~1% for transport properties at high pressure) permits the use of LITA for making fundamental measurements of gas properties and gas equations of state. Our experiments have focused mainly in this area. However, LITA may also be applied for making other measurements in systems of scientific or engineering interest. LITA measurements in the primary and reflected-shock regions at the exit of a combustion-driven shock tube are imminent. These tests will assist in designing a system for making LITA measurements on T5, a large free-piston shock tunnel. The ability to measure transport properties is important both for fundamental and applied studies, since transport is of primary importance in phenomena ranging from combustion to boundary-layer behavior in hypersonic flows.

2. Physics of LITA

LITA scattering involves *driving* of perturbations of a medium via opto-acoustic effects and *coherent scattering* of light off these perturbations into a coherent signal beam via acousto-optic effects. In the driving step, light from a pulsed laser is split into two phase-coherent beams that intersect at a shallow angle in the medium. The region of intersection of the beams defines a sample volume of the LITA measurement. Interference of the driver beams creates an electric field intensity grating in the sample volume. The medium responds to the intensity of these driver beams. In an effect called *thermalization*, light energy absorbed by the medium subsequently heats the medium locally over a short time. The medium expands thermally upon heating. If the thermal expansion speed is comparable to the sound speed, sound waves are generated. A second effect is called *electrostriction*, in which molecules of the medium are accelerated toward or away from regions of intense light, depending on molecular properties. If the driving is rapid, the acceleration is impulsive and sound waves are created. If the driving is slow, molecules diffuse into or away from the regions of intense light, creating species concentration gradients. Through these and other generalized opto-acoustic effects, the driver laser can perturb the bulk properties of a medium.

The perturbations induced by the driver laser create perturbations in the susceptibility of the medium via acousto-optic effects. These perturbations therefore have the appearance of volume diffraction gratings. In the coherent scattering step, a CW or long-pulse source laser is trained on the sample volume at the Bragg angle of the gratings. Light from the source laser is scattered off the gratings into a coherent signal beam that is detected through large f-number optics. The signal is modulated in time by the evolution of the laser-induced perturbations. From this modulation, accurate physical properties of the medium may be inferred, including the sound speed, thermal diffusivity, and other properties that affect the evolution of the laser-induced gratings.

An analytical expression for LITA signals has been derived (Cummings et al. 1995). This expression includes finite-beam-size and duration effects. The effects of electrostriction and finite-rate thermalization are treated using first- and second-order density-operator perturbation theory, respectively. The evolution of the laser-induced gratings is treated using the linearized equations of hydrodynamics. The equation for the far field of light scattered linearly from a density perturbation provides the distribution and magnitude of the electrical field of the signal beam scattered from these gratings. This field is integrated appropriately over the detection angle to provide the absolute amplitude and time history of the detected LITA signal. The expression for the LITA signal depends upon fluid properties, notably the sound speed, thermal diffusivity, acoustic damping rate, the characteristic thermalization rate, and the susceptibility of the medium at the driver- and source-laser wavelengths. In addition, the signal depends upon the beam geometries and time histories.

3. Experimental apparatus and data analysis

Figure 1 shows a schematic diagram of the experimental apparatus used in narrow-band and multiplex LITA experiments. The driver laser is a pulsed dye laser pumped by a 10-Hz, Q-switched, frequency-doubled Nd:YAG laser. The dye laser emission is typically ~20 mJ in ~5 ns pulses at ~590 nm. A 50% beam splitter (*bs*) evenly splits the driver laser emission. The split beams are synchronized in time and made parallel using mirrors $m2$, $m3$, and $m4$. The driver beams intersect at the focus of the 500-mm lens $\ell 1$. The beam-crossing angle is typically ~1°. The driver beams intersect inside a high-pressure test cell. The source laser used in narrow-band LITA experiments is a CW argon-ion laser producing ~1.3 W at 488 nm. A chopper wheel passes ~1-ms pulses from the CW beam at 10 Hz, synchronized with the driver laser pulses. A BOXCARS forward phase-matching geometry is used (Eckbreth 1988). A silicon PIN photodiode, $d2$, monitors the intensity of the source beam. The narrow-band LITA detection system consists of a fast photomultiplier tube (PMT) and optical filters to prevent signal contamination, including a 10-nm interference filter $f1$ at 488 nm and a 40-mm lens, $\ell 2$, and 20-μm pinhole spatial filter, *ph*. The LITA-signal path length to the detector is about 2.5 m. A 2-GS/s/channel, eight-bit, four-channel digital storage oscilloscope (DSO) records time histories of signals from the PMT and photodetectors.

Gas property measurements are extracted from the recorded signals by a process of optimal nonlinear filtering. The nonlinear filter is a global least-squares fit between theory and experiment with the gas properties as adjustable parameters. This filtering vastly reduces the effect of signal noise and improves measurement accuracy. However, systematic modeling errors can lead to skewed measurements (Cummings 1995).

4. Results

An initial set of LITA experiments was performed on NO_2-seeded static gases in a high-pressure cell with optical access to evaluate the fidelity of the theoretical model, to assess the practicality and range of applicability of LITA, and to identify experimental issues which may limit the accuracy of LITA measurements. Parts-per-million levels of NO_2 enhanced the absorption of the visible driver beams and thereby thermalization, providing strong LITA signals. The static-cell tests were conducted with air, He, and CO_2 around 300 K.

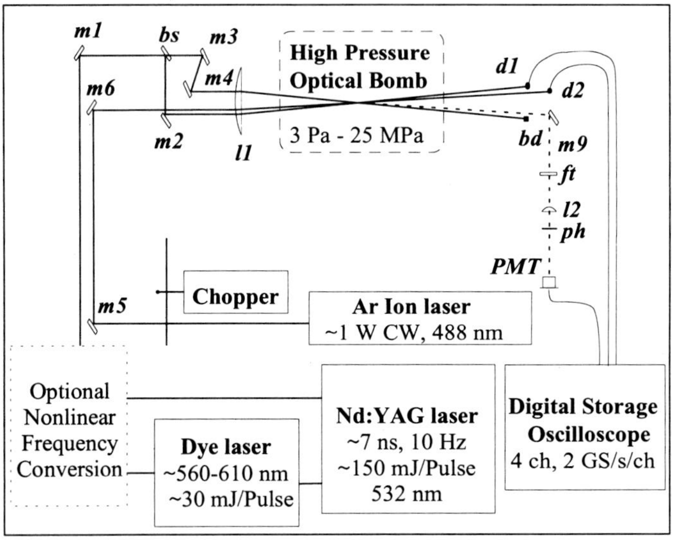

Figure 1. Schematic diagram of narrow-band LITA experimental arrangement.

Typical experimental LITA signal time histories and theoretical fits appear in Fig. 2. The nature of the signals is discussed in Cummings et al. (1995). The analytical expression accurately matches LITA signals over a wide range of pressures (Cummings 1995). The sound speed of air extracted through the nonlinear filtering of LITA signals taken over a range of pressures from ~0.1–100 atm appears in Fig. 3. The agreement with published data of Hilsenrath (1955) is ~0.1% and better above ~1 atm. Below atmospheric pressure, finite grating-Knudsen-number effects invalidate the hydrodynamics analysis and effects like rotational freezing contribute to significant errors in the measurements. Thermal diffusivity measurements of air appear in Fig. 4. At low pressures, details of the thermalization process that are ignored in the basic analysis cause measurements to be ~10% lower than published values. At higher pressures, accuracy is much better (~1%), although accurate measurements require accurate beam alignment. Because the theoretical model assumes perfect alignment, misalignment introduces a systematic modeling error. This error causes the nonlinear filter to skew the values that it extracts for the thermal diffusivity, typically by ~10%. In future experiments this sensitivity to beam alignment will be mitigated by the use of unfocused beams. Other issues that were identified include beam-steering, self-focusing, and self-defocusing effects that were apparent in high-pressure gases, particularly near-critical CO_2.

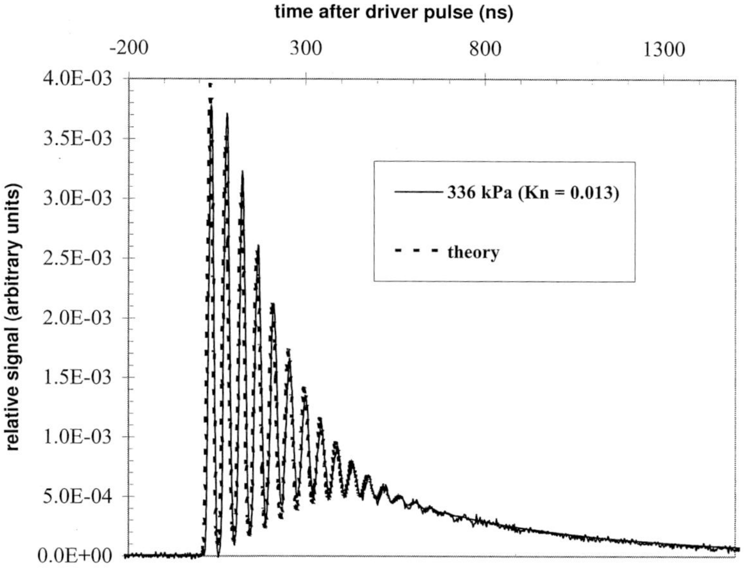

Figure 2. Experimental and theoretical LITA signals from thermalization in air at ~300 K.

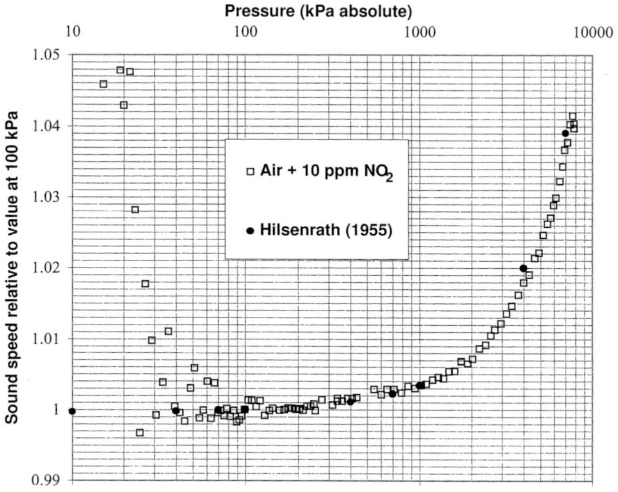

Figure 3. Variation of sound speed of air at ~300 K with pressure. Squares are LITA measurements; independent published data appear as solid circles.

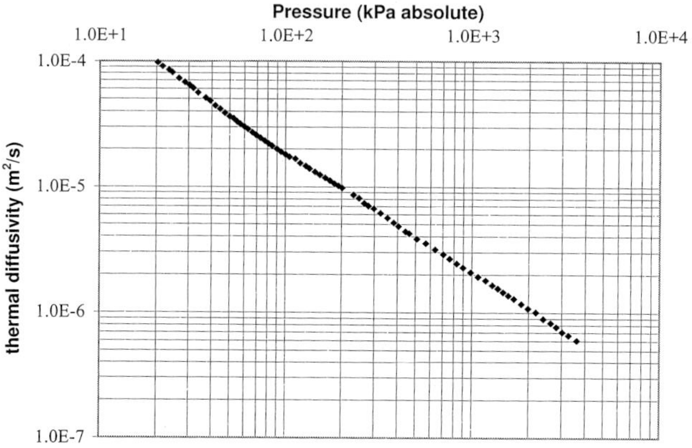

Figure 4. Variation of thermal diffusivity of air at ~300 K with pressure. Measurements are extracted from LITA signals.

5. Conclusions

LITA has been demonstrated to be an accurate tool for measuring gas properties. The analytical expression for the signal accurately models experimental signals. Over a wide range of pressure and in gases ranging from He to near-critical CO_2, LITA measurement of the sound speed featured ~0.1% accuracy, limited at low pressures by finite grating-Knudsen-number effects. Thermal diffusivity measurements at elevated pressures featured ~1% accuracy under ideal experimental conditions, although minor beam misalignments were a systematic error that reduced the accuracy typically by a factor of 10.

Acknowledgement. This work was supported by the Air Force Office of Scientific Research under Grant Nos. F 49620-92-J-0110 and F 49620-93-1-0338.

References

Cummings EB (1995) Laser-Induced Thermal Acoustics, Ph.D. Thesis, California Institute of Technology.

Cummings EB, Leyva IA, Hornung HG (1995) Laser-induced thermal acoustic (LITA) signals from finite beams. Appl Opt 34:3290.

Eckbreth AC (1988) Laser Diagnostics for Combustion Species and Temperature, Cambridge, MA, Abacus Press.

Hilsenrath J (1955) Tables of Thermal Properties of Gases, NBS Circular 564.

PLIF imaging of nitric oxide in high enthalpy flows

P. C. Palma, T. J. McIntyre*, A. F. P. Houwing
Aerophysics and Laser-based Diagnostics Research Laboratories
Department of Physics, Faculty of Science
Australian National University, Canberra, 0200 Australia

Abstract: Planar laser-induced fluorescence is performed in a free-piston shock tunnel. Nitric oxide, which occurs naturally in the flow, is excited with the 225-nm radiation from a Raman-shifted tunable excimer laser. The Mach 7 flow over a 25 mm diameter cylinder is imaged for three different rotational lines and a temperature map of the flow field is calculated. The maximum flow enthalpy was limited to 4.1 MJ/kg due to high flow luminosity which is produced by metallic contaminants in the flow. A reflective filter is used to reduce the influence of flow luminosity making these measurements feasible. Freestream temperature measurements are in excellent agreement with those predicted from numerical flow calculations. Several higher enthalpy shots (14 MJ/kg) with a wedge in the flow were also performed with an insignificant amount of contaminate emission being observed.

Key words: Planar laser-induced fluorescence, Flow visualization, Shock tunnel

1. Introduction

Planar laser-induced fluorescence (PLIF) has recently been applied to the study of shock-tunnel flows with great success (Palmer et al. 1994). This technique allows two-dimensional visualization of rotational and vibrational temperatures, velocities, and chemical-species number densities. In the past, shadowgraph, schlieren, and interferometry have been used to provide non-intrusive measurements of flow parameters and supplement surface-pressure and hea-transfer measurements. These techniques can usually only provide integrated measurements through the flow and therefore mask any three-dimensional behavior. PLIF can provide spatially resolved cross sections through the flow at virtually instantaneous moments of the flow development, hence overcoming limitations associated with the above-mentioned line-of-sight techniques.

Nitric oxide, NO, is naturally present in heated air flows such as the ones examined here. This situation is much preferable to seeding a chemical species

*Present address: Dept. of Physics, University of Queensland, Brisbane, 4072, Australia.

into the flow which can often be impractical (NO is toxic) or which may influence the dynamics of the flow. NO has some advantageous spectroscopic properties that make it an excellent choice for PLIF thermometry. These include collisional quenching rates and excited-state radiative lifetimes that are independent of rotational quantum number. It also exhibits strong non-resonant fluorescence which alleviates problems associated with radiative trapping and laser scatter. The spectroscopy of NO is also very well characterized, which significantly aids in data analysis and modeling.

Applying PLIF to free-piston shock tunnels provides a unique set of experimental problems not encountered in other flow facilities. Real-gas effects are prominent and the high temperature and pressure variations in the imaging region mean careful consideration of PLIF imaging parameters is necessary. Metallic contaminants which are eroded from the shock-tube walls produce strong emission in high-temperature regions in the flow field. This background emission can easily swamp fluorescence signals and so careful spectral filtering is necessary (Palma et al. 1993). Conventional dielectric filters were found to be unsuitable since they attenuate the signal by as much as 80%. By using a narrow-band dielectric mirror however, narrow-band filtering can be employed without signifcant loss of signal.

2. Theory

The theory of PLIF thermometry has been well described elsewhere and is not reviewed here (see Seitzman et al. 1994, Palmer et al. 1994). Essentially, PLIF gives a relative measure of the population of the probed state. If two different rotational states are probed and the ratio of the signals taken, the other signal dependencies (fluorescence yields, quenching rates, collection efficiencies, etc.) cancel out and the ratio is only dependent on the temperature. Different transition pairs have different temperature sensitivities and careful choice of the pair of transitions is necessary.

Collisional broadening and shifts of absorption lines are also a major concern in shock-tunnel flows due to the large pressure and temperature variations. Since these effects are J independent they can be eliminated from the signal ratio. However, if the overlap between the laser spectral profile and the absorption line is significantly reduced, the fluorescence signal may become too weak to detect against background flow luminosity.

For PLIF thermometry in unsteady or turbulent flows it is necessary to use two lasers and cameras to excite the two transitions in rapid sequence. However, for steady flows it suffices to use one laser and camera and assume a degree of shock-tunnel repeatability by taking images on successive shots of the tunnel. The latter method has been employed here.

3. Experiment

The experiments were performed on the T2 free-piston shock tunnel (Stalker 1967) at the Australian National University. The T2 nozzle is a 30-degree full-angle conical geometry with a 6.4-mm diameter throat and an area ratio of 144. The flow field studied was a Mach 7 flow over a 25-mm diameter cylinder. The primary shock speed was 1.90 km/s giving a flow enthalpy of 4.1 MJ/kg. The nozzle reservoir pressure was 28.4 MPa and the calculated temperature was 3320 K. The freestream pressure and temperature were calculated to be 3 kPa and 340 $\pm$ 30 K respectively and the freestream NO concentration was calculated to be 5%.

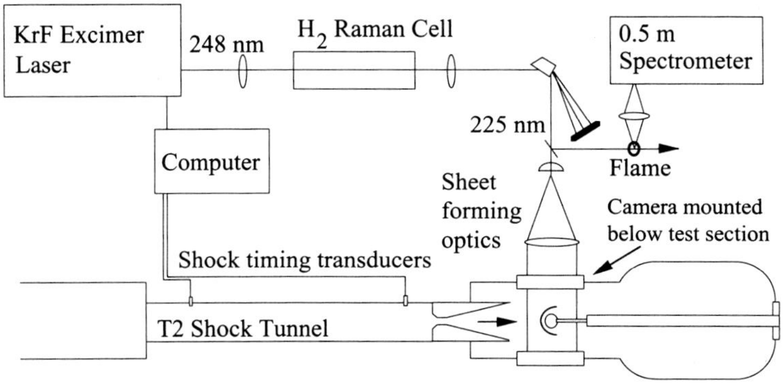

Figure 1. Schematic of experimental set up. The intensified CCD camera is mounted below the test section. The laser sheet monitor has not been included but would normally be placed immediately before the test section window.

A schematic of the experimental set up is shown in Fig. 1. A cell containing high pressure hydrogen (1 MPa) is used to Raman shift the output of a tunable excimer laser (Lambda Physik EMG150ETS) operating with KrF at 248 nm. The first anti-Stokes order is separated from the other orders with a Pellin-Broca prism and formed into a sheet using a 40-mm focal length cylindrical lens and a 1000-mm focal length spherical lens. This radiation is tunable between 224.9 and 225.4 nm which coincides with the $A^2\Sigma^+ \leftarrow X^2\Pi(0,0)$ band of NO. A small portion of the beam is split off and used for wavelength calibration by passing it through an H_2 + O_2 flame. The laser-induced fluorescence (LIF) from the flame is imaged onto a 0.5-m spectrometer (which acts as a 9-nm bandpass filter), allowing LIF from only the NO (0,1) vibrational band near 236 nm to be detected. The energy of the Raman-shifted beam was typically 4 mJ per pulse with a linewidth of approximately 0.25 cm^{-1}.

The laser is tuned to a particular rotational line prior to each firing of the shock tunnel. Immediately before firing, the tunnel operator stops the laser via a remote switch. After firing, the nozzle reservoir pressure transducer detects the shock reflection at the end of the shock tube and the laser is fired 450 μs later, this delay being chosen so as to coincide with the period of steady flow. An intensified CCD camera (250-ns gating period) is used to capture the fluorescence image at right angles to the laser sheet. The fluorescence is turned through 90 degrees by reflecting it off a dielectric mirror (normal incidence KrF excimer mirror) which had maximum reflectance between 210 and 240 nm. A 2-mm thick UG-5 Schott glass filter is used to block off elastic scatter at the laser wavelength, thus allowing only the (0,1) and higher vibrational bands through to the camera. The dielectric mirror and short camera gating time were able to efficiently filter out the broad-band background contaminant emission.

The variation in energy across the laser sheet is measured simultaneously by use of a dye cell and CCD camera (Seitzman et al. 1994). This energy-distribution measurement is then used to correct the PLIF image and it also takes account of any pulse-to-pulse fluctuations in the total laser energy.

4. Results

Three rotational lines were chosen within the tuning range of the laser to make a wider range of temperatures measurable. Figures 2a,b, and c show the PLIF images for the three different rotational lines. Each image is the average of five shots and each has been corrected for flow luminosity and the laser sheet profile. In Fig. 2a there is strong fluorescence in the freestream because the J = 18.5 level is strongly populated at the freestream temperature. The edges of the core flow are clearly visible in this image. Fig. 2b shows less freestream signal and in Fig. 2c the signal is not significantly above the background noise due to the increasing rotational quantum number (J = 27.5 and 38.5 respectively).

Images in Fig. 2d and e show temperature maps derived from the first three PLIF images of Fig. 2 and calibrated by calculating the appropriate constants. The image in Fig. 2d was derived from the images for the J = 18.5 and 27.5 lines. This line-pair ratio is accurate for temperatures between 200 and 1000 K which is suitable for the freestream temperature measurement. From this image the freestream temperature was determined to be 320 $\pm$ 30 K which agrees with the calculated value of 340 $\pm$ 30 K. The temperatures behind the shock in this image are not valid since this line-pair ratio is only accurate for temperatures below 1000 K. Likewise in image Fig. 2e the freestream temperature is not valid since this line-pair ratio (J = 18.5 and 38.5) is only accurate for temperatures between 1000 and 3200 K. Image Fig. 2e shows the temperature distribution behind the shock. Immediately in front of the body a temperature of about 3200 K is

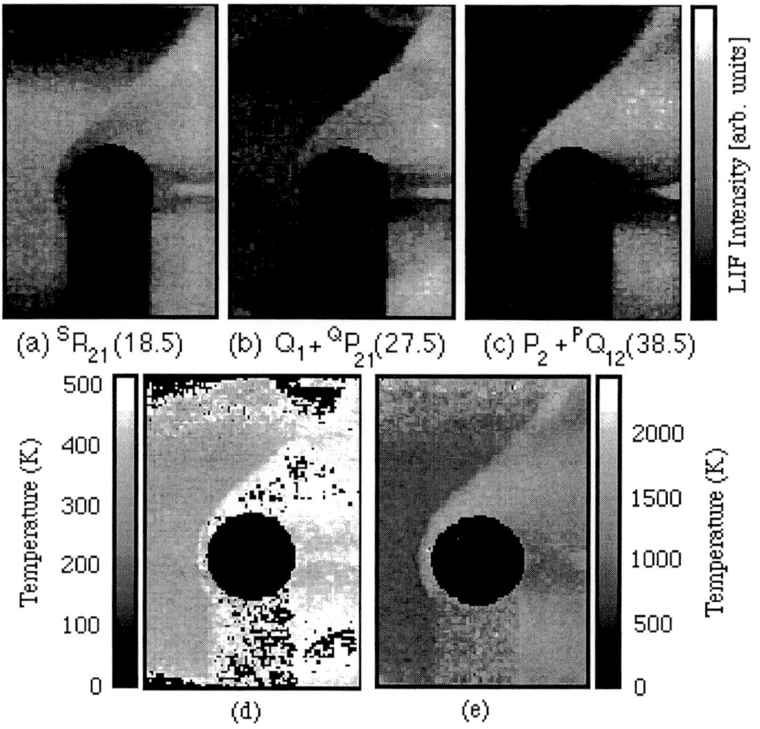

Figure 2. Mach 7 flow over a 25-mm diameter cylinder at 4 MJ/kg. The flow in each image is from left to right and the laser sheet is entering from the top producing the shadow beneath the cylinder: (a), (b), and (c) are PLIF images for three different NO rotational lines; (d) is the temperature calculated from (a) and (b); (e) is the temperature calculated from (a) and (c).

expected but only 1800 to 2000 K is measured. This discrepancy is undoubtedly due to the high collisional quenching in this region (pressure = 300 kPa) which, as can be seen in Fig. 2a and b images, has significantly reduced the fluorescence signal intensity to below the limit of sensitivity of the camera.

4.1. Higher enthalpy shots

Several shots were performed at a 14-MJ/kg condition to ascertain the effectiveness of the filtering used to eliminate the background contaminate emission. A 35-degree full-angle symmetric wedge was used since with the current filtering there was too much emission to go much above 4 MJ/kg with the cylinder.

Figure 3 shows a single-shot raw PLIF image of the flow over the wedge. The interaction between the corner expansion and the shock can be clearly seen, and emission near the wedge tip is just visible.

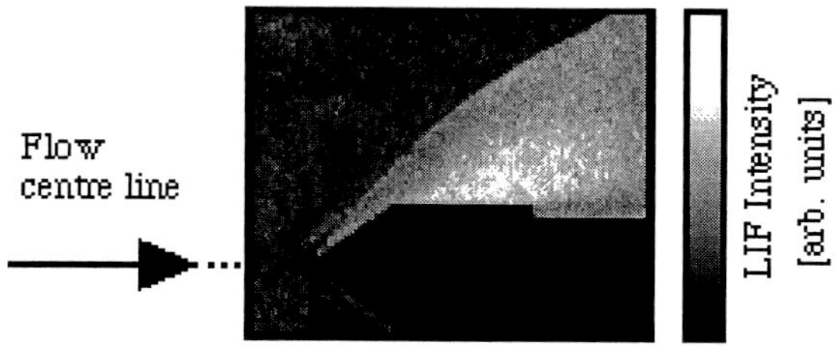

Figure 3. Raw PLIF image of flow over a 35 degree wedge at 14 MJ/kg.

5. Conclusions

PLIF imaging of NO in a shock-tunnel flow has been presented. A Raman-shifted tunable excimer laser has been used to probe naturally occuring NO in a Mach 7 flow over a cylinder. The measured freestream temperature shows excellent agreement with theory and the discrepancy behind the shock has been explained as being due to collision quenching of the excited state population. Shots performed at 14 MJ/kg on a wedge demonstrate the effectiveness of the filtering employed to remove background emission. However, the spectral filtering and camera gating times can still be significantly improved upon to increase the usefulness of this technique on free-piston shock tunnels.

Acknowledgement. The authors would like to thank Paul Walsh for his technical assistance. This work was funded by the Australian Research Council.

References

Palma PC, Houwing AFP, Sandeman RJ (1993) Absolute intensity measurements of impurity emissions in a shock tunnel and their consequences for laser induced fluorescence experiments. Shock Waves 3:49–53.

Palmer JL, Houwing AFP, Thurber MC, Wehe SD, Hanson RK (1994) PLIF Imaging of Transient Shock Phenomena in Hypersonic Flows. AIAA Paper No. 94-2642.

Seitzman JM, Hanson RK, DeBarber PA, Hess CF (1994) Application of quantitative two-line OH PLIF for temporally resolved planar thermometry in reacting flows. Appl Opt 33:4000–4012.

Stalker RJ (1967) A study of the free-piston shock tunnel. AIAA J 5:2160.

A skin-friction gauge for impulsive facilities

C. P. Goyne, A. Paull, R. J. Stalker
Department of Mechanical Engineering
The University of Queensland
Queensland 4072 Australia

Abstract: A new skin-friction gauge has been designed for use in impulsive facilities. Experiments were conducted in the T4 free-piston shock tunnel at the University of Queensland. Measurements were conducted in laminar and turbulent boundary layers using air as a test gas. The test-section free stream had a 4.7-MJ/kg stagnation enthalpy and a Mach number of 6.2. The measurements compared well with laminar and turbulent analytical theory.

Key words: Skin friction, Boundary layer, Shock tunnel

1. Introduction

The accurate prediction of skin friction is of vital importance to the operation of any vehicle traveling at hypersonic speeds. Skin-friction drag can become the most dominant factor affecting the limit of vehicle performance because of its strong velocity dependence. In order to cope with the encountered heat-transfer loads at such speeds, a vehicle has to rely on cooling by passive or active means. This results in a regime of high-enthalpy flow over highly cooled surfaces, particularly with high Reynolds numbers. Heat-transfer measurements are often used to predict wall-shear levels, however this method can not be conclusively used on flows within this regime. This problem is exacerbated by a lack of simultaneous measurements of skin-friction and heat-transfer under such conditions (Holden 1972).

In order to directly measure skin-friction in high-enthalpy flows, a new skin-friction gauge has been designed for use in the T4 free-piston shock tunnel at the University of Queensland. The work follows that of Kelly et al. (1992) and utilizes a different principle for strain-gauge-type sensors (Novean et al. 1995) because a faster response time is required. This paper presents research results with the new gauge including measurements in a laminar boundary layer and tentative turbulent boundary-layer measurements. The experiments were conducted in air with a nominal Mach number of 6.2 and stagnation enthalpy of 4.7 MJ/kg.

2. Gauge description and calibration

The gauge incorporated a piezoceramic measuring element and therefore possessed a suitably fast response time for use in impulsive facilities. The design included an acceleration-measuring element which was used for acceleration compensation. The compensation element was not exposed to the flow. Both elements consisted of a 10-mm diameter, 0.8-mm thick invar disk which covered a $7 \times 7 \times 1.5$-mm lead zirconate piezoceramic block. This low-thermal-expansion invar disk was of the appropriate thickness to protect the ceramic from heat-transfer loads until the end of the test time. The configuration resulted in a gauge with a natural frequency of 30 kHz.

The gauge also incorporated a brass housing, to act as an inertial mass, a felt mounting system which vibrationally isolated the gauge from the model, and internally mounted charge amplifiers.

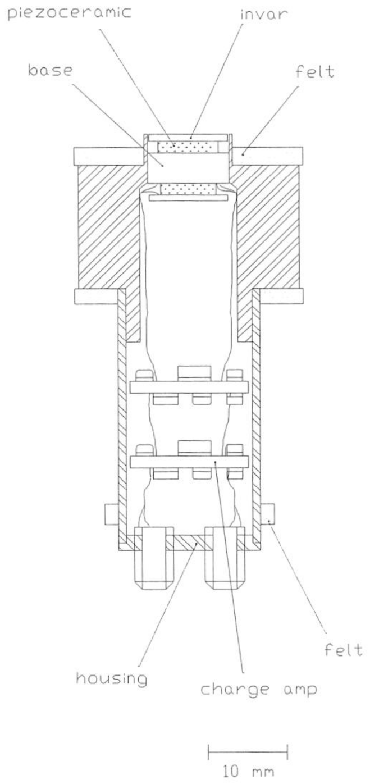

Figure 1. Skin-friction gauge schematic.

The gauge was calibrated for shear, pressure, and acceleration separately in bench tests. The technique used in each case was chosen to closely reflect the operational environment of the gauge.

The shear calibrations were obtained by mounting the gauge in a prone position and loading the measuring invar disk with an effective shear stress. A known weight was hung vertically by an adhered cotton thread and impulsively released with a sharp blow from beneath. Step inputs with a rise time of 2 ms were typically achieved. Figure 2a presents a typical shear calibration.

Pressure calibrations were obtained using a small shock tube positioned over a mounted gauge. The tube was 430 mm long and atmospherically driven by manually bursting a cellophane diaphragm. The rig was capable of step pressure changes near 130 μs. A typical pressure calibration is presented in Fig. 2b.

By mounting a gauge in the model and impulsively loading the model with a blow from a hammer, a calibration for acceleration between the measuring and acceleration elements was obtained. Ascertainment of this factor allowed for routine subtraction of the acceleration component during testing.

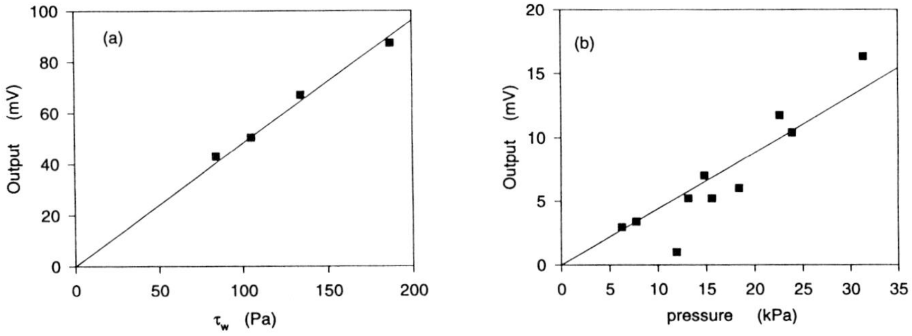

Figure 2. Typical skin-friction gauge calibration for (a) shear and (b) pressure.

3. Model and test conditions

The experimental measurements were conducted in T4 using an instrumented 1.5-m long flat plate that formed one of the inner walls of a rectangular duct, 120 by 60 mm at the inlet. In order to allow for boundary-layer displacement, the non-instrumented walls each had a half-degree divergence. Pressure and thin-film heat-transfer gauges were mounted every 50 mm along axial lines, 25 mm each from the plate centerline. The first transducer station was located at 370 mm from a sharp leading edge. Skin-friction gauge measurements were conducted along the plate centerline at 370, 770, 970, and 1170 mm from the leading edge. The measurements were made using a single gauge located at different positions in separate runs in T4. The flow conditions were nominally the same in each run.

The experiments were conducted in air. During the test time, the measured test-section static, Pitot, and stagnation pressures were 5 kPa, 370.1 kPa, and 26.6 Mpa, respectively. The static wall pressure along the instrumented plate was constant to within a 20% variation. Using ESTC (McIntosh 1968) and NENZF (Lordi et al. 1966), the free-stream stagnation enthalpy, Mach number, and static temperature were calculated to be 4.7 MJ/kg, 6.2, and 498.9 K, respectively.

4. Gauge measurements and analysis

The skin-friction gauge measurements were found to contain a spurious component that was believed to be due to heat-transfer between the gauge and the test-section flow. A method was devised to compensate for this effect (Goyne et al. 1995). This involved taking a measurement in the laminar part of the boundary-layer during one run in T4, and comparing this with a measurement at the same location during another run, but with the gauge rotated by 180 degrees relative to the flow. A compensated shear-signal time history for a measurement in the laminar part of the boundary-layer, at 370 mm from the leading edge, is presented in Fig. 3. During the test time, from 5.3 to 5.6 ms, the signal agrees well with the level predicted by laminar theory (van Driest 1952). For comparison, heat-transfer measurements near the skin-friction measurement location are presented in Fig. 4. As Reynolds analogy implies, the skin-friction does follow the heat-transfer trends.

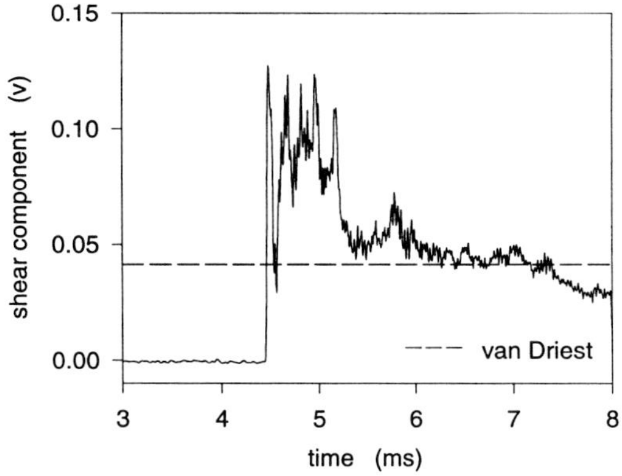

Figure 3. Compensated skin-friction gauge signal and theory of van Driest (1952). Time average of 50 μs.

Compensated skin-friction measurements in the turbulent part of the boundary layer, at 770, 970, and 1170 mm, are presented in Fig. 5. These mea-

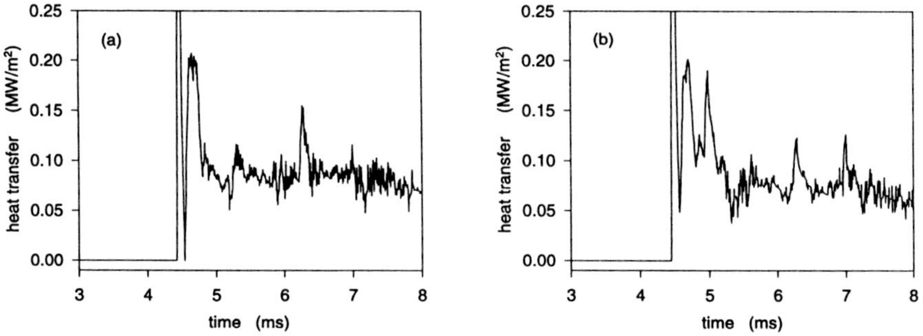

Figure 4. Wall heat-transfer at (a) 370 mm and (b) 420 mm from the leading edge. Time average of 40 μs.

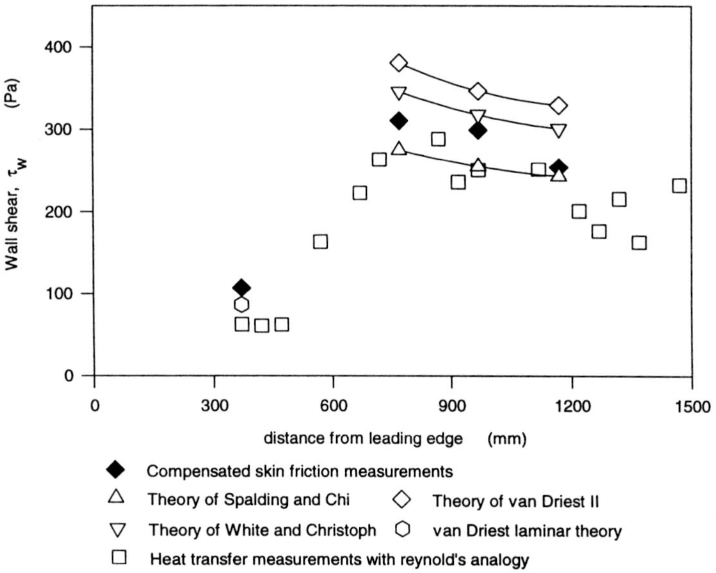

Figure 5. Skin-friction measurements and theory vs. distance from leading edge.

surements are considered tentative because of the compensation technique employed. Also presented are the predictions for turbulent wall shear by three compressible theories, White and Christoph (1972), van Driest (1956), and Spalding and Chi (1964). The calculations were performed with a turbulent virtual origin at 470 mm. It can be seen that the turbulent skin-friction measurements fall within the range predicted by the turbulent theories. The heat-transfer measurements, with Reynolds analogy applied, are also presented for comparison. These measurements confirm that transition to a turbulent boundary-layer did occur.

5. Conclusions

A new skin-friction gauge has been designed for use in impulsive facilities. The gauge had a natural frequency of 30 Hz and was calibrated for shear, pressure, and acceleration in separate bench tests. It has been tested in the T4 free-piston shock tunnel using air as a test gas with a stagnation enthalpy of 4.7 MJ/kg and Mach number of 6.2. Shear measurements in a laminar boundary layer and tentative measurements in a turbulent boundary layer compared well with theory.

Acknowledgement. The authors wish to thank the National Aeronautics and Space Administration for support under grant NAGW-674.

References

Goyne CP, Paull A, Stalker RJ (1995) A skin-friction gauge for impulsive flows. In: 31st Joint Propulsion Conference, AIAA 95-3152.

Holden MS (1972) An Experimental Investigation of Turbulent Boundary Layers at High Mach Number and Reynolds Numbers, NASA CR-1121147.

Kelly GM, Simmons JM, Paull A (1992) A skin-friction gauge for use in hypervelocity impulse facilities. AIAA J 30:844–845.

Lordi JA, Mates RE, Moselle RJ (1966) Computer Program for the Numerical Simulation of Non-Equilibrium Expansion of Reacting Gas Mixtures, NASA CR-472.

McIntosh MK (1968) Computer Program for the Numerical Calculation of Frozen and Equilibrium Conditions in Shock Tunnels, Department of Physics, Australian National University, Australia.

Novean M, Schetz J, Hazelton D, Bowersox R (1995) Skin-friction measurements in short duration, high enthalpy flows. In: 6th International Aerospace Planes and Hypersonics Technologies Conference, AIAA 95-6109.

Spalding DB, Chi SW (1964) The drag of a compressible turbulent boundary layer on a smooth flat plate with and without heat transfer. J Fluid Mech 18:1.

van Driest ER (1952) Investigation of Laminar Boundary-Layer in Compressible Fluids Using the Crocco Method, NACA TN-2597.

van Driest ER (1956) The problem of aerodynamic heating. Aero Eng Review, October.

White FM, Christoph GH (1972) A simple theory for the two-dimensional compressible turbulent boundary layer. J Basic Eng, September.

Imaging and quantitative measurement of transverse injection in supersonic flow using LIF

H. Shida, T. Nagayama, Y. Deguchi
Nagasaki Research and Development Center
Mitsubishi Heavy Industries Ltd.
1-1 Akunoura-Machi, Nagasaki 850-91 Japan

Abstract: NO-based PLIF methods of imaging and quantitative measurement of injectant concentration of transverse injection in supersonic flow are shown. The instantaneous and multi-frame average images of 3D and 2D gas jets and the consequent distribution of NO concentration were measured; such distributions were also measured by other optical methods such as schlieren and Mach-Zehnder interferometry.

Key words: Laser-induced fluorescence

1. Introduction

The laser-induced fluorescence (LIF) method of fluid-flow analysis features a high level of versatility including not only flow visualization, one of advantages typically offered by optical techniques, but also the quantitative, or even 3D, measurement of such flow state quantities as temperature, pressure, velocity, density, concentration, etc. and, in addition, the capturing of a large number of data at as many as about $\sim 10^5$ space points. Expectation is high that LIF will prove useful as a new engineering tool for analysis, particularly of combustion flow, supersonic flow, and hypersonic flow. Though the transition in research on LIF from the pioneering stage to the present status is outlined in the paper presented by Hanson et al. (1990), research conducted so far has focused principally on the flow visualization or imaging as exemplified by studies on the imaging of combustion flow for measurement (Kychakoff et al. 1982), the imaging of supersonic combustion flow for measurement (Lee et al. 1991), the imaging of mutual interference of two gas jets (Honma et al. 1993), etc. Attempts at LIF imaging of gas flow with an emphasis on quantitative measurement of flow state quantities are exemplified by such studies as the measurement of distribution of NO and OH concentrations in the gas turbine combustor (Deguchi et al. 1993), the measurement of concentration distribution of a gas (I_2) jet injected perpendicularly into supersonic transverse fluid flow (Hollo et al. 1988), etc. The authors conducted this time a series of tests in a shock wind tunnel, in which sonic jet of N_2 gas containing 0.1% of NO was injected perpendicularly into the supersonic transverse free flow of N_2 gas for the LIF imaging of the injectant with NO as well as for the quantitative measurement of the admixed NO concentration. Consequently the instantaneous and time-average phenomena of the sonic jet of N_2 gas mixing into the supersonic free N_2 gas flow of $M_\infty = 2.8$ were successfully visualized with NO followed by the quantitative measurement of NO concentration, as herein reviewed.

2. LIF method of fluid-flow analysis

LIF, representing one of the modern optical measurement techniques, allows one to obtain necessary information in visible form on invisible substances by exciting with a laser beam the atoms or molecules of substances and thus causing them to emit fluorescence. Desirable for this purpose, therefore, are molecules (or atoms) which emit fluorescence even when excited by laser beam of relatively low energy level and whose absorption wavelength is highly tunable with the laser-beam wavelength. For that reason, I_2, NO, OH, or O_2 is used as the tracer element. See Hanson et al. (1990) for the generalized working principle of the LIF method of imaging and measurement. Analyzing the supersonic flow field by LIF as in the research conducted by the authors requires the following technical considerations: (i) laser-beam irradiation with the excitation wavelengths which are not susceptible to the Doppler effect; (ii) avoidance of molecules of other gases exerting quenching effect on the NO molecule fluorescence intensity. Of the above technical considerations, (i) concerns the excitation with shift in wavelength due to the Doppler effect of molecule motions and the consequent decline in molecule excitation efficiency. The unwanted Doppler effect was avoided by using a laser beam of relatively broad spectral line width capable of exciting the NO molecules for measurement even in the event of shift in absorption line due to the Doppler effect. On the other hand, (ii) was dealt with by employing N_2 gas for the main free flow and thus excluding O_2 and other gases so that the quenching effect of their molecules could be avoided as much as possible.

3. Test apparatus employed

3.1. Shock wind tunnel and test model

The tests were conducted in the shock wind tunnel capable of generating the airstream of $M_\infty = 3$. Figure 1 shows a schematic drawing of the shock wind tunnel and associated test equipment. Principal components include a bellows-type air bag, a vacuum tank to produce and maintain the required pressure differential, a quick operating valve capable of instantaneous open/close operations located between the air bag and vacuum tank, a nozzle to produce the fluid flow high in velocity, and a test, or measuring, section. The nozzle for the

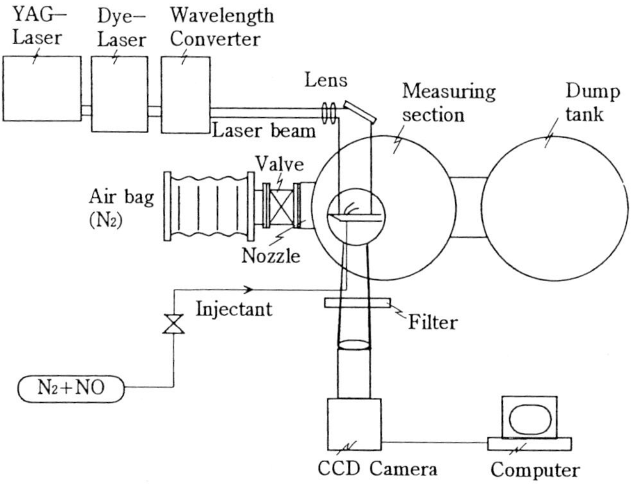

Figure 1. Schematic drawing of the shock wind tunnel and associated test equipment.

Mach number of $M_\infty = 3$ is rectangular in cross section measuring 300 mm × 300 mm at its discharge outlet connecting to the wind-tunnel test section. Three optical windows with quartz glass are provided 190 mm upstream of the nozzle discharge outlet end, one measuring 200 mm × 50 mm in the nozzle top wall for laser-beam irradiation in sheet form and the remaining two, each measuring 100 mm in diameter, in the nozzle side wall and the test section side wall for use as measurement windows in conjunction with the CCD camera. The injection of a sonic gas jet of 99.9% N_2 and 0.1% NO into the supersonic transverse flow of N_2 gas was conducted using an element test model fabricated to simulate the hydrogen fuel injection nozzle in the scramjet engine combustion chamber. A flat plate (270 mm wide × 370 mm long × 18 mm thick) with a leading edge of a sharp-angled wedge form and with an injection opening at 180 mm from the leading edge was fabricated. The injection opening in the plate was designed so that it could be used as a 1-mm wide × 100-mm long two-dimensional slit or as a circular hole 6 mm in diameter. Also, the associated piping was sized so that chocked flow of $M = 1.0$ could be produced at the injection opening. The 0.1% of NO mixed in the N_2 gas jet was meant to serve as the tracer element.

3.2. Optical instrumentation system

The optical instrumentation system shown in Fig. 1 was used in the tests. The YAG laser with 800 mJ/P in output was used as the light source to activate the dye laser, and then the dye laser beam was converted into ultraviolet light bringing the laser beam wavelength in tune with the NO-molecule absorption wavelength. The tests were conducted by exciting the NO molecules using the absorption line of R_{21} (8.5) (225.748 nm in wavelength) and measuring the fluorescence of NO molecules with absorption lines (0.0) to (0.6) in band spectrum (235 to 300 nm in wavelength). The laser-beam spectral line width was enlarged to $\sim 1.0\ cm^{-1}$ to neutralize the Doppler effect causing the shift in NO-molecule excitation wavelength. The laser beam was turned into sheet form by means of a cylindrical lens to irradiate the measuring section through the aforementioned 200-mm $\times$ 50-mm optical window. The signal light from the measuring section, with the background luminescence removed by means of a band-pass filter, was condensed by a high-efficiency condensing system equipped with the Maksutov-Newton telescope, for photographing with the CCD camera fitted with the image intensifier. The CCD camera employed has 284 $\times$ 384 picture cells. The signal from the CCD camera was transmitted to a computer for image processing and concentration analysis.

4. Test results and consideration

In the tests conducted this time, a three-dimensional gas jet ($d = 6$ mm, $M_j = 1.0$, and $P_{jo} = 32$ kPa) and a two-dimensional gas jet (1-mm width, $M_j = 1.0$, and $P_{jo} = 20$ kPa) were each injected into the main supersonic gas flow of $M_j = 2.8$ and $P_{jo} = 2.45$ kPa, and the NO-LIF-imaged gas jet injection as viewed sideways was photographed in each instance. The side-view imaging covered a rectangular region from about 10 mm upstream of the injection opening to 80 mm downstream and about 80 mm in height. Figure 2 shows the PLIF image of NO using a circular-hole injection opening with 15 shots of laser beam in total and at intervals of about 100 μs and the distribution of NO number density calculated from time-averaged, or 15-frame average, LIF images. The NO number density is defined as the number of NO molecules per unit volume of 1.0 cm^3 (with the values in the figures indicated in $\times\ 10^{-12}$). Evident from Fig. 2 are masses of NO molecules being carried downstream while mixing together with surrounding gas or mixing vigorously with one another and with surrounding gas at the barrel shock periphery and in the vicinity of the gas jet upper edge. Figure 3 shows the PLIF image of a gas jet with NO as obtained by locating the laser-sheet irradiation plane 1.5 mm ($Y/D = 0.25$) from the center of the gas jet; the distribution of NO number density is thereby calculated. The high-concentration region observed in the vicinity of $X/D = 4$ is deemed as representing the region present at the center of a pair of vortices generated in the gas jet, which is in good agreement with what was observed in the I_2-PLIF imaging of gas jet conducted by Hollo et al. (1988). Figure 4 shows examples of the two-dimensional gas-jet images obtained by the schlieren method and Mach-Zehnder interferometry.

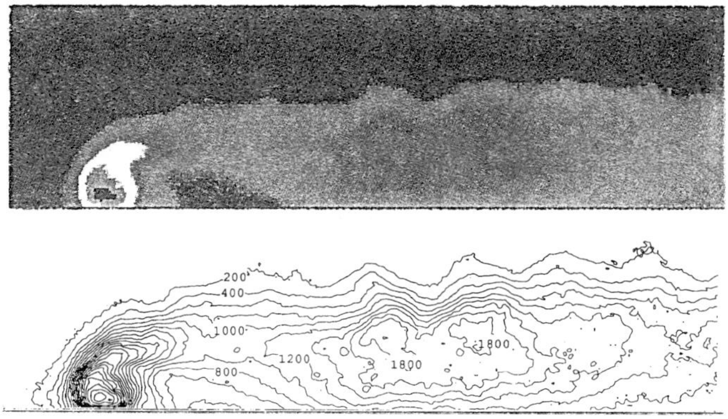

Figure 2. NO-PLIF image and NO number density distribution ($Y/D = 0$).

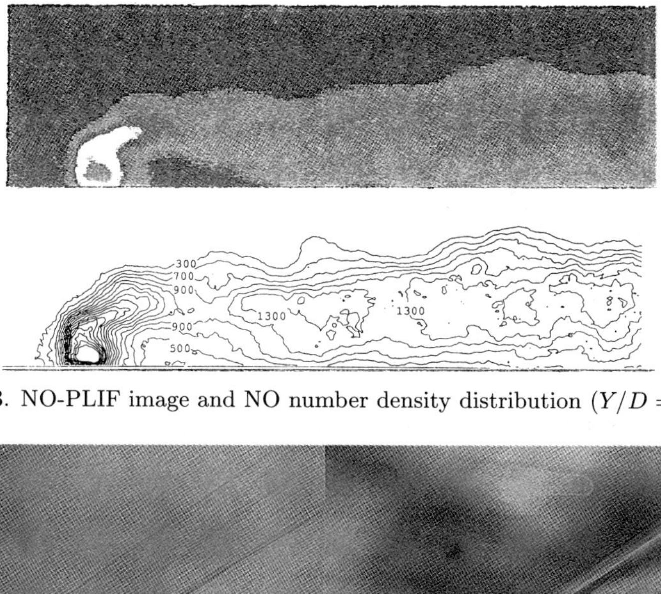

Figure 3. NO-PLIF image and NO number density distribution ($Y/D = 0.25$).

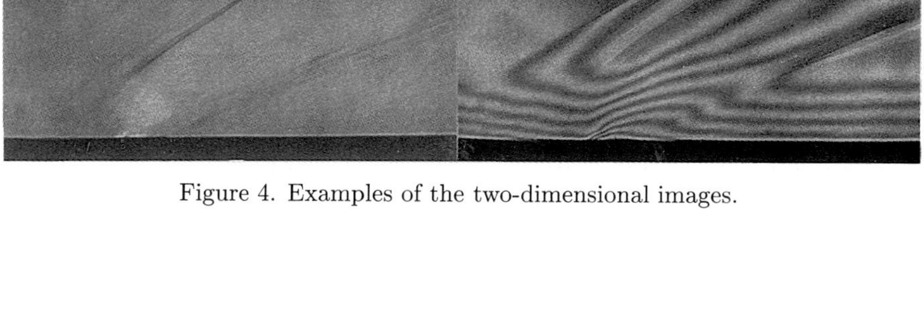

Figure 4. Examples of the two-dimensional images.

5. Concluding remarks

The authors conducted a series of tests in the shock wind tunnel for NO-PLIF imaging of N_2 gas jets injected perpendicularly into supersonic transverse N_2 gas flow as well as for the quantitative measurement of distribution of NO number density. Consequently the instantaneous and time-averaged NO-PLIF images of the gas jet mixing with the main gas flow were successfully obtained followed by the quantitative analysis of distributions of NO number density, all photographed by CCD camera and subject to computer image processing. The experimental research conducted by the authors this time clearly identified the details of gas-jet mixing behaviors and, in addition, provided data for verifying the accuracy of CFD code simulation of the mixing behaviors of supersonic fluid flows, the latter of which is deemed to carry with it significant technical implications as important as the development of the NO-LIF method of fluid-flow analysis.

References

Deguchi Y, Nada M, Ianada M, Nishida T (1993) OH-, NO-measurement in a gas turbine combustor using LIF. In: Proc 31st Combustion Symposium, pp 428–430 (in Japanese).

Hanson RK, Seitzman JM, Paul PH (1990) Planer laser-fluoresence imaging of combustion gases. Appl Physics B50:441.

Hollo SD, Hartfield Jr RJ, McDaniel JC (1988) Injectant mole fraction measurements of transverse injection in a nonreacting supersonic combustor. J Propulsion and Power 4(6):591–597.

Honma N, Tomita T, Fujiwara T, Saito A (1993) Three-dimensional LIF visualization of two supersonic jets interference. In: Proc Shock Wave Symposium, pp 1287–1290 (in Japanese).

Kychakoff G, Howe RD, Hanson RK, McDaniel JC (1982) Quantitative visualization of combustion species in a plane. Appl Optics 21:3225–3227.

Lee MP, McMillin BK, Palmer JL, Hanson RK (1991) Two-Dimensional Imaging of Combustion Phenomena in a Shock Tube Using Planar Laser-Induced Fluorescence, AIAA-91-0460, AIAA 29th Aerospace Sciences Meeting, Reno, NV.

On the role of laser absorption in planar laser-induced fluorescence imaging of high-enthalpy flows

W. M. Ruyten, M. S. Smith, W. D. Williams
Micro Craft Technology, Inc./AEDC Operations
Arnold Engineering Development Center
Arnold AFB, TN 37389-1300 USA*

Abstract: Planar laser-induced fluorescence measurements of hypersonic flows in the AEDC Impulse Facility are reported. Unlike in low-density flows, laser absorption and spectral hole burning must be accounted for. This is accomplished through the use of a comprehensive spectro-physical model for the absorption and signal-generation processes. Results of dual-wavelength PLIF measurements are compared with the results of theoretical predictions for the free stream and the shock flow around a conical test article.

Key words: Hypersonic flow, High-enthalpy flow, PLIF imaging, Flow visualization, Laser absorption, CFD validation

1. Introduction

One of the hypersonic test facilities at the Arnold Engineering Development Center (AEDC) is the Impulse Facility. This is a free-piston, light-gas-driven, reflected shock tunnel. During shakedown testing of this facility during FY94 and FY95, planar laser-induced fluorescence (PLIF) diagnostics were performed to assess the feasibility of their use in a hypersonic facility. Based on validation testing in a small laboratory shock tube (Smith et al. 1994), it was expected that quantitative data could be obtained in the larger facility as well. Initial results (Ruyten et al. 1994) were encouraging, although significant disagreements with results from CFD solutions (Maus and Rock 1995) were identified. Subsequent work has revealed that a significant shortcoming of the spectro-physical model used in Ruyten et al. (1994) was that laser absorption and spectral hole burning were not taken into account. Updated results are reported here.

*The research reported herein was performed by the Arnold Engineering Development Center (AEDC), Air Force Materiel Command. Work and analysis for this research were done by personnel of Micro Craft Technology, Inc./AEDC Operations, technical services contractor for the AEDC aerospace flight dynamics facilities. Further reproduction is authorized to satisfy needs of the U.S. Government.

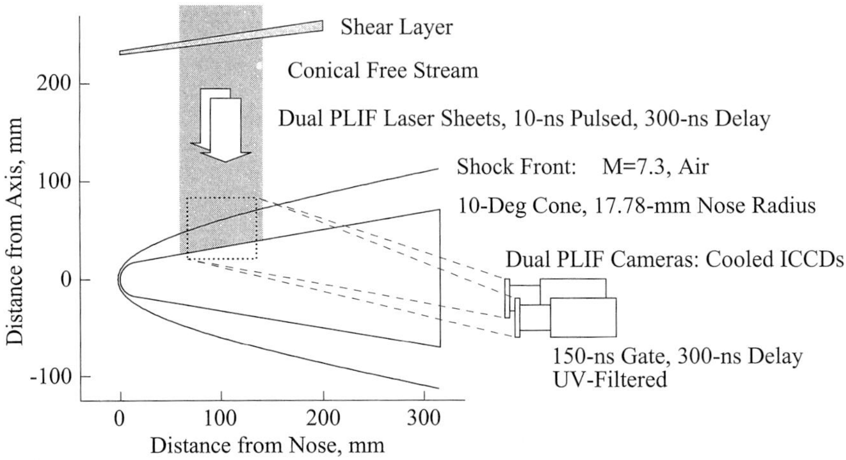

Figure 1. Test geometry and dual PLIF setup.

2. Experimental

Figure 1 shows the most important elements of the experimental setup. Two cooled, intensified CCD cameras are used to image the PLIF signals from a region above a conical test article. The two laser sheets are aligned along the vertical plane through the cone axis. The lasers are tuned to different A←X transitions of nitric oxide which occurs naturally in the flow. The PLIF images are corrected for camera dark-count and nonuniform energy distribution in the laser sheets. Also, they are scaled geometrically based on pre-run images of a calibration target. The pair of processed images for Impulse Facility run 21 is shown in Fig. 2. Also shown are the PLIF signals along a radial cut through the flow. Both show the effect of laser-beam attenuation in the free stream, which extends out to the free-jet shear layer at a distance of about 250 mm from the cone axis (see Fig. 1).

3. Spectro-physical model

In low-density flows, the calculation of flow temperature from a pair of energy-normalized PLIF images is relatively straightforward, as it depends only on the ratio of the two signals (Gross et al. 1987, McMillen et al. 1993). The situation is much more complicated in the presence of laser absorption and spectral hole burning (see Fig. 3a). Thus our initial approach toward performing a comparison against CFD results was to calculate synthetic PLIF images from a given CFD solution, and compare these to the measured signals (see

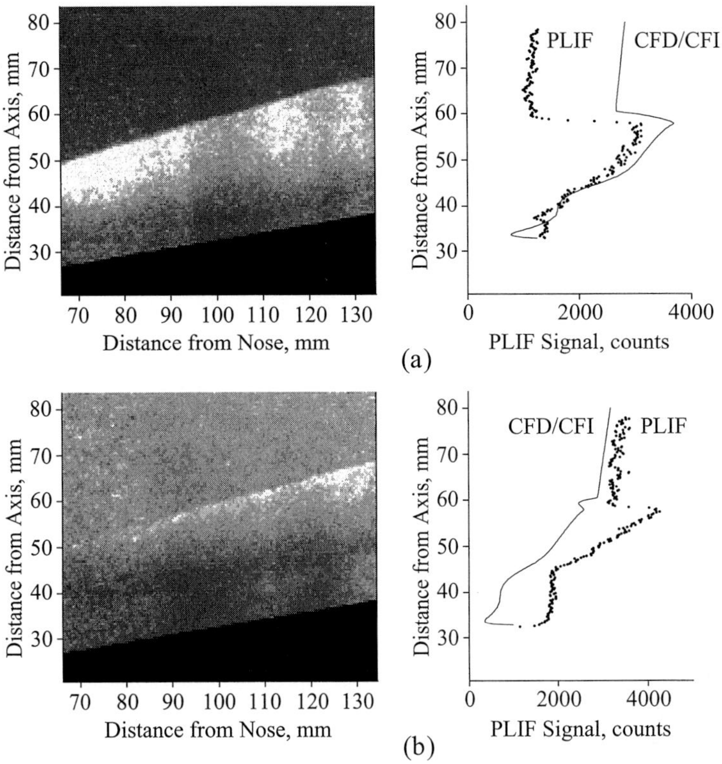

Figure 2. Energy-normalized dual PLIF images (left) and radial PLIF profiles 100 mm from the nose (right) for run 21: (a) $Q_{11} + P_{21}(28\frac{1}{2})$ line at 225.134 nm; (b) $R_{11} + Q_{21}(13\frac{1}{2})$ line at 225.716 nm. The same line-pair was used for run 24; for run 23, the 225.134 line was replaced by the $P_{22} + Q_{12}(34\frac{1}{2})$ line at 225.453 nm.

Fig. 2). The spectro-physical model used for this purpose includes the effects of Doppler and pressure broadening, Doppler and pressure shift, frequency-dependent laser absorption, and collisional quenching (using the rate constants of Gray et al. 1993). Spectral line positions and line strengths were calculated from Reisel et al. (1992). The model correctly reproduces the PLIF signal observed in a reference cell at room temperature over a range of NO densities.

The calculation of synthetic PLIF signals is referred to in Ruyten et al. (1994) as 'Computational Flow Imaging' (CFI). The strength of the technique is that it allows a comparison of calculated and measured signals to be made even when the number of variables upon which the synthetic PLIF signals depend (rotational and vibrational temperature, densities of each of the flow species, and, in the case of spectral hole burning, the radial flow velocity) is greater than the number of independent measurements. Figure 2 shows CFI results obtained from a CFD solution for run 21. Other than predicting the correct signal slope in the free stream, the discrepancy between the calculated and measured PLIF signals is considerable.

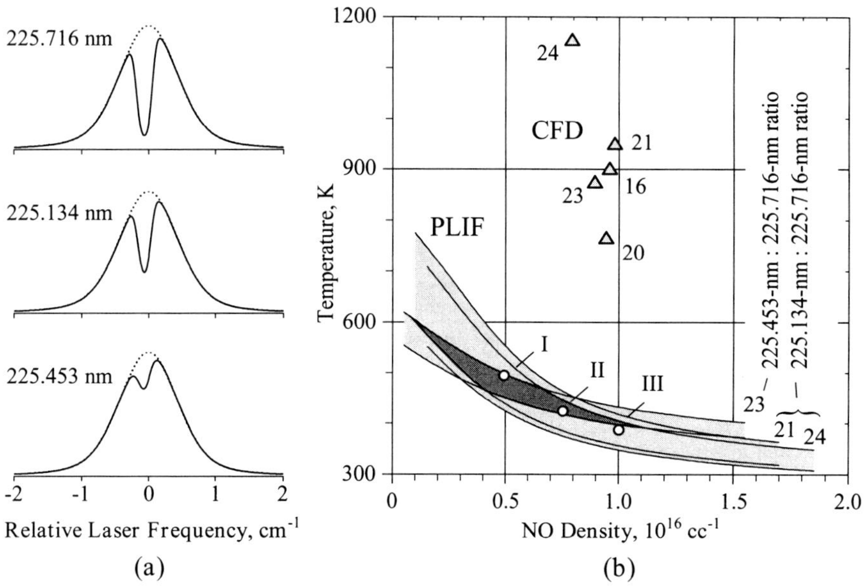

Figure 3. Effects of laser absorption: (a) calculated laser spectral profiles in the free stream just above the shock for the point marked "CFD 16" in (b); the unattenuated profiles are given by the dotted curves; (b) comparison of dual-PLIF-derived free-stream parameters and CFD values. Arabic numerals are run numbers. Roman numerals are referred to in Fig. 4.

4. Model-assisted PLIF analysis

Using the spectro-physical model, the dual PLIF data were analyzed further in two steps. In the first step, contours of constant PLIF signal ratios in the free stream were calculated as a function of NO density and (rotational) temperature. Thus each of the measured signal ratios from runs 21, 23, and 24 defines a curve in Fig. 3b (in the absence of spectral hole burning these would be straight lines). Assigning a (conservative) uncertainty of ± 25 percent to the measured signal ratios, the curves broaden into the lightly shaded bands in

Fig. 3b. Assuming that the conditions for the three runs were the same (this is true only approximately), the most probable combinations of NO density and temperature in the free stream are given by the overlap region of these three bands (dark shading). The size and position of this region are largely independent of the other flow variables. Not surprisingly, the signal ratios constrain the temperature more tightly than the density. But both the NO density and the temperature in the overlap region differ significantly from the CFD values. Notably, the measured temperature is much lower than the predicted one.

The second step involves the determination of the temperatures and densities inside the shock layer. This step is accomplished by an algorithm which iteratively adjusts the CFD solution to achieve a perfect replication of the measured PLIF signals. Starting in the free stream, the algorithm 'marches' radially inward along a vertical cut through the flow, all the while adjusting the NO density and temperature and accounting for the spectral hole burning, such that the experimental PLIF signals are reproduced exactly. The results, shown in Fig. 4, depend on the 'starting' values assumed in the free stream, as indicated by the Roman numerals in Figs. 3 and 4. Also, the results depend on the other flow variables. As such, this approach does not yield a unique answer. It demonstrates however, qualitatively as well as quantitatively, just *how* the NO density and temperature should differ from the CFD results to reproduce the experimental dual PLIF data, as governed by the spectro-physical model. Again, there is considerable disagreement between the CFD and experimental results, especially for the temperature profile close to the model.

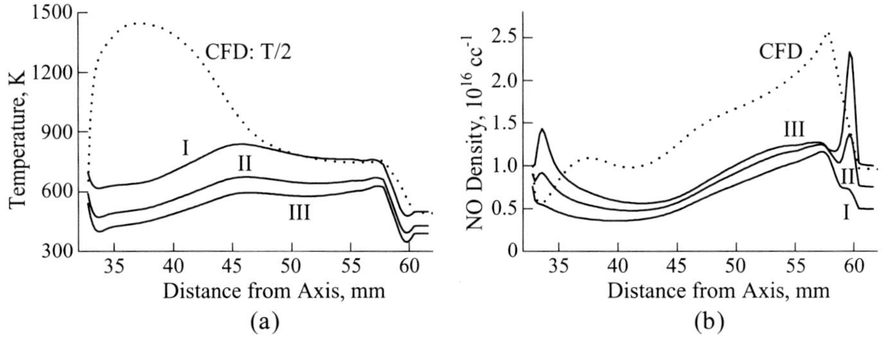

Figure 4. Flow parameters determined from the (smoothed) experimental PLIF profiles at 100 mm from the nose in Fig. 3: (a) temperature (the CFD temperature shown is only half of the actual value); (b) NO density. The curves marked 'I', 'II', and 'III' correspond to free-stream conditions as marked in Fig. 3b.

5. Discussion

One possible explanation for the significant discrepancies between the experimental data and the CFD data might lie in flow contamination. At burst pressures in excess of those employed for the data runs here, significant optical blocking of the flow was observed. The effect on the PLIF temperature measurement should be minimal, because PLIF signal ratios should be unaffected by nonresonant laser scatter (unlike spectral hole burning, nonresonant scatter is marked by a uniform attenuation of the laser spectral profile). Density measurements would be low in the presence of laser scatter. On the other hand, the effect of flow contaminants on the CFD solution are not known. Resolution of the discrepancies would best be addressed by a combined experimental-computational effort.

Most importantly, a procedure has been demonstrated which may be used to perform quantitative analysis of PLIF data in a high-enthalphy, hypersonic flow facility, even when laser absorption and spectral hole burning are present.

Acknowledgement. Discussions with J. R. Maus are gratefully acknowledged.

References

Gray JA, Paul PH, Durant JL, Thomann JW Jr (1993) Collisional Electronic Quenching of NO ($A^2\Sigma^+$) Measured at High Temperatures, AIAA Paper 93-0924.

Gross KP, McKenzie RL, Logan P (1987) Measurement of temperature, density, pressure, and their fluctuations in supersonic turbulence using laser-induced fluorescence. Exp Fluids 5:372.

Maus JR, Rock SG (1995) Computational Results for AEDC Free Piston Shock Tunnel and Comparison with Calibration Data, AIAA Paper 95-6039.

McMillen BK, Palmer JL, Hanson RK (1993) Temporally resolved, two-line fluorescence imaging of NO temperature in a transverse jet in a supersonic cross flow. Appl Opt 32:7532.

Reisel JR, Carter CD, Laurendeau NM (1992) Einstein coefficients for rotational lines of the $(0,0)$ band of the NO $A^2\Sigma^+ - X^2\Pi$ system. J Quant Spectrosc Radiat Transfer 47:43.

Ruyten WM, Williams WD, Heltsley FL (1994) Computational Flow Imaging for Planar Laser-Induced Fluorescence Applications (CFI-PLIF), AIAA Paper 94-2621.

Smith MS, Williams WD, Price LL, Jones JH (1994) Shocktube Planar Laser-Induced Fluorescence Measurements in Support of the AEDC Impulse Facility, AIAA Paper 94-2649.

High-speed image processing of interferograms in experiments on weak shock waves

Y. Kurokui, T. Aoyama, N. Watanabe, K. Maeno, H. Honma
Depertment of Mechanical Engineering, Chiba University
1-33 Yayoi-cho, Inage-ku, Chiba 263 Japan

Abstract: In order to capture small density variations in the flow field behind weak shock waves, an image-processing system for analyzing interferograms is developed. A Mach-Zehnder interferometer is used for visualization of the flow field, and the images are directly photographed by a CCD camera. The image is displayed immediately after each shot, and processed on a personal computer. Using this system, weak shock waves in carbon dioxide are observed. This paper presents the details of density distributions of flow fields which are generated by weak shock waves passing over a hemicylinder in carbon dioxide, and compares them with CFD results.

Key words: Image processing, Finite-fringe interferogram, Dispersed waves

1. Introduction

Interferometry is often used for visualization of flow fields. There are two methods: infinite- and finite-fringe interferometry. Since for the infinite-fringe interferograms each fringe indicates an equi-density line, it is easy to observe directly density distributions in flow fields. However, in case of weak shock waves, it becomes difficult to measure detailed density variations using the infinite-fringe interferograms, due to the decrease of the number of fringes. On the contrary, for the finite-fringe interferograms, it is not so easy to observe directly density distributions, but the detailed density variations can be measured by tracing the fringe shift if it is precisely detected. The measurement of fringe shifts for the whole flow field is laborious and impractical in the classical method of manual processing from standard photos of interferograms.

Recent advances in computer-aided image processing make it possible to analyze the images in digitized form. The present study aims at developing high-speed image processing of interferograms for experiments on weak shock waves by applying computer-aided image processing to interferometry. There are two methods for image processing of interferograms. One is fringe-pattern analysis using Fourier Transform (see Bone et al. 1986) and another is the fringe-trace method. The present study adopts the latter method which involves a simpler technique. The scheme is based on the work of Fujise et al. (1988). and was improved so as to be applicable to the present system. The present method is applied to dispersed waves passing over a hemicylinder in carbon dioxide.

2. Experiment

A schematic diagram of the experimental apparatus is shown in Fig. 1. A diaphragmless shock tube is used to generate weak shock waves. The hemicylinder model with a 20-mm diameter is inserted in the lower wall of the test section. Carbon dioxide is used as both the driver and driven gas to maintain the purity of the gases. Experiments are conducted with initial pressures $p_1 = 147$ to 196 kPa and incident-shock Mach numbers 1.02 to 1.06. The incident-shock Mach number is derived from the pressure ratio across the shock wave by using the shock relation for a perfect gas. The pressure is measured by a piezoelectric transducer.

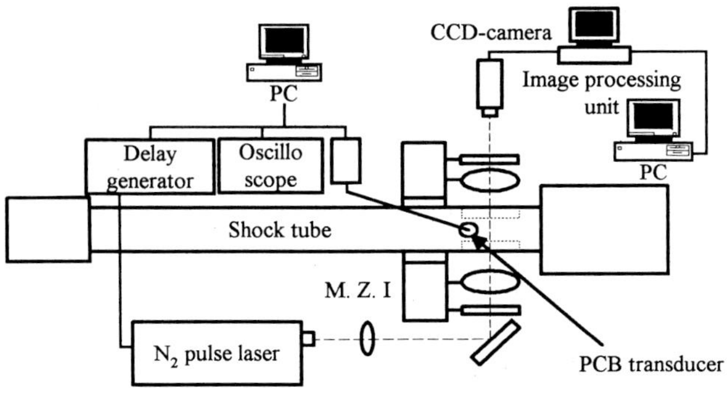

Figure 1. A schematic diagram of the experimental apparatus.

The shock waves are visualized by a Mach-Zehnder interferometer (MZ-4/Mizojiri). The light source is a pulsed nitrogen laser (UV14/Molectron) with the pulse width of 10 ns at the wavelength of 337.1 nm. The interferograms are photographed directly by a CCD camera (IK-536/Toshiba). The image is displayed immediately by an image-processing unit (PIP-4000/ADS), so that the optical setup and photography difficulties can be largely reduced in comparison with the conventional use of an ordinary camera. The image is forwarded to a personal computer (PC-9801RA/NEC), and recorded on its MO disk. The whole experimental process from the generation of the shock wave to the recording of images is controlled automatically by a personal computer.

3. Image processing

One image consists of 512 × 480 pixels with 8-bit (256 step) brightness levels. The image-processing routine is as follows: (1) Image reformation: As a preprocess of thresholding, the contrast of the image is adjusted, since the light intensity is not uniform all over the image. (2) Thresholding: The 8-bit brightness image data are transformed into 1-bit (two-value) data using a threshold value. The mode method is used to determine the threshold value in order to avoid such processing errors as disappearance and merging of fringes. (3) Thinning: Each fringe is transformed to one pixel line passing through the center of the fringe to make it easy to find its position. (4) Tracking operation: The fringe is traced to measure its shifted distance from initial position. (5) Calculation of density: The density distribution is calculated from the fringe shift. The densities between adjacent fringes are interpolated linearly.

The processing steps (1) to (5) are automatically carried out after a manual positioning of the hemicylinder in the coordinate system of the image. The processing time is about 30 seconds. A quick display of the density data can be made in various ways, i.e., equi-density contour or spatial density distribution.

It should be noted that the present method of image processing is applicable only to fully dispersed waves whose front structures exhibit continuous changes of density due to vibrational relaxation of gas molecules. Some additional techniques would be needed for obtaining the density data across shock waves with discontinuous fronts, such as simultaneous pressure measurements and the use of white light. In the present study, the image processing has been carried out also for the interferograms of partly dispersed waves with discontinuous fronts, with reference to the pressure data.

4. Results and discussion

In order to confirm the accuracy of the image processing, the procedure is applied to CFD data for a shock wave passing over a hemicylinder in carbon dioxide. The CFD data are obtained by solving the two-dimensional Euler equations with a relaxation equation by using an FCT-MacCormack method (see Honma et al. 1993). The resultant density distribution is transformed to a finite-fringe interferogram (exhibited in Fig. 2), and then the interferogram is processed by steps (1) – (5) as described.

Figure 3a shows the pseudo-infinite-fringe interferogram that is transformed directly from the CFD data, and Fig. 3b shows the one that is obtained from image processing of the CFD data. The density interval of each fringe is $\Delta\rho = 3.68 \times 10^{-2}$ kg/m^3. Figure 4 shows the density distribution along a line through the flow field, as marked in Fig. 2, for original and processed data. It can be seen in Figs. 3 or 4 that the agreement is fairly good in view of reproducibility.

Now that we are sure that the processing error is very small, the next step is to apply the present method to experimental images. Figure 5a shows a

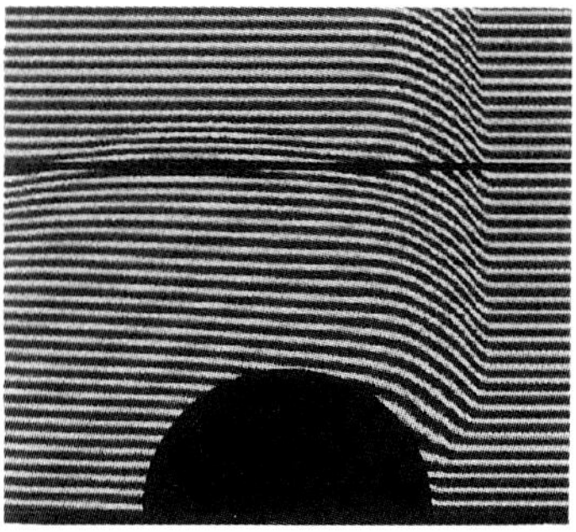

Figure 2. A pseudo-finite-fringe interferogram transformed directly from CFD data ($M_s = 1.027$, $p_1 = 196$ kPa, $\delta/d = 0.25$).

Figure 3a. A pseudo-infinite-fringe interferogram transformed directly from CFD data ($M_s = 1.027$, $p_1 = 196$ kPa, $\delta/d = 0.25$).

Figure 3b. A pseudo-infinite-fringe interferogram obtained from the image processing of Fig. 2 ($M_s = 1.027$, $p_1 = 196$ kPa, $\delta/d = 0.25$).

pseudo-infinite-fringe interferogram that is obtained from image processing of a finite-fringe interferogram photographed in an experiment. The density interval of each fringe is $\Delta\rho = 3.68 \times 10^{-2}$ kg/m^3. For comparison, an infinite-fringe interferogram which is photographed with the same conditions as Fig. 5a is shown in Fig. 5b. The incident-shock Mach number is $M_s = 1.027$ for both cases. The pseudo-infinite-fringe interferogram shows fairly good agreement with infinite-fringe interferograms in view of the reproducibility of the experimental images.

The digitized experimental data make it possible to compare the CFD data in detail, as shown in Figs. 6 and 7. Figure 6 shows the instantaneous density distributions on the surface of the hemicylinder plotted against the angle θ which is measured from the stagnation point of the hemicylinder as illustrated in the figure. The density is normalized by the density ahead of the incident wave. Its

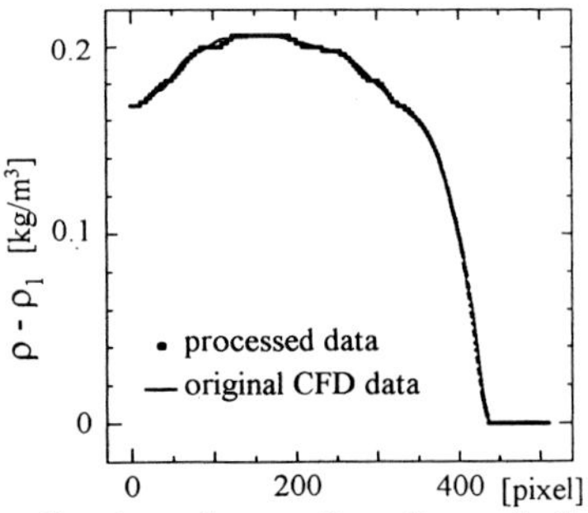

Figure 4. The density distribution along a line through the flow field, as marked in Fig. 2, for original and processed data.

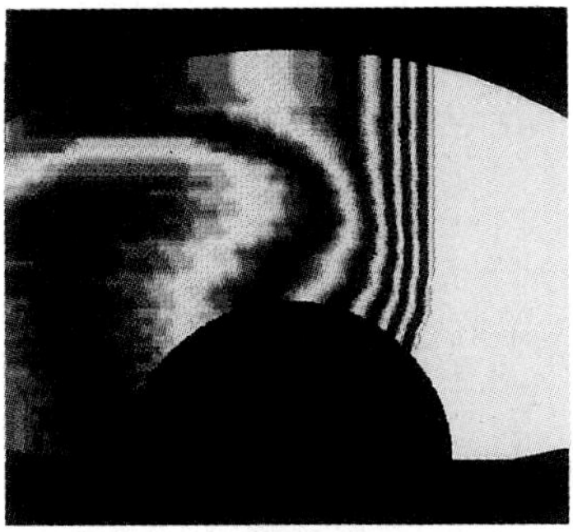

Figure 5a. A pseudo-infinite-fringe interferogram obtained from image processing ($M_s = 1.027$, $p_1 = 196$ kPa).

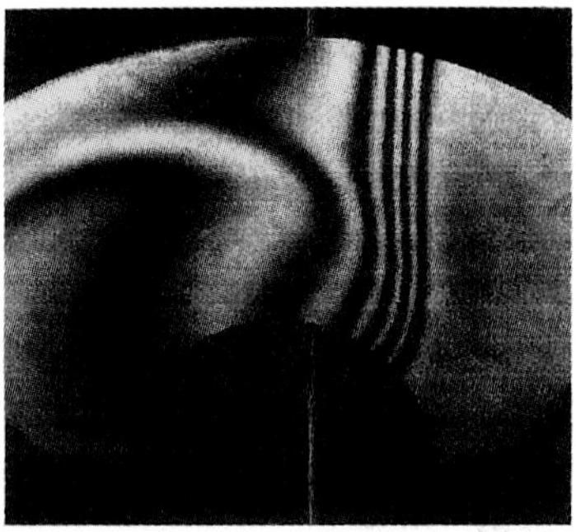

Figure 5b. An infinite-fringe interferogram photographed in an experiment ($M_s = 1.027$, $p_1 = 196$ kPa).

shock Mach number is $M_s = 1.040$. The solid and broken lines indicate, respectively, experimental and computational density distributions. The experimental distributions are obtained from image processing of interferograms. The curves A to E correspond to the density distributions for each position of advancement of the incident wave. A fairly good agreement between solid and broken lines is obtained for curves A and E, while there are some discrepancies for curves B through D. Near at the stagnation point ($\theta = 0$), the density values for the CFD results are 10 – 20% higher than those for the experimental ones.

Figure 7 shows the density distribution plotted against the radial distance from the center of the hemicylinder at the angle $\omega = 45°$. The density is normalized by the density ahead of the incident wave, and the radial distance is normalized by the radius of the hemicylinder. Its shock Mach number is $M_s = 1.040$. The solid and broken lines also indicate, respectively, experimental

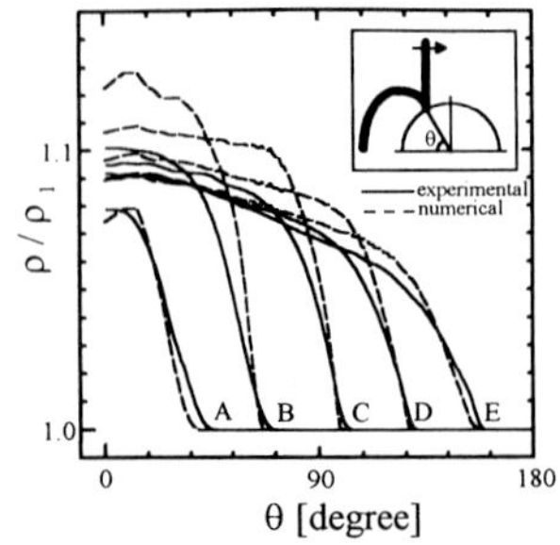

Figure 6. Density distributions along the surface ($M_s = 1.040$).

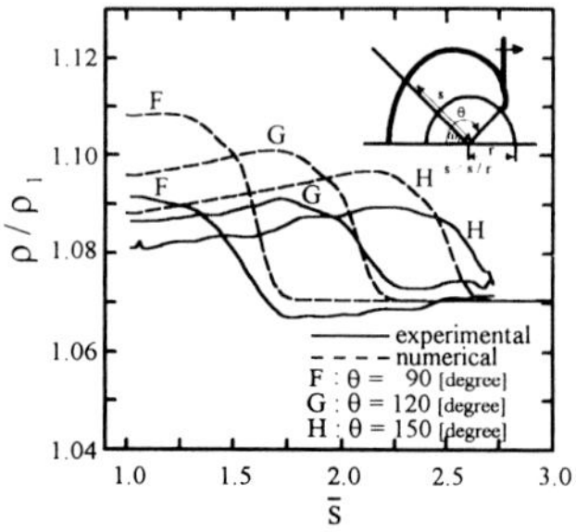

Figure 7. Density distributions along radial direction ($M_s = 1.040$, $\omega = 45°$).

and computational density distributions. The curves F to H correspond to the density distributions for each position of advancement of the reflected wave.

The characteristic features of a cylindrical wave appear in both computational and experimental cases. The peak values of density attenuate as the wave proceeds, and there is an expansion region behind the reflected wave. However, large discrepancies can be seen between solid and broken lines. The density values for the CFD results are higher than those for the experimental ones behind the reflected wave, and the attenuation rate is smaller for the latter than for the former. The reason for these discrepancies seems to be attributed to nonuniformities behind the incident dispersed wave in the experiment.

5. Conclusion

Experimental efficiency is improved by use of the described visualization system including a CCD camera. The detailed density distributions of flow fields, generated by weak shock waves passing over a hemicylinder in carbon dioxide are obtained from finite-fringe interferograms by image processing. The small processing error is verified by applying the image-processing system to CFD data. Some discrepancies between experimental and CFD data are clarified, but their exact reasons are still obscure.

References

Bone DJ, Bacher HA, Sandemann RJ (1986) Fringe-pattern analysis using a 2-d Fourier transform. Appl Opt 25(10):1653–1660.

Fujise K, Honma H, Yoshida H (1988) Reflection of weak shock waves in carbon dioxide. Theoretical and Applied Mechanics 36:113–120.

Honma H, Maeno K, Itabashi M, Watanabe N (1993) Weak shock waves passing over a hemicylinder in carbon dioxide. In: Shock Waves @ Marseille, Brun R, Dumitrescu LZ (eds) Springer, Vol. 2, pp 361–366.

Medical Applications

Experimental and mathematical model of living-tissue cell damage induced by plane shock waves

M. Tamagawa, T. Akamatsu, K. Yoshida
Department of Mechanical Engineering, Kyoto University
Yoshida-honmachi, Sakyo-ku, Kyoto, Japan

Abstract: This paper describes a model of living-tissue cell damage induced by plane shock waves. In this study, red blood cells and micro capsules are damaged by incident plane shock waves, and then evaluated to estimate the degree of damage by shock wave. Mathematical models of this damage are analyzed, and the results show that cell damage depends on (1) elastic modulus of cell, (2) bulk modulus of cell, and (3) width of cell membrane.

Key words: Living-tissue cell, Red blood cell, Micro capsule

1. Introduction

Recently, Extracorporeal Shock Wave Lithotripsy (ESWL) has become pervasive in medicine. Using lithotripsy, kidney stones are broken, but living tissues are injured as well (Chaussy 1982). Why living tissues are injured has not been not elucidated. To resolve this problem, Tamagawa and Akamatsu developed an experimental apparatus to generate saw-tooth single-pulse pressure waves rising from atmospheric conditions (Tamagawa and Akamatsu 1991, Tamagawa and Akamatsu 1993). Animal experiments were then executed to determine the threshold for injury to rats' kidneys (living tissue). It was found that damage to living tissue depends on the rise time of the incident shock wave, duration time, maximum pressure, impulse, and energy.

Since living tissue is composed of living-tissue cells, it is necessary to investigate the effects of shock waves on living-tissue cells. Threshold data thus obtained can be used to develop safety criteria for lithotripsy, but each type of living tissue will exhibit its own threshold. If universal experimental and mathematical models are constructed based on the properties of living tissues (elastic modulus, damping ratio, density, Poisson ratio), we will be able to predict the amount of damage various tissues will undergo.

In this study, injuries to living tissue (red blood cells) and micro capsules by plane shock waves are executed. Then the oscillation and transient response of an elastic shell model of living tissue cell in water are evaluated.

2. Nomenclature

Here, a, h, and r are radius of specified cell, width of cell membrane, and radial coordinate; ρ, ρ_s, and c are density of surrounding fluid, density of cell membrane, and sound velocity; δ_r and δ_θ are radius displacement and angular displacement; Π, Π_{int}, and Π^{sca} are total pressure, internal pressure, and scattering pressure; and E and ν are Young's Modulus and Poisson ratio.

3. Injury to living-tissue cells by plane shock waves

3.1. Red blood cells and micro capsules

In this study, red blood cells and micro capsules are used for injury experiments. The micro capsules are pressure-sensing particles pasted on pressure-sensor sheets (Fuji Film Co. Ltd). The dynamic properties of red blood cells and micro capsules are shown in Table 1. In the mathematical model, they also have different cell-membrane widths.

Table 1. Properties of micro capsules and red blood cells.

	Micro capsule (industrial material)	Red blood cell (living tissue)
diameter (μm)	3 - 40 (av. 20)	7.2-8.6
depth(μm)	0.08	1-2
config.	pseudo spherical	disk
material	Polyurethane	Protein, Phosphoric
contents	Lueco dye stuff Dially-ethan	Hemoglobin
life time	?	120 days
elastic modulas (Pa)	$4.2 - 5.4 \times 10^6$ (at 23 C -65%RH)	$10^5 - 10^7$

3.2. Experimental apparatus

Figure 1 shows the free-piston shock tube (Tamagawa and Akamatsu 1993) used for injury tests of red blood cells and micro capsules. The specimens are located at the end of the shock tube. To control the pressure wave at the test section, air, Ar, and He gases are used as driver gases in the compression tube.

Figure 2 shows a typical pressure wave. The pressure wave propagates between the end wall and the water surface, and has no effect on the cell because the cell's natural frequency is high compared with that of the wave. The energy the cells are subjected to is defined as $E = \int_0^{T0} \frac{p^2}{Z} dt$. Figure 3 shows part of the test section in detail; the test case is filled with blood and micro capsules in suspension to prevent air bubbles. The concentration of red blood cells and micro capsules in liquid is expressed as a volume ratio (hematocrit or Ht). In this injury experiment, two volume ratios are compared.

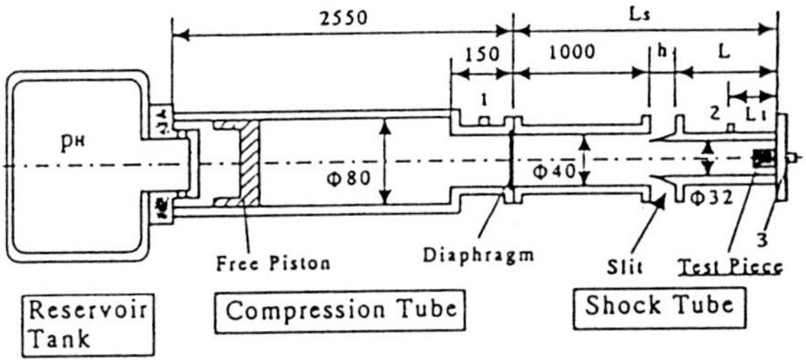

Figure 1. Free-piston shock-tube for bio tests.

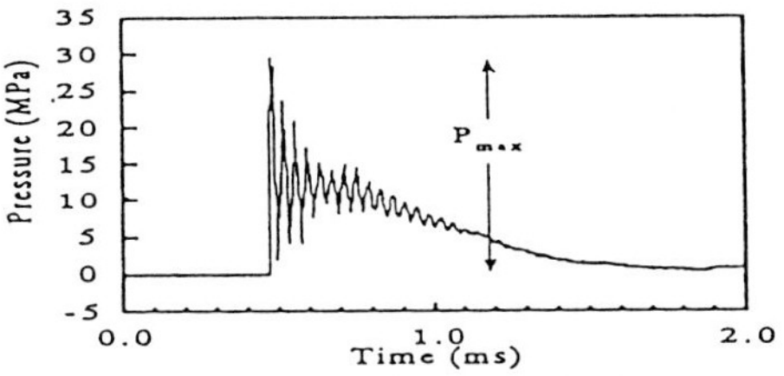

Figure 2. Pressure history of test case.

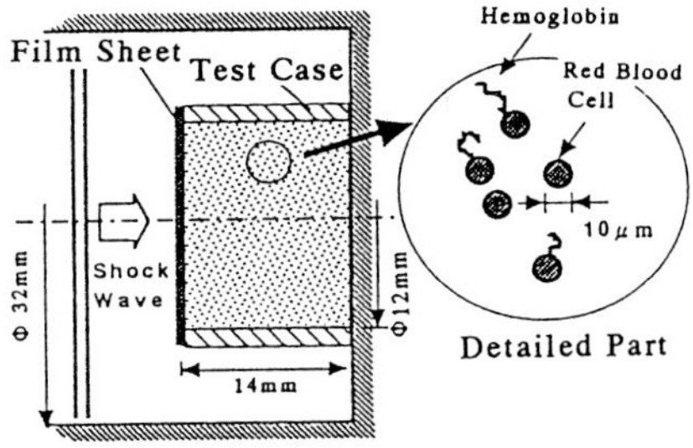

Figure 3. Detail of test section.

The degree of injury from shock waves is obtained from evaluating differences in the optical absorption ΔA of free-plasma hemoglobin for red blood cells and free-water dye stuff for micro capsules. In case of red blood cells, ΔA should be transduced to hemoglobin density ΔHb (mg/dl). The degree of injury (ΔD_b for red blood cells, ΔD_m for micro capsules) is modified by Ht as follows: $\Delta D_b = \Delta Hb(1 - Ht)/Ht$, $\Delta D_m = \Delta A(1 - Ht)/Ht$.

3.3. Results and discussion

Injury to red blood cells and micro capsules. Figure 4 shows the degree of injury with maximum pressure and energy applied. It is found that the larger the maximum pressure and energy are, the greater the degree of injury. It is also found that when the Ht of red blood cells is low (5% Ht) the degree of injury is larger than that for 45% Ht. In the case of high Ht, the frequency of the incident shock wave decreases before its arrival at the red blood cell because of the cell's elasticity. Therefore the effect of shock waves in high-Ht suspensions is smaller than that in low-Ht suspensions.

Figure 5 shows the injury to micro capsules by plane shock waves. The same trends as for red blood cells are found in this figure. It is also found that in the micro-capsule experiments, the effects of Ht difference are distinguishable; in the red blood cell this is not distinguishable, the reason being that there is aggregation red blood cells in blood.

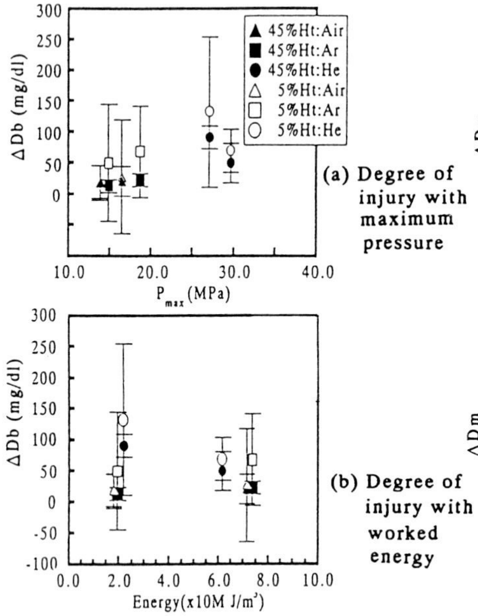

Figure 4. Injury experiment: red blood cells.

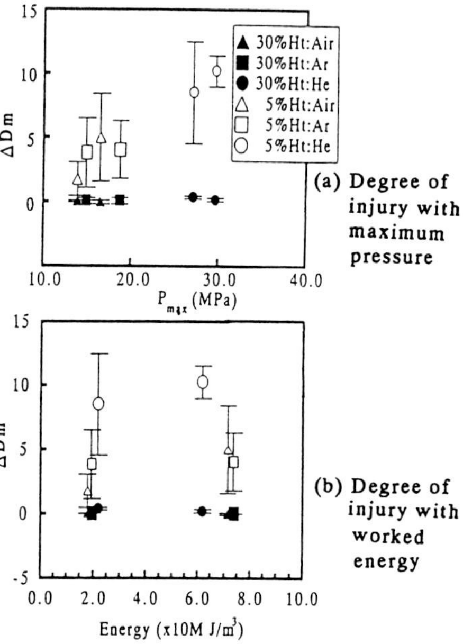

Figure 5. Injury experiment: micro capsules.

4. Deformation analysis and mathematical model

4.1. Construction of mathematical model

In the mathematical analysis, the red blood cells and the micro capsules are modeled as spherical elastic shells filled with liquid (see Fig. 6a). In this analysis, the Poisson ratio ν is 0.499.

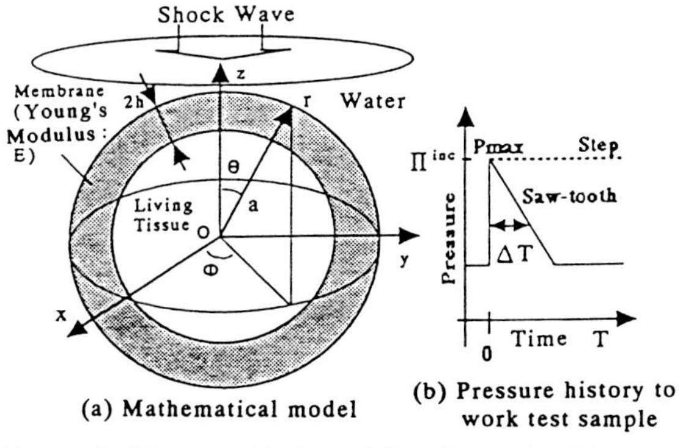

Figure 6. Mathematical model and pressure history.

4.2. Formulation

The spherical shells (living-tissue cells) are analyzed according to the method of Haung's analysis (Haung 1968). The non-dimensional physical parameters are defined as follows: $R = r/a$, $w = \delta_r/a$, $u = \delta_\theta/a$, $\Pi = \pi/(\rho c^2)$, and $T = t/(a/c)$. In a spherical coordinate system, the variables w, u, and Π are expanded with the Legendre function. And by using a Lagrange equation composed of kinematic energy (shell oscillations) and elastic energy, the equations of fluid structure are derived as follows:

$$\frac{d^2 w_m}{dT^2} + A_{1m} w_m - A_{2m} u_m = -M(\Pi_m(1, R, T) - \Pi_{int}) \quad (1)$$

$$\frac{d^2 u_m}{dT^2} + A_{3m} u_m - A_{4m} w_m = 0 \quad (2)$$

where A_{1m}, A_{2m}, A_{3m}, and A_{4m} are coefficients including Young's Modulus and the width of the cell membrane.

The internal pressure of the cells is obtained from the general definition of the bulk modulus, and the surrounding pressure is obtained from the spherical Helmholtz equation. Then the displacement of shells and the surrounding pressure are linked to satisfy the boundary conditions at the interface. Particularly, the solved solutions in Laplace space are shown as follows (for $m = 0$ and $m > 1$):

$$\bar{w_0}(s) = \frac{-Msf(s)}{Q_0(s)}, \qquad \bar{w_m}(s) = \frac{-sf(s)(2m+1)M(s^2 + C_m)s^{m-1}}{Q_m(s)} \quad (3)$$

4.3. Oscillation analysis

Stationary oscillation of the cell is analyzed using $\bar{w}_m(iw)$. In this paper, only the effects of internal bulk modulus are discussed. Figure 7 shows the relationship between frequency and amplitude in terms of the bulk modulus K. It is found that there are two natural frequency points in the case of $K = 1.0 \times 10^8$ (Pa) and there are no natural frequency points in the case of $K = 1.0 \times 10^{10}$ (Pa). As a result, it is seen the bulk modulus has a very large effect on how much damage to living-tissue cells occurs.

4.4. Transient analysis (deformation processes)

Figure 8 shows the deformation processes of red blood cells and micro capsules in time. In this analysis, $T = 2$ refers to the duration of one shock wave propagating through a cell. It is found that after $T = 10$ (corresponding to five times the shock duration) the cell displacement is extremely large. At this time, red blood cells expand toward the direction in which the shock wave was propagating, while micro capsules expand toward the perpendicular direction.

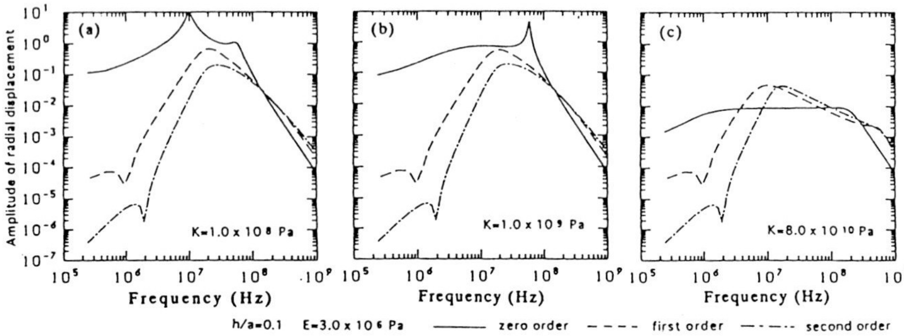

Figure 7. Radial displacement with frequency by changing bulk modulus.

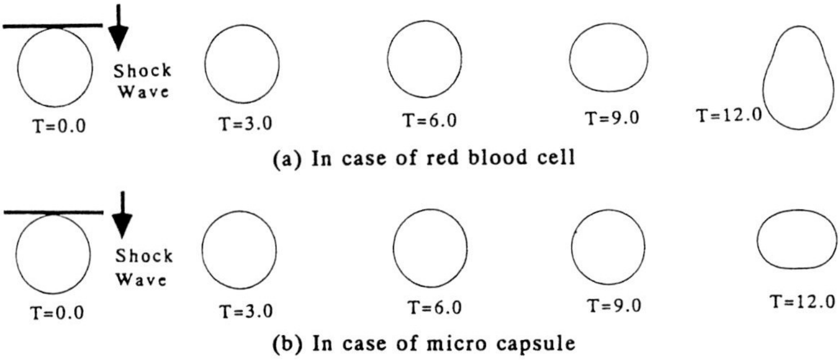

Figure 8. Deformation process of the cells by shock wave.

5. Conclusion

In this study, it is concluded that (1) the utility of micro capsules as an experimental model of living-tissue cells is established; and (2) the effect of the bulk modulus of the internal liquid in cells on shock-induced damage is large.

References

Chaussy CH (1982) Extracorporeal Shock Wave Lithotripsy, Karger, Munchen.

Haung HJ (1968) Transient interaction of plane acoustic waves with a spherical elastic shell. Accoust Soc Am 45(3)661–670.

Tamagawa M, Akamatsu T (1991) Development of test apparatus for the shock induced injury of living tissue. In: Proc 18th Int Shock Wave Symp, Sendai, Japan, pp 1177–1180.

Tamagawa M, Akamatsu T (1993) Effects of shock waves on living tissues using shock tubes for bio-tests. In: Shock Waves @ Marseille, R Brun and L Dumitrescu (eds), Springer, pp 430–434.

Application of extracorporeal shock wave on bone

K. Ikeda, A. Nakatani, K. Tomita
Department of Orthopaedic Surgery, School of Medicine
Kanazawa University, 13-1 Takaramachi, Kanazawa 920 Japan
K. Takayama
Shock Wave Research Center, Institute of Fluid Science, Tohoku University
2-1-1, Katahira, Aoba-ku, Sendai 980 Japan

Abstract: We developed a new extracorporeal shock wave (ESW) generator which generated three to six times higher peak pressures than the lithotriptor. Ten shots of ESW induced new bone formation on rabbit femurs six weeks after application. Application of ESW on bone is thus available.

Key words: Extracorporeal shock wave, Bone, Fracture

1. Introduction

Extracorporeal shock waves (ESW) are widely accepted as a therapeutic modality for urinary stones. Although the mechanism for the destruction of urinary stones is not completely worked out, ESW can crack urinary stones with little damage to the surrounding soft tissue. This study was based on the idea that bone is characteristically similar to urinary stones in the human body (Takayama 1987).

The values of acoustic impedance of both bones and urinary stones are nearly the same (bones: 4100 m/s and urinary stones: 4000–6000 m/s); and application of ESW to bones would be beneficial to the field of orthopaedics.

Treatment of the delayed union of fractures is still a challenging area. Delayed union is a delay of bone union in fracture healing, usually secondary to bone defect, infection, or instability of the fracture site. These factors could be treated with antibiotics, immobilization, or bone grafting. However, there are some delayed unions that occur with no obvious explanation. One of the proposed mechanisms is that poor blood circulation to the fracture site decreases the rate of new bone formation. In such cases, orthopaedic surgeons create new fractures around the original fracture site with chisels. This procedure stimulates osteogenic activities and enhances bone formation (we call this procedure "decortication"). However, this procedure carries certain morbidity risks to the patients. If we could create small fractures on the bone noninvasively, it will be extremely beneficial to the patients.

There were a few reports of applying ESW to delayed union of fractures (Valchanov and Michailov 1991; Haupt et al. 1992; Schleberger and Senge 1992). However, all these studies used the lithotriptor that was designed for urinary stones. These lithotriptors were made in consideration of the safety of other internal organs, and thus were not powerful enough for application on bones. Bones are not composed of crystals like urinary stones, but have a strong structure. In addition, all these studies did not have any in-vivo experimental models. Therefore, the effect of ESW on bone is still unclear. The purpose of our study is to develop a new ESW generator that has enough power to affect bones and create new bone formation in vivo.

2. Materials and methods

The ESW generator was custom-made for this study by Yachiyoda Kogyo Co. Ltd (Chiba, Japan). Shock waves were generated by exploding a 10-mg silver azide pellet placed at the first focus (f_1) of an ellipsoidal reflector. The shock wave was then reflected by the reflector and focused onto the second focus (f_2) aimed at the rabbit femur. The shock wave was transmitted through degased water placed within a water bag, similar to a dry-type lithotriptor. The ellipsoidal reflector has a major axis of 108 mm and a minor axis of 90 mm. The dimensions of the lithotriptor were 130 mm and 93 mm. The length of our second focus is shorter as bone is more superficial than urinary stones. The position of the explosion pins were moved from a perpendicular position to a parallel position with respect to the major axis. This modification improved the efficacy of reflecting the shock waves. Peak pressure at f_2 was about 150 – 300 MPa, three to six times higher than the usual lithotriptor (Ikeda et al. 1993; Nakatani et al. 1994). Our study includes both in-vivo and in-vitro studies of the effect of our ESW generator on rabbit femurs.

(1) Three Japanese white rabbit femurs were harvested after euthanasia with an overdose of Thiobarbital (150 mg/kg). ESW was applied to the femur immediately after extraction to preserve water content. ESW was applied until fractures were created. The femurs were then fixed with 10 % formalin, decalcified, and embedded in paraffin. Specimens were stained with hematoxylin and eosin and were examined under light microscopy.

(2) Nine Japanese white rabbits were anesthetized by intravenous administration of Thiobarbital (30–50 mg/kg) and placed on an operating table with the mid-point of the right femur located at the focus of the generator. ESW was applied from the lateral side of the right femur. The animals were divided into three groups. Group I consisted of three femurs that were given five shots of ESW and harvested immediately after treatment. Group II consisted of three femurs that were given ten shots of ESW and harvested immediately after treat-

ment. Group III consisted of three femurs that were given ten shots of ESW and harvested six weeks after treatment. Radiographs were taken immediately after treatment for all specimens and an additional set of radiographs were taken for group III six weeks after treatment. All harvested specimens were fixed and examined histologically in a manner similar to that described above.

3. Results

(1) Bicortical bone penetration was achieved with three and five shots of ESW with our custom-made generator in vitro (Fig. 1). The defect created at the opposite cortex was much smaller than the cortical defect on the same side of the generator. However, the intramedullary tissues (bone marrow and vessels) were not damaged under microscopic examination. (2) No fractures were

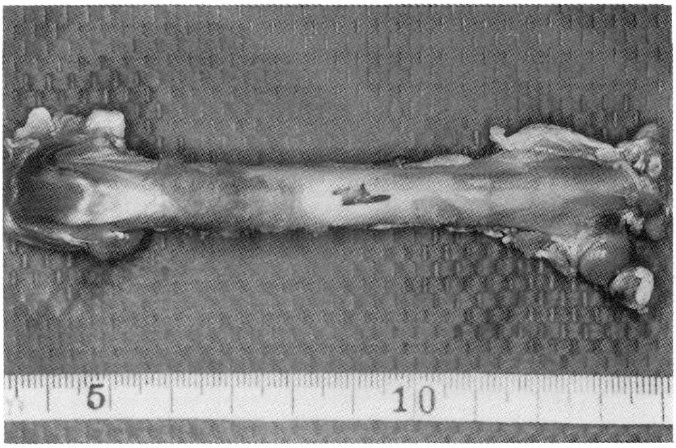

Figure 1. Rabbit femur is perforated by three shots of ESW in vitro.

observed for group I and group II radiographically. However, on histological examination, group I had stripping of the periosteum (osteomembrane) from the surface of the cortex but the periosteum was not torn. Hematoma was observed beneath the periosteum. The inner surface of the opposite cortex was smashed and saucerized along the path of the ESW (Fig. 2). However, no fractures were observed histologically. In group II, micro-fractures were observed histologically that were undetected radiographically. In group III, six weeks after treatment, new bone was seen histologically beneath the periosteum around the previous fracture lines (Fig. 3). There were more osteoblasts and osteoclasts in the lacunas than in the intact areas. These results all indicate active bone remodeling. Hypertrophy on both sides of the femur was noted on radiographic examination (Fig. 4).

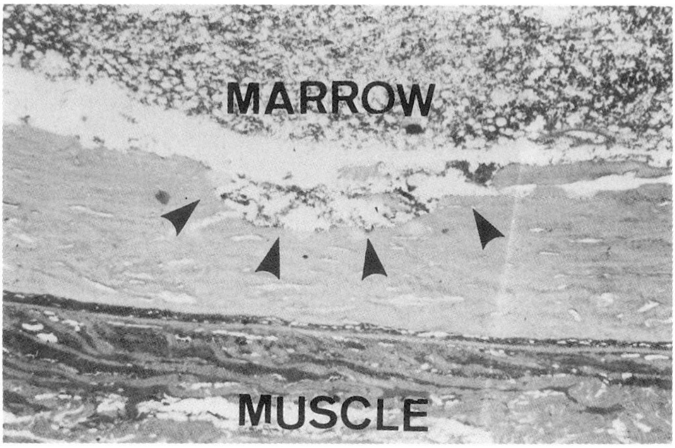

Figure 2. The inner surface of the opposite cortex is made smooth, shallow, and concave by ten shots of ESW in vivo.

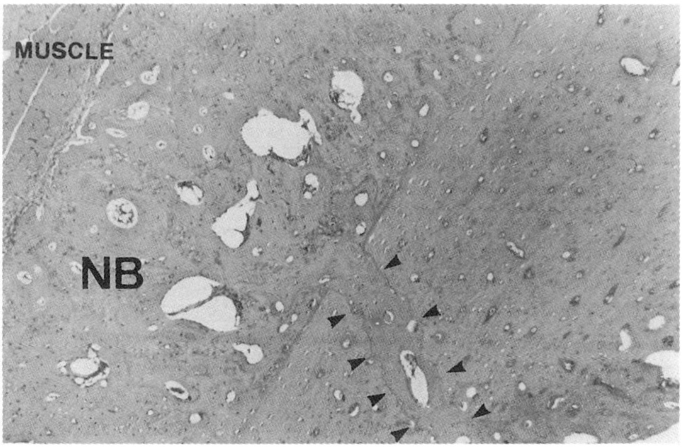

Figure 3. Six weeks after application of ten shots of ESW. The new bone formation (NB) is visible beneath the periosteum and between an original fracture line (arrows).

4. Discussion

This is the first study that showed *new bone formation* initiated by ESW. The former studies have assumed that micro-fractures and subperiosteal hematoma would stimulate bone formation but this was never demonstrated by any of them. We have also failed to create any bone defect in a rat's fibular bone using the less powerful lithotriptor. Therefore, we have some doubt that the

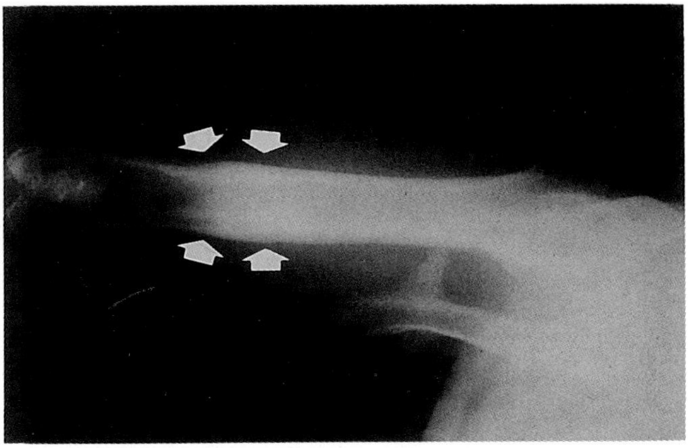

Figure 4. Six weeks after treatment, new bone formation is noted upon radiographic examination. Hypertrophy is observed on both sides of the femur (arrows).

lithotriptor would be able to affect delayed or nonunions in clinical cases. Our ESW generator can emit three to six times higher peak pressures than the lithotriptor. It has also proven to create micro-fractures in rabbit femurs and induce new bone formation in vivo.

The fracture pattern in vivo was interesting. The opposite cortex was affected but to a lesser degree. ESW might have been transmitted through the bone-marrow tissues and damaged the inner cortex on the opposite side. With ten shots of ESW, the inner cortex on the opposite side was made smooth, shallow, and concave in vivo, and we assume that we could penetrate the opposite cortex with more ESW treatment. For in-vitro studies, the lithotriptor of Yachiyoda Kogyo needed 118.0 shots to create bicortical penetration while the new generator only took 3.7 shots (average of three rabbit femur study) to create a similar defect. The first purpose of this study was achieved as our ESW generator is clearly more effective than the usual lithotriptor. For the in-vivo study, more shots of ESW were necessary to create fracture because surrounding tissues and blood flow damped the effect. Ten shots of ESW did not create any gross fracture that could be detected radiographically.

However, our main goal is to induce bone formation but not gross bone fractures. Six weeks after application of ESW, we could observe remarkable new bone formation by X-ray. Hypertrophy of the femoral shaft was seen. The fact that we could induce new bone formation without creating gross fracture is very useful when considering clinical application. Gross fractures would lead to severe complications and are difficult to control.

We named the induction of bone formation by extracorporeal shock wave ESWIB (ESW Induced Bone formation). We believe that ESWIB will be applicable to clinical cases in the near future and that ESWIB will enable us to treat delayed union of fracture without invasive surgery. Electric or electromagnetic fields have been used to stimulate bone union in clinical settings. However, the results of these treatments were not satisfying as the healing time is long and the success rate is low. Among these conservative and invasive treatment options, we strongly believe that ESWIB will be extremely valuable in the future. Moreover, ESWIB could be utilized in other clinical settings, e.g.: (1) treatment of delayed or nonunion (for stimulation of bone fusion); (2) stimulation to induce hypertrophy of vascularized fibular graft (increase strength of fibular grafts); (3) treatment of apophysitis (replacing drilling procedure for the closure of the epiphysis); (4) stimulation of ossification of callus during limb lengthening; and (5) osteotomy (early malunions of pediatric fractures could be weakened by ESW to enable realignment).

5. Conclusion

We developed a new ESW generator which generated three to six times higher peak pressures than the lithotriptor. Ten shots of ESW created micro-fractures on rabbit femurs in vivo. Six weeks after application of ESW, hypertrophy of the femoral shaft was observed. ESW Induced Bone formation (ESWIB) will be one of the valuable options for the treatment of delayed union and other orthopaedic problems.

Acknowledgement. We really appreciate Benjamin Ma's help in completing this study.

References

Haupt G, Ekkernkamp A, Gerety B, Chvapil M (1992) Influence of shock waves on fracture healing. Urology 39:529

Ikeda K, Nakatani A, Tajiri K, Tomita K, Takayama K (1993) Extracorporeal shock wave induced bone formation. JJME 31, suppl 2:28

Nakatani A, Ikeda K, Tomita K, Takayama K (1994) Experimental study of extracorporeal shock wave induced bone formation (ESWIB). JJME 32, suppl 2:59

Schleberger R, Senge T (1992) Non-invasive treatment of long-bone pseudoarthrosis by shock waves. Arch Orthop Trauma Surg 111:224.

Takayama K (1987) Shock Wave, Maruzen, Tokyo.

Valchanov VD, Michailov P (1991) High energy shock waves in the treatment of delayed and nonunion of fractures. International Orthopaedics 15:181.

Author Index

Abdul-Baki, I., 1113
Abgrall, R., 1545
Absil, L.H.J., 685
Achasov, O.V., 1131
Adachi, T., 429
Adam, S., 1225
Adomeit, G., 929, 935, 1027
Ahrens, T.J., 1401, 1449, 1455, 1461
Ahyi, A., 513
Aizik, F., 147
Akamatsu, T., 1649
Aleshin, A.N., 575, 611
Anderson, J., 801
Ando, Y., 355
Aoki, T., 723
Aoyama, T., 1641
Arai, T., 729
Archer, R.D., 1267
Aslam, T., 97
Aso, S., 355, 759, 765

Bélanger, J., 1119
Babinsky, H., 141, 495, 1599
Baillion, M., 869
Bakker, P.G., 1587
Ball, G.J., 173, 557
Ballmann, J., 741
Bauer, H.-J., 947
Baum, J.D., 679
Bazarov, S.B., 441, 477
Bazhenova, T.V., 441, 477, 1169
Bdzil, J., 97
Beck, W.H., 185, 281
Becker, M., 1231
Behnia, M., 1267
Bell, J.B., 617
Belov, V.Y., 1425
Ben-Dor, G., 147, 369, 393, 405, 447, 459, 1303, 1333, 1357
Benjamin, R.F., 587
Bershader, D., 165, 519, 911, 1605
Beylich, A.E., 471
Billiotte, M., 869
Blaha, J., 1249
Blumenthal, R., 935
Bonazza, R., 1189
Boris, J., 801
Bormotova, T.A., 441, 477
Boulton, D., 1033
Boyd, I., 293
Brück, S., 215
Braun, S., 1327
Braun-Unkhoff, M., 857
Brazier, J.P., 735
Breitung, W., 1101
Brennen, C.E., 1213
Britan, A., 1285
Brocher, E., 343
Brouillette, M., 507, 1089
Brown, M.S., 1611
Bruckner, A.P., 1063
Brun, R., 33, 227
Budil, K., 581
Budzinski, J.M., 587
Burcat, A., 1045
Burtschell, Y., 227, 1545, 1569

Cadman, P., 863, 1039
Candler, G.V., 323
Čanić, S., 435
Cardoso, M., 1545
Casey, R.T., 1267
Cates, J., 423
Causon, D.M., 1009, 1195
Chadwick, K., 179
Chaix, A., 227
Chambarel, A., 343
Chauvin, A.H., 983

Chebotareva, E.I., 575
Chen, H., 833
Chernin, A., 453
Chpoun, A., 369, 375, 405, 795
Christiansen, W.H., 135
Chue, R.S.M., 777
Cieslak, S., 299
Clifford, L.J., 1033
Colella, P., 617
Cordonnier, P., 343
Cramer, M.S., 783, 1163, 1327
Crutchfield, W.Y., 789
Cullis, I.G., 997
Cummings, E.B., 1611

Damion, J.P., 1527
Daru, V., 1527
Das, H.K., 1279
Davidson, D.F., 923, 941
Davies, G.A.O, 1207
Davis, J., 191, 203
Deguchi, Y., 1629
Delale, C.F., 1237
Demmig, F., 661
Dewey, J.M., 399
Di Rosa, M.D., 923
Diaconescu, G., 1351
Dihu, R., 1137
Dorofeev, S.B., 77, 673, 1101
Druguet, M.C., 227, 1545
Dumitrescu, L.Z., 227, 1569
Dumitrescu, M.-P., 227, 1487, 1581
Dupré, G., 893
Dvinyaninov, M., 1045
Dvoinishnikov, A.E., 673

East, R.A., 173, 557
Efimenko, A.A., 1101
Eitelberg, G., 185, 221, 257, 1575
El-Beshbeeshy, M., 1189
Elizarova, T.G., 795
Elperin, I., 1303
Elperin, T., 147, 1285, 1309
Emelyanov, A.V., 1169
Ende, H., 311
Eremin, A.V., 881, 917, 1169
Exberger, R., 911

Fabre, E., 299
Falcovitz, J., 489, 1309
Favorsky, A.P., 605
Felthun, L., 551
Fieweger, K., 929, 935, 1027
Figura, E.S., 997
Flanagan, O., 1207
Fletcher, D., 1605
Fortov, V.E., 1443
Frank, P., 857
Frohn, A., 905, 977
Frolov, S.M., 1015
Frost, D.L., 121, 1113, 1387
Fu, Y., 1243
Fuchs, J.W., 747
Fujikawa, S., 1219
Fujita, M., 1431
Fujiwara, T., 263, 765, 1051, 1071, 1157, 1175
Fukuda, M., 355
Fursenko, A., 691
Fuyuto, T., 1071

Gai, S.L., 233, 239, 245, 317
Galametz, I., 647
Gallagher, K.G., 1401
Gallis, M.A., 179
Galyukov, A., 691, 1593
Gao, D.Z., 1515
Garen, W., 1231
Gelfand, B.E., 935, 1027, 1321
George, A., 311
Gilmore, J., 911, 1605
Gimelshein, S.F., 471, 717
Goel, B., 1443
Golub, V.V., 441, 477

Gong, M., 1243
Gorbachev, Y.E., 717
Goto, T., 483
Gou, G.X., 1515
Goyne, C.P., 771, 1623
Grönig, H., 147, 153, 305, 501, 1143
Graur, I.A., 795
Greenough, J.A., 617
Griffin, D., 1497
Grove, J.W., 629
Gu, J.-H., 1509
Gvozdeva, L.G., 375; 1345, 1393

Hänel, D., 1077
Hébert, C., 507
Haas, J.-F., 647
Halat-Augier, C., 893
Han, Z., 1243
Hannemann, K., 221, 257, 1575
Hanson, R.K., 275, 923, 941
Harvey, J.K., 179
Hayashi, A.K., 1051, 1071, 1157
Hayashi, M., 765
He, Y., 953
Heberlein, D.C., 1137
Henderson, L.F., 789
Hertzberg, A., 135
Herzler, J., 839
Hidaka, Y., 899
Higashino, F., 361, 381, 1201
Higgins, A.J., 1063
Hill, L., 1413
Hillier, R., 699, 1207
Hindelang, F.J., 827
Hironde, N., 203
Hitchings, D., 1207
Hoeijmakers, H., 1363
Holden, M.S., 179
Holmes, R.L., 629
Hongo, S., 729
Honma, H., 159, 1181, 1255, 1641
Honore, D., 269
Hornung, H.G., 1611
Hornung, K., 1261
Horvath, T.J., 221
Houas, L., 635, 869
Houwing, A.F.P., 991, 1617
Hu, G., 387
Huq, N., 863

Igra, O., 147, 1285, 1303, 1309, 1339
Ikeda, K., 1655
Ingram, D.M., 1009
Inomata, T., 959
Inoue, A., 959
Itoh, K., 43, 777, 1503
Itoh, S., 1431
Ivanov, M.S., 471, 717
Iwamoto, N., 483
Iwamoto, R., 899

Jacobs, J.W., 587, 617
Jacobs, P.A., 1539
Jacobs, P.G., 1491
Jiang, B., 1009
Jiang, J.P., 1285, 1309
Jiang, Z.L., 465
Joubert, X., 287
Jourdan, G., 635

Kambe, T., 1021
Kameda, M., 1201
Kamel, M.R., 275
Kaneko, R., 1481
Kaneshige, M., 1119
Kao, C.T., 165, 519
Kapila, A., 1413
Karashima, K., 759
Kastell, D., 221
Kawase, T., 729
Kedrinskii, V.K., 107
Kellerer, H., 947
Kemmou, A., 513
Kendall, M.A., 1491

Kern, R.D., 833
Keyfitz, B.L., 435
Khomik, S.V., 1321
Kleine, H., 501, 1351
Klenk, W., 905
Klomfass, A., 741
Kluwick, A., 1327
Knowlen, C., 1063
Knowles, J.K., 1449
Kobayashi, S., 429
Kochurko, A.S., 1101
Kodama, T., 569
Koffi-Kpante, K., 869
Koike, T., 899
Komp, K.H., 935
Komurasaki, K., 263
Komuro, T., 1503
Kondoh, N., 723
Koppenwallner, G., 203
Koreeda, J., 1181
Kotani, M., 1219
Krek, R., 185, 1575
Kruse, T., 851, 875, 887
Kubota, S., 1431
Kudo, S., 361
Kumar, D., 711
Kurian, J., 349, 1279
Kurokui, Y., 1641

Löhner, R., 679
Labracherie, L., 869
Labuda, S.A., 1131
Lackner, K.S., 87
Lagutov, Y.P., 1345, 1393
Laskin, A., 971
Lazareva, E.V., 575, 611
Le, J., 387, 525
Leblanc, J.E., 1175
Lebo, I.G., 605
Lee, J.H.S., 1113, 1351, 1387
Lee, J.J., 1089
Lee, M.G., 545, 563, 705
Legoete, T., 1375
Lele, S.K., 653
Lenartz, M., 153
Lengrand, J.C., 369, 795
Levy, A., 1333, 1357
Li, H., 369, 393, 447, 1333
Li, K.Y., 1083
Li, Z.-F., 1509
Lifshitz, A., 971, 1045
Liu, J.J., 411
Lord, S.J., 1207
Luo, H., 679

Mölder, S., 337
Müller, S., 741
Müller, T., 281
Méolans, J.G., 983
Maekawa, S., 759
Maeno, K., 159, 1181, 1255, 1641
Mahesh, K., 653
Makhsud, A., 343
Makris, A., 1113, 1387
Mallinson, S.G., 233, 317
Manion, J.A., 839
Maršík, F., 1237, 1249
Markelov, G.N., 471, 717
Marsh, S.P., 1419
Martin, R., 227
Mashimo, S., 723
Mathieu, G., 311
Matsui, H., 965
Matsui, Y., 1481
Matsumoto, M., 1219
Matsumoto, Y., 1201
Matsuo, K., 723
Maul, J., 197
Maus, J.R., 1533
McIntyre, T.J., 991, 1617
McKenzie, N.R., 1533
Medvedev, S.P., 1321
Mee, D.J., 771
Meiron, D.I., 623

Menikoff, R., 87
Merlen, A., 299, 513
Messitt, D.G., 1551
Meyer, S., 911
Mick, H.-J., 875, 881
Miller, G.H., 1467
Miller, P., 581
Milne, A.M., 1033
Milthorpe, J., 239
Milton, B.E., 1267
Minota, T., 545, 563
Minucci, M.A.S., 1551
Mitomo, T., 765
Miura, Y., 1473
Moin, P., 653
Mond, M., 147, 667, 1437
Morgan, R.G., 251, 1491, 1563
Mori, S., 1051
Morris, C.I., 275
Mortimer, B., 551
Mouti, M., 983
Mudford, N.R., 173, 233, 245, 317
Muelder, S.A., 53
Muylaert, J., 191, 203, 1587
Myrabo, L.N., 1551

Nagamatsu, H.T., 1551
Nagano, S., 1431
Nagayama, T., 1629
Nagnibeda, E.A., 983
Nagoya, H., 489
Nakatani, A., 1655
Naumann, C., 857
Naumann, K.W., 311
Neely, A., 251
Nerenberg, J., 1387
Neuvazhayev, V.Y., 599
Nguyen, T., 753
Ni, H., 387
Nikishin, V.V., 605
Nishida, M., 545, 563, 705
Nishimori, Y., 1255
Norikan, K., 1137
Noskov, M.A., 1015
Nozawa, H., 1255

O'Keefe, J.D., 1455
Odagawa, M., 159
Ofengeim, D., 537
Ogawa, H., 1157
Oh, C., 753
Ohama, Y., 1181
Okamoto, M., 1473
Olejniczak, J., 323
Olivier, H., 305
Onishi, Y., 747
Onodera, O., 495, 1481, 1521
Oran, E., 753, 801
Oya, M., 965

Paál, G., 1231
Page, N., 1497
Paillard, C.E., 893
Palma, P.C., 991, 1617
Park, S., 783
Passerel, D., 369, 375
Patel, D.L., 1137
Patz, G., 1057
Paull, A., 1557, 1623
Penyazkov, O.G., 1131
Pernod, P., 513
Petersen, E.L., 923, 941
Petrie, P.J., 1539
Peyser, T., 581
Pfahl, U., 929, 1027
Piana, J., 753
Pichugin, S.V., 477
Polenov, A.N., 1321
Popov, I.V., 605
Pruiksma, J., 1369
Pu, Y.K., 1083
Puckett, E.G., 617, 1467
Pullin, D.I., 623

Quenett, C., 1003

Quirk, J.J., 1125

Röhrig, M., 923, 941
Rabbath, P.A.B., 991
Ragozin, D.S., 1131
Ramana Rao, T.V., 349
Ramesh Kumar, R., 349
Ray, J., 641
Rayer, C., 647
Redlinger, R., 1101
Reichenbach, H., 17
Resnyansky, A.D., 1425
Ritzerfeld, E., 501
Roberts, G.T., 173, 317
Robin, L., 269
Rodriguez, G., 647
Rom, J., 1149
Romensky, E.I., 1425
Rosenband, V., 1407
Rosenhauer, M., 281
Roth, P., 1, 845, 851, 875, 881, 887, 1077, 1273, 1315
Rozanov, V.B., 605
Rutkevich, I., 667, 1437
Ruyten, W., 1635

Séror, S., 735
Saint-Martin-Tillet, X., 753
Saito, T., 141, 691
Sakamura, Y., 417
Sakurai, A., 1021
Samtaney, R., 623, 641
Sanderson, S.R., 209
Sasoh, A., 1063, 1481
Sato, H., 159, 1219
Sato, K., 759, 899, 1503
Saunders, R., 1009
Schading, G.N., 1273
Schall, E., 735
Scheer, M., 281
Schnerr, G.H., 1225, 1237
Schotz, M., 393
Schuricht, P.H., 173
Schwendeman, D., 531
Seiler, F., 1057
Seitz, M., 1381
Sekine, N., 1521
Sellschop, R., 1375
Serov, A.O., 331
Shamoun, B., 1189
Sharma, S., 911, 1605
Sharov, Y.L., 1345, 1393
Sharp, D.H., 629
Shaw, L.L., 53
Shepherd, J.E., 1119
Sherbak, N.B., 375, 1393
Shida, H., 1629
Shimaura, N., 1201
Short, M., 1125
Shtemenko, L.S., 331
Shugaev, F.V., 331
Shulmeister, A.M., 441, 477
Shumova, V.V., 881, 917, 1169
Sidorov, V.P., 673, 1101
Simeonides, G., 203, 1587
Sislian, J.P., 141
Skews, B., 551, 1357, 1375, 1381
Smeets, G., 1003, 1057, 1363
Smeulders, D., 1363, 1369
Smith, M., 1635
Sommer, T., 887
Sopuch, P., 1249
Sorek, S., 1357
Srulijes, J., 1057
Stalker, R., 1497, 1623
Stewart, D., 97
Stollery, J.L., 711
Strecker, J.J.F., 1315
Stry, P., 581
Stuhler, H., 977
Sturtevant, B., 191, 209, 423, 593
Sudnitsin, O., 1563
Sugiyama, H., 729

Sumida, W.K., 1137
Sun, X.W., 159
Suzuki, T., 417, 429

Takahashi, K., 959
Takano, Y., 483
Takasugi, N., 1219
Takata, T., 1455
Takayama, F., 1021
Takayama, K., 141, 465, 489, 495, 569, 1481, 1521, 1599, 1655
Taki, T., 1021
Tamagawa, M., 1649
Tan, T.-H., 1419
Tanno, H., 1503
Tannou, M., 355
Tarkenton, G.M., 1163
Tatschl, R., 1015
Teodorczyk, A., 1095
Thibault, P. A., 1107
Thoma, M.L., 827
Thomas, G.O., 1095
Thorembey, M.-H., 647
Thurber, M.C., 275
Timnat, Y.M., 1407
Timofeev, E., 495, 537, 691, 1593
Tishkin, V.F., 605
Titov, S.N., 575
Toffin, E., 667
Tomita, K., 1655
Tsang, W., 839
Tsuchiya, K., 965
Tulis, A.J., 1137
Tumino, G., 203

Ueda, S., 257
Umii, K., 355
Uno, N., 729
Uphoff, U., 1077

van den Berg, A.C., 685
van Dongen, M., 1363, 1369
van Gool, C., 1369
van Netten A.A., 399
Vasiliev, E., 1303
Vervisch, P., 269
Vetter, M., 593
Virgona, R.J., 381, 789
Voinovich, P., 453, 495, 537, 691, 1593
von Ellenrieder, K., 165, 519
Vorobiev, O.Y., 1443
Vuillon, J., 459

Wallus, H., 661
Walpot, L.M.G., 1587
Walpot, W., 203
Wang, B., 153
Wang, B.-Y., 1291, 1297
Wang, D.-Z., 1297
Wang, R., 863
Wang, Y.-C., 1213
Ward, S.M., 1095
Watanabe, N., 1641
Watanabe, Y., 361
Watson, L.T., 783
Weber, Y., 801
Weerheijm, J., 685
Wehe, S.D., 275
Whiting, E., 911
Widdecke, N., 905
Williams, G.H., 1425
Williams, W., 1635
Wilson, G.J., 67
Winfree, N.A., 1449
Wintergerst, K., 857
Wittig, S., 947
Woiki, D., 845
Wolanski, P., 1015
Wollenhaupt, M., 281
Wu, Q.-S., 1291, 1297
Wu, X., 525

Xu, Y.-H., 1291, 1297

Xu, Y.-Q., 1291

Yamada, K., 489, 569
Yamane, Y., 355
Yang, H., 525
Yang, J.M., 141
Yang, W., 1461
Yano, S., 263
Yao, J., 833
Ye, X., 525
Yin, X.-.Z, 417
Yoshida, K., 1649
Yoshida, T., 1051
Yoshinaka, A., 1113
Yu, H.-R., 1509
Yu, Q., 501, 1143

Zabusky, N.J., 641
Zanchetta, M.A., 699
Zaretsky, E., 1339, 1437
Zaslonko, I.S., 917
Zaytsev, S.G., 575, 611
Zeitoun, D., 227, 459, 735, 1545, 1569
Zerf, G., 1261
Zhang, F., 1107
Zhang, X.-Y., 1509
Zhao, W., 1509
Zhong, X., 287
Ziborov, V.S., 881

Keyword Index

Ablation, 263
Absorption, 911
Absorption spectra, 923
Absorption spectroscopy, 857, 887
Acceleration measurement, 311
Acetone pyrolysis, 899
Acetylene, 851
Adaptive methods, 617
Advanced ceramics, 1407
Aerodynamic force measurement, 311
Aerodynamic heating, 765
Aerodynamics, 337
α-methylnaphthalene, 1027
Aluminum, 1137
Aluminum dust-air mixture, 1083
Angular momentum, 519
Animation, 141
ARAS, 881, 959, 965
Astrophysics, 453
Asymptotic expansion, 735
Atomic oxygen, 911
Atomization, 1027, 1267
Attenuation law, 147
Automobile engine, 1521

Backwards step, 239
Baffle system, 1291
Ballistic range, 263
Base flow, 179
Bifurcation, 801, 1225
Bimodal distribution, 1315
Blast damage assessment, 685
Blast interaction with structures, 685
Blast parameters, 673
Blast wave, 77, 453, 557, 685, 1321, 1387
Blast-wave propagation, 1351
Blunt body, 735
Blunt cone, 221, 239
Blunt protuberance, 759
Bluntness, 711
Bone, 1655
Boundary layer, 67, 245, 251, 735, 1497, 1533, 1581, 1623
Boundary-layer blockage, 1563
Boundary-layer formation, 1563
Break-up, 1027, 1261
Break-up time, 1027
Bubble, 1201
Bubble dynamics, 159

C_2, 887
Catalyticity, 317
Cavitation, 159, 1207
Cavitation noise, 1213
Cavitation structure, 107
Ceramic armor, 1449
CFD, 495, 789, 1143, 1225, 1569
CFD validation, 1635
Chemical in-tube accelerators, 1149
Chemical kinetics, 845, 851, 857, 875, 929
Chemical nonequilibrium, 735, 977, 1545
Chemical reaction, 765
Chemical reactions, finite-rate, 1063
Chemical separation, 1473
Chemically reacting hypersonic rarefied flows, 717
Cloud cavitation, 1213
Cloud of particles, 1279
CN emission, 869
CO_2 laser absorption, 635
Coflowing jets, 349
Collision broadening, 923
Combustible gas mixture, 1057
Combustion, 1003, 1057, 1551
Combustion synthesis, 1407

Combustion torch, 1051
Comet Shoemaker-Levy 9, 1455
Communication blackout, 1175
Compaction, 1413
Compaction waves, 1357
Comparative rate, 953
Comparative rate measurements, 839
Compositional change, 1473
Composites, fibrous and laminated, 1425
Compressibility, 519
Compressible flow, 581
Compressible foam, 1345, 1393
Compressible turbulence, 507
Compressible turbulent boundary layer, 1015
Compression corner, 233
Compression wave, 723, 1339
Computational model, 1533
Condensation, 1219, 1231, 1243, 1255
Condensation coefficient, 1219
Condensation shock, 1249
Cone, 417
Continuum equations, 795
Control flap, 711
Control jets, 173
Converging shock waves, 661
Corundum, 1401
Critical pressure ratio, 343
Cryogenic liquid, 159
Curvature effect, 87, 97
Cylinder flow, 215

Damkohler number, 251
Decay, 1339
Decomposition, 827, 953, 971
Defect formulation, 735
Deflagration, 77, 929, 1107
Deflagration-to-detonation transition (DDT), 929, 1015, 1107, 1413
Deformation, 1261
Density measurement, 647
Detection, 1557
Detonation, 77, 97, 153, 929, 1071, 1077, 1101, 1107, 1119, 1125, 1143, 1149
Detonation driver, 153, 1509
Detonation initiation, 107, 1131
Detonation, over driving, 1503
Detonation propagation limits, 1113
Deuterated ethylene, 833
DFWM, 1611
Diaphragm rupture, 1539
Diaphragmless shock tube, 141
Diesel engine, 929
Differential interferometry, 647
Diffraction, 141, 483, 513, 789
Direct numerical simulation, 653
Direct Simulation Monte Carlo (DSMC), 717, 753
Disk, 795
Dispersed waves, 1641
Dissipation, 519
Dissociation, 33, 245
Dissociation modeling, 293
Dissociation-recombination, 917
Disturbances on shock front, 331
Drag, 1261
Drag coefficient, 1285
Driver, 1491, 1497
Driver gas, 1557
Driver-gas contamination, 777, 1581
Droplet, 1027, 1267
Dry-powder dispersion, 1315
Dust deposit, 1321
Dust-gas suspensions, 147
Dust lofting, 1309
Dusty gas, 1297, 1303, 1309
Dusty shock, 1291

Edge interactions, 97
Effective drag coefficient, 1279
Effective stress, 1363
Ejecta, 1461

Elastic bodies, 513
Electro-optic shutter, 53
Embedded shock formation, 1237
Emission-absorption spectroscopy, 1169
Engine silencer, 1521
Entrainment, 1267
Entropy, 1491
Environmental, 947
Equation of state, 1437, 1443
Equilibrium gas dynamics, 1369
Ethylene combustion, 1033
Euler equations, 459, 531
Evaporation, 747, 1027, 1219
Excitation, 667
Exhaust gas-pipe system, 1521
Expansion tube, 67, 275, 1539, 1563
Explosions, 77, 685
Explosive compaction, 1407
Explosive, granular, 1413
Extracorporeal shock wave lithotripsy (ESWL), 107, 1655

Fay and Riddell theory, 305
Finite-compression waves, 411
Finite-volume code, 1539
Flame acceleration, 1083
Flame, spontaneous, 1015
Flow establishment times, 179
Flow imaging, 275, 991
Flow visualization, 165, 257, 545, 569, 593, 729, 759, 1189, 1225, 1617, 1635
Fluorocarbons, 551
Foam, 1375, 1381
Foam, flexible, 1363, 1369
Foam, permeable, 1351
Fourier transform, 1599
Fracture, 1655
Fragmentation wave system, 661
Free-Lagrange method, 557
Free piston, 1491, 1497
Free-piston driver, 43
Free-piston shock tunnel, 43, 227, 257, 1503, 1515
Friction, 1413
Fringe analysis, 1599
Front tracking, 587, 629, 1195
Fuel-coolant interaction, 121
Fullerene formation, 863
Furonitrile, 971

Gas bubble, 569
Gas phase, 851, 875
Gas-wall interaction, 197
Gaseous detonation, 673, 1113
Geometrical shock dynamics, 423, 531
Global simulation method, 1009
Godunov methods, 1467
Griceite, 1401
Grid, adaptive, 1143
Grid generation, 691
Grid, local refinement, 691
Grids, combined, 537
Grid, unstructured, 495
Growth rate, 623, 641

H_2S, 965
Halis, 191
Halocarbons, 1045
Halomethanes, 959
Heat engine, 135
Heat flux, 245, 257
Heat transfer, 251, 1491
HEG shock tunnel, 185, 221, 1575, 1587
Heterogeneous detonation, 1089, 1137
HFC-134a, 1255
HHK, 203
High enthalpy, 191, 251, 317, 777, 1587
High-enthalpy effects, 179
High-enthalpy flow, 269, 1515, 1635
High-enthalpy nozzle flow, 1575
High Knudsen-number flow, 753

High pressure, 923, 935, 941
High-speed photography, 53
High temperature, 833
Historical report, 17
Hollow projectile, 1063
Holographic interferometry, 141, 525, 1599
Homogeneous condensation, 1237
Hood, 1481
Hugoniot jump conditions, 87
Hydraulic shock, 1207
Hydrazine, 893
Hydrocarbons, 929
Hydrodynamic code, 997, 1009
Hydrodynamic instability, 575, 611
Hydrogen, 1101
Hydrogen-air, 935
Hydrogen combustion, 1009
Hyperboloid flare, 185
Hypersonic facility, 227, 1487, 1581
Hypersonic flow, 173, 179, 191, 221, 245, 263, 275, 281, 293, 299, 305, 323, 387, 699, 735, 1563, 1635
Hypersonic nozzle, 1569
Hypersonic receptivity, 287
Hypervelocity, 185, 191, 233, 239, 317, 337, 623, 1119, 1149
Hysteresis, 375, 405, 471

Ideal dissociating gas, 623
Ignition, 1027, 1057, 1413
Ignition-delay time, 929, 1027
Ignition, spontaneous, 1021
Ignition times, 941
Image-converter camera, 53
Image processing, 1641
Imaging spectroscopy, 1181
Impact, 1449, 1455, 1461
Improved tangential velocity gradient, 305
Impulsive jet, 441
Induction zone, 997
Integral method, 305
Interaction of vortices, 563
Interference heating, 173
Interferometry, 495, 647
Interferometry, finite-fringe, 1641
Intermittency, 699
Internal cavity flow, 361
Intersecting characteristics, 1237
Ionization, 667
IR diode laser absorption spectroscopy, 905
IR-laser absorption, 899
IR-spectroscopy, 1273
Iron, 1401
Isentropic compression, 1497

Jet interaction, 311, 1131
Jets, 453, 581
Jet, liquid, 569
Jetting, 1461, 1467
Jupiter, 1455

Kármán integral equations, 1563
Kinetic theory, 747
Kinetics, 893, 997
Kinetics, shock-tube, 1

Laminar separation, 711
Large-scale experiments, 1101
Laser absorption, 1635
Laser diagnostics, 923, 941, 1051
Laser differential interferometry, 723
Laser-flash photolysis, 881
Laser holography, 1261
Laser-illuminated photography, 53
Laser-induced fluorescence (LIF), 269, 281, 1629
Laser photolysis, 965
LDV, 729
Leading edge bluntness, 233
Light extinction, 1189

Linear analysis, 653
Liner implosion, 1419
Lineshape, 911
Liquid-spray combustion, 863
LITA, 1611
Living time, 1027
Living-tissue cell, 1649
Local similarity, 741
Low temperature, 935, 1255

Mach disc, 349
Mach reflection, 97, 369, 375, 387, 393, 399, 405, 411, 471, 1255
Magnetohydrodynamics, 1157
Mantle-core structure, 1473
Material interfaces, 1195
Measurements, 191
Melting, 1413
Methane ignition, 1045
Methyl halides, 959
MHD shocks, 1163
Micro capsule, 1649
Microchannel flows, 753
Microinhomogeneities, 107
Minimum entropy production, 447, 1249
Mixing, 581
Model computation, 661
Modeling, 857, 893
Molecular dynamics, 1219
Monte Carlo method, 293, 795
Moving shock waves, 465
Multi-component flows, 1195
Multi-fluid methods, 489, 617
Multi-pulsed ruby laser, 53
Multipass technique, 827
Multiphase flow, 1107

n-decane, 1027
Navier-Stokes equations, 537, 753, 789
Nd-B-Fe powder, 1431
NH_4Cl/Ar aerosols, 1273
Nitric oxide (NO), 269, 827, 905
Nitromethane, 1089
NO spectroscopy, 281
NO_x, 857
Non-deuterated ethylene, 833
Nonequilibrium, 33, 221, 337, 741, 765, 1587
Nonequilibrium condensation, 1225
Nonequilibrium Navier-Stokes, 203
Nonequilibrium radiation, 1169, 1181
Nonlinearity parameter, 551
Non-uniform distribution, 1201
Nova laser, 581
Nozzle starting process, 1545
Nozzles, 1297, 1303, 1563, 1587
Numerical blast simulation, 679
Numerical computation, 221, 343, 453, 459, 465, 501, 605, 1033, 1063, 1071, 1125, 1297, 1363, 1425, 1521, 1527, 1575, 1593

Off-axis loading, 1425
OH, 923, 941
Optical-flow visualization, 1521
Optical method, 197, 1345
Origin of the Moon, 1473
Oscillating shock wave, 759
Oxidation, 965
Oxy-hydrogen detonation, 1509
Ozone, 827

Particle coagulation, 1315
Pattern recognition, 661
Percolation, 1393
Perfluoropropylene, 953
Permeability, 1363
Phase transformation, 1449
Phase transition, 1249
Photolysis, 845
Piston-stopping device, 1487
Planar laser-induced fluorescence, 275, 991, 1617, 1635

Plasmas, 1163
Plasma, anisotropic, 1175
Plasma, nonideal, 1443
Plasma, waves in, 1175
PLIF, 275, 991, 1617, 1635
Polycyclic aromatic hydrocarbons, 863
Porous material, 1321, 1333, 1357, 1375, 1381
Porous medium, 1113, 1387
Porous wall, 1481
Precursor shock, 1369
Pressure, 947
Pressure amplification, 1375, 1381
Pressure fluctuations, 1527
Pressure measurement, 1057
Pressure-wave refrigerator, 1593
Projectile, 1119
Projectile acceleration, 1057
Pseudo-shock wave, 729
Pulsating instabilities, 1125
Pulse laser, 159
Pyrolysis, 845, 959

Quantum fluid, 1327

Radiation, 263
Ram accelerator, 1057, 1063
Raman scattering, 1605
Rankine-Hugoniot relations, 1369
Rapid phase transformation, 121
Rarefaction focusing, 107
Rarefaction waves, 1243
Rarefied gas dynamics, 795
Rate coefficients, 959
Rate constant, 953
RDX, 1137
Reaction mechanism, 899
Reaction rate, 899
Reactive flow, 735, 801, 1033
Real gas, 191
Real-gas effects, 203
Recombination, 251, 1169
Red blood cell, 1649
Reduced mechanism, 1009
Reduced models, 641
Reentry vehicle, 1175
Reflected-shock/boundary-layer interaction, 777, 801
Reflection, 141, 417, 429, 441, 447, 483, 551, 1291, 1375, 1381
Reflection, anomalous, 381
Reflection, regular, 375, 405, 1157, 1333
Relaxation models, 977
Retrograde gases, 1231
Reynolds shear stress, 653
Richtmyer-Meshkov instability, 581, 587, 593, 605, 629, 635, 647
Riemann problem, 435
Rotational-vibrational excitation, 905
RR to MR transition, 459

Safety distances, 685
Scaling, 77
Scavenger, 953
SCRAM jet engine, 355, 1003
Self-ignition, 929, 935
Self-similarity, 399
Separation, 185, 191
70-deg cone, 179
Shaped-charge jet, 53
Shear layer, 221, 361
Shock acceleration, 587
Shock acceleration, diffuse interfaces, 617
Shock attenuation, 1351
Shock capturing, 87, 531
Shock-cell structure, 343
Shock compression, 1419
Shock consolidation, 1431
Shock diffraction, 477, 501, 531, 1291
Shock diffraction, reactive, 1009
Shock focusing, 423, 495, 1131
Shock heating, 1419

Shock impingement, 209, 477
Shock-induced mixing, 587
Shock-induced separation, 783
Shock instability, 667
Shock interaction, 87, 323, 435
Shock interaction, boundary-layer, 191, 203, 705, 729
Shock interaction, bubble 489, 569
Shock interaction, disturbance, 287
Shock interaction, dust 1309
Shock interaction, foam, 1387
Shock interaction, interface, 635
Shock interaction, sphere, 1285
Shock interaction, turbulence, 653
Shock interaction, turbulent boundary-layer, 729
Shock interference, 209
Shock loading, 107
Shock polar, 1157
Shock reflection, 141, 417, 429, 441, 447, 483, 551, 1291, 1375, 1381
Shock refraction, 557
Shock-on-shock interaction, 209, 215
Shock structure, 349
Shock tunnel, 153, 191, 203, 281, 311, 771, 777, 991, 1533, 1539, 1551, 1605, 1617, 1623
Shock-vortex interaction, 165, 1021
Shock-wave diffraction, 477, 501, 531, 1291
Shock-wave impingement, 209, 477
Shock-wave propagation, 1195, 1297, 1425
Shock-wave reflection, 141, 417, 429, 441, 447, 483, 551, 1291, 1375, 1381
Shock wave, standing, 747
Shock wave, strong, 1181
Shocklet, formation of, 545, 563
Shocks in liquids, 513
Short-duration flow, 311
Silicon, 875
Single-pulse shock tube, 839
Skin friction, 1623
Soot, 863, 947
Spherical condensed phase, 747
Spontaneous acoustic emission, 1437
Stability, 471
Stability of shocks, 1437
Stagnation-point heating, 305
Stagnation streamline, 741
Starch, 1137
Steady flow, 369, 393, 477
Steady supersonic flow, 459
Stream-wise vortex, 355
Stress relaxation time, 1339
Sulfur, 845
Superfluid, 1327
Superorbital expansion tube, 251
Supersonic combustion, 991
Supersonic expansion, 1169
Supersonic flow, 361
Supersonic mixing, 355
Supersonic projectile, 465
Surface heat flux, 317

T5, 191, 203
Temperature measurements, 869
Temperature probe, 1497
Temperature, shock, 1401
Terrestrial planets, 1473
Test time, 777
Tetra-ethoxysilane decomposition, 839
Thermal conductivity, 1401
Thermal decomposition, 833, 887, 893, 959, 1273
Thermal grating, 1611
Thermal nonequilibrium, 977
Thermodynamic fluctuations, 653
Three-dimensional flows, 417
Three-dimensional simulation, 1101
Time-of-flight mass spectrometry, 833
Time-resolved visualization, 661

Titan, 869
TNT, 1137
TNT-equivalent, 673
Transition, 179, 429, 471, 699, 711, 723, 771
Transition criterion, 369
Transition to detonation, 1083
Transonic flow, 1225
Transonic shocks, 435
Transport terms, 33
Tuned operation, 43, 1503
Tunnel calibration, 257
Tunnel sonic boom, 1481
Turbulence, 575, 611, 729
Turbulent kinetic energy, 653
Turbulent mixing, 593, 599, 647
Turbulent mixing zone, 635
Turbulent spots, 771
TVD method, 215, 387, 1357, 1545
Two-phase analysis, 1357
Two-phase flow, 1071, 1077, 1201, 1225

Underwater shock wave, 1431
Unsteady boundary layer, 723
Unsteady flow, 67, 537, 1279
Unsteady reflection, 429
Unstructured Euler solver, 691

Vacuum ultraviolet, 911
Vapor explosion, 107, 121, 1189
Velocimetry, 1611
Vibration-dissociation coupling, 323, 983
Vibrational nonequilibrium, 735, 881
Vibrational populations, 33, 983
Vibrational relaxation, 917, 1605
Viscous nonequilibrium flows, 215
Visiometrics, 641
Visualization techniques, 17, 501
Void fraction, 1189
von Neumann reflection, 381, 435
Vortex, 441, 519
Vortex, resonant excitement of, 331
Vortex ring, 477, 545, 563, 1021
Vortex-wall interaction, 545
Vorticity, 507

Wake flow, 221
Wall pressure, 257
Water droplet, 1261
Wave configuration, 393
Wave machinery, 135
Wave processes, 1593
Wave splitting, 1369
Wave structure, 429
Wave transmission, 1089
Waveriders, 299
Waves in plasma, 1175
Weak shock wave, 483, 1231, 1521
World Trade Center, 679

X-rays, 647